1 BF52134

Handbook of
STRUCTURAL ENGINEERING

Handbook of
STRUCTURAL
ENGINEERING

Edited by
W.F. CHEN

CRC Press
Boca Raton   New York

Acquiring Editor: *Tim Pletscher*
Project Editor: *Susan Fox*
Cover Design: *Denise Craig*
PrePress: *Kevin Luong*

Library of Congress Cataloging-in-Publication Data

Handbook of structural engineering / W.F. Chen, editor-in-chief,
  p.   cm.
    Includes bibliographical references and index.
    ISBN 0-8493-2674-5 (alk. paper)
    1. Structural engineering.   I. Chen, Wai-Fah, 1936-
  TA633.H36 1997
  624.1—dc21   97-148
                                                             CIP

Preface

The *Handbook of Structural Engineering* is a comprehensive reference work and resource book covering the broad spectrum of structural engineering. It has been written with the practicing structural engineer in mind. The ideal reader will be a M.S.-level engineer with a need for a single reference source to keep abreast of new techniques and practices, as well as review standard practices.

The area of structural engineering includes such fundamental typical topics as strength of materials, structural analysis in both classical and numerical computer methods, structural design in both steel and concrete as well as composite construction and earthquake engineering. The types of special structures involved in a typical structural engineering work includes cold-formed steel, aluminum, and timber structures; bridge, building, and shell structures; as well as space and tower offshore structures. Special research topics in structural engineering include such typical topics as shock and seismic loading, connections and detailing, fatigue and fracture, construction and repairs, active and passive control as well as computer-aided analysis and design using advanced methods.

To this end, the Handbook is divided into three parts. It sets out initially to review the basic theory and analysis and design of engineering structures (Section I: Fundamentals), goes on to show how these processes are applied to the design, analysis, and evaluation of building, bridge, tower and shell structures (Section II: Special Structures), and finally in-depth treatments of some important topics in structural engineering are addressed and their recent developments evaluated and assessed (Section III: Special Topics).

The Handbook stresses professional applications; emphasis has been placed on ready-to-use material. It contains many formulas and tables that give immediate answers to questions arising from practical work. It also contains a brief description of the essential elements of each subject, omitting the derivations of formulas, thus enabling the reader to see the global picture and to understand the logical background of these results and to think beyond them. Traditional as well as new and innovative practices are covered including the three basic aspects of structural engineering: theoretical, experimental, and computing.

The subdivision of Sections II and III is made somewhat arbitrary, as the many subjects of the individual chapters are somewhat independent and are cross-linked to chapters in Section I in many ways. To this end, an overview of the structure, organization, and content of the book can be seen by examining the table of contents presented at the beginning of the book, while an in-depth view of a particular subject can be seen by examining the individual table of contents preceding each chapter. The references at the end of each chapter can be consulted for more detailed studies.

The chapters have been written by many authors in different countries covering structural engineering practices, research, and developments in North America, Europe, and Pacific Rim countries. This Handbook may provide a glimpse of the rapid trend in recent years toward international outsourcing of practice and competition in all dimensions of engineering. The way we practice structural engineering in North America and the way we educate students in the U.S. is rapidly changing. Thus, the book may also be used as a survey of the practice of structural engineering around the world. I wish to thank all the authors for their contributions and also to acknowledge at CRC Press, Tim Pletscher, Acquiring Editor, and Susan Fox, Senior Project Editor.

W.F. Chen
West Lafayette, IN

Editor

W.F. Chen is a George E. Goodwin Distinguished Professor of Civil Engineering and Head of the Department of Structural Engineering, School of Civil Engineering at Purdue University. He received his B.S. in civil engineering from the National Cheng-Kung University, Taiwan, in 1959, M.S. in structural engineering from Lehigh University, Bethlehem, Pennsylvania in 1963, and Ph.D. in solid mechanics from Brown University, Providence, Rhode Island in 1966.

Dr. Chen's research interests cover several areas, including constitutive modeling of engineering materials, soil and concrete plasticity, structural connections, and structural stability. He is the recipient of numerous engineering awards, including the AISC T.R. Higgins Lectureship Award, the ASCE Raymond C. Reese Research Prize, and the ASCE Shortridge Hardesty Award. He was elected to the National Academy of Engineering in 1995, and an Honorary Membership in the American Society of Civil Engineers in 1997.

Dr. Chen is a member of the Executive Committee of the Structural Stability Research Council, the Specification Committee of the American Institute of Steel Construction, and the editorial board of six technical journals. He has worked as a consultant for Exxon's Production and Research Division on offshore structures, for Skidmore, Owings and Merril on tall steel buildings, and for World Bank on the Chinese University Development Projects.

A widely respected author, Dr. Chen's works include *Limit Analysis and Soil Plasticity* (Elsevier, 1975), the two-volume *Theory of Beam-Columns* (McGraw-Hill, 1976–77), *Plasticity in Reinforced Concrete* (McGraw-Hill, 1982), *Plasticity for Structural Engineers* (Springer-Verlag, 1988), and *Stability Design of Steel Frames* (CRC Press, 1991). He is the editor of two book series, one in structural engineering and the other in civil engineering. He has authored or coauthored more than 500 papers in journals and conference proceedings. He is the author or coauthor of 18 books, has edited 12 books, and has contributed chapters to 28 other books. His more recent books are *Plastic Design and Second-Order Analysis of Steel Frames* (Springer-Verlag, 1994), the two-volume *Constitutive Equations for Engineering Materials* (Elsevier, 1994), *Stability Design of Semi-Rigid Frames* (Wiley-Interscience, 1995), and *LRFD Steel Design Using Advanced Analysis* (CRC Press, 1997). He is editor-in-chief of the *Civil Engineering Handbook* (CRC Press, 1995), winner of the Choice Outstanding Academic Book Award for 1996 – *Choice Magazine*.

Contributors

T. Balendra
Department of Civil Engineering, National
University of Singapore, Singapore

Reidar Bjorhovde
Department of Civil and Environmental
Engineering, University of Pittsburgh,
Pittsburgh, PA

O.W. Blodgett
The Lincoln Electric Company, Cleveland, OH

W. F. Chen
School of Civil Engineering, Purdue University,
West Lafayette, IN

Edoardo Cosenza
University of Naples, Napoli, Italy

D.F. Dargush
Department of Civil Engineering, State
University of New York at Buffalo, Buffalo, NY

Robert J. Dexter
Department of Civil Engineering, Lehigh
University, Bethlehem, PA

J. M. Doyle
Sargent & Lundy, LLC, Chicago, IL

Lian Duan
Division of Structures, California Department of
Transportation, Sacramento, CA

Jackson Durkee
Consulting Structural Engineer, Bethlehem, PA

Mohamed Elgaaly
Department of Civil & Architectural Engineering,
Drexel University, Philadelphia, PA

Shu-jin Fang
Sargent & Lundy, Chicago, IL

John W. Fisher
Department of Civil Engineering, Lehigh
University, Bethlehem, PA

Kenneth J. Fridley
Department of Civil & Environmental
Engineering, Washington State University,
Pullman, WA

Phillip L. Gould
Department of Civil Engineering, Washington
University, St. Louis, MO

Ami Grider
School of Civil Engineering, Purdue University,
West Lafayette, IN

Seung-Eock Kim
Department of Civil Engineering, Sejong
University, Seoul, South Korea

Jacob Kramer
Sargent & Lundy, Chicago, IL

Wilfried B. Krätzig
Ruhr-University, Bochum, Germany

Tien T. Lan
Department of Civil Engineering, Chinese
Academy of Building Research, Beijing, China

Roberto Leon
School of Civil and Environmental Engineering,
Georgia Institute of Technology, Atlanta, GA

J.Y. Richard Liew
Department of Civil Engineering, The National
University of Singapore, Singapore

E. M. Lui
Department of Civil and Environmental
Engineering, Syracuse University, Syracuse, NY

Peter W. Marshall
MHP Systems Engineering, Houston, Texas

Clarence D. Miller
Consulting Engineer, Bloomington, IN

D.K. Miller
The Lincoln Electric Company, Cleveland, OH

Julio A. Ramirez
School of Civil Engineering, Purdue University,
West Lafayette, IN

Mark Reno
Division of Structures, California Department of
Transportation, Sacramento, CA

D. V. Rosowsky
Department of Civil Engineering, Clemson
University, Clemson, SC

Subir Roy
Sargent & Lundy, Chicago, IL

Charles Scawthorn
EQE International, San Francisco, California and
Tokyo, Japan

N.E. Shanmugam
Department of Civil Engineering, The National
University of Singapore, Singapore

Maurice L. Sharp
Consultant—Aluminum Structures, Avonmore,
PA

T.T. Soong
Department of Civil Engineering, State
University of New York at Buffalo, Buffalo, NY

Shouji Toma
Department of Civil Engineering,
Hokkai-Gakuen University, Sapporo, Japan

Eiki Yamaguchi
Department of Civil Engineering, Kyushu
Institute of Technology, Kitakyusha, Japan

C.H. Yu
Department of Civil Engineering, The National
University of Singapore, Singapore

Wei-Wen Yu
Department of Civil Engineering, University of
Missouri-Rolla, Rolla, MO

Young Mook Yun
Department of Civil Engineering, Kyungpook
National University, Taegu, South Korea

Riccardo Zandonini
Department of Structural Mechanics and Design,
University of Trento, Povo, Italy

Contents

SECTION I   Fundamentals

1   Basic Theory of Plates and Elastic Stability   *Eiki Yamaguchi* *1-1*

2   Structural Analysis   *J.Y. Richard Liew, N.E. Shanmugam, and C.H. Yu* . .   *2-1*

3   Structural Steel Design   *E. M. Lui* *3-1*

4   Structural Concrete Design   *Amy Grider, Julio A. Ramirez, and
Young Mook Yun* *4-1*

5   Earthquake Engineering   *Charles Scawthorn* *5-1*

6   Composite Construction   *Edoardo Cosenza and Riccardo Zandonini* . . .   *6-1*

SECTION II   Special Structures

7   Cold-Formed Steel Structures   *Wei-Wen Yu* *7-1*

8   Aluminum Structures   *Maurice L. Sharp* *8-1*

9   Timber Structures   *Kenneth J. Fridley* *9-1*

10   Bridge Structures   *Shouji Toma, Lian Duan, and Wai-Fah Chen* *10-1*

11   Shell Structures   *Clarence D. Miller* *11-1*

12   Multistory Frame Structures   *J. Y. Richard Liew, T. Balendra, and
W. F. Chen* *12-1*

13   Space Frame Structures   *Tien T. Lan* *13-1*

14   Cooling Tower Structures   *Phillip L. Gould and Wilfried B. Krätzig* *14-1*

15   Transmission Structures   *Shu-jin Fang, Subir Roy, and Jacob Kramer* . . .   *15-1*

SECTION III   Special Topics

16   Performance-Based Seismic Design Criteria For Bridges   *Lian Duan
and Mark Reno* *16-1*

17   Effective Length Factors of Compression Members   *Lian Duan and
W.F. Chen* *17-1*

18 Stub Girder Floor Systems   *Reidar Bjorhovde* *18-1*

19 Plate and Box Girders   *Mohamed Elgaaly* *19-1*

20 Steel Bridge Construction   *Jackson Durkee* *20-1*

21 Basic Principles of Shock Loading   *O.W. Blodgett and D.K. Miller* *21-1*

22 Welded Connections   *O.W. Blodgett and D. K. Miller* *22-1*

23 Composite Connections   *Roberto Leon* *23-1*

24 Fatigue and Fracture   *Robert J. Dexter and John W. Fisher* *24-1*

25 Underground Pipe   *J. M. Doyle and S.J. Fang* *25-1*

26 Structural Reliability   *D. V. Rosowsky* *26-1*

27 Passive Energy Dissipation and Active Control   *T.T. Soong and
    G.F. Dargush* *27-1*

28 An Innnovative Design For Steel Frame Using Advanced Analysis
    Seung-Eock Kim and W. F. Chen *28-1*

29 Welded Tubular Connections—CHS Trusses   *Peter W. Marshall* *29-1*

Index *I-1*

Section I
FUNDAMENTALS

1

Basic Theory of Plates and Elastic Stability

1.1   Introduction 1-1
1.2   Plates 1-1
       Basic Assumptions • Governing Equations • Boundary Conditions • Circular Plate • Examples of Bending Problems
1.3   Stability 1-11
       Basic Concepts • Structural Instability • Columns • Thin-Walled Members • Plates

Eiki Yamaguchi
Department of Civil Engineering,
Kyushu Institute of Technology,
Kitakyusha, Japan

1.4   Defining Terms 1-26
References 1-27
Further Reading 1-27

1.1   Introduction

This chapter is concerned with basic assumptions and equations of plates and basic concepts of elastic stability. Herein, we shall illustrate the concepts and the applications of these equations by means of relatively simple examples; more complex applications will be taken up in the following chapters.

1.2   Plates

1.2.1   Basic Assumptions

We consider a continuum shown in Figure 1.1. A feature of the body is that one dimension is much smaller than the other two dimensions:

$$t << L_x, L_y \tag{1.1}$$

where t, L_x, and L_y are representative dimensions in three directions (Figure 1.1). If the continuum has this geometrical characteristic of Equation 1.1 and is flat before loading, it is called a plate. Note that a shell possesses a similar geometrical characteristic but is curved even before loading.

The characteristic of Equation 1.1 lends itself to the following assumptions regarding some stress and strain components:

$$\sigma_z = 0 \tag{1.2}$$

$$\varepsilon_z = \varepsilon_{xz} = \varepsilon_{yz} = 0 \tag{1.3}$$

We can derive the following displacement field from Equation 1.3:

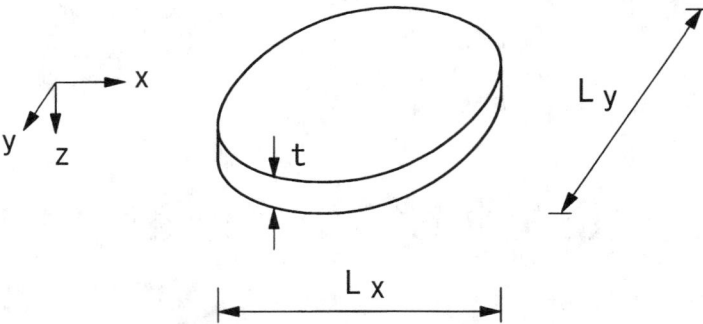

Figure 1.1   Plate.

$$u(x, y, z) = u_0(x, y) - z\frac{\partial w_0}{\partial x}$$

$$v(x, y, z) = v_0(x, y) - z\frac{\partial w_0}{\partial y} \qquad (1.4)$$

$$w(x, y, z) = w_0(x, y)$$

where u, v, and w are displacement components in the directions of x-, y-, and z-axes, respectively. As can be realized in Equation 1.4, u_0 and v_0 are displacement components associated with the plane of $z = 0$. Physically, Equation 1.4 implies that the linear filaments of the plate initially perpendicular to the middle surface remain straight and perpendicular to the deformed middle surface. This is known as the Kirchhoff hypothesis. Although we have derived Equation 1.4 from Equation 1.3 in the above, one can arrive at Equation 1.4 starting with the Kirchhoff hypothesis: the Kirchhoff hypothesis is equivalent to the assumptions of Equation 1.3.

1.2.2   Governing Equations

Strain-Displacement Relationships

Using the strain-displacement relationships in the continuum mechanics, we can obtain the following strain field associated with Equation 1.4:

$$\varepsilon_x = \frac{\partial u_0}{\partial x} - z\frac{\partial^2 w_0}{\partial x^2}$$

$$\varepsilon_y = \frac{\partial v_0}{\partial y} - z\frac{\partial^2 w_0}{\partial y^2} \qquad (1.5)$$

$$\varepsilon_{xy} = \frac{1}{2}\left(\frac{\partial u_0}{\partial y} + \frac{\partial v_0}{\partial x}\right) - z\frac{\partial^2 w_0}{\partial x \partial y}$$

This constitutes the strain-displacement relationships for the plate theory.

Equilibrium Equations

In the plate theory, equilibrium conditions are considered in terms of resultant forces and moments. This is derived by integrating the equilibrium equations over the thickness of a plate. Because of Equation 1.2, we obtain the equilibrium equations as follows:

$$\frac{\partial N_x}{\partial x} + \frac{\partial N_{xy}}{\partial y} + q_x = 0 \tag{1.6a}$$

$$\frac{\partial N_{xy}}{\partial x} + \frac{\partial N_y}{\partial y} + q_y = 0 \tag{1.6b}$$

$$\frac{\partial V_x}{\partial x} + \frac{\partial V_y}{\partial y} + q_z = 0 \tag{1.6c}$$

where N_x, N_y, and N_{xy} are in-plane stress resultants; V_x and V_y are shearing forces; and q_x, q_y, and q_z are distributed loads per unit area. The terms associated with τ_{xz} and τ_{yz} vanish, since in the plate problems the top and the bottom surfaces of a plate are subjected to only vertical loads.

We must also consider the moment equilibrium of an infinitely small region of the plate, which leads to

$$\frac{\partial M_x}{\partial x} + \frac{\partial M_{xy}}{\partial y} - V_x = 0$$

$$\frac{\partial M_{xy}}{\partial x} + \frac{\partial M_y}{\partial y} - V_y = 0 \tag{1.7}$$

where M_x and M_y are bending moments and M_{xy} is a twisting moment.

The resultant forces and the moments are defined mathematically as

$$N_x = \int_z \sigma_x dz \tag{1.8a}$$

$$N_y = \int_z \sigma_y dz \tag{1.8b}$$

$$N_{xy} = N_{yx} = \int_z \tau_{xy} dz \tag{1.8c}$$

$$V_x = \int_z \tau_{xz} dz \tag{1.8d}$$

$$V_y = \int_z \tau_{yz} dz \tag{1.8e}$$

$$M_x = \int_z \sigma_x z dz \tag{1.8f}$$

$$M_y = \int_z \sigma_y z dz \tag{1.8g}$$

$$M_{xy} = M_{yx} = \int_z \tau_{xy} z dz \tag{1.8h}$$

The resultant forces and the moments are illustrated in Figure 1.2.

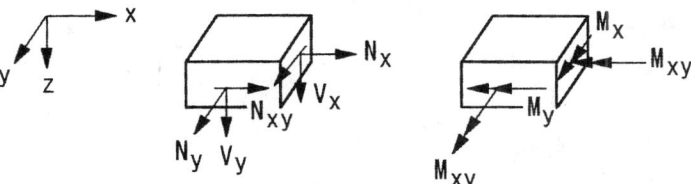

Figure 1.2 Resultant forces and moments.

Constitutive Equations

Since the thickness of a plate is small in comparison with the other dimensions, it is usually accepted that the constitutive relations for a state of plane stress are applicable. Hence, the stress-strain relationships for an isotropic plate are given by

$$
\left\{ \begin{array}{c} \sigma_x \\ \sigma_y \\ \tau_{xy} \end{array} \right\} = \frac{E}{1 - v^2} \left[\begin{array}{ccc} 1 & v & 0 \\ v & 1 & 0 \\ 0 & 0 & (1-v)/2 \end{array} \right] \left\{ \begin{array}{c} \varepsilon_x \\ \varepsilon_y \\ \gamma_{xy} \end{array} \right\}
\tag{1.9}
$$

where E and v are Young's modulus and Poisson's ratio, respectively. Using Equations 1.5, 1.8, and 1.9, the constitutive relationships for an isotropic plate in terms of stress resultants and displacements are described by

$$
N_x = \frac{Et}{1 - v^2} \left(\frac{\partial u_0}{\partial x} + v \frac{\partial v_0}{\partial y} \right)
\tag{1.10a}
$$

$$
N_y = \frac{Et}{1 - v^2} \left(\frac{\partial v_0}{\partial y} + v \frac{\partial u_0}{\partial x} \right)
\tag{1.10b}
$$

$$
N_{xy} = N_{yx} \frac{Et}{2(1 + v)} \left(\frac{\partial v_0}{\partial x} + \frac{\partial u_0}{\partial y} \right)
\tag{1.10c}
$$

$$
M_x = -D \left(\frac{\partial^2 w_0}{\partial x^2} + v \frac{\partial^2 w_0}{\partial y^2} \right)
\tag{1.10d}
$$

$$
M_y = -D \left(\frac{\partial^2 w_0}{\partial y^2} + v \frac{\partial^2 w_0}{\partial x^2} \right)
\tag{1.10e}
$$

$$
M_{xy} = M_{yx} = -(1 - v)D \frac{\partial^2 w_0}{\partial x \partial y}
\tag{1.10f}
$$

where t is the thickness of a plate and D is the flexural rigidity defined by

$$
D = \frac{Et^3}{12(1 - v^2)}
\tag{1.11}
$$

In the derivation of Equation 1.10, we have assumed that the plate thickness t is constant and that the initial middle surface lies in the plane of $Z = 0$. Through Equation 1.7, we can relate the shearing forces to the displacement.

Equations 1.6, 1.7, and 1.10 constitute the framework of a plate problem: 11 equations for 11 unknowns, i.e., N_x, N_y, N_{xy}, M_x, M_y, M_{xy}, V_x, V_y, u_0, v_0, and w_0. In the subsequent sections, we shall drop the subscript 0 that has been associated with the displacements for the sake of brevity.

In-Plane and Out-Of-Plane Problems

As may be realized in the equations derived in the previous section, the problem can be decomposed into two sets of problems which are uncoupled with each other.

1. In-plane problems
   The problem may be also called a stretching problem of a plate and is governed by

$$
\frac{\partial N_x}{\partial x} + \frac{\partial N_{xy}}{\partial y} + q_x = 0
$$

$$
\frac{\partial N_{xy}}{\partial x} + \frac{\partial N_y}{\partial y} + q_y = 0
\tag{1.6a,b}
$$

$$N_x = \frac{Et}{1-\nu^2}\left(\frac{\partial u}{\partial x} + \nu\frac{\partial v}{\partial y}\right)$$

$$N_y = \frac{Et}{1-\nu^2}\left(\frac{\partial v}{\partial y} + \nu\frac{\partial u}{\partial x}\right)$$

$$N_{xy} = N_{yx} = \frac{Et}{2(1+\nu)}\left(\frac{\partial v}{\partial x} + \frac{\partial u}{\partial y}\right) \qquad (1.10\text{a}\sim\text{c})$$

Here we have five equations for five unknowns. This problem can be viewed and treated in the same way as for a plane-stress problem in the theory of two-dimensional elasticity.

2. Out-of-plane problems
This problem is regarded as a bending problem and is governed by

$$\frac{\partial V_x}{\partial x} + \frac{\partial V_y}{\partial y} + q_z = 0 \qquad (1.6\text{c})$$

$$\frac{\partial M_x}{\partial x} + \frac{\partial M_{xy}}{\partial y} - V_x = 0$$

$$\frac{\partial M_{xy}}{\partial x} + \frac{\partial M_y}{\partial y} - V_y = 0 \qquad (1.7)$$

$$M_x = -D\left(\frac{\partial^2 w}{\partial x^2} + \frac{\partial^2 w}{\partial y^2}\right)$$

$$M_y = -D\left(\frac{\partial^2 w}{\partial y^2} + \frac{\partial^2 w}{\partial x^2}\right)$$

$$M_{xy} = M_{yx} = -(1-\nu)D\frac{\partial^2 w}{\partial x \partial y} \qquad (1.10\text{d}\sim\text{f})$$

Here are six equations for six unknowns.

Eliminating V_x and V_y from Equations 1.6c and 1.7, we obtain

$$\frac{\partial^2 M_x}{\partial x^2} + 2\frac{\partial^2 M_{xy}}{\partial x \partial y} + \frac{\partial^2 M_y}{\partial y^2} + q_z = 0 \qquad (1.12)$$

Substituting Equations 1.10d~f into the above, we obtain the governing equation in terms of displacement as

$$D\left(\frac{\partial^4 w}{\partial x^4} + 2\frac{\partial^4 w}{\partial x^2 \partial y^2} + \frac{\partial^4 w}{\partial y^4}\right) = q_z \qquad (1.13)$$

or

$$\nabla^4 w = \frac{q_z}{D} \qquad (1.14)$$

where the operator is defined as

$$\nabla^4 = \nabla^2\nabla^2$$

$$\nabla^2 = \frac{\partial^2}{\partial x^2} + \frac{\partial^2}{\partial y^2} \qquad (1.15)$$

1.2.3 Boundary Conditions

Since the in-plane problem of a plate can be treated as a plane-stress problem in the theory of two-dimensional elasticity, the present section is focused solely on a bending problem.

Introducing the n-s-z coordinate system alongside boundaries as shown in Figure 1.3, we define the moments and the shearing force as

$$
\begin{aligned}
M_n &= \int_z \sigma_n z\, dz \\
M_{ns} &= M_{sn} = \int_z \tau_{ns} z\, dz \\
V_n &= \int_z \tau_{nz}\, dz
\end{aligned}
\tag{1.16}
$$

In the plate theory, instead of considering these three quantities, we combine the twisting moment and the shearing force by replacing the action of the twisting moment M_{ns} with that of the shearing force, as can be seen in Figure 1.4. We then define the joint vertical as

$$
S_n = V_n + \frac{\partial M_{ns}}{\partial s}
\tag{1.17}
$$

The boundary conditions are therefore given in general by

$$
w = \overline{w} \ \text{or}\ S_n = \overline{S}_n
\tag{1.18}
$$

$$
-\frac{\partial w}{\partial n} = \overline{\lambda}_n \ \text{or}\ M_n = \overline{M}_n
\tag{1.19}
$$

where the quantities with a bar are prescribed values and are illustrated in Figure 1.5. These two sets of boundary conditions ensure the unique solution of a bending problem of a plate.

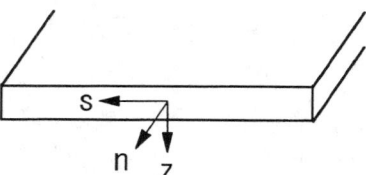

Figure 1.3 n-s-z coordinate system.

The boundary conditions for some practical cases are as follows:

1. Simply supported edge

$$
w = 0, \quad M_n = \overline{M}_n
\tag{1.20}
$$

2. Built-in edge

$$
w = 0, \quad \frac{\partial w}{\partial n} = 0
\tag{1.21}
$$

3. Free edge

$$
M_n = \overline{M}_n, \quad S_n = \overline{S}_n
\tag{1.22}
$$

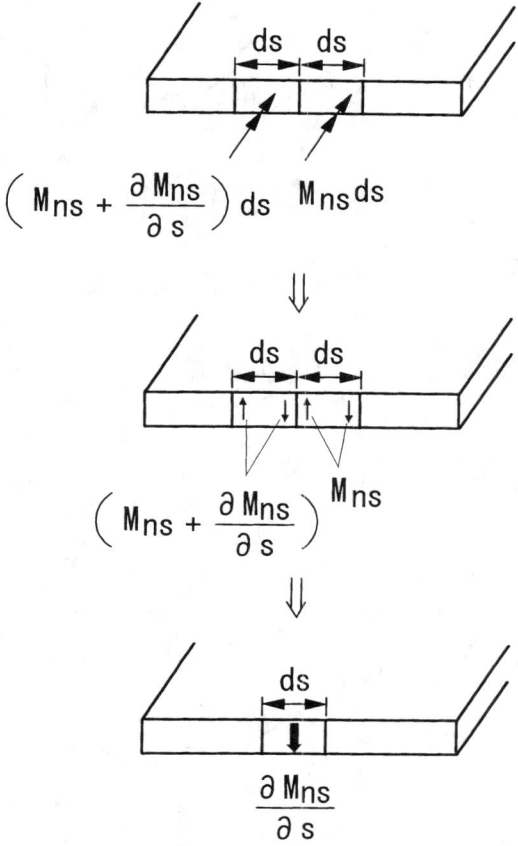

Figure 1.4 Shearing force due to twisting moment.

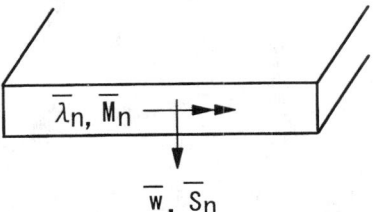

Figure 1.5 Prescribed quantities on the boundary.

4. Free corner (intersection of free edges)

At the free corner, the twisting moments cause vertical action, as can be realized is Figure 1.6. Therefore, the following condition must be satisfied:

$$-2M_{xy} = \overline{P} \tag{1.23}$$

where $\overline{P}$ is the external concentrated load acting in the Z direction at the corner.

1.2.4 Circular Plate

Governing equations in the cylindrical coordinates are more convenient when circular plates are dealt with. Through the coordinate transformation, we can easily derive the Laplacian operator in

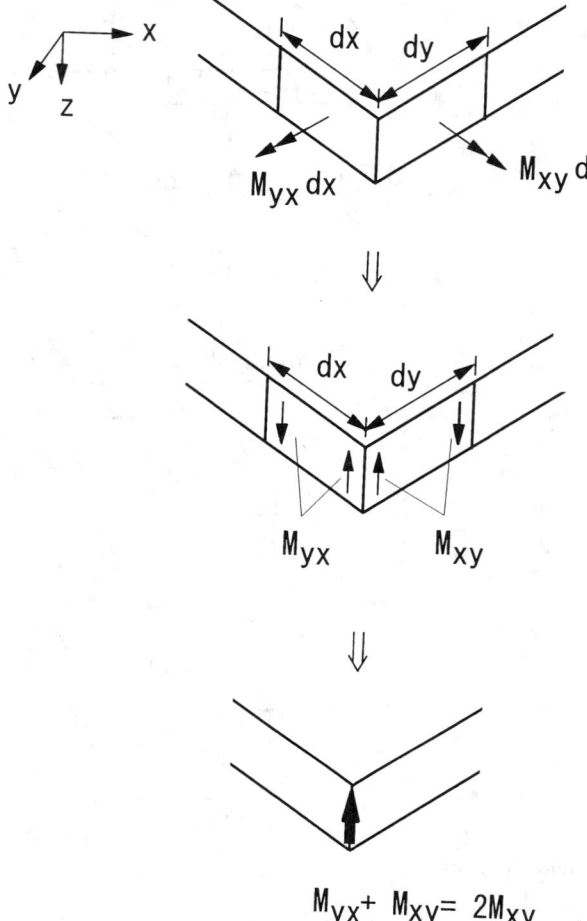

Figure 1.6   Vertical action at the corner due to twisting moment.

the cylindrical coordinates and the equation that governs the behavior of the bending of a circular plate:

$$\left(\frac{\partial^2}{\partial r^2} + \frac{1}{r}\frac{\partial}{\partial r} + \frac{1}{r^2}\frac{\partial^2}{\partial\theta^2}\right)\left(\frac{\partial^2}{\partial r^2} + \frac{1}{r}\frac{\partial}{\partial r} + \frac{1}{r^2}\frac{\partial^2}{\partial\theta^2}\right)w = \frac{q_z}{D} \qquad (1.24)$$

The expressions of the resultants are given by

$$M_r = -D\left[(1-v)\frac{\partial^2 w}{\partial r^2} + v\nabla^2 w\right]$$

$$M_\theta = -D\left[\nabla^2 w + (1-v)\frac{\partial^2 w}{\partial r^2}\right]$$

$$M_{r\theta} = M_{\theta r} = -D(1-v)\frac{\partial}{\partial r}\left(\frac{1}{r}\frac{\partial w}{\partial\theta}\right) \qquad (1.25)$$

$$S_r = V_r + \frac{1}{r}\frac{\partial M_{r\theta}}{\partial\theta}$$

$$S_\theta = V_\theta + \frac{\partial M_{r\theta}}{\partial r}$$

When the problem is axisymmetric, the problem can be simplified because all the variables are independent of θ. The governing equation for the bending behavior and the moment-deflection relationships then become

$$\frac{1}{r}\frac{d}{dr}\left[r\frac{d}{dr}\left\{\frac{1}{r}\frac{d}{dr}\left(r\frac{dw}{dr}\right)\right\}\right] = \frac{q_z}{D} \tag{1.26}$$

$$
\begin{aligned}
M_r &= D\left(\frac{d^2w}{dr^2} + \frac{v}{r}\frac{dw}{dr}\right) \\
M_\theta &= D\left(\frac{1}{r}\frac{dw}{dr} + v\frac{d^2w}{dr^2}\right) \\
M_{r\theta} &= M_{\theta r} = 0
\end{aligned} \tag{1.27}
$$

Since the twisting moment does not exist, no particular care is needed about vertical actions.

1.2.5 Examples of Bending Problems

Simply Supported Rectangular Plate Subjected to Uniform Load

A plate shown in Figure 1.7 is considered here. The governing equation is given by

$$\frac{\partial^4 w}{\partial x^4} + 2\frac{\partial^4 w}{\partial x^2 \partial y^2} + \frac{\partial^4 w}{\partial y^4} = \frac{q_0}{D} \tag{1.28}$$

in which q_0 represents the intensity of the load. The boundary conditions for the plate are

$$
\begin{aligned}
w &= 0, \quad M_x = 0 \quad \text{along } x = 0, a \\
w &= 0, \quad M_y = 0 \quad \text{along } y = 0, b
\end{aligned} \tag{1.29}
$$

Using Equation 1.10, we can rewrite the boundary conditions in terms of displacement. Furthermore, since $w = 0$ along the edges, we observe $\frac{\partial^2 w}{\partial x^2} = 0$ and $\frac{\partial^2 w}{\partial y^2} = 0$ for the edges parallel to the x and y axes, respectively, so that we may describe the boundary conditions as

$$
\begin{aligned}
w &= 0, \quad \frac{\partial^2 w}{\partial x^2} = 0 \quad \text{along } x = 0, a \\
w &= 0, \quad \frac{\partial^2 w}{\partial y^2} = 0 \quad \text{along } y = 0, b
\end{aligned} \tag{1.30}
$$

We represent the deflection in the double trigonometric series as

$$w = \sum_{m=1}^{\infty}\sum_{n=1}^{\infty} A_{mn} \sin\frac{m\pi x}{a} \sin\frac{n\pi y}{b} \tag{1.31}$$

It is noted that this function satisfies all the boundary conditions of Equation 1.30. Similarly, we express the load intensity as

$$q_0 = \sum_{m=1}^{\infty}\sum_{n=1}^{\infty} B_{mn} \sin\frac{m\pi x}{a} \sin\frac{n\pi y}{b} \tag{1.32}$$

where

$$B_{mn} = \frac{16q_0}{\pi^2 mn} \tag{1.33}$$

Substituting Equations 1.31 and 1.32 into 1.28, we can obtain the expression of A_{mn} to yield

$$w = \frac{16q_0}{\pi^6 D}\sum_{m=1}^{\infty}\sum_{n=1}^{\infty}\frac{1}{mn\left(\frac{m^2}{a^2} + \frac{n^2}{b^2}\right)^2}\sin\frac{m\pi x}{a}\sin\frac{n\pi y}{b} \tag{1.34}$$

We can readily obtain the expressions for bending and twisting moments by differentiation.

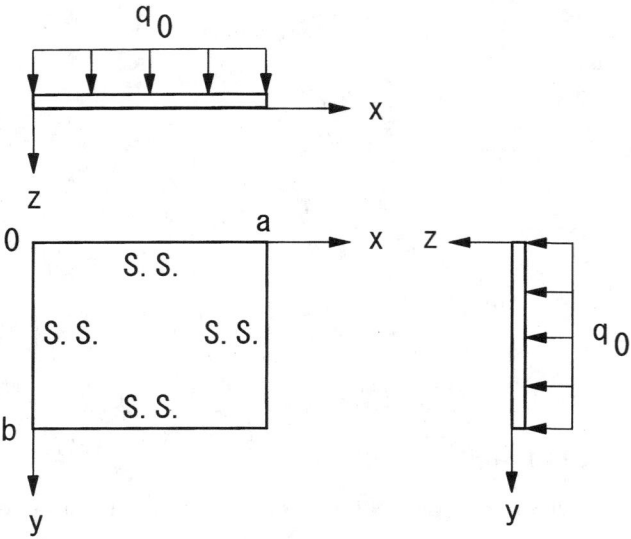

Figure 1.7   Simply supported rectangular plate subjected to uniform load.

Axisymmetric Circular Plate with Built-In Edge Subjected to Uniform Load

The governing equation of the plate shown in Figure 1.8 is

$$\frac{1}{r}\frac{d}{dr}\left[r\frac{d}{dr}\left\{\frac{1}{r}\frac{d}{dr}\left(r\frac{dw}{dr}\right)\right\}\right] = \frac{q_0}{D} \tag{1.35}$$

where q_0 is the intensity of the load. The boundary conditions for the plate are given by

$$w = \frac{dw}{dr} = 0 \text{ at } r = a \tag{1.36}$$

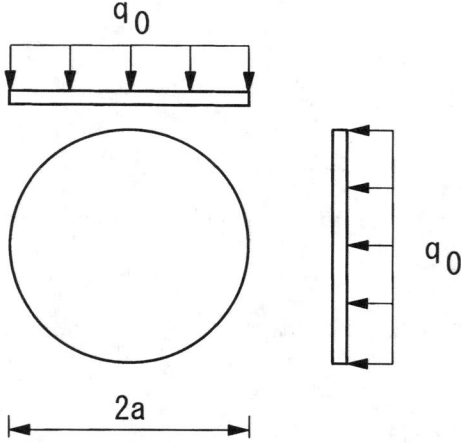

Figure 1.8   Circular plate with built-in edge subjected to uniform load.

We can solve Equation 1.35 without much difficulty to yield the following general solution:

$$w = \frac{q_0 r^4}{64D} + A_1 r^2 \ln r + A_2 \ln r + A_3 r^2 + A_4 \tag{1.37}$$

We have four constants of integration in the above, while there are only two boundary conditions of Equation 1.36. Claiming that no singularities should occur in deflection and moments, however, we can eliminate A_1 and A_2, so that we determine the solution uniquely as

$$w = \frac{q_0 a^4}{64D} \left(\frac{r^2}{a^2} - 1\right)^2 \tag{1.38}$$

Using Equation 1.27, we can readily compute the bending moments.

1.3 Stability

1.3.1 Basic Concepts

States of Equilibrium

To illustrate various forms of equilibrium, we consider three cases of equilibrium of the ball shown in Figure 1.9. We can easily see that if it is displaced slightly, the ball on the concave spherical surface will return to its original position upon the removal of the disturbance. On the other hand, the ball on the convex spherical surface will continue to move farther away from the original position if displaced slightly. A body that behaves in the former way is said to be in a state of stable equilibrium, while the latter is called unstable equilibrium. The ball on the horizontal plane shows yet another behavior: it remains at the position to which the small disturbance has taken it. This is called a state of neutral equilibrium.

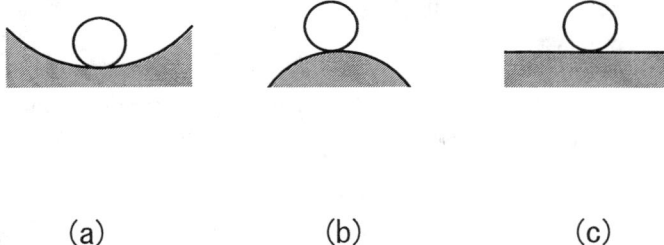

(a)   (b)   (c)

Figure 1.9   Three states of equilibrium.

For further illustration, we consider a system of a rigid bar and a linear spring. The vertical load P is applied at the top of the bar as depicted in Figure 1.10. When small disturbance θ is given, we can compute the moment about Point B M_B, yielding

$$\begin{aligned} M_B &= PL\sin\theta - (kL\sin\theta)(L\cos\theta) \\ &= L\sin\theta(P - kL\cos\theta) \end{aligned} \tag{1.39}$$

Using the fact that θ is infinitesimal, we can simplify Equation 1.39 as

$$\frac{M_B}{\theta} = L(P - kL) \tag{1.40}$$

We can claim that the system is stable when M_B acts in the opposite direction of the disturbance θ; that it is unstable when M_B and θ possess the same sign; and that it is in a state of neutral equilibrium when M_B vanishes. This classification obviously shares the same physical definition as that used in the first example (Figure 1.9). Mathematically, the classification is expressed as

$$(P - kL) \begin{cases} < 0 & : \text{stable} \\ = 0 & : \text{neutral} \\ > 0 & : \text{unstable} \end{cases} \tag{1.41}$$

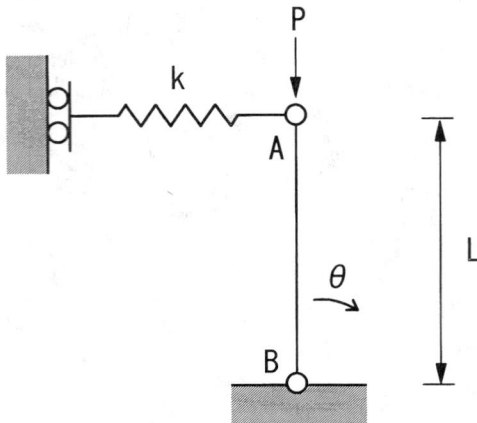

Figure 1.10   Rigid bar AB with a spring.

Equation 1.41 implies that as P increases, the state of the system changes from stable equilibrium to unstable equilibrium. The critical load is kL, at which multiple equilibrium positions, i.e., $\theta = 0$ and $\theta \neq 0$, are possible. Thus, the critical load serves also as a bifurcation point of the equilibrium path. The load at such a bifurcation is called the buckling load.

For the present system, the buckling load of kL is stability limit as well as neutral equilibrium. In general, the buckling load corresponds to a state of neutral equilibrium, but not necessarily to stability limit. Nevertheless, the buckling load is often associated with the characteristic change of structural behavior, and therefore can be regarded as the limit state of serviceability.

Linear Buckling Analysis

We can compute a buckling load by considering an equilibrium condition for a slightly deformed state. For the system of Figure 1.10, the moment equilibrium yields

$$PL \sin\theta - (kL \sin\theta)(L \cos\theta) = 0 \tag{1.42}$$

Since θ is infinitesimal, we obtain

$$L\theta(P - kL) = 0 \tag{1.43}$$

It is obvious that this equation is satisfied for any value of P if θ is zero: $\theta = 0$ is called the trivial solution. We are seeking the buckling load, at which the equilibrium condition is satisfied for $\theta \neq 0$. The trivial solution is apparently of no importance and from Equation 1.43 we can obtain the following buckling load P_C:

$$P_C = kL \tag{1.44}$$

A rigorous buckling analysis is quite involved, where we need to solve nonlinear equations even when elastic problems are dealt with. Consequently, the linear buckling analysis is frequently employed. The analysis can be justified, if deformation is negligible and structural behavior is linear before the buckling load is reached. The way we have obtained Equation 1.44 in the above is a typical application of the linear buckling analysis.

In mathematical terms, Equation 1.43 is called a characteristic equation and Equation 1.44 an eigenvalue. The linear buckling analysis is in fact regarded as an eigenvalue problem.

1.3.2   Structural Instability

Three classes of instability phenomenon are observed in structures: bifurcation, snap-through, and softening.

We have discussed a simple example of bifurcation in the previous section. Figure 1.11a depicts a schematic load-displacement relationship associated with the bifurcation: Point A is where the bifurcation takes place. In reality, due to imperfection such as the initial crookedness of a member and the eccentricity of loading, we can rarely observe the bifurcation. Instead, an actual structural behavior would be more like the one indicated in Figure 1.11a. However, the bifurcation load is still an important measure regarding structural stability and most instabilities of a column and a plate are indeed of this class. In many cases we can evaluate the bifurcation point by the linear buckling analysis.

In some structures, we observe that displacement increases abruptly at a certain load level. This can take place at Point A in Figure 1.11b; displacement increases from U_A to U_B at P_A, as illustrated by a broken arrow. The phenomenon is called snap-through. The equilibrium path of Figure 1.11b is typical of shell-like structures, including a shallow arch, and is traceable only by the finite displacement analysis.

The other instability phenomenon is the softening: as Figure 1.11c illustrates, there exists a peak load-carrying capacity, beyond which the structural strength deteriorates. We often observe this phenomenon when yielding takes place. To compute the associated equilibrium path, we need to resort to nonlinear structural analysis.

Since nonlinear analysis is complicated and costly, the information on stability limit and ultimate strength is deduced in practice from the bifurcation load, utilizing the linear buckling analysis. We shall therefore discuss the buckling loads (bifurcation points) of some structures in what follows.

1.3.3 Columns

Simply Supported Column

As a first example, we evaluate the buckling load of a simply supported column shown in Figure 1.12a. To this end, the moment equilibrium in a slightly deformed configuration is considered. Following the notation in Figure 1.12b, we can readily obtain

$$w'' + k^2 w = 0 \tag{1.45}$$

where

$$k^2 = \frac{P}{EI} \tag{1.46}$$

EI is the bending rigidity of the column. The general solution of Equation 1.45 is

$$w = A_1 \sin kx + A_2 \cos kx \tag{1.47}$$

The arbitrary constants A_1 and A_2 are to be determined by the following boundary conditions:

$$w = 0 \text{ at } x = 0 \tag{1.48a}$$
$$w = 0 \text{ at } x = L \tag{1.48b}$$

Equation 1.48a gives $A_2 = 0$ and from Equation 1.48b we reach

$$A_1 \sin kL = 0 \tag{1.49}$$

$A_1 = 0$ is a solution of the characteristic equation above, but this is the trivial solution corresponding to a perfectly straight column and is of no interest. Then we obtain the following buckling loads:

$$P_C = \frac{n^2 \pi^2 EI}{L^2} \tag{1.50}$$

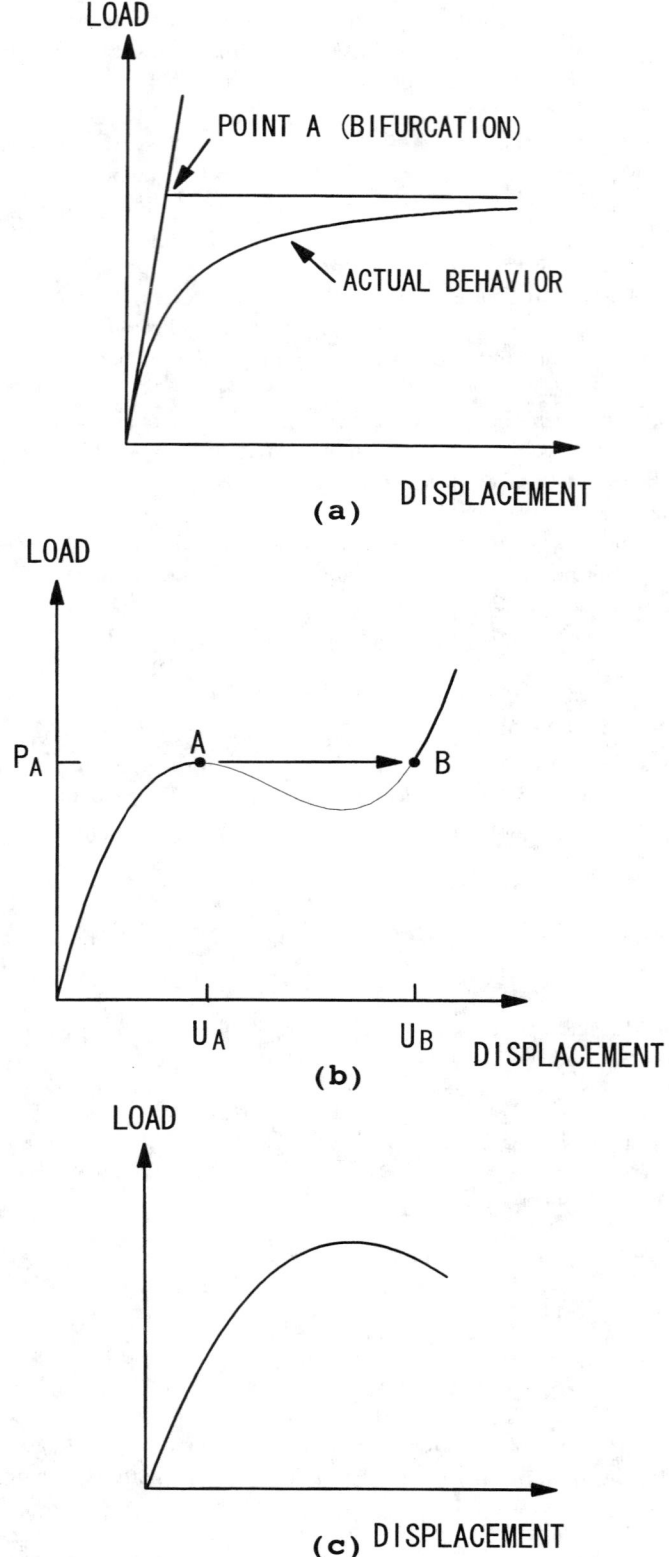

Figure 1.11   Unstable structural behaviors.

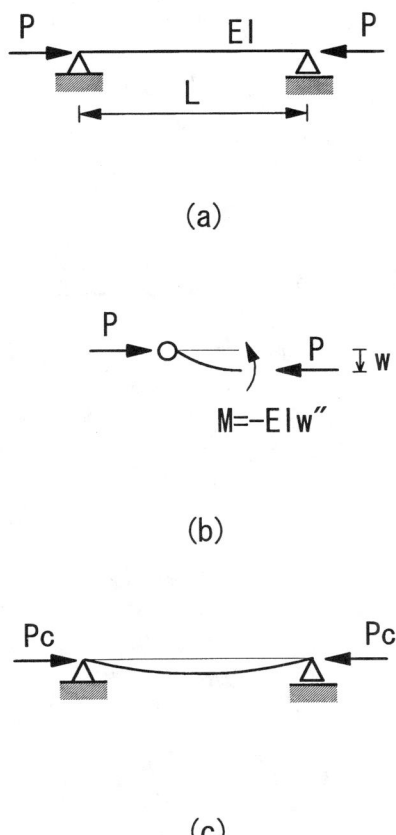

Figure 1.12 Simply-supported column.

Although n is any integer, our interest is in the lowest buckling load with $n = 1$ since it is the critical load from the practical point of view. The buckling load, thus, obtained is

$$P_C = \frac{\pi^2 EI}{L^2} \tag{1.51}$$

which is often referred to as the Euler load. From $A_2 = 0$ and Equation 1.51, Equation 1.47 indicates the following shape of the deformation:

$$w = A_1 \sin \frac{\pi x}{L} \tag{1.52}$$

This equation shows the buckled shape only, since A_1 represents the undetermined amplitude of the deflection and can have any value. The deflection curve is illustrated in Figure 1.12c.

The behavior of the simply supported column is summarized as follows: up to the Euler load the column remains straight; at the Euler load the state of the column becomes the neutral equilibrium and it can remain straight or it starts to bend in the mode expressed by Equation 1.52.

Cantilever Column

For the cantilever column of Figure 1.13a, by considering the equilibrium condition of the free body shown in Figure 1.13b, we can derive the following governing equation:

$$w'' + k^2 w = k^2 \delta \tag{1.53}$$

where δ is the deflection at the free tip. The boundary conditions are

$$
\begin{aligned}
w &= 0 \text{ at } x = 0 \\
w' &= 0 \text{ at } x = 0 \\
w &= \delta \text{ at } x = L
\end{aligned}
\tag{1.54}
$$

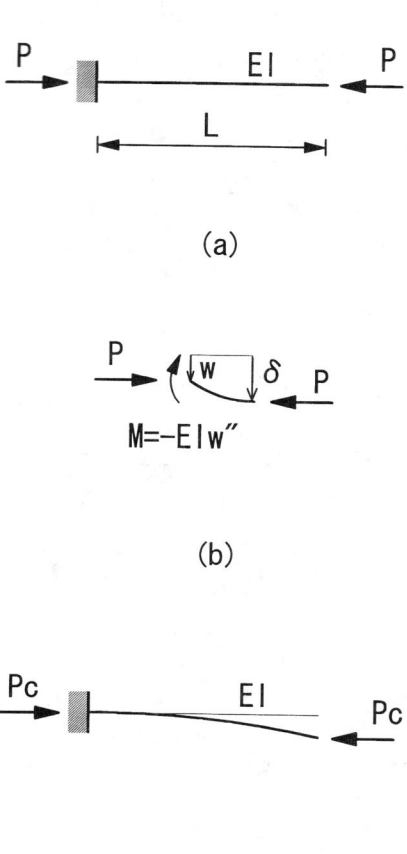

(a)

(b)

(c)

Figure 1.13   Cantilever column.

From these equations we can obtain the characteristic equation as

$$
\delta \cos kL = 0
\tag{1.55}
$$

which yields the following buckling load and deflection shape:

$$
P_C = \frac{\pi^2 EI}{4L^2}
\tag{1.56}
$$

$$
w = \delta \left(1 - \cos \frac{\pi x}{2L}\right)
\tag{1.57}
$$

The buckling mode is illustrated in Figure 1.13c. It is noted that the boundary conditions make much difference in the buckling load: the present buckling load is just a quarter of that for the simply supported column.

Higher-Order Differential Equation

We have thus far analyzed the two columns. In each problem, a second-order differential equation was derived and solved. This governing equation is problem-dependent and valid only for a particular problem. A more consistent approach is possible by making use of the governing equation for a beam-column with no laterally distributed load:

$$EIw^{IV} + Pw'' = q \tag{1.58}$$

Note that in this equation P is positive when compressive. This equation is applicable to any set of boundary conditions. The general solution of Equation 1.58 is given by

$$w = A_1 \sin kx + A_2 \cos kx + A_3 x + A_4 \tag{1.59}$$

where $A_1 \sim A_4$ are arbitrary constants and determined from boundary conditions.

We shall again solve the two column problems, using Equation 1.58.

1. Simply supported column (Figure 1.12a)
   Because of no deflection and no external moment at each end of the column, the boundary conditions are described as

$$
\begin{aligned}
w &= 0, \quad w'' = 0 \text{ at } x = 0 \\
w &= 0, \quad w'' = 0 \text{ at } x = L
\end{aligned}
\tag{1.60}
$$

From the conditions at $x = 0$, we can determine

$$A_2 = A_4 = 0 \tag{1.61}$$

Using this result and the conditions at $x = L$, we obtain

$$
\begin{bmatrix} \sin kL & L \\ -k^2 \sin kL & 0 \end{bmatrix}
\begin{Bmatrix} A_1 \\ A_3 \end{Bmatrix} =
\begin{Bmatrix} 0 \\ 0 \end{Bmatrix}
\tag{1.62}
$$

For the nontrivial solution to exist, the determinant of the coefficient matrix in Equation 1.62 must vanish, leading to the following characteristic equation:

$$k^2 L \sin kL = 0 \tag{1.63}$$

from which we arrive at the same critical load as in Equation 1.51. By obtaining the corresponding eigenvector of Equation 1.62, we can get the buckled shape of Equation 1.52.

2. Cantilever column (Figure 1.13a)
   In this column, we observe no deflection and no slope at the fixed end; no external moment and no external shear force at the free end. Therefore, the boundary conditions are

$$
\begin{aligned}
w &= 0, \quad w' = 0 \quad \text{at } x = 0 \\
w'' &= 0, \quad w''' + k^2 w' = 0 \quad \text{at } x = L
\end{aligned}
\tag{1.64}
$$

Note that since we are dealing with a slightly deformed column in the linear buckling analysis, the axial force has a transverse component, which is why P comes in the boundary condition at $x = L$.

The latter condition at $x = L$ eliminates A_3. With this and the second condition at $x = 0$, we can claim $A_1 = 0$. The remaining two conditions then lead to

$$\begin{bmatrix} 1 & 1 \\ k^2 \cos kL & 0 \end{bmatrix} \begin{Bmatrix} A_2 \\ A_4 \end{Bmatrix} = \begin{Bmatrix} 0 \\ 0 \end{Bmatrix} \tag{1.65}$$

The smallest eigenvalue and the corresponding eigenvector of Equation 1.65 coincide with the buckling load and the buckling mode that we have obtained previously in Section 1.3.3.

Effective Length

We have obtained the buckling loads for the simply supported and the cantilever columns. By either the second- or the fourth-order differential equation approach, we can compute buckling loads for a fixed-hinged column (Figure 1.14a) and a fixed-fixed column (Figure 1.14b) without much difficulty:

$$P_C = \frac{\pi^2 EI}{(0.7L)^2} \quad \text{for a fixed - hinged column}$$

$$P_C = \frac{\pi^2 EI}{(0.5L)^2} \quad \text{for a fixed - hinged column} \tag{1.66}$$

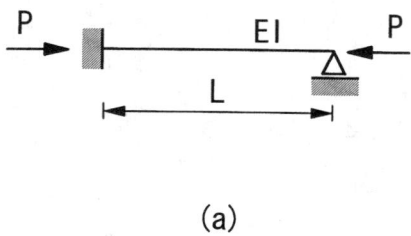

(a)

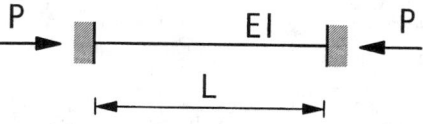

(b)

Figure 1.14   (a) Fixed-hinged column; (b) fixed-fixed column.

For all the four columns considered thus far, and in fact for the columns with any other sets of boundary conditions, we can express the buckling load in the form of

$$P_C = \frac{\pi^2 EI}{(KL)^2} \tag{1.67}$$

where KL is called the effective length and represents presumably the length of the equivalent Euler column (the equivalent simply supported column).

For design purposes, Equation 1.67 is often transformed into

$$\sigma_C = \frac{\pi^2 E}{(KL/r)^2} \tag{1.68}$$

where r is the radius of gyration defined in terms of cross-sectional area A and the moment of inertia I by

$$r = \sqrt{\frac{I}{A}} \tag{1.69}$$

For an ideal elastic column, we can draw the curve of the critical stress σ_C vs. the slenderness ratio KL/r, as shown in Figure 1.15a.

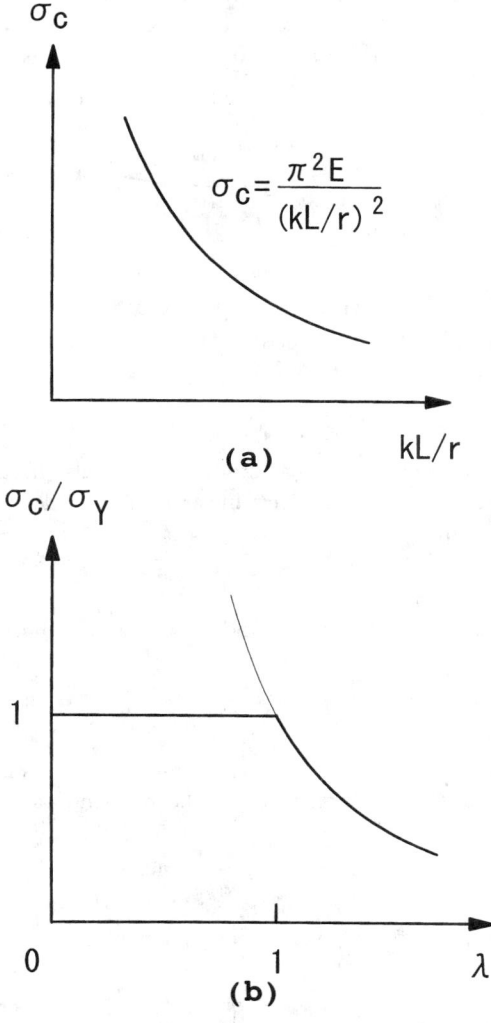

Figure 1.15   (a) Relationship between critical stress and slenderness ratio; (b) normalized relationship.

For a column of perfectly plastic material, stress never exceeds the yield stress σ_Y. For this class of column, we often employ a normalized form of Equation 1.68 as

$$\frac{\sigma_C}{\sigma_Y} = \frac{1}{\lambda^2} \tag{1.70}$$

where

$$\lambda = \frac{1}{\pi} \frac{KL}{r} \sqrt{\frac{\sigma_Y}{E}} \tag{1.71}$$

This equation is plotted in Figure 1.15b. For this column, with $\lambda < 1.0$, it collapses plastically; elastic buckling takes place for $\lambda > 1.0$.

Imperfect Columns

In the derivation of the buckling loads, we have dealt with the idealized columns; the member is perfectly straight and the loading is concentric at every cross-section. These idealizations help simplify the problem, but perfect members do not exist in the real world: minor crookedness of shape and small eccentricities of loading are always present. To this end, we shall investigate the behavior of an initially bent column in this section.

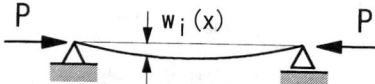

Figure 1.16   Initially bent column.

We consider a simply supported column shown in Figure 1.16. The column is initially bent and the initial crookedness w_i is assumed to be in the form of

$$w_i = a \sin \frac{\pi x}{L} \tag{1.72}$$

where a is a small value, representing the magnitude of the initial deflection at the midpoint. If we describe the additional deformation due to bending as w and consider the moment equilibrium in this configuration, we obtain

$$w'' + k^2 w = -k^2 a \sin \frac{\pi x}{L} \tag{1.73}$$

where k^2 is defined in Equation 1.46. The general solution of this differential equation is given by

$$w = A \sin \frac{\pi x}{L} + B \cos \frac{\pi x}{L} + \frac{P/P_E}{1 - P/P_E} a \sin \frac{\pi x}{L} \tag{1.74}$$

where P_E is the Euler load, i.e., $\pi^2 EI/L^2$. From the boundary conditions of Equation 1.48, we can determine the arbitrary constants A and B, yielding the following load-displacement relationship:

$$w = \frac{P/P_E}{1 - P/P_E} a \sin \frac{\pi x}{L} \tag{1.75}$$

By adding this expression to the initial deflection, we can obtain the total displacement w_t as

$$w_t = w_i + w = \frac{a}{1 - P/P_E} \sin \frac{\pi x}{L} \tag{1.76}$$

Figure 1.17 illustrates the variation of the deflection at the midpoint of this column w_m.

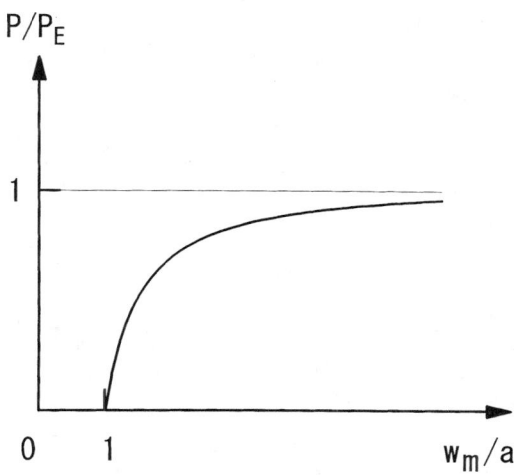

Figure 1.17 Load-displacement curve of the bent column.

Unlike the ideally perfect column, which remains straight up to the Euler load, we observe in this figure that the crooked column begins to bend at the onset of the loading. The deflection increases slowly at first, and as the applied load approaches the Euler load, the increase of the deflection is getting more and more rapid. Thus, although the behavior of the initially bent column is different from that of bifurcation, the buckling load still serves as an important measure of stability.

We have discussed the behavior of a column with geometrical imperfection in this section. However, the trend observed herein would be the same for general imperfect columns such as an eccentrically loaded column.

1.3.4 Thin-Walled Members

In the previous section, we assumed that a compressed column would buckle by bending. This type of buckling may be referred to as flexural buckling. However, a column may buckle by twisting or by a combination of twisting and bending. Such a mode of failure occurs when the torsional rigidity of the cross-section is low. Thin-walled open cross-sections have a low torsional rigidity in general and hence are susceptible of this type of buckling. In fact, a column of thin-walled open cross-section usually buckles by a combination of twisting and bending: this mode of buckling is often called the torsional-flexural buckling.

A bar subjected to bending in the plane of a major axis may buckle in yet another mode: at the critical load a compression side of the cross-section tends to bend sideways while the remainder is stable, resulting in the rotation and lateral movement of the entire cross-section. This type of buckling is referred to as lateral buckling. We need to use caution in particular, if a beam has no lateral supports and the flexural rigidity in the plane of bending is larger than the lateral flexural rigidity.

In the present section, we shall briefly discuss the two buckling modes mentioned above, both of which are of practical importance in the design of thin-walled members, particularly of open cross-section.

Torsional-Flexural Buckling

We consider a simply supported column subjected to compressive load P applied at the centroid of each end, as shown in Figure 1.18. Note that the x axis passes through the centroid of every cross-section. Taking into account that the cross-section undergoes translation and rotation as illustrated in Figure 1.19, we can derive the equilibrium conditions for the column deformed slightly by the torsional-flexural buckling

$$EI_y v^{IV} + Pv'' + Pz_s \phi'' = 0$$
$$EI_z w^{IV} + Pw'' - Py_s \phi'' = 0 \tag{1.77}$$
$$EI_w \phi^{IV} + \left(Pr_s^2 \phi'' - GJ\right)\phi'' + Pz_s v'' - Py_s w'' = 0$$

where

v, w	=	displacements in the y, z-directions, respectively
ϕ	=	rotation
EI_w	=	warping rigidity
GJ	=	torsional rigidity
y_s, z_s	=	coordinates of the shear center

and

$$
\begin{aligned}
EI_y &= \int_A y^2 \, dA \\
EI_z &= \int_A z^2 \, dA \\
r_s^2 &= \frac{I_s}{A}
\end{aligned}
\tag{1.78}
$$

where

I_s	=	polar moment of inertia about the shear center
A	=	cross-sectional area

We can obtain the buckling load by solving the eigenvalue problem governed by Equation 1.77 and the boundary conditions of

$$v = v'' = w = w'' = \phi = \phi'' = 0 \text{ at } x = 0, L \tag{1.79}$$

For doubly symmetric cross-section, the shear center coincides with the centroid. Therefore, y_s, z_s, and r_s vanish and the three equations in Equation 1.77 become independent of each other, if the cross-section of the column is doubly symmetric. In this case, we can compute three critical loads as follows:

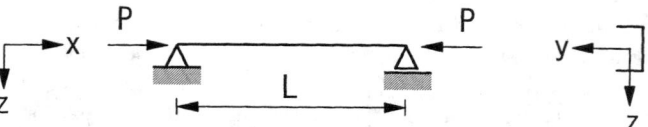

Figure 1.18   Simply-supported thin-walled column.

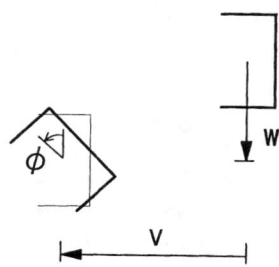

Figure 1.19   Translation and rotation of the cross-section.

$$P_{yC} = \frac{\pi^2 E I_y}{L^2} \qquad (1.80a)$$

$$P_{zC} = \frac{\pi^2 E I_z}{L^2} \qquad (1.80b)$$

$$P_{\phi C} = \frac{1}{r_s^2}\left(GJ + \frac{\pi^2 E I_w}{L^2}\right) \qquad (1.80c)$$

The first two are associated with flexural buckling and the last one with torsional buckling. For all cases, the buckling mode is in the shape of $\sin\frac{\pi x}{L}$. The smallest of the three would be the critical load of practical importance: for a relatively short column with low GJ and EI_w, the torsional buckling may take place.

When the cross-section of a column is symmetric with respect only to the y axis, we rewrite Equation 1.77 as

$$EI_y v^{IV} + Pv'' = 0 \qquad (1.81a)$$
$$EI_z w^{IV} + Pw'' - Py_s\phi'' = 0 \qquad (1.81b)$$
$$EI_w \phi^{IV} + \left(Pr_s^2 - GJ\right)\phi'' - Py_s w'' = 0 \qquad (1.81c)$$

The first equation indicates that the flexural buckling in the $x - y$ plane occurs independently and the corresponding critical load is given by P_{yC} of Equation 1.80a. The flexural buckling in the $x - z$ plane and the torsional buckling are coupled. By assuming that the buckling modes are described by $w = A_1 \sin\frac{\pi x}{L}$ and $\phi = A_2 \sin\frac{\pi x}{L}$, Equations 1.81b,c yields

$$\begin{bmatrix} P - P_{zC} & -Py_s \\ -Py_s & r_s^2\left(P - P_{\phi C}\right) \end{bmatrix}\left\{\begin{array}{c} A_1 \\ A_2 \end{array}\right\} = \left\{\begin{array}{c} 0 \\ 0 \end{array}\right\} \qquad (1.82)$$

This eigenvalue problem leads to

$$f(P) = r_s^2\left(P - P_{\phi C}\right)\left(P - P_{zC}\right) - (Py_s)^2 = 0 \qquad (1.83)$$

The solution of this quadratic equation is the critical load associated with torsional-flexural buckling. Since $f(0) = r_s^2 P_{\phi C} P_{zC} > 0$, $f(P_{\phi C}) = -(Py_s)^2 < 0$, and $f(P_{zC}) = -(Py_s)^2 < 0$, it is obvious that the critical load is lower than P_{zC} and $P_{\phi C}$. If this load is smaller than P_{yC}, then the torsional-flexural buckling will take place.

If there is no axis of symmetry in the cross-section, all the three equations in Equation 1.77 are coupled. The torsional-flexural buckling occurs in this case, since the critical load for this buckling mode is lower than any of the three loads in Equation 1.80.

Lateral Buckling

The behavior of a simply supported beam in pure bending (Figure 1.20) is investigated. The equilibrium condition for a slightly translated and rotated configuration gives governing equations for the bifurcation. For a cross-section symmetric with respect to the y axis, we arrive at the following equations:

$$EI_y v^{IV} + M\phi'' = 0 \qquad (1.84a)$$
$$EI_z w^{IV} = 0 \qquad (1.84b)$$
$$EI_w \phi^{IV} - (GJ + M\beta)\phi'' + Mv'' = 0 \qquad (1.84c)$$

where

$$\beta = \frac{1}{I_z} \int_A \left\{ y^2 + (z - z_s)^2 \right\} z \, dA \tag{1.85}$$

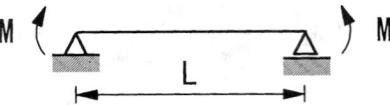

Figure 1.20   Simply supported beam in pure bending.

Equation 1.84b is a beam equation and has nothing to do with buckling. From the remaining two equations and the associated boundary conditions of Equation 1.79, we can evaluate the critical load for the lateral buckling. By assuming the bucking mode is in the shape of $\sin \frac{\pi x}{L}$ for both v and ϕ, we obtain the characteristic equation

$$M^2 - \beta P_{yC} M - r_s^2 P_{yC} P_{\phi C} = 0 \tag{1.86}$$

The smallest root of this quadratic equation is the critical load (moment) for the lateral buckling. For doubly symmetric sections where β is zero, the critical moment M_C is given by

$$M_C = \sqrt{r_s^2 P_{yC} P_{\phi C}} = \sqrt{\frac{\pi^2 E I_y}{L^2} \left(GJ + \frac{\pi^2 E I_w}{L^2} \right)} \tag{1.87}$$

1.3.5   Plates

Governing Equation

The buckling load of a plate is also obtained by the linear buckling analysis, i.e., by considering the equilibrium of a slightly deformed configuration. The plate counterpart of Equation 1.58, thus, derived is

$$D\nabla^4 w + \left(\overline{N}_x \frac{\partial^2 w}{\partial x^2} + 2\overline{N}_{xy} \frac{\partial^2 w}{\partial x \partial y} + \overline{N}_y \frac{\partial^2 w}{\partial y^2} \right) = 0 \tag{1.88}$$

The definitions of $\overline{N}_x$, $\overline{N}_y$, and $\overline{N}_{xy}$ are the same as those of N_x, N_y, and N_{xy} given in Equations 1.8a through 1.8c, respectively, except the sign; $\overline{N}_x$, $\overline{N}_y$, and $\overline{N}_{xy}$ are positive when compressive. The boundary conditions are basically the same as discussed in Section 1.2.3 except the mechanical condition in the vertical direction: to include the effect of in-plane forces, we need to modify Equation 1.18 as

$$S_n + N_n \frac{\partial w}{\partial n} + N_{ns} \frac{\partial w}{\partial s} = \overline{S}_n \tag{1.89}$$

where

$$N_n = \int_z \sigma_n dz$$

$$N_{ns} = \int_z \tau_{ns} dz \tag{1.90}$$

Simply Supported Plate

As an example, we shall discuss the buckling load of a simply supported plate under uniform compression shown in Figure 1.21. The governing equation for this plate is

$$D\nabla^4 w + \overline{N}_x \frac{\partial^2 w}{\partial x^2} = 0 \tag{1.91}$$

and the boundary conditions are

$$w = 0, \quad \frac{\partial^2 w}{\partial x^2} = 0 \quad \text{along } x = 0, a$$

$$w = 0, \quad \frac{\partial^2 w}{\partial y^2} = 0 \quad \text{along } y = 0, b \tag{1.92}$$

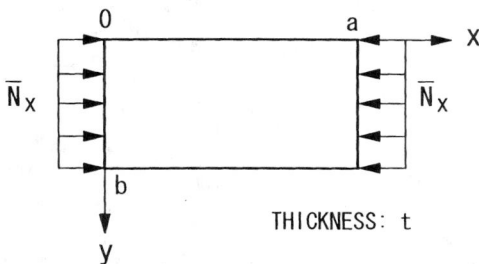

Figure 1.21   Simply supported plate subjected to uniform compression.

We assume that the solution is of the form

$$w = \sum_{m=1}^{\infty} \sum_{n=1}^{\infty} A_{mn} \sin \frac{m\pi x}{a} \sin \frac{n\pi x}{b} \tag{1.93}$$

where m and n are integers. Since this solution satisfies all the boundary conditions, we have only to ensure that it satisfies the governing equation. Substituting Equation 1.93 into 1.91, we obtain

$$A_{mn} \left[\pi^4 \left(\frac{m^2}{a^2} + \frac{n^2}{b^2} \right)^2 - \frac{\overline{N}_x}{D} \frac{m^2 \pi^2}{a^2} \right] = 0 \tag{1.94}$$

Since the trivial solution is of no interest, at least one of the coefficients a_{mn} must not be zero, the consideration of which leads to

$$\overline{N}_x = \frac{\pi^2 D}{b^2} \left(m \frac{b}{a} + \frac{n^2}{m} \frac{a}{b} \right)^2 \tag{1.95}$$

As the lowest $\overline{N}_x$ is crucial and $\overline{N}_x$ increases with n, we conclude $n = 1$: the buckling of this plate occurs in a single half-wave in the y direction and

$$\overline{N}_{xC} = \frac{k\pi^2 D}{b^2} \tag{1.96}$$

or

$$\sigma_{xC} = \frac{\overline{N}_{xC}}{t} = k \frac{\pi^2 E}{12(1 - v^2)} \frac{1}{(b/t)^2} \tag{1.97}$$

where

$$k = \left(m \frac{b}{a} + \frac{1}{m} \frac{a}{b} \right)^2 \tag{1.98}$$

Note that Equation 1.97 is comparable to Equation 1.68, and k is called the buckling stress coefficient.

The optimum value of m that gives the lowest $\overline{N}_{xC}$ depends on the aspect ratio a/b, as can be realized in Figure 1.22. For example, the optimum m is 1 for a square plate while it is 2 for a plate of $a/b = 2$. For a plate with a large aspect ratio, $k = 4.0$ serves as a good approximation. Since the aspect ratio of a component of a steel structural member such as a web plate is large in general, we can often assume k is simply equal to 4.0.

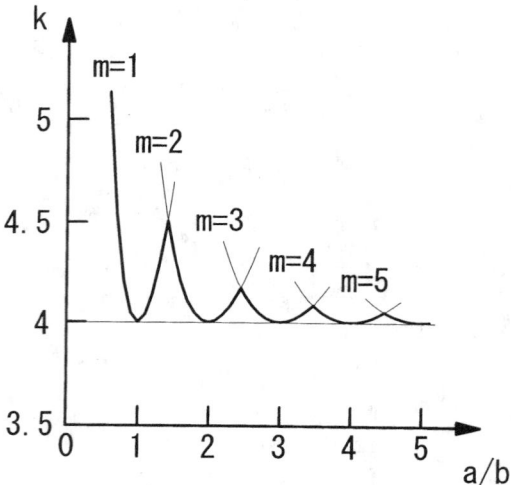

Figure 1.22   Variation of the buckling stress coefficient k with the aspect ratio a/b.

1.4 Defining Terms

The following is a list of terms as defined in the *Guide to Stability Design Criteria for Metal Structures,* 4th ed., Galambos, T.V., Structural Stability Research Council, John Wiley & Sons, New York, 1988.

Bifurcation: A term relating to the load-deflection behavior of a perfectly straight and perfectly centered compression element at critical load. Bifurcation can occur in the inelastic range only if the pattern of post-yield properties and/or residual stresses is symmetrically disposed so that no bending moment is developed at subcritical loads. At
   the critical load a member can be in equilibrium in either a straight or slightly deflected
   configuration, and a bifurcation results at a branch point in the plot of axial load vs.
   lateral deflection from which two alternative load-deflection plots are mathematically
   valid.

Braced frame: A frame in which the resistance to both lateral load and frame instability is
   provided by the combined action of floor diaphragms and structural core, shear walls,
   and/or a diagonal K brace, or other auxiliary system of bracing.

Effective length: The equivalent or effective length (KL) which, in the Euler formula for a
   hinged-end column, results in the same elastic critical load as for the framed member
   or other compression element under consideration at its theoretical critical load. The
   use of the effective length concept in the inelastic range implies that the ratio between
   elastic and inelastic critical loads for an equivalent hinged-end column is the same as
   the ratio between elastic and inelastic critical loads in the beam, frame, plate, or other
   structural element for which buckling equivalence has been assumed.

Instability: A condition reached during buckling under increasing load in a compression
   member, element, or frame at which the capacity for resistance to additional load is
   exhausted and continued deformation results in a decrease in load-resisting capacity.

Stability: The capacity of a compression member or element to remain in position and support
   load, even if forced slightly out of line or position by an added lateral force. In the elastic
   range, removal of the added lateral force would result in a return to the prior loaded
   position, unless the disturbance causes yielding to commence.

Unbraced frame: A frame in which the resistance to lateral loads is provided primarily by the
   bending of the frame members and their connections.

References

[1] Chajes, A. 1974. *Principles of Structural Stability Theory,* Prentice-Hall, Englewood Cliffs, NJ.

[2] Chen, W.F. and Atsuta, T. 1976. *Theory of Beam-Columns,* vol. 1: *In-Plane Behavior and Design,* and vol. 2: *Space Behavior and Design,* McGraw-Hill, NY.

[3] Thompson, J.M.T. and Hunt, G.W. 1973. *A General Theory of Elastic Stability,* John Wiley & Sons, London, U.K.

[4] Timoshenko, S.P. and Woinowsky-Krieger, S. 1959. *Theory of Plates and Shells,* 2nd ed., McGraw-Hill, NY.

[5] Timoshenko, S.P. and Gere, J.M. 1961. *Theory of Elastic Stability,* 2nd ed., McGraw-Hill, NY.

Further Reading

[1] Chen, W.F. and Lui, E.M. 1987. *Structural Stability Theory and Implementation,* Elsevier, New York.

[2] Chen, W.F. and Lui, E.M. 1991. *Stability Design of Steel Frames,* CRC Press, Boca Raton, FL.

[3] Galambos, T.V. 1988. *Guide to Stability Design Criteria for Metal Structures,* 4th ed., Structural Stability Research Council, John Wiley & Sons, New York.

2

Structural Analysis

2.1   Fundamental Principles 2-1
2.2   Flexural Members 2-5
2.3   Trusses 2-23
2.4   Frames 2-27
2.5   Plates 2-39
2.6   Shell 2-61
2.7   Influence Lines...................................... 2-70
2.8   Energy Methods in Structural Analysis 2-75
2.9   Matrix Methods 2-83
2.10 The Finite Element Method 2-97
2.11 Inelastic Analysis 2-118
2.12 Frame Stability...................................... 2-142
2.13 Structural Dynamic 2-165
2.14 Defining Terms 2-178
References.. 2-180
Further Reading 2-183

J.Y. Richard Liew,
N.E. Shanmugam, and
C.H. Yu
Department of Civil Engineering
The National University of
Singapore, Singapore

2.1   Fundamental Principles

Structural analysis is the determination of forces and deformations of the structure due to applied loads. *Structural design* involves the arrangement and proportioning of structures and their components in such a way that the assembled structure is capable of supporting the designed loads within the allowable limit states. An analytical model is an idealization of the actual structure. The structural model should relate the actual behavior to material properties, structural details, and loading and boundary conditions as accurately as is practicable.

All structures that occur in practice are three-dimensional. For building structures that have regular layout and are rectangular in shape, it is possible to idealize them into two-dimensional frames arranged in orthogonal directions. *Joints* in a structure are those points where two or more **members** are connected. A **truss** is a structural system consisting of members that are designed to resist only axial forces. Axially loaded members are assumed to be pin-connected at their ends. A structural system in which joints are capable of transferring end moments is called a *frame*. Members in this system are assumed to be capable of resisting **bending moment** axial force and **shear force**. A structure is said to be two dimensional or planar if all the members lie in the same plane. **Beams** are those members that are subjected to bending or flexure. They are usually thought of as being in horizontal positions and loaded with vertical loads. *Ties* are members that are subjected to axial tension only, while struts (columns or posts) are members subjected to axial compression only.

2.1.1   Boundary Conditions

A *hinge* represents a pin connection to a structural assembly and it does not allow translational
movements (Figure 2.1a). It is assumed to be frictionless and to allow rotation of a *member* with

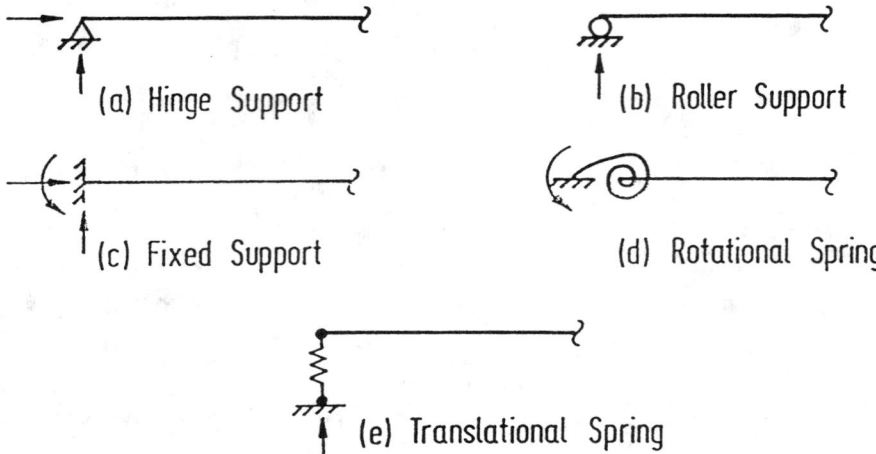

Figure 2.1   Various boundary conditions.

respect to the others. A *roller* represents a kind of support that permits the attached structural part
to rotate freely with respect to the foundation and to translate freely in the direction parallel to the
foundation surface (Figure 2.1b) No translational movement in any other direction is allowed. A
fixed support (Figure 2.1c) does not allow rotation or translation in any direction. A *rotational spring*
represents a support that provides some rotational restraint but does not provide any translational
restraint (Figure 2.1d). A *translational spring* can provide partial restraints along the direction of
deformation (Figure 2.1e).

2.1.2   Loads and Reactions

Loads may be broadly classified as *permanent loads* that are of constant magnitude and remain in
one position and *variable loads* that may change in position and magnitude. Permanent loads are
also referred to as *dead loads* which may include the self weight of the structure and other loads
such as walls, floors, roof, plumbing, and fixtures that are permanently attached to the structure.
Variable loads are commonly referred to as live or imposed loads which may include those caused by
construction operations, wind, rain, earthquakes, snow, blasts, and temperature changes in addition
to those that are movable, such as furniture and warehouse materials.

Ponding load is due to water or snow on a flat roof which accumulates faster than it runs off.
Wind loads act as pressures on windward surfaces and pressures or suctions on leeward surfaces.
Impact loads are caused by suddenly applied loads or by the vibration of moving or movable loads.
They are usually taken as a fraction of the live loads. *Earthquake loads* are those forces caused by the
acceleration of the ground surface during an earthquake.

A structure that is initially at rest and remains at rest when acted upon by applied loads is said
to be in a state of *equilibrium.* The resultant of the external loads on the body and the supporting
forces or **reactions** is zero. If a structure or part thereof is to be in equilibrium under the action of
a system of loads, it must satisfy the six static equilibrium equations, such as

$$\sum F_x = 0, \qquad \sum F_y = 0, \qquad \sum F_z = 0$$
$$\sum M_x = 0, \qquad \sum M_y = 0, \qquad \sum M_z = 0 \qquad (2.1)$$

The summation in these equations is for all the components of the forces (F) and of the moments (M) about each of the three axes x, y, and z. If a structure is subjected to forces that lie in one plane, say x-y, the above equations are reduced to:

$$\sum F_x = 0, \qquad \sum F_y = 0, \qquad \sum M_z = 0 \qquad (2.2)$$

Consider, for example, a beam shown in Figure 2.2a under the action of the loads shown. The

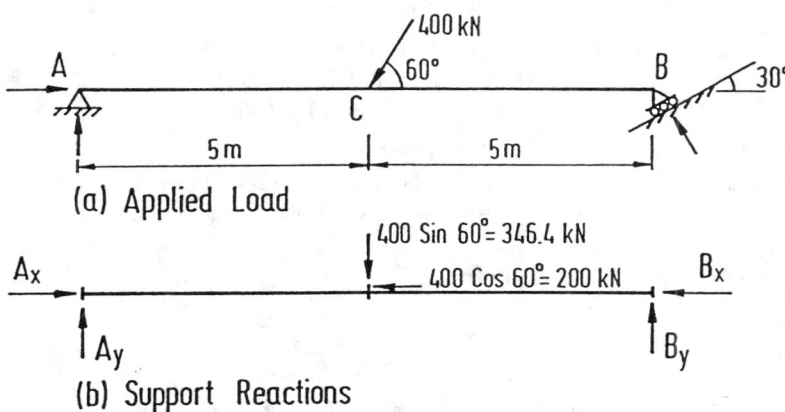

(a) Applied Load

(b) Support Reactions

Figure 2.2 Beam in equilibrium.

reaction at support B must act perpendicular to the surface on which the rollers are constrained to roll upon. The support reactions and the applied loads, which are resolved in vertical and horizontal directions, are shown in Figure 2.2b.

From geometry, it can be calculated that $B_y = \sqrt{3}B_x$. Equation 2.2 can be used to determine the magnitude of the support reactions. Taking moment about B gives

$$10A_y - 346.4x5 = 0$$

from which

$$A_y = 173.2 \text{ kN}.$$

Equating the sum of vertical forces, $\sum F_y$ to zero gives

$$173.2 + B_y - 346.4 = 0$$

and, hence, we get

$$B_y = 173.2 \text{ kN}.$$

Therefore,

$$B_x = B_y/\sqrt{3} = 100 \text{ kN}.$$

Equilibrium in the horizontal direction, $\sum F_x = 0$ gives,

$$A_x - 200 - 100 = 0$$

and, hence,

$$A_x = 300 \text{ kN}.$$

There are three unknown reaction components at a fixed end, two at a hinge, and one at a roller. If, for a particular structure, the total number of unknown reaction components equals the number of equations available, the unknowns may be calculated from the equilibrium equations, and the structure is then said to be *statically determinate externally*. Should the number of unknowns be greater than the number of equations available, the structure is *statically indeterminate externally*; if less, it is *unstable externally*. The ability of a structure to support adequately the loads applied to it is dependent not only on the number of reaction components but also on the arrangement of those components. It is possible for a structure to have as many or more reaction components than there are equations available and yet be unstable. This condition is referred to as *geometric instability*.

2.1.3 Principle of Superposition

The principle states that if the structural behavior is linearly elastic, the forces acting on a structure may be separated or divided into any convenient fashion and the structure analyzed for the separate cases. Then the final results can be obtained by adding up the individual results. This is applicable to the computation of structural responses such as moment, shear, **deflection**, etc.

However, there are two situations where the principle of superposition cannot be applied. The first case is associated with instances where the geometry of the structure is appreciably altered under load. The second case is in situations where the structure is composed of a material in which the stress is not linearly related to the strain.

2.1.4 Idealized Models

Any complex structure can be considered to be built up of simpler components called *members* or **elements**. Engineering judgement must be used to define an idealized structure such that it represents the actual structural behavior as accurately as is practically possible.

Structures can be broadly classified into three categories:

1. *Skeletal structures* consist of line elements such as a bar, beam, or column for which the length is much larger than the breadth and depth. A variety of skeletal structures can be obtained by connecting line elements together using hinged, rigid, or semi-rigid joints. Depending on whether the axes of these members lie in one plane or in different planes, these structures are termed as *plane structures* or *spatial structures*. The line elements in these structures under load may be subjected to one type of force such as axial force or a combination of forces such as shear, moment, torsion, and axial force. In the first case the structures are referred to as the truss-type and in the latter as frame-type.

2. *Plated structures* consist of elements that have length and breadth of the same order but are much larger than the thickness. These elements may be plane or curved in plane, in which case they are called plates or shells, respectively. These elements are generally used in combination with beams and bars. Reinforced concrete slabs supported on beams, box-girders, plate-girders, cylindrical shells, or water tanks are typical examples of plate and shell structures.

3. Three-dimensional solid structures have all three dimensions, namely, length, breadth, and depth, of the same order. Thick-walled hollow spheres, massive raft foundation, and dams are typical examples of solid structures.

Recent advancement in finite element methods of structural analysis and the advent of more powerful computers have enabled the economic analysis of skeletal, plated, and solid structures.

2.2 Flexural Members

One of the most common structural elements is a *beam*; it bends when subjected to loads acting transversely to its centroidal axis or sometimes by loads acting both transversely and parallel to this axis. The discussions given in the following subsections are limited to straight beams in which the centroidal axis is a straight line with shear center coinciding with the centroid of the cross-section. It is also assumed that all the loads and reactions lie in a simple plane that also contains the centroidal axis of the flexural member and the principal axis of every cross-section. If these conditions are satisfied, the beam will simply bend in the plane of loading without twisting.

2.2.1 Axial Force, Shear Force, and Bending Moment

Axial force at any transverse cross-section of a straight beam is the algebraic sum of the components acting parallel to the axis of the beam of all loads and reactions applied to the portion of the beam on either side of that cross-section. *Shear force* at any transverse cross-section of a straight beam is the algebraic sum of the components acting transverse to the axis of the beam of all the loads and reactions applied to the portion of the beam on either side of the cross-section. Bending moment at any transverse cross-section of a straight beam is the algebraic sum of the moments, taken about an axis passing through the centroid of the cross-section. The axis about which the moments are taken is, of course, normal to the plane of loading.

2.2.2 Relation Between Load, Shear, and Bending Moment

When a beam is subjected to transverse loads, there exist certain relationships between load, shear, and bending moment. Let us consider, for example, the beam shown in Figure 2.3 subjected to some arbitrary loading, p.

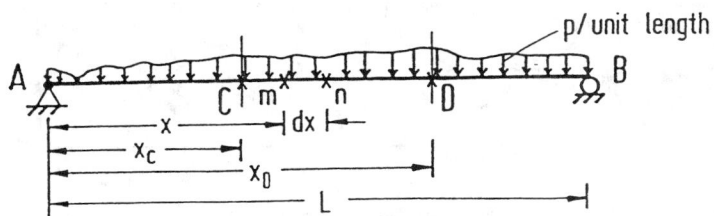

Figure 2.3   A beam under arbitrary loading.

Let S and M be the shear and bending moment, respectively, for any point 'm' at a distance x, which is measured from A, being positive when measured to the right. Corresponding values of shear and bending moment at point 'n' at a differential distance dx to the right of m are $S + dS$ and $M + dM$, respectively. It can be shown, neglecting the second order quantities, that

$$p = \frac{dS}{dx} \tag{2.3}$$

and

$$S = \frac{dM}{dx} \tag{2.4}$$

Equation 2.3 shows that the rate of change of shear at any point is equal to the intensity of load applied to the beam at that point. Therefore, the difference in shear at two cross-sections C and D is

$$S_D - S_C = \int_{x_C}^{x_D} p\,dx \tag{2.5}$$

We can write in the same way for moment as

$$M_D - M_C = \int_{x_C}^{x_D} S dx \qquad (2.6)$$

2.2.3 Shear and Bending Moment Diagrams

In order to plot the shear force and bending moment diagrams it is necessary to adopt a sign convention for these responses. A shear force is considered to be positive if it produces a clockwise moment
about a point in the free body on which it acts. A negative shear force produces a counterclockwise
moment about the point. The bending moment is taken as positive if it causes compression in the
upper fibers of the beam and tension in the lower fiber. In other words, **sagging moment** is positive
and hogging moment is negative. The construction of these diagrams is explained with an example
given in Figure 2.4.

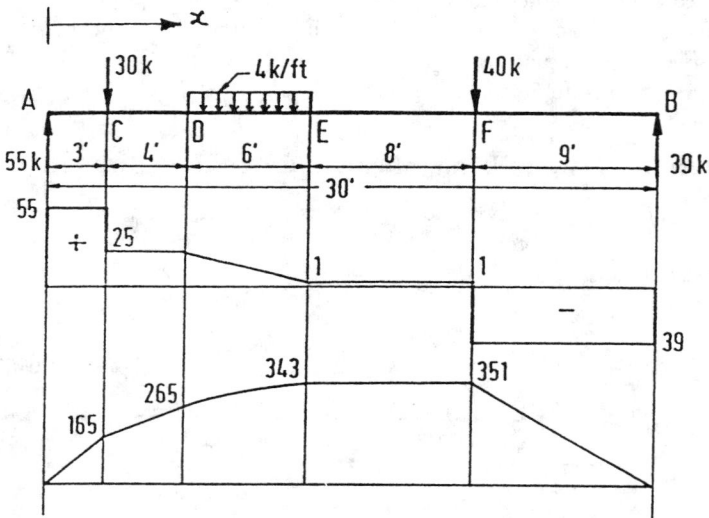

Figure 2.4   Bending moment and shear force diagrams.

The section at E of the beam is in equilibrium under the action of applied loads and internal forces
acting at E as shown in Figure 2.5. There must be an internal vertical force and internal bending

Figure 2.5   Internal forces.

moment to maintain equilibrium at Section E. The vertical force or the moment can be obtained
as the algebraic sum of all forces or the algebraic sum of the moment of all forces that lie on either
side of Section E.

The shear on a cross-section an infinitesimal distance to the right of point A is $+55$ k and, therefore, the shear diagram rises abruptly from 0 to $+55$ at this point. In the portion AC, since there is no additional load, the shear remains $+55$ on any cross-section throughout this interval, and the diagram is a horizontal as shown in Figure 2.4. An infinitesimal distance to the left of C the shear is $+55$, but an infinitesimal distance to the right of this point the 30 k load has caused the shear to be reduced to $+25$. Therefore, at point C there is an abrupt change in the shear force from $+55$ to $+25$. In the same manner, the shear force diagram for the portion CD of the beam remains a rectangle. In the portion DE, the shear on any cross-section a distance x from point D is

$$S = 55 - 30 - 4x = 25 - 4x$$

which indicates that the shear diagram in this portion is a straight line decreasing from an ordinate of $+25$ at D to $+1$ at E. The remainder of the shear force diagram can easily be verified in the same way. It should be noted that, in effect, a concentrated load is assumed to be applied at a point and, hence, at such a point the ordinate to the shear diagram changes abruptly by an amount equal to the load.

In the portion AC, the bending moment at a cross-section a distance x from point A is $M = 55x$. Therefore, the bending moment diagram starts at 0 at A and increases along a straight line to an ordinate of $+165$ k-ft at point C. In the portion CD, the bending moment at any point a distance x from C is $M = 55(x + 3) - 30x$. Hence, the bending moment diagram in this portion is a straight line increasing from 165 at C to 265 at D. In the portion DE, the bending moment at any point a distance x from D is $M = 55(x + 7) - 30(X + 4) - 4x^2/2$. Hence, the bending moment diagram in this portion is a curve with an ordinate of 265 at D and 343 at E. In an analogous manner, the remainder of the bending moment diagram can be easily constructed.

Bending moment and shear force diagrams for beams with simple boundary conditions and subject to some simple loading are given in Figure 2.6.

2.2.4 Fix-Ended Beams

When the ends of a beam are held so firmly that they are not free to rotate under the action of applied loads, the beam is known as a built-in or fix-ended beam and it is statically indeterminate. The bending moment diagram for such a beam can be considered to consist of two parts, namely the free bending moment diagram obtained by treating the beam as if the ends are simply supported and the fixing moment diagram resulting from the restraints imposed at the ends of the beam. The solution of a fixed beam is greatly simplified by considering Mohr's principles which state that:

1. the area of the fixing bending moment diagram is equal to that of the free bending moment diagram
2. the centers of gravity of the two diagrams lie in the same vertical line, i.e., are equidistant from a given end of the beam

The construction of bending moment diagram for a fixed beam is explained with an example shown in Figure 2.7. **P Q U T** is the free bending moment diagram, M_s, and **P Q R S** is the fixing moment diagram, M_i. The net bending moment diagram, M, is shown shaded. If A_s is the area of the free bending moment diagram and A_i the area of the fixing moment diagram, then from the first Mohr's principle we have $A_s = A_i$ and

$$\frac{1}{2} \times \frac{Wab}{L} \times L = \frac{1}{2}(M_A + M_B) \times L$$

$$M_A + M_B = \frac{Wab}{L} \tag{2.7}$$

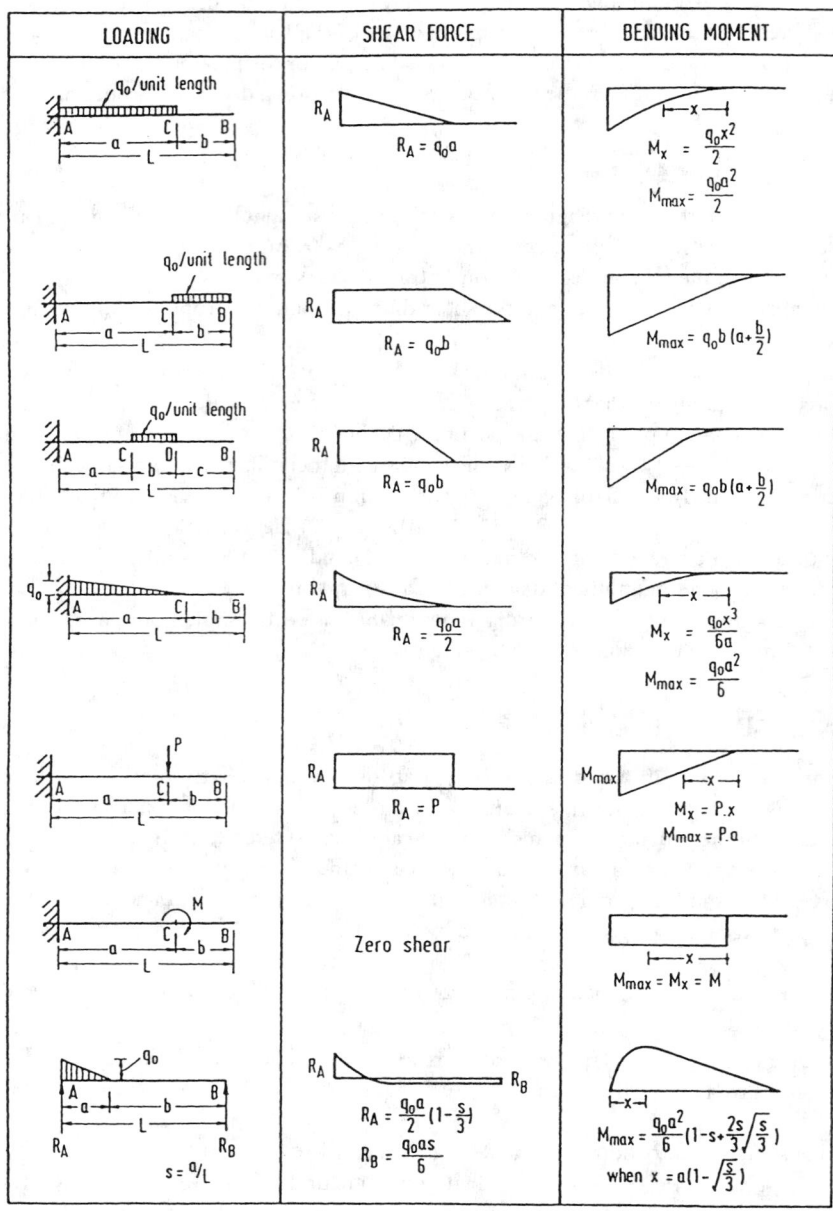

Figure 2.6   Shear force and bending moment diagrams for beams with simple boundary conditions subjected to selected loading cases.

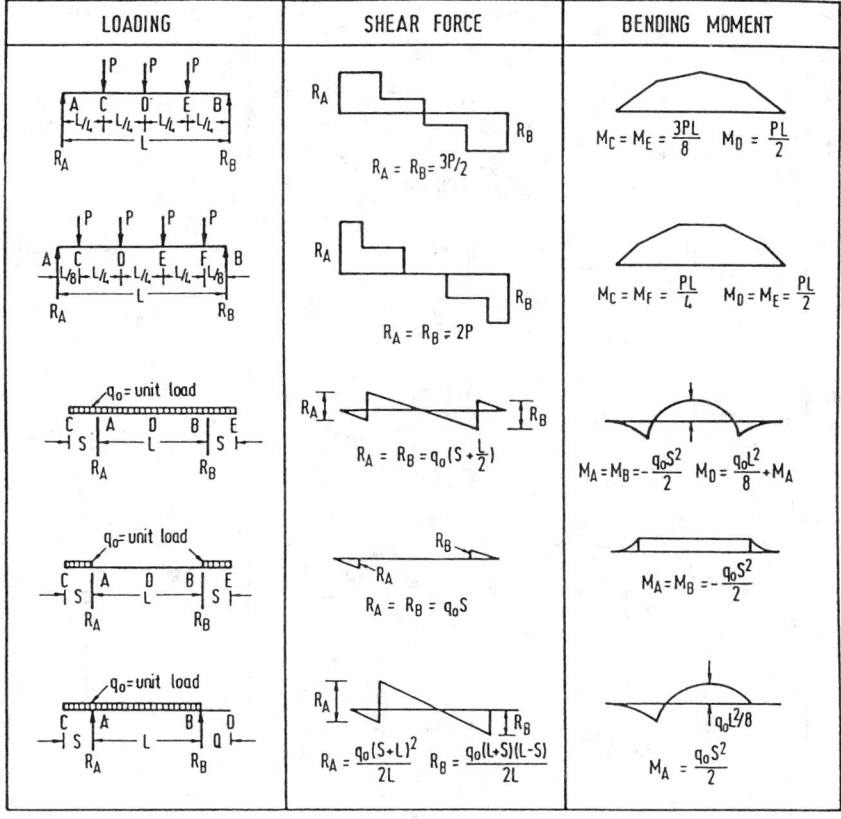

Figure 2.6 *(Continued)* Shear force and bending moment diagrams for beams with simple boundary conditions subjected to selected loading cases.

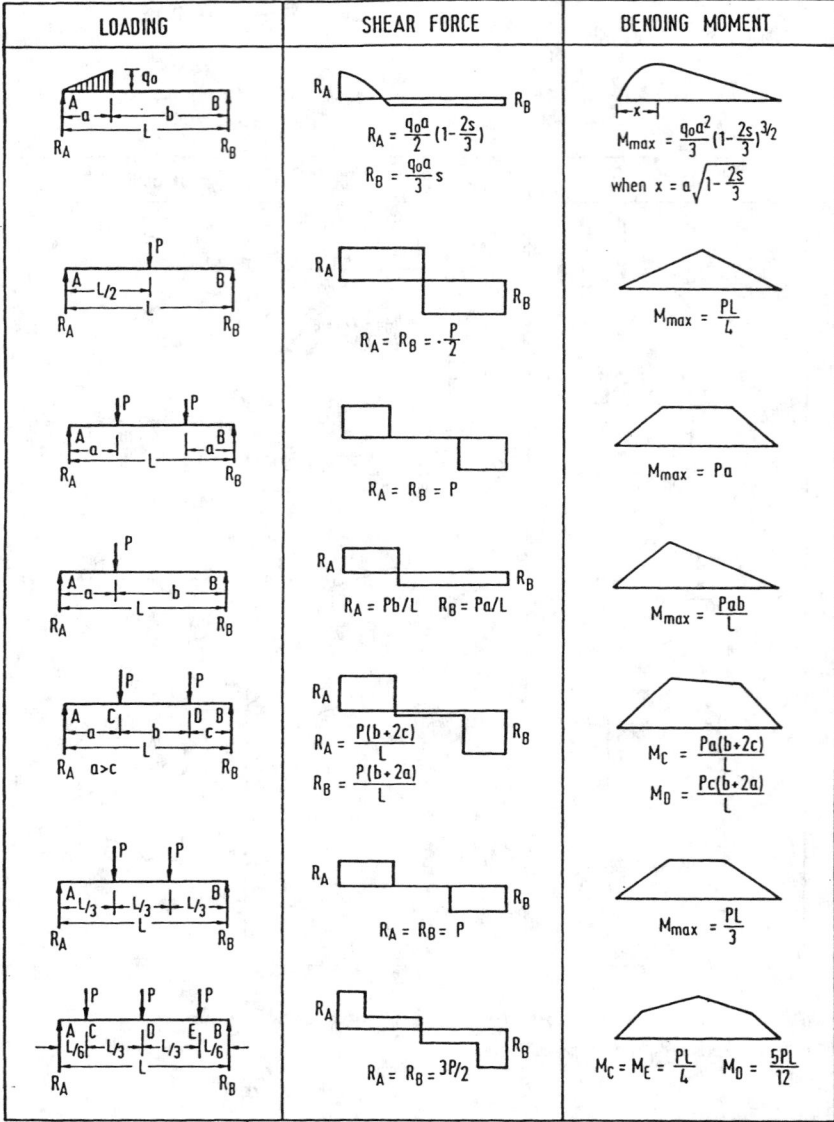

Figure 2.6   *(Continued)* Shear force and bending moment diagrams for beams with simple boundary conditions subjected to selected loading cases.

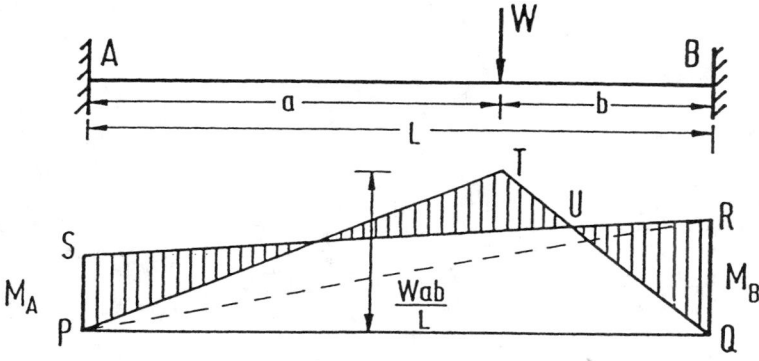

Figure 2.7 Fixed-ended beam.

From the second principle, equating the moment about A of A_s and A_i, we have,

$$M_A + 2M_B = \frac{Wab}{L^3}\left(2a^2 + 3ab + b^2\right) \tag{2.8}$$

Solving Equations 2.7 and 2.8 for M_A and M_B, we get

$$M_A = \frac{Wab^2}{L^2}$$

$$M_B = \frac{Wa^2b}{L^2}$$

Shear force can be determined once the bending moment is known. The shear force at the ends of the beam, i.e., at A and B are

$$S_A = \frac{M_A - M_B}{L} + \frac{Wb}{L}$$

$$S_B = \frac{M_B - M_A}{L} + \frac{Wa}{L}$$

Bending moment and shear force diagrams for some typical loading cases are shown in Figure 2.8.

2.2.5 Continuous Beams

Continuous beams, like fix-ended beams, are statically indeterminate. Bending moments in these beams are functions of the geometry, moments of inertia and modulus of elasticity of individual members besides the load and **span**. They may be determined by Clapeyron's Theorem of three moments, moment distribution method, or slope deflection method.

An example of a two-span continuous beam is solved by Clapeyron's Theorem of three moments. The theorem is applied to two adjacent spans at a time and the resulting equations in terms of unknown support moments are solved. The theorem states that

$$M_A L_1 + 2M_B(L_1 + L_2) + M_C L_2 = 6\left(\frac{A_1 x_1}{L_1} + \frac{A_2 x_2}{L_2}\right) \tag{2.9}$$

in which M_A, M_B, and M_C are the **hogging moment** at the supports A, B, and C, respectively, of two adjacent spans of length L_1 and L_2 (Figure 2.9); A_1 and A_2 are the area of bending moment diagrams produced by the vertical loads on the simple spans AB and BC, respectively; x_1 is the centroid of A_1 from A, and x_2 is the distance of the centroid of A_2 from C. If the beam section is

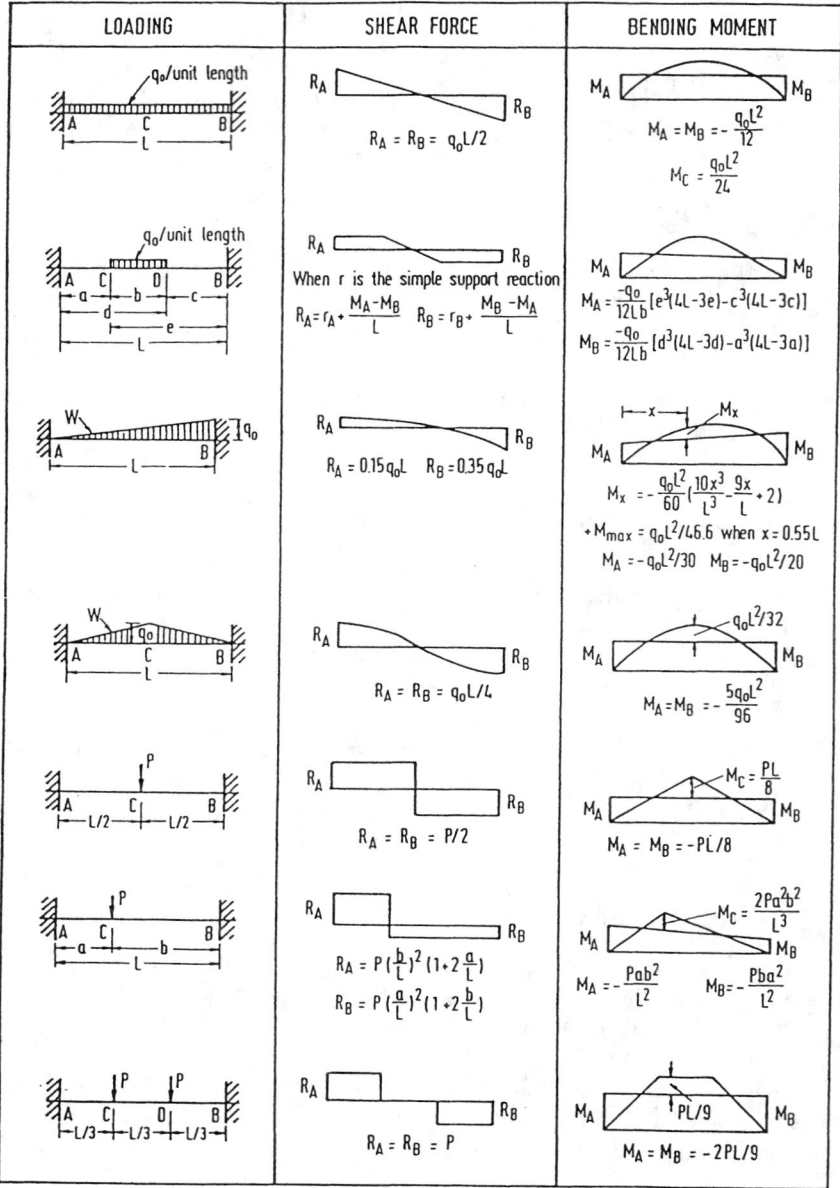

Figure 2.8   Shear force and bending moment diagrams for built-up beams subjected to typical loading cases.

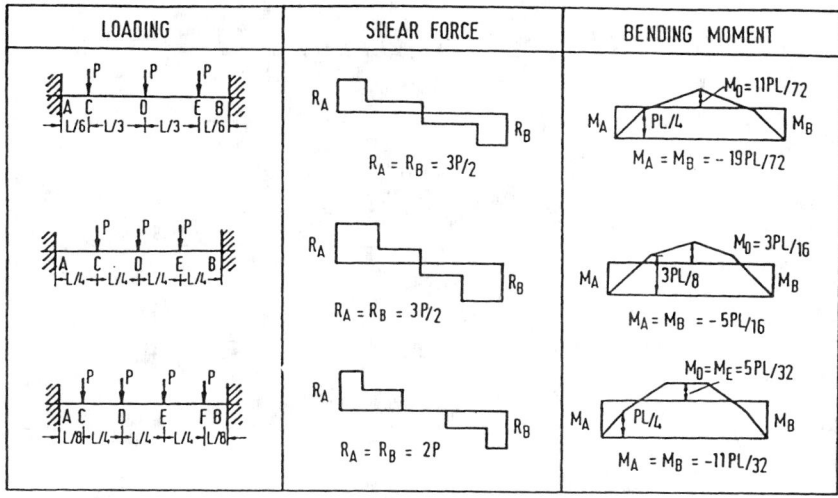

Figure 2.8 *(Continued)* Shear force and bending moment diagrams for built-up beams subjected to typical loading cases.

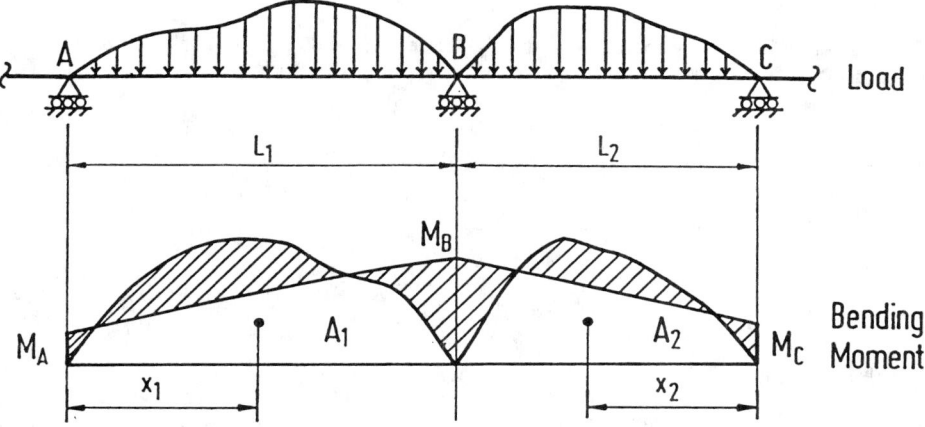

Figure 2.9 Continuous beams.

constant within a span but remains different for each of the spans, Equation 2.9 can be written as

$$M_A \frac{L_1}{I_1} + 2M_B \left(\frac{L_1}{I_1} + \frac{L_2}{I_2} \right) + M_C \frac{L_2}{I_2} = 6 \left(\frac{A_1 x_1}{L_1 I_1} + \frac{A_2 x_2}{L_2 I_2} \right) \tag{2.10}$$

in which I_1 and I_2 are the moments of inertia of beam section in span L_1 and L_2, respectively.

EXAMPLE 2.1:

The example in Figure 2.10 shows the application of this theorem. For spans AC and BC

$$M_A \times 10 + 2M_C(10 + 10) + M_B \times 10$$

$$= 6 \left[\frac{\frac{1}{2} \times 500 \times 10 \times 5}{10} + \frac{\frac{2}{3} \times 250 \times 10 \times 5}{10} \right]$$

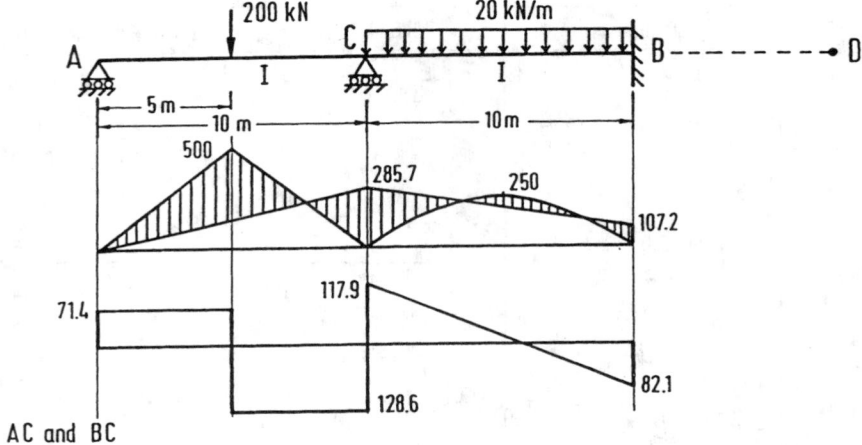

Figure 2.10   Example—continuous beam.

Since the support at A is simply supported, $M_A = 0$. Therefore,

$$4M_C + M_B = 1250 \qquad (2.11)$$

Considering an imaginary span BD on the right side of B, and applying the theorem for spans CB and BD

$$M_C \times 10 + 2M_B(10) + M_D \times 10 = 6 \times \frac{(2/3) \times 10 \times 5}{10} \times 2$$
$$M_C + 2M_B = 500 \text{ (because } M_C = M_D \text{)} \qquad (2.12)$$

Solving Equations 2.11 and 2.12 we get

$$M_B = 107.2 \text{ kNm}$$
$$M_C = 285.7 \text{ kNm}$$

Shear force at A is

$$S_A = \frac{M_A - M_C}{L} + 100 = -28.6 + 100 = 71.4 \text{ kN}$$

Shear force at C is

$$S_C = \left(\frac{M_C - M_A}{L} + 100\right) + \left(\frac{M_C - M_B}{L} + 100\right)$$
$$= (28.6 + 100) + (17.9 + 100) = 246.5 \text{ kN}$$

Shear force at B is

$$S_B = \left(\frac{M_B - M_C}{L} + 100\right) = -17.9 + 100 = 82.1 \text{ kN}$$

The bending moment and shear force diagrams are shown in Figure 2.10.

2.2.6   Beam Deflection

There are several methods for determining beam deflections: (1) moment-area method, (2) conjugate-beam method, (3) virtual work, and (4) Castigliano's second theorem, among others.

The elastic curve of a member is the shape the neutral axis takes when the member deflects under load. The inverse of the radius of curvature at any point of this curve is obtained as

$$\frac{1}{R} = \frac{M}{EI} \tag{2.13}$$

in which M is the bending moment at the point and EI is the flexural rigidity of the beam section. Since the deflection is small, $\frac{1}{R}$ is approximately taken as $\frac{d^2y}{dx^2}$, and Equation 2.13 may be rewritten as:

$$M = EI\frac{d^2y}{dx^2} \tag{2.14}$$

In Equation 2.14, y is the deflection of the beam at distance x measured from the origin of coordinate. The change in slope in a distance dx can be expressed as Mdx/EI and hence the slope in a beam is obtained as

$$\theta_B - \theta_A = \int_A^B \frac{M}{EI}dx \tag{2.15}$$

Equation 2.15 may be stated as the change in slope between the tangents to the elastic curve at two points is equal to the area of the M/EI diagram between the two points.

Once the change in slope between tangents to the elastic curve is determined, the deflection can be obtained by integrating further the slope equation. In a distance dx the neutral axis changes in direction by an amount $d\theta$. The deflection of one point on the beam with respect to the tangent at another point due to this angle change is equal to $d\delta = xd\theta$, where x is the distance from the point at which deflection is desired to the particular differential distance.

To determine the total deflection from the tangent at one point A to the tangent at another point B on the beam, it is necessary to obtain a summation of the products of each $d\theta$ angle (from A to B) times the distance to the point where deflection is desired, or

$$\delta_B - \delta_A = \int_A^B \frac{Mx\,dx}{EI} \tag{2.16}$$

The deflection of a tangent to the elastic curve of a beam with respect to a tangent at another point is equal to the moment of M/EI diagram between the two points, taken about the point at which deflection is desired.

Moment Area Method

Moment area method is most conveniently used for determining slopes and deflections for beams in which the direction of the tangent to the elastic curve at one or more points is known, such as **cantilever** beams, where the tangent at the fixed end does not change in slope. The method is applied easily to beams loaded with concentrated loads because the moment diagrams consist of straight lines. These diagrams can be broken down into single triangles and rectangles. Beams supporting uniform loads or uniformly varying loads may be handled by integration. Properties of some of the shapes of $\frac{M}{EI}$ diagrams designers usually come across are given in Figure 2.11.

It should be understood that the slopes and deflections that are obtained using the moment area theorems are with respect to tangents to the elastic curve at the points being considered. The theorems do not directly give the slope or deflection at a point in the beam as compared to the horizontal axis (except in one or two special cases); they give the change in slope of the elastic curve from one point to another or the deflection of the tangent at one point with respect to the tangent at another point. There are some special cases in which beams are subjected to several concentrated loads or the combined action of concentrated and uniformly distributed loads. In such cases it is advisable to separate the concentrated loads and uniformly distributed loads and the moment area method can be applied separately to each of these loads. The final responses are obtained by the principle of superposition.

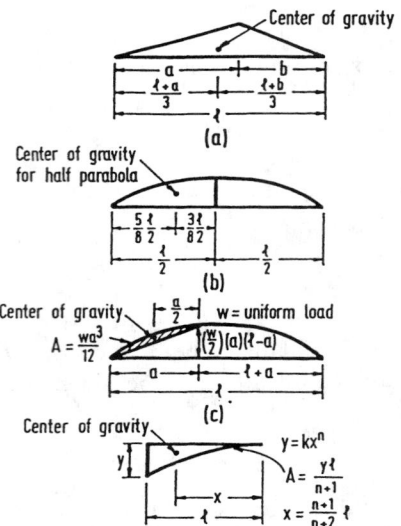

Figure 2.11   Typical M/EI diagram.

For example, consider a simply supported beam subjected to uniformly distributed load q as shown in Figure 2.12. The tangent to the elastic curve at each end of the beam is inclined. The

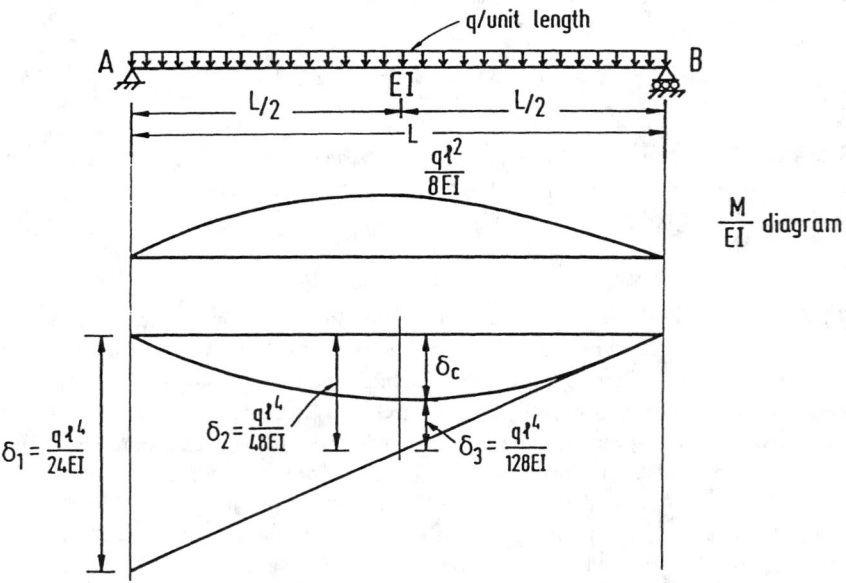

Figure 2.12   Deflection-simply supported beam under UDL.

deflection δ_1 of the tangent at the left end from the tangent at the right end is found as $ql^4/24EI$. The distance from the original chord between the supports and the tangent at right end, δ_2, can be computed as $ql^4/48EI$. The deflection of a tangent at the center from a tangent at right end, δ_3, is determined in this step as $\frac{ql^4}{128EI}$. The difference between δ_2 and δ_3 gives the centerline deflection as $\frac{5}{384}\frac{ql^4}{EI}$.

2.2.7 Curved Flexural Members

The flexural formula is based on the assumption that the beam to which bending moment is applied is initially straight. Many members, however, are curved before a bending moment is applied to them. Such members are called curved beams. It is important to determine the effect of initial curvature of a beam on the stresses and deflections caused by loads applied to the beam in the plane of initial curvature. In the following discussion, all the conditions applicable to straight-beam formula are assumed valid except that the beam is initially curved.

Let the curved beam DOE shown in Figure 2.13 be subjected to the loads Q. The surface in which the fibers do not change in length is called the neutral surface. The total deformations of the fibers between two normal sections such as AB and $A_1 B_1$ are assumed to vary proportionally with the distances of the fibers from the neutral surface. The top fibers are compressed while those at the bottom are stretched, i.e., the plane section before bending remains plane after bending.

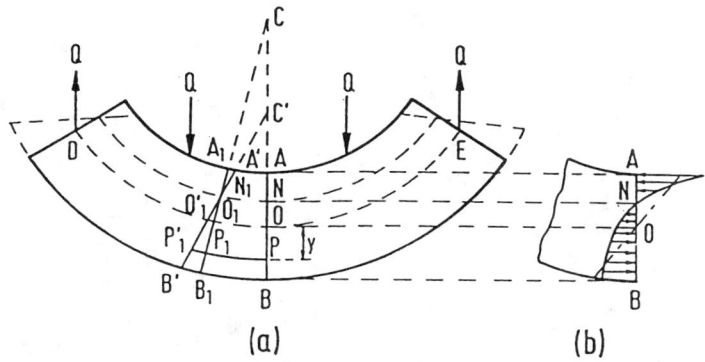

Figure 2.13   Bending of curved beams.

In Figure 2.13 the two lines AB and $A_1 B_1$ are two normal sections of the beam before the loads are applied. The change in the length of any fiber between these two normal sections after bending is represented by the distance along the fiber between the lines $A_1 B_1$ and $A'B'$; the neutral surface is represented by NN_1, and the stretch of fiber PP_1 is $P1P_1'$, etc. For convenience it will be assumed that the line AB is a line of symmetry and does not change direction.

The total deformations of the fibers in the curved beam are proportional to the distances of the fibers from the neutral surface. However, the strains of the fibers are not proportional to these distances because the fibers are not of equal length. Within the elastic limit the stress on any fiber in the beam is proportional to the strain of the fiber, and hence the elastic stresses in the fibers of a curved beam are not proportional to the distances of the fibers from the neutral surface. The resisting moment in a curved beam, therefore, is not given by the expression $\sigma I/c$. Hence, the neutral axis in a curved beam does not pass through the centroid of the section. The distribution of stress over the section and the relative position of the neutral axis are shown in Figure 2.13b; if the beam were straight, the stress would be zero at the centroidal axis and would vary proportionally with the distance from the centroidal axis as indicated by the dot-dash line in the figure. The stress on a normal section such as AB is called the circumferential stress.

Sign Conventions

The bending moment M is positive when it decreases the radius of curvature, and negative when it increases the radius of curvature; y is positive when measured toward the convex side of the beam, and negative when measured toward the concave side, that is, toward the center of curvature. With these sign conventions, σ is positive when it is a tensile stress.

Circumferential Stresses

Figure 2.14 shows a free body diagram of the portion of the body on one side of the section; the equations of equilibrium are applied to the forces acting on this portion. The equations obtained are

$$\sum F_z = 0 \text{ or } \int \sigma da = 0 \tag{2.17}$$

$$\sum M_z = 0 \text{ or } M = \int y\sigma da \tag{2.18}$$

Figure 2.15 represents the part ABB_1A_1 of Figure 2.13a enlarged; the angle between the two sections AB and A_1B_1 is $d\theta$. The bending moment causes the plane A_1B_1 to rotate through an angle $\Delta d\theta$, thereby changing the angle this plane makes with the plane BAC from $d\theta$ to $(d\theta + \Delta d\theta)$;

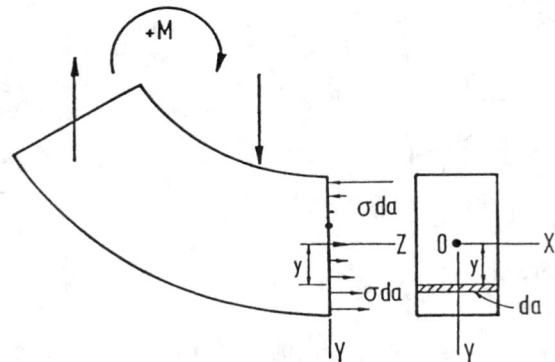

Figure 2.14   Free-body diagram of curved beam segment.

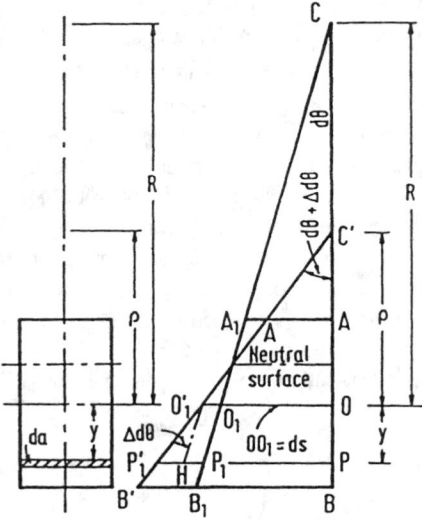

Figure 2.15   Curvature in a curved beam.

the center of curvature is changed from C to C', and the distance of the centroidal axis from the center of curvature is changed from R to ρ. It should be noted that y, R, and ρ at any section are measured from the centroidal axis and not from the neutral axis.

It can be shown that the bending stress σ is given by the relation

$$\sigma = \frac{M}{aR}\left(1 + \frac{1}{Z}\frac{y}{R+y}\right) \tag{2.19}$$

in which

$$Z = -\frac{1}{a}\int \frac{y}{R+y}da$$

σ is the tensile or compressive (circumferential) stress at a point at the distance y from the centroidal axis of a transverse section at which the bending moment is M; R is the distance from the centroidal axis of the section to the center of curvature of the central axis of the unstressed beam; a is the area of the cross-section; Z is a property of the cross-section, the values of which can be obtained from the expressions for various areas given in Table 2.1. Detailed information can be obtained from [51].

EXAMPLE 2.2:

The bent bar shown in Figure 2.16 is subjected to a load $P = 1780$ N. Calculate the circumferential

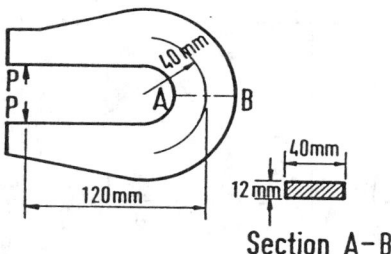

Figure 2.16   Bent bar.

stress at A and B assuming that the elastic strength of the material is not exceeded. We know from Equation 2.19

$$\sigma = \frac{P}{a} + \frac{M}{aR}\left(1 + \frac{1}{Z}\frac{y}{R+y}\right)$$

in which

a = area of rectangular section $= 40 \times 12 = 480$ mm^2
R = 40 mm
y_A = -20
y_B = $+20$
P = 1780 N
M = $-1780 \times 120 = -213600$ N mm

From Table 2.1, for rectangular section

$$Z = -1 + \frac{R}{h}\left[\log_e \frac{R+c}{R-c}\right]$$
$$h = 40 \text{ mm}$$
$$c = 20 \text{ mm}$$

TABLE 2.1   Analytical Expressions for Z

	$$Z = \frac{1}{4}\left(\frac{c}{R}\right)^2 + \frac{1}{8}\left(\frac{c}{R}\right)^4 + \frac{5}{64}\left(\frac{c}{R}\right)^6 + \frac{7}{128}\left(\frac{c}{R}\right)^8 + \cdots$$ $$Z = -1 + 2\left(\frac{R}{c}\right)^2 - 2\left(\frac{R}{c}\right)\sqrt{\left(\frac{R}{c}\right)^2 - 1}$$
	$$Z = \frac{1}{3}\left(\frac{c}{R}\right)^2 + \frac{1}{5}\left(\frac{c}{R}\right)^4 + \frac{1}{7}\left(\frac{c}{R}\right)^6 + \cdots$$ $$Z = -1 + \frac{R}{h}\left[\log_e\left(\frac{R+c}{R-c}\right)\right]$$
	$$Z = -1 + \frac{R}{ah}\left\{[b_1 h + (R + c_1)(b - b_1)]\log_e\left(\frac{R+c_1}{R-c_2}\right) - (b - b_1)h\right\}$$ $$Z = -1 + \frac{2R}{(b+b_1)h}\left\{\left[b_1 + \frac{b-b_1}{h}(R+c_1)\right]\log_e\left(\frac{R+c_1}{R-c_2}\right) - (b-b_1)\right\}$$
	$$Z = -1 + 2\frac{R}{h^2}\left[(R+c_1)\log_e\left(\frac{R+c_1}{R-c_2}\right) - h\right]$$
	$$Z = \frac{1}{4}\left(\frac{c}{R}\right)^2 + \frac{1}{8}\left(\frac{c}{R}\right)^4 + \frac{5}{64}\left(\frac{c}{R}\right)^6 + \frac{7}{128}\left(\frac{c}{R}\right)^8 + \cdots$$ $$Z = -1 + 2\left(\frac{R}{c}\right)^2 - 2\left(\frac{R}{c}\right)\sqrt{\left(\frac{R}{c}\right)^2 - 1}$$

TABLE 2.1   Analytical Expressions for Z *(continued)*

	$$Z = -1 + \frac{2R}{c_2^2 - c_1^2}\left[\sqrt{R^2 - c_1^2} - \sqrt{R^2 - c_2^2}\right]$$
	$$Z = -1 + \frac{1}{bc_2 - b_1 c_1}\left\{bc_2\left[2\left(\frac{R}{c_2}\right)^2 - 2\left(\frac{R}{c_2}\right)\sqrt{\left(\frac{R}{c_2}\right)^2 - 1}\right]\right.$$ $$\left. - b_1 c_1\left[2\left(\frac{R}{c_1}\right)^2 - 2\left(\frac{R}{c_1}\right)\sqrt{\left(\frac{R}{c_1}\right)^2 - 1}\right]\right\}$$
	$$Z = -1 + \frac{R}{a}\,[b_1\log_e(R + c_1) + (t - b_1)\log_e(R + c_4)$$ $$+ (b - t)\log_e(R - c_3) - b\log_e(R - c_2)]$$
	The value of Z for each of these three sections may be found from the expression above by making $$b_1 = b, \quad c_2 = c_1, \quad \text{and} \quad c_3 = c_4$$ $$Z = -1 + \frac{R}{a}\left[b\log_e\left(\frac{R + c_2}{R - c_2}\right) + (t - b)\log_e\left(\frac{R + c_1}{R - c_1}\right)\right]$$ $$\text{Area} = a = 2[(t - b)c_1 + bc_2]$$

TABLE 2.1   Analytical Expressions for Z *(continued)*

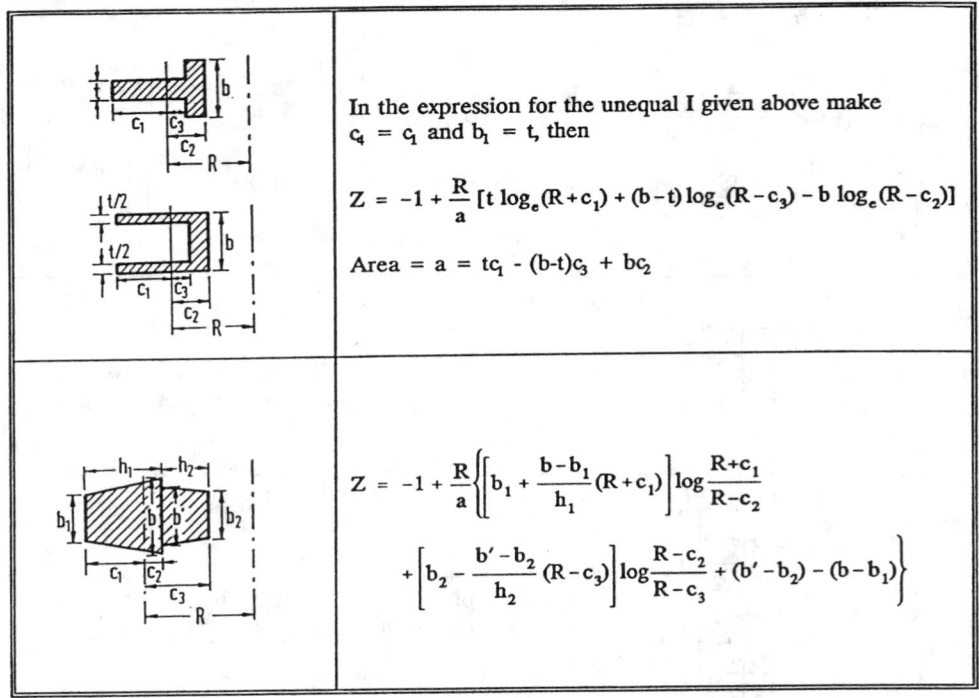

In the expression for the unequal I given above make $c_4 = c_1$ and $b_1 = t$, then

$$Z = -1 + \frac{R}{a} [t \log_e (R + c_1) + (b - t) \log_e (R - c_3) - b \log_e (R - c_2)]$$

Area $= a = tc_1 - (b-t)c_3 + bc_2$

$$Z = -1 + \frac{R}{a} \left\{ \left[b_1 + \frac{b - b_1}{h_1} (R + c_1) \right] \log \frac{R + c_1}{R - c_2} \right.$$

$$\left. + \left[b_2 - \frac{b' - b_2}{h_2} (R - c_3) \right] \log \frac{R - c_2}{R - c_3} + (b' - b_2) - (b - b_1) \right\}$$

From Seely, F.B. and Smith, J.O., *Advanced Mechanics of Materials,* John Wiley & Sons, New York, 1952. With permission.

Hence,

$$Z = -1 + \frac{40}{40}\left[\log_e \frac{40+20}{40-20}\right] = 0.0986$$

Therefore,

$$\sigma_A = \frac{1780}{480} + \frac{-213600}{480\times40}\left(1 + \frac{1}{0.0986}\frac{-20}{40-20}\right) = 105.4 \text{ N mm}^2 \text{ (tensile)}$$

$$\sigma_B = \frac{1780}{480} + \frac{-213600}{480\times40}\left(1 + \frac{1}{0.0986}\frac{20}{40+20}\right) = -45 \text{ N mm}^2 \text{ (compressive)}$$

2.3   Trusses

A structure that is composed of a number of bars pin connected at their ends to form a stable framework is called a truss. If all the bars lie in a plane, the structure is a planar truss. It is generally assumed that loads and reactions are applied to the truss only at the joints. The centroidal axis of each member is straight, coincides with the line connecting the joint centers at each end of the member, and lies in a plane that also contains the lines of action of all the loads and reactions. Many truss structures are three dimensional in nature and a complete analysis would require consideration of the full spatial interconnection of the members. However, in many cases, such as bridge structures and simple roof systems, the three-dimensional framework can be subdivided into planar components for analysis as planar trusses without seriously compromising the accuracy of the results. Figure 2.17 shows some typical idealized planar truss structures.

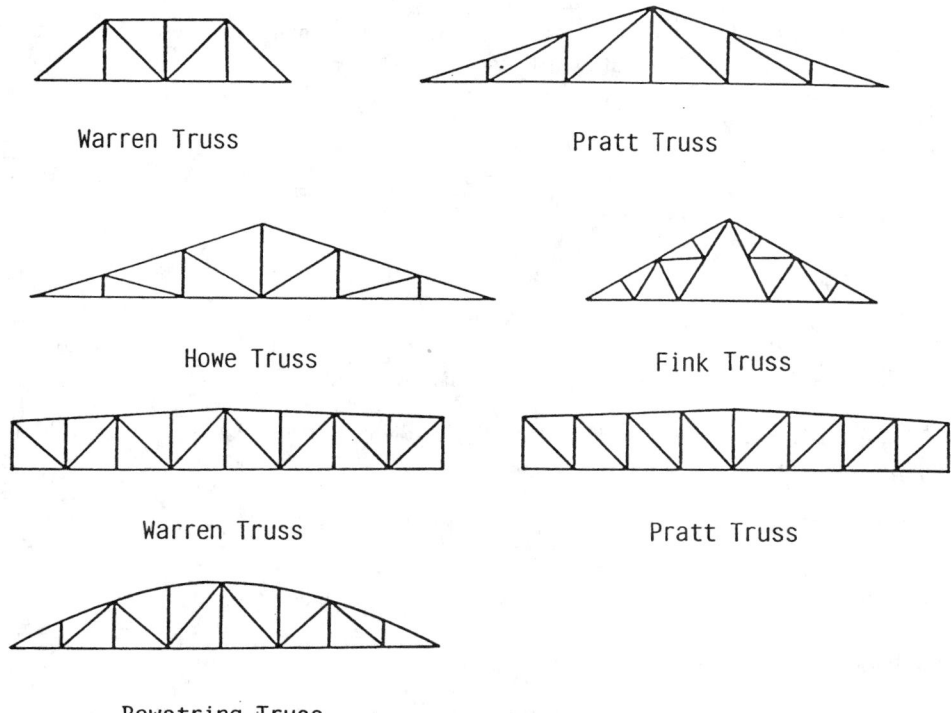

Figure 2.17   Typical planar trusses.

There exists a relation between the number of members, m, number of joints, j, and reaction components, r. The expression is

$$m = 2j - r \qquad (2.20)$$

which must be satisfied if it is to be statically determinate internally. The least number of reaction components required for external stability is r. If m exceeds $(2j - r)$, then the excess members are called redundant members and the truss is said to be statically indeterminate.

Truss analysis gives the bar forces in a truss; for a statically determinate truss, these bar forces can be found by employing the laws of statics to assure internal equilibrium of the structure. The process requires repeated use of free-body diagrams from which individual bar forces are determined. The *method of joints* is a technique of truss analysis in which the bar forces are determined by the sequential isolation of joints—the unknown bar forces at one joint are solved and become known bar forces at subsequent joints. The other method is known as *method of sections* in which equilibrium of a part of the truss is considered.

2.3.1 Method of Joints

An imaginary section may be completely passed around a joint in a truss. The joint has become a free body in equilibrium under the forces applied to it. The equations $\sum H = 0$ and $\sum V = 0$ may be applied to the joint to determine the unknown forces in members meeting there. It is evident that no more than two unknowns can be determined at a joint with these two equations.

EXAMPLE 2.3:

A truss shown in Figure 2.18 is symmetrically loaded, and it is sufficient to solve half the truss by considering the joints 1 through 5. At Joint 1, there are two unknown forces. Summation of the vertical components of all forces at Joint 1 gives

$$135 - F_{12} \sin 45 = 0$$

which in turn gives the force in the member 1-2, $F_{12} = 190.0$ kN (compressive). Similarly, summation of the horizontal components gives

$$F_{13} - F_{12} \cos 45° = 0$$

Substituting for F_{12} gives the force in the member 1-3 as

$$F_{13} = 135 \ \text{kN} \ \text{(tensile)}.$$

Now, Joint 2 is cut completely and it is found that there are two unknown forces F_{25} and F_{23}. Summation of the vertical components gives

$$F_{12} \cos 45° - F_{23} = 0.$$

Therefore,

$$F_{23} = 135 \ \text{kN} \ \text{(tensile)}.$$

Summation of the horizontal components gives

$$F_{12} \sin 45° - F_{25} = 0$$

and hence

$$F_{25} = 135 \ \text{kN} \ \text{(compressive)}.$$

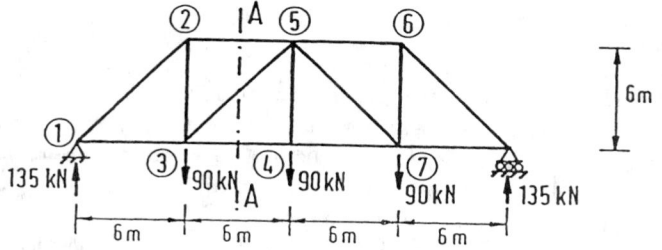

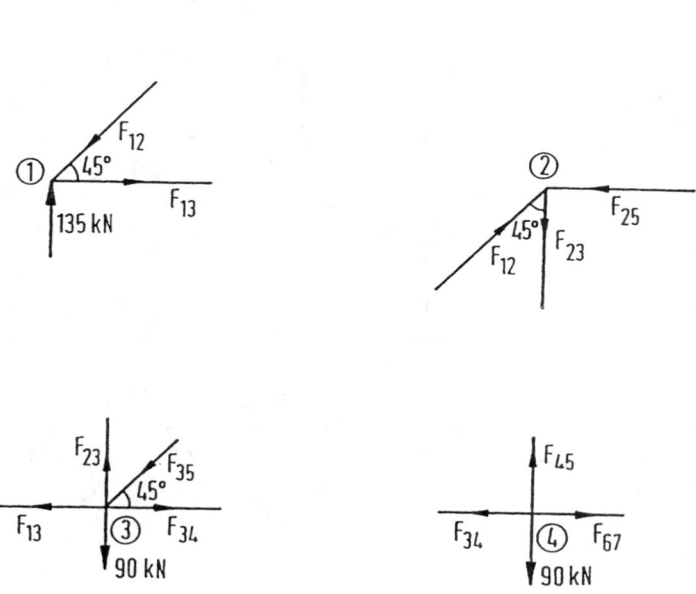

Figure 2.18 Example—methods of joints, planar truss.

After solving for Joints 1 and 2, one proceeds to take a section around Joint 3 at which there are now two unknown forces, namely, F_{34} and F_{35}. Summation of the vertical components at Joint 3 gives

$$F_{23} - F_{35} \sin 45° - 90 = 0$$

Substituting for F_{23}, one obtains $F_{35} = 63.6$ kN (compressive). Summing the horizontal components and substituting for F_{13} one gets

$$-135 - 45 + F_{34} = 0$$

Therefore,

$$F_{34} = 180 \text{ kN (tensile).}$$

The next joint involving two unknowns is Joint 4. When we consider a section around it, the summation of the vertical components at Joint 4 gives

$$F_{45} = 90 \text{ kN (tensile).}$$

Now, the forces in all the members on the left half of the truss are known and by symmetry the forces in the remaining members can be determined. The forces in all the members of a truss can also be determined by making use of the method of section.

2.3.2 Method of Sections

If only a few member forces of a truss are needed, the quickest way to find these forces is by the Method of Sections. In this method, an imaginary cutting line called a section is drawn through a stable and determinate truss. Thus, a section subdivides the truss into two separate parts. Since the entire truss is in equilibrium, any part of it must also be in equilibrium. Either of the two parts of the truss can be considered and the three equations of equilibrium $\sum F_x = 0$, $\sum F_y = 0$, and $\sum M = 0$ can be applied to solve for member forces.

The example considered in Section 2.3.1 (Figure 2.19) is once again considered. To calculate the

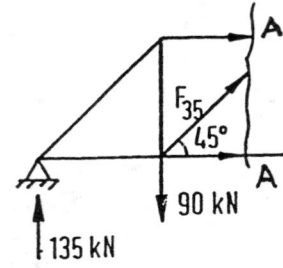

Figure 2.19 Example—method of sections, planar truss.

force in the member 3-5, F_{35}, a section AA should be run to cut the member 3-5 as shown in the figure. It is only required to consider the equilibrium of one of the two parts of the truss. In this case, the portion of the truss on the left of the section is considered. The left portion of the truss as shown in Figure 2.19 is in equilibrium under the action of the forces, namely, the external and internal forces. Considering the equilibrium of forces in the vertical direction, one can obtain

$$135 - 90 + F_{35} \sin 45° = 0$$

Therefore, F_{35} is obtained as

$$F_{35} = -45\sqrt{2} \text{ kN}$$

The negative sign indicates that the member force is compressive. This result is the same as the one obtained by the Method of Joints. The other member forces cut by the section can be obtained by considering the other equilibrium equations, namely, $\sum M = 0$. More sections can be taken in the same way so as to solve for other member forces in the truss. The most important advantage of this method is that one can obtain the required member force without solving for the other member forces.

2.3.3 Compound Trusses

A compound truss is formed by interconnecting two or more simple trusses. Examples of compound trusses are shown in Figure 2.20. A typical compound roof truss is shown in Figure 2.20a in which two simple trusses are interconnected by means of a single member and a common joint. The compound truss shown in Figure 2.20b is commonly used in bridge construction and in this case, three members are used to interconnect two simple trusses at a common joint. There are three simple trusses interconnected at their common joints as shown in Figure 2.20c.

The Method of Sections may be used to determine the member forces in the interconnecting members of compound trusses similar to those shown in Figure 2.20a and b. However, in the case of cantilevered truss, the middle simple truss is isolated as a free body diagram to find its reactions. These reactions are reversed and applied to the interconnecting joints of the other two simple trusses.

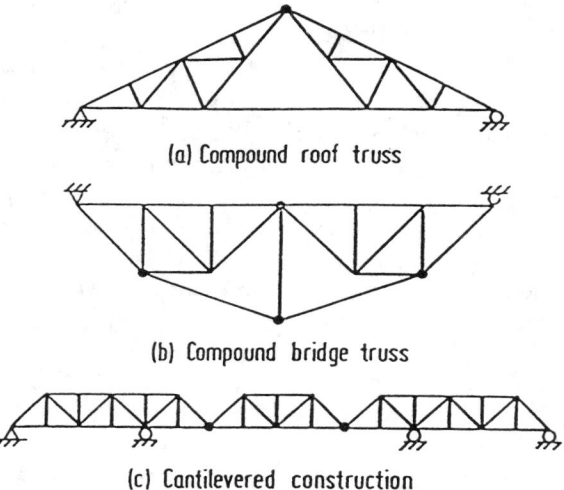

(a) Compound roof truss

(b) Compound bridge truss

(c) Cantilevered construction

Figure 2.20   Compound truss.

After the interconnecting forces between the simple trusses are found, the simple trusses are analyzed by the Method of Joints or the Method of Sections.

2.3.4   Stability and Determinacy

A stable and statically determinate plane truss should have at least three members, three joints, and three reaction components. To form a stable and determinate plane truss of 'n' joints, the three members of the original triangle plus two additional members for each of the remaining $(n - 3)$ joints are required. Thus, the minimum total number of members, m, required to form an internally stable plane truss is $m = 2n - 3$. If a stable, simple, plane truss of n joints and $(2n - 3)$ members is supported by three independent reaction components, the structure is stable and determinate when subjected to a general loading. If the stable, simple, plane truss has more than three reaction components, the structure is externally indeterminate. That means not all of the reaction components can be determined from the three available equations of statics. If the stable, simple, plane truss has more than $(2n - 3)$ members, the structure is internally indeterminate and hence all of the member forces cannot be determined from the $2n$ available equations of statics in the Method of Joints. The analyst must examine the arrangement of the truss members and the reaction components to determine if the simple plane truss is stable. Simple plane trusses having $(2n - 3)$ members are not necessarily stable.

2.4   Frames

Frames are statically indeterminate in general; special methods are required for their analysis. Slope deflection and moment distribution methods are two such methods commonly employed. Slope deflection is a method that takes into account the flexural displacements such as rotations and deflections and involves solutions of simultaneous equations. Moment distribution on the other hand involves successive cycles of computation, each cycle drawing closer to the "exact" answers. The method is more labor intensive but yields accuracy equivalent to that obtained from the "exact" methods. This method, however, remains the most important hand-calculation method for the analysis of frames.

2.4.1 Slope Deflection Method

This method is a special case of the stiffness method of analysis, and it is convenient for hand analysis of small structures. Moments at the ends of frame members are expressed in terms of the rotations and deflections of the joints. Members are assumed to be of constant section between each pair of supports. It is further assumed that the joints in a structure may rotate or deflect, but the angles between the members meeting at a joint remain unchanged.

The member force-displacement equations that are needed for the slope deflection method are written for a member AB in a frame. This member, which has its undeformed position along the x axis is deformed into the configuration shown in Figure 2.21. The positive axes, along with the positive member-end force components and displacement components, are shown in the figure.

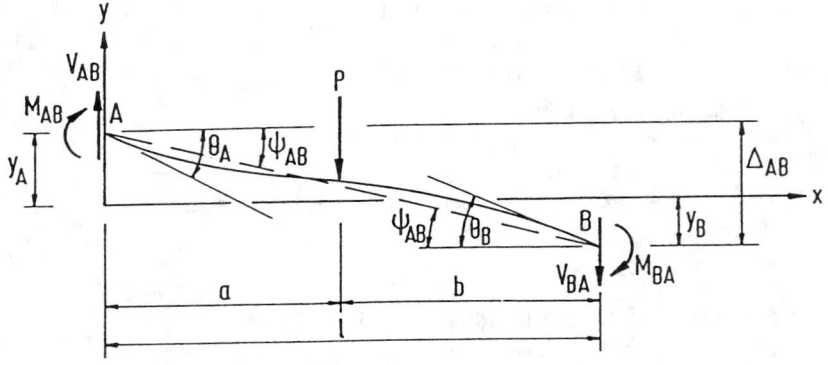

Figure 2.21   Deformed configuration of a beam.

The equations for end moments are written as

$$M_{AB} = \frac{2EI}{l}(2\theta_A + \theta_B - 3\psi_{AB}) + M_{FAB}$$

$$M_{BA} = \frac{2EI}{l}(2\theta_B + \theta_A - 3\psi_{AB}) + M_{FBA} \qquad (2.21)$$

in which M_{FAB} and M_{FBA} are fixed-end moments at supports A and B, respectively, due to the applied load. ψ_{AB} is the rotation as a result of the relative displacement between the member ends A and B given as

$$\psi_{AB} = \frac{\Delta_{AB}}{l} = \frac{y_A + y_B}{l} \qquad (2.22)$$

where Δ_{AB} is the relative deflection of the beam ends. y_A and y_B are the vertical displacements at ends A and B. Fixed-end moments for some loading cases may be obtained from Figure 2.8. The slope deflection equations in Equation 2.21 show that the moment at the end of a member is dependent on member properties EI, dimension l, and displacement quantities. The fixed-end moments reflect the transverse loading on the member.

2.4.2 Application of Slope Deflection Method to Frames

The slope deflection equations may be applied to statically indeterminate frames with or without sidesway. A frame may be subjected to sidesway if the loads, member properties, and dimensions of the frame are not symmetrical about the centerline. Application of slope deflection method can be illustrated with the following example.

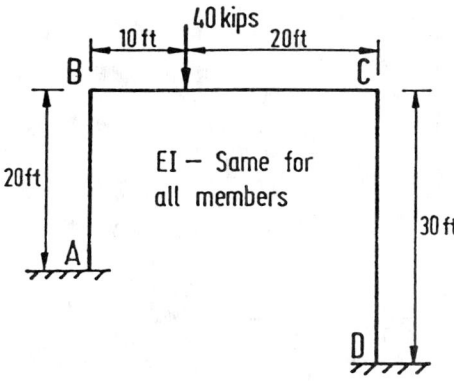

Figure 2.22 Example—slope deflection method.

EXAMPLE 2.4:

Consider the frame shown in Figure 2.22. subjected to sidesway Δ to the right of the frame. Equation 2.21 can be applied to each of the members of the frame as follows:

Member AB:

$$M_{AB} = \frac{2EI}{20}\left(2\theta_A + \theta_B - \frac{3\Delta}{20}\right) + M_{FAB}$$

$$M_{BA} = \frac{2EI}{20}\left(2\theta_B + \theta_A - \frac{3\Delta}{20}\right) + M_{FBA}$$

$$\theta_A = 0, \quad M_{FAB} = M_{FBA} = 0$$

Hence,

$$M_{AB} = \frac{2EI}{20}(\theta_B - 3\psi) \tag{2.23}$$

$$M_{BA} = \frac{2EI}{20}(2\theta_B - 3\psi) \tag{2.24}$$

in which

$$\psi = \frac{\Delta}{20}$$

Member BC:

$$M_{BC} = \frac{2EI}{30}(2\theta_B + \theta_C - 3 \times 0) + M_{FBC}$$

$$M_{CB} = \frac{2EI}{30}(2\theta_C + \theta_B - 3 \times 0) + M_{FCB}$$

$$M_{FBC} = -\frac{40 \times 10 \times 20^2}{30^2} = -178 \text{ ft-kips}$$

$$M_{FCB} = -\frac{40 \times 10^2 \times 20}{30^2} = 89 \text{ ft-kips}$$

Hence,

$$M_{BC} = \frac{2EI}{30}(2\theta_B + \theta_C) - 178 \tag{2.25}$$

$$M_{CB} = \frac{2EI}{30}(2\theta_C + \theta_B) + 89 \tag{2.26}$$

Member CD:

$$M_{CD} = \frac{2EI}{30}\left(2\theta_C + \theta_D - \frac{3\Delta}{30}\right) + M_{FCD}$$

$$M_{DC} = \frac{2EI}{30}\left(2\theta_D + \theta_C - \frac{3\Delta}{30}\right) + M_{FDC}$$

$$M_{FCD} = M_{FDC} = 0$$

Hence,

$$M_{DC} = \frac{2EI}{30}\left(\theta_C - 3 \times \frac{2}{3}\psi\right) = \frac{2EI}{30}(2\theta_C - 2\psi) \tag{2.27}$$

$$M_{DC} = \frac{2EI}{30}\left(\theta_C - 3 \times \frac{2}{3}\psi\right) = \frac{2EI}{30}(\theta_C - 2\psi) \tag{2.28}$$

Considering moment equilibrium at Joint B

$$\sum M_B = M_{BA} + M_{BC} = 0$$

Substituting for M_{BA} and M_{BC}, one obtains

$$\frac{EI}{30}(10\theta_B + 2\theta_C - 9\psi) = 178$$

or

$$10\theta_B + 2\theta_C - 9\psi = \frac{267}{K} \tag{2.29}$$

where $K = \frac{EI}{20}$.

Considering moment equilibrium at Joint C

$$\sum M_C = M_{CB} + M_{CD} = 0$$

Substituting for M_{CB} and M_{CD} we get

$$\frac{2EI}{30}(4\theta_C + \theta_B - 2\psi) = -89$$

or

$$\theta_B + 4\theta_C - 2\psi = -\frac{66.75}{K} \tag{2.30}$$

Summation of base shear equals to zero, we have

$$\sum H = H_A + H_D = 0$$

or

$$\frac{M_{AB} + M_{BA}}{1_{AB}} + \frac{M_{CD} + M_{DC}}{1_{CD}} = 0$$

Substituting for M_{AB}, M_{BA}, M_{CD}, and M_{DC} and simplifying

$$2\theta_B + 12\theta_C - 70\psi = 0 \tag{2.31}$$

Solution of Equations 2.29 to 2.31 results in

$$\theta_B = \frac{42.45}{K}$$

$$\theta_C = \frac{20.9}{K}$$

and

$$\psi = \frac{12.8}{K} \qquad (2.32)$$

Substituting for θ_B, θ_C, and ψ from Equations 2.32 into Equations 2.23 to 2.28 we get,

$$
\begin{aligned}
M_{AB} &= 10.10 \text{ ft-kips} \\
M_{BA} &= 93 \text{ ft-kips} \\
M_{BC} &= -93 \text{ ft-kips} \\
M_{CB} &= 90 \text{ ft-kips} \\
M_{CD} &= -90 \text{ ft-kips} \\
M_{DC} &= -62 \text{ ft-kips}
\end{aligned}
$$

2.4.3 Moment Distribution Method

The moment distribution method involves successive cycles of computation, each cycle drawing closer to the "exact" answers. The calculations may be stopped after two or three cycles, giving a very good approximate analysis, or they may be carried on to whatever degree of accuracy is desired. Moment distribution remains the most important hand-calculation method for the analysis of continuous beams and frames and it may be solely used for the analysis of small structures. Unlike the slope deflection method, this method does require the solution to simultaneous equations.

The terms constantly used in moment distribution are fixed-end moments, unbalanced moment, distributed moments, and carry-over moments. When all of the joints of a structure are clamped to prevent any joint rotation, the external loads produce certain moments at the ends of the members to which they are applied. These moments are referred to as *fixed-end moments*. Initially the joints in a structure are considered to be clamped. When the joint is released, it rotates if the sum of the fixed-end moments at the joint is not zero. The difference between zero and the actual sum of the end moments is the *unbalanced moment*. The unbalanced moment causes the joint to rotate. The rotation twists the ends of the members at the joint and changes their moments. In other words, rotation of the joint is resisted by the members and resisting moments are built up in the members as they are twisted. Rotation continues until equilibrium is reached—when the resisting moments equal the unbalanced moment—at which time the sum of the moments at the joint is equal to zero. The moments developed in the members resisting rotation are the *distributed moments*. The distributed moments in the ends of the member cause moments in the other ends, which are assumed fixed, and these are the *carry-over moments*.

Sign Convention

The moments at the end of a member are assumed to be positive when they tend to rotate the member clockwise about the joint. This implies that the resisting moment of the joint would be counter-clockwise. Accordingly, under gravity loading condition the fixed-end moment at the left end is assumed as counter-clockwise ($-ve$) and at the right end as clockwise ($+ve$).

Fixed-End Moments

Fixed-end moments for several cases of loading may be found in Figure 2.8. Application of moment distribution may be explained with reference to a continuous beam example as shown in Figure 2.23. Fixed-end moments are computed for each of the three spans. At Joint B the unbalanced moment is obtained and the clamp is removed. The joint rotates, thus distributing the unbalanced moment to the B-ends of spans BA and BC in proportion to their distribution factors. The values of these distributed moments are carried over at one-half rate to the other ends of the members. When equilibrium is reached, Joint B is clamped in its new rotated position and Joint C

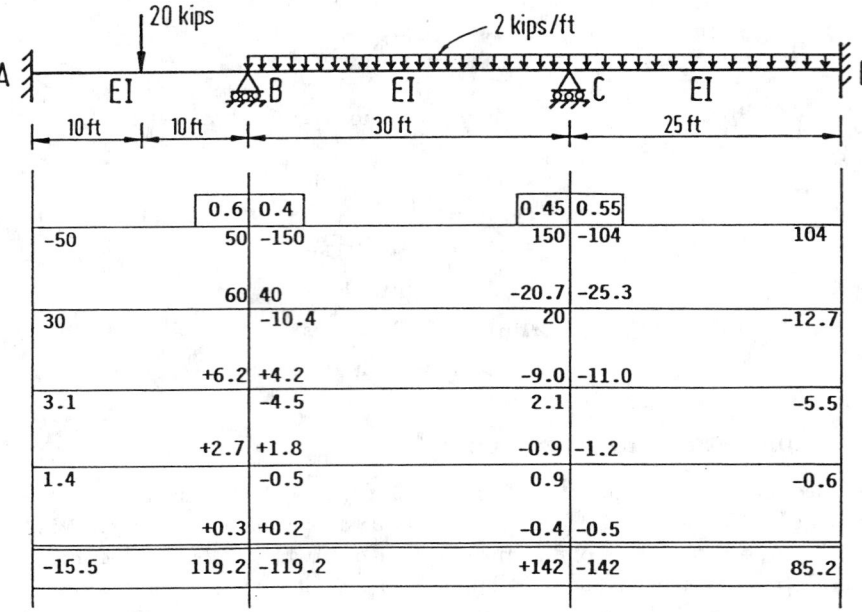

Figure 2.23   Example—continuous beam by moment distribution.

is released afterwards. Joint C rotates under its unbalanced moment until it reaches equilibrium, the rotation causing distributed moments in the C-ends of members CB and CD and their resulting carry-over moments. Joint C is now clamped and Joint B is released. This procedure is repeated again and again for Joints B and C, the amount of unbalanced moment quickly diminishing, until the release of a joint causes negligible rotation. This process is called moment distribution.

The stiffness factors and distribution factors are computed as follows:

$$DF_{BA} = \frac{K_{BA}}{\sum K} = \frac{I/20}{I/20 + I/30} = 0.6$$

$$DF_{BC} = \frac{K_{BC}}{\sum K} = \frac{I/30}{I/20 + I/30} = 0.4$$

$$DF_{CB} = \frac{K_{CB}}{\sum K} = \frac{I/30}{I/30 + I/25} = 0.45$$

$$DF_{CD} = \frac{K_{CD}}{\sum K} = \frac{I/25}{I/30 + I/25} = 0.55$$

Fixed-end moments

$$M_{FAB} = -50 \text{ ft-kips;} \quad M_{FBC} = -150 \text{ ft-kips;} \quad M_{FCD} = -104 \text{ ft-kips}$$
$$M_{FBA} = 50 \text{ ft-kips;} \quad M_{FCB} = 150 \text{ ft-kips;} \quad M_{FDC} = 104 \text{ ft-kips}$$

When a clockwise couple is applied at the near end of a beam, a clockwise couple of half the magnitude is set up at the far end of the beam. The ratio of the moments at the far and near ends is defined as *carry-over factor*, and it is $\frac{1}{2}$ in the case of a straight prismatic member. The carry-over factor was developed for carrying over to fixed ends, but it is applicable to simply supported ends, which must have final moments of zero. It can be shown that the beam simply supported at the far end is only three-fourths as stiff as the one that is fixed. If the stiffness factors for end spans that are simply supported are modified by three-fourths, the simple end is initially balanced to zero and no carry-overs are made to the end afterward. This simplifies the moment distribution process significantly.

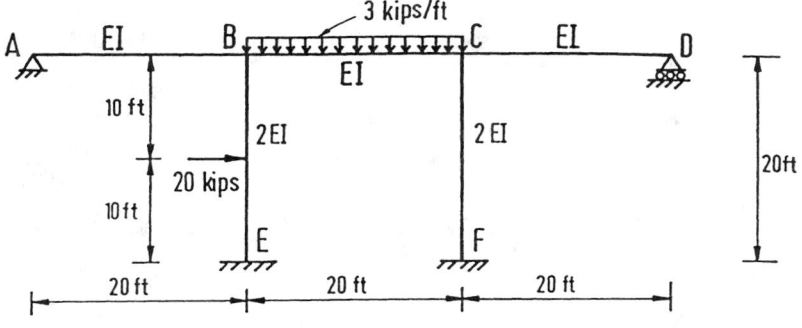

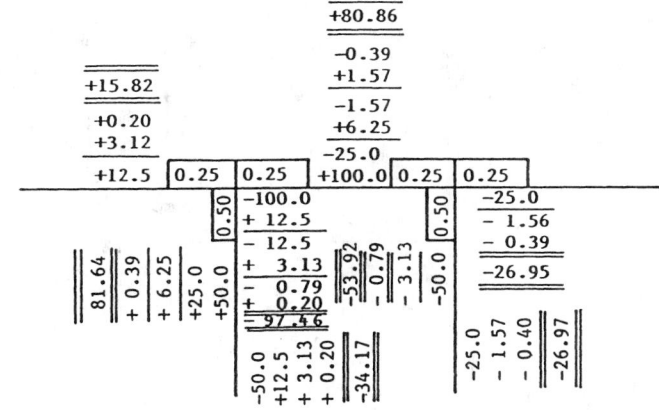

Figure 2.24   Example—non-sway frame by moment distribution.

Moment Distribution for Frames

Moment distribution for frames without sidesway is similar to that for continuous beams. The example shown in Figure 2.24 illustrates the applications of moment distribution for a frame without sidesway.

$$DF_{BA} = \frac{EI/20}{\frac{EI}{20} + \frac{EI}{20} + \frac{2EI}{20}} = 0.25$$

Similarly

$$
\begin{aligned}
DF_{BE} &= 0.50; & DF_{BC} &= 0.25 \\
M_{FBC} &= -100 \text{ ft-kips}; & M_{FCB} &= 100 \text{ ft-kips} \\
M_{FBE} &= 50 \text{ ft-kips}; & M_{FEB} &= -50 \text{ ft-kips}.
\end{aligned}
$$

Structural frames are usually subjected to sway in one direction or the other due to asymmetry of the structure and eccentricity of loading. The sway deflections affect the moments resulting in unbalanced moment. These moments could be obtained for the deflections computed and added to the originally distributed fixed-end moments. The sway moments are distributed to columns. Should a frame have columns all of the same length and the same stiffness, the sidesway moments will be the same for each column. However, should the columns have differing lengths and/or stiffness, this will not be the case. The sidesway moments should vary from column to column in proportion to their I/l^2 values.

The frame in Figure 2.25 shows a frame subjected to sway. The process of obtaining the final moments is illustrated for this frame.

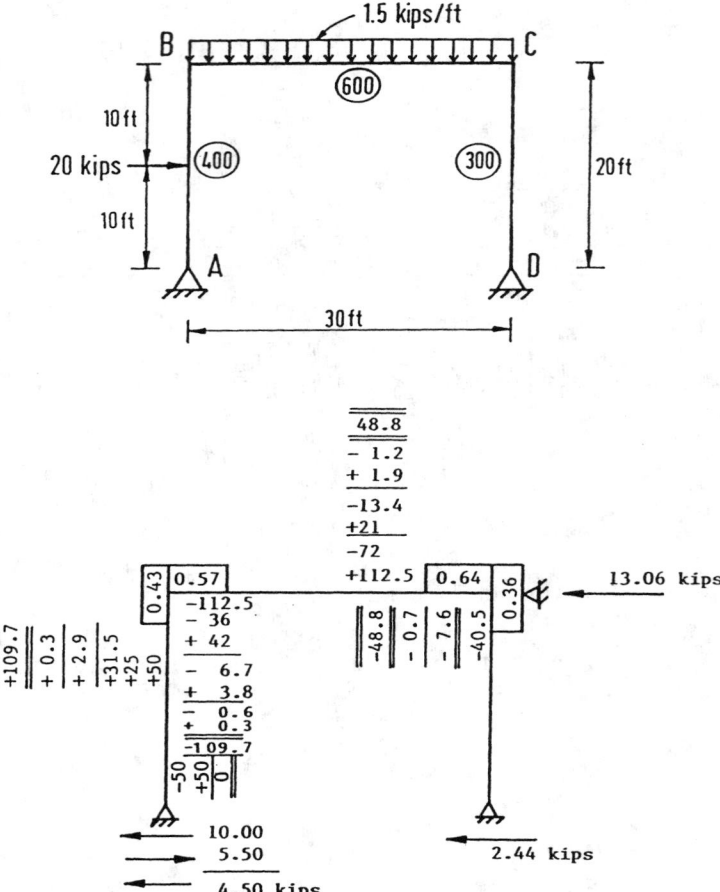

Figure 2.25   Example—sway frame by moment distribution.

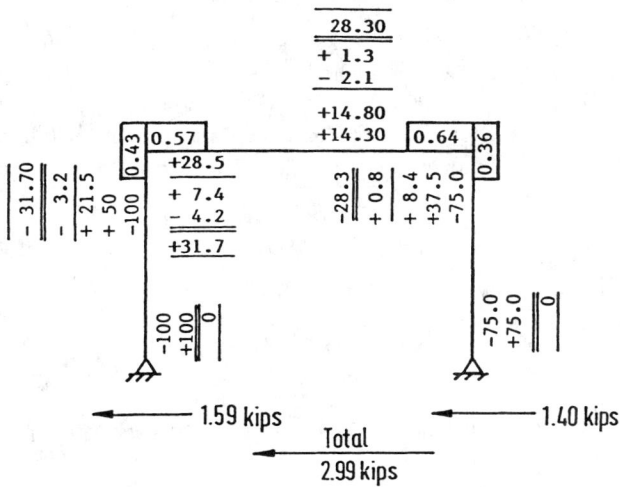

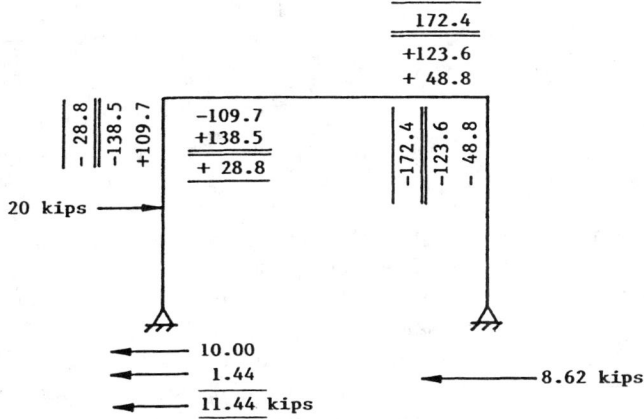

Figure 2.25 *(Continued)* Example—sway frame by moment distribution.

The frame sways to the right and the sidesway moment can be assumed in the ratio

$$\frac{400}{20^2} : \frac{300}{20^2} \quad \text{(or)} \quad 1 : 0.7$$

Final moments are obtained by adding distributed fixed-end moments and $\frac{13.06}{2.99}$ times the distributed assumed sidesway moments.

2.4.4   Method of Consistent Deformations

The method of consistent deformations makes use of the principle of deformation compatibility to analyze indeterminate structures. This method employs equations that relate the forces acting on the structure to the deformations of the structure. These relations are formed so that the deformations are expressed in terms of the forces and the forces become the unknowns in the analysis.

Let us consider the beam shown in Figure 2.26a. The first step, in this method, is to determine

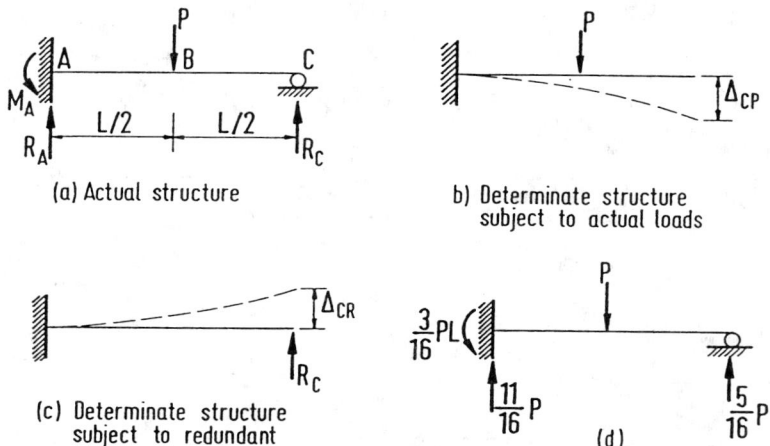

Figure 2.26   Beam with one redundant reaction.

the degree of indeterminacy or the number of redundants that the structure possesses. As shown in the figure, the beam has three unknown reactions, R_A, R_C, and M_A. Since there are only two equations of equilibrium available for calculating the reactions, the beam is said to be indeterminate to the first degree. Restraints that can be removed without impairing the load-supporting capacity of the structure are referred to as *redundants*.

Once the number of redundants is known, the next step is to decide which reaction is to be removed in order to form a determinate structure. Any one of the reactions may be chosen to be the redundant provided that a stable structure remains after the removal of that reaction. For example, let us take the reaction R_C as the redundant. The determinate structure obtained by removing this restraint is the cantilever beam shown in Figure 2.26b. We denote the deflection at end C of this beam, due to P, by Δ_{CP}. The first subscript indicates that the deflection is measured at C and the second subscript that the deflection is due to the applied load P. Using the moment area method, it can be shown that $\Delta_{CP} = 5PL^3/48EI$. The redundant R_C is then applied to the determinate cantilever beam, as shown in Figure 2.26c. This gives rise to a deflection Δ_{CR} at point C the magnitude of which can be shown to be $R_C L^3/3EI$.

In the actual indeterminate structure, which is subjected to the combined effects of the load P and the redundant R_C, the deflection at C is zero. Hence the algebraic sum of the deflection Δ_{CP} in

Figure 2.26b and the deflection Δ_{CR} in Figure 2.26c must vanish. Assuming downward deflections to be positive, we write

$$\Delta_{CP} - \Delta_{CR} = 0 \qquad (2.33)$$

or

$$\frac{5PL^3}{48EI} - \frac{R_C L^3}{3EI} = 0$$

from which

$$R_C = \frac{5}{16}P$$

Equation 2.33, which is used to solve for the redundant, is referred to as an equation of consistent of deformation.

Once the redundant R_C has been evaluated, one can determine the remaining reactions by applying the equations of equilibrium to the structure in Figure 2.26a. Thus, $\sum F_y = 0$ leads to

$$R_A = P - \frac{5}{16}P = \frac{11}{16}P$$

and $\sum M_A = 0$ gives

$$M_A = \frac{PL}{2} - \frac{5}{16}PL = \frac{3}{16}PL$$

A free body of the beam, showing all the forces acting on it, is shown in Figure 2.26d.

The steps involved in the method of consistent deformations are:

1. The number of redundants in the structure is determined.
2. Enough redundants are removed to form a determinate structure.
3. The displacements that the applied loads cause in the determinate structure at the points where the redundants have been removed are then calculated.
4. The displacements at these points in the determinate structure due to the redundants are obtained.
5. At each point where a redundant has been removed, the sum of the displacements calculated in Steps 3 and 4 must be equal to the displacement that exists at that point in the actual indeterminate structure. The redundants are evaluated using these relationships.
6. Once the redundants are known, the remaining reactions are determined using the equations of equilibrium.

Structures with Several Redundants

The method of consistent deformations can be applied to structures with two or more redundants. For example, the beam in Figure 2.27a is indeterminate to the second degree and has two redundant reactions. If we let the reactions at B and C be the redundants, then the determinate structure obtained by removing these supports is the cantilever beam shown in Figure 2.27b. To this determinate structure we apply separately the given load (Figure 2.27c) and the redundants R_B and R_C one at a time (Figures 2.27d and e).

Since the deflections at B and C in the original beam are zero, the algebraic sum of the deflections in Figures 2.27c, d, and e at these same points must also vanish.

Thus,

$$\begin{aligned} \Delta_{BP} - \Delta_{BB} - \Delta_{BC} &= 0 \\ \Delta_{CP} - \Delta_{CB} - \Delta_{CC} &= 0 \end{aligned} \qquad (2.34)$$

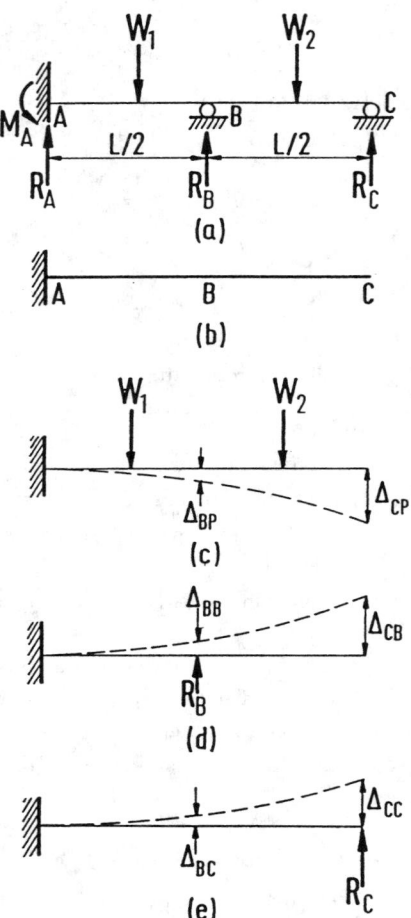

Figure 2.27   Beam with two redundant reactions.

It is useful in the case of complex structures to write the equations of consistent deformations in the form

$$\Delta_{BP} - \delta_{BB} R_B - \delta_{BC} R_C = 0$$
$$\Delta_{CP} - \delta_{CB} R_B - \delta_{CC} R_C = 0 \qquad (2.35)$$

in which δ_{BC}, for example, denotes the deflection at B due to a unit load at C in the direction of R_C. Solution of Equation 2.35 gives the redundant reactions R_B and R_C.

EXAMPLE 2.5:

Determine the reactions for the beam shown in Figure 2.28 and draw its shear force and bending moment diagrams.

It can be seen from the figure that there are three reactions, namely, M_A, R_A, and R_C one more than that required for a stable structure. The reaction R_C can be removed to make the structure determinate. We know that the deflection at support C of the beam is zero. One can determine the deflection δ_{CP} at C due to the applied load on the cantilever in Figure 2.28b. The deflection δ_{CR} at C due to the redundant reaction on the cantilever (Figure 2.28c) can be determined in the same way. The compatibility equation gives

$$\delta_{CP} - \delta_{CR} = 0$$

By moment area method,

$$
\begin{aligned}
\delta_{CP} &= \frac{20}{EI} \times 2 \times 1 + \frac{1}{2} \times \frac{20}{EI} \times 2 \times \frac{2}{3} \times 2 \\
&\quad + \frac{40}{EI} \times 2 \times 3 + \frac{1}{2} \times \frac{60}{EI} \times 2 \times \left(\frac{2}{3} \times 2 + 2 \right) \\
&= \frac{1520}{3EI} \\
\delta_{CR} &= \frac{1}{2} \times \frac{4R_C}{EI} \times 4 \times \frac{2}{3} \times 4 = \frac{64R_C}{3EI}
\end{aligned}
$$

Substituting for δ_{CP} and δ_{CR} in the compatibility equation one obtains

$$\frac{1520}{3EI} - \frac{64R_C}{3EI} = 0$$

from which

$$R_C = 23.75 \text{ kN} \uparrow$$

By using statical equilibrium equations we get

$$R_A = 6.25 \text{ kN} \uparrow$$

and

$$M_A = 5 \text{ kNm.}$$

The shear force and bending moment diagrams are shown in Figure 2.28d.

2.5 Plates

2.5.1 Bending of Thin Plates

When the thickness of an object is small compared to the other dimensions, it is called a **thin plate**. The plane parallel to the faces of the plate and bisecting the thickness of the plate, in the

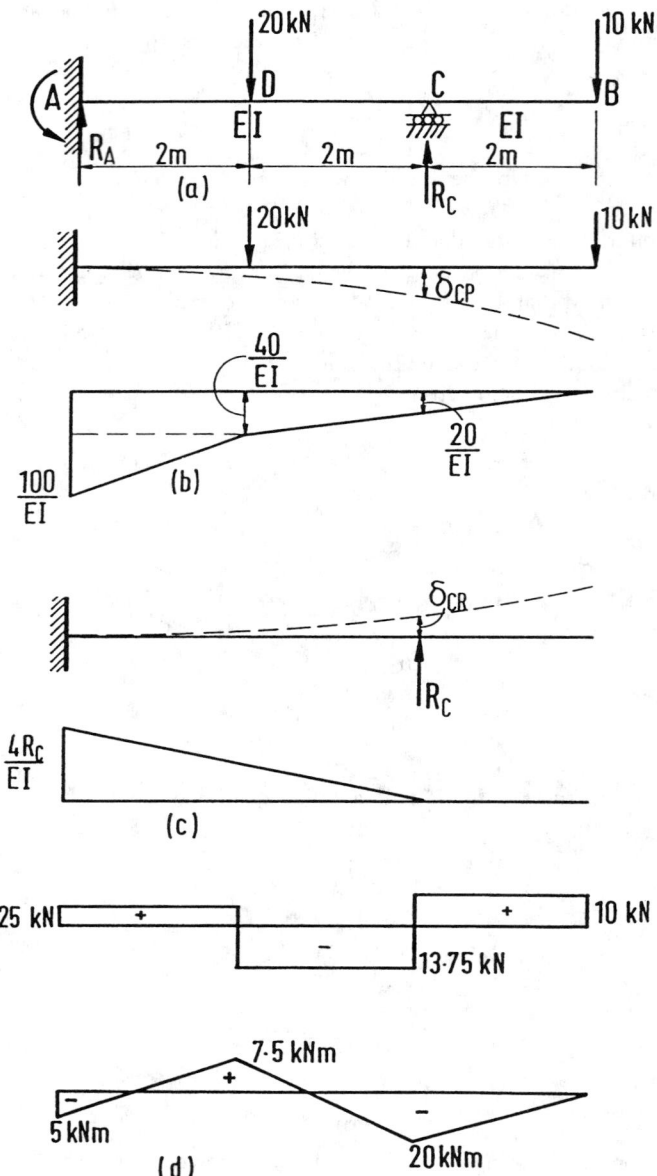

Figure 2.28   Example 2.5.

undeformed state, is called the middle plane of the plate. When the deflection of the middle plane is small compared with the thickness, h, it can be assumed that

1. There is no deformation in the middle plane.
2. The normal of the middle plane before bending is deformed into the normals of the middle plane after bending.
3. The normal stresses in the direction transverse to the plate can be neglected.

Based on these assumptions, all stress components can be expressed by deflection w' of the plate. w' is a function of the two coordinates (x, y) in the plane of the plate. This function has to satisfy a linear partial differential equation, which, together with the boundary conditions, completely defines w'.

Figure 2.29a shows a plate element cut from a plate whose middle plane coincides with the xy plane. The middle plane of the plate subjected to a lateral load of intensity 'q' is shown in Figure 2.29b. It can be shown, by considering the equilibrium of the plate element, that the stress

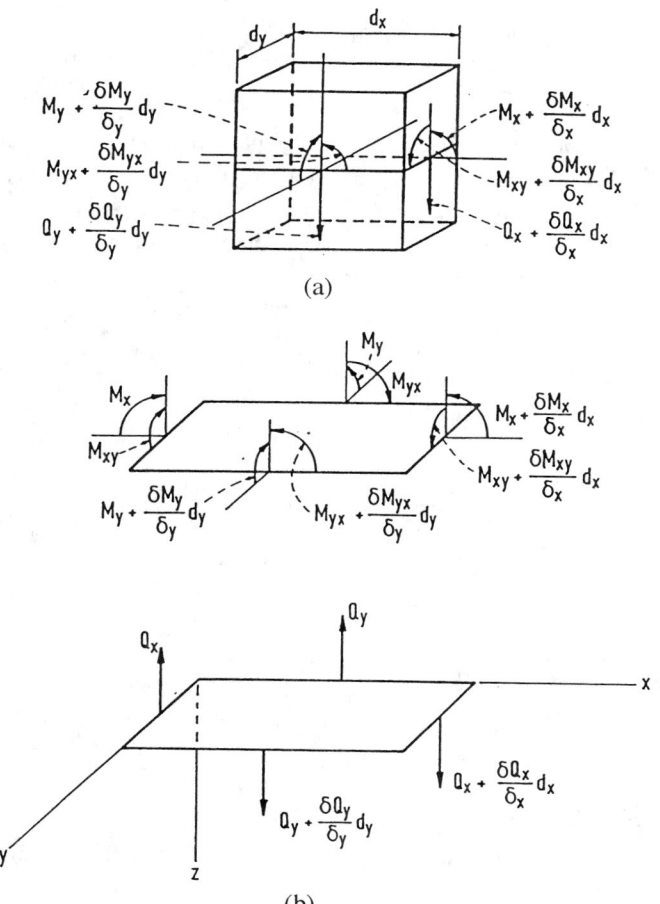

Figure 2.29   (a) Plate element; (b) stress resultants.

resultants are given as

$$M_x = -D\left(\frac{\partial^2 w}{\partial x^2} + \nu\frac{\partial^2 w}{\partial y^2}\right)$$

$$M_y = -D\left(\frac{\partial^2 w}{\partial y^2} + \nu\frac{\partial^2 w}{\partial x^2}\right)$$

$$M_{xy} = -M_{yx} = D(1-\nu)\frac{\partial^2 w}{\partial x\partial y} \tag{2.36}$$

$$V_x = \frac{\partial^3 w}{\partial x^3} + (2-\nu)\frac{\partial^3 w}{\partial x\partial y^2} \tag{2.37}$$

$$V_y = \frac{\partial^3 w}{\partial y^3} + (2-\nu)\frac{\partial^3 w}{\partial y\partial x^2} \tag{2.38}$$

$$Q_x = -D\frac{\partial}{\partial x}\left(\frac{\partial^2 w}{\partial x^2} + \frac{\partial^2 w}{\partial y^2}\right) \tag{2.39}$$

$$Q_y = -D\frac{\partial}{\partial y}\left(\frac{\partial^2 w}{\partial x^2} + \frac{\partial^2 w}{\partial y^2}\right) \tag{2.40}$$

$$R = 2D(1-\nu)\frac{\partial^2 w}{\partial x\partial y} \tag{2.41}$$

where

M_x and M_y	=	bending moments per unit length in the x and y directions, respectively
M_{xy} and M_{yx}	=	twisting moments per unit length
Q_x and Q_y	=	shearing forces per unit length in the x and y directions, respectively
V_x and V_y	=	supplementary shear forces in the x and y directions, respectively
R	=	corner force
D	=	$\frac{Eh^3}{12(1-\nu^2)}$, flexural rigidity of the plate per unit length
E	=	modulus of elasticity
ν	=	Poisson's Ratio

The governing equation for the plate is obtained as

$$\frac{\partial^4 w}{\partial x^4} + 2\frac{\partial^4 w}{\partial x^2\partial y^2} + \frac{\partial^4 w}{\partial y^4} = \frac{q}{D} \tag{2.42}$$

Any plate problem should satisfy the governing Equation 2.42 and boundary conditions of the plate.

2.5.2 Boundary Conditions

There are three basic boundary conditions for plate problems. These are the clamped edge, the simply supported edge, and the free edge.

Clamped Edge

For this boundary condition, the edge is restrained such that the deflection and slope are zero along the edge. If we consider the edge $x = a$ to be clamped, we have

$$(w)_{x=a} = 0 \qquad\qquad \left(\frac{\partial w}{\partial x}\right)_{x=a} = 0 \tag{2.43}$$

Simply Supported Edge

If the edge $x = a$ of the plate is simply supported, the deflection w along this edge must be zero. At the same time this edge can rotate freely with respect to the edge line. This means that

$$(w)_{x=a} = 0; \qquad \left(\frac{\partial^2 w}{\partial x^2}\right)_{x=a} = 0 \tag{2.44}$$

Free Edge

If the edge $x = a$ of the plate is entirely free, there are no bending and twisting moments or vertical shearing forces. This can be written in terms of w, the deflection as

$$\left(\frac{\partial^2 w}{\partial x^2} + v\frac{\partial^2 w}{\partial y^2}\right)_{x=a} = 0$$

$$\left(\frac{\partial^3 w}{\partial x^3} + (2 - v)\frac{\partial^3 w}{\partial x \partial y^2}\right)_{x=a} = 0 \tag{2.45}$$

2.5.3   Bending of Simply Supported Rectangular Plates

A number of the plate bending problems may be solved directly by solving the differential Equation 2.42. The solution, however, depends on the loading and boundary condition. Consider a simply supported plate subjected to a sinusoidal loading as shown in Figure 2.30. The differential

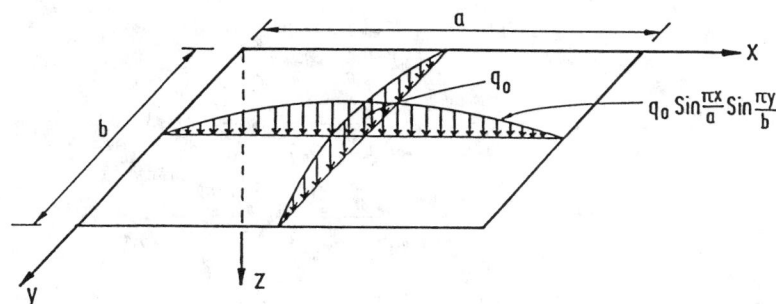

Figure 2.30   Rectangular plate under sinusoidal loading.

Equation 2.42 in this case becomes

$$\frac{\partial^4 w}{\partial x^4} + 2\frac{\partial^4 w}{\partial x^2 \partial y^2} + \frac{\partial^4 w}{\partial y^4} = \frac{q_o}{D} \sin\frac{\pi x}{a} \sin\frac{\pi y}{b} \tag{2.46}$$

The boundary conditions for the simply supported edges are

$$w = 0, \qquad \frac{\partial^2 w}{\partial x^2} = 0 \ \text{ for } \ x = 0 \ \text{ and } \ x = a$$

$$w = 0, \qquad \frac{\partial^2 w}{\partial y^2} = 0 \ \text{ for } \ y = 0 \ \text{ and } \ y = b \tag{2.47}$$

The deflection function becomes

$$w = w_0 \sin\frac{\pi x}{a} \sin\frac{\pi y}{b} \tag{2.48}$$

which satisfies all the boundary conditions in Equation 2.47. w_0 must be chosen to satisfy Equation 2.46. Substitution of Equation 2.48 into Equation 2.46 gives

$$\pi^4 \left(\frac{1}{a^2} + \frac{1}{b^2} \right)^2 w_0 = \frac{q_o}{D}$$

The deflection surface for the plate can, therefore, be found as

$$w = \frac{q_o}{\pi^4 D \left(\frac{1}{a^2} + \frac{1}{b^2} \right)^2} \sin \frac{\pi x}{a} \sin \frac{\pi y}{b} \tag{2.49}$$

Using Equations 2.49 and 2.36, we find expression for moments as

$$M_x = \frac{q_o}{\pi^2 \left(\frac{1}{a^2} + \frac{1}{b^2} \right)^2} \left(\frac{1}{a^2} + \frac{v}{b^2} \right) \sin \frac{\pi x}{a} \sin \frac{\pi y}{b}$$

$$M_y = \frac{q_o}{\pi^2 \left(\frac{1}{a^2} + \frac{1}{b^2} \right)^2} \left(\frac{v}{a^2} + \frac{1}{b^2} \right) \sin \frac{\pi x}{a} \sin \frac{\pi y}{b}$$

$$M_{xy} = \frac{q_o(1-v)}{\pi^2 \left(\frac{1}{a^2} + \frac{1}{b^2} \right)^2 ab} \cos \frac{\pi x}{a} \cos \frac{\pi y}{b} \tag{2.50}$$

Maximum deflection and maximum bending moments that occur at the center of the plate can be written by substituting $x = a/2$ and $y = b/2$ in Equation 2.50 as

$$w_{max} = \frac{q_o}{\pi^4 D \left(\frac{1}{a^2} + \frac{1}{b^2} \right)^2} \tag{2.51}$$

$$(M_x)_{max} = \frac{q_o}{\pi^2 \left(\frac{1}{a^2} + \frac{1}{b^2} \right)^2} \left(\frac{1}{a^2} + \frac{v}{b^2} \right)$$

$$(M_y)_{max} = \frac{q_o}{\pi^2 \left(\frac{1}{a^2} + \frac{1}{b^2} \right)^2} \left(\frac{v}{a^2} + \frac{1}{b^2} \right)$$

If the plate is square, then $a = b$ and Equation 2.51 becomes

$$w_{max} = \frac{q_o a^4}{4\pi^4 D'}$$

$$(M_x)_{max} = (M_y)_{max} = \frac{(1+v)}{4\pi^2} q_o a^2 \tag{2.52}$$

If the simply supported rectangular plate is subjected to any kind of loading given by

$$q = q(x, y) \tag{2.53}$$

the function $q(x, y)$ should be represented in the form of a double trigonometric series as

$$q(x, y) = \sum_{m=1}^{\infty} \sum_{n=1}^{\infty} q_{mn} \sin \frac{m\pi x}{a} \sin \frac{n\pi y}{b} \tag{2.54}$$

in which q_{mn} is given by

$$q_{mn} = \frac{4}{ab} \int_0^a \int_0^b q(x, y) \sin \frac{m\pi x}{a} \sin \frac{n\pi y}{b} dx dy \tag{2.55}$$

From Equations 2.46, 2.53, 2.54, and 2.55 we can obtain the expression for deflection as

$$w = \frac{1}{\pi^4 D} \sum_{m=1}^{\infty} \sum_{n=1}^{\infty} \frac{q_{mn}}{\left(\frac{m^2}{a^2} + \frac{n^2}{b^2}\right)^2} \sin\frac{m\pi x}{a} \sin\frac{n\pi y}{b} \tag{2.56}$$

If the applied load is uniformly distributed of intensity q_o, we have

$$q(x, y) = q_o$$

and from Equation 2.55 we obtain

$$q_{mn} = \frac{4q_o}{ab} \int_0^a \int_0^b \sin\frac{m\pi x}{a} \sin\frac{n\pi y}{b} dxdy = \frac{16q_o}{\pi^2 mn} \tag{2.57}$$

in which 'm' and 'n' are odd integers. $q_{mn} = 0$ if 'm' or 'n' or both of them are even numbers. We can, therefore, write the expression for deflection of a simply supported plate subjected to uniformly distributed load as

$$w = \frac{16q_o}{\pi^6 D} \sum_{m=1}^{\infty} \sum_{n=1}^{\infty} \frac{\sin\frac{m\pi x}{a} \sin\frac{n\pi y}{b}}{mn \left(\frac{m^2}{a^2} + \frac{n^2}{b^2}\right)^2} \tag{2.58}$$

where $m = 1, 3, 5, \ldots$ and $n = 1, 3, 5, \ldots$

The maximum deflection occurs at the center and it can be written by substituting $x = \frac{a}{2}$ and $y = \frac{b}{2}$ in Equation 2.58 as

$$w_{max} = \frac{16q_o}{\pi^6 D} \sum_{m=1}^{\infty} \sum_{n=1}^{\infty} \frac{(-1)^{\frac{m+n}{2}-1}}{mn \left(\frac{m^2}{a^2} + \frac{n^2}{b^2}\right)^2} \tag{2.59}$$

Equation 2.59 is a rapid converging series and a satisfactory approximation can be obtained by taking only the first term of the series; for example, in the case of a square plate,

$$w_{max} = \frac{4q_o a^4}{\pi^6 D} = 0.00416 \frac{q_o a^4}{D}$$

Assuming $\nu = 0.3$, we get for the maximum deflection

$$w_{max} = 0.0454 \frac{q_o a^4}{E h^3}$$

The expressions for bending and twisting moments can be obtained by substituting Equation 2.58 into Equation 2.36. Figure 2.31 shows some loading cases and the corresponding loading functions.

The above solution for uniformly loaded cases is known as Navier solution. If two opposite sides (say $x = 0$ and $x = a$) of a rectangular plate are simply supported, the solution taking the deflection function as

$$w = \sum_{m=1}^{\infty} Y_m \sin\frac{m\pi x}{a} \tag{2.60}$$

can be adopted. This solution was proposed by Levy [53]. Equation 2.60 satisfies the boundary conditions $w = 0$ and $\frac{\partial^2 w}{\partial x^2} = 0$ on the two simply supported edges. Y_m should be determined such that it satisfies the boundary conditions along the edges $y = \pm\frac{b}{2}$ of the plate shown in Figure 2.32 and also the equation of the deflection surface

No.	Load $q(x,y) = \sum_{m}\sum_{n} q_{mn} \sin\dfrac{m\pi x}{a} \sin\dfrac{n\pi y}{b}$	Expansion Coefficients q_{mn}
1	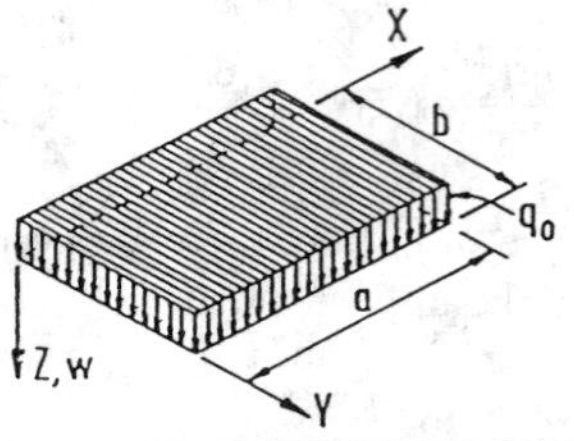	$q_{mn} = \dfrac{16 q_0}{\pi^2 mn}$ $(m, n = 1, 3, 5, \ldots)$
2	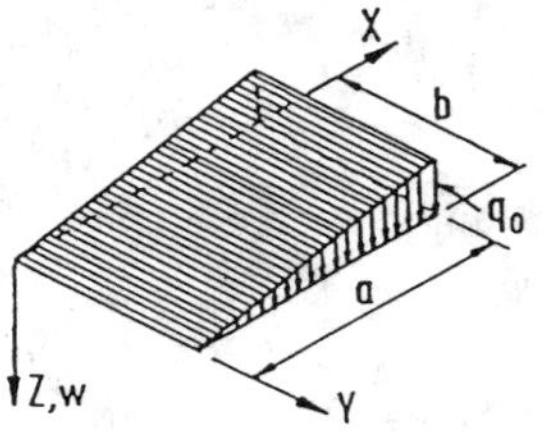	$q_{mn} = \dfrac{-8 q_0 \cos m\pi}{\pi^2 mn}$ $(m, n = 1, 3, 5, \ldots)$
3	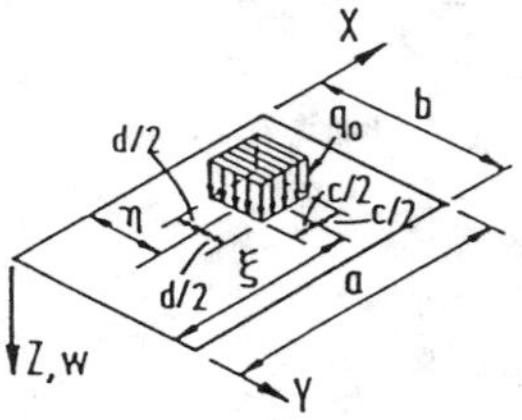	$P_{mn} = \dfrac{16 q_0}{\pi^2 mn} \sin\dfrac{m\pi \xi}{a} \sin\dfrac{n\pi \eta}{b}$ $\times \sin\dfrac{m\pi c}{2a} \sin\dfrac{n\pi d}{2b}$ $(m, n = 1, 3, 5, \ldots)$
4	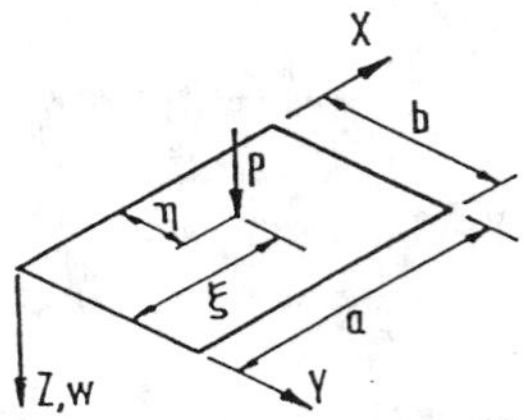	$q_{mn} = \dfrac{4 q_0}{ab} \sin\dfrac{m\pi \xi}{a} \sin\dfrac{n\pi \eta}{b}$ $(m, n = 1, 2, 3, \ldots)$

Figure 2.31   Typical loading on plates and loading functions.

No.	Load $q(x,y) = \sum_M \sum_a q_{mn} \sin\frac{m\pi x}{a} \sin\frac{n\pi y}{b}$	Expansion coefficients q_{mn}
5	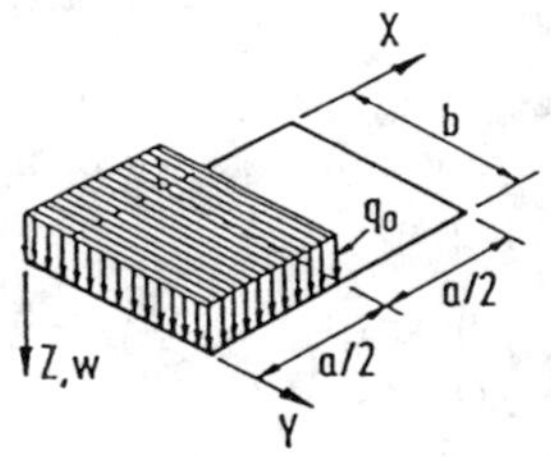 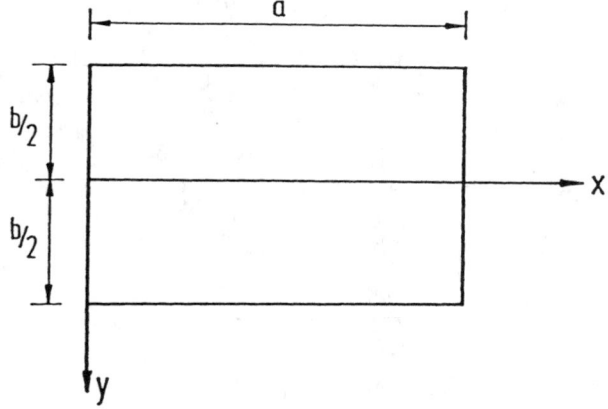	$q_{mn} = \dfrac{8q_0}{\pi^2 mn}$ for $m,n = 1,3,5,\ldots$ $q_{mn} = \dfrac{16q_0}{\pi^2 mn}$ for $\begin{cases} m = 2,6,10,\ldots \\ n = 1,3,5,\ldots \end{cases}$
6	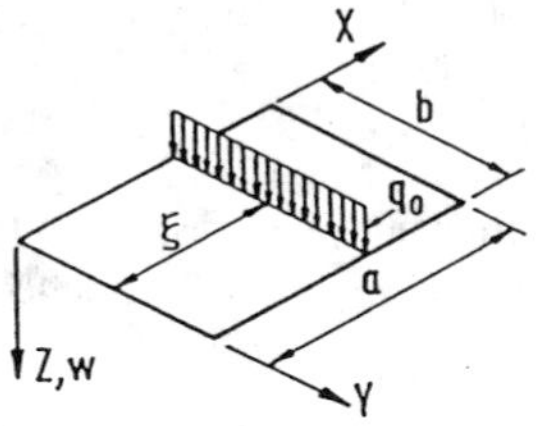	$q_{mn} = \dfrac{4q_0}{\pi an} \sin\dfrac{m\pi\xi}{a}$ $(m,n = 1,2,3,\ldots)$

Figure 2.31   *(Continued)* Typical loading on plates and loading functions.

Figure 2.32   Rectangular plate.

$$\frac{\partial^4 w}{\partial x^4} + 2\frac{\partial^4 w}{\partial x^2 \partial y^2} + \frac{\partial^4 w}{\partial y^4} = \frac{q_0}{D} \tag{2.61}$$

q_0 being the intensity of uniformly distributed load.

The solution for Equation 2.61 can be taken in the form

$$w = w_1 + w_2 \tag{2.62}$$

for a uniformly loaded simply supported plate. w_1 can be taken in the form

$$w_1 = \frac{q_0}{24D}\left(x^4 - 2ax^3 + a^3 x\right) \tag{2.63}$$

representing the deflection of a uniformly loaded strip parallel to the x axis. It satisfies Equation 2.61 and also the boundary conditions along $x = 0$ and $x = a$.

The expression w_2 has to satisfy the equation

$$\frac{\partial^4 w_2}{\partial x^4} + 2\frac{\partial^4 w_2}{\partial x^2 \partial y^2} + \frac{\partial^4 w_2}{\partial y^4} = 0 \tag{2.64}$$

and must be chosen such that Equation 2.62 satisfies all boundary conditions of the plate. Taking w_2 in the form of series given in Equation 2.60 it can be shown that the deflection surface takes the form

$$\begin{aligned}
w =\;& \frac{q_0}{24D}\left(x^4 - 2ax^3 + a^3 x\right) + \frac{q_0 a^4}{24D}\sum_{m=1}^{\infty}\left(A_m \cosh\frac{m\pi y}{a}\right. \\
& + B_m \frac{m\pi y}{a}\sinh\frac{m\pi y}{a} + C_m \sinh\frac{m\pi y}{a} \\
& \left. + D_m \frac{m\pi y}{a}\cosh\frac{m\pi y}{a}\right)\sin\frac{m\pi x}{a}
\end{aligned} \tag{2.65}$$

Observing that the deflection surface of the plate is symmetrical with respect to the x axis, we keep in Equation 2.65 only an even function of y; therefore, $C_m = D_m = 0$. The deflection surface takes the form

$$\begin{aligned}
w =\;& \frac{q_0}{24D}\left(x^4 - 2ax^3 + a^3 x\right) + \frac{q_0 a^4}{24D}\sum_{m=1}^{\infty}\left(A_m \cosh\frac{m\pi y}{a}\right. \\
& \left. + B_m \frac{m\pi y}{a}\sinh\frac{m\pi y}{a}\right)\sin\frac{m\pi x}{a}
\end{aligned} \tag{2.66}$$

Developing the expression in Equation 2.63 into a trigonometric series, the deflection surface in Equation 2.66 is written as

$$w = \frac{q_0 a^4}{D}\sum_{m=1}^{\infty}\left(\frac{4}{\pi^5 m^5} + A_m \cosh\frac{m\pi y}{a} + B_m \frac{m\pi y}{a}\sinh\frac{m\pi y}{a}\right)\sin\frac{m\pi x}{a} \tag{2.67}$$

Substituting Equation 2.67 in the boundary conditions

$$w = 0, \qquad \frac{\partial^2 w}{\partial y^2} = 0 \tag{2.68}$$

one obtains the constants of integration A_m and B_m and the expression for deflection may be written as

$$\begin{aligned}
w =\;& \frac{4q_0 a^4}{\pi^5 D}\sum_{m=1,3,5\ldots}^{\infty}\frac{1}{m^5}\left(1 - \frac{\alpha_m \tanh\alpha_m + 2}{2\cosh\alpha_m}\cosh\frac{2\alpha_m y}{b}\right. \\
& \left. + \frac{\alpha_m}{2\cosh\alpha_m}\frac{2y}{b}\sinh\frac{2\alpha_m y}{b}\right)\sin\frac{m\pi x}{a}
\end{aligned} \tag{2.69}$$

in which $\alpha_m = \frac{m\pi b}{2a}$.

Maximum deflection occurs at the middle of the plate, $x = \frac{a}{2}$, $y = 0$ and is given by

$$w = \frac{4q_0 a^4}{\pi^5 D}\sum_{m=1,3,5\ldots}^{\infty}\frac{(-1)^{\frac{m-1}{2}}}{m^5}\left(1 - \frac{\alpha_m \tanh\alpha_m + 2}{2\cosh\alpha_m}\right) \tag{2.70}$$

Solution of plates with arbitrary boundary conditions are complicated. It is possible to make some simplifying assumptions for plates with the same boundary conditions along two parallel edges in order to obtain the desired solution. Alternately, the energy method can be applied more efficiently to solve plates with complex boundary conditions. However, it should be noted that the accuracy of results depends upon the deflection function chosen. These functions must be so chosen that they satisfy at least the kinematics boundary conditions.

Figure 2.33 gives formulas for deflection and bending moments of rectangular plates with typical boundary and loading conditions.

Case No.	Structural System and Static Loading	Deflection and Internal Forces
1	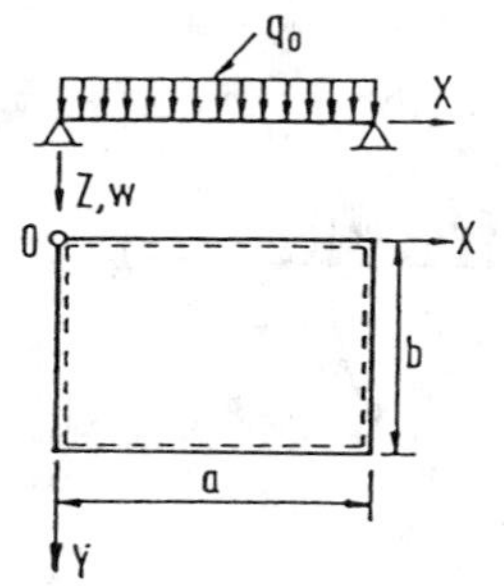	$$w = \frac{16q_0}{\pi^6 D} \sum_m \sum_n \frac{\sin\frac{m\pi x}{a}\sin\frac{n\pi y}{b}}{mn\left(\frac{m^2}{a^2}+\frac{n^2}{b^2}\right)^2}$$ $$m_x = \frac{16q_0 a^2}{\pi^4} \sum_m \sum_n \frac{\left(m^2+v\frac{n^2}{e^2}\right)\sin\frac{m\pi x}{a}\sin\frac{n\pi y}{b}}{mn\left(m^2+\frac{n^2}{e^2}\right)^2}$$ $$m_y = \frac{16q_0 a^2}{\pi^4} \sum_m \sum_n \frac{\left(\frac{n^2}{e^2}+vm^2\right)\sin\frac{m\pi x}{a}\sin\frac{n\pi y}{b}}{mn\left(m^2+\frac{n^2}{e^2}\right)^2}$$ $$\epsilon = \frac{b}{a}, \quad m = 1,3,5,\ldots,\infty; \quad n = 1,3,5,\ldots,\infty$$
2	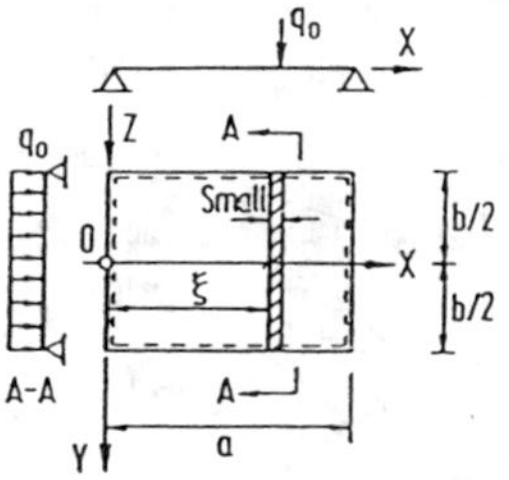	$$w = \frac{a^4}{D\pi^4}\sum_{m=1}^{\infty}\frac{P_m}{m^4}\left(1 - \frac{2+\alpha_m\tanh\alpha_m}{2\cosh\alpha_m}\cos\lambda_m y\right.$$ $$\left. + \frac{\lambda_m y\sinh\lambda_m y}{2\cosh\alpha_m}\right)\sin\lambda_m x$$ where $$P_m = \frac{2q_0}{a}\sin\frac{m\pi\xi}{a} \qquad \lambda_m = \frac{m\pi}{a}$$ $$m = 1,2,3,\ldots \qquad \alpha_m = \frac{m\pi b}{2a}$$
3	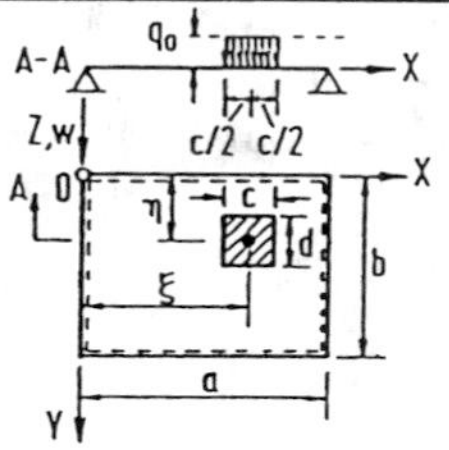	$$w = \frac{16q_0}{D\pi^6}\sum_m\sum_n \frac{\sin\frac{m\pi\xi}{a}\sin\frac{n\pi\eta}{b}\sin\frac{m\pi c}{b}\sin\frac{n\pi d}{2b}}{mn\left(\frac{m^2}{a^2}+\frac{n^2}{b^2}\right)^2}$$ $$\times\sin\frac{m\pi x}{a}\sin\frac{n\pi y}{b}$$ $$m = 1,2,3,\ldots$$ $$n = 1,2,3,\ldots$$
4	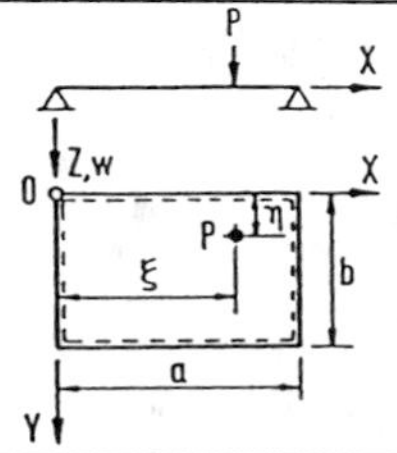	$$w = \frac{4P}{D\pi^4 ab}\sum_m\sum_n \frac{\sin\frac{m\pi\xi}{a}\sin\frac{n\pi\eta}{b}\sin\frac{m\pi x}{a}\sin\frac{n\pi y}{b}}{\left(\frac{m^2}{a^2}+\frac{n^2}{b^2}\right)^2}$$ $$m = 1,2,3,\ldots$$ $$n = 1,2,3,\ldots$$

Figure 2.33 Typical loading and boundary conditions for rectangular plates.

2.5.4 Bending of Circular Plates

In the case of symmetrically loaded circular plate, the loading is distributed symmetrically about the axis perpendicular to the plate through its center. In such cases, the deflection surface to which the middle plane of the plate is bent will also be symmetrical. The solution of circular plates can be conveniently carried out by using polar coordinates.

Stress resultants in a circular plate element are shown in Figure 2.34. The governing differential

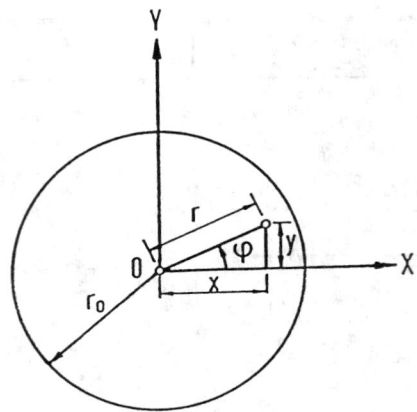

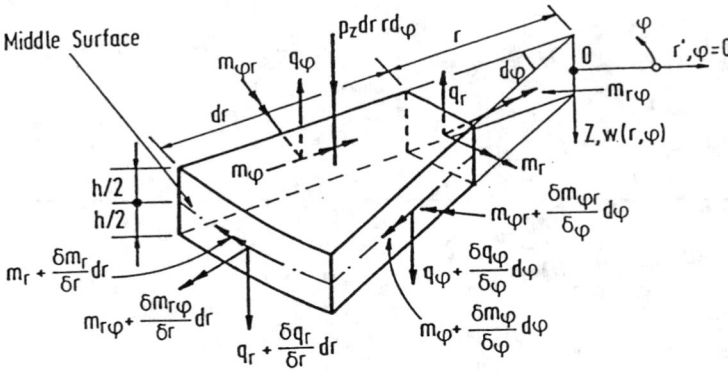

Figure 2.34   (a) Circular plate; (b) stress resultants.

equation is expressed in polar coordinates as

$$\frac{1}{r}\frac{d}{dr}\left\{r\frac{d}{dr}\left[\frac{1}{r}\frac{d}{dr}\left(r\frac{dw}{dr}\right)\right]\right\} = \frac{q}{D} \tag{2.71}$$

in which q is the intensity of loading.

In the case of uniformly loaded circular plates, Equation 2.71 can be integrated successively and the deflection at any point at a distance r from the center can be expressed as

$$w = \frac{q_0 r^4}{64D} + \frac{C_1 r^2}{4} + C_2 \log\frac{r}{a} + C_3 \tag{2.72}$$

in which q_o is the intensity of loading and a is the radius of the plate. C_1, C_2, and C_3 are constants of integration to be determined using the boundary conditions.

For a plate with clamped edges under uniformly distributed load q_o, the deflection surface reduces to

$$w = \frac{q_o}{64D} \left(a^2 - r^2 \right)^2 \tag{2.73}$$

The maximum deflection occurs at the center where $r = 0$, and is given by

$$w = \frac{q_o a^4}{64D} \tag{2.74}$$

Bending moments in the radial and tangential directions are respectively given by

$$\begin{aligned} M_r &= \frac{q_o}{16} \left[a^2(1+v) - r^2(3+v) \right] \\ M_t &= \frac{q_o}{16} \left[a^2(1+v) - r^2(1+3v) \right] \end{aligned} \tag{2.75}$$

The method of superposition can be applied in calculating the deflections for circular plates with simply supported edges. The expressions for deflection and bending moment are given as follows:

$$\begin{aligned} w &= \frac{q_o(a^2 - r^2)}{64D} \left(\frac{5+v}{1+v} a^2 - r^2 \right) \\ w_{max} &= \frac{5+v}{64(1+v)} \frac{q_o a^4}{D} \\ M_r &= \frac{q_o}{16}(3+v)(a^2 - r^2) \\ M_t &= \frac{q_o}{16} \left[a^2(3+v) - r^2(1+3v) \right] \end{aligned} \tag{2.76}$$
$$\tag{2.77}$$

This solution can be used to deal with plates with circular holes at the center and subjected to concentric moment and shearing forces. Plates subjected to concentric loading and concentrated loading also can be solved by this method. More rigorous solutions are available to deal with irregular loading on circular plates. Once again energy method can be employed advantageously to solve circular plate problems. Figure 2.35 gives deflection and bending moment expressions for typical cases of loading and boundary conditions on circular plates.

2.5.5 Strain Energy of Simple Plates

The strain energy expression for a simple rectangular plate is given by

$$\begin{aligned} U = \frac{D}{2} \int \int_{area} & \left\{ \left(\frac{\partial^2 w}{\partial x^2} + \frac{\partial^2 w}{\partial y^2} \right)^2 \right. \\ & \left. -2(1-v) \left[\frac{\partial^2 w}{\partial x^2} \frac{\partial^2 w}{\partial y^2} - \left(\frac{\partial^2 w}{\partial x \partial y} \right)^2 \right] \right\} dxdy \end{aligned} \tag{2.78}$$

Suitable deflection function $w(x, y)$ satisfying the boundary conditions of the given plate may be chosen. The strain energy, U, and the work done by the given load, $q(x, y)$,

$$W = -\int \int_{area} q(x, y)w(x, y)dxdy \tag{2.79}$$

can be calculated. The total potential energy is, therefore, given as $V = U + W$. Minimizing the total potential energy the plate problem can be solved.

Case No.	Structural System and Static Loading	Deflection and Internal Forces
1	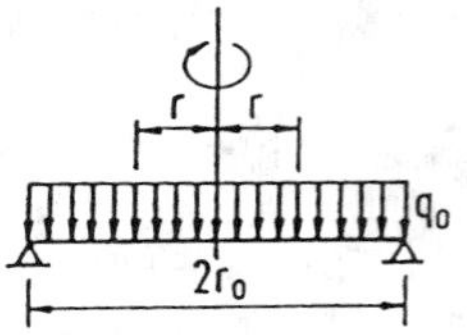 (circular plate, uniform load q_0, span $2r_0$)	$w = \dfrac{q_0 r_0^4}{64D(1+v)}[2(3+v)C_1 - (1+v)C_0]$ $m_r = \dfrac{q_0 r_0^2}{16}(3+v)C_1$ $m_\theta = \dfrac{q_0 r_0^2}{16}[2(1-v) - (1+3v)C_1]$ $q_r = \dfrac{q_0 r_0}{2}\rho$ $\rho = \dfrac{r}{r_0}$ $C_0 = 1-\rho^4$ $C_1 = 1-\rho^2$
2	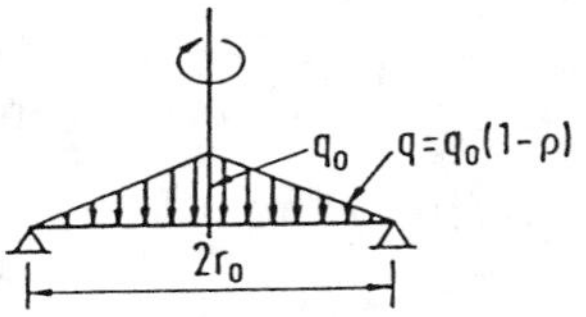 (triangular load q_0, $q = q_0(1-\rho)$, span $2r_0$)	$w = \dfrac{q_0 r_0^4}{14400D}\left[\dfrac{3(183+43v)}{1+v} - \dfrac{10(71+29v)}{1+v}\rho^2 + 225\rho^4 - 64\rho^5\right]$ $(m_r)_{\rho=0} = (m_\varphi)_{\rho=0} = \dfrac{q_0 r_0^4}{720}(71+29v);$ $(q_r)_{\rho=1} = -\dfrac{q_0 r_0}{6}$ $\rho = \dfrac{r}{r_0}$
3	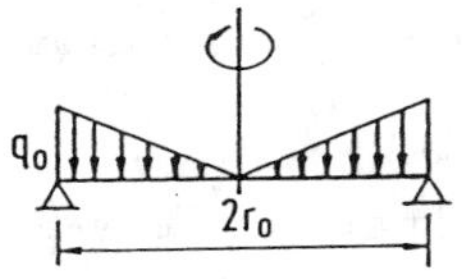 (load q_0, span $2r_0$)	$w = \dfrac{q_0 r_0^4}{450D}\left[\dfrac{3(6+v)}{1+v} - \dfrac{5(4+v)}{1+v}\rho^2 + 2\rho^5\right]$ $(m_r)_{\rho=0} = (m_\varphi)_{\rho=0} = \dfrac{q_0 r_0^2}{45}(4+v);$ $(q_r)_{\rho=1} = -\dfrac{q_0 r_0}{3}$ $\rho = \dfrac{r}{r_0}$
4	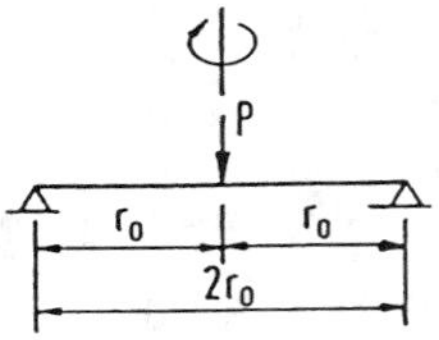 (central point load P, span $2r_0$)	$w = \dfrac{P r_0^2}{16\pi D}\left[\dfrac{3+v}{1+v}C_1 + 2C_2\right]$ $m_r = \dfrac{P}{4\pi}(1+v)C_3$ $m_\varphi = \dfrac{P}{4\pi}[(1-v) - (1+v)C_3]$ $q_r = \dfrac{P}{2\pi r_0 \rho}$ $C_1 = 1-\rho^2$ $C_2 = \rho^2 \ln\rho$ $C_3 = \ln\rho$ $\rho = \dfrac{r}{r_0}$

Figure 2.35   Typical loading and boundary conditions for circular plates.

Case No.	Structural System and Static Loading	Deflection and Internal Forces
5	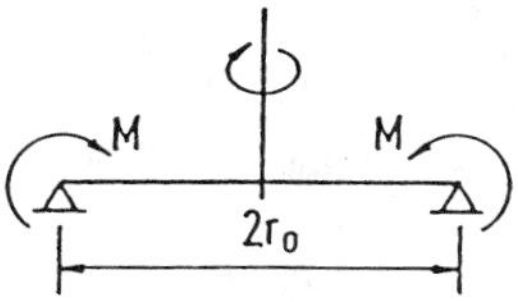	$w = \dfrac{M r_0^2}{2D(1+v)C_1}$ $m_r = m_\varphi = M$ $q_r = 0$ $C_1 = 1 - \rho^2, \quad \rho = \dfrac{r}{r_0}$
6	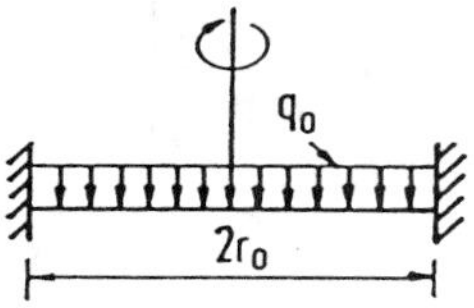	$w = \dfrac{q_0 r_0^4}{64D}(1-\rho^2)^2$ $q_r = -\dfrac{q_0 r_0}{2}\rho$ $m_r = \dfrac{q_0 r_0^2}{16}[1 + v - (3+v)\rho^2]$ $\rho = \dfrac{r}{r_0}$ $m_\varphi = \dfrac{q_0 r_0^2}{16}[1 + v - (1+3v)\rho^2]$
7	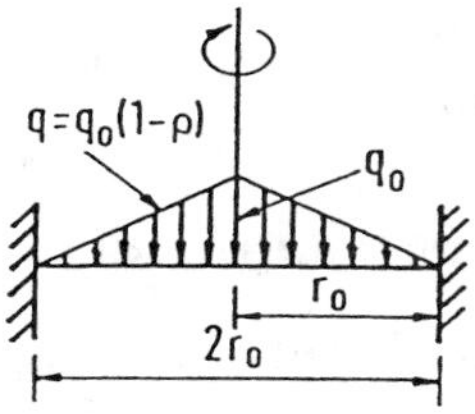	$w = \dfrac{q_0 r_0^4}{14400D}(129 - 290\rho^2 + 225\rho^4 - 64\rho^5)$ $(m_r)_{\rho=0} = (m_\varphi)_{\rho=0} = \dfrac{29 q_0 r_0^2}{720}(1+v)$ $(q_r)_{\rho=1} = -\dfrac{q_0 r_0}{6}$ $(m_r)_{\rho=1} = (m_\varphi)_{\rho=1} = -\dfrac{7 q_0 r_0^2}{120}$ $\rho = \dfrac{r}{r_0}$
8	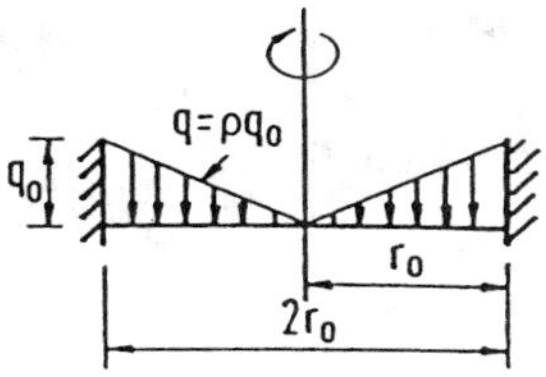	$w = \dfrac{q_0 r_0^4}{450D}(3 - 5\rho^2 + 2\rho^5)$ $q_r = -\dfrac{q_0 r_0}{3}\rho^2$ $m_r = \dfrac{q_0 r_0^2}{45}[1 + v - (4+v)\rho^3]$ $\rho = \dfrac{r}{r_0}$ $m_\varphi = \dfrac{q_0 r_0^2}{45}[1 + v - (1+4v)\rho^3]$
9	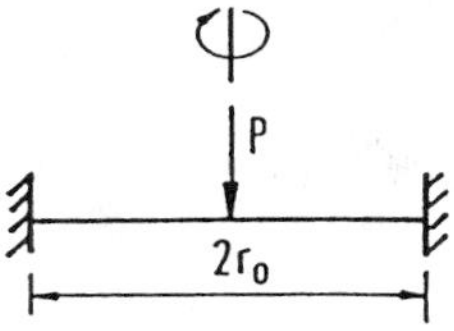	$w = \dfrac{P r_0^2}{16\pi D}(1 - \rho^2 + 2\rho^2 \ln\rho)$ $q_r = -\dfrac{P}{2\pi r_0 \rho}$ $m_r = -\dfrac{P}{4\pi}[1 + (1+v)\ln\rho]$ $\rho = \dfrac{r}{r_0}$ $m_\varphi = -\dfrac{P}{4\pi}[v + (1+v)\ln\rho]$

Figure 2.35 *(Continued)* Typical loading and boundary conditions for circular plates.

$$\left[\frac{\partial^2 w}{\partial x^2} \frac{\partial^2 w}{\partial y^2} - \left(\frac{\partial^2 w}{\partial x \partial y} \right)^2 \right]$$

The term is known as the Gaussian curvature.

If the function $w(x, y) = f(x) \cdot \phi(y)$ (product of a function of x only and a function of y only) and $w = 0$ at the boundary are assumed, then the integral of the Gaussian curvature over the entire plate equals zero. Under these conditions

$$U = \frac{D}{2} \int \int_{area} \left(\frac{\partial^2 w}{\partial x^2} + \frac{\partial^2 w}{\partial y^2} \right)^2 dx dy$$

If polar coordinates instead of rectangular coordinates are used and axial symmetry of loading and deformation is assumed, the equation for strain energy, U, takes the form

$$U = \frac{D}{2} \int \int_{area} \left\{ \left(\frac{\partial^2 w}{\partial r^2} + \frac{1}{r} \frac{\partial w}{\partial r} \right)^2 - \frac{2(1-v)}{r} \frac{\partial w}{\partial r} \frac{\partial^2 w}{\partial r^2} \right\} r dr d\theta \qquad (2.80)$$

and the work done, W, is written as

$$W = - \int \int_{area} q w r dr d\theta \qquad (2.81)$$

Detailed treatment of the Plate Theory can be found in [56].

2.5.6 Plates of Various Shapes and Boundary Conditions

Simply Supported Isosceles Triangular Plate Subjected to a Concentrated Load

Plates of shapes other than circle and rectangle are used in some situations. A rigorous solution of the deflection for a plate with a more complicated shape is likely to be very difficult. Consider, for example, the bending of an isosceles triangular plate with simply supported edges under concentrated load P acting at an arbitrary point (Figure 2.36). A solution can be obtained for this plate by considering a mirror image of the plate as shown in the figure. The deflection of

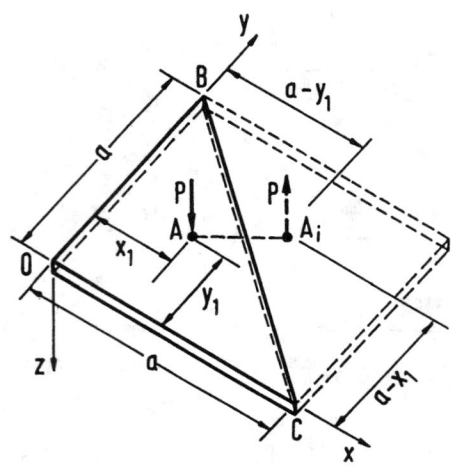

Figure 2.36   Isosceles triangular plate.

OBC of the square plate is identical with that of a simply supported triangular plate OBC. The deflection owing to the force P can be written as

$$w_1 = \frac{4Pa^2}{\pi^4 D} \sum_{m=1}^{\infty} \sum_{n=1}^{\infty} \frac{\sin(m\pi x_1/a)\sin(n\pi y_1/a)}{(m^2+n^2)^2} \sin\frac{m\pi x}{a} \sin\frac{n\pi y}{a} \qquad (2.82)$$

Upon substitution of $-P$ for P, $(a-y_1)$ for x_1, and $(a-x1)$ for $y1$ in Equation 2.82 we obtain the deflection due to the force $-P$ at A_i:

$$w_2 = -\frac{4Pa^2}{\pi^4 D} \sum_{m=1}^{\infty} \sum_{n=1}^{\infty} (-1)^{m+n} \frac{\sin(m\pi x_1/a)\sin(n\pi y_1/a)}{(m^2+n^2)^2} \sin\frac{m\pi x}{a} \sin\frac{n\pi y}{a} \qquad (2.83)$$

The deflection surface of the triangular plate is then

$$w = w_1 + w_2 \qquad (2.84)$$

Equilateral Triangular Plates

The deflection surface of a simply supported plate loaded by uniform moment M_o along its boundary and the surface of a uniformly loaded membrane, uniformly stretched over the same triangular boundary, are identical. The deflection surface for such a case can be obtained as

$$w = \frac{M_o}{4aD}\left[x^3 - 3xy^2 - a(x^2+y^2) + \frac{4}{27}a^3\right] \qquad (2.85)$$

If the simply supported plate is subjected to uniform load p_o the deflection surface takes the form

$$w = \frac{p_o}{64aD}\left[x^3 - 3xy^2 - a(x^2+y^2) + \frac{4}{27}a^3\right]\left(\frac{4}{9}a^2 - x^2 - y^2\right) \qquad (2.86)$$

For the equilateral triangular plate (Figure 2.37) subjected to uniform load and supported at the corners approximate solutions based on the assumption that the total bending moment along each

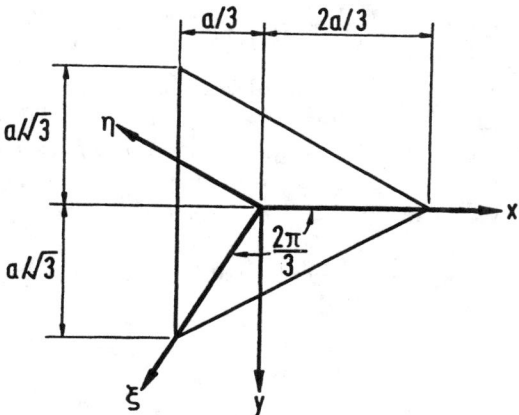

Figure 2.37 Equilateral triangular plate with coordinate axes.

side of the triangle vanishes were obtained by Vijakkhana et al. [58] who derived an equation for deflection surface as

$$
\begin{aligned}
w = \frac{qa^4}{144(1-v^2)D} \Bigg[& \frac{8}{27}(7+v)(2-v) - (7+v)(1-v) \\
& \left(\frac{x^2}{a^2} + \frac{y^2}{a^2}\right) - (5-v)(1+v)\left(\frac{x^3}{a^3} - 3\frac{xy^2}{a^3}\right) \\
& + \frac{9}{4}(1-v^2)\left(\frac{x^4}{a^4} + 2\frac{x^2 y^2}{a^4} + \frac{y^4}{a^4}\right)\Bigg]
\end{aligned}
\tag{2.87}
$$

The errors introduced by the approximate boundary condition, i.e., the total bending moment along each side of the triangle vanishes, are not significant because its influence on the maximum deflection and stress resultants is small for practical design purposes. The value of the twisting moment on the edge at the corner given by this solution is found to be exact.

The details of the mathematical treatment may be found in [58].

Rectangular Plate Supported at Corners

Approximate solutions for rectangular plates supported at the corners and subjected to uniformly distributed load were obtained by Lee and Ballesteros [36]. The approximate deflection surface is given as

$$
\begin{aligned}
w = \frac{qa^4}{48(1-v^2)D} \Bigg[& (10+v-v^2)\left(1+\frac{b^4}{a^4}\right) - 2(7v-1)\frac{b^2}{a^2} \\
& + 2\left((1+5v)\frac{b^2}{a^2} - (6+v-v^2)\right)\frac{x}{a} \\
& + 2\left((1+5v) - (6+v-v^2)\frac{b^2}{a^2}\right)\frac{y^2}{a^2} \\
& + (2+v-v^2)\frac{x^4+y^4}{a^4} - 6(1+v)\frac{x^2 y^2}{a^4}\Bigg]
\end{aligned}
\tag{2.88}
$$

The details of the mathematical treatment may be found in [36].

2.5.7 Orthotropic Plates

Plates of anisotropic materials have important applications owing to their exceptionally high bending stiffness. A nonisotropic or anisotropic material displays direction-dependent properties. Simplest among them are those in which the material properties differ in two mutually perpendicular directions. A material so described is orthotropic, e.g., wood. A number of manufactured materials are approximated as orthotropic. Examples include corrugated and rolled metal sheets, fillers in sandwich plate construction, plywood, fiber reinforced composites, reinforced concrete, and gridwork. The latter consists of two systems of equally spaced parallel ribs (beams), mutually perpendicular, and attached rigidly at the points of intersection.

The governing equation for orthotropic plates similar to that of isotropic plates (Equation 2.42) takes the form

$$
D_x \frac{\delta^4 w}{\delta x^4} + 2H \frac{\delta^4 w}{\delta x^2 \delta y^2} + D_y \frac{\delta^4 w}{\delta y^4} = q
\tag{2.89}
$$

In which

$$
D_x = \frac{h^3 E_x}{12}, \quad D_y = \frac{h^3 E_y}{12}, \quad H = D_{xy} + 2G_{xy}, \quad D_{xy} = \frac{h^3 E_{xy}}{12}, \quad G_{xy} = \frac{h^3 G}{12}
$$

The expressions for D_x, D_y, D_{xy}, and G_{xy} represent the flexural rigidities and the torsional rigidity of an orthotropic plate, respectively. E_x, E_y, and G are the orthotropic plate moduli. Practical considerations often lead to assumptions, with regard to material properties, resulting in approximate expressions for elastic constants. The accuracy of these approximations is generally the most significant factor in the orthotropic plate problem. Approximate rigidities for some cases that are commonly encountered in practice are given in Figure 2.38.

Geometry	Rigidities
A. Reinforced concrete slab with x and y directed reinforcement steel bars 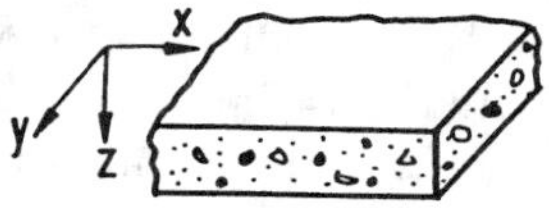	$D_x = \dfrac{E_c}{1-v_c^2}\left[I_{cx} + \left(\dfrac{E_s}{E_c} - 1\right)I_{sx}\right]$ $\qquad$ $D_y = \dfrac{E_c}{1-v_c^2}\left[I_{cy} + \left(\dfrac{E_s}{E_c} - 1\right)I_{sy}\right]$ $G_{xy} = \dfrac{1-v_c}{2}\sqrt{D_x D_y}$ $\qquad$ $H = \sqrt{D_x D_y}$ $\qquad$ $D_{xy} = v_c\sqrt{D_x D_y}$ v_c: Poisson's ratio for concrete E_c, E_s: Elastic modulus of concrete and steel, respectively $I_{cx}(I_{sx})$, $I_{cy}(I_{sy})$: Moment of ineretia of the slab (steel bars) about neutral axis in the section x = constant and y = constant, respectively
B. Plate reinforced by equidistant stiffeners 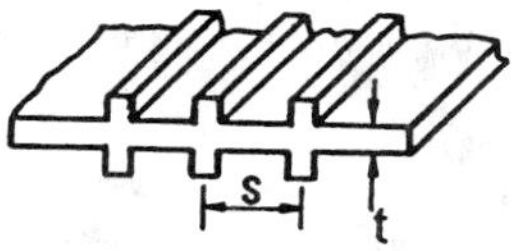	$D_x = H = \dfrac{E t^3}{12(1-v^2)}$ $\qquad$ $D_y = \dfrac{E t^3}{12(1-v^2)} + \dfrac{E'I}{s}$ E, E': Elastic modulus of plating and stiffeners, respectively v : Poisson's ratio of plating s : Spacing between centerlines of stiffeners I : Moment of inertia of the stiffener cross section with respect to midplane of plating
C. Plate reinforced by a set of equidistant ribs 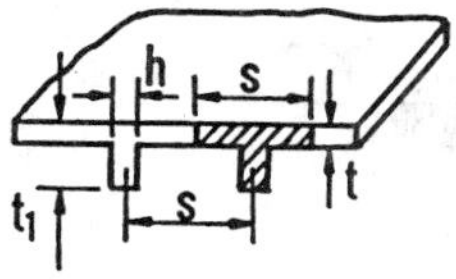	$D_x = \dfrac{Est^3}{12[s - h + h(t\,t_1)^3]}$ $\qquad$ $D_y = \dfrac{EI}{s}$ $H = 2G'_{xy} + \dfrac{C}{s}$ $\qquad$ $D_{xy} = 0$ C : Torsional rigidity of one rib I : Moment of inertia about neutral axis of a T-section of width s (shown as shaded) G'_{xy} : Torsional rigidity of the plating E : Elastic modulus of the plating
D. Corrugated plate 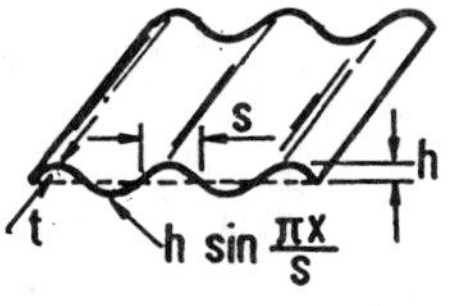	$D_x = \dfrac{s}{\lambda}\dfrac{Et^3}{12(1-v^2)}$ $\qquad$ $D_y = EI$, $H = \dfrac{\lambda}{a}\dfrac{Et^3}{12(1+v)}$ $\qquad$ $D_{xy} = 0$ where $\lambda = s\left(1 + \dfrac{\pi^2 h^2}{4s^2}\right)$ $\qquad$ $I = 0.5h^2 t\left[1 - \dfrac{0.81}{1 + 2.5(h/2s)^2}\right]$

Figure 2.38 Various orthotropic plates.

General solution procedures applicable to the case of isotropic plates are equally applicable to the orthotropic plates as well. Deflections and stress-resultants can thus be obtained for orthotropic plates of different shapes with different support and loading conditions. These problems have been researched extensively and solutions concerning plates of various shapes under different boundary and loading conditions may be found in the references, namely [37, 52, 53, 56, 57].

2.5.8 Buckling of Thin Plates

Rectangular Plates

Buckling of a plate involves bending in two planes and is therefore fairly complicated. From a mathematical point of view, the main difference between columns and plates is that quantities such as deflections and bending moments, which are functions of a single independent variable, in columns become functions of two independent variables in plates. Consequently, the behavior of plates is described by partial differential equations, whereas ordinary differential equations suffice for describing the behavior of columns. A significant difference between columns and plates is also apparent if one compares their buckling characteristics. For a column, buckling terminates the ability of the member to resist axial load, and the critical load is thus the failure load of the member. However, the same is not true for plates. These structural elements can, subsequently to reaching the critical load, continue to resist increasing axial force, and they do not fail until a load considerably in excess of the critical load is reached. The critical load of a plate is, therefore, not its failure load. Instead, one must determine the load-carrying capacity of a plate by considering its postbuckling behavior.

To determine the critical in-plane loading of a plate by the concept of neutral equilibrium, a governing equation in terms of biaxial compressive forces N_x and N_y and constant shear force N_{xy} as shown in Figure 2.39 can be derived as

$$D\left(\frac{\delta^4 w}{\delta x^4} + 2\frac{\delta^4 w}{\delta x^2 \delta y^2} + \frac{\delta^4 w}{\delta y^4}\right) + N_x \frac{\delta^2 w}{\delta x^2} + N_y \frac{\delta^2 w}{\delta y^2} + 2N_{xy}\frac{\delta^2 w}{\delta x \delta y} = 0 \qquad (2.90)$$

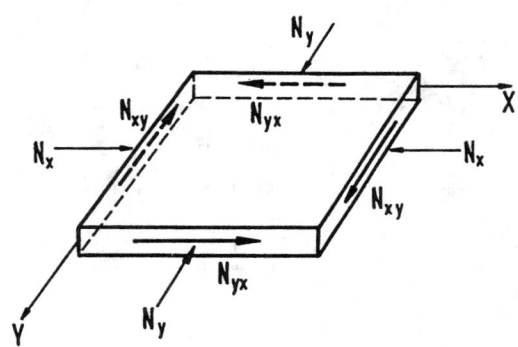

Figure 2.39   Plate subjected to in-plane forces.

The critical load for uniaxial compression can be determined from the differential equation

$$D\left(\frac{\delta^4 w}{\delta x^4} + 2\frac{\delta^4 w}{\delta x^2 \delta y^2} + \frac{\delta^4 w}{\delta y^4}\right) + N_x \frac{\delta^2 w}{\delta x^2} = 0 \qquad (2.91)$$

which is obtained by setting $N_y = N_{xy} = 0$ in Equation 2.90.

For example, in the case of a simply supported plate Equation 2.91 can be solved to give

$$N_x = \frac{\pi^2 a^2 D}{m^2}\left(\frac{m^2}{a^2} + \frac{n^2}{b^2}\right)^2 \qquad (2.92)$$

The critical value of N_x, i.e., the smallest value, can be obtained by taking n equal to 1. The physical meaning of this is that a plate buckles in such a way that there can be several half-waves in the direction of compression but only one half-wave in the perpendicular direction. Thus, the expression for the critical value of the compressive force becomes

$$(N_x)_{cr} = \frac{\pi^2 D}{a^2} \left(m + \frac{1}{m} \frac{a^2}{b^2} \right)^2 \tag{2.93}$$

The first factor in this expression represents the Euler load for a strip of unit width and of length a. The second factor indicates in what proportion the stability of the continuous plate is greater than the stability of an isolated strip. The magnitude of this factor depends on the magnitude of the ratio a/b and also on the number m, which gives the number of half-waves into which the plate buckles. If 'a' is smaller than 'b', the second term in the parenthesis of Equation 2.93 is always smaller than the first and the minimum value of the expression is obtained by taking $m = 1$, i.e., by assuming that the plate buckles in one half-wave. The critical value of N_x can be expressed as

$$N_{cr} = \frac{k \pi^2 D}{b^2} \tag{2.94}$$

The factor k depends on the aspect ratio a/b of the plate and m, the number of half-waves into which the plate buckles in the x direction. The variation of k with a/b for different values of m can be plotted, as shown in Figure 2.40. The critical value of N_x is the smallest value that is obtained

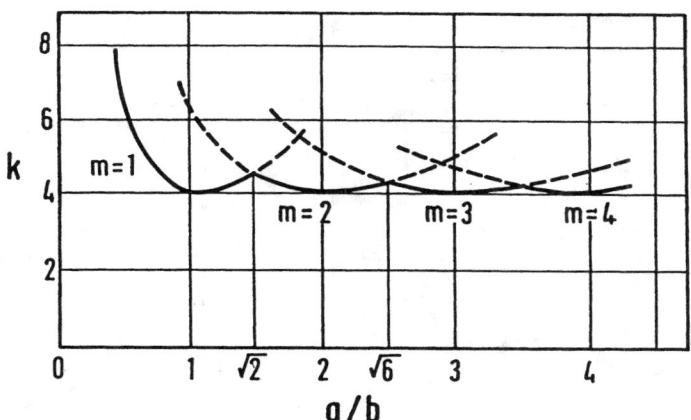

Figure 2.40   Buckling stress coefficients for unaxially compressed plate.

for $m = 1$ and the corresponding value of k is equal to 4.0. This formula is analogous to Euler's formula for buckling of a column.

In the more general case in which normal forces N_x and N_y and the shearing forces N_{xy} are acting on the boundary of the plate, the same general method can be used. The critical stress for the case of a uniaxially compressed simply supported plate can be written as

$$\sigma_{cr} = 4 \frac{\pi^2 E}{12(1 - v^2)} \left(\frac{h}{b} \right)^2 \tag{2.95}$$

The critical stress values for different loading and support conditions can be expressed in the form

$$f_{cr} = k \frac{\pi^2 E}{12(1 - v^2)} \left(\frac{h}{b} \right)^2 \tag{2.96}$$

in which f_{cr} is the critical value of different loading cases. Values of k for plates with several different boundary and loading conditions are given in Figure 2.41.

Case	Boundary condition	Type of stress	Value of k for long plate
(a)	S.S. / S.S. S.S. / S.S	Compression	4.0
(b)	Fixed / S.S. S.S. / Fixed	Compression	6.97
(c)	S.S. / S.S. S.S. / Free	Compression	0.425
(d)	Fixed / S.S. S.S. / Free	Compression	1.277
(e)	Fixed / S.S. S.S. / S.S.	Compression	5.42
(f)	S.S / S.S. S.S. / S.S.	Shear	5.34
(g)	Fixed / Fixed Fixed / Fixed	Shear	8.98
(h)	S.S. / S.S. S.S. / S.S.	Bending	23.9
(i)	Fixed / Fixed Fixed / Fixed	Bending	41.8

Figure 2.41   Values of K for plate with different boundary and loading conditions.

Circular Plates

The critical value of the compressive forces N_r uniformly distributed around the edge of a circular plate of radius r_o, clamped along the edge (Figure 2.42) can be determined by using the governing equation

$$r^2 \frac{d^2\phi}{dr^2} + r \frac{d\phi}{dr} - \phi = -\frac{Qr^2}{D} \qquad (2.97)$$

in which ϕ is the angle between the axis of revolution of the plate surface and any normal to the plate, r is the distance of any point measured from the center of the plate, and Q is the shearing force per unit of length. When there are no lateral forces acting on the plate, the solution of Equation 2.97 involves a Bessel function of the first order of the first and second kind and the resulting critical

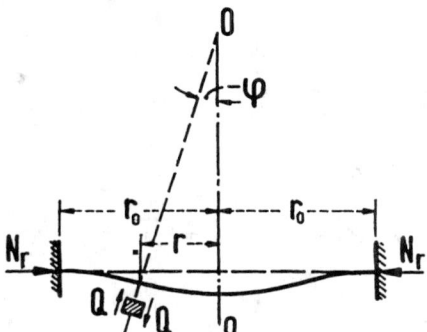

Figure 2.42 Circular plate under compressive loading.

value of N_r is obtained as

$$(N_r)_{cr} = \frac{14.68D}{r_0^2} \tag{2.98}$$

The critical value of N_r for the plate when the edge is simply supported can be obtained in the same way as

$$(N_r)_{cr} = \frac{4.20D}{r_0^2} \tag{2.99}$$

2.6 Shell

2.6.1 Stress Resultants in Shell Element

A **thin shell** is defined as a shell with a thickness that is relatively small compared to its other dimensions. Also, deformations should not be large compared to the thickness. The primary difference between a shell structure and a plate structure is that the former has a curvature in the unstressed state, whereas the latter is assumed to be initially flat. The presence of initial curvature is of little consequence as far as flexural behavior is concerned. The membrane behavior, however, is affected significantly by the curvature. Membrane action in a surface is caused by in-plane forces. These forces may be primary forces caused by applied edge loads or edge deformations, or they may be secondary forces resulting from flexural deformations.

In the case of the flat plates, secondary in-plane forces do not give rise to appreciable membrane action unless the bending deformations are large. Membrane action due to secondary forces is, therefore, neglected in small deflection theory. If the surface, as in the case of shell structures, has an initial curvature, membrane action caused by secondary in-plane forces will be significant regardless of the magnitude of the bending deformations.

A plate is likened to a two-dimensional beam and resists transverse loads by two dimensional bending and shear. A membrane is likened to a two-dimensional equivalent of the cable and resists loads through tensile stresses. Imagine a membrane with large deflections (Figure 2.43a), reverse the load and the membrane and we have the structural shell (Figure 2.43b) provided that the shell is stable for the type of load shown. The membrane resists the load through tensile stresses but the ideal thin shell must be capable of developing both tension and compression.

Consider an infinitely small shell element formed by two pairs of adjacent planes which are normal to the middle surface of the shell and which contain its principal curvatures as shown in Figure 2.44a. The thickness of the shell is denoted as h. Coordinate axes x and y are taken tangent at 'O' to the lines of principal curvature and the axis z normal to the middle surface. r_x and r_y are the principal radii of curvature lying in the xz and yz planes, respectively. The resultant forces per

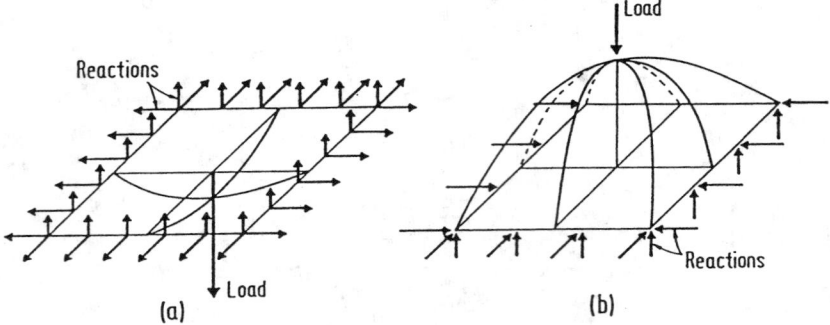

Figure 2.43   Membrane with large deflections.

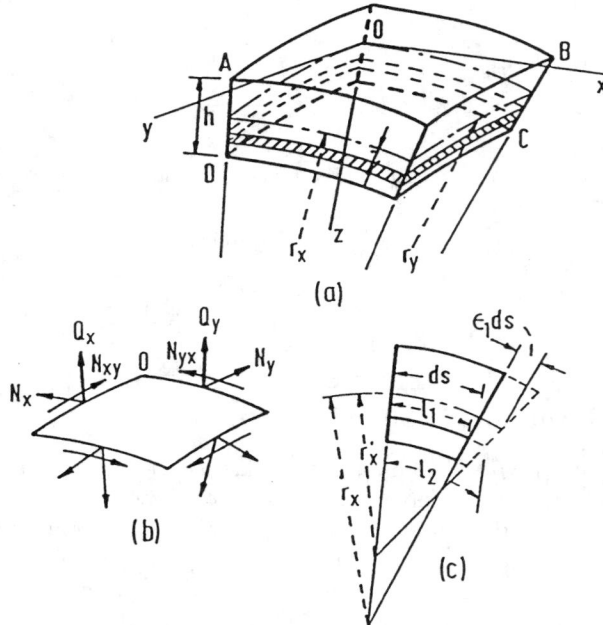

Figure 2.44   A shell element.

unit length of the normal sections are given as

$$N_x = \int_{-h/2}^{h/2} \sigma_x \left(1 - \frac{z}{r_y}\right) dz, \quad N_y = \int_{-h/2}^{h/2} \sigma_y \left(1 - \frac{z}{r_x}\right) dz$$

$$N_{xy} = \int_{-h/2}^{h/2} \tau_{xy} \left(1 - \frac{z}{r_y}\right) dz, \quad N_{yx} = \int_{-h/2}^{h/2} \tau_{yx} \left(1 - \frac{z}{r_x}\right) dz$$

$$Q_x = \int_{-h/2}^{h/2} \tau_{xz} \left(1 - \frac{z}{r_y}\right) dz, \quad Q_y = \int_{-h/2}^{h/2} \tau_{yz} \left(1 - \frac{z}{r_x}\right) dz \qquad (2.100)$$

The bending and twisting moments per unit length of the normal sections are given by

$$M_x = \int_{-h/2}^{h/2} \sigma_x z \left(1 - \frac{z}{r_y}\right) dz, \quad M_y = \int_{-h/2}^{h/2} \sigma_y z \left(1 - \frac{z}{r_x}\right) dz$$

$$M_{xy} = -\int_{-h/2}^{h/2} \tau_{xy} z \left(1 - \frac{z}{r_y}\right) dz, \quad M_{yx} = \int_{-h/2}^{h/2} \tau_{yx} z \left(1 - \frac{z}{r_x}\right) dz \qquad (2.101)$$

It is assumed, in bending of the shell, that linear elements as AD and BC (Figure 2.44), which are normal to the middle surface of the shell, remain straight and become normal to the deformed middle surface of the shell. If the conditions of a shell are such that bending can be neglected, the problem of stress analysis is greatly simplified because the resultant moments (Equation 2.101) vanish along with shearing forces Q_x and Q_y in Equation 2.100. Thus, the only unknowns are N_x, N_y, and $N_{xy} = N_{yx}$ and these are called membrane forces.

2.6.2 Membrane Theory of Shells of Revolution

Shells having the form of surfaces of revolution find extensive application in various kinds of containers, tanks, and domes. Consider an element of a shell cut by two adjacent meridians and two parallel circles as shown in Figure 2.45. There will be no shearing forces on the sides of the element

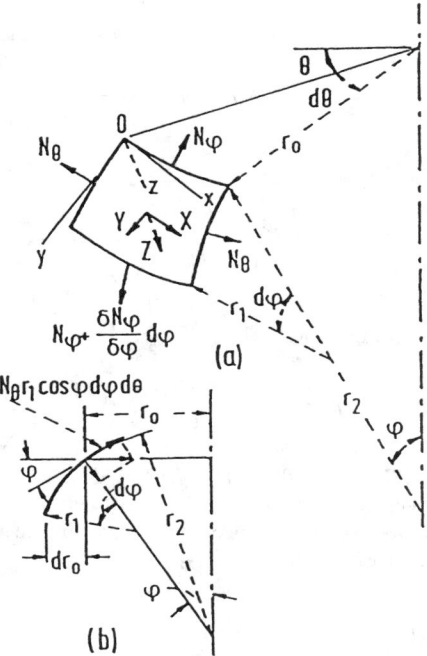

Figure 2.45   An element from shells of revolution—symmetrical loading.

because of the symmetry of loading. By considering the equilibrium in the direction of the tangent to the meridian and z, two equations of equilibrium are written, respectively, as

$$\frac{d}{d\phi}(N_\phi r_0) - N_\theta r_1 \cos\phi + Y r_1 r_0 = 0$$

$$N_\phi r_0 + N_\theta r_1 \sin\phi + Z r_1 r_0 = 0 \tag{2.102}$$

The force N_θ and N_ϕ can be calculated from Equation 2.102 if the radii r_0 and r_1 and the components Y and Z of the intensity of the external load are given.

2.6.3 Spherical Dome

The spherical shell shown in Figure 2.46 is assumed to be subjected to its own weight; the intensity of the self weight is assumed as a constant value q_o per unit area. Considering an element of the

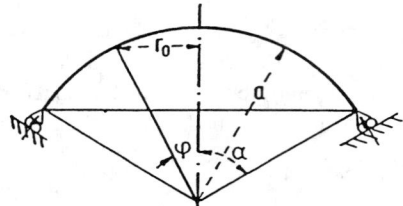

Figure 2.46   Spherical dome.

shell at an angle ϕ, the self weight of the portion of the shell above this element is obtained as

$$r = 2\pi \int_0^\phi a^2 q_o \sin\phi\, d\phi$$

$$= 2\pi a^2 q_o (1 - \cos\phi)$$

Considering the equilibrium of the portion of the shell above the parallel circle defined by the angle ϕ, we can write

$$2\pi r_0 N_\phi \sin\phi + R = 0 \tag{2.103}$$

Therefore,

$$N_\phi = -\frac{aq(1 - \cos\phi)}{\sin^2\phi} = -\frac{aq}{1 + \cos\phi}$$

We can write from Equation 2.102

$$\frac{N_\phi}{r_1} + \frac{N_\theta}{r_2} = -Z \tag{2.104}$$

Substituting for N_ϕ and R into Equation 2.104

$$N_\theta = -aq \left(\frac{1}{1 + \cos\phi} - \cos\phi \right)$$

It is seen that the forces N_ϕ are always negative. Thus, there is a compression along the meridians that increases as the angle ϕ increases. The forces N_θ are also negative for small angles ϕ. The stresses as calculated above will represent the actual stresses in the shell with great accuracy if the supports are of such a type that the reactions are tangent to meridians as shown in the figure.

2.6.4   Conical Shells

If a force P is applied in the direction of the axis of the cone as shown in Figure 2.47, the stress distribution is symmetrical and we obtain

$$N_\phi = -\frac{P}{2\pi r_0 \cos\alpha}$$

By Equation 2.104, one obtains $N_\theta = 0$.

In the case of a conical surface in which the lateral forces are symmetrically distributed, the membrane stresses can be obtained by using Equations 2.103 and 2.104. The curvature of the meridian in the case of a cone is zero and hence $r_1 = \infty$; Equations 2.103 and 2.104 can, therefore, be written as

$$N_\phi = -\frac{R}{2\pi r_0 \sin\phi}$$

and

$$N_\theta = -r_2 Z = -\frac{Z r_0}{\sin\phi}$$

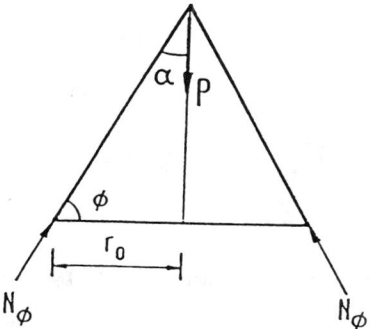

Figure 2.47 Conical shell.

If the load distribution is given, N_ϕ and N_θ can be calculated independently.

For example, a conical tank filled with a liquid of specific weight γ is considered as shown in Figure 2.48. The pressure at any parallel circle *mn* is

$$p = -Z = \gamma(d - y)$$

For the tank, $\phi = \alpha + \frac{\pi}{2}$ and $r_0 = y \tan \alpha$.

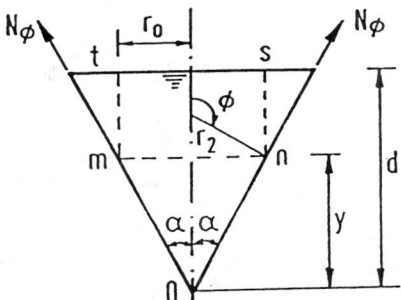

Figure 2.48 Inverted conical tank.

Therefore,

$$N_\theta = \frac{\gamma(d - y)y \tan \alpha}{\cos \alpha}$$

N_θ is maximum when $y = \frac{d}{2}$ and hence

$$(N_\theta)_{max} = \frac{\gamma d^2 \tan \alpha}{4 \cos \alpha}$$

The term R in the expression for N_ϕ is equal to the weight of the liquid in the conical part *mno* and the cylindrical part must be as shown in Figure 2.47. Therefore,

$$
\begin{aligned}
R &= -\left[\frac{1}{3}\pi y^3 \tan^2 \alpha + \pi y^2 (d - y) \tan^2 \alpha\right]\gamma \\
&= -\pi \gamma y^2 \left(d - \frac{2}{3}y\right)\tan^2 \alpha
\end{aligned}
$$

Hence,

$$N_\phi = \frac{\gamma y \left(d - \frac{2}{3}y\right)\tan \alpha}{2 \cos \alpha}$$

N_ϕ is maximum when $y = \frac{3}{4}d$ and

$$(N_\phi)_{max} = \frac{3}{16}\frac{d^2\gamma\tan\alpha}{\cos\alpha}$$

The horizontal component of N_ϕ is taken by the reinforcing ring provided along the upper edge of the tank. The vertical components constitute the reactions supporting the tank.

2.6.5 Shells of Revolution Subjected to Unsymmetrical Loading

Consider an element cut from a shell by two adjacent meridians and two parallel circles (Figure 2.49). In the general case, shear forces $N_{\varphi\theta} = N_{\theta\varphi}$ in addition to normal forces N_φ and N_θ will act on the

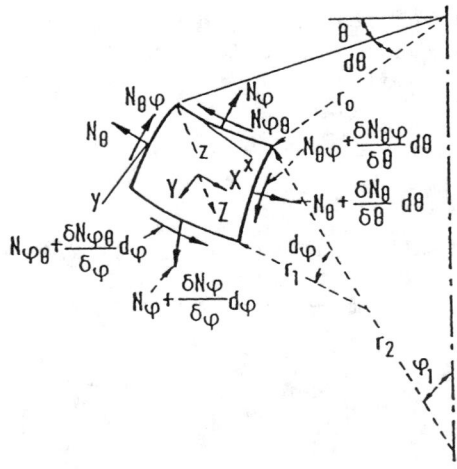

Figure 2.49   An element from shells of revolution—unsymmetrical loading.

sides of the element. Projecting the forces on the element in the y direction we obtain the equation

$$\frac{\partial}{\partial\varphi}(N_\varphi r_0) + \frac{\partial N_{\theta\varphi}}{\partial\theta}r_1 - N_\theta r_1\cos\varphi + Y r_1 r_0 = 0 \qquad (2.105)$$

Similarly the forces in the x direction can be summed up to give

$$\frac{\partial}{\partial\varphi}(r_0 N_{\varphi\theta}) + \frac{\partial N_\theta}{\partial\theta}r_1 + N_{\theta\varphi}r_1\cos\varphi + X r_0 r_1 = 0 \qquad (2.106)$$

Since the projection of shearing forces on the z axis vanishes, the third equation is the same as Equation 2.104. The problem of determining membrane stresses under unsymmetrical loading reduces to the solution of Equations 2.104, 2.105, and 2.106 for given values of the components X, Y, and Z of the intensity of the external load.

2.6.6 Membrane Theory of Cylindrical Shells

It is assumed that the generator of the shell is horizontal and parallel to the x axis. An element is cut from the shell by two adjacent generators and two cross-sections perpendicular to the x axis, and its position is defined by the coordinate x and the angle φ. The forces acting on the sides of the element are shown in Figure 2.50b.

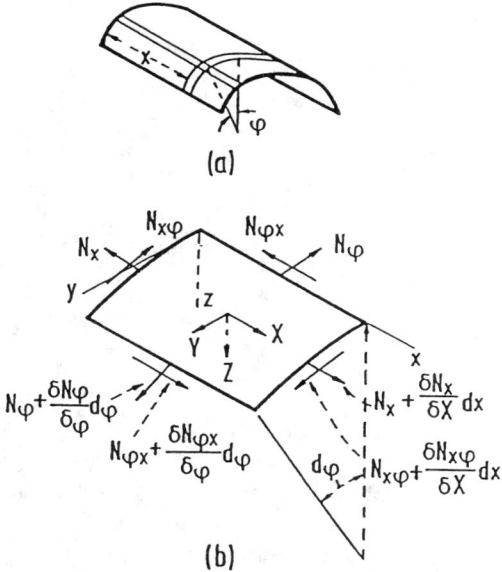

Figure 2.50   Membrane forces on a cylindrical shell element.

The components of the distributed load over the surface of the element are denoted as X, Y, and Z. Considering the equilibrium of the element and summing up the forces in the x direction, we obtain

$$\frac{\partial N_x}{\partial x} r\, d\varphi\, dx + \frac{\partial N_{\varphi x}}{\partial \varphi} d\varphi\, dx + X r\, d\varphi\, dx = 0$$

The corresponding equations of equilibrium in the y and z directions are given, respectively, as

$$\frac{\partial N_{x\varphi}}{\partial x} r\, d\varphi\, dx + \frac{\partial N_{\varphi}}{\partial \varphi} d\varphi\, dx + Y r\, d\varphi\, dx = 0$$

$$N_{\varphi}\, d\varphi\, dx + Z r\, d\varphi\, dx = 0$$

The three equations of equilibrium can be simplified and represented in the following form:

$$\frac{\partial N_x}{\partial x} + \frac{1}{r}\frac{\partial N_{x\varphi}}{\partial \varphi} = -X$$

$$\frac{\partial N_{x\varphi}}{\partial x} + \frac{1}{r}\frac{\partial N_{\varphi}}{\partial \varphi} = -Y$$

$$N_{\varphi} = -Z_r \tag{2.107}$$

In each particular case we readily find the value of N_{φ}. Substituting this value in the second of the equations, we then obtain $N_{x\varphi}$ by integration. Using the value of $N_{x\varphi}$ thus obtained we find N_x by integrating the first equation.

2.6.7   Symmetrically Loaded Circular Cylindrical Shells

In practical applications problems in which a circular shell is subjected to the action of forces distributed symmetrically with respect to the axis of the cylinder are common. To establish the equations required for the solution of these problems, we consider an element, as shown in Figures 2.50a and 2.51, and consider the equations of equilibrium. From symmetry, the membrane shearing forces $N_{x\varphi} = N_{\varphi x}$ vanish in this case; forces N_{φ} are constant along the circumference. From symmetry, only the forces Q_z do not vanish. Considering the moments acting on the element

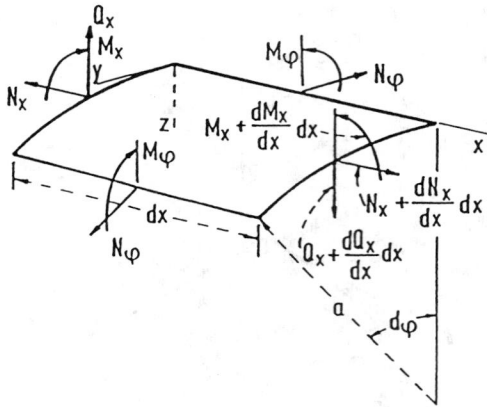

Figure 2.51   Stress resultants in a cylindrical shell element.

in Figure 2.51, from symmetry it can be concluded that the twisting moments $M_{x\varphi} = M_{\varphi x}$ vanish and the bending moments M_φ are constant along the circumference. Under such conditions of symmetry three of the six equations of equilibrium of the element are identically satisfied. We have to consider only the equations obtained by projecting the forces on the x and z axes and by taking the moment of the forces about the y axis. For example, consider a case in which external forces consist only of a pressure normal to the surface. The three equations of equilibrium are

$$\frac{dN}{dx}adxd\varphi = 0$$

$$\frac{dQ_x}{dx}adxd\varphi + N_\varphi dxd\varphi + Zadxd\varphi = 0$$

$$\frac{dM_x}{dx}adxd\varphi - Q_x adxd\varphi = 0 \tag{2.108}$$

The first one indicates that the forces N_x are constant, and they are taken equal to zero in the further discussion. If they are different from zero, the deformation and stress corresponding to such constant forces can be easily calculated and superposed on stresses and deformations produced by lateral load. The remaining two equations are written in the simplified form:

$$\frac{dQ_x}{dx} + \frac{1}{a}N_\varphi = -Z$$

$$\frac{dM_x}{dx} - Q_x = 0 \tag{2.109}$$

These two equations contain three unknown quantities: N_φ, Q_x, and M_x. We need, therefore, to consider the displacements of points in the middle surface of the shell.

The component v of the displacement in the circumferential direction vanishes because of symmetry. Only the components u and w in the x and z directions, respectively, are to be considered. The expressions for the strain components then become

$$\varepsilon_x = \frac{du}{dx} \quad \varepsilon_\varphi = -\frac{w}{a} \tag{2.110}$$

By Hooke's law, we obtain

$$N_x = \frac{Eh}{1-v^2}(\varepsilon_x + v\varepsilon_\varphi) = \frac{Eh}{1-v^2}\left(\frac{du}{dx} - v\frac{w}{a}\right) = 0$$

$$N_\varphi = \frac{Eh}{1-v^2}(\varepsilon_\varphi + v\varepsilon_x) = \frac{Eh}{1-v^2}\left(-\frac{w}{a} + v\frac{du}{dx}\right) = 0 \tag{2.111}$$

From the first of these equation it follows that

$$\frac{du}{dx} = v\frac{w}{a}$$

and the second equation gives

$$N_\varphi = -\frac{Ehw}{a} \tag{2.112}$$

Considering the bending moments, we conclude from symmetry that there is no change in curvature in the circumferential direction. The curvature in the x direction is equal to $-d^2w/dx^2$. Using the same equations as for plates, we then obtain

$$
\begin{aligned}
M_\varphi &= v M_x \\
M_x &= -D\frac{d^2w}{dx^2}
\end{aligned}
\tag{2.113}
$$

where

$$D = \frac{Eh^3}{12(1 - v^2)}$$

is the flexural rigidity per unit length of the shell.

Eliminating Q_x from Equation 2.109, we obtain

$$\frac{d^2 M_x}{dx^2} + \frac{1}{a}N_\varphi = -Z$$

from which, by using Equations 2.112 and 2.113, we obtain

$$\frac{d^2}{dx^2}\left(D\frac{d^2w}{dx^2}\right) + \frac{Eh}{a^2}w = Z \tag{2.114}$$

All problems of symmetrical deformation of circular cylindrical shells thus reduce to the integration of Equation 2.114.

The simplest application of this equation is obtained when the thickness of the shell is constant. Under such conditions, Equation 2.114 becomes

$$D\frac{d^4w}{dx^4} + \frac{Eh}{a^2}w = Z$$

Using the notation

$$\beta^4 = \frac{Eh}{4a^2 D} = \frac{3(1 - v^2)}{a^2 h^2} \tag{2.115}$$

Equation 2.115 can be represented in the simplified form

$$\frac{d^4w}{dx^4} + 4\beta^4 w = \frac{Z}{D} \tag{2.116}$$

The general solution of this equation is

$$
\begin{aligned}
w = \ & e^{\beta x}(C_1 \cos \beta x + C_2 \sin \beta x) \\
& + e^{-\beta x}(C_3 \cos \beta x + C_4 \sin \beta x) + f(x)
\end{aligned}
\tag{2.117}
$$

Detailed treatment of shell theory can be obtained from Timoshenko and Woinowsky-Krieger [56].

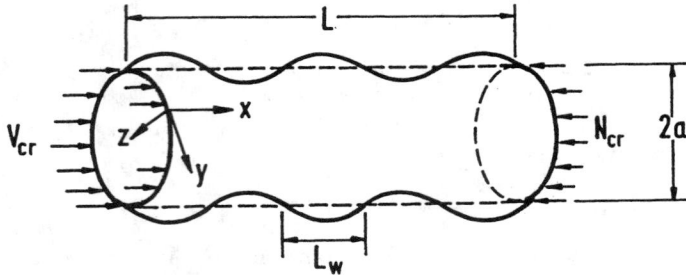

Figure 2.52   Buckling of a cylindrical shell.

2.6.8   Buckling of Shells

If a circular cylindrical shell is uniformly compressed in the axial direction, buckling symmetrical with respect to the axis of the cylinder (Figure 2.52) may occur at a certain value of the compressive load. The critical value of the compressive force N_{cr} per unit length of the edge of the shell can be obtained by solving the differential equation

$$D\frac{d^4w}{dx^4} + N\frac{d^2w}{dx^2} + Eh\frac{w}{a^2} = 0 \qquad (2.118)$$

in which a is the radius of the cylinder and h is the wall thickness.

Alternatively, the critical force per unit length may also be obtained by using the energy method. For a cylinder of length L simply supported at both ends one obtains

$$N_{cr} = D\left(\frac{m^2\pi^2}{L^2} + \frac{EhL^2}{Da^2m^2\pi^2}\right) \qquad (2.119)$$

For each value of m there is a unique buckling mode shape and a unique buckling load. The lowest value is of greatest interest and is thus found by setting the derivative of N_{cr} with respect to L equal to zero for $m = 1$. With Poisson's Ratio, $= 0.3$, the buckling load is obtained as

$$N_{cr} = 0.605\frac{Eh^2}{a} \qquad (2.120)$$

It is possible for a cylindrical shell be subjected to uniform external pressure or to the combined action of axial and uniform lateral pressure. In such cases the mathematical treatment is more involved and it requires special considerations.

More detailed treatment of such cases may be found in Timoshenko and Gere [55].

2.7   Influence Lines

Structures such as bridges, industrial buildings with travelling cranes, and frames supporting conveyor belts are subjected to moving loads. Each member of these structures must be designed for the most severe conditions that can possibly be developed in that member. Live loads should be placed at the position where they will produce these severe conditions. The critical positions for placing live loads will not be the same for every member. On some occasions it is possible to determine by inspection where to place the loads to give the most critical forces, but on many other occasions it is necessary to resort to certain criteria to find the locations. The most useful of these methods is influence lines.

An **influence line** for a particular response such as reaction, shear force, bending moment, and axial force is defined as a diagram the ordinate to which at any point equals the value of that response attributable to a unit load acting at that point on the structure. Influence lines provide a systematic procedure for determining how the force in a given part of a structure varies as the applied load moves about on the structure. Influence lines of responses of **statically determinate structures** consist only of straight lines whereas they are curves for **statically indeterminate structures**. They are primarily used to determine where to place live loads to cause maximum force and to compute the magnitude of those forces. The knowledge of influence lines helps to study the structural response under different moving load conditions.

2.7.1 Influence Lines for Shear in Simple Beams

Figure 2.53 shows influence lines for shear at two sections of a simply supported beam. It is assumed

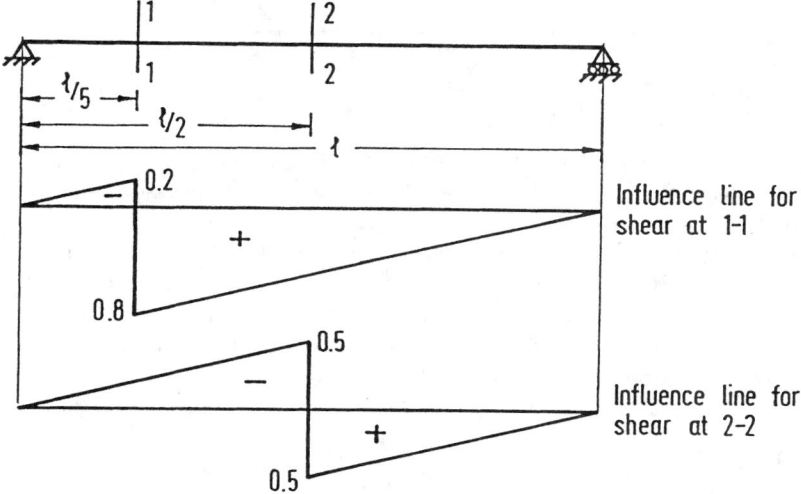

Figure 2.53 Influence line for shear force.

that positive shear occurs when the sum of the transverse forces to the left of a section is in the upward direction or when the sum of the forces to the right of the section is downward. A unit force is placed at various locations and the shear force at sections 1-1 and 2-2 are obtained for each position of the unit load. These values give the ordinate of influence line with which the influence line diagrams for shear force at sections 1-1 and 2-2 can be constructed. Note that the slope of the influence line for shear on the left of the section is equal to the slope of the influence line on the right of the section. This information is useful in drawing shear force influence line in other cases.

2.7.2 Influence Lines for Bending Moment in Simple Beams

Influence lines for bending moment at the same sections, 1-1 and 2-2 of the **simple beam** considered in Figure 2.53, are plotted as shown in Figure 2.54. For a section, when the sum of the moments of all the forces to the left is clockwise or when the sum to the right is counter-clockwise, the moment is taken as positive. The values of bending moment at sections 1-1 and 2-2 are obtained for various positions of unit load and plotted as shown in the figure.

It should be understood that a shear or bending moment diagram shows the variation of shear or moment across an entire structure for loads fixed in one position. On the other hand, an influence line for shear or moment shows the variation of that response at one particular section in the structure caused by the movement of a unit load from one end of the structure to the other.

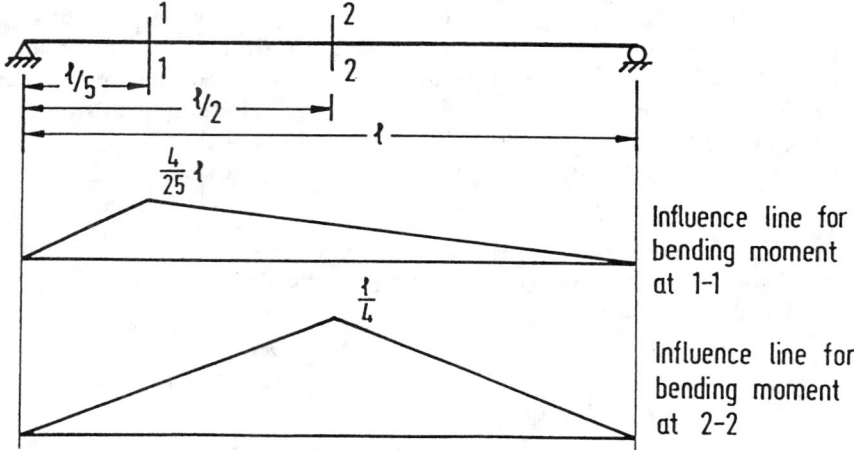

Figure 2.54   Influence line for bending moment.

Influence lines can be used to obtain the value of a particular response for which it is drawn when the beam is subjected to any particular type of loading. If, for example, a uniform load of intensity q_o per unit length is acting over the entire length of the simple beam shown in Figure 2.53, then the shear force at section 1-1 is given by the product of the load intensity, q_o, and the net area under the influence line diagram. The net area is equal to $0.3l$ and the shear force at section 1-1 is, therefore, equal to $0.3q_o l$. In the same way, the bending moment at the section can be found as the area of the corresponding influence line diagram times the intensity of loading, q_o. The bending moment at the section is, therefore, $(0.08l^2 \times q_o =)0.08q_o l^2$.

2.7.3   Influence Lines for Trusses

Influence lines for support reactions and member forces may be constructed in the same manner as those for various beam functions. They are useful to determine the maximum load that can be applied to the **truss**. The unit load moves across the truss, and the ordinates for the responses under consideration may be computed for the load at each panel point. Member force, in most cases, need not be calculated for every panel point because certain portions of influence lines can readily be seen to consist of straight lines for several panels. One method used for calculating the forces in a chord member of a truss is by the Method of Sections discussed earlier.

The truss shown in Figure 2.55 is considered for illustrating the construction of influence lines for trusses.

The member forces in $U_1 U_2$, $L_1 L_2$, and $U_1 L_2$ are determined by passing a section 1-1 and considering the equilibrium of the free body diagram of one of the truss segments. Unit load is placed at L_1 first and the force in $U_1 U_2$ is obtained by taking moment about L_2 of all the forces acting on the right-hand segment of the truss and dividing the resulting moment by the lever arm (the perpendicular distance of the force in $U_1 U_2$ from L_2). The value thus obtained gives the ordinate of the influence diagram at L_1 in the truss. The ordinate at L_2 obtained similarly represents the force in $U_1 U_2$ for unit load placed at L_2. The influence line can be completed with two other points, one at each of the supports. The force in the member $L_1 L_2$ due to unit load placed at L_1 and L_2 can be obtained in the same manner and the corresponding influence line diagram can be completed. By considering the horizontal component of force in the diagonal of the panel, the influence line for force in $U_1 L_2$ can be constructed. Figure 2.55 shows the respective influence diagram for member forces in $U_1 U_2$, $L_1 L_2$, and $U_1 L_2$. Influence line ordinates for the force in a chord member of a

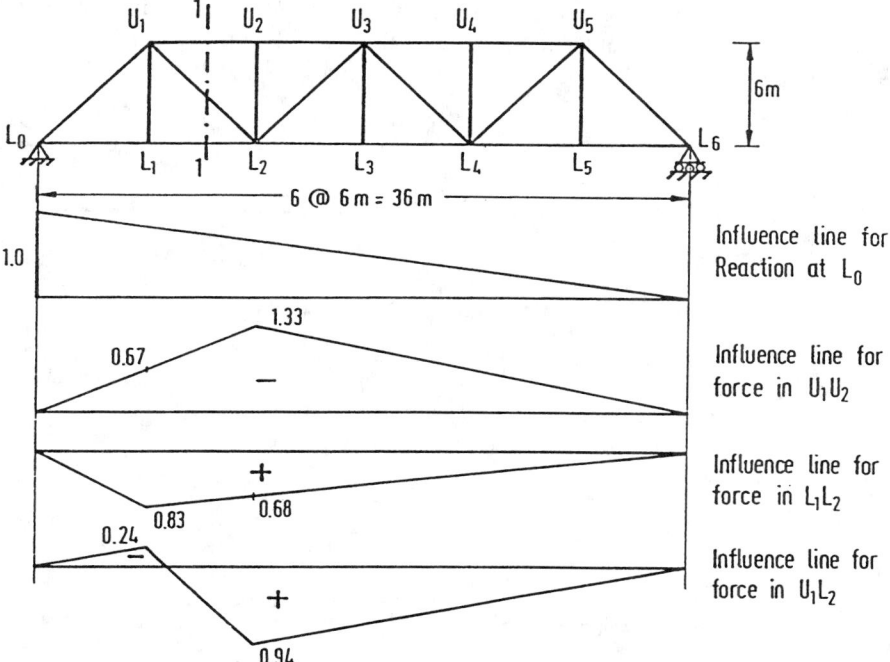

Figure 2.55   Influence line for truss.

"curved-chord" truss may be determined by passing a vertical section through the panel and taking moments at the intersection of the diagonal and the other chord.

2.7.4   Qualitative Influence Lines

One of the most effective methods of obtaining influence lines is by the use of Müller-Breslau's principle, which states that "the ordinates of the influence line for any response in a structure are equal to those of the deflection curve obtained by releasing the restraint corresponding to this response and introducing a corresponding unit displacement in the remaining structure". In this way, the shape of the influence lines for both statically determinate and indeterminate structures can be easily obtained especially for beams.

To draw the influence lines of

1. Support reaction: Remove the support and introduce a unit displacement in the direction of the corresponding reaction to the remaining structure as shown in Figure 2.56 for a symmetrical overhang beam.

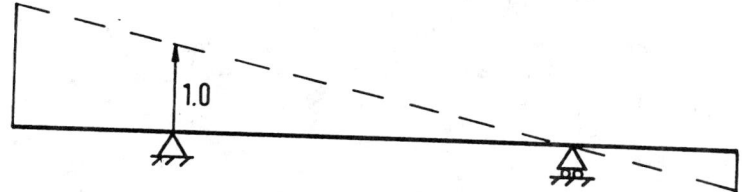

Figure 2.56   Influence line for support reaction.

2. Shear: Make a cut at the section and introduce a unit relative translation (in the direction of positive shear) without relative rotation of the two ends at the section as shown in Figure 2.57.

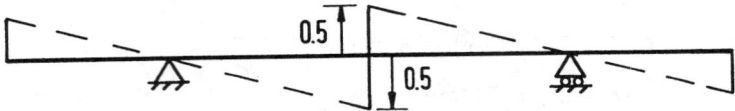

Figure 2.57   Influence line for midspan shear force.

3. Bending moment: Introduce a hinge at the section (releasing the bending moment) and apply bending (in the direction corresponding to positive moment) to produce a unit relative rotation of the two beam ends at the hinged section as shown in Figure 2.58.

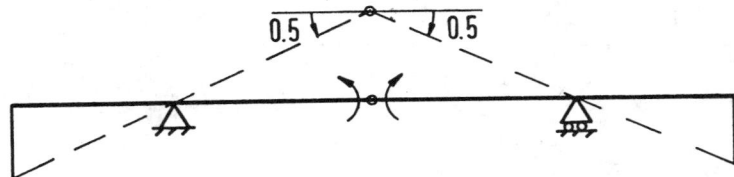

Figure 2.58   Influence line for midspan bending moment.

2.7.5   Influence Lines for Continuous Beams

Using Müller-Breslau's principle, the shape of the influence line of any response of a continuous beam can be sketched easily. One of the methods for beam deflection can then be used for determining the ordinates of the influence line at critical points. Figures 2.59 to 2.61 show the influence lines of bending moment at various points of two, three, and four span continuous beams.

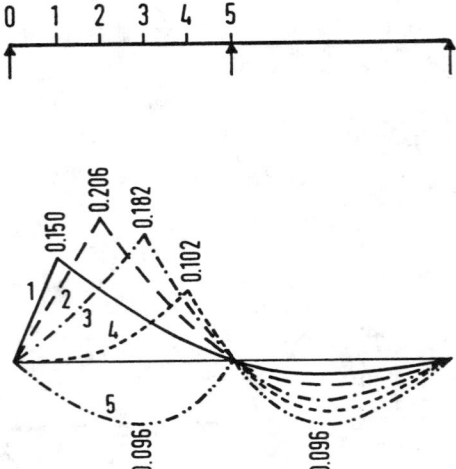

Figure 2.59   Influence lines for bending moments—two span beam.

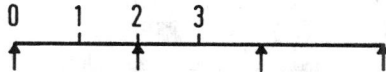

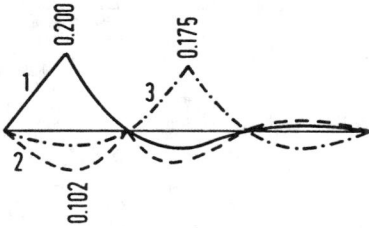

Figure 2.60   Influence lines for bending moments—three span beam.

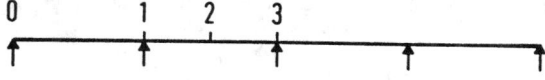

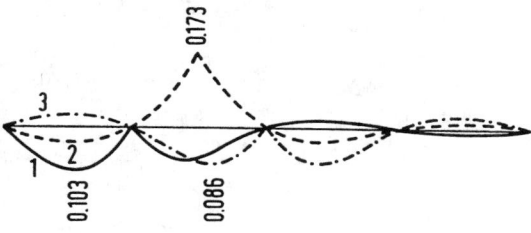

Figure 2.61   Influence lines for bending moments—four span beam.

2.8   Energy Methods in Structural Analysis

Energy methods are a powerful tool in obtaining numerical solutions of statically indeterminate problems. The basic quantity required is the *strain energy,* or work stored due to deformations, of the structure.

2.8.1   Strain Energy Due to Uniaxial Stress

In an axially loaded bar with constant cross-section, the applied load causes normal stress σ_y as shown in Figure 2.62. The tensile stress σ_y increases from zero to a value σ_y as the load is gradually applied. The original, unstrained position of any section such as $C - C$ will be displaced by an amount v. A section $D - D$ located a differential length below $C - C$ will have been displaced by an amount $v + \left(\frac{\partial v}{\partial y}\right) dy$. As σ_y varies with the applied load, from zero to σ_y, the work done by the

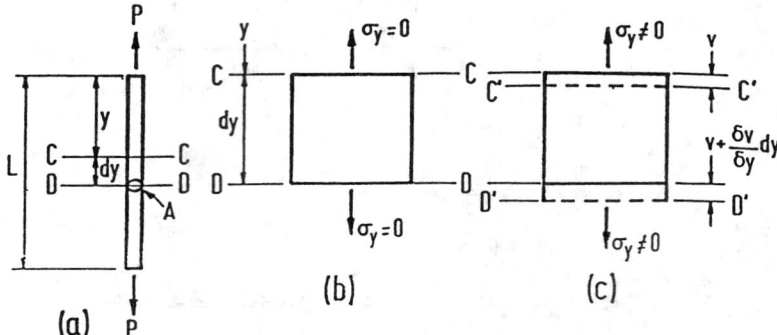

Figure 2.62   Axially loaded bar.

forces external to the element can be shown to be

$$dV = \frac{1}{2E}\sigma_y^2 A\,dy = \frac{1}{2}\sigma_y \varepsilon_y A\,dy \qquad (2.121)$$

in which A is the area of cross-section of the bar and ε_y is the strain in the direction of σ_y.

2.8.2   Strain Energy in Bending

It can be shown that the strain energy of a differential volume $dx\,dy\,dz$ stressed in tension or compression in the x direction only by a normal stress σ_x will be

$$dV = \frac{1}{2E}\sigma_x^2 dx\,dy\,dz = \frac{1}{2}\sigma_x \varepsilon_x dx\,dy\,dz \qquad (2.122)$$

When σ_x is the bending stress given by $\sigma_x = \frac{My}{I}$ (see Figure 2.63), then $dV = \frac{1}{2E}\frac{M^2 y^2}{I^2}dx\,dy\,dz$, where I is the **moment of inertia** of the cross-sectional area about the neutral axis.

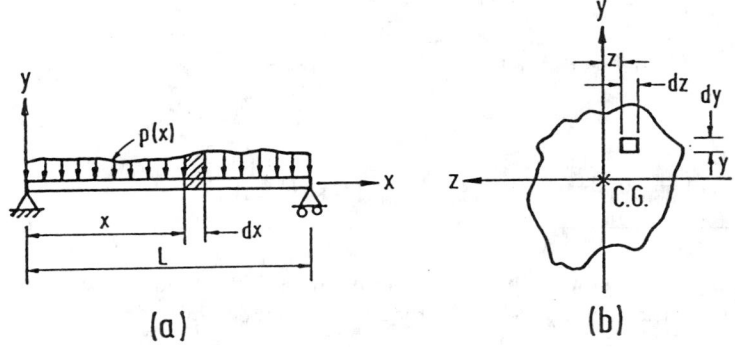

Figure 2.63   Beam under arbitrary bending load.

The total strain energy of bending of a beam is obtained as

$$V = \int \int \int_{\text{volume}} \frac{1}{2E}\frac{M^2}{I^2}y^2 dz\,dy\,dx$$

where

$$I = \int \int_{\text{area}} y^2 dz\,dy$$

Therefore,

$$V = \int_{length} \frac{M^2}{2EI} dx \tag{2.123}$$

2.8.3 Strain Energy in Shear

Figure 2.64 shows an element of volume $dx\,dy\,dz$ subjected to shear stress τ_{xy} and τ_{yx}.

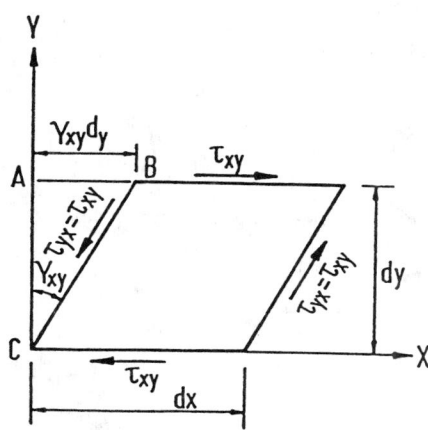

Figure 2.64 Shear loading.

For static equilibrium, it can readily be shown that

$$\tau_{xy} = \tau_{yx}$$

The shear strain, γ is defined as AB/AC. For small deformations, it follows that

$$\gamma_{xy} = \frac{AB}{AC}$$

Hence, the angle of deformation γ_{xy} is a measure of the shear strain. The strain energy for this differential volume is obtained as

$$dV = \frac{1}{2} \left(\tau_{xy} dz dx \right) \gamma_{xy} dy = \frac{1}{2} \tau_{xy} \gamma_{xy} dx dy dz \tag{2.124}$$

Hooke's Law for shear stress and strain is

$$\gamma_{xy} = \frac{\tau_{xy}}{G} \tag{2.125}$$

where G is the shear modulus of elasticity of the material. The expression for strain energy in shear reduces to

$$dV = \frac{1}{2G} \tau_{xy}^2 dx dy dz \tag{2.126}$$

2.8.4 The Energy Relations in Structural Analysis

The energy relations or laws such as (1) Law of Conservation of Energy, (2) Theorem of Virtual Work, (3) Theorem of Minimum Potential Energy, and (4) Theorem of Complementary Energy are of fundamental importance in structural engineering and are used in various ways in structural analysis.

The Law of Conservation of Energy

There are many ways of stating this law. For the purpose of structural analysis it will be sufficient to state it in the following way:

> If a structure and the external loads acting on it are isolated so that these neither receive nor give out energy, then the total energy of this system remains constant.

A typical application of the Law of Conservation of Energy can be made by referring to Figure 2.65 which shows a cantilever beam of constant cross-sections subjected to a concentrated load at its end. If only bending strain energy is considered,

$$
\begin{aligned}
\text{External work} \quad &= \quad \text{Internal work} \\
\frac{P\delta}{2} \quad &= \quad \int_0^L \frac{M^2 dx}{2EI}
\end{aligned}
$$

Substituting $M = -Px$ and integrating along the length gives

$$\delta = \frac{PL^3}{3EI} \tag{2.127}$$

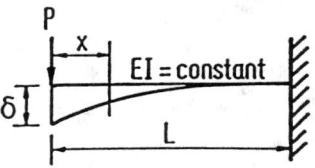

Figure 2.65   Cantilever beam.

The Theorem of Virtual Work

The Theorem of Virtual Work can be derived by considering the beam shown in Figure 2.66. The full curved line represents the equilibrium position of the beam under the given loads. Assume

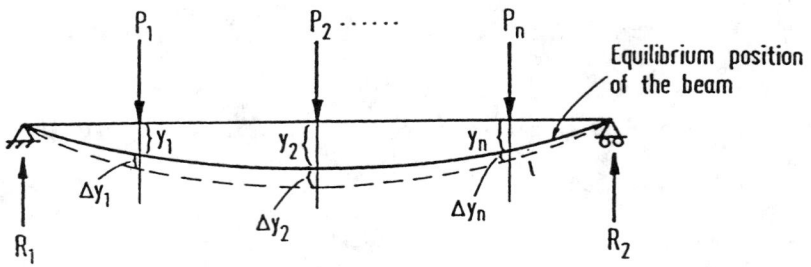

Figure 2.66   Equilibrium of a simply supported beam under loading.

the beam to be given an additional small deformation consistent with the boundary conditions. This is called a virtual deformation and corresponds to increments of deflection $\Delta_{y1}, \Delta_{y2}, ..., \Delta_{yn}$ at loads $P_1, P_2, ..., P_n$ as shown by the broken line.

The change in potential energy of the loads is given by

$$\Delta(P.E.) = \sum_{i=1}^{n} P_i \Delta y_i \qquad (2.128)$$

By the Law of Conservation of Energy this must be equal to the internal strain energy stored in the beam. Hence, we may state the Theorem of Virtual Work in the following form:

> If a body in equilibrium under the action of a system of external loads is given any small (virtual) deformation, then the work done by the external loads during this deformation is equal to the increase in internal strain energy stored in the body.

The Theorem of Minimum Potential Energy

Let us consider the beam shown in Figure 2.67. The beam is in equilibrium under the action

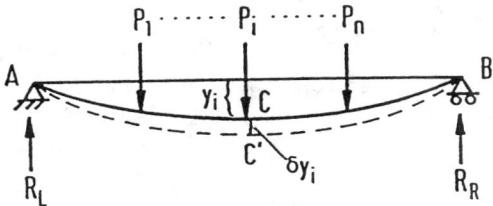

Figure 2.67 Simply supported beam under point loading.

of loads, $P_1, P_2, P_3, ..., P_i, ..., P_n$. The curve ACB defines the equilibrium positions of the loads and reactions. Now apply by some means an additional small displacement to the curve so that it is defined by $AC'B$. Let y_i be the original equilibrium displacement of the curve beneath a particular load P_i. The additional small displacement is called δ_{yi}. The potential energy of the system while it is in the equilibrium configuration is found by comparing the potential energy of the beam and loads in equilibrium and in the undeflected position. If the change in potential energy of the loads is W and the strain energy of the beam is V, the total energy of the system is

$$U = W + V \qquad (2.129)$$

If we neglect the second-order terms, then

$$\delta U = \delta(W + V) = 0 \qquad (2.130)$$

The above is expressed as the Principle or Theorem of Minimum Potential Energy which can be stated as

> Of all displacements satisfying given boundary conditions, those that satisfy the equilibrium conditions make the potential energy a minimum.

Castigliano's Theorem

An example of application of energy methods to the field of structural engineering is Castigliano's Theorem. The theorem applies only to structures stressed within the elastic limit. Also, all deformations must be linear homogeneous functions of the loads. Castigliano's Theorem can be derived using the expression for total potential energy as follows: For a beam in equilibrium loaded as in Figure 2.66, the total energy is

$$U = -[P_1 y_1 + P_2 y_2 + ...P_j y_j + ...P_n y_n] + V \qquad (2.131)$$

For an elastic system, the strain energy, V, turns out to be one half the change in the potential energy of the loads.

$$V = \frac{1}{2} \sum_{i=1}^{i=n} P_i y_i \qquad (2.132)$$

Castigliano's Theorem results from studying the variation in the strain energy, V, produced by a differential change in one of the loads, say P_j.

If the load P_j is changed by a differential amount δP_j and if the deflections y are linear functions of the loads, then

$$\frac{\partial V}{\partial P_j} = \frac{1}{2} \sum_{i=1}^{i=n} P_i \frac{\partial y_i}{\partial P_j} + \frac{1}{2} y_j = y_j \qquad (2.133)$$

Castigliano's Theorem is stated as follows:

> The partial derivatives of the total strain energy of any structure with respect to any one of the applied forces is equal to the displacement of the point of application of the force in the direction of the force.

To find the deflection of a point in a beam that is not the point of application of a concentrated load, one should apply a load $P = 0$ at that point and carry the term P into the strain energy equation. Finally, introduce the true value of $P = 0$ into the expression for the answer.

EXAMPLE 2.6:

For example, it is required to determine the bending deflection at the free end of a cantilever loaded as shown in Figure 2.68.

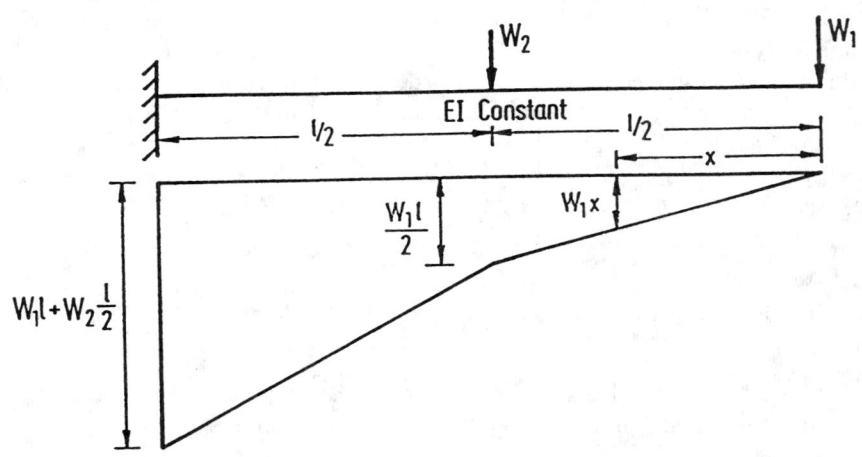

Figure 2.68   Example 2.6

Solution

$$V = \int_0^L \frac{M^2}{2EI} dx$$

$$\Delta = \frac{\partial V}{\partial W_1} = \int_0^L \frac{M}{EI} \frac{\partial M}{\partial W_1} dx$$

$$M = W_1 x \qquad 0 < x \frac{L}{2}$$

$$= W_1 x + W_2 \left(x - \frac{\ell}{2} \right) \quad \frac{L}{2} < x < L$$

$$\Delta = \frac{1}{EI} \int_0^{\ell/2} W_1 x \times x\, dx + \frac{1}{EI} \int_{\ell/2}^{\ell} \left[W_1 x + W_2 \left(x - \frac{\ell}{2} \right) \right] x\, dx$$

$$= \frac{W_1 \ell^3}{24EI} + \frac{7W_1 \ell^3}{24EI} + \frac{5W_2 \ell^3}{48EI}$$

$$= \frac{W_1 \ell^3}{3EI} + \frac{5W_2 \ell^3}{48EI}$$

Castigliano's Theorem can be applied to determine deflection of trusses as follows:
We know that the increment of strain energy for an axially loaded bar is given as

$$dV = \frac{1}{2E} \sigma_y^2 A\, dy$$

Substituting, $\sigma_y = \frac{S}{A}$, where S is the axial load in the bar and integrating over the length of the bar the total strain energy of the bar is given as

$$V = \frac{S^2 L}{2AE} \qquad (2.134)$$

The deflection component Δ_i of the point of application of a load P_i in the direction of P_i is given as

$$\Delta_i = \frac{\partial V}{\partial P_i} = \frac{\partial}{\partial P_i} \sum \frac{S^2 L}{2AE} = \sum \frac{S \frac{\partial S}{\partial P_i} L}{AE}$$

EXAMPLE 2.7:

Let us consider the truss shown in Figure 2.69. It is required to determine the vertical deflection at 'g' of the truss when loaded as shown in the figure. Let us first replace 20 k load at 'g' by P and carry out the calculations in terms of P. At the end, P will be replaced by the actual value, namely 20 k.

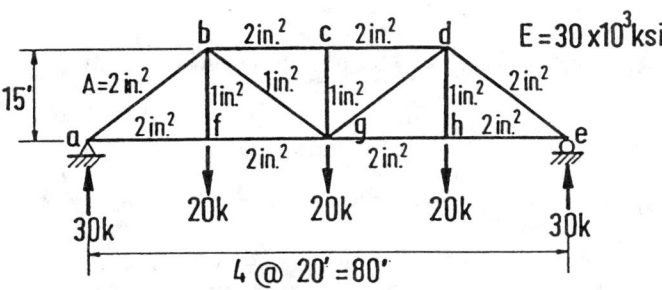

Figure 2.69 Example 2.7.

Member	A in.^{2}	L ft	S	$\frac{\delta S}{\delta P}$	n	$nS\frac{\delta S}{\delta P}\frac{L}{A}$
ab	2	25	$-(33.3 + 0.83P)$	-0.83	2	$(691 + 17.2P)$
af	2	20	$(26.7 + 0.67P)$	0.67	2	$(358 + 9P)$
fg	2	20	$(26.7 + 0.67P)$	0.67	2	$(358 + 9P)$
bf	1	15	20	0	2	0
bg	1	25	$0.83P$	0.83	2	$34.4P$
bc	2	20	$-26.7 - 1.33P$	-1.33	2	$(710 + 35.4P)$
cg	1	15	0	0	1	0

'n' indicates the number of similar members $\qquad \sum \frac{S\frac{\delta S}{\delta P}L}{A} \qquad 2117 + 105P$

With

$$P = 20k$$

$$\Delta_g = \sum \frac{S\frac{\delta S}{\delta P}L}{AE} = \frac{(2117 + 105 \times 20) \times 12}{30 \times 10^3} = 1.69 \text{ in.}$$

2.8.5   Unit Load Method

The unit load method is a versatile tool in the solution of deflections of both trusses and beams. Consider an elastic body in equilibrium under loads P_1, P_2, P_3, P_4, ...P_n and a load p applied at point O, as shown in Figure 2.70. By Castigliano's Theorem, the component of the deflection of

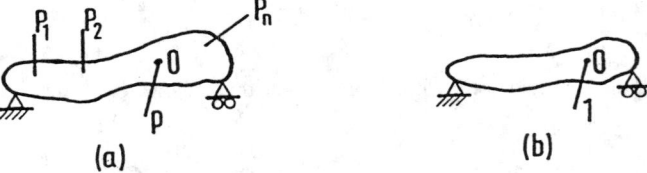

(a)   **(b)**

Figure 2.70   Elastic body in equilibrium under load.

point O in the direction of the applied force p is

$$\delta_{o_p} = \frac{\partial V}{\partial p} \tag{2.135}$$

in which V is the strain energy of the body. It has been shown in Equation 2.123, that the strain energy of a beam, neglecting shear effects, is given by

$$V = \int_0^L \frac{M^2}{2EI} dx$$

Also it was shown that if the elastic body is a truss, from Equation 2.134

$$V = \sum \frac{S^2 L}{2AE}$$

For a beam, therefore, from Equation 2.135

$$\delta_{o_p} = \int_L \frac{M\frac{\partial M}{\partial p}dx}{EI} \tag{2.136}$$

and for a truss,

$$\delta_{op} = \sum \frac{S \frac{\partial S}{\partial p} L}{AE} \tag{2.137}$$

The bending moments M and the axial forces S are functions of the load p as well as of the loads $P_1, P_2, ... P_n$. Let a unit load be applied at O on the elastic body and the corresponding moment be m if the body is a beam, and the forces in the members of the body be u if the body is a truss. For the body in Figure 2.70, the moments M and the forces S due to the system of forces $P_1, P_2, ... P_n$ and p at O applied separately can be obtained by superposition as

$$M = M_p + pm \tag{2.138}$$

$$S = S_p + pu \tag{2.139}$$

in which M_P and S_P are, respectively, moments and forces produced by $P_1, P_2, ... P_n$. Then

$$\frac{\partial M}{\partial p} = m = \text{moments produced by a unit load at } O \tag{2.140}$$

$$\frac{\partial S}{\partial p} = u = \text{stresses produced by a unit load at } O \tag{2.141}$$

Using Equations 2.140 and 2.141 in Equations 2.136 and 2.137, respectively,

$$\delta_{op} = \int_L \frac{Mmdx}{EI} \tag{2.142}$$

$$\delta_{op} = \sum \frac{SuL}{AE} \tag{2.143}$$

EXAMPLE 2.8:

Determine, using the unit load method, the deflection at C of a simple beam of constant cross-section loaded as shown in Figure 2.71a.

Solution   The bending moment diagram for the beam due to the applied loading is shown in Figure 2.71b. A unit load is applied at C where it is required to determine the deflection as shown in Figure 2.71c and the corresponding bending moment diagram is shown in Figure 2.71d. Now, using Equation 2.142, we have

$$
\begin{aligned}
\delta_c &= \int_0^L \frac{Mmdx}{EI} \\
&= \frac{1}{EI} \int_0^{\frac{L}{4}} (Wx) \left(\frac{3}{4} x \right) dx + \frac{1}{EI} \int_{\frac{L}{4}}^{\frac{3L}{4}} \left(\frac{WL}{4} \right) \frac{1}{4} (L - x) dx \\
&\quad + \frac{1}{EI} \int_{\frac{3L}{4}}^L W(L - x) \frac{1}{4} (L - x) dx \\
&= \frac{WL^3}{48EI}
\end{aligned}
$$

Further details on energy methods in structural analysis may be found in [10].

2.9   Matrix Methods

In this method of structural analysis, a set of simultaneous equations that describe the load-deformation characteristics of the structure under consideration are formed. These equations

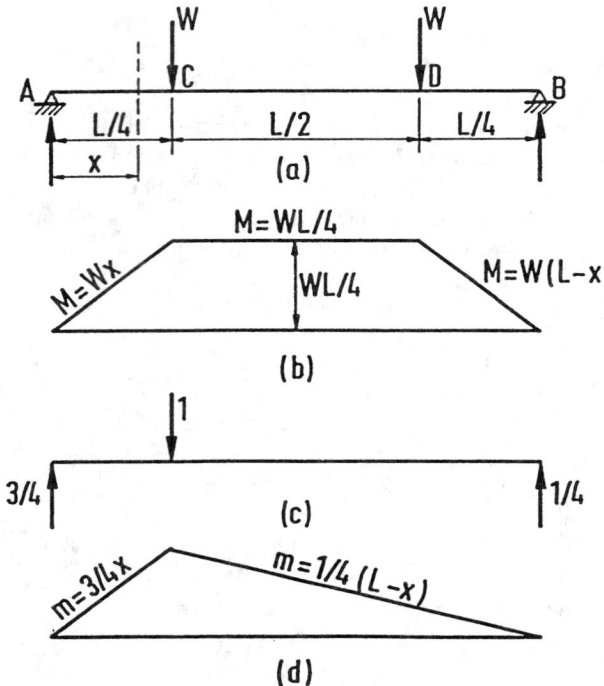

Figure 2.71   Example 2.8.

are solved using the matrix algebra to obtain the load-deformation characteristics of discrete or finite elements into which the structure has been subdivided. Matrix algebra is ideally suited for setting up and solving equations on the computer. Matrix structural analysis has two methods of approach. The first is called the flexibility method in which forces are used as independent variables and the second is called the stiffness method; the second method employs deformations as the independent variables. The two methods are also called the force method and the displacement method, respectively.

2.9.1   Flexibility Method

In a structure, the forces and displacements are related to one another by using stiffness influence coefficients. Let us consider, for example, a simple beam in which three concentrated loads W_1, W_2, and W_3 are applied at sections 1, 2, and 3, respectively as shown in Figure 2.72. Now, the deflection

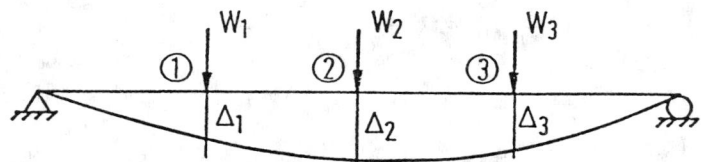

Figure 2.72   Simple beam under concentrated loads.

at section 1, Δ_1 can be expressed as

$$\Delta_1 = F_{11}W_1 + F_{12}W_2 + F_{13}W_3$$

in which F_{11}, F_{12}, and F_{13} are called flexibility coefficients and they are, respectively, defined as the deflection at section 1 due to unit loads applied at sections 1, 2, and 3. Deflections at sections 2 and 3 are similarly given as

$$\Delta_2 = F_{21}W_1 + F_{22}W_2 + F_{23}W_3$$

and

$$\Delta_3 = F_{31}W_1 + F_{32}W_2 + F_{33}W_3 \qquad (2.144)$$

These expressions are written in the matrix form as

$$\left\{ \begin{array}{c} \Delta_1 \\ \Delta_2 \\ \Delta_3 \end{array} \right\} = \left[\begin{array}{ccc} F_{11} & F_{12} & F_{13} \\ F_{21} & F_{22} & F_{23} \\ F_{31} & F_{32} & F_{33} \end{array} \right] \left\{ \begin{array}{c} W_1 \\ W_2 \\ W_3 \end{array} \right\}$$

(or)

$$\{\Delta\} = [F]\{W\} \qquad (2.145)$$

The matrix $[F]$ is called the flexibility matrix. It can be shown, by applying Maxwell's reciprocal theorem [10], that the matrix $\{F\}$ is a symmetric matrix.

The flexibility matrix for a cantilever beam loaded as shown in Figure 2.73 can be constructed as follows.

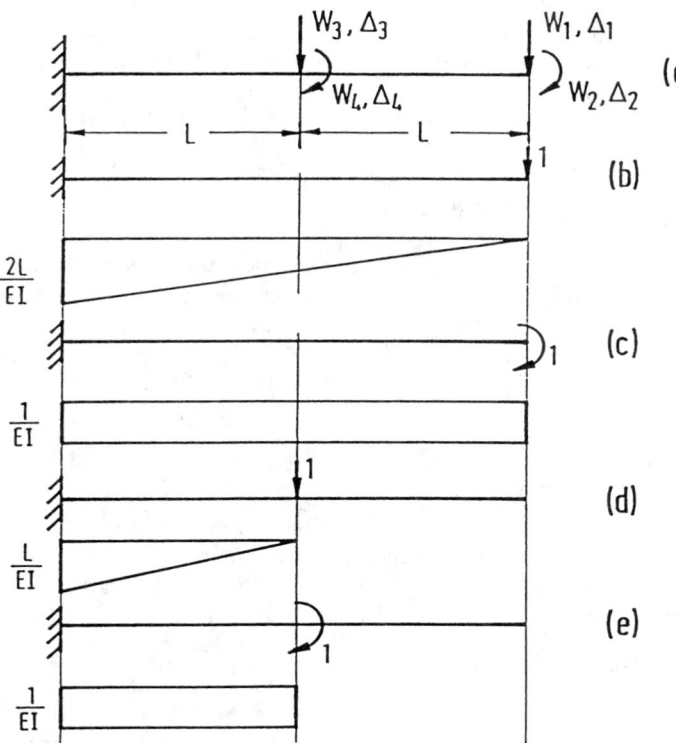

Figure 2.73   Cantilever beam.

The first column in the flexibility matrix can be generated by applying a unit vertical load at the free end of the cantilever as shown in Figure 2.73b and making use of the moment area method. We get

$$F_{11} = \frac{8L^3}{3EI}, \qquad F_{21} = \frac{2L^2}{EI}, \qquad F_{31} = \frac{5L^3}{6EI}, \qquad F_{41} = \frac{3L^2}{2EI}$$

Columns 2, 3, and 4 are, similarly, generated by applying unit moment at the free end and unit force and unit moment at the mid-span as shown in Figures 2.73c, d, and e, respectively. Combining the results thus obtained, one gets the flexibility matrix as

$$\begin{Bmatrix} \Delta_1 \\ \Delta_2 \\ \Delta_3 \\ \Delta_4 \end{Bmatrix} = \frac{1}{EI} \begin{bmatrix} \frac{8L^3}{3} & 2L^2 & \frac{5L^3}{6} & \frac{3L^2}{2} \\ 2L^2 & 2L & \frac{L^2}{2} & L \\ \frac{5L^3}{6} & \frac{L^2}{2} & \frac{L^3}{3} & \frac{L^2}{2} \\ \frac{3L^2}{2} & L & \frac{L^2}{2} & L \end{bmatrix} \begin{Bmatrix} W_1 \\ W_2 \\ W_3 \\ W_4 \end{Bmatrix} \qquad (2.146)$$

The above method to generate the flexibility matrix for a given structure is extremely impractical. It is therefore recommended to subdivide a given structure into several elements and to form the flexibility matrix for each of the elements. The flexibility matrix for the entire structure is then obtained by combining the flexibility matrices of the individual elements.

Force transformation matrix relates what occurs in these elements to the behavior of the entire structure. Using the conditions of equilibrium, it relates the element forces to the structure forces. The principle of conservation of energy may be used to generate transformation matrices.

2.9.2 Stiffness Method

Forces and deformations in a structure are related to one another by means of stiffness influence coefficients. Let us consider, for example, a simply supported beam subjected to end moments W_1 and W_2 applied at supports 1 and 2 and let the respective rotations be denoted as Δ_1 and Δ_2 as shown in Figure 2.74. We can now write the expressions for end moments W_1 and W_2 as

$$\begin{aligned} W_1 &= K_{11}\Delta_1 + K_{12}\Delta_2 \\ W_2 &= K_{21}\Delta_1 + K_{22}\Delta_2 \end{aligned} \qquad (2.147)$$

in which K_{11} and K_{12} are called stiffness influence coefficients defined as moments at 1 due to unit rotation at 1 and 2, respectively. The above equations can be written in matrix form as

$$\begin{Bmatrix} W_1 \\ W_2 \end{Bmatrix} = \begin{bmatrix} K_{11} & K_{12} \\ K_{21} & K_{22} \end{bmatrix} \begin{Bmatrix} \Delta_1 \\ \Delta_2 \end{Bmatrix}$$

or

$$\{W\} = [K]\{\Delta\} \qquad (2.148)$$

The matrix $[K]$ is referred to as stiffness matrix. It can be shown that the flexibility matrix of a structure is the inverse of the stiffness matrix and vice versa. The stiffness matrix of the whole structure is formed out of the stiffness matrices of the individual elements that make up the structure.

2.9.3 Element Stiffness Matrix

Axially Loaded Member

Figure 2.75 shows an axially loaded member of constant cross-sectional area with element forces q_1 and q_2 and displacements δ_1 and δ_2. They are shown in their respective positive directions. With unit displacement $\delta_1 = 1$ at node 1, as shown in Figure 2.75, axial forces at nodes 1 and 2 are obtained as

$$K_{11} = \frac{EA}{L}, \qquad K_{21} = -\frac{EA}{L}$$

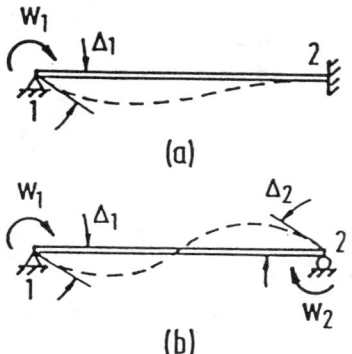

Figure 2.74 Simply supported beam.

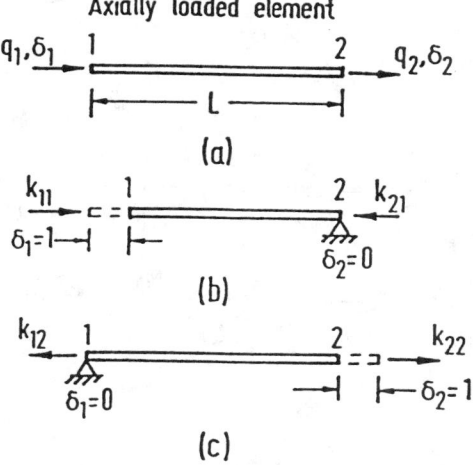

Figure 2.75 Axially loaded member.

In the same way by setting $\delta_2 = 1$ as shown in Figure 2.75 the corresponding forces are obtained as

$$K_{12} = -\frac{EA}{L}, \qquad K_{22} = \frac{EA}{L}$$

The stiffness matrix is written as

$$\left\{ \begin{array}{c} q_1 \\ q_2 \end{array} \right\} = \left[\begin{array}{cc} K_{11} & K_{12} \\ K_{21} & K_{22} \end{array} \right] \left\{ \begin{array}{c} \delta_1 \\ \delta_2 \end{array} \right\}$$

or

$$\left\{ \begin{array}{c} q_1 \\ q_2 \end{array} \right\} = \frac{EA}{L} \left[\begin{array}{cc} 1 & -1 \\ -1 & 1 \end{array} \right] \left\{ \begin{array}{c} \delta_1 \\ \delta_2 \end{array} \right\}. \tag{2.149}$$

Flexural Member

The stiffness matrix for the flexural element shown in Figure 2.76 can be constructed as follows. The forces and the corresponding displacements, namely the moments, the shears, and the corresponding rotations and translations at the ends of the member, are defined in the figure. The

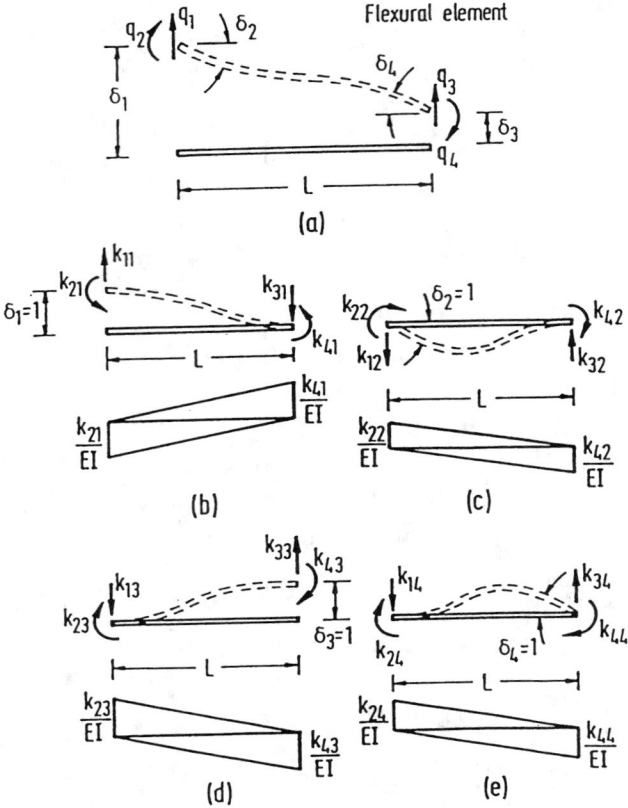

Figure 2.76   Beam element—stiffness matrix.

matrix equation that relates these forces and displacements can be written in the form

$$
\begin{bmatrix} q_1 \\ q_2 \\ q_3 \\ q_4 \end{bmatrix} = \begin{bmatrix} K_{11} & K_{12} & K_{13} & K_{14} \\ K_{21} & K_{22} & K_{23} & K_{24} \\ K_{31} & K_{32} & K_{33} & K_{34} \\ K_{41} & K_{42} & K_{43} & K_{44} \end{bmatrix} \begin{bmatrix} \delta_1 \\ \delta_2 \\ \delta_3 \\ \delta_4 \end{bmatrix}
$$

The terms in the first column consist of the element forces q_1 through q_4 that result from displacement $\delta_1 = 1$ when $\delta_2 = \delta_3 = \delta_4 = 0$. This means that a unit vertical displacement is imposed at the left end of the member while translation at the right end and rotation at both ends are prevented as shown in Figure 2.76. The four member forces corresponding to this deformation can be obtained using the moment-area method.

The change in slope between the two ends of the member is zero and the area of the M/EI diagram between these point must, therefore, vanish. Hence,

$$
\frac{K_{41}L}{2EI} - \frac{K_{21}L}{2EI} = 0
$$

and

$$
K_{21} = K_{41} \tag{2.150}
$$

The moment of the M/EI diagram about the left end of the member is equal to unity. Hence,

$$
\frac{K_{41}L}{2EI}\left(\frac{2L}{3}\right) - \frac{K_{21}L}{2EI}\left(\frac{L}{3}\right) = 1
$$

and in view of Equation 2.150,

$$K_{41} = K_{21} = \frac{6EI}{L^2}$$

Finally, moment equilibrium of the member about the right end leads to

$$K_{11} = \frac{K_{21} + K_{41}}{L} = \frac{12EI}{L^3}$$

and from equilibrium in the vertical direction we obtain

$$K_{31} = K_{11} = \frac{12EI}{L^3}$$

The forces act in the directions indicated in Figure 2.76b. To obtain the correct signs, one must compare the forces with the positive directions defined in Figure 2.76a. Thus,

$$K_{11} = \frac{12EI}{L^3}, \qquad K_{21} = -\frac{6EI}{L^2}, \qquad K_{31} = -\frac{12EI}{L^3}, \qquad K_{41} = -\frac{6EI}{L^2}$$

The second column of the stiffness matrix is obtained by letting $\delta_2 = 1$ and setting the remaining three displacements equal to zero as indicated in Figure 2.76c. The area of the M/EI diagram between the ends of the member for this case is equal to unity, and hence,

$$\frac{K_{22}L}{2EI} - \frac{K_{42}L}{2EI} = 1$$

The moment of the M/EI diagram about the left end is zero, so that

$$\frac{K_{22}L}{2EI}\left(\frac{L}{3}\right) - \frac{K_{42}L}{2EI}\left(\frac{2L}{3}\right) = 0$$

Therefore, one obtains

$$K_{22} = \frac{4EI}{L}, \qquad\qquad K_{42} = \frac{2EI}{L}$$

From vertical equilibrium of the member,

$$K_{12} = K_{32}$$

and moment equilibrium about the right end of the member leads to

$$K_{12} = \frac{K_{22} + K_{42}}{L} = \frac{6EI}{L^2}$$

Comparison of the forces in Figure 2.76c with the positive directions defined in Figure 2.76a indicates that all the influence coefficients except k_{12} are positive. Thus,

$$K_{12} = -\frac{6EI}{L^2}, \qquad K_{22} = \frac{4EI}{L}, \qquad K_{32} = \frac{6EI}{L^2}, \qquad K_{42} = \frac{2EI}{L}$$

Using Figures 2.76d and e, the influence coefficients for the third and fourth columns can be obtained. The results of these calculations lead to the following element-stiffness matrix:

$$
\begin{bmatrix} q_1 \\ q_2 \\ q_3 \\ q_4 \end{bmatrix}
=
\begin{bmatrix}
\frac{12EI}{L^3} & -\frac{6EI}{L^2} & -\frac{12EI}{L^3} & -\frac{6EI}{L^2} \\
-\frac{6EI}{L^2} & \frac{4EI}{L} & \frac{6EI}{L^2} & \frac{2EI}{L} \\
-\frac{12EI}{L^3} & \frac{6EI}{L^2} & \frac{12EI}{L^3} & \frac{6EI}{L^2} \\
-\frac{6EI}{L^2} & \frac{2EI}{L} & \frac{6EI}{L^2} & \frac{4EI}{L}
\end{bmatrix}
\begin{bmatrix} \delta_1 \\ \delta_2 \\ \delta_3 \\ \delta_4 \end{bmatrix}
\qquad (2.151)
$$

Note that Equation 2.150 defines the element-stiffness matrix for a flexural member with constant flexural rigidity EI.

If axial load in a frame member is also considered the general form of an element, then the stiffness matrix for an element shown in Figure 2.77 becomes

$$
\begin{bmatrix} q_1 \\ q_2 \\ q_3 \\ q_4 \\ q_5 \\ q_6 \end{bmatrix} = \begin{bmatrix} \frac{EA}{L} & 0 & 0 & -\frac{EA}{L} & 0 & 0 \\ 0 & \frac{12EI}{L^3} & -\frac{6EI}{L^2} & 0 & -\frac{12EI}{L^3} & -\frac{6EI}{L^2} \\ 0 & -\frac{6EI}{L^2} & \frac{4EI}{L} & 0 & \frac{6EI}{L^2} & \frac{2EI}{L} \\ -\frac{EI}{L} & 0 & 0 & \frac{EI}{L} & 0 & 0 \\ 0 & -\frac{12EI}{L^3} & \frac{6EI}{L^2} & 0 & \frac{12EI}{L^3} & \frac{6EI}{L^2} \\ 0 & -\frac{6EI}{L^2} & \frac{2EI}{L} & 0 & \frac{6EI}{L^2} & \frac{4EI}{L} \end{bmatrix} \begin{bmatrix} \delta_1 \\ \delta_2 \\ \delta_3 \\ \delta_4 \\ \delta_5 \\ \delta_6 \end{bmatrix}
$$

(or)

$$[q] = [k_c][\delta] \tag{2.152}$$

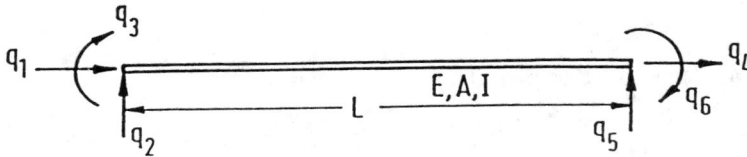

Figure 2.77   Beam element with axial force.

2.9.4   Grillages

Another common type of structure is one in which the members all lie in one plane with loads being applied in the direction normal to this plane. This type of structure is commonly adopted in building floor systems, bridge decks, ship decks, floors, etc. The grid floor, for purposes of analysis by using matrix method, can be treated as a **space frame**. However, the solution can be simplified by considering the grid member as a planar grid. A typical grid floor is shown in Figure 2.78a. The significant member forces in a member and the corresponding deformations are as shown in Figure 2.78b.

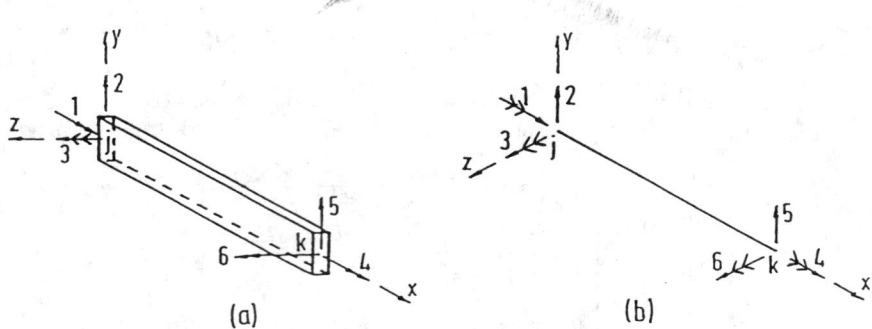

Figure 2.78   A grid member.

The member stiffness matrix can be written as

$$K = \begin{bmatrix} \frac{GJ}{L} & 0 & 0 & -\frac{GJ}{L} & 0 & 0 \\ 0 & \frac{12EI_z}{L^3} & \frac{6EI_z}{L^2} & 0 & -\frac{12EI_z}{L^3} & \frac{6EI_z}{L^2} \\ 0 & \frac{6EI_z}{L^2} & \frac{4EI_z}{L} & 0 & -\frac{6EI_z}{L^2} & \frac{2EI_z}{L} \\ -\frac{GJ}{L} & 0 & 0 & \frac{GJ}{L} & 0 & 0 \\ 0 & -\frac{12EI_z}{L^2} & -\frac{6EI_z}{L^2} & 0 & \frac{12EI_z}{L^3} & -\frac{6EI_z}{L^2} \\ 0 & \frac{6EI_z}{L^2} & \frac{2EI_z}{L} & 0 & -\frac{6EI_z}{L^2} & \frac{4EI_z}{L} \end{bmatrix} \qquad (2.153)$$

2.9.5 Structure Stiffness Matrix

Equation 2.152 has been expressed in terms of the coordinate system of the individual members. In a structure consisting of many members there would be as many systems of coordinates as the number of members. Before the internal actions in the members of the structure can be related, all forces and deflections must be stated in terms of one single system of axes common to all—the structure axes. The transformation from element to structure coordinates is carried out separately for each element and the resulting matrices are then combined to form the structure-stiffness matrix. A separate transformation matrix $[T]$ is written for each element and a relation of the form

$$[\delta]_n = [T]_n [\Delta]_n \qquad (2.154)$$

is written in which $[T]_n$ defines the matrix relating the element deformations of element 'n' to the structure deformations at the ends of that particular element. The element and structure forces are related in the same way as the corresponding deformations as

$$[q]_n = [T]_n [W]_n \qquad (2.155)$$

where $[q]_n$ contains the element forces for element 'n' and $[W]_n$ contains the structure forces at the extremities of the element. The transformation matrix $[T]_n$ can be used to transform element 'n' from its local coordinates to structure coordinates. We know, for an element n, the force-deformation relation is given as

$$[q]_n = [k]_n [\delta]_n$$

Substituting for $[q]_n$ and $[\delta]_n$ from Equations 2.153 and 2.154 one obtains

$$[T]_n [W]_n = [k]_n [T]_n [\Delta]_n$$

or

$$\begin{aligned} [W]_n &= [T]_n^{-1} [k]_n [T]_n [\Delta]_n \\ &= [T]_n^T [k]_n [T]_n [\Delta]_n \\ &= [K]_n [\Delta]_n \\ [K]_n &= [T]_n^T [k]_n [T]_n \end{aligned} \qquad (2.156)$$

$[K]_n$ is the stiffness matrix which transforms any element 'n' from its local coordinate to structure coordinates. In this way, each element is transformed individually from element coordinate to structure coordinate and the resulting matrices are combined to form the stiffness matrix for the entire structure.

Member stiffness matrix $[K]_n$ in structure coordinates for a truss member shown in Figure 2.79, for example, is given as

$$[K]_n = \frac{AE}{L} \begin{bmatrix} \lambda^2\mu & \lambda\mu & -\lambda^2 & -\lambda\mu \\ \lambda\mu & \mu^2 & -\lambda\mu & -\mu^2 \\ -\lambda^2 & -\lambda\mu & \lambda^2 & \lambda\mu \\ -\lambda\mu & -\mu^2 & \lambda\mu & \mu^2 \end{bmatrix} \begin{matrix} i \\ j \\ k \\ \ell \end{matrix} \qquad (2.157)$$

in which $\lambda = \cos\phi$ and $\mu = \sin\phi$.

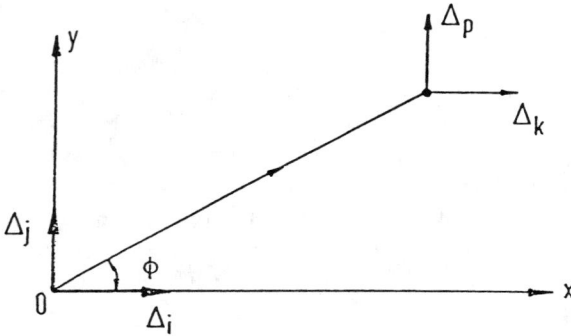

Figure 2.79   A grid member.

To construct $[K]_n$ for a given member it is necessary to have the values of λ and μ for the member. In addition, the structure coordinates i, j, k, and ℓ at the extremities of the member must be known.

Member stiffness matrix $[K]_n$ in structural coordinates for a flexural member shown in Figure 2.80 can be written as

$$[K]_n =$$

$$\begin{bmatrix} \left(\lambda^2\frac{AE}{L} + \mu^2\frac{12EI}{L^3}\right) & & & & & \text{(Symmetric)} \\ \mu\lambda\left(\frac{AE}{L} - \frac{12EI}{L^3}\right) & \left(\mu^2\frac{AE}{L} + \lambda^2\frac{12EI}{L^3}\right) & & & & \\ -\mu\left(\frac{6EI}{L^2}\right) & \lambda\left(\frac{6EI}{L^2}\right) & \frac{4EI}{L} & & & \\ \left(-\lambda^2\frac{AE}{L} - \mu^2\frac{12EI}{L^3}\right) & \mu\lambda\left(\frac{AE}{L} - \frac{12EI}{L^3}\right) & \mu\left(\frac{6EI}{L^2}\right) & \left(\lambda^2\frac{AE}{L} + \mu^2\frac{12EI}{L^3}\right) & & \\ -\mu\lambda\left(\frac{AE}{L} - \frac{12EI}{L^3}\right) & \left(-\mu^2\frac{AE}{L} - \lambda^2\frac{12EI}{L^3}\right) & -\lambda\left(\frac{6EI}{L^2}\right) & \mu\lambda\left(\frac{AE}{L} - \frac{12EI}{L^3}\right) & \left(\mu^2\frac{AE}{L} + \lambda^2\frac{12EI}{L^3}\right) & \\ -\mu\left(\frac{6EI}{L^2}\right) & \lambda\left(\frac{6EI}{L^2}\right) & \frac{2EI}{L} & \mu\left(\frac{6EI}{L^2}\right) & -\lambda\left(\frac{6EI}{L^2}\right) & \frac{4EI}{L} \end{bmatrix}$$

$$(2.158)$$

where $\lambda = \cos\phi$ and $\mu = \sin\phi$.

EXAMPLE 2.9:

Determine the displacement at the loaded point of the truss shown in Figure 2.81. Both members have the same area of cross-section $A = 3$ in.^{2} and $E = 30 \times 10^3$ ksi.

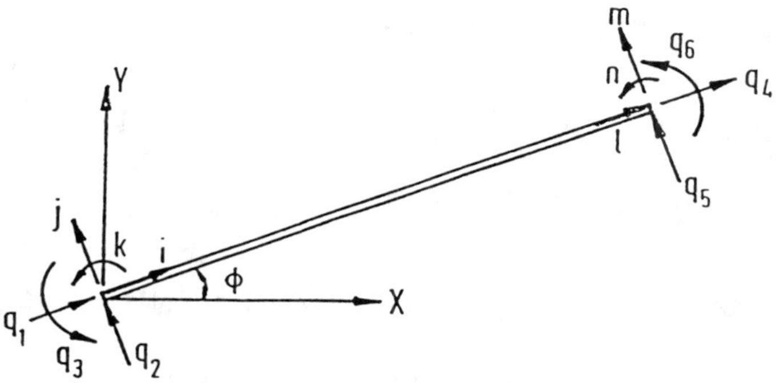

Figure 2.80 A flexural member in global coordinate.

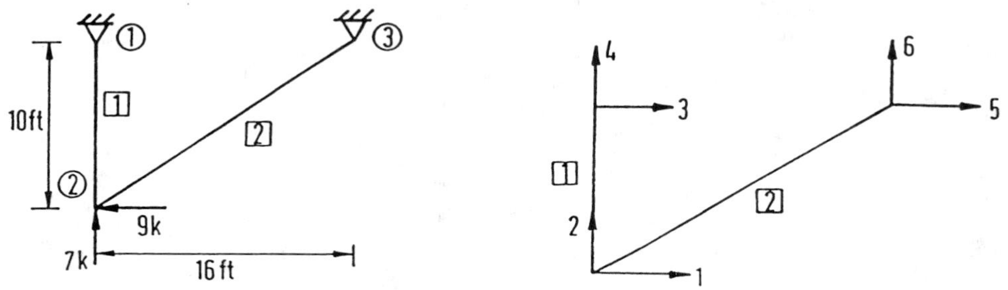

Figure 2.81 Example 2.9.

The details required to form the element stiffness matrix with reference to structure coordinates axes are listed below:

Member	Length	ϕ	λ	μ	i	j	k	l
1	10 ft	90°	0	1	1	2	3	4
2	18.9 ft	32°	0.85	0.53	1	2	5	6

We now use these data in Equation 2.157 to form $[K]_n$ for the two elements.
For member 1,

$$\frac{AE}{L} = \frac{3\times 30\times 10^3}{120} = 750$$

$$[K]_1 = \begin{array}{c} \begin{matrix} 1 & 2 & 3 & 4 \end{matrix} \\ \begin{bmatrix} 0 & 0 & 0 & 0 \\ 0 & 750 & 0 & -750 \\ 0 & 0 & 0 & 0 \\ 0 & -750 & 0 & 750 \end{bmatrix} \end{array} \begin{matrix} 1 \\ 2 \\ 3 \\ 4 \end{matrix}$$

For member 2,

$$\frac{AE}{L} = \frac{3\times 30\times 10^3}{18.9\times 12} = 397$$

$$[K]_2 = \begin{array}{c} \begin{matrix} 1 & 2 & 5 & 6 \end{matrix} \\ \begin{bmatrix} 286 & 179 & -286 & -179 \\ 179 & 111 & -179 & -111 \\ -286 & -179 & 286 & 179 \\ -179 & -111 & 179 & 111 \end{bmatrix} \end{array} \begin{matrix} 1 \\ 2 \\ 5 \\ 6 \end{matrix}$$

Combining the element stiffness matrices, $[K]_1$ and $[K]_2$, one obtains the structure stiffness matrix as follows:

$$
\begin{bmatrix} W_1 \\ W_2 \\ W_3 \\ W_4 \\ W_5 \\ W_6 \end{bmatrix} = \begin{bmatrix} 286 & 179 & 0 & 0 & -286 & -179 \\ 179 & 861 & 0 & -750 & -179 & -111 \\ 0 & 0 & 0 & 0 & 0 & 0 \\ 0 & -750 & 0 & 750 & 0 & 0 \\ -286 & -179 & 0 & 0 & 286 & 179 \\ -179 & -111 & 0 & 0 & 179 & 111 \end{bmatrix} \begin{bmatrix} \Delta_1 \\ \Delta_2 \\ \Delta_3 \\ \Delta_4 \\ \Delta_5 \\ \Delta_6 \end{bmatrix}
$$

The stiffness matrix can now be subdivided to determine the unknowns. Let us consider Δ_1 and Δ_2 the deflections at joint 2 which can be determined in view of $\Delta_3 = \Delta_4 = \Delta_5 = \Delta_6 = 0$ as follows:

$$
\begin{bmatrix} \Delta_1 \\ \Delta_2 \end{bmatrix} = \begin{bmatrix} 286 & 179 \\ 179 & 861 \end{bmatrix}^{-1} \begin{bmatrix} -9 \\ 7 \end{bmatrix}
$$

or

$$
\begin{aligned}
\Delta_1 &= 0.042 \text{ in. to the left} \\
\Delta_2 &= 0.0169 \text{ in. upward}
\end{aligned}
$$

EXAMPLE 2.10:

A simple triangular frame is loaded at the tip by 20 kips as shown in Figure 2.82. Assemble the structure stiffness matrix and determine the displacements at the loaded node.

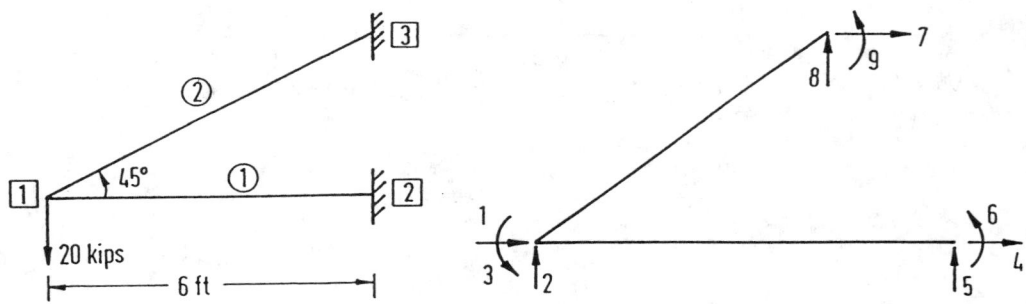

Figure 2.82   Example 2.10.

Member	Length (in.)	A (in.^{2})	I (in.^{4})	ϕ	λ	μ
1	72	2.4	1037	0	1	0
2	101.8	3.4	2933	45°	0.707	0.707

For members 1 and 2 the stiffness matrices in structure coordinates can be written by making use of Equation 2.158.

$$[K]_1 = 10^3 \times \begin{array}{c} \\ \\ \\ \\ \\ \\ \\ \end{array} \begin{array}{cccccc} 1 & 2 & 3 & 4 & 5 & 6 \\ 1 & 0 & 0 & -1 & 0 & 0 \\ 0 & 1 & 36 & 0 & -1 & 36 \\ 0 & 36 & 1728 & 0 & -36 & 864 \\ -1 & 0 & 0 & 1 & 0 & 0 \\ 0 & -1 & -36 & 0 & 1 & -36 \\ 0 & 36 & 864 & 0 & -36 & 1728 \end{array} \begin{array}{c} 1 \\ 2 \\ 3 \\ 4 \\ 5 \\ 6 \end{array}$$

and

$$[K]_2 = 10^3 \times \begin{array}{cccccc} 1 & 2 & 3 & 7 & 8 & 9 \\ 1 & 0 & -36 & -1 & 0 & -36 \\ 0 & 1 & 36 & 0 & 1 & 36 \\ -36 & 36 & 3457 & 36 & -36 & 1728 \\ -1 & 0 & 36 & 1 & 0 & 36 \\ 0 & 1 & -36 & 0 & 1 & -36 \\ -36 & 36 & 1728 & 36 & -36 & 3457 \end{array} \begin{array}{c} 1 \\ 2 \\ 3 \\ 7 \\ 8 \\ 9 \end{array}$$

Combining the element stiffness matrices $[K]_1$ and $[K]_2$, one obtains the structure stiffness matrix as follows:

$$[K] = 10^3 \times \begin{array}{ccccccccc} 2 & 0 & -36 & -1 & 0 & 0 & -1 & 0 & -36 \\ 0 & 2 & 72 & 0 & -1 & 36 & 0 & 1 & 36 \\ -36 & 72 & 5185 & 0 & -36 & 864 & 36 & -36 & 1728 \\ -1 & 0 & 0 & 1 & 0 & 0 & 0 & 0 & 0 \\ 0 & -1 & -36 & 0 & 1 & -36 & 0 & 0 & 0 \\ 0 & 36 & 864 & 0 & -36 & 1728 & 0 & 0 & 0 \\ -1 & 0 & 36 & 0 & 0 & 0 & 1000 & 0 & 36 \\ 0 & 1 & -36 & 0 & 0 & 0 & 0 & 1 & -36 \\ -36 & 36 & 1728 & 0 & 0 & 0 & 36 & 36 & 3457 \end{array} \begin{array}{c} 1 \\ 2 \\ 3 \\ 4 \\ 5 \\ 6 \\ 7 \\ 8 \\ 9 \end{array}$$

The deformations at joints 2 and 3 corresponding to Δ_5 to Δ_9 are zero since joints 2 and 4 are restrained in all directions. Cancelling the rows and columns corresponding to zero deformations in the structure stiffness matrix, one obtains the force deformation relation for the structure:

$$\begin{bmatrix} F_1 \\ F_2 \\ F_3 \end{bmatrix} = \begin{bmatrix} 2 & 0 & -36 \\ 0 & 2 & 72 \\ -36 & 72 & 5185 \end{bmatrix} \times 10^3 \begin{bmatrix} \Delta_1 \\ \Delta_2 \\ \Delta_3 \end{bmatrix}$$

Substituting for the applied load $F_2 = -20$ kips, the deformations are given as

$$\begin{bmatrix} \Delta_1 \\ \Delta_2 \\ \Delta_3 \end{bmatrix} = \begin{bmatrix} 2 & 0 & -36 \\ 0 & 2 & 72 \\ -36 & 72 & 5185 \end{bmatrix}^{-1} \times 10^3 \begin{bmatrix} 0 \\ -20 \\ 0 \end{bmatrix}$$

or

$$\begin{bmatrix} \Delta_1 \\ \Delta_2 \\ \Delta_3 \end{bmatrix} = \begin{bmatrix} 6.66 \text{ in.} \\ -23.334 \text{ in.} \\ 0.370 \text{ rad} \end{bmatrix} \times 10^3$$

2.9.6 Loading Between Nodes

The problems discussed thus far have involved concentrated forces and moments applied to nodes only. But real structures are subjected to distributed or concentrated loading between nodes as

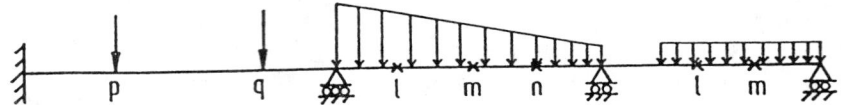

Figure 2.83   Loading between nodes.

shown in Figure 2.83. Loading may range from a few concentrated loads to an infinite variety of uniform or nonuniformly distributed loads. The solution method of matrix analysis must be modified to account for such load cases.

One way to treat such loads in the matrix analysis is to insert artificial nodes, such as p and q as shown in Figure 2.83. The degrees of freedom corresponding to the additional nodes are added to the total structure and the necessary additional equations are written by considering the requirements of equilibrium at these nodes. The internal member forces on each side of nodes p and q must equilibrate the external loads applied at these points. In the case of distributed loads, suitable nodes such as l, m, and n shown in Figure 2.83 are selected arbitrarily and the distributed loads are lumped as concentrated loads at these nodes. The degrees of freedom corresponding to the arbitrary and real nodes are treated as unknowns of the problem. There are different ways of obtaining equivalence between the lumped and the distributed loading. In all cases the lumped loads must be statically equivalent to the distributed loads they replace.

The method of introducing arbitrary nodes is not a very elegant procedure because the number of unknown degrees of freedom made the solution procedure laborious. The approach that is of most general use with the displacement method is one employing the related concepts of artificial joint restraint, fixed-end forces, and equivalent nodal loads.

2.9.7   Semi-Rigid End Connection

A rigid connection holds unchanged the original angles between intersecting members; a simple connection allows the member end to rotate freely under gravity load, a **semi-rigid connection** possesses a moment capacity intermediate between the simple and the rigid. A simplified linear relationship between the moment m acting on the connection and the resulting connection rotation ψ in the direction of m is assumed giving

$$M = R\frac{EI}{L}\psi \tag{2.159}$$

where EI and L are the flexural rigidity and length of the member, respectively. The non-dimensional quantity R, which is a measure of the degree of rigidity of the connection, is called the rigidity index. For a simple connection, R is zero and for a rigid connection, R is infinity. Considering the semi-rigidity of joints, the member flexibility matrix for flexure is derived as

$$\begin{bmatrix} \phi_1 \\ \phi_2 \end{bmatrix} = \frac{L}{EI} \begin{bmatrix} \frac{1}{3} + \frac{1}{R_1} & -\frac{1}{6} \\ -\frac{1}{6} & \frac{1}{3} + \frac{1}{R_2} \end{bmatrix} \begin{bmatrix} M_1 \\ M_2 \end{bmatrix} \tag{2.160}$$

or

$$[\phi] = [F][M] \tag{2.161}$$

where ϕ_1 and ϕ_2 are as shown in Figure 2.84.

For convenience, two parameters are introduced as follows:

$$p_1 = \frac{1}{1 + \frac{3}{R_1}}$$

and

$$p_2 = \frac{1}{1 + \frac{3}{R_2}}$$

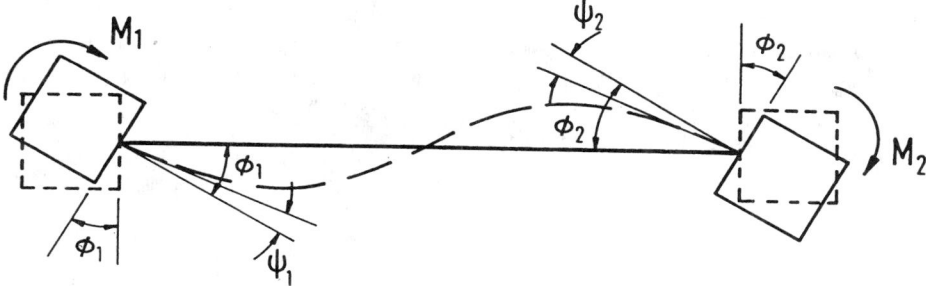

Figure 2.84 A flexural member with semi-rigid end connections.

where p_1 and p_2 are called the fixity factors. For hinged connections, both the fixity factors (p) and the rigidity index (R) are zero; but for rigid connections, the fixity factor is 1 and the rigidity index is infinity. Since the fixity factor can only vary from 0 to 1, its use is more convenient in the analyses of structures with semi-rigid connections.

Equation 2.160 can be rewritten to give

$$[F] = \frac{L}{EI} \begin{bmatrix} \frac{1}{3p_1} & -\frac{1}{6} \\ -\frac{1}{6} & \frac{1}{3p_2} \end{bmatrix} \tag{2.162}$$

From Equation 2.162, the modified member stiffness matrix $[K]$ for a member with semi-rigid and connections expresses the member end moments, M_1 and M_2, in terms of the member end rotations, ϕ_1 and ϕ_2, as

$$[K] = EI \begin{bmatrix} k_{11} & k_{12} \\ k_{21} & k_{22} \end{bmatrix} \tag{2.163a}$$

Expressions for k_{11}, and $k_{12} = k_{21}$ and k_{22} may be obtained by inverting the $[F]$ matrix. Thus,

$$k_{11} = \frac{12/p_2}{4/(p_1 p_2) - 1} \tag{2.163b}$$

$$k_{12} = k_{21} = \frac{6}{4(p_1 p_2) - 1} \tag{2.163c}$$

$$k_{22} = \frac{12/p_1}{4/(p_1 p_2) - 1} \tag{2.163d}$$

The modified member stiffness matrix $[K]$, as expressed by Equations 2.163 will be needed in the stiffness method of analysis of frames in which there are semi-rigid member-end connections.

2.10 The Finite Element Method

Many problems that confront the design analyst, in practice, cannot be solved by analytical methods. This is particularly true for problems involving complex material properties and boundary conditions. Numerical methods, in such cases, provide approximate but acceptable solutions. Of the many numerical methods developed before and after the advent of computers, the finite element method has proven to be a powerful tool. This method can be regarded as a natural extension of the matrix methods of structural analysis. It can accommodate complex and difficult problems such as nonhomogenity, nonlinear stress-strain behavior, and complicated boundary conditions. The finite element method is applicable to a wide range of boundary value problems in engineering and it dates back to the mid-1950s with the pioneering work by Argyris [4], Clough [21], and others. The method was first applied to the solution of plane stress problems and extended subsequently to the solution of plates, shells, and axisymmetric solids.

2.10.1   Basic Concept

The finite element method is based on the representation of a body or a structure by an assemblage of subdivisions called finite elements as shown in Figure 2.85. These elements are considered

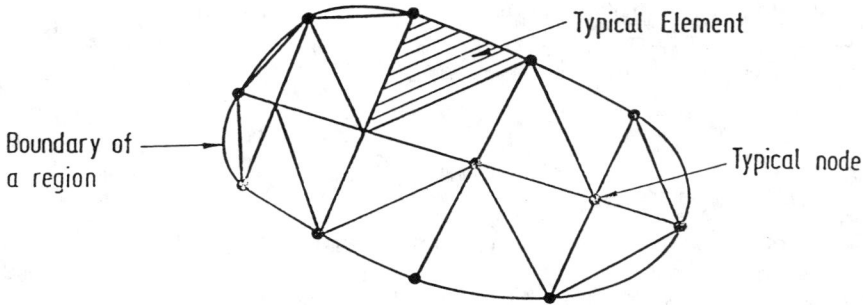

Figure 2.85   Assemblage of subdivisions.

to be connected at nodes. Displacement functions are chosen to approximate the variation of displacements over each finite element. Polynomials are commonly employed to express these functions. Equilibrium equations for each element are obtained by means of the principle of minimum potential energy. These equations are formulated for the entire body by combining the equations for the individual elements so that the continuity of displacements is preserved at the nodes. The resulting equations are solved satisfying the boundary conditions in order to obtain the unknown displacements.

The entire procedure of the finite element method involves the following steps: (1) the given body is subdivided into an equivalent system of finite elements, (2) suitable displacement function is chosen, (3) element stiffness matrix is derived using variational principle of mechanics such as the principle of minimum potential energy, (4) global stiffness matrix for the entire body is formulated, (5) the algebraic equations thus obtained are solved to determine unknown displacements, and (6) element strains and stresses are computed from the nodal displacements.

2.10.2   Basic Equations from Theory of Elasticity

Figure 2.86 shows the state of stress in an elemental volume of a body under load. It is defined in terms of three normal stress components σ_x, σ_y, and σ_z and three shear stress components τ_{xy}, τ_{yz}, and τ_{zx}. The corresponding strain components are three normal strains ε_x, ε_y, and ε_z and three shear strains γ_{xy}, γ_{yz}, and γ_{zx}. These strain components are related to the displacement components u, v, and w at a point as follows:

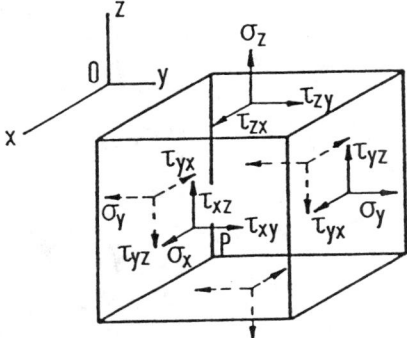

Figure 2.86   State of stress in an elemental volume.

$$\varepsilon_x = \frac{\partial u}{\partial x} \qquad \gamma_{xy} = \frac{\partial v}{\partial x} + \frac{\partial u}{\partial y}$$

$$\varepsilon_y = \frac{\partial v}{\partial y} \qquad \gamma_{yz} = \frac{\partial w}{\partial y} + \frac{\partial v}{\partial z}$$

$$\varepsilon_z = \frac{\partial w}{\partial z} \qquad \gamma_{zx} = \frac{\partial u}{\partial z} + \frac{\partial w}{\partial x} \qquad (2.164)$$

The relations given in Equation 2.164 are valid in the case of the body experiencing small deformations. If the body undergoes large or finite deformations, higher order terms must be retained.

The stress-strain equations for isotropic materials may be written in terms of the Young's modulus and Poisson's ratio as

$$\sigma_x = \frac{E}{1 - \nu^2}[\varepsilon_x + \nu(\varepsilon_y + \varepsilon_z)]$$

$$\sigma_y = \frac{E}{1 - \nu^2}[\varepsilon_y + \nu(\varepsilon_z + \varepsilon_x)]$$

$$\sigma_z = \frac{E}{1 - \nu^2}[\varepsilon_z + \nu(\varepsilon_x + \varepsilon_y)]$$

$$\tau_{xy} = G\gamma_{xy}, \quad \tau_{yz} = G\gamma_{yz}, \quad \tau_{zx} = G\gamma_{zx} \qquad (2.165)$$

2.10.3 Plane Stress

When the elastic body is very thin and there are no loads applied in the direction parallel to the thickness, the state of stress in the body is said to be plane stress. A thin plate subjected to in-plane loading as shown in Figure 2.87 is an example of a plane stress problem. In this case, $\sigma_z = \tau_{yz} = \tau_{zx} = 0$ and the constitutive relation for an isotropic continuum is expressed as

$$\begin{bmatrix} \sigma_x \\ \sigma_y \\ \sigma_{xy} \end{bmatrix} = \frac{E}{1 - \nu^2} \begin{bmatrix} 1 & \nu & 0 \\ \nu & 1 & 0 \\ 0 & 0 & \frac{1-\nu}{2} \end{bmatrix} \begin{bmatrix} \varepsilon_x \\ \varepsilon_y \\ \gamma_{xy} \end{bmatrix} \qquad (2.166)$$

2.10.4 Plane Strain

The state of plane strain occurs in members that are not free to expand in the direction perpendicular to the plane of the applied loads. Examples of some plane strain problems are retaining walls, dams, long cylinder, tunnels, etc. as shown in Figure 2.88. In these problems ε_z, γ_{yz}, and γ_{zx} will vanish and hence,

$$\sigma_z = \nu(\sigma_x + \sigma_y)$$

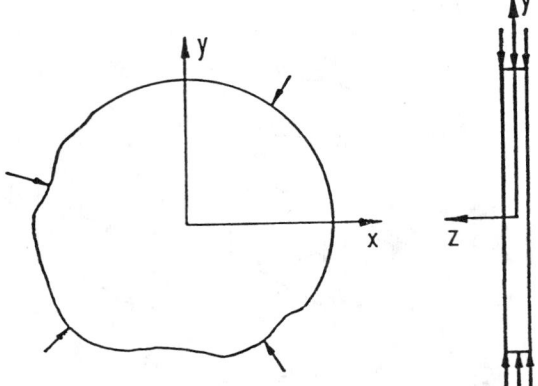

Figure 2.87   Plane-stress problem.

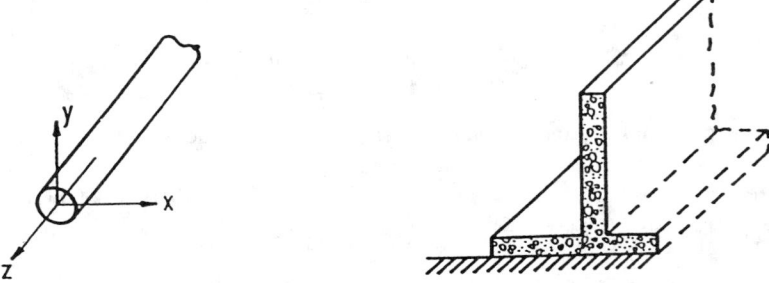

Figure 2.88   Practical examples of plane-strain problems.

The constitutive relations for an isotropic material are written as

$$
\begin{bmatrix} \sigma_x \\ \sigma_y \\ \tau_{xy} \end{bmatrix} = \frac{E}{(1+\nu)(1-2\nu)} \begin{bmatrix} (1-\nu) & \nu & 0 \\ \nu & (1-\nu) & 0 \\ 0 & 0 & \frac{1-2\nu}{2} \end{bmatrix} \begin{bmatrix} \varepsilon_x \\ \varepsilon_y \\ \gamma_{xy} \end{bmatrix} \qquad (2.167)
$$

2.10.5   Element Shapes and Discretization

The process of subdividing a continuum is an exercise of engineering judgement. The choice depends on the geometry of the body. A finite element generally has a simple one-, two-, or three-dimensional configuration. The boundaries of elements are often straight lines and the elements can be one, two, or three dimensional as shown in Figure 2.89. While subdividing the continuum, one has to decide the number, shape, size, and configuration of the elements in such a way that the original body is simulated as closely as possible. Nodes must be located in locations where abrupt changes in geometry, loading, and material properties occur. A node must be placed at the point of application of a concentrated load because all loads are converted into equivalent nodal-point loads.

It is easy to subdivide a continuum into a completely regular one having the same shape and size. But problems encountered in practice do not involve regular shape; they may have regions of steep gradients of stresses. A finer subdivision may be necessary in regions where stress concentrations are expected in order to obtain a useful approximate solution. Typical examples of mesh selection are shown in Figure 2.90.

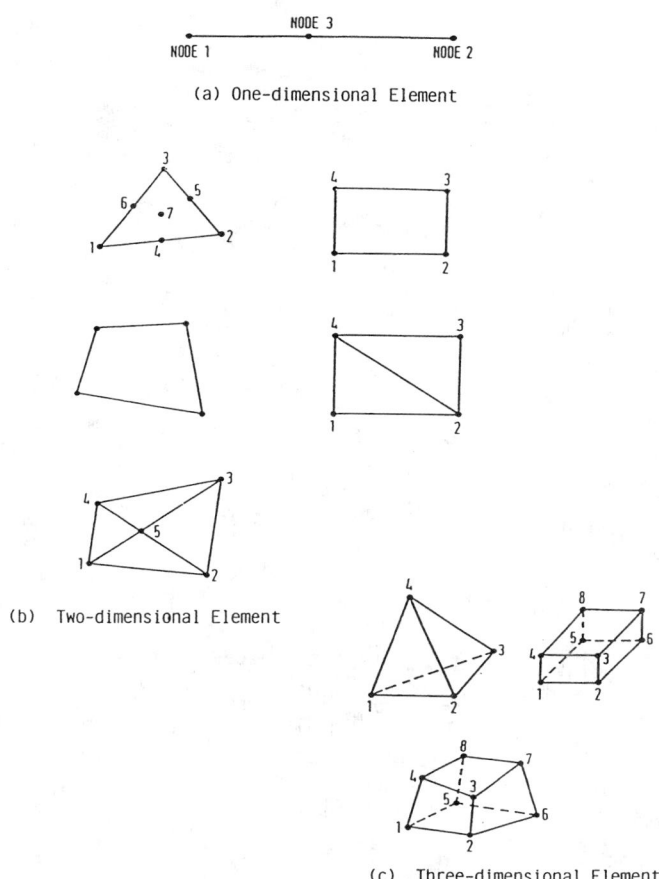

Figure 2.89   (a) One-dimensional element; (b) two-dimensional elements; (c) three-dimensional elements.

2.10.6   Choice of Displacement Function

Selection of displacement function is the important step in the finite element analysis because it determines the performance of the element in the analysis. Attention must be paid to select a displacement function which (1) has the number of unknown constants as the total number of degrees of freedom of the element, (2) does not have any preferred directions, (3) allows the element to undergo rigid-body movement without any internal strain, (4) is able to represent states of constant stress or strain, and (5) satisfies the compatibility of displacements along the boundaries with adjacent elements. Elements that meet both the third and fourth requirements are known as *complete elements*.

A polynomial is the most common form of displacement function. Mathematics of polynomials are easy to handle in formulating the desired equations for various elements and convenient in digital computation. The degree of approximation is governed by the stage at which the function is truncated. Solutions closer to exact solutions can be obtained by including more number of terms. The polynomials are of the general form

$$w(x) = a_1 + a_2 x + a_3 x^2 + \ldots a_{n+1} x^n \qquad (2.168)$$

The coefficients 'a's are known as *generalized displacement amplitudes*. The general polynomial form for a two-dimensional problem can be given as

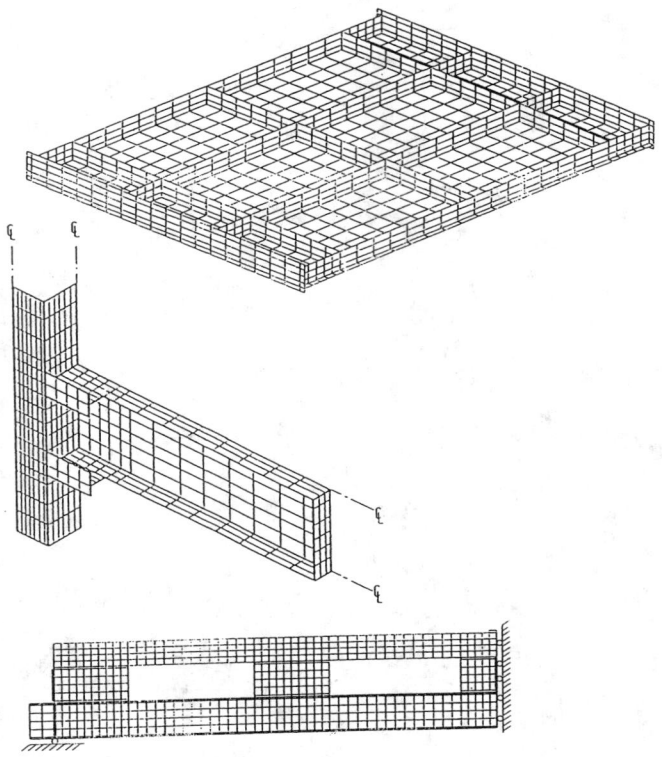

Figure 2.90   Typical examples of finite element mesh.

$$u(x, y) = a_1 + a_2 x + a_3 y + a_4 x^2 + a_5 xy + a_6 y^2 + \ldots a_m y^n$$
$$v(x, y) = a_{m+1} + a_{m+2} x + a_{m+3} y + a_{m+4} x^2 + a_{m+5} xy$$
$$+ a_{m+6} y^2 + \ldots + a_{2m} y^n$$

in which

$$m = \sum_{i=1}^{n+1} i \tag{2.169}$$

These polynomials can be truncated at any desired degree to give constant, linear, quadratic, or higher order functions. For example, a linear model in the case of a two-dimensional problem can be given as

$$u = a_1 + a_2 x + a_3 y$$
$$v = a_4 + a_5 x + a_6 y \tag{2.170}$$

A quadratic function is given by

$$u = a_1 + a_2 x + a_3 y + a_4 x^2 + a_5 xy + a_6 y^2$$
$$v = a_7 + a_8 x + a_9 y + a_{10} x^2 + a_{11} xy + a_{12} y^2 \tag{2.171}$$

The Pascal triangle shown below can be used for the purpose of achieving isotropy, i.e., to avoid displacement shapes that change with a change in local coordinate system.

$$
\begin{array}{cccccccccccc}
 & & & & & 1 & & & & & & & \text{Constant} \\
 & & & & x & & & y & & & & & \text{Linear} \\
 & & & x^2 & & & xy & & & y^2 & & & \text{Quadratic} \\
 & & x^3 & & & x^2y & & & xy^2 & & & y^3 & \text{Cubic} \\
 & x^4 & & & x^3y & & & x^2y^2 & & & xy^3 & & y^4 & \text{Quantic} \\
x^5 & & x^4y & & & x^3y^2 & & & x^2y^3 & & & xy^4 & & & y^5 & \text{Quintic}
\end{array}
$$

Line of Symmetry

2.10.7   Nodal Degrees of Freedom

The deformation of the finite element is specified completely by the nodal displacement, rotations, and/or strains which are referred to as *degrees of freedom*. Convergence, geometric isotropy, and potential energy function are the factors that determine the minimum number of degrees of freedom necessary for a given element. Additional degrees of freedom beyond the minimum number may be included for any element by adding secondary external nodes. Such elements with additional degrees of freedom are called higher order elements. The elements with more additional degrees of freedom become more flexible.

2.10.8   Isoparametric Elements

The scope of finite element analysis is also measured by the variety of element geometries that can be constructed. Formulation of element stiffness equations requires the selection of displacement expressions with as many parameters as there are node-point displacements. In practice, for planar conditions, only the four-sided (quadrilateral) element finds as wide an application as the triangular element. The simplest form of quadrilateral, the rectangle, has four node points and involves two displacement components at each point for a total of eight degrees of freedom. In this case one would choose four-term expressions for both u and v displacement fields. If the description of the element is expanded to include nodes at the mid-points of the sides, an eight-term expression would be chosen for each displacement component.

The triangle and rectangle can approximate the curved boundaries only as a series of straight line segments. A closer approximation can be achieved by means of *isoparametric* coordinates. These are non-dimensionalized curvilinear coordinates whose description is given by the same coefficients as are employed in the displacement expressions. The displacement expressions are chosen to ensure continuity across element interfaces and along supported boundaries, so that geometric continuity is ensured when the same forms of expressions are used as the basis of description of the element boundaries. The elements in which the geometry and displacements are described in terms of the same parameters and are of the same order are called *isoparametric elements*. The isoparametric concept enables one to formulate elements of any order which satisfy the completeness and compatibility requirements and which have isotropic displacement functions.

2.10.9   Isoparametric Families of Elements

Definitions and Justifications

For example, let u_i represent nodal displacements and x_i represent nodal x-coordinates. The interpolation formulas are

$$
u = \sum_{i=1}^{m} N_i u_i \qquad x = \sum_{i=1}^{n} N_i' x_i
$$

where N_i and N are shape functions written in terms of the intrinsic coordinates. The value of u and the value of x at a point within the element are obtained in terms of nodal values of u_i and x_i from the above equations when the (intrinsic) coordinates of the internal point are given. Displacement components v and w in the y and z directions are treated in a similar manner.

The element is *isoparametric* if $m = n$, $N_i = N$, and the same nodal points are used to define both element geometry and element displacement (Figure 2.91a); the element is *subparametric* if

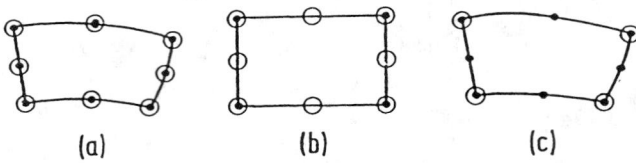

Figure 2.91   (a) Isoparametric element; (b) subparametric element; (c) superparametric element.

$m > n$ and the order of N_i is larger than N_i' (Figure 2.91b); the element is *superparametric* if $m < n$ and the order of N_i is smaller than N_i' (Figure 2.91c). The isoparametric elements can correctly display rigid-body and constant-strain modes.

2.10.10   Element Shape Functions

The finite element method is not restricted to the use of linear elements. Most finite element codes, commercially available, allow the user to select between elements with linear or quadratic interpolation functions. In the case of quadratic elements, fewer elements are needed to obtain the same degree of accuracy in the nodal values. Also, the two-dimensional quadratic elements can be shaped to model a curved boundary. Shape functions can be developed based on the following properties: (1) each shape function has a value of one at its own node and is zero at each of the other nodes, (2) the shape functions for two-dimensional elements are zero along each side that the node does not touch, and (3) each shape function is a polynomial of the same degree as the interpolation equation. Shape function for typical elements are given in Figures 2.92a and b.

2.10.11   Formulation of Stiffness Matrix

It is possible to obtain all the strains and stresses within the element and to formulate the stiffness matrix and a consistent load matrix once the displacement function has been determined. This consistent load matrix represents the equivalent nodal forces which replace the action of external distributed loads.

As an example, let us consider a linearly elastic element of any of the types shown in Figure 2.93. The displacement function may be written in the form

$$\{f\} = [P]\{A\} \tag{2.172}$$

in which $\{f\}$ may have two components $\{u, v\}$ or simply be equal to w, $[P]$ is a function of x and y only, and $\{A\}$ is the vector of undetermined constants. If Equation 2.172 is applied repeatedly to the nodes of the element one after the other, we obtain a set of equations of the form

$$\{D^*\} = [C]\{A\} \tag{2.173}$$

in which $\{D^*\}$ is the nodal parameters and $[C]$ is the relevant nodal coordinates. The undetermined constants $\{A\}$ can be expressed in terms of the nodal parameters $\{D^*\}$ as

$$\{A\} = [C]^{-1}\{D^*\} \tag{2.174}$$

Element name	Configuration	DOF	Shape functions
Two-node linear element	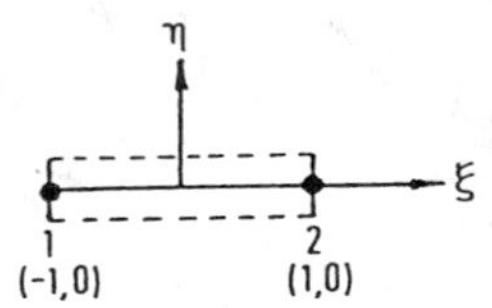	+	$N_i = \dfrac{1}{2}(1 + \xi_0);$ $i = 1, 2$
Three-node parabolic element	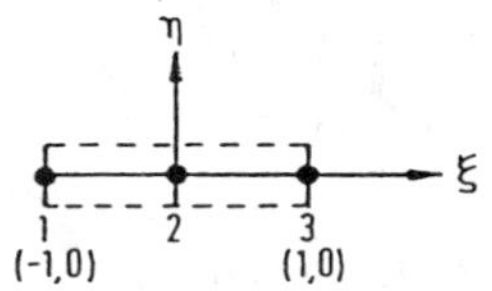	+	$N_i = \dfrac{1}{2}\xi_0(1 + \xi_0);\ i = 1, 3$ $N_i = (1 - \xi^2);\ i = 2$
Four-node cubic element	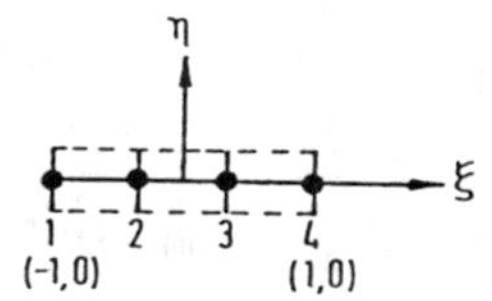	+	$N_i = \dfrac{1}{16}(1 + \xi_0)(9\xi^2 - 1)$ $i = 1, 4$ $N_i = \dfrac{9}{16}(1 + 9\xi_0)(1 - \xi^2)$ $i = 2, 3$
Five-node quartic element	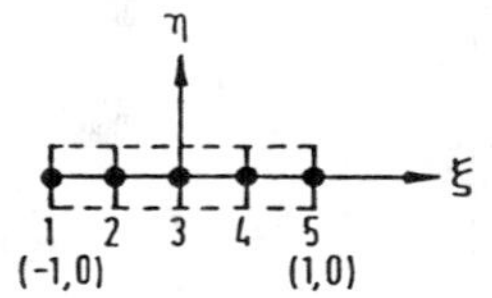	+	$N_i = \dfrac{1}{6}(1 + \xi_0)\ \{4\xi_0(1 - \xi^2)$ $\qquad + 3\xi_0\}$ $i = 1, 5$ $N_i = 4\xi_0(1 - \xi^2)(1 + 4\xi_0)$ $i = 2, 4$ $N_3 = (1 - 4\xi^2)(1 - \xi^2)$

Figure 2.92a   Shape functions for typical elements.

Element name	Configuration	DOF	Shape functions
Four-node plane quadrilateral	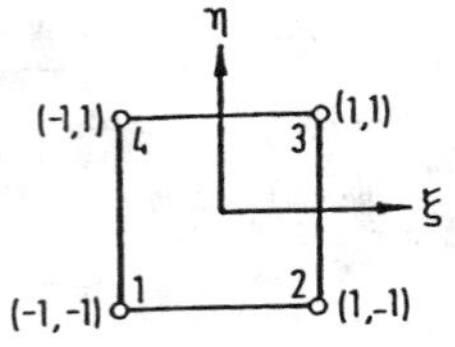	u, v	$N_i = \frac{1}{4}(1+\xi_0)(1+\eta_0)$; $i = 1, 2, 3, 4$
Eight-node plane quadrilateral	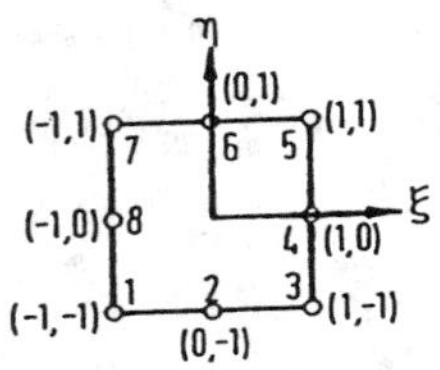	u, v	$N_i = \frac{1}{4}(1+\xi_0)(1+\eta_0)$ $(\xi_0 + \eta_0 - 1)$; $i = 1, 3, 5, 7$ $N_i = \frac{1}{2}(1-\xi^2)(1+\eta_0)$ $i = 2, 6$ $N_i = \frac{1}{2}(1-\eta^2)(1+\xi_0)$ $i = 4, 8$
Twelve-node plane quadrilateral	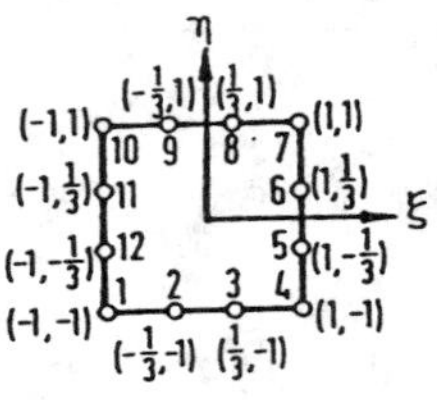	u, v	$N_i = \frac{1}{32}(1+\xi_0)(1+\eta_0)$ $(-10+9(\xi^2+\eta^2))$ $i = 1, 4, 7, 10$ $N_i = \frac{9}{32}(1+\xi_0)(1+\eta^2)$ $(1+9\eta_0)$ $i = 5, 6, 11, 12$ $N_i = \frac{9}{32}(1+\eta_0)(1-\xi^2)$ $(1+9\xi_0)$ $i = 2, 3, 8, 9$

Figure 2.92b   Shape functions for typical elements.

Serial no.	Element name	Configuration	DOF	Shape functions
4	Six-node linear Quadrilateral	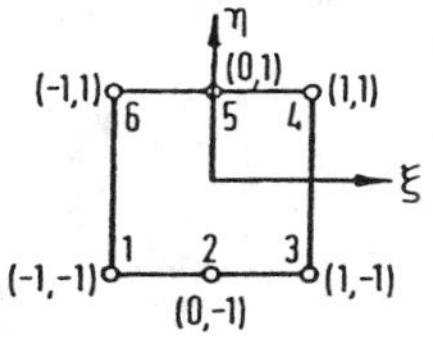 η axis with nodes: 6 at (-1,1), 5 at (0,1), 4 at (1,1); node labels 1 at (-1,-1), 2 at (0,-1), 3 at (1,-1); ξ axis	u, v	$N_i = \dfrac{\xi_0}{4}(1+\xi_0)(1+\eta_0)$ $i = 1, 3, 4, 6$ $N_i = \dfrac{1}{2}(1-\xi^2)(1+\eta_0)$ $i = 2, 5$
5	Eight-node plane quadrilateral	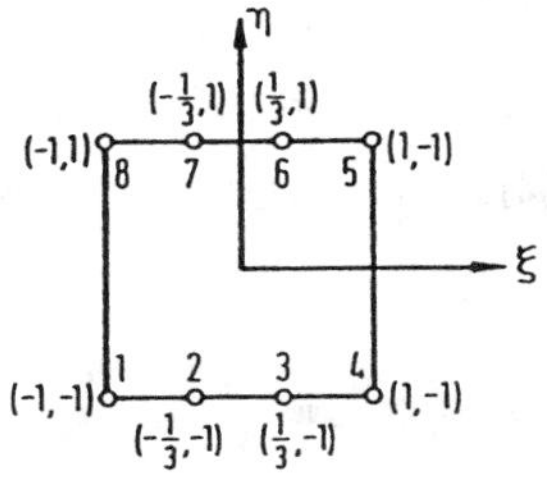 η axis; nodes 8,7,6,5 at top with $(-\tfrac{1}{3},1)$ $(\tfrac{1}{3},1)$, $(-1,1)$, $(1,-1)$; nodes 1,2,3,4 at bottom with $(-1,-1)$, $(-\tfrac{1}{3},-1)$, $(\tfrac{1}{3},-1)$, $(1,-1)$; ξ axis	u, v	$N_i = \dfrac{1}{32}(1+\xi_0)(-1+9\xi^2)$ $(1+\eta_0)$ $i = 1, 4, 5, 8$ $N_i = \dfrac{9}{32}(1-\xi^2)(1+9\xi_0)$ $(1+\eta_0)$ $i = 2, 3, 6, 7$
6	Seven-node plane quadrilateral	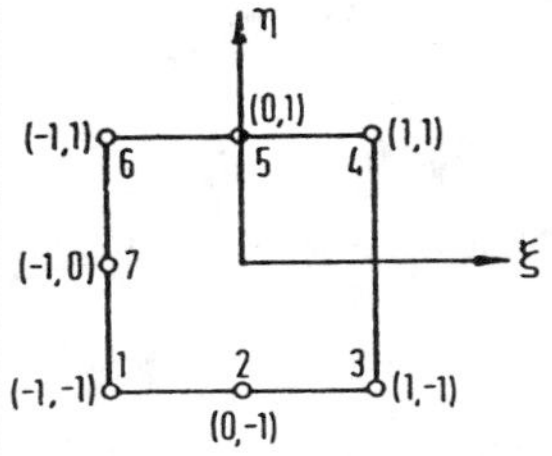 η axis; nodes 6 at (-1,1), 5 at (0,1), 4 at (1,1); node 7 at (-1,0); nodes 1 at (-1,-1), 2 at (0,-1), 3 at (1,-1); ξ axis	u, v	$N_1 = \dfrac{1}{4}(1-\xi)(1-\eta)$ $(1+\xi+\eta)$ $N_2 = \dfrac{1}{2}(1-\eta)(1-\xi^2)$ $N_3 = \dfrac{\xi}{4}(1+\xi)(1-\eta)$ $N_4 = \dfrac{\xi}{4}(1+\xi)(1+\eta)$ $N_5 = \dfrac{1}{2}(1+\eta)(1-\xi^2)$ $N_6 = -\dfrac{1}{4}(1-\xi)(1+\eta)$ $(1+\xi-\eta)$ $N_7 = \dfrac{1}{2}(1-\xi)(1-\eta^2)$

Figure 2.92b *(Continued)* Shape functions for typical elements.

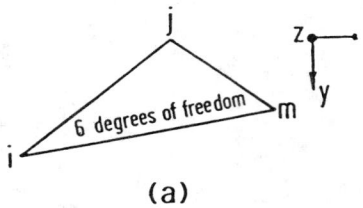

(a)

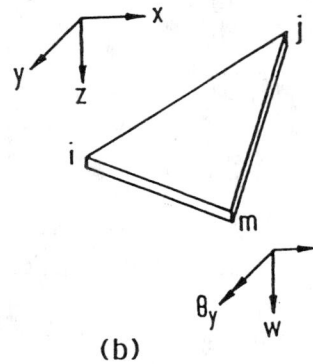

(b)

Figure 2.93   Degrees of freedom. (a) Triangular plane-stress element; (b) triangular bending element.

Substituting Equation 2.174 into Equation 2.172

$$\{f\} = [P][C]^{-1}\{D^*\} \tag{2.175}$$

Constructing the displacement function directly in terms of the nodal parameters one obtains

$$\{f\} = [L]\{D^*\} \tag{2.176}$$

where $[L]$ is a function of both (x, y) and $(x, y)_{i,j,m}$ given by

$$[L] = [P][C]^{-1} \tag{2.177}$$

The various components of strain can be obtained by appropriate differentiation of the displacement function. Thus,

$$\{\varepsilon\} = [B]\{D^*\} \tag{2.178}$$

$[B]$ is derived by differentiating appropriately the elements of $[L]$ with respect to x and y. The stresses $\{\sigma\}$ in a linearly elastic element are given by the product of the strain and a symmetrical elasticity matrix $[E]$. Thus,

$$\{\sigma\} \;=\; [E]\{\varepsilon\}$$
$$\text{or} \quad \{\sigma\} \;=\; [E][B]\{D^*\} \tag{2.179}$$

The stiffness and the consistent load matrices of an element can be obtained using the principle of minimum total potential energy. The potential energy of the external load in the deformed configuration of the element is written as

$$W = -\{D^*\}^T\{Q^*\} - \int_a \{f\}^T\{q\}da \tag{2.180}$$

In Equation 2.180 $\{Q^*\}$ represents concentrated loads at nodes and $\{q\}$ represents the distributed loads per unit area. Substituting for $\{f\}^T$ from Equation 2.176 one obtains

$$W = -\{D^*\}^T\{Q^*\} - \{D^*\}^T \int_a [L]^T\{q\}da \qquad (2.181)$$

Note that the integral is taken over the area a of the element. The strain energy of the element integrated over the entire volume v is given as

$$U = \frac{1}{2}\int_v \{\varepsilon\}^T\{\sigma\}dv$$

Substituting for $\{\varepsilon\}$ and $\{\sigma\}$ from Equations 2.178 and 2.179, respectively,

$$U = \frac{1}{2}\{D^*\}^T\left(\int_v [B]^T[E][B]dv\right)\{D^*\} \qquad (2.182)$$

The total potential energy of the element is

$$V = U + W$$

or

$$V = \frac{1}{2}\{D^*\}^T\left(\int_v [B]^T[E][B]dv\right)\{D^*\} - \{D^*\}^T\{Q^*\}$$
$$- \{D^*\}^T\int_a [L]^T\{q\}da \qquad (2.183)$$

Using the principle of minimum total potential energy, we obtain

$$\left(\int_v [B]^T[E][B]dv\right)\{D^*\} = \{Q^*\} + \int_a [L]^T\{q\}da$$

or

$$[K]\{D^*\} = \{F^*\} \qquad (2.184)$$

where

$$K] = \int_v [B]^T[E][B]dv \qquad (2.185a)$$

and

$$\{F^*\} = \{Q^*\} + \int_a [L]^T\{q\}da \qquad (2.185b)$$

2.10.12   Plates Subjected to In-Plane Forces

The simplest element available in two-dimensional stress analysis is the triangular element. The stiffness and consistent load matrices of such an element will now be obtained by applying the equation derived in the previous section.

Consider the triangular element shown in Figure 2.93a. There are two degrees of freedom per node and a total of six degrees of freedom for the entire element. We can write

$$u = A_1 + A_2x + A_3y$$

and

$$v = A_4 + A_5 x + A_6 y$$

expressed as

$$\{f\} = \left\{ \begin{array}{c} u \\ v \end{array} \right\} = \left[\begin{array}{cccccc} 1 & x & y & 0 & 0 & 0 \\ 0 & 0 & 0 & 1 & x & y \end{array} \right] \left\{ \begin{array}{c} A_1 \\ A_2 \\ A_3 \\ A_4 \\ A_5 \\ A_6 \end{array} \right\} \tag{2.186}$$

or

$$\{f\} = [P]\{A\} \tag{2.187}$$

Once the displacement function is available, the strains for a plane problem are obtained from

$$\varepsilon_x = \frac{\partial u}{\partial x} \qquad \varepsilon_y = \frac{\partial v}{\partial y}$$

and

$$\gamma_{xy} = \frac{\partial u}{\partial y} \frac{\partial v}{\partial x}$$

The matrix $[B]$ relating the strains to the nodal displacement $\{D^*\}$ is thus given as

$$[B] = \frac{1}{2\Delta} \left[\begin{array}{cccccc} b_i & 0 & b_j & 0 & b_m & 0 \\ 0 & c_i & 0 & c_j & 0 & c_m \\ c_i & b_j & c_j & b_j & c_m & b_m \end{array} \right] \tag{2.188}$$

b_i, c_i, etc. are constants related to the nodal coordinates only. The strains inside the element must all be constant and hence the name of the element.

For derivation of strain matrix, only isotropic material is considered. The plane stress and plane strain cases can be combined to give the following elasticity matrix which relates the stresses to the strains

$$[E] = \left[\begin{array}{ccc} C_1 & C_1 C_2 & 0 \\ C_1 C_2 & C_1 & 0 \\ 0 & 0 & C_{12} \end{array} \right] \tag{2.189}$$

where

$$C_1 = \bar{E}/(1 - v^2) \quad \text{and} \quad C_2 = v \quad \text{for plane stress}$$

and

$$C_1 = \frac{\bar{E}(1 - v)}{(1 + v)(1 - 2v)} \quad \text{and} \quad C_2 = \frac{v}{(1 - v)} \quad \text{for plane strain}$$

and for both cases,

$$C_{12} = C_1(1 - C_2)/2$$

and $\bar{E} = $ Modulus of elasticity.

The stiffness matrix can now be formulated according to Equation 2.185a

$$[E][B] = \frac{1}{2\Delta} \left[\begin{array}{ccc} C_1 & C_1 C_2 & 0 \\ C_1 C_2 & C_1 & 0 \\ 0 & 0 & C_{12} \end{array} \right] \left[\begin{array}{cccccc} b_i & 0 & b_j & 0 & b_m & 0 \\ 0 & c_i & 0 & c_j & 0 & c_m \\ c_i & b_j & c_j & b_j & c_m & b_m \end{array} \right]$$

where Δ is the area of the element.

The stiffness matrix is given by Equation 2.185a as

$$[K] = \int_v [B]^T [E][B] dv$$

The stiffness matrix has been worked out algebraically to be

$$[K] = \frac{h}{4\Delta} \begin{bmatrix}
\begin{matrix} C_1 b_i^2 \\ +C_{12}c_i^2 \end{matrix} & & & & & \\
\begin{matrix} C_1 C_2 b_i c_i \\ +C_{12} b_i c_i \end{matrix} & \begin{matrix} C_1 c_i^2 \\ +C_{12} b_i^2 \end{matrix} & & \text{Symmetrical} & & \\
\begin{matrix} C_1 b_i b_j \\ +C_{12} c_i c_j \end{matrix} & \begin{matrix} C_1 C_2 b_j c_i \\ +C_{12} b_i c_j \end{matrix} & \begin{matrix} C_1 b_j^2 \\ +C_{12} c_j^2 \end{matrix} & & & \\
\begin{matrix} C_1 C_2 b_i c_j \\ +C_{12} b_j c_i \end{matrix} & \begin{matrix} C_1 c_i c_j \\ +C_1 b_i b_j \end{matrix} & \begin{matrix} C_1 C_2 b_j c_j \\ +C_{12} b_j c_j \end{matrix} & \begin{matrix} C_1 c_j^2 \\ +C_{12} b_j^2 \end{matrix} & & \\
\begin{matrix} C_1 b_i b_m \\ +C_{12} c_i c_m \end{matrix} & \begin{matrix} C_1 C_2 b_m c_i \\ +C_{12} b_i c_m \end{matrix} & \begin{matrix} C_1 b_j b_m \\ +C_{12} c_j c_m \end{matrix} & \begin{matrix} C_1 C_2 b_m c_j \\ +C_{12} b_j c_m \end{matrix} & \begin{matrix} C_1 b_m^2 \\ +C_{12} c_m^2 \end{matrix} & \\
\begin{matrix} C_1 C_2 b_i c_m \\ +C_{12} b_m c_i \end{matrix} & \begin{matrix} C_1 c_i c_m \\ +C_{12} b_i b_m \end{matrix} & \begin{matrix} C_1 C_2 b_j c_m \\ +C_{12} b_m c_j \end{matrix} & \begin{matrix} C_1 c_j c_m \\ +C_{12} b_j b_m \end{matrix} & \begin{matrix} C_1 C_2 b_m c_m \\ +C_{12} b_m c_m \end{matrix} & \begin{matrix} C_1 c_m^2 \\ +C_{12} b_m^2 \end{matrix}
\end{bmatrix}$$

2.10.13 Beam Element

The stiffness matrix for a beam element with two degrees of freedom (one deflection and one rotation) can be derived in the same manner as for other finite elements using Equation 2.185a.

The beam element has two nodes, one at each end, and two degrees of freedom at each node, giving it a total of four degrees of freedom. The displacement function can be assumed as

$$f = w = A_1 + A_2 x + A_3 x^2 + A_4 x^3$$

i.e.,

$$f = \begin{bmatrix} 1 & x & x^2 & x^3 \end{bmatrix} \begin{Bmatrix} A_1 \\ A_2 \\ A_3 \\ A_4 \end{Bmatrix}$$

or

$$f = [P]\{A\}$$

With the origin of the x and y axis at the left-hand end of the beam, we can express the nodal-displacement parameters as

$$\begin{aligned}
D_1^* &= (w)_{x=0} = A_1 + A_2(0) + A_3(0)^2 + A_4(0)^3 \\
D_2^* &= \left(\frac{dw}{dx}\right)_{x=0} = A_2 + 2A_3(0) + 3A_4(0)^2 \\
D_3^* &= (w)_{x=l} = A_1 + A_2(l) + A_3(l)^2 + A_4(l)^3 \\
D_4^* &= \left(\frac{dw}{dx}\right)_{x=l} = A_2 + 2A_3(l) + 3A_4(l)^2
\end{aligned}$$

or

$$\{D^*\} = [C]\{A\}$$

where

$$\{A\} = [C]^{-1}\{D^*\}$$

and

$$[C]^{-1} = \begin{bmatrix} 1 & 0 & 0 & 0 \\ 0 & 1 & 0 & 0 \\ \frac{-3}{l^2} & \frac{-2}{l} & \frac{3}{l^2} & \frac{-1}{l} \\ \frac{2}{l^3} & \frac{1}{l^2} & \frac{-2}{l^3} & \frac{1}{l^2} \end{bmatrix} \tag{2.190}$$

Using Equation 2.190, we obtain

$$[L] = [P][C]^{-1}$$

or

$$[C]^{-1} = \left[\left(1 - \frac{3x^2}{l^2} + \frac{2x^3}{l^3} \right) \left(x - \frac{2x^2}{l} + \frac{x^3}{l^2} \right) \left(\frac{3x^2}{l^2} - \frac{2x^3}{l^3} \right) \left(-\frac{x^2}{l} + \frac{x^3}{l^2} \right) \right] \tag{2.191}$$

Neglecting shear deformation

$$\{\varepsilon\} = -\frac{d^2 y}{dx^2}$$

Substituting Equation 2.178 into Equation 2.191 and the result into Equation 2.179

$$\{\varepsilon\} = \left[\left| \frac{6}{l^2} - \frac{12x}{l^3} \right| \frac{4}{l} - \frac{6x}{l^2} \right| -\frac{6}{l^2} + \frac{12x}{l^3} \left| \frac{2}{l} - \frac{6x}{l^2} \right] \{D^*\}$$

or

$$\{\varepsilon\} = [B]\{D^*\}$$

Moment-curvature relationship is given by

$$M = \bar{E} I \left(-\frac{d^2 y}{dx^2} \right)$$

where $\bar{E}$ = modulus of elasticity.

We know that $\{\sigma\} = [E]\{\varepsilon\}$, so we have for the beam element

$$[E] = \bar{E} I$$

The stiffness matrix can now be obtained from Equation 2.185a written in the form

$$[K] = \int_0^l [B]^T [E][B]dx$$

with the integration over the length of the beam. Substituting for $[B]$ and $[E]$, we obtain

$$[K] = \bar{E} I \int_0^l$$

$$\begin{bmatrix} \frac{36}{\ell^4} - \frac{144x}{\ell^5} + \frac{144x^2}{\ell^6} & & \text{symmetrical} & \\ \frac{24}{\ell^3} - \frac{84x}{\ell^4} + \frac{72x^2}{\ell^5} & \frac{16}{\ell^2} - \frac{48x}{\ell^3} + \frac{36x^2}{\ell^4} & & \\ \frac{-36}{\ell^4} + \frac{144x}{\ell^5} - \frac{144x^2}{\ell^6} & \frac{-24}{\ell^3} + \frac{84x}{\ell^4} - \frac{72x^2}{\ell^5} & \frac{36}{\ell^4} - \frac{144x}{\ell^5} + \frac{144x^2}{\ell^6} & \\ \frac{12}{\ell^3} - \frac{60x}{\ell^4} + \frac{72x^2}{\ell^5} & \frac{8}{\ell^2} - \frac{36x}{\ell^3} + \frac{36x^2}{\ell^4} & \frac{-12}{\ell^3} + \frac{60x}{\ell^4} - \frac{72x^2}{\ell^5} & \frac{4}{\ell^2} - \frac{24x}{\ell^3} + \frac{36x^2}{\ell^4} \end{bmatrix} dx$$

$$\tag{2.192}$$

or

$$[K] = \bar{E} I \begin{bmatrix} \frac{12}{\ell^3} & & \text{symmetrical} & \\ \frac{6}{\ell^2} & \frac{4}{\ell} & & \\ \frac{-12}{\ell^3} & \frac{-6}{\ell^2} & \frac{12}{\ell^3} & \\ \frac{6}{\ell^2} & \frac{2}{\ell} & \frac{-6}{\ell^2} & \frac{4}{\ell} \end{bmatrix} \tag{2.193}$$

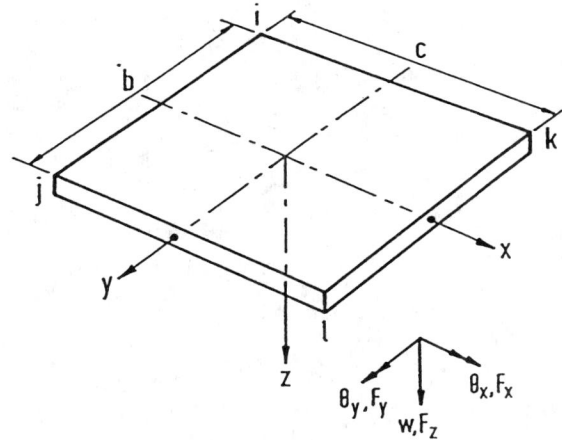

Figure 2.94 Rectangular bending element.

2.10.14 Plates in Bendings—Rectangular Element

For the rectangular bending element shown in Figure 2.94 with three degrees of freedom (one deflection and two rotations) at each node, the displacement function can be chosen as a polynomial with 12 undetermined constants as

$$
\begin{aligned}
\{f\} = w \;=\;& A_1 + A_2 x + A_3 y + A_4 x^2 + A_5 xy + A_6 y^2 + A_7 x^3 \\
&+ A_8 x^2 y + A_9 xy^2 + A_{10} y^3 + A_{11} x^3 y + A_{12} xy^3
\end{aligned}
\tag{2.194}
$$

or

$$
\{f\} = \{P\}\{A\}
$$

The displacement parameter vector is defined as

$$
\{D^*\} = \left\{ w_i, \theta_{xi}, \theta_{yi} \,\middle|\, w_j, \theta_{xj}, \theta_{yj} \,\middle|\, w_k, \theta_{xk}, \theta_{yk} \,\middle|\, w_\ell, \theta_{x\ell}, \theta_{y\ell} \right\}
$$

where

$$
\theta_x = \frac{\partial w}{\partial y} \quad \text{and} \quad \theta_y = -\frac{\partial w}{\partial x}
$$

As in the case of beam it is possible to derive from Equation 2.194 a system of 12 equations relating $\{D^*\}$ to constants $\{A\}$. The last equation

$$
w = \big[[L]_i | [L]_j | [L]_k | [L]_l \big] \{D^*\}
\tag{2.195}
$$

The curvatures of the plate element at any point (x, y) are given by

$$
\{\varepsilon\} = \left\{
\begin{array}{c}
\dfrac{-\partial^2 w}{\partial x^2} \\[4pt]
\dfrac{-\partial^2 w}{\partial y^2} \\[4pt]
\dfrac{2\partial^2 w}{\partial x \partial y}
\end{array}
\right\}
$$

By differentiating Equation 2.195, we obtain

$$
\{\varepsilon\} = \big[[B]_i | [B]_j | [B]_k | [B]_l \big] \{D^*\}
\tag{2.196}
$$

or

$$
\{\varepsilon\} = \sum_{r=i,j,k,l} [B]_r \{D^*\}_r
\tag{2.197}
$$

where

$$[B]_r = \begin{bmatrix} -\frac{\partial^2}{\partial x^2}[L]_r \\ \hline -\partial\frac{2}{\partial y^2}[L]_r \\ \hline 2\frac{\partial^2}{\partial x \partial y}[L]_r \end{bmatrix}$$

(2.198)

and

$$\{D^*\}_r = \{w_r, \theta_{xr}, \theta_{yr}\}$$

(2.199)

For an isotropic slab, the moment-curvature relationship is given by

$$\{\sigma\} = \{M_x M_y M_{xy}\}$$

(2.200)

$$[E] = N \begin{bmatrix} 1 & \nu & 0 \\ \nu & 1 & 0 \\ 0 & 0 & \frac{1-\nu}{2} \end{bmatrix}$$

(2.201)

and

$$N = \frac{\bar{E}h^3}{12(1-\nu^2)}$$

(2.202)

For orthotropic slabs with the principal directions of orthotropy coinciding with the x and y axes, no additional difficulty is experienced. In this case we have

$$[E] = \begin{bmatrix} D_x & D_1 & 0 \\ D_1 & D_y & 0 \\ 0 & 0 & D_{xy} \end{bmatrix}$$

(2.203)

where D_x, D_1, D_y, and D_{xy} are the orthotropic constants used by Timoshenko and Woinowsky-Krieger [56],

$$\left.\begin{array}{l} D_x = \frac{E_x h^3}{12(1-\nu_x \nu_y)} \\ D_y = \frac{E_y h^3}{12(1-\nu_x \nu_y)} \\ D_1 = \frac{\nu_x E_y h^3}{12(1-\nu_x \nu_y)} = \frac{\nu_y E_x h^3}{12(1-\nu_x \nu_y)} \\ D_{xy} = \frac{G h^3}{12} \end{array}\right\}$$

(2.204)

where E_x, E_y, ν_x, ν_y, and G are the orthotropic material constants, and h is the plate thickness.

Unlike the strain matrix for the plane stress triangle (see Equation 2.188), the stress and strain in the present element vary with x and y. In general we calculate the stresses (moments) at the four corners. These can be expressed in terms of the nodal displacements by Equation 2.179 which, for an isotropic element, takes the form

$$
\begin{Bmatrix}
\{\sigma\}_i \\ \hline
\{\sigma\}_j \\ \hline
\{\sigma\}_k \\ \hline
\{\sigma\}_r
\end{Bmatrix}
=
\frac{N}{cb}
\begin{bmatrix}
6p^{-1}+6vp & 4vc & -4b & -6vp & 2vc & -4b & -6p^{-1} & 0 & 0 & 0 & 0 & 0\\
6p+6vp^{-1} & 4c & -4vb & -6p & 2c & -4vb & -6vp^{-1} & 0 & 0 & 0 & 0 & 0\\
-(1-v) & -(1-v)b & (1-v)c & (1-v) & 0 & (1-v) & (1-v) & 0 & -(1-v)b & 0 & 0 & 0\\
-6vp & -2vc & 0 & 6p^{-1}+6vp & -4vc & 0 & 0 & 4vc & -2b & -6p^{-1} & 0 & 0\\
-6p & -2c & 0 & 6p+6vp^{-1} & -4c & 0 & 0 & 4c & -2vb & -6vp^{-1} & 0 & 0\\
-(1-v) & 0 & (1-v) & (1-v) & -(1-v)b & 0 & 0 & (1-v)b & 0 & (1-v) & -(1-v)c & -(1-v)\\
-6p^{-1} & 0 & 2b & 0 & 0 & 2b & 6p^{-1}+6vp & -6vp & 0 & 0 & 2vc & -4b\\
-6vp^{-1} & 0 & 2b & 0 & 0 & 2vb & 6p+6vp^{-1} & -6p & 0 & 0 & 2c & -4vb\\
-(1-v) & -(1-v)b & 0 & (1-v) & 0 & 0 & (1-v) & -(1-v) & -(1-v)c & (1-v)b & 0 & -(1-v)c\\
0 & 0 & 0 & -6p^{-1} & 0 & -2b & 0 & 0 & -4vc & 6p^{-1}+6vp & -4vc & 0\\
0 & 0 & 0 & -6vp^{-1} & 0 & -2vb & 0 & 0 & -4c & 6p+6vp^{-1} & -4c & 0\\
-(1-v) & 0 & 0 & (1-v) & -(1-v)b & 0 & (1-v) & -(1-v) & (1-v)b & -(1-v) & (1-v)b & -(1-v)c
\end{bmatrix}
\begin{Bmatrix}
\{D^*\}_i \\ \hline
\{D^*\}_j \\ \hline
\{D^*\}_k \\ \hline
\{D^*\}_r
\end{Bmatrix}
\tag{2.205}
$$

where $p = c/b$.

The stiffness matrix corresponding to the 12 nodal coordinates can be calculated by

$$[K] = \int_{-b/2}^{b/2} \int_{-c/2}^{c/2} [B]^T [E][B]dxdy \tag{2.206}$$

For an isotropic element, this gives

$$[K^*] = \frac{N}{15cb}[T][\bar{k}][T] \tag{2.207}$$

where

$$[T] = \begin{bmatrix} [T_s] & \text{Submatrices not} \\ & [T_s] & \text{shown are} \\ & & [T_s] & \text{zero} \\ & & & [T_s] \end{bmatrix} \tag{2.208}$$

$$[T_s] = \begin{bmatrix} 1 & 0 & 0 \\ 0 & b & 0 \\ 0 & 0 & c \end{bmatrix} \tag{2.209}$$

and $[\bar{K}]$ is given by the matrix shown in Equation 2.210.

$$
[\bar{K}] =
\begin{bmatrix}
60p^{-2}+60p^2-12\nu+42 & & & & & & & & & & & \\[2pt]
30p^2+12\nu+3 & 20p^2-4\nu+4 & & & & & & & & & & \\[2pt]
-(30p^{-2}+12\nu+3) & -15\nu & 20p^{-2}-4\nu+4 & & & & & & & & & \\[2pt]
+30p^{-2}-60p^2+12\nu-42 & -30p^2+3\nu-3 & -15p^{-2}+12\nu+3 & 60p^{-2}+60p^2-12\nu+42 & & & & & & & & \\[2pt]
30p^2-3\nu+3 & 10p^2+4\nu-4 & 0 & -(30p^2+12\nu+3) & 20p^2-4\nu+4 & & & & & & & \\[2pt]
-15p^{-2}+12\nu+3 & 0 & 10p^{-2}+4\nu-4 & -(30p^{-2}+12\nu+3) & 15\nu & 20p^{-2}-4\nu+4 & & & & & & \\[2pt]
-60p^{-2}+30p^2+12\nu-42 & 15p^{-2}-12\nu-3 & 15p^{-2}-3\nu+3 & -60p^{-2}+30p^2+12\nu-42 & 30p^2-3\nu+3 & -15p^{-2}+3\nu+3 & 60p^{-2}+60p^2-12\nu+42 & & & & & \\[2pt]
15p^{-2}-12\nu-3 & 10p^2+4\nu-4 & 0 & -30p^2+3\nu+3 & 10p^2+4\nu-4 & 0 & 30p^2+12\nu+3 & 20p^2-4\nu+4 & & & & \\[2pt]
-30p^{-2}+3\nu-3 & 0 & 5p^{-2}-\nu+1 & 15p^{-2}-12\nu-3 & 0 & 10p^{-2}+4\nu-4 & -(30p^{-2}+12\nu+3) & -15\nu & 20p^{-2}-4\nu+4 & & & \\[2pt]
-30p^{-2}+30p^2-12\nu+42+3 & -15p^{-2}-3\nu+3 & -15p^{-2}-3\nu+3 & -30p^{-2}-60p^2+12\nu-42+3 & -15p^{-2}-3\nu+3 & 15p^{-2}-3\nu+3 & 30p^{-2}-60p^2+12\nu-42+3 & -30p^2+3\nu+3 & 15p^{-2}-3\nu+3 & 60p^{-2}+60p^2-12\nu+42 & & \\[2pt]
15p^{-2}-3\nu-3 & 5p^{-2}-\nu+1 & 0 & -15p^{-2}+12\nu+3 & 10p^2+4\nu-4 & 0 & 30p^2+3\nu+3 & 10p^2+4\nu-4 & 0 & -(30p^2+12\nu+3) & 20p^2-4\nu+4 & \\[2pt]
-15p^{-2}-3\nu-3 & 0 & 10p^{-2}+\nu-1 & -30p^{-2}+3\nu-3 & 0 & 5p^{-2}-\nu+1 & 15p^{-2}-12\nu-3 & 0 & 10p^{-2}+\nu-1 & -(30p^{-2}+12\nu+3) & -15\nu & 20p^{-2}-4\nu+4
\end{bmatrix}
$$

(symmetrical)

(2.210)

If the element is subjected to a uniform load in the z direction of intensity q, the consistent load vector becomes

$$\{Q_q^*\} = q \int_{-b/2}^{b/2} \int_{-c/2}^{c/2} [L]^T \, dx \, dy \tag{2.211}$$

where $\{Q_q^*\}$ are 12 forces corresponding to the nodal displacement parameters. Evaluating the integrals in this equation gives

$$\{Q_q^*\} = qcb \begin{Bmatrix} 1/4 \\ b/24 \\ -c/24 \\ --- \\ 1/4 \\ -b/24 \\ -c/24 \\ --- \\ 1/4 \\ b/24 \\ c/24 \\ --- \\ 1/4 \\ -b/24 \\ c/24 \end{Bmatrix} \tag{2.212}$$

More details on the Finite Element Method can be found in [23] and [27].

2.11 Inelastic Analysis

2.11.1 An Overall View

Inelastic analysis can be generalized into two main approaches. The first approach is known as *plastic hinge analysis*. The analysis assumes that structural elements remain elastic except at critical regions where zero-length **plastic hinges** are allowed to form. The second approach is known as *plastic-zone analysis*. The analysis follows explicitly the gradual spread of yielding throughout the volume of the structure. Material yielding in the member is modeled by discretization of members into several beam-column elements, and subdivision of the cross-sections into many "fibers". The plastic-zone analysis can predict accurately the inelastic response of the structure. However, the plastic hinge analysis is considered to be more efficient than the plastic-zone analysis since it requires, in most cases, one beam-column element per member to capture the stability of column members subject to end loading.

If geometric nonlinear effect is not considered, the plastic hinge analysis predicts the maximum load of the structure corresponding to the formation of a plastic collapse mechanism [16]. First-order plastic hinge analysis is finding considerable application in continuous beams and low-rise building frames where members are loaded primarily in flexure. For tall building frames and for frames with slender columns subjected to sidesway, the interaction between structural inelasticity and instability may lead to collapse prior to the formation of a plastic mechanism [54]. If equilibrium equations are formulated based on deformed geometry of the structure, the analysis is termed *second order*. The need for a **second-order analysis** of steel frame is increasing in view of the American Specifications [3] which give explicit permission for the engineer to compute load effects from a direct second-order analysis.

This section presents the virtual work principle to explain the fundamental theorems of plastic hinge analysis. Simple and approximate techniques of practical **plastic analysis** methods are then introduced. The concept of hinge-by-hinge analysis is presented. The more advanced topics such as second-order elastic plastic hinge, refined plastic hinge analysis, and plastic zone analysis are covered in Section 2.12.

2.11.2 Ductility

Plastic analysis is strictly applicable for materials that can undergo large deformation without fracture. Steel is one such material with an idealized stress-strain curve as shown in Figure 2.95. When

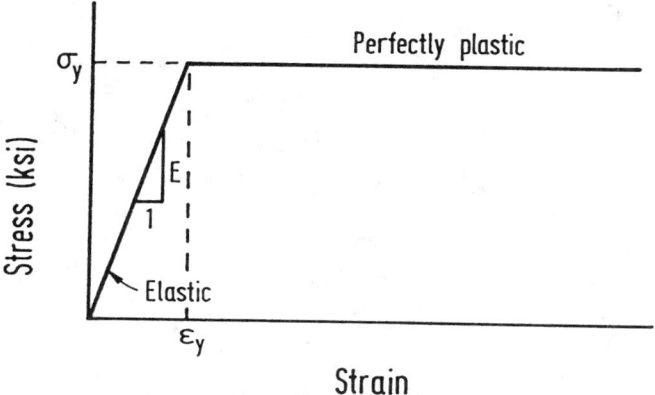

Figure 2.95 Idealized stress-strain curve.

steel is subjected to tensile force, it will elongate elastically until the yield stress is reached. This is followed by an increase in strain without much increase in stress. Fracture will occur at very large deformation. This material idealization is generally known as *elastic-perfectly plastic* behavior. For a compact section, the attainment of initial yielding does not result in failure of a section. The compact section will have reserved plastic strength that depends on the shape of the cross-section. The capability of the material to deform under constant load without decrease in strength is the *ductility* characteristic of the material.

2.11.3 Redistribution of Forces

The benefit of using a ductile material can be demonstrated from an example of a three-bar system shown in Figure 2.96. From the equilibrium condition of the system,

$$2T_1 + T_2 = P \qquad (2.213)$$

Assuming elastic stress-strain law, the displacement and force relationship of the bars may be written as:

$$\delta = \frac{T_1 L_1}{AE} = \frac{T_2 L_2}{AE} \qquad (2.214)$$

Since $L_2 = L_1/2 = L/2$, Equation 2.214 can be written as

$$T_1 = \frac{T_2}{2} \qquad (2.215)$$

where T_1 and T_2 are the tensile forces in the rods, L_1 and L_2 are length of the rods, A is a cross-section area, and $E =$ elastic modulus. Solving Equations 2.214 and 2.215 for T_2:

$$T_2 = \frac{P}{2} \qquad (2.216)$$

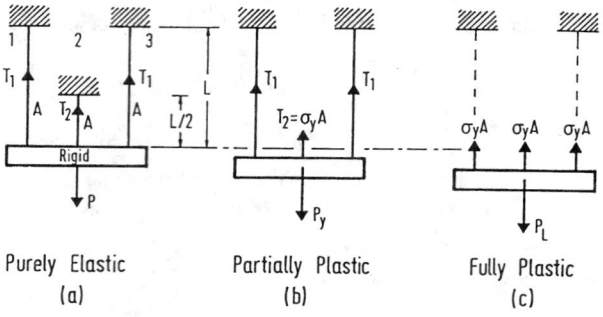

Purely Elastic (a)   Partially Plastic (b)   Fully Plastic (c)

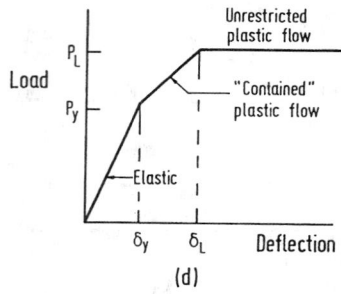

Figure 2.96   Force redistribution in a three-bar system: (a) elastic, (b) partially yielded, (c) fully plastic, (d) load-deflection curve.

The load at which the structure reaches the first yield (in Figure 2.96b) is determined by letting $T_2 = \sigma_y A$. From Equation 2.216:

$$P_y = 2T_2 = 2\sigma_y A \tag{2.217}$$

The corresponding displacement at first yield is

$$\delta_y = \varepsilon_y L = \frac{\sigma_y L}{2E} \tag{2.218}$$

After Bar 2 is yielded, the system continues to take additional load until all three bars reach their maximum strength of $\sigma_y A$, as shown in Figure 2.96c. The plastic limit load of the system is thus written as

$$P_L = 3\sigma_y A \tag{2.219}$$

The process of successive yielding of bars in this system is known as inelastic redistribution of forces. The displacement at the incipient of collapse is

$$\delta_L = \varepsilon_y L = \frac{\sigma_y L}{E} \tag{2.220}$$

Figure 2.96d shows the load-displacement behavior of the system when subjected to increasing force. As load increases, Bar 2 will reach its maximum strength first. As it yielded, the force in the member remains constant, and additional loads on the system are taken by the less critical bars. The system will eventually fail when all three bars are fully yielded. This is based on an assumption that material strain-hardening does not take place.

2.11.4   Plastic Hinge

A plastic hinge is said to form in a structural member when the cross-section is fully yielded. If material strain hardening is not considered in the analysis, a fully yielded cross-section can undergo indefinite rotation at a constant restraining **plastic moment** M_p.

Most of the plastic analyses assume that plastic hinges are concentrated at zero length plasticity. In reality, the yield zone is developed over a certain length, normally called the *plastic hinge length*, depending on the loading, boundary conditions, and geometry of the section. The hinge lengths of beams (ΔL) with different support and loading conditions are shown in Figures 2.97a, b, and c. Plastic hinges are developed first at the sections subjected to the greatest curvature. The possible

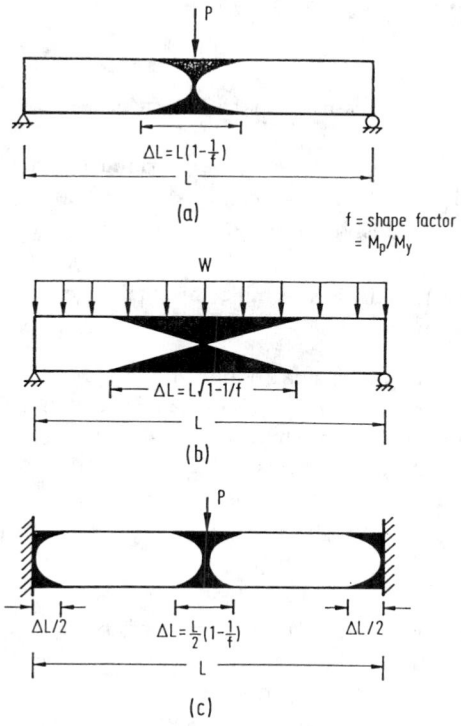

f = shape factor
= M_p/M_y

Figure 2.97   Hinge lengths of beams with different support and loading conditions.

locations for plastic hinges to develop are at the points of concentrated loads, at the intersections of members involving a change in geometry, and at the point of zero shear for member under uniform distributed load.

2.11.5   Plastic Moment

A knowledge of full plastic moment capacity of a section is important in plastic analysis. It forms the basis for limit load analysis of the system. Plastic moment is the moment resistance of a fully yielded cross-section. The cross-section must be fully compact in order to develop its plastic strength. The component plates of a section must not buckle prior to the attainment of full moment capacity.

The plastic moment capacity, M_p, of a cross-section depends on the material yield stress and the section geometry. The procedure for the calculation of M_p may be summarized in the following two steps:

1. The plastic neutral axis of a cross-section is located by considering equilibrium of forces normal to the cross-section. Figure 2.98a shows a cross-section of arbitrary shape subjected to increasing moment. The plastic neutral axis is determined by equating the force in compression (C) to that in tension (T). If the entire cross-section is made of the

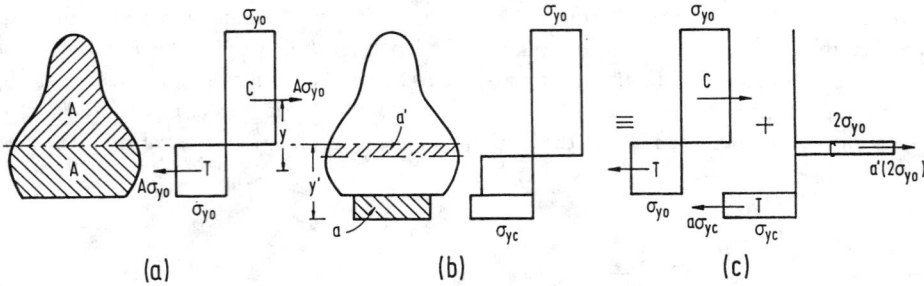

Figure 2.98   Cross-section of arbitrary shape subjected to bending.

same material, the plastic neutral axis can be determined by dividing the cross-sectional area into two equal parts. If the cross-section is made of more than one type of material, the plastic neutral axis must be determined by summing the normal force and letting the force equal zero.

2. The plastic moment capacity is determined by obtaining the moment generated by the tensile and compressive forces.

Consider an arbitrary section with area $2A$ and with one axis of symmetry of which the section is strengthened by a cover plate of area "a" as shown in Figure 2.98b. Further assume that the yield strength of the original section and the cover plate is σ_{yo} and σ_{yc}, respectively. At the full plastic state, the total axial force acting on the cover plate is $a\sigma_{yc}$. In order to maintain equilibrium of force in axial direction, the plastic neutral axis must shift down from its original position by a', i.e.,

$$a' = \frac{a\sigma_{yc}}{2\sigma_{yo}} \tag{2.221}$$

The resulting plastic capacity of the "built-up" section may be obtained by summing the full plastic moment of the original section and the moment contribution by the cover plate. The additional capacity is equal to the moment caused by the cover plate force $a\sigma_{yc}$ and a force due to the fictitious stress $2\sigma_{yo}$ acting on the area a' resulting from the shifting of plastic neutral axis from tension zone to compression zone as shown in Figure 2.98c.

Figure 2.99 shows the computation of plastic moment capacity of several shapes of cross-section. Based on the principle developed in this section, the plastic moment capacities of typical cross-sections may be generated. Additional information for sections subjected to combined bending, torsion, shear, and axial load can be found in Mrazik et al. [43].

2.11.6   Theory of Plastic Analysis

There are two main assumptions for first-order plastic analysis:

1. The structure is made of ductile material that can undergo large deformations beyond elastic limit without fracture or buckling.
2. The deflections of the structure under loading are small so that second-order effects can be ignored.

An "exact" plastic analysis solution must satisfy three basic conditions. They are *equilibrium*, *mechanism*, and *plastic moment* conditions. The plastic analysis disregards the continuity condition as required by the elastic analysis of indeterminate structures. The formation of plastic hinge in members leads to discontinuity of slope. If sufficient plastic hinges are formed to allow the structure to deform into a mechanism, it is called a *mechanism* condition. Since plastic analysis utilizes the limit of resistance of the member's plastic strength, the plastic moment condition is required to

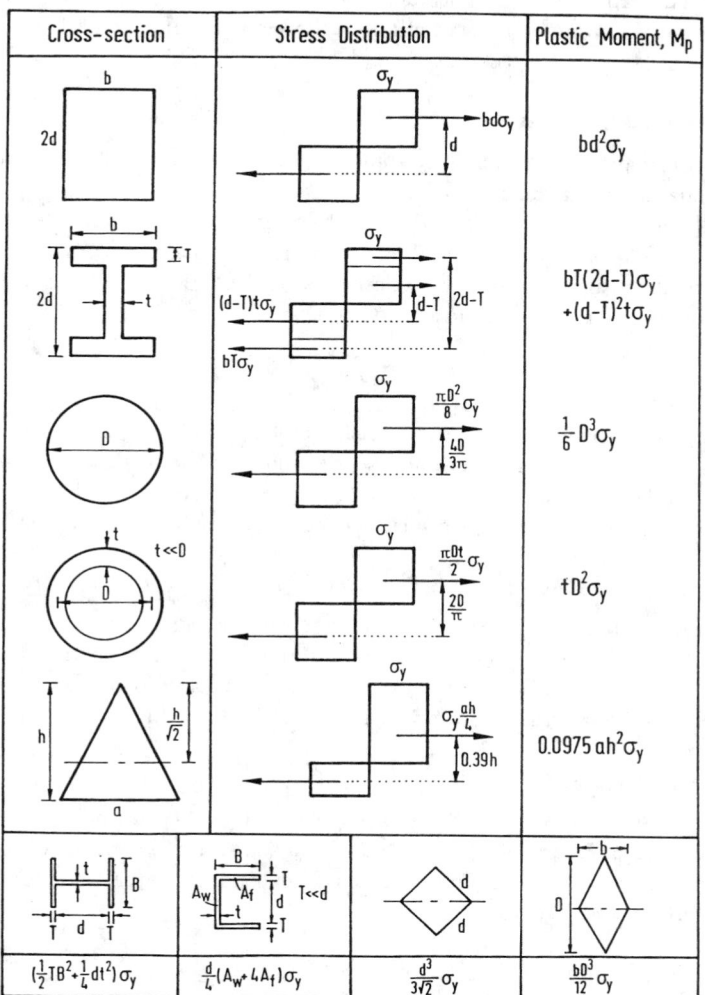

Figure 2.99 Plastic moment capacities of sections.

ensure that the resistance of the cross-sections is not violated anywhere in the structure. Lastly, the equilibrium condition, which is the same condition to be satisfied in elastic analysis, requires that the sum of all applied forces and reactions should be equal to zero, and all internal forces should be self-balanced.

When all the three conditions are satisfied, then the resulting plastic analysis for limiting load is the "correct" limit load. The collapse loads for simple structures such as beams and **portal frames** can be solved easily using a direct approach or through visualization of the formation of "correct" collapse mechanism. However, for more complex structures, the exact solution satisfying all three conditions may be difficult to predict. Thus, simple techniques using approximate methods of analysis are often used to assess these solutions. These techniques, called equilibrium and mechanism methods, will be discussed in the subsequent sections.

Principle of Virtual Work

The virtual work principle may be applied to relate a system of forces in equilibrium to a system of compatible displacements. For example, if a structure in equilibrium is given a set of small compatible displacement, then the work done by the external loads on these external displacements is equal to the work done by the internal forces on the internal deformation. In plastic analysis, internal deformations are assumed to be concentrated at plastic hinges. The virtual work equation for hinged structures can be written in an explicit form as

$$\sum P_i \delta_j = \sum M_i \theta_j \qquad (2.222)$$

where P_i is an external load and M_i is an internal moment at a hinge location. Both the P_i and M_i constitute an equilibrium set and they must be in equilibrium. δ_j are the displacements under the point loads P_i and in the direction of the loads. θ_j are the plastic hinge rotations under the moment M_i. Both δ_j and θ_j constitute a displacement set and they must be compatible with each other.

Lower Bound Theorem

For a given structure, if there exists any distribution of bending moments in the structure that satisfies both the equilibrium and plastic moment conditions, then the load factor, λ_L, computed from this moment diagram must be equal to or less than the collapse load factor, λ_c, of the structure. Lower bound theorem provides a safe estimate of the collapse limit load, i.e., $\lambda_L \leq \lambda_c$.

Upper Bound Theorem

For a given structure subjected to a set of applied loads, a load factor, λ_u, computed based on an assumed collapse mechanism must be greater than or equal to the true collapse load factor, λ_c. Upper bound theorem, which uses only the mechanism condition, over-estimates or is equal to the collapse limit load, i.e., $\lambda_u \geq \lambda_c$.

Uniqueness Theorem

A structure at collapse has to satisfy three conditions. First, a sufficient number of plastic hinges must be formed to turn the structure, or part of it, into a mechanism; this is called the mechanism condition. Second, the structure must be in equilibrium, i.e., the bending moment distribution must satisfy equilibrium with the applied loads. Finally, the bending moment at any cross-section must not exceed the full plastic value of that cross-section; this is called the plastic moment condition. The theorem simply implies that the collapse load factor, λ_c, obtained from the three basic conditions (mechanism, equilibrium, and plastic moment) has a unique value.

The proof of the three theorems can be found in Chen and Sohal [16]. A useful corollary of the lower bound theorem is that if at a load factor, λ, it is possible to find a bending moment diagram that satisfies both the equilibrium and moment conditions but not necessarily the mechanism condition, then the structure will not collapse at that load factor unless the load happens to be the collapse load. A corollary of the upper bound theorem is that the true load factor at collapse is the smallest possible one that can be determined from a consideration of all possible mechanisms of collapse. This concept is very useful in finding the collapse load of the system from various combinations of mechanisms. From these theorems, it can be seen that the lower bound theorem is based on the equilibrium approach while the upper bound technique is based on the mechanism approach. These two alternative approaches to an exact solution, called the *equilibrium method* and the *mechanism method,* will be discussed in the sections that follow.

2.11.7 Equilibrium Method

The equilibrium method, which employs the lower bound theorem, is suitable for the analysis of continuous beams and frames in which the structural redundancies are not exceeding two. The procedures for obtaining the equilibrium equations of a statically indeterminate structure and to evaluate its plastic limit load are as follows:

To obtain the equilibrium equations of a statically indeterminate structure:

1. Select the redundant(s).
2. Free the redundants and draw a moment diagram for the determinate structure under the applied loads.
3. Draw a moment diagram for the structure due to the redundant forces.
4. Superimpose the moment diagrams in Steps 2 and 3.
5. Obtain the maximum moment at critical sections of the structure utilizing the moment diagram in Step 4.

To evaluate the plastic limit load of the structure:

6. Select value(s) of redundant(s) such that the plastic moment condition is not violated at any section in the structure.
7. Determine the load corresponding to the selected redundant(s).
8. Check for the formation of a mechanism. If a collapse mechanism condition is met, then the computed load is the exact plastic limit load. Otherwise, it is a lower bound solution.
9. Adjust the redundant(s) and repeat Steps 6 to 9 until the exact plastic limit load is obtained.

EXAMPLE 2.11:   Continuous Beam

Figure 2.100a shows a two-span continuous beam which is analyzed using the equilibrium method. The plastic limit load of the beam is calculated based on the step-by-step procedure described in the previous section as follows:

1. Select the redundant force as M_1 which is the bending moment at the intermediate support, as shown in Figure 2.100b.
2. Free the redundants and draw the moment diagram for the determinate structure under the applied loads, as shown in Figure 2.100c.
3. Draw the moment diagram for the structure due to the redundant moment M_1 as shown in Figure 2.100d.

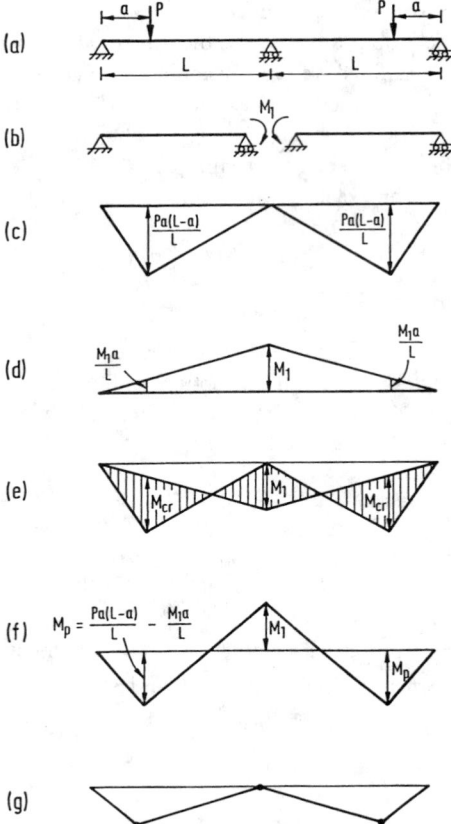

Figure 2.100   Analysis of a two-span continuous using the equilibrium method.

4. Superimpose the moment diagrams in Figures 2.100c and d. The results are shown in Figure 2.100e. The moment diagram in Figure 2.100e is redrawn on a single straight base line. The critical moment in the beam is

$$M_{cr} = \frac{Pa(L-a)}{L} - \frac{M_1 a}{L} \tag{2.223}$$

The maximum moment at the critical sections of the structure utilizing the moment diagram in Figure 2.100e is obtained. By letting $M_{cr} = M_p$, the resulting moment distribution is shown in Figure 2.100f.

5. A lower bound solution may be obtained by selecting a value of redundant moment M_1. For example, if $M_1 = 0$ is selected, the moment diagram is reduced to that shown in Figure 2.100c. By equating the maximum moment in the diagram to the plastic moment, M_p, we have

$$M_{cr} = \frac{Pa(L-a)}{L} = M_p \tag{2.224}$$

which gives $P = P_1$ as

$$P_1 = \frac{M_p L}{a(L-a)} \tag{2.225}$$

The moment diagram in Figure 2.100c shows a plastic hinge formed at each span. Since two plastic hinges in each span are required to form a plastic mechanism, the load P_1 is a lower bound solution. However, if the redundant moment M_1 is set equal to the plastic moment M_p, and letting the maximum moment in Figure 2.100f be equal to the plastic moment, we have

$$M_{cr} = \frac{Pa(L-a)}{L} - \frac{M_p a}{L} = M_p \tag{2.226}$$

which gives $P = P_2$ as

$$P_2 = \frac{M_p(L+a)}{a(L-a)} \tag{2.227}$$

6. Since a sufficient number of plastic hinges has formed in the beams (Figure 2.100g) to arrive at a collapse mechanism, the computed load, P_2, is the exact plastic limit load.

EXAMPLE 2.12: Portal Frame

A pinned-base rectangular frame subjected to vertical load V and horizontal load H is shown in Figure 2.101a. All the members of the frame, AB, BD, and DE are made of the same section with moment capacity M_p. The objective is to determine the limit value of H if the frame's width-to-height ratio L/h is 1.0.

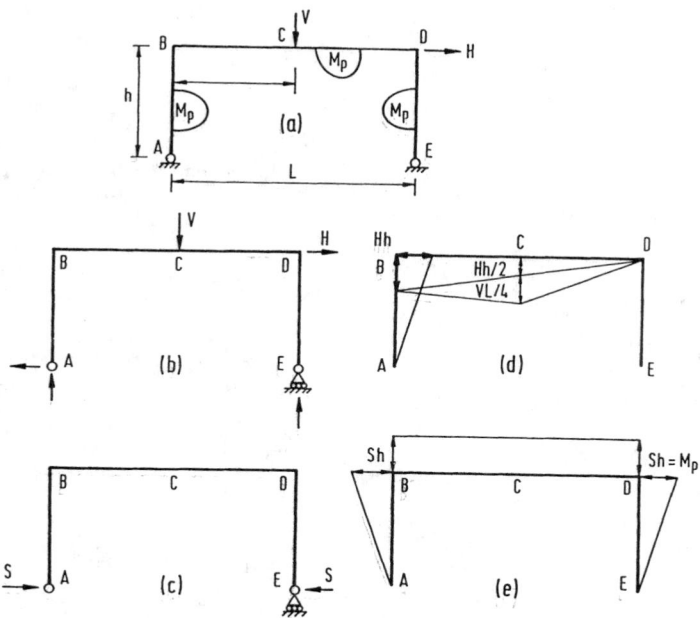

Figure 2.101 Analysis of portal frame using the equilibrium method.

Procedure: The frame has one degree of redundancy. The redundancy for this structure can be chosen as the horizontal reaction at E. Figures 2.101b and c show the resulting determinate frame loaded by the applied loads and redundant force. The moment diagrams corresponding to these two loading conditions are shown in Figures 2.101d and e.

The horizontal reaction S should be chosen in such a manner that all three conditions (equilibrium, plastic moment, and mechanism) are satisfied. Formation of two plastic hinges is necessary to form a mechanism. The plastic hinges may be formed at B, C, and D. Assuming that a plastic hinge is formed at D, as shown in Figure 2.101e, we have

$$S = \frac{M_p}{h} \tag{2.228}$$

Corresponding to this value of S, the moments at B and C can be expressed as

$$M_B = Hh - M_p \tag{2.229}$$

$$M_C = \frac{Hh}{2} + \frac{VL}{4} - M_p \tag{2.230}$$

The condition for the second plastic hinge to form at B is $|M_B| > |M_C|$. From Equations 2.229 and 2.230 we have

$$Hh - M_p > \frac{Hh}{2} + \frac{VL}{4} - M_p \tag{2.231}$$

and

$$\frac{V}{H} < \frac{h}{L} \tag{2.232}$$

The condition for the second plastic hinge to form at C is $|M_C| > |M_B|$. From Equations 2.229 and 2.230 we have

$$Hh - M_p < \frac{Hh}{2} + \frac{VL}{4} - M_p \tag{2.233}$$

and

$$\frac{V}{H} > \frac{h}{L} \tag{2.234}$$

For a particular combination of V, H, L, and h, the collapse load for H can be calculated.

(a) When $L/h = 1$ and $V/H = 1/3$, we have

$$M_B = Hh - M_p \tag{2.235}$$

$$M_C = \frac{Hh}{2} + \frac{Hh}{12} - M_p = \frac{7}{12}Hh - M_p \tag{2.236}$$

Since $|M_B| > |M_C|$, the second plastic hinge will form at B, and the corresponding value for H is

$$H = \frac{2M_p}{h} \tag{2.237}$$

(b) When $L/h = 1$ and $V/H = 3$, we have

$$M_B = Hh - M_p \tag{2.238}$$

$$M_C = \frac{Hh}{2} + \frac{3}{4}Hh - M_p = \frac{5}{4}Hh - M_p \tag{2.239}$$

Since $|M_C| > |M_B|$, the second plastic hinge will form at C, and the corresponding value for H is

$$H = \frac{1.6M_p}{h} \tag{2.240}$$

2.11.8 Mechanism Method

This method, which is based on the upper bound theorem, states that the load computed on the basis of an assumed failure mechanism is never less than the exact plastic limit load of a structure. Thus, it always predicts the upper bound solution of the collapse limit load. It can also be shown that the minimum upper bound is the limit load itself. The procedure for using the mechanism method has the following two steps:

1. Assume a failure mechanism and form the corresponding work equation from which an upper bound value of the plastic limit load can be estimated.
2. Write the equilibrium equations for the assumed mechanism and check the moments to see whether the plastic moment condition is met everywhere in the structure.

To obtain the true limit load using the mechanism method, it is necessary to determine every possible collapse mechanism of which some are the combinations of a certain number of independent mechanisms. Once the independent mechanisms have been identified, a work equation may be established for each combination and the corresponding collapse load is determined. The lowest load among those obtained by considering all the possible combination of independent mechanisms is the correct plastic limit load.

Number of Independent Mechanisms

The number of possible independent mechanisms n, for a structure, can be determined from

$$n = N - R \tag{2.241}$$

where N is the number of critical sections at which plastic hinges might form, and R is the degrees of redundancy of the structure.

Critical sections generally occur at the points of concentrated loads, at joints where two or more members are meeting at different angles, and at sections where there is an abrupt change in section geometries or properties. To determine the number of redundant Rs of a structure, it is necessary to free sufficient supports or restraining forces in structural members so that the structure becomes an assembly of several determinate sub-structures.

Figure 2.102 shows two examples. The cuts that are made in each structure reduce the structural

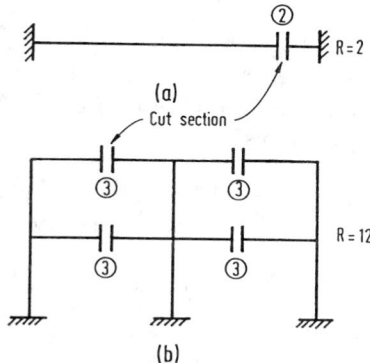

Figure 2.102   Number of redundants in (a) a beam and (b) a frame.

members to either cantilevers or simply supported beams. The fixed-end beam requires a shear force and a moment to restore continuity at the cut section, and thus $R = 2$. For the two-story frame, an axial force, shear, and moment are required at each cut section for full continuity, and thus $R = 12$.

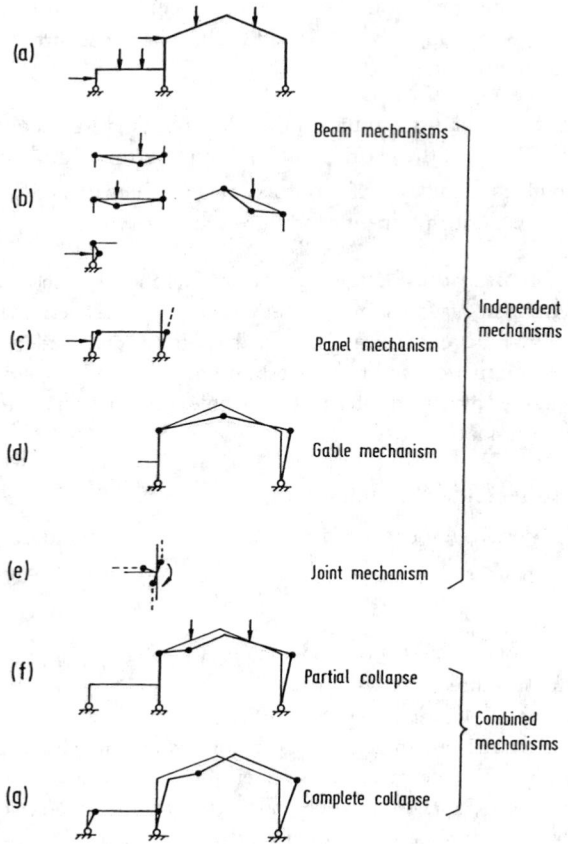

Figure 2.103   Typical plastic mechanisms.

Types of Mechanism

Figure 2.103a shows a frame structure subjected to a set of loading. The frame may fail by different types of collapse mechanisms dependent on the magnitude of loading and the frame's configurations. The collapse mechanisms are:

(a) *Beam Mechanism:* Possible mechanisms of this type are shown in Figure 2.103b.

(b) *Panel Mechanism:* The collapse mode is associated with sidesway as shown in Figure 2.103c.

(c) *Gable Mechanism:* The collapse mode is associated with the spreading of column tops with respect to the column bases as shown in Figure 2.103d.

(d) *Joint Mechanism:* The collapse mode is associated with the rotation of joints of which the adjoining members developed plastic hinges and deformed under an applied moment as shown in Figure 2.103e.

(e) *Combined Mechanism:* It can be a partial collapse mechanism as shown in Figure 2.103f or it may be a complete collapse mechanism as shown in Figure 2.103g.

The principal rule for combining independent mechanisms is to obtain a lower value of collapse load. The combinations are selected in such a way that the external work becomes a maximum and the internal work becomes a minimum. Thus, the work equation would require that the mechanism involve as many applied loads as possible and at the same time to eliminate as many plastic hinges as possible. This procedure will be illustrated in the following example.

EXAMPLE 2.13:   Rectangular Frame

A fixed-end rectangular frame has a uniform section with $M_p = 20$ and carries the load shown in Figure 2.104. Determine the value of load ratio λ at collapse.

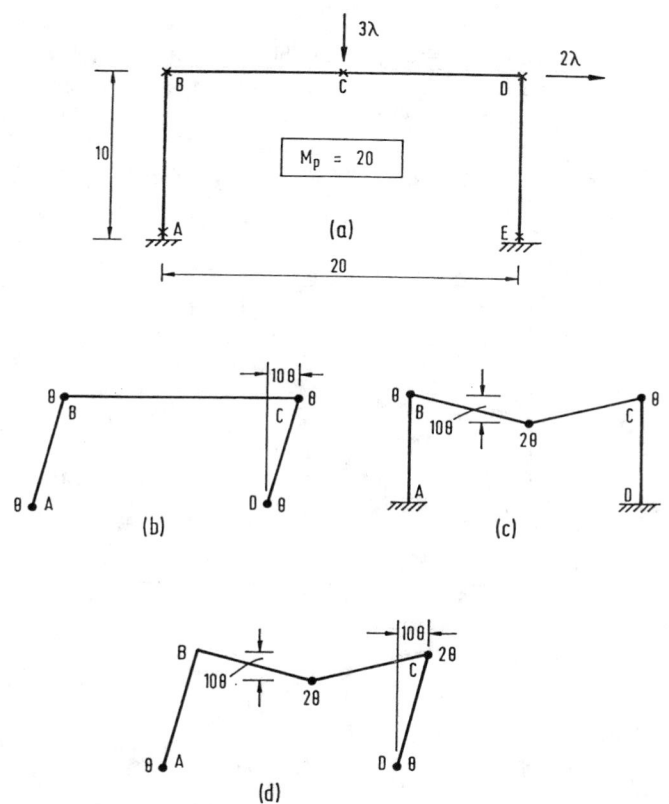

Figure 2.104   Collapse mechanisms of a fixed base portal frame.

Solution

Number of possible plastic hinges	$N = 5$
Number of redundancies	$R = 3$
Number of independent mechanisms	$N - R = 2.$

The two independent mechanisms are shown in Figures 2.104b and c, and the corresponding work equations are

Panel mechanism	$20\lambda = 4(20) = 80$	$\Rightarrow \lambda = 4$
Beam mechanism	$30\lambda = 4(20) = 80$	$\Rightarrow \lambda = 2.67$

The combined mechanisms are now examined to see whether they will produce a lower λ value. It is observed that only one combined mechanism is possible. The mechanism is shown in Figure 2.104c involving cancellation of a plastic hinge at B. The calculation of the limit load is described below:

Panel mechanism	$20\lambda = 4(20)$
Beam mechanism	$30\lambda = 4(20)$
Addition	$50\lambda = 8(20)$
Cancellation of plastic hinge	$-2(20)$
Combined mechanism	$50\lambda = 6(20)$
	$\Rightarrow \lambda = 2.4$

The combined mechanism results in a smaller value for λ and no other possible mechanism can produce a lower load. Thus, $\lambda = 2.4$ is the collapse load.

EXAMPLE 2.14:   Frame Subjected to Distributed Load

When a frame is subjected to distributed loads, the maximum moment and hence the plastic hinge location is not known in advance. The exact location of the plastic hinge may be determined by writing the work equation in terms of the unknown distance and then maximizing the plastic moment by formal differentiation.

Consider the frame shown in Figure 2.105a. The sidesway collapse mode in Figure 2.105b leads

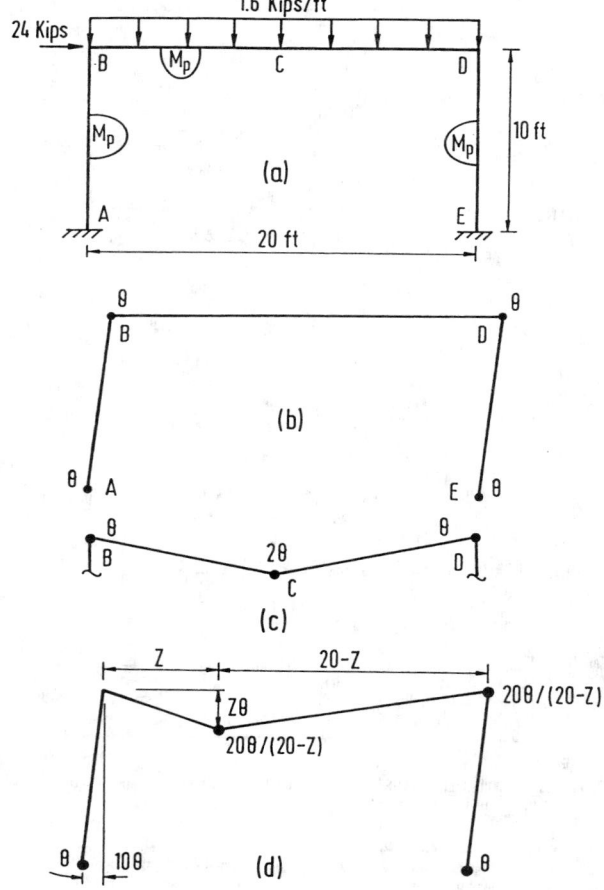

Figure 2.105   A portal frame subjected to a combined uniform distributed load and a horizontal load.

to the following work equation:

$$4M_p = 24(10\theta)$$

which gives

$$M_p = 60 \ \text{kip-ft}$$

The beam mechanism of Figure 2.105c gives

$$4M_p\theta = \frac{1}{2}(10\theta)32$$

which gives

$$M_p = 40 \ \text{kip-ft}$$

In fact, the correct mechanism is shown in Figure 2.105d in which the distance Z from the plastic hinge location is unknown. The work equation is

$$24(10\theta) + \frac{1}{2}(1.6)(20)(z\theta) = M_p\left(2 + 2\left(\frac{20}{20-z}\right)\right)\theta$$

which gives

$$M_p = \frac{(240 + 16z)(20 - z)}{80 - 2z}$$

To maximize M_p, the derivative of M_p is set to zero, i.e.,

$$(80 - 2z)(80 - 32z) + (4800 + 80z - 16z^2)(2) = 0$$

which gives

$$z = 40 - \sqrt{1100} = 6.83 \ \text{ft}$$

and

$$M_p = 69.34 \ \text{kip-ft}$$

In practice, uniform load is often approximated by applying several equivalent point loads to the member under consideration. Plastic hinges thus can be assumed to form only at the concentrated load points, and the calculations become simpler when the structural system becomes more complex.

2.11.9 Gable Frames

The mechanism method is used to determine the plastic limit load of the gable frame shown in Figure 2.106. The frame is composed of members with plastic moment capacity of 270 kip-in. The column bases are fixed. The frame is loaded by a horizontal load H and vertical concentrated load V. A graph from which V and H cause the collapse of the frame is to be produced.

    Solution Consider the three modes of collapse as follows:

1. **Plastic hinges form at** A, C, D, **and** E

   The mechanism is shown in Figure 2.106b. The instantaneous center O for member CD is located at the intersection of AC and ED extended. From similar triangles ACC_1 and OCC_2, we have

   $$\frac{OC_2}{CC_2} = \frac{C_1A}{C_1C}$$

   which gives

   $$OC_2 = \frac{C_1A}{C_1C}CC_2 = \frac{22.5(9)}{18} = 11.25 \ \text{ft}$$

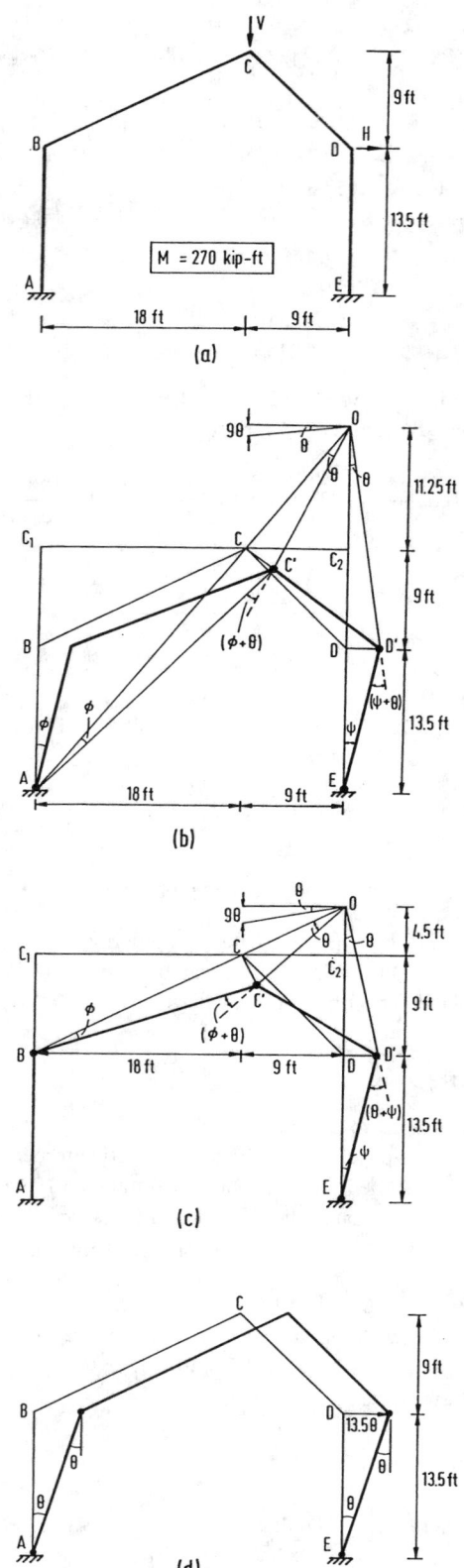

Figure 2.106   Collapse mechanisms of a fixed base gable frame.

From triangles ACC' and $CC'O$, we have

$$AC(\phi) = OC(\theta)$$

which gives

$$\phi = \frac{OC}{AC}\theta = \frac{CC_2}{C_1C}\theta = \frac{9}{8}\theta = \frac{1}{2}\theta$$

Similarly, from triangles ODD' and EDD', the rotation at E is given as

$$DE(\Psi) = OD(\theta)$$

which gives

$$\Psi = \frac{OD}{DE}\theta = 1.5\theta$$

From the hinge rotations and displacements, the work equation for this mechanism can be written as

$$V(9\theta) + H(13.5\Psi) = M_p[\phi + (\phi + \theta) + (\theta + \Psi) + \Psi]$$

Substituting values for ψ and ϕ and simplifying, we have

$$V + 2.25H = 180$$

2. **Mechanism with Hinges at** B, C, D, **and** E

Figure 2.106c shows the mechanism in which the plastic hinge rotations and displacements at the load points can be expressed in terms of the rotation of member CD about the instantaneous center O.

From similar triangles BCC_1 and OCC_2, we have

$$\frac{OC_2}{CC_2} = \frac{BC_1}{C_1C}$$

which gives

$$OC_2 = \frac{BC_1}{C_1C}CC_2 = \frac{9}{18}(9) = 4.5 \text{ ft}$$

From triangles BCC' and $CC'O$, we have

$$BC(\phi) = OC(\theta)$$

which gives

$$\phi = \frac{OC}{BC}\theta = \frac{OC_2}{BC_1}\theta = \frac{4.5}{9}\theta\frac{1}{2}\theta$$

Similarly, from triangles ODD' and EDD', the rotation at E is given as

$$DE(\Psi) = OD(\theta)$$

which gives

$$\Psi = \frac{OD}{DE}\theta = \theta$$

The work equation for this mechanism can be written as

$$V(9\theta) + H(13.5\Psi) = M_p[\phi + (\phi + \theta) + (\theta + \Psi) + \Psi]$$

Substituting values of ψ and ϕ and simplifying, we have

$$V + 1.5H = 150$$

3. **Mechanism with Hinges at A, B, D, and E**

   The hinge rotations and displacements corresponding to this mechanism are shown in Figure 2.106d. The rotation of all hinges is θ. The horizontal load moves by 13.5θ but the horizontal load has no vertical displacement. The work equation becomes

$$H(13.5\theta) = M_p(\theta + \theta + \theta + \theta)$$

or

$$H = 80 \text{ kips}$$

The interaction equations corresponding to the three mechanisms are plotted in Figure 2.107. By carrying out moment checks, it can be shown that Mechanism 1 is valid for portion AB of the curve, Mechanism 2 is valid for portion BC, and Mechanism 3 is valid only when $V = 0$.

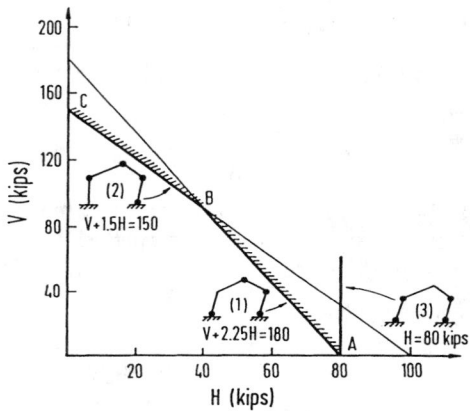

Figure 2.107   Vertical load and horizontal force interaction curve for collapse analysis of a gable frame.

2.11.10   Analysis Charts for Gable Frames

Pinned-Base Gable Frames

Figure 2.108a shows a pinned-end gable frame subjected to a uniform gravity load $\lambda w L$ and a horizontal load $\lambda 1 H$ at the column top. The collapse mechanism is shown in Figure 2.108b. The work equation is used to determine the plastic limit load. First, the instantaneousness of rotation O is determined by considering similar triangles,

$$\frac{OE}{CF} = \frac{L}{xL} \quad \text{and} \quad \frac{OE}{CF} = \frac{OE}{h_1 + 2xh_2} \tag{2.242}$$

and

$$OD = OE - h_1 = \frac{(1-x)h_1 + 2xh_2}{x} \tag{2.243}$$

From the horizontal displacement of D,

$$\theta h_1 = \phi OD \tag{2.244}$$

of which

$$\phi = \frac{x}{(1-x) + 2xk}\theta \tag{2.245}$$

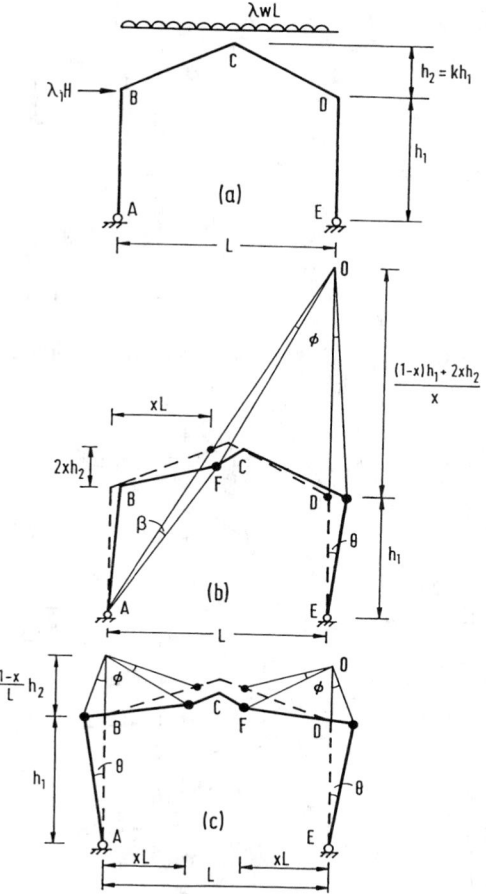

Figure 2.108 Pinned base gable frame subjected to a combined uniform distributed load and horizontal load.

where $k = h_2/h_1$. From the vertical displacement at C,

$$\beta = \frac{1-x}{(1-x) + 2xk}\theta \tag{2.246}$$

The work equation for the assumed mechanism is

$$\lambda_1 H h_1 \beta + \frac{\lambda w L^2}{2}(1-x)\phi = M_p(\beta + 2\phi + \theta) \tag{2.247}$$

which gives

$$M_p = \frac{(1-x)\lambda_1 H h_1 + (1-x)x\lambda w L^2/2}{2(1+kx)} \tag{2.248}$$

Differentiating M_p in Equation 2.248 with respect to x and solve for x,

$$x = \frac{A-1}{k} \tag{2.249}$$

where

$$A = \sqrt{(1+k)(1-Uk)} \quad \text{and} \quad U = \frac{2\lambda_1 H h_1}{\lambda w L^2} \tag{2.250}$$

Substituting for x in the expression for M_p gives

$$M_p = \frac{\lambda w L^2}{8}\left[\frac{U(2+U)}{A^2+2A-Uk^2+1}\right] \tag{2.251}$$

In the absence of horizontal loading, the gable mechanism, as shown in Figure 2.108c, is the failure mode. In this case, letting $H = 0$ and $U = 0$ gives [31]:

$$M_p = \frac{\lambda w L^2}{8}\left[\frac{1}{1+k+\sqrt{1+k}}\right] \tag{2.252}$$

Equation 2.251 can be used to produce a chart as shown in Figure 2.109 by which the value of M_p can be determined rapidly by knowing the values of

$$k = \frac{h_2}{h_1} \quad \text{and} \quad U = \frac{2\lambda_1 H h_1}{\lambda w L^2} \tag{2.253}$$

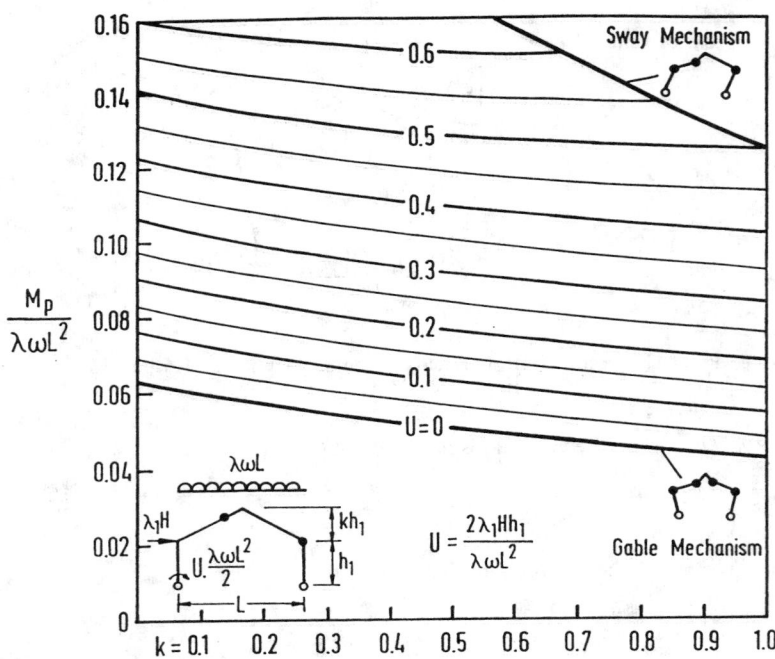

Figure 2.109   Analysis chart for pinned base gable frame.

Fixed-Base Gable Frames

Similar charts can be generated for fixed-base gable frames as shown in Figure 2.110. Thus, if the values of loading, λw and $\lambda_1 H$, and frame geometry, h_1, h_2, and L, are known, the parameters k and U can be evaluated and the corresponding value of $M_p/(\lambda w L^2)$ can be read directly from the appropriate chart. The required value of M_p is obtained by multiplying the value of $M_p/(\lambda w L^2)$ by $\lambda w L^2$.

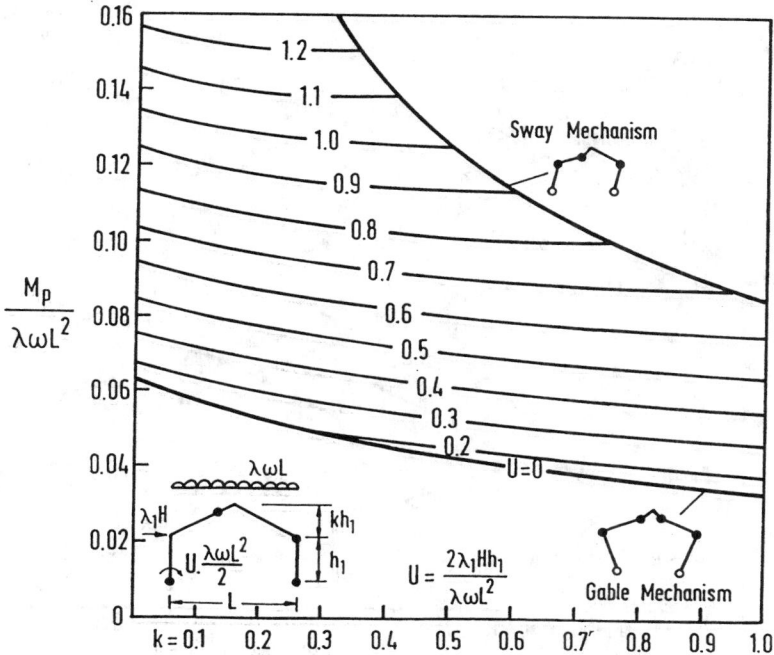

Figure 2.110 Analysis chart for fixed base gable frame.

2.11.11 Grillages

Grillage is a type of structure consisting of straight beams lying on the same plane, subjected to loads acting perpendicular to the plane. An example of such structure is shown in Figure 2.111. The grillage consists of two equal simply supported beams of span length $2L$ and full plastic moment M_p. The two beams are connected rigidly at their centers where a concentrated load W is carried.

The collapse mechanism consists of four plastic hinges formed at the beams adjacent to the point load as shown in Figure 2.111. The work equation is

$$WL\theta = 4M_p\theta$$

of which the collapse load is

$$W = \frac{4M_p}{L}$$

Six-Beam Grillage

A grillage consisting of six beams of span length $4L$ each and full plastic moment M_p is shown in Figure 2.112. A total load of 9W acts on the grillage, splitting into concentrated loads W at the nine nodes. Three collapse mechanisms are possible. Ignoring member twisting due to torsional forces, the work equations associated with the three collapse mechanisms are computed as follows:

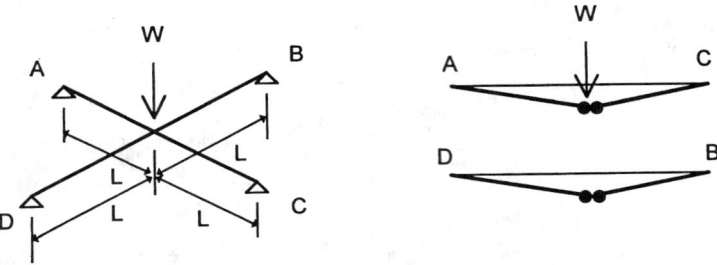

Figure 2.111   Two-beam grillage system.

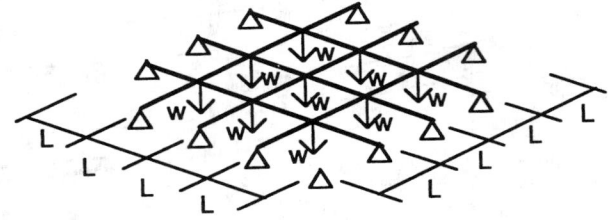

Figure 2.112   Six-beam grillage system.

Mechanism 1 (Figure 2.113a)	
Work equation	$9wL\theta = 12M_p\theta$
of which	$w = \frac{12}{9}\frac{M_p}{L} = \frac{4M_p}{3L}$
Mechanism 2 (Figure 2.113b)	
Work equation	$wL\theta = 8M_p\theta$
of which	$w = \frac{8M_p}{L}$
Mechanism 3 (Figure 2.113c)	
Work equation	$w2L2\theta + 4 \times w2L\theta = M_p(4\theta + 8\theta)$
of which	$w = \frac{M_p}{L}$

The lowest upper bound load corresponds to Mechanism 3. This can be confirmed by conducting a moment check to ensure that bending moments anywhere are not violating the plastic moment condition. Additional discussion of plastic analysis of grillages can be found in [6] and [29].

2.11.12   Vierendeel Girders

Figure 2.114 shows a simply supported girder in which all members are rigidly joined and have the same plastic moment M_p. It is assumed that axial loads in the members do not cause member instability. Two possible collapse mechanisms are considered as shown in Figures 2.114b and c. The work equation for Mechanism 1 is

$$W3\theta L = 20M_p\theta$$

so that

$$W = \frac{20M_p}{3L}$$

The work equation for Mechanism 2 is

$$W3\theta L = 16M_p\theta$$

or

$$W = \frac{16M_p}{3L}$$

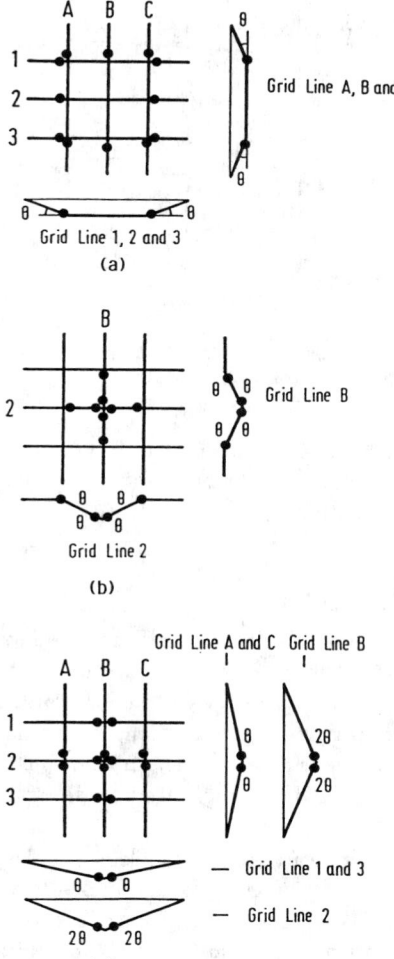

Figure 2.113 Six-beam grillage system. (a) Mechanism 1. (b) Mechanism 2. (c) Mechanism 3.

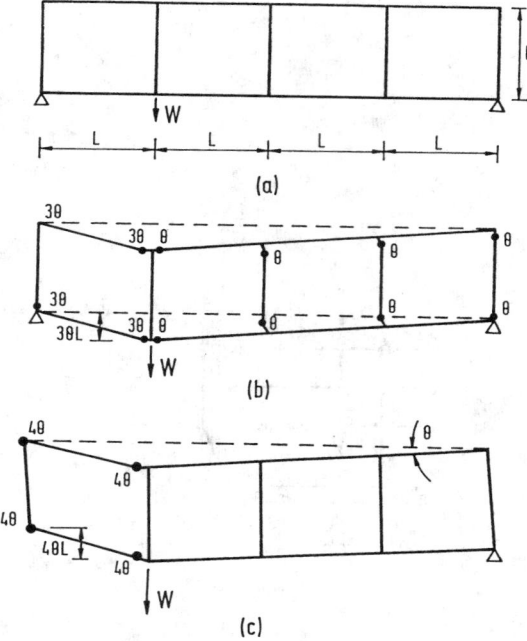

Figure 2.114 Collapse mechanisms of a Vierendeel girder.

It can be easily proved that the collapse load associated with Mechanism 2 is the correct limit load. This is done by constructing an equilibrium set of bending moments and checking that they are not violating the plastic moment condition.

2.11.13   First-Order Hinge-By-Hinge Analysis

Instead of finding the collapse load of the frame, it may be useful to obtain information about the distribution and redistribution of forces prior to reaching the collapse load. Elastic-plastic hinge analysis (also known as hinge-by-hinge analysis) determines the order of plastic hinge formation, the load factor associated with each plastic-hinge formation, and member forces in the frame between each hinge formation. Thus, the state of the frame can be defined at any load factor rather than only at the state of collapse. This allows for a more accurate determination of member forces at the design load level.

Educational and commercial software are now available for elastic-plastic hinge analysis [16]. The computations of deflections for simple beams and multi-story frames can be done using the virtual work method [5, 8, 16, 34]. The basic assumption of first-order elastic-plastic hinge analysis is that the deformations of the structure are insufficient to alter radically the equilibrium equations. This assumption ceases to be true for slender members and structures, and the method gives unsafe predictions of limit loads.

2.12   Frame Stability

2.12.1   Categorization of Analysis Methods

Several stability analysis methods have been utilized in research and practice. Figure 2.115 shows schematic representations of the load-displacement results of a sway frame obtained from each type of analysis to be considered.

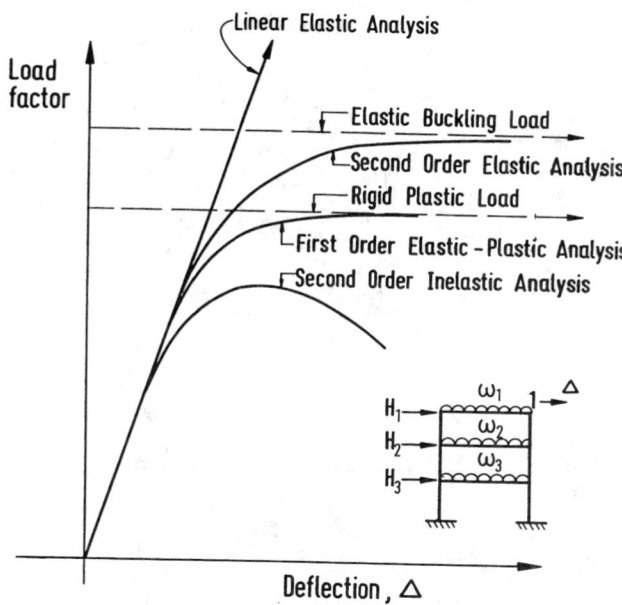

Figure 2.115   Categorization of stability analysis methods.

Elastic Buckling Analysis

The elastic buckling load is calculated by linear buckling or bifurcation (or eigenvalue) analysis. The buckling loads are obtained from the solutions of idealized elastic frames subjected to idealized loads which do not produce direct bending in the structure. The only displacements that occur before buckling occurs are those in the directions of the applied loads. When buckling (bifurcation) occurs, the displacements increase without bound, assuming linearized theory of elasticity and small displacement as shown by the horizontal straight line in Figure 2.115. The load at which these displacements occur is known as the buckling load, or commonly referred to as bifurcation load. For structural models that actually exhibit a bifurcation from the primary load path, the elastic buckling load is the largest load that the model can sustain, at least within the vicinity of the bifurcation point, provided that the post-buckling path is in unstable equilibrium. If the secondary path is in stable equilibrium, the load can still increase beyond the critical load value.

Buckling analysis is a common tool for calculations of column effective lengths. The effective length factor of a column member can be calculated using the procedure described in Section 2.12.2. The buckling analysis provides useful indices of the stability behavior of structures; however, it does not predict actual behavior of all but idealized structures with gravity loads applied only at the joints.

Second-Order Elastic Analysis

The analysis is formulated based on the deformed configuration of the structure. When derived rigorously, a second-order analysis can include both the member curvature ($P - \delta$) and the sidesway ($P - \Delta$) stability effects. The P-δ effect is associated with the influence of the axial force acting through the member displacement with respect to the rotated chord, whereas the $P - \Delta$ effect is the influence of axial force acting through the relative sidesway displacements of the member ends. It is interesting to note that a structural system will become stiffer when its members are subjected to tension. Conversely, the structure will become softer when its members are in compression. Such behavior can be illustrated for a simple model shown in Figure 2.116. There is a clear advantage for a designer to take advantage of the stiffer behavior for tension structures. However, the detrimental effects associated with second-order deformations due to the compression forces must be considered in designing structures subjected to predominant gravity loads.

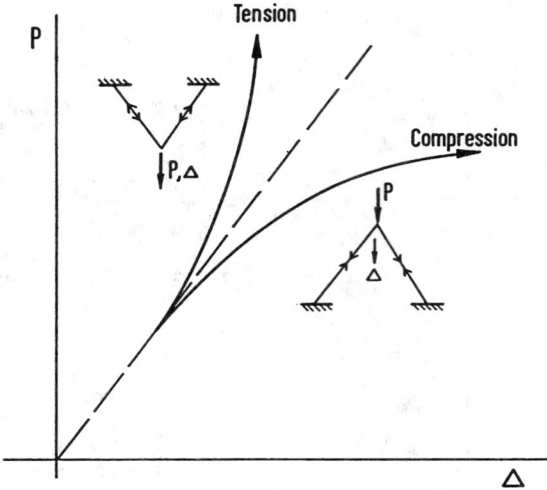

Figure 2.116   Behavior of frame in compression and tension.

Unlike the first-order analysis in which solutions can be obtained in a rather simple and direct manner, a second-order analysis often requires an iterative procedure to obtain solutions. Although second-order analysis can account for all the stability effects, it does not provide information on the actual inelastic strength of the structure. In design, the combined effects due to stability and inelasticity must be considered for a proper evaluation of member and system strengths. The load-displacement curve generated from a second-order analysis will gradually approach the horizontal straight line which represents the buckling load obtained from the elastic buckling analysis, as shown in Figure 2.115. Differences in the two loads may arise from the fact that the elastic stability limit is calculated for equilibrium based on the deformed configuration whereas the elastic critical load is calculated as a bifurcation from equilibrium on the undeformed geometry of the frame.

The load-displacement response of most frame structures usually does not involve any actual bifurcation or branch from one equilibrium solution to another equilibrium path. In some cases, the second-order elastic incremental response may not have any limit. The reader is referred to Chapter 1 of [14] for a basic discussion of these behavioral issues.

Recent works on second-order elastic analysis have been reported in Liew et al. [40]; White and Hajjar [64]; Chen and Lui [15], Liew and Chen [39], and Chen and Kim [17], among others. Second-order analysis programs which can take into consideration connection flexibility are also available [1, 24, 26, 28, 30, 41, 42].

Second-Order Inelastic Analysis

Second-order inelastic analysis refers to any method of analysis that can capture geometrical and material nonlinearities in the analysis. The most refined inelastic analysis method is called spread-of-plasticity analysis. It involves discretization of members into many line segments, and the cross-section of each segment into a number of finite elements. Inelasticity is captured within the cross-sections and along the member length. The calculation of forces and deformations in the structure after yielding requires iterative trial-and-error processes because of the nonlinearity of the load-deformation response, and the change in cross-section effective stiffness at inelastic regions associated with the increase in the applied loads and the change in structural geometry. Although most of the plastic-zone analysis methods have been developed for planar analysis [59, 62], three-dimensional plastic-zone techniques are also available involving various degrees of refinements [12, 20, 60, 63].

The simplest second-order inelastic analysis is the elastic-plastic hinge approach. The analysis assumes that the element remains elastic except at its ends where zero-length plastic hinges are allowed to form [15, 18, 19, 41, 42, 45, 66, 69, 70]. Second-order plastic hinge analysis allows efficient analysis of large scale building frames. This is particularly true for structures in which the axial forces in the component members are small and the predominate behavior is associated with bending actions. Although elastic-plastic hinge approaches can provide essentially the same load-displacement predictions as second-order plastic-zone methods for many frame problems, they cannot be classified as advanced analysis for use in frame design. Some modifications to the elastic-plastic hinge are required to qualify the methods as advanced analysis, and they are discussed in Section 2.12.8.

Figure 2.115 shows the load-displacement curve (a smooth curve with a descending branch) obtained from the second-order inelastic analysis. The computed limit load should be close to that obtained from the plastic-zone analysis.

2.12.2 Columns Stability

Stability Equations

The stability equation of a column can be obtained by considering an infinitesimal deformed segment of the column as shown in Figure 2.117. By summing the moment about point b, we obtain

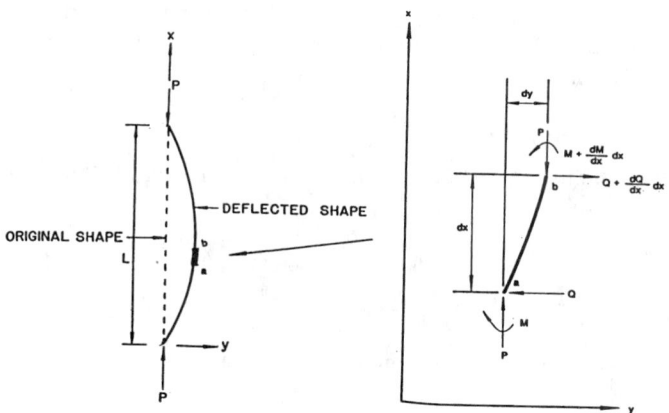

Figure 2.117   Stability equations of a column segment.

$$Qdx + Pdy + M - \left(M + \frac{dM}{dx}dx \right) = 0$$

or, upon simplification

$$Q = \frac{dM}{dx} - P\frac{dy}{dx} \qquad (2.254)$$

Summing force horizontally, we can write

$$-Q + \left(Q + \frac{dQ}{dx}dx \right) = 0$$

or, upon simplification

$$\frac{dQ}{dx} = 0 \qquad (2.255)$$

Differentiating Equation 2.254 with respect to x, we obtain

$$\frac{dQ}{dx} = \frac{d^2 M}{dx^2} - P\frac{d^2 y}{dx^2} \tag{2.256}$$

which, when compared with Equation 2.255, gives

$$\frac{d^2 M}{dx^2} - P\frac{d^2 y}{dx^2} = 0 \tag{2.257}$$

Since $M = -EI\frac{d^2 y}{dx^2}$, Equation 2.257 can be written as

$$EI\frac{d^4 y}{dx^4} + P\frac{d^2 y}{dx^2} = 0 \tag{2.258}$$

or

$$y^{IV} + k^2 y'' = 0 \tag{2.259}$$

Equation 2.259 is the general fourth-order differential equation that is valid for all support conditions. The general solution to this equation is

$$y = A\sin kx + B\cos kx + Cx + D \tag{2.260}$$

To determine the critical load, it is necessary to have four boundary conditions: two at each end of the column. In some cases, both geometric and force boundary conditions are required to eliminate the unknown coefficients (A, B, C, D) in Equation 2.260.

Column with Pinned Ends

For a column pinned at both ends as shown in Figure 2.118a, the four boundary conditions are

$$y(x = 0) = 0, \qquad M(x = 0) = 0 \tag{2.261}$$
$$y(x = L) = 0, \qquad M(x = L) = 0 \tag{2.262}$$

Since $M = -EIy''$, the moment conditions can be written as

$$y''(0) = 0 \text{ and } y''(x = L) = 0 \tag{2.263}$$

Using these conditions, we have
$$B = D = 0 \tag{2.264}$$

The deflection function (Equation 2.260) reduces to

$$y = A\sin kx + Cx \tag{2.265}$$

Using the conditions $y(L) = y''(L) = 0$, Equation 2.265 gives

$$A\sin kL + CL = 0 \tag{2.266}$$

and

$$-Ak^2 \sin kL = 0 \tag{2.267}$$

$$\begin{bmatrix} \sin kL & L \\ -k^2 \sin kL & 0 \end{bmatrix} \begin{bmatrix} A \\ C \end{bmatrix} = \begin{bmatrix} 0 \\ 0 \end{bmatrix} \tag{2.268}$$

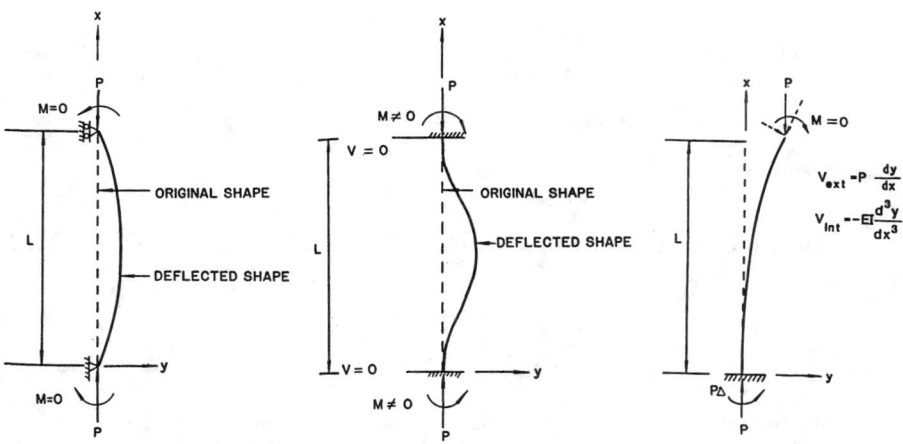

Figure 2.118 Column with (a) pinned ends, (b) fixed ends, and (c) fixed-free ends.

If $A = C = 0$, the solution is trivial. Therefore, to obtain a nontrivial solution, the determinant of the coefficient matrix of Equation 2.268 must be zero, i.e.,

$$\det \begin{vmatrix} \sin kL & L \\ -k^2 \sin kL & 0 \end{vmatrix} = 0 \qquad (2.269)$$

or

$$k^2 L \sin kL = 0 \qquad (2.270)$$

Since $k^2 L$ cannot be zero, we must have

$$\sin kL = 0 \qquad (2.271)$$

or

$$kL = n\pi, \quad n = 1, 2, 3, \ldots. \qquad (2.272)$$

The lowest buckling load corresponds to the first mode obtained by setting $n = 1$:

$$P_{cr} = \frac{\pi^2 EI}{L^2} \qquad (2.273)$$

Column with Fixed Ends

The four boundary conditions for a fixed-end column are (Figure 2.118b):

$$y(x = 0) = y'(x = 0) = 0 \qquad (2.274)$$
$$y(x = L) = y'''(x = L) = 0 \qquad (2.275)$$

Using the first two boundary conditions, we obtain

$$D = -B, \qquad C = -Ak \qquad (2.276)$$

The deflection function (Equation 2.260) becomes

$$y = A(\sin kx - kx) + B(\cos kx - 1) \qquad (2.277)$$

Using the last two boundary conditions, we have

$$\begin{bmatrix} \sin kL - kL & \cos kL - 1 \\ \cos kL - 1 & -\sin kL \end{bmatrix} \begin{bmatrix} A \\ B \end{bmatrix} = \begin{bmatrix} 0 \\ 0 \end{bmatrix} \qquad (2.278)$$

For a nontrivial solution, we must have

$$\det \begin{vmatrix} \sin kL - kL & \cos kL - 1 \\ \cos kL - 1 & -\sin kL \end{vmatrix} = 0 \tag{2.279}$$

or, after expanding

$$kL \sin kL + 2 \cos kL - 2 = 0 \tag{2.280}$$

Using trigonometrical identities $\sin kL = 2 \sin(kL/2) \cos(kL/2)$ and $\cos kL = 1 - 2 \sin^2(kL/2)$, Equation 2.280 can be written as

$$\sin \frac{kL}{2} \left(\frac{kL}{2} \cos \frac{kL}{2} - \sin \frac{kL}{2} \right) = 0 \tag{2.281}$$

The critical load for the symmetric buckling mode is $P_{cr} = 4\pi^2 EI/L^2$ by letting $\sin(kL/2) = 0$. The buckling load for the antisymmetric buckling mode is $P_{cr} = 80.766EI/L^2$ by letting the bracket term in Equation 2.281 equal zero.

Column with One End Fixed and One End Free

The boundary conditions for a fixed-free column are (Figure 2.118c): at the fixed end

$$y(x = 0) = y'(x = 0) = 0 \tag{2.282}$$

and, at the free end, the moment $M = EIy''$ is equal to zero

$$y''(x = L) = 0 \tag{2.283}$$

and the shear force $V = -dM/dx = -EIy'''$ is equal to Py' which is the transverse component of P acting at the free end of the column.

$$V = -EIy''' = Py' \tag{2.284}$$

It follows that the shear force condition at the free end has the form

$$y''' + k^2 y' = 0 \tag{2.285}$$

Using the boundary conditions at the fixed end, we have

$$B + D = 0, \quad \text{and} \quad Ak + C = 0 \tag{2.286}$$

The boundary conditions at the free end give

$$A \sin kL + B \cos kL = 0 \quad \text{and} \quad C = 0 \tag{2.287}$$

In matrix form, Equations 2.286 and 2.287 can be written as

$$\begin{bmatrix} 0 & 1 & 1 \\ k & 0 & 0 \\ \sin kL & \cos kL & 0 \end{bmatrix} \begin{bmatrix} A \\ B \\ C \end{bmatrix} = \begin{bmatrix} 0 \\ 0 \\ 0 \end{bmatrix} \tag{2.288}$$

For a nontrivial solution, we must have

$$\det \begin{vmatrix} 0 & 1 & 1 \\ k & 0 & 0 \\ \sin kL & \cos kL & 0 \end{vmatrix} = 0 \tag{2.289}$$

the characteristic equation becomes

$$k \cos kL = 0 \tag{2.290}$$

Since k cannot be zero, we must have $\cos kL = 0$ or

$$kL = \frac{n\pi}{2} \quad n = 1, 3, 5, \ldots \tag{2.291}$$

The smallest root ($n = 1$) gives the lowest critical load of the column

$$P_{cr} = \frac{\pi^2 EI}{4L^2} \tag{2.292}$$

The boundary conditions for columns with various end conditions are summarized in Table 2.2.

TABLE 2.2   Boundary Conditions for Various End Conditions

End conditions	Boundary conditions	
Pinned	$y = 0$	$y'' = 0$
Fixed	$y = 0$	$y' = 0$
Guided	$y' = 0$	$y''' = 0$
Free	$y'' = 0$	$y''' + k^2 y' = 0$

Column Effective Length Factor

The effective length factor, K, of columns with different end boundary conditions can be obtained by equating the P_{cr} load obtained from the buckling analysis with the Euler load of a pinned-ended column of effective length KL.

$$P_{cr} = \frac{\pi^2 EI}{(KL)^2}$$

The effective length factor can be obtained as

$$K = \sqrt{\frac{\pi^2 EI / L^2}{P_{cr}}} \tag{2.293}$$

The K factor is a factor that can be multiplied to the actual length of the end-restrained column to give the length of an equivalent pinned-ended column whose buckling load is the same as that of the end-restrained column. Table 2.3 [2, 3] summarizes the theoretical K factors for columns with different boundary conditions. Also shown in the table are the recommended K factors for design applications. The recommended values for design are equal or larger than the theoretical values to account for semi-rigid effects of the connections used in practice.

2.12.3   Beam-Column Stability

Figure 2.119a shows a beam-column subjected to an axial compressive force P at the ends, a lateral load w along the entire length and end moments M_A and M_B. The stability equation can be derived by considering the equilibrium of an infinitesimal element of length ds as shown in Figure 2.119b. The cross-section forces S and H act in the vertical and horizontal directions.

TABLE 2.3   Comparison of Theoretical and Design K Factors

	(a)	(b)	(c)	(d)	(e)	(f)
Buckled shape of column is shown by dashed line						
Theoretical K value	0.5	0.7	1.0	1.0	2.0	2.0
Recommended design value when ideal conditions are approximated	0.65	0.80	1.2	1.0	2.10	2.0
End condition code		Rotation fixed and translation fixed				
		Rotation free and translation fixed				
		Rotation fixed and translation free				
		Rotation free and translation free				

Considering equilibrium of forces

(a) Horizontal equilibrium

$$H + \frac{dH}{ds}ds - H = 0 \tag{2.294}$$

(b) Vertical equilibrium

$$S + \frac{dS}{ds}ds - S + wds = 0 \tag{2.295}$$

(c) Moment equilibrium

$$M + \frac{dM}{ds}ds - M - \left(S + \frac{dS}{ds} + S\right)\cos\theta\left(\frac{ds}{2}\right)$$
$$+ \left(H + \frac{dH}{ds}ds + H\right)\sin\theta\left(\frac{ds}{2}\right) = 0 \tag{2.296}$$

Since $(dS/ds)ds$ and $(dH/ds)ds$ are negligibly small compared to S and H, the above equilibrium equations can be reduced to

$$\frac{dH}{ds} = 0 \tag{2.297a}$$

$$\frac{dS}{ds} + w = 0 \tag{2.297b}$$

$$\frac{dM}{ds} - S\cos\theta + H\sin\theta = 0 \tag{2.297c}$$

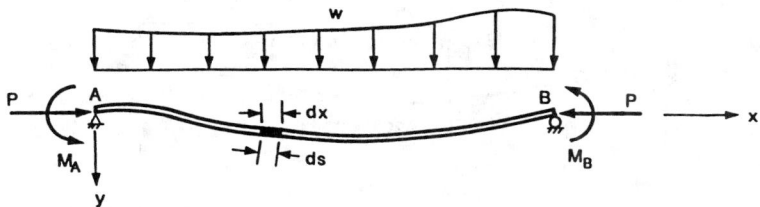

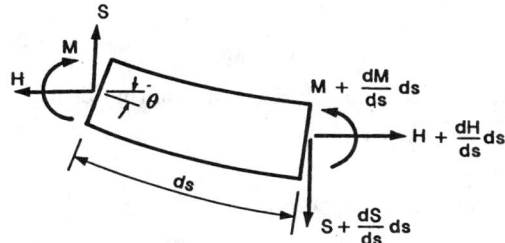

Figure 2.119 Basic differential equation of a beam-column.

For small deflections and neglecting shear deformations

$$ds \cong dx, \qquad \cos\theta \cong 1 \qquad \sin\theta \cong \theta \cong \frac{dy}{dx} \tag{2.298}$$

where y is the lateral displacement of the member. Using the above approximations, Equation 2.297 can be written as

$$\frac{dM}{dx} - S + H\frac{dy}{dx} = 0 \tag{2.299}$$

Differentiating Equation 2.299 and substituting Equations 2.297a and 2.297b into the resulting equation, we have

$$\frac{d^2M}{dx^2} + w + H\frac{d^2y}{dx^2} = 0 \tag{2.300}$$

From elementary mechanics of materials, it can easily be shown that

$$M = -EI\frac{d^2y}{dx^2} \tag{2.301}$$

Upon substitution of Equation 2.301 into Equation 2.300 and realizing that $H = -P$, we obtain

$$EI\frac{d^4y}{dx^4} + P\frac{d^2y}{dx^2} = w \tag{2.302}$$

The general solution to this differential equation has the form

$$y = A\sin kx + B\cos kx + Cx + D + f(x) \tag{2.303}$$

where $k = \sqrt{P/EI}$ and $f(x)$ is a particular solution satisfying the differential equation. The constants A, B, C, and D can be determined from the boundary conditions of the beam-column under investigation.

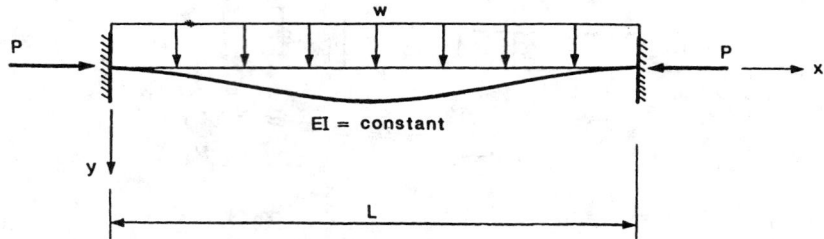

Figure 2.120   Beam-column subjected to uniform loading.

Beam-Column Subjected to Transverse Loading

Figure 2.120 shows a fixed-ended beam-column with uniformly distributed load w. The general solution to Equation 2.302 is

$$y = A \sin kx + B \cos kx + Cx + D + \frac{w}{2EIk^2}x^2 \tag{2.304}$$

Using the boundary conditions

$$y_{x=0} = 0 \qquad y'_{x=0} = 0 \qquad y_{x=L} = 0 \qquad y'_{x=L} = 0 \tag{2.305}$$

in which a prime denotes differentiation with respect to x, it can be shown that

$$A = \frac{wL}{2EIk^3} \tag{2.306a}$$

$$B = \frac{wL}{2EIk^3 \tan(kL/2)} \tag{2.306b}$$

$$C = -\frac{wL}{2EIk^2} \tag{2.306c}$$

$$D = -\frac{wL}{2EIk^3 \tan(kL/2)} \tag{2.306d}$$

Upon substitution of these constants into Equation 2.304, the deflection function can be written as

$$y = \frac{wL}{2EIk^3}\left[\sin kx + \frac{\cos kx}{\tan(kL/2)} - kx - \frac{1}{\tan(kL/2)} + \frac{kx^2}{L}\right] \tag{2.307}$$

The maximum moment for this beam-column occurs at the fixed ends and is equal to

$$M_{max} = -EIy''\,|_{x=0} = -EIy''\,|_{x=L} = -\frac{wL^2}{12}\left[\frac{3(\tan u - u)}{u^2 \tan u}\right] \tag{2.308}$$

where $u = kL/2$.

Since $wL^2/12$ is the maximum first-order moment at the fixed ends, the term in the bracket represents the theoretical moment amplification factor due to the P-δ effect.

For beam-columns with other transverse loading and boundary conditions, a similar approach can be followed to determine the moment amplification factor. Table 2.4 summarizes the expressions for the theoretical and design moment amplification factors for some loading conditions [2, 3].

TABLE 2.4   Theoretical and Design Moment Amplification Factor
$$\left(u = kL/2 = \tfrac{1}{2}\sqrt{(PL^2/EI)}\right)$$

Boundary conditions	P_{cr}	Location of M_{max}	Moment amplification factor
Hinged-hinged	$\dfrac{\pi^2 EI}{L^2}$	Mid-span	$\dfrac{2(\sec u - 1)}{u^2}$
Hinged-fixed	$\dfrac{\pi^2 EI}{(0.7L)^2}$	End	$\dfrac{2(\tan u - u)}{u^2(1/2u - 1/\tan 2u)}$
Fixed-fixed	$\dfrac{\pi^2 EI}{(0.5L)^2}$	End	$\dfrac{3(\tan u - u)}{u^2 \tan u}$
Hinged-hinged	$\dfrac{\pi^2 EI}{L^2}$	Mid-span	$\dfrac{\tan u}{u}$
Hinged-fixed	$\dfrac{\pi^2 EI}{(0.7L)^2}$	End	$\dfrac{4u(1 - \cos u)}{3u^2 \cos u(1/2u - 1/\tan 2u)}$
Fixed-fixed	$\dfrac{\pi^2 EI}{(0.5L)^2}$	Mid-span and end	$\dfrac{2(1 - \cos u)}{u \sin u}$

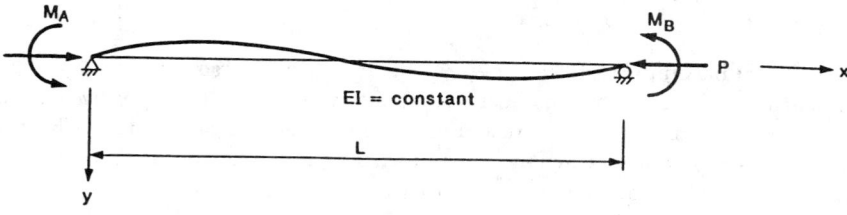

Figure 2.121   Beam-column subjected to end moments.

Beam-Column Subjected to End Moments

Consider the beam-column shown in Figure 2.121. The member is subjected to an axial force of P and end moments M_A and M_B. The differential equation for this beam-column can be obtained from Equation 2.302 by setting $w = 0$:

$$EI\frac{d^4 y}{dx^4} + P\frac{d^2 y}{dx^2} = 0 \tag{2.309}$$

The general solution is

$$y = A\sin kx + B\cos kx + Cx + D \tag{2.310}$$

The constants A, B, C, and D are determined by enforcing the four boundary conditions

$$y\,|_{x=0} = 0, \qquad y''\,|_{x=0} = \frac{M_A}{EI}, \qquad y\,|_{x=L} = 0, \qquad y''\,|_{x=L} = \frac{-M_B}{EI} \tag{2.311}$$

to give

$$A = \frac{M_A\cos kL + M_B}{EIk^2\sin kL} \tag{2.312a}$$

$$B = -\frac{M_A}{EIk^2} \tag{2.312b}$$

$$C = -\left(\frac{M_A + M_B}{EIk^2 L}\right) \tag{2.312c}$$

$$D = \frac{M_A}{EIk^2} \tag{2.312d}$$

Substituting Equations 2.312 into the deflection function Equation 2.310 and rearranging gives

$$
\begin{aligned}
y &= \frac{1}{EIk}\left[\frac{\cos kL}{\sin kL}\sin kx - \cos kx - \frac{x}{L} + 1\right]M_A \\
&+ \frac{1}{EIk^2}\left[\frac{1}{\sin kL}\sin kx - \frac{x}{L}\right]M_B
\end{aligned}
\tag{2.313}
$$

The maximum moment can be obtained by first locating its position by setting $dM/dx = 0$ and substituting the result into $M = -EIy''$ to give

$$
M_{\max} = \frac{\sqrt{(M_A^2 + 2M_A M_B \cos kL + M_B^2)}}{\sin kL}
\tag{2.314}
$$

Assuming that M_B is the larger of the two end moments, Equation 2.314 can be expressed as

$$
M_{\max} = M_B\left[\frac{\sqrt{\{(M_A/M_B)^2 + 2(M_A/M_B)\cos kL + 1\}}}{\sin kL}\right]
\tag{2.315}
$$

Since M_B is the maximum first-order moment, the expression in brackets is therefore the theoretical moment amplification factor. In Equation 2.315, the ratio (M_A/M_B) is positive if the member is bent in double (or reverse) curvature and the ratio is negative if the member is bent in single curvature. A special case arises when the end moments are equal and opposite (i.e., $M_B = -M_A$). By setting $M_B = -M_A = M_0$ in Equation 2.315, we have

$$
M_{\max} = M_0\left[\frac{\sqrt{\{2(1 - \cos kL)\}}}{\sin kL}\right]
\tag{2.316}
$$

For this special case, the maximum moment always occurs at mid-span.

2.12.4   Slope Deflection Equations

The slope-deflection equations of a beam-column can be derived by considering the beam-column shown in Figure 2.121. The deflection function for this beam-column can be obtained from Equation 2.313 in terms of M_A and M_B as:

$$
\begin{aligned}
y &= \frac{1}{EIk^2}\left[\frac{\cos kL}{\sin kL}\sin kx - \cos kx - \frac{x}{L} + 1\right]M_A \\
&+ \frac{1}{EIk^2}\left[\frac{1}{\sin kL}\sin kx - \frac{x}{L}\right]M_B
\end{aligned}
\tag{2.317}
$$

from which

$$
\begin{aligned}
y' &= \frac{1}{EIk}\left[\frac{\cos kL}{\sin kL}\cos kx + \sin kx - \frac{1}{kL}\right]M_A \\
&+ \frac{1}{EIk}\left[\frac{\cos kx}{\sin kL} - \frac{1}{kL}\right]M_B
\end{aligned}
\tag{2.318}
$$

The end rotations θ_A and θ_B can be obtained from Equation 2.318 as

$$
\begin{aligned}
\theta_A &= y'(x = 0) = \frac{1}{EIk}\left[\frac{\cos kL}{\sin kL} - \frac{1}{kL}\right]M_A + \frac{1}{EIk}\left[\frac{1}{\sin kL} - \frac{1}{kL}\right]M_B \\
&= \frac{L}{EI}\left[\frac{kL\cos kL - \sin kL}{(kL)^2 \sin kL}\right]M_A + \frac{L}{EI}\left[\frac{kL - \sin kL}{(kL)^2 \sin kL}\right]M_B
\end{aligned}
\tag{2.319}
$$

and

$$
\begin{aligned}
\theta_B &= y'(x = L) = \frac{1}{EIk}\left[\frac{1}{\sin kL} - \frac{1}{kL}\right]M_A + \frac{1}{EIk}\left[\frac{\cos kL}{\sin kL} - \frac{1}{kL}\right]M_B \\
&= \frac{L}{EI}\left[\frac{kL - \sin kL}{(kL)^2 \sin kL}\right]M_A + \frac{L}{EI}\left[\frac{kL \cos kL - \sin kL}{(kL)^2 \sin kL}\right]M_B
\end{aligned}
\tag{2.320}
$$

The moment rotation relationship can be obtained from Equations 2.319 and 2.320 by arranging M_A and M_B in terms of θ_A and θ_B as:

$$
M_A = \frac{EI}{L}(s_{ii}\theta_A + s_{ij}\theta_B)
\tag{2.321}
$$

$$
M_B = \frac{EI}{L}(s_{ji}\theta_A + s_{jj}\theta_B)
\tag{2.322}
$$

where

$$
s_{ii} = s_{jj} = \frac{kL \sin kL - (kL)^2 \cos kL}{2 - 2\cos kL - kL \sin kL}
\tag{2.323}
$$

$$
s_{ij} = s_{ji} = \frac{(kL)^2 - kL \sin kL}{2 - 2\cos kL - kL \sin kL}
\tag{2.324}
$$

are referred to as the *stability functions*.

Equations 2.321 and 2.322 are the slope-deflection equations for a beam-column that is not subjected to transverse loading and relative joint translation. It should be noted that when P approaches zero, $kL = \sqrt{(P/EI)}L$ approaches zero, and by using L'Hospital's rule, it can be shown that $s_{ij} = 4$ and $s_{ij} = 2$. Values for s_{ii} and s_{ij} for various values of kL are plotted as shown in Figure 2.122.

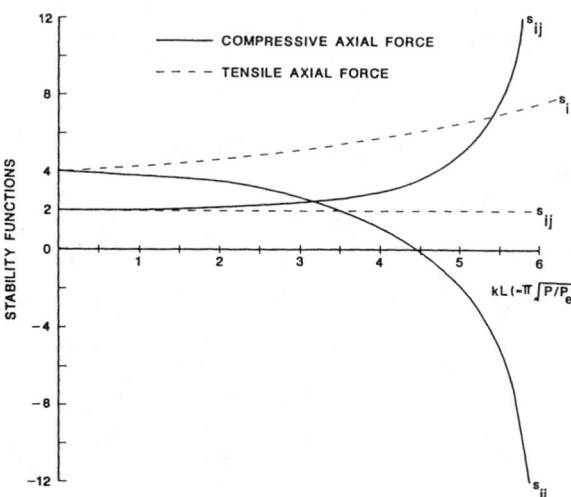

Figure 2.122 Plot of stability functions.

Equations 2.322 and 2.323 are valid if the following conditions are satisfied:

1. The beam is prismatic.
2. There is no relative joint displacement between the two ends of the member.

3. The member is continuous, i.e., there is no internal hinge or discontinuity in the member.
4. There is no in-span transverse loading on the member.
5. The axial force in the member is compressive.

If these conditions are not satisfied, some modifications to the slope-deflection equations are necessary.

Members Subjected to Sidesway

If there is a relative joint translation, Δ, between the member ends, as shown in Figure 2.123, the slope-deflection equations are modified as

$$
\begin{aligned}
M_A &= \frac{EI}{L}\left[s_{ii}\left(\theta_A - \frac{\Delta}{L}\right) + s_{ij}\left(\theta_b - \frac{\Delta}{L}\right)\right] \\
&= \frac{EI}{L}\left[s_{ii}\theta_A + s_{ij}\theta_B - (s_{ii} + s_{ij})\frac{\Delta}{L}\right] & (2.325)
\end{aligned}
$$

$$
\begin{aligned}
M_B &= \frac{EI}{L}\left[s_{ij}\left(\theta_A - \frac{\Delta}{L}\right) + s_{ii}\left(\theta_b - \frac{\Delta}{L}\right)\right] \\
&= \frac{EI}{L}\left[s_{ij}\theta_A + s_{ii}\theta_B - (s_{ii} + s_{ij})\frac{\Delta}{L}\right] & (2.326)
\end{aligned}
$$

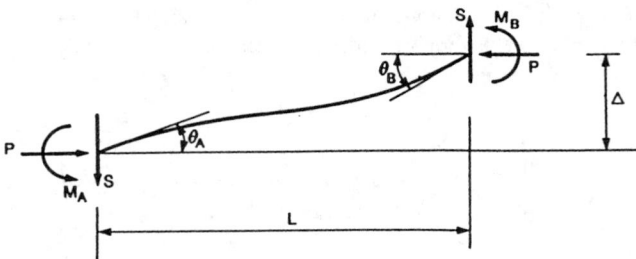

Figure 2.123 Beam-column subjected to end moments and sidesway.

Member with a Hinge at One End

If a hinge is present at B end of the member, the end moment there is zero, i.e.,

$$
M_B = \frac{EI}{L}\left(s_{ij}\theta_A + s_{ii}\theta_B\right) = 0 \tag{2.327}
$$

from which

$$
\theta_B = -\frac{s_{ij}}{s_{ii}}\theta_A \tag{2.328}
$$

Upon substituting Equation 2.328 into Equation 2.325, we have

$$
M_A = \frac{EI}{L}\left(s_{ii} - \frac{s_{ij}^2}{s_{ii}}\right)\theta_A \tag{2.329}
$$

If the member is hinged at A rather than at B, Equation 2.329 is still valid provided that the subscript A is changed to B.

Member with End Restraints

If the member ends are connected by two linear elastic springs, as in Figure 2.124, with spring constants, R_{kA} and R_{kB} at the A and B ends, respectively, the end rotations of the linear spring are M_A/R_{kA} and M_B/R_{kB}. If we denote the total end rotations at joints A and B by θ_A and θ_B,

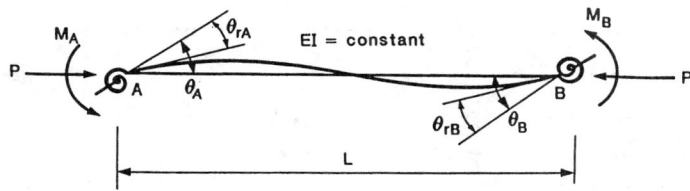

Figure 2.124   Beam-column with end springs.

respectively, then the member end rotations, with respect to its chord, will be $(\theta_A - M_A/R_{kA})$ and $(\theta_B - M_B/R_{kB})$. As a result, the slope-deflection equations are modified to

$$M_A = \frac{EI}{L}\left[s_{ii}\left(\theta_A - \frac{M_A}{R_{kA}}\right) + s_{ij}\left(\theta_B - \frac{M_B}{R_{kB}}\right)\right] \tag{2.330}$$

$$M_B = \frac{EI}{L}\left[s_{ij}\left(\theta_A - \frac{M_A}{R_{kA}}\right) + s_{jj}\left(\theta_B - \frac{M_B}{R_{kB}}\right)\right] \tag{2.331}$$

Solving Equations 2.330 and 2.331 simultaneously for M_A and M_B gives

$$M_A = \frac{EI}{LR^*}\left[\left(s_{ii} + \frac{EIs_{ii}^2}{LR_{kB}} - \frac{EIs_{ij}^2}{LR_{kB}}\right)\theta_A + s_{ij}\theta_B\right] \tag{2.332}$$

$$M_B = \frac{EI}{LR^*}\left[s_{ij}\theta_A + \left(s_{ii} + \frac{EIs_{ii}^2}{LR_{kA}} - \frac{EIs_{ij}^2}{LR_{kA}}\right)\theta_B\right] \tag{2.333}$$

where

$$R^* = \left(1 + \frac{EIs_{ii}}{LR_{kA}}\right)\left(1 + \frac{EIs_{ii}}{LR_{kB}}\right) - \left(\frac{EI}{L}\right)^2\frac{s_{ij}^2}{R_{kA}R_{kB}} \tag{2.334}$$

In writing Equations 2.332 to 2.333, the equality $s_{jj} = s_{ii}$ has been used. Note that as R_{kA} and R_{kB} approach infinity, Equations 2.332 and 2.333 reduce to Equations 2.321 and 2.322, respectively.

Member with Transverse Loading

For members subjected to transverse loading, the slope-deflection Equations 2.321 and 2.322 can be modified by adding an extra term for the fixed-end moment of the member.

$$M_A = \frac{EI}{L}\left(s_{ii}\theta_A + s_{ij}\theta_B\right) + M_{FA} \tag{2.335}$$

$$M_B = \frac{EI}{L}\left(s_{ij}\theta_A + s_{jj}\theta_B\right) + M_{FB} \tag{2.336}$$

Table 2.5 give the expressions for the fixed-end moments of five commonly encountered cases of transverse loading. Readers are referred to [14, 15] for more details.

TABLE 2.5 Beam-Column Fixed-End Moments [15] $\left(u = kL/2 = \dfrac{L}{2}\sqrt{\dfrac{P}{EI}} \right)$

Case	Fixed-End Moments
	$M_{FA} = \dfrac{QL}{8} \left[\dfrac{2(1-\cos u)}{u \sin u} \right]$ $M_{FB} = - M_{FA}$
	$M_{FA} = \dfrac{wL^2}{12} \left[\dfrac{3(\tan u - u)}{u^2 \tan u} \right]$ $M_{FB} = - M_{FA}$
	$M_{FA} = \dfrac{QL}{d} \left[\dfrac{2ub}{L} \cos 2u - 2u \cos \dfrac{2ub}{L} - \sin 2u \right.$ $\left. + \sin \dfrac{2ua}{L} + \sin \dfrac{2ub}{L} + \dfrac{2ua}{L} \right]$ $M_{FB} = - \dfrac{QL}{d} \left[\dfrac{2ua}{L} \cos 2u - 2u \cos \dfrac{2ua}{L} - \sin 2u \right.$ $\left. + \sin \dfrac{2ub}{L} + \sin \dfrac{2ua}{L} + \dfrac{2ub}{L} \right]$ where $d = 2u(2 - 2 \cos 2u - 2u \sin 2u)$
	$M_{FA} = \dfrac{wL^2}{8u^2 e} \left[(2u \cosec 2u - 1)\left(\dfrac{2ub}{L} - \sin \dfrac{2ub}{L}\right) \right.$ $\left. + (\tan u)\left(1 - \cos \dfrac{2ub}{L} - \dfrac{2u^2 b^2}{L^2}\right) \right]$ $M_{FB} = \dfrac{-wL^2}{8u^2 e} \left[(2u \cot 2u - 1)\left(\dfrac{2ub}{L} - \sin \dfrac{2ub}{L}\right) \right.$ $+ (\tan u)\left(1 - \cos \dfrac{2ub}{L} + \dfrac{2u^2 b^2}{L^2}\right)$ $\left. -2u \left(1 - \cos \dfrac{2ub}{L}\right) \right]$ where $e = \tan u - u$
	$M_{FA} = \dfrac{M_o}{2e} \left[(2u \cosec 2u - 1) \sin \dfrac{2ub}{L} \right.$ $\left. - (\tan u)\left(1 - \cos \dfrac{2ub}{L}\right) \right]$ $M_{FB} = \dfrac{M}{2e} \left[(1 - 2u \cot 2u) \sin \dfrac{2ub}{L} \right.$ $- (\tan u)\left(1 + \cos \dfrac{2ub}{L}\right)$ $\left. + 2u \cos \dfrac{2ub}{L} \right]$ where $e = \tan u - u$

Member with Tensile Axial Force

For members subjected to tensile force, Equations 2.321 and 2.322 can be used provided that the stability functions are redefined as

$$s_{ii} = s_{jj} = \frac{(kL)^2 \cosh kL - kL \sinh kL}{2 - 2\cosh kL + kL \sinh kL} \tag{2.337}$$

$$s_{ij} = s_{ji} = \frac{kL \sinh kL - (kL)^2}{2 - 2\cosh kL + kL \sinh kL} \tag{2.338}$$

Member Bent in Single Curvature with $\theta_B = -\theta_A$

For the member bent in single curvature in which $\theta_B = -\theta_A$, the slope-deflection equations reduce to

$$M_A = \frac{EI}{L}\left(s_{ii} - s_{ij}\right)\theta_A \tag{2.339}$$

$$M_B = -M_A \tag{2.340}$$

Member Bent in Double Curvature with $\theta_B = \theta_A$

For the member bent in double curvature such that $\theta_B = \theta_A$, the slope-deflection equations become

$$M_A = \frac{EI}{L}\left(s_{ii} + s_{ij}\right)\theta_A \tag{2.341}$$

$$M_B = M_A \tag{2.342}$$

2.12.5 Second-Order Elastic Analysis

The basis of the formulation is that the beam-column element is prismatic and initially straight. An update-Lagrangian approach [7] is assumed. There are two methods to incorporate second-order effects: (1) the stability function approach and (2) the geometric stiffness (or finite element) approach. The stability function approach is based directly on the governing differential equations of the problem as described in Section 2.12.4, whereas the stiffness approach is based on an assumed cubic polynomial variation of the transverse displacement along the element length. Therefore, the stability function approach is more exact in terms of representing the member stability behavior. However, the geometric stiffness approach is easier to implement for three-dimensional analysis.

For either of these approaches, the linearized element stiffness equations may be expressed in either incremental or total force and displacement forms as

$$[K]\{d\} + \{r_f\} = \{r\} \tag{2.343}$$

where $[K]$ is the element stiffness matrix, $\{d\} = \{d_1, d_2, ..., d_6\}^T$ is the element nodal displacement vector, $\{r_f\} = \{r_{f1}, r_{f2}, ..., r_{f6}\}^T$ is the element fixed-end force vector due to the presence of in-span loading, and $\{r\} = \{r_1, r_2, ..., r_6\}^T$ is the nodal force vector as shown in Figure 2.125. If the stability function approach is employed, the stiffness matrix of a two-dimensional beam-column element may be written as

$$[K] = \frac{EI}{L}\begin{bmatrix} \frac{A}{I} & 0 & 0 & -\frac{A}{I} & 0 & 0 \\ & \frac{2(S_{ii}+S_{ij})-(kL)^2}{L^2} & \frac{S_{ii}+S_{ij}}{L} & 0 & \frac{-2(S_{ii}+S_{ij})+(kL)^2}{L^2} & \frac{S_{ii}+S_{ij}}{L} \\ & & S_{ii} & 0 & \frac{-(S_{ii}+S_{ij})}{L} & S_{ij} \\ & & & \frac{A}{I} & 0 & 0 \\ & & & & \frac{2(S_{ii}+S_{ij})-(kL)^2}{L^2} & \frac{-(S_{ii}+S_{ij})}{L} \\ & \text{sym.} & & & & S_{ii} \end{bmatrix} \tag{2.344}$$

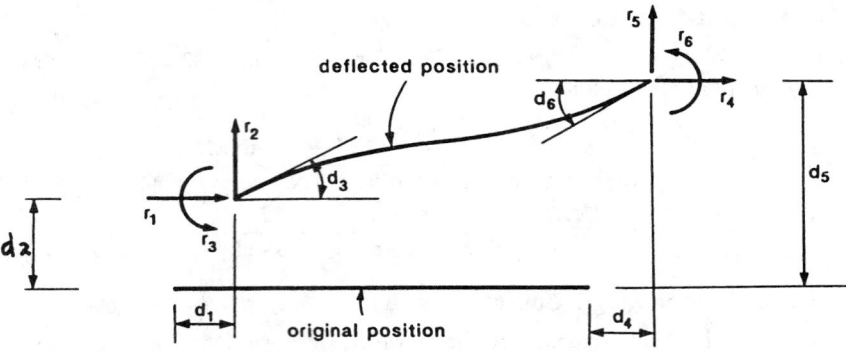

Figure 2.125   Nodal displacements and forces of a beam-column element.

where S_{ii} and S_{ij} are the member stiffness coefficients obtained from the elastic beam-column stability functions [14]. These coefficient may be expressed as

$$S_{ii} = \begin{cases} \frac{kL\sin(kL)-(kL)^2\cos(kL)}{2-2\cos(kL)-kL\sin(kL)} & \text{for } P < 0 \\ \frac{(kL)^2\cosh(kL)-kL\sinh(kL)}{2-2\cosh(kL)+kL\sin(kL)} & \text{for } P > 0 \end{cases} \tag{2.345}$$

$$S_{ij} = \begin{cases} \frac{(kL)^2-kL\sin(kL)}{2-2\cos(kL)-\rho\sin(kL)} & \text{for } P < 0 \\ \frac{kL\sinh(kL)-(kL)^2}{2-2\cosh(kL)+\rho\sinh(kL)} & \text{for } P > 0 \end{cases} \tag{2.346}$$

where $kL = L\sqrt{\frac{P}{EI}}$, and P is positive in compression and negative in tension.

The fixed-end force vector r_f is a 6×1 matrix which can be computed from the in-span loading in the beam-column. If curvature shortening is ignored, $r_{f1} = r_{f4} = 0$, $r_{f3} = M_{FA}$, and $r_{f6} = M_{FB}$. M_{FA} and M_{FB} can be obtained from Table 2.5 for different in-span loading conditions. r_{f2} and r_{f5} can be obtained from equilibrium of forces.

If the axial force in the member is small, Equation 2.344 can be simplified by ignoring the higher order terms of the power series expansion of the trigonometric functions. The resulting element stiffness matrix becomes:

$$[K] = \frac{EI}{L}\begin{bmatrix} \frac{A}{I} & 0 & 0 & -\frac{A}{I} & 0 & 0 \\ & \frac{12}{L^2} & \frac{6}{L} & 0 & \frac{-12}{L^2} & \frac{6}{L} \\ & & 4 & 0 & \frac{-6}{L} & 2 \\ & & & \frac{A}{I} & 0 & 0 \\ & & & & \frac{12}{L^2} & \frac{-6}{L} \\ & \text{sym.} & & & & 4 \end{bmatrix} + P\begin{bmatrix} 0 & 0 & 0 & 0 & 0 & 0 \\ & \frac{6}{5L} & \frac{1}{10} & 0 & \frac{-6}{5L} & \frac{1}{10} \\ & & \frac{2L}{15} & 0 & \frac{-1}{10} & \frac{-L}{30} \\ & & & 0 & 0 & 0 \\ & & & & \frac{6}{5L} & \frac{-1}{10} \\ & \text{sym.} & & & & \frac{2L}{15} \end{bmatrix} \tag{2.347}$$

The first term on the right is the first-order elastic stiffness matrix, and the second term is the geometric stiffness matrix, which accounts for the effect of axial force on the bending stiffness of the member. Detailed discussions on the limitation of the geometric stiffness approach vs. the stability function approach are given in White et al. [66].

2.12.6   Modifications to Account for Plastic Hinge Effects

There are two commonly used approaches for representing plastic hinge behavior in a second-order elastic-plastic hinge formulation [19]. The most basic approach is to model the plastic hinge behavior as a "real hinge" for the purpose of calculating the element stiffness. The change in moment capacity due to the change in axial force can be accommodated directly in the numerical formulation. The change in moment is determined in the force recovery at each solution step such that, for continued plastic loading, the new force point is positioned at the strength surface at the current value of the axial force. A detailed description of these procedures is given by Chen and Lui [15], Chen et al. [19], and Lee and Basu [35], among others.

Alternatively, the elastic-plastic hinge model may be formulated based on the "extending and contracting" plastic hinge model. The plastic hinge can rotate and extend/contract for plastic loading and axial force. The formulation can follow the force-space plasticity concept using the normality flow rule relative to the cross-section surface strength [13]. Formal derivations of the beam-column element based on this approach have been presented by Porter and Powell [46] and Orbison et al. [45], among others.

2.12.7 Modification for End Connections

The moment rotation relationship of the beam-column with end connections at both ends can be expressed as (Equations 2.332 and 2.333):

$$M_A = \frac{EI}{L}\left[s_{ii}^*\theta_A + s_{ij}^*\theta_B\right] \tag{2.348}$$

$$M_B = \frac{EI}{L}\left[s_{ij}^*\theta_A + s_{jj}^*\theta_B\right] \tag{2.349}$$

where

$$S_{ii}^* = \frac{S_{ii} + \frac{EIS_{ii}^2}{LR_{kB}} - \frac{EIS_{ij}^2}{LR_{kB}}}{\left[1+\frac{EIS_{ii}}{LR_{kA}}\right]\left[1+\frac{EIS_{jj}}{LR_{kB}}\right] - \left[\frac{EI}{L}\right]^2 \frac{S_{ij}^2}{R_{kA}R_{kB}}} \tag{2.350}$$

$$S_{jj}^* = \frac{S_{ii} + \frac{EIS_{ii}^2}{LR_{kA}} - \frac{EIS_{ij}^2}{LR_{kA}}}{\left[1+\frac{EIS_{ii}}{LR_{kA}}\right]\left[1+\frac{EIS_{jj}}{LR_{kB}}\right] - \left[\frac{EI}{L}\right]^2 \frac{S_{ij}^2}{R_{kA}R_{kB}}} \tag{2.351}$$

and

$$S_{ij}^* = \frac{S_{ij}}{\left[1+\frac{EIS_{ii}}{LR_{kA}}\right]\left[1+\frac{EIS_{jj}}{LR_{kB}}\right] - \left[\frac{EI}{L}\right]^2 \frac{S_{ij}^2}{R_{kA}R_{kB}}} \tag{2.352}$$

The member stiffness relationship can be written in terms of six degrees of freedom beam-column element shown in Figure 2.126 as

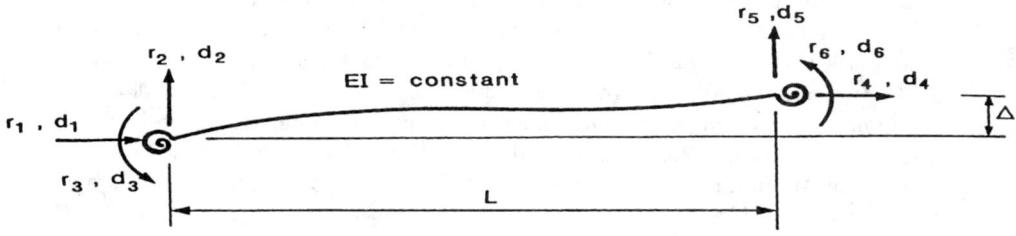

Figure 2.126   Nodal displacements and forces of a beam-column with end connections.

$$
\begin{pmatrix} r_1 \\ r_2 \\ r_3 \\ r_4 \\ r_5 \\ r_6 \end{pmatrix} = \frac{EI}{L}
$$

$$
\begin{bmatrix}
\frac{A}{I} & 0 & 0 & -\frac{A}{I} & 0 & 0 \\[4pt]
 & \frac{S_{ii}^{*}+2S_{ij}^{*}+S_{jj}^{*}-(kL)^2}{L^2} & \frac{S_{ii}^{*}+S_{ij}^{*}}{L} & 0 & \frac{-(S_{ii}^{*}+2S_{ij}^{*}+S_{jj}^{*})+(kL)^2}{L^2} & \frac{S_{ij}^{*}+S_{jj}^{*}}{L} \\[4pt]
 & & S_{ii}^{*} & 0 & \frac{-(S_{ii}^{*}+S_{ij}^{*})}{L} & S_{ij}^{*} \\[4pt]
 & & & \frac{A}{I} & 0 & 0 \\[4pt]
 & & & & \frac{S_{ii}^{*}+2S_{ij}^{*}+S_{jj}^{*}-(kL)^2}{L^2} & \frac{-(S_{ij}^{*}+S_{jj}^{*})}{L} \\[4pt]
 & \text{sym.} & & & & S_{jj}^{*}
\end{bmatrix}
$$

$$
\begin{pmatrix} d_1 \\ d_2 \\ d_3 \\ d_4 \\ d_5 \\ d_6 \end{pmatrix}
$$

(2.353)

2.12.8   Second-Order Refined Plastic Hinge Analysis

The main limitation of the conventional elastic-plastic hinge approach is that it over-predicts the strength of columns that fail by inelastic flexural buckling. The key reason for this limitation is the modeling of a member by a perfect elastic element between the plastic hinge locations. Furthermore, the elastic-plastic hinge model assumes that material behavior changes abruptly from the elastic state to the fully yielded state. The element under consideration exhibits a sudden stiffness reduction upon the formation of a plastic hinge. This approach, therefore, overestimates the stiffness of a member loaded into the inelastic range [42, 64, 65, 66]. This leads to further research and development of an alternative method called the *refined plastic hinge approach.* This approach is based on the following improvements to the elastic-plastic hinge model:

1. A column tangent-modulus model E_t is used in place of the elastic modulus E to represent the distributed plasticity along the length of a member due to axial force effects. The member inelastic stiffness, represented by the member axial and bending rigidities $E_t A$ and $E_t I$, is assumed to be the function of axial load only. In other words, $E_t A$ and $E_t I$ can be thought of as the properties of an effective core of the section, considering column action only. The tangent modulus captures the effect of early yielding in the cross-section due to residual stresses, which was believed to be the cause of the low strength of inelastic column buckling. The tangent modulus approach also has been utilized in previous work by Orbison et al. [45], Liew [38], and White et al. [66] to improve the accuracy of the elastic-plastic hinge approach for structures in which members are subjected to large axial forces.

2. Distributed plasticity effects associated with flexure are captured by gradually degrading the member stiffness at the plastic hinge locations as yielding progresses under increasing load as the cross-section strength is approached. Several models of this type have been proposed in recent literature based on extensions to the elastic-plastic hinge approach [47] as well as the tangent modulus inelastic hinge approach [41, 42, 66]. The rationale of modeling stiffness degradation associated with both axial and flexural actions is that the tangent modulus model represents the column strength behavior in the limit of pure axial compression, and the plastic hinge stiffness degradation model represents the beam behavior in pure bending, thus the combined effects of these two approaches should also satisfy the cases in which the member is subjected to combined axial compression and bending.

It has been shown that with the above two improvements, the refined plastic hinge model can be used with sufficient accuracy to provide a quantitative assessment of a member's performance up to failure. Detailed descriptions of the method and discussion of results generated by the method are given in White et al. [66] and Chen et al. [19].

2.12.9 Second-Order Plastic Zone Analysis

Plastic-zone analyses can be classified into two main types, namely 3-D shell element and 2-D beam-column approaches. In the 3-D plastic-zone analysis, the structure is modeled using a large number of finite 3-D shell elements, and the elastic constitutive matrix, in the usual incremental stress-strain relations, is replaced by an elastic-plastic constitutive matrix once yielding is detected. This analysis approach typically requires numerical integration for the evaluation of the stiffness matrix. Based on a deformation theory of plasticity, the combined effects of normal and shear stresses may be accounted for. The 3-D spread-of-plasticity analysis is computational intensive and best suited for analyzing small-scale structures.

The second-approach for plastic-zone analysis is based on the use of beam-column theory, in which the member is discretized into many beam-column segments, and the cross-section of each segment is further subdivided into a number of fibers. Inelasticity is typically modeled by the consideration of normal stress only. When the computed stresses at the centroid of any fibers reach the uniaxial normal strength of the material, the fiber is considered yielded. Compatibility is treated by assuming that full continuity is retained throughout the volume of the structure in the same manner as for elastic range calculations. Most of the plastic-zone analysis methods developed are meant for planar (2-D) analysis [18, 59, 62]. Three-dimensional plastic-zone techniques are also available involving various degrees of refinements [60, 63].

A plastic-zone analysis, which includes the spread of plasticity, residual stresses, initial geometric imperfections, and any other significant second-order behavioral effects, is often considered to be an exact analysis method. Therefore, when this type of analysis is employed, the checking of member interaction equations is not required. However, in reality, some significant behavioral effects such as joint and connection performances tend to defy precise numerical and analytical modeling. In such cases, a simpler method of analysis that adequately captures the inelastic behavior would be sufficient for engineering application. Second-order plastic hinge based analysis is still the preferred method for advanced analysis of large-scale steel frames.

2.12.10 Three-Dimensional Frame Element

The two-dimensional beam-column formulation can be extended to a three-dimensional space frame element by including additional terms due to shear force, bending moment, and torsion. The following stiffness equation for a space frame element has been derived by Yang and Kuo [67] by referring to Figure 2.127:

$$[k_e]\{d\} + [k_g]\{d\} = \{^2f\} - \{^1f\} \qquad (2.354)$$

where

$$\{d\}^T = \{d_1, d_2, ..., d_{12}\} \qquad (2.355)$$

is the displacement vector which consists of three translations and three rotations at each node, and

$$\{^if\}^T = \{^if_1, {}^if_2, ..., {}^if_{12}\} \qquad i = 1, 2 \qquad (2.356)$$

are the force vectors which consist of the corresponding nodal forces at configurations $i = 1$ and $i = 2$, respectively.

The physical interpretation of Equation 2.356 is as follows: By increasing the nodal forces acting on the element from $\{^1f\}$ to $\{^2f\}$, further deformations $\{d\}$ may occur with the element, resulting in the

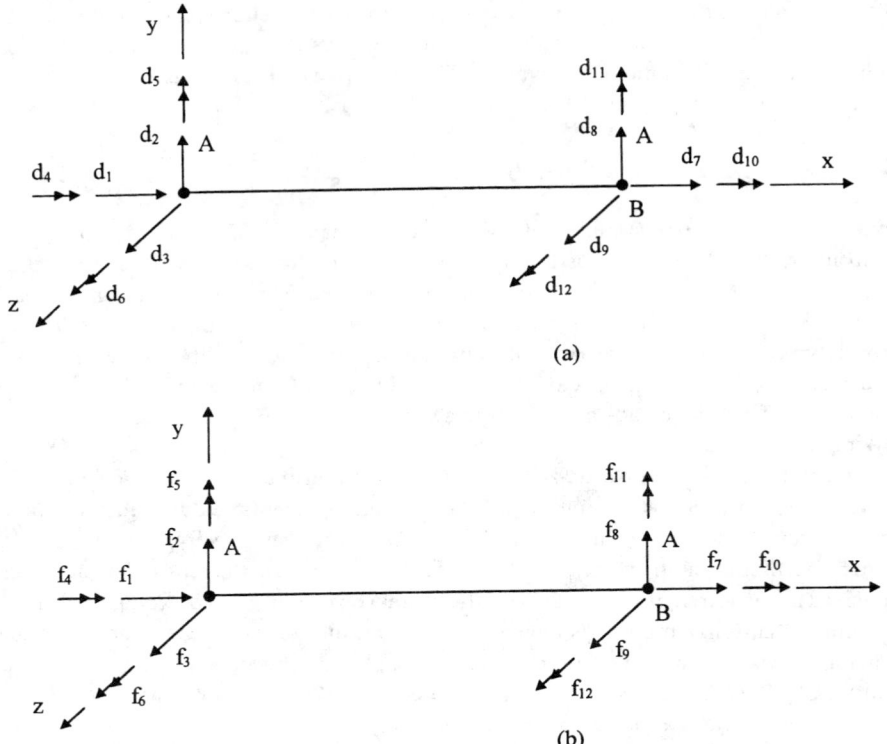

Figure 2.127 Three-dimensional frame element: (a) nodal degrees of freedom and (b) nodal forces.

motion of the element from configuration associated with the forces $\{^1 f\}$ to the new configuration associated with $\{^2 f\}$. During this process of deformation, the increments in the nodal forces, i.e., $\{^2 f\} - \{^1 f\}$, will be resisted not only by the elastic actions generated by the elastic stiffness matrix $[k_e]$ but also by the forces induced by the change in geometry as represented by the geometric stiffness matrix $[k_g]$.

The only assumption with the incremental stiffness equation is that the strains occurring with each incremental step should be small so that the approximations implied by the incremental constitutive law are not violated.

The elastic stiffness matrix $[K_e]$ for the space frame element, which has a 12 x 12 dimension, can be derived as follows:

$$[k] = \begin{bmatrix} [k_1] & [k_2] \\ [k_2]^T & [k_3] \end{bmatrix} \tag{2.357}$$

where the submatrices are

$$[k_1] = \begin{bmatrix} \frac{EA}{L} & 0 & 0 & 0 & 0 & 0 \\ 0 & \frac{12EI_z}{L^3} & 0 & 0 & 0 & \frac{6EI_z}{L^2} \\ 0 & 0 & \frac{12EI_y}{L^3} & 0 & -\frac{6EI_y}{L^2} & 0 \\ 0 & 0 & 0 & \frac{GJ}{L} & 0 & 0 \\ 0 & 0 & 0 & 0 & \frac{4EI_y}{L} & 0 \\ 0 & 0 & 0 & 0 & 0 & \frac{4EI_z}{L} \end{bmatrix} \tag{2.358}$$

$$[k_2] = \begin{bmatrix} -\frac{EA}{L} & 0 & 0 & 0 & 0 & 0 \\ 0 & -\frac{12EI_z}{L^3} & 0 & 0 & 0 & \frac{6EI_z}{L^2} \\ 0 & 0 & -\frac{12EI_y}{L^3} & 0 & -\frac{6EI_y}{L^2} & 0 \\ 0 & 0 & 0 & -\frac{GJ}{L} & 0 & 0 \\ 0 & 0 & \frac{6EI_y}{L^2} & 0 & \frac{2EI_y}{L} & 0 \\ 0 & -\frac{6EI_z}{L^2} & 0 & 0 & 0 & \frac{2EI_z}{L} \end{bmatrix} \tag{2.359}$$

$$[k_3] = \begin{bmatrix} \frac{EA}{L} & 0 & 0 & 0 & 0 & 0 \\ 0 & \frac{12EI_z}{L^3} & 0 & 0 & 0 & -\frac{6EI_z}{L^2} \\ 0 & 0 & \frac{12EI_y}{L^3} & 0 & \frac{6EI_y}{L^2} & 0 \\ 0 & 0 & 0 & \frac{GJ}{L} & 0 & 0 \\ 0 & 0 & 0 & 0 & \frac{4EI_y}{L} & 0 \\ 0 & 0 & 0 & 0 & 0 & \frac{4EI_z}{L} \end{bmatrix} \tag{2.360}$$

where I_x, I_y, and I_z are the moment of inertia about x-, y-, and z-axes; L = member length, E = modulus of elasticity, A = cross-sectional area, G = shear modulus, and J = torsional stiffness.

The geometric stiffness matrix for a three-dimensional space frame element can be given as follows:

$$[k_g] = \begin{bmatrix} a & 0 & 0 & 0 & -d & -e & -a & 0 & 0 & 0 & -n & -o \\ & b & 0 & d & g & k & 0 & -b & 0 & n & -g & k \\ & & c & e & h & g & 0 & 0 & -c & o & -h & -g \\ & & & f & i & l & 0 & -d & -e & -f & -i & -l \\ & & & & j & 0 & d & -g & h & -i & p & -q \\ & & & & & m & e & -k & -g & -l & q & r \\ & & & & & & a & 0 & 0 & 0 & n & o \\ & & & & & & & b & 0 & -n & g & -k \\ & & & & & & & & c & -o & h & g \\ & & & & & & & & & f & i & l \\ & \text{sym.} & & & & & & & & & j & o \\ & & & & & & & & & & & m \end{bmatrix} \tag{2.361}$$

where

$$a = -\frac{f_6 + f_{12}}{L^2}, \quad b = \frac{6f_7}{5L}, \quad c = -\frac{f_5 + f_{11}}{L^2}, \quad d = \frac{f_5}{L}, \quad e = \frac{f_6}{L}, \quad f = \frac{f_7 J}{AL},$$

$$g = \frac{f_{10}}{L}, \quad h = -\frac{f_7}{10}, \quad i = \frac{f_6 + f_{12}}{6}, \quad j = \frac{2f_7 L}{15}, \quad k = -\frac{f_5 + f_{11}}{6}, \quad l = \frac{f_{11}}{L},$$

$$m = \frac{f_{12}}{L}, \quad n = -\frac{f_7 L}{30}, \quad o = -\frac{f_{10}}{2}.$$

Further details can be obtained from [67].

2.13 Structural Dynamic

2.13.1 Equation of Motion

The essential physical properties of a linearly elastic structural system subjected to external dynamic loading are its mass, stiffness properties, and energy absorption capability or damping. The principle of dynamic analysis may be illustrated by considering a simple single-story structure as shown in Figure 2.128. The structure is subjected to a time-varying force $f(t)$. k is the spring constant that relates the lateral story deflection x to the story shear force, and the dash pot relates the damping force to the velocity by a damping coefficient c. If the mass, m, is assumed to concentrate at the

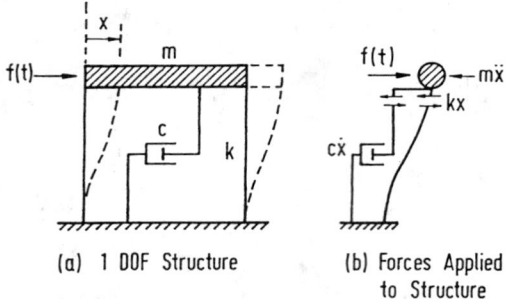

(a) 1 DOF Structure   (b) Forces Applied
                                   to Structure

Figure 2.128   (a) One DOF structure; (b) forces applied to structures.

beam, the structure becomes a single-degree-of-freedom (SDOF) system. The equation of motion
of the system may be written as:

$$m\ddot{x} + c\dot{x} + kx = f(t) \tag{2.362}$$

Various solutions to Equation 2.362 can give an insight into the behavior of the structure under
dynamic situation.

2.13.2   Free Vibration

In this case the system is set to motion and allowed to vibrate in the absence of applied force $f(t)$.
Letting $f(t) = 0$, Equation 2.362 becomes

$$m\ddot{x} + c\dot{x} + kx = 0 \tag{2.363}$$

Dividing Equation 2.363 by the mass m, we have

$$\ddot{x} + 2\xi\omega x + \omega^2 x = 0 \tag{2.364}$$

where

$$2\xi\omega = \frac{c}{m} \quad \text{and} \quad \omega^2 = \frac{k}{m} \tag{2.365}$$

The solution to Equation 2.364 depends on whether the vibration is damped or undamped.

Case 1: Undamped Free Vibration
   In this case, $c = 0$, and the solution to the equation of motion may be written as:

$$x = A \sin \omega t + B \cos \omega t \tag{2.366}$$

where $\omega = \sqrt{k/m}$ is the circular frequency. A and B are constants that can be determined by the
initial boundary conditions. In the absence of external forces and damping the system will vibrate
indefinitely in a repeated cycle of vibration with an amplitude of

$$X = \sqrt{A^2 + B^2} \tag{2.367}$$

and a natural frequency of

$$f = \frac{\omega}{2\pi} \tag{2.368}$$

The corresponding natural period is

$$T = \frac{2\pi}{\omega} = \frac{1}{f} \tag{2.369}$$

The undamped free vibration motion as described by Equation 2.366 is shown in Figure 2.129.

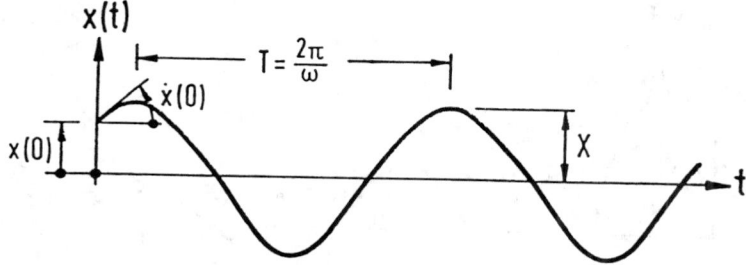

Figure 2.129 Response of undamped free vibration.

Case 2: Damped Free Vibration

If the system is not subjected to applied force and damping is presented, the corresponding solution becomes

$$x = A \exp(\lambda_1 t) + B \exp(\lambda_2 t) \tag{2.370}$$

where

$$\lambda_1 = \omega\left[-\xi + \sqrt{\xi^2 - 1}\right] \tag{2.371}$$

and

$$\lambda_2 = \omega\left[-\xi - \sqrt{\xi^2 - 1}\right] \tag{2.372}$$

The solution of Equation 2.370 changes its form with the value of ξ defined as

$$\xi = \frac{c}{2\sqrt{mk}} \tag{2.373}$$

If $\xi^2 < 1$ the equation of motion becomes

$$x = \exp(-\xi\omega t)(A \cos \omega_d t + B \sin \omega_d t) \tag{2.374}$$

where ω_d is the damped angular frequency defined as

$$\omega_d = \sqrt{(1 - \xi^2)}\omega \tag{2.375}$$

For most building structures ξ is very small (about 0.01) and therefore $\omega_d \approx \omega$. The system oscillates about the neutral position as the amplitude decays with time t. Figure 2.130 illustrates an example of such motion. The rate of decay is governed by the amount of damping present.

If the damping is large, then oscillation will be prevented. This happens when $\xi^2 > 1$ and the behavior is referred to as *overdamped*. The motion of such behavior is shown in Figure 2.131.

Damping with $\xi^2 = 1$ is called critical damping. This is the case where minimum damping is required to prevent oscillation and the critical damping coefficient is given as

$$c_{cr} = 2\sqrt{km} \tag{2.376}$$

where k and m are the stiffness and mass of the system.

The degree of damping in the structure is often expressed as a proportion of the critical damping value. Referring to Equations 2.373 and 2.376, we have

$$\xi = \frac{c}{c_{cr}} \tag{2.377}$$

ξ is called the critical damping ratio.

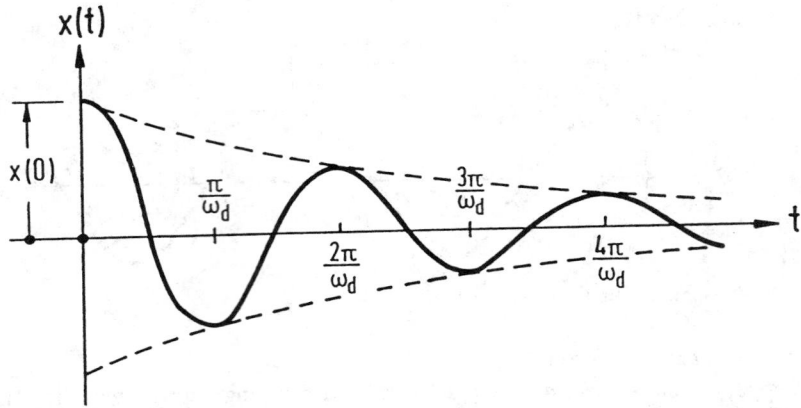

Figure 2.130   Response of damped free vibration.

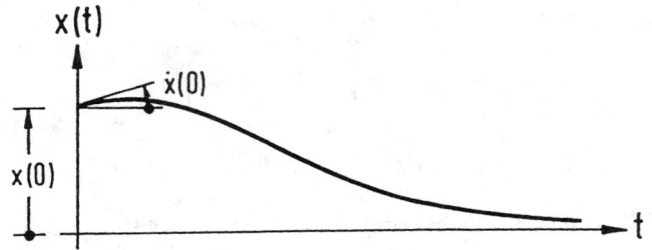

Figure 2.131   Response of free vibration with critical damping.

2.13.3 Forced Vibration

If a structure is subjected to a sinusoidal motion such as a ground acceleration of $\ddot{x} = F \sin \omega_f t$, it will oscillate and after some time the motion of the structure will reach a steady state. For example, the equation of motion due to the ground acceleration (from Equation 2.364) is

$$\ddot{x} + 2\xi \omega \dot{x} + \omega^2 x = -F \sin \omega_f t \tag{2.378}$$

The solution to the above equation consists of two parts; the complimentary solution given by Equation 2.366 and the particular solution. If the system is damped, oscillation corresponding to the complementary solution will decay with time. After some time, the motion will reach a steady state and the system will vibrate at a constant amplitude and frequency. This motion, which is called force vibration, is described by the particular solution expressed as

$$x = C_1 \sin \omega_f t + C_2 \cos \omega_f t \tag{2.379}$$

It can be observed that the steady force vibration occurs at the frequency of the excited force, ω_f, not the natural frequency of the structure, ω.

Substituting Equation 2.379 into Equation 2.378, the displacement amplitude can be shown to be

$$X = -\frac{F}{\omega^2} \frac{1}{\sqrt{\left[\left\{ 1 - \left(\frac{\omega_f}{\omega} \right)^2 \right\}^2 + \left(\frac{2\xi \omega_f}{\omega} \right)^2 \right]}} \tag{2.380}$$

The term $-F/\omega^2$ is the static displacement caused by the force due to the inertia force. The ratio of the response amplitude relative to the static displacement $-F/\omega^2$ is called the dynamic displacement amplification factor, D, given as

$$D = \frac{1}{\sqrt{\left[\left\{1 - \left(\frac{\omega_f}{\omega}\right)^2\right\}^2 + \left(\frac{2\xi\omega_f}{\omega}\right)^2\right]}} \tag{2.381}$$

The variation of the magnification factor with the frequency ratio ω_f/ω and damping ratio ξ is shown in Figure 2.132.

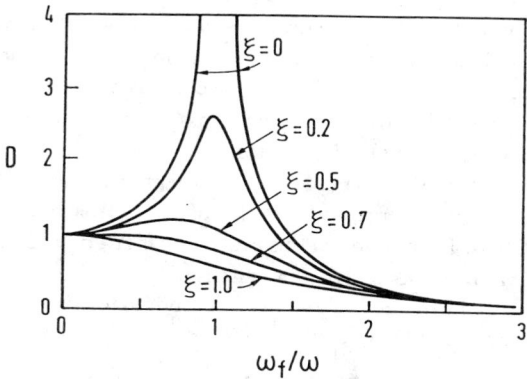

Figure 2.132 Variation of dynamic amplification factor with frequency ratio.

When the dynamic force is applied at a frequency much lower than the natural frequency of the system ($\omega_f/\omega \ll 1$), the response is quasi-static. The response is proportional to the stiffness of the structure, and the displacement amplitude is close to the static deflection.

When the force is applied at a frequency much higher than the natural frequency ($\omega_f/\omega \gg 1$), the response is proportional to the mass of the structure. The displacement amplitude is less than the static deflection ($D < 1$).

When the force is applied at a frequency close to the natural frequency, the displacement amplitude increases significantly. The condition at which $\omega_f/\omega = 1$ is known as resonance.

Similarly, the ratio of the acceleration response relative to the ground acceleration may be expressed as

$$D_a = \left|\frac{\ddot{x} + \ddot{x}_g}{\ddot{x}_g}\right| = \sqrt{\frac{1 + \left(\frac{2\xi\omega_f}{\omega}\right)^2}{\left[\left\{1 - \left(\frac{\omega_f}{\omega}\right)^2\right\}^2 + \left(\frac{2\xi\omega_f}{\omega}\right)^2\right]}} \tag{2.382}$$

D_a is called the dynamic acceleration magnification factor.

2.13.4 Response to Suddenly Applied Load

Consider the spring-mass damper system of which a load P_o is applied suddenly. The differential equation is given by

$$M\ddot{x} + c\dot{x} + kx = P_o \tag{2.383}$$

If the system is started at rest, the equation of motion is

$$x = \frac{P_o}{k}\left[1 - \exp(-\xi\omega t)\left\{\cos\omega_d t + \frac{\xi\omega}{\omega_d}\sin\omega_d t\right\}\right] \tag{2.384}$$

If the system is undamped, then $\xi = 0$ and $\omega_d = \omega$, and we have

$$x = \frac{P_o}{k}[1 - \cos \omega_d t] \tag{2.385}$$

The maximum displacement is $2(P_o/k)$ corresponding to $\cos \omega_d t = -1$. Since P_o/k is the maximum static displacement, the dynamic amplification factor is equal to 2. The presence of damping would naturally reduce the dynamic amplification factor and the force in the system.

2.13.5 Response to Time-Varying Loads

Some forces and ground motions that are encountered in practice are rather complex in nature. In general, numerical analysis is required to predict the response of such effects, and the finite element method is one of the most common techniques to be employed in solving such problems.

The evaluation of responses due to time-varying loads can be carried out using the Piecewise Exact Method. In using this method, the loading history is divided into small time intervals. Between these points, it is assumed that the slope of the load curve remains constant. The entire load history is represented by a piecewise linear curve, and the error of this approach can be minimize by reducing the length of the time steps. Description of this procedure is given in [21].

Other techniques employed include Fourier analysis of the forcing function followed by solution for Fourier components in the frequency domain. For random forces, random vibration theory and spectrum analysis may be used [25, 61].

2.13.6 Multiple Degree Systems

In multiple degree systems, an independent differential equation of motion can be written for each degree of freedom. The nodal equations of a multiple degree system consisting of n degrees of freedom may be written as

$$[m]\{\ddot{x}\} + [c]\{\dot{x}\} + [k]\{x\} = \{F(t)\} \tag{2.386}$$

where $[m]$ is a symmetrical $n \times n$ matrix of mass, $[c]$ is a symmetrical $n \times n$ matrix of damping coefficient, and $\{F(t)\}$ is the force vector which is zero in the case of free vibration.

Consider a system under free vibration without damping. The general solution of Equation 2.386 is assumed in the form of

$$\begin{Bmatrix} x_1 \\ x_2 \\ \vdots \\ x_n \end{Bmatrix} = \begin{bmatrix} \cos(\omega t - \phi) & 0 & 0 & 0 \\ 0 & \cos(\omega t - \phi) & 0 & 0 \\ \vdots & \vdots & \vdots & \vdots \\ 0 & 0 & 0 & \cos(\omega t - \phi) \end{bmatrix} \begin{Bmatrix} C_1 \\ C_2 \\ \vdots \\ C_n \end{Bmatrix} \tag{2.387}$$

where angular frequency ω and phase angle ϕ are common to all x's. In this assumed solution, ϕ and C_1, C_2, C_n are the constants to be determined from the initial boundary conditions of the motion and ω is a characteristic value (eigenvalue) of the system.

Substituting Equation 2.387 into Equation 2.386 yields

$$\begin{bmatrix} k_{11} - m_{11}\omega^2 & k_{12} - m_{12}\omega^2 & \cdots & k_{1n} - m_{1n}\omega^2 \\ k_{21} - m_{21}\omega^2 & k_{22} - m_{22}\omega^2 & \cdots & k_{2n} - m_{2n}\omega^2 \\ \vdots & \vdots & \vdots & \vdots \\ k_{n1} - m_{n1}\omega^2 & k_{n2} - m_{n2}\omega^2 & \cdots & k_{nn} - m_{nn}\omega^2 \end{bmatrix} \begin{Bmatrix} C_1 \\ C_2 \\ \vdots \\ C_n \end{Bmatrix} \cos(\omega t - \phi) = \begin{Bmatrix} 0 \\ 0 \\ \vdots \\ 0 \end{Bmatrix} \tag{2.388}$$

or

$$\left[[k] - \omega^2[m]\right]\{C\} = \{0\} \tag{2.389}$$

where $[k]$ and $[m]$ are the $n \times n$ matrices, ω^2 and $\cos(\omega t - \phi)$ are scalars, and $\{C\}$ is the amplitude vector. For non-trivial solutions, $\cos(\omega t - \phi) \neq 0$; thus, solution to Equation 2.389 requires the determinant of $\left[[k] - \omega^2[m]\right] = 0$. The expansion of the determinant yields a polynomial of n degree as a function of ω^2, the n roots of which are the eigenvalues $\omega_1, \omega_2, \omega_n$.

If the eigenvalue ω for a normal mode is substituted in Equation 2.389, the amplitude vector $\{C\}$ for that mode can be obtained. $\{C_1\}, \{C_2\}, \{C_3\}, \{C_n\}$ are therefore called the *eigenvectors*, the absolute values of which must be determined through initial boundary conditions. The resulting motion is a sum of n harmonic motions, each governed by the respective natural frequency ω, written as

$$\{x\} = \sum_{i=1}^{n} \{C_i\} \cos(\omega_i t - \phi) \tag{2.390}$$

2.13.7 Distributed Mass Systems

Although many structures may be approximated by lumped mass systems, in practice all structures are distributed mass systems consisting of an infinite number of particles. Consequently, if the motion is repetitive, the structure has an infinite number of natural frequencies and mode shapes. The analysis of a distributed-parameter system is entirely equivalent to that of a discrete system once the mode shapes and frequencies have been determined because in both cases the amplitudes of the modal response components are used as generalized coordinates in defining the response of the structure.

In principle an infinite number of these coordinates is available for a distributed-parameter system, but in practice only a few modes, usually those of lower frequencies, will provide a significant contribution to the overall response. Thus, the problem of a distributed-parameter system can be converted to a discrete system form in which only a limited number of modal coordinates is used to describe the response.

Flexural Vibration of Beams

The motion of the distributed mass system is best illustrated by a classical example of a uniform beam with a span length L and flexural rigidity EI and a self-weight of m per unit length, as shown in Figure 2.133a. The beam is free to vibrate under its self-weight. From Figure 2.133b, dynamic equilibrium of a small beam segment of length dx requires:

$$\frac{\partial V}{\partial x} dx = m dx \frac{\partial^2 y}{\partial t^2} \tag{2.391}$$

in which

$$\frac{\partial^2 y}{\partial x^2} = \frac{M}{EI} \tag{2.392}$$

and

$$V = -\frac{\partial M}{\partial x}, \qquad \frac{\partial V}{\partial x} = -\frac{\partial^2 M}{\partial x^2} \tag{2.393}$$

Substituting these equations into Equation 2.391 leads to the equation of motion of the flexural beam:

$$\frac{\partial^4 y}{\partial x^4} + \frac{m}{EI} \frac{\partial^2 y}{\partial t^2} = 0 \tag{2.394}$$

Equation 2.394 can be solved for beams with given sets of boundary conditions. The solution consists of a family of vibration modes with corresponding natural frequencies. Standard results

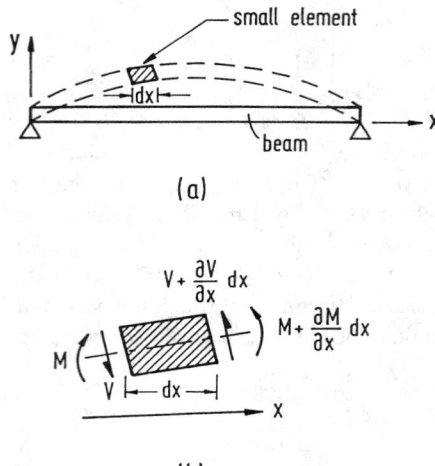

Figure 2.133   (a) Beam in flexural vibration; (b) equilibrium of beam segment in vibration.

are available in Table 2.6 to compute the natural frequencies of uniform flexural beams of different supporting conditions. Methods are also available for dynamic analysis of continuous beams [21].

Shear Vibration of Beams

Beams can deform by flexure or shear. Flexural deformation normally dominates the deformation of slender beams. Shear deformation is important for short beams or in higher modes of slender beams. Table 2.7 gives the natural frequencies of uniform beams in shear, neglecting flexural deformation. The natural frequencies of these beams are inversely proportional to the beam length L rather than L^2, and the frequencies increase linearly with the mode number.

Combined Shear and Flexure

The transverse deformation of real beams is the sum of flexure and shear deformations. In general, numerical solutions are required to incorporate both the shear and flexural deformation in the prediction of natural frequency of beams. For beams with comparable shear and flexural deformations, the following simplified formula may be used to estimate the beam's frequency:

$$\frac{1}{f^2} = \frac{1}{f_f^2} + \frac{1}{f_s^2} \tag{2.395}$$

where f is the fundamental frequency of the beam, and f_f and f_s are the fundamental frequencies predicted by the flexure and shear beam theory [50].

Natural Frequency of Multistory Building Frames

Tall building frames often deform more in the shear mode than in flexure. The fundamental frequencies of many multistory building frameworks can be approximated by [32, 48]

$$f = \alpha \frac{\sqrt{B}}{H} \tag{2.396}$$

where α is approximately equal to $11\sqrt{m}$/s, B is the building width in the direction of vibration, and H is the building height. This empirical formula suggests that a shear beam model with f inversely proportional to H is more appropriate than a flexural beam for predicting natural frequencies of buildings.

TABLE 2.6 Frequencies and Mode Shapes of Beams in Flexural Vibration

$f_n = \dfrac{k_n}{2\pi}\sqrt{\dfrac{EI}{mL^4}}$ HZ $n = 1, 2, 3...$		L = Length (m) EI = Flexural Rigidity (Nm^2) M = Mass per unit length (kg/m)	
Boundary Conditions	K_n; $n = 1,2,3$	Mode Shape $y_n\left(\dfrac{x}{L}\right)$	A_n; $n = 1,2,3...$
Pinned - Pinned	$(n\pi)^2$	$\sin\dfrac{n\pi x}{L}$	-
Fixed - Fixed	22.37 61.67 120.90 199.86 298.55 $(2n+1)\dfrac{\pi^2}{4}$; $n > 5$	$\cosh\dfrac{\sqrt{K_n}\,x}{L} - \cos\dfrac{\sqrt{K_n}\,x}{L}$ $- A_n\left(\sinh\dfrac{\sqrt{K_n}x}{L} - \sin\dfrac{\sqrt{K_n}x}{L}\right)$	0.98250 1.00078 0.99997 1.00000 0.99999 1.0; $n > 5$
Fixed - Pinned	15.42 49.96 104.25 178.27 272.03 $(4n+1)^2\dfrac{\pi^2}{4}$; $n > 5$	$\cosh\dfrac{\sqrt{K_n}\,x}{L} - \cos\dfrac{\sqrt{K_n}\,x}{L}$ $- A_n\left(\sinh\dfrac{\sqrt{K_n}x}{L} - \sin\dfrac{\sqrt{K_n}x}{L}\right)$	1.00078 1.00000 1.0; $n > 3$
Cantilever	3.52 22.03 61.69 120.90 199.86 $(2n-1)^2\dfrac{\pi^2}{4}$; $n > 5$	$\cosh\dfrac{\sqrt{K_n}\,x}{L} - \cos\dfrac{\sqrt{K_n}\,x}{L}$ $- A_n\left(\sinh\dfrac{\sqrt{K_n}x}{L} - \sin\dfrac{\sqrt{K_n}x}{L}\right)$	0.73410 1.01847 0.99922 1.00003 1.0; $n > 4$

TABLE 2.7   Frequencies and Mode Shapes of Beams in Shear Vibration

$f_n = \dfrac{K_n}{2\pi}\sqrt{\dfrac{KG}{\rho L^2}}$ HZ	L = Length K = Shear Coefficient (Cowper, 1966) G = Shear Modulus = $E/[2(1+\upsilon)]$ ρ = Mass Density	
Boundary Condition	K_n; n = 1,2,3...	Mode Shape $y_n\left(\dfrac{x}{L}\right)$
Fixed - Free	$n\pi$; n = 1,2,3...	$\cos\dfrac{n\pi x}{L}$; n = 1,2,3...
Fixed - Fixed	$n\pi$; n = 1,2,3...	$\sin\dfrac{n\pi x}{L}$; n = 1,2,3...

2.13.8 Portal Frames

A portal frame consists of a cap beam rigidly connected to two vertical columns. The natural frequencies of portal frames vibrating in the fundamental symmetric and asymmetric modes are shown in Tables 2.8 and 2.9, respectively.

The beams in these frames are assumed to be uniform and sufficiently slender so that shear, axial, and torsional deformations can be neglected. The method of analysis of these frames is given in [68]. The vibration is assumed to be in the plane of the frame, and the results are presented for portal frames with pinned and fixed bases.

If the beam is rigid and the columns are slender and uniform, but not necessarily identical, then the natural fundamental frequency of the frame can be approximated using the following formula [49]:

$$f = \frac{1}{2\pi} \left[\frac{12 \sum E_i I_i}{L^3 \left(M + 0.37 \sum M_i \right)} \right]^{1/2} \quad \text{Hz} \tag{2.397}$$

where M is the mass of the beam, M_i is the mass of the i-th column, and $E_i I_i$ are the flexural rigidity of the i-th column. The summation refers to the sum of all columns, and i must be greater or equal to 2. Additional results for frames with inclined members are discussed in [11].

2.13.9 Damping

Damping is found to increase with the increasing amplitude of vibration. It arises from the dissipation of energy during vibration. The mechanisms contributing to energy dissipation are material damping, friction at interfaces between components, and energy dissipation due to foundation interacting with soil, among others. Material damping arises from the friction at bolted connections and frictional interaction between structural and non-structural elements such as partitions and cladding.

The amount of damping in a building can never be predicted precisely, and design values are generally derived based on dynamic measurements of structures of a corresponding type. Damping can be measured based on the rate of decay of free vibration following an impact; by spectral methods based on analysis of response to wind loading; or by force excitation by mechanical vibrator at varying frequency to establish the shape of the steady state resonance curve. However, these methods may not be easily carried out if several modes of vibration close in frequency are presented.

Table 2.10 gives values of modal damping that are appropriate for use when amplitudes are low. Higher values are appropriate at larger amplitudes where local yielding may develop, e.g., in seismic analysis.

2.13.10 Numerical Analysis

Many less complex dynamic problems can be solved without much difficulty by hand methods. For more complex problems, such as determination of natural frequencies of complex structures, calculation of response due to time-varying loads, and response spectrum analysis to determine seismic forces, may require numerical analysis. The finite element method has been shown to be a versatile technique for this purpose.

The global equations of an undamped force-vibration motion, in matrix form, may be written as

$$[M]\{\ddot{x}\} + [K]\{\dot{x}\} = \{F(t)\} \tag{2.398}$$

where

$$[K] = \sum_{i=1}^{n} [k_i] \qquad [M] = \sum_{i=1}^{n} [m_i] \qquad [F] = \sum_{i=1}^{n} [f_i] \tag{2.399}$$

TABLE 2.8   Fundamental Frequencies of Portal Frames in Asymmetrical Mode of Vibration

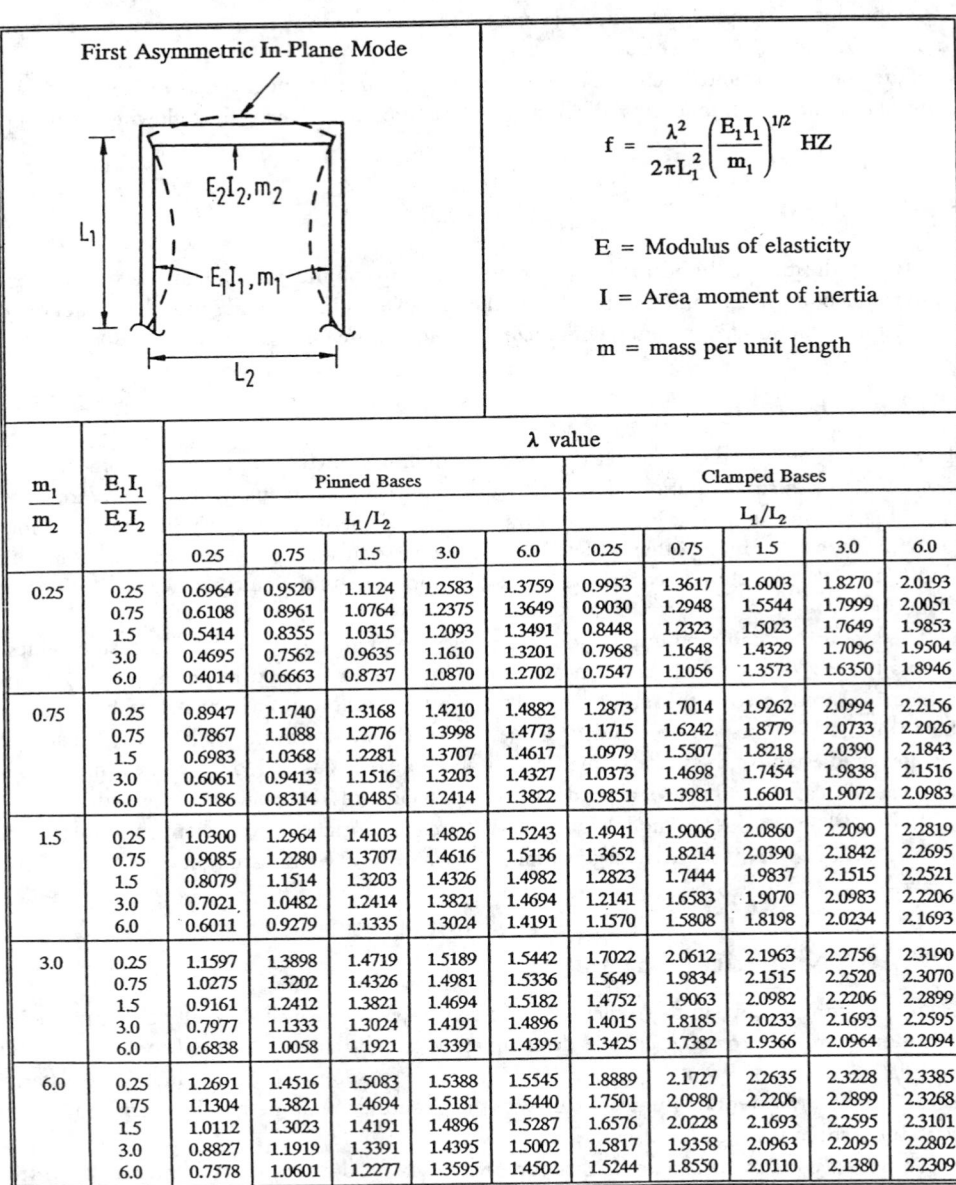

First Asymmetric In-Plane Mode

$$f = \frac{\lambda^2}{2\pi L_1^2}\left(\frac{E_1 I_1}{m_1}\right)^{1/2} \text{ HZ}$$

E = Modulus of elasticity

I = Area moment of inertia

m = mass per unit length

$\dfrac{m_1}{m_2}$	$\dfrac{E_1 I_1}{E_2 I_2}$	Pinned Bases L_1/L_2					Clamped Bases L_1/L_2				
		0.25	0.75	1.5	3.0	6.0	0.25	0.75	1.5	3.0	6.0
0.25	0.25	0.6964	0.9520	1.1124	1.2583	1.3759	0.9953	1.3617	1.6003	1.8270	2.0193
	0.75	0.6108	0.8961	1.0764	1.2375	1.3649	0.9030	1.2948	1.5544	1.7999	2.0051
	1.5	0.5414	0.8355	1.0315	1.2093	1.3491	0.8448	1.2323	1.5023	1.7649	1.9853
	3.0	0.4695	0.7562	0.9635	1.1610	1.3201	0.7968	1.1648	1.4329	1.7096	1.9504
	6.0	0.4014	0.6663	0.8737	1.0870	1.2702	0.7547	1.1056	1.3573	1.6350	1.8946
0.75	0.25	0.8947	1.1740	1.3168	1.4210	1.4882	1.2873	1.7014	1.9262	2.0994	2.2156
	0.75	0.7867	1.1088	1.2776	1.3998	1.4773	1.1715	1.6242	1.8779	2.0733	2.2026
	1.5	0.6983	1.0368	1.2281	1.3707	1.4617	1.0979	1.5507	1.8218	2.0390	2.1843
	3.0	0.6061	0.9413	1.1516	1.3203	1.4327	1.0373	1.4698	1.7454	1.9838	2.1516
	6.0	0.5186	0.8314	1.0485	1.2414	1.3822	0.9851	1.3981	1.6601	1.9072	2.0983
1.5	0.25	1.0300	1.2964	1.4103	1.4826	1.5243	1.4941	1.9006	2.0860	2.2090	2.2819
	0.75	0.9085	1.2280	1.3707	1.4616	1.5136	1.3652	1.8214	2.0390	2.1842	2.2695
	1.5	0.8079	1.1514	1.3203	1.4326	1.4982	1.2823	1.7444	1.9837	2.1515	2.2521
	3.0	0.7021	1.0482	1.2414	1.3821	1.4694	1.2141	1.6583	1.9070	2.0983	2.2206
	6.0	0.6011	0.9279	1.1335	1.3024	1.4191	1.1570	1.5808	1.8198	2.0234	2.1693
3.0	0.25	1.1597	1.3898	1.4719	1.5189	1.5442	1.7022	2.0612	2.1963	2.2756	2.3190
	0.75	1.0275	1.3202	1.4326	1.4981	1.5336	1.5649	1.9834	2.1515	2.2520	2.3070
	1.5	0.9161	1.2412	1.3821	1.4694	1.5182	1.4752	1.9063	2.0982	2.2206	2.2899
	3.0	0.7977	1.1333	1.3024	1.4191	1.4896	1.4015	1.8185	2.0233	2.1693	2.2595
	6.0	0.6838	1.0058	1.1921	1.3391	1.4395	1.3425	1.7382	1.9366	2.0964	2.2094
6.0	0.25	1.2691	1.4516	1.5083	1.5388	1.5545	1.8889	2.1727	2.2635	2.3228	2.3385
	0.75	1.1304	1.3821	1.4694	1.5181	1.5440	1.7501	2.0980	2.2206	2.2899	2.3268
	1.5	1.0112	1.3023	1.4191	1.4896	1.5287	1.6576	2.0228	2.1693	2.2595	2.3101
	3.0	0.8827	1.1919	1.3391	1.4395	1.5002	1.5817	1.9358	2.0963	2.2095	2.2802
	6.0	0.7578	1.0601	1.2277	1.3595	1.4502	1.5244	1.8550	2.0110	2.1380	2.2309

λ value

TABLE 2.9 Fundamental Frequencies of Portal Frames in Asymmetrical Mode of Vibration.

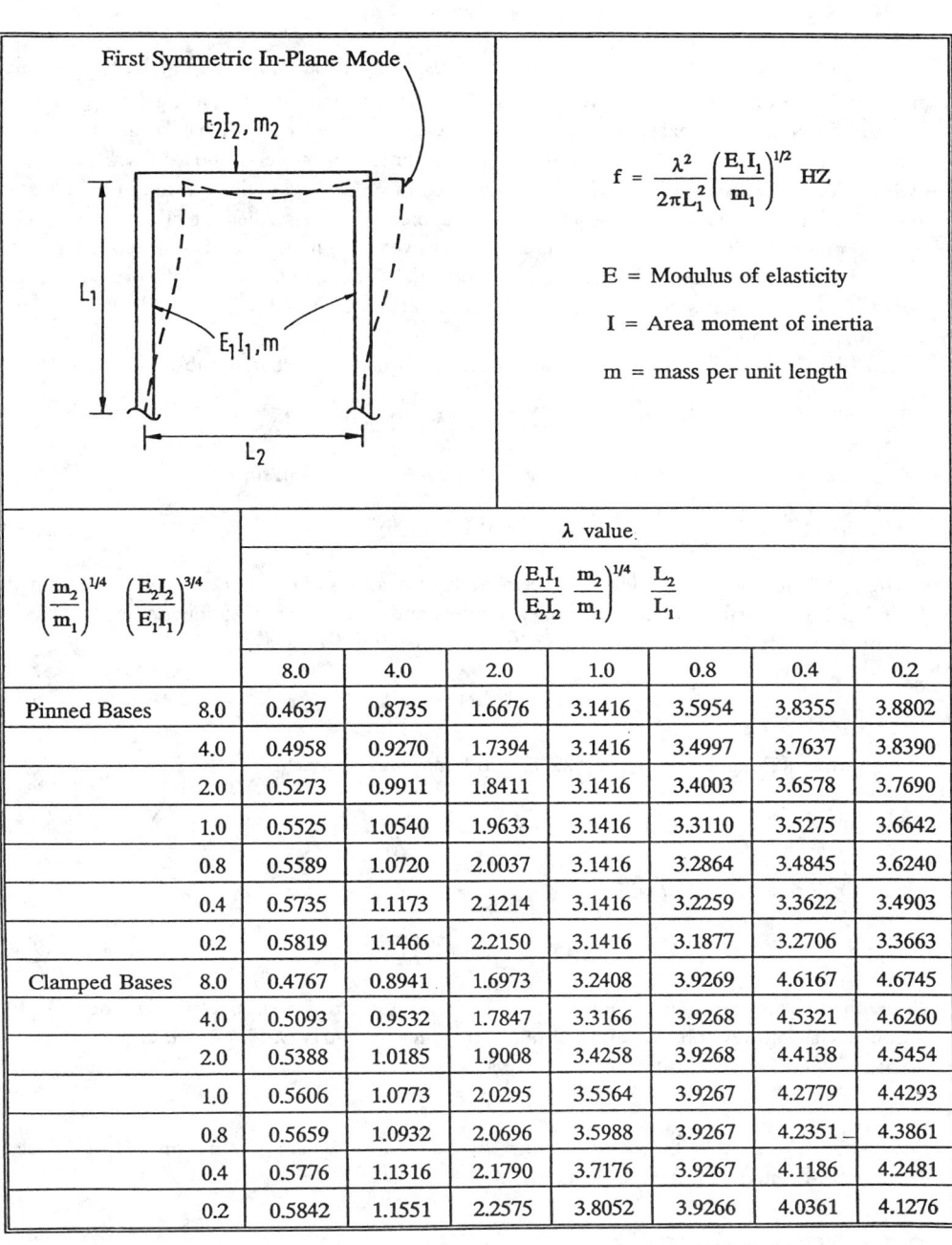

First Symmetric In-Plane Mode

$E_2 I_2, m_2$

$E_1 I_1, m$

$$f = \frac{\lambda^2}{2\pi L_1^2}\left(\frac{E_1 I_1}{m_1}\right)^{1/2} \text{ HZ}$$

E = Modulus of elasticity

I = Area moment of inertia

m = mass per unit length

$\left(\dfrac{m_2}{m_1}\right)^{1/4}\left(\dfrac{E_2 I_2}{E_1 I_1}\right)^{3/4}$		λ value $\left(\dfrac{E_1 I_1}{E_2 I_2}\dfrac{m_2}{m_1}\right)^{1/4}\dfrac{L_2}{L_1}$						
		8.0	4.0	2.0	1.0	0.8	0.4	0.2
Pinned Bases	8.0	0.4637	0.8735	1.6676	3.1416	3.5954	3.8355	3.8802
	4.0	0.4958	0.9270	1.7394	3.1416	3.4997	3.7637	3.8390
	2.0	0.5273	0.9911	1.8411	3.1416	3.4003	3.6578	3.7690
	1.0	0.5525	1.0540	1.9633	3.1416	3.3110	3.5275	3.6642
	0.8	0.5589	1.0720	2.0037	3.1416	3.2864	3.4845	3.6240
	0.4	0.5735	1.1173	2.1214	3.1416	3.2259	3.3622	3.4903
	0.2	0.5819	1.1466	2.2150	3.1416	3.1877	3.2706	3.3663
Clamped Bases	8.0	0.4767	0.8941	1.6973	3.2408	3.9269	4.6167	4.6745
	4.0	0.5093	0.9532	1.7847	3.3166	3.9268	4.5321	4.6260
	2.0	0.5388	1.0185	1.9008	3.4258	3.9268	4.4138	4.5454
	1.0	0.5606	1.0773	2.0295	3.5564	3.9267	4.2779	4.4293
	0.8	0.5659	1.0932	2.0696	3.5988	3.9267	4.2351	4.3861
	0.4	0.5776	1.1316	2.1790	3.7176	3.9267	4.1186	4.2481
	0.2	0.5842	1.1551	2.2575	3.8052	3.9266	4.0361	4.1276

TABLE 2.10   Typical Structural Damping Values

Structural type	Damping value, ξ (%)
Unclad welded steel structures	0.3
Unclad bolted steel structures	0.5
Floor, composite and non-composite	1.5-3.0
Clad buildings subjected to sidesway	1

are the global stiffness, mass, and force matrices, respectively. $[k_i]$, $[m_i]$, and $\{f_i\}$ are the stiffness, mass, and force of the ith element, respectively. The elements are assembled using the direct stiffness method to obtain the global equations such that intermediate continuity of displacements is satisfied at common nodes and, in addition, interelement continuity of acceleration is also satisfied.

Equation 2.398 is the matrix equation discretized in space. To obtain solution of the equation, discretization in time is also necessary. The general method used is called direct integration. There are two methods for direct integration: *implicit* or *explicit*. The first, and simplest, is an explicit method known as the central difference method [9]. The second, more sophisticated but more versatile, is an implicit method known as the Newmark method [44]. Other integration methods are also available in [7].

The natural frequencies are determined by solving Equation 2.398 in the absence of force $F(t)$ as

$$[M]\{\ddot{x}\} + [K]\{x\} = 0 \tag{2.400}$$

The standard solution for $x(t)$ is given by the harmonic equation in time

$$\{x(t)\} = \{X\}e^{i\omega t} \tag{2.401}$$

where $\{X\}$ is the part of the nodal displacement matrix called natural modes, which are assumed to be independent of time, i is the imaginary number, and ω is the natural frequency.

Differentiating Equation 2.401 twice with respect to time, we have

$$\ddot{x}(t) = \{X\}\left(-\omega^2\right)e^{i\omega t} \tag{2.402}$$

Substituting of Equations 2.401 and 2.402 into Equation 2.400 yields

$$e^{i\omega t}\left([K] - \omega^2[M]\right)\{X\} = 0 \tag{2.403}$$

Since $e^{i\omega t}$ is not zero, we obtain

$$\left([K] - \omega^2[M]\right)\{X\} = 0 \tag{2.404}$$

Equation 2.404 is a set of linear homogeneous equations in terms of displacement mode $\{X\}$. It has a non-trivial solution if the determinant of the coefficient matrix $\{X\}$ is non-zero; that is

$$[K] - \omega^2[M] = 0 \tag{2.405}$$

In general, Equation 2.405 is a set of n algebraic equations, where n is the number of degrees of freedom associated with the problem.

2.14   Defining Terms

Arch: Principal load-carrying member curved in elevation; resistance to applied loading developed by axial thrust and bending.

Beam: A straight or curved structural member, primarily supporting loads applied at right angles to the longitudinal axis.

Bending moment: Bending moment due to a force or a system of forces at a cross-section is computed as the algebraic sum of all moments to one side of the section.

Built-in beam: A beam restrained at its ends against vertical movement and rotation.

Cables: Flexible structures with no moment-carrying capacity.

Cantilever: A beam restrained against movement and rotation at one end and free to deflect at the other end.

Continuous beam: A beam that extends over several supports.

Deflection: Movement of a structure or parts of a structure under applied loads.

Element: Part of a cross-section forming a distinct part of the whole.

Grillage: Structures in which the members all lie in one plane with loads being applied in the direction normal to this plane.

Hogging moment: Bending moment causing upward deflection in a beam.

Influence line: An influence line indicates the effect at a given section of a unit load placed at any point on the structure.

Member: Any individual component of a structural frame.

Moment of inertia: The second moment of area of a section about the elastic neutral axis.

Plastic analysis: Analysis assuming redistribution of moments within the structure in a continuous construction.

Plastic hinge: Position at which a member has developed its plastic moment of resistance.

Plastic moment: Moment capacity allowing for redistribution of stress within a cross-section.

Plastic section: A cross-section that can develop a plastic hinge with sufficient rotational capacity to allow redistribution of bending moments within the section.

Portal frame: A single-story continuous plane frame deriving its strength from bending resistance and arch action.

Reaction: The load carried by each support.

Rigid frame: An indeterminate plane frame consisting of members with fixed end connections.

Sagging moment: An applied bending moment causing a sagging deflection in the beam.

Second-order analysis: Analysis considering the equilibrium formulated based on deformed structural geometry.

Semi-rigid connection: A connection that possesses a moment capacity intermediate between the simple and rigid connection.

Shear force: An internal force acting normal to the longitudinal axis; given by the algebraic sum of all forces to one side of the section chosen.

Simple beam: A beam restrained at its end only against vertical movement.

Space frame: A three-dimensional structure.

Span: The distance between the supports of a beam or a truss.

Static load: A noncyclic load that produces no dynamic effects.

Statically determinate structure: A structure in which support reactions may be found from the equations of equilibrium.

Statically indeterminate structure: A structure in which equations of equilibrium are not sufficient to determine the reactions.

Thin plate: A flat surface structure in which the thickness is small compared to the other dimensions.

Thin shell: A curved surface structure with a thickness relatively small compared to its other dimensions.

Truss: A coplanar system of structural members joined at their ends to form a stable framework.

References

[1] Ackroyd, M. H. 1985. *Design of Flexibly Connected Steel Frames,* Final Research Report to the American Iron and Steel Institute for Project No. 333, Rensselaer Polytechnic Institute, Troy, NY.

[2] AISC. 1989. *Allowable Stress Design Specification for Structural Steel Buildings,* American Institute of Steel Construction, Chicago, IL.

[3] AISC. 1993. *Load and Resistance Factor Design Specification for Structural Steel Buildings,* American Institute of Steel Construction, 2nd ed., Chicago, IL.

[4] Argyris, J. H. 1960. Energy Theorems and Structural Analysis, London, Butterworth. (Reprinted from *Aircraft Engineering,* Oct. 1954 - May 1955).

[5] ASCE. 1971. *Plastic Design in Steel—A Guide and Commentary, Manual 41,* American Society of Civil Engineers.

[6] Baker, L. and Heyman, J. 1969. *Plastic Design of Frames: 1. Fundamentals,* Cambridge University Press, 228 pp.

[7] Bathe, K. J. 1982. *Finite Element Procedures for Engineering Analysis,* Prentice Hall, Englewood Cliffs, NJ, 735 pp.

[8] Beedle, L. S. 1958. *Plastic Design of Steel Frames,* John Wiley & Sons, New York.

[9] Biggs, J. M. 1964. *Introduction to Structural Dynamics,* McGraw-Hill, New York.

[10] Borg, S. F. and Gennaro, J. J. 1959. *Advanced Structural Analysis,* D. Van Nostrand Company, Princeton, NJ.

[11] Chang, C. H. 1978. Vibration of frames with inclined members, *J. Sound Vib.,* 56, 201-214.

[12] Chen, W. F. and Atsuta, T. 1977. *Theory of Beam-Column, Vol. 2, Space Behavior and Design,* MacGraw-Hill, New York, 732 pp.

[13] Chen, W. F. and Han, D. J. 1988. *Plasticity for Structural Engineers,* Springer-Verlag, New York.

[14] Chen, W. F. and Lui, E. M. 1987. *Structural Stability—Theory and Implementation,* Prentice Hall, Englewood Cliffs, NJ, 490 pp.

[15] Chen W. F. and Lui, E. M. 1991. *Stability Design of Steel Frames,* CRC Press, Boca Raton, FL, 380 pp.

[16] Chen, W. F. and Sohal, I. S. 1995. *Plastic Design and Advanced Analysis of Steel Frames,* Springer-Verlag, New York.

[17] Chen, W. F. and Kim, S. E. 1997. *LRFD Steel Design Using Advanced Analysis,* CRC Press, Boca Raton, FL.

[18] Chen, W. F. and Toma, S. 1994. *Advanced Analysis in Steel Frames: Theory, Software and Applications,* CRC Press, Boca Raton, FL, 384 pp.

[19] Chen, W. F., Goto, Y., and Liew, J. Y. R. 1996. *Stability Design of Semi-Rigid Frames,* John Wiley & Sons, New York, 468 pp.

[20] Clark, M. J. 1994. Plastic-zone analysis of frames, in *Advanced Analysis of Steel Frames, Theory, Software, and Applications,* W.F. Chen and S. Toma, Eds., CRC Press, 195-319, chapt. 6.

[21] Clough, R. W. and Penzien, J. 1993. *Dynamics of Structures,* 2nd ed., McGraw-Hill, New York, 738 pp.

[22] Cowper, G. R. 1966. The shear coefficient in Timoshenko's beam theory, *J. Applied Mech.*, 33, 335-340.

[23] Desai, C. S. and Abel, J. F. 1972. *Introduction to the Finite Element Method*, Van Nostrand Reinhold and Company, New York.

[24] Dhillon, B. S. and Abdel-Majid, S. 1990. Interactive analysis and design of flexibly connected frames, *Computer & Structures*, 36(2), 189-202.

[25] Dowrick D. J. 1988. *Earthquake Resistant Design for Engineers and Architects*, 2nd ed., John Wiley & Sons, New York.

[26] Frye, M J. and Morris, G. A. 1976. Analysis of flexibly connected steel frames, *Can. J. Civ. Eng.*, 2(3), 280-291.

[27] Ghali, A. and Neville, A. M. 1978. *Structural Analysis*, Chapman and Hall, London.

[28] Goto, Y. and Chen, W. F. 1987. On the computer-based design analysis for the flexibly jointed frames, *J. Construct. Steel Res.*, Special Issue on Joint Flexibility in Steel Frames, (W. F. Chen, Ed.), 8, 203-231.

[29] Heyman, J. 1971. *Plastic Design of Frames: 2. Applications*, Cambridge University Press, 292 pp.

[30] Heish, S. H. 1990. *Analysis of Three-Dimensional Steel Frames with Semi-Rigid Connections*, Structural Engineering Report 90-1, School of Civil and Environmental Engineering, Cornell University, Ithaca, NY, 211 pp.

[31] Horne, M. R. 1964. *The Plastic Design of Columns*, BCSA Publication No. 23.

[32] Housner, G. W. and Brody, A. G. 1963. Natural periods of vibration of buildings, *J. Eng. Mech. Div.*, ASCE, 89, 31-65.

[33] Kardestuncer, H. and Norrie, D. H. 1988. *Finite Element Handbook*, McGraw-Hill, New York.

[34] Knudsen, K. E., Yang, C. H., Johnson, B. G., and Beedle, L. S. 1953. Plastic strength and deflections of continuous beams, *Welding J.*, 32(5), 240 pp.

[35] Lee, S. L. and Ballesteros, P. 1960. Uniformly loaded rectangular plate supported at the corners, *Int. J. Mech. Sci.*, 2, 206-211.

[36] Lee, S. L. and Basu, P. K. 1989. Secant method for nonlinear semi-rigid frames, *J. Construct. Steel Res.*, 14(2), 49-67.

[37] Lee, S. L., Karasudhi, P., Zakeria, M., and Chan, K. S. 1971. Uniformly loaded orthotropic rectangular plate supported at the corners, *Civil Eng. Trans.*, pp. 101-106.

[38] Liew J. Y. R. 1992. Advanced Analysis for Frame Design, Ph.D dissertation, School of Civil Engineering, Purdue University, West Lafayette, IN, May, 393 pp.

[39] Liew, J. Y. R. and Chen, W. F. 1994. Trends toward advanced analysis, in *Advanced Analysis in Steel Frames: Theory, Software and Applications*, Chen, W. F. and Toma, S., Eds., CRC Press, Boca Raton, FL, 1-45, chapt. 1.

[40] Liew, J. Y. R., White, D. W., and Chen, W. F. 1991. Beam-column design in steel frameworks—Insight on current methods and trends, *J. Construct. Steel Res.*, 18, 259-308.

[41] Liew, J. Y. R., White, D. W., and Chen, W. F. 1993a. Limit-states design of semi-rigid frames using advanced analysis, Part 1: Connection modelling and classification, Part 2: Analysis and design, *J. Construct. Steel Res.*, Elsevier Science Publishers, London, 26(1), 1-57.

[42] Liew, J. Y. R., White, D. W., and Chen, W. F. 1993b. Second-order refined plastic hinge analysis for frame design: Parts 1 & 2, *J. Structural Eng.*, ASCE, Nov., 119(11), 3196-3237.

[43] Mrazik, A., Skaloud, M., and Tochacek, M. 1987. *Plastic Design of Steel Structures*, Ellis Horwood Limited, 637 pp.

[44] Newmark, N. M. 1959. A method of computation for structural dynamic, *J. Eng. Mech.*, ASCE, 85(EM3), 67-94.

[45] Orbison, J. G. 1982. *Nonlinear Static Analysis of Three-Dimensional Steel Frames,* Department of Structural Engineering, Report No. 82-6, Cornell University, Ithaca, NY, 243 pp.

[46] Porter, F. L. and Powell, G. M. 1971. Static and dynamic analysis of inelastic frame structures, Report No. EERC 71-3, Earthquake Engineering Research Center, University of California, Berkeley.

[47] Powell, G. H. and Chen, P. F.-S. 1986. 3D beam column element with generalized plastic hinges, *J. Eng. Mech.,* ASCE, 112(7), 627-641.

[48] Rinne, J. E. 1952. Building code provisions for aseimic design, in *Proc. Symp. Earthquake and Blast Effects on Structures,* Los Angeles, CA, 291-305.

[49] Robert, D. B. 1979. *Formulas for Natural Frequency and Mode Shapes,* Van Nostrand Reinhold and Company, New York, 491 pp.

[50] Rutenberg, A. 1975. Approximate natural frequencies for coupled shear walls, *Earthquake Eng. Struct. Dynam.,* 4, 95-100.

[51] Seely, F. B. and Smith, J. O. 1952. *Advanced Mechanics of Materials,* John Wiley & Sons, New York, 137-187.

[52] Shanmugam, N. E, Rose, H., Yu, C. H., Lee, S. L. 1988. Uniformly loaded rhombic orthotropic plates supported at corners, *Computers Struct.,* 30(5), 1037-1045.

[53] Shanmugam, N. E., Rose, H., Yu, C. H., Lee, S. L. 1989. Corner supported isosceles triangular orthotropic plates, *Computers Struct.,* 32(5), 963-972.

[54] Structural Stability Research Council. 1988. *Guide to Stability Design Criteria for Metal Structures,* Galambos, T. V., Ed., Fourth ed., John Wiley & Sons, New York, 786 pp.

[55] Timoshenko, S. P. and Gere, J. M. 1961. *Theory of Elastic Stability,* McGraw-Hill, New York.

[56] Timoshenko, S. P. and Krieger, S. W. 1959. *Theory of Plates and Shells,* McGraw-Hill, New York.

[57] Tsai, S. W. and Cheron, T. 1968. *Anisotropic Plates* (translated from the Russian edition by S. G. Lekhnitskii), Gordon and Breach Science Publishers, New York, 1968.

[58] Vijakkhna, P., Karasudhi, P., and Lee, S. L. 1973. Corner supported equilateral triangular plates, *Int. J. Mech. Sci.,* 15, 123-128.

[59] Vogel, U. 1985. *Calibrating Frames,* Stahlbau, 10, Oct., 295-301.

[60] Wang, Y. C. 1988. Ultimate Strength Analysis of 3-D Beam Columns and Column Subassemblages with Flexible Connections, Ph.D thesis, University of Sheffield, England.

[61] Warburton, G. B. 1976. *The Dynamical Behavior of Structures,* 2nd ed., Pergamon Press, New York.

[62] White, D. W. 1985. Material and Geometric Nonlinear Analysis of Local Planar Behavior in Steel Frames Using Iterative Computer Graphics, M.S. thesis., Cornell University, Ithaca, NY, 281 pp.

[63] White, D. W. 1988. Analysis of Monotonic and Cyclic Stability of Steel Frame Subassemblages, Ph.D. dissertation, Cornell University, Ithaca, NY.

[64] White, D. W. and Hajjar, J. F. 1991. Application of second-order elastic analysis in LRFD: research to practice, *Engineering J.,* AISC, 28(4), 133-148.

[65] White, D. W., Liew, J. Y. R., and Chen, W. F. 1991. *Second-order Inelastic Analysis for Frame Design: A Report to SSRC Task Group 29 on Recent Research and the Perceived State-of-the-Art,* Structural Engineering Report, CE-STR-91-12, Purdue University, 116 pp.

[66] White, D. W., Liew, J. Y. R., and Chen, W. F. 1993. Toward advanced analysis in LRFD, in *Plastic Hinge Based Methods for Advanced Analysis and Design of Steel Frames—An Assessment of The State-Of-The-Art,* Structural Stability Research Council, Lehigh University, Bethlehem, PA, 95-173.

[67] Yang, Y. B. and Kuo, S. R. 1994. *Theory and Analysis of Nonlinear Framed Structures,* Prentice Hall, Singapore, 539-545.

[68] Yang, Y. T. and Sun C. T. 1973. Axial-flexural vibration of frameworks using finite element approach, *J. Acoust. Soc. Am.,* 53, 137-146.

[69] Ziemian, R. D., McGuire, W., and Deierlien, G. G. 1992. Inelastic Limit States Design: Part I - Plannar Frame Studies, *J. Struct. Eng.,* ASCE, 118(9).

[70] Ziemian, R. D., McGuire, W., and Deierlien, G.G. 1992. Inelastic Limit States Design: Part II - Three-Dimensional Frame Study, *J. Struct. Eng.,* ASCE, 118(9).

Further Reading

The *Structural Engineering Handbook* by E. H. Gaylord and C. N. Gaylord provides a reference work on structural engineering and deals with planning, design, and construction of a variety of engineering structures.

The *Finite Element Handbook* by H. Kardestuncer and D. H. Norrie presents the underlying mathematical principles, the fundamental formulations, and both commonly used and specialized applications of the finite element method.

3

Structural Steel Design[1]

3.1   Materials 3-2
Stress-Strain Behavior of Structural Steel • Types of Steel • Fire-proofing of Steel • Corrosion Protection of Steel • Structural Steel Shapes • Structural Fasteners • Weldability of Steel

3.2   Design Philosophy and Design Formats 3-7
Design Philosophy • Design Formats

3.3   Tension Members 3-9
Allowable Stress Design • Load and Resistance Factor Design • Pin-Connected Members • Threaded Rods

3.4   Compression Members 3-17
Allowable Stress Design • Load and Resistance Factor Design • Built-Up Compression Members

3.5   Flexural Members 3-26
Allowable Stress Design • Load and Resistance Factor Design • Continuous Beams • Lateral Bracing of Beams

3.6   Combined Flexure and Axial Force 3-44
Allowable Stress Design • Load and Resistance Factor Design

3.7   Biaxial Bending 3-47
Allowable Stress Design • Load and Resistance Factor Design

3.8   Combined Bending, Torsion, and Axial Force 3-48

3.9   Frames 3-49

3.10   Plate Girders 3-50
Allowable Stress Design • Load and Resistance Factor Design

3.11   Connections 3-58
Bolted Connections • Welded Connections • Shop Welded-Field Bolted Connections • Beam and Column Splices

3.12   Column Base Plates and Beam Bearing Plates (LRFD Approach) 3-84
Column Base Plates • Anchor Bolts • Beam Bearing Plates

3.13   Composite Members (LRFD Approach) 3-92
Composite Columns • Composite Beams • Composite Beam-Columns • Composite Floor Slabs

3.14   Plastic Design 3-98
Plastic Design of Columns and Beams • Plastic Design of Beam-Columns

3.15   Defining Terms 3-99

References 3-100

Further Reading 3-102

E. M. Lui
Department of Civil and Environmental Engineering, Syracuse University, Syracuse, NY

---

[1]The material in this chapter was previously published by CRC Press in *The Civil Engineering Handbook*, W.F. Chen, Ed., 1995.

0-8493-2674-5/97/$0.00+$.50
© 1997 by CRC Press LLC

3.1   Materials

3.1.1   Stress-Strain Behavior of Structural Steel

Structural steel is an important construction material. It possesses attributes such as *strength, stiffness, toughness,* and *ductility* that are very desirable in modern constructions. Strength is the ability of a material to resist stresses. It is measured in terms of the material's yield strength, F_y, and ultimate or tensile strength, F_u. For steel, the ranges of F_y and F_u ordinarily used in constructions are 36 to 50 ksi (248 to 345 MPa) and 58 to 70 ksi (400 to 483 MPa), respectively, although higher strength steels are becoming more common. Stiffness is the ability of a material to resist deformation. It is measured as the slope of the material's stress-strain curve. With reference to Figure 3.1 in which uniaxial engineering stress-strain curves obtained from coupon tests for various grades of steels are shown, it is seen that the modulus of elasticity, E, does not vary appreciably for the different steel grades. Therefore, a value of 29,000 ksi (200 GPa) is often used for design. Toughness is the ability

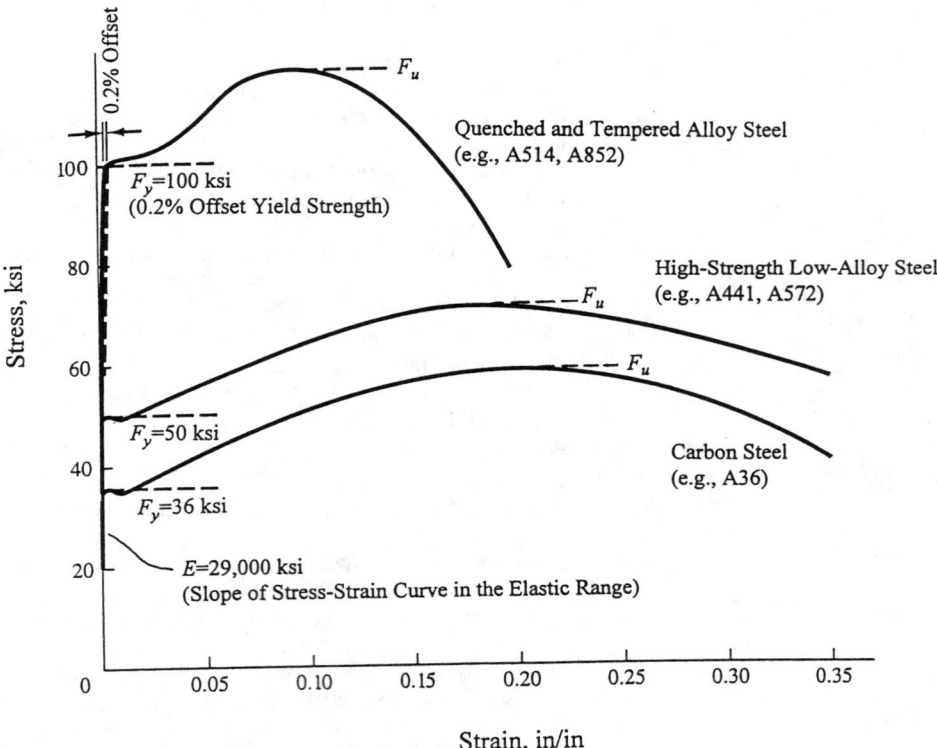

Figure 3.1   Uniaxial stress-strain behavior of steel.

of a material to absorb energy before failure. It is measured as the area under the material's stress-strain curve. As shown in Figure 3.1, most (especially the lower grade) steels possess high toughness which is suitable for both static and seismic applications. Ductility is the ability of a material to undergo large inelastic, or plastic, deformation before failure. It is measured in terms of percent elongation or percent reduction in area of the specimen tested in uniaxial tension. For steel, percent elongation ranges from around 10 to 40 for a 2-in. (5-cm) gage length specimen. Ductility generally decreases with increasing steel strength. Ductility is a very important attribute of steel. The ability of structural steel to deform considerably before failure by fracture allows an indeterminate structure to undergo stress redistribution. Ductility also enhances the energy absorption characteristic of the structure, which is extremely important in seismic design.

3.1.2 Types of Steel

Structural steels used for construction purpose are generally grouped into several major American Society of Testing and Materials (ASTM) classifications:

Carbon Steels (ASTM A36, ASTM A529, ASTM 709)

In addition to iron, the main ingredients of this category of steels are carbon (maximum content = 1.7%) and manganese (maximum content = 1.65%), with a small amount (< 0.6%) of silicon and copper. Depending on the amount of carbon content, different types of carbon steels can be identified:

Low carbon steel–carbon content < 0.15%

Mild carbon steel–carbon content varies from 0.15 to 0.29%

Medium carbon steel–carbon content 0.30 to 0.59%

High carbon steel–carbon content 0.60 to 1.70%

The most commonly used structural carbon steel has a mild carbon content. It is extremely ductile and is suitable for both bolting and welding. ASTM A36 is used mainly for buildings. ASTM A529 is occasionally used for bolted and welded building frames and trusses. ASTM 709 is used primarily for bridges.

High Strength Low Alloy Steels (ASTM A441, ASTM A572)

These steels possess enhanced strength as a result of the presence of one or more alloying agents such as chromium, copper, nickel, silicon, vanadium, and others in addition to the basic elements of iron, carbon, and manganese. Normally, the total quantity of all the alloying elements is below 5% of the total composition. These steels generally have higher corrosion-resistant capability than carbon steels. A441 steel was discontinued in 1989; it is superseded by A572 steel.

Corrosion-Resistant High Strength Low Alloy Steels (ASTM A242, ASTM A588)

These steels have enhanced corrosion-resistant capability because of the addition of copper as an alloying element. Corrosion is severely retarded when a layer of patina (an oxidized metallic film) is formed on the steel surfaces. The process of oxidation normally takes place within 1 to 3 years and is signified by a distinct appearance of a deep reddish-brown to black coloration of the steel. For the process to take place, the steel must be subjected to a series of wetting-drying cycles. These steels, especially ASTM 588, are used primarily for bridges and transmission towers (in lieu of galvanized steel) where members are difficult to access for periodic painting.

Quenched and Tempered Alloy Steels (ASTM A852, ASTM A514, ASTM A709, ASTM A852)

The quantities of alloying elements used in these steels are in excess of those used in carbon and low alloy steels. In addition, they are heat treated by quenching and tempering to enhance their strengths. These steels do not exhibit well-defined yield points. Their yield stresses are determined by the 0.2% offset strain method. These steels, despite their enhanced strength, have reduced ductility (Figure 3.1) and care must be exercised in their usage as the design **limit state** for the structure or structural elements may be governed by serviceability considerations (e.g., deflection, vibration) and/or local buckling (under compression).

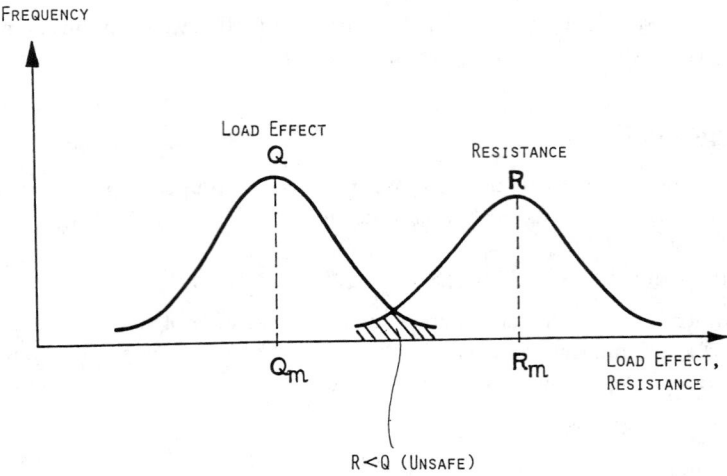

Figure 3.2   Frequency distribution of load effect and resistance.

In recent years, a new high strength steel produced using the *thermal-mechanical control process* (TMCP) has been developed. Compared with other high strength steels, TMCP steel has been shown to possess higher strength (for a given carbon equivalent value), enhanced toughness, improved weldability, and lower yield-to-tensile strength ratio, F_y/F_u. A low F_y/F_u value is desirable because there is an inverse relationship between F_y/F_u of the material and rotational capacity of the member. Research on TMCP steel is continuing and, as of this writing, TMCP steel has not been given an ASTM designation.

A summary of the specified minimum yield stresses, F_y, the specified minimum tensile strengths, F_u, and general usages for these various categories of steels are given in Table 3.1.

3.1.3   Fireproofing of Steel

Although steel is an incombustible material, its strength (F_y, F_u) and stiffness (E) reduce quite noticeably at temperatures normally reached in fires when other materials in a building burn. Exposed steel members that will be subjected to high temperature when a fire occurs should be fireproofed to conform to the fire ratings set forth in city codes. Fire ratings are expressed in units of time (usually hours) beyond which the structural members under a standard ASTM Specification (E119) fire test will fail under a specific set of criteria. Various approaches are available for fireproofing steel members. Steel members can be fireproofed by encasement in concrete if a minimum cover of 2 in. (51 mm) of concrete is provided. If the use of concrete is undesirable (because it adds weight to the structure), a lath and plaster (gypsum) ceiling placed underneath the structural members supporting the floor deck of an upper story can be used. In lieu of such a ceiling, spray-on materials such as mineral fibers, perlite, vermiculite, gypsum, etc. can also be used for fireproofing. Other means of fireproofing include placing steel members away from the source of heat, circulating liquid coolant inside box or tubular members and the use of insulative paints. These special paints foam and expand when heated, thus forming a shield for the members [26]. For a more detailed discussion of structural steel design for fire protection, refer to the latest edition of AISI publication No. FS3, *Fire-Safe Structural Steel-A Design Guide*. Additional information on fire-resistant standards and fire protection can be found in the AISI booklets on *Fire Resistant Steel Frame Construction, Designing Fire Protection for Steel Columns,* and *Designing Fire Protection for Steel Trusses* as well as in the *Uniform Building Code*.

TABLE 3.1 Types of Steels

ASTM designation	F_y (ksi)[a]	F_u (ksi)[a]	Plate thickness (in.)[b]	General usages
A36	36	58-80	To 8	Riveted, bolted, and welded buildings and bridges.
A529	42	60-85	To 0.5	Similar to A36. The higher yield
	50	70-100	To 1.5	stress for A529 steel allows for savings in weight. A529 supersedes A441.
A572 Grade 42	42	60	To 6	Similar to A441. Grades 60 and 65
Grade 50	50	65	To 4	not suitable for welded bridges.
Grade 60	60	75	To 1.25	
Grade 65	65	80	To 1.25	
A242	42	63	1.5 to 5	Riveted, bolted, and
	46	67	0.75 to 1.5	welded buildings and bridges.
	50	70	0.5 to 0.75	Used when weight savings and enhanced atmospheric corrosion resistance are desired. Specific instructions must be provided for welding.
A588	42	63	5 to 8	Similar to A242. Atmospheric
	46	67	4 to 5	corrosion resistance is about
	50	70	To 4	four times that of A36 steel.
A709 Grade 36	36	58-80	To 4	Primarily for use in bridges.
Grade 50	50	65	To 4	
Grade 50W	50	70	To 4	
Grade 70W	70	90-110	To 4	
Grade 100 & 100W	90	100-130	2.5 to 4	
Grade 100 & 100W	100	110-130	To 2.5	
A852	70	90-110	To 4	Plates for welded and bolted construction where atmospheric corrosion resistance is desired.
A514	90-100	100-130	2.5 to 6	Primarily for welded bridges. Avoid
		110-130		usage if ductility is important.

[a] 1 ksi = 6.895 MPa
[b] 1 in. = 25.4 mm

3.1.4 Corrosion Protection of Steel

Atmospheric corrosion occurs when steel is exposed to a continuous supply of water and oxygen. The rate of corrosion can be reduced if a barrier is used to keep water and oxygen from contact with the surface of bare steel. Painting is a practical and cost effective way to protect steel from corrosion. The Steel Structures Painting Council issues specifications for the surface preparation and the painting of steel structures for corrosion protection of steel. In lieu of painting, the use of other coating materials such as epoxies or other mineral and polymeric compounds can be considered. The use of corrosion resistance steel such as ASTM A242 and A588 steel or galvanized steel is another alternative.

3.1.5 Structural Steel Shapes

Steel sections used for construction are available in a variety of shapes and sizes. In general, there are three procedures by which steel shapes can be formed: hot-rolled, cold-formed, and welded. All steel shapes must be manufactured to meet ASTM standards. Commonly used steel shapes include the wide flange (W) sections, the American Standard beam (S) sections, bearing pile (HP) sections, American Standard channel (C) sections, angle (L) sections, and tee (WT) sections as well as bars, plates, pipes, and tubular sections. H sections which, by dimensions, cannot be classified as W or S shapes are designated as miscellaneous (M) sections, and C sections which, by dimensions, cannot be classified as American Standard channels are designated as miscellaneous channel (MC) sections.

Hot-rolled shapes are classified in accordance with their tensile property into five size groups by the American Society of Steel Construction (AISC). The groupings are given in the AISC Manuals [21, 22] Groups 4 and 5 shapes and group 3 shapes with flange thickness exceeding 1-1/2 in. are generally used for application as **compression members.** When weldings are used, care must be exercised to minimize the possibility of cracking in regions at the vicinity of the welds by carefully reviewing the material specification and fabrication procedures of the pieces to be joined.

3.1.6 Structural Fasteners

Steel sections can be fastened together by rivets, bolts, and welds. While rivets were used quite extensively in the past, their use in modern steel construction has become almost obsolete. Bolts have essentially replaced rivets as the primary means to connect nonwelded structural components.

Bolts

Four basic types of bolts are commonly in use. They are designated by ASTM as A307, A325, A490, and A449. A307 bolts are called unfinished or ordinary bolts. They are made from low carbon steel. Two grades (A and B) are available. They are available in diameters from 1/4 in. to 4 in. in 1/8 in. increments. They are used primarily for low-stress connections and for secondary members. A325 and A490 bolts are called high-strength bolts. A325 bolts are made from a heat-treated medium carbon steel. They are available in three types: Type 1—bolts made of medium carbon steel; Type 2—bolts made of low carbon martensite steel; and Type 3—bolts having atmospheric-corrosion resistance and weathering characteristics comparable to A242 and A588 steel. A490 bolts are made from quenched and tempered alloy steel and thus have a higher strength than A325 bolts. Like A325 bolts, three types (Types 1 to 3) are available. Both A325 and A490 bolts are available in diameters from 1/2 in. to 1-1/2 in. in 1/8 in. increments. They are used for general construction purposes. A449 bolts are made from quenched and tempered steel. They are available in diameters from 1/4 in. to 3 in. A449 bolts are used when diameters over 1-1/2 in. are needed. They are also used for anchor bolts and threaded rod.

High-strength bolts can be tightened to two conditions of tightness: snug-tight and fully tight. Snug-tight conditions can be attained by a few impacts of an impact wrench, or the full effort of a worker using an ordinary spud wrench. Snug-tight conditions must be clearly identified on the design drawing and are permitted only if the bolts are not subjected to tension loads, and loosening or fatigue due to vibration or load fluctuations are not design considerations. Bolts used in slip-critical conditions (i.e., conditions for which the integrity of the connected parts is dependent on the frictional force developed between the interfaces of the joint) and in conditions where the bolts are subjected to direct tension are required to be fully tightened to develop a pretension force equal to about 70% of the minimum tensile stress F_u of the material from which the bolts are made. This can be accomplished by using the turn-of-the-nut method, the calibrated wrench method, or by the use of alternate design fasteners or direct tension indicator [28].

Welds

Welding is a very effective means to connect two or more pieces of material together. The four most commonly used welding processes are *Shielded Metal Arc Welding* (SMAW), *Submerged Arc Welding* (SAW), *Gas Metal Arc Welding* (GMAW), and *Flux Core Arc Welding* (FCAW) [7]. Welding can be done with or without filler materials although most weldings used for construction utilized filler materials. The filler materials used in modern day welding processes are electrodes. Table 3.2 summarizes the electrode designations used for the aforementioned four most commonly used welding processes.

Finished welds should be inspected to ensure their quality. Inspection should be performed by qualified welding inspectors. A number of inspection methods are available for weld inspections. They include visual, the use of liquid penetrants, magnetic particles, ultrasonic equipment, and radiographic methods. Discussion of these and other welding inspection techniques can be found in the *Welding Handbook* [6].

TABLE 3.2   Electrode Designations

Welding processes	Electrode designations	Remarks
Shielded metal arc welding (SMAW)	E60XX E70XX E80XX E100XX E110XX	The 'E' denotes electrode. The first two digits indicate tensile strength in ksi.[a] The two 'X's represent numbers indicating the usage of the electrode.
Submerged arc welding (SAW)	F6X-EXXX F7X-EXXX F8X-EXXX	The 'F' designates a granular flux material. The digit(s) following the 'F' indicate the tensile strength in ksi (6 means 60 ksi, 10 means 100 ksi, etc.).
	F10X-EXXX F11X-EXXX	The digit before the hyphen gives the Charpy V-notched impact strength. The 'E' and the 'X's that follow represent numbers relating to the use of the electrode.
Gas metal arc welding (GMAW)	ER70S-X ER80S ER100S ER110S	The digits following the letters 'ER' represent the tensile strength of the electrode in ksi.
Flux cored arc welding (FCAW)	E6XT-X E7XT-X E8XT E10XT E11XT	The digit(s) following the letter 'E' represent the tensile strength of the electrode in ksi (6 means 60 ksi, 10 means 100 ksi, etc.).

[a] 1 ksi = 6.895 MPa

3.1.7   Weldability of Steel

Most ASTM specification construction steels are weldable. In general, the strength of the electrode used should equal or exceed the strength of the steel being welded [7]. The table below gives ranges of chemical elements in steel within which good weldability is assured [8].

Element	Range for good weldability	Percent requiring special care
Carbon	0.06-0.25	0.35
Manganese	0.35-0.80	1.40
Silicon	0.10 max.	0.30
Sulfur	0.035 max.	0.050
Phosphorus	0.030 max.	0.040

Weldability of steel is closely related to the amount of carbon in steel. Weldability is also affected by the presence of other elements. A quantity known as *carbon equivalent value*, giving the amount of carbon and other elements in percent composition, is often used to define the chemical requirements in steel. One definition of the carbon equivalent value C_{eq} is

$$C_{eq} = \text{Carbon} + \frac{(\text{Manganese} + \text{Silicon})}{6} + \frac{(\text{Copper} + \text{Nickel})}{15}$$
$$+ \frac{(\text{Chromium} + \text{Molybdenum} + \text{Vanadium} + \text{Columbium})}{5} \quad (3.1)$$

A steel is considered weldable if $C_{eq} \leq 0.50\%$ for steel in which the carbon content does not exceed 0.12%, and if $C_{eq} \leq 0.45\%$ for steel in which the carbon content exceeds 0.12%.

3.2   Design Philosophy and Design Formats

3.2.1   Design Philosophy

Structural design should be performed to satisfy three criteria: (1) strength, (2) serviceability, and (3) economy. *Strength* pertains to the general integrity and safety of the structure under extreme load conditions. The structure is expected to withstand occasional overloads without severe distress and damage during its lifetime. *Serviceability* refers to the proper functioning of the structure as related to its appearance, maintainability, and durability under normal, or **service load**, conditions. Deflection, vibration, permanent deformation, cracking, and corrosion are some design considerations associated with serviceability. *Economy* concerns the overall material and labor costs required for the design, fabrication, erection, and maintenance processes of the structure.

3.2.2 Design Formats

At present, steel design can be performed in accordance with one of the following three formats:

1. *Allowable Stress Design (ASD)*—**ASD** has been in use for decades for steel design of buildings and bridges. It continues to enjoy popularity among structural engineers engaged in steel building design. In allowable stress (or working stress) design, member stresses computed under the action of service (or working) loads are compared to some predesignated stresses called allowable stresses. The allowable stresses are usually expressed as a function of the yield stress (F_y) or tensile stress (F_u) of the material. To account for overload, understrength, and approximations used in structural analysis, a factor of safety is applied to reduce the nominal resistance of the structural member to a fraction of its tangible capacity. The general format for an allowable stress design has the form

$$\frac{R_n}{F.S.} \geq \sum_{i=1}^{m} Q_{ni} \qquad (3.2)$$

where R_n is the nominal resistance of the structural component expressed in a unit of stress; Q_{ni} is the service, or working stresses computed from the applied working load of type i; $F.S.$ is the factor of safety; i is the load type (dead, live, wind, etc.), and m is the number of load type considered in the design. The left-hand side of the equation, $R_n/F.S.$, represents the allowable stress of the structural component.

2. *Plastic Design (PD)*—**PD** makes use of the fact that steel sections have reserved strength beyond the first yield condition. When a section is under flexure, yielding of the cross-section occurs in a progressive manner, commencing with the fibers farthest away from the neutral axis and ending with the fibers nearest the neutral axis. This phenomenon of progressive yielding, referred to as *plastification,* means that the cross-section does not fail at first yield. The additional moment that a cross-section can carry in excess of the moment that corresponds to first yield varies depending on the shape of the cross-section. To quantify such reserved capacity, a quantity called *shape factor*, defined as the ratio of the *plastic moment* (moment that causes the entire cross-section to yield, resulting in the formation of a **plastic hinge**) to the *yield moment* (moment that causes yielding of the extreme fibers only) is used. The shape factor for hot-rolled I-shaped sections bent about the strong axes has a value of about 1.15. The value is about 1.50 when these sections are bent about their weak axes.

   For an indeterminate structure, failure of the structure will not occur after the formation of a plastic hinge. After complete yielding of a cross-section, force (or, more precisely, moment) redistribution will occur in which the unfailed portion of the structure continues to carry any additional loadings. Failure will occur only when enough cross-sections have yielded rendering the structure unstable, resulting in the formation of a *plastic collapse mechanism.*

   In plastic design, the factor of safety is applied to the applied loads to obtain **factored loads**. A design is said to have satisfied the strength criterion if the load effects (i.e., forces, shears, and moments) computed using these factored loads do not exceed the nominal plastic strength of the structural component. Plastic design has the form

$$R_n \geq \gamma \sum_{i=1}^{m} Q_{ni} \tag{3.3}$$

where R_n is the nominal plastic strength of the member; Q_{ni} is the nominal load effect from loads of type i; γ is the load factor; i is the load type; and m is the number of load types.

In steel building design, the load factor is given by the AISC Specification as 1.7 if Q_n consists of dead and live gravity loads only, and as 1.3 if Q_n consists of dead and live gravity loads acting in conjunction with wind or earthquake loads.

3. *Load and Resistance Factor Design (LRFD)*—**LRFD** is a probability-based limit state design procedure. In its development, both load effects and resistance were treated as random variables. Their variabilities and uncertainties were represented by frequency distribution curves. A design is considered satisfactory according to the strength criterion if the resistance exceeds the load effects by a comfortable margin. The concept of safety is represented schematically in Figure 3.2. Theoretically, the structure will not fail unless R is less than Q as shown by the shaded portion in the figure where the R and Q curves overlap. The smaller this shaded area, the less likely that the structure will fail. In actual design, a **resistance factor** ϕ is applied to the nominal resistance of the structural component to account for any uncertainties associated with the determination of its strength and a **load factor** γ is applied to each load type to account for the uncertainties and difficulties associated with determining its actual load magnitude. Different load factors are used for different load types to reflect the varying degree of uncertainty associated with the determination of load magnitudes. In general, a lower load factor is used for a load that is more predicable and a higher load factor is used for a load that is less predicable. Mathematically, the LRFD format takes the form

$$\phi R_n \geq \sum_{i=1}^{m} \gamma_i Q_{ni} \tag{3.4}$$

where ϕR_n represents the **design** (or usable) **strength**, and $\Sigma \gamma Q_{ni}$ represents the required strength or load effect for a given load combination. Table 3.3 shows the load combinations to be used on the right hand side of Equation 3.4. For a safe design, all load combinations should be investigated and the design is based on the worst case scenario.

LRFD is based on the limit state design concept. A **limit state** is defined as a condition in which a structure or structural component becomes unsafe (that is, a violation of the strength limit state) or unsuitable for its intended function (that is, a violation of the serviceability limit state). In a limit state design, the structure or structural component is designed in accordance to its limits of usefulness, which may be strength related or serviceability related.

3.3 Tension Members

Tension members are to be designed to preclude the following possible modes of failures under normal load conditions: Yielding in gross section, fracture in effective net section, block shear, shear rupture along plane through the fasteners, bearing on fastener holes, prying (for lap or hanger-type joints). In addition, the fasteners' strength must be adequate to prevent failure in the fasteners. Also, except for rods in tension, the slenderness of the tension member obtained by dividing the length of the member by its least radius of gyration should preferably not exceed 300.

TABLE 3.3   Load Factors and Load
Combinations

$1.4D$
$1.2D + 1.6L + 0.5(L_r \text{ or } S \text{ or } R)$
$1.2D + 1.6(L_r \text{ or } S \text{ or } R) + (0.5L \text{ or } 0.8W)$
$1.2D + 1.3W + 0.5L + 0.5(L_r \text{ or } S \text{ or } R)$
$1.2D \pm 1.0E + 0.5L + 0.2S$
$0.9D \pm (1.3W \text{ or } 1.0E)$

where

D	=	dead load
L	=	live load
L_r	=	roof live load
W	=	wind load
S	=	snow load
E	=	earthquake load
R	=	nominal load due to initial rainwater or ice exclusive of the ponding contribution

The load factor on L in the third, fourth, and fifth
load combinations shown above shall equal 1.0
for garages, areas occupied as places of public
assembly, and all areas where the live load is
greater than 100 psf (47.9 N/m^2).

3.3.1   Allowable Stress Design

The computed tensile stress, f_t, in a tension member shall not exceed the allowable stress for tension,
F_t, given by $0.60F_y$ for yielding on the gross area, and by $0.50F_u$ for fracture on the effective net
area. While the gross area is just the nominal cross-sectional area of the member, the *effective net
area* is the smallest cross-sectional area accounting for the presence of fastener holes and the effect
of **shear lag**. It is calculated using the equation

$$A_e = UA_n$$

$$= U\left[A_g - \sum_{i=1}^{m} d_{ni}t_i + \sum_{j=1}^{k}\left(\frac{s^2}{4g}\right)_j t_j\right] \tag{3.5}$$

where

U is a reduction coefficient given by [25]

$$U = 1 - \frac{\bar{x}}{l} \leq 0.90 \tag{3.6}$$

in which l is the length of the connection and $\bar{x}$ is the distance measured as shown in Figure 3.3. For
a given cross-section the largest $\bar{x}$ is used in Equation 3.6 to calculate U. This reduction coefficient
is introduced to account for the shear lag effect that arises when some component elements of the
cross-section in a joint are not connected, rendering the connection less effective in transmitting
the applied load. The terms in brackets in Equation 3.5 constitute the so-called net section A_n. The
various terms are defined as follows:

A_g	=	gross cross-sectional area
d_n	=	nominal diameter of the hole (bolt cutout), taken as the nominal bolt diameter plus 1/8 of an inch (3.2 mm)
t	=	thickness of the component element
s	=	longitudinal center-to-center spacing (pitch) of any two consecutive fasteners in a chain of staggered holes
g	=	transverse center-to-center spacing (gage) between two adjacent fasteners gage lines in a chain of staggered holes

The second term inside the brackets of Equation 3.5 accounts for loss of material due to bolt
cutouts, the summation is carried for all bolt cutouts lying on the failure line. The last term

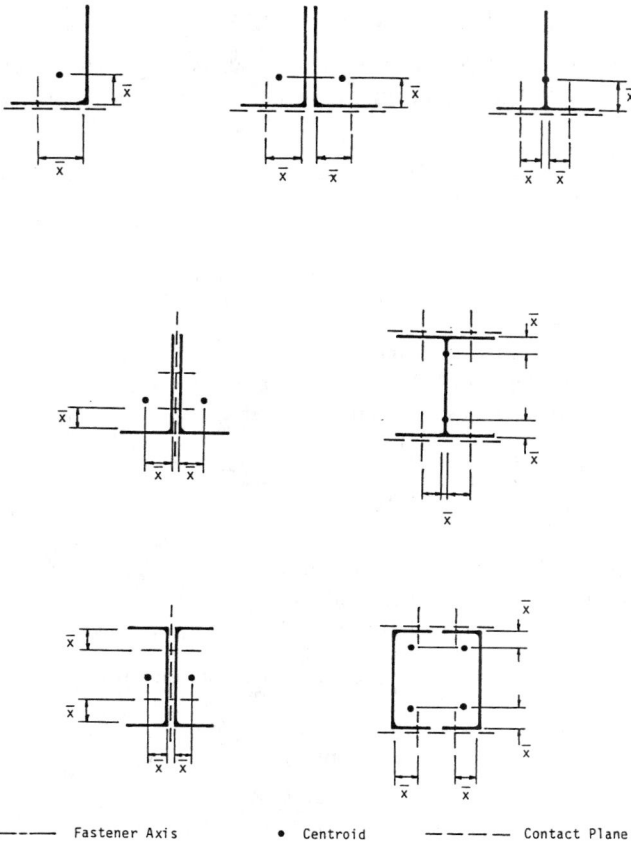

——·—— Fastener Axis   • Centroid   —— —— Contact Plane

Figure 3.3 Definition of $\bar{x}$ for selected cross-sections.

inside the brackets of Equation 3.5 indirectly accounts for the effect of the existence of a combined stress state (tensile and shear) along an inclined failure path associated with staggered holes. The summation is carried for all staggered paths along the failure line. This term vanishes if the holes are not staggered. Normally, it is necessary to investigate different failure paths that may occur in a connection, the critical failure path is the one giving the smallest value for A_e.

To prevent block shear failure and shear rupture, the allowable stresses for block shear and shear rupture are specified as follows.

Block shear:

$$R_{BS} = 0.30A_v F_u + 0.50A_t F_u \tag{3.7}$$

Shear rupture:

$$F_v = 0.30F_u \tag{3.8}$$

where

A_v = net area in shear
A_t = net area in tension
F_u = specified minimum tensile strength

The tension member should also be designed to possess adequate thickness and the fasteners should be placed within a specific range of spacings and edge distances to prevent failure due to bearing and failure by prying action (see section on Connections).

3.3.2 Load and Resistance Factor Design

According to the LRFD Specification [18], tension members designed to resist a factored axial force of P_u calculated using the load combinations shown in Table 3.3 must satisfy the condition of

$$\phi_t P_n \geq P_u \tag{3.9}$$

The design strength $\phi_t P_n$ is evaluated as follows.

Yielding on gross section:

$$\phi_t P_n = 0.90[F_y A_g] \tag{3.10}$$

where
0.90 = the resistance factor for tension
F_y = the specified minimum yield stress of the material
A_g = the gross cross-sectional area of the member

Fracture in effective net section:

$$\phi_t P_n = 0.75[F_u A_e] \tag{3.11}$$

where
0.75 = the resistance factor for fracture in tension
F_u = the specified minimum tensile strength
A_e = the effective net area given in Equation 3.5

Block shear: If $F_u A_{nt} \geq 0.6 F_u A_{nv}$ (i.e., shear yield-tension fracture)

$$\phi_t P_n = 0.75[0.60 F_y A_{gv} + F_u A_{nt}] \tag{3.12a}$$

If $F_u A_{nt} < 0.6 F_u A_{nv}$ (i.e., shear fracture-tension yield)

$$\phi_t P_n = 0.75[0.60 F_u A_{nv} + F_y A_{gt}] \tag{3.12b}$$

where
0.75 = the resistance factor for block shear
F_y, F_u = the specified minimum yield stress and tensile strength, respectively
A_{gv} = the gross area of the torn-out segment subject to shear
A_{nt} = the net area of the torn-out segment subject to tension
A_{nv} = the net area of the torn-out segment subject to shear
A_{gt} = the gross area of the torn-out segment subject to tension

EXAMPLE 3.1:

Using LRFD, select a double channel tension member shown in Figure 3.4a to carry a dead load D of 40 kips and a live load L of 100 kips. The member is 15 feet long. Six 1-in. diameter A325 bolts in standard size holes are used to connect the member to a 3/8-in. gusset plate. Use A36 steel ($F_y = 36$ ksi, $F_u = 58$ ksi) for all the connected parts.

Load Combinations:
From Table 3.3, the applicable load combinations are:

$$1.4D = 1.4(40) = 56 \text{ kips}$$
$$1.2D + 1.6L = 1.2(40) + 1.6(100) = 208 \text{ kips}$$

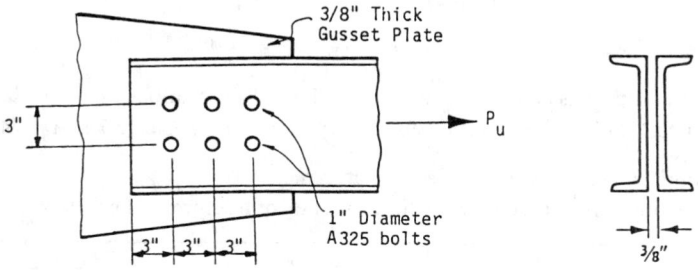

(a)   A Double Channel Tension Member

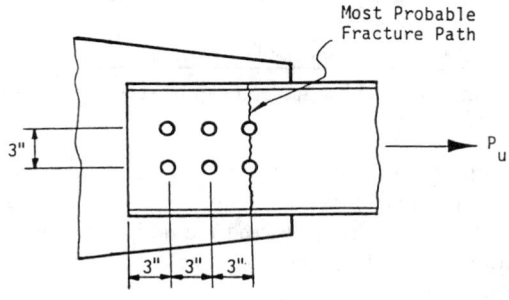

(b)   Fracture Failure

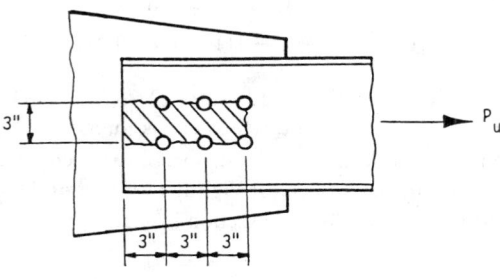

(c)   Block Shear Failure

Figure 3.4   Design of a double-channel tension member (1 in. = 25.4 mm).

The design of the tension member is to be based on the larger of the two, i.e., 208 kips and so *each* channel is expected to carry 104 kips.

Yielding in gross section:
    Using Equations 3.9 and 3.10, the gross area required to prevent cross-section yielding is

$$0.90[F_y A_g] \geq P_u$$
$$0.90[(36)(A_g)] \geq 104$$
$$(A_g)_{req'd} \geq 3.21 \text{ in}^2$$

From the section properties table contained in the AISC-LRFD Manual, one can select the following trial sections: C8x11.5 (A_g =3.38 in^2), C9x13.4 (A_g =3.94 in^2), C8x13.75 (A_g =4.04 in^2).

Check for the limit state of fracture on effective net section:
    The above sections are checked for the limiting state of fracture in the following table.

Section	A_g (in.^{2})	t_w (in.)	$\bar{x}$ (in.)	U^a	A_e^b (in.^{2})	$\phi_t P_n$ (kips)
C8x11.5	3.38	0.220	0.571	0.90	2.6	113.1
C9x13.4	3.94	0.233	0.601	0.90	3.07	133.5
C8x13.75	4.04	0.303	0.553	0.90	3.02	131.4

[a] Equation 3.6
[b] Equation 3.5, Figure 3.4b

From the last column of the above table, it can be seen that fracture is not a problem for any of the trial section.

Check for the limit state of block shear:
    Figure 3.4c shows a possible block shear failure mode. To avoid block shear failure the required strength of P_u =104 kips should not exceed the design strength, $\phi_t P_n$, calculated using Equation 3.12a or Equation 3.12b, whichever is applicable.
    For the C8x11.5 section:

$$A_{gv} = 2(9)(0.220) = 3.96 \text{ in.}^2$$
$$A_{nv} = A_{gv} - 5(1+1/8)(0.220) = 2.72 \text{ in.}^2$$
$$A_{gt} = (3)(0.220) = 0.66 \text{ in.}^2$$
$$A_{nt} = A_{gt} - 1(1+1/8)(0.220) = 0.41 \text{ in.}^2$$

Substituting the above into Equations 3.12b since $[0.6F_u A_{nv}$ =94.7 kips] is larger than $[F_u A_{nt} = 23.8$ kips], we obtain $\phi_t P_n$ =88.8 kips, which is less than P_u =104 kips. The C8x11.5 section is therefore not adequate. Significant increase in block shear strength is not expected from the C9x13.4 section because its web thickness t_w is just slightly over that of the C8x11.5 section. As a result, we shall check the adequacy of the C8x13.75 section instead.
    For the C8x13.75 section:

$$A_{gv} = 2(9)(0.303) = 5.45 \text{ in.}^2$$
$$A_{nv} = A_{gv} - 5(1+1/8)(0.303) = 3.75 \text{ in.}^2$$
$$A_{gt} = (3)(0.303) = 0.91 \text{ in.}^2$$
$$A_{nt} = A_{gt} - 1(1+1/8)(0.303) = 0.57 \text{ in.}^2$$

Substituting the above into Equations 3.12b since $[0.6F_u A_{nv}$ =130.5 kips] is larger than $[F_u A_{nt} = 33.1$ kips] we obtain $\phi_t P_n$ =122 kips, which exceeds the required strength P_u of 104 kips. Therefore, block shear will not be a problem for the C8x13.75 section.

Check for the limiting slenderness ratio:

Using the parallel axis theorem, the least radius of gyration of the double channel cross-section is calculated to be 0.96 in. Therefore, $L/r = (15)(12)/0.96 = 187.5$ which is less than the recommended maximum value of 300.

Check for the adequacy of the connection:

The calculations are shown in an example in the section on Connections.

Longitudinal spacing of connectors:

According to Section J3.5 of the LRFD Specification, the maximum spacing of connectors in built-up tension members shall not exceed:

- 24 times the thickness of the thinner plate or 12 in. for painted members or unpainted members not subject to corrosion.
- 14 times the thickness of the thinner plate or 7 in. for unpainted members of weathering steel subject to atmospheric corrosion.

Assuming the first condition applies, a spacing of 6 in. is to be used.

Use 2C8x13.75 Connected Intermittently at 6-in. Interval

3.3.3 Pin-Connected Members

Pin-connected members shall be designed to preclude the following modes of failure: (1) tension yielding on the gross area; (2) tension fracture on the effective net area; (3) longitudinal shear on the effective area; and (4) bearing on the projected pin area (Figure 3.5).

Allowable Stress Design

The allowable stresses for tension yield, tension fracture, and shear rupture are $0.60F_y$, $0.45F_y$, and $0.30F_u$, respectively. The allowable stresses for bearing are given in the section on Connections.

Load and Resistance Factor Design

The design tensile strength $\phi_t P_n$ for a pin-connected member is given as follows:

Tension on gross area: See Equation 3.10

Tension on effective net area:

$$\phi_t P_n = 0.75[2t b_{eff} F_u] \tag{3.13}$$

Shear on effective area:

$$\phi_{sf} P_n = 0.75[0.6 A_{sf} F_u] \tag{3.14}$$

Bearing on projected pin area: See section on Connections

The terms in the above equations are defined as follows:

a = shortest distance from edge of the pin hole to the edge of the member measured in the direction of the force

A_{pb} = projected bearing area $= dt$

A_{sf} = $2t(a + d/2)$

b_{eff} = $2t + 0.63$, but not more than the actual distance from the edge of the hole to the edge of the part measured in the direction normal to the applied force

d = pin diameter

t = plate thickness

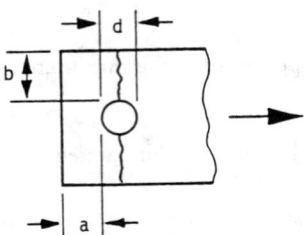

TENSION FRACTURE

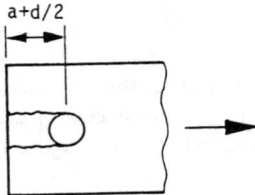

LONGITUDINAL SHEAR

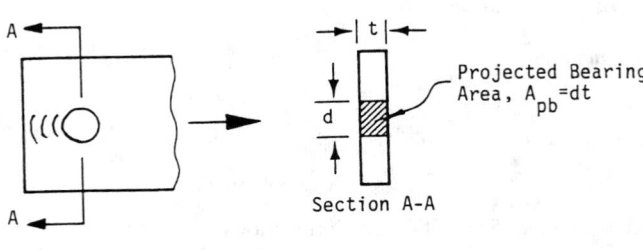

BEARING

Figure 3.5   Failure modes of pin-connected members.

3.3.4 Threaded Rods

Allowable Stress Design

Threaded rods under tension are treated as bolts subject to tension in allowable stress design. These allowable stresses are given in the section on Connections.

Load and Resistance Factor Design

Threaded rods designed as tension members shall have a gross area A_b given by

$$A_b \geq \frac{P_u}{\phi 0.75 F_u} \tag{3.15}$$

where

A_b = the gross area of the rod computed using a diameter measured to the outer extremity of the thread
P_u = the factored tensile load
ϕ = the resistance factor given as 0.75
F_u = the specified minimum tensile strength

3.4 Compression Members

Compression members can fail by yielding, inelastic buckling, or elastic buckling depending on the slenderness ratio of the members. Members with low slenderness ratios tend to fail by yielding while members with high slenderness ratios tend to fail by elastic buckling. Most compression members used in construction have intermediate slenderness ratios and so the predominant mode of failure is inelastic buckling. Overall member buckling can occur in one of three different modes: flexural, torsional, and flexural-torsional. Flexural buckling occurs in members with doubly symmetric or doubly antisymmetric cross-sections (e.g., I or Z sections) and in members with singly symmetric sections (e.g., channel, tee, equal-legged angle, double angle sections) when such sections are buckled about an axis that is *perpendicular* to the axis of symmetry. Torsional buckling occurs in members with doubly symmetric sections such as cruciform or built-up shapes with very thin walls. Flexural-torsional buckling occurs in members with singly symmetric cross-sections (e.g., channel, tee, equal-legged angle, double angle sections) when such sections are buckled about the axis of symmetry and in members with unsymmetric cross-sections (e.g., unequal-legged L). Normally, torsional buckling of symmetric shapes is not particularly important in the design of hot-rolled compression members. It either does not govern or its buckling strength does not differ significantly from the corresponding weak axis flexural buckling strengths. However, torsional buckling may become important for open sections with relatively thin component plates. It should be noted that for a given cross-sectional area, a closed section is much stiffer torsionally than an open section. Therefore, if torsional deformation is of concern, a closed section should be used. Regardless of the mode of buckling, the governing effective slenderness ratio (Kl/r) of the compression member preferably should not exceed 200.

In addition to the slenderness ratio and cross-sectional shape, the behavior of compression members is affected by the relative thickness of the component elements that constitute the cross-section. The relative thickness of a component element is quantified by the width-thickness ratio (b/t) of the element. The width-thickness ratios of some selected steel shapes are shown in Figure 3.6. If the width-thickness ratio falls within a limiting value (denoted by the LRFD specification [18] as λ_r) as shown in Table 3.4, the section will not experience local buckling prior to overall buckling of the member. However, if the width-thickness ratio exceeds this limiting width-thickness value, consideration of local buckling in the design of the compression member is required.

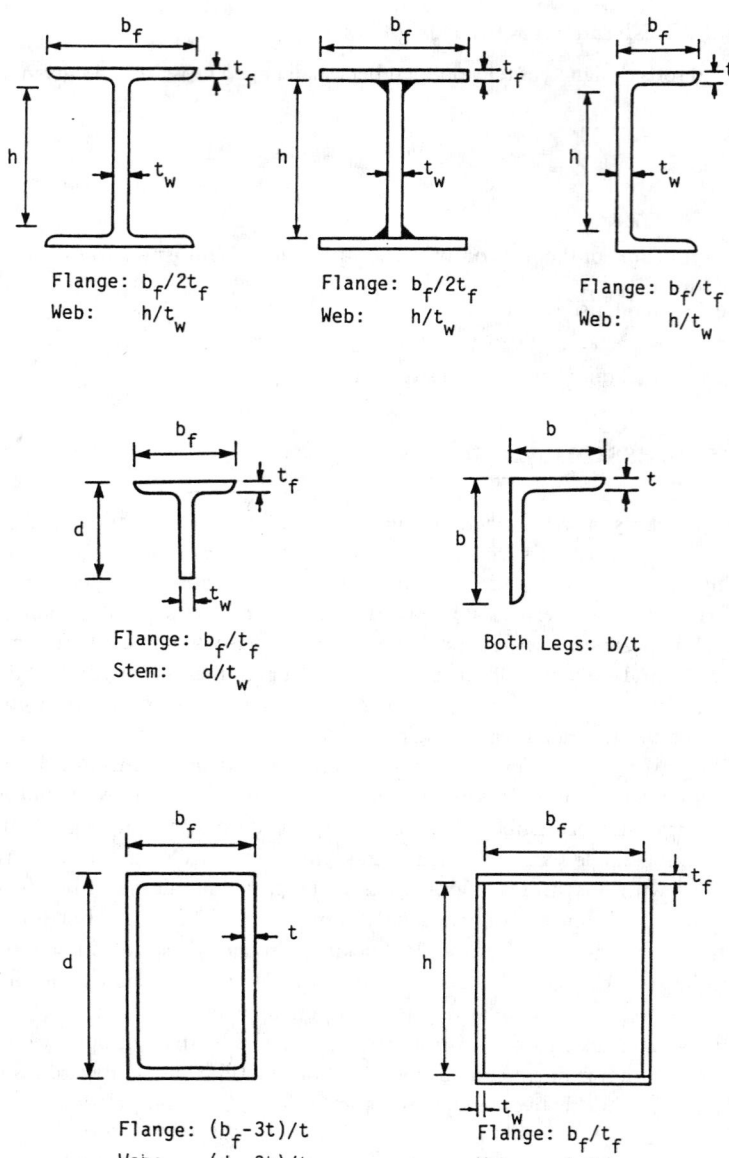

Figure 3.6   Definition of width-thickness ratio of selected cross-sections.

To facilitate the design of compression members, column tables for W, tee, double-angle, square/rectangular tubular, and circular pipe sections are available in the AISC Manuals for both allowable stress design [21] and load and resistance factor design [22].

TABLE 3.4 Limiting Width-Thickness Ratios for Compression Elements Under Pure Compression

Component element	Width-thickness ratio	Limiting value, λ_r
Flanges of I-shaped sections; plates projecting from compression elements; outstanding legs of pairs of angles in continuous contact; flanges of channels.	b/t	$95/\sqrt{f_y}$
Flanges of square and rectangular box and hollow structural sections of uniform thickness; flange cover plates and diaphragm plates between lines of fasteners or welds.	b/t	$238/\sqrt{f_y}$
Unsupported width of cover plates perforated with a succession of access holes.	b/t	$317/\sqrt{f_y}$
Legs of single angle struts; legs of double angle struts with separators; unstiffened elements (i.e., elements supported along one edge).	b/t	$76/\sqrt{f_y}$
Flanges projecting from built-up members.	b/t	$109/\sqrt{(F_y/k_c^a)}$
Stems of tees.	d/t	$127/\sqrt{F_y}$
All other uniformly compressed elements (i.e., elements supported along two edges). Circular hollow sections.	b/t h/t_w D/t D = outside diameter t = wall thickness	$253/\sqrt{F_y}$ $3,300/F_y$

$^a k_c$ = $4/\sqrt{(h/t_w)}$, and $0.35 \leq k_c \leq 0.763$ for I-shaped sections, $k_c = 0.763$ for other sections.
F_y = specified minimum yield stress, in ksi.

3.4.1 Allowable Stress Design

The computed compressive stress, f_a, in a compression member shall not exceed its allowable value given by

$$F_a = \begin{cases} \dfrac{\left[1-\frac{(Kl/r)^2}{2C_c^2}\right]f_y}{\frac{5}{3}+\frac{3(Kl/r)}{8C_c}-\frac{(Kl/r)^3}{8C_c^3}}, & \text{if } Kl/r \leq C_c \\ \dfrac{12\pi^2 E}{23(Kl/r)^2}, & \text{if } Kl/r > C_c \end{cases} \tag{3.16}$$

where Kl/r is the slenderness ratio, K is the effective length factor of the compression member (see Section 3.4.3), l is the unbraced member length, r is the radius of gyration of the cross-section, E is the modulus of elasticity, and $C_c = \sqrt{(2\pi^2 E/F_y)}$ is the slenderness ratio that demarcates between inelastic member buckling from elastic member buckling. Kl/r should be evaluated for both buckling axes and the larger value used in Equation 3.16 to compute F_a.

The first of Equation 3.16 is the allowable stress for inelastic buckling, and the second of Equation 3.16 is the allowable stress for elastic buckling. In ASD, no distinction is made between flexural, torsional, and flexural-torsional buckling.

3.4.2 Load and Resistance Factor Design

Compression members are to be designed so that the design compressive strength $\phi_c P_n$ will exceed the required compressive strength P_u. $\phi_c P_n$ is to be calculated as follows for the different types of overall buckling modes.

Flexural Buckling (with width-thickness ratio $< \lambda_r$):

$$\phi_c P_n = \begin{cases} 0.85 \left[A_g (0.658^{\lambda_c^2}) F_y \right], & \text{if } \lambda_c \leq 1.5 \\ 0.85 \left[A_g \left(\frac{0.877}{\lambda_c^2} \right) F_y \right], & \text{if } \lambda_c > 1.5 \end{cases} \qquad (3.17)$$

where

λ_c = $(KL/r\pi)\sqrt{(F_y/E)}$ is the slenderness parameter
A_g = gross cross-sectional area
F_y = specified minimum yield stress
E = modulus of elasticity
K = effective length factor
l = unbraced member length
r = radius of gyration of the cross-section

The first of Equation 3.17 is the design strength for inelastic buckling and the second of Equation 3.17 is the design strength for elastic buckling. The slenderness parameter $\lambda_c = 1.5$ is therefore the value that demarcates between inelastic and elastic behavior.

Torsional Buckling (with width-thickness ratio $< \lambda_r$):

$\phi_c P_n$ is to be calculated from Equation 3.17, but with λ_c replaced by λ_e given by

$$\lambda_e = \sqrt{(F_y/F_e)} \qquad (3.18)$$

where

$$F_e = \left[\frac{\pi^2 E C_w}{(K_z L)^2} + GJ \right] \frac{1}{I_x + I_y} \qquad (3.19)$$

in which

C_w = warping constant
G = shear modulus = 11,200 ksi (77,200 MPa)
I_x, I_y = moment of inertia about the major and minor principal axes, respectively
J = torsional constant
K_z = effective length factor for torsional buckling

The warping constant C_w and the torsional constant J are tabulated for various steel shapes in the AISC-LRFD Manual [22]. Equations for calculating approximate values for these constants for some commonly used steel shapes are shown in Table 3.5.

Flexural-Torsional Buckling (with width-thickness ratio $\leq \lambda_r$):

Same as for torsional buckling except F_e is now given by
For singly symmetric sections:

$$F_e = \frac{F_{es} + F_{ez}}{2H} \left[1 - \sqrt{1 - \frac{4 F_{es} F_{ez} H}{(F_{es} + F_{ez})^2}} \right] \qquad (3.20)$$

where

F_{es} = F_{ex} if the x-axis is the axis of symmetry of the cross-section, or F_{ey} if the y-axis is the axis of symmetry of the cross-section
F_{ex} = $\pi^2 E/(Kl/r)_x^2$
F_{ey} = $\pi^2 E/(Kl/r)_x^2$
H = $1 - (x_o^2 + y_o^2)/r_o^2$

in which

K_x, K_y = effective length factors for buckling about the x and y axes, respectively

TABLE 3.5 Approximate Equations for C_w and J

Structural shape	Warping constant, C_w	Torsional constant, J
I	$h'^2 I_c I_t/(I_c + I_t)$	$\sum C_i(b_i t_i^3/3)$

where
b_i = width of component element i
t_i = thickness of component element i
C_i = correction factor for component element i (see values below)

Structural shape	Warping constant, C_w
C	$(b' - 3E_o)h'^2 b'^2 t_f/6 + E_o^2 I_x$ where $E_o = b'^2 t_f/(2b't_f + h't_w/3)$
T	$(b_f^3 t_f^3/4 + h''^3 t_w^3)/36$ (≈ 0 for small t)
L	$(l_1^3 t_1^3 + l_2^3 t_2^3)/36$ (≈ 0 for small t)

b_i/t_i	C_i
1.00	0.423
1.20	0.500
1.50	0.588
1.75	0.642
2.00	0.687
2.50	0.747
3.00	0.789
4.00	0.843
5.00	0.873
6.00	0.894
8.00	0.921
10.00	0.936
∞	1.000

b' = distance measured from toe of flange to center line of web
h' = distance between centerline lines of flanges
h'' = distance from centerline of flange to tip of stem
l_1, l_2 = length of the legs of the angle
t_1, t_2 = thickness of the legs of the angle
b_f = flange width
t_f = average thickness of flange
t_w = thickness of web
I_c = moment of inertia of compression flange taken about the axis of the web
I_t = moment of inertia of tension flange taken about the axis of the web
I_x = moment of inertia of the cross-section taken about the major principal axis

l = unbraced member length
r_x, r_y = radii of gyration about the x and y axes, respectively
x_o, y_o = the shear center coordinates with respect to the centroid Figure 3.7
r_o^2 = $x_o^2 + y_o^2 + r_x^2 + r_y^2$

Numerical values for r_o and H are given for hot-rolled W, channel, tee, and single- and double-angle sections in the AISC-LRFD Manual [22].

For unsymmetric sections:
F_e is to be solved from the cubic equation

$$(F_e - F_{ex})(F_e - F_{ey})(F_e - F_{ez}) - F_e^2(F_e - F_{ey})\left(\frac{x_o}{r_o}\right)^2 - F_e^2(F_e - F_{ex})\left(\frac{y_o}{r_o}\right)^2 = 0 \quad (3.21)$$

The terms in the above equations are defined the same as in Equation 3.20.

Local Buckling (with width-thickness ratio $\geq \lambda_r$):
   Local buckling in a component element of the cross-section is accounted for in design by introducing a reduction factor Q in Equation 3.17 as follows:

$$\phi_c P_n = \begin{cases} 0.85\left[A_g Q\left(0.658^{Q\lambda^2}\right)F_y\right], & \text{if } \lambda\sqrt{Q} \leq 1.5 \\ 0.85\left[A_g\left(\frac{0.877}{\lambda^2}\right)F_y\right], & \text{if } \lambda\sqrt{Q} > 1.5 \end{cases} \quad (3.22)$$

where $\lambda = \lambda_c$ for flexural buckling, and $\lambda = \lambda_e$ for flexural-torsional buckling.
   The Q factor is given by

$$Q = Q_s Q_a \quad (3.23)$$

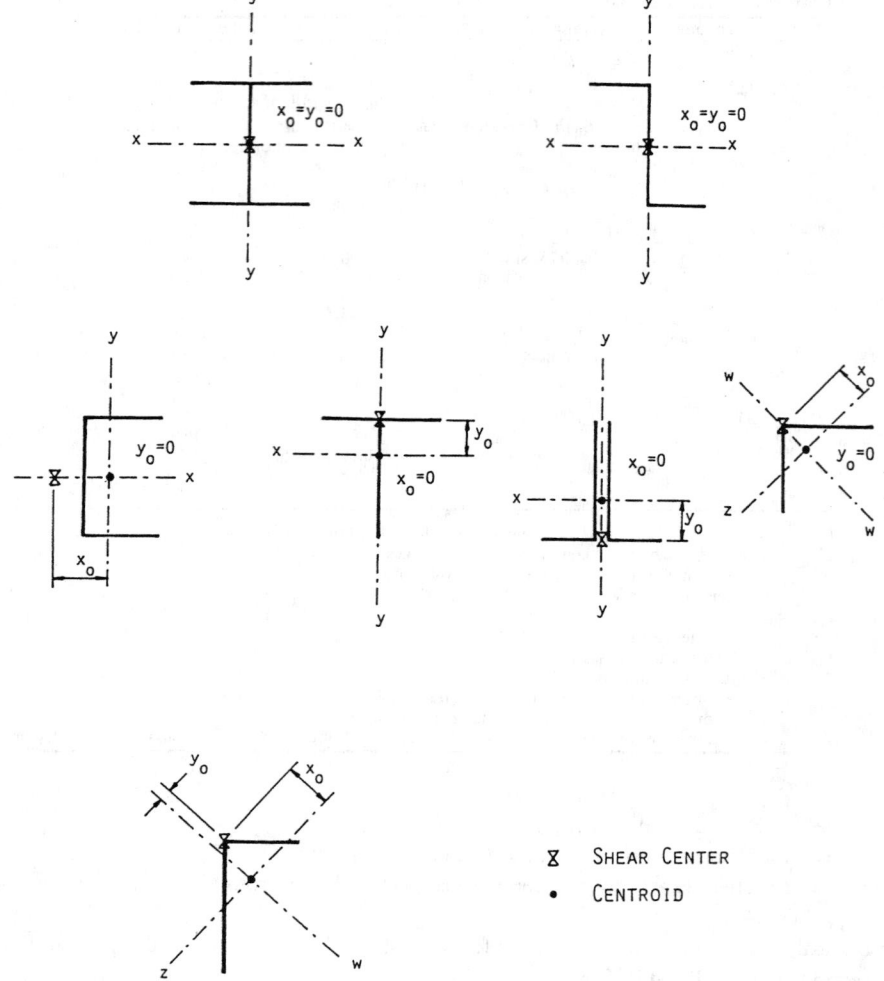

Figure 3.7   Location of shear center for selected cross-sections.

where

Q_s is the reduction factor for unstiffened compression elements of the cross-section (see Table 3.6); and Q_a is the reduction factor for stiffened compression elements of the cross-section (see Table 3.7)

3.4.3   Built-Up Compression Members

Built-up members are members made by bolting and/or welding together two or more standard structural shapes. For a built-up member to be fully effective (i.e., if all component structural shapes are to act as one unit rather than as individual units), the following conditions must be satisfied:

1. The ends of the built-up member must be prevented from slippage during buckling.
2. Adequate fasteners must be provided along the length of the member.
3. The fasteners must be able to provide sufficient gripping force on all the component shapes being connected.

TABLE 3.6 Formulas for Q_s

Structural element	Range of b/t	Q_s
Single angles	$76.0/\sqrt{F_y} < b/t < 155/\sqrt{F_y}$	$1.340 - 0.00447(b/t)\sqrt{f_y}$
	$b/t \geq 155/\sqrt{f_y}$	$15,500/[F_y(b/t)^2]$
Flanges, angles, and plates projecting from columns or other compression members	$95.0/\sqrt{F_y} < b/t < 176/\sqrt{f_y}$	$1.415 - 0.00437(b/t)\sqrt{f_y}$
	$b/t \geq 176/\sqrt{F_y}$	$20,000/[F_y(b/t)^2]$
Flanges, angles, and plates projecting from built-up columns or other compression members	$109/\sqrt{(F_y/k_c^a)} < b/t < 200/\sqrt{(F_y/k_c)}$	$1.415 - 0.00381(b/t)\sqrt{(F_y/k_c)}$
	$b/t \geq 200/\sqrt{(F_y/k_c)}$	$26,200k_c/[F_y(b/t)^2]$
Stems of tees	$127/\sqrt{F_y} < b/t < 176/\sqrt{F_y}$	$1.908 - 0.00715(b/t)\sqrt{F_y}$
	$b/t \geq 176/\sqrt{f_y}$	$20,000/[F_y(b/t)^2]$

^{a} see footnote a in Table 3.4
F_y = specified minimum yield stress, in ksi
b = width of the component element
t = thickness of the component element

TABLE 3.7 Formula for Q_a

$$Q_s = \frac{\text{effective area}}{\text{actual area}}$$

The effective area is equal to the summation of the effective areas of the stiffened elements of the cross-section. The effective area of a stiffened element is equal to the product of its thickness t and its effective width b_e given by:

For flanges of square and rectangular sections of uniform thickness: when $b/t \geq \frac{238^a}{\sqrt{f}}$

$$b_e = \frac{326t}{\sqrt{f}}\left[1 - \frac{64.9}{(b/t)\sqrt{f}}\right] \leq b$$

For other uniformly compressed elements: when $b/t \geq \frac{253^a}{\sqrt{f}}$

$$b_e = \frac{326t}{\sqrt{f}}\left[1 - \frac{57.2}{(b/t)\sqrt{f}}\right] \leq b$$

where
b = actual width of the stiffened element
f = computed elastic compressive stress in the stiffened elements, in ksi

$^a b_e$ = b otherwise.

Condition 1 is satisfied if all component shapes in contact at the ends of the member are connected by a weld having a length not less than the maximum width of the member or by fully tightened bolts spaced longitudinally not more than four diameters apart for a distance equal to 1-1/2 times the maximum width of the member.

Condition 2 is satisfied if continuous welds are used throughout the length of the built-up compression member.

Condition 3 is satisfied if either welds or fully tightened bolts are used as the fasteners.

While condition 1 is mandatory, conditions 2 and 3 can be violated in design. If condition 2 or 3 is violated, the built-up member is not fully effective and slight slippage among component shapes may occur. To account for the decrease in capacity due to slippage, a modified slenderness ratio is used for the computation of the design compressive strength when buckling of the built-up member is about an axis *coincide* or *parallel* to at least one plane of contact for the component shapes. The modified slenderness ratio $(KL/r)_m$ is given as follows:

If condition 2 is violated:

$$\left(\frac{KL}{r}\right)_m = \sqrt{\left(\frac{KL}{r}\right)_o^2 + \frac{0.82\alpha^2}{(1+\alpha^2)}\left(\frac{a}{r_{ib}}\right)^2} \qquad (3.24)$$

If conditions 2 and 3 are violated:

$$\left(\frac{KL}{r}\right)_m = \sqrt{\left(\frac{KL}{r}\right)_o^2 + \left(\frac{a}{r_i}\right)^2} \qquad (3.25)$$

In the above equations, $(KL/r)_o = (KL/r)_x$ if the buckling axis is the x-axis and at least one plane of contact between component shapes is parallel to that axis; $(KL/r)_o = (KL/r)_y$ if the buckling axis is the y axis and at least one plane of contact is parallel to that axis. a is the longitudinal spacing of the fasteners, r_i is the minimum radius of gyration of any component element of the built-up cross-section, r_{ib} is the radius of gyration of an individual component relative to its centroidal axis parallel to the axis of buckling of the member, h is the distance between centroids of component elements measured perpendicularly to the buckling axis of the built-up member.

No modification to (KL/r) is necessary if the buckling axis is perpendicular to the planes of contact of the component shapes. Modifications to both $(KL/r)_x$ and $(KL/r)_y$ are required if the built-up member is so constructed that planes of contact exist in both the x and y directions of the cross-section.

Once the modified slenderness ratio is computed, it is to be used in the appropriate equation to calculate F_a in allowable stress design, or $\phi_c P_n$ in load and resistance factor design.

An additional requirement for the design of built-up members is that the effective slenderness ratio, Ka/r_i, of each component shape, where K is the effective length factor of the component shape between adjacent fasteners, does not exceed 3/4 of the governing slenderness ratio of the built-up member. This provision is provided to prevent component shape buckling between adjacent fasteners from occurring prior to overall buckling of the built-up member.

EXAMPLE 3.2:

Using LRFD, determine the size of a pair of cover plates to be bolted, using snug-tight bolts, to the flanges of a W24x229 section as shown in Figure 3.8 so that its design strength, $\phi_c P_n$, will be increased by 15%. Also, determine the spacing of the bolts in the longitudinal direction of the built-up column. The effective lengths of the section about the major $(KL)_x$ and minor $(KL)_y$ axes are both equal to 20 ft. A36 steel is to be used.

Determine design strength for the W24x229 section:

Since $(KL)_x = (KL)_y$ and $r_x > r_y$, $(KL/r)_y$ will be greater than $(KL/r)_x$ and the design strength will be controlled by flexural buckling about the minor axis. Using section properties, $r_y = 3.11$ in. and $A = 67.2$ in.^{2}, obtained from the AISC-LRFD Manual [22], the slenderness parameter λ_c about the minor axis can be calculated as follows:

$$(\lambda_c)_y = \frac{1}{\pi}\left(\frac{KL}{r}\right)_y \sqrt{\frac{F_y}{E}} = \frac{1}{3.142}\left(\frac{20 \times 12}{3.11}\right)\sqrt{\frac{36}{29,000}} = 0.865$$

Substituting $\lambda_c = 0.865$ into Equation 3.17, the design strength of the section is

$$\phi_c P_n = 0.85\left[67.2\left(0.658^{0.865^2}\right)36\right] = 1503 \text{ kips}$$

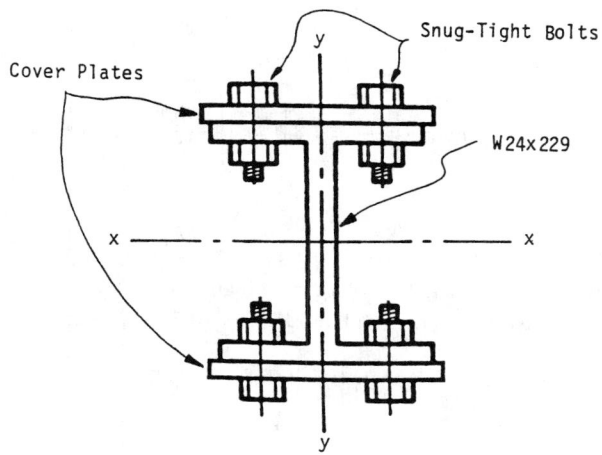

Figure 3.8 Design of cover plates for a compression member.

Alternatively, the above value of $\phi_c P_n$ can be obtained directly from the column tables contained in the AISC-LRFD Manual.

Determine design strength for the built-up section:

The built-up section is expected to possess a design strength which is 15% in excess of the design strength of the W24x229 section, so

$$(\phi_c P_n)_{req'd} = (1.15)(1503) = 1728 \text{ kips}$$

Determine size of the cover plates:

After cover plates are added, the resulting section is still doubly symmetric. Therefore, the overall failure mode is still flexural buckling. For flexural buckling about the minor axis (y-y), no modification to (KL/r) is required because the buckling axis is perpendicular to the plane of contact of the component shapes and no relative movement between the adjoining parts is expected. However, for flexural buckling about the major (x-x) axis, modification to (KL/r) is required because the buckling axis is parallel to the plane of contact of the adjoining structural shapes and slippage between the component pieces will occur. We shall design the cover plates assuming flexural buckling about the minor axis will control and check for flexural buckling about the major axis later.

A W24x229 section has a flange width of 13.11 in.; so, as a trial, use cover plates with widths of 13 in. as shown in Figure 3.8a. Denoting t as the thickness of the plates, we have

$$(r_y)_{\text{built-up}} = \sqrt{\frac{(I_y)_{\text{W-shape}} + (I_y)_{\text{plates}}}{A_{\text{W-shape}} + A_{\text{plates}}}} = \sqrt{\frac{651 + 183.1t}{67.2 + 26t}}$$

and

$$(\lambda_c)_{y,\text{built-up}} = \frac{1}{\pi}\left(\frac{KL}{r}\right)_{y,\text{built-up}}\sqrt{\frac{F_y}{E}} = 2.69\sqrt{\frac{67.2 + 26t}{651 + 183.1t}}$$

Assuming $(\lambda)_{y,\text{built-up}}$ is less than 1.5, one can substitute the above expression for λ_c in Equation 3.17. With $\phi_c P_n$ equals 1728, we can solve for t. The result is $t = 1/2$ in. Backsubstituting $t = 1/2$ into the above expression, we obtain $(\lambda)_{c,\text{built-up}} = 0.884$ which is indeed <1.5. So, try 13"x1/2" cover plates.

Check for local buckling:

For the I-section:

$$\text{Flange:} \quad \left[\frac{b_f}{2t_f} = 3.8\right] \quad < \quad \left[\frac{95}{\sqrt{F_y}} = 15.8\right]$$

$$\text{Web:} \quad \left[\frac{h_c}{t_w} = 22.5\right] \quad < \quad \left[\frac{253}{\sqrt{F_y}} = 42.2\right]$$

For the cover plates, if 3/4-in. diameter bolts are used and assuming an edge distance of 1-1/4 in., the width of the plate between fasteners will be $13-2.5 = 10.5$ in. Therefore, we have

$$\left[\frac{b}{t} = \frac{10.5}{1/2} = 21\right] < \left[\frac{238}{\sqrt{F_y}} = \frac{238}{\sqrt{36}} = 39.7\right]$$

Since the width-thickness ratios of all component shapes do not exceed the limiting width-thickness ratio for local buckling, local buckling is not a concern.

Check for flexural buckling about the major (x-x) axis:

Since the built-up section is doubly symmetric, the governing buckling mode will be flexural buckling regardless of the axes. Flexural buckling will occur about the major axis if the modified slenderness ratio $(KL/r)_m$ about the major axis exceeds $(KL/r)_y$. Therefore, as long as $(KL/r)_m$ is less than $(KL/r)_y$, buckling will occur about the minor axis and flexural buckling about the major axis will not be controlled. In order to arrive at an optimal design, we shall determine the longitudinal fastener spacing, a, such that the modified slenderness ratio $(KL/r)_m$ about the major axis will be equal to $(KL/r)_y$. That is, we shall solve for a from the equation

$$\left[\left(\frac{KL}{r}\right)_m = \sqrt{\left(\frac{KL}{r}\right)_x^2 + \left(\frac{a}{r_i}\right)^2}\right] = \left[\left(\frac{KL}{r}\right)_y = 78.9\right]$$

In the above equation, $(KL/r)_x$ is the slenderness ratio about the major axis of the built-up section, r_i is the least radius of gyration of the component shapes, which in this case is the cover plate. Substituting $(KL/r)_x = 21.56$, $r_i = r_{\text{cover plate}} = \sqrt{(I/A)_{\text{cover plate}}} = \sqrt{[(1/2)^2/112]} = 0.144$ into the above equation, we obtain $a = 10.9$ in. Since $(KL) = 20$ ft, we shall use $a = 10$ in. for the longitudinal spacing of the fasteners.

Check for component shape buckling between adjacent fasteners:

$$\left[\frac{Ka}{r_i} = \frac{1 \times 10}{0.144} = 69.44\right] > \left[\frac{3}{4}\left(\frac{KL}{r}\right)_y = \frac{3}{4}(78.9) = 59.2\right]$$

Since the component shape buckling criterion is violated, we need to decrease the longitudinal spacing from 10 in. to 8 in.

Use 13"x1/2" cover plates bolted to the flanges of the W24x229 section by 3/4-in. diameter fully tightened bolts spaced 8 in. longitudinally.

3.5 Flexural Members

Depending on the width-thickness ratios of the component elements, steel sections used for **flexural members** are classified as compact, noncompact, and slender element sections. Compact sections are sections that can develop the cross-section plastic moment (M_p) under flexure and sustain that moment through a large hinge rotation without fracture. Noncompact sections are sections that either cannot develop the cross-section full plastic strength or cannot sustain a large hinge rotation at M_p, probably due to local buckling of the flanges or web. Slender element sections are sections that fail by local buckling of component elements long before M_p is reached. A section is considered compact if all its component elements have width-thickness ratios less than a limiting value (denoted as λ_p in LRFD). A section is considered noncompact if one or more of its component elements have width-thickness ratios that fall in between λ_p and λ_r. A section is considered to be a slender element if one or more of its component elements have width-thickness ratios that exceed λ_r. Expressions for λ_p and λ_r are given in the Table 3.8

In addition to the compactness of the steel section, another important consideration for beam design is the lateral unsupported (unbraced) length of the member. For beams bent about their strong axes, the failure modes, or limit states, vary depending on the number and spacing of lateral supports provided to brace the compression flange of the beam. The compression flange of a beam behaves somewhat like a compression member. It buckles if adequate lateral supports are not provided in a phenomenon called lateral torsional buckling. *Lateral torsional buckling* may or may not be accompanied by yielding, depending on the lateral unsupported length of the beam. Thus, lateral torsional buckling can be inelastic or elastic. If the lateral unsupported length is large, the limit state is elastic lateral torsional buckling. If the lateral unsupported length is smaller, the limit state is inelastic lateral torsional buckling. For compact section beams with adequate lateral supports, the limit state is full yielding of the cross-section (i.e., plastic hinge formation). For noncompact section beams with adequate lateral supports, the limit state is flange or web local buckling.

For beams bent about their weak axes, lateral torsional buckling will not occur and so the lateral unsupported length has no bearing on the design. The limit states for such beams will be formation of a plastic hinge if the section is compact. The limit state will be flange or web local buckling if the section is noncompact.

Beams subjected to high shear must be checked for possible web shear failure. Depending on the width-thickness ratio of the web, failure by shear yielding or web shear buckling may occur. Short, deep beams with thin webs are particularly susceptible to web shear failure. If web shear is of concern, the use of thicker webs or web reinforcements such as stiffeners is required.

Beams subjected to concentrated loads applied in the plane of the web must be checked for a variety of possible flange and web failures. Failure modes associated with concentrated loads include local flange bending (for tensile concentrated load), local web yielding (for compressive concentrated load), web crippling (for compressive load), sidesway web buckling (for compressive load), and compression buckling of the web (for a compressive load pair). If one or more of these conditions is critical, transverse stiffeners extending at least one-half the beam depth (use full depth for compressive buckling of the web) must be provided adjacent to the concentrated loads.

Long beams can have deflections that may be too excessive, leading to problems in serviceability. If deflection is excessive, the use of intermediate supports or beams with higher flexural rigidity is required.

The design of flexural members should satisfy the following criteria: (1) flexural strength criterion, (2) shear strength criterion, (3) criteria for concentrated loads, and (4) deflection criterion. To facilitate beam design, a number of beam tables and charts are given in the AISC Manuals [21, 22] for both allowable stress and load and resistance factor design.

3.5.1 Allowable Stress Design

Flexural Strength Criterion

The computed flexural stress, f_b, shall not exceed the allowable flexural stress, F_b, given as follows (in all equations, the minimum specified yield stress, F_y, cannot exceed 65 ksi):

TABLE 3.8 λ_p and λ_r for Members Under Flexural Compression

Component element	Width-thickness ratio[a]	λ_p	λ_r
Flanges of I-shaped rolled beams and channels	b/t	$65/\sqrt{F_y}$	$141/\sqrt{(F_y - 10)}$[b]
Flanges of I-shaped hybrid or welded beams	b/t	$65/\sqrt{F_{yf}}$ (non-seismic) $52/\sqrt{F_{yf}}$ (seismic) F_{yf} = yield stress of flange	$162/\sqrt{(F_{yf} - 16.5)/k_c}$[c] F_{yw} = yield stress of web
Flanges of square and rectangular box and hollow structural sections of uniform thickness; flange cover plates and diaphragm plates between lines of fasteners or welds	b/t	$190/\sqrt{F_y}$	$238/\sqrt{F_y}$
Unsupported width of cover plates perforated with a succession of access holes	b/t	NA	$317/\sqrt{F_y}$
Legs of single angle struts; legs of double angle struts with separators; unstiffened elements	b/t	NA	$76/\sqrt{F_y}$
Stems of tees	d/t	NA	$127/\sqrt{F_y}$
Webs in flexural compression	h_c/t_w	$640/\sqrt{F_y}$ (non-seismic) $520/\sqrt{F_y}$ (seismic)	$970/\sqrt{F_y}$[d]
Webs in combined flexural and axial compression	h_c/t_w	For $P_u/\phi_b P_y \le 0.125$: $640(1 - 2.75 P_u/\phi_b P_y)/\sqrt{F_y}$ (non-seismic) $520(1 - 1.54 P_u/\phi_b P_y)/\sqrt{F_y}$ (seismic) For $P_u/\phi_b P_y > 0.125$: $191(2.33 - P_u/\phi_b P_y)/\sqrt{F_y}$ $\ge 253/\sqrt{F_y}$ $\phi_b = 0.90$ P_u = factored axial force; $P_y = A_g F_y$.	$970/\sqrt{F_y}$[d]
Circular hollow sections	D/t D = outside diameter; t = wall thickness	$2,070/F_y$ $1,300/F_y$ for plastic design	$8,970/F_y$

[a] See Figure 3.6 for definition of b, h_c, and t

[b] For ASD, this limit is $95/\sqrt{F_y}$

[c] For ASD, this limit is $95/\sqrt{(F_{yf}/k_c)}$, where $k_c = 4.05/(h/t)^{0.46}$ if $h/t > 70$, otherwise $k_c = 1.0$

[d] For ASD, this limit is $760/\sqrt{F_b}$

Note: All stresses have units of ksi.

Compact-Section Members Bent About Their Major Axes
For $L_b \leq L_c$,

$$F_b = 0.66F_y \tag{3.26}$$

where

L_c = smaller of $\{76b_f/\sqrt{F_y}, 20000/(d/A_f)F_y\}$, for I and channel shapes

= $[1950 + 1200(M_1/M_2)](b/F_y) \geq 1200(b/F_y)$, for box sections, rectangular and circular tubes

in which

b_f = flange width, in.

d = overall depth of section, ksi

A_f = area of compression flange, in.^{2}

b = width of cross-section, in.

M_1/M_2 = ratio of the smaller to larger moment at the ends of the unbraced length of the beam. M_1/M_2 is positive for reverse curvature bending and negative for single curvature bending.

For the above sections to be considered compact, in addition to having the width-thickness ratios of their component elements falling within the limiting value of λ_p shown in Table 3.8, the flanges of the sections must be continuously connected to the webs. For box-shaped sections, the following requirements must also be satisfied: the depth-to-width ratio should not exceed six, and the flange-to-web thickness ratio should exceed two.

For $L_b > L_c$, the allowable flexural stress in tension is given by

$$F_b = 0.60F_y \tag{3.27}$$

and the allowable flexural stress in compression is given by the larger value calculated from Equation 3.28 and Equation 3.29. Equation 3.28 normally controls for deep, thin-flanged sections where warping restraint torsional resistance dominates, and Equation 3.29 normally controls for shallow, thick-flanged sections where St. Venant torsional resistance dominates.

$$F_b = \begin{cases} \left[\dfrac{2}{3} - \dfrac{F_y(l/r_T)^2}{1530 \times 10^3 C_b}\right] F_y \leq 0.60F_y, & \text{if } \sqrt{\dfrac{102,000C_b}{F_y}} \leq \dfrac{l}{r_T} < \sqrt{\dfrac{510,000C_b}{F_y}} \\ \dfrac{170,000C_b}{(l/r_T)^2} \leq 0.60F_y, & \text{if } \dfrac{l}{r_T} \geq \sqrt{\dfrac{510,000C_b}{F_y}} \end{cases} \tag{3.28}$$

$$F_b = \frac{12,000C_b}{ld/A_f} \leq 0.60F_y \tag{3.29}$$

where

l = distance between cross-sections braced against twist or lateral displacement of the compression flange, in.

r_T = radius of gyration of a section comprising the compression flange plus 1/3 of the compression web area, taken about an axis in the plane of the web, in.

A_f = compression flange area, in.^{2}

C_b = $12.5M_{max}/(2.5M_{max} + 3M_A + 4M_B + 3M_C)$

M_{max}, M_A, M_B, M_C = maximum moment, quarter-point moment, midpoint moment, and three-quarter point moment along the unbraced length of the member, respectively.

For simplicity in design, C_b can conservatively be taken as unity.

It should be cautioned that Equations 3.28 and 3.29 are applicable only to I and channel shapes with an axis of symmetry in, and loaded in the plane of the web. In addition, Equation 3.29 is applicable only if the compression flange is solid and approximately rectangular in shape, and its area is not less than the tension flange.

Compact Section Members Bent About Their Minor Axes

Since lateral torsional buckling will not occur for bending about the minor axes, regardless of the value of L_b, the allowable flexural stress is

$$F_b = 0.75 F_y \tag{3.30}$$

Noncompact Section Members Bent About Their Major Axes

For $L_b \leq L_c$,

$$F_b = 0.60 F_y \tag{3.31}$$

where L_c is defined as for Equation 3.26.

For $L_b > L_c$, F_b is given in Equation 3.27, 3.28, or 3.29.

Noncompact Section Members Bent About Their Minor Axes

Regardless of the value of L_b,

$$F_b = 0.60 F_y \tag{3.32}$$

Slender Element Sections

Refer to the section on Plate Girders.

Shear Strength Criterion

For practically all structural shapes commonly used in constructions, the shear resistance from the flanges is small compared to the webs. As a result, the shear resistance for flexural members is normally determined on the basis of the webs only. The amount of web shear resistance is dependent on the width-thickness ratio h/t_w of the webs. If h/t_w is small, the failure mode is web yielding. If h/t_w is large, the failure mode is web buckling. To avoid web shear failure, the computed shear stress, f_v, shall not exceed the allowable shear stress, F_v, given by

$$F_v = \begin{cases} 0.40 F_y, & \text{if } \dfrac{h}{t_w} \leq \dfrac{380}{\sqrt{F_y}} \\[2ex] \dfrac{C_v}{2.89} F_y \leq 0.40 F_y, & \text{if } \dfrac{h}{t_w} > \dfrac{380}{\sqrt{F_y}} \end{cases} \tag{3.33}$$

where

$$
\begin{aligned}
C_v &= 45{,}000 k_v / F_y (h/t_w)^2, \text{ if } C_v \leq 0.8 \\
    &= 190 \sqrt{(k_v/F_y)}/(h/t_w), \text{ if } C_v > 0.8 \\
k_v &= 4.00 + 5.34/(a/h)^2, \text{ if } a/h \leq 1.0 \\
    &= 5.34 + 4.00/(a/h)^2, \text{ if } a/h > 1.0 \\
t_w &= \text{web thickness, in.} \\
a &= \text{clear distance between transverse stiffeners, in.} \\
h &= \text{clear distance between flanges at section under investigation, in.}
\end{aligned}
$$

Criteria for Concentrated Loads

Local Flange Bending

If the concentrated force that acts on the beam flange is tensile, the beam flange may experience excessive bending, leading to failure by fracture. To preclude this type of failure, transverse stiffeners are to be provided opposite the tension flange unless the length of the load when measured across the beam flange is less than 0.15 times the flange width, or if the flange thickness, t_f, exceeds

$$0.4 \sqrt{\frac{P_{bf}}{F_y}} \tag{3.34}$$

where

P_{bf} = computed tensile force multiplied by 5/3 if the force is due to live and dead loads only, or by 4/3 if the force is due to live and dead loads in conjunction with wind or earthquake loads, kips.

F_y = specified minimum yield stress, ksi.

Local Web Yielding

To prevent local web yielding, the concentrated compressive force, R, should not exceed $0.66R_n$, where R_n is the web yielding resistance given in Equation 3.52 or Equation 3.53, whichever applies.

Web Crippling

To prevent web crippling, the concentrated compressive force, R, should not exceed $0.50R_n$, where R_n is the web crippling resistance given in Equation 3.54, Equation 3.55, or Equation 3.56, whichever applies.

Sidesway Web Buckling

To prevent sidesway web buckling, the concentrated compressive force, R, should not exceed R_n, where R_n is the sidesway web buckling resistance given in Equation 3.57 or Equation 3.58, whichever applies, except the term $C_r t_w^3 t_f / h^2$ is replaced by $6,800 t_w^3 / h$.

Compression Buckling of the Web

When the web is subjected to a pair of concentrated forces acting on both flanges, buckling of the web may occur if the web depth clear of fillet, d_c, is greater than

$$\frac{4100 t_w^3 \sqrt{F_y}}{P_{bf}} \tag{3.35}$$

where t_w is the web thickness, F_y is the minimum specified yield stress, and P_{bf} is as defined in Equation 3.34.

Deflection Criterion

Deflection is a serviceability consideration. Since most beams are fabricated with a camber which somewhat offsets the dead load deflection, consideration is often given to deflection due to live load only. For beams supporting plastered ceilings, the service live load deflection preferably should not exceed $L/360$ where L is the beam span. A larger deflection limit can be used if due considerations are given to ensure the proper functioning of the structure.

EXAMPLE 3.3:

Using ASD, determine the amount of increase in flexural capacity of a W24x55 section bent about its major axis if two 7"x1/2" (178mmx13mm) cover plates are bolted to its flanges as shown in Figure 3.9. The beam is laterally supported at every 5-ft (1.52-m) interval. Use A36 steel. Specify the type, diameter, and longitudinal spacing of the bolts used if the maximum shear to be resisted by the cross-section is 100 kips (445 kN).

Section properties:

A W24x55 section has the following section properties:

$b_f = 7.005$ in. $t_f = 0.505$ in. $d = 23.57$ in. $t_w = 0.395$ in. $I_x = 1350$ in.[4] $S_x = 114$ in.[3]

Check compactness:

Refer to Table 3.8, and assuming that the transverse distance between the two bolt lines is 4 in.,

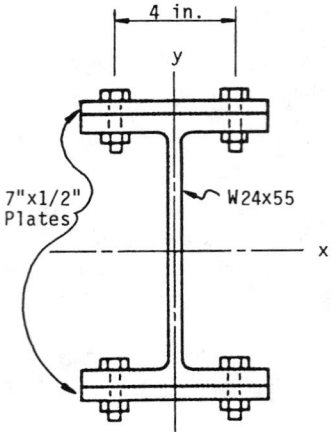

Figure 3.9   Cover-plated beam section.

we have

$$\text{Beam flanges} \qquad \left[\tfrac{b_f}{2t_f} = 6.94"\right] \quad < \quad \left[\tfrac{65}{\sqrt{F_y}} = 10.8"\right]$$

$$\text{Beam web} \qquad \left[\tfrac{d}{t_w} = 59.7"\right] \quad < \quad \left[\tfrac{640}{\sqrt{F_y}} = 107"\right]$$

$$\text{Cover plates} \qquad \left[\tfrac{4}{1/2} = 8"\right] \quad < \quad \left[\tfrac{190}{\sqrt{F_y}} = 31.7"\right]$$

Therefore, the section is compact.

Determine the allowable flexural stress, F_b:

Since the section is compact and the lateral unbraced length, $L_b = 60$ in., is less than $L_c = 83.4$ in., the allowable bending stress from Equation 3.26 is $0.66F_y = 24$ ksi.

Determine section modulus of the beam with cover plates:

$$\begin{aligned}
S_{x,\text{combination section}} &= \frac{I_{x,\text{combination section}}}{c} \\
&= \frac{1350 + 2\left[\left(\tfrac{1}{12}\right)(7)(1/2)^3 + (7)(1/2)(12.035)^2\right]}{(11.785 + 1/2)} = 192 \ \text{in.}^3
\end{aligned}$$

Determine flexural capacity of the beam with cover plates:

$$M_{x,\text{combination section}} = S_{x,\text{combination section}}F_b = (192)(24) = 4608 \ \text{k-in.}$$

Since the flexural capacity of the beam without cover plates is

$$M_x = S_x F_b = (114)(24) = 2736 \ \text{k-in.}$$

the increase in flexural capacity is 68.4%.

Determine diameter and longitudinal spacing of bolts:

From *Mechanics of Materials*, the relationship between the shear flow, q, the number of bolts per shear plane, n, the allowable bolt shear stress, F_v, the cross-sectional bolt area, A_b, and the longitudinal bolt spacing, s, at the interface of two component elements of a combination section is given by

$$\frac{n F_v A_b}{s} = q$$

Substituting $n = 2, q = VQ/I = (100)[(7)(1/2)(12.035)]/2364 = 1.78$ k/in. into the above equation, we have

$$\frac{F_v A_b}{s} = 0.9 \text{ k/in.}$$

If 1/2" diameter A325N bolts are used, we have $A_b = \pi(1/2)^2/4 = 0.196$ in.^{2}, and $F_v = 21$ ksi (from Table 3.12), from which s can be solved from the above equation to be 4.57 in. However, for ease of installation, use $s = 4.5$ in.

In calculating the section properties of the combination section, no deduction is made for the bolt holes in the beam flanges nor the cover plates. This is allowed provided that the following condition is satisfied:

$$0.5 F_u A_{fn} \geq 0.6 F_y A_{fg}$$

where F_y and F_u are the minimum specified yield strength and tensile strength, respectively. A_{fn} is the net flange area and A_{fg} is the gross flange area. For this problem

Beam flanges

$$\left[0.5 F_u A_{fn} = 0.5(58)(7.005 - 2 \times 1/2)(0.505) = 87.9 \text{ kips} \right]$$

$$> \left[0.6 F_y A_{fg} = 0.6(36)(7.005)(0.505) = 76.4 \text{ kips} \right]$$

Cover Plates

$$\left[0.5 F_u A_{fn} = 0.5(58)(7 - 2 \times 1/2)(1/2) = 87 \text{ kips} \right]$$

$$> \left[0.6 F_y A_{fg} = 0.6(36)(7)(1/2) = 75.6 \text{ kips} \right]$$

so the use of the gross cross-sectional area to compute section properties is justified. In the event that the condition is violated, cross-sectional properties should be evaluated using an effective tension flange area A_{fe} given by

$$A_{fe} = \frac{5}{6} \frac{F_u}{F_y} A_{fn}$$

Use 1/2" diameter A325N bolts spaced 4.5" apart longitudinally in two lines 4" apart to connect the cover plates to the beam flanges.

3.5.2 Load and Resistance Factor Design

Flexural Strength Criterion

Flexural members must be designed to satisfy the flexural strength criterion of

$$\phi_b M_n \geq M_u \tag{3.36}$$

where $\phi_b M_n$ is the design flexural strength and M_u is the required strength. The design flexural strength is determined as follows:

Compact Section Members Bent About Their Major Axes
For $L_b \leq L_p$, (Plastic hinge formation)

$$\phi_b M_n = 0.90 M_p \tag{3.37}$$

For $L_p < L_b \leq L_r$, (Inelastic lateral torsional buckling)

$$\phi_b M_n = 0.90 C_b \left[M_p - (M_p - M_r) \left(\frac{L_b - L_p}{L_r - L_p} \right) \right] \leq 0.90 M_p \tag{3.38}$$

For $L_b > L_r$, (Elastic lateral torsional buckling)

For I-shaped members and channels:

$$\phi_b M_n = 0.90 C_b \left[\frac{\pi}{L_b} \sqrt{EI_y GJ + \left(\frac{\pi E}{L_b}\right)^2 I_y C_w} \right] \le 0.90 M_p \qquad (3.39)$$

For solid rectangular bars and symmetric box sections:

$$\phi_b M_n = 0.90 C_b \frac{57,000\sqrt{JA}}{L_b/r_y} \le 0.90 M_p \qquad (3.40)$$

The variables used in the above equations are defined in the following.

L_b = lateral unsupported length of the member
L_p, L_r = limiting lateral unsupported lengths given in the following table

Structural shape	L_p	L_r
I-shaped sections, chanels	$300 r_y/\sqrt{F_{yf}}$	$\left[r_y X_1/F_L \right] \left\{ \sqrt{\left[1 + \sqrt{\left(1 + X_2 F_L^2\right)} \right]} \right\}$
	where	where
	r_y = radius of gyration about minor axis, in.	r_y = radius of gyration about minor axis, in.
	F_{yf} = flange yield stress, ksi	$X_1 = (\pi/S_x)\sqrt{(EGJA/2)}$
		$X_2 = (4C_w/I_y)(S_x/GJ)^2$
		F_L = smaller of $(F_{yf} - F_r)$ or F_{yw}
		F_{yf} = flange yield stress, ksi
		F_{yw} = web yield stress, ksi
		F_r = 10 ksi for rolled shapes, 16.5 ksi for welded shapes
		S_x = elastic section modulus about the major axis, in.^{3} (use S_{xc}, the elastic section modulus about the major axis with respect to the compression flange if the compression flange is larger than the tension flange)
		I_y = moment of inertia about the minor axis, in.^{4}
		J = torsional constant, in.^{4}
		C_w = warping constant, in.^{6}
		E = modulus of elasticity, ksi
		G = shear modulus, ksi
Solid rectangular bars, symmetric box sections	$\left[3,750 r_y\sqrt{(JA)} \right]/M_p$	$\left[57,000 r_y\sqrt{(JA)} \right]/M_r$
	where	where
	r_y = radius of gyration about minor axis, in.	r_y = radius of gyration about minor axis, in.
	J = torsional constant, in.^{4}	J = torsional constant, in.^{4}
	A = cross-sectional area, in.^{2}	A = cross-sectional area, in.^{2}
	M_p = plastic moment capacity = $F_y Z_x$	$M_r = F_y S_x$ for solid rectangular bar, $F_{yf} S_{eff}$ for box sections
	F_y = yield stress, ksi	F_y = yield stress, ksi
	Z_x = plastic section modulus about the major axis, in.^{3}	F_{yf} = flange yield stress, ksi
		S_x = plastic section modulus about the major axis, in.^{3}

Note: L_p given in this table are valid only if the bending coefficient C_b is equal to unity. If $C_b > 1$, the value of L_p can be increased. However, using the L_p expressions given above for $C_b > 1$ will give a conservative value for the flexural design strength.

and

$M_p = F_y Z_x$

$M_r = F_L S_x$ for I-shaped sections and channels, $F_y S_x$ for solid rectangular bars, $F_{yf} S_{eff}$ for box sections

$F_L =$ smaller of $(F_{yf} - F_r)$ or F_{yw}

$F_{yf} =$ flange yield stress, ksi

$F_{yw} =$ web yield stress

$F_r =$ 10 ksi for rolled sections, 16.5 ksi for welded sections

$F_y =$ specified minimum yield stress

$S_x =$ elastic section modulus about the major axis

$S_{eff} =$ effective section modular, calculated using effective width b_e, in Table 3.7

$Z_x =$ plastic section modulus about the major axis

$I_y =$ moment of inertia about the minor axis

$J =$ torsional constant

$C_w =$ warping constant

$E =$ modulus of elasticity

$G =$ shear modulus

$C_b = 12.5 M_{max}/(2.5 M_{max} + 3 M_A + 4 M_B + 3 M_C)$

$M_{max}, M_A, M_B, M_C =$ maximum moment, quarter-point moment, midpoint moment, and three-quarter point moment along the unbraced length of the member, respectively.

C_b is a factor that accounts for the effect of moment gradient on the lateral torsional buckling strength of the beam. Lateral torsional buckling strength increases for a steep moment gradient. The worst loading case as far as lateral torsional buckling is concerned is when the beam is subjected to a uniform moment resulting in single curvature bending. For this case $C_b = 1$. Therefore, the use of $C_b = 1$ is conservative for the design of beams.

Compact Section Members Bent About Their Minor Axes

Regardless of L_b, the limit state will be a plastic hinge formation

$$\phi_b M_n = 0.90 M_{py} = 0.90 F_y Z_y \tag{3.41}$$

Noncompact Section Members Bent About Their Major Axes

For $L_b \leq L'_p$, (Flange or web local buckling)

$$\phi_b M_n = \phi_b M'_n = 0.90 \left[M_p - (M_p - M_r) \left(\frac{\lambda - \lambda_p}{\lambda_r - \lambda_p} \right) \right] \tag{3.42}$$

where

$$L'_p = L_p + (L_r - L_p) \left(\frac{M_p - M'_n}{M_p - M_r} \right) \tag{3.43}$$

L_p, L_r, M_p, M_r are defined as before for compact section members, and

For flange local buckling:

$\lambda = b_f/2t_f$ for I-shaped members, b_f/t_f for channels

$\lambda_p = 65/\sqrt{F_y}$

$\lambda_r = 141/\sqrt{(F_y - 10)}$

For web local buckling:

$\lambda = h_c/t_w$

$$\lambda_p = 640/\sqrt{F_y}$$
$$\lambda_r = 970/\sqrt{F_y}$$

in which

b_f	=	flange width
t_f	=	flange thickness
h_c	=	twice the distance from the neutral axis to the inside face of the compression flange less the fillet or corner radius
t_w	=	web thickness

For $L_p' < L_b \le L_r$, (Inelastic lateral torsional buckling), $\phi_b M_n$ is given by Equation 3.38 except that the limit $0.90 M_p$ is to be replaced by the limit $0.90 M_n'$.

For $L_b > L_r$, (Elastic lateral torsional buckling), $\phi_b M_n$ is the same as for compact section members as given in Equation 3.39 or Equation 3.40.

Noncompact Section Members Bent About Their Minor Axes
Regardless of the value of L_b, the limit state will be either flange or web local buckling, and $\phi_b M_n$ is given by Equation 3.42.

Slender Element Sections
Refer to the section on Plate Girder.

Tees and Double Angle Bent About Their Major Axes
The design flexural strength for tees and double-angle beams with flange and web slenderness ratios less than the corresponding limiting slenderness ratios λ_r shown in Table 3.8 is given by

$$\phi_b M_n = 0.90 \left[\frac{\pi \sqrt{E I_y G J}}{L_b} (B + \sqrt{1 + B^2}) \right] \le 0.90 (C M_y) \tag{3.44}$$

where

$$B = \pm 2.3 \left(\frac{d}{L_b} \right) \sqrt{\frac{I_y}{J}} \tag{3.45}$$

$C = 1.5$ for stems in tension, and 1.0 for stems in compression.
Use the plus sign for B if the *entire* length of the stem along the unbraced length of the member is in tension. Otherwise, use the minus sign. The other variables in Equation 3.44 are defined as before in Equation 3.39.

Shear Strength Criterion

For a satisfactory design, the design shear strength of the webs must exceed the factored shear acting on the cross-section, i.e.,

$$\phi_v V_n \ge V_u \tag{3.46}$$

Depending on the slenderness ratios of the webs, three limit states can be identified: shear yielding, inelastic shear buckling, and elastic shear buckling. The design shear strength that corresponds to each of these limit states is given as follows:

For $h/t_w \le 418/\sqrt{F_{yw}}$, (Shear yielding of web)

$$\phi_v V_n = 0.90 [0.60 F_{yw} A_w] \tag{3.47}$$

For $418/\sqrt{F_{yw}} < h/t_w \le 523/\sqrt{F_{yw}}$, (Inelastic shear buckling of web)

$$\phi_v V_n = 0.90 \left[0.60 F_{yw} A_w \frac{418/\sqrt{F_{yw}}}{h/t_w} \right] \tag{3.48}$$

For $523/\sqrt{F_{yw}} < h/t_w \leq 260$, (Elastic shear buckling of web)

$$\phi_v V_n = 0.90 \left[\frac{132,000 A_w}{(h/t_w)^2} \right] \tag{3.49}$$

The variables used in the above equations are defined in the following:

h = clear distance between flanges less the fillet or corner radius, in.
t_w = web thickness, in.
F_{yw} = yield stress of web, ksi
A_w = dt_w, in.^{2}
d = overall depth of section, in.

Criteria for Concentrated Loads

When concentrated loads are applied normal to the flanges in planes parallel to the webs of flexural members, the flange(s) and web(s) must be checked to ensure that they have sufficient strengths ϕR_n to withstand the concentrated forces R_u, i.e.,

$$\phi R_n \geq R_u \tag{3.50}$$

The design strength for a variety of limit states are given below:

Local Flange Bending
The design strength for local flange bending is given by

$$\phi R_n \geq 0.90[6.25t_f^2 F_{yf}] \tag{3.51}$$

where
t_f = flange thickness of the loaded flange, in.
F_{yf} = flange yield stress, ksi

Local Web Yielding
The design strength for yielding of a beam web at the toe of the fillet under tensile or compressive loads acting on one or both flanges are:

If the load acts at a distance from the beam end which exceeds the depth of the member

$$\phi R_n = 1.00[(5k + N)F_{yw}t_w] \tag{3.52}$$

If the load acts at a distance from the beam end which does not exceed the depth of the member

$$\phi R_n = 1.00[(2.5k + N)F_{yw}t_w] \tag{3.53}$$

where
k = distance from outer face of flange to web toe of fillet
N = length of bearing on the beam flange
F_{yw} = web yield stress
t_w = web thickness

Web Crippling
The design strength for crippling of a beam web under compressive loads acting on one or both flanges are:

If the load acts at a distance from the beam end which exceeds half the depth of the beam

$$\phi R_n = 0.75 \left\{ 135 t_w^2 \left[1 + 3 \left(\frac{N}{d} \right) \left(\frac{t_w}{t_f} \right)^{1.5} \right] \sqrt{\frac{F_{yw} t_f}{t_w}} \right\} \tag{3.54}$$

If the load acts at a distance from the beam end which does not exceed half the depth of the beam and if $N/d \leq 0.2$

$$\phi R_n = 0.75 \left\{ 68 t_w^2 \left[1 + 3 \left(\frac{N}{d} \right) \left(\frac{t_w}{t_f} \right)^{1.5} \right] \sqrt{\frac{F_{yw} t_f}{t_w}} \right\} \tag{3.55}$$

If the load acts at a distance from the beam end which does not exceed half the depth of the beam and if $N/d > 0.2$

$$\phi R_n = 0.75 \left\{ 68 t_w^2 \left[1 + \left(\frac{4N}{d} - 0.2 \right) \left(\frac{t_w}{t_f} \right)^{1.5} \right] \sqrt{\frac{F_{yw} t_f}{t_w}} \right\} \tag{3.56}$$

where
d =   overall depth of the section, in.
t_f =   flange thickness, in.
The other variables are the same as those defined in Equations 3.52 and 3.53.

Sidesway Web Buckling
Sidesway web buckling may occur in the web of a member if a compressive concentrated load is applied to a flange which is not restrained against relative movement by stiffeners or lateral bracings. The sidesway web buckling design strength for the member is:

If the loaded flange is restrained against rotation about the longitudinal member axis and $(h_c/t_w)(l/b_f) \leq 2.3$

$$\phi R_n = 0.85 \left\{ \frac{C_r t_w^3 t_f}{h^2} \left[1 + 0.4 \left(\frac{h/t_w}{l/b_f} \right)^3 \right] \right\} \tag{3.57}$$

If the loaded flange is not restrained against rotation about the longitudinal member axis and $(h_c/t_w)(l/b_f) \leq 1.7$

$$\phi R_n = 0.85 \left\{ \frac{C_r t_w^3 t_f}{h^2} \left[0.4 \left(\frac{h/t_w}{l/b_f} \right)^3 \right] \right\} \tag{3.58}$$

where
t_f =   flange thickness, in.
t_w =   web thickness, in.
h =   clear distance between flanges less the fillet or corner radius for rolled shapes; distance between adjacent lines of fasteners or clear distance between flanges when welds are used for built-up shapes, in.
b_f =   flange width, in.
l =   largest laterally unbraced length along either flange at the point of load, in.
C_r =   960,000 if $M_u/M_y < 1$ at the point of load, ksi
      =   480,000 if $M_u/M_y \geq 1$ at the point of load, ksi

Compression Buckling of the Web
This limit state may occur in members with unstiffened webs when both flanges are subjected to compressive forces. The design strength for this limit state is

$$\phi R_n = 0.90 \left[\frac{4,100t_w^3 \sqrt{F_{yw}}}{h} \right] \tag{3.59}$$

This design strength shall be reduced by 50% if the concentrated forces are acting at a distance from the beam end which is half the beam depth. The variables in Equation 3.59 are the same as those defined in Equations 3.56 to 3.58.

Stiffeners shall be provided in pairs if any one of the above strength criteria is violated. If the local flange bending or the local web yielding criterion is violated, the stiffener pair to be provided to carry the excess R_u need not extend more than one-half the web depth. The stiffeners shall be welded to the loaded flange if the applied force is tensile. They shall either bear on or be welded to the loaded flange if the applied force is compressive. If the web crippling or the compression web buckling criterion is violated, the stiffener pair to be provided shall extend the full height of the web. They shall be designed as axially loaded compression members (see section on Compression Members) with an effective length factor $K = 0.75$, a cross-section A_g composed of the cross-sectional areas of the stiffeners plus $25t_w^2$ for interior stiffeners, and $12t_w^2$ for stiffeners at member ends.

Deflection Criterion

The deflection criterion is the same as that for ASD. Since deflection is a serviceability limit state, service (rather than factored) loads should be used in deflection computations.

3.5.3 Continuous Beams

Continuous beams shall be designed in accordance with the criteria for flexural members given in the preceding section. However, a 10% reduction in negative moments due to gravity loads is allowed at the supports provided that:

1. the maximum positive moment between supports is increased by 1/10 the average of the negative moments at the supports;

2. the section is compact;

3. the lateral unbraced length does not exceed L_c (for ASD), or L_{pd} (for LRFD) where L_c is as defined in Equation 3.26 and L_{pd} is given by

$$L_{pd} = \begin{cases} \frac{3,600+2,200(M_1/M_2)}{F_y}r_y, & \text{for I-shaped members} \\ \frac{5,000+3,000(M_1/M_2)}{F_y}r_y, & \text{for solid rectangular and box sections} \end{cases} \tag{3.60}$$

in which

F_y = specified minimum yield stress of the compression flange, ksi

r_y = radius of gyration about the minor axis, in.

M_1/M_2 = ratio of smaller to larger moment within the unbraced length, taken as positive if the moments cause reverse curvature and negative if the moments cause single curvature.

4. the beam is not a hybrid member;

5. the beam is not made of high strength steel;

6. the beam is continuous over the supports (i.e., not cantilevered).

EXAMPLE 3.4:

Using LRFD, select the lightest W section for the three-span continuous beam shown in Figure 3.10a to support a uniformly distributed dead load of 1.5 k/ft (22 kN/m) and a uniformly distributed live load of 3 k/ft (44 kN/m). The beam is laterally braced at the supports A,B,C, and D. Use A36 steel.

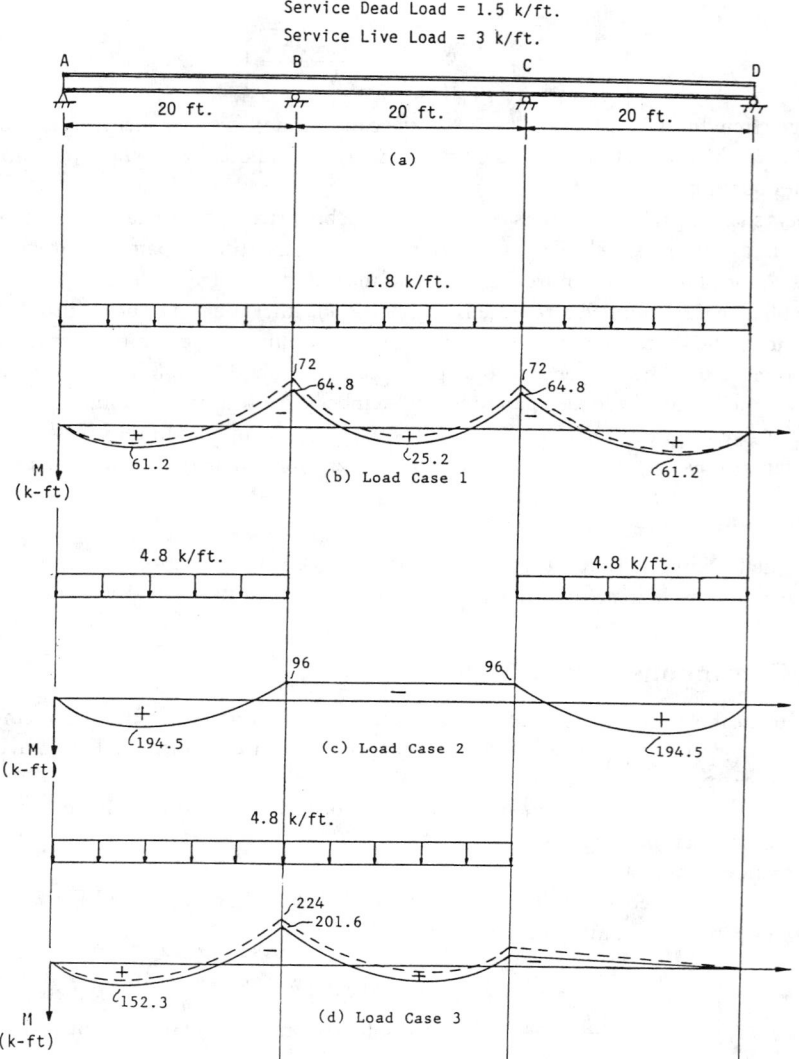

Figure 3.10   Design of a three-span continuous beam (1 k = 4.45 kN, 1 ft = 0.305 m).

Load combinations

The beam is to be designed based on the worst load combination of Table 3.3 By inspection, the load combination $1.2D + 1.6L$ will control the design. Thus, the beam will be designed to support a factored uniformly distributed dead load of $1.2 \times 1.5 = 1.8$ k/ft and a factored uniformly distributed live load of $1.6 \times 3 = 4.8$ k/ft.

Placement of loads

The uniform dead load is to be applied over the entire length of the beam as shown in Figure 3.10b. The uniform live load is to be applied to spans AB and CD as shown in Figure 3.10c to obtain the maximum positive moment and it is to be applied to spans AB and BC as shown in Figure 3.10d to obtain the maximum negative moment.

Reduction of negative moment at supports

Assuming the beam is compact and $L_b < L_{pd}$ (we shall check these assumptions later), a 10%

reduction in support moment due to gravity load is allowed provided that the maximum moment is increased by 1/10 the average of the negative support moments. This reduction is shown in the moment diagrams as solid lines in Figures 3.10b and 3.10d (The dotted lines in these figures represent the unadjusted moment diagrams). This provision for support moment reduction takes into consideration the beneficial effect of moment redistribution in continuous beams and it allows for the selection of a lighter section if the design is governed by negative moments. Note that no reduction in negative moments is made to the case when only spans AB and CD are loaded. This is because for this load case, the negative support moments are less than the positive in-span moments.

Determination of the required flexural strength, M_u

Combining load case 1 and load case 2, the maximum positive moment is found to be 256 kip-ft. Combining load case 1 and load case 3, the maximum negative moment is found to be 266 kip-ft. Thus, the design will be controlled by the negative moment and so $M_u = 266$ kip-ft.

Beam selection

A beam section is to be selected based on Equation 3.36. The critical segment of the beam is span BC. For this span, the lateral unsupported length, L_b, is equal to 20 ft. For simplicity, the bending coefficient, C_b, is conservatively taken as 1. The selection of a beam section is facilitated by the use of a series of beam charts contained in the AISC-LRFD Manual [22]. Beam charts are plots of flexural design strength $\phi_b M_n$ of beams as a function of the lateral unsupported length L_b based on Equations 3.37 to 3.39. A beam is considered satisfactory for the limit state of flexure if the beam strength curve envelopes the required flexural strength for a given L_b.

For the present example, $L_b = 20$ ft. and $M_u = 266$ kip-ft, the lightest section (the first solid curve that envelopes $M_u = 266$ kip-ft for $L_b = 20$ ft) obtained from the chart is a W16x67 section. Upon adding the factored dead weight of this W16x67 section to the specified loads, the required strength increases from 266 kip-ft to 269 kip-ft. Nevertheless, the beam strength curve still envelopes this required strength for $L_b = 20$ ft; therefore, the section is adequate.

Check for compactness

For the W16x67 section,

$$\text{Flange:} \quad \left[\frac{b_f}{2t_f} = 7.7\right] \quad < \quad \left[\frac{65}{\sqrt{F_y}} = 10.8\right]$$

$$\text{Web:} \quad \left[\frac{h_c}{t_w} = 35.9\right] \quad < \quad \left[\frac{640}{\sqrt{F_y}} = 106.7\right]$$

Therefore, the section is compact.

Check whether $L_b < L_{pd}$

Using Equation 3.64, with $M_1/M_2 = 0$, $r_y = 2.46$ in. and $F_y = 36$ ksi, we have $L_{pd} = 246$ in. (or 20.5 ft). Since $L_b = 20$ ft is less than $L_{pd} = 20.5$ ft, the assumption made earlier is validated.

Check for the limit state of shear

The selected section must satisfy the shear strength criterion of Equation 3.46. From structural analysis, it can be shown that maximum shear occurs just to the left of support B under load case 1 (for dead load) and load case 3 (for live load). It has a magnitude of 81.8 kips. For the W16x67 section, $h/t_w = 35.9$ which is less than $418/\sqrt{F_{yw}} = 69.7$, so the design shear strength is given by Equation 3.47. We have, for $F_{yw} = 36$ ksi and $A_w = dt_w = (16.33)(0.395)$,

$$[\phi_v V_n = 0.90(0.60 F_{yw} A_w) = 125 \text{ kips }] > [V_u = 81.8 \text{ kips }]$$

Therefore, shear is not a concern. Normally, the limit state of shear will not be controlled unless for short beams subjected to heavy loads.

Check for limit state of deflection

Deflection is a serviceability limit state. As a result, a designer should use service (not factored) loads, for deflection calculations. In addition, most beams are cambered to offset deflection caused by dead loads, so only live loads are considered in deflection calculations. From structural analysis, it can be shown that maximum deflection occurs in span AB and CD when (service) live loads are placed on those two spans. The magnitude of the deflection is 0.297 in. Assuming the maximum allowable deflection is $L/360$ where L is the span length between supports, we have an allowable deflection of $20 \times 12/360 = 0.667$ in. Since the calculated deflection is less than the allowable deflection, deflection is not a problem.

Check for the limit state of web yielding and web crippling at points of concentrated loads

From a structural analysis it can be shown that maximum support reaction occurs at support B when the beam is subjected to loads shown as load case 1 (for dead load) and load case 3 (for live load). The magnitude of the reaction R_u is 157 kips. Assuming point bearing, i.e., $N = 0$, we have, for $d = 16.33$ in., $k = 1.375$ in., $t_f = 0.665$ in., and $t_w = 0.395$ in.,

Web Yielding: $[\phi R_n =$ Equation 3.52 $= 97.8$ kips $]$ $<$ $[R_u = 157$ kips $]$
Web Crippling: $[\phi R_n =$ Equation 3.54 $= 123$ kips $]$ $<$ $[R_u = 157$ kips $]$

Thus, both the web yielding and web crippling criteria are violated. As a result, we need to provide web stiffeners or bearing plate at support B. Suppose we choose the latter, the size of the bearing plate is to be determined by solving Equation 3.52 and Equation 3.54 for N, given $R_u = 157$ kips. Solving Equation 3.52 and Equation 3.54 for N, we obtain $N = 4.2$ in. and 3.3 in., respectively. So, use $N = 4.25$ in. The width of the plate, B, should conform with the flange width, b_f, of the W-section. The W16x67 section has a flange width of 10.235 in., so use $B = 10.5$ in. For uniformity, use the same size plate at all the supports. The bearing plates are to be welded to the supporting flange of the W-section.

Use a W16x67 section. Provide bearing plates of size $10.5'' \times 4''$ at the supports.

3.5.4   Lateral Bracing of Beams

The design strength of beams that bent about their major axes depends on their lateral unsupported length, L_b. The manner a beam is braced against out-of-plane deformation affects its design. Bracing can be provided by various means such as cross beams, or diaphragms, or encasement of the compression flange of the beam in the floor slab. Although neither the ASD nor the LRFD specification addresses the design of braces, a number of methodologies have been proposed in the literature for the design of braces [30]. It is important to note that braces must be designed with sufficient strength and stiffness to prevent out-of-plane movement of the beam at the braced points. In what follows, the requirements for brace design as specified in the Canadian Standards Association [10] for the limit states design of steel structures will be given.

Stiffness Requirement:

The stiffness of the bracing assembly in a direction perpendicular to the longitudinal axis of the braced member, in the plane of buckling, must exceed

$$k_b = \frac{\beta C_f}{L_b}[1 + \frac{\Delta_o}{\Delta_b}] \tag{3.61}$$

where

β = bracing coefficient given by 2, 3, 3.41, and 3.63 for one, two, three, and four or more equally spaced braces, respectively.

C_f = compressive force in the braced member. In a limit state design, C_f for a doubly symmetric I-shaped beam can be calculated by dividing its design flexural strength, $\phi_b M_p$, by the distance between centroids of the flanges, $d - t_f$.

L_b = lateral unsupported length of the braced member.

Δ_o = initial misalignment of the braced member at the point of the brace. Δ_o may be taken as the sweep tolerance of the braced member.

Δ_b = displacement of the braced member and bracing assembly under force C_f at the point of the brace. For a trial design, Δ_b may be taken as Δ_o.

Strength Requirement:

In addition to the stiffness requirement as stipulated above, braces must be designed for strength that exceed

$$P_b = k_b \Delta_b \tag{3.62}$$

where P_b is the force in the bracing assembly under factored loads.

If a series of parallel members are being braced, a reduced initial misalignment can be used in Equation 3.61 to account for probable force redistribution when the bracing assembly transfers force from one braced member to another. This reduced misalignment is given by

$$\Delta_m = \left[0.2 + \frac{0.8}{\sqrt{n}} \right] \Delta_o \tag{3.63}$$

where n is the number of parallel braced members.

Finally, if the brace is under compression, a maximum slenderness, l/r, not exceeding 200 is recommended.

EXAMPLE 3.5:

Design an I-shaped cross beam 15 ft (4.6 m) in length to be used as lateral braces to brace a 30-ft (9.1 m) long W30x90 beam at every third point. The sweep of the W30x90 section is 0.36 in. (9 mm) in the plane of the brace. A36 steel is used.

If a brace is provided at every third point, L_b for the W30x90 section is 10 ft. Therefore, the design flexural strength, $\phi_b M_n$, is 8890 kip-in, from which $C_f = \phi_b M_n/(d-t_f) = 8890/(29.53-0.610) = 307$ kips. As a first trial, assume $\Delta_b = \Delta_o = 0.36$ in., we can calculate from Equation 3.61 and Equation 3.62 the minimum stiffness and strength requirements for the cross beam

$$k_b = \frac{3(307)}{10 \times 12} [1 + \frac{0.36}{0.36}] = 15.4 \text{ kips/in.}$$

$$P_b = (15.4)(0.36) = 5.5 \text{ kips}$$

Since the cross beam will be subject to compression, its slenderness ratio, l/r, should not exceed 200.

Try a W4x13 section ($A = 3.83$ in^2, $r_y = 1.00$ in., $\phi_c P_n = 25$ kips)

Stiffness, $\frac{EA}{l} = \frac{(29000)(3.83)}{15 \times 12} = 617$ kips/in. > 15.4 kips/in.

Strength, $\phi_c P_n = 25$ kips > 5.5 kips

Slenderness, $\frac{l}{r_y} = \frac{15 \times 12}{1.00} = 180$ < 200

Recalculate Δ_b using $P_b = 5.5$ kips and check adequacy of the W4x13 section.

$$\Delta_b = \frac{P_b l}{EA} = \frac{(5.5)(15 \times 12)}{(29000)(3.83)} = 0.0089 \text{ in.}$$

$$k_b = \frac{3(307)}{10 \times 12} \left[1 + \frac{0.36}{0.0089} \right] = 318 \text{ kips/in.} \quad < 617 \text{ kips/in.} \quad \text{OK}$$

$$P_b = (318)(0.0089) = 2.83 \text{ kips} \quad < 25 \text{ kips} \quad \text{OK}$$

Recalculate Δ_b using $P_b = 2.83$ kips and check adequacy of the W4x13 section.

$$\Delta_b = \frac{P_b l}{EA} = \frac{(2.83)(15 \times 12)}{(29000)(3.83)} = 0.0046 \text{ in.}$$

$$k_b = \frac{3(307)}{10 \times 12}\left[1 + \frac{0.36}{0.0046}\right] = 608 \text{ kips/in.} \quad < 617 \text{ kips/in.} \quad \text{OK}$$

$$P_b = (608)(0.0046) = 2.80 \text{ kips} \quad < 25 \text{ kips} \quad \text{OK}$$

Since P_b, and hence Δ_b, has converged, no more iterations are necessary.

Use W4x13 cross beams as braces.

3.6   Combined Flexure and Axial Force

When a member is subject to the combined action of bending and axial force, it must be designed to resist stresses and forces arising from both bending and axial actions. While a tensile axial force may induce a stiffening effect on the member, a compressive axial force tends to destabilize the member, and the instability effects due to member instability (P-δ effect) and frame instability (P-Δ effect) must be properly accounted for. P-δ effect arises when the axial force acts through the lateral deflection of the member relative to its chord. P-Δ effect arises when the axial force acts through the relative displacements of the two ends of the member. Both effects tend to increase member deflection and moment, and so they must be considered in the design. A number of approaches are available in the literature to handle these so-called P-delta effects (see for example [9, 13]). The design of members subject to combined bending and axial force is facilitated by the use of interaction equations. In these equations, the effects of bending and axial actions are combined in a certain manner to reflect the capacity demand on the member.

3.6.1   Allowable Stress Design

The interaction equations are:

If the axial force is tensile:

$$\frac{f_a}{F_t} + \frac{f_{bx}}{F_{bx}} + \frac{f_{by}}{F_{by}} \leq 1.0 \tag{3.64}$$

where

f_a	=	computed axial tensile stress
f_{bx}, f_{by}	=	computed bending tensile stresses about the major and minor axes, respectively
F_{bx}, F_{by}	=	allowable bending stresses about the major and minor axes, respectively (see section on Flexural Members)
F_t	=	allowable tensile stress (see section on Tension Members)

If the axial force is compressive:
Stability requirement

$$\frac{f_a}{F_a} + \left[\frac{C_{mx}}{\left(1 - \frac{f_a}{F'_{ex}}\right)}\right]\frac{f_{bx}}{F_{bx}} + \left[\frac{C_{my}}{\left(1 - \frac{f_a}{F'_{ey}}\right)}\right]\frac{f_{by}}{F_{by}} \leq 1.0 \tag{3.65}$$

Yield requirement

$$\frac{f_a}{0.60F_y} + \frac{f_{bx}}{F_{bx}} + \frac{f_{by}}{F_{by}} \leq 1.0 \tag{3.66}$$

However, if the axial force is small (when $f_a/F_a \leq 0.15$), the following interaction equation can be used in lieu of the above equations.

$$\frac{f_a}{F_a} + \frac{f_{bx}}{F_{bx}} + \frac{f_{by}}{F_{by}} \leq 1.0 \qquad (3.67)$$

The terms in the Equations 3.65 to 3.67 are defined as follows:

f_a, f_{bx}, f_{by} = computed axial compressive stress, computed bending stresses about the major and minor axes, respectively. These stresses are to be computed based on a *first-order analysis*

F_y = minimum specified yield stress

F'_{ex}, F'_{ey} = Euler stresses about the major and minor axes $(\pi^2 E/(Kl/r)_x, \pi^2 E/(Kl/r)_y)$ divided by a factor of safety of 23/12

C_m = a coefficient to account for the effect of moment gradient on member and frame instabilities (C_m is defined in the section on LRFD to follow)

The other terms are defined as in Equation 3.64.

The terms in brackets in Equation 3.65 are moment magnification factors. The computed bending stresses f_{bx}, f_{by} are magnified by these magnification factors to account for the P-delta effects in the member.

3.6.2 Load and Resistance Factor Design

Doubly or singly symmetric members subject to combined flexure and axial force shall be designed in accordance with the following interaction equations:

For $P_u/\phi P_n \geq 0.2$

$$\frac{P_u}{\phi P_n} + \frac{8}{9}\left(\frac{M_{ux}}{\phi_b M_{nx}} + \frac{M_{uy}}{\phi_b M_{ny}}\right) \leq 1.0 \qquad (3.68)$$

For $P_u/\phi P_n < 0.2$

$$\frac{P_u}{2\phi P_n} + \frac{8}{9}\left(\frac{M_{ux}}{\phi_b M_{nx}} + \frac{M_{uy}}{\phi_b M_{ny}}\right) \leq 1.0 \qquad (3.69)$$

where, if P is tensile

P_u = factored tensile axial force

P_n = design tensile strength (see section on Tension Members)

M_u = factored moment (preferably obtained from a second-order analysis)

M_n = design flexural strength (see section on Flexural Members)

$\phi = \phi_t$ = resistance factor for tension = 0.90

ϕ_b = resistance factor for flexure = 0.90

and, if P is compressive

P_u = factored compressive axial force

P_n = design compressive strength (see section on Compression Members)

M_u = required flexural strength (see discussion below)

M_n = design flexural strength (see section on Flexural Members)

$\phi = \phi_c$ = resistance factor for compression = 0.85

ϕ_b = resistance factor for flexure = 0.90

The required flexural strength M_u shall be determined from a second-order elastic analysis. In lieu of such an analysis, the following equation may be used

$$M_u = B_1 M_{nt} + B_2 M_{lt} \qquad (3.70)$$

where

M_{nt}	=	factored moment in member assuming the frame does not undergo lateral translation (see Figure 3.11)
M_{lt}	=	factored moment in a member as a result of lateral translation (see Figure 3.11)
B_1	=	$C_m/(1 - P_u/P_e) \geq 1.0$ is the P-δ moment magnification factor
P_e	=	$\pi^2 EI/(KL)^2$, with $K \leq 1.0$ in the plane of bending
C_m	=	a coefficient to account for moment gradient, determined from the following discussion
B_2	=	$1/[1 - (\Sigma P_u \Delta_{oh}/\Sigma HL)]$ or $B_2 = 1/[1 - (\Sigma P_u/\Sigma P_e)]$
ΣP_u	=	sum of all factored loads acting on and above the story under consideration
Δ_{oh}	=	first-order interstory translation
ΣH	=	sum of all lateral loads acting on and above the story under consideration
L	=	story height
P_e	=	$\pi^2 EI/(KL)^2$

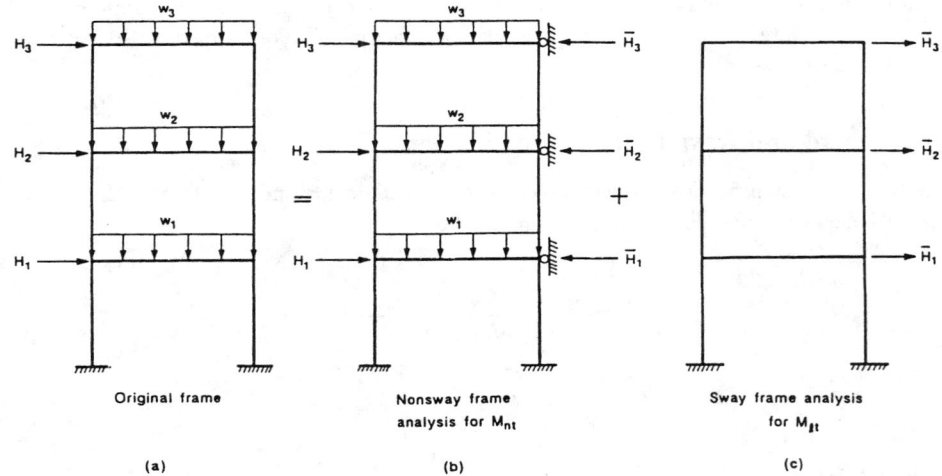

Original frame   Nonsway frame   Sway frame analysis
analysis for M_{nt} for M_{lt}

(a)   (b)   (c)

Figure 3.11   Calculation of M_{nt} and M_{lt}.

For end-restrained members which do not undergo relative joint translation and are not subject to transverse loading between their supports in the plane of bending, C_m is given by

$$C_m = 0.6 - 0.4 \left(\frac{M_1}{M_2} \right)$$

where M_1/M_2 is the ratio of the smaller to larger member end moments. The ratio is positive if the member bends in reverse curvature and negative if the member bends in single curvature.

For end-restrained members which do not undergo relative joint translation and are subject to transverse loading between their supports in the plane of bending

$$C_m = 0.85$$

For unrestrained members which do not undergo relative joint translation and are subject to transverse loading between their supports in the plane of bending

$$C_m = 1.00$$

The selection of trial sections for use as **beam-columns** is facilitated by rewriting the interaction equations of Equation 3.68 and 3.69 into the so-called equivalent axial load form:

For $P_u/\phi_c P_n > 0.2$

$$P_u + m_x M_{ux} + m_y U M_{uy} \le \phi_c P_n \qquad (3.71)$$

For $P_u/\phi_c P_n \le 0.2$

$$\frac{P_u}{2} + \frac{9}{8} m_x M_{ux} + \frac{9}{8} m_y U M_{uy} \le \phi_c P_n \qquad (3.72)$$

where

$$m_x = (8/9)(\phi_c P_n/\phi_b M_{nx})$$
$$m_y U = (8/9)(\phi_c P_n/\phi_b M_{ny})$$

Numerical values for m and U are provided in the AISC Manual [22]. The advantage of using Equations 3.71 and 3.72 for preliminary design is that the terms on the left-hand side of the inequality can be regarded as an equivalent axial load, $(P_u)_{eff}$, thus allowing the designer to take advantage of the column tables provided in the manual for selecting trial sections.

3.7 Biaxial Bending

Members subjected to bending about both principal axes (e.g., purlins on an inclined roof) should be designed for **biaxial bending**. Since both moment about the major axis M_{ux} and moment about the minor axis M_{uy} create flexural stresses over the cross-section of the member, the design must take into consideration this stress combination.

3.7.1 Allowable Stress Design

The following interaction equation is often used for the design of beams subject to biaxial bending

$$f_{bx} + f_{by} \le 0.60 F_y$$

or,

$$\frac{M_x}{S_x} + \frac{M_y}{S_y} \le 0.60 F_y \qquad (3.73)$$

where

$$
\begin{aligned}
M_x, M_y &= \text{service load moments about the major and minor axes, respectively} \\
S_x, S_y &= \text{elastic section moduli about the major and minor axes, respectively} \\
F_y &= \text{specified minimum yield stress}
\end{aligned}
$$

EXAMPLE 3.6:

Using ASD, select a W section to carry dead load moments $M_x = 20$ k-ft (27 kN-m) and $M_y = 5$ k-ft (6.8 kN-m), and live load moments $M_x = 50$ k-ft (68 kN-m) and $M_y = 15$ k-ft (20 kN-m). Use steel having $F_y = 50$ ksi (345 MPa).

Calculate service load moments:

$$M_x = M_{x,\text{dead}} + M_{x,\text{live}} = 20 + 50 = 70 \quad \text{k-ft}$$
$$M_y = M_{y,\text{dead}} = M_{y,\text{live}} = 5 + 15 = 20 \quad \text{k-ft}$$

Select section:

Substituting the above service load moments into Equation 3.73, we have

$$\frac{70 \times 12}{S_x} + \frac{20 \times 12}{S_y} \leq 0.60(50) \quad \text{or,} \quad 840 + 240\frac{S_x}{S_y} \leq 30 S_x$$

For W sections with depth below 14 in. the value of S_x/S_y normally falls in the range 3 to 8, and for W sections with depth above 14 in. the value of S_x/S_y normally falls in the range 5 to 12. Assuming $S_x/S_y = 10$, we have from the above equation, $S_x \geq 108$ in.^{3}. Using the Allowable Stress Design Selection Table in the AISC-ASD Manual, lets try a W24x55 section ($S_x = 114$ in.^{3}, $S_y = 8.30$ in.^{3}). For the W24x55 section

$$\left[840 + 240\frac{114}{8.30} = 4136\right] > [30S_x = 30(114) = 3420] \quad \therefore \quad \text{NG}$$

The next lightest section is W21x62 ($S_x = 127$ in.^{3}, $S_y = 13.9$ in.^{3}). For this section

$$\left[840 + 240\frac{127}{13.9} = 3033\right] < [30S_x = 30(127) = 3810] \quad \therefore \quad \text{OK}$$

Use a W21x62 section.

3.7.2 Load and Resistance Factor Design

To avoid distress at the most severely stressed point, the following equation for the yielding limit state must be satisfied:

$$f_{un} \leq \phi_b F_y \tag{3.74}$$

where

f_{un}	=	$M_{ux}/S_x + M_{uy}/S_y$ is the flexural stress under factored loads
S_x, S_y	=	are the elastic section moduli about the major and minor axes, respectively
ϕ_b	=	0.90
F_y	=	specified minimum yield stress

In addition, the limit state for lateral torsional buckling about the major axis should also be checked, i.e.,

$$\phi_b M_{nx} \geq M_{ux} \tag{3.75}$$

$\phi_b M_{nx}$ is the design flexural strength about the major axis (see section on Flexural Members). Note that lateral torsional buckling will not occur about the minor axis. Equation 3.74 can be rearranged to give:

$$S_x \leq \frac{M_{ux}}{\phi_b F_y} + \frac{M_{uy}}{\phi_b F_y}\left(\frac{S_x}{S_y}\right) \approx \frac{M_{ux}}{\phi_b F_y} + \frac{M_{uy}}{\phi_b F_y}\left(3.5\frac{d}{b_f}\right) \tag{3.76}$$

The approximation $(S_x/S_y) \approx (3.5d/b_f)$ where d is the overall depth and b_f is the flange width was suggested by Gaylord et al. [15] for doubly symmetric I-shaped sections. The use of Equation 3.74 greatly facilitates the selection of trial sections for use in biaxial bending problems.

3.8 Combined Bending, Torsion, and Axial Force

Members subjected to the combined effect of bending, torsion, and axial force should be designed to satisfy the following limit states:

Yielding under normal stress

$$\phi F_y \geq f_{un} \tag{3.77}$$

where

ϕ =   0.90

F_y =   specified minimum yield stress

f_{un} =   maximum normal stress determined from an elastic analysis under factored loads

Yielding under shear stress

$$\phi(0.6F_y) \geq f_{uv} \tag{3.78}$$

where

ϕ =   0.90

F_y =   specified minimum yield stress

f_{uv} =   maximum shear stress determined from an elastic analysis under factored loads

Buckling

$$\phi_c F_{cr} \geq f_{un} \ \text{ or } \ f_{uv}, \quad \text{whichever is applicable} \tag{3.79}$$

where

$\phi_c F_{cr}$ = $\phi_c P_n / A_g$, in which $\phi_c P_n$ is the design compressive strength of the member (see section on Compression Members) and A_g is the gross cross-section area

f_{un}, f_{uv} =   normal and shear stresses as defined in Equation 3.77 and 3.78

3.9 Frames

Frames are designed as a collection of structural components such as beams, beam-columns (columns), and connections. According to the restraint characteristics of the connections used in the construction, frames can be designed as Type I (rigid framing), Type II (simple framing), Type III (semi-rigid framing) in ASD, or fully restrained (rigid), partially restrained (semi-rigid) in LRFD. The design of rigid frames necessitates the use of connections capable of transmitting the full or a significant portion of the moment developed between the connecting members. The rigidity of the connections must be such that the angles between intersecting members should remain virtually unchanged under factored loads. The design of semi-rigid frames is permitted upon evidence of the connections to deliver a predicable amount of moment restraint. The main members joined by these connections must be designed to assure that their ultimate capacities will not exceed those of the connections. The design of simple frames is based on the assumption that the connections provide no moment restraint to the beam insofar as gravity loads are concerned but these connections should have adequate capacity to resist wind moments. Semi-rigid and simple framings often incur inelastic deformation in the connections. The connections used in these constructions must be proportioned to possess sufficient ductility to avoid overstress of the fasteners or welds.

Regardless of the types of constructions used, due consideration must be given to account for member and frame instability (P-δ and P-Δ) effects either by the use of a second-order analysis or by other means such as moment magnification factors. The end-restrained effect on members should also be accounted for by the use of the effective length factor (see Chapter 17).

Frames can be designed as **sidesway inhibited** (braced) or **sidesway uninhibited** (unbraced). In sidesway inhibited frames, frame **drift** is controlled by the presence of a bracing system (e.g., shear walls, diagonal or cross braces, etc.). In sidesway uninhibited frames, frame drift is limited by the flexural rigidity of the connected members and diaphragm action of the floors. Most sidesway uninhibited frames are designed as Type I or Type FR frames using moment connections. Under normal circumstances, the amount of interstory drift under service loads should not exceed $h/500$ to $h/300$ where h is the story height. Higher value of interstory drift is allowed only if it does not create serviceability concerns.

Beams in sidesway inhibited frames are often subject to high axial forces. As a result, they should be designed as beam-columns using beam-column interaction equations. Furthermore, vertical bracing systems should be provided for braced multistory frames to prevent vertical buckling of the frames under gravity loads.

3.10 Plate Girders

Plate girders are built-up beams. They are used as flexural members to carry extremely large lateral loads. A flexural member is considered as a plate girder if the width-thickness ratio of the web, h_c/t_w, exceeds $760/\sqrt{F_b}$ (F_b =allowable flexural stress) according to ASD, or $970/\sqrt{F_{yf}}$ (F_{yf} =minimum specified flange yield stress) according to LRFD. Because of the large web slenderness, plate girders are often designed with transverse stiffeners to reinforce the web and to allow for post-buckling (shear) strength (i.e., **tension field action**) to develop. Table 3.9 summarizes the requirements for transverse stiffeners for plate girders based on the web slenderness ratio h/t_w. Two types of transverse stiffeners are used for plate girders: bearing stiffeners and intermediate stiffeners. Bearing stiffeners are used at unframed girder ends and at concentrated load points where the web yielding or web crippling criterion is violated. Bearing stiffeners extend the full depth of the web from the bottom of the top flange to the top of the bottom flange. Intermediate stiffeners are used when the width-thickness ratio of the web, h/t_w, exceeds 260, or when the shear criterion is violated, or when tension field action is considered in the design. Intermediate stiffeners need not extend the full depth of the web but must be in contact with the compression flange of the girder.

Normally, the depths of plate girder sections are so large that simple beam theory which postulates that plane sections before bending remain plane after bending does not apply. As a result, a different set of design formulas for plate girders are required.

TABLE 3.9 Web Stiffeners Requirements

Range of web slenderness	Stiffeners requirements
$\frac{h}{t_w} \le 260$	Plate girder can be designed without web stiffeners.
$260 \le \frac{h}{t_w} \le \frac{14,000}{\sqrt{F_{yf}(F_{yf}+16.5)}}$	Plate girder must be designed with web stiffeners. The spacing of stiffeners, a, can exceed $1.5h$. The actual spacing is determined by the shear criterion.
$\frac{14,000}{\sqrt{F_{yf}(F_{yf}+16.5)}} < \frac{h}{t_w} \le \frac{2,000}{\sqrt{F_{yf}}}$	Plate girder must be designed with web stiffeners. The spacing of stiffeners, a, cannot exceed $1.5h$.

a = clear distance between stiffeners
h = clear distance between flanges when welds are used or the distance between adjacent lines of fasteners when bolts are used
t_w = web thickness
F_{yf} = compression flange yield stress, ksi

3.10.1 Allowable Stress Design

Allowable Bending Stress

The maximum bending stress in the compression flange of the girder computed using the flexure formula shall not exceed the allowable value, F_b', given by

$$F_b' = F_b R_{PG} R_e \tag{3.80}$$

where
F_b = applicable allowable bending stress as discussed in the section on Flexural Members
R_{PG} = plate girder stress reduction factor = $1 - 0.0005(A_w/A_f)(h/t_w - 760/\sqrt{F_b}) \le 1.0$

R_e = hybrid girder factor = $[12 + (A_w/A_f)(3\alpha - \alpha^3)]/[12 + 2(A_w/A_f)] \leq 1.0$, $R_e = 1$
for non-hybrid girder
A_w = area of web
A_f = area of compression flange
α = $0.60F_{yw}/F_b \leq 1.0$
F_{yw} = yield stress of web

Allowable Shear Stress

Without tension field action:
The allowable shear stress is the same as that for beams given in Equation 3.33.

With tension field action:
The allowable shear stress is given by

$$F_v = \frac{F_y}{2.89}\left[C_v + \frac{1 - C_v}{1.15\sqrt{1 + (a/h)^2}}\right] \leq 0.40F_y \tag{3.81}$$

Note that tension field action can be considered in the design only for non-hybrid girders. If tension field action is considered, transverse stiffeners must be provided and spaced at a distance such that the computed average web shear stress, f_v, obtained by dividing the total shear by the web area does not exceed the allowable shear stress, F_v, given by Equation 3.81. In addition, the computed bending tensile stress in the panel where tension field action is considered cannot exceed $0.60F_y$, nor $(0.825 - 0.375 f_v/F_v)F_y$ where f_v is the computed average web shear stress and F_v is the allowable web shear stress given in Equation 3.81. The shear transfer criterion given by Equation 3.84 must also be satisfied.

Transverse Stiffeners

Transverse stiffeners must be designed to satisfy the following criteria.

Moment of inertial criterion:
With reference to an axis in the plane of the web, the moment of inertia of the stiffeners, in cubic inches, shall satisfy the condition

$$I_{st} \geq \left(\frac{h}{50}\right)^4 \tag{3.82}$$

where h is the clear distance between flanges, in inches.

Area criterion:
The total area of the stiffeners, in square inches, shall satisfy the condition

$$A_{st} \geq \frac{1 - C_v}{2}\left[\frac{a}{h} - \frac{(a/h)^2}{\sqrt{1 + (a/h)^2}}\right]YDht_w \tag{3.83}$$

where
C_v = shear buckling coefficient as defined in Equation 3.33
a = stiffeners' spacing
h = clear distance between flanges
t_w = web thickness, in.
Y = ratio of web yield stress to stiffener yield stress
D = 1.0 for stiffeners furnished in pairs, 1.8 for single angle stiffeners, and 2.4 for single plate stiffeners

Shear transfer criterion:

If tension field action is considered, the total shear transfer, in kips/in., of the stiffeners shall not be less than

$$f_{vs} = h \sqrt{\left(\frac{F_{yw}}{340}\right)^3} \qquad (3.84)$$

where

F_{yw} = web yield stress, ksi

h = clear distance between flanges, in.

The value of f_{vs} can be reduced proportionally if the computed average web shear stress, f_v, is less than F_v given in Equation 3.81.

3.10.2 Load and Resistance Factor Design

Flexural Strength Criterion

Doubly or singly symmetric single-web plate girders loaded in the plane of the web should satisfy the flexural strength criterion of Equation 3.36. The plate girder design flexural strength is given by:

For the limit state of tension flange yielding

$$\phi_b M_n = 0.90[S_{xt} R_e F_{yt}] \qquad (3.85)$$

For the limit state of compression flange buckling

$$\phi_b M_n = 0.90[S_{xc} R_{PG} R_e F_{cr}] \qquad (3.86)$$

where

S_{xt} = section modulus referred to the tension flange = I_x/c_t

S_{xc} = section modulus referred to the compression flange = I_x/c_c

I_x = moment of inertia about the major axis

c_t = distance from neutral axis to extreme fiber of the tension flange

c_c = distance from neutral axis to extreme fiber of the compression flange

R_{PG} = plate girder bending strength reduction factor = $1 - a_r(h_c/t_w - 970/\sqrt{F_{cr}})/(1,200 + 300a_r) \le 1.0$

R_e = hybrid girder factor = $[12 + a_r(3m - m^3)]/(12 + 2a_r) \le 1.0$ ($R_e = 1$ for non-hybrid girder)

a_r = ratio of web area to compression flange area

m = ratio of web yield stress to flange yield stress or F_{cr}

F_{yt} = tension flange yield stress

F_{cr} = critical compression flange stress calculated as follows:

Limit state	Range of slenderness	F_{cr}
Flange local buckling	$\dfrac{b_f}{2t_f} \leq \dfrac{65}{\sqrt{F_{yf}}}$	F_{yf}
	$\dfrac{65}{\sqrt{F_{yf}}} < \dfrac{b_f}{2t_f} \leq \dfrac{230}{\sqrt{F_{yf}/k_c}}$	$F_{yf}\left[1 - \dfrac{1}{2}\left(\dfrac{\frac{b_f}{2t_f} - \frac{65}{\sqrt{F_{yf}}}}{\frac{230}{\sqrt{F_{yf}/k_c}} - \frac{64}{\sqrt{F_{yf}}}}\right)\right] \leq F_{yf}$
	$\dfrac{b_f}{2t_f} > \dfrac{230}{\sqrt{F_{yf}/k_c}}$	$\dfrac{26{,}200k_c}{\left(\frac{b_f}{2t_f}\right)^2}$
Lateral torsional buckling	$\dfrac{L_b}{r_T} \leq \dfrac{300}{\sqrt{F_{yf}}}$	F_{yf}
	$\dfrac{300}{\sqrt{F_{yf}}} < \dfrac{L_b}{r_T} \leq \dfrac{756}{\sqrt{F_{yf}}}$	$C_b F_{yf}\left[1 - \dfrac{1}{2}\left(\dfrac{\frac{L_b}{r_T} - \frac{300}{\sqrt{F_{yf}}}}{\frac{756}{\sqrt{F_{yf}}} - \frac{300}{\sqrt{F_{yf}}}}\right)\right] \leq F_{yf}$
	$\dfrac{L_b}{r_T} > \dfrac{756}{\sqrt{F_{yf}}}$	$\dfrac{286{,}000C_b}{\left(\frac{L_b}{r_T}\right)^2}$

$k_c \;=\; 4/\sqrt{(h/t_w)}, \; 0.35 \leq k_c \leq 0.763$

$b_f \;=\;$ compression flange width
$t_f \;=\;$ compression flange thickness
$L_b \;=\;$ lateral unbraced length of the girder

$r_T \;=\; \sqrt{[(t_f b_f^3/12 + h_c t_w^3/72)/(b_f t_f + h_c t_w/6)]}$

$h_c \;=\;$ twice the distance from the neutral axis to the inside face of the compression flange less the fillet
$t_w \;=\;$ web thickness
$F_{yf} \;=\;$ yield stress of compression flange, ksi
$C_b \;=\;$ Bending coefficient (see section on Flexural Members)

F_{cr} must be calculated for both flange local buckling and lateral torsional buckling. The smaller value of F_{cr} is used in Equation 3.86.

The plate girder bending strength reduction factor R_{PG} is a factor to account for the nonlinear flexural stress distribution along the depth of the girder. The hybrid girder factor is a reduction factor to account for the lower yield strength of the web when the nominal moment capacity is computed assuming a homogeneous section made entirely of the higher yield stress of the flange.

Shear Strength Criterion

Plate girders can be designed with or without the consideration of tension field action. If tension field action is considered, intermediate web stiffeners must be provided and spaced at a distance, a, such that a/h is smaller than 3 or $[260/(h/t_w)]^2$, whichever is smaller. Also, one must check the flexure-shear interaction of Equation 3.89, if appropriate. Consideration of tension field action is not allowed if (1) the panel is an end panel, (2) the plate girder is a hybrid girder, (3) the plate girder is a web tapered girder, or (4) a/h exceeds 3 or $[260/(h/t_w)]^2$, whichever is smaller.

The design shear strength, $\phi_v V_n$, of a plate girder is determined as follows:

If tension field action is not considered:
$\phi_v V_n$ are the same as those for beams as given in Equations 3.47 to 3.49.

If tension field action is considered and $h/t_w \leq 187/\sqrt{(k_v/F_{yw})}$:

$$\phi_v V_n = 0.90[0.60 A_w F_{yw}] \tag{3.87}$$

and, if $h/t_w > 187/\sqrt{(k_v/F_{yw})}$:

$$\phi_v V_n = 0.90 \left[0.60 A_w F_{yw} \left(C_v + \frac{1 - C_v}{1.15\sqrt{1 + (a/h)^2}} \right) \right] \tag{3.88}$$

where

k_v = $5 + 5/(a/h)^2$ (k_v shall be taken as 5.0 if a/h exceeds 3.0 or $[260/(h/t_w)]^2$, whichever is smaller)

A_w = dt_w

F_{yw} =   web yield stress, ksi

C_v =   shear coefficient, calculated as follows:

Range of h/t_w	C_v
$187\sqrt{\dfrac{k_v}{F_{yw}}} \le \dfrac{h}{t_w} \le 234\sqrt{\dfrac{k_v}{F_{yw}}}$	$\dfrac{187\sqrt{k_v/F_{yw}}}{h/t_w}$
$\dfrac{h}{t_w} > 234\sqrt{\dfrac{k_v}{F_{yw}}}$	$\dfrac{44{,}000 k_v}{(h/t_w)^2 F_{yw}}$

Flexure-Shear Interaction

Plate girders designed for tension field action must satisfy the flexure-shear interaction criterion in regions where $0.60\phi V_n \le V_u \le \phi V_n$ and $0.75\phi M_n \le M_u \le \phi M_n$

$$\frac{M_u}{\phi M_n} + 0.625\frac{V_u}{\phi V_n} \le 1.375 \tag{3.89}$$

where $\phi = 0.90$.

Bearing Stiffeners

Bearing stiffeners must be provided for a plate girder at unframed girder ends and at points of concentrated loads where the web yielding or the web crippling criterion is violated (see section on Concentrated Load Criteria). Bearing stiffeners shall be provided in pairs and extended from the upper flange to the lower flange of the girder. Denoting b_{st} as the width of one stiffener and t_{st} as its thickness, bearing stiffeners shall be portioned to satisfy the following limit states:

For the limit state of local buckling

$$\frac{b_{st}}{t_{st}} \le \frac{95}{\sqrt{F_y}} \tag{3.90}$$

For the limit state of compression

The design compressive strength, $\phi_c P_n$, must exceed the required compressive force acting on the stiffeners. $\phi_c P_n$ is to be determined based on an effective length factor K of 0.75 and an effective area, A_{eff}, equal to the area of the bearing stiffeners plus a portion of the web. For end bearing, this effective area is equal to $2(b_{st} t_{st}) + 12t_w^2$; and for interior bearing, this effective area is equal to $2(b_{st} t_{st}) + 25t_w^2$. t_w is the web thickness. The slenderness parameter, λ_c, is to be calculated using a radius of gyration, $r = \sqrt{(I_{st}/A_{eff})}$, where $I_{st} = t_{st}(2b_{st} + t_w)^3/12$.

For the limit state of bearing

The bearing strength, ϕR_n, must exceed the required compression force acting on the stiffeners. ϕR_n is given by

$$\phi R_n \ge 0.75[1.8 F_y A_{pb}] \tag{3.91}$$

where F_y is the yield stress and A_{pb} is the bearing area.

Intermediate Stiffeners

Intermediate stiffeners shall be provided if (1) the shear strength capacity is calculated based on tension field action, (2) the shear criterion is violated (i.e., when the V_u exceeds $\phi_v V_n$), or (3) the web slenderness h/t_w exceeds $418/\sqrt{F_{yw}}$. Intermediate stiffeners can be provided in pairs or on one side of the web only in the form of plates or angles. They should be welded to the compression flange and the web but they may be stopped short of the tension flange. The following requirements apply to the design of intermediate stiffeners:

Local Buckling

The width-thickness ratio of the stiffener must be proportioned so that Equation 3.90 is satisfied to prevent failure by local buckling.

Stiffener Area

The cross-section area of the stiffener must satisfy the following criterion:

$$A_{st} \geq \frac{F_{yw}}{F_y} \left[0.15 D h t_w (1 - C_v) \frac{V_u}{\phi_v V_n} - 18 t_w^2 \right] \geq 0 \tag{3.92}$$

where

F_y = yield stress of stiffeners

D = 1.0 for stiffeners in pairs, 1.8 for single angle stiffeners, and 2.4 for single plate stiffeners

The other terms in Equation 3.92 are defined as before in Equation 3.87 and Equation 3.88.

Stiffener Moment of Inertia

The moment of inertia for stiffener pairs taken about an axis in the web center or for single stiffeners taken in the face of contact with the web plate must satisfy the following criterion:

$$I_{st} \geq a t_w^3 \left[\frac{2.5}{(a/h)^2} - 2 \right] \geq 0.5 a t_w^3 \tag{3.93}$$

Stiffener Length

The length of the stiffeners, l_{st}, should fall within the range

$$h - 6 t_w < l_{st} < h - 4 t_w \tag{3.94}$$

where h is the clear distance between the flanges less the widths of the flange-to-web welds and t_w is the web thickness.

If intermittent welds are used to connect the stiffeners to the girder web, the clear distance between welds shall not exceed $16 t_w$, or 10 in. If bolts are used, their spacing shall not exceed 12 in.

Stiffener Spacing

The spacing of the stiffeners, a, shall be determined from the shear criterion $\phi_v V_n \geq V_u$. This spacing shall not exceed the smaller of $3h$ and $[260/(h/t_w)]^2 h$.

EXAMPLE 3.7:

Using LRFD, design the cross-section of an I-shaped plate girder shown in Figure 3.12a to support a factored moment M_u of 4600 kip-ft (6240 kN-m), dead weight of the girder is included. The girder is a 60-ft (18.3-m) long simply supported girder. It is laterally supported at every 20-ft (6.1-m) interval. Use A36 steel.

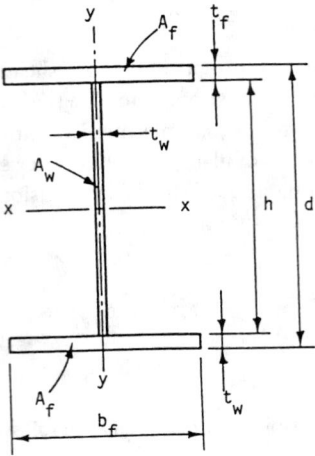

(a) Plate Girder Nomenclature

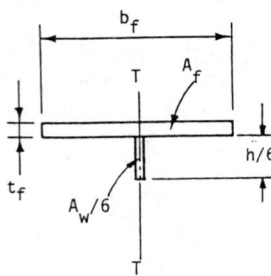

(b) Calculation of r_T

Figure 3.12   Design of a plate girder cross-section.

Proportion of the girder web

Ordinarily, the overall depth-to-span ratio d/L of a building girder is in the range 1/12 to 1/10. So, let us try $h = 70$ in. Also, knowing h/t_w of a plate girder is in the range $970/\sqrt{F_{yf}}$ and $2{,}000/\sqrt{F_{yf}}$, let us try $t_w = 5/16$ in.

Proportion of the girder flanges

For a preliminary design, the required area of the flange can be determined using the flange area method

$$A_f \approx \frac{M_u}{F_y h} = \frac{4600 \text{ kip-ft} \times 12 \text{ in./ft}}{(36 \text{ ksi })(70 \text{ in.})} = 21.7 \text{ in.}^2$$

So, let $b_f = 20$ in. and $t_f = 1\text{-}1/8$ in. giving $A_f = 22.5$ in.^{2}

Determine the design flexural strength $\phi_b M_n$ of the girder:
Calculate I_x:

$$I_x = \sum [I_i + A_i y_i^2]$$

$$= [8932 + (21.88)(0)^2] + 2[2.37 + (22.5)(35.56)^2]$$
$$= 65840 \text{ in.}^4$$

Calculate S_{xt}, S_{xc}:

$$S_{xt} = S_{xc} = \frac{I_x}{c_t} = \frac{I_x}{c_c} = \frac{65840}{35 + 1.125} = 1823 \text{ in.}^3$$

Calculate r_T: Refer to Figure 3.12b,

$$r_T = \sqrt{\frac{I_T}{A_f + \frac{1}{6}A_w}} = \sqrt{\frac{(1.125)(20)^3/12 + (11.667)(5/16)^3/12}{22.5 + \frac{1}{6}(21.88)}} = 5.36 \text{ in.}$$

Calculate F_{cr}:

For Flange Local Buckling (FLB),

$$\left[\frac{b_f}{2t_f} = \frac{20}{2(1.125)} = 8.89\right] < \left[\frac{65}{\sqrt{F_{yf}}} = \frac{65}{\sqrt{36}} = 10.8\right] \quad \text{so,} \quad F_{cr} = F_{yf} = 36 \text{ ksi}$$

For Lateral Torsional Buckling (LTB),

$$\left[\frac{L_b}{r_T} = \frac{20 \times 12}{5.36} = 44.8\right] < \left[\frac{300}{\sqrt{F_{yf}}} = \frac{300}{\sqrt{36}} = 50\right] \quad \text{so,} \quad F_{cr} = F_{yf} = 36 \text{ ksi}$$

Calculate R_{PG}:

$$R_{PG} = 1 - \frac{a_r(h_c/t_w - 970/\sqrt{F_{cr}})}{(1,200 + 300a_r)} = 1 - \frac{0.972[70/(5/16) - 970/\sqrt{36}]}{[1,200 + 300(0.972)]} = 0.96$$

Calculate $\phi_b M_n$:

$$\phi_b M_n = \text{smaller of} \begin{cases} 0.90 \, S_{xt} R_e F_{yt} = (0.90)(1823)(1)(36) = 59,065 \text{ kip-in.} \\ 0.90 \, S_{xc} R_{PG} R_e F_{cr} = (0.90)(1823)(0.96)(1)(36) = 56,700 \text{ kip-in.} \end{cases}$$
$$= 56,700 \text{ kip-in.}$$
$$= 4725 \text{ kip-ft.}$$

Since $[\phi_b M_n = 4725 \text{ kip-ft}] > [M_u = 4600 \text{ kip-ft}]$, the cross-section is acceptable.
Use web plate 5/16"x70" and two flange plates 1-1/8"x20" for the girder cross-section.

EXAMPLE 3.8:

Design bearing stiffeners for the plate girder of the preceding example for a factored end reaction of 260 kips.

Since the girder end is unframed, bearing stiffeners are required at the supports. The size of the stiffeners must be selected to ensure that the limit states of local buckling, compression, and bearing are not violated.

Limit state of local buckling

Refer to Figure 3.13, try $b_{st} = 8$ in. To avoid problems with local buckling, $b_{st}/2t_{st}$ must not exceed $95/\sqrt{F_y} = 15.8$. Therefore, try $t_{st} = 1/2$ in. So, $b_{st}/2t_{st} = 8$ which is less than 15.8.

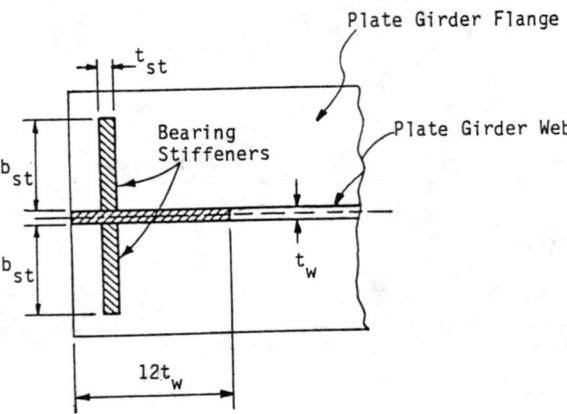

Figure 3.13   Design of bearing stiffeners.

Limit state of compression

$$
\begin{aligned}
A_{eff} &= 2(b_{st}t_{st}) + 12t_w^2 = 2(8)(0.5) + 12(5/16)^2 = 9.17 \text{ in.}^2 \\
I_{st} &= t_{st}(2b_{st} + t_w)^3/12 = 0.5[2(8) + 5/16]^3/12 = 181 \text{ in.}^4 \\
r_{st} &= \sqrt{(I_{st}/A_{eff})} = \sqrt{(181/9.17)} = 4.44 \text{ in.} \\
Kh/r_{st} &= 0.75(70)/4.44 = 11.8 \\
\lambda_c &= (Kh/\pi r_{st})\sqrt{(F_y/E)} = (11.8/3.142)\sqrt{(36/29,000)} = 0.132
\end{aligned}
$$

and from Equation 3.17

$$
\phi_c P_n = 0.85(0.658^{\lambda_c^2})F_y A_{st} = 0.85(0.658)^{0.132^2}(36)(9.17) = 279 \text{ kips}
$$

Since $\phi_c P_n > 260$ kips, the design is satisfactory for compression.

Limit state of bearing

Assuming there is a 1/4-in. weld cutout at the corners of the bearing stiffeners at the junction of the stiffeners and the girder flanges, the bearing area for the stiffener pairs is $A_{pb} = (8 - 0.25)(0.5)(2) = 7.75 \text{ in.}^2$. Substitute this into Equation 3.91, we have $\phi R_n = 0.75(1.8)(36)(7.75) = 377$ kips, which exceeds the factored reaction of 260 kips. So, bearing is not a problem.

Use two 1/2"x 8" plates for bearing stiffeners.

3.11   Connections

Connections are structural elements used for joining different members of a framework. Connections can be classified according to:

- the type of connecting medium used: bolted connections, welded connections, bolted-welded connections, riveted connections
- the type of internal forces the connections are expected to transmit: shear (semi-rigid, simple) connections, moment (rigid) connections
- the type of structural elements that made up the connections: single plate angle connections, double web angle connections, top and seated angle connections, seated beam connections, etc.

- the type of members the connections are joining: beam-to-beam connections (beam splices), column-to-column connections (column splices), beam-to-column connections, hanger connections, etc.

To properly design a connection, a designer must have a thorough understanding of the behavior of the joint under loads. Different modes of failure can occur depending on the geometry of the connection and the relative strengths and stiffnesses of the various components of the connection. To ensure that the connection can carry the applied loads, a designer must check for all perceivable modes of failure pertinent to each component of the connection and the connection as a whole.

3.11.1 Bolted Connections

Bolted connections are connections whose components are fastened together primarily by bolts. The four basic types of bolts commonly used for steel construction are discussed in the section on Structural Fasteners. Depending on the direction and line of action of the loads relative to the orientation and location of the bolts, the bolts may be loaded in tension, shear, or a combination of tension and shear. For bolts subjected to shear forces, the design shear strength of the bolts also depends on whether or not the threads of the bolts are excluded from the shear planes. A letter X or N is placed at the end of the ASTM designation of the bolts to indicate whether the threads are excluded or not excluded from the shear planes, respectively. Thus, A325X denotes A325 bolts whose threads are excluded from the shear planes and A490N denotes A490 bolts whose threads are not excluded from the shear planes. Because of the reduced shear areas for bolts whose threads are not excluded from the shear planes, these bolts have lower design shear strengths than their counterparts whose threads are excluded from the shear planes.

Bolts can be used in both bearing-type connections and slip-critical connections. Bearing-type connections rely on bearing between the bolt shanks and the connecting parts to transmit forces. Some slippage between the connected parts is expected to occur for this type of connection. Slip-critical connections rely on the frictional force developing between the connecting parts to transmit forces. No slippage between connecting elements is expected for this type of connection. Slip-critical connections are used for structures designed for vibratory or dynamic loads such as bridges, industrial buildings, and buildings in regions of high seismicity. Bolts used in slip-critical connections are denoted by the letter F after their ASTM designation, e.g., A325F, A490F.

Bolt Holes

Holes made in the connected parts for bolts may be standard size, oversized, short slotted, or long slotted. Table 3.10 gives the maximum hole dimension for ordinary construction usage.

TABLE 3.10 Nominal Hole Dimensions

Bolt diameter, d (in.)	Hole dimensions			
	Standard (dia.)	Oversize (dia.)	Short-slot (width × length)	Long-slot (width × length)
1/2	9/16	5/8	9/16×11/16	9/16×1-1/4
5/8	11/16	13/16	11/16×7/8	11/16×1-9/16
3/4	13/16	15/16	13/16×1	13/16×1-7/8
7/8	15/16	1-1/16	15/16×1-1/8	15/16×2-3/16
1	1-1/16	1-1/4	1-1/16×1-5/16	1-1/16×2-1/2
≥ 1-1/8	d+1/16	d+5/16	(d+1/16)×(d+3/8)	(d+1/16)×(2.5d)

Note: 1 in. = 25.4 mm.

Standard holes can be used for both bearing-type and slip-critical connections. Oversized holes shall be used only for slip-critical connections. Short- and long-slotted holes can be used for both bearing-type and slip-critical connections provided that when such holes are used for bearing, the direction of slot is transverse to the direction of loading.

Bolts Loaded in Tension

If a tensile force is applied to the connection such that the direction of the load is parallel to the longitudinal axes of the bolts, the bolts will be subjected to tension. The following condition must be satisfied for bolts under tensile stresses.

Allowable Stress Design:

$$f_t \leq F_t \tag{3.95}$$

where
f_t = computed tensile stress in the bolt, ksi
F_t = allowable tensile stress in bolt (see Table 3.11)

Load and Resistance Factor Design:

$$\phi_t F_t \geq f_t \tag{3.96}$$

where
ϕ_t = 0.75
f_t = tensile stress produced by factored loads, ksi
F_t = nominal tensile strength given in Table 3.11

TABLE 3.11 F_t of Bolts, ksi

	ASD		LRFD	
Bolt type	F_t, ksi (static loading)	F_t, ksi (fatigue loading)	F_t, ksi (static loading)	F_t, ksi (fatigue loading)
A307	20	Not allowed	45.0	Not allowed
A325	44.0	If $N \leq 20{,}000$: F_t = same as for static loading	90.0	If $N \leq 20{,}000$: F_t = same as for static loading
		If $20{,}000 < N \leq 500{,}000$: $F_t = 40$ (A325) $= 49$ (A490) If $N > 500{,}000$:		If $20{,}000 < N \leq 500{,}000$: $F_t = 0.30F_u$ (at service loads)
A490	54.0	$F_t = 31$(A325) $= 38$ (A490) where N = number of cycles F_u = minimum specified tensile strength, ksi	113	If $N > 500{,}000$: $F_t = 0.25F_u$ (at service loads) where N = number of cycles F_u = minimum specified tensile strength, ksi

Note: 1 ksi = 6.895 MPa.

Bolts Loaded in Shear

When the direction of load is perpendicular to the longitudinal axes of the bolts, the bolts will be subjected to shear. The condition that needs to be satisfied for bolts under shear stresses is as follows.

Allowable Stress Design:

$$f_v \leq F_v \tag{3.97}$$

where

f_v = computed shear stress in the bolt, ksi

F_v = allowable shear stress in bolt (see Table 3.12)

Load and Resistance Factor Design:

$$\phi_v F_v \geq f_v \tag{3.98}$$

where

ϕ_v = 0.75 (for bearing-type connections), 1.00 (for slip-critical connections when standard, oversized, short-slotted, or long-slotted holes with load perpendicular to the slots are used), 0.85 (for slip-critical connections when long-slotted holes with load in the direction of the slots are used)

f_v = shear stress produced by factored loads (for bearing-type connections), or by service loads (for slip-critical connections), ksi

F_v = nominal shear strength given in Table 3.12

TABLE 3.12 F_v of Bolts, ksi

Bolt type	F_v, ksi	
	ASD	LRFD
A307	10.0[a] (regardless of whether or not threads are excluded from shear planes)	24.0[a] (regardless of whether or not threads are excluded from shear planes)
A325N	21.0[a]	48.0[a]
A325X	30.0[a]	60.0[a]
A325F[b]	17.0 (for standard size holes) 15.0 (for oversized and short-slotted holes) 12.0 (for long-slotted holes when direction of load is transverse to the slots) 10.0 (for long-slotted holes when direction of load is parallel to the slots)	17.0 (for standard size holes) 15.0 (for oversized and short-slotted holes) 12.0 (for long-slotted holes)
A490N	28.0[a]	60.0[a]
A490X	40.0[a]	75.0[a]
A490F[b]	21.0 (for standard size holes) 18.0 (for oversized and short-slotted holes) 15.0 (for long-slotted holes when direction of load is transverse to the slots) 13.0 (for long-slotted holes when direction of load is parallel to the slots)	21.0 (for standard size holes) 18.0 (for oversized and short-slotted holes) 15.0 (for long-slotted holes)

[a] tabulated values shall be reduced by 20% if the bolts are used to splice tension members having a fastener pattern whose length, measured parallel to the line of action of the force, exceeds 50 in.

[b] tabulated values are applicable only to class A surface, i.e., clean mill surface and blast cleaned surface with class A coatings (with slip coefficient = 0.33). For design strengths with other coatings, see RCSC "Load and Resistance Factor Design Specification to Structural Joints Using ASTM A325 or A490 Bolts" [28]

Note: 1 ksi = 6.895 MPa.

Bolts Loaded in Combined Tension and Shear

If a tensile force is applied to a connection such that its line of action is at an angle with the longitudinal axes of the bolts, the bolts will be subjected to combined tension and shear. The conditions that need to be satisfied are given as follows.

Allowable Stress Design:

$$f_v \leq F_v \quad \text{and} \quad f_t \leq F_t \tag{3.99}$$

where

f_v, F_v = as defined in Equation 3.97

f_t = computed tensile stress in the bolt, ksi

F_t = allowable tensile stress given in Table 3.13

Load and Resistance Factor Design:

$$\phi_v F_v \geq f_v \quad \text{and} \quad \phi_t F_t \geq f_t \tag{3.100}$$

where

ϕ_v, F_v, f_v = as defined in Equation 3.98

ϕ_t = 1.0

f_t = tensile stress due to factored loads (for bearing-type connection), or due to service loads (for slip-critical connections), ksi

F_t = nominal tension stress limit for combined tension and shear given in Table 3.13

TABLE 3.13 F_t for Bolts Under Combined Tension and Shear, ksi

	Bearing-type connections			
	ASD		LRFD	
Bolt type	Threads not excluded from the shear plane	Threads excluded from the shear plane	Threads not excluded from the shear plane	Threads excluded from the shear plane
A307	$26 - 1.8 f_v \leq 20$		$59 - 1.9 f_v \leq 45$	
A325	$\sqrt{(44^2 - 4.39 f_v^2)}$	$\sqrt{(44^2 - 2.15 f_v^2)}$	$117 - 1.9 f_v \leq 90$	$117 - 1.5 f_v \leq 90$
A490	$\sqrt{(54^2 - 3.75 f_v^2)}$	$\sqrt{(54^2 - 1.82 f_v^2)}$	$147 - 1.9 f_v \leq 113$	$147 - 1.5 f_v \leq 113$

Slip-critical connections

For ASD:

F_t = values given above

F_v = $[1 - (f_t A_b / T_b)] \times$ (values of F_v given in Table 3.12)

where

f_t = computed tensile stress in the bolt, ksi

T_b = pretension load = $0.70 F_u A_b$, kips

F_u = minimum specified tensile strength, ksi

A_b = nominal cross-sectional area of bolt, in.^{2}

For LRFD:

F_t = values given above

F_v = $[1 - (T/T_b)] \times$ (values of F_v given in Table 3.12)

where

T = service tensile force, kips

T_b = pretension load = $0.70 F_u A_b$, kips

F_u = minimum specified tensile strength, ksi

A_b = nominal cross-sectional area of bolt, in.^{2}

Note: 1 ksi = 6.895 MPa.

Bearing Strength at Fastener Holes

Connections designed on the basis of bearing rely on the bearing force developed between the fasteners and the holes to transmit forces and moments. The limit state for bearing must therefore be checked to ensure that bearing failure will not occur. Bearing strength is independent of the type of fastener. This is because the bearing stress is more critical on the parts being connected than on the fastener itself. The AISC specification provisions for bearing strength are based on preventing excessive hole deformation. As a result, bearing capacity is expressed as a function of the type of holes (standard, oversized, slotted), bearing area (bolt diameter times the thickness of the connected parts), bolt spacing, edge distance (L_e), strength of the connected parts (F_u) and the number of fasteners in the direction of the bearing force. Table 3.14 summarizes the expressions used in ASD and LRFD for calculating the bearing strength and the conditions under which each expression is valid.

TABLE 3.14   Bearing Capacity

Conditions	ASD Allowable bearing stress, F_p, ksi	LRFD Design bearing strength, ϕR_n, ksi
1. For standard or short-slotted holes with $L_e \geq 1.5d$, $s \geq 3d$ and number of fasteners in the direction of bearing ≥ 2	$1.2F_u$	$0.75[2.4dtF_u]$
2. For long-slotted holes with direction of slot transverse to the direction of bearing and $L_e \geq 1.5d$, $s \geq 3d$ and the number of fasteners in the direction of bearing ≥ 2	$1.0F_u$	$0.75[2.0dtF_u]$
3. If neither condition 1 nor 2 above is satisfied	$L_eF_u/2d \leq 1.2F_u$	For the bolt hole nearest the edge: $0.75[L_etF_u]$ $\leq 0.75[2.4dtF_u]^a$ For the remaining bolt holes: $0.75[(s-d/2)tF_u]$ $\leq 0.75[2.4dtF_u]^a$
4. If hole deformation is not a design consideration and adequate spacing and edge distance is provided (see sections on Minimum Fastener Spacing and Minimum Edge Distance)	$1.5F_u$	For the bolt hole nearest the edge: $0.75[L_etF_u]$ $\leq 0.75[3.0dtF_u]$ For the remaining bolt holes: $0.75[(s-d/2)tF_u]$ $\leq 0.75[3.0dtF_u]$

^{a} For long-slotted bolt holes with direction of slot transverse to the direction of bearing, this limit is $0.75[2.0dtF_u]$

L_e = edge distance (i.e., distance measured from the edge of the connected part to the center of a standard hole or the center of a short- and long-slotted hole perpendicular to the line of force. For oversized holes and short- and long-slotted holes parallel to the line of force, L_e shall be increased by the edge distance increment C_2 given in Table 3.16)

s = fastener spacing (i.e., center to center distance between adjacent fasteners measured in the direction of bearing. For oversized holes and short- and long-slotted holes parallel to the line of force, s shall be increased by the spacing increment C_1 given in Table 3.15)

d = nominal bolt diameter, in.

t = thickness of the connected part, in.

F_u = specified minimum tensile strength of the connected part, ksi

TABLE 3.15   Values of Spacing Increment, C_1, in.

Nominal diameter of fastener (in.)	Standard holes	Oversized holes	Slotted Holes		
			Transverse to line of force	Parallel to line of force	
				Short-slots	Long-slots^a
$\leq 7/8$	0	1/8	0	3/16	$3d/2-1/16$
1	0	3/16	0	1/4	23/16
$\geq 1\text{-}1/8$	0	1/4	0	5/16	$3d/2-1/16$

^{a} When length of slot is less than the value shown in Table 3.10, C_1 may be reduced by the difference between the value shown and the actual slot length.
Note: 1 in. = 25.4 mm.

Minimum Fastener Spacing

To ensure safety, efficiency, and to maintain clearances between bolt nuts as well as to provide room for wrench sockets, the fastener spacing, s, should not be less than $3d$ where d is the nominal fastener diameter.

TABLE 3.16 Values of Edge Distance Increment, C_2, in.

Nominal diameter of fastener (in.)	Oversized holes	Slotted holes		Slot parallel to edge
		Slot transverse to edge		
		Short-slot	Long-slot[a]	
$\leq 7/8$	1/16	1/8	$3d/4$	0
1	1/8	1/8	$3d/4$	
$\leq 1\text{-}1/8$	1/8	3/16	$3d/4$	

[a] If the length of the slot is less than the maximum shown in Table 3.10, the value shown may be reduced by one-half the difference between the maximum and the actual slot lengths.
Note: 1 in. = 25.4 mm.

Minimum Edge Distance

To prevent excessive deformation and shear rupture at the edge of the connected part, a minimum edge distance L_e must be provided in accordance with the values given in Table 3.17 for standard holes. For oversized and slotted holes, the values shown must be incremented by C_2 given in Table 3.16.

TABLE 3.17 Minimum Edge Distance for Standard Holes, in.

Nominal fastener diameter (in.)	At sheared edges	At rolled edges of plates, shapes, and bars or gas cut edges
1/2	7/8	3/4
5/8	1-1/8	7/8
3/4	1-1/4	1
7/8	1-1/2	1-1/8
1	1-3/4	1-1/4
1-1/8	2	1-1/2
1-1/4	2-1/4	1-5/8
over 1-1/4	1-3/4 x diameter	1-1/4 x diameter

Note: 1 in. = 25.4 mm.

Maximum Fastener Spacing

A limit is placed on the maximum value for the spacing between adjacent fasteners to prevent the possibility of gaps forming or buckling from occurring in between fasteners when the load to be transmitted by the connection is compressive. The maximum fastener spacing measured in the direction of the force is given as follows.

For painted members or unpainted members not subject to corrosion: smaller of $24t$ where t is the thickness of the thinner plate and 12 in.

For unpainted members of weathering steel subject to atmospheric corrosion: smaller of $14t$ where t is the thickness of the thinner plate and 7 in.

Maximum Edge Distance

A limit is placed on the maximum value for edge distance to prevent prying action from occurring. The maximum edge distance shall not exceed the smaller of 12_t where t is the thickness of the connected part and 6 in.

EXAMPLE 3.9:

Check the adequacy of the connection shown in Figure 3.4a. The bolts are 1-in. diameter A325N bolts in standard holes.

Check bolt capacity

All bolts are subjected to double shear. Therefore, the design shear strength of the bolts will be twice that shown in Table 3.12. Assuming each bolt carries an equal share of the factored applied load, we have from Equation 3.98

$$[\phi_v F_v = 0.75(2 \times 48) = 72 \ \text{ksi}] > \left[f_v = \frac{208}{(6)\left(\frac{\pi 1^2}{4}\right)} = 44.1 \ \text{ksi} \right]$$

The shear capacity of the bolt is therefore adequate.

Check bearing capacity of the connected parts

With reference to Table 3.14, it can be seen that condition 1 applies for the present problem. Therefore, we have

$$\left[\phi R_n = 0.75(2.4)(1)\left(\frac{3}{8}\right)(58) = 39.2 \ \text{kips} \right] > \left[R_u = \frac{208}{6} = 34.7 \ \text{kips} \right]$$

and so bearing is not a problem. Note that bearing on the gusset plate is more critical than bearing on the webs of the channels because the thickness of the gusset plate is less than the combined thickness of the double channels.

Check bolt spacing

The minimum bolt spacing is $3d = 3(1) = 3$ in. The maximum bolt spacing is the smaller of $14t = 14(.303) = 4.24$ in. or 7 in. The actual spacing is 3 in. which falls within the range of 3 to 4.24 in., so bolt spacing is adequate.

Check edge distance

From Table 3.17, it can be determined that the minimum edge distance is 1.25 in. The maximum edge distance allowed is the smaller of $12t = 12(0.303) = 3.64$ in. or 6 in. The actual edge distance is 3 in. which falls within the range of 1.25 to 3.64 in., so edge distance is adequate.

The connection is adequate.

Bolted Hanger Type Connections

A typical hanger connection is shown in Figure 3.14. In the design of such connections, the designer must take into account the effect of *prying action*. Prying action results when flexural deformation occurs in the tee flange or angle leg of the connection (Figure 3.15). Prying action tends

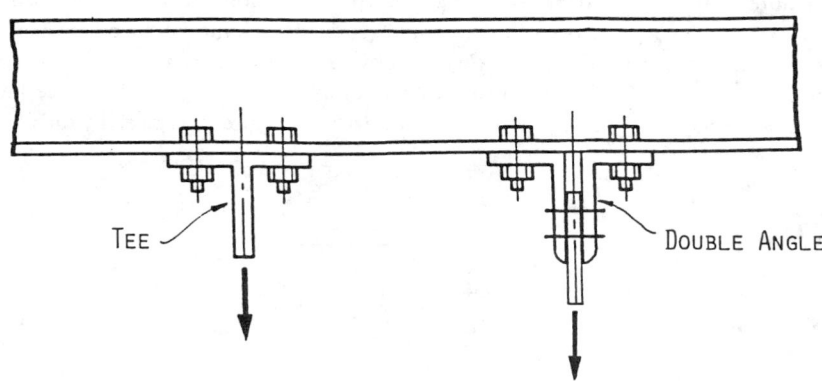

Figure 3.14   Hanger connections.

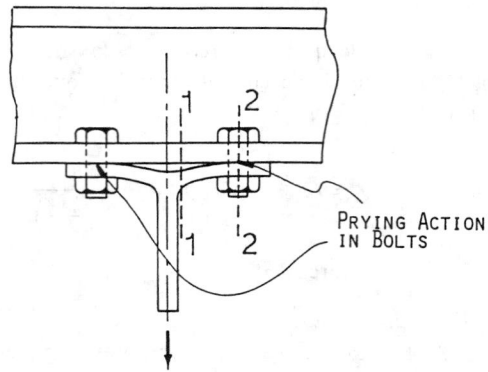

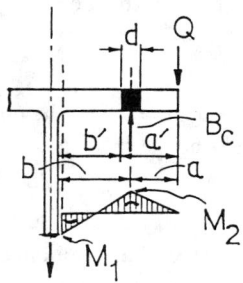

Figure 3.15   Prying action in hanger connections.

to increase the tensile force, called prying force, in the bolts. To minimize the effect of prying, the fasteners should be placed as close to the tee stem or outstanding angle leg as the wrench clearance will permit (see Tables on Entering and Tightening Clearances in Volume II-Connections of the AISC-LRFD Manual [22]). In addition, the flange and angle thickness should be proportioned so that the full tensile capacities of the bolts can be developed.

Two failure modes can be identified for hanger type connections: formation of plastic hinges in the tee flange or angle leg at cross-sections 1 and 2, and tensile failure of the bolts when the tensile force including prying action $B_c(= T + Q)$ exceeds the tensile capacity of the bolt B. Since the determination of the actual prying force is rather complex, the design equation for the required thickness for the tee flange or angle leg is semi-empirical in nature. It is given by the following.

If ASD is used:

$$t_{req'd} = \sqrt{\frac{8Tb'}{pF_y(1 + \delta\alpha')}}$$

(3.101)

where

T = tensile force per bolt due to service load exclusive of initial tightening and prying force, kips

The other variables are as defined in Equation 3.102 except that B in the equation for α' is defined as the allowable tensile force per bolt. A design is considered satisfactory if the thickness of the tee flange or angle leg t_f exceeds $t_{req'd}$ and $B > T$.

If LRFD is used:

$$t_{req'd} = \sqrt{\frac{4T_u b'}{\phi_b p F_y (1 + \delta \alpha')}} \tag{3.102}$$

where

ϕ_b	=	0.90
T_u	=	factored tensile force per bolt exclusive of initial tightening and prying force, kips
p	=	length of flange tributary to each bolt measured along the longitudinal axis of the tee or double angle section, in.
δ	=	ratio of net area at bolt line to gross area at angle leg or stem face $= (p - d')/p$
d'	=	diameter of bolt hole = bolt diameter $+1/8''$, in.
α'	=	$[(B/T_u - 1)(a'/b')]/\{\delta[1 - (B/T_u - 1)(a'/b')]\}$, but not larger than 1 (if α' is less than zero, use $\alpha' = 1$)
B	=	design tensile strength of one bolt $= \phi F_t A_b$, kips (ϕF_t is given in Table 3.11 and A_b is the nominal diameter of the bolt)
a'	=	$a + d/2$
b'	=	$b - d/2$
a	=	distance from bolt centerline to edge of tee flange or angle leg but not more than $1.25b$, in.
b	=	distance from bolt centerline to face of tee stem or outstanding leg, in.

A design is considered satisfactory if the thickness of the tee flange or angle leg t_f exceeds $t_{reg'd}$ and $B > T_u$.

Note that if t_f is much larger than $t_{reg'd}$, the design will be too conservative. In this case α' should be recomputed using the equation

$$\alpha' = \frac{1}{\delta}\left[\frac{4T_u b'}{\phi_b p t_f^2 F_y} - 1\right] \tag{3.103}$$

As before, the value of α' should be limited to the range $0 \le \alpha' \le 1$. This new value of α' is to be used in Equation 3.102 to recalculate $t_{reg'd}$.

Bolted Bracket Type Connections

Figure 3.16 shows three commonly used bracket type connections. The bracing connection shown in Figure 3.16a should be designed so that the line of action the force passes through is the centroid of the bolt group. It is apparent that the bolts connecting the bracket to the column flange are subjected to combined tension and shear. As a result, the capacity of the connection is limited to the combined tensile-shear capacities of the bolts in accordance with Equation 3.99 in ASD and Equation 3.100 in LRFD. For simplicity, f_v and f_t are to be computed assuming that both the tensile and shear components of the force are distributed evenly to all bolts. In addition to checking for the bolt capacities, the bearing capacities of the column flange and the bracket should also be checked. If the axial component of the force is significant, the effect of prying should also be considered.

In the design of the eccentrically loaded connections shown in Figure 3.16b, it is assumed that the neutral axis of the connection lies at the center of gravity of the bolt group. As a result, the bolts above the neutral axis will be subjected to combined tension and shear and so Equation 3.99 or Equation 3.100 needs to be checked. The bolts below the neutral axis are subjected to shear only and so Equation 3.97 or Equation 3.98 applies. In calculating f_v, one can assume that all bolts in the bolt group carry an equal share of the shear force. In calculating f_t, one can assume that the tensile force varies linearly from a value of zero at the neutral axis to a maximum value at the bolt

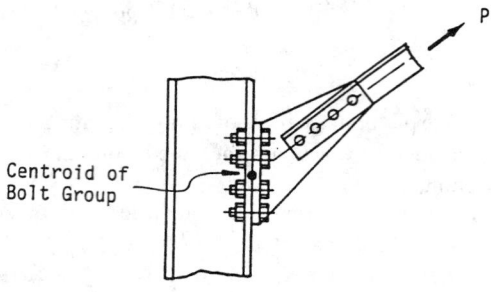

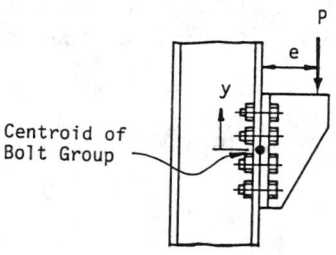

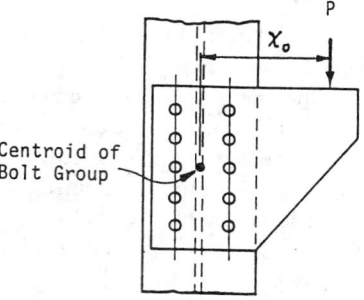

Figure 3.16   Bolted bracket-type connections.

farthest away from the neutral axis. Using this assumption, f_t can be calculated from the equation Pey/I where y is the distance from the neutral axis to the location of the bolt above the neutral axis and $I = \sum A_b y^2$ is the moment of inertia of the bolt areas with A_b equal to the cross-sectional area of each bolt. The capacity of the connection is determined by the capacities of the bolts and the bearing capacity of the connected parts.

For the eccentrically loaded bracket connection shown in Figure 3.16c, the bolts are subjected to shear. The shear force in each bolt can be obtained by adding vectorally the shear caused by the applied load P and the moment $P\chi_o$. The design of this type of connection is facilitated by the use of tables contained in the AISC Manuals for Allowable Stress Design and Load and Resistance Factor Design [21, 22].

In addition to checking for bolt shear capacity, one needs to check the bearing and shear rupture capacities of the bracket plate to ensure that failure will not occur in the plate.

Bolted Shear Connections

Shear connections are connections designed to resist shear force only. These connections are not expected to provide appreciable moment restraint to the connection members. Examples of these connections are shown in Figure 3.17. The framed beam connection shown in Figure 3.17a consists of two web angles which are often shop-bolted to the beam web and then field-bolted to the column flange. The seated beam connection shown in Figure 3.17b consists of two flange angles often shop-bolted to the beam flange and field-bolted to the column flange. To enhance the strength and stiffness of the seated beam connection, a stiffened seated beam connection shown in Figure 3.17c is sometimes used to resist large shear force. Shear connections must be designed to sustain appreciable deformation and yielding of the connections is expected. The need for ductility often limits the thickness of the angles that can be used. Most of these connections are designed with angle thickness not exceeding 5/8 in.

The design of the connections shown in Figure 3.17 is facilitated by the use of design tables contained in the AISC-ASD and AISC-LRFD Manuals. These tables give design loads for the connections with specific dimensions based on the limit states of bolt shear, bearing strength of the connection, bolt bearing with different edge distances, and block shear (for coped beams).

Bolted Moment-Resisting Connections

Moment-resisting connections are connections designed to resist both moment and shear. These connections are often referred to as rigid or fully restrained connections as they provide full continuity between the connected members and are designed to carry the full factored moments. Figure 3.18 shows some examples of moment-resisting connections. Additional examples can be found in the AISC-ASD and AISC-LRFD Manuals and Chapter 4 of the AISC Manual on Connections [20].

Design of Moment-Resisting Connections

An assumption used quite often in the design of moment connections is that the moment is carried solely by the flanges of the beam. The moment is converted to a couple F_f given by $F_f = M/(d - t_f)$ acting on the beam flanges as shown in Figure 3.19.

The design of the connection for moment is considered satisfactory if the capacities of the bolts and connecting plates or structural elements are adequate to carry the flange force F_f. Depending on the geometry of the bolted connection, this may involve checking: (a) the shear and/or tensile capacities of the bolts, (b) the yield and/or fracture strength of the moment plate, (c) the bearing strength of the connected parts, and (d) bolt spacing and edge distance as discussed in the foregoing sections.

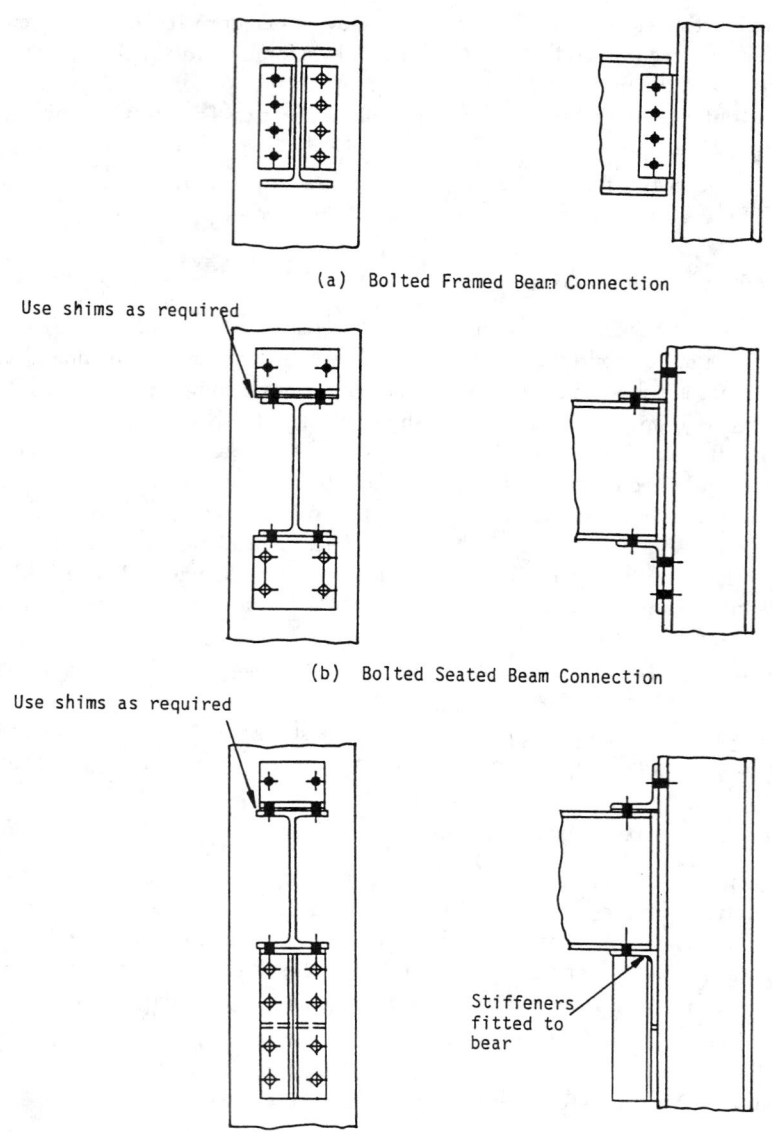

(a) Bolted Framed Beam Connection

Use shims as required

(b) Bolted Seated Beam Connection

Use shims as required

Stiffeners
fitted to
bear

(c) Bolted Stiffened Seated Beam Connection

Figure 3.17   Bolted shear connections. (a) Bolted frame beam connection. (b) Bolted seated beam connection. (c) Bolted stiffened seated beam connection.

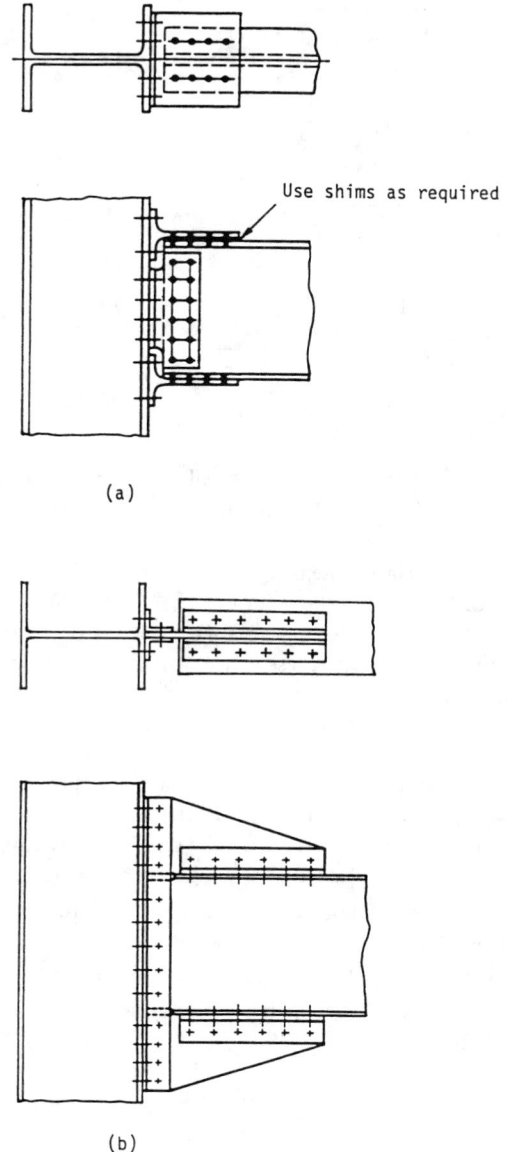

(a)

(b)

Figure 3.18 Bolted moment connections.

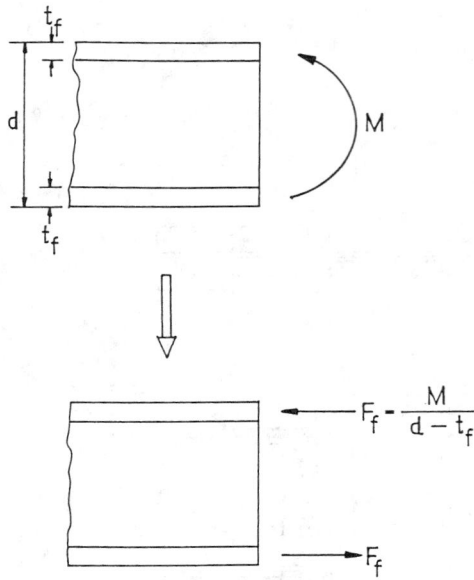

Figure 3.19   Flange forces in moment connections.

As for shear, it is common practice to assume that all the shear resistance is provided by the shear plates or angles. The design of the shear plates or angles is governed by the limit states of bolt shear, bearing of the connected parts, and shear rupture.

If the moment to be resisted is large, the flange force may cause bending of the column flange, or local yielding, crippling, or buckling of the column web. To prevent failure due to bending of the column flange or local yielding of the column web (for a tensile F_f) as well as local yielding, crippling or buckling of the column web (for a compressive F_f), column stiffeners should be provided if any one of the conditions discussed in the section on Criteria on Concentrated Loads is violated.

Following is a set of guidelines for the design of column web stiffeners [21, 22]:

1. If local web yielding controls, the area of the stiffeners (provided in pairs) shall be determined based on any excess force beyond that which can be resisted by the web alone. The stiffeners need not extend more than one-half the depth of the column web if the concentrated beam flange force F_f is applied at only one column flange.

2. If web crippling or compression buckling of the web controls, the stiffeners shall be designed as axially loaded compression members (see section on Compression Members). The stiffeners shall extend the entire depth of the column web.

3. The welds that connect the stiffeners to the column shall be designed to develop the full strength of the stiffeners.

In addition, the following recommendations are given:

1. The width of the stiffener plus one-half of the column web thickness should not be less than one-half the width of the beam flange nor the moment connection plate which applies the force.

2. The stiffener thickness should not be less than one-half the thickness of the beam flange.

3. If only one flange of the column is connected by a moment connection, the length of the stiffener plate does not have to exceed one-half the column depth.

4. If both flanges of the column are connected by moment connections, the stiffener plate should extend through the depth of the column web and welds should be used to connect

the stiffener plate to the column web with sufficient strength to carry the unbalanced moment on opposite sides of the column.

5. If column stiffeners are required on both the tension and compression sides of the beam, the size of the stiffeners on the tension side of the beam should be equal to that on the compression size for ease of construction.

In lieu of stiffener plates, a stronger column section could be used to preclude failure in the column flange and web.

For a more thorough discussion of bolted connections, the readers are referred to the book by Kulak et al. [16]. Examples on the design of a variety of bolted connections can be found in the AISC-LRFD Manual [22] and the AISC Manual on Connections [20]

3.11.2 Welded Connections

Welded connections are connections whose components are joined together primarily by welds. The four most commonly used welding processes are discussed in the section on Structural Fasteners. Welds can be classified according to:

- types of welds: groove, fillet, plug, and slot welds.
- positions of the welds: horizontal, vertical, overhead, and flat welds.
- types of joints: butt, lap, corner, edge, and tee.

Although fillet welds are generally weaker than groove welds, they are used more often because they allow for larger tolerances during erection than groove welds. Plug and slot welds are expensive to make and they do not provide much reliability in transmitting tensile forces perpendicular to the faying surfaces. Furthermore, quality control of such welds is difficult because inspection of the welds is rather arduous. As a result, plug and slot welds are normally used just for stitching different parts of the members together.

Welding Symbols

A shorthand notation giving important information on the location, size, length, etc. for the various types of welds was developed by the American Welding Society [6] to facilitate the detailing of welds. This system of notation is reproduced in Figure 3.20.

Strength of Welds

In ASD, the strength of welds is expressed in terms of allowable stress. In LRFD, the design strength of welds is taken as the smaller of the design strength of the base material ϕF_{BM} and the design strength of the weld electrode ϕF_W. These allowable stresses and design strengths are summarized in Table 3.18 [18, 21]. When a design uses ASD, the computed stress in the weld shall not exceed its allowable value. When a design uses LRFD, the design strength of welds should exceed the required strength obtained by dividing the load to be transmitted by the effective area of the welds.

Effective Area of Welds

The effective area of groove welds is equal to the product of the width of the part joined and the effective throat thickness. The effective throat thickness of a full-penetration groove weld is taken as the thickness of the thinner part joined. The effective throat thickness of a partial-penetration groove weld is taken as the depth of the chamfer for J, U, bevel, or V (with bevel $\geq 60°$) joints and it is taken as the depth of the chamfer minus 1/8 in. for bevel or V joints if the bevel is between 45° and 60°. For flare bevel groove welds the effective throat thickness is taken as $5R/16$ and for flare V-groove the effective throat thickness is taken as $R/2$ (or $3R/8$ for GMAW process when $R \geq 1$ in.). R is the radius of the bar or bend.

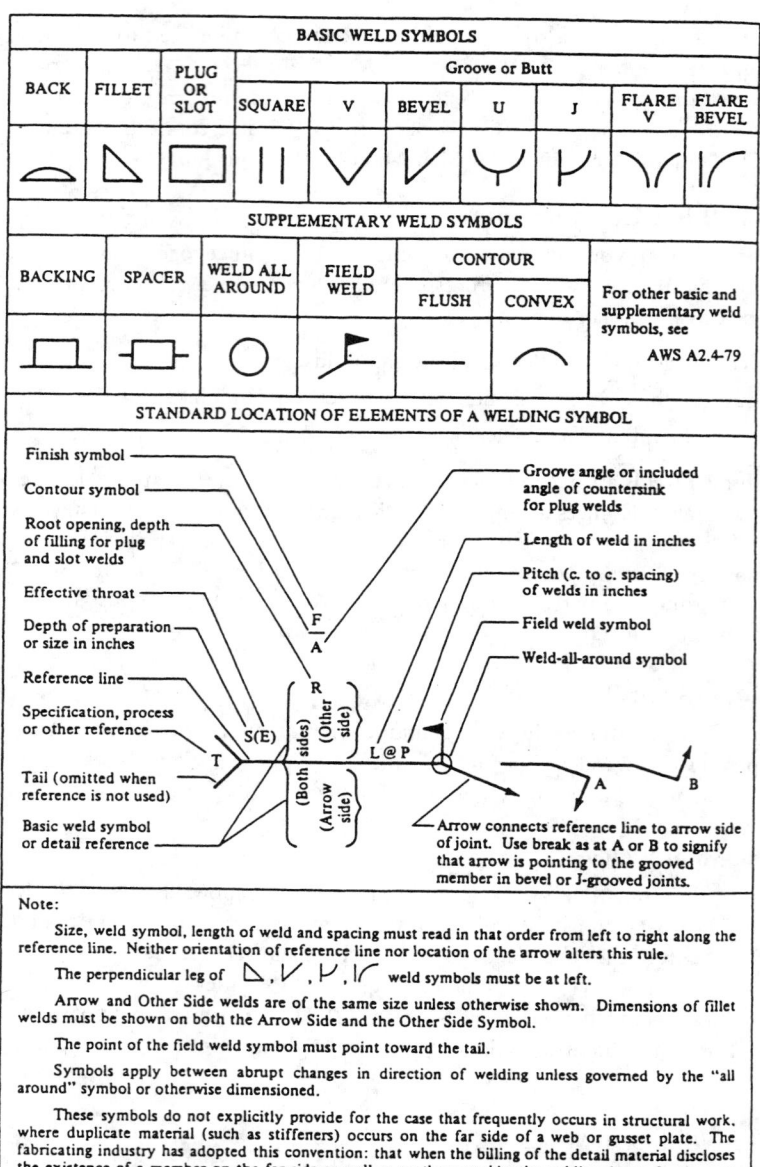

Figure 3.20   Basic weld symbols.

TABLE 3.18 Strength of Welds

Types of weld and stress[a]	Material	ASD allowable stress	LRFD ϕF_{BM} or ϕF_W	Required weld strength level[b,c]
		Full penetration groove weld		
Tension normal to effective area	Base	Same as base metal	$0.90 F_y$	"Matching" weld must be used
Compression normal to effective area	Base	Same as base metal	$0.90 F_y$	Weld metal with a strength level equal to or less than "matching" must be used
Tension of compression parallel to axis of weld	Base	Same as base metal	$0.90 F_y$	
Shear on effective area	Base weld electrode	$0.30\times$ nominal tensile strength of weld metal	$0.90[0.60 F_y]$ $0.80[0.60 F_{EXX}]$	
		Partial penetration groove welds		
Compression normal to effective area	Base	Same as base metal	$0.90 F_y$	Weld metal with a strength level equal to or less than "matching" weld metal may be used
Tension or compression parallel to axis of weld[d]				
Shear parallel to axis of weld	Base weld electrode	$0.30\times$ nominal tensile strength of weld metal	$0.75[0.60 F_{EXX}]$	
Tension normal to effective area	Base weld electrode	$0.30\times$ nominal tensile strength of weld metal $\leq 0.18\times$ yield stress of base metal	$0.90 F_y$ $0.80[0.60 F_{EXX}]$	
		Fillet welds		
Stress on effective area	Base weld electrode	$0.30\times$ nominal tensile strength of weld metal	$0.75[0.60 F_{EXX}]$ $0.90 F_y$	Weld metal with a strength level equal to or less than "matching" weld metal may be used
Tension or compression parallel to axis of weld[d]	Base	Same as base metal	$0.90 F_y$	
		Plug or slot welds		
Shear parallel to faying surfaces (on effective area)	Base weld electrode	$0.30\times$ nominal tensile strength of weld metal	$0.75[0.60 F_{EXX}]$	Weld metal with a strength level equal to or less than "matching" weld metal may be used

[a] see below for effective area
[b] see AWS D1.1 for "matching" weld material
[c] weld metal one strength level stronger than "matching" weld metal will be permitted
[d] fillet welds partial-penetration groove welds joining component elements of built-up members such as flange-to-web connections may be designed without regard to the tensile or compressive stress in these elements parallel to the axis of the welds

The effective area of fillet welds is equal to the product of length of the fillets including returns and the effective throat thickness. The effective throat thickness of a fillet weld is the shortest distance from the root of the joint to the face of the diagrammatic weld as shown in Figure 3.21. Thus, for an equal leg fillet weld, the effective throat is given by 0.707 times the leg dimension. For fillet weld made by the submerged arc welding process (SAW), the effective throat thickness is taken as the leg size (for 3/8-in. and smaller fillet welds) or as the theoretical throat plus 0.11-in. (for fillet weld over 3/8-in.). A larger value for the effective throat thickness is permitted for welds made by the SAW process to account for the inherently superior quality of such welds.

The effective area of plug and slot welds is taken as the nominal cross-sectional area of the hole or slot in the plane of the faying surface.

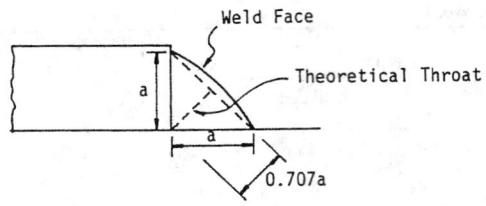

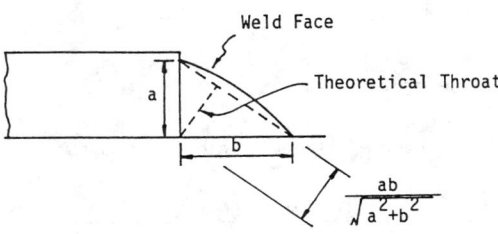

Figure 3.21   Effective throat of fillet welds.

Size and Length Limitations of Welds

To ensure effectiveness, certain size and length limitations are imposed for welds. For partial-penetration groove welds, minimum values for the effective throat thickness are given in Table 3.19.

TABLE 3.19   Minimum Effective Throat Thickness of Partial-Penetration Groove Welds

Thickness of the thicker part joined, t (in.)	Minimum effective throat thickness (in.)
$t \leq 1/4$	1/8
$1/4 < t \leq 1/2$	3/16
$1/2 < t \leq 3/4$	1/4
$3/4 < t \leq 1\text{-}1/2$	5/16
$1\text{-}1/2 < t \leq 2\text{-}1/4$	3/8
$2\text{-}1/4 < t \leq 6$	1/2
> 6	5/8

Note: 1 in. = 25.4 mm.

For fillet welds, the following size and length limitations apply:

Minimum Size of Leg—The minimum leg size is given in Table 3.20.

TABLE 3.20   Minimum Leg Size of Fillet Welds

Thickness of thicker part joined, t (in.)	Minimum leg size (in.)
$\leq 1/4$	1/8
$1/4 < t \leq 1/2$	3/16
$1/2 < t \leq 3/4$	1/4
$> 3/4$	5/16

Note: 1 in. = 25.4 mm.

Maximum Size of Leg—Along the edge of a connected part less than 1/4 thick, the maximum leg size is equal to the thickness of the connected part. For thicker parts, the maximum leg size is t minus 1/16 in. where t is the thickness of the part.

Minimum effective length of weld—The minimum effective length of a fillet weld is four times its nominal size. If a shorter length is used, the leg size of the weld shall be taken as 1/4 its effective length for purpose of stress computation. The length of fillet welds used for flat bar tension members shall not be less than the width of the bar if the welds are provided in the longitudinal direction only. The transverse distance between longitudinal welds should not exceed 8 in. unless the effect of shear lag is accounted for by the use of an effective net area.

Maximum effective length of weld—The maximum effective length of a fillet weld loaded by forces parallel to the weld shall not exceed 70 times the size of the fillet weld leg.

End returns—End returns must be continued around the corner and must have a length of at least two times the size of the weld leg.

Welded Connections for Tension Members

Figure 3.22 shows a tension angle member connected to a gusset plate by fillet welds. The

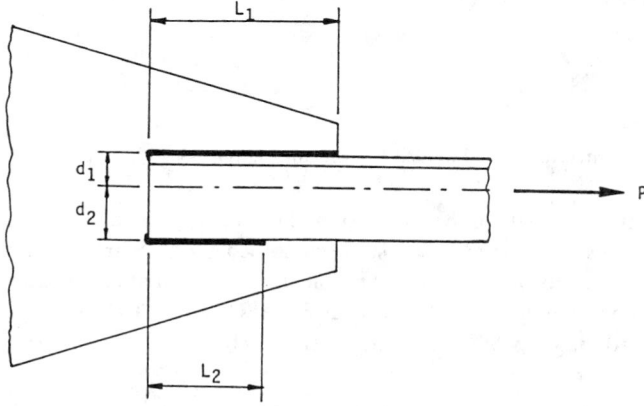

Figure 3.22   An eccentrically loaded welded tension connection.

applied tensile force P is assumed to act along the center of gravity of the angle. To avoid eccentricity, the lengths of the two fillet welds must be proportioned so that their resultant will also act along the center of gravity of the angle. For example, if LRFD is used, the following equilibrium equations can be written:

Summing force along the axis of the angle

$$(\phi F_M)t_{eff}L_1 + (\phi F_m)t_{eff}L_2 = P_u \tag{3.104}$$

Summing moment about the center of gravity of the angle

$$(\phi F_M)t_{eff}L_1d_1 = (\phi F_M)t_{eff}L_2d_2 \tag{3.105}$$

where P_u is the factored axial force, ϕF_M is the design strength of the welds as given in Table 3.18, t_{eff} is the effective throat thickness, L_1, L_2 are the lengths of the welds, and d_1, d_2 are the transverse distances from the center of gravity of the angle to the welds. The two equations can be used to solve for L_1 and L_2. If end returns are used, the added strength of the end returns should also be included in the calculations.

Welded Bracket Type Connections

A typical welded bracket connection is shown in Figure 3.23. Because the load is eccentric with

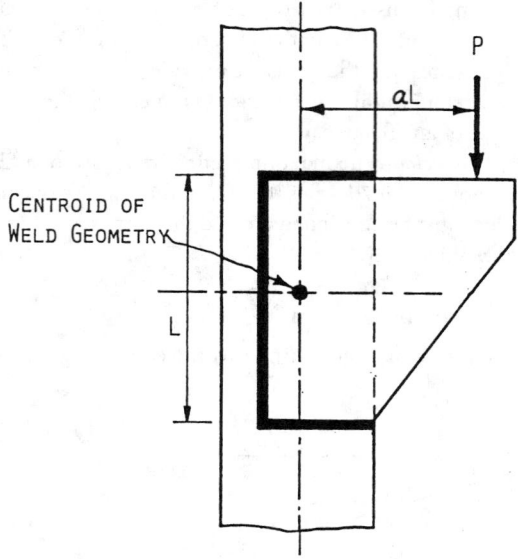

Figure 3.23   An eccentrically loaded welded bracket connection.

respect to the center of gravity of the weld group, the connection is subjected to both moment and shear. The welds must be designed to resist the combined effect of direct shear for the applied load and any additional shear from the induced moment. The design of the welded bracket connection is facilitated by the use of design tables in the AISC-ASD and AISC-LRFD Manuals. In both ASD and LRFD, the load capacity for the connection is given by

$$P = CC_1 Dl \tag{3.106}$$

where

P =   allowable load (in ASD), or factored load, P_u (in LRFD), kips
l =   length of the vertical weld, in.
D =   number of sixteenths of an inch in fillet weld size
C_1 =   coefficients for electrode used (see table below)
C =   coefficients tabulated in the AISC-ASD and AISC-LRFD Manuals. In the tables, values of
         C for a variety of weld geometries and dimensions are given

	Electrode	E60	E70	E80	E90	E100	E110
ASD	F_v (ksi)	18	21	24	27	30	33
	C_1	0.857	1.0	1.14	1.29	1.43	1.57
LRFD	F_{EXX} (ksi)	60	70	80	90	100	110
	C_1	0.857	1.0	1.03	1.16	1.21	1.34

Welded Connections with Welds Subjected to Combined Shear and Flexure

Figure 3.24 shows a welded framed connection and a welded seated connection. The welds for these connections are subjected to combined shear and flexure. For purpose of design, it is common practice to assume that the shear force per unit length, R_S, acting on the welds is a constant and is

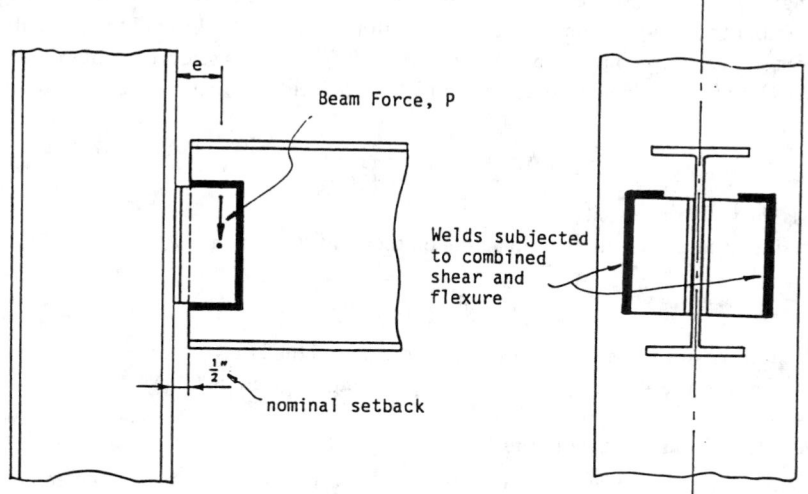

WELDED FRAMED CONNECTION

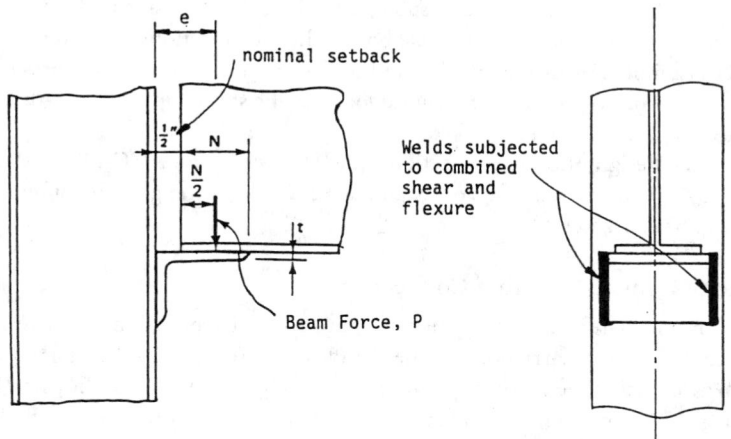

WELDED SEATED CONNECTION

Figure 3.24   Welds subjected to combined shear and flexure.

given by

$$R_S = \frac{P}{2l} \tag{3.107}$$

where P is the allowable load (in ASD), or factored load, P_u (in LRFD), and l is the length of the vertical weld.

In addition to shear, the welds are subjected to flexure as a result of load eccentricity. There is no general agreement on how the flexure stress should be distributed on the welds. One approach is to assume that the stress distribution is linear with half the weld subjected to tensile flexure stress and half is subjected to compressive flexure stress. Based on this stress distribution and ignoring the returns, the flexure tension force per unit length of weld, R_F, acting at the top of the weld can be written as

$$R_F = \frac{Mc}{I} = \frac{P_e(l/2)}{2l^3/12} = \frac{3P_e}{l^2} \tag{3.108}$$

where e is the load eccentricity.

The resultant force per unit length acting on the weld, R, is then

$$R = \sqrt{R_S^2 + R_F^2} \tag{3.109}$$

For a satisfactory design, the value R/t_{eff} where t_{eff} is the effective throat thickness of the weld should not exceed the allowable values or design strengths given in Table 3.18.

Welded Shear Connections

Figure 3.25 shows three commonly used welded shear connections: a framed beam connection, a seated beam connection, and a stiffened seated beam connection. These connections can be designed by using the information presented in the earlier sections on welds subjected to eccentric shear and welds subjected to combined tension and flexure. For example, the welds that connect the angles to the beam web in the framed beam connection can be considered as eccentrically loaded welds and so Equation 3.106 can be used for their design. The welds that connect the angles to the column flange can be considered as welds subjected to combined tension and flexure and so Equation 3.109 can be used for their design. Like bolted shear connections, welded shear connections are expected to exhibit appreciable ductility and so the use of angles with thickness in excess of 5/8 in. should be avoided. To prevent shear rupture failure, the shear rupture strength of the critically loaded connected parts should be checked.

To facilitate the design of these connections, the AISC-ASD and AISC-LRFD Manuals provide design tables by which the weld capacities and shear rupture strengths for different connection dimensions can be checked readily.

Welded Moment-Resisting Connections

Welded moment-resisting connections (Figure 3.26), like bolted moment-resisting connections, must be designed to carry both moment and shear. To simplify the design procedure, it is customary to assume that the moment, to be represented by a couple F_f as shown in Figure 3.19, is to be carried by the beam flanges and that the shear is to be carried by the beam web. The connected parts (e.g., the moment plates, welds, etc.) are then designed to resist the forces F_f and shear. Depending on the geometry of the welded connection, this may include checking: (a) the yield and/or fracture strength of the moment plate, (b) the shear and/or tensile capacity of the welds, and (c) the shear rupture strength of the shear plate.

If the column to which the connection is attached is weak, the designer should consider the use of column stiffeners to prevent failure of the column flange and web due to bending, yielding, crippling, or buckling (see section on Design of Moment-Resisting Connections).

Examples on the design of a variety of welded shear and moment-resisting connections can be found in the AISC Manual on Connections [20] and the AISC-LRFD Manual [22].

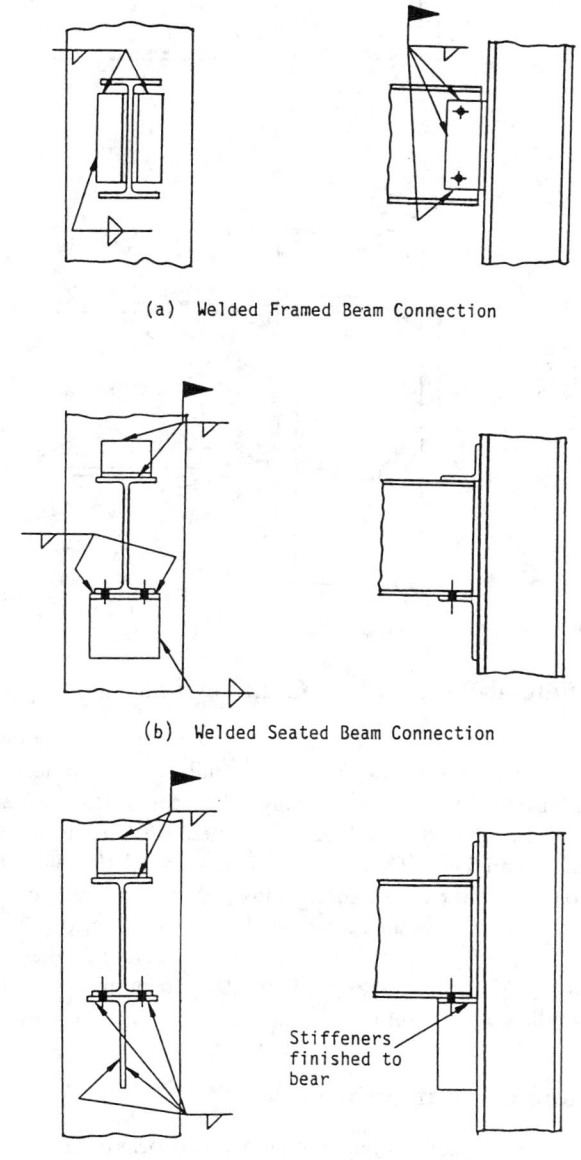

(a) Welded Framed Beam Connection

(b) Welded Seated Beam Connection

Stiffeners
finished to
bear

(c) Welded Stiffened Seated Beam Connection

Figure 3.25 Welded shear connections. (a) Framed beam connection, (b) seated beam connection, (c) stiffened beam connection.

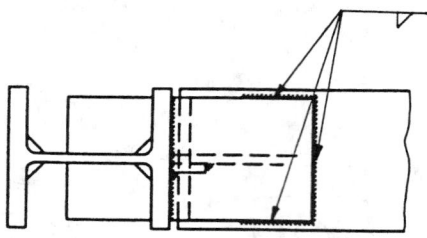

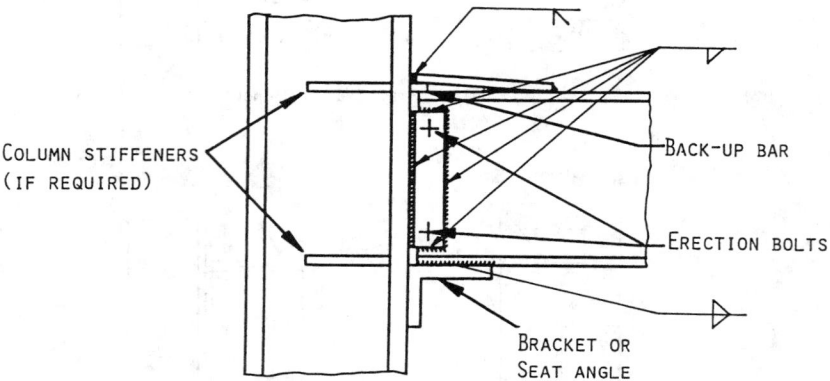

Figure 3.26   Welded moment connections.

3.11.3 Shop Welded-Field Bolted Connections

A large percentage of connections used for construction are shop welded and field bolted types. These connections are usually more cost effective than fully welded connections and their strength and ductility characteristics often rival those of fully welded connections. Figure 3.27 shows some of these connections. The design of shop welded–field bolted connections is also covered in the AISC Manual on Connections and the AISC-LRFD Manual. In general, the following should be checked: (a) Shear/tensile capacities of the bolts and/or welds, (b) bearing strength of the connected parts, (c) yield and/or fracture strength of the moment plate, and (d) shear rupture strength of the shear plate. Also, as for any other types of moment connections, column stiffeners shall be provided if any one of the following criteria is violated: column flange bending, local web yielding, crippling, and compression buckling of the column web.

3.11.4 Beam and Column Splices

Beam and column splices (Figure 3.28) are used to connect beam or column sections of different sizes. They are also used to connect beams or columns of the same size if the design calls for an extraordinarily long span. Splices should be designed for both moment and shear unless it is the intention of the designer to utilize the splices as internal hinges. If splices are used for internal hinges, provisions must be made to ensure that the connections possess adequate ductility to allow for large hinge rotation.

Splice plates are designed according to their intended functions. Moment splices should be designed to resist the flange force $F_f = M/(d - t_f)$ (Figure 3.19) at the splice location. In particular, the following limit states need to be checked: yielding of gross area of the plate, fracture

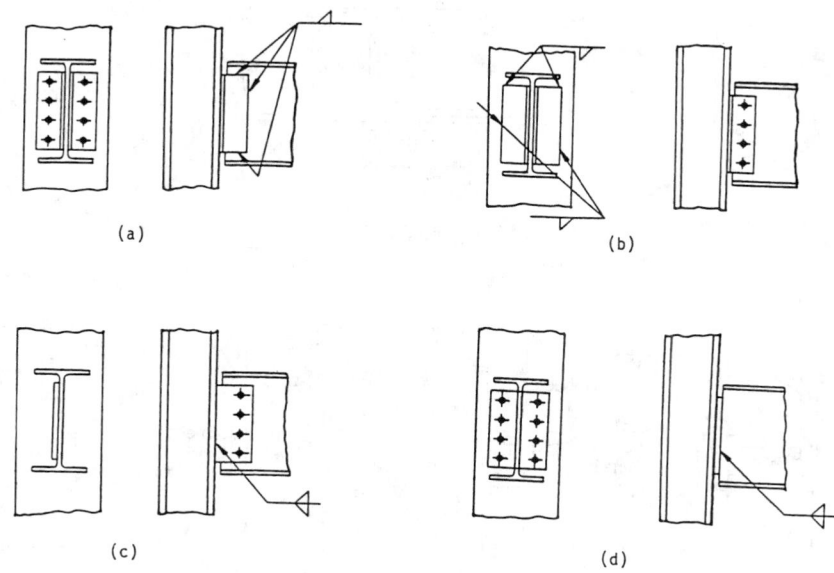

SHOP WELDED-FIELD BOLTED SHEAR CONNECTIONS

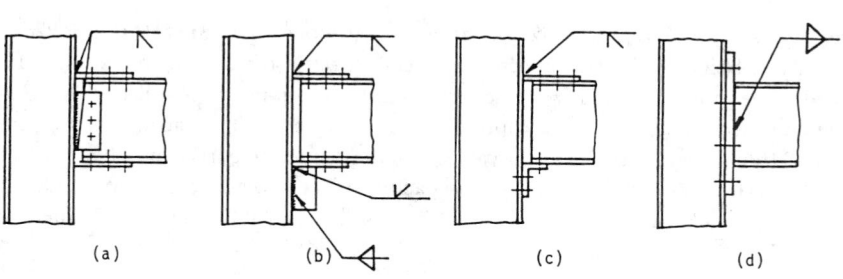

SHOP WELDED-FIELD BOLTED MOMENT CONNECTIONS

Figure 3.27 Shop-welded field-bolted connections.

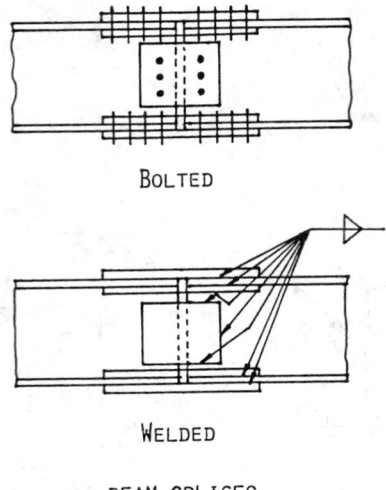

BOLTED

WELDED

BEAM SPLICES

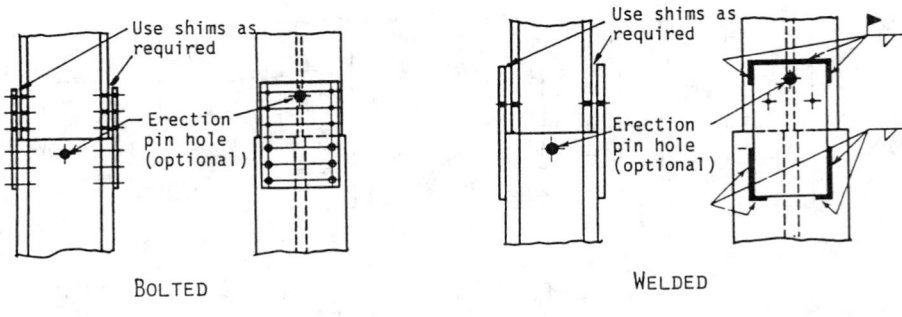

BOLTED   WELDED

COLUMN SPLICES

Figure 3.28   Bolted and welded beam and column splices.

of net area of the plate (for bolted splices), bearing strengths of connected parts (for bolted splices), shear capacity of bolts (for bolted splices), and weld capacity (for welded splices). Shear splices should be designed to resist the shear forces acting at the locations of the splices. The limit states that need to be checked include: shear rupture of the splice plates, shear capacity of bolts under an eccentric load (for bolted splices), bearing capacity of the connected parts (for bolted splices), shear capacity of bolts (for bolted splices), and weld capacity under an eccentric load (for welded splices). Design examples of beam and column splices can be found in the AISC Manual of Connections [20] and the AISC-LRFD Manuals [22].

3.12   Column Base Plates and Beam Bearing Plates (LRFD Approach)

3.12.1   Column Base Plates

Column base plates are steel plates placed at the bottom of columns whose function is to transmit column loads to the concrete pedestal. The design of column base plates involves two major steps: (1) determining the size $N \times B$ of the plate, and (2) determining the thickness t_p of the plate. Generally, the size of the plate is determined based on the limit state of bearing on concrete and the thickness of the plate is determined based on the limit state of plastic bending of critical sections in the plate. Depending on the types of forces (axial force, bending moment, shear force) the plate will be subjected to, the design procedures differ slightly. In all cases, a layer of grout should be placed between the base plate and its support for the purpose of leveling and anchor bolts should be provided to stabilize the column during erection or to prevent uplift for cases involving large bending moment.

Axially Loaded Base Plates

Base plates supporting concentrically loaded columns in frames in which the column bases are assumed pinned are designed with the assumption that the column factored load P_u is distributed uniformly to the area of concrete under the base plate. The size of the base plate is determined from the limit state of bearing on concrete. The design bearing strength of concrete is given by the equation

$$\phi_c P_p = 0.60 \left[0.85 f_c' A_1 \sqrt{\frac{A_2}{A_1}} \right] \tag{3.110}$$

where
f_c' = compressive strength of concrete
A_1 = area of base plate
A_2 = area of concrete pedestal that is geometrically similar to and concentric with the loaded area, $A_1 \le A_2 \le 4A_1$

From Equation 3.110, it can be seen that the bearing capacity increases when the concrete area is greater than the plate area. This accounts for the beneficial effect of confinement. The upper limit of the bearing strength is obtained when $A_2 = 4A_1$. Presumably, the concrete area in excess of $4A_1$ is not effective in resisting the load transferred through the base plate.

Setting the column factored load, P_u, equal to the bearing capacity of the concrete pedestal, $\phi_c P_p$, and solving for A_1 from Equation 3.110, we have

$$A_1 = \frac{1}{A_2} \left[\frac{P_u}{0.6(0.85 f_c')} \right]^2 \tag{3.111}$$

The length, N, and width, B, of the plate should be established so that $N \times B > A_1$. For an efficient design, the length can be determined from the equation

$$N \approx \sqrt{A_1} + 0.50(0.95d - 0.80b_f) \tag{3.112}$$

where $0.95d$ and $0.80b_f$ define the so-called effective load bearing area shown cross-hatched in Figure 3.29a. Once N is obtained, B can be solved from the equation

$$B = \frac{A_1}{N} \tag{3.113}$$

Both N and B should be rounded up to the nearest full inches.

The required plate thickness, $t_{reg'd}$, is to be determined from the limit state of yield line formation along the most severely stressed sections. A yield line develops when the cross-section moment capacity is equal to its plastic moment capacity. Depending on the size of the column relative to the plate and the magnitude of the factored axial load, yield lines can form in various patterns on the plate. Figure 3.29 shows three models of plate failure in axially loaded plates. If the plate is large compared to the column, yield lines are assumed to form around the perimeter of the effective load bearing area (the cross-hatched area) as shown in Figure 3.29a. If the plate is small and the column factored load is light, yield lines are assumed to form around the inner perimeter of the I-shaped area as shown in Figure 3.29b. If the plate is small and the column factored load is heavy, yield lines

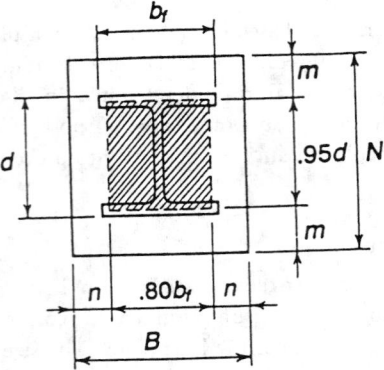

(a) Plate with Large m,n

(b) Lightly Loaded Plate with Small m, n

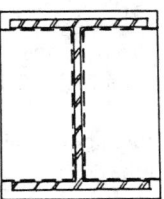

(c) Heavily Loaded Plate with Small m, n

Figure 3.29   Failure models for centrally loaded column base plates.

are assumed to form around the inner edge of the column flanges and both sides of the column web as shown in Figure 3.29c. The following equation can be used to calculate the required plate thickness

$$t_{req'd} = l \sqrt{\frac{2P_u}{0.90 F_y B N}} \tag{3.114}$$

where l is the larger of m, n, and $\lambda n'$ given by

$$m = \frac{(N - 0.95d)}{2}$$

$$n = \frac{(B - 0.80b_f)}{2}$$

$$n' = \frac{\sqrt{db_f}}{4}$$

and

$$\lambda = \frac{2\sqrt{X}}{1 + \sqrt{1-X}} \le 1$$

in which

$$X = \left(\frac{4db_f}{(d+b_f)^2}\right)\frac{P_u}{\phi_c P_p}$$

Base Plates for Tubular and Pipe Columns

The design concept for base plates discussed above for I-shaped sections can be applied to the design of base plates for rectangular tubes and circular pipes. The critical section used to determine the plate thickness should be based on 0.95 times the outside column dimension for rectangular tubes and 0.80 times the outside dimension for circular pipes [11].

Base Plates with Moments

For columns in frames designed to carry moments at the base, base plates must be designed to support both axial forces and bending moments. If the moment is small compared to the axial force, the base plate can be designed without consideration of the tensile force which may develop in the anchor bolts. However, if the moment is large, this effect should be considered. To quantify the relative magnitude of this moment, an eccentricity $e = M_u/P_u$ is used. The general procedures for the design of base plates for different values of e will be given in the following [11].

Small eccentricity, $e \le N/6$

If e is small, the bearing stress is assumed to distribute linearly over the entire area of the base plate (Figure 3.30). The maximum bearing stress is given by

$$f_{max} = \frac{P_u}{BN} + \frac{M_u c}{I} \tag{3.115}$$

where $c = N/2$ and $I = BN^3/12$.

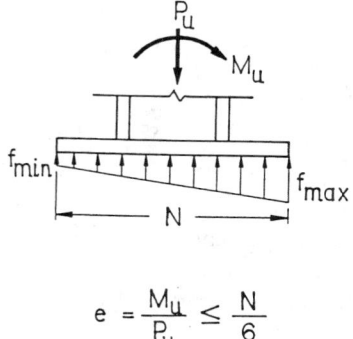

$$e = \frac{M_u}{P_u} \le \frac{N}{6}$$

Figure 3.30 Eccentrically loaded column base plate (small load eccentricity).

The size of the plate is to be determined by a trial and error process. The size of the base plate should be such that the bearing stress calculated using Equation 3.115 does not exceed $\phi_c P_p/A_1$, given by

$$0.60\left[0.85 f_c'\sqrt{\frac{A_2}{A_1}}\right] \le 0.60[1.7 f_c'] \tag{3.116}$$

The thickness of the plate is to be determined from

$$t_p = \sqrt{\frac{4M_{plu}}{0.90F_y}} \tag{3.117}$$

where M_{plu} is the moment per unit width of critical section in the plate. M_{plu} is to be determined by assuming that the portion of the plate projecting beyond the critical section acts as an inverted cantilever loaded by the bearing pressure. The moment calculated at the critical section divided by the length of the critical section (i.e., B) gives M_{plu}.

Moderate eccentricity, $N/6 < e \leq N/2$
   For plates subjected to moderate moments, only portions of the plate will be subjected to bearing stress (Figure 3.31). Ignoring the tensile force in the anchor bolt in the region of the plate where no

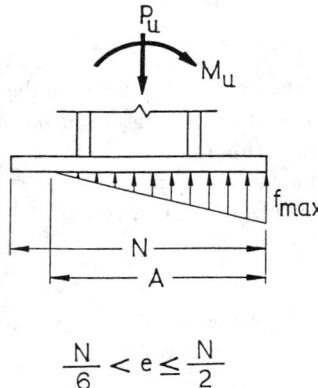

Figure 3.31   Eccentrically loaded column base plate (moderate load eccentricity).

bearing occurs and denoting A as the length of the plate in bearing, the maximum bearing stress can be calculated from force equilibrium consideration as

$$f_{\max} = \frac{2P_u}{AB} \tag{3.118}$$

where $A = 3(N/2 - e)$ is determined from moment equilibrium. The plate should be portioned such that $f_{\max}$ does not exceed the value calculated using Equation 3.116. t_p is to be determined from Equation 3.117.

Large eccentricity, $e > N/2$
   For plates subjected to large bending moments so that $e > N/2$, one needs to take into consideration the tensile force developing in the anchor bolts (Figure 3.32). Denoting T as the resultant force in the anchor bolts, force equilibrium requires that

$$T + P_u = \frac{f_{\max}AB}{2} \tag{3.119}$$

and moment equilibrium requires that

$$P_u\left(N' - \frac{N}{2}\right) + M = \frac{f_{\max}AB}{2}\left(N' - \frac{A}{3}\right) \tag{3.120}$$

The above equations can be used to solve for A and T. The size of the plate is to be determined using a trial-and-error process. The size should be chosen such that $f_{\max}$ does not exceed the value

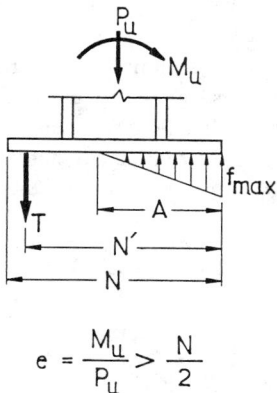

$$e = \frac{M_u}{P_u} > \frac{N}{2}$$

Figure 3.32 Eccentrically loaded column base plate (large load eccentricity).

calculated using Equation 3.116, A should be smaller than N' and T should not exceed the tensile capacity of the bolts.

Once the size of the plate is determined, the plate thickness t_p is to be calculated using Equation 3.117. Note that there are two critical sections on the plate, one on the compression side of the plate and the other on the tension side of the plate. Two values of M_{plu} are to be calculated and the larger value should be used to calculate t_p.

Base Plates with Shear

Under normal circumstances, the factored column base shear is adequately resisted by the frictional force developed between the plate and its support. Additional shear capacity is also provided by the anchor bolts. For cases in which exceptionally high shear force is expected, such as in a bracing connection or in which uplift occurs which reduces the frictional resistance, the use of shear lugs may be necessary. Shear lugs can be designed based on the limit states of bearing on concrete and bending of the lugs. The size of the lug should be proportioned such that the bearing stress on concrete does not exceed $0.60(0.85 f'_c)$. The thickness of the lug can be determined from Equation 3.117. M_{plu} is the moment per unit width at the critical section of the lug. The critical section is taken to be at the junction of the lug and the plate (Figure 3.33).

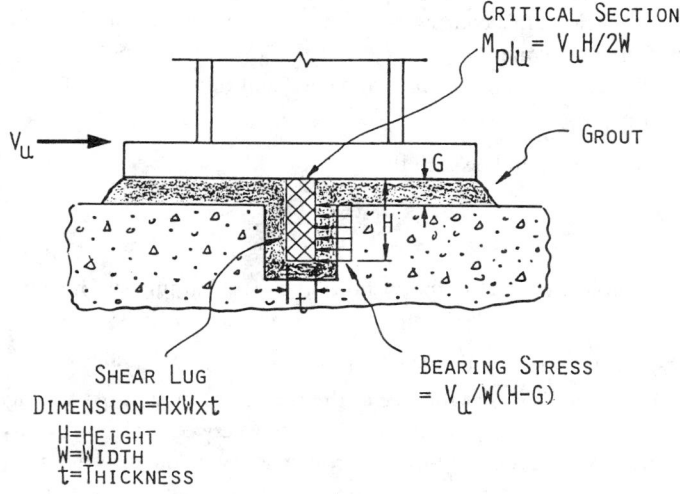

Figure 3.33 Column base plate subjected to shear.

3.12.2 Anchor Bolts

Anchor bolts are provided to stabilize the column during erection and to prevent uplift for cases involving large moments. Anchor bolts can be cast-in-place bolts or drilled-in bolts. The latter are placed after the concrete is set and are not too often used. Their design is governed by the manufacturer's specifications. Cast-in-place bolts are hooked bars, bolts, or threaded rods with nuts (Figure 3.34) placed before the concrete is set. Of the three types of cast-in-place anchors

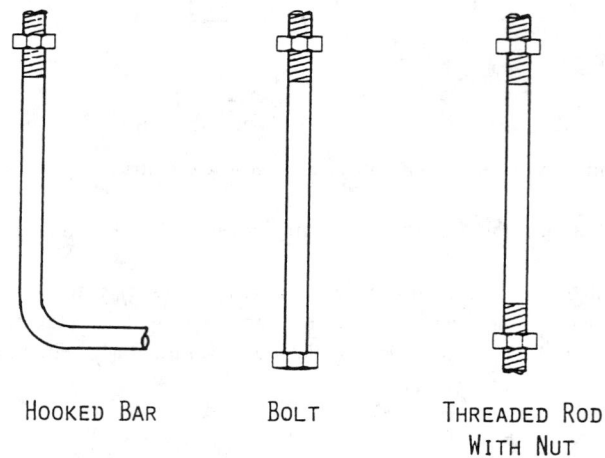

HOOKED BAR   BOLT   THREADED ROD
                                      WITH NUT

Figure 3.34 Base plate anchors.

shown in the figure, the hooked bars are recommended for use only in axially loaded base plates. They are not normally relied upon to carry significant tensile force. Bolts and threaded rods with nuts can be used for both axially loaded base plates or base plates with moments. Threaded rods with nuts are used when the length and size required for the specific design exceed those of standard size bolts. Failure of bolts or threaded rods with nuts occur when their tensile capacities are reached. Failure is also considered to occur when a cone of concrete is pulled out from the pedestal. This cone pull-out type of failure is depicted schematically in Figure 3.35. The failure cone is assumed to radiate out from the bolt head or nut at an angle of 45° with tensile failure occurring along the surface of the cone at an average stress of $4\sqrt{f'_c}$ where f'_c is the compressive strength of concrete in psi. The load that will cause this cone pull-out failure is given by the product of this average stress and the projected area the cone A_p [23, 24]. The design of anchor bolts is thus governed by the limit states of tensile fracture of the anchors and cone pull-out.

Limit State of Tensile Fracture

  The area of the anchor should be such that

$$A_g \geq \frac{T_u}{\phi_t 0.75 F_u} \tag{3.121}$$

where A_g is the required gross area of the anchor, F_u is the minimum specified tensile strength, and ϕ_t is the resistance factor for tensile fracture which is equal to 0.75.

Limit State of Cone Pull-Out

  From Figure 3.35, it is clear that the size of the cone is a function of the length of the anchor. Provided that there is sufficient edge distance and spacing between adjacent anchors, the amount of tensile force required to cause cone pull-out failure increases with the embedded length of the anchor. This concept can be used to determine the required embedded length of the anchor. Assuming that

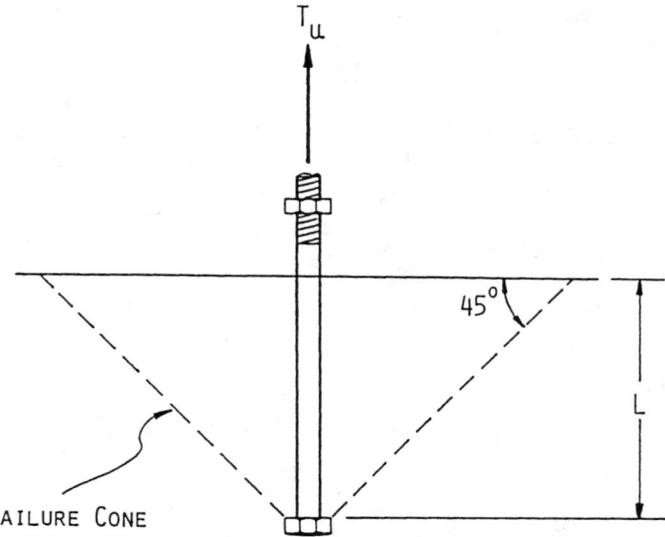

Figure 3.35   Cone pullout failure.

the failure cone does not intersect with another failure cone nor the edge of the pedestal, the required embedded length can be calculated from the equation

$$L \geq \sqrt{\frac{A_p}{\pi}} = \sqrt{\frac{(T_u/\phi_t 4\sqrt{f_c'})}{\pi}} \qquad (3.122)$$

where A_p is the projected area of the failure cone, T_u is the required bolt force in pounds, f_c' is the compressive strength of concrete in psi and ϕ_t is the resistance factor assumed to be equal to 0.75. If failure cones from adjacent anchors overlap one another or intersect with the pedestal edge, the projected area A_p must be adjusted according (see, for example [23, 24]).

The length calculated using the above equation should not be less than the recommended values given by [29]. These values are reproduced in the following table. Also shown in the table are the recommended minimum edge distances for the anchors.

Bolt type (material)	Minimum embedded length	Minimum edge distance
A307 (A36)	12d	5d > 4 in.
A325 (A449)	17d	7d > 4 in.
d = nominal diameter of the anchor		

3.12.3   Beam Bearing Plates

Beam bearing plates are provided between main girders and concrete pedestals to distribute the girder reactions to the concrete supports (Figure 3.36). Beam bearing plates may also be provided between cross beams and girders if the cross beams are designed to sit on the girders.

Beam bearing plates are designed based on the limit states of web yielding, web crippling, bearing on concrete, and plastic bending of the plate. The dimension of the plate along the beam axis, i.e., N, is determined from the web yielding or web crippling criterion (see section on Concentrated Load Criteria), whichever is more critical. The dimension B of the plate is determined from Equation 3.113 with A_1 calculated using Equation 3.111. P_u in Equation 3.111 is to be replaced by R_u, the factored reaction at the girder support.

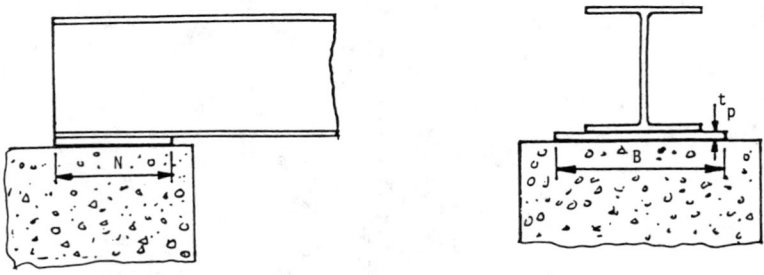

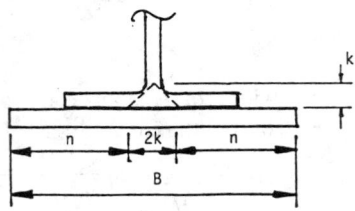

Figure 3.36   Beam bearing plate.

Once the size $B \times N$ is determined, the plate thickness t_p can be calculated using the equation

$$t_p = \sqrt{\frac{2 R_u n^2}{0.90 F_y B N}} \qquad (3.123)$$

where R_u is the factored girder reaction, F_y is the yield stress of the plate and $n = (B - 2k)/2$ in which k is the distance from the web toe of the fillet to the outer surface of the flange. The above equation was developed based on the assumption that the critical sections for plastic bending in the plate occur at a distance k from the centerline of the web.

3.13 Composite Members (LRFD Approach)

Composite members are structural members made from two or more materials. The majority of composite sections used for building constructions are made from steel and concrete. Steel provides strength and concrete provides rigidity. The combination of the two materials often results in efficient load-carrying members. Composite members may be concrete-encased or concrete-filled. For concrete-encased members (Figure 3.37a), concrete is casted around steel shapes. In addition to enhancing strength and providing rigidity to the steel shapes, the concrete acts as a fire-proofing material to the steel shapes. It also serves as a corrosion barrier shielding the steel from corroding under adverse environmental conditions. For concrete-filled members (Figure 3.37b), structural steel tubes are filled with concrete. In both concrete-encased and concrete-filled sections, the rigidity of the concrete often eliminates the problem of local buckling experienced by some slender elements of the steel sections.

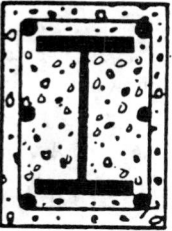

(a) CONCRETE ENCASED COMPOSITE SECTION

 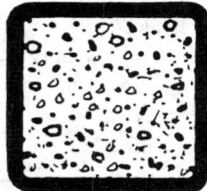 

(b) CONCRETE FILLED COMPOSITE SECTIONS

Figure 3.37 Composite columns.

Some disadvantages associated with composite sections are that concrete creeps and shrinks. Furthermore, uncertainties with regard to the mechanical bond developed between the steel shape and the concrete often complicate the design of beam-column joints.

3.13.1 Composite Columns

According to the LRFD Specification [18], a compression member is regarded as a composite column if (1) the cross-sectional area of the steel shape is at least 4% of the total composite area. If this condition is not satisfied, the member should be designed as a reinforced concrete column. (2) Longitudinal reinforcements and lateral ties are provided for concrete-encased members. The cross-sectional area of the reinforcing bars shall be 0.007 in.^{2} per inch of bar spacing. To avoid spalling, lateral ties shall be placed at a spacing not greater than 2/3 the least dimension of the composite cross-section. For fire and corrosion resistance, a minimum clear cover of 1.5 in. shall be provided. (3) The compressive strength of concrete f'_c used for the composite section falls within the range 3 to 8 ksi for normal weight concrete and not less than 4 ksi for light weight concrete. These limits are set because they represent the range of test data available for the development of the design equations. (4) The specified minimum yield stress for the steel shapes and reinforcing bars used in calculating the strength of the composite column does not exceed 55 ksi. This limit is set because this stress corresponds to a strain below which the concrete remains unspalled and stable. (5) The minimum wall thickness of the steel shapes for concrete filled members is equal to $b\sqrt{(F_y/3E)}$ for rectangular sections of width b and $D\sqrt{(F_y/8E)}$ for circular sections of outside diameter D.

Design Compressive Strength

The design compressive strength, $\phi_c P_n$, shall exceed the factored compressive force, P_u. The design compressive strength is given as follows:

For $\lambda_c \leq 1.5$

$$\phi_c P_n = \begin{cases} 0.85 \left[\left(0.658^{\lambda_c^2} \right) A_s F_{my} \right], & \text{if } \lambda_c \leq 1.5 \\ 0.85 \left[\left(\frac{0.877}{\lambda_c^2} \right) A_s F_{my} \right], & \text{if } \lambda_c > 1.5 \end{cases} \tag{3.124}$$

where

$$\lambda_c = \frac{KL}{r_m \pi} \sqrt{\frac{F_{my}}{E_m}} \tag{3.125}$$

$$F_{my} = F_y + c_1 F_{yr} \left(\frac{A_r}{A_s} \right) + c_2 f_c' \left(\frac{A_c}{A_s} \right) \tag{3.126}$$

$$E_m = E + c_3 E_c \left(\frac{A_c}{A_s} \right) \tag{3.127}$$

A_c = area of concrete, in.^{2}
A_r = area of longitudinal reinforcing bars, in.^{2}
A_s = area of steel shape, in.^{2}
E = modulus of elasticity of steel, ksi
E_c = modulus of elasticity of concrete, ksi
F_y = specified minimum yield stress of steel shape, ksi
F_{yr} = specified minimum yield stress of longitudinal reinforcing bars, ksi
f_c' = specified compressive strength of concrete, ksi
c_1, c_2, c_3 = coefficients given in table below

Type of composite section	c_1	c_2	c_3
Concrete encased shapes	0.7	0.6	0.2
Concrete-filled pipes and tubings	1.0	0.85	0.4

In addition to satisfying the condition $\phi_c P_n \geq P_u$, the bearing condition for concrete must also be satisfied. Denoting $\phi_c P_{nc} (= \phi_c P_{n,\text{composite section}} - \phi_c P_{n,\text{steel shape alone}})$ as the portion of compressive strength resisted by the concrete and A_B as the loaded area (the condition), then if the supporting concrete area is larger than the loaded area, the bearing condition that needs to be satisfied is

$$\phi_c P_{nc} \leq 0.60[1.7 f_c' A_B] \tag{3.128}$$

3.13.2 Composite Beams

For steel beams fully encased in concrete, no additional anchorage for shear transfer is required if (1) at least 1.5 in. concrete cover is provided on top of the beam and at least 2 in. cover is provided over the sides and at the bottom of the beam, and (2) spalling of concrete is prevented by adequate mesh or other reinforcing steel. The design flexural strength $\phi_b M_n$ can be computed using either an elastic or plastic analysis.

If an elastic analysis is used, ϕ_b shall be taken as 0.90. A linear strain distribution is assumed for the cross-section with zero strain at the neutral axis and maximum strains at the extreme fibers. The stresses are then computed by multiplying the strains by E (for steel) or E_c (for concrete). Maximum stress in steel shall be limited to F_y, and maximum stress in concrete shall be limited to $0.85 f_c'$. Tensile strength of concrete shall be neglected. M_n is to be calculated by integrating the resulting stress block about the neutral axis.

If a plastic analysis is used, ϕ_c shall be taken as 0.90, and M_n shall be assumed to be equal to M_p, the plastic moment capacity of the steel section alone.

3.13.3 Composite Beam-Columns

Composite beam-columns shall be designed to satisfy the interaction equation of Equation 3.68 or Equation 3.69, whichever is applicable, with $\phi_c P_n$ calculated based on Equations 3.124 to 3.127, P_e calculated using the equation $P_e = A_s F_{my}/\lambda_c^2$, and $\phi_b M_n$ calculated using the following equation [14]:

$$\phi_b M_n = 0.90 \left[Z F_y + \frac{1}{3}(h_2 - 2c_r) A_r F_{yr} + \left(\frac{h_2}{2} - \frac{A_w F_y}{1.7 f_c' h_1} \right) A_w F_y \right] \qquad (3.129)$$

where
Z = plastic section modulus of the steel section, in.^{3}
c_r = average of the distance measured from the compression face to the longitudinal reinforcement in that face and the distance measured from the tension face to the longitudinal reinforcement in that face, in.
h_1 = width of the composite section perpendicular to the plane of bending, in.
h_2 = width of the composite section parallel to the plane of bending, in.
A_r = cross-sectional area of longitudinal reinforcing bars, in.^{2}
A_w = web area of the encased steel shape (= 0 for concrete-filled tubes)

If $0 < (P_u/\phi_c P_n) \le 0.3$, a linear interpolation of $\phi_b M_n$ calculated using the above equation assuming $P_u/\phi_c P_n = 0.3$ and that for beams with $P_u/\phi_c P_n = 0$ (see section on Composite Beams) should be used.

3.13.4 Composite Floor Slabs

Composite floor slabs (Figure 3.38) can be designed as shored or unshored. In shored construction, temporary shores are used during construction to support the dead and accidental live loads until the concrete cures. The supporting beams are designed on the basis of their ability to develop composite action to support all factored loads after the concrete cures. In unshored construction, temporary shores are not used. As a result, the steel beams alone must be designed to support the dead and accidental live loads before the concrete has attained 75% of its specified strength. After the concrete is cured, the composite section should have adequate strength to support all factored loads.

Composite action for the composite floor slabs shown in Figure 3.38 is developed as a result of the presence of shear connectors. If sufficient shear connectors are provided so that the maximum flexural strength of the composite section can be developed, the section is referred to as fully composite. Otherwise, the section is referred to as partially composite. The flexural strength of a partially composite section is governed by the shear strength of the shear connectors. The horizontal shear force V_h, which should be designed for at the interface of the steel beam and the concrete slab, is given by:

In regions of positive moment

$$V_h = \min(0.85 f_c' A_c, A_s F_y, \sum Q_n) \qquad (3.130)$$

In regions of negative moment

$$V_h = \min(A_r F_{yr}, \sum Q_n) \qquad (3.131)$$

where
f_c' = compressive strength of concrete, ksi
A_c = effective area of the concrete slab $= t_c b_{eff}$, in.^{2}
t_c = thickness of the concrete slab, in.
b_{eff} = effective width of the concrete slab, in.

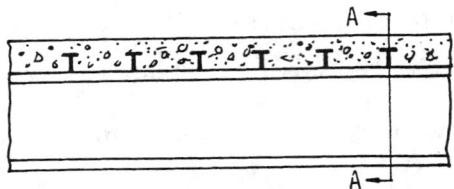

COMPOSITE FLOOR SLAB WITH STUD SHEAR CONNECTORS

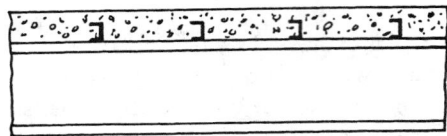

COMPOSITE FLOOR SLAB WITH CHANNEL SHEAR CONNECTORS

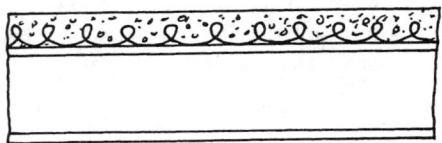

COMPOSITE FLOOR SLAB WITH SPIRAL SHEAR CONNECTORS

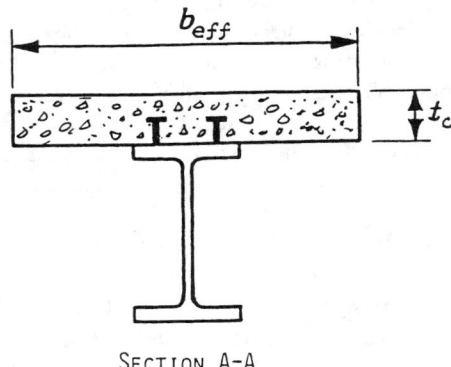

SECTION A-A

Figure 3.38   Composite floor slabs.

	$=$	$\min(L/4, s)$, for an interior beam
	$=$	$\min(L/8+$ distance from beam centerline to edge of slab, $s/2+$ distance from beam centerline to edge of slab), for an exterior beam
L	$=$	beam span measured from center-to-center of supports, in.
s	$=$	spacing between centerline of adjacent beams, in.
A_s	$=$	cross-sectional area of the steel beam, in.^{2}
F_y	$=$	yield stress of the steel beam, ksi
A_r	$=$	area of reinforcing steel within the effective area of the concrete slab, in.^{2}
F_{yr}	$=$	yield stress of the reinforcing steel, ksi
ΣQ_n	$=$	sum of nominal shear strengths of the shear connectors, kips

The nominal shear strength of a shear connector (used without a formed steel deck) is given by:

For a stud shear connector

$$Q_n = 0.5A_{sc}\sqrt{f'_c E_c} \le A_{sc} F_u \tag{3.132}$$

For a channel shear connector

$$Q_n = 0.3(t_f + 0.5t_w)L_c\sqrt{f'_c E_c} \tag{3.133}$$

where

A_{sc} = cross-sectional area of the shear stud, in.^{2}
f'_c = compressive strength of concrete, ksi
E_c = modulus of elasticity of concrete, ksi
F_u = minimum specified tensile strength of the shear stud, ksi
t_f = flange thickness of the channel, in.
t_w = web thickness of the channel, in.
L_c = length of the channel, in.

If a formed steel deck is used, Q_n must be reduced by a reduction factor. The reduction factor depends on whether the deck ribs are perpendicular or parallel to the steel beam. Expressions for the reduction factor are given in the AISC-LRFD Specification [18].

For full composite action, the number of connectors required between the *maximum* moment point and the *zero* moment point of the beam is given by

$$N = \frac{V_h}{Q_n} \tag{3.134}$$

For partial composite action, the number of connectors required is governed by the condition $\phi_b M_n \ge M_u$, where $\phi_b M_n$ is governed by the shear strength of the connectors.

The placement and spacing of the shear connectors should comply with the following guidelines:

1. The shear connectors shall be uniformly spaced with the region of maximum moment and zero moment. However, the number of shear connectors placed between a concentrated load point and the nearest zero moment point must be sufficient to resist the factored moment M_u.

2. Except for connectors installed in the ribs of formed steel decks, shear connectors shall have at least 1 in. of lateral concrete cover.

3. Unless located over the web, diameter of shear studs must not exceed 2.5 times the thickness of the beam flange.

4. The longitudinal spacing of the studs should fall in the range 6 times the stud diameter to 8 times the slab thickness if a solid slab is used or 4 times the stud diameter to 8 times the slab thickness if a formed steel deck is used.

The design flexural strength $\phi_b M_n$ of the composite beam with shear connectors is determined as follows:

In regions of positive moments

For $h_c/t_w \le 640/\sqrt{F_{yf}}$, $\phi_b = 0.85$, M_n = moment capacity determined using a plastic stress distribution assuming concrete crushes at a stress of $0.85 f'_c$ and steel yields at a stress of F_y. If a portion of the concrete slab is in tension, the strength contribution of that portion of concrete is ignored. The determination of M_n using this method is very similar to the technique used for computing the moment capacity of a reinforced concrete beam according to the ultimate strength method.

For $h_c/t_w > 640/\sqrt{F_{yf}}$, $\phi_b = 0.90$, M_n = moment capacity determined using superposition of elastic stress, considering the effect of shoring. The determination of M_n using this method is quite similar to the technique used for computing the moment capacity of a reinforced concrete beam according to the working stress method.

In regions of negative moments

$\phi_b M_n$ is to be determined for the steel section alone in accordance with the requirements discussed in the section on Flexural Members.

To facilitate design, numerical values of $\phi_b M_n$ for composite beams with shear studs in solid slabs are given in tabulated form by the AISC-LRFD Manual. Values of $\phi_b M_n$ for composite beams with formed steel decks are given in a publication by the Steel Deck Institute [19].

3.14 Plastic Design

Plastic analysis and design is permitted only for steels with yield stress not exceeding 65 ksi. The reason for this is that steels with high yield stress lack the ductility required for inelastic rotation at hinge locations. Without adequate inelastic rotation, moment redistribution (which is an important characteristic for plastic design) cannot take place.

In plastic design, the predominant limit state is the formation of plastic hinges. Failure occurs when sufficient plastic hinges have formed for a collapse mechanism to develop. To ensure that plastic hinges can form and can undergo large inelastic rotation, the following conditions must be satisfied:

1. Sections must be compact. That is, the width-thickness ratios of flanges in compression and webs must not exceed λ_p in Table 3.8.
2. For columns, the slenderness parameter λ_c (see section on Compression Members) shall not exceed $1.5K$ where K is the effective length factor, and P_u from gravity and horizontal loads shall not exceed $0.75A_g F_y$.
3. For beams, the lateral unbraced length L_b shall not exceed L_{pd} where

For doubly and singly symmetric I-shaped members loaded in the plane of the web

$$L_{pd} = \frac{3,600 + 2,200(M_1/M_2)}{F_y} r_y \qquad (3.135)$$

and for solid rectangular bars and symmetric box beams

$$L_{pd} = \frac{5,000 + 3,000(M_1/M_2)}{F_y} r_y \geq \frac{3,000 r_y}{F_y} \qquad (3.136)$$

In the above equations, M_1 is the smaller end moment within the unbraced length of the beam. $M_2 = M_p$ is the plastic moment ($= Z_x F_y$) of the cross-section. r_y is the radius of gyration about the minor axis, in inches, and F_y is the specified minimum yield stress, in ksi.

L_{pd} is not defined for beams bent about their minor axes nor for beams with circular and square cross-sections because these beams do not experience lateral torsional bucking when loaded.

3.14.1 Plastic Design of Columns and Beams

Provided that the above limitations are satisfied, the design of columns shall meet the condition $1.7F_a A \geq P_u$ where F_a is the allowable compressive stress given in Equation 3.16, A is the gross cross-sectional area, and P_u is the factored axial load.

The design of beams shall satisfy the conditions $M_p \geq M_u$ and $0.55F_y t_w d \geq V_u$ where M_u and V_u are the factored moment and shear, respectively. M_p is the plastic moment capacity F_y is the minimum specified yield stress, t_w is the beam web thickness, and d is the beam depth. For beams subjected to concentrated loads, all failure modes associated with concentrated loads (see section on Concentrated Load Criteria) should also be prevented.

Except at the location where the last hinge forms, a beam bending about its major axis must be braced to resist lateral and torsional displacements at plastic hinge locations. The distance between adjacent braced points should not exceed l_{cr} given by

$$l_{cr} = \begin{cases} \left(\frac{1375}{F_y} + 25\right) r_y, & \text{if } -0.5 < \frac{M}{M_p} < 1.0 \\ \left(\frac{1375}{F_y}\right) r_y, & \text{if } -1.0 < \frac{M}{M_p} \leq -0.5 \end{cases} \qquad (3.137)$$

where

r_y = radius of gyration about the weak axis
M = smaller of the two end moments of the unbraced segment
M_p = plastic moment capacity
M/M_p = is taken as positive if the unbraced segment bends in reverse curvature, and it is taken as negative if the unbraced segment bends in single curvature

3.14.2 Plastic Design of Beam-Columns

Beam-columns designed on the basis of plastic analysis shall satisfy the following interaction equations for stability (Equation 3.138) and for strength (Equation 3.139).

$$\frac{P_u}{P_{cr}} + \frac{C_m M_u}{\left(1 - \frac{P_u}{P_e}\right) M_m} \leq 1.0 \qquad (3.138)$$

$$\frac{P_u}{P_y} + \frac{M_u}{1.18 M_p} \leq 1.0 \qquad (3.139)$$

where

P_u = factored axial load
P_{cr} = $1.7 F_a A$, F_a is defined in Equation 3.16 and A is the cross-sectional area
P_y = yield load $= A F_y$
P_e = Euler buckling load $= \pi^2 E I / (Kl)^2$
C_m = coefficient defined in the section on Compression Members
M_u = factored moment
M_p = plastic moment $= Z F_y$
M_m = maximum moment that can be resisted by the member in the absence of axial load
   = M_{px} if the member is braced in the weak direction
   = $\{1.07 - [(l/r_y)\sqrt{F_y}]/3160\} M_{px} \leq M_{px}$ if the member is unbraced in the weak direction
l = unbraced length of the member
r_y = radius of gyration about the minor axis
M_{px} = plastic moment about the major axis $= Z_x F_y$
F_y = minimum specified yield stress

3.15 Defining Terms

ASD: Acronym for Allowable Stress Design.

Beam-columns: Structural members whose primary function is to carry loads both along and transverse to their longitudinal axes.

Biaxial bending: Simultaneous bending of a member about two orthogonal axes of the cross-section.

Built-up members: Structural members made of structural elements jointed together by bolts, welds, or rivets.

Composite members: Structural members made of both steel and concrete.

Compression members: Structural members whose primary function is to carry loads along their longitudinal axes

Design strength: Resistance provided by the structural member obtained by multiplying the nominal strength of the member by a resistance factor.

Drift: Lateral deflection of a building.

Factored load: The product of the nominal load and a load factor.

Flexural members: Structural members whose primary function is to carry loads transverse to their longitudinal axes.

Limit state: A condition in which a structural or structural component becomes unsafe (strength limit state) or unfit for its intended function (serviceability limit state).

Load factor: A factor to account for the unavoidable deviations of the actual load from its nominal value and uncertainties in structural analysis in transforming the applied load into a load effect (axial force, shear, moment, etc.)

LRFD: Acronym for Load and Resistance Factor Design.

PD: Acronym for Plastic Design.

Plastic hinge: A yielded zone of a structural member in which the internal moment is equal to the plastic moment of the cross-section.

Resistance factor: A factor to account for the unavoidable deviations of the actual resistance of a member from its nominal value.

Service load: Nominal load expected to be supported by the structure or structural component under normal usage.

Sidesway inhibited frames: Frames in which lateral deflections are prevented by a system of bracing.

Sidesway uninhibited frames: Frames in which lateral deflections are not prevented by a system of bracing.

Shear lag: The phenomenon in which the stiffer (or more rigid) regions of a structure or structural component attract more stresses than the more flexible regions of the structure or structural component. Shear lag causes stresses to be unevenly distributed over the cross-section of the structure or structural component.

Tension field action: Post-buckling shear strength developed in the web of a plate girder. Tension field action can develop only if sufficient transverse stiffeners are provided to allow the girder to carry the applied load using truss-type action after the web has buckled.

References

[1] AASHTO. 1992. *Standard Specification for Highway Bridges.* 15th ed., American Association of State Highway and Transportation Officials, Washington D.C.

[2] ASTM. 1988. *Specification for Carbon Steel Bolts and Studs, 60000 psi Tensile Strength (A307- 88a).* American Society for Testing and Materials, Philadelphia, PA.

[3] ASTM. 1986. *Specification for High Strength Bolts for Structural Steel Joints (A325-86).* American Society for Testing and Materials, Philadelphia, PA.

[4] ASTM. 1985. *Specification for Heat-Treated Steel Structural Bolts, 150 ksi Minimum Tensile Strength (A490-85).* American Society for Testing and Materials, Philadelphia, PA.

[5] ASTM. 1986. *Specification for Quenched and Tempered Steel Bolts and Studs (A449-86)*. American Society for Testing and Materials, Philadelphia, PA.

[6] AWS. 1987. *Welding Handbook*. 8th ed., 1, *Welding Technology*, American Welding Society, Miami, FL.

[7] AWS. 1996. *Structural Welding Code-Steel*. American Welding Society, Miami, FL.

[8] Blodgett, O.W. Distortion... How to Minimize it with Sound Design Practices and Controlled Welding Procedures Plus Proven Methods for Straightening Distorted Members. *Bulletin G261*, The Lincoln Electric Company, Cleveland, OH.

[9] Chen, W.F. and Lui, E.M. 1991. *Stability Design of Steel Frames*, CRC Press, Boca Raton, FL.

[10] CSA. 1994. *Limit States Design of Steel Structures*. CSA Standard CAN/CSA S16.1-94, Canadian Standards Association, Rexdale, Ontantio.

[11] Dewolf, J.T. and Ricker, D.T. 1990. *Column Base Plates*. Steel Design Guide Series 1, American Institute of Steel Construction, Chicago, IL.

[12] Disque, R.O. 1973. Inelastic K-factor in column design. *AISC Eng. J.*, 10(2):33-35.

[13] Galambos, T.V., Ed. 1988. *Guide to Stability Design Criteria for Metal Structures*. 4th ed., John Wiley & Sons, New York.

[14] Galambos, T.V. and Chapuis, J. 1980. *LRFD Criteria for Composite Columns and Beam Columns*. Washington University, Department of Civil Engineering, St. Louis, MO.

[15] Gaylord, E.H., Gaylord, C.N., and Stallmeyer, J.E. 1992. *Design of Steel Structures*, 3rd ed., McGraw-Hill, New York.

[16] Kulak, G.L., Fisher, J.W., and Struik, J.H.A. 1987. *Guide to Design Criteria for Bolted and Riveted Joints*, 2nd ed., John Wiley & Sons, New York.

[17] Lee, G.C., Morrel, M.L., and Ketter, R.L. 1972. Design of Tapered Members. *WRC Bulletin No. 173*.

[18] *Load and Resistance Factor Design Specification for Structural Steel Buildings*. 1993. American Institute of Steel Construction, Chicago, IL.

[19] *LRFD Design Manual for Composite Beams and Girders with Steel Deck*. 1989. Steel Deck Institute, Canton, OH.

[20] *Manual of Steel Construction-Volume II Connections*. 1992. ASD 1st ed./LRFD 1st ed., American Institute of Steel Construction, Chicago, IL.

[21] *Manual of Steel Construction-Allowable Stress Design*. 1989. 9th ed., American Institute of Steel Construction, Chicago, IL.

[22] *Manual of Steel Construction-Load and Resistance Factor Design*. 1994. Vol. I and II, 2nd ed., American Institute of Steel Construction, Chicago, IL.

[23] Marsh, M.L. and Burdette, E.G. 1985. Multiple bolt anchorages: Method for determining the effective projected area of overlapping stress cones. *AISC Eng. J.*, 22(1):29-32.

[24] Marsh, M.L. and Burdette, E.G. 1985. Anchorage of steel building components to concrete. *AISC Eng. J.*, 22(1):33-39.

[25] Munse, W.H. and Chesson E., Jr. 1963. Riveted and Bolted Joints: Net Section Design. *ASCE J. Struct. Div.*, 89(1):107-126.

[26] Rains, W.A. 1976. A new era in fire protective coatings for steel. *Civil Eng.*, ASCE, September:80-83.

[27] RCSC. 1985. *Allowable Stress Design Specification for Structural Joints Using ASTM A325 or A490 Bolts*. American Institute of Steel Construction, Chicago, IL.

[28] RCSC. 1988. *Load and Resistance Factor Design Specification for Structural Joints Using ASTM A325 or A490 Bolts*. American Institute of Steel Construction, Chicago, IL.

[29] Shipp, J.G. and Haninge, E.R. 1983. Design of headed anchor bolts. *AISC Eng. J.*, 20(2):58-69.

[30] SSRC. 1993. *Is Your Structure Suitably Braced?* Structural Stability Research Council, Bethlehem, PA.

Further Reading

The following publications provide additional sources of information for the design of steel structures:

General Information

[1] Chen, W.F. and Lui, E.M. 1987. *Structural Stability—Theory and Implementation,* Elsevier, New York.
[2] Englekirk, R. 1994. *Steel Structures—Controlling Behavior Through Design,* John Wiley & Sons, New York.
[3] *Stability of Metal Structures—A World View.* 1991. 2nd ed., Lynn S. Beedle (editor-in-chief), Structural Stability Research Council, Lehigh University, Bethlehem, PA.
[4] Trahair, N.S. 1993. *Flexural-Torsional Buckling of Structures,* CRC Press, Boca Raton, FL.

Allowable Stress Design

[5] Adeli, H. 1988. *Interactive Microcomputer-Aided Structural Steel Design,* Prentice-Hall, Englewood Cliffs, NJ.
[6] Cooper S.E. and Chen A.C. 1985. *Designing Steel Structures—Methods and Cases,* Prentice-Hall, Englewood Cliffs, NJ.
[7] Crawley S.W. and Dillon, R.M. 1984. *Steel Buildings Analysis and Design,* 3rd ed., John Wiley & Sons, New York.
[8] Fanella, D.A., Amon, R., Knobloch, B., and Mazumder, A. 1992. *Steel Design for Engineers and Architects,* 2nd ed., Van Nostrand Reinhold, New York.
[9] Kuzmanovic, B.O. and Willems, N. 1983. *Steel Design for Structural Engineers,* 2nd ed., Prentice-Hall, Englewood Cliffs, NJ.
[10] McCormac, J.C. 1981. *Structural Steel Design,* 3rd ed., Harper & Row, New York.
[11] Segui, W.T. 1989. *Fundamentals of Structural Steel Design,* PWS-KENT, Boston, MA.
[12] Spiegel, L. and Limbrunner, G.F. 1986. *Applied Structural Steel Design,* Prentice-Hall, Englewood Cliffs, NJ.

Plastic Design

[13] Horne, M.R. and Morris, L.J. 1981. *Plastic Design of Low-Rise Frames,* Constrado Monographs, Collins, London, England.
[14] *Plastic Design in Steel-A Guide and Commentary.* 1971. 2nd ed., ASCE Manual No. 41, ASCE-WRC, New York.
[15] Chen, W.F. and Sohal, I.S. 1995. *Plastic Design and Second-Order Analysis of Steel Frames,* Springer-Verlag, New York.

Load and Resistance Factor Design

[16] Geschwindner, L.F., Disque, R.O., and Bjorhovde, R. 1994. *Load and Resistance Factor Design of Steel Structures,* Prentice-Hall, Englewood Cliffs, NJ.
[17] McCormac, J.C. 1995. *Structural Steel Design—LRFD Method,* 2nd ed., Harper & Row, New York.
[18] Salmon C.G. and Johnson, J.E. 1990. *Steel Structures—Design and Behavior,* 3rd ed., Harper & Row, New York.
[19] Segui, W.T. 1994. *LRFD Steel Design,* PWS, Boston, MA.
[20] Smith, J.C. 1996. *Structural Steel Design—LRFD Approach,* 2nd ed., John Wiley & Sons, New York.

[21] Chen, W.F. and Kim, S.E. 1997. *LRFD Steel Design Using Advanced Analysis,* CRC Press, Boca Raton, FL.

[22] Chen, W.F., Goto, Y., and Liew, J.Y.R. 1996. *Stability Design of Semi-Rigid Frames,* John Wiley & Sons, New York.

4

Structural Concrete Design[1]

4.1 Properties of Concrete and Reinforcing Steel 4-2
Properties of Concrete • Lightweight Concrete • Heavyweight
Concrete • High-Strength Concrete • Reinforcing Steel

4.2 Proportioning and Mixing Concrete 4-6
Proportioning Concrete Mix • Admixtures • Mixing

4.3 Flexural Design of Beams and One-Way Slabs 4-8
Reinforced Concrete Strength Design • Prestressed Concrete
Strength Design

4.4 Columns under Bending and Axial Load 4-14
Short Columns under Minimum Eccentricity • Short Columns
under Axial Load and Bending • Slenderness Effects • Columns
under Axial Load and Biaxial Bending

4.5 Shear and Torsion 4-19
Reinforced Concrete Beams and One-Way Slabs Strength De-
sign • Prestressed Concrete Beams and One-Way Slabs Strength
Design

4.6 Development of Reinforcement........................ 4-25
Development of Bars in Tension • Development of Bars in Com-
pression • Development of Hooks in Tension • Splices, Bundled
Bars, and Web Reinforcement

4.7 Two-Way Systems..................................... 4-27
Definition • Design Procedures • Minimum Slab Thickness and
Reinforcement • Direct Design Method • Equivalent Frame
Method • Detailing

4.8 Frames 4-37
Analysis of Frames • Design for Seismic Loading

4.9 Brackets and Corbels 4-43

4.10 Footings 4-45
Types of Footings • Design Considerations • Wall Footings •
Single-Column Spread Footings • Combined Footings • Two-
Column Footings • Strip, Grid, and Mat Foundations • Footings
on Piles

4.11 Walls 4-60
Panel, Curtain, and Bearing Walls • Basement Walls • Partition
Walls • Shears Walls

4.12 Defining Terms 4-65
References.. 4-70
Further Reading 4-70

Amy Grider and
Julio A. Ramirez
School of Civil Engineering,
Purdue University,
West Lafayette, IN

Young Mook Yun
Department of Civil Engineering,
National University,
Taegu, South Korea

---

[1]The material in this chapter was previously published by CRC Press in *The Civil Engineering Handbook*, W.F. Chen, Ed., 1995.

0-8493-2674-5/97/$0.00+$.50
© 1997 by CRC Press LLC

At this point in the history of development of reinforced and prestressed concrete it is necessary to reexamine the fundamental approaches to design of these composite materials. Structural engineering is a worldwide industry. Designers from one nation or a continent are faced with designing a project in another nation or continent. The decades of efforts dedicated to harmonizing concrete design approaches worldwide have resulted in some successes but in large part have led to further differences and numerous different design procedures. It is this abundance of different design approaches, techniques, and code regulations that justifies and calls for the need for a unification of design approaches throughout the entire range of structural concrete, from plain to fully prestressed [5].

The effort must begin at all levels: university courses, textbooks, handbooks, and standards of practice. Students and practitioners must be encouraged to think of a single continuum of structural concrete. Based on this premise, this chapter on concrete design is organized to promote such unification. In addition, effort will be directed at dispelling the present unjustified preoccupation with complex analysis procedures and often highly empirical and incomplete sectional mechanics approaches that tend to both distract the designers from fundamental behavior and impart a false sense of accuracy to beginning designers. Instead, designers will be directed to give careful consideration to overall structure behavior, remarking the adequate flow of forces throughout the entire structure.

4.1 Properties of Concrete and Reinforcing Steel

The designer needs to be knowledgeable about the properties of concrete, reinforcing steel, and prestressing steel. This part of the chapter summarizes the material properties of particular importance to the designer.

4.1.1 Properties of Concrete

Workability is the ease with which the ingredients can be mixed and the resulting mix handled, transported, and placed with little loss in homogeneity. Unfortunately, workability cannot be measured directly. Engineers therefore try to measure the consistency of the concrete by performing a slump test.

The slump test is useful in detecting variations in the uniformity of a mix. In the slump test, a mold shaped as the frustum of a cone, 12 in. (305 mm) high with an 8 in. (203 mm) diameter base and 4 in. (102 mm) diameter top, is filled with concrete (ASTM Specification C143). Immediately after filling, the mold is removed and the change in height of the specimen is measured. The change in height of the specimen is taken as the slump when the test is done according to the ASTM Specification.

A well-proportioned workable mix settles slowly, retaining its original shape. A poor mix crumbles, segregates, and falls apart. The slump may be increased by adding water, increasing the percentage of fines (cement or aggregate), entraining air, or by using an admixture that reduces water requirements; however, these changes may adversely affect other properties of the concrete. In general, the slump specified should yield the desired consistency with the least amount of water and cement.

Concrete should withstand the weathering, chemical action, and wear to which it will be subjected in service over a period of years; thus, durability is an important property of concrete. Concrete resistance to freezing and thawing damage can be improved by increasing the watertightness, entraining 2 to 6% air, using an air-entraining agent, or applying a protective coating to the surface. Chemical agents damage or disintegrate concrete; therefore, concrete should be protected with a resistant coating. Resistance to wear can be obtained by use of a high-strength, dense concrete made with hard aggregates.

Excess water leaves voids and cavities after evaporation, and water can penetrate or pass through the concrete if the voids are interconnected. Watertightness can be improved by entraining air or reducing water in the mix, or it can be prolonged through curing.

Volume change of concrete should be considered, since expansion of the concrete may cause buckling and drying shrinkage may cause cracking. Expansion due to alkali-aggregate reaction can be avoided by using nonreactive aggregates. If reactive aggregates must be used, expansion may be reduced by adding pozzolanic material (e.g., fly ash) to the mix. Expansion caused by heat of hydration of the cement can be reduced by keeping cement content as low as possible; using Type IV cement; and chilling the aggregates, water, and concrete in the forms. Expansion from temperature increases can be reduced by using coarse aggregate with a lower coefficient of thermal expansion. Drying shrinkage can be reduced by using less water in the mix, using less cement, or allowing adequate moist curing. The addition of pozzolans, unless allowing a reduction in water, will increase drying shrinkage. Whether volume change causes damage usually depends on the restraint present; consideration should be given to eliminating restraints or resisting the stresses they may cause [8].

Strength of concrete is usually considered its most important property. The compressive strength at 28 d is often used as a measure of strength because the strength of concrete usually increases with time. The compressive strength of concrete is determined by testing specimens in the form of standard cylinders as specified in ASTM Specification C192 for research testing or C31 for field testing. The test procedure is given in ASTM C39. If drilled cores are used, ASTM C42 should be followed.

The suitability of a mix is often desired before the results of the 28-d test are available. A formula proposed by W. A. Slater estimates the 28-d compressive strength of concrete from its 7-d strength:

$$S_{28} = S_7 + 30\sqrt{S_7} \tag{4.1}$$

where

S_{28} = 28-d compressive strength, psi
S_7 = 7-d compressive strength, psi

Strength can be increased by decreasing water-cement ratio, using higher strength aggregate, using a pozzolan such as fly ash, grading the aggregates to produce a smaller percentage of voids in the concrete, moist curing the concrete after it has set, and vibrating the concrete in the forms. The short-time strength can be increased by using Type III portland cement, accelerating admixtures, and by increasing the curing temperature.

The stress-strain curve for concrete is a curved line. Maximum stress is reached at a strain of 0.002 in./in., after which the curve descends.

The modulus of elasticity, E_c, as given in ACI 318-89 (Revised 92), *Building Code Requirements for Reinforced Concrete* [1], is:

$$E_c = w_c^{1.5}33\sqrt{f_c'} \quad \text{lb/ft}^3 \text{ and psi} \tag{4.2a}$$

$$E_c = w_c^{1.5}0.043\sqrt{f_c'} \quad \text{kg/m}^3 \text{ and MPa} \tag{4.2b}$$

where

w_c = unit weight of concrete
f_c' = compressive strength at 28 d

Tensile strength of concrete is much lower than the compressive strength—about $7\sqrt{f_c'}$ for the higher-strength concretes and $10\sqrt{f_c'}$ for the lower-strength concretes.

Creep is the increase in strain with time under a constant load. Creep increases with increasing water-cement ratio and decreases with an increase in relative humidity. Creep is accounted for in design by using a reduced modulus of elasticity of the concrete.

4.1.2   Lightweight Concrete

Structural lightweight concrete is usually made from aggregates conforming to ASTM C330 that are usually produced in a kiln, such as expanded clays and shales. Structural lightweight concrete has a density between 90 and 120 lb/ft^3 (1440 to 1920 kg/m^3).

Production of lightweight concrete is more difficult than normal-weight concrete because the aggregates vary in absorption of water, specific gravity, moisture content, and amount of grading of undersize. Slump and unit weight tests should be performed often to ensure uniformity of the mix. During placing and finishing of the concrete, the aggregates may float to the surface. Workability can be improved by increasing the percentage of fines or by using an air-entraining admixture to incorporate 4 to 6% air. Dry aggregate should not be put into the mix because it will continue to absorb moisture and cause the concrete to harden before placement is completed. Continuous water curing is important with lightweight concrete.

No-fines concrete is obtained by using pea gravel as the coarse aggregate and 20 to 30% entrained air instead of sand. It is used for low dead weight and insulation when strength is not important. This concrete weighs from 105 to 118 lb/ft^3 (1680 to 1890 kg/m^3) and has a compressive strength from 200 to 1000 psi (1 to 7 MPa).

A porous concrete made by gap grading or single-size aggregate grading is used for low conductivity or where drainage is needed.

Lightweight concrete can also be made with gas-forming of foaming agents which are used as admixtures. Foam concretes range in weight from 20 to 110 lb/ft^3 (320 to 1760 kg/m^3). The modulus of elasticity of lightweight concrete can be computed using the same formula as normal concrete. The shrinkage of lightweight concrete is similar to or slightly greater than for normal concrete.

4.1.3   Heavyweight Concrete

Heavyweight concretes are used primarily for shielding purposes against gamma and x-radiation in nuclear reactors and other structures. Barite, limonite and magnetite, steel punchings, and steel shot are typically used as aggregates. Heavyweight concretes weigh from 200 to 350 lb/ft^3 (3200 to 5600 kg/m^3) with strengths from 3200 to 6000 psi (22 to 41 MPa). Gradings and mix proportions are similar to those for normal weight concrete. Heavyweight concretes usually do not have good resistance to weathering or abrasion.

4.1.4   High-Strength Concrete

Concretes with strengths in excess of 6000 psi (41 MPa) are referred to as high-strength concretes. Strengths up to 18,000 psi (124 MPa) have been used in buildings.

Admixtures such as superplasticizers, silica fume, and supplementary cementing materials such as fly ash improve the dispersion of cement in the mix and produce workable concretes with lower water-cement ratios, lower void ratios, and higher strength. Coarse aggregates should be strong fine-grained gravel with rough surfaces.

For concrete strengths in excess of 6000 psi (41 MPa), the modulus of elasticity should be taken as

$$E_c = 40,000\sqrt{f_c'} + 1.0 \times 10^6 \tag{4.3}$$

where

f_c' = compressive strength at 28 d, psi [4]

The shrinkage of high-strength concrete is about the same as that for normal concrete.

4.1.5   Reinforcing Steel

Concrete can be reinforced with welded wire fabric, deformed reinforcing bars, and prestressing tendons.

Welded wire fabric is used in thin slabs, thin shells, and other locations where space does not allow the placement of deformed bars. Welded wire fabric consists of cold drawn wire in orthogonal patterns—square or rectangular and resistance-welded at all intersections. The wire may be smooth (ASTM A185 and A82) or deformed (ASTM A497 and A496). The wire is specified by the symbol W for smooth wires or D for deformed wires followed by a number representing the cross-sectional area in hundredths of a square inch. On design drawings it is indicated by the symbol WWF followed by spacings of the wires in the two 90° directions. Properties for welded wire fabric are given in Table 4.1.

TABLE 4.1   Wire and Welded Wire Fabric Steels

AST designation	Wire size designation	Minimum yield stress,[a] f_y		Minimum tensile strength	
		ksi	MPa	ksi	MPa
A82-79 (cold-drawn wire) (properties apply when material is to be used for fabric)	W1.2 and larger[b] Smaller than W1.2	65 56	450 385	75 70	520 480
A185-79 (welded wire fabric)	Same as A82; this is A82 material fabricated into sheet (so-called "mesh") by the process of electric welding				
A496-78 (deformed steel wire) (properties apply when material is to be used for fabric)	D1-D31[c]	70	480	80	550
A497-79	Same as A82 or A496; this specification applies for fabric made from A496, or from a combination of A496 and A82 wires				

[a] The term "yield stress" refers to either *yield point,* the well-defined deviation from perfect elasticity, or *yield strength,* the value obtained by a specified offset strain for material having no well-defined yield point.

[b] The W number represents the nominal cross-sectional area in square inches multiplied by 100, for smooth wires.

[c] The D number represents the nominal cross-sectional area in square inches multiplied by 100, for deformed wires.

From Wang, C.-K. and Salmon, C.G. 1985. *Reinforced Concrete Design,* 4th ed., Harper Row, New York. With permission.

The deformations on a deformed reinforcing bar inhibit longitudinal movement of the bar relative to the concrete around it. Table 4.2 gives dimensions and weights of these bars. Reinforcing bar steel can be made of billet steel of grades 40 and 60 having minimum specific yield stresses of 40,000 and 60,000 psi, respectively (276 and 414 MPa) (ASTM A615) or low-alloy steel of grade 60, which is intended for applications where welding and/or bending is important (ASTM A706). Presently, grade 60 billet steel is the most predominantly used for construction.

Prestressing tendons are commonly in the form of individual wires or groups of wires. Wires of different strengths and properties are available with the most prevalent being the 7-wire low-relaxation strand conforming to ASTM A416. ASTM A416 also covers a stress-relieved strand, which is seldom used in construction nowadays. Properties of standard prestressing strands are given in Table 4.3. Prestressing tendons could also be bars; however, this is not very common. Prestressing bars meeting ASTM A722 have been used in connections between members.

The modulus of elasticity for non-prestressed steel is 29,000,000 psi (200,000 MPa). For prestressing steel, it is lower and also variable, so it should be obtained from the manufacturer. For 7-wires strands conforming to ASTM A416, the modulus of elasticity is usually taken as 27,000,000 psi (186,000 MPa).

TABLE 4.2   Reinforcing Bar Dimensions and Weights

Bar number	Diameter (in.)	Diameter (mm)	Area (in.^{2})	Area (cm^2)	Weight (lb/ft)	Weight (kg/m)
3	0.375	9.5	0.11	0.71	0.376	0.559
4	0.500	12.7	0.20	1.29	0.668	0.994
5	0.625	15.9	0.31	2.00	1.043	1.552
6	0.750	19.1	0.44	2.84	1.502	2.235
7	0.875	22.2	0.60	3.87	2.044	3.041
8	1.000	25.4	0.79	5.10	2.670	3.973
9	1.128	28.7	1.00	6.45	3.400	5.059
10	1.270	32.3	1.27	8.19	4.303	6.403
11	1.410	35.8	1.56	10.06	5.313	7.906
14	1.693	43.0	2.25	14.52	7.65	11.38
18	2.257	57.3	4.00	25.81	13.60	20.24

TABLE 4.3   Standard Prestressing Strands, Wires, and Bars

Tendon type	Grade f_{pu} ksi	Diameter in.	Area in.^{2}	Weight plf
Seven-wire strand	250	1/4	0.036	0.12
	270	3/8	0.085	0.29
	250	3/8	0.080	0.27
	270	1/2	0.153	0.53
	250	1/2	0.144	0.49
	270	0.6	0.215	0.74
	250	0.6	0.216	0.74
Prestressing wire	250	0.196	0.0302	0.10
	240	0.250	0.0491	0.17
	235	0.276	0.0598	0.20
Deformed prestressing bars	157	5/8	0.28	0.98
	150	1	0.85	3.01
	150	1 1/4	1.25	4.39
	150	1 3/8	1.58	5.56

From Collins, M.P. and Mitchell, D. 1991. *Prestressed Concrete Structures*, 1st ed., Prentice-Hall, Englewood Cliffs, NJ. With permission.

4.2 Proportioning and Mixing Concrete

4.2.1 Proportioning Concrete Mix

A concrete mix is specified by the weight of water, sand, coarse aggregate, and admixture to be used per 94-pound bag of cement. The type of cement (Table 4.4), modulus of the aggregates, and maximum size of the aggregates (Table 4.5) should also be given. A mix can be specified by the weight ratio of cement to sand to coarse aggregate with the minimum amount of cement per cubic yard of concrete.

In proportioning a concrete mix, it is advisable to make and test trial batches because of the many variables involved. Several trial batches should be made with a constant water-cement ratio but varying ratios of aggregates to obtain the desired workability with the least cement. To obtain results similar to those in the field, the trial batches should be mixed by machine.

When time or other conditions do not allow proportioning by the trial batch method, Table 4.6 may be used. Start with mix B corresponding to the appropriate maximum size of aggregate. Add just enough water for the desired workability. If the mix is undersanded, change to mix A; if oversanded, change to mix C. Weights are given for dry sand. For damp sand, increase the weight of sand 10 lb, and for very wet sand, 20 lb, per bag of cement.

TABLE 4.4   Types of Portland Cement[a]

Type	Usage
I	Ordinary construction where special properties are not required
II	Ordinary construction when moderate sulfate resistance or moderate heat of hydration is desired
III	When high early strength is desired
IV	When low heat of hydration is desired
V	When high sulfate resistance is desired

[a] According to ASTM C150.

TABLE 4.5   Recommended Maximum Sizes of Aggregate[a]

Minimum dimension of section, in.	Maximum size, in., of aggregate for:		
	Reinforced-concrete beams, columns, walls	Heavily reinforced slabs	Lightly reinforced or unreinforced slabs
5 or less	. . .	3/4 – 1 1/2	3/4 – 1 1/2
6–11	3/4 – 1 1/2	1 1/2	1 1/2 – 3
12–29	1 1/2 – 3	3	3 – 6
30 or more	1 1/2 – 3	3	6

[a] *Concrete Manual*. U.S. Bureau of Reclamation.

TABLE 4.6   Typical Concrete Mixes[a]

Maximum size of aggregate, in.	Mix designation	Bags of cement per yd^3 of concrete	Aggregate, lb per bag of cement		
			Sand		Gravel or crushed stone
			Air-entrained concrete	Concrete without air	
1/2	A	7.0	235	245	170
	B	6.9	225	235	190
	C	6.8	225	235	205
3/4	A	6.6	225	235	225
	B	6.4	225	235	245
	C	6.3	215	225	265
1	A	6.4	225	235	245
	B	6.2	215	225	275
	C	6.1	205	215	290
1 1/2	A	6.0	225	235	290
	B	5.8	215	225	320
	C	5.7	205	215	345
2	A	5.7	225	235	330
	B	5.6	215	225	360
	C	5.4	205	215	380

[a] *Concrete Manual*. U.S. Bureau of Reclamation.

4.2.2   Admixtures

Admixtures may be used to modify the properties of concrete. Some types of admixtures are set accelerators, water reducers, air-entraining agents, and waterproofers. Admixtures are generally helpful in improving quality of the concrete. However, if admixtures are not properly used, they could have undesirable effects; it is therefore necessary to know the advantages and limitations of the proposed admixture. The ASTM Specifications cover many of the admixtures.

Set accelerators are used (1) when it takes too long for concrete to set naturally; such as in cold weather, or (2) to accelerate the rate of strength development. Calcium chloride is widely used as a set accelerator. If not used in the right quantities, it could have harmful effects on the concrete and reinforcement.

Water reducers lubricate the mix and permit easier placement of the concrete. Since the workability of a mix can be improved by a chemical agent, less water is needed. With less water but the same cement content, the strength is increased. Since less water is needed, the cement content could also be decreased, which results in less shrinkage of the hardened concrete. Some water reducers also slow down the concrete set, which is useful in hot weather and integrating consecutive pours of the concrete.

Air-entraining agents are probably the most widely used type of admixture. Minute bubbles of air are entrained in the concrete, which increases the resistance of the concrete to freeze-thaw cycles and the use of ice-removal salts.

Waterproofing chemicals are often applied as surface treatments, but they can be added to the concrete mix. If applied properly and uniformly, they can prevent water from penetrating the concrete surface. Epoxies can also be used for waterproofing. They are more durable than silicone coatings, but they may be more costly. Epoxies can also be used for protection of wearing surfaces, patching cavities and cracks, and glue for connecting pieces of hardened concrete.

4.2.3 Mixing

Materials used in making concrete are stored in batch plants that have weighing and control equipment and bins for storing the cement and aggregates. Proportions are controlled by automatic or manually operated scales. The water is measured out either from measuring tanks or by using water meters.

Machine mixing is used whenever possible to achieve uniform consistency. The revolving drum-type mixer and the countercurrent mixer, which has mixing blades rotating in the opposite direction of the drum, are commonly used.

Mixing time, which is measured from the time all ingredients are in the drum, "should be at least 1.5 minutes for a 1-yd^3 mixer, plus 0.5 min for each cubic yard of capacity over 1 yd^3" [ACI 304-73, 1973]. It also is recommended to set a maximum on mixing time since overmixing may remove entrained air and increase fines, thus requiring more water for workability; three times the minimum mixing time can be used as a guide.

Ready-mixed concrete is made in plants and delivered to job sites in mixers mounted on trucks. The concrete can be mixed en route or upon arrival at the site. Concrete can be kept plastic and workable for as long as 1.5 hours by slow revolving of the mixer. Mixing time can be better controlled if water is added and mixing started upon arrival at the job site, where the operation can be inspected.

4.3   Flexural Design of Beams and One-Way Slabs

4.3.1   Reinforced Concrete Strength Design

The basic assumptions made in flexural design are:

1. Sections perpendicular to the axis of bending that are plane before bending remain plane after bending.

2. A perfect bond exists between the reinforcement and the concrete such that the strain in the reinforcement is equal to the strain in the concrete at the same level.

3. The strains in both the concrete and reinforcement are assumed to be directly proportional to the distance from the neutral axis (ACI 10.2.2) [1].

4. Concrete is assumed to fail when the compressive strain reaches 0.003 (ACI 10.2.3).

5. The tensile strength of concrete is neglected (ACI 10.2.5).

6. The stresses in the concrete and reinforcement can be computed from the strains using stress-strain curves for concrete and steel, respectively.

7. The compressive stress-strain relationship for concrete may be assumed to be rectangular, trapezoidal, parabolic, or any other shape that results in prediction of strength in substantial agreement with the results of comprehensive tests (ACI 10.2.6). ACI 10.2.7 outlines the use of a rectangular compressive stress distribution which is known as the Whitney rectangular stress block. For other stress distributions see *Reinforced Concrete Mechanics and Design* by James G. MacGregor [8].

Analysis of Rectangular Beams with Tension Reinforcement Only

Equations for M_n and ϕM_n: Tension Steel Yielding Consider the beam shown in Figure 4.1. The compressive force, C, in the concrete is

$$C = \left(0.85 f_c'\right) ba \tag{4.4}$$

The tension force, T, in the steel is

$$T = A_s f_y \tag{4.5}$$

For equilibrium, $C = T$, so the depth of the equivalent rectangular stress block, a, is

$$a = \frac{A_s f_y}{0.85 f_c' b} \tag{4.6}$$

Noting that the internal forces C and T form an equivalent force-couple system, the internal moment is

$$M_n = T(d - a/2) \tag{4.7}$$

or

$$M_n = C(d - a/2)$$

ϕM_n is then

$$\phi M_n = \phi T(d - a/2) \tag{4.8}$$

or

$$\phi M_n = \phi C(d - a/2)$$

where $\phi = 0.90$.

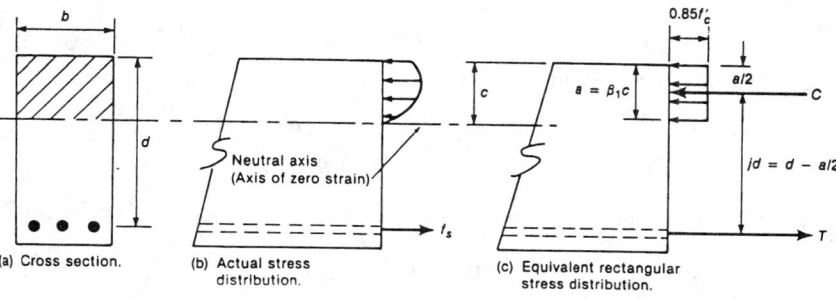

(a) Cross section.   (b) Actual stress distribution.   (c) Equivalent rectangular stress distribution.

Figure 4.1   Stresses and forces in a rectangular beam. (From MacGregor, J.G. 1992. *Reinforced Concrete Mechanics and Design,* 2nd ed., Prentice-Hall, Englewood Cliffs, N.J. With permission.)

Equation for M_n and ϕM_n: Tension Steel Elastic   The internal forces and equilibrium are given by:

$$
\begin{aligned}
C &= T \\
0.85 f_c' b a &= A_s f_s \\
0.85 f_c' b a &= \rho b d E_s \varepsilon_s
\end{aligned}
\tag{4.9}
$$

From strain compatibility (see Figure 4.1),

$$
\varepsilon_s = \varepsilon_{cu} \left(\frac{d-c}{c} \right)
\tag{4.10}
$$

Substituting ε_s into the equilibrium equation, noting that $a = \beta_1 c$, and simplifying gives

$$
\left(\frac{0.85 f_c'}{\rho E_s \varepsilon_{cu}} \right) a^2 + (d)a - \beta_1 d^2 = 0
\tag{4.11}
$$

which can be solved for a. Equations 4.7 and 4.8 can then be used to obtain M_n and ϕM_n.

Reinforcement Ratios   The reinforcement ratio, ρ, is used to represent the relative amount of tension reinforcement in a beam and is given by

$$
\rho = \frac{A_s}{bd}
\tag{4.12}
$$

At the balanced strain condition the maximum strain, ε_{cu}, at the extreme concrete compression fiber reaches 0.003 just as the tension steel reaches the strain $\varepsilon_y = f_y / E_s$. The reinforcement ratio in the balanced strain condition, ρ_b, can be obtained by applying equilibrium and compatibility conditions. From the linear strain condition, Figure 4.1,

$$
\frac{c_b}{d} = \frac{\varepsilon_{cu}}{\varepsilon_{cu} + \varepsilon_y} = \frac{0.003}{0.003 + \dfrac{f_y}{29,000,000}} = \frac{87,000}{87,000 + f_y}
\tag{4.13}
$$

The compressive and tensile forces are:

$$
\begin{aligned}
C_b &= 0.85 f_c' b \beta_1 c_b \\
T_b &= f_y A_{sb} = \rho_b b d f_y
\end{aligned}
\tag{4.14}
$$

Equating C_b to T_b and solving for ρ_b gives

$$
\rho_b = \frac{0.85 f_c' \beta_1}{f_y} \left(\frac{c_b}{d} \right)
\tag{4.15}
$$

which on substitution of Equation 4.13 gives

$$
\rho_b = \frac{0.85 f_c' \beta_1}{f_y} \left(\frac{87,000}{87,000 + f_y} \right)
\tag{4.16}
$$

ACI 10.3.3 limits the amount of reinforcement in order to prevent nonductile behavior:

$$
\max \rho = 0.75 \rho_b
\tag{4.17}
$$

ACI 10.5 requires a minimum amount of flexural reinforcement:

$$
\rho_{\min} = \frac{200}{f_y}
\tag{4.18}
$$

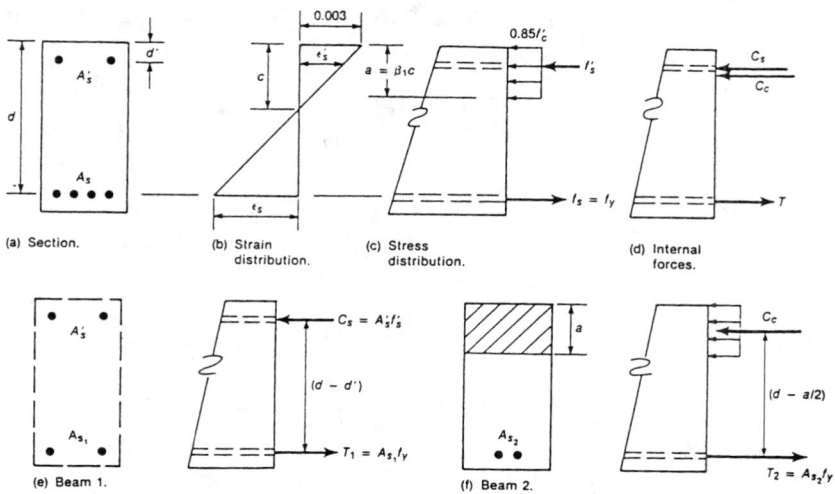

Figure 4.2 Strains, stresses, and forces in beam with compression reinforcement. (From MacGregor, J.G. 1992. *Reinforced Concrete Mechanics and Design,* 2nd ed., Prentice-Hall, Englewood Cliffs, NJ. With permission.)

Analysis of Beams with Tension and Compression Reinforcement

For the analysis of doubly reinforced beams, the cross-section will be divided into two beams. Beam 1 consists of the compression reinforcement at the top and sufficient steel at the bottom so that $T_1 = C_s$; beam 2 consists of the concrete web and the remaining tensile reinforcement, as shown in Figure 4.2

Equation for M_n: Compression Steel Yields The area of tension steel in beam 1 is obtained by setting $T_1 = C_s$, which gives $A_{s1} = A'_s$. The nominal moment capacity of beam 1 is then

$$M_{n1} = A'_s f_y \left(d - d'\right) \tag{4.19}$$

Beam 2 consists of the concrete and the remaining steel, $A_{s2} = A_s - A_{s1} = A_s - A'_s$. The compression force in the concrete is

$$C = 0.85 f'_c ba \tag{4.20}$$

and the tension force in the steel for beam 2 is

$$T = \left(A_s - A'_s\right) f_y \tag{4.21}$$

The depth of the compression stress block is then

$$a = \frac{\left(A_s - A'_s\right) f_y}{0.85 f'_c b} \tag{4.22}$$

Therefore, the nominal moment capacity for beam 2 is

$$M_{n2} = \left(A_s - A'_s\right) f_y (d - a/2) \tag{4.23}$$

The total moment capacity for a doubly reinforced beam with compression steel yielding is the summation of the moment capacity for beam 1 and beam 2; therefore,

$$M_n = A'_s f_y \left(d - d'\right) + \left(A_s - A'_s\right) f_y (d - a/2) \tag{4.24}$$

Equation for M_n: Compression Steel Does Not Yield Assuming that the tension steel yields, the internal forces in the beam are

$$
\begin{aligned}
T &= A_s f_y \\
C_c &= 0.85 f_c' b a \\
C_s &= A_s' \left(E_s \varepsilon_s' \right)
\end{aligned}
\tag{4.25}
$$

where

$$
\varepsilon_s' = \left(1 - \frac{\beta_1 d'}{a} \right)(0.003)
\tag{4.26}
$$

From equilibrium, $C_s + C_c = T$ or

$$
0.85 f_c' b a + A_s' E_s \left(1 - \frac{\beta_1 d'}{a} \right)(0.003) = A_s f_y
\tag{4.27}
$$

This can be rewritten in quadratic form as

$$
\left(0.85 f_c' b \right) a^2 + \left(0.003 A_s' E_s - A_s F_y \right) a - \left(0.003 A_s' E_s \beta_1 d' \right) = 0
\tag{4.28}
$$

where a can be calculated by means of the quadratic equation. Therefore, the nominal moment capacity in a doubly reinforced concrete beam where the compression steel does not yield is

$$
M_n = C_c \left(d - \frac{a}{2} \right) + C_s \left(d - d' \right)
\tag{4.29}
$$

Reinforcement Ratios The reinforcement ratio at the balanced strain condition can be obtained in a similar manner as that for beams with tension steel only. For compression steel yielding, the balanced ratio is

$$
(\rho - \rho')_b = \frac{0.85 f_c' \beta_1}{f_y} \left(\frac{87,000}{87,000 + f_y} \right)
\tag{4.30}
$$

For compression steel not yielding, the balanced ratio is

$$
\left(\rho - \frac{\rho' f_s'}{f_y} \right)_b = \frac{0.85 f_c' \beta_1}{f_y} \left(\frac{87,000}{87,000 + f_y} \right)
\tag{4.31}
$$

The maximum and minimum reinforcement ratios as given in ACI 10.3.3 and 10.5 are

$$
\begin{aligned}
\rho_{max} &= 0.75 \rho_b \\
\rho_{min} &= \frac{200}{f_y}
\end{aligned}
\tag{4.32}
$$

4.3.2 Prestressed Concrete Strength Design

Elastic Flexural Analysis

In developing elastic equations for prestress, the effects of prestress force, dead load moment, and live load moment are calculated separately, and then the separate stresses are superimposed, giving

$$
f = -\frac{F}{A} \pm \frac{F e y}{I} \pm \frac{M y}{I}
\tag{4.33}
$$

where $(-)$ indicates compression and $(+)$ indicates tension. It is necessary to check that the stresses in the extreme fibers remain within the ACI-specified limits under any combination of loadings that many occur. The stress limits for the concrete and prestressing tendons are specified in ACI 18.4 and 18.5 [1].

ACI 18.2.6 states that the loss of area due to open ducts shall be considered when computing section properties. It is noted in the commentary that section properties may be based on total area if the effect of the open duct area is considered negligible. In pretensioned members and in post-tensioned members after grouting, section properties can be based on gross sections, net sections, or effective sections using the transformed areas of bonded tendons and nonprestressed reinforcement.

Flexural Strength

The strength of a prestressed beam can be calculated using the methods developed for ordinary reinforced concrete beams, with modifications to account for the differing nature of the stress-strain relationship of prestressing steel compared with ordinary reinforcing steel.

A prestressed beam will fail when the steel reaches a stress f_{ps}, generally less than the tensile strength f_{pu}. For rectangular cross-sections the nominal flexural strength is

$$M_n = A_{ps} f_{ps} d - \frac{a}{2} \tag{4.34}$$

where

$$a = \frac{A_{ps} f_{ps}}{0.85 f_c' b} \tag{4.35}$$

The steel stress f_{ps} can be found based on strain compatibility or by using approximate equations such as those given in ACI 18.7.2. The equations in ACI are applicable only if the effective prestress in the steel, f_{se}, which equals P_e/A_{ps}, is not less than $0.5 f_{pu}$. The ACI equations are as follows.

(a) For members with bonded tendons:

$$f_{ps} = f_{pu} \left(1 - \frac{\gamma_p}{\beta_1} \left[\rho \frac{f_{pu}}{f_c'} + \frac{d}{d_p} (\omega - \omega')\right]\right) \tag{4.36}$$

If any compression reinforcement is taken into account when calculating f_{ps} with Equation 4.36, the following applies:

$$\left[\rho_p \frac{f_{pu}}{f_c'} + \frac{d}{d_p} (\omega - \omega')\right] \geq 0.17 \tag{4.37}$$

and

$$d' \leq 0.15 d_p$$

(b) For members with unbonded tendons and with a span-to-depth ratio of 35 or less:

$$f_{ps} = f_{se} + 10,000 + \frac{f_c'}{100\rho_p} \leq \left\{ \begin{array}{c} f_{py} \\ f_{se} + 60,000 \end{array} \right\} \tag{4.38}$$

(c) For members with unbonded tendons and with a span-to-depth ratio greater than 35:

$$f_{ps} = f_{se} + 10,000 + \frac{f_c'}{300\rho_p} \leq \left\{ \begin{array}{c} f_{py} \\ f_{se} + 30,000 \end{array} \right\} \tag{4.39}$$

The flexural strength is then calculated from Equation 4.34. The design strength is equal to ϕM_n, where $\phi = 0.90$ for flexure.

Reinforcement Ratios

ACI requires that the total amount of prestressed and nonprestressed reinforcement be adequate to develop a factored load at least 1.2 times the cracking load calculated on the basis of a modulus of rupture of $7.5\sqrt{f_c'}$.

To control cracking in members with unbonded tendons, some bonded reinforcement should be uniformly distributed over the tension zone near the extreme tension fiber. ACI specifies the minimum amount of bonded reinforcement as

$$A_s = 0.004A \tag{4.40}$$

where A is the area of the cross-section between the flexural tension face and the center of gravity of the gross cross-section. ACI 19.9.4 gives the minimum length of the bonded reinforcement.

To ensure adequate ductility, ACI 18.8.1 provides the following requirement:

$$\left\{ \begin{array}{c} \omega_p \\[4pt] \omega_p + \left(\dfrac{d}{d_p}\right)(\omega - \omega') \\[8pt] \omega_{pw} + \left(\dfrac{d}{d_p}\right)(\omega_w - \omega'_w) \end{array} \right\} \le 0.36\beta_1 \tag{4.41}$$

ACI allows each of the terms on the left side to be set equal to $0.85\, a/d_p$ in order to simplify the equation.

When a reinforcement ratio greater than $0.36\,\beta_1$ is used, ACI 18.8.2 states that the design moment strength shall not be greater than the moment strength based on the compression portion of the moment couple.

4.4   Columns under Bending and Axial Load

4.4.1   Short Columns under Minimum Eccentricity

When a symmetrical column is subjected to a concentric axial load, P, longitudinal strains develop uniformly across the section. Because the steel and concrete are bonded together, the strains in the concrete and steel are equal. For any given strain it is possible to compute the stresses in the concrete and steel using the stress-strain curves for the two materials. The forces in the concrete and steel are equal to the stresses multiplied by the corresponding areas. The total load on the column is the sum of the forces in the concrete and steel:

$$P_o = 0.85f_c'\left(A_g - A_{st}\right) + f_y A_{st} \tag{4.42}$$

To account for the effect of incidental moments, ACI 10.3.5 specifies that the maximum design axial load on a column be, for spiral columns,

$$\phi P_{n(\max)} = 0.85\phi\left[.85f_c'\left(A_g - A_{st}\right) + f_y A_{st}\right] \tag{4.43}$$

and for tied columns,

$$\phi P_{n(\max)} = 0.80\phi\left[.85f_c'\left(A_g - A_{st}\right) + f_y A_{st}\right] \tag{4.44}$$

For high values of axial load, ϕ values of 0.7 and 0.75 are specified for tied and spiral columns, respectively (ACI 9.3.2.2b) [1].

Short columns are sufficiently stocky such that slenderness effects can be ignored.

4.4.2 Short Columns under Axial Load and Bending

Almost all compression members in concrete structures are subjected to moments in addition to axial loads. Although it is possible to derive equations to evaluate the strength of columns subjected to combined bending and axial loads, the equations are tedious to use. For this reason, interaction diagrams for columns are generally computed by assuming a series of strain distributions, each corresponding to a particular point on the interaction diagram, and computing the corresponding values of P and M. Once enough such points have been computed, the results are summarized in an interaction diagram. For examples on determining the interaction diagram, see *Reinforced Concrete Mechanics and Design* by James G. MacGregor [8] or *Reinforced Concrete Design* by Chu-Kia Wang and Charles G. Salmon [11].

Figure 4.3 illustrates a series of strain distributions and the resulting points on the interaction diagram. Point A represents pure axial compression. Point B corresponds to crushing at one face and zero tension at the other. If the tensile strength of concrete is ignored, this represents the onset of cracking on the bottom face of the section. All points lower than this in the interaction diagram represent cases in which the section is partially cracked. Point C, the farthest right point, corresponds to the balanced strain condition and represents the change from compression failures for higher loads and tension failures for lower loads. Point D represents a strain distribution where the reinforcement has been strained to several times the yield strain before the concrete reaches its crushing strain.

The horizontal axis of the interaction diagram corresponds to pure bending where $\phi = 0.9$. A transition is required from $\phi = 0.7$ or 0.75 for high axial loads to $\phi = 0.9$ for pure bending. The change in ϕ begins at a capacity ϕP_a, which equals the smaller of the balanced load, ϕP_b, or 0.1 $f'_c A_g$. Generally, ϕP_b exceeds 0.1 $f'_c A_g$ except for a few nonrectangular columns.

ACI publication SP-17A(85), *A Design Handbook for Columns*, contains nondimensional interaction diagrams as well as other design aids for columns [2].

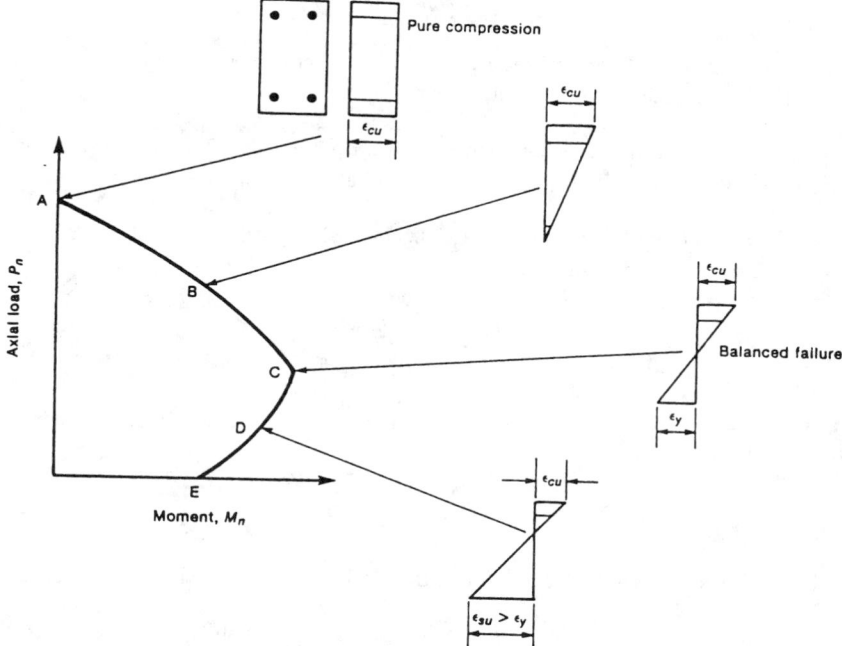

Figure 4.3   Strain distributions corresponding to points on interaction diagram.

4.4.3　Slenderness Effects

ACI 10.11 describes an approximate slenderness-effect design procedure based on the moment magnifier concept. The moments are computed by ordinary frame analysis and multiplied by a moment magnifier that is a function of the factored axial load and the critical buckling load of the column. The following gives a summary of the moment magnifier design procedure for slender columns in frames.

1. *Length of Column.* The unsupported length, l_u, is defined in ACI 10.11.1 as the clear distance between floor slabs, beams, or other members capable of giving lateral support to the column.

2. *Effective length.* The effective length factors, k, used in calculating δ_b shall be between 0.5 and 1.0 (ACI 10.11.2.1). The effective length factors used to compute δ_s shall be greater than 1 (ACI 10.11.2.2). The effective length factors can be estimated using ACI Fig. R10.11.2 or using ACI Equations (A)–(E) given in ACI R10.11.2. These two procedures require that the ratio, ψ, of the columns and beams be known:

$$\psi = \frac{\sum (E_c I_c / l_c)}{\sum (E_b I_b / l_b)} \tag{4.45}$$

In computing ψ it is acceptable to take the EI of the column as the uncracked gross $E_c I_g$ of the columns and the EI of the beam as $0.5 \, E_c I_g$.

3. *Definition of braced and unbraced frames.* The ACI Commentary suggests that a frame is braced if either of the following are satisfied:

(a) If the stability index, Q, for a story is less than 0.04, where

$$Q = \frac{\sum P_u \Delta_u}{H_u h_s} \leq 0.04 \tag{4.46}$$

(b) If the sum of the lateral stiffness of the bracing elements in a story exceeds six times the lateral stiffness of all the columns in the story.

4. *Radius of gyration.* For a rectangular cross-section r equals $0.3 \, h$, and for a circular cross-section r equals $0.25 \, h$. For other sections, r equals $\sqrt{I/A}$.

5. *Consideration of slenderness effects.* ACI 10.11.4.1 allows slenderness effects to be neglected for columns in braced frames when

$$\frac{k l_u}{r} < 34 - 12 \frac{M_{1b}}{M_{2b}} \tag{4.47}$$

ACI 10.11.4.2 allows slenderness effects to be neglected for columns in unbraced frames when

$$\frac{k l_u}{r} < 22 \tag{4.48}$$

If $k l_u / r$ exceeds 100, ACI 10.11.4.3 states that design shall be based on second-order analysis.

6. *Minimum moments.* For columns in a braced frame, M_{2b} shall be not less than the value given in ACI 10.11.5.4. In an unbraced frame ACI 10.11.5.5 applies for M_{2s}.

7. *Moment magnifier equation.* ACI 10.11.5.1 states that columns shall be designed for the factored axial load, P_u, and a magnified factored moment, M_c, defined by

$$M_c = \delta_b M_{2b} + \delta_s M_{2s} \tag{4.49}$$

where M_{2b} is the larger factored end moment acting on the column due to loads causing no appreciable sidesway (lateral deflections less than $l/1500$) and M_{2s} is the larger factored end moment due to loads that result in an appreciable sidesway. The moments are computed from a conventional first-order elastic frame analysis. For the above equation, the following apply:

$$\delta_b = \frac{C_m}{1 - P_u/\phi P_c} \geq 1.0 \tag{4.50}$$

$$\delta_s = \frac{1}{1 - \sum P_u/\phi \sum P_c} \geq 1.0$$

For members braced against sidesway, ACI 10.11.5.1 gives $\delta_s = 1.0$.

$$C_m = 0.6 + 0.4\frac{M_{1b}}{M_{2b}} \geq 0.4 \tag{4.51}$$

The ratio M_{1b}/M_{2b} is taken as positive if the member is bent in single curvature and negative if the member is bent in double curvature. Equation 4.51 applies only to columns in braced frames. In all other cases, ACI 10.11.5.3 states that $C_m = 1.0$.

$$P_c = \frac{\pi^2 EI}{(kl_u)^2} \tag{4.52}$$

where

$$EI = \frac{E_c I_g/5 + E_s I_{se}}{1 + \beta_d} \tag{4.53}$$

or, approximately

$$EI = \frac{E_c I_g/2.5}{1 + \beta_d} \tag{4.54}$$

When computing δ_b,

$$\beta_d = \frac{\text{Axial load due to factored dead load}}{\text{Total factored axial load}} \tag{4.55}$$

When computing δ_s,

$$\beta_d = \frac{\text{Factored sustained lateral shear in the story}}{\text{Total factored lateral shear in the story}} \tag{4.56}$$

If δ_b or δ_s is found to be negative, the column should be enlarged. If either δ_b or δ_s exceeds 2.0, consideration should be given to enlarging the column.

4.4.4 Columns under Axial Load and Biaxial Bending

The nominal ultimate strength of a section under biaxial bending and compression is a function of three variables, P_n, M_{nx}, and M_{ny}, which may also be expressed as P_n acting at eccentricities $e_y = M_{nx}/P_n$ and $e_x = M_{ny}/P_n$ with respect to the x and y axes. Three types of failure surfaces can be defined. In the first type, S_1, the three orthogonal axes are defined by P_n, e_x, and e_y; in the second type, S_2, the variables defining the axes are $1/P_n$, e_x, and e_y; and in the third type, S_3 the axes are P_n, M_{nx}, and M_{ny}. In the presentation that follows, the Bresler reciprocal load method makes use of the reciprocal failure surface S_2, and the Bresler load contour method and the PCA load contour method both use the failure surface S_3.

Bresler Reciprocal Load Method

Using a failure surface of type S_2, Bresler proposed the following equation as a means of approximating a point on the failure surface corresponding to prespecified eccentricities e_x and e_y:

$$\frac{1}{P_{ni}} = \frac{1}{P_{nx}} + \frac{1}{P_{ny}} - \frac{1}{P_0}$$

(4.57)

where

P_{ni}	=	nominal axial load strength at given eccentricity along both axes
P_{nx}	=	nominal axial load strength at given eccentricity along x axis
P_{ny}	=	nominal axial load strength at given eccentricity along y axis
P_0	=	nominal axial load strength for pure compression (zero eccentricity)

Test results indicate that Equation 4.57 may be inappropriate when small values of axial load are involved, such as when P_n/P_0 is in the range of 0.06 or less. For such cases the member should be designed for flexure only.

Bresler Load Contour Method

The failure surface S_3 can be thought of as a family of curves (load contours) each corresponding to a constant value of P_n. The general nondimensional equation for the load contour at constant P_n may be expressed in the following form:

$$\left(\frac{M_{nx}}{M_{ox}}\right)^{\alpha_1} + \left(\frac{M_{ny}}{M_{oy}}\right)^{\alpha_2} = 1.0$$

(4.58)

where

M_{nx}	=	$P_n e_y$; $M_{ny} = P_n e_x$
M_{ox}	=	M_{nx} capacity at axial load P_n when M_{ny} (or e_x) is zero
M_{oy}	=	M_{ny} capacity at axial load P_n when M_{nx} (or e_y) is zero

The exponents α_1 and α_2 depend on the column dimensions, amount and arrangement of the reinforcement, and material strengths. Bresler suggests taking $\alpha_1 = \alpha_2 = \alpha$. Calculated values of α vary from 1.15 to 1.55. For practical purposes, α can be taken as 1.5 for rectangular sections and between 1.5 and 2.0 for square sections.

PCA (Parme-Gowens) Load Contour Method

This method has been developed as an extension of the Bresler load contour method in which the Bresler interaction Equation 4.58 is taken as the basic strength criterion. In this approach, a point on the load contour is defined in such a way that the biaxial moment strengths M_{nx} and M_{ny} are in the same ratio as the uniaxial moment strengths M_{ox} and M_{oy},

$$\frac{M_{ny}}{M_{nx}} = \frac{M_{oy}}{M_{ox}} = \beta \tag{4.59}$$

The actual value of β depends on the ratio of P_n to P_0 as well as the material and cross-sectional properties, with the usual range of values between 0.55 and 0.70. Charts for determining β can be found in ACI Publication SP-17A(85), *A Design Handbook for Columns* [2].

Substituting Equation 4.59 into Equation 4.58,

$$\left(\frac{\beta M_{ox}}{M_{ox}}\right)^\alpha + \left(\frac{\beta M_{oy}}{M_{oy}}\right)^\alpha = 1$$
$$2\beta^\alpha = 1$$
$$\beta^\alpha = 1/2 \tag{4.60}$$
$$\alpha = \frac{\log 0.5}{\log \beta}$$

thus,

$$\left(\frac{M_{nx}}{M_{ox}}\right)^{\log 0.5/\log\beta} + \left(\frac{M_{ny}}{M_{oy}}\right)^{\log 0.5/\log\beta} = 1 \tag{4.61}$$

For more information on columns subjected to biaxial bending, see *Reinforced Concrete Design* by Chu-Kia Wang and Charles G. Salmon [11].

4.5 Shear and Torsion

4.5.1 Reinforced Concrete Beams and One-Way Slabs Strength Design

The cracks that form in a reinforced concrete beam can be due to flexure or a combination of flexure and shear. Flexural cracks start at the bottom of the beam, where the flexural stresses are the largest. Inclined cracks, also called *shear cracks* or *diagonal tension cracks,* are due to a combination of flexure and shear. Inclined cracks must exist before a shear failure can occur.

Inclined cracks form in two different ways. In thin-walled I-beams in which the shear stresses in the web are high while the flexural stresses are low, a web-shear crack occurs. The inclined cracking shear can be calculated as the shear necessary to cause a principal tensile stress equal to the tensile strength of the concrete at the centroid of the beam.

In most reinforced concrete beams, however, flexural cracks occur first and extend vertically in the beam. These alter the state of stress in the beam and cause a stress concentration near the tip of the crack. In time, the flexural cracks extend to become flexure-shear cracks. Empirical equations have been developed to calculate the flexure-shear cracking load, since this cracking cannot be predicted by calculating the principal stresses.

In the ACI Code, the basic design equation for the shear capacity of concrete beams is as follows:

$$V_u \le \phi V_n \tag{4.62}$$

where V_u is the shear force due to the factored loads, ϕ is the strength reduction factor equal to 0.85 for shear, and V_n is the nominal shear resistance, which is given by

$$V_n = V_c + V_s \tag{4.63}$$

where V_c is the shear carried by the concrete and V_s is the shear carried by the shear reinforcement.

The torsional capacity of a beam as given in ACI 11.6.5 is as follows:

$$T_u \leq \phi T_n \tag{4.64}$$

where T_u is the torsional moment due to factored loads, ϕ is the strength reduction factor equal to 0.85 for torsion, and T_n is the nominal torsional moment strength given by

$$T_n = T_c + T_c \tag{4.65}$$

where T_c is the torsional moment strength provided by the concrete and T_s is the torsional moment strength provided by the torsion reinforcement.

Design of Beams and One-Way Slabs Without Shear Reinforcement: for Shear

The critical section for shear in reinforced concrete beams is taken at a distance d from the face of the support. Sections located at a distance less than d from the support are designed for the shear computed at d.

Shear Strength Provided by Concrete   Beams without web reinforcement will fail when inclined cracking occurs or shortly afterwards. For this reason the shear capacity is taken equal to the inclined cracking shear. ACI gives the following equations for calculating the shear strength provided by the concrete for beams without web reinforcement subject to shear and flexure:

$$V_c = 2\sqrt{f_c'}b_w d \tag{4.66}$$

or, with a more detailed equation:

$$V_c = \left(1.9\sqrt{f_c'} + 2500\rho_w \frac{V_u d}{M_u}\right)b_w d \leq 3.5\sqrt{f_c'}b_w d \tag{4.67}$$

The quantity $V_u d/M_u$ is not to be taken greater than 1.0 in computing V_c where M_u is the factored moment occurring simultaneously with V_u at the section considered.

Combined Shear, Moment, and Axial Load   For members that are also subject to axial compression, ACI modifies Equation 4.66 as follows (ACI 11.3.1.2):

$$V_c = 2\left(1 + \frac{N_u}{2000A_k}\right)\sqrt{f_c'}b_w d \tag{4.68}$$

where N_u is positive in compression. ACI 11.3.2.2 contains a more detailed calculation for the shear strength of members subject to axial compression.

For members subject to axial tension, ACI 11.3.1.3 states that shear reinforcement shall be designed to carry total shear. As an alternative, ACI 11.3.2.3 gives the following for the shear strength of members subject to axial tension:

$$V_c = 2\left(1 + \frac{N_u}{500A_g}\right)\sqrt{f_c'}b_w d \tag{4.69}$$

where N_u is negative in tension. In Equation 4.68 and 4.69 the terms $\sqrt{f_c'}$, N_u/A_g, 2000, and 500 all have units of psi.

Combined Shear, Moment, and Torsion   For members subject to torsion, ACI 11.3.1.4 gives the equation for the shear strength of the concrete as the following:

$$V_c = \frac{2\sqrt{f_c'}b_w d}{\sqrt{1 + (2.5C_t T_u/V_u)^2}} \tag{4.70}$$

where

$$T_u \geq \phi\left(0.5\sqrt{f_c'}\sum x^2 y\right)$$

Design of Beams and One-Way Slabs Without Shear Reinforcements: for Torsion

ACI 11.6.1 requires that torsional moments be considered in design if

$$T_u \geq \phi \left(0.5\sqrt{f_c'} \sum x^2 y \right) \tag{4.71}$$

Otherwise, torsion effects may be neglected.

The critical section for torsion is taken at a distance d from the face of support, and sections located at a distance less than d are designed for the torsion at d. If a concentrated torque occurs within this distance, the critical section is taken at the face of the support.

Torsional Strength Provided by Concrete   Torsion seldom occurs by itself; bending moments and shearing forces are typically present also. In an uncracked member, shear forces as well as torques produce shear stresses. Flexural shear forces and torques interact in a way that reduces the strength of the member compared with what it would be if shear or torsion were acting alone. The interaction between shear and torsion is taken into account by the use of a circular interaction equation. For more information, refer to *Reinforced Concrete Mechanics and Design* by James G. MacGregor [8].

The torsional moment strength provided by the concrete is given in ACI 11.6.6.1 as

$$T_c = \frac{0.8\sqrt{f_c'}x^2 y}{\sqrt{1 + (0.4V_u/C_t T_u)^2}} \tag{4.72}$$

Combined Torsion and Axial Load   For members subject to significant axial tension, ACI 11.6.6.2 states that the torsion reinforcement must be designed to carry the total torsional moment, or as an alternative modify T_c as follows:

$$T_c = \frac{0.8\sqrt{f_c'}x^2 y}{\sqrt{1 + (0.4V_u/C_t T_u)^2}} \left(1 + \frac{N_u}{500 A_g} \right) \tag{4.73}$$

where N_u is negative for tension.

Design of Beams and One-Way Slabs without Shear Reinforcement:

Minimum Reinforcement   ACI 11.5.5.1 requires a minimum amount of web reinforcement to be provided for shear and torsion if the factored shear force V_u exceeds one half the shear strength provided by the concrete ($V_u \geq 0.5\phi V_c$) except in the following:

(a) Slabs and footings

(b) Concrete joist construction

(c) Beams with total depth not greater than 10 inches, 2 1/2 times the thickness of the flange, or 1/2 the width of the web, whichever is greatest

The minimum area of shear reinforcement shall be at least

$$A_{v(\min)} = \frac{50 b_w s}{f_y} \quad \text{for } T_u < \phi \left(0.5\sqrt{f_c'} \sum x^2 y \right) \tag{4.74}$$

When torsion is to be considered in design, the sum of the closed stirrups for shear and torsion must satisfy the following:

$$A_v + 2A_t \geq \frac{50 b_w s}{f_y} \tag{4.75}$$

where A_v is the area of two legs of a closed stirrup and A_t is the area of only one leg of a closed stirrup.

Design of Stirrup Reinforcement for Shear and Torsion

Shear Reinforcement　Shear reinforcement is to be provided when $V_u \geq \phi V_c$, such that

$$V_s \geq \frac{V_u}{\phi} - V_c \tag{4.76}$$

The design yield strength of the shear reinforcement is not to exceed 60,000 psi.

When the shear reinforcement is perpendicular to the axis of the member, the shear resisted by the stirrups is

$$V_s = \frac{A_v f_y d}{s} \tag{4.77}$$

If the shear reinforcement is inclined at an angle α, the shear resisted by the stirrups is

$$V_s = \frac{A_v f_y (\sin\alpha + \cos\alpha) d}{s} \tag{4.78}$$

The maximum shear strength of the shear reinforcement is not to exceed $8\sqrt{f_c'}b_w d$ as stated in ACI 11.5.6.8.

Spacing Limitations for Shear Reinforcement　ACI 11.5.4.1 sets the maximum spacing of vertical stirrups as the smaller of $d/2$ or 24 inches. The maximum spacing of inclined stirrups is such that a 45° line extending from midheight of the member to the tension reinforcement will intercept at least one stirrup.

If V_s exceeds $4\sqrt{f_c'}b_w d$, the maximum allowable spacings are reduced to one half of those just described.

Torsion Reinforcement　Torsion reinforcement is to be provided when $T_u \geq \phi T_c$, such that

$$T_s \geq \frac{T_u}{\phi} - T_c \tag{4.79}$$

The design yield strength of the torsional reinforcement is not to exceed 60,000 psi.

The torsional moment strength of the reinforcement is computed by

$$T_s = \frac{A_t \alpha_t x_1 y_1 f_y}{s} \tag{4.80}$$

where

$$\alpha_t = [0.66 + 0.33(y_t/x_t)] \geq 1.50 \tag{4.81}$$

where A_t is the area of one leg of a closed stirrup resisting torsion within a distance s. The torsional moment strength is not to exceed $4 T_c$ as given in ACI 11.6.9.4.

Longitudinal reinforcement is to be provided to resist axial tension that develops as a result of the torsional moment (ACI 11.6.9.3). The required area of longitudinal bars distributed around the perimeter of the closed stirrups that are provided as torsion reinforcement is to be

$$A_l \geq 2A_t \frac{(x_1 + y_1)}{s}$$

$$A_l \geq \left[\frac{400xs}{f_y}\left(\frac{T_u}{T_u + \frac{V_u}{3C_t}}\right) = 2A_t\right]\left(\frac{x_1 + y_1}{s}\right) \tag{4.82}$$

Spacing Limitations for Torsion Reinforcement　ACI 11.6.8.1 gives the maximum spacing of closed stirrups as the smaller of $(x_1 + y_1)/4$ or 12 inches.

The longitudinal bars are to spaced around the circumference of the closed stirrups at not more than 12 inches apart. At least one longitudinal bar is to be placed in each corner of the closed stirrups (ACI 11.6.8.2).

Design of Deep Beams

ACI 11.8 covers the shear design of deep beams. This section applies to members with $l_n/d < 5$ that are loaded on one face and supported on the opposite face so that compression struts can develop between the loads and the supports. For more information on deep beams, see *Reinforced Concrete Mechanics and Design*, 2nd ed. by James G. MacGregor [8].

The basic design equation for simple spans deep beams is

$$V_u \leq \phi \left(V_c + V_s \right) \tag{4.83}$$

where V_c is the shear carried by the concrete and V_s is the shear carried by the vertical and horizontal web reinforcement.

The shear strength of deep beams shall not be taken greater than

$$
\begin{aligned}
V_n &= 8\sqrt{f'_c}b_w d \text{ for } l_n/d < 2 \\
V_n &= \frac{2}{3}\left(10 + \frac{l_n}{d}\right)\sqrt{f'_c}b_w d \text{ for } 2 \leq l_n/d \leq 5
\end{aligned} \tag{4.84}
$$

Design for shear is done at a critical section located at $0.15 l_n$ from the face of support in uniformly loaded beams, and at the middle of the shear span for beams with concentrated loads. For both cases, the critical section shall not be farther than d from the face of the support. The shear reinforcement required at this critical section is to be used throughout the span.

The shear carried by the concrete is given by

$$V_c = 2\sqrt{f'_c}b_w d \tag{4.85}$$

or, with a more detailed calculation,

$$V_c = \left(3.5 - 2.5\frac{M_u}{V_u d}\right)\left(1.9\sqrt{f'_c} + 2500\rho_w \frac{V_u d}{M_u}\right)b_w d \leq 6\sqrt{f'_c}b_w d \tag{4.86}$$

where

$$\left(3.5 - 2.5\frac{M_u}{V_u d}\right) \leq 2.5 \tag{4.87}$$

In Equations 4.86 and 4.87 M_u and V_u are the factored moment and shear at the critical section.

Shear reinforcement is to be provided when $V_u \geq \phi V_c$ such that

$$V_s = \frac{V_u}{\phi} - V_c \tag{4.88}$$

where

$$V_s = \left[\frac{A_v}{s}\left(\frac{1 + l_n/d}{12}\right) + \frac{A_{vh}}{s_2}\left(\frac{11 - l_n/d}{12}\right)\right]f_y d \tag{4.89}$$

where A_v and s are the area and spacing of the vertical shear reinforcement and A_{vh} and s_2 refer to the horizontal shear reinforcement.

ACI 11.8.9 and 11.8.10 require minimum reinforcement in both the vertical and horizontal sections as follows:

(a) vertical direction

$$A_v \geq 0.0015 b_w s \tag{4.90}$$

where

$$s \leq \left\{ \begin{array}{c} d/5 \\ 18 \text{ in.} \end{array} \right\} \tag{4.91}$$

(b) horizontal direction

$$A_{vh} \geq 0.0025 b_w s_2 \qquad (4.92)$$

where

$$s_2 \leq \left\{ \begin{array}{c} d/3 \\ 18 \text{ in.} \end{array} \right\} \qquad (4.93)$$

4.5.2 Prestressed Concrete Beams and One-Way Slabs Strength Design

At loads near failure, a prestressed beam is usually heavily cracked and behaves similarly to an ordinary reinforced concrete beam. Many of the equations developed previously for design of web reinforcement for nonprestressed beams can also be applied to prestressed beams.

Shear design is based on the same basic equation as before,

$$V_u \leq \phi \left(V_c + V_s \right)$$

where $\phi = 0.85$.

The critical section for shear is taken at a distance $h/2$ from the face of the support. Sections located at a distance less than $h/2$ are designed for the shear computed at $h/2$.

Shear Strength Provided by the Concrete

The shear force resisted by the concrete after cracking has occurred is taken as equal to the shear that caused the first diagonal crack. Two types of diagonal cracks have been observed in tests of prestressed concrete.

1. Flexure-shear cracks, occurring at nominal shear V_{ci}, start as nearly vertical flexural cracks at the tension face of the beam, then spread diagonally upward toward the compression face.

2. Web shear cracks, occurring at nominal shear V_{cw}, start in the web due to high diagonal tension, then spread diagonally both upward and downward.

The shear strength provided by the concrete for members with effective prestress force not less than 40% of the tensile strength of the flexural reinforcement is

$$V_c = \left(0.6\sqrt{f_c'} + 700 \frac{V_u d}{M_u} \right) b_w d \leq 2\sqrt{f_c'} b_w d \qquad (4.94)$$

V_c may also be computed as the lesser of V_{ci} and V_{cw}, where

$$V_{ci} = 0.6\sqrt{f_c'} b_w d + V_d + \frac{V_i M_{cr}}{M_{\max}} \geq 1.7\sqrt{f_c'} b_w d \qquad (4.95)$$

$$M_{cr} = \left(\frac{I}{y_t} \right) \left(6\sqrt{f_c'} + f_{pc} - f_d \right) \qquad (4.96)$$

$$V_{cw} = \left(3.5\sqrt{f_c'} + 0.3 f_{pc} \right) b_w d + V_p \qquad (4.97)$$

In Equations 4.95 and 4.97 d is the distance from the extreme compression fiber to the centroid of the prestressing steel or $0.8h$, whichever is greater.

Shear Strength Provided by the Shear Reinforcement

Shear reinforcement for prestressed concrete is designed in a similar manner as for reinforced concrete, with the following modifications for minimum amount and spacing.

Minimum Reinforcement The minimum area of shear reinforcement shall be at least

$$A_{v(min)} = \frac{50b_w s}{f_y} \text{ for } T_u < \phi\left(0.5\sqrt{f'_c}\sum x^2 y\right) \tag{4.98}$$

or

$$A_{v(min)} = \frac{A_{ps} f_{pu} s}{80 f_{yd}}\sqrt{\frac{d}{b_w}} \tag{4.99}$$

Spacing Limitations for Shear Reinforcement ACI 11.5.4.1 sets the maximum spacing of vertical stirrups as the smaller of $(3/4)h$ or 24 in. The maximum spacing of inclined stirrups is such that a 45° line extending from midheight of the member to the tension reinforcement will intercept at least one stirrup.

If V_s exceeds $4\sqrt{f'_c}b_w d$, the maximum allowable spacings are reduced to one-half of those just described.

4.6 Development of Reinforcement

The development length, l_d, is the shortest length of bar in which the bar stress can increase from zero to the yield strength, f_y. If the distance from a point where the bar stress equals f_y to the end of the bar is less than the development length, the bar will pull out of the concrete. Development lengths are different for tension and compression.

4.6.1 Development of Bars in Tension

ACI Fig. R12.2 gives a flow chart for determining development length. The steps are outlined below.

The basic tension development lengths have been found to be (ACI 12.2.2). For no. 11 and smaller bars and deformed wire:

$$l_{db} = \frac{0.04A_b f_y}{\sqrt{f'_c}} \tag{4.100}$$

For no. 14 bars:

$$l_{db} = \frac{0.085 f_y}{\sqrt{f'_c}} \tag{4.101}$$

For no. 18 bars:

$$l_{db} = \frac{0.125 f_y}{\sqrt{f'_c}} \tag{4.102}$$

where $\sqrt{f'_c}$ is not to be taken greater than 100 psi.

The development length, l_d, is computed as the product of the basic development length and modification factors given in ACI 12.2.3, 12.2.4, and 12.2.5. The development length obtained from ACI 12.2.2 and 12.2.3.1 through 12.2.3.5 shall not be less than

$$\frac{0.03d_b f_y}{\sqrt{f'_c}} \tag{4.103}$$

as given ACI 12.2.3.6.

The length computed from ACI 12.2.2 and 12.2.3 is then multiplied by factors given in ACI 12.2.4 and 12.2.5. The factors given in ACI 12.2.3.1 through 12.2.3.3 and 12.2.4 are required, but the factors in ACI 12.2.3.4, 12.2.3.5, and 12.2.5 are optional.

The development length is not to be less than 12 inches (ACI 12.2.1).

4.6.2 Development of Bars in Compression

The basic compression development length is (ACI 12.3.2)

$$l_{db} = \frac{0.02d_b f_y}{\sqrt{f'_c}} \geq 0.003d_b f_y \tag{4.104}$$

The development length, l_d, is found as the product of the basic development length and applicable modification factors given in ACI 12.3.3.

The development length is not to be less than 8 inches (ACI 12.3.1).

4.6.3 Development of Hooks in Tension

The basic development length for a hooked bar with $f_y = 60,000$ psi is (ACI 12.5.2)

$$l_{db} = \frac{1200d_b}{\sqrt{f'_c}} \tag{4.105}$$

The development length, l_{dh}, is found as the product of the basic development length and applicable modification factors given in ACI 12.5.3.

The development length of the hook is not to be less than 8 bar diameters or 6 inches (ACI 12.5.1). Hooks are not to be used to develop bars in compression.

4.6.4 Splices, Bundled Bars, and Web Reinforcement

Splices

Tension Lap Splices   ACI 12.15 distinguishes between two types of tension lap splices depending on the amount of reinforcement provided and the fraction of the bars spliced in a given length—see ACI Table R12.15.2. The splice lengths for each splice class are as follows:

Class A splice: 1.0 l_d

Class B splice: 1.3 l_d

where l_d is the tensile development length as computed in ACI 12.2 without the modification factor for excess reinforcement given in ACI 12.2.5. The minimum splice length is 12 inches.

Lap splices are not to be used for bars larger than no. 11 except at footing to column joints and for compression lap splices of no. 14 and no. 18 bars with smaller bars (ACI 12.14.2.1). The center-to-center distance between two bars in a lap splice cannot be greater than one-fifth the required lap splice length with a maximum of 6 inches (ACI 12.14.2.3). ACI 21.3.2.3 requires that tension lap splices of flexural reinforcement in beams resisting seismic loads be enclosed by hoops or spirals.

Compression Lap Splices   The splice length for a compression lap splice is given in ACI 12.16.1 as

$$l_s = 0.0005 f_y d_b \text{ for } f_y \leq 60,000 \text{ psi} \tag{4.106}$$
$$l_s = (0.0009 f_y - 24) d_b \text{ for } f_y > 60,000 \text{ psi} \tag{4.107}$$

but not less than 12 inches. For f'_c less than 3000 psi, the lap length must be increased by one-third.

When different size bars are lap spliced in compression, the splice length is to be the larger of:

1. Compression splice length of the smaller bar, or

2. Compression development length of larger bar.

Compression lap splices are allowed for no. 14 and no. 18 bars to no. 11 or smaller bars (ACI 12.16.2).

End-Bearing Splices   End-bearing splices are allowed for compression only where the compressive stress is transmitted by bearing of square cut ends held in concentric contact by a suitable device. According to ACI 12.16.4.2 bar ends must terminate in flat surfaces within 1 1/2° of right angles to the axis of the bars and be fitted within 3° of full bearing after assembly. End-bearing splices are only allowed in members containing closed ties, closed stirrups, or spirals.

Welded Splices or Mechanical Connections   Bars stressed in tension or compression may be spliced by welding or by various mechanical connections. ACI 12.14.3, 12.15.3, 12.15.4, and 12.16.3 govern the use of such splices. For further information see *Reinforced Concrete Design,* by Chu-Kia Wang and Charles G. Salmon [11].

Bundled Bars

The requirements of ACI 12.4.1 specify that the development length for bundled bars be based on that for the individual bar in the bundle, increased by 20% for a three-bar bundle and 33% for a four-bar bundle. ACI 12.4.2 states that "a unit of bundled bars shall be treated as a single bar of a diameter derived from the equivalent total area" when determining the appropriate modification factors in ACI 12.2.3 and 12.2.4.3.

Web Reinforcement

ACI 12.13.1 requires that the web reinforcement be as close to the compression and tension faces as cover and bar-spacing requirements permit. The ACI Code requirements for stirrup anchorage are illustrated in Figure 4.4.

(a) ACI 12.13.3 requires that each bend away from the ends of a stirrup enclose a longitudinal bar, as seen in Figure 4.4(a).

(b) For no. 5 or D31 wire stirrups and smaller with any yield strength and for no. 6, 7, and 8 bars with a yield strength of 40,000 psi or less, ACI 12.13.2.1 allows the use of a standard hook around longitudinal reinforcement, as shown in Figure 4.4(b).

(c) For no. 6, 7, and 8 stirrups with f_y greater than 40,000 psi, ACI 12.13.2.2 requires a standard hook around a longitudinal bar plus an embedment between midheight of the member and the outside end of the hook of at least $0.014 d_b f_y / \sqrt{f_c'}$.

(d) Requirements for welded wire fabric forming U stirrups are given in ACI 12.13.2.3.

(e) Pairs of U stirrups that form a closed unit shall have a lap length of $1.3 l_d$ as shown in Figure 4.4(c). This type of stirrup has proven unsuitable in seismic areas.

(f) Requirements for longitudinal bars bent to act as shear reinforcement are given in ACI 12.13.4.

4.7   Two-Way Systems

4.7.1   Definition

When the ratio of the longer to the shorter spans of a floor panel drops below 2, the contribution of the longer span in carrying the floor load becomes substantial. Since the floor transmits loads in two directions, it is defined as a *two-way system*, and flexural reinforcement is designed for both directions. Two-way systems include *flat plates, flat slabs, two-way slabs,* and *waffle slabs* (see Figure 4.5). The choice between these different types of two-way systems is largely a matter of the architectural layout, magnitude of the design loads, and span lengths. A flat plate is simply a slab of uniform thickness supported directly on columns, generally suitable for relatively light loads. For larger loads and spans, a flat slab becomes more suitable with the column capitals and drop panels

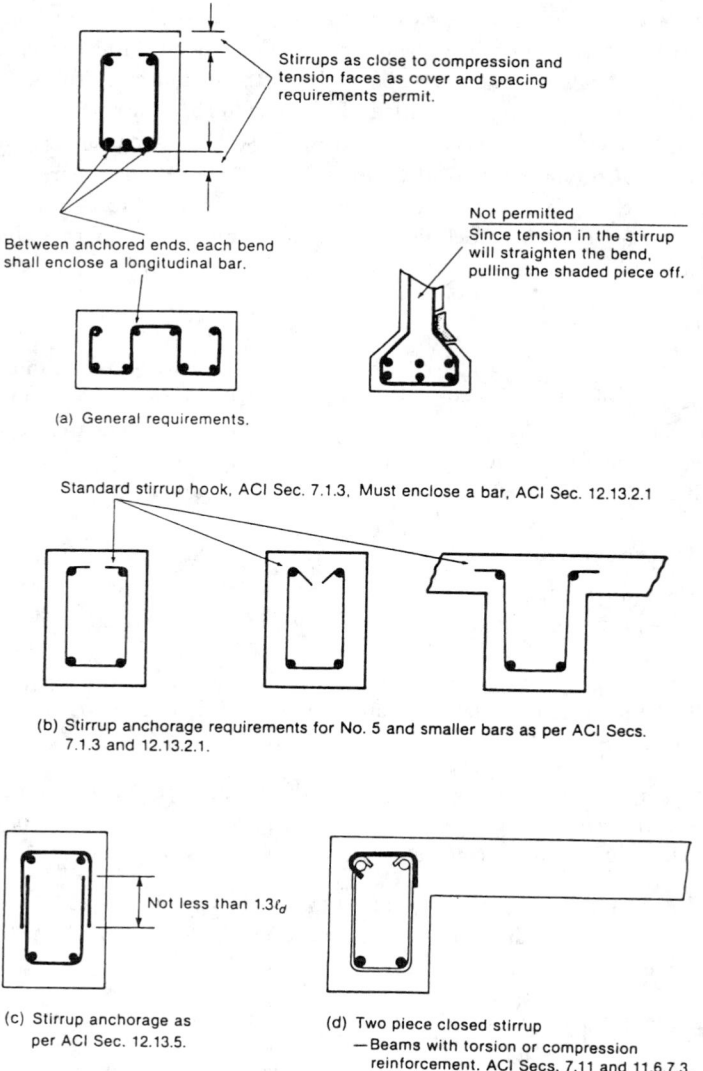

Figure 4.4   Stirrup detailing requirements. (From Wang, C.-K. and Salmon, C.G. 1985. *Reinforced Concrete Design*, 4th ed., Harper Row, New York. With permission.)

providing higher shear and flexural strength. A slab supported on beams on all sides of each floor panel is generally referred to as a two-way slab. A waffle slab is equivalent to a two-way joist system or may be visualized as a solid slab with recesses in order to decrease the weight of the slab.

4.7.2   Design Procedures

The ACI code [1] states that a two-way slab system "may be designed by any procedure satisfying conditions of equilibrium and geometric compatibility if shown that the design strength at every section is at least equal to the required strength ... and that all serviceability conditions, including specified limits on deflections, are met" (p.204). There are a number of possible approaches to the analysis and design of two-way systems based on elastic theory, limit analysis, finite element analysis, or combination of elastic theory and limit analysis. The designer is permitted by the

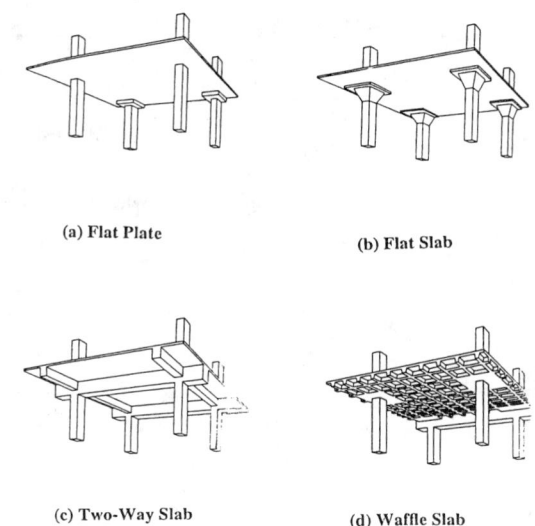

(a) Flat Plate

(b) Flat Slab

(c) Two-Way Slab

(d) Waffle Slab

Figure 4.5 Two-way systems.

ACI Code to adopt any of these approaches provided that all safety and serviceability criteria are satisfied. In general, only for cases of a complex two-way system or unusual loading would a finite element analysis be chosen as the design approach. Otherwise, more practical design approaches are preferred. The ACI Code details two procedures—the *direct design method* and the *equivalent frame method*—for the design of floor systems with or without beams. These procedures were derived from analytical studies based on elastic theory in conjunction with aspects of limit analysis and results of experimental tests. The primary difference between the direct design method and equivalent frame method is in the way moments are computed for two-way systems.

The *yield-line theory* is a limit analysis method devised for slab design. Compared to elastic theory, the yield-line theory gives a more realistic representation of the behavior of slabs at the ultimate limit state, and its application is particularly advantageous for irregular column spacing. While the yield-line method is an upper-bound limit design procedure, *strip method* is considered to give a lower-bound design solution. The strip method offers a wide latitude of design choices and it is easy to use; these are often cited as the appealing features of the method.

Some of the earlier design methods based on moment coefficients from elastic analysis are still favored by many designers. These methods are easy to apply and give valuable insight into slab behavior; their use is especially justified for many irregular slab cases where the preconditions of the direct design method are not met or when column interaction is not significant. Table 4.7 lists the moment coefficients taken from method 2 of the 1963 ACI Code.

As in the 1989 code, two-way slabs are divided into column strips and middle strips as indicated by Figure 4.6, where l_1 and l_2 are the center-to-center span lengths of the floor panel. A column strip is a design strip with a width on each side of a column centerline equal to $0.25l_2$ or $0.25l_1$, whichever is less. A middle strip is a design strip bounded by two column strips. Taking the moment coefficients from Table 4.7, bending moments per unit width M for the middle strips are computed from the formula

$$M = (\text{Coef.})wl_s^2 \tag{4.108}$$

where w is the total uniform load per unit area and l_s is the shorter span length of l_1 and l_2. The average moments per unit width in the column strip is taken as two-thirds of the corresponding moments in the middle strip.

TABLE 4.7   Elastic Moment Coefficients for Two-Way Slabs

	Short span						Long span, all
	Span ratio, short/long						
Moments	1.0	0.9	0.8	0.7	0.6	0.5 and less	span ratios
Case 1—Interior panels							
Negative moment at:							
Continuous edge	0.033	0.040	0.048	0.055	0.063	0.083	0.033
Discontinuous edge	—	—	—	—	—	—	—
Positive moment at midspan	0.025	0.030	0.036	0.041	0.047	0.062	0.025
Case 2—One edge discontinuous							
Negative moment at:							
Continuous edge	0.041	0.048	0.055	0.062	0.069	0.085	0.041
Discontinuous edge	0.021	0.024	0.027	0.031	0.035	0.042	0.021
Positive moment at midspan	0.031	0.036	0.041	0.047	0.052	0.064	0.031
Case 3—Two edges discontinuous							
Negative moment at:							
Continuous edge	0.049	0.057	0.064	0.071	0.078	0.090	0.049
Discontinuous edge	0.025	0.028	0.032	0.036	0.039	0.045	0.025
Positive moment at midspan:	0.037	0.043	0.048	0.054	0.059	0.068	0.037
Case 4—Three edges discontinuous							
Negative moment at:							
Continuous edge	0.058	0.066	0.074	0.082	0.090	0.098	0.058
Discontinuous edge	0.029	0.033	0.037	0.041	0.045	0.049	0.029
Positive moment at midspan:	0.044	0.050	0.056	0.062	0.068	0.074	0.044
Case 5—Four edges discontinuous							
Negative moment at:							
Continuous edge	—	—	—	—	—	—	—
Discontinuous edge	0.033	0.038	0.043	0.047	0.053	0.055	0.033
Positive moment at midspan	0.050	0.057	0.064	0.072	0.080	0.083	0.050

4.7.3   Minimum Slab Thickness and Reinforcement

ACI Code Section 9.5.3 contains requirements to determine minimum slab thickness of a two-way system for deflection control. For slabs without beams, the thickness limits are summarized by Table 4.8, but thickness must not be less than 5 in. for slabs without drop panels or 4 in. for slabs with drop panels. In Table 4.8 l_n is the length of clear span in the long direction and α is the ratio of flexural stiffness of beam section to flexural stiffness of a width of slab bounded laterally by centerline of adjacent panel on each side of beam.

For slabs with beams, it is necessary to compute the minimum thickness h from

$$h = \frac{l_n \left(0.8 + \dfrac{f_y}{200,000} \right)}{36 + 5\beta \left[\alpha_m - 0.12 \left(1 + \dfrac{1}{\beta} \right) \right]} \tag{4.109}$$

but not less than

$$h = \frac{l_n \left(0.8 + \dfrac{f_y}{200,000} \right)}{36 + 9\beta} \tag{4.110}$$

and need not be more than

$$h = \frac{l_n \left(0.8 + \dfrac{f_y}{200,000} \right)}{36} \tag{4.111}$$

where β is the ratio of clear spans in long-to-short direction and α_m is the average value of α for all beams on edges of a panel. In no case should the slab thickness be less than 5 in. for $\alpha_m < 2.0$ or less than 3 1/2 in. for $\alpha_m \geq 2.0$.

Minimum reinforcement in two-way slabs is governed by shrinkage and temperature controls to minimize cracking. The minimum reinforcement area stipulated by the ACI Code shall not be

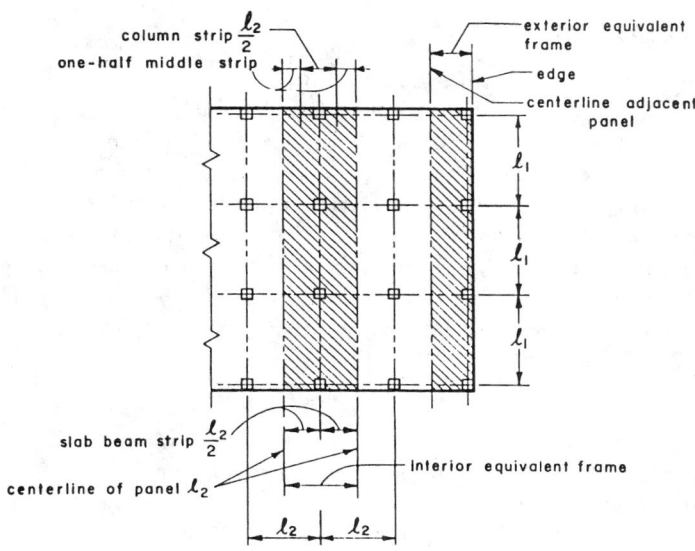

Figure 4.6 Definitions of equivalent frame, column strip, and middle strip. (From ACI Committee 318. 1992. *Building Code Requirements for Reinforced Concrete and Commentary, ACI 318-89 (Revised 92) and ACI 318R-89 (Revised 92)*, Detroit, MI. With permission.)

less than 0.0018 times the gross concrete area when grade 60 steel is used (0.0020 when grade 40 or grade 50 is used). The spacing of reinforcement in two-way slabs shall exceed neither two times the slab thickness nor 18 in.

TABLE 4.8 Minimum Thickness of Two-Way Slabs without Beams

Yield stress f_y, psi[a]	Exterior panels		Interior panels
	Without edge beams	With edge beams[b]	
Without drop panels			
40,000	$l_n/33$	$l_n/36$	$l_n/36$
60,000	$l_n/30$	$l_n/33$	$l_n/33$
With drop panels			
40,000	$l_n/36$	$l_n/40$	$l_n/40$
60,000	$l_n/33$	$l_n/36$	$l_n/36$

[a] For values of reinforcement yield stress between 40,000 and 60,000 psi minimum thickness shall be obtained by linear interpolation.

[b] Slabs with beams between columns along exterior edges. The value of α for the edge beam shall not be less than 0.8.

From ACI Committee 318. 1992. *Building Code Requirements for Reinforced Concrete and Commentary, ACI 318-89 (Revised 92) and ACI 318R-89 (Revised 92)*, Detroit, MI. With permission.

4.7.4   Direct Design Method

The direct design method consists of a set of rules for the design of two-ways slabs with or without beams. Since the method was developed assuming simple designs and construction, its application is restricted by the code to two-way systems with a minimum of three continuous spans, successive span lengths that do not differ by more than one-third, columns with offset not more than 10% of the span, and all loads are due to gravity only and uniformly distributed with live load not exceeding three times dead load. The direct design method involves three fundamental steps: (1) determine the total factored static moment; (2) distribute the static moment to negative and positive sections; and (3) distribute moments to column and middle strips and to beams, if any. The total factored static moment M_o for a span bounded laterally by the centerlines of adjacent panels (see Figure 4.6) is given by

$$M_o = \frac{w_u l_2 l_n^2}{8} \tag{4.112}$$

In an interior span, $0.65\,M_o$ is assigned to each negative section and $0.35\,M_o$ is assigned to the positive section. In an end span, M_o is distributed according to Table 4.9. If the ratio of dead load to live load is less than 2, the effect of pattern loading is accounted for by increasing the positive moment following provisions in ACI Section 13.6.10. Negative and positive moments are then proportioned to the column strip following the percentages in Table 4.10, where β_t is the ratio of the torsional stiffness of edge beam section to flexural stiffness of a width of slab equal to span length of beam. The remaining moment not resisted by the column strip is proportionately assigned to the corresponding half middle strip. If beams are present, they are proportioned to resist 85% of column strip moments. When $(\alpha l_2/l_1)$ is less than 1.0, the proportion of column strip moments resisted by beams is obtained by linear interpolation between 85% and zero. The shear in beams is determined from loads acting on tributary areas projected from the panel corners at 45 degrees.

TABLE 4.9   Direct Design Method—Distribution of Moment in End Span

	(1)	(2)	(3)	(4)	(5)
		Slab with beams between all supports	Slab without beams between interior supports		Exterior edge fully restrained
	Exterior edge unrestrained		Without edge beam	With edge beam	
Interior negative-factored moment	0.75	0.70	0.70	0.70	0.65
Positive-factored moment	0.63	0.57	0.52	0.50	0.35
Exterior negative-factored moment	0	0.16	0.26	0.30	0.65

From ACI Committee 318. 1992. *Building Code Requirements for Reinforced Concrete and Commentary, ACI 318-89 (Revised 92) and ACI 318R-89 (Revised 92)*, Detroit, MI. With permission.

4.7.5   Equivalent Frame Method

For two-way systems not meeting the geometric or loading preconditions of the direct design method, design moments may be computed by the equivalent frame method. This is a more general method and involves the representation of the three-dimensional slab system by dividing it into a series of two-dimensional "equivalent" frames (Figure 4.6). The complete analysis of a two-way system consists of analyzing the series of equivalent interior and exterior frames that span longitudinally and transversely through the system. Each equivalent frame, which is centered on a column line and bounded by the center lines of the adjacent panels, comprises a horizontal slab-beam strip and equivalent columns extending above and below the slab beam (Figure 4.7). This structure

TABLE 4.10 Proportion of Moment to Column Strip in Percent

Interior negative-factored moment			
ℓ_2/ℓ_1	0.5	1.0	2.0
$(\alpha_1\ell_2/\ell_1) = 0$	75	75	75
$(\alpha_1\ell_2/\ell_1) \geq 1.0$	90	75	45
Positive-factored moment				
---	---	---	---	---
$(\alpha_1\ell_2/\ell_1) = 0$	$B_t = 0$	100	100	100
	$B_t \geq 2.5$	75	75	75
$(\alpha_1\ell_2/\ell_1) \geq 1.0$	$B_t = 0$	100	100	100
	$B_t = 2.5$	90	75	45
Exterior negative-factored moment			
$(\alpha_1\ell_2/\ell_1) = 0$	60	60	60
$(\alpha_1\ell_2/\ell_1) \geq 1.0$	90	75	45

From ACI Committee 318. 1992. *Building Code Requirements for Reinforced Concrete and Commentary, ACI 318-89 (Revised 92) and ACI 318R-89 (Revised 92)*, Detroit, MI. With permission.

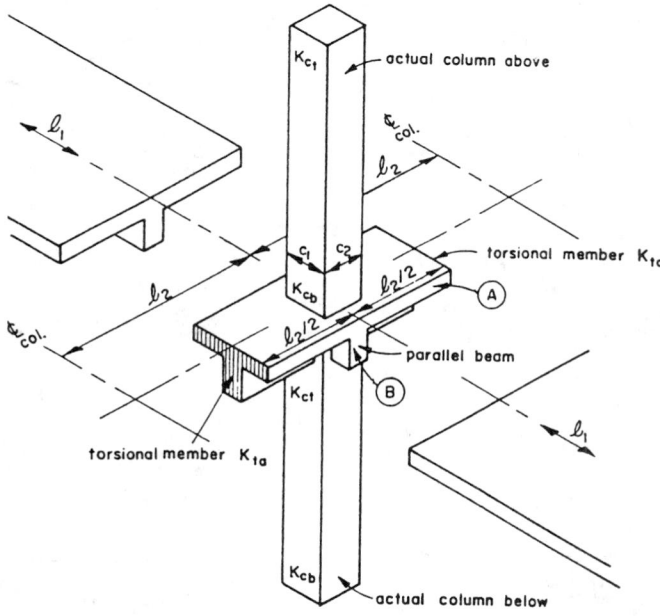

Figure 4.7 Equivalent column (columns plus torsional members).

is analyzed as a frame for loads acting in the plane of the frame, and the moments obtained at critical sections across the slab-beam strip are distributed to the column strip, middle strip, and beam in the same manner as the direct design method (see Table 4.10). In its original development, the equivalent frame method assumed that analysis would be done by moment distribution. Presently, frame analysis is more easily accomplished in design practice with computers using general purpose programs based on the direct stiffness method. Consequently, the equivalent frame method is now often used as a method for modeling a two-way system for computer analysis.

For the different types of two-way systems, the moment of inertias for modeling the slab-beam element of the equivalent frame are indicated in Figure 4.8. Moments of inertia of slab beams are based on the gross area of concrete; the variation in moment of inertia along the axis is taken into account, which in practice would mean that a node would be located on the computer model where a change of moment of inertia occurs. To account for the increased stiffness between the center

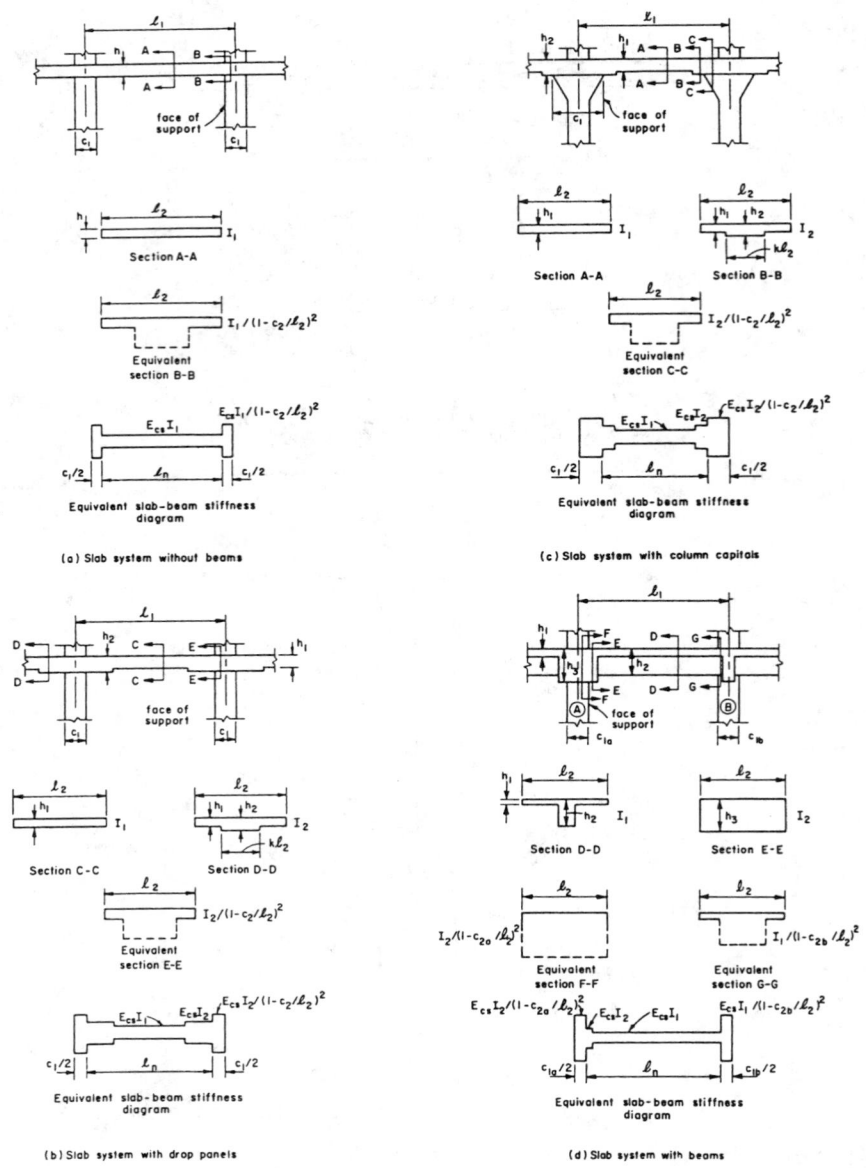

Figure 4.8   Slab-beam stiffness by equivalent frame method. (From ACI Committee 318. 1992. *Building Code Requirements for Reinforced Concrete and Commentary, ACI 318-89 (Revised 92) and ACI 318R-89 (Revised 92),* Detroit, MI. With permission.)

of the column and the face of column, beam, or capital, the moment of inertia is divided by the quantity $(1 - c_2/l_2)^2$, where c_2 and l_2 are measured transverse to the direction of the span. For column modeling, the moment of inertia at any cross-section outside of joints or column capitals may be based on the gross area of concrete, and the moment of inertia from the top to bottom of the slab-beam joint is assumed infinite.

Torsion members (Figure 4.7) are elements in the equivalent frame that provide moment transfer between the horizontal slab beam and vertical columns. The cross-section of torsional members are assumed to consist of the portion of slab and beam having a width according to the conditions

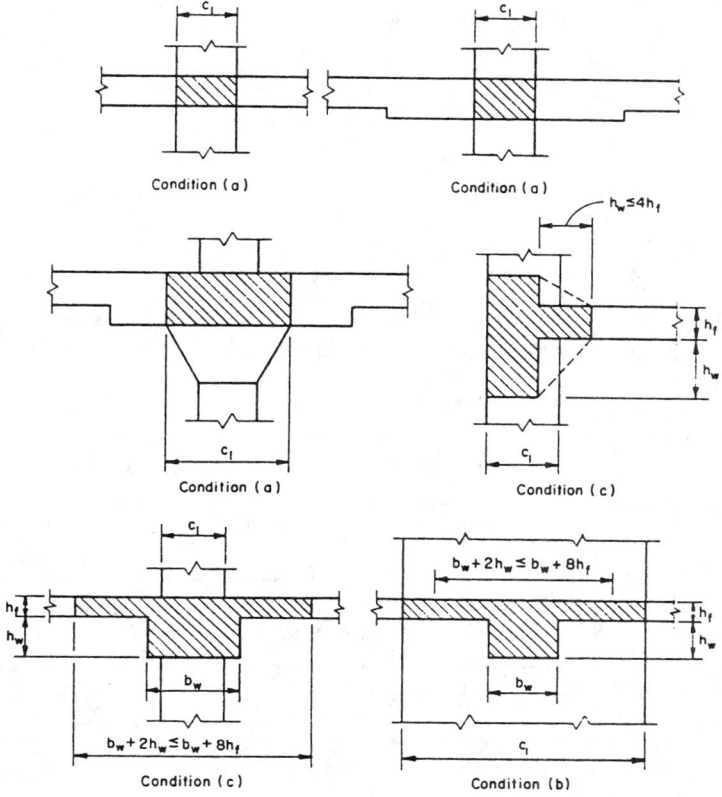

Figure 4.9 Torsional members. (From ACI Committee 318. 1992. *Building Code Requirements for Reinforced Concrete and Commentary, ACI 318-89 (Revised 92) and ACI 318R-89 (Revised 92)*, Detroit, MI. With permission.)

depicted in Figure 4.9. The stiffness K_t of the torsional member is calculated by the following expression:

$$K_t = \sum \frac{9E_{cs}C}{l_2\left(1 - \frac{c_2}{l_2}\right)^3} \tag{4.113}$$

where E_{cs} is the modulus of elasticity of the slab concrete and torsional constant C may be evaluated by dividing the cross-section into separate rectangular parts and carrying out the following summation:

$$C = \sum\left(1 - 0.63\frac{x}{y}\right)\frac{x^3 y}{3} \tag{4.114}$$

where x and y are the shorter and longer dimension, respectively, of each rectangular part. Where beams frame into columns in the direction of the span, the increased torsional stiffness K_{ta} is obtained by multiplying the value K_t obtained from Equation 4.113 by the ratio of (a) moment inertia of slab with such beam, to (b) moment of inertia of slab without such beam. Various ways have been suggested for incorporating torsional members into a computer model of an equivalent frame. The model implied by the ACI Code is one that has the slab beam connected to the torsional members, which are projected out of the plane of the columns. Others have suggested that the torsional members be replaced by rotational springs at column ends or, alternatively, at the slab-beam ends. Or, instead of rotational springs, columns may be modeled with an equivalent value of the moment of inertia modified by the equivalent column stiffness K_{ec} given in the commentary of the code. Using Figure 4.7, K_{ec} is computed as

$$K_{ec} = \frac{K_{ct} + K_{cb}}{1 + \dfrac{K_{ct} + K_{cb}}{K_{ta} + K_{ta}}} \qquad (4.115)$$

where K_{ct} and K_{cb} are the top and bottom flexural stiffnesses of the column.

4.7.6   Detailing

The ACI Code specifies that reinforcement in two-way slabs without beams have minimum extensions as prescribed in Figure 4.10. Where adjacent spans are unequal, extensions of negative moment

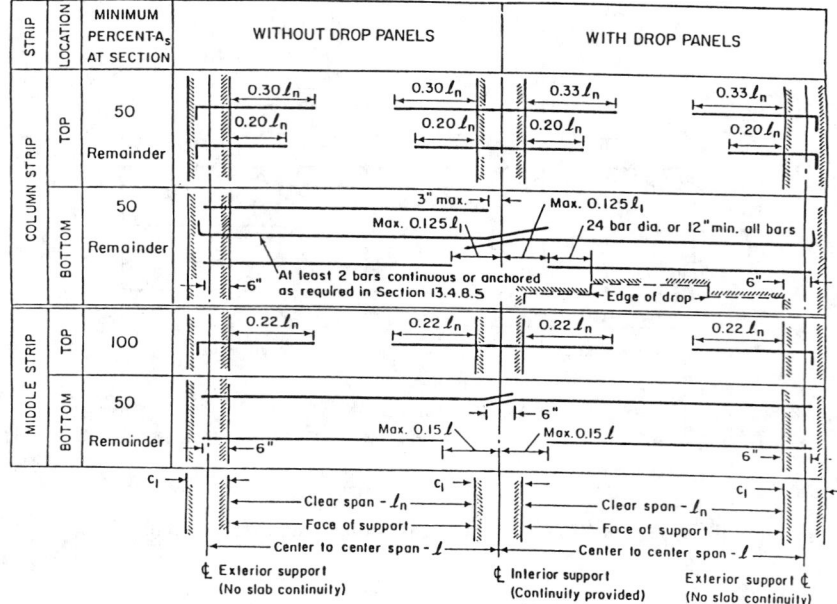

Figure 4.10   Minimum extensions for reinforcement in two-way slabs without beams. (From ACI Committee 318. 1992. *Building Code Requirements for Reinforced Concrete and Commentary, ACI 318-89 (Revised 92) and ACI 318R-89 (Revised 92)*, Detroit, MI. With permission.)

reinforcement shall be based on the longer span. Bent bars may be used only when the depth-span ratio permits use of bends 45 degrees or less. And at least two of the column strip bottom bars in each direction shall be continuous or spliced at the support with Class A splices or anchored within support. These bars must pass through the column and be placed within the column core. The purpose of this "integrity steel" is to give the slab some residual capacity following a single punching shear failure.

The ACI Code requires drop panels to extend in each direction from centerline of support a distance not less than one-sixth the span length, and the drop panel must project below the slab at least one-quarter of the slab thickness. The effective support area of a column capital is defined by the intersection of the bottom surface of the slab with the largest right circular cone whose surfaces are located within the column and capital and are oriented no greater than 45 degrees to the axis of the column.

4.8 Frames

A structural frame is a three-dimensional structural system consisting of straight members that are built monolithically and have rigid joints. The frame may be one bay long and one story high—such as portal frames and gable frames—or it may consist of multiple bays and stories. All members of frame are considered continuous in the three directions, and the columns participate with the beams in resisting external loads.

Consideration of the behavior of reinforced concrete frames at and near the ultimate load is necessary to determine the possible distributions of bending moment, shear force, and axial force that could be used in design. It is possible to use a distribution of moments and forces different from that given by linear elastic structural analysis if the critical sections have sufficient ductility to allow redistribution of actions to occur as the ultimate load is approached. Also, in countries that experience earthquakes, a further important design is the ductility of the structure when subjected to seismic-type loading, since present seismic design philosophy relies on energy dissipation by inelastic deformations in the event of major earthquakes.

4.8.1 Analysis of Frames

A number of methods have been developed over the years for the analysis of continuous beams and frames. The so-called classical methods—such as application of the theorem of three moments, the method of least work, and the general method of consistent deformation—have proved useful mainly in the analysis of continuous beams having few spans or of very simple frames. For the more complicated cases usually met in practice, such methods prove to be exceedingly tedious, and alternative approaches are preferred. For many years the closely related methods of slope deflection and moment distribution provided the basic analytical tools for the analysis of indeterminate concrete beams and frames. In offices with access to high-speed digital computers, these have been supplanted largely by matrix methods of analysis. Where computer facilities are not available, moment distribution is still the most common method. Approximate methods of analysis, based either on an assumed shape of the deformed structure or on moment coefficients, provide a means for rapid estimation of internal forces and moments. Such estimates are useful in preliminary design and in checking more exact solutions, and in structures of minor importance may serve as the basis for final design.

Slope Deflection

The method of slope deflection entails writing two equations for each member of a continuous frame, one at each end, expressing the end moment as the sum of four contributions: (1) the restraining moment associated with an assumed fixed-end condition for the loaded span, (2) the moment associated with rotation of the tangent to the elastic curve at the near end of the member, (3) the moment associated with rotation of the tangent at the far end of the member, and (4) the moment associated with translation of one end of the member with respect to the other. These equations are related through application of requirements of equilibrium and compatibility at the joints. A set of simultaneous, linear algebraic equations results for the entire structure, in which the structural displacements are unknowns. Solution for these displacements permits the calculation of all internal forces and moments.

This method is well suited to solving continuous beams, provided there are not very many spans. Its usefulness is extended through modifications that take advantage of symmetry and antisymmetry, and of hinge-end support conditions where they exist. However, for multistory and multibay frames in which there are a large number of members and joints, and which will, in general, involve translation as well as rotation of these joints, the effort required to solve the correspondingly large number of simultaneous equations is prohibitive. Other methods of analysis are more attractive.

Moment Distribution

The method of moment distribution was developed to solve problems in frame analysis that involve many unknown joint displacements. This method can be regarded as an iterative solution of the slope-deflection equations. Starting with fixed-end moments for each member, these are modified in a series of cycles, each converging on the precise final result, to account for rotation and translation of the joints. The resulting series can be terminated whenever one reaches the degree of accuracy required. After obtaining member-end moments, all member stress resultants can be obtained by use of the laws of statics.

Matrix Analysis

Use of matrix theory makes it possible to reduce the detailed numerical operations required in the analysis of an indeterminate structure to systematic processes of matrix manipulation which can be performed automatically and rapidly by computer. Such methods permit the rapid solution of problems involving large numbers of unknowns. As a consequence, less reliance is placed on special techniques limited to certain types of problems; powerful methods of general applicability have emerged, such as the matrix displacement method. Account can be taken of such factors as rotational restraint provided by members perpendicular to the plane of a frame. A large number of alternative loadings may be considered. Provided that computer facilities are available, highly precise analyses are possible at lower cost than for approximate analyses previously employed.

Approximate Analysis

In spite of the development of refined methods for the analysis of beams and frames, increasing attention is being paid to various approximate methods of analysis. There are several reasons for this. Prior to performing a complete analysis of an indeterminate structure, it is necessary to estimate the proportions of its members in order to know their relative stiffness upon which the analysis depends. These dimensions can be obtained using approximate analysis. Also, even with the availability of computers, most engineers find it desirable to make a rough check of results—using approximate means—to detect gross errors. Further, for structures of minor importance, it is often satisfactory to design on the basis of results obtained by rough calculation.

Provided that points of inflection (locations in members at which the bending moment is zero and there is a reversal of curvature of the elastic curve) can be located accurately, the stress resultants for a framed structure can usually be found on the basis of static equilibrium alone. Each portion of the structure must be in equilibrium under the application of its external loads and the internal stress resultants. The use of approximate analysis in determining stress resultants in frames is illustrated using a simple rigid frame in Figure 4.11.

ACI Moment Coefficients

The ACI Code [1] includes moment and shear coefficients that can be used for the analysis of buildings of usual types of construction, span, and story heights. They are given in ACI Code Sec. 8.3.3. The ACI coefficients were derived with due consideration of several factors: a maximum allowable ratio of live to dead load (3:1); a maximum allowable span difference (the larger of two adjacent spans not exceed the shorter by more than 20%); the fact that reinforced concrete beams are never simply supported but either rest on supports of considerable width, such as walls, or are built monolithically like columns; and other factors. Since all these influences are considered, the ACI coefficients are necessarily quite conservative, so that actual moments in any particular design are likely to be considerably smaller than indicated. Consequently, in many reinforced concrete structures, significant economy can be effected by making a more precise analysis.

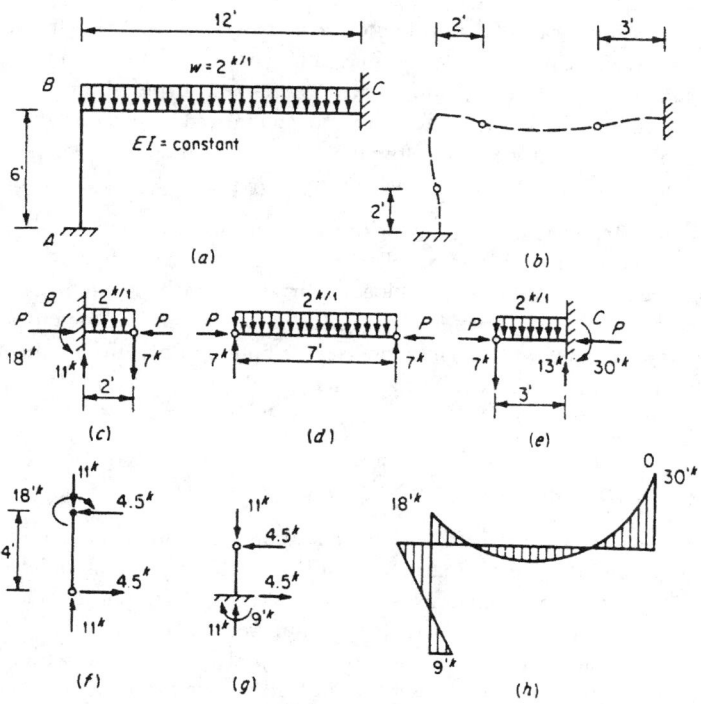

Figure 4.11 Approximate analysis of rigid frame. (From Nilson, A.H. and Winter, G. 1992. *Design of Concrete Structures,* 11th ed., MacGraw-Hill, New York. With permission.)

Limit Analysis

Limit analysis in reinforced concrete refers to the redistribution of moments that occurs through-out a structure as the steel reinforcement at a critical section reaches its yield strength. Under working loads, the distribution of moments in a statically indeterminate structure is based on elastic theory, and the whole structure remains in the elastic range. In limit design, where factored loads are used, the distribution of moments at failure when a mechanism is reached is different from that distribution based on elastic theory. The ultimate strength of the structure can be increased as more sections reach their ultimate capacity. Although the yielding of the reinforcement introduces large deflections, which should be avoided under service, a statically indeterminate structure does not collapse when the reinforcement of the first section yields. Furthermore, a large reserve of strength is present between the initial yielding and the collapse of the structure.

In steel design the term *plastic design* is used to indicate the change in the distribution of moments in the structure as the steel fibers, at a critical section, are stressed to their yield strength. Limit analysis of reinforced concrete developed as a result of earlier research on steel structures. Several studies had been performed on the principles of limit design and the rotation capacity of reinforced concrete plastic hinges.

Full utilization of the plastic capacity of reinforced concrete beams and frames requires an exten-sive analysis of all possible mechanisms and an investigation of rotation requirements and capacities at all proposed hinge locations. The increase of design time may not be justified by the limited gains obtained. On the other hand, a restricted amount of redistribution of elastic moments can safely be made without complete analysis and may be sufficient to obtain most of the advantages of limit analysis.

A limited amount of redistribution is permitted under the ACI Code, depending upon a rough measure of available ductility; without explicit calculation of rotation requirements and capacities.

The ratio ρ/ρ_b—or in the case of doubly reinforced members, $(\rho - \rho')/\rho_b$—is used as an indicator of rotation capacity, where ρ_b is the balanced steel ratio. For singly reinforced members with $\rho = \rho_b$, experiments indicate almost no rotation capacity, since the concrete strain is nearly equal to ε_{cu} when steel yielding is initiated. Similarly, in a doubly reinforced member, when $\rho - \rho' = \rho_b$, very little rotation will occur after yielding before the concrete crushes. However, when ρ or $\rho - \rho'$ is low, extensive rotation is usually possible. Accordingly, ACI Code Sec. 8.3 provides as follows:

> Except where approximate values for moments are used, it is permitted to increase or decrease negative moments calculated by elastic theory at supports of continuous flexural members for any assumed loading arrangement by not more than 20 $[1 - (\rho - \rho')/\rho_b]$ percent. The modified negative moments shall be used for calculating moments at sections within the spans. Redistribution of negative moments shall be made only when the section at which moment is reduced is so designed that ρ or $\rho - \rho'$ is not greater than 0.5 ρ_b [1992].

4.8.2 Design for Seismic Loading

The ACI Code contains provisions that are currently considered to be the minimum requirements for producing a monolithic concrete structure with adequate proportions and details to enable the structure to sustain a series of oscillations into the inelastic range of response without critical decay in strength. The provisions are intended to apply to reinforced concrete structures located in a seismic zone where major damage to construction has a high possibility of occurrence, and are designed with a substantial reduction in total lateral seismic forces due to the use of lateral load-resisting systems consisting of ductile moment-resisting frames. The provisions for frames are divided into sections on flexural members, columns, and joints of frames. Some of the important points stated are summarized below.

Flexural Members

Members having a factored axial force not exceeding $A_g f'_c/10$, where A_g is gross section of area (in.^{2}), are regarded as flexural members. An upper limit is placed on the flexural steel ratio ρ. The maximum value of ρ should not exceed 0.025. Provision is also made to ensure that a minimum quantity of top and bottom reinforcement is always present. Both the top and the bottom steel are to have a steel ratio of a least $200/f_y$, with the steel yield strength f_y in psi throughout the length of the member. Recommendations are also made to ensure that sufficient steel is present to allow for unforeseen shifts in the points of contraflexure. At column connections, the positive moment capacity should be at least 50% of the negative moment capacity, and the reinforcement should be continuous through columns where possible. At external columns, beam reinforcement should be terminated in the far face of the column using a hook plus any additional extension necessary for anchorage.

The design shear force V_e should be determined from consideration of the static forces on the portion of the member between faces of the joints. It should be assumed that moments of opposite sign corresponding to probable strength M_{pr} act at the joint faces and that the member is loaded with the factored tributary gravity load along its span. Figure 4.12 illustrates the calculation. Minimum web reinforcement is provided throughout the length of the member, and spacing should not exceed $d/4$ in plastic hinge zones and $d/2$ elsewhere, where d is effective depth of member. The stirrups should be closed around bars required to act as compression reinforcement and in plastic hinge regions, and the spacing should not exceed specified values.

Columns

Members having a factored axial force exceeding $A_g f'_c/10$ are regarded as columns of frames serving to resist earthquake forces. These members should satisfy the conditions that the shortest

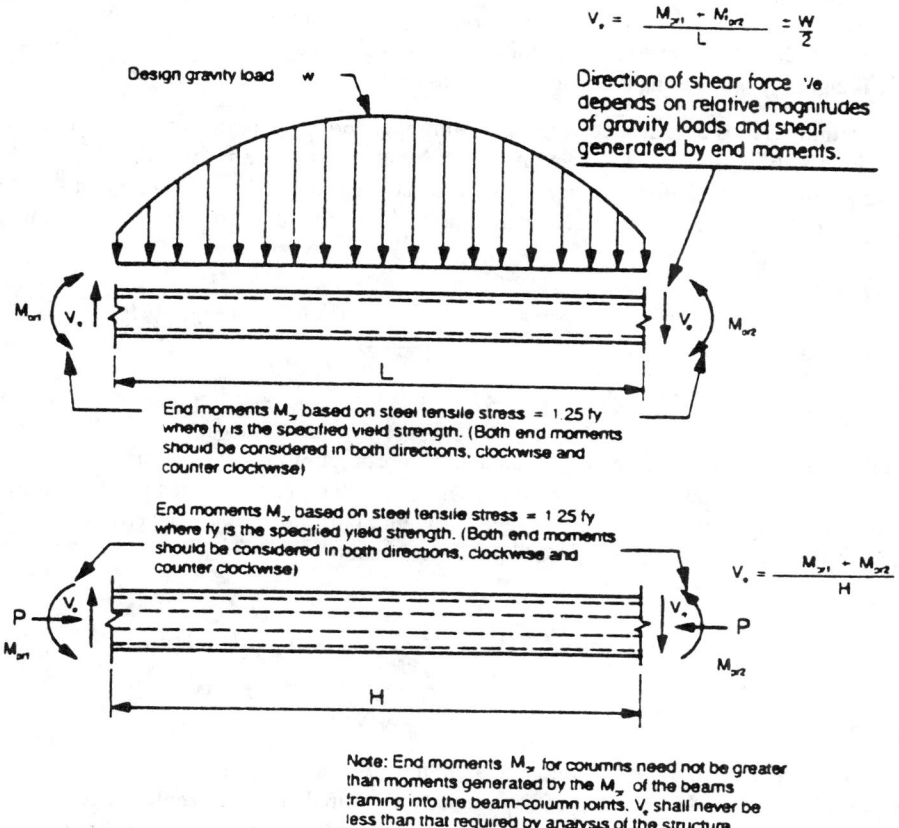

Figure 4.12 Design shears for girders and columns. (From ACI Committee 318. 1992. *Building Code Requirements for Reinforced Concrete and Commentary, ACI 318-89 (Revised 92) and ACI 318R-89 (Revised 92),* Detroit, MI. With permission.)

cross-sectional dimension—measured on a straight line passing through the geometric centroid—should not be less than 12 in. and that the ratio of the shortest cross-sectional dimension to the perpendicular dimension should not be less than 0.4. The flexural strengths of the columns should satisfy

$$\sum M_e \geq (6/5) \sum M_g \tag{4.116}$$

where $\sum M_e$ is sum of moments, at the center of the joint, corresponding to the design flexural strength of the columns framing into that joint and where $\sum M_g$ is sum of moments, at the center of the joint, corresponding to the design flexural strengths of the girders framing into that joint. Flexural strengths should be summed such that the column moments oppose the beam moments. Equation 4.116 should be satisfied for beam moments acting in both directions in the vertical plane of the frame considered. The requirement is intended to ensure that plastic hinges form in the girders rather than the columns.

The longitudinal reinforcement ratio is limited to the range of 0.01 to 0.06. The lower bound to the reinforcement ratio refers to the traditional concern for the effects of time-dependent deformations of the concrete and the desire to have a sizable difference between the cracking and yielding moments. The upper bound reflects concern for steel congestion, load transfer from floor elements to column in low-rise construction, and the development of large shear stresses. Lap splices are permitted only within the center half of the member length and should be proportioned as tension splices.

Welded splices and mechanical connections are allowed for splicing the reinforcement at any section, provided not more than alternate longitudinal bars are spliced at a section and the distance between splices is 24 in. or more along the longitudinal axis of the reinforcement.

If Equation 4.116 is not satisfied at a joint, columns supporting reactions from that joint should be provided with transverse reinforcement over their full height to confine the concrete and provide lateral support to the reinforcement. Where a spiral is used, the ratio of volume of spiral reinforcement to the core volume confined by the spiral reinforcement, ρ_s, should be at least that given by

$$\rho_s = 0.45 \frac{f_c'}{f_y} \left(\frac{A_g}{A_c} - 1 \right) \tag{4.117}$$

but not less than $0.12\, f_c'/f_{yh}$, where A_c is the area of core of spirally reinforced compression member measured to outside diameter of spiral in in.^{2} and f_{yh} is the specified yield strength of transverse reinforcement in psi. When rectangular reinforcement hoop is used, the total cross-sectional area of rectangular hoop reinforcement should not be less than that given by

$$A_{sh} = 0.3 \left(s h_c f_c'/f_{yh} \right) \left[\left(A_g/A_{ch} \right) - 1 \right] \tag{4.118}$$

$$A_{sh} = 0.09 s h_c f_c'/f_{yh} \tag{4.119}$$

where s is the spacing of transverse reinforcement measured along the longitudinal axis of column, h_c is the cross-sectional dimension of column core measured center-to-center of confining reinforcement, and A_{sh} is the total cross-sectional area of transverse reinforcement (including crossties) within spacing s and perpendicular to dimension h_c. Supplementary crossties, if used, should be of the same diameter as the hoop bar and should engage the hoop with a hook. Special transverse confining steel is required for the full height of columns that support discontinuous shear walls.

The design shear force V_e should be determined from consideration of the maximum forces that can be generated at the faces of the joints at each end of the column. These joint forces should be determined using the maximum probable moment strength M_{pr} of the column associated with the range of factored axial loads on the column. The column shears need not exceed those determined from joint strengths based on the probable moment strength M_{pr}, of the transverse members framing into the joint. In no case should V_e be less than the factored shear determined by analysis of the structure Figure 4.12.

Joints of Frames

Development of inelastic rotations at the faces of joints of reinforced concrete frames is associated with strains in the flexural reinforcement well in excess of the yield strain. Consequently, joint shear force generated by the flexural reinforcement is calculated for a stress of $1.25\, f_y$ in the reinforcement.

Within the depth of the shallowed framing member, transverse reinforcement equal to at least one-half the amount required for the column reinforcement should be provided where members frame into all four sides of the joint and where each member width is at least three-fourths the column width. Transverse reinforcement as required for the column reinforcement should be provided through the joint to provide confinement for longitudinal beam reinforcement outside the column core if such confinement is not provided by a beam framing into the joint.

The nominal shear strength of the joint should not be taken greater than the forces specified below for normal weight aggregate concrete:

$$20\sqrt{f'_c A_j} \quad \text{for joints confined on all four faces}$$

$$15\sqrt{f'_c A_j} \quad \text{for joints confined on three faces or on two opposite faces}$$

$$12\sqrt{f'_c A_j} \quad \text{for others}$$

where A_j is the effective cross-sectional area within a joint in a plane parallel to plane of reinforcement generating shear in the joint (Figure 4.13). A member that frames into a face is considered to provide

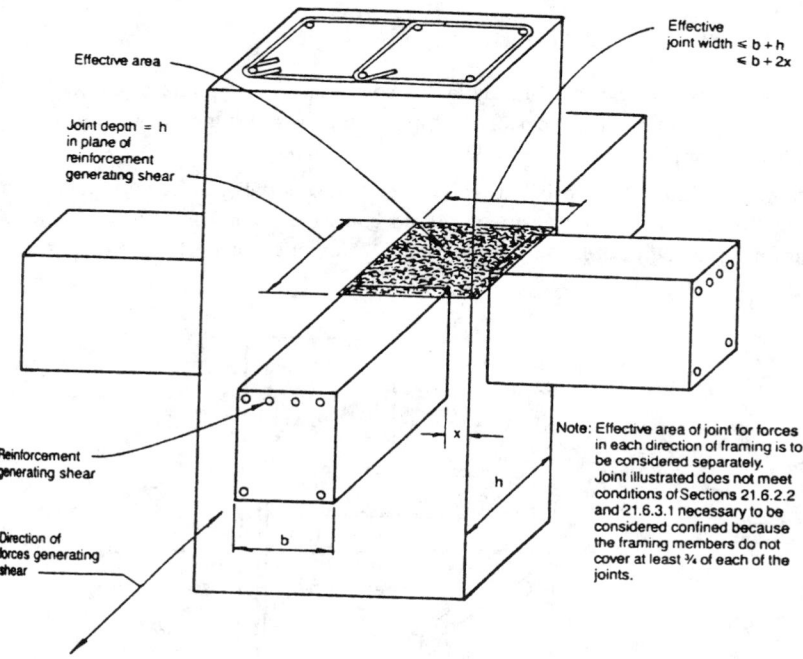

Figure 4.13 Effective area of joint. (From ACI Committee 318. 1992. *Building Code Requirements for Reinforced Concrete and Commentary, ACI 318-89 (Revised 92) and ACI 318R-89 (Revised 92)*, Detroit, MI. With permission.)

confinement to the joint if at least three-quarters of the face of the joint is covered by the framing member. A joint is considered to be confined if such confining members frame into all faces of the joint. For lightweight-aggregate concrete, the nominal shear strength of the joint should not exceed three-quarters of the limits given above.

Details of minimum development length for deformed bars with standard hooks embedded in normal and lightweight concrete and for straight bars are contained in ACI Code Sec. 21.6.4.

4.9 Brackets and Corbels

Brackets and corbels are cantilevers having shear span depth ratio, a/d, not greater than unity. The shear span a is the distance from the point of load to the face of support, and the distance d shall be measured at face of support (see Figure 4.14).

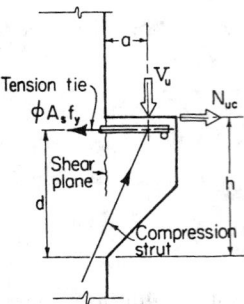

Figure 4.14   Structural action of a corbel. (From ACI Committee 318. 1992. *Building Code Requirements for Reinforced Concrete and Commentary, ACI 318-89 (Revised 92) and ACI 318R-89 (Revised 92)*, Detroit, MI. With permission.)

The corbel shown in Figure 4.14 may fail by shearing along the interface between the column and the corbel by yielding of the tension tie, by crushing or splitting of the compression strut, or by localized bearing or shearing failure under the loading plate.

The depth of a bracket or corbel at its outer edge should be less than one-half of the required depth d at the support. Reinforcement should consist of main tension bars with area A_s and shear reinforcement with area A_h (see Figure 4.15 for notation). The area of primary tension

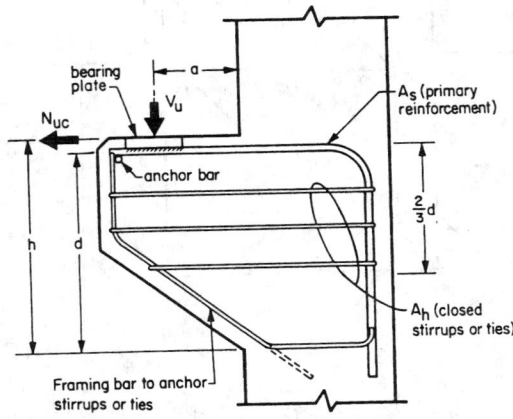

Figure 4.15   Notation used. (From ACI Committee 318. 1992. *Building Code Requirements for Reinforced Concrete and Commentary, ACI 318-89 (Revised 92) and ACI 318R-89 (Revised 92)*, Detroit, MI. With permission.)

reinforcement A_s should be made equal to the greater of $(A_f + A_n)$ or $(2A_{vf}/3 - A_n)$, where A_f is the flexural reinforcement required to resist moment $[V_u a + N_{uc}(h - d)]$, A_n is the reinforcement required to resist tensile force N_{uc}, and A_{vf} is the shear-friction reinforcement required to resist shear V_u:

$$A_f = \frac{M_u}{\phi f_y jd} = \frac{V_u a + N_{uc}(h - d)}{\phi f_y jd} \tag{4.120}$$

$$A_n = \frac{N_{uc}}{\phi f_y} \tag{4.121}$$

$$A_{vf} = \frac{V_u}{\phi f_y \mu} \qquad\qquad (4.122)$$

In the above equations, f_y is the reinforcement yield strength; ϕ is 0.9 for Equation 4.120 and 0.85 for Equations 4.121 and 4.122. In Equation 4.120, the lever arm jd can be approximated for all practical purposes in most cases as $0.85d$. Tensile force N_{uc} in Equation 4.121 should not be taken less than $0.2 V_u$ unless special provisions are made to avoid tensile forces. Tensile force N_{uc} should be regarded as a live load even when tension results from creep, shrinkage, or temperature change. In Equation 4.122, $V_u/\phi (= V_n)$ should not be taken greater than $0.2 f_c' b_w d$ nor $800 b_w d$ in pounds in normal-weight concrete. For "all-lightweight" or "sand-lightweight" concrete, shear strength V_n should not be taken greater than $(0.2 - 0.07a/d)f_c' b_w d$ nor $(800 - 280a/d)b_w d$ in pounds. The coefficient of friction μ in Equation 4.122 should be 1.4λ for concrete placed monolithically, 1.0λ for concrete placed against hardened concrete with surface intentionally roughened, 0.6λ for concrete placed against hardened concrete not intentionally roughened, and 0.7λ for concrete anchored to as-rolled structural steel by headed studs or by reinforcing bars, where λ is 1.0 for normal weight concrete, 0.85 for "sand-lightweight" concrete, and 0.75 for "all-lightweight" concrete. Linear interpolation of λ is permitted when partial sand replacement is used.

The total area of closed stirrups or ties A_h parallel to A_s should not be less than $0.5(A_s - A_n)$ and should be uniformly distributed within two-thirds of the depth of the bracket adjacent to A_s.

At front face of bracket or corbel, primary tension reinforcement A_s should be anchored in one of the following ways: (a) by a structural weld to a transverse bar of at least equal size; weld to be designed to develop specified yield strength f_y of A_s bars; (b) by bending primary tension bars A_s back to form a horizontal loop, or (c) by some other means of positive anchorage. Also, to ensure development of the yield strength of the reinforcement A_s near the load, bearing area of load on bracket or corbel should not project beyond straight portion of primary tension bars A_s, nor project beyond interior face of transverse anchor bar (if one is provided). When corbels are designed to resist horizontal forces, the bearing plate should be welded to the tension reinforcement A_s.

4.10 Footings

Footings are structural members used to support columns and walls and to transmit and distribute their loads to the soil in such a way that (a) the load bearing capacity of the soil is not exceeded, (b) excessive settlement, differential settlement, and rotations are prevented, and (c) adequate safety against overturning or sliding is maintained. When a column load is transmitted to the soil by the footing, the soil becomes compressed. The amount of settlement depends on many factors, such as the type of soil, the load intensity, the depth below ground level, and the type of footing. If different footings of the same structure have different settlements, new stresses develop in the structure. Excessive differential settlement may lead to the damage of nonstructural members in the buildings, even failure of the affected parts.

Vertical loads are usually applied at the centroid of the footing. If the resultant of the applied loads does not coincide with the centroid of the bearing area, a bending moment develops. In this case, the pressure on one side of the footing will be greater than the pressure on the other side, causing higher settlement on one side and a possible rotation of the footing.

If the bearing soil capacity is different under different footings—for example, if the footings of a building are partly on soil and partly on rock—a differential settlement will occur. It is customary in such cases to provide a joint between the two parts to separate them, allowing for independent settlement.

4.10.1 Types of Footings

Different types of footings may be used to support building columns or walls. The most commonly used ones are illustrated in Figure 4.16(a–g). A simple file footing is shown in Figure 4.16(h).

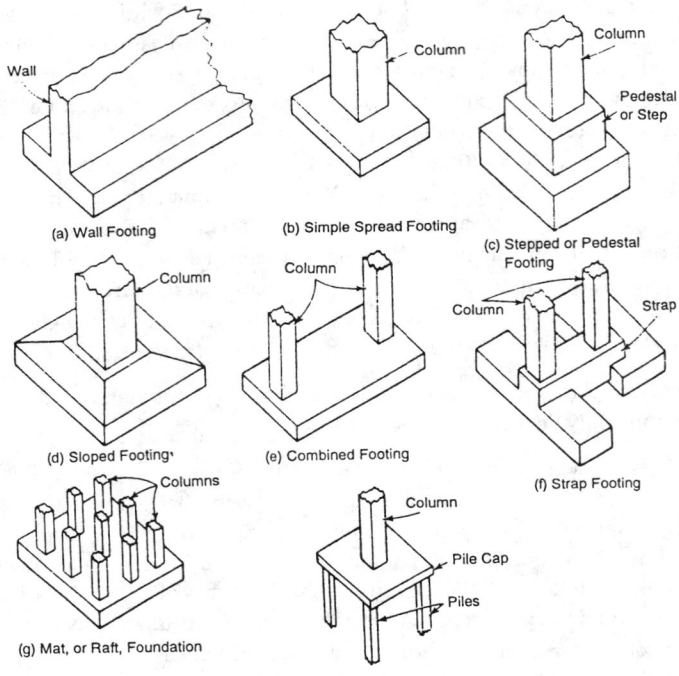

Figure 4.16   Common types of footings for walls and columns. (From ACI Committee 340. 1990. *Design Handbook in Accordance with the Strength Design Method of ACI 318-89.* Volume 2, SP-17. With permission.)

   For walls, a spread footing is a slab wider than the wall and extending the length of the wall [Figure 4.16(a)]. Square or rectangular slabs are used under single columns [Figure 4.16(b–d)]. When two columns are so close that their footings would merge or nearly touch, a combined footing [Figure 4.16(e)] extending under the two should be constructed. When a column footing cannot project in one direction, perhaps because of the proximity of a property line, the footing may be helped out by an adjacent footing with more space; either a combined footing or a strap (cantilever) footing [Figure 4.16(f)] may be used under the two.

   For structures with heavy loads relative to soil capacity, a mat or raft foundation [Figure 4.16(g)] may prove economical. A simple form is a thick, two-way-reinforced-concrete slab extending under the entire structure. In effect, it enables the structure to float on the soil, and because of its rigidity it permits negligible differential settlement. Even greater rigidity can be obtained by building the raft foundation as an inverted beam-and-girder floor, with the girders supporting the columns. Sometimes, also, inverted flat slabs are used as mat foundations.

4.10.2 Design Considerations

Footings must be designed to carry the column loads and transmit them to the soil safely while satisfying code limitations. The design procedure must take the following strength requirements into consideration:

- The area of the footing based on the allowable bearing soil capacity

- Two-way shear or punching shear

- One-way shear

- Bending moment and steel reinforcement required

- Dowel requirements

- Development length of bars

- Differential settlement

These strength requirements will be explained in the following sections.

Size of Footings

The required area of concentrically loaded footings is determined from

$$A_{\text{req}} = \frac{D + L}{q_a} \qquad (4.123)$$

where q_a is allowable bearing pressure and D and L are, respectively, unfactored dead and live loads. Allowable bearing pressures are established from principles of soil mechanics on the basis of load tests and other experimental determinations. Allowable bearing pressures q_a under service loads are usually based on a safety factor of 2.5 to 3.0 against exceeding the ultimate bearing capacity of the particular soil and to keep settlements within tolerable limits. The required area of footings under the effects of wind W or earthquake E is determined from the following:

$$A_{\text{req}} = \frac{D + L + W}{1.33 q_a} \quad \text{or} \quad \frac{D + L + E}{1.33 q_a} \qquad (4.124)$$

It should be noted that footing sizes are determined for unfactored service loads and soil pressures, in contrast to the strength design of reinforced concrete members, which utilizes factored loads and factored nominal strengths.

A footing is eccentrically loaded if the supported column is not concentric with the footing area or if the column transmits—at its juncture with the footing—not only a vertical load but also a bending moment. In either case, the load effects at the footing base can be represented by the vertical load P and a bending moment M. The resulting bearing pressures are again assumed to be linearly distributed. As long as the resulting eccentricity $e = M/P$ does not exceed the kern distance k of the footing area, the usual flexure formula

$$q_{\text{max, min}} = \frac{P}{A} + \frac{M_c}{I} \qquad (4.125)$$

permits the determination of the bearing pressures at the two extreme edges, as shown in Figure 4.17(a). The footing area is found by trial and error from the condition $q_{\text{max}} \leq q_a$. If the eccentricity falls outside the kern, Equation 4.125 gives a negative value for q along one edge of the footing. Because no tension can be transmitted at the contact area between soil and footing, Equation 4.125 is no longer valid and bearing pressures are distributed as in Figure 4.17(b).

Once the required footing area has been determined, the footing must then be designed to develop the necessary strength to resist all moments, shears, and other internal actions caused by the applied loads. For this purpose, the load factors of the ACI Code apply to footings as to all other structural components.

Depth of footing above bottom reinforcement should not be less than 6 in. for footings on soil, nor less than 12 in. for footings on piles.

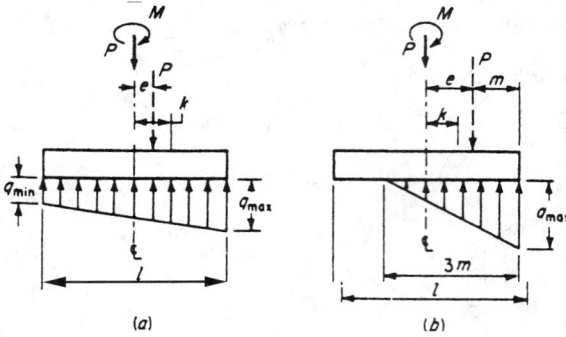

Figure 4.17   Assumed bearing pressures under eccentric footings. (From Wang, C.-K. and Salmon, C.G. 1985. *Reinforced Concrete Design,* 4th ed., Harper Row, New York. With permission.)

Two-Way Shear (Punching Shear)

ACI Code Sec. 11.12.2 allows a shear strength V_c of footings without shear reinforcement for two-way shear action as follows:

$$V_c = \left(2 + \frac{4}{\beta_c}\right)\sqrt{f_c'}b_o d \leq 4\sqrt{f_c'}b_o d \tag{4.126}$$

where β_c is the ratio of long side to short side of rectangular area, b_o is the perimeter of the critical section taken at $d/2$ from the loaded area (column section), and d is the effective depth of footing. This shear is a measure of the diagonal tension caused by the effect of the column load on the footing. Inclined cracks may occur in the footing at a distance $d/2$ from the face of the column on all sides. The footing will fail as the column tries to punch out part of the footing, as shown in Figure 4.18.

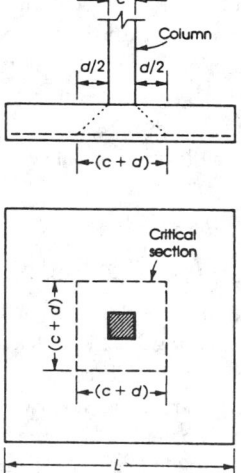

Figure 4.18   Punching shear (two-way). (From MacGregor, J.G. 1992. *Reinforced Concrete Mechanics and Design,* 2nd ed., Prentice-Hall, Englewood Cliffs, NJ. With permission.)

One-Way Shear

For footings with bending action in one direction, the critical section is located at a distance d from the face of the column. The diagonal tension at section m-m in Figure 4.19 can be checked as is done in beams. The allowable shear in this case is equal to

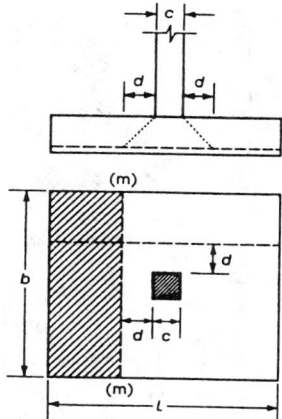

Figure 4.19 One-way shear. (From MacGregor, J.G. 1992. *Reinforced Concrete Mechanics and Design,* 2nd ed., Prentice-Hall, Englewood Cliffs, NJ. With permission.)

$$\phi V_c = 2\phi\sqrt{f_c'}bd \qquad (4.127)$$

where b is the width of section m-m. The ultimate shearing force at section m-m can be calculated as follows:

$$V_u = q_u b \left(\frac{L}{2} - \frac{c}{2} - d \right) \qquad (4.128)$$

where b is the side of footing parallel to section m-m.

Flexural Reinforcement and Footing Reinforcement

The theoretical sections for moment occur at face of the column (section n-n, Figure 4.20). The bending moment in each direction of the footing must be checked and the appropriate reinforcement must be provided. In square footings the bending moments in both directions are equal. To determine the reinforcement required, the depth of the footing in each direction may be used. As the bars in one direction rest on top of the bars in the other direction, the effective depth d varies with the diameter of the bars used. The value of d_{min} may be adopted.

The depth of footing is often controlled by the shear, which requires a depth greater than that required by the bending moment. The steel reinforcement in each direction can be calculated in the case of flexural members as follows:

$$A_s = \frac{M_u}{\phi f_y (d - a/2)} \qquad (4.129)$$

The minimum steel percentage requirement in flexural member is equal to $200/f_y$. However, ACI Code Sec. 10.5.3 indicates that for structural slabs of uniform thickness, the minimum area and maximum spacing of steel in the direction of bending should be as required for shrinkage and temperature reinforcement. This last minimum steel reinforcement is very small and a higher minimum reinforcement ratio is recommended, but not greater than $200/f_y$.

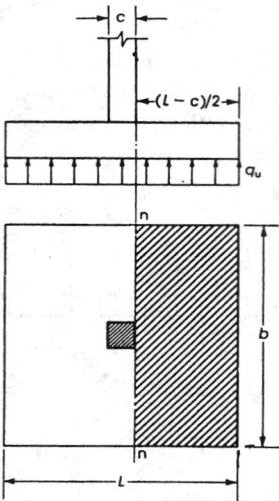

Figure 4.20   Critical section of bending moment. (From MacGregor, J.G. 1992. *Reinforced Concrete Mechanics and Design,* 2nd ed., Prentice-Hall, Englewood Cliffs, NJ. With permission.)

The reinforcement in one-way footings and two-way footings must be distributed across the entire width of the footing. In the case of two-way rectangular footings, ACI Code Sec 15.4.4 specifies that in the long direction the total reinforcement must be placed uniformly within a band width equal to the length of the short side of the footing according to

$$\frac{\text{Reinforcement band width}}{\text{Total reinforcement in short direction}} = \frac{2}{\beta + 1} \tag{4.130}$$

where β is the ratio of the long side to the short side of the footing. The band width must be centered on the centerline of the column (Figure 4.21). The remaining reinforcement in the short direction must be uniformly distributed outside the band width. This remaining reinforcement percentage should not be less than required for shrinkage and temperature.

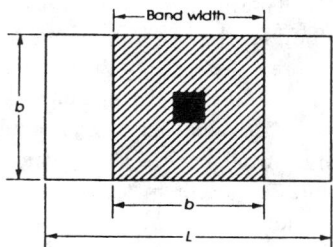

Figure 4.21   Band width for reinforcement distribution. (From MacGregor, J.G. 1992. *Reinforced Concrete Mechanics and Design,* 2nd ed., Prentice-Hall, Englewood Cliffs, NJ. With permission.)

When structural steel columns or masonry walls are used, the critical sections for moments in footing are taken at halfway between the middle and the edge of masonry walls, and halfway between the face of the column and the edge of the steel base place (ACI Code Sec. 15.4.2).

Bending Capacity of Column at Base

The loads from the column act on the footing at the base of the column, on an area equal to the area of the column cross-section. Compressive forces are transferred to the footing directly by bearing on the concrete. Tensile forces must be resisted by reinforcement, neglecting any contribution by concrete.

Forces acting on the concrete at the base of the column must not exceed the bearing strength of concrete as specified by the ACI Code Sec.10.15:

$$N = \phi \left(0.85 f'_c A_1\right) \qquad (4.131)$$

where ϕ is 0.7 and A_1 is the bearing area of the column. The value of the bearing strength given in Equation 4.131 may be multiplied by a factor $\sqrt{A_2/A_1} \leq 2.0$ for bearing on footings when the supporting surface is wider on all sides other than the loaded area. Here A_2 is the area of the part of the supporting footing that is geometrically similar to and concentric with the load area (Figure 4.22). Since $A_2 > A_1$, the factor $\sqrt{A_2/A_1}$ is greater than unity, indicating that the allowable

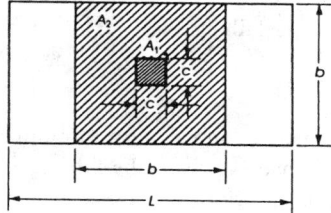

Figure 4.22 Bearing areas on footings. $A_1 = c^2$, $A_2 = b^2$. (From MacGregor, J.G. 1992. *Reinforced Concrete Mechanics and Design*, 2nd ed., Prentice-Hall, Englewood Cliffs, NJ. With permission.)

bearing strength is increased because of the lateral support from the footing area surrounding the column base. If the calculated bearing force is greater than N or the modified one with $r\sqrt{A_2/A_1}$, reinforcement must be provided to transfer the excess force. This is achieved by providing dowels or extending the column bars into the footing. If the calculated bearing force is less than either N or the modified one with $r\sqrt{A_2/A_1}$, then minimum reinforcement must be provided. ACI Code Sec. 15.8.2 indicates that the minimum area of the dowel reinforcement is at least $0.005A_g$ but not less than 4 bars, where A_g is the gross area of the column section of the supported member. The minimum reinforcement requirements apply to the case in which the calculated bearing forces are greater than N or the modified one with $r\sqrt{A_2/A_1}$.

Dowels on Footings

It was explained earlier that dowels are required in any case, even if the bearing strength is adequate. The ACI Code specifies a minimum steel ratio $\rho = 0.005$ of the column section as compared to $\rho = 0.01$ as minimum reinforcement for the column itself. The minimum number of dowel bars needed is four; these may be placed at the four corners of the column. The dowel bars are usually extended into the footing, bent at their ends, and tied to the main footing reinforcement.

ACI Code Sec. 15.8.2 indicates that #14 and #18 longitudinal bars, in compression only, may be lap-spliced with dowels. Dowels should not be larger than #11 bar and should extend (1) into supported member a distance not less than the development length of #14 or 18" bars or the splice length of the dowels—whichever is greater, and (2) into the footing a distance not less than the development length of the dowels.

Development Length of the Reinforcing Bars

The critical sections for checking the development length of reinforcing bars are the same as those for bending moments. Calculated tension or compression in reinforcement at each section should be developed on each side of that section by embedment length, hook (tension only) or mechanical device, or a combination thereof. The development length for compression bar is

$$l_d = 0.02 f_y d_b \sqrt{f_c'} \tag{4.132}$$

but not less than $0.0003 f_y d_b \geq 8$ in. For other values, refer to ACI Code, Chapter 12. Dowel bars must also be checked for proper development length.

Differential Settlement

Footings usually support the following loads:

- Dead loads from the substructure and superstructure

- Live loads resulting from materials or occupancy

- Weight of materials used in backfilling

- Wind loads

Each footing in a building is designed to support the maximum load that may occur on any column due to the critical combination of loadings, using the allowable soil pressure.

The dead load, and maybe a small portion of the live load, may act continuously on the structure. The rest of the live load may occur at intervals and on some parts of the structure only, causing different loadings on columns. Consequently, the pressure on the soil under different loadings will vary according to the loads on the different columns, and differential settlement will occur under the various footings of one structure. Since partial settlement is inevitable, the problem is defined by the amount of differential settlement that the structure can tolerate. The amount of differential settlement depends on the variation in the compressibility of the soils, the thickness of compressible material below foundation level, and the stiffness of the combined footing and superstructure. Excessive differential settlement results in cracking of concrete and damage to claddings, partitions, ceilings, and finishes.

For practical purposes it can be assumed that the soil pressure under the effect of sustained loadings is the same for all footings, thus causing equal settlements. The sustained load (or the usual load) can be assumed equal to the dead load plus a percentage of the live load, which occurs very frequently on the structure. Footings then are proportioned for these sustained loads to produce the same soil pressure under all footings. In no case is the allowable soil bearing capacity to be exceeded under the dead load plus the maximum live load for each footing.

4.10.3 Wall Footings

The spread footing under a wall [Figure 4.16(a)] distributes the wall load horizontally to preclude excessive settlement. The wall should be so located on the footings as to produce uniform bearing pressure on the soil (Figure 4.23), ignoring the variation due to bending of the footing. The pressure is determined by dividing the load per foot by the footing width.

The footing acts as a cantilever on opposite sides of the wall under downward wall loads and upward soil pressure. For footings supporting concrete walls, the critical section for bending moment is at the face of the wall; for footings under masonry walls, halfway between the middle and edge of the wall. Hence, for a one-foot-long strip of symmetrical concrete-wall footing, symmetrically

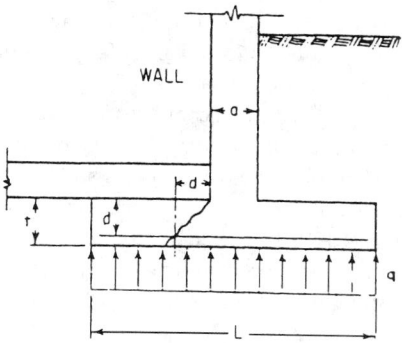

Figure 4.23 Reinforced-concrete wall footing. (From Wang, C.-K. and Salmon, C.G. 1985. *Reinforced Concrete Design*, 4th ed., Harper Row, New York. With permission.)

loaded, the maximum moment, ft-lb, is

$$M_u = \frac{1}{8} q_u (L - a)^2 \qquad (4.133)$$

where q_u is the uniform pressure on soil (lb/ft^2), L is the width of footing (ft), and a is wall thickness (ft).

For determining shear stresses, the vertical shear force is computed on the section located at a distance d from the face of the wall. Thus,

$$V_u = q_u \left(\frac{L - a}{2} - L \right) \qquad (4.134)$$

The calculation of development length is based on the section of maximum moment.

4.10.4 Single-Column Spread Footings

The spread footing under a column [Figure 4.16(b–d)] distributes the column load horizontally to prevent excessive total and differential settlement. The column should be located on the footing so as to produce uniform bearing pressure on the soil, ignoring the variation due to bending of the footing. The pressure equals the load divided by the footing area.

In plan, single-column footings are usually square. Rectangular footings are used if space restrictions dictate this choice or if the supported columns are of strongly elongated rectangular cross-section. In the simplest form, they consist of a single slab [Figure 4.16(b)]. Another type is that of Figure 4.16(c), where a pedestal or cap is interposed between the column and the footing slab; the pedestal provides for a more favorable transfer of load and in many cases is required in order to provide the necessary development length for dowels. This form is also known as a *stepped footing*. All parts of a stepped footing must be poured in a single pour in order to provide monolithic action. Sometimes sloped footings like those in Figure 4.16(d) are used. They requires less concrete than stepped footings, but the additional labor necessary to produce the sloping surfaces (formwork, etc.) usually makes stepped footings more economical. In general, single-slab footings [Figure 4.16(b)] are most economical for thicknesses up to 3 ft.

The required bearing area is obtained by dividing the total load, including the weight of the footing, by the selected bearing pressure. Weights of footings, at this stage, must be estimated and usually amount to 4 to 8% of the column load, the former value applying to the stronger types of soils.

Once the required footing area has been established, the thickness h of the footing must be determined. In single footings the effective depth d is mostly governed by shear. Two different types of shear strength are distinguished in single footings: two-way (or punching) shear and one-way (or beam) shear. Based on the Equations 4.126 and 4.127 for punching and one-way shear strength, the required effective depth of footing d is calculated.

Single-column footings represent, as it were, cantilevers projecting out from the column in both directions and loaded upward by the soil pressure. Corresponding tension stresses are caused in both these directions at the bottom surface. Such footings are therefore reinforced by two layers of steel, perpendicular to each other and parallel to the edge. The steel reinforcement in each direction can be calculated using Equation 4.129. The critical sections for development length of footing bars are the same as those for bending. Development length may also have to be checked at all vertical planes in which changes of section or of reinforcement occur, as at the edges of pedestals or where part of the reinforcement may be terminated.

When a column rests on a footing or pedestal, it transfers its load to only a part of the total area of the supporting member. The adjacent footing concrete provides lateral support to the directly loaded part of the concrete. This causes triaxial compression stresses that increase the strength of the concrete, which is loaded directly under the column. The design bearing strength of concrete must not exceed the one given in Equation 4.131 for forces acting on the concrete at the base of column and the modified one with $r\sqrt{A_2/A_1}$ for supporting area wider than the loaded area. If the calculated bearing force is greater than the design bearing strength, reinforcement must be provided to transfer the excess force. This is done either by extending the column bars into the footing or by providing dowels, which are embedded in the footing and project above it.

4.10.5 Combined Footings

Spread footings that support more than one column or wall are known as *combined footings*. They can be divided into two categories: those that support two columns, and those that support more than two (generally large numbers of) columns.

In buildings where the allowable soil pressure is large enough for single footings to be adequate for most columns, two-column footings are seen to become necessary in two situations: (1) if columns are so close to the property line that single-column footings cannot be made without projecting beyond that line, and (2) if some adjacent columns are so close to each other that their footings would merge.

When the bearing capacity of the subsoil is low so that large bearing areas become necessary, individual footings are replaced by continuous strip footings, which support more than two columns and usually all columns in a row. Mostly, such strips are arranged in both directions, in which case a grid foundation is obtained, as shown in Figure 4.24. Such a grid foundation can be done by single footings because the individual strips of the grid foundation represent continuous beams whose moments are much smaller than the cantilever moments in large single footings that project far out from the column in all four directions.

For still lower bearing capacities, the strips are made to merge, resulting in a mat foundation, as shown in Figure 4.25. That is, the foundation consists of a solid reinforced concrete slab under the entire building. In structural action such a mat is very similar to a flat slab or a flat plate, upside down—that is, loaded upward by the bearing pressure and downward by the concentrated column reactions. The mat foundation evidently develops the maximum available bearing area under the building. If the soil's capacity is so low that even this large bearing area is insufficient, some form of deep foundation, such as piles or caissons, must be used.

Grid and mat foundations may be designed with the column pedestals—as shown in Figures 4.24 and 4.25—or without them, depending on whether or not they are necessary for shear strength and the development length of dowels. Apart from developing large bearing areas, another advantage of grid and mat foundations is that their continuity and rigidity help in reducing differential settlements of individual columns relative to each other, which may otherwise be caused by local variations in the quality of subsoil, or other causes. For this purpose, continuous spread foundations are frequently

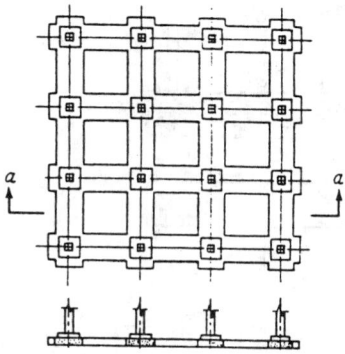

Figure 4.24 Grid foundation. (From Wang, C.-K. and Salmon, C.G. 1985. *Reinforced Concrete Design,* 4th ed., Harper Row, New York. With permission.)

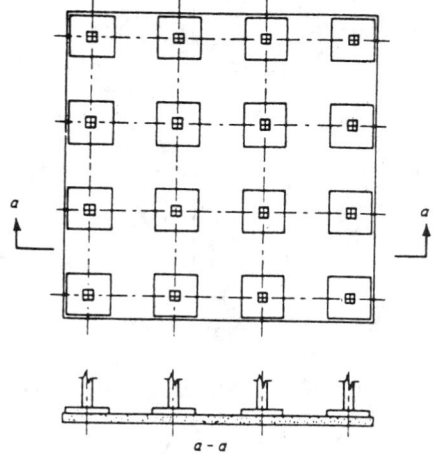

Figure 4.25 Mat foundation. (From Wang, C.-K. and Salmon, C.G. 1985. *Reinforced Concrete Design,* 4th ed., Harper Row, New York. With permission.)

used in situations where the superstructure or the type of occupancy provides unusual sensitivity to differential settlement.

4.10.6   Two-Column Footings

The ACI Codes does not provide a detailed approach for the design of combined footings. The design, in general, is based on an empirical approach. It is desirable to design combined footings so that the centroid of the footing area coincides with the resultant of the two column loads. This produces uniform bearing pressure over the entire area and forestalls a tendency for the footings to tilt. In plan, such footings are rectangular, trapezoidal, or T shaped, the details of the shape being arranged to produce coincidence of centroid and resultant. The simple relationships of Figure 4.26 facilitate the determination of the shapes of the bearing area [7]. In general, the distances m and n are given, the former being the distance from the center of the exterior column to the property line and the latter the distance from that column to the resultant of both column loads.

Another expedient, which is used if a single footing cannot be centered under an exterior column, is to place the exterior column footing eccentrically and to connect it with the nearest interior column by a beam or strap. This strap, being counterweighted by the interior column load, resists the tilting

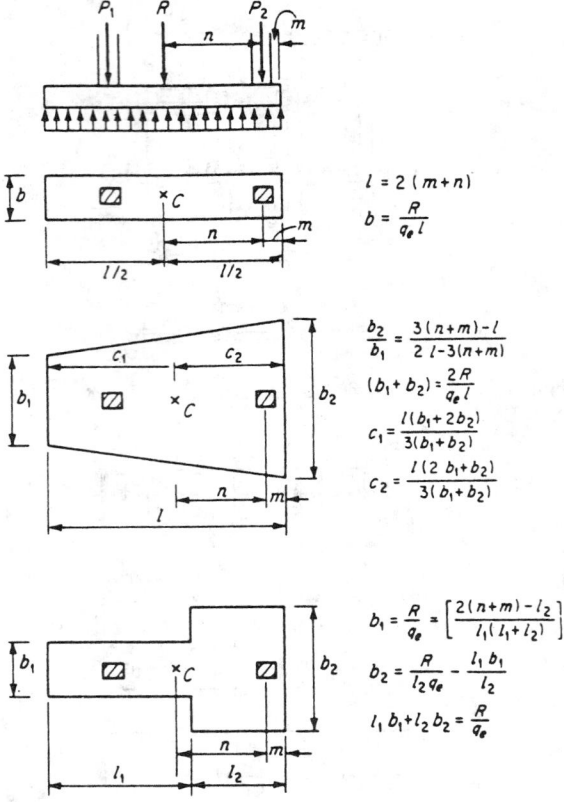

Figure 4.26 Two-column footings. (From Fintel, M. 1985. *Handbook of Concrete Engineering,* 2nd ed., Van Nostrand Reinhold, New York. With permission.)

tendency of the eccentric exterior footings and equalizes the pressure under it. Such foundations are known as *strap, cantilever,* or *connected footings.*

The strap may be designed as a rectangular beam spacing between the columns. The loads on it include its own weight (when it does not rest on the soil) and the upward pressure from the footings. Width of the strap usually is selected arbitrarily as equal to that of the largest column plus 4 to 8 inches so that column forms can be supported on top of the strap. Depth is determined by the maximum bending moment. The main reinforcing in the strap is placed near the top. Some of the steel can be cut off where not needed. For diagonal tension, stirrups normally will be needed near the columns (Figure 4.27). In addition, longitudinal placement steel is set near the bottom of the strap, plus reinforcement to guard against settlement stresses.

The footing under the exterior column may be designed as a wall footing. The portions on opposite sides of the strap act as cantilevers under the constant upward pressure of the soil. The interior footing should be designed as a single-column footing. The critical section for punching shear, however, differs from that for a conventional footing. This shear should be computed on a section at a distance $d/2$ from the sides and extending around the column at a distance $d/2$ from its faces, where d is the effective depth of the footing.

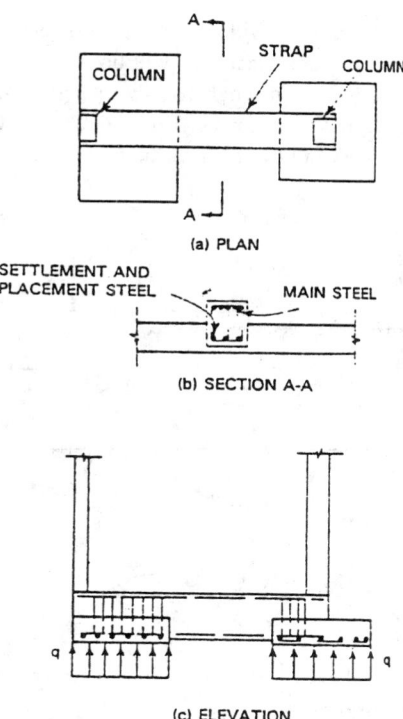

Figure 4.27   Strap (cantilever) footing. (From Fintel, M. 1985. *Handbook of Concrete Engineering,* 2nd ed., Van Nostrand Reinhold, New York. With permission.)

4.10.7   Strip, Grid, and Mat Foundations

In the case of heavily loaded columns, particularly if they are to be supported on relatively weak or uneven soils, continuous foundations may be necessary. They may consist of a continuous strip footing supporting all columns in a given row or, more often, of two sets of such strip footings intersecting at right angles so that they form one continuous grid foundation (Figure 4.24). For even larger loads or weaker soils the strips are made to merge, resulting in a mat foundation (Figure 4.25).

For the design of such continuous foundations it is essential that reasonably realistic assumptions be made regarding the distribution of bearing pressures, which act as upward loads on the foundation. For compressible soils it can be assumed in first approximation that the deformation or settlement of the soil at a given location and the bearing pressure at that location are proportional to each other. If columns are spaced at moderate distances and if the strip, grid, or mat foundation is very rigid, the settlements in all portions of the foundation will be substantially the same. This means that the bearing pressure, also known as *subgrade reaction,* will be the same provided that the centroid of the foundation coincides with the resultant of the loads. If they do not coincide, then for such rigid foundations the subgrade reaction can be assumed as linear and determined from statics in the same manner as discussed for single footings. In this case, all loads—the downward column loads as well as the upward-bearing pressures—are known. Hence, moments and shear forces in the foundation can be found by statics alone. Once these are determined, the design of strip and grid foundations is similar to that of inverted continuous beams, and design of mat foundations is similar to that of inverted flat slabs or plates.

On the other hand, if the foundation is relatively flexible and the column spacing large, settlements will no longer be uniform or linear. For one thing, the more heavily loaded columns will cause larger settlements, and thereby larger subgrade reactions, than the lighter ones. Also, since the continuous strip or slab midway between columns will deflect upward relative to the nearby columns, soil settlement—and thereby the subgrade reaction—will be smaller midway between columns than directly at the columns. This is shown schematically in Figure 4.28. In this case the subgrade reaction can no longer be assumed as uniform. A reasonably accurate but fairly complex analysis can then be made using the theory of beams on elastic foundations.

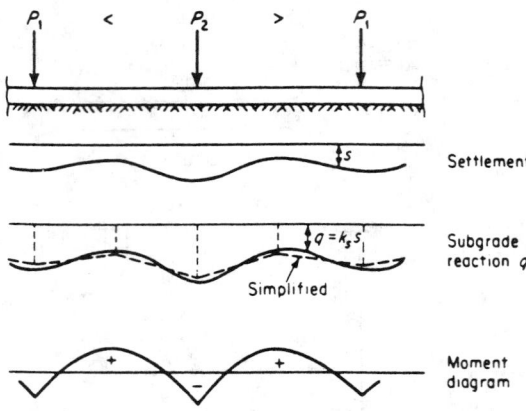

Figure 4.28   Strip footing. (From Fintel, M. 1985. *Handbook of Concrete Engineering*, 2nd ed., Van Nostrand Reinhold, New York. With permission.)

A simplified approach has been developed that covers the most frequent situations of strip and grid foundations [4]. The method first defines the conditions under which a foundation can be regarded as rigid so that uniform or overall linear distribution of subgrade reactions can be assumed. This is the case when the average of two adjacent span lengths in a continuous strip does not exceed $1.75/\lambda$, provided also that the adjacent span and column loads do not differ by more than 20% of the larger value. Here,

$$\lambda = 4\sqrt{\frac{k_s b}{3 E_c I}} \tag{4.135}$$

where

$k_s = Sk'_s$

$k'_s = $ coefficient of subgrade reaction as defined in soils mechanics, basically force per unit area required to produce unit settlement, kips/ft^3

$b = $ width of footing, ft

$E_c = $ modulus of elasticity of concrete, kips/ft^2

$I = $ moment of inertia of footing, ft^4

$S = $ shape factor, being $[(b + 1)/2b]^2$ for granular soils and $(n + 0.5)/1.5n$ for cohesive soils, where n is the ratio of longer to shorter side of strip

   If the average of two adjacent spans exceeds $1.75/\lambda$, the foundation is regarded as flexible. Provided that adjacent spans and column loads differ by no more than 20%, the complex curvelinear distribution of subgrade reaction can be replaced by a set of equivalent trapezoidal reactions, which are also shown in Figure 4.28. The report of ACI Committee 436 contains fairly simple equations for determining the intensities of the equivalent pressures under the columns and at the middle of the spans and also gives equations for the positive and negative moments caused by these equivalent subgrade reactions. With this information, the design of continuous strip and grid footings proceeds similarly to that of footings under two columns.

Mat foundations likewise require different approaches, depending on whether they can be classified as rigid or flexible. As in strip footings, if the column spacing is less than $1/\lambda$, the structure may be regarded as rigid, soil pressure can be assumed as uniformly or linearly distributed, and the design is based on statics. On the other hand, when the foundation is considered flexible as defined above, and if the variation of adjacent column loads and spans is not greater than 20%, the same simplified procedure as for strip and grid foundations can be applied to mat foundations. The mat is divided into two sets of mutually perpendicular strip footings of width equal to the distance between midspans, and the distribution of bearing pressures and bending moments is carried out for each strip. Once moments are determined, the mat is in essence treated the same as a flat slab or plate, with the reinforcement allocated between column and middle strips as in these slab structures.

This approach is feasible only when columns are located in a regular rectangular grid pattern. When a mat that can be regarded as rigid supports columns at random locations, the subgrade reactions can still be taken as uniform or as linearly distributed and the mat analyzed by statics. If it is a flexible mat that supports such randomly located columns, the design is based on the theory of plates on elastic foundation.

4.10.8 Footings on Piles

If the bearing capacity of the upper soil layers is insufficient for a spread foundation, but firmer strata are available at greater depth, piles are used to transfer the loads to these deeper strata. Piles are generally arranged in groups or clusters, one under each column. The group is capped by a spread footing or cap that distributes the column load to all piles in the group. Reactions on caps act as concentrated loads at the individual piles, rather than as distributed pressures. If the total of all pile reactions in a cluster is divided by area of the footing to obtain an equivalent uniform pressure, it is found that this equivalent pressure is considerably higher in pile caps than for spread footings.

Thus, it is in any event advisable to provide ample rigidity—that is, depth for pile caps—in order to spread the load evenly to all piles.

As in single-column spread footings, the effective portion of allowable bearing capacities of piles, R_a, available to resist the unfactored column loads is the allowable pile reaction less the weight of footing, backfill, and surcharge per pile. That is,

$$R_e = R_a - W_f \tag{4.136}$$

where W_f is the total weight of footing, fill, and surcharge divided by the number of piles.

Once the available or effective pile reaction R_e is determined, the number of piles in a concentrically loaded cluster is the integer next larger than

$$n = \frac{D + L}{R_e} \tag{4.137}$$

The effects of wind and earthquake moments at the foot of the columns generally produce an eccentrically loaded pile cluster in which different piles carry different loads. The number and location of piles in such a cluster is determined by successive approximation from the condition that the load on the most heavily loaded pile should not exceed the allowable pile reaction R_a. Assuming a linear distribution of pile loads due to bending, the maximum pile reaction is

$$R_{\max} = \frac{P}{n} + \frac{M}{I_{pg}/c} \tag{4.138}$$

where P is the maximum load (including weight of cap, backfill, etc.), M is the moment to be resisted by the pile group, both referred to the bottom of the cap, I_{pg} is the moment of inertia of the entire pile group about the centroidal axis about which bendings occurs, and c is the distance from that axis to the extreme pile.

Piles are generally arranged in tight patterns, which minimizes the cost of the caps, but they cannot be placed closer than conditions of deriving and of undisturbed carrying capacity will permit. AASHTO requires that piles be spaced at least 2 ft 6 in. center to center and that the distance from the side of a pile to the nearest edge of the footing be 9 in. or more.

The design of footings on piles is similar to that of single-column spread footings. One approach is to design the cap for the pile reactions calculated for the factored column loads. For a concentrically loaded cluster this would give $R_u = (1.4D + 1.7L)/n$. However, since the number of piles was taken as the next larger integer according to Equation 4.138, determining R_u in this manner can lead to a design where the strength of the cap is less than the capacity of the pile group. It is therefore recommended that the pile reaction for strength design be taken as

$$R_u = R_e \times \text{Average load factor} \qquad (4.139)$$

where the average load factor is $(1.4D + 1.7L)/(D + L)$. In this manner the cap is designed to be capable of developing the full allowable capacity of the pile group.

As in single-column spread footings, the depth of the pile cap is usually governed by shear. In this regard both punching and one-way shear need to be considered. The critical sections are the same as explained earlier under "Two-Way Shear (Punching Shear)" and "One-Way Shear." The difference is that shears on caps are caused by concentrated pile reactions rather than by distributed bearing pressures. This poses the question of how to calculate shear if the critical section intersects the circumference of one or more piles. For this case the ACI Code accounts for the fact that pile reaction is not really a point load, but is distributed over the pile-bearing area. Correspondingly, for piles with diameters d_p, it stipulates as follows:

Computation of shear on any section through a footing on piles shall be in accordance with the following:

(a) The entire reaction from any pile whose center is located $d_p/2$ or more outside this section shall be considered as producing shear on that section.

(b) The reaction from any pile whose center is located $d_p/2$ or more inside the section shall be considered as producing no shear on that section.

(c) For intermediate portions of the pile center, the portion of the pile reaction to be considered as producing shear on the section shall be based on straight-line interpolation between the full value at $d_p/2$ outside the section and zero at $d_p/2$ inside the section [1].

In addition to checking punching and one-way shear, punching shear must be investigated for the individual pile. Particularly in caps on a small number of heavily loaded piles, it is this possibility of a pile punching upward through the cap which may govern the required depth. The critical perimeter for this action, again, is located at a distance $d/2$ outside the upper edge of the pile. However, for relatively deep caps and closely spaced piles, critical perimeters around adjacent piles may overlap. In this case, fracture, if any, would undoubtedly occur along an outward-slanting surface around both adjacent piles. For such situations the critical perimeter is so located that its length is a minimum, as shown for two adjacent piles in Figure 4.29.

4.11 Walls

4.11.1 Panel, Curtain, and Bearing Walls

As a general rule, the exterior walls of a reinforced concrete building are supported at each floor by the skeleton framework, their only function being to enclose the building. Such walls are called *panel walls*. They may be made of concrete (often precast), cinder concrete block, brick, tile blocks, or

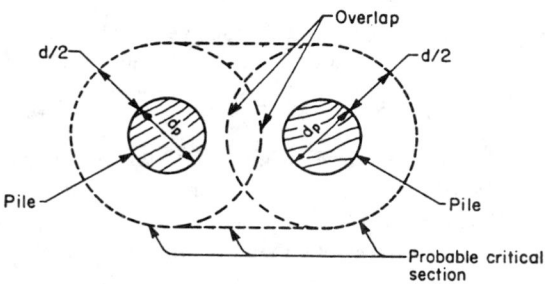

Figure 4.29 Modified critical section for shear with overlapping critical perimeters.

insulated metal panels. The thickness of each of these types of panel walls will vary according to the material, type of construction, climatological conditions, and the building requirements governing the particular locality in which the construction takes place. The pressure of the wind is usually the only load that is considered in determining the structural thickness of a wall panel, although in some cases exterior walls are used as diaphragms to transmit forces caused by horizontal loads down to the building foundations.

Curtain walls are similar to panel walls except that they are not supported at each story by the frame of the building; rather, they are self supporting. However, they are often anchored to the building frame at each floor to provide lateral support.

A bearing wall may be defined as one that carries any vertical load in addition to its own weight. Such walls may be constructed of stone masonry, brick, concrete block, or reinforced concrete. Occasional projections or pilasters add to the strength of the wall and are often used at points of load concentration. Bearing walls may be of either single or double thickness, the advantage of the latter type being that the air space between the walls renders the interior of the building less liable to temperature variation and makes the wall itself more nearly moistureproof. On account of the greater gross thickness of the double wall, such construction reduces the available floor space.

According to ACI Code Sec. 14.5.2 the load capacity of a wall is given by

$$\phi P_{nw} = 0.55\phi f_c' A_g \left[1 - \left(\frac{kl_c}{32h} \right)^2 \right] \qquad (4.140)$$

where
ϕP_{nw} = design axial load strength
A_g = gross area of section, in.^{2}
l_c = vertical distance between supports, in.
h = thickness of wall, in.
ϕ = 0.7

and where the effective length factor k is taken as 0.8 for walls restrained against rotation at top or bottom or both, 1.0 for walls unrestrained against rotation at both ends, and 2.0 for walls not braced against lateral translation.

In the case of concentrated loads, the length of the wall to be considered as effective for each should not exceed the center-to-center distance between loads; nor should it exceed the width of the bearing plus 4 times the wall thickness. Reinforced concrete bearing walls should have a thickness of a least 1/25 times the unsupported height or width, whichever is shorter. Reinforced concrete bearing walls of buildings should be not less than 4 in. thick.

Minimum ratio of horizontal reinforcement area to gross concrete area should be 0.0020 for deformed bars not larger than #5—with specified yield strength not less than 60,000 psi or 0.0025 for other deformed bars—or 0.0025 for welded wire fabric not larger than W31 or D31. Minimum ratio of vertical reinforcement area to gross concrete area should be 0.0012 for deformed bars not larger than #5—with specified yield strength not less than 60,000 psi or 0.0015 for other deformed bars—or 0.0012 for welded wire fabric not larger than W31 or D31. In addition to the minimum reinforcement, not less than two #5 bars shall be provided around all window and door openings. Such bars shall be extended to develop the bar beyond the corners of the openings but not less than 24 in.

Walls more than 10 in. thick should have reinforcement for each direction placed in two layers parallel with faces of wall. Vertical and horizontal reinforcement should not be spaced further apart than three times the wall thickness, or 18 in. Vertical reinforcement need not be enclosed by lateral ties if vertical reinforcement area is not greater than 0.01 times gross concrete area, or where vertical reinforcement is not required as compression reinforcement.

Quantity of reinforcement and limits of thickness mentioned above are waived where structural analysis shows adequate strength and stability. Walls should be anchored to intersecting elements such as floors, roofs, or to columns, pilasters, buttresses, and intersecting walls, and footings.

4.11.2 Basement Walls

In determining the thickness of basement walls, the lateral pressure of the earth, if any, must be considered in addition to other structural features. If it is part of a bearing wall, the lower portion may be designed either as a slab supported by the basement and floors or as a retaining wall, depending upon the type of construction. If columns and wall beams are available for support, each basement wall panel of reinforced concrete may be designed to resist the earth pressure as a simple slab reinforced in either one or two directions. A minimum thickness of 7.5 in. is specified for reinforced concrete basement walls. In wet ground a minimum thickness of 12 in. should be used. In any case, the thickness cannot be less than that of the wall above.

Care should be taken to brace a basement wall thoroughly from the inside (1) if the earth is backfilled before the wall has obtained sufficient strength to resist the lateral pressure without such assistance, or (2) if it is placed before the first-floor slab is in position.

4.11.3 Partition Walls

Interior walls used for the purpose of subdividing the floor area may be made of cinder block, brick, precast concrete, metal lath and plaster, clay tile, or metal. The type of wall selected will depend upon the fire resistance required; flexibility of rearrangement; ease with which electrical conduits, plumbing, etc. can be accommodated; and architectural requirements.

4.11.4 Shears Walls

Horizontal forces acting on buildings—for example, those due to wind or seismic action—can be resisted by a variety of means. Rigid-frame resistance of the structure, augmented by the contribution of ordinary masonry walls and partitions, can provide for wind loads in many cases. However, when heavy horizontal loading is likely—such as would result from an earthquake—reinforced concrete shear walls are used. These may be added solely to resist horizontal forces; alternatively, concrete walls enclosing stairways or elevator shafts may also serve as shear walls.

Figure 4.30 shows a building with wind or seismic forces represented by arrows acting on the edge of each floor or roof. The horizontal surfaces act as deep beams to transmit loads to vertical resisting elements A and B. These shear walls, in turn, act as cantilever beams fixed at their base to carry loads down to the foundation. They are subjected to (1) a variable shear, which reaches maximum at the base, (2) a bending moment, which tends to cause vertical tension near the loaded edge and compression at the far edge, and (3) a vertical compression due to ordinary gravity loading from the structure. For the building shown, additional shear walls C and D are provided to resist loads acting in the log direction of the structure.

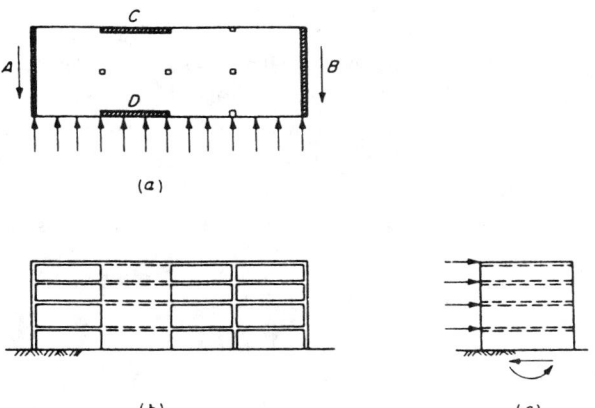

Figure 4.30   Building with shear walls subject to horizontal loads: (a) typical floor; (b) front elevation; (c) end elevation.

The design basis for shear walls, according to the ACI Code, is of the same general form as that used for ordinary beams:

$$V_u \le \phi V_n \qquad (4.141)$$

$$V_n = V_c + V_s \qquad (4.142)$$

Shear strength V_n at any horizontal section for shear in plane of wall should not be taken greater than $10\sqrt{f_c'}hd$. In this and all other equations pertaining to the design of shear walls, the distance of d may be taken equal to $0.8l_w$. A larger value of d, equal to the distance from the extreme compression face to the center of force of all reinforcement in tension, may be used when determined by a strain compatibility analysis.

The value of V_c, the nominal shear strength provided by the concrete, may be based on the usual equations for beams, according to ACI Code. For walls subjected to vertical compression,

$$V_c = 2\sqrt{f_c'}hd \qquad (4.143)$$

and for walls subjected to vertical tension N_u,

$$V_c = 2\left(1 + \frac{N_u}{500A_g}\right)\sqrt{f_c'}hd \qquad (4.144)$$

where N_u is the factored axial load in pounds, taken negative for tension, and A_g is the gross area of horizontal concrete section in square inches. Alternatively, the value of V_c may be based on a more detailed calculation, as the lesser of

$$V_c = 3.3\sqrt{f_c'}hd + \frac{N_u d}{4l_w} \qquad (4.145)$$

or

$$V_c = \left[0.6\sqrt{f_c'} + \frac{l_w\left(1.25\sqrt{f_c'} + 0.2N_u/l_w h\right)}{M_u/V_u - l_w/2}\right]hd \qquad (4.146)$$

Equation 4.145 corresponds to the occurrence of a principal tensile stress of approximately $4\sqrt{f_c'}$ at the centroid of the shear-wall section. Equation 4.146 corresponds approximately to the occurrence of a flexural tensile stress of $6\sqrt{f_c'}$ at a section $l_w/2$ above the section being investigated. Thus the two equations predict, respectively, web-shear cracking and flexure-shear cracking. When the quantity $M_u / V_u - l_w/2$ is negative, Equation 4.146 is inapplicable. According to the ACI Code, horizontal sections located closer to the wall base than a distance $l_w/2$ or $h_w/2$, whichever less, may be designed for the same V_c as that computed at a distance $l_w/2$ or $h_w/2$.

When the factored shear force V_u does not exceed $\phi V_c/2$, a wall may be reinforced according to the minimum requirements given in Sec. 12.1. When V_u exceeds $\phi V_c/2$, reinforcement for shear is to be provided according to the following requirements.

The nominal shear strength V_s provided by the horizontal wall steel is determined on the same basis as for ordinary beams:

$$V_s = \frac{A_v f_y d}{s_2} \tag{4.147}$$

where A_v is the area of horizontal shear reinforcement within vertical distance s_2, (in.^{2}), s_2 is the vertical distance between horizontal reinforcement, (in.), and f_y is the yield strength of reinforcement, psi. Substituting Equation 4.147 into Equation 4.142, then combining with Equation 4.141, one obtains the equation for the required area of horizontal shear reinforcement within a distance s_2:

$$A_v = \frac{(V_u - \phi V_c) s_2}{\phi f_y d} \tag{4.148}$$

The minimum permitted ratio of horizontal shear steel to gross concrete area of vertical section, ρ_n, is 0.0025 and the maximum spacing s_2 is not exceed $l_w/5$, $3h$, or 18 in.

Test results indicate that for low shear walls, vertical distributed reinforcement is needed as well as horizontal reinforcement. Code provisions require vertical steel of area A_h within a spacing s_1, such that the ratio of vertical steel to gross concrete area of horizontal section will not be less than

$$\rho_n = 0.0025 + 0.5 \left(2.5 - \frac{h_w}{l_w} \right) (\rho_h - 0.0025) \tag{4.149}$$

nor less than 0.0025. However, the vertical steel ratio need not be greater than the required horizontal steel ratio. The spacing of the vertical bars is not to exceed $l_w/3$, $3h$, or 18 in.

Walls may be subjected to flexural tension due to overturning moment, even when the vertical compression from gravity loads is superimposed. In many but not all cases, vertical steel is provided, concentrated near the wall edges, as in Figure 4.31. The required steel area can be found by the usual methods for beams.

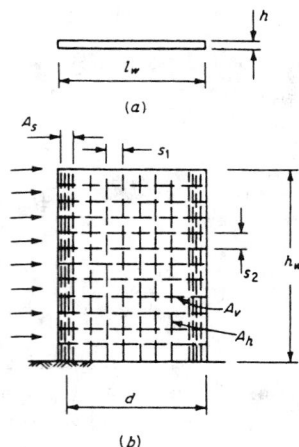

Figure 4.31   Geometry and reinforcement of typical shear wall: (a) cross section; (b) elevation.

The ACI Code contains requirements for the dimensions and details of structural walls serving as part of the earthquake-force resisting systems. The reinforcement ratio, $\rho_v (= A_{sv}/A_{cv}$; where A_{cv} is the net area of concrete section bounded by web thickness and length of section in the direction of shear force considered, and A_{sv} is the projection on A_{cv} of area of distributed shear reinforcement crossing the plane of A_{cv}), for structural walls should not be less than 0.0025 along the longitudinal and transverse axes. Reinforcement provided for shear strength should be continuous and should be distributed across the shear plane. If the design shear force does not exceed $A_{cv}\sqrt{f'_c}$, the shear reinforcement may conform to the reinforcement ratio given in Sec. 12.1. At least two curtains of reinforcement should be used in a wall if the in-plane factored shear force assigned to the wall exceeds $2A_{cv}\sqrt{f'_c}$. All continuous reinforcement in structural walls should be anchored or spliced in accordance with the provisions for reinforcement in tension for seismic design.

Proportioning and details of structural walls that resist shear forces caused by earthquake motion is contained in the ACI Code Sec. 21.7.3.

4.12 Defining Terms

The terms common in concrete engineering as defined in and selected from the Cement and Concrete Terminology Report of ACI Committee 116 are given below [1, Further Reading].

Allowable stress: Maximum permissible stress used in design of members of a structure and based on a factor of safety against yielding or failure of any type.

Allowable stress design (ASD): Design principle according to which stresses resulting from service or working loads are not allowed to exceed specified allowable values.

Balanced load: Combination of axial force and bending moment that causes simultaneous crushing of concrete and yielding of tension steel.

Balanced reinforcement: An amount and distribution of flexural reinforcement such that the tensile reinforcement reaches its specified yield strength simultaneously with the concrete in compression reaching its assumed ultimate strain of 0.003.

Beam: A structural member subjected primarily to flexure; depth-to-span ratio is limited to 2/5 for continuous spans, or 4/5 for simple spans, otherwise the member is to be treated as a deep beam.

Beam-column: A structural member that is subjected simultaneously to bending and substantial axial forces.

Bond: Adhesion and grip of concrete or mortar to reinforcement or to other surfaces against which it is placed; to enhance bond strength, ribs or other deformations are added to reinforcing bars.

Camber: A deflection that is intentionally built into a structural element or form to improve appearance or to offset the deflection of the element under the effects of loads, shrinkage, and creep.

Cast-in-place concrete: Concrete poured in its final or permanent location; also called *in situ* concrete; opposite of precast concrete.

Column: A member used to support primarily axial compression loads with a height of at least three times its least lateral dimensions; the capacity of short columns is controlled by strength; the capacity of long columns is limited by buckling.

Column strip: The portion of a flat slab over a row of columns consisting of the two adjacent quarter panels on each side of the column centerline.

Composition construction: A type of construction using members made of different materials (e.g., concrete and structural steel), or combining members made of cast-in-place concrete and precast concrete such that the combined components act together as a single member; strictly speaking, reinforced concrete is also composite construction.

Compression member: A member subjected primarily to longitudinal compression; often synonymous with "column".

Compressive strength: Strength typically measured on a standard 6×12 in. cylinder of concrete in an axial compression test, 28 d after casting.

Concrete: A composite material that consists essentially of a binding medium within which are embedded particles or fragments of aggregate; in portland cement concrete, the binder is a mixture of portland cement and water.

Confined concrete: Concrete enclosed by closely spaced transverse reinforcement, which restrains the concrete expansion in directions perpendicular to the applied stresses.

Construction joint: The surface where two successive placements of concrete meet, across which it may be desirable to achieve bond, and through which reinforcement may be continuous.

Continuous beam or slab: A beam or slab that extends as a unit over three or more supports in a given direction and is provided with the necessary reinforcement to develop the negative moments over the interior supports; a redundant structure that requires a statically indeterminant analysis (opposite of simple supported beam or slab).

Cover: In reinforced concrete, the shortest distance between the surface of the reinforcement and the outer surface of the concrete; minimum values are specified to protect the reinforcement against corrosion and to assure sufficient bond strength.

Cracks: Results of stresses exceeding concrete's tensile strength capacity; cracks are ubiquitous in reinforced concrete and needed to develop the strength of the reinforcement, but a design goal is to keep their widths small (hairline cracks).

Cracked section: A section designed or analyzed on the assumption that concrete has no resistance to tensile stress.

Cracking load: The load that causes tensile stress in a member to be equal to the tensile strength of the concrete.

Deformed bar: Reinforcing bar with a manufactured pattern of surface ridges intended to prevent slip when the bar is embedded in concrete.

Design strength: Ultimate load-bearing capacity of a member multiplied by a strength reduction factor.

Development length: The length of embedded reinforcement to develop the design strength of the reinforcement; a function of bond strength.

Diagonal crack: An inclined crack caused by a diagonal tension, usually at about 45 degrees to the neutral axis of a concrete member.

Diagonal tension: The principal tensile stress resulting from the combination of normal and shear stresses acting upon a structural element.

Drop panel: The portion of a flat slab in the area surrounding a column, column capital, or bracket which is thickened in order to reduce the intensity of stresses.

Ductility: Capability of a material or structural member to undergo large inelastic deformations without distress; opposite of brittleness; very important material property, especially for earthquake-resistant design; steel is naturally ductile, concrete is brittle but it can be made ductile if well confined.

Durability: The ability of concrete to maintain its qualities over long time spans while exposed to weather, freeze-thaw cycles, chemical attack, abrasion, and other service load conditions.

Effective depth: Depth of a beam or slab section measured from the compression face to the centroid of the tensile reinforcement.

Effective flange width: Width of slab adjoining a beam stem assumed to function as the flange of a T-section.

Effective prestress: The stress remaining in the prestressing steel or in the concrete due to pre-stressing after all losses have occurred.

Effective span: The lesser of the distance between centers of supports and the clear distance between supports plus the effective depth of the beam or slab.

Flat slab: A concrete slab reinforced in two or more directions, generally without beams or girders to transfer the loads to supporting members, but with drop panels or column capitals or both.

High-early strength cement: Cement producing strength in mortar or concrete earlier than regular cement.

Hoop: A one-piece closed reinforcing tie or continuously wound tie that encloses the longitudinal reinforcement.

Interaction diagram: Failure curve for a member subjected to both axial force and the bending moment, indicating the moment capacity for a given axial load and vice versa; used to develop design charts for reinforced concrete compression members.

Lightweight concrete: Concrete of substantially lower unit weight than that made using normal-weight gravel or crushed stone aggregate.

Limit analysis: See **Plastic analysis.**

Limit design: A method of proportioning structural members based on satisfying certain strength and serviceability limit states.

Load and resistance factor design (LRFD): See **Ultimate strength design.**

Load factor: A factor by which a service load is multiplied to determine the factored load used in ultimate strength design.

Modulus of elasticity: The ratio of normal stress to corresponding strain for tensile of compressive stresses below the proportional limit of the material; for steel, $E_s = 29,000$ ksi; for concrete it is highly variable with stress level and the strength f_c'; for normal-weight concrete and low stresses, a common approximation is $E_c = 57,000\sqrt{f_c'}$.

Modulus of rupture: The tensile strength of concrete as measured in a flexural test of a small prismatic specimen of plain concrete.

Mortar: A mixture of cement paste and fine aggregate; in fresh concrete, the material occupying the interstices among particles of coarse aggregate.

Nominal strength: The strength of a structural member based on its assumed material properties and sectional dimensions, before application of any strength reduction factor.

Plastic analysis: A method of structural analysis to determine the intensity of a specified load distribution at which the structure forms a collapse mechanism.

Plastic hinge: Region of flexural member where the ultimate moment capacity can be developed and maintained with corresponding significant inelastic rotation, as main tensile steel is stressed beyond the yield point.

Post-tensioning: A method of prestressing reinforced concrete in which the tendons are tensioned after the concrete has hardened (opposite of pretensioning).

Precast concrete: Concrete cast elsewhere than its final position, usually in factories or factory-like shop sites near the final site (opposite of cast-in-place concrete).

Prestressed concrete: Concrete in which internal stresses of such magnitude and distribution are introduced that the tensile stresses resulting from the service loads are counteracted to a desired degree; in reinforced concrete the prestress is commonly introduced by tensioning embedded tendons.

Prestressing steel: High strength steel used to prestress concrete, commonly seven-wire strands, single wires, bars, rods, or groups of wires or strands.

Pretensioning: A method of prestressing reinforced concrete in which the tendons are tensioned before the concrete has hardened (opposite of post-tensioning).

Ready-mixed concrete: Concrete manufactured for delivery to a purchaser in a plastic and unhardened state; usually delivered by truck.

Rebar: Short for reinforcing bar; see **Reinforcement**.

Reinforced concrete: Concrete containing adequate reinforcement (prestressed or not) and designed on the assumption that the two materials act together in resisting forces.

Reinforcement: Bars, wires, strands, and other slender members that are embedded in concrete in such a manner that the reinforcement and the concrete act together in resisting forces.

Safety factor: The ratio of a load producing an undesirable state (such as collapse) and an expected or service load.

Service loads: Loads on a structure with high probability of occurrence, such as dead weight supported by a member or the live loads specified in building codes and bridge specifications.

Shear key: A recess or groove in a joint between successive lifts or placements of concrete, which is filled with concrete of the adjacent lift, giving shear strength to the joint.

Shear span: The distance from a support of a simply supported beam to the nearest concentrated load.

Shear wall: See **Structural wall**.

Shotcrete: Mortar or concrete pneumatically projected at high velocity onto a surface.

Silica fume: Very fine noncrystalline silica produced in electric arc furnaces as a by-product of the production of metallic silicon and various silicon alloys (also know as condensed silica fume); used as a mineral admixture in concrete.

Slab: A flat, horizontal (or neatly so) molded layer of plain or reinforced concrete, usually of uniform thickness, either on the ground or supported by beams, columns, walls, or other frame work. See also **Flat slab**.

Slump: A measure of consistency of freshly mixed concrete equal to the subsidence of the molded specimen immediately after removal of the slump cone, expressed in inches.

Splice: Connection of one reinforcing bar to another by lapping, welding, mechanical couplers, or other means.

Split cylinder test: Test for tensile strength of concrete in which a standard cylinder is loaded to failure in diametral compression applied along the entire length (also called Brazilian test).

Standard cylinder: Cylindric specimen of 12-in. height and 6-in. diameter, used to determine standard compressive strength and splitting tensile strength of concrete.

Stiffness coefficient: The coefficient k_{ij} of stiffness matrix **K** for a multi-degree of freedom structure is the force needed to hold the ith degree of freedom in place, if the jth degree of freedom undergoes a unit of displacement, while all others are locked in place.

Stirrup: A reinforcement used to resist shear and diagonal tension stresses in a structural member; typically a steel bar bent into a U or rectangular shape and installed perpendicular to or at an angle to the longitudinal reinforcement, and properly anchored; the term "stirrup" is usually applied to lateral reinforcement in flexural members and the term "tie" to lateral reinforcement in compression members. See **Tie**.

Strength design: See **Ultimate strength design**.

Strength reduction factor: Capacity reduction factor (typically designated as ϕ) by which the nominal strength of a member is to be multiplied to obtain the design strength; specified by the ACI Code for different types of members.

Structural concrete: Concrete used to carry load or to form an integral part of a structure (opposite of, for example, insulating concrete).

T-beam: A beam composed of a stem and a flange in the form of a "T", with the flange usually provided by a slab.

Tension stiffening effect: The added stiffness of a single reinforcing bar due to the surrounding uncracked concrete between bond cracks.

Tie: Reinforcing bar bent into a loop to enclose the longitudinal steel in columns; tensile bar to hold a form in place while resisting the lateral pressure of unhardened concrete.

Ultimate strength design (USD): Design principle such that the actual (ultimate) strength of a member or structure, multiplied by a strength factor, is no less than the effects of all service load combinations, multiplied by respective overload factors.

Unbonded tendon: A tendon that is not bonded to the concrete.

Under-reinforced beam: A beam with less than balanced reinforcement such that the reinforcement yields before the concrete crushes in compression.

Water-cement ratio: Ratio by weight of water to cement in a mixture; inversely proportional to concrete strength.

Water-reducing admixture: An admixture capable of lowering the mix viscosity, thereby allowing a reduction of water (and increase in strength) without lowering the workability (also called superplasticizer).

Whitney stress block: A rectangular area of uniform stress intensity $0.85f'_c$, whose area and centroid are similar to that of the actual stress distribution in a flexural member near failure.

Workability: General property of freshly mixed concrete that defines the ease with which it can be placed into forms without honeycombs; closely related to slump.

Working stress design: See **Allowable stress design**.

Yield-line theory: Method of structural analysis of plate structures at the verge of collapse under factored loads.

References

[1] ACI Committee 318. 1992. *Building Code Requirements for Reinforced Concrete and Commentary, ACI 318-89 (Revised 92) and ACI 318R-89 (Revised 92)* (347pp.). Detroit, MI.
[2] ACI Committee 340. 1990. *Design Handbook in Accordance with the Strength Design Method of ACI 318-89*. Volume 2, SP-17 (222 pp.).
[3] ACI Committee 363. 1984. State-of-the-art report on high strength concrete. *ACI J. Proc.* 81(4):364-411.
[4] ACI Committee 436. 1996. Suggested design procedures for combined footings and mats. *J. ACI.* 63:1041-1057.
[5] Breen, J.E. 1991. Why structural concrete? *IASE Colloq. Struct. Concr.* Stuttgart, pp.15-26.
[6] Collins, M.P. and Mitchell, D. 1991. *Prestressed Concrete Structures*, 1st ed., Prentice Hall, Englewood Cliffs, N.J.
[7] Fintel, M. 1985. *Handbook of Concrete Engineering*. 2nd ed., Van Nostrand Reinhold, New York.
[8] MacGregor, J.G. 1992. *Reinforced Concrete Mechanics and Design*, 2nd ed., Prentice Hall, Englewood Cliffs, N.J.
[9] Nilson, A.H. and Winter, G. 1992. *Design of Concrete Structures*, 11th ed., McGraw-Hill, New York.
[10] *Standard Handbook for Civil Engineers*, 2nd ed., McGraw-Hill, New York.
[11] Wang, C.-K. and Salmon, C. G. 1985. *Reinforced Concrete Design*, 4th ed., Harper Row, New York.

Further Reading

[1] ACI Committee 116. 1990. Cement and Concrete Terminology, Report 116R-90, American Concrete Institute, Detroit, MI.
[2] Ferguson, P.M., Breen, J.E., and Jirsa, J.O. 1988. *Reinforced Concrete Fundamentals*, 5th ed., John Wiley & Sons, New York.
[3] Lin, T-Y. and Burns, N.H. 1981. *Design of Prestressed Concrete Structures*, 3rd ed., John Wiley & Sons, New York.
[4] Meyer, C. 1996. *Design of Concrete Structures*, Prentice-Hall, Upper Saddle River, NJ.

5

Earthquake Engineering

5.1	Introduction	5-1
5.2	Earthquakes	5-3
	Causes of Earthquakes and Faulting • Distribution of Seismicity • Measurement of Earthquakes • Strong Motion Attenuation and Duration • Seismic Hazard and Design Earthquake • Effect of Soils on Ground Motion • Liquefaction and Liquefaction-Related Permanent Ground Displacement	
5.3	Seismic Design Codes	5-54
	Purpose of Codes • Historical Development of Seismic Codes • Selected Seismic Codes	
5.4	Earthquake Effects and Design of Structures	5-61
	Buildings • Non-Building Structures	
5.5	Defining Terms	5-72
	References	5-75
	Further Reading	5-82

Charles Scawthorn
EQE International, San Francisco, California and Tokyo, Japan

5.1 Introduction

Earthquakes are naturally occurring broad-banded vibratory ground motions, caused by a number of phenomena including tectonic ground motions, volcanism, landslides, rockbursts, and human-made explosions. Of these various causes, tectonic-related earthquakes are the largest and most important. These are caused by the fracture and sliding of rock along **faults** within the Earth's crust. A fault is a zone of the earth's crust within which the two sides have moved — faults may be hundreds of miles long, from 1 to over 100 miles deep, and not readily apparent on the ground surface. Earthquakes initiate a number of phenomena or agents, termed **seismic hazards**, which can cause significant damage to the built environment — these include fault rupture, vibratory ground motion (i.e., shaking), inundation (e.g., tsunami, seiche, dam failure), various kinds of permanent ground failure (e.g., liquefaction), fire or hazardous materials release. For a given earthquake, any particular hazard can dominate, and historically each has caused major damage and great loss of life in specific earthquakes. The expected damage given a specified value of a hazard parameter is termed **vulnerability**, and the product of the hazard and the vulnerability (i.e., the expected damage) is termed the **seismic risk**. This is often formulated as

$$E(D) = \int_H E(D \mid H) p(H) dH \tag{5.1}$$

where

H	=	hazard
$p(\cdot)$	=	refers to probability
D	=	damage
$E(D\mid H)$	=	vulnerability
$E(\cdot)$	=	the expected value operator

Note that damage can refer to various parameters of interest, such as casualties, economic loss, or temporal duration of disruption. It is the goal of the earthquake specialist to reduce seismic risk. The probability of having a specific level of damage (i.e., $p(D) = d$) is termed the **fragility**.

For most earthquakes, shaking is the dominant and most widespread agent of damage. Shaking near the actual earthquake rupture lasts only during the time when the fault ruptures, a process that takes seconds or at most a few minutes. The seismic waves generated by the rupture propagate long after the movement on the fault has stopped, however, spanning the globe in about 20 minutes. Typically earthquake ground motions are powerful enough to cause damage only in the near field (i.e., within a few tens of kilometers from the causative fault). However, in a few instances, long period motions have caused significant damage at great distances to selected lightly damped structures. A prime example of this was the 1985 Mexico City earthquake, where numerous collapses of mid- and high-rise buildings were due to a Magnitude 8.1 earthquake occurring at a distance of approximately 400 km from Mexico City.

Ground motions due to an earthquake will vibrate the base of a structure such as a building. These motions are, in general, three-dimensional, both lateral and vertical. The structure's mass has inertia which tends to remain at rest as the structure's base is vibrated, resulting in deformation of the structure. The structure's load carrying members will try to restore the structure to its initial, undeformed, configuration. As the structure rapidly deforms, energy is absorbed in the process of material deformation. These characteristics can be effectively modeled for a single degree of freedom (SDOF) mass as shown in Figure 5.1 where m represents the mass of the structure, the elastic spring (of stiffness k = force / displacement) represents the restorative force of the structure, and the dashpot **damping** device (damping coefficient c = force/velocity) represents the force or energy lost in the process of material deformation. From the equilibrium of forces on mass m due

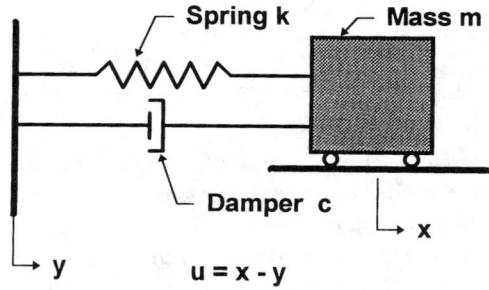

Figure 5.1   Single degree of freedom (SDOF) system.

to the spring and dashpot damper and an applied load $p(t)$, we find:

$$m\ddot{u} + c\dot{u} + ku = p(t) \tag{5.2}$$

the solution of which [32] provides relations between circular frequency of vibration ω, the natural frequency f, and the natural period T:

$$\omega^2 = \frac{k}{m} \tag{5.3}$$

$$f = \frac{1}{T} = \frac{\omega}{2\pi} = \frac{1}{2\pi}\sqrt{\frac{k}{m}} \tag{5.4}$$

Damping tends to reduce the amplitude of vibrations. **Critical damping** refers to the value of damping such that free vibration of a structure will cease after one cycle ($c_{crit} = 2m\omega$). Damping is conventionally expressed as a percent of critical damping and, for most buildings and engineering structures, ranges from 0.5 to 10 or 20% of critical damping (increasing with displacement amplitude). Note that damping in this range will not appreciably affect the natural period or frequency of vibration, but does affect the amplitude of motion experienced.

5.2   Earthquakes

5.2.1   Causes of Earthquakes and Faulting

In a global sense, tectonic earthquakes result from motion between a number of large plates comprising the earth's crust or lithosphere (about 15 in total), (see Figure 5.2). These plates are driven by the convective motion of the material in the earth's mantle, which in turn is driven by heat generated at the earth's core. Relative plate motion at the fault interface is constrained by friction and/or **asperities** (areas of interlocking due to protrusions in the fault surfaces). However, strain energy accumulates in the plates, eventually overcomes any resistance, and causes slip between the two sides of the fault. This sudden slip, termed **elastic rebound** by Reid [101] based on his studies of regional deformation following the 1906 San Francisco earthquake, releases large amounts of energy, which constitutes the earthquake. The location of initial radiation of seismic waves (i.e., the first location of dynamic rupture) is termed the **hypocenter**, while the projection on the surface of the earth directly above the hypocenter is termed the **epicenter**. Other terminology includes **near-field** (within one source dimension of the epicenter, where source dimension refers to the length or width of faulting, whichever is less), **far-field** (beyond near-field), and **meizoseismal** (the area of strong shaking and damage). Energy is radiated over a broad spectrum of frequencies through the earth, in **body waves** and **surface waves** [16]. Body waves are of two types: P waves (transmitting energy via push-pull motion), and slower S waves (transmitting energy via shear action at right angles to the direction of motion). Surface waves are also of two types: horizontally oscillating **Love waves** (analogous to S body waves) and vertically oscillating **Rayleigh waves**.

While the accumulation of strain energy within the plate can cause motion (and consequent release of energy) at faults at any location, earthquakes occur with greatest frequency at the boundaries of the tectonic plates. The boundary of the Pacific plate is the source of nearly half of the world's great earthquakes. Stretching 40,000 km (24,000 miles) around the circumference of the Pacific Ocean, it includes Japan, the west coast of North America, and other highly populated areas, and is aptly termed the Ring of Fire. The interiors of plates, such as ocean basins and continental shields, are areas of low seismicity but are not inactive — the largest earthquakes known to have occurred in North America, for example, occurred in the New Madrid area, far from a plate boundary. Tectonic plates move very slowly and irregularly, with occasional earthquakes. Forces may build up for decades or centuries at plate interfaces until a large movement occurs all at once. These sudden, violent motions produce the shaking that is felt as an earthquake. The shaking can cause direct damage to buildings, roads, bridges, and other human-made structures as well as triggering fires, landslides, tidal waves (tsunamis), and other damaging phenomena.

Faults are the physical expression of the boundaries between adjacent tectonic plates and thus may be hundreds of miles long. In addition, there may be thousands of shorter faults parallel to or branching out from a main fault zone. Generally, the longer a fault the larger the earthquake it can generate. Beyond the main tectonic plates, there are many smaller sub-plates ("platelets") and simple blocks of crust that occasionally move and shift due to the "jostling" of their neighbors and/or the major plates. The existence of these many sub-plates means that smaller but still damaging earthquakes are possible almost anywhere, although often with less likelihood.

Faults are typically classified according to their sense of motion (Figure 5.3). Basic terms include **transform** or **strike slip** (relative fault motion occurs in the horizontal plane, parallel to the strike of the fault), **dip-slip** (motion at right angles to the strike, up- or down-slip), **normal** (dip-slip motion, two sides in tension, move away from each other), **reverse** (dip-slip, two sides in compression, move towards each other), and **thrust** (low-angle reverse faulting).

Figure 5.2   Global seismicity and major tectonic plate boundaries.

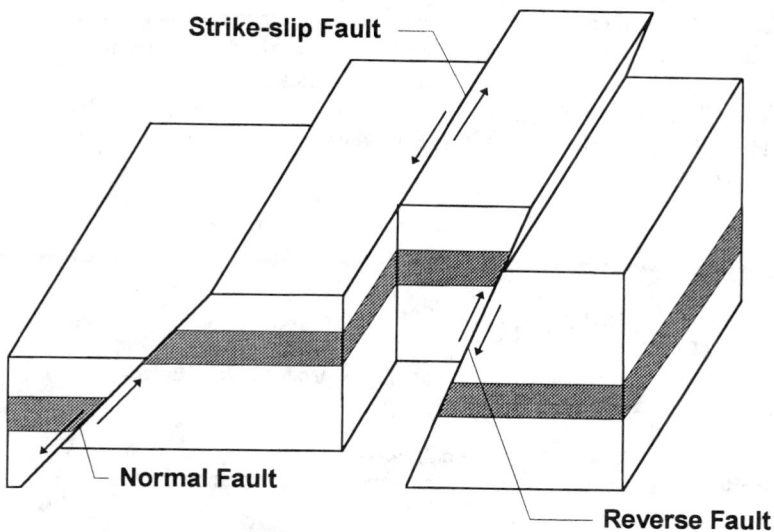

Figure 5.3   Fault types.

Generally, earthquakes will be concentrated in the vicinity of faults. Faults that are moving more rapidly than others will tend to have higher rates of seismicity, and larger faults are more likely than others to produce a large event. Many faults are identified on regional geological maps, and useful information on fault location and displacement history is available from local and national geological surveys in areas of high seismicity. Considering this information, areas of an expected large earthquake in the near future (usually measured in years or decades) can be and have been identified. However, earthquakes continue to occur on "unknown" or "inactive" faults. An important development has been the growing recognition of **blind thrust faults**, which emerged as a result of several earthquakes in the 1980s, none of which were accompanied by surface faulting [120]. Blind thrust faults are faults at depth occurring under anticlinal folds — since they have only subtle surface expression, their seismogenic potential can be evaluated by indirect means only [46]. Blind thrust faults are particularly worrisome because they are hidden, are associated with folded topography in general, including areas of lower and infrequent seismicity, and therefore result in a situation where the potential for an earthquake exists in any area of anticlinal geology, even if there are few or no earthquakes in the historic record. Recent major earthquakes of this type have included the 1980 M_w 7.3 El- Asnam (Algeria), 1988 M_w 6.8 Spitak (Armenia), and 1994 M_w 6.7 Northridge (California) events.

Probabilistic methods can be usefully employed to quantify the likelihood of an earthquake's occurrence, and typically form the basis for determining the **design basis earthquake**. However, the earthquake generating process is not understood well enough to reliably predict the times, sizes, and locations of earthquakes with precision. In general, therefore, communities must be prepared for an earthquake to occur at any time.

5.2.2   Distribution of Seismicity

This section discusses and characterizes the distribution of seismicity for the U.S. and selected areas.

Global

It is evident from Figure 5.2 that some parts of the globe experience more and larger earthquakes than others. The two major regions of seismicity are the circum-Pacific **Ring of Fire** and the **Trans-Alpide** belt, extending from the western Mediterranean through the Middle East and the northern India sub-continent to Indonesia. The Pacific plate is created at its South Pacific extensional boundary — its motion is generally northwestward, resulting in relative strike-slip motion in California and New Zealand (with, however, a compressive component), and major compression and **subduction** in Alaska, the Aleutians, Kuriles, and northern Japan. Subduction refers to the plunging of one plate (i.e., the Pacific) beneath another, into the mantle, due to convergent motion, as shown in Figure 5.4.

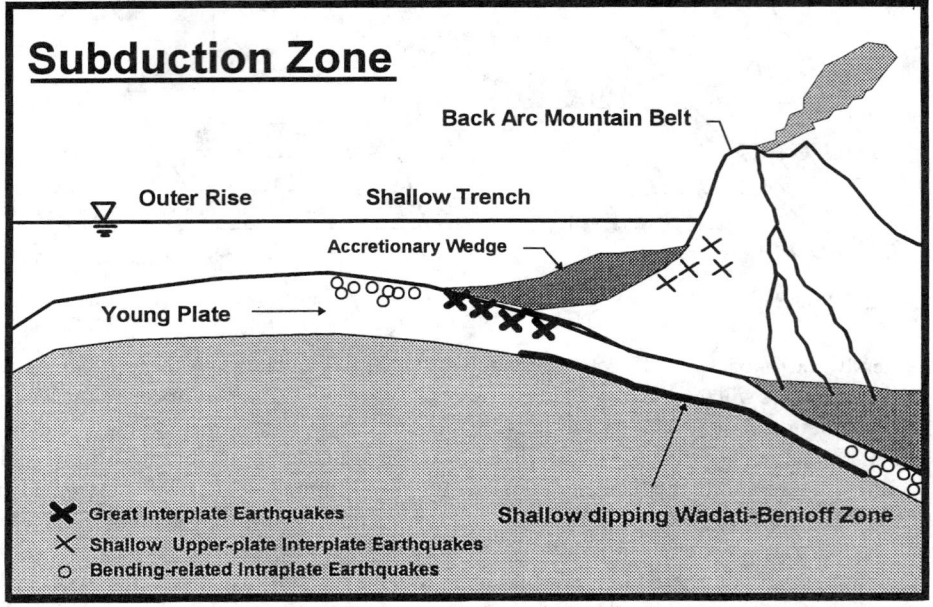

Figure 5.4   Schematic diagram of subduction zone, typical of west coast of South America, Pacific Northwest of U.S., or Japan.

Subduction zones are typically characterized by volcanism, as a portion of the plate (melting in the lower mantle) re-emerges as volcanic lava. Subduction also occurs along the west coast of South America at the boundary of the Nazca and South American plate, in Central America (boundary of the Cocos and Caribbean plates), in Taiwan and Japan (boundary of the Philippine and Eurasian plates), and in the North American Pacific Northwest (boundary of the Juan de Fuca and North American plates). The Trans-Alpide seismic belt is basically due to the relative motions of the African and Australian plates colliding and subducting with the Eurasian plate.

U.S.

Table 5.1 provides a list of selected U.S. earthquakes. The San Andreas fault system in California and the Aleutian Trench off the coast of Alaska are part of the boundary between the North American and Pacific tectonic plates, and are associated with the majority of U.S. seismicity (Figure 5.5 and Table 5.1). There are many other smaller fault zones throughout the western U.S. that are also helping to release the stress that is built up as the tectonic plates move past one another, (Figure 5.6). While California has had numerous destructive earthquakes, there is also clear evidence that the potential exists for great earthquakes in the Pacific Northwest [11].

On the east coast of the U.S., the cause of earthquakes is less well understood. There is no plate boundary and very few locations of active faults are known so that it is more difficult to assess where

TABLE 5.1　Selected U.S. Earthquakes

Yr	M	D	Lat.		Long.		M	MMI	Fat.	USD mills	Locale
1755	11	18						8			Nr Cape Ann, MA (MMI from STA)
1774	2	21						7			Eastern VA (MMI from STA)
1791	5	16						8			E. Haddam, CT (MMI from STA)
1811	12	16	36	N	90	W	8.6	-			New Madrid, MO
1812	1	23	36.6	N	89.6	W	8.4	12			New Madrid, MO
1812	2	7	36.6	N	89.6	W	8.7	12			New Madrid, MO
1817	10	5						8			Woburn, MA (MMI from STA)
1836	6	10	38	N	122	W	-	10	-		California
1838	6	0	37.5	N	123	W	-	10	-		California
1857	1	9	35	N	119	W	8.3	7	-		San Francisco, CA
1865	10	8	37	N	122	W	-	9			San Jose, Santa Cruz, CA
1868	4	3	19	N	156	W	-	10	81		Hawaii
1868	10	21	37.5	N	122	W	6.8	10	3		Hayward, CA
1872	3	26	36.5	N	118	W	8.5	10	50		Owens Valley, CA
1886	9	1	32.9	N	80	W	7.7	9	60	5	Charleston, SC, Ms from STA
1892	2	24	31.5	N	117	W	-	10	-		San Diego County, CA
1892	4	19	38.5	N	123	W	-	9	-		Vacaville, Winters, CA
1892	5	16	14	N	143	E	-	-	-		Agana, Guam
1897	5	31					5.8	8			Giles County, VA (mb from STA)
1899	9	4	60	N	142	W	8.3	-	-		Cape Yakataga, AK
1906	4	18	38	N	123	W	8.3	11	700?	400	San Francisco, CA (deaths more?)
1915	10	3	40.5	N	118	W	7.8	-			Pleasant Valley, NV
1925	6	29	34.3	N	120	W	6.2	-	13	8	Santa Barbara, CA
1927	11	4	34.5	N	121	W	7.5	9			Lompoc, Port San Luis, CA
1933	3	11	33.6	N	118	W	6.3	-	115	40	Long Beach, CA
1934	12	31	31.8	N	116	W	7.1	10			Baja, Imperial Valley, CA
1935	10	19	46.6	N	112	W	6.2	-	2	19	Helena, MT
1940	5	19	32.7	N	116	W	7.1	10	9	6	SE of Elcentro, CA
1944	9	5	44.7	N	74.7	W	5.6	-		2	Massena, NY
1949	4	13	47.1	N	123	W	7	8	8	25	Olympia, WA
1951	8	21	19.7	N	156	W	6.9	-			Hawaii
1952	7	21	35	N	119	W	7.7	11	13	60	Central, Kern County, CA
1954	12	16	39.3	N	118	W	7	10			Dixie Valley, NV
1957	3	9	51.3	N	176	W	8.6	-		3	Alaska
1958	7	10	58.6	N	137	W	7.9	-	5		Lituyabay, AK—Landslide
1959	8	18	44.8	N	111	W	7.7	-			Hebgen Lake, MT
1962	8	30	41.8	N	112	W	5.8	-		2	Utah
1964	3	28	61	N	148	W	8.3	-	131	540	Alaska
1965	4	29	47.4	N	122	W	6.5	7	7	13	Seattle, WA
1971	2	9	34.4	N	118	W	6.7	11	65	553	San Fernando, CA
1975	3	28	42.1	N	113	W	6.2	8	-	1	Pocatello Valley, ID
1975	8	1	39.4	N	122	W	6.1	-	-	6	Oroville Reservoir, CA
1975	11	29	19.3	N	155	W	7.2	9	2	4	Hawaii
1980	1	24	37.8	N	122	W	5.9	7	1	4	Livermore, CA
1980	5	25	37.6	N	119	W	6.4	7	-	2	Mammoth Lakes, CA
1980	7	27	38.2	N	83.9	W	5.2	-	-	1	Maysville, KY
1980	11	8	41.2	N	124	W	7	7	5	3	N Coast, CA
1983	5	2	36.2	N	120	W	6.5	8	-	31	Central, Coalinga, CA
1983	10	28	43.9	N	114	W	7.3	-	2	13	Borah Peak, ID
1983	11	16	19.5	N	155	W	6.6	8	-	7	Kapapala, HI
1984	4	24	37.3	N	122	W	6.2	7	-	8	Central Morgan Hill, CA
1986	7	8	34	N	117	W	6.1	7	-	5	Palm Springs, CA
1987	10	1	34.1	N	118	W	6	8	8	358	Whittier, CA
1987	11	24	33.2	N	116	W	6.3	6	2	-	Superstition Hills, CA
1989	6	26	19.4	N	155	W	6.1	6			Hawaii
1989	10	18	37.1	N	122	W	7.1	9	62	6,000	Loma Prieta, CA
1990	2	28	34.1	N	118	W	5.5	7	-	13	Claremont, Covina, CA
1992	4	23	34	N	116	W	6.3	7			Joshua Tree, CA
1992	4	25	40.4	N	124	W	7.1	8		66	Humboldt, Ferndale, CA
1992	6	28	34.2	N	117	W	6.7	8			Big Bear Lake, Big Bear, CA
1992	6	28	34.2	N	116	W	7.6	9	3	92	Landers, Yucca, CA
1992	6	29	36.7	N	116	W	5.6	-			Border of NV and CA
1993	3	25	45	N	123	W	5.6	7			Washington-Oregon
1993	9	21	42.3	N	122	W	5.9	7	2	-	Klamath Falls, OR
1994	1	16	40.3	N	76	W	4.6	5			PA, Felt, Canada
1994	1	17	34.2	N	119	W	6.8	9	57	30,000	Northridge, CA
1994	2	3	42.8	N	111	W	6	7			Afton, WY
1995	10	6	65.2	N	149	W	6.4	-			AK (Oil pipeline damaged)

Note: STA refers to [3]. From NEIC, Database of Significant Earthquakes Contained in *Seismicity Catalogs*, National Earthquake Information Center, Goldon, CO, 1996. With permission.

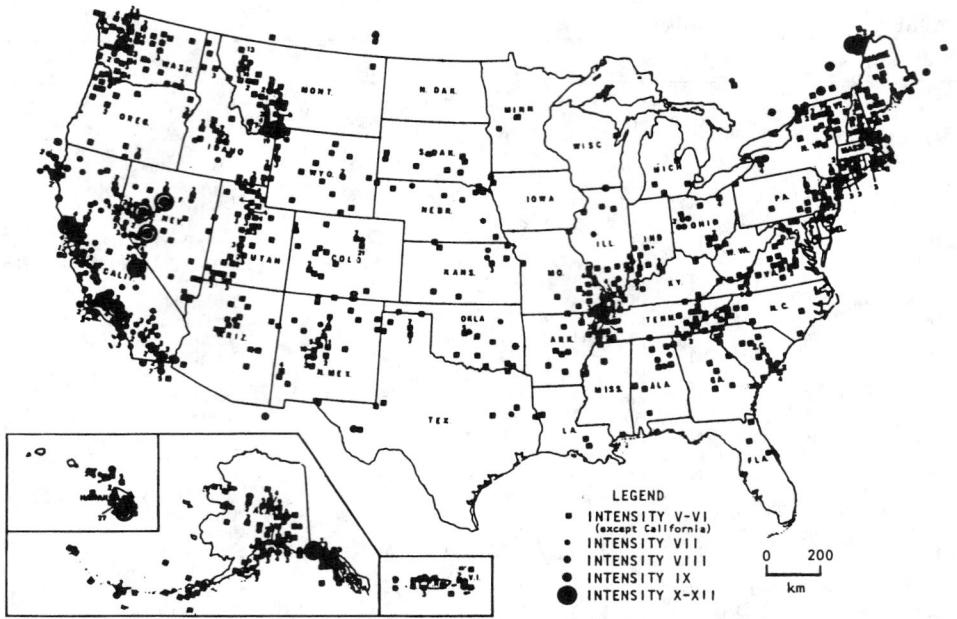

Figure 5.5   U.S. seismicity. (From Algermissen, S. T., *An Introduction to the Seismicity of the United States,* Earthquake Engineering Research Institute, Oakland, CA, 1983. With permission. Also after Coffman, J. L., von Hake, C. A., and Stover, C. W., *Earthquake History of the United States,* U.S. Department of Commerce, NOAA, Pub. 41-1, Washington, 1980.)

earthquakes are most likely to occur. Several significant historical earthquakes have occurred, such as in Charleston, South Carolina, in 1886, and New Madrid, Missouri, in 1811 and 1812, indicating that there is potential for very large and destructive earthquakes [56, 131]. However, most earthquakes in the eastern U.S. are smaller magnitude events. Because of regional geologic differences, eastern and central U.S. earthquakes are felt at much greater distances than those in the western U.S., sometimes up to a thousand miles away [58].

Other Areas

Table 5.2 provides a list of selected 20th-century earthquakes with fatalities of approximately 10,000 or more. All the earthquakes are in the Trans-Alpide belt or the circum-Pacific ring of fire, and the great loss of life is almost invariably due to low-strength masonry buildings and dwellings. Exceptions to this rule are the 1923 Kanto (Japan) earthquake, where most of the approximately 140,000 fatalities were due to fire; the 1970 Peru earthquake, where large landslides destroyed whole towns; and the 1988 Armenian earthquake, where large numbers were killed in Spitak and Leninakan due to poor quality pre-cast concrete construction. The Trans-Alpide belt includes the Mediterranean, which has very significant seismicity in North Africa, Italy, Greece, and Turkey due to the Africa plate's motion relative to the Eurasian plate; the Caucasus (e.g., Armenia) and the Middle East (Iran, Afghanistan), due to the Arabian plate being forced northeastward into the Eurasian plate by the African plate; and the Indian sub-continent (Pakistan, northern India), and the subduction boundary along the southwestern side of Sumatra and Java, which are all part of the Indian-Australian plate. Seismicity also extends northward through Burma and into western China. The Philippines, Taiwan, and Japan are all on the western boundary of the Philippines sea plate, which is part of the circum-Pacific ring of fire.

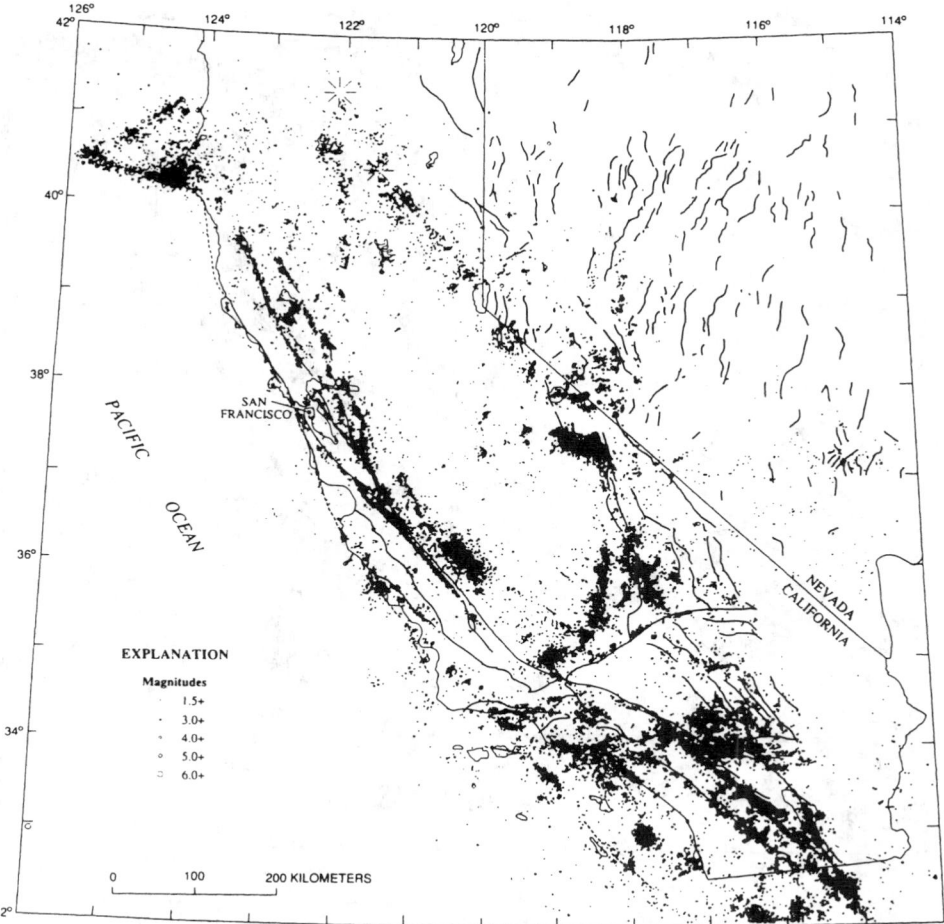

Figure 5.6 Seismicity for California and Nevada, 1980 to 1986. $M > 1.5$ (Courtesy of Jennings, C. W., Fault Activity Map of California and Adjacent Areas, Department of Conservation, Division of Mines and Geology, Sacramento, CA, 1994.)

Japan is an island archipelago with a long history of damaging earthquakes [128] due to the interaction of four tectonic plates (Pacific, Eurasian, North American, and Philippine) which all converge near Tokyo. Figure 5.7 indicates the pattern of Japanese seismicity, which is seen to be higher in the north of Japan. However, central Japan is still an area of major seismic risk, particularly Tokyo, which has sustained a number of damaging earthquakes in history. The Great Kanto earthquake of 1923 (M7.9, about 140,000 fatalities) was a great subduction earthquake, and the 1855 event (M6.9) had its epicenter in the center of present-day Tokyo. Most recently, the 1995 MW 6.9 Hanshin (Kobe) earthquake caused approximately 6,000 fatalities and severely damaged some modern structures as well as many structures built prior to the last major updating of the Japanese seismic codes (ca. 1981).

The predominant seismicity in the Kuriles, Kamchatka, the Aleutians, and Alaska is due to subduction of the Pacific Plate beneath the North American plate (which includes the Aleutians and extends down through northern Japan to Tokyo). The predominant seismicity along the western boundary of North American is due to transform faults (i.e., strike-slip) as the Pacific Plate displaces northwestward relative to the North American plate, although the smaller Juan de Fuca plate offshore

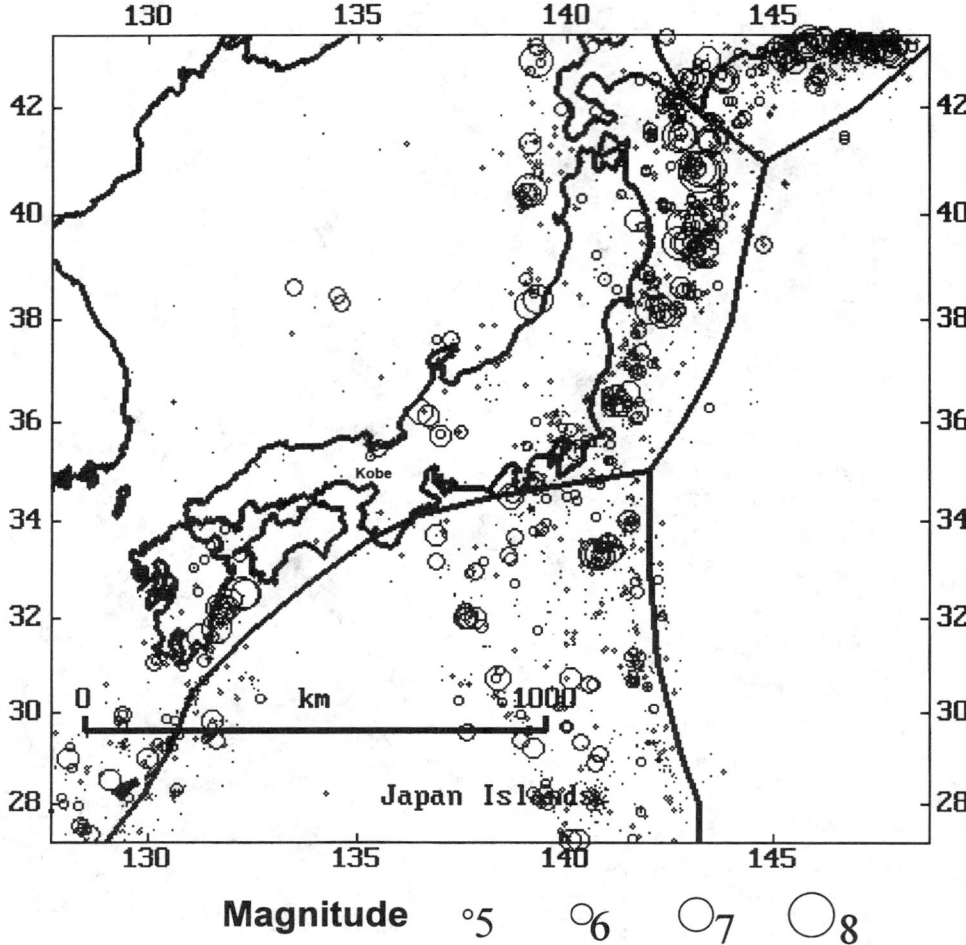

Figure 5.7   Japanese seismicity (1960 to 1965).

Washington and Oregon, and the still smaller Gorda plate offshore northern California, are driven into subduction beneath North American by the Pacific Plate. Further south, the Cocos plate is similarly subducting beneath Mexico and Central America due to the Pacific Plate, while the Nazca Plate lies offshore South America. Lesser but still significant seismicity occurs in the Caribbean, primarily along a series of trenches north of Puerto Rico and the Windward islands. However, the southern boundary of the Caribbean plate passes through Venezuela, and was the source of a major earthquake in Caracas in 1967. New Zealand's seismicity is due to a major plate boundary (Pacific with Indian-Australian plates), which transitions from thrust to transform from the South to the North Island [108]. Lesser but still significant seismicity exists in Iceland where it is accompanied by volcanism due to a spreading boundary between the North American and Eurasian plates, and through Fenno-Scandia, due to tectonics as well as glacial rebound. This very brief tour of the major seismic belts of the globe is not meant to indicate that damaging earthquakes cannot occur elsewhere — earthquakes can and have occurred far from major plate boundaries (e.g., the 1811-1812 New Madrid intraplate events, with several being greater than magnitude 8), and their potential should always be a consideration in the design of a structure.

TABLE 5.2   Selected 20th Century Earthquakes with Fatalities Greater than 10,000

Yr	M	D	Lat.	Long.	M	MMI	Deaths	Damage USD millions	Locale
1976	7	27	39.5 N	118 E	8	10	655,237	$2,000	China: NE: Tangshan
1920	12	16	36.5 N	106 E	8.5	—	200,000		China: Gansu and Shanxi
1923	9	1	35.3 N	140 E	8.2	—	142,807	$2,800	Japan: Toyko, Yokohama, Tsunami
1908	2	0	38.2 N	15.6 E	7.5	—	75,000		Italy: Sicily
1932	12	25	39.2 N	96.5 E	7.6	—	70,000		China: Gansu Province
1970	5	31	9.1 S	78.8 W	7.8	9	67,000	$500	Peru
1990	6	20	37 N	49.4 E	7.7	7	50,000		Iran: Manjil
1927	5	22	37.6 N	103 E	8	—	40,912		China: Gansu Province
1915	1	13	41.9 N	13.6 E	7	11	35,000		Italy: Abruzzi, Avezzano
1935	5	30	29.5 N	66.8 E	7.5	10	30,000		Pakistan: Quetta
1939	12	26	39.5 N	38.5 E	7.9	12	30,000		Turkey: Erzincan
1939	1	25	36.2 S	72.2 W	8.3	—	28,000	$100	Chile: Chillan
1978	9	16	33.4 N	57.5 E	7.4	—	25,000	$11	Iran: Tabas
1988	12	7	41 N	44.2 E	6.8	10	25,000	$16,200	CIS: Armenia
1976	2	4	15.3 N	89.2 W	7.5	9	22,400	$6,000	Guatemala: Tsunami
1974	5	10	28.2 N	104 E	6.8	—	20,000		China: Yunnan and Sichuan
1948	10	5	37.9 N	58.6 E	7.2	—	19,800		CIS: Turkmenistan: Aschabad
1905	4	4	33 N	76 E	8.6	—	19,000		India: Kangra
1917	1	21	8 S	115 E	—	—	15,000		Indonesia: Bali, Tsunami
1968	8	31	33.9 N	59 E	7.3	—	15,000		Iran
1962	9	1	35.6 N	49.9 E	7.3	—	12,225		Iran: NW
1907	10	21	38.5 N	67.9 E	7.8	9	12,000		CIS: Uzbekistan: SE
1960	2	29	30.4 N	9.6 W	5.9	—	12,000		Morocco: Agadir
1980	10	10	36.1 N	1.4 E	7.7	—	11,000		Algeria: Elasnam
1934	1	15	26.5 N	86.5 E	8.4	—	10,700		Nepal-India
1918	2	13	23.5 N	117 E	7.3	—	10,000		China: Guangdong Province
1933	8	25	32 N	104 E	7.4	—	10,000		China: Sichuan Province
1975	2	4	40.6 N	123 E	7.4	10	10,000		China: NE: Yingtao

From NEIC, Database of Significant Earthquakes Contained in *Seismicity Catalogs*, National Earthquake Information Center, Goldon, CO, 1996. With permission.

5.2.3   Measurement of Earthquakes

Earthquakes are complex multi-dimensional phenomena, the scientific analysis of which requires measurement. Prior to the invention of modern scientific instruments, earthquakes were qualitatively measured by their effect or **intensity**, which differed from point-to-point. With the deployment of seismometers, an instrumental quantification of the entire earthquake event — the unique **magnitude** of the event — became possible. These are still the two most widely used measures of an earthquake, and a number of different scales for each have been developed, which are sometimes confused.[1] Engineering design, however, requires measurement of earthquake phenomena in units such as force or displacement. This section defines and discusses each of these measures.

Magnitude

An individual earthquake is a unique release of strain energy. Quantification of this energy has formed the basis for measuring the earthquake event. Richter [103] was the first to define earthquake magnitude as

$$M_L = \log A - \log A_o \tag{5.5}$$

---

[1]Earthquake magnitude and intensity are analogous to a lightbulb and the light it emits. A particular lightbulb has only one energy level, or wattage (e.g., 100 watts, analogous to an earthquake's magnitude). Near the lightbulb, the light intensity is very bright (perhaps 100 ft-candles, analogous to MMI IX), while farther away the intensity decreases (e.g., 10 ft-candles, MMI V). A particular earthquake has only one magnitude value, whereas it has many intensity values.

where M_L is **local magnitude** (which Richter only defined for Southern California), A is the maximum trace amplitude in microns recorded on a standard Wood-Anderson short-period torsion seismometer,[2] at a site 100 km from the epicenter, log A_o is a standard value as a function of distance, for instruments located at distances other than 100 km and less than 600 km. Subsequently, a number of other magnitudes have been defined, the most important of which are **surface wave magnitude** M_S, **body wave magnitude** m_b, and **moment magnitude** M_W. Due to the fact that M_L was only locally defined for California (i.e., for events within about 600 km of the observing stations), surface wave magnitude M_S was defined analogously to M_L using teleseismic observations of surface waves of 20-s period [103]. Magnitude, which is defined on the basis of the amplitude of ground displacements, can be related to the total energy in the expanding wave front generated by an earthquake, and thus to the total energy release. An empirical relation by Richter is

$$\log_{10} E_s = 11.8 + 1.5 M_s \tag{5.6}$$

where E_s is the total energy in ergs.[3] Note that $10^{1.5} = 31.6$, so that an increase of one magnitude unit is equivalent to 31.6 times more energy release, two magnitude units increase is equivalent to 1000 times more energy, etc. Subsequently, due to the observation that deep-focus earthquakes commonly do not register measurable surface waves with periods near 20 s, a body wave magnitude m_b was defined [49], which can be related to M_s [38]:

$$m_b = 2.5 + 0.63 M_s \tag{5.7}$$

Body wave magnitudes are more commonly used in eastern North America, due to the deeper earthquakes there. A number of other magnitude scales have been developed, most of which tend to *saturate* — that is, asymptote to an upper bound due to larger earthquakes radiating significant amounts of energy at periods longer than used for determining the magnitude (e.g., for M_s, defined by measuring 20 s surface waves, saturation occurs at about $M_s > 7.5$). More recently, **seismic moment** has been employed to define a **moment magnitude** M_w([53]; also denoted as bold-face *M*) which is finding increased and widespread use:

$$\log M_o = 1.5 M_w + 16.0 \tag{5.8}$$

where seismic moment M_o (dyne-cm) is defined as [74]

$$M_o = \mu A \bar{u} \tag{5.9}$$

where μ is the material shear modulus, A is the area of fault plane rupture, and $\bar{u}$ is the mean relative displacement between the two sides of the fault (the averaged fault slip). Comparatively, M_w and M_s are numerically almost identical up to magnitude 7.5. Figure 5.8 indicates the relationship between moment magnitude and various magnitude scales.

For lay communications, it is sometimes customary to speak of great earthquakes, large earthquakes, etc. There is no standard definition for these, but the following is an approximate categorization:

Earthquake Magnitude*	Micro Not felt	Small < 5	Moderate 5 ∼ 6.5	Large 6.5 ∼ 8	Great > 8

* Not specifically defined.

---

[2]The instrument has a natural period of 0.8 s, critical damping ration 0.8, magnification 2,800.
[3]Richter [104] gives 11.4 for the constant term, rather than 11.8, which is based on subsequent work. The uncertainty in the data make this difference, equivalent to an energy factor = 2.5 or 0.27 magnitude units, inconsequential.

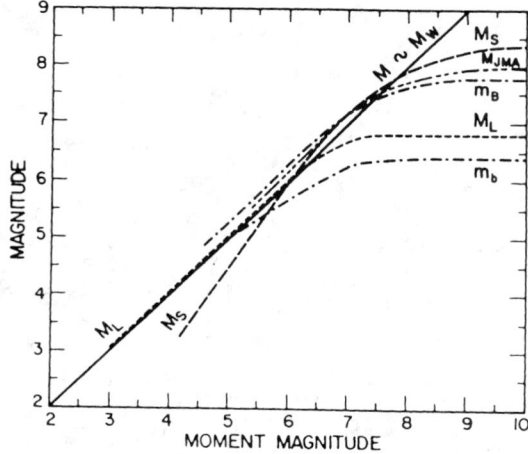

Figure 5.8 Relationship between moment magnitude and various magnitude scales. (From Campbell, K. W., Strong Ground Motion Attenuation Relations: A Ten-Year Perspective, *Earthquake Spectra*, 1(4), 759-804, 1985. With permission.)

From the foregoing discussion, it can be seen that magnitude and energy are related to fault rupture length and slip. Slemmons [114] and Bonilla et al. [17] have determined statistical relations between these parameters, for worldwide and regional data sets, segregated by type of faulting (normal, reverse, strike-slip). The worldwide results of Bonilla et al. for all types of faults are

$$M_s = 6.04 + 0.708 \log_{10} L \quad s = .306 \tag{5.10}$$

$$\log_{10} L = -2.77 + 0.619 M_s \quad s = .286 \tag{5.11}$$

$$M_s = 6.95 + 0.723 \log_{10} d \quad s = .323 \tag{5.12}$$

$$\log_{10} d = -3.58 + 0.550 M_s \quad s = .282 \tag{5.13}$$

which indicates, for example that, for $M_s = 7$, the average fault rupture length is about 36 km (and the average displacement is about 1.86 m). Conversely, a fault of 100 km length is capable of about a $M_s = 7.5$[4] event. More recently, Wells and Coppersmith [130] have performed an extensive analysis of a dataset of 421 earthquakes. Their results are presented in Table 5.3a and b.

Intensity

In general, seismic intensity is a measure of the effect, or the strength, of an earthquake hazard at a specific location. While the term can be applied generically to engineering measures such as peak ground acceleration, it is usually reserved for qualitative measures of location-specific earthquake effects, based on observed human behavior and structural damage. Numerous intensity scales were developed in pre-instrumental times. The most common in use today are the **Modified Mercalli Intensity** (**MMI**) [134], Rossi-Forel (**R-F**), Medvedev-Sponheur-Karnik (**MSK**) [80], and the Japan Meteorological Agency (**JMA**) [69] scales.

MMI is a subjective scale defining the level of shaking at specific sites on a scale of I to XII. (MMI is expressed in Roman numerals to connote its approximate nature). For example, moderate shaking that causes few instances of fallen plaster or cracks in chimneys constitutes MMI VI. It is difficult

---

[4]Note that $L = g(M_s)$ should not be inverted to solve for $M_s = f(L)$, as a regression for $y = f(x)$ is different than a regression for $x = g(y)$.

Table 5.3a Regressions of Rupture Length, Rupture Width, Rupture Area and Moment Magnitude

Equation[a]	Slip type[b]	Number of events	Coefficients and standard errors a(sa)	Coefficients and standard errors b(sb)	Standard deviation s	Correlation coefficient r	Magnitude range	Length/width range (km)
$M = a + b*\log(SRL)$	SS	43	5.16(0.13)	1.12(0.08)	0.28	0.91	5.6 to 8.1	1.3 to 432
	R	19	5.00(0.22)	1.22(0.16)	0.28	0.88	5.4 to 7.4	3.3 to 85
	N	15	4.86(0.34)	1.32(0.26)	0.34	0.81	5.2 to 7.3	2.5 to 41
	All	77	5.08(0.10)	1.16(0.07)	0.28	0.89	5.6 to 8.1	1.3 to 432
$\log(SRL) = a + b*M$	SS	43	−3.55(0.37)	0.74(0.05)	0.23	0.91	5.6 to 8.1	1.3 to 432
	R	19	−2.86(0.55)	0.63(0.08)	0.20	0.88	5.4 to 7.4	3.3 to 85
	N	15	−2.01(0.65)	0.50(0.10)	0.21	0.81	5.2 to 7.3	2.5 to 41
	All	77	−3.22(0.27)	0.69(0.04)	0.22	0.89	5.2 to 8.1	1.3 to 432
$M = a + b*\log(RLD)$	SS	93	4.33(0.06)	1.49(0.05)	0.24	0.96	4.8 to 8.1	1.5 to 350
	R	50	4.49(0.11)	1.49(0.09)	0.26	0.93	4.8 to 7.6	1.1 to 80
	N	24	4.34(0.23)	1.54(0.18)	0.31	0.88	5.2 to 7.3	3.8 to 63
	All	167	4.38(0.06)	1.49(0.04)	0.26	0.94	4.8 to 8.1	1.1 to 350
$\log(RLD) = a + b*M$	SS	93	−2.57(0.12)	0.62(0.02)	0.15	0.96	4.8 to 8.1	1.5 to 350
	R	50	−2.42(0.21)	0.58(0.03)	0.16	0.93	4.8 to 7.6	1.1 to 80
	N	24	−1.88(0.37)	0.50(0.06)	0.17	0.88	5.2 to 7.3	3.8 to 63
	All	167	−2.44(0.11)	0.59(0.02)	0.16	0.94	4.8 to 8.1	1.1 to 350
$M = a + b*\log(RW)$	SS	87	3.80(0.17)	2.59(0.18)	0.45	0.84	4.8 to 8.1	1.5 to 350
	R	43	4.37(0.16)	1.95(0.15)	0.32	0.90	4.8 to 7.6	1.1 to 80
	N	23	4.04(0.29)	2.11(0.28)	0.31	0.86	5.2 to 7.3	3.8 to 63
	All	153	4.06(0.11)	2.25(0.12)	0.41	0.84	4.8 to 8.1	1.1 to 350
$\log(RW) = a + b*M$	SS	87	−0.76(0.12)	0.27(0.02)	0.14	0.84	4.8 to 8.1	1.5 to 350
	R	43	−1.61(0.20)	0.41(0.03)	0.15	0.90	4.8 to 7.6	1.1 to 80
	N	23	−1.14(0.28)	0.35(0.05)	0.12	0.86	5.2 to 7.3	3.8 to 63
	All	153	−1.01(0.10)	0.32(0.02)	0.15	0.84	4.8 to 8.1	1.1 to 350
$M = a + b*\log(RA)$	SS	83	3.98(0.07)	1.02(0.03)	0.23	0.96	4.8 to 7.9	3 to 5,184
	R	43	4.33(0.12)	0.90(0.05)	0.25	0.94	4.8 to 7.6	2.2 to 2,400
	N	22	3.93(0.23)	1.02(0.10)	0.25	0.92	5.2 to 7.3	19 to 900
	All	148	4.07(0.06)	0.98(0.03)	0.24	0.95	4.8 to 7.9	2.2 to 5,184
$\log(RA) = a + b*M$	SS	83	−3.42(0.18)	0.90(0.03)	0.22	0.96	4.8 to 7.9	3 to 5,184
	R	43	−3.99(0.36)	0.98(0.06)	0.26	0.94	4.8 to 7.6	2.2 to 2,400
	N	22	−2.87(0.50)	0.82(0.08)	0.22	0.92	5.2 to 7.3	19 to 900
	All	148	−3.49(0.16)	0.91(0.03)	0.24	0.95	4.8 to 7.9	2.2 to 5,184

[a] SRL—surface rupture length (km); RLD—subsurface rupture length (km); RW—downdip rupture width (km); RA—rupture area (km^2).

[b] SS—strike slip; R—reverse; N—normal.

From Wells, D. L. and Coopersmith, K. J., Empirical Relationships Among Magnitude, Rupture Length, Rupture Width, Rupture Area and Surface Displacements, *Bull. Seis. Soc. Am.*, 84(4), 974-1002, 1994. With permission.

Table 5.3b Regressions of Displacement and Moment Magnitude

Equation[a]	Slip type[b]	Number of events	Coefficients and standard errors		Standard deviation s	Correlation coefficient r	Magnitude range	Displacement range (km)
			a(sa)	b(sb)				
M = a + b * log(MD)	SS	43	6.81(0.05)	0.78(0.06)	0.29	0.90	5.6 to 8.1	0.01 to 14.6
	[R[c]	21	*6.52(0.11)*	*0.44(0.26)*	*0.52*	*0.36*	*5.4 to 7.4*	*0.11 to 6.5]*
	N	16	6.61(0.09)	0.71(0.15)	0.34	0.80	5.2 to 7.3	0.06 to 6.1
	All	80	6.69(0.04)	0.74(0.07)	0.40	0.78	5.2 to 8.1	0.01 to 14.6
log(MD) = a + b∗M	SS	43	−7.03(0.55)	1.03(0.08)	0.34	0.90	5.6 to 8.1	0.01 to 14.6
	[R	21	*−1.84(1.14)*	*0.29(0.17)*	*0.42*	*0.36*	*5.4 to 7.4*	*0.11 to 6.5}*
	N	16	−5.90(1.18)	0.89(0.18)	0.38	0.80	5.2 to 7.3	0.06 to 6.1
	All	80	−5.46(0.51)	0.82(0.08)	0.42	0.78	5.2 to 8.1	0.01 to 14.6
M = a + b ∗ log(AD)	SS	29	7.04(0.05)	0.89(0.09)	0.28	0.89	5.6 to 8.1	0.05 to 8.0
	[R	15	*6.64(0.16)*	*0.13(0.36)*	*0.50*	*0.10*	*5.8 to 7.4*	*0.06 to 1.5}*
	N	12	6.78(0.12)	0.65(0.25)	0.33	0.64	6.0 to 7.3	0.08 to 2.1
	All	56	6.93(0.05)	0.82(0.10)	0.39	0.75	5.6 to 8.1	0.05 to 8.0
log(AD) = a + b∗ M	SS	29	−6.32(0.61)	0.90(0.09)	0.28	0.89	5.6 to 8.1	0.05 to 8.0
	[R	15	*−0.74(1.40)*	*0.08(0.21)*	*0.38*	*0.10*	*5.8 to 7.4*	*0.06 to 1.5}*
	N	12	−4.45(1.59)	0.63(0.24)	0.33	0.64	6.0 to 7.3	0.08 to 2.1
	All	56	−4.80(0.57)	0.69(0.08)	0.36	0.75	5.6 to 8.1	0.05 to 8.0

[a] MD—maximum displacement (m); AD—average displacement (M).

[b] SS—strike slip; R—reverse; N—normal.

[c] Regressions for reverse-slip relationships shown in italics and brackets are not significant at a 95% probability level.

From Wells, D. L. and Coopersmith, K. J., Empirical Relationships Among Magnitude, Rupture Length, Rupture Width, Rupture Area and Surface Displacements, *Bull. Seis. Soc. Am.*, 84(4), 974-1002, 1994. With permission.

to find a reliable relationship between magnitude, which is a description of the earthquake's total energy level, and intensity, which is a subjective description of the level of shaking of the earthquake at specific sites, because shaking severity can vary with building type, design and construction practices, soil type, and distance from the event.

TABLE 5.4   Modified Mercalli Intensity Scale of 1931

I	Not felt except by a very few under especially favorable circumstances.
II	Felt only by a few persons at rest, especially on upper floors of buildings. Delicately suspended objects may swing.
III	Felt quite noticeably indoors, especially on upper floors of buildings, but many people do not recognize it as an earthquake. Standing motor cars may rock slightly. Vibration like passing truck. Duration estimated.
IV	During the day felt indoors by many, outdoors by few. At night some awakened. Dishes, windows, and doors disturbed; walls make creaking sound. Sensation like heavy truck striking building. Standing motor cars rock noticeably.
V	Felt by nearly everyone; many awakened. Some dishes, windows, etc. broken; a few instances of cracked plaster; unstable objects overturned. Disturbance of trees, poles, and other tall objects sometimes noticed. Pendulum clocks may stop.
VI	Felt by all; many frightened and run outdoors. Some heavy furniture moved; a few instances of fallen plaster or damaged chimneys. Damage slight.
VII	Everybody runs outdoors. Damage negligible in buildings of good design and construction slight to moderate in well built ordinary structures; considerable in poorly built or badly designed structures. Some chimneys broken. Noticed by persons driving motor cars.
VIII	Damage slight in specially designed structures; considerable in ordinary substantial buildings, with partial collapse; great in poorly built structures. Panel walls thrown out of frame structures. Fall of chimneys, factory stacks, columns, monuments, walls. Heavy furniture overturned. Sand and mud ejected in small amounts. Changes in well water. Persons driving motor cars disturbed.
IX	Damage considerable in specially designed structures; well-designed frame structures thrown out of plumb; great in substantial buildings, with partial collapse. Buildings shifted off foundations. Ground cracked conspicuously. Underground pipes broken.
X	Some well-built wooden structures destroyed; most masonry and frame structures destroyed with foundations; ground badly cracked. Rails bent. Landslides considerable from river banks and steep slopes. Shifted sand and mud. Water splashed over banks.
XI	Few, if any (masonry), structures remain standing. Bridges destroyed. Broad fissures in ground. Underground pipelines completely out of service. Earth slumps and land slips in soft ground. Rails bent greatly.
XII	Damage total. Waves seen on ground surfaces. Lines of sight and level distorted. Objects thrown upward into the air.

After Wood, H. O. and Neumann, Fr., Modified Mercalli Intensity Scale of 1931, *Bull. Seis. Soc. Am.*, 21, 277-283, 1931.

Note that MMI X is the maximum considered physically possible due to "mere" shaking, and that MMI XI and XII are considered due more to permanent ground deformations and other geologic effects than to shaking.

Other intensity scales are defined analogously (see Table 5.5, which also contains an approximate conversion from MMI to acceleration a [PGA, in cm/s^2, or gals]). The conversion is due to Richter [103] (other conversions are also available [84].

$$\log a = MMI/3 - 1/2 \qquad (5.14)$$

Intensity maps are produced as a result of detailed investigation of the type of effects tabulated in Table 5.4, as shown in Figure 5.9 for the 1994 M_W 6.7 Northridge earthquake. Correlations have been developed between the area of various MMIs and earthquake magnitude, which are of value for seismological and planning purposes.

Figure 10 correlates A_{felt} vs. M_W. For pre-instrumental historical earthquakes, A_{felt} can be estimated from newspapers and other reports, which then can be used to estimate the event magnitude, thus supplementing the seismicity catalog. This technique has been especially useful in regions with a long historical record [4, 133].

Time History

Sensitive strong motion seismometers have been available since the 1930s, and they record actual ground motions specific to their location (Figure 5.11). Typically, the ground motion records,

TABLE 5.5   Comparison of Modified Mercalli (MMI) and Other Intensity Scales

a^a	MMI[b]	R-F[c]	MSK[d]	JMA[e]
0.7	I	I	I	0
1.5	II	I to II	II	I
3	III	III	III	II
7	IV	IV to V	IV	II to III
15	V	V to VI	V	III
32	VI	VI to VII	VI	IV
68	VII	VIII-	VII	IV to V
147	VIII	VIII+ to IX−	VIII	V
316	IX	IX+	IX	V to VI
681	X	X	X	VI
$(1468)^f$	XI	—	XI	VII
$(3162)^f$	XII	—	XII	

[a] gals
[b] Modified Mercalli Intensity
[c] Rossi-Forel
[d] Medvedev-Sponheur-Karnik
[e] Japan Meteorological Agency
[f] a values provided for reference only. MMI > X are due more to geologic effects.

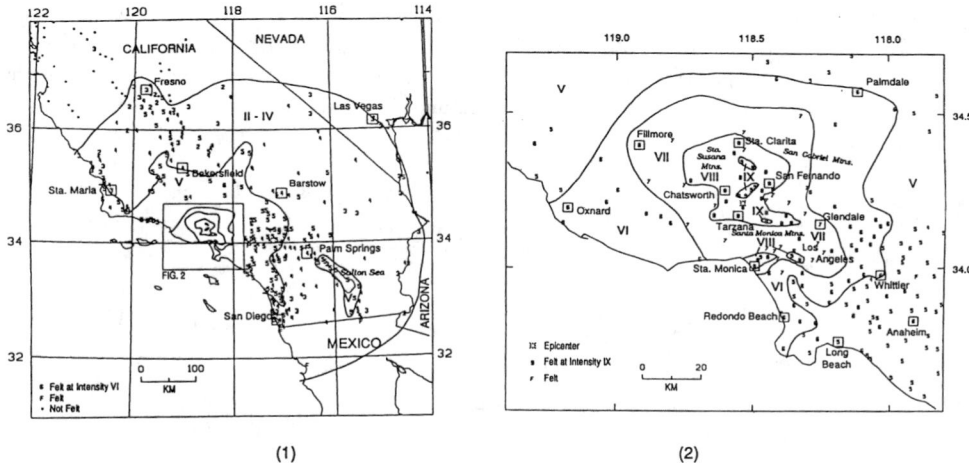

(1)   (2)

Figure 5.9   MMI maps, 1994 M_W 6.7 Northridge Earthquake. (1) Far-field isoseismal map. Roman numerals give average MMI for the regions between isoseismals; arabic numerals represent intensities in individual communities. Squares denote towns labeled in the figure. Box labeled "FIG. 2" identifies boundaries of that figure. (2) Distribution of MMI in the epicentral region. (Courtesy of Dewey, J.W. et al., Spacial Variations of Intensity in the Northridge Earthquake, in Woods, M.C. and Seiple, W.R., Eds., The Northridge California Earthquake of 17 January 1994, California Department of Conservation, Division of Mines and Geology, Special Publication 116, 39-46, 1995.)

termed *seismographs* or *time histories*, have recorded acceleration (these records are termed *accelerograms*), for many years in analog form on photographic film and, more recently, digitally. Analog records required considerable effort for correction due to instrumental drift, before they could be used.

Time histories theoretically contain complete information about the motion at the instrumental location, recording three *traces* or orthogonal records (two horizontal and one vertical). Time histories (i.e., the earthquake motion at the site) can differ dramatically in duration, frequency content, and amplitude. The maximum amplitude of recorded acceleration is termed the **peak ground acceleration**, PGA (also termed the ZPA, or **zero period acceleration**). Peak ground

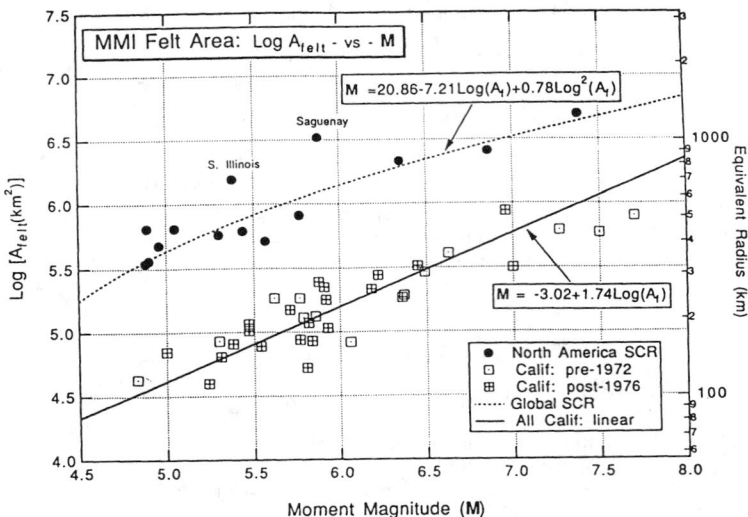

Figure 5.10 $\log A_{felt}$ (km^2) vs. M_W. Solid circles denote ENA events and open squares denote California earthquakes. The dashed curve is the $M_W - A_{felt}$ relationship of an earlier study, whereas the solid line is the fit determined by Hanks and Johnston, for California data. (Courtesy of Hanks J. W. and Johnston A. C., Common Features of the Excitation and Propagation of Strong Ground Motion for North American Earthquakes, *Bull. Seis. Soc. Am.*, 82(1), 1-23, 1992.)

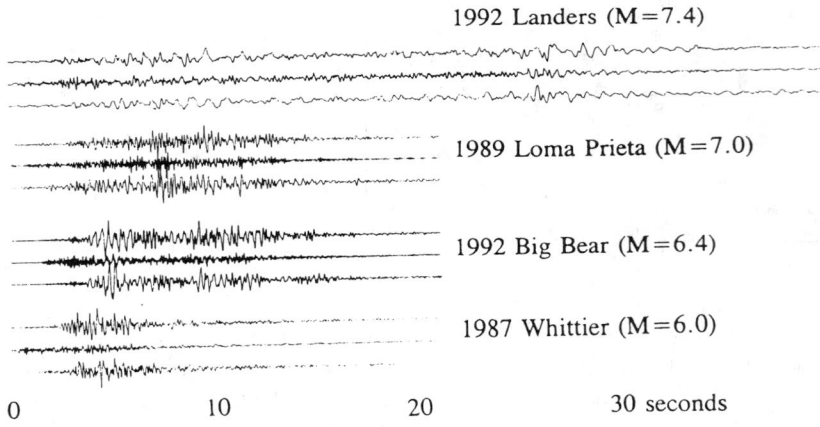

Figure 5.11 Typical earthquake accelerograms. (Courtesy of Darragh, R. B., Huang, M. J., and Shakal, A. F., Earthquake Engineering Aspects of Strong Motion Data from Recent California Earthquakes, *Proc. Fifth U.S. Natl. Conf. Earthquake Eng.*, 3, 99-108, 1994, Earthquake Engineering Research Institute. Oakland, CA.)

velocity (PGV) and peak ground displacement (PGD) are the maximum respective amplitudes of velocity and displacement. Acceleration is normally recorded, with velocity and displacement being determined by numerical integration; however, velocity and displacement meters are also deployed, to a lesser extent. Acceleration can be expressed in units of cm/s^2 (termed **gals**), but is often also expressed in terms of the fraction or percent of the acceleration of gravity (980.66 gals, termed 1g). Velocity is expressed in cm/s (termed **kine**). Recent earthquakes (1994 Northridge, M_w 6.7 and 1995 Hanshin [Kobe] M_w 6.9) have recorded PGA's of about 0.8g and PGV's of about 100 kine — almost 2g was recorded in the 1992 Cape Mendocino earthquake.

Elastic Response Spectra

If the SDOF mass in Figure 5.1 is subjected to a time history of ground (i.e., base) motion similar to that shown in Figure 5.11, the elastic **structural response** can be readily calculated as a function of time, generating a **structural response time history**, as shown in Figure 5.12 for several oscillators with differing natural periods. The response time history can be calculated by direct

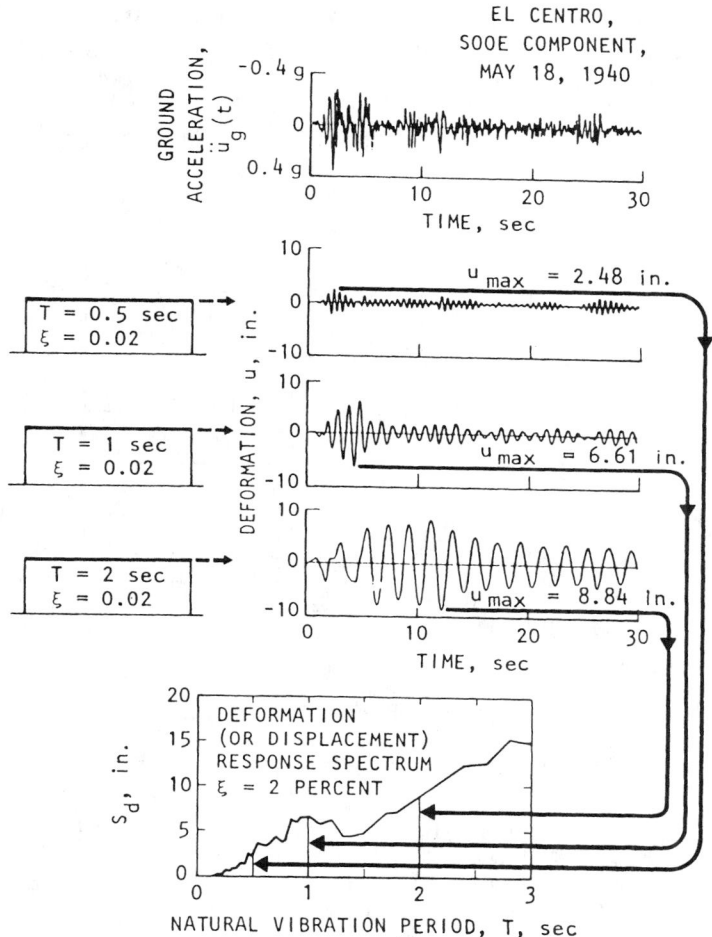

Figure 5.12   Computation of deformation (or displacement) response spectrum. (From Chopra, A. K., *Dynamics of Structures, A Primer,* Earthquake Engineering Research Institute, Oakland, CA, 1981. With permission.)

integration of Equation 5.1 in the **time domain**, or by solution of the **Duhamel** integral [32]. However, this is time-consuming, and the elastic response is more typically calculated in the **frequency domain**

$$v(t) = \frac{1}{2\pi} \int_{\varpi=-\infty}^{\infty} H(\varpi)c(\varpi)\exp(i\varpi t)d\varpi \tag{5.15}$$

where

$v(t)$ = the elastic structural displacement response time history
ϖ = frequency

$H(\varpi) = \dfrac{1}{-\varpi^2 m + ic + k}$ is the complex frequency response function

$c(\varpi) = \int_{\varpi=-\infty}^{\infty} p(t)\exp(-i\varpi t)dt$ is the Fourier transform of the input motion (i.e., the Fourier transform of the ground motion time history)

which takes advantage of computational efficiency using the Fast Fourier Transform.

For design purposes, it is often sufficient to know only the maximum amplitude of the response time history. If the natural period of the SDOF is varied across a spectrum of engineering interest (typically, for natural periods from .03 to 3 or more seconds, or frequencies of 0.3 to 30+ Hz), then the plot of these maximum amplitudes is termed a **response spectrum**. Figure 5.12 illustrates this process, resulting in S_d, the *displacement response spectrum*, while Figure 5.13 shows (a) the S_d, displacement response spectrum, (b) S_v, the *velocity response spectrum* (also denoted PSV, the

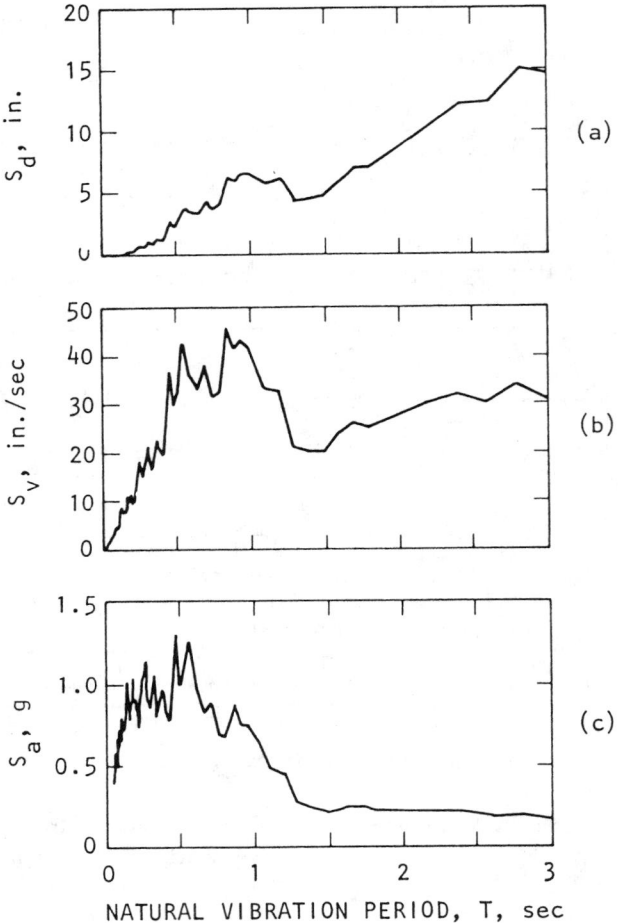

Figure 5.13 Response spectra spectrum. (From Chopra, A. K., *Dynamics of Structures, A Primer*, Earthquake Engineering Research Institute, Oakland, CA, 1981. With permission.)

pseudo spectral velocity, pseudo to emphasize that this spectrum is not exactly the same as the relative velocity response spectrum [63], and (c) S_a, the acceleration *response spectrum*. Note that

$$S_v = \frac{2\pi}{T} S_d = \varpi S_d \qquad (5.16)$$

and

$$S_a = \frac{2\pi}{T} S_v = \varpi S_v = \left(\frac{2\pi}{T}\right)^2 S_d = \varpi^2 S_d \tag{5.17}$$

Response spectra form the basis for much modern earthquake engineering structural analysis and design. They are readily calculated *if* the ground motion is known. For design purposes, however, response spectra must be estimated. This process is discussed below. Response spectra may be plotted in any of several ways, as shown in Figure 5.13 with arithmetic axes, and in Figure 5.14 where the velocity response spectrum is plotted on tripartite logarithmic axes, which equally enables reading of displacement and acceleration response. Response spectra are most normally presented for 5% of critical damping.

While actual response spectra are irregular in shape, they generally have a concave-down arch or trapezoidal shape, when plotted on tripartite log paper. Newmark observed that response spectra tend to be characterized by three regions: (1) a region of constant acceleration, in the high frequency portion of the spectra; (2) constant displacement, at low frequencies; and (3) constant velocity, at intermediate frequencies, as shown in Figure 5.15. If a **spectrum amplification factor** is defined as the ratio of the spectral parameter to the ground motion parameter (where parameter indicates acceleration, velocity or displacement), then response spectra can be estimated from the data in Table 5.6, provided estimates of the ground motion parameters are available. An example spectra using these data is given in Figure 5.15.

TABLE 5.6 Spectrum Amplification Factors for Horizontal Elastic Response

Damping,	One sigma (84.1%)			Median (50%)		
% Critical	A	V	D	A	V	D
0.5	5.10	3.84	3.04	3.68	2.59	2.01
1	4.38	3.38	2.73	3.21	2.31	1.82
2	3.66	2.92	2.42	2.74	2.03	1.63
3	3.24	2.64	2.24	2.46	1.86	1.52
5	2.71	2.30	2.01	2.12	1.65	1.39
7	2.36	2.08	1.85	1.89	1.51	1.29
10	1.99	1.84	1.69	1.64	1.37	1.20
20	1.26	1.37	1.38	1.17	1.08	1.01

From Newmark, N. M. and Hall, W. J., *Earthquake Spectra and Design*, Earthquake Engineering Research Institute, Oakland, CA, 1982. With permission.

A standardized response spectra is provided in the Uniform Building Code [126] for three soil types. The spectra is a smoothed average of normalized 5% damped spectra obtained from actual ground motion records grouped by subsurface soil conditions at the location of the recording instrument, and are applicable for earthquakes characteristic of those that occur in California [111]. If an estimate of ZPA is available, these normalized shapes may be employed to determine a response spectra, appropriate for the soil conditions. Note that the maximum amplification factor is 2.5, over a period range approximately 0.15 s to 0.4 - 0.9 s, depending on the soil conditions. Other methods for estimation of response spectra are discussed below.

Inelastic Response Spectra

While the foregoing discussion has been for elastic response spectra, most structures are not expected, or even designed, to remain elastic under strong ground motions. Rather, structures are expected to enter the *inelastic* region — the extent to which they behave inelastically can be defined

RESPONSE SPECTRUM

IMPERIAL VALLEY EARTHQUAKE

MAY 18, 1940 — 2037 PST

IIIA001 40.001.0 EL CENTRO SITE
IMPERIAL VALLEY IRRIGATION DISTRICT COMP S00E
DAMPING VALUES ARE 0, 2, 5, 10, AND 20 PERCENT OF CRITICAL

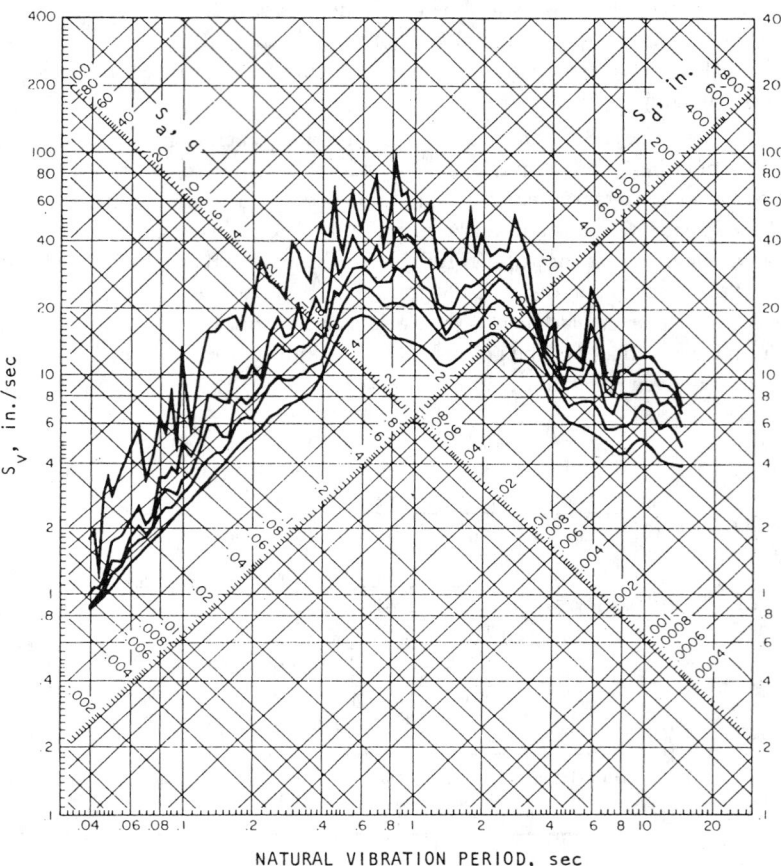

NATURAL VIBRATION PERIOD, sec

Figure 5.14 Response spectra, tri-partite plot (El Centro S 0° E component). (From Chopra, A. K., *Dynamics of Structures, A Primer,* Earthquake Engineering Research Institute, Oakland, CA, 1981. With permission.)

by the **ductility factor,** μ

$$\mu = \frac{u_m}{u_y} \tag{5.18}$$

where u_m is the maximum displacement of the mass under actual ground motions, and u_y is the displacement at yield (i.e., that displacement which defines the extreme of elastic behavior). Inelastic response spectra can be calculated in the time domain by direct integration, analogous to elastic response spectra but with the structural stiffness as a non-linear function of displacement, $k = k(u)$. If elastoplastic behavior is assumed, then elastic response spectra can be readily modified to reflect inelastic behavior [90] on the basis that (a) at low frequencies (0.3 Hz <) displacements are the same; (b) at high frequencies (> 33 Hz), accelerations are equal; and (c) at intermediate frequencies,

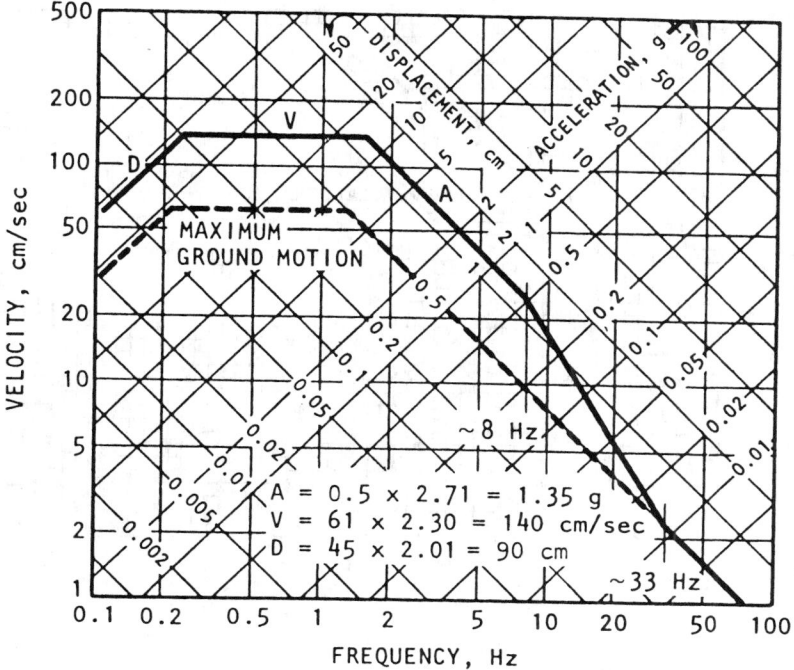

Figure 5.15 Idealized elastic design spectrum, horizontal motion (ZPA = 0.5g, 5% damping, one sigma cumulative probability. (From Newmark, N. M. and Hall, W. J., *Earthquake Spectra and Design,* Earthquake Engineering Research Institute, Oakland, CA, 1982. With permission.)

the absorbed energy is preserved. Actual construction of inelastic response spectra on this basis is shown in Figure 5.17, where $DVAA_o$ is the elastic spectrum, which is reduced to D' and V' by the ratio of $1/\mu$ for frequencies less than 2 Hz, and by the ratio of $1/(2\mu - 1)^{1/2}$ between 2 and 8 Hz. Above 33 Hz there is no reduction. The result is the inelastic acceleration spectrum ($D'V'A'A_o$), while $A''A_o'$ is the inelastic displacement spectrum. A specific example, for ZPA = 0.16g, damping = 5% of critical, and $\mu = 3$ is shown in Figure 5.18.

Response Spectrum Intensity and Other Measures

While the elastic response spectrum cannot directly define damage to a structure (which is essentially inelastic deformation), it captures in one curve the amount of elastic deformation for a wide variety of structural periods, and therefore may be a good overall measure of ground motion intensity. On this basis, Housner defined a **response spectrum intensity** as the integral of the elastic response spectrum velocity over the period range 0.1 to 2.5 s.

$$SI(h) = \int_{T=0.1}^{2.5} Sv(h, T)dT \tag{5.19}$$

where h = damping (as a percentage of c_{crit}). A number of other measures exist, including Fourier amplitude spectrum [32] and Arias Intensity [8]:

$$I_A = \frac{\pi}{g} \int_0^t a^2(t)dt \tag{5.20}$$

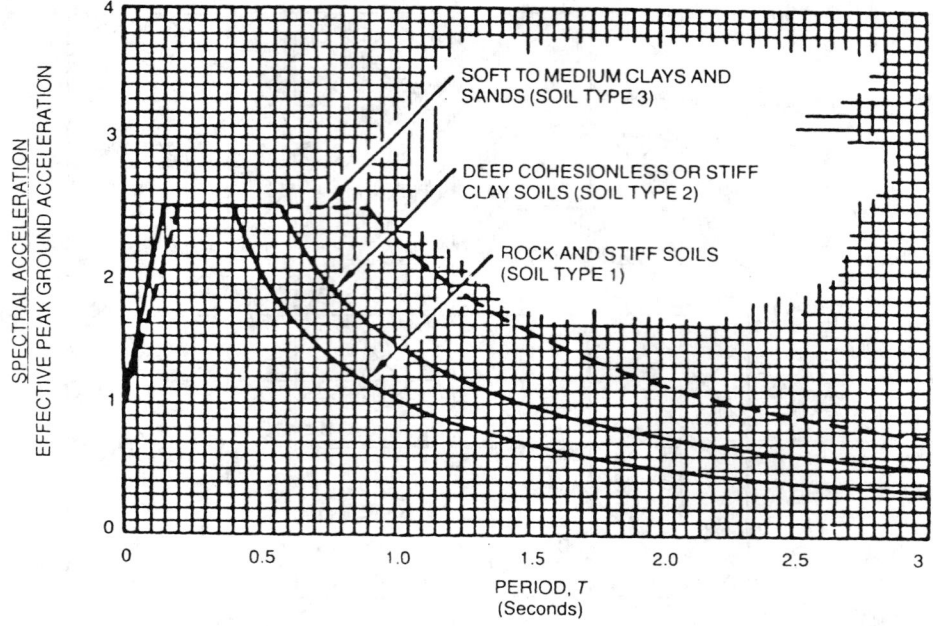

Figure 5.16 Normalized response spectra shapes. (From Uniform Building Code, Structural Engineering Design Provisions, vol. 2, Intl. Conf. Building Officials, Whittier, 1994. With permission.)

Engineering Intensity Scale

Lastly, Blume [14] defined a measure of earthquake intensity, the Engineering Intensity Scale (EIS), which has been relatively underutilized but is worth noting as it attempts to combine the engineering benefits of response spectra with the simplicity of qualitative intensity scales, such as MMI. The EIS is simply a 10x9 matrix which characterizes a 5% damped elastic response spectra (Figure 5.19). Nine period bands (0.01-.1, -.2, -.4, -.6, -1.0, -2.0, -4.0, -7.0, -10,0 s), and ten S_v levels (0.01-0.1, -1.0, -4.0, -10.0, -30.0, -60.0, -100., -300., -1000. kine) are defined. As can be seen, since the response spectrum for the example ground motion in period band II (0.1-0.2 s) is predominantly in S_v level 5 (10-30 kine), it is assigned EIS 5 (X is assigned where the response spectra does not cross a period band). In this manner, a nine-digit EIS can be assigned to a ground motion (in the example, it is X56,777,76X), which can be reduced to three digits (5,7,6) by averaging, or even to one digit (6, for this example). Numerically, single digit EIS values tend to be a unit or so lower than the equivalent MMI intensity value.

5.2.4 Strong Motion Attenuation and Duration

The rate at which earthquake ground motion decreases with distance, termed **attenuation**, is a function of the regional geology and inherent characteristics of the earthquake and its source. Three major factors affect the severity of ground shaking at a site: (1) **source** — the size and type of the earthquake, (2) **path** — the distance from the source of the earthquake to the site and the geologic characteristics of the media earthquake waves pass through, and (3) **site-specific effects** — type of soil at the site. In the simplest of models, if the seismogenic source is regarded as a point, then from considering the relation of energy and earthquake magnitude and the fact that the volume of a hemisphere is proportion to R^3 (where R represents radius), it can be seen that energy per unit volume is proportional to $C 10^{aM} R^{-3}$, where C is a constant or constants dependent on the earth's crustal properties. The constant C will vary regionally — for example, it has long been observed that

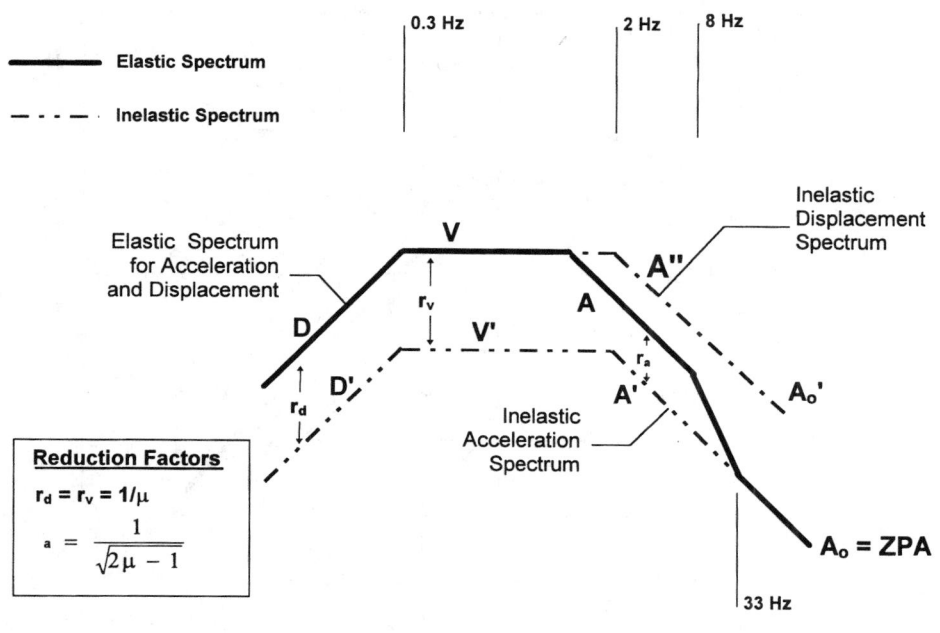

Inelastic Response Spectra for Earthquakes
(elasto-plastic)

Figure 5.17   Inelastic response spectra for earthquakes. (After Newmark, N. M. and Hall, W. J., *Earthquake Spectra and Design,* Earthquake Engineering Research Institute, Oakland, CA, 1982.)

attenuation in eastern North America (ENA) varies significantly from that in western North America (WNA) — earthquakes in ENA are felt at far greater distances. Therefore, attenuation relations are regionally dependent. Another regional aspect of attenuation is the definition of terms, especially magnitude, where various relations are developed using magnitudes defined by local observatories.

A very important aspect of attenuation is the definition of the distance parameter; because attenuation is the change of ground motion with location, this is clearly important. Many investigators use differing definitions; as study has progressed, several definitions have emerged: (1) hypocentral distance (i.e., straight line distance from point of interest to hypocenter, where hypocentral distance may be arbitrary or based on regression rather than observation), (2) epicentral distance, (3) closest distance to the causative fault, and (4) closest horizontal distance from the station to the point on the earth's surface that lies directly above the seismogenic source. In using attenuation relations, it is critical that the correct definition of distance is consistently employed.

Methods for estimating ground motion may be grouped into two major categories: empirical and methods based on seismological models. Empirical methods are more mature than methods based on seismological models, but the latter are advantageous in explicitly accounting for source and path, therefore having explanatory value. They are also flexible, they can be extrapolated with more confidence, and they can be easily modified for additional factors. Most seismological model-based methods are stochastic in nature — Hanks and McGuire's [54] seminal paper has formed the basis for many of these models, which "assume that ground acceleration is a finite-duration segment of a stationary random process, completely characterized by the assumption that acceleration follows Brune's [23] source spectrum (for California data, typically about 100 bars), and that the duration of strong shaking is equal to reciprocal of the source **corner frequency**" f_o (the frequency above which earthquake radiation spectra vary with ϖ^{-3} - below f_o, the spectra are proportional to seismic moment [108]). Since there is substantial ground motion data in WNA, seismological model-based

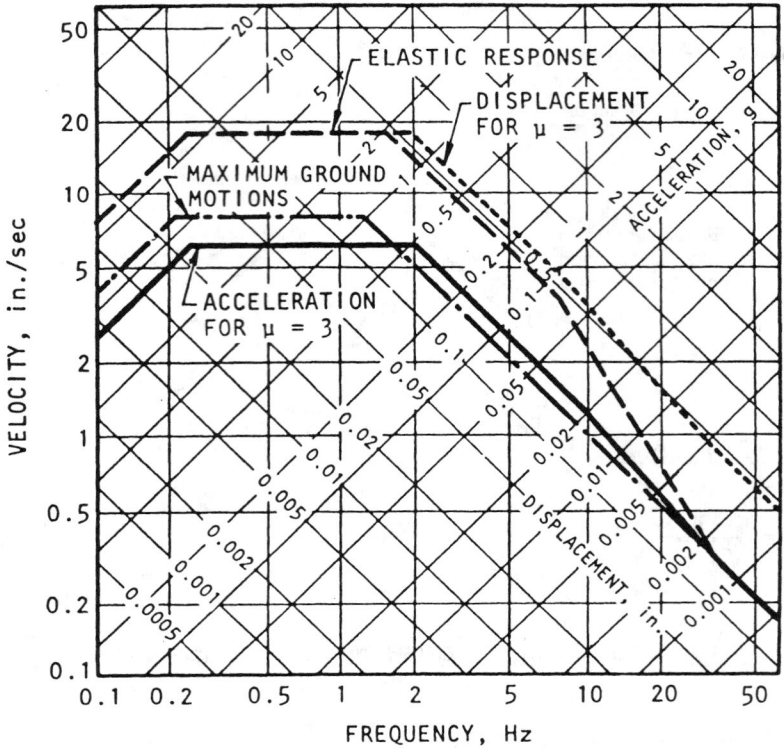

Figure 5.18   Example inelastic response spectra. (From Newmark, N. M. and Hall, W. J., *Earthquake Spectra and Design,* Earthquake Engineering Research Institute, Oakland, CA, 1982. With permission.)

relations have had more value in ENA, where few records exist. The Hanks-McGuire method has, therefore, been usefully applied in ENA [123] where Boore and Atkinson [18] found, for hard-rock sites, the relation:

$$\log y = c_0 + c_1 r - \log r \tag{5.21}$$

where

y	$=$	a ground motion parameter (PSV, unless c_i coefficients for a_{max} are used)
r	$=$	hypocentral distance (km)
c_i	$=$	$\xi_o^i + \sum \xi_n^i (M_W - 6)^n$ $I = 0, 1$ summation for $n = 1, 2, 3$ (see Table 5.7)

Similarly, Toro and McGuire [123] furnish the following relation for rock sites in ENA:

$$\ln Y = c_0 + c_1 M + c_2 \ln(R) + c_3 R \tag{5.22}$$

where the c_0 - c_3 coefficients are provided in Table 5.8, M represents m_{Lg}, and R is the closest distance between the site and the causative fault at a minimum depth of 5 km.

These results are valid for hypocentral distances of 10 to 100 km, and m_{Lg} 4 to 7.

More recently, Boore and Joyner [19] have extended their hard-rock relations to deep soil sites in ENA:

$$\log y = a'' + b(m - 6) + c(m - 6)^2 + d(m - 6)^3 - \log r + kr \tag{5.23}$$

where a'' and other coefficients are given in Table 5.9, m is moment magnitude (M_W), and r is hypocentral distance (km) although the authors suggest that, close to long faults, the distance should be the nearest distance to seismogenic rupture. The coefficients in Table 5.9 should not be used outside the ranges $10 < r < 400$ km, and $5.0 < M_W < 8.5$.

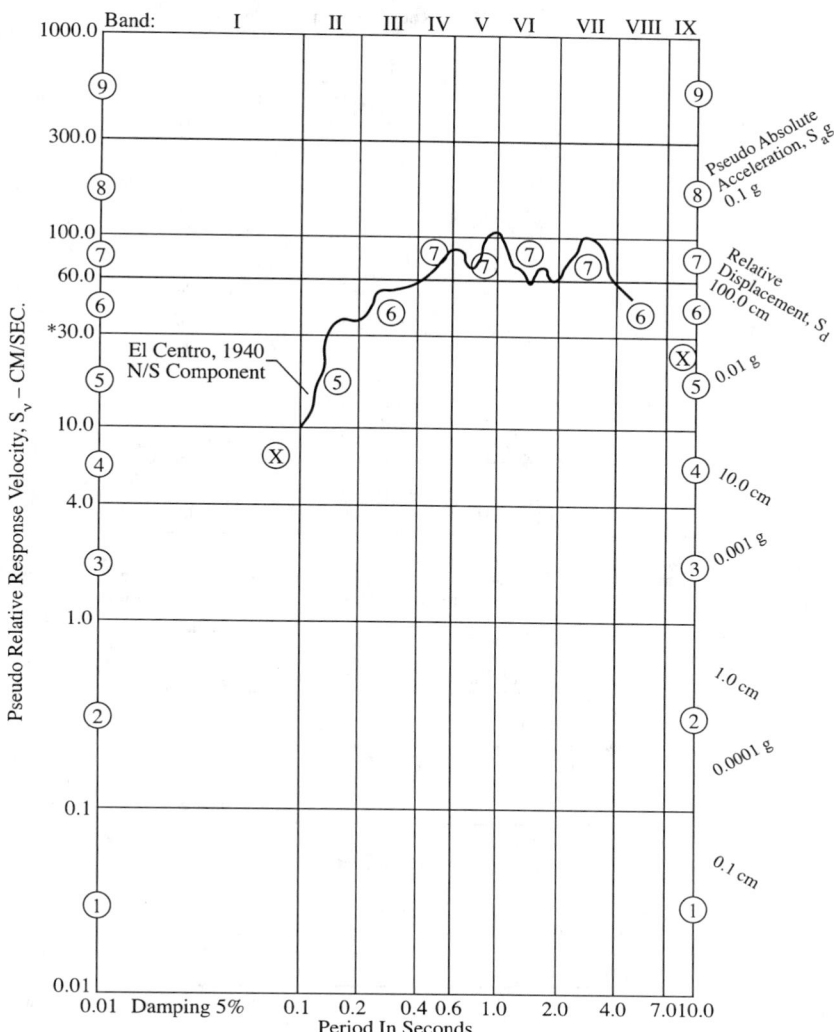

Figure 5.19 Engineering intensity scale (EIS) matrix with example. (From Blume, J. A., An Engineering Intensity Scale for Earthquakes and Other Ground Motions, *Bull. Seis. Soc. Am.,* 60(1), 217-229, 1970. With permission.)

In WNA, due to more data, empirical methods based on regression of the ground motion parameter vs. magnitude and distance have been more widely employed, and Campbell [28] offers an excellent review of North American relations up to 1985. Initial relationships were for PGA, but regression of the amplitudes of response spectra at various periods is now common, including consideration of fault type and effects of soil.

Some current favored relationships are:

Campbell and Bozorgnia [29] (PGA - Worldwide Data)

$$\begin{aligned}
\ln(PGA) = \ &-3.512 + 0.904M - 1.328\ln\sqrt{\{R_s^2 + [0.149\exp(0.647M)]^2\}} \\
&+ [1.125 - 0.112\ln(R_s) - 0.0957M]F \\
&+ [0.440 - 0.171\ln(R_s)]S_{sr} + [0.405 - 0.222\ln(R_s)]S_{hr} + \varepsilon \quad (5.24)
\end{aligned}$$

TABLE 5.7 Eastern North America Hard-Rock Attenuation Coefficients[a]

Response frequency (Hz)		ξ_0	ξ_1	ξ_2	ξ_3
0.2	c_0:	$1.743E+00$	$1.064E+00$	$-4.293E-02$	$-5.364E-02$
	c_1:	$-3.130E-04$	$1.415E-03$	$-1.028E-03$	
0.5	c_0:	$2.141E+00$	$8.521E-01$	$-1.670E-01$	
	c_1:	$-2.504E-04$		$-2.612E-04$	
1.0	c_0:	$2.300E+00$	$6.655E-01$	$-1.538E-01$	
	c_1:	$-1.024E-03$	$-1.144E-04$	$1.109E-04$	
2.0	c_0:	$2.317E+00$	$5.070E-01$	$-9.317E-02$	
	c_1:	$-1.683E-03$	$1.492E-04$	$1.203E-04$	
5.0	c_0:	$2.239E+00$	$3.976E-01$	$-4.564E-02$	
	c_1:	$-2.537E-03$	$5.468E-04$	$7.091E-05$	
10.0	c_0:	$2.144E+00$	$3.617E-01$	$-3.163E-02$	
	c_1:	$-3.094E-03$	$7.640E-04$		
20.0	c_0:	$2.032E+00$	$3.438E-01$	$-2.559E-02$	
	c_1:	$-3.672E-03$	$8.956E-04$	$-4.219E-05$	
a_{max}	c_0:	$3.763E+00$	$3.354E-01$	$-2.473E-02$	
	c_1:	$-3.885E-03$	$1.042E-03$	$-9.169E-05$	

[a] See Equation 5.21.
From Boore, D.M. and Atkinson, G.M., Stochastic Prediction of Ground Motion and Spectral Response Parameters at Hard-Rock Sites in Eastern North America, *Bull. Seis. Soc. Am.*, 77, 440-487, 1987. With permission.

TABLE 5.8 ENA Rock Attenuation Coefficients[a]

Y	c_0	c_1	c_2	c_3
PSRV (1 Hz)	-9.283	2.289	-1.000	$-.00183$
PSRV (5 Hz)	-2.757	1.265	-1.000	$-.00310$
PSRV (10 Hz)	-1.717	1.069	-1.000	$-.00391$
PGA (cm/s^2)	2.424	0.982	-1.004	$-.00468$

[a] See Equation 5.22.
Spectral velocities are given in cm/s; peak acceleration is given in cm/s^2
From Toro, G.R. and McGuire, R.K., An Investigation Into Earthquake Ground Motion Characteristics in Eastern North America, *Bull. Seis. Soc. Am.*, 77, 468-489, 1987. With permission.

where

PGA	=	the geometric mean of the two horizontal components of peak ground acceleration (g)
M	=	moment magnitude (M_W)
R_s	=	the closest distance to seismogenic rupture on the fault (km)
F	=	0 for strike-slip and normal faulting earthquakes, and 1 for reverse, reverse-oblique, and thrust faulting earthquakes
S_{sr}	=	1 for soft-rock sites
S_{hr}	=	1 for hard-rock sites
$S_{sr} = S_{hr}$	=	0 for alluvium sites
ε	=	a random error term with zero mean and standard deviation equal to $\sigma_{\ln}(PGA)$, the standard error of estimate of $\ln(PGA)$

This relation is intended for meizoseismal applications, and should not be used to estimate PGA at distances greater than about 60 km (the limit of the data employed for the regression). The relation is based on 645 near-source recordings from 47 worldwide earthquakes (33 of the 47 are California records — among the other 14 are the 1985 M_W 8.0 Chile, 1988 M_W 6.8 Armenia, and 1990 M_W 7.4 Manjil Iran events). R_s should not be assigned a value less than the depth of the top

TABLE 5.9 Coefficients for Ground-Motion Estimation at Deep-Soil Sites in Eastern North America in Terms of M_W^a

T (sec)	a'	a''	b	c	d	k	M at max[b]
S_V							
0.05	0.020	1.946	0.431	−0.028	−0.018	−0.00350	8.35
0.10	0.040	2.267	0.429	−0.026	−0.018	−0.00240	8.38
0.15	0.015	2.377	0.437	−0.031	−0.017	−0.00190	8.38
0.20	0.015	2.461	0.447	−0.037	−0.016	−0.00168	8.38
0.30	0.010	2.543	0.472	−0.051	−0.012	−0.00140	8.47
0.40	0.015	2.575	0.499	−0.066	−0.009	−0.00110	8.50
0.50	0.010	2.588	0.526	−0.080	−0.007	−0.00095	8.48
0.75	0.000	2.586	0.592	−0.111	−0.001	−0.00072	8.58
1.00	0.000	2.567	0.655	−0.135	0.002	−0.00058	8.57
1.50	0.000	2.511	0.763	−0.165	0.004	−0.00050	8.55
2.00	0.000	2.432	0.851	−0.180	0.002	−0.00039	8.47
3.00	0.000	2.258	0.973	−0.176	−0.008	−0.00027	8.38
4.00	0.000	2.059	1.039	−0.145	−0.022	−0.00020	8.34
a_{max}	0.030	3.663	0.448	−0.037	−0.016	−0.00220	8.38
$S_{V\,max}$	0.020	2.596	0.608	−0.038	−0.022	−0.00055	8.51
$S_{A\,max}$	0.040	4.042	0.433	−0.029	−0.017	−0.00180	8.40

[a] The distance used is generally the hypocentral distance; we suggest that, close to long faults, the distance should be the nearest distance to seismogenic rupture. The response spectra are for random horizontal components and 5% damping. The units of a_{max} and S_A are cm/s^2; the units of S_V are cm/s. The coefficients in this table should not be used outside the ranges $10 \lesssim r \lesssim 400$ km and $5.0 \lesssim M \lesssim 8.5$. See also Equation 5.23.

[b] "M at max" is the magnitude at which the cubic equation attains its maximum value; for larger magnitudes, we recommend that the motions be equated to those for "M at max". From Boore, D.M. and Joyner, W.B., Estimation of Ground Motion at Deep-Soil Sites in Eastern North America, *Bull. Seis. Soc. Am.*, 81(6), 2167-2185, 1991. With permission.

of the seismogenic crust, or 3 km. Regarding the uncertainty, ε was estimated as:

$$\sigma_{\ln}(PGA) = \begin{vmatrix} 0.55 & \text{if } PGA < 0.068 \\ 0.173 - 0.140 \ln(PGA) & \text{if } 0.068 \leq PGA \leq 0.21 \\ 0.39 & \text{if } PGA > 0.21 \end{vmatrix}$$

Figure 5.20 indicates, for alluvium, median values of the attenuation of peak horizontal acceleration with magnitude and style of faulting.

Joyner and Boore (PSV - WNA Data) [20, 67]

Similar to the above but using a two-step regression technique in which the ground motion parameter is first regressed against distance and then amplitudes regressed against magnitude, Boore, Joyner, and Fumal [20] have used WNA data to develop relations for PGA and PSV of the form:

$$\log Y = b_1 + b_2(M - 6) + b_3(M - 6)^2 + b_4 r + b_5 \log_{10} r + b_6 G_B + b_7 G_C + \varepsilon_r + \varepsilon_e \quad (5.25)$$

where

Y = the ground motion parameter (in cm/s for PSV, and g for PGA)

M = moment magnitude (M_W)

r = $(d^2 + h^2)^{(1/2)}$ = distance (km), where h is a fictitious depth determined by regression, and d is the closest horizontal distance from the station to the point on the earth's surface that lies directly above the rupture

G_B, G_C = site classification indices ($G_B = 1$ for class B site, $G_C = 1$ for class C site, both zero otherwise), where Site Class A has shear wave velocities (averaged over the upper 30 m) > 750 m/s, Site Class B is 360 to 750 m/s, and Site Class C is 180 to 360 m/s (class D sites, < 180 m/s, were not included). In effect, class A are rock, B are firm soil sites, C are deep alluvium/soft soils, and D would be very soft sites

$\varepsilon_r + \varepsilon_e$ = independent random variable measures of uncertainty, where ε_r takes on a specific value for each record, and ε_e for each earthquake

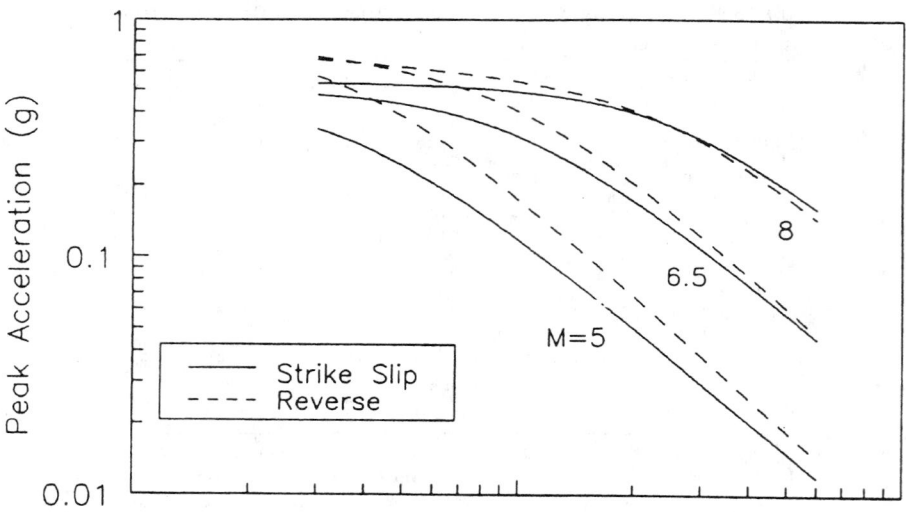

Figure 5.20   Campbell and Bozorgnia worldwide attenuation relationship showing (for alluvium) the scaling of peak horizontal acceleration with magnitude and style of faulting. (From Campbell, K.W. and Bozorgnia, Y., Near-Source Attenuation of Peak Horizontal Acceleration from Worldwide Accelerograms Recorded from 1957 to 1993, Proc. Fifth U.S. National Conference on Earthquake Engineering, Earthquake Engineering Research Institute, Oakland, CA, 1994. With permission.)

b_i, h =   coefficients (see Table 5.10 and Table 5.11)

The relation is valid for magnitudes between 5 and 7.7, and for distances (d) $\leq$ 100 km. The coefficients in Equation 5.25 are for 5% damped response spectra — Boore et al. [20] also provide similar coefficients for 2%, 10%, and 20% damped spectra, as well as for the random horizontal coefficient (i.e., both horizontal coefficients, not just the larger, are considered). Figure 5.21 presents curves of attenuation of PGA and PSV for Site Class C, using these relations, while Figure 5.22 presents a comparison of this, the Campbell and Bozorgnia [29] and Sadigh et al. [105] attenuation relations, for two magnitude events on alluvium.

The foregoing has presented attenuation relations for PGA (Worldwide) and response spectra (ENA and WNA). While there is some evidence [136] that meizoseismal strong ground motion may not differ as much regionally as previously believed, regional attenuation in the far-field differs significantly (e.g., ENA vs. WNA). One regime that has been treated in a special class has been large subduction zone events, such as those that occur in the North American Pacific Northwest (PNW), in Alaska, off the west coast of Central and South America, off-shore Japan, etc. This is due to the very large earthquakes that are generated in these zones, with long duration and a significantly different path. A number of relations have been developed for these events [10, 37, 81, 115, 138] which should be employed in those regions. A number of other investigators have developed attenuation relations for other regions, such as China, Japan, New Zealand, the Trans-Alpide areas, etc., which should be reviewed when working in those areas (see the References).

In addition to the seismologically based and empirical models, there is another method for attenuation or ground motion modeling, which may be termed **semi-empirical methods** (Figure 5.23) [129]. The approach discretizes the earthquake fault into a number of subfault elements, finite rupture on each of which is modeled with radiation therefrom modeled via Green's functions. The resulting wave-trains are combined with empirical modeling of scattering and other factors to generate time-histories of ground motions for a specific site. The approach utilizes a rational framework with powerful explanatory features, and offers useful application in the very near-field of large earthquakes, where it is increasingly being employed.

TABLE 5.10 Coefficients for 5% Damped PSV, for the Larger Horizontal Component

T(s)	B1	B2	B3	B4	B5	B6	B7	H
.10	1.700	.321	−.104	.00000	−.921	.039	.128	6.18
.11	1.777	.320	−.110	.00000	−.929	.065	.150	6.57
.12	1.837	.320	−.113	.00000	−.934	.087	.169	6.82
.13	1.886	.321	−.116	.00000	−.938	.106	.187	6.99
.14	1.925	.322	−.117	.00000	−.939	.123	.203	7.09
.15	1.956	.323	−.117	.00000	−.939	.137	.217	7.13
.16	1.982	.325	−.117	.00000	−.939	.149	.230	7.13
.17	2.002	.326	−.117	.00000	−.938	.159	.242	7.10
.18	2.019	.328	−.115	.00000	−.936	.169	.254	7.05
.19	2.032	.330	−.114	.00000	−.934	.177	.264	6.98
.20	2.042	.332	−.112	.00000	−.931	.185	.274	6.90
.22	2.056	.336	−.109	.00000	−.926	.198	.291	6.70
.24	2.064	.341	−.105	.00000	−.920	.208	.306	6.48
.26	2.067	.345	−.101	.00000	−.914	.217	.320	6.25
.28	2.066	.349	−.096	.00000	−.908	.224	.333	6.02
.30	2.063	.354	−.092	.00000	−.902	.231	.344	5.79
.32	2.058	.358	−.088	.00000	−.897	.236	.354	5.57
.34	2.052	.362	−.083	.00000	−.891	.241	.363	5.35
.36	2.045	.366	−.079	.00000	−.886	.245	.372	5.14
.38	2.038	.369	−.076	.00000	−.881	.249	.380	4.94
.40	2.029	.373	−.072	.00000	−.876	.252	.388	4.75
.42	2.021	.377	−.068	.00000	−.871	.255	.395	4.58
.44	2.013	.380	−.065	.00000	−.867	.258	.401	4.41
.46	2.004	.383	−.061	.00000	−.863	.261	.407	4.26
.48	1.996	.386	−.058	.00000	−.859	.263	.413	4.16
.50	1.988	.390	−.055	.00000	−.856	.265	.418	3.97
.55	1.968	.397	−.048	.00000	−.848	.270	.430	3.67
.60	1.949	.404	−.042	.00000	−.842	.275	.441	3.43
.65	1.932	.410	−.037	.00000	−.837	.279	.451	3.23
.70	1.917	.416	−.033	.00000	−.833	.283	.459	3.08
.75	1.903	.422	−.029	.00000	−.830	.287	.467	2.97
.80	1.891	.427	−.025	.00000	−.827	.290	.474	2.89
.85	1.881	.432	−.022	.00000	−.826	.294	.481	2.85
.90	1.872	.436	−.020	.00000	−.825	.297	.486	2.83
.95	1.864	.440	−.018	.00000	−.825	.301	.492	2.84
1.00	1.858	.444	−.016	.00000	−.825	.305	.497	2.87
1.10	1.849	.452	−.014	.00000	−.828	.312	.506	3.00
1.20	1.844	.458	−.013	.00000	−.832	.319	.514	3.19
1.30	1.842	.464	−.012	.00000	−.837	.326	.521	3.44
1.40	1.844	.469	−.013	.00000	−.843	.334	.527	3.74
1.50	1.849	.474	−.014	.00000	−.851	.341	.533	4.08
1.60	1.857	.478	−.016	.00000	−.859	.349	.538	4.46
1.70	1.866	.482	−.019	.00000	−.868	.357	.543	4.86
1.80	1.878	.485	−.022	.00000	−.878	.365	.547	5.29
1.90	1.891	.488	−.025	.00000	−.888	.373	.551	5.74
2.00	1.905	.491	−.028	.00000	−.898	.381	.554	6.21

The equations are to be used for $5.0 <= M <= 7.7$ and $d <= 100.0$ km.
From Boore, D. M., Joyner, W. B., and Fumal, T. E., Estimation of Response Spectra and Peak Acceleration from Western North American Earthquakes: An Interim Report, U.S.G.S. Open-File Report 93-509, Menlo Park, CA, 1993. With permission.

TABLE 5.11 Coefficients for the Random and Larger Horizontal Components of Peak Acceleration

Comp.	b_1	b_2	b_3	b_4	b_5	b_6	b_7	h
Random	−.105	.229	0	0	−.778	.162	.251	5.57
Larger	−.038	.216	0	0	−.777	.158	.254	5.48

Courtesy of Boore, D.M., Joyner, W.B., and Fumal, T.E., Estimation of Response Spectra and Peak Acceleration from Western North American Earthquakes: An Interim Report, U.S.G.S. Open-File Report 93-509, Menlo Park, CA, 1993.

The foregoing has also dealt exclusively with *horizontal* ground motions, yet **vertical ground motions** can be very significant. The common practice for many years has been to take the ratio (V/H) as one half or two thirds (practice varies). Recent work [22] has found that response spectra V/H ratio is a function of period, distance to source, and magnitude, with the ratio being larger than 2/3 in the near-field and having a peak at about 0.1 s, and less than 2/3 at long periods.

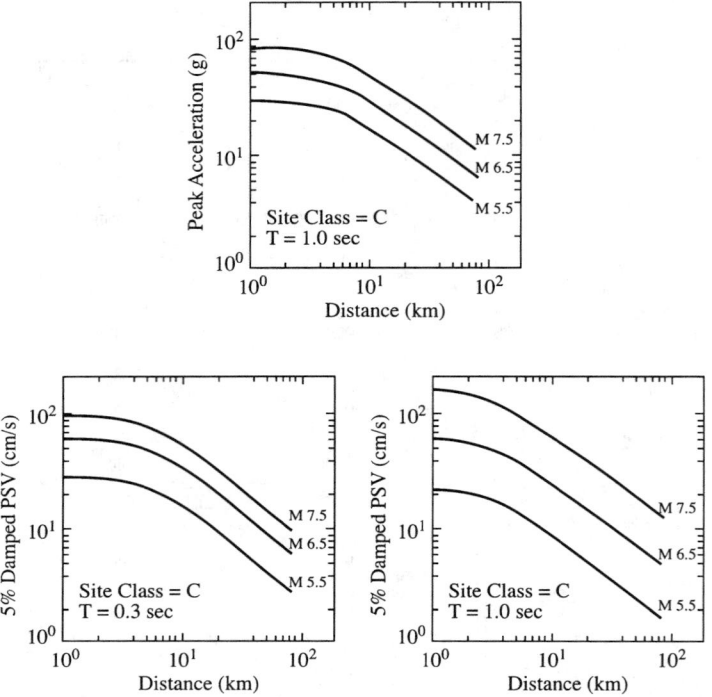

Figure 5.21 Attenuation of PGA and PSV for Site Class C. (From Boore, D.M., Joyner, W.B., and Fumal, T.E., Estimation of Response Spectra and Peak Acceleration from Western North American Earthquakes: An Interim Report, U.S.G.S. Open-File Report 93-509, Menlo Park, CA, 1993. With permission.)

An important aspect of ground motion is **duration** — larger earthquakes shake longer, forcing structures through more (typically inelastic) cycles and thus tending to cause more damage. Since a typical fault rupture velocity [16] may be on the order of 2.5 km/s, it can be readily seen from Equation 5.11 that a magnitude 7 event will require about 14 s for fault rupture, and a magnitude 7.5 event about 40 s (and note that the radiated wave train will increase in duration due to scattering). Thus, strong ground motion can be felt for several seconds to significantly longer than a minute. Because the duration of strong ground motion is very significant, there have been a number of attempts at quantifying, and therefore a number of definitions of, strong ground motion. These have included *bracketed duration* D_B (time interval between the first and the last time when the acceleration exceeds some level, usually taken [15] to be 0.05g), *fractional or normalized duration* D_F (elapsed time between the first and the last acceleration excursion greater than α times PGA [70]), and D_{TB} (the time interval during which 90% of the total energy is recorded at the station [124] equal to the time interval between attainment of 5% and 95% of the total Arias intensity of the record). McGuire and Barnhard [78] have used these definitions to examine 50 strong motion records (3 components each), and found:

$$\ln D = c_1 + c_2 M + c_3 S + c_4 V + c_5 \ln R \qquad (5.26)$$

where D is D_B, D_F, or D_{TB}, (for D_F, $\alpha = 0.5$ in this case), M is earthquake magnitude, $S = 0,1$ for rock or alluvium, $V = 0,1$ for horizontal or vertical component, R typically closest distance to the rupture surface (km), and c_i are coefficients in Table 5.12. However, McGuire and Barnhard note that there is large uncertainty in these estimates due to varying source effects, travel paths, etc.

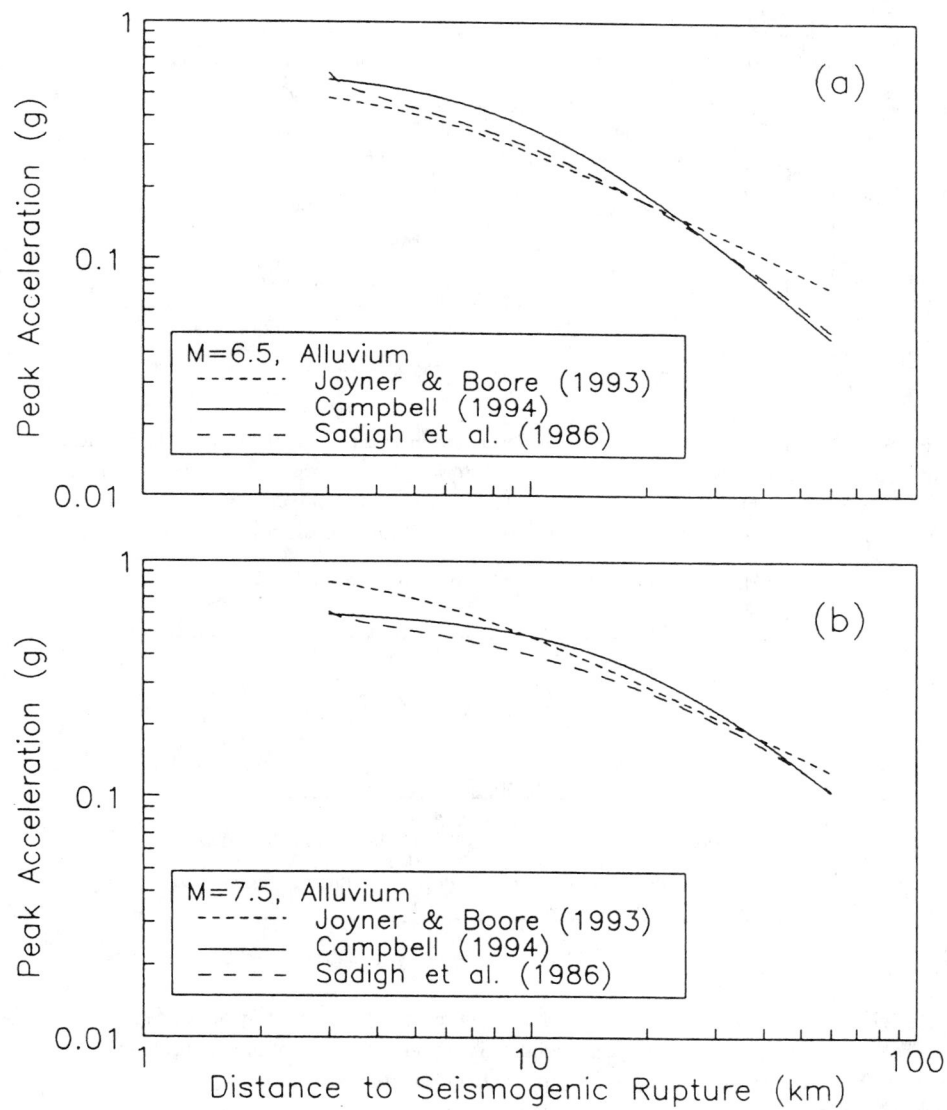

Figure 5.22   Comparison of PGA on alluvium, for various relations. (From Campbell, K.W. and Bozorgnia, Y., Near-Source Attenuation of Peak Horizontal Acceleration from Worldwide Accelerograms Recorded from 1957 to 1993, Proc. Fifth U.S. National Conference on Earthquake Engineering, Earthquake Engineering Research Institute, Oakland, CA, 1994. With permission.)

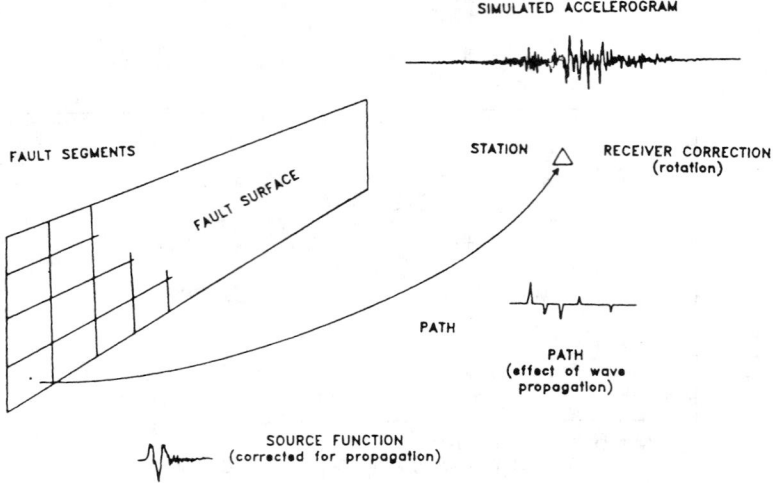

SEMI−EMPIRICAL SIMULATION PROCEDURE

Figure 5.23   Semi-empirical ground motion simulation procedure. (From Wald, D.J., Burdick, L.J., and Somerville, P.G., Simulation of Acceleration Time Histories Close to Large Earthquakes, in *Earthquake Engineering and Soil Dynamics II—Recent Advances in Ground-Motion Evaluation,* Thun, J. L. V., Ed., Geotechnical Spec. Publ. No. 20, Am. Soc. Civil Engrs., New York, 1988. With permission.)

TABLE 5.12   Coefficients for Strong Ground Motion Duration Estimation

D	Duration	C_1	C_2	C_3	C_4
D_B	Bracketed	2.3	2.0	0.20	−0.16
D_F	Fractional (normalized)	−4.5	0.74	0.36	0.31
D_{TB}	Trifunac-Brady (time for energy absorption)	0.19	0.15	0.73	0.23

$\ln D = c_1 + c_2 M + c_3 S + c_4 V + c_5 \ln R$
From McGuire, R.K. and Barnhard, T.P., The Usefulness of Ground Motion Duration in Predicting the Severity of Seismic Shaking, *Proc. 2nd U.S. Natl. Conf. on Earthquake Eng.,* Earthquake Engineering Research Institute, Oakland, CA, 1979. With permission.

As Trifunac and Novikova [125] discuss, strong motion duration may be represented by the sum of three terms: $dur = t_0 + t_\Delta + t_{region}$, where t_0 is the duration of the source fault rupture, t_Δ is the increase in duration due to propagation path effects (scattering), and t_{region} is prolongation effects caused by the geometry of the regional geologic features and of the local soil. This approach will be increasingly useful, but requires additional research. Note that response spectra are not strongly correlated with, or good measures of, duration — that is, an elastic SDOF oscillator will reach its maximum amplitude within several cycles of harmonic motion, and two earthquakes (one of long, the other of short duration, but both with similar PGA) may have similar elastic response spectra.

Lastly, it should be noted that the foregoing has dealt exclusively with attenuation of engineering measures of ground motions — there are also a number of attenuation relations available for MMI, R-F, MSK, and other qualitative measures of ground motion, specific to various regions. However, the preferred method is to employ attenuation relations for engineering measures, and then convert the results to MMI or other intensity measures, using various conversions [84, 103].

5.2.5 Seismic Hazard and Design Earthquake

The foregoing sections provide an overview of earthquake measures and occurrence. If an earthquake location and magnitude are specified, attenuation relations may be employed to estimate the PGA or response spectra at a site, which can then be employed for design of a structure. However, since earthquake occurrence is a random process, the specification of location and magnitude is not a simple matter. The basic question facing the designer is, *what is the earthquake which the structure should be designed to withstand?* Note that this is termed the **design earthquake**, although in actuality hazard parameters (e.g., PGA, response spectra) are the specific parameters in question. Basically, three approaches may be employed in determining a design earthquake: they can be characterized as (1) code approach, (2) upper-bound approach, or (3) Probabilistic Seismic Hazard Analysis approach. This section briefly describes these approaches.

Code Approach

The code approach is to simply employ the lateral force coefficients as specified in the applicable design code. Most countries and regions have macro-zoned their jurisdiction [97], and have regional maps available which provide a lateral force coefficient. Figure 5.24 for example is the seismic zonation map of the U.S. which provides a zone factor Z as part of the determination of the lateral force coefficient. This mapping is based on probabilistic methods [3] such that the ground

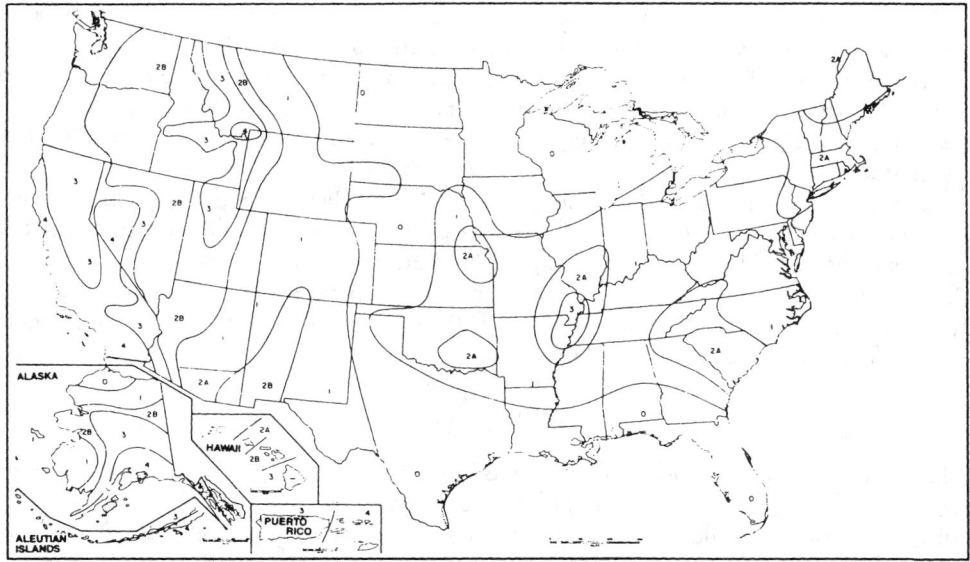

Figure 5.24 Seismic zone map of the U.S. (From Uniform Building Code, Vol. 2, Structural Engineering Design Provisions, Intl. Conf. Building Officials, Whittier, 1994. With permission.)

motion parameters are intended to have about a 10% probability of being exceeded during any 50-year period (this is discussed further below). The advantages of this approach are simplicity and ease, and obvious compliance with local requirements. The disadvantages are inappropriateness for unusual structures, and that the methods employed in the mapping have been regional in nature and may have overlooked local geology.

Upper-Bound Approach

The upper-bound approach consists of reviewing the geology and historic seismicity of the region, to determine the largest event that is physically capable of occurring in the vicinity and affecting the site. In high seismicity areas, this approach is feasible because very large faults may be readily identifiable. Using historic data and/or fault length-magnitude relations, a maximum magnitude event can be assigned to the fault and, using attenuation relations, a PGA or other engineering measure can be estimated for the site, based on the distance. This approach has a number of drawbacks including lack of understanding of the degree of conservatism and potentially excessive design requirements, so that it is rarely employed, and then only for critical structures.

Probabilistic Seismic Hazard Analysis

The Probabilistic Seismic Hazard Analysis (PSHA) approach entered general practice with Cornell's [35] seminal paper, and basically employs the theorem of total probability to formulate:

$$P(Y) = \sum_F \sum_M \sum_R p(Y|M, R) p(M) \qquad (5.27)$$

where

Y	=	a measure of intensity, such as PGA, response spectral parameters PSV, etc.
$p(Y|M, R)$	=	the probability of Y given earthquake magnitude M and distance R (i.e., attenuation)
$p(M)$	=	the probability of occurrence of a given earthquake magnitude M
F	=	indicates seismic sources, whether discrete such as faults, or distributed

This process is illustrated in Figure 5.25, where various seismic sources (faults modeled as line sources and dipping planes, and various distributed or area sources, including a background source to account for miscellaneous seismicity) are identified, and their seismicity characterized on the basis of historic seismicity and/or geologic data. The effects at a specific site are quantified on the basis of strong ground motion modeling, also termed attenuation. These elements collectively are the **seismotectonic model** — their integration results in the seismic hazard.

There is extensive literature on this subject [86, 102], so only key points will be discussed here. Summation is indicated, as integration requires closed form solutions, which are usually precluded by the empirical form of the attenuation relations. The $p(Y|M, R)$ term represents the full probabilistic distribution of the attenuation relation — summation must occur over the full distribution, due to the significant uncertainty in attenuation. The $p(M)$ term is referred to as the **magnitude-frequency** relation, which was first characterized by Gutenberg and Richter [48] as

$$\log N(m) = a_N - b_N m \qquad (5.28)$$

where $N(m) =$ the number of earthquake events equal to or greater than magnitude m occurring on a seismic source per unit time, and a_N and b_N are regional constants ($10^{a_N} =$ the total number of earthquakes with magnitude > 0, and b_N is the rate of seismicity; b_N is typically 1 ± 0.3). Gutenberg and Richter's examination of the seismicity record for many portions of the earth indicated this relation was valid, for selected magnitude ranges. That is, while this relation appears as a straight line when plotted on semi-log paper, the data is only linear for a selected middle range of magnitudes, typically falling below the line for both the smaller and larger magnitudes, as shown in Figure 5.26a for Japan earthquake data for the period 1885 to 1990. The fall-off for smaller magnitudes is usually attributed to lack of instrumental sensitivity. That is, some of the smaller events are not detected. Typically, some improved instruments, better able to detect distant small earthquakes, are introduced during any observation period. This can be seen in Figure 5.26(b), where the number of detected earthquakes are relatively few in the early decades of the record. The fall-off for larger magnitudes is usually attributed to two reasons: (1) the observation period is shorter than the return period of

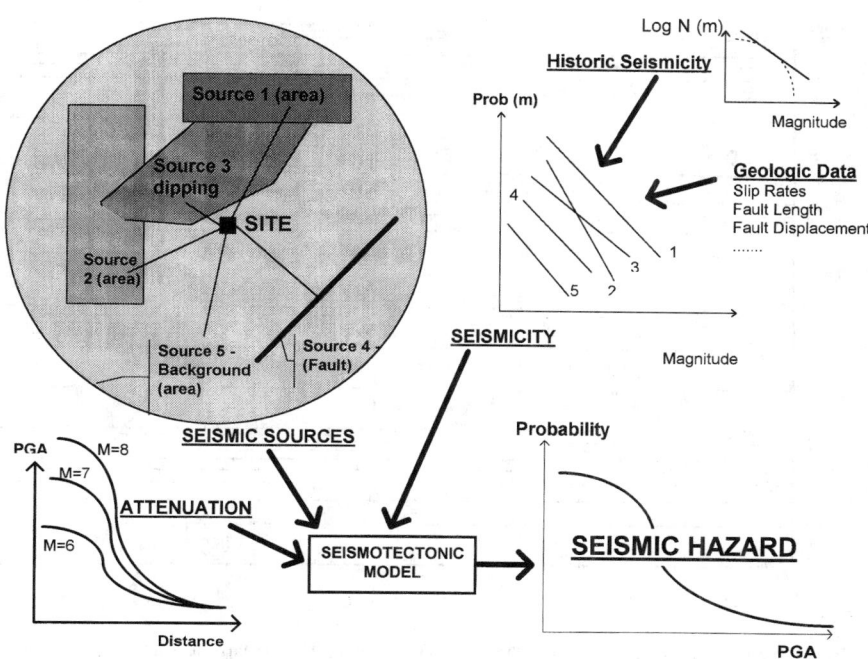

Figure 5.25 Elements of seismic hazard analysis — seismotectonic model is composed of seismic sources, whose seismicity is characterized on the basis of historic seismicity and geologic data, and whose effects are quantified at the site via strong motion attenuation models.

the largest earthquakes, and (2) there is some physical limit to the size of earthquakes, so that the Gutenberg-Richter relation cannot be indefinitely extrapolated to larger and larger magnitudes.

The Gutenberg-Richter relation can be normalized to

$$F(m) = 1. - \exp[-B_M(m - M_o)] \tag{5.29}$$

where $F(m)$ is the cumulative distribution function (CDF) of magnitude, B_M is a regional constant, and M_o is a small enough magnitude such that lesser events can be ignored. Combining this with a Poisson distribution to model large earthquake occurrence [44] leads to the CDF of earthquake magnitude per unit time

$$F(m) = \exp[-\exp\{-a_M(m - \mu_M)\}] \tag{5.30}$$

which has the form of a Gumbel [47] extreme value type I (largest values) distribution (denoted $EX_{I,L}$), which is an unbounded distribution (i.e., the variate can assume any value). The parameters a_M and μ_M can be evaluated by a least squares regression on historical seismicity data, although the probability of very large earthquakes tends to be overestimated. Several attempts have been made to account for this (e.g., Cornell and Merz [36]). Yegulalp and Kuo [137] have used Gumbel's Type III (largest value, denoted $EX_{III,L}$) to successfully account for this deficiency. This distribution

$$F(m) = \exp\left[-\left(\frac{w - m}{w - u}\right)^k\right] \tag{5.31}$$

has the advantage that w is the largest possible value of the variate (i.e., earthquake magnitude), thus permitting (when w, u, and k are estimated by regression on historical data) an estimate of the source's largest possible magnitude. It can be shown [137] that estimators of w, u, and k can be

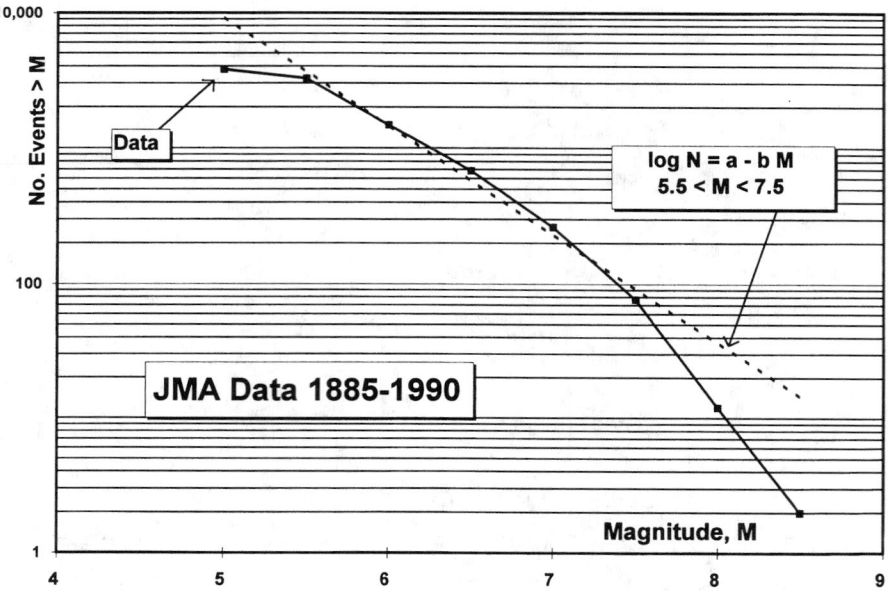

Figure 5.26   (a) Plot of seismicity data for Japan, 1885 to 1990, from Japan Meteorological Agency Catalog. Note actual data falls below Gutenberg-Richter relation at smaller and larger magnitudes.

obtained by satisfying Kuhn-Tucker conditions although if the data is too incomplete, the $EX_{III,L}$ parameters approach those of the $EX_{I,L}$:

$$u \to \mu_M \qquad k/(w - u) \to a_M$$

Determination of these parameters requires careful analysis of historical seismicity data (which is highly complex and something of an art [40], and the merging of the resulting statistics with estimates of maximum magnitude and seismicity made on the basis of geological evidence (i.e., as discussed above, maximum magnitude can be estimated from fault length, fault displacement data, time since last event and other evidence, and seismicity can be estimated from fault slippage rates combined with time since last event; see Schwartz [109] for an excellent discussion of these aspects). In a full probabilistic seismic hazard analysis, many of these aspects are treated fully or partially probabilistically, including the attenuation, magnitude-frequency relation, upper- and lower-bound magnitudes for each source zone, geographical bounds of source zones, fault rupture length, and many other aspects. The full treatment requires complex specialized computer codes, which incorporate uncertainty via use of multiple alternative source zonations, attenuation relations, and other parameters [13, 43] often using a logic tree format (Figure 5.27). A number of codes have been developed using the public domain FRISK (Fault Risk) code first developed by McGuire [77].

Several topics are worth briefly noting:

- While analysis of the seismicity of a number of regions indicates that the Gutenberg-Richter relation $\log N(M) = a - bM$ is a good overall model for the magnitude-frequency or probability of occurrence relation, studies of late Quaternary faults during the 1980s indicated that the exponential model is not appropriate for expressing earthquake recurrence on individual faults or fault segments [109]. Rather, it was found that many individual faults tend to generate essentially the same size or **characteristic earthquake** [109], having a relatively narrow range of magnitudes at or near the maximum that can be produced by the geometry, mechanical properties, and state of stress of the fault. This implies that, relative to the Gutenberg-Richter magnitude-frequency

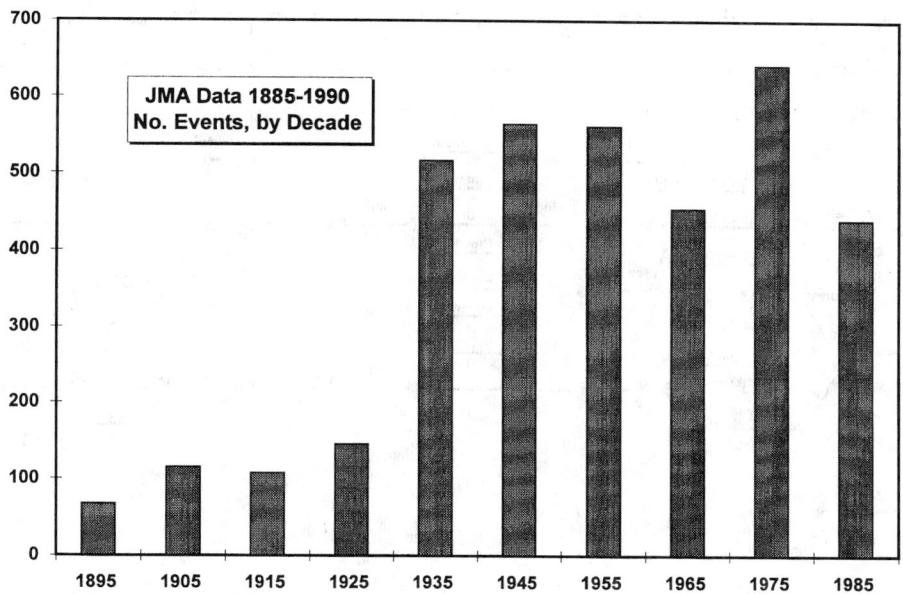

Figure 5.26   (b) Number of events by decade, for Japan, 1885 to 1990, from Japan Meteorological Agency Catalog.

relation, faults exhibiting characteristic earthquake behavior will have relatively less seismicity (i.e., higher b value) at low and moderate magnitudes, and more near the characteristic earthquake magnitude (i.e., lower b value).

- Most probabilistic seismic hazard analysis models assume the Gutenberg-Richter exponential distribution of earthquake magnitude, and that earthquakes follow a **Poisson process**, occurring on a seismic source zone randomly in time and space. This implies that the time between earthquake occurrences is exponentially distributed, and that the time of occurrence of the next earthquake is independent of the elapsed time since the prior earthquake.[5] The CDF for the exponential distribution is:

$$F(t) = 1 - \exp(-\lambda t) \qquad (5.32)$$

Note that this forms the basis for many modern building codes, in that the probabilistic seismic hazard analysis results are selected such that the seismic hazard parameter (e.g., PGA) has a "10% probability of exceedance in 50 years" [126]; that is, if $t = 50$ years and $F(t) = 0.1$ (i.e., only 10% probability that the event has occurred in t years), then $\lambda = .0021$ per year, or 1 per 475 years. A number of more sophisticated models of earthquake occurrence have been investigated, including time- predictable models [5], and renewal models [68, 91]

- As was seen from the data for Japan in Figure 5.26, historic seismicity observations vary with event size — large magnitude events will have been noted and recorded for perhaps the full length of the historic record, while small magnitude events will only have been recorded with the advent of instruments, with quality of the instrumental record usually improving in more recent times. Thus, in analyzing any historic seismicity record, the

---

[5]For this aspect, the Poisson model is often termed a *memoryless* model.

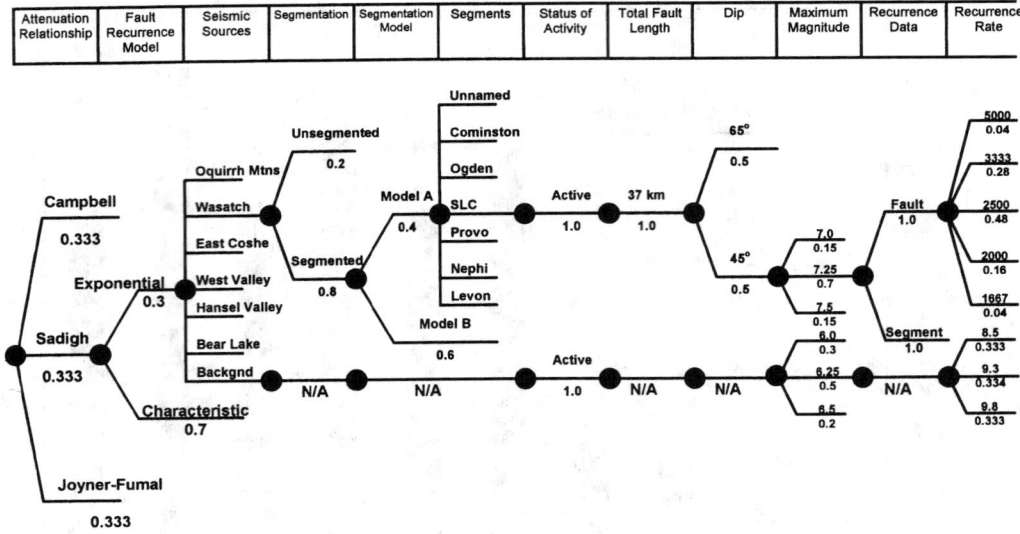

Attenuation Relationship	Fault Recurrence Model	Seismic Sources	Segmentation	Segmentation Model	Segments	Status of Activity	Total Fault Length	Dip	Maximum Magnitude	Recurrence Data	Recurrence Rate

Figure 5.27   Hazard model logic tree. Parameter values are shown above the branches and assessed probabilities are shown in parentheses. (From Schwartz, D.P., Geologic Characterization of Seismic Sources: Moving into the 1990s, in *Earthquake Engineering and Soil Dynamics II — Recent Advances in Ground-Motion Evaluation*, Thun, J. L. V., Ed., Geotechnical Spec. Publ. No 20, Am. Soc. Civil Engrs., New York, 1988. With permission. Also after Youngs, R. R. and Coopersmith, K. J., Implication of Fault Slip Rates and Earthquake Recurrence Models to Probabilistic Seismic Hazard Estimates, *Bull. Seis. Soc. Am.*, 75, 939-964, 1987.)

issue of **completeness** is important and should be analyzed. Stepp [121] has developed a method assuming the earthquake sequence can be modeled as a Poisson distribution. If $k_1, k_2, k_3 \ldots k_n$ are the number of events per unit time interval, then $\lambda = \frac{1}{n} \sum_{i=1}^{n} k_i$ and its variance is $\sigma^2 = \lambda/n$ where n equals the number of unit time intervals. Taking the unit time interval to be one year gives $\sigma_\lambda = \sqrt{\lambda}/\sqrt{T}$ as the standard deviation of the estimate of the mean, where T is the sample length (in years). That is, this provides a test for stationarity of the observational quality. If the data for a magnitude interval (say $5.6 < M < 6.5$, to test for quality of observation for events in this magnitude range) is plotted as $\log(\sigma_\lambda)$ vs. $\log(T)$, then the portion of the record with slope $T^{-1/2}$ can be considered homogeneous (Figure 5.28) and used with data for other magnitude ranges (but for different observational periods) similarly tested for homogeneity, to develop estimates of magnitude-frequency parameters. A more recent method is also provided by Habermann [50].

- Equation 5.27 is quite general, and used to develop estimates of MMI, PGA, response spectra, or other measures of seismic hazard. Construction of response spectra is usually performed in one of two ways:

    1. Using probabilistic seismic hazard analysis to obtain an estimate of the PGA, and using this to scale a normalized response spectral shape. Alternatively, estimating PGA and PSV (also perhaps PSD) and using these to fit a normalized response spectral shape for each portion of the spectra. Since probabilistic response spectra are a composite of the contributions of varying earthquake magnitudes at varying distances, the ground motions of which attenuate differently at different periods, this method has the drawback that the resulting spectra have varying (and unknown) probabilities of exceedance at different periods. Because of this

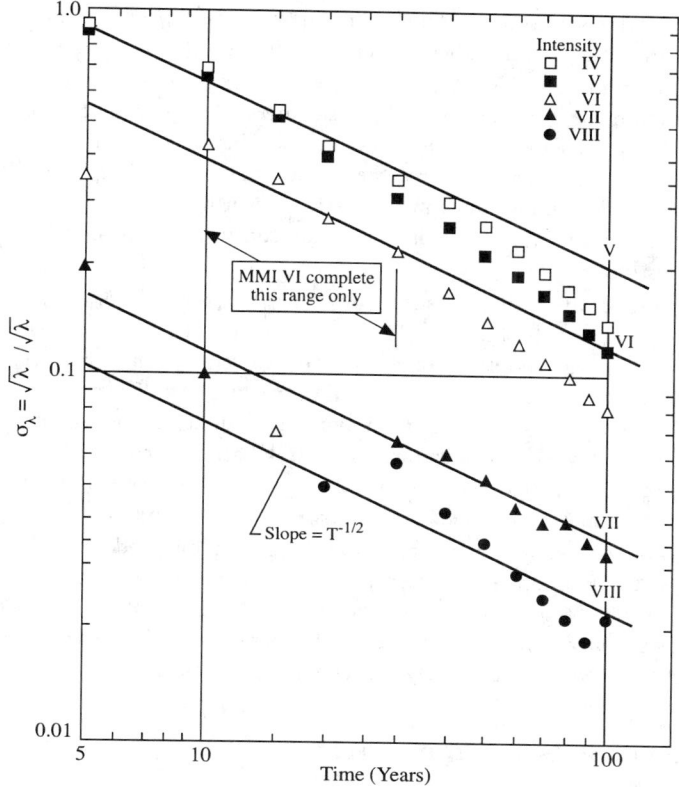

Figure 5.28 Standard deviation of the estimate of the man of the annual number of events as a function of sample length and intensity class. (After Stepp, J.C., Analysis of Completeness of the Earthquake Sample in the Puget Sound Area and Its Effect on Statistical Estimates of Earthquake Hazard, Proc. First Conf. Microzonation, Seattle, 1972.)

drawback, this method is less favored at present, but still offers the advantage of economy of effort.

2. An alternative method results in the development of **uniform hazard spectra** [7], and consists of performing the probabilistic seismic hazard analysis for a number of different periods, with attenuation equations appropriate for each period (e.g., those of Boore, Joyner, and Fumal). This method is currently preferred, as the additional effort is not prohibitive and the resulting response spectra has the attribute that the probability of exceedance is independent of frequency. However, the resulting spectra do not represent the response from any one event.

Selection of Design Earthquake

The foregoing discussion has outlined the methods for probabilistic seismic hazard analysis. However, the question still remains, *given the results of a probabilistic seismic hazard analysis, what values of PGA, response spectra, etc. should be employed or equivalently, what is the appropriate probability of exceedance?* This is a complex question; for ordinary building structures, regulated by the local authorities, the default value for the probability of exceedance would be that comparable with the local building code — usually about 10% probability of exceedance in 50 years. The value of probabilistic seismic hazard analysis is that the results are site-specific and utilize the latest and most detailed data on local seismic sources. Those results may indicate, for code-comparable probabilities of exceedance, design parameters either greater or lesser than specified in the code.

Many structures, however, are either (a) atypical, such as large bridges, civil works (e.g., water supply pump stations or treatment plants), data centers, emergency operations centers, etc. or (b) not ordinarily regulated by local authorities, such as industrial structures in large manufacturing or process complexes. For these structures, the choice of the appropriate probability of exceedance is more difficult — their importance may demand design for less frequent events (i.e., smaller probability of exceedance) or, if their failure does not constitute a life-safety hazard, economies in design may be justified.

One approach for addressing this issue involves careful assessment of the total costs of damage associated with varying levels of hazard. This should be undertaken in a **cost-benefit framework**. The total costs of damage should include not only the value of the structure, but the value of potential losses to equipment and inventory in the structure, and the associated business costs of lost operations while repairs are required. In many cases, design levels exceeding that of the local building code may be found to be warranted, when total costs are considered. It should be noted, however, that assessment of structural damage is not easy, and assessment of damage to equipment and contents, and business costs of lost operations, is even more complex.

It should always be kept in mind that the first goal of the building code and the structural designer is the maintenance of public safety; therefore, designs using parameter values lower than the local building code (i.e., parameters with probabilities of exceedance exceeding those of the local building code) may carry undue risk to safety. Therefore, the parameters of the local building code should normally be taken as indicative of society's criteria for appropriate risk.

5.2.6   Effect of Soils on Ground Motion

The effect of different types of soils on earthquake ground motion and damage has long been noted. By observation of damage and mapping of seismic intensities, such as following the 1906 San Francisco [73] and other earthquakes, it was observed that the greatest damage generally correlated with geologically recent alluvial deposits and non-engineered fills, so that softer soils were generally considered more damaging. However, an S or soil factor to account for site-specific soil effects was only introduced into the base shear formula prevailing in the U.S. in 1976. Prior to that time there had been no variation in the base shear coefficients for different site conditions although it had been pointed out that "the absence of a soil factor... should not be interpreted as meaning that the effect of soil conditions on building response is not important" [111]. The current Uniform Building Code defines four site coefficients, as shown in Table 5.13. Quantification of the effects of soils on ground motion has generally been by either analytical or empirical methods.

Analytical Methods

Analysis of dynamic ground response is usually accomplished using SHAKE [106] or similar programs, which are based on the vertical propagation of shear waves in a layered half-space. The approach involves using equivalent linear properties taking into account non-linear soil behavior — soil properties are unit weight, maximum shear modulus, variation of shear modulus, and damping ratio with shear strain. A one-dimensional vertical shear wave propagation is assumed, and the shear modulus and damping ratio are adjusted iteratively based on strain compatibility. The solution is in the frequency domain to take advantage of the computational efficiency of the Fast Fourier Transform method.

In application, a time history is assumed at a nearby rock or firm soil surface outcrop, and SHAKE or a similar program *deconvolves* the motion, that is, it is used to determine the corresponding time history at some depth (below the layer of interest, usually at the underlying "basement rock"). Using this analytically derived time history at depth then, the same program is used for the profile

TABLE 5.13 Site Coefficients[a]

Type	Description	S Factor
S_1	A soil profile with either: (a) A rock-like material characterized by a shear-wave velocity greater than 2,500 ft/s (763 m/s) or by other suitable means of classification, or (b) Medium-dense or medium-stiff to stiff soil conditions, where soil depth is less than 200 ft (60 960 mm).	1.0
S_2	A soil profile with predominantly medium-dense to dense or medium-stiff to stiff soil conditions, where the soil depth exceeds 200 ft (60 960 mm).	1.2
S_3	A soil profile containing more than 200 ft (60 960 mm) of soft to medium-stiff clay but not more than 40 ft (12 192 mm) of soft clay.	1.5
S_4	A soil profile containing more than 40 ft (12 192 mm) of soft clay characterized by a shear wave velocity less than 500 ft/s (152.4 m/s).	2.0

[a] The site factor shall be established from properly substantiated geotechnical data. In locations where the soil properties are not known in sufficient detail to determine the soil profile type, soil profile S_3 shall be used. Soil profile S_4 need not be assumed unless the building official determines that soil profile S_4 may be present at the site, or in the event that soil profile S_4 is established by geotechnical data.
After Uniform Building Code, Vol. 2, Structural Engineering Design Provisions, Intl. Conf. Building Officials, Whittier, 1994.

of interest to determine time histories at the surface or any intervening layer. This procedure is typically employed to determine both (a) the effect of soils on surface ground motions and (b) time histories or equivalent number of cyclic shear stresses for the analysis of liquefaction potential for a layer of interest.

Empirical Methods

Empirical methods generally take the form of a modification factor for the site-specific soil profile to be applied to a "base" ground motion estimate arrived at by attenuation relation or other methods. These modifiers were required previously, when attenuation relations had insufficient data for correlation against varying soil profiles. More recently, with additional data (as seen above), attenuation relations are now available for several different soil types so that this approach is receding, although it is still required at times. Evernden [45] has provided a series of modifiers for MMI, based on a study of a number of U.S. intensity distributions, correlated against surficial geology maps. Seed and Idriss developed a well-known relationship for PGA (Figure 5.29), for four different types of soil deposit:

1. Rock

2. Stiff soil deposits involving cohesionless soils or stiff clays to about 200 ft in depth

3. Deep cohesionless soil deposits with depths greater than about 250 ft

4. Deposits of soft to medium stiff clays and sands

   Note that these type-definitions differ from those employed in the model building codes (see Table 5.13).

As Seed and Idriss [113] note, "apart from deposits involving soft to medium stiff clay, values of PGA developed on different types of soil do not differ appreciably, particularly at acceleration levels less than about 0.3 to 0.4g". However, PGV do differ significantly with a general pattern as indicated in Table 5.14, the significance of which was first noted by Newmark [89].

Similar but more detailed ratios form the basis for development of response spectra [83, 98] and future lateral force provisions are tending to be based on the velocity portion of the spectrum as well as PGA [21, 24].

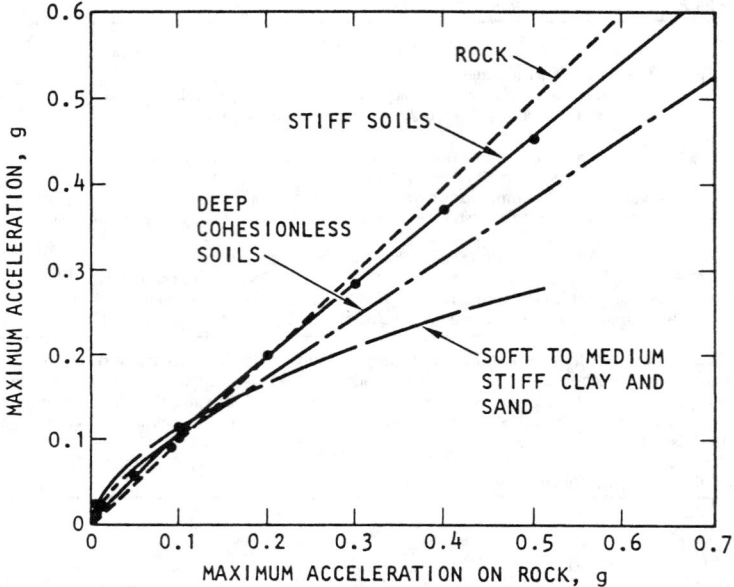

Figure 5.29 Approximate relationships between maximum acceleration on rock and other local site conditions. (From Seed, H.B. and Idriss, I.M., Ground Motions and Soil Liquefaction During Earthquakes, Earthquake Engineering Research Institute, Oakland, CA, 1982. With permission.)

TABLE 5.14   Typical PGV/PGA Ratios

Geologic condition	PGV / PGA
Rock	55 cm/s/g
Stiff soils (< 200 ft)	110 cm/s/g
Deep stiff soils (> 200 ft)	135 cm/s/g

From Seed, H.B. and Idriss, I.M., Ground Motions and Soil Liquefaction during Earthquakes, Earthquake Engineering Research Institute, Oakland, CA, 1982. With permission.

5.2.7 Liquefaction and Liquefaction-Related Permanent Ground Displacement

Liquefaction refers to a process resulting in a soil's loss of shear strength, due to a transient excess of pore water pressure. The process is shown in Figure 5.30 and typically consists of loose granular soil with a high water table being strongly shaken during an earthquake; that is, cyclically sheared. The soil particles initially have large voids between them. Due to shaking, the particles are displaced relative to each other, and tend to more tightly pack,[6] decreasing the void volume. The water, which had occupied the voids (and being incompressible), comes under increased pressure and migrates upward towards or to the surface, where the pressure is relieved. The water usually carries soil with it, and the resulting ejecta are variously termed **sand boils** or **mud volcanoes** (Figure 5.31).

Seismic liquefaction of soils was noted in numerous earthquakes but the phenomenon was first well understood following the 1964 Alaska and, particularly, 1964 Niigata (Japan) earthquakes,

---

[6]Analogous to a can of coffee grounds, which "tamps" down when struck against the tabletop.

Liquefaction

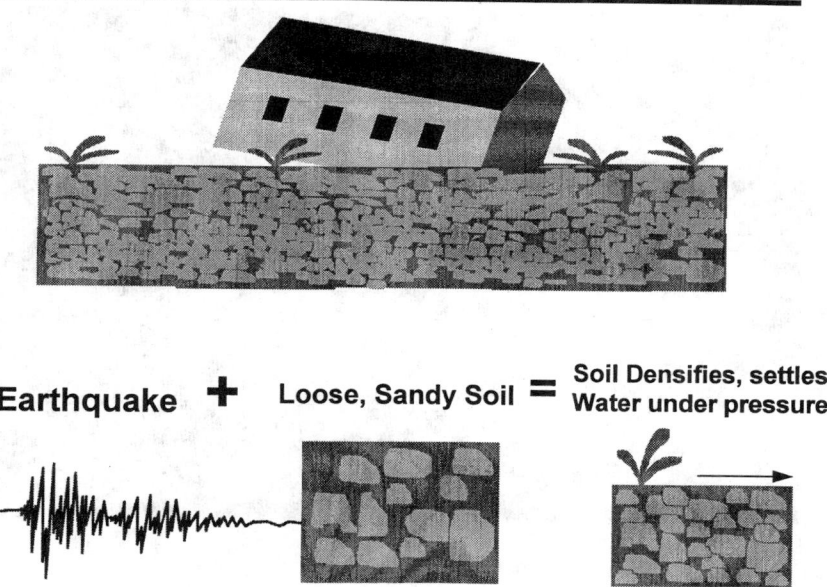

Earthquake + Loose, Sandy Soil = Soil Densifies, settles Water under pressure

Figure 5.30   Soil Liquefaction — loose granular soil density under shaking, settling, and putting ground water under pressure. The pressure is relieved when the water migrates to the surface, where it produces sand boils.

where dramatic effects were observed [51, 93, 95, 111]. Liquefaction is a major source of damage in earthquakes, since (a) the soil's loss of shear strength results in partial or total loss of bearing capacity, resulting in foundation failure unless the structure is founded below the liquefying layer; (b) liquefaction may result in large lateral spreads and permanent ground displacements, often measured in meters and occasionally resulting in catastrophic slides, such as occurred at Turnagain Heights in the 1964 Alaska earthquake [112]; and (c) for both of these reasons, various lifelines, particularly buried water, wastewater, and gas pipes, typically sustain numerous breaks which can result in system failure and lead to major secondary damage, such as fire following earthquake.

Recent earthquakes where these effects occurred and liquefaction was a significant agent of damage have included:

- 1989 M 7.1 Loma Prieta (San Francisco) — major liquefaction and resulting disruption at Port of Oakland; significant liquefaction in Marina and other districts of San Francisco, causing building damage and numerous underground pipe failures [93].

- 1990 M 7.7 Philippines — major liquefaction in Dagupan City with many low-and mid-rise buildings collapsed or left tilting; many bridge failures due to lateral spreading of embankments causing piers to fail.

- 1991 M 7.4 Valle de la Estrella (Costa Rica) — numerous bridge failures due to lateral spreading of embankments causing piers to fail.

- 1994 M 6.9 Hanshin (Kobe, Japan) — major liquefaction in port areas, resulting in widespread ground failure, port disruption, and a portion of approximately 2,000 underground pipe breaks.

Evaluating liquefaction potential requires consideration of a number of factors, including **grain-size distribution** — generally, poorly graded sands (i.e., most particles about the same size) are

Figure 5.31   Photograph of sand boils.

Figure 5.31 *(Continued)* Photograph of sand boils.

much more susceptible to liquefaction than well- graded (i.e., particles of many differing sizes), since good grading results in better natural packing and better grain-grain contact. Silts and clays are generally much less susceptible to liquefaction, although the potential should not be ignored, and large-grained sands and gravels have such high permeability that pore water pressures usually dissipate before liquefaction can occur. **Relative density** is basically a measure of the packing — higher relative density means better packing and more grain-grain contact, while lower relative density, or looseness, indicates a higher potential for liquefaction. **Water table depth** is critical, as only submerged deposits are susceptible to liquefaction. A loose sandy soil above the water table is not liquefiable by itself, although upward-flowing water from a lower liquefying layer can initiate liquefaction even above the pre-event water table. Note also that water table depths can fluctuate significantly, seasonally, or over longer periods and for natural reasons or due to human intervention. Earthquake **acceleration and duration** are also critical, as these are the active causative agents for liquefaction. Acceleration must be high enough and duration long enough to cause sufficient shear strains to mobilize grain redistribution.

The basic evaluation procedure for liquefaction involves a comparison of the deposit's cyclic shear *capacity* with the *demand* imposed by an earthquake (Figure 5.32), where capacity and demand refer to the number of cycles of shear stress of a given amplitude, under similar confining pressures (comparable to the deposit's *in situ* conditions).

Determination of a soil's capacity against liquefaction can be determined by laboratory tests of undisturbed soil samples, in dynamic triaxial or cyclic simple shear tests for the appropriate confining pressures, or by correlation of these properties with some measurable *in situ* characteristic, such as blow count (**N value**) from a **Standard Penetration Test** (SPT). Demand can be determined via

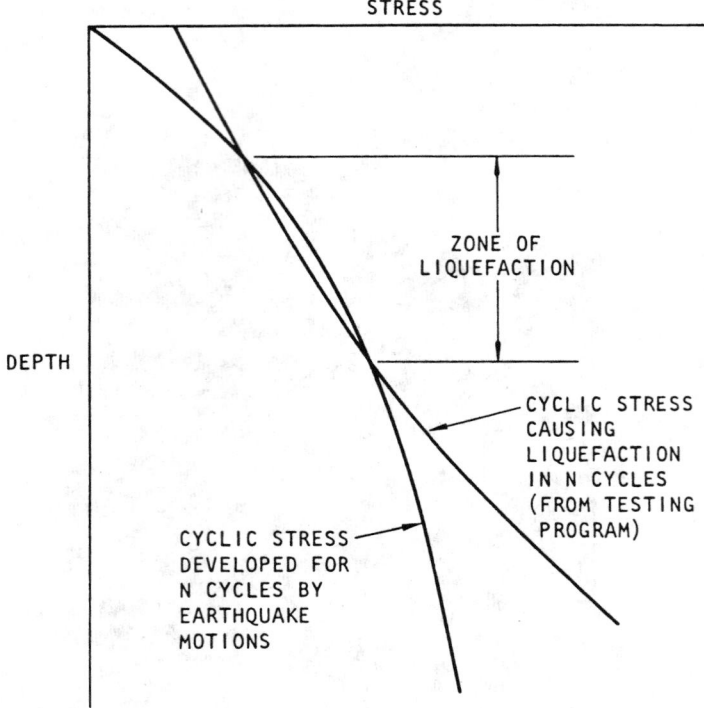

Figure 5.32   Method of evaluating liquefaction potential. (From Seed, H.B. and Idriss, I.M., Ground Motions and Soil Liquefaction During Earthquakes, Earthquake Engineering Research Institute, Oakland, CA, 1982. With permission.)

analysis of dynamic ground response using SHAKE or a similar program (see previous discussion), or by Seed's simplified procedure [113].

Simplified Procedure for Evaluation of Liquefaction Potential

The procedure developed by Seed is based on the assumption of the shear stresses induced at any point in a soil deposit during an earthquake being

   a) the mass of the soil column above the point (equal to $\gamma h/g$, where γ is the unit weight of soil, h is the depth of the point, and g is gravity)

   b) times the maximum ground acceleration, a_{max}

   c) then being corrected by a factor r_d to account for the soil column's deformable behavior, so that the maximum shear stress is $\tau_{max} = \gamma h/g \cdot a_{max} \cdot r_d$. The factor r_d is indicated in Figure 5.33. Note that while a range is indicated, most liquefiable soils are in the top 40 ft or so of a profile, where the range is relatively narrow. If the average value is used, the error is on the order of 5%.

Finally, Seed reduces the maximum shear stress by a factor of 0.65, to an "equivalent uniform average shear stress", so that τ_{av}, the average cyclic shear stress, is:

$$\tau_{av} = 0.65 \cdot \gamma h/g \cdot a_{max} \cdot r_d \tag{5.33}$$

τ_{av} and the above values provide a simple procedure for evaluation of the average cyclic shear stresses induced at different depths by a given earthquake for which the maximum ground surface

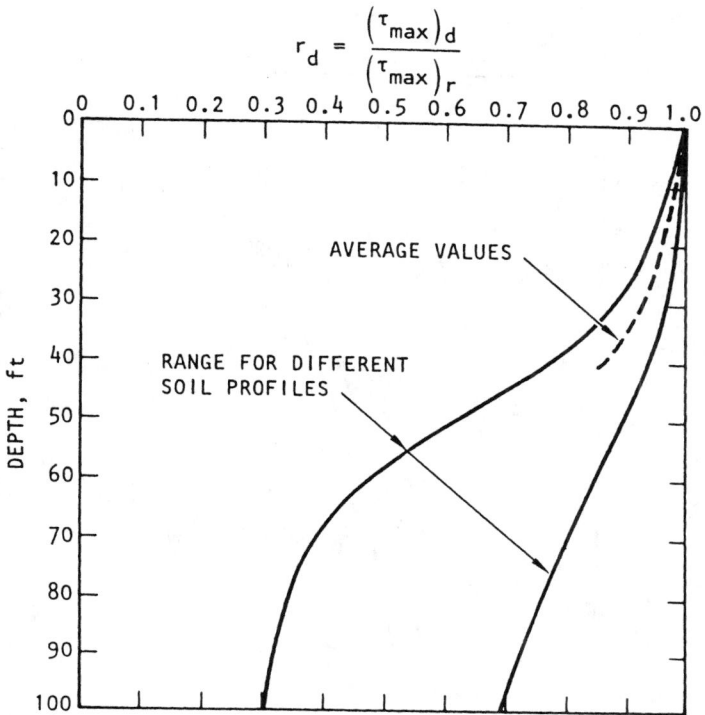

$$r_d = \frac{(\tau_{max})_d}{(\tau_{max})_r}$$

Figure 5.33   Range of values of r_d for different soil profiles. (From Seed, H.B. and Idriss, I.M., Ground Motions and Soil Liquefaction During Earthquakes, Earthquake Engineering Research Institute, Oakland, CA, 1982. With permission.)

acceleration is known [113]. In order to determine a soil's capacity against liquefaction, Seed studied SPT data from a number of sites where liquefaction had and had not occurred. He found that if the SPT data were

> a) normalized to an effective overburden pressure of 1 ton per square foot by a factor C_N
> (Figure 5.34, C_N is a function of the effective overburden pressure at the depth where the penetration test was conducted) to obtain N_1 (measured penetration of the soil under an effective overburden pressure of 1 ton per square foot, so that $N_1 = C_N \cdot N$)
>
> b) plotted against the cyclic stress ratio, τ_{av}/σ'_v, where σ'_v is the effective overburden pressure (effective meaning buoyant weight is used for soils below the water table)

then a reasonable boundary could be drawn between sites where liquefaction had and had not occurred, and could also be extended to various magnitudes (Figure 5.35).   The extension to various magnitudes was based on the number of cycles of τ_{av} caused by an earthquake, termed **significant stress cycles** N_c, which depends on the duration of ground shaking, and thus on the magnitude of the event. Representative numbers are indicated in the following table:

Magnitude	5-1/4	6	6-3/4	7-1/2	8-1/2
N_c (number of significant stress cycles)	2 ~ 3	5	10	15	26

In Figure 5.35, the region to the left of the curve (for the appropriate magnitude) is the region of liquefaction. Figure 5.35 is appropriate for sandy soils ($D_{50} > 0.25$ mm) and can be used for silty sands provided N_1 for the silty sand is increased by 7.5 before entering the chart [113].

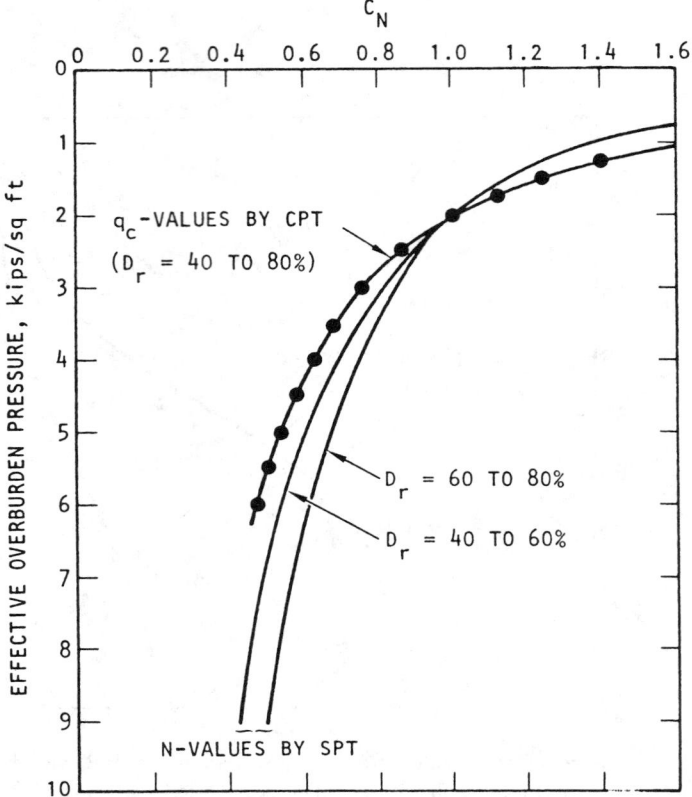

Figure 5.34   Chart for values of C_N. (From Seed, H.B. and Idriss, I.M., Ground Motions and Soil Liquefaction During Earthquakes, Earthquake Engineering Research Institute, Oakland, CA, 1982. With permission.)

As an example of Seed's simplified procedure for evaluation of liquefaction potential, consider a site located such that 0.3g maximum ground surface is expected, due to a magnitude 8 earthquake. If the water table is at 5 ft below the ground surface, then the induced cyclic stress ratio is found to be:

$$\tau_{av}/\sigma_v' = [0.65 \cdot \gamma h/g \cdot a_{\max} \cdot r_d]/\sigma_v'$$
$$= (0.65)(15 \times 110)(0.3)(0.95)/(5 \times 110 + 10 \times 47.5) = 0.3$$

If the SPT blow count at 15 ft depth is $N = 13$, then $N_1 = 16$. From Figure 5.35 we see that for a magnitude 8 earthquake and $N_1 = 16$, liquefaction is expected if the induced cyclic stress ratio is 0.16 or greater. Given its value of 0.3, liquefaction is to be expected.

Normally, results of an evaluation of liquefaction potential are expressed in terms of a factor of safety against liquefaction, or:

$$\text{Factor of Safety} = \tau_l/\tau_d \qquad\qquad (5.34)$$

where

τ_l = average cyclic stress required to cause liquefaction in N cycles (i.e., the soil's *capacity*)
τ_d = average cyclic stress induced by an earthquake for N cycles (i.e., the earthquake's *demand*)
    In the example above, the factor of safety is 0.53 ($= 0.16 / 0.30$).

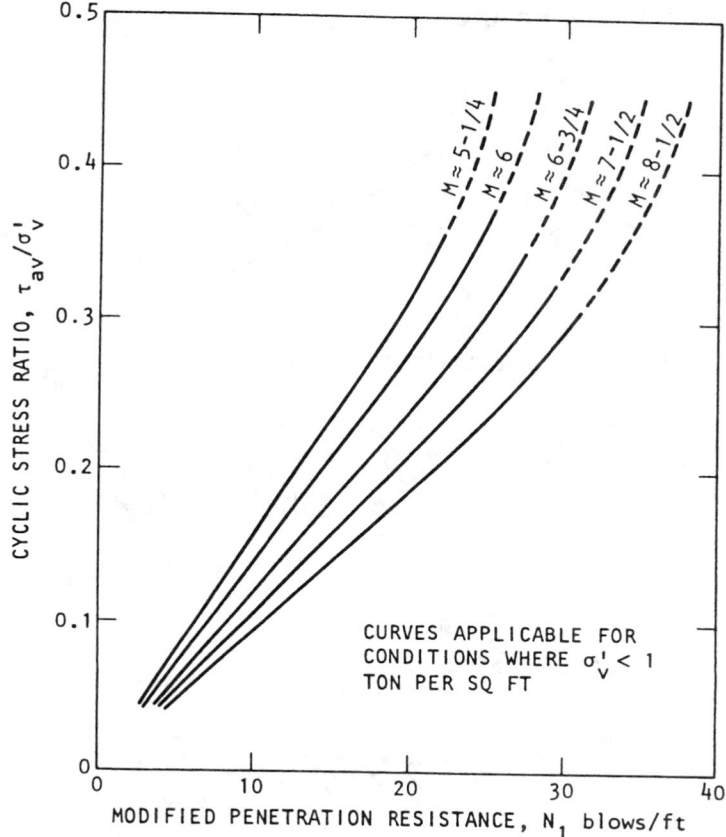

Figure 5.35 Chart for evaluation of liquefaction potential for sands for different magnitude earthquakes. (From Seed, H.B. and Idriss, I.M., Ground Motions and Soil Liquefaction During Earthquakes, Earthquake Engineering Research Institute, Oakland, CA, 1982. With permission.)

The foregoing has outlined the estimation of the occurrence of liquefaction. An important aspect (particularly for lifelines) is, given liquefaction, "What are the **permanent ground displacements** (PGD)?". Liquefaction-related PGD can be vertical, lateral, or a combination. Only lateral PGD are discussed here. Lateral PGD may be of three general types [92] (Figure 5.36):

- Flow failure on steep slopes, characterized by large displacements.
- Ground oscillation, n flat ground with liquefaction at depth decoupling surface soil layers from the underlying unliquefied ground. This decoupling allows large transient ground oscillations, although the residual PGD are usually small and chaotic.
- Lateral spread lies between flow failure and ground oscillation, occurring on horizontal ground or flat slopes, but resulting in large PGD, typically on the order of meters.

There are several methods available for quantification of PGD, including finite element analysis [100], sliding block analysis [27, 88], and empirical procedures [12]. Empirical procedures offer significant advantages when high confidence or precision is not required, and are the only feasible methods for estimation of lateral PGD in certain situations, such as extensive lifelines (e.g., buried pipelines).

Regarding empirical relations, Bartlett and Youd [12] analyzed nearly 500 horizontal displacement vectors from U.S. and Japanese earthquakes, ranging in magnitude from 6.4 to 9.2. In consideration

(a) Flow Failure

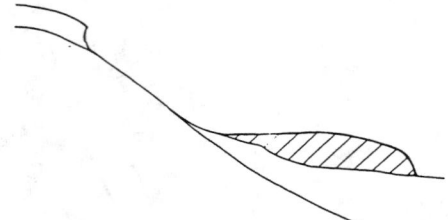

(b) Ground Oscillation

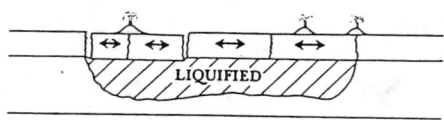

(c) Lateral Spread

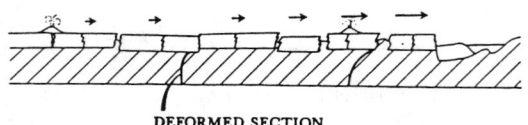

DEFORMED SECTION

Figure 5.36   Modes of lateral permanent ground displacement due to liquefaction. (After NRC, *Liquefaction of Soils During Earthquakes,* National Research Council, National Academy Press, Washington, 1985.)

of the influence of free-face (i.e., near steep banks), two empirical relations were developed:

For free-face conditions:

$$
\begin{aligned}
\log D_H \ = \ & -16.3658 + 1.1782M - 0.9275 \log R - 0.0133R \\
& + 0.6572 \log W + 0.3483 \log T_{15} + 4.527 \log(100 - F_{15}) \\
& - 0.9224 D50_{15}
\end{aligned}
\tag{5.35}
$$

For ground slope conditions:

$$
\begin{aligned}
\log D_H \ = \ & -15.7870 + 1.1782M - 0.9275 \log R - 0.0133R \\
& + 0.4293 \log S + 0.3483 \log T_{15} + 4.527 \log(100 - F_{15}) \\
& - 0.9224 D50_{15}
\end{aligned}
\tag{5.36}
$$

where
D_H = estimated lateral ground displacement in meters
$D50_{15}$ = average mean grain size in granular layers included in T_{15}, in millimeters
F_{15} = average fines content (fraction of sediment sample passing a No. 200 sieve) for granular layers included in T_{15}, in percent
M = earthquake magnitude (moment magnitude)

R = horizontal distance from the site to the nearest point on a surface project of the seismic source zone, in kilometers

S = ground slope, in percent

T_{15} = cumulative thickness of saturated granular layers with corrected blow counts (N_1) less than 15, in meters

W = ratio of the height (H) of the free-face to the distance (L) from the base of the free-face to the point in question, in percent

Allowable ranges for these parameters are indicated in Table 5.15. Displacements estimated using these relations are considered generally valid within a factor of 2 (note that doubling the estimated displacement would provide an estimate with a high likelihood of not being exceeded).

TABLE 5.15   Ranges of Input Values for Independent Variables for Which Predicted Results are Verified by Case-History Observations

Variable	Range of values in case history database
Magnitude	$6.0 < M < 8.0$
Free-face ratio	$1\% < W < 20\%$
Ground slope	$0.1\% < S < 6\%$
Thickness of loose layer	$0.3\ \text{m} < T_{15} < 12\ \text{m}$
Fines content	$0\% < F_{15} < 50\%$
Mean grain size	$0.1\ \text{mm} < d50_{15} < 1\ \text{mm}$
Depth to bottom of section	Depth to bottom of liquefied zone $< 15\ \text{m}$

After Bartlett, S.F. and Youd, T.L., Empirical Analysis of Horizontal Ground Displacement Generated by Liquefaction-Induced Lateral Spread, Tech Rept. NCEER-92-0021, National Center of Earthquake Engineering Research, Buffalo, NY, 1992.

Mitigation of liquefaction is accomplished by a number of methods, including:

- **Excavation and replacement**, that is, if the structure is sufficiently large or important, and the liquefiable layer sufficiently shallow, it may be cost- effective to place the foundation of the structure below the liquefiable layer. Note, however, that liquefaction may still occur around the structure, with possible disruption of adjacent streets, entrances, and utilities.

- **Compaction** can be accomplished by a number of methods, more complete details being given in Mitchell [82] and Hausmann [57]. Methods include:

  - Vibrostabilization:

    Vibro-compaction, in which a vibrating pile or head accompanied by water jetting is dynamically injected into the ground and then withdrawn. Vibration of the head as it is withdrawn compacts an annulus of soil, and this is repeated on a closely spaced grid. Sand may be backfilled to compensate for volume reduction. This technique achieves good results in clean granular soils with less than about 20% fines.

    Vibro-replacement (stone columns) is used in soils with higher fines content ($> 20\%$) or even in clayey soils which cannot be satisfactorily vibrated. The stone columns act as vertical drains (see below).

  - Dynamic compaction via the use of heavy dropped weights or small explosive charges. Weights up to 40 tons are dropped from heights up to 120 ft on a grid pattern, attaining significant compaction to depths of a maximum of 40 ft. Best used in large open areas due to vibrations, noise, and flying debris.

  - Compaction piles

- *Grouting*, where grout is injected at high pressure, filling voids in the soil. This technique offers the advantage of being able to be used in small, difficult to access areas (e.g., in basements of buildings, under bridges). Grouting can be used in several ways:

  - Compaction — a very stiff soil-cement-water mixture is injected and compacts an annulus around the borehole, this being repeated on a closely spaced grid.

  - Chemical — low-viscosity chemical gels are injected, forming a strong sandstone-like material. Long-term stability of the grout should be taken into account.

  - Jet-grouting — very high pressure water jets are used to cut a cylindrical hole to the desired depth, and the material is replaced or mixed with admixtures to form a stabilized column.

- *Soil-mixing* [65] — large rotary augers are used to churn up and mix the soil with admixtures and form stabilized columns or walls.

- *Permeability* can be enhanced by a number of methods, including:

  - Placement of stone columns — rather than by vibro-replacement, a hole is augured or clamshell-excavated, and filled with gravel or cobbles, on a closely spaced grid. If liquefaction occurs, any increase in pore water pressure is dissipated via the high permeability of the stone columns.

  - Soil wicks — similar to stone columns, but *wicks* consisting of geotextiles are inserted into the ground on a very closely spaced grid. Their high permeability similarly dissipates any increase in pore water pressure.

- *Grouting* — many materials are available to cement or otherwise create adhesion between soil grains, thus decreasing their cyclic mobility and any packing due to shaking.

Figure 5.37 illustrates the general soil particle size ranges for applicability of various stabilization techniques.

5.3 Seismic Design Codes

5.3.1 Purpose of Codes

The purpose of design codes in general is to develop a better built environment and improve public safety. The Uniform Building Code (UBC) is typical of many design codes, and states as its purpose:

> The purpose of this code is to provide minimum standards to safeguard life or limb, health, property and public welfare by regulating .....

In this and many other codes, it needs to be emphasized that these are **minimums standards**; that is, codes define design requirements, procedures, and other aspects that are the minimum considered necessary to achieve the code's purposes. Seismic design codes are a subset, and are generally a section or portion of many building and other design codes. The UBC sections on earthquake design are based on the SEAOC "Blue Book" (i.e., the *Recommended Lateral Force Requirements* of the Structural Engineers Association of California, SEAOC [111]), which states:

> The primary purpose of these recommendations is to provide minimum standards for use in building design regulation to maintain public safety in the extreme earth-quakes likely to occur at the building's site. The SEAOC recommendations primarily are intended to safeguard against major failures and loss of life, not to limit damage, maintain functions or provide for easy repair....Structures designed in conformance with these recommendations should, in general, be able to:

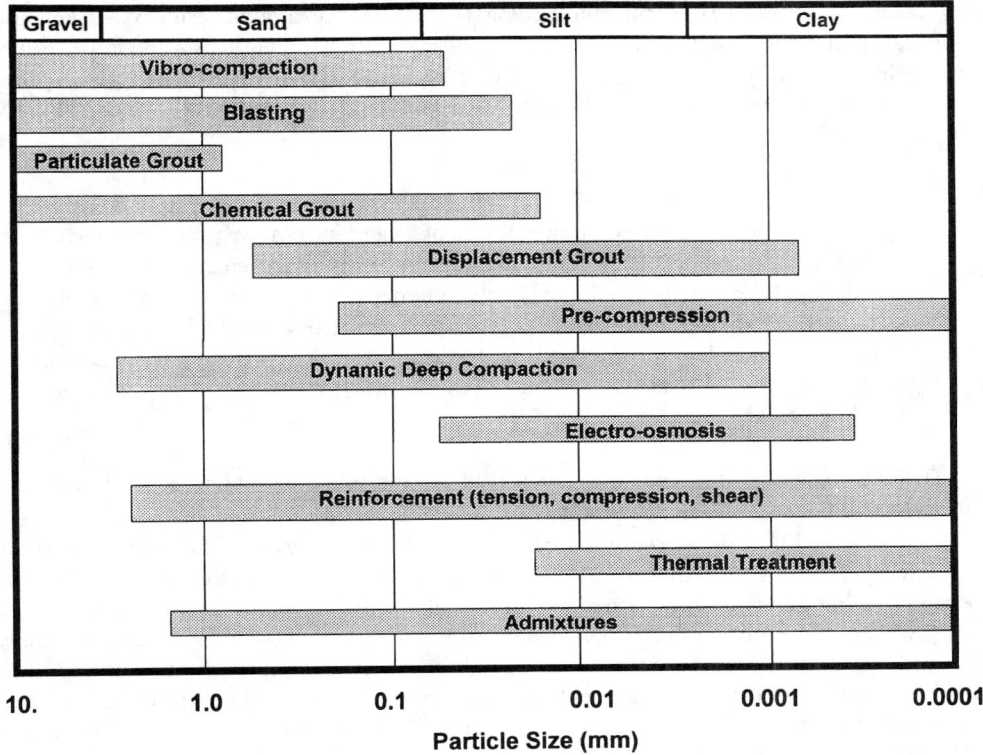

Figure 5.37 Applicable grain size ranges for different stabilization methods. (From Mitchell, J.K., Soil Improvement: State-of-the-Art, *Proc. 10th Intl. Conf. Soil Mech. Found. Eng.*, 4, 509-565, 1981. With permission.)

1. Resist minor levels of earthquake ground motion without damage;

2. Resist moderate levels of earthquake ground motion without structural damage, but possibly experience some nonstructural damage;

3. Resist major levels of earthquake ground motion having an intensity equal to the strongest either experienced or forecast for the building site, without collapse, but possibly with some structural as well as nonstructural damage.

It is expected that structural damage, even in a major earthquake, will be limited to a repairable level for structures that meet these requirements.

The general public, and many decision-makers (i.e., building owners, public officials, investors, etc.), are often not aware that substantial damage can be incurred by a structure in an earthquake, even though the structure is in conformance with the latest design codes (and has been approved as such by local building officials).

Designers should make it a practice to inform owners and other decision-makers of this aspect of earthquake design, and that exceeding minimum requirements may be cost-effective, when costs of damage and disruption are considered.

5.3.2 Historical Development of Seismic Codes

While there is some evidence that efforts were made to codify apparently successful building aspects following earthquakes in earlier times, the first codifications of the concept of **equivalent lateral** force (ELF) in earthquake resistant building design occurred in the early 20th century. San Francisco was rebuilt after the earthquake and fire of 1906 under provisions that a 30 psf wind force, to affect both wind and earthquake resistance, would be adequate for a building designed with a proper system of bracing [111]. Following the 1908 M7.5 Messina (Italy) earthquake, which killed over 80,000:

> The government of Italy responded to the Messina earthquake by appointing a special committee composed of nine practicing engineers and five professors of engineering... The report of this committee appears to be the first engineering recommendation that earthquake resistant structures be designed by means of the equivalent static method (the percent g method)...M. Panetti, Professor of Applied Mechanics in Turin...recommended that the first story be designed for a horizontal force equal to 1/12 the weight above and the second and third stories to be designed for 1/8 of the building weight above [59].

The ELF concept was proposed in Japan in 1914 by Prof. Riki [96] but not required. It was also used by leading structural engineers in the U.S., but without clear codification.

Following the 1923 Tokyo earthquake, the Japanese Urban Building Law Enforcement Regulations were revised in 1924 to introduce a 0.10 seismic coefficient. Following the 1925 Santa Barbara earthquake in California, the International Conference of Building Officials in 1927 adopted the UBC which "suggested" an ELF ranging from 0.075 to 0.10g for cities located in an area "subject to earthquake shocks" [111]. In 1933, as a result of the Long Beach earthquake, the Field Act in California required all public school buildings to be designed per an ELF (0.10g for masonry, 0.02~0.05 for other types), and the Riley Act required all buildings be designed per an ELF of 0.02g. The 1935 UBC required 0.08g. Concepts of dynamic response of structures were perhaps first adopted by Los Angeles and, in 1952, the ASCE *Lateral Forces of Earthquake and Wind* report [6] appeared, which related lateral force coefficients to the fundamental period of the structure and provided a framework for codification of dynamic analysis of structures. In Japan, the Building Standard Law was proclaimed in 1950, and required an ELF of 0.20g.

In 1959 the first edition of the SEAOC "Blue Book" appeared. Since that time, the Blue Book (written entirely on a volunteer basis by members of SEAOC) has traditionally been adopted for the seismic provisions of the UBC and other model codes in the U.S., and in many other countries, with modifications appropriate to local conditions. Following a series of earthquakes in the 1960s and 1970s,[7] the UBC was significantly revised in 1976 to generally increase the ELF and, perhaps more importantly, change a number of other requirements, such as concrete detailing. Also introduced was the "R_W" factor, intended to account for and take advantage of overstrength and ductility of LFRS. However, the R_W values are based on professional judgement and experience and their qualification, on a rational basis, has proven challenging. Similarly, the Japanese Building Standard Law was revised in 1971 and 1981 to significantly decrease spacing of hoops or other transverse concrete reinforcement and other respects.

In 1978 the Applied Technology Council's ATC3-06 [9] was issued, marking a major step forward in seismic hazard and dynamic response analysis. That effort has since been continued by the Building Seismic Safety Council (BSSC) founded in 1979 as a result of the passage in 1977 by the U.S. Congress of the Earthquake Hazards Reduction Act (PL 95-124) and creation of the National Earthquake Hazards Reduction Program (NEHRP). BSSC's NEHRP Recommended Provisions for Seismic Regulations for New Buildings were last updated in 1994 [25].

---

[7] For example, 1964 Alaska, 1964 Niigata, 1967 Santa Rosa, 1968 Tokachi-Oki, 1971 San Fernando, and 1978 Miyagiken-oki.

This brief history has only discussed, in very abbreviated form, the development of building seismic design codes, primarily for the U.S. and, to a lesser extent, for Japan. It is summarized, to some extent, in Figure 5.38 which shows a number of earthquakes that have been significant for code development, as well as showing selected other developments, and the general range in the increase in the ELF seismic coefficient. Note also the growth in the number of strong motion records, from the first in the 1933 Long Beach earthquake, to 1971 where the approximately 100 records obtained in that event effectively doubled the number of available records worldwide, to the thousands available today.

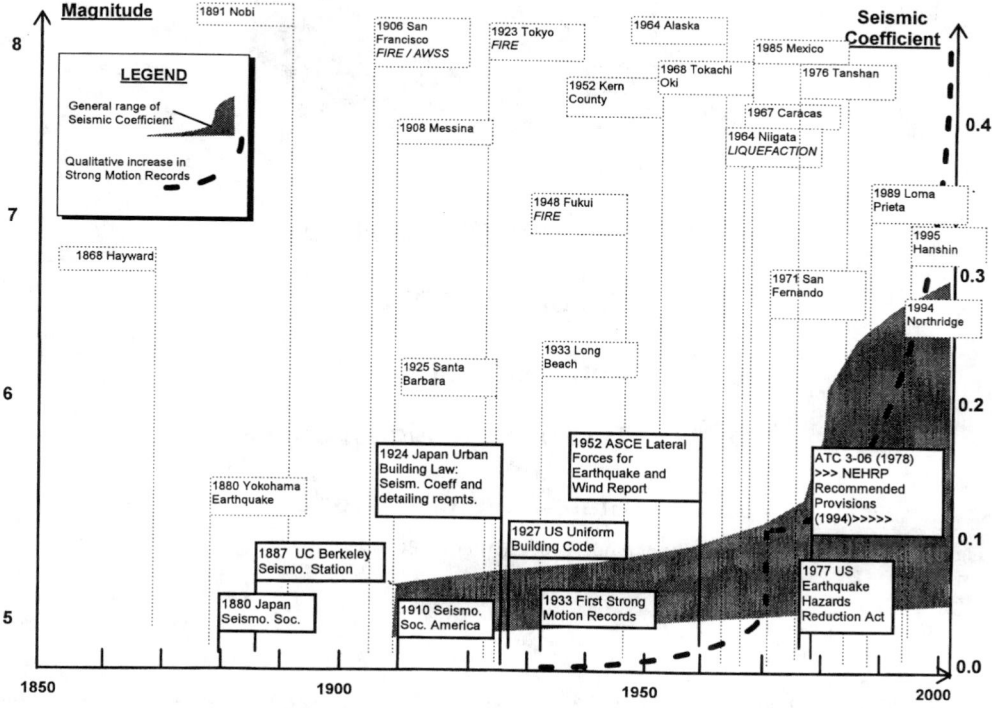

Figure 5.38 Schematic development of seismic design codes.

5.3.3 Selected Seismic Codes

In the U.S., building code enforcement derives from the police power of the state, which is typically administered either at the state or local level (which grants this power to cities and other jurisdictions), unless pre-empted by state or federal regulation. Each jurisdiction can thus choose to write its own building ordinances and codes but, practically, many jurisdictions choose to adopt a so-called **model code**, of which three exist in the U.S. — the **UBC** written by the International Conference of Building Officials (ICBO, founded about 1921), the **BOCA National Code**, written by the Building Officials & Code Administrators International (BOCA, founded in 1915), and the Standard Building Code (**SBC**) written by the Southern Building Code Congress International (SBCCI, founded more than 50 years ago). The UBC serves as the model code primarily for jurisdictions in the western U.S., BOCA in the central and northeast, and the SBC in the southeast portion of the U.S. (see Figure 5.39). Seismic provisions in these three model codes follow similar formats, derived from the SEAOC Blue Book as discussed above, but adoption of specific provisions varies. In 1994 these three organizations adopted a common format, and will join to write a single International Building Code (**IBC**) by the year 2000, under the umbrella of the Council of American Building Officials (**CABO**). The seismic provisions of the IBC will draw on the NEHRP Recommended Provisions [25] as well as a new performance-based *Vision 2000* code format currently under development by SEAOC.

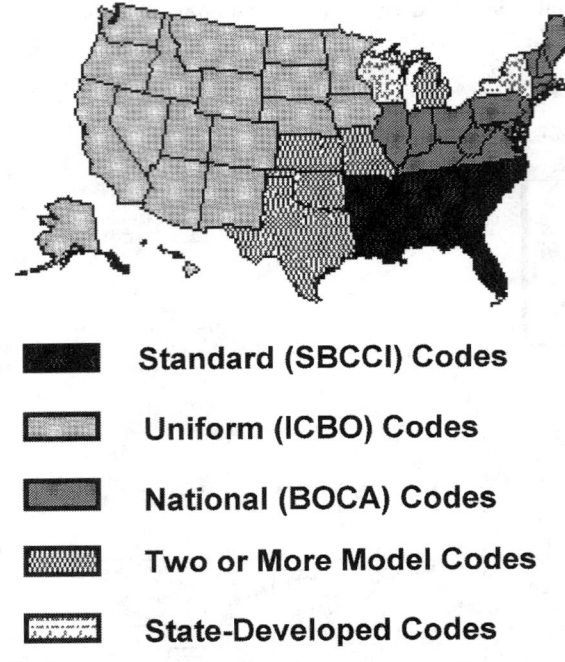

■ **Standard (SBCCI) Codes**

□ **Uniform (ICBO) Codes**

▦ **National (BOCA) Codes**

▨ **Two or More Model Codes**

▨ **State-Developed Codes**

Figure 5.39   Geographic influence of model codes. (After CABO www.cabo.org.)

Seismic design of monolithic reinforced concrete structures in the U.S. is guided by the provisions of ACI-318 [1] of the American Concrete Institute (ACI) — the commentary to ACI-318 is a valuable guide to understanding seismic effects on reinforced concrete.

Seismic design codes for other countries and for non-building structures cannot be covered in detail here. A few selected resources include Earthquake Resistant Regulations [64], which is a compilation of earthquake building design codes and regulations used in seismically active regions of the world (seismic regulations for 37 countries are included); the International Handbook of Earthquake Engineering [97], which presents a good explanation of seismic codes and some design practices for 34 countries, as well as a summary of the main developments in seismic code activity for each country; and the so-called Tri-services Manual [127], which provides criteria and guidance for the design of structures to resist the effects of earthquakes, including architectural components, mechanical and electrical equipment supports, some structures other than buildings, and utility systems. Another useful resource is the Practice of Earthquake Hazard Assessment [79], which summarizes probabilistic seismic hazard assessment as it is practiced in 88 seismically active countries throughout the world. Table 5.16 summarizes the countries that are treated by each of these resources.

TABLE 5.16   Table of Countries and Related Seismic Design Code Information

Country	E	I	P	S	U	Country	E	I	P	S	U
Afghanistan			X	X		Ecuador			X	X	
Albania			X	X		Egypt	X	X	X	X	
Algeria	X	X		X		El Salvador	X	X	X		
Angola				X		Equatorial Guinea				X	
Argentina	X	X	X	X		Ethiopia	X	X	X		
Australia	X	X	X	X		Fiji			X	X	
Austria	X		X			Finland			X	X	
Azores				X	X	France	X	X	X	X	
Bahama Islands				X	X	French West Indies				X	
Bahrain				X		Gabon				X	
Bangladesh			X	X		Gambia				X	
Belgium			X	X		Germany	X		X	X	
Belize				X		Ghana				X	
Benin				X		Greece	X	X	X	X	
Bermuda				X	X	Greenland				X	X
Bolivia			X	X		Grenada				X	
Botswana				X		Guatemala				X	
Brazil			X	X		Guinea				X	
Brunei				X		Haiti				X	
Bulgaria	X	X	X	X		Honduras				X	
Burma				X		Hong Kong				X	
Burundi				X		Hungary		X	X	X	
Cameroon				X		Iceland				X	X
Canada	X	X	X	X		India	X	X	X	X	
Canal Zone				X	X	Indonesia	X	X		X	
Cape Verde				X		Iran	X	X	X	X	
Caroline Islands				X	X	Iraq				X	
Central African Rep.				X		Ireland			X	X	
Chad				X		Israel	X	X	X	X	
Chile	X	X	X	X		Italy	X	X	X	X	
China	X	X	X	X		Ivory Coast			X	X	
Colombia	X	X	X	X		Jamaica			X	X	
Congo				X		Japan	X	X	X	X	X
Costa Rica	X	X	X	X		Johnston Island				X	X
Cuba	X	X	X			Jordan		X		X	
Cyprus			X	X		Kenya				X	
Czech Republic			X	X		Korea				X	
Denmark				X		Kuwait				X	
Djibouti				X		Kwajalein				X	
Dominican Rep.				X		Laos				X	
Eastern Caribbean			X	X		Lebanon				X	
						Leeward Islands				X	X
						Lesotho			X		

TABLE 5.16   Table of Countries and Related Seismic Design Code Information
(continued)

Country	E	I	P	S	U	Country	E	I	P	S	U
Liberia				X		South Yemen				X	
Libya				X		Spain	X	X	X	X	
Luxembourg				X		Sri Lanka				X	
Malagasy Rep.				X		Swaziland				X	
Malawi				X		Sweden				X	
Malaysia			X	X		Switzerland	X		X	X	
Mali				X		Syria				X	
Malta				X		Taiwan		X	X	X	
Marcus Island					X	Tanzania				X	
Mariana				X	X	Thailand		X	X	X	
Islands				X		Togo				X	
Marshall				X		Trinidad &			X	X	X
Islands						Tobago					
Mauritania				X		Tunisia			X	X	
Mauritius				X		Turkey	X	X	X	X	X
Mexico	X	X		X		Uganda				X	
Morocco				X		United Arab				X	
Mozambique				X		Emir's					
Nepal			X	X		United			X	X	
Netherlands			X	X		Kingdom					
New Zealand	X	X	X	X		Upper Volta				X	
Nicaragua	X			X		Uruguay				X	
Niger				X		USA	X	X	X	X	X
Nigeria			X	X		USSR, Former	X	X	X	X	
Norway			X	X		Venezuela	X	X		X	
Oman				X		Vietnam			X	X	
Pakistan				X		Wake Island				X	X
Panama				X		Yemen Arab				X	
Papua New			X	X		Rep.					
Guinea						Yugoslavia	X	X	X	X	
Paraguay				X		Fmr					
Peru	X	X		X		Zaire			X	X	
Philippines	X			X	X	Zambia				X	
Poland			X	X		Zimbabwe				X	
Portugal	X	X			X						
Puerto Rico		X		X	X	E: Earthquake Resistant					
Quatar				X		Regulations: A World List–					
Rep. of				X		1992					
Rwanda											
Romania	X	X	X	X		P: Practice of Earthquake					
Samoa				X	X	Hazard Assessment					
Saudi Arabia				X							
Senegal				X		I: International Handbook					
Seychelles				X		of Earthquake					
Sierra Leone				X		Engineering					
Singapore				X							
Slovakia			X			S: Seismic Design for Buildings					
Somalia				X							
South Africa			X	X		U: Uniform Building Code					

After NCEER http://nceer.eng.buffalo.edu/bibs/intcodes.html

5.4 Earthquake Effects and Design of Structures

Many different types of earthquake damage occur in structures. This section discusses general earthquake performance of buildings and selected transportation and industrial structures, with the emphasis more towards buildings, especially those typically built in the western U.S. Specific aspects of structural analysis and design of buildings, other structures, steel, concrete, wood, masonry, and other topics are discussed in other chapters.

5.4.1 Buildings

In buildings, earthquake damage can be divided into two categories: structural damage and non-structural damage, both of which can be hazardous to building occupants. Structural damage means degradation of the building's structural support systems (i.e., vertical and lateral force resisting systems), such as the building frames and walls. Non-structural damage refers to any damage that does not affect the integrity of the structural support system. Examples of non-structural damage are a chimney collapsing, windows breaking or ceilings falling, piping damage, disruption of pumps, control panels, telecommunications equipment, etc. Non-structural damage can still be life-threatening and costly. The type of damage to be expected is a complex issue that depends on the structural type and age of the building, its configuration, construction materials, the site conditions, the proximity of the building to neighboring buildings, and the type of non-structural elements.

How Earthquake Forces are Resisted

Buildings experience horizontal distortion when subjected to earthquake motion (Figure 5.40). When these distortions become large, the damage can be catastrophic. Therefore, most buildings are designed with **lateral force resisting** systems (LFRS) to resist the effects of earthquake forces and maintain displacements within specified limits. LFRS are usually capable of resisting only forces that result from ground motions parallel to them. However, the combined action of LFRS along the width and length of a building can typically resist earthquake motion from any direction. LFRS differ from building to building because the type of system is controlled to some extent by the basic layout and structural elements of the building. Basically, LFRS consist of axial- (tension and/or compression), shear- and/or bending-resistant elements.

In wood frame stud-wall buildings, the resistance to lateral loads is typically provided by (a) for older buildings, especially houses, wood diagonal "let-in" bracing, and (b) for newer (primarily post-WW2) buildings, plywood siding "shear walls". Without the extra strength provided by the bracing or plywood, walls would distort excessively or "rack", resulting in broken windows, stuck doors, cracked plaster, and, in extreme cases, collapse.

The earthquake resisting systems in modern steel buildings take many forms. Many types of bracing configurations have been used (diagonal, "X", "V", "K", etc). Moment-resisting steel frames are also capable of resisting lateral loads. In this type of construction, the connections between the beams and the columns are designed to resist the rotation of the column relative to the beam. Thus, the beam and the column work together and resist lateral movement by bending. This is contrary to the braced frame, where loads are resisted through tension and compression forces in the braces. Steel buildings are sometimes constructed with moment resistant frames in one direction and braced frames in the other, or with integral concrete or masonry shear walls.

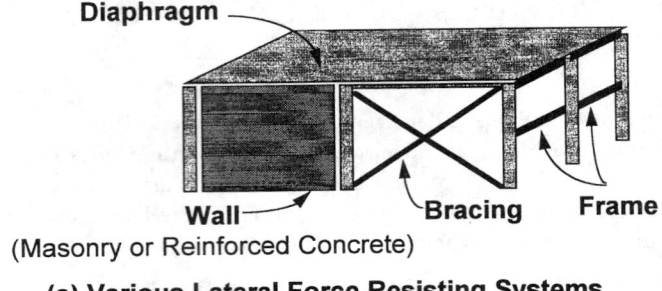

(a) Various Lateral Force Resisting Systems

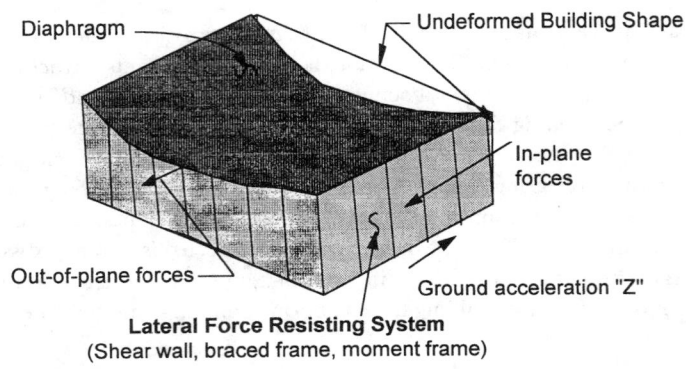

(b) Effect of Ground Acceleration on Building

Figure 5.40   Resistance of earthquake forces: (a) variety of LFRS and (b) effect of ground acceleration on building.

In concrete structures, **shear walls**[8] are sometimes used to provide lateral resistance, in addition to moment-resisting frames. Ideally, these shear walls are continuous reinforced-concrete walls extending from the foundation to the roof of the building, and can be exterior or interior walls. They are interconnected with the rest of the concrete frame, and thus resist the motion of one floor relative to another. Shear walls can also be constructed of reinforced brick, or reinforced concrete masonry units.

Certain problems in earthquake resistiveness are independent of building type and include the following:

- **Configuration**, or the general vertical and/or horizontal shape of buildings, is an important factor in earthquake performance and damage. Buildings that have simple, regular, symmetric configurations generally display the best performance in earthquakes. The reasons for this are (1) non-symmetric buildings tend to have twist (i.e., have significant torsional modes) in addition to shaking laterally, and (2) the various "wings" of a building tend to act independently, resulting in differential movements, cracking, and other damage. Rotational motion introduces additional damage, especially at re-entrant or

---

[8]Termed shear walls because the depth-to-length ratio is so large that, deformation is primarily due to shear rather than bending.

"internal" corners of the building. The term "configuration" also refers to the geometry of lateral load resisting systems as well as the geometry of the building. Asymmetry can exist in the placement of bracing systems, shear walls, or moment-resisting frames that are used to provide earthquake resistance in a building. This type of asymmetry of the LFRS can result in twisting or differential motion, with the same consequences as asymmetry in the building plan. An important aspect of configuration is **soft story**, which is a story of a building significantly less stiff than adjacent stories (that is, a story in which the lateral stiffness is 70% or less than that in the story above, or less than 80% of the average stiffness of the three stories above [25]). Soft stories often (but not always) occur on the ground floor, where commercial or other reasons require a greater story height, and large windows or openings for ingress or commercial display (e.g., the building might have masonry curtain walls for the full height, except at the ground floor, where these are replaced with large windows, for a store's display). Due to inadequate stiffness, a disproportionate amount of the entire building's drift is concentrated at the soft story, resulting in non-structural and potential structural damage. Many older buildings with soft stories, built prior to recognition of this aspect, collapse due to excessive ductility demands at the soft story.

- **Pounding** is the collision of adjacent buildings during an earthquake due to insufficient lateral clearance. Such collision can induce very high and unforeseen accelerations and story shears in the overall structure. Additionally, if adjacent buildings have varying story heights, a relatively rigid floor or roof diaphragm may impact an adjacent building at or near mid-column height, causing bending or shear failure in the columns, and subsequent story collapse. Under earthquake lateral loading, buildings deflect significantly. These deflections or **drift** are limited by code, and adjacent buildings must be separated by a **seismic gap** equal to the sum of their actual calculated drifts (i.e., ideally, each building set back from its property line by the drift). Pounding has been the cause of a number of building collapses, most notably in the 1985 Mexico City earthquake.

Estimation of Earthquake Forces

As discussed above, estimation of the forces an earthquake may impose on a building may be accomplished by use of an Equivalent Lateral Force (ELF) procedure, or by development of a design basis earthquake using probabilistic methods involving seismicity magnitude-frequency relations, attenuation, etc. For many years, the UBC and other codes determined the ELF according to varients on the equation $V = ZICKSW$ (parameters defined below with the exception of K, varied by type of structure). The current UBC [126] at present determined the minimum ELF, as follows:

$$V = \frac{ZIC}{R_W} W \qquad (5.37)$$

where

V = total design lateral force or shear at the base

Z = a seismic zone factor given in Table 16-I of the UBC, and varying between 0.075 (Zone 1, low seismicity areas) and 0.40 (Zone 4, high seismicity areas)

I = importance factor given in Table 16-K of the UBC, varying between $1.0 \sim 1.25$

C = $\frac{1.25S}{T^{2/3}} \leq 2.75$ (and C/R_W shall be ≥ 0.075)

R_W = a numerical coefficient defined in UBC Tables 16-N and 16-P, and varying between $4 \sim 12$ for buildings, and $3 \sim 5$ for selected nonbuilding structures

S = site coefficient for soil characteristics given in UBC Table 16-J (see Table 5.13 of this section)

T = fundamental period of vibration, in seconds, of the structure in the direction under consideration

W = total seismic dead load

In 1997, the UBC revised the determination of total design base shear to be determined according to:

$$V = \frac{C_v I}{RT} W$$

except that the total design base shear need not exceed the following:

$$V = \frac{2.5 C_a I}{R} W$$

nor be less than $V = 0.11 C_a I W$. In this, note that R is not the same as the previous R_W. The new term, R, still accounts for inherent overstrength and global ductility of LFRS, but is somewhat less (R varies from a minimum of 2.8 to a maximum of 8.5) such that base shear levels are increased by about 40% on average. C_a and C_v are seismic coefficients depending on soil type and seismic zone, effectively replacing the previous S factor. In Seismic Zone 4 (the highest), total design base shear is further limited to not be less than $V = 0.8 Z N_v (I/R) W$, where N_v is a near-source factor used in the determination of C_v, related to the proximity of the structure to known faults (and depending on slip rate and maximum magnitude, for the fault; note that there is an analogous N_a). Another important addiction is a Reliability/Redundancy Factor $1.0 \le \rho \le 1.5$, which can significantly increase the total design base shear for non-redundant structures.

A number of other requirements and conditions are detailed in the UBC and other seismic codes, and the reader is referred to them for details.

Types of Buildings and Typical Earthquake Performance

The typical earthquake performances of different types of common construction systems are described in this section, to provide insights into the type of damage that may be expected and the strengthening that may be necessary.

Wood Frame　Wood frame structures tend to be mostly low rise (one to three stories, occasionally four stories). Vertical framing may be of several types: stud wall, braced post and beam, or timber pole:

- Stud walls are typically constructed of 2 in. by 4 in. wood members set vertically about 16 in. apart — multiple story buildings may have 2x6 or larger studs. These walls are braced by plywood sheathing or by diagonals made of wood or steel. Most detached single and low-rise multiple family residences in the U.S. are of stud wall wood frame construction.
- Post and beam construction is not very common in the U.S., although it is the basis of the traditional housing in other countries (e.g., Europe, Japan), and in the U.S. is found mostly in older housing and larger buildings (i.e., warehouses, mills, churches, and theaters). This type of construction consists of larger rectangular (6 in. by 6 in. and larger) or sometimes round wood columns framed together with large wood beams or trusses.
- Timber pole buildings are a less common form of construction found mostly in sub-urban/rural areas. Generally adequate seismically when first built, they are more often subject to wood deterioration due to the exposure of the columns, particularly near the ground surface. Together with an often found "soft story" in this building type, this deterioration may contribute to unsatisfactory seismic performance.

Stud wall buildings have performed very well in past U.S. earthquakes for ground motions of about 0.5g or less, due to inherent qualities of the structural system and because they are lightweight and low rise. Cracking in plaster and stucco may occur, and these act to degrade the strength of the building to some extent (i.e., the plaster and stucco may in fact form part of the LFRS, sometimes by design). This is usually classified as non-structural damage but, in fact, dissipates a lot of the earthquake induced energy. However, the most common type of structural damage in older wood frame buildings results from a lack of connection between the superstructure and the foundation — so-called "cripple wall" construction. This kind of construction is common in the milder climes of the western U.S., where full basements are not required, and consists of an air space (typically $2 \sim 3$ ft) left under the house. The short stud walls under the first floor (termed by carpenters a **cripple wall** because of their less than full height) were usually built without bracing, so that there is no adequate LFRS for this short height. Plywood sheathing nailed to the cripple studs may be used to strengthen the cripple walls. Additionally, the mud sill in these older (typically pre-WW2) housings may not be bolted to the foundation. As a result, houses can slide off their foundations when not properly bolted to the foundation, resulting in major damage to the building as well as to plumbing and electrical connections. Overturning of the entire structure is usually not a problem because of the low-rise geometry. In many municipalities, modern codes require wood structures to be bolted to their foundations. However, the year that this practice was adopted will differ from community to community and should be checked.

Garages often have a very large door opening in one wall with little or no bracing. This wall has almost no resistance to lateral forces, which is a problem if a heavy load such as a second story sits on top of the garage (so-called house over garage, or HOGs). Homes built over garages have sustained significant amounts of damage in past earthquakes, with many collapses. Therefore, the house-over-garage configuration, which is found commonly in low-rise apartment complexes and some newer suburban detached dwellings, should be examined more carefully and perhaps strengthened.

Unreinforced masonry chimneys also present a life-safety problem. They are often inadequately tied to the building and therefore fall when strongly shaken. On the other hand, chimneys of reinforced masonry generally perform well.

Some wood frame structures, especially older buildings in the eastern U.S., have masonry veneers that may represent another hazard. The veneer usually consists of one wythe of brick (a wythe is a term denoting the width of one brick) attached to the stud wall. In older buildings, the veneer is either insufficiently attached or has poor quality mortar, which often results in peeling off of the veneer during moderate and large earthquakes.

Post and beam buildings tend to perform well in earthquakes, if adequately braced. However, walls often do not have sufficient bracing to resist horizontal motion and thus they may deform excessively.

The 1994 M_W 6.7 Northridge earthquake was the largest earthquake to occur directly within an urbanized area since the 1971 San Fernando earthquake; ground motions were as high as 0.9g, and substantial numbers of modern wood-frame dwellings sustained significant damage, including major cracking of veneers, gypsum board walls, and splitting of wood wall studs. It may be inferred from this, as well as the performance observed in the more sparsely populated epicentral regions of the 1989 M_W 7.1 Loma Prieta earthquake, that the U.S. single family dwelling design begins to sustain substantial non-structural and structural damage for PGA in excess of about 0.5g.

Steel Frame Buildings   Steel frame buildings generally may be classified as either moment resisting frames (**MRF**) or **braced frames**, based on their lateral force resisting systems. In concentric braced frames, the lateral forces or loads are resisted by the tension only, or tensile and compressive strength of the bracing, which can assume a number of different configurations including diagonal, 'V', inverted "V" (also termed chevron), "K", etc. A recent development in seismic bracing is the eccentric brace frame (**EBF**). Here the bracing is slightly offset from the main beam to column connection, and a short section of beam is expected to deform significantly under major seismic forces and thereby dissipate a considerable portion of the energy. Moment resisting frames resist lateral loads and deformations by the bending stiffness of the beams and columns (there is no bracing).

Steel frame buildings have tended to perform satisfactorily in earthquakes with ground motions less than about 0.5g because of their strength, flexibility, and lightness. Collapse in earthquakes has been very rare, although steel frame buildings did collapse, for example, in the 1985 Mexico City earthquake. More recently, following the 1994 M_W 6.7 Northridge earthquake, a number of MRF were found to have sustained serious cracking in the beam column connection (see Figure 5.41, which shows one of a number of different types of cracking that were found following the Northridge earthquake). The cracking typically initiated at the lower beam flange location and propagated upward into the shear panel. Similar cracking was also observed following the 1995 M_W 6.9 Hanshin (Kobe) earthquake, which experienced similar levels of ground motion as Northridge.

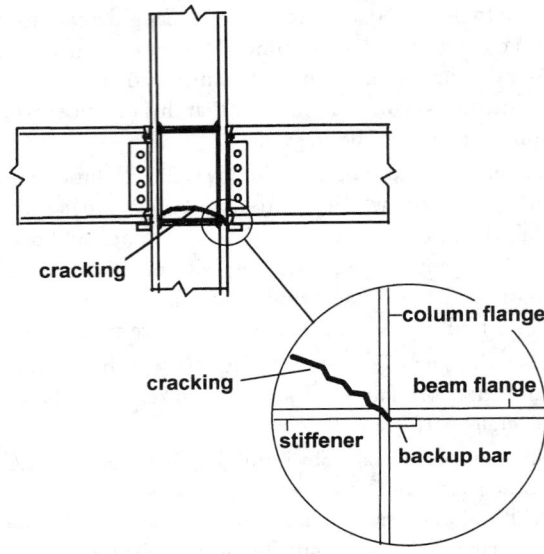

Figure 5.41   Example steel moment-frame connection cracking.

More worrisome is that, as of this writing, some steel buildings in the San Francisco Bay Area have been found to have similar cracking, presumably as a result of the 1989 M_W 7.1 Loma Prieta earthquake. As a result, there is an on-going effort by a consortium of research organizations, termed SAC (funded by the Federal Emergency Management Agency) to better understand and develop solutions for this problem.

Light gage steel buildings are used for agricultural structures, industrial factories, and warehouses. They are typically one story in height, sometimes without interior columns, and often enclose a large floor area. Construction is typically of steel frames spanning the short dimension of the building and resisting lateral forces as moment frames. Forces in the long direction are usually resisted by diagonal steel rod bracing. These buildings are usually clad with lightweight metal or reinforced concrete siding, often corrugated. Because these buildings are low-rise, lightweight, and constructed of steel members, they usually perform relatively well in earthquakes. Collapses do not usually occur. Some typical problems are (a) insufficient capacity of tension braces can lead to their elongation and, in turn, building damage, and (b) inadequate connection to the foundation can allow the building columns to slide.

Concrete Buildings   Several construction subtypes fall under this category: (a) moment resisting frames (nonductile or ductile), (b) shear wall structures, and (c) pre-cast, including tilt-up structures. The most prevalent of these is **nonductile reinforced concrete frame** structures with or without infill walls built in the U.S. between about 1920 and (in the western U.S.) about 1972. In many others portions of the U.S. this type of construction continues to the present. This group includes large multistory commercial, institutional, and residential buildings constructed using flat slab frames, waffle slab frames, and the standard girder-column type frames. These structures generally are more massive than steel frame buildings, are under-reinforced (i.e., have insufficient reinforcing steel embedded in the concrete) and display low ductility. Some typical problems are: (a) large tie spacings in columns can lead to a lack of concrete confinement and/or shear failure; (b) placement of inadequate rebar splices at the same location can lead to column failure; (c) insufficient shear strength in columns can lead to shear failure prior to the development of moment hinge capacity; (d) insufficient shear tie anchorage can prevent the column from developing its full shear capacity; (e) lack of continuous beam reinforcement can result in hinge formation during load reversal ; (f) inadequate reinforcing of beam-column joints or location of beam bar splices at columns can lead to failures; and (g) the relatively low stiffness of the frame can lead to substantial non-structural damage.

Ductile RC frames where special reinforcing details are required in order to furnish satisfactory load-carrying performance under large deflections (termed ductility) have usually only been required in the highly seismic portions of the U.S. since the late 1960s (first provisions appeared in the 1967 UBC). ACI-318 [1] provides comprehensive treatment for the **ductile detailing**, which involves a number of special requirements including close spacing of lateral reinforcement in order to attain **confinement** of a concrete core, appropriate relative dimensioning of beams and columns, 135° hooks on lateral reinforcement, hooks on main beam reinforcement within the column, etc.

Concrete shear wall buildings consist of a concrete box or frame structural system with walls constituting the main LFRS. The entire structure, along with the usual concrete diaphragm, is typically cast in place. Shear walls in buildings can be located along the perimeter, as interior walls, or around the service or elevator core. This building type generally tends to perform better than concrete frame buildings. They are heavier than steel frame buildings but they are also rigid due to the shear walls. Some types of damage commonly observed in taller buildings are caused by vertical discontinuities, pounding, and/or irregular configuration. Other damages specific to this building type are (a) shear cracking and distress occurring around openings in concrete shear walls during large seismic events; (b) shear failure occurring at wall construction joints usually at a load level below the expected capacity; and (c) bending failures resulting from insufficient chord steel lap lengths.

Tilt-up buildings are a common type of construction in the western U.S., and consist of concrete wall panels cast on the ground and then tilted upward into their final positions. More recently, wall panels are fabricated off-site and trucked in. The wall panels are welded together at embedments, or held in place by cast-in-place columns or steel columns, depending on the region. The floor and roof beams are often glue-laminated wood or steel open webbed joists that are attached to the tilt-up wall panels; these panels may be load bearing or non-load bearing, depending on the region. These buildings tend to be low-rise industrial or office buildings. Before 1973 in the western U.S., many tilt-up buildings did not have sufficiently strong connections or anchors between the walls and the roof and floor diaphragms. During an earthquake, weak anchors pulled out of the walls, causing the floors or roofs to collapse. The connections between concrete panels are also vulnerable to failure. Without these, the building loses much of its lateral force-resisting capacity. For these reasons, many tilt-up buildings were damaged in the 1971 San Fernando earthquake. Since 1973, tilt-up construction practices have changed in California and other high seismicity regions, requiring positive wall-diaphragm connection and prohibiting cross-grain bending in wall ledgers. (Such requirements may not have yet been made in other regions of the country.) However, a large number of these older, pre-1970s vintage tilt-up buildings still exist and have not been retrofitted to correct this wall-anchor defect. These buildings are a prime source of seismic risk. In areas of low or moderate seismicity, inadequate wall anchor details continue to be employed. Damage to tilt-up buildings was observed again in the 1994 M_W 6.7 Northridge earthquake, where the primary problems were poor wall anchorage into the concrete, and excessive forces due to flexible roof diaphragms amplifying ground motion to a greater extent than anticipated in the code.

Precast concrete frame construction, first developed in the 1930s, was not widely used until the 1960s. The precast frame is essentially a post and beam system in concrete where columns, beams, and slabs are prefabricated and assembled on site. Various types of members are used: vertical load carrying elements may be T's, cross shapes, or arches and are often more than one story in height. Beams are often T's and double T's, or rectangular sections. Prestressing of the members, including pre-tensioning and post-tensioning, is often employed. The LFRS is often concrete CIP shear walls. The earthquake performance of this structural type varies greatly and is sometimes poor. This type of building can perform well if the details used to connect the structural elements have sufficient strength and ductility (toughness). Because structures of this type often employ cast-in-place concrete shear walls for lateral load resistance, they experience the same types of damage as other shear wall building types. Some of the problem areas specific to precast frames are (a) failure of poorly designed connections between prefabricated elements; (b) accumulated stresses due to shrinkage and creep and due to stresses incurred in transportation; (c) loss of vertical support due to inadequate bearing area and/or insufficient connection between floor elements and columns; and (d) corrosion of metal connectors between prefabricated elements. A number of precast parking garages failed in the 1994 M_W 6.7 Northridge earthquake, including a large structure at the Cal State Northridge campus which sustained a progressive failure. This structure had a perimeter pre-cast MRF and interior non-ductile columns — the MRF sustained large but tolerable deflections; however, interior non-ductile columns failed under these deflections, resulting in an interior collapse, which then pulled the exterior MRF's over.

Masonry **Reinforced masonry** buildings are mostly low-rise perimeter bearing wall structures, often with wood diaphragms although precast or cast-in-place concrete is sometimes used. Floor and roof assemblies usually consist of timber joists and beams, glue-laminated beams, or light steel joists. The bearing walls consist of grouted and reinforced hollow or solid masonry units. Interior supports, if any, are often wood or steel columns, wood stud frames, or masonry walls. Generally, they are less than five stories in height although many mid-rise masonry buildings exist. Reinforced masonry buildings can perform well in moderate earthquakes if they are adequately reinforced and grouted and if sufficient diaphragm anchorage exists.

Most **unreinforced masonry** (URM) bearing wall structures in the western U.S. were built before 1934, although this construction type was permitted in some jurisdictions having moderate or high seismicity until the late 1940s or early 1950s (in low-seismicity jurisdictions, URM may still be a common type of construction even today). These buildings usually range from one to six stories in height and construction typically varies according to the type of use, although wood floor and roof diaphragms are common. Smaller commercial and residential buildings usually have light wood floor/roof joists supported on the typical perimeter URM wall and interior wood load bearing partitions. Larger buildings, such as industrial ware- houses, have heavier floors and interior columns, usually of wood. The bearing walls of these industrial buildings tend to be thick, often as much as 24 in. or more at the base. Wall thicknesses of residential buildings range from 9 in. at upper floors to 17 in. at lower floors. Unreinforced masonry structures are recognized as perhaps the most hazardous structural type. They have been observed to fail in many modes during past earthquakes. Typical problems are

1. **Insufficient anchorage**—Because the walls, parapets, and cornices were not positively anchored to the floors, they tend to fall out. The collapse of bearing walls can lead to major building collapses. Some of these buildings have anchors as a part of the original construction or as a retrofit. These older anchors exhibit questionable performance.

2. **Excessive diaphragm deflection**—Because most of the floor diaphragms are constructed of wood sheathing, they are very flexible and permit large out-of-plane deflection at the wall transverse to the direction of the force. The large drift, occurring at the roof line, can cause the masonry wall to collapse under its own weight.

3. **Low shear resistance**—The mortar used in these older buildings is often made of lime and sand, with little or no cement, and has very little shear strength. The bearing walls will be heavily damaged and collapse under large loads.

4. **Wall slenderness**—Some of these buildings have tall story heights and thin walls. This condition, especially in non-load bearing walls, will result in out-of-plane failure, under severe lateral load. Failure of a non-load-bearing wall represents a falling hazard, whereas the collapse of a load- bearing wall will lead to partial or total collapse of the structure.

Innovative techniques can be divided into two broad categories: passive control (base isolation, energy dissipation) and active control, which increasingly are being applied to the design of new structures or to the retrofit of existing structures against wind, earthquakes, and other external loads [117]. The distinction between passive and active control is that passive systems require no active intervention or energy source, while active systems typically monitor the structure and incoming ground motion, and seek to actively control masses or forces in the structure (via moving weights, variable tension tendons, etc.) so as to develop a structural response (ideally) equal and opposite to the structural response due to the incoming ground motion. Recently developed semi-active control systems appear to combine the best features of both approaches, offering the reliability of passive devices, yet maintaining the versatility and adaptability of fully active systems. Magnetorheological (MR) dampers, for example, are new semi-active control devices that use MR fluids to create controllable damper. Initial results indicate that these devices are quite promising for civil engineering applications [30, 41, 42, 118].

Passive Control

Base isolation consists of softening of the shear capacity of a structure's connection with the ground, while maintaining vertical load carrying capacity, so as to reduce the earthquake ground motion input to the structure [71]. This has mostly been accomplished to date via the use of various types of rubber, lead-rubber, rubber-steel composite, or other types of bearings, beneath columns. Key aspects of most of the base isolation systems developed to date are: (1) they are economically limited to selected classes of structures (not too tall or short); (2) they require additional foundation expense, including special treatment of incoming utility lines; and (3) they require a certain amount of "rattlespace" around the structure, to accomodate the additional displacements that the bearings will undergo. For new structures, these requirements are not especially onerous, and a number of new structures in Japan, and a few in other countries, have been designed for base isolation. Most applications of this technology in the U.S., however, have been for the retrofit of existing (usually historic [26, 72]) structures, where the technology permitted increased seismic capacity without major modification to the architectural features. The technique has been applied to a number of highway bridges in the U.S., as well as some industrial structures.

Supplemental Damping — If damping can be significantly increased, then structural response (and therefore forces and displacements) are greatly reduced [34, 55]. Supplemental damping systems include *friction* systems (e.g., Sumitomo, Pall, and Friction-Slip) based on Coulomb friction, *self-centering* friction resistance that is proportional to displacement (e.g., Fluor-Daniel Energy Dissipating Restraint), or various *energy dissipation* mechanisms: ADAS (added damping and stiffness) elements, which utilize the yielding of mild-steel X-plates; viscoelastic shear dampers using a 3M acrylic copolymer as the dissipative element; or Nickel-Titanium alloy shape-memory devices that take advantage of reversible, stress-induced phase changes in the alloy to dissipate energy [2]. These systems are generally still in the developmental stage, although there are no special obstacles for implementation of the ADAS system [85, 132, 135], which has seen one application to date in the U.S. [99, 107] and more in other countries.

Active Control

Active control depends on actively modifying a structure's mass, stiffness, or geometric properties during its dynamic response, in such a manner as to counteract and reduce excessive displacements [76]. Tuned mass dampers, reliance on liquid sloshing [75], and active tensioning of tendons are methods currently under investigation [61, 62, 116]. Most methods of active control are real-time, relying on measurement of structural response, rapid structural computation, and fast-acting energy sources. A number of issues of reliability remain to be resolved [119].

5.4.2 Non-Building Structures

Bridges

The most vulnerable components of **conventional bridges** have been support bearings, abutments, piers, footings, and foundations. A common deficiency is that unrestrained expansion joints are not equipped to handle large relative displacements (inadequate support length), and simple bridge spans fall. Skewed bridges, in particular, have performed poorly in past earthquakes because they respond partly in rotation, resulting in an unequal distribution of forces to bearings and supports. Rocker bearings have proven most vulnerable. Roller bearings generally remain stable in earthquakes, except they may become misaligned and horizontally displaced. Elastomeric bearing pads are relatively stable although they have been known to "walk out" under severe shaking. Failure of backfill near abutments is common and can lead to tilting, horizontal movement or settlement of abutments, spreading and settlement of fills, and failure of foundation members. Abutment damage rarely leads to bridge collapse. Liquefaction of saturated soils in river channels and floodplains and subsequent loss of support have caused many bridge failures in past earthquakes, notably in the 1990 M_W 7.7 Philippines and 1991 M_W 7.1 Costa Rica earthquakes. Pounding of adjacent, simply supported spans can cause bearing damage and cracking of the girders and deck slab. Piers have failed in a number of earthquakes including most recently the 1994 M_W 6.7 Northridge and 1995 MW 6.9 Hanshin (Kobe) earthquakes, primarily because of insufficient transverse confining steel, and inadequate longitudinal steel splices and embedment into the foundation. Bridge superstructures have not exhibited any particular weaknesses other than being dislodged from their bearings.

Regarding aseismic design, bridge behavior during an earthquake can be very complex. Unlike buildings, which generally are connected to a single foundation through the diaphragm action of the base slab, bridges have multiple supports with varying foundation and stiffness characteristics. In addition, longitudinal forces are resisted by the abutments through a combination of passive backwall pressures and foundation embedment when the bridge moves toward an abutment, but by the abutment foundation only as the bridge moves away from an abutment. Significant movement must occur at bearings before girders impact abutments and bear against them, further complicating the response. To accurately assess the dynamic response of all but the simplest bridges, a three-dimensional dynamic analysis should be performed. Special care is required for design of hinges for continuous bridges. Restraint for spans or adequate bearing lengths to accommodate motions are the most effective ways to mitigate damage. In order to prevent damage to piers, proper confinement, splices, and embedment into the foundation should be provided. Similarly, sufficient steel should be provided in footings. Loads resisted by bridges may be reduced through use of energy absorption features including ductile columns, lead-filled elastomeric bearings, and restrainers. Foundation failure can be prevented by ensuring sufficient bearing capacity, proper foundation embedment, and sufficient consolidation of soil behind retaining structures.

In general, **railway bridges** may be steel, concrete, wood, or masonry construction, and their spans may be any length. Included are open and ballasted trestles, drawbridges, and fixed bridges. Bridge components include a bridge deck, stringers and girder, ballast, rails and ties, truss members, piers, abutments, piles, and caissons. Railroads sometimes share major bridges with highways (suspension bridges), but most railway bridges are older and simpler in configuration than highway bridges. Bridges that cross streams or narrow drainage passages typically have simple-span deck plate girders or beams. Longer spans use simple trusses supported on piers. Only a few of the more recently constructed bridges have continuous structural members.

The major cause of damage to trestles is displacement of unconsolidated sediments on which the substructures are supported, resulting in movement of pile-supported piers and abutments. Resulting superstructure damage has consisted of compressed decks and stringers, as well as collapsed spans. Shifting of the piers and abutments may shear anchor bolts. Girders can also shift on their piers. Failures of approaches or fill material behind abutments can result in bridge closure. Movable bridges are more vulnerable than fixed bridges; slight movement of piers supporting drawbridges can result in binding so that they cannot be opened without repairs.

Industrial Structures

Most large water, oil, and other storage **tanks** are unanchored, cylindrical tanks supported directly on the ground. Smaller tanks may be anchored, and tanks can also be elevated or buried. Older oil tanks are either fixed or floating roof, while more modern larger tanks are almost exclusively floating-roofed. Potable water tanks are invariably fixed roof. Diameters range from approximately 40 ft to more than 250 ft. Tank height on larger tanks is nearly always less than the diameter. Construction materials include welded, bolted, or riveted steel (for water), concrete and sometimes (for water or chemicals, and in smaller sizes) fiberglass. Tank foundations may consist of sand or gravel, or a concrete ring wall supporting the shell, supported on piles in poor soils. On-ground storage tanks are subject to a variety of earthquake damage mechanisms, generally due to sliding or rocking. Rocking is typically due to fluid sloshing, which must be considered for most tanks [60]. Specific failure modes include:

1. failure of weld between baseplate and wall
2. buckling of tank wall (termed elephant foot buckling)
3. rupture of attached rigid piping due to sliding or rocking of the tank
4. implosion of tank resulting from rapid loss of contents and negative internal pressure
5. differential settlement
6. anchorage failure or tearing of tank wall
7. failure of roof-to-shell connection or damage to roof seals for floating roofs (and loss of contents)
8. failure of shell at bolts or rivets because of tensile hoop stresses
9. total collapse

Torsional rotations of floating roofs may damage attachments such as guides, ladders, etc. Aseismic design practices for ground storage tanks include the use of flexible piping, pressure relief valves, and well-compacted foundations and concrete ring walls that prevent differential settlement. Adequate freeboard to prevent sloshing against the roof should be maintained. Positive attachment between the roof and shell should be provided for fix-roofed tanks. The bottom plate and its connection to the shell should be stiffened to resist uplift forces, and the baseplate should be protected against corrosion. Abrupt changes in thickness between adjacent courses should be avoided. Properly detailed ductile anchor bolts may be feasible on smaller steel tanks. Maintaining a height-to-diameter ratio of between 0.3 and 0.7 for tanks supported on the ground controls seismic loading.

In general, **ports/cargo handling equipment** comprise buildings (predominantly warehouses), waterfront structures, cargo handling equipment, paved aprons, conveyors, scales, tanks, silos, pipelines, railroad terminals, and support services. Building type varies, with steel frame being a common construction type. Waterfront structures include quay walls, sheet-pile bulkheads, and pile-supported piers. Quay walls are essentially waterfront masonry or caisson walls with earth fills behind them. Piers are commonly wood or concrete construction and often include batter piles to resist lateral transverse loads. Cargo handling equipment for loading and unloading ships includes cranes for containers, bulk loaders for bulk goods, and pumps for liquid fuels. Additional handling equipment is used for transporting goods throughout port areas. By far the most significant source of earthquake-induced damage to port and harbor facilities has been porewater pressure buildup in the saturated cohesionless soils that prevail at these facilities. This pressure buildup can lead to application of excessive lateral pressures to quay walls by backfill materials, liquefaction, and massive submarine sliding. This has occurred in a number of earthquakes including the 1985 M_W 7.8 Chile and 1989 M_W 7.1 Loma Prieta earthquakes but most notably and on a massive scale in the 1995 M_W 6.9 Hanshin (Kobe) earthquake, where several kilometers of deep-water caisson rotated outward, rendering dozens of ship berths unusable.

Electric transmission substations generally receive power at high voltages (often 220 kV or more) and step it down to lower voltages for distribution. The substations generally consist of one or more control buildings, steel towers, conductors, ground wires, underground cables, and extensive electrical equipment including banks of circuit breakers, switches, wave traps, buses, capacitors, voltage regulators, and massive transformers. Circuit breakers (oil or gas) protect transformers against power surges due to short circuits. Buses provide transmission linkage of the many and varied components within the substation. Transformers and voltage regulators serve to maintain the predetermined voltage or to step down or step up from one voltage to another. Porcelain lightning arresters are used to protect the system from voltage spikes caused by lightning. Long, cantilevered porcelain components (e.g., bushings and lightning arresters) are common on many electrical equipment items.

In earthquakes, unanchored or improperly anchored equipment may slide or topple, experiencing damage or causing attached piping and conduit to fail. In the yard, steel towers are typically damaged only by soil failures. Porcelain bushings, insulators, and lightning arresters are brittle and vulnerable to shaking and are frequently damaged, especially at the highest voltages. Transformers are large, heavy pieces of equipment that are frequently unanchored or inadequately anchored, so that they may shift, tear the attached conduit, break bushings, damage radiators, and spill insulating oil. Transformers in older substations that are mounted on rails frequently have fallen off their rails unless anchored. Frequently, inadequate slack in conductors or rigid bus bars result in porcelain damage due to differential motion of supports. Aseismic design practice includes the use of damping devices for porcelain; proper anchorage for equipment (avoid the use of friction clips); provision of conductor slack between equipment in the substation; use of breakaway connectors to reduce loads on porcelain bushings and insulators; and replacement of single cantilever-type insulator supports with those having multiple supports. Transformer radiators that cantilever from the body of a transformer can be braced. Adequate spacing between equipment can reduce the likelihood of secondary damage resulting from adjacent equipment falling.

5.5 Defining Terms

Attenuation: the rate at which earthquake ground motion decreases with distance.

Body waves: vibrational waves transmitted through the body of the earth, which are of two types: P waves (transmitting energy via dilatational or push-pull motion), and slower S waves (transmitting energy via shear action at right angles to the direction of motion).

Characteristic earthquake: a relatively narrow range of magnitudes at or near the maximum that can be produced by the geometry, mechanical properties, and state of stress of a fault [110].

Completeness: homogeneity of the seismicity record.

Corner frequency, f_o **:** the frequency above which earthquake radiation spectra vary with ϖ^{-3} - below f_o, the spectra are proportional to seismic moment.

Cripple wall: a carpenter's term indicating a wood frame wall of less than full height, usually built without bracing.

Critical damping: the value of damping such that free vibration of a structure will cease after one cycle ($c_{crit} = 2m\omega$).

Damping: represents the force or energy lost in the process of material deformation (damping coefficient c = force per velocity).

Design (basis) earthquake: the earthquake (as defined by various parameters, such as PGA, response spectra, etc.) for which the structure will be, or was, designed.

Dip-slip: motion at right angles to the strike, up- or down-slip.

Dip: the angle between a plane, such as a fault, and the earth's surface.

Ductile detailing: special requirements such as for reinforced concrete and masonry, close spacing of lateral reinforcement to attain **confinement** of a concrete core, appropriate relative dimensioning of beams and columns, 135° hooks on lateral reinforcement, hooks on main beam reinforcement within the column, etc.

Ductile frames: frames required to furnish satisfactory load-carrying performance under large deflections (i.e., ductility). In reinforced concrete and masonry this is achieved by **ductile detailing**.

Ductility factor: the ratio of the total displacement (elastic plus inelastic) to the elastic (i.e., yield) displacement.

Epicenter: the projection on the surface of the earth directly above the hypocenter.

Far-field: beyond near-field, also termed **teleseismic**.

Fault: a zone of the earth's crust within which the two sides have moved — faults may be hundreds of miles long, from 1 to over 100 miles deep, and not readily apparent on the ground surface.

Fragility: the probability of having a specific level of damage given a specified level of hazard.

Hypocenter: the location of initial radiation of seismic waves (i.e., the first location of dynamic rupture).

Intensity: a metric of the effect, or the strength, of an earthquake hazard at a specific location, commonly measured on qualitative scales such as MMI, MSK, and JMA.

Lateral force resisting system: a structural system for resisting horizontal forces due, for example, to earthquake or wind (as opposed to the vertical force resisting system, which provides support against gravity).

Liquefaction: a process resulting in a soil's loss of shear strength, due to a transient excess of pore water pressure.

Magnitude-frequency relation: the probability of occurrence of a selected magnitude — the most common is $\log_{10} n(m) = a - bm$ [48].

Magnitude: a unique measure of an individual earthquake's release of strain energy, measured on a variety of scales, of which the **moment magnitude** M_W (derived from seismic moment) is preferred.

Meizoseismal: the area of strong shaking and damage.

MMI (modified mercalli intensity) scale: see Table 5.4.

MSK: see table 5.5.

Near-field: within one source dimension of the epicenter, where source dimension refers to the length or width of faulting, whichever is less. (Alternatively termed near-source, this region is also defined as limited to about 10 km from the fault rupture surface, where the large velocity pulse dominated spectra are common.)

Non-ductile frames: frames lacking ductility or energy absorption capacity due to lack of ductile detailing — ultimate load is sustained over a smaller deflection (relative to ductile frames) and for fewer cycles.

Normal fault: a fault that exhibits dip-slip motion, where the two sides are in tension and move away from each other.

Peak ground acceleration (PGA): the maximum amplitude of recorded acceleration (also termed the **ZPA**, or **zero period acceleration**).

Pounding: the collision of adjacent buildings during an earthquake due to insufficient lateral clearance.

Response spectrum: a plot of maximum amplitudes (acceleration, velocity, or displacement) of a single degree of freedom oscillator (SDOF), as the natural period of the SDOF is varied across a spectrum of engineering interest (typically, for natural periods from .03 to 3 or more seconds, or frequencies of 0.3 to 30+ hz).

Reverse fault: a fault that exhibits dip-slip motion, where the two sides are in compression and move away towards each other.

Sand boils or **mud volcanoes:** ejecta of solids (i.e., sand, silt) carried to the surface by water, due to liquefaction.

Seismic hazard: a generic term for the expected seismic intensity at a site, more commonly measured using engineering measures, such as PGA, response spectra, etc.

Seismic hazards: the phenomena and/or expectation of an earthquake-related agent of damage, such as fault rupture, vibratory ground motion (i.e., shaking), inundation (e.g., tsunami, seiche, dam failure), various kinds of permanent ground failure (e.g., liquefaction), fire, or hazardous materials release.

Seismic moment: the moment generated by the forces generated on an earthquake fault during slip.

Seismic risk: the product of the hazard and the vulnerability (i.e., the expected damage or loss, or associated full probability distribution).

Seismotectonic model: a mathematical model representing the seismicity, attenuation, and related environment.

Soft story: a story of a building significantly less stiff than adjacent stories (that is, the lateral stiffness is 70% or less than that in the story above, or less than 80% of the average stiffness of the three stories above [25]).

Spectrum amplification factor: the ratio of a response spectral parameter to the ground motion parameter (where parameter indicates acceleration, velocity, or displacement).

Strike: the intersection of a fault and the surface of the earth, usually measured from north (e.g., the fault strike is N 60° W).

Subduction: refers to the plunging of a tectonic plate (e.g., the Pacific) beneath another plate (e.g., the North American) down into the mantle, due to convergent motion.

Surface waves: vibrational waves transmitted within the surficial layer of the earth, which are of two types: horizontally oscillating **Love waves** (analogous to s body waves) and vertically oscillating **Rayleigh waves.**

Thrust fault: low-angle reverse faulting (**blind thrust faults** are faults at depth occurring under anticlinal folds — they have only subtle surface expression).

Transform or **strike slip fault:** a fault where relative fault motion occurs in the horizontal plane, parallel to the strike of the fault.

Uniform hazard spectra: response spectra with the attribute that the probability of exceedance is independent of frequency.

Vulnerability: the expected damage given a specified value of a hazard parameter.

References

[1] ACI. 1995. *Building Code Requirements for Structural Concrete (ACI 318-95) and Commentary (ACI 318R-95)*, American Concrete Institute, Farmington Hills, MI.

[2] Aiken, I.D., Nims, D.K., Whittaker, A.S., and, Kelly, J.M. 1993. Testing of Passive Energy Dissipation Systems. *Earthquake Spectra*, 9(3), 336-370.

[3] Algermissen, S.T. 1983. *An Introduction to the Seismicity of the United States*, Earthquake Engineering Research Institute, Oakland, CA.

[4] Ambrayses, N.N. and Melville, C.P. 1982. *A History of Persian Earthquakes*, Cambridge University Press, Massachusetts.

[5] Anagnos, T. and Kiremidjian, A. S. 1984. Temporal Dependence in Earthquake Occurrence, in *Proc. Eighth World Conf. Earthquake Eng.*, 1, 255-262, Earthquake Engineering Research Institute, Oakland, CA.

[6] Anderson, Blume, Degenkolb et al. 1952. Lateral Forces of Earthquake and Wind, *Trans. Am. Soc. Civil Engrs.*, 117, *Am. Soc. Civil Engrs.*, New York.

[7] Anderson, J. G. and Trifunac, M D. 1977. Uniform Risk Absolute Acceleration Spectra, *Advances In Civil Engineering Through Engineering Mechanics: Proceedings Of The Second Annual Engineering Mechanics Division Specialty Conference*, Raleigh, NC, May 23-25, 1977, American Society of Civil Engineers, New York, pp. 332-335.

[8] Arias, A. 1970. A Measure of Earthquake Intensity, in *Seismic Design of Nuclear Power Plants*, R. Hansen, Ed., MIT Press, Massachusetts .

[9] ATC 3-06. 1978. *Tentative Provisions For The Development Of Seismic Regulations For Buildings*, Applied Technology Council, Redwood City, CA.

[10] Atkinson, G. M. 1995. Attenuation and Source Parameters of Earthquakes in the Cascadia Region, *Bull. Seis. Soc. Am.*, 85(5), 1327-1342.

[11] Atwater, B.F. et al. 1995. Summary of Coastal Geologic Evidence for Past Great Earthquakes at the Cascadia Subduction Zone, *Earthquake Spectra*, 11(1), 1-18, 1995.

[12] Bartlett, S.F. and Youd, T.L. 1992. *Empirical Analysis of Horizontal Ground Displacement Generated by Liquefaction-Induced Lateral Spread*, Tech. Rept. NCEER-92-0021, National Center for Earthquake Engineering Research, Buffalo, NY.

[13] Bernreuter, D.L. et al. 1989. *Seismic Hazard Characterization of 69 Nuclear Power Plant Sites East of the Rocky Mountains*, U.S. Nuclear Regulatory Commission, NUREG/CR-5250.

[14] Blume, J.A. 1970. An Engineering Intensity Scale for Earthquakes and other Ground Motions, *Bull. Seis. Soc. Am.*, 60(1), 217-229.

[15] Bolt, B.A. 1973. Duration of Strong Ground Motion, *Proc. 5th World Conf. On Earthquake Engineering*, 1, 1304-1313.

[16] Bolt, B.A. 1993. *Earthquakes*, W.H. Freeman and Co., New York.

[17] Bonilla, M.G. et al. 1984. Statistical Relations Among Earthquake Magnitude, Surface Rupture Length, and Surface Fault Displacement, *Bull. Seis. Soc. Am.*, 74(6), 2379-2411.

[18] Boore, D.M. and Atkinson, G.M. 1987. Stochastic Prediction of Ground Motion and Spectral Response Parameters at Hard-Rock Sites in Eastern North America, *Bull. Seis. Soc. Am.*, 77, 440-487.

[19] Boore, D.M. and Joyner, W.B. 1991. Estimation of Ground Motion at Deep-Soil Sites in Eastern North America, *Bull. Seis. Soc. Am.*, 81(6), 2167-2185.

[20] Boore, D.M., Joyner, W.B. and Fumal, T.E. 1993. Estimation of Response Spectra and Peak Acceleration from Western North American Earthquakes: An Interim Report, U.S.G.S. Open-File Report 93-509, Menlo Park, CA.

[21] Borcherdt, R.D. 1994. New developments in estimating site effects on ground motion, in *ATC-35 Seminar on New Developments in Earthquake Ground Motion Estimation and Implications for Engineering Design Practice*, Applied Technology Council, Redwood City, CA.

[22] Bozorgnia, Y. and Campbell, K.W. 1995. Spectral Characteristics of Vertical Ground Motion in the Northridge and other Earthquakes, *Proc. 4th U.S. Conf. On Lifeline Earthquake Eng.*, pp. 660-667, Am. Soc. Civil Engrs., New York.

[23] Brune, J.N. 1971, 1972. Tectonic Stress and the Spectra of Seismic Shear Waves from Earthquakes, *J. Geophys. Res.*, 75, 4997-5002 (and Correction, 76, 5002).

[24] BSSC. 1994. *NEHRP Recommended Provisions for Seismic Regulations for New Buildings, Part 1: Provisions*, prepared by the Building Seismic Safety Council for the Federal Emergency Management Agency, Building Seismic Safety Council, Washington, D.C.

[25] BSSC. 1994. *NEHRP Recommended Provisions for Seismic Regulations for New Buildings, Part 1: Provision, Part 2: Commentary*, 1994 Edition, Building Seismic Safety Council, Washington, D.C.

[26] Buckle, I G. 1995. Application of Passive Control Systems to the Seismic Retrofit of Historical Buildings in the United States of America. Construction for Earthquake Hazard Mitigation: Seismic Isolation Retrofit of Structures, Proceedings of the 1995 Annual Seminar of the ASCE Metropolitan Section, American Society of Civil Engineers, New York, February 6-7.

[27] Byrne, P.M. et al. 1992. Earthquake Induced Displacement of Soil-Structure Systems, *Proc. 10th World Conf. On Earthquake Eng.*, p. 1407-1412, Earthquake Engineering Research Institute, Oakland, CA.

[28] Campbell, K.W. 1985. Strong Ground Motion Attenuation Relations: A Ten-Year Perspective, *Earthquake Spectra*, 1(4), 759-804.

[29] Campbell, K.W. and Bozorgnia, Y. 1994. Near-Source Attenuation of Peak Horizontal Acceleration from Worldwide Accelerograms Recorded from 1957 to 1993, Proc. Fifth U.S. National Conference on Earthquake Engineering, Earthquake Engineering Research Institute, Oakland, CA.

[30] Carlson, J.D. and Spencer, Jr., B.F. 1996. Magneto-Rheological Fluid Dampers for Semi-Active Seismic Control, Proceedings of the 3rd International Conference on Motion and Vibration Control, September 1-6, Chiba, Japan, 3, 35-40.

[31] Chopra, A.K. 1981. *Dynamics of Structures, A Primer*, Earthquake Engineering Research Institute, Oakland CA.

[32] Clough, R.W. and Penzien, J. 1975. *Dynamics of Structures*, McGraw-Hill, New York.

[33] Coffman, J.L., von Hake, C.A., and Stover, C.W. 1980. *Earthquake History of the United States*, U.S. Dept. of Commerce, NOAA, Pub. 41-1, Washington, D.C.

[34] Constantinou, M C. and Symans, M D. 1993. Seismic Response of Structures with Supplemental Damping. Structural Design of Tall Buildings, 2(2), 77-92.

[35] Cornell, C.A. 1968. Engineering Seismic Risk Analysis, *Bull. Seis. Soc. Am.*, 58(5), 1583-1606.

[36] Cornell, C.A. and Merz, H.A. 1973. Seismic Risk Analysis Based on a Quadratic Magnitude Frequency Law, *Bull. Seis. Soc. Am.*, 63(6) 1992-2006.

[37] Crouse, C. B. 1991. Ground-Motion Attenuation Equations For Earthquakes On The Cascadia Subduction Zone, *Earthquake Spectra*, 7(2) 201-236.

[38] Darragh, R.B. Huang, M.J., and Shakal, A.F. 1994. Earthquake Engineering Aspects of Strong Motion Data from Recent California Earthquakes, Proc. Fifth U.S. National Conf. Earthquake Engineering, v. III, 99-108, Earthquake Engineering Research Institute, Oakland, CA.

[39] Dewey, J.W. et al. 1995. Spatial Variations of Intensity in the Northridge Earthquake, in Woods, M.C. and Seiple. W.R., Eds., *The Northridge California Earthquake of 17 January 1994*, Calif. Dept. Conservation, Div. Mines and Geology, Special Publ. 116, p. 39-46.

[40] Donovan, N.C. and Bornstein, A.E. 1978. Uncertainties in Seismic Risk Procedures, *J. Geotech. Div.*, 104(GT7), 869-887.

[41] Dyke, S.J., Spencer, Jr., B.F., Sain, M.K., and Carlson, J.D. In press. Modeling and Control of Magnetorheological Dampers for Seismic Response Reduction, *Smart Mater. Struc.*

[42] Dyke, S.J., Spencer, Jr., B.F., Sain, M.K., and Carlson, J.D. 1996. Experimental Verification of Semi-Active Structural Control Strategies Using Acceleration Feedback, Proceedings of the 3rd International Conference on Motion and Vibration Control, September 1-6, Chiba, Japan, 3, 291-296.

[43] Electric Power Research Institute. 1986. *Seismic Hazard Methodology for the Central and Eastern United States*, EPRI NP-4726, Menlo Park, CA.

[44] Esteva, L. 1976. *Seismicity*, in Lomnitz, C. and Rosenblueth, E., Eds., *Seismic Risk and Engineering Decisions*, Elsevier, New York.

[45] Evernden, J. F. 1991. Computer programs and data bases useful for prediction of ground motion resulting from earthquakes in California and the conterminous United States, U.S. Geological Survey Open-file report 91-338, Menlo Park, CA.

[46] Greenwood, R.B. 1995. Characterizing blind thrust fault sources — an overview, in Woods, M.C. and Seiple. W.R., Eds., *The Northridge California Earthquake of 17 January 1994*, Calif. Dept. Conservation, Div. Mines and Geology, Special Publ. 116, p. 279-287.

[47] Gumbel, E.J. 1958. *Statistics of Extremes*, Columbia University Press, New York.

[48] Gutenberg, B. and Richter, C.F. 1954. *Seismicity of the Earth and Associated Phenomena*, Princeton University Press, Princeton, NJ.

[49] Gutenberg, B. and Richter, C. F. 1956. Magnitude and Energy of Earthquakes, *Annali de Geofisica*, 9(1), 1-15.

[50] Habermann, R. E. 1987. Man-Made Changes of Seismicity Rates, *Bull. Seis. Soc. Am.*, 77(1), 141-159.

[51] Hamada, M., O'Rourke, T D., Ed. 1992. *Case Studies of Liquefaction and Lifeline Performance During Past Earthquakes, Volume 1: Japanese Case Studies*. National Center for Earthquake Engineering Research, State University of New York, Buffalo, NY.

[52] Hanks, T.C. and Johnston, A.C. 1992. Common Features of the Excitation and Propagation of Strong Ground Motion for North American Earthquakes, *Bull. Seis. Soc. Am.*, 82(1), 1-23.

[53] Hanks, T.C. and Kanamori, H. 1979. A Moment Magnitude Scale, *J. Geophys. Res.*, 84, 2348-2350.

[54] Hanks, T.C. and McGuire, R.K. 1981. The Character of High Frequency Strong Ground Motion, *Bull. Seis. Soc. Am.*, 71, 2071-2095.

[55] Hanson, R.D., Aiken, I.D., Nims, D.K., Richter, P.J., and Bachman, R.E. 1993. State-of-the-Art and State-of-the-Practice in Seismic Energy Dissipation, Proceedings of ATC-17-1 Seminar on Seismic Isolation, Passive Energy Dissipation, and Active Control, San Francisco, CA, March 11-12, Volume 2; Passive Energy Dissipation, Active Control, and Hybrid Control Systems, Applied Technology Council, Redwood City, CA, 449-471.

[56] Harlan, M.R. and Lindbergh, C. 1988. An Earthquake Vulnerability Analysis of the Charleston, South Carolina, Area, Rept. No. CE-88-1, Dept. of Civil Eng., The Citadel, Charleston, SC.

[57] Hausmann, M.R. 1990. *Engineering Principles of Ground Modification,* McGraw-Hill, New York.

[58] Hopper, M.G. 1985. Estimation of Earthquake Effects Associated with Large Earthquakes in the New Madrid Seismic Zone, U.S.G.S. Open File Report 85-457, Washington, D.C.

[59] Housner, G. 1984. Historical View of Earthquake Engineering, *Proc. Post-Conf. Volume, Eighth World Conf. On Earthquake Engineering,* Earthquake Engineering Research Institute, Oakland, CA., as quoted by S. Otani [96].

[60] Housner, G.W. 1963. The Dynamic Behaviour of Water Tanks, *Bull. Seis. Soc. Am.,* 53(2), 381-387.

[61] Housner, G.W. and Masri, S.F., Eds. 1990. Proceedings of the U.S. National Workshop on Structural Control Research, 25-26 October, University of Southern California, USC Publication No. M9013.

[62] Housner, G.W., Masri, S.F., and Soong, T.T. 1992. Recent Developments in Active Structural Control Research in the U.S.A., Invited Presentation at the First European Conference on Smart Structures & Materials, University of Strathclyde, Glasgow, UK, 12-14 May, B. Culshaw, P.T. Gardiner, and A. McDonach, Eds., Institute of Physics Publishing, Bristol, U.K., pp 201-206.

[63] Hudson, D.E. 1979. *Reading and Interpreting Strong Motion Accelerograms,* Earthquake Engineering Research Institute, Oakland, CA.

[64] IAEE. 1992. Earthquake *Resistant Regulations: A World List-1992.* Rev. ed., International Association for Earthquake Engineering, Tokyo.

[65] Jasperse, B.H. and Ryan, C.R. 1992. Stabilization and Fixation Using Soil Mixing: Grouting, Soil Improvement, and Geosynthetics, *ASCE Geotechnical Special Publication No. 30,* p. 1273-1284.

[66] Jennings, C.W. 1994. Fault Activity Map of California and Adjacent Areas, Dept. of Conservation, Div. Mines and Geology, Sacramento, CA.

[67] Joyner, W.B. and Boore, D.M. 1988. Measurements, Characterization and Prediction of Strong Ground Motion, in *Earthquake Engineering and Soil Dynamics II - Recent Advances in Ground-Motion Evaluation,* J.L. v. Thun, Ed., Geotechnical Spec. Publ. No. 20., American Soc. Civil Engrs., New York.

[68] Kameda, H. and Takagi, H. 1981. Seismic Hazard Estimation Based on Non-Poisson Earthquake Occurrences, *Mem. Fac. Eng., Kyoto University,* v. XLIII, Part 3, July, Kyoto.

[69] Kanai, K. 1983. *Engineering Seismology,* University of Tokyo Press, Tokyo.

[70] Kawashima, K. and Aizawa, K. 1989. Bracketed and Normalized Duration of Earthquake Ground Acceleration, *Earthquake Eng. Struct. Dynam.,* 18, 1041-1051.

[71] Kelly, J. and Michell, C. 1990. Earthquake Simulator Testing of a Combined Sliding Bearing and Rubber Bearing Isolation System, Report No. UCB/EERC-87/04 December, University of California at Berkeley.

[72] Kelly, J.M. and Way, D. 1991. Economic Feasibility and Seismic Rehabilitation of Existing Buildings Via Base Isolation. Seismic Retrofit of Historic Buildings Conference Workbook, Proceedings of the Second Technical Conference on the Seismic Retrofit of Historic Buildings, San Francisco, November 17-19. Look, D.L., Ed., Western Chapter of the Association of Preservation Technology, San Francisco.

[73] Lawson, A.C. and Reid, H.F. 1908-10. *The California Earthquake Of April 18, 1906. Report Of The State Earthquake Investigation Commission, California.* State Earthquake Investigation Commission, Carnegie Institution of Washington, Washington, D.C.

[74] Lomnitz, C. 1974. *Global Tectonics and Earthquake Risk,* Elsevier, New York.

[75] Lou, J.Y.K., Lutes, L.D., and Li, J.J. 1994. Active Tuned Liquid Damper for Structural Control, Proceedings of the First World Conference on Structural Control: International Association for Structural Control, Los Angeles, 3-5 August, Housner, G.W., et al., Eds., International Association for Structural Control, Los Angeles, 2, TP1-70–TP1-79.

[76] Luco, J.E., Mita, A. 1992. Active Control of The Seismic Response of Tall Non-Uniform Buildings, Proceedings of the Tenth World Conference on Earthquake Engineering, 19-24 July, Madrid; A A Balkema, Rotterdam, 4, 2143-2148.

[77] McGuire, R. K. 1978. FRISK: *Computer Program for Seismic Risk Analysis Using Faults as Earthquake Sources.* U.S. Geological Survey, Reports — United States Geological Survey Open File 78-1007.

[78] McGuire, R.K. and Barnhard, T.P. 1979. The Usefulness of Ground Motion Duration in Predicting the Severity of Seismic Shaking, *Proc. 2nd U.S. Natl. Conf. on Earthquake Engineering,* Earthquake Engineering Research Institute, Oakland, CA.

[79] McGuire, R.K., Ed. 1993. *Practice of Earthquake Hazard Assessment.* International Association of Seismology and Physics of the Earth's Interior.

[80] Meeting on Up-dating of MSK-64. 1981. Report on the Ad-hoc Panel Meeting of Experts on Up-dating of the MSK-64 Seismic Intensity Scale, Jene, 10-14 March 1980, *Gerlands Beitr. Geophys.,* Leipzeig 90, 3, 261-268.

[81] Midorikawa, S. 1991. Attenuation of Peak Ground Acceleration and Velocity From Large Subduction Earthquakes, *Proc. Fourth Intl. Conf. On Seismic Zonation,* Stanford University, August 25-29, 1991, Earthquake Engineering Research Institute, Oakland, CA, 2, 179-186.

[82] Mitchell, J.K. 1981. Soil Improvement: State-of-the-Art, *Proc. 10th Intl. Conf. Soil Mech. Found. Eng.,* 4, 509-565, Stockholm.

[83] Mohraz, B. 1976. A Study of Earthquake Response Spectra for Different Geological Conditions, *Bull. Seis. Soc. Am.,* 66, 915-935.

[84] Murphy J.R. and O'Brien, L.J. 1977. The Correlation of Peak Ground Acceleration Amplitude with Seismic Intensity and Other Physical Parameters, *Bull. Seis. Soc. Am.,* 67(3), 877-915.

[85] Nacer, N. and Hanson, R.D. 1994. Inelastic Behavior of Supplemental Steel Damping Devices. Proceedings of the Second International Conference on Earthquake Resistant Construction and Design, Berlin, 15-17 June, Savidis, S.A., Ed., A A Balkema, Rotterdam, 2, 805-812.

[86] National Academy Press. 1988. *Probabilistic Seismic Hazard Analysis,* National Academy of Sciences, Washington, D.C.

[87] NEIC. 1996. Database of Significant Earthquakes Contained in *Seismicity Catalogs,* National Earthquake Information Center, Golden, CO.

[88] Newmark, N.M. 1965. Effects of Earthquakes on Dams and Embankments, *Geotechnique,* 15(2), 129-160.

[89] Newmark, N.M. 1973. *A Study of Vertical and Horizontal Earthquake Spectra,* Dir. Of Licensing, U.S. Atomic Energy Commission, Washington, D.C.

[90] Newmark, N.M. and Hall, W.J. 1982. *Earthquake Spectra and Design,* Earthquake Engineering Research Institute, Oakland, CA.

[91] Nishenko, S.P. and Buland, R. 1987. A Generic Recurrence Interval Distribution For Earthquake Forecasting, *Bull. Seis. Soc. Am.,* 77, 1382-1399.

[92] NRC. 1985. *Liquefaction of Soils During Earthquakes,* National Research Council, National Academy Press, Washington, D.C.

[93] O'Rourke, T D. and Hamada, M., Ed. 1992. Case Studies of Liquefaction and Lifeline Performance During Past Earthquakes: Volume 2, United States Case Studies. National Center for Earthquake Engineering Research, State University of New York, Buffalo, NY.

[94] O'Rourke, T.D., Pease, J.W., and Stewart, H. E. 1992. Lifeline Performance and Ground Deformation During the Earthquake, in O'Rourke, T., Ed. Geological Survey (U.S.) *Loma Prieta, California, Earthquake Of October 17, 1989: Marina District.* U.S. Government Printing Office, Washington, D.C.

[95] Ohsaki, Y. 1966. Niigata Earthquakes, 1964 Building Damage and Condition, Soils and Foundations, Japanese Soc. Of Soil Mechanics and Foundation Engineering, 6(2), 14-37. March.

[96] Otani, S. 1995. A Brief History of Japanese Seismic Design Requirements, *Concrete International.*

[97] Paz, M., Ed. 1994. *International Handbook of Earthquake Engineering: Codes, Programs, and Examples,* Chapman & Hall, New York.

[98] Peng, M.H., Elghadamsi, F., and Mohraz, B. 1989. A Simplified Procedure for Constructing Probabilistic Response Spectra, *Earthquake Spectra,* 5(2), Earthquake Engineering Research Institute, Oakland.

[99] Perry, C.L., Fierro, E.A., Sedarat, H., and Scholl, R.E. 1993. Seismic Upgrade in San Francisco Using Energy Dissipation Devices, *Earthquake Spectra,* 9(3), 559-580.

[100] Prevost, J.H. 1981. *DYNA-FLOW: A Nonlinear Transient Finite Element Analysis Program,* Rept. No. 81-SM-1, Dept. Civil Eng., Princeton University, Princeton, NJ.

[101] Reid, H.F. 1910. The Mechanics of the Earthquake, the California Earthquake of April 18, 1906, Report of the State Investigation Committee, v. 2, Carnegie Institution of Washington, Washington, D.C.

[102] Reiter, L. 1990. *Earthquake Hazard Analysis, Issues and Insights,* Columbia University Press, NY.

[103] Richter, C. F. 1935. An Instrumental Earthquake Scale, *Bull. Seis. Soc. Am.* 25, 1- 32.

[104] Richter, C.F. 1958. *Elementary Seismology,* W.H. Freeman and Co., San Francisco, CA.

[105] Sadigh, K., Egan, J., and Youngs, R. 1986. Specification of Ground Motion for Seismic Design of Long Period Structures, *Earthquake Notes,* 57, 13.

[106] Schnabel, P.B., Lysmer, J., and Seed, H.B. 1972. SHAKE — *A Computer Program for Earthquake Response Analysis of Horizontally Layered Sites,* Earthquake Engineering Research Center, Rept. No. EERC-72-12, University of California at Berkeley, CA.

[107] Scholl, R.E. 1994. Comparison of Conventional and Damped Seismic Retrofits of a Two-Story Steel Frame Building, Proceedings of the Fifth U.S. National Conference on Earthquake Engineering, July 10-14, Chicago, IL, Earthquake Engineering Research Institute, Oakland, CA, I, pages 725-733.

[108] Scholz, C.H. 1990. *The Mechanics of Earthquakes and Faulting,* Cambridge University Press, MA.

[109] Schwartz, D.P. 1988. Geologic Characterization of Seismic Sources: Moving into the 1990s, in *Earthquake Engineering and Soil Dynamics II — Recent Advances in Ground-Motion Evaluation,* J.L. v. Thun, Ed., Geotechnical Spec. Publ. No. 20, *Am. Soc. Civil Eng.,* New York.

[110] Schwartz, D.P. and Coppersmith, K.J. 1984. Fault Behavior and Characteristic Earthquakes: Examples from the Wasatch and San Andreas Faults, *J. Geophys. Res.,* 89, 5681-5698.

[111] SEAOC. 1980. *Recommended Lateral Force Requirements and Commentary,* Seismology Committee, Structural Engineers of California, San Francisco, CA.

[112] Seed, H.B. and Wilson, S.D. 1967. The Turnagain Heights Landslide, Anchorage, Alaska, *J. Soil Mech. Found. Div., Am. Soc. Civil Eng.,* 93, SM4, 325-353.

[113] Seed, H.B. and Idriss, I.M. 1982. Ground Motions and Soil Liquefaction during Earthquakes, Earthquake Engineering Research Institute, Oakland, CA.

[114] Slemmons, D.B. 1977. State-of-the-Art for Assessing Earthquake Hazards in the United States, Report 6: Faults and Earthquake Magnitude, U.S. Army Corps of Engineers, Waterways Experiment Station, Misc. Paper s-73-1, 129 pp.

[115] Somerville, P. 1989. Strong Ground Motion Attenuation in the Puget Sound-Portland Region, *Proceedings Of Conference XLVIII: 3rd Annual Workshop On Earthquake Hazards*

in the Puget Sound, Portland Area; March 28-30, 1989, Portland, Oregon; Hays-Walter, W., Ed. U.S. Geological Survey, Reston, VA.

[116] Soong, T.T., Masri, S.F., and Housner, G.W. 1991. An Overview of Active Structural Control Under Seismic Loads, *Earthquake Eng. Res. Inst. Spectra,* 7(3), 483-506.

[117] Soong, T.T and Costantinou, M.C., Eds. 1994. Passive and Active Structural Vibration Control in Civil Engineering, CISM International Centre for Mechanical Sciences, Vol. 345, Springer-Verlag, New York.

[118] Spencer, Jr., B.F., Dyke, S.J., Sain M.K., and Carlson, J.D. In press. Phenomenological Model of a Magnetorheological Damper, *J. Eng. Mech.,* ASCE.

[119] Spencer, Jr., B.F., Sain, M.K., Won, C.-H., Kaspari, Jr., D.C., and Sain, P.M. 1994. Reliability-Based Measures of Structural Control Robustness, *Structural Safety,* 15, 111-129.

[120] Stein, R.S. and Yeats, R.S. 1989. Hidden Earthquakes, *Sci. Am.,* June.

[121] Stepp, J.C. 1972. Analysis of Completeness of the Earthquake Sample in the Puget Sound Area and Its Effect on Statistical Estimates of Earthquake Hazard, *Proc. First Conf. Microzonation,* Seattle, WA.

[122] Structural Engineers Association of California. 1988. *Recommended Lateral Force Requirements and Tentative Commentary,* Structural Engineers Association of California, San Francisco, CA.

[123] Toro, G.R. and McGuire, R.K. 1987. An Investigation into Earthquake Ground Motion Characteristics in Eastern North America, *Bull. Seis. Soc. Am.,* 77, 468-489.

[124] Trifunac, M.D. and Brady, A.G. 1975. A Study on the Duration of Strong Earthquake Ground Motion, *Bull. Seis. Soc. Am.,* 65, 581-626.

[125] Trifunac, M.D. and Novikova, E.I. 1995. State of the Art Review on Strong Motion Duration, *Prof. 10th European Conf. On Earthquake Engineering,* pp. 131-140.

[126] Uniform Building Code. 1994. Volume 2, Structural Engineering Design Provisions, Intl. Conf. Building Officials, Whittier.

[127] U.S. Army. 1992. *Seismic Design for Buildings* (Army: TM 5-809-10; Navy: NAVFAC P-355; USAF: AFM-88-3, Chapter 13). Washington D.C.: Departments of the Army, Navy, and Air Force, 1992. 407 pages. Available from the National Technical Information Service (NTIS), Military Publications Division, Washington.

[128] Usami, T. 1981. *Nihon Higai Jishin Soran* (List of Damaging Japanese Earthquakes), University of Tokyo Press (in Japanese).

[129] Wald, D.J., Burdick, L.J., and Somerville, P.G. 1988. Simulation of Acceleration Time Histories Close to Large Earthquakes, in *Earthquake Engineering and Soil Dynamics II — Recent Advances in Ground-Motion Evaluation,* J.L. v. Thun, Ed., Geotechnical Spec. Publ. No. 20, *Am. Soc. Civil Eng.,* New York.

[130] Wells, D.L. and Coppersmith, K.J. 1994. Empirical Relationships Among Magnitude, Rupture Length, Rupture Width, Rupture Area and Surface Displacement, *Bull. Seis. Soc. Am.,* 84(4), 974-1002.

[131] Wheeler, R.L. et al. 1994. Elements of Infrastructure and Seismic Hazard in the Central United States, U.S.G.S. Prof. Paper 1538-M, Washington, D.C.

[132] Whittaker, A., Bertero, V., Alonso, J., and Thompson, C. 1989. Earthquake Simulator Testing of Steel Plate Added Damping and Stiffness Elements, Earthquake Engineering Research Center, University of California, Berkeley, SEL TA654.6.R37 no.89/02. Report: UCB-EERC-89-02., 201 pages.

[133] Woo, G., Wood, R., and Muir. 1984. *British Seismicity and Seismic Hazard,* in *Proc. Eighth World Conf. Earthquake Eng.,* 1, 39-44, Earthquake Engineering Research Institute, Oakland, CA.

[134] Wood, H.O. and Neumann, Fr. 1931. Modified Mercalli Intensity Scale of 1931, *Bull. Seis. Soc. Am.,* 21, 277-283.

[135] Xia, C. and Hanson, R.D. 1992. Influence of Adas Element Parameters on Building Seismic Response. *J. Struc. Eng.*, 118(7), 1903-1918.

[136] Yamazaki, F. 1993. Comparative Study of Attenuation Characteristics of Ground Motion in Europe, North America and Japan, *Bull. Earthquake Resistant Structure Res. Center*, University of Tokyo, 26, 39-56.

[137] Yegulalp, T.M. and Kuo, J.T. 1974. Statistical Prediction of the Occurrence of Maximum Magnitude Earthquakes, *Bull. Seis. Soc. Am.*, 64(2), 393-414.

[138] Youngs, R. R. and Coppersmith, K J. 1989. Attenuation Relationships for Evaluation of Seismic Hazards From Large Subduction Zone Earthquakes, *Proc. Conference XLVIII: 3rd Annual Workshop On Earthquake Hazards in the Puget Sound, Portland Area;* March 28-30, 1989, Portland, OR, Hays-Walter, W., Ed., U.S. Geological Survey, Reston, VA.

[139] Youngs, R.R. and Coppersmith, K.J. 1987. Implication of Fault Slip Rates and Earthquake Recurrence Models to Probabilistic Seismic Hazard Estimates, *Bull. Seis. Soc. Am.*, 75, 939-964.

Further Reading

A number of attenuation relations have been developed for other regions, such as China, Japan, New Zealand, the Trans-Alpide areas, etc. including the following.

[1] Ambrayses, N. N. 1995. Prediction of Earthquake Peak Ground Acceleration In Europe. *Earthquake Engineering and Structural Dynamics*, 24(4), 467-490.

[2] Caillot, V. and Bard, P.Y. 1993. Magnitude, Distance and Site Dependent Spectra From Italian Accelerometric Data, *European Earthquake Engineering*, 6(1), 37-48.

[3] Capuano, P., Gasparini, P., Peronaci, M., and Scarpa, R. 1992. Strong Ground Motion and Source Parameters for Earthquakes in the Apennines, Italy, *Earthquake Spectra*, 8(4), 529-554.

[4] Dan, K., Ishii, T., and Ebihara, M. 1993. Estimation of Strong Ground Motions in Meizoseismal Region of the 1976 Tangshan, China, Earthquake. *Bull. Seis. Soc. Am.*, 83, 6, 1756-1777.

[5] Dowrick, D J. and Sritharan, S. 1993. Attenuation of Peak Ground Accelerations in Some Recent New Zealand Earthquakes, *Bull. New Zealand Natl. Soc. Earthquake Eng.* 26(1), 3-13.

[6] Dowrick, D.J. 1992. Attenuation of Modified Mercalli Intensity in New Zealand Earthquakes, *Earthquake Engineering and Structural Dynamics*, 21(3), 181-196.

[7] *Earthquake Motion and Ground Conditions: In Commemoration of the 20th Anniversary of the Research Subcommittee on Earthquake Ground Motion, the Architectural Institute of Japan.* 1993. Architectural Institute of Japan, Tokyo.

[8] Fukushima, Y. and Tanaka, T. 1991. New Attenuation Relation for Peak Horizontal Acceleration of Strong Earthquake Ground Motion in Japan, *Shimizu Tech. Res. Bull.*, 10, 1-11.

[9] Grandori, G. 1991. Macroseismic Intensity Versus Epicentral Distance: The Case of Central Italy, *Tectonophysics*, 193, 1-3, 165-171.

[10] Hatzidimitriou, P., Papazachos, C., Kiratzi, A., and Theodulidis, N. 1993. Estimation of Attenuation Structure and Local Earthquake Magnitude Based on Acceleration Records in Greece, *Tectonophysics*, 217(3/4), 243-253.

[11] Matuschka, T. and Davis, B. K. 1991. Derivation of an Attenuation Model in Terms of Spectral Acceleration for New Zealand. *Proceedings [Of The] Pacific Conference on Earthquake Engineering*, Auckland, New Zealand, 20-23 November, 1991, New Zealand National Society For Earthquake Engineering, Wellington, 2, 123-134.

[12] Molas, G. L. and Yamazaki, F. 1995. Attenuation of Earthquake Ground Motion in Japan Including Deep Focus Events, *Bull. Seis. Soc. Am.,* 85, 5, 1343-1358.

[13] Monachesi, G. 1994. Seismic Hazard Assessment Using Intensity Point Data, *Soil Dynamics and Earthquake Engineering,* 13, 3, 219-226.

[14] Niazi, M. and Bozorgnia, Y. 1992. 1990 Manjil, Iran, Earthquake: Geology and Seismology Overview, Pga Attenuation, and Observed Damage, *Bull. Seis. Soc. Am.,* 82(2), 774-799.

[15] *Proceedings of the International Workshop on Strong Motion Data.* 1994. Japan Port and Harbour Research Institute, Yokosuka, Japan.

[16] Ruan, A. and Sun, C. 1992. A Study on Attenuation Law of Seismic Ground Motion in the Loess Region of Northwestern China, *Earthquake Research In China,* 6, 4, 431-442.

[17] Tao, X. and Zheng, G. 1991. Intensity Attenuation Relationship For Seismic Zonation of China, *Proceedings of the Fourth International Conference On Seismic Zonation,* Stanford University, August 25-29, 1991; Earthquake Engineering Research Institute, Oakland, CA.

6

Composite Construction

6.1 Introduction 6-1
Historical Overview • Scope • Design Codes
6.2 Materials 6-5
Concrete • Reinforcing Steel • Structural Steel • Steel Decking
• Shear Connectors
6.3 Simply-Supported Composite Beams 6-8
Beam Response and Failure Modes • The Effective Width of
Concrete Flange • Elastic Analysis • Plastic Analysis • Vertical
Shear • Serviceability Limit States • Worked Examples[1]
6.4 Continuous Beams 6-25
Introduction • Effective Width • Local Buckling and Classifi-
cation of Cross-Sections • Elastic Analysis of the Cross-Section
• Plastic Resistance of the Cross-Section • Serviceability Limit
States • Ultimate Limit State • The Lateral-Torsional Buckling
• Worked Examples
6.5 The Shear Connection................................ 6-44
The Shear Transfer Mechanisms • The Shear Strength of Mechanical Shear Connectors • Steel-Concrete Interface Separation • Shear Connectors Spacing • Shear Connection Detailing
• Transverse Reinforcement • The Shear Connection in Fully
and Partially Composite Beams • Worked Examples
6.6 Composite Columns 6-69
Types of Sections and Advantages • Failure Mechanisms • The
Elastic Behavior of the Section • The Plastic Behavior of the Section • The Behavior of the Members • Influence of Local Buckling • Shear Effects • Load Introduction Region • Restrictions
for the Application of the Design Methods • Worked Examples
6.7 Composite Slabs 6-91
The Steel Deck • The Composite Slab • Worked Examples
Notations 6-108
References 6-113
Codes and Standards 6-115
Further Reading 6-116

Edoardo Cosenza
*University of Naples,
Napoli, Italy*

Riccardo Zandonini
*Department of Structural
Mechanics and Design,
University of Trento,
Povo, Italy*

6.1 Introduction

6.1.1 Historical Overview

The history of structural design may be explained in terms of a continuous progress toward optimal constructional systems with respect to aesthetic, engineering, and economic parameters. If the attention is focused on the structure, optimality is mainly sought through improvement of the form and of the materials. Moreover, creative innovation of the form combined with advances of material properties and technologies enables pursuit of the human challenge to the "natural" limitations to the height (buildings) and span (roofs and bridges) of the structural systems.

Advances may be seen to occur as a step-by-step process of development. While the enhancement of the properties of already used materials contributes to the "in-step" continuous advancement, new materials as well as the synergic combination of known materials permit structural systems to make a step forward in the way to optimality.

Use of composite or hybrid material solutions is of particular interest, due to the significant potential in overall performance improvement obtained through rather modest changes in manufacturing and constructional technologies. Successful combinations of materials may even generate a new material, as in the case of reinforced concrete or, more recently, fiber-reinforced plastics. However, most often the synergy between structural components made of different materials has shown to be a fairly efficient choice. The most important example in this field is represented by the steel-concrete composite construction, the enormous potential of which is not yet fully exploited after more than one century since its first appearance.

"Composite bridges" and "composite buildings" appeared in the U.S. in the same year, 1894 [34, 46]:

1. The Rock Rapids Bridge in Rock Rapids, Iowa, made use of curved steel I-beams embedded in concrete.
2. The Methodist Building in Pittsburgh had concrete-encased floor beams.

The composite action in these cases relied on interfacial bond between concrete and steel. Efficiency and reliability of bond being rather limited, attempts to improve concrete-to-steel joining systems were made since the very beginning of the century, as shown by the shearing tabs system patented by Julius Kahn in 1903 (Figure 6.1a). Development of efficient mechanical shear connectors

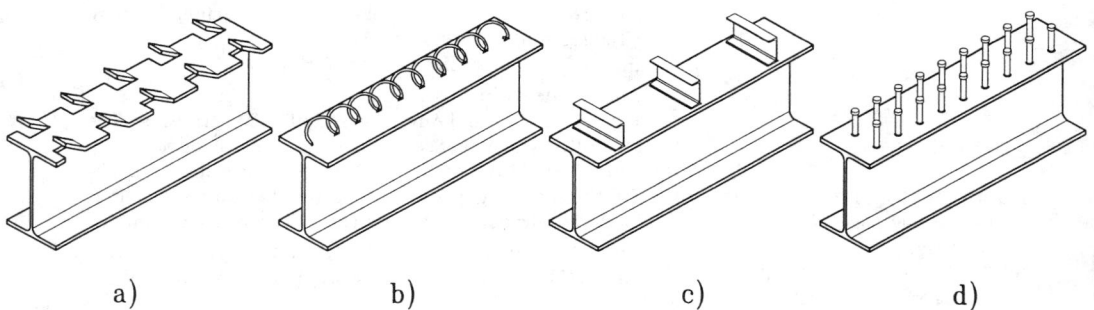

| a)   b)   c)   d)

Figure 6.1   Historical development of shear connectors. (a) Shearing tabs system (Julius Kahn 1903). (b) Spiral connectors. (c) Channels. (d) Welded studs.

progressed quite slowly, despite the remarkable efforts both in Europe (spiral connectors and rigid connectors) and North America (flexible channel connectors). The use of welded headed studs (in 1956) was hence a substantial breakthrough. By coincidence, welded studs were used the same year in a bridge (Bad River Bridge in Pierre, South Dakota) and a building (IBM's Education Building in Poughkeepsie, New York). Since then, the metal studs have been by far the most popular shear transferring device in steel-concrete composite systems for both building and bridge structures.

The significant interest raised by this "new material" prompted a number of studies, both in Europe and North America, on composite members (columns and beams) and connecting devices. The increasing level of knowledge then enabled development of Code provisions, which first appeared for buildings (the New York City Building Code in 1930) and subsequently for bridges (the AASHO specifications in 1944).

In the last 50 years extensive research projects made possible a better understanding of the fairly complex phenomena associated with composite action, codes evolved significantly towards acceptance of more refined and effective design methods, and constructional technology progressed at a brisk pace. However, these developments may be considered more a consequence of the increasing popularity of composite construction than a cause of it. In effect, a number of advantages with respect to structural steel and reinforced concrete were identified and proven, as:

- high stiffness and strength (beams, girders, columns, and moment connections)
- inherent ductility and toughness, and satisfactory damping properties (e.g., encased columns, beam-to-column connections)
- quite satisfactory performance under fire conditions (all members and the whole system)
- high constructability (e.g., floor decks, tubular infilled columns, moment connections)

Continuous development toward competitive exploitation of composite action was first concentrated on structural elements and members, and was based mainly on technological innovation as in the use of steel-concrete slabs with profiled steel sheeting and of headed studs welded through the metal decking, which successfully spread composite slab systems in the building market since the 1960s. Innovation of types of structural forms is a second important factor on which more recent advances (in the 1980s) were founded: composite trusses and stub girders are two important examples of novel systems permitting fulfillment of structural requirements and easy accommodation of air ducts and other services.

A very recent trend in the design philosophy of tall buildings considers the whole structural system as a body where different materials can cohabit in a fairly beneficial way. Reinforced concrete, steel, and composite steel-concrete members and subsystems are used in a synergic way, as in the cases illustrated in Figure 6.2. These mixed systems often incorporate composite superframes, whose columns, conveniently built up by taking advantage of the steel erection columns (Figure 6.3), tend to become more and more similar to highly reinforced concrete columns. The development of such systems stresses again the vitality of composite construction, which seems to increase rather than decline.

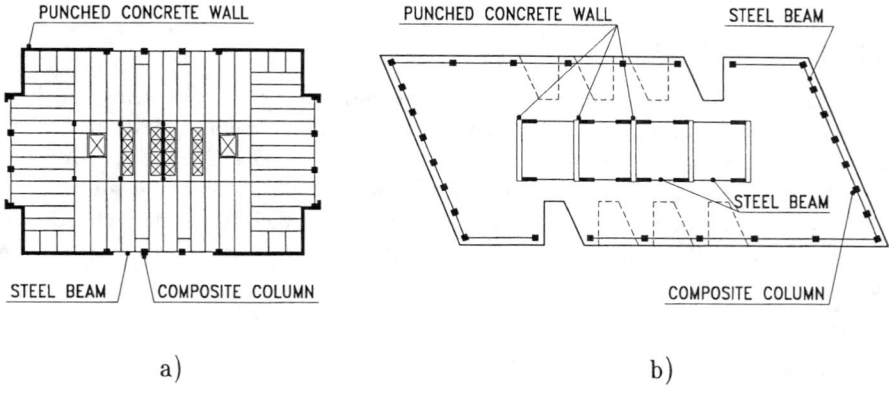

a)   b)

Figure 6.2 Composite systems in buildings. (a) Momentum Place, Dallas, Texas. (b) First City Tower, Houston, Texas. (After Griffis, L.G. 1992. Composite Frame Construction, *Constructional Steel Design. An International Guide,* P.J. Dowling, et al., Eds., Elsevier Applied Science, London.)

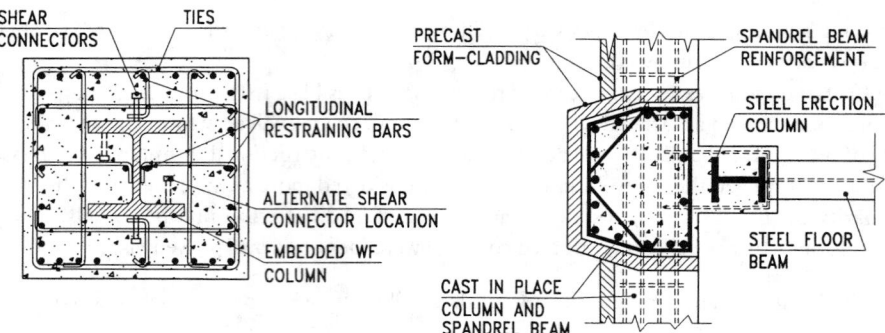

Figure 6.3   Columns in composite superframes. (After Griffis, L.G. 1992. Composite Frame Construction, *Constructional Steel Design. An International Guide,* P.J. Dowling, et al., Eds., Elsevier Applied Science, London.)

6.1.2   Scope

The variety of structural forms and the continuous evolution of composite systems precludes the possibility of comprehensive coverage. This chapter has the more limited goal of providing practicing structural engineers with a reference text dealing with the key features of the analysis and design of composite steel-concrete members used in building systems. The attention is focused on the response and design criteria under static loading of individual components (members and elements) of traditional forms of composite construction. Recent developments in floor systems and composite connections are dealt with in Chapters 18 and 23, respectively.

Emphasis is given to the behavioral aspects and to the suitable criteria to account for them in the design process. Introduction to the practical usage of these criteria requires that reference is made to design codes. This is restricted to the main North American and European Specifications and Standards, and has the principal purpose of providing general information on the different application rules. A few examples permit demonstration of the general design criteria.

Problems related to members in special composite systems as composite superframes are not included, due to the limited space. Besides, their use is restricted to fairly tall buildings, and their construction and design requires rather sophisticated analysis methods, often combined with "creative" engineering understanding [21].

6.1.3   Design Codes

The complexity of the local and global response of composite steel-concrete systems, and the number of possible different situations in practice led to the use of design methods developed by empirical processes. They are based on, and calibrated against, a set of test data. Therefore, their applicability is limited to the range of parameters covered by the specific experimental background.

This feature makes the reference to codes, and in particular to their application rules, of substantial importance for any text dealing with design of composite structures. In this chapter reference is made to two codes:

1. AISC-LRFD Specifications [1993]
2. Eurocode 4 [1994]

Besides, the ASCE Standards [1991] for the design of composite slabs are referred to, as this subject is not covered by the AISC-LRFD Specifications. These codes may be considered representative of the design approaches of North America and Europe, respectively. Moreover, they were issued or revised very recently, and hence reflect the present state of knowledge. Both codes are based on the limit states methodology and were developed within the framework of first order approaches to probabilistic design. However, the format adopted is quite different. This operational difference, together with the general scope of the chapter, required a "simplified" reference to the codes. The key features of the formats of the two codes are highlighted here, and the way reference is made to the code recommendations is then presented.

The Load and Resistance Factor Design (LRFD) specifications adopted a design criterion, which expresses reliability requirements in terms of the general formula

$$\phi R_n \geq E_m \left(\sum \gamma_{Fi} F_{i.m} \right) \tag{6.1}$$

where on the resistance side R_n represents the nominal resistance and ϕ is the "resistance factor", while on the loading side E_m is the "mean load effect" associated to a given load combination $\sum \gamma_{Fi} F_{i.m}$ and γ_{Fi} is the "load factor" corresponding to mean load $F_{i.m}$. The nominal resistance is defined as the resistance computed according to the relevant formula in the Code, and relates to a specific limit state. This "first-order" simplified probabilistic design procedure was calibrated with reference to the "safety index" β expressed in terms of the mean values and the coefficients of variation of the relevant variables only, and assumed as a measure of the degree of reliability. Application of this procedure requires that (1) the nominal strength be computed using the nominal specified strengths of the materials, (2) the relevant resistance factor be applied to obtain the "design resistance", and (3) this resistance be finally compared with the corresponding mean load effect (Equation 6.1).

In Eurocode 4, the fundamental reliability equation has the form

$$R_d \left(f_{i.k} / \gamma_{m.i} \right) \geq E_d \left(\sum \gamma_{Fi} F_{i.k} \right) \tag{6.2}$$

where on the resistance side the design value of the resistance, R_d, appears, determined as a function of the characteristic values of the strengths $f_{i.k}$ of the materials of which the member is made. The factors $\gamma_{m.i}$ are the "material partial safety factors"; Eurocode 4 adopts the following material partial safety factors:

$$\gamma_c = 1.5; \quad \gamma_s = 1.10; \quad \gamma_{sr} = 1.15$$

for concrete, structural steel, and reinforcing steel, respectively.

On the loading side the design load effect, E_d, depends on the relevant combination of the characteristic factored loads $\gamma_{Fi} F_{i.k}$. Application of this checking format requires the following steps: (1) the relevant resistance factor be applied to obtain the "design strength" of each material, (2) the design strength R_d be then computed using the factored materials' strengths, and (3) the resistance R_d be finally compared with the corresponding design load effect E_d (Equation 6.2).

Therefore, the two formats are associated with two rather different resistance parameters (R_n and R_d), and design procedures. A comprehensive and specific reference to the two codes would lead to a uselessly complex text. It seemed consistent with the purpose of this chapter to refer in any case to the "unfactored" values of the resistances as explicitly (LRFD) or implicitly (Eurocode 4) given in code recommendations, i.e., to resistances based on the nominal and characteristic values of material strengths, respectively. Factors (ϕ or $\gamma_{m.i}$) to be applied to determine the design resistance are specified only when necessary. Finally, in both codes considered, an additional reduction factor equal to 0.85 is introduced in order to evaluate the design strength of concrete.

6.2 Materials

Figure 6.4 shows stress-strain curves typical of concrete, and structural and reinforcing steel. The properties are covered in detail in Chapters 3 and 4 of this Handbook, which deal with steel and reinforced concrete structures, respectively. The reader will hence generally refer to these sections. However, some data are provided specific to the use of these materials in composite construction, which include limitations imposed by the present codes to the range of material grades that can be

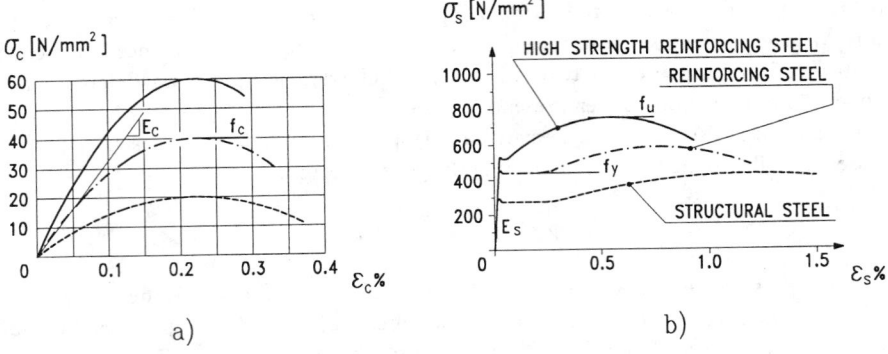

Figure 6.4   Stress-strain curves. (a) Typical compressive stress-strain curves for concrete. (b) Typical stress-strain curves for steel.

selected, in view of the limited experience presently available. Moreover, the main characteristics of the materials used for elements or components typical of composite construction, like stud connectors and metal steel decking, are given.

6.2.1   Concrete

Composite action implies that forces are transferred between steel and concrete components. The transfer mechanisms are fairly complex. Design methods are supported mainly by experience and test data, and their use should be restricted to the range of concrete grades and strength classes sufficiently investigated. It should be noted that concrete strength significantly affects the local and overall performance of the shear connection, due to the inverse relation between the resistance and the strain capacity of this material. Therefore, the capability of redistribution of forces within the shear connection is limited by the use of high strength concretes, and consequently the applicability of plastic analysis and of design methods based on full redistribution of the shear forces supported by the connectors (as the partial shear connection design method discussed in Section 6.7.2) is also limited.

The LRFD specifications [AISC, 1993] prescribe for composite flexural elements that concrete meet quality requirements of ACI [1989], made with ASTM C33 or rotary-kiln produced C330 aggregates with concrete unit weight not less than 14.4 kN/m^3 (90 pcf)[1]. This allows for the development of the full flexural capacity according to test results by Olgaard et al. [38]. A restriction is also imposed on the concrete strength in composite compressed members to ensure consistency of the specifications with available experimental data: the strength upper limit is 55 N/mm^2 (8 ksi) and the lower limit is 20 N/mm^2 (3 ksi) for normal weight concrete, and 27 N/mm^2 (4 ksi) for lightweight concrete.

The recommendations of Eurocode 4 [CEN, 1994] are applicable for concrete strength classes up the C50/60 (see Table 6.1), i.e., to concretes with cylinder characteristic strength up to 50 N/mm^2. The use of higher classes should be justified by test data. Lightweight concretes with unit weight not less than 16 kN/m^3 can be used.

Compression tests permit determination of the immediate concrete strength f_c. The strength under sustained loads is obtained by applying to f_c a reduction factor 0.85.

Time dependence of concrete properties, i.e., shrinkage and creep, should be considered when

---

[1]The Standard International (S.I.) system of units is used in this chapter. Quantities are also expressed (in parenthesis) in American Inch-Pound units, when reference is made to American Code specified values.

TABLE 6.1 Values of Characteristic Compressive strength (f_c), Characteristic tensile strength (f_{ct}), and Secant Modulus of Elasticity (E_c) proposed by Eurocode 4

		Class of concrete[a]						
		C 20/25	C 25/30	C 30/37	C 35/45	C 40/50	C 45/55	C 50/60
f_c	(N/mm^2)	20	25	30	35	40	45	50
f_{ct}	(N/mm^2)	2.2	2.6	2.9	3.2	3.5	3.8	4.1
E_c	(kN/mm^2)	29	30.5	32	33.5	35	36	37

[a] Classification refer to the ratio of cylinder to cube strength.

determining the response of composite structures under sustained loads, with particular reference to member stiffness. Simple design methods can be adopted to treat them.

Stiffness and stress calculations of composite beams may be based on the transformed cross-section approach first developed for reinforced concrete sections, which uses the modular ratio $n = E_s/E_c$ to reduce the area of the concrete component to an equivalent steel area. A value of the modular ratio may be suitably defined to account for the *creep effect* in the analysis:

$$n_{ef} = \frac{E_s}{E_{c.ef}} = \frac{E_s}{[E_c/(1+\phi)]}$$ (6.3)

where

$E_{c.ef}$ = an effective modulus of elasticity for the concrete
ϕ = a creep coefficient approximating the ratio of creep strain to elastic strain for sustained compressive stress

This coefficient may generally be assumed as 1 leading to a reduction by half of the modular ratio for short term loading; a value $\phi = 2$ (i.e., a reduction by a factor 3) is recommended by Eurocode 4 when a significant portion of the live loads is likely to be on the structure quasi-permanently. The effects of shrinkage are rarely critical in building design, except when slender beams are used with span to depth ratio greater than 20.

The total long-term drying *shrinkage* strains ε_{sh} varies quite significantly, depending on concrete, environmental characteristics, and the amount of restraint from steel reinforcement. The following design values are provided by the Eurocode 4 for ordinary cases:

1. Dry environments

   - 325×10^{-6} for normal weight concrete
   - 500×10^{-6} for lightweight concrete

2. Other environments and infilled members

   - 200×10^{-6} for normal weight concrete
   - 300×10^{-6} for lightweight concrete

Finally, the same value of the *coefficient of thermal expansion* may be conveniently assumed as for the steel components (i.e., 10×10^{-6} per °C), even for lightweight concrete.

6.2.2 Reinforcing Steel

Rebars with yield strength up to 500 N/mm^2 (72 ksi) are acceptable in most instances. The reinforcing steel should have adequate ductility when plastic analysis is adopted for continuous beams. This factor should hence be carefully considered in the selection of the steel grade, in particular when high strength steels are used.

A different requirement is implied by the limitation of 380 N/mm^2 (55 ksi) specified by AISC for the yield strength of the reinforcement in encased composite columns; this is aimed at ensuring that buckling of the reinforcement does not occur before complete yielding of the steel components.

6.2.3   Structural Steel

Structural steel alloys with yield strength up to 355 N/mm^2 (50 ksi for American grades) can be used in composite members, without any particular restriction. Studies of the performance of composite members and joints made of high strength steel are available covering a yield strength range up to 780 N/mm^2 (113 ksi) (see also [47]). However, significant further research is needed to extend the range of structural steels up to such levels of strength. Rules applicable to steel grades Fe420 and Fe460 (with $f_y = 420$ and 460 N/mm^2, respectively) have been recently included in the Eurocode 4 as Annex H [1996]. Account is taken of the influence of the higher strain at yielding on the possibility to develop the full plastic sagging moment of the cross-section, and of the greater importance of buckling of the steel components.

The AISC specification applies the same limitation to the yield strength of structural steel as for the reinforcement (see the previous section).

6.2.4   Steel Decking

The increasing popularity of composite decking, associated with the trend towards higher flexural stiffnesses enabling possibility of greater unshored spans, is clearly demonstrated by the remarkable variety of products presently available. A wide range of shapes, depths (from 38 to 200 mm [15 to 79 in.]), thicknesses (from 0.76 to 1.52 mm [5/24 to 5/12 in.]), and steel grades (with yield strength from 235 to 460 N/mm^2 [34 to 67 ksi]) may be adopted. Mild steels are commonly used, which ensure satisfactory ductility.

The minimum thickness of the sheeting is dictated by protection requirements against corrosion. Zinc coating should be selected, the total mass of which should depend on the level of aggressiveness of the environment. A coating of total mass 275 g/m^2 may be considered adequate for internal floors in a non-aggressive environment.

6.2.5   Shear Connectors

The steel quality of the connectors should be selected according to the method of fixing (usually welding or screwing). The welding techniques also should be considered for welded connectors (studs, anchors, hoops, etc.).

Design methods implying redistribution of shear forces among connectors impose that the connectors do possess adequate deformation capacity. A problem arises concerning the mechanical properties to be required to the stud connectors. Standards for material testing of welded studs are not available. These connectors are obtained by cold working the bar material, which is then subject to localized plastic straining during the heading process. The Eurocode hence specifies requirements for the ultimate-to-yield strength ratio ($f_u/f_y \geq 1.2$) and to the elongation at failure (not less than 12% on a gauge length of $5.65 \sqrt{A_o}$, with A_o cross-sectional area of the tensile specimen) to be fulfilled by the finished (cold drawn) product. Such a difficulty in setting an appropriate definition of requirements in terms of material properties leads many codes to prescribe, for studs, cold bending tests after welding as a means to check "ductility".

6.3   Simply-Supported Composite Beams

Composite action was first exploited in flexural members, for which it represents a "natural" way to enhance the response of structural steel. Many types of composite beams are currently used in building and bridge construction. Typical solutions are presented in Figures 6.5, 6.6, and 6.7. With reference to the steel member, either rolled or welded I sections are the preferred solution in building systems (Figure 6.5a); hollow sections are chosen when torsional stiffness is a critical design factor (Figure 6.5b). The trend towards longer spans (higher than 10 m) and the need of freedom in accommodating services made the composite truss become more popular (Figure 6.5c). In bridges, multi-girder (Figure 6.6a) and box girder can be adopted; box girders may have either a closed (Figure 6.6b) or an open (Figure 6.6c) cross-section. With reference to the concrete element, use of traditional solid slabs are now basically restricted to bridges. Composite decks with steel

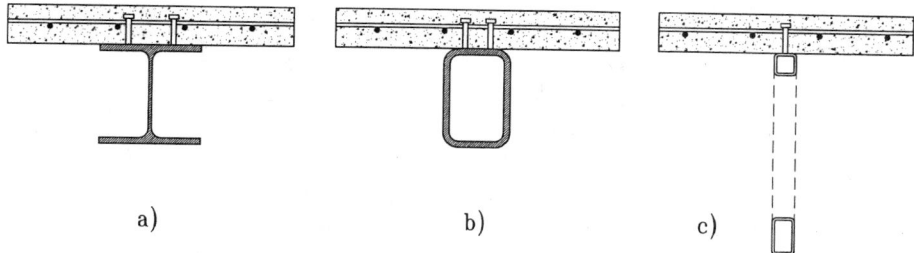

Figure 6.5 Typical composite beams. (a) I-shape steel section. (b) Hollow steel section. (c) Truss system.

profiled sheetings are the most popular solution (Figure 6.7a,b) in building structures because their use permits elimination of form-works for concrete casting and also reduction of the slab depth, as for example in the recently developed "slim floor" system shown in Figure 6.7c. Besides, full or partial encasement of the steel section significantly improves the performance in fire conditions (Figures 6.7d and 6.7e).

The main features of composite beam behavior are briefly presented, with reference to design. Due to the different level of complexity, and the different behavioral aspects involved in the analysis and design of simply supported and continuous composite beams, separate chapters are devoted to these two cases.

6.3.1 Beam Response and Failure Modes

Simply supported beams are subjected to positive (sagging) moment and shear. Composite steel-concrete systems are advantageous in comparison with both reinforced concrete and structural steel members:

- With respect to reinforced concrete beams, concrete is utilized in a more efficient way, i.e., it is mostly in compression. Concrete in tension, which may be a significant portion of the member in reinforced concrete beams, does not contribute to the resistance, while it increases the dead load. Moreover, cracking of concrete in tension has to be controlled, to avoid durability problems as reinforcement corrosion. Finally, construction methods can be chosen so that form-work is not needed.
- With respect to structural steel beams, a large part of the steel section, or even the entire steel section, is stressed in tension. The importance of local and flexural-torsional buckling is substantially reduced, if not eliminated, and plastic resistance can be achieved in most instances. Furthermore, the sectional stiffness is substantially increased, due to the contribution of the concrete flange deformability problems are consequently reduced, and tend not to be critical.

In summary, it can be stated that simply supported composite beams are characterized by an efficient use of both materials, concrete and steel; low sensitivity to local and flexural-torsional buckling; and high stiffness.

The design analyses may focus on few critical phenomena and the associated limit states. For the usual uniform loading pattern, typical failure modes are schematically indicated in Figure 6.8:

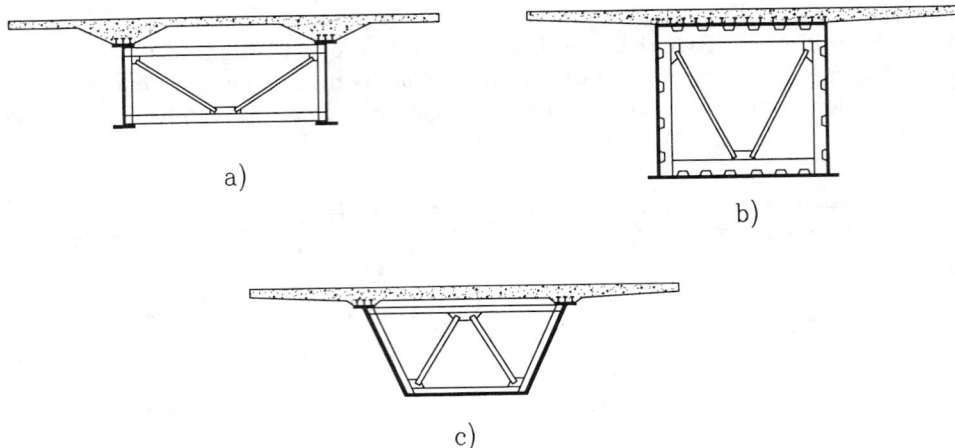

Figure 6.6   Typical system for composite bridges. (a) Multi-girder. (b) Box girder with closed cross-section. (c) Box girder with open cross-section.

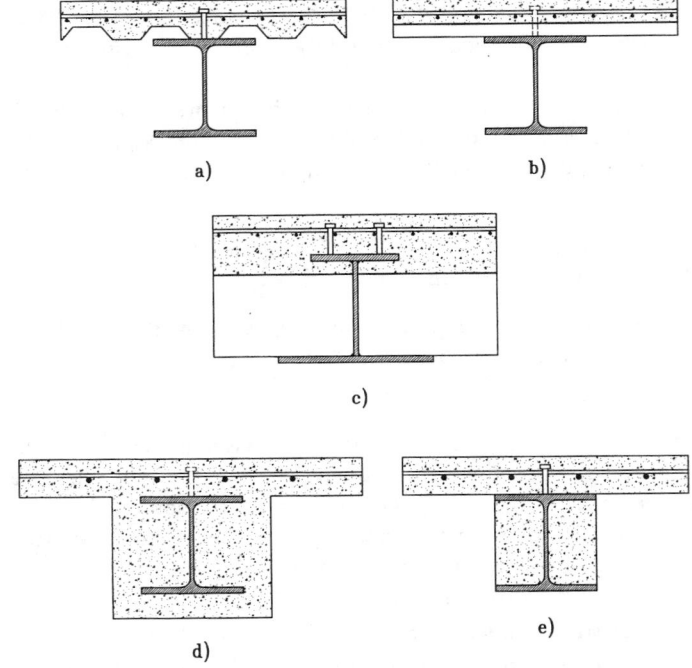

Figure 6.7   Typical system for composite floors. (a) Deck rib parallel to the steel beam. (b) Deck rib normal to the steel beam. (c) Slim-floor system. (d) Fully encased steel section. (e) Partially encased steel section.

mode I is by attainment of the ultimate moment of resistance in the midspan cross-section, mode II is by shear failure at the supports, and mode III is by achievement of the maximum strength of the shear connection between steel and concrete in the vicinity of the supports. A careful design of the structural details is necessary in order to avoid local failures as the longitudinal shear failure of the slab along the planes shown in Figure 6.9, where the collapse under longitudinal shear does not involve the connectors, or the concrete flange failure by splitting due to tensile transverse forces.

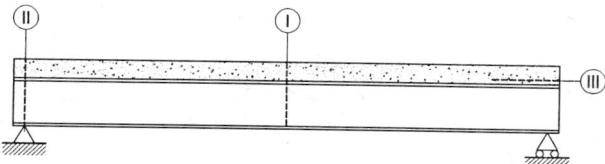

Figure 6.8 Typical failure modes for composite beam: critical sections.

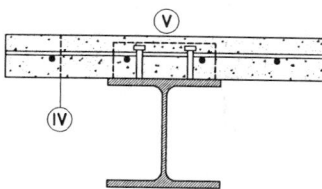

Figure 6.9 Potential shear failure planes.

The behavioral features and design criteria for the shear connection and the slabs are dealt with in Chapters 5 and 7, respectively. In the following the main concepts related to the design analysis of simply supported composite beams are presented, under the assumption that interface slip can be disregarded and the strength of the shear connection is not critical. In the following, the behavior of the elements is examined in detail, analyzing at first the evaluation of the concrete flange effective width.

During construction the member can be temporarily supported (i.e., shored construction) at intermediate points, in order to reduce stresses and deformation of the steel section during concrete casting. The construction procedures can affect the structural behavior of the composite beam. In the case of the unshored construction, the weight of fresh concrete and constructional loads are supported by the steel member alone until concrete has achieved at least 75% of its strength and the composite action can develop, and the steel section has to be checked for all the possible loading condition arising during construction. In particular, the verification against lateral-torsional buckling can become important because there is not the benefit of the restraint provided by concrete slab, and the steel section has to be suitably braced horizontally.

In the case of shored constructions, the overall load, including self weight, is resisted by the composite member. This method of construction is advantageous from a stactical point of view, but it may lead to significant increase of cost. The props are usually placed at the half and the quarters of the span, so that full shoring is obtained. The effect of the construction method on the stress state and deformation of the members generally has to be accounted for in design calculations. It is interesting to observe that if the composite section does possess sufficient ductility, the method of construction does not influence the ultimate capacity of the structure. The different responses of shored and unshored "ductile" members are shown in Figure 6.10: the behavior under service loading is very different but, if the elements are ductile enough, the two structures attain the same ultimate capacity. More generally, the composite member ductility permits a number of phenomena, such as shrinkage of concrete, residual stresses in the steel sections, and settlement of supports, to be neglected at ultimate. On the other hand, all these actions can substantially influence the performance in service and the ultimate capacity of the member in the case of slender cross-sections susceptible to local buckling in the elastic range.

6.3.2   The Effective Width of Concrete Flange

The traditional form of composite beam (Figure 6.7) can be modeled as a T-beam, the flange of which is the concrete slab. Despite the inherent in plane stiffness, the geometry, characterized by a significant width for which the shear lag effect is non-negligible, and the particular loading

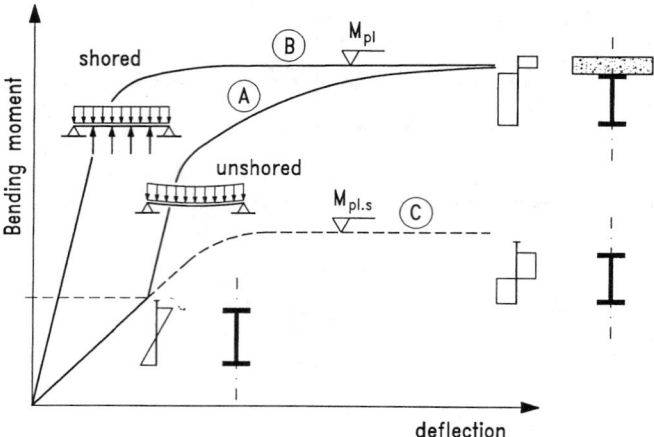

Figure 6.10 Bending moment relationship for unshored (curve A) and shored (curve B) composite beams and steel beam (curve C).

condition (through concentrated loads at the steel-concrete interface), make the response of the concrete "flange" truly bi-dimensional in terms of distribution of strains and stresses. However, it is possible to define a suitable breadth of the concrete flange permitting analysis of a composite beam as a mono-dimensional member by means of the usual beam theory. The definition of such an "effective width" may be seen as the very first problem in the analysis of composite members in bending. This width can be determined by the equivalence between the responses of the beam computed via the beam theory, and via a more refined model accounting for the actual bi-dimensional behavior of the slab. In principle, the equivalence should be made with reference to the different parameters characterizing the member performance (i.e., the elastic limit moment, the ultimate moment of resistance, the maximum deflections), and to different loading patterns. A number of numerical studies of this problem are available in the literature based on equivalence of the elastic or inelastic response [1, 2, 9, 23], and rather refined approaches were developed to permit determination of elastic effective widths depending on the various design situations and related limit states. Some codes provide detailed, and quite complex, rules based on these studies. However, recent parametric numerical analyses, the findings of which were validated by experimental results, indicated that simple expressions for effective width calculations can be adopted, if the effect of the non-linear behavior of concrete and steel is taken into account. Moreover, the assumption, in design global analyses, of a constant value for the effective width b_{eff} leads to satisfactorily accurate results. These outcomes are reflected by recent design codes. In particular, both the Eurocode 4 and the AISC specifications assume, in the analysis of simply supported beams, a constant effective width b_{eff} obtained as the sum of the effective widths $b_{e.i}$ at each side of the beam web, determined via the following expression (Figure 6.11):

$$b_{e.i} = \frac{l_o}{8} \tag{6.4}$$

where l_o is the beam span. The values of $b_{e.i}$ should be lower than one-half the distance between center-lines of adjacent beams or the distance to the slab free edge, as shown in Figure 6.11.

6.3.3 Elastic Analysis

When the interface slip can be neglected as assumed here, a similar procedure for the analysis of reinforced concrete sections can be adopted for composite members subject to bending. In fact, the cross-sections remain plane and then the strains vary linearly along the section depth. The stress diagram is also linear if the concrete stress is multiplied by the modular ratio $n = E_s/E_c$

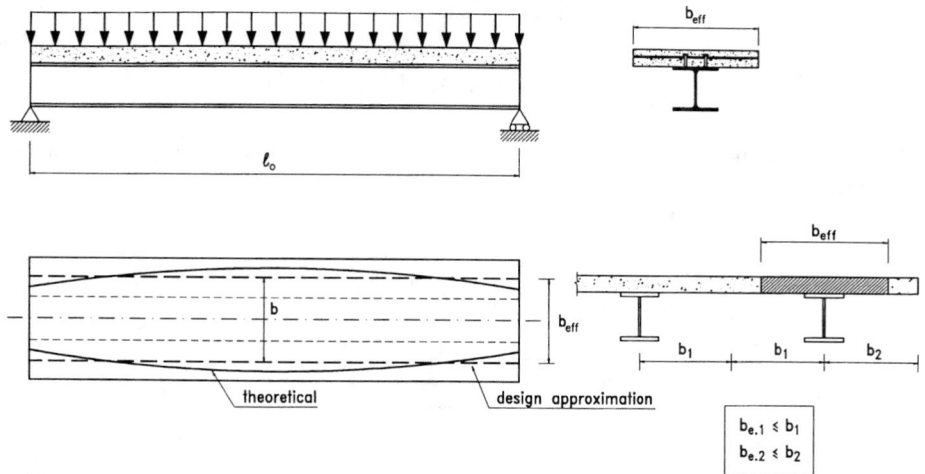

Figure 6.11 Effective width of slab.

between the elastic moduli E_s and E_c of steel and concrete, respectively. As further assumptions, the concrete tensile strength is neglected, as it is the presence of reinforcement placed in the concrete compressive area in view of its modest contribution. The theory of the transformed sections can be used, i.e., the composite section is replaced by an equivalent all-steel section[2], the flange of which has a breadth equal to b_{eff}/n. The translational equilibrium of the section requires the centroidal axis

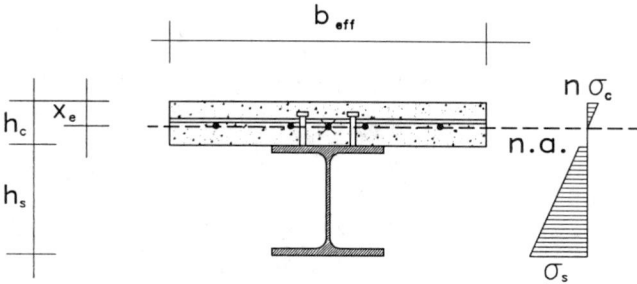

Figure 6.12 Elastic stress distribution with neutral axis in slab.

to be coincident with the neutral axis. Therefore, the position of the neutral axis can be determined by imposing that the first moment of the effective area of the cross-section is equal to zero. In the case of a solid concrete slab, and if the elastic neutral axis lies in the slab (Figure 6.12), this condition leads to the equation:

$$S = \frac{1}{n} \frac{b_{eff} \cdot x_e^2}{2} - A_s \cdot \left(\frac{h_s}{2} + h_c - x_e \right) = 0 \qquad (6.5)$$

that is quadratic in the unknown x_e (which is the distance of elastic neutral axis to the top fiber of the concrete slab). Once the value of x_e is calculated, the second moment of area of the transformed

---

[2]Transformation to an equivalent all-concrete section is a viable alternative.

cross-section can be evaluated by the following expression:

$$I = \frac{1}{n} \frac{b_{eff} \cdot x_e^3}{3} + I_s + A_s \cdot \left(\frac{h_s}{2} + h_c - x_e \right)^2 \tag{6.6}$$

The same procedure (Figure 6.13) is used if the whole cross-section is effective, i.e., if the elastic neutral axis lies in the steel profile. In this case it results:

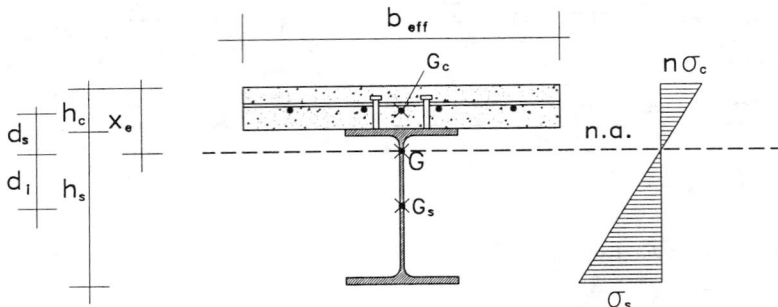

Figure 6.13   Elastic stress distribution with neutral axis in steel beam.

$$x_e = d_s + \frac{h_c}{2} \tag{6.7}$$

where

$$d_s = \frac{A_s}{A_s + b_{eff} \cdot h_c / n} \cdot \frac{h_c + h_s}{2}$$

where d_s is the distance between the centroid of the slab and the centroid of the transformed section;

$$I = I_s + \frac{1}{n} \frac{b_{eff} \cdot h_c^3}{12} + A^* \cdot \frac{(h_s + h_c)^2}{4} \tag{6.8}$$

where

$$A^* = \frac{A_s \cdot b_{eff} \cdot h_c / n}{A_s + b_{eff} \cdot h_c / n}$$

Extension of Equations 6.5 to 6.8 to beams with composite steel-concrete slabs is straightforward. The application to this case is provided by Example 6.2.

When the neutral axis depth and the second moment of area of the composite section are known, the maximum stress of concrete in compression and of structural steel in tension associated with a bending moment M are evaluated by the following expressions:

$$\sigma_c = \frac{1}{n} \frac{M}{I} \cdot x_e \tag{6.9}$$

$$\sigma_s = \frac{M}{I} \cdot (h_s + h_c - x_e) \tag{6.10}$$

These stresses must be lower than the relevant maximum design stresses allowed at the elastic limit condition. In the case of unshored construction, determination of the elastic stress distribution should take into account that the steel section alone resists all the permanent loads acting on the steelwork before composite action can develop.

In many instances, it is convenient to refer, in cross-sectional verifications, to the applied moment rather than to the stress distribution. Therefore, it is useful to define an "elastic moment of resistance" as the moment at which the strength of either structural steel or concrete is achieved. This elastic limit moment can be determined as the lowest of the moments associated with the attainment of the elastic limit condition, and obtained from Equations 6.9 and 6.10 by imposing the maximum stress equal to the design limit stress values of the relevant material (i.e., that $\sigma_c = f_{c.d}$ and $\sigma_s = f_{y.s.d}$). As the nominal resistances are assumed as in the AISC specifications

$$M_{el} = \min \left\{ f_{c.d} \cdot \frac{n \cdot I}{x_e}, \, f_{y.s.d} \cdot \frac{I}{h_s + h_c - x_e} \right\} \tag{6.11}$$

The stress check is then indirectly satisfied if (and only if) it results:

$$M \leq M_{el} \tag{6.12}$$

where M is the maximum value of the bending moment for the load combination considered.

The elastic analysis approach, based on the transformed section concept, requires the evaluation of the modular coefficient n. Through an appropriate definition of this coefficient it is possible to compute the stress distribution under sustained loads as influenced by creep of concrete. In particular, the reduction of the effective stiffness of the concrete due to creep is reflected by a decrease of the modular ratio, and consequently the stress in the concrete slab decreases, while the stress in the steel section increases. Values can be obtained for the reduced effective modulus of elasticity $E_{c.ef}$ of concrete, accounting for the relative proportion of long- to short-term loads. Codes may suggest values of $E_{c.ef}$ defined accordingly to common load proportions in practice (see Section 6.2.1 for Eurocode 4 specifications). Selection of the appropriate modular ratio n would permit, in principle, the variation of the stress distribution in the cross-section to be checked at different times during the life of the structure.

6.3.4 Plastic Analysis

Refined non-linear analysis of the composite beam can be carried out accounting for yielding of the steel section and inelasticity of the concrete slab. However, the stress state typical of composite beams under sagging moments usually permits the plastic moment of the composite section to be achieved. In most instances the plastic neutral axis lies in the slab and the whole of the steel section is in tension, which results in:

- local buckling not being a critical phenomenon
- concrete strains being limited, even when the full yielding condition of the steel beam is achieved

Therefore, the plastic method of analysis is applicable to most simply supported composite beams. Such a tool is so practically advantageous that it is the non-linear design method for these members. In particular, this approach is based on equilibrium equations at ultimate, and does not depend on the constitutive relations of the materials and on the construction method. The plastic moment can be computed by application of the rectangular stress block theory. Moreover, the concrete may be assumed, in composite beams, to be stressed uniformly over the full depth x_{pl} of the compression side of the plastic neutral axis, while for the reinforced concrete sections usually the stress block depth is limited to $0.8\,x_{pl}$. The evaluation of the plastic moment requires calculation of the following quantities:

$$F_{c.max} = 0.85 b_{eff} \cdot h_c \cdot f_c \tag{6.13}$$

$$F_{s.max} = A_s \cdot f_{y.s} \tag{6.14}$$

These are, respectively, the maximum compression force that the slab can resist and the maximum tensile force that the steel profile can resist. If $F_{c.max}$ is greater than $F_{s.max}$, the plastic neutral axis lies in the slab; in this case (Figure 6.14) the interaction force between slab and steel profile is $F_{s.max}$ and the plastic neutral axis depth is defined by a simple first order equation:

$$F_{c.max} > F_{s.max} \Rightarrow F_c = F_s = A_s \cdot f_{y.s} \tag{6.15}$$

$$0.85b_{eff} \cdot x_{pl} \cdot f_c = A_s \cdot f_{y.s} \tag{6.16}$$

$$x_{pl} = \frac{A_s \cdot f_{y.s}}{0.85b_{eff} \cdot f_c} \tag{6.17}$$

It can be observed that using stress block, the plastic analysis allows evaluation of the neutral

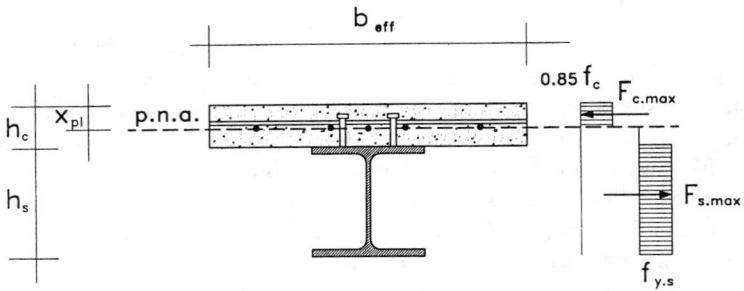

Figure 6.14   Plastic stress distribution with neutral axis in slab.

axis depth by means of an equation of lower degree than in the elastic analysis: in this last case Equation 6.5 the stresses have a linear distribution and the equation is of the second order. The internal bending moment lever arm (distance between line of action of the compression and tension resultants) is then evaluated by the following expression:

$$h^* = \frac{h_s}{2} + h_c - \frac{x_{pl}}{2} \tag{6.18}$$

The plastic moment can then be determined as:

$$M_{pl} = A_s \cdot f_{y.s} \cdot h^* \tag{6.19}$$

If $F_{s.max}$ is greater than $F_{c.max}$, the neutral axis lies in the steel profile (Figure 6.15); in this case it results:

$$F_{c.max} < F_{s.max} \Rightarrow F_c = F_s = 0.85b_{eff} \cdot h_c \cdot f_c \tag{6.20}$$

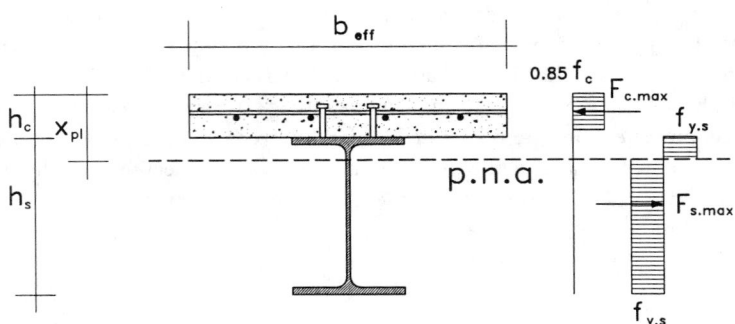

Figure 6.15   Plastic stress distribution with neutral axis in steel beam.

Two different cases can take place; in the first case:

$$F_c > F_w = d \cdot t_w \cdot f_{y.s} \tag{6.21}$$

where
t_w = the web thickness
d = the clear distance between the flanges
and:

$$M_{pl} \simeq F_{s.max} \cdot \frac{h_s}{2} + F_c \cdot \frac{h_c}{2} \tag{6.22}$$

In the second case:

$$F_c < F_w = d \cdot t_w \cdot f_{y.s} \tag{6.23}$$

and:

$$M_{pl} = M_{pl.s} + F_c \cdot \frac{h_s + h_c}{2} - \frac{F_c^2}{4 \cdot t_w \cdot f_{y.s}} \tag{6.24}$$

where
$M_{pl.s}$ = the plastic moment of the steel profile

The design value of the plastic moment of resistance has to be computed in accordance to the format assumed in the reference code. If the Eurocode 4 provisions are used, in Equations 6.13 and 6.14, the "design values" of strength $f_{c.d}$ and $f_{y.s.d}$ shall be used (see Section 6.1.3) instead of the "unfactored" strength f_c and $f_{y.s}$; i.e., the design plastic moment given by Equation 6.19 is evaluated in the following way:

$$M_{pl.d} = A_s \cdot f_{y.s.d} \cdot h^* \tag{6.25a}$$

where h^* has to be computed with reference to the plastic neutral axis x_{pl} associated with design values of the material strengths.

If the AISC specification are considered, the nominal values of the material strengths shall be used and the safety factor $\phi_b = 0.9$ affects the nominal value of the plastic moment:

$$M_{pl.d} = \phi_b \left(A_s \cdot f_{y.s} \cdot h^* \right) \tag{6.25b}$$

6.3.5 Vertical Shear

In composite elements shear is carried mostly by the web of the steel profile; the contributions of concrete slab and steel flanges can be neglected in the design due to their width. The design shear strength can be determined by the same expression as for steel profiles:

$$V_{pl} = A_v \cdot f_{y.s.V} \tag{6.26}$$

where
A_v = the shear area of the steel section
$f_{y.s.V}$ = the shear strength of the structural steel

With reference to the usual case of I steel sections, and considering the different values assumed for $f_{y.s.V}$, the AISC and Eurocode specifications provide the same shear resistance; in fact:

$$\text{AISC} \quad V_{pl} = h_s \cdot t_w \cdot (0.6 \cdot f_{y.s}) \tag{6.27a}$$

$$\text{Eurocode} \quad V_{pl} = 1.04 \cdot h_s \cdot t_w \cdot \frac{f_{y.s}}{\sqrt{3}} \tag{6.27b}$$

and $(1.04 / \sqrt{3}) = 0.60$.

The design value of the plastic shear capacity is obtained either by multiplying the value of V_{pl} from Equation 6.27a by a Φ_v factor equal to 0.90 (AISC), or by using in Equation 6.27b the design value of $f_{y.s.d}$ (Eurocode).

For slender beam webs (i.e., when their depth-to-thickness ratio is lower than $69 / \sqrt{235/f_{y.s}}$ with $f_{y.s}$ in N/mm^2) the shear resistance should be suitably determined by taking into account web buckling in shear.

The shear-moment interaction is not important in simply supported beams (in fact for usual loading conditions where the moment is maximum the shear is zero; where the shear is maximum the moment is zero). The situation in continuous beams is different (see Chapter 4).

6.3.6 Serviceability Limit States

The adequacy of the performance under service loads requires that the use, efficiency, or appearance of the structure are not impaired. Besides, the stress state in concrete also needs to be limited due to the possible associated durability problems. Micro-cracking of concrete (when stressed over 0.5 f_c) may allow development of rebars corrosion in aggressive environments. This aspect has to be addressed with reference to the specific design conditions.

As to the member deformability, the stiffness of a composite beams in sagging bending is far higher than in the case of steel members of equal depth, due to the significant contribution of the concrete flange (see Equations 6.6 and 6.8). Therefore, deflection limitation is less critical than in steel systems. However, the effect of concrete creep and shrinkage has to be evaluated, which may significantly increase the beam deformation as computed for short-term loads. In service the beam should behave elastically. Under the assumption of full interaction the usual formulae for beam deflection calculation can be used. As an example, the deflection under a uniformly distributed load, p, is obtained as:

$$\delta = \frac{5}{384} \frac{p \cdot l^4}{E_s \cdot I} \tag{6.28}$$

For unshored beams, the construction sequence and the deflection of the steel section under the permanent loads has to be taken into account before development of composite action is added to the deflections of the composite beam under the relevant applied loads.

The value of the moment of inertia I of the transformed section, and hence the value of δ, depends on the modular ratio, n. Therefore, the effective modulus (EM) theory enables the effect of concrete creep to be incorporated in design calculations without any additional complexity. Determination of the deflection under sustained loads simply requires that an effective modular ration $n_{ef} = E_s/E_{c.ef}$ is used when computing I via Equation 6.6 or 6.8.

The effect of the shrinkage strain ε_{sh} could be evaluated considering that the compatibility of the composite beam requires a tension force N_{sh} to develop in the slab equal to:

$$N_{sh} = \varepsilon_{sh} \cdot E_{c.ef} \cdot b \cdot h_c \tag{6.29}$$

This force is applied in the centroid of the slab and, due to equilibrium, produces a positive moment M_{sh} equal to:

$$M_{sh} = N_{sh} \cdot d_s \tag{6.30}$$

where d_s is given by Equation 6.7. This moment is constant along the beam. The additional deflection can be determined as:

$$\delta_{sh} = \frac{M_{sh} \cdot l^2}{8 \cdot E_s \cdot I} = 0.125 \cdot \varepsilon_{sh} \cdot \frac{b \cdot h_c \cdot d_s}{n_{ef} \cdot I} \cdot l^2 \qquad (6.31)$$

Typical values of ε_{sh} are given in Section 6.2.1. The influence of shrinkage on the deflection is usually important in a dry environment and for span-to-beam depth ratios greater than 20.

In partially composite beams, the deflection associated with the interface slip has also to be accounted for. A simplified treatment is presented in the Section on page **6**-63.

The total deflection should be lower than limit values compatible with the serviceability requirements specific to the building system considered. Reference values given by the Eurocode 4 are presented in Table 6.2.

TABLE 6.2   Eurocode 4 Limiting Values for Vertical Deflections

Conditions	Limits	
	δ_{max}	δ_2
roofs generally	L/200	L/250
roofs frequently carrying personnel other than for maintenance	L/250	L/300
floors generally	L/250	L/300
floors and roofs supporting brittle finish on non-flexible partitions	L/250	L/350
floors supporting columns (unless the deflection has been included in the global analysis for the ultimate limit state)	L/400	L/500
* where δ_{max} can impair the appearance of the buildings	L/250	-
For cantilever: L= twice cantilever span		

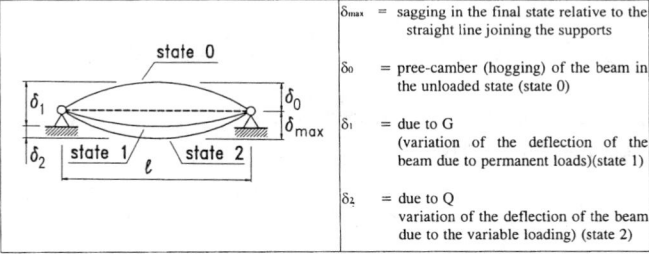

δ_{max}	= sagging in the final state relative to the straight line joining the supports
δ_0	= pree-camber (hogging) of the beam in the unloaded state (state 0)
δ_1	= due to G (variation of the deflection of the beam due to permanent loads)(state 1)
δ_2	= due to Q variation of the deflection of the beam due to the variable loading) (state 2)

The vibration control is strongly correlated to the deflection control. In fact it can be shown that the fundamental frequency of a simply supported floor beam is given by:

$$f = \frac{18}{\sqrt{\delta}} \qquad (6.32)$$

where
δ = the immediate deflection (mm) due to the self weight. A value of f equal to 4 Hertz (cycles for second) may be considered acceptable for the comfort of people in buildings

6.3.7 Worked Examples[3]

EXAMPLE 6.1:   Composite beam with solid concrete slab

In Figure 6.16 it is reported that the cross-section of a simply supported beam with a span length l equals 10 m; the steel profile (IPE 500) is characterized by $A_s = 11600$ mm^2 and $I_s = 4.82 \cdot 10^8$ mm^4. The solid slab has a thickness of 120 mm; the spacing with adjacent beams is 5000 mm. The beam is subjected to a uniform load p =40 kN/m (16 kN/m of dead load, 24 kN/m of live load) at the serviceability condition. A shored construction is considered.

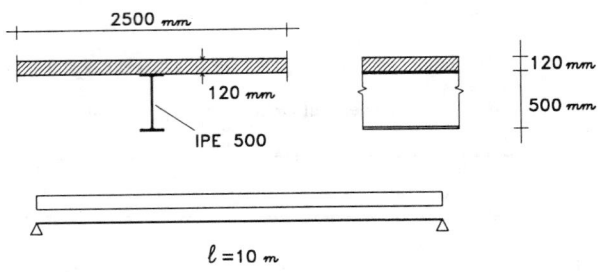

Figure 6.16   Cross-section of a simply supported beam with a span length l equals 10 m.

1. Determination of design moment:
   The design moment for service conditions is

   $$M = \frac{40 \cdot 10^2}{8} = 500 \text{ kNm}$$

   For the ultimate limit state, considering a factor of 1.35 for the dead load and 1.5 for the live load, it is

   $$M = \frac{(1.35 \cdot 16 + 1.5 \cdot 24) \cdot 10^2}{8} = 720 \text{ kNm}$$

2. Evaluation of the effective width (Section 6.3.2):
   Applying Equation 6.4 it results:

   $$b_{\text{eff}} = 2 \cdot \frac{l}{8} = 2 \cdot \frac{10000}{8} = 2500 \text{ mm}$$

   Thus, only 2500 mm of 5000 mm are considered as effective.

3. Elastic analysis of the cross-section (Section 6.3.3):
   The following assumptions are made:

   $$
   \begin{aligned}
   E_c &= 30500 \text{ N / mm}^2 \\
   E_s &= 210000 \text{ N / mm}^2
   \end{aligned}
   $$

---

[3] All examples refer to Eurocode rules, with which the authors are more familiar.

Therefore, at the short time the modular ratio results:

$$n = \frac{210000}{30500} = 6.89$$

Equation 6.5 becomes:

$$\frac{1}{6.89}\frac{2500}{2} \cdot x_e^2 - 11600 \cdot \left(\frac{500}{2} + 120 - x_e\right) = 0$$

then:

$$181.4x_e^2 + 11600 \cdot x_e - 4292000 = 0$$
$$x_e = 125.1 \text{ mm} > 120 \text{ mm}$$

i.e., the entire slab is under compression and Equation 6.7 shall be considered:

$$x_e = \frac{120}{2} + \frac{11600}{11600 + 120 \cdot 2500/6.89}\frac{120 + 500}{2} = 125.2 \text{ mm}$$

From Equation 6.8 it results:

$$A^* = \frac{11600 \cdot 2500 \cdot 120/6.89}{11600 + 2500 \cdot 120/6.89} = 9160 \text{ mm}^2$$

$$I = \frac{1}{6.89}\frac{2500 \cdot 120^3}{12} + 4.82 \cdot 10^8 + 9160\frac{(500 + 120)^2}{4} = 1.41 \cdot 10^9 \text{ mm}^4$$

The maximum stress in concrete and steel are the following (see Equations 6.9 and 6.10):

$$\sigma_c = \frac{500 \cdot 10^6}{6.89 \cdot 1.41 \cdot 10^9} \cdot 125.2 = 6.4 \text{ N / mm}^2$$

$$\sigma_s = \frac{500 \cdot 10^6}{1.41 \cdot 10^9} \cdot (500 + 120 - 125.2) = 175.5 \text{ N / mm}^2$$

For the long term calculation, a creep coefficient $\phi = 2$ is assumed obtaining the following modular ratio:

$$n_{ef} = 6.89 \cdot 3 = 20.67$$

Equation 6.6 gives:

$$\frac{2500}{20.67 \cdot 2} \cdot x_e^2 - 11600 \cdot \left(\frac{500}{2} + 120 - x_e\right) = 0$$
$$60.47 \cdot x_e^2 + 11600 \cdot x_e - 4292000 = 0$$
$$x_e = 187.2 \text{ mm} > 120 \text{ mm}$$

The elastic neutral axis lies in the steel profile web and the slab is entirely compressed. Therefore, it is (Equations 6.7, and 6.8):

$$x_e = \frac{120}{2} + \frac{11600}{11600 + 120 \cdot 2500/20.67} \cdot \frac{120 + 500}{2} = 197.7 \text{ mm}$$

$$A^* = \frac{11600 \cdot 2500 \cdot 120/20.67}{11600 + 2500 \cdot 120/20.67} = 6447 \text{ mm}^2$$

$$I = \frac{2500 \cdot 120^3}{20.67 \cdot 12} + 4.82 \cdot 10^8 + 6447 \cdot \frac{(500 + 120)^2}{4} = 1.12 \cdot 10^9 \text{ mm}^4$$

The stresses result:

$$\sigma_c = \frac{500 \cdot 10^6}{20.67 \cdot 1.12 \cdot 10^9} \cdot 197.7 = 4.3 \text{ N} / \text{mm}^2$$

$$\sigma_s = \frac{500 \cdot 10^6}{1.12 \cdot 10^9} \cdot (620 - 197.7) = 188.5 \text{ N} / \text{mm}^2$$

The concrete stress decreases 33% while the steel stress increases 7%.

4. Plastic analysis of the cross-section (Section 6.3.4):
   The design strength of materials are assumed as $f_{c.d} = 14.2 \text{ N/mm}^2$ (including the coefficient 0.85) for concrete and $f_{y.s.d} = 213.6 \text{ N/mm}^2$ for steel. By means Equations 6.13, and 6.14 it is:

$$F_{c.\text{max}} = 2500 \cdot 120 \cdot 14.2 = 4260000 \text{N} = 4260 \text{ kN}$$

$$F_{s.\text{max}} = 11600 \cdot 213.6 = 2477760 \text{N} = 2478 \text{ kN}$$

Consequently, it is assumed:

$$F_c = F_s = 2478 \text{ kN}$$

and, considering the design strength in Equation 6.17:

$$x_{pl} = \frac{11600 \cdot 213.6}{2500 \cdot 14.2} = 69.8 \text{ mm}$$

The internal arm is (Equation 6.18):

$$h^* = \frac{500}{2} + 120 - \frac{69.8}{2} = 335 \text{ mm}$$

and the plastic moment results (Equation 6.19):

$$M_{pl} = 2478 \cdot 0.335 = 830 \text{ kNm}$$

Thus, the value of the plastic resistance is greater than the design moment at the ultimate conditions ($M = 720 \text{ kNm}$).

5. Ultimate shear of the section (Section 6.3.5):

$$A_v = 1.04 \cdot 500 \cdot 10.2 = 5304 \text{ mm}^2$$

$$V_{pl} = 5304 \frac{213.6}{\sqrt{3}} = 654100 \text{ N} = 654 \text{ kN}$$

6. Control of deflection (Section 6.3.6):

The deflection at short term is

$$\delta(t=0) = \frac{5}{384} \cdot \frac{40 \cdot 10000^4}{210000 \cdot 1.41 \cdot 10^9} = 17.6 \text{ mm} = \frac{1}{568} \cdot l$$

The deflection at long term is increased by the creep and shrinkage effects. The 50% of the live load is considered as long term load; thus, in this verification the load is $16 + 0.5 \cdot 24 = 28$ kN/m. If only the creep effect is taken into account, the following result is obtained:

$$\delta(t=\infty) = \frac{5}{384} \cdot \frac{28 \cdot 10000^4}{210000 \cdot 1.12 \cdot 10^9} = 15.5 \text{ mm}$$

As regards the shrinkage effect, a final value of the strain is assumed equal to

$$\varepsilon_{sh} = 200 \cdot 10^{-6}$$

and the increment of deflection due to shrinkage results (Equation 6.32):

$$\delta_{sh} = 0.125 \cdot 200 \cdot 10^{-6} \cdot \frac{120 \cdot 5000 \cdot (197.7 - 60)}{20.67 \cdot 1.12 \cdot 10^9} \cdot 10000^2 = 8.9 \text{ mm}$$

The final value of the deflection is

$$\delta(t=\infty) = 15.5 + 8.9 = 24.4 \text{ mm} = \frac{l}{410}$$

EXAMPLE 6.2: Composite beam with concrete slab with metal decking.

In Figure 6.17 it is reported that the cross-section of a simple supported beam with a span length l equals 10 m; the difference with the previous example consists in the use of a profiled steel sheeting. The structural steel (IPE 500) is characterized by $A_s = 11600$ mm^2 and $I_s = 4.82 \cdot 10^8$ mm^4. The slab thickness is $55 + 65$ mm. The spacing of beams is 5000 mm. The beam is subjected to a uniform load $p = 40$ kN/m (16 kN/m of dead load, 24 kN/m of live load) at serviceability condition.

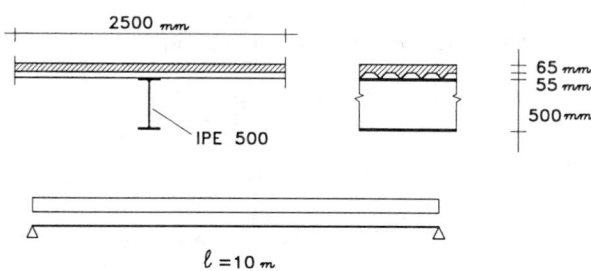

Figure 6.17 Cross-section of a simple supported beam with a span length l equals 10 m.

1. Determination of design moment:

The design moment for service conditions is

$$M = \frac{40 \cdot 10^2}{8} = 500 \text{ kNm}$$

For the ultimate limit state, considering a factor of 1.35 for the dead load and of 1.5 for the live load, it is

$$M = \frac{(1.35 \cdot 16 + 1.5 \cdot 24) \cdot 10^2}{8} = 720 \text{ kNm}$$

2. Evaluation of the effective width (Section 6.3.2):

$$b_{\text{eff}} = 2 \cdot \frac{10000}{8} = 2500 \text{ mm}$$

only 2500 of 5000 mm are considered as effective.

3. Elastic analysis of the cross-section (Section 6.3.3):
At a short time, the results are the following:

$$181.4x_e^2 + 11600 \cdot x_e - 4292000 = 0$$
$$x_e = 125.1 \text{ mm} > 65 \text{ mm}$$

i.e., the entire slab is under compression and Equation 6.7 shall be considered:

$$x_e = \frac{65}{2} + \frac{11600}{11600 + 65 \cdot 2500/6.89} \cdot \left(120 + \frac{500}{2} - \frac{65}{2}\right) = 143.8 \text{ mm}$$

$$I = \frac{2500 \cdot 65}{6.89} \cdot \left(143.8 - \frac{65}{2}\right) + \frac{2500 \cdot 65^3}{12 \cdot 6.89} + 4.82 \cdot 10^8$$
$$+ 11600 \cdot \left(\frac{500}{2} + 120 - 143.8\right) = 1.38 \cdot 10^9 \text{ mm}^4$$

$$\sigma_c = 7.6 \text{ N / mm}^2$$
$$\sigma_s = 172.5 \text{ N / mm}^2$$

For long term calculation it is

$$x_e = \frac{65}{2} + \frac{11600}{11600 + 65 \cdot 2500/20.67} \cdot \left(120 + \frac{500}{2} - \frac{65}{2}\right) = 233.7 \text{ mm}$$

The elastic neutral axis lies in the steel profile web and the slab is entirely compressed.

$$I = 1.02 \cdot 10^9 \text{ mm}^4$$

The stresses result:

$$\sigma_c = 5.5 \text{ N / mm}^2$$
$$\sigma_s = 189.4 \text{ N / mm}^2$$

The concrete stress decreases 28% while the steel stress increases 10%.

4. Plastic analysis of the section (Section 6.3.4):

$$F_{c.\text{max}} = 2500 \cdot 65 \cdot 14.2 = 2308000 \text{ N} = 2308 \text{ kN}$$
$$F_{s.\text{max}} = 11600 \cdot 213.6 = 2477760 \text{ N} = 2478 \text{ kN}$$

Consequently, it is

$$F_c = F_s = 2308 \text{ kN}$$

In this case

$$
\begin{aligned}
t_w &= 10.2 \text{ mm} \qquad t_f = 16 \text{ mm} \\
F_w &= (500 - 32) \cdot 10.2 \cdot 213.6 = 1019641/\text{N} = 1020 \text{ kN} \\
F_c &> F_w
\end{aligned}
$$

Therefore, the plastic neutral axis lies in the web of the steel profile:

$$
M_{pl} \simeq 2308 \cdot \left(120 - \frac{65}{2}\right) + 2478 \cdot \frac{500}{2} = 821 \text{ kNm}
$$

5. Ultimate shear of the section (Section 6.3.5):

$$
\begin{aligned}
A_v &= 1.04 \cdot 500 \cdot 10.2 = 5304 \text{ mm}^2 \\
V_{pl} &= 654 \text{ kN}
\end{aligned}
$$

6. Control of deflection (Section 6.3.6):
The control at short term provides:

$$
\delta(t = 0) = 18 \text{ mm} = \frac{1}{556} \cdot l
$$

The deflection at long term is increased by the creep and shrinkage effects. The 50% of the live load is considered as long term load; thus, in this verification the load is $16 + 0.5 \cdot 24 = 28$ kN/m. If only the creep effect is taken into account, the following result is obtained:

$$
\delta(t = \infty) = \frac{5}{384} \cdot \frac{28 \cdot 10000^4}{210000 \cdot 1.02 \cdot 10^9} = 17.0 \text{ mm}
$$

As regards the shrinkage effect, a final value of the strain is assumed equal to:

$$
\begin{aligned}
\varepsilon_{sh} &= 200 \cdot 10^{-6} \\
\delta_{sh} &= 0.125 \cdot 200 \cdot 10^{-6} \cdot \frac{65 \cdot 5000 \cdot (233.7 - 32.5)}{20.67 \cdot 1.02 \cdot 10^9} \cdot 10000^2 = 7.8 \text{ mm}
\end{aligned}
$$

The final value of the deflection is

$$
\delta(t = \infty) = 17.0 + 7.8 = 24.8 \text{ mm} = \frac{1}{403} \cdot l
$$

6.4 Continuous Beams

6.4.1 Introduction

Beam continuity may represent an efficient stactical solution with reference to both load capacity and stiffness. In composite buildings, different kinds of continuity may, in principle, be achieved, as indicated by Puhali et al. [40], between the beams and the columns and, possibly, between adjacent beams. Furthermore, the degree of continuity can vary significantly in relation to the performance of joints as to both strength and stiffness: joints can be designed to be full or partial strength

(strength) and rigid, semi-rigid, or pinned (stiffness). Despite the growing popularity of semi-rigid partial strength joints (see Chapter 23), rigid joints may still be considered the solution most used in building frames. Structural solutions for the flooring system were also proposed (see for example Brett et al., [7]), which allow an efficient use of beam continuity without the burden of costly joints.

In bridge structures, the use of continuous beams is very advantageous for it enables joints along the beams to be substantially reduced, or even eliminated. This results in a remarkable reduction in design work load, fabrication and construction problems, and structural cost.

From the structural point of view, the main benefits of continuous beams are the following:

- at the serviceability limit state: deformability is lower than that of simply supported beams, providing a reduction of deflections and vibrations problems
- at the ultimate limit state: moment redistribution may allow an efficient use of resistance capacity of the sections under positive and negative moment.

However, the continuous beam is subjected to hogging (negative) bending moments at intermediate supports, and its response in these regions is not efficient as under sagging moments, for the slab is in tension and the lower part of the steel section is in compression. The first practical consequence is the necessity of an adequate reinforcement in the slab. Besides, the following problems arise:

- at the serviceability limit state: concrete in tension cracks and the related problems such as control of the cracks width, the need of a minimum reinforcement, etc., have to be accounted for in the design. Moreover, deformability increases reducing the beneficial effect of the beam continuity
- at the ultimate limit state: compression in steel could cause buckling problems either locally (in the bottom flange in compression and/or in the web) or globally (distortional lateral-torsional buckling)

Other problems can arise as well; i.e., in simply supported beams, the shear-moment interaction is usually negligible, while at the intermediate supports of continuous beams both shear and bending can simultaneously attain high values, and shear-moment interaction becomes critical.

In the following, the various aspects described above are discussed. In this Section the assumption of full shear-concrete interaction is still maintained, i.e., the shear connection is assumed to be a "full" shear connection (see Section 6.5). Problems related to use of the partial shear connection are discussed in detail in Section 6.5.7.

6.4.2   Effective Width

The general definition of the effective width, b_{eff}, is the same for the simply supported beam (Section 6.3.2). The determination of the effective width along a continuous beam is certainly a more complex problem. Besides the type of loading and geometrical characteristics, several other parameters are involved, which govern the strain (stress) state in the slab in the hogging moment regions. This complexity results in different provisions in the various national codes. However, it should be noted that the variability of b_{eff} along the beam would imply, if accounted for, a substantial burden for design analysis. For a continuous composite beam, it was shown that the selection in the global analysis of a suitable effective width constant within each span allows us to obtain internal forces with satisfactory accuracy. On the other hand, sectional verification should be performed with reference to the "local" value of b_{eff}. The effective width in the moment negative zone allows evaluation of the reinforcement area that is effective in the section.

The AISC provisions suggest use of Equation 6.4, considering the full span length and center-to-center support for the analysis of continuous beams. No recommendations are provided for sectional verification.

Eurocode 4 also recommends that in the global analysis b_{eff} is assumed to be constant over the whole length of each span, and equal to the value at midspan. The resistance of the critical cross-sections is determined using the values of b_{eff} computed via Equation 6.4, where the length l is replaced by the length l_0 defined as in Figure 6.18. The effective width depends on the type of applied moment (hogging or sagging) and span (external, internal, cantilever). The value of b_{eff} in the hogging moment enables determination of the effective area of steel reinforcement to be considered in design calculations.

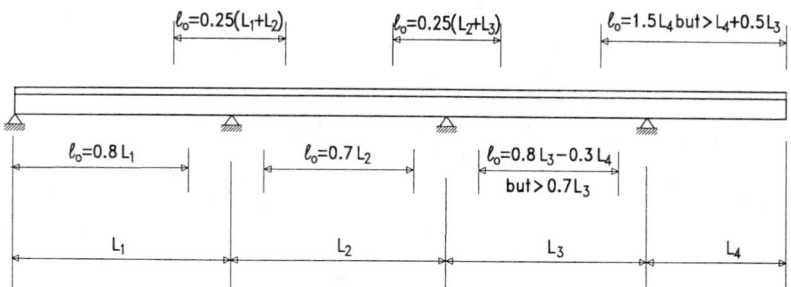

Figure 6.18 Equivalent span for effective width of concrete flange.

6.4.3 Local Buckling and Classification of Cross-Sections

Local buckling has to be accounted for in the very preliminary phase of design; due to the occurrence of local buckling, sections subjected to negative moment may not attain their plastic moment of resistance or develop the plastic rotation required for the full moment redistribution, associated with the formation of a beam plastic mechanism. In order to enable a preliminary assessment of strength and rotation capacity, steel sections can be classified according to the slendernesses of the flanges and of the web [10]. Four different member behaviors could be identified, according to the importance of local buckling effects:

1. members that develop the full plastic moment capacity and also possess a rotation capacity sufficient to make, in most practical cases, a beam plastic mechanism
2. members that can develop their plastic moment of resistance, but then have limited rotation capacity
3. members that achieve the elastic moment of resistance associated with yielding of steel in the more stressed fiber, but not the plastic moment of resistance
4. members for which local buckling occurs still in the elastic range, so that even the elastic limit moment cannot be developed and elastic local buckling govern resistance

In Figure 6.19, the four behaviors described above are schematically presented.

The definition of reliable limitations for the flange and web slenderness that take into account the different performances is a very complex problem both theoretically and experimentally. Besides, the wide range of possible geometries, the influence of the loading conditions and the mechanical characteristics of the steel material have to be considered. Moreover, the interaction with the lateral-torsional buckling mode in the distortional form has to be considered. The buckling problems depend not only on the flange and/or web slenderness but also on the mechanical ratio $f_{y.s}/E_s$; since the elasticity modulus E_s is constant for all steel grades, the yield strength $f_{y.s}$ can be assumed as the reference parameter taking into account steel grade. As a rule, the higher the yield strength, the lower the upper limit slenderness for a given class.

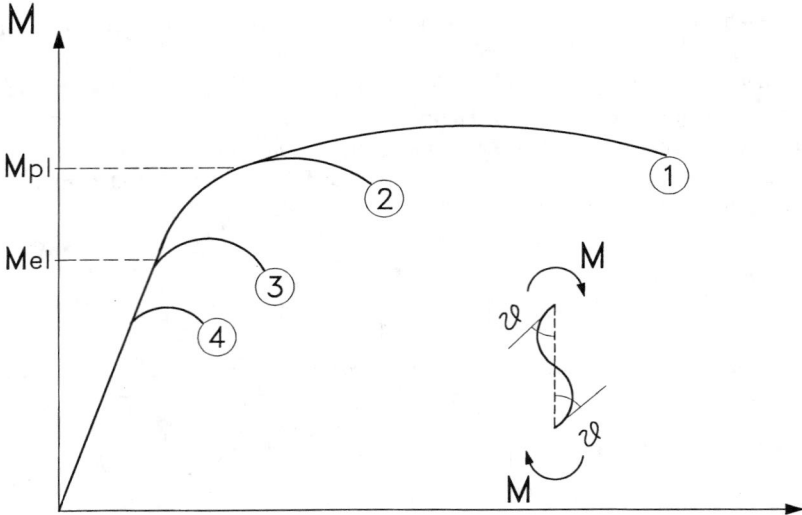

Figure 6.19   Different behaviors of composite members expressed in terms of moment-rotation (M-θ) relationships.

The complexity of the problem and the still limited knowledge available force code specifications to be based on rather conservative assumptions.

The classification limits specified by Eurocode 4 are reported in Tables 6.3 and 6.4. The four different behaviors are defined as "plastic – class 1", "compact – class 2", "semi-compact – class 3", and "slender – class 4". The limitations accounting for the fabrication processes are different for rolled and welded shapes; also, the presence of web encasement is allowed for by different limitations, which take into account the beneficial restraint offered by the encasing concrete.

The AISC specifications define only three classes. However, a fourth class is suggested for seismic use, due to the higher overall structural ductility required to dissipate seismic energy. The cross-sections are then classified as "seismic", "compact", "non compact", and "slender". The seismic sections guarantee a plastic rotation capacity (i.e., a rotational ductility defined as the ratio between the ultimate plastic rotation and the rotation at the onset of yield) in the range of 7 to 9, while the compact sections have a rotational ductility of at least 3. With reference to steel rolled I sections in pure bending moment, and using the same symbols as in Tables 6.3 and 6.4, the AISC slenderness limitations are given in Table 6.5, where $f_{y,s}$ represents the nominal strength of steel in ksi. The values associated with $f_{y,s}$ in N/mm^2 are also provided in brackets. The values provided by Eurocode 4 and AISC provisions show a good agreement when both Class 1 and Class 2 sections are compared to seismic and compact sections, respectively. The comparison between Class 3 and non-compact sections highlights that the values provided by Eurocode 4 are more restrictive.

6.4.4   Elastic Analysis of the Cross-Section

Cross-sectional behavior in sagging bending has been treated in Section 6.3.3 to which reference can be made.

In the negative moment regions, where the concrete slab is subject to tensile stresses, two main states of the composite beam can be identified with reference to the value of moment M_{cr} at which cracks start to develop. When the bending moment is lower than M_{cr}, the cross-section is in the "state 1 uncracked" and its uncracked moment of inertia I_1 can be evaluated by the same procedure of the section subjected to positive moment (see Section 6.3.3). When M is greater than M_{cr} the cross-section enters the "state 2 cracked", characterized by the moment of inertia I_2. In this phase,

TABLE 6.3 Eurocode 4 Maximum Width-to-Thickness Ratios for Steel Webs

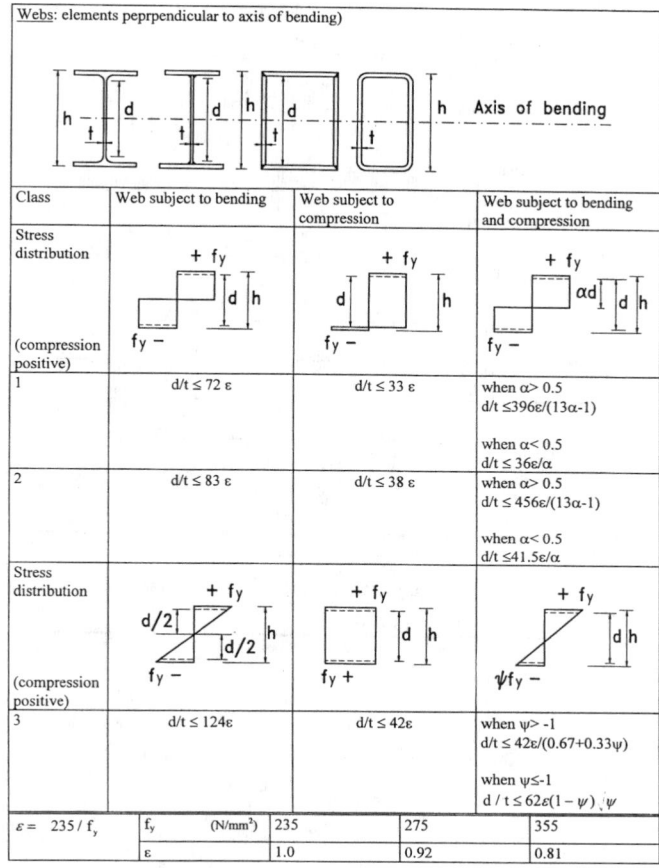

	Webs: elements peprpendicular to axis of bending)		

Class	Web subject to bending	Web subject to compression	Web subject to bending and compression
Stress distribution (compression positive)	$+ f_y$ ... d h ... $f_y -$	$+ f_y$ ... d h ... $f_y -$	$+ f_y$ ... αd d h ... $f_y -$
1	$d/t \le 72\,\varepsilon$	$d/t \le 33\,\varepsilon$	when $\alpha > 0.5$ $d/t \le 396\varepsilon/(13\alpha-1)$ when $\alpha < 0.5$ $d/t \le 36\varepsilon/\alpha$
2	$d/t \le 83\,\varepsilon$	$d/t \le 38\,\varepsilon$	when $\alpha > 0.5$ $d/t \le 456\varepsilon/(13\alpha-1)$ when $\alpha < 0.5$ $d/t \le 41.5\varepsilon/\alpha$
Stress distribution (compression positive)	$+ f_y$ $d/2$ $d/2$ h $f_y -$	$+ f_y$ d h $f_y +$	$+ f_y$ d h $\psi f_y -$
3	$d/t \le 124\varepsilon$	$d/t \le 42\varepsilon$	when $\psi > -1$ $d/t \le 42\varepsilon/(0.67+0.33\psi)$ when $\psi \le -1$ $d/t \le 62\varepsilon(1-\psi)\sqrt{\psi}$
$\varepsilon = \sqrt{235/f_y}$	f_y (N/mm²) 235	275	355
	ε 1.0	0.92	0.81

the elastic neutral axis x_e usually lies within the steel section, so that concrete does not collaborate to the stiffness and strength of the composite section. As a consequence, the effective cross-section of the composite beams consists only of steel (reinforcement bars and steel section). The moment of inertia I_2 and the stresses can be computed straightforwardly. The same general considerations apply to elastic verification of cracked composite beams, already discussed in Section 6.3.3 with reference to beams in sagging bending.

6.4.5 Plastic Resistance of the Cross-Section

In most cases, as already discussed, sections in positive bending have the neutral axis within the slab. The steel section is hence fully (or predominantly) in tension and plastic analysis can be applied, i.e., sections are in class 1 or 2 (compact). The stress block model, presented in Section 6.3.4 may be adopted for determining the plastic moment of resistance of the cross-section. Plastic analysis under hogging moment requires a preliminary classification of the cross-section as plastic or compact. The fully plastic stress distribution of the composite cross-section under hogging moments is shown in Figure 6.20: the location of the plastic neutral axis (i.e., the depth x_{pl}) is determined by imposing

TABLE 6.4   Eurocode 4 Maximum Width-to-Thickness Ratios for Steel Outstanding Flanges in Compression

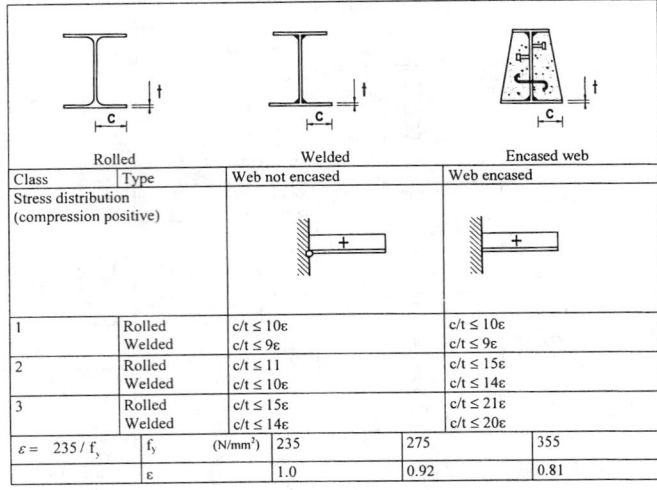

Class	Type	Welded Web not encased	Encased web Web encased
Stress distribution (compression positive)			
1	Rolled	c/t ≤ 10ε	c/t ≤ 10ε
	Welded	c/t ≤ 9ε	c/t ≤ 9ε
2	Rolled	c/t ≤ 11	c/t ≤ 15ε
	Welded	c/t ≤ 10ε	c/t ≤ 14ε
3	Rolled	c/t ≤ 15ε	c/t ≤ 21ε
	Welded	c/t ≤ 14ε	c/t ≤ 20ε
$\varepsilon = \sqrt{235/f_y}$	f_y (N/mm²)	235	275	355
	ε	1.0	0.92	0.81

TABLE 6.5   Limitation to the Local Slenderness of AISC: Rolled I Sections

	c/t (flange)	d/t (web)
Seismic	$52/\sqrt{f_{y.s}}$	$520/\sqrt{f_{y.s}}$
	$(137/\sqrt{f_{y.s}})$	$(1365/\sqrt{f_{y.s}})$
Compact	$65/\sqrt{f_{y.s}}$	$640/\sqrt{f_{y.s}}$
	$(171/\sqrt{f_{y.s}})$	$(1680/\sqrt{f_{y.s}})$
Non-compact	$141/\sqrt{f_{y.s}-10}$	$970/\sqrt{f_{y.s}}$
	$(370/\sqrt{f_{y.s}-10})$	$(2547/\sqrt{f_{y.s}})$

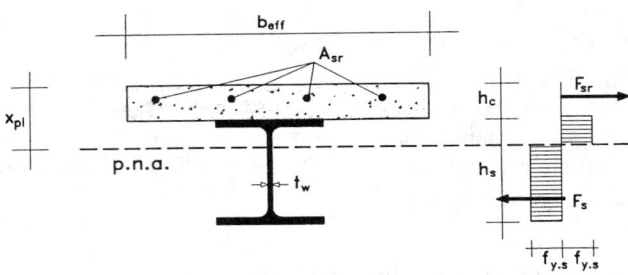

Figure 6.20   Plastic stress distribution under hogging moment.

the equilibrium to the translation in the direction of the beam axis. Usually the neutral axis lies in the steel web, and the value of x_{pl} is given by the following expression:

$$x_{pl} = \frac{h_s}{2} + h_c - \frac{F_{sr}/2}{t_w f_{y.s}} \tag{6.33}$$

where F_{sr} is the plastic strength of the reinforcement:

$$F_{sr} = A_{sr} \cdot f_{y.sr} \tag{6.34}$$

The evaluation of the plastic moment is then carried out by imposing the equilibrium of the cross-section to the rotation respect to the neutral axis:

$$M_{pl} = M_{pl.s} + F_{sr} \cdot \left(\frac{h_s}{2} + h_c - c \right) - \frac{F_{sr}^2}{4 t_w f_{y.s}} \tag{6.35}$$

where c is the concrete cover. It can be observed that the form of Equation 6.35 giving the hogging moment resistance M_{pl} is very similar to one Equation 6.25a obtained for the section in sagging bending, i.e., the plastic bending capacity of the composite section may be seen as the sum of the two contributions: the plastic moment of the steel section and the moment of the steel reinforcement.

The design value of M_{pl} can be obtained as:

- Eurocode 4:

$$M_{pl.d} = \frac{M_{pl.s}}{\gamma_s} + \frac{F_{sr}}{\gamma_{sr}} \cdot \left(\frac{h_s}{2} + h_c - c \right) - \frac{\gamma_s}{\gamma_{sr}^2} \frac{F_{sr}^2}{4 t_w f_{y.s}} \tag{6.36}$$

- AISC

$$M_{pl.d} = 0.85 \cdot M_{pl} \tag{6.37}$$

The AISC specifications also cover the case of continuous members, where composite action is developed only in sagging moment, while the steel section alone resists in negative moment regions. In this case, consistently with specifications for steel members, the hogging design plastic resistance would be computed as:

$$M_{pl.d} = 0.90 M_{pl.s} \tag{6.38}$$

In the region of negative moment, the ultimate bending resistance might be reduced by "high" vertical shear. In the usual design model, which neglects the contribution of the slab to the vertical shear resistance (see Equation 6.26), the same upper limit value may be assumed for the applied shear as in steel sections, which exceeded the ultimate moment of resistance reduced by shear force. When shear buckling of the steel section web is not critical, this limit value is defined as a percentage of the plastic shear resistance V_{pl}.

The Eurocode 4 specifies it as $0.5 V_{pl}$, with V_{pl} defined as in Equation 6.26: if the design shear V is higher than $0.5 V_{pl}$, a part of the web is assumed to carry the shear force, therefore, a fictitious reduced yield strength $f_{y.s.r}$ is used in the determination of the web contribution to the bending resistance:

$$f_{y.s.r} = f_{y.s} \cdot \left[1 - \left(\frac{2 \cdot V}{V_{pl}} - 1 \right)^2 \right] \tag{6.39}$$

where
$f_{y.s}$ = the yield strength of the web material
When $V = V_{pl}$, the bending resistance of the cross-section is equal to the plastic moment capacity of the part of the cross-section remaining after deduction of the web.

6.4.6 Serviceability Limit States

Global Analysis

Elastic calculation of the bending moment distribution under service load combinations is the preliminary step of the design analysis aimed at checking the member against serviceability limit states. As already mentioned, this verification, for continuous composite beams, has to consider many different aspects. A problem arises in performing the elastic analysis, due to change in stiffness of the hogging moment region caused by slab cracking. This problem is different than in reinforced concrete members: the stiffness reduction of a composite member, due to concrete cracking, takes place only in the hogging moment zone and it is very important resulting in a significant redistribution of moments, while in reinforced concrete continuous beam cracking occurs in positive moment zones as well; hence, the associated moment redistribution is usually not so remarkable.

In order to determine the bending moment distribution, the following three procedures may be adopted, which are presented in order of decreasing difficulty:

1. A non-linear analysis accounting for the tension stiffening effect in the cracked zone, and the consequent contribution to the section stiffness of the concrete between two adjacent cracks due to transferring of forces between reinforcement and concrete by means of bond. The effect of the slip between steel and concrete should also be taken into account in the case of partial shear connection.

2. An elastic analysis that assumes the beam flexural stiffness varies as schematically shown in model "a" of Figure 6.21: in the negative moment zone of the beam, where the moment is higher than the cracking one, the "cracked" stiffness EI_2 is used, while the "uncracked" stiffness EI_1 characterizes the remainder of the beam. In order to further simplify the procedure, the length of the cracked zone can be pre-defined as a percentage of α of the span l. Eurocode 4 recommends a cracking length equal to $0.15\, l$.

3. An elastic uncracked analysis based on model "b" shown in Figure 6.21, which considers for the whole beam the "uncracked" stiffness EI_1 and accounts for the effect of cracking by redistributing the internal forces between the negative and positive moment regions; the redistribution allowed by the codes ranges from 10 to 15%.

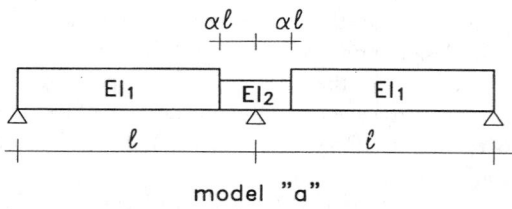

model "a"

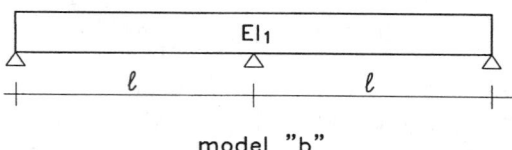

model "b"

Figure 6.21   Stiffness distribution models for composite continuous beam design analysis.

The choice of the stiffness distribution model is a key design issue; in fact, a model that underestimates the beam stiffness over-estimates both the deflections and the moment redistribution, i.e., the procedure is on the safe side for the deformability control, but on the unsafe side for the resistance verification of the cross-sections.

The refined non-linear analysis (method 1 in the above) is particularly complex, and it is not covered in this chapter. Reference can be made to the literature (see for example [11, 12, 31]).

If method 2 is adopted, the analysis can be performed in a simple way by usual computer programs: an intermediate node should be added at the location of the cross-section where the moment is equal to the cracking moment and the beam state changes from the uncracked state (inertia I_1) to the cracked state (inertia I_2). When the value of the cracked length is not pre-defined, the position of this intermediate node varies with the applied load, due to the redistribution of moments.

As an alternative, a beam element can be used, the stiffness matrix of which takes into account the influence of cracking [13].

If this last approach is used, it is possible to develop simplified formulations for assessing the design value of moment redistribution to be adopted in associated uncracked analysis. By means of an example based on a continuous beam with two equal spans, Cosenza and Pecce [13] pointed out that the moment redistribution on the central support could be calculated by means of the following expression:

$$\frac{M_r}{M_e} = \frac{0.890 + 0.110i}{0.614 + 0.386i} \tag{6.40}$$

where

M_r = the moment after the redistribution
M_e = the elastic moment computed for the "uncracked" beam (Figure 6.22)
i = the ratio I_1/I_2

Figure 6.22 Bending moment diagram before and after redistribution.

This moment redistribution ratio, which is based on a value of α equal to 0.15, can be compared to the following formula:

$$\frac{M_r}{M_e} = i^{-0.35} \tag{6.40a}$$

provided by Eurocode 4.

The good agreement between Equation 6.40 and 6.40a is shown in Table 6.6.

TABLE 6.6 Evaluation of the Moment Redistribution
Due to Cracking

		I_1/I_2			
		1	2	3	4
M_r/M_e	Equation 6.40	1	0.801	0.688	0.616
	Equation 6.40a	1	0.785	0.681	0.616

Stress Limitation

As already mentioned in Section 6.3.6, high stresses in the materials under service loads have to be prevented. High compression in concrete could cause microcracking and, consequently, durability problems; moreover, the creep effect can be very high, and even exceed the range of applicability of the linear theory with an unexpected increase of deflections. Analogously, yielding of steel in tension may lead to excessive beam deformation and increase crack widths in the hogging moment regions, resulting in a greater importance of rebars corrosion. Some codes provide limit stress levels for steel and concrete[4], which should be considered as reference values to be appraised against specific design conditions.

Deflections

The most advantageous feature of composite continuous beams is the lower deformability with respect to simply supported beams. The greatest overall stiffness enables use of more slender floor systems, which meet serviceability deflection requirements due to continuity effect. The problem arises as to the accuracy of the determination of the beam deflection, which depends on the model adopted in the analysis. Some indication on the opposite influences of continuity and slab cracking on maximum deflections is useful. Some results are presented here with reference to a two span beam subjected to uniform loading. Spans may be different. By means of simple calculations, the following expressions of the midspan deflection can be obtained:

$$\frac{\delta}{\delta_{ss}} = \frac{0.129 + 0.285i + 0.001i^2}{0.614 + 0.386i} \tag{6.41}$$

$$\frac{\delta}{\delta_{i=1}} = \frac{0.311 + 0.687i + 0.002i^2}{0.614 + 0.386i} \tag{6.42}$$

which link the midspan deflection δ of the continuous beam with δ_{ss}; which is, the deflection of a simply supported beam with the same mechanical and geometric characteristics; and $\delta_{i=1}$, which is the deflection evaluated considering the inertia I_1 constant along the whole of the beam. The first expression represents the reduction of deformability due to the continuity effect with respect to the simply supported beam; the second provides the increase of deformability due to cracking. Equation 6.42 can be written also in a simplified form: Cosenza and Pecce [13] proposed the following formula:

$$\frac{\delta}{\delta_{i=1}} = i^{0.29} \tag{6.43}$$

The comparison between Equations 6.41, 6.42, and 6.43 is reported in Table 6.7. Despite the significant adverse effect of the concrete cracking (the associated increase of deflection ranges from 27 to 43%), the reduction of deflection due to continuity remains substantial and greater for all cases considered to 40%.

TABLE 6.7 Calculation of Deflections in Continuous Beams

	I_1/I_2			
	1	2	3	4
δ/δ_{ss}	0.415	0.507	0.560	0.588
$\delta/\delta_{i=1}$	1	1.273	1.349	1.432
$i^{0.29}$	1	1.222	1.375	1.494

The influence of cracking should be considered as well in the case of long term effects (creep and shrinkage, see Section 6.3.6). A comparison between different methodologies is provided by Dezi and Tarantino [16, 17].

---

[4]Concrete stress is restricted to 50 to 60% of f_c, and stresses in the structural steel to 0.85 to 1.0 $f_{y.s}$.

On the other hand, another effect that reduces the deflection is "tension stiffening", i.e., the stiffening effect given by the concrete in tension between the cracks. This effect was analyzed in several studies (see for example [12, 26]), and it may be advantageously included in design when the associated additional complexity of calculations is justified by the importance of the design project.

Control of Cracking and Minimum Reinforcement

The width of the cracks caused in the slab by negative moments has to be checked if one or both of the following conditions are present:

- durability problems as in an aggressive environment
- aesthetic problems

Usually in buildings both problems are negligible because the environment internal to the building is rarely aggressive and, furthermore, the finishing of the floor somehow protects the slab against, and hides, the cracks. However, if durability and/or aesthetic problems exist, control is necessary. The crack width can be reduced by the following design criteria:

- using reinforcing bars with small diameters and spaced relatively closely
- restricting the stress in the reinforcement
- choosing the amount of reinforcement adequately, in order to avoid a very critical situation where the cracking moment is greater than the moment that leads to the yielding of the reinforcement

The nominal design value of the crack width is usually restrained to values ranging from 0.1 to 0.5 mm according to the various environmental situations. Direct evaluation of the crack width is obtained by multiplying the crack distance by the average strain in the reinforcement. This approach it is recommended in aggressive environments. For its application, reference can be made to Eurocode 2 [1992]. The direct evaluation can be avoided if a minimum reinforcement between 0.4 and 0.6% of the concrete slab is employed. Moreover, the crack width, w, can be restricted to 0.3 (moderately aggressive environment) or to 0.5 mm (little aggressive environment) if the stress limitations under service loads, reported in Table 6.8 as function of the rebars diameter (in mm), are fulfilled.

TABLE 6.8   Maximum Stress in the Steel
Reinforcement to Limit Crack Width

σ_s N/mm^2	Rebar diameter (mm)	
	$w = 0.3$ mm	$w = 0.5$ mm
160	32	36
200	25	36
240	20	36
280	16	30
320	12	22
360	10	18
400	8	14
450	6	12

Additional Techniques to Limit Serviceability Problems

In some cases, the flexural capacity under negative moments cannot be sufficient with reference to serviceability limit states. Consequently, appropriate constructional methods can be adopted to solve these problems. In particular:

- Pre-imposed deformation at the intermediate supports can be applied to modify the moment distribution values of the hogging moment region. As a result, the maximum positive moment in the spans increases and the cross-section under positive moment has to be checked.
- The slab in the negative moment zone of the beam can be prestressed, either by the usual post-tension reinforcement included in a duct or by external prestressing cables.

In both construction techniques, which are generally adopted in bridge construction, creep and shrinkage of concrete reduce the effects of imposed deformations. Therefore, these phenomena should be taken into account in design, also by means of simplified calculation procedures [16, 17]. It is evident that the aforementioned techniques provide benefits with reference to the ultimate limit states as well.

6.4.7 Ultimate Limit State

Different methods can be adopted to analyze the structure in ultimate conditions, the main features of which are summarized in the following.

Plastic Analysis

The requirement that all relevant cross-sections are plastic or compact may not be sufficient for a composite beam to achieve the plastic collapse condition. It has been proven [27] that the rotational capacity is sufficient to develop the collapse mechanism only if further particular limitations are meet as to the structural regularity, the loading pattern, and the lateral restraint. The limitations, more in detail, are the following:

- Adjacent spans do not differ in length by more than 50% of the shorter span, and end spans do not exceed 15% of the length of the adjacent span.
- In any span in which more than half of the total design load is concentrated within a length of one-fifth of the span, at any hinge location under sagging moment, no more than 15% of the overall depth of the member is in compression.
- The steel flange under compression at a plastic hinge location is laterally restrained.

If these requirements are fulfilled, the limit design approach can be applied in design analysis. In this case, at the ultimate limit condition in the external spans, there is the following relation between the total applied load, p, and the negative and positive plastic moment of resistance of the beam (Figure 6.23a):

$$M_{pl}^{(+)} + \frac{1}{2}M_{pl}^{(-)} \cong \frac{pl^2}{8} \tag{6.44}$$

Whereas in the intermediate spans, it is (Figure 6.23b)

$$M_{pl}^{(+)} + M_{pl}^{(-)} = \frac{pl^2}{8} \tag{6.45}$$

The static advantages are remarkable with respect to the case of a simply supported beam. For example, Equation 6.44 suggests for a beam with one fixed end and one supported end that has the same type of behavior of a continuous beam with two symmetrical spans, an increment of the ultimate load capacity of approximately $0.5\,M_{pl}^{(-)}$. However, the redistribution of bending moment required to cause the formation of the plastic mechanism is very large; the degree of redistributions is defined as the reduction of the elastic negative moment necessary to obtain the final moment in the mechanism situation, i.e., $M_{pl}^{(-)}$ divided by the initial value of elastic moment:

$$r = \frac{pl^2/8 - M_{pl}^{(-)}}{pl^2/8} = 1 - \frac{M_{pl}^{(-)}}{pl^2/8} \tag{6.46}$$

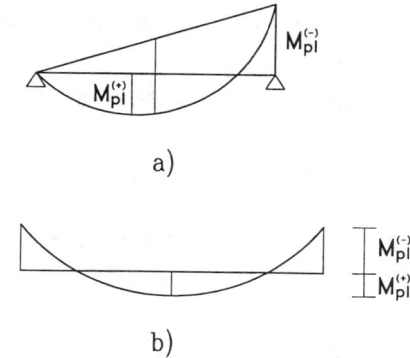

Figure 6.23   Structural model for plastic analysis. (a) External span. (b) Internal span.

The degree of redistribution also depends on the ratio, m, between the negative and positive plastic moments of resistance of the beam:

$$m = \frac{M_{pl}^{(-)}}{M_{pl}^{(+)}}$$ (6.47)

In the case of a symmetrical two-span continuous beam under uniform loading, the use of Equations 6.44 through 6.46 enables the following relation to be established between r and m:

$$r = \frac{2 - m}{2 + m}$$ (6.48)

This relation provides the value of the redistribution, r, necessary to achieve the full mechanism condition in a continuous beam as a function of the ratio, m, between the plastic moments. The values of r associated with selected values of m are presented in Table 6.9. Since m is always less than 1,

TABLE 6.9   Relationship Between Redistribution Degree and Plastic Moment Ratio to Achieve the Plastic Mechanism Condition

m	0.1	0.2	0.3	0.4	0.5	0.6	0.7	0.8	0.9	1
r	90%	82%	74%	67%	60%	54%	48%	43%	38%	33%

except if very large amounts of reinforcement are placed in the slab (the practical values of m vary between 0.4 and 0.7), r results between 50% and 70%. Thus, in many instances, the development of a full plastic mechanism requires such a high moment redistribution that the corresponding service loads would almost cause problems of excessive stress, deflections, cracking, i.e., the serviceability limit states would govern design.

Non-Linear Analysis

If the limitation on the regularity and loading conditions required for applying the plastic analysis are not met, a large redistribution can still be considered up to the formation of a mechanism, if a refined non-linear analysis is performed. As already mentioned, refined methods of analysis are available in the literature [11, 14, 31], while detailed recommendations for use in practice are not yet available. As a general step-by-step procedure, the following approach can be outlined:

1. Cracking of the negative moment zone is accounted for and inelasticity is concentrated in the relevant locations. The plastic rotations associated with the ultimate design load combination, or the mechanism formation, are then determined.

2. The required plastic rotations are compared to the allowable ones.

A refined analysis also allows performance in the serviceability conditions to be checked, since the response of the beam is followed by a step-by-step procedure.

Linear Analysis with Redistribution

The simplified design approach, which combines elastic linear analysis with redistribution of internal forces, can be adopted also for the verifications at ultimate. The amount of redistribution allowed depends on:

- The type of linear analysis: if an uncracked analysis has been performed, the allowable redistribution is higher, since it also has to take into account the effect of cracking.
- The class of the sections: the available plastic rotations are different for plastic, compact, or semi-compact sections, so the allowable redistribution is also different.

Eurocode 4 specifies maximum allowed degrees of the redistribution shown in Table 6.10. The differences between the redistribution accepted in uncracked analysis and the redistribution accepted in cracked analysis is the estimated redistribution due to cracking. This difference varies between 10 and 15%. It is worth noticing that the maximum redistribution is 40% in the uncracked analysis and 25% in the cracked analysis. These values may appear very high, but as already observed in Section 6.4.7 for attaining the plastic sagging moment at midspan (which is usually much higher than the negative one), an even larger redistribution is necessary; on the other hand, higher values of the redistribution are not compatible with the verifications at the serviceability condition.

TABLE 6.10   Allowable Redistribution in Linear Analysis

	class 1	class 2	class 3	class 4
Cracked analysis	25%	15%	10%	-
Uncracked analysis	40%	30%	20%	10%

6.4.8   The Lateral-Torsional Buckling

Under hogging moments, a substantial part of the steel cross-section is stressed in compression, and lateral-torsional buckling can occur. The restraint offered by the concrete slab, which is usually very stiff and continuously connected to the steel section through the shear connectors, prevents this type of buckling from developing in the usual mode characterized by the lateral and torsional displacement of the member. Lateral-torsional buckling in composite beams takes the form of "lateral-distortional" buckling, as described in Figure 6.24.

This very complex problem received attention by a few researchers, who developed sophisticated methods of analysis [6, 22, 37, 48].

From a practical point of view, the usual formulations for lateral-torsional buckling can be applied with appropriate modifications to account for the restraint offered by the slab. For example, Eurocode 4 provisions suggest the same approach of the usual lateral-torsional buckling varying the buckling length of the beam in a suitable way. Such a simplified approach is not applicable to sections with high depth for which a numerical analysis is necessary. In most instances the critical load associated with distortional buckling is usually higher than the "lateral-torsional" one and this problem is often negligible for the shapes commonly used in building structures. A parametric analysis [ECCS, 1993] showed that IPE and HE types, characterized by depths higher than the values provided by Table 6.11, have to be checked against distortional buckling.

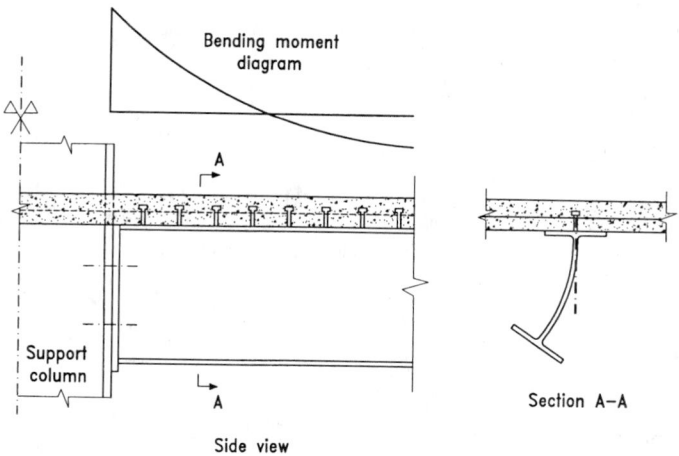

Figure 6.24 Distortional buckling of composite beam in a hogging moment zone.

TABLE 6.11 Limit Values of Steel Section Depth to Avoid Distortional Buckling

Steel member	$d_{s,\lim}$ (mm)	
	$f_{y,s} < 240$ N/mm²	$f_{y,s} < 360$ N/mm²
IPE	600	400
HEA	800	650
HEB	900	700
Encased beam	1000	1000

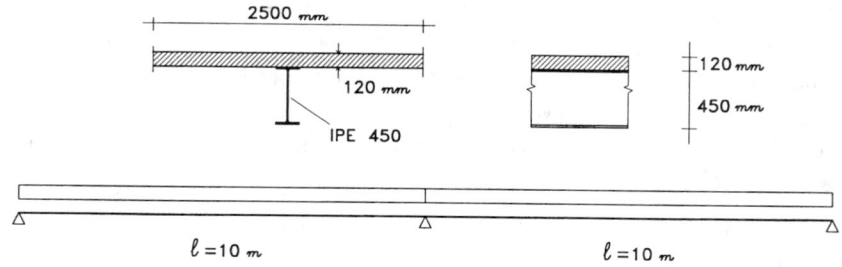

Figure 6.25 Cross-section of a continuous beam with two equal spans.

6.4.9 Worked Examples

EXAMPLE 6.3:

In Figure 6.25 it is reported that the cross-section of a continuous beam with two equal spans has a span length of l equal to 10 m; the steel profile (IPE 450) is characterized by $A_s = 9880$ mm², $I_s = 3.37 \cdot 10^8$ mm⁴, $t_w = 9.4$ mm, $t_f = 14.6$ mm. The solid slab has a thickness of 120 mm; the spacing with adjacent beams is 5000 mm. The reinforcement of slab in the negative moment zone is realized by bars of diameter 14 mm with interspacing 250 mm. The beam is subjected to a uniform load p = 40 kN/m (16 kN/m of dead load, 24 kN/m of live load) at the serviceability condition. The material has the following design strength:

- concrete 14.2 N/mm^2 (including the coefficient 0.85)
- structural steel 213.6 N/mm^2
- reinforcement steel 365.2 N/mm^2

For the service state, the design load is

$$p = 16 + 24 = 40 \text{ kN} / \text{m}$$

For the ultimate limit state, considering a factor of 1.35 for the dead load and 1.5 for the live load, it is

$$p = 1.35 \cdot 16 + 1.5 \cdot 24 = 57.6 \text{ kN} / \text{m}$$

1. Determination of design moment:
   Considering the "uncracked analysis", the negative design moment for service conditions is

$$M^{(-)} = \frac{40 \cdot 10^2}{8} = 500 \text{ kNm}$$

   At the ultimate state it is

$$M^{(-)} = \frac{57.6 \cdot 10^2}{8} = 720 \text{ kNm}$$

- Ultimate limit state

1. Evaluation of the effective width (Section 6.4.2):
   Applying Equation 6.4, it results:

$$b_{\text{eff}} = 2 \cdot \frac{1}{8} = 2 \cdot \frac{10000}{8} = 2500 \text{ mm}$$

   Thus, only 2500 of 5000 mm are considered as effective and $A_{sr} = 1540 \text{ mm}^2$.

2. Plastic negative moment resistance:
   The plastic resistance of the section in the negative moment is evaluate according to Section 6.4.5

$$
\begin{aligned}
F_{sr} &= 1540 \cdot 365.2 = 5.62 \cdot 10^5 \text{ N} \\
x_{pl} &= \frac{450}{2} + 120 - \frac{5.62 \cdot 10^5}{2 \cdot 9.4 \cdot 213.6} = 205 \text{ mm} \\
M_{pl.s} &= 363.5 \cdot 10^6 \text{ N} = 363.5 \text{ kNm} \\
M_{pl}^{(-)} &= 363.5 \cdot 10^6 + 5.62 \cdot 10^5 \cdot \left(\frac{450}{2} + 120 - 30 \right) \\
&\quad - \frac{\left(5.62 \cdot 10^5 \right)^2}{4 \cdot 9.4 \cdot 213.6} = 501 \cdot 10^6 \text{ Nm} = 501 \text{ kNm}
\end{aligned}
$$

3. Plastic shear resistance of the cross-section:

$$
\begin{aligned}
A_v &= 1.04 \cdot 450 \cdot 9.4 = 4399 \text{ mm}^2 \\
V_{pl} &= 4399 \frac{213.6}{\sqrt{3}} = 542494 \text{N} = 542 \text{ kN}
\end{aligned}
$$

Safely the maximum shear at the ultimate state is evaluated by the uncracked analysis without considering the redistribution of bending moment; it is

$$V = \frac{57.6 \cdot 10}{2} + \frac{720}{10} = 360 \text{ kN}$$

Thus, the shear verify is satisfied.

4. Bending-shear interaction:
   V results greater than $V_{pl}/2$; thus, the strength of the web is reduced by the following factor:

$$-\left(2 \cdot \frac{V}{V_{pl}} - 1\right)^2 = 1 - \left(2 \cdot \frac{360}{542} - 1\right)^2 = 0.892$$

thus leading to a design reduced strength:

$$f_{y.s.d.r} = 0.892 \cdot 213.6 = 190 \text{ N} / \text{mm}^2$$

The evaluation of the reduced moment resistance results:

$$x_{pl} = \frac{450}{2} + 120 - \frac{5.62 \cdot 10^5}{2 \cdot 9.4 \cdot 191} = 188 \text{ mm}$$

$$M_{pl}^{(-)} = 363.5 \cdot 10^6 + 5.62 \cdot 10^5 \cdot \left(\frac{450}{2} + 120 - 30\right) - \frac{(5.62 \cdot 10^5)^2}{4 \cdot 9.4 \cdot 190} = 497 \text{ kNm}$$

5. Check of the section class (Section 6.4.3):

   - for the flange

$$c/t_f = 69.3/14.6 = 4.7 < 10 \cdot 1 \rightarrow \text{ class 1}$$

   - for the web

$$\alpha = \frac{450 - 205 - 35.6}{378.8} = 0.55$$

$$d/t_w = 378.8/9.4 = 40.3 < 396/(13 \cdot 0.55 - 1) = 64.4 \rightarrow \text{ class 1}$$

   Thus, the steel profile is in class 1 (Tables 6.3 and 6.4).

6. Plastic positive moment resistance:
   By means of Equations 6.13 and 6.14 it is

$$F_{c.max} = 2500 \cdot 120 \cdot 14.2 = 4260000 \text{ N} = 4260 \text{ kN}$$

$$F_{s.max} = 9880 \cdot 213.6 = 2110368 \text{ N} = 2110 \text{ kN}$$

Consequently, it is assumed:

$$F_c = F_s = 2110 \text{ kN}$$

and, considering the design strength in Equation 6.17:

$$x_{pl} = \frac{9880 \cdot 213.6}{2500 \cdot 14.2} = 59.4 \text{ mm}$$

the neutral axis lies on the slab. The internal arm is (Equation 6.18):

$$h^* = \frac{450}{2} + 120 - \frac{59.4}{2} = 315 \text{ mm}$$

and the design plastic moment results (Equation 6.19):

$$M_{pl}^{(+)} = 2110 \cdot 0.315 = 665 \text{ kNm}$$

7. Required redistribution:
   To satisfy the verify to plastic negative resistance, the following redistribution is required:

$$r = 1 - \frac{497}{720} = 31\%$$

that is, less than the value 40% that is possible for class 1 sections and uncraked analysis (Table 6.10). In the midspan the maximum positive moment, by equilibrium, is

$$M_d^{(+)} = 238 \cdot 4.1 - \frac{57.6 \cdot 4.1^2}{2} = 492 \text{ kNm}$$

assuming the reaction of the support and the abscissa of maximum moment the following values:

$$R = \frac{57.6 \cdot 10}{2} - \frac{497}{10} = 238 \text{ kN} \qquad z_{M_{max}} = \frac{238}{57.6} = 4.1 \text{ m}$$

It results:

$$M_d^{(+)} = 492 \text{ kNm} < M_{pl}^{(+)} = 665 \text{ kNm}$$

8. Flexural-torsional buckling:
   In this case the control is not necessary because the steel profile depth is lower than the limit depth provided in Table 6.11 for IPE sections ($h_s = 450$ mm $< d_{s.\lim} = 600$ mm).

- Serviceability limit state.
  The following assumptions are made:

$$E_c = 30500 \text{ N / mm}^2$$
$$E_s = 210000 \text{ N / mm}^2$$

Therefore, at the short time the modular ratio results:

$$n = \frac{210000}{30500} = 6.89$$

Neglecting the contribution of the reinforcement, the state 1 uncracked is characterized by the following parameters:

$$A^* = \frac{9880 \cdot 2500 \cdot 120/6.89}{9880 + 2500 \cdot 120/6.89} = 8053 \text{ mm}^2$$

$$I_1 = \frac{1}{6.89} \frac{2500 \cdot 120^3}{12} + 3.37 \cdot 10^8 + 8053 \frac{(450 + 120)^2}{4} = 1.04 \cdot 10^9 \text{ mm}^4$$

After concrete cracking in the negative moment zone, it is necessary to consider the cracked state, leading to:

$$x_e = 30 + \frac{9880 \cdot (450/2 + 120 - 30)}{9880 + 1540} = 302 \text{ mm}$$

$$I_2 = 1540 \cdot (302 - 30)^2 + 9880 \cdot (302 - 120 - 450/2)^2 + 3.37 \cdot 10^8$$
$$= 4.69 \cdot 10^8 \text{ mm}^4$$

Thus, the parameter i given in Section 6.4.6 is:

$$i = \frac{1.04 \cdot 10^9}{4.69 \cdot 10^8} = 2.2$$

For the long term calculation, a creep coefficient $f = 2$ is assumed obtaining the following modular ratio:

$$n_{ef} = 6.89 \cdot 3 = 20.67$$

and then, in the uncracked state:

$$x_e = \frac{120}{2} + \frac{9880}{9880 + 120 \cdot 2500/20.67} \frac{450 + 120}{2} = 175 \text{ mm}$$

$$A^* = \frac{9880 \cdot 2500 \cdot 120/20.67}{9880 + 2500 \cdot 120/20.67} = 5878 \text{ mm}^2$$

$$I_1 = \frac{1}{20.67} \frac{2500 \cdot 120^3}{12} + 3.37 \cdot 10^8 + 5878 \frac{(450 + 120)^2}{4} = 8.32 \cdot 10^8 \text{ mm}^4$$

In the cracked state, the section has no variation, thus, leading to:

$$i = \frac{8.32 \cdot 10^8}{4.69 \cdot 10^8} = 1.8$$

1. Redistribution due to cracking

   At short term ($t = 0$), according to Equation 6.40, it results:

   $$\frac{M_r}{M_e} = \frac{0.890 + 0.110 \cdot 2.2}{0.614 + 0.386 \cdot 2.2} = 0.77$$

   Using the simplified Equation 6.40a it is

   $$\frac{M_r}{M_e} = i^{-0.35} = 2.2^{-0.35} = 0.76$$

   obtaining very similar results. In the case of long term condition ($t = \infty$), the results are

   $$\frac{M_r}{M_e} = \frac{0.890 + 0.110 \cdot 1.8}{0.614 + 0.386 \cdot 1.8} = 0.83$$

   $$\frac{M_r}{M_e} = i^{-0.35} = 1.8^{-0.35} = 0.81$$

2. Stress control

   In the negative moment zone, considering the redistribution due to cracking, it is $t = 0$

   $$\sigma_s = \frac{500 \cdot 10^6 \cdot 0.77}{4.69 \cdot 10^8} \cdot (450 + 120 - 302) = 220 \text{ N} / \text{mm}^2$$

   $$\sigma_{sr} = \frac{500 \cdot 10^6 \cdot 0.77}{4.69 \cdot 10^8} \cdot (302 - 30) = 223 \text{ N} / \text{mm}^2$$

$t = \infty$

$$\sigma_s = \frac{500 \cdot 10^6 \cdot 0.83}{4.69 \cdot 10^8} \cdot (450 + 120 - 302) = 237 \, \text{N} / \text{mm}^2$$

$$\sigma_{sr} = \frac{500 \cdot 10^6 \cdot 0.83}{4.69 \cdot 10^8} \cdot (302 - 30) = 240 \, \text{N} / \text{mm}^2$$

3. Control of deflections:
   Considering Equation 6.41, it follows: $t = 0$

$$\frac{\delta}{\delta_{ss}} = \frac{0.129 + 0.285 \cdot 2.2 + 0.001 \cdot 2.2^2}{0.614 + 0.386 \cdot 2.2} = 0.52$$

$$\delta_{ss} = \frac{5}{384} \cdot \frac{40 \cdot 10000^4}{210000 \cdot 1.04 \cdot 10^9} = 24 \, \text{mm}$$

and then:

$$\delta(t = 0) = 0.52 \cdot 24 = 12 \, \text{mm} \quad \delta = \frac{1}{833} \cdot l$$

The deflection at long term is increased by the creep; the effect of shrinkage in the hyperstatic system is considered negligible. Safely the entire live load is considered as long term load. The following result is obtained ($i = 1.8$):

$$\frac{\delta}{\delta_{ss}} = \frac{0.129 + 0.285 \cdot 1.8 + 0.001 \cdot 1.8^2}{0.614 + 0.386 \cdot 1.8} = 0.49$$

$$\delta_{ss} = \frac{5}{384} \cdot \frac{40 \cdot 10000^4}{210000 \cdot 8.32 \cdot 10^8} = 30 \, \text{mm}$$

$$\delta(t = \infty) = 0.49 \cdot 30 = 15 \, \text{mm} = \frac{1}{680} l$$

4. Control of cracking and minimum reinforcement:
   Using Table 6.8, the crack width results is less than 0.3 mm; the interpolation of the limitations referred to diameter 16 and 12 leads to a stress limit of 300 N/mm^2 that is higher than the reinforcement stress at the serviceability condition.

6.5 The Shear Connection

The mutual transfer of forces between the steel and concrete components is the key mechanism, which makes composite action possible. This mechanism generally involves a complex combination of forces (or stresses) acting at the steel concrete interface. In design, the main attention is focused on the forces (or stresses) parallel to interface, i.e., on the longitudinal shear forces (and stresses). The components of the interface forces perpendicular to the interface, which may play a significant role in the transfer mechanism, are principally considered through the selection of suitable detailing.

It can be stated that the shear connection is a factor of substantial importance and, in many instances, it permits achievement of the required performance. Therefore, a substantial research effort has been devoted to the development of the fundamental knowledge of the response and performance of the different shear connections available or proposed for practical use. The basics of this knowledge are presented in this chapter, with main reference to provide design oriented information.

Preliminary to any treatment of the behavioral features and design criteria of the shear connection, it seems convenient to give some useful definitions and classifications based on the key behavioral parameters of stiffness, strength, and ductility:

- Stiffness: a shear connection realizes either full interaction (the connection is "rigid" and no slip occurs under stress at the steel-concrete interface) or partial interaction (the connection is flexible and interface slip occurs).
- Resistance: when the overall resistance of the connection can be conveniently considered as in plastic design, a full connection has the shear strength sufficient to make the composite structural element (beam or slab) to develop its ultimate flexural resistance before collapse is achieved. If this condition is not fulfilled, the connection is a partial connection. A structural element with full shear connection is a fully composite structural element. A structural element with partial shear connection is a partially composite structural element. The ratio $F_c/F_{c.f}$ between the resistance of the shear connection and the minimum resistance required by the full connection condition defines the degree of shear connection.
- Finally, a connection is ductile if its deformation (slip) capacity is adequate for a complete redistribution of the forces acting on the individual connectors. The ductility demand depends on the span and the degree of shear connection.

The behavioral parameters relevant to the type of analysis adopted in design (i.e., elastic, inelastic, or plastic analysis) have to be considered. In particular, connection flexibility should be accounted for in elastic and inelastic analyses, which would make design rather complex. However, the simplified assumption of full interaction is satisfactory for most shear connections used in practice: the effect of slip is in fact negligible.

6.5.1 The Shear Transfer Mechanisms

Various forms of shear transfer can be identified for nature and effectiveness, namely:

- adhesion and chemical bond
- interface friction
- mechanical interlock
- dowel action

Shear transfer via adhesion and bond has the non-negligible advantage of being associated with no steel-concrete slip. However, tests show a rather low maximum shear resistance, which decreases rapidly and, remarkably, in the post-ultimate range of response. Moreover, this form of shear strength is highly dependent on factors, such as the quality of the steel surface and the concrete shrinkage, the control and quantification of which is difficult, if at all possible. Therefore, low values have to be assumed in design for the bond strength. Nevertheless, bond might be sufficient when the demand of interface shear capacity is limited as in composite columns (Figure 6.26a) or in fully encased beams (Figure 6.7d), at least in the elastic range (see also the AISC specifications at clause I1 and I3).

Eurocode 4 specifies the following values of the bond stress (including the effect of friction) to be considered when checking the connection effectiveness of composite columns:

- completely encased sections   0.6 N/mm^2
- concrete filled hollow sections   0.4 N/mm^2
- flanges in partially encased sections   0.2 N/mm^2
- webs in partially encased sections   zero

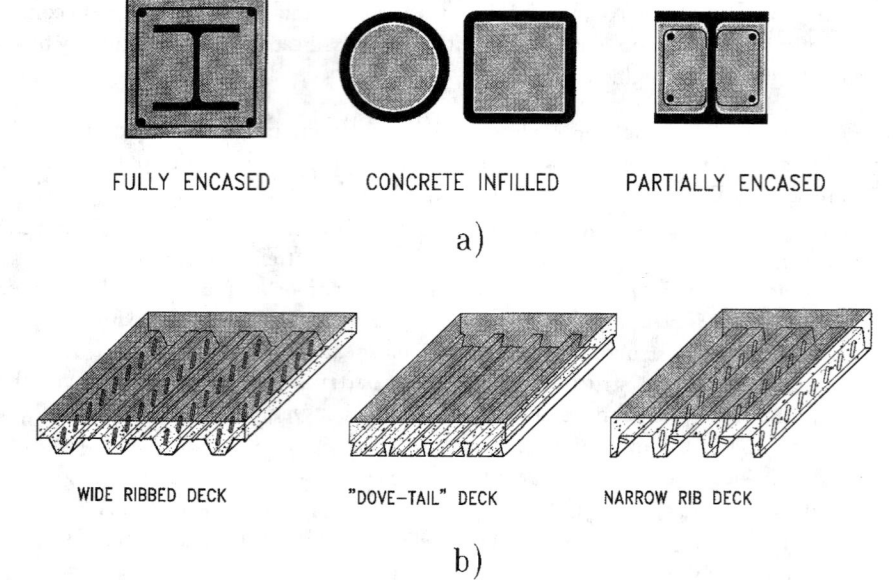

FULLY ENCASED   CONCRETE INFILLED   PARTIALLY ENCASED

a)

WIDE RIBBED DECK   "DOVE–TAIL" DECK   NARROW RIB DECK

b)

Figure 6.26   Composite system. (a) Composite columns. (b) Composite slab.

Friction is often associated to bond in resisting shear. In flexural members, the tendency of steel and concrete elements to separate usually makes frictional action rapidly deteriorate. A suitable geometry of the composite element, as in the composite slab of Figure 6.26b with "dove-tailed" profiled sheeting, prevents separation and allows frictional interlock to develop throughout the response.

Mechanical interlock is obtained by embossing the metal decking so that slip at the interface is resisted by bearing between the steel ribs and the concrete indentations. The effectiveness of the embossments depends on their geometrical dimensions (mainly the height and depth) and shape. Enhancement of the shear transfer capacity in composite slabs is achieved if frictional and mechanical interlock are combined. The complexity of these interlock shear transfer mechanisms dictates that the response of the shear connection is determined by appropriate tests (see Figure 6.27). As discussed in Section 6.7, standard testing procedures are specified in most relevant codes, as in the Eurocode 4 and the ASCE Standard for the Structural Design of Composite Slabs (1991).

The transfer of large shear forces dictates that suitable mechanical connectors are used. New types of connectors have been continuously developing since the early stages of composite construction (see also Figure 6.1). Therefore, an increasing variety of forms of shear connectors is available for practical use. Despite possible significant differences, they all act as steel dowels embedded in concrete, and hence apply a concentrated load to the concrete slab, the diffusion of which requires careful consideration in the design of the slab detailing (Figure 6.28). Figure 6.29 shows some of the connectors more frequently used, for most of which design rules are provided by the relevant codes. The headed stud is by far the most popular connector.

The behavior and modes of failure of each type of connector highly depend on the local interaction with the concrete, and can only be determined via ad hoc tests: the so-called *push-out tests* schematically shown in Figure 6.30a. Experimental shear force vs. slip curves are plotted in Figure 6.30b for a few connector types. All types of mechanical shear connectors possess a limited deformation capacity. However, in many instances, the associated slip is sufficient to make the design flexural resistance and rotation capacity of the composite section to be developed. If this condition is fulfilled, the connectors (and the connection) can be classified as ductile. As mentioned in the introduction to this section, the ductility requirements depend on the span and the degree

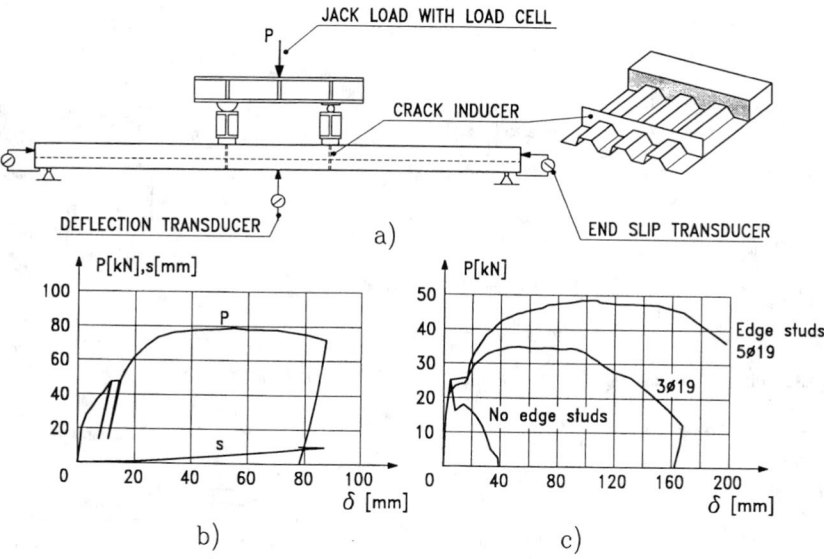

Figure 6.27   Slab responses after Bode and Sauerborn [5].

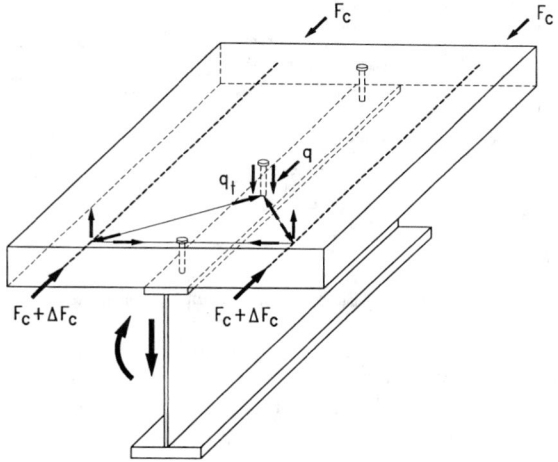

Figure 6.28   Shear transfer mechanism.

of shear connection: the classification of a connector as ductile should hence be associated with a definition of a combination of:

1. a range for these parameters
2. a characteristic value of slip capacity

Eurocode 4 assumes a characteristic slip capacity of 6 mm as the reference parameter in the calibration of the recommendations related to partial composite beams with ductile connectors. On the basis of an assessment of the available experimental data [29], it then classifies as *ductile* for the given ranges of span and degree of shear connection (see Section 6.5.7), only the friction grip bolts and the welded stud connectors meeting the following requirements:

- overall length H_{sc} after welding not less than 4 times the diameter d_{sc}
- $16 \text{ mm} \le d_{sc} \le 22 \text{ mm}$

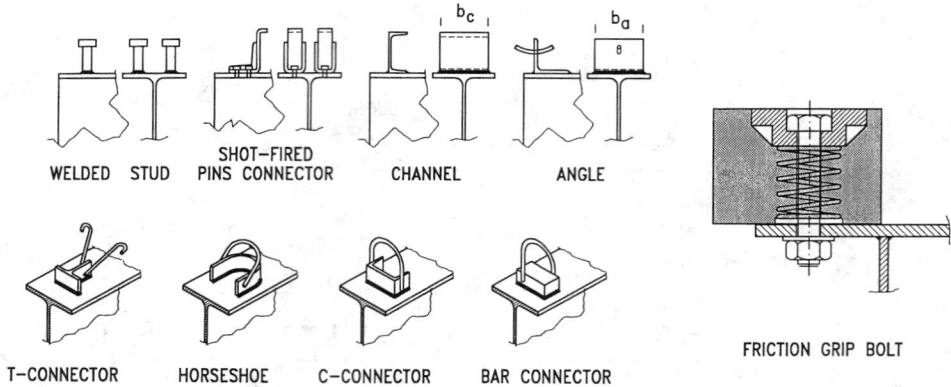

Figure 6.29   Typical shear connectors.

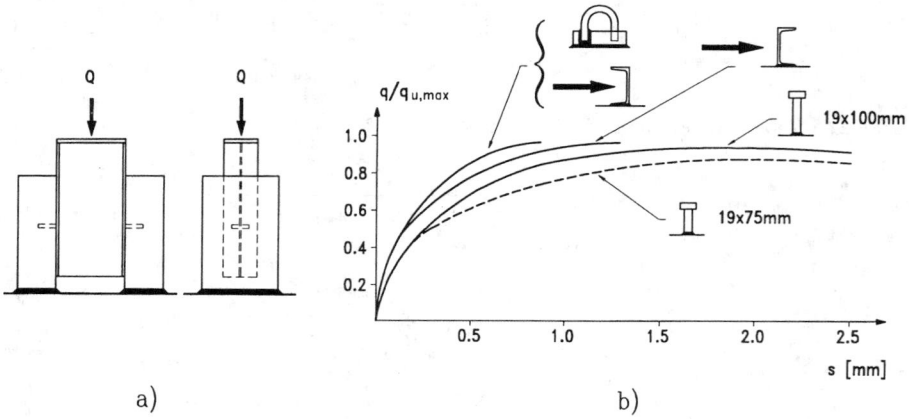

Figure 6.30   Test on shear connectors. (a) Push-out specimen. (b) Non dimensional load-slip relationship.

Despite the assumed limitations appearing to be rather strict, most of the shear connections in buildings fall into the ductile category. Therefore, current practice should not be largely affected.

Stud shear connectors may also be used to provide end anchorage in composite slabs: their effect is significant on both the resistance and the ductility of the shear connection as a whole (see Figure 6.27c).

6.5.2   The Shear Strength of Mechanical Shear Connectors

The performance of a mechanical shear connector depends on the numerous factors governing the mutually interactive response of the connector and the surrounding concrete. Despite the number of attempts, a numerical model enabling satisfactory simulation of the many facets of this complex problem has not yet been developed. Design should hence rely on the available sets of experimental data. Besides, due to the inherent difficulty of deducing the resistance of shear connectors from beam tests, most of the usable data come from the push-out tests, which do not entirely reproduce, in terms of loading and confinement from the slab, the more favorable connector's condition in a composite beam. A further problem arises from the lack of consistency among the geometrical and detailing features of push-out specimens designed differently in different research studies. This situation demands for a careful selection of "comparable" data, in the background studies to code specifications [30].

Only the stud connectors were so extensively investigated that a sufficient body of information has been built up to enable a truly statistical determination of the nominal, characteristic, and design values of the resistance. The knowledge of the load-slip response of other types of connector is far more limited and design rules are based also on successful practice.

This situation is well reflected by present codes, which cover shear connectors at a rather different extension. The AISC specifications provide design rules only for studs and channel shear connectors, i.e., to the most popular connectors utilized in buildings. Besides, even when a wide set of shear connectors is covered by a code, as in the case of Eurocode 4, recent types of connectors such as, for example, the shot-fired steel pins in Figure 6.29, cannot be treated. Suitable test programs should be carried out in order to define their performance in terms of design strength and slip capacity. Procedures for conducting and assessing the results of push-out tests are given in many standards as in Eurocode 4.

Stud Connectors Used in Solid Concrete Slabs

Both Eurocode 4 and AISC associate the resistance of a stud with the failure either of the concrete, which crushes in the zone at the lower part of the stud shank, or of the stud shank, which fractures directly above the weld collar under shear, flexure, and tension. Interaction between these two modes of failure is not explicitly accounted for in order to maintain design simplicity. The format of the design rules is the same, although the design values of the resistances are quite different, due partly to the different philosophies on which the two codes are based and partly to the different sets of data on which the adopted strength model was calibrated. The shear resistance q_u of an individual stud is the lesser of

$$q_{u.c} = k_c A_{sc} (f_c E_c)^{0.5} \tag{6.49}$$

$$q_{u.s} = k_s A_{sc} f_{u.sc} \tag{6.50}$$

where A_{sc} represents the cross-sectional area of a stud shear connector. The AISC specifications assume $k_c = 0.5$ and $k_s = 1.0$, while in Eurocode 4 $k_c = 0.36$ and $k_s = 0.8$. Moreover, the Eurocode limits to 500 N/mm^2 the value of the ultimate tensile strength $f_{u.sc}$ to be assumed in Equation 6.50, and it restricts the application of the resistance Equations 6.49 and 6.50 to studs with diameter not greater than 22 mm (7/8 in.).

Experimental results had proven that the height-to-diameter ratio for the stud (H_{sc}/d_{sc}) affects the resistance $q_{u.c}$: the full resistance (Equation 6.49) is developed only when $H_{sc}/d_{sc} \geq 4$. All studs in the study by Ollgaard et al. [38], on which AISC rules are based, satisfy this requirement. The AISC specifications apply only to studs with $H_{sc}/d_{sc} \geq 4$, even if this requirement is not explicitly stated. The European code aims at permitting use of a wider range of studs (studs with lower height may be conveniently used in shallow floor systems). Therefore, it specifies a reduction coefficient of the resistance $q_{u.c}$ expressed as $\alpha = 0.2[(H_{sc}/d_{sc}) + 1] \leq 1$. In any case, studs with $H_{sc}/d_{sc} < 3$ cannot be used.

A comparison between the two specifications is presented in Figure 6.31 with reference to a 19 mm (3/4 in.) stud with $H_{sc}/d_{sc} \geq 4$. Figure 6.32 illustrates the influence of the concrete density as by the Eurocode relation between this parameter and the modulus of elasticity E_c.

The response and strength of a stud connector welded to a rather thin flange may be affected by the deformation of the flange; a restriction to the stud diameter related to the flange thickness t_f is needed when the influence of this factor is not appraised experimentally. Both the AISC specifications and the Eurocode restrict d_{sc} to 2.5 times t_f, unless the stud is placed in correspondence to the beam web.

The Eurocode obtains the design resistance q_d by dividing q_u by a safety factor $\gamma_V = 1.25$. In the AISC-LRFD format, the uncertainty related to the stud resistance is included in the resistance factor ϕ_b affecting the beam moment capacity.

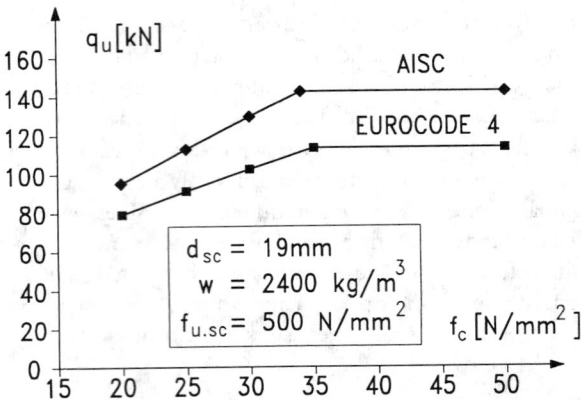

Figure 6.31 Stud connector shear resistance–concrete resistance relationship in accordance with AISC and Eurocode 4 provisions.

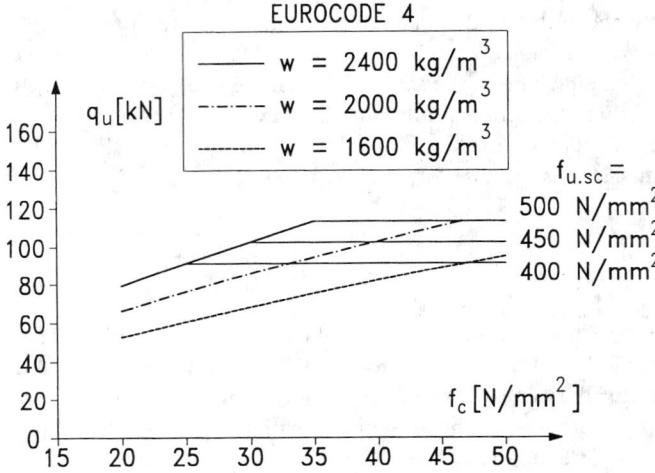

Figure 6.32 Influence of the concrete density on the stud connector shear resistance in accordance with Eurocode 4 provisions.

Stud Connectors Used with Profiled Steel Decking

The most popular solution for floor systems in composite framed construction makes use of decks where the profiled steel sheeting acts compositely with a concrete "ribbed" slab. The studs are placed within a rib, and their performance is fairly different than in the previous case of solid concrete slab. The concrete stiffness, degree of confinement, and the resistance mechanism of the studs which are loaded by a highly eccentrical longitudinal shear force are all different.

The prime parameters affecting the stud behavior are

- the orientation of the ribs relative to the beam span
- the rib geometry as characterized by the b_r / h_r ratio
- the stud height H_{sc} relative to the rib height h_r.

Proposals have been made to account for the influence of the relevant parameters on stud ultimate resistance. However, the available data do not enable a comprehensive design method to be developed. In codes, the effects of the main factors are accounted for via a suitable reduction factor.

Deck Ribs Oriented Parallel to Steel Beams

Studies conducted at Lehigh University in the 1970s [20] provide the sole background to this problem. The results indicated that the resistance of the stud in a rib parallel to the supporting beam can be determined by reducing the resistance in a solid slab Equations 6.49 and 6.50 by the factor

$$k_{rp} = 0.6 \cdot \frac{b_r}{h_r} \left[\frac{H_{sc}}{h_r} - 1.0 \right] \le 1.0 \qquad (6.51)$$

which mainly accounts for the limited restraint provided to the concrete by the sheeting side walls. This restraint is even negligible when the deck is split longitudinally at the beam, as shown in Figure 6.33. Good practice in this case would suggest meeting the requirements set for concrete haunches.

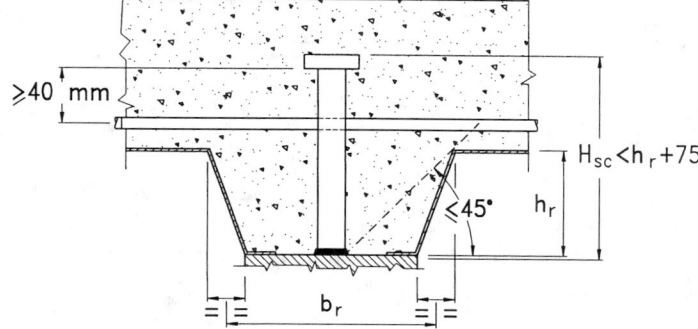

Figure 6.33   Geometrical restraint provided to the concrete by sheeting side walls.

Equation 6.51 appears in both codes. However, the AISC specifies that it applies only to very narrow ribs ($b_r/h_r < 1.5$), while the Eurocode does not give any range of application, but it limits H_{sc} to $h_r + 75$mm.

Deck Ribs Oriented Perpendicular to Steel Beams

The efficiency of the floor system may require the steel sheeting to be placed with the ribs transverse to the supporting beam. Figure 6.34 schematically illustrates the possible failure modes of the shear connection either by collapse of the stud connector in flexure, tension, and shear or by fracture of the concrete rib associated with local concrete crushing. This deck arrangement apparently involves

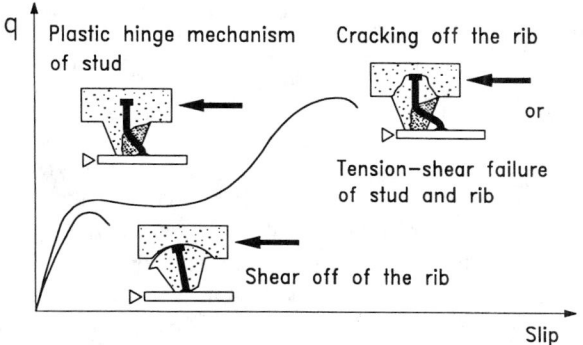

Figure 6.34   Failure modes for a stud connector in a deck with transverse ribs.

a concrete rib that is significantly stressed, as it acts as the transfer medium of the longitudinal shear between the concrete slab above the sheeting and the base of the stud. Moreover, the stud connector, subject to a highly eccentric load, tends to be less effective than in solid slabs. Its performance in terms of strength and ductility may be adversely affected by the interaction with other connectors in the same rib (i.e., when the number of studs in a rib, N_r, increases) and/or the reduced efficiency of longitudinal restraint offered by the concrete (i.e., when the studs are placed off center).

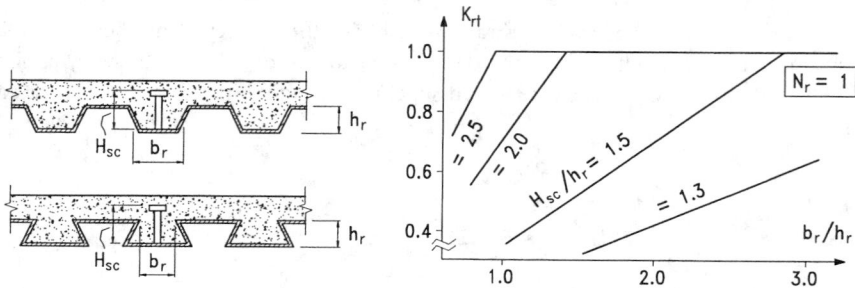

Figure 6.35   Reduction factor k_{rt} vs. geometrical parameters.

A comprehensive investigation carried out at Lehigh University [20], which also accounted for the results of previous studies, permitted definition of the general form of the relationship (see figure 6.35) between the main parameters governing the shear connection performance and the reduction factor k_{rt} to be applied to stud resistance in solid concrete slabs:

$$k_{rt} = \frac{c}{\sqrt{N_r}} \frac{b_r}{h_r} \left[\frac{H_{sc}}{h_r} - 1.0 \right] \le 1.0 \tag{6.52}$$

The same study proposed for the coefficient c the value 0.85, which was adopted by the AISC-LRFD code.

Further research work conducted mainly to investigate decks with European types of steel sheeting [24, 35] indicated that, in order to avoid unsafe assessment of the shear strength:

1. a lower value should be assumed for coefficient c
2. a range of application of Equation 6.52 should be defined

Therefore, Eurocode 4 assumed c = 0.7 and limited application of Equation 6.52 to

- stud diameter $d_{sc} \le 20$ mm
- studs welded through-deck
- studs with ultimate tensile strength not greater than 450 N/mm^2
- ribs with $b_r / h_r \ge 1.$, and $h_r \le 85$ mm
- $k_{rt} \le 0.8$ when $N_r \ge 2$

The limited knowledge still available does not allow coverage of the practice, common in a few countries, of welding studs through holes cut in the sheeting.

Neither code takes into account the effect of off center placement of the studs, resulting from the presence of stiffening ribs in the sheeting. This effect may be significant. It would be advisable to locate the stud in the favorable position as shown in Figure 6.36. In the case of symmetrically loaded simply supported beams, the "strong" side is the one nearest to the closest support.

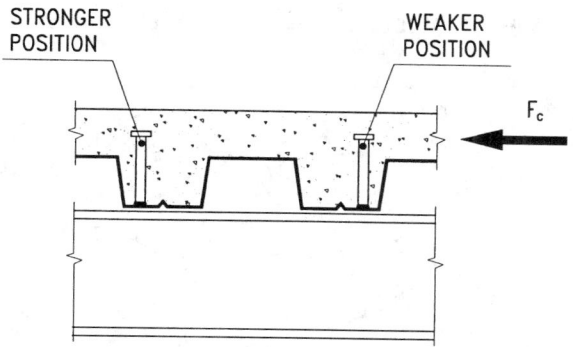

Figure 6.36   Different locations of the stud in the rib.

Other Types of Connectors

The knowledge of the performance of other types of connectors is far more restricted, and it is mainly based on research work conducted in the 1970s. This reflects the rather limited use of connectors other than studs both in building and bridge practice. Some new types of shear connector, such as the cold formed "seat element" connected to the steel beam by means of shot fired pins shown in Figure 6.29, are increasingly employed. Their response was found satisfactory in a few studies of composite beams and joints [4, 15, 33], and comparable to equivalent welded studs. However, no specific rules are currently provided in codes, and their use requires suitable testing to be carried out.

The main code recommendations are reported here, with a few explanatory notes.

Channel Connectors

The Eurocode does not cover channel connectors, while channel connectors are the only connectors other than studs included in the AISC specifications, which are based on the work by Slutter and Driscoll [44]. The strength equation proposed there is modified in order to cover the case of lightweight concrete. Their capacity is then determined as

$$q_u = 0.3 \cdot (t_f + 0.5t_w) \cdot b_c \sqrt{f_c E_c} \tag{6.53}$$

where t_f, t_w, and b_c are the flange thickness, the web thickness, and the length of the channel, respectively.

The effect of the position of the channel flange relative to the direction of the shear force on the connector's strength (see figure 6.30) is not considered, and Equation 6.53 implicitly refers to the weakest position where the flanges are "looking" at the shear force, which was the only one covered by Slutter and Driscoll.

Angle Connectors

Angle connectors were investigated mainly in France, and the Eurocode design formula is based on the French studies:

$$q_u = 10b_a h_{ac}^{3/4} f_c^{2/3} \tag{6.54}$$

where b_a and h_{ac} are the length and the height of the outstanding leg of the connector. The design resistance $q_{u.d}$ is then obtained by applying to q_u a partial safety factor equal to 1.25. Further design requirements relate to:

- the welds fastening the angle to the beam flange, the design of which should be based on a conventional eccentricity of the shear force of $h_{ac}/4$
- the reinforcement bar necessary to resist uplift, whose area should be $\geq 0.1 \, q_{u.d} \gamma_{sr} / f_{y.sr}$, where γ_{sr} is the partial safety factor for steel reinforcement

The recommended position of the angles with respect to the thrust in the slab is shown in Figure 6.37.

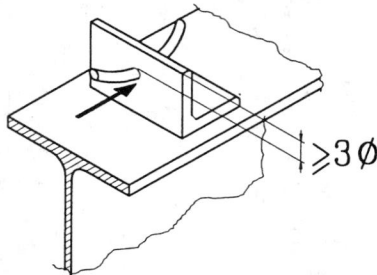

Figure 6.37   Recommended position of angle connectors with respect to the thrust in the slab.

Friction Grip Bolts

Friction grip bolts may be advantageously employed to connect precast concrete slabs. The Eurocode 4 recommendations are consistent with the rules for preloaded bolts in Eurocode 3. These should be accounted for, even if not explicitly mentioned: e.g., the standard nominal clearance should be met by the holes in the beam flange, otherwise the reduction factor for the friction resistance as defined in Eurocode 3 must be applied.

The friction resistance is taken as

$$q_u = \mu \frac{F_{b.pr}}{\gamma_V} \qquad (6.55)$$

where

μ = the friction coefficient

$F_{b.pr}$ = the preload in the bolt

γ_V = the partial safety factor ($\gamma_V = 1.0$ for serviceability checks and $\gamma_V = 1.25$ for checks at ultimate)

The value of the friction coefficient μ strongly depends on the flexural stiffness of the steel flange. The Eurocode specifies $\mu = 0.50$ for steel flanges with thickness $t_f \geq 10$ mm and 0.55 for $t_f \geq 15$ mm. The surface of the flange should be prepared by blasting and removing loose rust; besides, pitting should not be present.

The bolt preload force is significantly reduced by shrinkage and creep. This effect should be taken into account when determining $F_{b.pr}$. In absence of an accurate determination (by testing or by an evaluation of existing data) a reduction of 40% can be assumed. In many instances, it may not be convenient to ensure that the bolted connection resists by friction the shear force at ultimate. In this case, the ultimate strength of the connector can be by the bolt's shear resistance or by "bearing" against the concrete. The latter resistance can be assumed to be equal to the value specified for studs (see Equation 6.49). Suitable testing may enable use of a further possible design criterion, which combines friction and shear in the bolt to achieve the required ultimate resistance.

Block Connectors, Anchors, and Hoops in Solid Slabs

Block connectors are those connectors with a flat surface opposing the shear force, and whose stiffness is so high that pressure on the concrete may be considered uniformly distributed at least at failure. The connector resistance is determined as

$$q_u = cA_{f1}f_c \qquad (6.56)$$

where the coefficient $c = \sqrt{A_{f2}/A_{f1}}$ is ≤ 2.5 for normal weight concrete and ≤ 2.0 for lightweight concrete, and the areas A_{f1} and A_{f2} are defined in Figure 6.38a. Only the part of the area, A_{f2}, which lies within the concrete section, should be taken into account.

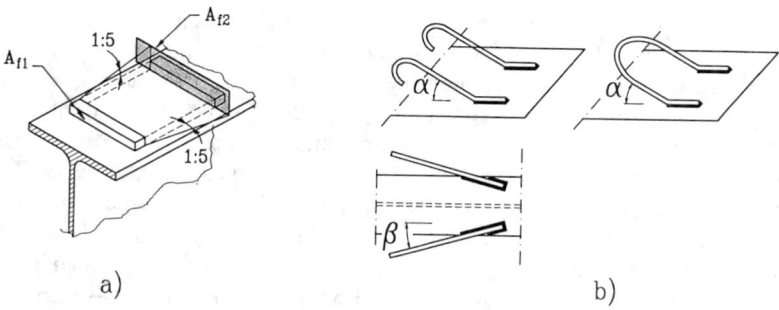

Figure 6.38 (a) Definition of areas A_{f1} and A_{f2}. (b) Definition of angles α and β.

If the welds fastening the connector to the beam flange are satisfactorily designed, failure is attained by concrete crushing. The limit to the coefficient c accounts for the effect of the triaxial stress state on the concrete strength, and simultaneously imposes that the lateral restraint is adequate (through its link with the area A_{f2}) and that the distance of the connectors is sufficient to ensure that their full resistance capability is developed.

The design resistance is obtained by dividing q_u by the partial safety factor for concrete γ_c. Ties and hoops should also be present, and designed to resist the uplift force to be assumed as equal to $1/10$ $q_{u.d}$.

The resistance of each leg of anchors and hoops to longitudinal shear is defined as

$$q_u = A_{sr} \cdot f_{y.sr} \cdot \frac{\cos\beta}{\sqrt{1+\sin^2\alpha}} \tag{6.57}$$

where A_{sr} and $f_{y.sr}$ refer to the bar area and yield strength, respectively, and the angles α and β are defined in Figure 6.38b.

The failure is associated to the anchor or hoop. The design resistance is obtained by dividing q_u by the appropriate material partial safety factor: either γ_s or γ_{sr}.

As indicated in the above, block connectors require combination with anchors or hoops in order to ensure adequate resistance to uplift. It may be convenient to use these types of connectors to resist the longitudinal shear force as well. Due to the different stiffness and different failure modes of these connectors, the total resistance $q_{u.tot}$ is less than the sum of the resistances of each of the connectors considered individually; the following expressions may be used in absence of accurate design resistance determination through testing:

$$q_{u.tot} = q_{u.block} + 0.5 q_{u.anchors} \tag{6.58}$$

$$q_{u.tot} = q_{u.block} + 0.7 q_{u.hoop} \tag{6.59}$$

The welds that connect the system to the beam flange should be designed for a total shear force equal to $1.2\, q_{u.block} + q_{u.anchors}$ (or $q_{u.hoop}$).

6.5.3 Steel-Concrete Interface Separation

The analytical model of the composite beam assumes that the curvatures of the concrete and steel elements are the same: i.e., that there is no separation between the concrete flanges and the steel section. However, the shear transfer mechanism implies that forces also arise acting perpendicularly to the interface, which, inter alias, tend to cause steel concrete separation (see also Figure 6.28). Therefore, care should be taken to provide the shear connection with the capacity of resisting such uplift forces. The most efficient solution assigns this additional role to the same shear connectors.

It has been shown that the forces required to maintain full contact are substantially lower than the corresponding shear forces. A ratio of 1 to 10 may usually be assumed in design. As a consequence, no specific strength check is usually required for they can be resisted by connectors with suitable shapes such as headed studs and channels. On the other hand, additional elements resisting uplift should be placed when inadequate connectors are used: for example, anchor bars or hooks are needed in case of angle or block connectors (see Figure 6.29). Moreover, attention should be paid to the detailing. In particular, the spacing of connectors should be limited (the Eurocode recommends the lower value between 800 mm and six times the slab thickness), and, in the case of composite slabs with the ribs transverse to the beam, some form of anchorage should be present in every rib.

In several circumstances it should be considered that other factors may lead to an increased importance of vertical separation, as the loading condition (e.g., suspended loads), the stabilizing role of the concrete flange (typically against lateral buckling), and the cross-sectional variation of the member along its span.

6.5.4 Shear Connectors Spacing

The shear connectors should be suitably distributed among, and spaced along, the various shear spans in which the beam is divided so that the sectional beam capacity is always greater than the design moment. Shear spans, and the correspondent *critical lengths of the steel-concrete interface* to be considered in design verifications can be conveniently defined as the distances between two adjacent *critical cross-sections*, which to that purpose are

- supports
- free ends of cantilever
- sections of maximum bending moment
- sections where heavy concentrated loads are applied
- sections at which a sudden change of the cross-section occurs

The most viable criterion of spacing of connectors distributes them consistently with the longitudinal shear flow determined through an elastic analysis. On the other hand, uniform spacing would commonly be preferred, whenever possible, for it is practically advantageous. This alternative relies on interface slip to fully redistribute the forces within the shear connection, and hence requires an adequate slip capacity of the connectors.

Available information on slip capacity is quite limited, and certainly not sufficient for allowing a reliable assessment of this parameter to be obtained for most types of connector. Therefore, the Eurocode 4 restrains uniform spacing along a critical length l_{cr} to studs and friction grip bolts. Furthermore, this applies to beams that:

- have all critical sections in the span considered in Class 1 or Class 2 (see Section 6.4.3)
- meet the specified limitations as to the degree of shear connection in each critical length
- have a ratio between the plastic moment of the composite section and the plastic moment of the steel section not greater than 2.5

The AISC specifications seem less severe. They cover only studs and channel connectors, and consider their "ductility" sufficient for uniform spacing along the shear span in all cases.

Shear connectors may also ensure the stability of the steel flange. Therefore, an upper limit to their spacing may be associated with their effectiveness as stabilizing elements. The following values are recommended in the Eurocode:

- center-to-center spacing not greater than

    slab in contact over the full length (e.g., solid slab)

$$22 t_f \cdot \sqrt{\frac{235}{f_{y.s}}}$$

slab not in contact over the full length (e.g., slab with ribs transverse to the beam)

$$15t_f \cdot \sqrt{\frac{235}{f_{y.s}}}$$

- distance to the edge of the steel compression flange not greater than

$$9t_f \cdot \sqrt{\frac{235}{f_{y.s}}}$$

where
t_f = thickness of the flange
$f_{y.s}$ = nominal yield strength of the flange in N/mm^2

6.5.5 Shear Connection Detailing

Quality of a structure always implies a careful selection and sizing of details in design. In composite structures, the large variety of practical situations is combined with the lack of tools enabling a sufficiently accurate appraisal of the three-dimensional complex stress state deriving locally from the composite action. The detailing rules, which are provided by codes, do not have any scientific basis. Rather, they stem from previous successful experience in practice or adequate experimental data. These recommendations should be considered, for their very nature, just as applications rules deemed to satisfy the design principle. Moreover, the practical range of possibilities in composite structures is so broad that no set of rules can cover it in a comprehensive way. Therefore, engineering judgment should be used when applying them in design.

6.5.6 Transverse Reinforcement

Mechanical connectors apply concentrated loads to the concrete slab which disperse into this element through shear and transverse tensile stresses. This complex stress state may result in longitudinal splitting of the slab (tensile stresses) or formation of inclined (herring bone) cracks. Therefore, transverse reinforcement is needed to resist these forces. In most instances, reinforcing bars are already present transversally to the beam as longitudinal rebars of the slab spanning between adjacent beams. However, the capability to resist forces associated with the shear transfer should also be checked.

Transverse rebars play basically the same role as stirrups in reinforced concrete beams. Therefore, background to current design approach are the studies on shear transfer in reinforced concrete by Mattock and Hawkins [32]. Distinction should be made between the shear transfer mechanism in the uncracked slab and in the presence of a longitudinal crack. In the former case, the behavior may be schematically modeled via a strut and tie truss analogy, as shown in Figure 6.28. When the shear is transferred through a cracked plane, three contributions to the resistance can be identified:

1. dowel action of the rebars
2. aggregate interlock
3. friction

Development of the last two contributions requires that transverse rebars are placed, which prevents the crack width from increasing.

The shear transfer capacity in cracked situations is significantly lower than in uncracked slabs. However, the former condition is assumed in design. Besides conservativeness, this approach is simpler, for in this case the shear strength does not depend on the longitudinal stress in the slab. The steel reinforcement resists by bending across the surface of the crack, and its maximum "shear capacity" depends on the steel yield strength and on the relative stiffness and strength of the steel and concrete. Nevertheless, the steel yield resistance remains the main factor. Interlock and friction contributions are limited by fracture of interface protrusions, which is a function of the concrete tensile strength.

The shear transfer capacity per unit length of a shear plane may be written in the following form, which accounts for both resistance contributions

$$v_l = k_1 A_{c.v} f_{ct} + k_2 A_{ef.v} f_{y.sr} \tag{6.60}$$

where

k_1 and k_2 =   factors accounting indirectly for the other parameters affecting v_l
$A_{c.v}$ =   the area per unit length of the concrete shear surface in the failure plane considered
$A_{ef.v}$ =   the area per unit length of the reinforcing bars which can be considered effective
f_{ct} =   the tensile strength of concrete

The experimental data suggest that a lower limit to the reinforcement strength per unit length should be imposed, in order to ensure applicability of Equation 6.60. An upper bound is also necessary because higher reinforcement ratios lead to different failure modes; this limit may be written in the form

$$v_l = k_3 A_{c.v} f_c \tag{6.61}$$

In Eurocode 4, Equations 6.60 and 6.61 are written, respectively, as

$$v_l = 2.5 \eta A_{c.v} \tau_u + A_{ef.v} f_{y.sr} + v_{sd} \tag{6.60a}$$

and

$$v_1 = 0.2 \eta A_{c.v} f_c + \frac{v_{sd}}{\sqrt{3}} \tag{6.61a}$$

where the coefficient η is a factor accounting for concrete density: $\eta = 1$ for normal weight concrete and $\eta = 0.3 + 0.7$ (w/24) for lightweight concrete (with w expressed in kN/m^3), τ_u is the shear strength to be taken as $0.25 f_{ct}$ and v_{sd} is the possible contribution of the steel sheeting of the composite slab. The lowest of the resistances computed by Equations 6.60a and 6.61a has to be taken.

It should be noted that the steel decking can contribute only when its ribs are transverse to the beam, and its value, v_{sd}, should be:

- assumed as $v_{sd} = A_{sd} f_{y.sd}$, if the sheeting is continuous across the beam
- limited by the bearing resistance of the stud against the sheeting, when the sheeting is not continuous, but anchored by studs welded through the sheeting

The shear transfer capacity has to be checked for all the relevant critical shear surfaces shown in Figure 6.39. A value of the total longitudinal shear equal to the design shear force for the connectors is usually considered. However, the variation of the longitudinal shear along the width of the concrete slab may be considered (see [8]) and different values determined for the various shear failure surfaces.

A minimum area of transverse reinforcement equal to $0.002 A_{c.v}$ is prescribed in the AISC specifications, which, on the other hand, do not provide any specific criterion for transverse reinforcement design.

Spacing of the transverse rebars should be consistent with spacing of the shear connectors.

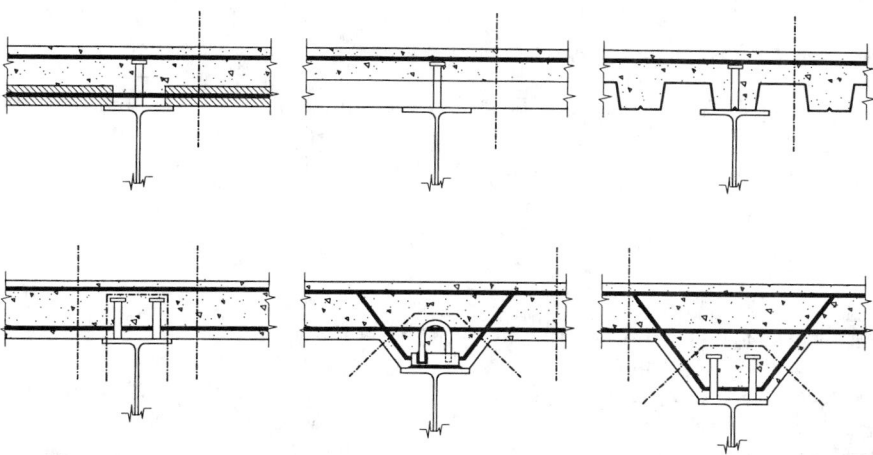

Figure 6.39   Relevant shear surface for the evaluation of the shear transfer capacity.

6.5.7   The Shear Connection in Fully and Partially Composite Beams

The methods of analysis and design of composite beams presented in Sections 6.3 and 6.4 are based on the assumption of full interaction. As already mentioned, the effect of interface slip on the performance of a composite beam may generally be neglected in composite beams designed elastically, when the shear connection has adequate resistance in all the critical lengths of the member, and shear connectors are used which proved to possess satisfactory stiffness, as for the ones covered by design codes.

In the case of plastic design, a distinction has to be made between fully and partially composite beams as defined in the introduction to this section. In fully composite beams, i.e., if the requirement of full shear connection is fulfilled, the deformation of the connectors can be neglected in design analyses. The shear connection design does not differ remarkably from the elastic case, however, the level of strength to be considered in each critical length does differ. The spacing and ductility requirements of the connectors are the same. In many instances, fully composite beams are not the optimal solution, and partially composite beams may better suit the requirements in service and at ultimate, and at a lower cost. The high redistribution in partial shear connection implies that the ductility demand to connectors becomes significant. Furthermore, it sharply increases with the decrease of the degree of shear connection. The performance of partially composite beams depends dramatically on the ductility of the connectors and this is a key design parameter. The main aspects of partially composite beam design are presented in this section. The design of the shear connection in fully composite beams is also treated, as a necessary preliminary step.

Fully Composite Beams

Fully composite beams can develop by definition the full plastic moment in positive and negative bending. The collapse mode is by formation of a plastic beam mechanism. The design of a full shear connection refers to this collapse condition, and is based on the overall longitudinal shear to be transferred at the steel-concrete interface in each beam length between two adjacent plastic hinges or between a plastic hinge and the nearest support (i.e., generally in a half span).

Such an overall longitudinal shear V_l can be determined straightforwardly by imposing the equilibrium to longitudinal translation to the slab:

$$V_l = F_{c.f} + F_{t.f} \tag{6.62}$$

where

$F_{c.f}$ = the compression force in the slab in the section where the sagging moment is maximum (see Equation 6.13)

$F_{t.f}$ = the tensile force in the slab in the section where the negative bending moment is maximum (see Equation 6.34; $F_{t.f}$ is zero in a simply supported beam)

The minimum number of connectors required to transfer V_l is obtained as

$$N_f = \frac{\left(F_{c.f} + F_{t.f}\right)}{q_u} \qquad (6.63)$$

where

q_u = the ultimate shear resistance of the chosen shear connector type

The connectors may be uniformly spaced along the half span. However, a critical cross-section may be located between the plastic hinges considered (or between the plastic hinge in sagging moment and a support), and a further check is needed to suitably distribute the total number of connectors between the critical length. To this purpose, the force in the slab has to be computed at the intermediate critical cross-section (usually by an elastic analysis), and the total change in longitudinal force V_l can then be subdivided between the two critical lengths and the total number of connectors shared accordingly.

Partially Composite Beams

Fully composite beams are not always the most efficient solution. Moreover, the condition of full shear connection is normally difficult to achieve when composite slabs are used with metal decking with the ribs parallel to the beam axis. As to the efficiency, it may be noted, among other aspects, that:

- The design of the slab and the steel section is often governed by criteria other than strength.
- In a building, only a limited number of steel sections will commonly be selected and used. As a result, several beams will be oversized, if fully composite.
- This possible beam over-strength is paid in terms of cost of placement of connectors and detailing (mainly the required amount of transverse reinforcement in the slab).

Therefore, the use of connectors in numbers lower than that required by the condition of full shear connection appears advantageous, for it enables the ultimate capacity of the beam to be "tailored" to the design value of the moment. This approach, the so-called partial shear connection design, was recently developed and refined through numerous experimental and numerical studies.

The influence of the reduction of the number N of connectors (i.e., of the allowable compressive force F_c in the slab) on the moment resistance of the beam is schematically illustrated in Figure 6.40a, where representative stress distributions are also shown for the different ranges of behavior. The reduction in moment capacity is moderate until the mode of failure is associated with concrete crushing (line BC in the figure). A further decrease of shear connectors makes the shear connection to fail first, and the moment capacity reduces sharply (line A'B). Failure of the shear connection may happen even before the plastic moment $Mpl.s$ of the steel section is attained (line AA'). There is no particular reason to consider differently, in design practice, the two modes of failure, at least when ductile shear connectors are used. The shear connection failure in this case is not more sudden than the flexural mode. On the contrary, a greater conservativeness is necessary in case of non-ductile connectors.

The relationships M − N (M − F_c) and, therefore, the location of points A' and B depend on the type of loading and the beam span, and are affected by the strain hardening of the structural steel. Accurate determination of these relations is too burdensome in routine design. Approximate approaches were then developed, and are included in several codes. They are briefly presented here.

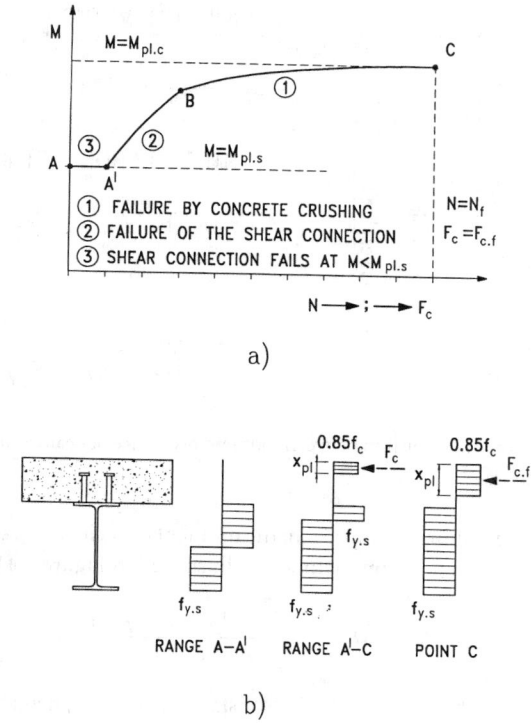

a)

b)

Figure 6.40 Influence of the reduction of shear connectors. (a) Moment resistance of beam-shear capacity. (b) Possible stress redistribution in the cross-section.

Besides, deflections of partially composite beams have to be computed allowing for the effect of interface slip. These calculations represent an additional design difficulty; simplified methods are covered as well.

The Moment Resistance

As already underlined, the redistribution of forces on connectors required to exploit the flexural resistance of a partially composite beam is significant. The deformation capacity of the shear connectors (ductile or non-ductile connectors) is the key parameter for the selection of the design method.

A. Ductile Connectors

If plastic theory can be applied (i.e., all critical cross-sections may be classified as compact) and the deformation capacity of the connectors is sufficient to assume that a complete redistribution can take place within the shear connection in the shear span considered, then the ultimate strength of the partially composite beam at a critical cross-section can be computed rather easily via the same equilibrium approach based on simple plastic theory as for fully composite beams. Due to the occurrence of slip, the slab and the steel beam will have different neutral axes, as shown in Figure 6.40b. The depth $x_{pl} < h_c$ of the concrete slab contributing to the flexural strength is determined by equating the stress block and the shear connection resistance; that is, by posing

$$0.85 f_c x_{pl} b_{\text{eff}} = \sum q_u \qquad (6.64)$$

The equilibrium of the full composite section to the translation enables calculation of the plastic neutral axis of the steel section, which may lie within the top flange or the web. The determination of the moment capacity M_u is then straightforward (see Section 6.3.4).

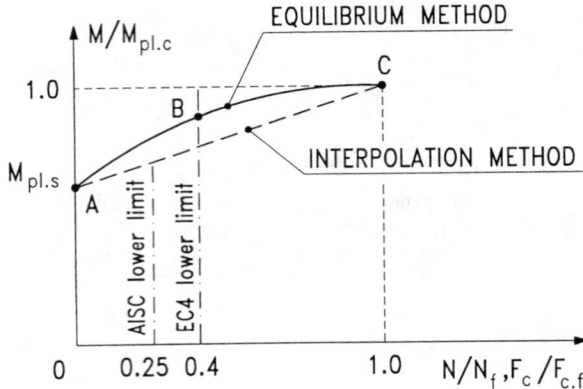

Figure 6.41   Non-dimensional relationship between moment resistance of beam and shear connector capacity for partially composite beams.

As an alternative to the equilibrium approach, the moment capacity of a partially composite beam can be determined via the interpolation method, as illustrated in Figure 6.41:

$$M_w = M_{pl.s} + \frac{F_c \left(M_{pl.c} - M_{pl.s} \right)}{F_{c.f}} \tag{6.65}$$

In most instances, the accuracy of this method is satisfactory. Besides, it enables a direct assessment of the value of F_c required to resist the design moment, M, being

$$F_c = F_{c.f} \frac{\left(M_w - M_{pl.s} \right)}{\left(M_{pl.c} - M_{pl.s} \right)} \tag{6.66}$$

Several studies were recently devoted to the appraisal of the "ductility" requirement of the connectors to be associated with partial connection plastic design [3, 29]. They confirmed the substantial deformation capacity needed to allow redistribution for low degrees of shear connection and long spans. The main outcomes provided the background to the Eurocode relations between the beam span and the minimum degree of shear connection allowed for application of Equations 6.65 and 6.66, which are illustrated in Figure 6.42. The cases specifically covered are

- steel sections with equal flanges (line A)
- steel section with unequal flanges (line B valid for bottom flanges with area not greater than 3 times the upper flange area), for which a more strict limitation is specified
- composite beams with studs connectors in composite slabs with profiled steel sheeting and ribs transverse to the beam (line C), for which tests by Mottram and Johnson [35] indicated the possible availability of higher slip capacities, at least when certain conditions are met by the connectors, the steel section, and the deck sheeting

These relationships are consistent with the definition of ductile connectors given in the same code, which classifies as ductile stud connectors, friction grip bolts, and all connectors having a characteristic slip capacity, determined from push tests, not lower than 6 mm.

B. "Non-Ductile" Connectors

When the condition of full redistribution within the shear span cannot be attained, elastic-plastic calculations should, in principle, be performed to determine the M vs. F_c curve, accounting for the influence of slip (curve A' B' C in Figure 6.43). Such calculations are too complex for usual design practice, even if full interaction is assumed in the consideration of the need to limit the slip and the

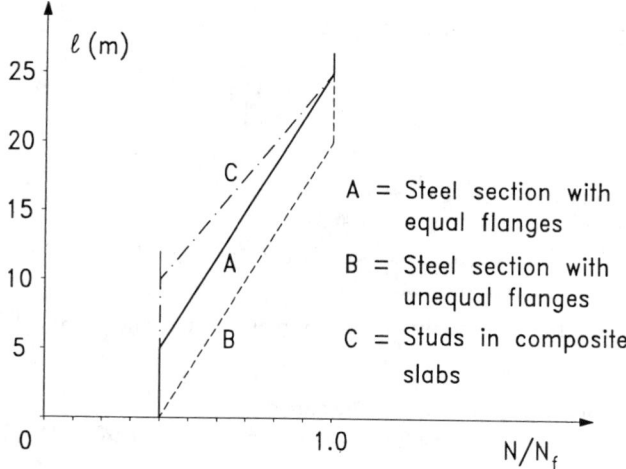

Figure 6.42   Relationship between the beam span and the minimum degree of shear connection.

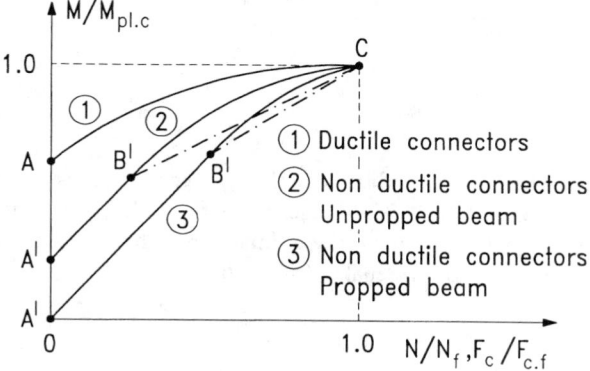

Figure 6.43   Non-dimensional relationship between moment resistance of beam and shear connector capacity for different connector behaviors.

fact that slip reduces the longitudinal shear force for a given moment distribution. A substantial simplification is made possible by the relatively easy determination of the three key points A′, B′, and C and by the nature of the curve linear up to point B′ and convex-upwards in the range B′C, which enables use of a straight line also in this range. Point A′ is associated with the condition of no composite action ($F_c = 0$); in the case of unshored construction, the coordinate of point A′ is the bending moment M_s resisted by the steel section alone. Point B′ is associated with the moment M_{el} for which the steel bottom flange achieves its yield strength. Computation of M_{el} should account for the method of construction, i.e., M_{el} should include, for unshored beams, the moment M_s. At point C, the full moment capacity of the composite section is achieved.

The Deflections

The effect of interface slip on beam deflections may be neglected in many instances, even in case of partial shear connection, as shown by several experimental studies [28]. Besides, design criteria usually aim to reduce the importance of longitudinal slip, and shear connections are designed accordingly. However, such an effect tends to become significant as the degree of shear connection N/N_f decreases and/or the forces on the shear connectors increase, and should be considered when N/N_f is "low".

The recent parametric studies conducted in view of the preparation of Eurocode 4 supported the proposal by Johnson and May [28]. The approach, included in the Eurocode, accounts for the effect of incomplete interaction in the calculation of the beam deflection δ_p by the simple relation

$$\delta_p = \delta_f + k\left(\delta_s - \delta_f\right)\left[1 - \frac{N}{N_f}\right] \tag{6.67}$$

where
δ_f = the deflection for the fully composite beam
δ_s = the deflection of steel beam acting alone
k = a factor allowing for the influence of construction type: $k = 0.5$ for shored construction and $k = 0.3$ for unshored construction

The Eurocode also defines the requirements to be met in order to neglect the slip effect in unshored beams. They are

- $N/N_f \geq 0.5$, or force q on a shear connector $\leq 0.7\, q_u$
  and in the case of slabs with profiled steel decking transverse to the beam
- height of the ribs $h_r \leq 80$ mm

6.5.8 Worked Examples

EXAMPLE 6.4:

With reference to Example 6.1, the verification of the shear connection is proposed in the following. It is assumed that the connection is full interaction, with headed stud connectors (Figure 6.44), having $H_{sc} = 100$ mm and $d_{sc} = 19$ mm. The ultimate strength of the connectors is $f_{u.sc.B} = 450$ N/mm^2 and a partial safety factor γ_v equal to 1.25 is considered.

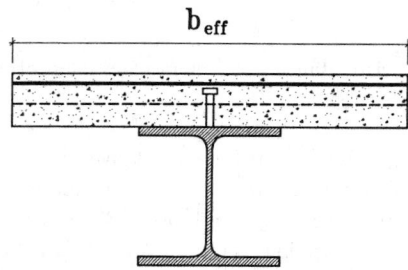

Figure 6.44   Cross-section of the composite beam.

1. Longitudinal shear force — In accordance with the hypothesis of full shear connection, the total design longitudinal shear V_1 to transfer by shear connectors, spaced between the point of maximum sagging moment and the end support is:

$$V_1 = \min\ (F_{s.\text{max}},\ F_{c.\text{max}})$$

   where $F_{s.\text{max}}$ and $F_{c.\text{max}}$ are the steel beam and the concrete slab limit resistance, respectively. Therefore, neglecting the contribution offered by the longitudinal slab reinforcement and with reference to the steel beam (see Equation 6.14):

$$F_{s.\text{max}} = \frac{A_a \cdot f_{y.s.k}}{\gamma_s} = \frac{11600 \cdot 235}{1.1} \cdot 10^{-3} = 2478.2\ \text{kN}$$

and with reference to the concrete (see Equation 6.13):

$$F_{c.max} = \frac{0.85 A_c \cdot f_{c.k}}{\gamma_c} = \frac{0.85 \cdot b_{eff} \cdot h_c \cdot f_{c.k}}{\gamma_c}$$

$$= \frac{0.85 \cdot 2500 \cdot 120 \cdot 25}{1.5} \cdot 10^{-3} = 4250 \text{ kN}$$

Then, $V_1 = 2478.2$ kN

2. Design resistance of shear connection (see Section 6.5.2) — It is necessary to make reference to the lower value between

   - the resistance of concrete (Equation 6.49):

   $$q_{u.c} = \frac{k_c \cdot A_{sc} \cdot \sqrt{f_{c.d} \cdot E_c}}{\gamma_v} = \frac{0.36 \cdot \frac{\pi \cdot 19^2}{4} \sqrt{25 \cdot 30500}}{1.25} \cdot 10^{-3} = 71.27 \text{ kN}$$

   - the resistance of stud connectors (Equation 6.50):

   $$q_{u.s} = k_s \cdot f_{u.sc.d} \cdot \frac{A_{sc}}{\gamma_v} = 0.8 \cdot 450 \cdot \frac{\pi \cdot 19^2}{4} \frac{1}{1.25} \cdot 10^3 = 81.7 \text{ kN}$$

   Then, $q_{u.d} = q_{u.c} = 71.27$ kN

3. Design of the connection

   - Minimum number of connectors:

   $$N_f = 2 \cdot \frac{V_1}{q_{u.d}} = 2 \cdot \frac{2478.2}{71.27} = 69.54$$

   It is assumed
   $$N = 2 \cdot 35 = 70 \text{ headed stud connectors}$$

   - Spacing of the connectors (see Section 6.5.4)
   The stud connectors are spaced uniformly, being:

     - all critical sections are in class 1
     - $\frac{N}{N_f} = \frac{70}{69.54} = 1.001 \geq 0.25 + 0.03 \cdot l = 0.25 + 0.03 \cdot 10 = 0.55$
     - $\frac{M_{pl}}{M_{pl.s}} = \frac{M_{pl.c}}{W_s \cdot f_{y.s.k}/\gamma_s} = \frac{830}{2194 \cdot 235/1.10} \cdot 10^{-6} = 1.77 < 2.5$

   It is assumed a uniform spacing between studs:

   $$i = \frac{l}{N} = \frac{10000}{70} = 143 \text{ mm}$$

   It is hence assumed 80 headed stud connectors, spacing of 125 mm.

4. Detailing of the shear connection (see Section 6.5.4)

   - Spacing:

   $$i \leq 22 \cdot t_f \sqrt{\frac{235}{f_{y.s.k}}} = 22 \cdot 16 \cdot \sqrt{\frac{235}{235}} = 352 \text{ mm}$$

   - Overall height:

   $$6 H_{sc} < i < 5 d_{sc} \qquad 6 \cdot 100 < 125 < 5 \cdot 19 = 95 \text{ mm}$$

5. Reinforcement in the slab — The area of the reinforcement in a solid slab must not be less than 0.002 times the concrete area being reinforced and should be uniformly spaced. Referring to a width of 1000 mm, minimum amount of reinforcement is:

$$0.002 \cdot A_c = 0.002 \cdot 120 \cdot 1000 = 240 \text{ mm}^2/\text{m}$$

It is assumed a double layers of reinforcement ϕ 10 mm/200 mm

$$A_{ef.v} = 2 \cdot 392.7 = 785.4 \text{ mm}^2/\text{m}$$

6. Longitudinal shear in the slab (see Section 6.5.6) — In accordance with Eurocode 4, the shear surfaces are presented in Figure 6.45.

- Section a-a — The design resistance is determined from the smaller value between (Equations 6.60a and 6.61a):

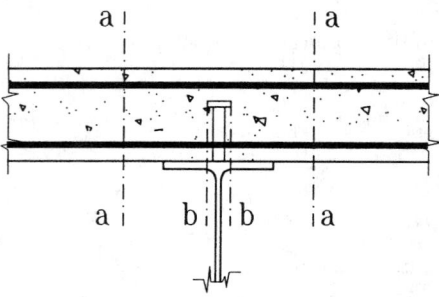

Figure 6.45   Longitudinal shear surfaces.

$$
\begin{aligned}
v_1 &= 2.5\eta A_{cv}\tau_{u.d} + A_{ef.v} f_{y.sr.d} + v_{sd} \\
&= 2.5 \cdot (1000 \cdot 120) \cdot 1 \cdot 0.30 + 2 \cdot 392.7 \cdot \frac{420}{1.15} = 376842 \text{ N/m} \cong 377 \text{ kN/m}
\end{aligned}
$$

and

$$
\begin{aligned}
v_1 &= 0.2\eta A_{cv} 0.85 \frac{f_{c.k}}{\gamma_c} + \frac{v_{sd}}{\sqrt{3}} \\
&= 0.2 \cdot (1000 \cdot 120) \cdot \frac{0.85 \cdot 25}{1.5} = 340000 \text{ N/m} \cong 340 \text{ kN / m}
\end{aligned}
$$

where

$$
\begin{aligned}
\tau_{u.d} &= 0.30 \text{ N/mm}^2 \\
A_{cv} &= 120 * 1000 = 120000 \text{ mm}^2 \\
\eta &= 1.0 \qquad \text{(normal concrete)} \\
v_{sd} &= 0 \qquad \text{(contribution of the profiled steel sheeting)}
\end{aligned}
$$

The design longitudinal shear force per unit length is:

$$v = \frac{q_u}{i} = \frac{71.27}{125} = 0.570 \text{ kN / mm} = 570 \text{ kN / m}$$

Referring to only one section of the headed stud connector it is assumed:

$$\frac{v}{2} = 285 \text{ kN / m} < 340 \text{ kN / m}$$

- Section b-b — It is preliminary to calculate the length of the shear surface. In accordance with Eurocode 4 and assuming a width of 1000 mm, it is assumed:

$$A_{cv} = (2 \cdot H_{sc} + d_{sc}) \cdot 1000 = (2 \cdot 100 + 19) \cdot 1000 = 219000 \text{ mm}^2 / \text{m}$$

then:

$$
\begin{aligned}
v_1 &= 2.5\eta A_{cv}\tau_{u.d} + A_{ef.v}\frac{f_{y.sr.k}}{\gamma_{sr}} + v_{sd} \\
&= 2.5 \cdot 219000 \cdot 1 \cdot \frac{0.30}{1.10} + 2 \cdot 392.7 \cdot \frac{420}{1.15} = 451092 \text{ N / m} \simeq 451 \text{ kN/m}
\end{aligned}
$$

Being v less then v_1, the section b-b will be safe.

EXAMPLE 6.5:

With reference to Example 6.1, the design of the shear connection is proposed. It is assumed Partial shear connection with ductile connectors. In particular, 19 mm diameter stud connectors with ultimate tensile strength $f_{u.sc.k} = 450 \text{ N/mm}^2$ are used.

It is also assumed that the headed stud connectors have an overall length after welding not less than 4 times the diameter and with a shank of diameter not less than 16 mm and not exceeding 22 mm (see Eurocode 4 – 6.1.2).

1. Connection design — The connection is considered ductile if:

$$\frac{N}{N_f} \geq 0.25 + 0.03l$$

where

l = total length of the beam
N = number of shear connectors
N_f = number of shear connectors for total shear connection

Then

$$\frac{N}{N_f} \geq 0.25 + 0.03 \cdot 10.00 = 0.55$$

The stud design shear resistance is evaluated as the lesser value between

- the resistance of the concrete (Equation 6.49):

$$q_{u.c} = \frac{k_c \cdot A_{sc} \cdot \sqrt{f_{c.d} \cdot E_c}}{\gamma_v} = 71.27 \text{ kN}$$

- the resistance of the connectors (Equation 6.50):

$$q_{u.s} = k_s \cdot A_{sc} \frac{f_{u.sc.k}}{\gamma_v} = 0.8 \cdot \frac{\pi \cdot 19^2}{4} \frac{450}{1.25} \cdot 10^{-3} = 81.7 \text{ kN}$$

Then, $q_u = q_{u.c} = 71.27$ kN

According to Examples 6.1 and 6.2, the shear force between the section at midspan and the one at the support is

$$V_1 = F_{smax} = 2478.2 \text{ kN}$$

In the hypothesis of total shear connection the minimum number of stud connectors is:

$$N_f = \frac{V_1}{q_u} = \frac{2478.2}{71.27} = 34.97$$

Therefore, it is possible to evaluate N.

$$N = 0.55 \cdot N_f = 0.55 \cdot 34.77 = 19.12$$

Hence, 20 connectors are selected for the half beam. The corresponding force F_c is

$$F_c = N \cdot q_u = 20 \cdot 71.27 = 1425.4 \text{ kN}$$

As a consequence, the connection system on the beam is made with 40 headed stud connectors.

2. Connection check — In accordance with the interpolation method, it is:

$$
\begin{aligned}
M_{pl.s} &= W_s \cdot \frac{f_{y.s.k}}{\gamma_s} = 2194 \cdot \frac{235}{1.1} = 468.7 \text{ kNm} \\
M_{pl.c} &= 830 \text{ kNm} \\
M &= 720 \text{ kNm} \\
\frac{N}{N_f} &= \frac{F_c}{F_{c.f}} = \frac{M - M_{pl.s}}{M_{pl.c} - M_{pl.s}} = \frac{720 - 468.7}{830 - 468.7} = 0.696
\end{aligned}
$$

The number of headed stud connectors is then evaluated.

$$N = 0.696 \cdot N_f = 0.696 \cdot 67.8 = 47.2$$

Hence, 50 headed stud connectors along the beam are required.

3. Detailing of the connection (see Section 6.5.4)

- Spacing of the connectors
  The stud connectors are spaced uniformly, being:
  - all critical sections are in class 1
  - $\frac{N}{N_f} = \frac{40}{67.8} = 0.59 \geq 0.25 + 0.03 \cdot 10 = 0.55$
  - $\frac{M_{pl.c}}{M_{pl.s}} = \frac{M_{pl.c}}{W_s \cdot f_{y.s.k}/\gamma_s} = \frac{830}{2194 \cdot 235/1.10} \cdot 10^{-6} = 1.77 < 2.5$

  It is assumed a uniform spacing between studs:

  $$i = \frac{l}{N} = \frac{10000}{40} = 250 \text{ mm} < 22 t_f \sqrt{\frac{235}{f_{y.s.k}}} = 22 \cdot 16 \cdot \sqrt{\frac{235}{235}} = 352 \text{ mm}$$

  It is assumed 40 headed stud connectors, spacing of 250 mm.

- Overall height:

$$6H_{sc} < i < 5d_{sc} \qquad 600\text{ mm} = 6 \cdot 100 < 250 < 5 \cdot 19 = 95\text{ mm}$$

The evaluation of the minimum reinforcement in the slab and the check of the longitudinal shear in the slab must be carried out as indicated in the previous example.

4. Maximum beam deflection — In accordance with Eurocode 4, it is

$$\frac{\delta}{\delta_f} = 1 + 0.3 \cdot \left(1 - \frac{N}{N_f}\right) \cdot \left(\frac{\delta_s}{\delta_f} - 1\right)$$

$\delta_f = 24.4$ mm (see Example 6.1 in Section 6.3.7)

$$\delta_s = \frac{5 \cdot p \cdot l^4}{384 \cdot E \cdot I} = 51.46\text{ mm}$$

Hence:

$$\frac{\delta}{\delta_f} = 1 + 0.3 \cdot \left(1 - \frac{40}{70}\right) \cdot \left(\frac{51.46}{24.1} - 1\right) = 1.1426$$

$$\delta = 27.88\text{ mm} \cong \frac{1}{359}l$$

6.6 Composite Columns

6.6.1 Types of Sections and Advantages

The most common types of steel-concrete composite columns are shown in Figure 6.46. The cross-sections can be classified in 3 groups:

- fully encased
- partially encased
- concrete filled

The first type of cross-section, which is illustrated in Figure 6.46a, is characterized by I-steel profiles fully encased in concrete. In the second type of cross-section (See Figure 6.46b), the steel profile is partially encased in concrete while the external surface of the steel flanges is uncovered. In the third type, the concrete completely fills a steel hollow section. In this case, the column behavior is different when rectangular steel sections (Figure 6.46c) or circular steel sections (Figure 6.46d) are considered.

Composite columns embody several advantages. In general, the two limit cases of a composite section are the reinforced concrete section, when the steel area is small, and the steel section, when concrete is not introduced. Thereby, the composite system is a more complete structural system than simple reinforced concrete or steel elements.

When adopting a composite section, the amount of structural steel, reinforcing steel and concrete area, and the geometry as well as the position of the three materials represent relevant design parameters. Indeed, a number of different combinations is possible thus leading to a flexible design. From a technical viewpoint, the fully encased and partially encased columns offer good fire and corrosion resistance properties, owing to the protection offered by concrete. On the other hand, the most important benefit of the concrete filled type owing to the cross-section is that the steel tube

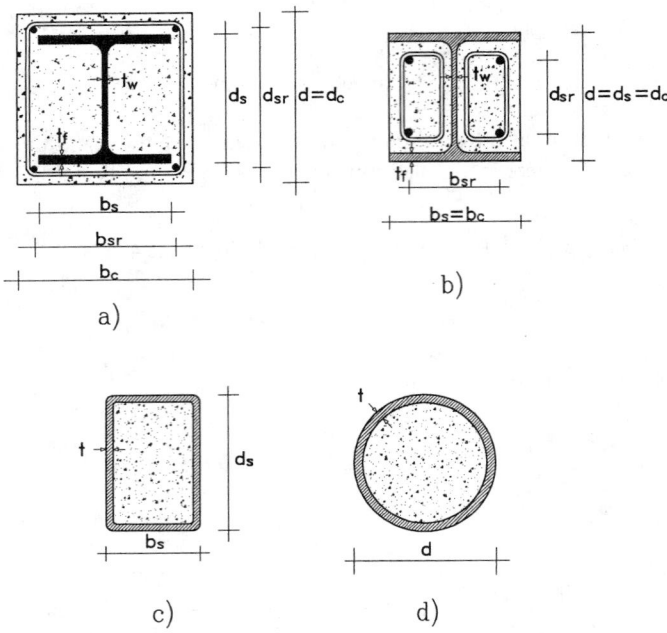

Figure 6.46 Typical steel concrete composite columns. (a) I steel profile fully encased in concrete. (b) I steel profile partially encased in concrete. (c) Hollow steel section filled of concrete. (d) Circular steel section filled of concrete.

also serves as form-work for concrete even if adequate additives have to be used for reducing the concrete separation from the steel section due to shrinkage.

Further advantages are associated with the constructional techniques: for instance, it is possible to set up entirely the steel part of the structure and then to complete it with concrete at alternate levels reducing erection time. It is also possibly a convenient precast of partially encased columns. In particular, the steel profile can be filled with concrete in a horizontal position and then the column can be turned 180° and completed with the remaining concrete.

Moreover, structural benefits can be identified. One important aspect is that concrete prevents local buckling, more effectively in fully encased sections but also in partially encased ones. Also, for concrete filled sections this problem is reduced. Indeed, concrete represents an effective bound for steel in order to prevent or delay the critical warping. Thereby, the elements are generally characterized by a compact behavior while the section reaches full plastic state. In the concrete filled type, the steel provides benefits to concrete. In detail, the confinement effect due to steel is high for the rectangular sections and very high for circular sections, resulting in the increasing of strength with a great enhancement of ductility. The aforementioned last advantage appears to be more relevant to seismic countries, in which composite columns are largely used (see, as an example, Japan).

Clearly, the performances of composite columns require a proper design of connections, both between the two materials (i.e., concrete and steel) and between the two elements (i.e., beams and/or columns and base columns). The problem of steel-concrete connection is analyzed in this section while connections between elements are analyzed in Chapter 23 of this Handbook.

6.6.2 Failure Mechanisms

Composite columns are characterized by several typical failure mechanisms. Collapse due to combined compression and bending could occur, together with the phenomena that characterizes the behavior of slender beam-columns (i.e., geometrical imperfections, erection imperfections, and residual stresses). The shear interaction mechanism could also be present, especially for stocky elements. Local buckling is usually prevented.

A problem relevant to composite systems is represented by the force transfer mechanism between the two components. The force transferring and/or the load introduction have to be developed along a short part of the column in order to consider the element as composite, i.e., full interaction between concrete and steel arises. Reference to bond is possible, but in most cases mechanical connectors are necessary to produce mechanical interlock. The evaluation of connector strength can be made by means of the rules in Section 6.5.

6.6.3 The Elastic Behavior of the Section

All the sections reported in Figure 6.46 are characterized by the centroid of steel profile, reinforcement, and concrete that are coincident, owing to the symmetry about both axes; other cases are more complex [42] and are out of the scope of this chapter.

Due to the aforementioned symmetry, the geometrical characteristics can be evaluated in a simple manner. If concrete is uncracked, the overall full section must be considered, and as a result, the area, A, and the inertia, I, of the composite section can be evaluated as the sum of the area and the inertia of the two components by introducing the modular ratio, n:

$$A = A_s + A_{sr} + A_c/n \tag{6.68}$$

$$I = I_s + I_{sr} + I_c/n \tag{6.69}$$

The long term effects can be taken into account by the EM (Effective Modulus) method. In detail, Equation 6.9 can be adopted to introduce the creep effects caused by dead loads. Both the maximum and minimum stress in concrete, steel profile, and steel rebars can be computed by means of the following expressions:

$$\sigma_{c.\max(\min)} = \frac{N}{n \cdot A} \pm \frac{M}{nI} \cdot \frac{d_c}{2} \tag{6.70}$$

$$\sigma_{s.\max(\min)} = \frac{N}{A} \pm \frac{M}{I} \cdot \frac{d_s}{2} \tag{6.71}$$

$$\sigma_{sr.\max(\min)} = \frac{N}{A} \pm \frac{M}{I} \cdot \frac{d_{sr}}{2} \tag{6.72}$$

where M and N represent the design values (see Figure 6.46 and 6.47a). If the section is in a cracked condition, only the concrete in compression has to be considered (Figure 6.47b), and the approach is similar to the one used for reinforced concrete sections. The elastic neutral axis, x_e, can be evaluated by means of the following equation:

$$I - S\left(\frac{M}{N} - \frac{d}{2} + x_e\right) = 0 \tag{6.73}$$

where
S = the first moment of the cross-section effective area (steel and concrete assumed to be under compression) with respect to the elastic neutral axis
I = the inertia of the effective section respect to the same line
d = the overall dimension of the cross-section
Finally, concrete contributions are divided by the modular ratio, n.

Moreover, stresses can be computed by means of similar expressions that are typical of reinforced concrete sections:

$$\sigma_c = \frac{N}{n \cdot S} \cdot x_e \tag{6.74}$$

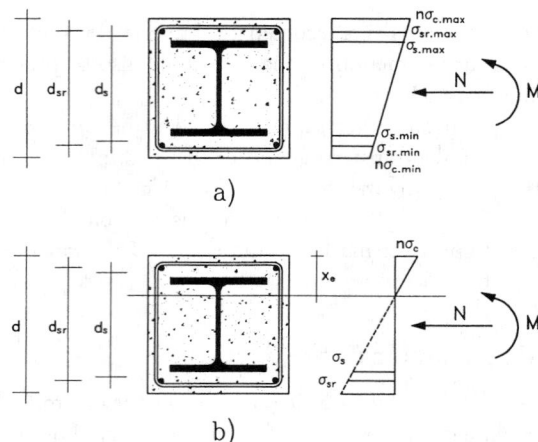

Figure 6.47   Beam-column composite element. (a) Composite section fully effective. (b) Cracked composite section.

$$\sigma_s = \frac{N}{S} \cdot \left(\frac{d + d_s}{2} - x_e \right) \tag{6.75}$$

$$\sigma_{sr} = \frac{N}{S} \cdot \left(\frac{d + d_{sr}}{2} - x_e \right) \tag{6.76}$$

The stress control follows the same indications as the ones adopted for composite beams (see Section 6.3.4).

6.6.4   The Plastic Behavior of the Section

Resistance of the Section Under Compression

First, the uniaxial compression case is considered in order to highlight the main aspects of the problem. Experimental results showed that the resistance of the section can be evaluated as the sum of the strength of the three components (i.e., concrete, structural steel, and reinforcing steel). This approach is allowed if the maximum stress in concrete is reached when steel is yielded (Figure 6.48). Since the concrete strain at the maximum stress is about 0.2%, and by assuming the elastic modulus of steel is about 210000 N/mm^2, maximum stress is reached simultaneously in the two materials

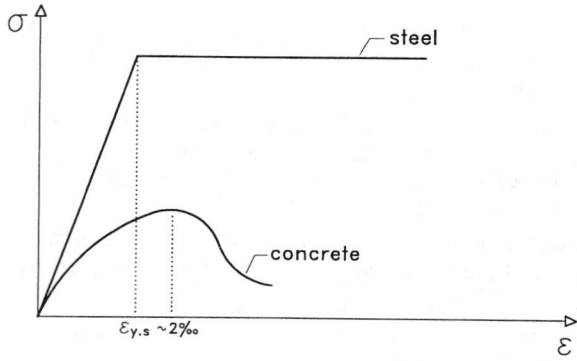

Figure 6.48   Stress-strain relationship for concrete and steel.

if the steel stress at yielding is lower than $2/1000 \times 210000 = 420$ N/mm^2. On the basis of the aforementioned consideration, AISC provisions specify a maximum yield stress of 55 ksi (about 380 N/mm^2). Thereby, high strength steel is excluded as steel yield strain could be higher than the peak strain of concrete. As a result, the yield stress of steel could be reached when concrete behaves in the softening range so as to get the section resistance lower than the sum of the resistance of the two components. However, the sum of the two resistances can still be obtained only if concrete is well confined: in such conditions, concrete is ductile and stress remains practically constant even for high strain values.

According to Eurocode 4 provisions, the "design" plastic axial resistance of the section N_{pl} is evaluated by dividing the "characteristic" strength of material $f_{y.s}$, $f_{y.sr}$, and f_c by means of the partial safety factors as follows:

- encased sections:

$$N_{pl} = \frac{f_{y.s} \cdot A_s}{1.10} + \frac{f_{y.sr} \cdot A_{sr}}{1.15} + \frac{0.85 \cdot f_c \cdot A_c}{1.50} \qquad (6.77)$$

- rectangular concrete filled sections:

$$N_{pl} = \frac{f_{y.s} \cdot A_s}{1.10} + \frac{f_{y.sr} \cdot A_{sr}}{1.15} + \frac{f_c \cdot A_c}{1.50} \qquad (6.78)$$

Additional corrective factors are introduced for circular concrete-filled sections to take into account the confinement action, which is both very effective and beneficial in concrete and reduces the normal bearing capacity of steel that is subjected to biaxial tension and compression.

According to the AISC provisions, the resistance of the section is computed as the product of the "nominal strength" of the materials $f_{y.s}$, $f_{y.sr}$, and f_c times the resistance factor $\phi_c = 0.85$. As a result,

$$N_{pl} = 0.85 \cdot \left(f_{y.s} \cdot A_s + c_1 \cdot f_{y.sr} \cdot A_{sr} + c_2 \cdot f_c \cdot A_c \right) \qquad (6.79)$$

in which the strength of materials has to be considered as a "nominal value", while the numerical factors, c_i, assume the following values:

- encased sections

$$c_1 = 0.70; \quad c_2 = 0.60 \qquad (6.80a)$$

- concrete-filled sections

$$c_1 = 1.00; \quad c_2 = 0.85 \qquad (6.80b)$$

A comparison between the code approaches can be performed by deriving AISC formulation in the EC4 format. One obtains the following quantities:

- encased sections:

$$N_{pl} = \frac{f_{y.s} \cdot A_s}{1.18} + \frac{f_{y.sr} \cdot A_{sr}}{1.68} + \frac{0.85 \cdot f_c \cdot A_c}{1.67} \qquad (6.81)$$

- concrete-filled sections:

$$N_{pl} = \frac{f_{y.s} \cdot A_s}{1.18} + \frac{f_{y.sr} \cdot A_{sr}}{1.18} + \frac{f_c \cdot A_c}{1.38} \qquad (6.82)$$

A comparison between Equations 6.77 and 6.78 and Equations 6.81 and 6.82 mirrors differences that are not very remarkable, especially if the different procedures are considered, which are then applied to evaluate the strength of the material (characteristic values vs. nominal values). It should be noted that AISC provisions remarkably reduce the reinforcement contribution in encased sections.

Resistance of a Section to Combined Compression and Bending

The behavior of composite cross-sections under compression and bending actions is determined by the interaction curves M–N. The relevant codes provide methods quite different for defining the interaction curves. For example, the Japanese provisions (based on the method of Wakabaiashi) are reported in [AIJ, 1987] and the AISC provisions are reported in [AISC, 1994]. The Eurocode 4 approach is based on the study of Roik and Bergman [41, 42].

The general procedure consists of defining some points of the interaction curve by means of the solution of both translational and rotational equilibrium equations of the section. These are based on Bernoulli's hypothesis and by introducing the constitutive relationships of materials, which are described in Section 6.2. The method is theoretically simple, but it requires a considerable effort and it is not used in practice. Thereby, simplified methods are required.

The ductility of the materials allows a full plastic analysis to be used. With reference to the interaction curve of Figure 6.49, point A defines the uniaxial plastic resistance (N = N_{pl}, M = 0) and it can be determined by means of the formulation reported in the previous paragraph. Point B represents the plastic moment resistance (M = M_{pl}, N = 0).

AISC code suggests a simple procedure in order to draw the interaction curve, which consists of approximating the curve by means of two linear segments with the evaluation of N_{pl} and M_{pl} only.

The studies of Roik and Bergman [41, 42] suggest a piecewise linear curve on the safe side with respect to the actual interaction curve to be drawn. The minimum number of points necessary to obtain a realistic multi-linear curve ranges between four or five. These points can be determined by assuming a full plastic stress distribution (i.e., stress block) in all the materials (concrete, structural steel, and reinforcing steel). The result of the method is schematically shown in Figure 6.50. It can be observed that point C is sufficient to obtain a simplified, but conservative, approximation. Point C, which can be readily defined without any additional evaluation, is determined by the same moment M_{pl} that characterizes the condition N = 0. This consideration allows $N_{pl.c}$ to be easily identified as the resistance of the concrete area, by comparing the stress pattern to the one of point B:

$$N_{pl.c} = \frac{c_3 \cdot f_c \cdot A_c}{1.5} \tag{6.83}$$

($c_3 = 1.0$ for filled and $c_3 = 0.85$ for encased sections, respectively)

Point D is identified through the coordinates (N = $N_{pl.c}/2$ and M = M_{max}). Point D can be safely neglected, leading to a slightly conservative interaction curve. In general, the interaction curve A – C – D is convex and, as a result, on the safe side.

In order to evaluate M_{max} and M_{pl}, by means of the stress pattern of Figure 6.50, the following equations can be defined:

$$M_{max} = W_s \cdot f_{y.s} + W_{sr} \cdot f_{y.sr} + \frac{W_c}{2} \cdot 0.85 \cdot f_c \tag{6.84}$$

$$M_{pl} = (W_s - W_{s.n}) \cdot f_{y.s} + (W_{sr} - W_{sr.n}) \cdot f_{y.sr}$$
$$+ \frac{1}{2} \cdot (W_c - W_{c.n}) \cdot 0.85 \cdot f_c \tag{6.85}$$

where

$W_s, W_{sr}, $ and W_c = the plastic moduli of the steel profile, reinforcement, and concrete

$W_{s.n}, W_{sr.n},$ and $W_{c.n}$ = the plastic moduli of the part of steel profile, reinforcement, and concrete in the height of section $\pm h_n$, where h_n is the distance between the plastic neutral axis line and the centroid line in correspondence of $M = M_{pl}$

In Example 6.6, applications of Equations 6.84 and 6.85 will be analyzed. In Equations 6.84 and 6.85, the strength of materials should be considered as "unfactored" strength. In order to obtain

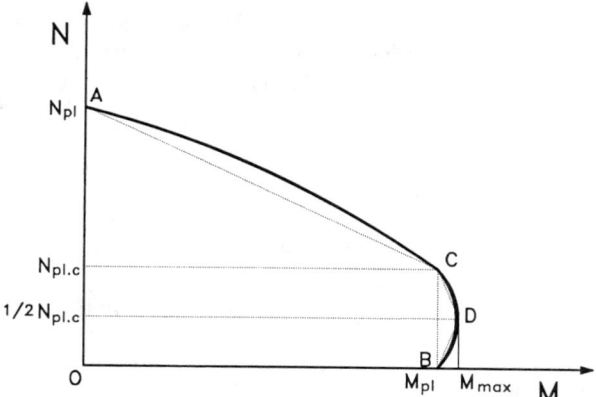

Figure 6.49 M–N interaction curve.

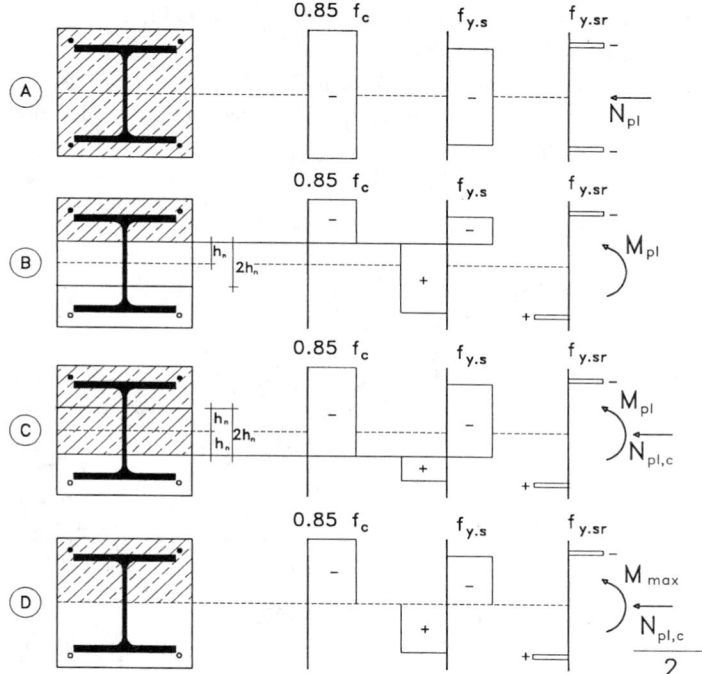

Figure 6.50 Stress distributions corresponding to an interaction curve.

the design values of Eurocode 4, it is necessary to introduce the partial safety factors. Clearly, the interaction curve characterizes the sole section behavior or the behavior of a very stocky element. The actual behavior of members is analyzed in the next paragraph.

6.6.5 The Behavior of the Members

Resistance of Members to Compression

As far as buckling problems are concerned, both AISC and Eurocode 4 extend the approach of the steel columns to the composite ones

According to Eurocode 4, the design ultimate bearing capacity of the composite column N_u has to be determined by considering the imperfection effects and residual stresses. The influence of these effects on the axial resistance of the section N_{pl}, evaluated by means of Equation 6.77 or 6.78, are introduced by means of the factor χ of the buckling curves, in order to evaluate the ultimate axial load of column N_u:

$$N_u = \chi \cdot N_{pl} \tag{6.86}$$

$$\chi = \frac{1}{\phi + \sqrt{\phi^2 - \lambda^2}} \tag{6.87}$$

$$\phi = 0.5 \cdot \left[1 + \alpha \cdot (\lambda - 0.2) + \lambda^2 \right] \tag{6.88}$$

where

λ = the relative slenderness of the column that shall be defined in the following

α = the imperfection factor that is equal to 0.21, 0.34, and 0.49 for three different curves named a, b, and c, respectively (Figure 6.51)

In particular, these three buckling curves a, b, and c refer to concrete-filled cross-sections, encased cross-sections loaded along the strong axis, and encased sections loaded along the weak axis, respectively.

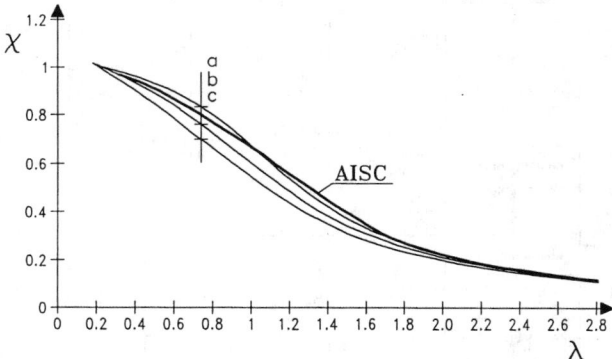

Figure 6.51   Buckling curves for Eurocode 4 and AISC provisions.

AISC code allows the critical stress σ_{cr} (F_{cr} according to the symbols of AISC) to be evaluated, which has to be multiplied by the steel area A_s as a function of a conventional value of steel yield stress f_{my}, modified in order to account for the other components of the section. The buckling curve comprises two branches represented by the following equations:

$$\sigma_{cr} = \left(0.658^{\lambda^2} \right) \cdot f_{my} \ \ \text{if } \lambda \leq 1.5 \tag{6.89a}$$

$$\sigma_{cr} = \left(\frac{0.877}{\lambda^2} \right) \cdot f_{my} \ \ \text{if } \lambda \geq 1.5 \tag{6.89b}$$

The comparison among the three buckling curves of Eurocode 4 and the one of AISC is shown in Figure 6.51. In order to draw the AISC curve, the following relations are adopted:

$$\chi = \left(0.658^{\lambda^2}\right) \quad \text{for } \lambda \leq 1.5 \tag{6.90a}$$

$$\chi = \left(\frac{0.877}{\lambda^2}\right) \quad \text{for } \lambda \geq 1.5 \tag{6.90b}$$

Eurocode 4 defines the "relative slenderness" by means of the following expression:

$$\lambda = \sqrt{\frac{N_{pl}}{N_{cr}}} = \frac{k \cdot l}{\pi} \cdot \sqrt{\frac{f_{y.s} \cdot A_s + f_{y.sr} \cdot A_{sr} + 0.85 \cdot f_c \cdot A_c}{E_s \cdot I_s + E_{sr} \cdot I_{sr} + 0.8 \cdot E_c \cdot I_c / 1.35}} \tag{6.91}$$

Since N_{pl} is expressed by means of Equations 6.77 or 6.78, by assuming the characteristic values of the strength without the partial safety factors:

$$N_{pl} = f_{y.s} \cdot A_s + f_{y.sr} \cdot A_{sr} + c_3 \cdot f_c \cdot A_c \tag{6.92}$$
$$(c_3 = 1.0 \text{ filled} \qquad c_3 = 0.85 \text{ encased})$$

The critical bearing capacity N_{cr} is calculated as follows:

$$N_{cr} = \frac{\pi^2 \cdot (EI)_e}{(kl)^2} \tag{6.93}$$

where

kl = the effective length
$(EI)_e$ = the effective flexural stiffness of the composite section

$$(EI)_e = E_s \cdot I_s + E_{sr} \cdot I_{sr} + 0.59 E_c \cdot I_c \tag{6.94}$$

The factor $0.8/1.35 = 0.59$ that reduces the elasticity modulus of concrete considers a secant modulus owing to the non-linear behavior of concrete and it allows a good agreement of the theoretical and experimental results to be achieved [41]. The creep influence can be remarkable for slender columns; therefore, in such a case, an additional reduction of the elasticity modulus of concrete that leads to an effective elasticity modulus:

$$E_{c.ef} = E_c \cdot \left(1 - 0.5 \cdot \frac{N_p}{N}\right) \tag{6.95}$$

where the ratio between the axial load owing to the dead load, N_p, and the total axial load, N, is considered to weight the creep effect. As an example, if the dead load is 2/3 and the live load is 1/3 of the total axial load, the effective modulus of concrete is obtained by a reduction factor equal to $1 - 0.5 \cdot 2/3 = 0.67$.

The slenderness definition of AISC is quite different since the following expression is introduced:

$$\lambda = \frac{kl}{\pi \cdot r_m} \cdot \sqrt{\frac{f_{y.s} \cdot A_s + c_1 \cdot f_{y.sr} \cdot A_{sr} + c_2 \cdot f_c \cdot A_c}{E_s \cdot A_s + c_4 \cdot E_c \cdot A_c}} \tag{6.96}$$

where

$r_m = (I_s/A_s)^{0.5}$ = the gyration radius of the steel part

The factors c_1, c_2 have already been used to define Equation 6.79, while c_4 assumes the following values:

- encased cross-sections:

$$c_4 = 0.20$$

- concrete cross-filled:

$$c_4 = 0.40$$

Also, in this case the comparison between the two codes can be affected by changing the format of AISC slenderness according to the Eurocode 4 expression:

$$\lambda = \frac{kl}{\pi} \cdot \sqrt{\frac{f_{y.s} \cdot A_s + c_1 \cdot f_{y.sr} \cdot A_{sr} + c_2 \cdot f_c \cdot A_c}{E_s \cdot I_s + c_4 \cdot E_c \cdot I_s \cdot A_c / A_s}} \tag{6.97}$$

The conceptual difference between Equations 6.91 and 6.97 is that the reinforcement contribution to the effective inertia is neglected and the concrete contribution is evaluated in a simplified manner in Equation 6.97.

The reduction factor χ allows the interaction curve of the section to be reduced by obtaining the interaction curve of the member. Under a pure compression regime, the column design is satisfied if:

$$N < \chi \cdot N_{pl} \tag{6.98}$$

Resistance of Members to Combined Compression and Bending

The check of a column subject to combined compression and bending has to be carried out by means of the following steps:

- The interaction curve of the section has to be evaluated.
- The interaction curve of the member shall be drawn reducing the interaction curve of the section to take into account the geometrical non-linearity related both to the axial compression (axial buckling) and to bending (flexural-torsional buckling). Moreover, it has to be considered the influence of geometrical imperfections (by the fabrication of the elements as well as the erection procedure), mechanical imperfections, (i.e., residual stresses), bending moment pattern along the element, the presence of lateral restraints, etc.
- The external actions have to be increased in a simplified way (in particular, the bending moment owing to the load has to be increased) to evaluate the stress introduced by the geometrical non-linearity that can influence the slender elements.

By considering the procedure of Eurocode 4, one assumes that no bending moment, M, can be applied in correspondence of $N_u = \chi N_{pl}$. However, when the axial compression is null, the moment M_{pl} can be applied entirely. As a result, for a generic value of N, Eurocode 4 suggests referral to the line between these two limiting values. For the design value of the axial force, N_d, the plastic moment, M_{pl}, reported in Figure 6.52 can be carried by the column. Moreover, what follows can be observed:

- The use of stress blocks for the materials is unsafe to a certain extent because it implies infinite ductility of concrete. As a result, Eurocode 4 suggests reduction of M_{pl} by a factor of 0.9.
- The geometrical imperfections of the column which are taken into account in the buckling curves are considered in the most unfavorable combination: constant along all the element. Thus, the method previously explained is referred to as a constant distribution of the bending moment. If the moment is variable along the column, the procedure is much safer. It is possible to consider the interaction curve marked by the dotted line in Figure 6.52 that intersects the points M = 0; N = $\chi_n N_{pl}$ where

$$\chi_n = \chi \cdot \frac{1 - r}{4} \tag{6.99}$$

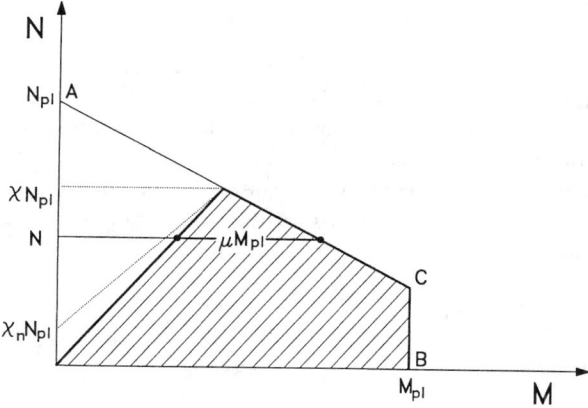

Figure 6.52 M–N design interaction diagram.

where r is limited between -1 and 1 and represents the ratio between the values of bending moments at the member ends. The design bending moment, M, that has to be considered in the check is an equivalent moment, owing to the variability along the element.

A number of studies dealt with this aspect; in particular, Eurocode 4 takes into account the second order moment multiplying the first order bending moment by means of a factor, a:

$$a = \frac{c}{1 - N/N_{cr}} \geq 1.0 \quad ; \quad c = 0.66 + 0.44 \cdot r \geq 0.44 \tag{6.100}$$

where r is already defined in Figure 6.53. In the factor, a, the term above takes into account the moment distribution along the column. In detail, c equals 1 if the moment is constant, the term

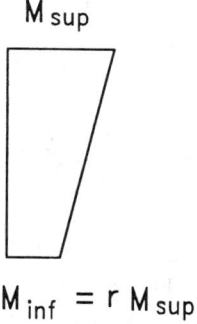

Figure 6.53 Definition of r factor in accordance with Eurocode 4.

below is lower than 1, and increases the equivalent moment introducing the effect of the second order moment (this term approaches 1 if N is much lower than N_{cr}, while it approaches 0 if N tends to N_{cr}). Finally, the check of the column requires that the moment M is lower than the ultimate value M_u equal to 0.9 μM_{pl}; i.e.:

$$M \leq M_u = 0.9 \cdot \mu \cdot M_{pl} \tag{6.101}$$

The aforementioned numerical procedure is developed by means of the examples to follow regarding fully encased columns, partially encased columns, and concrete-filled columns reported in Section 6.6.9.

6.6.6 Influence of Local Buckling

The elastic approach of Section 6.6.3 has to be applied to evaluate the stress condition under the serviceability loading; conversely, the full plastic bearing capacity of the section can be reached almost always at failure. In fact, as mentioned previously, the presence of the concrete reduces the local buckling phenomena; in particular, in fully encased sections local buckling is definitely prevented if a certain minimum concrete cover is used. In the other cases, local buckling can be excluded if limit ratios of depth-to-thickness for the steel section are respected. By considering Eurocode 4, it follows that:

- partially encased I-sections:

$$b_s/t_f \leq 44 \cdot \sqrt{235/f_{y.s}} \tag{6.102}$$

- circular hollow steel sections:

$$d_s/t \leq 90 \cdot \sqrt{235/f_{y.s}} \tag{6.103}$$

- rectangular hollow steel sections:

$$b_s/t \leq 52 \cdot \sqrt{235/f_{y.s}} \tag{6.104}$$

where $f_{y.s}$ is the characteristic strength in N/mm^2. These restrictions are loose with respect to the case of simple profiles analyzed in Section 6.4.3.

The AISC approach is slightly different by considering the following restrictions:

- circular hollow steel sections:

$$t \geq d_s\sqrt{\frac{f_{y.s}}{8 \cdot E_s}} \tag{6.105}$$

- rectangular hollow steel sections:

$$t \geq b_s\sqrt{\frac{f_{y.s}}{3 \cdot E_s}} \tag{6.106}$$

where
$f_{y.s}$ = the nominal strength in N/mm^2

In order to compare different codes, the AISC provisions can be expressed in the same format of Eurocode 4 by assuming $E_s = 210000$ N/mm^2 and are reported in the following:

- circular hollow steel sections:

$$d_s/t \leq 85 \cdot \sqrt{\frac{235}{f_{y.s}}} \tag{6.107}$$

- rectangular hollow steel sections:

$$b_s/t \leq 52 \cdot \sqrt{\frac{235}{f_{y.s}}} \tag{6.108}$$

One can observe that the restrictions of the two codes are similar.

6.6.7 Shear Effects

In stocky members, or in the case of high horizontal loads, the shear influence can be remarkable. In this case, the interaction curve N–M–V has to be drawn both for the section and for the member. To analyze this problem in a general manner, the constitutive relationships of materials and appropriate resistance criteria are required. Moreover, all the geometrical and mechanical non-linearities and imperfections of the element have to be considered in the computation.

Usually, the provisions suggested for the continuous beam (see Section 6.4.5), where interaction between shear and bending takes place, apply to the columns. In detail, the approximate procedure assumes that shear is carried out by means of the web of the steel profile and only a part of the web is considered able to resist the combined compression and bending moment.

6.6.8 Load Introduction Region

The connection between horizontal (beams) and vertical (beam columns) usually requires complex systems that are analyzed in Chapter 23 of this Handbook. However, when the joints are set, it is necessary to ensure the load introduction along a short length of the members out of the nodal zone. As an example, Eurocode 4 suggests that the load introduction length should not exceed twice the overall dimensions of the cross-section. The transfer of the shear forces can be guaranteed by steel-concrete bond only if the bond stress is lower than prescribed limits. As already stated in Section 6.2 bond stress has to satisfy the following limits:

- 0.6 N/mm^2 for encased cross-sections
- 0.4 N/mm^2 for concrete-filled cross-sections

Otherwise, mechanical shear connectors must be used. Analogous rules of the connections to beams have to be used. Moreover, in the case of fully encased beams, Eurocode 4 provisions allow an increase of the shear capacity of stud connectors owing to the friction between concrete and steel flange, using a frictional coefficient of 0.5.

Another aspect of the problem regards the load introduction to the members. When the supporting concrete area is wider than the loaded area, the restrained lateral expansion offered by the remaining concrete provides a strength increase; AISC provisions consider this phenomenon increasing of the design strength of the concrete with a factor of 1.7.

6.6.9 Restrictions for the Application of the Design Methods

The rules explained in the previous paragraphs according to both Eurocode 4 and AISC are effective if some restrictions are fulfilled. Indeed, the simplified procedures are based on large experimental and numerical analyses that, however, cannot take into account all design conditions. Thereby, though it is always possible to apply a general procedure introducing the constitutive relationships of materials, the rules suggested by Eurocode 4 can be used only if the following restrictions are satisfied:

- the cross-sections are symmetric about the two axes and the cross-section is constant along the member
- the factor δ_s

$$\delta_s = \frac{A_s \cdot f_{y.s}/1.1}{N_{pl}} \tag{6.109}$$

that represents the contribution of structural steel in the plastic axial load capacity varies between 0.2 and 0.9; otherwise, the member has to be designed as a reinforced concrete element corresponding to a lower restriction or as a steel element corresponding to a higher restriction.

- The relative slenderness λ shall be lower than 2.
- If the longitudinal reinforcement is considered in the design, the minimum share of 0.3% and maximum share of 4% of the concrete area shall be provided.
- In the fully encased cross-sections, a minimum concrete clear cover of 40 mm shall be provided.

Moreover, the additional restrictions must be considered: maximum clear cover along the axis direction of 0.3 d; maximum clear cover along the weak axis of 0.4 b_s. Likewise, the AISC provisions require the following restrictions:

- The steel profile area has to be at least 4% of the composite cross-section area.
- In the case of encased sections, longitudinal reinforcement shall be continuous at framed levels; the spacing of ties shall be not greater than 2/3 the minimum dimension of the cross-section.
- The cross-sectional area of the transverse and longitudinal reinforcement shall be at least 1.8 mm^2 per millimeter of bar spacing (0.007 in.^{2} per inch of bar spacing). The clear cover of both longitudinal and transversal reinforcement shall be at least 38.1 mm (1.5 in.).
- The concrete, reinforcing steel, and structural steel shall follow the requirements of par 2.1, 2.2, and 2.3.
- In the case of concrete-filled sections, the restrictions expressed by Equations 6.107 and 6.108 reported in Section 6.6.6 have to be taken into account.
- The gyration radius of the steel profile r_m (see Section 6.6.5) shall not be less than 0.3 times the overall dimension of the cross-section in the plane of buckling.

6.6.10 Worked Examples

In what follows, the Eurocode 4 procedure is developed with reference to two beam columns characterized by the same length, bending moment pattern, and axial load, but different type of cross-section. In Example 6.6, the case of a fully encased cross-section is analyzed; in Example 6.7, a rectangular concrete-filled cross-section is analyzed. In both cases, the column is characterized by the following conditions:

- length $l = 4000$ mm
- restraints: fixed at the lower end and simply supported at the top end in both vertical planes
- axial load: N = 850 kN
- bending moment: linear pattern as illustrated in Figure 6.54 characterized by M = 140 kNm at the top end and M = −70 kNm at the bottom end ($r = -0.5$ as shown in Figure 6.53)

Materials are characterized by the following characteristic strength, design strength, and elasticity modulus:

- structural steel: $f_{y.s} = 355$ N/mm^2, $f_{y.s.d} = 322.7$ N/mm^2; $E_s = 210000$ N/mm^2
- reinforcing steel: $f_{y.sr} = 420$ N/mm^2, $f_{y.sr.d} = 365.2$ N/mm^2; $E_{sr} = 210000$ N/mm^2
- concrete: $f_c = 20$ N/mm^2; $f_{c.d} = 13.3$ N/mm^2; $E_c = 29000$ N/mm^2

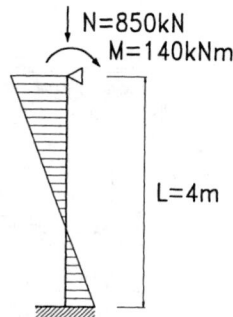

Figure 6.54 Distribution of bending moment in the column.

EXAMPLE 6.6: Fully encased cross-section

The section has dimension 300 mm · 300 mm; the steel profile is HEA 200; the reinforcement is provided by four bars with diameter 12 mm.

The geometric characteristics of the cross-section are (see Figure 6.55):

$$b_c = 300 \text{ mm}; d_c = 300 \text{ mm};$$
$$b_{rs} = 230 \text{ mm}, d_{rs} = 230 \text{ mm};$$
$$b_s = 200 \text{ mm}, d_s = 190 \text{ mm}:$$

thus leading to the following area and second order moment of the cross-section:

- steel

$$A_s = 5380 \text{ mm}^2,$$
$$I_s \text{ (strong axis)} = 3692.10^4 \text{ mm}^4, I_s \text{ (weak axis)} = 1336.10^4 \text{ mm}^4$$

- rebars:

$$A_{sr} = 4 \cdot 113 = 452 \text{ mm}^2; I_{sr} = 4 \cdot 113 \cdot 115^2 = 598 \cdot 10^4 \text{ mm}^4$$

- concrete

$$A_c = 300 \cdot 300 - 5380 - 452 = 84,168 \text{ mm}^2$$
$$I_c = \frac{300^4}{12} - 3692 \cdot 10^4 - 598 \cdot 10^4 = 6321 \cdot 10^4 \text{ mm}^4 \quad \text{(strong axis)}$$
$$I_c = \frac{300^4}{12} - 1336 \cdot 10^4 - 598 \cdot 10^4 = 6557 \cdot 10^4 \text{ mm}^4 \quad \text{(weak axis)}$$

1. Application of simplified rules and shear interaction — The plastic axial resistance of the cross-section according to Equation 6.77 is

$$N_{pl} = \frac{355 \cdot 5380}{1.10} + \frac{420 \cdot 452}{1.15} + \frac{0.85 \cdot 20 \cdot 84168}{1.50} = 2855 \cdot 10^3 \text{ N} = 2855 \text{ kN}$$

The steel contribute as given by Equation 6.109 is

$$\delta_s = \frac{5380 \cdot 355/1.1}{2855 \cdot 10^3} = 0.608$$

The other limitations given by Section 6.6.9 are satisfied; thereby the simplified method of Eurocode 4 can be applied. It should be observed that there are no problems of local buckling that is typical of fully encased elements (see Section 6.6.6). It should also be observed that:

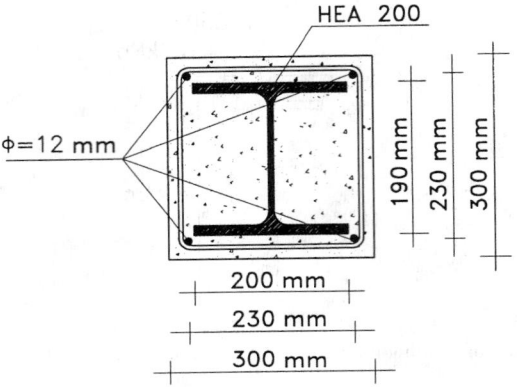

Figure 6.55   Geometric characteristics of the cross-section.

- the design value of shear is

$$V = \frac{140 + 70}{4} = 52.5 \text{ kN}$$

- the plastic shear resistance is

$$V_{pl} = 1.04 \cdot 6.5 \cdot 200 \cdot \frac{355}{1.1\sqrt{3}} = 252 \cdot 10^3 \text{ N} = 252 \text{ kN}$$

thus not leading to interaction between bending moment and shear ($V < 0.5\ V_{pl}$; Section 6.6.7).

2. Determination of the *M–N* interaction curve — With reference to Section 6.6.4, the general evaluation of the interaction N–M curve requires a number of couples (N,M) using the two equations of equilibrium and Bernoulli assumption to be evaluated. The full interaction curve of the cross-section defined in this example, evaluated point by point by means of a numerical procedure, is illustrated in Figure 6.56. It is evident that point A, B, C, and D provide a quite good approximation of the actual curve.

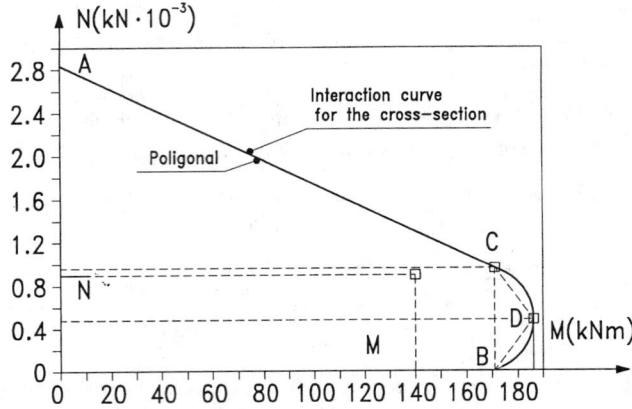

Figure 6.56   Cross-section interaction curve.

The points are defined as follows:

- point A: $(N = N_{pl},\ M = 0)$
- point B: $(N = 0,\ M = M_{pl})$
- point C: $(N = N_{pl.c},\ M = M_{pl})$
- point D: $(N = N_{pl.c}/2,\ M = M_{max})$

By means of Equation 6.83, one obtains what follows:

$$N_{pl.c} = 0.85 \cdot 13.3 \cdot 8416 = 953 \cdot 10^3 \text{ N} = 953 \text{ kN}$$

M_{max} and M_{pl} are evaluated by Equations 6.84 and 6.85:

$$M_{max} = W_s \cdot f_{y.s.d} + W_{sr} \cdot f_{yy.sr.d} + \frac{W_c}{2} 0.85 \cdot f_{c.d}$$

$$M_{pl} = (W_s - W_{s.n}) \cdot f_{y.s.d} + (W_s - W_{sr.n}) \cdot f_{y.sr.d} + \frac{1}{2} \cdot (W_c - W_{c.n}) \cdot 0.85 \cdot f_{c.d}$$

where:

$$W_s = 407 \cdot 10^3 \text{ mm}^3$$
$$W_{sr} = 4 \cdot 113 \cdot 115 = 52 \cdot 10^3 \text{ mm}^3$$
$$W_c = \frac{b_c h_c^2}{4} - W_s - W_{sr} = \frac{300 \cdot 300^2}{4} - 407 \cdot 10^3 - 52 \cdot 10^3 = 6291 \cdot 10^3 \text{ mm}^3$$

and

$$h_n = \frac{953 \cdot 10^3/2}{300 \cdot 0.85 \cdot 13.33 + 6.5 \cdot (2 \cdot 322.7 - 0.85 \cdot 13.33)} = 63.4 \text{ mm}$$
$$W_{s.n} = t_w \cdot h_n^2 = 6.5 \cdot 63.4^2 = 26127 \text{ mm}^3$$
$$W_{sr.n} = 0$$
$$W_{c.n} = b_c \cdot h_n^2 - W_{s.n} = 300 \cdot 63.4^2 - 26127 = 1180 \cdot 10^3 \text{ mm}^3$$

It follows that:

$$
\begin{aligned}
M_{max} &= 407 \cdot 10^3 \cdot 322.7 + 52 \cdot 10^3 \cdot 365.2 \\
&\quad + \frac{6291 \cdot 10^3}{2} 0.85 \cdot 13.33 = 186 \cdot 10^3 \text{ N} = 186 \text{ kN} \\
M_{pl} &= (407 - 26) \cdot 10^3 \cdot 322.7 + (52) \cdot 10^3 \cdot 365.2 \\
&\quad + \frac{1}{2}(6291 - 1180) \cdot 10^3 \cdot 0.85 \cdot 13.3 = 171 \cdot 10^5 \text{ Nmm} = 171
\end{aligned}
$$

3. Ultimate axial load evaluation — In order to evaluate the ultimate axial load, it is necessary to compute the plastic axial resistance, without partial safety factors, i.e.:

$$N_{pl} = 355 \cdot 5380 + 420 \cdot 452 + 0.85 \cdot 20 \cdot 84168 = 3530 \cdot 10^3 \text{ N} = 3530 \text{ kN}$$

It is also necessary to evaluate the effective elastic flexural stiffness, by means of Equation 6.94, both in the planes of the strong axis of the steel profile and of the weak axis of the steel profile:

- strong axis:

$$(EI)_e = 210000 \cdot 3692 \cdot 10^4 + 210000 \cdot 598 \cdot 10^4$$
$$+ \frac{0.8 \cdot 29000 \cdot 6557 \cdot 10^5}{1.35} = 1.987 \cdot 10^{13} \text{ N} \cdot \text{mm}^2$$

- weak axis:

$$(EI)_e = 210000 \cdot 1336 \cdot 10^4 + 210000 \cdot 598 \cdot 10^4$$
$$+ \frac{0.8 \cdot 29000 \cdot 6321 \cdot 10^5}{1.35} = 1.492 \cdot 10^{13} \text{ N} \cdot \text{mm}^2$$

By considering a value of the effective length factor k = 0.7 in both planes, the slenderness and the buckling curve factor are according to Equation 6.91:

- strong axis:

$$\lambda = \frac{0.7 \cdot 4000}{\pi} \cdot \sqrt{\frac{3530 \cdot 10^3}{1.987 \cdot 10^{13}}} = 0.376$$

In this plane it shall be considered the stability curve b and the value of the imperfection factor α is 0.34. It follows by means of Equation 6.88:

$$\phi = 0.5 \left[1 + 0.34 \cdot (0.376 - 0.2) + 0.376^2 \right] = 0.60$$

The stability curve factor is given by means of Equation 6.87:

$$\chi = \frac{1}{0.601 + \sqrt{0.601^2 - 0.376^2}} = 0.935$$

- weak axis

$$\lambda = \frac{0.7 \cdot 4000}{\pi} \cdot \sqrt{\frac{3530 \cdot 10^3}{1.492 \cdot 10^{13}}} = 0.433$$

In this plane it shall be considered the stability curve c and the value of the imperfection factor $\alpha = 0.49$; it follows by means of Equation 6.88:

$$\phi = 0.5 \left[1 + 0.49 \cdot (0.433 - 0.2) + 0.433^2 \right] = 0.65$$

The stability curve factor is given by means of Equation 6.87:

$$\chi = \frac{1}{0.651 + \sqrt{0.651^2 - 0.433^2}} = 0.879$$

It is evident that the minimum value of the stability curve factor is given by the weak axis, thus leading to (Equation 6.86):

$$N_u = 0.879 \cdot 2855 = 2510 \text{ kN}$$

4. Member check — By means of Equation 6.98, we get:

$$\chi_n = 0.879 \cdot \frac{1 + 0.5}{4} = 0.330$$

Being:

$$\frac{N}{N_{pl}} = 0.298 < 0.330$$

In Figure 6.57, the procedure for the evaluation of the factor μ (Equation 6.101 and Figure 6.52) is shown; the check point is under the point C; thus, it is on the safe side to consider a value of $\mu = 1$. Consequently, the value of the ultimate moment of the column is the following:

$$M_u = 0.9 \cdot 1 \cdot 171 = 154 \text{ kNm}$$

It is evident that the magnification factor, a, of Equation 6.100 is equal to 1. Thus, it follows that:

$$M = 140 \text{ kNm} < M_u = 154 \text{ kNm}$$

and the column check is satisfied.

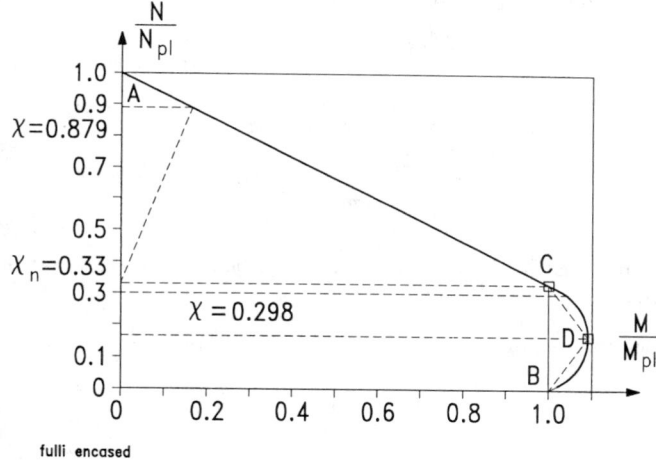

Figure 6.57 Procedure for the evaluation of the factor μ.

EXAMPLE 6.7: Square steel section filled of concrete

The section is characterized by the dimension 220 mm · 220 mm; the steel section has a square shape with a constant thickness t equal to 7.1 (Figure 6.58). Thus, the area and second order moment of the cross-section are

• steel:

$$A_s = t \cdot 2 \cdot (b - t) = 7.1 \cdot 2 \cdot (220 - 7.1) = 6046 \text{ mm}^2$$

$$I_s = \frac{b^4}{12} - \frac{(b - 2 \cdot t)^4}{12} = \frac{220^4}{12} - \frac{(220 - 2 \cdot 7.1)^4}{12} = 4573 \cdot 10^4 \text{ mm}^4$$

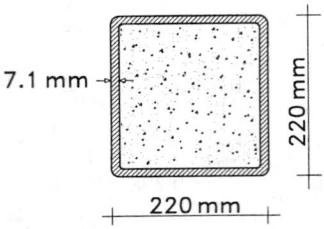

Figure 6.58 Geometric characteristics of the cross-section.

- concrete

$$A_c = (b - 2 \cdot t)^2 = (220 - 2 \cdot 7.1)^2 = 42354 \text{ mm}^2$$

$$I_c = \frac{(b - 2 \cdot t)^4}{12} = \frac{(220 - 2 \cdot 7.1)^4}{12} = 14949 \cdot 10^4 \text{ mm}^4$$

1. Application of the simplified rules and shear interaction — The plastic axial resistance of the cross-section is according to Equation 6.77:

$$N_{pl} = \frac{355 \cdot 6046}{1.10} + \frac{20 \cdot 42354}{1.50} = 2515 \cdot 10^3 \text{ N} = 2515 \text{ kN}$$

The steel contribution is given by Equation 6.109 and it is equal to:

$$\delta_s = \frac{6046 \cdot 355/1.1}{2515 \cdot 10^3} = 0.776$$

The local slenderness is according to (Section 6.6.6):

$$\frac{d}{t} = \frac{220}{7.1} = 30.98 < 52\sqrt{\frac{235}{355}} = 42.12$$

Thus, the simplified method of Eurocode 4 can be applied. However, it should be observed that:

- the design value of shear is

$$V = \frac{140 + 70}{4} = 52.5 \text{ kN}$$

- while the plastic shear resistance is

$$V_{pl} = 1.04 \cdot 2 \cdot 7.1 \cdot 220 \cdot \frac{355}{1.1\sqrt{3}} = 605 \cdot 10^3 \text{ N} = 605 \text{ kN}$$

thus not leading to interaction between bending moment and shear ($V_d < 0.5$ V_{pl}; see Section 6.6.7).

2. Determination of the M–N interaction curve — With reference to Section 6.6.4, the evaluation of interaction N–M curve requires the definition of a number of couple (N,M) by using the two equations of equilibrium and the Bernoulli assumption. The full interaction curve of the cross-section defined in this example, evaluated point by point by means of a numerical procedure, is illustrate in Figure 6.59. In this case, the linear curve by means of the points A, B, C, and D provide an approximation less accurate; therefore, additional couple of points E and F should be considered. The points are defined as follows:

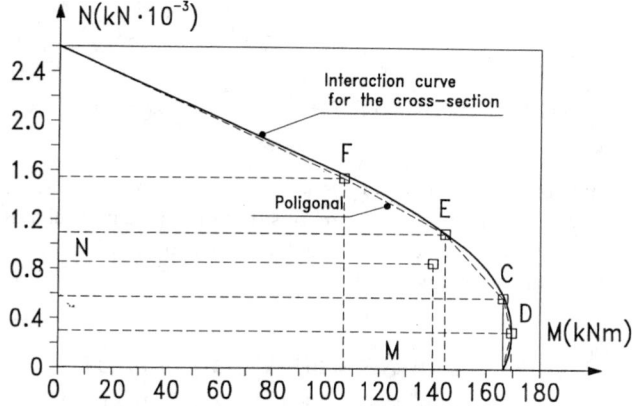

Figure 6.59 Interaction curve of the cross-section.

- point A: $(N = N_{pl}, \; M = 0)$
- point B: $(N = 0, \; M = M_{pl})$
- point C: $(N = N_{pl.c}, \; M = M_{pl})$
- point D: $(N = N_{pl.c}/2, \; M = M_{max})$
- point E: $(N = N_E, \; M = M_E)$
- point F: $(N = N_F, \; M = M_F)$

By means of Equation 6.83:

$$N_{pl.c} = f_{c.d} \cdot A_c = 13.3 \cdot 42354 = 563 \cdot 10^3 \text{ N} = 563 \text{ N}$$

M_{max} and M_{pl} are evaluated by means of Equations 6.84 and 6.85 where:

$$
\begin{aligned}
W_s &= 2\left[b \cdot t \cdot \left(\frac{b-t}{2}\right) + \left(\frac{b}{2} - t\right)^2 \cdot t\right] \\
&= 2\left[220 \cdot 7.1 \cdot \left(\frac{220 - 7.1}{2}\right) + \left(\frac{220}{2} - 7.1\right)^2 \cdot 7.1\right] = 483 \cdot 10^3 \text{ mm}^3 \\
W_{sr} &= 0 \\
W_c &= \frac{(b - 2 \cdot t)^3}{4} = \frac{(220 - 2 \cdot 7.1)^3}{4} = 2179 \cdot 10^3 \text{ mm}^3
\end{aligned}
$$

and

$$
\begin{aligned}
h_n &= \frac{563 \cdot 10^3 / 2}{206 \cdot 13.3 + 2 \cdot 7.1 \cdot (2 \cdot 322.7 - 13.3)} = 24 \text{ mm} \\
W_{c.n} &= (b - 2 \cdot t) \cdot h_n^2 = (220 - 2 \cdot 7.1) \cdot 24^2 = 118.5 \cdot 10^3 \text{ mm}^3 \\
W_{s.n} &= b \cdot h_n^2 - W_{c.n} = 220 \cdot 24^2 - 118.5 \cdot 10^3 = 8220 \text{ mm}^3 \\
W_{sr.n} &= 0
\end{aligned}
$$

It follows that:

$$M_{max} = 483 \cdot 10^3 \cdot 322.7 + 2179 \cdot 10^3 \cdot \frac{13.3}{2} = 170 \cdot 10^3 \, N = 170 \, kN$$

$$M_{pl} = \left(483 \cdot 10^3 - 8220\right) \cdot 322.7 + \frac{1}{2}(2179 - 118.5) \cdot 10^3 \cdot 13.3 = 167 \, kNm$$

By using the same procedure, it is also possible to define points E and F, i.e., it is possible to determine the values h_E and h_F of the position of plastic neutral axis, and than to evaluate N and M by means of the equilibrium equations. The points illustrated in Figure 6.59 are characterized by:

$$h_E = 66.84; \; N_E = 1077 \, kN, \; M_E = 143.3 \, kNm$$
$$h_F = 102.9; \; N_F = 1507 \, kN, \; M_F = 107.0 \, kNm$$

3. Ultimate axial load evaluation — In order to evaluate the ultimate axial load it is necessary to compute the plastic axial resistance, without partial safety factors, i.e.:

$$N_{pl} = 355 \cdot 6046 + 20 \cdot 42354 = 2993 \cdot 10^3 \, N = 2993 \, kN$$

It is also necessary to evaluate the effective elastic flexural stiffness, by means of Equation 6.94, than in this particular case is equal in both planes:

$$(EI)_e = 210000 \cdot 4573 \cdot 10^4 + \frac{0.8 \cdot 29000 \cdot 14949 \cdot 10^4}{1.35} = 1.217 \cdot 10^{13} \, N \cdot mm^2$$

By considering a values of the effective length factor $k = 0.7$, the slenderness and the buckling curve factor are equal to (see Equation 6.91):

$$\lambda = \frac{0.7 \cdot 4000}{\pi} \cdot \sqrt{\frac{2993 \cdot 10^3}{1.217 \cdot 10^{13}}} = 0.442$$

In this plane, it shall be considered the stability curve a and the value of the imperfection factor α is 0.21; it follows by means of Equation 6.88 that:

$$\phi = 0.5\left[1 + 0.21 \cdot (0.442 - 0.2) + 0.442^2\right] = 0.623$$

The stability curve factor is given by Equation 6.87:

$$\chi = \frac{1}{0.623 + \sqrt{0.623^2 - 0.442^2}} = 0.941$$
$$N_u = 0.941 \cdot 2515 = 2367 \, kN$$

4. Member check — By means of Equation 6.98, one gets:

$$\chi_n = 0.941 \cdot \frac{1 + 0.5}{4} = 0.353$$

Being:

$$\frac{N}{N_{pl}} = 0.338 < 0.353$$

In Figure 6.60, the check point is above the point C and as a result $\mu = 0.94$. Consequently, the value of the ultimate moment of the column is

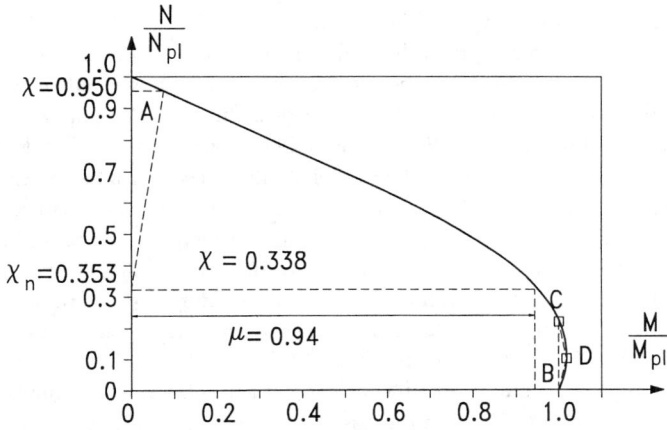

Figure 6.60 Procedure for the evaluation of factor μ.

$$M_u = 0.9 \cdot 0.94 \cdot 167 = 141 \text{ kNm}$$

It is evident that the magnification factor, a, of Equation 6.99 is equal to 1. Thus, it follows that:

$$M = 140 \text{ kNm} \ < \ M_u = 141 \text{ kNm}$$

and the column check is satisfied.

6.7   Composite Slabs

Since its development in North America in the late 1950s, composite floor systems using light gauge metal sheeting proved to be a very efficient solution, which became increasingly popular worldwide (Figure 6.26b). The steel deck serves

- first as a working platform and safety netting system,
- then as a shuttering for the *in situ* casting of concrete,
- and finally as the bottom tensile "reinforcement" of the composite slab.

This capability of efficiently fulfilling different roles during construction and in service conditions is certainly one of the main factors of the success of composite flooring. As already mentioned in the introduction to this section, a second key factor was related to the technological breakthrough provided by the possibility of welding stud connectors through the sheeting by means of a convenient and reliable process.

   The designer should suitably consider the diverse aspects and potential problems associated with the different phases of erection and service. These are discussed here, and design criteria and specifications are then provided, which include the recent methodologies also enabling treatment of slabs with partial shear connection theory. Eurocode 4 includes detailed rules for the design of composite slabs. The AISC-LRFD specifications, on the contrary, do not cover this subject; reference is hence made to the ASCE Standards (1991).

6.7.1   The Steel Deck

The profiled steel sheeting may be seen as a mono-directional structural system whose geometry, depth, and thickness are dictated by the types of load imposed during construction, and by the economical requirement of maximizing the span without need of shoring. This purpose led to an increase of the deck depth from values lower than 50 mm (2") to 70/75 mm (3") and more. Even sheetings 200 mm (8") deep are presently available. The rib width is also important in relation to the composite beam performance: when the composite slab acts together with the steel beam and the ribs are transverse to the beam axis, narrow deck flutes would penalize the stud resistance, often resulting in use of more studs. Wider flutes tend to characterize present deck profiles. Wide-rib profiles have a ratio h_r/b_r between the rib height and average width greater than 2. Sheeting thicknesses range from 0.76 to 1.52 mm (0.3 to 0.6 in.). Deeper decks require greater thickness in order not to have significant out-of-plane deformability, which would reduce the shear transfer capacity in the composite slab. The necessary protection against corrosion is provided by zinc coating. In many instances, the deflection under fresh concrete is the parameter governing deck selection and design. Therefore, the steel resistance is not fully exploited, and use of high strength steels would generally not be advantageous.

The steel sheeting is a structural system consisting of a number of plates subject to in plane and out of plane bending and shear. Even in the elastic range, only a rather refined method, such as the folded plate analysis, would enable an accurate determination of the stresses and deformations [18]. However, when loads are low, the response of the deck under uniform loading is close to that of a beam, and this model can be adopted also when, under increasing loads, local buckling may occur in the compressed parts. This phase of the response requires effective sectional properties to be utilized in design calculations. Buckling is not leading to abrupt failure, which is associated either to extensive sectional yielding or further development of buckles involving folded lines as well.

In the construction stage the sheeting acts as a working platform and shuttering system for the fresh concrete. The concrete is in a liquid state and applies a load normal to each of the plate components. As a result, the sheeting bends transversally due to the variation in lateral restraint from the center to the edge. However, it is acceptable, for design purposes, to model the wet concrete load as a uniform load. Besides, the "ponding" effect of concrete due to the deck deflection imposes an additional load. As a working platform, the deck supports different construction loads, including the ones related to concreting (heaping, pipelines, and pumping). Local overloads as well as vibration and impact effects may be significant, and should be considered, depending on site equipment and operations.

Codes specify minimum construction loads to be used, in addition to the weight of the fresh concrete, for the design checking of the steel deck. In several instances, the designer should assess the construction loads in order to better approximate the actual conditions. Uniform and concentrated live loads are given to simulate the overall and local effects. Differences in value and distribution also reflect the different constructional practices.

The ASCE standards prescribe to consider either a uniform load of 1.0 kN/m^2 (20 psf) or a concentrated load 2.2 kN/m (150 lb per foot of width). The sequence of construction should be accounted for, and several possible loading conditions should be defined for determination of moments, support reactions, and deflections.

Eurocode 4 allows for the local nature of the construction loads and applies a characteristic load of 1.5 kN/m^2 (30 psf) distributed on any area 3 m × 3 m, while the remaining area should be subject to a load of 0.75 kN/m^2 (15 psf). Furthermore, the sheeting should be able to resist, in absence of concrete, a concentrated load of 1 kN (0.22 kip) on a square area of side 300 mm (11.8 in.), so that a sufficient resistance against crushing of the profile is ensured.

The loads specified by ASCE are nominal loads to be directly utilized in safety and deflection checks. These should be made in accordance to the AISI specifications [1986], which are in the allowable stress format. Partial safety factors should be applied to the characteristic load values given by Eurocode 4 in order to obtain the design load combination.

Elastic methods of analysis should be used to compute the internal forces. The slenderness ratios of the component plates are usually so high (typically about 50) that local buckling governs the resistance of the deck, even when flats are stiffened. However, the effect of local buckling may be neglected in many instances, and the design analysis performed assuming uniform stiffness. A rough account for the loss of effectiveness of some parts of a continuous sheeting may then be obtained via partial moment redistribution. This approach is rather conservative. However, more accurate calculations should involve iterative procedures to determine the effective cross-sectional properties (see for example the methods provided in the AISI specifications or in the part of Eurocode 3 [1994]).

In consideration of the fairly complex response and the many parameters involved (some of which, as the variation of the yield strength in the cross-section due to the forming process and the presence of embossments, are difficult to be accounted for in a simple yet reliable way), a number of design aids were developed and are available to practitioners, mainly providing values of stiffness and resistance based on tests commissioned by the manufacturers.

The verification in service is based on a check of the midspan deflection, δ, under the wet concrete weight: the deflection limit is assumed as $l/180$ or 20 mm, whichever the minimum. The Eurocode prescribes that when this limit is exceeded, the effect of concrete ponding should be allowed for in the design. A uniform load corresponding to an additional concrete thickness of 0.7δ may be assumed for that purpose.

6.7.2   The Composite Slab

When concrete has achieved its full strength, the deck acts as a composite slab, and the steel sheeting serves as the bottom reinforcement under sagging moments. The concrete is continuous over the whole floor span. However, the amount of bottom tensile reinforcement provided by the sheeting is sufficient to make it advantageous to consider and design the slab as simply supported. A design method based on the elastic uncracked analysis with limited redistribution would lead, in fact, to maximum allowable loads for a continuous slab lower than those determined assuming the slab as simply supported. Top reinforcement is present anyway for shrinkage and temperature effects as well as for crack control over the intermediate supports. A minimum amount of reinforcement is specified by Eurocode 4 as 0.2% of the cross-section of the concrete above the steel ribs for unshored construction, and 0.4% for shored construction. Further requirements relate to the minimum values of the total depth, h_t, of the slab and of the thickness of the concrete cover, h_c. These values are similar in the Eurocode and the ASCE specifications (see Table 6.12), for they basically reflect past satisfactory performances and satisfy the need for consistency with other detailing rules. The analysis usually considers a slab strip of unit width, which depends on the system of units adopted: 1 m (SI units) or 1 ft (American units system).

TABLE 6.12   Minimum Values of the Slab Depth

Depth	EUROCODE 4	ASCE
h_t [mm]	80 (90)[a]	90
h_c [mm]	40 (50)[a]	50

[a] For slabs acting compositely with the beam

Simply Supported Composite Slabs

The same modes of failure already identified for composite beams (see Section 6.3) may be associated with ultimate conditions of a composite slab: I) flexural resistance, II) longitudinal shear, and III) vertical shear. Critical sections for each of the possible modes are schematically shown in Figure 6.61a with reference to the simply supported case. The type of mechanism for longitudinal shear transfer, which involves bond and frictional interlock as discussed in Section 6.5.1, makes the condition of complete shear connection difficult to achieve for the slab geometries and spans typical of current design practice. Therefore, collapse is primarily due to the loss of shear transfer capacity at the steel deck — concrete interface (failure mode II). However, the bending capacity may become the critical parameter for slabs with full shear connection (either long slabs or slabs with efficient end anchorage), and vertical shear may govern design of slabs with fairly low span-to-depth ratios.

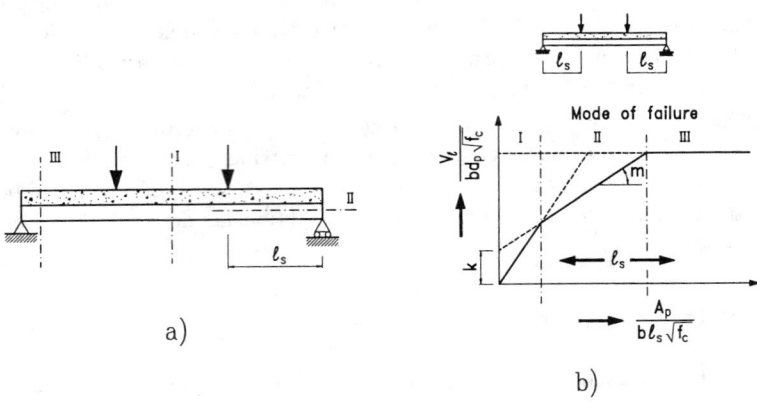

a)

b)

Figure 6.61   Composite beams. (a) Typical failure modes: critical sections. (b) Shear-bond line.

The Longitudinal Shear Capacity

Several experimental studies permitted identification of the main features of the shear bond failure with formation of major diagonal cracks at approximately one-quarter to one-third of the span from the supports accompanied by significant slip and vertical separation at the steel-concrete interface. The shear transfer mechanism in composite slabs is fairly complex. Besides material properties, its efficiency depends on many parameters, in particular to those related to the sheeting and to its deformations, such as the geometry (height, shape, and orientation) and spacing of the embossments and the out-of-plane flexibility of the sheeting component plates. Reliable models for the various phenomena involved (i.e., chemical bond, friction, and mechanical interlock) are not yet available. Moreover, end anchorage may be provided over the supports by shear connectors (typically welded studs) or suitable deformation of the sheeting, which prevent slip between the concrete slab and steel deck and enhance the slab resistance to shear bond failure (see Figure 6.27c). In some instances, combined action of in span shear bond and end anchorages should be taken into account. As a consequence of the complexity of the phenomenon and the lack of comprehensive analysis models, the determination of the longitudinal shear capacity of a composite slab is still based on performance tests.

In 1976, Porter and Ekberg [39] proposed the empirical method, on which most design code recommendations are based: the so-called m–k method. This approach conveniently relates the vertical shear resistance, V_u, at shear bond failure and the shear span, l_s, in which that failure occurs. The factors m and k are the slope and the ordinate intercept of the shear–bond line (Figure 6.62) obtained by linear regression analysis of the test results. Tests are performed on assemblies of the type presented in Figure 6.27, with two concentrated line loads applied at a distance, l_s, from the supports. This "shear span" is specified to be equal to one quarter of the total span l in the Eurocode, which also prescribes the use of crack inducers to improve the accuracy of determination of the effective shear span to be utilized in the evaluation of test results. The test program should include two groups of specimens: the first with rather long and shallow slabs (region A in Figure 6.62) and the second with larger depths and smaller shear lengths (region B). Detailing, construction, loading procedure, and assessment of test data are specified in the codes.

The method requires that the maximum vertical shear not exceed the longitudinal shear bond

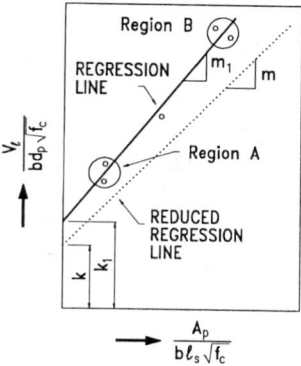

Figure 6.62 Shear bond line.

capacity. The unfactored shear bond capacity may be expressed as

$$V_{l.u} = bd_p \left[\left(\frac{m A_p \sqrt{f_c}}{b l_s} \right) + k\sqrt{f_c} \right] \tag{6.110}$$

where

A_p = the cross-sectional area of the steel deck for unit width
b = the specimen width
d_p = the distance of the top fiber of the composite slab to the centroid of the steel deck

Test results by Evans and Wright [18] showed that the influence of the concrete strength is modest and may be neglected. Based on this outcome, formula 6.110 in Eurocode 4 becomes

$$V_{l.u} = bd_p \left[\left(\frac{m A_p}{b l_s} \right) + k \right] \tag{6.111}$$

The moderate effect of the concrete strength is also reflected by the relation provided in the ASCE standards to permit approximation of the shear bond capacity, $V_{l.u2}$, of a composite slab with concrete strength, $f_{c.2}$, when an identical slab was tested but with concrete strength $f_{c.1}$:

$$V_{l.u.2} = V_{l.u.1} \left(\frac{f_{c.2}}{f_{c.1}} \right)^{1/4} \tag{6.112}$$

The design shear bond strength is obtained by reducing the unfactored resistance:

- by multiplying for the resistance factor $\phi = 0.75$ the ASCE value of $V_{l.u}$ (Equation 6.110)
- by dividing for the safety factor $\gamma_{V_l} = 1.25$ the Eurocode value of $V_{l.u}$ (Equation 6.111).

Application of Equations 6.110 and 6.111 requires preliminary determination of the value of shear span l_s, accounting for the actual loading on the slab, which is, in most instances, different from the two-point loading condition of the test and implies a variation along the shear span of the longitudinal shear force per unit length. A suitable criterion for determining an equivalent l_s to be assumed in design calculations is presented in Figure 6.63 with reference to the uniform loading case: the equivalence of the two cases is imposed by assuming that the support reactions and the areas of the shear diagrams are the same.

Due to its empirical nature, the m and k factors should be strictly applied only to the slab configuration tested. A limited extension is possible to cover slabs with the same steel deck type, but different for the thickness of the sheeting, slab depth, and steel yield strength. However, some conservativeness is necessary in defining the range of applicability. Furthermore, the method cannot take into account the very different behaviors that may be associated with slab failure, which range

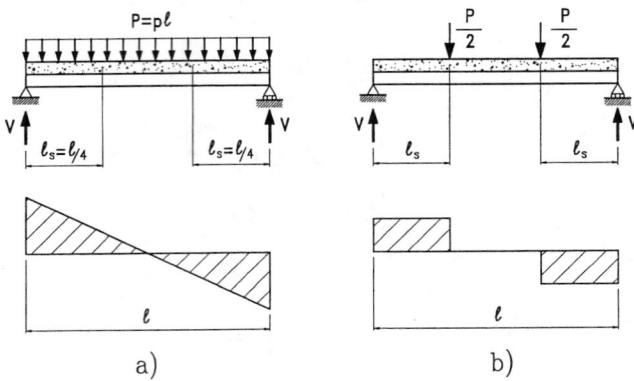

Figure 6.63   Suitable criterion for determining the equivalent length l_s. (a) Uniform load. (b) Point loads.

from very brittle to fairly ductile. Figure 6.27 shows typical composite slab responses, and points out the effect of end anchorage also on the type of failure.

The conceptual model of flexural composite members with partial shear connection and incomplete interaction can be satisfactorily applied to composite slabs with ductile behavior as demonstrated by recent German studies [5]. This type of approach enables better exploitation of the resistance capacity of composite slabs made with the types of sheeting profiles presently available. Eurocode 4 includes partial connection design, which utilizes the results of the same type of testing specified for the m–k method; for the sake of clarity, this design procedure will be illustrated after having discussed the flexural resistance of partially composite slabs.

The Flexural Capacity

Traditional approaches to the analysis and design of composite slabs adopted and adapted the methods and design criteria developed for reinforced concrete elements. The steel sheeting is hence considered and modeled as the tensile reinforcement, and limitations are imposed to ensure that failure is associated with a "ductile" mode: i.e., that crushing of the concrete in compression is avoided while the "steel reinforcement" may achieve its full plasticity strength. Restrictions are made on the depth of the concrete in compression or a range is defined, as in the ASCE standards (1991), within which only the "plastic" stress block analysis can be applied. Consistently with the reinforced concrete analogy, the parameter adopted to define the upper boundary of such a range is a suitable "reinforcement" ratio, obtained by modifying the relevant expression in the ACI standards, and hence assumed as the ratio of the steel deck area to the effective concrete area in the unit slab width b:

$$\rho = \frac{A_p}{bd_p} \tag{6.113}$$

The balanced value of it, ρ_b, also defined in accordance to the ACI Standards (1989), is

$$\rho_b = \left(\frac{0.85\beta f_c}{f_{y.p}}\right) \cdot \left[\frac{\varepsilon_c E_s \cdot (h_t - h_p)}{(\varepsilon_c E_s + f_{y.p}) d_p}\right] \tag{6.114}$$

where β is a stress block depth factor depending on concrete strength (β is equal to 0.85 for $f_c \leq 28$ N/mm², and decreases by 0.05 for each increase of f_c equal to 7 N/mm² down to a minimum value of 0.65), ε_c is the maximum allowable concrete strain ($\varepsilon_c = 0.003$ mm/mm in the ASCE standards), and h_t is the overall thickness of the slab.

The ratio ρ_b defined by Equation 6.114 refers to the balanced cross-sectional strain condition involving simultaneous achievement of the maximum concrete strain, ε_c, and full plastification of the steel sheeting. A slab with a reinforcement ratio lower than ρ_b is under-reinforced and the

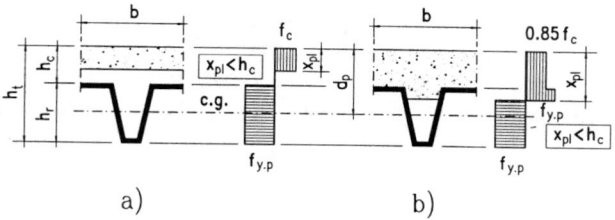

Figure 6.64   Simplified calculation method for plastic analysis of slab.

nominal moment resistance can be determined via a stress block analysis (Figure 6.64a), then

$$M_{pl.cs} = A_p f_{y.p} \left(d_p - \frac{x_{pl}}{2} \right) \qquad (6.115)$$

where
$x_{pl} =$ the depth of the concrete stress block

$$x_{pl} = \frac{A_p \cdot f_{y.p}}{0.85 \cdot b \cdot f_c} \qquad (6.116)$$

$b \;=\;$ the unit slab width

Equations 6.115 and 6.116 assume that the plastic neutral axis is in the slab and the whole steel deck yields in tension. Therefore, it is not applicable to deep decks for which the plastic neutral axis would lie within the deck profile, and for decks made of steel grades with low ductility. The latter aspect is covered in the ASCE standards by imposing that the ratio f_u/f_y for the steel deck shall not be lower than 1.08.

Determination of the flexural resistance of over-reinforced slabs (for which $\rho > \rho_b$) requires use of a general strain analysis in order to take account of all the various phenomena that possibly affect the slab performance. Besides including material nonlinearities, a refined analytical model should enable simulation of several events such as fracture of the deck in tension, buckling of the deck parts in compression, presence of additional reinforcing bars, crushing of concrete, and interface slip between steel and concrete. Furthermore, the influence of shoring on strains and stresses should also be accounted for.

The value of the strength reduction factor, ϕ, for the different types of slab recognizes the characteristics of the mode of failure: ϕ is assumed equal to 0.85 for under-reinforced slabs, and decreased to 0.70 for over-reinforced slabs and to 0.65 for under-reinforced slabs for which $f_u/f_y \leq 1.08$ in consideration of the possibility of battle failure.

Recent research studies [45] pointed out that several features are peculiar of the steel deck acting as "reinforcement", which cause composite slabs to perform differently from reinforced concrete members. In particular, they are less sensitive to concrete failure.
These features are

- the bending stiffness and strength of the steel deck, which becomes significant for deep decks

- the yield strength of the sheeting, usually substantially lower than that of the reinforcing bars

- the fact that the selfweight of the slab is resisted by the sheeting alone, which is subject to important stresses before acting composite with the concrete

As a consequence, traditional approaches, based on the behavioral analogy with reinforced concrete members, were found to be rather conservative, in particular with respect to the range of application of plastic analysis. A more general procedure for determining the ultimate flexural resistance of the slab was then proposed. Equations 6.115 and 6.116 implicitly assume that the shear bond is sufficient to cause the full flexural capacity of the composite slab to develop (i.e., the case of a fully composite slab), and that the neutral axis lies in the concrete ($x_{pl} \leq h_c$, Figure 6.64a). For unusually deep decks, the neutral axis lies within the steel section ($x_{pl} > h_c$, Figure 6.64b). In this case, a simplified approach can be used to compute $M_{pl.cs}$ illustrated in Figure 6.65 which neglects the concrete in the rib. The tensile force in the sheeting can be decomposed in two forces: one at the bottom, $N_{pl.pr}$,

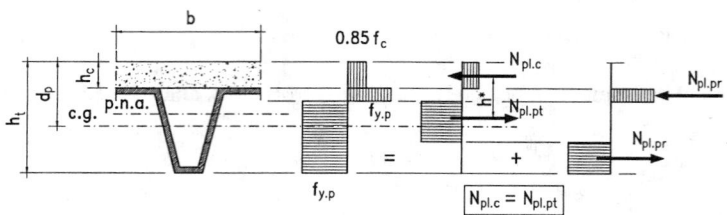

Figure 6.65   Additional simplified calculation method for plastic analysis of slab.

while the other, $N_{pl.pt}$, is equal to the compression force $0.85\,bh_c\,f_c$. The contribution of the forces, $N_{pl.pc}$, defines a moment, $M_{pl.pr}$, that may be considered as the plastic moment of the steel deck, $M_{pl.p}$, reduced by the presence of the axial force, $N_{pl.pt}$. The reduced plastic moment, $M_{pl.pr}$, can be obtained from the interaction diagram of the sheeting (see Figure 6.66). A good approximation

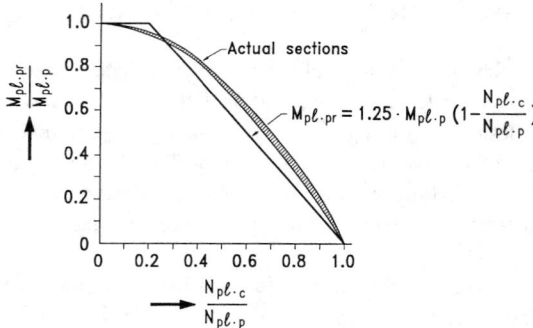

Figure 6.66   Interaction curve N–M for profiled sheetings.

is provided by the following expression:

$$M_{pl.pr} = 1.25 M_{pl.p}\left(1 - \frac{N_{pl.c}}{N_{pl.p}}\right) \leq M_{pl.p} \qquad (6.117)$$

where $M_{pl.p}$ and $N_{pl.p}$ are the plastic moment and the full plastic axial resistance of the sheeting, respectively.

The plastic moment of resistance of the composite slab is then obtained as

$$M_{pl.cs} = M_{pl.pr} + 0.85 bh_c f_c h^* = M_{pl.pc} + N_{pl.c} \cdot h^* \qquad (6.118)$$

The lever arm h^* can be satisfactorily approximated by the relationship [45]

$$h^* = h_t - 0.5 h_c - e_p + \left[(e_p - e) \cdot \frac{N_{pl.c}}{N_{pl.p}}\right] \qquad (6.119)$$

where e and e_p are the distances from the bottom of the slab to the centroid and the plastic neutral axis of the steel sheet, respectively (Figure 6.67).

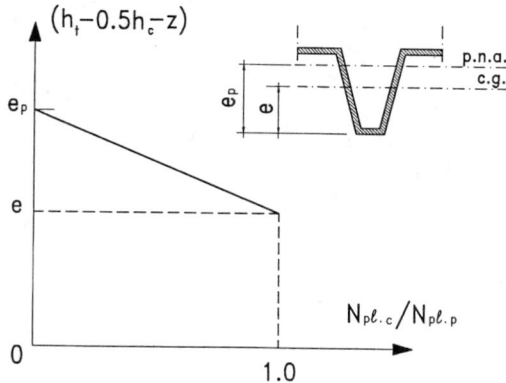

Figure 6.67   Plot of the relationship Equation 6.119.

The extension of the method to the case of partially composite slabs is straightforward. The compression force in the concrete, F_c, is lower than the value, $F_{c.f}$, associated with the condition of full shear connection, and two neutral axes are present in the cross-section: i.e., the first lying in the concrete and the second within the steel sheeting. The depth of the concrete stress block is given by

$$x_{pl} = \frac{F_c}{0.85 \cdot b \cdot f_c} \leq h_c \tag{6.120}$$

By replacing in Equations 6.117 to 6.119 h_c with x_{pl} and $N_{pl.pt}$ with F_c, the moment resistance can be computed as:

$$h^* = h_t - 0.5x_{pl} - e_p + \left[(e_p - e) \frac{F_c}{N_{pl.p}} \right] \tag{6.121}$$

$$M_{pl.pr} = 1.25 M_{pl.p} \cdot \left[1 - \frac{F_c}{N_{pl.p}} \right] \leq M_{pl.p} \tag{6.122}$$

$$M_{w.cs} = M_{pl.pr} + F_c h^* \tag{6.123}$$

Slab Design Based on Partial Interaction Theory

A composite slab for which the shear bond failure is ductile may be treated as a partially composite slab. In this condition, the flexural and longitudinal shear capacity of the section are linked, in the sense that the bending resistance of the slab is limited by the maximum shear resistance. A design procedure based on the analytical model of partially composite members was included in Annex E of Eurocode 4. This approach enables the effect of end anchorages to be accounted for in a rather straightforward and efficient way. The procedure, which makes use of the results of standard tests for the determination of the factors m and k presented before, is briefly described here. Ductile behavior is defined as that of a composite slab for which the failure experimental load is at least 10% higher than the load at which end slip is observed. If collapse is associated with a midspan deflection δ greater than $l/50$, the load at $\delta = l/50$ has to be assumed as a failure load.

The partial-interaction design method requires that the mean ultimate shear stress is determined. As a first step, for the slab considered, the curve expressing the relationship between the unfactored moment of resistance and the degree of interaction $\eta = F_c/F_{c.f}$ has to be defined (Figure 6.68). The following must occur:

1. The moment capacity, $M_{pl.cs}$, of the fully composite slab is computed and the corresponding value of the compressive force, $F_{c.f}$, in the concrete slab is determined.

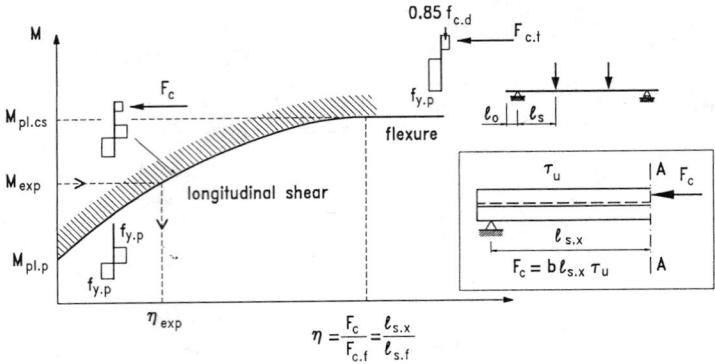

Figure 6.68   Composite slab partial interaction diagram.

2. A value of h^* is selected, to which the partial-interaction compression force $F_c = \eta F_{c.f}$ corresponds, and the value of the associated moment capacity $M_{pl.cs}$ is computed via Equations 6.121 to 6.123.

3. The calculations at Step 2 are repeated for a number of values of η sufficient to define the $M_{w.cs} - \eta$ curve.

This curve is then used in combination with the experimental ultimate moment to obtain the degree of interaction of the considered slab and then the value of F_c. Finally, under the assumption of uniform shear bond resistance along the shear span l_s and the overhang l_o, the mean ultimate shear stress is obtained as

$$\tau_u = \frac{F_c}{b \cdot (l_s + l_o)} \tag{6.124}$$

A value of τ_u is determined for each test. A statistical treatment of these values then enables definition of a design value $\tau_{u.d}$. The Eurocode first defines a characteristic ultimate shear $\tau_{u.k}$ as the mean value reduced by 10%, and then the design value as $\tau_{u.d} = \tau_{u.k}/1.25$.

It should be noted that while a reduced number of additional tests is sufficient for assessing the influence of the steel deck thickness, the thickness of the slab remarkably affects the partial interaction curve, and results cannot be applied to slabs with higher slab thicknesses than the specimens tested.

In the verification procedure, the partial-interaction diagram is computed using the design values of slab dimensions and material strengths. If $l_{s.x}$ is the distance of a cross-section from the closer support, the assumption of uniform shear strength distribution implies that $F_c = b\tau_{u.d}l_{s.x}$ and allows substitution of the ratio $F_c/F_{c.f}$ with the ratio $l_{s.x}/l_{s.f}$, with $l_{s.f}$ the value of the shear span for which the full shear interaction is achieved. Verification that the design bending moment is lower than the moment of resistance is simply obtained by superimposition as shown in Figure 6.69.

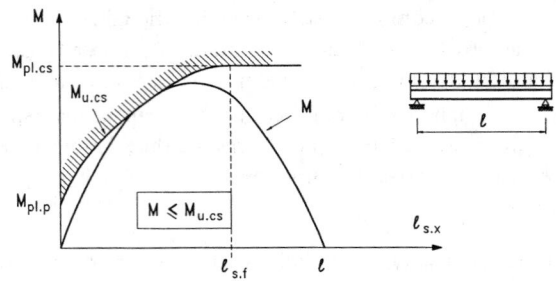

Figure 6.69   Verification procedure for simply supported slab under uniform loading.

If end anchorages are used, three additional tests are required to determine their contribution V_{ea} to the longitudinal shear strength. From each test, this contribution can be computed as

$$V_{ea} = \eta F_{c.f} - b\,(l_s - l_o)\,\tau_{um} \tag{6.125}$$

where τ_{um} is the mean value of c_u from the tests carried out without end anchorage.

The mean value of V_{ea} reduced by 10% is assumed to be a characteristic value, $V_{ea.k}$. The design resistance is then obtained as $V_{ea.k}/1.25$. When defining the design partial-interaction diagram, the compressive force, F_c, should account for the contribution of the end anchorage:

$$F_c = bl_{s.x}c_{u.d} + V_{ea.d} \tag{6.126}$$

This will result in a shifting of the design diagram as illustrated in Figure 6.70.

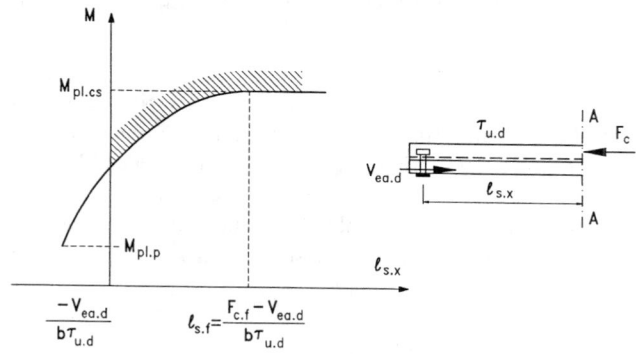

Figure 6.70   Design partial interaction diagram for a slab with end anchorage.

Reinforcing bars are often provided in the slab ribs to enhance the fire response. The contribution of rebars to the flexural resistance of the slab may also be taken advantage of. The moment capacity can be obtained by adding up the plastic resistances associated with the relevant reinforcing bars (see Figure 6.71).

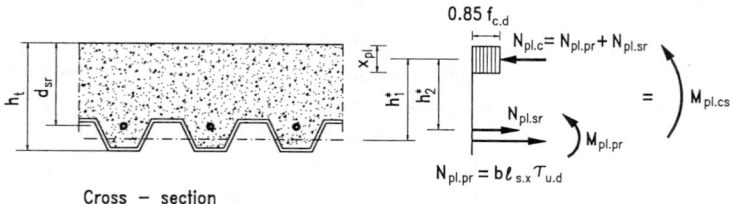

Figure 6.71   Contribution of additional longitudinal reinforcement.

Continuous Composite Slabs

Elastic analysis with limited moment redistribution and plastic analysis can both be adopted. In most instances, the latter approach is advantageous. However, it imposes that rotation capacities are checked in the hogging moment regions, which are substantially affected by the ductility of the reinforcing bars. If high ductility rebars are selected, the method can be applied to commonly used slabs for spans up to 5 m [45].

The hogging ultimate moment of resistance can be computed by the stress block theory, also accounting for the contribution of the sheeting when continuous over the support. The possible buckling of plate components should be considered. The restraint offered by the concrete allows for relaxation of the related rules. The Eurocode recommends that the effective widths be taken as twice the values given for class 1 steel webs. When the deck is not continuous over the support, the slab in the hogging moment region should be modeled as a reinforced concrete element.

If elastic design analysis is adopted, the large available sagging moment of resistance makes modeling the slab as independent simple spans the most convenient approach. The elastic analysis with limited redistribution of moments (up to 30% is allowed by the Eurocode) is less advantageous in terms of load carrying capacity. On the other hand, continuity may become beneficial to reduce deflections and meet serviceability requirements.

When checking the shear bond resistance of the slab portions in sagging moment, an effective simple span equal to the distance between points of contraflexure may be assumed for internal spans, while for end spans the full exterior span length has to be used. The regions in the hogging moment provide a constraint to shear slippage, which is modest for end spans and should be neglected.

Vertical Shear

The resistance to vertical shear is mainly provided by the ribs, and formulae for reinforced concrete T-beams can be applied, if suitably adjusted. Moreover, the shear stresses in the sheeting consequent to it functioning as shuttering during concrete casting can be neglected and the total shear force can be considered as resisted by the composite cross-section.

Reference can be made to a slab width equal to the distance between the centers of adjacent ribs, and the unfactored vertical shear resistance expressed as (Eurocode 4):

$$V_v = b_r d_p \tau_u k_v (1.2 + 40\rho) \tag{6.127}$$

where
τ_u = the shear strength of concrete
b_r = mean width of concrete rib
k_v = $(1.6 - d_p) \geq 1$ (with d_p in m)
ρ = $A_p / b_r d_p$
A_p = the effective area in tension within width b_o
The shear strength for Eurocode should be taken as equal to $0.25\, f_{ct}$.

Punching Shear and Two-Way Action

Heavy concentrated loads may be applied to the slab (i.e., by the wheels of fork-lift trucks), which make the slab subject to two-way action and may cause failure by punching shear.

The limited experimental knowledge available is not sufficient to allow for an appraisal of the sheeting contribution to the resistance to punching shear. This resistance is hence generally determined as for reinforced concrete sections. An effective area can be defined accounting for the different stiffnesses of the slab in the two directions. The critical perimeter, C_p, can be obtained by a 45° dispersion of the load down to the centroidal axis of the sheeting in the longitudinal direction and to the top of the sheeting in the transverse direction (see Figure 6.72):

$$C_p = 2\pi h_c + 2\left(2d_p + a_p - 2h_c\right) + 2b_p + 8h_f \tag{6.128}$$

where a_p and b_p define the loaded area, and h_f is the height of finishes.

In analogy with vertical shear, and considering the height of concrete over the deck, h_c, as the effective depth, the punching shear resistance can be written as

$$V_p = C_p d_p \tau_u k_v (1.2 + 40\rho) \tag{6.129}$$

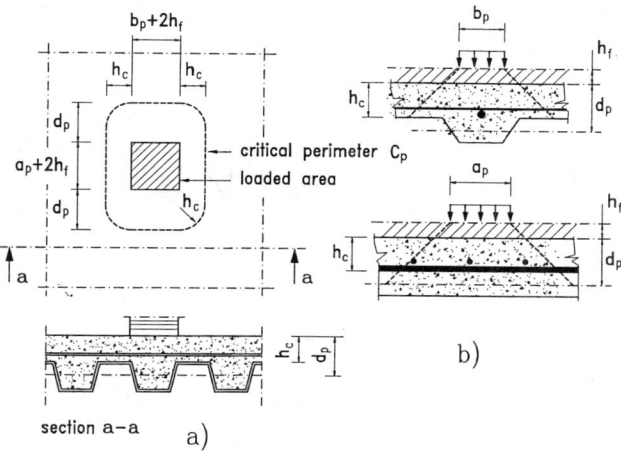

Figure 6.72 Critical perimeter for punching shear.

The ASCE standards limit the nominal shear stress to $2\sqrt{f_c}$ in inch-pound units ($0.166\sqrt{f_c}$ if f_c in N/mm^2) in view of the incomplete two-way action.

The load distribution requires the slab to possess adequate flexural strength in the transverse direction, and suitable transverse reinforcement should be placed in consideration of the negligible bending strength of the steel sheeting transverse to the deck ribs.

Serviceability Limit State

The performance in service is verified mainly with reference to cracking of concrete and to the flexural stiffness (through a limitation of the midspan deflections).

Cracking of Concrete

Cracking of concrete may occur in the regions over the supports where some degree of continuity develops due to the intrinsic continuity of the concrete slab, also when the composite slab is conceived and designed as a series of simply supported elements. Control of the crack width would require that the criteria are used, which were developed and codified for reinforced concrete members. When the environment is not aggressive and the width of the cracks is not critical for the functioning of the structure, placement of nominal anti-crack reinforcement would be sufficient to satisfy serviceability requirements. The Eurocode specifies a minimum amount of reinforcement equal to 0.2% of the concrete area over the deck for unpropped slabs, and 0.4 for propped slabs.

Deflections

The floor deflection has to be limited to the values that ensure that no damage is induced by floor deformation in partitions and other nonstructural elements. Values of maximum deflections are provided in the codes, which can usually be allowed in buildings. Table 6.13 presents the limit values given in the ASCE standards, while the ones in Eurocode 4 are in Table 6.2. Besides, both codes provide limitations to the span-to-depth ratios (Table 6.14), which are related, though in a different way, to in service conditions. The ASCE span-to-depth ratios refer to the total depth, h_t, of the slab, and intend to provide guidance to obtain satisfactory in-service deflections. Slab deformation should be computed and checked. On the contrary, fulfillment of the Eurocode limitation, which refers to the effective depth of the slab, d_p, allows for deflection calculations to be omitted, at least when slip does not significantly affect the slab response. In some instances, a more accurate assessment is necessary, accounting for the expected type of behavior of nonstructural elements chosen in the specific design project.

In the deflection calculation different components have to be taken into account that are associated with the various facets of the response. In particular, the deformation under short and long term

TABLE 6.13 Limiting Values for Vertical Deflections Recommended by ASCE Standards

Type of composite slab	Type of deflection	Deflection limitation (in.)
Flat roofs	Immediate deflection due to live load	$\frac{l}{180}$
Floors	Immediate deflection due to live load	$\frac{l_i}{360}$
Roofs or floor slabs supporting or attached to non-structural elements likely to be damaged by large deflections	That part of the total deflection which occurs after attachment of the non-structural elements, the sum of the long-time deflection to immediate deflection due to any additional live load	$\frac{l_i}{480}$
Roofs or floor slabs supporting or attached to non-structural elements not likely to be damaged by large deflections		$\frac{l_i}{240}$

TABLE 6.14 Recommended Limiting Values for Span-to-Depth Ratios

	ASCE: l/h_t	EUROCODE 4: l/d_p
Simply supported slabs	22	25
External span of continuous slabs	27	32
Internal spans of continuous slabs	32	35

loading and the effect of interface slip need to be considered. Immediate slab deformation, δ_{st}, can be determined via a linear elastic analysis. In continuous slabs, calculations can assume that the slab has a uniform stiffness characterized by a moment of inertia equal to the average of those of the cracked and uncracked section.

The effect of slip may be important in external spans, and it should accounted for, with reference to the results of the relevant performance tests. It generally may be neglected when the experimental data indicate that end slip greater than 0.5 mm does not occur at loads equal to 1.2 times the service loads. If this condition is not fulfilled, there are two possible alternatives to the calculation of the deflection including slip:

- suitable end anchorages can be provided, or
- the design service loads are reduced so that the previous limit on end slip is met.

Additional deflections under long-term loading, δ_{lt}, may be approximated, as for reinforced concrete members, as a quota of the elastic deflection under short-term loads, i.e.,

$$\delta_{lt} = k_\delta \delta_{st} \qquad (6.130)$$

The ASCE standards specify as k_d the same factor as in the ACI 318 Code:

$$k_\delta = \left[2 - 1.2 \left(\frac{A_{s.c}}{A_{s.t}} \right) \right] \leq 0.6 \qquad (6.131)$$

where

$A_{s.c}$ and $A_{s.t}$ =   the areas of steel in compression and tension under service loads, respectively

The area, $A_{s.c}$, includes reinforcement and the possible portion of the steel deck in compression. The Eurocode enables a simplified appraisal of the total deflection under sustained loads to be obtained, for slabs with normal density concrete, via a linear elastic analysis performed with a slab stiffness based on an average modular ratio for long and short term effects. In the calculations of total deflections for serviceability checks, loads should be carefully selected. The immediate deformations induced by all the dead loads applied before placement of the relevant nonstructural elements can be neglected. However, the applied forces simulating the effect of shore removal, which contributes to long-term deflections, should be accounted for as well. Finally, in some instances, the shear-bond slip may also cause significant additional deflections under sustained loads. Test data are needed in these cases to provide appropriate input to design calculations.

6.7.3   Worked Examples

EXAMPLE 6.8:   Design of a composite slab

With reference to Example 6.2 the design of the composite slab is proposed in the following, which is made by using HiBond 55/100 steel profiled sheeting. A reinforcing steel mesh is placed in the concrete slab with 25 mm top cover, providing 188 mm²/m of reinforcement (ϕ 6 mm p = 150 mm). For the check of the longitudinal shear, the m and k parameters for the HiBond 55/100 profiled sheeting are equal to 86 and 0.69, respectively. The analysis is related to a strip of 1000 mm. The stactical system and typical cross-section are given in Figure 6.73.

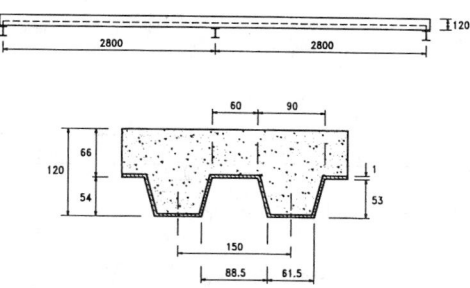

Figure 6.73   Stactical system and typical cross-section.

- Characteristic of materials:
  The design strength of materials is assumed.

Concrete:		$f_{c.k} = 25$ N/mm²	$\gamma_c = 1.5$
Steel:	reinforcement	$f_{y.sr.k} = 235$ N/mm²	$\gamma_s = 1.15$
	steel sheeting	$f_{y.p.k} = 320$ N/mm²	$\gamma_p = 1.10$

A. Verification of the profiled sheeting (see Section 6.7.1)
Loads:

Self-weight of the sheet	g_p	0.13	kNm/m²	($\gamma_F = 1.35$; 1)
Weight of the wet concrete construction load	g_c	2.27	kNm/m²	($\gamma_F = 1.35$; 1)
	q_{m1}	1.5	kNm/m²	($\gamma_F = 1.5$; 0)
	q_{m2}	0.75	kNm/m²	($\gamma_F = 1.5$; 0)

Stactical system and loading cases are given in Figures 6.74 and 6.75.

1. Ultimate limit states — By elastic calculation of the beam, it follows:

   - Maximum positive bending moment $M^{(+)} = 3.62$ kNm/m
   - Maximum negative bending moment $M^{(-)} = 6.12$ kNm/m
   - Maximum support reaction $F = 19.10$ kN/m

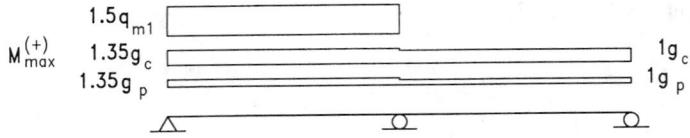

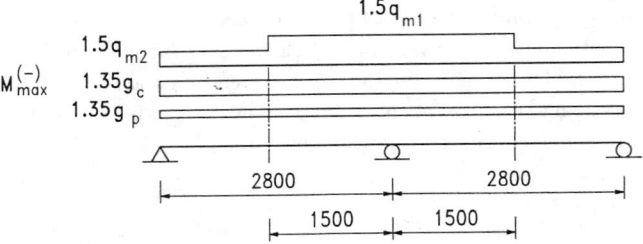

Figure 6.74   Stactical system and loading cases for ultimate limit states.

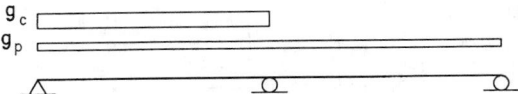

Figure 6.75   Loading case for serviceability limit state.

- Cross-section resistance (trade literature):

$$
\begin{aligned}
\gamma_p & & = & \ 1.1 \\
M_{pl.p}^{(+)} & \quad 7.41/\gamma_p & = & \ 6.74 \quad \text{kNm/m} \\
M_{pl.p}^{(-)} & \quad 7.48/\gamma_p & = & \ 6.8 \quad \ \text{kNm/m} \\
F_p & \quad 68.1/\gamma_p & = & \ 61.9 \quad \text{kN/m}
\end{aligned}
$$

- Safety check:

$$
\begin{aligned}
M^{(+)} & \quad 3.62 \ \text{kNm/m} & < & \ 6.74 \quad \text{kNm/m} \\
M^{(-)} & \quad 6.12 \ \text{kNm/m} & < & \ 6.8 \quad \ \text{kNm/m} \\
F & \quad 19.10 \ \text{kN/m} & < & \ 61.9 \quad \text{kN/m}
\end{aligned}
$$

$$
\frac{M^{(-)}}{\dfrac{M_{pl.p}^{(-)}}{\gamma_p}} + \frac{F}{\dfrac{F_p}{\gamma_p}} = \frac{6.12}{6.8} + \frac{19.10}{61.9} = 0.9 + 0.31 = 1.21 < 1.25
$$

2. Serviceability limit states — The analysis makes with reference to the stactical system in Figure 6.75.
   The maximum moment is

$$
M = 1.773 \ \text{kNm/m}
$$

   - Deflection control
     By means of an elastic analysis (Figure 6.75), it is $\delta = 7.19$ mm.
     Check:

$$
\delta = 7.19 \ \text{mm} \ < l/180 \qquad ((2800/180) = 15.56 \ \text{mm} \ < 20 \ \text{mm})
$$

B. Verification of the composite slab.

It is assumed that the composite slab is designed as a series of a simply supported beams (Figure 6.76).

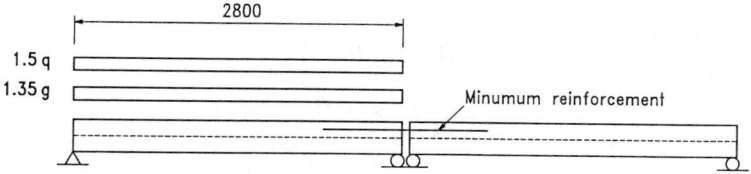

Figure 6.76 Composite slab as a series of a simply supported beams.

Loads:

Self-weight of the slab and floor finishes $\quad g = 3.05 \text{ kNm/m}^2 \quad (\gamma_g = 1.35)$

Live load $\qquad\qquad\qquad\qquad\qquad\quad q = 4.8 \text{ kNm/m}^2 \quad (\gamma_q = 1.5)$

- Design load

$$p = \left(\gamma_g g + \gamma_q q\right) \cdot b = (1.35 \cdot 3.05 + 1.5 \cdot 4.8) \cdot 1.00 = 11.32 \text{ kN/m}$$

- Ultimate limit state
  Design bending moment

$$M^{(+)} = \frac{p \cdot l^2}{8} = \frac{11.32 \cdot 2.80^2}{8} = 11.09 \text{ kN/m}$$

Determination of the design bending resistance.
Calculation of the plastic neutral axis

$$x_{pl} = \frac{A_p \cdot f_{y.p.k}/\gamma_p}{b \cdot 0.85 f_{c.k}/\gamma_c} = \frac{1482 \cdot 320/1.10}{1000 \cdot 0.85 \cdot 25/1.5} = 30.43 \text{ mm} \; < 120 \text{ mm}$$

Design bending resistance:

$$
\begin{aligned}
M_{pl.cs}^{(+)} &= A_p \cdot \frac{f_{y.p.k}}{\gamma_p} \left(d_p - \frac{x_{pl}}{2} \right) \\
&= 1482 \cdot \frac{320}{1.1} \left(120 - 27.5 - \frac{30.43}{2} \right) = 33.3 \text{ kNm/m}
\end{aligned}
$$

Check

$$M^{(+)} = 11.09 \text{ kNm/m} \le 33.3 \text{ kNm/m}$$

Longitudinal shear: m-k method (see Section 6.7.2). Calculation of design shear force

$$V_1 = p \cdot \frac{l}{2} = 11.32 \cdot \frac{2.8}{2} = 15.85 \text{ kN}$$

Design shear resistance (Equation 6.111):

$$m = 86, \quad k = 0.069$$

$$V_{1.u} = b \cdot d_p \cdot \frac{\left(m A_p/b l_s + k \right)}{\gamma} = 1000 \cdot 92.5 \cdot \frac{\left(\frac{86 \cdot 1482}{1000 \cdot 2800/4} + 0.069 \right)}{1.25} = 18.6 \text{ kN}$$

Check

$$V_1 = 15.85 \text{ kN} \; \le 18.6 \text{ kN} \; = V_{1.u}$$

- Vertical shear

  Calculation of design vertical shear resistance:

$$V_v = b_0 d_p \tau_{w.d} k_v (1.2 + 40\rho).$$

where

$$\tau_{w.d} = 0.30 \text{ N/mm}^2$$

$$k_v = 1.6 - d_p = 1.6 - 0.0925 = 1.51 \text{ mm}$$

$$1.2 + 40\rho_0 = 1.2 + 40 \cdot \frac{A_p}{b_0 \cdot d_p} = 1.2 + 40 \cdot \frac{188}{500 \cdot 92.5} = 1.36$$

Hence,

$$V_v = 28675 \text{ N} \cong 28.7 \text{ kN}$$

Check

$$V_1 = 15.85 \text{ kN} < 28.7 \text{ kN} = V_v$$

- Serviceability limit state — The deflection of the sheeting under its selfweight and the weight of the concrete is

$$\delta = 7.19 \text{ mm}$$

For the evaluation of the deflection of the composite slab, it is assumed the hypothesis of continuous beam. The deflection under the weight of floor finishes is:

$$\delta_g = 0.132 \text{ mm}$$

The deflection of the composite slab under variable loading of long duration has the value:

$$\delta_p = 1.661 \text{ mm}$$

where the deflection evaluation has been evaluated as the average of the value for the cracked an uncracked section, calculated with an average coefficient n $= \frac{E_s}{E_c} = 15$.
Checks

$$\delta_{tot} = \delta + \delta_g + \delta_p = 7.19 + 0.132 + 1.661 = 8.983 \text{ mm} < \frac{l}{250} = \frac{2800}{250} = 11.26 \text{ mm}$$

$$\delta_2 = \delta_g + \delta_p = 0.132 + 1.661 = 1.793 \text{ mm} < \frac{l}{300} = 9.3 \text{ mm}$$

- Cracking of concrete — The amount of reinforcement an intermediate support is:

$$\rho = \frac{A_s}{bh_c} = \frac{188}{1000 \cdot 65} = 0.29\% > 0.2\%$$

This percentage is greater than the minimum recommended by Eurocode 4.

Notations

a	=	amplificaton factor
b	=	slab width of a composite beam, width of a composite column section, or unit width of a composite slab
b_a	=	length of the outstanding leg of connection
b_c	=	base width of concrete component of a composite column section
b_e	=	effective width of the portion of the concrete flange on each side of the steel web

b_{eff}	$=$	effective width of the concrete flange of a composite beam
b_r	$=$	average width of concrete rib or haunch
b_s	$=$	base width of structural steel component of a composite column section
b_{sr}	$=$	base width of reinforcement component of a composite column section
c	$=$	outstanding flange of the steel profile, numerical factor
c_i	$=$	numerical factors
d	$=$	overall depth of member, clear distance between flanges of the steel profile
d_b	$=$	beam depth
d_c	$=$	depth of concrete component of a composite column section
d_p	$=$	distance between the top fiber of a composite slab to the centroid of the steel deck
d_{inf}	$=$	distance between centroid of structural steel profile and centroid of composite beam
d_{sc}	$=$	diameter of stud shear connector
d_s	$=$	depth of structural steel component of a composite column section
d_s, d_i	$=$	distances between centroid of concrete slab and centroid of composite beam
d_{sr}	$=$	depth of reinforcement component of a composite column section
f	$=$	fundamental frequency of a beam
f_c	$=$	nominal compressive strength of concrete
$f_{c.d}$	$=$	design cylinder strength of concrete under compression
$f_{c.k}$	$=$	characteristic cylinder strength of concrete under compression
f_{ct}	$=$	tensile strength of concrete
f_{my}	$=$	modified yielding stress of structural steel
f_y	$=$	yield strength
f_k	$=$	characteristic strength
$f_{y.s}$	$=$	specified yield strength
$f_{y.s.d}$	$=$	design yielding stress of structural steel
$f_{y.s.k}$	$=$	characteristic yield strength of structural steel
$f_{y.s.v}$	$=$	design yielding shear resistance of structural steel
$f_{y.sr.d}$	$=$	design yielding stress of reinforcing steel
$f_{y.sr.k}$	$=$	characteristic yielding stress of reinforcing steel
$f_{y.p}$	$=$	specified yield strength of steel deck
$f_{y.sr}$	$=$	specified yield strength of steel reinforcement
$f_{y.s.r}$	$=$	reduced specified yield strength of steel
f_u	$=$	specified tensile strength of steel
$f_{u.sc}$	$=$	ultimate tensile resistance of shear studs
h	$=$	height, thickness
h_{ac}	$=$	height of the outstanding leg of an angle connector
h_c	$=$	thickness of the solid concrete slab, thickness of the concrete above steel deck
h_p	$=$	overall depth of steel deck
h_n	$=$	distance between the line of plastic neutral axis and the line of centroid of cross-section
h_r	$=$	rib height
h_s	$=$	height of structural steel profiles of a composite beam
h_t	$=$	total thickness of composite slab
h^*	$=$	internal arm distance (distance between line of action of compression resultant and line of action of tension resultant)
k	$=$	factor, ordinate intercept of shear-bond regression line
k_c, k_s	$=$	coefficient for evaluation of shear resistance (for concrete and steel, respectively)
k_{rp}	$=$	reduction factor for resistance of stud in composite deck with rib parallel to the beam
k_{rt}	$=$	reduction factor for resistance of stud in composite deck with rib transverse to the beam

k_l	=	effective length of a column for buckling evaluation
i	=	ratio between the uncracked and cracked inertia of composite cross-section
l	=	length of a beam or column
l_0	=	approximate distance between the points of zero moment
l_s	=	shear span
$l_{s.f}$	=	shear span at which full interaction is achieved in a composite slab
$l_{s.x}$	=	distance from a cross-section and the closer support
m	=	slope of shear-bond regression line, ratio between moments
n	=	modular ratio: E_s/E_c
n_{ef}	=	modular ratio for creep analysis $E_s/E_{c.ef}$
p	=	vertical load per unit length
q	=	force on a shear connector
q_u	=	maximum resistance of shear connector
$q_{u.c}$	=	shear resistance associated with concrete failure
$q_{u.s}$	=	shear resistance associated with connector failure
$q_{u.tot}$	=	total shear resistance
$q_{u.block}$	=	shear resistance of the block
$q_{u.anchors}$	=	shear resistance of the anchors
$q_{u.hoop}$	=	shear resistance of the hoop
q_{Rd}	=	design resistance of shear connector
r	=	numerical factor for buckling evaluation, redistribution degree, ratio between moments
r_m	=	radius of gyration of steel part of a composite column
s	=	slip
t	=	thickness
t_c	=	thickness of the concrete slab
t_f	=	flange thickness
t_w	=	web thickness
v	=	shear force per unit length
v_l	=	longitudinal shear force per unit length
v_{sd}	=	shear resistance of steel decking
w	=	unit weight of concrete
x_e	=	elastic neutral axis depth
x_{pl}	=	plastic neutral axis depth
A	=	area of the overall full composite section
A_0	=	original cross-sectional area of steel specimen
A_c	=	cross-sectional area of concrete
$A_{c.v}$	=	area of concrete per unit length of a shear plane
$A_{ef.v}$	=	effective area of transverse reinforcement per unit length
A_{f1}	=	area of the front of the block connector
A_{f2}	=	enlarged area of the front surface of the block connector
A_p	=	cross-sectional area of steel deck for unit width
A_s	=	cross-sectional area of structural steel
A_{sc}	=	cross-sectional area of stud shear connector
A_{sd}	=	area of steel decking
A_{sr}	=	area of steel reinforcement
A_v	=	vertical shear area of structural steel
A^*	=	effective cross-section area
E_c	=	modulus of elasticity of concrete
$E_{c.ef}$	=	effective modulus of elasticity of concrete for creep analysis

E_d	=	design load effect
E_m	=	mean load effect
E_k	=	characteristic load effect
E_s	=	modulus of elasticity of steel
$(EI)_e$	=	global flexural stiffness of column cross-section for buckling evaluation
F	=	action, force, load
F_m	=	mean load
$F_{b.pr}$	=	preload force in bolt
F_c	=	compression force in the concrete slab
$F_{c.f}$	=	value of F_c associated with the plastic moment of the composite beam section
$F_{c.max}$	=	maximum possible stress resultant of concrete cross-sectional component
F_k	=	characteristic load value
F_s	=	tensile force in the steel profile
$F_{s.max}$	=	maximum possible stress resultant of structural steel cross-sectional component
F_{sr}	=	tensile force in reinforcing steel
F_t	=	tensile force in the concrete slab
$F_{t.f}$	=	tensile force in the concrete slab when bending moment is maximum
F_w	=	plastic axial load resistance of steel web
H_{sc}	=	overall height of a stud
I	=	inertia of the composite cross-section
I_c	=	inertia of the concrete cross-sectional component
I_s	=	inertia of the structural steel cross-sectional component
I_{sr}	=	inertia of the reinforcing steel cross-sectional component
I_1	=	inertia of the uncracked composite cross-section
I_2	=	inertia of the cracked composite cross-section
L	=	span, length
M	=	bending moment
M_{cr}	=	first cracking moment
M_e	=	bending moment before redistribution
M_{el}	=	elastic limit moment
M_{pl}	=	plastic moment
$M_{pl.d.c}$	=	design plastic moment of the composite cross-section
$M_{pl.c}$	=	plastic moment of the composite beam
$M_{pl.cs}$	=	plastic moment of the composite slab
$M_{pl.s}$	=	plastic moment of the steel section alone
$M_{pl.p}$	=	plastic moment of the steel deck alone
$M_{pl.pr}$	=	reduced plastic moment of the steel deck alone
M_r	=	bending moment after redistribution
M_s	=	bending moment resistant of the steel section alone
M_{sh}	=	bending moment induced by shrinkage
M_u	=	ultimate moment capacity of a partially composite beam
$M_{u.cs}$	=	ultimate moment capacity of a partially composite slab
M_{unf}	=	unfactored value of the moment capacity
N	=	axial load, number of connectors in shear span
N_{cr}	=	buckling load of a composite column
N_{el}	=	elastic axial resistance
N_f	=	minimum number of connectors required to transfer F_{cf}
N_r	=	number of studs in a rib of a composite slab
N_p	=	permanent part of the axial load
N_{pl}	=	plastic axial resistance of a section
$N_{pl.c}$	=	plastic axial resistance of concrete component

$N_{pl.p}$	=	plastic axial resistance of steel deck alone
$N_{pl.pc}$	=	plastic axial resistance of steel deck in compression
N_{sh}	=	axial load induced by shrinkage
$N_{pl.sr}$	=	plastic tensile resistance of steel reinforcement
N_u	=	ultimate axial load of a column
P	=	applied load
Q_c	=	resistance of the shear connection on a given beam length
Q_{cf}	=	resistance of a full connection on a given beam length
R_d	=	design value of the resistance
R_k	=	characteristic value of the resistance
R_n	=	nominal resistance
S	=	first moment of area of the composite cross-section
T	=	tensile force
V	=	shear force
V_l	=	longitudinal shear force
V_{ea}	=	shear strength of an end anchorage
V_{pl}	=	plastic vertical shear resistance
W_c	=	plastic resistance modulus of concrete component
W_s	=	plastic resistance modulus of steel component
W_{sr}	=	plastic resistance modulus of reinforcing steel component
$W_{c.n}$	=	plastic resistance modulus of concrete component in the range $\pm h_n$
$W_{s.n}$	=	plastic resistance modulus of steel component in the range $\pm h_n$
$W_{sr.n}$	=	plastic resistance modulus of reinforcing steel component in the range $\pm h_n$
α	=	angle, shear reduction coefficient, imperfection factor
β	=	angle, safety angle
δ	=	beam deflection
δ_c	=	structural steel contribution
δ_f	=	beam deflection in case of full shear connection
δ_p	=	beam deflection in case of partial shear connection
δ_s	=	deflection of the steel section acting alone
δ_{sh}	=	deflection due to shrinkage
δ_{ss}	=	deflection of the simply supported beam
ε_c	=	strain of concrete
ε_s	=	strain of steel
$\varepsilon_{y.s}$	=	strain of steel at welding
ε_{sh}	=	shrinkage strain
ε_{sr}	=	reinforcing steel strain
γ	=	partial safety factor
γ_c	=	partial safety factor for concrete
γ_F	=	load factor
γ_m	=	partial safety factor for material strength
γ_s	=	partial safety factor for structural steel
γ_{sr}	=	partial safety factor for steel reinforcement
γ_V	=	partial safety factor for shear connectors
λ	=	relative slenderness of a composite column
μ	=	coefficient for the evaluation of ultimate moment, friction coefficient
σ_c	=	concrete stress
σ_s	=	structural steel stress
σ_{sr}	=	reinforcing steel stress
σ_{cr}	=	critical stress (F_{cr} following AISC symbols)

$\sigma_{c.max(min)}$	$=$	maximum (minimum) concrete compression stress
$\sigma_{s.max(min)}$	$=$	maximum (minimum) structural steel stress
$\sigma_{sr.max(min)}$	$=$	maximum (minimum) reinforcing steel stress
χ	$=$	stability curve factor
χ_n	$=$	effects of imperfection factor for buckling evaluation
η	$=$	degree of shear connection
μ	$=$	friction coefficient
τ_c	$=$	shear resistance of concrete
τ_u	$=$	mean ultimate shear stress in a composite slab
$\tau_{u.d}$	$=$	design value of τ_u
$\tau_{u.k}$	$=$	characteristic value of τ_u
ϕ	$=$	creep coefficient, resistance factor, non-dimensional term to define buckling curves, diameter of reinforcing bars
ϕ_c	$=$	resistance factor of column
ϕ_b	$=$	resistance factor of beam

References

[1] Allern, D.N. de G. and Severn, R.T. 1961. Composite Action Between Beams and Slabs Under Transverse Load, *Struct. Eng.*, 39, 149-154.

[2] Ansourian, P. 1975. An Application of the Method of Finite Elements to the Analysis of Composite Floor System, *Proc. Inst. Civil Eng.*, London, 59, 699-726.

[3] Aribert, J.M. 1990. Design of Composite Beams with a Partial Shear Connection. *Proc. IABSE Symp. Mixed Structures Including New Materials*, 60, 215-220 (in French).

[4] Aribert, J.M. and Abdel Aziz, K. 1990. Calcul de poutres mextes jusqu'à l'état ultime avec un effect de soulèvement à l'interface aciier-bèton, *Construction Mètallique*, 22(4), 3-36.

[5] Bode, H. and Sauerborn, I. 1992. Modern Design Concept for Composite Slabs with Ductile Behavior, in *Composite Construction in Steel and Concrete*, 2nd ed. W.S. Easterling and W.M.K. Morris, ASCE, New York, 125-141.

[6] Bradford, M.A. and Gao, Z. 1992. Distorsional Buckling Solutions for Continuous Composite Beams, *ASCE J. Struct. Eng.*, 118,(1), 73-89.

[7] Brett, P.R., Nethercot, D.A., and Owens, G.W. 1997. Continuous Construction in Steel for Roofs and Composite Floors, *Struct. Eng.*, 65A, 355-363.

[8] Buckner, C.D., Deville, D.J., and McKee, D.C. 1981. Shear Strength of Slab in Stub-Girders, *ASCE J. Sruct. Div.*, February.

[9] Carbone, V. and Nascè, V. 1985. On the Effective Widht and Other Interaction Aspects in Composite Steel-Concrete Beams, in *Proc. of the 10th Italian Conf. Constructional Steel*, Montecatini, 121-135.

[10] Climenhaga, J.J. and Johnson, R.P. 1972. Local Buckling in Continuous Composite Beams, *Struct. Eng.*, 50 (9), 367-374.

[11] Cosenza, E., Mazzolani, S.M., and Pecce, M. 1992. Ultimate Limit State Checking of Continuous Composite Beams Designed to Eurocode Recommendations, *Costruzioni Metalliche*, 5, 273-287.

[12] Cosenza, E. and Pecce, M. 1991. Deflections and Redistribution of Moments Due to the Cracking in Steel-Concrete Composite Continuous Beams Designed to Eurocode 4, in *Proc. ICSAS 91*, Elsevier Applied Science, 3, 52-61, Singapore, May.

[13] Cosenza, E. and Pecce, M. 1993. Composite Steel-Concrete Structures and Eurocode 4; New Research Results: Cracked Analysis of Continuous Composite Beams, *Lecture notes, International Advanced School, Eurocode 3 and 4*, Budapest, Hungary, December.

[14] Couchman, G. and Lebet, J-P. 1993. Design Rules for Continuous Composite Beams using Class 1 and 2 Steel Sections — Applicability of EC 4. *EPFL-ICOM Publication 290*, Lausanne, November.

[15] Crisinel, M. 1990. Partial-Interaction Analysis of Composite Beams with Profiled Sheeting and Nonweld Shear Connectors, *J. Construct. Steel Res.*, Special Issue on Composite Construction, R. Zandonini, Ed., 15 (1&2), 65-98.

[16] Dezi, L. and Tarantino, A. M. 1993. Creep in Composite Continuous Beams, I: Theoretical Treatment, *ASCE J. Struct. Eng.*, 7 (119), 2095-2111.

[17] Dezi, L. and Tarantino, A. M. 1993. Creep in Composite Continuous Beams. I: Parametric Study. *ASCE J. Struct. Eng.*, 7 (119), 2112-2133.

[18] Evans, H.R. and Wright, H.D. 1988. Steel-Concrete Composite Flooring Deck Structures, in *Steel-Concrete Composite Structures, Stability and Strength*, Narayanan, R., Ed., pp.21-52, Elsevier Applied Science, London.

[19] Goble, G.G. 1968. Shear Strength of Thin Flange Composite Specimens, *AISC Eng. J.*, 5 (2).

[20] Grant, J.A., Fisher, J.W., and Slutter, R.G. 1977. Composite Beams with Formed Steel Deck, *AISC Eng. J.*, 14 (1). 24-42.

[21] Griffis, L.G. 1992. Composite Frame Construction, *Constructional Steel Design, An International Guide*, Dowling, P.J. et al., Eds., p. 523-553, Elsevier Applied Science, London.

[22] Hancock, G.J. 1978. Local, Distorsional and Lateral Buckling on I-beams, *ASCE J. Struct. Div.*, 106 (ST7), 1557-1571.

[23] Heins, C.P. and Fain, H.H. 1976. Effective Composite Beams Width at Ultimate Load, *ASCE J. Struct. Div.*, 102 (ST11), 2163-2179.

[24] Lawson, R.M. 1992. Shear connection in Composite Beams, in *Composite Construction in Steel and Concrete*, 2nd ed. W.S. Easterling and W.M.K. Morris, ASCE, New York, pp. 81-97.

[25] Luttrel, L.D. and Prassanan, S. 1986. Method for Predicting Strengths in Composite Slabs, in *Proc. 8th Intl. Speciality Conference on Cold Formed Structures*, pp. 419-431, University of Missouri-Rolla, St. Louis, Missouri.

[26] Johnson, R.P. and Allison, R.W. 1981. Shrinkage and Tension Stiffening in Negative Moment Regions of Composite Beams, *Struct. Eng.*, 59b (1), 10-15.

[27] Johnson, R.P. and Hope-Gill, M.C. 1976. Applicability of Simple Plastic Theory to Continuous Composite Beams, in *Proc. Inst. Civ. Eng.*, 61 (2), 127-143.

[28] Johnson, R.P. and May, I.M. 1978. Tests on Restrained Composite Columns, *Struct. Eng.*, 56B, 21-8, June.

[29] Johnson, R.P. and Molenstra, N. 1991. Partial Shear Connection in Composite Beams for Buildings, in *Proc. Inst. Civ. Eng.*, 91 (2), 679 - 704.

[30] Johnson, R.P. and Anderson, D. 1993. Designer's Handbook to Eurocode 4: Part 1.1, Design of Steel and Composite Structures, British Standards Institution, London.

[31] Kemp, A.R. and Dekker, N.W. 1991. Available Rotation Capacity in Steel and Composite Beams, *Struct. Eng.*, 69 (5), 88-97.

[32] Mattock, A.H. and Hawkins, N.M. 1972. Shear Transfer in Reinforced Concrete — Recent Research, *J. Prestressed Concrete Inst.*, March/April, 55-75.

[33] Mele, M. and Puhali, R. 1985. Experimental Analysis of Cold Formed Shear Connectors in Steel-Concrete Composite Beams, *Costruzioni Metalliche*, 36(5/6), 239-251, 291-302.

[34] Moore, W.P. Jr. 1988. An Overview of Composite Construction in the United States, *Composite Construction in Steel and Concrete*, C.D. Buckner and I.M. Viest, Eds., ASCE, New York p.1-17.

[35] Mottram, J.T. and Johnson, R.P. 1990. Push Tests on Studs Welded Through Profiled Steel Sheeting, *Struct. Eng.*, 68, 187-193.

[36] Oehlers, D.J. 1989. Splitting Induced by Shear Connectors in Composite Beams, *ASCE J. Struct. Eng. Div.*, 115, 341-362.

[37] Oehlers, D.J. and Bradford, M.A. 1995. *Composite Steel and Concrete Structural Members, Fundamental Behavior*, Pergamon Press.

[38] Ollgaard, J.G., Slutter, R.G. and Fisher, J.W. 1971. Shear Strength of Stud Shear Connections in Lightweight and Normal Weight Concrete, *AISC Eng. J.*, 8 (2).

[39] Porter, M.L. and Ekberg, C.E. 1976. Design Recommendations for Steel Deck Floor Slabs, *ASCE J. Struct. Div.*, 102 (ST11), 2121-2136.

[40] Puhali, R., Smotlak, I. and Zandonini, R. 1990. Semi-Rigid Composite Action: Experimental Analysis and a Suitable Model, *J. Constructional Steel Res.*, Special Issue on Composite Construction, Zandonini R., Ed., 15 (1&2), 121-151.

[41] Roik, K. and Bergmann, R. 1989. Report on Eurocode 4, Clause 4.8 and 4.9, Composite Columns, Harmonization of European Construction Codes, Report EC4/6/89, Minister fur Raumordnung, Bauwesen und Studtebau der Bundesrepublik Deutschland, reference number RSII 1-6741028630, July.

[42] Roik, K. and Bergmann, R. 1992. Composite Columns, in *Constructional Steel Design, An International Guide*, Dowling, P.J., Harding, J.E., and Bjorhovde, R. Eds., Elsevier Applied Science, p.443-470

[43] Schuster, R. M. and Ekberg, C.E. 1970. Commentary on the Tentative Recommendations for the Design of Cold-Formed Steel Decking as Reinforcement for Concrete Floor Slabs, Research Report, Iowa State University, Ames, Iowa.

[44] Slutter, R.G. and Driscoll, G.C. 1965. Flexural Strength of Steel-Concrete Composite Beams, *ASCE J. Struct. Div.*, 91 (ST2).

[45] Stark, J.W.B. and Brekelmans, J.W.P.M. 1990. Plastic Design of Continuous Composite Slabs, *J. Constructional Steel Res.*, 15, 23-47.

[46] Viest, I.M. 1992. Composite Construction: Recent Past, Present and Near Future, in *Composite Construction in Steel and Concrete*, 2nd ed., Easterling W.S. and Morris, W.M.K. Eds., ASCE, New York, p. 1-16.

[47] Wakabayashi, M. and Minami, K. 1990. Application of High Strength Steel to Composite Structures, in *Proc. IABSE Symp. Mixed Struct. Including New Mater.*, 60, 59-64.

[48] Trahair, N.S. 1993. Flexural-Torsional Buckling of Structures, Chapman and Hall, London, 1993.

Codes and Standards

American Concrete Institute. 1989. Building Code Requirements for Reinforced Concrete. ACI 318-89. Detroit, Michigan.

American Institute of Steel Construction. 1993. Load and Resistance Factor Design Specifications for Structural Steel Buildings, Chicago, Illinois.

American Iron and Steel Institute. 1986. Specifications for the Design of Cold-Formed Steel Structural Members, Washington, D.C.

American Society of Civil Engineers. 1991. Standard for the Structural Design of Composite Slabs, ANSI/ASCE 3-91, New York.

American Society of Civil Engineers. 1991. Standard Practice for Construction and Inspection of Composite Slabs, ANSI/ASCE 9-91, New York.

Architectural Institute of Japan. 1987. AIJ Standard for Steel Reinforced Concrete Structures.

European Committee for Standardisation (CEN). 1992. ENV 1992-1-1 Eurocode 2, Common Unified Rules for Reinforced Concrete Structures, Brussels, Belgium.

European Committee for Standardisation (CEN). 1992. ENV 1993-1-1 Eurocode 3, Design of Steel Structures, Part 1.1, General Rules and Rules for Buildings, Brussels, Belgium.

European Committee for Standardisation (CEN). 1994. ENV 1993-1-3 Eurocode 3, Design of Steel Structures, Part 1.3 Cold Formed Thin Gauge Members and Sheeting, Brussels, Belgium.

European Committee for Standardisation (CEN). 1994. ENV 1994-1-1 Eurocode 4, Design of Composite Steel and Concrete Structures, Part 1.1 General Rules and Rules for Buildings, Brussels, Belgium.

European Convention for Constructional Steelwork. 1981. Composite Structures (Model Code for Eurocode 4: Part 1), The Construction Press, London.

European Convention for Constructional Steelwork. 1993. Composite Structures, Composite Beams and Columns to Eurocode 4, publication n. 72, TC 11.

Further Reading

[1] Jonhson R.P. and Anderson, D. 1993. *Designer's Handbook to Eurocode 4 — Part 1.1: Design of Composite Steel and Concrete Structures,* Telford Press, pp. 182.

[2] Jonhson, R.P. 1994. *Composite Structures of Steel and Concrete, Vol. 1 Beams, Slabs, Columns, and Frames for Buildings,* Blackwell, pp. 210.

[3] Oehlers, D.J. and Bradford, M.A. 1995. *Composite Steel and Concrete Structural Members — Fundamental Behavior,* Pergamon Press, pp. 549.

[4] Viest, I.M., Colaco, J.P., Furlong, R.W., Griffis, L.G., Leon, R.T., and Wyllies, L.A., Eds. 1996. *Composite Construction: Design for Buildings,* McGraw-Hill/ASCE, New York, pp. 416.

Section II
SPECIAL STRUCTURES

7

Cold-Formed Steel Structures

7.1   Introduction 7-1
7.2   Design Standards 7-3
7.3   Design Bases 7-4
      Allowable Stress Design (ASD) • Limit States Design or Load
      and Resistance Factor Design (LRFD)
7.4   Materials and Mechanical Properties 7-7
      Yield Point, Tensile Strength, and Stress-Strain Relationship •
      Strength Increase from Cold Work of Forming • Modulus of
      Elasticity, Tangent Modulus, and Shear Modulus • Ductility
7.5   Element Strength 7-11
      Maximum Flat-Width-to-Thickness Ratios • Stiffened Ele-
      ments under Uniform Compression • Stiffened Elements with
      Stress Gradient • Unstiffened Elements under Uniform Com-
      pression • Uniformly Compressed Elements with an Edge Stiff-
      ener • Uniformly Compressed Elements with Intermediate
      Stiffeners
7.6   Member Design 7-19
      Sectional Properties • Linear Method for Computing Sectional
      Properties • Tension Members • Flexural Members • Concen-
      trically Loaded Compression Members • Combined Axial Load
      and Bending • Cylindrical Tubular Members
7.7   Connections and Joints 7-42
      Welded Connections • Bolted Connections • Screw Connec-
      tions
7.8   Structural Systems and Assemblies 7-46
      Metal Buildings • Shear Diaphragms • Shell Roof Structures •
      Wall Stud Assemblies • Residential Construction • Composite
      Construction
7.9   Defining Terms 7-53
References 7-54
Further Reading 7-56

Wei-Wen Yu
Department of Civil Engineering,
University of Missouri-Rolla,
Rolla, MO

7.1   Introduction

Cold-formed steel members as shown in Figure 7.1 are widely used in building construction, bridge construction, storage racks, highway products, drainage facilities, grain bins, transmission towers, car bodies, railway coaches, and various types of equipment. These sections are cold-formed from carbon or low alloy steel sheet, strip, plate, or flat bar in cold-rolling machines or by press brake or bending brake operations. The **thicknesses** of such members usually range from 0.0149 in. (0.378 mm) to about 1/4 in. (6.35 mm) even though steel plates and bars as thick as 1 in. (25.4 mm) can be cold-formed into structural shapes.

Figure 7.1   Various shapes of cold-formed steel sections. (From Yu, W.W. 1991. *Cold-Formed Steel Design,* John Wiley & Sons, New York. With permission.)

The use of cold-formed steel members in building construction began in the 1850s in both the U.S. and Great Britain. However, such steel members were not widely used in buildings in the U.S. until the 1940s. At the present time, cold-formed steel members are widely used as construction materials worldwide.

Compared with other materials such as timber and concrete, cold-formed steel members can offer the following advantages: (1) lightness, (2) high strength and stiffness, (3) ease of prefabrication and mass production, (4) fast and easy erection and installation, and (5) economy in transportation and handling, just to name a few.

From the structural design point of view, cold-formed steel members can be classified into two major types: (1) individual structural framing members (Figure 7.2) and (2) panels and decks (Figure 7.3).

In view of the fact that the major function of the individual framing members is to carry load, structural strength and stiffness are the main considerations in design. The sections shown in Figure 7.2 can be used as primary framing members in buildings up to four or five stories in height. In tall multistory buildings, the main framing is typically of heavy hot-rolled shapes and the secondary elements such as wall studs, joists, decks, or panels may be of cold-formed steel members. In this case, the heavy hot-rolled steel shapes and the cold-formed steel sections supplement each other.

The cold-formed steel sections shown in Figure 7.3 are generally used for roof decks, floor decks, wall panels, and siding material in buildings. Steel decks not only provide structural strength to carry loads, but they also provide a surface on which flooring, roofing, or concrete fill can be applied as shown in Figure 7.4. They can also provide space for electrical conduits. The cells of cellular panels can also be used as ducts for heating and air conditioning. For **composite slabs,** steel decks

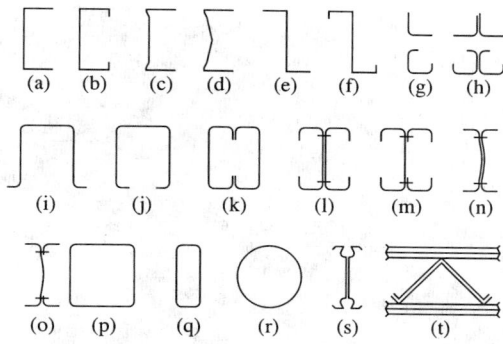

Figure 7.2 Cold-formed steel sections used for structural framing. (From Yu, W.W. 1991. *Cold-Formed Steel Design*, John Wiley & Sons, New York. With permission.)

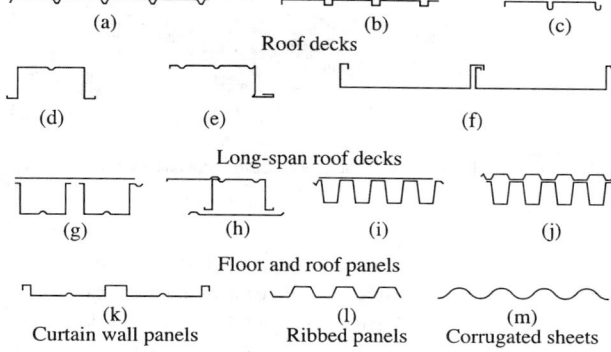

Figure 7.3 Decks, panels, and corrugated sheets. (From Yu, W.W. 1991. *Cold-Formed Steel Design*, John Wiley & Sons, New York. With permission.)

are used not only as formwork during construction, but also as reinforcement of the composite system after the concrete hardens. In addition, load-carrying panels and decks not only withstand loads normal to their surface, but they can also act as shear diaphragms to resist forces in their own planes if they are adequately interconnected to each other and to supporting members.

During recent years, cold-formed steel sections have been widely used in residential construction and pre-engineered metal buildings for industrial, commercial, and agricultural applications. Metal building systems are also used for community facilities such as recreation buildings, schools, and churches. For additional information on cold-formed steel structures, see Yu [49], Rhodes [36], and Hancock [28].

7.2 Design Standards

Design standards and recommendations are now available in Australia [39], Austria [31], Canada [19], Czechoslovakia [21], Finland [26], France [20], Germany [23], India [30], Japan [14], The Netherlands [27], New Zealand [40], The People's Republic of China [34], The Republic of South Africa [38], Sweden [44], Romania [37], U.K. [17], U.S. [7], USSR [41], and elsewhere. Since 1975, the European Convention for Constructional Steelwork [24] has prepared several documents for the design and testing of cold-formed sheet steel used in buildings. In 1989, Eurocode 3 provided design information for cold-formed steel members.

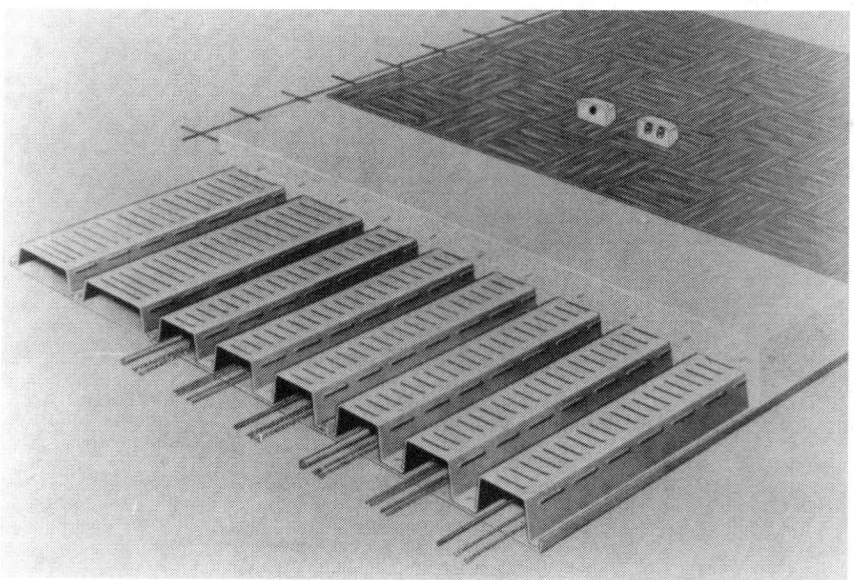

Figure 7.4　Cellular floor decks. (From Yu, W.W. 1991. *Cold-Formed Steel Design,* John Wiley & Sons, New York. With permission.)

This chapter presents discussions on the design of cold-formed steel structural members for use in buildings. It is mainly based on the current AISI combined specification [7] for allowable stress design (**ASD**) and load and resistance factor design (**LRFD**). It should be noted that in addition to the AISI specification, in the U.S., many trade associations and professional organizations have issued special design requirements for using cold-formed steel members as floor and roof decks [42], roof trusses [6], open web steel joists [43], transmission poles [10], storage racks [35], shear diaphragms [7, 32], composite slabs [11], metal buildings [33], light framing systems [15], guardrails, structural supports for highway signs, luminaries, and traffic signals [4], automotive structural components [5], and others. For the design of cold-formed stainless steel structural members, see ASCE Standard 8-90 [12].

7.3　Design Bases

For cold-formed steel design, two design approaches are being used. They are: (1) ASD and (2) LRFD. Both methods are briefly discussed in this section.

7.3.1　Allowable Stress Design (ASD)

In the ASD approach, the **required strengths** (moments, axial forces, and shear forces) in structural members are computed by accepted methods of structural analysis for the specified **nominal** or working loads for all applicable load combinations listed below [7].

1. D
2. $D + L + (L_r \text{ or } S \text{ or } R_r)$
3. $D + (W \text{ or } E)$
4. $D + L + (L_r \text{ or } S \text{ or } R_r) + (W \text{ or } E)$

where

D	=	dead load
E	=	earthquake load
L	=	live load due to intended use and occupancy
L_r	=	roof live load
R_r	=	rain load, except for ponding
S	=	snow load
W	=	wind load

In addition, due consideration should also be given to the loads due to (1) fluids with well-defined pressure and maximum heights, (2) weight and lateral pressure of soil and water in soil, (3) ponding, and (4) contraction or expansion resulting from temperature, shrinkage, moisture changes, creep in component materials, movement due to different settlement, or combinations thereof.

The required strengths should not exceed the allowable **design strengths** permitted by the applicable design standard. The allowable design strength is determined by dividing the **nominal strength** by a safety factor as follows:

$$R_a = R_n / \Omega \qquad (7.1)$$

where

R_a	=	allowable design strength
R_n	=	nominal strength
Ω	=	**safety factor**

For the design of cold-formed steel structural members using the AISI ASD method [7], the safety factors are given in Table 7.1.

When wind or earthquake loads act in combination with dead and/or live loads, it has been a general practice to permit the allowable design strength to be increased by a factor of one-third because the action of wind or earthquake on a structure is highly localized and of very short duration. This can also be accomplished by permitting a 25% reduction in the combined load effects without the increase of the allowable design strength.

7.3.2 Limit States Design or Load and Resistance Factor Design (LRFD)

Two types of **limit states** are considered in the LRFD method. They are: (1) the limit state of strength required to resist the extreme loads during the life of the structure and (2) the limit state of serviceability for a structure to perform its intended function..

For the limit state of strength, the general format of the LRFD method is expressed by the following equation:

$$\Sigma \gamma_i Q_i \leq \phi R_n \qquad (7.2)$$

where

$\Sigma \gamma_i Q_i$	=	required strength
ϕR_n	=	design strength
γ_i	=	**load factors**
Q_i	=	load effects
ϕ	=	**resistance factor**
R_n	=	nominal strength

The load factors and load combinations are specified in various standards. According to the AISI Specification [7], the following load factors and load combinations are used for cold-formed steel design:

1. $1.4D + L$
2. $1.2D + 1.6L + 0.5(L_r$ or S or $R_r)$
3. $1.2D + 1.6(L_r$ or S or $R_r) + (0.5L$ or $0.8W)$

TABLE 7.1 Safety Factors, Ω, and Resistance Factors, ϕ, used in the AISI Specification [7]

Type of strength	ASD safety factor, Ω	LRFD resistance factor, ϕ
(a) Stiffeners		
Transverse stiffeners	2.00	0.85
Shear stiffeners[a]	1.67	0.90
(b) Tension members (see also bolted connections)	1.67	0.95
(c) Flexural members		
Bending strength		
For sections with stiffened or partially stiffened compression flanges	1.67	0.95
For sections with unstiffened compression flanges	1.67	0.90
Laterally unbraced beams	1.67	0.90
Beams having one flange through-fastened to deck or sheathing (C- or Z-sections)	1.67	0.90
Beams having one flange fastened to a standing seam roof system	1.67	0.90
Web design		
Shear strength[a]	1.67	0.90
Web crippling		
For single unreinforced webs	1.85	0.75
For I-sections	2.00	0.80
For two nested Z-sections	1.80	0.85
(d) Concentrically loaded compression members	1.80	0.85
(e) Combined axial load and bending		
For tension	1.67	0.95
For compression	1.80	0.85
For bending	1.67	0.90-0.95
(f) Cylindrical tubular members		
Bending strength	1.67	0.95
Axial compression	1.80	0.85
(g) Wall studs and wall assemblies		
Wall studs in compression	1.80	0.85
Wall studs in bending	1.67	0.90-0.95
(h) Diaphragm construction	2.00-3.00	0.50-0.65
(i) Welded connections		
Groove welds		
Tension or compression	2.50	0.90
Shear (welds)	2.50	0.80
Shear (base metal)	2.50	0.90
Arc spot welds		
Welds	2.50	0.60
Connected part	2.50	0.50-0.60
Minimum edge distance	2.00-2.22	0.60-0.70
Tension	2.50	0.60
Arc seam welds		
Welds	2.50	0.60
Connected part	2.50	0.60
Fillet welds		
Longitudinal loading (connected part)	2.50	0.55-0.60
Transverse loading (connected part)	2.50	0.60
Welds	2.50	0.60
Flare groove welds		
Transverse loading (connected part)	2.50	0.55
Longitudinal loading (connected part)	2.50	0.55
Welds	2.50	0.60
Resistance Welds	2.50	0.65
(j) Bolted connections		
Minimum spacing and edge distance	2.00-2.22	0.60-0.70
Tension strength on net section		
With washers		
Double shear connection	2.00	0.65
Single shear connection	2.22	0.55
Without washers	2.22	0.65
Bearing strength	2.22	0.55-0.70
Shear strength of bolts	2.40	0.65
Tensile strength of bolts	2.00-2.25	0.75
(k) Screw connections	3.00	0.50
(l) Shear rupture	2.00	0.75
(m) Connections to other materials (Bearing)	2.50	0.60

[a] When $h/t \leq 0.96\sqrt{Ek_v/F_y}$, $\quad \Omega = 1.50$, $\quad \phi = 1.0$

4. $1.2D + 1.3W + 0.5L + 0.5(L_r$ or S or $R_r)$
5. $1.2D + 1.5E + 0.5L + 0.2S$
6. $0.9D - (1.3W$ or $1.5E)$

All symbols were defined previously.
Exceptions:

1. The load factor for E in combinations (5) and (6) should be equal to 1.0 when the seismic load model specified by the applicable code or specification is limit state based.
2. The load factor for L in combinations (3), (4), and (5) should be equal to 1.0 for garages, areas occupied as places of public assembly, and all areas where the live load is greater than 100 psf.
3. For wind load on individual purlins, girts, wall panels, and roof decks, multiply the load factor for W by 0.9.
4. The load factor for L_r in combination (3) should be equal to 1.4 in lieu of 1.6 when the roof live load is due to the presence of workmen and materials during repair operations.

In addition, the following LRFD criteria apply to roof and floor composite construction using cold-formed steel:

$$1.2D_s + 1.6C_w + 1.4C$$

where
D_s = weight of steel deck
C_w = weight of wet concrete during construction
C = construction load, including equipment, workmen, and formwork, but excluding the weight of the wet concrete.

Table 7.1 lists the ϕ factors, which are used for the AISI LRFD method for the design of cold-formed steel members and connections [7]. It should be noted that different load factors and resistance factors may be used in different standards. These factors are selected for the specific nominal strength equations adopted by the given standard or specification.

7.4 Materials and Mechanical Properties

In the AISI Specification [7], 14 different steels are presently listed for the design of cold-formed steel members. Table 7.2 lists steel designations, ASTM designations, **yield points**, tensile strengths, and elongations for these steels.

From a structural standpoint, the most important properties of steel are as follows:

1. Yield point or yield strength, F_y
2. Tensile strength, F_u
3. Stress-strain relationship
4. Modulus of elasticity, tangent modulus, and shear modulus
5. Ductility
6. Weldability
7. Fatigue strength

In addition, formability, durability, and toughness are also important properties for cold-formed steel.

TABLE 7.2   Mechanical Properties of Steels Referred to in the AISI 1996 Specification

Steel designation	ASTM designation	Yield point, F_y (ksi)	Tensile strength, F_u (ksi)	Elongation (%) In 2-in. gage length	Elongation (%) In 8-in. gage length
Structural steel	A36	36	58-80	23	—
High-strength low-alloy structural steel	A242 (3/4 in. and under)	50	70	—	18
	(3/4 in. to 1-1/2 in.)	46	67	21	18
Low and intermediate	A283 Gr. A	24	45-60	30	27
tensile strength	B	27	50-65	28	25
carbon plates, shapes	C	30	55-75	25	22
and bars	D	33	60-80	23	20
Cold-formed welded	A500				
and seamless carbon	Round tubing				
steel structural tubing	A	33	45	25	—
in rounds and shapes	B	42	58	23	—
	C	46	62	21	—
	D	36	58	23	—
	Shaped tubing				
	A	39	45	25	—
	B	46	58	23	—
	C	50	62	21	—
	D	36	58	23	—
Structural steel with 42 ksi	A529 Gr. 42	42	60-85	—	19
minimum yield point	50	50	70-100	—	18
Hot-rolled carbon steel	A570 Gr. 30	30	49	21-25	—
sheets and strips of	33	33	52	18-23	—
structural quality	36	36	53	17-22	—
	40	40	55	15-21	—
	45	45	60	13-19	—
	50	50	65	11-17	—
High-strength low-alloy	A572 Gr. 42	42	60	24	20
columbium-vanadium	50	50	65	21	18
steels of structural	60	60	75	18	16
quality	65	65	80	17	15
High-strength low-alloy structural steel with 50 ksi minimum yield point	A588	50	70	21	18
Hot-rolled and cold-rolled high-strength low-alloy steel sheet and strip with improved corrosion resistance	A606 Hot-rolled as rolled coils; annealed, or normalized; and cold-rolled	45	65	22	—
	Hot-rolled as rolled cut lengths	50	70	22	—
Hot-rolled and cold-rolled	A607 Gr. 45	45	60 (55)	Hot-rolled 23-25 Cold-rolled 22	
high-strength low-alloy columbium and/or vanadium steel sheet and strip	50	50	65 (60)	Hot-rolled 20-22 Cold-rolled 20	—
	55	55	70 (65)	Hot-rolled 18-20 Cold-rolled 18	—
	60	60	75 (70)	Hot-rolled 16-18 Cold-rolled 16	—
	65	65	80 (75)	Hot-rolled 14-16 Cold-rolled 15	—
	70	70	85 (80)	Hot-rolled 12-14 Cold-rolled 14	—
Cold-rolled carbon	A611 Gr. A	25	42	26	—
structural steel sheet	B	30	45	24	—
	C	33	48	22	—
	D	40	52	20	—
	E	80	82	—	—

TABLE 7.2 Mechanical Properties of Steels Referred to in the AISI 1996 Specification
(continued)

Steel designation	ASTM designation	Yield point, F_y (ksi)	Tensile strength, F_u (ksi)	Elongation (%) In 2-in. gage length	Elongation (%) In 8-in. gage length
Zinc-coated steel sheets	A653 SQ Gr. 33	33	45	20	—
of structural quality	37	37	52	18	—
	40	40	55	16	—
	50 (class 1)	50	65	12	—
	50 (class 3)	50	70	12	—
	80	80	82	—	—
	HSLA Gr. 50	50	60	20	—
	60	60	70	16	—
	70	70	80	12(14)	—
	80	80	90	10(12)	—
Hot-rolled high-strength	A715 Gr. 50	50	60	22-24	—
low-alloy steel sheets	60	60	70	20-22	—
and strip with improved	70	70	80	18	—
formability	80	80	90	14	—
Aluminum-zinc	A792 Gr. 33	33	45	20	—
alloy-coated by the	37	37	52	18	—
hot-dip process	40	40	55	16	—
general requirements	50	50	65	12	—
	80	80	82	—	—

Notes:
1. The tabulated values are based on ASTM Standards.
2. 1 in. = 25.4 mm; 1 ksi = 6.9 MPa.
3. A653 Structural Quality Grade 80, Grade E of A611, and Structural Quality Grade 80 of A792 are allowed in the AISI Specification under special conditions. For these grades, $F_y = 80$ ksi, $F_u = 82$ ksi, elongations are unspecified. See AISI Specification for reduction of yield point and tensile strength.
4. For A653 steel, HSLA Grades 70 and 80, the elongation in 2-in. gage length given in the parenthesis is for Type II. The other value is for Type I.
5. For A607 steel, the tensile strength given in the parenthesis is for Class 2. The other value is for Class 1.

7.4.1   Yield Point, Tensile Strength, and Stress-Strain Relationship

As listed in Table 7.2, the yield points or yield strengths of all 14 different steels range from 24 to 80 ksi (166 to 552 MPa). The tensile strengths of the same steels range from 42 to 100 ksi (290 to 690 MPa). The ratios of the tensile strength-to-yield point vary from 1.12 to 2.22. As far as the stress-strain relationship is concerned, the stress-strain curve can either be the sharp-yielding type (Figure 7.5a) or the gradual-yielding type (Figure 7.5b).

7.4.2   Strength Increase from Cold Work of Forming

The mechanical properties (yield point, tensile strength, and ductility) of cold-formed steel sections, particularly at the corners, are sometimes substantially different from those of the flat steel sheet, strip, plate, or bar before forming. This is because the cold-forming operation increases the yield point and tensile strength and at the same time decreases the ductility. The effects of cold-work on the mechanical properties of corners usually depend on several parameters. The ratios of tensile strength-to-yield point, F_u/F_y, and inside bend radius-to-thickness, R/t, are considered to be the most important factors to affect the change in mechanical properties of cold-formed steel sections. Design equations are given in the AISI Specification [7] for computing the tensile yield strength of corners and the average full-section tensile yield strength for design purposes.

7.4.3   Modulus of Elasticity, Tangent Modulus, and Shear Modulus

The strength of cold-formed steel members that are governed by buckling depends not only on the yield point but also on the modulus of elasticity, E, and the tangent modulus, E_t. A value of $E = 29,500$ ksi (203 GPa) is used in the AISI Specification for the design of cold-formed steel structural

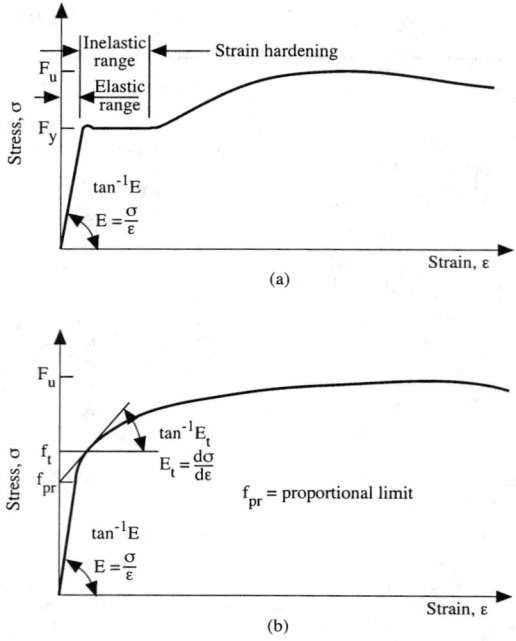

Figure 7.5   Stress-strain curves of steel sheet or strip. (a) Sharp-yielding. (b) Gradual-yielding. (From Yu, W.W. 1991. *Cold-Formed Steel Design,* John Wiley & Sons, New York. With permission.)

members. This E value is slightly larger than the value of 29,000 ksi (200 GPa), which is being used in the AISC Specification for the design of hot-rolled shapes. The tangent modulus is defined by the slope of the stress-strain curve at any given stress level as shown in Figure 7.5b. For sharp-yielding steels, $E_t = E$ up to the yield, but with gradual-yielding steels, $E_t = E$ only up to the proportional limit, f_{pr} (Figure 7.5b). Once the stress exceeds the proportional limit, the tangent modulus E_t becomes progressively smaller than the initial modulus of elasticity. For cold-formed steel design, the shear modulus is taken as $G = 11,300$ ksi (77.9 GPa) according to the AISI Specification.

7.4.4   Ductility

According to the AISI Specification, the ratio of F_u/F_y for the steels used for structural framing members should not be less than 1.08, and the total elongation should not be less than 10% for a 2-in. (50.8 mm) gage length. If these requirements cannot be met, an exception can be made for purlins and girts for which the following limitations should be satisfied when such a material is used: (1) local elongation in a 1/2-in. (12.7 mm) gage length across the fracture should not be less than 20% and (2) uniform elongation outside the fracture should not be less than 3%. It should be noted that the required ductility for cold-formed steel structural members depends mainly on the type of application and the suitability of the material. The same amount of ductility that is considered necessary for individual framing members may not be needed for roof panels, siding, and similar applications. For this reason, even though Structural Grade 80 of ASTM A653 steel, Grade E of A611 steel, and Grade 80 of A792 steel do not meet the AISI requirements of the F_u/F_y ratio and the elongation, these steels can be used for roofing, siding, and similar applications provided that (1) the yield strength, F_y, used for design is taken as 75% of the specified minimum yield point or 60 ksi (414 MPa), whichever is less, and (2) the tensile strength, F_u, used for design is taken as 75% of the specified minimum tensile stress or 62 ksi (427 MPa), whichever is less.

7.5 Element Strength

For cold-formed steel members, the width-to-thickness ratios of individual elements are usually large. These thin elements may buckle locally at a stress level lower than the yield point of steel when they are subject to compression in flexural bending and axial compression as shown in Figure 7.6. Therefore, for the design of such thin-walled sections, **local buckling** and postbuckling strength of thin elements have often been the major design considerations. In addition, shear buckling and web crippling should also be considered in the design of beams.

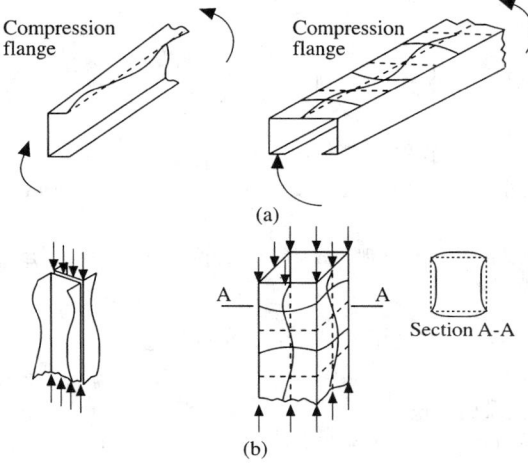

Figure 7.6 Local buckling of compression elements. (a) Beams. (b) Columns. (From Yu, W.W. 1991. *Cold-Formed Steel Design,* John Wiley & Sons, New York. With permission.)

7.5.1 Maximum Flat-Width-to-Thickness Ratios

In cold-formed steel design, the maximum **flat-width-to-thickness ratio**, w/t, for flanges is limited to the following values in the AISI Specification:

1. **Stiffened compression element** having one longitudinal edge connected to a web or flange element, the other stiffened by

   Simple lip 60

   Any other kind of stiffener 90

2. Stiffened compression element with both longitudinal edges connected to other stiffened element 500

3. **Unstiffened compression element** and elements with an inadequate edge stiffener 60

For the design of beams, the maximum depth-to-thickness ratio, h/t, for webs are:

1. For unreinforced webs: $(h/t)_{max} = 200$

2. For webs that are provided with transverse stiffeners:

   Using bearing stiffeners only: $(h/t)_{max} = 260$

   Using bearing stiffeners and intermediate stiffeners: $(h/t)_{max} = 300$

7.5.2   Stiffened Elements under Uniform Compression

The strength of a stiffened compression element such as the compression flange of a hat section is governed by yielding if its w/t ratio is relatively small. It may be governed by local buckling as shown in Figure 7.7 at a stress level less than the yield point if its w/t ratio is relatively large.

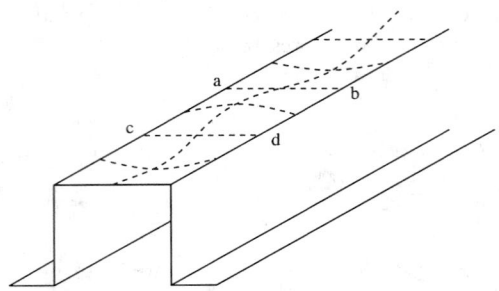

Figure 7.7   Local buckling of stiffened compression flange of hat-shaped beam.

The elastic local buckling stress, f_{cr}, of simply supported square plates and long plates can be determined as follows:

$$f_{cr} = \frac{k\pi^2 E}{12(1 - \mu^2)(w/t)^2} \tag{7.3}$$

where

k = local buckling coefficient
E = modulus of elasticity of steel $= 29.5 \times 10^3$ ksi (203 GPa)
w = width of the plate
t = thickness of the plate
μ = Poisson's ratio

It is well known that stiffened compression elements will not collapse when the local buckling stress is reached. An additional load can be carried by the element after buckling by means of a redistribution of stress. This phenomenon is known as postbuckling strength and is most pronounced for elements with large w/t ratios.

The mechanism of the postbuckling action can be easily visualized from a square plate model as shown in Figure 7.8 [48]. It represents the portion *abcd* of the compression flange of the hat section illustrated in Figure 7.7. As soon as the plate starts to buckle, the horizontal bars in the grid of the model will act as tie rods to counteract the increasing deflection of the longitudinal struts.

In the plate, the stress distribution is uniform prior to its buckling. After buckling, a portion of the prebuckling load of the center strip transfers to the edge portion of the plate. As a result, a nonuniform stress distribution is developed, as shown in Figure 7.9. The redistribution of stress continues until the stress at the edge reaches the yield point of steel and then the plate begins to fail.

For cold-formed steel members, a concept of "effective width" has been used for practical design. In this approach, instead of considering the nonuniform distribution of stress over the entire width of the plate, w, it is assumed that the total load is carried by a fictitious effective width, b, subjected to a uniformly distributed stress equal to the edge stress, f_{max}, as shown in Figure 7.9. The width, b, is selected so that the area under the curve of the actual nonuniform stress distribution is equal to the sum of the two parts of the equivalent rectangular shaded area with a total width, b, and an intensity of stress equal to the edge stress, f_{max}. Based on the research findings of von Karman, Sechler, and Donnell [45], and Winter [47], the following equations have been developed in the AISI Specification for computing the **effective design width**, b, for stiffened elements under uniform compression [7]:

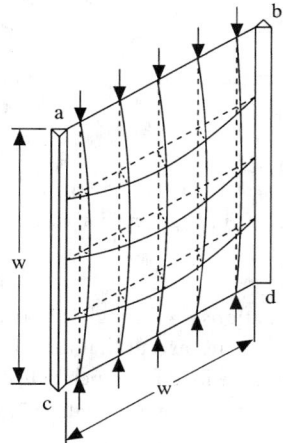

Figure 7.8   Postbuckling strength model. (From Yu, W.W. 1991. *Cold-Formed Steel Design,* John Wiley & Sons, New York. With permission.)

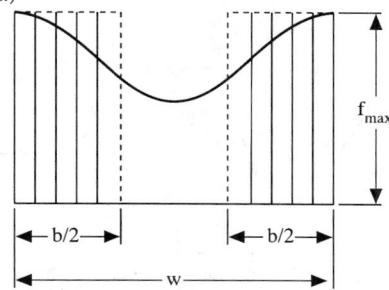

Figure 7.9   Stress distribution in stiffened compression elements.

(a) Strength Determination

$$\text{1. When } \lambda \leq 0.673, b = w \tag{7.4}$$

$$\text{2. When } \lambda > 0.673, b = \rho w \tag{7.5}$$

where

b = effective design width of uniformly compressed element for strength determination (Figure 7.10)

w = flat width of compression element

ρ = reduction factor determined from Equation 7.6:

$$\rho = (1 - 0.22/\lambda)/\lambda \leq 1 \tag{7.6}$$

where λ = plate slenderness factor determined from Equation 7.7:

$$\lambda = (1.052/\sqrt{k})(w/t)(\sqrt{f/E}) \tag{7.7}$$

where

k = plate buckling coefficient = 4.0 for stiffened elements supported by a web on each longitudinal edge as shown in Figure 7.10

t = thickness of compression element

E = modulus of elasticity

f = maximum compressive edge stress in the element without considering the safety factor

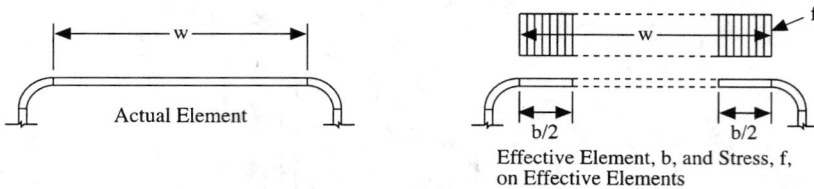

Figure 7.10   Effective design width of stiffened compression elements.

(b) Deflection Determination

For deflection determination, Equations 7.4 through 7.7 can also be used for computing the effective design width of compression elements, except that the compressive stress should be computed on the basis of the effective section at the load for which deflection is calculated.

The relationship between ρ and λ according to Equation 7.6 is shown in Figure 7.11.

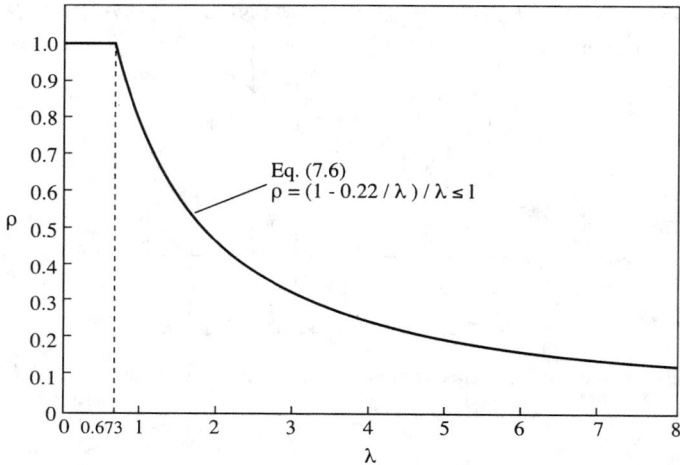

Figure 7.11   Reduction factor, ρ, vs. slenderness factor, λ. (From Yu, W.W. 1991. *Cold-Formed Steel Design*, John Wiley & Sons, New York. With permission.)

EXAMPLE 7.1:

Calculate the effective width of the compression flange of the box section (Figure 7.12) to be used as a beam bending about the x-axis. Use $F_y = 33$ ksi. Assume that the beam webs are fully effective and that the bending moment is based on initiation of yielding.

   Solution   Because the compression flange of the given section is a uniformly compressed stiffened element, which is supported by a web on each longitudinal edge, the effective width of the flange for strength determination can be computed by using Equations 7.4 through 7.7 with $k = 4.0$.

Assume that the bending strength of the section is based on Initiation of Yielding, $\bar{y} \geq 2.50$ in. Therefore, the slenderness factor λ for $f = F_y$ can be computed from Equation 7.7, i.e.,

$$k = 4.0$$

$$w = 6.50 - 2(R + t) = 6.192 \text{ in.}$$

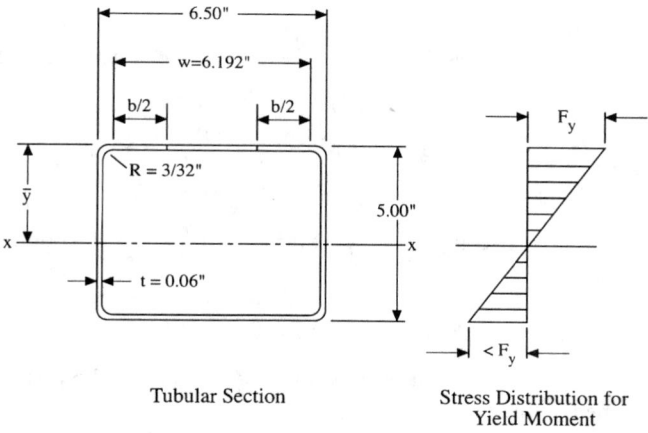

Tubular Section

Stress Distribution for Yield Moment

Figure 7.12 Example 7.1. (From Yu, W.W. 1991. *Cold-Formed Steel Design*, John Wiley & Sons, New York. With permission.)

$$w/t = 103.2$$
$$f = 33 \text{ ksi}$$
$$\lambda = (1.052/\sqrt{k})(w/t)\sqrt{f/E}$$
$$= (1.052/\sqrt{4.0})(103.2)\sqrt{33/29,500} = 1.816$$

Since $\lambda > 0.673$, use Equations 7.5 and 7.6 to compute the effective width, b, as follows:

$$b = \rho w = [(1 - 0.22/\lambda)/\lambda]w$$
$$= [(1 - 0.22/1.816)/1.816](6.192) = 3.00 \text{ in.}$$

7.5.3 Stiffened Elements with Stress Gradient

When a flexural member is subject to bending moment, the beam web is under the stress gradient condition (Figure 7.13), in which the compression portion of the web may buckle due to the compressive stress caused by bending. The effective width of the beam web can be determined from the following AISI provisions:

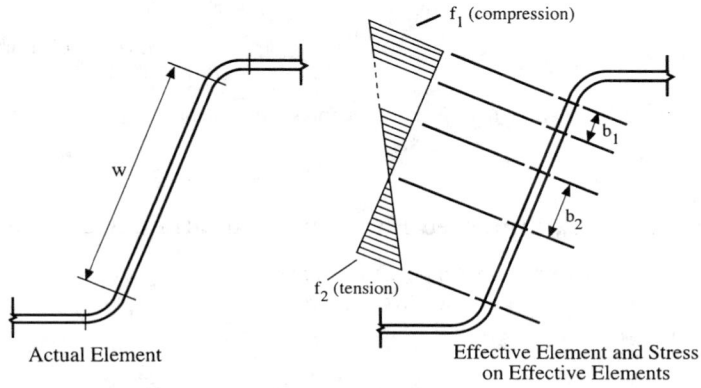

Actual Element

Effective Element and Stress on Effective Elements

Figure 7.13 Stiffened elements with stress gradient.

(a) Strength Determination

The effective widths, b_1 and b_2, as shown in Figure 7.13, should be determined from the following equations:

$$b_1 = b_e/(3 - \psi) \tag{7.8}$$

For $\psi \leq -0.236$

$$b_2 = b_e/2 \tag{7.9}$$

$b_1 + b_2$ should not exceed the compression portion of the web calculated on the basis of effective section.

For $\psi > -0.236$

$$b_2 = b_e - b_1 \tag{7.10}$$

where b_e = effective width b determined by Equation 7.4 or Equation 7.5 with f_1 substituted for f and with k determined as follows:

$$k = 4 + 2(1 - \psi)^3 + 2(1 - \psi) \tag{7.11}$$

$$\psi = f_2/f_1 \tag{7.12}$$

f_1, f_2 = stresses shown in Figure 7.13 calculated on the basis of effective section. f_1 is compression (+) and f_2 can be either tension (−) or compression. In case f_1 and f_2 are both compression, $f_1 \geq f_2$

(b) Deflection Determination

The effective widths used in computing deflections should be determined as above, except that f_{d1} and f_{d2} are substituted for f_1 and f_2, where f_{d1} and f_{d2} are the computed stresses f_1 and f_2 as shown in Figure 7.13 based on the effective section at the load for which deflection is determined.

7.5.4 Unstiffened Elements under Uniform Compression

The effective width of unstiffened elements under uniform compression as shown in Figure 7.14 can also be computed by using Equations 7.4 through 7.7, except that the value of k should be taken as 0.43 and the flat width w is measured as shown in Figure 7.14.

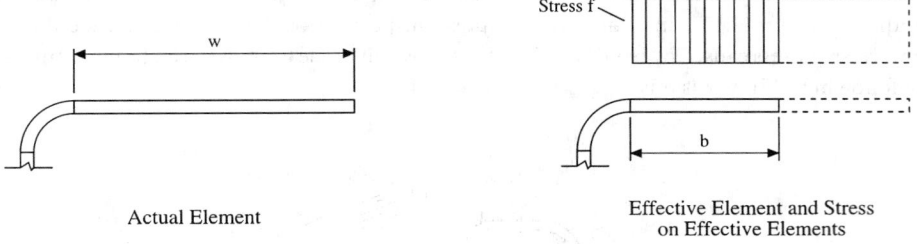

| | |
| Actual Element | Effective Element and Stress
on Effective Elements |

Figure 7.14   Effective design width of unstiffened compression elements.

7.5.5 Uniformly Compressed Elements with an Edge Stiffener

The following equations can be used to determine the effective width of the uniformly compressed elements with an edge stiffener as shown in Figure 7.15.

Case I: For $w/t \leq S/3$

$$I_a = 0 \quad \text{(no edge stiffener needed)}$$

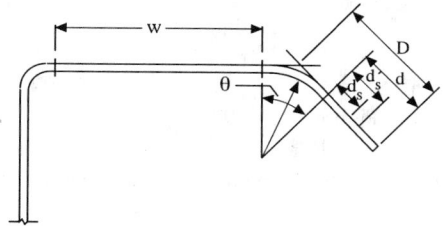

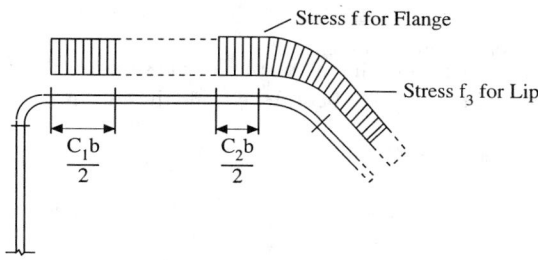

Figure 7.15 Compression elements with an edge stiffener.

$$b = w$$
$$d_s = d'_s \quad \text{for simple lip stiffener}$$
$$A_s = A'_s \quad \text{for other stiffener shapes} \tag{7.13}$$

Case II: For $S/3 < w/t < S$

$$I_a/t^4 = 399\{[(w/t)/S] - \sqrt{k_u/4}\}^3$$
$$n = 1/2$$
$$C_2 = I_s/I_a \leq 1$$
$$C_1 = 2 - C_2 \tag{7.14}$$

b should be calculated according to Equations 7.4 through 7.7, where

$$k = C_2^n(k_a - k_u) + k_u$$
$$k_u = 0.43$$

For simple lip stiffener with $140° \geq \theta \geq 40°$ and $D/w \leq 0.8$ where θ is as shown in Figure 7.15:

$$k_a = 5.25 - 5(D/w) \leq 4.0$$
$$d_s = C_2 d'_s$$

For a stiffener shape other than simple lip:

$$k_a = 4.0$$
$$A_s = C_2 A'_s$$

Case III: For $w/t \geq S$
$$I_a/t^4 = [115(w/t)/S] + 5 \tag{7.15}$$

C_1, C_2, b, k, d_s, and A_s are calculated per Case II with $n = 1/3$

where

S	=	$1.28\sqrt{E/f}$
k	=	buckling coefficient
d, w, D	=	dimensions shown in Figure 7.15
d_s	=	reduced effective width of the stiffener
d_s'	=	effective width of the stiffener calculated as unstiffened element under uniform compression
C_1, C_2	=	coefficients shown in Figure 7.15
A_s	=	reduced area of the stiffener
I_a	=	adequate moment of inertia of the stiffener, so that each component element will behave as a stiffened element
I_s, A_s'	=	moment of inertia of the full section of the stiffener about its own centroidal axis parallel to the element to be stiffened, and the effective area of the stiffener, respectively

For the stiffener shown in Figure 7.15,

$$I_s = (d^3 t \sin^2 \theta)/12$$
$$A_s' = d_s' t$$

7.5.6 Uniformly Compressed Elements with Intermediate Stiffeners

The effective width of uniformly compressed elements with intermediate stiffeners can also be determined from the AISI Specification, which includes separate design rules for compression elements with only one intermediate stiffener and compression elements with more than one intermediate stiffener.

Uniformly Compressed Elements with One Intermediate Stiffener

The following equation can be used to determine the effective width of the uniformly compressed elements with one intermediate stiffener as shown in Figure 7.16.

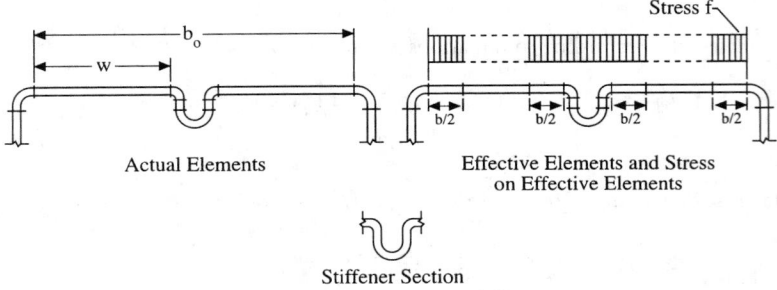

Figure 7.16   Compression elements with one intermediate stiffener.

Case I: For $b_0/t \leq S$

$$I_a = 0 \quad \text{(no intermediate stiffener needed)}$$
$$b = w$$
$$A_s = A_s'$$

Case II: For $S < b_0/t < 3S$

$$I_a/t^4 = [50(b_0/t)/S] - 50$$

b and A_s are calculated according to Equations 7.4 through 7.7, where

$$k = 3(I_s/I_a)^{1/2} + 1 \le 4$$
$$A_s = A'_s(I_s/I_a) \le A'_s$$

Case III: For $b_0/t \ge 3S$

$$I_a/t^4 = [128(b_0/t)/S] - 285$$

b and A_s are calculated according to Equations 7.4 through 7.7, where

$$k = 3(I_s/I_a)^{1/3} + 1 \le 4$$
$$A_s = A'_s(I_s/I_a) \le A'_s$$

In the above equations, all symbols were defined previously.

Uniformly Compressed Elements with More Than One Intermediate Stiffener

For the determination of the effective width of sub-elements, the stiffeners of a stiffened element with more than one stiffener should be disregarded unless each intermediate stiffener has the minimum I_s as follows:

$$I_{\min}/t^4 = 3.66\sqrt{(w/t)^2 - (0.136E)/F_y} \ge 18.4$$

where

w/t = width-thickness ratio of the larger stiffened sub-element
I_s = moment of inertia of the full stiffener about its own centroidal axis parallel to the element to be stiffened

For additional requirements, see the AISI Specification.

7.6 Member Design

This chapter deals with the design of the following cold-formed steel structural members: (a) tension members, (b) **flexural members**, (c) concentrically loaded compression members, (d) combined axial load and bending, and (e) cylindrical tubular members. The nominal strength equations with safety factors (Ω) and resistance factors (ϕ) are provided in the Specification [7] for the given limit states.

7.6.1 Sectional Properties

The sectional properties of a member such as area, moment of inertia, section modulus, and radius of gyration are calculated by using the conventional methods of structural design. These properties are based on either full cross-section dimensions, effective widths, or net section, as applicable.

For the design of tension members, the nominal tensile strength is presently based on the net section. However, for flexural members and axially loaded compression members, the full dimensions are used when calculating the critical moment or load, while the effective dimensions, evaluated at the stress corresponding to the critical moment or load, are used to calculate the nominal strength.

7.6.2 Linear Method for Computing Sectional Properties

Because the thickness of cold-formed steel members is usually uniform, the computation of sectional properties can be simplified by using a "linear" or "midline" method. In this method, the material of each element is considered to be concentrated along the centerline or midline of the steel sheet and the area elements are replaced by straight or curved "line elements". The thickness dimension, t, is introduced after the linear computations have been completed. Thus, the total area is $A = Lt$, and the moment of inertia of the section is $I = I't$, where L is the total length of all line elements and I' is the moment of inertia of the centerline of the steel sheet. The moments of inertia of straight line elements and circular line elements are shown in Figure 7.17.

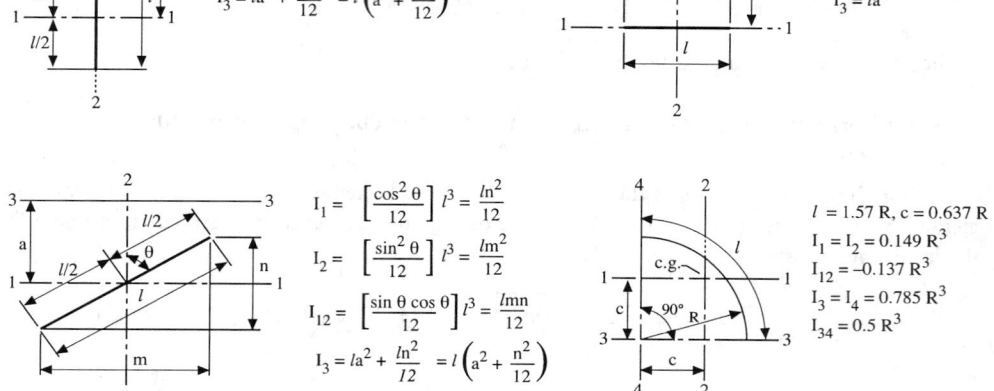

Figure 7.17   Properties of line elements.

7.6.3   Tension Members

The nominal tensile strength of axially loaded cold-formed steel tension members is determined by the following equation:

$$T_n = A_n F_y \qquad (7.16)$$

where

T_n =   nominal tensile strength

A_n =   net area of the cross-section

F_y =   design yield stress

When tension members use bolted connections or circular holes, the nominal tensile strength is also limited by the tensile capacity of connected parts treated separately by the AISI Specification [7] under the section title of Bolted Connections.

7.6.4   Flexural Members

For the design of flexural members, consideration should be given to several design features: (a) bending strength and deflection, (b) shear strength of webs and combined bending and shear, (c) web crippling strength and combined bending and web crippling, and (d) bracing requirements. For some cases, special consideration should also be given to shear lag and flange curling due to the use of thin materials.

Bending Strength

Bending strengths of flexural members are differentiated according to whether or not the member is laterally braced. If such members are laterally supported, they are designed according to the nominal section strength. Otherwise, if they are laterally unbraced, then the bending strength may be governed by the lateral buckling strength. For channels or Z-sections with tension flange attached to deck or sheathing and with compression flange laterally unbraced, and for such members having one flange fastened to a standing seam roof system, the nominal bending strength should be reduced according to the AISI Specification.

Nominal Section Strength

Two design procedures are now used in the AISI Specification for determining the nominal bending strength. They are: (I) *Initiation of Yielding* and (II) *Inelastic Reserve Capacity*.

According to Procedure I on the basis of initiation of yielding, the nominal moment, M_n, of the cross-section is the effective yield moment, M_y, determined for the effective areas of flanges and the beam web. The effective width of the compression flange and the effective depth of the web can be computed from the design equations given in Section 7.5. The yield moment of a cold-formed steel flexural member is defined as the moment at which an outer fiber (tension, compression, or both) first attains the yield point of the steel. Figure 7.18 shows three types of stress distribution for yield

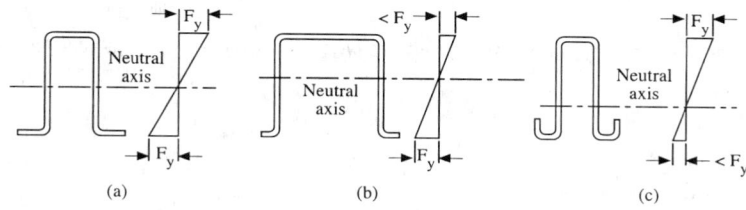

Figure 7.18   Stress distribution for yield moment (based on initiation of yielding).

moment based on different locations of the neutral axis. Accordingly, the nominal section strength for initiation of yielding can be computed as follows:

$$M_n = M_y = S_e F_y \qquad (7.17)$$

where

S_e = elastic section modulus of the effective section calculated with the extreme compression or tension fiber at F_y

F_y = design yield stress

For cold-formed steel design, S_e is usually computed by using one of the following two cases:

1. If the neutral axis is closer to the tension than to the compression flange (Case c), the maximum stress occurs in the compression flange, and therefore the plate slenderness ratio λ (Equation 7.7) and the effective width of the compression flange are determined by the w/t ratio and $f = F_y$. This procedure is also applicable to those beams for which the neutral axis is located at the mid-depth of the section (Case a).

2. If the neutral axis is closer to the compression than to the tension flange (Case b), the maximum stress of F_y occurs in the tension flange. The stress in the compression flange depends on the location of the neutral axis, which is determined by the effective area of the section. The latter cannot be determined unless the compressive stress is known. The closed-form solution of this type of design is possible but would be a very tedious and complex procedure. It is, therefore, customary to determine the sectional properties of the section by successive approximation.

See Examples 7.2 and 7.3 for the calculation of nominal bending strengths.

EXAMPLE 7.2:

Use the ASD and LRFD methods to check the adequacy of the I-section with unstiffened flanges as shown in Figure 7.19. The nominal moment is based on the initiation of yielding using $F_y = 50$ ksi. Assume that lateral bracing is adequately provided. The dead load moment $M_D = 30$ in.-kips and the live load moment $M_L = 150$ in.-kips.

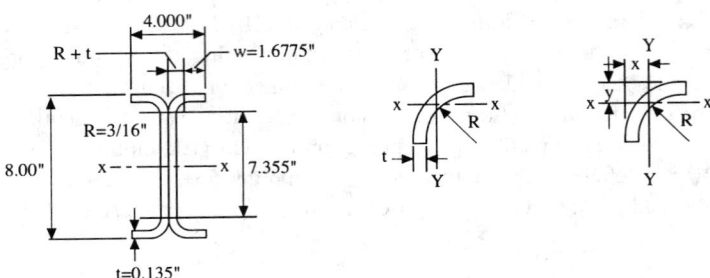

Figure 7.19   Example 7.2. (From Yu, W.W. 1991. *Cold-Formed Steel Design*, John Wiley & Sons, New York. With permission.)

Solution
(A) ASD Method

1. Location of Neutral Axis. For $R = 3/16$ in. and $t = 0.135$ in., the sectional properties of the corner element are as follows:

$$
\begin{aligned}
I_x &= I_y = 0.0003889 \ \text{in.}^4 \\
A &= 0.05407 \ \text{in.}^2 \\
x &= y = 0.1564 \ \text{in.}
\end{aligned}
$$

For the unstiffened compression flange,

$$w = 1.6775 \ \text{in.}, \ w/t = 12.426$$

Using $k = 0.43$ and $f = F_y = 50$ ksi,

$$
\begin{aligned}
\lambda &= (1.052/\sqrt{k})(w/t)\sqrt{f/E} = 0.821 > 0.673 \\
b &= [(1 - 0.22/0.821)/0.821](1.6775) = 1.496 \ \text{in.}
\end{aligned}
$$

Assuming the web is fully effective, the neutral axis is located at $y_{cg} = 4.063$ in. as shown in Figure 7.20. Since $y_{cg} > d/2$, initial yield occurs in the compression flange. Therefore, $f = F_y$.

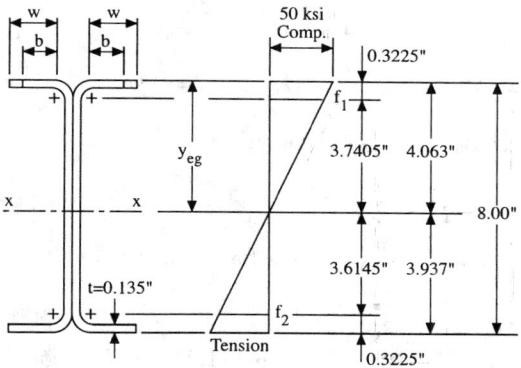

Figure 7.20 Stress distribution in webs. (From Yu, W.W. 1991. *Cold-Formed Steel Design*, John Wiley & Sons, New York. With permission.)

2. Check the web for full effectiveness as follows (Figure 7.20):

$$f_1 = 46.03 \text{ ksi (compression)}$$
$$f_2 = -44.48 \text{ ksi (tension)}$$
$$\psi = f_2/f_1 = -0.966.$$

Using Equation 7.11,

$$k = 4 + 2(1 - \psi)^3 + 2(1 - \psi) = 23.13$$
$$h = 7.355 \text{ in.}$$
$$h/t = 54.48$$
$$\lambda = (1.052/\sqrt{k})(54.48)\sqrt{46.03/29,500}$$
$$= 0.471 < 0.673$$
$$b_e = h = 7.355 \text{ in.}$$
$$b_1 = b_e/(3 - \psi) = 1.855 \text{ in.}$$
$$b_2 = b_e/2 = 3.6775 \text{ in.}$$

Since $b_1 + b_2 = 5.5325$ in. > 3.7405 in., the web is fully effective.

3. The moment of inertia I_x is

$$I_x = \Sigma(Ay^2) + 2I_{web} - (\Sigma A)(y_{cg})^2$$
$$= 25.382 \text{ in.}^4$$

The section modulus for the top fiber is

$$S_e = I_x/y_{cg} = 6.247 \text{ in.}^3$$

4. Based on initiation of yielding, the nominal moment for section strength is

$$M_n = S_e F_y = 312.35 \text{ in.-kips}$$

5. The allowable moment or design moment is

$$M_a = M_n/\Omega = 312.35/1.67 = 187.04 \text{ in.-kips}$$

Based on the given data, the required moment is

$$M = M_D + M_L = 30 + 150 = 180 \text{ in.-kips}$$

Since $M < M_a$, the I-section is adequate for the ASD method.

(B) LRFD Method

1. Based on the nominal moment M_n computed above, the design moment is

$$\phi_b M_n = 0.90(312.35) = 281.12 \text{ in.-kips}$$

2. The required moment for combined dead and live moments is

$$
\begin{aligned}
M_u &= 1.2M_D + 1.6M_L \\
&= (1.2 \times 30) + (1.6 \times 150) \\
&= 276.00 \text{ in.-kips}
\end{aligned}
$$

Since $\phi_b M_n > M_u$, the I-section is adequate for bending strength according to the LRFD approach.

EXAMPLE 7.3:

Determine the nominal moment about the x-axis for the hat section with stiffened compression flange as shown in Figure 7.21. Assume that the yield point of steel is 50 ksi. Use the linear method. The nominal moment is determined by initiation of yielding.

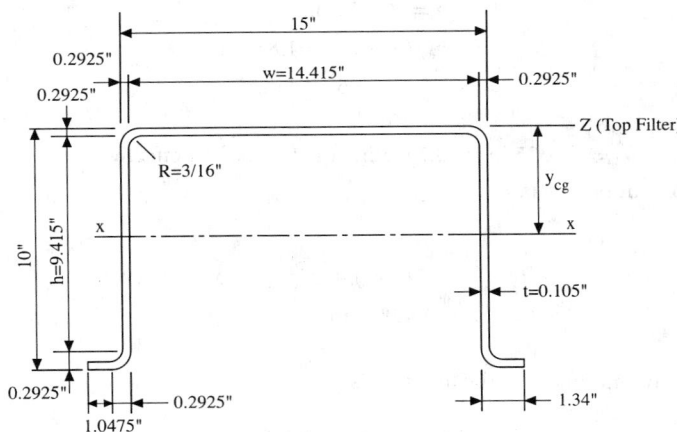

Figure 7.21   Example 7.3. (From Yu, W.W. 1991. *Cold-Formed Steel Design,* John Wiley & Sons, New York. With permission.)

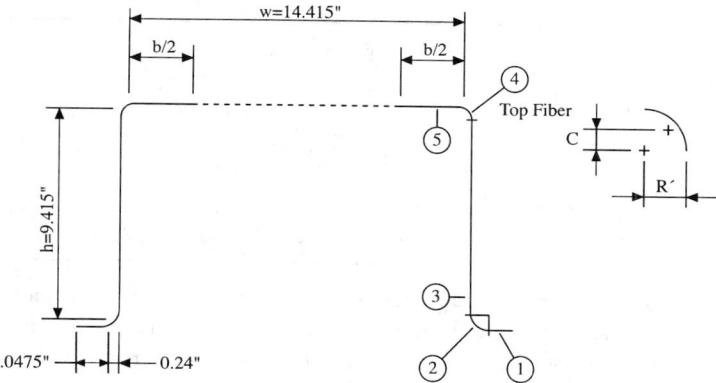

Figure 7.22 Line elements. (From Yu, W.W. 1991. *Cold-Formed Steel Design,* John Wiley & Sons, New York. With permission.)

Solution

1. Calculation of Sectional Properties. In order to use the linear method, midline dimensions are shown in Figure 7.22.

   A. Corner element (Figures 7.17 and 7.22)

   $$R' = R + t/2 = 0.240 \text{ in.}$$

   Arc length

   $$\begin{aligned} L &= 1.57R' = 0.3768 \text{ in.} \\ c &= 0.637R' = 0.1529 \text{ in.} \end{aligned}$$

   B. Location of neutral axis

   a. First approximation. For the compression flange,

   $$\begin{aligned} w &= 15 - 2(R + t) = 14.415 \text{ in.} \\ w/t &= 137.29 \end{aligned}$$

   Using Equations 7.4 through 7.7 and assuming $f = F_y = 50$ ksi,

   $$\begin{aligned} \lambda &= \frac{1.052}{\sqrt{4}}(137.29)\sqrt{\frac{50}{29500}} = 2.973 > 0.673 \\ \rho &= \left(1 - \frac{0.22}{2.973}\right)/2.973 = 0.311 \\ b &= \rho w = 0.311(14.415) = 4.483 \text{ in.} \end{aligned}$$

   By using the effective width of the compression flange and assuming the web is fully effective, the neutral axis can be located as follows:

Element	Effective length L (in.)	Distance from top fiber y (in.)	Ly (in.^{2})
1	$2 \times 1.0475 = 2.0950$	9.9475	20.8400
2	$2 \times 0.3768 = 0.7536$	9.8604	7.4308
3	$2 \times 9.4150 = 18.8300$	5.0000	94.1500
4	$2 \times 0.3768 = 0.7536$	0.1396	0.1052
5	4.4830	0.0525	0.2354
Total	26.9152		122.7614

$$y_{cg} = \frac{\Sigma(Ly)}{\Sigma L} = \frac{122.7614}{26.9152} = 4.561 \text{ in.}$$

Because the distance y_{cg} is less than the half-depth of 5.0 in., the neutral axis is closer to the compression flange and, therefore, the maximum stress occurs in the tension flange. The maximum compressive stress can be computed as follows:

$$f = 50 \left(\frac{4.561}{10 - 4.561} \right) = 41.93 \text{ ksi}$$

Since the above computed stress is less than the assumed value, another trial is required.

b. Second approximation. Assuming that

$$f = 40.70 \text{ ksi}$$
$$\lambda = 2.682 > 0.673$$
$$b = 4.934 \text{ in.}$$

Element	Effective length L (in.)	Distance from top fiber y (in.)	Ly (in.^{2})	Ly^2 (in.^{3})
1	2.0950	9.9475	20.8400	207.3059
2	0.7536	9.8604	7.4308	73.2707
3	18.8300	5.0000	94.1500	470.7500
4	0.7536	0.1396	0.1052	0.0147
5	4.9340	0.0525	0.2590	0.0136
Total	27.3662		122.7850	751.3549

$$y_{cg} = \frac{122.7850}{27.3662} = 4.487 \text{ in.}$$

$$f = \left(\frac{4.487}{10 - 4.487} \right) = 40.69 \text{ ksi}$$

Since the above computed stress is close to the assumed value, it is O.K.

C. Check the effectiveness of the web. Use the AISI Specification to check the effectiveness of the web element. From Figure 7.23,

$$f_1 = 50(4.1945/5.513) = 38.04 \text{ ksi (compression)}$$
$$f_2 = -50(5.2205/5.513) = -47.35 \text{ ksi (tension)}$$
$$\psi = f_2/f_1 = -1.245. \text{ Using Equation 7.11,}$$
$$k = 4 + 2(1 - \psi)^3 + 2(1 - \psi)$$
$$= 4 + 2(2.245)^3 + 2(2.245) = 31.12$$

$$h/t = 9.415/0.105 = 89.67 < 200 \quad \text{O.K.}$$

$$\lambda = \frac{1.052}{\sqrt{31.12}}(89.67)\sqrt{\frac{38.04}{29,500}} = 0.607 < 0.673$$

$$b_e = h = 9.415 \text{ in.}$$

$$b_1 = b_e/(3 - \psi) = 2.218 \text{ in.}$$

Since $\psi < -0.236$,

$$b_2 = b_e/2 = 4.7075 \text{ in.}$$

$$b_1 + b_2 = 6.9255 \text{ in.}$$

Because the computed value of $(b_1 + b_2)$ is greater than the compression portion of the web (4.1945 in.), the web element is fully effective.

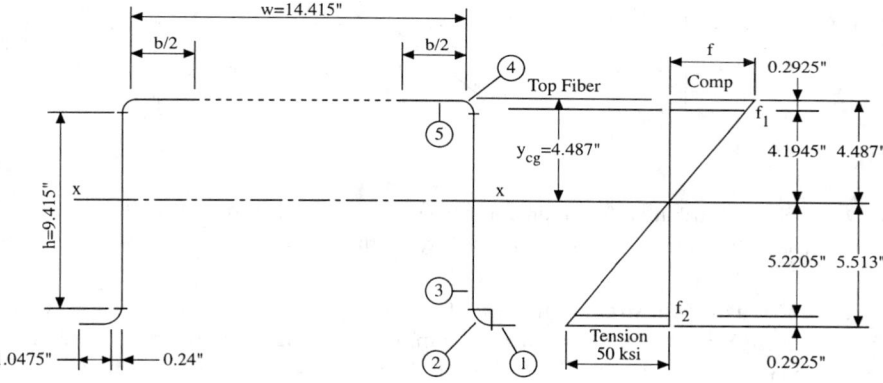

Figure 7.23 Effective lengths and stress distribution using fully effective webs. (From Yu, W.W. 1991. *Cold-Formed Steel Design*, John Wiley & Sons, New York. With permission.)

D. Moment of inertia and section modulus. The moment of inertia based on line elements is

$$2I_3' = 2\left(\frac{1}{12}\right)(9.415)^3 = 139.0944$$

$$\Sigma(Ly^2) = 751.3549$$

$$I_z' = 2I_3' + \Sigma(Ly^2) = 890.4493 \text{ in.}^3$$

$$(\Sigma L)(y_{cg})^2 = 27.3662(4.487)^2 = 550.9683 \text{ in.}^3$$

$$I_x' = I_z' - (\Sigma L)(y_{cg})^2 = 339.4810 \text{ in.}^3$$

The actual moment of inertia is

$$I_x = I_x't = (339.4810)(0.105) = 35.646 \text{ in.}^4$$

The section modulus relative to the extreme tension fiber is

$$S_x = 35.646/5.513 = 6.466 \ \ \text{in.}^3$$

2. Nominal Moments. The nominal moment for section strength is

$$M_n = S_e F_y = S_x F_y = (6.466)(50) = 323.30 \ \ \text{in.-kips}$$

Once the nominal moment is computed, the design moments for the ASD and LRFD methods can be determined as illustrated in Example 7.2.

According to Procedure II of the AISI Specification, the nominal moment, M_n, is the maximum bending capacity of the beam by considering the inelastic reserve strength through partial plastification of the cross-section as shown in Figure 7.24. The inelastic stress distribution in the cross-section

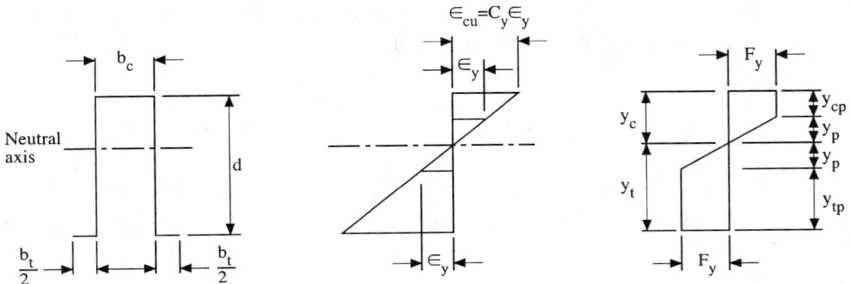

Figure 7.24   Stress distribution for maximum moment (inelastic reserve strength). (From Yu, W.W. 1991. *Cold-Formed Steel Design*, John Wiley & Sons, New York. With permission.)

depends on the maximum strain in the compression flange, which is limited by the Specification for the given width-to-thickness ratio of the compression flange. On the basis of the maximum compression strain allowed in the Specification, the neutral axis can be located by Equation 7.18 and the nominal moment, M_n, can be determined by using Equation 7.19:

$$\int \sigma \, dA = 0 \tag{7.18}$$

$$\int \sigma y \, dA = M \tag{7.19}$$

where σ is the stress in the cross-section. For additional information, see Yu [49].

Lateral Buckling Strength

The nominal lateral buckling strength of unbraced segments of singly-, doubly-, and point-symmetric sections subjected to lateral buckling, M_n, can be determined as follows:

$$M_n = S_c \frac{M_c}{S_f} \tag{7.20}$$

where
S_f = elastic section modulus of the full unreduced section for the extreme compression fiber
S_c = elastic section modulus of the effective section calculated at a stress M_c/S_f in the extreme compression fiber
M_c = critical moment for singly-, doubly-, and point-symmetric sections calculated as follows:

1. For $M_e \geq 2.78 M_y$:

$$M_c = M_y \tag{7.21}$$

2. For $2.78 M_y > M_e > 0.56 M_y$:

$$M_c = \frac{10}{9} M_y \left(1 - \frac{10 M_y}{36 M_e} \right) \tag{7.22}$$

3. For $M_e \leq 0.56 M_y$:

$$M_c = M_e \tag{7.23}$$

where

M_y = moment causing initial yield at the extreme compression fiber of the full section

= $S_f F_y$

M_e = elastic critical moment calculated according to (a) or (b) below:

(a) For singly-, doubly-, and point-symmetric sections:

M_e = $C_b r_0 A \sqrt{\sigma_{ey} \sigma_t}$ for bending about the symmetry axis. For singly-symmetric sections, x-axis is the axis of symmetry oriented such that the shear center has a negative x-coordinate. For point-symmetric sections, use $0.5 M_e$. Alternatively, M_e can be calculated using the equation for doubly-symmetric I-sections or point-symmetric sections given in (b)

M_e = $C_s A \sigma_{ex} [j + C_s \sqrt{j^2 + r_0^2 (\sigma_t / \sigma_{ex})}] / C_{TF}$ for bending about the centroidal axis perpendicular to the symmetry axis for singly-symmetric sections only

C_s = $+1$ for moment causing compression on the shear center side of the centroid

C_s = -1 for moment causing tension on the shear center side of the centroid

σ_{ex} = $\pi^2 E / (K_x L_x / r_x)^2$

σ_{ey} = $\pi^2 E / (K_y L_y / r_y)^2$

σ_t = $[G J + \pi^2 E C_w / (K_t L_t)^2] / A r_0^2$

A = full cross-sectional area

$$C_b = 12.5 M_{max} / (2.5 M_{max} + 3 M_A + 4 M_B + 3 M_C) \tag{7.24}$$

In Equation 7.24,

M_{max} = absolute value of maximum moment in the unbraced segment

M_A = absolute value of moment at quarter point of unbraced segment

M_B = absolute value of moment at centerline of unbraced segment

M_C = absolute value of moment at three-quarter point of unbraced segment

C_b is permitted to be conservatively taken as unity for all cases. For cantilevers or overhangs where the free end is unbraced, C_b shall be taken as unity. For members subject to combined axial load and bending moment, C_b shall be taken as unity.

E = modulus of elasticity

C_{TF} = $0.6 - 0.4 (M_1 / M_2)$

where

M_1 is the smaller and M_2 the larger bending moment at the ends of the unbraced length in the plane of bending, and where M_1 / M_2, the ratio of end moments, is positive when M_1 and M_2 have the same sign (reverse curvature bending) and negative when they are of opposite sign (single curvature bending). When the bending moment at any point within an unbraced length is larger than that at both ends of this length, and for members subject to combined compressive axial load and bending moment, C_{TF} shall be taken as unity.

r_0 = Polar radius of gyration of the cross-section about the shear center

= $\sqrt{r_x^2 + r_y^2 + x_0^2}$

r_x, r_y = radii of gyration of the cross-section about the centroidal principal axes

G = shear modulus

K_x, K_y, K_t = effective length factors for bending about the x- and y-axes, and for twisting

L_x, L_y, L_t = unbraced length of compression member for bending about the x- and y-axes, and for twisting

x_0 = distance from the shear center to the centroid along the principal x-axis, taken as negative

J = St. Venant torsion constant of the cross-section

C_w = torsional warping constant of the cross-section

j = $[\int_A x^3 dA + \int_A xy^2 dA]/(2I_y) - x_0$

(b) For I- or Z-sections bent about the centroidal axis perpendicular to the web (x-axis):
In lieu of (a), the following equations may be used to evaluate M_e:

$$M_e = \pi^2 E C_b d I_{yc}/L^2 \quad \text{for doubly-symmetric I-sections} \tag{7.25}$$

$$M_e = \pi^2 E C_b d I_{yc}/(2L^2) \quad \text{for point-symmetric Z-sections} \tag{7.26}$$

In Equations 7.25 and 7.26,

d = depth of section

E = modulus of elasticity

I_{yc} = moment of inertia of the compression portion of a section about the gravity axis of the entire section parallel to the web, using the full unreduced section

L = unbraced length of the member

EXAMPLE 7.4:

Determine the nominal moment for lateral buckling strength for the I-beam used in Example 7.2. Assume that the beam is braced laterally at both ends and midspan. Use $F_y = 50$ ksi.

   Solution

1. Calculation of Sectional Properties

   Based on the dimensions given in Example 7.2 (Figures 7.19 and 7.20), the moment of inertia, I_x, and the section modulus, S_f, of the full section can be computed as shown in the following table.

Element	Area A (in.^{2})	Distance from mid-depth y (in.)	Ay^2 (in.^{4})
Flanges	4(1.6775)(0.135) = 0.9059	3.9325	14.0093
Corners	4(0.05407) = 0.2163	3.8436	3.1955
Webs	2(7.355)(0.135) = 1.9859	0	0
Total	3.1081		17.2048

$$2 I_{web} = 2(1/12)(0.135)(7.355)^3 = 8.9522$$
$$I_x = 26.1570 \text{ in.}^4$$
$$S_f = I_x/(8/2) = 6.54 \text{ in.}^3$$

The value of I_{yc} can be computed as shown below.

Element	Area A (in.^{2})	Distance from y axis, x (in.)	Ax^2 (in.^{4})
Flanges	4(1.6775)(0.135) = 0.9059	1.1613	1.2217
Corners	4(0.05407) = 0.2163	0.1564	0.0053
Webs	2(7.355)(0.135) = 1.9859	0.0675	0.0090
Total	3.1081		1.2360

$$I_{flanges} = 4(1/12)0.135(1.6775)^3 = 0.2124$$
$$I_y = 1.4484 \text{ in.}^4$$
$$I_{yc} = I_y/2 = 0.724 \text{ in.}^4$$

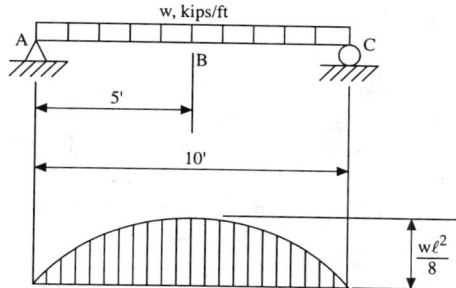

Figure 7.25 Example 7.4. (From Yu, W.W. 1991. *Cold-Formed Steel Design,* John Wiley & Sons, New York. With permission.)

Considering the lateral supports at both ends and midspan, and the moment diagram shown in Figure 7.25, the value of C_b for the segment AB or BC is 1.30 according to Equation 7.24. Using Equation 7.25,

$$
\begin{aligned}
M_e &= \pi^2 E C_b \frac{d I_{yc}}{L^2} \\
&= \pi^2 (29,500)(1.30)\frac{(8)(0.724)}{(5 \times 12)^2} = 608.96 \ \text{in.-kips} \\
M_y &= S_f F_y = (6.54)(50) = 327.0 \ \text{in.-kips} \\
0.56 M_y &= 183.12 \ \text{in.-kips} \\
2.78 M_y &= 909.06 \ \text{in.-kips}
\end{aligned}
$$

Since $2.78\,M_y > M_e > 0.56 M_y$, from Equation 7.22,

$$
\begin{aligned}
M_c &= \frac{10}{9} M_y \left(1 - \frac{10 M_y}{36 M_e}\right) \\
&= \frac{10}{9}(327.0)\left[1 - \frac{10(327.0)}{36(608.96)}\right] \\
&= 309.14 \ \text{in.-kips}
\end{aligned}
$$

Based on Equation 7.20, the nominal moment for lateral buckling strength is

$$
M_n = S_c \frac{M_c}{S_f}
$$

in which S_c is the elastic section modulus of the effective section calculated at a compressive stress of $f = M_c/S_f = 309.14/6.54 = 47.27$ ksi. By using the same procedure illustrated in Example 7.2, $S_c = 6.295$ in.^{3}. Therefore, the nominal moment for lateral buckling strength is

$$
M_n = (6.295)\left(\frac{309.14}{6.54}\right) = 297.6 \ \text{in.-kips}
$$

For channels or Z-sections having the tension flange through-fastened to deck or sheathing with the compression flange laterally unbraced and loaded in a plane parallel to the web, the nominal flexural strength is determined by $M_n = R S_e F_y$, where R is a reduction factor [7]. A similar approach is used for beams having one flange fastened to a standing seam roof system.

Unusually Wide Beam Flanges and Short Span Beams

When beam flanges are unusually wide, special consideration should be given to the possible effects of shear lag and flange curling. Shear lag depends on the type of loading and the span-to-width ratio and is independent of the thickness. Flange curling is independent of span length but depends on the thickness and width of the flange, the depth of the section, and the bending stresses in both tension and compression flanges.

In order to consider the shear lag effects, the effective widths of both tension and compression flanges should be used according to the AISI Specification.

When a straight beam with unusually wide and thin flanges is subject to bending, the portion of the flange most remote from the web tends to deflect toward the neutral axis due to the effect of longitudinal curvature of the beam and the applied bending stresses in both flanges. For the purpose of controlling the excessive flange curling, the AISI Specification provides an equation to limit the flange width.

Shear Strength

The shear strength of beam webs is governed by either yielding or buckling of the web element, depending on the depth-to-thickness ratio, h/t, and the mechanical properties of steel. For beam webs having small h/t ratios, the nominal shear strength is governed by shear yielding. When the h/t ratio is large, the nominal shear strength is controlled by elastic shear buckling. For beam webs having moderate h/t ratios, the shear strength is based on inelastic shear buckling.

For the design of beam webs, the AISI Specification provides the following equations for determining the nominal shear strength:

For $h/t \leq 0.96\sqrt{Ek_v/F_y}$:

$$V_n = 0.60F_y ht \tag{7.27}$$

For $0.96\sqrt{Ek_v/F_y} < h/t \leq 1.415\sqrt{Ek_v/F_y}$:

$$V_n = 0.64t^2\sqrt{k_v F_y E} \tag{7.28}$$

For $h/t > 1.415\sqrt{Ek_v/F_y}$:

$$V_n = \pi^2 Ek_v t^3/[12(1-\mu^2)h] = 0.905Ek_v t^3/h \tag{7.29}$$

where

V_n = nominal shear strength of beam
h = depth of the flat portion of the web measured along the plane of the web
t = web thickness
k_v = shear buckling coefficient determined as follows:

1. For unreinforced webs, $k_v = 5.34$
2. For beam webs with transverse stiffeners satisfying the AISI requirements when $a/h \leq 1.0$:

$$k_v = 4.00 + \frac{5.34}{(a/h)^2}$$

when $a/h > 1.0$:

$$k_v = 5.34 + \frac{4.00}{(a/h)^2}$$

where

a = the shear panel length for unreinforced web element
  = the clear distance between transverse stiffeners for reinforced web elements

For a web consisting of two or more sheets, each sheet should be considered as a separate element carrying its share of the shear force.

Combined Bending and Shear

For continuous beams and cantilever beams, high bending stresses often combine with high shear stresses at the supports. Such beam webs must be safeguarded against buckling due to the combination of bending and shear stresses. Based on the AISI Specification, the moment and shear should satisfy the interaction equations listed in Table 7.3.

TABLE 7.3   Interaction Equations Used for Combined Bending and Shear

	ASD	LRFD
Beams with unreinforced webs	$\left(\frac{M}{M_{axo}}\right)^2 + \left(\frac{V}{V_a}\right)^2 \le 1.0$ (7.30)	$\left(\frac{M_u}{\phi_b M_{nxo}}\right)^2 + \left(\frac{V_u}{\phi_v V_n}\right)^2 \le 1.0$ (7.32)
Beams with transverse web stiffeners	$M \le M_a$ and $V \le V_a$ $0.6\left(\frac{M}{M_{axo}}\right) + \left(\frac{V}{V_a}\right) \le 1.3$ (7.31)	$M_u \le \phi_b M_n$ and $V_u \le \phi_v V_n$ $0.6\left(\frac{M_u}{\phi_b M_{nxo}}\right) + \left(\frac{V_u}{\phi_v V_n}\right) \le 1.3$ (7.33)

M	=	bending moment
M_a	=	allowable moment when bending alone exists
M_{axo}	=	allowable moment about the centroidal x-axis determined in accordance with the specification excluding the consideration of lateral buckling
V	=	unfactored shear force
V_a	=	allowable shear force when shear alone exists
ϕ_b	=	resistance factor for bending
ϕ_v	=	resistance factor for shear
M_n	=	nominal flexural strength when bending alone exists
M_{nxo}	=	nominal flexural strength about the centroidal x-axis determined in accordance with the specification excluding the consideration of lateral buckling
M_u	=	required flexural strength
V_n	=	nominal shear strength when shear alone exists
V_u	=	required shear strength

Web Crippling

For cold-formed steel beams, transverse stiffeners are not frequently used for beam webs. The webs may cripple due to the high local intensity of the load or reaction as shown in Figure 7.26. Because the theoretical analysis of web crippling is rather complex due to the involvement of many

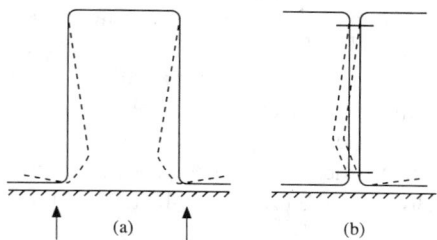

(a)   (b)

Figure 7.26   Web crippling of cold-formed steel beams.

factors, the present AISI design equations are based on extensive experimental investigations under four loading conditions: (1) end one-flange (EOF) loading, (2) interior one-flange (IOF) loading, (3) end two-flange (ETF) loading, and (4) interior two-flange (ITF) loading [29, 46, 50]. The loading conditions used for the tests are illustrated in Figure 7.27.

The nominal web crippling strength for a given loading condition can be determined from the AISI equations [7] on the basis of the thickness of web element, design yield stress, the bend radius-

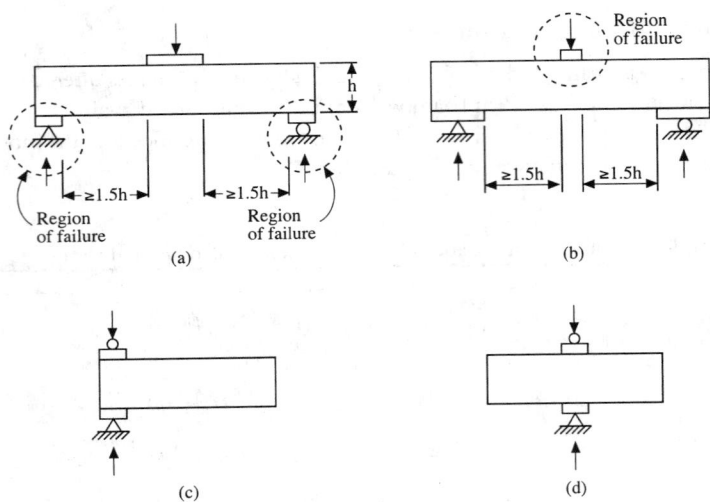

Figure 7.27 Loading conditions for web crippling tests. (a) EOF loading. (b) IOF loading. (c) ETF loading. (d) ITF loading. (From Yu, W.W. 1991. *Cold-Formed Steel Design*, John Wiley & Sons, New York. With permission.)

to-thickness ratio, the depth-to-thickness ratio, the bearing length-to-thickness ratio, and the angle between the plane of the web and the plane of the bearing surface. Tables 7.4a and Table 7.4b list the equations for determining the nominal web crippling strengths of one- and two-flange loading conditions, respectively.

Combined Bending and Web Crippling

For combined bending and web crippling, the design of beam webs should be based on the interaction equations provided in the AISI Specification [7]. These equations are presented in Table 7.5.

Bracing Requirements

In cold-formed steel design, braces should be designed to restrain lateral bending or twisting of a loaded beam and to avoid local crippling at the points of attachment. When channels and Z-shaped sections are used as beams and loaded in the plane of the web, the AISI Specification [7] provides design requirements to restrain twisting of the beam under the following two conditions: (1) the top flange is connected to deck or sheathing material in such a manner as to effectively restrain lateral deflection of the connected flange, and (2) neither flange is connected to sheathing. In general, braces should be designed to satisfy the strength and stiffness requirements. For beams using symmetrical cross sections, such as I-beams, the AISI Specification does not provide specific requirements for braces. However, the braces may be designed for a capacity of 2% of the force resisted by the compression portion of the beam. This is a frequently used rule of thumb but is a conservative approach, as proven by a rigorous analysis.

7.6.5 Concentrically Loaded Compression Members

Axially loaded cold-formed steel **compression members** should be designed for the following limit states: (1) yielding, (2) overall column buckling (flexural buckling, torsional buckling, or **torsional-flexural buckling**), and (3) local buckling of individual elements. The governing failure mode depends on the configuration of the cross-section, thickness of material, unbraced length, and end restraint.

Table 7.4a Nominal Web Crippling Strength for One-Flange Loading, per Web, P_n

	Shapes having single webs	Shapes having single webs	I-Sections or similar sections
	Stiffened or partially stiffened flanges	Unstiffened flanges	Stiffened, partially stiffened, and unstiffened flanges
End reaction opposing loads spaced $> 1.5h$	$t^2 k C_3 C_4 C_\theta [331 - 0.61(h/t)]$ $\times [1 + 0.01(N/t)] C_9$ [a]	$t^2 k C_3 C_4 C_\theta [217 - 0.28(h/t)]$ $\times [1 + 0.01(N/t)] C_9$ [a,b]	$t^2 F_y C_6 (10.0 + 1.25\sqrt{N/t})$
Interior reactions opposing loads spaced $> 1.5h$	$t^2 k C_1 C_2 C_\theta [538 - 0.74(h/t)]$ $\times [1 + 0.007(N/t)] C_9$ [c]	$t^2 k C_1 C_2 C_\theta [538 - 0.74(h/t)]$ $\times [1 + 0.007(N/t)] C_9$ [c]	$t^2 F_y C_5 (0.88 + 0.12m)$ $\times (15.0 + 3.25\sqrt{N/t})$

[a] When $F_y \geq 66.5$ ksi (459 MPa), the value of $K C_3$ shall be taken as 1.34.
[b] When $N/t > 60$, the factor $[1 + 0.01(N/t)]$ may be increased to $[0.71 + 0.015(N/t)]$
[c] When $N/t > 60$, the factor $[1 + 0.007(N/t)]$ may be increased to $[0.75 + 0.011(N/t)]$

$C_1 = (1.22 - 0.22k)$
$C_2 = (1.06 - 0.06R/t) \leq 1.0$
$C_3 = (1.33 - 0.33k)$
$C_4 = (1.15 - 0.15R/t) \leq 1.0$ but not less than 0.50
$C_5 = (1.49 - 0.53k) \geq 0.6$
$C_6 = 1 + (h/t)/750$ when $h/t \leq 150$
     $= 1.20$, when $h/t > 150$
$C_9 = 1.0$ for U.S. customary units, kips and in.
     $= 6.9$ for SI units, N and mm
$C_\theta = 0.7 + 0.3(\theta/90)^2$
$F_y = $ design yield stress of the web
$h = $ depth of the flat portion of the web measured along the plane of the web
$k = 894 F_y / E$
$m = t/0.075$, when t is in in.
$m = t/1.91$, when t is in mm
$t = $ web thickness
$N = $ actual length of bearing
$R = $ inside bend radius
$\theta = $ angle between the plane of the web and the plane of the bearing surface $\geq 45°$, but not more than $90°$

Table 7.4b Nominal Web Crippling Strength for Two-Flange Loading, per Web, P_n

	Shapes having single webs	Shapes having single webs	I-Sections or similar sections
	Stiffened or partially stiffened flanges	Unstiffened flanges	Stiffened, partially stiffened, and unstiffened flanges
End reaction opposing loads spaced $\leq 1.5h$	$t^2 k C_3 C_4 C_\theta [244 - 0.57(h/t)]$ $\times [1 + 0.01(N/t)] C_9$ [a]	$t^2 k C_3 C_4 C_\theta [244 - 0.57(h/t)]$ $\times [1 + 0.01(N/t)] C_9$ [a]	$t^2 F_y C_8 (0.64 + 0.31m)$ $\times (10.0 + 1.25\sqrt{N/t})$
Interior reaction opposing loads spaced $\leq 1.5h$	$t^2 k C_1 C_2 C_\theta [771 - 2.26(h/t)]$ $\times [1 + 0.0013(N/t)] C_9$	$t^2 k C_1 C_2 C_\theta [771 - 2.26(h/t)]$ $\times [1 + 0.0013(N/t)] C_9$	$t^2 F_y C_7 (0.82 + 0.15m)$ $\times (15.0 + 3.25\sqrt{N/t})$

[a] When $F_y \geq 66.5$ ksi (459 MPa), the value of $K C_3$ shall be taken as 1.34.
$C_7 = 1/k$, when $h/t \leq 66.5$
     $= [1.10 - (h/t)/665]/k$, when $h/t > 66.5$
$C_8 = [0.98 - (h/t)/865]/k$
$C_1, C_2, C_3, C_4, C_9, C_\theta, F_y, h, k, m, t, N, R$, and θ are defined in Table 7.4a.

TABLE 7.5 Interaction Equations for Combined Bending and Web Crippling

	ASD	LRFD
Shapes having single unrein-forced webs	$1.2\left(\frac{P}{P_a}\right)+\left(\frac{M}{M_{axo}}\right)\leq 1.5$ (7.34)	$1.07\left(\frac{P_u}{\phi_w P_n}\right)+\left(\frac{M_u}{\phi_b M_{nxo}}\right)\leq 1.42$ (7.37)
Shapes having multiple un-reinforced webs such as I-sections	$1.1\left(\frac{P}{P_a}\right)+\left(\frac{M}{M_{axo}}\right)\leq 1.5$ (7.35)	$0.82\left(\frac{P_u}{\phi_w P_n}\right)+\left(\frac{M_u}{\phi_b M_{nxo}}\right)\leq 1.32$ (7.38)
Support point of two nested Z-shapes	$\frac{M}{M_{no}}+\frac{P}{P_n}\leq 1.0$ (7.36)	$\frac{M_u}{M_{no}}+\frac{P_u}{P_n}\leq 1.68\phi$ (7.39)

Note: The AISI Specification includes some exception clauses, under which the effect of combined bending and web crippling need not be checked.

P = concentrated load or reaction in the presence of bending moment
P_a = allowable concentrated load or reaction in the absence of bending moment
P_n = nominal web crippling strength for concentrated load or reaction in the absence of bending moment (for Equations 7.34, 7.35, 7.37, and 7.38)
P_n = nominal web crippling strength assuming single web interior one-flange loading for the nested Z-sections, i.e., sum of the two webs evaluated individually (for Equations 7.36 and 7.39)
P_u = required strength for the concentrated load or reaction in the presence of bending moment
M = applied bending moment at, or immediately adjacent to, the point of application of the concentrated load or reaction
M_{axo} = allowable moment about the centroidal x-axis determined in accordance with the specification excluding the consideration of lateral buckling
M_{no} = nominal flexural strength for the nested Z-sections, i.e., the sum of the two sections evaluated individually, excluding lateral buckling
M_{nxo} = nominal flexural strength about the centroidal x-axis determined in accordance with the specification excluding the consideration of lateral buckling
M_u = required flexural strength at, or immediately adjacent to, the point of application of the concentrated load or reaction P_u
ϕ = resistance factor = 0.9
ϕ_b = resistance factor for bending
ϕ_w = resistance factor for web crippling

Yielding

A very short, compact column under axial load may fail by yielding. For this case, the nominal axial strength is the yield load, i.e.,

$$P_n = P_y = AF_y \tag{7.40}$$

where A is the full cross-sectional area of the column and F_y is the yield point of steel.

Overall Column Buckling

Overall column buckling may be one of the following three types:
1. Flexural buckling — bending about a principal axis. The elastic flexural buckling stress is

$$F_e = \frac{\pi^2 E}{(KL/r)^2} \tag{7.41}$$

where
E = modulus of elasticity
K = effective length factor for flexural buckling (Figure 7.28)
L = unbraced length of member for flexural buckling
r = radius of gyration of the full section

2. Torsional buckling — twisting about shear center. The elastic torsional buckling stress is

$$F_e = \frac{1}{Ar_0^2}\left[GJ + \frac{\pi^2 E C_w}{(K_t L_t)^2}\right] \tag{7.42}$$

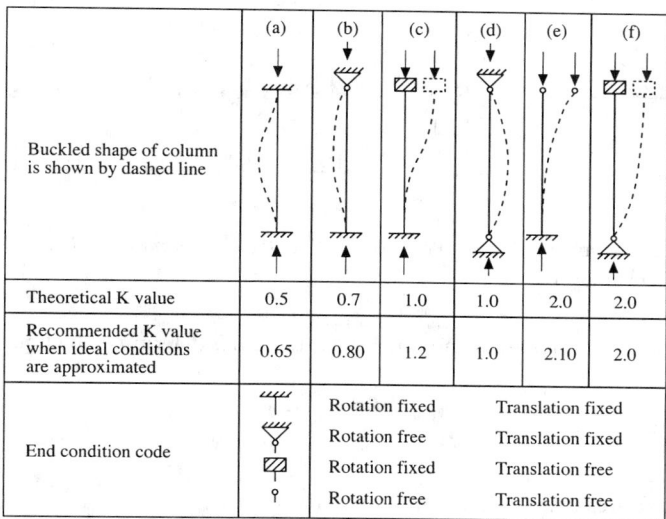

Figure 7.28 Effective length factor K for concentrically loaded compression members.

where

A = full cross-sectional area
C_w = torsional warping constant of the cross-section
G = shear modulus
J = St. Venant torsion constant of the cross-section
K_t = effective length factor for twisting
L_t = unbraced length of member for twisting
r_0 = polar radius of gyration of the cross-section about shear center

3. Torsional-flexural buckling — bending and twisting simultaneously. The elastic torsional-flexural buckling stress is

$$F_e = [(\sigma_{ex} + \sigma_t) - \sqrt{(\sigma_{ex} + \sigma_t)^2 - 4\beta\sigma_{ex}\sigma_t}]/2\beta \qquad (7.43)$$

where

β = $1 - (x_0/r_0)^2$
σ_{ex} = $\pi^2 E/(K_x L_x/r_x)^2$
σ_t = the same as Equation 7.42
x_0 = distance from shear center to the centroid along the principal x-axis

For doubly-symmetric and **point-symmetric shapes** (Figure 7.29), the overall column buckling can be either flexural type or torsional type. However, for singly-symmetric shapes (Figure 7.30), the overall column buckling can be either flexural buckling or torsional-flexural buckling.

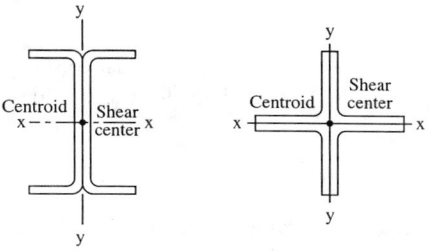

Figure 7.29 Doubly-symmetric shapes.

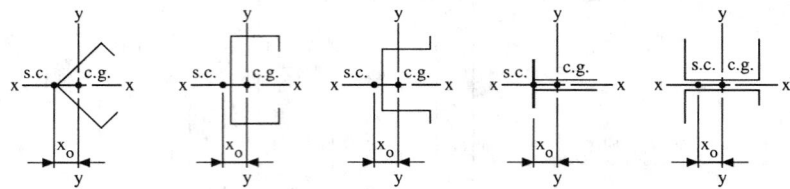

Figure 7.30 Singly-symmetric shapes. (From Yu, W.W. 1991. *Cold-Formed Steel Design*, John Wiley & Sons, New York. With permission.)

For overall column buckling, the nominal axial strength is determined by Equation 7.44:

$$P_n = A_e F_n \tag{7.44}$$

where

A_e = effective area determined for the stress F_n

F_n = nominal buckling stress determined as follows:

For $\lambda_c \leq 1.5$:

$$F_n = (0.658^{\lambda_c^2}) F_y \tag{7.45}$$

For $\lambda_c > 1.5$:

$$F_n = \left[\frac{0.877}{\lambda_c^2} \right] F_y \tag{7.46}$$

The use of the effective area A_e in Equation 7.44 is to reflect the effect of local buckling on the reduction of column strength. In Equations 7.45 and 7.46,

$$\lambda_c = \sqrt{F_y / F_e}$$

in which F_e is the least of elastic flexural buckling stress (Equation 7.41), torsional buckling stress (Equation 7.42), and torsional-flexural buckling stress (Equation 7.43), whichever is applicable.

For the design of compression members, the slenderness ratio should not exceed 200, except that during construction, KL/r preferably should not exceed 300.

For nonsymmetric shapes whose cross-sections do not have any symmetry, either about an axis or about a point, the elastic torsional-flexural buckling stress should be determined by rational analysis or by tests. See AISI Design Manual [8].

In addition to the above design provisions for the design of axially loaded columns, the AISI Specification also provides design criteria for compression members having one flange through-fastened to deck or sheathing.

EXAMPLE 7.5:

Determine the allowable axial load for the square tubular column shown in Figure 7.31. Assume that $F_y = 40$ ksi, $K_x L_x = K_y L_y = 10$ ft, and the dead-to-live load ratio is 1/5. Use the ASD and LRFD methods.

> *Solution*

(A) ASD Method

Since the square tube is a doubly-symmetric closed section, it will not be subject to torsional-flexural buckling. It can be designed for flexural buckling.

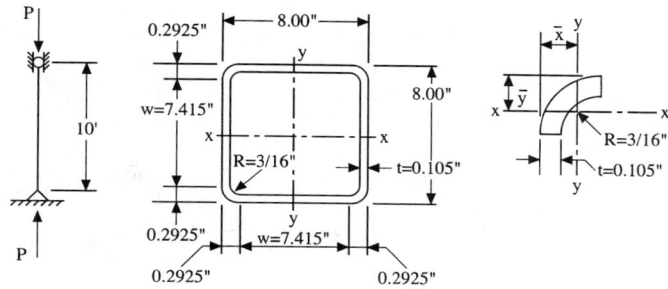

Figure 7.31 Example 7.5. (From Yu, W.W. 1991. *Cold-Formed Steel Design,* John Wiley & Sons, New York. With permission.)

1. Sectional Properties of Full Section

$$w = 8.00 - 2(R + t) = 7.415 \text{ in.}$$
$$A = 4(7.415 \times 0.105 + 0.0396) = 3.273 \text{ in.}^2$$
$$I_x = I_y = 2(0.105)[(1/12)(7.415)^3 + 7.415(4 - 0.105/2)^2] + 4(0.0396)(4.0 - 0.1373)^2$$
$$= 33.763 \text{ in.}^4$$
$$r_x = r_y = \sqrt{I_x/A} = \sqrt{33.763/3.273} = 3.212 \text{ in.}$$

2. Nominal Buckling Stress, F_n. According to Equation 7.41, the elastic flexural buckling stress, F_e, is computed as follows:

$$\frac{KL}{r} = \frac{10 \times 12}{3.212} = 37.36 < 200 \text{ O.K.}$$
$$F_e = \frac{\pi^2 E}{(KL/r)^2} = \frac{\pi^2(29500)}{(37.36)^2} = 208.597 \text{ ksi}$$
$$\lambda_c = \sqrt{\frac{F_y}{F_e}} = \sqrt{\frac{40}{208.597}} = 0.438 < 1.5$$
$$F_n = (0.658^{\lambda_c^2})F_y = (0.658^{0.438^2})40 = 36.914 \text{ ksi}$$

3. Effective Area, A_e. Because the given square tube is composed of four stiffened elements, the effective width of stiffened elements subjected to uniform compression can be computed from Equations 7.4 through 7.7 by using $k = 4.0$:

$$w/t = 7.415/0.105 = 70.619$$
$$\lambda = \frac{1.052}{\sqrt{k}} \left(\frac{w}{t}\right) \sqrt{\frac{F_n}{E}}$$
$$\lambda = 1.052/\sqrt{4}(70.619)\sqrt{36.914/29,500} = 1.314$$

Since $\lambda > 0.673$, from Equation 7.5,

$$b = \rho w$$

where

$$\rho = (1 - 0.22/\lambda)/\lambda = (1 - 0.22/1.314)/1.314 = 0.634$$

Therefore, $b = (0.634)(7.415) = 4.701$ in.
The effective area is

$$A_e = 3.273 - 4(7.415 - 4.701)(0.105) = 2.133 \text{ in.}^2$$

4. **Nominal and Allowable Loads.** Using Equation 7.44, the nominal load is

$$P_n = A_e F_n = (2.133)(36.914) = 78.738 \text{ kips}$$

The allowable load is

$$P_a = P_n/\Omega_c = 78.738/1.80 = 43.74 \text{ kips}$$

(B) LRFD Method

In Item (A) above, the nominal axial load, P_n, was computed to be 78.738 kips. The design axial load for the LRFD method is

$$\phi_c P_n = 0.85(78.738) = 66.93 \text{ kips.}$$

Based on the load combination of dead and live loads, the required axial load is

$$P_u = 1.2P_D + 1.6P_L = 1.2P_D + 1.6(5P_D) = 9.2P_D$$

where
P_D = axial load due to dead load
P_L = axial load due to live load
By using $P_u = \phi_c P_n$, the values of P_D and P_L are computed as follows:

$$P_D = 66.93/9.2 = 7.28 \text{ kips}$$
$$P_L = 5P_D = 36.40 \text{ kips}$$

Therefore, the allowable axial load is

$$P_a = P_D + P_L = 43.68 \text{ kips}$$

It can be seen that the allowable axial loads determined by the ASD and LRFD methods are practically the same.

7.6.6 Combined Axial Load and Bending

The AISI Specification provides interaction equations for combined axial load and bending.

Combined Tensile Axial Load and Bending

For combined tensile axial load and bending, the required strengths should satisfy the interaction equations presented in Table 7.6. These equations are to prevent yielding of the tension flange and to prevent failure of the compression flange of the member.

Combined Compressive Axial Load and Bending

Cold-formed steel members under combined compressive axial load and bending are usually referred to as **beam-columns.** Such members are often found in framed structures, trusses, and exterior wall studs. For the design of these members, the required strengths should satisfy the AISI interaction equations presented in Table 7.7.

TABLE 7.6   Interaction Equations for Combined Tensile Axial Load and Bending

	ASD	LRFD
Check tension flange	$\dfrac{\Omega_b M_x}{M_{nxt}} + \dfrac{\Omega_b M_y}{M_{nyt}} + \dfrac{\Omega_t T}{T_n} \le 1.0$ (7.47)	$\dfrac{M_{ux}}{\phi_b M_{nxt}} + \dfrac{M_{uy}}{\phi_b M_{nyt}} + \dfrac{T_u}{\phi_t T_n} \le 1.0$ (7.49)
Check compression flange	$\dfrac{\Omega_b M_x}{M_{nx}} + \dfrac{\Omega_b M_y}{M_{ny}} + \dfrac{\Omega_t T}{T_n} \le 1.0$ (7.48)	$\dfrac{M_{ux}}{\phi_b M_{nx}} + \dfrac{M_{uy}}{\phi_b M_{ny}} - \dfrac{T_u}{\phi_t T_n} \le 1.0$ (7.50)

M_{nx}, M_{ny} = nominal flexural strengths about the centroidal x- and y-axes
M_{nxt}, M_{nyt} = $S_{ft} F_y$
M_{ux}, M_{uy} = required flexural strengths with respect to the centroidal axes
M_x, M_y = required moments with respect to the centroidal axes of the section
S_{ft} = section modulus of the full section for the extreme tension fiber about the appropriate axis
T = required tensile axial load
T_n = nominal tensile axial strength
T_u = required axial strength
ϕ_b = resistance factor for bending
ϕ_t = resistance factor for tension
Ω_b = safety factor for bending
Ω_t = safety factor for tension

TABLE 7.7   Interaction Equations for Combined Compressive Axial Load and Bending

ASD	LRFD
when $\Omega_c P / P_n \le 0.15$,	when $P_u / \phi_c P_n \le 0.15$,
$\dfrac{\Omega_c P}{P_n} + \dfrac{\Omega_b M_x}{M_{nx}} + \dfrac{\Omega_b M_y}{M_{ny}} \le 1.0$ (7.51)	$\dfrac{P_u}{\phi_c P_n} + \dfrac{M_{ux}}{\phi_b M_{nx}} + \dfrac{M_{uy}}{\phi_b M_{ny}} \le 1.0$ (7.54)
when $\Omega_c P / P_n > 0.15$,	when $P_u / \phi_c P_n > 0.15$,
$\dfrac{\Omega_c P}{P_n} + \dfrac{\Omega_b C_{mx} M_x}{M_{nx} \alpha_x} + \dfrac{\Omega_b C_{my} M_y}{M_{ny} \alpha_y} \le 1.0$ (7.52)	$\dfrac{P_u}{\phi_c P_n} + \dfrac{C_{mx} M_{ux}}{\phi_b M_{nx} \alpha_x} + \dfrac{C_{my} M_{uy}}{\phi_b M_{ny} \alpha_y} \le 1.0$ (7.55)
$\dfrac{\Omega_c P}{P_{no}} + \dfrac{\Omega_b M_x}{M_{nx}} + \dfrac{\Omega_b M_y}{M_{ny}} \le 1.0$ (7.53)	$\dfrac{P_u}{\phi_c P_{no}} + \dfrac{M_{ux}}{\phi_b M_{nx}} + \dfrac{M_{uy}}{\phi_b M_{ny}} \le 1.0$ (7.56)

M_x, M_y = required moments with respect to the centroidal axes of the effective section determined for the required axial strength alone
M_{nx}, M_{ny} = nominal flexural strengths about the centroidal axes
M_{ux}, M_{uy} = required flexural strengths with respect to the centroidal axes of the effective section determined for the required axial strength alone
P = required axial load
P_n = nominal axial strength determined in accordance with Equation 7.44
P_{no} = nominal axial strength determined in accordance with Equation 7.44, for $F_n = F_y$
P_u = required axial strength
α_x = $1 - \Omega_c P / P_{EX}$ (for Equation 7.52)
α_y = $1 - \Omega_c P / P_{EY}$ (for Equation 7.52)
α_x = $1 - P_u / P_{EX}$ (for Equation 7.55)
α_y = $1 - P_u / P_{EY}$ (for Equation 7.55)
P_{EX} = $\pi^2 E I_x / (K_x L_x)^2$
P_{EY} = $\pi^2 E I_y / (K_y L_y)^2$
Ω_b = safety factor for bending
Ω_c = safety factor for concentrically loaded compression
C_{mx}, C_{my} = coefficients whose value shall be taken as follows:

1. For compression members in frames subject to joint translation (sidesway) $C_m = 0.85$

2. For restrained compression members in frames braced against joint translation and not subject to transverse loading between their supports in the plane of bending $C_m = 0.6 - 0.4 (M_1/M_2)$, where M_1/M_2 is the ratio of the smaller to the larger moment at the ends of that portion of the member under consideration which is unbraced in the plane of bending. M_1/M_2 is positive when the member is bent in reverse curvature and negative when it is bent in single curvature

3. For compression members in frames braced against joint translation in the plane of loading and subject to transverse loading between their supports, the value of C_m may be determined by rational analysis. However, in lieu of such analysis, the following values may be used: (a) for members whose ends are restrained, $C_m = 0.85$; (b) for members whose ends are unrestrained, $C_m = 1.0$

I_x, I_y, L_x, L_y, K_x, and K_y were defined previously.

7.6.7 Cylindrical Tubular Members

Thin-walled cylindrical tubular members are economical sections for compression and torsional members because of their large ratio of radius of gyration to area, the same radius of gyration in all directions, and the large torsional rigidity. The AISI design provisions are limited to the ratio of outside diameter-to-wall thickness, D/t, not greater than $0.441 E/F_y$.

Bending Strength

For cylindrical tubular members subjected to bending, the nominal flexural strengths are as follows according to the D/t ratio:

1. For $D/t \leq 0.070E/F_y$:

$$M_n = 1.25 F_y S_f \tag{7.57}$$

2. For $0.070E/F_y < D/t \leq 0.319E/F_y$:

$$M_n = [0.970 + 0.020(E/F_y)/(D/t)]F_y S_f \tag{7.58}$$

3. For $0.319E/F_y < D/t \leq 0.441E/F_y$:

$$M_n = [0.328E/(D/t)]S_f \tag{7.59}$$

where
S_f = elastic section modulus of the full, unreduced cross-section
Other symbols were defined previously.

Compressive Strength

When cylindrical tubes are used as concentrically loaded compression members, the nominal axial strength is determined by Equation 7.44, except that (1) the elastic buckling stress, F_e, is determined for flexural buckling by using Equation 7.41 and (2) the effective area, A_e, is calculated by Equation 7.60.

$$A_e = [1 - (1 - R^2)(1 - A_0/A)]A \tag{7.60}$$

where
R = $\sqrt{F_y/2F_e}$
A_0 = $\{0.037/[(DF_y)/(tE)] + 0.667\}A \leq A$
A = area of the unreduced cross-section
In the above equations, the value A_0 is the reduced area due to the effect of local buckling [8, 49].

7.7 Connections and Joints

Welds, bolts, screws, rivets, and other special devices such as metal stitching and adhesives are generally used for cold-formed steel connections. The AISI Specification contains only the design provisions for welded connections, bolted connections, and screw connections. These design equations are based primarily on the experimental data obtained from extensive test programs.

7.7.1 Welded Connections

Welds used for cold-formed steel construction may be classified as arc welds (or fusion welds) and resistance welds. Arc welding is usually used for connecting cold-formed steel members to each other as well as connecting such thin members to heavy, hot-rolled steel framing members. It is used for groove welds, arc spot welds, arc seam welds, fillet welds, and flare groove welds. The AISI design provisions for welded connections are applicable only for cold-formed steel structural members, in which the thickness of the thinnest connected part is 0.18 in. (4.57 mm) or less. Otherwise, when the thickness of connected parts is thicker than 0.18 in. (4.57 mm), the welded connection should be designed according to the AISC Specifications [1, 2]. Additional design information on structural welding of sheet steels can also be found in the AWS Code [16].

Arc Welds

According to the AISI Specification, the nominal strengths of arc welds can be determined from the equations given in Table 7.8. The design strengths can then be computed by using the safety factor or resistance factor provided in Table 7.1.

Resistance Welds

The nominal shear strengths of resistance welds are provided in the AISI Specification [7] according to the thickness of the thinnest outside sheet. They are applicable for all structural grades of low-carbon steel, uncoated or galvanized with 0.9 oz/ft^2 of sheet or less, and medium carbon and low-alloy steels.

7.7.2 Bolted Connections

Due to the thinness of the connected parts, the design of bolted connections in cold-formed steel construction is somewhat different from that in hot-rolled heavy construction. The AISI design provisions are applicable only to cold-formed members or elements less than 3/16 in. (4.76 mm) in thickness. For materials not less than 3/16 in. (4.76 mm), the bolted connection should be designed in accordance with the AISC Specifications [1, 2].

In the AISI Specification, five types of bolts (A307, A325, A354, A449, and A490) are used for connections in cold-formed steel construction, in which A449 and A354 bolts should be used as an equivalent of A325 and A490 bolts, respectively, whenever bolts with smaller than 1/2-in. diameters are required.

On the basis of the failure modes occurring in the tests of bolted connections, the AISI criteria deal with three major design considerations for the connected parts: (1) longitudinal shear failure, (2) tensile failure, and (3) bearing failure. The nominal strength equations are given in Table 7.9.

In addition, design strength equations are provided for shear and tension in bolts. Accordingly, the AISI nominal strength for shear and tension in bolts can be determined as follows:

$$P_n = A_b F$$

where

A_b = gross cross-sectional area of bolt
F = nominal shear or tensile stress given in Table 7.10.

For bolts subjected to the combination of shear and tension, the reduced nominal tension stress is given in Table 7.11.

7.7.3 Screw Connections

Screws can provide a rapid and effective means to fasten sheet metal siding and roofing to framing members and to connect individual siding and roofing panels. Design equations are presently given in the AISI Specification for determining the nominal shear strength and the nominal tensile strength of connected parts and screws. These design requirements should be used for self-tapping screws with diameters larger than or equal to 0.08 in. (2.03 mm) but not exceeding 1/4 in. (6.35 mm). The screw can be thread-forming or thread-cutting, with or without drilling point. The spacing

TABLE 7.8   Nominal Strength Equations for Arc Welds

Type of weld	Type of strength	Nominal strength P_n (kips)
Groove welds (Figure 7.32)	• Tension or compression • Shear strength of weld • Shear strength of connected part	$Lt_e F_y$ $Lt_e(0.6F_{xx})$ $Lt_e(F_y/\sqrt{3})$
Arc spot welds (Figure 7.33)	• Shear strength Strength of weld Strength of connected part	 $0.589\,d_e^2 F_{xx}$
	1. $d_a/t \le 0.815\sqrt{E/F_u}$	$2.20 t d_a F_u$
	2. $0.815\sqrt{E/F_u} < (d_a/t) < 1.397\sqrt{E/F_u}$	$0.28[1+(5.59\sqrt{E/F_u})/(d_a/t)]t d_a F_u$
	3. $d_a/t \ge 1.397\sqrt{E/F_u}$	$1.40 t d_a F_u$
	Shear strength of connected part based on end distance • Tensile strength Strength of weld Strength of connected part	$e F_u t$ $0.785 d_e^2 F_{xx}$
	1. $F_u/E < 0.00187$	$[6.59 - 3150(F_u/E)](t d_a F_u)$ $\le 1.46 t d_a F_u$
	2. $F_u/E \ge 0.00187$	$0.70 t d_a F_u$
Arc seam welds (Figure 7.34)	• Shear strength • Strength of connected part	$[\pi d_e^2/4 + L d_e]0.75 F_{xx}$ $2.5 t F_u(0.25L + 0.96 d_a)$
Fillet welds (Figure 7.35)	• Shear strength of weld (for $t > 0.15$ in.) • Strength of connected part 1. Longitudinal loading	$0.75 t_w L F_{xx}$
	$\quad L/t < 25$:	$[1-(0.01L/t)]t L F_u$
	$\quad L/t \ge 25$:	$0.75 t L F_u$
	2. Transverse loading	$t L F_u$
Flare groove welds (Figure 7.36)	• Shear strength of weld (for $t > 0.15$ in.) • Strength of connected part 1. Transverse loading	$0.75 t_w L F_{xx}$ $0.833 t L F_u$
	2. Longitudinal loading	
	For $t \le t_w < 2t$ or if lip height $< L$	$0.75 t L F_u$
	For $t_w \ge 2t$ and lip height $\ge L$	$1.50 t L F_u$

d = visible diameter of outer surface of arc spot weld
d_a = average diameter of the arc spot weld at mid-thickness of t
d_a = $(d - t)$ for single sheet
d_a = $(d - 2t)$ for multiple sheets
d_e = effective diameter of fused area at plane of maximum shear transfer
d_e = $0.7d - 1.5t \le 0.55d$
e = distance measured in the line of force from the centerline of a weld to the nearest edge of an adjacent weld or to the end of the connected part toward which the force is directed
F_u = tensile strength of the connected part
F_y = yield point of steel
F_{xx} = filler metal strength designation in AWS electrode classification
L = length of weld
P_n = nominal strength of weld
t = thickness of connected sheet
t_w = effective throat dimension for groove weld, see AISI specification
t_w = effective throat = $0.707\,w_1$ or $0.707\,w_2$, whichever is smaller
w_1 = leg of weld
w_2 = leg of weld

See AISI Specification for additional design information.

between the centers of screws and the distance from the center of a screw to the edge of any part in the direction of the force should not be less than three times the diameter.

According to the AISI Specification, the nominal strength per screw is determined from Table 7.12. See Figures 7.37 and 7.38 for t_1, t_2, F_{u1}, and F_{u2}.

TABLE 7.9 Nominal Strength Equations for Bolted Connections

Type of strength	Nominal strength, P_n
Shear strength based on spacing and edge distance	$te F_u$
Tensile strength	
1. With washers under bolt head and nuts	$(1 - 0.9\,r + 3rd/s)\,F_u A_n \leq F_u A_n$
2. No washers or only one washer under bolt head and nuts	$(1 - r + 2.5\,rd/s)\,F_u A_n \leq F_u A_n$
Note: The tensile strength computed above should not exceed $A_n F_y$.	
Bearing strength	
1. With washers under bolt head and nut	
• Inside sheet of double shear connection	
$\quad F_u/F_y \geq 1.08$	$3.33 F_u dt$
$\quad F_u/F_y < 1.08$	$3.00 F_u dt$
• Single shear and outside sheets of double shear connection	$3.00 F_u dt$
2. Without washers under bolt head and nut or with only one washer	
• Inside sheet of double shear connection	$3.00 F_u dt$
• Single shear and outside sheets of double shear connection	$2.22 F_u dt$

A_n	=	net area of the connected part
d	=	diameter of bolt
e	=	distance measured in the line of force from the center of bolt to the nearest edge of an adjacent hole or to the end of the connected part
F_u	=	tensile strength of the connected part
F_y	=	specified yield point of steel
r	=	force transmitted by the bolt or bolts at the section considered, divided by the tension force in the member at that section. If r is less than 0.2, it may be taken equal to zero
s	=	spacing of bolts perpendicular to line of force
t	=	thickness of thinnest connected part

For the convenience of designers, the following table gives the correlation between the common number designation and the nominal diameter for screws.

Number designation	Nominal diameter, d (in.)
0	0.060
1	0.073
2	0.086
3	0.099
4	0.112
5	0.125
6	0.138
7	0.151
8	0.164
10	0.190
12	0.216
1/4	0.250

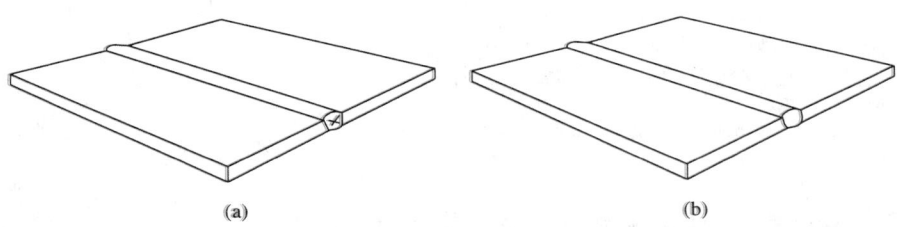

(a)   (b)

Figure 7.32 Groove welds.

In addition to the design requirements discussed above, the AISI Specification also includes some provisions for spacing of connectors when two channels are connected to form an I-section or when compression elements are joined to other parts of built-up members by intermittent connections.

TABLE 7.10   Nominal Tensile and Shear Stresses for Bolts

Description of bolts	Nominal tensile stress F_{nt}, ksi	Nominal shear stress F_{nv}, ksi
A307 Bolts, Grade A, 1/4 in. $\leq d <$ 1/2 in.	40.5	24.0
A307 Bolts, Grade A, $d \geq$ 1/2 in.	45.0	27.0
A325 Bolts, when threads are not excluded from shear planes	90.0	54.0
A325 Bolts, when threads are excluded from shear planes	90.0	72.0
A354 Grade BD Bolts, 1/4 in. $\leq d <$ 1/2 in., when threads are not excluded from shear planes	101.0	59.0
A354 Grade BD Bolts, 1/4 in. $\leq d <$ 1/2 in. when threads are excluded from shear planes	101.0	90.0
A449 Bolts, 1/4 in. $\leq d <$ 1/2 in., when threads are not excluded from shear planes	81.0	47.0
A449 Bolts, 1/4 in. $\leq d <$ 1/2 in., when threads are excluded from shear planes	81.0	72.0
A 490 Bolts, when threads are not excluded from shear planes	112.5	67.5
A490 Bolts, when threads are excluded from shear planes	112.5	90.0

TABLE 7.11   Nominal Tension Stresses, F'_{nt} (ksi), for Bolts Subjected to the Combination of Shear and Tension

(A) ASD Method		
Description of bolts	Threads not excluded from shear planes	Threads excluded from shear planes
A325 Bolts	$110 - 3.6f_v \leq 90$	$110 - 2.8f_v \leq 90$
A354 Grade BD Bolts	$122 - 3.6f_v \leq 101$	$122 - 2.8f_v \leq 101$
A449 Bolts	$100 - 3.6f_v \leq 81$	$100 - 2.8f_v \leq 81$
A490 Bolts	$136 - 3.6f_v \leq 112.5$	$136 - 2.8f_v \leq 112.5$
A307 Bolts, Grade A		
When 1/4 in. $\leq d <$ 1/2 in.	$52 - 4f_v \leq 40.5$	$52 - 4f_v \leq 40.5$
When $d \geq$ 1/2 in.	$58.5 - 4f_v \leq 45$	$58.5 - 4f_v \leq 45$
(B) LRFD Method		
A325 Bolts	$113 - 2.4f_v \leq 90$	$113 - 1.9f_v \leq 90$
A354 Grade BD Bolts	$127 - 2.4f_v \leq 101$	$127 - 1.9f_v \leq 101$
A449 Bolts	$101 - 2.4f_v \leq 81$	$101 - 1.9f_v \leq 81$
A490 Bolts	$141 - 2.4f_v \leq 112.5$	$141 - 1.9f_v \leq 112.5$
A307 Bolts, Grade A		
When 1/4 in. $\leq d <$ 1/2 in.	$47 - 2.4f_v \leq 40.5$	$47 - 2.4f_v \leq 40.5$
When $d \geq$ 1/2 in.	$52 - 2.4f_v \leq 45$	$52 - 2.4f_v \leq 45$

d = diameter of bolt
f_v = shear stress based on gross cross-sectional area of bolt

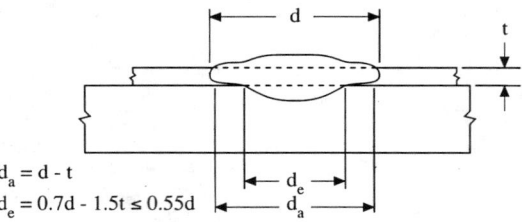

$d_a = d - t$
$d_e = 0.7d - 1.5t \leq 0.55d$

Figure 7.33   Arc spot weld — single thickness of sheet.

7.8 Structural Systems and Assemblies

In the past, cold-formed steel components have been used in different structural systems and assemblies such as metal buildings, shear diaphragms, shell roof structures, wall stud assemblies, residential construction, and composite construction.

TABLE 7.12 Nominal Strength Equations for Screws

Type of strength	Nominal strength
Shear strength	
1. Connection shear	
• For $t_2/t_1 \leq 1.0$:	a. $P_{ns} = 4.2(t_2^3 d)^{1/2} F_{u2}$
use smallest of three considerations	b. $P_{ns} = 2.7\, t_1 d\, F_{u1}$
	c. $P_{ns} = 2.7\, t_2 d\, F_{u2}$
• For $t_2/t_1 \geq 2.5$:	a. $P_{ns} = 2.7\, t_1 d\, F_{u1}$
use smaller of two considerations	b. $P_{ns} = 2.7\, t_2 d\, F_{u2}$
• For $1.0 < t_2/t_1 < 2.5$:	
use linear interpolation	
2. Shear in screws	$\geq 1.25\, P_{ns}$
Tensile strength	
1. Connection tension	
Pull-out strength	$P_{not} = 0.85\, t_c d\, F_{u2}$
Pull-over strength	$P_{nov} = 1.5\, t_1 d_w\, F_{u1}$
2. Tension in screws	$P_{nt} \geq 1.25$
	(lesser of P_{not} and P_{nov}

d	=	diameter of screw
d_w	=	larger of the screw head diameter or the washer diameter, and should be taken not larger than 1/2 in. (12.7 mm)
F_{u1}	=	tensile strength of member in contact with the screw head
F_{u2}	=	tensile strength of member not in contact with the screw head
P_{ns}	=	nominal shear strength per screw
P_{nt}	=	nominal tension strength per screw
P_{not}	=	nominal pull-out strength per screw
P_{nov}	=	nominal pull-over strength per screw
t_1	=	thickness of member in contact with the screw head
t_2	=	thickness of member not in contact with the screw head

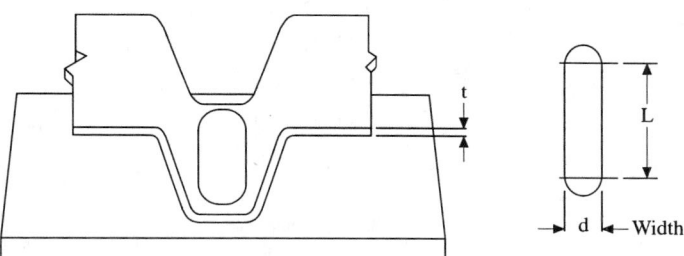

Figure 7.34 Arc seam weld.

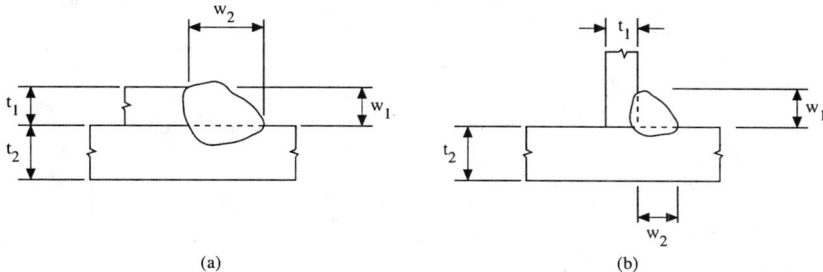

(a)   (b)

Figure 7.35 Fillet welds.

7.8.1 Metal Buildings

Standardized metal buildings have been widely used in industrial, commercial, and agricultural applications. This type of metal building has also been used for community facilities because it can provide attractive appearance, fast construction, low maintenance, easy extension, and lower long-term cost.

In general, metal buildings are made of welded rigid frames with cold-formed steel sections used for purlins, girts, roofs, and walls. In the U.S., the design of standardized metal buildings is often based on the Low Rise Building Systems published by the Metal Building Manufacturers Associ-

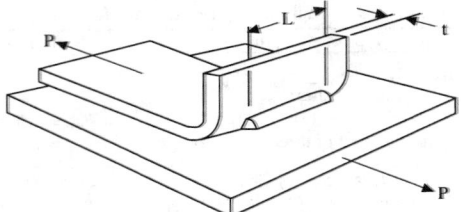

Flare-Bevel Groove Weld

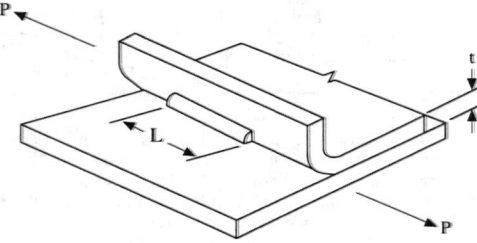

Flare Bevel Groove Weld

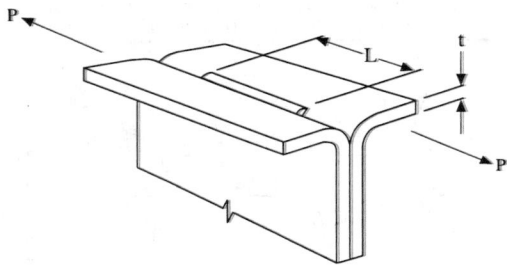

Flare V-Groove Weld

Figure 7.36   Flare groove welds.

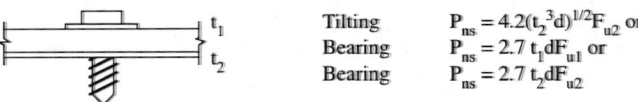

Tilting $\quad P_{ns} = 4.2(t_2^3 d)^{1/2} F_{u2}$ or
Bearing $\quad P_{ns} = 2.7\, t_1 d F_{u1}$ or
Bearing $\quad P_{ns} = 2.7\, t_2 d F_{u2}$

Figure 7.37   Screw connection for $t_2/t_1 \leq 1.0$.

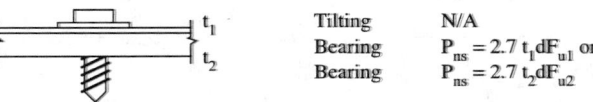

Tilting $\quad$ N/A
Bearing $\quad P_{ns} = 2.7\, t_1 d F_{u1}$ or
Bearing $\quad P_{ns} = 2.7\, t_2 d F_{u2}$

Figure 7.38   Screw connection for $t_2/t_1 \geq 2.5$.

ation [33]. This document contains design practices, commentary, common industry practices, guide specifications, and nomenclature for metal building systems. In other countries, many design concepts and building systems have been developed.

7.8.2   Shear Diaphragms

In building construction, it has been a common practice to provide a separate bracing system to resist horizontal loads due to wind load or earthquake. However, steel floor and roof panels, with or without concrete fill, are capable of resisting horizontal loads in addition to the beam strength for gravity loads if they are adequately interconnected to each other and to the supporting frame. For the same reason, wall panels can provide not only enclosure surfaces and support normal loads, but they can also provide diaphragm action in their own planes.

The structural performance of a diaphragm construction can be evaluated by either calculations or tests. Several analytical procedures exist, and are summarized in the literature [3, 18, 22, 32]. Tested performance can be measured by the procedures of the Standard Method for Static Load Testing of Framed Floor, Roof and Wall Diaphragm Construction for Buildings, ASTM E455 [13]. A general discussion of structural diaphragm behavior is given by Yu [49].

Shear diaphragms should be designed for both strength and stiffness. After the nominal shear strength is established by calculations or tests, the design strength can be determined on the basis of the safety factor or resistance factor given in the Specification. Six cases are classified in the AISI Specification for the design of shear diaphragms according to the type of failure mode, connection, and loading. Because the quality of mechanical connectors is easier to control than welded connections, a relatively smaller safety factor or larger resistance factor is used for mechanical connections. As far as the loading is concerned, the safety factors for earthquake are slightly larger than those for wind due to the ductility demands required by seismic loading.

7.8.3 Shell Roof Structures

Shell roof structures such as folded-plate and hyperbolic paraboloid roofs have been used in building construction for churches, auditoriums, gymnasiums, schools, restaurants, office buildings, and airplane hangars. This is because the effective use of steel panels in roof construction is not only to provide an economical structure but also to make the building architecturally attractive and flexible for future extension. The design methods used in engineering practice are mainly based on the successful investigation of shear diaphragms and the structural research on shell roof structures.

A folded-plate roof structure consists of three major components. They are (1) steel roof panels, (2) fold line members at ridges and valleys, and (3) end frame or end walls as shown in Figure 7.39. Steel roof panels can be designed as simply supported slabs in the transverse direction between fold lines. The reaction of the panels is then applied to fold lines as a line loading, which can be resolved into two components parallel to the two adjacent plates. These load components are carried by an

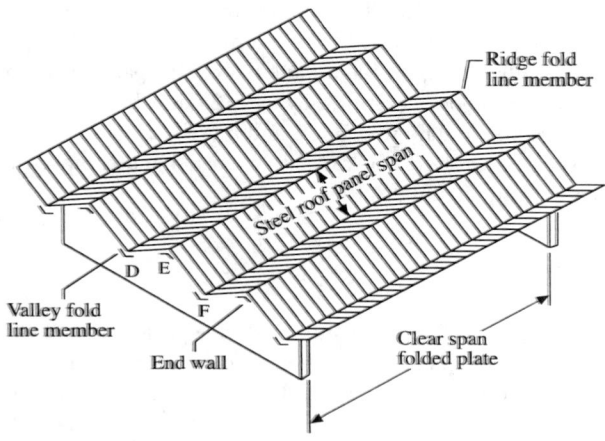

Figure 7.39 Folded-plate structure. (From Yu, W.W. 1991. *Cold-Formed Steel Design*, John Wiley & Sons, New York. With permission.)

inclined deep girder spanned between end frames or end walls. These deep girders consist of fold line members as flanges and steel panels as a web element. The longitudinal flange force in fold line members can be obtained by dividing the bending moment of the deep girder by its depth. The shear force is resisted by the diaphragm action of the steel roof panels. In addition to the strength, the deflection characteristics of the folded-plate roof should also be investigated, particularly for long-span structures. In the past, it has been found that a method similar to the Williot diaphragm for determining truss deflections can also be used for the prediction of the deflection of a steel folded-plate roof. The in-plane deflection of each plate should be computed as a sum of the deflections due to flexure, shear, and seam slip, considering the plate temporarily separated from the adjacent plates. The true displacement of the fold line can then be determined analytically or graphically by a Williot diagram. The above discussion deals with a simplified method. The finite-element method can provide a more detailed analysis for various types of loading, support, and material.

The hyperbolic paraboloid roof has also gained popularity due to the economical use of materials and its appearance. This type of roof can be built easily with either single- or double-layer standard steel roof deck panels because hyperbolic paraboloid has straight line generators. Figure 7.40 shows four common types of hyperbolic paraboloid roofs which may be modified or varied in other ways to achieve a striking appearance. The method of analysis depends on the curvature of the shell used

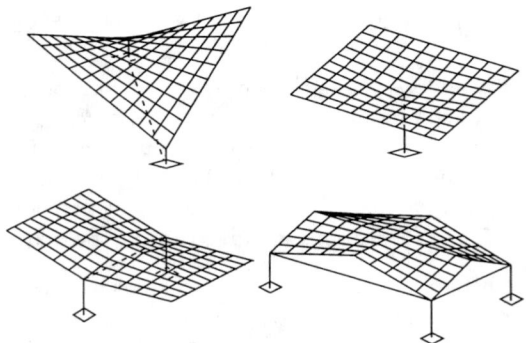

Figure 7.40   Types of hyperbolic paraboloid roofs. (From Yu, W.W. 1991. *Cold-Formed Steel Design*, John Wiley & Sons, New York. With permission.)

for the roof. If the uniformly loaded shell is deep, the membrane theory may be used. For the case of a shallow shell or a deep shell subjected to unsymmetrical loading, the finite-element method will provide accurate results. Using the membrane theory, the panel shear for a uniformly loaded hyperbolic paraboloid roof can be determined by $wab/2h$, in which w is the applied load per unit surface area, a and b are horizontal projections, and h is the amount of corner depression of the surface. This panel shear force should be carried by tension and compression framing members. For additional design information, see Yu [49].

7.8.4   Wall Stud Assemblies

Cold-formed steel I-, C-, Z-, or box-type studs are widely used in walls with their webs placed perpendicular to the wall surface. The walls may be made of different materials, such as fiber board, lignocellulosic board, plywood, or gypsum board. If the wall material is strong enough and there is adequate attachment provided between wall material and studs for lateral support of the studs, then the wall material can contribute to the structural economy by increasing the usable strength of the studs substantially.

The AISI Specification provides the requirements for two types of stud design. The first type is "All Steel Design", in which the wall stud is designed as an individual compression member neglecting the structural contribution of the attached sheathing. The second type is "Sheathing Braced Design", in which consideration is given to the bracing action of the sheathing material due to the shear rigidity and the rotational restraint provided by the sheathing. Both solid and perforated webs are permitted. The subsequent discussion deals with the sheathing braced design of wall studs.

Wall Studs in Compression

The AISI design provisions are used to prevent three possible modes of failure. The first requirement is for column buckling between fasteners in the plane of the wall (Figure 7.41). For this case, the limit state may be either (1) flexural buckling, (2) torsional buckling, or (3) torsional-flexural buckling depending on the geometric configuration of the cross-section and the spacing of fasteners. The nominal compressive strength is based on the stud itself without considering any interaction with the sheathing material.

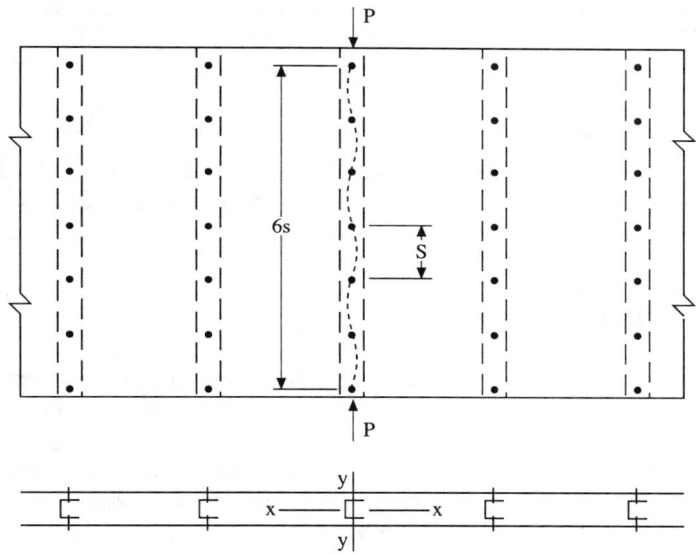

Figure 7.41   Buckling of studs between fasteners. (From Yu, W.W. 1991. *Cold-Formed Steel Design,* John Wiley & Sons, New York. With permission.)

The second requirement is for overall column buckling of wall studs braced by shear diaphragms on both flanges (Figure 7.42). For this case, the AISI Specification provides equations for calculating the critical stresses in order to determine the nominal axial strength by considering the shear rigidity of the sheathing material. These lengthy equations can be found in Section D4 of the Specification [7].

The third requirement is to prevent shear failure of the sheathing by limiting the shear strain within the permissible value for a given sheathing material.

Wall Studs in Bending

The nominal flexural strength of wall studs is determined by the nominal section strength by using the "All Steel Design" approach and neglecting the structural contribution of the attached sheathing material.

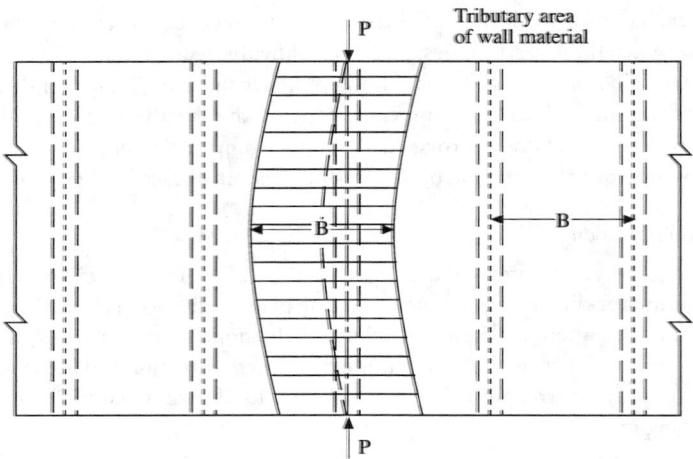

Figure 7.42   Overall column buckling of studs. (From Yu, W.W. 1991. *Cold-Formed Steel Design*, John Wiley & Sons, New York. With permission.)

Wall Studs with Combined Axial Load and Bending

The AISI interaction equations presented in Table 7.7 are also applicable to wall studs subjected to combined axial load and bending with the exception that the nominal flexural strength be evaluated by excluding lateral buckling considerations.

7.8.5   Residential Construction

During recent years, cold-formed steel members have been increasingly used in residential construction as roof trusses, wall framing, and floor systems (Figure 7.43). Because of the lack of standard sections and design tables, prescriptive standards have recently been developed by the National Association of Home Builders (NAHB) Research Center and the Housing and Urban Development (HUD). The sectional properties and load-span design tables for a selected group of C-sections are calculated in accordance with the AISI Specification [9].

For the design of cold-formed steel trusses and shear walls using steel studs, design guides have been published by the American Iron and Steel Institute [6, 9].

7.8.6   Composite Construction

Cold-formed steel decks have been used successfully in composite roof and floor construction. For this type of application, the steel deck performs the dual role of serving as a form for the wet concrete during construction and as positive reinforcements for the slab during service.

As far as the design method for the composite slab is concerned, many designs have been based on the SDI Specification for composite steel floor deck [42]. This document contains requirements and recommendations on materials, design, connections, and construction practice. Since 1984, the American Society of Civil Engineers has published a standard specification for the design and construction of composite slabs [11].

When the composite construction is composed of steel beams or girders with cold-formed steel deck, the design should be based on the AISC Specification [1, 2].

Figure 7.43 Steel house using cold-formed members for walls, joists, and trusses.

7.9 Defining Terms

ASD (allowable stress design): A method of proportioning structural components such that the allowable stress, allowable force, or allowable moment is not exceeded when the structure is subjected to all appropriate combinations of nominal loads.

Beam-column: A structural member subjected to combined compressive axial load and bending.

Buckling load: The load at which a compressed element, member, or frame assumes a deflected position.

Cold-formed steel members: Shapes that are manufactured by press-braking blanks sheared from sheets, cut lengths of coils or plates, or by roll forming cold- or hot-rolled coils or sheets.

Composite slab: A slab in which the load-carrying capacity is provided by the composite action of concrete and steel deck (as reinforcement).

Compression members: Structural members whose primary function is to carry concentric loads along their longitudinal axes.

Design strength: R_n/Ω for ASD or ϕR_n for LRFD (force, moment, as appropriate), provided by the structural component.

Effective design width: Reduced flat width of an element due to local buckling for design purposes. The reduced width is termed the effective width or effective design width.

Effective length: The equivalent length KL used in design equations.

Flexural members (beams): Structural members whose primary function is to carry transverse loads and/or moments.

Flat-width-to-thickness ratio: The flat width of an element measured along its plane, divided by its thickness.

Limit state: A condition at which a structure or component becomes unsafe (strength limit state) or no longer useful for its intended function (serviceability limit state).

Load factor: A factor that accounts for unavoidable deviations of the actual load from the nominal load.

Local buckling: Buckling of elements only within a section, where the line junctions between elements remain straight and angles between elements do not change.

LRFD (load and resistance factor design): A method of proportioning structural components such that no applicable limit state is exceeded when the structure is subjected to all appropriate load combinations.

Multiple-stiffened elements: An element that is stiffened between webs, or between a web and a stiffened edge, by means of intermediate stiffeners that are parallel to the direction of stress. A sub-element is the portion between adjacent stiffeners or between web and intermediate stiffener or between edge and intermediate stiffener.

Nominal loads: The loads specified by the applicable code not including load factors.

Nominal strength: The capacity of a structure or component to resist the effects of loads, as determined by computations using specified material strengths and dimensions with equations derived from accepted principles of structural mechanics or by tests of scaled models, allowing for modeling effects, and differences between laboratory and field conditions.

Point-symmetric section: A point-symmetric section is a section symmetrical about a point (centroid) such as a Z-section having equal flanges.

Required strength: Load effect (force, moment, as appropriate) acting on the structural component determined by structural analysis from the factored loads for LRFD or nominal loads for ASD (using most appropriate critical load combinations).

Resistance factor: A factor that accounts for unavoidable deviations of the actual strength from the nominal value.

Safety factor: A ratio of the stress (or strength) at incipient failure to the computed stress (or strength) at design load (or service load).

Stiffened or partially stiffened compression elements: A stiffened or partially stiffened compression element is a flat compression element with both edges parallel to the direction of stress stiffened either by a web, flange, stiffening lip, intermediate stiffener, or the like.

Stress: Stress as used in this chapter means force per unit area and is expressed in ksi (kips per square inch) for U.S. customary units or MPa for SI units.

Thickness: The thickness of any element or section should be the base steel thickness, exclusive of coatings.

Torsional-flexural buckling: A mode of buckling in which compression members can bend and twist simultaneously without change in cross-sectional shape.

Unstiffened compression elements: A flat compression element which is stiffened at only one edge parallel to the direction of stress.

Yield point: Yield point as used in this chapter means either yield point or yield strength of steel.

References

[1] American Institute of Steel Construction. 1989. *Specification for Structural Steel Buildings — Allowable Stress Design and Plastic Design*, Chicago, IL.

[2] American Institute of Steel Construction. 1993. *Load and Resistance Factor Design Specification for Structural Steel Buildings,* Chicago, IL.

[3] American Iron and Steel Institute. 1967. *Design of Light-Gage Steel Diaphragms,* New York.

[4] American Iron and Steel Institute. 1983. *Handbook of Steel Drainage and Highway Construction Products,* Washington, D.C.

[5] American Iron and Steel Institute. 1996. *Automotive Steel Design Manual,* Washington, D.C.

[6] American Iron and Steel Institute. 1995. *Design Guide for Cold-Formed Steel Trusses.* Publication RG-9518, Washington, D.C.

[7] American Iron and Steel Institute. 1996. *Specification for the Design of Cold-Formed Steel Structural Members,* Washington, D.C.

[8] American Iron and Steel Institute. 1996. *Cold-Formed Steel Design Manual,* Washington, D.C.

[9] American Iron and Steel Institute. 1996. *Residential Steel Framing Manual for Architects, Engineers, and Builders,* Washington, D.C.

[10] American Society of Civil Engineers. 1978. *Design of Steel Transmission Pole Structures,* New York.

[11] American Society of Civil Engineers. 1984. *Specification for the Design and Construction of Composite Steel Deck Slabs,* ASCE Standard, New York.

[12] American Society of Civil Engineers. 1991. *Specification for the Design of Cold-Formed Stainless Steel Structural Members,* ANSI/ASCE-8-90, New York.

[13] American Society for Testing and Materials. 1993. *Standard Method for Static Load Testing of Framed Floor, Roof and Wall Diaphragm Construction for Buildings,* ASTM E455, Philadelphia, PA.

[14] Architectural Institute of Japan. 1985. *Recommendations for the Design and Fabrication of Light Weight Steel Structures,* Japan.

[15] Association of the Wall and Ceiling Industries—International and Metal Lath/Steel Framing Association. 1979. *Steel Framing Systems Manual,* Chicago, IL.

[16] American Welding Society. 1989. *Structural Welding Code — Sheet Steel,* AWS D1.3-89, Miami, FL.

[17] British Standards Institution. 1987. *British Standard: Structural Use of Steelwork in Building. Part 5. Code of Practice for Design of Cold-Formed Sections,* BS 5950: Part 5: 1987.

[18] Bryan, E.R. and Davies, J.M. 1981. *Steel Diaphragm Roof Decks — A Design Guide With Tables for Engineers and Architects,* Granada Publishing, New York.

[19] Canadian Standards Association. 1994. *Cold-Formed Steel Structural Members,* CAN3-S136-M94, Ottawa, Canada.

[20] Centre Technique Industriel de la Construction Metallique. 1978. *Recommendations pourle Calcul des Constructions a Elements Minces en Acier.*

[21] Czechoslovak State Standard. 1987. *Design of Light Gauge Cold-Formed Profiles in Steel Structures,* CSN 73 1402.

[22] Department of Army. 1985. *Seismic Design for Buildings,* U.S. Army Technical Manual 5-809-10, Washington, D.C.

[23] DIN 18807. 1987. *Trapezprofile im Hochbau* (Trapezoidal Profiled Sheeting in Building). Deutsche Norm (German Standard).

[24] European Convention for Constructional Steelwork. 1987. *European Recommendations for the Design of Light Gauge Steel Members,* CONSTRADO, London.

[25] Eurocode 3. 1989. *Design of Steel Structures.*

[26] Finnish Ministry of Environment. 1988. *Building Code Series of Finland: Specification for Cold-Formed Steel Structures.*

[27] Groep Stelling Fabrikanten — GSF. 1977. *Richtlijnen Voor de Berekening van Stalen Industriele Magezijnstellingen*, RSM.

[28] Hancock, G.J. 1994. *Cold-Formed Steel Structures*, Australian Institute of Steel Construction, North Sydney, NSW, Australia.

[29] Hetrakul, N. and Yu, W.W. 1978. Structural Behavior of Beam Webs Subjected to Web Crippling and a Combination of Web Crippling and Bending, *Final Report*, University of Missouri-Rolla, Rolla, MO.

[30] Indian Standards Institution. 1975. *Indian Standard Code of Practice for Use of Cold-Formed Light-Gauge Steel Structural Members in General Building Construction*, IS:801-1975.

[31] Lagereinrichtungen. 1974. ONORM B 4901, Dec.

[32] Luttrell, L.D. 1988. *Steel Deck Institute Diaphragm Design Manual*, 2nd ed., Steel Deck Institute, Canton, OH.

[33] Metal Building Manufacturers Association. 1996. *Low Rise Building Systems Manual*, Cleveland, OH.

[34] People's Republic of China National Standard. 1989. *Technical Standard for Thin-Walled Steel Structures*, GBJ 18-87, Beijing, China.

[35] Rack Manufacturers Institute. 1990. *Specification for the Design, Testing, and Utilization of Industrial Steel Storage Racks*, Charlotte, NC.

[36] Rhodes, J. 1991. *Design of Cold-Formed Steel Structures*, Elsevier Publishing, New York.

[37] *Romanian Specification for Calculation of Thin-Walled Cold-Formed Steel Members*, STAS 10108/2-83.

[38] South African Institute of Steel Construction. 1995. *Code of Practice for the Design of Structural Steelwork*.

[39] Standards Australia. 1996. *Cold-Formed Steel Structures*, Australia.

[40] Standards New Zealand. 1996. *Cold-Formed Steel Structures*, New Zealand.

[41] State Building Construction of USSR. 1988. *Building Standards and Rules: Design Standards — Steel Construction*, Part II, Moscow.

[42] Steel Deck Institute. 1995. *Design Manual for Composite Decks, Form Decks, Roof Decks, and Cellular Floor Deck with Electrical Distribution*, Publication No. 29. Canton, OH.

[43] Steel Joist Institute. 1995. *Standard Specification and Load Tables for Open Web Steel Joists*, 40th ed., Myrtle Beach, SC.

[44] Swedish Institute of Steel Construction. 1982. *Swedish Code for Light Gauge Metal Structures*, Publication 76.

[45] von Karman, T., Sechler, E.E., and Donnell, L.H. 1932. The Strength of Thin Plates in Compression, *Trans. ASME*, Vol. 54.

[46] Winter, G. and Pian, R.H.M. 1946. Crushing Strength of Thin Steel Webs, *Bulletin* 35. Pt. 1, Cornell University Engineering Experiment Station, Ithaca, NY.

[47] Winter, G. 1947. Strength of Thin Steel Compression Flanges, *Trans. ASCE*, Vol. 112.

[48] Winter, G. 1970. *Commentary on the Specification for the Design of Cold-Formed Steel Structural Members*, American Iron and Steel Institute, New York.

[49] Yu, W.W. 1991. *Cold-Formed Steel Design*, John Wiley & Sons, New York.

[50] Zetlin, L. 1955. Elastic Instability of Flat Plates Subjected to Partial Edge Loads. *J. Structural Div.*, ASCE, Vol. 81, New York.

Further Reading

Guide to Stability Design Criteria for Metal Structures, edited by T.V. Galambos, presents general information, interpretation, new ideas, and research results on a full range of structural stability concerns. It was published by John Wiley & Sons in 1988.

Cold-Formed Steel in Tall Buildings, edited by W.W. Yu, R. Baehre, and T. Toma, provides readers with information needed for the design and construction of tall buildings, using cold-formed steel for structural members and/or architectural components. It was published by McGraw-Hill in 1993.

Thin-Walled Structures, edited by J. Rhodes and K.P. Chong, is an international journal which publishes papers on theory, experiment, design, etc. related to cold-formed steel sections, plate and shell structures, and others. It was published by Elsevier Applied Science. A special issue of the Journal on Cold-Formed Steel Structures was edited by J. Rhodes and W.W. Yu, Guest Editor, and published in 1993.

Proceedings of the International Specialty Conference on Cold-Formed Steel Structures, edited by W.W. Yu, J.H. Senne, and R.A. LaBoube, has been published by the University of Missouri-Rolla since 1971. This publication contains technical papers presented at the International Specialty Conferences on Cold-Formed Steel Structures.

"Cold-Formed Steel Structures", by J. Rhodes and N.E. Shanmugan, in *The Civil Engineering Handbook* (W.F. Chen, Editor-in-Chief), presents discussions of cold-formed steel sections, local buckling of plate elements, and the design of cold-formed steel members and connections. It was published by CRC Press in 1995.

8

Aluminum Structures

8.1	Introduction	8-1
	The Material • Alloy Characteristics • Codes and Specifications	
8.2	Structural Behavior	8-4
	General • Component Behavior • Joints • Fatigue	
8.3	Design	8-20
	General Considerations • Design Studies	
8.4	Economics of Design	8-29
8.5	Defining Terms	8-30
References		8-30
Further Reading		8-30

Maurice L. Sharp
Consultant—Aluminum Structures,
Avonmore, PA

8.1 Introduction

8.1.1 The Material

Background

Of the structural materials used in construction, aluminum was the latest to be introduced into the market place even though it is the most abundant of all metals, making up about 1/12 of the earth's crust. The commercial process was invented simultaneously in the U.S. and Europe in 1886. Commercial production of the metal started thereafter using an electrolytic process that economically separated aluminum from its oxides. Prior to this time aluminum was a precious metal. The initial uses of aluminum were for cooking utensils and electrical cables. The earliest significant structural use of aluminum was for the skins and members of a dirigible called the *Shenendoah* completed in 1923. The first structural design handbook was developed in 1930 and the first specification was issued by the industry in 1932 [4].

Product Forms

Aluminum is available in all the common product forms: flat-rolled, extruded, cast, and forged. Fasteners such as bolts, rivets, screws, and nails are also manufactured. The available thicknesses of flat-rolled products range from 0.006 in. or less for foil to 7.0 in. or more for plate. Widths to 17 ft are possible. Shapes in aluminum are extruded. Some presses can extrude sections up to 31 in. wide. The extrusion process allows the material to be placed in areas that maximize structural properties and joining ease. Because the cost of extrusion dies is relatively low, most extruded shapes are designed for specific applications. Castings of various types and forgings are possibilities for three-dimensional shapes and are used in some structural applications. The design of castings is not covered in detail in structural design books and specifications primarily because there can be a wide range of quality depending on the casting process. The quality of the casting affects structural performance.

Alloy and Temper Designation

The four-digit number used to designate alloys is based on the main alloying ingredients. For example, magnesium is the principal alloying element in alloys whose designation begins with a 5(5083, 5456, 5052, etc.). Cast designations are similar to wrought designations but a decimal is placed between the third and fourth digit(356.0). The second part of the designation is the temper which defines the fabrication process. If the term starts with T, e.g., -T651, the alloy has been subjected to a thermal heat treatment. These alloys are often referred to as heat-treatable alloys. The numbers after the T show the type of treatment and any subsequent mechanical treatment such as a controlled stretch. The temper of alloys that harden with mechanical deformation starts with H, e.g., -H116. These alloys are referred to as non-heat-treatable alloys. The type of treatment is defined by the numbers in the temper designation. A 0 temper is the fully annealed temper. The full designation of an alloy has the two parts that define both chemistry and fabrication history, e.g., 6061-T651.

8.1.2 Alloy Characteristics

Physical Properties

Physical properties usually vary only by a few percent depending on alloy. Some nominal values are given in Table 8.1.

TABLE 8.1 Some Nominal Properties of Aluminum Alloys

Property	Value
Weight	0.1 lb/in.^3
Modulus of elasticity	
Tension and compression	10,000 ksi
Shear	3,750 ksi
Poisson's ratio	1/3
Coefficient of thermal expansion (68 to 212 °F)	0.000013 per °F

Data from Gaylord and Gaylord, *Structural Engineering Handbook*, McGraw-Hill, New York, 1990.

The density of aluminum is low, about 1/3 that of steel, which results in lightweight structures. The modulus of elasticity is also low, about 1/3 of that of steel, which affects design when deflection or buckling controls.

Mechanical Properties

Mechanical properties for a few alloys used in general purpose structures are given in Table 8.2. The stress-strain curves for aluminum alloys do not have an abrupt break when yielding but rather have a gradual bend (see Figure 8.1). The yield strength is defined as the stress corresponding to a 0.002 in./in. permanent set. The alloys shown in Table 8.2 have moderate strength, excellent resistance to corrosion in the atmosphere, and are readily joined by mechanical fasteners and welds. These alloys often are employed in outdoor structures without paint or other protection. The higher strength aerospace alloys are not shown. They usually are not used for general purpose structures because they are not as resistant to corrosion and normally are not welded.

Toughness

The accepted measure of toughness of aluminum alloys is fracture toughness. Most high strength aerospace alloys can be evaluated in this manner; however, the moderate strength alloys

TABLE 8.2   Minimum Mechanical Properties

Alloy and temper	Product	Thickness range, in.	Tension		Compression	Shear		Bearing	
			TS	YS	YS	US	YS	US	YS
3003-H14	Sheet and plate	0.009–1.000	20	17	14	12	10	40	25
5456-H116	Sheet and plate	0.188–1.250	46	33	27	27	19	87	56
6061-T6	Sheet and plate	0.010–4.000	42	35	35	27	20	88	58
6061-T6	Shapes	All	38	35	35	24	20	80	56
6063-T5	Shapes	to 0.500	22	16	16	13	9	46	26
6063-T6	Shapes	All	30	25	25	19	14	63	40

Data from The Aluminum Association, *Structural Design Manual*, 1994.
Note: All properties are in ksi. TS is tensile strength, YS is yield strength, and US is ultimate strength.

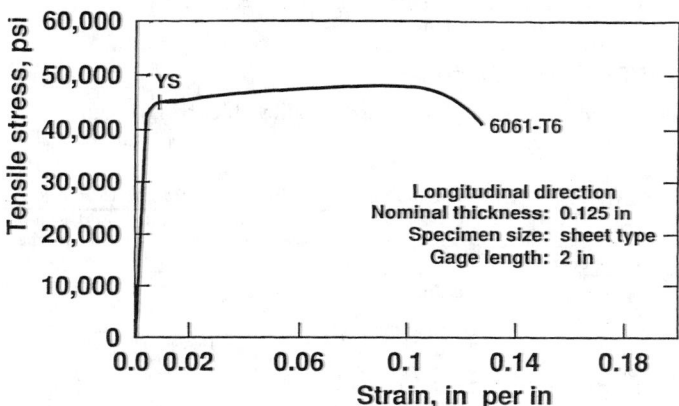

Figure 8.1   Stress-strain curve.

employed for general purpose structures cannot because they are too tough to get valid results in the test. Aluminum alloys also do not exhibit a transition temperature; their strength and ductility actually increase with decrease in temperature. Some alloys have a high ratio of yield strength to tensile strength (compared to mild steel) and most alloys have a lower elongation than mild steel, perhaps 8 to 10%, both considered to be negative factors for toughness. However, these alloys do have sufficient ductility to redistribute stresses in joints and in sections in bending to achieve full strength of the components. Their successful use in various types of structures (bridges, bridge decks, tractor-trailers, railroad cars, building structures, and automotive frames) has demonstrated that they have adequate toughness. Thus far there has not been a need to modify the design based on toughness of aluminum alloys.

8.1.3   Codes and Specifications

Allowable stress design (ASD) for building, bridge, and other structures that need the same factor of safety, and Load and Resistance Factor Design (LRFD)for building and similar type structures have been published by the Aluminum Association [1]. These specifications are included in a design manual that also has design guidelines, section properties of shapes, design examples, and numerous other aids for the designer.

The American Association of State Highway and Transportation Officials has published LRFD Specifications that cover bridges of aluminum and other materials [2]. The equations for strength and behavior of aluminum components are essentially the same in all of these specifications. The margin of safety for design differs depending on the type of specification and the type of structure.

Codes and standards are available for other types of aluminum structures. Lists and summaries are provided elsewhere [1, 3].

8.2 Structural Behavior

8.2.1 General

Compared to Steel

The basic principles of design for aluminum structures are the same as those for other ductile metals such as steel. Equations and analysis techniques for global structural behavior such as load-deflection behavior are the same. Component strength, particularly buckling, post buckling, and fatigue, are defined specifically for aluminum alloys. The behavior of various types of components are provided below. Strength equations are given. The designer needs to incorporate appropriate factors of safety when these equations are used for practical designs.

Safety and Resistance Factors

Table 8.3 gives factors of safety as utilized for allowable stress design.

TABLE 8.3 Factors of Safety for Allowable Stress Design

Component	Failure mode	Buildings and similar type structures	Bridges and similar type structures
Tension	Yielding	1.95	2.20
	Ultimate strength	1.65	1.85
Columns	Yielding (short col.)	1.65	1.85
	Buckling	1.95	2.20
Beams	Tensile yielding	1.65	1.85
	Tensile ultimate	1.95	2.20
	Compressive yielding	1.65	1.85
	Lateral buckling	1.65	1.85
Thin plates in	Ultimate in columns	1.95	2.20
compression	Ultimate in beams	1.65	1.85
Stiffened flat	Shear yield	1.65	1.85
webs in shear	Shear buckling	1.20	1.35
Mechanically	Bearing yield	1.65	1.85
fastened	Bearing ultimate	2.34	2.64
joints	Shear str./rivets, bolts	2.34	2.64
Welded joints	Shear str./fillet welds	2.34	2.64
	Tensile str./butt welds	1.95	2.20
	Tensile yield/ butt welds	1.65	1.85

Data from The Aluminum Association, *Structural Design Manual*, 1994.

The calculated strength of the part is divided by these factors. This allowable stress must be less than the stress calculated using the total load applied to the part. In LRFD, the calculated strength of the part is multiplied by the resistance factors given in Table 8.4. This calculated stress must be less than that calculated using factored loads. Equations for determining the factored loads are given in the appropriate specifications discussed previously.

Buckling Curves for Alloys

The equations for the behavior of aluminum components apply to all thicknesses of material and to all aluminum alloys. Equations for buckling in the elastic and inelastic range are provided. Figure 8.2 shows the format generally used for both component and element behavior. Strength of the component is normally considered to be limited by the yield strength of the material. For buckling behavior, coefficients are defined for two classes of alloys, those that are heat treated with temper designations -T5 or higher and those that are not heat treated or are heat treated with temper

TABLE 8.4 Resistance Factors for LRFD

Component	Limit state	Buildings	Bridges
Tension	Yielding	0.95	0.90
	Ultimate strength	0.85	0.75
Columns	Buckling	Varies with slenderness ratio	Varies with slenderness ratio
Beams	Tensile yielding	0.95	0.90
	Tensile ultimate	0.85	0.80
	Compressive yielding	0.95	0.90
	Lateral buckling	0.85	0.80
Thin plates in compression	Yielding	0.95	0.90
	Ultimate strength	0.85	0.80
Stiffened flat webs in shear	Yielding	0.95	0.90
	Buckling	0.90	0.80

Buildings data from The Aluminum Association, *Structural Design Manual,* 1994.
Bridges data from American Association of State Highway and Transportation Officials,
AASHTO LRFD Bridge Design Specifications, 1994.

designations -T4 or lower. Different coefficients are needed because of the differences in the shapes of the stress-strain curves for the two classes of alloys.

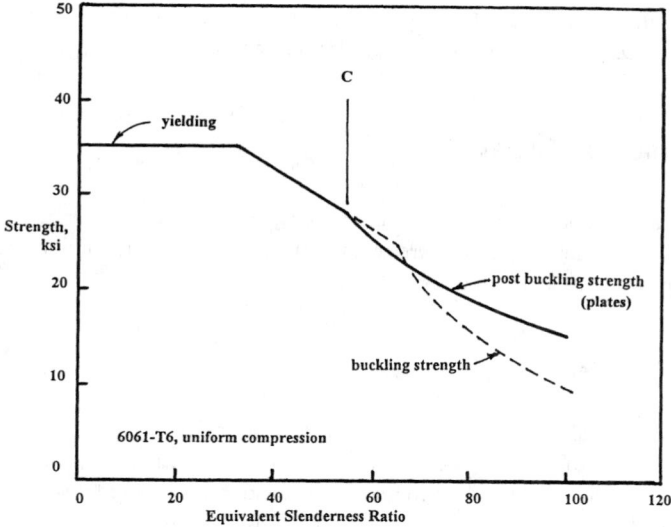

Figure 8.2 Buckling of components.

Effects of Welding

In most applications some efficiency is obtained by using alloys that have been thermally treated or strain hardened to achieve higher strength. The alloys are readily welded. However, welding partially anneals a narrow band of material (about 1.0 in. on either side of the weld) and thus this *heat affected* material has a lower strength than the rest of the member. The lower strength is accounted for in the design equations presented later.

If the strength of the heat affected material is less than the yield strength of the parent material, the plastic deformation of the component at failure loads will be confined to that narrow band of lower strength material. In this case the component fails with only a small total deformation, thus exhibiting low structural toughness. For good structural toughness the strength of the heat-affected material should be well above the yield strength of the parent material. In the case of liquid natural gas containers, an annealed temper of the plate, 5083-0, has been employed to achieve maximum toughness. The strength of the welded material is the same as that of the parent material and there is essentially no effect of welding on structural behavior.

Effects of Temperature

All of the properties important to structural behavior (static strength, elongation, fracture toughness, and fatigue strength) increase with decrease in temperature. Elongation increases but static and fatigue strength decrease at elevated temperatures. Alloys behave differently but significant changes in mechanical properties can occur at temperatures over 300°F.

Effects of Strain Rate

Aluminum alloys are relatively insensitive to strain rate. There is some increase in mechanical properties at high strain rates. Thus, published strength data based on conventional tests are normally used for calculations for cases of rapidly applied loads.

8.2.2 Component Behavior

This section presents equations and discussion for determining the strength of various types of aluminum components. These equations are consistent with those employed in current specifications and publications for aluminum structures.

Members in Tension

Because various alloys have different amounts of strain hardening both yielding and fracture strength of members should be checked. The net section of the member is used in the calculation. The calculated net section stress is compared to the yield strength and tensile strength of the alloy as given in Table 8.2. Larger factors of safety are applied for ultimate rather than yield strength in both ASD and LRFD specifications as noted in Tables 8.3 and 8.4.

The strengths across a groove weld are given for some alloys in Table 8.5. These properties are used for the design of tension members with transverse welds that affect the entire cross-section. For members with longitudinal welds in which only part of the cross-section is affected by welds, the tensile or yield strength may be calculated using the following equation:

$$F_{pw} = F_n - \frac{A_w}{A}(F_n - F_w) \qquad (8.1)$$

TABLE 8.5 Minimum Strengths of Groove Welds

Parent material	Filler metal	Tension TS[a]	Tension YS[b]	Compression YS[b]	Shear US
3003-H14	1100	14	7	7	10
5456-H116	5556	42	26	24	25
6061-T6	5356	24	20	20	15
6061-T6	4043	24	15	15	15
6063-T5,-T6	4043	17	11	11	11

Data from The Aluminum Association, *Structural Design Manual,* 1994.
Note: All strengths are in ksi. TS is tensile strength, YS is yield strength, and US is ultimate strength.
[a] ASME weld-qualification values. The design strength is considered to be 90% of these values.
[b] Corresponds to a 0.2% set on a 10-in. gage length.

where

F_{pw} = strength of member with a portion of cross-section affected by welding
F_n = strength of unaffected parent metal
F_w = strength of material affected by welding
A = area of cross-section
A_w = area that lies within 1 in. of a weld

Columns Under Flexural Buckling

The Euler column formula is employed for the elastic region and straight line equations in the inelastic region. The straight line equations are a close approximation to the tangent modulus column curve. The equations for column strength are as follows:

$$F_c = B_c - D_c \frac{KL}{r} \qquad \frac{KL}{r} \leq C_c \tag{8.2}$$

$$F_c = \frac{\pi^2 E}{(KL/r)^2} \qquad \frac{KL}{r} > C_c \tag{8.3}$$

where

F_c = column strength, ksi
L = unsupported length of column, in.
r = radius of gyration, in.
K = effective-length factor
E = modulus of elasticity, ksi
B_c, D_c, C_c = constants depending on mechanical properties (see below)

For wrought products with tempers starting with -O, -H, -T1, -T2, -T3, and -T4, and cast products,

$$B_c = F_{cy} \left[1 + \left(\frac{F_{cy}}{1000} \right)^{1/2} \right] \tag{8.4}$$

$$D_c = \frac{B_c}{20} \left(\frac{6B_c}{E} \right)^{1/2} \tag{8.5}$$

$$C_c = \frac{2B_c}{3D_c} \tag{8.6}$$

For wrought products with tempers starting with -T5, -T6, -T7, -T8, and -T9,

$$B_c = F_{cy} \left[1 + \left(\frac{F_{cy}}{2250} \right)^{1/2} \right] \tag{8.7}$$

$$D_c = \frac{B_c}{10} \left(\frac{B_c}{E} \right)^{1/2} \tag{8.8}$$

$$C_c = 0.41 \frac{B_c}{D_c} \tag{8.9}$$

where F_{cy} = compressive yield strength, ksi

The column strength of a welded member is generally less than that of a member with the same cross-section but without welds. If the welds are longitudinal and affect part of the cross-section, the column strength is given by Equation 8.1. The strengths in this case are column buckling values assuming all parent metal and all heat-affected metal. If the member has transverse welds that affect the entire cross-section, and occur away from the ends, the strength of the column is calculated assuming that the entire column is heat-affected material. Note that the constants for the heat-affected material are given by Equations 8.4, 8.5, and 8.6. If transverse welds occur only at the ends, then the equations for parent metal are used but the strength is limited to the yield strength across the groove weld.

Columns Under Flexural-Torsional Buckling

Thin, open sections that are unsymmetrical about one or both principal axes may fail by combined torsion and flexure. This strength may be estimated using a previously developed equation that relates the combined effects to pure flexural and pure torsional buckling of the section. The equation below is in the form of effective and equivalent slenderness ratios and is in good agreement with test data [3]. The equation must be solved by trial for the general case.

$$\left[1 - \left(\frac{\lambda_c}{\lambda_y} \right)^2 \right] \left[1 - \left(\frac{\lambda_c}{\lambda_x} \right)^2 \right] \left[1 - \left(\frac{\lambda_c}{\lambda_\Phi} \right)^2 \right]$$
$$- \left(\frac{y_o}{r_o} \right)^2 \left[1 - \left(\frac{\lambda_c}{\lambda_x} \right)^2 \right] - \left(\frac{x_o}{r_o} \right)^2 \left[1 - \left(\frac{\lambda_c}{\lambda_y} \right)^2 \right] = 0 \tag{8.10}$$

where

λ_c	=	equivalent slenderness ratio for flexural-torsional buckling
λ_x, λ_y	=	slenderness ratios for flexural buckling in the x and y directions, respectively
x_o, y_o	=	distances between centroid and shear center, parallel to principal axes
r_o	=	$[(I_{xo} + I_{yo})/A]^{1/2}$
I_{xo}, I_{yo}	=	moments of inertia about axes through shear center
λ_Φ	=	equivalent slenderness ratio for torsional buckling

$$\lambda_\phi = \sqrt{ \frac{I_x + I_y}{ \frac{3J}{8\pi^2} + \frac{C_w}{(K_\Phi L)^2} } } \tag{8.11}$$

where

J	=	torsion constant
C_w	=	warping constant
K_ϕ	=	effective length coefficient for torsional buckling
L	=	length of column
I_x, I_y	=	moments of inertia about the centroid (principal axes)

Beams

Beams that are supported against lateral-torsional buckling fail by excessive yielding or fracture of the tension flange at bending strengths above that corresponding to stresses reaching the tensile or yield strength at the extreme fiber. This additional strength may be accounted for by applying a shape factor to the tensile or yield strength of the alloy. Nominal shape factors for some aluminum shapes are given in Table 8.6. These factors vary slightly with alloy because they are affected by the shape of the stress-strain curve but the values shown are reasonable for all aluminum alloys.

This higher bending strength can be developed provided that the cross-section is compact enough so that local buckling does not occur at a lower stress. Limitations on various types of elements are given in Table 8.7. The bending moment for compact sections is as follows.

$$M = ZSF \qquad (8.12)$$

where

M = moment corresponding to yield or ultimate strength of the beam
S = section modulus of the section
F = yield or tensile strength of the alloy
Z = shape factor

TABLE 8.6 Shape Factors for Aluminum Beams

Cross-section	Yielding, K_y	Ultimate, K_u
I and channel (major axis)	1.07	1.16
I (minor axis)	1.30	1.42
Rectangular tube	1.10	1.22
Round tube	1.17	1.24
Solid rectangle	1.30	1.42
Solid round	1.42	1.70

Data from Gaylord and Gaylord, *Structural Engineering Handbook,* McGraw-Hill, New York, 1990, and The Aluminum Association, *Structural Design Manual,* 1994.

TABLE 8.7 Limiting Ratios of Elements for Plastic Bending

Element	Limiting ratio
Outstanding flange of I or channel	$b/t \le 0.30(E/F_{cy})^{1/2}$
Lateral buckling of I or channel	
Uniform moment	$L/r_y \le 1.2(E/F_{cy})^{1/2}$
Moment gradient	$L/r_y \le 2.2(E/F_{cy})^{1/2}$
Web of I or rectangular tube	$b/t \le 0.45(E/F_{cy})^{1/2}$
Flange of rectangular tube	$b/t \le 1.13(E/F_{cy})^{1/2}$
Round tube	$D/t \le 2.0(E/F_{cy})^{1/2}$

Data from Gaylord and Gaylord, *Structural Engineering Handbook,* McGraw-Hill, New York, 1990.

Effects of Joining

If there are holes in the tension flange, the net section should be used for calculating the section modulus. Welding affects beam strength in the same way as it does tensile strength. The groove weld strength is used when the entire cross-section is affected by welds. Beams may not develop the bending strength as given by Equation 8.12 at the locations of the transverse welds. In these locations it is reasonable to use a shape factor equal to 1.0. If only part of the section is affected by welds, Equation 8.1 is used to calculate strength and compact sections can develop the moment as given by Equation 8.12. In the calculation the flange is considered to be the area that lies farther than 2/3 of the distance between the neutral axis and the extreme fiber.

Lateral Buckling

Beams that do not have continuous support for the compression flange may fail by lateral buckling. For aluminum beams an equivalent slenderness is defined and substituted in column formulas in place of KL/r. The slenderness ratios for buckling of I-sections, WF-shapes, and channels are as follows.

For beams with end moments only or transverse loads applied at the neutral axis:

$$\lambda_b = 1.4 \frac{L_b}{\sqrt{\frac{I_y d C_b}{S_c} \sqrt{1.0 + 0.152 \frac{J}{I_y} \left(\frac{L_b}{d}\right)^2}}} \tag{8.13}$$

For beams with loads applied to top and bottom flanges where the load is free to move laterally with the beam:

$$\lambda_b = 1.4 \frac{L_b}{\sqrt{\frac{I_y d C_b}{S_c} \left[\pm 0.5 + \sqrt{1.25 + 0.152 \frac{J}{I_y} \left(\frac{L_b}{d}\right)^2}\right]}} \tag{8.14}$$

where

λ_b	=	equivalent slenderness ratio for beam buckling (to be used in place of KL/r in the column formula)
C_b	=	coefficient depending on loading and beam supports = $12.5 M_{max}/(2.5 M_{max} + 3 M_A + 4 M_B + 3 M_C)$ for simple supports
M_{max}	=	absolute value of maximum moment in the unbraced beam segment
M_A	=	absolute value of moment at quarter-point of the unbraced beam segment
M_B	=	absolute value of moment at midpoint of the unbraced beam segment
M_C	=	absolute value of moment at three-quarter-point of the unbraced beam segment
d	=	depth of beam
I_y	=	moment of inertia about axis parallel to web
S_c	=	section modulus for compression flange
J	=	torsion constant

The plus sign is to be used in Equation 8.14 if the load acts on the bottom (tension) flange, the minus sign if it acts on the top (compression) flange.

Equations 8.13 and 8.14 may also be used for cantilever beams of the specified cross-section by the use of the appropriate factor C_b. For a concentrated load at the end, the factor is 1.28 and for a uniform lateral load, the factor is 2.04.

These equations also may be applied to I-sections in which the tension and compression flanges are of somewhat different size. In this case the beam properties are calculated as though the tension flange is the same size as the compression flange. The depth of the section is maintained.

Lateral buckling strengths of welded beams are affected similarly to that of flexural buckling of columns. For cases in which part of the compression flange has heat-affected material, Equation 8.1 is used. The total flange area is that farther than 2/3 the distance from the neutral axis to the extreme fiber. If the beam has transverse welds away from the supports, the strength of the beam is calculated as though the entire beam is of heat-affected material.

For other types of cross-sections and loadings not provided for above, and for cases in which the loads cause torsional stresses in the beam, other equations and analysis are needed. Some cases are covered elsewhere [1, 3].

Members Under Combined Bending and Axial Loads

The same interaction equations may be used for aluminum as for steel members. The following equations are for bending in one direction. Both formulas must be checked.

$$\frac{f_a}{F_{ao}} + \frac{f_b}{F_b} \leq 1.0 \tag{8.15}$$

$$\frac{f_a}{F_a} + \frac{C_b f_b}{F_b(1.0 - f_a/F_e)} \leq 1.0 \tag{8.16}$$

where

f_a	=	average compressive stress from axial load
f_b	=	maximum compressive bending stress
F_a	=	strength of member as a column
F_b	=	strength of member as a beam
F_{ao}	=	strength of member as a short column
F_e	=	$\pi^2 E/(KL/r)^2$

Buckling of Thin, Flat Elements of Columns and Beams Under Uniform Compression

The elastic buckling of plates is calculated using classical plate buckling theory. For inelastic stresses, straight line formulas that approximate a secant-tangent modulus combination are used. These straight line formulas give higher stresses than those for columns that use tangent modulus, and they are in close agreement with test data. An equivalent slenderness ratio, $K_p b/t$, is utilized in the equations.

$$F_p = B_p - \frac{D_p K_p b}{t} \qquad \frac{K_p b}{t} \leq C_p \tag{8.17}$$

$$F_p = \frac{\pi^2 E}{(K_p b/t)^2} \qquad K_p \frac{b}{t} > C_p \tag{8.18}$$

where

F_p	=	buckling stress of plate, ksi
b	=	clear width of plate
t	=	thickness of plate
K_p	=	coefficient depending on conditions of edge restraint of plate (see Table 8.8)
B_p, D_p, C_p	=	alloy constants defined below

TABLE 8.8 Values of K_p for Plate Elements

Type of member	Stress distribution	Edge support	K_p
Column	Uniform compression	One edge free, one edge supported	5.1
		Both edges supported	1.6
Beam (flange)	Uniform compression	One edge free, one edge supported	5.1
		Both edges supported	1.6
Beam (web)	Varying from compression on one edge to tension on the other edge	Compression edge free, tension edge with partial restraint	3.5
		Both edges supported	0.67

Data from Gaylord and Gaylord, *Structural Engineering Handbook*, MacGraw-Hill, New York, 1990.

For wrought products with tempers starting with -0, -H, -T1, -T2, -T3, and -T4, and cast products,

$$B_p = F_{cy}\left[1 + \frac{(F_{cy})^{1/3}}{7.6}\right] \tag{8.19}$$

$$D_p = \frac{B_p}{20}\left(\frac{6B_p}{E}\right)^{1/2} \tag{8.20}$$

$$C_p = \frac{2B_p}{3D_p} \tag{8.21}$$

For wrought products with tempers starting with -T5, -T6, -T7, -T8, and -T9,

$$B_p = F_{cy}\left[1 + \frac{(F_{cy})^{1/3}}{11.4}\right] \tag{8.22}$$

$$D_p = \frac{B_p}{10}\left(\frac{B_p}{E}\right)^{1/2} \tag{8.23}$$

$$C_p = 0.41\frac{B_p}{D_p} \tag{8.24}$$

Buckling of Thin, Flat Elements of Beams Under Bending

For webs under bending loads, Equations 8.17 and 8.18 apply for buckling in the inelastic and elastic ranges. C_p is given by Equation 8.21. However, the values of B_p and D_p are higher than those of elements under uniform compression because they include a shape factor effect, the same as that defined for beams. The constants for the straight-line equation are as follows. They apply to all alloys and tempers.

$$B_p = 1.3F_{cy}\left[1 + \frac{(F_{cy})^{1/3}}{7}\right] \tag{8.25}$$

$$D_p = \frac{B_p}{20}\left(\frac{6B_p}{E}\right)^{1/2} \tag{8.26}$$

Post-Buckling Strength of Thin Elements of Columns and Beams

Most thin elements can develop strengths much higher than the elastic buckling strength as given by Equation 8.18. This higher strength is used in design. Elements of angle, cruciform, and channel (flexural buckling about the weak axis) columns may not develop post-buckling strength. Thus, the buckling strength should be used for these cases. For other cases, the post-buckling strength (in the elastic buckling region) is given as follows:

$$F_{cr} = k_2\frac{\sqrt{B_p E}}{K_p b/t} \quad \text{for} \quad b/t > \frac{k_1 B_p}{K_p D_p} \tag{8.27}$$

where

F_{cr}	=	ultimate strength of plate in compression, ksi
B_p, D_p	=	coefficients defined in Equations 8.19 through 8.26
k_1, k_2	=	coefficients. $k_1 = 0.5$ and $k_2 = 2.04$ for wrought products whose temper starts with -0, -H.-T1, -T2, -T3, and -T4, and castings. $k_1 = 0.35$ and $k_2 = 2.27$ for wrought products whose temper starts with -T5, -T6, -T7, -T8 and -T9

Weighted Average Strength of Thin Sections

In many cases a component will have elements with different calculated buckling strengths. An estimate of the component strength is obtained by equating the ultimate strength of the section multiplied by the total area to the sum of the strength of each element times its area, and solving for the ultimate strength. This weighted average approach gives a close estimate of strength for columns and for beam flanges.

Effect of Local Buckling on Column and Beam Strength

If local buckling occurs at a stress below that for overall buckling of a column or beam, the strength of the component will be reduced. Thus, the elements should be proportioned such that they are stable at column or beam buckling strengths. There are methods for taking into account local buckling on column or beam strength provided elsewhere [1, 3].

Shear Buckling of Plates

The same equations apply to stiffened and unstiffened webs. Equivalent slenderness ratios are defined for each case. Straight line equations are employed in the inelastic range and the Euler formula in the elastic range. The equations are as follows:

$$F_s = B_s - D_s \lambda_s \qquad \lambda_s \leq C_s \tag{8.28}$$

$$F_s = \frac{\pi^2 E}{\lambda_s^2} \qquad \lambda_s > C_s \tag{8.29}$$

For tempers -0, -H, -T1, -T2, -T3, and -T4

$$B_s = F_{sy}\left(1 + \frac{F_{sy}^{1/3}}{6.2}\right) \tag{8.30}$$

$$D_s = \frac{B_s}{20}\left(\frac{6B_s}{E}\right)^{1/2} \tag{8.31}$$

$$C_s = \frac{2B_s}{3D_s} \tag{8.32}$$

For tempers -T5, -T6, -T7, -T8, and -T9

$$B_s = F_{sy}\left(1 + \frac{F_{sy}^{1/3}}{9.3}\right) \tag{8.33}$$

$$D_s = \frac{B_s}{10}\left(\frac{B_s}{E}\right)^{1/2} \tag{8.34}$$

$$C_s = 0.41\frac{B_s}{D_s} \tag{8.35}$$

where
F_{sy} = shear yield strength, ksi
λ_s = 1.25h/t for unstiffened webs
    = $1.25a_1/t[1 + 0.7(a_1/a_2)^2]^{1/2}$ for stiffened webs
h = clear depth of web
t = web thickness
a_1 = smallest dimension of shear panel
a_2 = largest dimension of shear panel

Web Crushing

One of the design limitations of formed sheet members in bending and thin-webbed beams is local failure of the web under concentrated loads. Also, there is interaction between the effects of the concentrated load and the bending strength of the web.

For interior loads

$$P = \frac{t^2(N + 5.4)(\sin \Theta)(0.46F_{cy} + 0.02\sqrt{EF_{cy}})}{0.4 + r(1 - \cos \Theta)} \tag{8.36}$$

where
P = maximum load on one web, kips
N = length of load, in.
F_{cy} = compressive yield strength, ksi
E = modulus of elasticity, ksi
r = radius between web and top flange, in.

t =   thickness, in.

Θ =   angle between plane of web and plane of loading (flange)

    For loads at the end of the beam

$$P = \frac{1.2t^2(N+1.3)(\sin \Theta)(0.46F_{cy} + 0.02\sqrt{EF_{cy}}}{0.4 + r(1 - \cos \Theta)} \tag{8.37}$$

If there is significant bending stresses at the point of the concentrated load, the interaction may be calculated using the following equation:

$$\left(\frac{M}{M_u}\right)^{1.5} + \left(\frac{P}{P_u}\right)^{1.5} \leq 1.0 \tag{8.38}$$

where

M =   applied moment, in.-kips

P =   applied concentrated load, kips

M_u =   maximum moment in bending, in.-kip

P_u =   web crippling load, kips

Stiffeners for Flat Plates

The addition of stiffeners to thin elements greatly improves efficiency of material. This is especially important for aluminum components because there usually is no need for a minimum thickness based on corrosion, and thus parts can be thin compared to those of steel, for example.

Stiffening Lips for Flanges

The buckling strength of the combined lip and flange is calculated using the equations for column buckling given previously and the following equivalent slenderness ratio that replaces the effective slenderness ratio. Element buckling of the flange plate and lip must also be considered.

$$\lambda = \pi \sqrt{\frac{I_p}{3/8J + 2\sqrt{C_w K_\phi / E}}} \tag{8.39}$$

where

λ =   equivalent slenderness ratio (to be used in the column buckling equations)

I_p = $I_{xo} + I_{yo}$ = polar moment of inertia of lip and flange about center of rotation, in.[4]

I_{xo}, I_{yo} =   moments of inertia of lip and flange about center of rotation, in.[4]

K_ϕ =   elastic restraint factor (the torsional restraint against rotation as calculated from the application of unit outward forces at the centroid of the combined lip and flange of a one unit long strip of the section), in.-lb/in.

J =   torsion constant, in.[4]

C_w = $b^2(I_{yc} - bt^3/12)$ = warping term for lipped flange about center of rotation, in.[6]

I_{yc} =   moment of inertia of flange and lip about their combined centroidal axis (the flange is considered to be parallel to the y-axis), in.[4]

E =   modulus of elasticity, ksi

b =   flange width, in.

t =   flange thickness, in.

Intermediate Stiffeners for Plates in Compression

Longitudinal stiffeners (oriented parallel to the direction of the compressive stress) are often used to stabilize the compression flanges of formed sheet products and can be effective for any thin element of a column or the compression flange of a beam. The buckling strength of a plate supported on both edges with intermediate stiffeners is calculated using the column buckling equations and an equivalent slenderness ratio. The strength of the individual elements, plate between stiffeners and stiffener elements, also must be evaluated.

$$\lambda = \frac{4Nb}{\sqrt{3}t} \sqrt{\frac{1 + A_s/bt}{1 + \sqrt{1 + 32I_e/3t^3b}}} \tag{8.40}$$

where

λ = equivalent slenderness ratio for stiffener that replaces the effective slenderness ratio in the column formulas

N = total number of panels into which the longitudinal stiffeners divide the plate, in.

b = stiffener spacing, in.

t = thickness of plate, in.

I_e = moment of inertia of plate-stiffener combination about neutral axis using an effective width of plate equal to b, in.[4]

A_s = area of stiffener (not including any of the plate), in.[2]

Intermediate Stiffeners for Plates in Shear (Girder Webs)

Transverse stiffeners on girder webs must be stiff enough so that they remain straight during buckling of the plate between stiffeners. The following equations are proposed for design.

For $s/h \leq 0.4$

$$I_s = \frac{0.46Vh^2}{E}(s/h) \tag{8.41}$$

For $s/h > 0.4$

$$I_s = \frac{0.073Vh^2}{E}(h/s) \tag{8.42}$$

where

I_s = moment of inertia of stiffener (about face of web plate for stiffeners on one side of web only), in.[4]

s = stiffener spacing, in.

h = clear height of web, in.

V = shear force on web at stiffener location, kips

E = modulus of elasticity, ksi

Corrugated Webs

Corrugated sheet is highly efficient in carrying shear loads. Webs of girders and roofs and side walls of buildings are practical applications. The behavior of these panels, particularly the shear stiffness, is dependant not only on the type and size of corrugation but also on the manner in which it is attached to edge members. Test information and design suggestions are published elsewhere [3]. Some of the failure modes to consider are as follows.

1. Overall shear buckling. Primarily this is a function of the size of corrugations, the length of the panel parallel to the corrugations, the attachment, and the alloy.
2. Local buckling. Individual flat or curved elements of the corrugations must be checked for buckling strength.
3. Failure of corrugations and/or fastening at the attachment to the edge framing. If not completely attached at the ends, the corrugation may roll or collapse at the supports or fastening may fail.
4. Excessive deformation. This characteristic is difficult to calculate and the best guidelines are based on test data. The shear deformation can be many times that of flat webs particularly for those cases in which the fastening at the supports is not continuous.

Local Buckling of Tubes and Curved Panels

The strength of these members for each type of loading is defined by an equation for elastic buckling that applies to all alloys and two equations for the inelastic region which are dependent on alloy and temper. The members also need to be checked for overall buckling.

Round Tubes Under Uniform Compression

The local buckling strength is given by the following equations.

$$F_t = B_t - D_t \sqrt{\frac{R}{t}} \qquad \frac{R}{t} \le C_t \tag{8.43}$$

$$F_t = \frac{\pi^2 E}{16(R/t)\left(1 + \frac{\sqrt{R/t}}{35}\right)^2} \qquad \frac{R}{t} > C_t \tag{8.44}$$

where
F_t = buckling stress for round tube in end compression, ksi
R = mean radius of tube, in.
t = thickness of tube, in.
C_t = intersection of equations for elastic and inelastic buckling (determined by charting or trial and error)

The values of the constants, B_t and D_t, are given by the following formulas.

For wrought products with tempers starting with -O, -H, -T1, -T2, -T3, and -T4, and cast products,

$$B_t = F_{cy}\left(1 + \frac{F_{cy}^{1/5}}{5.8}\right) \tag{8.45}$$

$$D_t = \frac{B_t}{3.7}\left(\frac{B_t}{E}\right)^{1/3} \tag{8.46}$$

For wrought products with tempers starting with -T5, -T6, -T7, -T8, and -T9,

$$B_t = F_{cy}\left(1 + \frac{F_{cy}^{1/5}}{8.7}\right) \tag{8.47}$$

$$D_t = \frac{B_t}{4.5}\left(\frac{B_t}{E}\right)^{1/3} \tag{8.48}$$

where
F_{cy} = compressive yield strength, ksi

For welded tubes, Equations 8.45 and 8.46 are used along with the yield strength for welded material. The accuracy of these equations has been verified for tubes with circumferential welds and R/t ratios equal to or less than 20. For tubes with much thinner walls, limited tests show that much lower buckling strengths may occur [3].

Round Tubes and Curved Panels Under Bending

For curved elements of panels under bending, such as corrugated sheet, the local buckling strength of the compression flange may be determined using the same equations as given in the preceding section for tubes under uniform compression.

In the case of round tubes under bending a higher compressive buckling strength is available for low R/t ratios due to the shape factor effect. (Tests have indicated that this higher strength is not developed in curved panels.) The equations for tubes in bending for low R/t are given later. Note that the buckling of tubes in bending is provided by two equations in the inelastic region, that defined below and that for intermediate R/t ratios, which is the same as that for uniform compression, and the equation for elastic behavior, which also is the same as that for tubes under uniform compression.

$$F_{tb} = B_{tb} - D_{tb}\sqrt{R/t} \qquad R/t \leq C_{tb} \qquad (8.49)$$

where

F_{tb} = buckling stress for round tube in bending, ksi
R = mean radius of tube, in.
t = thickness of tube, in.
C_{tb} = $[(B_{tb} - B_t)/(D_{tb} - D_t)]^2$, intersection of curves, Equations 8.43 and 8.49

The values of the constants, B_{tb} and D_{tb}, are given by the following formulas.

For wrought products with tempers starting with -0, -H, -T1, -T2, -T3, and -T4, and cast products,

$$B_{tb} = 1.5F_y \left(1 + \frac{F_y^{1/5}}{5.8}\right) \qquad (8.50)$$

$$D_{tb} = \frac{B_{tb}}{2.7}\left(\frac{B_{tb}}{E}\right)^{1/3} \qquad (8.51)$$

For wrought products with tempers starting with -T5, -T6, -T7, -T8, and -T9,

$$B_{tb} = 1.5F_y \left(1 + \frac{F_y^{1/5}}{8.7}\right) \qquad (8.52)$$

$$D_{tb} = \frac{B_{tb}}{2.7}\left(\frac{B_{tb}}{E}\right)^{1/3} \qquad (8.53)$$

where
F_y = tensile or compressive yield strength, whichever is lower, ksi

Round Tubes and Curved Panels Under Torsion and Shear

Thin walled curved members can buckle under torsion. Long tubes are covered in specifications [1] and provisions for stiffened and unstiffened cases are provided elsewhere [3].

8.2.3 Joints

Mechanical Connections

Aluminum components are joined by aluminum rivets, aluminum and steel (galvanized, aluminized, or stainless) bolts, and clinches. The joints are normally designed as bearing-type connections because the as-received surfaces of aluminum products have a low coefficient of friction and slip often occurs at working loads. Some information has been developed for the amount of roughening of the surfaces and the limiting thicknesses of material for designing a friction-type joint, although current U.S. specifications do not cover this type of design.

Table 8.9 presents strength data for a few of the rivet and bolt alloys available. Rivets are not recommended for applications that introduce large tensile forces on the fastener. The joints are proportioned based on the shear strength of the fastener and the bearing strength of the elements being joined. The bearing strengths apply to edge distances equal to at least twice the fastener diameter, otherwise reduced values apply. Steel bolts are often employed in aluminum structures. They are generally stronger than the aluminum bolts, and may be required for pulling together parts during assembly. They also have high fatigue strength, which is important in applications in which the fastener is subject to cyclic tension. The steel bolts must be properly coated or be of the 300 series stainless to avoid galvanic corrosion between the aluminum elements and the fastener.

TABLE 8.9   Strengths of Aluminum Bolts and Rivets

Alloy and temper	Minimum expected strength, ksi	
	Shear	Tension on net area
Rivets		
6053-T61	20	—
6061-T6	25	—
Bolts		
2024-T4	37	62
6061-T6	25	42
7075-T73	41	68

Data from The Aluminum Association, *Structural Design Manual*, 1994.

Thin aluminum roofing and siding products are commonly used in the building industry. One failure mode is the pulling of the sheathing off of the fastener due to uplift forces from wind. The pull-through strength is a function of the strength of the sheet, the geometry of the product, the location of the fastener, the hole diameter, and the size of the head of the fastener [3].

Welded Connections

The aluminum alloys employed in most nonaerospace applications are readily welded, and many structures are fabricated with this method of joining. Transverse groove weld strengths and appropriate filler alloys for a few of the alloys are given in Table 8.5. Because the weld strengths are usually less than those of the base material, the design of aluminum welded structures is somewhat different from that of steel structures. Techniques for designing aluminum components with longitudinal and transverse welds are provided in the preceding section. If the welds are inclined to the direction of stress, neither purely longitudinal or transverse, the strength of the connection more closely approximates that of the transversely welded case.

Fillet weld strengths are given in Table 8.10. Two categories are defined, longitudinal and transverse. These strengths are based on tests of specimens in which the welds were symmetrically placed and had no large bending component. In the case of longitudinal fillets, the welds were subjected to primarily shear stresses. The transverse fillet welds carried part of the load in tension. The difference in stress states accounts for the higher strengths for transverse welds. Aluminum specifications utilize the values for longitudinal welds for all orientations of welds because many types of transverse fillet welds cannot develop the strengths shown due to having a more severe stress state than the test specimens, e.g., more bending stress. Proportioning of complex fillet weld configurations is done using structural analysis techniques appropriate for steel and other metals.

TABLE 8.10   Minimum Shear Strengths of Fillet Welds

Filler alloy	Shear strength, ksi	
	Longitudinal	Transverse
4043[a]	11.5	15
5356	17	26
5554	17	23
5556	20	30

[a] Naturally aged (2 to 3 months)
Data from Sharp, *Behavior and Design of Aluminum Structures*, McGraw-Hill, New York, 1993.

Adhesive Bonded Connections

Adhesive bonding is not used as the only joining method for main structural components of nonaerospace applications. It is employed in combination with other joining methods and for secondary members. Although there are many potential advantages in the performance of adhesive joints compared to those for mechanical and welded joints, particularly in fatigue, there are too many uncertainties in design to use them in primary structures. Some of the problems in design are as follows.

1. There are no specific adhesives identified for general structures. The designer needs to work with adhesive experts to select the proper one for the application.
2. In order to achieve long-term durability proper pretreatment of the metal is required. There are little data available for long-term behavior, so the designer should supplement the design with durability tests.
3. There is no way to inspect for the quality of the joint. Proper quality control of the joining process should result in good joints. However, a mistake can result in very low strengths, and the bad joint cannot be detected by inspection.
4. There are calculation procedures for proportioning simple joints in thin materials. Techniques for designing complex joints of thicker elements are under development, but are not adequate for design at this time.

8.2.4   Fatigue

Fatigue is a major design consideration for many aluminum applications, e.g., aircraft, cars, trucks, railcars, bridges, and bridge decks. Most field failures of metal structures are by fatigue. The current design method used for all specifications, aluminum and steel structures, is to define categories of details that have essentially the same fatigue strength and fatigue curves for each of these categories. Smooth components, bolted and riveted joints, and welded joints are covered in the categories. For a new detail the designer must select the category that has a similar local stress. Chapter 24 of this Handbook provides details of this method of design.

Many of the unique characteristics related to the fatigue behavior of aluminum components have been summarized [5]. Some general comments from this reference follow.

1. Some cyclic loads, such as wind induced vibration and dynamic effects in forced vibration, are nearly impossible to design for because stresses are high and the number of cycles build up quickly. These loads must be reduced or eliminated by design.
2. Good practice to eliminate known features of structures causing fatigue, such as sharp notches and high local stresses due to concentrated loads, should be employed in all cases. In some applications the load spectrum is not known, e.g., light poles, and fatigue resistant joints must be employed.
3. The fatigue strength of aluminum parts is higher at low temperature and lower at elevated temperature compared to that at room temperature.
4. Corrosion generally does not have a large effect on the fatigue strength of welded and mechanically fastened joints but considerably lowers that of smooth components. Protective measures, such as paint, improve fatigue strength in most cases.
5. Many of the joints for aluminum structures are unusual in that they are quite different from those of the fatigue categories provided in the specifications. Stress analysis to define the critical local stress is useful in these cases. Test verification is desirable if practical.

8.3   Design

Aluminum should be considered for applications in which life cycle costs are favorable compared to competing materials. The costs include:

1. Acquisition, refining, and manufacture of the metal
2. Fabrication of the metal into a useful configuration
3. Assembly and erection of the components in the final structure
4. Maintenance and operation of the structure over its useful life
5. Disposal after the useful life

The present markets for aluminum have developed because of life cycle considerations. Transportation vehicles, one of the largest markets, with aerospace applications, aircraft, trucks, cars, and railcars, are light weight thus saving fuel costs and are corrosion resistant thus minimizing maintenance costs. Packaging, another large market, makes use of close loop recycling that returns used cans to rolling mills that produce sheet for new cans. Building and infrastructure uses were developed because of the durability of aluminum in the atmosphere without the need for painting, thus saving maintenance costs.

8.3.1   General Considerations

Product Selection

Most aluminum structures are constructed of flat rolled products, sheet and plate, and extrusions because they provide the least cost solution. The properties and quality of these products are guaranteed by producers. The flat rolled products may be bent or formed into shapes and joined to make the final structure. Extrusions should be considered for all applications requiring constant section members. Most extruders can supply shapes whose cross-section fits within a 10-in. circle. Larger shapes are made by a more limited number of manufacturers and are more costly. Extrusions are attractive for use because the designer can incorporate special features to facilitate joining, place material in the section to optimize efficiency, and consolidate number of parts (compared with fabricated sheet parts). Because die costs are low the designer should develop unique shapes for most applications.

Forgings are generally more expensive than extrusions and plate, and are employed in aerospace applications and wheels, where the three-dimensional shape and high performance and quality are essential. Castings are also used for three-dimensional shapes, but the designer must work with the supplier for design assistance.

Alloy Selection

For extrusions alloy 6061 is best for higher strength applications and 6063 is preferred if the strength requirements are less. 5XXX alloys have been extruded and have higher as-welded strength and ductility in structures but they are generally much more expensive to manufacture, compared to the 6XXX alloys.

6061 sheet and plate are also available and are used for many applications. For the highest as-welded strength, the 5XXX alloys are employed.

Table 8.11 shows alloys that have been employed in some applications. Choice of specific alloy depends on cost, strength, formability, weldability, and finishing characteristics.

Corrosion Resistance

Alloys shown in Table 8.11, 3XXX, 5XXX, and 6XXX, have high resistance to general atmospheric corrosion and can be employed without painting. Tests of small, thin specimens of these

TABLE 8.11 Selection of Alloy

Application	Specific use	Alloys
Architecture		
Sheet	Curtain walls, roofing and siding, mobile homes	3003, 3004, 3105
extrusions	Window frames, railings, building frames	6061, 6063
Highway		
Plate	Signs, bridge decks	5086, 5456, 6061
extrusions	Sign supports, lighting standards, bridge railings	6061, 6063
Industrial		
Plate	Tanks, pressure vessels, pipe	3003, 3004, 5083, 5086, 5456, 6061
Transportation		
Sheet/plate	Automobiles, trailers, railcars, shipping containers, boats	5052, 5083, 5086, 6061, 6009
extrusions	Stiffeners/framing	6061
Miscellaneous extrusions	Scaffolding, towers, ladders	6061, 6063

Data from Gaylord and Gaylord, *Structural Engineering Handbook,* McGraw-Hill, New York, 1990.

alloys in a seacoast or industrial environment for over 50 years of exposure have shown that the depth of attack is small and self-limiting. A hard oxide layer forms on the surface of the component, which prevents significant additional corrosion.

If aluminum components are attached to steel components, protective measures must be employed to prevent galvanic corrosion. These measures include painting the steel components and placing a sealant in the joint. Stainless steel or galvanized fasteners are also required.

Some of the 5XXX alloys with magnesium content over about 3% may be sensitized by sustained elevated temperatures and lose their resistance to corrosion. For these applications alloys 5052 and 5454 may be used.

Metal Working

All of the usual fabrication processes can be used with aluminum. Forming capabilities vary with alloy. Special alloys are available for automotive applications in which high formability is required. Aluminum parts may be machined, cut, or drilled and the operations are much easier to accomplish compared to steel parts.

Finishing

Aluminum structures may be painted or anodized to achieve a color of choice. These finishes have excellent long-term durability. Bright surfaces also may be accomplished by mechanical polishing and buffing.

8.3.2 Design Studies

Some specific design examples follow in which product form, alloy selection, and joining method are discussed. The Aluminum Association Specifications are used for calculations of component strength.

EXAMPLE 8.1: Lighting Standard

Design Requirements

1. Withstand wind loads for area
2. Fatigue and vibration resistant

3. Heat treatable after welding to achieve higher strength

4. Base that breaks away under vehicle impact

Alloy and Product

Round, extruded tubes of 6063-T4 are selected for the shaft. This alloy is easily extruded and has low cost and excellent corrosion resistance. The -T4 temper is required so that the pole can be tapered by a spinning operation, and so that the structure can be heat treated and aged after welding. A permanent mold casting of 356-T6 is selected for the base. The shaft extends through the base. This base may be acceptable for break away characteristics. If not, a break away device must be employed.

Joining

MIG circumferential welds are made at the top and bottom of the base using filler alloy 4043. This filler alloy must be employed because of the heat treat operation after welding. The corrosion resistance of a 5XXX filler alloy may be lowered by the heat treatment and aging.

Design Considerations

Wind induced vibration occasionally can be a problem. The vibration involves both the standard and luminare. Currently there is no accurate way to predict whether or not these structures will vibrate. Light pole manufacturers have dampers that they can use if necessary.

Calculation Example-Bending of Welded Tube

Determine the bending strength of a 8 in. diameter(outside) X 0.313 in. wall tube of 6063-T4, heat treated and aged after welding using ASD. Factors of safety corresponding to building type structures apply.

For this special case of fabrication, the specifications allow the use of allowable stresses for the welded construction equal to 0.85 times those for 6063-T6. Also, the allowable stresses can be increased 1/3 for wind loading.

1. The allowable tensile stress (tensile properties are given in Table 8.2, shape factors in Table 8.6, and factors of safety in Table 8.3) is as follows.
   Tensile strength: $F_{tu} = 0.85(1.24)(1.33)(30)/(1.95) = 21.6$ ksi
   Yield strength: $F_{ty} = 0.85(1.17)(1.33)(25)/(1.65) = 20.0$ ksi

2. The allowable compressive strength is given by Equations 8.43 to 8.53.
   $R/t = (4.0 - 0.313)/(0.313) = 11.8$. Equation 8.49 applies because R/t is less than $69.6(C_b t)$. Constants are determined from Equations 8.52 and 8.53.
   $F_{tb} = (0.85)(1.33)(45.7 - 2.8\sqrt{(R/t)})/1.65 = 24.7$ ksi

3. The lower of the three values, 20.0 ksi is used for design. This bending stress must be less than that calculated from all loads.

EXAMPLE 8.2:   Overhead Sign Truss

Design Requirements

1. Withstand wind loads for locality (signs and truss are considered).

2. Prevent wind induced vibration of truss and members.

3. Provide structure that does not need painting.

Alloy and Product

Extruded tubes of 6061-T651 are selected for the truss and end supports. This alloy is readily welded and has excellent corrosion resistance. It also is one of the lower cost extrusion alloys.

Joining

The individual members will be machined at the ends to fit closely with other parts and welded together using the MIG process. 5356 filler wire is specified to provide higher fillet weld strength compared to that for 4043 filler.

Design Considerations

Wind induced vibration must be prevented in these structures. The trusses are particularly susceptible to the wind when they do not have signs installed. Vibration of the entire truss can be controlled by the addition of a suitable damper (at midspan) and individual members must be designed to prevent vibration by limiting their slenderness ratio [3].

Calculation Example-Buckling of a Tubular Column with Welds at Ends

The diagonal member of the truss is a 4-in. diameter tube (outside diameter) of 6061-T651 with a wall thickness of 0.125 in. The radius of gyration is 1.37. Its length is 48 in. and it is welded at each end to chords using filler 5356. Use ASD factors of safety corresponding to bridge structures (Table 8.3). Assume that the effective length factor is 1.0. Allow 1/3 increase in stress because of wind loading.

$$KL/r = (1.0)(48)/1.37 = 35.0$$

For column buckling, Equation 8.2 applies ($KL/r \leq C_c$). The constants are calculated from Equations 8.7, 8.8, and 8.9 and parent metal properties (Table 8.2).

$$F_c = (1.33)(39.4 - 0.246(35.0))/2.20 = 18.6 \text{ ksi}$$

For yielding at the welds (the entire cross-section is affected at the ends), the properties in Table 8.5 are employed.

$$F_c = (1.33)(20)/1.85 = 14.4 \text{ ksi}$$

The allowable stress is the lower of these values, 14.4 ksi.

Calculation Example-Tubular Column with Welds at Ends and Midlength

This is the same construction as described above, except that the designer has specified that a bracket be circumferentially welded to the tube at midlength. This weld lowers the column buckling strength. The column is now designed as though all the material is heat-affected. Equation 8.2 still applies but constants are now calculated using Equations 8.4, 8.5, and 8.6 and properties from Table 8.5.

$$F_c = (1.33)[22.8 - (1.0)(35.0)(0.133)]/2.20 = 11.0 \text{ ksi}$$

This stress is less than that calculated previously for yielding(14.4 ksi) and now governs. This stress must be higher than that calculated using the total load on the structure.

EXAMPLE 8.3: Built-Up Highway Girder

Design Requirements

The loads to be used for static and fatigue strength calculations are provided in AASHTO specifications. Long time maintenance-free construction is also specified. The size of the girder is larger than the largest extrudable section.

Riveted Construction

Alloy 6061-T6 is selected for the web plate, flanges, and web stiffeners. This alloy has excellent corrosion resistance, is readily available, and has the highest strength for mechanical joining. Rivets of 6061-T6 are used for the joining because they are a good match for the parts of the girder from strength and corrosion considerations. The extrusions for the flanges and stiffeners are special sections designed to facilitate fabrication and to achieve maximum efficiency of material. A sealant is placed in the faying surfaces to enhance fatigue strength and to prevent ingress of detrimental substances.

Welded Construction

Alloy 5456-H116 is selected for the web plate and flange plate. This alloy has high as-welded strength compared to 6061-T6 and excellent corrosion resistance. Alloys 5083 and 5086 would also be satisfactory selections. The filler wire selected for MIG welding is 5556, to have high fillet weld strength.

Calculation Example: Strength of Riveted Joint

The 6061-T6 parts are assembled as received from the supplier, so that the joint must be designed as a bearing connection. Use ASD for design. The thickness of the web plate is 1/2 in. and it is attached to the legs of angle flanges (two angles) that are 3/4 in. thick. Rivets 1 in. in diameter (area is 0.785 in.^{2}) are used.

Allowable bearing load on the web (bearing area is $0.50 \times 1.0 = 0.50$ in.^{2}) for one fastener is (see Tables 8.2 and 8.3):

Based on yielding: $P = (58)(0.50)/1.85 = 15.7$ kips

Based on ultimate: $P = (88)(0.50)/2.64 = 16.7$ kips

Allowable shear load on one rivet with double shear:

Based on ultimate (see Tables 8.3 and 8.9):

$$P = (2)(.785)(25)/2.64 = 14.9 \text{ kips}$$

The allowable load per rivet is the smaller of the three values or 14.9 kips.

Calculation Example: Fatigue Life of Welded Girder with Longitudinal Fillet Welds

The allowable tensile strength for a 5456-H116 girder is (see Tables 8.2 and 8.3):

Based on yield: $F = 33/1.85 = 17.8$ ksi (governs)

Based on ultimate: $F = 46/2.2 = 20.9$ ksi

Calculate the number of cycles that the girder can sustain at a stress range corresponding to a stress of 1/2 the static design value (8.9 ksi). Category B applies to a connection with the fillet weld parallel to the direction of stress. For this category, the fatigue strength is:

The stress range: $S = 130N^{-.207} = 8.9$ ksi

The number of cycles: $N = 423,000$ cycles

(Fatigue equations from [1].)

Calculation Example: Intermediate Stiffeners

Stiffeners on girder webs must be of sufficient size to remain straight when the web buckles. Stiffener sizes are given by Equations 8.41 and 8.42.

EXAMPLE 8.4:   Roofing or Siding for a Building

Design Requirements

1. Withstand wind loads(uplift as well as downward pressure)
2. Withstand concentrated loads from foot pressure or from reactions at supports
3. Corrosion resistant so that painting is not needed

Alloy and Product

Sheet of alloy 3004-H14 is selected. This alloy and temper have sufficient formability to roll-form the trapezoidal shape desired. They also have excellent corrosion and reasonable strength. Other 3XXX alloys would also be good choices.

Design Considerations

Attachment of the sheet panels to the supporting structure must be strong enough to resist uplift forces. The pull through strength of the sheet product as well as the fastener strength are considered. Sufficient overlap of panels and fasteners at laps are needed for watertightness.

Calculation Example: Web Crushing Load at an Intermediate Support

Consider the shape shown in Figure 8.3. The bearing length is 2 in. (width of flange of support). Use LRFD specifications for buildings. The material properties for 3004-H14 are given

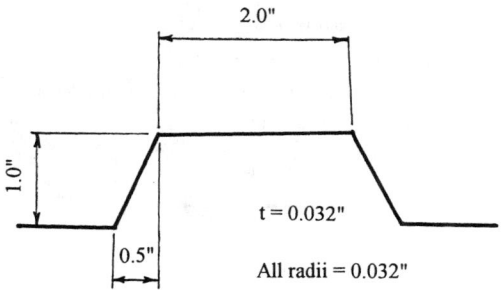

Figure 8.3   Example 8.4.

in Table 8.2 and the resistance factors are in Table 8.4. For an interior load use Equation 8.36. $\phi = 0.90$ for this case, the same as that for web buckling.

$$\phi P = \frac{(0.90)(0.032)^2(2.0 + 5.4)(0.866)(0.46 \times 14.0 + 0.02(10100 \times 14.0)^{1/2})}{0.4 + 0.032(1 - 0.5)}$$

$$= 0.198 \text{ kips per web}$$

This load must be higher than that calculated using the factored loads. (Equations for factored loads are given in the [1].)

Calculation Example: Bending Strength of Section

To calculate the section strength, the strength of the flange under uniform compression and the strength of the web under bending are calculated separately, and then combined using a weighted average calculation. The area of the web used in the calculation is that area beyond 2/3 of the distance from the neutral axis. The resistance factor for the strength calculations from Table 8.4 is 0.85. The radii are neglected in subsequent calculations so that plate widths are to the intersection point of elements. The width to points of tangency of the corner radii is more accurate.

Strength of Flange

Equation 8.27 governs because the b/t ratio (62.5) is larger than the b/t limit given for that equation. Values for B_p and D_p are given by Equations 8.19 and 8.20; the value of K_p is in Table 8.8.

$$\phi F_{cr} = (0.85)(2.04)(18.4 \times 10100)^{1/2}/(1.6)(62.5) = 7.5 \ \text{ksi}$$

Strength of Web

The web is in bending and has a h/t ratio of 35 so Equation 8.17 governs. Values for B_p and D_p are given by Equations 8.25 and 8.26, and the value of K_p is in Table 8.8.

$$\phi F_p = (0.85)(24.5 - (0.67)(.147)(35)) = 17.9 \ \text{ksi}$$

Strength of Section

The bending strength of the section is between that calculated for the flange and web. An accurate estimate of the strength is obtained from a weighted average calculation, which depends on the areas of the elements and the strength of each element. The area of the webs is that portion further than 2/3 of the distance from the neutral axis.

$$\phi F = [(2.0)t(7.5) + (2)(1.12)t(0.187)(17.9)]/(2.0 + 0.374)t = 9.5 \ \text{ksi}$$

This stress must be higher than that calculated using factored loads.

Calculation Example: Intermediate Stiffener

The bending strength of the section in Figure 8.3 can be increased significantly, with a small increase in material, by the addition of a formed stiffener at midwidth as illustrated in Figure 8.4. The strength of the stiffened panel is calculated using an equivalent slenderness ratio as given by

Figure 8.4   Example 8.4.

Equation 8.40 and column buckling equations. The addition of a few percent more material as illustrated in Figure 8.4 can increase section strength by over 25%.

Calculation Example: Combined Bending and Concentration Loads

The formed sheet product can experience high longitudinal compressive stresses and a high normal concentrated load at the same location such as an intermediate support. These stresses interact and must be limited as defined by Equation 8.38.

EXAMPLE 8.5:   Orthotropic Bridge Deck

Design Requirements

1. Withstand the static and impact loads as provided in an appropriate bridge design specification.
2. Withstand the cyclic loads provided in the specifications.

3. Fabricated by the use of welding.

4. Corrosion resistant so that painting is not needed.

5. Large prefabricated panels to shorten erection time.

Alloy and Product

The selection depends on the type of construction desired. Figure 8.5 is a plate reinforced by an extruded closed stiffener. This construction has been used successfully. The plate is 5456-

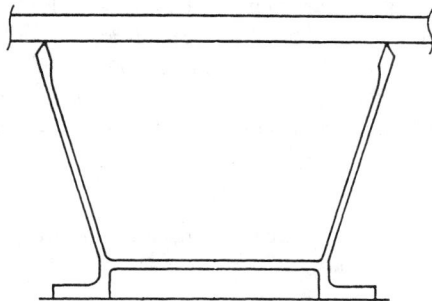

Figure 8.5   Example 8.5.

H116, chosen because of it's high as-welded strength. The extrusion is of 6061-T651 which has high strength and reasonable cost. The extrusion is designed to accommodate welding and attachment to supports to minimize fabrication costs. Both alloys have excellent corrosion resistance and will not need to be painted.

All extruded decks with segments either bolted or welded together, to achieve a shape similar to that in Figure 8.5, have also been used. 6061-T651 extrusions for all the segments are the choice in this case.

Joining

MIG welding with filler alloy 5556 is selected for attaching the extrusions to the plate. Fixturing is required to control the final shape of the panel.

Design Considerations

Large panels, 11 ft × 28 ft or larger, complete with wearing surface have been fabricated. The panels must be attached tightly to the supporting structure to avoid fatigue failures of the fasteners. Galvanized A356 bolts are suggested for the attachment to obtain high static and fatigue strengths.

Calculation Example: Bending Stresses in Plate and Section

Fatigue is the major design concern in a metal bridge deck. Wheel pressures cause bending stresses in the deck plate transverse to the direction of the stiffeners. These loads also cause longitudinal bending stresses in the stiffened panel. Fatigue evaluations are needed for both stresses. Deflection and static strength requirements of specifications also must be met.

EXAMPLE 8.6:   Ship Hull

Design Requirements

1. Withstand pressures from operation in seas, including dynamic pressures from storms.

2. Withstand stresses from bending and twisting of entire hull from storm conditions.
3. Hull is of welded construction.
4. Hull must plastically deform without fracture when impacting with another object.
5. Employ joints that are proven to be fatigue resistant in other metal ship structures.
6. Corrosion resistant so that painting is not required even for salt water exposure.

Alloy and Product

The hull is constructed of stiffened plate. Alloy 5456-H116 plate and 5456-H111 extruded stiffeners are selected. Main girders are fabricated using 5456-H116 plate. Other lower strength 5XXX alloys are also suitable choices. This alloy is readily welded and has high as-welded strength. The 5456-H111 extrusions are more costly than those of 6061-T6 but the welded 5XXX construction is much tougher using 5XXX stiffening, and thus would better accommodate damage without failure. This alloy has excellent resistance to corrosion in a salt water environment.

Joining

MIG welding using 5556 filler is specified. This is a high strength filler and is appropriate for joining parts of this high strength alloy.

Design Considerations

Loadings for hull or component design are difficult to obtain. The American Bureau of Shipping has requirements for the size of some of the components. Fireproofing is required in some areas.

Calculation Example: Buckling of Stiffened Panel

For a longitudinally framed vessel, the hull plate and stiffeners will be under compression from bending of the ship. The stiffened panel must be checked for column buckling between major transverse members using Equations 8.2 and 8.3. For hull construction that is subjected to normal pressures, the stiffened panel will have lateral bending as well as longitudinal compression. Equations 8.15 and 8.16 are needed in this case.

Elements of the stiffened panel must be checked for strength under the compression loads. In addition, an angle or T stiffener can fail by a torsion about an enforced axis of rotation, the point of attachment of the stiffener to the plate. Equation 8.39 may be used for the calculation.

EXAMPLE 8.7:   Latticed Tower or Space Frame

Design Requirements

1. Withstand wind, earthquake, and other imposed loads.
2. Corrosion resistant so that painting is not required.
3. Prevent wind induced vibration.

Alloy and Product

Extrusions of 6061-T651 are selected for the members because of their corrosion resistance, strength, and economy. 6063 extrusions can be more economical if the higher strength of 6061 is not needed. The designer should make full use of the extrusion process by designing features in the cross-section that will facilitate joining and erection, and that will result in optimal use of the material.

Joining

Mechanical fasteners are selected. Galvanized A325 or stainless steel fasteners are best for major structures because of their higher strength compared to those of aluminum.

Design Considerations

Overall buckling of the system as well as the buckling of components must be considered. The manner in which the members are attached at their ends can affect both component and overall strength.

Special extrusions in the form of angles, Y-sections and hat-sections have been used in these structures. Some of these sections can fail by flexural-torsional buckling under compressive loads. Equation 8.10 covers this case. These sections, because they are relatively flexible in torsion, can vibrate in the wind in torsion as well as flexure.

8.4   Economics of Design

There are two considerations that can affect the economy of aluminum structures: efficiency of design and life cycle costs. These considerations will be summarized briefly here.

Most structural designers are schooled in and are comfortable with design in steel. Although the design of aluminum structures is very similar to that in steel, there are differences in their basic characteristics that should be recognized.

1. The density of aluminum alloys is about 1/3 that of steel. Efficiently designed aluminum structures will weigh about 1/3 to 1/2 of those of efficiently designed steel structures, depending on failure mode. The lighter structures are governed by tensile or yield strength of the material and the heavier ones by fatigue, deflection, or buckling.
2. Modulus of elasticity of aluminum alloys is about 1/3 that of steel. The size/shape of efficient aluminum components will need to be larger for aluminum structures as compared to those of steel, for the same performance.
3. The fatigue strength of a joint of aluminum is 1/3 to 1/2 that of steel, with identical geometry. The size/shape of the aluminum component will need to be larger than that of steel to have the same performance.
4. The resistance to corrosion from the atmosphere of aluminum is much higher than that of steel. The thickness of aluminum parts can be much thinner than those of steel and painting is not needed for most aluminum structures.
5. Extrusions are used for aluminum shapes, not rolled sections as used for steel. The designer has much flexibility in the design to (1) consolidate parts, (2) include features for welding to eliminate machining, (3) include features to snap together parts or to accommodate mechanical fasteners, and (4) to include stiffeners, nonuniform thickness, and other features to provide the most efficient placement of metal. Because die costs are low, most extrusions are uniquely designed for the application.

Aluminum applications are economical generally because of life cycle considerations. In some cases, e.g., castings, aluminum can be competitive on a first cost basis compared to steel. Light weight and corrosion resistance are important in transportation applications. In this case the higher initial cost of the aluminum structure is more than offset by lower fuel costs and higher pay loads. Closed loop recycling is possible for aluminum and scrap has high value. The used beverage can is converted into sheet to make additional cans with no deterioration of properties.

8.5   Defining Terms

Alloy: Aluminum in which a small percentage of one or more other elements have been added primarily to improve strength.

Foil: Flat-rolled product that is less than 0.006 in. thick.

Heat-affected zone: Reduced strength material from welding measured 1 in. from centerline of groove weld or 1 in. from toe or heel of fillet weld.

Plate: Flat-rolled product that is greater than 0.25 in. thick.

Sheet: Flat-rolled product between 0.006 in. and 0.25 in. thick.

Temper: The measure of the characteristic of the alloy as established by the fabrication process.

References

[1] Aluminum Association. 1994. *Aluminum Design Manual,* Washington D.C.
[2] American Association of State Highway and Transportation Officials. 1994. *AASHTO LRFD Bridge Design Specifications,* Washington D.C.
[3] Sharp, M. L. 1993. *Behavior and Design of Aluminum Structures,* McGraw-Hill, New York.
[4] Sharp, M. L. 1994. Development of Aluminum Structural Technology in the United States, *Proceedings, 50th Anniversary Conference,* Structural Stability Research Council, Bethlehem, PA, 21-22 June 1994.
[5] Sharp, M. L., Nordmark, G. E., and Menzemer, C. C. 1996. *Fatigue Design of Aluminum Components and Structures,* McGraw-Hill, New York.

Further Reading

[1] Aluminum Association. 1987. *The Aluminum Extrusion Manual,* Washington D.C.
[2] American Bureau of Shipping. 1975. *Rules for Building and Classing Aluminum Vessels,* New York.
[3] American Welding Society. 1990. *ANSI/AWS D1.2-90 Structural Welding Code Aluminum,* Miami, FL.
[4] Gaylord and Gaylord. 1990. *Structural Engineering Handbook,* McGraw-Hill, New York.
[5] Kissell, J. R. and Ferry, R. L. 1995. *Aluminum Structures, A Guide to Their Specifications and Design,* John Wiley & Sons, New York.

9

Timber Structures

9.1   Introduction 9-1
         Types of Wood Products • Types of Structures • Design Specifications and Industry Resources
9.2   Properties of Wood 9-3
9.3   Preliminary Design Considerations 9-5
         Loads and Load Combinations • Design Values • Adjustment of Design Values
9.4   Beam Design 9-8
         Moment Capacity • Shear Capacity • Bearing Capacity • NDS® Provisions
9.5   Tension Member Design 9-14
9.6   Column Design 9-15
         Solid Columns • Spaced Columns • Built-Up Columns • NDS® Provisions
9.7   Combined Load Design 9-19
         Combined Bending and Axial Tension • Biaxial Bending or Combined Bending and Axial Compression • NDS® Provisions
9.8   Fastener and Connection Design 9-22
         Nails, Spikes, and Screws • Bolts, Lag Screws, and Dowels • Other Types of Connections • NDS® Provisions
9.9   Structural Panels 9-30
         Panel Section Properties • Panel Design Values • Design Resources
9.10   Shear Walls and Diaphragms.......................... 9-32
         Required Resistance • Shear Wall and Diaphragm Resistance • Design Resources
9.11   Trusses 9-33
9.12   Curved Beams and Arches 9-34
         Curved Beams • Arches • Design Resources
9.13   Serviceability Considerations 9-37
         Deflections • Vibrations • NDS® Provisions • Non-Structural Performance
9.14   Defining Terms 9-38
References 9-40
Further Reading 9-41

Kenneth J. Fridley
Department of Civil &
Environmental Engineering,
Washington State University,
Pullman, WA

9.1 Introduction

Wood is one of the earliest building materials, and as such its use often has been based more on tradition than principles of engineering. However, the structural use of wood and wood-based

0-8493-2674-5/97/$0.00+$.50
© 1997 by CRC Press LLC

materials has increased steadily in recent times. The driving force behind this increase in use is the ever-increasing need to provide economical housing for the world's population. Supporting this need, though, has been an evolution of our understanding of wood as a structural material and ability to analyze and design safe and functional timber structures. This evolution is evidenced by the recent industry-sponsored development of the *Load and Resistance Factor Design* (LRFD) *Standard for Engineered Wood Construction* [1, 5].

An accurate and complete understanding of any material is key to its proper use in structural applications, and structural timber and other wood-based materials are no exception to this requirement. This section introduces the fundamental mechanical and physical properties of wood that govern its structural use, then presents fundamental considerations for the design of timber structures. The basics of beam, column, connection, and structural panel design are presented. Then, issues related to shear wall and diaphragm, truss, and arch design are presented. The section concludes with a discussion of current serviceability design code provisions and other serviceability considerations relevant to the design of timber structures. The use of the new LRFD provisions for timber structures [1, 5] is emphasized in this section; however, reference is also made to existing allowable stress provisions [2] due to their current popular use.

9.1.1   Types of Wood Products

There are a wide variety of wood and wood-based structural building products available for use in most types of structures. The most common products include solid lumber, **glued laminated timber**, **plywood**, and orientated strand board (OSB). Solid sawn lumber was the mainstay of timber construction and is still used extensively; however, the changing resource base and shift to plantation-grown trees has limited the size and quality of the raw material. Therefore, it is becoming increasingly difficult to obtain high quality, large dimension **timbers** for construction. This change in raw material, along with a demand for stronger and more cost effective material, initiated the development of alternative products that can replace solid lumber. Engineered products such as wood composite **I-joists** and **structural composite lumber** (SCL) were the result of this evolution. These products have steadily gained popularity and now are receiving wide-spread use in construction.

9.1.2   Types of Structures

By far, the dominate types of structures utilizing wood and wood-based materials are residential and light commercial buildings. There are, however, numerous examples available of larger wood structures, such as gymnasiums, domes, and multistory office buildings. Light-frame construction is the most common type used for residential structures. Light-frame consists of nominal "2-by" lumber such as 2 × 4s (38 mm × 89 mm) up to 2 × 12s (38 mm × 286 mm) as the primary framing elements. Post-and-beam (or timber-frame) construction is perhaps the oldest type of timber structure, and has received renewed attention in specialty markets in recent years. Prefabricated panelized construction has also gained popularity in recent times. Reduced cost and shorter construction time have been the primary reasons for the interest in panelized construction. Both framed (similar to light-frame construction) and insulated (where the core is filled with a rigid insulating foam) panels are used. Other types of construction include glued-laminated construction (typically for longer spans), pole buildings (typical in so-called "agricultural" buildings, but making entry into commercial applications as well), and shell and folded plate systems (common for gymnasiums and other larger enclosed areas). The use of wood and wood-based products as only a part of a complete structural system is also quite common. For example, wood roof systems supported by masonry walls or wood floor systems supported by steel frames are common in larger projects.

Wood and wood-based products are not limited to building structures, but are also used in transportation structures as well. Timber bridges are not new, as evidenced by the number of covered bridges throughout the U.S. Recently, however, modern timber bridges have received renewed attention, especially for short-span, low-volume crossings.

9.1.3 Design Specifications and Industry Resources

The *National Design Specification for Wood Construction,* or NDS® [2], is currently the primary design specification for engineered wood construction. The NDS® is an allowable stress design (ASD) specification. As with the other major design specifications in the U.S., a *Load and Resistance Factor Design* (LRFD) *Standard for Engineered Wood Construction* [1, 5] has been developed and is recognized by all model building codes as an alternate to the NDS®. In this section, the LRFD approach to timber design will be emphasized; however, ASD requirements as provided by the NDS®, as well as other wood design specifications, also will be presented due to its current popularity and acceptance. Additionally, most provisions in the NDS® are quite similar to those in the LRFD except that the NDS® casts design requirements in terms of allowable stresses and loads and the LRFD utilizes nominal strength values and factored load combinations.

In addition to the NDS® and LRFD Standard, other design manuals, guidelines, and specifications are available. For example, the *Timber Construction Manual* [3] provides information related to engineered wood construction in general and glued laminated timber in more detail, and the *Plywood Design Specification* (PDS®)[6] and its supplements present information related to plywood properties and design of various panel-based structural systems. Additionally, various industry associations such as the APA–The Engineered Wood Association, American Institute of Timber Construction (AITC), American Forest & Paper Association–American Wood Council (AF&PA – AWC), Canadian Wood Council (CWC), Southern Forest Products Association (SFPA), Western Wood Products Association (WWPA), and Wood Truss Council of America (WTCA), to name but a few, provide extensive technical information.

One strength of the LRFD Specification is its comprehensive coverage of engineered wood construction. While the NDS® governs the design of solid-sawn members and connections, the *Timber Construction Manual* primarily provides procedures for the design of glued-laminated members and connections, and the PDS® addresses the design of plywood and other panel-based systems, the LRFD is complete in that it combines information from these and other sources to provide the engineer a comprehensive design specification, including design procedures for lumber, connections, I-joists, metal plate connected trusses, glued laminated timber, SCL, wood-base panels, timber poles and piles, etc. To be even more complete, the AF&PA has developed the *Manual of Wood Construction: Load & Resistance Factor Design* [1]. The Manual includes design value supplements, guidelines to design, and the formal LRFD Specification [5].

9.2 Properties of Wood

It is important to understand the basic structure of wood in order to avoid many of the pitfalls relative to the misuse and/or misapplication of the material. Wood is a natural, cellular, anisotropic, hyrgothermal, and viscoelastic material, and by its natural origins contains a multitude of inclusions and other defects.[1] The reader is referred to any number of basic texts that present a description of the fundamental structure and physical properties of wood as a material (e.g., [8, 11, 20]).

---

[1]The term "defect" may be misleading. Knots, grain characteristics (e.g., slope of grain, spiral grain, etc.), and other naturally occurring irregularities do reduce the effective strength of the member, but are accounted for in the grading process and in the assignment of design values. On the other hand, splits, checks, dimensional warping, etc. are the result of the drying process and, although they are accounted for in the grading process, may occur after grading and may be more accurately termed "defects".

One aspect of wood that deserves attention here, however, is the affect of moisture on the physical and mechanical properties and performance of wood. Many problems encountered with wood structures can be traced to moisture. The amount of moisture present in wood is described by the *moisture content* (MC), which is defined by the weight of the water contained in the wood as a percentage of the weight of the oven-dry wood. As wood is dried, water is first evaporated from the cell cavities. Then, as drying continues, water from the cell walls is drawn out. The **moisture content** at which *free* water in the cell cavities is completely evaporated, but the cell walls are still saturated, is termed the **fiber saturation point** (FSP). The FSP is quite variable among and within species, but is on the order of 24 to 34%. The FSP is an important quantity since most physical and mechanical properties are dependent on changes in MC below the FSP, and the MC of wood in typical structural applications is below the FSP. Finally, wood releases and absorbs moisture to and from the surrounding environment. When the wood equilibrates with the environment and moisture is not transferring to or from the material, the wood is said to have reached its **equilibrium moisture content** (EMC). Tables are available (see [20]) that provide the EMC for most species as a function of dry-bulb temperature and relative humidity. These tables allow designers to estimate in-service moisture contents that are required for their design calculations.

In structural applications, wood is typically dried to a MC near that expected in service prior to dimensioning and use. A major reason for this is that wood shrinks as its MC drops below the FSP. Wood machined to a specified size at a MC higher than that expected in service will therefore shrink to a smaller size in use. Since the amount any particular piece of wood will shrink is difficult to predict, it would be very difficult to control dimensions of wood if it was not machined after it was dried. Estimates of dimensional changes can be made with the use of published values of shrinkage coefficients for various species (see [20]).

In addition to simple linear dimensional changes in wood, drying of wood can cause warp of various types. Bow (distortion in the weak direction), crook (distortion in the strong direction), twist (rotational distortion), and cup (cross-sectional distortion similar to bow) are common forms of warp and, when excessive, can adversely affect the structural use of the member. Finally, drying stresses (internal stress resulting from differential shrinkage) can be quite significant and lead to checking (cracks formed along the growth rings) and splitting (cracks formed across the growth rings).

The mechanical properties of wood also are functions of the MC. Above the FSP, most properties are invariant with changes in MC, but most properties are highly affected by changes in the MC below the FPS. For example, the modulus of rupture of wood increases by nearly 4% for a 1% decrease in moisture content below the FSP. For structural design purposes, design values are typically provided for a specific maximum MC (e.g., 19%).

Load history can also have a significant effect on the mechanical performance of wood members. The load that causes failure is a function of the duration and/or rate the load is applied to the member; that is, a member can resist higher magnitude loads for shorter durations or, stated differently, the longer a load is applied, the less able a wood member is to support that load. This response is termed "load duration" effects in wood design. Figure 9.1 illustrates this effect by plotting the time-to-failure as a function of the applied stress expressed in terms of the short term (static) strength. There are many theoretical models proposed to represent this response, but the line shown in Figure 9.1 was developed at the U.S. Forest Products Laboratory in the early 1950s [20] and is the basis for design provisions (i.e., design adjustment factors) in both the LRFD and NDS®.

The design factors derived from the relationship illustrated in Figure 9.1 are appropriate only for stresses and not for stiffness or, more precisely, the modulus of elasticity. Very much related to load duration effects, the deflection of a wood member under sustained load increases over time. This response, termed creep effect, must be considered in design when deflections are critical from either a safety or serviceability standpoint. The main parameters that significantly affect the creep response of wood are stress level, moisture content, and temperature. In broad terms, a 50% increase in deflection after a year or two is expected in most situations, but can easily be upwards of 100%

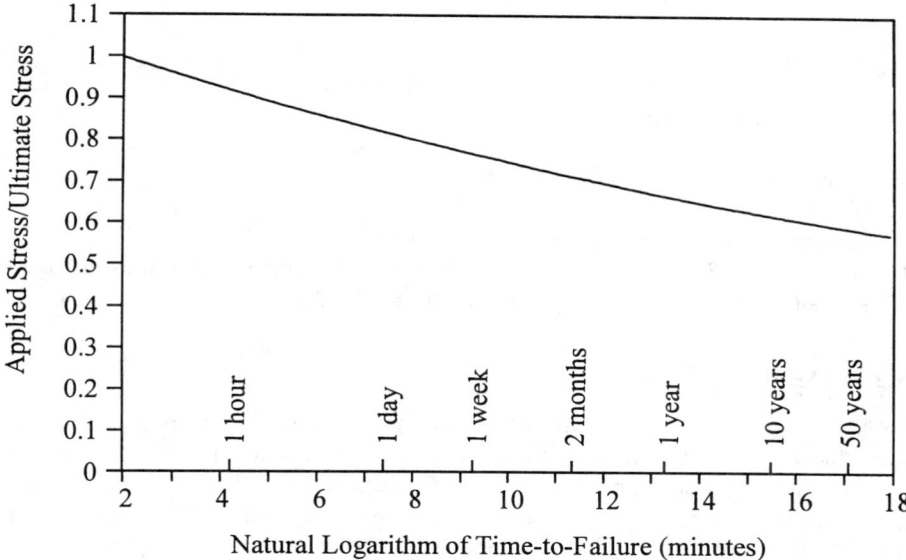

Figure 9.1   Load duration behavior of wood.

given the right conditions. In fact, if a member is subjected to continuous moisture cycling, a 100 to 150% increase in deflection could occur in a matter of a few weeks. Unfortunately, the creep response of wood, especially considering the effects of moisture cycling, is poorly understood and little guidance is available to the designer.

9.3   Preliminary Design Considerations

One of the first issues a designer must consider is determining the types of wood materials and/or wood products that are available for use. For smaller projects, it is better to select materials readily available in the region; for larger projects, a wider selection of materials may be possible since shipping costs may be offset by the volume of material required. One of the strengths of wood construction is its economics; however, the proper choice of materials is key to an efficient and economical wood structure. In this section, preliminary design considerations are discussed including loads and load combinations, design values and adjustments to the design values for in-use conditions.

9.3.1   Loads and Load Combinations

As with all structures designed in the U.S., nominal loads and load combinations for the design of wood structures are prescribed in the ASCE load standard [4]. The following basic factored load combinations must be considered in the design of wood structures when using the LRFD specification:

$$1.4D \tag{9.1}$$

$$1.2D + 1.6L + 0.5(L_r \text{ or } S \text{ or } R) \tag{9.2}$$

$$1.2D + 1.6(L_r \text{ or } S \text{ or } R) + (0.5L \text{ or } 0.8W) \tag{9.3}$$

$$1.2D + 1.3W + 0.5L + 0.5(L_r \text{ or } S \text{ or } R) \tag{9.4}$$

$$1.2D + 1.0E + 0.5L + 0.2S \tag{9.5}$$

$$0.9D - (1.3W \text{ or } 1.0E) \tag{9.6}$$

where

D = dead load
L = live load excluding environmental loads such as snow and wind
L_r = roof live load during maintenance
S = snow load
R = rain or ice load excluding ponding
W = wind load
E = earthquake load (determined in accordance in with [4])

For ASD, the ASCE load standard provides four load combinations that must be considered: D, $D + L + (L_r$ or S or $R)$, $D + (W$ or $E)$, and $D + L + (L_r$ or S or $R) + (W$ or $E)$.

9.3.2 Design Values

The AF&PA [1] *Manual of Wood Construction: Load and Resistance Factor Design* provides nominal design values for visually and mechanically graded lumber, glued laminated timber, and connections. These values include reference bending strength, F_b; reference tensile strength parallel to the grain, F_t; reference shear strength parallel to the grain, F_v; reference compressive strength parallel and perpendicular to the grain, F_c and $F_{c\perp}$, respectively; reference bearing strength parallel to the grain, F_g; and reference modulus of elasticity, E; and are appropriate for use with the LRFD provisions. In addition, the *Manual* provides design values for metal plate connections and trusses, structural composite lumber, structural panels, and other pre-engineered structural wood products. (It should be noted that the LRFD Specification [5] provides only the design provisions, and design values for use with the LRFD Specification are provided in the AF&PA *Manual*.)

Similarly, the Supplement to the NDS® [2] provides tables of design values for visually graded and machine stress rated lumber and glued laminated timber. The basic quantities are the same as with the LRFD, but are in the form of allowable stresses and are appropriate for use with the ASD provisions of the NDS®. Additionally, the NDS® provides tabulated allowable design values for many types of mechanical connections. Allowable design values for many proprietary products (e.g., SCL, I-joist, etc.) are provided by producers in accordance with established standards. For structural panels, design values are provided in the PDS® [6] and by individual product producers.

One main difference between the NDS® and LRFD design values, other than the NDS® prescribing allowable stresses and the LRFD prescribing nominal strengths, is the treatment of duration of load effects. Allowable stresses (except compression perpendicular to the grain) are tabulated in the NDS® and elsewhere for an assumed 10-year load duration in recognition of the duration of load effect discussed previously. The allowable compressive stress perpendicular to the grain is not adjusted since a deformation definition of failure is used for this mode rather than fracture as in all other modes; thus, the adjustment has been assumed unnecessary. Similarly, the modulus of elasticity is not adjusted to a 10-year duration since the adjustment is defined for strength, not stiffness. For the LRFD, short-term (i.e., 20 min) nominal strengths are tabulated for all strength values. In the LRFD, design strengths are reduced for longer duration design loads based on the load combination being considered. Conversely, in the NDS®, allowable stresses are increased for shorter load durations and decreased only for permanent (i.e., greater than 10 years) loading.

9.3.3 Adjustment of Design Values

In addition to providing *reference* design values, both the LRFD and the NDS® specifications provide adjustment factors to determine final *adjusted* design values. Factors to be considered include load duration (termed "time effect" in the LRFD), **wet service**, temperature, stability, size, volume, repetitive use, curvature, orientation (form), and bearing area. Each of these factors will be discussed further; however, it is important to note not all factors are applicable to all design values, and the designer must take care to properly apply the appropriate factors.

LRFD reference strengths and NDS® allowable stresses are based on the following specified reference conditions: (1) dry use in which the maximum EMC does not exceed 19% for solid wood and 16% for glued wood products; (2) continuous temperatures up to 32°C, occasional temperatures up to 65°C (or briefly exceeding 93°C for structural-use panels); (3) untreated (except for poles and piles); (4) new material, not reused or recycled material; and (5) single members without load sharing or composite action. To adjust the reference design value for other conditions, adjustment factors are provided which are applied to the published reference design value:

$$R' = R \cdot C_1 \cdot C_2 \cdots C_n \tag{9.7}$$

where R' = adjusted design value (resistance), R = reference design value, and $C_1, C_2, \ldots C_n$ = applicable adjustment factors. Adjustment factors, for the most part, are common between the LRFD and the NDS®. Many factors are functions of the type, **grade**, and/or species of material while other factors are common across the broad spectrum of materials. For solid sawn lumber, glued laminated timber, piles, and connections, adjustment factors are provided in the NDS® and the LRFD *Manual*. For other products, especially proprietary products, the adjustment factors are provided by the product producers. The LRFD and NDS® list numerous factors to be considered, including wet service, temperature, preservative treatment, fire-retardant treatment, composite action, load sharing (repetitive-use), size, beam stability, column stability, bearing area, form (i.e., shape), time effect (load duration), etc. Many of these factors will be discussed as they pertain to specific designs; however, some of the factors are unique for specific applications and will not be discussed further. The four factors that are applied across the board to all design properties are the wet service factor, C_M; temperature factor, C_t; preservative treatment factor, C_{pt}; and fire-retardant treatment factor, C_{rt}. The two treatment factors are provided by the individual treaters, but the wet service and temperature factors are provided in the LRFD *Manual*. For example, when considering the design of solid sawn lumber members, the adjustment values given in Table 9.1 for wet service, which is defined as the maximum EMC exceeding 19%, and Table 9.2 for temperature, which is applicable when continuous temperatures exceed 32°C, are applicable to all design values.

TABLE 9.1 Wet Service Adjustment Factors, C_M

| Thickness | Size adjusted[a] F_b | | F_t | Size adjusted[a] F_c | | F_v | $F_{c\perp}$ | E, E_{05} |
	≤ 20 MPa	> 20 MPa		≤ 12.4 MPa	>12.4 MPa			
≤ 90 mm	1.00	0.85	1.00	1.00	0.80	0.97	0.67	0.90
> 90 mm	1.00	1.00	1.00	0.91	0.91	1.00	0.67	1.00

[a] Reference value adjusted for size only.

TABLE 9.2 Temperature Adjustment Factors, C_t

| Sustained temperature (°C) | Dry use | | Wet use | |
	E, E_{05}	All other prop.	E, E_{05}	All other prop.
32 < T ≤ 48	0.9	0.8	0.9	0.7
48 < T ≤ 65	0.9	0.7	0.9	0.5

Since, as discussed, the LRFD and the NDS® handle time (duration of load) effects so differently and since duration of load effects are somewhat unique to wood design, it is appropriate to elaborate on it here. Whether using the NDS® or LRFD, a wood structure is designed to resist all appropriate load combinations — unfactored combinations for the NDS® and factored combinations for the LRFD. The time effects (LRFD) and load duration (NDS®) factors are meant to recognize the fact that the failure of wood is governed by a creep-rupture mechanism; that is, a wood member may fail at a load less than its short term strength if that load is held for an extended period of time. In the LRFD, the time effect factor, λ, is based on the load combination being considered as given in Table 9.3. In the NDS®, the load duration factor, C_D, is given in terms of the assumed cumulative duration of the design load. Table 9.4 provides commonly used load duration factors with the associated load combination.

TABLE 9.3   Time Effects Factors for Use in LRFD

Load combination	Time effect factor, λ
$1.4D$	0.6
$1.2D + 1.6L + 0.5(L_r$ or S or $R)$	0.7 when L from storage
	0.8 when L from occupancy
	1.25 when L from impact[a]
$1.2D + 1.6(L_r$ or S or $R) + (0.5L$ or $0.8W)$	0.8
$1.2D + 1.3W + 0.5L + 0.5(L_r$ or S or $R)$	1.0
$1.2D + 1.0E + 0.5L + 0.2S$	1.0
$0.9D - (1.3W$ or $1.0E)$	1.0

[a] For impact loading on connections, $\lambda = 1.0$ rather than 1.25.
From *Load and Resistance Factor Design (LRFD) for Engineered Wood Construction*, American Society of Civil Engineers (ASCE), AF&PA/ASCE 16-95. ASCE, New York, 1996. With permission.

TABLE 9.4   Load Duration Factors for Use in NDS®

Load duration	Load type	Load combination	Load duration factor, C_D
Permanent	Dead	D	0.9
Ten years	Occupancy live	$D + L$	1.0
Two months	Snow load	$D + L + S$	1.15
Seven days	Construction live	$D + L + L_r$	1.25
Ten minutes	Wind and	$D + (W$ or $E)$ and	1.6
	earthquake	$D + L + (L_r$ or S or $R) + (W$ or $E)$	
Impact	Impact loads	$D + L$ (L from impact)	2.0[a]

[a] For impact loading on connections, $\lambda = 1.6$ rather than 2.0.
From *National Design Specification for Wood Construction and Supplement*, American Forest and Paper Association (AF&PA), Washington, D.C., 1991. With permission.

Adjusted design values, whether they are allowable stresses or nominal strengths, are established in the same basic manner: the reference value is taken from an appropriate source (e.g., the LRFD *Manual* [1] or manufacture product literature) and is adjusted for various end-use conditions (e.g., wet use, load sharing, etc.). Additionally, depending on the design load combination being considered, a time effect factor (LRFD) or a load duration factor (NDS®) is applied to the adjusted resistance. Obviously, this rather involved procedure is critical, and somewhat unique, to wood design.

9.4   Beam Design

Bending members are perhaps the most common structural element. The design of wood beams follows traditional beam theory but, as mentioned previously, allowances must be made for the conditions and duration of loads expected for the structure. Additionally, many times bending members are not used as single elements, but rather as part of integrated systems such as a floor or roof system. As such, there exists a degree of member interaction (i.e., load sharing) which can be accounted for in the design. Wood bending members include sawn lumber, timber, glued laminated timber, SCL, and I-joists.

9.4.1 Moment Capacity

The flexural strength of a beam is generally the primary concern in a beam design, but consideration of other factors such as horizontal shear, bearing, and deflection are also crucial for a successful design. Strength considerations will be addressed here while serviceability design (i.e., deflection, etc.) will be presented in Section 9.13. In terms of moment, the LRFD [5] design equation is

$$M_u \leq \lambda \phi_b M' \tag{9.8}$$

where

M_u = moment caused by factored loads
λ = time effect factor applicable for the load combination under consideration
ϕ_b = resistance factor for bending = 0.85
M' = adjusted moment resistance

The moment caused by the factored load combination, M_u, is determined through typical methods of structural analysis. The assumption of linear elastic behavior is acceptable, but a nonlinear analysis is acceptable if supporting data exists for such an analysis. The resistance values, however, involve consideration of factors such as lateral support conditions and whether the member is part of a larger assembly.

Published design values for bending are given for use in the LRFD by AF&PA [1] in the form of a reference bending strength (stress), F_b. This value assumes strong axis orientation; an adjustment factor for flat-use, C_{fu}, can be used if the member will be used about the weak axis. Therefore, for strong $(x - x)$ axis bending, the moment resistance is

$$M' = M'_x = S_x \cdot F'_b \tag{9.9}$$

and for weak $(y - y)$ axis bending

$$M' = M'_y = S_y \cdot C_{fu} \cdot F'_b \tag{9.10}$$

where

M' = M'_x = adjusted strong axis moment resistance
M' = M'_y = adjusted weak axis moment resistance
S_x = section modulus for strong axis bending
S_y = section modulus for weak axis bending
F'_b = adjusted bending strength

For bending, typical adjustment factors to be considered include wet service, C_M; temperature, C_t; beam stability, C_L; size, C_F; volume (for glued laminated timber only), C_V; load sharing, C_r; form (for non-rectangular sections), C_f; and curvature (for glued laminated timber), C_c; and, of course, flat-use, C_{fu}. Many of these factors, including the flat-use factor, are functions of specific product types and species of materials, and therefore are provided with the reference design values. The two factors worth discussion here are the beam stability factor, which accounts for possible lateral-torsional buckling of a beam, and the load sharing factor, which accounts for system effects in repetitive assemblies.

The beam stability factor, C_L, is only used when considering strong axis bending since a beam oriented about its weak axis is not susceptible to lateral instability. Additionally, the beam stability factor and the volume effects factor for glued laminated timber are not used simultaneously. Therefore, when designing an unbraced, glued laminated beam, the lessor of C_L and C_V is used to determine the adjusted bending strength. The beam stability factor is taken as 1.0 for members with continuous lateral bracing or meeting limitations set forth in Table 9.5.

When the limitations in Table 9.5 are not met, C_L is calculated from

$$C_L = \frac{1 + \alpha_b}{2c_b} - \sqrt{\left(\frac{1 + \alpha_b}{2c_b}\right)^2 - \frac{\alpha_b}{c_b}} \tag{9.11}$$

TABLE 9.5   Conditions Defining Full Lateral Bracing

Depth to width (d/b)	Support conditions
≤ 2	No lateral support required.
> 2 and < 5	Ends supported against rotation.
≥ 5 and < 6	Compression edge continuously supported.
≥ 6 and < 7	Bridging, blocking, or X-bracing spaced no more than 2.4 m, or compression edge supported throughout its length and ends supported against rotation (typical in a floor system).
≥ 7	Both edges held in line throughout entire length.

where

$$\alpha_b = \frac{\phi_s M_e}{\lambda \phi_b M_x^*} \tag{9.12}$$

and

c_b = beam stability coefficient = 0.95
ϕ_s = resistance factor for stability = 0.85
M_e = elastic buckling moment
M_x^* = moment resistance for strong axis bending including all adjustment factors except C_{fu}, C_V, and C_L.

The elastic buckling moment can be determined for most rectangular timber beams through a simplified method where

$$M_e = 2.40 E_{05}' \frac{I_y}{l_e} \tag{9.13}$$

where

E_{05}' = adjusted fifth percentile modulus of elasticity
I_y = moment of inertia about the weak axis
l_e = effective length between bracing points of the compression side of the beam

The adjusted fifth percentile modulus of elasticity is determined from the published reference modulus of elasticity, which is a mean value meant for use in deflection serviceability calculations, by

$$E_{05}' = 1.03 E'(1 - 1.645 \cdot COV_E) \tag{9.14}$$

where E' = adjusted modulus of elasticity and COV_E = coefficient of variation of E. The factor 1.03 recognizes that E is published to include a 3% shear component. For glued laminated timber, values of E include a 5% shear component, so it is acceptable to replace the 1.03 factor by 1.05 for the design of glued laminated timber beams. The COV of E can be assumed as 0.25 for **visually graded lumber**, 0.11 for **machine stress rated (MSR)** lumber, and 0.10 for glued laminated timber [2]. For other products, $COVs$ or values of E_{05}' can be obtained from the producer. Also, the only adjustments needed to be considered for E are the wet service, temperature, and any preservative/fire-retardant treatment factors. The effective length, l_e, accounts for both the lateral motion and torsional phenomena and is given in the LRFD specification [1, 5] for numerous combinations of span types, end conditions, loading, bracing conditions, and actual unsupported span to depth ratios (l_u/d). Generally, for $l_u/d < 7$, the effective unbraced length, l_e, ranges from $1.33l_u$ to $2.06l_u$; for $7 \leq l_u/d \leq 14.3$, l_e ranges from $1.11l_u$ to $1.84l_u$; and for $l_u/d > 14.3$, l_e ranges from $0.9l_u + 3d$ to $1.63l_u + 3d$ where d = depth of the beam.

The load sharing factor, C_r, is a multiplier that can be used when a bending member is part of an assembly, such as the floor system illustrated in Figure 9.2, consisting of three or more members spaced no more than 610 mm on center and connected together by a load-distributing element, such as typical floor and roof **sheathing**. The factors recognize the beneficial effects of the sheathing in distributing loads away from less stiff members and are only applicable when considering uniformly applied loads. Assuming a strong correlation between strength and stiffness, this implies the load is distributed away from the weaker members as well, and that the value of C_r is dependent of the

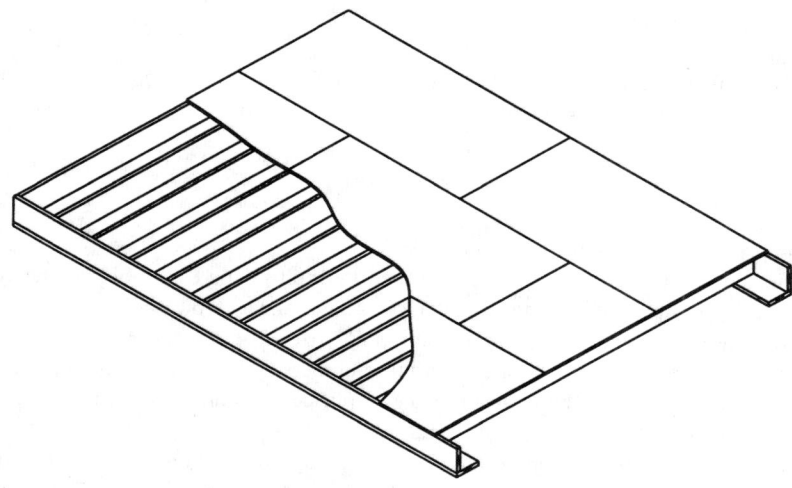

Figure 9.2 Typical wood floor assembly.

inherent variability of the system members. Table 9.6 provides values of C_r for various common framing materials.

TABLE 9.6 Load Sharing Factor, C_r

Assembly type	C_r
Solid sawn lumber framing members	1.15
I-joists with visually graded lumber flanges	1.15
I-joists with MSR lumber flanges	1.07
Glued laminated timber and SCL framing members	1.05
I-joists with SCL flanges	1.04

9.4.2 Shear Capacity

Similar to bending, the basic design equation for shear is given by

$$V_u \leq \lambda \phi_v V' \tag{9.15}$$

where

V_u = shear caused by factored loads

λ = time effect factor applicable for the load combination under consideration

ϕ_v = resistance factor for shear = 0.75

V' = adjusted shear resistance

Except in the design of I-joists, V_u is determined at a distance d (depth of the member) away from the face of the support if the loads acting on the member are applied to the face opposite the bearing area of the support. For other loading conditions and for I-joists, V_u is determined at the face of the support.

The adjusted shear resistance is computed from

$$V' = \frac{F_v' I b}{Q} \tag{9.16}$$

where

F_v' = adjusted shear strength parallel to the grain

I =   moment of inertia
b =   member width
Q =   statical moment of an area about the neutral axis

For rectangular sections, this equation simplifies to

$$V' = \frac{2}{3}F'_v bd \tag{9.17}$$

where d = depth of the rectangular section.

The adjusted shear strength, F'_v, is determined by multiplying the published reference shear strength, F_v, by all appropriate adjustment factors. For shear, typical adjustment factors to be considered include wet service, C_M; temperature, C_t; size, C_F; and shear stress, C_H. The shear stress factor allows for increased shear strength in members with limited splits, checks, and shakes and ranges from $C_H = 1.0$ implying the presence of splits, checks, and shakes to $C_H = 2.0$ implying no splits, checks, or shakes.

In wood construction, notches are often made at the support to allow for vertical clearances and tolerances as illustrated in Figure 9.3; however, stress concentrations resulting from these notches significantly affect the shear resistance of the section. At sections where the depth is reduced due to

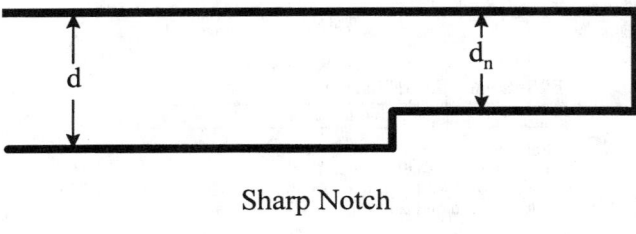

Sharp Notch

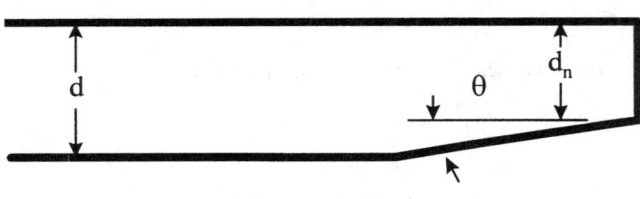

Tapered Notch

Figure 9.3   Notched beam: (a) sharp notch and (b) angled notch.

the presence of a notch, the shear resistance of the notched section is determined from

$$V' = \left(\frac{2}{3}F'_v bd_n\right)\left(\frac{d_n}{d}\right) \tag{9.18}$$

where d = depth of the unnotched section and d_n = depth of the member after the notch. When the notch is made such that it is actually a gradual tapered cut at an angle θ from the longitudinal axis of the beam, the stress concentrations resulting from the notch are reduced and the above equation becomes

$$V' = \left(\frac{2}{3}F'_v bd_n\right)\left(1 - \frac{(d - d_n)\sin\theta}{d}\right) \tag{9.19}$$

Similar to notches, connections too can produce significant stress concentrations resulting in reduced shear capacity. Where a connection produces at least one-half the member shear force on either side of the connection, the shear resistance is determined by

$$V' = \left(\frac{2}{3}F'_v b d_e\right)\left(\frac{d_e}{d}\right) \tag{9.20}$$

where d_e = effective depth of the section at the connection which is defined as the depth of the member less the distance from the unloaded edge (or nearest unloaded edge if both edges are unloaded) to the center of the nearest fastener for **dowel-type fasteners** (e.g., bolts). For additional information regarding connector design, see Section 9.8.

9.4.3 Bearing Capacity

The last aspect of beam design to be covered in this section is bearing at the supports. The governing design equation for bearing is

$$P_u \leq \lambda \phi_c P'_\perp \tag{9.21}$$

where

P_u = the compression force due to factored loads
λ = time effects factor corresponding to the load combination under consideration
ϕ_c = resistance factor for compression = 0.90
$P'_\perp$ = adjusted compression resistance perpendicular to the grain
 The adjusted compression resistance, $P'_\perp$, is determined by

$$P'_\perp = A_n F'_{c\perp} \tag{9.22}$$

where

A_n = net bearing area
$F'_{c\perp}$ = adjusted compression strength perpendicular to the grain

The adjusted compression strength, $F'_{c\perp}$, is determined by multiplying the reference compression strength perpendicular to the grain, $F_{c\perp}$, by all applicable adjustment factors, including wet service, C_M; temperature, C_t; and bearing area, C_b. The bearing area factor, C_b, allows an increase in the compression strength when the bearing length, l_b, is no more than 150 mm along the length of the member and is at least 75 mm from the end of the member, and is given by

$$C_b = (l_b + 9.5)/l_b \tag{9.23}$$

where l_b is in mm.

9.4.4 NDS® Provisions

In the ASD format provided by the NDS®, the design checks are in terms of allowable stresses and unfactored loads. The determined bending, shear, and bearing stresses in a member due to unfactored loads are required to be less than the adjusted allowable bending, shear, and bearing stresses, respectively, including load duration effects. The basic approach to the design of a beam element, however, is quite similar between the LRFD and NDS® and is based on the same principles of mechanics. One major difference between the two specifications, though, is the treatment of load duration effects with respect to bearing. In the LRFD, the design equation for bearing (Equation 9.21) includes the time effect factor, λ; however, the NDS® does not require any adjustment for load duration for bearing. The allowable compressive stress perpendicular to the grain as presented in the NDS® is not adjusted because the compressive stress perpendicular to the grain follows a deformation definition of failure rather than fracture as in all other modes; thus, the adjustment is considered unnecessary. Conversely, the LRFD specification assumes time effects to occur in all modes, whether it is strength- (fracture) based or deformation-based.

9.5 Tension Member Design

The design of tension members, either by LRFD or NDS®, is relatively straightforward. The basic design checking equation for a tension member as given by the LRFD Specification [5] is

$$T_u \leq \lambda \phi_t T' \tag{9.24}$$

where
T_u = the tension force due to factored loads
λ = time effects factor corresponding to the load combination under consideration
ϕ_t = resistance factor for tension = 0.80
T' = adjusted tension resistance parallel to the grain
  The adjusted compression resistance, T', is determined by

$$T' = A_n F'_t \tag{9.25}$$

where A_n = net cross-sectional area and F'_t = adjusted tension strength parallel to the grain. The adjusted compression strength, F'_t, is determined by multiplying the reference tension strength parallel to the grain, F_t, by all applicable adjustment factors, including wet service, C_M; temperature, C_t; and size, C_F.

  It should be noted that tension forces are typically transferred to a member through some type of mechanical connection. When, for example as illustrated in Figure 9.4, the centroid of an unsymmetric net section of a group of three or more connectors differs by 5% or more from the centroid of the gross section, then the tension member must be designed as a combined tension and bending member (see Section 9.7).

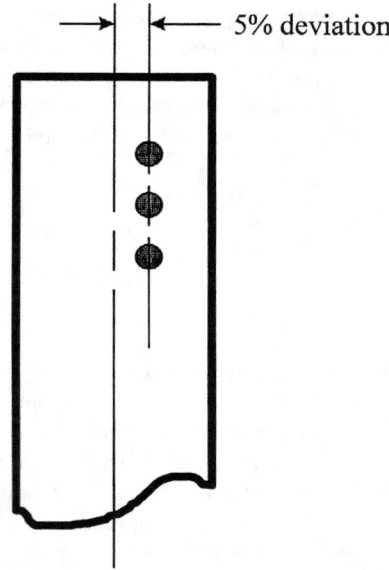

Figure 9.4   Eccentric bolted connection.

9.6   Column Design

The term *column* is typically considered to mean any compression member, including compressive members in trusses and posts as well as traditional columns. Three basic types of wood columns as illustrated in Figure 9.5 are (1) simple solid or traditional columns, which are single members such as sawn lumber, posts, timbers, poles, glued laminated timber, etc.; (2) spaced columns, which are two or more parallel single members separated at specific locations along their length by blocking and rigidly tied together at their ends; and (3) built-up columns, which consist of two or more members joined together by mechanical fasteners such that the assembly acts as a single unit.

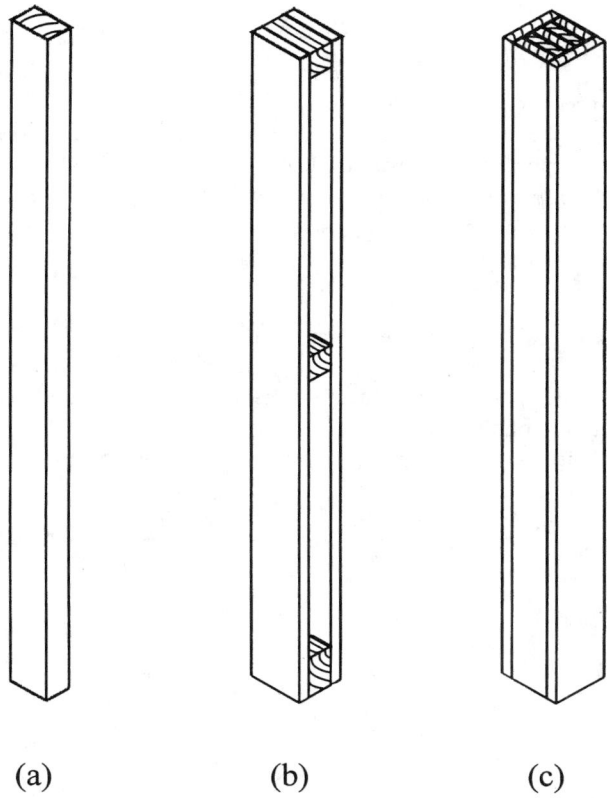

(a)   (b)   (c)

Figure 9.5   Typical wood columns: (a) simple wood column, (b) spaced column, and (c) built-up column.

Depending on the relative dimensions of the column as defined by the *slenderness ratio,* the design of wood columns is limited by the material's stiffness and strength parallel to the grain. The slenderness ratio is defined as the ratio of the effective length of the column, l_e, to the least radius of gyration, $r = \sqrt{I/A}$, where I = moment of inertia of the cross-section about the weak axis and A = cross-sectional area. The effective length is defined by $l_e = K_e l$, where K_e = effective length factor or buckling length coefficient and l = unbraced length of the column. The unbraced length, l, is measured as center to center distance between lateral supports. K_e is dependent on the column end support conditions and on whether sidesway is allowed or restrained. Table 9.7 provides values of K_e for various typical column configurations. Regardless of the column type of end conditions, the slenderness ratio, $K_e l/r$, is not permitted to exceed 175.

TABLE 9.7   Effect Length Factors for Wood Columns

Support conditions	Sideway restrained	Theoretical K_e	Recommended K_e^a
Fixed–fixed	Yes	0.50	0.65
Fixed–pinned	Yes	0.70	0.80
Fixed–fixed	No	1.00	1.20
Pinned–pinned	Yes	1.00	1.00
Fixed–free	No	2.00	2.10
Fixed–pinned	No	2.00	2.40

^{a} Values recommended by [5].

9.6.1   Solid Columns

The basic design equation for an axially loaded member as given by the LRFD Specification [5] is given as

$$P_u \le \lambda \phi_c P'_c \tag{9.26}$$

where

P_u =   the compression force due to factored loads
λ =   time effects factor corresponding to the load combination under consideration
ϕ_c =   resistance factor for compression = 0.90
P'_c =   adjusted compression resistance parallel to the grain.

   The adjusted compression resistance, P'_c, is determined by

$$P'_c = A F'_c \tag{9.27}$$

where A = gross area and F'_c = adjusted compression strength parallel to the grain. The adjusted compression strength, F'_c, is determined by multiplying the reference compression strength parallel to the grain, F_c, by all applicable adjustment factors, including wet service, C_M; temperature, C_t; size, C_F; and column stability, C_P.

   The column stability factor, C_P, accounts for partial lateral support for a column and is given by

$$C_P = \frac{1 + \alpha_c}{2c} - \sqrt{\left(\frac{1 + \alpha_c}{2c}\right)^2 - \frac{\alpha_c}{c}} \tag{9.28}$$

where

$$\alpha_c = \frac{\phi_s P_e}{\lambda \phi_c P'_0} \tag{9.29}$$

$$P_e = \frac{\pi^2 E'_{05} A}{\left(\frac{K_e l}{r}\right)^2} \tag{9.30}$$

and c = coefficient based on member type, ϕ_s = resistance factor for stability = 0.85, ϕ_b = resistance factor for compression = 0.90, λ = time effect factor for load combination under consideration, P_e = Euler buckling resistance, P'_0 = adjusted resistance of a fully braced (or so-called "zero-length") column, E'_{05} = adjusted fifth percentile modulus of elasticity, and A = gross cross-sectional area. The coefficient $c = 0.80$ for solid sawn members, 0.85 for round poles and piles, and 0.90 for glued laminated members and SCL. E'_{05} is determined as presented for beam stability using Equation 9.14, and P'_0 is determined using Equation 9.27, except that the reference compression strength, F_c, is *not* adjusted for stability (i.e., assume $C_P = 1.0$).

   Two common conditions occurring in solid columns are notches and tapers. When notches or holes are present in the middle half of the effective length (between inflection points), and the net moment of inertia at the notch or hole is less than 80% of the gross moment of inertia, or the length of the notch or hole is greater than the largest cross-sectional dimension of the column, then P'_c

(Equation 9.27) and C_P (Equation 9.28) are computed using the net area, A_n, rather than gross area, A. When notches or holes are present outside this region, the column resistance is taken as the lesser of that determined without considering the notch or hole (i.e., using gross area) and

$$P'_c = A_n F^*_c \qquad (9.31)$$

where F^*_c = the compression strength adjusted by all applicable factors *except* for stability (i.e., assume $C_P = 1.0$).

Two basic types of uniformly tapered solid columns exist: circular and rectangular. For circular tapered columns, the design diameter is taken as either (1) the diameter of the small end or (2) when the diameter of the small end, D_1, is at least one-third of the large end diameter, D_2,

$$D = D_1 + X(D_2 - D_1) \qquad (9.32)$$

where D = design diameter and X = a factor dependent on support conditions as follows:

1. Cantilevered, large end fixed: $X = 0.52 + 0.18(D_1/D_2)$ $\qquad$ (9.33a)
2. Cantilevered, small end fixed: $X = 0.12 + 0.18(D_1/D_2)$ $\qquad$ (9.33b)
3. Singly tapered, simple supports: $X = 0.32 + 0.18(D_1/D_2)$ $\qquad$ (9.33c)
4. Doubly tapered, simple supports: $X = 0.52 + 0.18(D_1/D_2)$ $\qquad$ (9.33d)
5. All other support conditions: $X = 0.33$ $\qquad$ (9.33e)

For uniformly tapered rectangular columns with constant width, the design depth of the member is handled in a manner similar to circular tapered columns, except that buckling in two directions must be considered. The design depth is taken as either (1) the depth of the small end or (2) when the depth of the small end, d_1, is at least one-third of the large end depth, d_2,

$$d = d_1 + X(d_2 - d_1) \qquad (9.34)$$

where d = design depth and X = a factor dependent on support conditions as follows:
For buckling in the tapered direction:

1. Cantilevered, large end fixed: $X = 0.55 + 0.15(d_1/d_2)$ $\qquad$ (9.35a)
2. Cantilevered, small end fixed: $X = 0.15 + 0.15(d_1/d_2)$ $\qquad$ (9.35b)
3. Singly tapered, simple supports: $X = 0.35 + 0.15(d_1/d_2)$ $\qquad$ (9.35c)
4. Doubly tapered, simple supports: $X = 0.55 + 0.15(d_1/d_2)$ $\qquad$ (9.35d)
5. All other support conditions: $X = 0.33$ $\qquad$ (9.35e)

For buckling in the non-tapered direction:

1. Cantilevered, large end fixed: $X = 0.63 + 0.07(d_1/d_2)$ $\qquad$ (9.35f)
2. Cantilevered, small end fixed: $X = 0.23 + 0.07(d_1/d_2)$ $\qquad$ (9.35g)
3. Singly tapered, simple supports: $X = 0.43 + 0.07(d_1/d_2)$ $\qquad$ (9.35h)
4. Doubly tapered, simple supports: $X = 0.63 + 0.07(d_1/d_2)$ $\qquad$ (9.35i)
5. All other support conditions: $X = 0.33$ $\qquad$ (9.35j)

In addition to these provisions, the design resistance of a tapered circular or rectangular column cannot exceed

$$P'_c = A_n F^*_c \qquad (9.36)$$

where A_n = net area of the column at any cross-section and F^*_c = the compression strength adjusted by all applicable factors *except* for stability (i.e., assume $C_P = 1.0$).

9.6.2   Spaced Columns

Spaced columns consist of two or more parallel single members separated at specific locations along their length by blocking and rigidly tied together at their ends. As defined in Figure 9.5b, $L_1 =$ overall length in the spaced column direction, $L_2 =$ overall length in the solid column direction, $L_3 =$ largest distance from the centroid of an end block to the center of the mid-length spacer, $L_{ce} =$ distance from the centroid of end block connectors to the nearer column end, $d_1 =$ width of individual components in the spaced column direction, and $d_2 =$ width of individual components in the solid column direction. Typically, the individual components of a spaced column are considered to act individually in the direction of the wide face of the members. The blocking, however, effectively reduces the unbraced length in the weak direction. Therefore, the following L/d ratios are imposed on spaced columns:

$$\text{1. In the spaced column direction:} \quad L_1/d_1 \leq 80 \qquad (9.37a)$$
$$L_3/d_1 \leq 40 \qquad (9.37b)$$
$$\text{2. In the solid column direction:}^2 \quad L_2/d_2 \leq 50 \qquad (9.37c)$$

Depending on the length L_{ce} relative to L_1, one of two effective length factors can be assumed for design in the spaced column direction. If sidesway is not allowed and $L_{ce} \leq 0.05L_1$, then the effective length factor is assumed as $K_e = 0.63$; or if there is no sidesway and $0.05L_1 < L_{ce} \leq 0.10L_1$, then assume $K_e = 0.53$. For columns with sidesway in the spaced column direction, an effective length factor greater than unity is determined as given in Table 9.7.

9.6.3   Built-Up Columns

Built-up columns consist of two or more members joined together by mechanical fasteners such that the assembly acts as a single unit. Conservatively, the capacity of a built-up member can be taken as the sum of resistances of the individual components. Conversely, if information regarding the rigidity and overall effectiveness of the fasteners is available, the designer can incorporate such information into the analysis and take advantage of the composite action provided by the fasteners; however, no codified procedures are available for the design of built-up columns. In either case, the fasteners must be designed appropriately to resist the imposed shear and tension forces (see Section 9.8 for fastener design).

9.6.4   NDS® Provisions

For rectangular columns, which are common in wood construction, the slenderness ratio can be expressed as the ratio of the unbraced length to the least cross-sectional dimension of the column, or L/d where d is the least cross-sectional dimension. This is the approach offered by the NDS® [2] which differs from the more general approach of the LRFD [5] and is identical to that used in the LRFD for spaced columns. Often, the unbraced length of a column is not the same about both the strong and weak axes and the slenderness ratios in both directions should be considered (e.g., $r_1 = L_1/d_1$ in the strong direction and $r_2 = L_2/d_2$ in the weak direction). One common example of such a case is wood studs in a load bearing wall where, if adequately fastened, the sheathing provides continuous lateral support in the weak direction and only the slenderness ratio about the strong axis needs to be determined. The slenderness ratio is not permitted to exceed 50^2 for single solid

---

^{2}For rectangular columns, the provision $L/d \leq 50$ is equivalent to the provision $KL/r \leq 175$.

columns or built-up columns, and is not permitted to exceed 80 for individual members of spaced columns; however, when used for temporary construction bracing, the allowable slenderness ratio is increased from 50 to 75 for single or built-up columns. All other provisions related to column design are equivalent between the NDS® and LRFD.

9.7 Combined Load Design

Often, structural wood members are subjected to bending about both principal axes and/or bending combined with axial loads. The bending can come from eccentric axial loads and/or laterally applied loads. The adjusted member resistances for moment, M', tension, T', and compression, P'_c, defined in Sections 9.4, 9.5, and 9.6 are used for combined load design in conjunction with an appropriate interaction equation. All other factors (e.g., the resistance factors ϕ_b, ϕ_t, and ϕ_c, and the time effect factor, λ) are also the same in combined load design as defined previously.

9.7.1 Combined Bending and Axial Tension

When a tension load acts simultaneously with bending about one or both principal axes, the following interaction equations must be satisfied:

1. Tension face:
$$\frac{T_u}{\lambda \phi_t T'} + \frac{M_{ux}}{\lambda \phi_b M'_s} + \frac{M_{uy}}{\lambda \phi_b M'_y} \leq 1.0 \qquad (9.38)$$

2. Compression face:
$$\frac{\left(M_{ux} - \frac{d}{6} T_u\right)}{\lambda \phi_b M'_x} + \frac{M_{uy}}{\lambda \phi_b \left(1 - \frac{M_{ux}}{\phi_b M_e}\right)^2} \leq 1.0 \qquad (9.39)$$

where

T_u	=	tension force due to factored loads
M_{ux} and M_{uy}	=	moment due to factored loads about the strong and weak axes, respectively
M'_x and M'_y	=	adjusted moment resistance about the strong and weak axes, respectively
M_e	=	elastic lateral buckling moment (Equation 9.13)
M'_s	=	M'_x computed assuming the beam stability factor $C_L = 1.0$ but including all other appropriate adjustment factors, including the volume factor C_V
d	=	depth of the member

Equations 9.38 and 9.39 assume rectangular sections. If a non-rectangular section is being designed, the quantity $d/6$ appearing in Equation 9.38 should be replaced by S_x/A where $S_x =$ the section modulus about the strong axis and $A =$ gross area of the section.

9.7.2 Biaxial Bending or Combined Bending and Axial Compression

When a member is being designed for either biaxial bending or for combined axial compression and bending about one or both principal axes, the following interaction equation must be satisfied:

$$\left(\frac{P_u}{\lambda \phi_c P'_c}\right)^2 + \frac{M_{mx}}{\lambda \phi_b M'_x} + \frac{M_{my}}{\lambda \phi_b M'_y} \leq 1.0 \qquad (9.40)$$

where

P_u	=	axial load due to factored loads
P'_c	=	adjusted compression resistance assuming the compression acts alone (i.e., no moments) for the axis of buckling providing the lower resistance value
M_{mx} and M_{my}	=	moments due to factored loads, including any magnification resulting from second-order moments, about the strong and weak axes, respectively

M'_x and M'_y =   adjusted strong and weak axes moment resistances, respectively, assuming the beam stability factor $C_L = 1.0$

The moments due to factored loads, M_{mx} and M_{my}, can be determined either of two ways: (1) using an appropriate second-order analysis procedure or (2) using a simplified magnification method. The moment magnification method recommended in the LRFD is given as follows:

$$M_{mx} = B_{bx}M_{bx} + B_{sx}M_{sx} \qquad (9.41)$$

$$M_{my} = B_{by}M_{by} + B_{sy}M_{sy} \qquad (9.42)$$

where M_{bx} and M_{by} = factored strong and weak axis moments, respectively, from loads producing no lateral translation or sidesway determined using an appropriate first-order analysis; M_{sx} and M_{sy} = factored strong and weak axis moments, respectively, from loads producing lateral translation or sidesway determined using an appropriate first-order analysis; and B_{bx}, B_{sx}, B_{by}, and B_{sy} = moment magnification factors to account for second-order effects and associated with M_{bx}, M_{sx}, M_{by}, and M_{sy}, respectively, and are determined as follows:

$$B_{bx} = \frac{C_{mx}}{\left(1 - \frac{P_u}{\phi_c P_{ex}}\right)} \geq 1.0 \qquad (9.43)$$

$$B_{by} = \frac{C_{mx}}{\left[1 - \frac{P_u}{\phi_c P_{ex}} - \left(\frac{M_{ux}}{\phi_b M_e}\right)^2\right]} \geq 1.0 \qquad (9.44)$$

$$B_{sx} = \frac{1}{\left(1 - \frac{\sum P_u}{\phi_c \sum P_{ex}}\right)} \geq 1.0 \qquad (9.45)$$

$$B_{sy} = \frac{1}{\left(1 - \frac{\sum P_u}{\phi_c \sum P_{ey}}\right)} \geq 1.0 \qquad (9.46)$$

where

P_{ex} and P_{ey} =   Euler buckling resistance about the strong and weak axes, respectively, as determined by Equation 9.30

$\sum P_u$ =   sum of all compression forces due to factored loads for all columns in the sidesway mode under consideration

$\sum P_{ex}$ and $\sum P_{ey}$ =   sum of all Euler buckling resistances for columns in the sidesway mode under consideration about its strong and weak axes, respectively

C_{mx} and C_{my} =   factor relating the actual moment diagram shape to an equivalent uniform moment diagram for moment applied about the strong and weak axes, respectively

All other terms are as defined previously. The factors C_{mx} and C_{my} are determined for one of two conditions:

1. For members braced against lateral joint translation with only end moments applied:

$$C_{mx} \text{ or } C_{my} = 0.60 - 0.40\frac{M_1}{M_2} \qquad (9.47)$$

where M_1/M_2 = ratio of the smaller magnitude end moment to the larger end moment in the plane of bending ($x - x$ or $y - y$) under consideration, with the ratio defined as being negative for single curvature and positive for double curvature.

2. For members braced against joint translation in the plane of bending under consideration with lateral loads applied between the joints:

a) C_{mx} or $C_{my} = 0.85$ for members with ends restrained against rotation, or

b) C_{mx} or $C_{my} = 1.00$ for members with ends unrestrained against rotation.

For members not braced against sidesway, all four moment magnification factors need to be determined; however, for members braced against sidesway, only B_{bx} and B_{sx} need to be determined.

9.7.3 NDS® Provisions

The primary difference between the LRFD approach for members subjected to combined load and that of the NDS® is in the design of members for combined axial and bending loads. The LRFD combines moments from all sources, including moments from eccentrically applied axial loads and moments from transversely applied loads. The NDS® provides the following general interaction equation:

$$\left(\frac{f_c}{F_c'}\right)^2 + \frac{f_{bx} + f_c\left(\frac{6e_y}{d_y}\right)\left(1 + 0.234\frac{f_c}{F_{cEx}}\right)}{F_{bx}'\left(1 - \frac{f_c}{F_{cEx}}\right)}$$

$$+ \frac{f_{by} + f_c\left(\frac{6e_x}{d_x}\right)\left(1 + 0.234\frac{f_c}{F_{cEy}}\right)}{F_{by}'\left[1 - \frac{f_c}{F_{cEy}} - \left(\frac{f_{bx}}{F_{bE}}\right)^2\right]} \le 1.0 \qquad (9.48)$$

where

f_c	=	compression stress due to unfactored loads
f_{bx} and f_{by}	=	bending stress about the strong and weak axes, respectively, due to unfactored loads
F_c'	=	adjusted allowable compression stress
F_{bx}' and F_{by}	=	adjusted allowable bending stress about the strong and weak axes, respectively
F_{cEx} and F_{cEy}	=	allowable Euler buckling stress about the strong and weak axes, respectively
F_{bE}	=	allowable buckling stress for bending
e_x and e_y	=	eccentricity in the x and y directions, respectively
d_x and d_y	=	cross-sectional dimension in the x (narrow dimension) and y (wide dimension) directions, respectively

The allowable buckling values are determined from

$$F_{cEx} = \frac{K_{cE}E'}{(l_{ex}/d_x)^2} \qquad (9.49)$$

$$F_{cEy} = \frac{K_{cE}E'}{(l_{ey}/d_y)^2} \qquad (9.50)$$

$$F_{bE} = \frac{K_{bE}E'}{(R_b)^2} \qquad (9.51)$$

where

K_{ce}	=	0.3 for visually graded and machine evaluated lumber, or 0.418 for products with a coefficient of variation on E less than or equal to 11% (e.g., MSR lumber and glued laminated timber)
K_{bE}	=	0.438 for visually graded and machine evaluated lumber, or 0.609 for products with a coefficient of variation on E less than or equal to 11%
E'	=	adjusted modulus of elasticity
R_b	=	slenderness ratio for bending as given by

$$R_b = \sqrt{\frac{l_{ey}d_y}{d_x^2}} \tag{9.52}$$

Note that l_{ey} is used in Equation 9.52 since lateral buckling in beams is only possible about the weak axis.

9.8   Fastener and Connection Design

The design of fasteners and connections for wood has undergone significant changes in recent years. Typical fastener and connection details for wood include nails, staples, screws, lag screws, dowels, and bolts. Additionally, **split rings, shear plates, truss plate connectors,** joist hangers, and many other types of connectors are available to the designer. The general LRFD design checking equation for connections is given as follows:

$$Z_u \leq \lambda \phi_z Z' \tag{9.53}$$

where

Z_u = connection force due to factored loads
λ = applicable time effect factor
ϕ_z = resistance factor for connections = 0.65
Z' = connection resistance adjusted by the appropriate adjustment factors

It should be noted that, for connections, the moisture adjustment is based on both in service condition and on conditions at the time of fabrication; that is, if a connection is fabricated in the wet condition but is to be used in service under a dry condition, the wet condition should be used for design purposes due to potential drying stresses which may occur. Also, C_M does not account for corrosion of metal components in a connection. Other adjustments specific to connection type (e.g., diaphragm factor, C_{di}; end grain factor, C_{eg}; group action factor, C_g; geometry factor, C_Δ; penetration depth factor, C_d; toe-nail factor, C_{tn}; etc.) will be discussed with their specific use. It should also be noted that the time effects factor, λ, is not allowed to exceed unity for connections as noted in Table 9.3. Additionally, when failure of a connection is controlled by a non-wood element (e.g., fracture of a bolt), then the time-effects factor is taken as unity since time effects are specific to wood and not applicable to non-wood components.

In both the LRFD Manual [1] and the NDS® [2], tables of reference resistances (LRFD) and allowable loads (NDS®) are available which significantly reduce the tedious calculations required for a simple connection design. In this section, the basic design equations and calculation procedures are presented, but design tables such as those given in the LRFD Manual and the NDS® are not provided here.

The design of general dowel-type connections (i.e., nails, spikes, screws, bolts, etc.) for lateral loading are currently based on possible yield modes. Formerly (i.e., all previous editions of the NDS®), empirical behavior equations were the basis for the design provisions. Figure 9.6 illustrates the various yield modes that must be considered for single and double shear connections. Based on these possible yield modes, lateral resistances are determined for the various dowel-type connections. Specific equations are presented in the following sections for nails and spikes, screws, bolts, and lag screws. In general, though, the dowel bearing strength, F_e, is required to determine the lateral resistance of a dowel-type connection. Obviously, this property is a function of the orientation of the applied load to the grain, and values of F_e are available for parallel to the grain, $F_{e\|}$, and perpendicular to the grain, $F_{e\perp}$. The dowel bearing strength or other angles to the grain, $F_{e\theta}$, is determined by

$$F_{e\theta} = \frac{F_{e\|}F_{e\perp}}{F_{e\|}\sin^2\theta + F_{e\perp}\cos^2\theta} \tag{9.54}$$

where θ = angle of load with respect to a direction parallel to the grain.

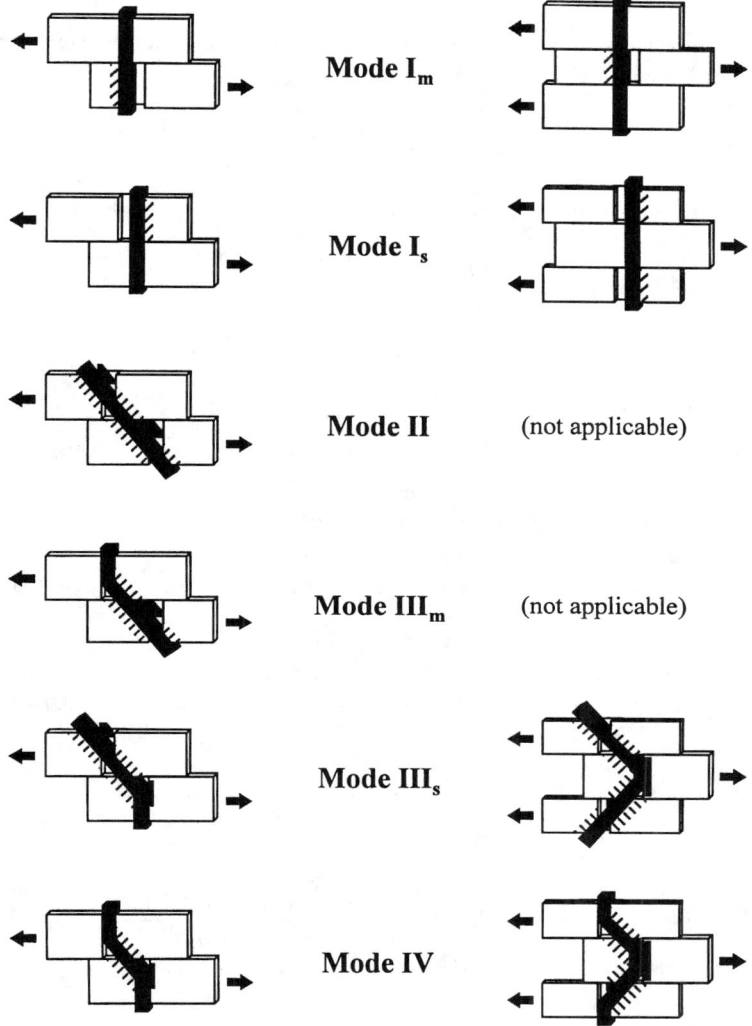

Figure 9.6 Yield modes for dowel-type connections. (Courtesy of American Forest & Paper Association, Washington, D.C.)

9.8.1 Nails, Spikes, and Screws

Nails and spikes are perhaps the most commonly used fasteners in wood construction. Nails are generally used when loads are light such as in the construction of diaphragms and shear walls; however, they are susceptible to working loose under vibration or withdrawal loads. Common wire nails and spikes are quite similar, except that spikes have larger diameters than nails. Both a 12d (i.e., 12-penny) nail and spike are 88.9 mm in length; however, a 12d nail has a diameter of 3.76 mm while a spike has a diameter of 4.88 mm. Many types of nails have been developed to provide better withdrawal resistance, such as deformed shank and coated nails. Nonetheless, nails and spikes should be designed to carry laterally applied load and not withdrawal.

Lateral Resistance

The reference lateral resistance of a single nail or spike in single shear is taken as the least value determined by the four governing modes:

$$I_s: \quad Z = \frac{3.3 D t_s F_{es}}{K_D} \tag{9.55}$$

$$III_m: \quad Z = \frac{3.3 k_1 D p F_{em}}{K_D (1 + 2R_e)} \tag{9.56}$$

$$III_s: \quad Z = \frac{3.3 k_2 D t_s F_{em}}{K_D (2 + R_e)} \tag{9.57}$$

$$IV: \quad Z = \frac{3.3 D^2}{K_D} \sqrt{\frac{2 F_{em} F_{yb}}{3(1 + R_e)}} \tag{9.58}$$

where

D = shank diameter

t_s = thickness of the side member

F_{es} = dowel bearing strength of the side member

p = shank penetration into member (see Figure 9.7)

R_e = ratio of dowel bearing strength of the main member to that of the side member
       = F_{em}/F_{es}

F_{yb} = bending yield strength of the dowel fastener (i.e., nail or spike in this case)

K_D = factor related to the shank diameter as follows: $K_D = 2.2$ for $D \leq 4.3$ mm, $K_D = 0.38D + 0.56$ for 4.3 mm $< D \leq 6.4$ mm, and $K_D = 3.0$ for $D > 6.4$ mm

k_1 and k_2 = factors related to material properties and connection geometry as follows:

$$K_1 = -1 + \sqrt{2(1 + R_e) + \frac{2 F_{yb}(1 + 2R_e) D^2}{3 F_{em} p^2}} \tag{9.59}$$

$$K_2 = -1 + \sqrt{\frac{2(1 + R_e)}{R_e} + \frac{2 F_{yb}(1 + 2R_e) D^2}{3 F_{em} t_s^2}} \tag{9.60}$$

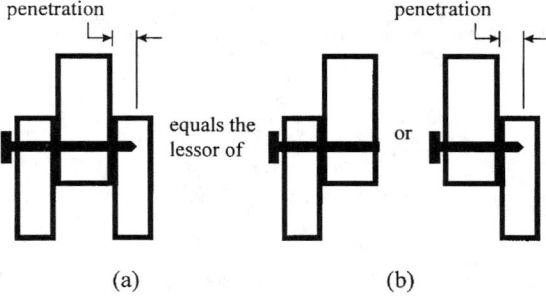

Figure 9.7   Double shear connection: (a) complete connection and (b) left and right shear planes.

Similarly, the reference lateral resistance of a single wood screw in single shear is taken as the least value determined by the three governing modes:

$$I_s: \quad Z = \frac{3.3 D t_s F_{es}}{K_D} \tag{9.61}$$

$$III_s: \quad Z = \frac{3.3K_3 Dt_s F_{em}}{K_D(2 + R_e)} \tag{9.62}$$

$$IV: \quad Z = \frac{3.3D^2}{K_D} \sqrt{\frac{1.75 F_{em} F_{yb}}{3(1 + R_e)}} \tag{9.63}$$

where K_D is defined for wood screws as it was for nails and spikes, and $K_3 = $ a factor related to material properties and connection geometry as follows:

$$K_3 = -1 + \sqrt{\frac{2(1 + R_e)}{R_e} + \frac{F_{yb}(2 + R_e)D^2}{2F_{em}t_s^2}} \tag{9.64}$$

For nail, spike, or wood screw connections with steel side plates, the above equations for yield mode I_s are not appropriate. Rather, the resistance for that mode should be computed as the bearing resistance of the fastener on the steel side plate. Also, when double shear connections are designed (Figure 9.7a), the reference lateral resistance is taken as twice the resistance of the weaker single shear representation of the left and right shear planes (Figure 9.7b).

For multi-nail, spike, or wood screw connections, the least resistance, as determined from Equations 9.55 through 9.58 for nails and spikes or Equations 9.61 through 9.63 for wood screws, is simply multiplied by the number of fasteners, n_f, in the connection detail. When multiple fasteners are used, the minimum spacing between fasteners in a row is $10D$ for wood side plates and $7D$ for steel side plates, and the minimum spacing between rows of fasteners is $5D$. Whether a single or a multiple nail, spike, or wood screw connection is used, the minimum distance from the end of a member to the nearest fastener is $15D$ with wood side plates and $10D$ with steel side plates for tension members, and $10D$ with wood side plates and $5D$ with steel side plates for compression members. Additionally, the minimum distance from the edge of a member to the nearest fastener is $5D$ for an unloaded edge, and $10D$ for a loaded edge.

The reference lateral resistance must be multiplied by all the appropriate adjustment factors. It is necessary to consider penetration depth, C_d, and end grain, C_{eg}, for nails, spikes, and wood screws. For nails and spikes, the minimum penetration allowed is $6D$, while for wood screws this minimum is $4D$. The penetration depth factor, $C_d = p/12D$, is applied to nails and spikes when the penetration depth is greater than the minimum but less than $12D$. Nails and spikes with a penetration depth greater than $12D$ assume $C_d = 1.0$. The penetration depth factor, $C_d = p/7D$, is applied to wood screws when the penetration depth is greater than the minimum but less than $7D$. Wood screws with a penetration depth greater than $7D$ assume $C_d = 1.0$. Whenever a nail, spike, or wood screw is driven into the end grain of a member, the end grain factor, $C_{eg} = 0.68$, is applied to the reference resistance. Finally, in addition to C_d and C_{eg}, a toe-nail factor, $C_{tn} = 0.83$, is applied to nails and spikes for "toe-nail" connections. A toe-nail is typically driven at an angle of approximately $30°$ to the member.

Axial Resistance

For connections loaded axially, tension is of primary concern and is governed by either fastener capacity (e.g., yielding of the nail) or fastener withdrawal. The tensile resistance of the fastener (i.e., nail, spike, or screw) is determined using accepted metal design procedure. The reference withdrawal resistance for nails and spikes with undeformed shanks in the side grain of the member is given by

$$Z_w = 31.6 DG^{2.5} pn_f \tag{9.65}$$

where $Z_w = $ reference withdrawal resistance in Newtons and $G = $ specific gravity of the wood. For nails and spikes with deformed shanks, design values are determined from tests and supplied by fastener manufactures, or Equation 9.65 can be used conservatively with $D = $ least shank diameter. For wood screws in the side grain,

$$Z_w = 65.3 DG^2 pn_f \tag{9.66}$$

A minimum wood screw depth of penetration of at least 25 mm or one-half the nominal length of the screw is required for Equation 9.66 to be applicable. No withdrawal resistance is assumed for nails, spikes, or wood screws used in end grain applications.

The end grain adjustment factor, C_{eg}, and the toe-nail adjustment factor, C_{tn}, as defined for lateral resistance, are applicable to the withdrawal resistances. The penetration factor is not applicable, however, to withdrawal resistances.

Combined Load Resistance

The adequacy of nail, spike, and wood screw connections under combined axial tension and lateral loading is checked using the following interaction equation:

$$\frac{Z_u \cos \alpha}{\lambda \phi_z Z'} + \frac{Z_u \sin \alpha}{\lambda \phi_z Z'_w} \le 1.0 \tag{9.67}$$

where α = angle between the applied load and the wood surface (i.e., $0°$ = lateral load and $90°$ = withdrawal/tension).

9.8.2   Bolts, Lag Screws, and Dowels

Bolts, lag screws, and dowels are commonly used to connect larger dimension members and when larger connection capacities are required. The provisions presented here are valid for bolts, lag screws, and dowels with diameters in the range of 6.3 mm $\le D \le$ 25.4 mm.

Lateral Resistance

The reference lateral resistance of a bolt or dowel in single shear is taken as the least value determined by the six governing modes:

$$I_m: \quad Z = \frac{0.83 D t_m F_{em}}{K_\theta} \tag{9.68}$$

$$I_s: \quad Z = \frac{0.83 D t_s F_{es}}{K_\theta} \tag{9.69}$$

$$II: \quad Z = \frac{0.93 K_1 D F_{es}}{K_\theta} \tag{9.70}$$

$$III_m: \quad Z = \frac{1.04 K_2 D t_m F_{em}}{K_\theta (1 + 2 R_e)} \tag{9.71}$$

$$III_s: \quad Z = \frac{1.04 K_3 D t_s F_{em}}{K_\theta (2 + R_e)} \tag{9.72}$$

$$IV: \quad Z = \frac{1.04 D^2}{K_\theta} \sqrt{\frac{2 F_{em} F_{yb}}{3(1 + R_e)}} \tag{9.73}$$

where

D	$=$	shank diameter
t_m and t_s	$=$	thickness of the main and side member, respectively
F_{em}	$=$	F_{es} = dowel bearing strength of the main and side member, respectively
R_e	$=$	ratio of dowel bearing strength of the main member to that of the side member $= F_{em}/F_{es}$
F_{yb}	$=$	bending yield strength of the dowel fastener (i.e., nail or spike in this case)
K_θ	$=$	factor related to the angle between the load and the main axis (parallel to the grain) of the member $= 1 + 0.25(\theta/90)$
$K_1, K_2,$ and K_3	$=$	factors related to material properties and connection geometry as follows:

$$K_1 = \frac{\sqrt{R_e + 2R_e^2(1 + R_t + R_t^2) + R_t^2 R_e^3} - R_e(1 + R_t)}{1 + R_e} \tag{9.74}$$

$$K_2 = -1 + \sqrt{2(1 + R_e) + \frac{2F_{yb}(1 + 2R_e)D^2}{3F_{em}t_m^2}} \tag{9.75}$$

$$K_3 = -1 + \sqrt{\frac{2(1 + R_e)}{R_e} + \frac{2F_{yb}(1 + 2R_e)D^2}{3F_{em}t_s^2}} \tag{9.76}$$

where R_t = ratio of the thickness of the main member to that of the side member = t_m/t_s.

The reference lateral resistance of a bolt or dowel in double shear is taken as the least value determined by the four governing modes:

$$I_m : \quad Z = \frac{0.83 D t_m F_{em}}{K_\theta} \tag{9.77}$$

$$I_s : \quad Z = \frac{1.66 D t_s F_{es}}{K_\theta} \tag{9.78}$$

$$III_s : \quad Z = \frac{2.08 K_3 D t_s F_{em}}{K_\theta(2 + R_e)} \tag{9.79}$$

$$IV : \quad Z = \frac{2.08 D^2}{K_\theta} \sqrt{\frac{2F_{em} F_{yb}}{3(1 + R_e)}} \tag{9.80}$$

where K_3 is defined by Equation 9.76.

Similarly, the reference lateral resistance of a single lag screw in single shear is taken as the least value determined by the three governing modes:

$$I_s : \quad Z = \frac{0.83 D t_s F_{es}}{K_\theta} \tag{9.81}$$

$$III_s : \quad Z = \frac{1.19 K_4 D t_s F_{em}}{K_\theta(2 + R_e)} \tag{9.82}$$

$$IV : \quad Z = \frac{1.11 D^2}{K_\theta} \sqrt{\frac{1.75 F_{em} F_{yb}}{3(1 + R_e)}} \tag{9.83}$$

where K_4 = a factor related to material properties and connection geometry as follows:

$$K_4 = -1 + \sqrt{\frac{2(1 + R_e)}{R_e} + \frac{F_{yb}(2 + R_e)D^2}{2F_{em}t_s^2}} \tag{9.84}$$

When double shear lag screw connections are designed, the reference lateral resistance is taken as twice the resistance of the weaker single shear representation of the left and right shear planes as was described for nail and wood screw connections.

Wood members are often connected to non-wood members with bolt and lag screw connections (e.g., wood to concrete, masonry, or steel). For connections with concrete or masonry main members, the dowel bear strength, F_{em}, for the concrete or masonry can be assumed the same as the wood side members with an effective thickness of twice the thickness of the wood side member. For connections with steel side plates, the equations for yield modes I_s and I_m are not appropriate. Rather, the resistance for that mode should be computed as the bearing resistance of the fastener on the steel side plate.

For multi-bolt, lag screw, and dowel connections, the least resistance is simply multiplied by the number of fasteners, n_f, in the connection detail. When multiple fasteners are used, the minimum spacings, edge distances, and end distances are dependent on the direction of loading. When loading is dominantly parallel to the grain, the minimum spacing between fasteners in a row (parallel to the grain) is $4D$, and the minimum spacing between rows (perpendicular to the grain) of fasteners is $1.5D$ but not greater than 127 mm.[3] The minimum edge distance is dependent on l_m = length of the fastener in the main member for spacing in the main member or total fastener length in the side members for side member spacing relative to the diameter of the fastener. For shorter fasteners ($l_m/D \le 6$), the minimum edge distance is $1.5D$, while for longer fasteners ($l_m/D > 6$), the minimum edge distance is the greater of $5D$ or one-half the spacing between rows (perpendicular to the grain). The minimum end distance is $7D$ for tension members and $4D$ for compression members. When loading is dominantly perpendicular to the grain, the minimum spacing within a row (perpendicular to the grain) is typically limited by the attached member but not to exceed 127 mm,[3] and the minimum spacing between rows (parallel to the grain) is dependent on l_m. For shorter fastener lengths ($l_m/D \le 2$), the spacing between rows is limited to $2D$; for medium fastener lengths ($2 < l_m/D < 6$), the spacing between rows is limited to $(5l_m + 10D)/8$; and for longer fastener lengths ($l_m/D \ge 6$), the spacing is limited to $5D$; but never should the spacing exceed than 127 mm.[3] The minimum edge distance is $4D$ for loaded edges and $1.5D$ for unloaded edges. Finally, the minimum end distance for a member loaded dominantly perpendicular to the grain is $4D$.

The reference lateral resistance must be multiplied by all the appropriate adjustment factors. It is necessary to consider group action, C_g, and geometry, C_Δ, for bolts, lag screws, and dowels. In addition, penetration depth, C_d, and end grain, C_{eg}, need to be considered for lag screws. The group action factor accounts for load distribution between bolts, lag screw, or dowels when one or more rows of fasteners are used and is defined by

$$C_g = \frac{1}{n_f} \sum_{i=1}^{n_r} a_i \qquad (9.85)$$

where n_f = number of fasteners in the connection, n_r = number of rows in the connection, and a_i = effective number of fasteners in row i due to load distribution in a row and is defined by

$$a_i = \left(\frac{1 + R_{EA}}{1 - m} \right) \left[\frac{m(1 - m^{2n_i})}{(1 + R_{EA}m^{n_i})(1 + m) - 1 + m^{2n_i}} \right] \qquad (9.86)$$

where

$$m = u - \sqrt{u^2 - 1} \qquad (9.87a)$$

$$u = 1 + \gamma \frac{s}{2} \left(\frac{1}{(EA)_m} + \frac{1}{(EA)_s} \right) \qquad (9.87b)$$

---

[3]The limit of 127 mm can be violated if allowances are made for dimensional changes of the wood.

and where γ = load/slip modulus for a single fastener; s = spacing of fasteners within a row; $(EA)_m$ and $(EA)_s$ = axial stiffness of the main and side member, respectively; R_{EA} = ratio of the smaller of $(EA)_m$ and $(EA)_s$ to the larger of $(EA)_m$ and $(EA)_s$. The load/slip modulus, γ, is either determined from testing or assumed as $\gamma = 0.246D^{1.5}$ kN/mm for bolts, lag screws, or dowels in wood-to-wood connections or $\gamma = 0.369D^{1.5}$ kN/mm for bolts, lag screws, or dowels in wood-to-steel connections.

The geometry factor, C_Δ, is used to adjust for connections in which either end distances and/or spacing within a row does not meet the limitations outlined previously. Defining a = actual minimum end distance, a_{min} = minimum end distance as specified previously, s = actual spacing of fasteners within a row, and s_{min} = minimum spacing as specified previously, the lessor of the following geometry factors are used to reduce the connection's adjusted resistance:

1. End distance:

$$\text{for,} \quad a \geq a_{min}, \qquad C_\Delta = 1.0$$
$$\text{for,} \quad a_{min}/2 \leq a < a_{min}, \quad C_\Delta = a/a_{min}$$

2. Spacing:

$$\text{for,} \quad s \geq s_{min}, \qquad C_\Delta = 1.0$$
$$\text{for,} \quad 3D \leq s < s_{min}, \quad C_\Delta = s/s_{min}$$

In addition to group action and geometry, the penetration depth factor, C_d, and end grain factor, C_{eg}, are applicable to lag screws (not bolts and dowels). The penetration of a lag screw, including the shank and thread less the threaded tip, is required to be at least $4D$. For penetrations of at least $4D$ but not more than $8D$, the connection resistance is multiplied by $C_d = p/8D$, where p = depth of penetration. For penetrations of at least $8D$, $C_d = 1.0$. The end grain factor, C_{eg}, is applied when a lag screw is driven in the end grain of a member and is given as $C_{eg} = 0.67$.

Axial Resistance

Again, the tensile resistance of the fastener (i.e., bolt, lag screw, or dowel) is determined using accepted metal design procedure. Withdrawal resistance is only appropriate for lag screws since bolts and dowels are "through-member" fasteners. For the purposes of lag screw withdrawal, the penetration depth, p, is assumed as the threaded length of the screw less the tip length, and the minimum penetration depth for withdrawal is the lessor of 25 mm or one-half the threaded length. The reference withdrawal resistance of a lag screw connection is then given by

$$Z_w = 92.6D^{0.75}G^{1.5}pn_f \tag{9.88}$$

where Z_w = reference withdrawal resistance in Newtons and G = specific gravity of the wood.

The end grain adjustment factor, C_{eg}, is applicable to the withdrawal resistance of lag screws and is defined as $C_{eg} = 0.75$.

Combined Load Resistance

The resistance of a bolt, dowel, or lag screw connection to combined axial and lateral load is given by:

$$Z'_\alpha = \frac{Z'Z'_w}{Z' \sin^2 \alpha + Z'_w \cos^2 \alpha} \tag{9.89}$$

where Z'_α = adjusted resistance at an angle α = angle between the applied load and the wood surface (i.e., $0°$ = lateral load and $90°$ = withdrawal/tension).

9.8.3   Other Types of Connections

A multitude of other connection types are available for design, including split rings, shear plates, truss plate connectors, joist hangers, and many other types of connectors. Many of the connection types are proprietary (e.g., truss plates, joist hangers, etc.), and as such their design resistances are provided by the fastener manufacture/producer.

9.8.4   NDS® Provisions

The basic approach used for the design of dowel type connections is identical between that presented here based on the LRFD Specification and that of the NDS®. The NDS® is less restrictive and, perhaps, helpful with respect to minimum edge distances, end distances, and spacings. For nails, spikes, and wood screws, the NDS® stipulates that the minimum edge distances, end distances, and spacings must be such that splitting of the wood is avoided. The other notable difference between the LRFD procedure and that of the NDS® is the group action factor for bolts, lag screws, and dowels. The group action factor, C_g, prescribed in the NDS® is given by

$$C_g = \left(\frac{1 + R_{EA}}{1 - m}\right)\left[\frac{m(1 - m^{2n})}{n[(1 + R_{EA}m^n)(1 + m) - 1 + m^{2n}]}\right] \tag{9.90}$$

where n = number of fasteners in the row, and all other factors are as defined previously. This equation is essentially equivalent to the factor a_i used to calculate the LRFD group action factor. The difference between this group action factor and that presented for the LRFD is that the LRFD accounts for load sharing in all rows, while the NDS® bases the adjustment of load sharing within one row of fasteners.

9.9   Structural Panels

Structural-use panels are wood-based panel products bonded with waterproof adhesives. Currently, structural-use panels include plywood, **oriented strand board (OSB)**, and composite panels (reconstituted wood-based material with wood veneer faces). The intended use for structural-use panels is primarily for floor, roof, and wall sheathing in residential, commercial, and industrial applications; therefore, these products typically must resist bending and shear stress in the panel without excessive deformation, and resist racking shear (i.e., diaphragm behavior). Due to the numerous types and formulations of structural-use panels, a performance-based system was developed [21]. Structural-use panels are qualified based on performance specifications for specified **span ratings.** As such, panels of various thickness and composition may be qualified for the same span rating; therefore, the designer is required to specify both a panel thickness and span rating. A span rating is a set of two numbers separated by a slash (e.g., 24/0, 32/16, 48/24, etc.) or a single number (e.g., 16 o.c., 24 o.c., etc.). When two numbers are provided, the first number indicates the allowable span (in inches) if the panel is used as a roof sheathing with the primary axis perpendicular to the rafters, and the second number indicates the allowable span (in inches) if the panel is used as a floor sheathing with the primary axis perpendicular to the joists. When only one number is provided, it is the maximum allowed spacing of the supporting members (in inches), and use is typically specific to either floor or wall applications.

The APA qualifies structural-use, performance rated panels into four types: Rated Sheathing, Structural I Rated Sheathing, Rated Sturd-I-Floor, and Rated Siding. Rated Sheathing is intended for use as subflooring and wall and roof sheathing, and will carry a span rating such as 24/16, 40/20, etc. Structural I Rated Sheathing is intended for use where shear and cross-panel properties are of importance, such as in diaphragms and shearwalls. A very common span rating for Structural I sheathing is 32/16. Common thicknesses for both Rated and Structural I Rated Sheathing range from nominally 8 mm to 19 mm. Sturd-I-Floor is specifically designed as a floor sheathing that can be used as both subflooring and underlayment for carpeted floors. Since the intended use is for floor applications only, one span number is provided in the rating (e.g., 16 o.c., 20 o.c., etc.). Nominal thicknesses range from 15 mm to 29 mm. Rated Siding is produced for exterior siding and can be manufactured to included textured surfaces for visual appearance. Since the intended use is for exterior siding applications only, one span number is provided in the rating (e.g., 16 o.c., 24 o.c., etc.) which indicates the maximum **stud** spacing allowed for use in an APA Sturd-I-Wall application. Nominal thicknesses range from 9 mm to 16 mm. It should be noted that the span ratings can be modified (i.e., decreased or increased) by local codes.

Although the performance-based product standards provide a simplified method for selecting an appropriate sheathing material, in many instances the designer is required to calculate resistances directly and/or the specific design is not well-suited or covered by the limited performance-based applications. In these cases, the LRFD Manual [1] provides reference properties for various structural-use panels, and the APA [6] provides similar properties for allowable stress design. Additionally, due to the composite nature of structural-use panels, design section properties are provided by the LRFD Manual and APA.

9.9.1 Panel Section Properties

Structural-use panels are composite, non-homogeneous, anisotropic panels, and as such would be difficult to analyze for routine application. To simplify the design process, effective section properties are provided for the primary and secondary directions on a per-unit-width basis. When the normal stress is parallel to the primary axis, the section properties for that direction should be used. Such would be the case when a structural-use panel is used to sheath a floor or roof assembly by laying the primary axis perpendicular to the joists or rafters (see Figure 9.2). As an example, the effective moment of inertia for plywood is calculated using transformed sections assuming a ratio of 35:1 for the modulus of elasticity parallel vs. perpendicular to the grain.

9.9.2 Panel Design Values

Design stresses (nominal in the case of LRFD and allowable for ASD), are based on two basic factors: the Grade Stress Level and Species Group. The Grade Stress Level is the grade of the panel and is designated by S-1 (highest grade), S-2, and S-3. The Species Group is used to classify the panel into one of four groups, designated by 1, 2, 3, or 4. Often, panels are made from more than one species, thus the necessity of creating Species Groups. If span-rated panels are used, then a relationship between the performance type (e.g., Structural I), panel thickness, span-rating, and the appropriate Grade Stress Level and Species Group is provided to determine applicable design stresses.

To further simplify the derivation of design stresses, tables of nominal resistances (i.e., M', V', T', $E'I$, etc.) are provided for the basic panel types based on span rating and thickness. This alleviates the need to determine both effective section properties and design stresses, both of which are complicated by (1) the composite, anisotropic nature of the products; (2) the diversity of structural-use panel products; as well as (3) the proprietary nature of the panel industry.

Design values and nominal resistances for structural-use panels, as with all structural wood products, are provided at specified reference conditions. In addition to the standard adjustments for wet service, C_M, and temperature, C_t, adjustments for width effects, C_w, and grade/construction, C_G, must be considered. The width factor, C_w, accounts for increased panel resistance for narrow sections. The published values are for 610 mm or wider. The grade/construction factor, C_G, is used when the properties of a particular panel differ from that of the reference grade or when panel materials have layups different than that for which the reference values are published. In both cases, the factors are specified by the product producer.

9.9.3 Design Resources

The NDS® does not cover the design of structural panels. The APA and Structural Board Association (SBA) are the primary groups responsible for providing design information to the designer for structural-use panels. The LRFD *Manual* [1] has included significant information relevant to load and resistance factor design of panels, but traditional ASD procedures are maintained by APA and SBA (e.g., see [6, 7, 17]). Additionally, a complete presentation of the allowable stress design of panels is given by McLain [15].

9.10 Shear Walls and Diaphragms

The primary lateral force-resisting systems used in timber structures are shear walls and diaphragms. Shear walls are structural wall assemblies designed to resist lateral forces applied in the plane of the wall and transmit those forces down to the base of the wall. A typical wood stud shear wall is illustrated in Figure 9.8. Diaphragms are horizontal (or nearly horizontal) systems which transmit lateral forces to the shear walls or other lateral force-resisting systems. A floor (see Figure 9.2) or roof system is often designed as a diaphragm as well as a gravity-load resisting system.

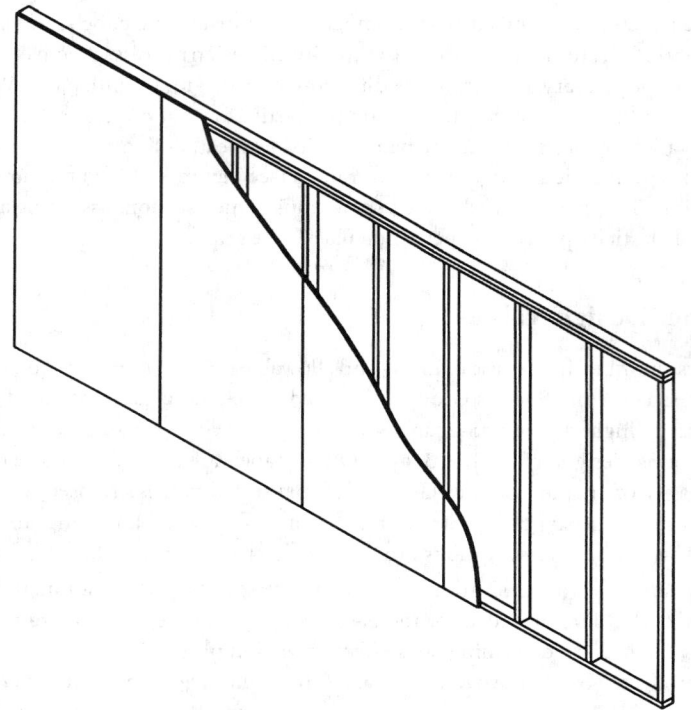

Figure 9.8   Typical shear wall assembly.

Shear walls and diaphragms are designed such that

$$D_u \leq \lambda \phi_z D'$$

(9.91)

where

D_u =   shear wall or diaphragm force due to factored loads

D' = adjusted shear wall or diaphragm resistance
λ = time effect factor
ϕ_z = resistance factor for shear walls and diaphragms = 0.65

Typically, D_u and D' are taken as the force applied and resisted, respectively, per unit length of wall or diaphragm. The LRFD Specification allows a simplified design procedure based on a beam analogy; however, more refined and accurate analysis and design procedures are allowed if the designer wishes to use one (e.g., finite element analysis). In the beam analog, the shear wall and/or diaphragm and its individual components are considered as thin, deep beams with the sheathing resisting the in-plane shear and the boundary members resisting the axial forces. This equates the sheathing to the web of an I-beam section and the boundary elements to the flanges of the I-beam. Boundary elements (e.g., studs, sills, rim-joists, etc.) must be included at all shear wall and diaphragm perimeters and around interior openings or other discontinuities.

9.10.1   Required Resistance

The required resistance of a shear wall or diaphragm, D_u, comes directly from the governing factored lateral load combination considering wind and/or seismic loads. Consideration must be made of loads acting both along each of the structure's principal axes and orthogonal effects. For more on the governing load combinations and their applications, refer to the ASCE Load Standard [4].

9.10.2   Shear Wall and Diaphragm Resistance

The reference resistance of a shear wall and/or diaphragm is generated one of several ways: (1) through experimental tests for specific assemblies, (2) using the beam analogy, or (3) through the use of more rigorous analysis procedures (e.g., finite element analysis). To determine the resistance of a shear wall or diaphragm, consideration should be given to sheathing, framing connection resistance and spacing, sheathing capacity and configuration, blocking, and framing capacity and spacing. For many typical assemblies (e.g., specific type and layout of structural sheathing, nail and nail spacing, and size, species, and spacing framing members), tables are available which provide a reference resistance per unit length (e.g., APA has numerous technical and design publications related to diaphragms and shear walls). Finally, a complete and adequate load transfer path from the shear wall or diaphragm to the supporting system must be designed.

The adjusted resistance of a shear wall and/or diaphragm, D', is determined by adjusting the reference resistance by all the adjustment factors previously discussed in this section. In addition, a diaphragm factor, $C_{di} = 1.10$, is used to account for the increase in resistance of nails used in diaphragms over that specified for single nail connections, recalling that no group action factor was available for nail connections.

9.10.3   Design Resources

Guidelines for shear wall and diaphragm design is provided in a number of publications by APA and is actually prescribed in many model building codes, but is not covered in the NDS®. The LRFD brings information from APA and other sources into a single document and covers shear wall and diaphragm design; but, additional information, such as that provided by the APA, may be helpful. Additionally, a good discussion of diaphragm and shear wall design is presented by Diekmann [9].

9.11   Trusses

One of the most popular structural uses of structural lumber is in wood trusses. This includes both trusses designed for field construction with bolted or nailed gusset plates, as well as the pre-engineered truss. Individual members of both truss types are sized following the procedures outlined previously in this section for bending, tension, and compression members. The connections for pre-engineered trusses, however, typically involve proprietary metal plate connectors (MPC). MPCs are light gauge galvanized steel and come in a variety of sizes, shapes, and configurations, depending on the plate producer. The sheet stock is punched to produce projections (teeth) that are pressed into the truss members to form virtually any planer joint configuration required for a truss. Since MPCs are proprietary and not interchangeable, metal plate connected trusses, including the connections, are designed by the truss supplier per design data provided by the building designer. The responsibilities of each party are outlined by the WTCA [22]. The building designer is required to design the structure suitable for supporting the truss and specifications for all truss, including orientations, spans, profiles, and minimum design loads and deflection performance. The building designer is also required to insure adequate permanent bracing and connection of the truss(es) to the structure. The truss supplier is required to design the truss per accepted standards (e.g., [5, 19]) to meet the requirements set forth by building designer, temporary and permanent bracing requirements, permitted loads, and calculated deflections. A good overview of the truss industry and the design and construction of wood truss systems is provided by Trus Joist MacMillan [18].

9.12 Curved Beams and Arches

The provisions discussed in this section are applicable to the design of glued laminated timber (as well as solid sawn lumber and SCL); however, the advantage of the laminating process, beyond providing large dimensioned members, is the ability to fabricate curved and arched sections. Such members are quite popular for medium to large structures requiring large open spans, such as churches, school gymnasiums, etc. Curved beams and arches of constant cross-section are discussed here; however, design provisions and procedures are available in the LRFD Specification and Manual [1, 5] for pitched and tapered curved beams and arches.

9.12.1 Curved Beams

The design of curved beams is identical to that of straight members, except that a curvature adjustment factor, C_c, is used to adjust the reference moment resistance and radial stresses need to be checked. A curved beam is defined as a member with a radius of curvature of the inside face of a lamination, R_f, defined such that $R_f \geq 100t$ for hardwoods and southern pine, and $R_f \geq 125t$ for other softwoods.

Moment Resistance

The adjusted moment resistance of a straight section was given by Equation 9.9. This basic equation is still valid for curved beams and arches, and the curvature factor, C_c, is used to modify the published resistance of a straight, glued, laminated section of constant cross-section to account for curvature on the bending capacity. The curvature factor is defined by

$$C_c = 1 - 2000 \left(\frac{t}{R_f} \right)^2 \tag{9.92}$$

where t = thickness of individual laminations. This factor is applied to the published reference bending strength or reference moment resistance just as any other adjustment factor. The remainder of the flexural design is then identical to that of straight beams.

Radial Stresses

Radial stresses occur in curved members in the radial direction (or transverse to the bending stresses). Tension occurs when the bending moments act to flatten the beam (or increase the radius of curvature), and compression occurs when moments tend to increase curvature (or decrease the radius of curvature). The result is commonly a tension stress perpendicular to the grain.

To account for radial stresses induced by bending moments in curved beams, the bending resistance limited by radial stress can be calculated using:

$$M' = \frac{2}{3} R_m b d F'_r \tag{9.93}$$

where
M' = adjusted moment resistance of the curved beam limited by radial stress
R_m = radius of curvature at the mid-depth of the cross-section
b = width of the section
d = depth of the section
F'_r = adjusted strength in the radial direction

The value of F'_r depends on the type of material, whether the stress is tension or compression, and whether radial reinforcement is provided. For example, F'_r = adjusted radial tension (perpendicular to the grain) strength when the radial stress is tension and no reinforcement is provided; $F'_r = F'_v/3$ when the radial stress is tension, Douglas-fir-Larch, Douglas-fir-South, Hem-Fir, Western Woods, or Canadian softwood species are used, and the design load is either wind or earthquake, or if reinforcement is provided to carry the radial force; or F'_r = adjusted radial compression (perpendicular to the grain) strength when the radial stress is compression. The radial strengths should be adjusted for other factors, such as temperature and moisture content.

9.12.2 Arches

The primary difference between curved beams and arches is the degree of arching; that is, when the radius of curvature is small enough, arch action is induced and axial forces become an integral part of the structural resistance. Arches can be assumed when $R_f < 100t$ for hardwoods and southern pine, and $R_f < 125t$ for other softwoods. The two basic types of glued laminated arches are three-hinged and two-hinged arches (See Figure 9.9), and include a variety of styles such as tudor, gothic, radial, parabolic, etc. The three-hinged arch is statically determinate owing to the moment release at the peak. The design of three-hinged arches are, therefore, relatively straightforward and should consider the combined bending moment and compression parallel to the grain along the length of the arch, and shear near the member ends. Two-hinged arches are statically indeterminate, and an appropriate method of analysis must be used to determine design moments, shears, and axial loads along the length of the arch. Once these forces and moments are determined, the design is again straightforward and should consider the combined bending moment and compression parallel to the grain along the length of the arch, and shear near the member ends. Two-hinged arches are slightly more efficient than three-hinged arches; however, constructability and transportation issues often dictate the use of three-hinged arches over two-hinged arches.

Moment Resistance

Again, the adjusted moment resistance of a straight section was given by Equation 9.9. This basic equation is still valid for arches, except now, in addition to the curvature factor defined for curved beams, the volume effect factor, C_V, is for arches as follows:

$$\text{1. For } F'_b(1 - C_V) \le f_c: \qquad C'_V = 1.0 \tag{9.94}$$

$$\text{2. For } F'_b(1 - C_V) > f_c: \qquad C'_V = C_V + f_c/F'_b \tag{9.95}$$

where
F'_b = adjusted bending strength of the arch, including all applicable adjustment factors except C_V

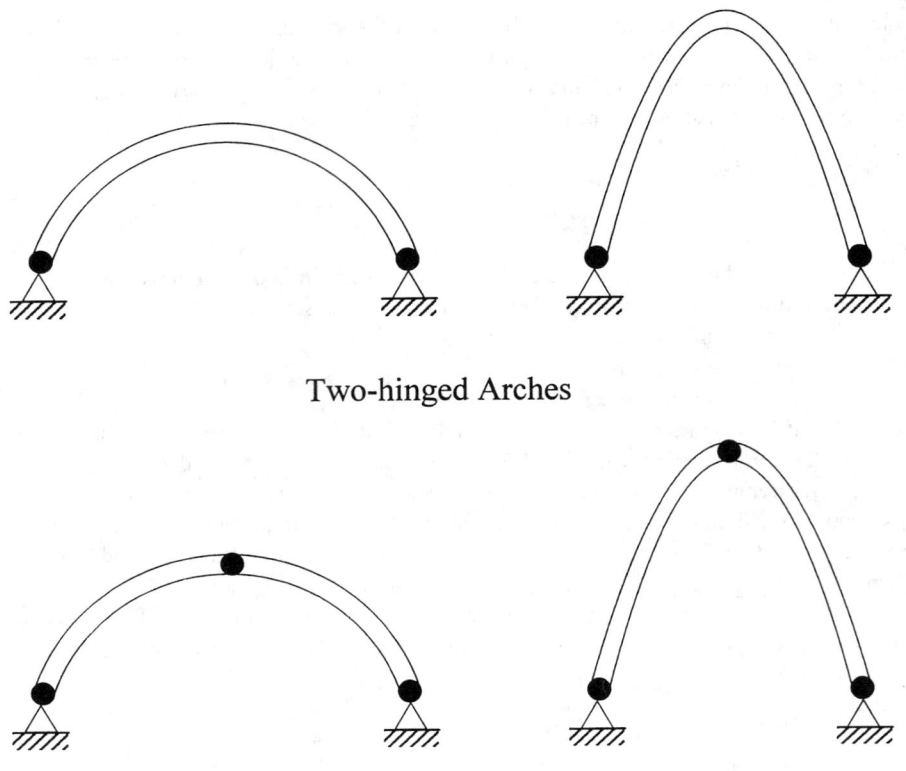

Two-hinged Arches

Three-hinged Arches

Figure 9.9   Two- and three-hinge arches.

C_V' = modified or corrected volume effect factor
f_c = compressive stress in the arch considering factored loads

   The remainder of the flexural design is then identical to that of straight and curved beams.

Axial Compression Resistance

   Again, the main difference between a curved beam and an arch is that the curvature has been increased to a point where axial forces become significant in arches. The procedures outlined previously for axial resistance are considered valid for arches. Typically, arches are considered to be fully braced against buckling about the weak axis and unbraced about the strong axis; however, design for buckling about the strong axis is typically not required due to arch action. Regardless, the bracing requirements presented for compression and lateral torsional buckling are considered applicable to arches.

Interaction Equation for Arches

   The interaction of bending moments and axial forces is accounted for in the same manner as presented for straight columns. However, since arches are often fully braced against buckling in the weak direction, the factored moment need not be magnified. Therefore, the interaction equation given by Equation 9.40 reduces to

$$\left(\frac{P_u}{\lambda \phi_c P_c'} \right)^2 + \frac{M_{bx}}{\lambda \phi_b M_x'} \leq 1.0 \tag{9.96}$$

for arches where P_u = factored axial load; P'_c = adjusted axial compression resistance; M_{bx} = factored moment about the strong axis; M'_x = adjusted moment capacity about the strong axis; λ = time effect factor; and ϕ_c and ϕ_b = resistance factors for axial compression and bending, respectively.

Radial Stresses

Due to the arch action and geometry, radial stresses are induced in arches and must be checked. The procedure to determine a moment resistance is limited by radial stress and is identical to that presented for curved beams.

9.12.3 Design Resources

The NDS® [2] provides design procedures essentially identical to that used in the LRFD Specification [5]. A good overview of glued laminated curved beam and arch deign is provided by Kasaguma [14]. Additional and more detailed design information and guidelines for curved, pitched, and tapered beams and arches are provide by the American Institute of Timber Construction (AITC) in numerous design aids as well as in the *Timber Construction Manual* [3].

9.13 Serviceability Considerations

Previous discussion in this section has focused entirely on strength design (i.e., strength resistance), which is intended to insure adequate safeguard against structural collapse. As with all types of structure, the serviceability performance of a timber structure is critical, not from a safety viewpoint, but rather for occupant satisfaction. Structural serviceability considerations (e.g., deflections, vibrations, etc.) often govern the design of a timber structure. Additionally, the lack of designer attention to detailing, especially detailing which protects wood from moisture, ultraviolet, and/or insect attack, often leads to performance and even structural problems.

9.13.1 Deflections

The most common serviceability design check is to compare the static (short-term) deflection resulting from service loads to some limiting quantity. Historically, this limit has been *span*/360 for floors subjected to a full uniform floor live load and *span*/240 for roof members subjected to roof live load. Model and local building codes may delineate more specific limits for static deflections.

To calculate the deflection used to compare to the limiting value, the appropriate design load, span, and material properties must be used. The following unfactored load combinations are used to determine the deflection:

$$D + L \tag{9.97}$$
$$D + 0.5S \tag{9.98}$$

where D, L, and S are the nominal design dead, live, and snow loads. The design span is taken as the clear span plus one-half the required bearing length at each support, and the adjusted modulus of elasticity, E', is used for serviceability calculations. The reference modulus of elasticity, E, for lumber is a mean value and includes a 3% reduction to account for shear effects; therefore, if shear deflections are calculated separately, it is appropriate to increase the reference value by 3%. For glued laminated timber, E includes a 5% decrease for shear. In some critical applications, it may be appropriate to use a fifth-percentile value of E rather than a mean value. In this case, E_{05} can be determined as presented in Equation 9.14.

In addition to short-term deflections, the effects of creep must be considered in the design of timber structures. The LRFD suggests using the following load combination when considering creep:

$$D + 0.5L \tag{9.99}$$

where it is assumed approximately half the live load has a sustained component which may attribute to creep. In addition, the LRFD suggests increasing the calculated deflection (using the above load combination) by 50 to 100%. A creep multiplier of 1.5 is suggested for glued laminated timber and seasoned sawn lumber, and a factor of 2.0 is used for unseasoned lumber. Recent research by Fridley [12] and Philpot et al. [16] provide a more complete treatment of creep for design, including various creep factors dependent on the load combination and reference period (time) under consideration. Regardless, the multiplier is used to magnify the calculated elastic deflection to account for creep effects.

9.13.2   Vibrations

Current serviceability design of wood joist floor systems is based on static deflection checks as discussed previously. In the U.S., there are no set guidelines to account for floor vibrations. This often leads to objectionable vibrations due to occupant loading. With the use of I-joists and other new, lightweight products, increasing spans can be achieved along with a decrease in the weight of the floor. These aspects tend to cause an increase in the number of complaints about annoying vibrations. Manufactures of proprietary products (e.g., wood I-joists) have moved toward providing span tables for serviceability which address vibrations. Often, this is as simple as limiting deflection to *span*/480 rather than *span*/360. A number of more rigorous methods for vibration design have been presented in the literature (e.g., see Kalkert et al. [13]). The LRFD, in its commentary, reviews several of these approaches, but makes no formal recommendation as to a preferred method of analysis or design.

9.13.3   NDS® Provisions

The LRFD adopted the same serviceability deflection design methodology as that provided by NDS® with respect to static (or short-term) and creep (or long-term) deflection. Being an ASD specification, the NDS® does not, however, provide separate load combinations for serviceability design. Rather, the same working/service loads are used for strength considerations. The NDS® offers no recommendations for vibrations either.

9.13.4   Non-Structural Performance

Often out of the control of the structural designer, the misconstruction and mis-installation of wood, wood materials, and building envelops can effectively shorten the life of a timber structure. From moisture intrusion to insect attack and decay, wood must be protected, or at least guarded from detrimental environments and conditions. A complete presentation of detailing to avoid problems is beyond the scope of this section, but Dost and Botsai [10] provide a comprehensive guide to detailing timber structures for performance with specific attention to protecting the structure and wood-based materials from deterioration. With proper detailing, and attention to details, a timber structure can enjoy a long and useful life, even in exposed environments.

9.14   Defining Terms

Dimensioned lumber: Solid members from 38 to 89 mm in thickness and 33 mm or greater in width (depth).

Dowel-type fasteners: Fasteners such as bolts, lag screws, wood screws, nails, spikes, etc.

Dry service: Structures wherein the maximum equilibrium moisture content of the wood does not exceed 19%.

Equilibrium moisture content: The moisture content at which wood neither gains moisture from nor loses moisture to the surrounding air.

Fiber saturation point: The point at which free water in the cell cavities is completely evaporated, but the cell walls are still saturated.

Glued laminated timber (Glulam): A composite wood member with wood laminations bonded together with adhesive and with the grain of all laminations oriented parallel longitudinally.

Grade: Classification of structural wood products with reference to strengh and use.

Green lumber: Lumber that has a moisture content exceeding 19%.

I-joists: Prefabricated manufactured members using solid or composite lumber flanges and structural panel webs.

Laminated veneer lumber (LVL): Composite lumber products comprised of wood veneer sheets oriented with the grain parallel to the lenght of the member.

Load-duration (time-effect): The strenght of wood is a function of the duration and/or rate the load is applied to the member. That is, a member can resist higher magnitude loads for shorter durations or, stated differently, the longer a load is applied, the less able a wood member is to support that load.

Machine-evaluated lumber (MEL): Lumber non-destructively evaluated and classified into strength classifications.

Machine stress-rated (MSR) lumber: Lumber non-destructively evaluated and assigned bending strength and modulus of elasticity design values.

Moisture content: The amount of moisture present in wood defined by the weight of water contained in the wood as a percentage of the weight of the oven-dry wood.

Oriented strandboard (OSB): Structural panel comprised of thin flakes oriented in cross-aligned layers.

Parallel strand lumber (PSL): Composite lumber products comprised of wood strands oriented with the grain parallel to the lenght of the member.

Plywood: Structural panel comprised of thin veneers oriented in cross-aligned layers.

Pressure-treated wood: Products treated to make them more resistant to biological attack or environmental conditions.

Repetitive member systems: A system such as a floor, roof, or wall assembly which consists of closely spaced parallel members exhibiting load sharing among the members.

Seasoned lumber: Lumber that has been dried to a maximum of 19% moisture content.

Shear plate: Circular plate embedded between two wood members or a wood member and a steel side plate with a single bolt to transfer shear from the wood to the bolt.

Sheathing: Lumber or structural panel attached to framing members, typically forming a wall, floor, or roof.

Span rating: Index used for structural panels identifying the maximum spacing of supporting framing members for roof, floor, and/or wall applications.

Split ring: Metal ring embedded into adjacent wood members to transmit shear force between them.

Stressed skin panel: Type of construction wherein the framing members and sheathing act as a composite system to resist load.

Structural composite lumber (SCL): Composite lumber products typically comprised of wood strands or veneers oriented with the grain parallel to the length of the member.

Stud: Vertical framing member used in walls, typically 2 × 4 and 2 × 6.

Timbers: Lumber of greater than 114 mm in its least dimension.

Truss-plate connectors: Light gage steel plates with punched teeth typically used in prefabricated wood truss systems.

Visually graded lumber: Lumber graded visually and assigned engineering design values.

Wet service: Structures wherein in the maximum equilibrium moisture content of the wood exceeds 19%.

References

[1] American Forest & Paper Association (AF&PA). 1996. *Manual of Wood Construction: Load and Resistance Factor Design,* AF&PA, Washington, D.C.

[2] American Forest & Paper Association (AF&PA). 1991. *National Design Specification for Wood Construction and Supplement,* AF&PA, Washington, D.C.

[3] American Institute of Timber Construction (AITC). 1994. *Timber Construction Manual,* 4th ed., John Wiley & Sons, New York.

[4] American Society of Civil Engineers (ASCE). 1996. *Minimum Design Loads for Buildings and Other Structures,* ASCE 7-75, ASCE, New York.

[5] American Society of Civil Engineers (ASCE). 1996. *Load and Resistance Factor Design (LRFD) for Engineered Wood Construction,* AF&PA/ASCE 16-95, ASCE, New York.

[6] APA—The Engineered Wood Association (APA). 1989. *The Plywood Design Specification and Supplements,* APA, Tacoma, WA.

[7] APA—The Engineered Wood Association (APA). 1993. *Design/Construction Guide, Residential and Commercial Construction,* APA, Tacoma, WA.

[8] Bodig, J. and Jayne, B. 1982. *Mechanics of Wood and Wood Composites,* Van Nostrand Reinhold, New York.

[9] Diekmann, E.F. 1995. Diaphragms and Shearwalls, in *Wood Engineering and Construction Handbook,* 2nd ed., K.F. Faherty and T.G. Williamson, Eds., McGraw Hill, New York.

[10] Dost, W.A. and Botsai, E.E. 1990. *Wood: Detailing for Performance,* GRDA Publications, Mill Valley, CA.

[11] Freas, A.D. 1995. Wood Properties, in *Wood Engineering and Construction Handbook,* 2nd ed., K.F. Faherty and T.G. Williamson, Eds., McGraw-Hill, New York.

[12] Fridley, K.J. 1992. Designing for Creep in Wood Structures, *Forest Products J.,* 42(3):23-28.

[13] Kalkert, R.E., Dolan, J.D., and Woeste, F.E. 1993. The Current Status of Analysis and Design for Annoying Wooden Floor Vibrations, *Wood Fiber Sci.,* 25(3):305-314.

[14] Kasaguma, R.K. 1995. Arches and Domes, in *Wood Engineering and Construction Handbook,* 2nd ed., K.F. Faherty and T.G. Williamson, Eds., McGraw-Hill, New York.

[15] McLain, T.E. 1995. Structural-Use Panels, in *Wood Engineering and Construction Handbook,* 2nd ed., K.F. Faherty and T.G. Williamson, Eds., McGraw-Hill, New York.

[16] Philpot, T.A., Rosowsky, D.V., and Fridley, K.J. 1993. Serviceability Design in LRFD for Wood, *J. Structural Eng.,* 119(12):3649-3667.

[17] Structural Board Association (SBA). 1995. *SpecRite:* Industrial, Residential, Codes and Grades, and Specifications on Disk. SBA, Onterio, Canada.

[18] Trus Joist MacMillan. 1995. Trusses, in *Wood Engineering and Construction Handbook,* 2nd ed., K.F. Faherty and T.G. Williamson, Eds., McGraw-Hill, New York.

[19] Truss Plate Institute (TPI). 1995. *National Design Specification for Metal Plate Connected Wood Truss Construction,* ANSI/TPI 1-1995, TPI, Madison, WI.

[20] United States Department of Agriculture (USDA). 1987. *Wood Handbook: Wood as an Engineering Material,* Agriculture Handbook 72, Forest Products Laboratory, USDA, Madison, WI.

[21] United States Department of Commerce. 1992. *Performance Standard for Wood-Based Structural-Use Panels,* PS 2-92, Washington D.C.

[22] Wood Truss Council of America (WTCA). 1995. *Standard Responsibilities in the Design Process Involving Metal Plate Connected Wood Trusses,* WTCA 1-1995, WTCA, Madison, WI.

Further Reading

[1] Breyer, D. 1993. *Design of Wood Structures,* 3rd ed., McGraw-Hill, New York.

[2] Faherty, K. and Williamson, T. 1995. *Wood Engineering and Construction Handbook,* 2nd ed., McGraw-Hill, New York.

[3] Gurfinkle, G. 1981. *Wood Engineering,* 2nd ed., Kendall/Hunt Publishing (available from Southern Forest Products Association, Box 524468, New Orleans, LA 70152).

[4] Hoyle, R. and Woeste, F. *Wood Technology in the Design of Structures,* 5th ed., Iowa State University Press, Ames, IA.

[5] Madsen, B. 1992. *Structural Behaviour of Timber,* Timber Engineering Ltd., N. Vancouver, British Columbia, Canada.

[6] Ritter, M. 1990. *Timber Bridges: Design, Construction, Inspection, and Maintenance,* United States Department of Agriculture, Forest Service, Washington, D.C.

[7] Somayaji, S. 1990. *Structural Wood Design,* West Publishing Company, New York.

[8] Stahlnaker, J. and Harris, E. 1989. *Structural Design in Wood,* Van Nostrand Reinhold, New York.

[9] *Western Woods Use Book,* 4th ed., 1996. Western Wood Products Association, Portland, OR.

[10] *Wood Design Manual.* 1990. Canadian Wood Council, Ottawa, Canada.

[11] *Wood as a Structural Material.* 1982. Clark C. Heritage Memorial Series on Wood, Vol. 2, Pennsylvania State University, University Park, PA.

10

Bridge Structures

Shouji Toma

Department of Civil Engineering,
Hokkai-Gakuen University,
Sapporo, Japan

Lian Duan

Division of Structures, California
Department of Transportation,
Sacramento, CA

Wai-Fah Chen

School of Civil Engineering,
Purdue University,
West Lafayette, IN

10.1 General 10-1
10.2 Steel Bridges 10-8
10.3 Concrete Bridges 10-13
10.4 Concrete Substructures 10-25
10.5 Floor System 10-28
10.6 Bearings, Expansion Joints, and Railings 10-31
10.7 Girder Bridges 10-34
10.8 Truss Bridges 10-42
10.9 Rigid Frame Bridges (Rahmen Bridges) 10-45
10.10 Arch Bridges 10-47
10.11 Cable-Stayed Bridges 10-49
10.12 Suspension Bridges 10-53
10.13 Defining Terms 10-59
Acknowledgment 10-60
References 10-60
Further Reading 10-61
Appendix: Design Examples 10-61

10.1 General

10.1.1 Introduction

A **bridge** is a structure that crosses over a river, bay, or other obstruction, permitting the smooth and safe passage of vehicles, trains, and pedestrians. An elevation view of a typical bridge is shown in Figure 10.1. A bridge structure is divided into an upper part (the **superstructure**), which consists of the slab, the **floor system**, and the main truss or **girders**, and a lower part (the **substructure**), which are columns, piers, towers, footings, piles, and **abutments**. The superstructure provides horizontal spans such as deck and girders and carries traffic loads directly. The substructure supports the horizontal spans, elevating above the ground surface. In this chapter, main structural features of common types of steel and concrete bridges are discussed. Two design examples, a two-span continuous, cast-in-place, **prestressed concrete** box **girder bridge** and a three-span continuous, composite plate girder bridge, are given in the Appendix.

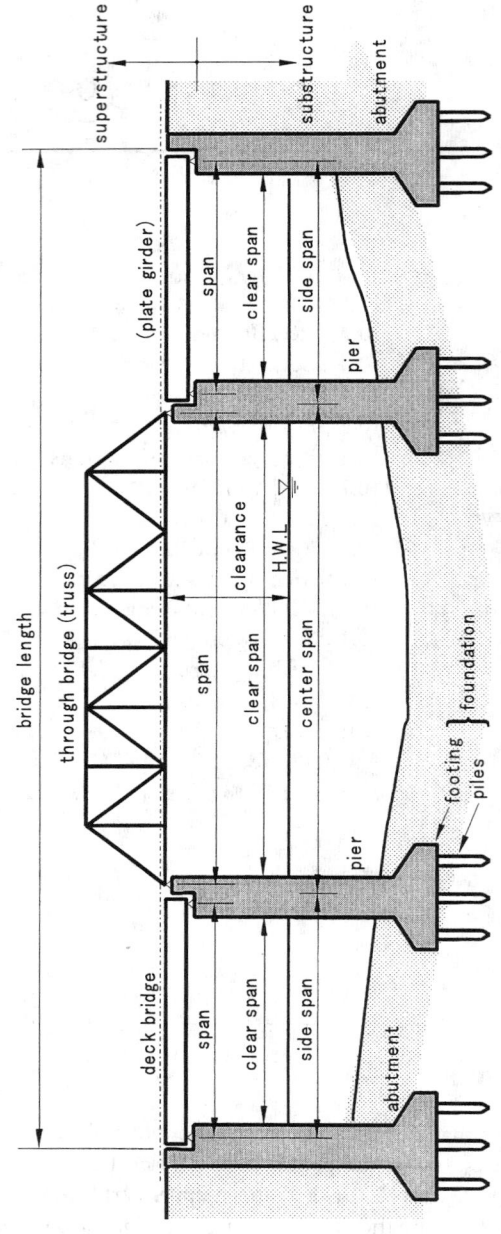

Figure 10.1   Elevation view of a typical bridge.

10.1.2 Classification

1. **Classification by Materials**

   Steel bridges: A steel bridge may use a wide variety of structural steel components and systems: girders, frames, trusses, arches, and suspension cables.

   Concrete bridges: There are two primary types of concrete bridges: reinforced and prestressed.

   Timber bridges: Wooden bridges are used when the span is relatively short.

   Metal alloy bridges: Metal alloys such as aluminum alloy and stainless steel are also used in bridge construction.

2. **Classification by Objectives**

   Highway bridges: bridges on highways.

   Railway bridges: bridges on railroads.

   Combined bridges: bridges carrying vehicles and trains.

   Pedestrian bridges: bridges carrying pedestrian traffic.

   Aqueduct bridges: bridges supporting pipes with channeled waterflow.

   Bridges can alternatively be classified into movable (for ships to pass the river) or fixed and permanent or temporary categories.

3. **Classification by Structural System (Superstructures)**

   Plate girder bridges: The main girders consist of a plate assemblage of upper and lower flanges and a web. H- or I-cross-sections effectively resist bending and shear.

   Box girder bridges: The single (or multiple) main girder consists of a box beam fabricated from steel plates or formed from concrete, which resists not only bending and shear but also torsion effectively.

   T-beam bridges: A number of reinforced concrete T-beams are placed side by side to support the live load.

   Composite girder bridges: The concrete **deck slab** works in conjunction with the steel girders to support loads as a united beam. The steel girder takes mainly tension, while the concrete slab takes the compression component of the bending moment.

   Grillage girder bridges: The main girders are connected transversely by floor beams to form a grid pattern which shares the loads with the main girders.

   Truss bridges: Truss bar members are theoretically considered to be connected with pins at their ends to form triangles. Each member resists an axial force, either in compression or tension. Figure 10.1 shows a Warren **truss bridge** with vertical members, which is a "trough bridge", i.e., the deck slab passes through the lower part of the bridge. Figure 10.2 shows a comparison of the four design alternatives evaluated for Minato Oh-Hasshi in Osaka, Japan. The truss frame design was selected.

   Arch bridges: The arch is a structure that resists load mainly in axial compression. In ancient times stone was the most common material used to construct magnificent **arch bridges**. There is a wide variety of arch bridges as will be discussed in Section 10.10

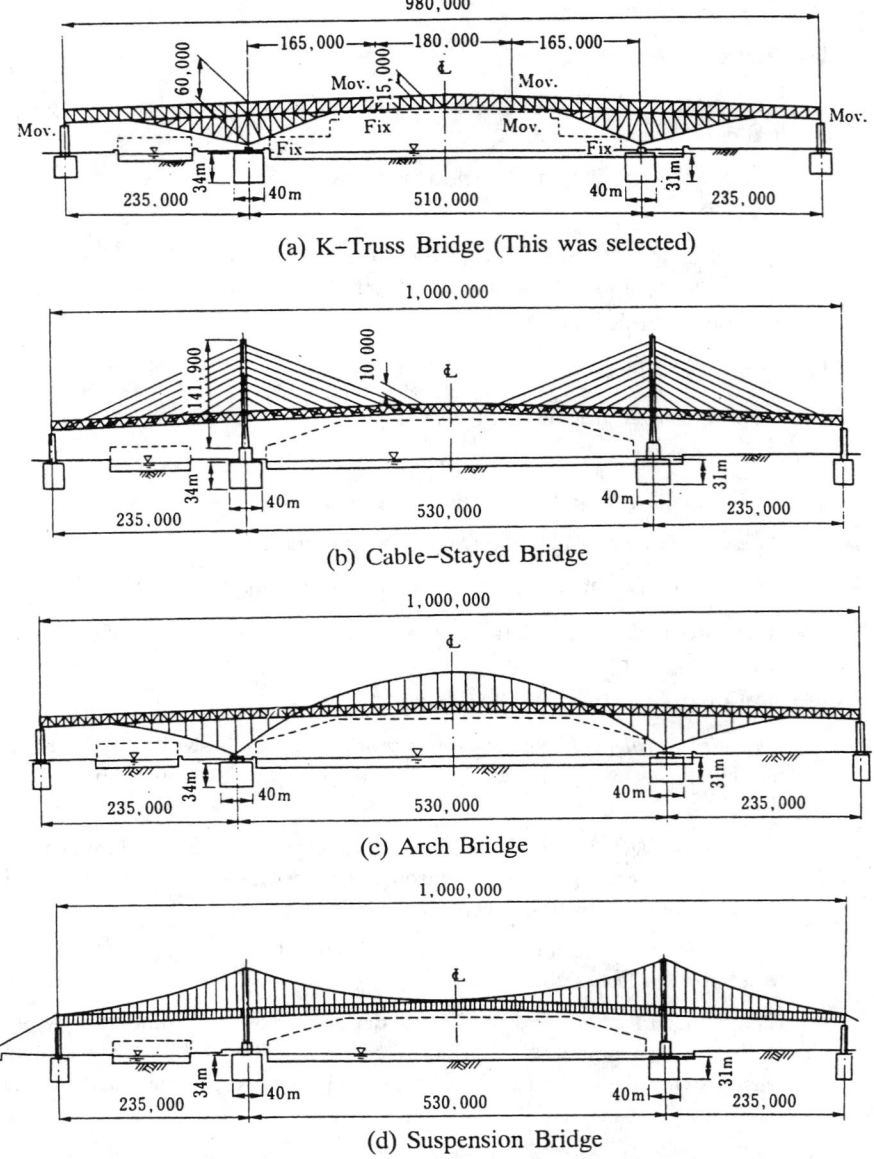

Figure 10.2   Design comparison for Minato Oh-Hashi, Japan. (From Hanshin Expressway Public Corporation, *Construction Records of Minato Oh-Hashi,* Japan Society of Civil Engineers, Tokyo [in Japanese], 1975. With permission.)

Cable-stayed bridges: The girders are supported by highly strengthened cables (often composed of tightly bound steel strands) which stem directly from the tower. These are most suited to bridge long distances.

Suspension bridges: The girders are suspended by hangers tied to the main cables which hang from the towers. The load is transmitted mainly by tension in cable. This design is suitable for very long span bridges.

Table 10.1 shows the span lengths appropriate to each type of bridge.

TABLE 10.1 Types of Bridges and Applicable Span Lengths

Bridge Type	Span range (m) / span scale: 10 20 30 40 50 60 70 80 90 100 110 120 130 140 150 160 170 180 190 200 250	Remarks
Simple Composite H-Beam		
Simple Composite Plate Girder		
Simple Composite Box Girder		
Continuous Non-Composite Plate Girder		
Continuous Non-Composite Box Girder		Zoo Br. 259. Costa e Silva Br. 300
Continuous Composite Plate Girder		Charlotte Br. 234
Steel Deck Plate Girder	Shorenji River Br. ●235	
Rigid Frame (Rahmen)	Hamana Lake Br. ●140	
Simple Truss	Mitsuishi Br. ●120	Minato-Oh-hashi 510. Quebec Br. 549
Continuous and Gerber Truss		Port-Mann Br. 366
Langer Girder	Makigi Br. ●150	
Reversed Langer Girder	New Benten Br. ●140	
Lohse Girder	Kobe-Oh-hashi ●217	
Reversed Lohse Girder	Saigo-Oh-hashi ●260	
Langer Truss		
Trussed Langer		
Nielsen Bridge		New Hamadera Br. 254. Fehmarnsund Br. 248.4
Arch Bridge		Omichima Br. 297. Bayonne Br. 504
		New River George Br. 518
Cable-Stayed Bridge		Iguchi Br. 490. Tatara Br. 890. Normandy Br. 856
Suspension Bridge		North and South Bisan SetoOhashi Br. 1,100, 990
		Akashi Br. 1,990. Humber Br. 1,410

Legend: ▬ commonly applicable range · ☐ fairly applicable range · ● special Cases

From JASBC, *Manual Design Data Book*, Japan Association of Steel Bridge Construction, Tokyo (in Japanese), 1981. With permission.

4. **Classification by Support Condition**
   Figure 10.3 shows three different support conditions for girder bridges.

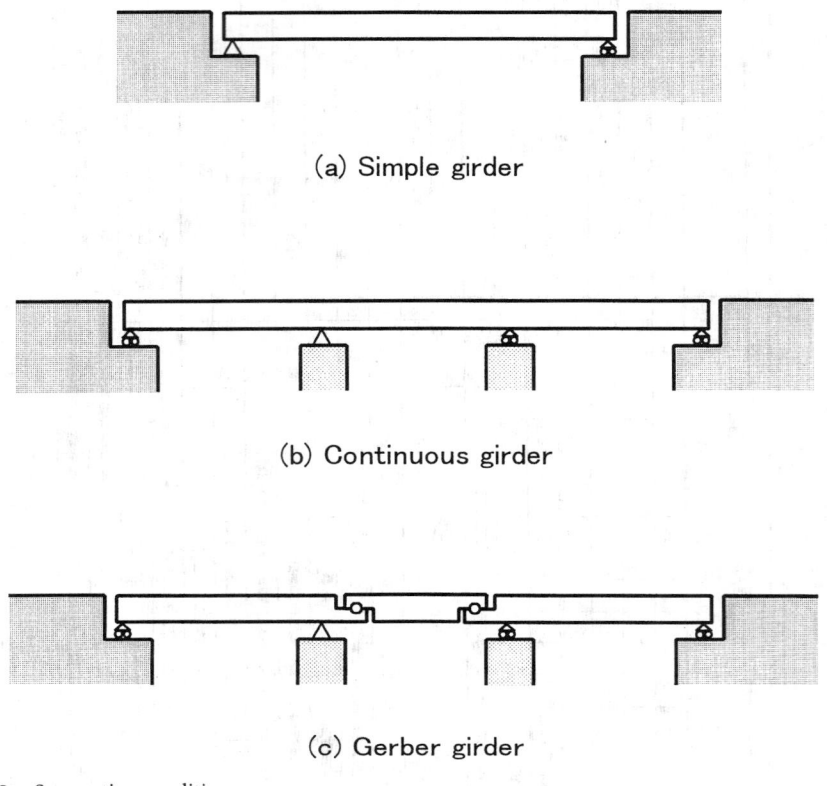

(a) Simple girder

(b) Continuous girder

(c) Gerber girder

Figure 10.3   Supporting conditions.

Simply supported bridges: The main girders or trusses are supported by a movable hinge at one end and a fixed hinge at the other (simple support); thus they can be analyzed using only the conditions of equilibrium.

Continuously supported bridges: Girders or trusses are supported continuously by more than three supports, resulting in a structurally indeterminate system. These tend to be more economical since fewer expansion joints, which have a common cause of service and maintenance problems, are needed. Sinkage at the supports must be avoided.

Gerber bridges (cantilever bridge): A continuous bridge is rendered determinate by placing intermediate hinges between the supports. Minato Oh-Hashi's bridge, shown in Figure 10.2a, is an example of a Gerber truss bridge.

10.1.3   Plan

Before the structural design of a bridge is considered, a bridge project will start with planning the fundamental design conditions. A bridge plan must consider the following factors:

1. **Passing Line and Location**

   A bridge, being a continuation of a road, does best to follow the line of the road. A right angle bridge is easy to design and construct but often forces the line to be bent. A skewed bridge or a curved bridge is commonly required for expressways or railroads where the road line must be kept straight or curved, even at the cost of a more difficult design (see Figure 10.4).

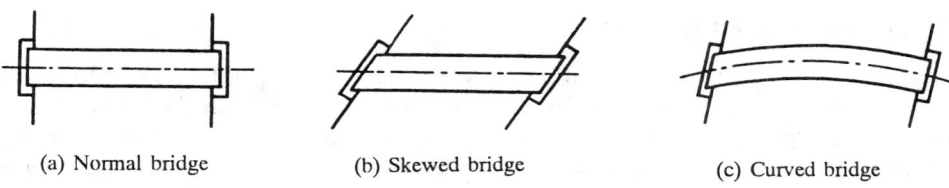

| (a) Normal bridge | (b) Skewed bridge | (c) Curved bridge |

Figure 10.4   Bridge lines.

2. **Width**

   The width of a highway bridge is usually defined as the width of the roadway plus that of the sidewalk, and often the same dimension as that of the approaching road.

3. **Type of Structure and Span Length**

   The types of substructures and superstructures are determined by factors such as the surrounding geographical features, the soil foundation, the passing line and its width, the length and span of the bridge, aesthetics, the requirement for clearance below the bridge, transportation of the construction materials and erection procedures, construction cost, period, and so forth.

4. **Aesthetics**

   A bridge is required not only to fulfill its function as a thoroughfare, but also to use its structure and form to blend, harmonize, and enhance its surroundings.

10.1.4   Design

The bridge design includes selection of a bridge type, structural analysis and member design, and preparation of detailed plans and drawings. The size of members that satisfy the requirements of design codes are chosen [1, 17]. They must sustain prescribed loads. Structural analyses are performed on a model of the bridge to ensure safety as well as to judge the economy of the design. The final design is committed to drawings and given to contractors.

10.1.5   Loads

Designers should consider the following loads in bridge design:

1. Primary loads exert constantly or continuously on the bridge.

   Dead load: weight of the bridge.

   Live load: vehicles, trains, or pedestrians, including the effect of impact. A vehicular load is classified into three parts by AASHTO [1]: the truck axle load, a tandem load, and a uniformly distributed lane load.

   Other primary loads may be generated by prestressing forces, the creep of concrete, the shrinkage of concrete, soil pressure, water pressure, buoyancy, snow, and centrifugal actions or waves.

2. Secondary loads occur at infrequent intervals.

> *Wind load:* a typhoon or hurricane.

> *Earthquake load:* especially critical in its effect on the substructure.

Other secondary loads come about with changes in temperature, acceleration, or temporary loads during erection, collision forces, and so forth.

10.1.6 Influence Lines

Since the live loads by definition move, the worst case scenario along the bridge must be determined. The maximum live load bending moment and shear envelopes are calculated conveniently using influence lines. The **influence line** graphically illustrates the maximum forces (bending moment and shear), reactions, and deflections over a section of girder as a load travels along its length. Influence lines for the bending moment and shear force of a simply supported beam are shown in Figure 10.5. For a concentrated load, the bending moment or shear at section A can be calculated by multiplying the load and the influence line scalar. For a uniformly distributed load, it is the product of the load intensity and the net area of the corresponding influence line diagram.

10.2 Steel Bridges

10.2.1 Introduction

The main part of a steel bridge is made up of steel plates which compose main girders or frames to support a concrete deck. Gas flame cutting is generally used to cut steel plates to designated dimensions. Fabrication by welding is conducted in the shop where the bridge components are prepared before being assembled (usually bolted) on the construction site. Several members for two typical steel bridges, plate girder and truss bridges, are given in Figure 10.6. The composite plate girder bridge in Figure 10.6a is a deck type while the truss bridge in Figure 10.6b is a through-deck type.

Steel has higher strength, ductility, and toughness than many other structural materials such as concrete or wood, and thus makes an economical design. However, steel must be painted to prevent rusting and also stiffened to prevent a local buckling of thin members and plates.

10.2.2 Welding

Welding is the most effective means of connecting steel plates. The properties of steel change when heated and this change is usually for the worse. Molten steel must be shielded from the air to prevent oxidization. Welding can be categorized by the method of heating and the shielding procedure. Shielded metal arc welding (SMAW), submerged arc welding (SAW), CO_2 gas metal arc welding (GMAW), tungsten arc inert gas welding (TIG), metal arc inert gas welding (MIG), electric beam welding, laser beam welding, and friction welding are common methods.

The first two welding procedures mentioned above, SMAW and SAW, are used extensively in bridge construction due to their high efficiency. Both use an electric arc, which is generally considered the most efficient method of applying heat. SMAW is done by hand and is suitable for welding complicated joints but is less efficient than SAW. SAW is generally automated and can be very effective for welding simple parts such as the connection between the flange and web of plate girders. A typical placement of these welding methods is shown in Figure 10.7. TIG and MIG use an electric arc for heat source and inert gas for shielding.

An electric beam weld must not be exposed to air, and therefore must be laid in a vacuum chamber. A laser beam weld can be placed in air but is less versatile than other types of welding. It cannot be

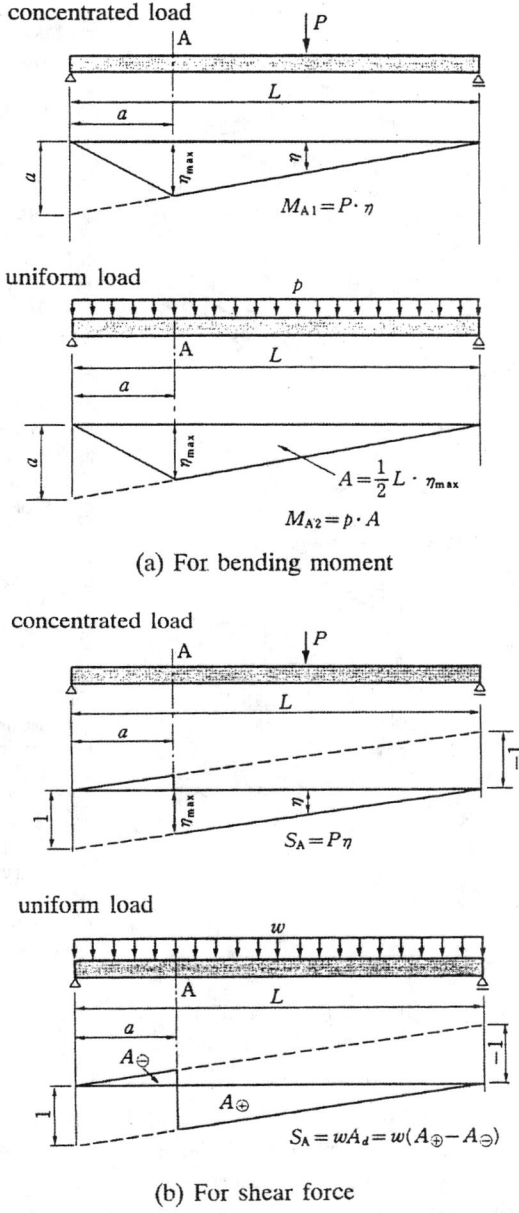

concentrated load

uniform load

(a) For bending moment

concentrated load

uniform load

(b) For shear force

Figure 10.5 Influence lines.

used on thick plates but is ideal for minute or artistic work. Since the welding equipment necessary for heating and shielding is not easy to handle on a construction site, all welds are usually laid in the fabrication shop.

The heating and cooling processes during welding induce residual stresses to the connected parts. The steel surfaces or parts of the cross section at some distance from the hot weld, cool first. When the area close to the weld then cools, it tries to shrink but is restrained by the more solidified and cooler parts. Thus, tensile residual stresses are trapped in the vicinity of the weld while the outer parts are put into compression.

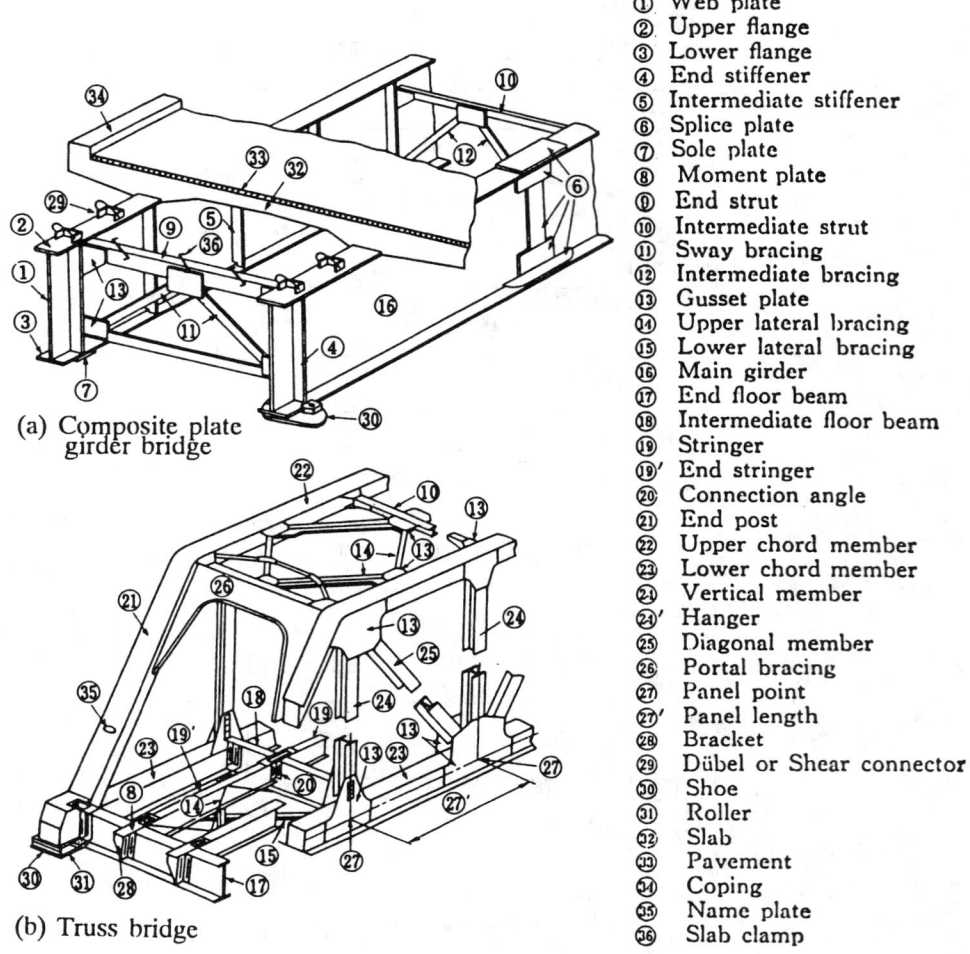

① Web plate
② Upper flange
③ Lower flange
④ End stiffener
⑤ Intermediate stiffener
⑥ Splice plate
⑦ Sole plate
⑧ Moment plate
⑨ End strut
⑩ Intermediate strut
⑪ Sway bracing
⑫ Intermediate bracing
⑬ Gusset plate
⑭ Upper lateral bracing
⑮ Lower lateral bracing
⑯ Main girder
⑰ End floor beam
⑱ Intermediate floor beam
⑲ Stringer
⑲′ End stringer
⑳ Connection angle
㉑ End post
㉒ Upper chord member
㉓ Lower chord member
㉔ Vertical member
㉔′ Hanger
㉕ Diagonal member
㉖ Portal bracing
㉗ Panel point
㉗′ Panel length
㉘ Bracket
㉙ Dübel or Shear connector
㉚ Shoe
㉛ Roller
㉜ Slab
㉝ Pavement
㉞ Coping
㉟ Name plate
㊱ Slab clamp

(a) Composite plate girder bridge

(b) Truss bridge

Figure 10.6   Member names of steel bridges. (From Tachibana, Y. and Nakai, H., *Bridge Engineering*, Kyoritsu Publishing Co., Tokyo, Japan [in Japanese], 1996. With permission.)

There are two types of welded joints: groove and fillet welds (Figure 10.8). The fillet weld is placed at the junction of two plates, often between a web and flange. It is a relatively simple procedure with no machining required. The groove weld, also called a butt weld, is suitable for joints requiring greater strength. Depending on the thickness of adjoining plates, the edges are beveled in preparation for the weld to allow the metal to fill the joint. Various groove weld geometries for full penetration welding are shown in Figure 10.8b.

Inspection of welding is an important task since an imperfect weld may well have catastrophic consequences. It is difficult to find faults such as an interior crack or a blow hole by observing only the surface of a weld. Many nondestructive testing procedures are available which use various devices, such as x-ray, ultrasonic waves, color paint, or magnetic particles. These all have their own advantages and disadvantages. For example, the x-ray and the ultrasonic tests are suitable for interior faults but require expensive equipment. Use of color paint or magnetic particles, on the other hand, is a cheap alternative but only detects surface flaws. The x-ray and ultrasonic tests are used in common bridge construction, but ultrasonic testing is becoming increasingly popular for both its "high tech" and its economical features.

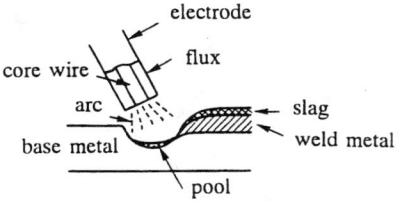

(a) Shielded Metal Arc Welding (SMAW)

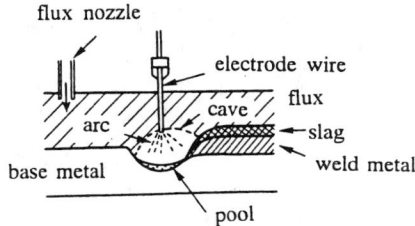

(b) Submerged Arc Welding (SAW)

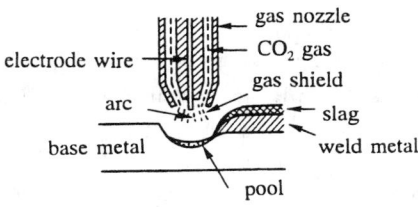

(c) Gas Metal Arc Welding (GMAW)

Figure 10.7 Welding methods. (From Nagai, N., *Bridge Engineering,* Kyoritsu Publishing Co., Tokyo, Japan [in Japanese], 1994. With permission.)

10.2.3 Bolting

Bolting does not require the skilled workmanship needed for welding, and is thus a simpler alternative. It is applied to the connections worked on construction site. Some disadvantages, however, are incurred: (1) splice plates are needed and the force transfer is indirect; (2) screwing-in of the bolts creates noise; and (3) aesthetically bolts are less appealing. In special cases that need to avoid these disadvantages, the welding may be used even for site connections.

There are three types of high-tensile strength-bolted connections: the slip-critical connection, the bearing-type connection (Figure 10.9), and the tensile connection (Figure 10.10). The slip-critical (friction) bolt is most commonly used in bridge construction as well as other steel structures because it is simpler than a bearing-type bolt and more reliable than a tension bolt. The force is transferred by the friction generated between the base plates and the splice plates. The friction resistance is induced by the axial compression force in the bolts.

The bearing-type bolt transfers the force by bearing against the plate as well as making some use of friction. The bearing-type bolt can transfer larger force than the friction bolts but is less forgiving with respect to the clearance space often existing between the bolt and the plate. These require that precise holes be drilled and at exact spacings. The force transfer mechanism for these connections is

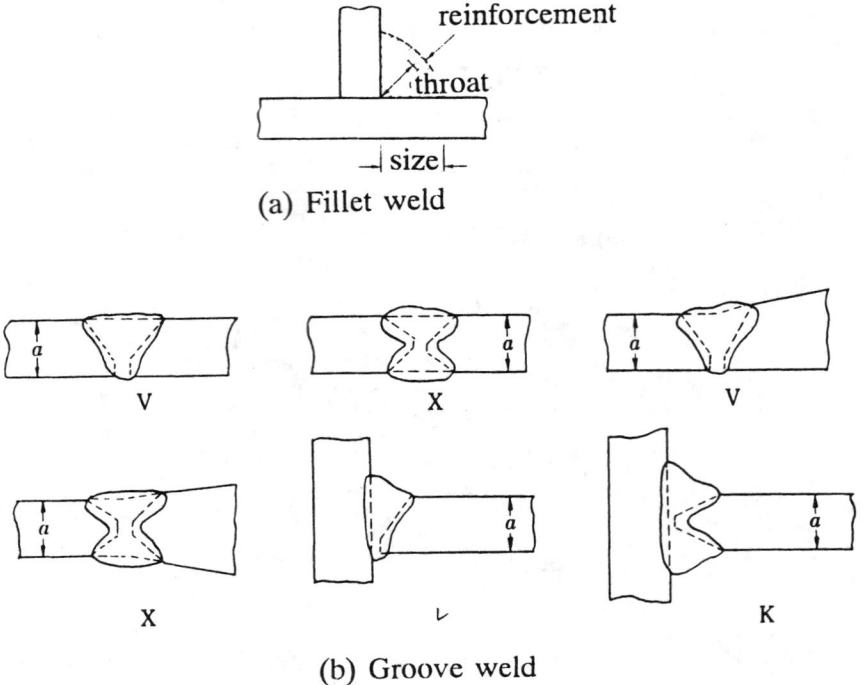

(a) Fillet weld

(b) Groove weld

Figure 10.8　Types of welding joints. (From Tachibana, Y. and Nakai, H., *Bridge Engineering*, Kyoritsu Publishing Co., Tokyo, Japan [in Japanese], 1996. With permission.)

shown in Figure 10.9. In the beam-to-column connection shown in Figure 10.10, the bolts attached to the column are tension bolts while the bolts on the beam are slip-critical bolts.

The tension bolt transfers force in the direction of the bolt axis. The tension type of bolt connection is easy to connect on site, but difficulties arise in distributing forces equally to each bolt, resulting in reduced reliability. Tension bolts may also be used to connect box members of the towers of **suspension bridges** where compression forces are larger than the tension forces. In this case, the compression is shared with butting surfaces of the plates and the tension is carried by the bolts.

10.2.4　Fabrication in Shop

Steel bridges are fabricated into members in the shop yard and then transported to the construction site for assembly. Ideally all constructional work would be completed in the shop to get the highest quality in the minimum construction time. The larger and longer the members can be, the better, within the restrictions set by transportation limits and erection tolerances. When crane ships for erection and barges for transportation can be used, one block can weigh as much as a thousand tons and be erected as a whole on the quay. In these cases the bridge is made of a single continuous block and much of the hassle usually associated with assembly and erection is avoided.

10.2.5　Construction on Site

The designer must consider the loads that occur during construction, generally different from those occurring after completion. Steel bridges are particularly prone to buckling during construction. The erection plan must be made prior to the main design and must be checked for every possible load case that may arise during erection, not only for strength but also for stability. Truck crane and

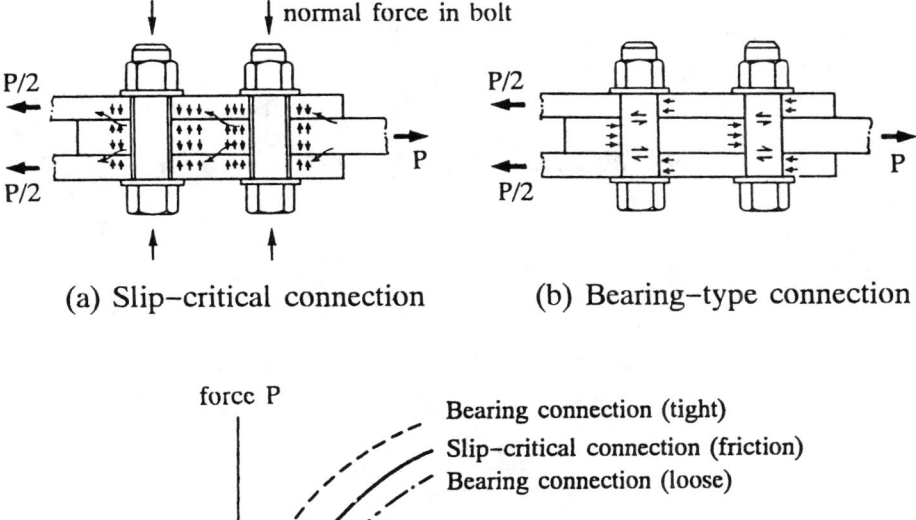

(a) Slip–critical connection   (b) Bearing–type connection

force P

Bearing connection (tight)
Slip–critical connection (friction)
Bearing connection (loose)

displacement

(c) Behavior of bolt connections

Figure 10.9   Slip-critical and bearing-type connections. (From Nagai, N., *Bridge Engineering*, Kyoritsu Publishing Co., Tokyo, Japan [in Japanese], 1994. With permission.)

bent erection (or staging erection); launching erection; cable erection; cantilever erection; and large block erection (or floating crane erection) are several techniques (see Figure 10.11). An example of the large block erection is shown in Figure 10.43, in which a 186-m, 4500-ton center block is transported by barge and lifted.

10.2.6   Painting

Steel must be painted to protect it from rusting. There is a wide variety of paints, and the life of a steel structure is largely influenced by its quality. In areas near the sea, the salty air is particularly harmful to exposed steel. The cost of painting is high but is essential to the continued good condition of the bridge. The color of the paint is also an important consideration in terms of its public appeal or aesthetic quality.

10.3   Concrete Bridges

10.3.1   Introduction

For modern bridges, both structural concrete and steel give satisfactory performance. The choice between the two materials depends mainly upon the cost of construction and maintenance. Generally, concrete structures require less maintenance than steel structures, but since the relative cost of steel and concrete is different from country to country, and may even vary throughout different parts of the same country, it is impossible to put one definitively above the other in terms of "economy".

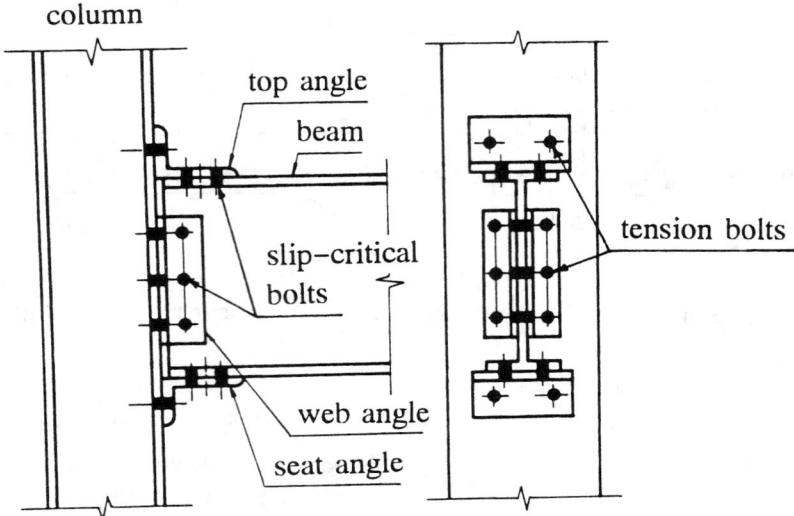

Figure 10.10   Tension-type connection.

In this section, the main features of common types of concrete bridge superstructures are briefly discussed. Concrete bridge substructures will be discussed in Section 10.4. A design example of a two-span continuous, cast-in-place, prestressed concrete box girder bridge is given in the Appendix. For a more detailed look at design procedures for concrete bridges, reference should be made to the recent books of Gerwick [7], Troitsky [24], Xanthakos [26, 27], and Tonias [23].

10.3.2   Reinforced Concrete Bridges

Figure 10.12 shows the typical reinforced concrete sections commonly used in highway bridge superstructures.

1. **Slab**

   A reinforced concrete slab (Figure 10.12a) is the most economical bridge superstructure for spans of up to approximately 40 ft (12.2 m). The slab has simple details and standard formwork and is neat, simple, and pleasing in appearance. Common spans range from 16 to 44 ft (4.9 to 13.4 m) with structural depth-to-span ratios of 0.06 for simple spans and 0.045 for continuous spans.

2. **T-Beam (Deck Girder)**

   The T-beams (Figure 10.12b) are generally economic for spans of 40 to 60 ft (12.2 to 18.3 m), but do require complicated formwork, particularly for skewed bridges. Structural depth-to-span ratios are 0.07 for simple spans and 0.065 for continuous spans. The spacing of girders in a T-beam bridge depends on the overall width of the bridge, the slab thickness, and the cost of the formwork and may be taken as 1.5 times the structural depth. The most commonly used spacings are between 6 and 10 ft (1.8 to 3.1 m).

3. **Cast-in-Place Box Girder**

   Box girders like the one shown in Figure 10.12c, are often used for spans of 50 to 120 ft (15.2 to 36.6 m). Its formwork for skewed structures is simpler than that required for the T-beam. Due to excessive dead load deflections, the use of reinforced concrete box girders over simple spans of 100 ft (30.5 m) or more may not be economical. The depth-to-span ratios are typically 0.06 for simple spans and 0.055 for continuous spans with the girders spaced at 1.5 times the structural depth. The high torsional resistance

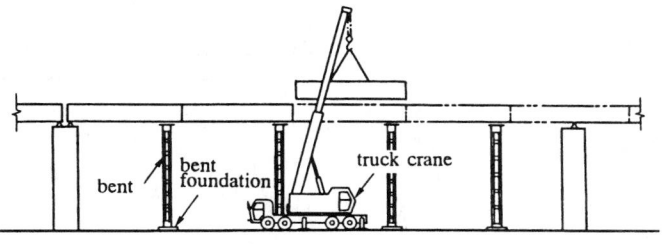

(a) Truck crane and bent erection

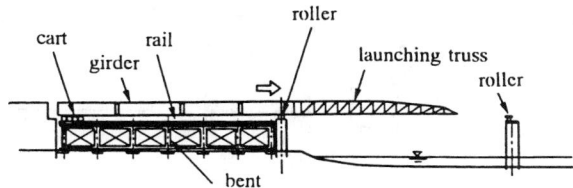

(b) Launching erection

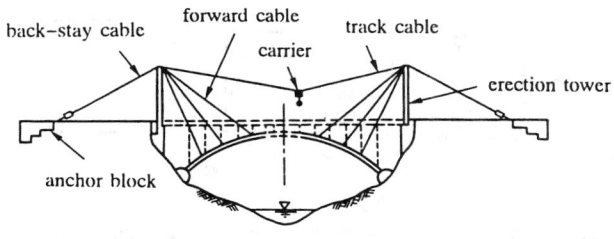

(c) Cable erection

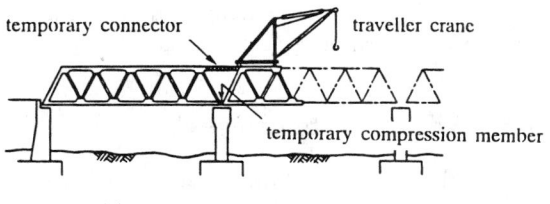

(d) Cantilever erection

Figure 10.11 Erections methods. (From Japan Construction Mechanization Association, *Cost Estimation of Bridge Erection,* Tokyo, Japan [in Japanese], 1991. With permission.)

of the box girder makes it particularly suitable for curved alignments, such as the ramps onto freeways. Its smooth flowing lines are appealing in metropolitan cities.

4. **Design Consideration**

A reinforced concrete highway bridge should be designed to satisfy the specification or code requirements, such as the AASHTO-**LRFD** [1] requirements (American Association of State Highway and Transportation Officials—Load and Resistance Factor Design) for all appropriate service, fatigue, strength, and extreme event limit states. In the AASHTO-LRFD [1], service limit states include cracking and deformation effects,

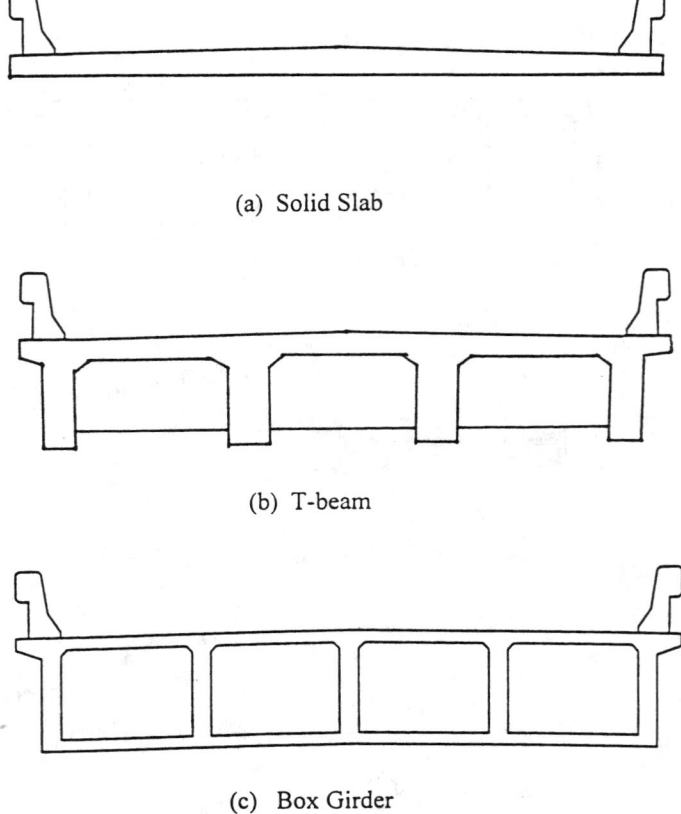

(a) Solid Slab

(b) T-beam

(c) Box Girder

Figure 10.12   Typical reinforced concrete sections in bridge superstructures.

and strength limit states consider the strength and stability of a structure. A bridge structure is usually designed for the strength limit states and is then checked against the appropriate service and extreme event limit states.

10.3.3   Prestressed Concrete Bridges

Prestressed concrete, using high-strength materials, makes an attractive alternative for long-span bridges. It has been widely used in bridge structures since the 1950s.

1. **Slab**
   Figure 10.13 shows Federal Highway Administration (FHWA) [6] standard types of precast, prestressed, voided slabs and their sectional properties. While cast-in-place, prestressed slab is more expensive than reinforced concrete slab, precast, prestressed slab is economical when many spans are involved. Common spans range from 20 to 50 ft (6.1 to 15.2 m). Structural depth-to-span ratios are 0.03 for both simple and continuous spans.

2. **Precast I Girder**
   Figure 10.14 shows AASHTO [6] standard types of I-beams. These compete with steel girders and generally cost more than reinforced concrete with the same depth-to-span ratios. The formwork is complicated, particularly for skewed structures. These sections

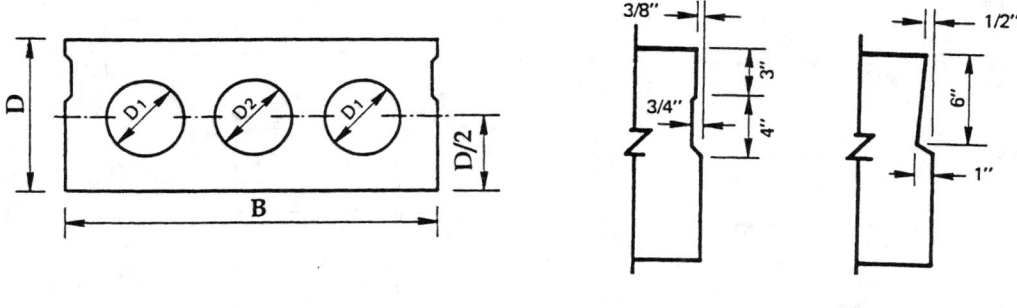

(a)   Typical Section   (b)   Alternative Shear Key

Span	Section Dimensions				Section Properties		
Range ft. (m)	width B in (mm)	Depth D in (mm)	$D1$ in (mm)	$D2$ in (mm)	A in^2 $(mm^2\ 10^6)$	I_x in^4 $(mm^4\ 10^9)$	S_x in^3 $(mm^3\ 10^6)$
25 (7.6)	48 (1,219)	12 (305)	0 (0)	0 (0)	576 (0.372)	6,912 (2.877)	1,152 (18.878)
30~35 (9.1~10.7)	48 (1,219)	15 (381)	8 (203)	8 (203)	569 (0.362)	12,897 (5.368)	1,720 (28.185)
40~45 (12.2~13.7)	48 (1,219)	18 (457)	10 (254)	10 (254)	628 (0.405)	21,855 (9.097)	2,428 (39.788)
50 (15.2)	48 (,1219)	21 (533)	12 (305)	10 (254)	703 (0.454)	34,517 (1.437)	3,287 (53.864)

Figure 10.13   Federal Highway Administration (FHWA) precast, prestressed, voided slab sections. (From Federal Highway Administration, *Standard Plans for Highway Bridges, Vol. 1, Concrete Superstructures,* U.S. Department of Transportation, Washington, D.C., 1990. With permission.)

are applicable to spans 30 to 120 ft (9.1 to 36.6 m). Structural depth-to-span ratios are 0.055 for simple spans and 0.05 for continuous spans.

3. **Box Girder**

   Figure 10.15 shows FHWA [6] standard types of precast box sections. The shape of a cast-in-place, prestressed concrete box girder is similar to the conventional reinforced concrete box girder (Figure 10.12c). The spacing of the girders can be taken as twice the structural depth. It is used mostly for spans of 100 to 600 ft (30.5 to 182.9 m). Structural depth-to-span ratios are 0.045 for simple spans and 0.04 for continuous spans. These sections are used frequently for simple spans of over 100 ft (30.5 m) and are particularly suitable for widening in order to control deflections. About 70 to 80% of California's highway bridge system is composed of prestressed concrete box girder bridges.

4. **Segmental Bridge**

   The segmentally constructed bridges have been successfully developed by combining the concepts of prestressing, box girder, and the cantilever construction [2, 20]. The first prestressed segmental box girder bridge was built in Western Europe in 1950. California's Pine Valley Bridge, as shown in Figure 10.16 (composed of three spans of 340 ft [103.6 m], 450 ft [137.2 m], and 380 ft [115.8 ft] with the pier height of 340 ft [103.6 m]), was the first cast-in-place segmental bridge built in the U.S., in 1974.

   The prestressed segmental bridges with precast or cast-in-place segmental can be classi-fied by the construction methods: (1) balanced cantilever, (2) span-by-span, (3) incre-mental launching, and (4) progressive placement. The selection between cast-in-place and precast segmental, and among various construction methods, is dependent on

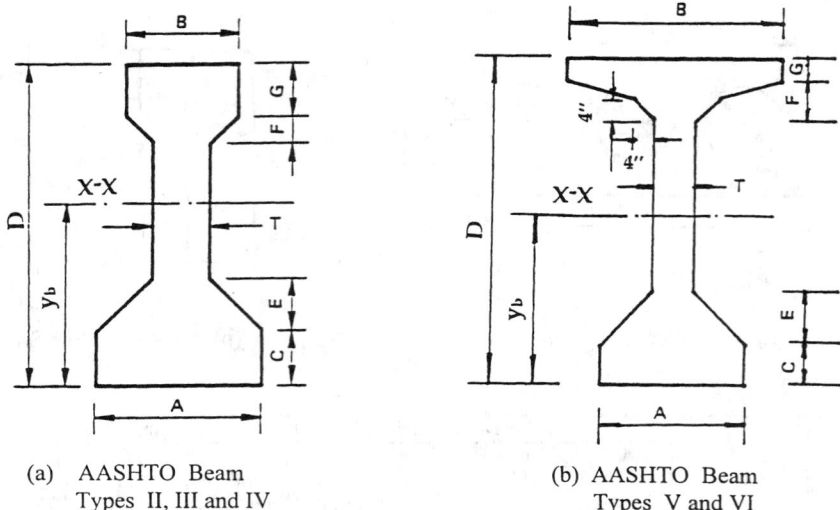

(a)   AASHTO Beam   (b) AASHTO Beam
      Types II, III and IV   Types V and VI

AASHTO	Section Dimensions in (mm)							
BEAM TYPE	Depth D	bot width A	web width T	top width B	C	E	F	G
II	36 (914)	18 (457)	6 (152)	12 (305)	6 (152)	6 (152)	3 (76)	6 (152)
III	45 (1143)	22 (559)	7 (178)	16 (406)	7 (178)	7.5 (191)	4.5 (114)	7 (178)
IV	54 (1372)	26 (660)	8 (203)	20 (508)	8 (203)	9 (229)	6 (152)	8 (203)
V	65 (1651)	28 (711)	8 (203)	42 (1067)	8 (203)	10 (254)	3 (76)	5 (127)
VI	72 (1829)	28 (711)	8 (203)	42 (1067)	8 (203)	10 (254)	3 (76)	5 (127)

AASHTO	Section Properties					
BEAM TYPE	A in^2 (mm^2 10^6)	Y_b in (mm)	I_x in^4 (mm^4 10^9)	S_b in^3 (mm^4 10^6)	S_t in^3 (mm^4 10^6)	Span Ranges ft (m)
II	369 (0.2381)	15.83 (402.1)	50,980 (21.22)	3220 (52.77)	2528 (41.43)	40 ~ 45 (12.2 ~ 13.7)
III	560 (0.3613)	20.27 (514.9)	125,390 (52.19)	6186 (101.38)	5070 (83.08)	50 ~ 65 (15.2 ~ 19.8)
IV	789 (0.5090)	24.73 (628.1)	260,730 (108.52)	10543 (172.77)	8908 (145.98)	70 ~ 80 (21.4 ~ 24.4)
V	1013 (0.6535)	31.96 (811.8)	521,180 (216.93)	16307 (267.22)	16791 (275.16)	90 ~ 100 (27.4 ~ 30.5)
VI	1085 (0.7000)	36.38 (924.1)	733,340 (305.24)	20158 (330.33)	20588 (337.38)	110 ~ 120 (33.5 ~ 36.6)

Figure 10.14   Precast, prestressed AASHTO (American Association of State Highway and Transportation Officials) I-beam sections. (From Federal Highway Administration, *Standard Plans for Highway Bridges, Vol. 1, Concrete Superstructures,* U.S. Department of Transportation, Washington, D.C., 1990. With permission.)

project features, site conditions, environmental and public constraints, construction time for the project, and equipment available. Table 10.2 lists the range of application of segmental bridges by span lengths [20].

5. **Design Consideration**

   Compared to reinforced concrete, the main design features of prestressed concrete are that stresses for concrete and prestressing steel and deformation of structures at each stage (i.e., during construction, stressing, handling, transportation, and erection as well as during the service life) and stress concentrations need to be investigated. In the following, we shall briefly discuss the AASHTO-LRFD [1] requirements for stress limits,

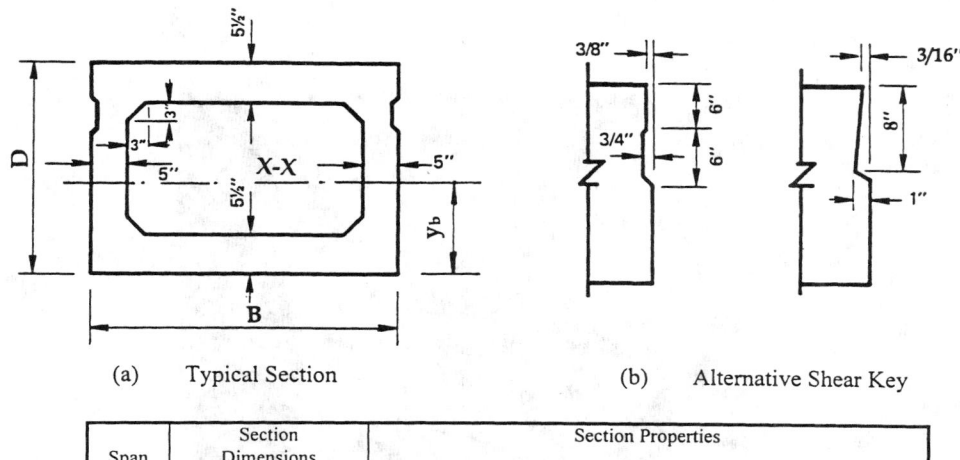

(a)   Typical Section   (b)   Alternative Shear Key

Span	Section Dimensions		Section Properties				
ft. (m)	width B in (mm)	Depth D in (mm)	A in² (mm² 10⁶)	Y_b in (mm)	I_x in⁴ (mm⁴ 10⁹)	S_b in³ (mm³ 10⁶)	S_t in³ (mm³ 10⁶)
50 (15.2)	48 (1,219)	27 (686)	693 (0.4471)	13.37 (339.6)	65,941 (27.447)	4,932 (80.821)	4,838 (79.281)
60 (18.3)	48 (1,219)	33 (838)	753 (0.4858)	16.33 (414.8)	110,499 (45.993)	6,767 (110.891)	6,629 (108.630)
70 (21.4)	48 (1,219)	39 (991)	813 (0.5245)	19.29 (490.0)	168,367 (70.080)	8,728 (143.026)	8,524 (139.683)
80 (24.4)	48 (1,219)	42 (1,067)	843 (0.5439)	20.78 (527.8)	203,088 (84.532)	9,773 (160.151)	9,571 (156.841)

Figure 10.15 Federal Highway Administration (FHWA) precast, pretensioned box sections. (From Federal Highway Administration, *Standard Plans for Highway Bridges, Vol. 1, Concrete Superstructures,* U.S. Department of Transportation, Washington, D.C., 1990. With permission.)

TABLE 10.2   Range of Application of Segmental Bridge Type by Span Length

Span ft (m)		Bridge types
0–150	(0–45.7)	I-type pretensioned girder
100–300	(30.5–91.4)	Cast-in-place post-tensioned box girder
100–300	(30.5–91.4)	Precast-balanced cantilever segmental, constant depth
200–600	(61.0–182.9)	Precast-balanced cantilever segmental, variable depth
200–1000	(61.0–304.8)	Cast-in-place cantilever segmental
800–1500	(243.8–457.2)	Cable-stay with balanced cantilever segmental

From Podolny, W. and Muller, J. M., *Construction and Design of Prestressed Concrete Segmental Bridges,* John Wiley & Sons, New York, 1982. With permission.

nominal flexural resistance, and shear resistance in designing a prestressed member.

a) *Stress Limits*

Calculations of stresses for concrete and prestressing steel are based mainly on the elastic theory. Tables 10.3 to 10.5 list the AASHTO-LRFD [1] stress limits for concrete and prestressing tendons.

b) *Nominal Flexural Resistance, M_n*

Flexural strength is based on the assumptions that (1) the strain is linearly distributed across a cross-section (except for deep flexural member); (2) the maximum usable strain at extreme compressive fiber is equal to 0.003; (3) the tensile strength of concrete is neglected; and (4) a concrete stress of 0.85 f_c' is uniformly distributed over an equivalent compression zone. For a member with a flanged section (Figure 10.17) subjected to uniaxial bending, the equations of equilibrium are used to give a nominal moment resistance of:

Figure 10.16a Pine Valley Bridge, California. Construction state. (From California Department of Transportation. With permission.)

Figure 10.16b Pine Valley Bridge, California. Construction completed. (From California Department of Transportation. With permission.)

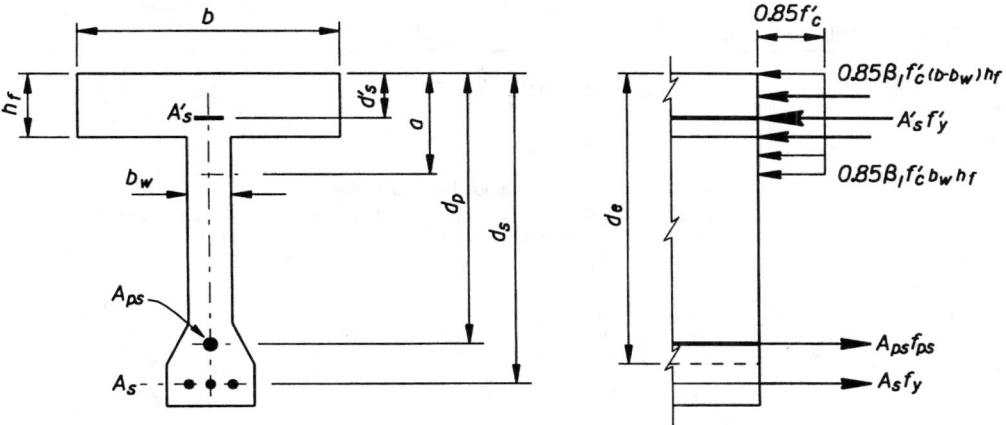

Figure 10.17 A flanged section at nominal moment capacity state.

TABLE 10.3 Stress Limits for Prestressing Tendons

Stress type	Prestressing method	Prestressing tendon type		
		Stress-relieved strand and plain high-strength bars	Low Relaxation strand	Deformed high-strength bars
At jacking (f_{pj})	Pretensioning	$0.72 f_{pu}$	$0.78 f_{pu}$	—
	Post-tensioning	$0.76 f_{pu}$	$0.80 f_{pu}$	$0.75 f_{pu}$
After transfer (f_{pt})	Pretensioning	$0.70 f_{pu}$	$0.74 f_{pu}$	—
	Post-tensioning			
	At anchorages and couplers immediately after anchor set	$0.70 f_{pu}$	$0.70 f_{pu}$	$0.66 f_{pu}$
	General	$0.70 f_{pu}$	$0.74 f_{pu}$	$0.66 f_{pu}$
At service limit state (f_{pe})	After all losses	$0.80 f_{py}$	$0.80 f_{py}$	$0.80 f_{py}$

From American Association of State Highway and Transportation Officials, *AASHTO LRFD Bridge Design Specifications*, First Edition, Washington, D.C., 1994. With permission.

$$M_n = A_{ps} f_{ps} \left(d_p - \frac{a}{2}\right) + A_s f_y \left(d_s - \frac{a}{2}\right)$$

$$- A'_s f'_y \left(d'_s - \frac{a}{2}\right) + 0.85 f'_c (b - b_w)\beta_1 h_f \left(\frac{a}{2} - \frac{h_f}{2}\right) \tag{10.1}$$

$$a = \beta c \tag{10.2}$$

$$c = \frac{A_{ps} f_{pu} + A_s f_y - A'_s f'_y - 0.85\beta_1 f'_c (b - b_w)h_f}{0.85\beta_1 f'_c b_w + k A_{ps} \frac{f_{pu}}{d_p}} \tag{10.3}$$

$$f_{ps} = f_{pu}\left(1 - k\frac{c}{d_p}\right) \tag{10.4}$$

$$k = 2\left(1.04 - \frac{f_{py}}{f_{pu}}\right) \tag{10.5}$$

where A represents area; f is stress; b is the width of the compression face of member; b_w is the web width of a section; h_f is the compression flange depth of a cross-section; d_p and d_s are distances from extreme compression fiber to the centroid of prestressing tendons and to centroid of

TABLE 10.4   Temporary Concrete Stress Limits at Jacking State Before Losses Due to Creep and Shrinkage—Fully Prestressed Components

Stress type		Area and condition	Stress ksi (MPa)
Compressive		Pretensioned	$0.60 f'_{ci}$
		Post-tensioned	$0.55 f'_{ci}$
		Precompressed tensile zone without bonded reinforcement	N/A
		Area other than the precompressed tensile zones and without bonded auxiliary reinforcement	$0.0948 \sqrt{f'_{ci}} \leq 0.2$ $\left(0.25 \sqrt{f'_{ci}} \leq 1.38 \right)$
Tensile	Nonsegmental bridges	Area with bonded reinforcement which is sufficient to resist 120% of the tension force in the cracked concrete computed on the basis of uncracked section	$0.22 \sqrt{f'_{ci}}$ $\left(0.58 \sqrt{f'_{ci}} \right)$
		Handling stresses in prestressed piles	$\left(0.158 \sqrt{f'_{ci}} \right)$ $\left(0.415 \sqrt{f'_{ci}} \right)$
	Longitudinal stress through joint in precompressed tensile zone	Type A joints with minimum bonded auxiliary reinforcement through the joints which is sufficient to carry the calculated tensile force at a stress of $0.5 f_y$ with internal tendons	$0.0948 \sqrt{f'_{ci}}$ max. tension $\left(0.25 \sqrt{f'_{ci}} \text{ max. tension} \right)$
		Type A joints without the minimum bonded auxiliary reinforcement through the joints with internal tendons	No tension
		Type B with external tendons	0.2 min. compression (1.38 min. compression)
Segmental bridges	Transverse stress through joints	For any type of joint	$0.0948 \sqrt{f'_{c}}$ max. tension $\left(0.25 \sqrt{f'_{c}} \text{ max. tension} \right)$
		Without bonded non-prestressed reinforcement	No tension
	Other area	Bonded reinforcement is sufficient to carry the calculated tensile force in the concrete on the assumption of an uncracked section at a stress of $0.5 f_{sy}$	$0.19 \sqrt{f'_{ci}}$ $\left(0.50 \sqrt{f'_{ci}} \right)$

Note: Type A joints are cast-in-place joints of wet concrete and/or epoxy between precast units. Type B joints are dry joints between precast units.
From American Association of State Highway and Transportation Officials, *AASHTO LRFD Bridge Design Specifications*, First Edition, Washington, D.C., 1994. With permission.

tension reinforcement, respectively; subscripts c and y indicate specified strength for concrete and steel, respectively; subscripts p and s signify prestressing steel and reinforcement steel, respectively; subscripts ps, py, and pu correspond to states of nominal moment capacity, yield, and specified tensile strength of prestressing steel, respectively; superscript prime ($'$) represents compression; and β_1 is the concrete stress block factor, equal to 0.85 $f'_c \leq 4000$ psi and 0.05 less for each 1000 psi of f'_c in excess of 4000 psi, and minimum $\beta_1 = 0.65$. The above equations also can be used for a rectangular section in which $b_w = b$ is taken.

Maximum reinforcement limit:

$$\frac{c}{d_e} \leq 0.42 \tag{10.6}$$

$$d_e = \frac{A_{ps} f_{ps} d_p + A_s f_y d_s}{A_{ps} f_{ps} + A_s f_y} \tag{10.7}$$

Minimum reinforcement limit:

$$\phi M_n \geq 1.2 M_{cr} \tag{10.8}$$

TABLE 10.5 Concrete Stress Limits at Service Limit State After All Losses—Fully Prestressed Components

Stress type	Area and condition		Stress ksi (MPa)
Compressive	Nonsegmental bridge at service state		$0.45 f_c'$
	Nonsegmental bridge during shipping and handling		$0.60 f_c'$
	Segmental bridge during shipping and handling		$0.45 f_c'$
Tensile	Precompressed tensile zone assuming uncracked sections	With bonded prestressing tendons other than piles	$0.19 \sqrt{f_c'}$ $(0.50 \sqrt{f_c'})$
		Subjected to severe corrosive conditions	$0.0948 \sqrt{f_c'}$
Nonsegmental bridges		With unbonded prestressing tendon	$\left(0.25\sqrt{f_c'}\right)$ No tension
	Longitudinal stress in precompressed tensile zone	Type A joints with minimum bonded auxiliary reinforcement through the joints which is sufficient to carry the calculated tensile force at a stress of $0.5 f_y$ with internal tendons	$0.0948\sqrt{f_c'}$ $(0.25\sqrt{f_c'})$
		Type A joints without the minimum bonded auxiliary reinforcement through the joints	No tension
		Type B with external tendons	0.2 min. compression (1.38 min. compression)
Segmental bridges	Transverse stress in precompressed tensile zone	For any type of joint	$0.0948 \sqrt{f_c'}$
		Type A joint without minimum bonded auxiliary reinforcement through joints	$\left(0.25\sqrt{f_c'}\right)$ No tension
	Other area (without bonded reinforcement)	Bonded reinforcement is sufficient to carry the calculated tensile force in the concrete on the assumption of an uncracked section at a stress of 0.5 f_{sy}	$0.19\sqrt{f_c'}$ $\left(0.50\sqrt{f_c'}\right)$

Note: Type A joints are cast-in-place joints of wet concrete and/or epoxy between precast units. Type B joints are dry joints between precast units.
From American Association of State Highway and Transportation Officials, *AASHTO LRFD Bridge Design Specifications*, First Edition, Washington, D.C., 1994. With permission.

in which ϕ is the flexural resistance factor 1.0 for prestressed concrete and 0.9 for reinforced concrete, and M_{cr} is the cracking moment strength given by the elastic stress distribution and the modulus of rupture of concrete.

c) *Nominal Shear Resistance, V_n*

The nominal shear resistance shall be determined by the following formulas:

$$V_n = \text{the lesser of} \begin{cases} V_c + V_s + V_p \\ 0.25 f_c' b_v d_v + V_p \end{cases} \tag{10.9}$$

where

$$V_c = \begin{cases} 0.0316\beta\sqrt{f_c'}b_v d_v & \text{(ksi)} \\ 0.083\beta\sqrt{f_c'}b_v d_v & \text{(MPa)} \end{cases} \tag{10.10}$$

$$V_s = \frac{A_v f_y d_v (\cos\theta + \cos\alpha)\sin\alpha}{s} \tag{10.11}$$

where b_v is the effective web width determined by subtracting the diameters of ungrouted ducts or one-half the diameters of grouted ducts; d_v is the effective depth between the resultants of the tensile and compressive forces due to flexure, but not less than the greater of $0.9\,d_e$ or $0.72h$; A_v is the area of transverse reinforcement within distance s; s is the spacing of the stirrups; α is the angle of inclination of transverse reinforcement to the longitudinal axis; β is a factor indicating the ability of diagonally cracked concrete to transmit tension; and θ is the angle of inclination of diagonal compressive stresses (Figure 10.18). The values of β and θ for sections with transverse reinforcement are given in Table 10.6. In this table, the shear stress, v, and strain, ε_x, in the reinforcement on the flexural tension side of the member are determined by:

$$v = \frac{V_u - \phi V_p}{\phi b_v d_v} \tag{10.12}$$

$$\varepsilon_x = \frac{\frac{M_u}{d_v} + 0.5 N_u + 0.5 V_u \cot\theta - A_{ps} f_{po}}{E_s A_s + E_p A_{ps}} \leq 0.002 \tag{10.13}$$

where M_u and N_u are the factored moment and axial force (taken as positive if compressive),

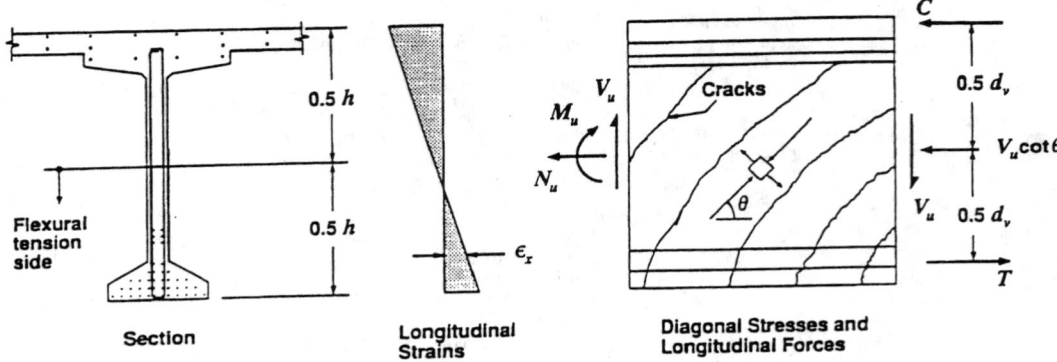

Figure 10.18   Illustration of A_c for shear strength calculation. (From American Association of State Highway and Transportation Officials, *AASHTO LRFD Bridge Design Specifications*, First Edition, Washington, D.C., 1994. With permission.)

respectively, associated with V_u, and f_{po} is the stress in prestressing steel when the stress in the surrounding concrete is zero and can be conservatively taken as the effective stress after losses, f_{pe}. When the value of ε_x calculated from the above equation is negative, its absolute value shall be reduced by multiplying by the factor F_ε, taken as:

$$F_\varepsilon = \frac{E_s A_s + E_p A_{ps}}{E_c A_c + E_s A_s + E_p A_{ps}} \tag{10.14}$$

where E_s, E_p, and E_c are modules of elasticity for reinforcement, prestressing steel, and concrete, respectively, and A_c is the area of concrete on the flexural tension side of the member, as shown in Figure 10.18.

Minimum transverse reinforcement:

$$A_{v\,min} = \begin{cases} 0.0316\sqrt{f_c'}\,\dfrac{b_v S}{f_y} & \text{(ksi)} \\[2mm] 0.083\sqrt{f_c'}\,\dfrac{b_v S}{f_y} & \text{(MPa)} \end{cases} \tag{10.15}$$

TABLE 10.6 Values of θ and β for Sections with Transverse Reinforcement

$\frac{v}{f'_c}$	Angle (degree)	−0.2	−0.15	−0.1	0	0.125	0.25	0.50	0.75	1.00	1.50	2.00
						$\varepsilon_x \times 1000$						
≤ 0.05	θ	27.0	27.0	27.0	27.0	27.0	28.5	29.0	33.0	36.0	41.0	43.0
	β	6.78	6.17	5.63	4.88	3.99	3.49	2.51	2.37	2.23	1.95	1.72
0.075	θ	27.0	27.0	27.0	27.0	27.0	27.5	30.0	33.5	36.0	40.0	42.0
	β	6.78	6.17	5.63	4.88	3.65	3.01	2.47	2.33	2.16	1.90	1.65
0.100	θ	23.5	23.5	23.5	23.5	24.0	26.5	30.5	34.0	36.0	38.0	39.0
	β	6.50	5.87	5.31	3.26	2.61	2.54	2.41	2.28	2.09	1.72	1.45
0.125	θ	20.0	21.0	22.0	23.5	26.0	28.0	31.5	34.0	36.0	37.0	38.0
	β	2.71	2.71	2.71	2.60	2.57	2.50	2.37	2.18	2.01	1.60	1.35
0.150	θ	22.0	22.5	23.5	25.0	27.0	29.0	32.0	34.0	36.0	36.5	37.0
	β	2.66	2.61	2.61	2.55	2.50	2.45	2.28	2.06	1.93	1.50	1.24
0.175	θ	23.5	24.0	25.0	26.5	28.0	30.0	32.5	34.0	35.0	35.5	36.0
	β	2.59	2.58	2.54	2.50	2.41	2.39	2.20	1.95	1.74	1.35	1.11
0.200	θ	25.0	25.5	26.5	27.5	29.0	31.0	33.0	34.0	34.5	35.0	36.0
	β	2.55	2.49	2.48	2.45	2.37	2.33	2.10	1.82	1.58	1.21	1.00
0.225	θ	26.5	27.0	27.5	29.0	30.5	32.0	33.0	34.0	34.5	36.5	39.0
	β	2.45	2.38	2.43	2.37	2.33	2.27	1.92	1.67	1.43	1.18	1.14
0.250	θ	28.0	28.5	29.0	30.0	31.0	32.0	33.0	34.0	35.5	38.5	41.5
	β	2.36	2.32	2.36	2.30	2.28	2.01	1.64	1.52	1.40	1.30	1.25

From American Association of State Highway and Transportation Officials, *AASHTO LRFD Bridge Design Specifications*, First Edition, Washington, D.C., 1994. With permission.

Maximum spacing of transverse reinforcement:

$$\text{For } V_u < 0.1 f'_c b_v d_v \quad s_{\max} = \text{ the smaller of } \begin{cases} 0.8 d_v \\ 24 \text{ in. } (600 \text{ mm}) \end{cases} \quad (10.16)$$

$$\text{For } V_u \geq 0.1 f'_c b_v d_v \quad s_{\max} = \text{ the smaller of } \begin{cases} 0.4 d_v \\ 12 \text{ in. } (300 \text{ mm}) \end{cases} \quad (10.17)$$

10.4 Concrete Substructures

10.4.1 Introduction

Bridge substructures transfer traffic loads from the superstructure to the footings and foundations. Vertical intermediate supports (piers or bents) and end supports (abutments) are included.

10.4.2 Bents and Piers

1. **Pile Bents**
   Pile extension, as shown in Figure 10.19a, is used for slab and T-beam bridges. It is usually used to cross streams when debris is not a problem.

2. **Solid Piers**
   Figure 10.19b shows a typical solid pier, used mostly when stream debris or fast currents are present. These are used for long spans and can be supported by spread footings or pile foundations.

3. **Column Bents**
   Column bents (Figure 10.19c) are generally used on dry land structures and are supported by spread footings or pile foundations. Multi-column bents are desirable for bridges in seismic zones. The single-column bent, such as a T bent (Figure 10.19d), modified T bent (C bent) (Figure 10.19e), or outrigger bent (Figure 10.19f), may be used when the location of the columns is restricted and changes of the alignment are impossible. To achieve a pleasing appearance at the minimum cost using standard column shapes, Caltrans [3] developed "standard architectural columns" (Figure 10.20). Prismatic sections of column types 1 and 1W, with one-way flares of column types 2

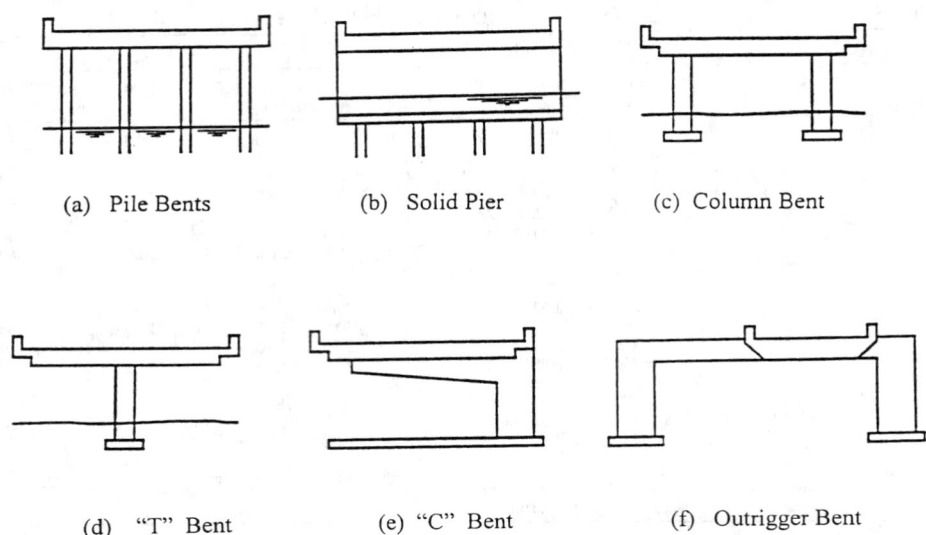

(a) Pile Bents   (b) Solid Pier   (c) Column Bent

(d) "T" Bent   (e) "C" Bent   (f) Outrigger Bent

Figure 10.19 Bridge substructures—piers and bents. (From California Department of Transportation, *Bridge Design Aids Manual,* Sacramento, CA, 1990. With permission.)

and 2W, and with two-way flares of column types 3 and 3W may be used for various highway bridges.

10.4.3 Abutments

Abutments are the end supports of a bridge. Figure 10.21 shows the typical abutments used for highway bridges. The seven types of abutments can be divided into two categories: open and closed ends. Selection of an abutment type depends on the requirements for structural support, movement, drainage, road approach, and earthquakes.

1. **Open-End Abutments**

   Open-end abutments include diaphragm abutments and short-seat abutments. These are the most frequently used abutments and are usually the most economical, adaptable, and attractive. The basic structural difference between the two types is that seat abutments permit the superstructure to move independently from the abutment while the diaphragm abutment does not. Since open-end abutments have lower abutment walls, there is less settlement in the road approaches than that experienced by higher backfilled closed abutments. They also provide for more economical widening than closed abutments.

2. **Closed-End Abutments**

   Closed-end abutments include cantilever, strutted, rigid frame, bin, and closure abutments. These are less commonly used, but for bridge widenings of the same kind, unusual sites, or in tightly constrained urban locations. Rigid frame abutments are generally used with tunnel-type single-span connectors and overhead structures which permit passage through a roadway embankment. Because the structural supports are adjacent to traffic these have a high initial cost and present a closed appearance to approaching traffic.

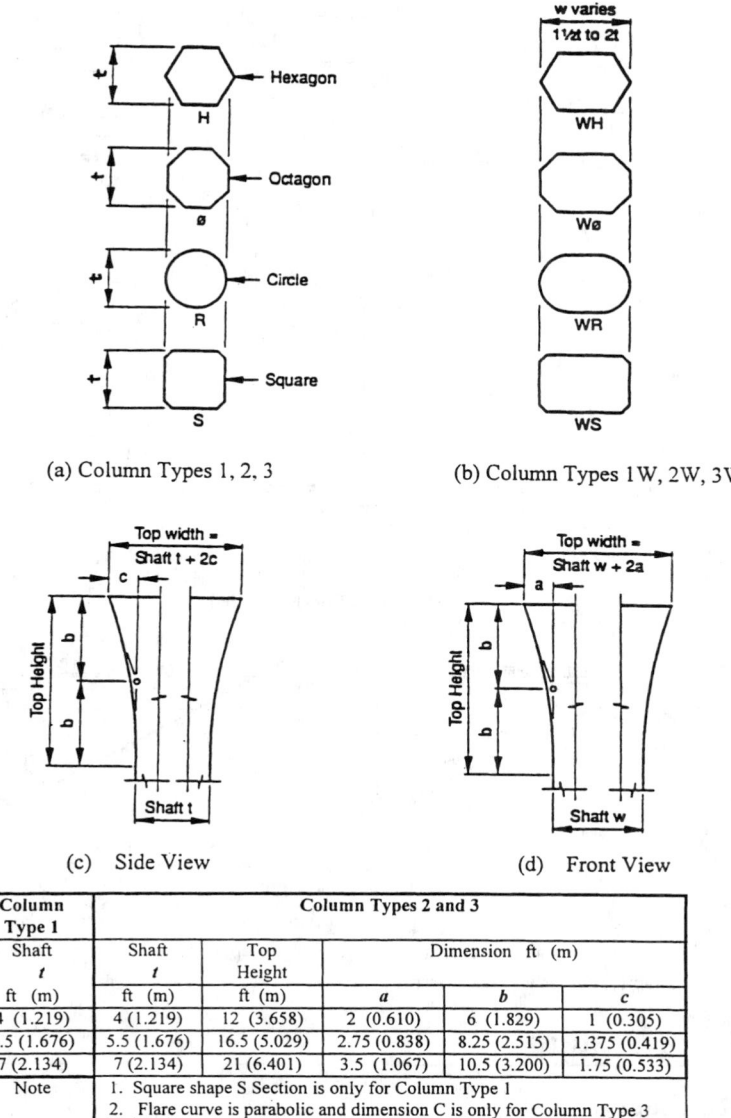

(a) Column Types 1, 2, 3   (b) Column Types 1W, 2W, 3W

(c) Side View   (d) Front View

Column Type 1	Column Types 2 and 3				
Shaft *t*	Shaft *t*	Top Height	Dimension ft (m)		
ft (m)	ft (m)	ft (m)	*a*	*b*	*c*
4 (1.219)	4 (1.219)	12 (3.658)	2 (0.610)	6 (1.829)	1 (0.305)
5.5 (1.676)	5.5 (1.676)	16.5 (5.029)	2.75 (0.838)	8.25 (2.515)	1.375 (0.419)
7 (2.134)	7 (2.134)	21 (6.401)	3.5 (1.067)	10.5 (3.200)	1.75 (0.533)
Note	1. Square shape S Section is only for Column Type 1 2. Flare curve is parabolic and dimension C is only for Column Type 3				

Figure 10.20 Caltrans (California Department of Transportation) standard architectural columns. (From California Department of Transportation, *Bridge Design Aids Manual,* Sacramento, CA, 1990. With permission.)

10.4.4 Design Consideration

After the recent 1989 Loma Prieta and the 1994 Northridge Earthquakes in the U.S. and the 1995 Kobe earthquake in Japan, major damages were found in substructures. Special attention, therefore, must be paid to seismic effects and the detailing of the ductile structures. Boundary conditions and soil–foundation–structure interaction in seismic analyses should also be carefully considered.

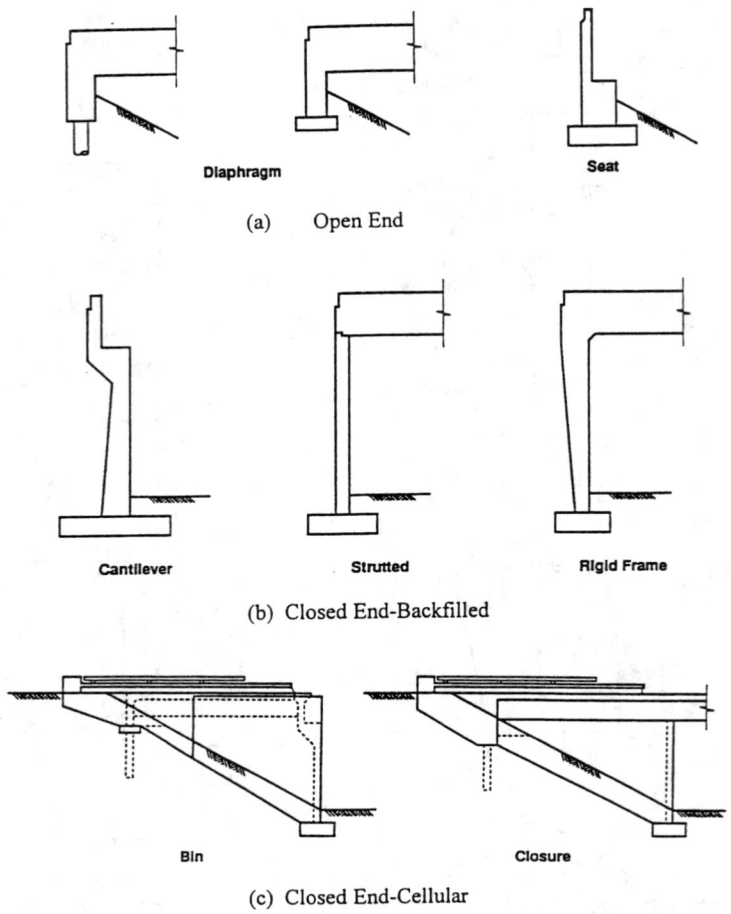

Figure 10.21 Typical types of abutments. (From California Department of Transportation, *Bridge Design Aids Manual*, Sacramento, CA, 1990. With permission.)

10.5 Floor System

10.5.1 Introduction

The floor system of a bridge usually consists of a deck, floor beams, and stringers. The deck directly supports the live load. Floor beams as well as stringers, shown in Figure 10.22, form a grillage and transmit the load from the deck to the main girders. The floor beams and stringers are used for framed bridges, i.e., truss, rahmen, and arch bridges (see Figures 10.40, 10.45, and 10.47), in which the spacing of the main girders or trusses is large. In an upper deck type of plate girder bridge the deck is directly supported by the main girders, and often there is no floor system because the main girders run in parallel and close together.

The floor system is classified as suitable for either highway or railroad bridges. The deck of a highway bridge is designed for the wheel loads of trucks using plate bending theory in two dimensions. Often in design practice, however, this plate theory is reduced to equivalent one-dimensional beam theory. The materials used are also classified into concrete, steel, or wood.

The recent influx of traffic flow has severely fatigued existing floor systems. Cracks in concrete decks and connections of floor system are often found in old bridges that have been in service for many years.

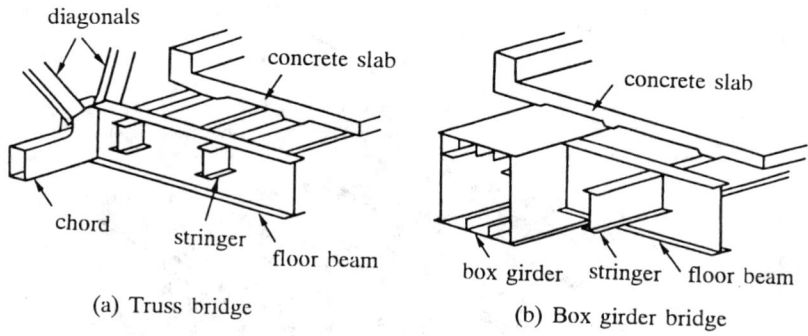

Figure 10.22   Floor system. (From Nagai, N., *Bridge Engineering*, Kyoritsu Publishing Co., Tokyo, Japan [in Japanese], 1994. With permission.)

10.5.2   Decks

1. **Concrete Deck**

   A reinforced concrete deck slab is most commonly used in highway bridges. It is the deck that is most susceptible to damage caused by the flow of traffic, which continues to increase. Urban highways are exposed to heavy traffic and must be repaired frequently. Recently, a composite deck slab was developed to increase the strength, ductility, and durability of decks without increasing their weight or affecting the cost and duration of construction. In a composite slab, the bottom steel plate serves both as a part of the slab and the formwork for pouring the concrete. There are many ways of combining the steel plate and the reinforcement. A typical example is shown in Figure 10.23. This slab is prefabricated in the yard and then the concrete is poured on site after girders have been placed. A precast, prestressed deck may reduce the time required to complete construction.

2. **Steel Deck**

   For long spans, the steel deck is used to minimize the weight of the deck. The steel deck plate is stiffened with longitudinal and transverse ribs as shown in Figure 10.24. The steel deck also works as the upper flange of the supporting girders. The pavement on the steel deck should be carefully finished to prevent water from penetrating through the pavement and causing the steel deck to rust.

10.5.3   Pavement

The pavement on the deck provides a smooth driving surface and prevents rain water from seeping into the reinforcing bars and steel deck below. A layer of waterproofing may be inserted between the pavement and the deck. Asphalt is most commonly used to pave highway bridges. Its thickness is usually 5 to 10 cm on highways and 2 to 3 cm on pedestrian bridges.

10.5.4   Stringers

The stringers support the deck directly and transmit the loads to floor beams, as can be seen in Figure 10.22. They are placed in the longitudinal direction just like the main girders are in a plate girder bridge and thus provide much the same kind of support.

   The stringers must be sufficiently stiff in bending to prevent cracks from forming in the deck or on the pavement surface. The design codes usually limit the vertical displacement caused by the weight of a truck.

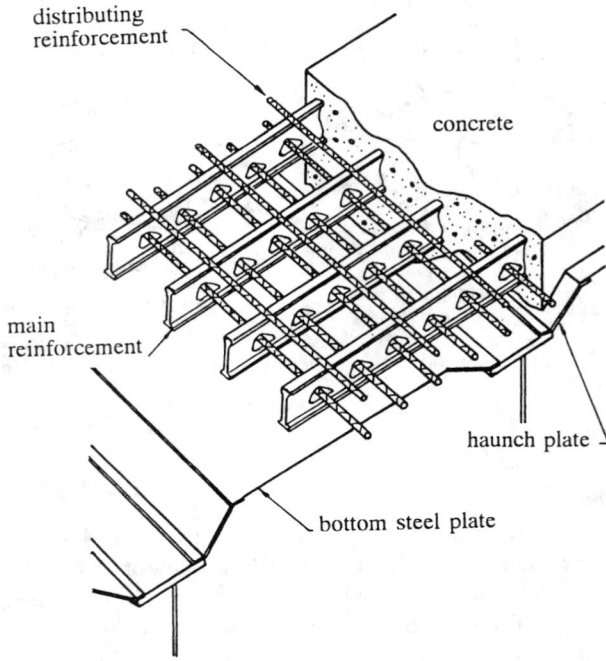

Figure 10.23   Composite deck. (From Japan Association of Steel Bridge Construction, *Planning of Steel Bridges*, Tokyo [in Japanese], 1988. With permission.)

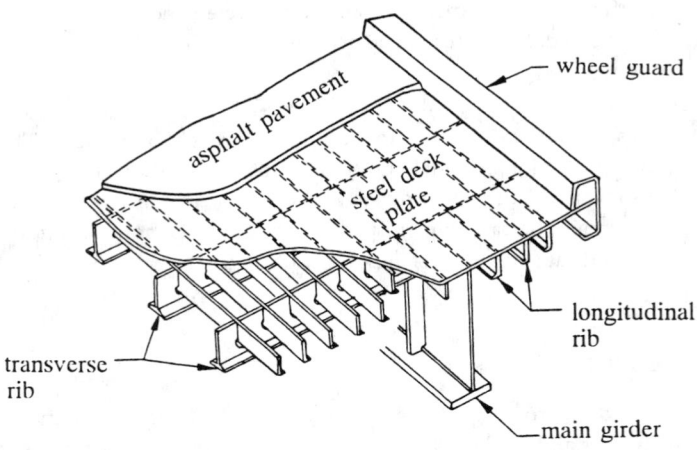

Figure 10.24   Steel plate deck. (From Japan Association of Steel Bridge Construction, *Outline of Steel Bridges*, Tokyo [in Japanese], 1985. With permission.)

10.5.5   Floor Beams

The floor beams are placed in the transverse direction and connected by high-tension bolts to the truss frame or arch, as shown in Figure 10.22. The floor beams support the stringers and transmit the loads to main girders, trusses, or arches. In other words, the main truss or arch receives the loads indirectly via the floor beams. The floor beams also provide transverse stiffness to bridges and thus improve the overall torsional resistance.

10.6 Bearings, Expansion Joints, and Railings

10.6.1 Introduction

Aside from the main components, such as the girders or the floor structure, other parts such as bearings (shoes), expansion joints, guardrailings, drainage paths, lighting, and sound-proofing walls also make up the structure of a bridge. Each plays a minor part but provides an essential function. Drains flush rain water off and wash away dust. Guardrailings and lights add to the aesthetic quality of the design as well as providing their obvious original functions. A sound-proofing wall may take away from the beauty of the structure but might be required by law in urban areas to isolate the sound of traffic from the surrounding residents. In the following section, bearings, expansion joints, and guardrailings are discussed.

10.6.2 Bearings (Shoes)

Bearings support the superstructure (the main girders, trusses, or arches) and transmit the loads to the substructure (abutments or piers). The bearings connect the upper and lower structures and carry the whole weight of the superstructure. The bearings are designed to resist these reaction forces by providing support conditions that are fixed or hinged. The hinged bearings may be movable or immovable; horizontal movement is restrained or unrestrained, i.e., horizontal reaction is produced or not. The amount of the horizontal movement is determined by calculating the elongation due to a temperature change.

Many bearings were found to have sustained extensive damage during the 1995 Kobe Earthquake in Japan, due to stress concentrations, which are the weak spots along the bridge. The bearings may play the role of a fuse to keep damage from occurring at vital sections of the bridge, but the risk of the superstructure falling down goes up. The girder-to-girder or girder-to-abutment connections prevent the girders from collapsing during strong earthquakes.

Many types of bearings are available. Some are shown in Figure 10.25 and briefly explained in the following:

Line bearings: The contacting line between the upper plate and the bottom round surface provides rotational capability as well as sliding. These are used in small bridges.

Plate bearings: The bearing plate has a plane surface on the top side which allows sliding and a spherical surface on the bottom allowing rotation. The plate is placed between the upper and lower shoes.

Hinged bearings (pin bearings): A pin is inserted between the upper and lower shoes allowing rotation but no translation in longitudinal direction.

Roller bearings: Lateral translation is unrestrained by using single or multiple rollers for hinged bearings or spherical bearings.

Spherical bearings (pivot bearings): Convex and concave spherical surfaces allow rotation in all directions and no lateral movement. The two types are: a point contact for large differences in the radii of each sphere and a surface contact for small differences in their radii.

Pendel bearings: An eye bar connects the superstructure and the substructure by a pin at each end. Longitudinal movement is permitted by inclining the eye bar; therefore, the distance of the pins at ends should be properly determined. These are used to provide a negative reaction in **cable-stayed bridges**. There is no resistance in the transverse direction.

Wind bearings: This type of bearing provides transverse resistance for wind and is often used with pendel bearings.

Elastomeric bearings: The flexibility of elastomeric or lead rubber bearings allows both ro-

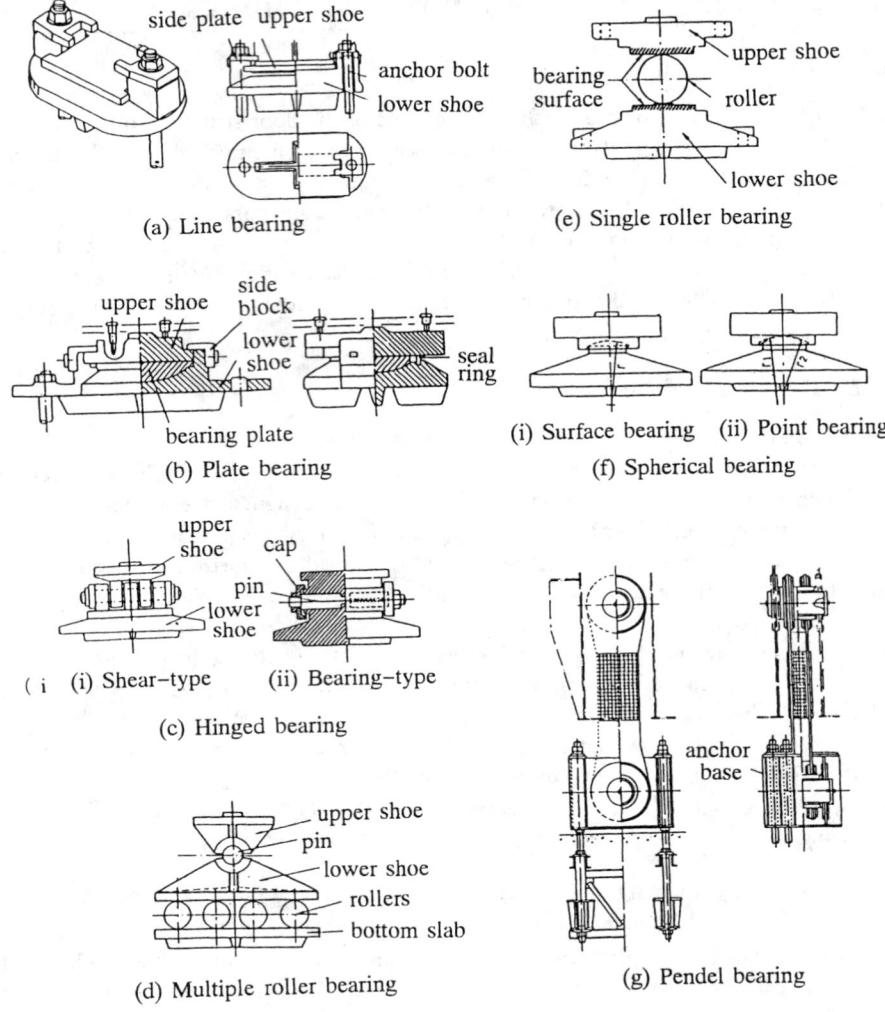

Figure 10.25 Types of bearings. (From Japan Association of Steel Bridge Construction, *A Guide Book of Bearing Design for Steel Bridges,* Tokyo [in Japanese], 1984. With permission.)

tation and horizontal movement. Figure 10.26 explains a principle of rubber-layered bearings by comparing with a unit rubber. A layered rubber is stiff, unlike a unit rubber, for vertical compression because the steel plates placed between the rubber restrain the vertical deformation of the rubber, but flexible for horizontal shear force like a unit rubber. The flexibility absorbs horizontal seismic energy and is ideally suited to resist earthquake actions. Since the disaster of the 1995 Kobe Earthquake in Japan, elastomeric rubber bearings have become more and more popular, but whether they effectively sustain severe vertical actions without damage is not certified.

Oil damper bearings: The oil damper bearings move under slow actions (such as temperature changes) but do not move under quick movements (such as those of an earthquake). They are used in continuous span bridges to distribute seismic forces.

A selection from these types of bearings is made according to the size of the bridge and the magnitude of predicted downward or upward reaction forces.

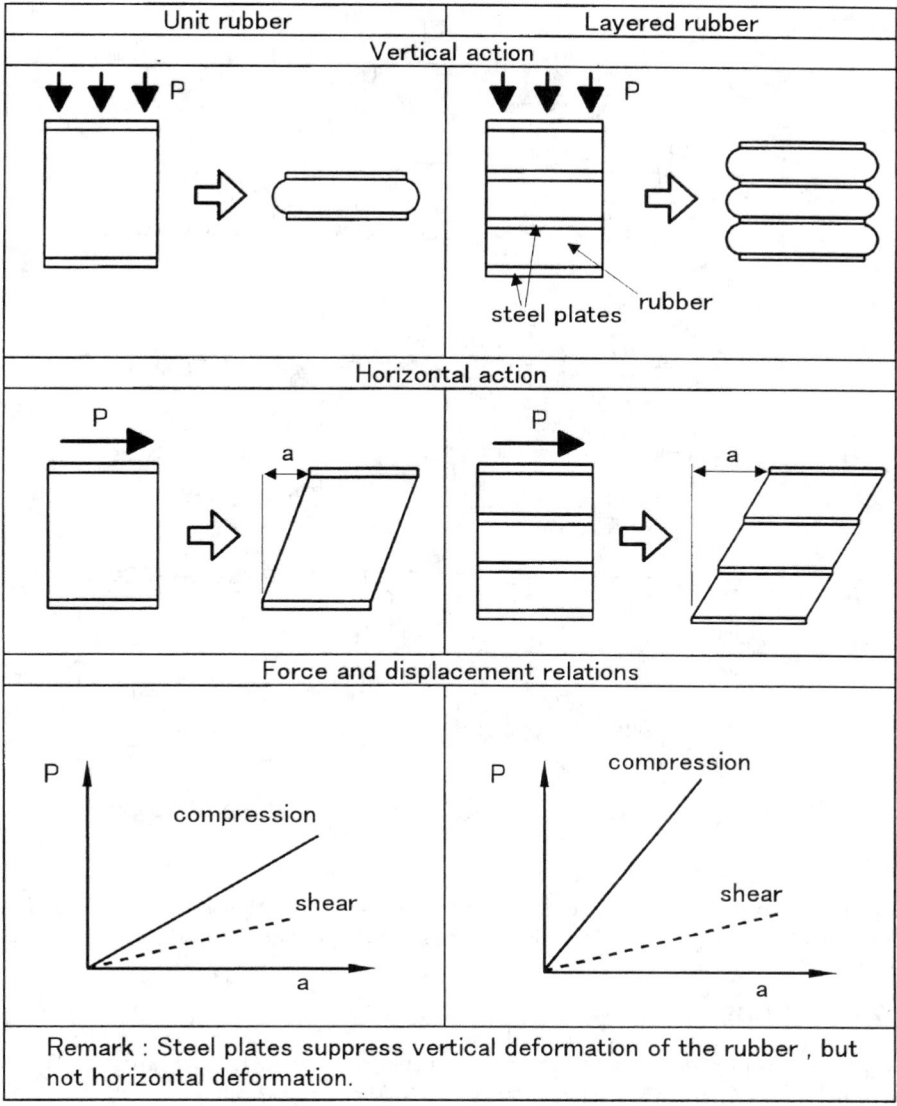

Figure 10.26 Properties of elastomeric bearings.

10.6.3 Expansion Joints

Expansion joints are provided to allow a bridge to adjust its length under changes in temperature or deformation by external loads. They are designed according to expanding length and material as classified in Figure 10.27. Steel expansion joints are most commonly used. A defect is often found at the boundary between the steel and the concrete slab where the disturbing jolt is given to drivers as they pass over the junction. To solve this problem, rubber joints are used on the road surface to provide a smooth transition for modern bridge construction (see Figure 10.27e), or continuous girders are more commonly adopted than simple girders.

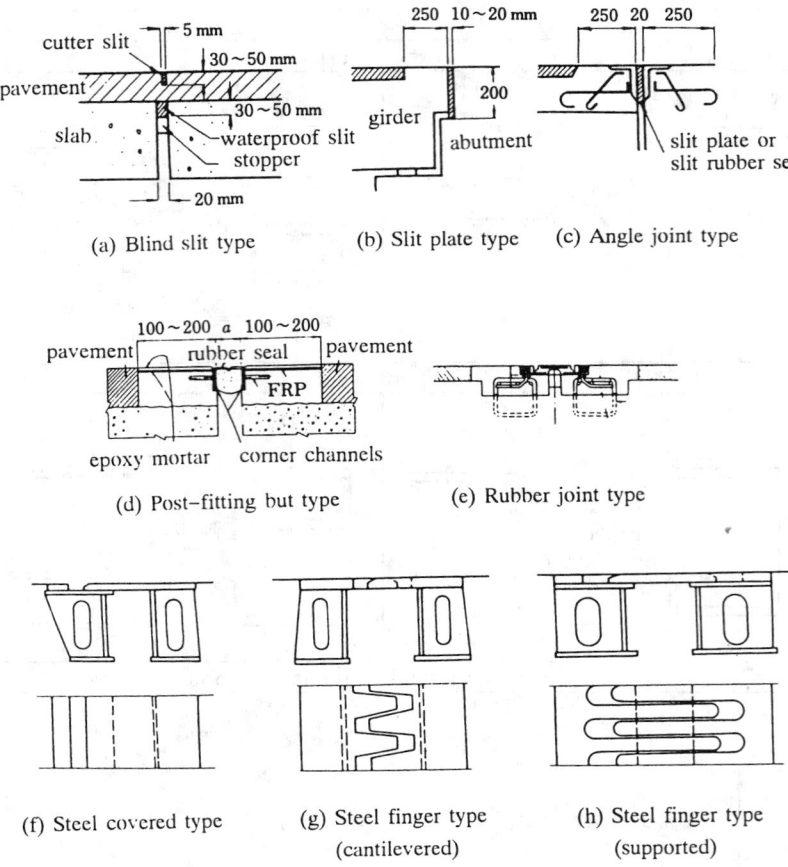

Figure 10.27   Types of expansion joints. (From Japan Association of Steel Bridge Construction, *A Guide Book of Expansion Joint Design for Steel Bridges,* Tokyo [in Japanese], 1984. With permission.)

10.6.4   Railings

Guardrailings are provided to ensure vehicles and pedestrians do not fall off the bridge. They may be a handrail for pedestrians, a heavier guard for vehicles, or a common railing for both. These are made from materials such as concrete, steel, or aluminum. The guardrailings are located prominently and are thus open to the critical eye of the public. It is important that they not only keep traffic within boundaries but also add to the aesthetic appeal of the whole bridge (Figure 10.28).

10.7   Girder Bridges

10.7.1   Structural Features

Girder bridges are structurally the simplest and the most common. They consist of a floor slab, girders, and the bearings which support and transmit gravity loads to the substructure. Girders resist bending moments and shear forces and are used to span short distances. Girders are classified by material into steel plate and box girders, reinforced or prestressed concrete T-beams, and **composite girders**. The box girder is also used often for prestressed concrete continuous bridges. The steel girder bridges are explained in this section; the concrete bridges were described in Section 10.3.

Kanazawa Hakkei Oh–Hashi,

Kanagawa, Japan (1982)

(The railing images waves.)

Figure 10.28 Pedestrian railing. (From Japan Association of Steel Bridge Construction, *Outline of Steel Bridges*, Tokyo [in Japanese], 1985. With permission.)

Figure 10.29 shows the structural composition of plate and box girder bridges and the load transfer path. In plate girder bridges, the live load is directly supported by the slab and then by the main girders. In box girder bridges the forces are taken first by the slab, then supported by the stringers and floor beams in conjunction with the main box girders, and finally taken to the substructure and foundation through the bearings.

Girders are classified as noncomposite or composite, that is, whether the steel girders act in tandem with the concrete slab (using shear connectors) or not. Since composite girders make use of the best properties of both steel and concrete, they are often the rational and economic choice. Less frequently H or I shapes are used for the main girders in short-span noncomposite bridges.

10.7.2   Plate Girder (Noncomposite)

The plate girder is the most economical shape designed to resist bending and shear; the moment of inertia is greatest for a relatively low weight per unit length. Figure 10.30 shows a plan of a typical plate girder bridge with four main girders spanning 30 m and a width of 8.5 m.

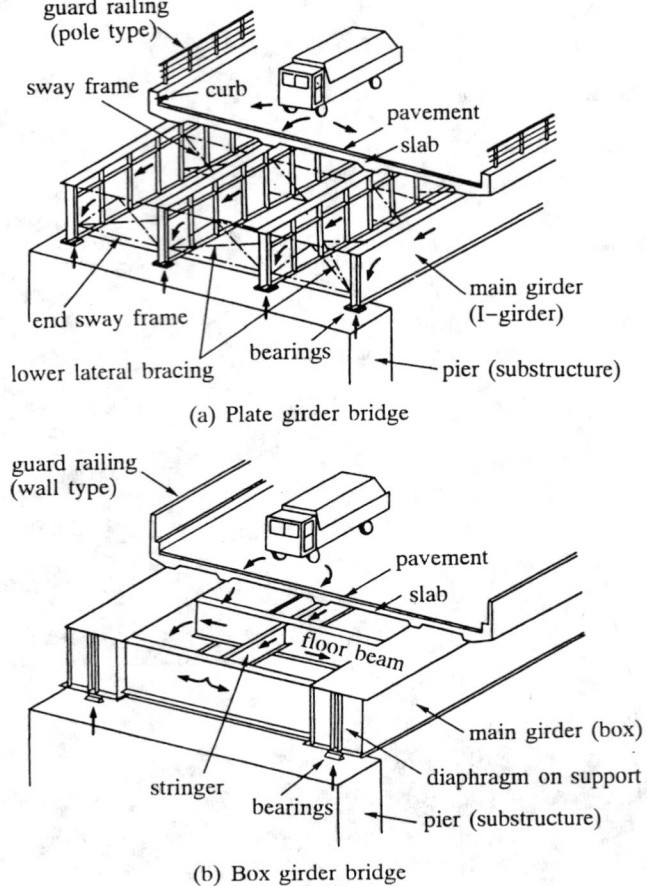

Figure 10.29   Steel girder bridges. (From Nagai, N., *Bridge Engineering*, Kyoritsu Publishing Co., Tokyo, Japan [in Japanese], 1994. With permission.)

The gravity loads are supported by several main plate girders, each manufactured by welding three plates: an upper and lower flange and a web. Figure 10.31 shows a block of plate girder and its fabrication process. The web and the flanges are cut from steel plate and welded. The block is fabricated in the shop and transported to the construction site for erection.

The design procedure for plate girders, primarily the sizing of the three plates, is as follows:

1. **Web height:** The web height is the fundamental design factor affecting the weight and cost of the bridge. If the height is too small, the flanges need to be large and the dead weight increases. The height (h) is determined empirically by dividing the span length (L) by a "reasonable" factor. Common ratios are $h/L = 1/18$ to $1/20$ for highway bridges and a little smaller for railway bridges. The web height also influences the stiffness of the bridge. Greater heights generally produce greater stiffness. However, if the height is too great, the web becomes unstable and must have its thickness supplemented or stiffeners added. These measures increase the weight and the cost. In addition, plate girders with excessively deep web and small flanges are liable to buckle laterally.

2. **Web thickness:** The web primarily resists shear forces, which are not usually significant when the web height is properly designed. The shear force is generally assumed to be distributed uniformly across the web instead of using the exact equation of beam theory.

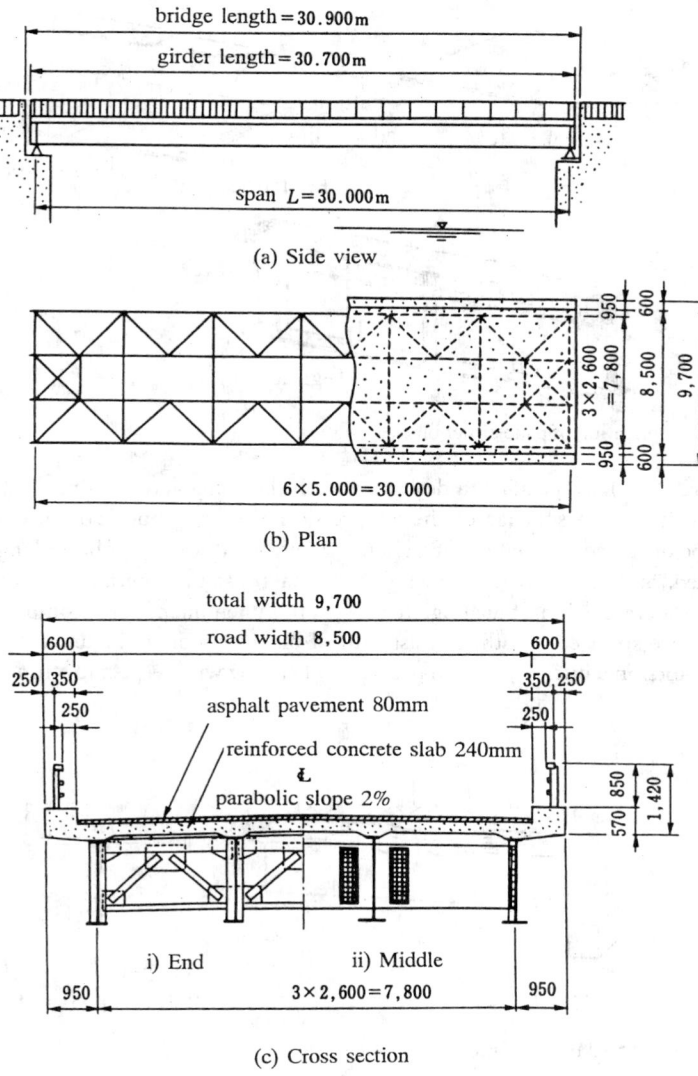

Figure 10.30 General plans of a typical plate girder bridge. (From Tachibana, Y. and Nakai, H., *Bridge Engineering*, Kyoritsu Publishing Co., Tokyo, Japan [in Japanese], 1996. With permission.)

The web thickness (t) is determined such that thinner is better as long as buckling is prevented. Since the web does not contribute much to the bending resistance, thin webs are most economical but the possibility of buckling increases. Therefore, the web is usually stiffened by horizontal and vertical stiffeners, which will be discussed later (see Figure 10.34). It is not primarily strength but rather stiffness that controls the design of webs.

3. **Area of flanges:** After the sizes of web are determined, the flanges are designed. The flanges work mostly in bending and the required area is calculated using equilibrium conditions imposed on the internal and external bending moment. A selection of strength for the steel material is principally made at this stage in the design process.

4. **Width and thickness of flanges:** The width and thickness can be determined by ensuring that the area of the flanges falls under the limiting width-to-thickness ratio, b/t

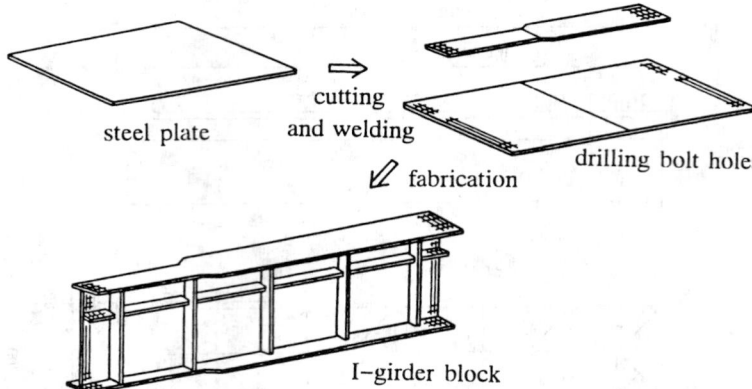

Figure 10.31   Fabrication of plate girder block.

(Figure 10.32), as specified in design codes. If the flanges are too thin (i.e., the width-to-thickness ratio is too large), the compression flange may buckle or the tension flange may be distorted by the heat of welding. Thus, the thickness of both flanges must be checked. Since plate girders have little torsional resistance, special attention should be paid to lateral torsional buckling. To prevent this phenomenon, the compression flange must have sufficient width to resist "out-of-plane" bending. Figure 10.33 shows the lateral torsional buckling that may occur by bending with respect to strong axis.

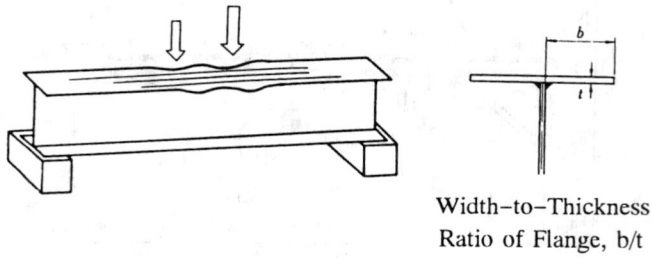

Width–to–Thickness
Ratio of Flange, b/t

Figure 10.32   Local buckling of compression flange.

After determining the member sizes, calculations of the resisting moment capacity are made to ensure code requirements are satisfied. If these fail, the above steps must be repeated until the specifications are met.

A few other important factors in the design of girder bridges will be explained in the following:

Design of web stiffeners: The horizontal and vertical stiffeners should be attached to the web (Figure 10.34) when it is relatively thin. Bending moment produces compression and tension in the web, separated by a neutral axis. The horizontal stiffener prevents buckling due to bending and is therefore attached to the compression side (the top half for a simply supported girder). Since the bending moment is largest near the midspan of a simply supported girder, the horizontal stiffeners are usually located there. If the web is not too deep nor its thickness too small, no stiffeners are necessary and fabrication costs are reduced. Vertical stiffeners, on the other hand, prevent shear buckling, which is produced by the tension and compression fields in diagonal directions. The compression field causes shear buckling. Since the shear force is largest near the supports, the most vertical stiffeners are needed there. Bearing stiffeners, which are designed independently

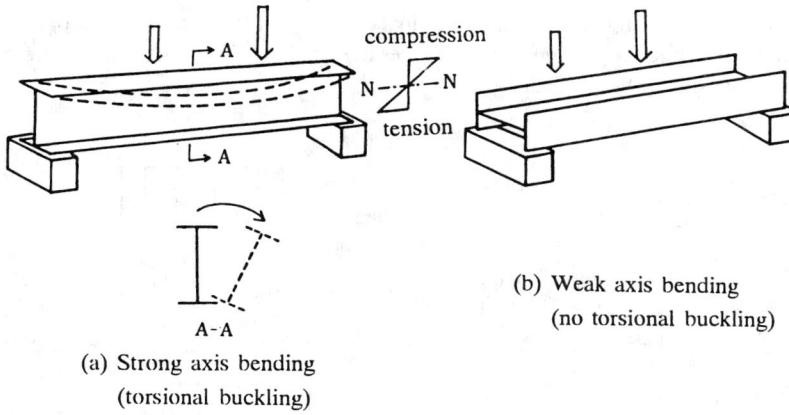

compression

N ---- N

tension

(b) Weak axis bending
(no torsional buckling)

A-A

(a) Strong axis bending
(torsional buckling)

Figure 10.33   Lateral torsional buckling.

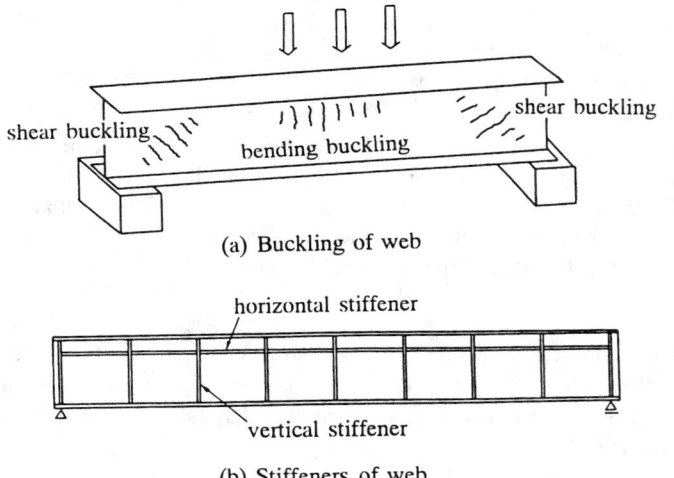

shear buckling   shear buckling
                          bending buckling

(a) Buckling of web

horizontal stiffener

vertical stiffener

(b) Stiffeners of web

Figure 10.34   Buckling and stiffeners of web.

just as any other compression member would be, are also required at the supports to combat large reaction forces. Buckling patterns of a web are shown in Figure 10.34.

Variable sections: The variable cross-sections may be used to save material and cost where the bending moment is smaller, that is, near the end of the span (see Figure 10.31). However, this reduction increases the labor required for welding and fabrication. The cost of labor and material must be balanced and traded off. In today's industrial climate, labor is more important and costly than the material. Therefore, the change of girder section is avoided. Likewise, thick plates are often specified to eliminate the number of stiffeners needed, thus to reduce the necessary labor.

10.7.3   Composite Girder

If two beams are simply laid one upon the other, as shown in Figure 10.35a, they act separately and only share the load depending on their relative flexural stiffness. In this case, slip occurs along the boundary between the beams. However, if the two beams are connected and slip prevented as shown in Figure 10.35b, they act as a unit, i.e., a composite girder. For composite plate girder bridges, the steel girder and the concrete slab are joined by shear connectors. In this way, the concrete slab becomes integral with the girder and usually takes most of the compression component of the bending moment while the steel plate girder takes the tension. Composite girders are much more effective than the simply tiered girder.

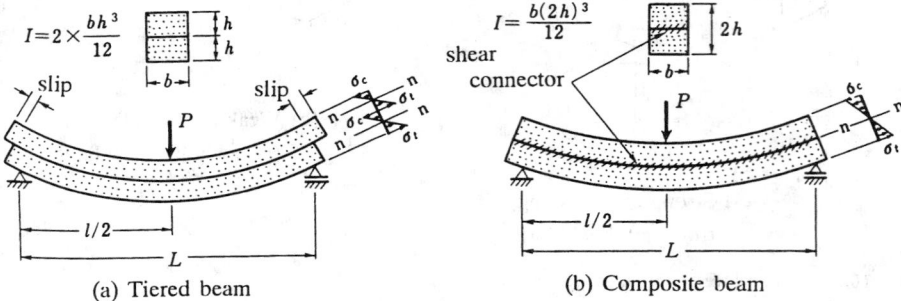

(a) Tiered beam   (b) Composite beam

Figure 10.35 Principle of tiered beam and composite beam. (From Tachibana, Y. and Nakai, H., *Bridge Engineering*, Kyoritsu Publishing Co., Tokyo, Japan [in Japanese], 1996. With permission.)

Let us consider the two cases shown in Figure 10.35 and note the difference between tiered beams and composite beams. Both have the same cross-sections and are subjected to a concentrated load at midspan. The moment of inertia for the composite beam is four times that of the tiered beams, thus the resulting vertical deflection is one-fourth. The maximum bending stress in the extreme (top or bottom) fiber is half that of the tiered beam configuration.

The corresponding stress distributions are shown in Figure 10.36. Points "S" and "V" are the center of area of the steel section and the composite section, respectively. According to beam theory, the strain distribution is linear but the stress distribution has a step change at the boundary between the steel and concrete.

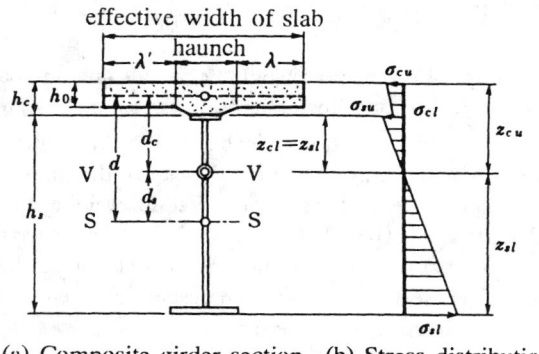

(a) Composite girder section   (b) Stress distribution

Figure 10.36 Section of composite girder. (From Tachibana, Y. and Nakai, H., *Bridge Engineering*, Kyoritsu Publishing Co., Tokyo, Japan [in Japanese], 1996. With permission.)

Three types of shear connectors—studs, horse shoes, and steel blocks—are shown in Figure 10.37. Studs are most commonly used since they are easily welded to the compression flange by the electric resistance welding, but the weld inspection is a cumbersome task. If the weld on a certain stud is

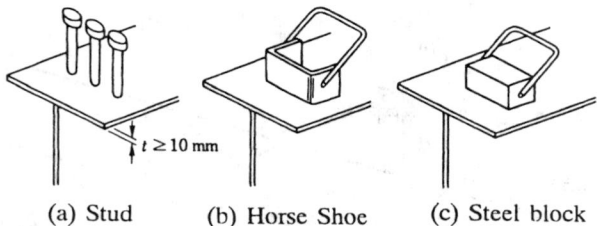

(a) Stud   (b) Horse Shoe   (c) Steel block

Figure 10.37   Types of shear connectors. (From Nagai, N., *Bridge Engineering,* Kyoritsu Publishing Co., Tokyo, Japan [in Japanese], 1994. With permission.)

poor, the stud may shear off and trigger a totally unforeseen failure mode. Other types are considered to maintain more reliability.

Shear connectors are needed most near the ends of the span, where the shear force is largest. This region is illustrated in Figure 10.35a, which shows the maximum shift due to slip occurs at the ends of tiered beams. It is this slip that is restrained by the shear connectors.

10.7.4   Grillage Girder

When girders are placed in a row and connected transversely by floor beams, the truck loads are distributed by the floor beams to the girders. This system is called a grillage of girders. If the main girders are plate girders, no stiffness in torsion is considered. On the other hand, box girders and concrete girders can be analyzed assuming stiffness is available to resist torsion. Floor beams increase the torsional resistance of the whole structural system of the bridge.

Let us consider the structural system shown in Figure 10.38a to observe the load distribution in a grillage system. This grillage has three girders with one floor beam at midspan. In this case, there are three nodal forces at the intersections of the girders and the floor beam but only two equilibrium equations ($V = 0$ and $M = 0$). Thus, it becomes one degree statically indeterminate. If we disconnect the intersection between main girder B and the floor beam and apply a pair of indeterminate forces, X, at point b, as shown in Figure 10.38b, X can be obtained using the compatibility condition at point b. Once the force, X, is found, the sectional forces in the girders can be calculated. This structural system is commonly applied to the practical design of plate girder bridges.

10.7.5   Box Girder

Structural configuration of box girders is illustrated in Figure 10.39. Since the box girder is a closed section, its resistance to torsion is high with no loss of strength in bending and shear. On the other hand, plate girders are open sections generally only considered effective in resisting bending and shear. Steel plates with longitudinal and transverse stiffeners are often used for decks on box girder or thin-walled structures instead of a concrete slab (Figure 10.39b) although a concrete slab is permissible.

Torsion is resisted in two parts: pure torsion (St. Venant torsion) and warping torsion. The pure torsional resistance of I-plate girders is negligible. However, for closed sections such as a box girder, the pure torsional resistance is considerable, making them particularly suited for curved bridges or long-span bridges. On the other hand, the warping torsion for box sections is negligible. The I-section girder has some warping resistance but it is not large compared to the pure torsion of closed sections.

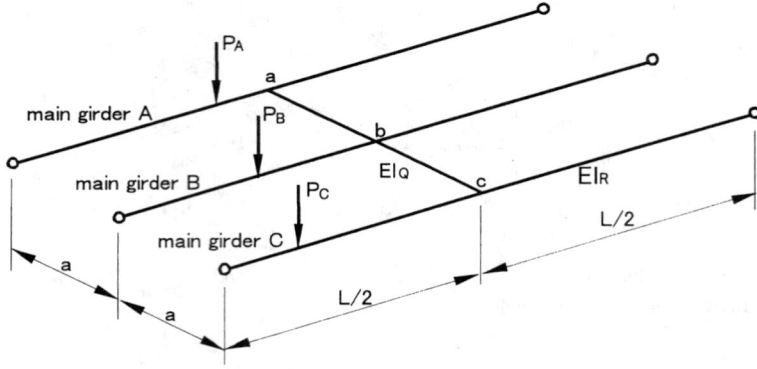

(a) One-Degree Indeterminate System

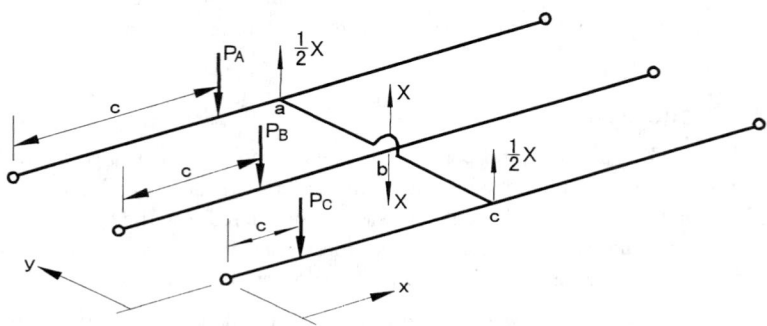

(b) Statically Determinate System

Figure 10.38   Grillage girders. (From Tachibana, Y. and Nakai, H., *Bridge Engineering*, Kyoritsu Publishing Co., Tokyo, Japan [in Japanese], 1996. With permission.)

10.8   Truss Bridges

10.8.1   Structural Features

The structural layout of a truss bridge is shown in Figure 10.40 for a through bridge with the deck located at the level of lower chords. The floor slab, which carries the live load, is supported by the floor system of stringers and cross beams. The load is transmitted to the main trusses at nodal connections, one on each side of the bridge, through the floor system and finally to the bearings. Lateral braces, which also are a truss frame, are attached to the upper and lower chords to resist horizontal forces such as wind and earthquake loads as well as torsional moments. The portal frame at the entrance provides transition of horizontal forces from the upper chords to the substructure.

Truss bridges can take the form of a deck bridge as well as a through bridge. In this case, the concrete slab is mounted on the upper chords and the sway bracing is placed between the vertical members of two main trusses to provide lateral stability.

A truss is composed of upper and lower chords, joined by diagonal and vertical members (web members). This frame action corresponds to beam action in that the upper and lower chords perform like flanges and the diagonal braces behave in much the same way as the web plate. The

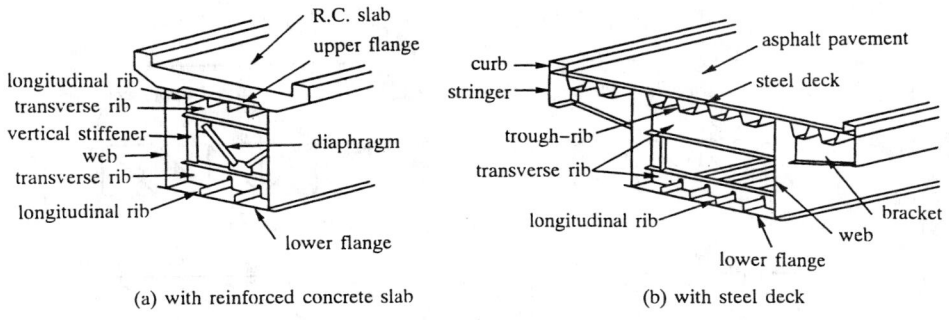

Figure 10.39 Box girders. (From Nagai, N., *Bridge Engineering*, Kyoritsu Publishing Co., Tokyo, Japan [in Japanese], 1994. With permission.)

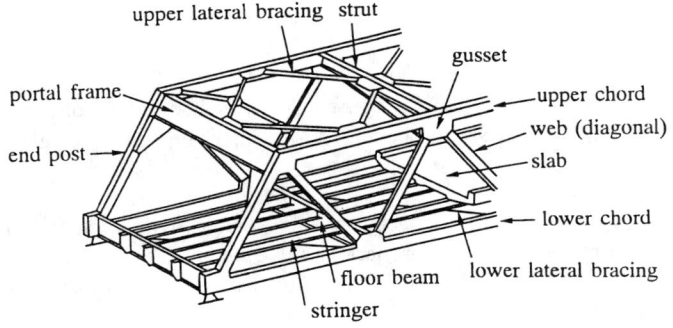

Figure 10.40 Truss bridge. (From Nagai, N., *Bridge Engineering*, Kyoritsu Publishing Co., Tokyo, Japan [in Japanese], 1994. With permission.)

chords are mainly in charge of bending moment while the web members take the shear force. Trusses are an assembly of bars, not plates, and thus are comparatively easier to erect on site and are often the choice for long bridges.

10.8.2 Types of Trusses

Figure 10.41 shows some typical trusses. A Warren truss is the most common and is a frame composed of isosceles triangles, where the web members are either in compression or tension. The web members of a Pratt truss are vertical and diagonal members where the diagonals are inclined toward the center and resist only tension. The Pratt truss is suitable for steel bridges since it is tension that is most effectively resisted. It should be noted, however, that vertical members of Pratt truss are in compression. A Howe truss is similar to the Pratt except that the diagonals are inclined toward the ends, leading to axial compression forces, and the vertical members resist tension. Wooden bridges often make use of the Howe truss since the connections of the diagonals in wood tend to compress. A K-truss, so named since the web members form a "K", is most economical in large bridges because the short member lengths reduce the risk of buckling.

10.8.3 Structural Analysis and Secondary Stress

The truss is a framed structure of bars, theoretically connected by hinges, forming stable triangles. Trusses contain triangle framed units to keep it stable. Its members are assumed to resist only tensile or compressive axial forces. A statically determinate truss can be analyzed using equilibrium conditions only. If more than the least number of members required for stability are provided, the

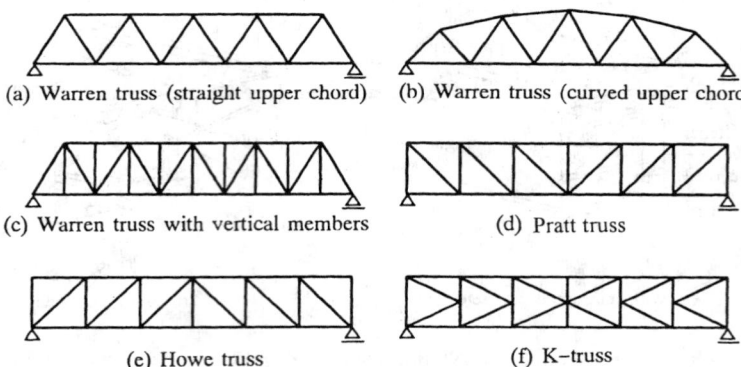

Figure 10.41   Types of trusses.

truss becomes indeterminate and can no longer be solved using only the conditions of equilibrium. The displacement compatibility should be added. An internally and/or externally indeterminate truss is best solved using computer software.

In practice, truss members are connected to gusset plates with high-tension bolts (see Figure 10.42), not rotation-free hinges, simply because these are much easier to fabricate. The "pinned" condition of theory is not reflected in the field. This discrepancy results in "secondary stresses"

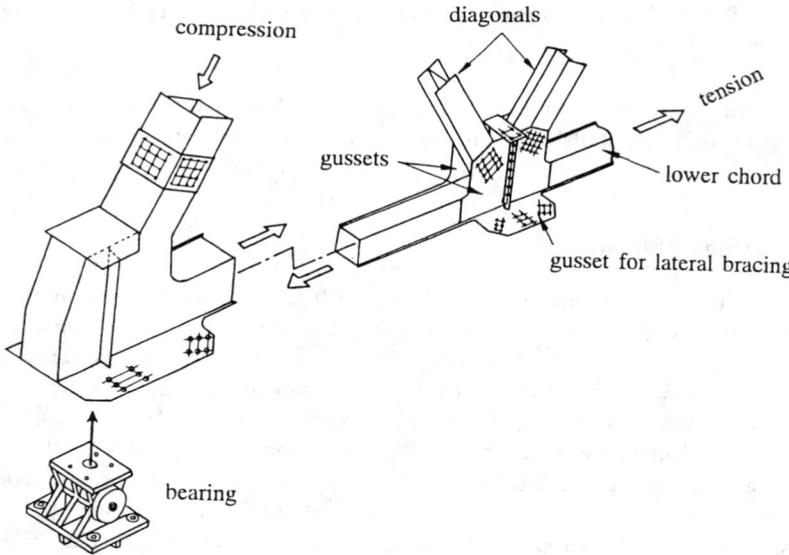

Figure 10.42   Nodal joints of a truss bridge. (From Japan Association of Steel Bridge Construction, *Outline of Steel Bridges*, Tokyo [in Japanese], 1985. With permission.)

(bending stresses) in the members. Secondary stresses are given by a computer analysis of a rigid frame and are usually found to be less than 20% of the primary (axial) stresses. If the truss members are properly designed, that is, the slenderness ratios of the truss bars are sufficiently large with no buckling, then secondary stresses can conveniently and reliably be disregarded.

10.8.4   Gerber Truss Bridge

Figure 10.43 is a photo of a Gerber truss bridge during the erection of the central part, which is the Minato Oh-Hashi in Japan. Its plan view is shown in Figure 10.2. A Gerber truss has intermediate hinges between the supports to create a statically determinate structural system. In the case of Minato Oh-Hashi, the soil condition at the bottom of the harbor was found to be not stiff and solid; thus the Gerber truss proved the wisest choice.

Lifting the center girder of the Minato Bridge into place.

Figure 10.43   Lifting erection of the Minato Oh-Hashi, Japan. (Gerber bridge, 1974). (From Hanshin Expressway Public Corporation, *Techno Gallery*, Osaka, Japan, 1994. With permission.)

10.9   Rigid Frame Bridges (Rahmen Bridges)

10.9.1   Structural Features

The members are rigidly connected in "rahmen" structures or "rigid frames". Unlike the truss and the arch bridges, which will be discussed in the following subsection, all the members are subjected to both an axial force and bending moments. Figure 10.44 shows various types of rahmen bridges.

The members of a rigid frame bridge are much larger than those in a typical building. Consequently, stress concentrations occur at the junctions of beams and columns which must be carefully designed using finite element analyses or experimental verification. The supports of rahmen bridges are either hinged or fixed, making it an externally indeterminate structure, and it is therefore not suitable when the foundation is likely to sink. The reactions at supports are horizontal and vertical forces at hinges, with the addition of a bending moment at a fixed base.

10.9.2   Portal Frame

A portal frame is the simplest design (Figure 10.44a) and is widely used for the piers of elevated highway bridges because the space underneath can be effectively used for other roads or parking lots. These piers were proved, in the 1995 Kobe Earthquake in Japan, to be more resilient, that is, to retain more strength and absorb more energy than single-legged piers.

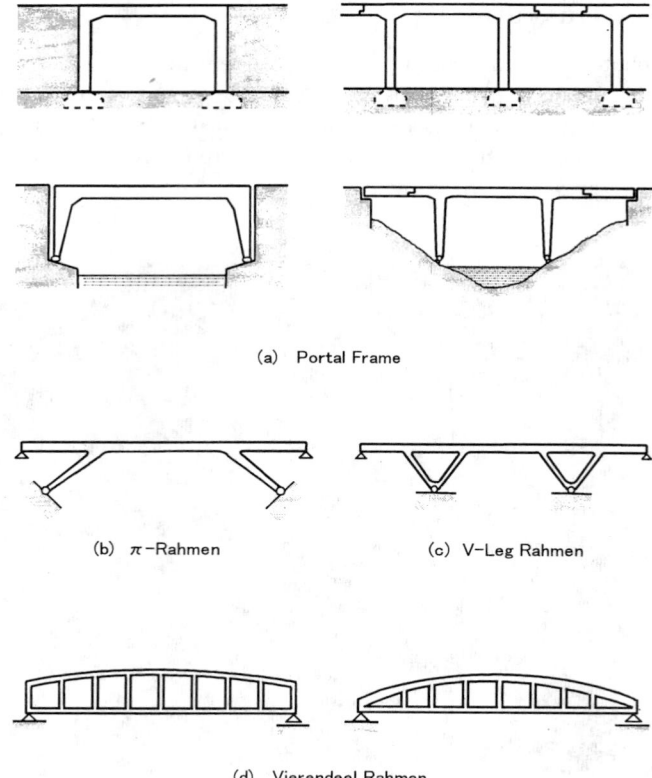

(a)   Portal Frame

(b) π-Rahmen   (c) V-Leg Rahmen

(d)   Vierendeel Rahmen

Figure 10.44   Types of rahmen bridges.

10.9.3 π-Rahmen (Strutted Beam Bridge)

The π-rahmen design is usually used for bridges in mountainous regions where the foundation is firm, passing over deep valleys with a relatively long span, or for bridges crossing over expressways (Figure 10.44b). As shown in the structural layout of a π-rahmen bridge in Figure 10.45, the two legs support the main girders, inducing axial compression in the center span of the girder. Live load on the deck is transmitted to the main girders through the floor system. Intermediate hinges may be inserted in the girders to make Gerber girders. A V-leg rahmen bridge is similar to a π-rahmen bridge but can span longer distances with no axial force in the center span of the girder (Figure 10.44c).

10.9.4   Vierendeel Bridge

The Vierendeel bridge is a rigid frame whose upper and lower chords are connected rigidly to the vertical members (Figure 10.44d). All the members are subjected to axial and shear forces as well as bending moments. This is internally a highly indeterminate system. Analysis of the Vierendeel frame must consider secondary stresses (see Section 10.8.3). It is more stiff than Langer or Lohse arch bridges in which some members take only axial forces.

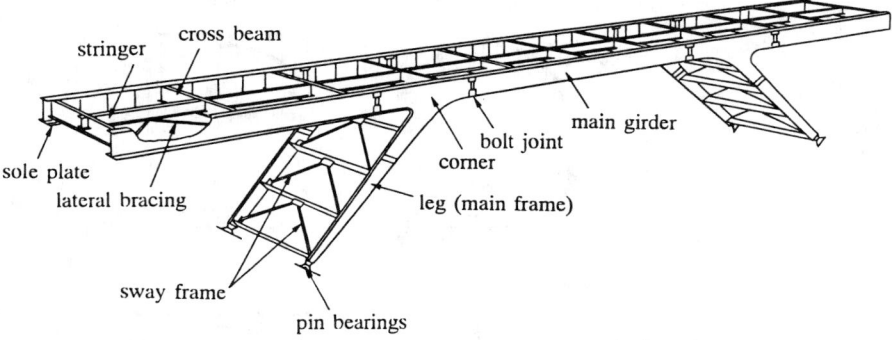

Figure 10.45 π-rahmen bridge. (From Japan Association of Steel Bridge Construction, *Outline of Steel Bridges,* Tokyo [in Japanese], 1985. With permission.)

10.10 Arch Bridges

10.10.1 Structural Features

An arch rib acts like a circular beam restrained not only vertically but also horizontally at both ends, and thus results in vertical and horizontal reactions at the supports. The horizontal reaction causes axial compression in addition to bending moments in the arch rib. The bending moments caused by the horizontal force balances those due to gravity loads. Compared with the axial force, the effect of the bending moment is usually small. That is why the arch is often made of materials that have high compressive strength, such as concrete, stone, or brick.

10.10.2 Types of Arches

An arch bridge includes the road deck and the supporting arch. Various types of arches are shown in Figure 10.46. In the figure, the thick line represents the members carrying bending moment, shear, and axial forces. The thin line represents members taking axial forces only. Arch bridges are classified into the deck and the through-deck types according to the location of the road surface, as shown in Figure 10.46. Since the deck in both types of bridges is sustained by either vertical columns or hangers to the arch, structurally the same axial force action, either compression or tension, is in effect in the members. The difference is that the vertical members of deck bridges take compressive forces and the hangers of through-deck bridges take tension. The live load acts on the arch only indirectly.

A basic structural type for an arch is a two-hinge arch (see Figure 10.46a). The two-hinge arch has one degree of indeterminacy externally because there are four end reactions. If one hinge is added at the crown of the arch, creating a three-hinge arch, it is rendered determinate. If the ends are clamped, turning it into a fixed arch, it becomes indeterminate to the third degree. The tied arch is subtended by two hinges by a tie and simply supported (Figure 10.46b). The tied arch is externally determinate but internally has one degree of indeterminacy. The floor structures hang from the arch and are isolated from the tie. Other types of arch bridges will be discussed later in more detail.

10.10.3 Structural Analysis

Almost all bridge design analyses, in this age of super computing power, use finite element methods. The analysis of an arch is basically the same as that for a frame. The web members are analyzed as truss bars which take only axial forces. The arch rib and the girders are analyzed as either trusses or beam-columns depending on the type of arch considered. Beam-columns take axial and shear

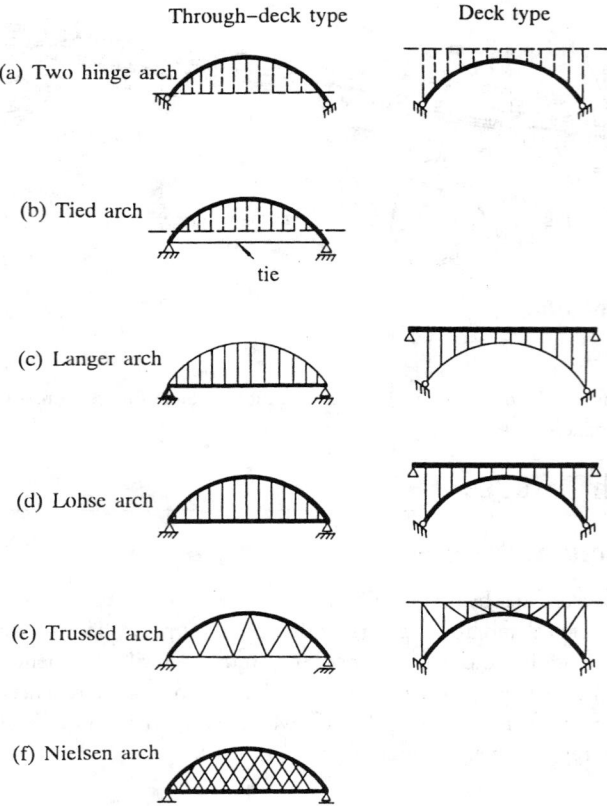

Figure 10.46   Types of arch bridges. (From Shimada, S., *Journal of Bridge and Foundation Engineering,* 25(8), 1991 [in Japanese]. With permission.)

forces and bending moments. An arch rib is usually made up of straight piece-wise components, not curved segments, and it is so analyzed.

10.10.4   Langer Bridge

The Langer arch is analyzed by assuming that the arch rib takes only axial compression (Figure 10.46c). The arch rib is thin, but the girders are deep and resist moment and shear as well as axial tension. The girders of the Langer bridge are regarded as being strengthened by the arch rib. Figure 10.47 shows the structural components of a Langer bridge.

    If diagonals are used in the web, it is called a trussed Langer. The difference between a trussed Langer and a standard truss is that the lower chord is a girder instead of just a bar. The Langer bridge is also determinate externally and indeterminate internally. The deck-type bridge of the Langer is often called a reversed Langer.

10.10.5   Lohse Bridge

The Lohse bridge is very similar to the Langer bridge except the Lohse bridge carries its resistance to bending in the arch rib as well as the girder (Figure 10.46d). By this assumption, the Lohse bridge is stiffer than the Langer. The distribution of bending moments in the arch rib and the girder depends on the stiffness ratio of the two members, which is the designer's decision. The Lohse arch bridges may be thought of as tiered beams (see Figure 10.35) connected by vertical members. The vertical

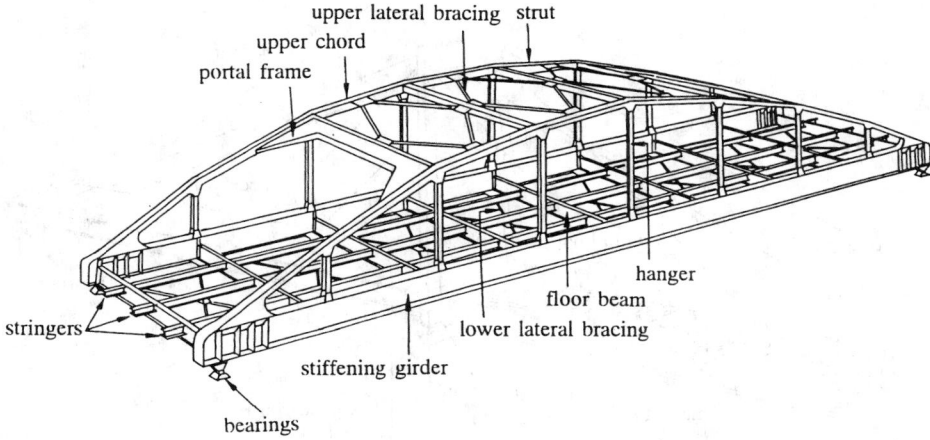

Figure 10.47   Langer arch bridge. (From Japan Association of Steel Bridge Construction, *Outline of Steel Bridges,* Tokyo [in Japanese], 1985. With permission.)

members are assumed to take only axial forces. Aesthetically, the Lohse is more imposing than the Langer, and is therefore suited to urban areas while the Langer fits into mountain areas.

10.10.6   Trussed Arch and Nielsen Arch Bridges

Generally diagonal members are not used in arch bridges, thus avoiding difficulty in structural analysis. However, recent advancements in computer technology have changed this outlook. New types of arch bridges, such as the trussed arch in which diagonal truss bars are used instead of vertical members or the Nielsen Lohse design in which tension rods are used for diagonals, have now been introduced (see Figure 10.46e, f). Diagonal web members increase the stiffness of a bridge more so than vertical members.

   All the members of the truss bridge take only axial forces. On the other hand, the trussed arch bridge may resist bending in either the arch rib or the girder, or both. Since the diagonals of the Nielsen Lohse bridge carry only axial tension, they are prestressed by the dead load to compensate for the compression force due to the live load.

10.11   Cable-Stayed Bridges

10.11.1   Structural Features

A cable-stayed bridge hangs the girders from diagonal cables that are tensioned from the tower, as shown in Figure 10.48. The cables of cable-stayed bridges are anchored in the girders. The girders are most often supported by movable or fixed hinges. Due to the diagonally tensioned cables, axial forces and bending moments are imposed on the girder and the tower. The bending moment in the girder is reduced when supported by the cables, and spans can be longer than conventional girder bridges (as long as 300 to 500 m). The maximum span length is the 890 m of the Tatara Bridge in Japan (see Table 10.1). Because of the wonder and beauty of this type, its design has been copied in even relatively small bridges including ones carrying only pedestrians. For long-span bridges, stability under strong wind currents should be carefully considered in the design. The dynamic effects of wind and earthquakes must be studied analytically and experimentally. Wind tunnel tests may be necessary to ensure excessive oscillation does not occur along the length of the bridge or in the tower. The cables also may resonate in the wind if they are thin and flexible. In this case, devices

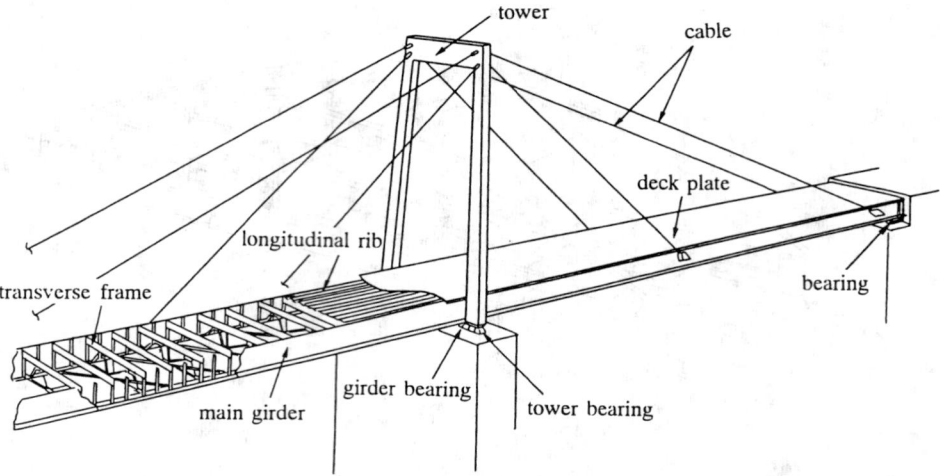

Figure 10.48   Cable-stayed bridge. (From Japan Association of Steel Bridge Construction, *Outline of Steel Bridges,* Tokyo [in Japanese], 1985. With permission.)

are necessary to curb the vibration. The stability of bridges under wind loads will be discussed in more detail in Section 10.12 (see Figure 10.61).

10.11.2   Types of Cable-Stayed Bridges

Cable-stayed bridges may be classified by the hanging formation of the cable and the shape of the tower. Figure 10.49 illustrates three typical cable formations. Structurally, the radial cable most effectively decreases the axial force in the tower and girders; however, difficulty in construction arises due to the structural complexity at the top of the tower. The fan type is more common because the cable connections at the tower are distributed. The harp type is aesthetically the most pleasing.

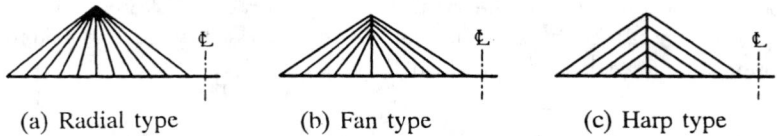

(a) Radial type   (b) Fan type   (c) Harp type

Figure 10.49   Types of cable formation. (From Nagai, N., *Bridge Engineering,* Kyoritsu Publishing Co., Tokyo, Japan [in Japanese], 1994. With permission.)

Figure 10.50 shows various tower designs. As the span length becomes large, columns such as the A, the H, or the upside-down Y shape are selected; these have significant torsional resistance.

10.11.3   Structural Analysis

The cable-stayed bridge is usually analyzed using linear elastic frame analysis. The cable is modeled as a bar element with hinged ends. Figure 10.51 shows the flow of gravity loads. Most of the load is transmitted to the substructure through the cables and the tower, but some goes to the girder directly. The smaller the bending stiffness of the girder, the less the load is taken by the girder. As the tower becomes higher, the tension force of the cable can be reduced.

Because of the sag in the cable due to its own weight, a reduced elastic modulus may be used in analysis. This reduced modulus is slightly lower than the actual elastic modulus of the cable material.

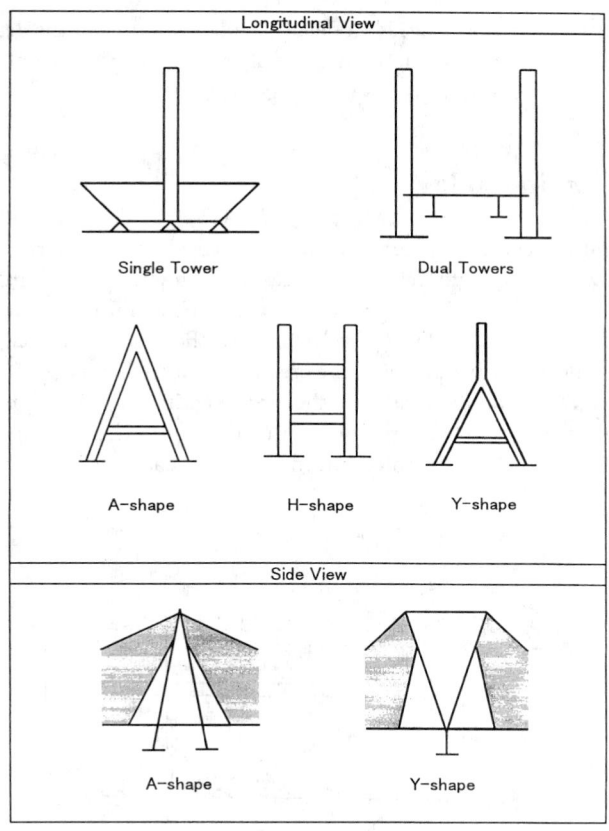

Figure 10.50   Types of towers. (From Nagai, N., *Bridge Engineering*, Kyoritsu Publishing Co., Tokyo, Japan [in Japanese], 1994. With permission.)

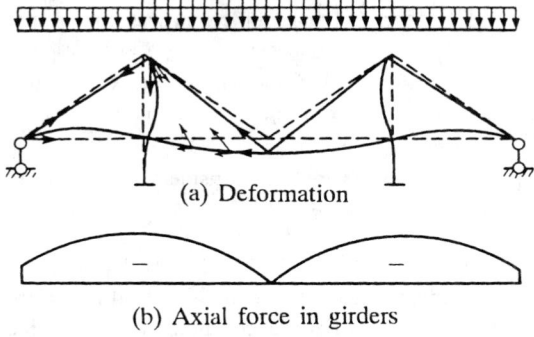

Figure 10.51   Force flow in cable-stayed bridges. (From Nagai, N., *Bridge Engineering*, Kyoritsu Publishing Co., Tokyo, Japan [in Japanese], 1994. With permission.)

The girder and the tower are designed to take axial compression, bending, and shear. Since the large force in the cable is concentrated on the girder and tower, stress concentration at those connections should be carefully checked using finite element analysis. Taking into the consideration the fact that the supports are subjected to large negative reactions (uplift), Pendel bearings are used. These, as mentioned previously (see Figure 10.25), are composed of an eye-bar and two end hinges, which may move horizontally and rotate freely.

In the preliminary design, the bridge is modeled as a plane frame. For the details, however, more precise analyses such as three-dimensional stress analyses may be used. Nonlinear effects may be taken into consideration for flexible long-span bridges.

10.11.4 Tension in Cable

One of the important aspects in the design of a cable-stayed bridge is the determination of the tension force in the cable, which is directly related to forces in the tower and the girder. Control on the tension force in the cables is critical. The pre-tension of the cables must be known because it changes the stresses in the girder and the tower. Figure 10.52 shows the bending moment distribution under dead loads along the bridge before and after the prestressing force is applied. It can be seen that the proper prestress reduces bending moments in the girder significantly. If the vertical component of the tension is selected to be equal to the reaction of the continuous girder (supported at the junction of the cable and girder), the bending moment in the girder can be reduced to match that of the continuous girder.

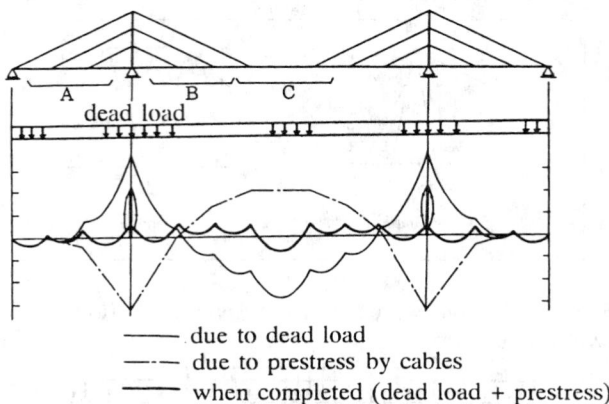

——— due to dead load
—·— due to prestress by cables
——— when completed (dead load + prestress)

Figure 10.52 Bending moment distribution. (From Japan Society of Civil Engineers, *Cable-Stayed Bridges—Technology and its Change*, Tokyo, Japan [in Japanese], 1990. With permission.)

The following three general principles are to be considered in determining cable tension [19]:

1. Avoid having any bending moments (generated by dead loads) in the tower. This is accomplished by balancing the horizontal components of the cable tension in the left and right ends of the tower.
2. Keep the bending moments in the girder small. It depends on the location and the distance between joints to the cable. Small distances (such as a multi-cable) will result in small bending moments in the girders.
3. Close the girder by connecting the center block lastly without using any compelling forces. The cable tension is selected such that zero sectional force exists at the center of the girder.

10.12 Suspension Bridges

10.12.1 Structural Features

Suspension bridges use two main cables suspended between two towers and anchored to blocks at the ends. Figure 10.53 shows the structural components of a suspension bridge. Stiffening girders

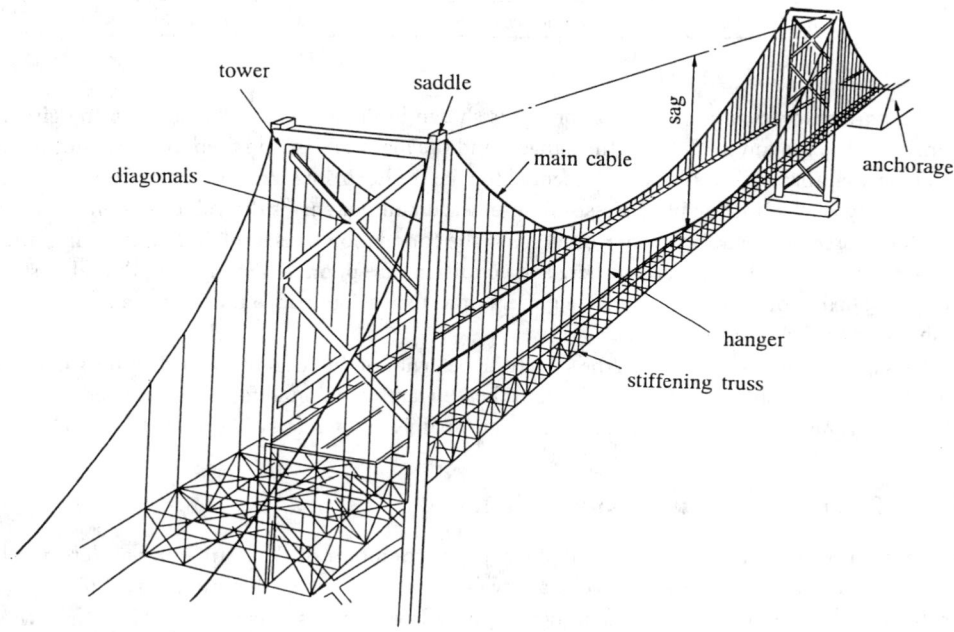

Figure 10.53 Suspension bridge. (From Japan Association of Steel Bridge Construction, *Outline of Steel Bridges*, Tokyo [in Japanese], 1985. With permission.)

are either truss or box type (see Figure 10.54) and hung from the main cables using hangers. The suspension bridge is most suitable for long spans. Table 10.7 is a list of the world's ten longest bridges, all of which are suspension bridges. The longest is the Akashi Kaikyo Bridge, which has a main span of 1990.8 m, in Japan. It was originally designed with a main span of 1990 m (Figure 10.55), but was extended by 0.8 m when the Kobe Earthquake came close to this mark in 1995.

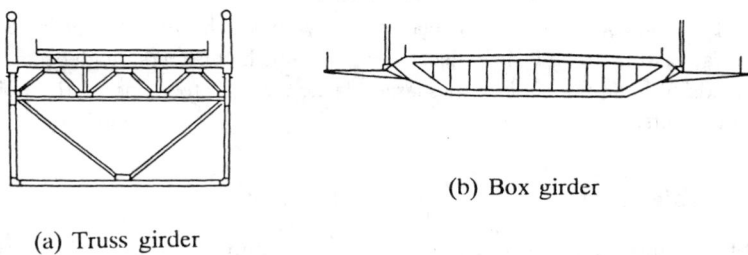

(a) Truss girder

(b) Box girder

Figure 10.54 Types of stiffening girders. (From Japan Association of Steel Bridge Construction, *Outline of Steel Bridges*, Tokyo [in Japanese], 1985. With permission.)

TABLE 10.7 The World's 10 Longest Bridges

Rank	Name	Center span (m)	Country	Year completed
1	Akashi Kaikyo Bridge	1990	Japan	1998 (est.)
2	Great Belt East Bridge	1624	Denmark	1997 (est.)
3	Humber Bridge	1410	England	1981
4	Tsing Ma Bridge	1377	China	1997 (est.)
5	Verrazano Gate Bridge	1298	U.S.	1964
6	Golden Gate Bridge	1280	U.S.	1937
7	Mackinac Straits Bridge	1158	U.S.	1957
8	Minami Bisan-Seto Bridge	1100	Japan	1988
9	Faith Sulton Mehmet Bridge	1090	Turkey	1988
10	Bosporus Bridge	1074	Turkey	1973

From Honshu Shikoku Bridge Authority, *Booklet and Brochures,* Japan. With permission.

The flow of forces in a suspension bridge is shown in Figure 10.56. The load on the girder is transmitted to the towers through the hangers and the main cables, and then to the anchor blocks. It can be seen that anchor blocks are essential to take the horizontal reaction force from the cables. The gravity of the anchor blocks resists the upward component of the cable tension force, and the shear force between the anchor blocks and the foundation resists the horizontal component. Construction difficulty may arise where soil conditions are poor. Different from the cable-stayed bridge, no axial force is induced in the girders of a suspension bridge unless it is a self-anchored suspension bridge (see Figure 10.57d).

The sag in the main cable affects the structural behavior of the suspension bridge: the smaller the sag, the larger the stiffness of the bridge and thereby large horizontal forces are applied to anchor blocks. In general the ratio of the sag to the main span is selected to be about 1:10.

10.12.2 Types of Suspension Bridges

Suspension bridges can be classified by the support condition of their stiffening girders and the main cable (Figure 10.57). The three-span, two-hinge type is most commonly used for highway bridges. The continuous girder is often adopted for railroad bridges to avoid "knuckle points", which adversely affect the trains. When the side span is short, the single-span type is selected. The main cables of self-anchored bridges are fixed to the girders instead of to the anchor blocks, making the construction of anchor blocks unnecessary; instead the axial compression is carried in the girders as in the cable-stayed bridge. There are special cases (such as the Severn Bridge in England) where diagonal hangers have been used.

10.12.3 Structural Analysis

If the dead load of the cable and the stiffening girders is assumed to be uniformly distributed along the bridge length, the deflection of the cable is parabolic and all dead loads are supported by the cable. In this case, only live loads act on the girder.

There are two analytical procedures: elastic theory, in which linear elastic material and small displacement are assumed, and the deflection theory, which considers the deflection of the cable due to live loads. When the span becomes large, elastic theory is too conservative in its estimation of bending moments.

10.12.4 Cable Design

For the cable, the high-strength steel wire, i.e., usually 5 mm in diameter with a strength of 160 to 180 kg/mm^2 (1760 N/mm^2) and zinc-galvanized, is used. There are several types of cables (Figure 10.58): strand rope, spiral rope, locked coil rope (LCR), and parallel wire strand (PWS).

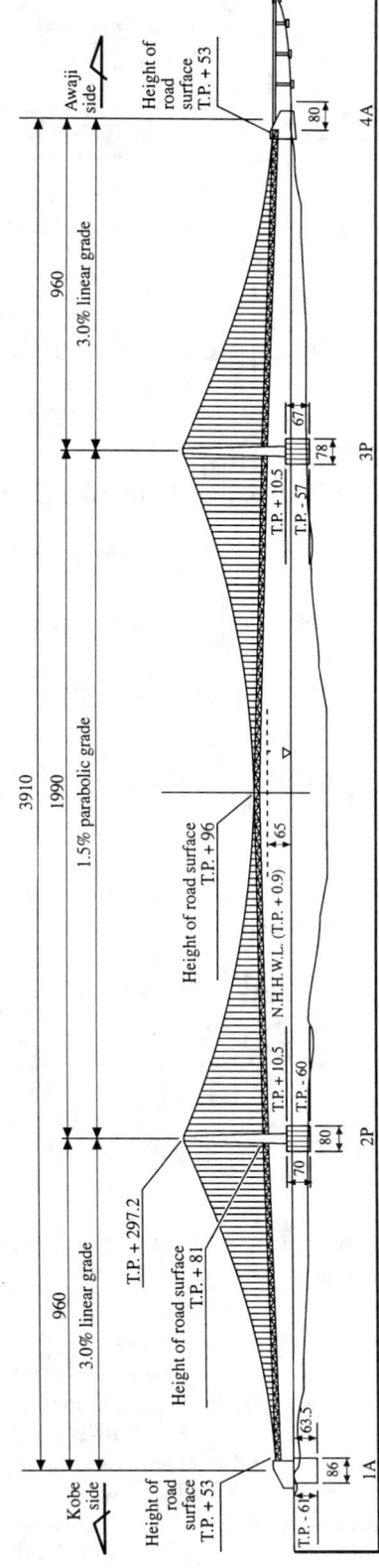

Figure 10.55 Side view of Akashi Kaikyo Bridge, Japan (1998 expected). (From Honshu Shikoku Bridge Authority, Technology of Akashi Kaikyo Bridge, Japan [in Japanese]. With permission.)

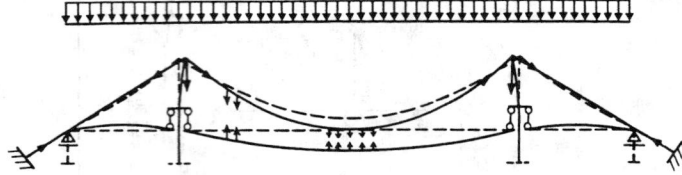

Figure 10.56 Force flow in a suspension bridge. (From Nagai, N., *Bridge Engineering*, Kyoritsu Publishing Co., Tokyo, Japan [in Japanese], 1994. With permission.)

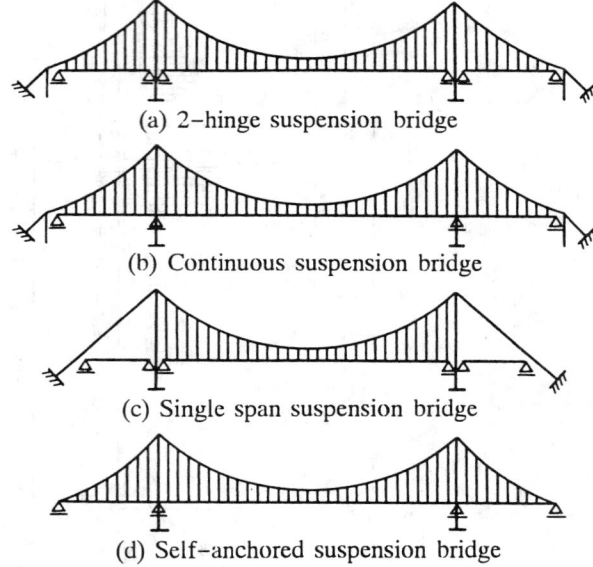

(a) 2–hinge suspension bridge

(b) Continuous suspension bridge

(c) Single span suspension bridge

(d) Self–anchored suspension bridge

Figure 10.57 Types of suspension bridges. (From Nagai, N., *Bridge Engineering*, Kyoritsu Publishing Co., Tokyo, Japan [in Japanese], 1994. With permission.)

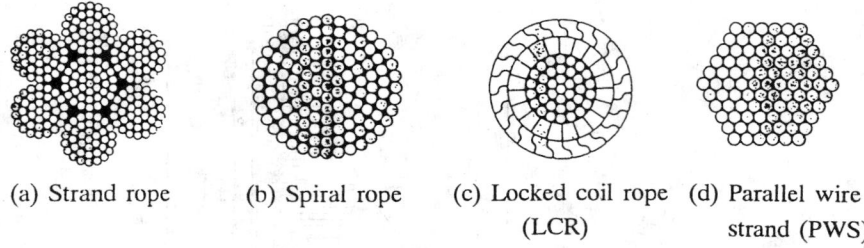

(a) Strand rope   (b) Spiral rope   (c) Locked coil rope   (d) Parallel wire
                                               (LCR)   strand (PWS)

Figure 10.58 Types of cables. (From Nagai, N., *Bridge Engineering*, Kyoritsu Publishing Co., Tokyo, Japan [in Japanese], 1994. With permission.)

The PWS is used most commonly for suspension bridges; thousands of parallel wire elements are bundled into a circle by a squeezing machine, then wrapped with steel wire and painted.

The wire is treated by an air spinning (AS) method or the prefabricated parallel wire strand method (PWSS). In the AS method, the 5-mm wire is elected by rounding between anchor blocks one by one until the prescribed number of wires is obtained. In the PWSS method, a strand that bundles 100 to 200 wire elements is suspended between the anchor blocks by fixing with a socket. In this method, the construction period can be short because more wires are elected at one time than in the AS method. The thick strand is more stable to wind but harder to handle during construction.

A foothold (or catwalk) must be provided under the cable for the workers to attach the cable band to the main cable.

10.12.5 Stiffening Girder

Truss or box type girders are used to stiffen suspension bridges. The girder must be carefully designed to have sufficient stiffness for wind stability. For very long spans trusses are most effective in improving the stiffness and stability (see Figure 10.54). The box girder is also often adopted due to its ease of fabrication.

10.12.6 Tower

The tower is designed to be subjected to large axial compression and bending moment. It is designed to have smaller bending stiffness in the longitudinal direction since the horizontal forces coming from both sides of the tower keep it balanced. Figure 10.59 shows a comparison of several towers used for various structures. The Sears Tower in Chicago, known as the tallest building, has a height of 443 m.

Figure 10.59 Comparison of towers. (From Honshu Shikoku Bridge Authority, Technology of Akashi Kaikyo Bridge, Japan [in Japanese]. With permission.)

A bridge tower usually consists of more than three cells inside, each having adequate resistance to torsion and local buckling under large axial forces. Mechanical dampers such as the TMD (tuned mass damper) or the TLD (tuned liquid damper) are often used during construction to control tower oscillations caused by wind forces. Figure 10.60 shows a typical construction procedure adopted for the Akashi Kaikyo Bridge, in which a climbing tower crane is used. An alternative method is to use a creeper crane, which clambers up along the tower.

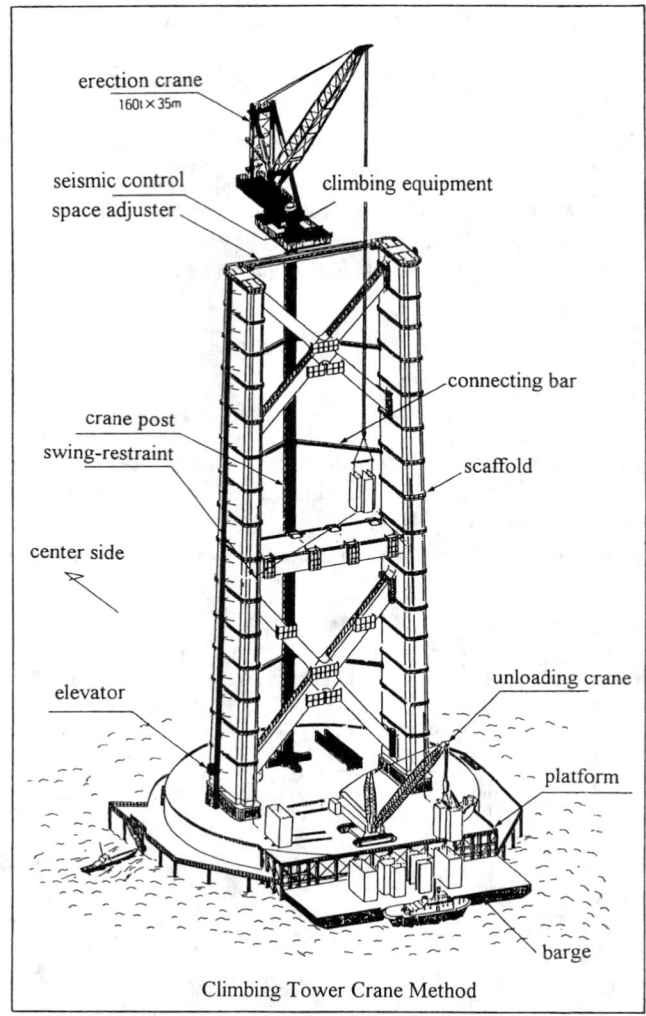

erection crane
160t × 35m

seismic control
space adjuster

climbing equipment

connecting bar

crane post
swing-restraint

scaffold

center side

unloading crane

elevator

platform

barge

Climbing Tower Crane Method

Figure 10.60 Construction of a tower. (From Honshu Shikoku Bridge Authority, Technology of Akashi Kaikyo Bridge, Japan [in Japanese]. With permission.)

10.12.7 Stability for Wind

Suspension bridges are so flexible that the dynamic stability under wind effects should be investigated using a wind tunnel. The dynamic responses may be categorized into three types, of which response behaviors are shown in Figure 10.61: vortex-induced oscillations, buffeting, and torsional flutters. The flutter, also called "galloping", is torsional oscillation and is especially dangerous since it is a self-diverging resonance and may incite failure quickly and easily. The flow of air increases the amplitude of oscillations under certain combinations of wind speed and structural characteristics (natural frequency), as illustrated in Figure 10.61. Flexible bridges, such as suspension or cable-stayed bridges, must be carefully designed if the wind speeds are likely to incite flutter.

Vortex-induced oscillations were once thought to be caused by the Karman vortex. Now it is understood to be the air flow coming from the surface or edge of the girders that yields vibrations which resonate with the natural frequency of the structure. This vibration occurs at a low and relatively narrow range of wind speeds and does not develop dangerous degrees of amplitude am-

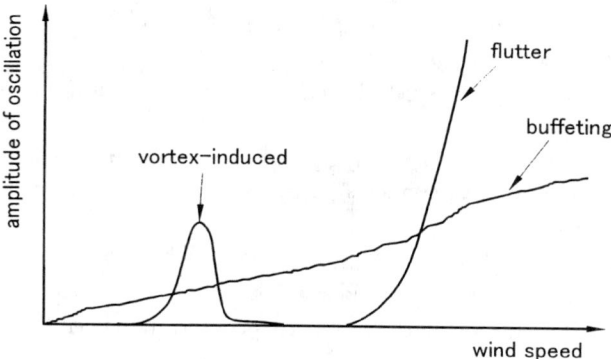

Figure 10.61 Dynamic response of a tower against wind. (From Nagai, N., *Bridge Engineering*, Kyoritsu Publishing Co., Tokyo, Japan [in Japanese], 1994. With permission.)

plification. Buffeting is a random vibration caused by turbulence in the air flow or spontaneous gusts. Horizontal movements are dominant and the amplitudes increase proportionally with the square of wind speed.

10.13 Defining Terms

Abutment: An end support for a bridge structure.

Arch bridge: A bridge that includes the road deck and the supporting arch.

Bridge: A structure that crosses over a river, bay, or other obstruction, permitting the smooth and safe passage of vehicles, trains, and pedestrians.

Cable-stayed bridge: A bridge in which the superstructure is hung from the diagonal cables that are tensioned from the tower.

Cast-in-place concrete: Concrete placed in its final position in the structure while still in a plastic state.

Composite girder: A stell girder connected to a concrete deck so that they respond to force effects as a unit.

Deck (slab): A component, with or without wearing surface, directly supporting wheel loads.

Floor system: A superstructure in which the deck is integral with its supporting components, such as floor beams and stringers.

Girder: A structural component whose primary function is to resist loads in flexure and shear. Generally, this term is used for fabricated sections.

Girder bridge: A bridge superstructure that consists of a floor slab, girders, and bearings.

Influence line: A continuous or discretized function over a section of girder whose value at a point, multiplied by a load acting normal to the girder at that point, yields the force effect being sought.

Lever rule: The static summation of moments about one point to calculate the reaction at a second point.

LRFD (Load and Resistance Factor Design): A method of proportioning structural components (members, connectors, connecting elements, and assemblages) such that no applicable limit state is exceeded when the structure is subjected to all appropriate load combinations.

Precast member: Concrete element cast in a location other than its final position.

Prestressed concrete: Concrete components in which the stresses and deformations are introduced by application of prestressing forces.

Rigid frame bridge: A bridge in which the superstructure and substructure members are rigidly connected.

Segmental bridge: A bridge in which primary load-supporting members are composed of individual members called segments post-tensioned together to act as a monolithic unit under loads.

Substructure: Structural parts of the bridge which provide the horizontal span.

Superstructure: Structural parts of the bridge which support the horizontal span.

Suspension bridge: A bridge in which the superstructure is suspended by two main cables and anchored to end blocks.

Truss bridge: A bridge superstructure which consists of a floor system and main trusses.

Acknowledgment

Many of the figures in this chapter are copied from other books and journals. The authors would like to express sincere gratitude to the original authors. Special thanks go to Prof. N. Nagai of the Nagaoka Institute of Science and Technology, Profs. Y. Tachibana and H. Nakai of Osaka City University, the Japan Association of Steel Bridge Construction, American Association of State Highway and Transportation Officials, and California Department of Transportation for their generosity.

References

[1] American Association of State Highway and Transportation Officials. 1994. *AASHTO LRFD Bridge Design Specifications,* 1st ed., AASHTO, Washington, D.C.

[2] American Association of State Highway and Transportation Officials. 1989. *Guide Specifications for Design and Construction of Segmental Concrete Bridges,* AASHTO, Washington, D.C.

[3] Caltrans. 1990. *Bridge Design Details Manual.* California Department of Transportation, Sacramento, CA.

[4] Caltrans. 1993. Bridge Design Practice Manual, vol. 2, California Department of Transportation, Sacramento, CA.

[5] Caltrans. 1990. *Bridge Design Aids Manual.* California Department of Transportation, Sacramento, CA.

[6] Federal Highway Administration. 1990. *Standard Plans for Highway Bridges, Vol. I, Concrete Superstructures,* U.S. Department of Transportation, FHWA, Washington, D.C.

[7] Gerwick, B.C., Jr. 1993. *Construction of Prestressed Concrete Structures,* 2nd ed., John Wiley & Sons, New York.

[8] Hanshin Expressway Public Corporation. 1975. *Construction Records of Minato Oh-Hashi,* HEPC, Japan Society of Civil Engineers, Tokyo, Japan (in Japanese).

[9] Hanshin Expressway Public Corporation. 1994. *Techno Gallery,* HEPC, Osaka, Japan.

[10] Honshu Shikoku Bridge Authority. Technology of Akashi Kaikyo Bridge, HSBA, Japan (in Japanese).

[11] Japan Association of Steel Bridge Construction. 1981. *Manual Design Data Book,* JASBC, Tokyo, Japan (in Japanese).

[12] Japan Association of Steel Bridge Construction. 1984. *A Guide Book of Bearing Design for Steel Bridges,* JASBC, Tokyo, Japan (in Japanese).

[13] Japan Association of Steel Bridge Construction. 1984. *A Guide Book of Expansion Joint Design for Steel Bridges,* JASBC, Tokyo, Japan (in Japanese).

[14] Japan Association of Steel Bridge Construction. 1985. *Outline of Steel Bridges,* JASBC, Tokyo, Japan (in Japanese).

[15] Japan Association of Steel Bridge Construction. 1988. *Planning of Steel Bridges,* JASBC, Tokyo, Japan (in Japanese).

[16] Japan Construction Mechanization Association. 1991. *Cost Estimation of Bridge Erection,* JCMA, Tokyo, Japan (in Japanese).

[17] Japan Road Association. 1993. *Specifications for Highway Bridges, Part I Common Provisions, Part II Steel Bridges, and Part III Concrete Bridges,* JRA, Tokyo, Japan (in Japanese).

[18] Japan Society of Civil Engineers. 1990. *Cable-Stayed Bridges—Technology and its Change,* JSCE, Tokyo, Japan (in Japanese).

[19] Nagai, N. 1994. *Bridge Engineering,* Kyoritsu Publishing Co., Tokyo, Japan (in Japanese).

[20] Podolny, W. and Muller, J.M. 1982. *Construction and Design of Prestressed Concrete Segmental Bridges,* John Wiley & Sons, New York.

[21] Shimada, S. 1991. Basic theory of arch structures, *Journal of Bridge and Foundation Engineering,* Kensetsu-Tosho, 25(8), 48-52, Tokyo, Japan (in Japanese).

[22] Tachibana, Y. and Nakai, H. 1996. *Bridge Engineering,* Kyoritsu Publishing Co., Tokyo, Japan (in Japanese).

[23] Tonias, D.E. 1995. *Bridge Engineering,* McGraw-Hill, New York.

[24] Troitsky, M.S. 1994. *Planning and Design of Bridges,* John Wiley & Sons, New York.

[25] VSL. 1994. VSL Post-Tensioning System, VSL Corporation, Campbell, CA.

[26] Xanthakos, P.P. 1994. *Theory and Design of Bridges,* John Wiley & Sons, New York.

[27] Xanthakos, P.P. 1995. *Bridge Substructure and Foundation Design,* Prentice-Hall, Upper Saddle River, NJ.

Further Reading

[1] Billington, D.P. 1983. *The Tower and the Bridge,* Basic Books, Inc., New York.

[2] Leonhardt, F. 1984. *Bridges, Aesthetics and Design,* MIT Press, Cambridge, MA.

[3] Chen, W. F. and Duan, L. 1998. *Handbook of Bridge Engineering,* CRC Press, Boca Raton, FL.

Appendix: Design Examples

10.A.1 Two-Span, Continuous, Cast-in-Place, Prestressed Concrete Box Girder Bridge

Given: A two-span, continuous, cast-in-place, prestressed concrete box girder bridge has two equal spans of length 157 ft (47.9 m) with a column bent. The superstructure is 34 ft (10.4 m) wide. The elevation of the bridge is shown in Figure 10.62a.

Material:

Initial concrete $f'_{ci} = 3500$ psi (24.13 MPa), $E_{ci} = 3372$ ksi (23,250 MPa)

Final concrete $f'_c = 4000$ psi (27.58 MPa), $E_c = 3600$ ksi (24,860 MPa)

Prestressing steel $f_{pu} = 270$ ksi (1860 MPa) low relaxation strand, $E_p = 28,500$ ksi (197,000 MPa)

Mild steel $f_y = 60$ ksi (414 MPa), $E_s = 29,000$ ksi (200,000 MPa)

Prestressing:

Anchorage set thickness $= 0.375$ in. (9.5 mm)

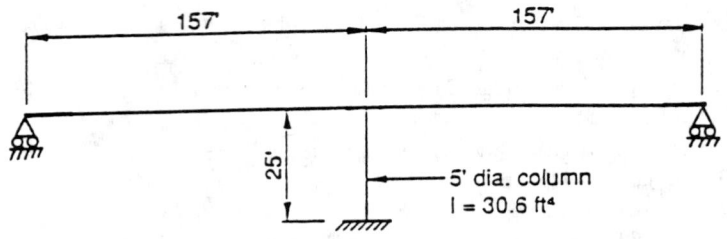

(a) Elevation View

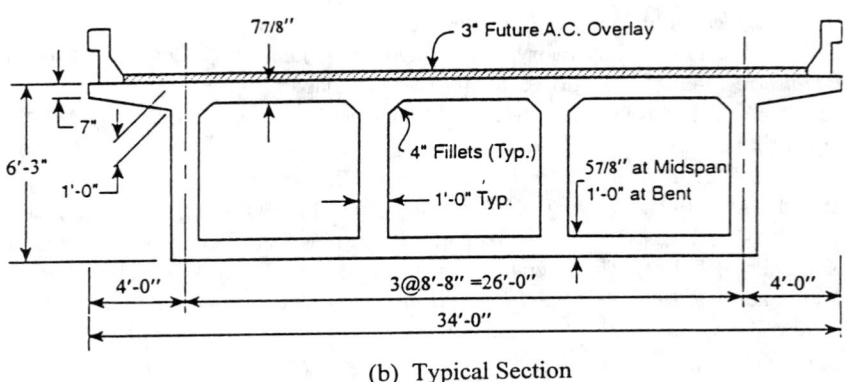

(b) Typical Section

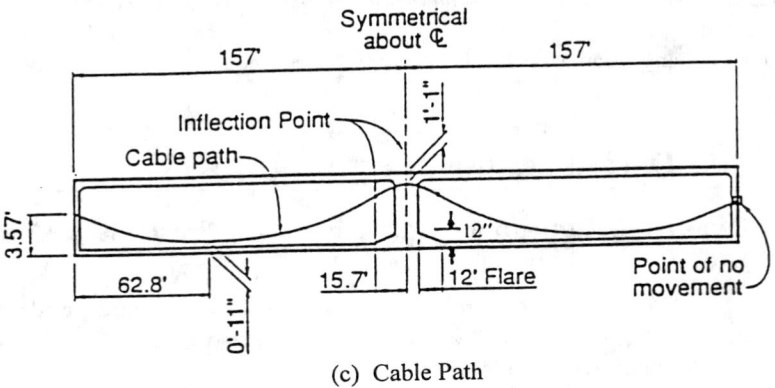

(c) Cable Path

Figure 10.62   A two-span, continuous, prestressed concrete box girder bridge.

Prestressing stress at jacking $f_{pj} = 0.8$

$f_{pu} = 216$ ksi (1489 MPa)

The secondary moments due to prestressing at the bent are $M_{DA} = 1.118 \, P_j$ (kips-ft)

$M_{DG} = 1.107 \, P_j$ (kips-ft)

Loads:

Dead load = self-weight + barrier rail + future wearing 3 in AC overlay

Live load = AASHTO HS20-44 + dynamic load allowance

Specification:

AASHTO-LRFD [1] (referred to as AASHTO in this example)

Requirements:

1. Determine cross-section geometry
2. Determine longitudinal section and cable path
3. Calculate loads
4. Calculate live load distribution factors
5. Calculate unfactored moments and shear demands for interior girder
6. Determine load factors for strength limit state I and service limit state I
7. Calculate section properties for interior girder
8. Calculate prestress losses
9. Determine prestressing force, P_j, for interior girder
10. Check concrete strength for interior girder, service limit state I
11. Flexural strength design for interior girder, strength limit state I
12. Shear strength design for interior girder, strength limit state I

   Solution
1. **Determine Cross-Section Geometry**
   1.1) Structural Depth, d
   For prestressed continuous spans, the structural depth, d, can be determined using a depth-to-span ratio (d/L) of 0.04.

   $$d = 0.04L = 0.04(157) = 6.28 \text{ ft } (1.91 \text{ m})$$
   $$\text{Use } d = 6.25 \text{ ft } (1.91 \text{ m})$$

   1.2) Girder Spacing, S
   To provide effective torsional resistance and a sufficient number of girders for prestress paths, the spacing of girders should not be larger than twice their depth.

   $$S_{\max} < 2d = 2(6.25) = 12.5 \text{ ft } (3.81 \text{ m})$$

   Using an overhang of 4 ft (1.22 m), the center-to-center distance between two exterior girders is 26 ft (7.92 m).

   Try three girders and two bays, $S = 26/2 = 13$ ft > 12.5 ft N.G.

   Try four girders and three bays, $S = 26/3 = 8.67$ ft < 12.5 ft O.K.

   Use a girder spacing, $S = 8.67$ ft (2.64 m)

   1.3) Typical Section
   From past experience and design practice, we select a thickness of 7 in. (178 mm) at the edge and 12 in. (305 mm) at the face of exterior girder for the overhang, the width of 12 in. (305 mm) for girders with the exterior girder flaring to 18 in. (457 mm) at the anchorage end. The length of this flare is usually taken as one-tenth of the span length 15.7 ft (4.79 m). The deck and soffit thicknesses depend on the clear distance between adjacent girders. We choose 7.875 in. (200 mm) and 5.875 in. (149 mm) for the deck and soffit thicknesses, respectively. A typical section for this example is shown in Figure 10.62b. The section properties of the box girder are :

Properties	Midspan	Bent (face of support)
A ft^2 (m^2)	57.25 (5.32)	68.98 (6.41)
I ft^4 (m^4)	325.45 (2.81)	403.56 (3.48)
y_b ft (m)	3.57 (1.09)	3.09 (0.94)

2. **Determine Longitudinal Section and Cable Path**

    To lower the center of gravity of the superstructure at the face of a bent cap in a cast-in-place post-tensioned box girder, the thickness of soffit is flared to 12 in., as shown in Figure 10.62c. A cable path is generally controlled by the maximum dead load moments and the position of the jack at the end section. Maximum eccentricities should occur at points of maximum dead load moment and almost no eccentricity should be present at the jacked end section. For this example, the maximum dead load moments occur at the bent cap, close to $0.4L$ for span 1 and $0.6L$ for span 2. A parabolic cable path is chosen as shown in Figure 10.62c.

3. **Calculate Loads**

    3.1) *Component Dead Load, DC*

    The component dead load, DC, includes all structural dead loads with the exception of the future wearing surface and specified utility loads. For design purposes, two parts of the DC are defined as:

    $DC1$: girder self-weight (150 lb/ft^3) acting at the prestressing state

    $DC2$: barrier rail weight (784 kips/ft) acting at service state after all losses

    3.2) *Wearing Surface Load, DW*

    The future wearing surface of 3 in. (76 mm) with a unit weight of 140 lb/ft^3 is designed for this bridge.

$$
\begin{aligned}
DW &= \text{(deck width–barrier width) (thickness of wearing surface) (unit weight)} \\
&= [34 - 2(1.75)](0.25)(140) = 1067.5 \ \text{lb/ft}
\end{aligned}
$$

    3.3) *Live Load, LL, and Dynamic Load Allowance, IM*

    The design live load, LL, is the AASHTO HS20-44 vehicular live load. To consider the wheel load impact from moving vehicles, the dynamic load allowance, $IM = 33\%$ (AASHTO-LRFD Table 3.6.2.1-1), is used.

4. **Calculate Live Load Distribution Factors**

    AASHTO-LRFD [1] recommends that approximate methods be used to distribute live load to individual girders. The dimensions relevant to this prestressed box girder are: depth, $d = 6.25$ ft (1.91 m); number of cells, $N_c = 3$; spacing of girders, $S = 8.67$ ft (2.64 m); span length, $L = 157$ ft (47.9 m); half of the girder spacing plus the total overhang, $W_e = 8.334$ ft (2.54 m); and the distance between the center of an exterior girder and the interior edge of a barrier, $d_e = 4 - 1.75 = 2.25$ ft (0.69 m). This box girder is within the range of applicability of the AASHTO approximate formulas. The live load distribution factors are calculated as follows.

    4.1) *Live Load Distribution Factor for Bending Moments*

    (a) Interior girder (AASHTO Table 4.6.2.2.2b-1)

        One lane loaded:

$$LD_M = \left(1.75 + \frac{S}{3.6}\right)\left(\frac{1}{L}\right)^{0.35}\left(\frac{1}{N_c}\right)^{0.45}$$

$$= \left(1.75 + \frac{8.67}{3.6}\right)\left(\frac{1}{157}\right)^{0.35}\left(\frac{1}{3}\right)^{0.45} = 0.432 \text{ lanes}$$

Two or more lanes loaded:

$$LD_M = \left(\frac{13}{N_c}\right)^{0.3}\left(\frac{S}{5.8}\right)\left(\frac{1}{L}\right)^{0.25}$$

$$= \left(\frac{13}{3}\right)^{0.3}\left(\frac{8.67}{5.8}\right)\left(\frac{1}{157}\right)^{0.25} = 0.656 \text{ lanes (controls)}$$

(b) Exterior girder (AASHTO Table 4.6.2.2.2d-1)

$$LD_M = \frac{W_e}{14} = \frac{8.334}{14} = 0.595 \text{ lanes (controls)}$$

4.2) Live Load Distribution Factor for Shear

(a) Interior girder (AASHTO Table 4.62.2.3a-1)

One lane loaded:

$$LD_V = \left(\frac{S}{9.5}\right)^{0.6}\left(\frac{d}{12L}\right)^{0.1}$$

$$= \left(\frac{8.67}{9.5}\right)^{0.6}\left(\frac{6.25}{12(157)}\right)^{0.1} = 0.535 \text{ lanes}$$

Two or more lanes loaded:

$$LD_V = \left(\frac{S}{7.3}\right)^{0.9}\left(\frac{d}{12L}\right)^{0.1}$$

$$= \left(\frac{8.67}{7.3}\right)^{0.9}\left(\frac{6.25}{12(157)}\right)^{0.1} = 0.660 \text{ lanes (controls)}$$

(b) Exterior girder (AASHTO Table 4.62.2.3b-1)

One lane loaded: Lever rule

The lever rule assumes that the deck in its transverse direction is simply supported by the girders and uses statics to determine the live load distribution to the girders. AASHTO-LRFD also requires that when the lever rule is used, the multiple presence factor, m, should apply. For a one loaded lane, $m = 1.2$. The lever rule model for the exterior girder is shown in Figure 10.63. From static equilibrium:

$$R = \frac{5.92}{8.67} = 0.683$$

$$LD_v = mR = 1.2(0.683) = 0.820 \text{ (controls)}$$

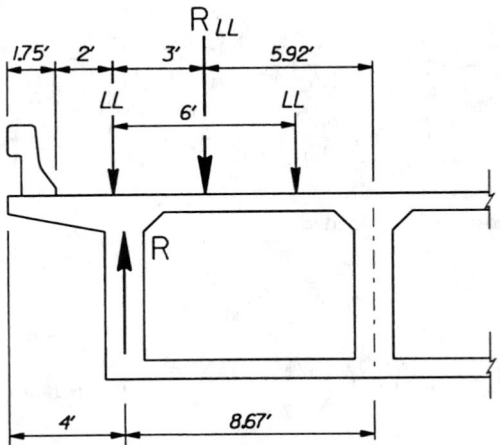

Figure 10.63   Live load distribution for exterior girder—lever rule.

Two or more lanes loaded: Modify interior girder factor by e

$$LD_V = e(LD_v)_{\text{interior girder}} = \left(0.64 + \frac{d_e}{12.5}\right)(LD_v)_{\text{interior girder}}$$

$$= \left(0.64 + \frac{2.25}{12.5}\right)(0.66) = 0.541$$

The live load distribution factors at the strength limit state:

Strength limit state I	Interior girder	Exterior girder
Bending moment	0.656 lanes	0.595 lanes
Shear	0.660 lanes	0.820 lanes

5. **Calculate Unfactored Moments and Shear Demands for Interior Girder**
   It is practically assumed that all dead loads are carried by the box girder and equally distributed to each girder. The live loads take forces to the girders according to live load distribution factors (AASHTO Article 4.6.2.2.2). Unfactored moment and shear demands for an interior girder are shown in Figures 10.64 and 10.65. Details are listed in Tables 10.8 and 10.9. Only the results for span 1 are shown in these tables and figures since the bridge is symmetrical about the bent.

6. **Determine Load Factors for Strength Limit State I and Service Limit State I**
   6.1) *General Design Equation* (AASHTO Article 1.3.2)

$$\eta \sum \gamma_i Q_i \leq \phi R_n \tag{10.18}$$

where γ_i are load factors, ϕ is a resistance factor, Q_i represents force effects, R_n is the nominal resistance, and η is a factor related to ductility, redundancy, and operational importance of that being designed. η is defined as:

$$\eta = \eta_D \eta_R \eta_I \geq 0.95 \tag{10.19}$$

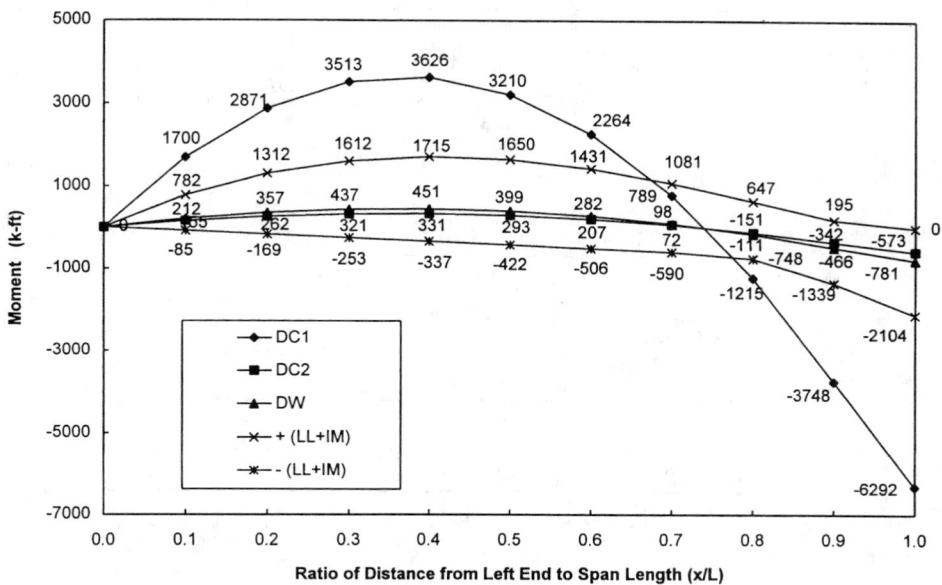

Figure 10.64   Moment envelopes for span 1.

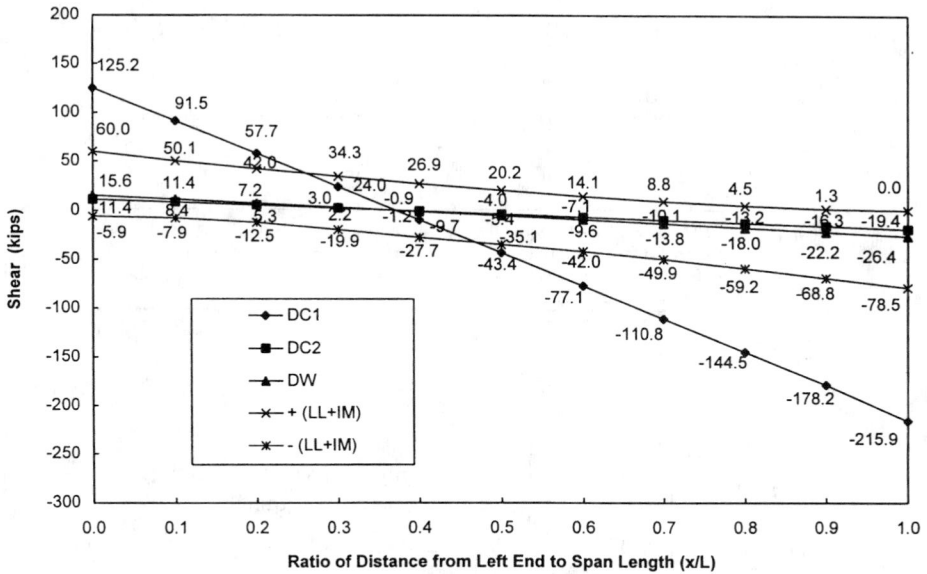

Figure 10.65   Shear envelopes for span 1.

TABLE 10.8   Moment and Shear Due to Unfactored Dead Load
for the Interior Girder (Span 1)

| | Unfactored dead load | | | | | |
| | $DC1^a$ | | $DC2^b$ | | DW^c | |
Location (x/L)	M_{DC1} (k-ft)	V_{DC1} (kips)	M_{DC2} (k-ft)	V_{DC2} (kips)	M_{DW} (k-ft)	V_{DW} (kips)
0.0	0	125.2	0	11.4	0	15.6
0.1	1700	91.5	155	8.4	212	11.4
0.2	2871	57.7	262	5.3	357	7.2
0.3	3513	24.0	321	2.2	437	3.0
0.4	3626	−9.7	331	−0.9	451	−1.2
0.5	3210	−43.4	293	−4.0	399	−5.4
0.6	2264	−77.1	207	−7.1	282	−9.6
0.7	789	−111	72	−10.1	98	−13.8
0.8	−1215	−145	−111	−13.2	−151	−18.0
0.9	−3748	−178	−342	−16.3	−466	−22.2
1.0	−6833	−216	−622	−19.4	−847	−26.4
	(−6292)		(−573)		(−781)	

Note: Moments in brackets are for face of support at the bent. Moments
in span 2 are symmetrical about the bent. Shear in span are antisymmetrical about the bent.
$^a DC1$, interior girder self-weight.
$^b DC2$, barrier self-weight.
$^c DW$, wearing surface load.

TABLE 10.9   Moment and Shear Envelopes and Associated Forces for the
Interior Girder Due to AASHTO HS20-44 Live Load (Span 1)

| | Positive moment and associated shear | | Negative moment and associated shear | | Shear and associated moment | |
| Location | M_{LL+IM} | V_{LL+IM} | M_{LL+IM} | V_{LL+IM} | V_{LL+IM} | M_{LL+IM} |
(x/L)	(k-ft)	(kips)	(k-ft)	(kips)	(kips)	(k-ft)
0.0	0	0	0	0	60.0	0
0.1	782	49.8	−85	−5.4	50.1	787
0.2	1312	41.8	−169	−5.4	42.0	1320
0.3	1612	29.3	−253	−5.4	34.3	1614
0.4	1715	21.8	−337	−5.4	−27.7	1650
0.5	1650	−30.0	−422	−5.4	−35.1	1628
0.6	1431	−36.7	−506	−5.4	−42.0	1424
0.7	1081	−42.6	−590	−5.4	−49.9	852
0.8	647	−47.8	−748	−8.3	−59.2	216
0.9	196	−32.9	−1339	−50.1	−68.8	−667
1.0	0	0	−2266	−67.8	−78.5	1788
			−(2104)			

Note: $LL + IM$ = AASHTO HS20-44 live load plus dynamic load allowance. Moments in
brackets are for face of support at the bent. Moments in span 2 are symmetrical about
the bent. Shear in span 2 is antisymmetrical about the bent. Live load distribution
factors are considered.

where

$$\eta_D = \begin{cases} 1.05 & \text{for nonductile components and connections} \\ 0.95 & \text{for ductile components and connections} \end{cases} \tag{10.20}$$

$$\eta_R = \begin{cases} 1.05 & \text{for nonredundant members} \\ 0.95 & \text{for redundant members} \end{cases} \tag{10.21}$$

$$\eta_I = \begin{cases} 1.05 & \text{operationally important bridge} \\ 0.95 & \text{general bridge} \\ & \text{only apply to strength and extreme} \\ & \text{event limit states} \end{cases} \tag{10.22}$$

For this bridge, the following values are assumed:

Limit states	Ductility η_D	Redundancy η_R	Importance η_I	η
Strength limit state	0.95	0.95	1.05	0.95
Service limit state	1.0	1.0	1.0	1.0

6.2) *Load Factors and Load Combinations*
The load factors and combinations are specified as (AASHTO Table 3.4.1-1):

Strength limit state I:	$1.25(DC1 + DC2) + 1.5(DW) + 1.75(LL + M)$
Service limit state I:	$DC1 + DC2 + DW + (LL + IM)$

7. **Calculate Section Properties for Interior Girder**
For an interior girder as shown in Figure 10.66, the effective flange width, b_{eff}, is determined (AASHTO Article 4.6.2.6) by

$$b_{eff} = \text{ the lesser of } \begin{cases} \dfrac{L_{eff}}{4} \\ 12h_f + b_w \\ S \end{cases} \qquad (10.23)$$

where L_{eff} is the effective span length and may be taken as the actual span length for simply supported spans and the distance between points of permanent load inflection for continuous spans; h_f is the compression flange depth; and b_w is the web width. The effective flange width and the section properties are shown in Table 10.10 for the interior girder.

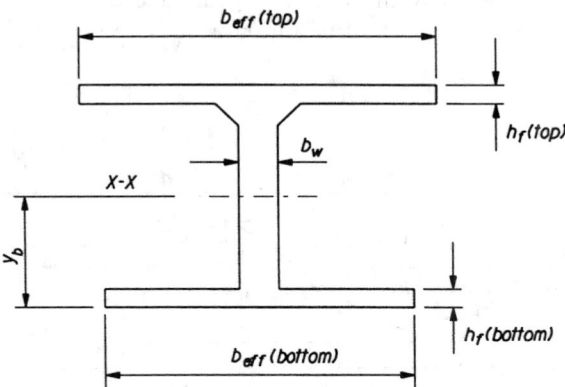

Figure 10.66 Effective flange width of interior girder.

8. **Calculate Prestress Losses**
For a cast-in-place post-tensioned box girder, two types of losses, instantaneous losses (friction, anchorage set, and elastic shortening) and time-dependent losses (creep and shrinkage of concrete and relaxation of prestressing steel) are significant. Since the prestress losses are not symmetrical about the bent for this bridge, the calculation is performed for both spans.
8.1) *Frictional Loss*, Δf_{pF}

TABLE 10.10 Effective Flange Width and Section Properties for Interior Girder

Location	Dimension	Midspan	Bent (face of support)
Top flange	h_f in. (mm)	7.875 (200)	7.875 (200)
	$L_{eff}/4$ in. (mm)	353 (8966)	235.5(11963)
	$12h_f + b_w$ in. (mm)	106.5 (2705)	106.5 (2705)
	S in. (mm)	104 (2642)	104 (2642)
	b_{eff} in. (mm)	104 (2642)	104 (2642)
Bottom flange	h_f in. (mm)	5.875 (149)	12 (305)
	$L_{eff}/4$ in. (mm)	353 (8966)	235.5 (11963)
	$12h_f + b_w$ in. (mm)	82.5 (2096)	156 (2096)
	S in. (mm)	104 (2642)	104 (2642)
	b_{eff} in. (mm)	82.5 (2096)	104 (2096)
Area	A ft^2 (m^2)	14.38 (1.336)	19.17 (1.781)
Moment of inertia	I ft^4 (m^4)	81.85 (0.706)	112.21 (0.968)
C.G.	y_b ft (m)	3.55 (1.082)	2.82 (0.860)

Note:
L_{eff} = 117.8 ft (35.9 m) for midspan,
L_{eff} = 78.5 ft (23.9 m) for the bent,
b_w = 12 in. (305 mm).

$$\Delta f_{pF} = f_{pj}\left(1 - e^{-(Kx+\mu\alpha)}\right) \tag{10.24}$$

where K is the wobble friction coefficient = 0.0002 1/ft (6.6 × 10^{-7} 1/mm); μ is the coefficient of friction = 0.25 (AASHTO Article 5.9.5.2.2a); x is the length of a prestressing tendon from the jacking end to the point considered; and α is the sum of the absolute values of angle change in the prestressing steel path from the jacking end. For a parabolic cable path (Figure 10.67), the angle change is $\alpha = 2e_p/L_p$, where e_p is the vertical distance between two control points and L_p is the horizontal distance between two control points. The details are given in Table 10.11.

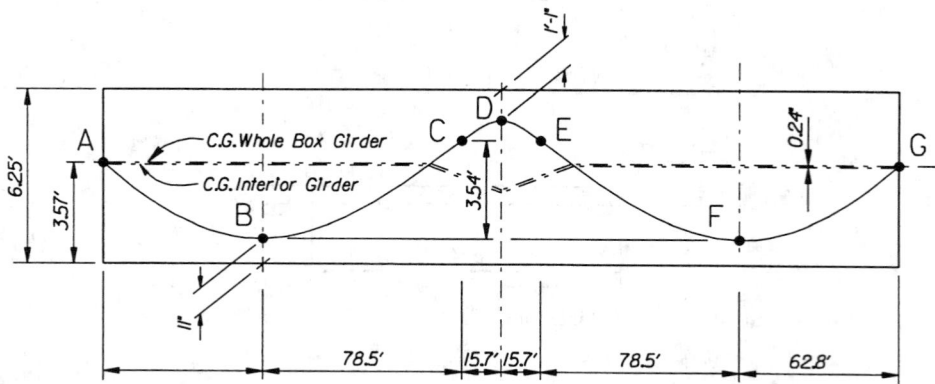

Figure 10.67 Parabolic cable path.

8.2) *Anchorage Set Loss*, Δf_{pA}

The effect of anchorage set on the cable stress can be approximated by the Caltrans procedure [4], as shown in Figure 10.68. It is assumed that the anchorage set loss changes linearly within the length, L_{pA}.

TABLE 10.11 Prestress Frictional Loss

Segment	e_p (in.)	L_p (ft)	α (rad)	$\sum \alpha$ (rad)	$\sum L_p$ (ft)	Point	Δf_{pF} (ksi)
A	31.84	0	0	0	0	A	0.0
AB	31.84	62.8	0.0845	0.0845	62.8	B	7.13
BC	42.50	78.5	0.0902	0.1747	141.3	C	14.90
CD	8.50	15.7	0.0902	0.2649	157.0	D	20.09
DE	8.50	15.7	0.0902	0.3551	172.7	E	25.06
EF	42.50	78.5	0.0902	0.4453	251.2	F	32.18
FG	31.84	62.8	0.0845	0.5298	314.0	G	38.23

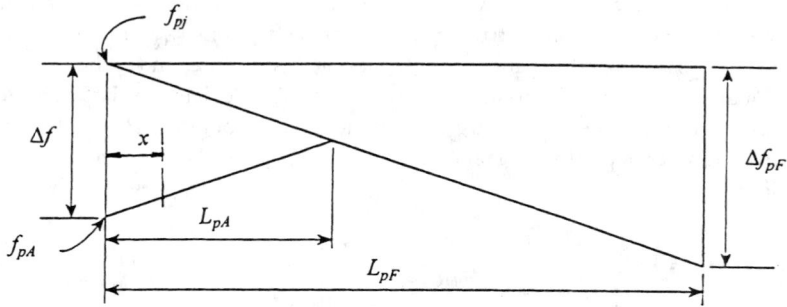

Figure 10.68 Anchorage set loss model. (From California Department of Transportation, *Bridge Design Practice*, Copyright 1983 (Figure 3–10, pages 3–46, updated March, 1993), Sacramento, CA, 1993. With permission.)

$$L_{pA} = \sqrt{\frac{E(\Delta L)L_{pF}}{\Delta f_{pF}}} \qquad (10.25)$$

$$\Delta f = \frac{2\Delta f_{pF} x}{L_{pF}} \qquad (10.26)$$

$$\Delta f_{pA} = \Delta f \left(1 - \frac{x}{L_{pA}}\right) \qquad (10.27)$$

where ΔL is the thickness of the anchorage set; E is the modulus of elasticity of the anchorage set; Δf is the change in stress due to the anchor set; L_{pA} is the length influenced by the anchor set; L_{pF} is the length to a point where loss is known; and x is the horizontal distance from the jacking end to the point considered.

For an anchor set thickness of $\Delta L = 0.375$ in. and $E = 29,000$ ksi, consider the point B where $L_{pF} = 141.3$ ft and $\Delta f_{pF} = 14.9$ ksi:

$$L_{pA} = \sqrt{\frac{E(\Delta L)L_{pF}}{\Delta f_{pF}}} = \sqrt{\frac{29,000(3/8)(141.3)}{12(14.90)}} = 92.71 \text{ ft } < 141.3 \text{ ft O.K}$$

$$\Delta f = \frac{2\Delta f_{pF} x}{L_{pF}} = \frac{2(14.90)(92.71)}{141.3} = 19.55 \text{ ksi}$$

$$\Delta f_{pA} = \Delta f \left(1 - \frac{x}{L_{pA}}\right) = 19.55\left(1 - \frac{x}{92.71}\right)$$

8.3) *Elastic Shortening Loss,* Δf_{pES}

The loss due to elastic shortening in post-tensioned members is calculated using the following formula (AASHTO Article 5.9.5.2.3b):

$$\Delta f_{pES} = \frac{N-1}{2N} \frac{E_p}{E_{ci}} f_{cgp} \tag{10.28}$$

where N is the number of identical prestressing tendons and f_{cgp} is the sum of the concrete stress at the center of gravity of the prestressing tendons due to the prestressing force after jacking and the self-weight of member at the section with the maximum moment. For post-tensioned structures with bonded tendons, f_{cgp} may be calculated at the center section of the span for simply supported structures and at the section with the maximum moment for continuous structures. To calculate the elastic shortening loss, we assume that the prestressing jack force for an interior girder $P_j = 1800$ kips and the total number of prestressing tendons $N = 4$. f_{cgp} is calculated for the mid-support section:

$$
\begin{aligned}
f_{cgp} &= \frac{P_j}{A} + \frac{P_j e^2}{I_x} + \frac{M_{DC1}e}{I_x} \\[2mm]
&= \frac{1800}{19.17(12)^2} + \frac{1800(28.164)^2}{112.21(12)^4} + \frac{(-6292)(12)(28.164)}{112.21(12)^4} \\[2mm]
&= 0.652 + 0.614 - 0.914 = 0.352 \ \text{ksi} \ (2448 \ \text{MPa}) \\[2mm]
\Delta f_{pES} &= \frac{N-1}{2N} \frac{E_p}{E_{ci}} f_{cgp} = \frac{4-1}{2(4)} \frac{28,500}{3370}(0.352) = 1.12 \ \text{ksi} \ (7.7 \ \text{MPa})
\end{aligned}
$$

8.4) *Time-Dependent Losses,* Δf_{pTM}

AASHTO provides a table to estimate the accumulated effect of time-dependent losses resulting from the creep and shrinkage of concrete and the relaxation of the steel tendons. From AASHTO Table 5.9.5.3-1:

$$\Delta f_{pTM} = 21 \ \text{ksi} \ (145 \ \text{MPa}) \ (\text{upper bound})$$

8.5) *Total Losses,* Δf_{pT}

$$\Delta f_{pT} = \Delta f_{pF} + \Delta f_{pA} + \Delta f_{pES} + \Delta f_{pTM}$$

Details are given in Table 10.12.

9. **Determine Prestressing Force,** P_j**, for Interior Girder**

Since the live load is not in general equally distributed to the girders, the prestressing force, P_j, required for each girder may differ. To calculate prestress jacking force, P_j, the initial prestress force coefficient, F_{pCI}, and final prestress force coefficient, F_{pCF}, are defined as:

$$F_{pCI} = 1 - \frac{\Delta f_{pF} + \Delta f_{pA} + \Delta f_{pES}}{f_{pj}} \tag{10.29}$$

$$F_{pCF} = 1 - \frac{\Delta f_{pT}}{f_{pj}} \tag{10.30}$$

TABLE 10.12 Cable Path and Prestress Losses

Span	(x/L)	Δf_{pF}	Δf_{pA}	Δf_{pES}	Δf_{pTM}	Δf_{pT}	F_{pCI}	F_{pCF}
				Prestress losses (ksi)			Force coeff.	
	0.0	0.00	19.55			41.67	0.904	0.807
	0.1	1.78	16.24			40.14	0.911	0.814
	0.2	3.56	12.93			38.61	0.918	0.821
	0.3	5.35	9.93			37.40	0.924	0.827
	0.4	7.13	6.31			35.56	0.933	0.835
1	0.5	8.68	3.00	1.12	21	33.79	0.941	0.844
	0.6	10.24				32.36	0.947	0.850
	0.7	11.79				33.91	0.940	0.843
	0.8	13.35	0.00			35.47	0.933	0.836
	0.9	14.90				37.02	0.926	0.829
	1.0	20.09				42.21	0.902	0.805
	0.0	20.09				42.21	0.902	0.805
	0.1	25.06				47.18	0.879	0.782
	0.2	26.49				48.61	0.872	0.775
	0.3	27.91				50.03	0.866	0.768
	0.4	29.34				51.46	0.859	0.762
	0.5	30.76	0.00	1.12	21	52.88	0.852	0.755
2	0.6	32.18				54.30	0.846	0.749
	0.7	33.69				55.81	0.839	0.742
	0.8	35.21				57.33	0.832	0.735
	0.9	36.72				58.84	0.825	0.728
	1.0	38.23				60.35	0.818	0.721

Note:

$$F_{pCI} = 1 - \frac{\Delta f_{pF} + \Delta f_{pA} + \Delta f_{pES}}{f_{pj}}$$

$$F_{pCF} = 1 - \frac{\Delta f_{pT}}{f_{pj}}$$

The secondary moment coefficients are defined as:

$$M_{sC} = \begin{cases} \frac{x}{L} \frac{M_{DA}}{P_j} & \text{for span 1} \\ (1 - \frac{x}{L}) \frac{M_{DG}}{P_j} & \text{for span 2} \end{cases} \tag{10.31}$$

where x is the distance from the left end for each span. The combined prestressing moment coefficients are defined as:

$$M_{psCI} = F_{pCI}(e) + M_{sC} \tag{10.32}$$

$$M_{psCF} = F_{pCF}(e) + M_{sC} \tag{10.33}$$

where e is the distance between the cable and the center of gravity of a cross-section; positive values of e indicate that the cable is above the center of gravity, and negative ones indicate the cable is below it.

The prestress force coefficients and the combined moment coefficients are calculated and listed in Table 10.13. According to AASHTO, the prestressing force, P_j, can be determined using the concrete tensile stress limit in the precompression tensile zone (see Table 10.5):

$$f_{DC1} + f_{DC2} + f_{DW} + f_{LL+IM} + f_{psF} \geq -0.19\sqrt{f_c'} \tag{10.34}$$

TABLE 10.13   Prestress Force and Moment Coefficients

Span	Location (x/L)	Cable path e (in.)	Force coeff. F_{pCI}	F_{pCF}	$F_{pCI}e$	$F_{pCF}e$	Moment coefficients (ft) M_{sC}	M_{psCI}	M_{psCF}
	0.0	0.240	0.904	0.807	0.018	0.016	0.000	0.018	0.016
	0.1	−13.692	0.911	0.814	−1.040	−0.929	0.112	−0.928	−0.817
	0.2	−23.640	0.918	0.821	−1.809	−1.618	0.224	−1.586	−1.394
	0.3	−29.136	0.924	0.827	−2.244	−2.008	0.335	−1.908	−1.672
	0.4	−31.596	0.933	0.835	−2.456	−2.200	0.447	−2.008	−1.752
1	0.5	−29.892	0.941	0.844	−2.344	−2.101	0.559	−1.785	−1.542
	0.6	−24.804	0.947	0.850	−1.958	−1.757	0.671	−1.287	−1.087
	0.7	−13.608	0.940	0.843	−1.278	−1.146	0.783	−0.495	−0.363
	0.8	−4.404	0.933	0.836	−0.342	−0.307	0.894	0.552	−0.588
	0.9	10.884	0.926	0.829	0.840	0.752	1.006	1.846	1.758
	1.0	28.164	0.902	0.805	2.117	1.888	1.118	3.235	3.006
	0.0	−28.164	0.902	0.805	2.117	1.888	1.107	3.224	2.995
	0.1	10.884	0.879	0.782	0.797	0.709	0.996	1.793	1.705
	0.2	−4.404	0.872	0.775	−0.320	−0.284	0.886	0.566	0.601
	0.3	−16.308	0.866	0.768	−1.176	−1.044	0.775	−0.401	−0.269
	0.4	−24.804	0.859	0.762	−1.776	−1.575	0.664	−1.111	−0.910
2	0.5	−29.892	0.852	0.755	−2.123	−1.881	0.554	−1.570	−1.328
	0.6	−31.596	0.846	0.749	−2.227	−1.971	0.443	−1.784	−1.528
	0.7	−29.136	0.839	0.742	−2.037	−1.801	0.332	−1.705	−1.469
	0.8	−23.640	0.832	0.735	−1.639	−1.447	0.221	−1.417	−1.226
	0.9	−13.692	0.825	0.728	−0.941	−0.830	0.111	−0.830	−0.719
	1.0	0.240	0.818	0.721	0.016	0.014	0.000	0.016	0.014

Note: e is the distance between the cable path and central gravity of the interior girder cross-section; positive means cable is above the central gravity and negative indicates cable is below the central gravity.

in which

$$f_{DC1} = \frac{M_{DC1}C}{I_x} \qquad (10.35)$$

$$f_{DC2} = \frac{M_{DC2}C}{I_x} \qquad (10.36)$$

$$f_{DW} = \frac{M_{DW}C}{I_x} \qquad (10.37)$$

$$f_{LL+IM} = \frac{M_{LL+IM}C}{I_x} \qquad (10.38)$$

$$f_{psF} = \frac{P_{pe}}{A} + \frac{(P_{pe}e)C}{I_x} + \frac{M_sC}{I_x} = \frac{F_{pCF}P_j}{A} + \frac{M_{psCF}P_jC}{I_x} \qquad (10.39)$$

where $C(= y_b$ or $y_t)$ is the distance from the extreme fiber to the center of gravity of the cross-section; f_c' is in ksi; and P_{pe} is the effective prestressing force after all losses have been incurred. From Equations 10.34 and 10.39, we have:

$$P_j = \frac{-f_{DC1} - f_{DC2} - f_{DW} - f_{LL+IM} - 0.19\sqrt{f_c'}}{\frac{F_{pCF}}{A} + \frac{M_{psCF}C}{I_x}} \qquad (10.40)$$

Detailed calculations are given in Table 10.14. Most critical points coincide with locations of maximum eccentricity: 0.4L in span 1, 0.6L in span 2, and at the bent. For this bridge, the controlling section is through the right face of the bent. Herein, $P_j = 1823$ kips (8109 kN). Rounding P_j up to 1830 kips (8140 kN) gives a required area of prestressing steel of $A_{ps} = P_j/f_{pj} = 1830/216 = 8.47$ in.^{2} (5465 mm^2).

10. **Check Concrete Strength for Interior Girder, Service Limit State I**
    Two criteria are imposed on the level of concrete stresses when calculating required concrete strength (AASHTO Article 5.9.4.2):

TABLE 10.14 Determination of Prestressing Jacking Force for an Interior Girder

Span	Location (x/L)	\multicolumn{5}{c}{Top fiber}					\multicolumn{5}{c}{Bottom fiber}				
		\multicolumn{4}{c}{Stress (psi)}	Jacking force	\multicolumn{4}{c}{Stress (psi)}	Jacking force
		f_{DC1}	f_{DC2}	f_{DW}	f_{LL+IM}	P_j (kips)	f_{DC1}	f_{DC2}	f_{DW}	f_{LL+IM}	P_j (kips)
1	0.0	0	0	0	0	—	0	0	0	0	—
	0.1	389	36	48	179	—	−512	−47	−64	−236	749
	0.2	658	60	82	301	—	−865	−79	−108	−395	1307
	0.3	805	73	100	369	—	−1058	−97	−132	−485	1542
	0.4	831	76	103	393	—	−1092	−100	−136	−517	1573
	0.5	735	67	91	378	—	−967	−88	−120	−497	1482
	0.6	519	47	64	328	—	−682	−62	−85	−431	1193
	0.7	181	16	22	248	—	−238	−22	−30	−326	455
	0.8	−278	−25	−35	−171	242	366	33	46	225	—
	0.9	−859	−78	−107	−307	1210	1129	103	140	403	—
	1.0	−1336	−122	−166	−447	1818	1098	100	136	367	—
2	0.0	−1336	−122	−166	−447	1823	1098	100	136	367	—
	0.1	−859	−78	−107	−307	1264	1129	103	140	403	—
	0.2	−278	−25	−35	−171	254	366	33	46	225	—
	0.3	181	16	22	248	—	−238	−22	−30	−326	520
	0.4	519	47	64	328	—	−682	−62	−85	−431	1371
	0.5	735	67	91	378	—	−967	−88	−120	−497	1691
	0.6	831	76	103	393	—	−1092	−100	−136	−517	1782
	0.7	805	73	100	369	—	−1058	−97	−132	−485	1739
	0.8	658	60	82	301	—	−865	−79	−108	−395	1474
	0.9	389	36	48	179	—	−512	−47	−64	−236	843
	1.0	0	0	0	0	—	0	0	0	0	—

Note: Positive stress indicates compression and negative stress indicates tension. P_j are obtained by Equation 10.40.

$$\begin{cases} f_{DC1} + f_{psI} \leq 0.55 f'_{ci} & \text{at prestressing state} \\ f_{DC1} + f_{DC2} + f_{DW} + f_{LL+IM} + f_{psF} \leq 0.45 f'_c & \text{at service state} \end{cases} \quad (10.41)$$

$$f_{psI} = \frac{P_{jI}}{A} + \frac{(P_{jI}e)\,C}{I_x} + \frac{M_{sI}C}{I_x} = \frac{F_{pCI}P_j}{A} + \frac{M_{psCI}P_jC}{I_x} \quad (10.42)$$

The concrete stresses in the extreme fibers (after instantaneous losses and final losses) are given in Tables 10.15 and 10.16. For the initial concrete strength in the prestressing state, the controlling location is the bottom fiber at 0.9L section in span 1. From Equation 10.41 we have:

$$f'_{ci,\text{req}} \geq \frac{f_{DC1} + f_{psI}}{0.55} = \frac{930}{0.55} = 1691 \text{ psi} \quad < 3500 \text{ psi}$$

$$\therefore \text{ choose } f'_{ci} = 3500 \text{ psi (24.13 MPa) O.K.}$$

For the final concrete strength at the service limit state, the controlling location is again in the bottom fiber at 0.9L section in span 1. From Equation 10.41 we have:

$$f'_{c,\text{req}} \geq \frac{f_{DC1} + f_{DC2} + f_{DW} + f_{LL+IM} + f_{psF}}{0.45}$$

$$= \frac{1539}{0.45} = 3420 \text{ psi} \quad < 4000 \text{ psi}$$

$$\therefore \text{ choose } f'_c = 4000 \text{ psi (27.58 MPa) O.K.}$$

11. **Flexural Strength Design for Interior Girder, Strength Limit State I**
    AASHTO requires that for the strength limit state I

TABLE 10.15 Concrete Stresses after Instantaneous Losses for the Interior Girder

		Top fiber stress (psi)					Bottom fiber stress (psi)				
Span	Location (x/L)	f_{DC1}	F^*_{pCI} P_j/A	M^*_{psCI} $P^*_j Y_t/I$	f_{psI}	Total initial stress	f_{DC1}	F^*_{pCI} P_j/A	M^*_{psCI} $P^*_j y_h/I$	f_{psI}	Total initial stress
1	0.0	0	799	8	807	807	0	799	−10	789	789
	0.1	389	805	−389	416	806	−512	805	512	1317	805
	0.2	658	812	−665	147	805	−865	812	874	1686	821
	0.3	805	817	−800	17	821	−1058	817	1052	1868	810
	0.4	831	824	−842	−18	813	−1092	824	1107	1931	839
	0.5	735	831	−748	83	819	−967	831	984	1815	848
	0.6	519	837	−540	298	816	−682	837	710	1547	865
	0.7	181	831	−208	623	804	−238	831	723	1104	866
	0.8	−278	825	231	1056	778	366	825	−304	520	866
	0.9	−859	818	774	1592	733	1129	818	−1017	−199	930
	1.0	−1336	598	1257	1854	519	1098	598	−1033	−435	663
2	0.0	−1336	598	1252	1850	514	1098	598	−1030	−432	666
	0.1	−859	777	752	1528	670	1129	777	−988	−212	917
	0.2	−278	771	237	1008	730	366	771	−312	−459	825
	0.3	181	765	−168	597	777	−238	765	221	986	749
	0.4	519	759	−466	293	812	−682	759	613	1372	690
	0.5	735	753	−658	95	830	−967	753	865	1619	652
	0.6	831	748	−748	0	830	−1092	748	983	1731	639
	0.7	805	741	−715	27	831	−1058	741	940	1681	623
	0.8	658	735	−594	141	799	−865	735	781	1516	651
	0.9	389	729	−348	381	770	−512	729	458	1187	675
	1.0	0	723	7	730	730	0	723	−9	714	714

Note: Positive stress indicates compression and negative stress indicates tension.

$$M_u \leq \phi M_n$$
$$M_u = \eta \sum \gamma_i M_i = 0.95\,[1.25(M_{DC1} + M_{DC2})$$
$$+ 1.5 M_{DW} + 1.75 M_{LLH}] + M_{ps}$$

where ϕ is the flexural resistance factor 1.0 and M_{ps} is the secondary moment due to prestress. Factored moment demands, M_u, for the interior girder in span 1 are calculated in Table 10.17. Although the moment demands are not symmetrical about the bent (due to different secondary prestress moments), the results for span 2 are similar and the differences will not be considered in this example. The detailed calculations for the flexural resistance, ϕM_n, are shown in Table 10.18. It is clear that no additional mild steel is required.

12. **Shear Strength Design for Interior Girder, Strength Limit State I**
   AASHTO requires that for the strength limit state *I*

$$V_u \leq \phi V_n$$
$$V_u = \eta \sum \gamma_i V_i$$
$$= 0.95\,[1.25(V_{DC1} + V_{DC2}) + 1.5 V_{DW} + 1.75 V_{LL+IM}] + V_{ps}$$

where ϕ is shear resistance factor 0.9 and V_{ps} is the secondary shear due to prestress. Factored shear demands, V_u, for the interior girder are calculated in Table 10.19. To determine the effective web width, assume that the VSL post-tensioning system of 5 to 12 tendon units [25] will be used with a grouted duct diameter of 2.88 in. In this example, $b_v = 12 - 2.88/2 = 10.56$ in. (268 mm). Detailed calculations of the shear resistance, ϕV_n (using two-leg #5 stirrups, $A_v = 0.62$ in.^{2} [419 mm^2]) for span 1, are shown in Table 10.20. The results for span 2 are similar to span 1 and the calculations are not repeated for this example.

TABLE 10.16 Concrete Stresses after Total Losses for the Interior Girder

		Top fiber stress (psi)				Bottom fiber stress (psi)				
Span	Location (x/L)	f_{LOAD}	F^*_{pCF} P^*_j/A	M^*_{psCF} $P^*_j Y_t/I$	f_{psF}	Total final stress	f_{LOAD}	F^*_{pCF} P^*_j/A	M^*_{psCF} $P^*_j y_b/I$	f_{psF}	Total final stress
	0.0	0	713	7	720	720	0	713	−9	704	704
	0.1	653	720	−343	377	1030	−858	720	450	1170	312
	0.2	1100	726	−584	141	1241	−1466	726	768	1494	48
	0.3	1348	731	−701	30	1377	−1772	731	922	1652	−119
	0.4	1403	738	−735	4	1407	−1844	738	966	1704	−140
1	0.5	1272	746	−647	99	1371	−1672	746	850	1596	−76
	0.6	958	751	−455	296	1254	−1260	751	599	1350	90
	0.7	467	745	−152	593	1060	−614	745	200	945	331
	0.8	−510	739	−246	985	475	670	739	−324	415	1085
	0.9	−1351	732	737	1469	119	1776	732	−969	−237	1539
	1.0	−2070	533	1168	1701	−368	1702	533	−960	−427	1275
	0.0	−2070	533	1164	1697	−373	1702	533	−957	−423	1278
	0.1	−1351	691	715	1406	55	1776	691	−940	−249	1527
	0.2	−510	685	252	937	427	670	685	−331	353	1024
	0.3	467	679	−113	566	1033	−614	679	148	828	213
	0.4	958	673	−382	292	1250	−1260	673	502	1175	−85
2	0.5	1272	667	−557	111	1383	−1672	667	732	1399	−273
	0.6	1403	662	−641	21	1424	−1844	662	842	1504	−340
	0.7	1348	655	−616	40	1387	−1772	655	809	1465	−307
	0.8	1100	649	−514	135	1235	−1466	649	676	1325	−122
	0.9	653	643	−302	341	994	−858	643	397	1040	181
	1.0	0	637	6	643	643	0	637	−8	629	629

Note: $f_{LOAD} = f_{DC1} + f_{DC2} + F_{DW} + f_{LL+IM}$. Positive stress indicates compression and negative stress indicates tension.

TABLE 10.17 Factored Moments for an Interior Girder (Span 1)

Location	M_{DC1} (kips-ft) Dead	M_{DC2} (kips-ft) Dead	M_{DW} (kips-ft) Wearing	M_{LL+IM} (kips-ft)		M_{ps} (kips-ft)	M_u (kips-ft)	
(x/L)	load-1	load-2	surface	Positive	Negative	P/S	Positive	Negative
0.0	0	0	0	0	0	0	0	0
0.1	1700	155	212	782	−85	205	4009	2569
0.2	2871	262	357	1312	−169	409	6820	4358
0.3	3513	321	437	1612	−253	614	8469	5368
0.4	3626	331	451	1715	−337	818	9012	5599
0.5	3210	293	399	1650	−422	1023	8494	5050
0.6	2264	207	282	1431	−506	1228	6942	3721
0.7	789	72	98	1081	−590	1432	4392	1613
0.8	−1215	−111	−151	647	−748	1637	922	−1397
0.9	−3748	−342	−466	196	−1339	1841	−3355	−5906
1.0	−6292	−573	−781	0	−2104	2046	−7219	−10716

Note: $M_u = 0.95[1.25(M_{DC1} + M_{MDC2}) + 1.5M_{DW} + 1.75M_{LL+IM}] + M_{ps}$

10.A.2 Three-Span, Continuous, Composite Plate Girder Bridge

Given: A three-span, continuous, composite plate girder bridge has two equal spans of length 160 ft (48.8 m) and one midspan of 210 ft (64 m). The superstructure is 44 ft (13.4 m) wide. The elevation, plan, and typical cross-section are shown in Figure 10.69.

Structural steel: A709 Grade 50 for web and flanges,

$$F_{yw} = F_{yt} = F_{yc} = F_y = 50 \text{ ksi (345 MPa)}$$

A709 Grade 36 for stiffeners, etc.,

$$F_{ys} = 36 \text{ ksi (248 MPa)}$$

TABLE 10.18 Flexural Strength Design for Interior Girder, Strength Limit State I (Span 1)

Location (x/L)	A_{ps} (in.^{2})	d_p (in.)	A_s (in.^{2})	d_s (in.)	b (in.)	c (in.)	f_{ps} (ksi)	d_e (in.)	a (in.)	ϕM_n (k-ft)	M_u (k-ft)
0.0		32.16	0	72.06	104	7.14	253.2	32.16	6.07	5206	0
0.1		46.09	0	72.06	104	7.27	258.1	46.09	6.18	7833	4009
0.2		56.04	0	72.06	104	7.33	260.1	56.04	6.23	9717	6820
0.3		61.54	0	72.06	104	7.35	261.0	61.54	6.25	10759	8469
0.4		64.00	0	72.06	104	7.36	261.3	64.00	6.26	11226	9012
0.5	8.47	62.29	0	72.06	104	7.36	261.1	62.29	6.25	10903	8494
0.6		57.20	0	72.06	104	7.34	260.3	57.20	6.24	9937	6942
0.7		48.71	0	72.06	104	7.29	258.7	48.71	6.20	8328	4392
0.8		38.20	0	71.06	82.5	21.19	228.1	38.20	18.01	−4965	−1397
0.9		53.48	0	71.06	82.5	23.36	237.0	53.48	19.86	−7822	−5906
1.0		62.00	0	71.06	104	8.13	261.0	62.00	6.25	−10848	−10716

Note:

1. Prestressing steel, $f_{ps} = f_{pu}(1 - k\frac{c}{d_p})$, $k = 2(1.04 - \frac{f_{py}}{f_{pu}})$

2. For flanged section, $c/d_e \leq 0.42$,

$$M_n = A_{ps}f_{ps}(d_p - \tfrac{a}{2}) + A_s f_y(d_s - \tfrac{a}{2})$$

$$-A'f'_y(d'_s - \tfrac{a}{2}) + 0.85f'_c(b - b_w)\beta_1 h_f(\tfrac{a}{2} - \tfrac{h_f}{2})$$

$$a = \beta_1 C$$

$$c = \frac{A_{ps}f_{pu} + A_s f_y - A'_s f'_y - 0.85\beta_1 f'_c(b - b_w)h_f}{0.85\beta_1 f'_c b_w + k A_{ps}\frac{f_{pu}}{d_p}}$$

3. For flanged section, $c/d_e > 0.42$—over-reinforced,

$$M_n = (0.36\beta_1 - 0.08\beta_1^2)f'_c b_w d_e^2 + 0.85\beta_1 f'_c(b - b_w)h_f(d_e - 0.5h_f)$$

$$d_e = \frac{A_{ps}f_{ps}d_p + A_s f_y d_s}{A_{ps}f_{ps} + A_s f_y}$$

4. For rectangular section, i.e., when $c < h_f$ take $b = b_w$ in the above formulas.

TABLE 10.19 Factored Shear for an Interior Girder (Span 1)

Location (x/L)	V_{DC1} (kips) Dead load-1	V_{DC2} (kips) Dead load-2	V_{DW} (kips) Wearing surface	V_{LL+IM} (kips) Envelopes	M_{LL+IM} (k-ft) Associated	V_{ps} (kips) P/S	V_u (kips)	M_u (k-ft) Associated
0.0	125.2	11.4	15.6	60.0	0	13.03	297.1	0
0.1	91.5	8.4	11.4	50.1	787	13.03	231.0	4017
0.2	57.7	5.3	7.2	42.0	1320	13.03	168.0	6883
0.3	24.0	2.2	3.0	34.3	1614	13.03	105.4	8472
0.4	−9.7	−0.9	−1.2	−27.7	1650	130.3	−47.3	8903
0.5	−43.4	−4.0	−5.4	−35.1	1628	13.03	−109.2	8457
0.6	−77.1	−7.1	−9.6	−42.0	1424	13.03	−170.3	6929
0.7	−111	−10.1	−13.8	−49.9	852	13.03	−233.1	4011
0.8	−145	−13.2	−18.0	−59.2	216	13.03	−298.3	205.4
0.9	−178	−16.3	−22.2	−68.8	−667	13.03	−364	−4790
1.0	−216	−19.4	−26.4	−78.5	−1788	13.03	−434.3	−10191

Note: $V_u = 0.95[1.25(V_{DC1} + V_{DC2}) + 1.5V_{DW} + 1.75V_{LL+IM}] + V_{ps}$

Concrete:

$$f'_c = 3250 \text{ psi } (22.4 \text{ MPa}),$$

$$E_c = 3250 \text{ ksi } (22,400 \text{ MPa}),$$

modular ratio, $n = 9$

Loads:

Dead load = self-weight + barrier rail + future wearing 3 in AC overlay

Live load = AASHTO HS20-44 + dynamic load allowance

Single-lane average daily truck traffic (ADTT) = 3600 (one way)

TABLE 10.20 Shear Strength Design for Interior Girder, Strength Limit State I (Span 1)

Location (x/L)	d_v (in.)	y' (rad)	V_p (kips)	v/f'_c	ε_x (1000)	θ (degree)	β	V_c (kips)	S (in.)	ϕV_n (kips)	$\|V_u\|$ (kips)
0.0	54.00	0.084	124.1	0.090	−0.028	23.5	6.50	234.4	12	460.7	297.1
0.1	54.00	0.063	93.9	0.071	−0.093	27	5.60	201.8	12	400.4	231.0
0.2	52.90	0.042	63.1	0.055	0.733	33	2.37	83.7	24	194.0	168.0
0.3	58.87	0.021	31.8	0.034	1.167	38	2.10	82.5	24	167.6	105.4
0.4	60.84	0.000	0.0	0.020	1.078	36	2.23	90.6	24	150.2	47.3
0.5	59.14	0.018	27.8	0.037	1.026	36	2.23	88.0	24	171.0	109.2
0.6	54.06	0.036	56.2	0.058	0.539	30	2.48	89.5	24	196.4	170.3
0.7	54.00	0.054	83.5	0.077	−0.106	27	5.63	202.9	12	392.0	233.1
0.8	54.00	0.072	110.4	0.097	−0.287	23.5	6.50	234.3	12	448.4	298.3
0.9	54.00	0.090	136.8	0.117	−0.137	23.5	3.49	125.8	9	420.5	364.0
1.0	57.42	0.000	0.0	0.199	2.677	36	1.0	38.3	3.5	478.9	434.3

Note:
1. $b_v = 10.56$ in. and y' is slope of the prestressing cable.
2. $A_v = 0.62$ in.^{2} (2#5)

$$V_n = \text{ the lesser of } \begin{cases} V_c + V_s + V_p \\ 0.25 f'_c b_v d_v + V_p \end{cases}$$

$$V_c = 0.0316\beta\sqrt{f'_c}b_v d_v, \quad V_s = \frac{A_v f_y d_v \cos\theta}{s}$$

$$v = \frac{V_u - \phi V_p}{\phi b_v d_v}, \quad \varepsilon_x = \frac{\frac{M_u}{d_v}+0.5N_u+0.5V_u\cot\theta - A_{ps}f_{po}}{E_s A_s + E_p A_{ps}} \leq 0.002$$

$$F_\varepsilon = \frac{E_s A_s + E_p A_{ps}}{E_c A_c + E_s A_s + E_p A_{ps}} \quad \text{(when } \varepsilon_x \text{ is negative, multiply by } F_\varepsilon)$$

$$A_{v\,min} = 0.0316\sqrt{f'_c}\frac{b_v S}{f_y}$$

For $V_u < 0.1 f'_c b_v d_v$, $S_{max} = \text{ smaller of } \begin{cases} 0.8 d_v \\ 24 \text{ in.} \end{cases}$

For $V_u \geq 0.1 f'_c b_v d_v$, $S_{max} = \text{ smaller of } \begin{cases} 0.4 d_v \\ 12 \text{ in.} \end{cases}$

Deck: Concrete slab deck with thickness of 10.875 in. (276 mm) has been designed
Construction: Unshored; unbraced length for compression flange, $L_b = 20$ ft (6.1 m)
Specification: AASHTO-LRFD [1] (referred to as AASHTO in this example)
Requirements: Design the following portions of an interior girder for maximum positive flexure region at span 1:

1. Calculate loads
2. Calculate live load distribution factors
3. Calculate unfactored moments and shear demands
4. Determine load factors for strength limit state I and fatigue limit state
5. Calculate composite section properties for positive flexure region
6. Flexural strength design, strength limit state I
7. Shear strength design, strength limit state I
8. Fatigue design, fatigue and fracture limit state
9. Intermediate transverse stiffener design
10. Shear connector design
11. Constructability check

Solution

1. **Calculate Loads**

   1.1) *Component Dead Load, DC for an Interior Girder*
   The component dead load, DC, includes all structural dead loads with the exception of the further wearing surface and specified utility loads. For design purposes, the two parts of DC are defined as:

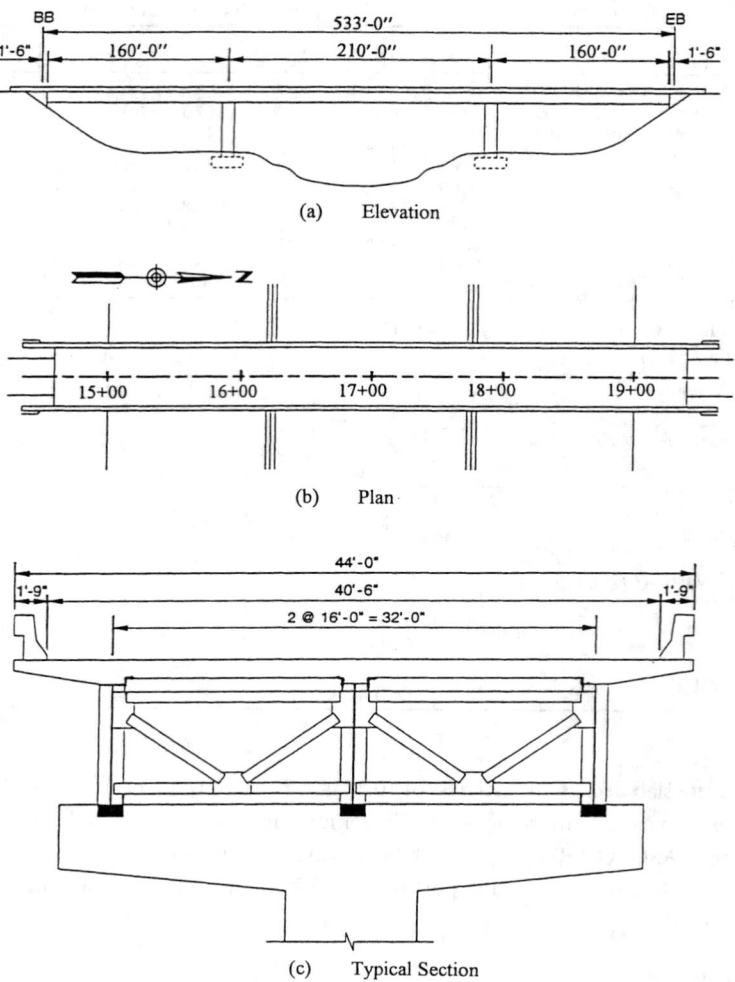

Figure 10.69   A three-span, continuous plate girder bridge.

*DC*1: Deck concrete (self-weight, 150 lb/ft³) and steel girder including bracing system and details (estimated weight, 300 lb/ft for each girder). Assume that *DC*1 is acting on the noncomposite section and is distributed to each girder by the tributary area. The tributary width for the interior girder is 16 ft (4.9 m).

$$DC1 = [(10.875/12)(16) + (1.5)(15.25 - 10.975)/12(1.5)](0.15) + 0.3$$
$$= 2.557 \text{ kips/ft} \quad (37.314 \text{ kN/m})$$

*DC*2: Barrier rail weight (784 kips/ft). Assume that *DC*2 is acting on the long-term composite section and is equally distributed to each girder.

$$DC2 = 0.784/3 = 0.261 \text{ kips/ft} \quad (3.809 \text{ kN/m})$$

1.2) *Wearing Surface Load, DW*
A future wearing surface of 3 in. (76 mm) with a unit weight of 140 lb/ft³ is assumed to be carried by the long-term composite section and equally distributed to each girder.

$$DW = \text{(deck width–barrier width) (thickness of wearing surface)}$$
$$\text{(unit weight)}/3$$
$$= [44 - 2(1.75)] (0.25)(0.14)/3 = 0.473 \text{ kips/ft} \quad (6.903 \text{ kN/m})$$

1.3) Live Load, LL, and Dynamic Load Allowance, IM

The design live load, LL, is the AASHTO HS20-44 vehicular live load. To consider the wheel load impact from moving vehicles, the dynamic load allowance, $IM = 33\%$ for the strength limit state and 15% for the fatigue limit state are used [AASHTO Table 3.6.2.1-1].

2. **Calculate Live Load Distribution Factors**

   2.1) Range Applicability of AASHTO Approximate Formulas

   AASHTO-LRFD [1] recommends that approximate methods be used to distribute live load to individual girders. For concrete deck on steel girders, live load distribution factors are dependent on the girder spacing, S, span length, L, concrete slab depth, t_s, longitudinal stiffness parameter, K_g, and number of girders, N_b. The range of applicability of AASHTO approximate formulas are 3.5 ft $\leq S \leq$ 16 ft; 4.5 in. $\leq t_s \leq$ 12 in.; 20 ft $\leq L \leq$ 240 ft; and $N_b \geq 4$. For this design example, $S = 16$ ft, $L_1 = L_3 = 160$ ft, $L_2 = 210$ ft, $t_s = 10.875$ in., and $N_b = 3 < 4$. It is obvious that this bridge is out of the range of applicability of AASHTO formulas. The conventional level rule is used to determine live load distribution factors.

   2.2) Level Rule

   The level rule assumes that the deck in its transverse direction is simply supported by the girders and uses statics to determine the live load distribution to the girders. AASHTO also requires that when the level rule is used, the multiple presence factor, m (1.2 for one loaded lane, 1.0 for two loaded lanes, 0.85 for three loaded lanes, and 0.65 for more than three loaded lanes), should apply.

   2.3) Live Load Distribution Factors for Strength Limit State

   Figure 10.70 shows locations of traffic lanes for the interior girder. For a 12-ft (3.6-m) traffic lane width, the number of traffic lanes for this bridge is three.

   (a) One lane loaded (Figure 10.70a)

   $$R = \frac{13}{16} = 0.8125 \text{ lanes}$$
   $$LD = mR = 1.2(0.8125) = 0.975 \text{ lanes}$$

   (b) Two lanes loaded (Figure 10.70b)

   $$R = \frac{13}{16} + \frac{9}{16} = 1.375 \text{ lanes}$$
   $$LD = mR = 1.0(1.375) = 1.375 \text{ lanes} \quad \text{(controls)}$$

   (c) Three lanes loaded (Figure 10.70c)

   $$R = \frac{(13 + 3)}{16} + \frac{7}{16} = 1.4375 \text{ lanes}$$
   $$LD = mR = 0.85(1.4375) = 1.222 \text{ lanes}$$

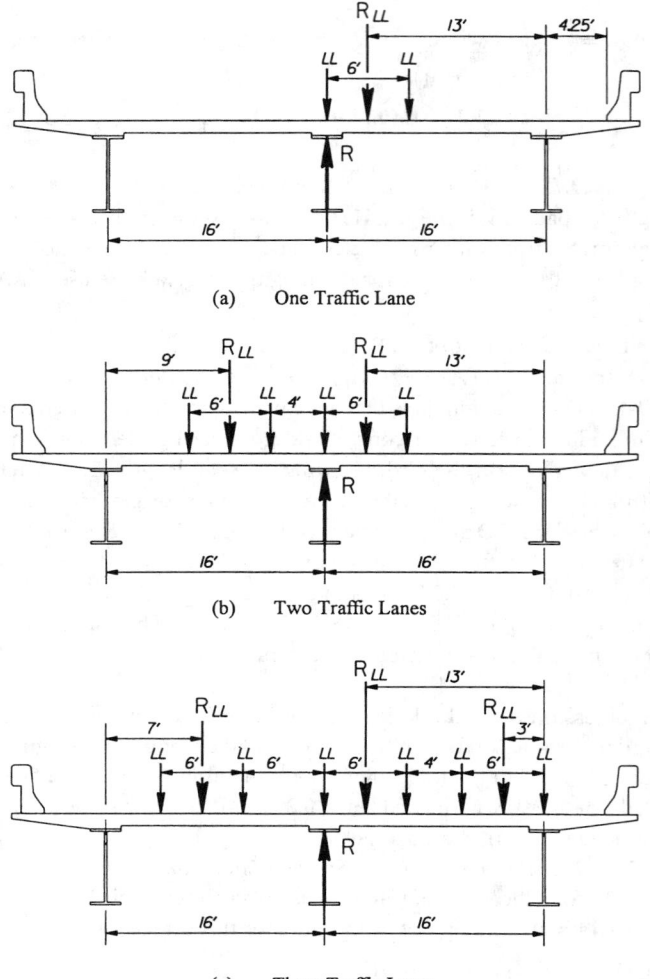

Figure 10.70   Live load distribution—lever rule.

2.4) Live Load Distribution Factors for Fatigue Limit State
AASHTO requires that one traffic lane load be used and multiple presence factors not be applied to the fatigue limit state. The live load distribution factor for the fatigue limit state, therefore, is obtained by one lane loaded without a multiple presence factor of 1.2.

$$LD = 0.813$$

3. **Calculate Unfactored Moments and Shear Demands**
   For an interior girder, unfactored moment and shear demands are shown in Figures 10.71 and 10.72 for the strength limit state and Figures 10.73 and 10.74 for the fatigue limit state. The details are listed in Tables 10.21 to 10.23. Only the results for span 1 and one half of span 2 are shown in these tables and figures since the bridge is symmetrical about the centerline of span 2.

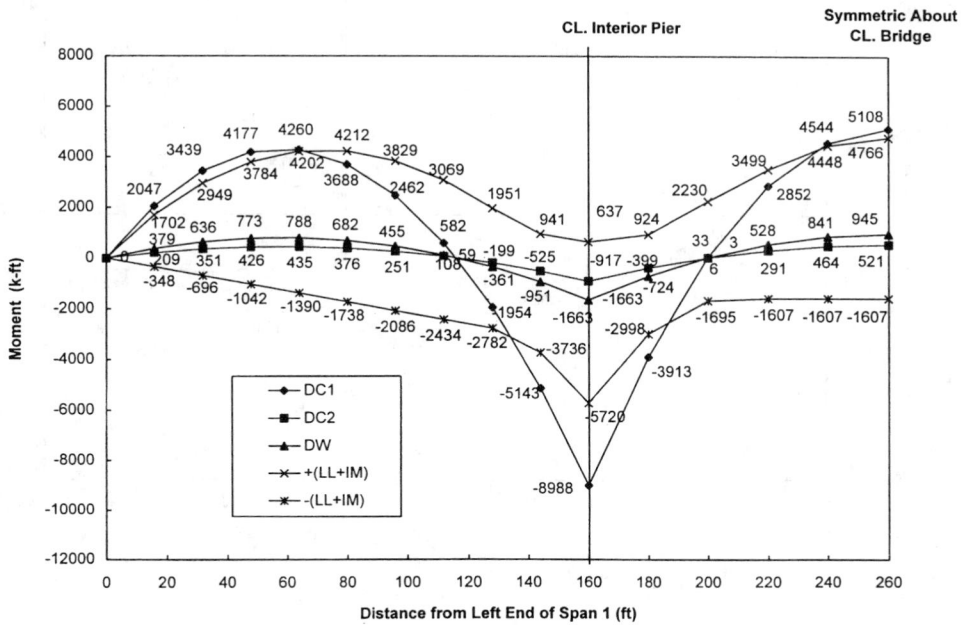

Figure 10.71 Moment envelopes due to unfactored loads.

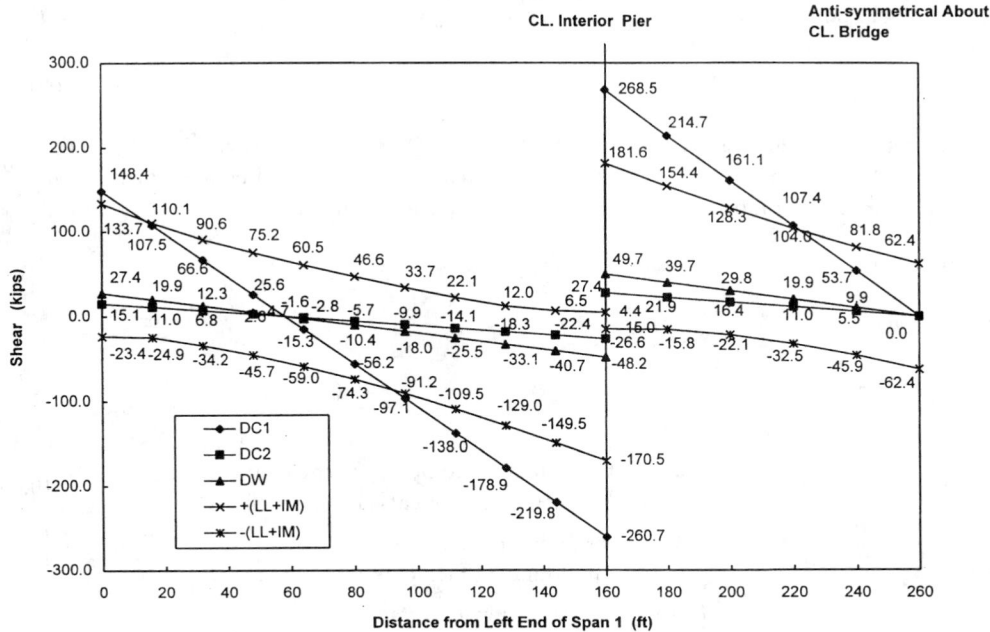

Figure 10.72 Shear envelopes due to unfactored loads.

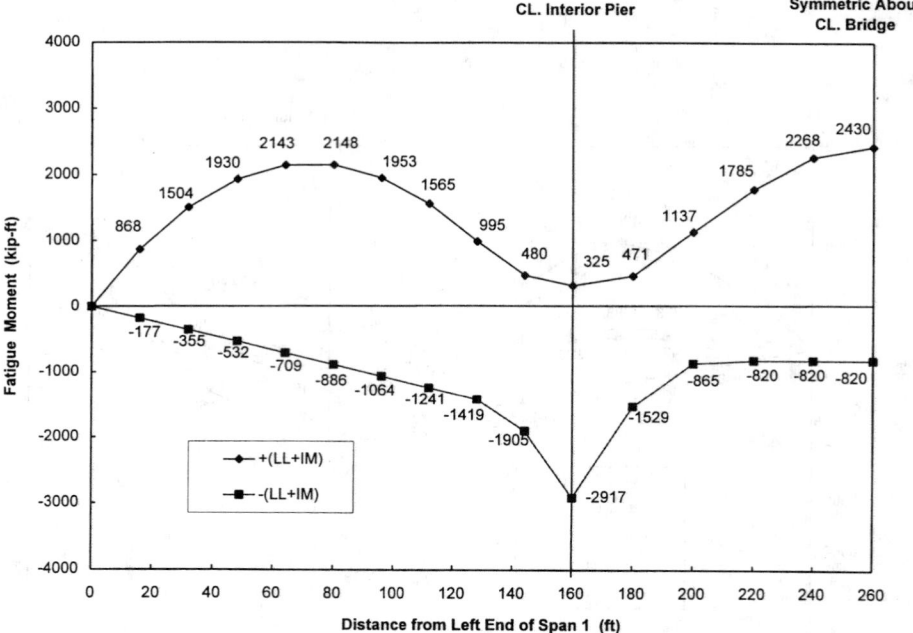

Figure 10.73   Unfactored moment due to fatigue loads.

4. **Determine Load Factors for Strength Limit State I and Fatigue Limit State**

4.1) *General Design Equation* (AASHTO Article 1.3.2)

$$\eta \sum \gamma_i Q_i \leq \phi R_n \tag{10.43}$$

where γ_i are load factors and ϕ resistance factors; Q_i represents force effects or demands; and R_n is the nominal resistance. η is a factor related to ductility, redundancy, and operational importance of that being designed and is defined as:

$$\eta = \eta_D \eta_R \eta_I \geq 0.95 \tag{10.44}$$

where

$$\eta_D = \begin{cases} 1.05 & \text{for nonductile components and connections} \\ 0.95 & \text{for ductile components and connections} \end{cases} \tag{10.45}$$

$$\eta_R = \begin{cases} 1.05 & \text{for nonredundant members} \\ 0.95 & \text{for redundant members} \end{cases} \tag{10.46}$$

$$\eta_I = \begin{cases} 1.05 & \text{operationally important bridge} \\ 0.95 & \text{general bridge} \\ & \text{only apply to strength} \\ & \text{and extreme event limit states} \end{cases} \tag{10.47}$$

TABLE 10.21 Moment Envelopes for Strength Limit State I

Span	Location (x/L)	M_{DC1} (kips-ft) Dead load-1	M_{DC2} (kips-ft) Dead load-2	M_{DW} (kips-ft) Wearing surface	M_{LL+IM} (kips-ft) Positive	M_{LL+IM} (kips-ft) Negative	M_u (kips-ft) Positive	M_u (kips-ft) Negative
	0.0	0	0	0	0	0	0	0
	0.1	2047	209	379	1702	−348	6049	2641
	0.2	3439	351	636	2949	−696	10310	4250
	0.3	4177	426	773	3784	−1042	12858	4835
	0.4	4260	435	788	4202	−1390	13684	4387
1	0.5	3688	376	682	4212	−1738	12800	2908
	0.6	2462	251	455	3829	−2086	10236	402
	0.7	582	59	108	3069	−2434	6017	−3131
	0.8	−1954	−199	−361	1951	−2782	173	−7696
	0.9	−5143	−525	−951	941	−3736	−6522	−14297
	1.0	−8988	−917	−1663	637	−5720	−13074	−23641
	0.0	−8988	−917	−1663	637	−5720	−13074	−23641
	0.1	−3913	−399	−724	924	−2998	−4616	−11136
	0.2	33	3	6	2230	−1695	3759	−2767
2	0.3	2852	291	528	3499	−1607	10302	1812
	0.4	4444	464	841	4448	−1607	14540	4473
	0.5	5108	521	945	4766	−1607	15954	5359

Note: Live load distribution factor, $LD = 1.375$. Dynamic load allowance, $IM = 33\%$. $M_u = 0.95\,[1.25\,(M_{DC1} + M_{DC2}) + 1.5 M_{DW} + 1.75 M_{LL+IM}]$

TABLE 10.22 Shear Envelopes for Strength Limit State I

Span	Location (x/L)	V_{DC1} (kips-ft) Dead load-1	V_{DC2} (kips-ft) Dead load-2	V_{DW} (kips-ft) Wearing surface	V_{LL+IM} (kips-ft) Positive	V_{LL+IM} (kips-ft) Negative	V_u (kips-ft) Positive	V_u (kips-ft) Negative
	0.0	148.4	15.1	27.4	133.7	−23.4	455.4	194.3
	0.1	107.5	11.0	19.9	110.1	−24.9	352.2	127.7
	0.2	66.6	6.8	12.3	90.6	−34.2	255.3	47.8
	0.3	25.6	2.6	4.7	75.2	−45.7	165.2	−35.7
	0.4	−15.3	−1.6	−2.8	60.5	−59.0	76.5	−122.1
1	0.5	−56.2	−5.7	−10.4	46.6	−74.3	−10.8	−211.8
	0.6	−97.1	−9.9	−18.0	33.7	−91.2	−96.7	−304.3
	0.7	−138.0	−14.1	−25.5	22.1	−109.5	−180.2	−398.9
	0.8	−178.9	−18.3	−33.1	12.0	−129.0	−261.5	−495.8
	0.9	−219.8	−22.4	−40.7	6.5	−149.5	−334.9	−594.1
	1.0	−260.7	−26.6	−48.2	4.4	−170.5	−402.5	−693.3
	0.0	268.5	27.4	49.7	181.6	−15.0	724.2	397.3
	0.1	214.7	21.9	39.7	154.4	−15.8	594.2	311.2
	0.2	161.1	16.4	29.8	128.3	−22.1	466.5	216.4
2	0.3	107.4	11.0	19.9	104.0	−32.5	341.8	115.0
	0.4	53.7	5.5	9.9	81.8	−45.9	220.4	8.1
	0.5	0	0	0	62.4	−62.4	103.8	−103.8

Note: Live load distribution factor, $LD = 1.375$. Dynamic load allowance, $IM = 33\%$. $V_u = 0.95\,[1.25\,(V_{DC1} + V_{DC2}) + 1.5 V_{DW} + 1.75 V_{LL+IM}]$

For this bridge, the following values are assumed:

Limit states	Ductility η_D	Redundancy η_R	Importance η_I	η
Strength limit state	0.95	0.95	1.05	0.95
Fatigue limit state	1.0	1.0	1.0	1.0

4.2) Load Factors and Load Combinations
The load factors and combinations are specified as (AASHTO Table 3.4.1-1):

Strength limit state I:	$1.25(DC1 + DC2) + 1.5(DW) + 1.75(LL + IM)$
Service limit state:	$0.75(LL + IM)$

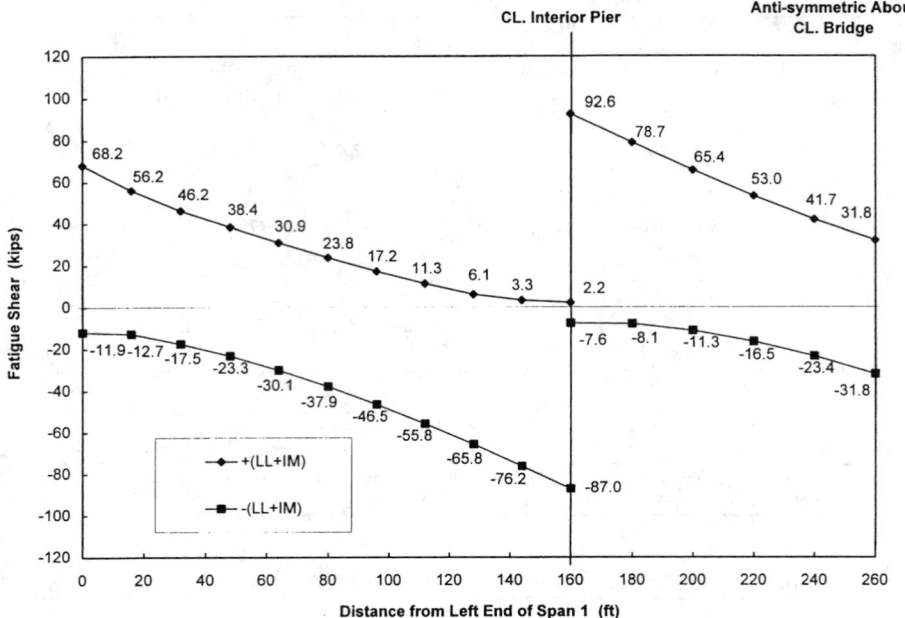

Figure 10.74   Unfactored shear due to fatigue loads.

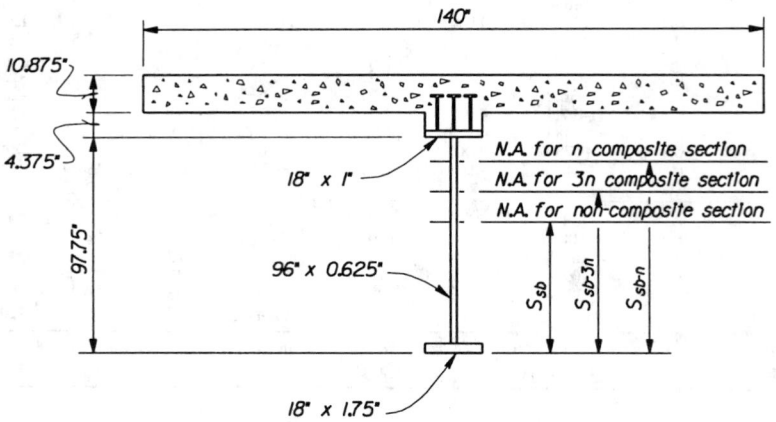

Figure 10.75   Typical cross-section in positive flexure region.

5. **Calculate Composite Section Properties for Positive Flexure Region**
   Try steel section (Figure 10.75) as:

Top flange:	$b_{fc} = 18$ in.	$t_{fc} = 1$ in.
Web:	$D = 96$ in.	$t_w = 0.625$ in.
Bottom flange:	$b_{ft} = 18$ in.	$t_{ft} = 1.75$ in.

TABLE 10.23　Moment and Shear Envelopes for Fatigue Limit State

		M_{LL+IM} (kips-ft)		V_{LL+IM} (kips)		$(M_{LL+IM})_u$ (kips-ft)		$(V_{LL+IM})_u$ (kips)	
Span	(x/L)	Positive	Negative	Positive	Negative	Positive	Negative	Positive	Negative
	0.0	0	0	68.2	−11.9	0	0	51.1	−8.9
	0.1	868	−177	56.2	−12.7	651	−133	42.1	−9.5
	0.2	1504	−355	46.2	−17.5	1128	−266	34.7	−13.1
	0.3	1930	−532	38.4	−23.3	1447	−399	28.8	−17.5
	0.4	2143	−709	30.9	−30.1	1607	−532	23.1	−22.6
1	0.5	2148	−886	23.8	−37.9	1611	−665	17.8	−28.4
	0.6	1953	−1064	17.2	−46.5	1465	−798	12.9	−34.9
	0.7	1565	−1241	11.3	−55.8	1174	−931	8.5	−41.9
	0.8	995	−1419	6.1	−65.8	746	−1064	4.6	−49.3
	0.9	480	−1905	3.3	−76.2	360	−1429	2.5	−57.2
	1.0	325	−2917	2.2	−87.0	243	−2188	1.7	−65.2
	0.0	325	−2917	92.6	−7.6	243	−2188	69.5	−5.7
	0.1	471	−1529	78.7	−8.1	353	−1146	59.1	−6.0
2	0.2	1137	−865	65.4	−11.3	853	−648	49.1	−8.5
	0.3	1785	−820	53.0	−16.5	1338	−615	39.8	−12.4
	0.4	2268	−820	41.7	−23.4	1701	−615	31.3	−17.6
	0.5	2430	−820	31.8	−31.8	1823	−615	23.9	−23.9

Note: Live load distribution factor, $LD = 0.813$. Dynamic load allowance, $IM = 15\%$. $(M_{LL+IM})_u = 0.75\,(M_{LL+IM})_u$ and $(V_{LL+IM})_u = 0.75\,(V_{LL+IM})_u$

5.1) *Effective Flange Width* (AASHTO Article 4.6.2.6)
For an interior girder, the effective flange width

$$b_{eff} = \text{the lesser of} \begin{cases} \dfrac{L_{eff}}{4} = \dfrac{115(12)}{4} = 345 \text{ in.} \\ 12t_s + \dfrac{b_f}{2} = (12)(10.875) + 18/2 = 140 \text{ in. (controls)} \\ S = (16)(12) = 192 \text{ in.} \end{cases}$$

where L_{eff} is the effective span length and may be taken as the actual span length for simply supported spans and the distance between the points of permanent load inflection for continuous spans; b_f is the top flange width of the steel girder.

5.2) *Elastic Composite Section Properties*
For the typical section (Figure 10.75) in the positive flexure region of span 1, the elastic section properties for the noncomposite, the short-term composite ($n = 9$), and the long-term composite ($3n = 27$), respectively, are calculated in Tables 10.24 to 10.26.

TABLE 10.24　Noncomposite Section Properties for Positive Flexure Region

Component	A (in.^{2})	y_i (in.)	$A_i y_i$ (in.^{3})	$y_i - y_{sb}$ (in.)	$A_i (y_i - y_{sb})^2$ (in.^{4})	I_o (in.^{4})
Top flange, 18 x 1	18	98.25	1,768.5	54.587	53,636	1.5
Web, 96 x 0.625	60	49.75	2,985.0	6.087	2,223	46,080
Bottom flange, 18 x 1.75	31.5	0.875	27.6	−42.788	57,670	8.04
$\sum$	109.5	—	4,781.1		113,529	46,090

$$y_{sb} = \frac{\sum A_i y_i}{\sum A_i} = \frac{4{,}781.1}{109.5} = 43.663 \text{ in.}$$

$$y_{st} = (1.75 + 96 + 1) - 43.663 = 55.087 \text{ in.}$$

$$I_{girder} \sum I_o + \sum A_i (y_i - y_{sb})^2 = 46{,}090 + 113{,}529 = 159{,}619 \text{ in.}^4$$

$$S_{sb} = \frac{I_{girder}}{y_{sb}} = \frac{159{,}619}{43.663} = 3{,}656 \text{ in.}^3$$

$$S_{st} = \frac{I_{girder}}{y_{st}} = \frac{159{,}619}{55.087} = 2{,}898 \text{ in.}^3$$

TABLE 10.25 Short-Term Composite Section Properties ($n = 9$)

Component	A (in.^{2})	y_i (in.)	$A_i y_i$ (in.^{3})	$y_i - y_{sb-n}$ (in.)	$A_i (y_i - y_{sb-n})^2$ (in.^{4})	I_o (in.^{4})
Steel section	109.5	43.663	4,781.1	−38.791	164,768	159,619
Concrete slab 140/9 x 10.875	169.17	107.563	18,196	25.109	106,653	1,667
$\sum$	278.67	—	22,977	—	271,421	161,286

$$y_{sb-n} \frac{\sum A_i y_i}{\sum A_i} = \frac{22,977}{278.67} = 82.454 \text{ in.}$$

$$y_{st-n} = (1.75 + 96 + 1) - 82.454 = 16.296 \text{ in.}$$

$$I_{com-n} = \sum I_o + \sum A_i (y_i - y_{sb-n})^2$$

$$= 161,286 + 271,421 = 432,707 \text{ in.}^4$$

$$S_{sb-n} = \frac{I_{con-n}}{y_{sb-n}} = \frac{432,707}{82.454} = 5,248 \text{ in.}^3$$

$$S_{st-n} = \frac{I_{com-n}}{y_{st-n}} = \frac{432,707}{16.296} = 26,553 \text{ in.}^3$$

TABLE 10.26 Long-Term Composite Section Properties ($3n = 27$)

Component	A (in.^{2})	y_i (in.)	$A_i y_i$ (in.^{3})	$y_i - y_{sb-3n}$ (in.)	$A_i \left(y_i - y_{sb-3n}\right)^2$ (in.^{4})	I_o (in.^{4})
Steel section	109.5	43.663	4,781.1	−21.72	51,661	159,619
Concrete slab 140/9 x 10.875	56.39	107.563	6,065.4	42.18	100,320	556
$\sum$	165.89	—	10,846.4	—	151,981	160,174

$$y_{sb-3n} = \frac{\sum A_i y_i}{\sum A_i} = \frac{10,846.4}{165.89} = 65.383 \text{ in.}$$

$$y_{st-3n} = (1.75 + 96 + 1) - 65.383 = 33.367 \text{ in.}$$

$$I_{com-3n} = \sum I_o + \sum A_i \left(y_i - y_{sb-3n}\right)^2$$

$$= 160,174 + 151,981 = 312,155 \text{ in.}^4$$

$$S_{sb-3n} = \frac{I_{con-3n}}{y_{sb-3n}} = \frac{312,155}{65.383} = 4,774 \text{ in.}^3$$

$$S_{st-3n} = \frac{I_{com-3n}}{y_{st-3n}} = \frac{312,155}{33.367} = 9,355 \text{ in.}^3$$

5.3) Plastic Moment Capacity, M_p

The plastic moment capacity, M_p, is determined using equilibrium equations. The reinforcement in the concrete slab is neglected in this example.

(a) Determine the location of the plastic neutral axis (PNA)

Assuming that the PNA is within the top flange of the steel girder (Figure 10.76) and that y_{PNA} is the distance from the top of the compression flange to the PNA, we obtain:

$$P_s + P_{c1} = P_{c2} + P_w + P_t \tag{10.48}$$

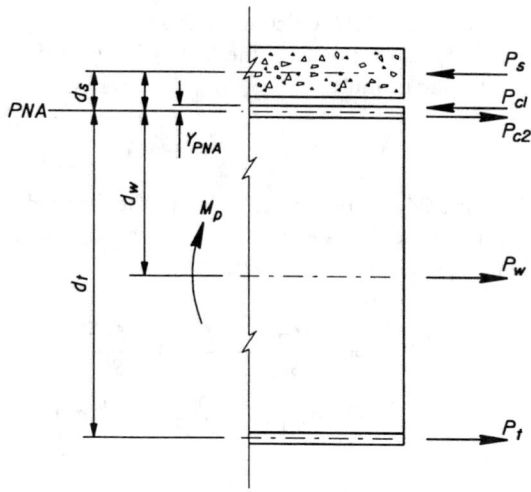

Figure 10.76 Plastic moment capacity state.

where

P_s = $0.85 f'_c b_{eff} t_s = 0.85(3.25)(140)(10.875) = 4206$ kips (18,708 kN)

P_{c1} = $y_{PNA} b_{fc} F_{yc}$

P_{c2} = $A_{fc} F_{yc} - P_{c1} = (t_{fc} - y_{PNA}) b_{fc} F_{yc}$

P_c = $P_{c1} + P_{c2} = A_{fc} F_{yc} = (18)(1)(50) = 900$ kips (4,003 kN)

P_w = $A_w F_{yw} = (96)(0.625)(50) = 3,000$ kips (13,344 kN)

P_t = $A_{ft} F_{yt} = (18)(1.75)(50) = 1,575$ kips (7,006 kN)

Substituting the above expressions into Equation (10.48) and solving for y_{PNA}, we obtain

$$y_{PNA} = \frac{t_{fc}}{2}\left(\frac{P_w + P_t - P_s}{P_c} + 1\right) \tag{10.49}$$

$$y_{PNA} = \frac{1}{2}\left(\frac{3000 + 1575 - 4206}{900} + 1\right) = 0.705 \text{ in.} \quad < t_{cf} = 1.0 \text{ in.} \quad \text{O.K.}$$

(b) Calculate M_p

Summing all forces about the PNA, we obtain:

$$M_p = \sum M_{PNA} = P_s d_s + P_{c1}\left(\frac{y_{PNA}}{2}\right) + P_{c2}\left(\frac{t_{fc} - y_{PNA}}{2}\right) + P_w d_w + P_t d_t \tag{10.50}$$

where

d_s = $\frac{10.875}{2} + 4.375 - 1 + 0.705 = 9.518$ in. (242 mm)

d_w = $\frac{96}{2} + 1 - 0.705 = 48.295$ in. (1,227 mm)

d_t = $\frac{1.75}{2} + 96 + 1 - 0.705 = 97.17$ in. (2,468 mm)

M_p = $(4,206)(9.518) + (18)(50)\frac{(0.705)^2}{2} + (18)(50)\frac{(1-0.705)^2}{2}$

$+ (3,000)(48.295) + (1,575)(97.17)$

= $338,223$ kips-in. = $28,185$ kips-ft (38,212 kN-m)

5.4) *Yield Moment*, M_y (AASHTO Article 6.10.5.1.2)

The yield moment, M_y, corresponds to the first yielding of either steel flange. It is obtained by the following formula:

$$M_y = M_{D1} + M_{D2} + M_{AD} \tag{10.51}$$

where M_{D1}, M_{D2}, and M_{AD} are moments due to the factored loads applied to the steel, the long-term and the short-term composite section, respectively. M_{AD} can be obtained by solving equation:

$$F_y = \frac{M_{D1}}{S_s} + \frac{M_{D2}}{S_{3n}} + \frac{M_{AD}}{S_n} \tag{10.52}$$

$$M_{AD} = S_n \left(F_y - \frac{M_{D1}}{S_s} - \frac{M_{D2}}{S_{3n}} \right) \tag{10.53}$$

$$\tag{10.54}$$

where S_s, S_n, and S_{3n} (see Tables 10.24 to 10.26) are section moduli for the noncomposite steel, the short-term and the long-term composite section, respectively. From Table 10.21, maximum factored positive moments M_{D1} and M_{D2} in span 1 are obtained at the location of $0.4L_1$.

$$
\begin{aligned}
M_{D1} &= (0.95)(1.25)(M_{DC1}) = (0.95)(1.25)(4260) = 5{,}059 \text{ kips-ft} \\
M_{D2} &= (0.95)\,(1.25M_{DC2} + 1.5M_{DW}) \\
&= (.095)\,[1.25(435) + 1.5(788)] = 1{,}640 \text{ kips-ft}
\end{aligned}
$$

For the top flange:

$$
\begin{aligned}
M_{AD} &= (26{,}553) \left(50 - \frac{5{,}059(12)}{2{,}898} - \frac{1{,}640(12)}{9{,}355} \right) \\
&= 715{,}552 \text{ kips-in.} = 59{,}629 \text{ kips-ft } (80{,}842 \text{ kN-m})
\end{aligned}
$$

For the bottom flange:

$$
\begin{aligned}
M_{AD} &= (5{,}248) \left(50 - \frac{5{,}059(12)}{3{,}656} - \frac{1{,}640(12)}{4{,}774} \right) \\
&= 153{,}623 \text{ kips-in.} = 12{,}802 \text{ kips-ft } (17{,}356 \text{ kN-m}) \text{ (controls)} \\
\therefore M_y &= 5{,}059 + 1{,}640 + 12{,}802 = 19{,}501 \text{ kips-ft } (26{,}438 \text{ kN-m})
\end{aligned}
$$

6. **Flexural Strength Design, Strength Limit State I**

   6.1) *Compactness of Steel Girder Section*

   The steel section is first checked to meet the requirements of a compact section (AASHTO Article 6.10.5.2.2).

   (a) Ductility requirement:

   $$D_p \leq \frac{d + t_s + t_h}{7.5}$$

   where D_p is the depth from the top of the concrete deck to the PNA, d is the depth of the steel girder, and t_h is the thickness of the concrete haunch above the top flange of the steel girder. The purpose of this requirement is to prevent permanent crashing of the concrete slab when the composite section approaches its plastic moment capacity. For this example, referring to Figure 10.75 and 10.76, we obtain:

$$
\begin{aligned}
D_p &= 10.875 + 4.375 - 1 + 0.705 = 14.955 \text{ in. } (381 \text{ mm}) \\
D_p &= 14.955 \text{ in. } < \frac{d + t_s + t_h}{7.5} = \frac{98.75 + 10.875 + 3.375}{7.5} \\
&= 15.067 \text{ in. O.K.}
\end{aligned}
$$

(b) Web slenderness requirement, $\frac{2D_{cp}}{t_w} \leq 3.76\sqrt{\frac{E}{F_{yc}}}$

where D_{cp} is the depth of the web in compression at the plastic moment state. Since the PNA is within the top flange, D_{cp} is equal to zero. The web slenderness requirement is satisfied.

(c) Compression flange slenderness and compression flange bracing requirement

It is usually assumed that the top flange is adequately braced by the hardened concrete deck; there are, therefore, no requirements for the compression flange slenderness and bracing for compact composite sections at the strength limit state.

∴ the section is a compact composite section.

6.2) *Moment of Inertia Ratio Limit* (AASHTO Article 6.10.1.1)
The flexural members shall meet the following requirement:

$$
0.1 \leq \frac{I_{yc}}{I_y} \leq 0.9
$$

where I_{yc} and I_y are the moments of inertia of the compression flange and steel girder about the vertical axis in the plane of web, respectively. This limit ensures that the lateral torsional bucking formulas are valid.

$$
\begin{aligned}
I_{yc} &= \frac{(1)(18)^3}{12} = 486 \text{ in.}^4 \\
I_y &= 486 + \frac{(96)(0.625)^3}{12} + \frac{(1.75)(18)^3}{12} + 1338 \text{ in.}^4 \\
0.1 &< \frac{I_{yc}}{I_y} = \frac{486}{1338} = 0.36 < 0.9 \quad \text{O.K.}
\end{aligned}
$$

6.3) *Nominal Flexure Resistance, M_n* (AASHTO Article 10.5.2.2a)
It is assumed that the adjacent interior-pier section is noncompact. For continuous spans with the noncompact interior support section, the nominal flexure resistance of a compact composite section is taken as:

$$
M_n = 1.3R_h M_y \leq M_p \tag{10.55}
$$

where R_h is a flange stress reduction factor taken as 1.0 for this homogeneous girder.

$$
M_n = 1.3(1.0)(19,501) = 25,351 \text{ kips-ft } < M_p = 28,185 \text{ kips-ft}
$$

6.4) Strength Limit State I

AASHTO-LRFD [1] requires that for strength limit state I

$$M_u \leq \phi_f M_n \tag{10.56}$$

where ϕ_f is the flexural resistance factor $= 1.0$. For the composite section in the positive flexure region in span 1, the maximum moment occurs at $0.4L_1$ (see Table 10.21).

$$M_u = 13,684 \text{ kips-ft} \ < \phi_f M_n = (1.0)(25,351) = 25,351 \text{ kips-ft} \ \text{O.K.}$$

7. **Shear Strength Design, Strength Limit State I**

   7.1) Nominal Shear Resistance, V_n

   (a) V_n for an unstiffened web (AASHTO Article 6.10.7.2)

$$V_n = \begin{cases} V_p = 0.58 F_{yw} D t_w & \text{For} \quad \frac{D}{t_w} \leq 2.46 \sqrt{\frac{E}{F_{yw}}} \\[2mm] 1.48 t_w^2 \sqrt{E F_{yw}} & \text{For } 2.46 \sqrt{\frac{E}{F_{yw}}} < \frac{D}{t_w} \leq 3.07 \sqrt{\frac{E}{F_{yw}}} \\[2mm] \frac{4.55 t_w^3 E}{D} & \text{For} \quad \frac{D}{t_w} > 3.07 \sqrt{\frac{E}{F_{yw}}} \end{cases} \tag{10.57}$$

   where D is depth of web and t_w is thickness of web.

$$\because \quad \frac{D}{t_w} = \frac{96}{0.625} = 153.6 > 3.07 \sqrt{\frac{E}{F_{yw}}} 3.07 \sqrt{\frac{29,000}{50}} = 73.9$$

$$\therefore \quad V_n = \frac{4.55 t_w^3 E}{D} = \frac{4.55 (0.625)^3 (29,000)}{96} = 335.6 \text{ kips } (1,493 \text{ kN})$$

   (b) V_n for an end-stiffened web panel (AASHTO Article 6.10.7.3.3c)

$$V_n = C V_p \tag{10.58}$$

$$C = \begin{cases} 1.0 & \text{For} \quad \frac{D}{t_w} < 1.10 \sqrt{\frac{Ek}{F_{yw}}} \\[2mm] \frac{1.10}{(D/t_w)} \sqrt{\frac{Ek}{F_{yw}}} & \text{For } 1.10 \sqrt{\frac{Ek}{F_{yw}}} \leq \frac{D}{t_w} \leq 1.38 \sqrt{\frac{Ek}{F_{yw}}} \\[2mm] \frac{1.52}{(D/t_w)^2} \sqrt{\frac{Ek}{F_{yw}}} & \text{For} \quad \frac{D}{t_w} > 1.38 \sqrt{\frac{Ek}{F_{yw}}} \end{cases} \tag{10.59}$$

$$k = 5 + \frac{5}{(d_o/D)^2} \tag{10.60}$$

   in which d_o is the spacing of transverse stiffeners (Figure 10.77).

   For $d_o = 240$ in. and $k = 5 + \frac{5}{(240/96)^2} = 5.80$

$$\because \quad \frac{D}{t_w} = 153.6 > 1.38 \sqrt{\frac{Ek}{F_{yw}}} = 1.38 \sqrt{\frac{29,000(5.8)}{50}} = 80$$

$$\therefore \quad C = \frac{152}{(153.6)^2} = \sqrt{\frac{29,000(5.80)}{50}} = 0.374$$

$$V_p = 0.58 F_{yw} D t_w = 0.58(50)(96)(0.625) = 1,740 \text{ kips } (7,740 \text{ kN})$$

$$V_n = C V_p = 0.374(1740) = 650.8 \text{ kips } (2,895 \text{ kN})$$

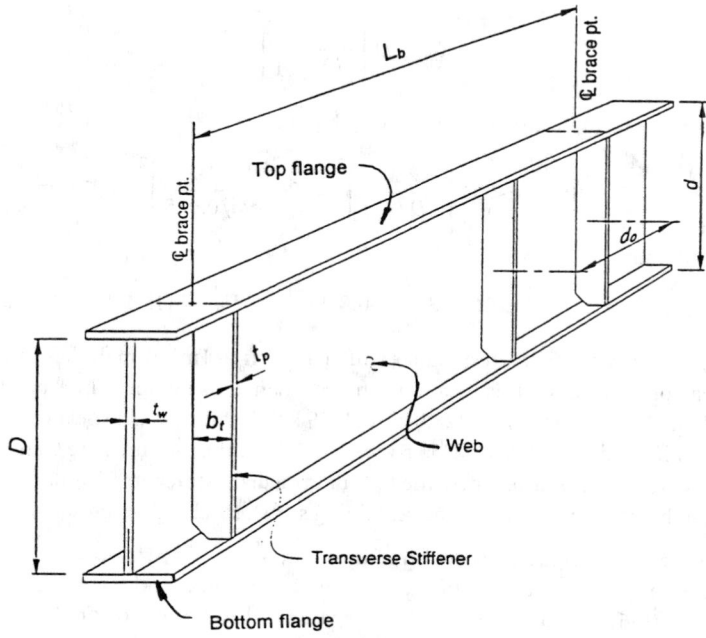

Figure 10.77 Typical steel girder dimensions.

(c) V_n for interior-stiffened web panel (AASHTO Article 6.10.7.3a)

$$V_n = \begin{cases} V_p \left(C + \dfrac{0.87(1-C)}{\sqrt{1+(d_o/D)^2}} \right) & \text{For } M_u \le 0.5\phi_f M_p \\[3mm] RV_p \left(C + \dfrac{0.87(1-C)}{\sqrt{1+(d_o/D)^2}} \right) \ge CV_p & \text{For } M_u > 0.5\phi_f M_p \end{cases} \tag{10.61}$$

where

$$R = \left[0.6 + 0.4 \left(\frac{\phi_f M_n - M_u}{\phi_f M_n - 0.75\phi_f M_y} \right) \right] \le 1.0 \tag{10.62}$$

7.2) Strength Limit State I
AASHTO-LRFD [1] requires that for strength limit state I

$$V_u \le \phi_v V_n \tag{10.63}$$

where ϕ_v is the shear resistance factor $= 1.0$.

(a) Left end of span 1:

$\because$ $V_u = 445.4$ kips $>$ $\phi_v V_n$ (for unstiffened web) $= 335.6$ kips

$\therefore$ Stiffeners are needed to increase shear capacity.

In order to facilitate handling of web panel sections, the spacing of transverse stiffeners shall meet (AASHTO Article 6.10.7.3.2) the following requirement:

$$d_o \leq D \left[\frac{260}{(D/t_w)} \right]^2 \tag{10.64}$$

Try $d_o = 240$ in. for end-stiffened web panel

$$d_o = 240 \text{ in.} \ < D \left[\frac{260}{(D/t_w)} \right]^2 = 96 \left[\frac{260}{96/0.625} \right]^2 = 275 \text{ in.} \quad \text{O.K.}$$

and then

$$\phi_v V_n = (1.0)650.8 = 650.8 \text{ kips} \ > V_u = 445.4 \text{ kips} \quad \text{O.K.}$$

(b) Location of the first intermediate stiffeners, 20 ft (6.1m) from the left end in span 1: Factored shear for this location can be obtained using linear interpolation from Table 10.22. Since $V_u = 328.0$ kips (1459 kN) is less than the shear capacity of the unstiffened web, $\phi_v V_n = 335.5$ kips (1492 kN), the intermediate transverse stiffeners may be omitted after the first intermediate stiffeners. Similar calculations can be used to determine the remaining stiffeners along the girder.

8. **Fatigue Design, Fatigue and Fracture Limit State**

The base metal at the connection plate welds to flanges, and webs located at 96 ft (29.26 m) (0.6L1) from the left end of span 1 will be checked for the fatigue load combination.

8.1) *Load-Induced Fatigue* (AASHTO Article 6.6.1.2)

The design requirements for load-induced fatigue apply only to (1) details subjected to a net applied tensile stress and (2) regions where the unfactored permanent loads produce compression, and only if the compressive stress is less than twice the maximum tensile stress resulting from the fatigue load combination. In the fatigue limit state, all stresses are calculated using the elastic section properties (Tables 10.24 to 10.26).

(a) Top-flange weld

The compressive stress at the top-flange weld due to unfactored permanent loads is obtained:

$$
\begin{aligned}
f_{DC} &= \frac{M_{DC1}(y_{st} - t_{fc})}{I_{girder}} + \frac{(M_{DC2} + M_{DW})(y_{st} - t_{fc})}{I_{com-3n}} \\
&= \frac{2462(12)(55.087 - 1.0)}{159,619} + \frac{(251 + 455)(12)(33.367 - 1.0)}{312,155} \\
&= 10.89 \text{ ksi (75.09 MPa)}
\end{aligned}
$$

Assume that the negative fatigue moments are carried by the steel section only in the positive flexure region. The maximum tensile stress at the top-flange weld at this location due to factored fatigue moment is

$$
\begin{aligned}
f_{LL+IM} &= \frac{\left| -(M_{LL+IM})_u \right| (y_{st} - t_{fc})}{I_{girder}} = \frac{798(12)(54.087)}{159,619} \\
&= 3.25 \text{ ksi (22.41 MPa)}
\end{aligned}
$$

∵ $f_{DC} = 10.89$ ksi $> 2f_{LL+IM} = 6.49$ ksi

∴   no need to check fatigue for the top-flange weld

(b) Bottom-flange weld

- Factored fatigue stress range, $(\Delta f)_u$

  For the positive flexure region, we assume that positive fatigue moments are applied to the short-term composite section and negative fatigue moments are applied to the noncomposite steel section only.

$$
\begin{aligned}
(\Delta f)_u &= \frac{(M_{LL+IM})_u \left(y_{sb-n} - t_{ft}\right)}{I_{com-n}} + \frac{\left|-(M_{LL+IM})_u\right| \left(y_{sb} - t_{ft}\right)}{I_{girder}} \\
&= \frac{1465(12)(82.454 - 1.75)}{432{,}707} + \frac{798(12)(43.663 - 1.75)}{159{,}619} \\
&= 5.79 \ \text{ksi} \ (39.92 \ \text{MPa})
\end{aligned}
$$

- Nominal fatigue resistance range, $(\Delta F)_n$

  For filet-welded connections with weld lines normal to the direction of stress, the base metal at transverse stiffeners to flange welds is fatigue detail category C' (AASHTO Table 6.6.1.2.3.-1).

$$
(\Delta F)_n = \left(\frac{A}{N}\right)^{1/3} \geq \frac{1}{2}(\Delta F)_{TH} \tag{10.65}
$$

where A is a constant dependent on detail category $= 44(10)^8$ for category C' and

$$
N = (365)(75)n(ADTT)_{ST} \tag{10.66}
$$
$$
ADTT_{ST} = p(ADTT) \tag{10.67}
$$

where p is a fraction of a truck in a single lane (AASHTO Table 3.6.1.4.2-1) $= 0.8$ for three-lane traffic, and n is the number of stress-range cycles per truck passage (AASHTO Table 6.6.1.2.5-2) $= 1.0$ for the positive flexure region.

$$
N = (365)(75)(1.0)(0.8)(3600) = 7.844(10)^7
$$

For category C' detail, $(\Delta F)_{TH} = 12$ ksi (AASHTO Table 6.6.1.2.5-3).

$$
\therefore \left(\frac{A}{N}\right)^{1/3} = \left(\frac{44(10)^8}{7.844(10)^7}\right)^{1/3} = 3.83 \ \text{ksi} \ < \frac{1}{2}(\Delta F)_{TH} = 6 \ \text{ksi}
$$
$$
\therefore (\Delta F)_n = \frac{1}{2}(\Delta F)_{TH} = 6 \ \text{ksi} \ (41.37 \ \text{MPa})
$$

- Fatigue limit state

  AASHTO requires that each detail shall satisfy:

$$
(\Delta f)_u \leq (\Delta F)_n \tag{10.68}
$$

For top-flange weld

$$
(\Delta f)_u = 5.79 \ \text{ksi} \ < (\Delta F)_n = 6 \ \text{ksi} \quad \text{O.K.}
$$

8.2) *Fatigue Requirements for Web* (AASHTO Article 6.10.4)

The purpose of these requirements is to control out-of-plane flexing of the web due to flexure and shear under repeated live loadings. The repeated live load is taken as twice the factored fatigue load.

(a) Flexure requirement

$$
f_{cf} \leq
\begin{cases}
R_h F_{yc} & \text{For } \quad \frac{2D_c}{t_w} \leq 5.76\sqrt{\frac{E}{F_{yc}}} \\[2mm]
R_h F_{yc}\left(3.58 - 0448\frac{2D_c}{t_w}\sqrt{\frac{F_{yc}}{E}}\right) & \text{For } 5.76\sqrt{\frac{E}{F_{yc}}} \leq \frac{2D_c}{t_w} \leq 6.43\sqrt{\frac{E}{F_{yc}}} \\[2mm]
28.9 R_h E\left(\frac{t_w}{2D_c}\right)^2 & \text{For } \quad \frac{2D_c}{t_w} > 6.43\sqrt{\frac{E}{F_{yc}}}
\end{cases}
\tag{10.69}
$$

where f_{cf} is the maximum elastic flexural stress in the compression flange due to the unfactored permanent loads and repeated live loadings; F_{yc} is the yield strength of the compression flange; and D_c is the depth of the web in compression.

- Depth of web in compression, Dc

  Considering the algebraic sum of stresses acting on different sections based on elastic section properties, D_c can be obtained by the following formula:

$$
D_c = \frac{f_{DC1} + f_{DC2} + f_{DW} + f_{LL+IM}}{\frac{f_{DC1}}{y_{st}} + \frac{f_{DC2}+f_{DW}}{y_{st-3n}} + \frac{f_{LL+IM}}{y_{st-n}}} - t_{fc}
$$

$$
= \frac{\frac{M_{DC1}}{S_{st}} + \frac{M_{DC2}+M_{DW}}{S_{st-3n}} + \frac{2(M_{LL+IM})u}{S_{st-n}}}{\frac{M_{DC1}}{I_{girder}} + \frac{M_{DC2}+M_{DW}}{I_{com-3n}} + \frac{2(M_{LL+IM})u}{I_{com-n}}} - t_{fc}
\tag{10.70}
$$

Substituting moments (Tables 10.21 and 10.23) and section properties (Tables 10.24 and 10.26) into Equation 10.70, we obtain:

$$
D_c = \frac{\frac{4260(12)}{2,898} + \frac{(435+788)(12)}{9,355} + \frac{2(1607)(12)}{26,553}}{\frac{4260(12)}{159,629} + \frac{(435+788)(12)}{312,155} + \frac{2(1607)(12)}{432,707}} - 1
$$

$$
= \frac{17.640 + 1.569 + 1.452}{0.320 + 0.047 + 0.089} - 1 = 44.29 \text{ in. } (1,125 \text{ mm})
$$

$$
\frac{2D_c}{t_w} = \frac{2(44.29)}{0.625} = 141.7 < 5.76\sqrt{\frac{E}{F_{yc}}} = 183.7
$$

- Maximum compressive stress in flange, f_{cf} (at location $0.4L_1$)

$$
\begin{aligned}
f_{cf} &= f_{DC1} + f_{DC2} + f_{DW} + f_{LL+IM} \\
&= \frac{M_{DC1}}{S_{st}} + \frac{M_{DC2} + M_{DW}}{S_{st-3n}} + \frac{2(M_{LL+IM})u}{S_{st-n}} \\
&= 17.64 + 1.57 + 1.45 = 20.66 < R_h F_{yc} = 50 \text{ ksi}
\end{aligned}
$$

(b) Shear (AASHTO Article 10.6.10.4.4)

The left end of span 1 is checked as follows:

- Fatigue load

$$
\begin{aligned}
V_u &= V_{DC1} + V_{DC2} + V_{DW} + 2(V_{LL+IM})_u \\
&= 148.4 + 15.1 + 27.4 + 2(51.1) = 293.1 \text{ kips } (1304 \text{ kN})
\end{aligned}
$$

- Fatigue shear stress

$$
v_{cf} = \frac{V_u}{Dt_w} = \frac{293.1}{96(0.625)} = 4.89 \text{ ksi } (33.72 \text{ MPa})
$$

- Fatigue shear resistance

$$
\begin{aligned}
C &= 0.374 \quad \text{(see Step 7)} \\
v_n &= 0.58 C F_{yw} = 0.58(0.374)(50) \\
&= 10.85 \text{ ksi } > v_{cf} = 4.89 \text{ ksi O.K.}
\end{aligned}
$$

8.3) Distortion-Induced Fatigue (AASHTO Article 6.6.1.3)

All transverse connection plates will be welded to both the tension and compression flanges to provide rigid load paths so distortion-induced fatigue (the development of significant secondary stresses) can be prevented.

8.4) Fracture Limit State (AASHTO Article 6.6.2)

Materials for main load-carrying components subjected to tensile stresses will meet the Charpy V-notch fracture toughness requirement (AASHTO Table 6.6.2-2) for temperature zone 2 (AASHTO Table 6.6.2-1).

9. **Intermediate Transverse Stiffener Design**

The intermediate transverse stiffener consists of two plates welded to both sides of the web. The design of the first intermediate transverse stiffener is discussed in the following.

9.1) Projecting Width, b_t, Requirements (AASHTO Article 6.10.8.1.2)

To prevent local buckling of the transverse stiffeners, the width of each projecting stiffener shall satisfy these requirements:

$$
\left\{ \begin{array}{c} 2.0 + \frac{d}{30} \\ 0.25 b_f \end{array} \right\} \le b_t \le \left\{ \begin{array}{c} 0.48 t_p \sqrt{\frac{E}{F_{ys}}} \\ 16 t_p \end{array} \right\} \tag{10.71}
$$

where b_f is the full width of the steel flange and F_{ys} is the specified minimum yield strength of the stiffener. To allow adequate space for cross-frame connections, try stiffener width $b_t = 6$ in. (152 mm):

$$
b_t = 6 \text{ in. } > \left\{ \begin{array}{l} 2.0 + \frac{d}{30} = 2.0 + \frac{98.75}{30} = 5.3 \text{ in.} \\ 0.25 b_f = 0.25(18) = 4.5 \text{ in. O.K.} \end{array} \right.
$$

Try $t_p = 0.5$ in. (13 mm) and obtain:

$$
b_t = 6 \text{ in. } < \left\{ \begin{array}{l} 0.48 t_p \sqrt{\frac{E}{F_{ys}}} = 0.48(0.5)\sqrt{\frac{29,000}{36}} = 6.8 \text{ in.} \\ 16 t_p = 16(0.5) = 8 \text{ in. O.K.} \end{array} \right.
$$

Use two 6 in. x 0.5 in. (152 mm x 13 mm) transverse stiffener plates.

9.2) *Moment of Inertia Requirement* (AASHTO Article 6.10.8.1.3)

The purpose of this requirement is to ensure sufficient rigidity of transverse stiffeners to adequately develop a tension field in the web.

$$I_t \geq d_o t_w^2 J \tag{10.72}$$

$$J = 2.5 \left(\frac{D_p}{d_o} \right)^2 - 2.0 \geq 0.5 \tag{10.73}$$

where I_t is the moment of inertia for the transverse stiffener taken about the edge in contact with the web for single stiffeners and about the mid-thickness of the web for stiffener pairs (Figure 10.78); D_p is the web depth for webs without longitudinal stiffeners.

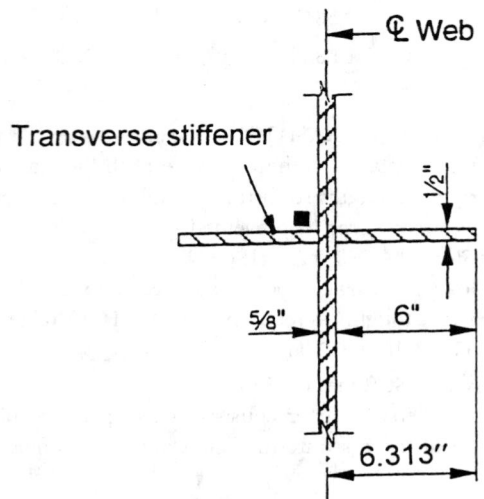

Figure 10.78   Cross-section of web and transverse stiffener.

$$\therefore \quad J = 2.5 \left(\frac{96}{240} \right)^2 - 2.0 = -1.6 < 0.5 \qquad \therefore \text{ Use } J = 0.5$$

$$I_t = 2 \left(\frac{6^3 (0.5)}{12} + (6)(0.5)(3.313)^2 \right) = 83.86 \text{ in.}^4$$

$$> \quad d_o t_w^2 J = (240)(0.625)^2 (0.5) = 46.88 \text{ in.}^4 \quad \text{O.K.}$$

9.3) *Area Requirement* (AASHTO Article 6.10.8.1.4)

This requirement ensures that transverse stiffeners have sufficient area to resist the vertical component of the tension field, and is only applied to transverse stiffeners required to carry the forces imposed by tension-field action.

$$A_s \geq A_{s\,\min} = \left(0.15 B D t_w (1 - C) \frac{V_u}{\phi_v V_n} - 18 t_w^2 \right) \left(\frac{F_{yw}}{F_{ys}} \right) \tag{10.74}$$

where $B = 1.0$ for stiffener pairs. From the previous calculation:

$$C = 0.374, \qquad\qquad F_{yw} = 50 \text{ ksi}, \qquad F_{ys} = 36 \text{ ksi}$$

$$V_u = 328.0 \text{ kips}, \qquad\qquad \phi_f V_n = 335.5 \text{ kips}, \quad t_w = 0.625 \text{ in.}$$

$$A_s = 2(6)(0.5) = 6 \text{ in.}^2 > A_{s\,min}$$

$$= \left(0.15(1.0)(96)(0.625)(1 - 0374)\frac{328.0}{335.5} - 18(0.625)^2 \right) \left(\frac{50}{36} \right)$$

$$= -0.635 \text{ in.}^2$$

The negative value of $A_{s\,min}$ indicates that the web has sufficient area to resist the vertical component of the tension field.

10. **Shear Connector Design**

In a composite girder, stud or channel shear connectors must be provided at the interface between the concrete deck slab and the steel section to resist the interface shear. For a composite bridge girder, the shear connectors should be normally provided throughout the length of the bridge (AASHTO Article 6.10.7.4.1). Stud shear connectors are chosen in this example and will be designed for the fatigue limit state and then checked against the strength limit state. The detailed calculations of the shear stud connectors for the positive flexure region of span 1 are given in the following. A similar procedure can be used to design the shear studs for other portions of the bridge.

10.1) *Stud Size* (AASHTO Article 6.10.7.4.1a)

To meet the limits for cover and penetration for shear connectors specified in AASHTO Article 6.10.7.4.1d, try:

$$\text{Stud height, } H_{stud} = 7 \text{ in.} \; > t_h + 2 = 3.375 + 2 = 5.375 \text{ in.} \;\; \text{O.K.}$$

$$\text{Stud diameter, } d_{stud} = 0.875 \text{ in.} \; < H_{stud}/4 = 1.75 \text{ in.} \qquad \text{O.K.}$$

10.2) *Pitch of Shear Stud, p, for Fatigue Limit State*

(a) Basic requirements (AASHTO Article 6.10.7.4.1b)

$$6d_{stud} \leq p = \frac{n_{stud} Z_r I_{com-n}}{V_{sr} Q} \leq 24 \text{ in.} \qquad\qquad (10.75)$$

where n_{stud} is the number of shear connectors in a cross-section; Q is the first moment of transformed section (concrete deck) about the neutral axis of the short-term composite section; V_{sr} is the shear force range in the fatigue limit state; and Z_r is the shear fatigue resistance of an individual shear connector.

(b) Fatigue resistance, Z_r (AASHTO Article 6.10.7.4.2)

$$Z_r = \alpha d_{stud}^2 \geq 5.5 d_{stud}^2 \qquad\qquad (10.76)$$

$$\alpha = 34.5 - 4.28 \log N \qquad\qquad (10.77)$$

where N is the number of cycles specified in AASHTO Article 6.6.1.2.5, $N = 7.844(10)^7$ cycle (see Step 8).

$$\alpha = 34.5 - 4.28 \log(7.844 \times 10^7) = 0.711 < 5.5$$

$$Z_r = 5.5 d_{stud}^2 = 5.5(0.875)^2 = 4.211 \text{ ksi}$$

(c) First moment, Q, and moment of initial, I_{com-n} (see Table 10.25)

$$
\begin{aligned}
Q &= \left(\frac{b_{eff} t_s}{9}\right)\left(y_{st-n} - t_h + \frac{t_s}{2}\right) \\
&= \left(\frac{140(10.875)}{9}\right)\left(16.296 + 3.375 + \frac{10.875}{2}\right) = 4247.52 \text{ in.}^3 \\
I_{com-n} &= 432,707 \text{ in.}^4
\end{aligned}
$$

(d) Required pitch for the fatigue limit state

Assume that shear studs are spaced at 6 in. transversely across the top flange of a steel section (Figure 10.75) and, using $n_{stud} = 3$ for this example, obtain

$$P_{required} = \frac{3(4.211)(432,707)}{V_{sr}(4,247.52)} = \frac{1,286.96}{V_{sr}}$$

The detailed calculations for the positive flexure region of span 1 are shown in Table 10.27.

TABLE 10.27   Shear Connector Design for the Positive Flexure Region in Span 1

Location (x/L)	V_{sr} (kips)	$P_{required}$ (in.)	P_{final} (in.)	$n_{total-stud}$
0.0	60.1	21.4	12	3
0.1	51.6	24.9	12	51
0.2	47.8	26.9	18	99
0.3	46.2	27.9	18	132
0.4	45.7	28.2	18	165
0.4	45.7	28.2	12	162
0.5	46.2	27.9	12	114
0.6	47.8	26.9	12	66
0.7	50.3	25.6	9	3

Note:

$$V_{sr} = \left|+ (V_{LL+IM})_u\right| + \left|- (V_{LL+IM})_u\right|$$

$$P_{required} = \frac{n_{stud} Z_r I_{com-n}}{V_{sr} Q} = \frac{1286.96}{V_{sr}}$$

$n_{total-stud}$ is the summation of number of shear studs between the locations of the zero moment and that location.

10.3) *Strength Limit State Check*

(a) Basic requirement (AASHTO Article 6.10.7.4.4a)

The resulting number of shear connectors provided between the section of maximum positive moment and each adjacent point of zero moment shall satisfy the following requirement:

$$n_{total-stud} \geq \frac{V_h}{\phi_{sc} Q_n} \tag{10.78}$$

where ϕ_{sc} is the resistance factor for shear connectors, 0.85; V_h is the nominal horizontal shear force; and Q_n is the nominal shear resistance of one stud shear connector.

(b) Nominal horizontal shear force (AASHTO Article 6.10.7.4.4b)

$$V_h = \text{the lesser of} \begin{cases} 0.85 f_c' b_{eff} t_s \\ F_{yw} D t_w + f_{yt} b_{ft} t_{ft} + F_{yc} b_{fc} t_{fc} \end{cases} \qquad (10.79)$$

$$
\begin{aligned}
V_{h-\text{concrete}} &= 0.5 f_c' b_{eff} t_s = 0.85(3.25)(140)(10.875) = 4,206 \text{ kips} \\
V_{h-\text{steel}} &= F_{yw} D t_w + F_{yt} b_{ft} t_{ft} + F_{yc} b_{fc} t_{fc} \\
&= 50\left[(18)(1.0) + (96)(0.625) + (18)(1.75)\right] = 5,475 \text{ kips} \\
\therefore V_h &= 4,206 \text{ kips} \quad (18,708 \text{ kN})
\end{aligned}
$$

(c) Nominal shear resistance (AASHTO Article 6.10.7.4.4c)

$$Q_n = 0.5 A_{sc} \sqrt{f_c' E_c} \leq A_{sc} F_u \qquad (10.80)$$

where A_{sc} is a cross-sectional area of a stud shear connector and F_u is the specified minimum tensile strength of a stud shear connector $= 60$ ksi (420 MPa).

$$
\begin{aligned}
0.5\sqrt{f_c' E_c} &= 0.5\sqrt{3.25(3,250)} = 51.4 \text{ kips} \quad < \quad F_u = 60 \text{ kips} \\
\therefore Q_n &= 0.5 A_{sc}\sqrt{f_c' E_c} = 51.4\left(\frac{\pi(0.875)^2}{4}\right) = 30.9 \text{ kips}
\end{aligned}
$$

(d) Check resulting number of shear stud connectors (see Table 10.27)

$$
\begin{aligned}
n_{\text{total-stud}} &= \begin{cases} 165 \text{ from left end } 0.4L_1 \\ 162 \text{ from } 0.4L_1 \text{ to } 0.7L_1 \end{cases} \\
&> \frac{V_h}{\phi_{sc} Q_n} = \frac{4206}{0.85(30.9)} = 160 \text{ O.K.}
\end{aligned}
$$

11. **Constructability Check**

For unshored construction, AASHTO requires that all I-section bending members be investigated for strength and stability during construction stages using appropriate load combinations given in AASHTO Table 3.4.1-1. The following checks are made for the steel girder section only under factored dead load, $DC1$. It is assumed that the final total dead load, $DC1$, produces the controlling maximum moments.

11.1) *Web Slenderness Requirement* (AASHTO 6.10.10.2.2)

$$\frac{2D_c}{t_w} \leq 6.77\sqrt{\frac{E}{f_c}} \qquad (10.81)$$

where f_c is the stress in compression flange due to the factored dead load, $DC1$, and D_c is the depth of the web in compression in the elastic range.

$$D_c = y_{st} - t_{fc} = 55.087 - 1 = 54.087 \text{ in. (1,374 mm)}$$

$$f_c = \frac{(0.95)(1.25)M_{DC1}}{S_{st}} = \frac{0.95(1.25)(4260)(12)}{2,898} = 20.95 \text{ ksi (145 MPa)}$$

$$\frac{2D_c}{t_w} = \frac{2(54.087)}{0.625} = 173.1 \le 6.77\sqrt{\frac{E}{f_c}} = 6.77\sqrt{\frac{29,000}{20.95}} = 251.9 \quad \text{O.K.}$$

$$\therefore \quad \text{no longitudinal stiffener is required}$$

11.2) *Compression Flange Slenderness Requirement* (AASHTO Article 6.10.10.2.3)
This requirement prevents the local buckling of the top flange before the concrete deck hardens.

$$\frac{b_f}{2t_f} \le 1.38 \sqrt{\frac{E}{f_c\sqrt{\frac{2D_c}{t_w}}}} \tag{10.82}$$

$$\frac{b_f}{2t_f} = \frac{18}{2(1.0)} = 9 \le 1.38 \sqrt{\frac{E}{f_c\sqrt{\frac{2D_c}{t_w}}}} = 1.38 \sqrt{\frac{29,000}{20.95\sqrt{173.1}}} = 14.2 \quad \text{O.K.}$$

11.3) *Compression Flange Bracing Requirement* (AASHTO Article 6.10.10.2.4)

(a) Flexure (AASHTO Article 6.10.6.4.1)

To ensure that a noncomposite steel girder has sufficient flexural resistance during construction, the moment capacity should be calculated considering lateral torsional buckling with an unbraced length, L_b (Figure 10.77).

For a steel girder without longitudinal stiffeners and $(2D_c/t_w) > \lambda_b\sqrt{E/F_{yc}}$, the nominal flexural resistance is

$$M_n = \begin{cases} 1.3R_h M_y \le M_p & \text{For } L_b \le L_p \\ C_b R_b R_h M_y \left[1 - 0.5\left(\frac{L_b - L_p}{L_p - L_r}\right)\right] \le R_b R_h M_y & \text{For } L_p < L_b \le L_r \\ C_b R_b R_h \frac{M_y}{2}\left(\frac{L_r}{L_b}\right)^2 \le R_b R_h M_y & \text{For } L_b > L_r \end{cases} \tag{10.83}$$

$$L_p \le 1.76r_t \sqrt{\frac{E}{F_{yc}}} \tag{10.84}$$

$$L_r = \sqrt{\frac{19.71 I_{yc} d}{S_{xc}} \frac{E}{F_{yc}}} \tag{10.85}$$

where λ_b equals 4.64 for a member with a compression flange area less than the tension flange area and 5.76 for members with a compression flange area equal to or greater than the tension flange area; r_t is the minimum radius of gyration of the compression flange of the steel section about the vertical axis; S_{xc} is the section modulus about the horizontal axis of the section to the compression flange (equal to S_{st} in Table 10.24); C_b is the moment gradient correction factor; and R_b is a flange stress reduction factor considering local buckling of a slender web (AASHTO Article 6.10.5.4.2).

$$C_b = 1.75 - 1.95\left(\frac{P_l}{P_h}\right) + 0.3\left(\frac{P_l}{P_h}\right)^2 \le 2.3 \qquad (10.86)$$

where P_l is the force in the compression flange at the braced point with the lower force due to the factored loading, and P_h is the force in the compression flange at the braced point with higher force due to the factored loading. C_b is conservatively taken as 1.0 in this example.

$$r_t = \sqrt{\frac{I_{yf}}{A_f}} = \sqrt{\frac{(18)^3(1.0)/12}{(18)(1.0)}} = 5.20 \text{ in. } (132 \text{ mm})$$

$$L_p = 1.76r_t\sqrt{\frac{E}{F_{yc}}} = 1.76(5.2)\sqrt{\frac{29{,}000}{50}} = 220 \text{ in. } < L_b = 240 \text{ in.}$$

$$L_r = \sqrt{\frac{19.71(486)(98.75)}{2{,}898}\frac{29{,}000}{50}} = 435 \text{ in. } (11{,}049 \text{ mm})$$

$$\frac{2D_c}{t_w} = 173.1 < \lambda_b\sqrt{\frac{E}{f_c}} = 4.64\sqrt{\frac{29{,}000}{20.95}} = 172.6$$

Since these two values are very close, take $R_b = 1.0$ (AASHTO Article 6.10.5.4.2).

$$\begin{aligned}
M_y &= S_{st}F_y = 2{,}898(50) = 144{,}900 \text{ kips-in. } = 12{,}075 \text{ kips-ft} \\
\therefore L_p &= 220 \text{ in. } < L_b = 240 \text{ in. } < L_r = 435 \text{ in.} \\
\therefore M_n &= (1.0)(1.0)(1.0)(12{,}075)\left[1 - 0.5\left(\frac{240-220}{435-220}\right)\right] \\
&= 11{,}513 \text{ kips-ft } < R_n R_h M_y = 12{,}075 \text{ kips-ft} \\
M_u &= 0.95(1.25)(4{,}260) = 5{,}059 \text{ kips-ft } (6{,}859 \text{ kN-m}) \\
&< \phi_f M_n + (1.0)(11{,}513) \\
&= 11{,}513 \text{ kips-ft } (15{,}609 \text{ kN-m}) \quad \text{O.K.}
\end{aligned}$$

(b) Shear (AASHTO Article 6.10.10.3)

Check the section at the first intermediate transverse stiffener, 20 ft (6.10 m) from the left end of span 1. V_u is taken conservatively from the location of $0.1L_1$.

$$V_u = 0.95(125)V_{DC1} = 0.95(1.25)(107.5) = 120.9 \text{ kips } (538 \text{ kN})$$

For an unstiffened web, $V_n = 335.5$ kips (1,492 kN); therefore, we obtain:

$$\phi_v V_n = (1.0)(335.5) = 335.5 \text{ kips } > V_u = 120.9 \text{ kips } \quad \text{O.K.}$$

11

Shell Structures

11.1 Introduction 11-1
   Overview • Production Practice • Scope • Limitations • Stress
   Components for Stability Analysis and Design • Materials • Ge-
   ometries, Failure Modes, and Loads • Buckling Design Method
   • Stress Factor • Nomenclature
11.2 Allowable Compressive Stresses for Cylindrical Shells . 11-10
   Uniform Axial Compression • Axial Compression Due to Bend-
   ing Moment • External Pressure • Shear • Sizing of Rings (Gen-
   eral Instability)
11.3 Allowable Compressive Stresses For Cones 11-14
   Uniform Axial Compression and Axial Compression
   Due to Bending • External Pressure • Shear • Local Stiffener
   Buckling
11.4 Allowable Stress Equations For Combined Loads 11-16
   For Combination of Uniform Axial Compression and Hoop
   Compression • For Combination of Axial Compression Due to
   Bending Moment, M, and Hoop Compression • For Combi-
   nation of Hoop Compression and Shear • For Combination of
   Uniform Axial Compression, Axial Compression Due to Bend-
   ing Moment, M, and Shear, in the Presence of Hoop Com-
   pression, $(f_h \neq 0)$ • For Combination of Uniform Axial Com-
   pression, Axial Compression Due to Bending Moment, M, and
   Shear, in the Absence of Hoop Compression, $(f_h = 0)$
11.5 Tolerances for Cylindrical and Conical Shells 11-18
   Shells Subjected to Uniform Axial Compression and Axial
   Compression Due to Bending Moment • Shells Subjected to
   External Pressure • Shells Subjected to Shear
11.6 Allowable Compressive Stresses 11-20
   Spherical Shells • Toroidal and Ellipsoidal Heads
11.7 Tolerances for Formed Heads 11-21
References 11-22
Further Reading 11-22

Clarence D. Miller
Consulting Engineer,
Bloomington, IN

11.1 Introduction

11.1.1 Overview

Many steel structures, such as elevated water tanks, oil and water storage tanks, offshore structures, and pressure vessels, are comprised of shell elements that are subjected to compression stresses. The shell elements are subject to instability resulting from the applied loads. The theoretical buckling strength based on linear elastic bifurcation analysis is well known for stiffened as well as unstiffened cylindrical and conical shells and unstiffened spherical and torispherical shells. Simple formulas have been determined for many geometries and types of loads. Initial geometric imperfections and residual stresses that result from the fabrication process, however, reduce the buckling strength of fabricated shells. The amount of reduction is dependent on the geometry of the shell, type of loading (axial compression, bending, external pressure, etc.), size of imperfections, and material properties.

11.1.2   Production Practice

The behavior of a cylindrical shell is influenced to some extent by whether it is manufactured in a pipe or tubing mill or fabricated from plate material. The two methods of production will be referred to as manufactured cylinders and fabricated cylinders. The distinction is important primarily because of the differences in geometric imperfections and residual stress levels that may result from the two different production practices. In general, fabricated cylinders may be expected to have considerably larger magnitudes of imperfections (in out-of-roundness and lack of straightness) than the mill manufactured products. Similarly, fabricated heads are likely to have larger shape imperfections than those produced by spinning. Spun heads, however, typically have a greater variation in thickness and greater residual stresses due to the cold working. The design rules given in this chapter apply to fabricated steel shells.

Fabricated shells are produced from flat plates by rolling or pressing the plates to the desired shape and welding the edges together. Because of the method of construction, the mechanical properties of the shells will vary along the length and around the circumference. Misfit of the edges to be welded together may result in unintentional eccentricities at the joints. In addition, welding tends to introduce out-of-roundness and out-of-straightness imperfections that must be taken into account in the design rules.

11.1.3   Scope

Rules are given for determining the allowable compressive stresses for unstiffened and ring stiffened circular cylinders and cones and unstiffened spherical, ellipsoidal, and torispherical heads. The allowable stress equations are based on theoretical buckling equations that have been reduced by knockdown factors and by plasticity reduction factors that were determined from tests on fabricated shells. The research leading to the development of the allowable stress equations is given in [2, 7, 8, 9, 10].

Allowable compressive stress equations are presented for cylinders and cones subjected to uniform axial compression, bending moment applied over the entire cross-section, external pressure, loads that produce in-plane shear stresses, and combinations of these loads. Allowable compressive stress equations are presented for formed heads that are subjected to loads that produce unequal biaxial stresses as well as equal biaxial stresses.

11.1.4   Limitations

The allowable stress equations are based on an assumed axisymmetric shell with uniform thickness for unstiffened cylinders and formed heads and with uniform thickness between rings for stiffened cylinders and cones. All shell penetrations must be properly reinforced. The results of tests on reinforced openings and some design guidance are given in [6]. The stability criteria of this chapter may be used for cylinders that are reinforced in accordance with the recommendations of this reference if the openings do not exceed 10% of the cylinder diameter or 80% of the ring spacing. Special consideration must be given to the effects of larger penetrations.

The proposed rules are applicable to shells with D/t ratios up to 2000 and shell thicknesses of 3/16 in. or greater. The deviations from true circular shape and straightness must satisfy the requirements stated in this chapter.

Special consideration must be given to ends of members or areas of load application where stress distribution may be nonlinear and localized stresses may exceed those predicted by linear theory. When the localized stresses extend over a distance equal to one half the wave length of the buckling mode, they should be considered as a uniform stress around the full circumference. Additional thickness or stiffening may be required.

Failure due to material fracture or fatigue and failures caused by dents resulting from accidental loads are not considered. The rules do not apply to temperatures where creep may occur.

11.1.5 Stress Components for Stability Analysis and Design

The internal stress field that controls the buckling of a cylindrical shell consists of the longitudinal, circumferential, and in-plane shear membrane stresses. The stresses resulting from a dynamic analysis should be treated as equivalent static stresses.

11.1.6 Materials

Steel

The allowable stress equations apply directly to shells fabricated from carbon and low alloy steel plate materials such as those given in Table 11.1 or Table UCS-23 of [3]. The steel materials in Table 11.1 are designated by group and class. Steels are grouped according to strength level and welding characteristics. *Group I* designates mild steels with specified minimum yield stresses ≤ 40 ksi and these steels may be welded by any of the processes as described in [5]. *Group II* designates intermediate strength steels with specified minimum yield stresses > 40 ksi and ≤ 52 ksi. These steels require the use of low hydrogen welding processes. *Group III* designates high strength steels with specified minimum yield stresses > 52 ksi. These steels may be used provided that each application is investigated with respect to weldability and special welding procedures that may be required. Consideration should be given to fatigue problems that may result from the use of higher working stresses, and notch toughness in relation to other elements of fracture control such as fabrication, inspection procedures, service stress, and temperature environment.

The steels in Table 11.1 have been classified according to their notch toughness characteristics. *Class C* steels are those that have a history of successful application in welded structures at service temperatures above freezing. Impact tests are not specified. *Class B* steels are suitable for use where thickness, cold work, restraint, stress concentration, and impact loading indicate the need for improved notch toughness. When impact tests are specified, Class B steels should exhibit Charpy V-notch energy of 15 ft-lbs for Group 1 and 25 ft-lbs for Group II at the lowest service temperature. The Class B steels given in Table 11.1 can generally meet the Charpy requirements at temperatures ranging from 50° to 32°F. *Class A* steels are suitable for use at subfreezing temperatures and for critical applications involving adverse combinations of the factors cited above. The steels given in Table 11.1 can generally meet the Charpy requirements for Class B steels at temperatures ranging from −4° to −40°F.

Other Materials

The design equations can also be applied to other materials for which a chart or table is provided in Subpart 3 of [4] by substituting the tangent modulus E_t for the elastic modulus E in the allowable stress equations. The method for finding the allowable stresses for shells constructed from these materials is determined by the following procedure.

TABLE 11.1   Steel Plate Materials

Group	Class	Specification	Specified minimum yield stress (ksi)[a]	Specified minimum tensile stress (ksi)[a]
I	C	ASTM A36 (to 2 in. thick)	36	58
		ASTM A131 Grade A (to 1/2 in. thick)	34	58
		ASTM A285 Grade C (to 3/4 in. thick)	30	55
I	B	ASTM A131 Grades B, D	34	58
		ASTM A516 Grade 65	35	65
		ASTM A573 Grade 65	35	65
		ASTM A709 Grade 36T2	36	58
I	A	ASTM A131 Grades CS, E	34	58
II	C	ASTM A572 Grade 42 (to 2 in. thick) ASTM A591 required over 1/2 in. thick	42	60
		ASTM A572 Grade 50 (to 2 in. thick) ASTM A591 required over 1/2 in. thick	50	65
II	B	ASTM A709 Grades 50T2, 50T3	50	65
		ASTM A131 Grade AH32	45.5	68
		ASTM A131 Grade AH36	51	71
II	A	API Spec 2H Grade 42	42	62
		API Spec 2H Grade 50 (to 2 1/2 in. thick)	50	70
		API Spec 2H Grade 50 (over 2 1/2 in. thick)	47	70
		API Spec 2W Grade 42 (to 1 in. thick)	42	62
		API Spec 2W Grade 42 (over 1 in. thick)	42	62
		API Spec 2W Grade 50 (to 1 in. thick)	50	65
		API Spec 2W Grade 50 (over 1 in. thick)	50	65
		API Spec 2W Grade 50T (to 1 in. thick)	50	70
		API Spec 2W Grade 50T (over 1 in. thick)	50	70
		API Spec 2Y Grade 42 (to 1 in. thick)	42	62
		API Spec 2Y Grade 42 (over 1 in. thick)	42	62
		API Spec 2Y Grade 50 (to 1 in. thick)	50	65
		API Spec 2Y Grade 50 (over 1 in. thick)	50	65
		API Spec 2Y Grade 50T (to 1 in. thick)	50	70
		API Spec 2Y Grade 50T (over 1 in. thick)	50	70
		ASTM A131 Grades DH32, EH32	45.5	68
		ASTM A131 Grades DH36, EH36	51	71
		ASTM A537 Class I (to 2 1/2 in. thick)	50	70
		ASTM A633 Grade A	42	63
		ASTM A633 Grades C, D	50	70
		ASTM A678 Grade A	50	70
III	A	ASTM A537 Class II (to 2 1/2 in. thick)	60	80
		ASTM A678 Grade B	60	80
		API Spec 2W Grade 60 (to 1 in. thick)	60	75
		API Spec 2W Grade 60 (over 1 in. thick)	60	75
		ASTM A710 Grade A Class 3 (to 2 in. thick)	75	85
		ASTM A710 Grade A Class 3 (2 in. to 4 in. thick)	65	75
		ASTM A710 Grade A Class 3 (over 4 in. thick)	60	70

[a] 1 ksi = 6.895 MPa

Step 1.   Calculate the value of factor A using the following equations.   The terms F_{xe}, F_{he}, and F_{ve} are defined in subsequent paragraphs.

$$A = \frac{F_{xe}}{E} \qquad A = \frac{F_{he}}{E} \qquad A = \frac{F_{ve}}{E}$$

Step 2. Using the value of A calculated in Step 1, enter the applicable material chart in Subpart 3 of [4] for the material under consideration. Move vertically to an intersection with the material temperature line for the design temperature. Use interpolation for intermediate temperature values.

Step 3. From the intersection obtained in Step 2, move horizontally to the right to obtain the value of B. E_t is given by the following equation:

$$E_t = \frac{2B}{A}$$

When values of A fall to the left of the applicable material/temperature line in Step 2, $E_t = E$.

Step 4. Calculate the allowable stresses from the following equations:

$$F_{xa} = \frac{F_{xe}}{FS} \frac{E_t}{E} \quad F_{ba} = F_{xa} \quad F_{ha} = \frac{F_{he}}{FS} \frac{E_t}{E} \quad F_{va} = \frac{F_{ve}}{FS} \frac{E_t}{E}$$

11.1.7 Geometries, Failure Modes, and Loads

Allowable stress equations are given for the following geometries and load conditions.

Geometries

1. Unstiffened Cylindrical, Conical, and Spherical Shells
2. Ring Stiffened Cylindrical and Conical Shells
3. Unstiffened Spherical, Ellipsoidal, and Torispherical Heads

The cylinder and cone geometries are illustrated in Figures 11.1 and 11.3 and the stiffener geometries are illustrated in Figure 11.4. The effective sections for ring stiffeners are shown in Figure 11.2. The maximum cone angle α shall not exceed 60°.

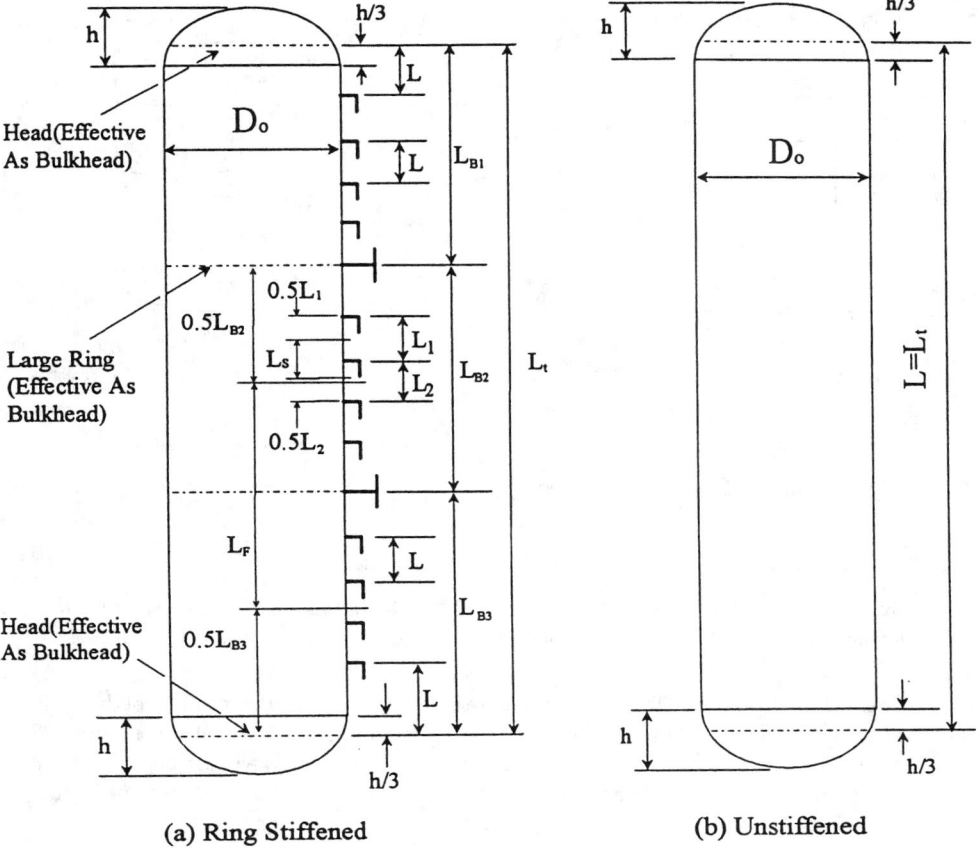

(a) Ring Stiffened   (b) Unstiffened

Figure 11.1   Geometry of cylinders.

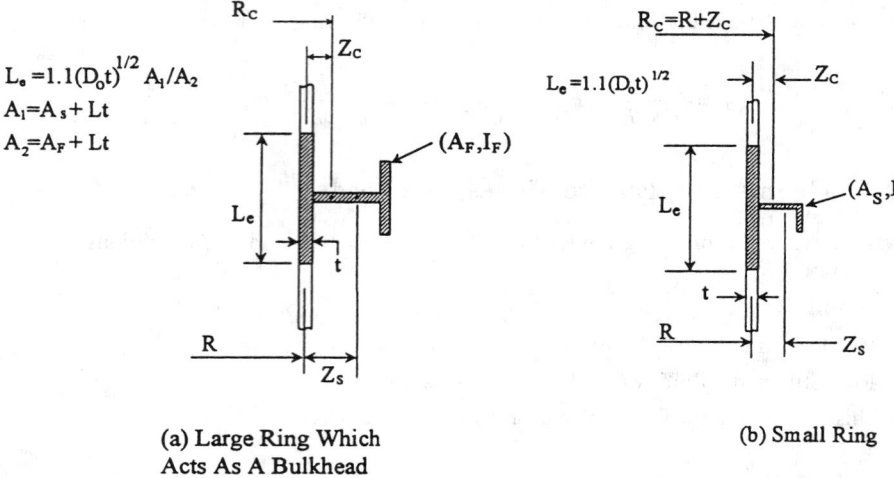

(a) Large Ring Which
Acts As A Bulkhead

(b) Small Ring

Figure 11.2   Sections through rings.

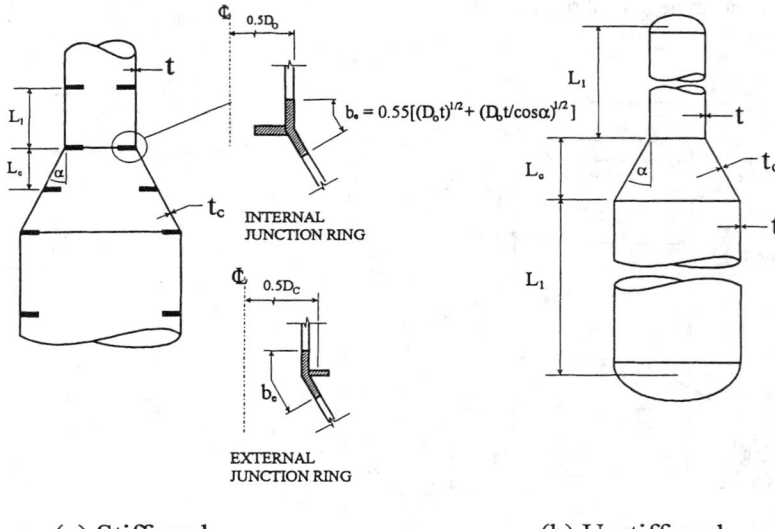

(a) Stiffened

(b) Unstiffened

Figure 11.3   Geometry of conical sections.

Failure Modes

Buckling stress equations are given herein for four failure modes that are defined below. The buckling patterns are both load and geometry dependent.

1. **Local Shell Buckling**—This mode of failure is characterized by the buckling of the shell in a radial direction. One or more waves will form in the longitudinal and circumferential directions. The number of waves and the shape of the waves are dependent on the geometry of the shell and the type of load applied. For ring stiffened shells, the stiffening rings are presumed to remain round prior to buckling.

2. **General Instability**—This mode of failure is characterized by the buckling of one or more rings together with the shell into a circumferential wave pattern with two or more waves.

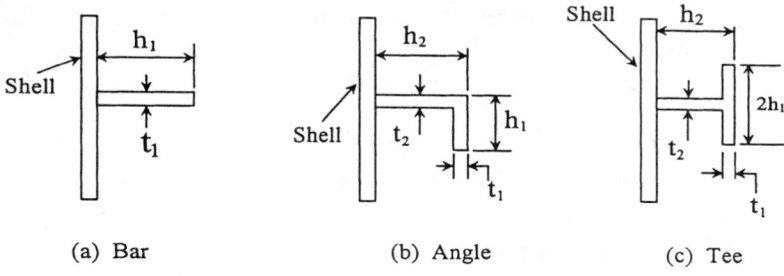

(a) Bar   (b) Angle   (c) Tee

Figure 11.4 Stiffener geometry.

3. **Column Buckling**—This mode of failure is characterized by out-of-plane buckling of the cylinder with the shell remaining circular prior to column buckling. The interaction between shell buckling and column buckling is taken into account by substituting the shell buckling stress for the yield stress in the column buckling formula.

4. **Local Buckling of Rings**—This mode of failure relates to the buckling of the stiffener elements such as the web and flange of a tee type stiffener. Most design rules specify requirements for compact sections to preclude this mode of failure. Very little analytical or experimental work has been done for this mode of failure in association with shell buckling.

Loads and Load Combinations

Allowable stress equations are given for the following types of stresses.

a. Cylinders and Cones

1. Uniform longitudinal compressive stresses
2. Longitudinal compressive stresses due to a bending moment acting across the full circular cross-section
3. Circumferential compressive stresses due to external pressure or other applied loads
4. In-plane shear stresses
5. Any combination of 1, 2, 3, and 4

b. Spherical Shells and Formed Heads

1. Equal biaxial stresses—both stresses are compressive
2. Unequal biaxial stresses—both stresses are compressive
3. Unequal biaxial stresses—one stress is tensile and the other is compressive

11.1.8 Buckling Design Method

The buckling strength formulations presented in this report are based on classical linear theory which is modified by reduction factors that account for the effects of imperfections, boundary conditions, nonlinearity of material properties, and residual stresses. The reduction factors are determined from approximate lower bound values of test data of shells with initial imperfections representative of the tolerance limits specified in this chapter. The validation of the knockdown factors is given in [7], [8], [9], and [10].

11.1.9 Stress Factor

The allowable stresses are determined by applying a stress factor, FS, to the predicted buckling stresses. The recommended values of FS are 2.0 when the buckling stress is elastic and 5/3 when the buckling stress equals the yield stress. A linear variation shall be used between these limits. The equations for FS are given below.

$$FS = 2.0 \quad \text{if } F_{ic} \leq 0.55F_y \tag{11.1a}$$

$$FS = 2.407 - 0.741F_{ic}/F_y \text{ if } 0.55F_y < F_{ic} < F_y \tag{11.1b}$$

$$FS = 1.667 \quad \text{if } F_{ic} = F_y \tag{11.1c}$$

F_{ic} is the predicted buckling stress, which is determined by letting $FS = 1$ in the allowable stress equations. For combinations of earthquake load or wind load with other loads, the allowable stresses may be increased by a factor of 4/3.

11.1.10 Nomenclature

Note: The terms not defined here are uniquely defined in the sections in which they are first used.

A = cross-sectional area of cylinder $A = \pi(D_o - t)t$, in.^{2}
A_S = cross-sectional area of a ring stiffener, in.^{2}
A_F = cross-sectional area of a large ring stiffener which acts as a bulkhead, in.^{2}
D_i = inside diameter of cylinder, in.
D_o = outside diameter of cylinder, in.
D_L = outside diameter at large end of cone, in.
D_S = outside diameter at small end of cone, in.
E = modulus of elasticity of material at design temperature, ksi
E_t = tangent modulus of material at design temperature, ksi
f_a = axial compressive membrane stress resulting from applied axial load, Q, ksi
f_b = axial compressive membrane stress resulting from applied bending moment, M, ksi
f_h = hoop compressive membrane stress resulting from applied external pressure, P, ksi
f_q = axial compressive membrane stress resulting from pressure load, Q_p, on the end of a cylinder, ksi.
f_v = shear stress from applied loads, ksi
f_x = $f_a + f_q$, ksi
F_{ba} = allowable axial compressive membrane stress of a cylinder due to bending moment, M, in the absence of other loads, ksi
F_{ca} = allowable compressive membrane stress of a cylinder due to axial compression load with $\lambda_c > 0.15$, ksi
F_{bha} = allowable axial compressive membrane stress of a cylinder due to bending in the presence of hoop compression, ksi
F_{hba} = allowable hoop compressive membrane stress of a cylinder in the presence of longitudinal compression due to a bending moment, ksi
F_{he} = elastic hoop compressive membrane failure stress of a cylinder or formed head under external pressure alone, ksi
F_{ha} = allowable hoop compressive membrane stress of a cylinder or formed head under external pressure alone, ksi
F_{hva} = allowable hoop compressive membrane stress in the presence of shear stress, ksi
F_{hxa} = allowable hoop compressive membrane stress of a cylinder in the presence of axial compression, ksi
F_{ta} = allowable tension stress, ksi
F_{va} = allowable shear stress of a cylinder subjected only to shear stress, ksi
F_{ve} = elastic shear buckling stress of a cylinder subjected only to shear stress, ksi
F_{vha} = allowable shear stress of a cylinder subjected to shear stress in the presence of hoop compression, ksi

F_{xa} = allowable compressive membrane stress of a cylinder due to axial compression load with $\lambda_c \leq 0.15$, ksi

F_{xc} = inelastic axial compressive membrane failure (local buckling) stress of a cylinder in the absence of other loads, ksi

F_{xe} = elastic axial compressive membrane failure (local buckling) stress of a cylinder in the absence of other loads, ksi

F_{xha} = allowable axial compressive membrane stress of a cylinder in the presence of hoop compression, ksi

F_y = minimum specified yield stress of material, ksi

F_u = minimum specified tensile stress of material, ksi

FS = stress factor

I_s' = moment of inertia of ring stiffener plus effective length of shell about centroidal axis of combined section, in.[4]

$$I_s' = I_s + A_s Z_s^2 \frac{L_e t}{A_s + L_e t} + \frac{L_e t^3}{12}$$

K = effective length factor for column buckling

I_s = moment of inertia of ring stiffener about its centroidal axis, in.[4]

L = design length of a vessel section between lines of support, in. A line of support is:

  1. a circumferential line on a head (excluding conical heads) at one-third the depth of the head from the head tangent line as shown in Figure 11.1

  2. a stiffening ring that meets the requirements of Equation 11.17

L_B = length of cylinder between bulkheads or large rings designed to act as bulkheads, in.

L_c = unbraced length of member, in.

L_e = effective length of shell, in. (see Figure 11.2)

L_F = one-half of the sum of the distances, L_B, from the center line of a large ring to the next large ring or head line of support on either side of the large ring, in. (see Figure 11.1)

L_s = one-half of the sum of the distances from the center line of a stiffening ring to the next line of support on either side of the ring, measured parallel to the axis of the cylinder, in. A line of support is described in the definition for L (see Figure 11.1).

L_t = overall length of vessel as shown in Figure 11.1, in.

M = applied bending moment across the vessel cross-section, in.-kips

M_s = $L_s/\sqrt{R_o t}$

M_x = $L/\sqrt{R_o t}$

P = applied external pressure, ksi

P_a = allowable external pressure in the absence of other loads, ksi

Q = applied axial compression load, kips

Q_p = axial compression load on end of cylinder resulting from applied external pressure, kips

R = radius to centerline of shell, in.

R_c = radius to centroid of combined ring stiffener and effective length of shell, in. $R_c = R + Z_c$

R_o = radius to outside of shell, in.

t = thickness of shell, less corrosion allowance, in.

t_c = thickness of cone, less corrosion allowance, in.

Z_c = radial distance from centerline of shell to centroid of combined section of ring and effective length of shell, in. $Z_c = \frac{A_s Z_s}{A_s + L_e t}$

Z_s = radial distance from center line of shell to centroid of ring stiffener (positive for outside rings), in.

S = elastic section modulus of full shell cross-section, in.[3]

$$S = \frac{\pi \left(D_o^4 - D_i^4 \right)}{32 D_o}$$

r =   radius of gyration of cylinder, in.

$$r = \frac{(D_o^2 + D_i^2)^{1/2}}{4}$$

λ_c =   slenderness factor

$$\lambda_c = \frac{KL_c}{\pi r} \left(\frac{F_{xa} \cdot FS}{E} \right)^{1/2}$$

11.2   Allowable Compressive Stresses for Cylindrical Shells

The maximum allowable stresses for cylindrical shells subjected to loads that produce compressive stresses are given by the following equations.

11.2.1   Uniform Axial Compression

Allowable longitudinal stress for a cylindrical shell under uniform axial compression is given by F_{xa} for values of $\lambda_c \le 0.15$ and by F_{ca} for values of $\lambda_c > 0.15$. F_{xa} is the smaller of the values given by Equations 11.3 and Equation 11.4.

$$\lambda_c = \frac{KL_c}{\pi r} \left(\frac{F_{xa} \cdot FS}{E} \right)^{1/2} \tag{11.2}$$

where KL_c is the effective length. L_c is the unbraced length. Recommended values for K [1] are 2.1 for members with one end free and the other end fixed, 1.0 for members with both ends pinned, 0.8 for members with one end pinned and the other end fixed, and 0.65 for members with both ends fixed.

Local Buckling (For $\lambda_c \le 0.15$)

$$F_{xa} = \frac{F_y}{FS} \quad \text{for} \quad \frac{D_o}{t} \le 135 \tag{11.3a}$$

$$F_{xa} = \frac{466 F_y}{\left(331 + \frac{D_o}{t} \right) FS} \quad \text{for} \quad 135 < \frac{D_o}{t} < 600 \tag{11.3b}$$

$$F_{xa} = \frac{0.5 F_y}{FS} \quad \text{for} \quad \frac{D_o}{t} \ge 600 \tag{11.3c}$$

or

$$F_{xa} = \frac{F_{xe}}{FS} \tag{11.4}$$

where

$$F_{xe} = \frac{C_x E_t}{D_o} \tag{11.5}$$

$$C_x = \frac{409\bar{c}}{389 + \frac{D_o}{t}} \quad \text{not to exceed } 0.9 \text{ for} \quad \frac{D_o}{t} < 1247$$

$$C_x = 0.25\bar{c} \quad \text{for} \quad \frac{D_o}{t} \ge 1247$$

$$\bar{c} = 2.64 \quad \text{for} \quad M_x \le 1.5$$

$$\bar{c} = \frac{3.13}{M_x^{0.42}} \text{ for } 1.5 < M_x < 15$$

$$\bar{c} = 1.0 \text{ for } M_x \geq 15$$

$$M_x = \frac{L}{(R_o t)^{1/2}} \tag{11.6}$$

Column Buckling (For $\lambda_c > 0.15$)

$$F_{ca} = F_{xa} \text{ for } \lambda_c \leq 0.15 \tag{11.7a}$$

$$F_{ca} = F_{xa} [1 - 0.74 (\lambda_c - 0.15)]^{0.3} \text{ for } 0.15 < \lambda_c < \sqrt{2} \tag{11.7b}$$

$$F_{ca} = \frac{0.88 F_{xa}}{\lambda_c^2} \text{ for } \lambda_c \geq \sqrt{2} \tag{11.7c}$$

11.2.2 Axial Compression Due to Bending Moment

Allowable longitudinal stress for a cylinder subjected to a bending moment acting across the full circular cross-section is given by F_{ba}.

$$F_{ba} = F_{xa} \text{ for } \frac{D_o}{t} \geq 135 \tag{11.8a}$$

$$F_{ba} = \frac{466 F_y}{FS \left(331 + \frac{D_o}{t}\right)} \text{ for } 100 \leq \frac{D_o}{t} < 135 \tag{11.8b}$$

$$F_{ba} = \frac{1.081 F_y}{FS} \text{ for } \frac{D_o}{t} < 100 \text{ and } \gamma \geq 0.11 \tag{11.8c}$$

$$F_{ba} = \frac{(1.4 - 2.9\gamma) F_y}{FS} \text{ for } \frac{D_o}{t} < 100 \text{ and } \gamma < 0.11 \tag{11.8d}$$

where F_{xa} is the smaller of the values given by Equations 11.3 and 11.4 and $\gamma = \frac{F_y D_o}{Et}$.

11.2.3 External Pressure

The allowable circumferential compressive stress for a cylinder under external pressure is given by F_{ha} and the allowable external pressure is given by the following equations:

$$P_a = 2 F_{ha} \frac{t}{D_o} \tag{11.9}$$

$$F_{ha} = \frac{F_y}{FS} \text{ for } \frac{F_{he}}{F_y} \geq 2.439 \tag{11.10a}$$

$$F_{ha} = \frac{0.7 F_y}{FS} \left(\frac{F_{he}}{F_y}\right)^{0.4} \text{ for } 0.552 < \frac{F_{he}}{F_y} < 2.439 \tag{11.10b}$$

$$F_{ha} = \frac{F_{he}}{FS} \text{ for } \frac{F_{he}}{F_y} \leq 0.552 \tag{11.10c}$$

where

$$F_{he} = \frac{1.6 C_h E t}{D_o} \tag{11.11}$$

$$C_h = 0.55\frac{t}{D_o} \quad \text{for} \quad M_x \geq 2\left(\frac{D_o}{t}\right)^{0.94}$$

$$C_h = 1.12M_x^{-1.058} \quad \text{for} \quad 13 < M_x < 2\left(\frac{D_o}{t}\right)^{0.94}$$

$$C_h = \frac{0.92}{M_x - 0.579} \quad \text{for} \quad 1.5 < M_x \leq 13$$

$$C_h = 1.0 \quad \text{for} \quad M_x \leq 1.5$$

11.2.4   Shear

Allowable in-plane shear stress for a cylindrical shell is given by F_{va}.

$$F_{va} = \frac{\eta_v F_{ve}}{FS} \tag{11.12}$$

where

$$F_{ve} = \frac{\alpha_v C_v E t}{D_o} \tag{11.13}$$

$$C_v = 4.454 \quad \text{for} \quad M_x \leq 1.5 \tag{11.14a}$$

$$C_v = \left(\frac{9.64}{M_x^2}\right)\left(1 + 0.0239M_x^3\right)^{1/2} \quad \text{for} \quad 1.5 < M_x < 26 \tag{11.14b}$$

$$C_v = \frac{1.492}{M_x^{1/2}} \quad \text{for} \quad 26 \leq M_x < 4.347\frac{D_o}{t} \tag{11.14c}$$

$$C_v = 0.716\left(\frac{t}{D_o}\right)^{1/2} \quad \text{for} \quad M_x \geq 4.347\frac{D_o}{t} \tag{11.14d}$$

$$\alpha_v = 0.8 \quad \text{for} \quad \frac{D_o}{t} \leq 500$$

$$\alpha_v = 1.389 - 0.218\log_{10}\left(\frac{D_o}{t}\right) \quad \text{for} \quad 500 < \frac{D_o}{t} \leq 1000$$

$$\eta_v = 1.0 \quad \text{for} \quad \frac{F_{ve}}{F_y} \leq 0.48$$

$$\eta_v = 0.43\frac{F_y}{F_{ve}} + 0.1 \quad \text{for} \quad 0.48 < \frac{F_{ve}}{F_y} < 1.7$$

$$\eta_v = 0.6\frac{F_y}{F_{ve}} \quad \text{for} \quad \frac{F_{ve}}{F_y} \geq 1.7$$

11.2.5   Sizing of Rings (General Instability)

Uniform Axial Compression and Axial Compression Due to Bending

When ring stiffeners are used to increase the allowable longitudinal compressive stress, the following equations must be satisfied. If $M_x \geq 15$, stiffener spacing is too large to be effective.

$$A_s \geq \left[\frac{0.334}{M_s^{0.6}} - 0.063\right]L_s t \quad \text{and} \quad A_s \geq 0.06L_s t \tag{11.15}$$

$$\text{also} \quad I'_s \geq \frac{5.33 L_s t^3}{M_s^{1.8}} \tag{11.16}$$

External Pressure

(a) Small Rings

$$I'_s \geq \frac{1.5 F_{he} L_s R_c^2 t}{E(n^2 - 1)} \tag{11.17}$$

F_{he} = stress determined from Equation 11.11 with $M_x = M_s$.

$$n^2 = \frac{2 D_o^{3/2}}{3 L_B t^{1/2}} \quad \text{and} \quad 4 \leq n^2 \leq 100$$

(b) Large Rings Which Act As Bulkheads

$$I'_s \geq I_F \quad \text{where} \quad I_F = \frac{F_{heF} L_F R_c^2 t}{2E} \tag{11.18}$$

I_F = the value of I'_s which makes a large stiffener act as a bulkhead. The effective length of shell is $L_e = 1.1\sqrt{D_o t}(A_1/A_2)$

A_1 = cross-sectional area of small ring plus shell area equal to $L_s t$, in.^{2}

A_2 = cross-sectional area of large ring plus shell area equal to $L_s t$, in.^{2}

R_c = radius to centroid of combined large ring and effective width of shell, in.

F_{heF} = average value of the hoop buckling stresses, F_{he}, over length L_F where F_{he} is determined from Equation 11.11, ksi

Shear

$$I'_s \geq 0.184 C_v M_s^{0.8} t^3 L_s \tag{11.19}$$

C_v = value determined from Equation 11.14 with $M_x = M_s$.

Local Stiffener Buckling

To preclude local buckling of the stiffener prior to shell buckling, the following stiffener properties shall be met. See Figure 11.4 for stiffener geometry.

(a) Flat Bar Stiffener, Flange of a Tee Stiffener, and Outstanding Leg of an Angle Stiffener

$$\frac{h_1}{t_1} \leq 0.375 \left(\frac{E}{F_y}\right)^{1/2} \tag{11.20}$$

where h_1 is the full width of a flat bar stiffener or outstanding leg of an angle stiffener and one-half of the full width of the flange of a tee stiffener and t_1 is the thickness of the bar, leg of angle, or flange of tee.

(b) Web of Tee Stiffener or Leg of Angle Stiffener Attached to Shell

$$\frac{h_2}{t_2} \leq 1.0 \left(\frac{E}{F_y}\right)^{1/2} \tag{11.21}$$

where h_2 is the full depth of a tee section or full width of an angle leg and t_2 is the thickness of the web or angle leg.

11.3   Allowable Compressive Stresses For Cones

Unstiffened conical transitions or cone sections between rings of stiffened cones with an angle $\alpha \leq 60°$ shall be designed for local buckling as an equivalent cylinder according to the following procedure. See Figure 11.3 for cone geometry.

11.3.1   Uniform Axial Compression and Axial Compression Due to Bending

Allowable Longitudinal and Bending Stresses

Assume an equivalent cylinder with diameter $D_e = D / \cos \alpha$, where D is the outside diameter of the cone at the cross-section under consideration and length equal to L_c. D_e is substituted for D_o in Equations 11.3 to Equations 11.8 to find F_{xa} and F_{ba} and L_c for L in Equation 11.6. The allowable stress must be satisfied at all cross-sections along the length of the cone.

Unstiffened Cone-Cylinder Junctions

Cone-cylinder junctions are subject to unbalanced radial forces (due to axial load and bending moment) and to localized bending stresses caused by the angle change. The longitudinal and hoop stresses at the junction may be evaluated as follows:

Longitudinal Stress—In lieu of detailed analysis, the localized bending stress at an unstiffened cone-cylinder junction may be estimated by the following equation.

$$f_b' = \frac{0.6t\sqrt{D\,(t + t_c)}}{t_e^2}\,(f_x + f_b)\tan \alpha \qquad (11.22)$$

where
$$
\begin{aligned}
D &= \text{outside diameter of cylinder at junction to cone} \\
t &= \text{thickness of cylinder} \\
t_c &= \text{thickness of cone} \\
t_e &= t \text{ to find stress in cylinder section} \\
t_e &= t_c \text{ to find stress in cone section} \\
\alpha &= \text{cone angle as defined in Figure 11.3} \\
f_x &= \text{uniform longitudinal stress in cylinder section at junction resulting from axial loads} \\
f_b &= \text{longitudinal stress in cylinder section at junction resulting from bending moment}
\end{aligned}
$$

For strength requirements, the total stress $(f_x + f_b + f_b')$ shall be limited to the minimum tensile strength given in Table 11.1 or Table U, Subpart 1 of [4] for the cone and cylinder material and $f_x + f_b$ shall be less than the allowable tensile stress F_t, where F_t is the smaller of $0.6F_y$ or $F_u/3$.

Hoop Stress—The hoop stress caused by the unbalanced radial line load may be estimated from:

$$f_h' = 0.45\sqrt{D/t}\,(f_x + f_b)\tan \alpha \qquad (11.23)$$

For hoop tension, f_h' shall be limited to the tensile allowable. For hoop compression, f_h' shall be limited to F_{ha} where F_{ha} is computed from Equation 11.10 with $F_{he} = 0.4Et/D$.

A cone-cylinder junction that does not satisfy the above criteria may be strengthened either by increasing the cylinder and cone wall thicknesses at the junction, or by providing a stiffening ring at the junction.

Cone-Cylinder Junction Rings

If stiffening rings are required, the section properties shall satisfy the following requirements:

$$A_c \;\geq\; \frac{tD}{F_y}(f_x + f_b)\tan \alpha \qquad (11.24)$$

$$I_c \geq \frac{tDD_c^2}{8E}(f_x + f_b)\tan\alpha \tag{11.25}$$

where

D = cylinder outside diameter at junction
D_c = diameter to centroid of composite ring section for external rings
D_c = D for internal rings
A_c = cross-sectional area of composite ring section
I_c = moment of inertia of composite ring section

In computing A_c and I_c the effective length of the shell wall acting as a flange for the composite ring section shall be computed from:

$$b_e = 0.55\left(\sqrt{D/t} + \sqrt{Dt_c/\cos\alpha}\right) \tag{11.26}$$

11.3.2 External Pressure

Allowable Circumferential Compression Stresses

Assume an equivalent cylinder with diameter $D_e = 0.5(D_L+D_S)$ and length $L_e = L_c/\cos\alpha$. This length and diameter shall be substituted into Equations 11.10 and 11.11 to determine F_{ha}.

Intermediate Stiffening Rings

If required, circumferential stiffening rings within cone transitions shall be sized using Equation 11.17 with $R_c = D/2$ where D is the cone diameter at the ring, t is the cone thickness, L_s is the average distance to adjacent rings along the cone axis, and F_{he} is the average of the elastic hoop buckling stress values computed for the two adjacent bays by the method given in the preceding paragraph.

Cone-Cylinder Junction Rings

A junction ring is not required for buckling due to external pressure if $f_h < F_{ha}$ where F_{ha} is determined from Equation 11.10 with F_{he} computed using C_h equal to $0.55\,(\cos\alpha)(t/D)$ in Equation 11.11. D is the cylinder diameter at the junction.

Circumferential stiffening rings required at the cone-cylinder junctions shall be sized such that the moment of inertia of the composite ring section satisfies the following equation:

$$I_c \geq \frac{D^2}{16E}\left\{tL_1F_{he} + \frac{t_cL_cF_{hec}}{\cos^2\alpha}\right\} \tag{11.27}$$

where

D = cylinder outside diameter at junction
L_c = distance to first stiffening ring in cone section along cone axis
L_1 = distance to first stiffening ring in cylinder section or line of support
F_{he} = elastic hoop buckling stress for cylinder (see Equation 11.11)
F_{hec} = F_{he} for cone section treated as an equivalent cylinder
t = cylinder thickness
t_c = cone thickness

11.3.3 Shear

Allowable In-Plane Shear Stress

Assume an equivalent cylinder with a length equal to the slant length of the cone between rings $(L_c/\cos\alpha)$ and a diameter $D_e = D/\cos\alpha$, where D is the outside diameter of the cone at the cross-section under consideration. This length and diameter shall be substituted into Equations 11.12 to 11.14 to determine F_{va}.

Intermediate Stiffening Rings

If required, circumferential stiffening rings within cone transition shall be sized using Equation 11.19 where L_s is the average distance to adjacent rings along the cone axis.

11.3.4 Local Stiffener Buckling

To preclude local buckling of a stiffener, the requirements of Equations 11.20 and 11.21 must be met.

11.4 Allowable Stress Equations For Unstiffened and Ring-Stiffened Cylinders and Cones Under Combined Loads

11.4.1 For Combination of Uniform Axial Compression and Hoop Compression

For $\lambda_c \leq 0.15$

The allowable stress in the longitudinal direction is given by F_{xha} and the allowable stress in the circumferential direction is given by F_{hxa}.

$$F_{xha} = \left(\frac{1}{F_{xa}^2} - \frac{C_1}{C_2 F_{xa} F_{ha}} + \frac{1}{C_2^2 F_{ha}^2} \right)^{-0.5} \tag{11.28}$$

where

$$C_1 = \frac{F_{xa} \cdot FS + F_{ha} \cdot FS}{F_y} - 1.0 \text{ and } C_2 = \frac{f_x}{f_h}$$

$$f_x = f_a + f_q = \frac{Q}{A} + \frac{Q_p}{A} \text{ and } f_h = \frac{PD_o}{2t}$$

$F_{xa} \cdot FS$ is given by the smaller of Equation 11.3 or 11.4, and $F_{ha} \cdot FS$ is given by Equation 11.10.

$$F_{hxa} = \frac{F_{xha}}{C_2} \tag{11.29}$$

For $0.15 < \lambda_c < 1.2$

F_{xha} is the smaller of F_{ah1} and F_{ah2} where $F_{ah1} = F_{xha}$ given by Equation 11.28 with $f_x = f_a$ and F_{ah2} is given by the following equation.

$$F_{ah2} = F_{ca} \left(1 - \frac{f_q}{F_y} \right) \tag{11.30}$$

F_{ca} is given by Equation 11.7.

11.4.2 For Combination of Axial Compression Due to Bending Moment, M, and Hoop Compression

The allowable stress in the longitudinal direction is given by F_{bha} and the allowable stress in the circumferential direction is given by F_{hba}.

$$F_{bha} = C_3 C_4 F_{ba} \tag{11.31}$$

where C_3 and C_4 are given by the following equations and F_{ba} is given by Equation 11.8.

$$C_4 = \frac{f_b}{f_h} \frac{F_{ha}}{F_{ba}}$$

$$C_3^2 \left(C_4^2 + 0.6 C_4 \right) + C_3^{2n} - 1 = 0 \qquad (11.32)$$

$$f_b = \frac{M}{S} \qquad f_h = \frac{P D_o}{2t} \qquad n = 5 - 4 \frac{F_{ha} \cdot FS}{F_y}$$

Solve for $C3$ from Equation 11.31 by iteration. $F_{ha} \cdot FS$ is given by Equation 11.10.

$$F_{hba} = F_{bha} \frac{f_h}{f_b} \qquad (11.33)$$

11.4.3 For Combination of Hoop Compression and Shear

The allowable shear stress is given by F_{vha} and the allowable circumferential stress is given by F_{hva}.

$$F_{vha} = \left[\left(\frac{F_{va}^2}{2 C_5 F_{ha}} \right)^2 + F_{va}^2 \right]^{1/2} - \frac{F_{va}^2}{2 C_5 F_{ha}} \qquad (11.34)$$

where $C_5 = \frac{f_v}{f_h}$ and F_{va} is given by Equation 11.12 and F_{ha} is given by Equation 11.10.

$$F_{hva} = \frac{F_{vha}}{C_5} \qquad (11.35)$$

11.4.4 For Combination of Uniform Axial Compression, Axial Compression Due to Bending Moment, M, and Shear, in the Presence of Hoop Compression, $(f_h \neq 0)$

$$\text{Let } K_s = 1 - \left(\frac{f_v}{F_{va}} \right)^2 \qquad (11.36)$$

For $\lambda_c \leq 0.15$

$$\left(\frac{f_a}{K_s F_{xha}} \right)^{1.7} + \frac{f_b}{K_s F_{bha}} \leq 1.0 \qquad (11.37)$$

F_{xha} is given by Equation 11.28, F_{bha} is given by Equation 11.30 and F_{va} is given by Equation 11.12.
For $0.15 < \lambda_c < 1.2$

$$\frac{f_a}{F_{xha}} + \frac{8}{9} \frac{\Delta f_b}{F_{bha}} \leq 1.0 \quad \text{for} \quad \frac{f_a}{F_{xha}} \geq 0.2 \qquad (11.38)$$

where

$$\Delta = \frac{C_m}{1 - f_a \cdot FS/F_e} \qquad F_e = \frac{\pi^2 E}{(K L_c / r)^2}$$

See Equation 11.2 for $K L_c$ and Equation 11.30 for F_{xha}. F_{bha} is given by Equation 11.31. FS is determined from Equation 11.1 where $F_{ic} = F_{xa} \cdot FS$ (see Equations 11.3 and 11.4). C_m is a coefficient whose value shall be taken as follows [1]:

1. For compression members in frames subject to joint translation (sidesway),

$$C_m = 0.85.$$

2. For rotationally restrained compression members in frames braced against joint translation and not subject to transverse loading between their supports in the plane of bending,

$$C_m = 0.6 - 0.4(M_1/M_2)$$

where M_1/M_2 is the ratio of the smaller to larger moments at the ends of that portion of the member that is unbraced in the plane of bending under consideration. M_1/M_2 is positive when the member is bent in reverse curvature and negative when bent in single curvature.

3. For compression members in frames braced against joint translation and subjected to transverse loading between their supports the following apply:

   a. for members whose ends are restrained against rotation in the plane of bending,

$$C_m = 0.85$$

   b. for members whose ends are unrestrained against rotation in the plane of bending,

$$C_m = 1.0$$

11.4.5 For Combination of Uniform Axial Compression, Axial Compression Due to Bending Moment, M, and Shear, in the Absence of Hoop Compression, $(f_h = 0)$

For $\lambda_c \leq 0.15$

$$\left(\frac{f_a}{K_s F_{xa}}\right)^{1.7} + \frac{f_b}{K_s F_{ba}} \leq 1.0 \tag{11.39}$$

F_{xa} is given by the smaller of Equations 11.3 or 11.4, F_{ba} is given by Equation 11.8 and K_s is given by Equation 11.36.

For $0.15 < \lambda_c < 1.2$

$$\frac{f_a}{K_s F_{ca}} + \frac{8}{9}\frac{\Delta f_b}{K_s F_{ba}} \leq 1.0 \text{ for } \frac{f_a}{K_s F_{ca}} \geq 0.2 \tag{11.40}$$

$$\frac{f_a}{2 K_s F_{ca}} + \frac{\Delta f_b}{K_s F_{ba}} \leq 1.0 \text{ for } \frac{f_a}{K_s F_{ca}} < 0.2 \tag{11.41}$$

F_{ca} is given by Equation 11.7, F_{ba} is given by Equation 11.31, and K_s is given by Equation 11.36. See Equation 11.38 for definition of Δ.

11.5 Tolerances for Cylindrical and Conical Shells

11.5.1 Shells Subjected to Uniform Axial Compression and Axial Compression Due to Bending Moment

The difference between the maximum and minimum diameters at any cross-section shall not exceed 1% of the nominal diameter at the cross-section under consideration. Additionally, the local deviation from a straight line, e, measured along a meridian over a gauge length L_x shall not exceed the maximum permissible deviation e_x given below.

$$e_x = 0.002R$$
$$L_x = 4\sqrt{Rt} \text{ but not greater than } L \text{ for cylinders}$$

L_x = $4\sqrt{Rt/\cos\alpha}$ but not greater than $L_c/\cos\alpha$ for cones

L_x = $25t$ across circumferential welds

Also L_x is not greater than 95% of the meridianal distance between circumferential welds.

11.5.2 Shells Subjected to External Pressure

The difference between the maximum and minimum diameters at any cross-section shall not exceed 1% of the nominal diameter at the cross-section under consideration. Additionally, the maximum deviation from a true circular form, e, shall not exceed the value given by Figure 11.5 or by the following equations.

$$e = 0.0165t\,(M_x + 3.25)^{1.069} \qquad 0.1t \le e \le 0.0242R \tag{11.42}$$

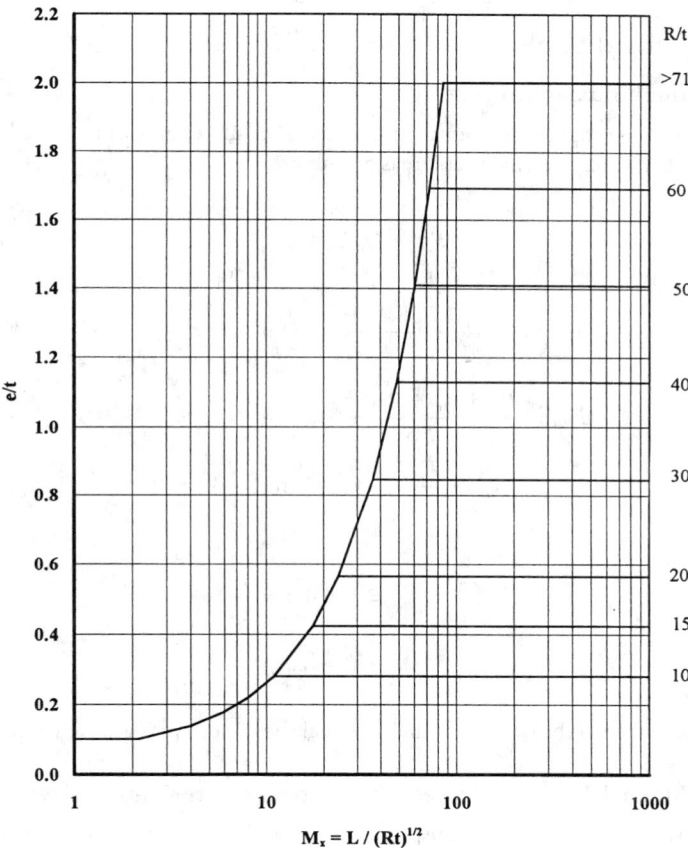

Figure 11.5 Values of e/t which give a buckling pressure of 80% of the theoretical buckling pressure.

Also, e shall not exceed $2t$. Measurements to determine e are made with a gauge or template with the chord length L_c given by the following equation.

$$L_c = 2R\sin(\pi/2n) \tag{11.43}$$

$$n = c\left(\frac{\sqrt{R/t}}{L/R}\right)^d \qquad 2 \le n \le 1.41(R/t)^{0.5} \tag{11.44}$$

where

$$c = 2.28(R/t)^{0.54} \leq 2.80$$
$$d = 0.38(R/t)^{0.044} \leq 0.485$$

11.5.3 Shells Subjected to Shear

The difference between the maximum and minimum diameters at any cross-section shall not exceed 1% of the nominal diameter at the cross-section under consideration.

11.6 Allowable Compressive Stresses for Spherical Shells and Formed Heads, With Pressure on Convex Side

11.6.1 Spherical Shells

With Equal Biaxial Stresses

The allowable compressive stress for a spherical shell under uniform external pressure is given by F_{ha} and the allowable external pressure is given by P_a.

$$F_{ha} = \frac{F_y}{FS} \qquad \text{for } \frac{F_{he}}{F_y} \geq 6.25 \tag{11.45a}$$

$$F_{ha} = \frac{1.31 F_y}{FS\left(1.15 + \frac{F_y}{F_{he}}\right)} \qquad \text{for } 1.6 < \frac{F_{he}}{F_y} < 6.25 \tag{11.45b}$$

$$F_{ha} = \frac{0.18 F_{he} + 0.45 F_y}{FS} \qquad \text{for } 0.55 < \frac{F_{he}}{F_y} \leq 1.6 \tag{11.45c}$$

$$F_{ha} = \frac{F_{he}}{FS} \qquad \text{for } \frac{F_{he}}{F_y} \leq 0.55 \tag{11.45d}$$

$$F_{he} = 0.075 E \frac{t}{R_o} \tag{11.46}$$

$$P_a = 2 F_{ha} \frac{t}{R_o} \tag{11.47}$$

where R_o is the radius to the outside of the spherical shell and F_{ha} is given by Equation 11.45.

With Unequal Biaxial Stresses—Both Stresses Are Compressive

The allowable compressive stresses for a spherical shell subjected to unequal biaxial stresses, σ_1 and σ_2, where both σ_1 and σ_2 are compression stresses resulting from applied loads, are given by the following equations.

$$F_{1a} = \frac{0.6}{1 - 0.4k} F_{ha} \tag{11.48}$$

$$F_{2a} = k F_{1a} \tag{11.49}$$

where $k = \sigma_2/\sigma_1$ and F_{ha} is given by Equation 11.45. F_{1a} is the allowable stress in the direction of σ_1 and F_{2a} is the allowable stress in the direction of σ_2. The larger of the compression stresses is σ_1.

With Unequal Biaxial Stresses—One Stress Is Compressive and the Other Is Tensile

The allowable compressive stress for a spherical shell subjected to unequal biaxial stresses σ_1 and σ_2, where σ_1 is a compression stress and σ_2 is a tensile stress, is given by F_{1a} where F_{1a} is the value of F_{ha} determined from Equation 11.45 with F_{he} given by Equation 11.50.

$$F_{he} = \left(C_o + C_p\right) E \frac{t}{R_o} \tag{11.50}$$

$$C_o = \frac{102.2}{195 + R_o/t} \quad \text{for} \quad \frac{R_o}{t} < 622$$

$$C_o = 0.125 \quad \text{for} \quad \frac{R_o}{t} \geq 622$$

$$C_p = \frac{1.06}{3.24 + \frac{1}{\bar{p}}}$$

$$\bar{p} = \frac{\sigma_2}{E}\frac{R_o}{t}$$

Shear

When shear is present, the principal stresses shall be calculated and used for σ_1 and σ_2.

11.6.2 Toroidal and Ellipsoidal Heads

The allowable compressive stresses for formed heads is determined by the equations given for spherical shells where R_o is defined below.

R_o = the outside radius of the crown portion of the head for torispherical heads, in.
R_o = the equivalent outside spherical radius taken as $K_o D_o$ for ellipsoidal heads, in.
K_o = factor depending on the ellipsoidal head proportions $D_o/2h_o$ (see Table 11.2)
h_o = outside height of the ellipsoidal head measured from the tangent line (head-bend line), in.

TABLE 11.2   Factor K_o

$D_o/2h_o$	...	3.0	2.8	2.6	2.4	2.2
K_o	...	1.36	1.27	1.18	1.08	0.99
$D_o/2h_o$	2.0	1.8	1.6	1.4	1.2	1.0
K_o	0.90	0.81	0.73	0.65	0.57	0.50

Note: Use interpolation for intermediate values.

11.7   Tolerances for Formed Heads

The inner surface of a spherical shell or formed head shall not deviate from the specified shape more than 1.25% of the nominal diameter of the vessel. Such deviations shall be measured perpendicular to the specified shape. Additionally, the maximum local deviation from a true circular form, e, for spherical shells and any spherical portion of a formed head designed for external pressure shall not exceed the shell thickness. Measurements to determine e are made with a gauge or template with the chord length L_c given by the following equation:

$$L_c = 3.72\sqrt{Rt}$$

References

[1] AISC. 1989. *Manual of Steel Construction, Allowable Stress Design,* 9th ed., Section C-C2, American Institute of Steel Construction, Chicago, IL.

[2] API 2U. 1987. API Bulletin 2U, "Bulletin on Stability Design of Cylindrical Shells," 1st ed., American Petroleum Institute, Washington, D.C.

[3] ASME. 1992. ASME Boiler and Pressure Vessel Code, Section VIII, Rules for Construction of Pressure Vessels, Division 1, American Society of Mechanical Engineers, New York.

[4] ASME. 1992. ASME Boiler and Pressure Vessel Code, Section II, Materials, Part D-Properties, American Society of Mechanical Engineers, New York.

[5] AWS. 1992, Structural Welding Code, AWS D1.1-92, American Welding Society.

[6] Miller, C.D. 1982. "Experimental Study of the Buckling of Cylindrical Shells With Reinforced Openings," ASME/ANS Nuclear Engineering Conference, Portland, OR.

[7] Miller, C.D. 1991. "ASME Code Case N-284: Metal Containment Shell Buckling Design Methods," Revision 1, in *Code Cases: Nuclear Components, ASME Boiler and Pressure Vessel Code,* American Society of Mechanical Engineers, New York, March 14, 1995.

[8] Miller, C.D. 1995. *An Evaluation of Codes and Standards Related to Buckling of Cylindrical Shells Subjected to Axial Compression, Bending and External Pressure,* UMI Dissertation Services, Ann Arbor, MI.

[9] Miller, C.D. and Saliklis, E.P. 1993. "Analysis of Cylindrical Shell Database and Validation of Design Formulations," API Project 90-56, Chicago Bridge & Iron Technical Services Co., Plainfield, IL.

[10] Miller, C.D. and Saliklis, E.P. 1995. "Analysis of Cylindrical Shell Database and Validation of Design Formulations, Phase 2: For D/t Values > 300," API Project 92-56, Chicago Bridge & Iron Technical Services Co., Plainfield, IL.

Further Reading

Additional information on the design of shell structures can be found in the following references:

[1] American Iron and Steel Institute, 1992. Steel Plate Engineering Data, Volume 1—Steel Tanks for Liquid Storage and Volume 2—Useful Information on the Design of Plate Structures.

[2] Wozniak, R.S. 1990. Steel Tanks, in *Structural Engineering Handbook,* Gaylord, E.H. and Gaylord, C.S. Eds., 3rd ed., McGraw-Hill, New York, 27-1 to 27-29.

The following is a list of codes, specifications, and standards that provide rules for the design of shell structures subject to instability from loads which produce compressive stresses in the shell elements. A comparison was made by Miller and Saliklis [8, 9, 10] of the predicted failure stresses given by each of these sets of rules with the test data obtained from over 600 tests on steel models representative of fabricated shells. The best fit equations were determined for each shell type and load. These equations were then modified to obtain a better fit with the test database. The equations given in this chapter are the results of these studies.

[3] API BUL 2U. 1987. *Bulletin on Stability Design of Cylindrical Shells,* 1st ed., American Petroleum Institute, Washington, D.C.

[4] API RP 2A-LRFD. 1993. *Recommended Practice for Planning, Designing and Constructing Fixed Offshore Platforms—Load and Resistance Factor Design,* 1st ed., American Petroleum Institute, Washington, D.C.

[5] API RP 2A-WSD. 1993. *Recommended Practice for Planning, Designing and Constructing Fixed Offshore Platforms—Working Stress Design*, 20th ed., American Petroleum Institute, Washington, D.C.

[6] API STD 620. 1990. *Design and Construction of Large, Welded Low-Pressure Storage Tanks*, 8th ed., American Petroleum Institute, Washington, D.C.

[7] API STD 650. 1993. *Welded Steel Tanks for Oil Storage*, 9th ed., American Petroleum Institute, Washington, D.C.

[8] ASME VIII. 1992. *Pressure Vessels, Division 2, ASME Boiler and Pressure Code*, American Society of Mechanical Engineers, Washington, D.C.

[9] AWWA D100. 1984. *AWWA Standard for Welded Steel Tanks for Water Storage*, American Water Works Association, Denver, CO.

[10] DIN 18800. 1990. *Stability of Shell Type Steel Structures*, German Code DIN 18800, Part 4.

[11] ECCS No. 56. 1988. *Buckling of Steel Shells, European Recommendations*, European Convention for Constructional Steelwork, Publication No. 56, 4th ed., Brussels, Belgium.

[12] NPD. 1990. *Buckling Criteria for Cylindrical Shells*, Norwegian Petroleum Directorate, Oslo, Norway.

12

Multistory Frame Structures

J. Y. Richard Liew
and T. Balendra
Department of Civil Engineering,
National University of Singapore,
Singapore, Singapore

W. F. Chen
School of Civil Engineering,
Purdue University,
West Lafayette, IN

12.1 Classification of Building Frames 12-1
Rigid Frames • Simple Frames (Pin-Connected Frames) • Brac-
ing Systems • Braced Frames vs. Unbraced Frames • Sway
Frames vs. Non-Sway Frames • Classification of Tall Building
Frames

12.2 Composite Floor Systems 12-8
Floor Structures in Multistory Buildings • Composite Floor
Systems • Composite Beams and Girders • Long-Span Floor-
ing Systems • Comparison of Floor Spanning Systems • Floor
Diaphragms

12.3 Design Concepts and Structural Schemes 12-23
Introduction • Gravity Frames • Bracing Systems • Moment-
Resisting Frames • Tall Building Framing Systems • Steel-
Concrete Composite Systems

12.4 Wind Effects on Buildings............................ 12-48
Introduction • Characteristics of Wind • Wind Induced Dynamic Forces • Response Due to Along Wind • Response Due
to Across Wind • Torsional Response • Response by Wind Tunnel Tests

12.5 Defining Terms 12-69
References 12-70
Further Reading 12-73

12.1 Classification of Building Frames

For building frame design, it is useful to define various frame systems in order to simplify models of analysis. For example, in the case of a braced frame, it is not necessary to separate frame and bracing behavior because both can be analyzed with a single model. On the other hand, for more complicated three-dimensional structures involving the interaction of different structural systems, simple models are useful for preliminary design and for checking computer results. These models should be able to capture the behavior of individual subframes and their effects on the overall structures.

The remainder of this section attempts to describe what a framed system represents, define when a framed system can be considered to be braced by another system, what is meant by a bracing system, and the difference between sway and non-sway frames. Various structural schemes for tall building construction are also given.

12.1.1 Rigid Frames

A rigid frame derives its lateral **stiffness** mainly from the bending rigidity of frame members interconnected by rigid joints. The joints shall be designed in such a manner that they have adequate strength and stiffness and negligible deformation. The deformation must be small enough to have any significant influence on the distribution of internal forces and moments in the structure or on the overall frame deformation.

A rigid unbraced frame should be capable of resisting lateral loads without relying on an additional bracing system for stability. The frame, by itself, has to resist all the design forces, including gravity as well as lateral forces. At the same time, it should have adequate lateral stiffness against sidesway when it is subjected to horizontal wind or earthquake loads. Even though the detailing of the rigid connections results in a less economic structure, rigid unbraced frame systems have the following benefits:

1. Rigid connections are more ductile and therefore the structure performs better in load reversal situations or in earthquakes.
2. From the architectural and functional points of view, it can be advantageous not to have any triangulated bracing systems or solid wall systems in the building.

12.1.2 Simple Frames (Pin-Connected Frames)

A simple frame refers to a structural system in which the beams and columns are pinned connected and the system is incapable of resisting any lateral loads. The stability of the entire structure must be provided for by attaching the simple frame to some form of bracing system. The lateral loads are resisted by the bracing systems while the gravity loads are resisted by both the simple frame and the bracing system.

In most cases, the lateral load response of the bracing system is sufficiently small such that second-order effects may be neglected for the design of the frames. Thus, the simple frames that are attached to the bracing system may be classified as non-sway frames. Figure 12.1 shows the principal components—simply frame and bracing system—of such a structure.

There are several reasons of adopting pinned connections in the design of steel multistory frames:

1. Pin-jointed frames are easier to fabricate and erect. For steel structures, it is more convenient to join the webs of the members without connecting the flanges.
2. Bolted connections are preferred over welded connections, which normally require weld inspection, weather protection, and surface preparation.
3. It is easier to design and analyze a building structure that can be separated into system resisting vertical loads and system resisting horizontal loads. For example, if all the girders are simply supported between the columns, the sizing of the simply supported girders and the columns is a straightforward task.
4. It is more cost effective to reduce the horizontal drift by means of bracing systems added to the simple framing than to use unbraced frame systems with rigid connections.

Actual connections in structures do not always fall within the categories of pinned or rigid connections. Practical connections are semi-rigid in nature and therefore the pinned and rigid conditions are only idealizations. Modern design codes allow the design of semi-rigid frames using the concept of wind moment design (type 2 connections). In wind moment design, the connection is assumed to be capable of transmitting only part of the bending moments (those due to the wind only). Recent development in the analysis and design of semi-rigid frames can be obtained from Chen et al. [15]. Design guidance is given in Eurocode 3 [22].

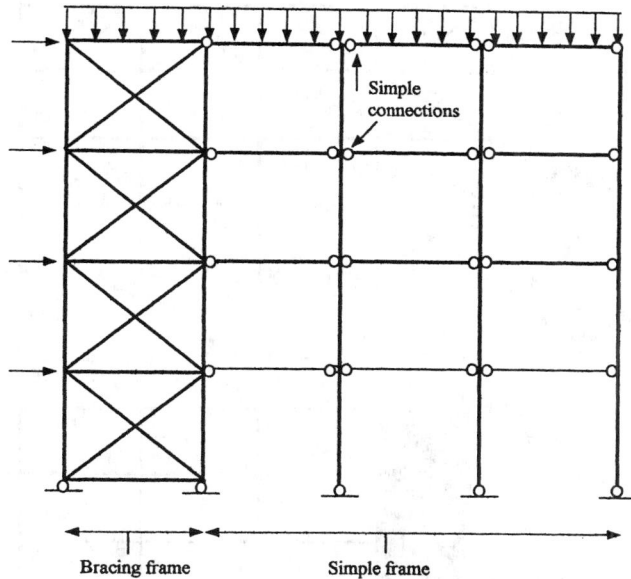

Figure 12.1   Simple braced frame.

12.1.3   Bracing Systems

Bracing systems refer to structures that can provide lateral stability to the overall framework. It may be in the form of triangulated frames, shear wall/cores, or rigid-jointed frames. It is common to find bracing systems represented as shown in Figure 12.2. They are normally located in buildings to accommodate lift shafts and staircases.

In steel structures, it is common to represent a *bracing system* by a triangulated truss because, unlike concrete structures where all the joints are naturally continuous, the most immediate way of making connections between steel members is to hinge one member to the other. As a result, common steel building structures are designed to have bracing systems in order to provide sidesway resistance. Therefore, bracing can only be obtained by use of triangulated trusses (Figure 12.2a) or, exceptionally, by a very stiff structure such as shear wall or core wall (Figure 12.2b). The efficiency of a building to resist lateral forces depends on the location and the types of the bracing systems employed, and the presence or absence of shear walls and cores around lift shafts and stair wells.

12.1.4   Braced Frames vs. Unbraced Frames

The main function of a bracing system is to resist lateral forces. Building frame systems can be separated into vertical load-resistance and horizontal load-resistance systems. In some cases, the vertical load-resistance system also has some capability to resist horizontal forces. It is necessary, therefore, to identify the two sources of resistance and to compare their behavior with respect to the horizontal actions. However, this identification is not that obvious since the bracing is integral within the structure. Some assumptions need to be made in order to define the two structures for the purpose of comparison.

Figures 12.3 and 12.4 represent the structures that are easy to define within one system: two sub-assemblies identifying the bracing system and the system to be braced. For the structure shown in Figure 12.3, there is a clear separation of functions in which the gravity loads are resisted by the hinged subassembly (Frame B) and the horizontal load loads are resisted by the braced assembly (Frame A). In contrast, for the structure in Figure 12.4, since the second sub-assembly (Frame B) is able to resist horizontal actions as well as vertical actions, it is necessary to assume that practically all

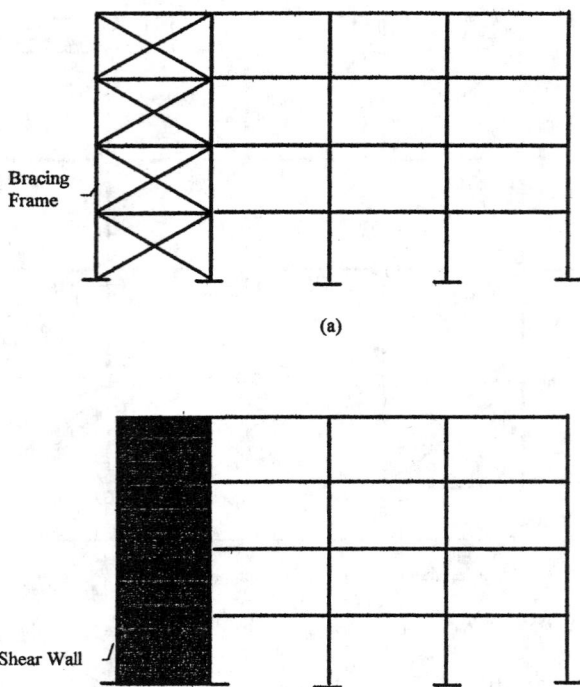

Figure 12.2   Common bracing systems: (a) vertical truss system and (b) shear wall.

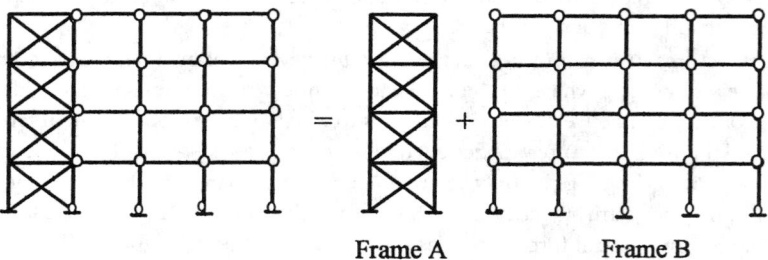

Figure 12.3   Pinned connected frames split into two subassemblies.

the horizontal actions are carried by the first sub-assembly (Frame A) in order to define this system as braced.

Eurocode 3 [22] gives a clear guidance in defining braced and unbraced frames. A frame may be classified as braced if its sway resistance is supplied by a bracing system in which its response to lateral loads is sufficiently stiff for it to be acceptably accurate to assume all horizontal loads are resisted by the bracing system. The frame can be classified as braced if the bracing system reduces its horizontal displacement by at least 80%.

For the frame shown in Figure 12.3, the hinged frame (Frame B) has no lateral stiffness, and Frame A (truss frame) resists all lateral load. In this case, Frame B is considered to be braced by Frame A. For the frame shown in Figure 12.4, Frame B may be considered to be a braced frame if the following deflection criterion is satisfied:

$$\left(1 - \frac{\Delta_A}{\Delta_B}\right) \geq 0.8 \qquad (12.1)$$

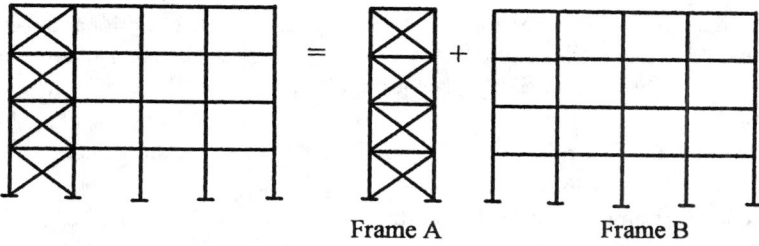

Frame A   Frame B

Figure 12.4   Mixed frames split into two subassemblies.

where

Δ_A = lateral deflection calculated from the truss frame (Frame A) alone

Δ_B = lateral deflection calculated from Frame B alone

Alternatively, the lateral stiffness of Frame A under the applied lateral load should be at least five times larger than that of Frame B:

$$K_A \geq 5K_B \tag{12.2}$$

where

K_A = lateral stiffness of Frame A

K_B = lateral stiffness of Frame B

12.1.5   Sway Frames vs. Non-Sway Frames

The identification of sway frames and non-sway frames in a building is useful for evaluating safety of structures against instability. In the design of multi-story building frame, it is convenient to isolate the columns from the frame and treat the stability of columns and the stability of frames as independent problems. For a column in a braced frame, it is assumed that the columns are restricted at their ends from horizontal displacements and therefore are only subjected to end moments and axial loads as transferred from the frame. It is then assumed that the frame, possibly by means of a bracing system, satisfies global stability checks and that the global stability of the frame does not affect the column behavior. This gives the commonly assumed *non-sway frame*. The design of columns in non-sway frames follows the conventional beam-column capacity check approach, and the column effective length may be evaluated based on the column end restraint conditions. Interaction equations for various cross-section shapes have been developed through years of research spent in the field of beam-column design [12].

Another reason for defining "sway" and "non-sway frames" is the need to adopt conventional analysis in which all the internal forces are computed on the basis of the undeformed geometry of the structure. This assumption is valid if second-order effects are negligible. When there is an interaction between overall frame stability and column stability, it is not possible to isolate the column. The column and the frame have to act interactively in a "sway" mode. The design of sway frames has to consider the frame subassemblage or the structure as a whole. Moreover, the presence of "inelasticity" in the columns will render some doubts on the use of the familiar concept of "elastic effective length" [45, 46].

On the basis of the above considerations, a definition can be established for sway and non-sway frames as:

> *A frame can be classified as non-sway if its response to in-plane horizontal forces is sufficiently stiff for it to be acceptably accurate to neglect any additional internal forces or moments arising from horizontal displacements of its nodes.*

British Code: BS5950:Part 1 [11] provides a procedure to distinguish between sway and non-sway frames as follows:

1. Apply a set of notional horizontal loads to the frame. These notional forces are to be taken as 0.5% of the factored dead plus vertical imposed loads and are applied in isolation, i.e., without the simultaneous application of actual vertical or horizontal loading.
2. Carry out a first-order linear elastic analysis and evaluate the individual relative sway deflection δ for each story.
3. If the actual frame is uncladed, the frame may be considered to be non-sway if the inter-story deflection of every story satisfies the following limit:

$$\delta < \frac{h}{4000}$$

where h = story height.
4. If the actual frame is cladded but the analysis is carried out on the bare frame, then in recognition of the fact that the cladding will substantially reduce deflections, the condition is reflected and the frame may be considered to be non-sway if

$$\delta < \frac{h}{2000}$$

where h = story height.
5. All frames not complying with the criteria in (3) or (4) are considered to be sway frames.

Eurocode 3 [22] also provides some guidelines to distinguish between sway and non-sway frames. It states that a frame may be classified as non-sway for a given load case if the elastic buckling load ratio P_{cr}/P for that load case satisfies the criterion:

$$P_{cr}/P \geq 10$$

where P_{cr} is the elastic critical buckling value for sway buckling and P is the design value of the total vertical load. When the system buckling load is 10 times the design load, the frame is said to be stiff enough to resist lateral load, and it is unlikely to be sensitive to sidesway deflections. AISC LRFD [3] does not give specific guidance on frame classification. However, for frames to be classified as non-sway in AISC LRFD format, the moment amplification factor, B_2, has to be small (a possible range is $B_2 < 1.10$) so that sway deflection would have negligible influence on the final value obtained from the beam-column capacity check.

12.1.6 Classification of Tall Building Frames

A tall building is defined uniquely as a building whose structure creates different conditions in its design, construction, and use than those for common buildings. From the structural engineer's view point, the selection of appropriate structural systems for tall buildings must satisfy two important criteria: strength and stiffness. The structural system must be adequate to resist lateral and gravity loads that cause horizontal shear deformation and overturning deformation. Other important issues that must be considered in planning the structural schemes and layout are the requirements for architectural details, building services, vertical transportation, and fire safety, among others. The efficiency of a structural system is measured in terms of its ability to resist higher lateral loads which increase with the height of the frame [30]. A building can be considered as tall when the effect of lateral loads is reflected in the design. Lateral deflections of tall buildings should be limited to prevent damage to both structural and non-structural elements. The accelerations at the top of the building during frequent windstorms should be kept within acceptable limits to minimize discomfort to the occupants (see Section 12.4).

Figure 12.5 shows a chart that defines, in general, the limits to which a particular system can be used efficiently for multi-story building projects. The various structural systems in Figure 12.5 can be broadly classified into two main types: (1) medium-height buildings with shear-type deformation predominant and (2) high-rise cantilever structures, such as framed tubes, diagonal tubes, and braced trusses. This classification of system forms is based primarily on their relative effectiveness in resisting lateral loads. At one end of the spectrum in Figure 12.5 is the moment resisting frames, which are efficient for buildings of 20 to 30 stories, and at the other end is the tubular systems with high cantilever efficiency. Other systems were placed with the idea that the application of any particular form is economical only over a limited range of building heights.

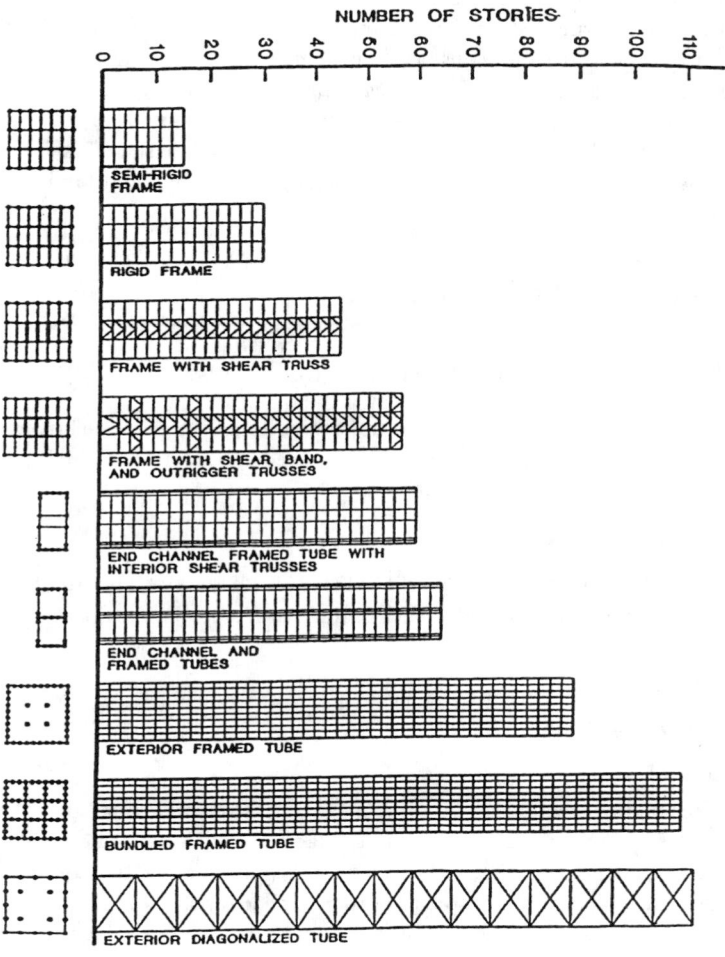

Figure 12.5   Various structural schemes.

An attempt has been made to develop a rigorous methodology for the cataloging of tall buildings with respect to their structural systems [16]. The classification scheme involves four levels of framing division: (1) primary framing system, (2) bracing subsystem, (3) floor framing, and (4) configuration and load transfer. While any cataloging scheme must address the pre-eminent focus on lateral load resistance, the load-carrying function of the tall building subsystems is rarely independent. An efficient high-rise system must engage vertical gravity load resisting elements in the lateral load subsystem in order to reduce the overall structural premium for resisting lateral loads. Further readings on design concepts and structural schemes for steel multi-story buildings can be found in Liew [41], and the design calculations and procedures for building frame structures using the AISC LRFD procedure are given in Liew and Chen [44].

Some degree of independence can be distinguished between the floor framing systems and the lateral load resisting systems, but the integration of these subassemblies into the overall structural scheme is crucial. Section 12.2 provides some advice for selecting composite floor systems to achieve the required stiffness and strength, and also highlights the ways where building services can be accommodated within normal floor zones. Several practical options for long-span construction are discussed, and their advantages and limitations are compared and contrasted. Design considerations for floor diaphragms are discussed. Section 12.3 provides some advice on the general principles to be applied when preparing a structural scheme for multistory steel and composite frames. The design procedure and construction considerations that are specific to steel gravity frames, braced frames, moment resisting frames, and the design approaches to be adopted for sizing multistory building frames are given. The potential use of steel-concrete composite material for high-rise construction is presented. Section 12.4 deals with the issues related to wind-induced effects on multistory frames. Dynamic effects due to along wind, across wind, and **torsional response** are considered with examples.

12.2   Composite Floor Systems

12.2.1   Floor Structures in Multistory Buildings

Tall building floor structures generally do not differ substantially from those in low-rise buildings; however, there are certain aspects and properties that need to be considered in design:

1. Floor weight to be minimized.
2. Floor should be able to resist construction loads during the erection process.
3. Integration of mechanical services (such as ducts and pipes) in the floor zone.
4. Fire resistance of the floor system.
5. Buildability of structures.
6. Long spanning capability.

Modern office buildings require large floor spans in order to create greater space flexibility for the accommodation of a greater variety of tenant floor plans. For tall building design, it is necessary to reduce the weight of the floors so as to reduce the size of columns and foundations and thus permit the use of larger space. Floors are required to resist vertical loads and they are usually supported by secondary beams. The spacing of the supporting beams must be compatible with the resistance of the floor slabs.

The floor systems can be made buildable by using prefabricated or precasted elements of steel and reinforced concrete in various combinations. Floor slabs can be precasted concrete slab, *in situ* concrete slab, or composite slabs with metal decking. Typical precast slabs are 4 to 7 m, thus avoiding the need of secondary beams. For composite slabs, metal deck spans ranging from 2 to 7 m may be used depending on the depth and shape of the deck profile. However, the permissible spans for steel decking are influenced by the method of construction; in particular, it depends on whether shoring is provided. Shoring is best avoided as the speed of construction is otherwise diminished for the construction of tall buildings.

Sometimes openings in the webs of beams are required to permit passage of horizontal services, such as pipes (for water and gas), cables (for electricity and tele and electronic communication), ducts (air-conditioning), etc.

In addition to strength, floor spanning systems must provide adequate stiffness to avoid large deflections due to live load which could lead to damage of plaster and slab finishers. Where the deflection limit is too severe, pre-cambering with an appropriate initial deformation equal and opposite to that due to the permanent loads can be employed to offset part of the deflection. In steel construction, steel members can be partially or fully encased in concrete for fire protection. For longer periods of fire resistance, additional reinforcement bars may be required.

12.2.2   Composite Floor Systems

Composite floor systems typically involve structural steel beams, joists, girders, or trusses linked via shear connectors with a concrete floor slab to form an effective T-beam flexural member resisting primarily gravity loads. The versatility of the system results from the inherent strength of the concrete floor component in compression and the tensile strength of the steel member. The main advantages of combining the use of steel and concrete materials for building construction are:

1. Steel and concrete may be arranged to produce ideal combinations of strength, with concrete efficient in compression and steel in tension.

2. Composite system is lighter in weight (about 20 to 40% lighter than concrete construction). This leads to savings in the foundation cost. Because of its light weight, site erection and installation are easier and thus labor cost can be minimized. Foundation cost can also be reduced.

3. The construction time is reduced because casting of additional floors may proceed without having to wait for the previously casted floors to gain strength. The steel decking system provides positive-moment reinforcement for the composite floor and requires only small amounts of reinforcement to control cracking and for fire resistance.

4. The construction of composite floors does not require highly skilled labor. The steel decking acts as a permanent formwork. Composite beams and slabs can accommodate raceways for electrification, communication, and an air distribution system. The slab serves as a ceiling surface to provide easy attachment of a suspended ceiling.

5. The composite slabs, when they are fixed in place, can act as an effective in-plane diaphragm that may provide effective lateral bracing to beams.

6. Concrete provides corrosion and thermal protection to steel at elevated temperatures. Composite slabs of 2-h fire rating can be achieved easily for most building requirements. Composite floor systems are advantageous because of the formation of the floor slab. The floor slab can be formed by the following methods:

   (a) a flat-soffit reinforced concrete slab (Figure 12.6a)

   (b) precast concrete planks with cast *in situ* concrete topping (Figure 12.6b)

   (c) precast concrete slab with *in situ* grouting at the joints (Figure 12.6c)

   (d) a metal steel deck, either composite or non-composite (Figure 12.6d)

The composite action of the beam or truss is due to shear studs welded directly through the metal deck, whereas the composite action of the metal deck results from side embossments incorporated into the steel sheet profile. The slab and beam arrangement typical in composite floor systems produces a rigid horizontal diaphragm, providing stability to the overall building system while distributing wind and seismic shears to the lateral load resisting systems.

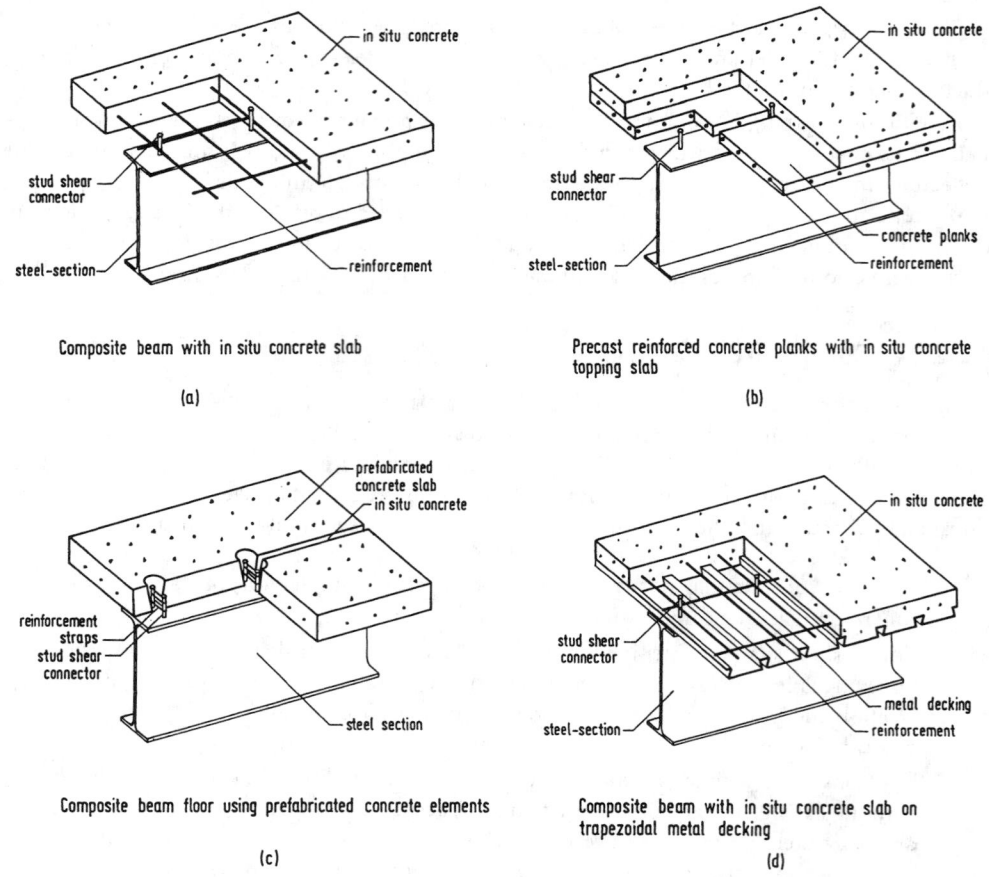

Figure 12.6   Composite beams with (a) flat-soffit reinforced concrete slab, (b) precast concrete planks and cast *in situ* concrete topping, (c) precast concrete slab and *in situ* concrete at the joints, and (d) metal steel deck supporting concrete slab.

12.2.3   Composite Beams and Girders

Steel and concrete composite beams may be formed by completely encasing a steel member in concrete with the composite action depending on the shear connectors connecting the concrete floor to the top flange of the steel member. Concrete encasement will provide fire resistance to the steel member. Alternatively, instead of using concrete encasement, direct sprayed-on cementitious and board-type fireproofing materials may be used economically to replace the concrete insulation on the steel members. The most common arrangement found in composite floor systems is a rolled or built-up steel beam connected to a formed steel deck and concrete slab (Figure 12.6d). The metal deck typically spans unsupported between steel members while also providing a working platform for concreting work. The metal decks may be oriented parallel or perpendicular to the composite beam span.

Figure 12.7a shows a typical building floor plan using composite steel beams. The stress distribution at working loads in a composite section is shown schematically in Figure 12.7b. The neutral axis is normally located very near to the top flange of the steel section. Therefore, the top flange is lightly stressed. Built-up beams or hybrid composite beams can be good choices in an attempt to use the structural steel material more efficiently (see Section 12.2.4). Also, composite beams of tapered flanges are possible. For a construction point of view, a relatively wide and thick top

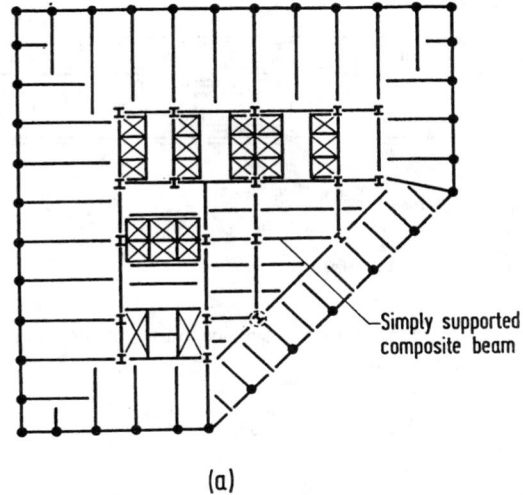

(a)

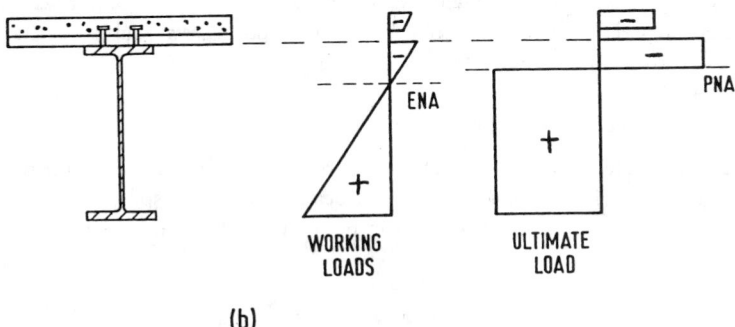

(b)

Figure 12.7   (a) Composite floor plan and (b) stress distribution in a composite cross-section.

flange must be provided for proper installation of shear stud and metal decking. However, in all of these cases, the increased fabrication costs must be evaluated, which tend to offset the saving from material efficiency.

A prismatic composite steel beam has two fundamental disadvantages over other types of composite floor framing types.

1. The member must be designed for the maximum bending moment near midspan and thus is often under stressed near the supports.
2. Building-services ductwork and piping must pass beneath the beam, or the beam must be provided with web openings (normally reinforced with plates or angles leading to higher fabrication costs) to allow access for this equipment as shown in Figure 12.8.

For this reason, a number of composite girder forms allowing the free passage of mechanical ducts and related services through the depth of the girder have been developed. Successful composite beam design requires the consideration of various serviceability issues such as long-term (creep) deflections and floor vibrations. Of particular concern is the occupant-induced floor vibrations. The relatively high flexural stiffness of most composite floor framing systems results in relatively low vibration amplitudes and therefore is effective in reducing perceptibility. Studies have shown that short to medium span (6 to 12 m) composite floor beams perform quite well and are rarely found to transmit annoying vibrations to the occupants. Particular care is required for long span

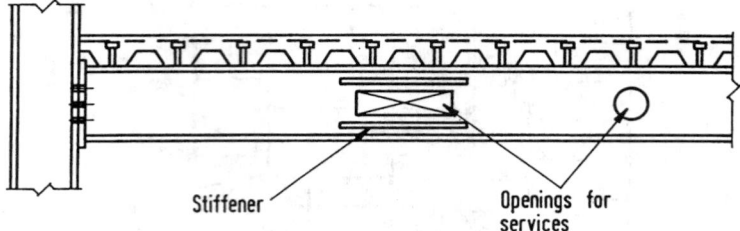

Figure 12.8   Web opening with horizontal reinforcements.

beams more than 12 m in range. Issues related to serviceability problems at various deflection or drift indices are discussed in Section 12.4.

12.2.4   Long-Span Flooring Systems

Long spans impose a burden on the beam design in terms of larger required flexural stiffness for limit-state designs. Besides satisfying both serviceability and ultimate strength limit states, the proposed system must also accommodate the incorporation of mechanical services within normal floor zones. Several practical options for long-span construction are available and they are discussed in the following subsections.

Beams With Web Openings

Standard castellated beams can be fabricated from hot-rolled beams by cutting along a zigzag line through the web. The top and bottom half-beams are then displaced to form castellations (Figure 12.9). Castellated composite beams can be used effectively for lightly serviced buildings.

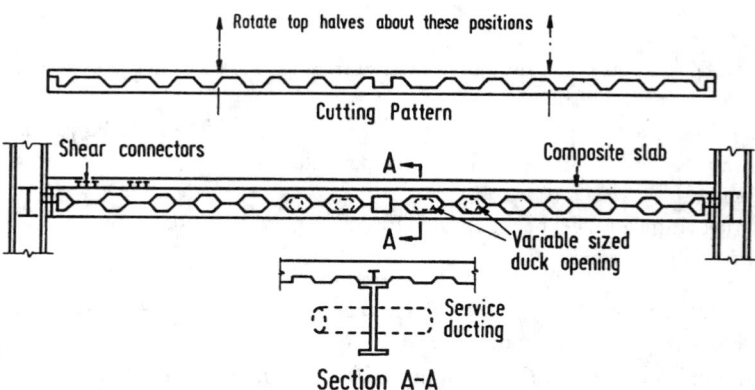

Figure 12.9   Composite castellated beams.

Although composite action does not increase the strength significantly, it increases the stiffness, and hence reduces deflection and the problem associated with vibration. Castellated beams have limited shear capacity and are best used as long span secondary beams where loads are low or where concentrated loads can be avoided. Their use may be limited due to the increased fabrication cost and the fact that the standard castellated openings are not large enough to accommodate the large mechanical ductwork common in modern high-rise buildings.

Horizontal stiffeners may be required to strengthen the web opening, and they are welded above and below the opening. The height of the opening should not be more than 70% of the beam depth, and the length should not be more than twice the beam depth. The best location of the openings is in the low shear zone of the beams, i.e., where the bending moment is high. This is because the webs do not contribute much to the moment resistance of the beam.

Fabricated Tapered Beams

The economic advantage of fabricated beams is that they can be designed to provide the required moment and shear resistance along the beam span in accordance with the loading pattern along the beam. Several forms of tapered beams are possible. A simply supported beam design with a maximum bending moment at the mid-span would require that they all effectively taper to a minimum at both ends (Figure 12.10), whereas a rigidly connected beam would have a minimum depth towards the mid-span. To make best use of this system, services should be placed towards the smaller depth of the beam cross-sections. The spaces created by the tapered web can be used for running services of modest size (Figure 12.10).

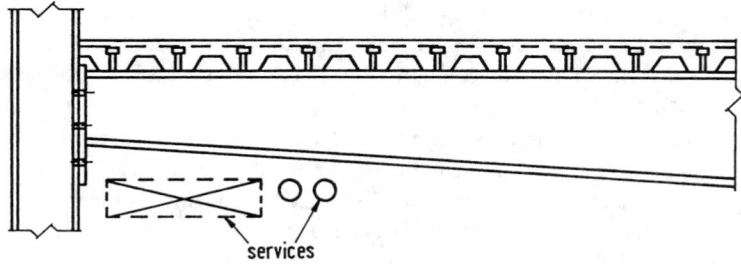

Figure 12.10   Tapered composite beams.

A hybrid girder can be formed with the top flange made of lower-strength steel in comparison with the steel grade for the bottom flange. The web plate can be welded to the flanges by double-sided fillet welds. Web stiffeners may be required at the change of section when taper slope exceeds approximately $6°$. Stiffeners are also required to enhance the shear resistance of the web especially when the web slenderness ratio is too high. Tapered beam is found to be economical for spans of 13 to 20 m. Further information on the design of fabricated beams with tapered webs can be found in Owens [51].

Haunched Beams

The span length of a composite beam can be increased by providing haunches or local stiffening of the beam-to-column connections as shown in Figure 12.11. Haunched beams are designed by

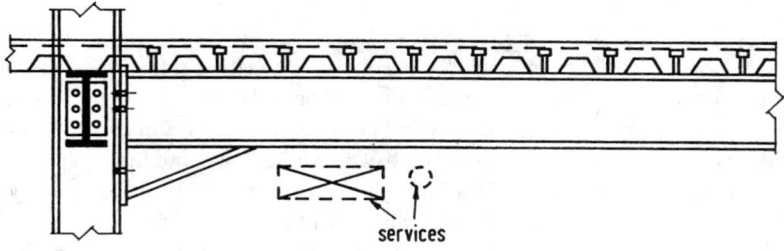

Figure 12.11   Haunched composite beam.

forming a rigid moment connection between the beams and columns. The haunch connections offer restraints to the beam and it helps to reduce mid-span moment and deflection. The beams are designed in a manner similar to continuous beams. Considerable economy can be gained in sizing the beams using continuous design which may lead to a reduction in beam depth up to 30% and deflection up to 50%.

The haunch may be designed to develop the required moment which is larger than the plastic moment resistance of the beam. In this case, the critical section is shifted to the tip of the haunch. The depth of the haunch is selected based on the required moment at the beam-to-column connections. The length of haunch is typically 5 to 7% the span length for non-sway frames or 7 to 15% for sway frames. Service ducts can pass below the beams as in conventional construction (Figure 12.11).

Haunched composite beams are usually used in the case where the beams frame directly into the major axis of the columns. This means that the columns must be designed to resist the moment transferred from the beam to the column. Thus, a heavier column and a more complex connection would be required in comparison with a structure design based on the assumption that the connections are pinned. The rigid frame action derived from the haunched connections can resist lateral loads due to wind without the need of vertical bracing. Haunched beams do offer higher strength and stiffness during the steel erection stage thus making this type of system particularly attractive for long span construction. However, haunched connections behave differently under positive and negative moments, as the connection configuration is not symmetrical about the bending axis.

The rationale of using the haunched beam approach is explained as follows. In continuous beam design, the moment distribution of a continuous beam would show that the support moment is generally larger than the mid-span moment up to the ratio of 1.8 times. The effective cross-sections of typical steel-concrete composite beam under hogging and sagging moment can be determined according to the usual stress block method of design. It can be observed that the hogging moment capacity of the composite section at the support is smaller than the sagging moment capacity near the mid-span. Therefore, there is a mismatch between the required greater support resistance and the much larger available sagging moment capacity.

When elastic analysis is used in the design of continuous composite beams, the potential large sagging moment capacities available from composite action can never be realized. One way to overcome this problem is to increase the moment resistance at the support (and hence utilize the full potential of larger sagging moment) by providing haunches at the supports. An optimum design can be achieved by designing the haunched section to develop the required moment at the support and the composite section to develop the required sagging moment. If this can be achieved in practice, the design does not require inelastic force redistribution and hence elastic analysis is adequate. However, analysis of haunched composite beams is more complicated because the member is non-prismatic (i.e., cross-section property varies along the length). The analysis of such beams requires the evaluation of section properties such as the beam's stiffness (EI) at different cross-sections. The analysis/design process is more involved because it requires the evaluation of serviceability deflection and ultimate strength limit state of non-prismatic members. Some guides on haunched beam design can be found in Lawson and Rackham [36].

Parallel Beam System

The system consists of two main beams with secondary beams run over the top of the main beams (see Figure 12.12). The main beams are connected to either side of the column. They can be made continuous over two or more spans supported on stubs attached to the columns. This will help in reducing the construction depth, and thus avoiding the usual beam-to-column connections. The secondary beams are designed to act compositely with the slab and may also be made to span continuously over the main beams. The need to cut the secondary beams at every junction is thus avoided. The parallel beam system is ideally suited for accommodating large service ducts in orthogonal directions (Figure 12.12). Small savings in steel weight are expected from the continuous

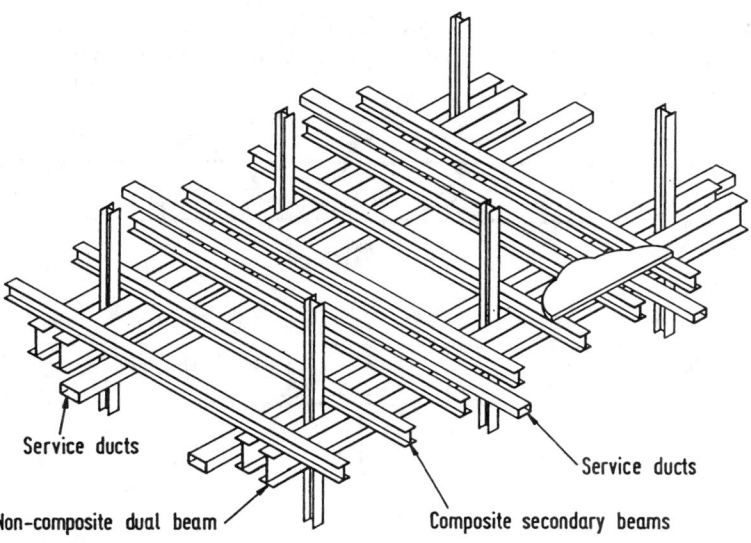

Figure 12.12 Parallel composite beam system.

construction because the primary beams are non-composite. However, the main beam can be made composite with the slab by welding beam stubs to the top flange of the main beam and connecting to the concrete slab through the use of shear studs (see the stud-girder system in Section 12.2.4). The simplicity of connections and ease of fabrication make this long-span beam option particularly attractive. Competitive pricing can be obtained from the fabricator. Further details on the parallel beam approach can be found in Brett and Rushton [10].

Composite Trusses

Trusses are frequently used in multistory buildings for very long span supports. The openings created in the truss braces can be used to accommodate large services. Although the cost of fabrication is higher in relation to the material cost, truss construction can be cost-effective for very long spans when compared to other structural schemes. An additional disadvantage other than fabrication cost is that truss configuration creates difficulty for fire protection. Fire protection wrapping is labor intensive and sprayed-protection systems cause a substantial mess to the services that pass through the web opening (see Figure 12.13). From a structural point of view, the benefit of using a composite truss is due to the increase in stiffness rather than strength.

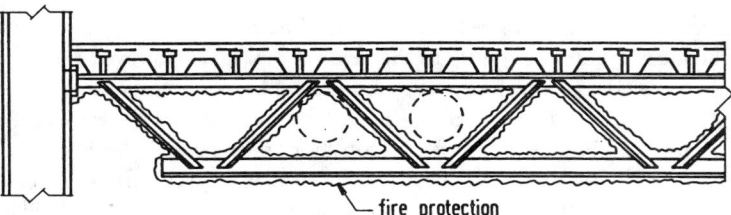

Figure 12.13 Composite truss.

Several forms of truss arrangement are possible. The three most common web framing configurations in floor truss and joist designs are: (1) Warren Truss, (2) Modified Warren Truss, and (3) Pratt Truss as shown in Figure 12.14. The efficiency of various web members in resisting vertical shear forces may be affected by the choice of a web-framing configuration. For example, the selection of Pratt web over Warren web may effectively shorten compression diagonals resulting in a more efficient use of these members.

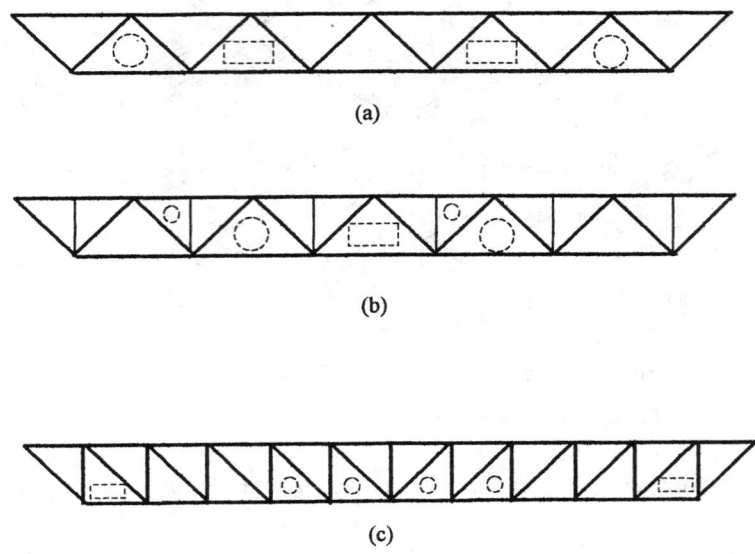

Figure 12.14   Truss configuration: (a) Warren truss, (b) Modified Warren truss, and (c) Pratt truss.

Experience has shown that both Pratt and Warren configurations of web framing are suitable for short span trusses with shallow depths. For truss with spans greater than 10 m, or effective depths larger than 700 mm, a modified Warren configuration is generally preferred. The Warren and modified Warren trusses are more popular for building construction since they offer larger web openings for services between bracing members.

The resistance of a composite truss is governed by (1) yielding of the bottom chord, (2) crushing of the concrete slab, (3) failure of the shear connectors, (4) buckling of top chord during construction, (5) buckling of web members, and (6) instability occurring during and after construction. To avoid brittle failures, ductile yielding of the bottom chord is the preferred failure mechanism. Thus, the bottom chord should be designed to yield prior to crushing of the concrete slab. The shear connectors should have sufficient capacity to transfer the horizontal shear between the top chord and the slab. During construction, adequate plan bracing should be provided to prevent top chord buckling. When composite action is considered, the top steel chord is assumed not to participate in the moment resistance of the truss because it is located very near to the neutral axis of the composite truss and, thus, contributes very little to the flexural capacity. However, the top chord has two functions: (1) it provides an attachment surface for the shear connectors, and (2) it resists the forces in the end panel without reliance on composite action unless shear connectors are placed over the seat or along a top chord extension. Thus, the top chord must be designed to resist the compressive force equilibrating the horizontal force component of the first web member. In addition, the top chord also transfers the factored shear force to the support, and must be designed accordingly.

The bottom chord shall be continuous and may be designed as an axially loaded tension member. The bottom chord shall be proportioned to yield before the concrete slab, web members, or the shear connectors fail.

The shear capacity of the steel top and bottom chords and concrete slab can be ignored in the evaluation of the shear resistance of a composite truss. The web members should be designed to resist vertical shear. Further references on composite trusses can be found in ASCE Task Committee [7] and Neals and Johnson [50].

Stub Girder System

The stub girder system involves the use of short beam stubs that are welded to the top flange of a continuous, heavier bottom girder member, and connected to the concrete slab through the use of shear studs. Continuous transverse secondary beams and ducts can pass through the openings formed by the beam stub. The natural openings in the stub girder system allow the integration of structural and service zones in two directions (Figure 12.15), permitting story-height reduction when compared with some other structural framing systems.

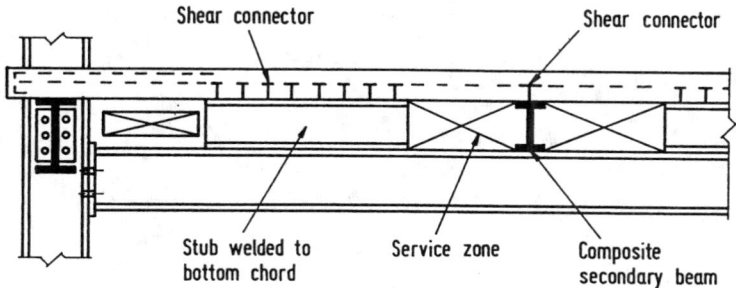

Figure 12.15 Stub girder system.

Ideally, stub-girders span about 12 to 15 m (usually from the center core wall to the exterior columns in a conventional office building) with the secondary framing or floor beams spanning about 6 to 9 m. The system is very versatile, particularly with respect to secondary framing spans with beam depths being adjusted to the required structural configuration and mechanical requirements. Overall girder depths vary only slightly, by varying the beam and stub depths. The major disadvantage of the stub girder system is that it requires temporary props at the construction stage, and these props have to remain until the concrete has gained adequate strength for composite action. However, it is possible to introduce an additional steel top chord, such as a T-section, which acts in compression to develop the required bending strength during construction. For span length greater than 15 m, stub-girders become impractical because the slab design becomes critical.

In the stub girder system, the floor beams are continuous over the main girders and splice at the locations near the points of inflection. The sagging moment regions of the floor beams are usually designed compositely with the deck-slab system, to produce savings in structural steel as well as to provide stiffness. The floor beams are bolted to the top flange of the steel bottom chord of the stub-girder, and two shear studs are usually specified on each floor beam, over the beam-girder connection, for anchorage to the deck-slab system. The stub-girder may be analyzed as a vierendeel girder, with the deck-slab acting as a compression top-chord, the full length steel girder as a tensile bottom-chord, and the steel stubs as vertical web members or shear panels.

Prestressed Composite Beams

Prestressing of the steel girders is carried out such that the concrete slab remains uncracked under the working loads and the steel is utilized fully in terms of stress in the tension zone of the girder.

Prestressing of a steel beam can be carried out using a precambering technique as depicted in Figure 12.16. First a steel girder member is prebent (Figure 12.16a), and is then subjected to preloading in the direction against the bending curvature until the required steel strength is reached

(Figure 12.16b). Second, the lower flange of the steel member, which is under tension, is encased in a reinforced concrete chord (Figure 12.16c). The composite action between the steel beam and the concrete slab is developed by providing adequate shear connectors at the interface. When the concrete gains adequate strength, the steel girder is prestressed by stress-relieving the precompressed tension chord (Figure 12.16d). Further composite action can be achieved by supplementing the girder with *in situ* or prefabricated reinforcement concrete slabs, and this will produce a double composite girder (Figure 12.16e).

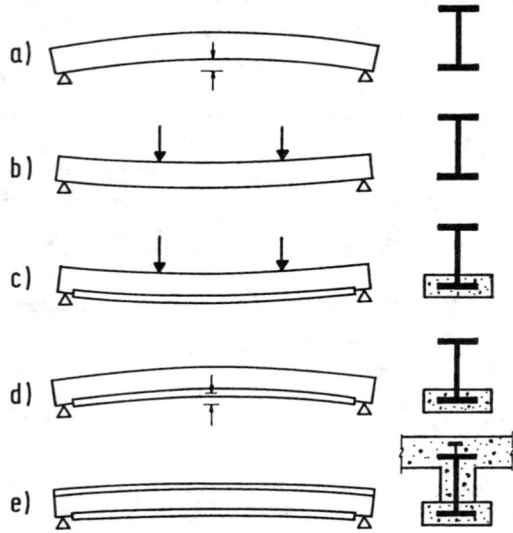

Figure 12.16   Process of prestressing using precambering technique.

The major advantages of this system is that the steel girders are encased in concrete on all sides and no corrosion and fire protection are required on the sections. The entire process of precambering and prestressing can be performed and automated in a factory. During construction, the lower concrete chord cast in the works can act as a formwork. If the distance between two girders is large, precast planks can be supported by the lower concrete chord as permanent formwork.

Prestressing can also be achieved by using tendons that can be attached to the bottom chord of a steel composite truss or the lower flange of a composite girder to enhance the load-carrying capacity and stiffness of long-span structures (Figure 12.17). This technique has been found to be popular for bridge construction in Europe and the U.S., although it is less common for building construction.

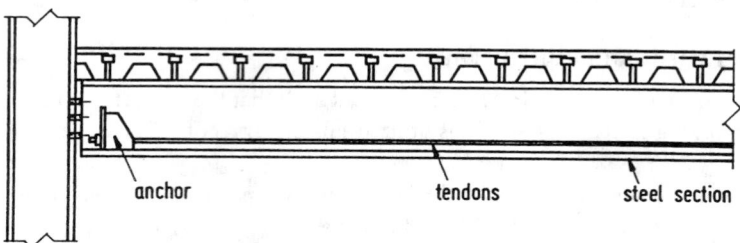

Figure 12.17   Prestressing of composite steel girders with tendons.

12.2.5 Comparison of Floor Spanning Systems

The conventional composite beams are the most common forms of floor construction for a large number of building projects. Typically they are highly efficient and economic with bay sizes in the range of 6 to 12 m. There is, however, much demand for larger column free areas where, with a traditional composite approach, the beams tend to become excessively deep, thus unnecessarily increasing the overall building height, with the consequent increases in cladding costs, etc. Spans exceeding 12 m are generally achieved by choosing an appropriate structural form that integrates the services within the floor structure, thereby reducing the overall floor zone depths. Although a long span solution may entail a small increase in structural costs, the advantages of greater flexibility and adaptability in service and the creation of column-free space often represent the most economic option over the design life of the building. Figure 12.18 compares the various structural options of a typical range of span lengths used in practice.

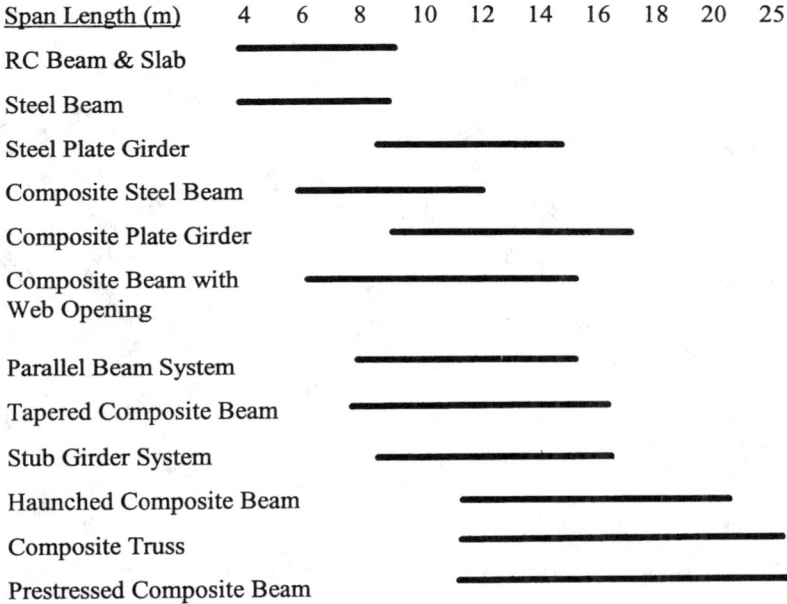

Figure 12.18 Comparison of composite floor systems.

12.2.6 Floor Diaphragms

Typically, beams and columns rigidly connected for moment resistance are placed in orthogonal directions to resist lateral loads. Each plane frame would assume to resist a portion of the overall wind shear which is determined from the individual frame stiffness in proportion to the overall stiffness of all frames in that direction. This is based on the assumption that the lateral loads are distributed to the various frames by the floor diaphragm which, for building structures, are normally assumed to have adequate in-plane stiffness. In order to develop proper diaphragm action, the floor slab must be attached to all columns and beams that participate in lateral-force resistance. For a building relying on bracing systems to resist all lateral load, the stability of the building depends on a rigid floor diaphragm to transfer wind shears from their point of application to the bracing systems such as lattice frames, shear walls, or core walls.

The use of composite floor diaphragms in place of in-plane steel bracing has become an accepted practice. The connection between slab and beams is often through shear studs that are welded directly through the metal deck to the beam flange. The connection between seams of adjacent deck panels is crucial and often through interlocking of panels overlapping each other. The diaphragm stresses are generally low and can be resisted by floor slabs that have adequate thickness for most buildings. Plan bracing is necessary when diaphragm action is not adequate. Figure 12.19a shows a triangulated plan bracing system that resists lateral load on one side and spans between the vertical walls. Figure 12.19b illustrates the case where the floor slab has adequate thickness and it can act as diaphragm resisting lateral loads and transmitting the forces to the vertical walls. However, if there is an abrupt change in lateral stiffness or where the shear must be transferred from one frame to the other due to the termination of a lateral bracing system at a certain height, large diaphragm stresses may be encountered and they must be accounted for through proper detailing of slab reinforcement. Also, diaphragm stresses may be high where there are large openings in the floor, in particular at the corners of the openings.

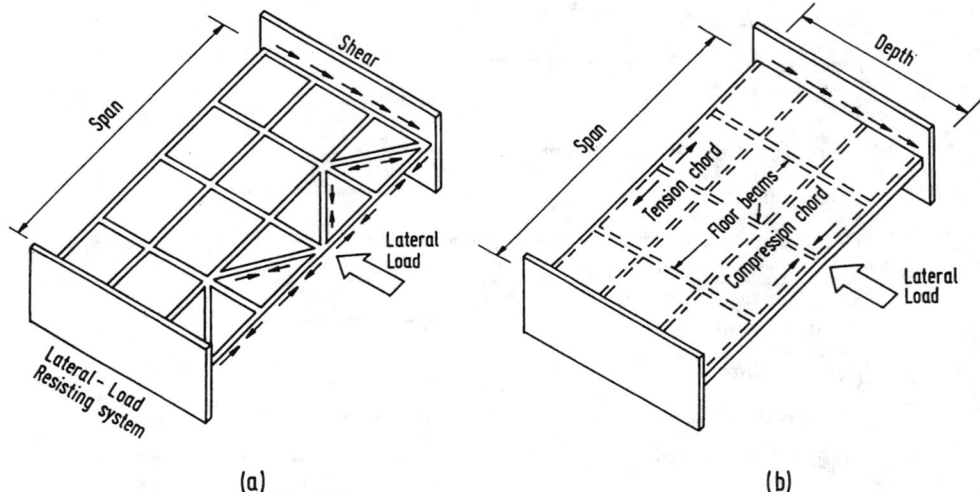

Figure 12.19   (a) Triangulated plan bracing system and (b) concrete floor diaphragm.

Diaphragms may be classified into three types, namely (1) flexible diaphragm, (2) semi-rigid diaphragm, and (3) rigid diaphragm. Common types of floor diaphragms that can be classified as rigid are (1) reinforced concrete slab, (2) composite slab with reinforced concrete slab supported by metal decking, and (3) precasted concrete slabs that are properly attached to one another. Floors that are classified as semi-rigid or flexible are steel deck without concrete fill or deck that is partially filled with concrete. However, the rigidity of a floor system must be comparable to the stiffness of the lateral-load resistance system. A rigid diaphragm will distribute lateral forces to the lateral-load resisting elements in proportion to their relative rigidities. Therefore, a vertical bracing system with high lateral stiffness will resist a greater proportion of the lateral force than a system with lower lateral stiffness. A flexible diaphragm behaves more like a beam spanning between the lateral-load resistance elements. It distributes lateral forces to the lateral systems on a tributary load basis, and it cannot resist any torsional forces. Semi-rigid diaphragms deflect like a beam under load, but possess some stiffness to distribute the loads to the lateral-load resistance systems in proportion to their rigidities. The load distribution process is a function of the floor stiffness and the vertical bracing stiffness.

The rigid diaphragm assumption is generally valid for most high-rise buildings (Figure 12.20a); however, as the plan aspect ratio (b/a) of the diaphragm linking two lateral systems exceeds 3 in 1 (see the illustration in Figure 12.20b), the diaphragm may become semi-rigid or flexible. For such cases, the wind shears must be allocated to the parallel shear frames according to the attributed area rather than relative stiffness of the frames.

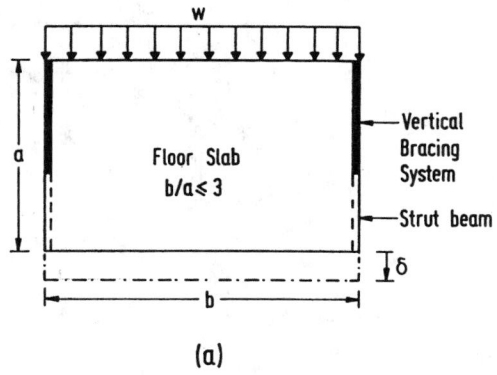

(a)

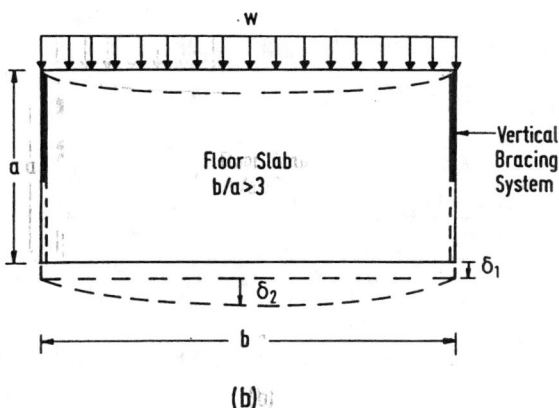

(b)

Figure 12.20 Diaphragm rigidity: (a) plan aspect ratio ≤ 3 and (b) plan aspect ratio > 3.

From the analysis point of view, a diaphragm is analogous to a deep beam with the slab forming the web and the peripheral members serving as the flanges as shown in Figure 12.19b. It is stressed principally in shear, but tension and compression forces must be accounted for in design.

A rigid diaphragm is useful to transmit torsional forces to the lateral-load resistance systems to maintain lateral stability. Figure 12.21a shows a building frame consisting of three shear walls resisting lateral forces acting in the direction of Wall A. The lateral load is assumed to act as a concentrated load with a magnitude F on each story. Figure 12.21b and 12.21c show the building plan having dimensions of L_1 and L_2. The lateral load resisting systems are represented in the plan by the solid lines which represent Wall A, Wall B, and Wall C. Since there is only one lateral resistance system (Wall A) in the direction of the applied load, the loading condition creates a torsion (F_e), and the diaphragm tends to rotate as shown by the dashed lines in Figure 12.21a. The lateral load resistance systems in Wall B and Wall C will provide the resistance forces to stabilize the torsional

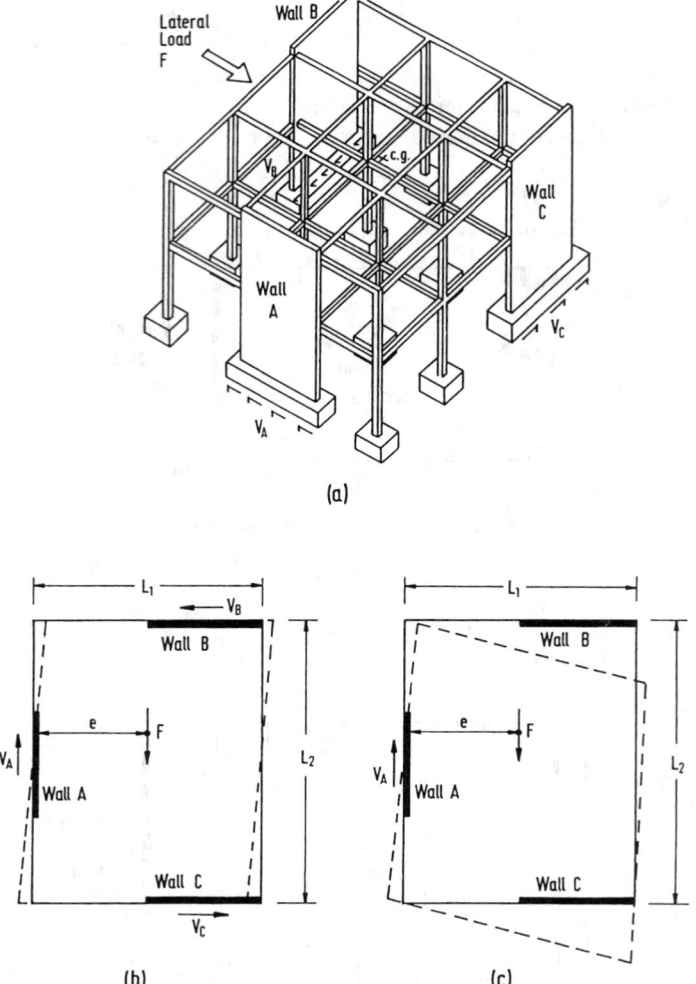

Figure 12.21   (a) Lateral force resisting systems in a building, (b) rigid diaphragm, and (c) flexible diaphragm.

force by generating a couple of shear resistances as:

$$V_B = V_C = \frac{Fe}{L_2}$$

Figure 12.21c illustrates the same condition except that a flexible diaphragm is used. The same torsional tendency exists, but the flexible diaphragm is unable to generate a resisting couple in Wall B and Wall C, and the structure will collapse as shown by the dashed lines. To maintain stability, a minimum of two vertical bracings in the direction of the applied force is required to eliminate the possibility of any torional effects.

The adequacy of the floor to act as a diaphragm depends very much on its type. Pre-cast concrete floor planks without any prestressing offer limited resistance to the racking effects of diaphragm action. In such cases, supplementary bracing systems in the plan, such as those shown in Figure 12.19a, are required for resistance of lateral forces. Where precast concrete floor units are employed, suffi-cient diaphragm action can be achieved by using a reinforced structural concrete topping, so that all individual floor planks are combined to form a single floor diaphragm. Composite concrete floors, incorporating permanent metal decking, provide excellent diaphragm action provided that the connections between the diaphragm and the peripheral members are adequate. When composite beams or girders are used, shear connectors will usually serve as boundary connectors and intermediate diaphragm-to-beam connectors. By fixing the metal decking to the floor beams, an adequate floor diaphragm can be achieved during the construction stage. It is essential at the start of the design of structural steelworks to consider the details of the flooring system to be used because these have a significant effect on the design of the structure. Table 12.1 summarizes the salient features of the various types of flooring systems in terms of their diaphragm actions.

Floor diaphragms may also be designed to provide lateral restraint to columns of multi-story buildings. In such cases, the shear required to be resisted by the floor diaphragm can be computed from the second-order forces caused by the vertical load acting on the story deflection of the column at the floor level under consideration. The stability force for the column may be transmitted directly to the deck-slab through bearing and gradually transferred into the floor framing connections through shear studs.

If metal decking is used, the metal deck provides column stability during erection, prior to concrete slab placement. Column loads are much lower during construction; hence, this condition may not be too critical. Special precaution must be given to limit the number of stories of steel erected ahead of the concrete floor construction. Overall building stability becomes important, possibly requiring the steel deck diaphragm to be supplemented with a concrete cover slab at various height levels in the structure.

TABLE 12.1   Details of Typical Flooring Systems and Their Relative Merits

Floor system	Typical span length (m)	Typical depth (mm)	Construction time	Degree of lateral restraint to beams	Degree of diaphragm action	Usage
In situ concrete	3–6	150–250	Medium	Very good	Very good	All categories but not often used in multistory buildings
Steel deck with *in situ* concrete	2.5–3.6 unshore > 3.6 shore	110–150	Fast	Very good	Very good	All categories especially in multistory office buildings
Pre-cast concrete	3–6	110–200	Fast	Fair-good	Fair-good	All categories with cranage requirements
Pre-stressed concrete	6–9	110–200	Medium	Fair-good	Fair-good	Multistory buildings and bridges

12.3   Design Concepts and Structural Schemes

12.3.1   Introduction

Multistory steel frames consist of a column and a beam interconnected to form a three-dimensional structure. A building frame can be stabilized either by some form of bracing system (braced frames) or can be stabilized by itself (unbraced frames). All building frames must be designed to resist lateral load to ensure overall stability. A common approach is to provide a gravity framing system with one or more lateral bracing system attached to it. This type of framing system, which is generally referred to as simple braced frames, is found to be cost-effective for multistory buildings of moderate height (up to 20 stories).

For gravity frames, the beams and columns are pinned connected and the frames are not capable of resisting any lateral loads. The stability of the entire structure is provided by attaching the gravity frames to some form of bracing system. The lateral loads are resisted mainly by the bracing systems, while the gravity loads are resisted by both the gravity frame and the bracing system. For buildings of moderate height, the bracing system's response to lateral forces is sufficiently stiff such that second-order effects may be neglected for the design of such frames.

In moment resisting frames, the beams and columns are rigidly connected to provide moment resistance at joints, which may be used to resist lateral forces in the absence of any bracing system. However, moment joints are rather costly to fabricate. In addition, it takes a longer time to erect a moment frame than a gravity frame.

A cost-effective framing system for multistory buildings can be achieved by minimizing the number of moment joints, replacing field welding by field bolting, and combining various framing schemes with appropriate bracing systems to minimize frame drift. A multistory structure is most economical and efficient when it can transmit the applied loads to the foundation by the shortest and most direct routes. For ease of construction, the structural schemes should be simple enough, which implies repetition of member and joints, adoption of standard structural details, straightforward temporary works, and minimal requirements for inter-related erection procedures to achieve the intended behavior of the completed structure.

Sizing of structural members should be based on the longest spans and largest attributed roof and/or floor areas. The same sections should be used for similar but less onerous cases. Simple structural schemes are quick to design and easy to erect. It also provides a good "benchmark" for further refinement. Many building structures have to accommodate extensive services within the floor zone. It is important that the engineer chooses a structural scheme (see Section 12.2) which can accommodate the service requirements within the restricted floor zone to minimize overall cost.

Scheme drawings for multistory building designs should include the following:

1. General arrangement of the structure including column and beam layout, **bracing frames**, and floor systems.
2. Critical and typical member sizes.
3. Typical cladding and bracing details.
4. Typical and unusual connection details.
5. Proposals for fire and corrosion protection.

This section offers advice on the general principles to be applied when preparing a structural scheme for multistory steel and composite frames. The aim is to establish several structural schemes that are practicable, sensibly economic, and functional to the changes that are likely to be encountered as the overall design develops. The section begins by examining the design procedure and construction considerations that are specific to steel gravity frames, braced frames, and moment resisting frames, and the design approaches to be adopted for sizing tall building frames. The potential use of steel-concrete composite material for high-rise construction is then presented. Finally, the design issues related to braced and unbraced composite frames are discussed, and future directions for research are highlighted.

12.3.2 Gravity Frames

Gravity frames refer to structures that are designed to resist only gravity loads. The bases for designing gravity frames are as follows:

1. The beam and girder connections transfer only vertical shear reactions without developing bending moment that will adversely affect the members and the structure as a whole.

2. The beams may be designed as a simply supported member.

3. Columns must be fully continuous. The columns are designed to carry axial loads only. Some codes of practice (e.g., [11]) require the column to carry nominal moments due to the reaction force at the beam end, applied at an appropriate eccentricity.

4. Lateral forces are resisted entirely by bracing frames or by shear walls, lift, or staircase closures, through floor diaphragm action.

General Guides

The following points should be observed in the design of gravity frames:

1. Provide lateral stability to gravity framing by arranging suitable braced bays or core walls deployed symmetrically in orthogonal directions, or wherever possible, to resist lateral forces.

2. Adopt a simple arrangement of slabs, beams, and columns so that loads can be transmitted to the foundations by the shortest and most direct load paths.

3. Tie all the columns effectively in orthogonal directions at every story. This may be achieved by the provision of beams or ties that are placed as close as practicable to the columns.

4. Select a flooring scheme that provides adequate lateral restraint to the beams and adequate diaphragm action to transfer the lateral load to the bracing system.

5. For tall building construction, choose a profiled-steel-decking composite floor construction if uninterrupted floor space is required and/or height is at a premium. As a guide, limit the span of the floor slab to 2.5 to 3.6 m; the span of the secondary beams to 6 to 12 m; and the span of the primary beams to 5 to 7 m. Otherwise, choose a precast or an *in situ* reinforced concrete floor, limiting its span to 5 to 6 m, and the span of the beams to 6 to 8 m approximately.

Structural Layout

In building construction, greater economy can be achieved through a repetition of similarly fabricated components. A regular column grid is less expensive than a non-regular grid for a given floor area. Orthogonal arrangements of beams and columns, as opposed to skewed arrangements, provide maximum repetition of standard details. In addition, greater economies can be achieved when the column grids in the plan are rectangular in which the secondary beams should span in the longer direction and the primary beams in the shorter, as shown in Figures 12.22a and b. This arrangement reduces the number of beam-to-beam connections and the number of individual members per unit area of the supported floor [52].

In gravity frames, the beams are assumed to be simply supported between columns. The effective beam span to depth ratio (L/D) is about 12 to 15 for steel beams and 18 to 22 for **composite beams**. The design of the beam is often dependent on the applied load, the type of beam system employed, and the restrictions on structural floor depth. The floor-to-floor height in a multistory building is influenced by the restrictions on overall building height and the requirements for services above and/or below the floor slab. Naturally, flooring systems involving the use of structural steel members that act compositely with the concrete slab achieve the longest spans (see Section 12.2.5).

Analysis and Design

The analysis and design of a simple braced frame must recognize the following points:

1. The members intersecting at a joint are pin connected.

2. The columns are not subjected to any direct moment transferred through the connection (nominal moments due to eccentricity of the beam reaction forces may be considered).

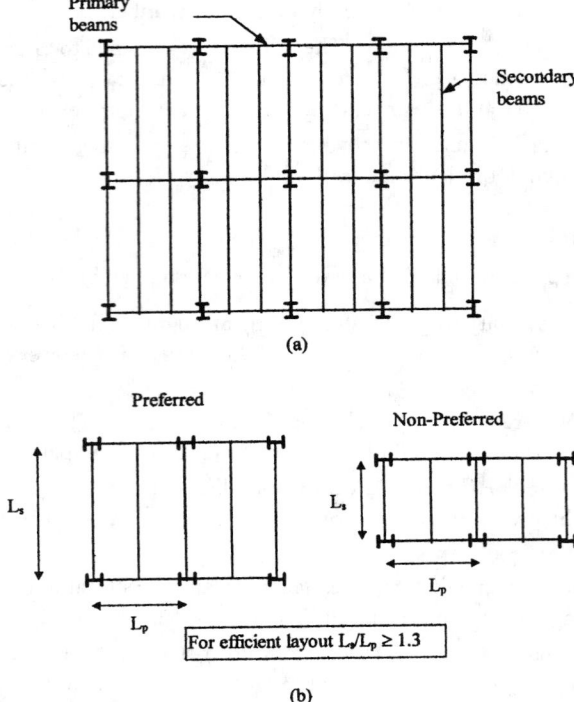

Figure 12.22   (a) Retangular grid layout and (b) preferred and non-preferred grid layout.

The design axial force in the column is predominately governed by floor loading and the tributary areas.

3. The structure is statically determinate. The internal forces and moments are therefore determined from a consideration of statics.

4. Gravity frames must be attached to a bracing system so as to provide lateral stability to the part of the structure resisting gravity load. The frame can be designed as a non-sway frame and the second-order moments associated with frame drift can be ignored.

5. The leaning column effects due to column sidesway must be considered in the design of the frames that are participating in sidesway resistance.

Since the beams are designed as simply supported between their supports, the bending moments and shear forces are independent of beam size. Therefore, initial sizing of beams is a straightforward task. Beam or girder members supporting more than 40 m^2 of floor at one story should be designed for a reduced live load in accordance with ASCE [6].

Most conventional types of floor slab construction will provide adequate lateral restraint to the compression flange of the beam. Consequently, the beams may be designed as laterally restrained beams without the moment resistance being reduced by lateral-torsional buckling.

Under the service loading, the total central deflection of the beam or the deflection of the beam due to unfactored live load (with proper precambering for dead load) should satisfy the deflection limits as given in Table 12.2.

In some occasions, it may be necessary to check the dynamic sensitivity of the beams. When assessing the deflection and dynamic sensitivity of secondary beams, the deflection of the supporting beams must also be included. Whether it is the strength, deflection, or dynamic sensitivity that controls the design will depend on the span-to-depth ratio of the beam. Figure 12.18 gives typical span ranges for beams in office buildings for which the design would be optimized for strength and serviceability. For beams with their span lengths exceeding those shown in Figure 12.18, serviceability limits due to deflection and vibration will most likely be the governing criteria for design.

TABLE 12.2   Recommended Deflection Limits for Steel Building Frames

Beam deflections from unfactored imposed loads	
Beams carrying plaster or brittle finish	Span/360 (with maximum of 1/4 to 1 in.)
Other beams	Span/240
Columns deflections from unfactored imposed and wind loads	
Column in single story frames	Height/300
Column in multistory frames	Height of story/300
For column supporting cladding which is sensitive to large movement	Height of story/500
Frame drift under 50 years wind load	
Frame drift	Frame height/450 ~ frame height/600

The required axial forces in the columns can be derived from the cumulative reaction forces from those beams that frame into the columns. Live load reduction should be considered in the design of columns in a multistory frame [6]. If the frame is braced against sidesway, the column node points are prevented from lateral translation. A conservative estimate of column effective length, KL, for buckling considerations is $1.0L$, where L is the story height. However, in cases where the columns above and below the story under consideration are underutilized in terms of load resistance, the restraining effects offered by these members may result in an effective length of less than $1.0L$ for the column under consideration. Such a situation arises where the column is continuous through the restraint points and the columns above and/or below the restraint points are of different length.

An example of such cases is the continuous column shown in Figure 12.23 in which Column AB is longer than Column BC and hence Column AB is restrained by Column BC at the restraint point B. A buckling analysis shows that the critical buckling load for the continuous column is $P_{cr} = 5.89EI/L^2$, which gives rise to an effective length factor of $K = 0.862$ for Column AB and $K = 1.294$ for Column BC. Column BC has a larger effective length factor because it provides restraint to Column AB, whereas Column AB has a smaller effective length factor because it is restrained by Column BC during buckling. Figure 12.24 summaries the reductions in effective length which may be considered for columns in a frame with different story heights having various values of a/L ratios [52].

Simple Shear Connections

Simple shear connections should be designed and detailed to allow free rotation and to prevent excessive transfer of moment between the beams and columns. Such connections should comply with the classification requirement for a "nominally pinned connection" in terms of both strength and stiffness. A computer program for connection classification has been made available in a book by Chen et al. [15], and their design implications for semi-rigid frames are discussed in Liew et al. [47].

Simple connections are designed to resist vertical shear at the beam end. Depending on the connection details adopted, it may also be necessary to consider an additional bending moment resulting from the eccentricity of the bolt line from the supporting face. Often the fabricator is told to design connections based on the beam end reaction for one-half uniformed distributed load (UDL). Unless the concentrated load is located very near to the beam end, UDL reactions are generally conservative. Because of the large reaction, the connection becomes very strong which may require a large number of bolts. Thus, it would be a good practice to design the connections for the actual forces used in the design of the beam. The engineer should give the design shear

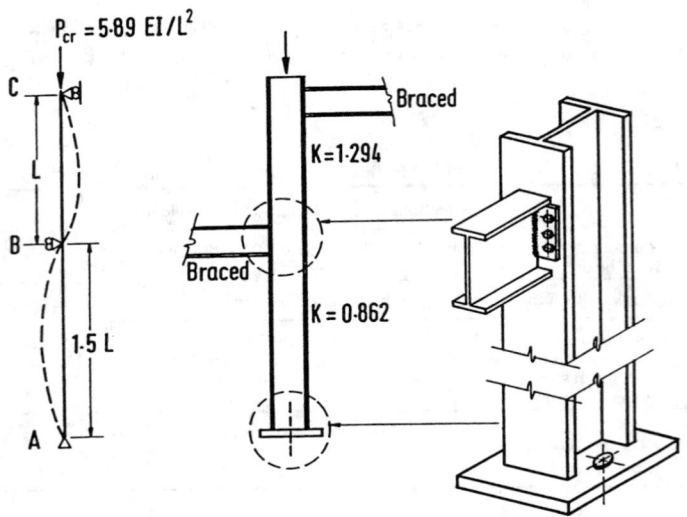

Figure 12.23   Buckling of a continuous column with intermediate restrain.

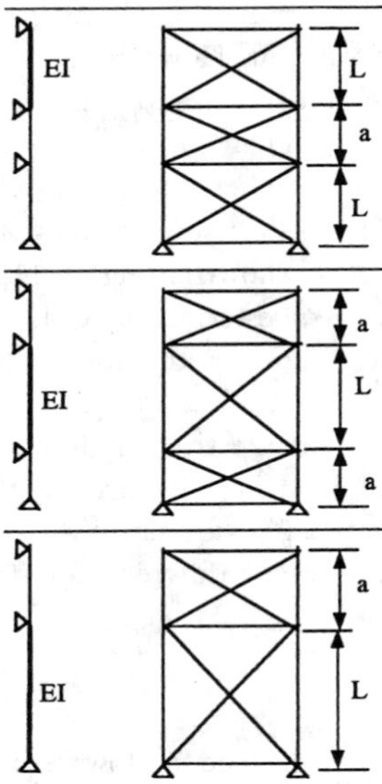

Column	Frame		a/L				
			0.2	0.4	0.6	0.8	1.0
EI		L, a, L	0.76	0.82	0.88	0.94	1.0
EI		a, L, a	0.57	0.65	0.75	0.87	1.0
EI		a, L	0.74	0.79	0.84	0.91	1.0

Figure 12.24   Effective length factors of continuous braced columns.

force for every beam to the steel fabricator so that a more realistic connection can be designed, instead of requiring all connections to develop the shear capacity of the beam. Figure 12.25 shows the typical connections that can be designed as simple connections. When the beam reaction is known, capacity tables developed for simple standard connections can be used for detailing such connections [2].

12.3.3   Bracing Systems

The main purpose of a bracing system is to provide the lateral stability to the entire structure. It has to be designed to resist all possible kinds of lateral loading due to external forces, e.g., wind forces, earthquake forces, and "leaning forces" from the gravity frames. The wind or the equivalent

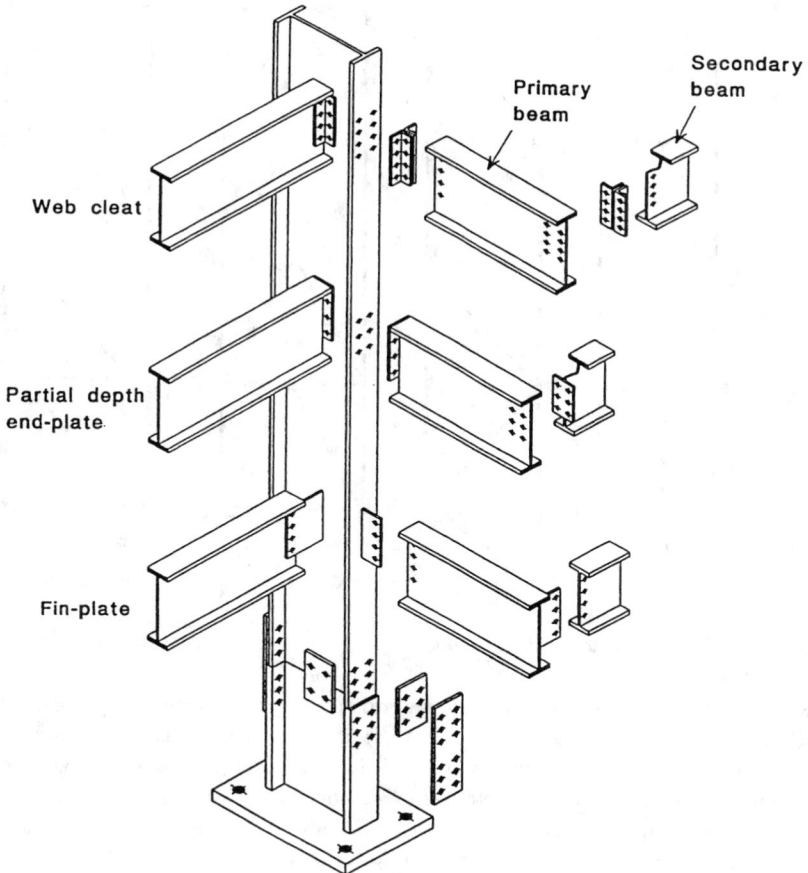

Figure 12.25 Typical beam-to-column connections to be considered as shear connections.

earthquake forces on the structure, whichever are greater, should be assessed and divided into the number of bracing bays resisting the lateral forces in each direction.

Structural Forms

Steel braced systems are often in a form of a vertical truss which behaves like cantilever elements under lateral loads developing tension and compression in the column chords. Shear forces are resisted by the bracing members. The truss diagonalization may take various forms, as shown in Figure 12.26. The design of such structures must take into account the manner in which the frames are erected, the distribution of lateral forces, and their sidesway resistance.

In the single braced forms, where a single diagonal brace is used (Figure 12.26a), it must be capable of resisting both tensile and compressive axial forces caused by the alternate wind load. Hollow sections may be used for the diagonal braces as they are stronger in compression. In the design of diagonal braces, gravity forces may tend to dominate the axial forces in the members and due consideration must be given in the design of such members. It is recommended that the slenderness ratio of the bracing member (L/r) not be greater than 200 to prevent the self-weight deflection of the brace limiting its compressive resistance.

In a cross-braced system (Figure 12.26b), the brace members are usually designed to resist tension only. Consequently, light sections such as structural angles and channels, or tie rods can be used to provide a very stiff overall structural response. The advantage of the cross-braced system is that

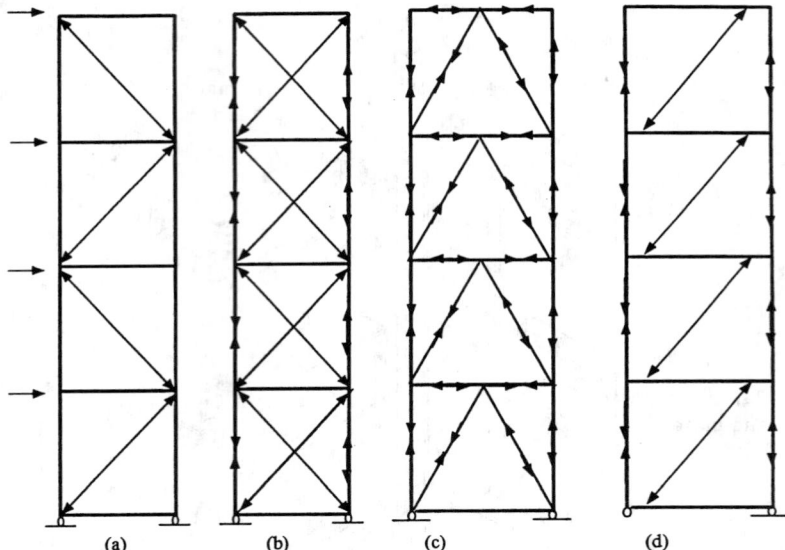

Figure 12.26   (a) Diagonal bracing, (b) cross-bracing, (c) K-bracing, and (d) eccentric bracing.

the beams are not subjected to significant axial force, as the lateral forces are mostly taken up by the bracing members.

The K trusses are common since the diagonals do not participate extensively in carrying column load, and can thus be designed for wind axial forces without gravity axial force being considered as a major contribution. A K-braced frame is more efficient in preventing sidesway than a cross-braced frame for equal steel areas of braced members used. This type of system is preferred for longer bay width because of the shorter length of the braces. A K-braced frame is found to be more efficient if the apexes of all the braces are pointing in the upward direction (Figure 12.26c).

For eccentrically braced frames, the center line of the brace is positioned eccentrically to the beam-column joint, as shown in Figure 12.26d. The system relies, in part, on flexure of the short segment of the beam between the brace-beam joint and the beam-column joint. The forces in the braces are transmitted to the column through shear and bending of the short beam segment. This particular arrangement provides a more flexible overall response. Nevertheless, it is more effective against seismic loading because it allows for energy dissipation due to flexural and shear yielding of the short beam segment.

Drift Assessment

The story drift Δ of a single story diagonally braced frame, as shown in Figure 12.27, can be approximated by the following equation:

$$\begin{aligned}
\Delta &= \Delta_s + \Delta_f \\
&= \frac{HL_d^3}{A_dEL^2} + \frac{Hh^3}{A_cEL^2}
\end{aligned} \qquad (12.3)$$

where
$$\begin{aligned}
\Delta &= \text{inter-story drift} \\
\Delta_s &= \text{story drift due to shear component} \\
\Delta_f &= \text{story drift due to flexural component} \\
A_c &= \text{area of the chord} \\
A_d &= \text{area of the diagonal brace} \\
E &= \text{modulus of elasticity}
\end{aligned}$$

H = horizontal force in the story
h = story height
L = length of braced bay
L_d = length of the diagonal brace

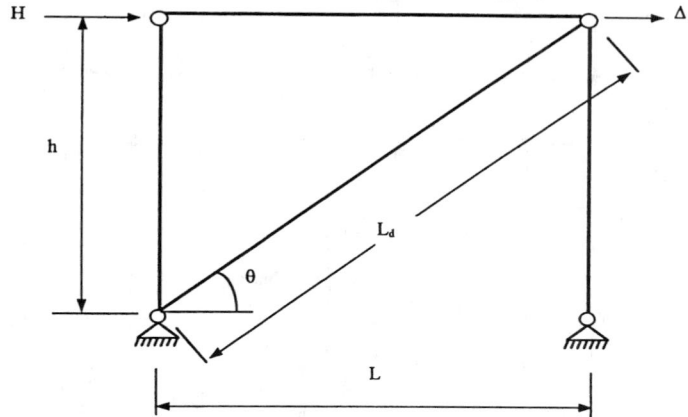

Figure 12.27   Lateral displacement of a diagonally braced frame.

The shear component Δ_s in Equation 12.3 is caused mainly by the straining of the diagonal brace. The deformation associated with girder compression has been neglected in the calculation of Δ_s because the axial stiffness of the girder is very much larger than the stiffness of the brace. The elongation of the diagonal braces gives rise to shear deformation of the frame, which is a function of the brace length, L_d, and the angle of the brace (L_d/L). A shorter brace length with a smaller brace angle will produce a lower story drift.

The flexural component of the frame drift is due to tension and compression of the windward and leeward columns. The extension of the windward column and shortening of the leeward column cause flexural deformation of the frame, which is a function of the area of the column and the ratio of the height-to-bay length (h/L). For a slender bracing frame with a large h/L ratio, the flexural component can contribute significantly to the overall story drift.

A low-rise braced frame deflects predominately in shear mode while high-rise braced frames tend to deflect more in flexural mode.

Design Considerations

Frames with braces connecting columns may obstruct locations of access openings such as windows and doors; thus, they should be placed where such access is not required, e.g., around elevators and service and stair wells. The location of the bracing systems within the structure will influence the efficiency with which the lateral forces can be resisted. The most appropriate position for the bracing systems is at the periphery of the building (Figure 12.28a) because this arrangement provides greater torsional resistance. Bracing frames should be situated where the center of lateral resistance is approximately equal to the center of shear resultant on the plan. Where this is not possible, torsional forces will be induced, and they must be considered when calculating the load carried by each braced system.

When core braced systems are used, they are normally located in the center of the building (Figure 12.28b). The torsional stability is then provided by the torsional rigidity of the core brace. For tall building frames, a minimum of three braced bents are required to provide transitional and torsional stability. These bents should be carefully arranged so that their planes of action do not meet

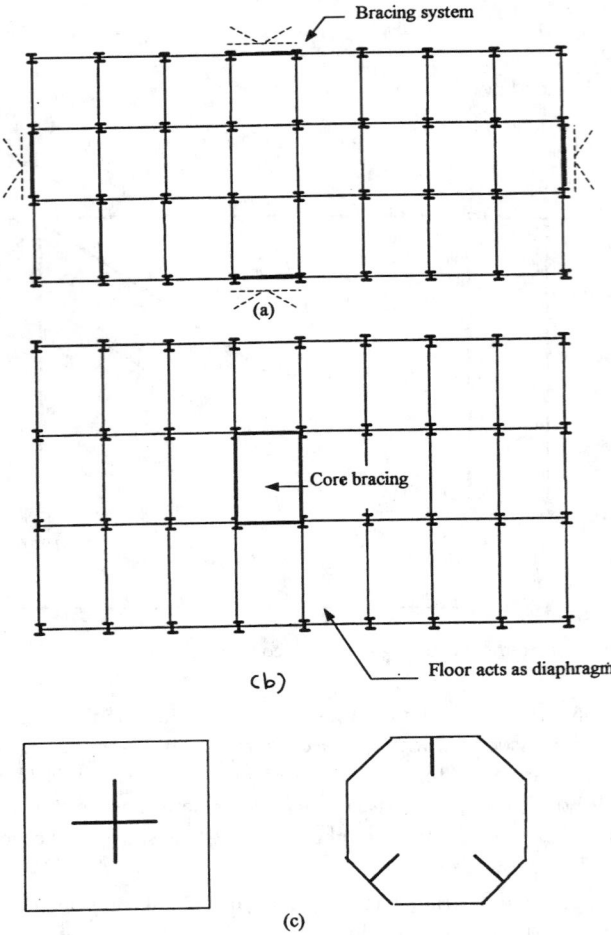

Figure 12.28   Locations of bracing systems: (a) exterior braced frames, (b) internal braced core, and (c) bracing arrangements to be avoided.

at one point so as to form a center of rotation. The bracing arrangement shown in Figure 12.28c should be avoided.

The flexibility of different bracing systems must be taken into account in the analysis because the stiffer braces will attract a larger share of the applied lateral load. For tall and slender frames, the bracing system itself can be a sway frame, and a second-order analysis is required to evaluate the required forces for ultimate strength and serviceability checks.

Lateral loads produce transverse shears, overturning moments, and sidesway. The stiffness and strength demands on the lateral system increase dramatically with height. The shear increases linearly, the overturning moment as a second power, and the sway as a fourth power of the height of the building. Therefore, apart from providing the strength to resist lateral shear and overturning moments, the dominant design consideration (especially for tall building) is to develop adequate lateral stiffness to control sway.

For serviceability verification, it requires that both the inter-story drifts and the lateral deflections of the structure as a whole must be limited. The limits depend on the sensitivity of the structural elements to shear deformations. Recommended limits for typical multistory frames are given in Table 12.2. When considering the ultimate limit state, the bracing system must be capable of transmitting the factored lateral loads safely down to the foundations. Braced bays should be effective throughout the full height of the building. If it is essential for bracing to be discontinuous at one level, provision must be made to transfer the forces to other braced bays. Where this is not possible, torsional forces may be induced, and they should be allowed for in the design (see Section 12.2.6).

The design of the internal bracing members in a steel bracing system is similar to the design of lattice trusses. The horizontal member in a latticed bracing system serves also as a floor beam. This member will be subjected to bending due to gravity loads and axial compression due to wind. The columns must be designed for additional forces due to leaning column effects from adjacent gravity frames. The resistance of the members should therefore be checked as a beam-column based on the appropriate load combinations.

Figure 12.29 shows an example of a building that illustrates the locations of vertical braced trusses provided at the four corners to achieve lateral stability. Diaphragm action is provided by 130 mm lightweight aggregate concrete slab which acts compositely with metal decking and floor beams. The floor beam-to-column connections are designed to resist shear force only as shown in the figure.

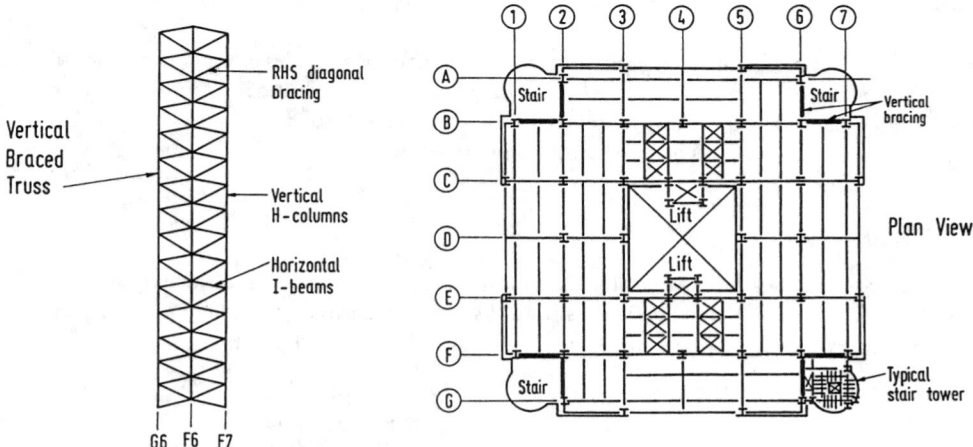

Figure 12.29   Simple building frame with vertical braced trusses located at the corners.

12.3.4   Moment-Resisting Frames

In cases where bracing systems would disturb the functioning of the building, rigidly jointed moment resisting frames can be used to provide lateral stability to the building, as illustrated in Figure 12.30a. The efficiency of development of lateral stiffness is dependent on bay span, number of bays in the frame, number of frames, and the available depth in the floors for the frame girders. For building with heights not more than three times the plan dimension, the moment frame system is an efficient form. Bay dimensions in the range of 6 to 9 m and structural height up to 20 to 30 stories are commonly used. However, as the building height increases, deeper girders are required to control drift; thus, the design becomes uneconomical.

When a rigid unbraced frame is subjected to lateral load, the horizontal shear in a story is resisted predominantly by the bending of columns and beams. These deformations cause the frame to deform in a shear mode. The design of these frames is controlled, therefore, by the bending stiffness of individual members. The deeper the member, the more efficiently the bending stiffness can be developed. A small part of the frame sidesway is caused by the overturning of the entire frame

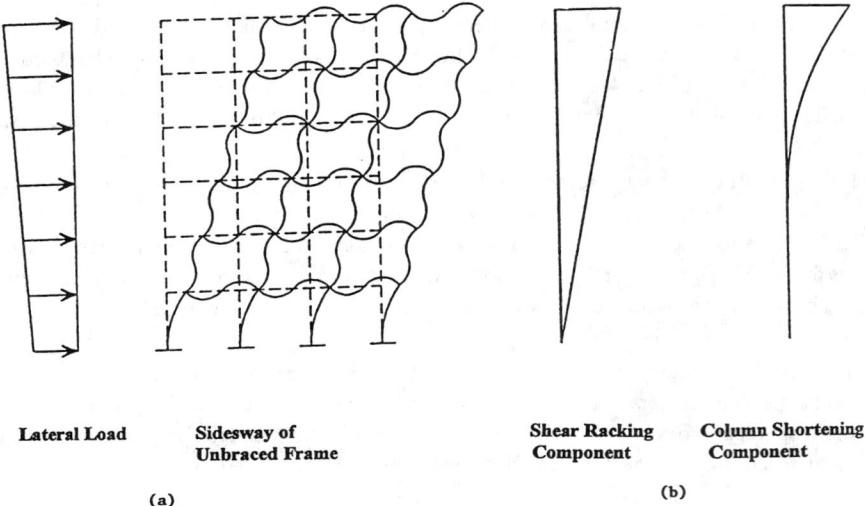

<table>
<tr><td>**Lateral Load**</td><td>**Sidesway of Unbraced Frame**</td><td>**Shear Racking Component**</td><td>**Column Shortening Component**</td></tr>
</table>

(a)   (b)

Figure 12.30   Sidesway resistance of a rigid unbraced frame.

resulting in shortening and elongation of the columns at opposite sides of the frame. For unbraced rigid frames up to 20 to 30 stories, the overturning moment contributes for about 10 to 20% of the total sway, whereas shear racking accounts for the remaining 80 to 90% (Figure 12.30b). However, the story drift due to overall bending tends to increase with height, while that due to shear racking tends to decrease.

Drift Assessment

Since shear racking accounts for most of the lateral sway, the design of such frames should be directed towards minimizing the sidesway due to shear. The shear displacement Δ in a typical story in a multistory frame, as shown in Figure 12.31, can be approximated by the equation:

$$\Delta_i = \frac{V_i h_i^2}{12E}\left(\frac{1}{\Sigma(I_{ci}/h_i)} + \frac{1}{\Sigma(I_{gi}/L_i)}\right) \tag{12.4}$$

where

Δ_i	=	the shear deflection of the i-th story
E	=	modulus of elasticity
I_c, I_g	=	second moment of area for columns and girders, respectively
h_i	=	height of the i-th story
L_i	=	length of girder in the i-th story
V_i	=	total horizontal shear force in the i-th story
$\Sigma(I_{ci}/h_i)$	=	sum of the column stiffness in the i-th story
$\Sigma(I_{gi}/L_i)$	=	sum of the girder stiffness in the i-th story

Examination of Equation 12.4 shows that sidesway deflection caused by story shear is influenced by the sum of the column and beam stiffness in a story. Since for multistory construction, span lengths are generally larger than the story height, the moment of inertia of the girders needs to be larger to match the column stiffness, as both of these members contribute equally to the story drift. As the beam span increases, considerably deeper beam sections will be required to control frame drift.

Since the gravity forces in columns are cumulative, larger column sizes are needed in lower stories as the frame height increases. Similarly, story shear forces are cumulative and, therefore, larger beam properties in lower stories are required to control lateral drift. Because of limitations in available

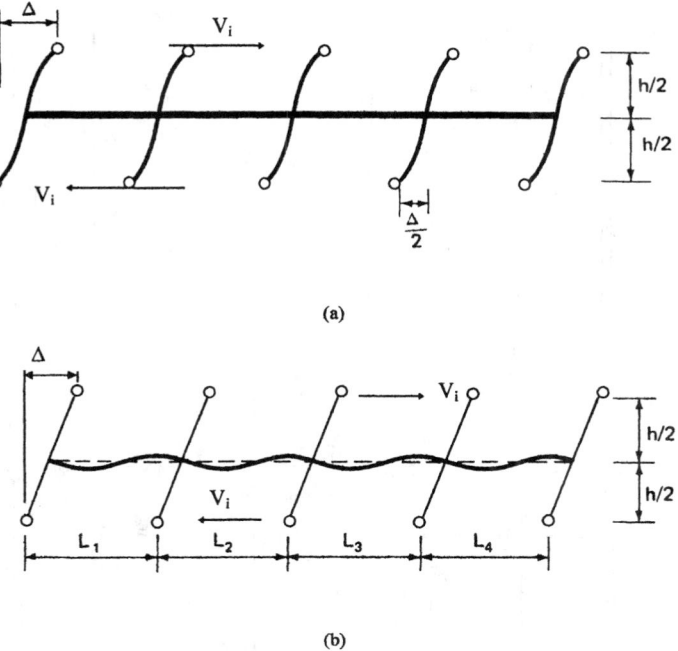

(a)

(b)

Figure 12.31   Story drift due to (a) bending of columns and (b) bending of girders.

depth, heavier beam members will need to be provided at lower floors. This is the major shortcoming of unbraced frames because considerable premium for steel weight is required to control lateral drift as building height increases.

Apart from the beam span, height-to-width ratios of the building play an important role in the design of such structures. Wider building frames allow a larger number of bays (i.e., larger values for story summation terms $\Sigma(I_{ci}/h_i)$ and $\Sigma(I_{gi}/L_i)$ in Equation 12.4) with consequent reduction in frame drift. Moment frames with closed spaced columns that are connected by deep beams are very effective in resisting sidesway. This kind of framing system is suitable for use in the exterior planes of the building.

Moment Connections

Fully welded moment joints are expensive to fabricate. To minimize labor cost and to speed up site erection, field bolting instead of field welding should be used. Figure 12.32 shows several types of bolted or welded moment connections that are used in practice. Beam-to-column flange connections can be shop-fabricated by welding of a beam stub to an end plate or directly to a column. The beam can then be erected by field bolting the end plate to the column flanges or splicing beams (Figures 12.32c and d).

An additional parameter to be considered in the design of columns of an unbraced frame is the "panel zone" between the column and the transverse framing beams. When an unbraced frame is subjected to lateral load, additional shear forces are induced in the column web panel as shown in Figure 12.33. The shear force is induced by the unbalanced moments from the adjoining beams causing the joint panel to deform in shear. The deformation is attributed to the large flexibility of the unstiffened column web. To prevent shear deformation so as to maintain the moment joint assumption as assumed in the global analysis, it may be necessary to stiffen the panel zone using either a doubler plate or a diagonal stiffener as shown in the joint details in Figures 12.32a and b. Otherwise, a heavier column with a larger web area is required to prevent excessive shear

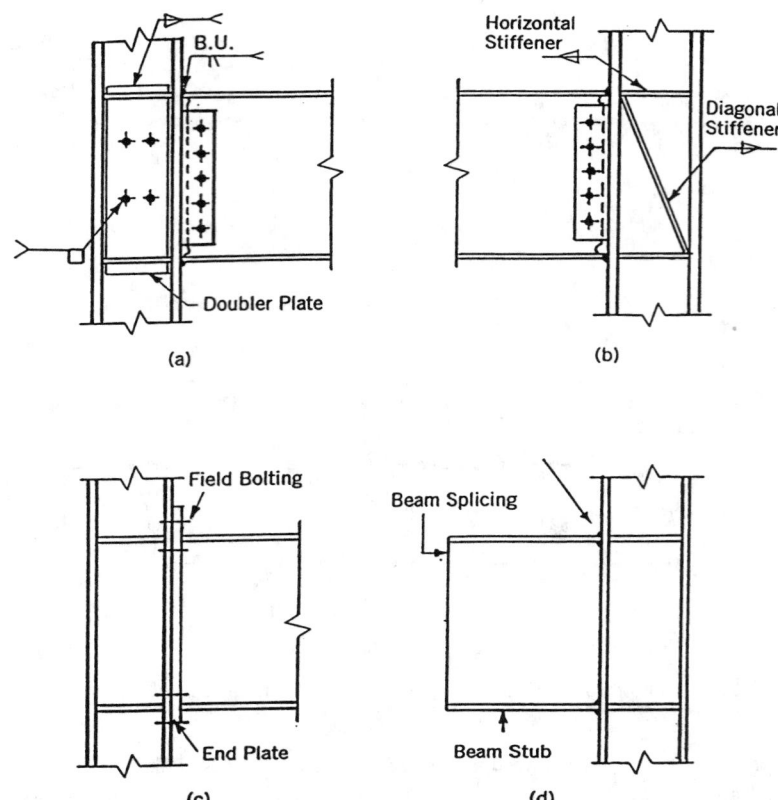

Figure 12.32   Rigid connections: (a) bolted and welded connection with doubler plate, (b) bolted and welded connection with diagonal stiffener, (c) bolted end-plate connection, and (d) beam-stub welded to column.

deformation, and this is often the preferred method as stiffeners and doublers can add significant costs to fabrication.

The engineer should not specify full strength moment connections unless they are required for ductile frame design for high seismic loads. For wind loads and for conventional moment frames where beams and columns are sized for stiffness (drift control) instead of strength, full strength moment connections are not required. Even so, many designers will specify full strength moment connections, adding to the cost of fabrication. Designing for actual loads has the potential to reduce column weight or the stiffener and doubler plate requirements.

If the panel zone is stiffened to prevent inelastic shear deformation, the conventional structural analysis based on the member center-line dimension will generally overestimate the frame displacement. If the beam-column joint sizes are relatively small compared to the member spans, the increase in frame stiffness using member center-line dimension will be offset by the increase in frame deflection due to panel-joint shear deformation. If the joint sizes are large, a more rigorous second-order analysis, which considers panel zone deformations, may be required for an accurate assessment of the frame response [43].

Analysis and Design of Unbraced Frames

Multistory moment frames are statically indeterminate. The required design forces can be determined using either: (1) elastic analysis or (2) plastic analysis. While elastic methods of analysis can be used for all kind of steel sections, plastic analysis is only applicable for frames whose members

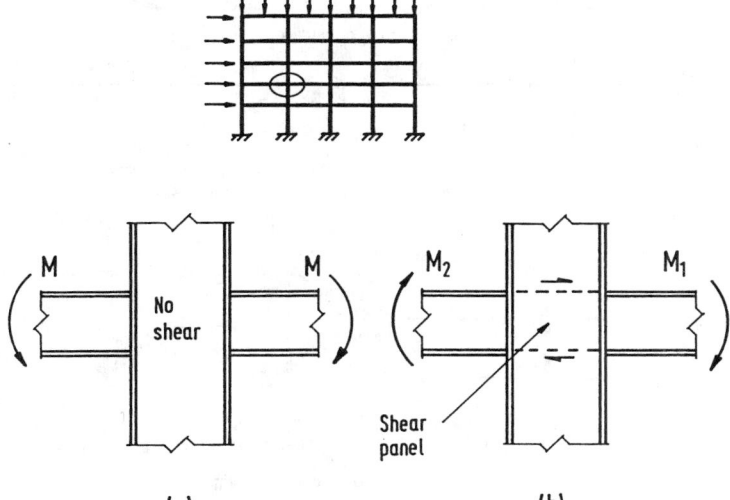

Figure 12.33 Forces acting on a panel joint: (a) balanced moment due to gravity load and (b) unbalanced moment due to lateral load.

are of plastic sections so as to enable the development of plastic hinges and to allow for inelastic redistribution of forces.

First-order elastic analysis can be used only in the following cases:

1. Where the frame is braced and not subjected to sidesway.

2. Where an indirect allowance for second-order effects is made through the use of moment amplification factors and/or the column effective length. Eurocode 3 requires only second-order moment or effective length factor to be used in the beam-column capacity checks. However, column and frame imperfections need to be modeled explicitly in the analysis. In AISC LRFD [3], both factors need to be computed for checking the member strength and stability, and the analysis is based on structures without initial imperfections.

The first-order elastic analysis is a convenient approach. Most design offices possess computer software capable of performing this method of analysis on large and highly indeterminate structures. As an alternative, hand calculations can be performed on appropriate sub-frames within the structure (see Figure 12.34) comprising a significantly reduced number of members. However, when conducting the analysis of an isolated sub-frame it is important that:

1. the sub-frame is indeed representative of the structure as a whole

2. the selected boundary conditions are appropriate

3. account is taken of the possible interaction effects between adjacent sub-frames

4. allowance is made for second-order effects through the use of column effective length or moment amplification factors

Plastic analysis generally requires more sophisticated computer programs, which enable second-order effects to be taken into account. Computer software is now available through recent publications made available by Chen and Toma [14] and Chen et al. [15]. For building structures in which the required rotations are not calculated, all members containing plastic hinges must have plastic cross-sections.

A basic procedure for the design of an unbraced frame is as follows:

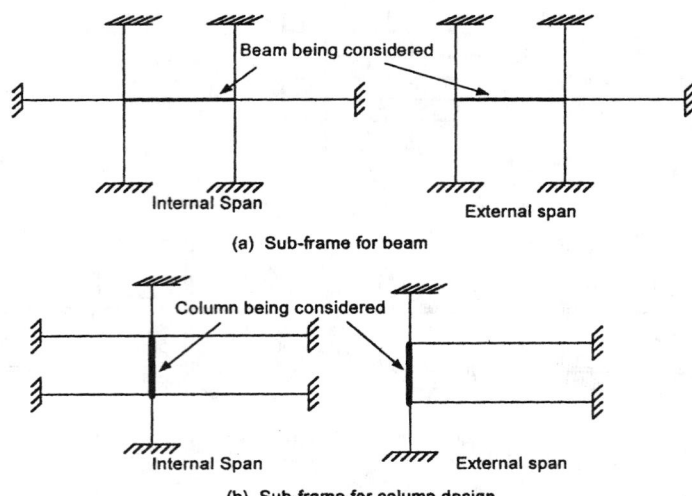

Figure 12.34   Sub-frame analysis for gravity loads.

1. Obtain approximate member size based on gravity load analysis of sub-frames shown in Figure 12.34. If sidesway deflection is likely to control (e.g., slender frames) use Equation 12.4 to estimate the member sizes.
2. Determine wind moments from the analysis of the entire frame subjected to lateral load. A simple portal wind analysis may be used in lieu of the computer analysis.
3. Check member capacity for the combined effects of factored lateral load plus gravity loads.
4. Check beam deflection and frame drift.
5. Redesign the members and perform final analysis/design check (a second-order elastic analysis is preferable at the final stage).

   The need to repeat the analysis to correspond to changed section sizes is unavoidable for highly redundant frames. Iteration of Steps 1 to 5 gives results that will converge to an economical design satisfying the various design constraints imposed on the analysis.

12.3.5   Tall Building Framing Systems

The following subsections discuss four classical systems that have been adopted for tall building constructions, namely (1) core braced, (2) moment-truss, (3) outriggle and belt, and (4) tube. Tall frames that utilize cantilever action will have higher efficiencies, but the overall structural efficiency depends on the height-to-width ratio. Interactive systems involving moment frame and vertical truss or core are effective up to 40 stories and represent most building forms for tall structures. Outrigger truss and belt truss help to further enhance the lateral stiffness by engaging the exterior frames with the core braces to develop cantilever actions. Exterior framed tube systems with closely spaced exterior columns connected by deep girders mobilize the three-dimensional action to resist lateral and torsional forces. Bundled tubes improve the efficiency of exterior frame tubes by providing internal stiffening to the exterior tube concept. Finally, by providing diagonal braces to the exterior framework, a superframe is formed and can be used for ultra-tall magastructures.

Core Braced Systems

This type of structural system relies entirely on the internal core for lateral load resistance. The basic concept is to provide internal shear wall core to resist the lateral forces (Figure 12.35). The surrounding steel framing is designed to carry gravity load only if simple framing is adopted.

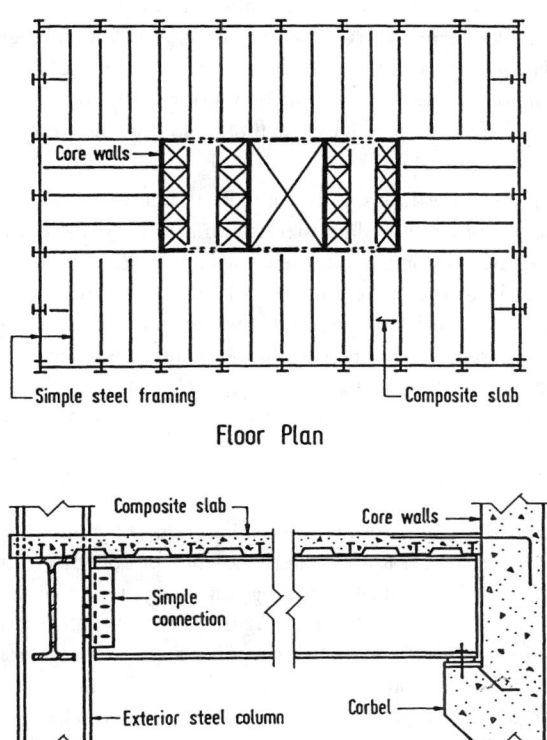

Figure 12.35 Core-brace frame. (a) Internal core walls with simple exterior framing and (b) beam-to-wall and beam-to-exterior column connections.

Otherwise, a rigid framing surrounding the core will enhance the overall lateral-force resistance of the structure. The steel beams can be simply connected to the core walls using a typical corbel detail, or by bearing in a wall pocket or by shear plate embedded in the core wall through studs. If rigid connection is required, the steel beams should be rigidly connected to steel columns embedded in the core wall. Rigid framing surrounding the cores is particularly useful in high seismic areas, and for very tall buildings that tend to attract stronger wind loads. They act as moment frames and provide resistance to some part of the lateral loads by engaging the core walls in the building.

The core generally provides all torsional and flexural rigidity and strength with no participation from the steel system. Conceptually, the core system should be treated as a cantilever wall system with punched openings for access. The floor-framing should be arranged in such a way that it distributes enough gravity loads to the core walls so that their design is controlled by compressive stresses even under wind loads. The geometric location of the core should be selected so as to minimize eccentricities for lateral load. The core walls need to have adequate torsional resistance for possible asymmetry of the core system where the center of the resultant shear load is acting at an eccentricity from the center of the lateral-force resistance.

A simple cantilever model should be adequate to analyze a core wall structure. However, if the structural form is a tube with openings for access, it may be necessary to perform a more accurate analysis to include the effect of openings. The walls can be analyzed by a finite element analysis using thin-walled plate elements. An analysis of this type may also be required to evaluate torsional stresses when the vertical profile of the core-wall assembly is asymmetrical.

The concrete core walls can be constructed using slip-form techniques, where the core walls could be advanced several floors (typically 4 to 6 story) ahead of the exterior steel framing. A core wall system represents an efficient type of structural system up to a certain height premium because of its cantilever action. However, when it is used alone, the massiveness of the wall structure increases with height, thereby inhibiting the free planning of interior spaces, especially in the core. The space occupied by the shear walls leads to loss of overall floor area efficiency, as compared to a tube system which could otherwise be used.

In commercial buildings where floor space is valuable, the large area taken up by a concrete column can be reduced by the use of an embedded steel column to resist the extreme loads encountered in tall buildings. Sometimes, particularly at the bottom open floors of a high rise structure where large open lobbies or atriums are utilized as part of the architectural design, a heavy embedded steel section as part of a composite column is necessary to resist high load and due to the large unbraced length. A heavy steel section in a composite column is often utilized where the column size is restricted architecturally and where reinforcing steel percentages would otherwise exceed the maximum code allowed values for the deign of reinforced concrete columns.

Moment-Truss Systems

Vertical shear trusses located around the inner core in one or both directions can be combined with perimeter moment-resisting frames in the facade of a building to form an efficient structure for lateral load resistance. An example of a building consisting of moment frames with shear trusses located at the center of the building is shown in Figure 12.36a. For the vertical trusses arranged in the North-South direction, either the K- or X-form of bracing is acceptable since access to lift-shafts is not required. However, K trusses are often preferred because in the case of X or single brace form bracings, the influence of gravity loads is rather significant. In the East-West direction, only the Knee bracing is effective in resisting lateral load.

In some cases, internal bracing can be provided using concrete shear walls as shown in Figure 12.36b. The internal core walls substitute the steel trusses in K, X, or a single brace form which may interfere with openings that provide access to, for example, elevators.

The interaction of shear frames and vertical trusses produces a combination of two deflection curves with the effect of more efficient stiffness. These moment frame-truss interacting systems are considered to be the most economical steel systems for buildings up to 40 stories. Figure 12.37 compares the sway characteristic of a 20-story steel frame subjected to the same lateral forces, but with different structural schemes, namely (1) unbraced moment frame, (2) simple-truss frame, and (3) moment-truss frame. The simple-truss frame helps to control lateral drift at the lower stories, but the overall frame drift increases toward the top of the frame. The moment frame, on the other hand, shows an opposite characteristic for sidesway in comparison with the simple braced frame. The combination of moment and truss frame provides overall improvement in reducing frame drift; the benefit becomes more pronounced towards the top of the frame. The braced truss is restrained by the moment frame at the upper part of the building while at the lower part, the moment frame is restrained by the truss frame. This is because the slope of frame sway displacement is relatively smaller than that of the truss at the top while the proportion is reversed at the bottom. The interacting forces between the truss frame and moment frame, as shown in Figure 12.38, enhance the combined moment-truss frame stiffness to a level larger than the summation of individual moment frame and truss stiffnesses.

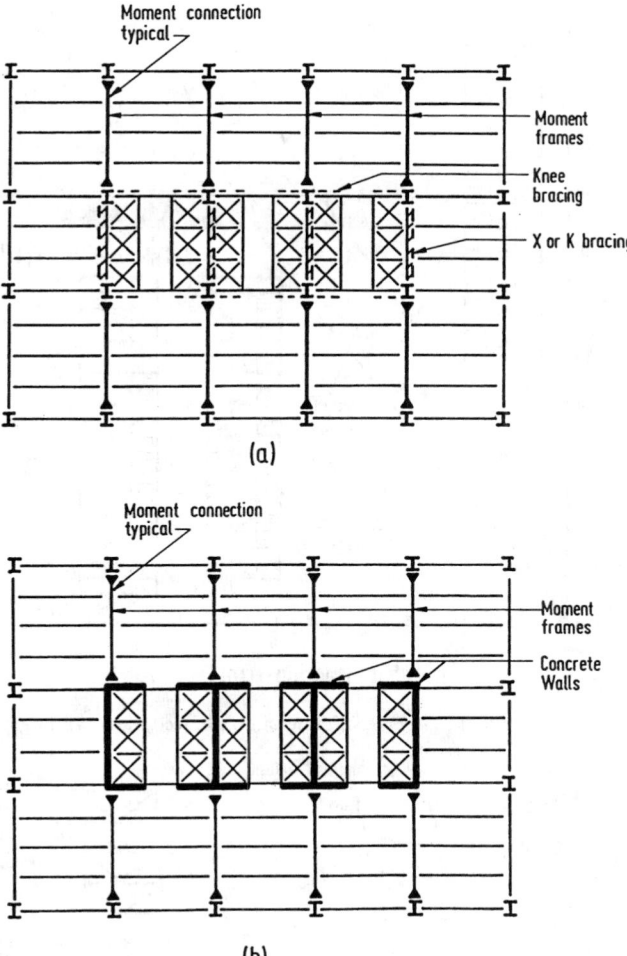

Figure 12.36 (a) Moment-frames with internal braced trusses. (b) Moment-frames with internal core walls.

Outrigger and Belt Truss Systems

Another significant improvement of lateral stiffness can be obtained if the vertical truss and the perimeter shear frame are connected on one or more levels by a system of outrigger and belt trusses. Figure 12.39 shows a typical example of such a system. The outrigger truss leads the wind forces of the core truss to the exterior columns providing cantilever behavior of the total frame system. The belt truss in the facade improves the cantilever participation of the exterior frame and creates a three-dimensional frame behavior.

Figure 12.40 shows a schematic diagram that demonstrates the sway characteristic of the overall building under lateral load. Deflection is significantly reduced by the introduction of the outrigger-belt trusses. Two kinds of stiffening effects can be observed; one is related to the participation of the external columns together with the internal core to act in a cantilever mode; the other is related to the stiffening of the external facade frame by the belt truss to act as a three-dimensional tube. The overall stiffness can be increased up to 25% as compared to the shear truss and frame system without such outrigger-belt trusses.

The efficiency of this system is related to the number of trussed levels and the depth of the truss. In some cases the outrigger and belt trusses have a depth of two or more floors. They are located

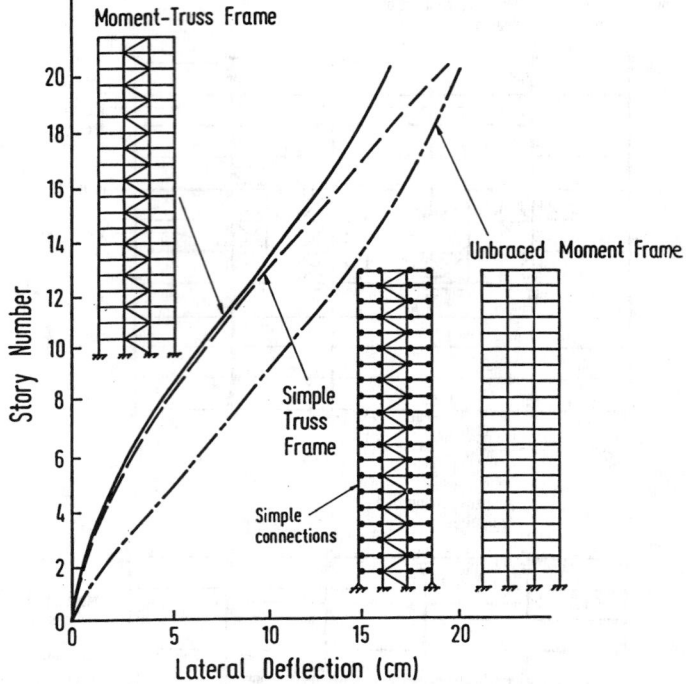

Figure 12.37   Sway characteristics of rigid braced frame, simple braced frame, and rigid unbraced frame.

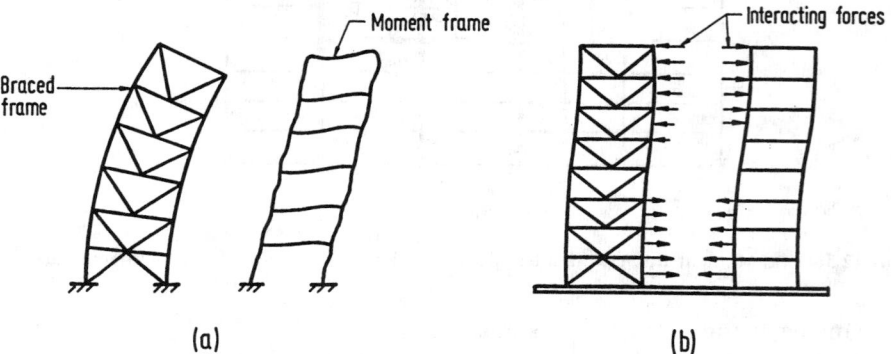

Figure 12.38   Behavior of frames subjected to lateral load: (a) independent behavior and (b) interactive behavior.

in services floors where there are no requirements for wide open spaces. These trusses are often pleasingly integrated into the architectural conception of the facade.

Frame Tube Systems

Figure 12.41 shows a typical frame tube system, which consists of a frame tube at the exterior of the building and gravity steel framing at the interior. The framed tube is constructed from wide columns placed at close centers connected by deep beams creating a punched wall appearance. The exterior frame tube structure resists all lateral loads of wind or earthquake whereas the gravity steel framing in the interior resists only its share of gravity loads. The behavior of the exterior frame tube is similar to a hollow perforated tube. The overturning moment under the action of lateral load is resisted by compression and tension of the leedward and windward columns, which are called the

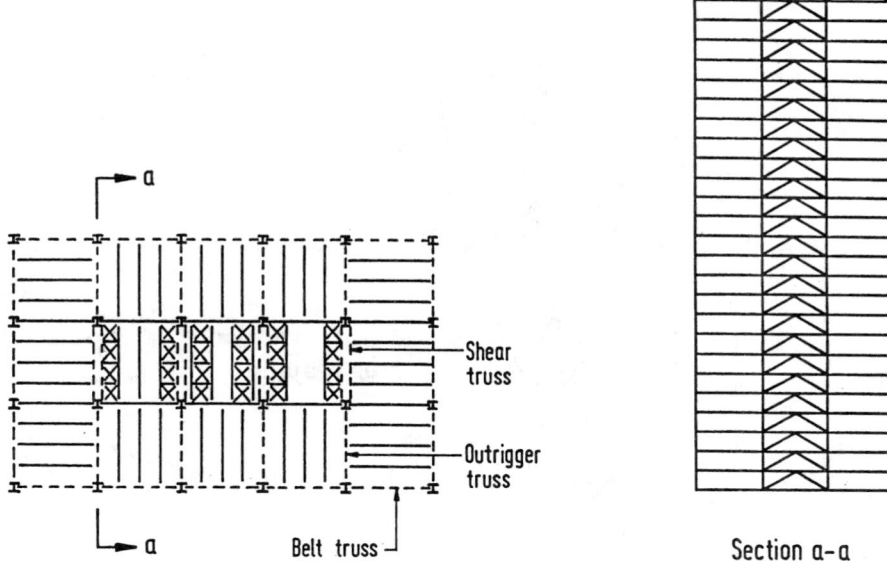

Figure 12.39   Outrigger and belt-truss system.

flange columns. The shear is resisted by bending of the columns and beams at the two sides of the building parallel to the direction of the lateral load, which are called the web frames.

Deepening on the shear rigidity of the frame tube, there may exist a shear lag across the windward and leeward sides of the tube. As a result of this, not all the flange columns resist the same amount of axial force. An approximate approach is to assume an equivalent column model as shown in Figure 12.42. In the calculation of the lateral deflection of the frame tube it is assumed that only the equivalent flange columns on the windward and leeward sides of the tube and the web frames would contribute to the moment of inertia of the tube.

The use of an exterior framed tube has three distinct advantages: (1) it develops high rigidity and strength for torsional and lateral-load resistance because the structural components are effectively placed at the exterior of the building forming a three-dimensional closed section; (2) massiveness of the frame tube system eliminates potential uplift difficulties and produces better dynamic behavior; and (3) the use of gravity steel framing in the interior has the advantages of flexibility and enables rapid construction. If a composite floor with metal decking is used, electrical and mechanical services can be incorporated in the floor zone.

Composite columns are frequently used in the perimeter of the building where the closely spaced columns work in conjunction with the spandrel beam (either steel or concrete) to form a three-dimensional cantilever tube rather than an assembly of two-dimensional plane frames. The exterior frame tube significantly enhances the structural efficiency in resisting lateral loads and thus reduces the shear wall requirements. However, in cases where a higher magnitude of lateral stiffness is required (such as for very tall buildings), internal wall cores and interior columns with floor framing can be added to transform the system into a tube-in-tube system. The concrete core may be strategically located to recapture elevator space and to provide transmission of mechanical ducts from shafts and mechanical rooms.

12.3.6   Steel-Concrete Composite Systems

Steel-concrete composite construction has gained wide acceptance as an alternative to pure steel and pure concrete construction. Composite building systems can be broadly categorized into two forms: one utilizes the core-braced system by means of interior shear walls, and the other utilizes

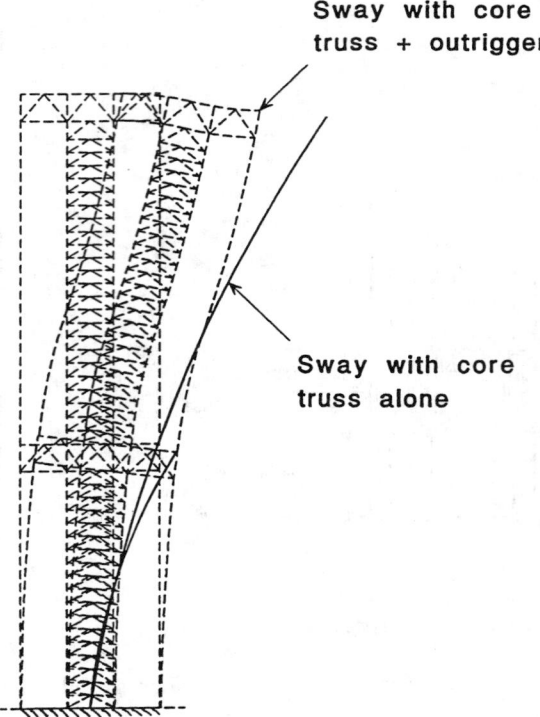

Figure 12.40   Improvement of lateral stiffness using outrigger-belt truss system.

exterior framing to form a tube for lateral load resistance. Combining these two structural forms will enable taller buildings to be constructed.

Braced Composite Frames Subjected to Gravity Loads

For composite frames resisting gravity load only, the beam-to-column connections behave as pinned before the placement of concrete. During construction, the beam is designed to resist concrete dead load and the construction load (to be treated as temporary live load). At the composite stage, the composite strength and stiffness of the beam should be utilized to resist the full design loads. For gravity frames consisting of bare steel columns and composite beams, there is now sufficient knowledge available for the designer to use composite action in the structural element as well as the semi-rigid composite joints to increase design choices, leading to more economical solutions [38, 39].

Figures 12.43a and b show the typical beam-to-column connections, one using a flushed end-plate bolted to the column flange and the other using a bottom angle with double web cleats. Composite action in the joint is acquired based on the tensile forces developed in the rebars acting with the balancing compression forces transmitted by the lower portion of the steel section that bears against the column flange to form a couple. Properly designed and detailed composite connections are capable of providing moment resistance up to the hogging resistance of the connecting members.

In designing the connections, slab reinforcements placed within 7 column flange widths are assumed to be effective in resisting the hogging moment. Reinforcement steels that fall outside this width should not be considered in calculating the resisting moment of the connection. The connections to edge columns should be carefully detailed to ensure adequate anchorage of re-bars. Otherwise, they shall be designed and detailed as simply supported. In a braced frame, a moment connection to the exterior column will increase the moments in the column, resulting in an increase of column size. Although the moment connections restrain the column from buckling by reducing

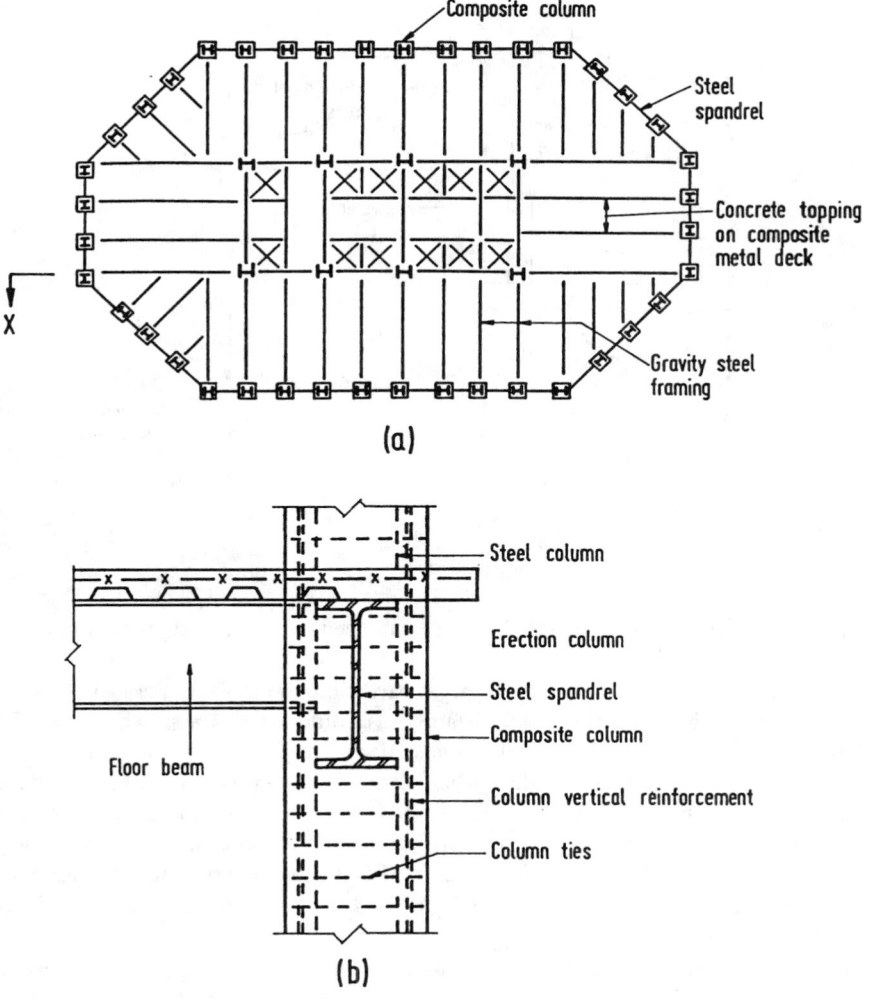

Figure 12.41 Composite tubular system.

the effective length, this is generally not adequate to offset the strength required to resist this moment.

The moment of inertia of the composite beam I_{cp} may be estimated using a weighted average of moment of inertia in the positive moment region (I_p) and negative moment region (I_n). For interior spans, approximately 60% of the span is experiencing positive moment; it is suggested that [37]:

$$I_{cp} = 0.6I_p + 0.4I_n \qquad (12.5)$$

where I_p is the lower bound moment of inertia for positive moment and I_n is the lower bound moment of inertia for negative moment. However, if the connections at both ends of the beam are designed and detailed for a simply supported beam, the beam will bend in single curvature under the action of gravity loads, and I_p should be used throughout.

Unbraced Composite Frames

If reinforcements are provided in the concrete encasement, composite design of members may be utilized for strength and stiffness assessment of the overall structure. The composite bending

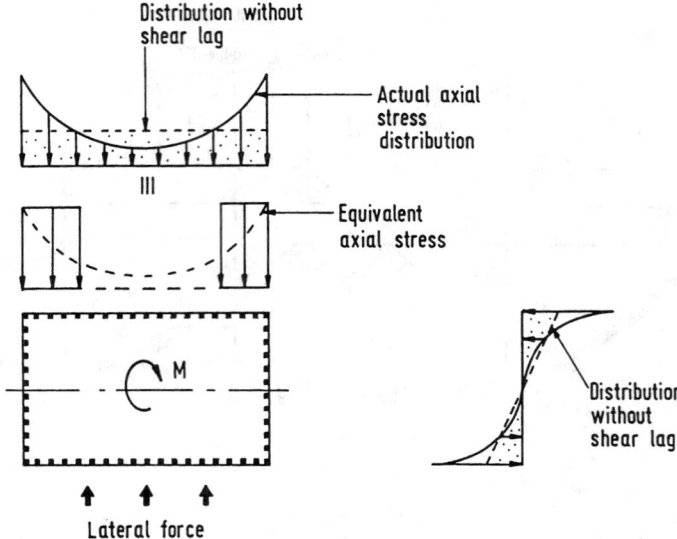

Figure 12.42   Equivalent column model for frame tube.

stiffness of the girder incorporating the slab may be utilized to reduce steel premium in controlling drift for high-rise frame design.

One approach is to use the composite beams as part of the frame. Since the slab element already exists, the composite flexural stiffness of the beams can be utilized with the steel beam alone resisting all negative moments. For an unbraced frame subjected to gravity and lateral loads, the beam typically bends in double curvature with a negative moment at one end of the beam and a positive moment at the other end. The concrete is assumed to be ineffective in tension; therefore, only the steel beam stiffness on the negative moment region and the composite stiffness on the positive moment region can be utilized for frame action. The frame analysis can be performed with a variable moment of inertia for the beams (see Figure 12.44). Further research is still needed in order to provide tangible guidance for design.

If semi-rigid composite joints are used in unbraced frames, the flexibility of the connections will contribute to additional drift over that of a fully rigid frame. In general, semi-rigid connections do not require the column size to be increased significantly over an equivalent rigid frame. This is because the design of frames with semi-rigid composite joints takes advantage of the additional stiffness in the beams provided by the composite action. The increase in beam stiffness would partially offset the additional flexibility introduced by the semi-rigid connections.

The story shear displacement Δ in an unbraced frame can be estimated using a modified expression from Equation 12.4 to account for the connection flexibility:

$$\Delta_i = \frac{V_i h_i^2}{12E}\left(\frac{1}{\Sigma(I_{ci}/h_i)} + \frac{1}{\Sigma(I_{cpi}/L_i)}\right) + \frac{V_i h_i^2}{\Sigma K_{con}} \tag{12.6}$$

where

Δ_i	=	the shear deflection of the i-th story
E	=	modulus of elasticity
I_c	=	moment of inertia for columns
I_g	=	moment of inertia of composite girder based on weighted average method
h_i	=	height of the i-th story
L_i	=	length of girder in the i-th story
V_i	=	total horizontal shear force in the i-th story

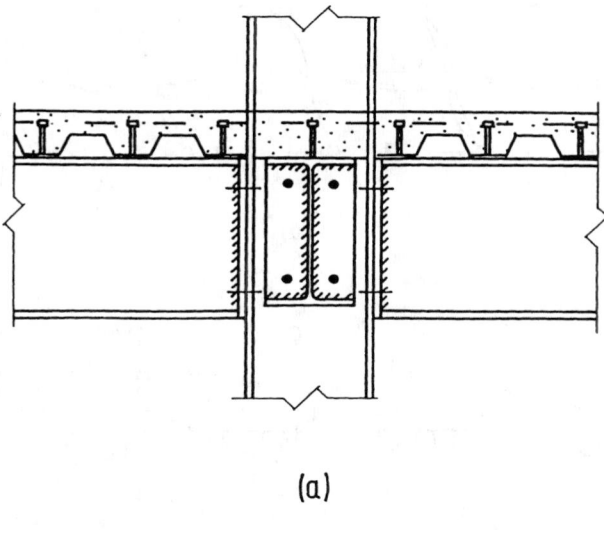

(a)

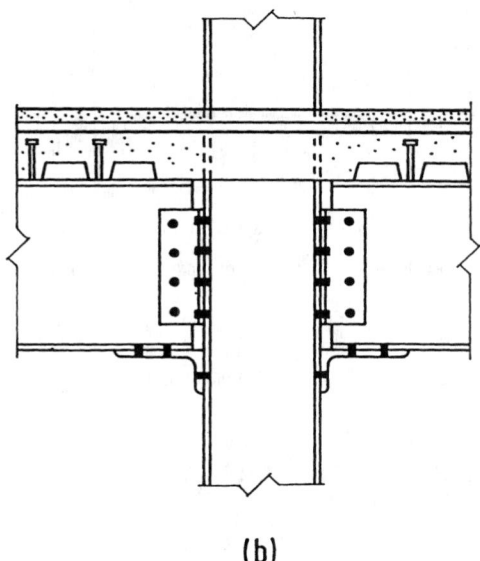

(b)

Figure 12.43 Composite beam-to-column connections with (a) flushed end plate and (b) seat and double web angles.

$\Sigma(I_{ci}/h_i)$ = sum of the column stiffness in the i-th story

$\Sigma(I_{gi}/L_i)$ = sum of the girder stiffness in the i-th story

ΣK_{con} = sum of the connection rotational stiffness in the i-th story

Further research is required to assess the performance of various types of composite connections used in building construction. Issues related to accurate modeling of effective stiffness of composite members and joints in unbraced frames for the computation of second-order effects and drifts need to be addressed.

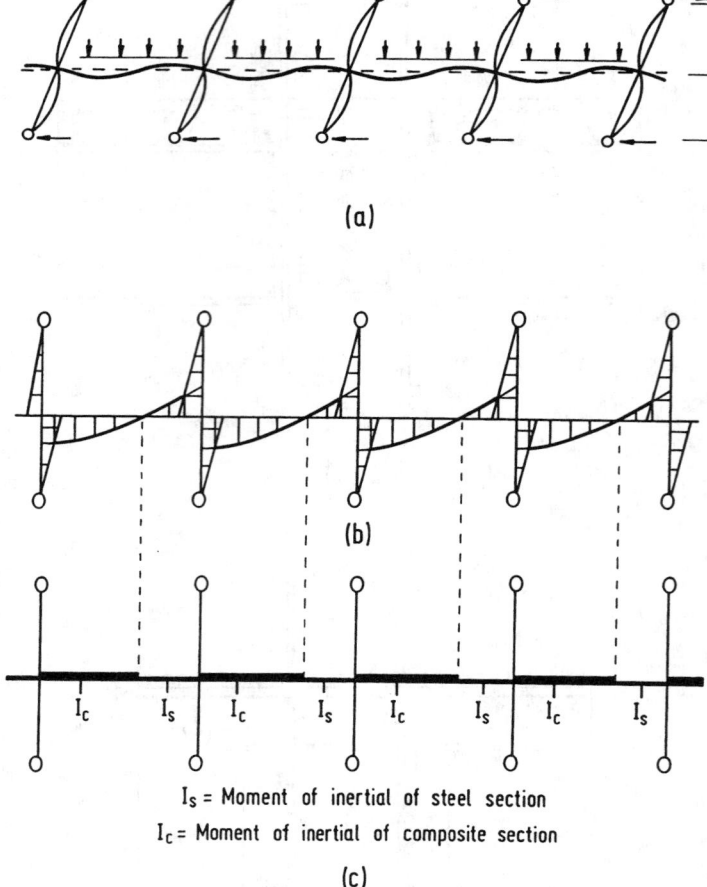

I_s = Moment of inertial of steel section

I_c = Moment of inertial of composite section

(c)

Figure 12.44   Composite unbraced frames: (a) story loads and idealization, (b) bending moment diagrams, and (c) composite beam stiffness.

12.4   Wind Effects on Buildings

12.4.1   Introduction

With the development of lightweight high strength materials, the recent trend is to build tall and slender buildings. The design of such buildings in non-seismic areas is often governed by the need to limit the wind-induced accelerations and drift to acceptable levels for human comfort and integrity of non-structural components, respectively. Thus, to check for serviceability of tall buildings, the peak resultant horizontal acceleration and displacement due to the combination of along wind, across wind, and torsional loads are required. As an approximate estimation, the peak effects due to along wind, across wind, and torsional responses may be determined individually and then combined vectorally. A reduction factor of .8 may be used on the combined value to account for the fact that in general the individual peaks do not occur simultaneously. If the calculated combined effect is less than any of the individual effects, then the latter should be considered for the design.

The effects of acceleration on human comfort is given in Table 12.3. The factors affecting the human response are:

1. Period of building—tolerence to acceleration tends to increase with **period**.

2. Women are more sensitive than men.
3. Children are more sensitive than adults.
4. Perception increases as you go from sitting on the floor, to sitting on a chair, to standing.
5. Perception threshold level decreases with prior knowledge that motion will occur.
6. Human body is more sensitive to fore-and-aft motion than to side-to-side motion.
7. Perception threshold is higher while walking than standing.
8. Visual cue—very sensitive to rotation of the building relative to fixed landmarks outside.
9. Acoustic cue—buildings make sounds while swaying due to rubbing of contact surfaces. These sounds, and sounds of the wind whistling, focus the attention on building motion even before motion is perceived, and thus lower the perception threshold.
10. The resultant translational acceleration due to the combination of longtitudinal, lateral, and torsional motions causes human discomfort. In addition, angular (torsional) motion appears to be more noticeable.

TABLE 12.3 Acceleration Limits for Different Perception Levels

Perception	Acceleration limits
Imperceptible	$a < 0.005$ g
Perceptible	0.005 g $< a < 0.015$ g
Annoying	0.015 g $< a < 0.05$ g
Very annoying	0.05 g $< a < 0.15$ g
Intolerable	$a > 0.15$ g

TABLE 12.4 Serviceability Problems at Various Deflection or Drift Indices

Deformation as a fraction of span or height	Visibility of deformation	Typical behavior
1/500	Not visible	Cracking of partition walls
1/300	Visible	General architectural damage Cracking in reinforced walls Cracking in secondary members Damage to ceiling and flooring Facade damage Cladding leakage Visual annoyance
1/200 - 1/300	Visible	Improper drainage
1/100 - 1/200	Visible	Damage to lightweight partitions, windows, finishes Impaired operation of removable components such as doors, windows, sliding partitions

Since the tolerable acceleration levels increase with period of building, the recommended design standard for peak acceleration for 10-year wind in commercial and residential buildings is as depicted in Figure 12.45 [28]. Lower acceleration levels are used for residential buildings for the following reasons:

1. Residential buildings are occupied for longer hours of the day and night and are therefore more likely to experience the design wind storm.
2. People are less sensitive to motion when they are occupied with their work than when they relax at home.
3. People are more tolerant of their work environment than of their home environment.

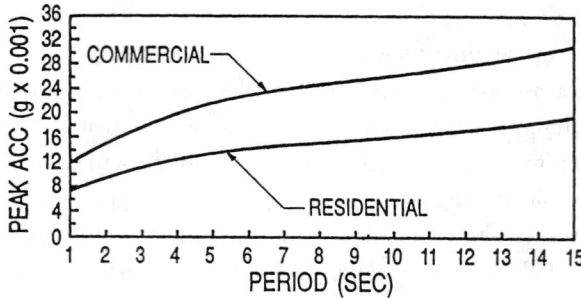

Figure 12.45   Design standard on peak acceleration for a 10-year return period.

4. Occupancy turnover rates are higher in commercial buildings than in residential buildings.

5. People can be evacuated easily from commercial buildings than residential buildings in the event of a peak storm.

The effects of excessive deflection on building components is described in Table 12.4. Thus, the allowable drift, defined as the resultant peak displacement at the top of the building divided by the height of the building, is generatly taken to be in the range 1/450 to 1/600.

Figure 12.46 depicts schematically the procedure of estimating the wind-induced accelerations and displacements in a building. The steps involved in this design procedure are described below with numerical examples for situations where the motion of the building does not affect the loads acting on the building. Finally, the situations when a wind tunnel study is required are listed at the end of this section.

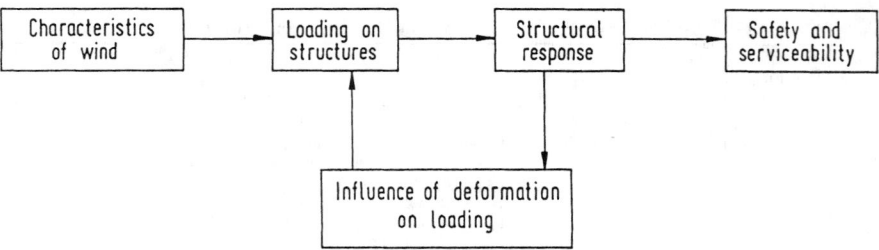

Figure 12.46   Schematic diagram for wind resistant design of structures.

12.4.2   Characteristics of Wind

Mean Wind Speed

The velocity of wind (wind speed) at great heights above the ground is constant and it is called the **gradient wind** speed. As shown in Figure 12.47, closer to the ground surface the wind speed is affected by frictional forces caused by the terrain and thus there is a **boundary layer** within which the wind speed varies from zero to the gradient wind speed. The thickness of the boundary layer (**gradient height**) depends on the ground roughness. For example, the gradient height is 457 m for large cities, 366 m for suburbs, 274 m for open terrain, and 213 m for open sea.

The velocity of wind averaged over 1 h is called the hourly mean wind speed, $\bar{U}$. The mean wind velocity profile within the atmospheric boundary layer is described by a power law

$$\bar{U}(z) = \bar{U}(z_{\text{ref}}) \left(\frac{z}{z_{\text{ref}}} \right)^{\alpha} \tag{12.7}$$

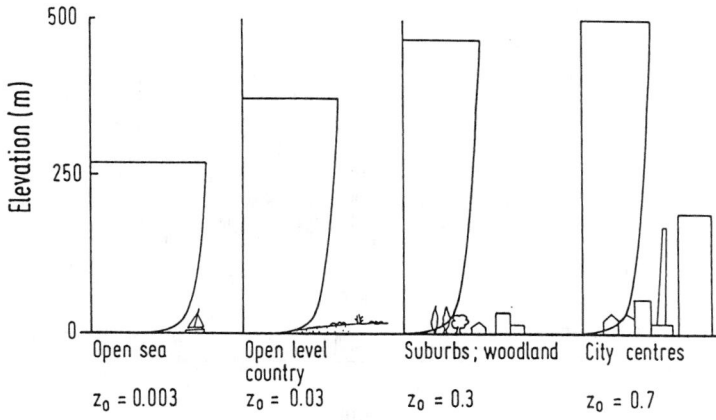

Figure 12.47 Mean wind profiles for different terrains.

in which $\bar{U}(z)$ is the mean wind speed at height z above the ground, z_{ref} is the reference height normally taken to be 10 m, and α is the power law exponent.

An alternative description of the mean wind velocity is by the logarithmic law, namely,

$$\bar{U}(z) = \frac{1}{K}u_* \ln\left(\frac{z-d}{z_o}\right) \tag{12.8}$$

in which u_* is the friction velocity, k is the von Karmon's constant equal to 0.4, z_o is the roughness length, and d is the height of zero-plane above the ground where the velocity is zero. Generally, zero plane is about 1 or 2 m below the average height of buildings and trees providing the roughness. Typical values of α, z_o, and d are given in Table 12.5 [4, 21].

TABLE 12.5 Typical Values of Terrain Parameters z_o, α, and d

	z_o	α	d (m)
City centers	.7	.33	15 to 25
Suburban terrain	.3	.22	5 to 10
Open terrain	.03	.14	0
Open sea	.003	.10	0

The roughness affects both the thickness of the boundary layer and the power law exponent. The thickness of the boundary layer and the power law exponent increase with the roughness of the surface. Consequently the velocity at any height decreases as the surface roughness increases. However, the gradient velocity will be the same for all surfaces. Thus, if the velocity of wind for a particular terrain is known, using Equation 12.7 and Table 12.5, the velocity at some other terrain can be computed.

Turbulence

The variation of wind velocity with time is shown in Figure 12.48. The eddies generated by the action of wind blowing over obstacles cause the turbulence. In general, the velocity of wind may be represented in a vector form as

$$U(z,t) = \bar{U}(z)\underline{i} + u(z,t)\underline{i} + v(z,t)\underline{j} + w(z,t)\underline{k} \tag{12.9}$$

where u, v, and w are the fluctuating components of the gust in x, y, z (longitudinal, lateral, and vertical axes) as shown in Figure 12.49 and $\bar{U}(z)$ is the mean wind along the x axis. The fluctuating

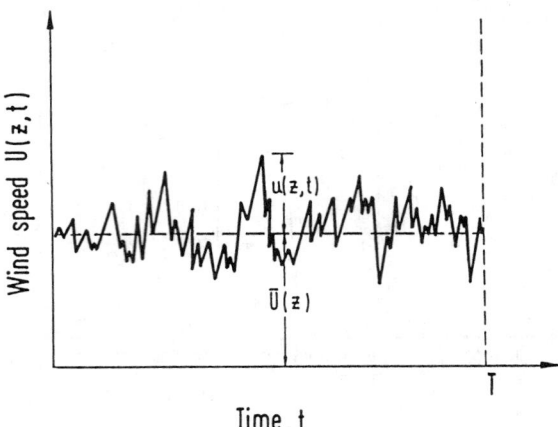

Figure 12.48   Variation of longitudinal component of turbulent wind with time.

component along the mean wind direction, u, is the largest and is therefore the most important for the vertical structures such as tall buildings which are flexible in the along wind direction. The vertical component w is important for horizontal structures that are flexible vertically, such as long span bridges.

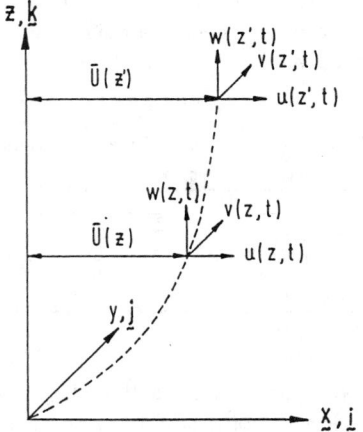

Figure 12.49   Velocity components of turbulent wind.

An overall measure of the intensity of turbulence is given by the root mean square value (r.m.s.). Thus, for the longitudinal component of the turbulence

$$\sigma_u(z) = \left[\frac{1}{T_o} \int_o^{T_o} \{u(z,t)^2\}dt \right]^{1/2} \tag{12.10}$$

where T_o is the averaging period. For the statistical properties of the wind to be independent on the part of the record that is being used, T_o is taken to be 1 h. Thus, the fluctuating wind is a stationary random function over 1 h.

The value of $\sigma_u(z)$ divided by the mean velocity $\bar{U}(z)$ is called the turbulence intensity

$$I_u(z) = \frac{\sigma_u(z)}{\bar{U}(z)} \tag{12.11}$$

which increases with ground roughness and decreases with height.

The variance of longitudinal turbulence can be determined from

$$\sigma_u^2 = \beta u_*^2 \tag{12.12}$$

where u_* is the friction velocity determined from Equation 12.8 and β which is independent of the height is given in Table 12.6 for various roughness lengths.

TABLE 12.6   Values of β for Various Roughness Lengths

z_o (m)	.005	0.7	.30	1.0	2.5
β	6.5	6.0	5.25	4.85	4.0

Integral Scales of Turbulence

The fluctuation of wind velocity at a point is due to eddies transported by the mean wind $\bar{U}$. Each eddy may be considered to be causing a periodic fluctuation at that point with a **frequency** n. The average size of the turbulent eddies are measured by **integral length scales**. For eddies associated with longitudinal velocity fluctuation, u, the integral length scales are L_u^x, L_u^y, and L_u^z describing the size of the eddies in longitudinal, lateral, and vertical directions, respectively. If L_u^y and L_u^z are comparable to the dimension of the structure normal to the wind, then the eddies will envelope the structure and give rise to well-correlated pressures and thus the effect is significant. On the other hand, if L_u^y and L_u^z are small, then the eddies produce uncorrelated pressures at various parts of the structure and the overall effect of the longitudinal turbulence will be small. Thus, the dynamic loading on a structure depends on the size of eddies.

Spectrum of Turbulence

The frequency content of the turbulence is represented by the power spectrum, which indicates the power or kinetic energy per unit time associated with eddies of different frequencies. An expression for the power spectrum is [60]:

$$\frac{n S_u(z, n)}{u_*^2} = \frac{200 f}{(1 + 50 f)^{5/3}} \tag{12.13}$$

where $f = nz/\bar{U}(z)$ is the reduced frequency. A typical spectrum of wind turbulence is shown in Figure 12.50. The spectrum has a peak value at a very low frequency around .04 Hz. As the typical range for the fundamental frequency of a tall building is .1 to 1 Hz, the buildings are affected by high-frequncy small eddies characterizing the decending part of the power spectrum.

Cross-Spectrum of Turbulence

The cross-spectrum of two continuous records is a measure of the degree to which the two records are correlated. If the records are taken at two points, M_1 and M_2, separated by a distance, r, then the cross-spectrum of longitudinal turbulent component is defined as

$$S_{u_1 u_2}(r, n) = S_{u_1 u_2}^c(r, n) + i S_{u_1 u_2}^q(r, n) \tag{12.14}$$

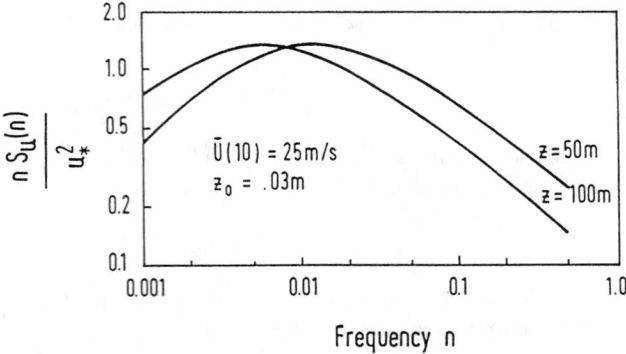

Figure 12.50   Power spectrum of longitudinal turbulence.

where the real and imaginary parts of the cross-spectrum are known as the co-spectrum and the quadrature spectrum, respectively. However, the latter is small enough to be neglected. Thus, the co-spectrum may be expressed non-dimensionally as the coherence and is given by

$$\gamma^2(r, n) = \frac{[S_{u_1 u_2}(r, n)]^2}{S_{u_1}(n) S_{u_2}(n)} \tag{12.15}$$

where $S_{u_1}(n)$ and $S_{u_2}(n)$ are the longitudinal velocity spectra at M_1 and M_2, respectively.

The square root of the coherence is given by the following expression [19]:

$$\gamma(r, n) = e^{-\hat{f}} \tag{12.16}$$

where

$$\hat{f} = \frac{n[c_z^2(z_1 - z_2)^2 + c_y^2(y_1 - y_2)^2]^{1/2}}{1/2[\bar{U}(z_1) + \bar{U}(z_2)]} \tag{12.17}$$

in which y_1, z_1 and y_2, z_2 are the coordinates of points M_1 and M_2. The line joining M_1 and M_2 is assumed to be perpendicular to the direction of the mean wind. The suggested values of c_y and c_z for the engineering calculation are 16 and 10, respectively [62].

12.4.3   Wind Induced Dynamic Forces

Forces Due to Uniform Flow

A bluff body in a two-dimensional flow, as shown in Figure 12.51, is subjected to a nett force in the direction of flow (**drag force**), and a force perpendicular to the flow (**lift force**). Furthermore, when the resultant force is eccentric to the elastic center, the body will be subjected to torsional moment. For uniform flow these forces and moment per unit height of the object are determined from

$$F_D = \frac{1}{2}\rho C_D B \bar{U}^2 \tag{12.18}$$

$$F_L = \frac{1}{2}\rho C_L B \bar{U}^2 \tag{12.19}$$

$$T = \frac{1}{2}\rho C_T B^2 \bar{U}^2 \tag{12.20}$$

where $\bar{U}$ is the mean velocity of the wind, ρ is the density of air, C_D and C_L are the drag and lift coefficients, C_T is the moment coefficient, and B is the characteristic length of the object such as the projected length normal to the flow.

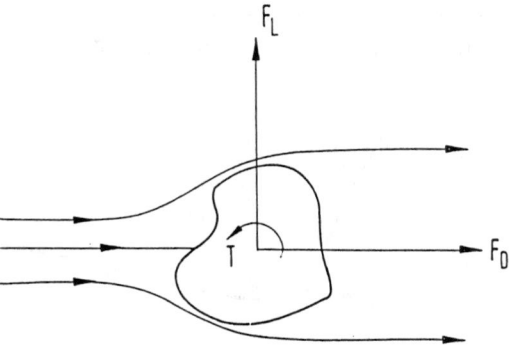

Figure 12.51   Drag and lift forces and torsional moment on a bluff body.

The drag coefficient for a rectangular building in the plan is shown in Figure 12.52 for various depth-to-breadth ratios [5]. The shear layers originating from the separation points at the windward corners surround a region known as the **wake**. For elongated sections, the stream lines that separate

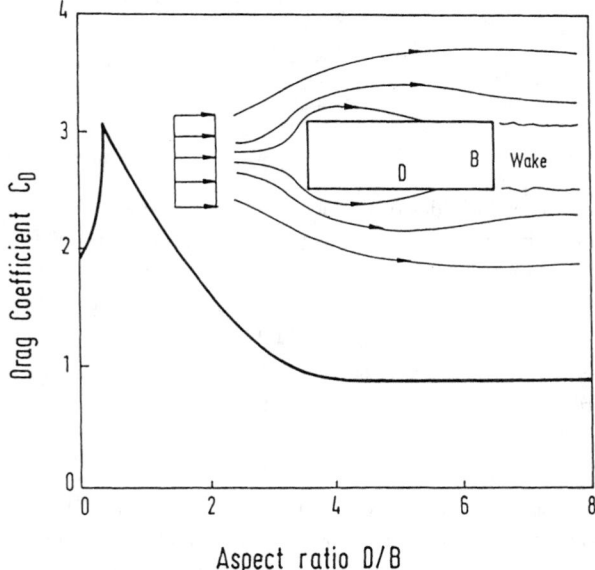

Aspect ratio D/B

Figure 12.52   Drag coefficient for a retangular section with different aspect ratios.

at the windward corners reattach to the body to form a narrower wake. This is attributed to the reduction in the drag for larger aspect ratios. For cylindrical buildings in the plan, the drag coefficient is dependent on Reynolds number as indicated in Figure 12.53.

Unlike the drag force, the lift force and torsional moment do not have a mean value for a symmetric object with a symmetric flow around it, as the symmetrical distribution of mean forces acting in the across wind direction cannot cause a net force. If the direction of wind is not parallel to the axes of symmetries or if the object is asymmetrical, then there will be a mean lift force and torsional moment. However, due to vortex shedding, fluctuating lift force and torsional moment will be present in both symmetric and non-symmetric structures. Figure 12.54 shows the mechanism of vortex shedding. Near the separation zones, strong shear stresses impart rotational motions to the fluid particles. Thus, discrete vortices are produced in the separation layers. These vortices

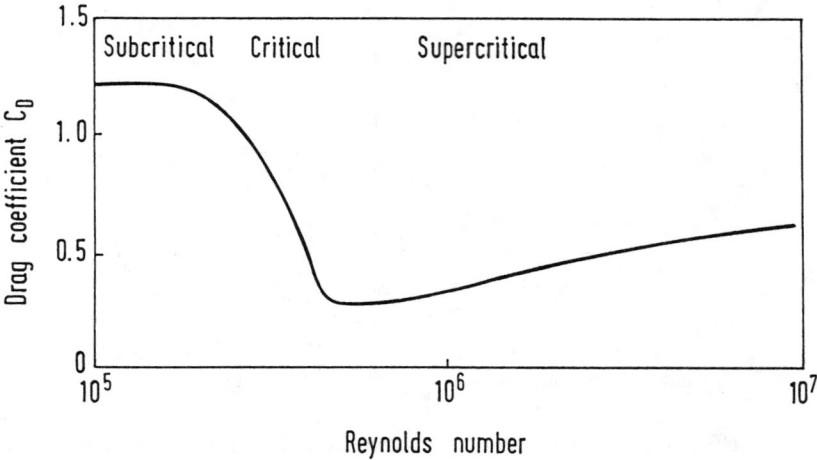

Figure 12.53   Effects of Reynolds number on drag coefficient of a circular cylinder.

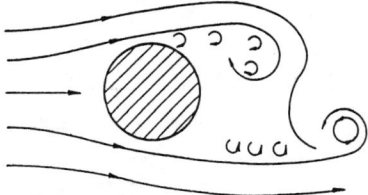

Figure 12.54   Vortex shedding in the wake of a bluff body.

are shed alternatively from the sides of the object. The asymmetric pressure distribution created by the vortices around the cross-section leads to an alternating transverse force (lift force) on the object. The vortex shedding frequency in Hz, n_s, is related to a non-dimensional parameter called the Strouhal number, S, defined as

$$S = n_s B / \bar{U} \tag{12.21}$$

where $\bar{U}$ is the mean wind speed and B is the width of the object normal to the wind. For objects with rounded profiles such as circular cylinders, the Strouhal number varies with the Reynolds number "Re" defined as

$$Re = \rho \bar{U} B / \mu \tag{12.22}$$

where ρ is the density of air and μ is the dynamic viscosity of the air. The vortex shedding becomes random in the transition region of $4 \times 10^5 < Re < 3 \times 10^6$ where the boundary layer at the surface of the cylinder changes from laminar to turbulent. Outside this transition range, the vortex shedding is regular producing a periodic lift force. For cross-sections with sharp corners, the Strouhal number is independent of the Reynolds number. The variation of the Strouhal number with length-to-breadth ratio of a rectangular cross-section is shown in Figure 12.55.

Forces Due to Turbulent Flow

If the wind is turbulent, then the velocity of the wind in the along wind direction is described as follows:

$$U(t) = \bar{U} + u(t) \tag{12.23}$$

where $\bar{U}$ is the mean wind and $u(t)$ is the turbulent component in the along wind direction. The time dependent drag force per unit height is obtained from Equation 12.18 by replacing $\bar{U}$ by $U(t)$. As the ratio $u(t)/\bar{U}$ is small, the time dependent drag force can be expressed as

$$f_D(t) = \bar{f}_D + f'_D(t) \tag{12.24}$$

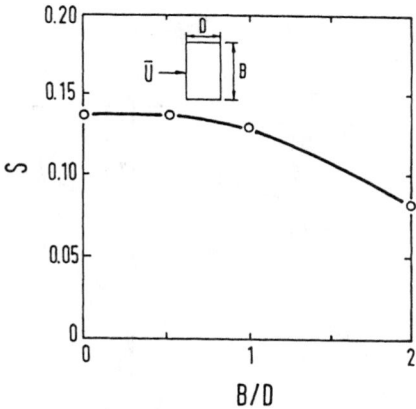

Figure 12.55 Strouhal number for a rectangular section.

where $\bar{f}_D$ and f'_D are the mean and the fluctuating parts of the drag force per unit height which are given by

$$\bar{f}_D = \frac{1}{2}\rho\bar{U}^2 C_D B \tag{12.25}$$

$$f'_D = \rho\bar{U}u C_D B \tag{12.26}$$

The spectral density of the fluctuating part of the drag force is obtained from the Fourier transformation of the auto correlation function as

$$S_{fD}(n) = \rho^2\bar{U}^2 B^2 C_D^2 S_u(n) \tag{12.27}$$

where $S_u(n)$ is the spectral density of the turbulent velocity, which may be obtained from Equation 12.13.

In practice, the presence of the structure distorts the turbulent flow, particularly the small high-frequency eddies. A correction factor known as the aerodynamic admittance function $\chi(n)$ may be introduced [17] to account for these effects. The following emprical formula has been suggested for $\chi(n)$ [62]:

$$\chi(n) = \frac{1}{1 + \left[\frac{2n\sqrt{A}}{\bar{U}(z)}\right]^{4/3}} \tag{12.28}$$

where A is the frontal area of the structure. Now with the introduction of the aerodynamic admittance function, Equation 12.27 may be rewritten as

$$S_{fD}(n) = \rho^2\bar{U}^2 B^2 C_D^2 \chi^2(n) S_u(n) \tag{12.29}$$

It is evident from Equation 12.26 that the fluctuating drag force varies linearly with the turbulence. Thus, large integral length scale and high **turbulent intensities** will cause strong buffeting and consequently increase the **along wind response** of the structure. However, the regularity of vortex shedding is affected by the presence of turbulence in the along wind and, hence, the across wind motion and torsional motion due to vortex shedding decrease as the level of turbulence increases.

12.4.4 Response Due to Along Wind

Tall slender buildings, where the breadth of the structure is small compared to the height, can be idealized as a line-like structure as shown in Figure 12.56. Modeling the building as a continuous

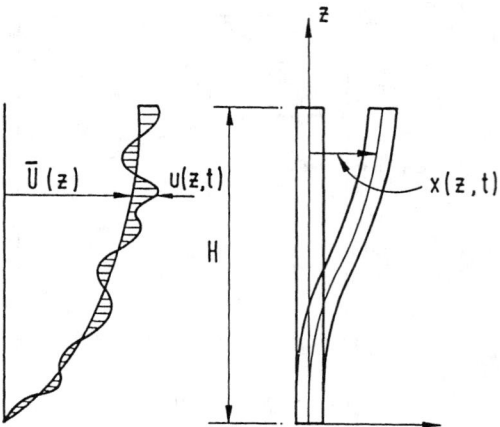

Figure 12.56   Typical deflection mode of a shear wall-frame building.

system, the governing equation of motion for along wind displacement $x(z, t)$ can be written as [29]:

$$m(z)\ddot{x}(z, t) + c(z)\dot{x}(z, t) + EI(z)x''''(z, t) - GA(z)x''(z, t) = f(z, t) \qquad (12.30)$$

where m, c, EI, and GA are, respectively, the mass, damping coefficient, flexural rigidity, and shear rigidity per unit height. Furthermore, $f(z, t)$ is the fluctuating wind load per unit height given in Equation 12.26.

Expressing the displacement in terms of the **normal coordinates,**

$$x(z, t) = \sum_{i=1}^{N} \phi_i(z)q_i(t) \qquad (12.31)$$

where ϕ_i is the i-th vibration mode shape and q_i is the i-th normal coordinate. Using the orthogonality conditions of **mode shapes,** Equation 12.30 can be expressed as [9]

$$m_i^* \ddot{q}_i + c_i^* \dot{q}_i + k_i^* q_i = p_i^* \quad i = 1 \text{ to } N \qquad (12.32)$$

where m_i^*, c_i^*, k_i^*, and p_i^* are the **generalized mass,** damping, stiffness, and force in the i-th mode of vibration. The generalized mass and force are determined from

$$m_i^* = \int_o^H m(z)\phi_i^2(z)dz \qquad (12.33)$$

$$p_i^* = \int_o^H f(z, t)\phi_i(z)dz$$

$$= \rho C_D B \int_o^H \bar{U}(z)u(z, t)\phi_i(z)dz \qquad (12.34)$$

Equation 12.32 consists of a set of uncoupled equations, each representing a single degree of freedom system. Using the random vibration theory [54], the power spectrum of the response in each normal coordinate is given by

$$S_{q_i}(n) = |H_i(n)|^2 S_{p_i^*}(n) \frac{1}{(k_i^*)^2} \qquad (12.35)$$

where

$$|H_i(n)| = \frac{1}{\left(\left[1 - \left(\frac{n}{n_i}\right)^2\right]^2 + 4\zeta_i^2 \left(\frac{n}{n_i}\right)^2\right)^{1/2}} \qquad (12.36)$$

and

$$k_i^* = 4\pi^2 n_i^2 m_i^* \tag{12.37}$$

in which n_i and ζ_i are the frequency and damping ratio in the i-th mode. The spectral density of the **generalized force** takes the form

$$
\begin{aligned}
S_{p_i^*}(n) &= \rho^2 C_D^2 B^2 \chi^2(n) \int_0^H \int_0^H \bar{U}(z_1) \bar{U}(z_2) S_{u_1 u_2}(r, n) \\
&\quad \phi_i(z_1) \phi_i(z_2) dz_1 \cdot dz_2
\end{aligned} \tag{12.38}
$$

where $S_{u_1 u_2}(r, n)$ is the cross-spectral density defined in Equation 12.14 with r being the distance between the coordinates z_1 and z_2. In Equation 12.38, the aerodynamic admittance has been incorporated to account for the distortion caused by the structure to the turbulent velocity.

In view of Equation 12.15, Equation 12.38 may be expressed as

$$
\begin{aligned}
S_{p_i^*}(n) &= \rho^2 C_D^2 B^2 \chi^2(n) \int_0^H \int_0^H \phi_i(z_1) \phi_i(z_2) \bar{U}(z_1) \bar{U}(z_2) \\
&\quad \sqrt{S_{u_1}(n)} \sqrt{S_{u_2}(n)} \gamma(r, n) dz_1 dz_2
\end{aligned} \tag{12.39}
$$

where $\gamma(r, n)$ is the square root of the coherence given in Equation 12.16, and $S_u(n)$ is the spectral density of the turbulent velocity.

The variance of the i-th normal coordinate is obtained from

$$\sigma_{q_i}^2 = \int_0^\infty S_{q_i}(n) dn = \frac{1}{(k_i^*)^2} \int_0^\infty |H_i \cdot (n)|^2 S_{p_i^*}(n) dn \tag{12.40}$$

The calculation of the above integral is very much simplified by observing the plot of the two components of the integrant shown in Figure 12.57. The mechanical admittance function is either 1.0 or 0 for most of the frequency range. However, over a relatively small range of frequencies around the natural frequency of the system, it attains very high values if the damping is small. As a result, the integrant takes the shape shown in Figure 12.57c. It has a sharp spike around the natural frequency of the system. The broad hump is governed by the shape of the turbulent velocity spectrum which is modified slightly by the aerodynamic admittance function. The area under the broad hump is the broad band or non-resonant response, whereas the area in the vicinity of the natural frequency gives the narrow band or **resonant response**. Thus, Equation 12.40 can be rewritten as

$$\sigma_{q_i}^2 = \frac{1}{(k_i^*)^2} \left[\int_o^{n_i - \Delta n} S_{p_i^*}(n) dn + \frac{\pi n_i}{4\zeta_i} S_{p_i^*}(n_i) \right] = \sigma_{Bq_i} + \sigma_{Dq_i}^2 \tag{12.41}$$

in which σ_{Bq_i} and σ_{Dq_i} are the non-resonating and resonating root mean square response of the i-th normal coordinate. As the responses due to various modes of vibration are statistically uncorrelated, the response of the system is given by

$$\sigma_x^2(z) = \sum_{i=1}^N \phi_i^2(z) \sigma_{Bq_i}^2 + \sum_{i=1}^N \phi_i^2(z) \sigma_{Dq_i}^2 \tag{12.42}$$

which gives the variance and, hence, the root mean square displacement at various heights.

The total displacement is obtained by including the static deflection due to the mean drag load, which is determined conveniently as follows:

In view of Equation 12.25, the mean generalized force is given by

$$
\begin{aligned}
\bar{f}_i &= \int_o^H \frac{1}{2} \rho C_D \bar{U}^2(z) B \phi_i(z) dz \\
&= \frac{1}{2} \rho C_D B \int_o^H \bar{U}^2(z) \phi_i(z) dz
\end{aligned} \tag{12.43}
$$

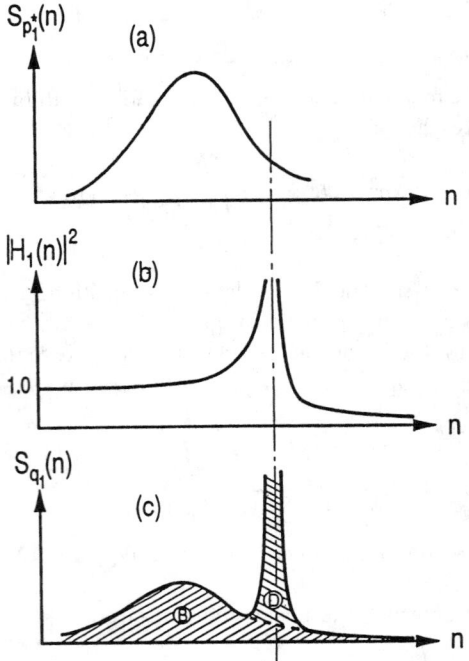

Figure 12.57   Schematic diagram for computation of response.

Then the mean displacement is determined from

$$\bar{x}(z) = \sum_{i=1}^{N} \phi_i(z) \left[\frac{\bar{f}_i}{(2\pi n_i)^2 m_i^*} \right] \tag{12.44}$$

The root mean square acceleration is obtained from

$$\sigma_{\ddot{x}}(z) = \left[\sum_{i=1}^{N} (2\pi n_i)^4 \phi_i^2(z) \sigma_{Dq_i}^2 \right]^{1/2} \tag{12.45}$$

The dynamic shear and bending moment at any height is obtained from the vibratory inertia forces in each mode and then by summing the modal contributions.

The probability of the response exceeding certain magnitude is determined using a **peak factor** on the root mean square response. Davenport [18] recommended the following expression for 50% probability of exceedence:

$$g_D = \sqrt{[2\ln(\nu T_0)]} + \frac{.577}{\sqrt{[2\ln(\nu T_0)]}} \tag{12.46}$$

where g_D is the peak factor, ν is the expected frequency at which the fluctuating response crosses the zero axis with a positive slope, and T_0 is the period (usually 3600 s) during which the peak response is assumed to occur.

For resonant response, ν is equal to the natural frequency and, thus, the peak factor for the resonant response g_D is obtained by setting $\nu = n$. For the non-resonating or broad band response the peak factor g_B has been evaluated to be 3.5 [20].

Using these peak factors, the most probable maximum value of the load effect, E, such as displacement, shear, bending moment, etc. are determined as follows:

$$E_{\max} = \bar{E} + \left[(g_B \sigma_{BE})^2 + (g_D \sigma_{DE})^2 \right]^{1/2} \tag{12.47}$$

where σ_{BE} and σDE are the non-resonating and resonating components of the load effect and $\bar{E}$ is the load effect due to mean wind.

EXAMPLE 12.1:

A rectangular building of height $H = 194$ m is situated in a suburban terrain. The breadth B and width D of the building are 56 m and 32 m, respectively. The period of the building corresponding to the fundamental sway mode is 5.15 s. The values of the mode shape at various heights are given below:

H(m)	0	20	40	75	95	135	150	170	194
ϕ	0	.032	.096	.248	.365	.611	.746	.849	1.0

The generalized mass and damping ratio corresponding to this mode are 18×10^6 kg and 2%, respectively.

Assuming that the mean wind profile follows the power law with a power law coefficient $\alpha = .22$, determine the maximum drift for a 50-year wind storm of 21 m/s at 10 m height, blowing normal to the breadth of the building. Given that the friction velocity is 2.96 m/s, the drag coefficient C_D is 1.3 and density of air $\rho = 1.2$ kg/m^3.

Solution

The mean height of the building $\bar{H} = 97$ m

$$\bar{U}(97) = \bar{U}(10)\left(\frac{97}{10}\right)^{.22} = 21\left(\frac{97}{10}\right)^{.22} = 34.6 \text{ m/s}$$

At mid-height, the reduced frequency

$$f = \frac{n\bar{H}}{\bar{U}(\bar{H})} = \frac{97n}{34.6} = 2.8n$$

From Equation 12.13, the spectrum of turbulent wind is given by

$$S_u(\bar{H}, n) = \frac{2.96^2 \times 200 \times 2.8}{(1 + 50 \times 2.8n)^{5/3}} = \frac{4906}{(1 + 140n)^{5/3}}$$

Resonant displacement

$$n_1 = \frac{1}{5.15} = .194 \text{ Hz},$$

$$S_u(\bar{H}, n_1) = \frac{4906}{(1 + 140 \times .194)^{5/3}} = 18.8 \text{ m}^2/\text{s}$$

The admittance function, from Equation 12.28, becomes

$$\chi(n) = \frac{1}{1 + \left(\frac{2n\sqrt{56 \times 194}}{34.6}\right)^{4/3}} = \frac{1}{1 + 10.96n^{4/3}}$$

$$\chi(n_1) = .45$$

From Equation 12.39,

$$S_{p_i^*}(n) = \rho^2 C_D^2 B^2 \chi^2(n) S_u(\bar{H}, n)\frac{[\bar{U}(\bar{H})]^2}{\bar{H}^{2\alpha}}$$

$$\int_0^H \int_0^H \phi_1(z_1)\phi_1(z_2)z_1^\alpha z_2^\alpha \gamma(z_1, z_2, n)dz_1dz_2$$

The square root of coherence γ is determined from Equation 12.16, considering only the vertical correlation. Thus,

$$S_{p_i^*}(n_1) = 1.2^2 \times 1.3^2 \times 56^2 \times (.45)^2 \times 18.8 \times \frac{(34.6)^2}{97.44} \times 16{,}900$$

$$= 7.85 \times 10^{10} \ \text{N}^2/\text{Hz}$$

From Equations 12.41 and 12.42, the variance of resonant displacement at the top of the building is obtained as

$$\sigma_D^2 = \phi_1^2(H)\sigma_{Dq_1}^2 = \frac{1}{(k_1^*)^2} \frac{\pi n_1}{4\zeta_1} S_{p_1^*}(n_1)$$

$$\sigma_D^2 = \left(\frac{1}{26.8 \times 10^6}\right)^2 \left(\frac{\pi(.194)}{4(.02)}\right)(7.85 \times 10^{10})10^6 \ \text{mm}^2$$

$$\sigma_D = 28.9 \ \text{mm}$$

Non-resonant displacement

The variance of non-resonant displacement at the top of the building is determined from Equations 12.41 and 12.42 as

$$\sigma_B^2 = \phi_1^2(H)\sigma_{Bq_1}^2 = \frac{1}{(k_1^*)^2} \int_0^{n_1 - \Delta n} S_{p_1^*}(n)\,dn$$

$$= \frac{565 \times 10^9}{(26.8 \times 10^6)^2} \times 10^6 \ \text{mm}$$

$$\sigma_B = 28 \ \text{mm}$$

Response to mean wind

From Equation 12.43, the mean generalized force

$$\bar{f}_1 = \frac{1}{2}\rho C_D B \int_o^H \bar{U}^2(z)\phi_1(z)\,dz$$

$$= \frac{1}{2}\rho C_D B[\bar{U}(\bar{H})]^2 \left(\frac{1}{\bar{H}}\right)^{2\alpha} \int_o^H z^{2\alpha}\phi_1(z)\,dz$$

$$= \frac{1}{2} \times 1.2 \times 1.3 \times 56 \times (34.6)^2 \left(\frac{1}{97}\right)^{.44} \times 684$$

$$= 4.8 \times 10^6 \ \text{N}$$

The generalized stiffness

$$k_1^* = \left(\frac{2\pi}{5.15}\right)^2 \times 18 \times 10^6$$

$$= 26.8 \times 10^6 \ \text{N/m}$$

Thus, the mean displacement

$$\bar{X} = \frac{4.8 \times 10^6}{26.8 \times 10^6} \times 10^3 = 179 \ \text{mm}$$

The peak factor g_D for resonant response is determined from Equation 12.46 as 3.78. Using a peak factor of 3.5 for non-resonant response, the most probable maximum displacement is

$$X_{\max} = \bar{X} + \sqrt{(g_B\sigma_B)^2 + (g_D\sigma_D)^2}$$

$$= 179 + \sqrt{(3.5 \times 28)^2 + (3.78 \times 28.9)^2}$$

$$= 326 \ \text{mm}$$

The most probable maximum drift would be

$$= \frac{326}{194,000} = \frac{1}{595}$$

The peak acceleration would be

$$
\begin{aligned}
&= .0289 \times 3.78 \times (2\pi)^2 \times (.194)^2 \\
&= 0.16 \text{ m/s}^2 \quad (1.6\% \text{ g})
\end{aligned}
$$

12.4.5   Response Due to Across Wind

For most modern tall buildings, the across wind response is more significant than the along wind response. Across wind vibration of a building is caused by the combination of forces from three sources: (1) buffeting by the turbulence in the across wind direction, (2) **wake excitation** due to vortex shedding, and (3) **lock-in**, a displacement dependent excitation.

The across wind force due to lateral turbulence in the approaching flow is generally small compared to the effects due to other mechanisms. Lock-in is the term used to describe large amplitude across wind motion that occurs when the vortex shedding frequency is close to the natural frequency. If the across wind response exceeds a certain critical value, the across wind response causes an increase in the excitation force, which in turn increases the response. The vortex shedding frequency tends to couple with the natural frequency of the structure for a range of wind velocities, and the large amplitude response will persist. Lock-in is likely to occur only in the case of structures with relatively low stiffness and low damping, operating near the critical wind velocity given by

$$\bar{U}_{\text{crit}} = \frac{n_o B}{S} \tag{12.48}$$

in which $\bar{U}_{\text{crit}}$ is the critical wind speed, B is the breadth of the structure normal to the wind stream, n_o (in Hz) is the fundamental natural frequency of the structure in the across-wind direction, and S is the Strouhal number.

Buildings should be designed so that lock-in effects do not occur during their anticipated life. If the root mean square displacement at the top of the structure is less than a certain critical value, then lock-in will not occur. For square tall buildings, the critical root mean square displacements σ_{yc} expressed as a ratio with respect to the breadth (σ_{yc}/B) are approximately .015, .025, and .045, respectively [55], for open terrain $(z_o = .07 \text{ m})$, suburban terrain $(z_o = 1.0 \text{ m})$, and city centers $(z_o = 2.5 \text{ m})$. For circular sections with diameter D, the value of σ_{yc}/D is approximately .006 for suburban terrain.

Thus, for buildings, the most common cause for across wind motion is the wake excitation. Although the turbulence in the atmospheric boundary layer affects the regularity of vortex shedding, the shed vortices have a predominant period which could be determined from an appropriate Strouhal number. Because the vortex shedding is random, the fluctuating across wind force is effectively broad-band as shown in Figure 12.58. The band width and the energy concentration near the vortex shedding frequency depends on the geometry of the building and the characteristics of the approach flow.

The response due to this across wind random excitation can be determined using the random vibration theory. Idealizing the tall building as a line-like structure, the across wind displacement $y(z, t)$ may be expressed in terms of the normal coordinates $r_i(t)$ as

$$y(z, t) = \sum_{i=1}^{N} \psi_i(z) r_i(t) \tag{12.49}$$

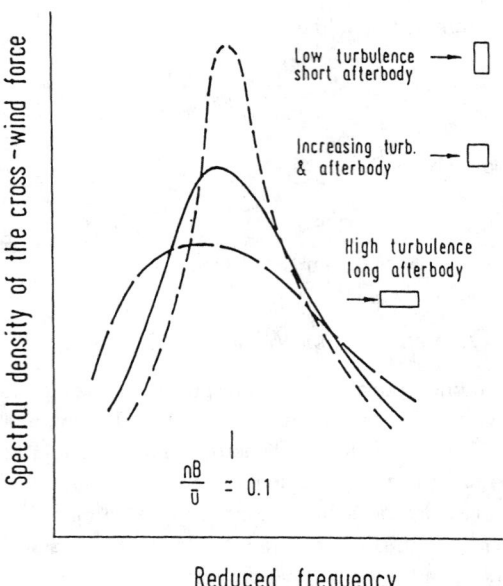

Figure 12.58   Effects of turbulence intensity and after body length on across wind force spectra.

where $\psi_i(z)$ is the i-th vibration mode in the across wind direction and N is the total number of modes considered to be significant. The governing equation of motion in terms of generalized mass m_i^*, generalized damping c_i^*, and generalized stiffness k_i^*, takes the form

$$m_i^* \ddot{r}_i + c_i^* \dot{r}_i + k_i^* r_i = f_i^*(t) \qquad i = 1 \text{ to } N \tag{12.50}$$

in which

$$
\begin{aligned}
m_i^* &= \int_o^H m(z)\psi_i^2(z)dz \\
k_i^* &= (2\pi n_i)^2 m_i^* \\
c_i^* &= 2\zeta_i \sqrt{m_i^* k_i^*} \\
f_i^*(t) &= \int_o^H f(z,t)\psi_i(z)dz
\end{aligned}
\tag{12.51}
$$

where H is the height of the building, $m(z)$ is the mass per unit length, n_i is the frequency of the i-th mode in the across wind direction, ζ_i is the damping ratio in the i-th mode, $f(z, t)$ is the across wind force per unit height, and $f_i^*(t)$ is the generalized across wind force in the i-th mode. The spectral density of each normal coordinate can be determined from

$$S_{r_i}(n) = \frac{|H_i(n)|^2}{(k_i^*)^2} S_{f_i^*}(n) \tag{12.52}$$

where $|H_i(n)|$ is the mechanical admittance function and $S_{f_i^*}$ is the **power spectral density** of the generalized across wind force.

The variance of the normal coordinate r_i is given by

$$\sigma_{r_i}^2 = \int_0^\infty S_{r_i}(n)dn \tag{12.53}$$

Hence, the variance of the across wind displacement is obtained from

$$\sigma_y^2(z) = \sum_{i=1}^{N} \psi_i^2(z)\sigma_{r_i}^2 \tag{12.54}$$

In Equation 12.53, if the contribution from the non-resonating component is neglected, then the root mean square response of the across wind displacement is given by

$$\sigma_y^2(z) = \left[\sum_{i=1}^{N} \frac{\psi_i^2(z)}{(2\pi n_i)^4 (m_i^*)^2} \left(\frac{\pi n_i}{4\zeta_i} \right) S_{f_i^*}(n_i) \right]^{1/2} \tag{12.55}$$

For convenient use of the above equation, the generalized force spectra obtained experimentally by Kwok and Melbourne [32] and Saunders and Melbourne [56] are presented in Figure 12.59 for various aspect ratios of square and rectangular buildings deflecting in a linear mode.

EXAMPLE 12.2:

Consider the building of Example 12.1. If the period of vibration in the across wind direction is 5.2 s, assuming a linear mode, determine the acceleration in the across wind direction for a 10-year wind storm of 14 m/s at 10 m height, given that the generalized mass corresponding to the linear mode is 17.5×10^6 kg and the damping in this mode of oscillation is 2%.
   Solution

The building is rectangular with an aspect ratio of

$$H : B : D = 6 : 1.75 : 1$$

Since the building is in a suburban terrain, the generalized cross wind force can be determined from Figure 12.59e.
   The wind speed at the tip of the building

$$\bar{U}(H) = \bar{U}(10) \times \left(\frac{194}{10} \right)^{.22} = 26.9 \text{ m/s}$$

The reduced frequency,

$$\frac{n_1 B}{\bar{U}(H)} = \frac{.192 \times 56}{26.9} = .40$$

then from Figure 12.59e.

$$S_{f_i^*}(n_1) = \left(\frac{.00018}{.192} \right) \left(\frac{1}{2} \times 1.2 \times 26.9^2 \times 56 \times 194 \right)^2$$
$$= 2.09 \times 10^{10} \text{ N}^2/\text{Hz}$$

From Equation 12.55,

$$\sigma_y = \left[\left(\frac{\pi \times .192}{4 \times .02} \right) (2.09 \times 10^{10}) \frac{1}{(2\pi \times .192)^4 (17.5 \times 10^6)^2} \right]^{1/2}$$
$$= .016 \text{ m}$$

Assuming a peak factor of 4, the peak acceleration in the cross wind direction

$$4 \times .016(2\pi)^2(.192)^2 = .093 \text{ m/s}^2$$

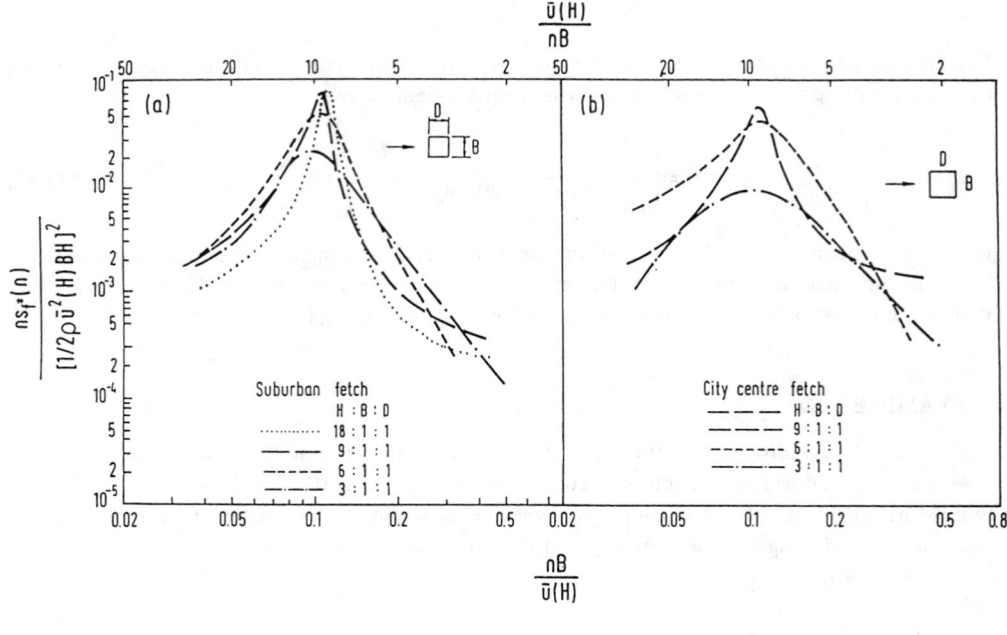

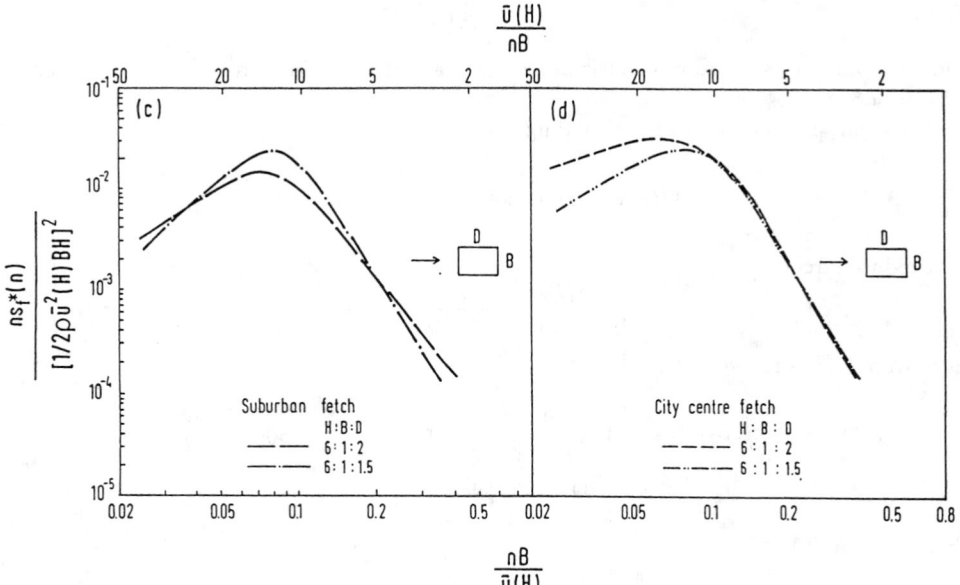

Figure 12.59   Generalized force spectra for a square and a rectangular building in suburban and city center fetch.

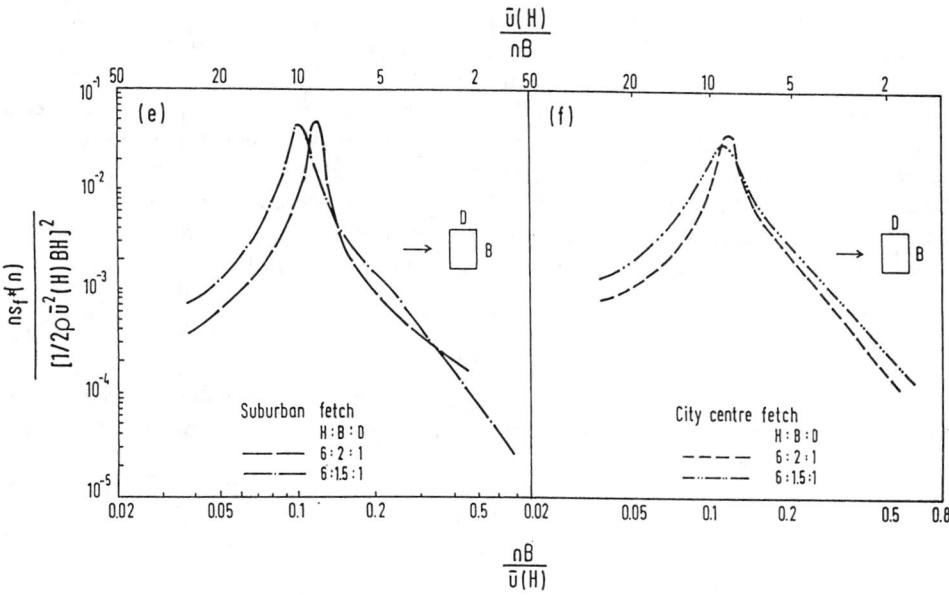

Figure 12.59 *(Continued)* Generalized force spectra for a square and a rectangular building in suburban and city center fetch.

12.4.6   Torsional Response

A building will be subjected to torsional motion when the instantaneous point of application of resultant aerodynamic load does not coincide with the center of mass and/or the elastic center. The major source for dynamic torque is the flow induced asymmetries in the lift force and the pressure fluctuation on the leeward side caused by the vortex shedding. Any eccentricities between the center of mass and center of stiffness present in asymmetrical buildings can amplify the torsional effects.

Balendra, Nathan, and Kang [8] have presented a time domain approach to estimate the coupled lateral-torsional motion of buildings due to along wind turbulence and across wind forces and torque due to wake excitation. The experimentally measured power spectra of across wind force and torsional moment [53] were used in this analysis. This method is useful at the final stages of design as specific details that are unique for a particular building can be easily incorporated in the analytical model. A useful method to assess the torsional effects at the preliminary design stage is given by the following empirical relation [58] which yields the peak base torque induced by wind speed $\bar{U}(H)$ at the top of the building as:

$$T_{\text{peak}} = \Psi(\bar{T} + g_T T_{\text{rms}}) \tag{12.56}$$

where Ψ is a reduction coefficient, g_T is the torsional peak factor equal to 3.8, and $\bar{T}$ and T_{rms} are the mean and root mean square base torques which are given by

$$T_{\text{rms}} = .00167 \frac{1}{\sqrt{\zeta_T}} \rho L^4 H n_T^2 U_r^{2.68} \tag{12.57}$$

$$\bar{T} = .038 \rho L^4 H n_T^2 U_r^2 \tag{12.58}$$

in which

$$L = \frac{\int |r| \, ds}{\sqrt{A}} \tag{12.59}$$

$$U_r = \frac{\bar{U}(H)}{n_T L} \tag{12.60}$$

where ρ is the density, H is the height of the building, n_T and ζ_T are the frequency and damping ratio in the fundamental torsional mode of vibration, $|r|$ is the distance between the elastic center and the normal to an element ds on the boundary of the building, and A is the cross-sectional area of the building. The expressions for $\bar{T}$ and T_{rms} are obtained for the most unfavorable directions for the mean and root mean square values of the base torque. In general, these directions do not coincide and furthermore will not be along the direction of the extreme winds expected to occur at the site. As such, a reduction coefficient Ψ ($.75 < \Psi \leq 1$) is incorporated in Equation 12.56.

For a linear fundamental mode shape, the peak torsional induced horizontal acceleration at the top of the building at a distance "a" from the elastic center is given by [27]

$$a\ddot{\theta} = \frac{2ag_T T_{\mathrm{rms}}}{\rho_b BDH r_m^2} \qquad (12.61)$$

where $\ddot{\theta}$ is the peak angular acceleration, ρ_b is the mass density of the building, B and D are the breadth and depth of the building, and r_m is the radius of gyration. For a rectangular building with uniform mass density,

$$r_m^2 = \frac{1}{12}(B^2 + D^2) \qquad (12.62)$$

EXAMPLE 12.3:

If the torsional frequency of the building in Example 12.2 is .8 Hz, assuming a linear mode and 2% damping ratio, determine the peak acceleration at the corner of the building due to torsional motion for a 10-year wind storm of 14 m/s, given that the center of rigidity is at the geometric center of the building.

     Solution

For a rectangular building

$$\int |r|\, ds = \frac{1}{2}(B^2 + D^2)$$

Thus, from Equations 12.59 and 12.60,

$$
\begin{aligned}
L &= \frac{1}{\sqrt{BD}}(B^2 + D^2)\frac{1}{2} = 49.1 \text{ m} \\
U_r &= \frac{U(H)}{n_T L} \frac{26.9}{.8 \times 49.1} = .685
\end{aligned}
$$

From Equation 12.58

$$
\begin{aligned}
T_{\mathrm{rms}} &= .00167\left(\frac{1}{\sqrt{(.02)}}\right)(1.2)(49.1)^4(194)(.8)^2(.685)^{2.68} \\
&= 3.71 \times 10^6 \text{ N.m}
\end{aligned}
$$

The average density of the building is determined as

$$\rho_b = \frac{3m_1^*}{AH} = \frac{3 \times 17.5 \times 10^6}{56 \times 32 \times 194} = 151 \text{ kg/m}^3$$

Thus, the peak torsional acceleration of the corner for which $a = 32.2$ m, is

$$a\ddot{\theta} = \frac{2 \times 32.2 \times 3.8 \times 3.71 \times 10^6}{151 \times 56 \times 32 \times 194 \times 346.7} = .05 \text{ m/s}^2$$

12.4.7 Response by Wind Tunnel Tests

There are many situations where analytical methods cannot be used to estimate certain types of wind loads and associated structural response. For example, the aerodynamic shape of the building is rather uncommon or the building is very flexible so that its motion affects the aerodynamic forces acting on the building. In such situations, a more accurate estimate of wind effects on buildings are obtained through **aeroelastic model** tests in a boundary-layer wind tunnel [9].

The aeroelastic model studies would provide the overall mean and dynamic loads, displacements, rotations, and accelerations. The aeroelastic model studies may be required under the following situations:

1. when the height-to-width ratio exceeds 5
2. when the structure is light with a density in the order of 1.5 kN/m^3
3. the fundamental period is long in the order of 5 to 10 s
4. when the natural frequency of the building in the cross wind direction is in the neighborhood of the shedding frequency
5. when the building is torsionally flexible
6. when the building is expected to execute strongly coupled lateral-torsional motion.

12.5 Defining Terms

Aeroelastic model: The model which simulates the dynamic properties of buildings to capture the motion dependent loads.

Along wind response: Response in the direction of wind.

Boundary layer: The layer within which the velocity varies because of ground roughness.

Bracing frames: Frames that provide lateral stability to the overall framework.

Composite beams: Steel beam acting compositely with part of the concrete slab through shear connectors.

Cross wind response: Response perpendicular to the direction of wind.

Drag force: Force in the direction of wind.

Frequency: Number of cycles per second.

Gradient height: Thickness of the boundary layer.

Gradient wind: Wind velocity above the boundary layer.

Generalized force: Force associated with a particular mode of vibration.

Generalized mass: Participating mass in a particular mode of vibration.

Integral length scale: A measure of average size of the eddies.

Lift force: Force perpendicular to the flow.

Lock-in: Situation where the vortex shedding frequency tends to couple with the frequency of the structure.

Long span systems: Structural systems that span a long distance. The design is likely to be governed by serviceability limit states.

Mode shapes: Free vibration deflection configurations in each frequency of the structure.

Non-resonating response: Response due to eddies whose frequencies are remote from the structural frequency.

Normal coordinates: Coordinates associated with modes of vibration.

Peak factor: Ratio between the peak and rms values.

Period: Duration of one complete cycle.

Power spectral density: Kinetic energy per unit time associated with eddies of different frequencies.

Resonant response: Response due to eddies whose frequencies are in the neighborhood of structural frequency.

Rigid frames: Frames resisting lateral load by bending of members which are rigidly connected.

Simple frames: Frames that have no lateral resistance and whose members are pinned connected.

Stiffness: Force required to produce unit displacement.

Sway frames: Frames in which the second-order effects due to gravity load acting on the deformed geometry can influence the force distribution in the structure.

Torsional response: Response causing twisting motion.

Turbulent intensity: Overall measure of intensity of turbulence.

Wake: Region surrounded by the shear layers originating from separation points.

Wake excitation: Excitation caused by the vortices in the wake.

References

[1] AISC. 1989. *Allowable Stress Design and Plastic Design Specifications for Structural Steel Buildings,* 9th ed., American Institute of Steel Construction, Chicago, IL.

[2] AISC. 1990. *LRFD-Simple Shear Connections,* American Institute of Steel Construction, Chicago, IL.

[3] AISC. 1993. *Load and Resistance Factor Design Specification for Structural Steel Buildings,* American Institute of Steel Construction, 2nd ed., Chicago, IL.

[4] ANSI. 1982. *American National Standard Building Code Requirements for Minimum Design Loads in Buildings and Other Structures,* A 58.1, New York.

[5] ASCE. 1987. Wind loading and wind induced structural response, *State-of-the-Art Report,* Committee on Wind Effects, New York.

[6] ASCE. 1990. Minimum design loads for buildings and other structures, ASCE Standard, *ASCE 7-88,* American Society of Civil Engineers.

[7] ASCE Task Committee. 1996. Proposed specification and commentary for composite joints and composite trusses, ASCE Task Committee on Design Criteria for Composite in Steel and Concrete, *J. Structural Eng., ASCE,* April, 122(4), 350-358.

[8] Balendra, T., Nathan, G. K., and Kang, K. H. 1989. Deterministic model for wind induced oscillations of buildings. *J. Eng. Mech., ASCE,* 115, 179-199.

[9] Balendra, T. 1993. *Vibration of Buildings to Wind and Earthquake Loads,* Springer-Verlag.

[10] Brett, P. and Rushton J. 1990. *Parallel Beam Approach—A Design Guide,* The Steel Construction Institute, U.K.

[11] BS5950:Part 1. 1990. Structural Use of Steelwork in Building. Part 1: Code of Practice for Design in Simple and Continuous Construction: Hot Rolled Section, *British Standards Institution,* London.

[12] Chen, W.F. and Atsuta, T. 1976. *Theory of Beam-Column, Vol. 1, In-Plane Behavior and Design,* MacGraw-Hill, New York.

[13] Chen W. F. and Lui, E. M. 1991. *Stability Design of Steel Frames,* CRC Press, Boca Raton, FL.

[14] Chen, W. F. and Toma, S. 1994. *Advanced Analysis in Steel Frames: Theory, Software and Applications,* CRC Press, Boca Raton, FL.

[15] Chen, W. F., Goto, Y., and Liew, J.Y.R. 1996. *Stability Design of Semi-Rigid Frames,* John Wiley & Sons, New York.

[16] Council On Tall Buildings and Urban Habitat. 1995. *Architecture of Tall Buildings,* Armstrong, P. J., Ed., McGraw-Hill, New York.

[17] Davenport, A. G. 1961. The application of statistical concepts to the wind loading of structures. *Proc. Inst. Civil Eng.,* 19, 449-472.

[18] Davenport, A. G. 1964. Note on the distribution of the largest value of a random function with application to gust loading. *Proc. Inst. Civil Eng.,* 28, 187-196.

[19] Davenport, A. G. 1968. The dependence of wind load upon meteorological parameters. *Proc. Intl. Res. Sem. Wind Effects on Buildings and Structures,* University of Toronto Press, Toronto, 19-82.

[20] ESDU. 1976. *The Response of Flexible Structures to Atmospheric Turbulence.* Item 76001, Engineering Sciences Data Unit, London.

[21] ESDU. 1985. *Characteristics of Atmospheric Turbulence Near the Ground, Part II: Single Point Data for Strong Winds* (Neutral Atmosphere). Item 85020, Engineering Sciences Data Unit, London.

[22] Eurocode 3. 1992. *Design of Steel Structures: Part 1.1—General Rules and Rules for Buildings,* National Application Document for use in the UK with ENV1993-1-1:1991, Draft for Development.

[23] Eurocode 3. 1992. *Design of Steel Structures: Part 1.1—General Rules and Rules for Buildings,* National Application Document for use in the UK with ENV1993-1-1:1991, Draft for Development.

[24] Eurocode 4. 1994. *Design of Composite Steel and Concrete Structures: General Rules for Buildings,* preENV 1994-1-1, European Committee for Standardization.

[25] Fishers, J.M. and West, M.A. 1990. *Serviceability Design Considerations for Low-Rise Buildings,* American Institute of Steel Construction, Chicago, IL.

[26] Geschwindner, L.F., Disque, R.O., and Bjorhovde, R. 1994. *Load and Resistance Factored Design of Steel Structures,* Prentice Hall, Englewood Cliffs, NJ.

[27] Greig, L. 1980. *Toward an Estimate of Wind Induced Dynamic Torque on Tall Buildings.* M.Sc. thesis, Department of Engineering, University of Western Ontario, London, Ontario.

[28] Griffis, L. G. 1993. Serviceability limit states under wind load. *Eng. J., AISC,* pp. 1-16.

[29] Heidebrecht, A. C. and Smith, B. S. 1973. Approximate analysis of tall wall-frame structures. *J. Structural Div., ASCE,* 99, 199-221.

[30] Iyengar, S.H., Baker, W.F., and Sinn, R. 1992. Multi-Story Buildings, in *Constructional Steel Design, An International Guide,* Chapter 6.2, Dowling, P. J., et al., Eds., Elsevier, England, 645-670.

[31] Knowles, P. R. 1985. *Design of Castellated Beams,* The Steel Construction Institute, U.K.

[32] Kwok, K. C. S. and Melbourne, W. H. 1981. Wind induced lock-in excitation of tall structures, *J. Structural Div., ASCE,* 107, 57-72.

[33] Lawson, R. M. 1987. *Design for Openings in Webs of Composite Beams CIRIA,* The Steel Construction Institute, U.K.

[34] Lawson, R. M. 1993. *Comparative Structure Cost of Modern Commercial Buildings,* The Steel Construction Institute, U.K.

[35] Lawson, R. M. and McConnel, R. E. 1993. *Design of Stub Girders,* The Steel Construction Institute, U.K.

[36] Lawson, R. M. and Rackham, J. W. 1989. *Design of Haunched Composite Beams in Buildings,* The Steel Construction Institute, U.K.

[37] Leon, R.T. 1990. Semi-rigid composite construction, *J. Constructional Steel Res.,* 15(1&2), 99-120.

[38] Leon, R.T. 1994. *Composite Semi-Rigid Construction, Steel Design: An International Guide,* R. Bjorhovde, J. Harding and P. Dowling, Eds., Elsevier, 501-522.

[39] Leon, R.T. and Ammerman, D.J. 1990. Semi-rigid composite connections for gravity loads, *Eng. J.*, AISC, 1st Qrt., 1-11.

[40] Leon, R. T., Hoffman, J.J., and Staeger, T. 1996. Partially restrained composite connections, *AISC Steel Design Guide Series 8, AISC.*

[41] Liew, J. Y. R. 1995. Design concepts and structural schemes for steel multi-story buildings, *J. Singapore Structural Steel Soc., Steel Structures*, 6(1), 45-59.

[42] Liew, J.Y.R. and Chen, W. F. 1994. Implications of using refined plastic hinge analysis for load and resistance factor design, *J. Thin-Walled Structures*, Elsevier Applied Science, London, UK, 20(1-4), 17-47.

[43] Liew, J.Y.R. and Chen, W.F. 1995. Analysis and design of steel frames considering panel joint deformations, *J. Structural Eng., ASCE*, 121(10), 1531-1540.

[44] Liew, J. Y. R. and Chen, W. F. 1997. LRFD - Limit Design of Frames, in *Steel Design Handbook*, Tamboli, A., Ed., McGraw-Hill, New York, Chapt. 6.

[45] Liew, J.Y.R., White, D. W., and Chen, W. F. 1991. Beam-column design in steel frameworks—Insight on current methods and trends. *J. Constructional Steel Res.*, 18, 259-308.

[46] Liew, J.Y.R., White, D. W., and Chen, W. F. 1992. Beam-Columns, in *Constructional Steel Design, An International Guide*, Dowling, P. J. et al., Eds., Elsevier, England, 105-132, Chapt. 5.1.

[47] Liew, J.Y.R., White, D. W., and Chen, W. F. 1993. Limit-states design of semi-rigid frames using advanced analysis. Part 1: Connection modelling and classification. Part II: Analysis and design, *J. Constructional Steel Res.*, Elsevier Science Publishers, London, 26(1), 1-57.

[48] Liew, J.Y.R., White, D. W., and Chen, W. F. 1993. Second-order refined plastic hinge analysis for frame design: Parts 1 & 2, *J. Structural Eng., ASCE*, 119(11), 3196-3237.

[49] Liew, J.Y.R., White, D. W., and Chen, W. F. 1994. Notional load plastic hinge method for frame design, *J. Structural Eng., ASCE*, 120(5), 1434-1454.

[50] Neals, S. and Johnson, R. 1992. *Design of Composite Trusses*, The Steel Construction Institute, U.K.

[51] Owens, G. 1989. *Design of Fabricated Composite Beams in Buildings*, The Steel Construction Institute, U.K.

[52] Owens, G.W. and Knowles, P.R. 1992. *Steel Designers' Manual*, 5th ed., Blackwell Scientific Publications, London.

[53] Reinhold, T. A. 1977. *Measurements of Simultaneous Fluctuating Loads at Multiple Levels on a Model of Tall Building in a Simulated Urban Boundary Layer*, Ph.D. thesis, Department of Civil Engineering, Virginia Polytechnic Institute and State University.

[54] Robson, J. D. 1963. *An Introduction to Random Vibration*, Edinburgh University Press, Scotland.

[55] Rosati, P. A. 1968. *An Experimental Study of the Response of a Square Prism to Wind Load*, Faculty of Graduate Studies, BLWT II-68, University of Western Ontario, London, Ontario, Canada.

[56] Saunders, J. W. and Melbourne, W. H. 1975. Tall rectangular building response to cross-wind excitation, *Proceedings of the 4th International Conference on Wind Effects on Building Structures*, Cambridge University Press.

[57] SCI. 1995. *Plastic Design of Single-Story Pitched-Roof Portal Frames to Eurocode 3*, Technical Report, SCI Publication 147, The Steel Construction Institute, U.K.

[58] Simiu, E. and Scanlan, R. H. 1986. *Wind Effects on Structures*, 2nd ed., John Wiley & Sons, New York.

[59] Simiu, E. 1974. Wind spectra and dynamic along wind response, *J. Structural Div., ASCE*, 100, 1897-1910.

[60] Taranath, B.S. 1988. *Structural Analysis and Design of Tall Buildings*, McGraw-Hill, New York.

[61] Vickery, B. J. 1965. *On the Flow Behind a Coarse Grid and Its Use as a Model of Atmospheric Turbulence in Studies Related to Wind Loads on Buildings,* Nat. Phys. Lab. Aero. Report 1143.

[62] Vickery, B. J. 1970. On the reliability of gust loading factors, *Proc. Tech. Meet. Concerning Wind Loads on Buildings and Structures,* National Bureau of Standards, *Building Science Series 30,* Washington D.C.

Further Reading

[1] Chen, W.F. and Kim, S.E. 1997. *LRFD Steel Design using Advanced Analysis,* CRC Press, Boca Raton, FL.

[2] Chen, W.F. and Sohal, I. 1995. *Plastic Design and Second-Order Analysis of Steel Frames,* Springer-Verlag, New York.

[3] Lawson, T.V. 1980. *Wind Effects on Buildings,* Applied Science Publishers.

[4] Smith, J.W. 1988. *Vibration of Structures — Application in Civil Engineering Design,* Chapman & Hall.

13

Space Frame Structures

13.1 Introduction to Space Frame Structures 13-1
General Introduction • Definition of the Space Frame • Basic Concepts • Advantages of Space Frames • Preliminary Planning Guidelines

13.2 Double Layer Grids 13-6
Types and Geometry • Type Choosing • Method of Support • Design Parameters • Cambering and Slope • Methods of Erection

13.3 Latticed Shells 13-17
Form and Layer • Braced Barrel Vaults • Braced Domes • Hyperbolic Paraboloid Shells • Intersection and Combination

13.4 Structural Analysis 13-25
Design Loads • Static Analysis • Earthquake Resistance • Stability

13.5 Jointing Systems 13-39
General Description • Proprietary System • Bearing Joints

13.6 Defining Terms 13-54

References 13-55

Further Reading 13-56

Tien T. Lan
Department of Civil Engineering,
Chinese Academy of
Building Research,
Beijing, China

13.1 Introduction to Space Frame Structures

13.1.1 General Introduction

A growing interest in space frame structures has been witnessed worldwide over the last half century. The search for new structural forms to accommodate large unobstructed areas has always been the main objective of architects and engineers. With the advent of new building techniques and construction materials, space frames frequently provide the right answer and satisfy the requirements for lightness, economy, and speedy construction. Significant progress has been made in the process of the development of the space frame. A large amount of theoretical and experimental research programs was carried out by many universities and research institutions in various countries. As a result, a great deal of useful information has been disseminated and fruitful results have been put into practice.

In the past few decades, the proliferation of the space frame was mainly due to its great structural potential and visual beauty. New and imaginative applications of space frames are being demonstrated in the total range of building types, such as sports arenas, exhibition pavilions, assembly halls, transportation terminals, airplane hangars, workshops, and warehouses. They have been used not only on long-span roofs, but also on mid- and short-span enclosures as roofs, floors, exterior walls, and canopies. Many interesting projects have been designed and constructed all over the world using a variety of configurations.

0-8493-2674-5/97/$0.00+$.50
© 1997 by CRC Press LLC

Some important factors that influence the rapid development of the space frame can be cited as follows. First, the search for large indoor space has always been the focus of human activities. Consequently, sports tournaments, cultural performances, mass assemblies, and exhibitions can be held under one roof. The modern production and the needs of greater operational efficiency also created demand for large space with a minimum interference from internal supports. The space frame provides the benefit that the interior space can be used in a variety of ways and thus is ideally suited for such requirements.

Space frames are highly statically indeterminate and their analysis leads to extremely tedious computation if by hand. The difficulty of the complicated analysis of such systems contributed to their limited use. The introduction of electronic computers has radically changed the whole approach to the analysis of space frames. By using computer programs, it is possible to analyze very complex space structures with great accuracy and less time involved.

Lastly, the space frame also has the problem of connecting a large number of members (sometimes up to 20) in space through different angles at a single point. The emergence of several connecting methods of proprietary systems has made great improvement in the construction of the space frame, which offered simple and efficient means for making connection of members. The exact tolerances required by these jointing systems can be achieved in the fabrication of the members and joints.

13.1.2 Definition of the Space Frame

If one looks at technical literature on structural engineering, one will find that the meaning of the **space frame** has been very diverse or even confusing. In a very broad sense, the definition of the space frame is literally a three-dimensional structure. However, in a more restricted sense, space frame means some type of special structure action in three dimensions. Sometimes structural engineers and architects seem to fail to convey with it what they really want to communicate. Thus, it is appropriate to define here the term space frame as understood throughout this section. It is best to quote a definition given by a Working Group on Spatial Steel Structures of the International Association [11].

> A space frame is a structure system assembled of linear elements so arranged that forces are transferred in a three-dimensional manner. In some cases, the constituent element may be two-dimensional. Macroscopically a space frame often takes the form of a flat or curved surface.

It should be noted that virtually the same structure defined as a space frame here is referred to as **latticed structures** in a State-of-the-Art Report prepared by the ASCE Task Committee on Latticed Structures [2] which states:

> A latticed structure is a structure system in the form of a network of elements (as opposed to a continuous surface). Rolled, extruded or fabricated sections comprise the member elements. Another characteristic of latticed structural system is that their load-carrying mechanism is three dimensional in nature.

The ASCE Report also specifies that the three-dimensional character includes flat surfaces with loading perpendicular to the plane as well as curved surfaces. The Report excludes structural systems such as common trusses or building frames, which can appropriately be divided into a series of planar frameworks with loading in the plane of the framework. In this section the terms *space frames* and *latticed structures* are considered synonymous.

A space frame is usually arranged in an array of single, double, or multiple layers of intersecting members. Some authors define space frames only as double layer grids. A single layer space frame that has the form of a curved surface is termed as **braced vault**, **braced dome**, or **latticed shell**.

Occasionally the term **space truss** appears in the technical literature. According to the structural analysis approach, a space frame is analyzed by assuming rigid joints that cause internal torsions and moments in the members, whereas a space truss is assumed as hinged joints and therefore has no internal member moments. The choice between space frame and space truss action is mainly determined by the joint-connection detailing and the member geometry is no different for both. However, in engineering practice, there is no absolutely rigid or hinged joints. For example, a double layer flat surface space frame is usually analyzed as hinged connections, while a single layer curved surface space frame may be analyzed either as hinged or rigid connections. The term *space frame* will be used to refer to both space frames and space trusses.

13.1.3 Basic Concepts

The space frame can be formed either in a flat or a curved surface. The earliest form of space frame structures is a single layer grid. By adding intermediate grids and including rigid connecting to the joist and girder framing system, the single layer grid is formed. The major characteristic of grid construction is the omni-directional spreading of the load as opposed to the linear transfer of the load in an ordinary framing system. Since such load transfer is mainly by bending, for larger spans, the bending stiffness is increased most efficiently by going to a double layer system. The load transfer mechanism of curved surface space frame is essentially different from the grid system that is primarily membrane-like action. The concept of a space frame can be best explained by the following example.

EXAMPLE 13.1:

It is necessary to design a roof structure for a square building. Figure 13.1a and b show two different ways of roof framing. The roof system shown in Figure 13.1a is a complex roof comprised of planar latticed trusses. Each truss will resist the load acting on it independently and transfer the load to the columns on each end. To ensure the integrity of the roof system, usually purlins and bracings are used between trusses. In Figure 13.1b, latticed trusses are laid orthogonally to form a system of space latticed grids that will resist the roof load through its integrated action as a whole and transfer the loads to the columns along the perimeters.Since the loads can be taken by the members in three dimensions, the corresponding forces in space latticed grids are usually less than that in planar trusses, and hence the depth can be decreased in a space frame.

The same concept can be observed in the design of a circular dome. Again, there are two different ways of framing a dome. The dome shown in Figure 13.2a is a complex dome comprised of elements such as arches, primary and secondary beams, and purlins, which all lie in a plane. Each of these elements constitutes a system that is stable by itself. In contrast, the dome shown in Figure 13.2b is an assembly of a series of longitudinal, meridional, and diagonal members, which is a certain form of latticed shell. It is a system whose resisting capacity is ensured only through its integral action as a whole.

The difference between planar structures and space frames can be understood also by examining the sequence of flow of forces. In a planar system, the force due to the roof load is transferred successively through the secondary elements, the primary elements, and then finally the foundation. In each case, loads are transferred from the elements of a lighter class to the elements of a heavier class. As the sequence proceeds, the magnitude of the load to be transferred increases, as does the span of the element. Thus, elements in a planar structure are characterized by their distinctive ranks, not only judging by the size of their cross-sections, but also by the importance of the task assigned to them. In contrast, in a space frame system, there is no sequence of load transfer and all elements contribute to the task of resisting the roof load in accordance with the three-dimensional geometry of the structure. For this reason, the ranking of the constituent elements similar to planar structures is not observed in a space frame.

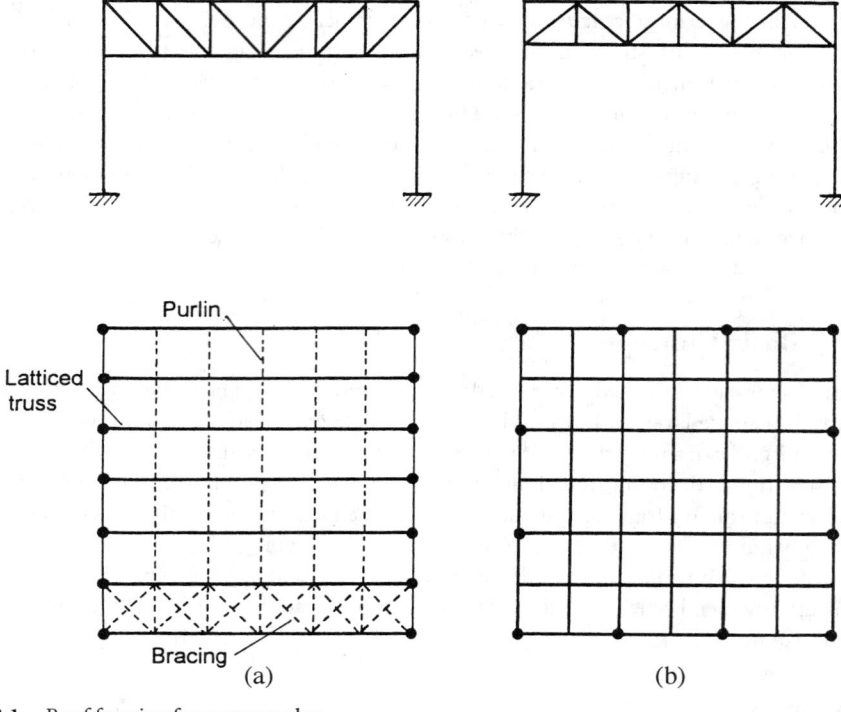

Figure 13.1   Roof framing for a square plan.

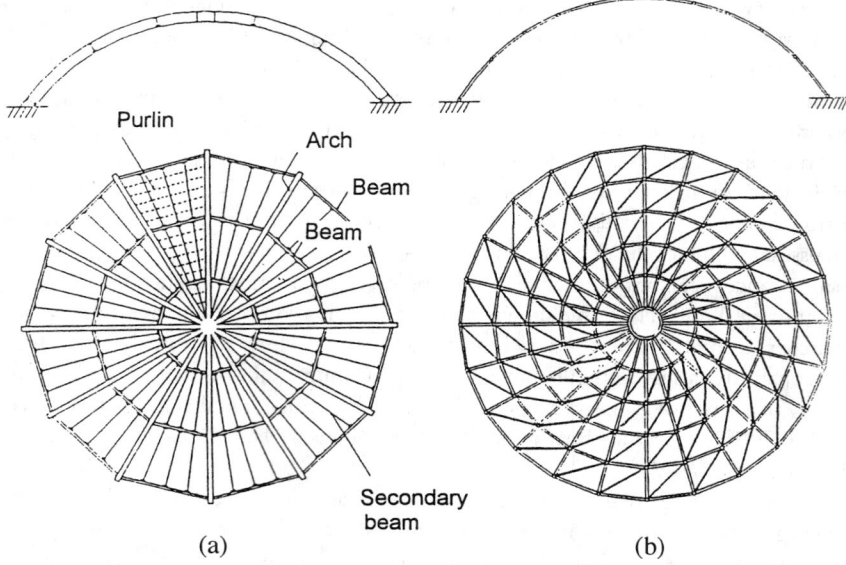

Figure 13.2   Roof framing for a circular dome.

13.1.4   Advantages of Space Frames

1. One of the most important advantages of a space frame structure is its light weight.
   It is mainly due to fact that material is distributed spatially in such a way that the
   load transfer mechanism is primarily axial—tension or compression. Consequently,
   all material in any given element is utilized to its full extent. Furthermore, most space

frames are now constructed with steel or aluminum, which decreases considerably their self-weight. This is especially important in the case of long span roofs that led to a number of notable examples of applications.

2. The units of space frames are usually mass produced in the factory so that they can take full advantage of an industrialized system of construction. Space frames can be built from simple prefabricated units, which are often of standard size and shape. Such units can be easily transported and rapidly assembled on site by semi-skilled labor. Consequently, space frames can be built at a lower cost.

3. A space frame is usually sufficiently stiff in spite of its lightness. This is due to its three-dimensional character and to the full participation of its constituent elements. Engineers appreciate the inherent rigidity and great stiffness of space frames and their exceptional ability to resist unsymmetrical or heavy concentrated load. Possessing greater rigidity, the space frames also allow greater flexibility in layout and positioning of columns.

4. Space frames possess a versatility of shape and form and can utilize a standard module to generate various flat space grids, latticed shell, or even free-form shapes. Architects appreciate the visual beauty and the impressive simplicity of lines in space frames. A trend is very noticeable in which the structural members are left exposed as a part of the architectural expression. Desire for openness for both visual impact as well as the ability to accommodate variable space requirements always calls for space frames as the most favorable solution.

13.1.5 Preliminary Planning Guidelines

In the preliminary stage of planning a space frame to cover a specific building, a number of factors should be studied and evaluated before proceeding to structural analysis and design. These include not only structural adequacy and functional requirements, but also the aesthetic effect desired.

1. In its initial phase, structural design consists of choosing the general form of the building and the type of space frame appropriate to this form. Since a space frame is assembled from straight, linear elements connected at nodes, the geometrical arrangement of the elements—surface shape, number of layers, grid pattern, etc.—needs to be studied carefully in the light of various pertinent requirements.

2. The geometry of the space frame is an important factor to be planned which will influence both the bearing capacity and weight of the structure. The **module** size is developed from the overall building dimensions, while the **depth** of the grid (in case of a double layer), the size of cladding, and the position of supports will also have a pronounced effect upon it. For a curved surface, the geometry is also related to the curvature or, more specifically, to the rise of the span. A compromise between these various aspects usually has to be made to achieve a satisfactory solution.

3. In a space frame, connecting joints play an important role, both functional and aesthetic, which is derived from their rationality during construction and after completion. Since joints have a decisive effect on the strength and stiffness of the structure and compose around 20 to 30% of the total weight, joint design is critical to space frame economy and safety. There are a number of proprietary systems that are used for space frame structures. A system should be selected on the basis of quality, cost, and erection efficiency. In addition, custom-designed space frames have been developed, especially for long span roofs. Regardless of the type of space frame, the essence of any system is the jointing system.

4. At the preliminary stage of design, choosing the type of space frame has to be closely related to the constructional technology. The space frames do not have such sequential

order of erection for planar structures and require special consideration on the method of construction. Usually a complete falsework has to be provided so that the structure can be assembled in the high place. Alternatively, the structure can be assembled on the ground, and certain techniques can be adopted to lift the whole structure, or its large part, to the final position.

13.2 Double Layer Grids

13.2.1 Types and Geometry

Double layer grids, or flat surface space frames, consist of two planar networks of members forming the top and bottom layers parallel to each other and interconnected by vertical and inclined web members. Double layer grids are characterized by the hinged joints with no moment or torsional resistance; therefore, all members can only resist tension or compression. Even in the case of connection by comparatively rigid joints, the influence of bending or torsional moment is insignificant.

Double layer grids are usually composed of basic elements such as:

- a planar latticed truss
- a pyramid with a square base that is essentially a part of an octahedron
- a pyramid with a triangular base (tetrahedron)

These basic elements used for various types of double-layer grids are shown in in Figure 13.3.

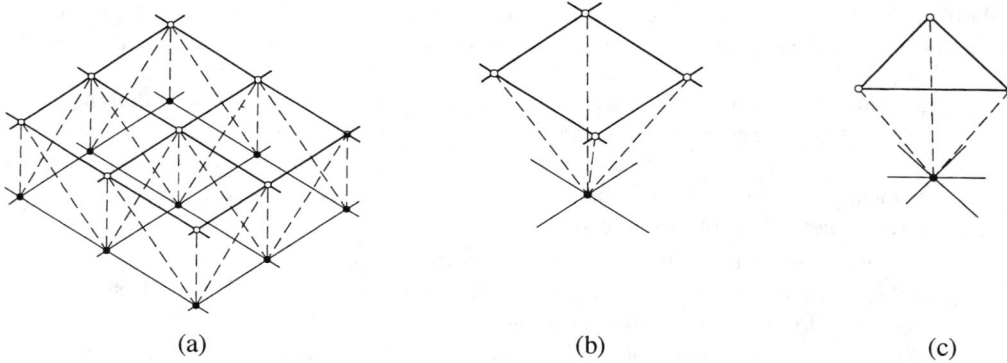

(a)   (b)   (c)

Figure 13.3   Basic elements of double layer grids.

A large number of types of double layer grids can be formed by these basic elements. They are developed by varying the direction of the top and bottom layers with respect to each other and also by the positioning of the top layer nodal points with respect to the bottom layer nodal points. Additional variations can be introduced by changing the size of the top layer grid with respect to the bottom layer grid. Thus, internal openings can be formed by omitting every second element in a normal configuration. According to the form of basic elements, double layer grids can be divided in two groups, i.e., **latticed grids** and **space grids**. The latticed grids consist of intersecting vertical latticed trusses and form a regular grid. Two parallel grids are similar in design, with one layer directly over the top of another. Both top and bottom grids are directionally the same. The space grids consist of a combination of square or triangular pyramids. This group covers the so-called offset grids, which consist of parallel grids having an identical layout with one grid offset from the other in plane but remaining directionally the same, as well as the so-called differential grids in which two parallel top and bottom grids are of a different layout but are chosen to coordinate and form a regular pattern [20].

The type of double layer grid can be chosen from the following most commonly used framing systems that are shown in Figure 13.4a through j. In Figure 13.4, top chord members are depicted with heavy solid lines, bottom chords are depicted with light solid lines and web members with dashed lines, while the upper joints are depicted by hollow circles and bottom joints by solid circles. Different types of double layer grids are grouped and named according to their composition and the names in the parenthesis indicate those suggested by other authors.

Group 1. Composed of latticed trusses

1. Two-way orthogonal latticed grids (square on square) (Figure 13.4a). This type of latticed grid has the advantage of simplicity in configuration and joint detail. All chord members are of the same length and lie in two planes that intersect at 90° to each other. Because of its weak torsional strength, horizontal bracings are usually established along the perimeters.

2. Two-way diagonal latticed grids (Figure 13.4b). The layout of the latticed grids is exactly the same as Type 1 except it is offset by 45° from the edges. The latticed trusses have different spans along two directions at each intersecting joint. Since the depth is all the same, the stiffness of each latticed truss varies according to its span. The latticed trusses of shorter spans may be considered as a certain kind of support for latticed trusses of longer span, hence more spatial action is obtained.

3. Three-way latticed grids (Figure 13.4c). All chord members intersect at 60° to each other and form equilateral triangular grids. It is a stiff and efficient system that is adaptable to those odd shapes such as circular and hexagonal plans. The joint detail is complicated by numerous members intersecting at one point, with 13 members in an extreme case.

4. One-way latticed grids (Figure 13.4d). It is composed of a series of mutually inclined latticed trusses to form a folded shape. There are only chord members along the spanning direction; therefore, one-way action is predominant. Like Type 1, horizontal bracings are necessary along the perimeters to increase the integral stiffness.

Group 2A. Composed of square pyramids

5. Orthogonal square pyramid space grids (square on square offset) (Figure 13.4e). This is one of the most commonly used framing patterns with top layer square grids offset over bottom layer grids. In addition to the equal length of both top and bottom chord members, if the angle between the diagonal and chord members is 45°, then all members in the space grids will have the same length. The basic element is a square pyramid that is used in some proprietary systems as prefabricated units to form this type of space grid.

6. Orthogonal square pyramid space grids with openings (square on square offset with internal openings, square on larger square) (Figure 13.4f). The framing pattern is similar to Type 5 except the inner square pyramids are removed alternatively to form larger grids in the bottom layer. Such modification will reduce the total number of members and consequently the weight. It is also visually affective as the extra openness of the space grids network produces an impressive architectural effect. Skylights can be used with this system.

7. Differential square pyramid space grids (square on diagonal) (Figure 13.4g). This is a typical example of differential grids. The two planes of the space grids are at 45° to each other which will increase the torsional stiffness effectively. The grids are arranged orthogonally in the top layer and diagonally in the bottom layer. It is one of the most

efficient framing systems with shorter top chord members to resist compression and longer bottom chords to resist tension. Even with the removal of a large number of members, the system is still structurally stable and aesthetically pleasing.

8. Diagonal square pyramid space grids (diagonal square on square with internal openings, diagonal on square) (Figure 13.4h). This type of space grid is also of the differential layout, but with a reverse pattern from Type 7. It is composed with square pyramids connected at their apices with fewer members intersecting at the node. The joint detail is relatively simple because there are only six members connecting at the top chord joint and eight members at the bottom chord joint.

Group 2B. Composed of triangular pyramids

9. Triangular pyramid space grids (triangle on triangle offset) (Figure 13.4i). Triangular pyramids are used as basic elements and are connected at their apices, thus forming a pattern of top layer triangular grids offset over bottom layer grids. If the depth of the space grids is equal to $\sqrt{2/3}$ chord length, then all members will have the same length.

10. Triangular pyramid space grids with openings (triangle on triangle offset with internal openings) (Figure 13.4j). Like Type 6, the inner triangular pyramids may also be removed alternatively. As the figure shown, triangular grids are formed in the top layer while triangular and hexagonal grids are formed in the bottom layer. The pattern in the bottom layer may be varied depending on the ways of removal. Such types of space grids have a good open feeling and the contrast of the patterns is effective.

13.2.2   Type Choosing

In the preliminary stage of design, it is most important to choose an appropriate type of double layer grid that will have direct influence on the overall cost and speed of construction. It should be determined comprehensively by considering the shape of the building plan, the size of the span, supporting conditions, magnitude of loading, roof construction, and architectural requirements. In general, the system should be chosen so that the space grid is built of relatively long tension members and short compression members.

In choosing the type, the steel weight is one of the important factors for comparison. If possible, the cost of the structure should also be taken into account, which is complicated by the different costs of joints and members. By comparing the steel consumption of various types of double layer grids with rectangular plans and supported along perimeters, it was found that the **aspect ratio** of the plan, defined here as the ratio of a longer span to a shorter span, has more influence than the span of the double layer grids. When the plan is square or nearly square (aspect ratio = 1 to 1.5), two-way latticed grids and all space grids of Group 2A, i.e., Type 1, 2, and 5 through 8, could be chosen. Of these types, the diagonal square pyramid space grids or differential square pyramid space grids have the minimum steel weight. When the plan is comparatively narrow (aspect ratio = 1.5 to 2), then those double layer grids with orthogonal gird systems in the top layer will consume less steel than those with a diagonal grid system. Therefore, two-way orthogonal latticed grids, orthogonal square pyramid space grids, and also those with openings and differential square pyramid space grids, i.e., Types 1, 5, 6, and 7, could be chosen. When the plan is long and narrow, the type of one-way latticed grid is the only selection. For square or rectangular double layer grids supported along perimeters on three sides and free on the other side, the selection of the appropriate types for different cases is essentially the same. The boundary along the free side should be strengthened either by increasing the depth or number of layers. Individual supporting structures such as trusses or girders along the free side are not necessary.

In case the double layer grids are supported on intermediate columns, type could be chosen from two-way orthogonal latticed grids, orthogonal square pyramid space grids, and also those with

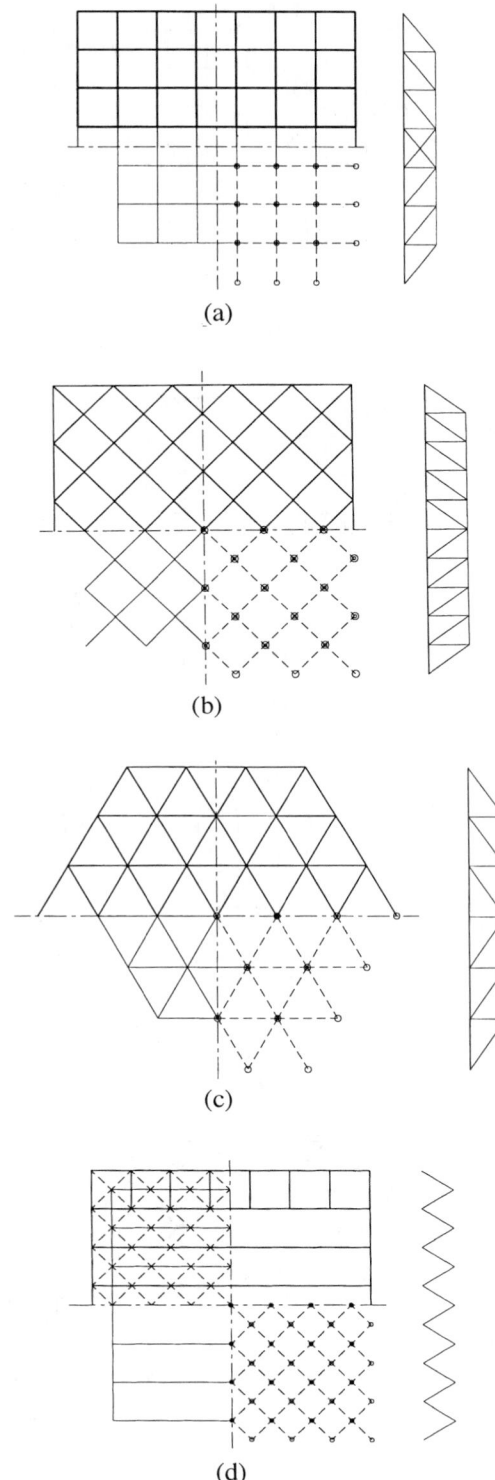

Figure 13.4 Framing system of double layer grids.

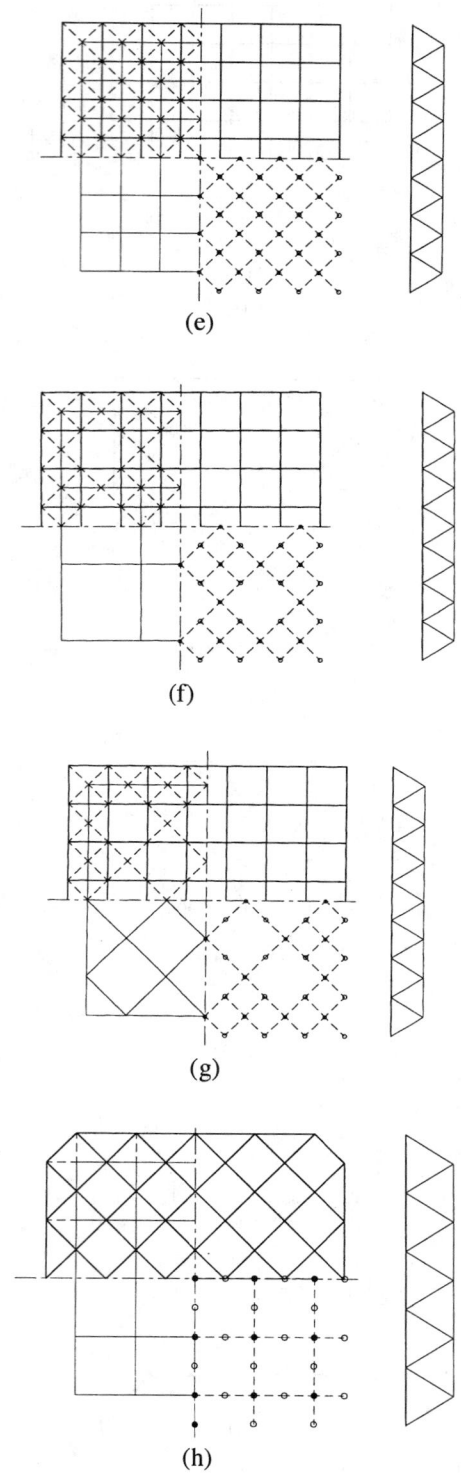

(e)

(f)

(g)

(h)

Figure 13.4   *(Continued)* Framing system of double layer grids.

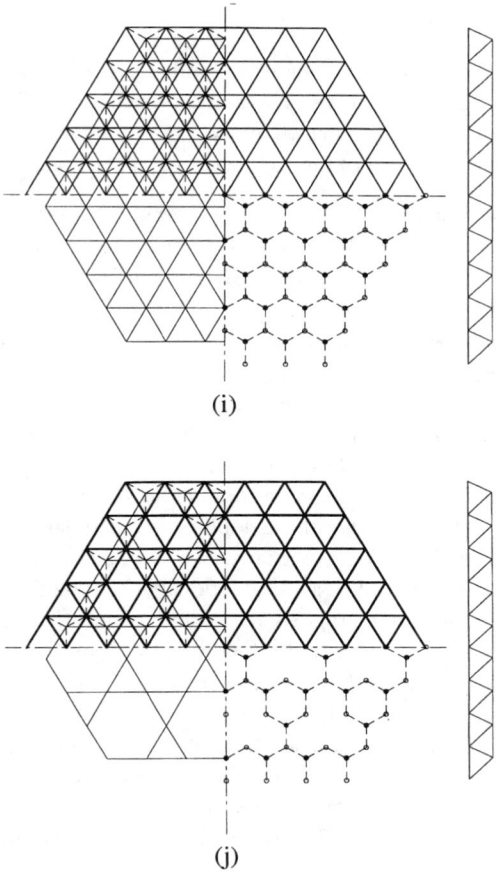

(i)

(j)

Figure 13.4 *(Continued)* Framing system of double layer grids.

openings, i.e., Types 1, 5, and 6. If the supports for multi-span double layer grids are combined with those along perimeters, then two-way diagonal latticed grids and diagonal square pyramid space grids, i.e., Types 2 and 8, could also be used.

For double layer grids with circular, triangular, hexagonal, and other odd shapes supporting along perimeters, types with triangular grids in the top layer, i.e., Types 3, 9, and 10, are appropriate for use.

The recommended types of double layer grids are summarized in Table 13.1 according to the shape of the plan and their supporting conditions.

TABLE 13.1 Type Choosing for Double Layer Grids

Shape of the plan	Supporting condition	Recommended types
Square, rectangular (aspect ratio = 1 to 1.5)	Along perimeters	1, 2, 5, 6, 7, 8
Rectangular (aspect ratio = 1.5 to 2)	Along perimeters	1, 5, 6, 7
Long strip (aspect ratio > 2)	Along perimeters	4
Square, rectangular	Intermediate support	1, 5, 6
Square, rectangular	Intermediate support combined with support along perimeters	1, 2, 5, 6, 8
Circular, triangular, hexagonal, and other odd shapes	Along perimeters	3, 9, 10

13.2.3 Method of Support

Ideal double layer grids would be square, circular, or other polygonal shapes with overhanging and continuous supports along the perimeters. This will approach more of a plate type of design which minimizes the maximum bending moment. However, the configuration of the building has a great number of varieties and the support of the double layer grids can take the following locations:

1. **Support along perimeters**—This is the most commonly used support location. The supports of double layer grids may directly rest on the columns or on ring beams connecting the columns or exterior walls. Care should be taken that the module size of grids matches the column spacing.

2. **Multi-column supports**—For single-span buildings, such as a sports hall, double layer grids can be supported on four intermediate columns as shown in Figure 13.5a. For buildings such as workshops, usually multi-span columns in the form of grids as shown in Figure 13.5b are used. Sometimes the column grids are used in combination with supports along perimeters as shown in Figure 13.5c. Overhangs should be employed where possible in order to provide some amount of stress reversal to reduce the interior chord forces and deflections. For those double layer grids supported on intermediate columns, it is best to design with overhangs, which are taken as 1/4 to 1/3 of the mid-span. Corner supports should be avoided if possible because they cause large forces in the edge chords. If only four supports are to be provided, then it is more desirable to locate them in the middle of the sides rather than at the corners of the building.

3. **Support along perimeters on three sides and free on the other side**—For buildings of a rectangular shape, it is necessary to have one side open, such as in the case of an airplane hanger or for future extension. Instead of establishing the supporting girder or truss on the free side, triple layer grids can be formed by simply adding another layer of several module widths (Figure 13.6). For shorter spans, it can also be solved by increasing the depth of the double layer grids. The sectional area of the members along the free side will increase accordingly.

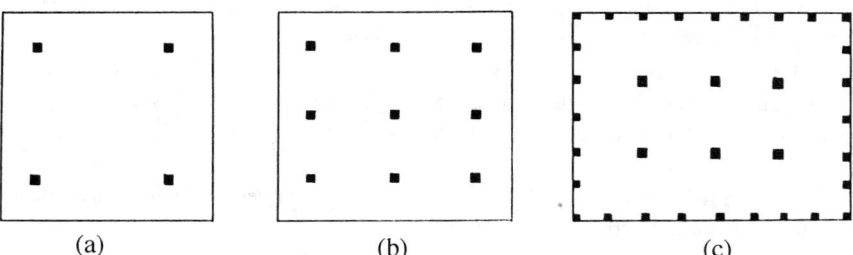

(a)	(b)	(c)

Figure 13.5   Multi-column supports.

The columns for double layer grids must support gravity loads and possible lateral forces. Typical types of support on multi-columns are shown in Figure 13.7. Usually the member forces around the support will be excessively large, and some means of transferring the loads to columns are necessary. It may carry the space grids down to the column top by an inverted pyramid as shown in Figure 13.7a or by triple layer grids as shown in Figure 13.7b, which can be employed to carry skylights. If necessary, the inverted pyramids may be extended down to the ground level as shown in Figure 13.7c. The spreading out of the concentrated column reaction on the space grids reduces the maximum chord and web member forces adjacent to the column supports and reduces the effective spans. The use of a vertical strut on column tops as shown in Figure 13.7d enables the space grids to

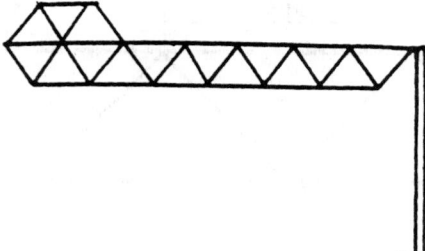

Figure 13.6   Triple layer grids on the free side.

be supported on top chords, but the vertical strut and the connecting joint have to be very strong. The use of crosshead beams on column tops as shown in Figure 13.7e produces the same effect as the inverted pyramid, but usually costs more in material and special fabrication.

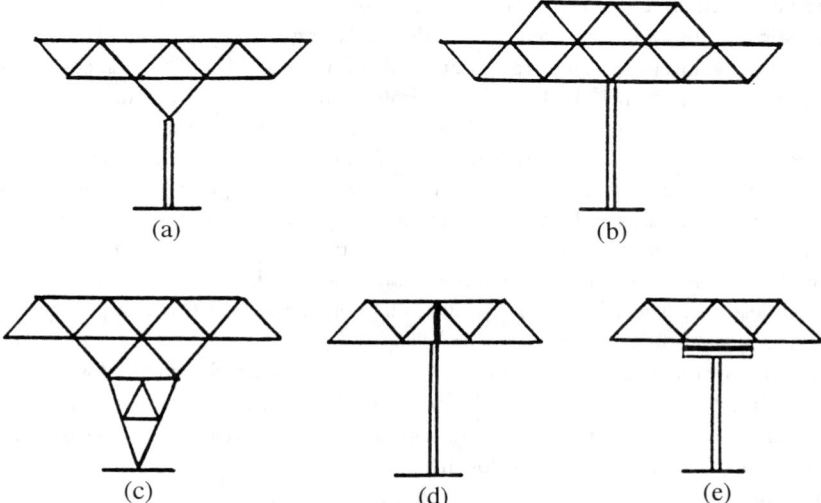

Figure 13.7   Supporting columns.

13.2.4   Design Parameters

Before any work can proceed on the analysis of a double layer grid, it is necessary to determine the depth and the module size. The depth is the distance between the top and bottom layers and the module is the distance between two joints in the layer of the grid (see Figure 13.8). Although these two parameters seem simple enough to determine, they will play an important role on the economy of the roof design. There are many factors influencing these parameters, such as the type of double layer grid, the span between the supports, the roof cladding, and also the proprietary system used. In fact, the depth and module size are mutually dependent which is related by the permissible angle between the center line of web members and the plane of the top and bottom chord members. This should be less than 30° or the forces in the web members and the length will be relatively excessive, but not greater than 60° or the density of the web members in the grid will become too high. For some of the proprietary systems, the depth and/or module are all standardized.

The depth and module size of double layer grids are usually determined by practical experience. In some of the paper and handbooks, figures on these parameters are recommended and one may

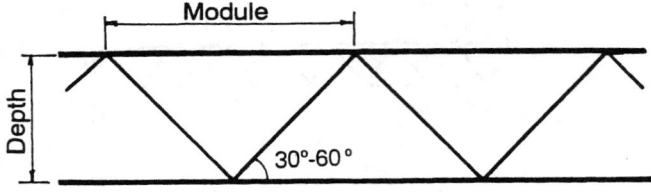

Figure 13.8 Depth and module.

find the difference is quite large. For example, the span-depth ratio varies from 12.5 to 25, or even more. It is usually considered that the depth of the space frame can be relatively small when compared with more conventional structures. This is generally true because double layer grids produce smaller deflections under load. However, depths that are small in relation to span will tend to use smaller modules and hence a heavier structure will result. In the design, almost unlimited possibilities exist in practice for the choice of geometry. It is best to determine these parameters through structural optimization.

Works have been done on the optimum design of double layer grids supported along perimeters. In an investigation by Lan [14], seven types of double layer grids were studied. The module dimension and depth of the space frame are chosen as the design variables. The total cost is taken as the objective function which includes the cost of members and joints as well as the roofing systems and enclosing walls. Such assumption makes the results realistic to a practical design. A series of double layer grids of different types spanning from 24 to 72 m was analyzed by optimization. It was found that the optimum design parameters were different for different types of roof systems. The module number generally increases with the span, and the steel purlin roofing system allows larger module sizes than that of reinforced concrete. The optimum depth is less dependent on the span and smaller depth can be used for a steel purlin roofing system. It should be observed that a smaller member density will lead to a grid with relatively few nodal points and thus the least possible production costs for nodes, erection expense, etc.

Through regression analysis of the calculated values by optimization method where the costs are within 3% optimum, the following empirical formulas for optimum span-depth ratios are obtained. It was found that the optimum depths are distributed in a belt and all the span-depth ratios within such range will give optimum effect in construction.

For a roofing system composed of reinforced concrete slabs

$$L/d = 12 \pm 2 \tag{13.1}$$

For a roofing system composed of steel purlins and metal decks

$$L/d = (510 - L)/34 \pm 2 \tag{13.2}$$

where L is the short span and d is the depth of the double layer grids.

Few data could be obtained from the past works. Regarding the optimum depth for steel purlin roofing systems, Geiger suggested the span-depth ratio to be varied from 10 to 20 with less than 10% variation in cost. Motro recommended a span-depth ratio of 15. Curves for diagonal square pyramid space grids (diagonal on square) were given by Hirata et al. and an optimum ratio of 10 was suggested. In the earlier edition of the *Specifications for the Design and Construction of Space Trusses* issued in China, the span-depth ratio is specified according to the span. These figures were obtained through the analysis of the parameters used in numerous design projects. A design handbook for double layer grids also gives graphs for determining upper and lower bounds of module dimension and depth. The relation between depth and span obtained from Equation 13.2 and relevant source is shown in Figure 13.9. For short and medium spans, the optimum values are in good agreement with those obtained from experience. It is noticeable that the span-depth ratio should decrease with

the span, yet an increasing tendency is found from experience which gives irrationally large values
for long spans.

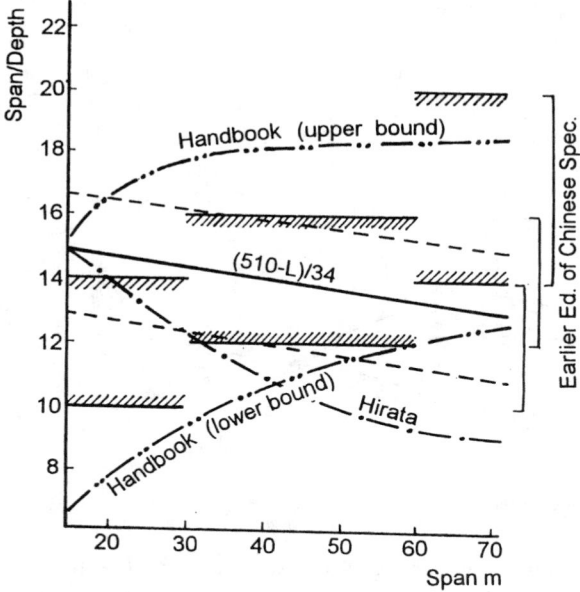

Figure 13.9   Relation between depth and span of double layer grids.

In the revised edition of the *Specification for the Design and Construction of Space Trusses* issued in
China, appropriate values of module size and depth for commonly used double layer grids simply
supported along the perimeters are given. Table 13.2 shows the range of module numbers of the
top chord and the span-depth ratios prescribed by the *Specifications*.

TABLE 13.2   Module Number and Span-Depth Ratio

Type of double layer grids	R.C. slab roofing system		Steel purlin roofing system	
	Module number	Span-depth ratio	Module number	Span-depth ratio
1, 5, 6	$(2 - 4) + 0.2L$			
		$10 - 14$	$(6-8)+0.7L$	$(13-17)-0.03L$
2, 7, 8	$(6-8)+0.08L$			

Note: 1. L Denotes the shorter span in meters. 2. When the span is less than 18 m, the
number of the module may be decreased.

13.2.5   Cambering and Slope

Most double layer grids are sufficiently stiff, so cambering is often not required. Cambering is
considered when the structure under load appears to be sagging and the deflection might be visually
undesirable. It is suggested that the cambering be limited to 1/300 of the shorter span. As shown
in Figure 13.10, cambering is usually done in (a) cylindrical, (b) ridge or (c, d) spherical shape. If
the grid is being fabricated on site by welding, then almost any type of camber can be obtained as

this is just a matter of setting the joint nodes at the appropriate levels. If the grid components are fabricated in the factory, then it is necessary to standardize the length of the members. This can be done by keeping either the top or bottom layer chords at the standard length, and altering the other either by adding a small amount to the length of each member or subtracting a small amount from it to generate the camber required.

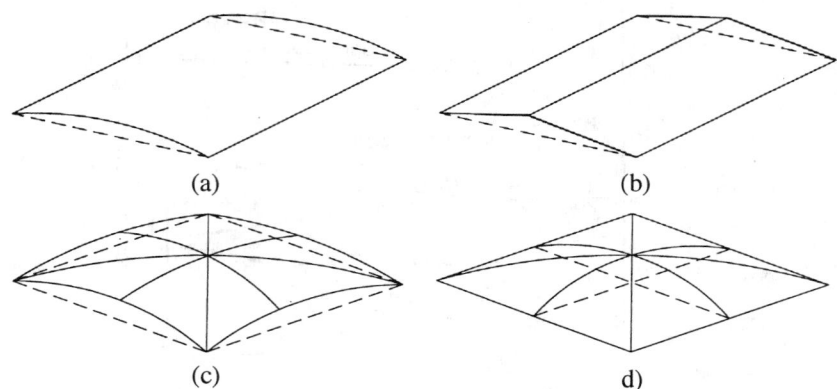

Figure 13.10   Ways of cambering.

Sometimes cambering is suggested so as to ensure that the rainwater drains off the roof quickly to avoid ponding. This does not seem to be effective especially when cambering is limited. To solve the water run-off problem in those locations with heavy rains, it is best to form a roof slope by the following methods (Figure 13.11):

1. Establishing short posts of different height on the joints of top layer grids.
2. Varying the depth of grids.
3. Forming a slope for the whole grid.
4. Varying the height of supporting columns.

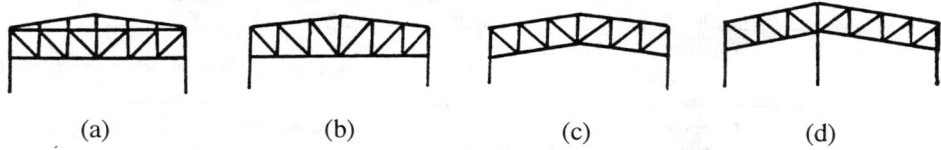

Figure 13.11   Ways of forming roof slope.

13.2.6   Methods of Erection

The method chosen for erection of a space frame depends on its behavior of load transmission and constructional details, so that it will meet the overall requirements of quality, safety, speed of construction, and economy. The scale of the structure being built, the method of jointing the individual elements, and the strength and rigidity of the space frame until its form is closed must all be considered. The general methods of erecting double layer grids are as follows. Most of them can also be applied to the construction of latticed shells.

1. **Assembly of space frame elements in the air**—Members and joints or prefabricated sub-assembly elements are assembled directly on their final position. Full scaffoldings are usually required for such types of erection. Sometimes only partial scaffoldings are used if cantilever erection of a space frame can be executed. The elements are fabricated at the shop and transported to the construction site and no heavy lifting equipment is required. It is suitable for all types of space frame with bolted connections.

2. **Erection of space frames by strips or blocks**—The space frame is divided on its plane into individual strips or blocks. These units are fabricated on the ground level, then hoisted up into the final position and assembled on the temporary supports. With more work being done on the ground, the amount of assembling work at high elevation is reduced. This method is suitable for those double layer grids where the stiffness and load-resisting behavior will not change considerably after dividing into strips or blocks, such as two-way orthogonal latticed grids, orthogonal square pyramid space grids, and the those with openings. The size of each unit will depend on the hoisting capacity available.

3. **Assembly of space frames by sliding element in the air**—Separate strips of space frame are assembled on the roof level by sliding along the rails established on each side of the building. The sliding units may either slide one after another to the final position and then assembled together or assembled successively during the process of sliding. Thus, the erection of a space frame can be carried out simultaneously with the construction work underneath, which leads to savings of construction time and cost of scaffoldings. The sliding technique is relatively simple, requiring no special lifting equipment. It is suitable for orthogonal grid systems where each sliding unit will remain geometrically non-deferrable.

4. **Hoisting of whole space frames by derrick masts or cranes**—The whole space frame is assembled on the ground level so that most of the assembling work can be done before hoisting. This will result in an increased efficiency and better quality. For short and medium spans, the space frame can be hoisted up by several cranes. For long-span space frames, derrick masts are used as the support and electric winches as the lifting power. The whole space frame can be translated or rotated in the air and then seated on its final position. This method can be employed to all types of double layer grids.

5. **Lifting-up the whole space frame**—This method also has the benefit or assembling space frames on the ground level, but the structure cannot move horizontally during lifting. Conventional equipment used is hydraulic jacks or lifting machines for lift-slab construction. An innovative method has been developed by using the center hole hydraulic jacks for slipforming.The space frame is lifted up simultaneously with the slipforms for r.c. columns or walls. This lifting method is suitable for double layer grids supported along perimeters or on multi-point supports.

6. **Jacking-up the whole space frame**—Heavy hydraulic jacks are established on the position of columns that are used as supports for jacking-up. Occasionally roof claddings, ceilings, and mechanical installations are also completed with the space frame on the ground level. It is appropriate for use in space frames with multi-point supports, the number of which is usually limited.

13.3 Latticed Shells

13.3.1 Form and Layer

The main difference between double layer grids and latticed shells is the form. For a double layer grid, it is simply a flat surface. For latticed shell, the variety of forms is almost unlimited. A common approach to the design of latticed shells is to start with the consideration of the form—a surface curved in space. The geometry of basic surfaces can be identified, according to the method of generation, as the surface of translation and the surface of rotation. A number of variations of form can be obtained by taking segments of the basic surfaces or by combining or adding them. In general, the geometry of surface has a decisive influence on essentially all characteristics of the structure: the manner in which it transfers loads, its strength and stiffness, the economy of construction, and finally the aesthetic quality of the completed project.

Latticed shells can be divided into three distinct groups forming singly curved, synclastic, and anticlastic surfaces. A barrel vault (cylindrical shell) represents a typical developable surface, having a zero curvature in the direction of generatrices. A spherical or elliptical dome (spheroid or elliptic paraboloid) is a typical example of a synclastic shell. A hyperbolic paraboloid is a typical example of an anticlastic shell.

Besides the mathematical generation of surface systems, there are other methods for finding shapes of latticed shells. Mathematically the surface can be defined by a high degree polynomial with the unknown coefficients determined from the known shape of the boundary and the known position of certain points at the interior required by the functional and architectural properties of the space. Experimentally the shape can be obtained by loading a net of chain wires, a rubber membrane, or a soap membrane in the desired manner. In each case the membrane is supported along a predetermined contour and at predetermined points. The resulting shape will produce a minimal surface that is characterized by a least surface area for a given boundary and also constant skin stress. Such experimental models help to develop an understanding about the nature of structural forms.

The inherent curvature in a latticed shell will give the structure greater stiffness. Hence, latticed shells can be built in single layer grids, which is a major difference from double layer grid. Of course, latticed shells may also be built in double layer grids. Although single layer and double layer latticed shells are similar in shape, the structural analysis and connecting detail are quite different. The single layer latticed shell is a structural system with rigid joints, while the double layer latticed shell has hinged joints. In practice, single layer latticed shells of short span with lightweight roofing may also be built with hinged joints. The members and connecting joints in a single layer shell of large span will resist not only axial forces as in a double layer shell, but also the internal moments and torsions. Since the single layer latticed shells are easily liable to buckling, the span should not be too large. There is no distinct limit between single and double layer, which will depend on the type of shell, the geometry and size of the framework, and the section of members.

13.3.2 Braced Barrel Vaults

The braced barrel vault is composed of member elements arranged on a cylindrical surface. The basic curve is a circular segment; however, occasionally a parabola, ellipse, or funicular line may also be used. Figure 13.12 shows the typical arrangement of a braced barrel vault. Its structural behavior depends mainly on the type and location of supports, which can be expressed as L/R, where L is the distance between the supports in longitudinal direction and R is the radius of curvature of the transverse curve.

If the distance between the supports is long and usually edge beams are used in the longitudinal direction (Figure 13.12a), the primary response will be beam action. For $1.67 < L/R < 5$, the barrel vaults are called long shells, which can be visualized as beams with curvilinear cross-sections. The beam theory with the assumption of linear stress distribution may be applied to barrel vaults that are of symmetrical cross-section and under uniform loading if $L/R > 3$. This class of barrel vault will have longitudinal compressive stresses near the crown of the vault, longitudinal tensile stresses towards the free edges, and shear stresses towards the supports.

As the distance between transverse supports becomes closer, or as the dimension of the longitudinal span becomes smaller than the dimension of the shell width such that $0.25 < L/R < 1.67$, then the primary response will be arch action in the transverse direction (Figure 13.12b). The barrel vaults are called short shells. Their structural behavior is rather complex and dependent on their geometrical proportions. The force distribution in the longitudinal direction is no longer linear, but in a curvilinear manner, trusses or arches are usually used as the transverse supports.

When a single braced barrel vault is supported continuously along its longitudinal edges on foundation blocks, or the ratio of L/R becomes very small, i.e., < 0.25 (Figure 13.12c), the forces are carried directly in the transverse direction to the edge supports. Its behavior may be visualized as the response of parallel arches. Displacement in the radial direction is resisted by cicumferential bending stiffness. Such type of barrel vault can be applied to buildings such as airplane hangars or gymnasia where the wall and roof are combined together.

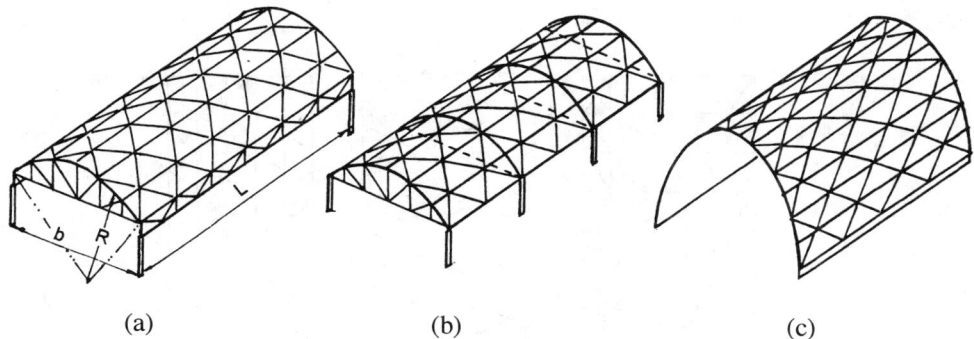

(a)   (b)   (c)

Figure 13.12   Braced barrel vaults.

There are several possible types of bracing that have been used in the construction of single layer braced barrel vaults. Figure 13.13 shows five principle types:

1. Orthogonal grid with single bracing of Warren truss (a)
2. Orthogonal grid with single bracing of Pratt truss (b)
3. Orthogonal grid with double bracing (c)
4. Lamella (d)
5. Three way (e)

The first three types of braced barrel vaults can be formed by composing latticed trusses with the difference in the arrangement of bracings (Figures 13.13a, b, and c). In fact, the original barrel vault was introduced by Foppl. It consists of several latticed trusses, spanning the length of the barrel and supported on the gables. After connection of the longitudinal booms of the latticed trusses, they became a part of the braced barrel vault of the single layer type.

The popular diamond-patterned lamella type of braced barrel vault consists of a number of interconnected modular units forming a rhombus shaped grid pattern (Figure 13.13d). Each unit, which is twice the length of the side of a diamond, is called a **lamella.** Lamella roofs proved ideal for prefabricated construction as all the units are of standard size. They were originally constructed of timber, but with the increase of span, steel soon became the most frequently used material.

To increase the stability of the structure and to reduce the deflections under unsymmetrical loads, purlins were employed for large span lamella barrel vaults. This created the three-way grid type of bracing and became very popular (Figure 13.13e). The three-way grid enables the construction of such systems using equilateral triangles composed of modular units, which are of identical length and can be connected with simple nodes.

Research investigations have been carried out on braced barrel vaults. One aspect of this research referred to the influence of different types of bracing on the resulting stress distribution. The

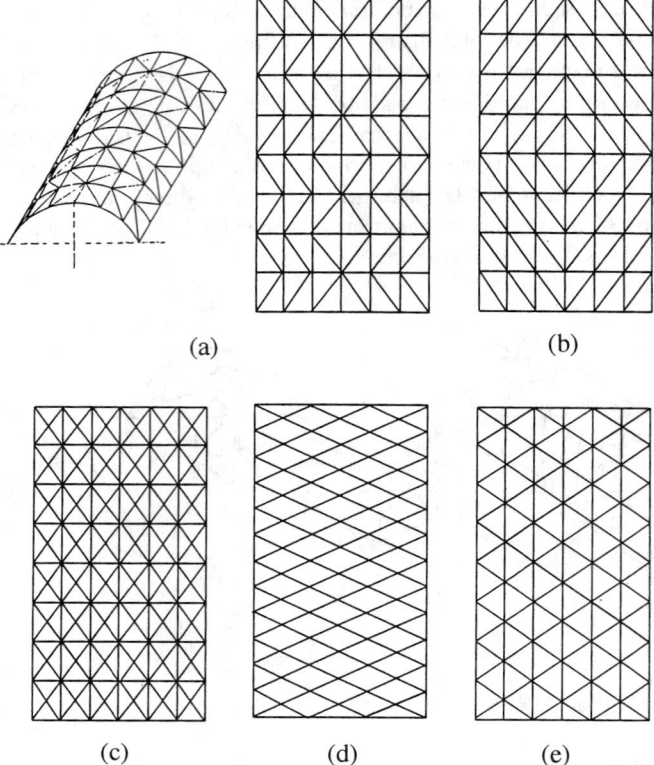

(a)   (b)

(c)   (d)   (e)

Figure 13.13   Types of bracing for braced barrel vaults.

experimental tests on the models proved that there are significant differences in the behavior of the structures, and the type of bracing has a fundamental influence upon the strength and load-carrying capacity of the braced barrel vaults. The three-way single layer barrel vaults exhibited a very uniform stress distribution under uniformly distributed load, and much smaller deflections in the case of unsymmetrical loading than for any of the other types of bracing. The experiments also showed that large span single layer braced barrel vaults are prone to instability, especially under the action of heavy unsymmetrical loads and that the rigidity of joints can exert an important influence on the overall stability of the structure.

   For double layer braced barrel vaults, if two- or three-way latticed trusses are used to form the top and bottom layers of the latticed shell, the grid pattern is identical as shown in Figure 13.13 for single layer shells. If square or triangular pyramids are used, either the top or bottom layer grid may follow the same pattern as shown in Figure 13.13.

   The usual height-to-width ratio for long shells varies from 1/5 to 1/7.5. When the barrel vault is supported along the longitudinal edges, then the height can be increased to 1/3 chord width. For long shells, if the longitudinal span is larger than 30 m, or for barrel vaults supported along longitudinal edges with a transverse span larger than 25 m, double layer grids are recommended. The thickness of the double layer barrel vault is usually taken from 1/20 to 1/40 of the chord width.

13.3.3   Braced Domes

Domes are one of the oldest and well-established structural forms and have been used in architecture since the earliest times. They are of special interest to engineers as they enclose a maximum amount of space with a minimum surface and have proved to be very economical in terms of consumption of constructional materials. The stresses in a dome are generally membrane and compressive in the most part of the shell except circumferential tensile stresses near the edge and small bending moments at the junction of the shell and the ring beam. Most domes are surfaces of revolution. The curves used to form the synclastic shell are spherical, parabolic, or elliptical covering circular or polygonal areas. Out of a large variety of possible types of braced domes, only four or five types proved to be frequently used in practice. They are shown in Figure 13.14.

1. Ribbed domes (a)
2. Schwedler domes (b)
3. Three-way grid domes (c)
4. Lamella domes (d, e)
5. Geodesic domes (f)

Ribbed domes are the earliest type of braced domes that were constructed (Figure 13.14a). A ribbed dome consists of a number of identical meridional solid girders or trusses, interconnected at the crown by a compression ring. The ribs are also connected by concentric rings to form grids in a trapezium shape. The ribbed dome is usually stiffened by a steel or reinforced concrete tension ring at its base.

A Schwedler dome also consists of meridional ribs connected together to a number of horizontal polygonal rings to stiffen the resulting structure so that it will be able to take unsymmetrical loads (Figure 13.14b). Each trapezium formed by intersecting meridional ribs with horizontal rings is subdivided into two triangles by a diagonal member. Sometimes the trapezium may also be subdivided by two cross-diagonal members. This type of dome was introduced by a German engineer, J.W. Schwedler, in 1863. The great popularity of Schwedler domes is due to the fact that, on the assumption of pin-connected joints, the structure can be analyzed as statically determinate. In practice, in addition to axial forces, all the members are also under the action of bending and torsional moments. Many attempts have been made in the past to simplify their analysis, but precise methods of analysis using computers have finally been applied to find the actual stress distribution.

The construction of a three-way grid dome is self-explanatory. It may be imagined as a curved form of three-way double layer grids (Figure 13.14c). It can also be constructed in single layer for the dome. The Japanese "Diamond Dome" system by Tomoegumi Iron Works belongs to this category. The theoretical analysis of three-way grid domes shows that even under unsymmetrical loading the forces in this configuration are very evenly distributed leading to economy in material consumption.

A Lamella dome is formed by intersecting two-way ribs diagonally to form a rhombus-shaped grid pattern. As in a lamella braced barrel vault, each lamella element has a length that is twice the length of the side of a diamond. The lamella dome can be distinguished further from parallel and curved domes. For a parallel lamella as shown in Figure 13.14d, the circular plan is divided into several sectors (usually six or eight), and each sector is subdivided by parallel ribs into rhombus grids of the same size. This type of lamella dome is very popular in the U.S. It is sometimes called a Kiewitt dome, named after its developer. For a curved lamella as shown in Figure 13.14e, rhombus grids of different size, gradually increasing from the center of the dome, are formed by diagonal ribs along the radial lines. Sometimes, for the purpose of establishing purlins for roof decks, concentric rings are introduced and a triangular network is generated.

The geodesic dome was developed by the American designer Buckminster Fuller, who turned architects' attention to the advantages of braced domes in which the elements forming the framework of the structure are lying on the great circle of a sphere. This is where the name "geodesic" came from (Figure 13.14f). The framework of these intersecting elements forms a three-way grid comprising virtually equilateral spherical triangles. In Fuller's original geodesic domes, he used an icosahedron as the basis for the geodesic subdivision of a sphere, then the spherical surface is divided into 20 equilateral triangles as shown in Figure 13.15a. This is the maximum number of equilateral

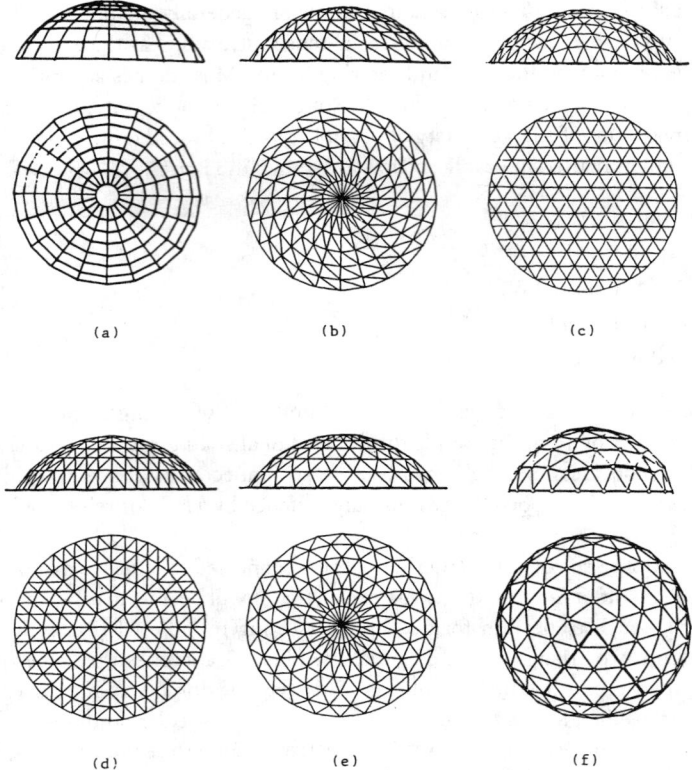

(a)   (b)   (c)

(d)   (e)   (f)

Figure 13.14   Braced domes.

triangles into which a sphere can be divided. For domes of larger span, each of these triangles can be subdivided into six triangles by drawing medians and bisecting the sides of each triangle. It is therefore possible to form 15 complete great circles regularly arranged on the surface of a sphere (see Figure 13.15b). Practice shows that the primary type of bracing, which is truly geodesic, is not sufficient because it would lead to an excessive length for members in a geodesic dome. Therefore, a secondary bracing has to be introduced. To obtain a more or less regular network of the bracing bars, the edges of the basic triangle are divided modularly. The number of modules into which each edge of the spherical icosahedron is divided depends mainly on the size of the dome, its span, and the type of roof cladding. This subdivision is usually referred to as "frequency" as depicted in Figure 13.15c. It must be pointed out that during such a subdivision, the resulting triangles are no longer equilateral. The members forming the skeleton of the dome show slight variation in their length. As the frequency of the subdivision increases, the member length reduces, and the number of components as well as the types of connecting joints increases. Consequently, this reflects in the increase of the final price of the geodesic dome, and is one of the reasons why geodesic domes, in spite of their undoubted advantages for smaller spans, do not compare equally well with other types of braced domes for larger spans.

The rise of a braced dome can be as flat as 1/6 of the diameter or as high as 3/4 of the diameter which will constitute a greater part of a sphere. For diameter of braced domes larger than 60 m, double layer grids are recommended. The ratio of the depth to the diameter is in the range of 1/30 to 1/50. For long spans, the depth can be taken as small as 1/100 of diameter.

The subdivision of the surface of a braced dome can also be considered by using one of the following three methods. The first method is based on the surface of revolution. The first set of lines of division is drawn as the meridional lines from the apex. Next, circumferential rings are

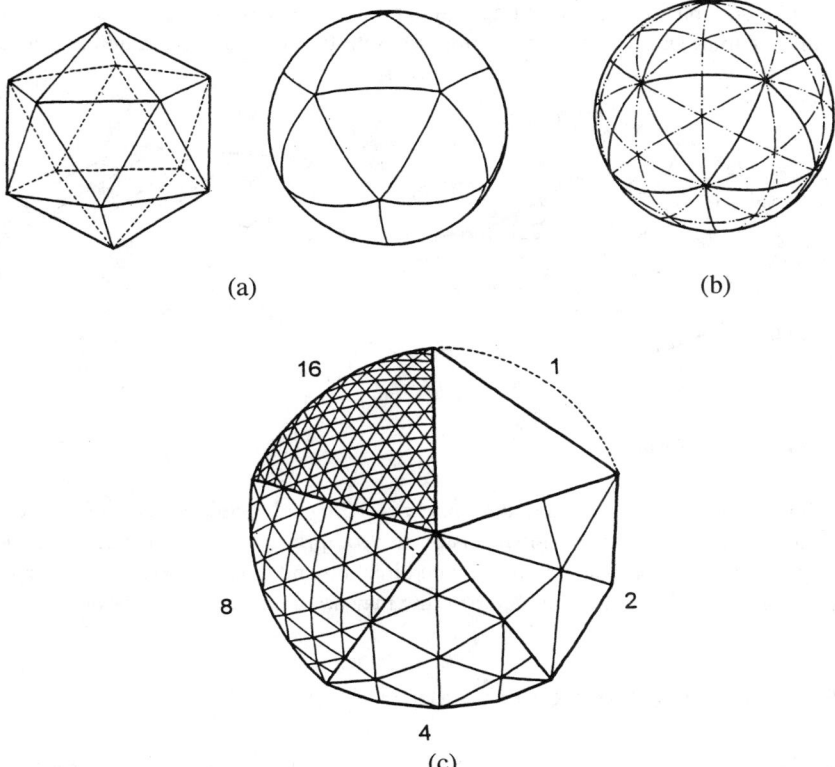

Figure 13.15 Geodesic subdivision.

added. This results in a ribbed dome and further a Schwedler dome. Alternately, the initial set may be taken as a series of spiral arcs, resulting in a division of the surface into triangular units as uniform as possible. This is achieved by drawing great circles in three directions as show in the case of a grid dome. A noteworthy type of division of a braced dome is the parallel lamella dome which is obtained by combining the first and second methods described above. The third method of subdivision results from projecting the edges of in-polyhedra onto the spherical surface, and then inscribing a triangular network of random frequency into this basic grid. A geodesic dome represents an application of this method, with the basic field derived from the isosahedron further subdivided with equilateral triangles.

13.3.4 Hyperbolic Paraboloid Shells

The hyperbolic paraboloid or hypar is a translational surface formed by sliding a concave paraboloid, called a generatrix, parallel to itself along a convex parabola, called a directrix, which is perpendicular to the generatrix (Figure 13.16a). By cutting the surface vertically, parabolas can be obtained and by cutting horizontally hyperbolas can be obtained. Such surfaces can also be formed by sliding a straight line along two other straight lines skewed with respect to each other (Figure 13.16b). The hyperbolic paraboloid is a doubly ruled surface; it can be defined by two families of intersecting straight lines that form in plan projection a rhombic grid. This is one of the main advantages of a hyperbolic paraboloid shell. Although it has a double curvature anticlastic surface, it can be built by using linear structural members only. Thus, single layer hypar shells can be fabricated from straight beams and double layer hypar shells from linear latticed trusses. The single hypar unit shown in

Figure 13.16 is suitable for use in building of square, rectangular, or elliptic plan. In practice, there exist an infinite number of ways of combining hypar units to enclose a given building space.

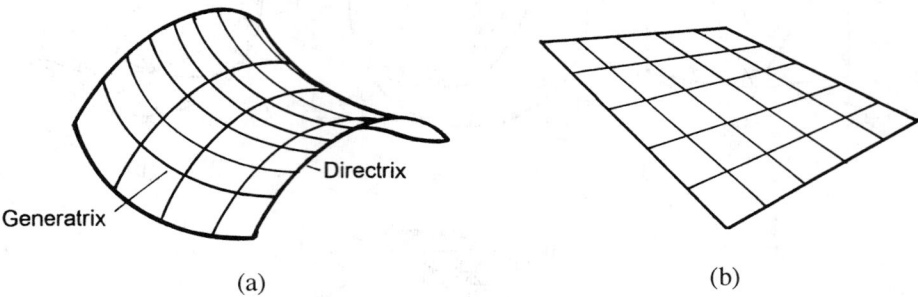

(a)   (b)

Figure 13.16   Hyperbolic paraboloid shells.

A shallow hyperbolic paraboloid under uniform loading acts primarily as a shear system, where the shear forces, in turn, causes diagonal tension and compression. The behavior of the surface can be visualized as thin compression arches in one direction and tension cables in the perpendicular direction. In reality, additional shear and bending may occur along the vicinity of the edges.

13.3.5   Intersection and Combination

The basic forms of latticed shells are single-curvature cylinders, double-curvature spheres, and hyperbolic paraboloids. Many interesting new shapes can be generated by intersecting and combining these basic forms. The art of intersection and combination is one of the important tools in the design of latticed shells. In order to fulfill the architectural and functional requirements, the load-resisting behavior of the structure as a whole and also its relation to the supporting structure should be taken into consideration.

For cylindrical shells, a simply way is to intersect through the diagonal as shown in Figure 13.17a. Two types of groined vaults on a square plane can be formed by combining the corresponding intersected curve surfaces as shown in Figures 13.17b and c. Likewise, combination of curved surfaces intersected from a cylinder produce a latticed shell on a hexagonal plan as shown in Figure 13.17d.

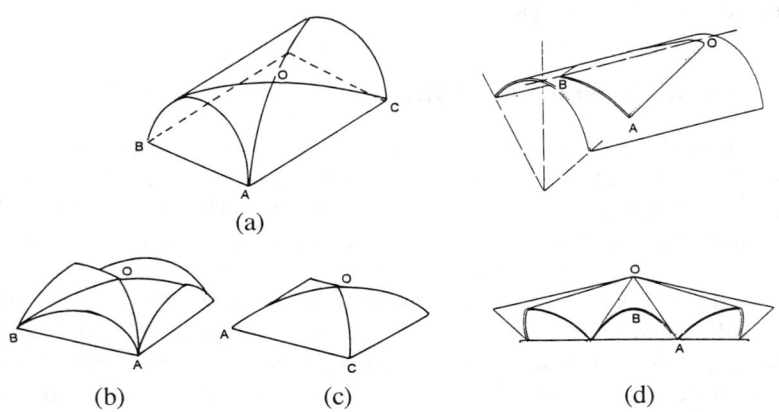

(a)

(b)   (c)   (d)

Figure 13.17   Intersection and combination of cylindrical shells.

For spherical shells, segments of the surface are used to cover planes other than circular, such as triangular, square, and polygonal as shown in Figure 13.18a, b, and c, respectively. Figure 13.18d shows a latticed shell on a square plane by combining the intersected curved surface from a sphere.

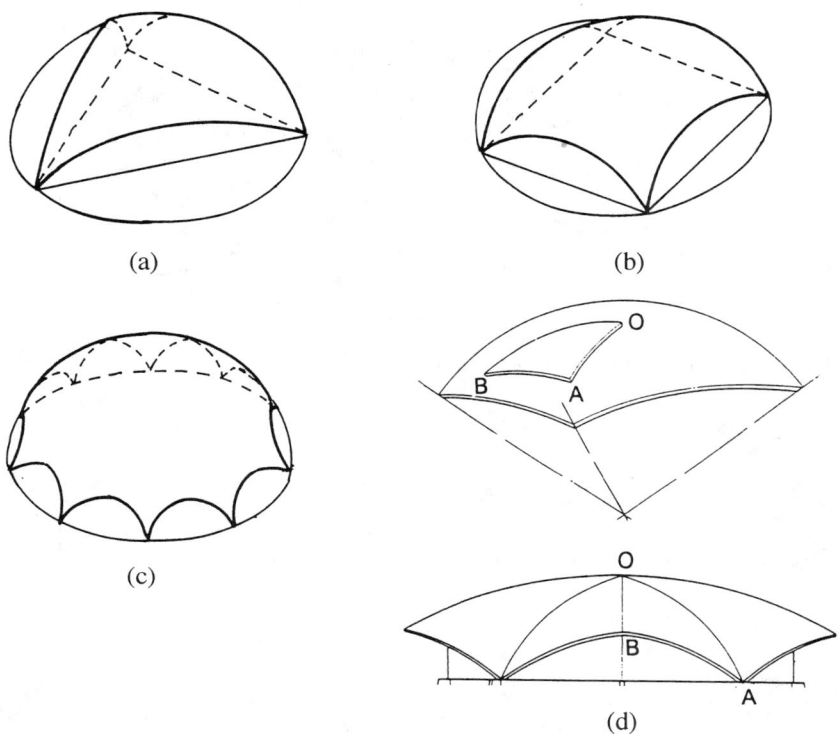

Figure 13.18   Intersection and combination of spherical shells.

It is usual to combine a segment of a cylindrical shell with hemispherical shells at two ends as shown in Figure 13.19. This form of latticed shell is an ideal plan for indoor track fields and ice skating rinks.

Different solutions for assembling single hyperbolic paraboloid units to cover a square plane are shown in Figure 13.20. The combination of four equal hypar units produces different types of latticed shells supported on a central column as well as two or four columns along the outside perimeter. These basic blocks, in turn, can be added in various ways to form the multi-bay buildings.

13.4   Structural Analysis

13.4.1   Design Loads

1. **Dead load**—The design dead load is established on the basis of the actual loads which may be expected to act on the structure of constant magnitude. The weight of various accessories—cladding, supported lighting, heat and ventilation equipment—and the weight of the space frame comprise the total dead load. An empirical formula is suggested to estimate the dead weight g of double layer grids.

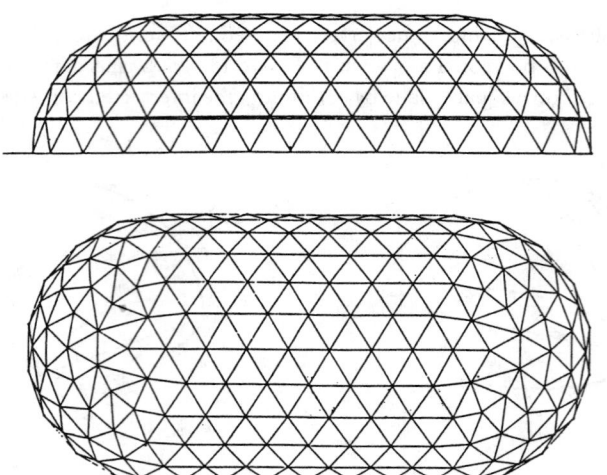

Figure 13.19   Combination of cylindrical and spherical shells.

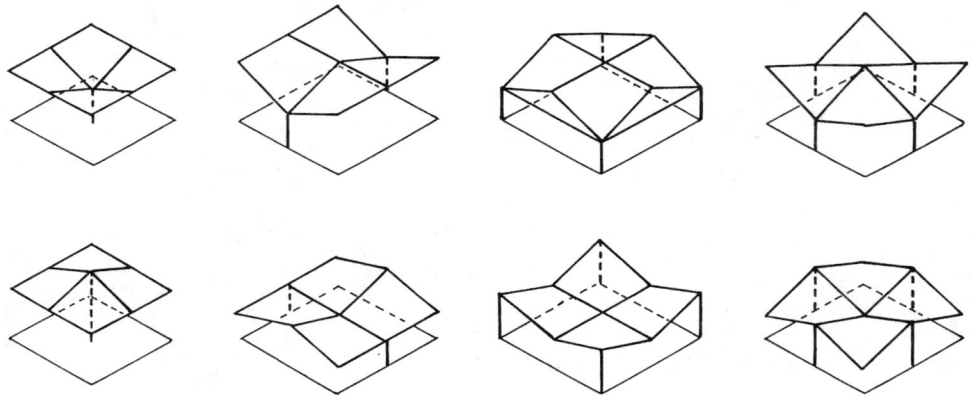

Figure 13.20   Combination of hyperbolic paraboloids.

$$g = 1/200 \left(\zeta \sqrt{q_w} L \right) \text{ kN/m}^2 \qquad (13.3)$$

where

q_w = all dead and live loads acting on a double layer grid except its self-weight in kN/m^2

L = shorter span in m

ζ = coefficient, 1.0 for steel tubes, 1.2 for mill sections

2. **Live load, snow or rain load**—Live load is specified by the local building code and compared with the possible snow or rain load. The larger one should be used as the design load. Each space frame is designed with a uniformly distributed snow load and further allowed for drifting depending upon the shape and slope of the structure. Often more than one assumed distribution of snow load is considered. Very little information can be found on this subject although a proposal was given by ISO for the determination of snow loads on simple curved roofs. The intensity of snow load as specified in *Basis for Design of Structures: Determination of Snow Loads on Roofs* [12] is reproduced as Figure 13.21. Rain load may be important in a tropical climate especially if the drainage

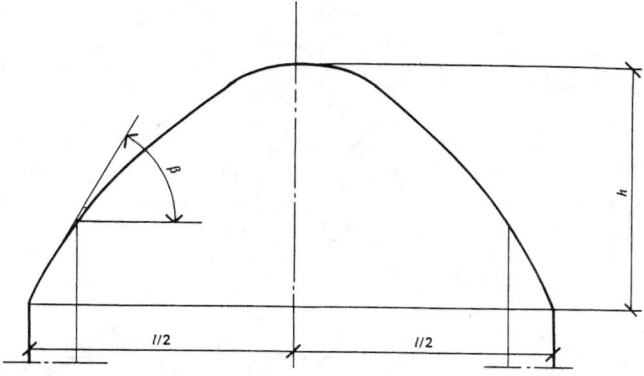

The following cases, 1 and 2, must be examined :

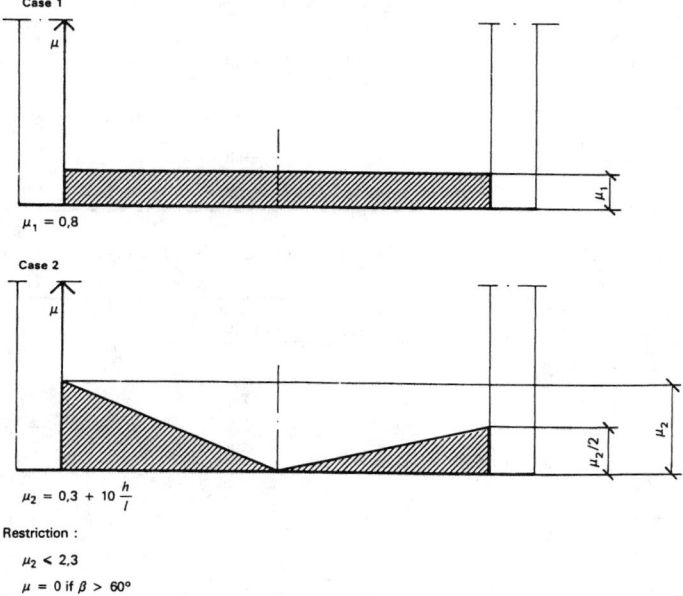

Case 1

$\mu_1 = 0,8$

Case 2

$\mu_2 = 0,3 + 10 \frac{h}{l}$

Restriction :

$\mu_2 < 2,3$

$\mu = 0$ if $\beta > 60°$

Figure 13.21 Snow loads on simple curved roof.

provisions are insufficient. Ponding results when water on a double layer grid flat roof accumulates faster than it runs off, thus causing excessive load on the roof.

3. **Wind load**—The wind loads usually represent a significant proportion of the overall forces acting on barrel vaults and domes. A detailed comparison of the available codes concerning wind loads has revealed quite a large difference between the practices adopted by various countries. Pressure coefficients for an arched roof springing from a ground surface that can be used for barrel vault designs are shown in Figure 13.22 and Table 13.3. For an arched roof resting on an elevated structure such as enclosure walls, the pressure coefficients are shown in Table 13.4.

The wind pressure distribution on buildings is also recommended by the European Convention for Constructional Steelwork. The pressure coefficients for an arched roof and spherical domes, either resting on the ground or on an elevated structure are presented in graphical forms as shown in Figure 13.23 and 13.24, respectively.

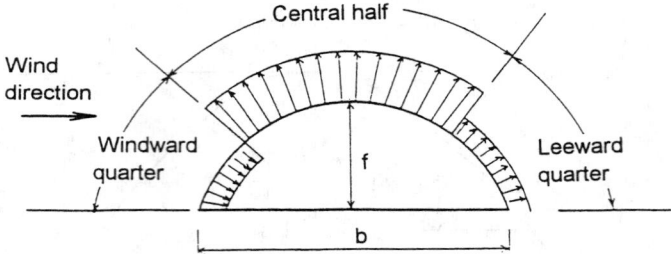

Figure 13.22   Wind pressure on an arched roof.

TABLE 13.3   Pressure Coefficient for an Arched Roof on the Ground

Country code	Windward quarter	Central half	Leeward quarter	Rise/span r
U.S.				
ANSI A 58.1-1982	1.4r	$-0.7 - r$	-0.5	$0 < r < 0.6$
U.S.S.R.				
BC&R 2.01.07-85	0.1	-0.8	-0.4	0.1
	0.3	-0.9	-0.4	0.2
	0.4	-1.0	-0.4	0.3
	0.6	-1.1	-0.4	
				0.4
	0.7	-1.2	-0.4	0.5
China				
GBJ 9-87-1987	0.1	-0.8	-0.5	0.1
	0.2	-0.8	-0.4	0.2
	0.6	-0.8	-0.4	0.5

TABLE 13.4   Pressure Coefficient for an Arched Roof on an Elevated Structure

Country code	Windward quarter	Central half	Leeward quarter	Rise/span r
U.S.				
ANSI A 58.1-1982	-0.9	$-0.7 - r$	-0.5	$0 < r < 0.2$
	1.5r $- 0.3^a$	$-0.7 - r$	-0.5	$0.2 \leq r < 0.3$
	2.75r $- 0.7$	$-0.7 - r$	-0.5	$0.3 \leq r \leq 0.6$
U.S.S.R.				
BC&R 2.01.07-85	$h_e/b = 0.2^b$			
	-0.2	-0.8	-0.4	0.1
	-0.1	-0.9	-0.4	0.2
	0.2	-1.0	-0.4	0.3
	0.5	-1.1	-0.4	0.4
	0.7	-1.2	-0.4	0.5
	$h_e/b > 1$			
	-0.8	-0.8	-0.4	0.1
	-0.7	-0.9	-0.4	0.2
	-0.3	-1.0	-0.4	0.3
	0.3	-1.1	-0.4	0.4
	0.7	-1.2	-0.4	0.5
China				
GBJ 9-87-1987	-0.8	-0.8	-0.5	0.1
	0	-0.8	-0.5	0.2
	0.6	-0.8	-0.5	0.5

^{a} Alternate coefficient 6r $- 2.1$ shall also be used.
^{b} h_e = height of the elevated structure.

It can be seen that significant variations in pressure coefficients from different codes of practice exist for three-dimensional curved space frames. This is due to the fact that these coefficients are highly dependent on Reynolds number, surface roughness, wind velocity profile, and turbulence. It may be concluded that the codes of practice are only suitable for preliminary design purposes, especially for those important long span space structures and latticed shells with peculiar shapes. It is therefore necessary

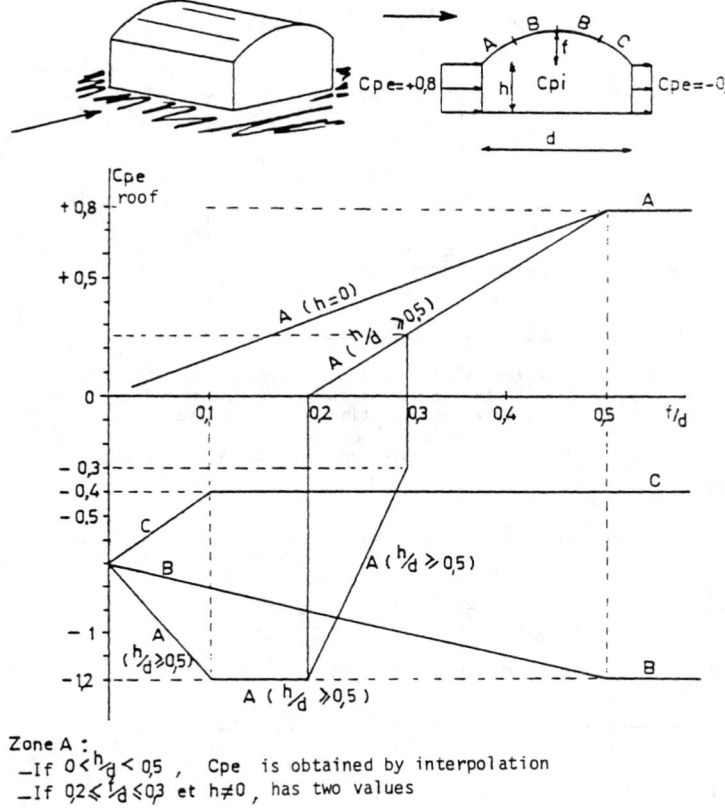

Zone A :
— If $0 < \frac{h}{d} < 0,5$, Cpe is obtained by interpolation
— If $0,2 \leqslant \frac{f}{d} \leqslant 0,3$ et $h \neq 0$, has two values

Figure 13.23 Wind pressure coefficients for an arched roof.

to undertake further wind tunnel tests in an attempt to more accurately establish the pressure distribution over the roof surface. For such tests, it is essential to simulate the velocity profile and turbulence of the natural wind and the Reynolds number effects associated with the curved surface.

4. **Temperature effect**—Most space frames are subject to thermal expansion and contraction due to changes in temperature, and thus may be subject to axial loads if restrained. Potential temperature effect must be considered in the design especially when the span is comparatively large. The choice of support locations—perimeter, intermediate columns—and types of support—fixed, slid or free rotation and translation—as well as the geometry of members adjacent to the support, all contribute to minimizing the effect of thermal expansion. The temperature effect of a space frame may be calculated by the ordinary matrix displacement method of analysis and most computer programs provide such a function.

For a double layer grid, if it satisfies one of the following requirements, the calculation for temperature effect may be exempted.

(a) The joints on supports allow the double layer grid to move horizontally.

(b) Double layer grids of less than 40 m span are supported along perimeters by independent reinforced concrete columns or brick pilasters.

(c) The displacement at the top of the column due to a unit force is greater or equal to the value calculated according to the following formula:

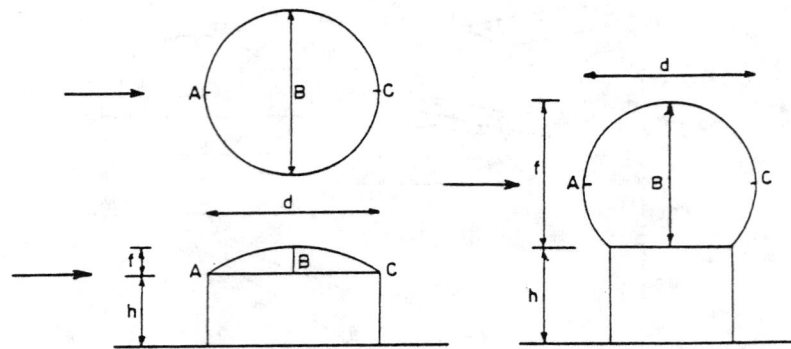

Cpe is constant along the intersection between the dome and planes
perpendicular to the wind : it can be determined as a first approximation by linear interpolation between the values at A, B and C.
In the same way, values of *Cpe* at A if $0 < h/d < h$ and at B or C if
$0 < h/d < 0,5$ can be obtained by linear interpolation.

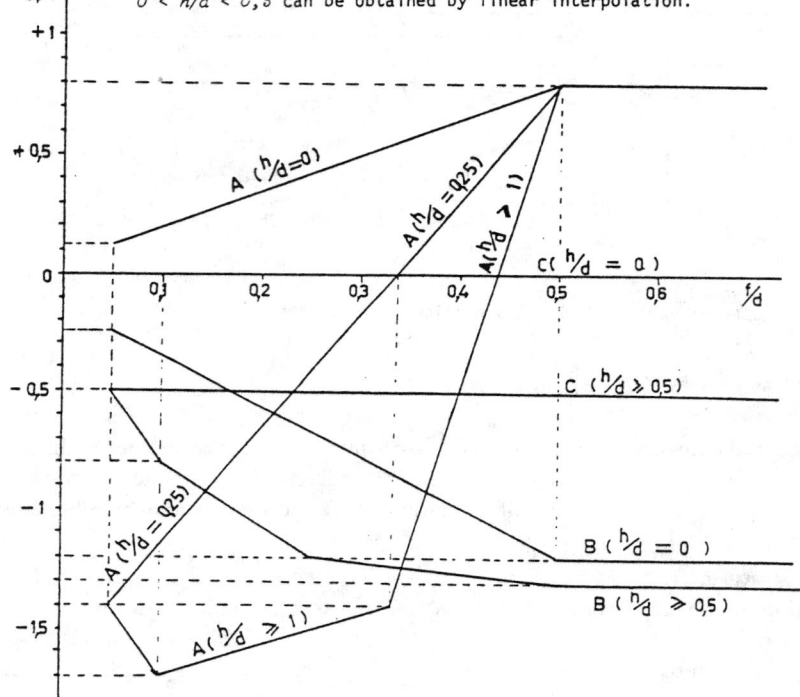

Figure 13.24   Wind pressure coefficients for spherical domes.

$$\delta = \frac{L}{2\xi E A}\left(\frac{E\alpha\Delta_t}{0.05\,[\sigma]} - 1\right)$$
(13.4)

where

L	=	span of double layer grid in the direction of checking temperature effect
E	=	modulus of elasticity
A	=	arithmetic mean value of the cross-sectional area of members in the supporting plane (top or bottom layer)
α	=	coefficient of thermal expansion

Δ_t =   temperature difference

$[\sigma]$ =   allowable stress of steel

ξ =   coefficient, when the chords in the supporting plane are arranged in orthogonal grids $\xi = 1$, in diagonal grids $\xi = 2$, and in three-way grids $\xi = 2$

5. **Construction loads**—During construction, structures may be subjected to loads different from the design loads after completion, depending on the sequence of construction and method of scaffoldings. For example, a space frame may be lifted up at points different from the final supports, or it may be constructed in blocks or strips. Therefore, the whole structure, or a portion of it, should be checked during various stages of construction.

13.4.2   Static Analysis

There are generally two different approaches in use for the analysis of space frames. In the first approach, the structure is analyzed directly as a general assembly of discrete members, i.e., **discrete method**. In the second approach, the structure is represented by an equivalent continuum like a plate or shell, i.e., **continuum analogy method**.

The advent of computers has radically changed the whole approach to the analysis and design of space frames. It has also been realized that matrix methods of analysis provide an extremely efficient means for rapid and accurate treatment of many types of space structures. In the matrix analysis, a structure is represented as a discrete system and all the usual equations of structural mechanics are written conveniently in matrix form. Thus, matrix analysis is particularly suitable to computer formulation, with an automatic sequence of operations. A number of general purpose computer programs, such as STRUDL and SAP, have been developed and are available to designers.

The two common formulations of the matrix analysis are the stiffness method and flexibility method. The stiffness method is also referred to as the displacement method because the displacements of the redundant members are treated as unknowns. The flexibility method (or force method) treats the forces in the members as unknowns. Of these two methods, the displacement method is widely used in most computer programs.

In the *displacement method,* the stiffness matrix of the whole structure is obtained by adding appropriately the stiffness matrixes of the individual elements. Supports are then introduced because the displacements at these points are known. A set of simultaneous equations are solved for displacements. From the joint displacements the member elongation can be found and hence the member forces and reaction at supports.

The matrix displacement method is by far the most accurate method for the analysis of space frames. It can be used without any limit on the type and shape of the structure, the loadings, the supporting conditions, or the variation of stiffness. The effect of temperature or uneven settlement of supports also can be analyzed conveniently by this method. For design work, a special purpose computer program for space frames is preferred; otherwise the input of generating nodal coordinates and member connectivity plus loading information will be a tremendous amount of work. Some sophisticated computer programs provide the functions of automatic design, optimization, and drafting.

Double layer grids can be analyzed as pin connected and rigidity of the joints does not change the stress by more than 10 to 15%. In the displacement method, bar elements are used with three unknown displacements in x, y, and z directions at each end. For single layer reticulated shells with rigid joints, bar elements are used and the unknowns are doubled, i.e., three displacements and three rotations. Under specific conditions, single layer braced domes may be analyzed as pin connection joints with reasonable accuracy if the rise of the dome is comparatively large and under symmetric loading.

When using a computer, the engineer must know the assumptions on which the program is based, the particular conditions for its use (boundary conditions for example), and the manner of introducing the input data. In the static analysis of the space frames, care should be taken on the following issues:

1. **Support conditions**—A fixed support (bolted or welded) in construction should not be treated literally as a completely fixed node in analysis. As a matter of fact, most space frames are supported on columns or walls that have a lateral flexibility. Upon the acting of external loads, there will be lateral displacements on the top of columns. Therefore, it is more reasonable to assume the support as horizontally movable rather than fixed, or as an elastic support by considering the stiffness of the supporting column.

2. **Criterion for the number of reanalyzes**— Usually a set of sectional areas are assumed for members and the computer will proceed to analyze the structure to obtain a set of member forces. Then the members are checked to see if the assumed areas are appropriate. If not, the structure should be reanalyzed until the forces and stiffness completely match each other. However, such extended reanalyzes by the stiffness method will induce a high concentration of stiffness and, hence, a great difference of member sections which is unacceptable for practical use. Therefore, it is necessary to limit the number of reanalyzis. In practice, certain criteria are specified such that the reanalysis will terminate automatically. One of the criterion is suggested as the number of the modified members less than 5% of the total number of members. Usually three or four runs will produce a satisfactory result.

3. **Checking of computer output**—It is dangerous for an engineer to rely on the computer output as being infallible. Always try to estimate and anticipate results. A simple manual calculation by approximate method and comparing it with computer output will be beneficial. By doing so, an order of magnitude for the results can be obtained. In this operation, intuition also plays an important role. At the same time, simple checks should be done to test the reliability of the computer program, such as the equilibrium of forces at nodes and the equilibrium of total loading with the summation of reactions. A check on the deflections along certain axes of the structure would also be helpful. The size and location of any large deflection should be noted. All deflections should be scanned to look for possible bad solutions caused by improper modeling of the structure. This check is made easily if the program has the ability to produce a deformed geometry plot.

A *continuum analogy method* may also be used for the static analysis of space frames. This is to replace a latticed structure by an equivalent continuum which exhibits equivalent behavior with respect to strength and stiffness. The equivalent rigidity is used for the stress and displacement analysis in the elastic range, and particularly so for stability and dynamic analysis. It is useful as well in order to provide an understanding of the overall behavior of the structure by large. By using equivalent rigidity, the thickness, elastic moduli, and Poisson's ratio are determined for the equivalent continuum, and the fundamental equations that govern the behavior of the equivalent continuum are established as in the usual continuum theory. Therefore, the methods of solution and the results of the theory of plates and shells are directly applicable. Thus, certain types of latticed shells and double layer grids can be analyzed by treating them as a continuum and applying the shell or plate analogy. This method has been found to be satisfactory where the loading is uniform and the load transfer is predominantly through membrane action.

Some difficulties may occur in the application of the continuum analogy method. The boundary conditions of the continuum cannot be entirely analogous to the boundary condition of the discrete prototype. Also, some of the effects that are relatively unimportant in the case of continua may be significant in the case of space frames. Two of these merit mention. The effect of shear deformation in elastic plates and shells is essentially negligible, whereas the contributions of web members connecting the layers of a space frame can be significant to the total deformation. Similarly, the correct continuum model of a rigidly connected space frame must allow for the possibility of rotation of joints independent of the rotations of normal sections. Such models are more complex than the usual ones, and few solutions of the governing equations exist.

It is useful to compare the discrete method and continuum analogy method. The continuum analogy method can only be applied to regular structures while the discrete method can handle arbitrary structural configuration. The computational time is much less for an equivalent shell analysis than a stiffness method analysis. The work involved in a continuum analogy method includes calculating the equivalent rigidity, the forces in the equivalent continuum, and finally the forces in the members. This will go through a discrete-continuum-discrete process and, hence, involves further approximation. To summarize, the continuum analogy method is most valuable at the stage of conceptual and preliminary design while the discrete method should be used for a working design.

13.4.3 Earthquake Resistance

One of the important issues that must be taken into consideration in the analysis and design of space frames is the earthquake excitation in case the structure is located in a seismic area. The response of the structure to earthquake excitation is dynamic in nature and usually a dynamic analysis is necessary. The analysis is complicated due to the fact that the amplitude of ground accelerations, velocities, and motions is not clearly determined. Furthermore, the stiffness, mass distribution, and damping characteristics of the structure will have a profound effect on its response: the magnitude of internal forces and deformations.

The dynamic behavior of a space frame can be studied first through the vibration characteristics of the structure that is represented by its natural frequencies. The earthquake effects can be reflected in response amplification through interaction with the natural dynamic characteristics of the structures. Thus, double layer grids can be treated as a pin-connected space truss system and their free vibration is formulated as an equation of motion for a freely vibrating undamped multi-degree-of-freedom system. By solving the generalized eigenvalue problems, the frequencies and vibration modes are obtained.

A series of double layer grids of different types and spans were taken for dynamic analysis [23]. The calculation results show some interesting features of the free vibration characteristics of the space frames. The difference between the frequencies of the first 10 vibration modes is so small that the frequency spectrums of space frames are rather concentrated. The variation of any design parameter will lead to the change of frequency. For instance, the boundary restraint has a significant influence on the fundamental period of the space frame: the stronger the restraint, the smaller the fundamental period.

The fundamental periods of most double layer grids range from 0.37 to 0.62 s which are less than that of planar latticed trusses of comparable size. This fact shows clearly that the space frames have relatively higher stiffness. Investigating into the relation of fundamental periods of different types of double layer grids with span, it is found that the fundamental period increases with the span, i.e., the space frames will be more flexible for longer span. The response of space frames with shorter span will be stronger.

The vibration modes of double layer grids could be classified mainly as vertical modes and horizontal modes that appear alternately. In most cases, the first vibration modes are vertical. The vertical modes of different types of double layer grids demonstrate essentially the same shape and the vertical frequencies for different space frames of equal span are very close to each other. It was found that the forces in the space frame due to vertical earthquake are mainly contributed by the first three symmetrical vertical modes. Certain relations could be established between the first three frequencies of the vertical mode as follows:

$$\omega_{v2} = (2 - 3.5)\,\omega_{v1} \tag{13.5}$$

$$\omega_{v3} = (4 - 4.6)\,\omega_{v2} \tag{13.6}$$

where ω_{v1}, ω_{v2}, and ω_{v3} are the first, second, and third vertical frequencies, respectively.

The simplest way to estimate the earthquake effect is a quasi-static model in which the dynamic action of the ground motion is simulated by a static action of equivalent loads. The manner in which the equivalent static loads are established is introduced in many seismic design codes of different countries. In the region where the maximum vertical acceleration is 0.05 g, usually the earthquake effect is not the governing factor in design and it is not necessary to check the forces induced by vertical or horizontal earthquake. In the area where the maximum vertical acceleration is 0.1 g or greater, a factor of 0.08 to 0.2, depending on different codes, is used to multiply the gravitational loads to represent the equivalent vertical earthquake load. It should be noticed that in certain seismic codes, the live load that forms a part of gravitational loads is reduced by 50%. The values of vertical seismic forces in the members of double layer grids are higher near the central region and decrease gradually towards the perimeters. Thus, the ratio between the forces in each member due to vertical earthquake and static load is not constant over the whole structure. The method of employing equivalent static load serves only as an estimation of the vertical earthquake effect and provides an adequate level of safety.

Due to the inherent horizontal stiffness of double layer grids, the forces induced by horizontal earthquake can be resisted effectively. In the region of 0.1 g maximum acceleration, if the space grids are supported along perimeters with short or medium span, it is not required to check the horizontal earthquake. However, for double layer grids of longer span or if the supporting structure underneath is rather flexible, seismic analysis in the horizontal direction should be taken. In the case of latticed shells with a curved surface, the response to horizontal earthquake is much stronger than double layer grids depending on the shape and supporting condition. Even in the region with maximum vertical acceleration of 0.05 g, the horizontal earthquake effect on latticed shells should be analyzed. In such analysis, coordinating the action of the space frame and the supporting structure should be considered. A simple way of coordination is to include the elastic effect of the supporting structure. This is represented by the elastic stiffness provided by the support in the direction of restraint. The space frame is analyzed as if the supports have horizontally elastic restraints. For a more accurate analysis, the supporting columns are taken as member with bending and axial stiffness and analyzed together with the space frame. In the analysis, it is also important to include the inertial effect of the supporting structure, which has influence on the horizontal earthquake response of the space frame.

In the case of more complex structures or large spans, dynamic analysis, such as the response spectrum method for modal analysis, should be used. Such method gives a good estimate of the maximum response during which the structure behaves elastically. For space frames, the vertical seismic action should be considered. However, few recorded data on the behavior of such structures under vertical earthquake exist. In some aseismic design codes, the magnitude of the vertical component may be taken as 50 to 65% of the horizontal motions. Use of 10 to 20 vibration modes is recommended for the space frames when applying the response spectrum method.

For space frames with irregular and complicated configurations or important long span structures, the time-history analysis method should be used. The number of acceleration records or synthesized acceleration curves for the time-history analysis is selected according to intensity, location of earthquake, and site category. In usual practice, at least three records are used for comparison. Such a method is an effective tool to calculate the earthquake response when large, inelastic deformations are expected.

The behavior of latticed structures under dynamic loads or, more specifically, the performance of latticed structures due to earthquake, was the main concern of structural engineers. An ASCE

Task Committee on Latticed Structures Under Extreme Dynamic Loads was formed to investigate this problem. One of the objectives was to determine if dynamic conditions have historically been the critical factor in failure of lattice structures. A short report on "Dynamic Considerations in Latticed Structures" was submitted by the Task Committee in 1984. Eight major failures of latticed roof structures were reported but notably none of them was due to earthquake.

Since the ASCE report was published, valuable information on the behavior of space frames during earthquake has been obtained through two seismic events. In 1995, the Hanshin area of Japan suffered a strong earthquake and many structures were heavily damaged or destroyed. However, when compared with other types of structures, most of the damage to space frames located in that area was relatively minor [13]. It is worthwhile to mention that two long span sport arenas of space frame construction were built on an artificial island in Kobe and no major structural damage was found. On the other hand, serious damage to a latticed shell was found on the roof structure of a hippodrome stand where many members were buckled. The cause of the damage is not due to the strength of the space frame itself but the failure of the supporting structure. Another example of serious damage to double layer grids for the roof structure of a theater occurred in 1985 when a strong earthquake struck the Kashigor District of Sinkiang Uygur Autonomous Region in China [15]. Failure was caused by a flaw in the design as the elastic stiffness and inertial effect of supporting structures were completely ignored. Behavior of space frame structures under a strong earthquake has generally been satisfactory from a strength point of view. Experiences gained from strong earthquakes shows the space frames demonstrate an effective spatial action and consequently a reasonably good aseismic behavior.

13.4.4 Stability

Although a great amount of research has been carried out to determine the buckling load of latticed shells, the available solutions are not satisfactory for practical use. The problem is complicated by the effect of geometric nonlinearity of the structure and also the influence of the joint system according to which the members can be considered as pin-connected or partially or completely restrained at the nodes.

The following points are important in the buckling analysis of latticed shells [7]:

1. Decision on which kind of nonlinearity is necessary to be used—only geometrical nonlinearity with the elastic analysis, or geometrical and material nonlinearities with the elastic-plastic analysis.
2. Choosing the physical model—equivalent continuum or discrete structure.
3. Choosing the computer model and numerical procedure for tracing the non-linear response for precritical behavior, collapse range, and post-critical behavior.
4. Study of factors influence load carrying capacity—buckling modes, density of network, geometrical and mechanical imperfections, plastic deformations, rigidity of joints, load distributions, etc.
5. Experimental investigations to provide data for analysis (rigidity of joints, postbuckling behavior of individual member, etc.) and confirmation of theoretical values.

Generally speaking, there are three types of buckling that may occur in latticed shells:

1. member buckling (Figure 13.25a)
2. local or dimple buckling at a joint (Figure 13.25b)
3. general or overall buckling of the whole structure (Figure 13.25c)

Member buckling occurs when an individual member becomes unstable, while the rest of the space frame (members and nodes) remain unaffected. The buckling load P_{cr} of a straight prismatic bar

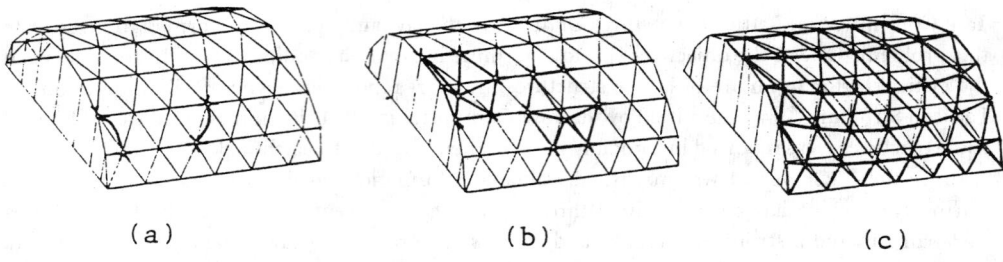

(a)   (b)   (c)

Figure 13.25   Different types of buckling.

under axial compression is given by

$$P_{cr} = \alpha \frac{\pi^2 E_e I}{l^2} \quad \alpha = \alpha\left(c_i, c_j, w_o, e, m\right) \tag{13.7}$$

where
E_e = effective modulus of elasticity that coincides with Young's modulus in the elastic range
I = moment of inertia of member
l = length of the member

The coefficient α takes different values depending on the parameter in the parentheses. The quantities c_i and c_j characterize the rotational stiffness of the joints, w_o is the initial imperfection, e is the eccentricity of the end compressive forces, and m is the end shear forces and moments. A reduced length l_o should be used in place of l when the ratio of the joint diameter to member length is relatively large. On the basis of Equation 13.7, the design code for steel structures in different countries provides methods for estimating member buckling, usually by introducing the slenderness ratio $\lambda = l/r$, where r is the radius of gyration of the member's section.

The *local buckling* of a space frame consists of a snap-through buckling which takes place at one joint. Snap-through buckling is characterized by a strong geometrical non-linearity. Local buckling is apt to occur when the ratio of t/R (where t is the equivalent shell thickness and R is the radius of curvature) is small. Similarly, local buckling of a space frame is likely to occur in single layer latticed shells.

Local buckling is greatly affected by the stiffness of and the loads on the adjacent members. Consider the pin-connected structure shown in Figure 13.26. Buckling load q_{cr} in terms of uniform normal load per unit area can be expressed as

$$\frac{AEl}{12R^3} \le q_{cr} \le \frac{AEl}{6R^3} \tag{13.8}$$

where
A = cross-sectional area of the member
E = modulus of elasticity
R = radius of an equivalent spherical shell through points B-A-B

In practice, different types of joints used in the design will possess different flexural strength; thus, the actual behavior of the joint and member assembly should be incorporated in determining the local buckling load. An approximate formula was proposed by Lind [16] and is applicable to triangular networks having all elements of the same cross-sections. For the uniform load, the critical load is

$$Q_{cr} = \frac{E_t}{1 + \alpha^2/8\pi^2}\left(0.47\frac{Al^3}{R^3} + 3\frac{BI}{lR}\right)$$

$$\alpha = l^2/rR \tag{13.9}$$

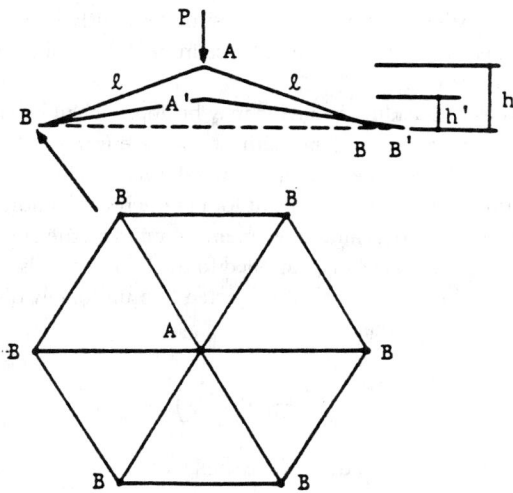

Figure 13.26 Local buckling of a pin-connected structure.

where

E_t = tangent modulus of elasticity
R = radius of curvature of the framework mid-surface
r = radius of gyration
B = non-dimensional bending stiffness of the grid given in Table 13.5

For the concentrated load, the following two formulae are presented

$$W_{cr} = \frac{3EAh^3}{l^3} \left[\frac{8B}{\alpha^2} + 0.241 \left(1 - 5.95 \frac{8B}{\alpha^2} \right) \right] \tag{13.10}$$

which is valid for $\alpha > 9$, and

$$W_{cr} = 0.0905 EA \left(\frac{l}{R} \right)^3 \tag{13.11}$$

for a regular pin-jointed triangular network.

TABLE 13.5 Equivalent Bending Stiffness B

α	1/32	1/16	1/8	1/4	1/2	1	2	4	8	16	32	64
B	0.868	0.873	0.886	0.950	1.176	1.85	3.15	4.83	6.48	7.35	7.80	7.90

The *overall buckling* occurs when a relatively large area of the space frame becomes unstable, and a relatively large number of joints is involved in the buckle. For most cases, in overall buckling of a space frame, the wave length is significantly greater than the member length. Local buckling often plays the role of trigger for overall buckling.

The type of buckling collapse of a space frame is greatly influenced by the following factors: its Gaussian curvature, whether it is a single or double layer system, the degree of statical indeterminacy, and the manner of supporting and loading. Generally speaking, a shallow shell of positive Gaussian curvature, like a dome, is more prone to overall buckling than a cylindrical shell of zero Gaussian curvature. Recent research reveals that a hyperbolic paraboloid shell is less vulnerable to overall buckling and the arrangement of the grids has a considerable influence on the stability and stiffness of the shell. It is best to arrange the members along the direction of compressive forces. A single layer space frame exhibits greater sensitivity to buckling than a double layer structure. Moreover, various types of buckling behavior may take place simultaneously in a complicated relation. For double layer grids, in most cases, it is sufficient to examine the member collapse which may occur in the compressive chord members.

The theoretical analysis of buckling behavior may be approached by two methods: continuum analogy analysis and discrete analysis. Since almost all space frames are constructed from nearly identical units arranged in a regular pattern, it is generally accepted that the analysis on the basis of the equivalent continuum serves as an important tool in the investigation of the buckling behavior of space frames. Numerous analytical and experimental studies on the buckling of continuous shells have been performed and the results can be applied to the latticed shells.

The buckling formula for a spherical shell subjected to a uniformly distributed load normal to the middle surface can be expressed as

$$q_{cr} = kE \left(\frac{t}{R} \right)^2 \qquad (13.12)$$

where t is the thickness and R is the radius of the shell.

Different values of the coefficient k were obtained by various investigators.

$$
\begin{aligned}
k \quad &= \quad 1.21 \text{ (Zoelly [1915], based on classical linear theory)} \\
&= \quad 0.7 \text{ (experiments on very carefully prepared models)} \\
&= \quad 0.366 \text{ (Karman and Tsien [1939], based on nonlinear elastic theory)} \\
&= \quad 0.228, 0.246 \text{ (del Pozo [1979], for } \mu = 0 \text{ and } 0.3, \text{ respectively)}
\end{aligned}
$$

For a triangulated dome where an equivalent thickness is used, Wright [22] derived the formula by using

$$
\begin{aligned}
E \quad &= \quad AE/3rl \quad t = 2\sqrt{3r} \quad k = 0.4 \\
q_{cr} \quad &= \quad 1.6E \frac{Ar}{lR^2} \qquad (13.13)
\end{aligned}
$$

The critical load for overall buckling may also be expressed as the following formula for comparison

$$q_{cr} = k' \frac{E}{R^2} t_m^{1/2} t_b^{3/2} \qquad (13.14)$$

where

$$
\begin{aligned}
t_m \quad &= \quad \text{effective in-plane thickness} \\
t_b \quad &= \quad \text{effective bending thickness} \\
k' \quad &= \quad 0.377 \text{ [22]} \\
&= \quad 0.365 \\
&= \quad 0.247 \\
&= \quad 0.294 \text{ [8]}
\end{aligned}
$$

Discrete analysis is a more powerful tool to study the whole process of instability for space frames. As shown in Figure 13.27, a structure may lose its stability when it has reached a "limit point", where the stiffness is lost completely. On the other hand, a structure such as a dome may lose its stability by a sudden buckling into a mode of deformation before the limit point, which occurs at a distinct **critical point**—"bifurcation point" on the load path. It should be noted that the initial imperfection of the structure will greatly reduce the value of critical load, and certain types of space frame are very sensitive to the present of imperfection.

In the stability analysis, usually the characteristic at certain special states is investigated, i.e., the stability mode and critical load are analyzed as an eigenvalue problem. Researchers are now more interested in studying the whole process of nonlinear stability. As a result of the development of the computer matrix method, numerical analyses of large systems have become straightforward. Therefore, the discrete analysis of a space frame, itself a discrete structure, is very suitable for the study of stability problems. Major problems encountered in the nonlinear stability process are:

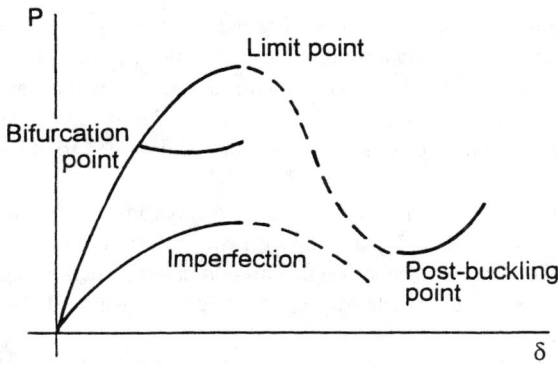

Figure 13.27   Instability points.

the mathematical and mechanical modeling of the structure, the numerical technique for solving nonlinear equations, and the tracing method for the nonlinear equilibrium path. Much research has been carried out in the above area.

The Newton-Raphson method or the modified Newton-Raphson method is the fundamental method for solving the nonlinear equilibrium equations and has proved to be one of the most effective methods. The purpose of tracing the nonlinear equilibrium path is as follows: (1) to provide equilibrium analysis for the pre-buckling state, (2) to determine the critical point, such as the limit point or bifurcation point on the load path and its critical load, (3) to trace the post-buckling response. On the basis of the increment-iteration process for the finite element method, techniques for the analysis of nonlinear equilibrium path and its tracing tactics have made significant progress in recent years. Numerical methods used for the construction of equilibrium paths associated with nonlinear problems, such as the load incremental method, constant arc-length method, displacement control method, etc. were developed by different authors. Because each of the techniques has its advantages and disadvantages in the derivation of fundamental equations, accuracy of solution, computing time, etc. the selection of the appropriate method has a profound influence on the efficiency of the computation. In the present stage of development, complicated equilibrium paths can be traced with the aid of the above technique. Computer programs have been developed for the whole process of nonlinear stability and can be used for the design of various types of latticed shells.

13.5   Jointing Systems

13.5.1   General Description

The jointing system is an extremely important part of a space frame design. An effective solution of this problem may be said to be fundamental to successful design and construction. The type of jointing depends primarily on the connecting technique, whether it is bolting, welding, or applying special mechanical connectors. It is also affected by the shape of the members. This usually involves a different connecting technique depending on whether the members are circular or square hollow sections or rolled steel sections. The effort expended on research and development of jointing systems has been enormous and many different types of connectors have been proposed in the past decades.

The joints for the space frame are more important than the ordinary framing systems because more members are connected to a single joint. Furthermore, the members are located in a three-dimensional space, and hence the force transfer mechanism is more complex. The role of the joints in a space frame is so significant that most of the successful commercial space frame systems utilize proprietary jointing systems. Thus, the joints in a space frame are usually more sophisticated than the joints in planar structures, where simple gusset plates will suffice.

In designing the jointing system, the following requirements should be considered. The joints must be strong and stiff, simple structurally and mechanically, and yet easy to fabricate without recourse to more advanced technology. The eccentricity at a joint should be kept to a minimum, yet the joint detailing should provide for the necessary tolerances that may be required during the construction. Finally, joints of space frames must be designed to allow for easy and effective maintenance.

The cost of the production of joints is one of the most important factors affecting the final economy of the finished structure. Usually the steel consumption of the connectors will constitute 15 to 30% of the total. Therefore, a successful prefabricated system requires joints that must be repetitive, mass produced, simple to fabricate, and able to transmit all the forces in the members interconnected at the node.

All connectors can be divided into two main categories: the purpose-made joint and the proprietary joint used in the industrialized system of construction. The purpose-made joints are usually used for long span structures where the application of standard proprietary joints is limited. An example of such types of joints is the cruciform gusset plate for connecting rolled steel sections as shown in Figure 13.28.

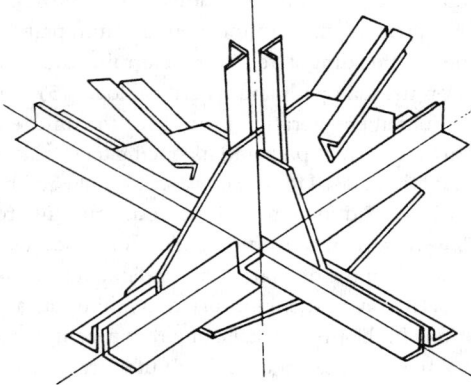

Figure 13.28   Connecting joint with cruciform gusset plate.

A survey around the world will reveal that there are over 250 different types of jointing systems suggested or used in practice, and there are some 50 commercial firms trying to specialize in the manufacture of proprietary jointing systems for space frames. Unfortunately, many of these systems have not proved attainment of great success mainly because of the complexity of the connecting method. Tables 13.6 through 13.8 give a comprehensive survey of the jointing systems all over the world. All the connection techniques can be divided into three main groups: (1) with a node, (2) without a node, and (3) with prefabricated units.

13.5.2   Proprietary System

Some of the most successful prefabricated jointing systems are summarized in Table 13.9. This is followed by a further description of each system.

1. **Mero**

    The Mero connector, introduced some 50 years ago by Dr. Mengeringhausen, proved to be extremely popular and has been used for numerous temporary and permanent buildings. Its joint consists of a node that is a spherical hot-pressed steel forging with flat

TABLE 13.6   Connection Types with a Node

Node		Connector	Member	Cross-section	Examples	Code
Sphere	Solid			○	Mero KK, Germany Montal, Germany Uzay, Italy	A11
				○	Steve Baer, U.S. Van Tel, NL KT space truss, Japan	
				●	Mero MT, Germany	
	Hollow			○	Spherobat, France	A12
				○	NS space truss, Japan Tuball, NL Orbik, U.K.	
				○	NS space truss, Japan Tuball, NL Orbik, U.K.	
	Hollow			○	SDC, France	A13
	Hollow			○	Oktaplatte, Germany	A14
				○	WHSJ, China	
	Hollow			○	Vestrut, Italy	A15
Cylinder	Solid			○	Triodetic, Canada nameless, East Germany	A21
	Hollow			○	Octatube Plus, NL nameless, Singapore	A22
				○	Pieter Huybers, NL	
				▢	nameless system, U.K.	
Disc	Flat			▢ ▢	Palc, Spain Power strut, U.S.	A31
				○	Pieter Huybers, NL	
				○	Tridimatec, France	A32
				▢ ▢	Moduspan, U.S. (former Unistrut) Space-Frame System VI, U.S. (Unistrut)	A33
	Welded			○	Boyd Auger, U.S. Octatube, NL	A34
				○	Piramodul large span, NL	
				○ ▢	Nodus, U.K.	A35

TABLE 13.6   Connection Types with a Node *(continued)*

Node	Connector	Member	Cross-section	Examples	Code
Solid				Montal, Germany	A41
				Mero BK, Germany	
Prism				Mero TK and ZK, Germany	
				Mero NK, Germany	A42
Hollow				Satterwhite, U.S.	

From Gerrits, J.M., The architectural impact of space frame systems, *Proc. Asia-Pacific Conf. on Shell and Spatial Structures 1996,* China Civil Engineering Society, Beijing, China, 1996. With permission.

facets and tapped holes. Members are circular hollow sections with cone-shaped steel forgings welded at the ends which accommodate connecting bolts. Bolts are tightened by means of a hexagonal sleeve and dowel pin arrangement, resulting in a completed joint such as that shown in Figure 13.29. Up to 18 members can be connected at a joint with no eccentricity. The manufacturer can produce modes of different size with diameter ranging from 46.5 mm to 350 mm, the corresponding bolts ranging from M12 to M64 with a maximum permissible force of 1413 kN. A typical space-module of a Mero system is a square pyramid (1/2 Octahedron) with both chord and diagonal members of the same length "a", angles extended are 90° or 60°. Thus, the depth of the space-module is $a/\sqrt{2}$ and the vertical angle between diagonal and chord member is 54.7°.

The Mero connector has the advantage that the axes of all members pass through the center of the node, eliminating eccentricity loading at the joint. Thus, the joint is only under the axial forces. Then tensile forces are carried along the longitudinal axis of the bolts and resisted by the tube members through the end cones. The compressive forces do not produce any stresses in the bolts; they are distributed to the node through the hexagonal sleeves. The size of the connecting bolt of compression members based on the diameter calculated from its internal forces may be reduced by 6 to 9 mm.

The diameter of a steel node may be determined by the following equations (Figure 13.29).

$$D \geq \sqrt{\left(\frac{d_2}{\sin\theta} + (d_1 ctg\theta + 2\xi d_1)\right)^2 + \eta^2 d_1^2} \qquad (13.15)$$

However, in order to satisfy the requirements of the connecting face of the sleeve, the diameter should be checked by the following equation:

$$D \geq \sqrt{\left(\frac{\eta d_2}{\sin\theta} + \eta d_1 ctg\theta\right)^2 + \eta^2 d_1^2} \qquad (13.16)$$

TABLE 13.7   Connection Types without a Node

Node		Connector	Member	Cross-section	Examples	Code
Form of member	Forming				Buckminster Fuller	B11
					Nonadome, NL	
	Flattened and bending				Radial, Australia	B12
					Harley, Australia	
Addition of member	Plate(s)				Mai Sky, U.S.	B21
	Member end				Pieter Huybers, NL	B22
					Pierce, U.S.	
					Buckminster Fuller	

From Gerrits, J.M., The architectural impact of space frame systems, *Proc. Asia-Pacific Conf. on Shell and Spatial Structures 1996,* China Civil Engineering Society, Beijing, China, 1996. With permission.

where

D	=	diameter of steel ball (mm)
θ	=	the smaller intersecting angle between two bolts (rad)
$d_1 d_2$	=	diameter of bolts (mm)
ξ	=	ratio between the inserted length of the bolt into the steel ball and the diameter of the bolt
η	=	ratio between the diameter of the circumscribed circle of the sleeve and the diameter of the bolt

ξ and η may be determined, respectively, by the design tension values or compression strength of bolt. Normally $\xi = 1.1$ and $\eta = 1.8$.

The diameter of a steel ball should be taken as the larger value calculated from the above two equations.

The Mero connector was originally developed for double layer grids. Due to the increasing usage of non-planar roof forms, it is required to construct the load-bearing space frame integrated with cladding element. A new type of jointing system called Mero Plus System was developed so that a variety of curved and folded structures are possible. Square or rectangular hollow sections are used to match the particular requirements of the cladding so that a flush transition from member to connecting node can be executed. The connector can transmit shear force, resist torsion, and in special cases can resist bending moment. There are four groups in this system which are described as follows.

(a) **Disc Node (Type TK)** (Figure 13.31)—This is a planar ring-shaped node connecting 5 to 10 members of square or rectangular sections. A single bolt is used

TABLE 13.8   Connection Types with Prefabricated Units

Node	Prefabricated unit	Member cross-section top / bracing / bottom	Examples	Code
Geometrical solid		⌐ O O ⊏ O O ⌐⊏ O O	Space deck, U.K. Mero DE, Germany Unistrut, France	C11
		⊓ ⌐ ⊔	nameless system, Italy	C12
2D components		□ □ □	Mai Sky, U.S.	C21
		⊏ O ⊏	nameless system, Italy	C22
3D components		I □ I	Cubic, U.K.	C31

From Gerrits, J.M., The architectural impact of space frame systems, *Proc. Asia-Pacific Conf. on Shell and Spatial Structures 1996*, China Civil Engineering Society, Beijing, China, 1996. With permission.

TABLE 13.9   Commonly Used Proprietary Systems

Name	Country	Period of development	Material	Connecting method
Mero	Germany	1940–1950	Steel Aluminum	Bolting
Space Deck	U.K.	1950–1960	Steel	Bolting
Triodetic	Canada	1950–1960	Aluminum Steel	Inserting member ends into hub
Unistrut (Moduspan)	U.S.	1950–1960	Steel	Bolting
Oktaplatte	Germany	1950–1960	Steel	Welding
Unibat	France	1960–1970	Steel	Bolting
Nodus	U.K.	1960–1970	Steel	Bolting and using pins
NS	Japan	1970–1980	Steel	Bolting

to connect the node and member and depth of the node is equal to the member section depth. Such jointing systems can transmit shear force and resist rotation. In the following discussion, the U-angle is designated as the angle between two members connected to the same node. Also, the V-angle is the angle between the member axis and the normal in the plane of the node which is a measure of curvature. For a disc node, the U-angle varies from 30° to 80° and the V-angle varies from 0° to 10°. This type of jointing system is essentially pin-jointed connections and is suitable for latticed shells made of triangular meshes.

(b) **Bowl Node (Type NK)** (Figure 13.32)—This is a hemispherical node connecting top chord and diagonal members. Single bolted connection from node to member is used. The top chord members of square or rectangular sections can be loaded in shear and are fitted flush to the nodes. Bowl nodes are used for double layer

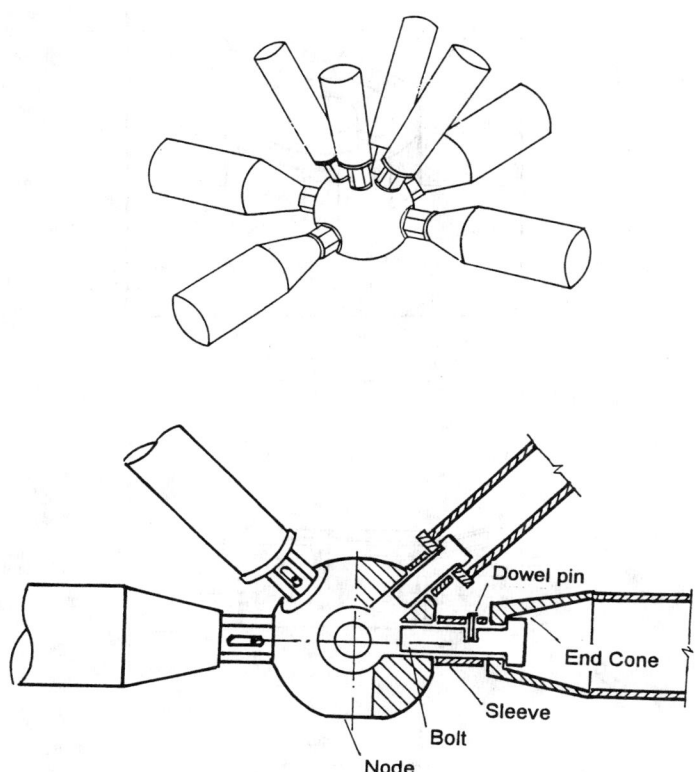

Figure 13.29 Mero system.

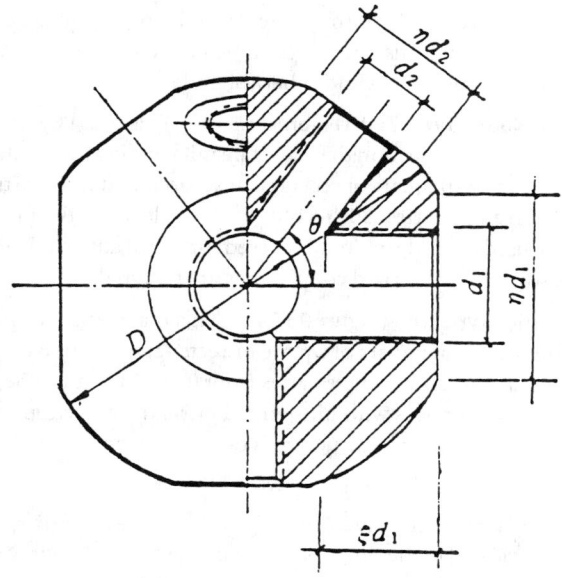

Figure 13.30 Dimensions of spherical node.

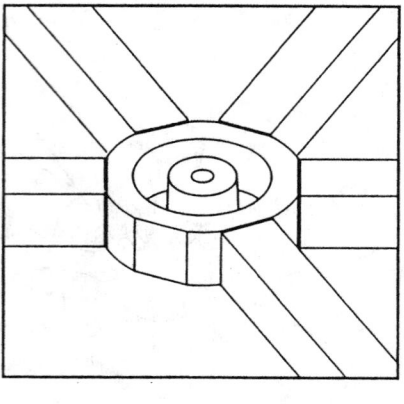

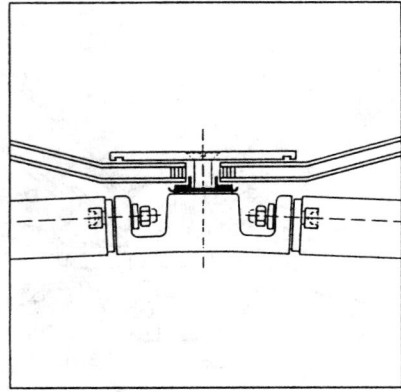

Figure 13.31   Disk node (Type TK).

planar and curved surfaces, in particular buildings irregular in plan or pyramid
in shape. The diagonals and lower chords are constructed in an ordinary Mero
system with circular tubes and spherical nodes.

(c) **Cylinder Node (Type ZK)** (Figure 13.33) —This is a a cylindrical node with a
multiple bolted connection that can transmit bending moment. Usually the node
can connect 5 to 10 square or rectangular sections that can take transverse loading.
Connection angle varies: 30° to 100° for V-angle, 0° to 10° for V-angle. Cylinder
nodes are used in singly or doubly curved surface of latticed shells with trapezoidal
meshes where flexural rigid connections are required.

(d) **Block Node (Type BK)** (Figure 13.34)—This is a a block- or prism-shaped solid
node connecting members of square or rectangular sections. The U-angle varies
from 70° to 120° and V-angle varies from 0° to 10°. It can be used for singly or
doubly curved surfaces with pin-jointed or rigid connections where the number
of members is small. The structure is of simple geometry and small dimensions.

2. **Space Deck**
   The Space Deck system, introduced in England in the early fifties, utilizes pyramidal
   units that are fabricated in the shop, as shown in Figure 13.35. The four diagonals made
   of rods or bars are welded to the corners of the angle frame and joined to a fabricated
   boss at the apex. It is based on square pyramid units that form a configuration of square
   on square offset double layer space grids. The units are field-bolted together through

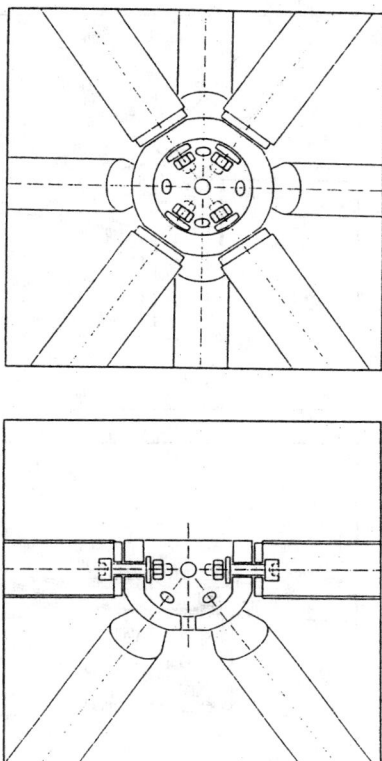

Figure 13.32   Bowl node (type NK).

the angle frames. The apexes of the units are connected in the field by using tie bars made from high-tensile steel bars. Camber can be achieved by adjusting the tie bar lengths, since right-hand and left-hand threading is provided in the boss. The Space Deck system is usually used for buildings of span less than 40 m with a standard module and depth of 1.2 m. A minimum structural depth of 0.75 m is also provided. For higher design loading and larger spans, alternative production modules of 1.5 m and 2.0 m with the same depth as the module are also available.

3. **Triodetic**

    The joint for the Triodetic system, developed in Canada, consists of an extruded aluminum connector hub with serrated keyways. Each member end is pressed in order to form a coined edge that fits into the hub keyway. The joint is completed when the members are inserted into the hub, washers are placed at each end of the hub, and a screw bolt is passed through the center of hub, as shown in Figure 13.36. The Triodetic connector can be used for any type of three-dimensional space frame. Originally only aluminum structures were built in this system, but later space frames were erected using galvanized steel tubes and aluminum hubs. Triodetic double layer grids have been used up to 33 m clear span. The basic module can be almost any size up to approximately 2.7 m in square. The depth is usually 70% of the module size.

4. **Unistrut**

    The Unistrut system was developed in the U.S. in the early fifties. Its joint consists of a connector plate that is press-formed from steel plate. The members are channel-shape

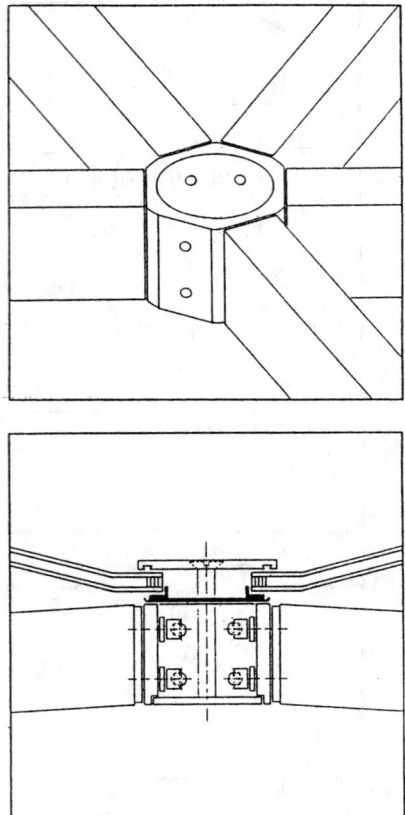

Figure 13.33   Cylinder node (type ZK).

cold-formed sections and are fastened to the connector plate by using a single bolt at
each end. The connectors for the top and bottom layers are identical and therefore the
Unistrut double layer grids consist of four components only, i.e., the connector plate,
the strut, the bolt, and the nut (see Figure 13.37). The maximum span for this system is
approximately 40 m with standard modules of 1.2 m and 1.5 m. The name of Moduspan
has also been used for this system.

5. **Oktaplatte**

The Oktaplatte system utilizes hollow steel spheres and circular tube members that are
connected by welding. The node is formed by welding two hemispherical shells together
which are made from steel plates either by hot or cold pressing. The hollow sphere may
be reinforced with an annular diaphragm. This type of node was popular at the early
stage of development of space frames. It is also useful for the long span structures where
other proprietary systems are limited by their bearing capacity. Hollow spheres with
diameter up to 500 mm have been used. It can be applied to single layer latticed shells
as the joint can be considered as semi-or fully rigid. The whole jointing system and the
hollow sphere with its parts are shown in Figure 13.38.

The allowable bearing strength of hollow spheres can be calculated by the following
empirical formulas:

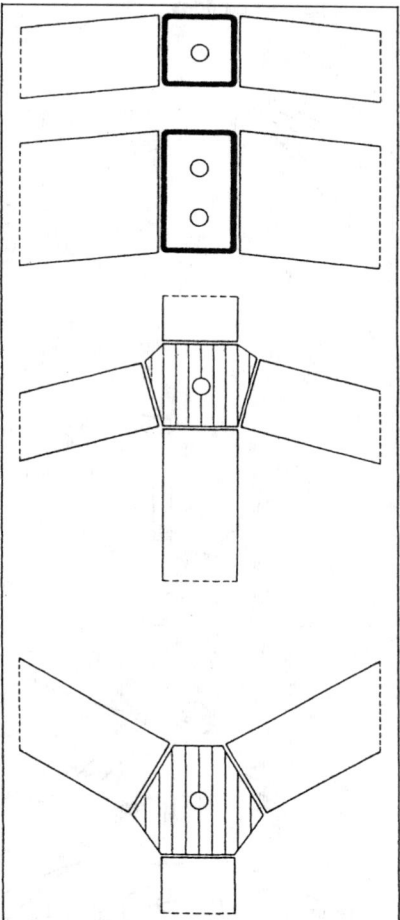

Figure 13.34   Block node (Type BK).

Under compression

$$N_c = \eta_c \left(6.6td - 2.2\frac{t^2d^2}{D} \right) \frac{1}{K} \text{ (tons)} \tag{13.17}$$

Under tension

$$N_t = \eta_t \left(0.6td\pi \right) [\sigma] \tag{13.18}$$

where

D	=	diameter of hollow sphere (cm)
t	=	wall thickness of hollow sphere (cm)
d	=	diameter of the tubular member (cm)
$[\sigma]$	=	allowable tensile stress
η_c, η_t	=	amplification factors due to the strengthening effect of the diaphragm, taken as 1.4 and 1.1, respectively
K	=	factor of safety

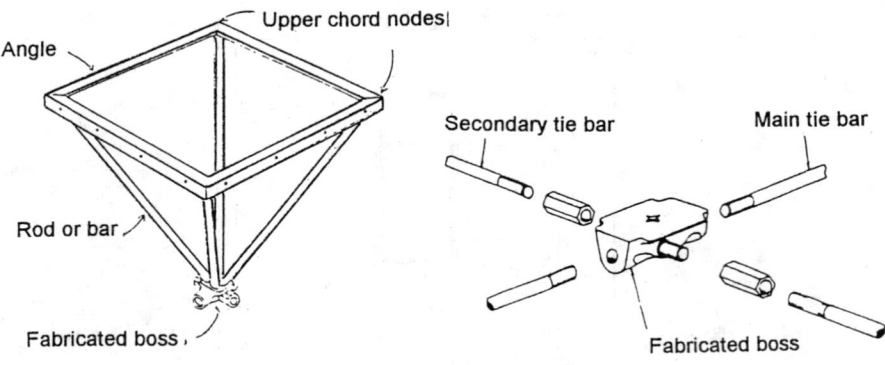

Figure 13.35   Space deck system.

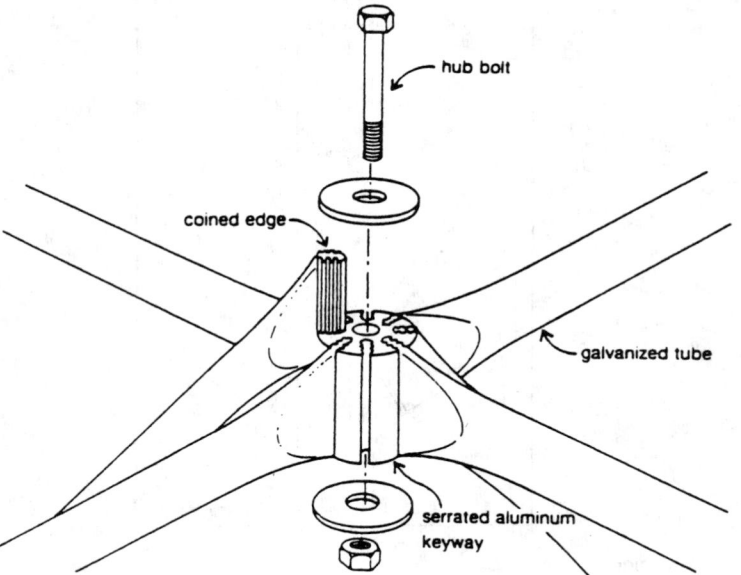

Figure 13.36   Triodetic system.

6. **Unibat**

The Unibat system, developed in France, consists of pyramidal units by arranging the top layer set on a diagonal grid relative to the bottom layer. The short length of the top chord members results in less material being required in these members to resist applied compressive and bending stresses. The standard units are connected to the adjacent units by means of a single high-tensile bolt at each upper corner. The apex and corners of the pyramidal unit may be forgings, to which the top chord and web members are welded. The units may employ any combination of rolled steel or structural sections. As shown in Figure 13.39, the top chords are rolled I sections and web members are square hollow sections. The bottom layer is formed by a two-way grid of circular hollow sections which are interconnected with the apex by a single vertical bolt. Numerous multi-story buildings, as well as large span roofs over sports buildings have been built using the Unibat system since 1970.

7. **Nodus**

The Nodus system was developed in England in the early seventies. Its joint consists of half-casings which are made of cast steel and have machined grooves and drilled

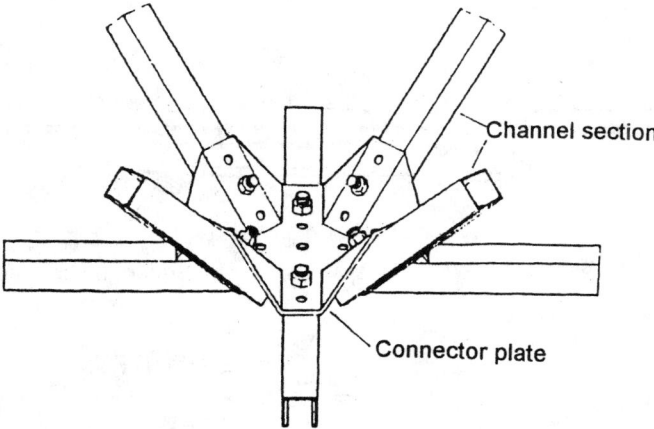

Figure 13.37   Unistrut system.

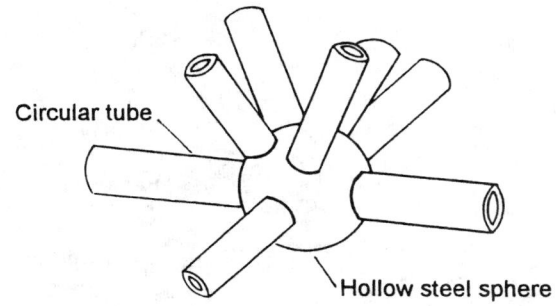

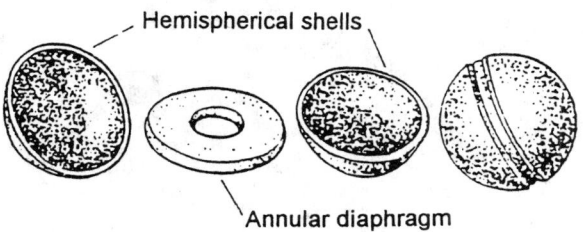

Figure 13.38   Oktaplatte system.

holes, as shown in Figure 13.40. The chord connections are made of forged steel and have machined teeth, and are full-strength welded to the member ends. The teeth and grooves have an irregular pitch in order to ensure proper engagement. The forked connectors are made of cast steel and are welded to the diagonal members. For the completed joint, the centroidal axes of the diagonals intersect at a point that generally does not coincide with the corresponding intersecting points of the chord members. This eccentricity produces some amount of local bending in the chord members and the joint components. Destructive load tests performed on typical joints usually result in failures due to bending of the teeth in the main half-casing. The main feature of the Nodus jointing system is that all fabrication is carried out in the workshop so that only the simplest erection techniques are necessary for the assembly of the structure on-site.

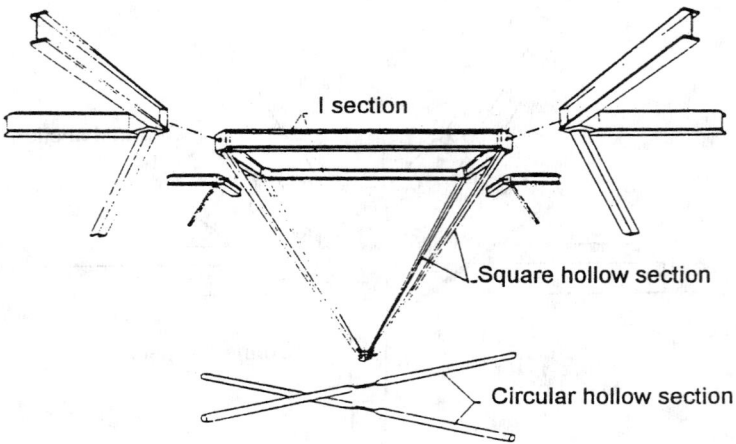

Figure 13.39   Unibat system.

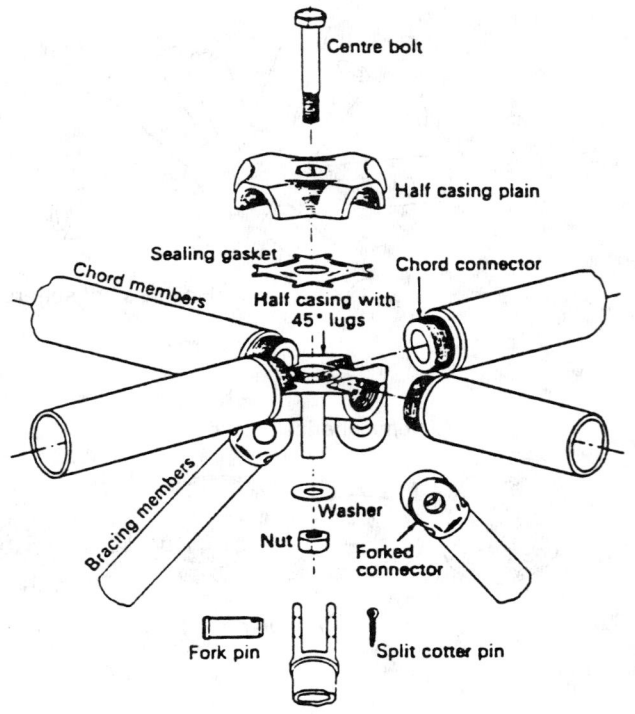

Figure 13.40   Nodus system.

8. NS Space Truss

The NS Space Truss system was introduced around 1970 by the Nippon Steel Corporation. It originates from the space truss technology developed for the construction of the huge roof at the symbol zone for Expo '70 in Japan. The NS Space Truss system has a joint consisting of thick spherical steel shell connectors open at the bottom for bolt insertion. The structural members are steel hollow sections having specially shaped end cones welded to both ends of the tube. End cones have threaded bolt holes. Special high strength bolts are used to join the tubular members to the spherical shell connector. The NS nodes enable several members to be connected to one node from any direction without any eccentricity of internal forces. The NS Space Truss system has been used successfully for many large span double and triple layer grids, domes, and other space structures. The connection detail of the NS node is shown in Figure 13.41.

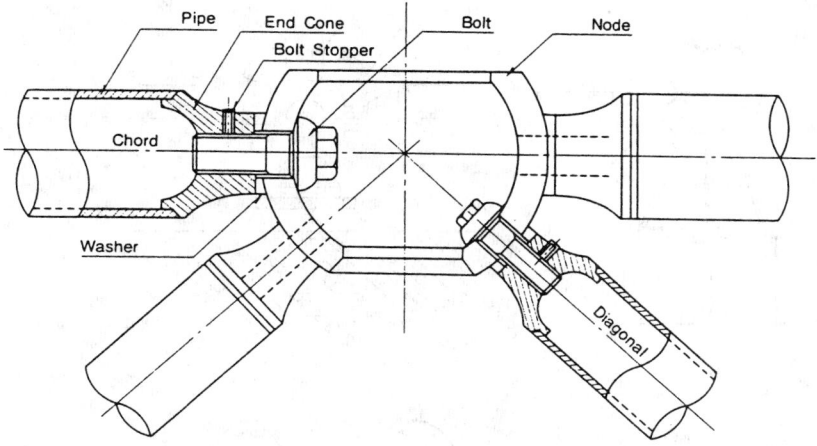

Figure 13.41   NS space truss.

13.5.3   Bearing Joints

Space frames are supported on columns or ring beams through bearing joints. These joints should posses enough strength and stiffness to transmit the reactions at the support safely. Under the vertical loading, bearing joints are usually under compression. In some double layer grids with diagonal layout, bearing joints at corners may resist tension. In latticed shells, both vertical and horizontal reactions are acting on the bearing joints.

The restraint of a bearing joint has a distinct influence on the joint displacement and member forces. The construction detail of a bearing support should conform to the restraint assumed in the design as near as possible. If such requirement is not satisfied, the magnitude or even the sign of the member forces may be changed.

The axes of all connecting members and the reaction should be intersected at one point at the support where a hinged joint is used. This will allow a free rotation of the joint. From an engineering standpoint, the space frame may be fixed in the vertical direction. While in the horizontal direction, it may be fixed either tangential or normal to the boundary or both. The way that the space frame is fixed often depends on the temperature effect. If the bearing support can allow a horizontal motion normal to the boundary, then the member forces due to the temperature variation can be neglected. In such case, the bearing should be constructed so that it can slide horizontally. For those space frames with large spans or complicated configurations, especially curved surface structures supported on sloped base, care should be exercised to ensure a reliable bearing support.

Typical details for bearing joints are shown in Figure 13.42. The simplest form of bearings is to establish the joint on a flat plate and anchored by bolts as shown in Figure 13.42a or b. This joint seems to be fixed at the support, but in structural analysis it has to be incorporated with the supporting structure, such as columns or walls that have a lateral flexibility. Figure 13.42c shows the joint is resting on a curved bearing block which allows rotation along the curved surface. Such type of construction can be considered as a hinged joint. If a laminated elastomeric pad is used under the joint as shown in Figure 13.42d, a new type of bearing joint is formed. Due to the shear deformation of the elastomeric pad, the joint can produce both rotation and horizontal movements. It is very effective to accommodate the horizontal deformation caused by temperature variation or earthquake action.

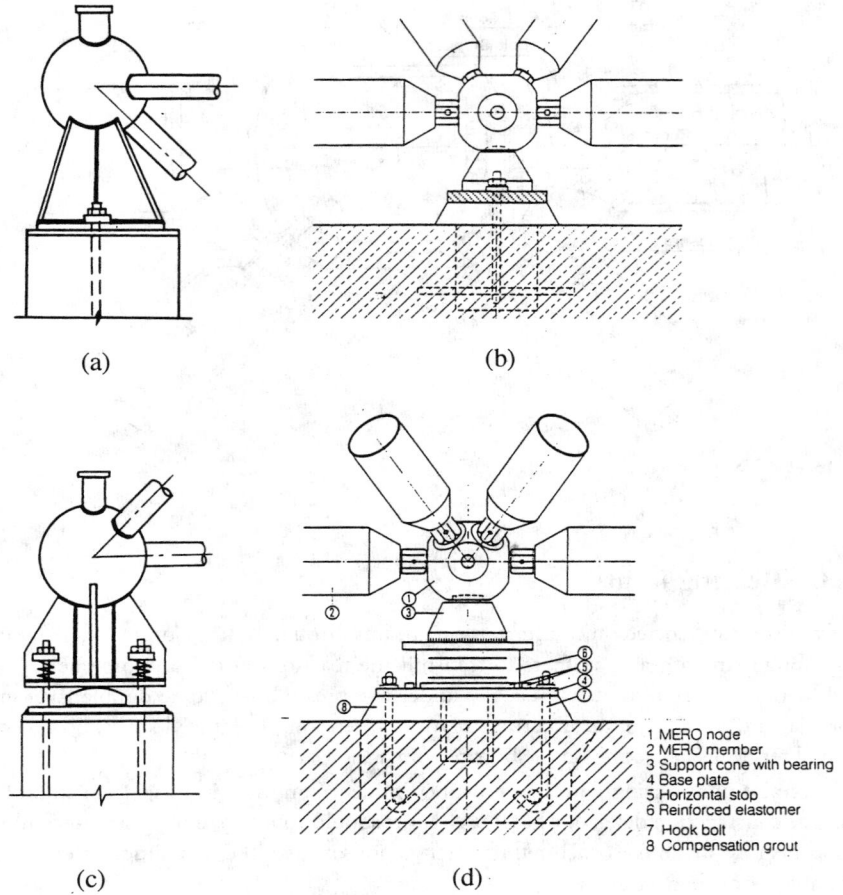

(a)   (b)

(c)   (d)

1 MERO node
2 MERO member
3 Support cone with bearing
4 Base plate
5 Horizontal stop
6 Reinforced elastomer
7 Hook bolt
8 Compensation grout

Figure 13.42   Bearing joints.

13.6   Defining Terms

Aspect ratio: Ratio of longer span to shorter span of a rectangular space frame.

Braced (barrel) vault: A space frame composed of member elements arranged on a cylindrical surface.

Braced dome: A space frame composed of member elements arranged on a spherical surface.

Continuum analogy method: A method for the analysis of a space frame where the structure is analyzed by assuming it as an equivalent continuum.

Depth: Distance between the top and bottom layer of a double layer space frame.

Discrete method: A method for the analysis of a space frame where the structure is analyzed directly as a general assembly of discrete members.

Double layer grids: A space frame consisting of two planar networks of members forming the top and bottom layers parallel to each other and interconnected by vertical and inclined members.

Geodesic dome: A braced dome in which the elements forming the network are lying on the great circle of a sphere.

Lamella: A unit used to form diamond shaped grids, the size being twice the length of the side of the diamond.

Latticed grids: Double layer grids consisting of intersecting vertical latticed trusses to form regular grids.

Latticed shell: A space frame consisting of curved networks of members built either in single or double layers.

Latticed structure: A structural system in the form of a network of elements whose load-carrying mechanism is three-dimensional in nature.

Local buckling: A snap-through buckling that takes place at one point.

Module: Distance between two joints in the layer of grid.

Overall buckling: Buckling that takes place at a relatively large area where a large number of joints is involved

Space frame: A structural system in the form of a flat or curved surface assembled of linear elements so arranged that forces are transferred in a three-dimensional manner.

Space grids: Double layer grids consisting of a combination of square or triangular pyramids to form offset or differential grids.

Space truss: A three-dimensional structure assembled of linear elements and assumed as hinged joints in structural analysis.

References

[1] ASCE Subcommittee on Latticed Structures of the Task Committee on Special Structures of the Committee on Metals of the Structural Division. 1972. Bibliography on latticed structures. *J. Struct. Div., Proc. ASCE.* 98(ST7):1545-1566.

[2] ASCE Task Committee on Latticed Structures of the Committee on Special Structures of the Committee on Metals of the Structural Division. 1976. Latticed structures: state-of-the-art report. *J. Struct. Div., Proc. ASCE.* 102 (ST11):2197-2230.

[3] ASCE Task Committee on Latticed Structures under Extreme Dynamic Loads of the Committee on Special Structures of the Committee on Metals of the Structural Division. 1984. Dynamic considerations in latticed structures. *J. Struct. Eng.*, 110(10):2547-2550.

[4] Chinese Academy of Building Research. 1981. *Specifications for the Design and Construction of Space Trusses (JGJ 7-80)*. China Building Industry Press, Beijing, China.

[5] European Convention for Constructional Steelwork (ECCS). 1980. *Recommendations for the Calculation of Wind Effects on Buildings and Structures.*

[6] Gerrits, J. M. 1996. The architectural impact of space frame systems. *Proc. Asia-Pacific Conf. on Shell and Spatial Structures 1996.* China Civil Engineering Society, Beijing, China.

[7] Gioncu, V. 1995. Buckling of reticulated shells: state-of-the-art. *Int. J. Space Struc.*, 10(1):1-46.

[8] Hangai, Y. and Tsuboi, Y. 1985. Buckling loads of reticulated single-layer space frames. *Theory and Experimental Investigation of Spatial Structures. Proc. IASS Congress 1985.* Moscow.

[9] Heki, K. 1993. Buckling of lattice domes—state of the art report. *Nonlinear Analysis and Design for Shell and Spatial Structures. Proc. Seiken-IASS Symp. 1993.* pp.159-166. Tokyo.

[10] Iffland, J. S. B. 1987. Preliminary design of space trusses and frames, in *Building Structural Design Handbook,* White, R. N. and Salmon, C. G., Eds., John Wiley & Sons, New York, 403-423.

[11] International Association for Shell and Spatial Structures (IASS) Working Group on Spatial Steel Structures. 1984. Analysis, design and realization of space frames. *Bull. IASS* No. 84/85, XXV(1/2):1-114.

[12] ISO 4355. 1981. *Basis for Design of Structures: Determination of Snow Loads on Roofs.*

[13] Kato, S., Kawaguchi, K., and Saka, T. 1995. Preliminary report on Hanshin earthquake. *Spatial Structures: Heritage, Present and Future. Proc. IASS Symp. 1995.* 2, 1059-1066. S. G. Editoriali, Padova, Italy.

[14] Lan, T. T. and Qian, R. 1986. A study on the optimum design of space trusses—optimal geometrical configuration and selection of type. *Shells, Membranes and Space Frames. Proc. IASS Symp. 1986.* 3, 191-198. Elsevier, Amsterdam.

[15] Lan, T. T. 1994. Structural failure and quality assurance of space frames. *Spatial, Lattice and Tension Structures. Proc. IASS-ASCE Intl. Symp. 1994.* 123-132. ASCE, New York.

[16] Lind, N. C. 1969. Local instability analysis of triangulated dome framework. *Structural Eng.,* 47(8).

[17] Makowski, Z. S., Ed. 1981. *Analysis, Design and Construction of Double Layer Grids.* Applied Science, London.

[18] Makowski, Z. S., Ed. 1984. *Analysis, Design and Construction of Braced Domes.* Granada, London.

[19] Makowski, Z. S., Ed. 1985. *Analysis, Design and Construction of Braced Barrel Vaults.* Elsevier Applied Science, London.

[20] Makowski, Z. S. 1992. Space frames and trusses, in *Constructional Steel Design. An International Guide.* Dowling, P. J. et al., Eds., Elsevier Applied Science, London, 791-843.

[21] Saitoh, M., Hangai, Y., Todu, I., and Okuhara, T. 1987. Design procedure for stability of reticulated single-layer domes. *Building Structures. Proc. Structure Congress 1987.* 368-376. ASCE New York.

[22] Wright, D. T. 1965. Membrane forces and buckling in reticulated shells. *J. Struct. Div. Proc. ASCE,* 91(ST1):173-201.

[23] Zhang, Y. G. and Lan, T. T. 1984. A practical method for the analysis of space frames under vertical earthquake loads. *Final Report, IABSE 12th Congress 1984.* pp.169-176. IABSE, Zurich.

Further Reading

An introduction to the practical design of space structures is presented in *Horizontal-Span Building Structures* by W. Schueller. It covers a wide range of topics, including the development, structural behavior, simplified analysis, and application of different types of space structures.

For further study of continuum analogy method of space frames, *Analysis and Design of Space Frames by the Continuum Method* by L. Kollar and I. Hegedus provides a good reference in this topic.

The quarterly journal *International Journal of Space Structures* reports advances in the theory and practice of space structures. Special issues treating individual topics of interest were published, such as *Stability of Space Structures*, V. Gioncu, Ed., 7(4), 1992 and *Prefabricated Spatial Frame Systems* by A. Hanaor, 10(3), 1995.

Conferences and symposiums are organized annually by the International Association for Shell and Spatial Structures (IASS). The proceedings document the latest developments in this field and provide a wealth of information on theoretical and practical aspects of space structures. The proceedings of conferences recently held are as follows:

[1] *Spatial Structures at the Turn of Millennium,* IASS Symposium 1991, Copenhagen.

[2] *Innovative Large Span Structures*, IASS-CSCE International Congress 1992, Toronto.

[3] *Public Assembly Structures from Antiquity to the Present*, IASS Symposium 1993, Istanbul.

[4] *Nonlinear Analysis and Design for Shell and Spatial Structures*, Seiken-IASS Symposium 1993, Tokyo.

[5] *Spatial, Lattice and Tension Structures*, IASS-ASCE International Symposium 1994, Atlanta.

[6] *Spatial Structures: Heritage, Present and Future*, IASS Symposium 1995, Milan.

[7] *Conceptual Design of Structures*, IASS International Symposium 1996, Stuttgart.

Journal of IASS is published three times a year and it covers design, analysis, construction, and other aspects of technology of all types of shell and spatial structures.

14

Cooling Tower Structures

14.1 Introduction 14-1
14.2 Components of a Natural Draft Cooling Tower 14-2
14.3 Damage and Failures 14-4
14.4 Geometry 14-6
14.5 Loading 14-7
14.6 Methods of Analysis 14-12
14.7 Design and Detailing of Components 14-24
14.8 Construction 14-29
References 14-31
Further Reading 14-32

Phillip L. Gould
Department of Civil Engineering,
Washington University,
St. Louis, MO

Wilfried B. Krätzig
Ruhr-University,
Bochum, Germany

14.1 Introduction

Hyperbolic cooling towers are large, thin shell reinforced concrete structures which contribute to environmental protection and to power generation efficiency and reliability. As shown in Figure 14.1, they may dominate the landscape but they possess a certain aesthetic .eloquence due to their doubly curved form. The operation of a cooling tower is illustrated in Figure 14.2. In a thermal power station, heated steam drives the turbogenerator which produces electric energy. To create an efficient heat sink at the end of this process, the steam is condensed and recycled into the boiler. This requires a large amount of cooling water, whose temperature is raised and then recooled in the tower.

In a so-called "wet" natural draft cooling tower, the heated water is distributed evenly through channels and pipes above the fill. As the water flows and drops through the fill sheets, it comes into contact with the rising cooler air. Evaporative cooling occurs and the cooled water is then collected in the water basin to be recycled into the condenser. The difference in density of the warm air inside and the colder air outside creates the natural draft in the interior. This upward flow of warm air, which leads to a continuous stream of fresh air through the air inlets into the tower, is protected against atmospheric turbulence by the reinforced concrete shell. The cooling tower shell is supported by a truss or framework of columns bridging the air inlet to the tower foundation.

There are also "dry" cooling towers that operate simply on the basis of convective cooling. In this case the water distribution, the fill, and the water basin are replaced by a closed piping system around the air inlet, resembling, in fact, a gigantic automobile radiator. While dry cooling towers are doubtless superior from the point of view of environmental protection, their thermal efficiency is only about 30% of comparable wet towers. If the flue gas is cleaned by a washing technology, it is frequently discharged into the atmosphere by the cooling tower upward flow. This saves reheating of the cleaned flue gas and the construction of a smoke stack (see Figure 14.2).

Figure 14.3 summarizes the historical development of natural draft cooling towers. Technical cooling devices first came into use at the end of the 19th century. The well-known hyperbolic shape of cooling towers was introduced by two Dutch engineers, Van Iterson and Kuyper, who in 1914

Figure 14.1   A group of hyperbolic cooling towers.

constructed the first hyperboloidal towers which were 35 m high. Soon, capacities and heights increased until around 1930, when tower heights of 65 m were achieved. The first such structures to reach higher than 100 m were the towers of the High Marnham Power Station in Britain.

Today's tallest cooling towers, located at several EDF nuclear power plants in France, reach heights of about 170 m. The key dimensions of one of the largest modern towers are shown in Figure 14.4. In relative proportions, the shell is thinner than an egg, and it is predicted that 200 m high towers will be constructed in the early 21st century.

14.2   Components of a Natural Draft Cooling Tower

The most prominent component of a natural draft cooling tower is the huge, towering shell. This shell is supported by diagonal, meridional, or vertical columns bridging the air inlet. The columns, made of high-strength reinforced concrete, are either prefabricated or cast *in situ* into moveable steel forms (Figure 14.5). After the erection of the ring of columns and the lower edge member, the climbing formwork is assembled and the stepwise climbing construction of the cooling tower shell begins (Figure 14.6a). Fresh concrete and reinforcement steel are supplied to the working site by a central crane anchored to the completed parts of the shell, and are placed in lifts up to 2 m high (Figure 14.6b). After sufficient strength has been gained, the complete forms are raised for the next lift.

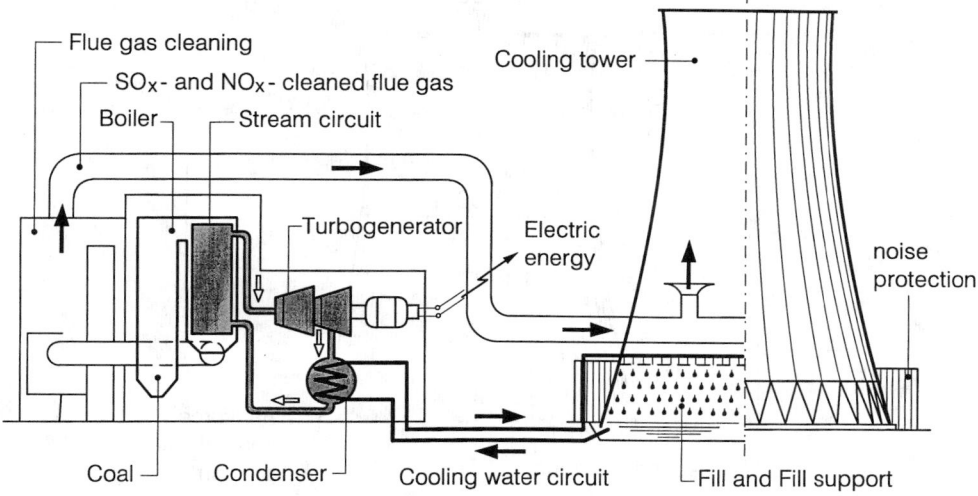

Figure 14.2 Thermal power plant with cleaned flue gas injection.

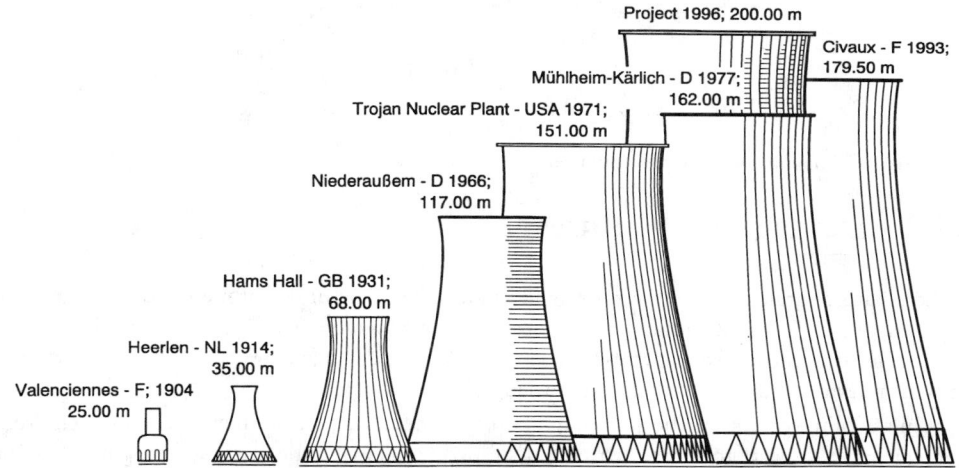

Figure 14.3 Historical development of natural draft cooling tower.

To enhance the durability of the concrete and to provide sufficient cover for the reinforcement, the cooling tower shell thickness should not be less than 16 to 18 cm. The shell itself should be sufficiently stiffened by upper and lower edge members. In order to achieve sufficient resistance against instability, large cooling tower shells may be stiffened by additional internal or external rings. These stiffeners may also serve as a repair or rehabilitation tool.

Wet cooling towers have a water basin with a cold water outlet at the base. These are both large engineered structures, able to handle up to 50 m^3/s of water circulation, as indicated in Figure 14.7. The fill construction inside the tower is a conventional frame structure, always prefabricated. It carries the water distribution, a large piping system, the spray nozzles, and the fill-package. Often dripping traps are applied on the upper surfaces of the fill to keep water losses through the uplift stream under 1%. Finally, noise protection elements around the inlet decrease the noise caused by the continuously dripping water, as illustrated in Figure 14.2.

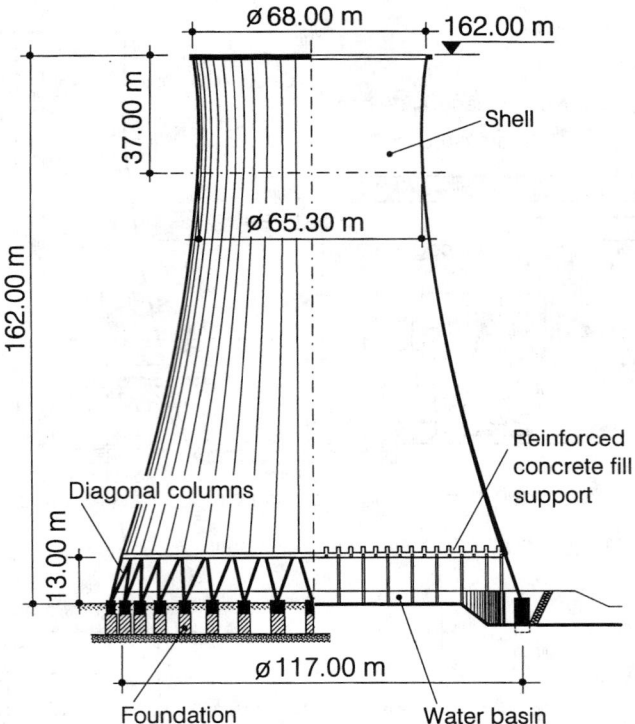

Figure 14.4   Cooling tower: Gundremmingen, Germany.

14.3   Damage and Failures

Today's natural draft cooling towers are safe and durable structures if properly designed and constructed. Nevertheless, it should be recognized that this high quality level has been achieved only after the lessons learned from a series of collapsed or heavily damaged towers have been incorporated into the relevant body of engineering knowledge.

While cooling towers have been the largest existing shell structures for many decades, their design and construction were formerly carried out simply by following the existing "recognized rules of craftsmanship", which had never envisaged constructions of this type and scale. This changed radically, however, in the wake of the Ferrybridge failures in 1965 [7]. On November 1st, 1965, three of eight 114 m high cooling towers collapsed during a Beaufort 12 gale in an obviously identical manner (Figure 14.8). Within a few years of this spectacular accident, the response phenomena of cooling towers had been studied in detail, and safety concepts with improved design rules were developed. These international research activities gained further momentum after the occurrence of failures in Ardeer (Britain) in 1973, Bouchain (France) in 1979, and Fiddler's Ferry (Britain) in 1984, the latter case clearly displaying the influence of dynamic and stability effects.

In surveying these failures, one can recognize at least four common circumstances:

1. The maximum design wind speed was often underestimated, so that the safety margin for the wind load was insufficient.
2. Group effects leading to higher wind speeds and increased vortex shedding influence on downstream towers were neglected.
3. Large regions of the shell were reinforced only in one central layer (in two orthogonal directions), or the double layer reinforcement was insufficient.

Figure 14.5   Fabrication of supporting columns.

4. The towers had no upper edge members or the existing members were too weak for stiffening the structure against dynamic wind actions.

Two towers in the U.S., namely at Willow Island, West Virginia, and at Port Gibson, Mississippi, were heavily damaged during their construction stage, the latter by a tornado. The Port Gibson tower was repaired partly by adding intermediate ring stiffeners [5]. Another tower in Poland collapsed without any definitive explanation having been published up to now, but probably because of considerable imperfections.

In addition to these cases, cracking of many cooling towers has been observed, often due to ground motions following underground coal mining, or just because of faulty design and construction. Obviously, any visible crack in a cooling tower shell is an indication of deterioration of its safety and reliability. It is thus imperative to conform to a design concept that guarantees sufficiently safe and reliable structures over a predetermined lifetime.

Although power plant construction over much of the industrialized west has slowed in the last decade, research and development on the structural aspects of hyperbolic cooling towers has continued [4, 9] and a new wave of construction for these impressive structures seems to be approaching. Engineers face this challenge with confidence in their improved analytical tools, in their ability to employ improved materials, and in their valuable experience in construction.

Figure 14.6a Climbing construction of the shell.

14.4 Geometry

The main elements of a cooling tower shell in the form of a hyperboloid of revolution are shown in Figure 14.9. This form falls into the class of structures known as thin shells. The cross-section as shown depicts the ideal profile of a shell generated by rotating the hyperboloid $R = f(Z)$ about the vertical (Z) axis. The coordinate Z is measured from the throat while z is measured from the base. All dimensions in the R-Z plane are specified on a reference surface, theoretically the middle surface of the shell but possibly the inner or outer surface. Dimensions through the thickness are then referred to this surface. There are several variations possible on this idealized geometry such as a cone-toroid with an upper and lower cone connected by a toroidal segment, two hyperboloids with different curves meeting at the throat, and an offset of the curve describing the shell wall from the axis of rotation.

Important elements of the shell include the *columns* at the base, which provide the necessary opening for the air; the *lintel*, either a discrete member or more often a thickened portion of the shell, which is designed to distribute the concentrated column reactions into the shell wall; the *shell wall* or *veil*, which may be of varying thickness and provides the enclosure; and the *cornice*, which like the lintel may be discrete or a thickened portion of the wall designed to stiffen the top against ovaling.

Figure 14.6b Steel reinforcement of shell wall.

Referring to Figure 14.9, the equation of the generating curve is given by

$$4R^2/d_T^2 - Z^2/b^2 = 1 \tag{14.1}$$

where b is a characteristic dimension of the shell that may be evaluated by

$$b = d_H Z_H / \sqrt{\left(d_H^2 - d_T^2\right)} \tag{14.2}$$

or by

$$b = d_U Z_U / \sqrt{\left(d_U^2 - d_T^2\right)} \tag{14.3}$$

if the upper and lower curves are different. The dimension b is related to the slope of the asymptote of the generating hyperbola (see Figure 14.9) by

$$b = 2cd_T \tag{14.4}$$

14.5   Loading

Hyperbolic cooling towers may be subjected to a variety of loading conditions. Most commonly, these are dead load (D), wind load (W), earthquake load (E), temperature variations (T), construction loads (C), and settlement (S). For the proportioning of the elements of the cooling tower, the

Figure 14.7   Water basin.

effects of the various loading conditions should be factored and combined in accordance with the applicable codes or standards. If no other codes or standards specifically apply, the factors and combinations given in ASCE 7 [11] are appropriate.

Dead load consists of the self-weight of the shell wall and the ribs, and the superimposed load from attachments and equipment.

Wind loading is extremely important in cooling tower design for several reasons. First of all, the amount of reinforcement, beyond a prescribed minimum level, is often controlled by the *net difference* between the *tension* due to wind loading and the dead load *compression*, and is therefore especially sensitive to variations in the tension. Second, the quasistatic velocity pressure on the shell wall is sensitive to the vertical variation of the wind, as it is for most structures, and also to the circumferential variation of the wind around the tower, which is peculiar to cylindrical bodies. While the vertical variation is largely a function of the regional climatic conditions and the ground surface irregularities, the circumferential variation is strongly dependent on the roughness properties of the shell wall surface. There are also additional wind effects such as internal suction, dynamic amplification, and group configuration.

The external wind pressure acting at any point on the shell surface is computed as [2, 9]

$$q(z, \theta) = q(z)H(\theta)(1 + g) \qquad (14.5)$$

in which

$q(z)$ =   effective velocity pressure at a height z above the ground level (Figure 14.9)
$H(\theta)$ =   coefficient for circumferential distribution of external wind pressure
$1 + g$ =   gust response factor
g =   peak factor

As mentioned above, $q(z)$ should be obtained from applicable codes or standards such as Reference [11].

The circumferential distribution of the wind pressure is denoted by $H(\theta)$ and is shown in Figure 14.10. The key regions are the windward meridian, $\theta = 0°$, the maximum side suction, $\theta \simeq 70°$,

Figure 14.8 Collapse of Ferrybridge Power Station shell.

and the back suction, $\theta \geq 90°$. These curves were determined by laboratory and field measurements as a function of the roughness parameter k/a as shown in Figure 14.11, in which k is the height of the rib and a is the mean distance between the ribs measured at about 1/3 of the height of the tower. Note that the coefficient along the windward meridian $H(0)$ reflects the so-called *stagnation pressure* while the side-suction is, remarkably, significantly affected by the surface roughness k/a. As will be discussed in a later section, the meridional forces in the shell wall and hence the required reinforcing steel are very sensitive to $H(\theta)$. In turn, the costs of construction are affected. Thus, the design of the ribs, or of alternative roughness elements, are an important consideration. For quantitative purposes, the equations of the various curves are given in Table 14.1 and tabulated values at 5° intervals are available [13].

TABLE 14.1 Functions of Pressure Curves $H(\Theta)$ and Corresponding Drag Coefficients c_W

Curve	Minimum pressure	Area I	Area II	Area III	c_W
K1.0	-1.0	$1 - 2.0\left(\sin\frac{90}{70}\Theta\right)^{2.267}$	$-1.0 + 0.5\left[\sin\left[\frac{90}{21}(\Theta - 70)\right]\right]^{2.395}$	-0.5	0.66
K1.1	-1.1	$1 - 2.1\left(\sin\frac{90}{71}\Theta\right)^{2.239}$	$-1.1 + 0.6\left[\sin\left[\frac{90}{22}(\Theta - 71)\right]\right]^{2.395}$	-0.5	0.64
K1.2	-1.2	$1 - 2.2\left(\sin\frac{90}{72}\Theta\right)^{2.205}$	$-1.2 + 0.7\left[\sin\left[\frac{90}{23}(\Theta - 72)\right]\right]^{2.395}$	-0.5	0.60
K1.3	-1.3	$1 - 2.3\left(\sin\frac{90}{73}\Theta\right)^{2.166}$	$-1.3 + 0.8\left[\sin\left[\frac{90}{24}(\Theta - 73)\right]\right]^{2.395}$	-0.5	0.56

The circumferential distribution of the external wind pressure may be presented in another manner which accents the importance of the asymmetry. If the distribution $H(\theta)$ is represented in

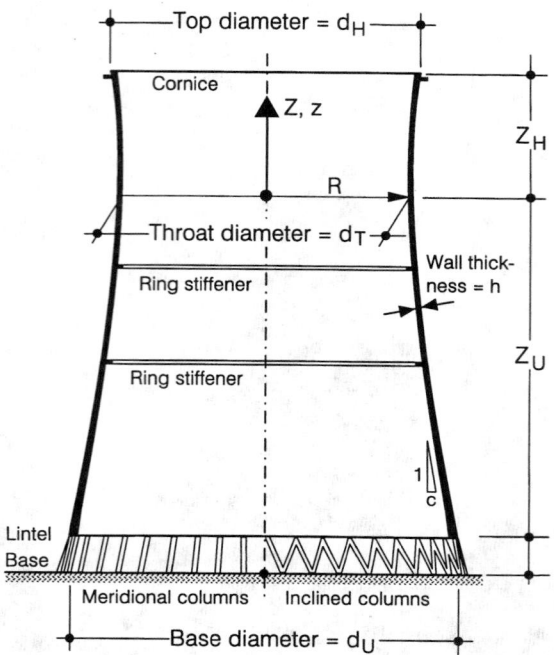

Figure 14.9   Hyperbolic cooling tower.

a Fourier cosine series of the form

$$H(\theta) = \sum_{n=0}^{n=\infty} A_n \cos n\theta \qquad (14.6)$$

the Fourier coefficients A_n for a distribution most similar to the curve for K1.3 are as follows [13]:

n	A_n
0	−0.3922
1	0.2602
2	0.6024
3	0.5046
4	0.1064
5	−0.0948
6	−0.0186
7	0.0468

Representative modes are shown in Figure 14.12. The $n = 0$ mode represents uniform expansion and contraction of the circumference, while $n = 1$ corresponds to beam-like bending about a diametrical axis resulting in translation of the cross-section. The higher modes $n > 1$ are peculiar to shells in that they produce undulating deformations around the cross-section with no net translation. The relatively large Fourier coefficients associated with $n = 2,3,4,5$ indicate that a significant portion of the loading will cause shell deformations in these modes. In turn, the corresponding local forces are significantly higher than a beam-like response would produce.

To account for the internal conditions in the tower during operation, it is common practice to add an axisymmetric internal suction coefficient $H = 0.5$ to the external pressure coefficients $H(\theta)$. In terms of the Fourier series representation, this would increase A_0 to −0.8922.

The dynamic amplification of the effective velocity pressure is represented by the parameter g in Equation 14.5. This parameter reflects the resonant part of the response of the structure and may be as much as 0.2 depending on the dynamic characteristics of the structure. However, when the basis of $q(z)$ includes some dynamic portion, such as the fastest-mile-of-wind, $(1 + g)$ is commonly taken as 1.0.

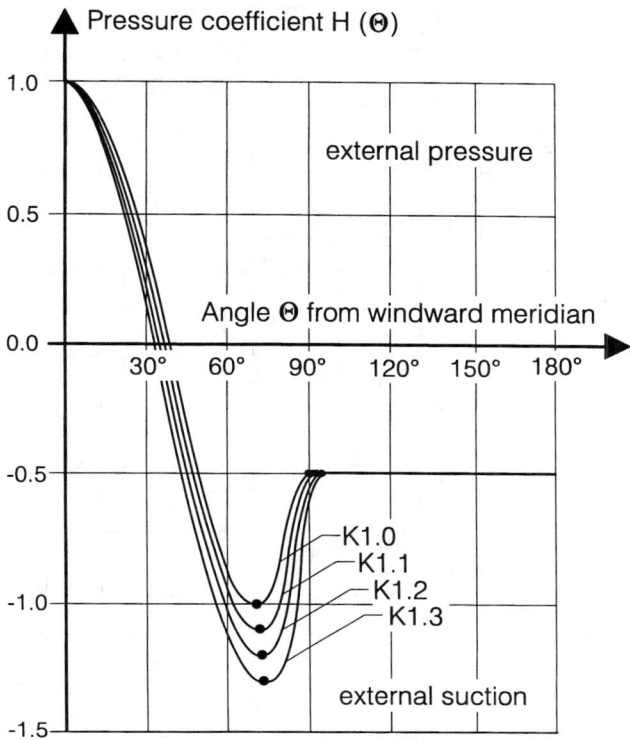

Figure 14.10   Types of circumferential pressure distribution.

Cooling towers are often constructed in groups and close to other structures, such as chimneys or boiler houses, which may be higher than the tower itself. When the spacing of towers is closer than 1.5 times the base diameter or 2 times the throat diameter, or when other tall structures are nearby, the wind pressure on any single tower may be altered in shape and intensity. Such effects should be studied carefully in boundary-layer wind tunnels in order not to overlook dramatic increases in the wind loading.

Earthquake loading on hyperbolic cooling towers is produced by ground motions transmitted from the foundation through the supporting columns and the lintel into the shell. If the base motion is assumed to be uniform vertically and horizontally, the circumferential effects are axisymmetrical ($n = 0$) and antisymmetrical ($n = 1$), respectively (see Figure 14.12). In the meridional direction, the magnitude and distribution of the earthquake-induced forces is a function of the mass of the tower and the dynamic properties of the structure (natural frequencies and damping) as well as the acceleration produced by the earthquake at the base of the structure. The most appropriate technique for determining the loads applied by a design earthquake to the shell and components is the response spectrum method which, in turn, requires a free vibration analysis to evaluate the natural frequencies [2, 3, 4]. It is common to use elastic spectra with 5% of critical damping. The supporting columns and foundation are critical for this loading condition and should be modeled in appropriate detail [3, 4].

Temperature variations on cooling towers arise from two sources: operating conditions and sunshine on one side. Typical operating conditions are an external temperature of $-15°C$ and internal temperature of $+30°C$. This is an axisymmetrical effect, $n = 0$ on Figure 14.12. For sunshine, a temperature gradient of 25°C constant over the height and distributed as a half-wave around one half of the circumference is appropriate. This loading would require a Fourier expansion in the form of Equation 14.6 and higher harmonic components, $n > 1$, to be considered.

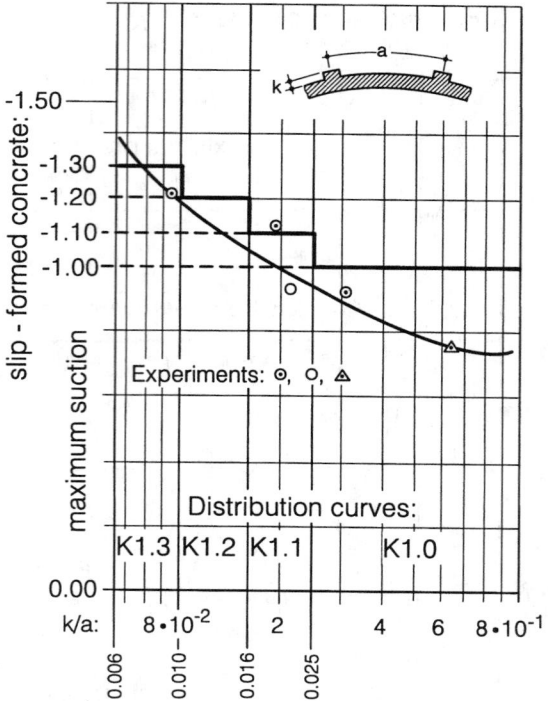

Figure 14.11   Surface roughness k/a and maximum side-suction.

Construction loads are generally caused by the fixing devices of climbing formwork, by tower crane anchors, and by attachments for material transport equipment as shown in Figure 14.13. These loads must be considered on the portion of the shell extant at the phase of construction.

Non-uniform settlement due to varying subsoil stiffness may be a consideration. Such effects should be modeled considering the interaction of the foundation and the soil.

14.6   Methods of Analysis

Thin shells may resist external loading through forces acting parallel to the shell surface, forces acting perpendicular to the shell surface, and moments. While the analysis of such shells may be formulated within the three-dimensional theory of elasticity, there are reduced theories which are two-dimensional and are expressed in terms of force and moment *intensities*. These intensities are traditionally based on a reference surface, generally the middle surface, and are forces and moments per unit length of the middle surface element upon which they act. They are called *stress resultants* and *stress couples*, respectively, and are associated with the three directions: circumferential, θ^1; meridional, θ^2; and normal, θ^3. In Figure 14.14, the extensional stress resultants, n_{11} and n_{22}, the in-plane shearing stress resultants, $n_{12} = n_{21}$, and the transverse shear stress resultants, $q_{12} = q_{21}$, are shown in the left diagram along with the components of the applied loading in the circumferential, meridional, and normal directions, p_1, p_2, and p_3, respectively. The bending stress couples, m_{11} and m_{22}, and the twisting stress couples, $m_{12} = m_{21}$, are shown in the right diagram along with the displacements v_1, v_2, and v_3 in the respective directions.

Historically, doubly curved thin shells have been designed to resist applied loading primarily through the extensional and shearing forces in the "plane" of the shell surface, as opposed to the transverse shears and bending and twisting moments which predominate in flat plates loaded normally to their surface. This is known as *membrane* action, as opposed to *bending* action, and

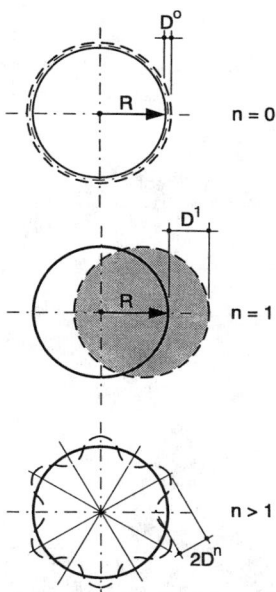

Figure 14.12 Harmonic components of the radial displacement.

is consistent with an accompanying theory and calculation methodology which has the advantage of being statically determinate. This methodology was well-suited for the pre-computer age and enabled many large thin shells, including cooling towers, to be rationally designed and economically constructed [9]. Because the conditions that must be provided at the shell boundaries in order to insure membrane action are not always achievable, shell bending should be taken into account even for shells designed by membrane theory. Remarkably, the accompanying bending often is confined to narrow regions in the vicinity of the boundaries and other discontinuities and may have only a minor effect on the shell design, such as local thickening and/or additional reinforcement. Many clever and insightful techniques have been developed over the years to approximate the effects of local bending in shells designed by the membrane theory.

As we have passed into and advanced in the computer age, it is no longer appropriate to use the membrane theory to analyze such extraordinary thin shells, except perhaps for preliminary design purposes. The finite element method is widely accepted as the standard contemporary technique and the attention shifts to the level of sophistication to be used in the finite element model. As is often the case, the greater the level of sophistication specified, the more data required. Consequently, a model may evolve through several stages, starting with a relatively simple version that enables the structure to be sized, to the most sophisticated version that may depict such phenomena as the sequence of progressive collapse of the as-built shell under various static and dynamic loading scenarios, the incremental effects of the progressive stages of construction, the influence of the operating environment, aging and deterioration on the structure, etc. The techniques described in the following paragraphs form a hierarchical progression from the relatively simple to the very complex, depending on the objective of the analysis.

In modeling cooling tower shells using the finite element method, there are a number of options. For the shell wall, ring elements, triangular elements, or quadrilateral elements have been used. Earlier, flat elements adapted from the two-dimensional elasticity and plate formulations were used to approximate the doubly curved surface. Such elements present a number of theoretical and computational problems and are *not* recommended for the analysis of shells. Currently, shell elements degenerated from three-dimensional solid elements are very popular. These elements have been utilized in both the ring and quadrilateral form.

Figure 14.13   Attachments on shell wall.

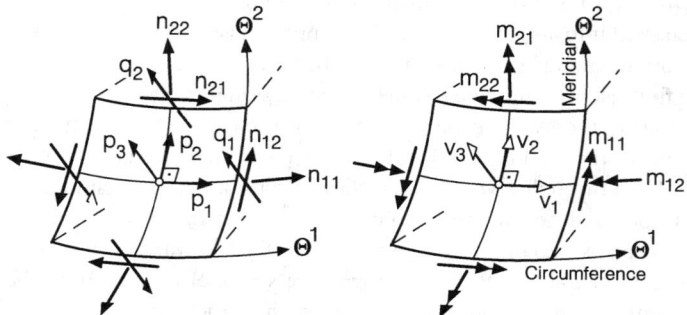

Figure 14.14   Surface loads, stress resultants, stress couples, and displacements.

The column region at the base of the shell presents a special modeling challenge. For static analysis, the lower boundary is often idealized as a uniform support at the lintel level. Then, a portion of the lower shell and the columns is considered in a subsequent analysis to account for the concentrated actions of the columns, which may penetrate only a relatively short distance into the shell wall. For dynamic analysis, it is important to include the column region along with the veil in the model. An equivalent shell element has proved useful in this regard if ring elements are used to model the shell [3, 4]. It may also be desirable to include some of the foundation elements, such as a ring beam at the base and even the supporting piles in a dynamic or settlement model.

The linear static analysis method is based on the classical bending theory of thin shells. While this theory has been formulated for many years, solutions for doubly curved shells have not been readily achievable until the development of computer-based numerical methods, most notably the finite element method. The outputs of such an analysis are the stress resultants and couples, defined on Figure 14.14, over the entire shell surface and the accompanying displacements. The analysis is based on the initial geometry, linear elastic material behavior, and a linear kinematic law. Some representative results of such analyses for a large cooling tower (Figure 14.15) are shown in Figures 14.16 through 14.24 for some of the important loading conditions discussed in the preceding section. The finite element model used considers the shell to be fixed at the top of the columns and,

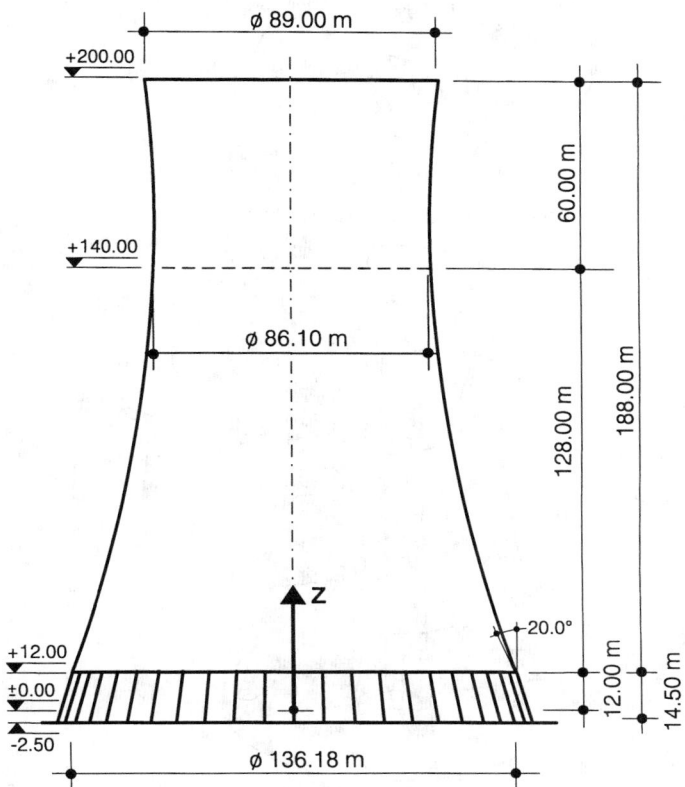

Figure 14.15   Design project for a 200-m high cooling tower: geometry.

thus, does not account for the effect of the concentrated column reactions. Also, in considering the analyses under the individual loading conditions, it should be remembered that the effects are to be factored and combined to produce design values.

The dead load analysis results in Figures 14.16 and 14.17 indicate that the shell is always under compression in both directions, except for a small circumferential tension near the top. This is a very desirable feature of this geometrical form.

In Figures 14.18 through 14.20, the results of an analysis for a quasistatic wind load using the K1.0 distribution from Figure 14.10 are shown. Large tensions in both the meridional and circumferential directions are present. The regions of tension may extend a considerable distance along the

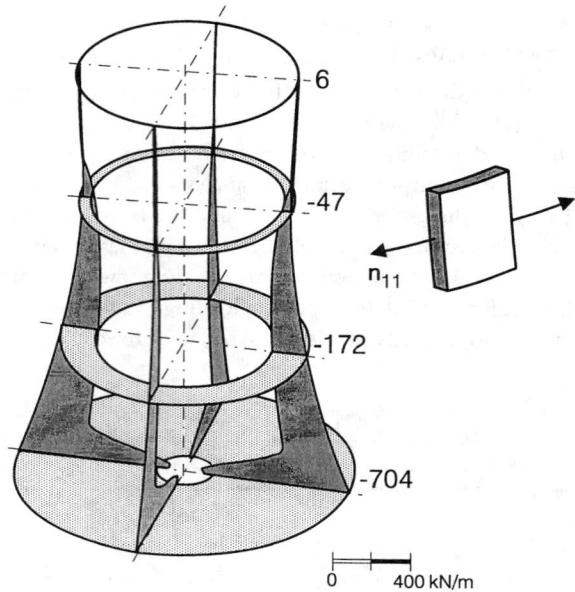

Figure 14.16   Circumferential forces n_{11G} under deadweight.

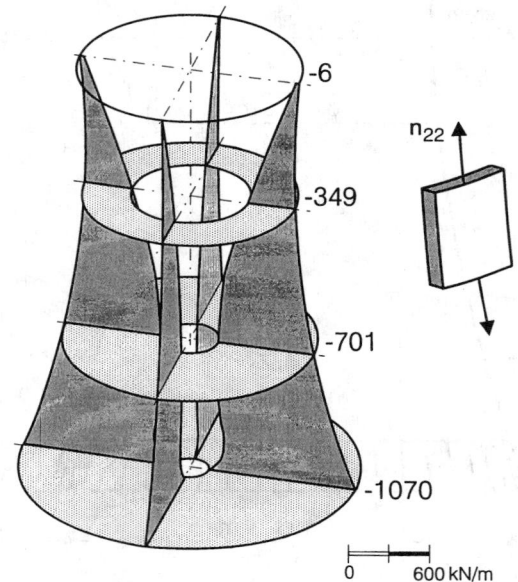

Figure 14.17   Meridional forces n_{22G} under deadweight.

circumference from the windward meridian, and the magnitude of the forces is strongly dependent on the distribution selected. In contrast to bluff bodies, where the magnitude of the extensional force along the meridian would be essentially a function of the overturning moment, the cylindrical-type body is also strongly influenced by the circumferential distribution of the applied pressure, a function of the surface roughness. The major effect of the shearing forces is at the level of the lintel where they are transferred into the columns. The internal suction effects (Figures 14.21 and 14.22) are significant only in the circumferential direction.

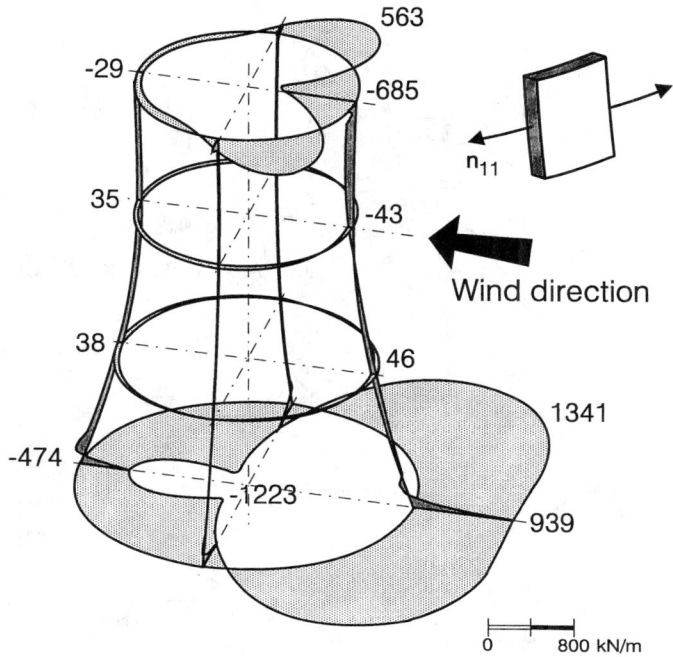

Figure 14.18 Circumferential forces n_{11W} under wind load.

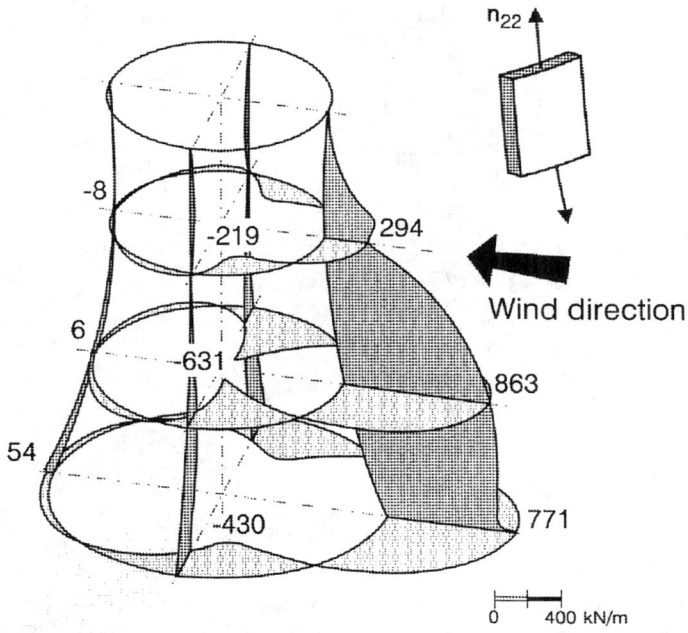

Figure 14.19 Meridional forces n_{22W} under wind load.

For the service temperature case shown in Figures 14.23 and 14.24, the main effects are bending in the lower region of the shell wall.

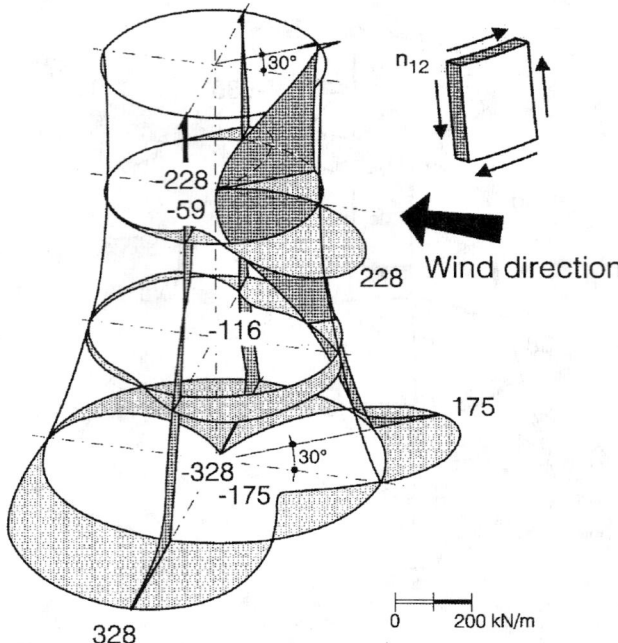

Figure 14.20   Shear forces n_{12W} under wind load.

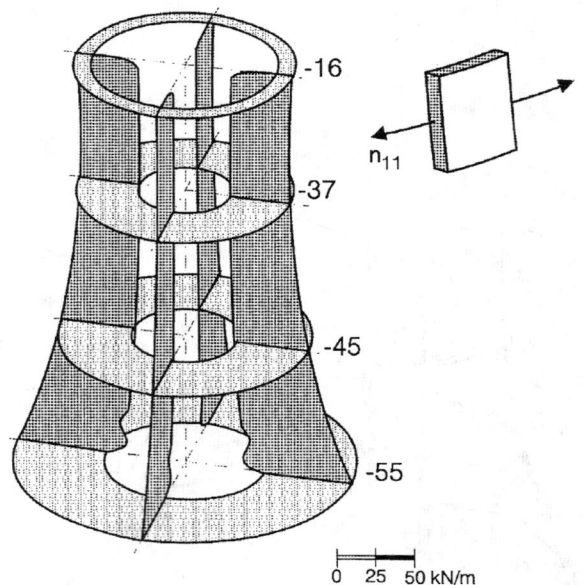

Figure 14.21   Circumferential forces n_{11S} under internal suction.

The analysis of hyperbolic cooling towers for instability or buckling is a subject that has been investigated for several decades [1]. Shell buckling is a complex topic to treat analytically in any case due to the influence of imperfections; for reinforced concrete, it is even more difficult. While the governing equations may be generalized to treat instability by using nonlinear strain-displacement relations and thereby introducing the geometric stiffness matrix, the correlation between the resulting analytical solutions and the possible failure of a reinforced concrete cooling tower is questionable.

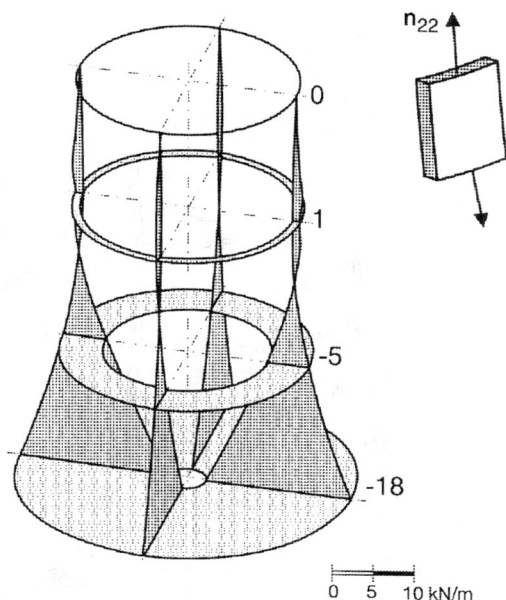

Figure 14.22 Meridional forces n_{22S} under internal suction.

Nevertheless, it has been common to analyze cooling tower shells under an unfactored combination of dead load plus wind load plus internal suction. The corresponding buckling pattern is shown in Figure 14.25.

Interaction diagrams calibrated from experimental studies based on bifurcation buckling are also available [9, 12, 13]. Additionally, there are empirical methods based on wind tunnel tests that consider a snap-through buckle at the upper edge at each stage of construction [13]. These formulas are proportional to h/R and are convenient for establishing an appropriate shell thickness. If buckling safety is evaluated based on such a linear buckling analysis or an experimental investigation, the buckling safety factor for realistic material parameters should exceed 5.0. Presently, however, the use of bifurcation buckling analyses should be confined to preliminary proportioning since more rational procedures based on nonlinear analysis have been developed to predict the collapse of reinforced concrete shells, as discussed in the following paragraphs.

Advances in the analyses of reinforced concrete have produced the capability to analyze shells taking into account the layered composition of the cross-section as shown in Figure 14.26. Using realistic material properties for steel and for concrete, including tension stiffening in the form shown in Figures 14.27 through 14.29, load-deflection relationships may be constructed for appropriate load combinations. These relationships progress from the linear elastic phase to initial cracking of the concrete through spreading of the cracks until collapse.

Results from a nonlinear study are presented in Figures 14.30 through 14.33. The geometry of the shell is given in Figure 14.30, the wind load factor λ is plotted against the maximum lateral displacement at the top of the shell in Figure 14.31, and the deformed shape for the collapse load is shown in Figure 14.32. Also, the pattern of cracking corresponding to the initial yielding of the reinforcement is indicated in Figure 14.33. For reinforced concrete shells, this type of analysis represents the state-of-the-art and provides a realistic evaluation of the capacity of such shells against extreme loading [8]. Also durability assessments can be performed by this concept, from which particularly weak and crack-endangered regions of the shell can be identified and further reinforced [10].

It is possible to obtain an estimate of the wind load factor, λ, from the results of a linear elastic analysis, even from a calculation based on membrane theory. This estimate is computed as the

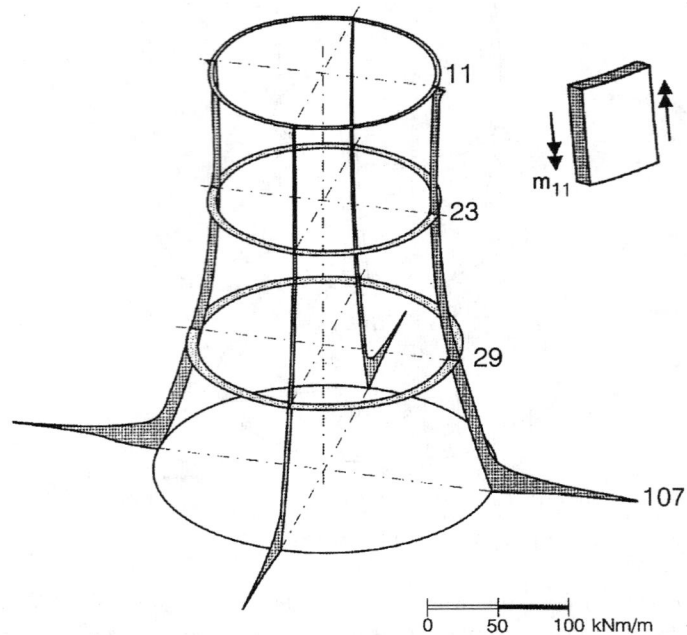

Figure 14.23 Circumferential bending moments m_{11T} under service temperature.

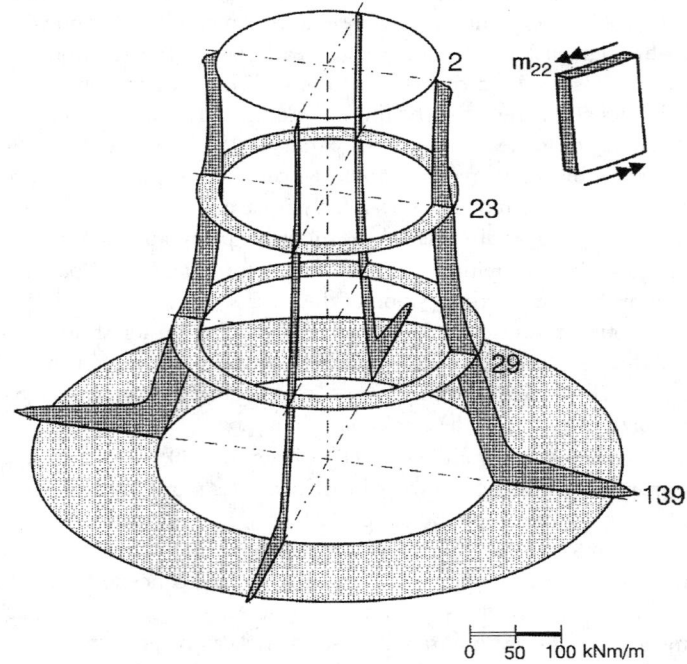

Figure 14.24 Meridional bending moments m_{22T} under service temperature.

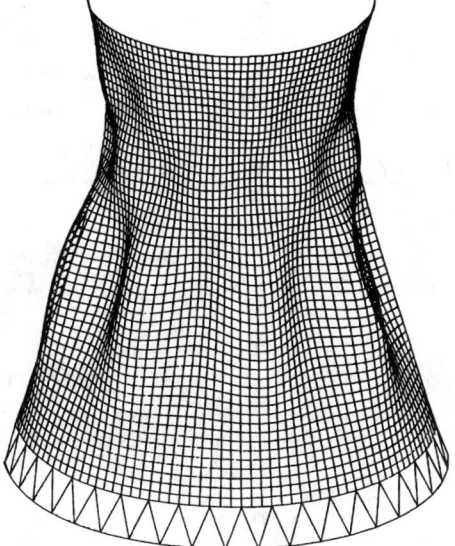

Figure 14.25 Buckling pattern of tower shell with upper ring beam: $D + W + S$.

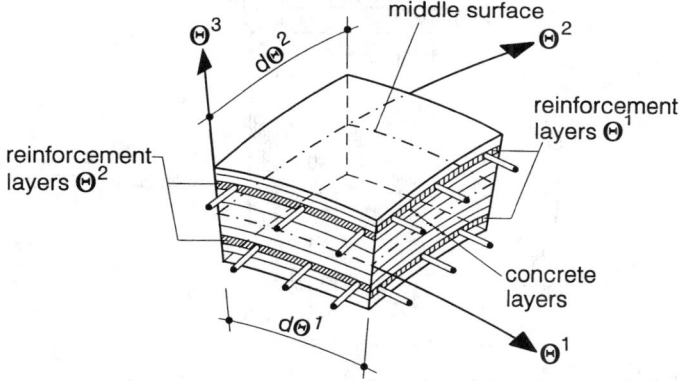

Figure 14.26 Layered model for reinforced concrete shell.

cracking load for the shell under a combination of $D + \lambda W$ and is predicated on the notion that the reinforcement may add only a modest amount of capacity to the tower beyond the cracking load [6]. The amount of reinforcement in the wall is often controlled by a specified minimum percentage augmented by that required to resist the net tension due to the factored load combinations. The steel provided is often less than the capacity of the concrete in tension, which is presumed to be lost when the concrete cracks. Therefore, the cracking load represents most of the ultimate capacity of the tower.

The maximum meridional tension location under the wind loading is identified, for example, as the value of $n_{22} = 863$ kN/m in Figure 14.19. The dead load at this location is obtained from Figure 14.17 as -701 kN/m. Taking the concrete tensile capacity as 2,400 kN/m^2 and the wall thickness as 16 cm, the tensile strength is 384 kN/m. Therefore, we have

$$- 701 + \lambda 863 = 384 \tag{14.7}$$

giving $\lambda = 1.26$ as the lower bound on the ultimate strength of the tower. Note that the tower used for the linear elastic analysis is much taller than the one shown in Figure 14.30.

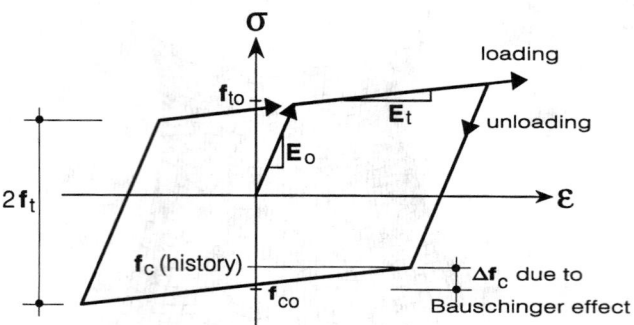

f_{to} : yield stress in tension
f_{co} : yield stress in compression
E_o : modulus of elasticity
E_t : tangent modulus of elasticity

Figure 14.27   Elasto-plastic material law for steel.

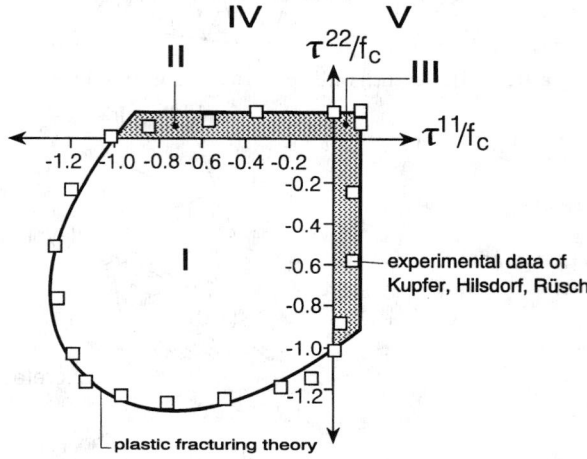

Figure 14.28   Biaxial failure envelope of Kupfter/Hilsdorf/Rüsch.

The dynamic analysis of cooling towers is usually associated with design for earthquake-induced forces. The most efficient approach is the response spectrum method, but a time history analysis may be appropriate if nonlinearities are to be included [2, 7]. For large shells the dynamic response due to wind is often investigated, at least to determine the positions of the nodal lines and areas of particularly intensive vibrations. In any case the first step is to carry out a free vibration analysis. This analysis represents the modes of free vibration associated with each natural frequency, f, or its inverse the natural period T, as the product of a circumferential mode proportional to $\sin n\theta$ or $\cos n\theta$ and a longitudinal mode along the z axis [3, 4]. Some representative results are shown on Figures 14.34 and 14.35, as discussed below.

As an illustration, the cooling tower from Figure 14.4 is again considered. Some key circumferential and longitudinal modes for a fixed-base boundary condition are shown in Figure 14.35. Also, the effects of different cornice stiffnesses are demonstrated. This model may be regarded as preliminary in that the relatively soft column supports are not properly represented, but it illustrates the salient characteristics of the modes of vibration. Most interesting are the frequency curves on Figure 14.34 for the first 10 harmonics, also demonstrating the influence of different cornice stiffnesses. Note that the natural frequencies *decrease* with increasing n until a minimum is reached whereupon they increase, a very typical behavior for cylindrical-type shells. Also, the stiffening of the cornice tends

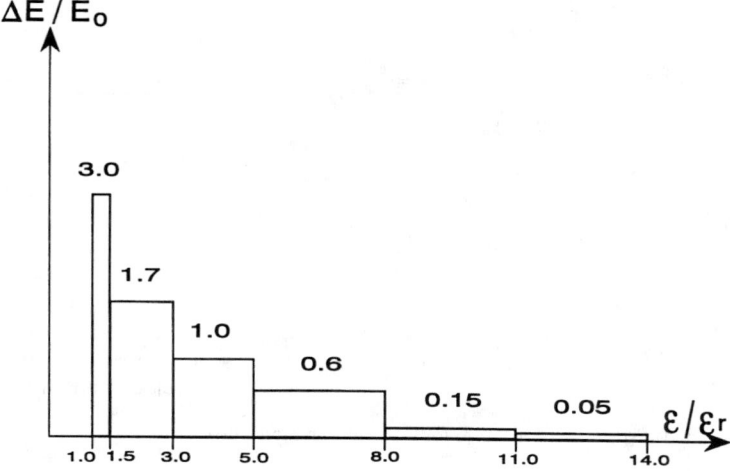

Figure 14.29   Additional modulus of elasticity due to tension stiffening.

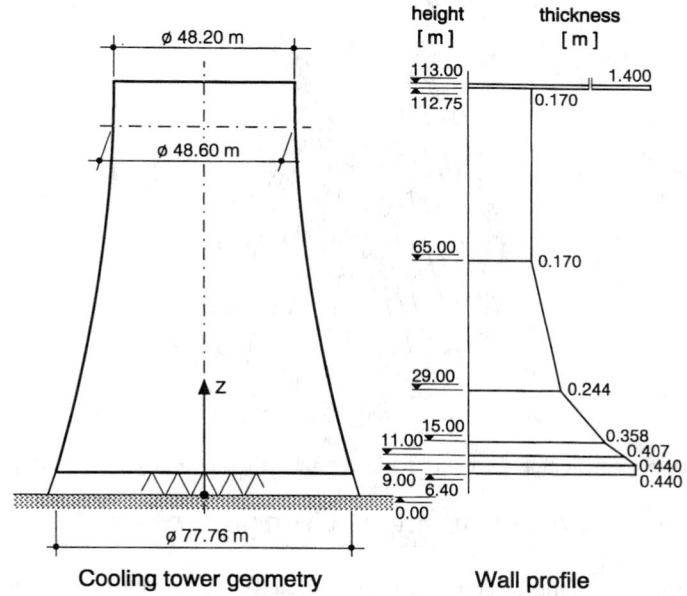

Cooling tower geometry   Wall profile

Figure 14.30   Shell geometry and wall thickness.

to raise the minimum frequency, which is desirable for resistance to dynamic wind. Longitudinally, the cornice stiffness effect is significant for odd modes only.

Specifically for earthquake effects, only the first mode participates in a linear analysis for uniform horizontal base motion and the respective values for $n = 1$ should be entered into the design response spectrum.

Results from a seismic analysis of a cooling tower are presented in Figures 14.36 to 14.39. The cooling tower of Figure 14.4 is subjected to a horizontal base excitation based on Figure 14.36, leading to a first circumferential mode ($n = 1$) participation. A response spectrum analysis provides the lateral displacements w of the tower axis, the meridional forces n_{22}, and the shear forces n_{12} as shown on the indicated figures. In general, cooling tower shells have proven to be reasonably resistant against seismic excitations, but obviously the most critical region is the connection between the columns and the lintel as portrayed in Figure 14.40.

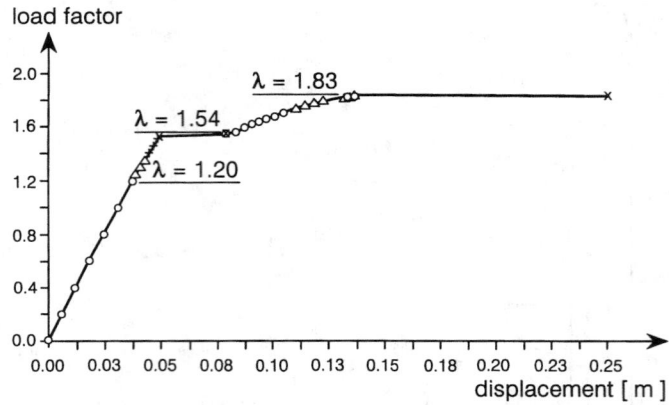

Figure 14.31   Load-displacement diagram for load combination $D + \lambda \bullet W$.

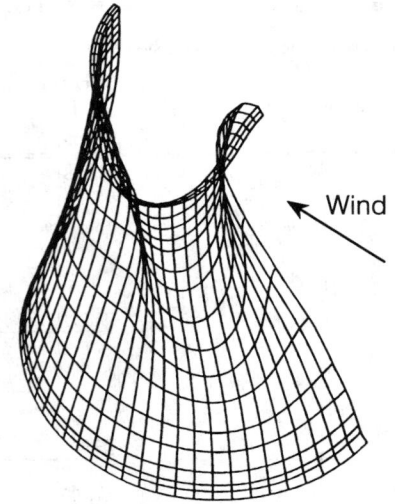

Figure 14.32   Displacement plot of the shell for load combination $D + \lambda \bullet W$: load factor $= 1.83$.

14.7   Design and Detailing of Components

The structural elements of the tower should be constructed with a suitable grade of concrete following the provisions of applicable codes and standards. The design of the mixture should reflect the conditions for placement of the concrete and the external and internal environment of the tower.

The shell wall should be of a thickness which will permit two layers of reinforcement in two perpendicular directions to be covered by a minimum of 3 cm of concrete, and should be no less than 16 cm thick [7, 13]. The buckling considerations mentioned in the previous section have proven to be a convenient and evidently acceptable criteria for setting the minimum wall thickness, subject to a nonlinear analysis. The formula

$$q_c = 0.052E(h/R)^{2.3} \tag{14.8}$$

where E = modulus of elasticity, has been used to estimate the critical shell buckling pressure q_c [1, 13]. Then, $h(z)$ is selected to provide a factor of safety of at least 5.0 with respect to the maximum velocity pressure along the windward meridian, $q(z)(1 + g)$. Also, the cornice should have a minimum stiffness of

$$I_x/d_H = 0.0015m^3 \tag{14.9}$$

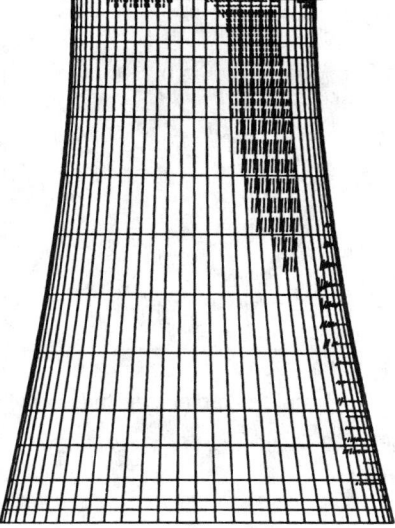

Figure 14.33   Crack pattern of the outer face of the shell for load combination $D + \lambda \bullet W$: load factor $= 1.54$.

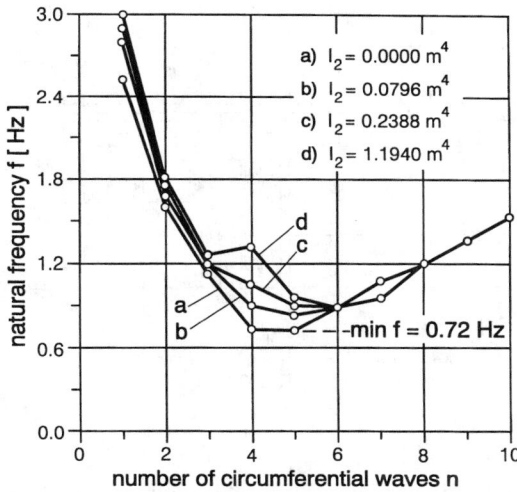

Figure 14.34   Natural frequencies for different cornice stiffnesses for the cooling tower on Figure 14.4.

where I_x is the moment of inertia of the uncracked cross-section about the vertical axis [13]. Some typical forms of the cornice cross-section are shown in Figure 14.41.

The elements of the cooling tower should be reinforced with deformed steel bars so as to provide for the tensile forces and moments arising from the controlling combination of factored loading cases. The shell walls may be proportioned as rectangular cross-sections subjected to axial forces and bending. As mentioned above, a mesh of two orthogonal layers of reinforcement should be provided in the shell walls, generally in the meridional and circumferential directions [2]. In each direction, the inner and outer layers should generally be the same, except near the edges where the bending may require an unsymmetrical mesh. It is preferable to locate the circumferential reinforcement outside of the meridional reinforcement except near the lintel, where the meridional reinforcement should be on the outside to stabilize the circumferential bars [13]. A typical heavily

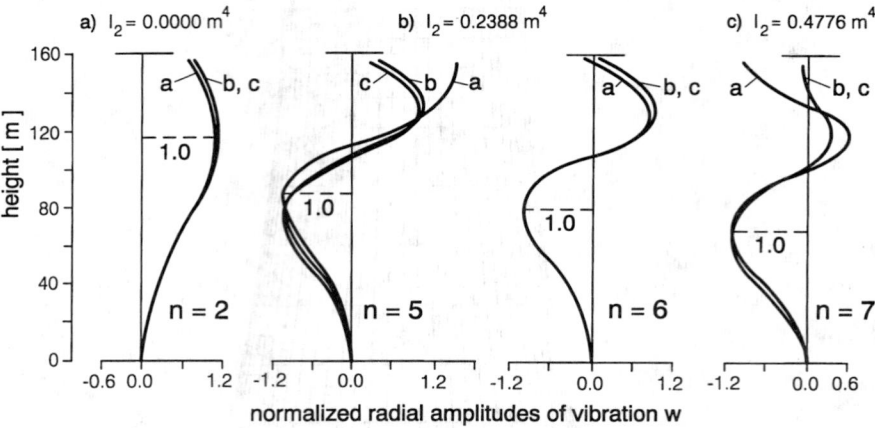

Figure 14.35   Normalized natural vibration modes for the cooling tower on Figure 14.4.

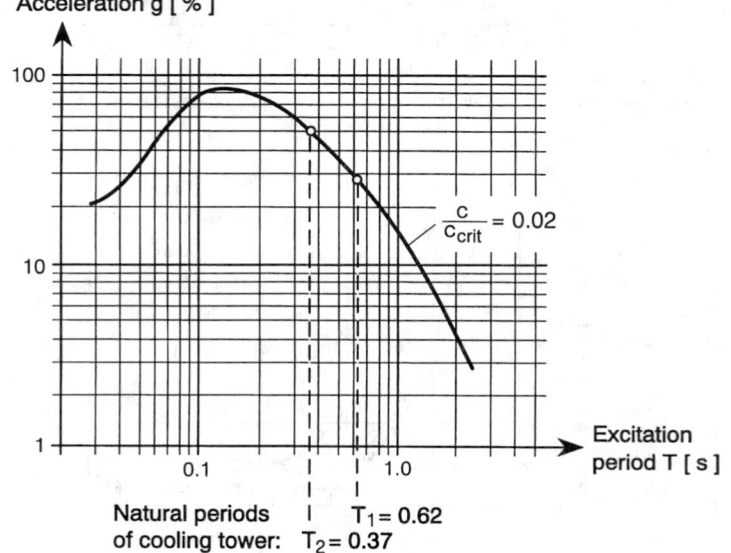

Figure 14.36   Seismic response spectrum.

reinforced segment of the lintel, also showing the anchorage of the column reinforcement into the shell, is depicted in Figure 14.42.

A summary of the most important minimum construction values for the shell wall is given in Figure 14.43 [13]. The bars should not be smaller than 8 mm diameter and, for meridional bars, not smaller than 10 mm. Further, a minimum of 0.35 to 0.45%, depending on the admissible cracking, should be used in each direction. The minimum cover, as mentioned above, should be 3 cm, the maximum spacing of the bars should be 20 cm, and the splices should be staggered as specified for the construction of walls in the applicable codes or standards. Particular attention should be given to splices in tensile zones.

The supporting columns should ideally be proportioned for the forces and moments computed from an analysis in which they are represented as discrete members, using the appropriate factored loading combinations [3]. If the column region has not been modeled discretely, but rather by a continuum approximation, the columns may be proportioned to resist the tributary factored forces and moments at the interface with the lintel, as computed from the shell analysis. The effective length may be taken as unity. Particular attention should be directed toward splices of the column

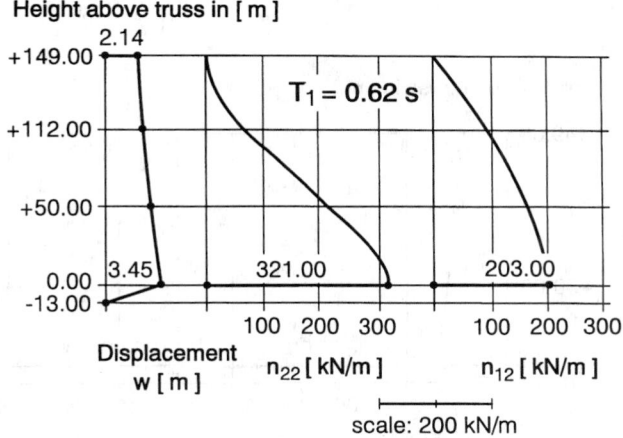

Figure 14.37 First axial mode seismic response.

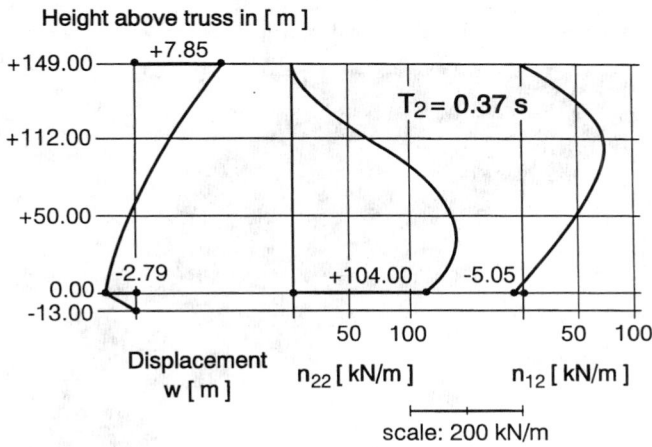

Figure 14.38 Second axial mode seismic response.

bars when net tension is present. Since large bars will be involved, welded splices are recommended in such regions.

It is possible to add discrete circumferential stiffeners to the shell to increase the stability or to restore capacity that may have been lost due to cracking or other deterioration [5] (see Figure 14.9). Such stiffeners can generally be included in a finite element model of the shell wall and should be proportioned for the forces computed from such an analysis. The eccentricity of the stiffeners with respect to the circumferential axis should be considered when the stiffeners are only on one side of the shell.

The foundations should be proportioned for the factored forces induced by the column reactions, or from the computed forces if the foundation is included in the model with the shell and columns. Reinforcement detailing and cover should be in accordance with the applicable codes or standards. Several improved forms for cooling tower foundations have been suggested. Figure 14.44 shows a flat ring footing suitable for uniform soil conditions, while Figure 14.45 portrays a stiff ring beam foundation appropriate for soil conditions that are non-uniform around the circumference. An example of an individual pier on bedrock is given in Figure 14.46.

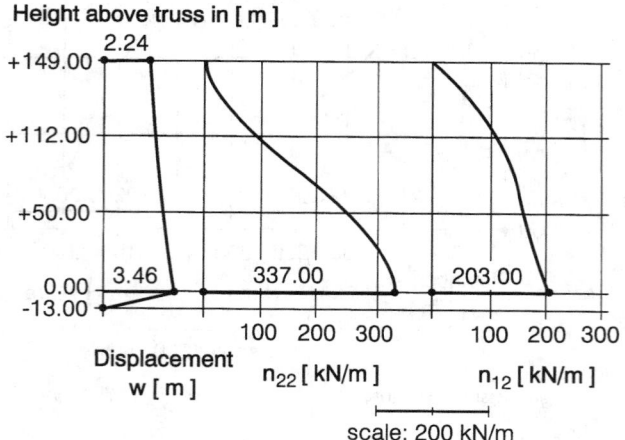

Figure 14.39   SRSS superposition of first and second axial modes.

Figure 14.40   Column to lintel connection.

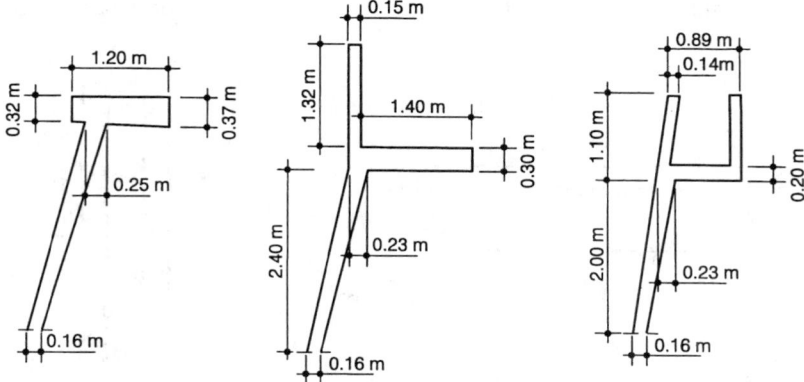

Figure 14.41   Suitable forms of the cornice.

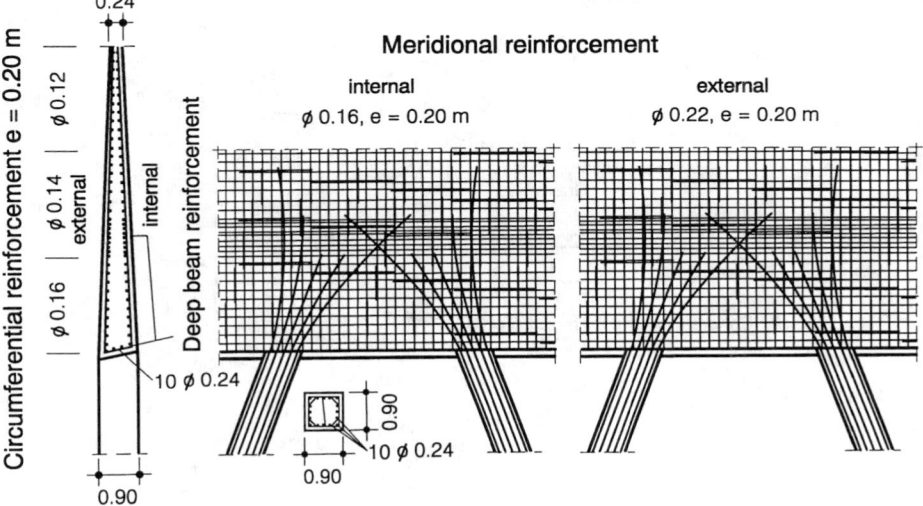

Figure 14.42   Lintel reinforcement.

14.8   Construction

Tolerances for tall concrete cooling tower shells have been debated for many years and reasonable values should take into consideration what is achievable and what is measurable. It should be noted that state-of-the-art finite element models are capable of analyzing the as-built shell as well as the design configuration, so that the effects of those irregularities arising during construction, or even those discovered later, may be quantitatively studied and sometimes corrected.

It is recommended that the actual wall thickness be no less than the design thickness and exceed this thickness by not more than 10%. The imperfections of the shell wall middle surface should not exceed one-half of the wall thickness or 10 cm. Deviations from the design geometry occurring during the construction should be corrected gradually, limiting the angular change in either direction to 1.5%. The column heads should be within 0.005 times the column height or ±6.0 cm of the design position, and foundation structures should also be within ±6.0 cm of the design location [13].

Formwork and scaffolding systems are generally proprietary and are provided by the constructor. Nevertheless, their influence on the shell quality is of utmost importance and diligent attention of the engineer is required. In general, the system should be designed to provide safety to operating personnel and to produce a sound structure. The working platforms should be designed for realistic

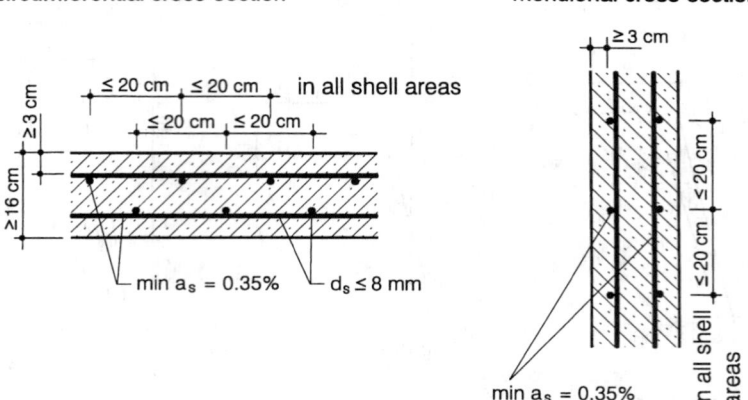

Figure 14.43   Important minimum construction values.

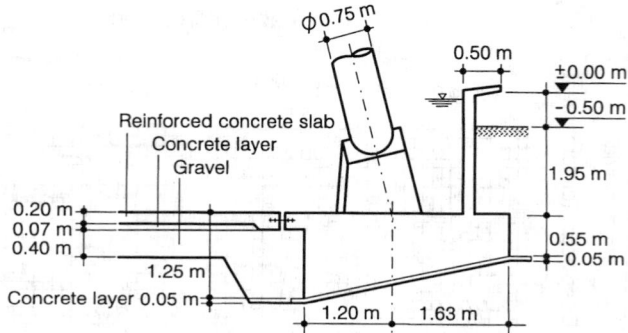

Figure 14.44   Flat ring foundation.

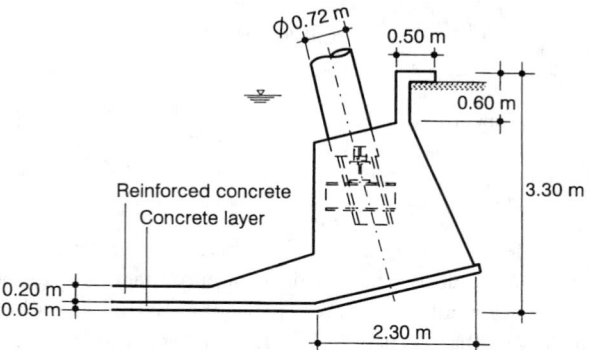

Figure 14.45   Ring beam foundation.

loading, and scaffolding systems used for continuous material transport should be designed and built taking into account the resulting loads.

The connections and joints between individual scaffolding units should be designed and built to act independently in case of collapse, so that the loss of one unit would not affect the adjacent units. Furthermore, at least two independent safety devices should be in place to prevent collapse.

The shell wall should be designed to resist the anchor loads of the scaffolding, based on the strength of the concrete which is expected to be available when the anchors are loaded. Continuous monitoring of the concrete strength during the climbing process is essential.

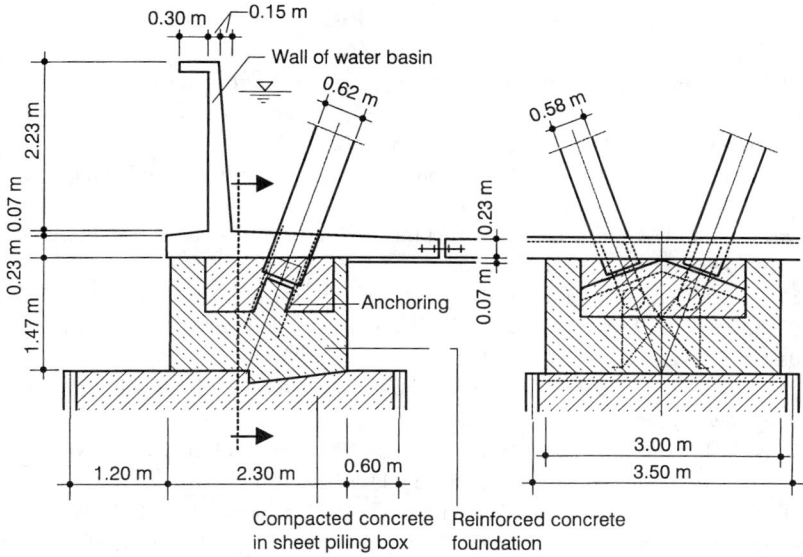

Figure 14.46 Individual reinforced concrete foundation on concrete base.

Cooling tower shells are subjected to a relatively severe environment over their lifetime, which may span several decades, and special care must be taken in order to provide a durable structure. The tower is subjected to the physical loads produced by wind, temperature, and moisture acting on concrete which may still be drying and hardening. Over the lifetime of the structure, it may be exposed to severe frost action in a saturated state, chemical attacks due to noxious substances in the atmosphere and in the water and water vapor, biological attacks due to microorganisms, and possibly additional chemical attacks due to reintroduced cleaned flue gases.

The concrete should be of high-quality approved materials including fly-ash. It should have the following properties:

- High resistance against chemical attacks
- High early strength
- High structural density
- High resistance against frost

The surface finish should be of high quality and the surface should be smooth and essentially free of shrink holes. Air bubbles deeper than 4 mm and unintended surface irregularities at joints should be avoided. The shell should be coated with a curing agent providing a high blocking effect and long durability.

Several single component (acrylate or polyurethene-based) or double component (epoxy resin-based) coating systems are approved worldwide and are in a process of continual improvement. Of utmost importance for any coating is the homogeneity of the applied film between $\geq 200 \mu m$ for single and $\geq 300 \mu$ for double component systems, since the durability of the complete coating is determined by the thinnest film spots.

References

[1] Abel, J.F. and Gould, P.L. 1981. *Buckling of Concrete Cooling Towers Shells,* ACI SP-67, American Concrete Institute, Detroit, Michigan, pp. 135-160.

[2] ACI-ASCE Committee 334. 1977. Recommended Practice for the Design and Construction of Reinforced Concrete Cooling Towers, *ACI J.*, 74(1), 22-31.

[3] Gould, P.L. 1985. *Finite Element Analysis of Shells of Revolution,* Pitman.

[4] Gould, P.L., Suryoutomo, H., and Sen, S.K. 1974. Dynamic analysis of column-supported hyperboloidal shells, *Earth. Eng. Struct. Dyn.*, 2, 269-279.

[5] Gould, P.L. and Guedelhoefer, O.C. 1988. Repair and Completion of Damaged Cooling Tower, *J. Struct. Eng.*, 115(3), 576-593.

[6] Hayashi, K. and Gould, P.L. 1983. Cracking load for a wind-loaded reinforced concrete cooling tower, *ACI J.*, 80(4), 318-325.

[7] IASS-Recommendations for the Design of Hyperbolic or Other Similarly Shaped Cooling Towers. 1977. Intern. Assoc. for Shell and Space Structures, Working Group No. 3, Brussels.

[8] Krätzig, W.B. and Zhuang, Y. 1992. Collapse simulation of reinforced natural draught cooling towers, *Eng. Struct.*, 14(5), 291-299.

[9] Krätzig, W.B. and Meskouris, K. 1993. Natural draught cooling towers: An increasing need for structural research, *Bull. IASS*, 34(1), 37-51.

[10] Krätzig, W.B. and Gruber, K.P. 1996. Life-Cycle Damage Simulations of Natural Draught Cooling Towers in *Natural Draught Cooling Towers,* Wittek, U. and Krätzig, W., Eds., A.A. Balkema, Rotterdam, 151-158.

[11] Minimum Design Loads for Buildings and Other Structures. 1994. ASCE Standard 7-93, ASCE, New York.

[12] Mungan, I. 1976. Buckling stress states of hyperboloidal shells, *J. Struct. Div.*, ASCE, 102, 2005-2020.

[13] VGB Guideline. 1990. Structural Design of Cooling Towers, VGB-Technical Committee, "Civil Engineering Problems of Cooling Towers", Essen, Germany.

Further Reading

[1] *Proceedings (First) International Symposium on Very Tall Reinforced Concrete Cooling Towers.* 1978. I.A.S.S., E.D.F., Paris, France, November.

[2] Gould, P.L., Krätzig, W.B., Mungan, I., and Wittek, U., Eds. 1984. *Natural Draught Cooling Towers.* Proceedings of the 2nd International Symposium on Natural Draught Cooling Towers, Springer-Verlag, Heidelberg.

[3] *Proceedings Third International Symposium on Natural Draught Cooling Towers.* 1989. I.A.S.S., E.D.F., Paris, France, April.

[4] Wittek, U. and Krätzig, W.B., Eds. 1996. *Natural Draught Cooling Towers.* Proceedings of the 4th International Symposium on Natural Draught Cooling Towers, A.A. Balkema, Rotterdam.

[5] British Standard Institution. 1996. BS 4485, Part 4: British Standard for Water Cooling Towers. Document 96/17117 DC 22.

[6] Syndicat National du Béton Armé et des Techniques Industrialisées. 1996. Regles de conception et de realisation des refrigerants atmospheriques en beton armé, Paris.

15

Transmission Structures

15.1   Introduction and Application 15-1
Application • Structure Configuration and Material • Con-
structibility • Maintenance Considerations • Structure Families
• State of the Art Review

15.2   Loads on Transmission Structures 15-5
General • Calculation of Loads Using NESC Code • Calculation
of Loads Using the ASCE Guide • Special Loads • Security Loads
• Construction and Maintenance Loads • Loads on Structure •
Vertical Loads • Transverse Loads • Longitudinal Loading

15.3   Design of Steel Lattice Tower 15-10
Tower Geometry • Analysis and Design Methodology • Allow-
able Stresses • Connections • Detailing Considerations • Tower
Testing

15.4   Transmission Poles 15-17
General • Stress Analysis • Tubular Steel Poles • Wood Poles •
Concrete Poles • Guyed Poles

15.5   Transmission Tower Foundations 15-23
Geotechnical Parameters • Foundation Types—Selection and
Design • Anchorage • Construction and Other Considerations
• Safety Margins for Foundation Design • Foundation Move-
ments • Foundation Testing • Design Examples

15.6   Defining Terms 15-30

References 15-31

Shu-jin Fang, Subir Roy,
and Jacob Kramer
Sargent & Lundy, Chicago, IL

15.1   Introduction and Application

Transmission structures support the phase conductors and shield wires of a transmission line. The
structures commonly used on transmission lines are either lattice type or pole type and are shown
in Figure 15.1. Lattice structures are usually composed of steel angle sections. Poles can be wood,
steel, or concrete. Each structure type can also be self-supporting or guyed. Structures may have
one of the three basic configurations: horizontal, vertical, or delta, depending on the arrangement
of the phase conductors.

15.1.1   Application

Pole type structures are generally used for voltages of 345-kV or less, while lattice steel structures
can be used for the highest of voltage levels. Wood pole structures can be economically used for
relatively shorter spans and lower voltages. In areas with severe climatic loads and/or on higher
voltage lines with multiple subconductors per phase, designing wood or concrete structures to meet
the large loads can be uneconomical. In such cases, steel structures become the cost-effective option.

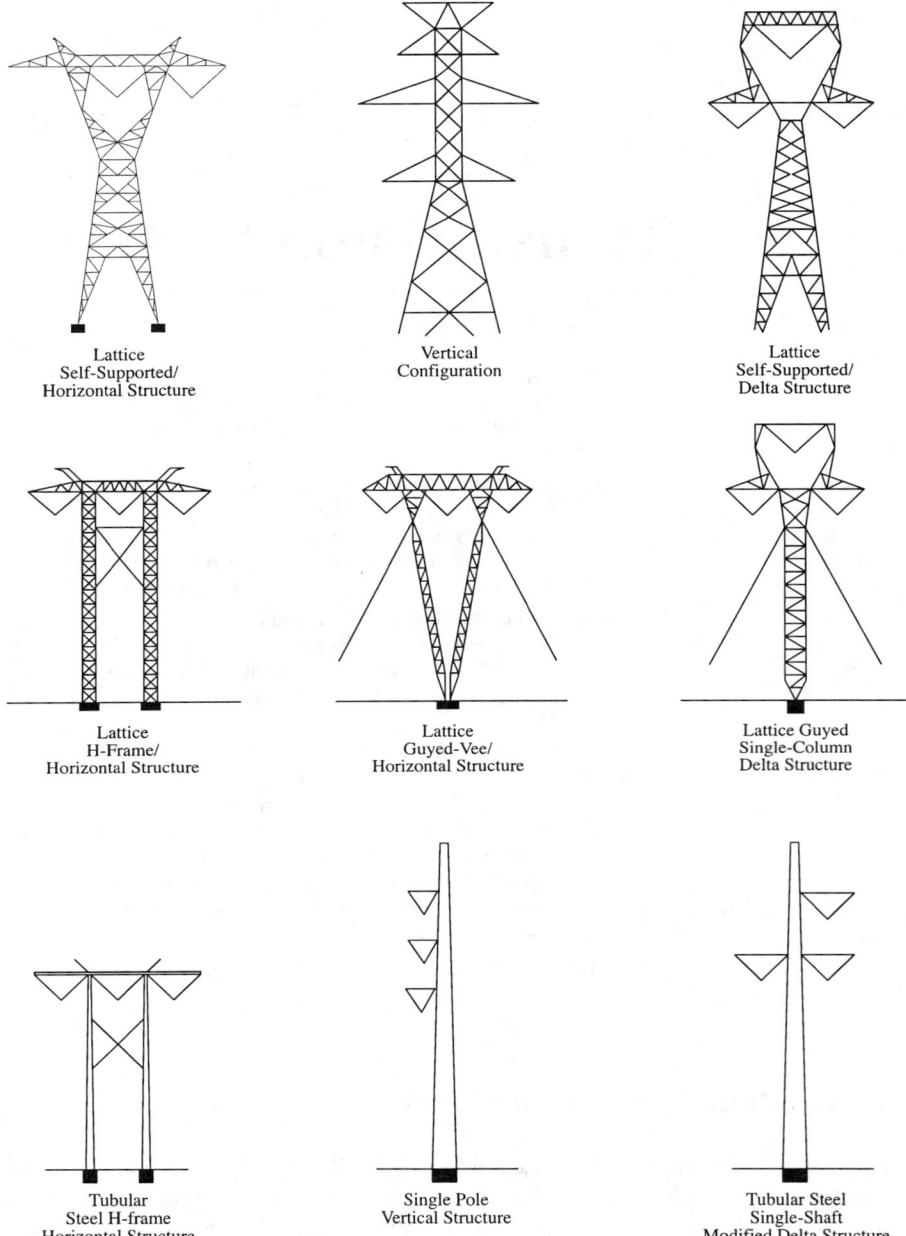

Figure 15.1   Transmission line structures.

Also, if greater longitudinal loads are included in the design criteria to cover various unbalanced loading contingencies, H-frame structures are less efficient at withstanding these loads. Steel lattice towers can be designed efficiently for any magnitude or orientation of load. The greater complexity of these towers typically requires that full-scale load tests be performed on new tower types and at least the tangent tower to ensure that all members and connections have been properly designed and detailed. For guyed structures, it may be necessary to proof-test all anchors during construction to ensure that they meet the required holding capacity.

15.1.2 Structure Configuration and Material

Structure cost usually accounts for 30 to 40% of the total cost of a transmission line. Therefore, selecting an optimum structure becomes an integral part of a cost-effective transmission line design. A structure study usually is performed to determine the most suitable structure configuration and material based on cost, construction, and maintenance considerations and electric and magnetic field effects. Some key factors to consider when evaluating the structure configuration are:

- A horizontal phase configuration usually results in the lowest structure cost.
- If right-of-way costs are high, or the width of the right-of-way is restricted or the line closely parallels other lines, a vertical configuration may be lower in total cost.
- In addition to a wider right-of-way, horizontal configurations generally require more tree clearing than vertical configurations.
- Although vertical configurations are narrower than horizontal configurations, they are also taller, which may be objectionable from an aesthetic point of view.
- Where electric and magnetic field strength is a concern, the phase configuration is considered as a means of reducing these fields. In general, vertical configurations will have lower field strengths at the edge of the right-of-way than horizontal configurations, and delta configurations will have the lowest single-circuit field strengths and a double-circuit with reverse or low-reactance phasing will have the lowest possible field strength.

Selection of the structure type and material depends on the design loads. For a single circuit 230-kV line, costs were estimated for single-pole and H-frame structures in wood, steel, and concrete over a range of design span lengths. For this example, wood H-frames were found to have the lowest installed cost, and a design span of 1000 ft resulted in the lowest cost per mile. As design loads and other parameters change, the relative costs of the various structure types and materials change.

15.1.3 Constructibility

Accessibility for construction of the line should be considered when evaluating structure types. Mountainous terrain or swampy conditions can make access difficult and use of helicopter may become necessary. If permanent access roads are to be built to all structure locations for future maintenance purposes, all sites will be accessible for construction.

To minimize environmental impacts, some lines are constructed without building permanent access roads. Most construction equipment can traverse moderately swampy terrain by use of wide-track vehicles or temporary mats. Transporting concrete for foundations to remote sites, however, increases construction costs.

Steel lattice towers, which are typically set on concrete shaft foundations, would require the most concrete at each tower site. Grillage foundations can also be used for these towers. However, the cost of excavation, backfill and compaction for these foundations is often higher than the cost of a drilled shaft. Unless subsurface conditions are poor, most pole structures can be directly embedded. However, if unguyed pole structures are used at medium to large line angles, it may be necessary to use drilled shaft foundations.

Guyed structures can also create construction difficulties in that a wider area must be accessed at each structure site to install the guys and anchors. Also, careful coordination is required to ensure that all guys are tensioned equally and that the structure is plumb.

Hauling the structure materials to the site must also be considered in evaluating constructibility. Transporting concrete structures, which weigh at least five times as much as other types of structures, will be difficult and will increase the construction cost of the line. Heavier equipment, more trips to transport materials, and more matting or temporary roadwork will be required to handle these heavy poles.

15.1.4   Maintenance Considerations

Maintenance of the line is generally a function of the structure material. Steel and concrete structures should require very little maintenance, although the maintenance requirements for steel structures depends on the type of finish applied. Tubular steel structures are usually galvanized or made of weathering steel. Lattice structures are galvanized. Galvanized or painted structures require periodic inspection and touch-up or reapplication of the finish while weathering steel structures should have relatively low maintenance. Wood structures, however, require more frequent and thorough inspections to evaluate the condition of the poles. Wood structures would also generally require more frequent repair and/or replacement than steel or concrete structures. If the line is in a remote location and lacks permanent access roads, this can be an important consideration in selecting structure material.

15.1.5   Structure Families

Once the basic structure type has been established, a family of structures is designed, based on the line route and the type of terrain it crosses, to accommodate the various loading conditions as economically as possible. The structures consist of tangent, angle, and deadend structures.

Tangent structures are used when the line is straight or has a very small line angle, usually not exceeding 3°. The line angle is defined as the deflection angle of the line into adjacent spans. Usually one tangent type design is sufficient where terrain is flat and the span lengths are approximately equal. However, in rolling and mountainous terrain, spans can vary greatly. Some spans, for example, across a long valley, may be considerably larger than the normal span. In such cases, a second tangent design for long spans may prove to be more economical. Tangent structures usually comprise 80 to 90% of the structures in a transmission line.

Angle towers are used where the line changes direction. The point at which the direction change occurs is generally referred to as the point of intersection (P.I.) location. Angle towers are placed at the P.I. locations such that the transverse axis of the cross arm bisects the angle formed by the conductor, thus equalizing the longitudinal pulls of the conductors in the adjacent spans. On lines where large numbers of P.I. locations occur with varying degrees of line angles, it may prove economical to have more than one angle structure design: one for smaller angles and the other for larger angles.

When the line angle exceeds 30°, the usual practice is to use a deadend type design. Deadend structures are designed to resist wire pulls on one side. In addition to their use for large angles, the deadend structures are used as terminal structures or for sectionalizing a long line consisting of tangent structures. Sectionalizing provides a longitudinal strength to the line and is generally recommended every 10 miles. Deadend structures may also be used for resisting uplift loads. Alternately, a separate strain structure design with deadend insulator assemblies may prove to be more economical when there is a large number of structures with small line angle subjected to uplift. These structures are not required to resist the deadend wire pull on one side.

15.1.6   State of the Art Review

A major development in the last 20 years has been in the area of new analysis and design tools. These include software packages and design guidelines [12, 6, 3, 21, 17, 14, 9, 8], which have greatly improved design efficiency and have resulted in more economical structures. A number of these tools have been developed based on test results, and many new tests are ongoing in an effort to refine the current procedures. Another area is the development of the reliability based design concept [6]. This methodology offers a uniform procedure in the industry for calculation of structure loads and strength, and provides a quantified measure of reliability for the design of various transmission line components.

Aside from continued refinements in design and analysis, significant progress has been made in the manufacturing technology in the last two decades. The advance in this area has led to the increasing usage of cold formed shapes, structures with mixed construction such as steel poles with lattice arms or steel towers with FRP components, and prestressed concrete poles [7].

15.2 Loads on Transmission Structures

15.2.1 General

Prevailing practice and most state laws require that transmission lines be designed, as a minimum, to meet the requirements of the current edition of the National Electrical Safety Code (NESC) [5]. NESC's rules for the selection of loads and overload capacity factors are specified to establish a minimum acceptable level of safety. The ASCE Guide for Electrical Transmission Line Structural Loading (ASCE Guide) [6] provides loading guidelines for extreme ice and wind loads as well as security and safety loads. These guidelines use reliability based procedures and allow the design of transmission line structures to incorporate specified levels of reliability depending on the importance of the structure.

15.2.2 Calculation of Loads Using NESC Code

NESC code [5] recognizes three loading districts for ice and wind loads which are designated as heavy, medium, and light loading. The radial thickness of ice and the wind pressures specified for the loading districts are shown in Table 15.1. Ice build-up is considered only on conductors and shield wires, and is usually ignored on the structure. Ice is assumed to weigh 57 lb/ft^3. The wind pressure applies to cylindrical surfaces such as conductors. On the flat surface of a lattice tower member, the wind pressure values are multiplied by a force coefficient of 1.6. Wind force is applied on both the windward and leeward faces of a lattice tower.

TABLE 15.1 Ice, Wind, and Temperature

	Loading districts		
	Heavy	Medium	Light
Radial thickness of ice (in.)	0.50	0.25	0
Horizontal wind pressure (lb/ft^2)	4	4	9
Temperature (°F)	0	+15	+30

NESC also requires structures to be designed for extreme wind loading corresponding to 50 year fastest mile wind speed with no ice loads considered. This provision applies to all structures without conductors, and structures over 60 ft supporting conductors. The extreme wind speed varies from a basic speed of 70 mph to 110 mph in the coastal areas.

In addition, NESC requires that the basic loads be multiplied by overload capacity factors to determine the design loads on structures. Overload capacity factors make it possible to assign relative importance to the loads instead of using various allowable stresses for different load conditions. Overload capacity factors specified in NESC have a larger value for wood structures than those for steel and prestressed concrete structures. This is due to the wide variation found in wood strengths and the aging effect of wood caused by decay and insect damage. In the 1990 edition, NESC introduced an alternative method, where the same overload factors are used for all the materials but a strength reduction factor is used for wood.

15.2.3 Calculation of Loads Using the ASCE Guide

The ASCE Guide [6] specifies extreme ice and extreme wind loads, based on a 50-year return period, which are assigned a reliability factor of 1. These loads can be increased if an engineer wants to use a higher reliability factor for an important line, for example a long line, or a line which provides the only source of load. The load factors used to increase the ASCE loads for different reliability factors are given in Table 15.2.

TABLE 15.2   Load Factor to Adjust Line Reliability

Line reliability factor, LRF	1	2	4	8
Load return period, RP	50	100	200	400
Corresponding load factor, $\bar{a}$	1.0	1.15	1.3	1.4

In calculating wind loads, the effects of terrain, structure height, wind gust, and structure shape are included. These effects are explained in detail in the ASCE Guide. ASCE also recommends that the ice loads be combined with a wind load equal to 40% of the extreme wind load.

15.2.4 Special Loads

In addition to the weather related loads, transmission line structures are designed for special loads that consider security and safety aspects of the line. These include security loads for preventing cascading type failures of the structures and construction and maintenance loads that are related to personnel safety.

15.2.5 Security Loads

Longitudinal loads may occur on the structures due to accidental events such as broken conductors, broken insulators, or collapse of an adjacent structure in the line due to an environmental event such as a tornado. Regardless of the triggering event, it is important that a line support structure be designed for a suitable longitudinal loading condition to provide adequate resistance against cascading type failures in which a larger number of structures fail sequentially in the longitudinal direction or parallel to the line. For this reason, longitudinal loadings are sometimes referred to as "anticascading", "failure containment", or "security loads".

There are two basic methods for reducing the risk of cascading failures, depending on the type of structure, and on local conditions and practices. These methods are: (1) design all structures for broken wire loads and (2) install stop structures or guys at specified intervals.

Design for Broken Conductors

Certain types of structures such as square-based lattice towers, 4-guyed structures, and single shaft steel poles have inherent longitudinal strength. For lines using these types of structures, the recommended practice is to design every structure for one broken conductor. This provides the additional longitudinal strength for preventing cascading failures at a relatively low cost.

Anchor Structures

When single pole wood structures or H-frame structures having low longitudinal strength are used on a line, designing every structure for longitudinal strength can be very expensive. In such cases, stop or anchor structures with adequate longitudinal strength are provided at specific intervals to limit the cascading effect. The Rural Electrification Administration [19] recommends a maximum interval of 5 to 10 miles between structures with adequate longitudinal capacity.

15.2.6 Construction and Maintenance Loads

Construction and maintenance (C&M) loads are, to a large extent, controllable and are directly related to construction and maintenance methods. A detailed discussion on these types of loads is included in the ASCE Loading Guide, and Occupation Safety and Health Act (OSHA) documents. It should be emphasized, however, that workers can be seriously injured as a result of structure overstress during C&M operations; therefore, personnel safety should be a paramount factor when establishing C&M loads. Accordingly, the ASCE Loading Guide recommends that the specified C&M loads be multiplied by a minimum load factor of 1.5 in cases where the loads are "static" and well defined; and by a load factor of 2.0 when the loads are "dynamic", such as those associated with moving wires during stringing operations.

15.2.7 Loads on Structure

Loads are calculated on the structures in three directions: vertical, transverse, and longitudinal. The transverse load is perpendicular to the line and the longitudinal loads act parallel to the line.

15.2.8 Vertical Loads

The vertical load on supporting structures consists of the weight of the structure plus the superimposed weight, including all wires, ice coated where specified.

Vertical load of wire V_w in. (lb/ft) is given by the following equations:

$$V_w = \text{wt. of bare wire } (lb/ft) + 1.24(d + I)I \qquad (15.1)$$

where
d = diameter of wire (in.)
I = ice thickness (in.)
Vertical wire load on structure (lb)

$$= Vw \times \text{ vertical design span } \times \text{ load factor} \qquad (15.2)$$

Vertical design span is the distance between low points of adjacent spans and is indicated in Figure 15.2.

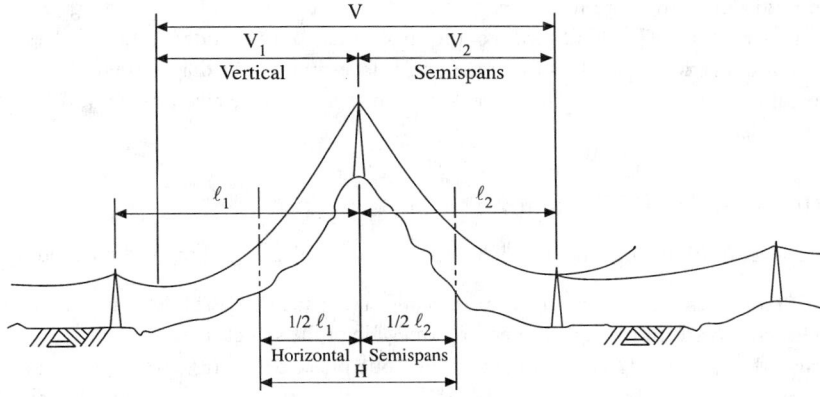

Figure 15.2 Vertical and horizontal design spans.

15.2.9 Transverse Loads

Transverse loads are caused by wind pressure on wires and structure, and the transverse component of the line tension at angles.

Wind Load on Wires

The transverse load due to wind on the wire is given by the following equations:

$$W_h = p \times d/12 \times \text{Horizontal Span} \times OCF \quad \text{(without ice)} \tag{15.3}$$

$$= p \times (d + 2I)/12 \times \text{Horizontal Span} \times OCF \quad \text{(with ice)} \tag{15.4}$$

where
W_h = transverse wind load on wire in lb
p = wind pressure in lb/ft^2
d = diameter of wire in in.
I = radial thickness of ice in in.
OCF = Overload Capacity Factor
Horizontal span is the distance between midpoints of adjacent spans and is shown in Figure 15.2.

Transverse Load Due to Line Angle

Where a line changes direction, the total transverse load on the structure is the sum of the transverse wind load and the transverse component of the wire tension. The transverse component of the tension may be of significant magnitude, especially for large angle structures. To calculate the total load, a wind direction should be used which will give the maximum resultant load considering the effects on the wires and structure.

The transverse component of wire tension on the structure is given by the following equation:

$$H = 2T \sin \theta/2 \tag{15.5}$$

where
H = transverse load due to wire tension in pounds
T = wire tension in pounds
θ = Line angle in degrees

Wind Load on Structures

In addition to the wire load, structures are subjected to wind loads acting on the exposed areas of the structure. The wind force coefficients on lattice towers depend on shapes of member sections, solidity ratio, angle of incidence of wind (face-on wind or diagonal wind), and shielding. Methods for calculating wind loads on transmission structures are given in the ASCE Guide as well the NESC code.

15.2.10 Longitudinal Loading

There are several conditions under which a structure is subjected to longitudinal loading:

Deadend Structures—These structures are capable of withstanding the full tension of the conductors and shield wires or combinations thereof, on one side of the structure.
Stringing— Longitudinal load may occur at any one phase or shield wire due to a hang-up in the blocks during stringing. The longitudinal load is taken as the stringing tension for the complete phase (i.e., all subconductors strung simultaneously) or a shield wire. In order to avoid any prestressing of the conductors, stringing tension is typically limited to the minimum tension required to keep

the conductor from touching the ground or any obstructions. Based on common practice and according to the IEEE "Guide to the Installation of Overhead Transmission Line Conductors" [4], stringing tension is generally about one-half of the sagging tension. Therefore, the longitudinal stringing load is equal to 50% of the initial, unloaded tension at 60°F.

Longitudinal Unbalanced Load—Longitudinal unbalanced forces can develop at the structures due to various conditions on the line. In rugged terrain, large differentials in adjacent span lengths, combined with inclined spans, could result in significant longitudinal unbalanced load under ice and wind conditions. Non-uniform loading of adjacent spans can also produce longitudinal unbalanced loads. This load is based on an ice shedding condition where ice is dropped from one span and not the adjacent spans. Reference [12] includes a software that is commonly used for calculating unbalanced loads on the structure.

EXAMPLE 15.1:   Problem

Determine the wire loads on a small angle structure in accordance with the data given below. Use NESC medium district loading and assume all intact conditions.
Given Data:

> Conductor: 954 kcm 45/7 ACSR
> Diameter = 1.165 in.
> Weight = 1.075 lb/ft
> Wire tension for NESC medium loading = 8020 lb

> Shield Wire: 3 No.6 Alumoweld
> Diameter = 0.349 in.
> Weight = 0.1781 lb/ft
> Wire tension for NESC medium loading = 2400 lb

> Wind Span = 1500 ft
> Weight Span = 1800 ft
> Line angle = 5°
> Insulator weight = 170 lb

Solution
NESC Medium District Loading

> 4 psf wind, 1/4-in. ice
> Ground Wire Iced Diameter = 0.349 + 2 × 0.25 = 0.849 in.
> Conductor Ice Diameter = 1.165 + 2 × 0.25 = 1.665 in.

Overload Capacity Factors for Steel

> Transverse Wind = 2.5
> Wire Tension = 1.65
> Vertical = 1.5

Conductor Loads On Tower

Transverse

$$\text{Wind} = 4\,\text{psf} \times 1.665''/12 \times 1500 \times 2.5 = 2080\,\text{lb}$$
$$\text{Line Angle} = 2 \times 8020 \times \sin 2.5° \times 1.65 = 1150\,\text{lb}$$
$$\text{Total} = 3230\,\text{lb}$$

Vertical

$$\text{Bare Wire} = 1.075 \times 1800 \times 1.5 = 2910\,\text{lb}$$
$$\text{Ice} = \{1.24(d+I)I\}1800 \times 1.5 = 1.24(1.165 + .25).25$$
$$\times\, 1800 \times 1.5 = 1185\,\text{lb}$$
$$\text{Insulator} = 170 \times 1.5 = 255\,\text{lb}$$
$$\text{Total} = 4350\,\text{lb}$$

Ground Wire Loads on Tower

Transverse

$$\text{Wind} = 4\,\text{psf} \times 0.849/12 \times 1500 \times 2.5 = 1060\,\text{lb}$$
$$\text{Line Angle} = 2 \times 2400 \times \sin 2.5 \times 1.65 = 350\,\text{lb}$$
$$\text{Total} = 1410\,\text{lb}$$

15.3   Design of Steel Lattice Tower

15.3.1   Tower Geometry

A typical single circuit, horizontal configuration, self-supported lattice tower is shown in Figure 15.3. The design of a steel lattice tower begins with the development of a conceptual design, which establishes the geometry of the structure. In developing the geometry, structure dimensions are established for the tower window, crossarms and bridge, shield wire peak, bracing panels, and the slope of the tower leg below the waist. The most important criteria for determining structure geometry are the minimum phase to phase and phase to steel clearance requirements, which are functions of the line voltage. Spacing of phase conductors may sometimes be dictated by conductor galloping considerations. Height of the tower peak above the crossarm is based on shielding considerations for lightning protection. The width of the tower base depends on the slope of the tower leg below the waist . The overall structure height is governed by the span length of the conductors between structures.

The lattice tower is made up of a basic body, body extension, and leg extensions. Standard designs are developed for these components for a given tower type. The basic body is used for all the towers regardless of the height. Body and leg extensions are added to the basic body to achieve the desired tower height.

The primary members of a tower are the leg and the bracing members which carry the vertical and shear loads on the tower and transfer them to the foundation. Secondary or redundant bracing members are used to provide intermediate support to the primary members to reduce their unbraced length and increase their load carrying capacity. The slope of the tower leg from the waist down has a significant influence on the tower weight and should be optimized to achieve an economical tower

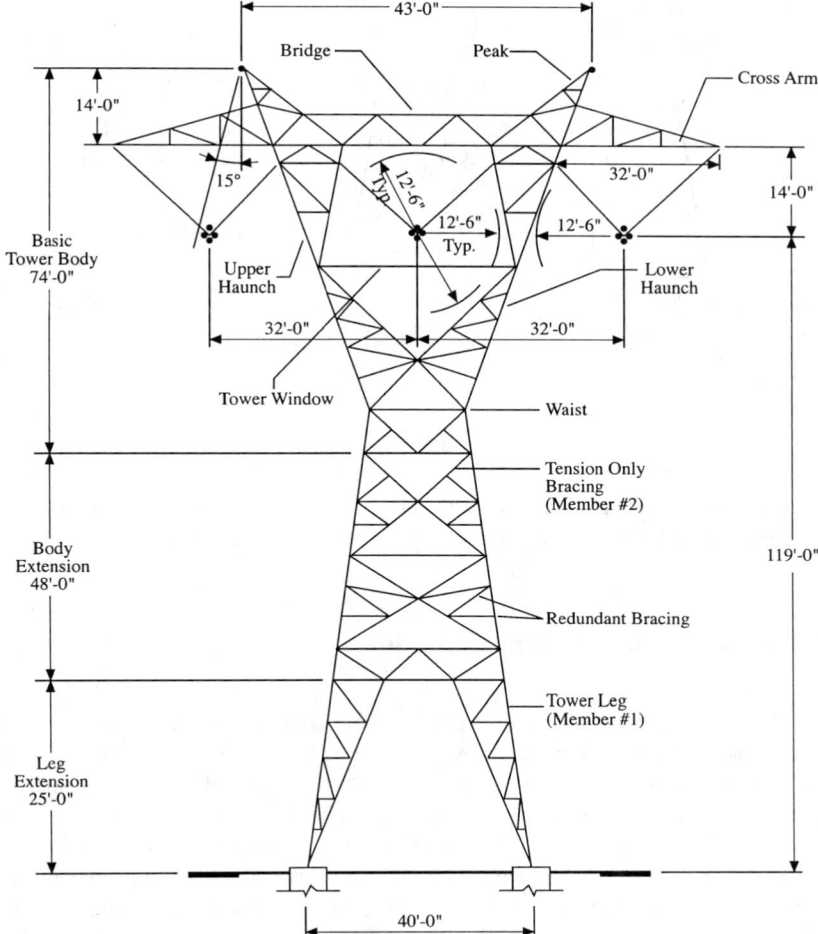

Figure 15.3   Single circuit lattice tower.

design. A flatter slope results in a wider tower base which reduces the leg size and the foundation size, but will increase the size of the bracing. Typical leg slopes used for towers range from 3/4 in. 12 for light tangent towers to 2 1/2 in. 12 for heavy deadend towers.

The minimum included angle ∞ between two intersecting members is an important factor for proper force distribution. Reference [3] recommends a minimum included angle of 15°, intended to develop a truss action for load transfer and to minimize moment in the member. However, as the tower loads increase, the preferred practice is to increase the included angle to 20° for angle towers and 25° for deadend towers [23].

Bracing members below the waist can be designed as a tension only or tension compression system as shown in Figure 15.4. In a tension only system shown in (a), the bracing members are designed to carry tension forces only, the compression forces being carried by the horizontal strut. In a tension/compression system shown in (b) and (c), the braces are designed to carry both tension and compression. A tension only system may prove to be economical for lighter tangent towers. But for heavier towers, a tension/compression system is recommended as it distributes the load equally to the tower legs.

A staggered bracing pattern is sometimes used on the adjacent faces of a tower for ease of connections and to reduce the number of bolt holes at a section. Tests [23] have shown that staggering of main bracing members may produce significant moment in the members especially for heavily

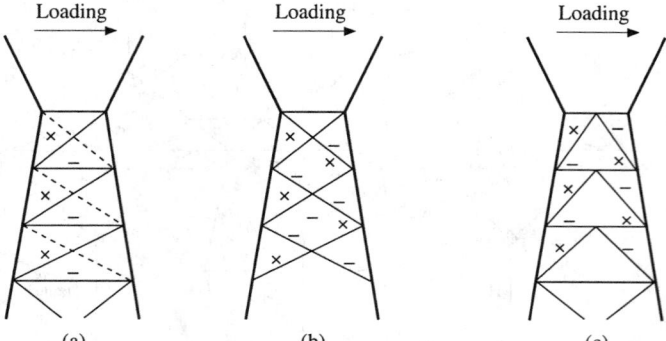

Figure 15.4   Bracing systems.

loaded towers. For heavily loaded towers, the preferred method is to stagger redundant bracing members and connect the main bracing members on the adjacent faces at a common panel point.

15.3.2   Analysis and Design Methodology

The ASCE Guide for Design of Steel Transmission Towers [3] is the industry document governing the analysis and design of lattice steel towers. A lattice tower is analyzed as a space truss. Each member of the tower is assumed pin-connected at its joints carrying only axial load and no moment. Today, finite element computer programs [12, 21, 17] are the typical tools for the analysis of towers for ultimate design loads. In the analytical model the tower geometry is broken down into a discrete number of joints (nodes) and members (elements). User input consists of nodal coordinates, member end incidences and properties, and the tower loads. For symmetric towers, most programs can generate the complete geometry from a part of the input. Loads applied on the tower are ultimate loads which include overload capacity factors discussed in Section 15.2. Tower members are then designed to the yield strength or the buckling strength of the member. Tower members typically consist of steel angle sections, which allow ease of connection. Both single- and double-angle sections are used. Aluminum towers are seldom used today due to the high cost of aluminum. Steel types commonly used on towers are ASTM A-36 ($Fy = 36$ ksi) or A-572 ($Fy = 50$ ksi). The most common finish for steel towers is hot-dipped galvanizing. Self-weathering steel is no longer used for towers due to the "pack-out" problems experienced in the past resulting in damaged connections.

Tower members are designed to carry axial compressive and tensile forces. Allowable stress in compression is usually governed by buckling, which causes the member to fail at a stress well below the yield strength of the material. Buckling of a member occurs about its weakest axis, which for a single angle section is at an inclination to the geometric axes. As the unsupported length of the member increases, the allowable stress in buckling is reduced.

Allowable stress in a tension member is the full yield stress of the material and does not depend on the member length. The stress is resisted by a net cross-section, the area of which is the gross area minus the area of the bolt holes at a given section. Tension capacity of an angle member may be affected by the type of end connection [3]. For example, when one leg of the angle is connected, the tension capacity is reduced by 10%. A further reduction takes place when only the short leg of an unequal angle is connected.

15.3.3 Allowable Stresses

Compression Member

The allowable compressive stress in buckling on the gross cross-sectional area of axially loaded compression members is given by the following equations [3]:

$$Fa = \left[1 - (KL/R)^2/(2Cc^2)\right] Fy \quad \text{if } KL/R = Cc \text{ or less} \tag{15.6}$$

$$Fa = 286000/(kl/r)^2 \quad \text{if } KL/R > Cc \tag{15.7}$$

$$Cc = (3.14)(2E/Fy)^{1/2} \tag{15.8}$$

where

Fa =   allowable compressive stress (ksi)
Fy =   yield strength (ksi)
E =   modulus of elasticity (ksi)
L/R =   maximum slenderness ratio = unbraced length /radius of gyration
K =   effective length co-efficient

The angle member must also be checked for local buckling considerations. If the ratio of the angle effective width to angle thickness (w/t) exceeds $80/(Fy)^{1/2}$, the value of Fa will be reduced in accordance with the provisions of Reference [3].

The above formulas indicate that the allowable buckling stress is largely dependent on the effective slenderness ratio (kl/r) and the material yield strength (Fy). It may be noted, however, that Fy influences the buckling capacity for short members only $(kl/r < Cc)$. For long members $(kl/r > Cc)$, the allowable buckling stress is unaffected by the material strength.

The slenderness ratio is calculated for different axes of buckling and the maximum value is used for the calculation of allowable buckling stress. In some cases, a compression member may have an intermediate lateral support in one plane only. This support prevents weak axis and in-plane buckling but not the out-of-plane buckling. In such cases, the slenderness ratio in the member geometric axis will be greater than in the member weak axis, and will control the design of the member.

The effective length coefficient K adjusts the member slenderness ratio for different conditions of framing eccentricity and the restraint against rotation provided at the connection. Values of K for six different end conditions, curves one through six, have been defined in Reference [3]. This reference also specifies maximum slenderness ratios of tower members, which are as follows:

Type of Member	Maximum KL/R
Leg	150
Bracing	200
Redundant	250

Tests have shown that members with very low L/R are subjected to substantial bending moment in addition to axial load. This is especially true for heavily loaded towers where members are relatively stiff and multiple bolted rigid joints are used [22]. A minimum L/R of 50 is recommended for compression members.

Tension Members

The allowable tensile force on the net cross-sectional area of a member is given by the following equation [3]:

$$P_t = Fy \cdot An \cdot K \tag{15.9}$$

where

P_t =   allowable tensile force (kips)
Fy =   yield strength of the material (ksi)

An = net cross-sectional area of the angle after deducting for bolt holes (in.^{2}). For unequal angles, if the short leg is connected, An is calculated by considering the unconnected leg to be the same size as the connected leg

K = 1.0 if both legs of the angle connected

   = 0.9 if one leg connected

The allowable tensile force must also meet the block shear criteria at the connection in accordance with the provisions of Reference [3].

Although the allowable force in a tension member does not depend on the member length, Reference [3] specifies a maximum L/R of 375 for these members. This limit minimizes member vibration under everyday steady state wind, and reduces the risk of fatigue in the connection.

15.3.4 Connections

Transmission towers typically use bearing type bolted connections. Commonly used bolt sizes are 5/8", 3/4", and 7/8" in diameter. Bolts are tightened to a snug tight condition with torque values ranging from 80 to 120 ft-lb. These torques are much smaller than the torque used in friction type connections in steel buildings. The snug tight torque ensures that the bolts will not slip back and forth under everyday wind loads thus minimizing the risk of fatigue in the connection. Under full design loads, the bolts would slip adding flexibility to the joint, which is consistent with the truss assumption.

Load carrying capacity of the bolted connections depends on the shear strength of the bolt and the bearing strength of the connected plate. The most commonly used bolt for transmission towers is A-394, Type 0 bolt with an allowable shear stress of 55.2 ksi across the threaded part. The maximum allowable stress in bearing is 1.5 times the minimum tensile strength of the connected part or the bolt. Use of the maximum bearing stress requires that the edge distance from the center of the bolt hole to the edge of the connected part be checked in accordance with the provisions of Reference [3].

15.3.5 Detailing Considerations

Bolted connections are detailed to minimize eccentricity as much as possible. Eccentric connections give rise to a bending moment causing additional shear force in the bolts. Sometimes small eccentricities may be unavoidable and should be accounted for in the design. The detailing specification should clearly specify the acceptable conditions of eccentricity.

Figure 15.5 shows two connections, one with no eccentricity and the second with a small eccentricity. In the first case the lines of force passing through the center of gravity (c.g.) of the members

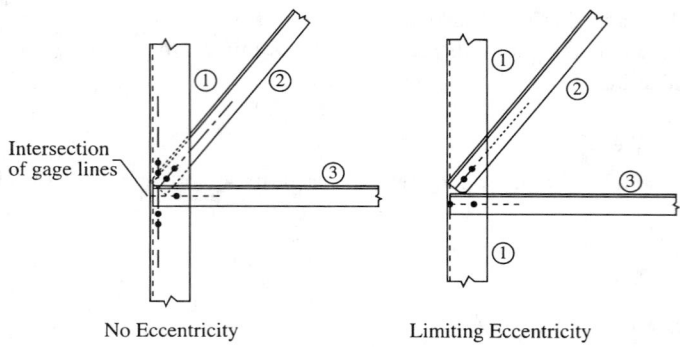

No Eccentricity   Limiting Eccentricity

Figure 15.5   Brace details.

intersect at a common point. This is the most desired condition producing no eccentricity. In the second case, the lines of force of the two bracing members do not intersect with that of the leg member thus producing an eccentricity in the connection. It is common practice to accept a small eccentricity as long as the intersection of the lines of force of the bracing members does not fall outside the width of the leg member. In some cases it may be necessary to add gusset plates to avoid large eccentricities.

In detailing double angle members, care should be taken to avoid a large gap between the angles that are typically attached together by stitch bolts at specified intervals. Tests [23] have shown that a double angle member with a large gap between the angles does not act as a composite member. This results in one of the two angles carrying significantly more load than the other angle. It is recommended that the gap between the two angles of a double angle member be limited to 1/2 in.

The minimum size of a member is sometimes dictated by the size of the bolt on the connected leg. The minimum width of members that can accommodate a single row of bolts is as follows:

Bolt diameter	Minimum width of member
5/8"	1 3/4"
3/4"	2"
7/8"	2 1/2"

Tension members are detailed with draw to facilitate erection. Members 15 ft in length, or less, are detailed 1/8 in. short, plus 1/16 in. for each additional 10 ft. Tension members should have at least two bolts on one end to facilitate the draw.

15.3.6 Tower Testing

Full scale load tests are conducted on new tower designs and at least the tangent tower to verify the adequacy of the tower members and connections to withstand the design loads specified for that structure. Towers are required to pass the tests at 100% of the ultimate design loads. Tower tests also provide insight into actual stress distribution in members, fit-up verification and action of the structure in deflected positions. Detailed procedures of tower testing are given in Reference [3].

EXAMPLE 15.2:

Description

Check the adequacy of the following tower components shown in Figure 15.3.

Member 1 (compressive leg of the leg extension)

$$\text{Member force} = 132 \text{ kips (compression)}$$
$$\text{Angle size} = L5 \times 5 \times 3/8"$$
$$F_y = 50 \text{ ksi}$$

Member 2 (tension member)

$$\text{Tensile force} = 22 \text{ kips}$$
$$\text{Angle size} = L2\ 1/2 \times 2 \times 3/16 \text{ (long leg connected)}$$
$$Fy = 36 \text{ ksi}$$

Bolts at the splice connection of Member 1

$$\text{Number of 5/8" bolts} = 6 \text{ (Butt Splice)}$$
$$\text{Type of bolt} = \text{A-394, Type O}$$

Solution

Member 1

$$\text{Member force} = 132 \text{ kips (compression)}$$
$$\text{Angle size} = L5 \times 5 \times 3/8"$$
$$F_y = 50 \text{ ksi}$$
$$\text{Find maximum } L/R$$
$$\text{Properties of } L \, 5 \times 5 \times 3/8"$$
$$\text{Area} = 3.61 \text{ in.}^2$$
$$r_x = r_y = 1.56 \text{ in.}$$
$$r_z = 0.99 \text{ in.}$$

Member 1 has the same bracing pattern in adjacent planes. Thus, the unsupported length is the same in the weak $(z - z)$ axis and the geometric axes $(x - x$ and $y - y)$.

$$l_z = l_x = l_y = 61"$$

Maximum $L/R = 61/0.99 = 61.6$
Allowable Compressive Stress:
Using Curve 1 for leg member (no framing eccentricity), per Reference [3], $k = 1.0$

$$
\begin{aligned}
KL/R &= L/R = 61.6 \\
Cc &= (3.14)(2E/Fy)^{1/2} \\
&= (3.14)(2 \times 29000/50)^{1/2} \\
&= 107.0 \text{ which is } > KL/R \\
Fa &= \left[1 - (KL/R)^2/(2Cc^2)\right] Fy \\
&= \left[1 - (61.6)^2/(2 \times 107.0^2)\right] 50.0 \\
&= 41.7 \text{ ksi}
\end{aligned}
$$

Allowable compressive load $= 41.7$ ksi $\times 3.61$ in.

$$= 150.6 \text{ kips} > 132 \text{ kips} \rightarrow \text{ O.K.}$$

Check local buckling:

$$
\begin{aligned}
w/t &= (5.0 - 7/8)/(3/8) = 11.0 \\
80/(Fy)^{1/2} &= 80/(50)^{1/2} = 11.3 > 11.0 \text{ O.K.}
\end{aligned}
$$

Member 2

$$\text{Tensile force} = 22 \text{ kips}$$
$$\text{Angle size} = L \, 2 - 1/2 \times 2 \times 3/16$$
$$\text{Area} = 0.81 \text{ in.}^2$$
$$Fy = 36 \text{ ksi}$$

Find tension capacity

$$P_t = Fy \cdot An \cdot K$$

$$\text{Diameter of bolt hole} = 5/8" + 1/16" = 11/16"$$

Assuming one bolt hole deduction in $2 - 1/2"$ leg width,

$$
\begin{aligned}
\text{Area of bolt hole} \quad &= \quad \text{angle th.} \times \text{hole diam.} \\
&= \quad (3/16)(11/16) = 0.128 \text{ in.}^2
\end{aligned}
$$

$$
\begin{aligned}
An \quad &= \quad \text{gross area} - \text{bolt hole area} \\
&= \quad 0.81 - 0.128 = 0.68 \text{ in.}^2 \\
K \quad &= \quad 0.9, \text{ since member end is connected by one leg} \\
P_t \quad &= \quad (36)(0.68)(0.9) = 22.1 \text{ kips} > 22.0 \text{ kips, O.K.}
\end{aligned}
$$

Bolts for Member 1

$$
\begin{aligned}
\text{Number of 5/8" bolts} &= 6 \text{ (Butt Splice)} \\
\text{Type of bolt} &= \text{A-394, Type O} \\
\text{Shear Strength } Fv &= 55.2 \text{ ksi} \\
\text{Root area thru threads} &= 0.202 \text{ in.}^2
\end{aligned}
$$

Shear capacity of bolts:

$$
\begin{aligned}
&\text{Bolts act in double shear at butt splice} \\
&\text{Shear capacity of 6 bolts in double shear} \\
&\quad = 2 \times (\text{Root area}) \times 55.2 \text{ ksi} \times 6 \\
&\quad = 133.8 \text{ kips} > 132 \text{ kips} \Rightarrow \text{ O.K.}
\end{aligned}
$$

Bearing capacity of connected part:

$$
\begin{aligned}
&\text{Thickness of connected angle} = 3/8" \\
&F_y \text{ of angle} = 50 \text{ ksi} \\
&\text{Capacity of bolt in bearing} \\
&\quad = 1.5 \times Fu \times \text{th. of angle} \times \text{dia. of bolt} \\
&Fu \text{ of 50 ksi material} = 65 \text{ ksi} \\
&\text{Capacity of 6 bolts in bearing} = 1.5 \times 65 \times 3/8 \times 5/8 \times 6 \\
&\quad = 137.1 \text{ kips} > 132 \text{ kips, O.K.}
\end{aligned}
$$

15.4 Transmission Poles

15.4.1 General

Transmission poles made of wood, steel, or concrete are used on transmission lines at voltages up to 345-kv. Wood poles can be economically used for relatively shorter spans and lower voltages whereas steel poles and concrete poles have greater strength and are used for higher voltages. For areas where severe climatic loads are encountered, steel poles are often the most cost-effective choice.

Pole structures have two basic configurations: single pole and H-frame (Figure 15.1). Single pole structures are used for lower voltages and shorter spans. H-frame structures consist of two poles connected by a framing comprised of the cross arm, the V-braces, and the X-braces. The use of X-braces significantly increases the load carrying capacity of H-frame structures.

At line angles or deadend conditions, guying is used to decrease pole deflections and to increase their transverse or longitudinal structural strength. Guys also help prevent uplift on H-frame structures. Large deflections would be a hindrance in stringing operations.

15.4.2 Stress Analysis

Transmission poles are flexible structures and may undergo relatively large lateral deflections under design loads. A secondary moment (or $P - \Delta$ effect) will develop in the poles due to the lateral deflections at the load points. This secondary moment can be a significant percent of the total moment. In addition, large deflections of poles can affect the magnitude and direction of loads caused by the line tension and stringing operations. Therefore, the effects of pole deflections should be included in the analysis and design of single and multi-pole transmission structures.

To properly analyze and design transmission structures, the standard industry practice today is to use nonlinear finite element computer programs. These computer programs allow efficient evaluation of pole structures considering geometric and/or material nonlinearities. For wood poles, there are several popular computer software programs available from EPRI [15]. They are specially developed for design and analysis of wood pole structures. Other general purpose commercial programs auch as SAP-90 and STAAD [20, 10] are available for performing small displacement $P - \Delta$ analysis.

15.4.3 Tubular Steel Poles

Steel transmission poles are fabricated from uniformly tapered hollow steel sections. The cross-sections of the poles vary from round to 16-sided polygonal with the 12-sided dodecagonal as the most common shape. The poles are formed into design cross-sections by braking, rolling, or stretch bending.

For these structures the usual industry practice is that the analysis, design, and detailing are performed by the steel pole supplier. This facilitates the design to be more compatible with fabrication practice and available equipment.

Design of tubular steel poles is governed by the ASCE Manual # 72 [9]. The Manual provides detailed design criteria including allowable stresses for pole masts and connections and stability considerations for global and local buckling. It also defines the requirements for fabrication, erection, load testing, and quality assurance.

It should be noted that steel transmission pole structures have several unique design features as compared to other tubular steel structures. First, they are designed for ultimate, or maximum anticipated loads. Thus, stress limits of the Manual #72 are not established for working loads but for ultimate loads.

Second, Manual #72 requires that stability be provided for the structure as a whole and for each structural element. In other words, the effects of deflected structural shape on structural stability should be considered in the evaluation of the whole structure as well as the individual element. It relies on the use of the large displacement nonlinear computer analysis to account for the $P - \Delta$ effect and check for stability. To prevent excessive deflection effects, the lateral deflection under factored loads is usually limited to 5 to 10% of the pole height. Pre-cambering of poles may be used to help meet the imposed deflection limitation on angle structures.

Lastly, due to its polygonal cross-sections combined with thin material, special considerations must be given to calculation of member section properties and assessment of local buckling.

To ensure a polygonal tubular member can reach yielding on its extreme fibers under combined axial and bending compression, local buckling must be prevented. This can be met by limiting the width to thickness ratio, w/t, to $240/(Fy)^{1/2}$ for tubes with 12 or fewer sides and $215/(Fy)^{1/2}$ for hexdecagonal tubes. If the axial stress is 1 ksi or less, the w/t limit may be increased to $260/(Fy)^{1/2}$ for tubes with 8 or fewer sides [9].

Special considerations should be given in the selection of the pole materials where poles are to be subjected to subzero temperatures. To mitigate potential brittle fracture, use of steel with good impact toughness in the longitudinal direction of the pole is necessary. Since the majority of pole structures are manufactured from steels of a yield strength of 50 to 65 ksi (i.e., ASTM A871 and A572), it is advantageous to specify a minimum Charpy-V-notch impact energy of 15 ft-lb at 0°F for plate thickness of 1/2 in. or less and 15 ft-lb at −20°F for thicker plates. Likewise, high strength anchor bolts made of ASTM A615-87 Gr.75 steel should have a minimum Charpy V-notch of 15 ft-lbs at −20°F.

Corrosion protection must be considered for steel poles. Selection of a specific coating or use of weathering steel depends on weather exposure, past experience, appearance, and economics. Weathering steel is best suited for environments involving proper wetting and drying cycles. Surfaces that are wet for prolonged periods will corrode at a rapid rate. A protective coating is required when such conditions exist. When weathering steel is used, poles should also be detailed to provide good drainage and avoid water retention. Also, poles should either be sealed or well ventilated to assure the proper protection of the interior surface of the pole. Hot-dip galvanizing is an excellent alternate means for corrosion protection of steel poles above grade. Galvanized coating should comply with ASTM A123 for its overall quality and for weight/thickness requirements.

Pole sections are normally joined by telescoping or slip splices to transfer shears and moments. They are detailed to have a lap length no less than 1.5 times the largest inside diameter. It is important to have a tight fit in slip joint to allow load transfer by friction between sections. Locking devices or flanged joints will be needed if the splice is subjected to uplift forces.

15.4.4 Wood Poles

Wood poles are available in different species. Most commonly used are Douglas Fir and Southern Yellow Pine, with a rupture bending stress of 8000 psi, and Western Red Cedar with a rupture bending stress of 6000 psi. The poles are usually treated with a preservative (pentachlorophenol or creosote). Framing materials for crossarm and braces are usually made of Douglas Fir or Southern Yellow Pine. Crossarms are typically designed for a rupture bending stress of 7400 psi.

Wood poles are grouped into a wide range of classes and heights. The classification is based on minimum circumference requirements specified by the American National Standard (ANSI) specification 05.1 for each species, each class, and each height [2]. The most commonly used pole classes are class 1, 2, 3, and H-1. Table 15.3 lists the moment capacities at groundline for these common classes of wood poles. Poles of the same class and length have approximately the same capacity regardless of the species.

The basic design principle for wood poles, as in steel poles, is to assure that the applied loads with appropriate overload capacity factors do not exceed the specified stress limits.

In the design of a single unguyed wood pole structure, the governing criteria is to keep the applied moments below the moment capacity of wood poles, which are assumed to have round solid sections. Theoretically the maximum stress for single unguyed poles under lateral load does not always occur at the ground line. Because all data have been adjusted to the ground line per ANSI 05.1 pole dimensions, only the stress or moment at the ground line need to be checked against the moment capacity. The total ground line moment is the sum of the moment due to transverse wire loads, the moment due to wind on pole, and the secondary moment. The moment due to the eccentric vertical load should also be included if the conductors are not symmetrically arranged.

TABLE 15.3   Moment Capacity at Ground Line for 8000 psi Douglas Fir and Southern Pine Poles

Class		H-1	1	2	3
Minimum circumference at top (in.)		29	27	25	23
Length of pole (ft)	Ground line distance from butt (ft)	Ultimate moment capacity, ft-lb			
50	7	220.3	187.2	152.1	121.7
55	7.5	246.4	204.2	167.1	134.7
60	8	266.8	222.3	183.0	148.7
65	8.5	288.4	241.5	200.0	163.5
70	9	311.2	261.9	218.1	179.4
75	9.5	335.3	283.4	230.3	190.2
80	10	360.6	306.2	250.2	201.5
85	10.5	387.2	321.5	263.7	213.3
90	11	405.2	337.5	285.5	225.5
95	11	438.0	357.3	303.2	—
100	11	461.5	387.3	321.5	
105	12	461.5	387.3	321.5	
110	12	514.2	424.1	354.1	

Design guidelines for wood pole structures are given in the REA (Rural Electrification Administration) Bulletin 62-1 [18] and IEEE Wood Transmission Structural Design Guide [15]. Because of the use of high overload factors, the REA and NESC do not require the consideration of secondary moments in the design of wood poles unless the pole is very flexible. It also permits the use of rupture stress. In contrast, IEEE requires the secondary moments be included in the design and recommends lower overload factors and use of reduction factors for computing allowable stresses. Designers can use either of the two standards to evaluate the allowable horizontal span for a given wood pole. Conversely, a wood pole can be selected for a given span and pole configuration.

For H-frames with X-braces, maximum moments may not occur at ground line. Sections at braced location of poles should also be checked for combined moments and axial loads.

15.4.5 Concrete Poles

Prestressed concrete poles are more durable than wood or steel poles and they are aesthetically pleasing. The reinforcing of poles consists of a spiral wire cage to prevent longitudinal cracks and high strength longitudinal strands for prestressing. The pole is spinned to achieve adequate concrete compaction and a dense smooth finish. The concrete pole typically utilizes a high strength concrete (around 12000 psi) and 270 ksi prestressing strands. Concrete poles are normally designed by pole manufacturers. The guideline for design of concrete poles is given in Reference [8]. Standard concrete poles are limited by their ground line moment capacity.

Concrete poles are, however, much heavier than steel or wood poles. Their greater weight increases transportation and handling costs. Thus, concrete poles are used most cost-effectively when there is a manufacturing plant near the project site.

15.4.6 Guyed Poles

At line angles and deadends, single poles and H-frames are guyed in order to carry large transverse loads or longitudinal loads. It is a common practice to use bisector guys for line angles up to 30° and in-line guys for structures at deadends or larger angles. The large guy tension and weight of conductors and insulators can exert significant vertical compression force on poles. Stability is therefore a main design consideration for guyed pole structures.

Structural Stability

The overall stability of guyed poles under combined axial compression and bending can be assessed by either a large displacement nonlinear finite element stress analysis or by the use of simplified approximate methods.

The rigorous stability analysis is commonly used by steel and concrete pole designers. The computer programs used are capable of assessing the structural stability of the guyed poles considering the effects of the stress-dependent structural stiffness and large displacements. But, in most cases, guys are modeled as tension-only truss elements instead of geometrically nonlinear cable elements. The effect of initial tension in guys is neglected in the analysis.

The simplified stability method is typically used in the design of guyed wood poles. The pole is treated as a strut carrying axial loads only and guys are to carry the lateral loads. The critical buckling load for a tapered guyed pole may be estimated by the Gere and Carter method [13].

$$Pcr = P(Dg/Da)^e \tag{15.10}$$

where P is the Euler buckling load for a pole with a constant diameter of Da at guy attachment and is equal to $9.87\ EI/(kl)^2$; Dg is the pole diameter at groundline; kl is the effective column length depending on end condition; e is an exponent constant equal to 2.7 for fixed-free ends and 2.0 for other end conditions. It should be noted that the exact end condition at the guyed attachment is difficult to evaluate. Common practice is to assume a hinged-hinged condition with k equal to 1.0. A higher k value should be chosen when there is only a single back guy.

For a pole guyed at multiple levels, the column stability may be checked as follows by comparing the maximum axial compression against the critical buckling load, Pcr, at the lowest braced location of the pole [15]:

$$[P1 + P2 + P3 + \cdots]/Pcr < 1/OCF \tag{15.11}$$

where OCF is the overload capacity factor and $P1$, $P2$, and $P3$ are axial loads at various guy levels.

Design of Guys

Guys are made of strands of cable attached to the pole and anchor by shackles, thimbles, clips, or other fittings. In the tall microwave towers, initial tension in the guys is normally set between 8 to 15% of the rated breaking strength (RBS) of the cable. However, there is no standard initial tension specified for guyed transmission poles. Guys are installed before conductors and ground wires are strung and should be tightened to remove slack without causing noticeable pole deflections. Initial tension in guys are normally in the range of 5 to 10% of RBS. For design of guys, the maximum tension under factored loads per NESC shall not exceed 90% of the cable breaking strength. Note that for failure containment (broken conductors) the guy tension may be limited to 0.85 RBS. A lower allowable of 65% of RBS would be needed if a linear load-deformation behavior of guyed poles is desired for extreme wind and ice conditions per ASCE Manual #72.

Considerations should be given to the range of ambient temperatures at the site. A large temperature drop may induce a significant increase of guy tension. Guys with an initial tension greater than 15% of RBS of the guy strand may be subjected to aeolian vibrations.

EXAMPLE 15.3:

Description

Select a Douglas Fir pole unguyed tangent structure shown below to withstand the NESC heavy district loads. Use an OCF of 2.5 for wind and 1.5 for vertical loads and a strength reduction factor of 0.65. Horizontal load span is 400 ft and vertical load span is 500 ft. Examine both cases with and without the $P - \Delta$ effect. The NESC heavy loading is 0.5 in. ice, 4 psf wind, and 0°F.

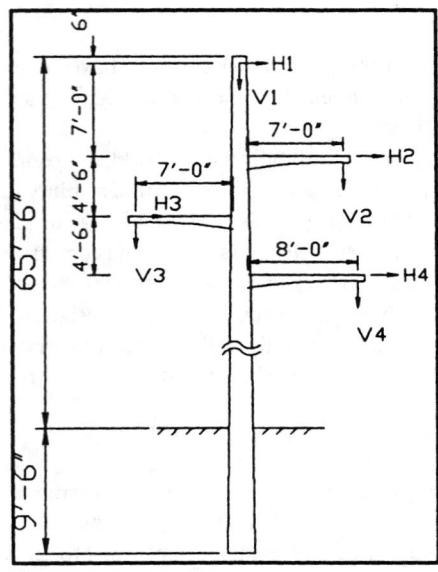

Ground Wire Loads

$$H1 = 0.453\#/\text{ft} \quad V1 = 0.807\#/\text{ft}$$

Conductor Loads

$$H2 = H3 = H4 = 0.732\#/\text{ft}$$
$$V2 = V3 = V4 = 2.284\#/\text{ft}$$
$$\text{Horizontal Span} = 400 \text{ ft}$$
$$\text{Vertical Span} = 500 \text{ ft}$$
$$\text{Line Angle} = 0°$$

Solution   A 75-ft class 1 pole is selected as the first trial. The pole will have a length of 9.5 ft buried below the groundline. The diameter of the pole is 9.59 in. at the top (Dt) and 16.3 in. at the groudline (Dg). Moment at groundline due to transverse wind on wire loads is

$$Mh = (0.732)(2.5)(400)(58 + 53.5 + 49) + (0.453)(2.5)(400)(65) = 146930 \text{ ft-lbs}$$

Moment at groundline due to vertical wire loads

$$Mv = (2.284)(1.5)(500)(8 + 7 - 7) = 13700 \text{ ft-lbs}$$

Moment due to 4 psf wind on pole

$$
\begin{aligned}
Mw &= \text{(wind pressure)} \ (OCF)H^2(Dg + 2Dt)/72 \\
   &= (4)(2.5)(65.5)^2(16.3 + 9.59 \times 2)/72 = 21140 \text{ ft-lbs}
\end{aligned}
$$

The total moment at groundline

$$Mt = 146930 + 13700 + 21140 = 181770 \text{ ft-lbs or } 181.7 \text{ ft-kips}$$

This moment is less than the moment capacity of the 75-ft class 1 pole, 184.2 ft-kips (i.e., 0.65 × 283.4, refer to Table 15.3). Thus, the 75-ft class 1 pole is adequate if the $P - \Delta$ effect is ignored.

To include the effect of the pole displacement, the same pole was modeled on the SAP-90 computer program using a modulus of elasticity of 1920 ksi. Under the factored NESC loading, the maximum displacement at the top of the pole is 67.9 in. The associated secondary moment at the groundline is 28.5 ft-kips, which is approximately 15.7% of the primary moment. As a result, a 75-ft class H1 Douglas Fir pole with an allowable moment of 217.9 ft-kips is needed when the $P - \Delta$ effect is considered.

15.5 Transmission Tower Foundations

Tower foundation design requires competent engineering judgement. Soil data interpretation is critical as soil and rock properties can vary significantly along a transmission line. In addition, construction procedures and backfill compaction greatly influence foundation performance.

Foundations can be designed for site specific loads or for a standard maximum load design. The best approach is to use both a site specific and standardized design. The selection should be based on the number of sites that will have a geotechnical investigation, inspection, and verification of soil conditions.

15.5.1 Geotechnical Parameters

To select and design the most economical type of foundation for a specific location, soil conditions at the site should be known through existing site knowledge or new explorations. Inspection should also be considered to verify that the selected soil parameters are within the design limits. The subsurface investigation program should be consistent with foundation loads, experience in the right-of-way conditions, variability of soil conditions, and the desired level of reliability.

In designing transmission structure foundations, considerations must be given to frost penetration, expansive or shrinking soils, collapsing soils, black shales, sinkholes, and permafrost. Soil investigation should consider the unit weight, angle of internal friction, cohesion, blow counts, and modulus of deformation. The blow count values are correlated empirically to the soil value. Lab tests can measure the soil properties more accurately especially in clays.

15.5.2 Foundation Types—Selection and Design

There are many suitable types of tower foundations such as steel grillages, pressed plates, concrete footings, precast concrete, rock foundations, drilled shafts with or without bells, direct embedment, pile foundations, and anchors. These foundations are commonly used as support for lattice, poles, and guyed towers. The selected type depends on the cost and availability [14, 24].

Steel Grillages

These foundations consist entirely of steel members and should be designed in accordance with Reference [3]. The surrounding soil should not be considered as bracing the leg. There are pyramid arrangements that transfer the horizontal shear to the base through truss action. Other types transfer the shear through shear members that engage the lateral resistance of the compacted backfill. The steel can be purchased with the tower steel and concrete is not required at the site.

Cast in Place Concrete

Cast in place concrete foundation consists of a base mat and a square of cylindrical pier. Most piers are kept in vertical position. However, the pier may be battered to allow the axial loads in the tower legs to intersect the mat centroid. Thus, the horizontal shear loads are greatly reduced for deadends and large line angles. Either stub angles or anchor bolts are embedded in the top of the pier so that the upper tower section can be spliced directly to the foundation. Bolted clip angles, welded stud shear connectors, or bottom plates are added to the stub angle. This type can also be precast elsewhere and delivered to the site. The design is accomplished by Reference [1].

Drilled Concrete Shafts

The drilled concrete shaft is the most common type of foundation now being used to support transmission structures. The shafts are constructed by power auguring a circular excavation, placing the reinforcing steel and anchor, and pouring concrete. Tubular steel poles are attached to the shafts using base plates welded to the pole with anchor bolts embedded in the foundation (Figure 15.6a). Lattice towers are attached through the use of stub angles or base plates with anchor bolts. Loose

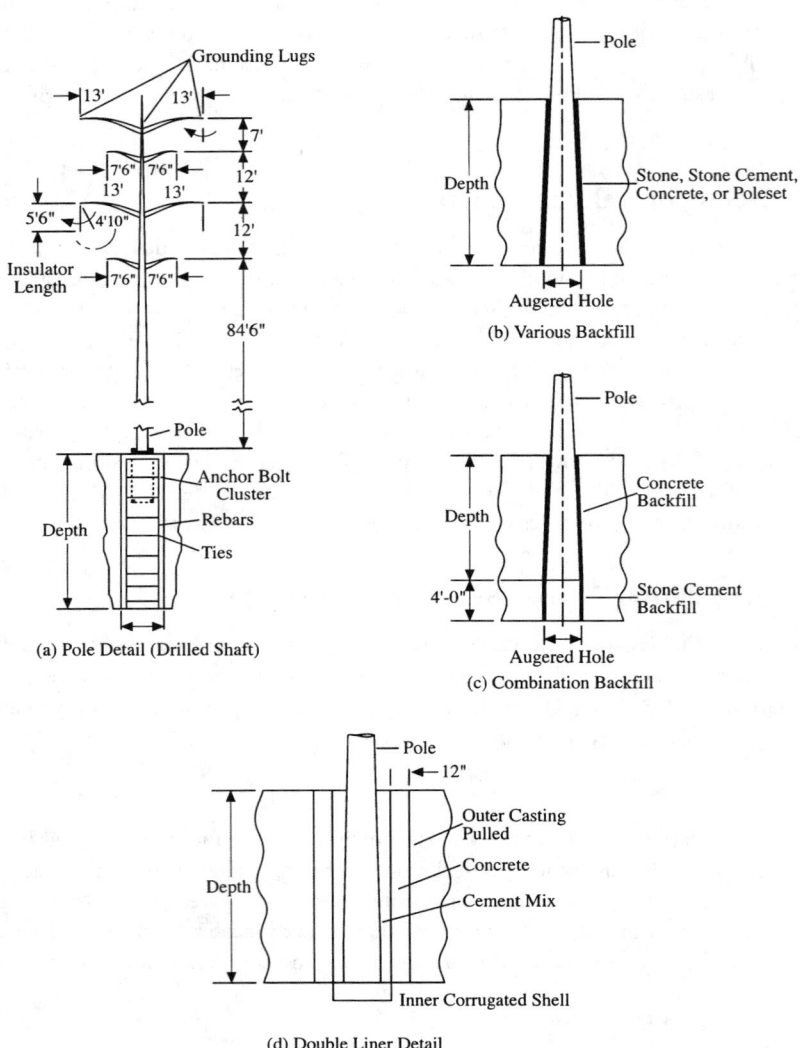

Figure 15.6   Direct embedment.

granular soil may require a casing or a slurry. If there is a water level, tremi concrete is required. The casing, if used, should be pulled as the concrete is poured to allow friction along the sides.

A minimum 4" slump should allow good concrete flow. Belled shafts should not be attempted in granular soil.

If conditions are right, this foundation type is the fastest and most economical to install as there is no backfilling required with dependency on compaction. Lateral procedures for design of drilled shafts under lateral and uplift loads are given in References [14] and [25].

Rock Foundations

If bedrock is close to the surface, a rock foundation can be installed. The rock quality designation (RQD) is useful in evaluating rock. Uplift capacity can be increased with drilled anchor rods or by shaping the rock. Blasting may cause shatter or fracture to rock. Drilling or power hammers are therefore preferred. It is also helpful to wet the hole before placing concrete to ensure a good bond.

Direct Embedment

Direct embedment of structures is the oldest form of foundation as it has been used on wood pole transmission lines since early times. Direct embedment consists of digging a hole in the ground, inserting the structure into the hole, and backfilling. Thus, the structure acts as its own foundation transferring loads to the *in situ* soil via the backfill. The backfill can be a stone mix, stone-cement mix, excavated material, polyurethane foam, or concrete (see Figure 15.6b and c). The disadvantage of direct embedment is the dependency on the quality of backfill material. To accurately get deflection and rotation of direct embedded structures, the stiffness of the embedment must be considered. Rigid caisson analysis will not give accurate results. The performance criteria for deflection should be for the combined pole and foundation. Instability of the augured hole and the presence of water may require a liner or double liners (see Figure 15.6d). The design procedure for direct embedment is similar to drilled shafts [14, 25, 16].

Vibratory Shells

Steel shells are installed by using a vibratory hammer. The top 6 or 8 ft (similar to slip joint requirements) of soil inside the shell is excavated and the pole is inserted. The annulus is then filled with a high strength non-shrink grout. The pole can also be attached through a flange connection which eliminates excavating and grouting. The shell design is similar to drilled shafts.

Piles

Piles are used to transmit loads through soft soil layers to stiffer soils or rock. The piles can be of wood, prestressed concrete, cast in place concrete, concrete filled shells, steel H piles, steel pipes filled with concrete, and prestressed concrete cylinder piles. The pipe selection depends on the loads, materials, and cost. Pile foundations are normally used more often for lattice towers than for H-framed structures or poles because piles have high axial load capacity and relatively low shear and bending capacity.

Besides the external loading, piles can be subjected to the handling, drying, and soil stresses. If piles are not tested, the design should be conservative. Reference [14] should be consulted for bearing, uplift, lateral capacity, and settlement. Driving formulas can be used to estimate dynamic capacity of the pile or group. Timber piles are susceptible to deterioration and should be treated with a preservative.

Anchors

Anchors are usually used to support guyed structures. The uplift capacity of rock anchors depends on the quality of the rock, the bond of the grout and rock with steel, and the steel strength. The uplift capacity of soil anchors depends on the resistance between grout and soil and end bearing if applicable. Multi-belled anchors in cohesive soil depend on the number of bells. The capacity of Helix anchors can be determined by the installation torque developed by the manufacturer. Spread anchor plate anchors depend on the soil weight plus the soil resistance.

Anchors provide resistance to upward forces. They may be prestressed or deadman anchors. Deadmen anchors are not loaded until the structure is loaded, while prestressed anchors are loaded when installed or proof loaded.

Helix soil anchors have deformed plates installed by rotating the anchor into the ground with a truck-mounted power auger. The capacity of the anchor is correlated to the amount of torque. Anchors are typically designed in accordance with the procedure given in Reference [14].

15.5.3 Anchorage

Anchorage of the transmission tower can consist of anchor bolts, stub angles with clip angles, or shear connectors and designed by Reference [3]. The anchor bolts can be smooth bars with a nut or head at the bottom, or deformed reinforcing bars with the embedment determined by Reference [1]. If the anchor bolt base plate is in contact with the foundation, the lateral or shear load is transferred to the foundation by shear friction. If there is no contact between the base plate and the concrete (anchor bolts with leveling nuts), the lateral load is transferred to the concrete by the side bearing of the anchor bolt. Thus, anchor bolts should be designed for a combination of tension (or compression), shear, and bending by linear interaction.

15.5.4 Construction and Other Considerations

Backfill

Excavated foundations require a high level of compaction that should be inspected and tested. During the original design the degree of compaction that may actually be obtained should be considered. This construction procedure of excavation and compaction increases the foundation costs.

Corrosion

The type of soil, moisture, and stray electric currents could cause corrosion of metals placed below the ground. Obtaining resistivity measurements would determine if a problem exists. Consideration could then be given to increasing the steel thickness, a heavier galvanizing coat, a bituminous coat, or in extreme cases a cathodic protection system. Hard epoxy coatings can be applied to steel piles. In addition, concrete can deteriorate in acidic or high sulfate soils.

15.5.5 Safety Margins for Foundation Design

The NESC requires the foundation design loads to be taken the same as NESC load cases used for design of the transmission structures. The engineer must use judgement in determining safety factors depending on the soil conditions, importance of the structures, and reliability of the transmission line. Unlike structural steel or concrete, soil does not have well-defined properties. Large variations exist in the geotechnical parameters and construction techniques. Larger safety margins should be provided where soil conditions are less uniform and less defined.

Although foundation design is based on ultimate strength design, there is no industry standard on strength reduction factors at present. The latest research [11] shows that uplift test results differed significantly from analytical predictions and uplift capacity. Based on a statistical analysis of 48 uplift tests on drilled piers and 37 tests on grillages and plates, the coefficients of variation were found to be approximately 30%. To achieve a 95% reliability, which is a 5% exclusion limit, an uplift strength reduction factor of 0.8 to 0.9 is recommended for drilled shafts and 0.7 to 0.8 for backfilled types of foundations.

15.5.6 Foundation Movements

Foundation movements may change the structural configuration and cause load redistribution in lattice structures and framed structures. For pole structures a small foundation movement can induce a large displacement at the top of the pole which will reduce ground clearance or cause problems in wire stringing. The amount of tolerable foundation settlements depends on the structure type and load conditions. However, there is no industry standard at the present time. For lattice structures, it is suggested that the maximum vertical foundation movement be limited to 0.004 times the base dimensions. If larger movements are expected, foundations can be designed to limit their movements or the structures can be designed to withstand the specified foundation movements.

15.5.7 Foundation Testing

Transmission line foundations are load tested to verify the foundation design for specific soils, adequacy of the foundation, research investigation, and to determine strength reduction factors. The load tests will refine foundation selection and verify the soil conditions and construction techniques. The load tests may be in uplift, download, lateral loads, overturning moment, or any necessary combination.

There should also be a geotechnical investigation at the test site to correlate the soil data with other locations. There are various test set-ups, depending on what type of loading is to be applied and what type of foundation is to be tested. The results should compare the analytical methods used to actual behaviors. The load vs. the foundation movements should be plotted in order to evaluate the foundation performance.

15.5.8 Design Examples

EXAMPLE 15.4: Spread Footing

Problem—Determine the size of a square spread footing for a combined moment (175 ft-k) and axial load (74 kips) using two alternate methods. In the first method, the minimum factor of safety against overturning is 1.7 and the maximum soil pressure is kept below an allowable soil bearing of 4000 psf. In the second method, no factor of safety against overturning is specified. Instead, the spread footing is designed so that the resultant reaction is within the middle third. This example shows that keeping the resultant in the middle third is a conservative design.

Solution

Method 1

Try a 8 ft x 8 ft footing

$$P = 74 \text{ kips}$$
$$Mo = 175 \text{ kip-ft}$$

P increase for footing size increase $= 0.3$ kips/ft^2

$$e = 175 \text{ k-ft}/74 \text{ kips} = 2.4 \text{ ft} > 8 \text{ ft}/6 = 1.33 \text{ ft}$$

Therefore, resultant is outside the middle third of the mat.

$$(4' - 2.4') \times 3 = 4.8 \text{ ft}$$
$$\text{S.P.} = (74 \text{ k})(2)/(4.8 \text{ ft})(8 \text{ ft}) = 3850 \text{ psf} < 4000 \text{ psf}$$
$$M_R = (74 \text{ k})(4 \text{ ft}) = 296 \text{ k-ft}$$
$$M_R/Mo = 296/175 = 1.7$$

FOS against overturning, O.K.

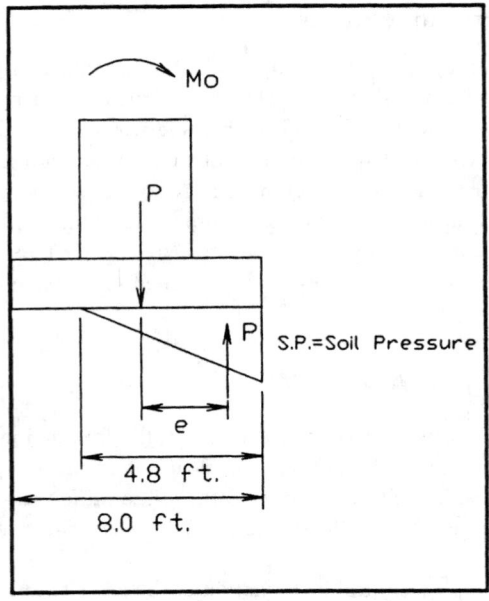

Method 2 (increase mat size to keep the resultant in the middle third) Try a 11.3 ft x 11.3 ft mat

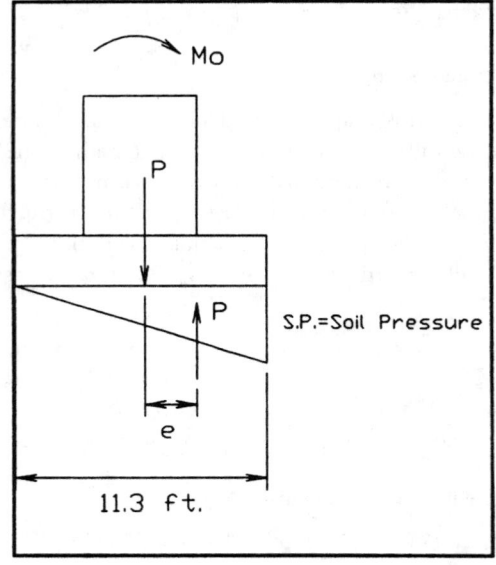

$$P \text{ increase } = \left[(11.3 \text{ ft})^2 - (8 \text{ ft})^2\right] \times 0.3 \text{ k/ft}^2 = 19.1 \text{ kips}$$
$$e = 175 \text{ k-ft}/(74 + 19.1) \text{ kips} = 1.88 \text{ ft} = 11.3 \text{ ft}/6$$

Resultant is within middle third.

$$\text{S.P.} = (93.1 \text{ k})(2)/(11.3)^2 = 1460 \text{ lbs/ft}^2 < 4000 \text{ lbs/ft}^2$$

Therefore, O.K.

Increase in mat size $= (11.3/8)^2 = 1.99$

Therefore, mat size has doubled, assuming that the mat thickness remains the same.

EXAMPLE 15.5:   Design of a Drilled Shaft

Problem—Determine the depth of a 5-ft diameter drilled shaft in cohesive soil with a cohesion of 1.25 ksf by both Broms and modified Broms methods. The foundation is subjected to a combined moment of 2000 ft-k and a shear of 20 kips under extreme wind loading. Manual calculation by Broms method is shown herein while the modified Broms method is made by the use of a computer program (CADPRO) [25], which determines the depth required, lateral displacement, and rotation of the foundation top. Calculations are made for various factors of safety (or strength reduction factor). The equations used in this example are based on Reference [25].

Foundation in Cohesive Soil:

$$M = 2000 \text{ ft-kips}$$
$$V = 20 \text{ kips}$$

Cohesion:

$$C = 1.25 \text{ ksf}$$
$$D = 5"$$

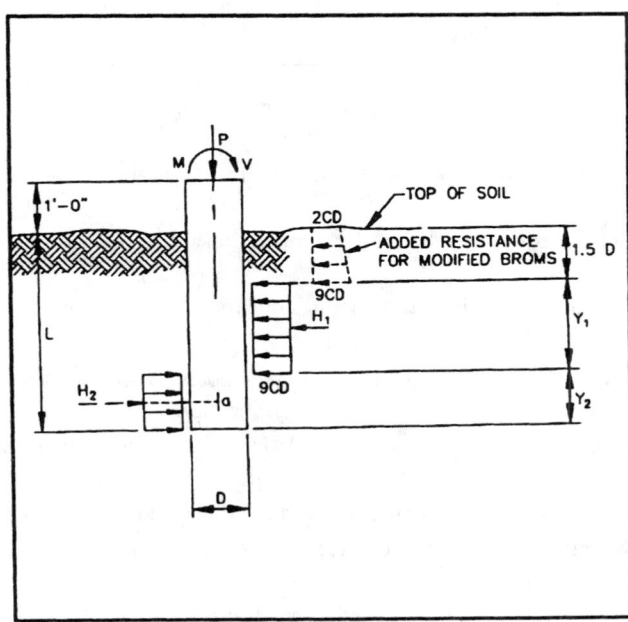

Solution

1. Use Broms Method [14]

$$M = 2000 + 20 \times 1$$
$$= 2020 \text{ ft-k}$$
$$H = M/V = 2020/20 = 101$$
$$q = V/(9\,CD) = 20/(9)(1.25)(5) = 0.356$$
$$L = 1.5D + q\left[1 + (2 + (4H + 6D)/q)^{0.5}\right]$$
$$= (1.5)(5) + .356\left[1 + (2 + (4)(101) + (6)(5))/0.356)^{0.5}\right] = 20.3 \text{ ft}$$

2. Comparison of Results of Broms Method and Modified Broms Method.

			Depth from	Modified Brom		
FOS	Φ	C used	broms (ft)	D (ft)	Δ	θ
1.0	1.0	1.25	20.3	19	.935	.457
1.33	0.75	.9375	22.3	19.5	.89	.474
1.5	0.667	.833	23.2	20.5	.81	.366
1.75	0.575	.714	24.6	23.0	.653	.262
2.0	0.5	.625	25.8	24.0	.603	.23

where

FOS	=	factor of safety
Φ	=	strength reduction factor
Δ	=	displacement, in.
θ	=	rotation, degrees

3. Conclusions

This example demonstrates that the modified Broms method provides a more economical design than the Broms method. It also shows that as the depth increases by 26%, the factor of safety increased from 1.0 to 2.0. The cost will also increase proportionally.

15.6   Defining Terms

Bearing connection: Shear resistance is provided by bearing of bolt against the connected part.

Block shear: A combination of shear and tensile failure through the end connection of a member.

Buckling: Mode of failure of a member under compression at stresses below the material yield stress.

Cascading effect: Progressive failure of structures due to an accident event.

Circuit: A system of usually three phase conductors.

Eccentric connection: Lines of force in intersecting members do not pass through a common work point, thus producing moment in the connection.

Galloping: High amplitude, low frequency oscillation of snow covered conductors due to wind on uneven snow formation.

Horizontal span: The horizontal distance between the midspan points of adjacent spans.

Leg and bracing members: Tension or compression members which carry the loads on the structure to the foundation.

Line angle: Denotes the change in the direction of a transmission line.

Line tension: The longitudinal tension in a conductor or shield wire.

Longitudinal load: Load on the supporting structure in a direction parallel to the line.

Overload capacity factor: A multiplier used with the unfactored load to establish the design factored load.

Phase conductors: Wires or cables intended to carry electric currents, extending along the route of the transmission line, supported by transmission structures.

Redundant member: Members that reduce the enbraced length of leg or brace members by providing intermediate support.

Sag: The distance measured vertically from a conductor to the straight line joining its two points of support.

Self supported structure: Unguyed structure supported on its own foundation.

Shear friction: A mechanism to transfer the shear force at anchor bolts to the concrete through wedge action and tension of anchor.

Shield wires: Wires installed on transmission structures intended to protect phase conductors against lightning strokes.

Slenderness ratio: Ratio of the member unsupported length to its least radius of gyration.

Span length: The horizontal distance between two adjacent supporting structures.

Staggered bracing: Brace members on adjoining faces of a lattice tower are not connected to a common point on the leg.

Stringing: Installation of conductor or shield wire on the structure.

Transverse load: Load on the supporting structure in a direction perpendicular to the line.

Uplift load: Vertically upward load at the wire attachment to the structure.

Vertical span: The horizontal distance between the maximum sag points of adjacent spans.

Voltage: The effective potential difference between any two conductors or between a conductor and ground.

References

[1] ACI Committee 318, 1995, *Building Code Requirements for Reinforced Concrete with Commentary,* American Concrete Institute (ACI), Detroit, MI.

[2] ANSI, 1979, *Specification and Dimensions for Wood Poles,* ANSI 05.1, American National Standard Institute, New York.

[3] ANSI/ASCE, 1991, *Design of Steel Latticed Transmission Structures,* Standard 10-90, American National Standard Institute and American Society of Civil Engineers, New York. (Former ASCE Manual No. 52).

[4] ANSI/IEEE, 1992, *IEEE Guide to the Installation of Overhead Transmission Line Conductors,* Standard 524, American National Structure Institute and Institute of Electrical and Electronic Engineers, New York.

[5] ANSI/IEEE, 1993, *National Electrical Safety Code,* Standard C2, American National Standard Institute and Institute of Electrical and Electronic Engineers, New York.

[6] ASCE, 1984, *Guideline for Transmission Line Structural Loading,* Committee on Electrical Transmission Structures, American Society of Civil Engineers, New York.

[7] ASCE, 1986, Innovations in the Design of Electrical Transmission Structures, *Proc. Conf. Struct. Div. Am. Soc. Civil Eng.,* New York.

[8] ASCE, 1987, *Guide for the Design and Use of Concrete Pole,* American Society of Civil Engineers, New York.

[9] ASCE, 1990, *Design of Steel Transmission Pole Structures,* ASCE Manual No. 72, Second ed., American Society of Civil Engineers, New York.

[10] CSI, 1992, *SAP90—A Series of Computer Programs for the Finite Element Analysis of Structures—Structural Analysis User's Manual,* Computer and Structures, Berkeley, CA.

[11] EPRI, 1983, Transmission Line Structure Foundations for Uplift-Compression Loading: Load Test Summaries, EPRI Report EL-3160, Electric Power Research Institute, Palo Alto, CA.

[12] EPRI, 1990, *T.L. Workstation Code,* EPRI (Electric Power Research Institute), Report EL-6420, Vol. 1-23, Palo Alto, CA.

[13] Gere, J.M. and Carter, W.O., 1962, Critical Buckling Loads for Tapered Columns, *J. Struct. Div.,* ASCE, 88(ST1), 112.

[14] IEEE, 1985, *IEEE Trial-Use Guide for Transmission Structure Foundation Design,* Standard 891, Institute of Electrical and Electronics Engineers, New York.

[15] IEEE, 1991, *IEEE Trial-Use Guide for Wood Transmission Structures,* IEEE Standard 751, Institute of Electrical and Electronic Engineers, New York.

[16] Kramer, J. M., 1978, Direct Embedment of Transmission Structures, Sargent & Lundy Transmission and Substation Conference, Chicago, IL.

[17] Peyrot, A.H., 1985, Microcomputer Based Nonlinear Structural Analysis of Transmission Line Systems, *IEEE Trans. Power Apparatus and Systems,* PAS-104 (11).

[18] REA, 1980, *Design Manual for High Voltage Transmission Lines,* Rural Electrification Administration (REA) Bulletin 62-1.

[19] REA, 1992, Design Manual for High Voltage Transmission Lines, Rural Electrification Administration (REA), Bulletin 1724E-200.

[20] REI, 1993, *Program STAAD-III—Structural Analysis and Design—User's Manual,* Research Engineers, Orange, CA.

[21] Rossow, E.C., Lo, D., and Chu, S.L, 1975, Efficient Design-Analysis of Physically Nonlinear Trusses, *J. Struct. Div.,* 839-853, ASCE, New York.

[22] Roy, S., Fang, S., and Rossow, E.C., 1982, Secondary Effects of Large Defection in Transmission Tower Structures, *J. Energy Eng.,* ASCE, 110-2, 157-172.

[23] Roy, S. and Fang, S., 1993, Designing and Testing Heavy Dead-End Towers, *Proc. Am. Power Conf.,* 55-I, 839-853, ASCE, New York.

[24] Simpson, K.D. and Yanaga, C.Y., 1982, Foundation Design Considerations for Transmission Structure, Sargent & Lundy Transmission and Distribution Conference, Chicago, IL.

[25] Simpson, K.D., Strains, T.R., et. al., 1992, Transmission Line Computer Software: The New Generation of Design Tool, Sargent & Lundy Transmission and Distribution Conference, Chicago, IL.

Section III
SPECIAL TOPICS

16

Performance-Based Seismic Design Criteria For Bridges

Notations 16-1
16.1 Introduction 16-5
      Damage to Bridges in Recent Earthquakes • No-Collapse-Based
      Design Criteria • Performance-Based Design Criteria • Back-
      ground of Criteria Development
16.2 Performance Requirements 16-10
      General • Safety Evaluation Earthquake • Functionality Evalu-
      ation Earthquake • Objectives of Seismic Design
16.3 Loads and Load Combinations 16-12
      Load Factors and Combinations • Earthquake Load • Wind
      Load • Buoyancy and Hydrodynamic Mass
16.4 Structural Materials 16-13
      Existing Materials • New Materials
16.5 Determination of Demands............................. 16-14
      Analysis Methods • Modeling Considerations
16.6 Determination of Capacities 16-16
      Limit States and Resistance Factors • Effective Length of Com-
      pression Members • Nominal Strength of Steel Structures •
      Nominal Strength of Concrete Structures • Structural Defor-
      mation Capacity • Seismic Response Modification Devices
16.7 Performance Acceptance Criteria 16-26
      General • Structural Component Classifications • Steel Struc-
      tures • Concrete Structures • Seismic Response Modification
      Devices
Defining Terms 16-35
Acknowledgments 16-36
References... 16-36
Further Reading 16-38
Appendix A 16-38
16.A.1 Section Properties for Latticed Members 16-38
16.A.2 Buckling Mode Interaction For Compression Built-up
       members 16-43
16.A.3 Acceptable Force D/C Ratios and Limiting Values 16-45
16.A.4 Inelastic Analysis Considerations 16-51

Lian Duan and
Mark Reno
Division of Structures,
California Department of
Transportation,
Sacramento, CA

Notations

The following symbols are used in this chapter. The section number in parentheses after definition of a symbol refers to the section where the symbol first appears or is defined.

0-8493-2674-5/97/$0.00+$.50
© 1997 by CRC Press LLC

A	=	cross-sectional area (Figure 16.9)
A_b	=	cross-sectional area of batten plate (Section 17.A.1)
A_{close}	=	area enclosed within mean dimension for a box (Section 17.A.1)
A_d	=	cross-sectional area of all diagonal lacings in one panel (Section 17.A.1)
A_e	=	effective net area (Figure 16.9)
A_{equiv}	=	cross-sectional area of a thin-walled plate equivalent to lacing bars considering shear transferring **capacity** (Section 17.A.1)
A_f	=	flange area (Section 17.A.1)
A_g	=	gross section area (Section 16.6.3)
A_{gt}	=	gross area subject to tension (Figure 16.9)
A_{gv}	=	gross area subject to shear (Figure 16.9)
A_i	=	cross-sectional area of individual component i (Section 17.A.1)
A_{nt}	=	net area subject to tension (Figure 16.9)
A_{nv}	=	net area subject to shear (Figure 16.9)
A_p	=	cross-sectional area of pipe (Section 16.6.3)
A_r	=	nominal area of rivet (Section 16.6.3)
A_s	=	cross-sectional area of steel members (Figure 16.8)
A_w	=	cross-sectional area of web (Figure 16.12)
A_i^*	=	cross-sectional area above or below plastic neutral axis (Section 17.A.1)
A_{equiv}^*	=	cross-sectional area of a thin-walled plate equivalent to lacing bars or battens assuming full section integrity (Section 17.A.1)
$\overline{B}$	=	ratio of width to depth of steel box section with respect to bending axis (Section 17.A.4)
C	=	distance from elastic neutral axis to extreme fiber (Section 17.A.1)
C_b	=	bending coefficient dependent on moment gradient (Figure 16.10)
C_w	=	warping constant, in.^{6} (Table 16.2)
D_Δ	=	**damage index** defined as ratio of elastic displacement demand to ultimate displacement (Section 17.A.3)
DC_{accept}	=	Acceptable force demand/capacity ratio (Section 16.7.1)
E	=	modulus of elasticity of steel (Figure 16.8)
E_c	=	modulus of elasticity of concrete (Section 16.4.2)
E_s	=	modulus of elasticity of reinforcement (Section 16.4.2)
E_t	=	tangent modulus (Section 17.A.4)
$(EI)_{eff}$	=	effective flexural stiffness (Section 17.A.4)
F_L	=	smaller of $(F_{yf} - F_r)$ or F_{yw}, ksi (Figure 16.10)
F_r	=	compressive residual stress in flange; 10 ksi for rolled shapes, 16.5 ksi for welded shapes (Figure 16.10)
F_u	=	specified minimum tensile strength of steel, ksi (Section 16.4.2)
F_{umax}	=	specified maximum tensile strength of steel, ksi (Section 16.4.2)
F_y	=	specified minimum yield stress of steel, ksi (Section 16.4.2)
F_{yf}	=	specified minimum yield stress of the flange, ksi (Figure 16.10)
F_{ymax}	=	specified maximum yield stress of steel, ksi (Section 16.4.2)
F_{yw}	=	specified minimum yield stress of the web, ksi (Figure 16.10)
G	=	shear modulus of elasticity of steel (Table 16.2)
I_b	=	moment of inertia of a batten plate (Section 17.A.1)
I_f	=	moment of inertia of one solid flange about weak axis (Section 17.A.1)
I_i	=	moment of inertia of individual component i (Section 17.A.1)
I_s	=	moment of inertia of the stiffener about its own centroid (Section 16.6.3)
I_{x-x}	=	moment of inertia of a section about x-x axis (Section 17.A.1)
I_{y-y}	=	moment of inertia of a section about y-y axis considering shear transferring capacity (Section 17.A.1)

I_y	=	moment of inertia about minor axis, in.[4] (Table 16.2)
J	=	torsional constant, in.[4] (Figure 16.10)
K_a	=	effective length factor of individual components between connectors (Figure 16.8)
K	=	effective length factor of a compression member (Section 16.6.2)
L	=	unsupported length of a member (Figure 16.8)
L_g	=	free edge length of gusset plate (Section 16.6.3)
M	=	bending moment (Figure 16.26)
M_1	=	larger moment at end of unbraced length of beam (Table 16.2)
M_2	=	smaller moment at end of unbraced length of beam (Table 16.2)
M_n	=	nominal flexural strength (Figure 16.10)
M_n^{FLB}	=	nominal flexural strength considering flange local buckling (Figure 16.10)
M_n^{LTB}	=	nominal flexural strength considering lateral torsional buckling (Figure 16.10)
M_n^{WLB}	=	nominal flexural strength considering web local buckling (Figure 16.10)
M_p	=	plastic bending moment (Figure 16.10)
M_r	=	elastic limiting buckling moment (Figure 16.10)
M_u	=	factored bending moment demand (Section 16.6.3)
M_y	=	yield moment (Figure 16.10)
$M_{p-batten}$	=	plastic moment of a batten plate about strong axis (Figure 16.12)
$M_{\varepsilon c}$	=	moment at which compressive strain of concrete at extreme fiber equal to 0.003 (Section 16.6.4)
N_s	=	number of shear planes per rivet (Section 16.6.3)
P	=	axial force (Section 17.A.4)
P_{cr}	=	elastic buckling load of a **built-up member** considering buckling mode interaction (Section 17.A.2)
P_L	=	elastic buckling load of an individual component (Section 17.A.2)
P_G	=	elastic buckling load of a global member (Section 17.A.2)
P_n	=	nominal axial strength (Figure 16.8)
P_u	=	factored axial load **demands** (Figure 16.13)
P_y	=	yield axial strength (Section 16.6.3)
P_n^*	=	nominal compressive strength of column (Figure 16.8)
P_n^{LG}	=	nominal compressive strength considering buckling mode interaction (Figure 16.8)
P_n^b	=	nominal tensile strength considering block shear rupture (Figure 16.9)
P_n^f	=	nominal tensile strength considering fracture in net section (Figure 16.9)
P_n^s	=	nominal compressive strength of a solid web member (Figure 16.8)
P_n^y	=	nominal tensile strength considering yielding in gross section (Figure 16.9)
P_n^{comp}	=	nominal compressive strength of lacing bar (Figure 16.12)
P_n^{ten}	=	nominal tensile strength of lacing bar (Figure 16.12)
Q	=	full reduction factor for slender compression elements (Figure 16.8)
Q_i	=	force effect (Section 16.3.1)
R_e	=	hybrid girder factor (Figure 16.10)
R_n	=	nominal shear strength (Section 16.6.3)
S	=	elastic section modulus (Figure 16.10)
S_{eff}	=	effective section modulus (Figure 16.10)
S_x	=	elastic section modulus about major axis, in.[3] (Figure 16.10)
T_n	=	nominal tensile strength of a rivet (Section 16.6.3)
V_c	=	nominal shear strength of concrete (Section 16.6.4)
V_n	=	nominal shear strength (Figure 16.12)
V_p	=	plastic shear strength (Section 16.6.3)
V_s	=	nominal shear strength of transverse reinforcement (Section 16.6.4)
V_t	=	shear strength carried bt truss mechanism (Section 16.6.4)
V_u	=	factored shear demand (Section 16.6.3)

X_1	=	beam buckling factor defined by AISC-LRFD [4] (Figure 16.11)
X_2	=	beam buckling factor defined by AISC-LRFD [4] (Figure 16.11)
Z	=	plastic section modulus (Figure 16.10)
a	=	distance between two connectors along member axis (Figure 16.8)
b	=	width of compression element (Figure 16.8)
b_i	=	length of particular segment of (Section 17.A.1)
d	=	effective depth of (Section 16.6.4)
f_c'	=	specified compressive strength of concrete (Section 16.6.5)
f_{cmin}	=	specified minimum compressive strength of concrete (Section 16.4.2)
f_r	=	modulus of rupture of concrete (Section 16.4.2)
f_{yt}	=	probable yield strength of transverse steel (Section 16.6.4)
h	=	depth of web (Figure 16.8) or depth of member in lacing plane (Section 17.A.1)
k	=	buckling coefficient (Table 16.3)
k_v	=	web plate buckling coefficient (Figure 16.12)
l	=	length from the last rivet (or bolt) line on a member to first rivet (or bolt) line on a member measured along the centerline of member (Section 16.6.3)
m	=	number of panels between point of maximum moment to point of zero moment to either side [as an approximation, half of member length ($L/2$) may be used] (Section 17.A.1)
m_{batten}	=	number of batten planes (Figure 16.12)
m_{lacing}	=	number of lacing planes (Figure 16.12)
n	=	number of equally spaced longitudinal compression flange stiffeners (Table 16.3)
n_r	=	number of rivets connecting lacing bar and main component at one joint (Figure 16.12)
r	=	radius of gyration, in. (Figure 16.8)
r_i	=	radius of gyration of local member, in. (Figure 16.8)
r_y	=	radius of gyration about minor axis, in. (Figure 16.10)
t	=	thickness of unstiffened element (Figure 16.8)
t_i	=	average thickness of segment b_i (Section 17.A.1)
t_{equiv}	=	thickness of equivalent thin-walled plate (Section 17.A.1)
t_w	=	thickness of the web (Figure 16.10)
v_c	=	permissible shear stress carried by concrete (Section 16.6.4)
x	=	subscript relating symbol to strong axis or x-x axis (Figure 16.13)
x_i	=	distance between y-y axis and center of individual component i (Section 17.A.1)
x_i^*	=	distance between center of gravity of a section A_i^* and plastic neutral y-y axis (Section 17.A.1)
y	=	subscript relating symbol to strong axis or y-y axis (Figure 16.13)
y_i^*	=	distance between center of gravity of a section A_i^* and plastic neutral x-x axis (Section 17.A.1)
Δ_{ed}	=	elastic displacement demand (Section 17.A.3)
Δ_u	=	ultimate displacement (Section 17.A.3)
α	=	separation ratio (Section 17.A.2)
α_x	=	parameter related to biaxial loading behavior for x-x axis (Section 17.A.4)
α_y	=	parameter related to biaxial loading behavior for y-y axis (Section 17.A.4)
β	=	0.8, reduction factor for connection (Section 16.6.3)
β_m	=	reduction factor for moment of inertia specified by Equation 16.28 (Section 17.A.1)
β_t	=	reduction factor for torsion constant may be determined Equation 16.38 (Section 17.A.1)
β_x	=	parameter related to uniaxial loading behavior for x-x axis (Section 17.A.4)
β_y	=	parameter related to uniaxial loading behavior for y-y axis (Section 17.A.4)
δ_o	=	imperfection (out-of-straightness) of individual component (Section 17.A.2)

γ_{LG} = buckling mode interaction factor to account for **buckling model interaction** (Figure 16.8)

λ = width-thickness ratio of compression element (Figure 16.8)

λ_b = $\frac{L}{r_y}$ (slenderness parameter of flexural moment dominant members) (Figure 16.10)

λ_{bp} = limiting beam slenderness parameter for plastic moment for **seismic design** (Figure 16.10)

λ_{br} = limiting beam slenderness parameter for elastic lateral torsional buckling (Figure 16.10)

λ_{bpr} = limiting beam slenderness parameter determined by Equation 16.25 (Table 16.2)

λ_c = $\left(\frac{KL}{r\pi}\right)\sqrt{\frac{F_y}{E}}$ (slenderness parameter of axial load dominant members) (Figure 16.8)

λ_{cp} = 0.5 (limiting column slenderness parameter for 90% of the axial yield load based on AISC-LRFD [4] column curve) (Table 16.2)

λ_{cpr} = limiting column slenderness parameter determined by Equation 16.24 (Table 16.2)

λ_{cr} = limiting column slenderness parameter for elastic buckling (Table 16.2)

λ_p = limiting width-thickness ratio for plasticity development specified in Table 16.3 (Figure 16.10)

λ_{pr} = limiting width-thickness ratio determined by Equation 16.23 (Table 16.2)

λ_r = limiting width-thickness ratio (Figure 16.8)

$\lambda_{p-\text{Seismic}}$ = limiting width-thickness ratio for seismic design (Table 16.2)

μ_Δ = displacement **ductility**, ratio of ultimate displacement to yield displacement (Section 16.6.4)

μ_ϕ = curvature ductility, ratio of ultimate curvature to yield curvature (Section 17.A.3)

ρ'' = ratio of transverse reinforcement volume to volume of confined core (Section 16.6.4)

ϕ = resistance factor (Section 16.6.1)

ϕ = angle between diagonal lacing bar and the axis perpendicular to the member axis (Figure 16.12)

ϕ_b = resistance factor for flexure (Figure 16.13)

ϕ_{bs} = resistance factor for block shear (Section 16.6.1)

ϕ_c = resistance factor for compression (Figure 16.13)

ϕ_t = resistance factor for tension (Figure 16.9)

ϕ_{tf} = resistance factor for tension fracture in net (section 16.6.1)

ϕ_{ty} = resistance factor for tension yield (Figure 16.9)

σ_c^{comp} = maximum concrete stress under uniaxial compression (Section 16.6.5)

σ_c^{ten} = maximum concrete stress under uniaxial tension (Section 16.6.5)

σ_s = maximum steel stress under uniaxial tension (Section 16.6.5)

τ_u = shear strength of a rivet (Section 16.6.3)

ε_s = maximum steel strain under uniaxial tension (Section 16.6.5)

ε_{sh} = strain hardening strain of steel (Section 16.4.2)

$\varepsilon_c^{\text{comp}}$ = maximum concrete strain under uniaxial compression (Section 16.6.5)

γ_i = load factor corresponding to Q_i (Section 16.3.1)

η = a factor relating to ductility, redundancy, and operational importance (Section 16.3.1)

16.1 Introduction

16.1.1 Damage to Bridges in Recent Earthquakes

Since the beginning of civilization, earthquake disasters have caused both death and destruction — the structural collapse of homes, buildings, and **bridges**. About 20 years ago, the 1976 Tangshan earthquake in China resulted in the tragic death of 242,000 people, while 164,000 people were severely injured, not to mention the entire collapse of the industrial city of Tangshan [39]. More recently, the 1989 Loma Prieta and the 1994 Northridge earthquakes in California [27, 28] and the 1995 Kobe earthquake in Japan [29] have exacted their tolls in the terms of deaths, injuries, and the collapse of the infrastructure systems which can in turn have detrimental effects on the economies. The damage and collapse of bridge structures tend to have a more lasting image on the public.

Figure 16.1 shows the collapsed elevated steel conveyor at Lujiatuo Mine following the 1976 Tangshan earthquake in China. Figures 16.2 and 16.3 show damage from the 1989 Loma Prieta earthquake: the San Francisco-Oakland Bay Bridge east span drop off and the collapsed double deck portion of the Cypress freeway, respectively. Figure 16.4 shows a portion of the R-14/I-5 interchange following the 1994 Northridge earthquake, which also collapsed following the 1971 San Fernando earthquake in California while it was under construction. Figure 16.5 shows a collapsed 500-m section of the elevated Hanshin Expressway during the 1995 Kobe earthquake in Japan. These examples of bridge damage, though tragic, have served as full-scale laboratory tests and have forced bridge engineers to reconsider their design principles and philosophies. Since the 1971 San Fernando earthquake, it has been a continuing challenge for bridge engineers to develop a safe seismic design procedure so that the structures are able to withstand the sometimes unpredictable devastating earthquakes.

Figure 16.1 Collapsed elevated steel conveyor at Lujiatuo Mine following the 1976 Tangshan earthquake in China. (From California Institute of Technology, The Greater Tangshan Earthquake, California, 1996. With permission.)

16.1.2 No-Collapse-Based Design Criteria

For seismic design and retrofit of ordinary bridges, the primary philosophy is to prevent collapse during severe earthquakes [13, 24, 25]. The structural survival without collapse has been a basis of seismic design and retrofit for many years [13]. To prevent the collapse of bridges, two alternative

Figure 16.2 Aerial view of collapsed upper and lower decks of the San Francisco-Oakland Bay Bridge (I-80) following the 1989 Loma Prieta earthquake in California. (Photo by California Department of Transportation. With permission.)

Figure 16.3 Collapsed Cypress Viaduct (I-880) following the 1989 Loma Prieta earthquake in California.

design approaches are commonly in use. First is the conventional force-based approach where the adjustment factor Z for ductility and risk assessment [12], or the response modification factor R [1], is applied to elastic member force levels obtained by acceleration spectra analysis. The second approach is the newer displacement-based design approach [13] where displacements are a major consideration in design. For more detailed information, reference is made to a comprehensive and state-of-the-art book by Prietley et al. [35]. Much of the information in this book is backed by California Department of Transportation (Caltrans)-supported research, directed at the seismic performance of bridge structures.

Figure 16.4 Collapsed SR-14/I-5 south connector overhead following the 1994 Northridge earthquake in California. (Photo by James MacIntyre. With permission.)

Figure 16.5 Collapsed Hanshin Expressway following the 1995 Kobe earthquake in Japan. (Photo by Mark Yashinsky. With permission.)

16.1.3 Performance-Based Design Criteria

Following the 1989 Loma Prieta earthquake, bridge engineers recognized the need for site-specific and project-specific design criteria for important bridges. A bridge is defined as "important" when one of the following criteria is met:

- The bridge is required to provide secondary life safety.
- Time for restoration of functionality after closure creates a major economic impact.
- The bridge is formally designated as critical by a local emergency plan.

Caltrans, in cooperation with various emergency agencies, has designated and defined the various important routes throughout the state of California. For important bridges, such as I-880 replacement [23] and R-14/I-5 interchange replacement projects, the design criteria [10, 11] including site-specific Acceleration Response Spectrum (ARS) curves and specific design procedures to reflect the desired performance of these structures were developed.

In 1995, Caltrans, in cooperation with engineering consulting firms, began the task of seismic retrofit design for the seven major toll bridges including the San Francisco-Oakland Bay Bridge (SFOBB) in California. Since the traditional seismic design procedures could not be directly applied to these toll bridges, various analysis and design concepts and strategies have been developed [7]. These differences can be attributed to the different post-earthquake performance requirements. As shown in Figure 16.6, the performance requirements for a specific project or bridge must be the first item to be established. Loads, materials, analysis methods and approaches, and detailed acceptance criteria are then developed to achieve the expected performance. The **no-collapse-based design** criteria shall be used unless performance-based design criteria is required.

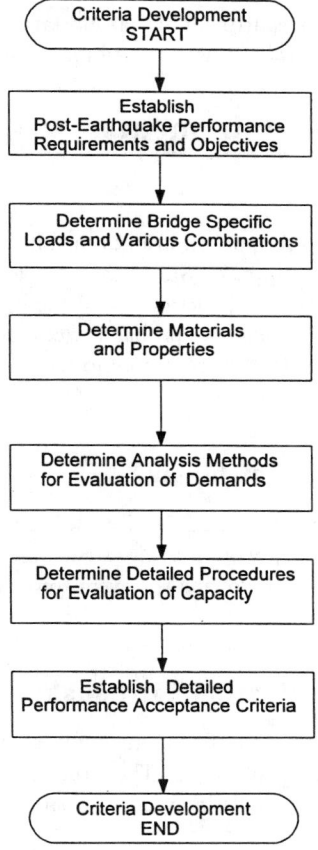

Figure 16.6   Development procedure of performance-based seismic design criteria for important bridges.

16.1.4 Background of Criteria Development

It is the purpose of this chapter to present performance-based criteria that may be used as a guideline for seismic design and retrofit of important bridges. More importantly, this chapter provides concepts for the general development of performance-based criteria. The appendices, as an integral part of the criteria, are provided for background and information of criteria development. However, it must be recognized that the desired performance of the structure during various earthquakes ultimately defines the design procedures.

Much of this chapter was primarily based on the Seismic Retrofit Design Criteria (*Criteria*) which was developed for the SFOBB West Spans [17]. The SFOBB *Criteria* was developed and based on past successful experience, various codes, specifications, and state-of-the-art knowledge.

The SFOBB, one of the national engineering wonders, provides the only direct highway link between San Francisco and the East Bay Communities. SFOBB (Figure 16.7) carries Interstate Highway 80 approximately 8-1/4 miles across San Francisco Bay since it first opened to traffic in 1936. The west spans of SFOBB, consisting of twin, end-to-end double-deck suspension bridges and a three-span double-deck continuous truss, crosses the San Francisco Bay from the city of San Francisco to Yerba Buena Island. The seismic retrofit design of SFOBB West Spans, as the top priority project of the California Department of Transportation, is a challenge to bridge engineers. A performance-based design *Criteria* [17] was, therefore, developed for SFOBB West Spans.

16.2 Performance Requirements

16.2.1 General

The seismic design and retrofit of important bridges shall be performed by considering both the higher level **Safety Evaluation Earthquake (SEE),** which has a mean return period in the range of 1000 to 2000 years, and the lower level **Functionality Evaluation Earthquake (FEE),** which has a mean return period of 300 years with a 40% probability of exceedance during the expected life of the bridge. It is important to note that the return periods of both the SEE and FEE are dictated by the engineers and seismologists.

16.2.2 Safety Evaluation Earthquake

The bridge shall remain serviceable after a SEE. Serviceable is defined as sustaining repairable damage with minimum impact to functionality of the bridge structure. In addition, the bridge will be open to emergency vehicles immediately following the event, provided bridge management personnel can provide access.

16.2.3 Functionality Evaluation Earthquake

The bridge shall remain fully operational after a FEE. Fully operational is defined as full accessibility to the bridge by current normal daily traffic. The structure may suffer repairable damage, but repair operations may not impede traffic in excess of what is currently required for normal daily maintenance.

16.2.4 Objectives of Seismic Design

The objectives of seismic design are as follows:

1. To keep the *Critical* structural components in the essentially elastic range during the SEE.

(a) West crossing spans.

(b) East crossing spans.

Figure 16.7 San Francisco-Oakland Bay Bridge. (Photo by California Department of Transportation. With permission.)

2. To achieve safety, reliability, serviceability, constructibility, and maintainability when the **Seismic Response Modification Devices (SRMDs)**, i.e., energy dissipation and isolation devices, are installed in bridges.

3. To devise expansion joint assemblies between bridge frames that either retain traffic support or, with the installation of deck plates, are able to carry the designated traffic after being subjected to SEE displacements.

4. To provide ductile load paths and detailing to ensure bridge safety in the event that future demands might exceed those demands resulting from current SEE ground motions.

16.3 Loads and Load Combinations

16.3.1 Load Factors and Combinations

New and retrofitted bridge components shall be designed for the applicable load combinations in accordance with the requirements of AASHTO-LRFD [1].

The load effect shall be obtained by

$$\text{Load effect} = \eta \sum \gamma_i Q_i \tag{16.1}$$

where
Q_i = force effect
η = a factor relating to ductility, redundancy, and operational importance
γ_i = load factor corresponding to Q_i

The AASHTO-LRFD load factors or load factors $\eta = 1.0$ and $\gamma_i = 1.0$ may be used for seismic design.

The live load on the bridge shall be determined by ADTT (Average Daily Truck Traffic) value for the project. The bridge shall be analyzed for the worst case with or without live load. The mass of the live loads shall not be included in the dynamic calculations. The intent of the live load combination is to include the weight effect of the vehicles only.

16.3.2 Earthquake Load

The earthquake load – ground motions and response spectra shall be considered at two levels: SEE and FEE. The ground motions and response spectra may be generated in accordance with Caltrans Guidelines [14, 15].

16.3.3 Wind Load

1. Wind load on structures — Wind loads shall be applied as a static equivalent load in accordance with AASHTO-LRFD [1] .

2. Wind load on live load — Wind pressure on vehicles shall be represented by a uniform load of 0.100 kips/ft (1.46 kN/m) applied at right angles to the longitudinal axis of the structure and 6.0 ft (1.85 m) above the deck to represent the wind load on vehicles.

3. Wind load dynamics — The expansion joints, SRMDs, and wind locks (tongues) shall be evaluated for the dynamic effects of wind loads.

16.3.4 Buoyancy and Hydrodynamic Mass

The buoyancy shall be considered to be an uplift force acting on all components below design water level. Hydrodynamic mass effects [26] shall be considered for bridges over water.

16.4   Structural Materials

16.4.1   Existing Materials

For seismic retrofit design, aged concrete with specified strength of 3250 psi (22.4 MPa) can be considered to have a compressive strength of 5000 psi (34.5 MPa). If possible, cores of existing concrete should be taken. Behavior of structural steel and reinforcement shall be based on mill certificate or tensile test results. If they are not available in bridge archives, a nominal strength of 1.1 times specified yield strength may be used [13].

16.4.2   New Materials

Structural Steel

New structural steel used shall be AASHTO designation M270 (ASTM designation A709) Grade 36 and Grade 50.

Welds shall be as specified in the Bridge Welding Code ANSI/AASHTO/AWS D1.5-95 [8].

Partial penetration welds shall not be used in regions of structural components subjected to possible inelastic deformation.

High strength bolts conforming to ASTM designation A325 shall be used for all new connections and for upgrading strengths of existing riveted connections. New bolted connections shall be designed as bearing-type for seismic loads and shall be slip-critical for all other load cases.

All bolts with a required length under the head greater than 8 in. shall be designated as ASTM A449 threaded rods (requiring nuts at each end) unless a verified source of longer bolts can be identified.

New anchor bolts shall be designated as ASTM A449 threaded rods.

Structural Concrete

All concrete shall be normal weight concrete with the following properties:

Specified compressive strength:	$f_{cmin} = 4,000$	psi (27.6MPa)
Modulus of elasticity:	$E_c = 57,000\sqrt{f_c'}$	psi
Modulus of rupture:	$f_r = 5\sqrt{f_c'}$	psi

Reinforcement

All reinforcement shall use ASTM A706 (Grade 60) with the following specified properties:

Specified minimum yield stress:	$F_y = 60$ ksi	(414 MPa)
Specified minimum tensile strength:	$F_u = 90$ ksi	(621 MPa)
Specified maximum yield stress:	$F_{ymax} = 78$ ksi	(538 MPa)
Specified maximum tensile strength:	$F_{umax} = 107$ ksi	(738 MPa)
Modulus of elasticity:	$E_s = 29,000$ ksi	(200,000 MPa)

$$\text{Strain hardening strain:} \quad \varepsilon_{sh} = \begin{cases} 0.0150 & \text{for \#8 and smallers bars} \\ 0.0125 & \text{for \#9} \\ 0.0100 & \text{for \#10 and \#11} \\ 0.0075 & \text{for \#14} \\ 0.0050 & \text{for \#18} \end{cases}$$

16.5 Determination of Demands

16.5.1 Analysis Methods

Static Linear Analysis

Static linear analysis shall be used to determine member forces due to self weight, wind, water currents, temperature, and live load.

Dynamic Response Spectrum Analysis

1. Dynamic response spectrum analysis shall be used for the local and regional stand alone models and the simplified global model described in Section 16.5.2 to determine mode shapes, structure periods, and initial estimates of seismic force and displacement demands.
2. Dynamic response spectrum analysis may be used on global models prior to time history analysis to verify model behavior and eliminate modeling errors.
3. Dynamic response spectrum analysis may be used to identify initial regions or members of likely inelastic behavior which need further refined analysis using inelastic nonlinear elements.
4. Site specific ARS curves shall be used, with 5% damping.
5. Modal responses shall be combined using the Complete Quadratic Combination (CQC) method and the resulting orthogonal responses shall be combined using either the Square Root of the Sum of the Squares (SRSS) method or the "30%" rule, e.g., $R_H = \text{Max}(R_x + 0.3R_y, R_y + 0.3R_x)$ [13].
6. Due to the expected levels of inelastic structural response in some members and regions, dynamic response spectrum analysis shall not be used to determine final design demand values or to assess the performance of the retrofitted structures.

Dynamic Time History Analysis

Site specific multi-support dynamic time histories shall be used in a dynamic time history analysis. All analyses incorporating significant nonlinear behavior shall be conducted using nonlinear inelastic dynamic time history procedures.

1. Linear elastic dynamic time history analysis — Linear elastic dynamic time history analysis is defined as dynamic time history analysis with considerations of geometrical linearity (small displacement), linear boundary conditions, and elastic members. It shall only be used to check regional and global models.
2. Nonlinear elastic dynamic time history analysis — Nonlinear elastic time history analysis is defined as dynamic time history analysis with considerations of geometrical nonlinearity, linear boundary conditions, and elastic members. It shall be used to determine areas of inelastic behavior prior to incorporating inelasticity into the regional and global models.
3. Nonlinear inelastic dynamic time history analysis – Level I — Nonlinear inelastic dynamic time history analysis – Level I is defined as dynamic time history analysis with considerations of geometrical nonlinearity, nonlinear boundary conditions, other inelastic elements (for example, dampers) and elastic members. It shall be used for the final determination of force and displacement demands for existing structures in combination with static gravity, wind, thermal, water current, and live load as specified in Section 16.3.

4. Nonlinear inelastic dynamic time history analysis – Level II — Nonlinear inelastic dynamic time history analysis – Level II is defined as dynamic time history analysis with considerations of geometrical nonlinearity, nonlinear boundary conditions, other inelastic elements (for example, dampers) and inelastic members. It shall be used for the final evaluation of response of the structures. Reduced material and section properties, and the yield surface equation suggested in the Appendix may be used for inelastic considerations.

16.5.2   Modeling Considerations

Global, Regional, and Local Models

The global models focus on the overall behavior and may include simplifications of complex structural elements. Regional models concentrate on regional behavior. Local models emphasize the localized behavior, especially complex inelastic and nonlinear behavior. In regional and global models where more than one foundation location is included in the model, multi-support time history analysis shall be used.

Boundary Conditions

Appropriate boundary conditions shall be included in the regional models to represent the interaction between the regional model and the adjacent portion of the structure not explicitly included. The adjacent portion not specifically included may be modeled using simplified structural combinations of springs, dashpots, and lumped masses.

Appropriate nonlinear elements such as gap elements, nonlinear springs, SRMDs, or specialized nonlinear finite elements shall be included where the behavior and response of the structure is determined to be sensitive to such elements.

Soil-Foundation-Structure-Interaction

Soil-Foundation-Structure-Interaction may be considered using nonlinear or hysteretic springs in the global and regional models. Foundation springs at the base of the structure which reflect the dynamic properties of the supporting soil shall be included in both regional and global models.

Section Properties of Latticed Members

For latticed members, the procedure proposed in the

Appendix may be used for member characterization.

Damping

When nonlinear member properties are incorporated in the model, Rayleigh damping shall be reduced, for example by 20%, compared with analysis with elastic member properties.

Seismic Response Modification Devices

The SRMDs, i.e., energy dissipation and isolation devices, shall be modeled explicitly using their hysteretic characteristics as determined by tests.

16.6 Determination of Capacities

16.6.1 Limit States and Resistance Factors

Limit States

The limit states are defined as those conditions of a structure at which it ceases to satisfy the provisions for which it was designed. Two kinds of limit states corresponding to SEE and FEE specified in Section 16.2 apply for seismic design and retrofit.

Resistance Factors

To account for unavoidable inaccuracies in the theory, variation in the material properties, workmanship, and dimensions, nominal strength of structural components should be modified by a resistance factor ϕ to obtain the design capacity or strength (resistance). The following resistance factors shall be used for seismic design:

- For tension fracture in net section ϕ_{tf} $=$ 0.8
- For block shear ϕ_{bs} $=$ 0.8
- For bolts and welds ϕ $=$ 0.8
- For all other cases ϕ $=$ 1.0

16.6.2 Effective Length of Compression Members

The **effective length factor** K for compression members shall be determined in accordance with Chapter 17 of this Handbook.

16.6.3 Nominal Strength of Steel Structures

Members

1. General — Steel members include rolled members and built-up members, such as latticed, battened, and perforated members. The design strength of those members shall be according to applicable provisions of AISC-LRFD [4]. Section properties of latticed members shall be determined in accordance with the Appendix.

2. Compression members — For compression members, the nominal strength shall be determined in accordance with Section E2 and Appendix B of AISC-LRFD [4]. For built-up members, effects of interaction of buckling modes shall be considered in accordance with the Appendix. A detailed procedure in a flowchart format is shown in Figure 16.8.

3. Tension Members — For tension members, the design strength shall be determined in accordance with Sections D1 and J4 of AISC-LRFD [4]. It is the smallest value obtained according to (i) yielding in gross section, (ii) fracture in net section, and (iii) block shear rupture. A detailed procedure is shown in Figure 16.9.

4. Flexural members — For flexural members, the nominal flexural strength shall be determined in accordance with Section F1 and Appendices B, F, and G of AISC-LRFD [4].

   - For *critical* members, the nominal flexural strength is the smallest value according to (i) initial yielding, (ii) lateral-torsional buckling, (iii) flange local buckling, and (iv) web local buckling.

   - For *other* members, the nominal flexural strength is the smallest value according to (i) plastic moment, (ii) lateral-torsional buckling, (iii) flange local buckling, and (iv) web local buckling.

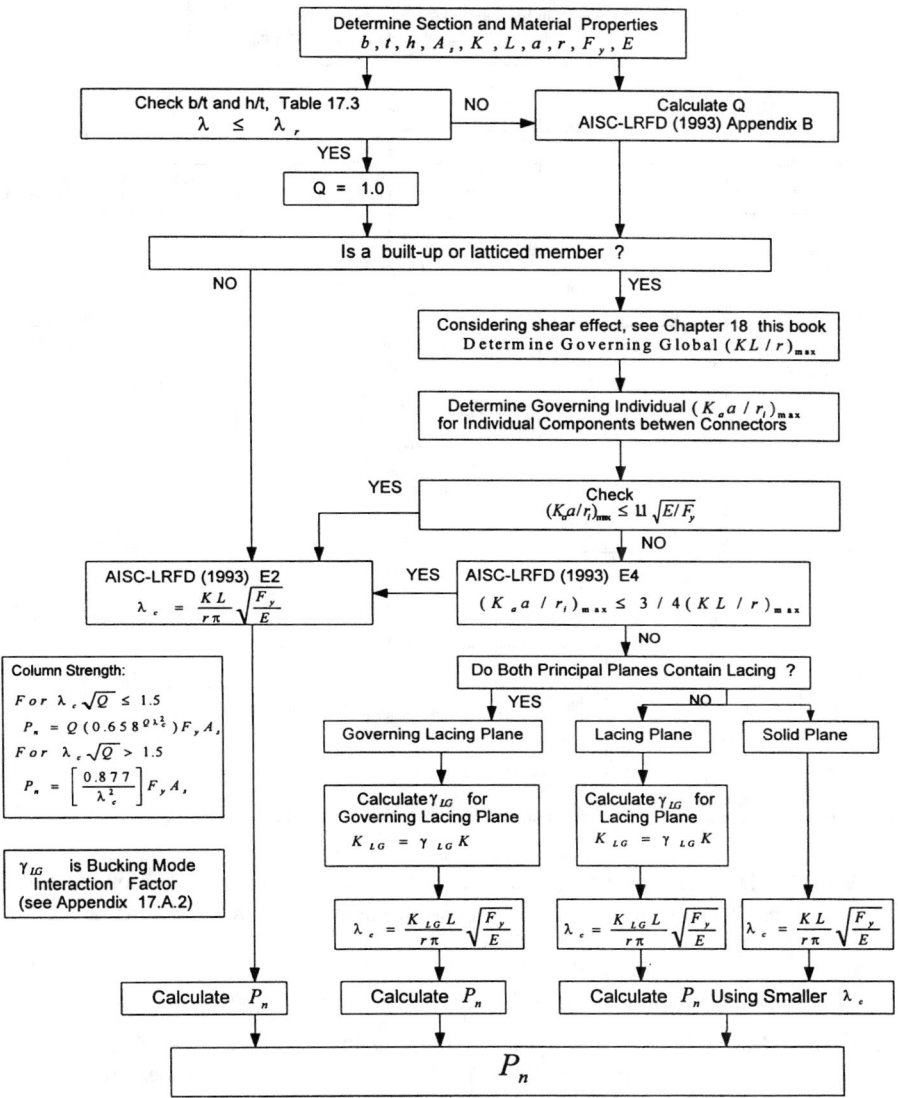

Figure 16.8   Evaluation procedure for nominal compressive strength of steel members.

Detailed procedures for flexural strength of box- and I-shaped members are shown in Figures 16.10 and 16.11, respectively.

5. Nominal shear strength — For solid-web steel members, the nominal shear strength shall be determined in accordance with Appendix F2 of AISC-LRFD [4]. For latticed members, the shear strength shall be based on shear-flow transfer-capacity of lacing bar, battens, and connectors as discussed in the Appendix. A detailed procedure for shear strength is shown in Figure 16.12.

6. Members subjected to bending and axial force — For members subjected to bending and axial force, the evaluation shall be according to Section H1 of AISC-LRFD [4], i.e., the bi-linear interaction equation shall be used. The recent study on "Cyclic Testing of Latticed Members for San Francisco-Oakland Bay Bridge" at UCSD [37] recommends that the

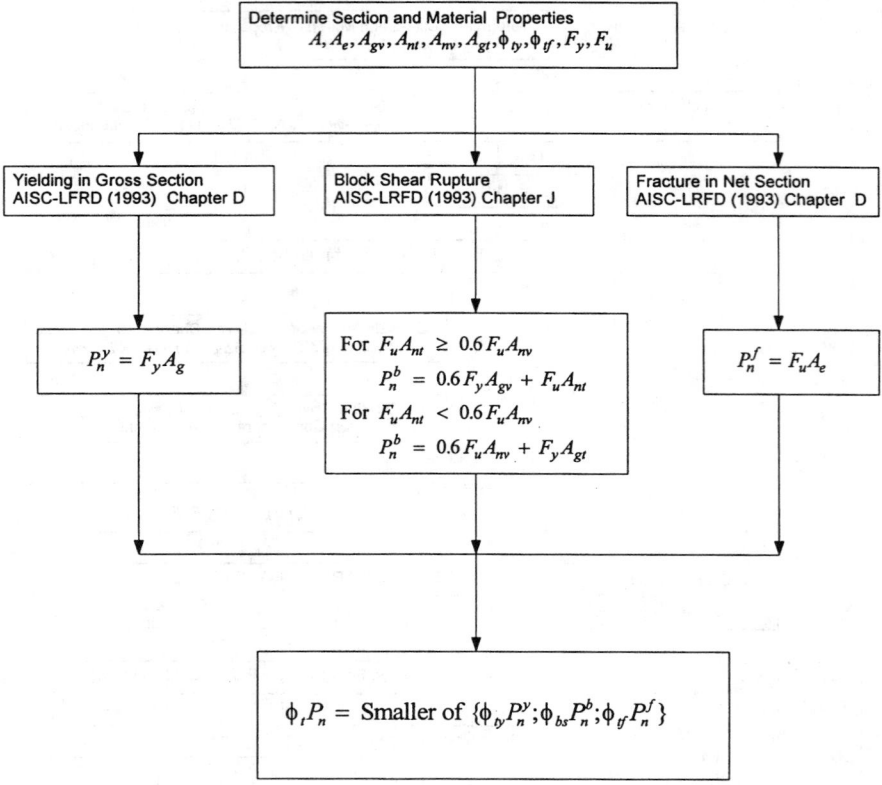

Figure 16.9   Evaluation procedure for tensile strength of steel members.

AISC-LRFD interaction equation can be used directly for seismic evaluation of latticed members. A detailed procedure for steel beam-columns is shown in Figure 16.13.

Gusset Plate Connections

1. General description — Gusset plates shall be evaluated for shear, bending, and axial forces according to Article 6.14.2.8 of AASHTO-LRFD [1]. The internal stresses in the gusset plate shall be determined according to Whitmore's method in which the *effective* area is defined as the width bound by two 30° lines drawn from the first row of the bolt or rivet group to the last bolt or rivet line. The stresses in the gusset plate may be determined by more rational methods or refined computer models.

2. Tension strength — The tension capacity of the gusset plates shall be calculated according to Article 6.13.5.2 of AASHTO-LRFD [1].

3. Compressive strength — The compression capacity of the gusset plates shall be calculated according to Article 6.9.4.1 of AASHTO-LRFD [1]. In using the AASHTO-LRFD Equations (6.9.4.1-1) and (6.9.4.1-2), symbol l is the length from the last rivet (or bolt) line on a member to first rivet (or bolt) line on a chord measured along the centerline of the member; K is effective length factor = 0.65; A_s is average *effective* cross-section area defined by Whitmore's method.

4. Limit of free edge to thickness ratio of gusset plate — When the free edge length to thickness ratio of a gusset plate $L_g/t > 1.6\sqrt{E/F_y}$, the compression stress of a gusset plate shall be less than $0.8\,F_y$; otherwise the plate shall be stiffened. The free edge length

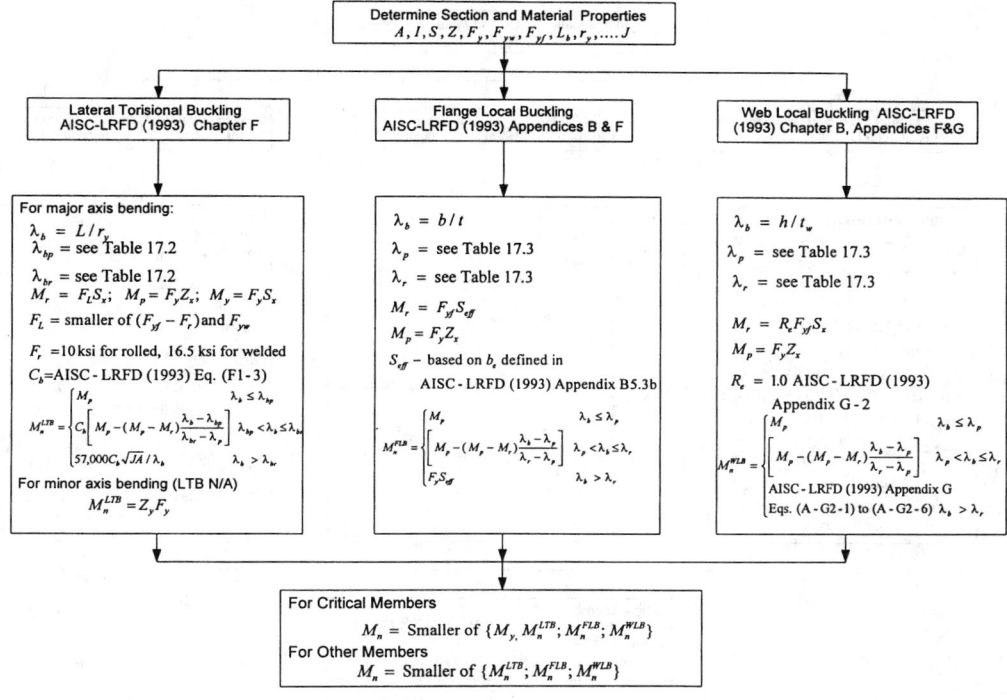

Figure 16.10   Evaluation procedure for nominal flexural strength of box-shaped steel members.

to thickness ratio of a gusset plate shall satisfy the following limit specified in Article 6.14.2.8 of AASHTO-LRFD [1].

$$\frac{L_g}{t} \le 2.06 \sqrt{\frac{E}{F_y}} \tag{16.2}$$

When the free edge is stiffened, the following requirements shall be satisfied:

- The stiffener plus a width of $10t$ of gusset plate shall have an l/r ratio less than or equal to 40.

- The stiffener shall have an l/r ratio less than or equal to 40 between fasteners.

- The stiffener moment of inertia shall satisfy [38]:

$$I_s \ge \begin{cases} 1.83t^4\sqrt{(b/t)^2 - 144} \\ 9.2t^4 \end{cases} \tag{16.3}$$

where

I_s = the moment of inertia of the stiffener about its own centroid

b = the width of the gusset plate perpendicular to the edge

t = the thickness of the gusset plate

5. In-plane moment strength of gusset plate (strong axis) — The nominal moment strength of a gusset plate shall be calculated by the following equation in Article 6.14.2.8 of AASHTO-LRFD [1]:

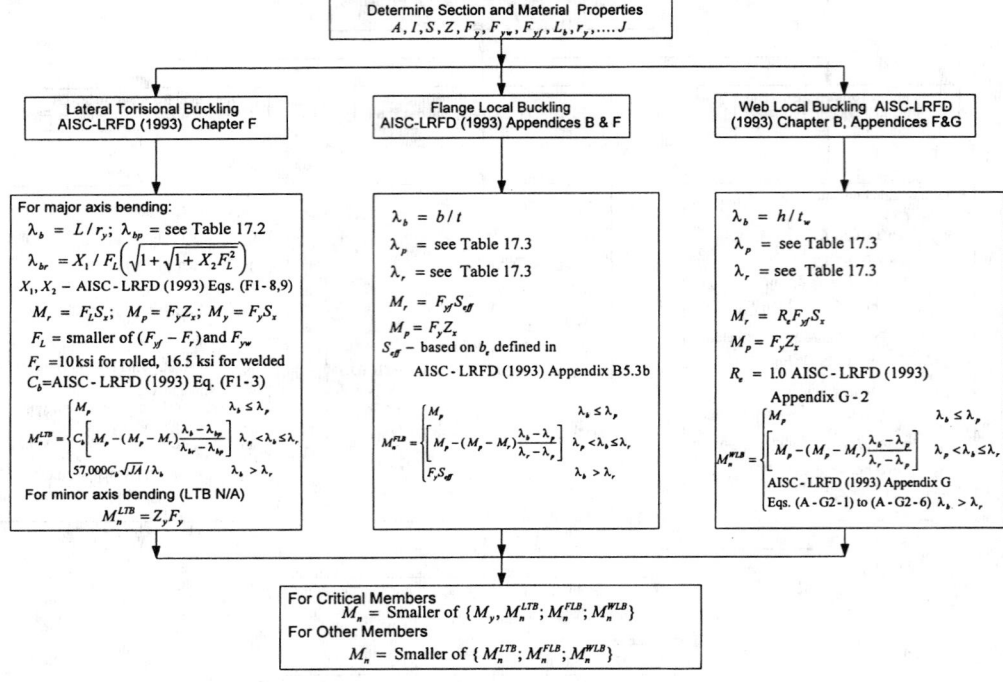

Figure 16.11   Evaluation procedure for nominal flexural strength of I-shaped steel members.

$$M_n = SF_y \qquad (16.4)$$

where
S = elastic section modulus about the strong axis

6. In-plane shear strength for a gusset plate — The nominal shear strength of a gusset plate shall be calculated by the following equations:

Based on gross section:

$$V_n = \text{smaller} \begin{cases} 0.4F_y A_{gv} & \text{for flexural shear} \\ 0.6F_y A_{gv} & \text{for uniform shear} \end{cases} \qquad (16.5)$$

Based on net section:

$$V_n = \text{smaller} \begin{cases} 0.4F_u A_{nv} & \text{for flexural shear} \\ 0.6F_u A_{nv} & \text{for uniform shear} \end{cases} \qquad (16.6)$$

where
A_{gv} = gross area subject to shear
A_{nv} = net area subject to shear
F_u = minimum tensile strength of the gusset plate

7. Initial yielding of gusset plate in combined in-plane moment, shear, and axial load — The initial yielding strength of a gusset plate subjected to a combined in-plane moment, shear, and axial load shall be determined by the following equations:

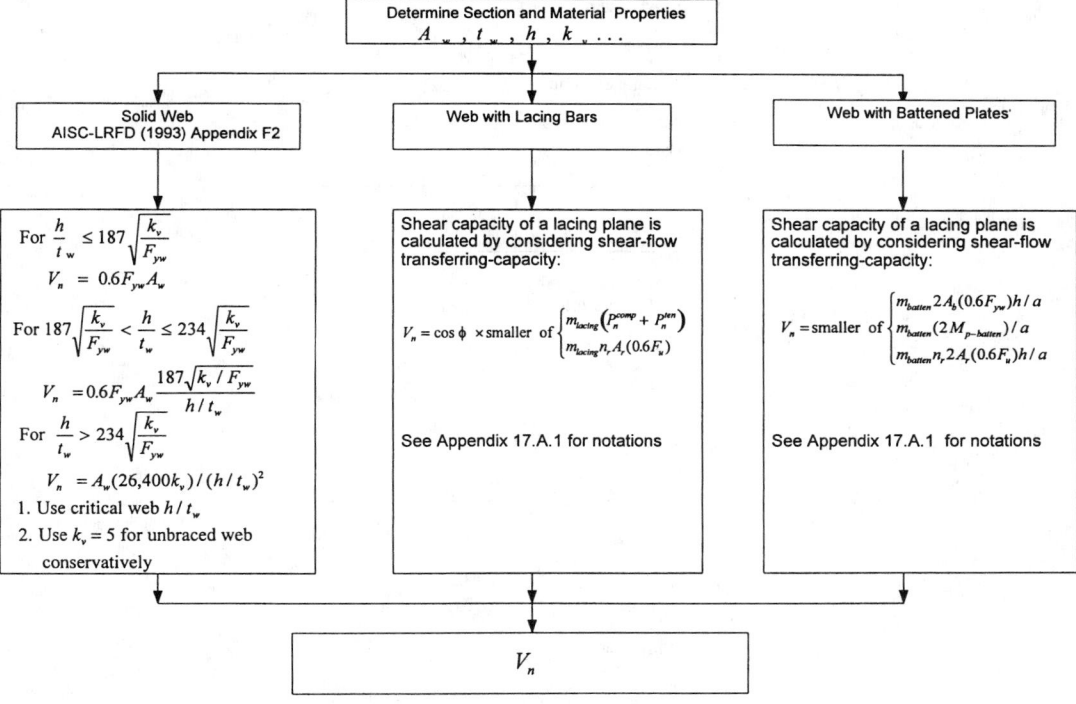

Figure 16.12   Evaluation procedure for nominal shear strength of steel members.

$$\frac{M_u}{M_n} + \frac{P_u}{P_y} \leq 1 \tag{16.7}$$

or

$$\left(\frac{V_u}{V_n}\right)^2 + \left(\frac{P_u}{P_y}\right)^2 \leq 1 \tag{16.8}$$

where

V_u = factored shear

M_u = factored moment

P_u = factored axial load

M_n = nominal moment strength determined by Equation 16.4

V_n = nominal shear strength determined by Equation 16.5

P_y = yield axial strength $(A_g F_y)$

A_g = gross section area of gusset plate

8. Full yielding of gusset plate in combined in-plane moment, shear, and axial load — Full yielding strength for a gusset plate subjected to combined in-plane moment, shear, and axial load has the form [6]:

$$\frac{M_u}{M_p} + \left(\frac{P_u}{P_y}\right)^2 + \frac{\left(\frac{V_u}{V_p}\right)^4}{\left[1 - \left(\frac{P_u}{P_y}\right)^2\right]} = 1 \tag{16.9}$$

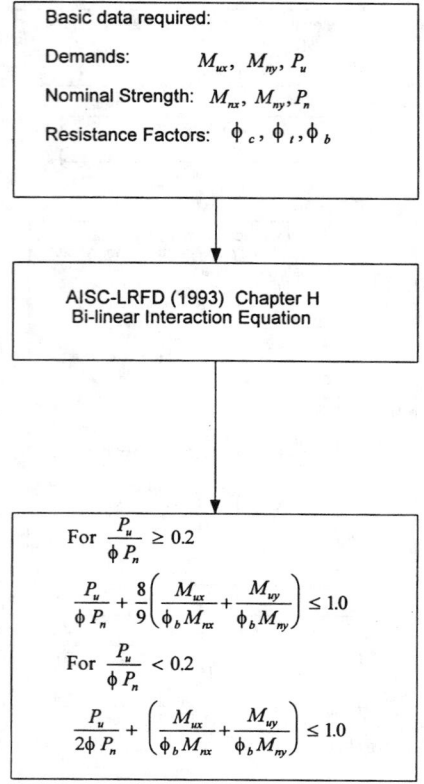

Figure 16.13   Evaluation procedure for steel beam-columns.

where
M_p =   plastic moment of pure bending ($Z F_y$)
V_p =   shear capacity of gusset plate ($0.6 A_g F_y$)
Z =   plastic section modulus

9. Block shear capacity — The block shear capacity shall be calculated according to Article 6.13.4 of AASHTO-LRFD [1].

10. Out-of-plane moment and shear consideration — Moment will be resolved into a couple acting on the near and far side gusset plates. This will result in tension or compression on the respective plates. This force will produce weak axis bending of the gusset plate.

Connections Splices

The splice section shall be evaluated for axial tension, flexure, and combined axial and flexural loading cases according to AISC-LRFD [4]. The member splice capacity shall be equal to or greater than the capacity of the smaller of the two members being spliced.

Eyebars

The tensile capacity of the eyebars shall be calculated according to Article D3 of AISC-LRFD [4].

Anchor Bolts (Rods) and Anchorage Assemblies

1. Anchorage assemblies for nonrocking mechanisms shall be anchored with sufficient capacity to develop the lesser of the seismic force demand and plastic strength of the columns. Anchorage assemblies may be designed for rocking mechanisms where yield is permitted — at which point rocking commences. Shear keys shall be provided to prevent excess lateral movement. The nominal shear strength of pipe guided shear keys shall be calculated by:

$$R_n = 0.6 F_y A_p \tag{16.10}$$

where
A_p = cross-section area of pipe

2. Evaluation of anchorage assemblies shall be based on reinforced concrete structure behavior with bonded or unbonded anchor rods under combined axial load and bending moment. All anchor rods outside of the compressive region may be taken to full minimum tensile strength.

3. The nominal strength of anchor bolts (rods) for shear, tension, and combined shear and tension shall be calculated according to Article 6.13.2 of AASHTO-LRFD [1].

4. Embedment length of anchor rods shall be such that a ductile failure occurs. Concrete failure surfaces shall be based on a shear stress of $2\sqrt{f_c'}$ and account for edge distances and overlapping shear zones. In no case should edge distances or embedments be less than those shown in Table 8-26 of the AISC-LRFD Manual [3]. New anchor rods shall be threaded to assure development.

Rivets and Holes

1. The bearing capacity on rivet holes shall be calculated according to Article 6.13.2.9 of AASHTO-LRFD [1].

2. Nominal shear strength of a rivet shall be calculated by the following formula:

$$R_n = 0.75 \beta F_u A_r N_s \tag{16.11}$$

where
β = 0.8, reduction factor for connections with more than two rivets and to account for deformation of connected material which causes nonuniform rivet shear force (see Article C6.13.2.7 of AASHTO-LRFD [1])
F_u = minimum tensile strength of the rivet
A_r = the nominal area of the rivet (before driving)
N_s = number of shear planes per rivet
It should be pointed out that the 0.75 factor is the ratio of the shear strength τ_u to the tensile strength F_u of a rivet. The research work by Kulak et al. [31] found that this ratio is independent of the rivet grade, installation procedure, diameter, and grip length and is about 0.75.

3. Tension capacity of a rivet shall be calculated by the following formula:

$$T_n = A_r F_u \tag{16.12}$$

4. Tensile capacity of a rivet subjected to combined tension and shear shall be calculated by the following formula:

$$T_n = A_r F_u \sqrt{1 - \frac{V_u}{R_n}} \qquad (16.13)$$

where

V_u = factored shear force
R_n = nominal shear strength of a rivet determined by Equation 16.11

Bolts and Holes

1. The bearing capacity on bolt holes shall be calculated according to Article 6.13.2.9 of AASHTO-LRFD [1].
2. The nominal strength of a bolt for shear, tension, and combined shear and tension shall be calculated according to Article 6.13.2 of AASHTO-LRFD [1].

Prying Action

Additional tension forces resulting from prying action must be accounted for in determining applied loads on rivets or bolts. The connected elements (primarily angles) must also be checked for adequate flexural strength. Prying action forces shall be determined from the equations presented in AISC-LRFD Manual Volume 2, Part 11 [3].

16.6.4 Nominal Strength of Concrete Structures

Nominal Moment Strength

The nominal moment strength M_n shall be calculated by considering combined biaxial bending and axial loads. It is defined as:

$$M_n = \text{smaller} \begin{cases} M_y \\ M_{\varepsilon c} \end{cases} \qquad (16.14)$$

where

M_y = moment corresponding to first steel yield
$M_{\varepsilon c}$ = moment at which compressive strain of concrete at extreme fiber equal to 0.003

Nominal Shear Strength

The nominal shear strength V_n shall be calculated by the following equations [12, 13].

$$\begin{aligned} V_n &= V_c + V_s & (16.15a) \\ \text{or} \quad V_n &= V_c + V_t & (16.15b) \end{aligned}$$

$$V_c = 0.8 v_c A_g \qquad (16.16)$$

$$v_c = \text{larger} \begin{cases} 2\left(1 + \frac{P_u}{2{,}000 A_g}\right)\sqrt{f_c'} \leq 3\sqrt{f_c'} \\ \text{Factor 1} \times \left(1 + \frac{P_u}{2{,}000 A_g}\right)\sqrt{f_c'} \leq 4\sqrt{f_c'} \end{cases} \qquad (16.17)$$

$$\text{Factor 1} = \frac{\rho'' f_{yt}}{150} + 3.67 - \mu_\Delta \leq 3.0$$

$$\rho'' = \frac{\text{volume of transverse reinforcement}}{\text{volume of confined core}} \qquad (16.18)$$

$$V_s = \begin{cases} A_v f_{yt} d/s & \text{for rectangular sections} \\ \frac{A_s f_{yt} D'}{2s} & \text{for circular sections} \end{cases} \qquad (16.19)$$

where
A_g = gross section area of concrete member
A_s = cross-sectional area of transverse reinforcement within space s
V_t = shear strength carried by truss mechanism
D' = hoop or spiral diameter
P_u = factored axial load associated with design shear V_u and P_u/A_g is in psi
d = effective depth of section
s = space of transverse reinforcement
f_{yt} = probable yield strength of transverse steel (psi)
μ_Δ = ductility demand ratio (1.0 will be used)

16.6.5 Structural Deformation Capacity

Steel Structures

Displacement capacity shall be evaluated by considering both material and geometrical non-linearity. Proper boundary conditions for various structures shall be carefully adjusted. The ultimate available displacement capacity is defined as the displacement corresponding to a load that drops a maximum of 20% from the peak load.

Reinforced Concrete Structures

Displacement capacity shall be evaluated using stand-alone push-over analysis models. Both the geometrical and material nonlinearities, as well as the foundation (nonlinear soil springs) shall be taken into account. The ultimate available displacement capacity is defined as the displacement corresponding to a maximum of 20% load reduction from the peak load, or to a specified stress-strain failure limit (surface), whichever occurs first.

The following parameters shall be used to define stress-strain failure limit (surface):

$$\sigma_c^{comp} = 0.85 f_c'$$
$$\sigma_c^{ten} = f_r = 5\sqrt{f_c'}$$
$$\varepsilon_c^{comp} = 0.003$$
$$\sigma_s = F_u$$
$$\varepsilon_s = 0.12$$

where
σ_c^{comp} = maximum concrete stress under uniaxial compression
σ_c^{ten} = maximum concrete stress under uniaxial tension
f_c' = specified compressive concrete strength
σ_s = maximum steel stress under uniaxial tension
ε_s = maximum steel strain under uniaxial tension
ε_c^{comp} = maximum concrete strain under uniaxial compression

16.6.6 Seismic Response Modification Devices

General

The SRMDs include the energy dissipation and seismic isolation devices. The basic purpose of energy dissipation devices is to increase the effective damping of the structure by adding dampers to the structure thereby reducing forces, deflections, and impact effects. The basic purpose of isolation devices is to change the fundamental mode of vibration so that the structure is subjected to lower earthquake forces. However, the reduction in force may be accompanied by an increase in displacement demand that shall be accommodated within the isolation system and any adjacent structures.

Determination of SRMDs Properties

The properties of SRMDs shall be determined by the specified testing program. References are made to AASHTO-Guide [2], Caltrans [18], and JMC [30]. The following items shall be addressed rigorously in the testing specification:

- Scales of specimens; at least two full-scale tests are required
- Loading (including lateral and vertical) history and rate
- Durability — design life
- Expected levels of strength and stiffness deterioration

16.7   Performance Acceptance Criteria

16.7.1   General

To achieve the performance objectives stated in Section 16.2, the various structural components shall satisfy the acceptable demand/capacity ratios, DC_{accept}, specified in this section. The general design format is given by the formula:

$$\frac{\text{Demand}}{\text{Capacity}} \leq DC_{accept} \qquad (16.20)$$

where demand, in terms of various factored forces (moment, shear, axial force, etc.), and deformations (displacement, rotation, etc.) shall be obtained by the nonlinear inelastic dynamic time history analysis – Level I defined in Section 16.5; and capacity, in terms of factored strength and deformations, shall be obtained according to the provisions set forth in Section 16.6. For members subjected to combined loadings, the definition of force **D/C ratios** is given in the Appendix.

16.7.2   Structural Component Classifications

Structural components are classified into two categories: critical and other. It is the aim that **other components** may be permitted to function as "fuses" so that the **critical components** of the bridge system can be protected during FEE and SEE. As an example, Table 16.1 shows structural component classifications and their definition for SFOBB West Span components.

16.7.3   Steel Structures

General Design Procedure

Seismic design of steel members shall be in accordance with the procedure shown in Figure 16.14. Seismic retrofit design of steel members shall be in accordance with the procedure shown in Figure 16.15.

Connections

Connections shall be evaluated over the length of the seismic event. For connecting members with force D/C ratios larger than one, 25% greater than the nominal capacities of the connecting members shall be used for connection design.

TABLE 16.1 Structural Component Classification

Component classification	Definition	Example (SFOBB West Spans)
Critical	Components on a critical path that carry bridge gravity load directly. The loss of capacity of these components would have serious consequences on the structural integrity of the bridge	Suspension cables Continuous trusses Floor beams and stringers Tower legs Central anchorage A-Frame Piers W-1 and W2 Bents A and B Caisson foundations Anchorage housings Cable bents
Other	All components other than *critical*	All other components

Note: Structural components include members and connections.

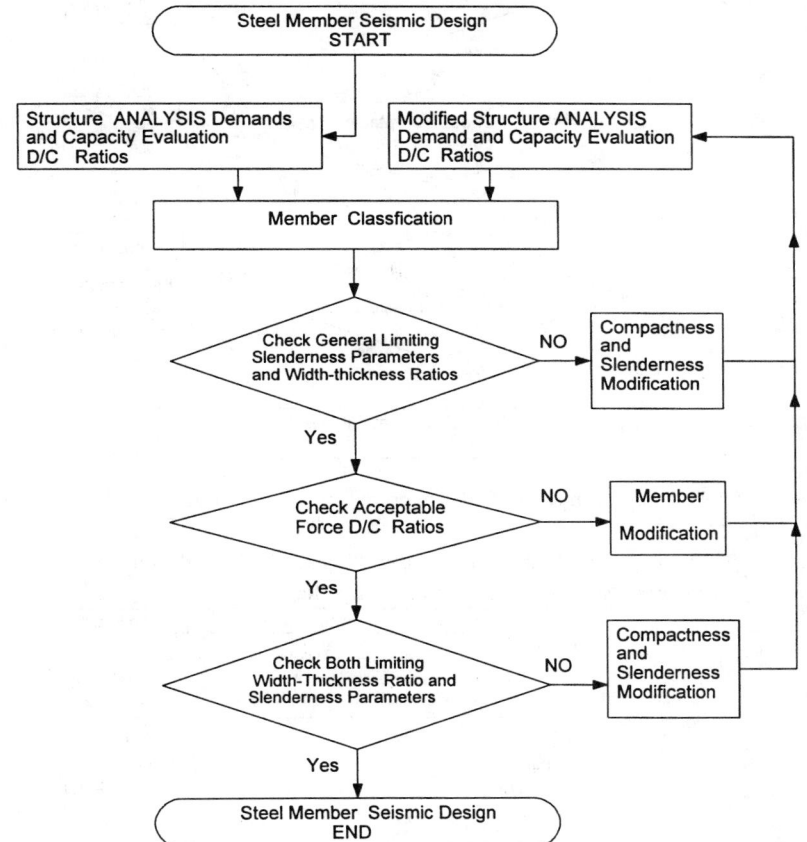

Figure 16.14 Steel member seismic design procedure.

General Limiting Slenderness Parameters and Width-Thickness Ratios

For all steel members regardless of their force D/C ratios, slenderness parameters λ_c for axial load dominant members, and λ_b for flexural dominant members shall not exceed the limiting values ($0.9\lambda_{cr}$ or $0.9\lambda_{br}$ for *critical*, λ_{cr} or λ_{br} for *other*) shown in Table 16.2.

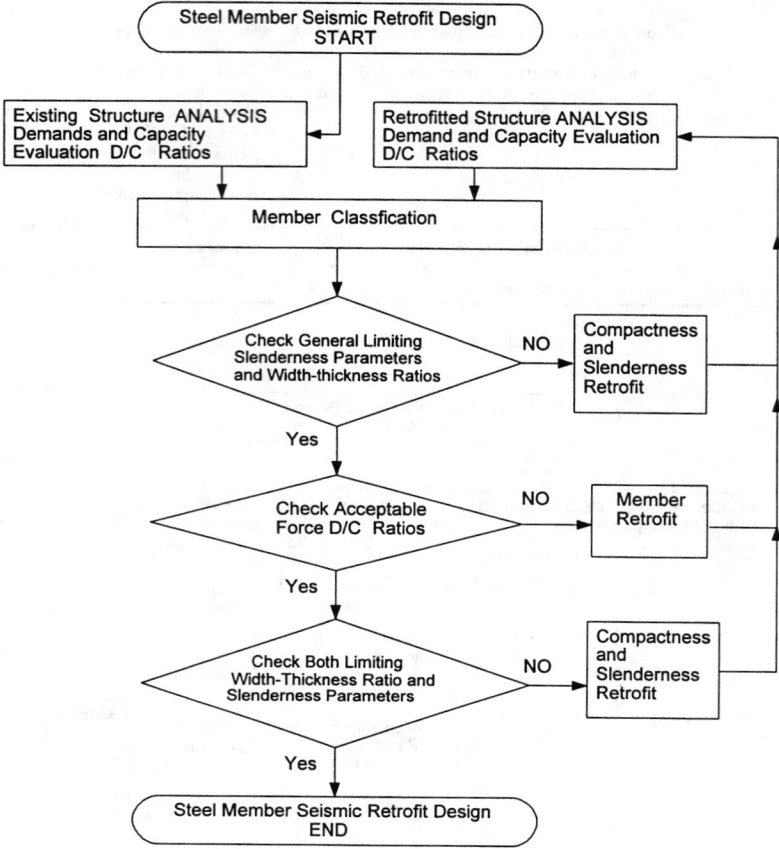

Figure 16.15   Steel member seismic retrofit design procedure.

TABLE 16.2   Acceptable Force Demand/Capacity Ratios and Limiting
Slenderness Parameters and Width/Thickness Ratios

| | | Limiting ratios | | Acceptable |
	Member classification	Slenderness parameter $(\lambda_c \text{ and } \lambda_b)$	Width/thickness λ $(b/t \text{ or } h/t_w)$	force D/C ratio D/C_{accept}
	Axial load	$0.9\,\lambda_{cr}$	λ_r	$DC_r = 1.0$
	dominant	λ_{cpr}	λ_{pr}	$1.0 \sim 1.2$
	$P_u/P_n \geq M_u/M_n$	λ_{cp}	λ_p	$DC_p = 1.2$
Critical	Flexural moment	$0.9\,\lambda_{br}$	λ_r	$DC_r = 1.0$
	dominant	λ_{bpr}	λ_{pr}	$1.2 \sim 1.5$
	$M_u/M_n > P_u/P_n$	λ_{bp}	λ_p	$DC_p = 1.5$
	Axial load	λ_{cr}	λ_r	$DC_r = 1.0$
	dominant	λ_{cpr}	λ_{pr}	$1.0 \sim 2.0$
	$P_u/P_n \geq M_u/M_n$	λ_{cp}	$\lambda_{p-\text{Seismic}}$	$DC_p = 2$
Other	Flexural moment	λ_{br}	λ_r	$DC_r = 1.0$
	dominant	λ_{bpr}	λ_{pr}	$1.0 \sim 2.5$
	$M_u/M_n > P_u/P_n$	λ_{bp}	$\lambda_{p-\text{Seismic}}$	$DC_p = 2.5$

Acceptable Force D/C Ratios and Limiting Values

Acceptable force D/C ratios, DC_{accept} and associated limiting slenderness parameters and width-thickness ratios for various members are specified in Table 16.2.

For all members with D/C ratios larger than one, slenderness parameters and width-thickness ratios shall not exceed the limiting values specified in Table 16.2. For existing steel members with D/C ratios less than one, width-thickness ratios may exceed λ_r specified in Table 16.3 and AISC-LRFD [4].

The following symbols are used in Table 16.2:

M_n = nominal moment strength of a member determined by Section 16.6
P_n = nominal axial strength of a member determined by Section 16.6
λ = width-thickness (b/t or h/t_w) ratio of compressive elements
λ_c = $(KL/r\pi)\sqrt{F_y/E}$, slenderness parameter of axial load dominant members
λ_b = L/r_y, slenderness parameter of flexural moment dominant members
λ_{cp} = 0.5, limiting column slenderness parameter for 90% of the axial yield load based on AISC-LRFD [4] column curve
λ_{bp} = limiting beam slenderness parameter for plastic moment for seismic design
λ_{cr} = 1.5, limiting column slenderness parameter for elastic buckling based on AISC-LRFD [4] column curve
λ_{br} = limiting beam slenderness parameter for elastic lateral torsional buckling

$$\lambda_{br} = \begin{cases} \dfrac{57,000\sqrt{JA}}{M_r} & \text{for solid rectangular bars and box sections} \\ \dfrac{X_1}{F_L}\sqrt{1+\sqrt{1+X_2 F_L^2}} & \text{for doubly symmetric I-shaped members and channels} \end{cases}$$

$$M_r = \begin{cases} F_L S_x & \text{for I-shaped member} \\ F_{yf} S_x & \text{for solid rectangular and box section} \end{cases}$$

$$X_1 = \frac{\pi}{S_x}\sqrt{\frac{EGJA}{2}}; \qquad X_2 = \frac{4C_w}{I_y}\left(\frac{S_x}{GJ}\right)^2; \qquad F_L = \text{smaller} \begin{cases} F_{yw} \\ F_{yf} - F_r \end{cases}$$

where

A = cross-sectional area, in.^{2}
L = unsupported length of a member
J = torsional constant, in.^{4}
r = radius of gyration, in.
r_y = radius of gyration about minor axis, in.
F_{yw} = yield stress of web, ksi
F_{yf} = yield stress of flange, ksi
E = modulus of elasticity of steel (29,000 ksi)
G = shear modulus of elasticity of steel (11,200 ksi)
S_x = section modulus about major axis, in.^{3}
I_y = moment of inertia about minor axis, in.^{4}
C_w = warping constant, in.^{6}

For doubly symmetric and singly symmetric I-shaped members with compression flange equal to or larger than the tension flange, including hybrid members (strong axis bending):

$$\lambda_{bp} = \begin{cases} \dfrac{[3,600+2,200(M_1/M_2)]}{F_y} & \text{for } \textit{other} \text{ members} \\ \dfrac{300}{\sqrt{F_{yf}}} & \text{for } \textit{critical} \text{ members} \end{cases} \qquad (16.21)$$

TABLE 16.3   Limiting Width-Thickness Ratio

No	Description of elements	Examples	Width-thickness ratios	λ_r	λ_p	$\lambda_{p\text{-}Seismic}$
	UNSTIFFENED ELEMENTS					
1	Flanges of I-shaped rolled beams and channels in flexure	Figures 16.16a, c	b/t	$\dfrac{141}{\sqrt{F_y-10}}$	$\dfrac{65}{\sqrt{F_y}}$	$\dfrac{52}{\sqrt{F_y}}$
2	Outstanding legs of pairs of angles in continuous contact; flanges of channels in axial compression; angles and plates projecting from beams or compression members	Figures 16.16d, e, f	b/t	$\dfrac{95}{\sqrt{F_y}}$	$\dfrac{65}{\sqrt{F_y}}$	$\dfrac{52}{\sqrt{F_y}}$
	STIFFENED ELEMENTS					
3	Flanges of square and rectangular box and hollow structural section of uniform thickness subject to bending or compression; flange cover plates and diaphragm plates between lines of fasteners or welds.	Figure 16.16b	b/t	$\dfrac{238}{\sqrt{F_y}}$	$\dfrac{190}{\sqrt{F_y}}$	$110/\sqrt{F_y}$ (tubes) $150/\sqrt{F_y}$ (others)
4	Unsupported width of cover plates perforated with a succession of access holes	Figure 16.16d	b/t	$\dfrac{317}{\sqrt{F_y}}$	$\dfrac{253}{\sqrt{F_y}}$	$\dfrac{152}{\sqrt{F_y}}$
5	All other uniformly compressed stiffened elements, i.e., supported along two edges.	Figures 16.16a, c, d, f	b/t h/t_w	$\dfrac{253}{\sqrt{F_y}}$	$\dfrac{190}{\sqrt{F_y}}$	$110/\sqrt{F_y}$ (w/lacing) $150/\sqrt{F_y}$ (others)
6	Webs in flexural compression	Figures 16.16a, c, d, f	h/t_w	$\dfrac{970}{\sqrt{F_y}}$	$\dfrac{640}{\sqrt{F_y}}$	$\dfrac{520}{\sqrt{F_y}}$
7	Webs in combined flexural and axial compression	Figures 16.16a, c, d, f	h/t_w	$\dfrac{970}{\sqrt{F_y}}\times\left(1-\dfrac{0.74P}{\phi_b P_y}\right)$	For $P_u \le 0.125\phi_b P_y$ $\dfrac{640}{\sqrt{F_y}}\left(1-\dfrac{2.75P}{\phi_b P_y}\right)$ For $P_u > 0.125\phi_b P_y$ $\dfrac{191}{\sqrt{F_y}}\left(2.33-\dfrac{P}{\phi_b P_y}\right)$ $\ge\dfrac{253}{\sqrt{F_y}}$	For $P_u \le 0.125\phi_b P_y$ $\dfrac{520}{\sqrt{F_y}}\left(1-\dfrac{1.54P}{\phi_b P_y}\right)$ For $P_u > 0.125\phi_b P_y$ $\dfrac{191}{\sqrt{F_y}}\left(2.33-\dfrac{P}{\phi_b P_y}\right)$ $\ge\dfrac{253}{\sqrt{F_y}}$
8	Longitudinally stiffened plates in compression	Figure 16.16e	b/t	$\dfrac{113\sqrt{k}}{\sqrt{F_y}}$	$\dfrac{95\sqrt{k}}{\sqrt{F_y}}$	$\dfrac{75\sqrt{k}}{\sqrt{F_y}}$

Notes:
1. Width-thickness ratios shown in **bold** are from AISC-LRFD [4] and AISC-Seismic Provisions [5].
2. k = buckling coefficient specified by Article 6.11.2.1.3a of AASHTO-LRFD [1]

for $n=1$, $k = (8I_s/bt^3)^{1/3} \le 4.0$ for $n=2,3,4$ and 5, $k = (14.3I_s/bt^3n^4)^{1/3} \le 4.0$

n = number of equally spaced longitudinal compression flange stiffeners

I_s = moment of inertia of a longitudinal stiffener about an axis parallel to the bottom flange and taken at the base of the stiffener

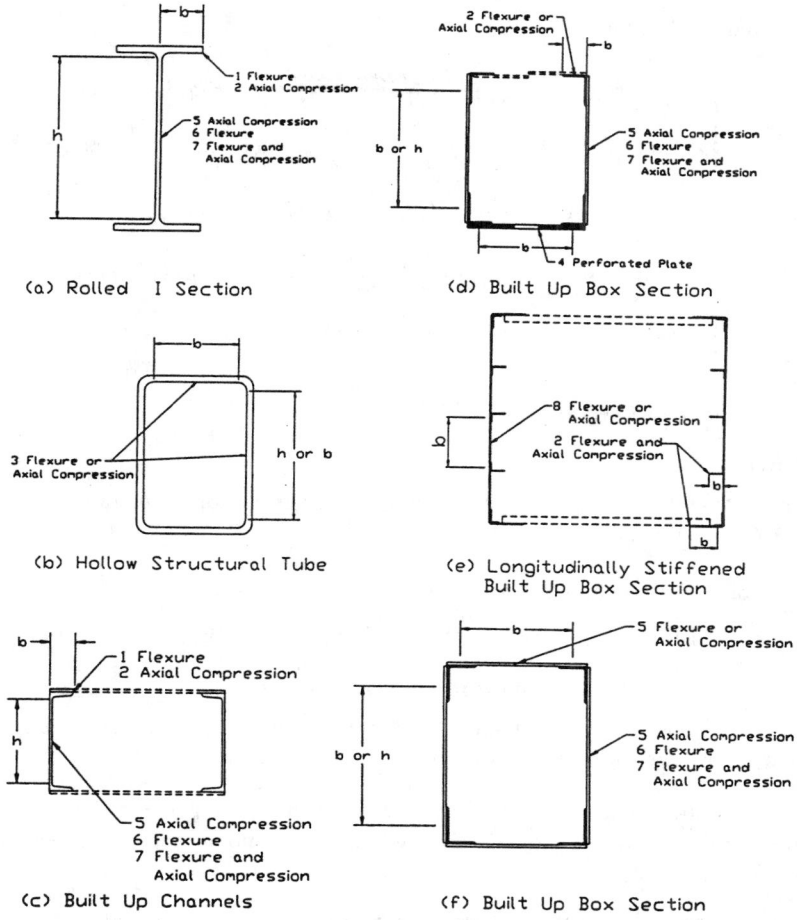

Figure 16.16 Typical cross-sections for steel members (SFOBB west spans).

in which

M_1 =   larger moment at end of unbraced length of beam
M_2 =   smaller moment at end of unbraced length of beam
(M_1/M_2) =   positive when moments cause reverse curvature and negative for single curvature

For solid rectangular bars and symmetric box beam (strong axis bending):

$$\lambda_{bp} = \begin{cases} \frac{[5,000+3,000(M_1/M_2)]}{F_y} \geq \frac{3,000}{F_y} & \text{for } \textit{other} \text{ members} \\[2ex] \frac{3,750}{M_p}\sqrt{J\,A} & \text{for } \textit{critical} \text{ members} \end{cases} \qquad (16.22)$$

in which

M_p =   plastic moment $(Z_x F_y)$
Z_x =   plastic section modulus about major axis
$\lambda_r, \lambda_p, \lambda_{p-\text{Seismic}}$ are limiting width thickness ratios specified by Table 16.3

$$\lambda_{pr} = \begin{cases} \left[\lambda_p + (\lambda_r - \lambda_p)\left(\frac{DC_p - DC_{\text{accept}}}{DC_p - DC_r}\right)\right] & \text{for } \textit{critical} \text{ members} \\[2ex] \left[\lambda_{p-\text{Seismic}} + (\lambda_r - \lambda_{p-\text{Seismic}})\left(\frac{DC_p - DC_{\text{accept}}}{DC_p - DC_r}\right)\right] & \text{for } \textit{other} \text{ members} \end{cases} \qquad (16.23)$$

For axial load dominant members $(P_u/P_n \geq M_u/M_n)$

$$\lambda_{cpr} = \begin{cases} \lambda_{cp} + \left(0.9\lambda_{cr} - \lambda_{cp}\right)\left(\frac{DC_p - DC_{\text{accept}}}{DC_p - DC_r}\right) & \text{for } \textit{critical} \text{ members} \\ \lambda_{cp} + \left(\lambda_{cr} - \lambda_{cp}\right)\left(\frac{DC_p - DC_{\text{accept}}}{DC_p - DC_r}\right) & \text{for } \textit{other} \text{ members} \end{cases} \quad (16.24)$$

For flexural moment dominant members $(M_u/M_n > P_u/P_n)$

$$\lambda_{bpr} = \begin{cases} \lambda_{bp} + \left(0.9\lambda_{br} - \lambda_{bp}\right)\left(\frac{DC_p - DC_{\text{accept}}}{DC_p - DC_r}\right) & \text{for } \textit{critical} \text{ members} \\ \lambda_{bp} + \left(\lambda_{br} - \lambda_{bp}\right)\left(\frac{DC_p - DC_{\text{accept}}}{DC_p - DC_r}\right) & \text{for } \textit{other} \text{ members} \end{cases} \quad (16.25)$$

16.7.4 Concrete Structures

General

For all concrete compression members regardless of their force D/C ratios, slenderness parameters KL/r shall not exceed 60.

> For *critical* components, force $DC_{\text{accept}} = 1.2$ and deformation $DC_{\text{accept}} = 0.4$.
>
> For *other* components, force $DC_{\text{accept}} = 2.0$ and deformation $DC_{\text{accept}} = 0.67$.

Beam-Column (Bent Cap) Joints

For concrete box girder bridges, the beam-column (bent cap) joints shall be evaluated and designed in accordance with the following guidelines [16, 40]:

1. Effective Superstructure Width — The effective width of a superstructure (box girder) on either side of a column to resist longitudinal seismic moment at bent (support) shall not be taken as larger than the superstructure depth.

   - The immediately adjacent girder on either side of a column within the effective superstructure width is considered effective.

   - Additional girders may be considered effective if refined bent-cap torsional analysis indicates that the additional girders can be mobilized.

2. Minimum Bent-Cap Width — Minimum cap width outside the column shall not be less than $D/4$ (D is column diameter or width in that direction) or 2 ft (0.61 m).

3. Acceptable Joint Shear Stress

   - For existing unconfined joints, acceptable principal tensile stress shall be taken as $3.5\sqrt{f_c'}$ psi ($0.29\sqrt{f_c'}$ MPa). If the principal tensile stress demand exceeds this value, the joint shear reinforcement specified in (4) shall be provided.

   - For new joints, acceptable principal tensile stress shall be taken as $12\sqrt{f_c'}$ psi ($1.0\sqrt{f_c'}$ MPa).

   - For existing and new joints, acceptable principal compressive stress shall be taken as $0.25 f_c'$.

4. Joint Shear Reinforcement

   - Typical flexure and shear reinforcement (see Figures 16.17 and 16.18) in bent caps shall be supplemented in the vicinity of columns to resist joint shear. All joint shear reinforcement shall be well distributed and provided within D/2 from the face of column.

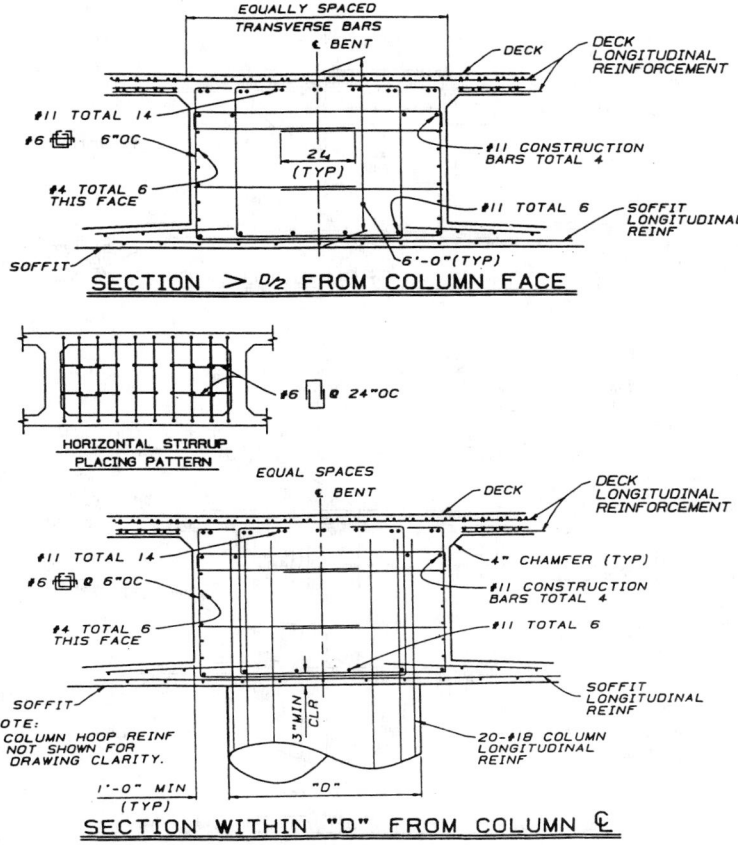

Figure 16.17 Example cap joint shear reinforcement — skews 0° to 20°.

- Vertical reinforcement including cap stirrups and added bars shall be 20% of the column reinforcement anchored into the joint. Added bars shall be hooked around main longitudinal cap bars. Transverse reinforcement in the joint region shall consist of hoops with a minimum reinforcement ratio of 0.4 (column steel area)/(embedment length of column bar into the bent cap)^2.

- Horizontal reinforcement shall be stitched across the cap in two or more intermediate layers. The reinforcement shall be shaped as hairpins, spaced vertically at not more than 18 in. (457 mm). The hairpins shall be 10% of column reinforcement. Spacing shall be denser outside the column than that used within the column.

- Horizontal side face reinforcement shall be 10% of the main cap reinforcement including top and bottom steel.

- For bent caps skewed greater than 20°, the vertical J-bars hooked around longitudinal deck and bent cap steel shall be 8% of the column steel (see Figure 16.18). The J-bars shall be alternatively 24 in. (600 mm) and 30 in. (750 mm) long and placed within a width of the column dimension on either side of the column centerline.

- All vertical column bars shall be extended as high as practically possible without interfering with the main cap bars.

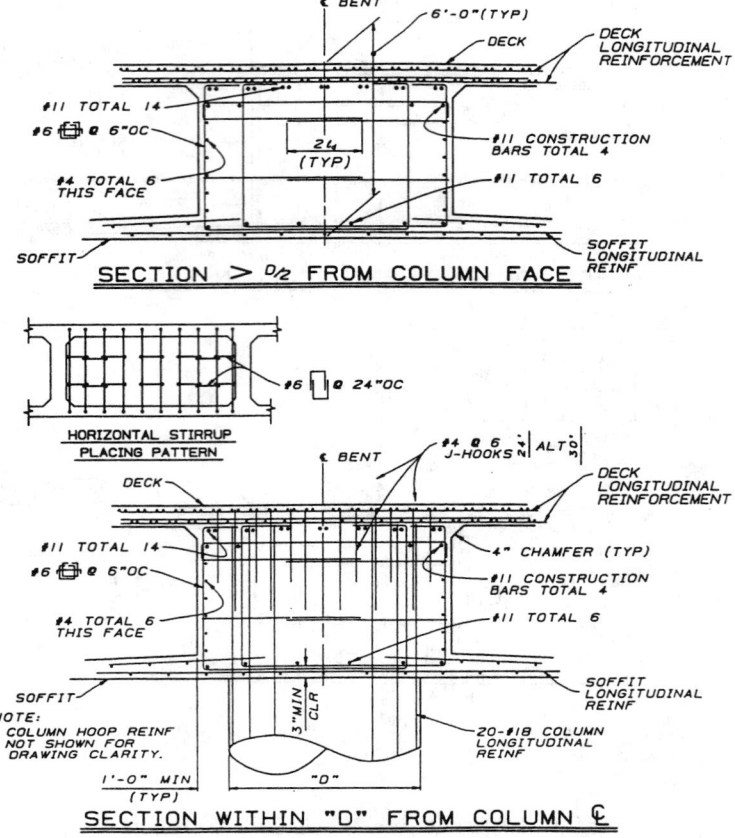

Figure 16.18   Example cap joint shear reinforcement — skews > 20°.

16.7.5   Seismic Response Modification Devices

General

Analysis methods specified in Section 16.5 shall apply for determining seismic design forces and displacements on SRMDs. Properties or capacities of SRMDs shall be determined by specified tests.

Acceptance Criteria

SRMDs shall be able to perform their intended function and maintain their design parameters for the design life (for example, 40 years) and for an ambient temperature range (for example from 30° to 125°F). The devices shall have accessibility for periodic inspections, maintenance, and exchange. In general, the SRMDs shall satisfy at least the following requirements:

- To remain stable and provide increasing resistance with the increasing displacement. Stiffness degradation under repeated cyclic load is unacceptable.

- To dissipate energy within the design displacement limits. For example: provisions may be made to limit the maximum total displacement imposed on the device to prevent device displacement failure, or the device shall have a displacement capacity 50% greater than the design displacement.

- To withstand or dissipate the heat build-up during reasonable seismic displacement time history.

- To survive for the number of cycles of displacement expected under wind excitation during the life of the device and to function at maximum wind force and displacement levels for at least, for example, five hours.

Defining Terms

Bridge: A structure that crosses over a river, bay, or other obstruction, permitting the smooth and safe passage of vehicles, trains, and pedestrians.

Buckling model interaction: A behavior phenomenon of compression built-up member; that is, interaction between the individual (or local) buckling mode and the global buckling mode.

Built-up member: A member made of structural metal elements that are welded, bolted, and/or riveted together.

Capacity: factored strength and deformation capacity obtained according to specified provisions.

Critical components: Structural components on a critical path that carry bridge gravity load directly. The loss of capacity of these components would have serious consequences on the structural integrity of the bridge.

Damage index: A ratio of elastic displacement demand to ultimate displacement.

D/C ratio: A ratio of demand to capacity.

Demands: In terms of various forces (moment, shear, axial force, etc.) and deformation (displacement, rotation, etc.) obtained by structural analysis.

Ductility: A nondimensional factor, i.e., ratio of ultimate deformation to yield deformation.

Effective length factor K: A factor that when multiplied by actual length of the end-restrained column gives the length of an equivalent pin-ended column whose elastic buckling load is the same as that of the end-restrained column

Functionality evaluation earthquake (FEE): An earthquake that has a mean return period of 300 years with a 40% probability of exceedance during the expected life of the bridge.

Latticed member: A member made of metal elements that are connected by lacing bars and batten plates.

LRFD (Load and Resistance Factor Design): A method of proportioning structural components (members, connectors, connecting elements, and assemblages) such that no applicable limit state is exceeded when the structure is subjected to all appropriate load combinations.

Limit states: Those conditions of a structure at which it ceases to satisfy the provision for which it was designed.

No-collapse-based design: Design that is based on survival limit state. The overall design concern is to prevent the bridge from catastrophic collapse and to save lives.

Other components: All components other than *critical.*

Performance-based seismic design: Design that is based on bridge performance requirements. The design philosophy is to accept some repairable earthquake damage and to keep bridge functional performance after earthquakes.

Safety evaluation earthquake (SEE): An earthquake that has a mean return period in the range of 1000 to 2000 years.

Seismic design: Design and analysis considering earthquake loads.

Seismic response modification devices (SRMDs): Seismic isolation and energy dissipation devices including isolators, dampers, or isolation/dissipation (I/D) devices.

Ultimate deformation: Deformation refers to a loading state at which structural system or a structural member can undergo change without losing significant load-carrying capacity. The ultimate deformation is usually defined as the deformation corresponding to a load that drops a maximum of 20% from the peak load.

Yield deformation: Deformation corresponds to the points beyond which the structure starts to respond inelastically.

Acknowledgments

First, we gratefully acknowledge the support from Professor Wai-Fah Chen, Purdue University. Without his encouragement, drive, and review, this chapter would not have been done in this timely manner.

Much of the material presented in this chapter was taken from the *San Francisco – Oakland Bay Bridge West Spans Seismic Retrofit Design Criteria (Criteria)* which was developed by Caltrans engineers. Substantial contribution to the *Criteria* came from the following: Lian Duan, Mark Reno, Martin Pohll, Kevin Harper, Rod Simmons, Susan Hida, Mohamed Akkari, and Brian Sutliff.

We would like to acknowledge the careful review of the SFOBB *Criteria* by Caltrans engineers: Abbas Abghari, Steve Altman, John Fujimoto, Don Fukushima, Richard Heninger, Kevin Keady, John Kung, Mike Keever, Rick Land, Ron Larsen, Brian Maroney, Steve Mitchell, Ramin Rashedi, Jim Roberts, Bob Tanaka, Vinacs Vinayagamoorthy, Ray Wolfe, Ray Zelinski, and Gus Zuniga.

The SFOBB *Criteria* was also reviewed by the Caltrans Peer Review Committee for the Seismic Safety Review of the Toll Bridge Retrofit Designs: Chuck Seim (Chairman), T.Y. Lin International; Professor Frieder Seible, University of California at San Diego; Professor Izzat M. Idriss, University of California at Davis; and Gerard Fox, Structural Consultant in New York. We are thankful for their input.

We are also appreciative of the review and suggestions of I-Hong Chen, Purdue University; Professor Ahmad Itani, University of Nevada at Reno; Professor Dennis Mertz, University of Delaware; and Professor Chia-Ming Uang, University of California at San Diego.

We express our sincere thanks to Enrico Montevirgen and Jerry Helm for their careful preparation of figures.

Finally, we gratefully acknowledge the continuous support of the California Department of Transportation.

References

[1] AASHTO. 1994. *LRFD Bridge Design Specifications,* 1st ed., American Association of State Highway and Transportation Officials, Washington, D.C.

[2] AASHTO. 1997. *Guide Specifications for Seismic Isolation Design,* American Association of State Highway and Transportation Officials, Washington, D.C.

[3] AISC. 1994. *Manual of Steel Construction — Load and Resistance Factor Design, Vol. 1-2,* 2nd ed., American Institute of Steel Construction, Chicago, IL.

[4] AISC. 1993. *Load and Resistance Factor Design Specification for Structural Steel Buildings,* 2nd ed., American Institute of Steel Construction, Chicago, IL.

[5] AISC. 1992. *Seismic Provisions for Structural Steel Buildings,* American Institute of Steel Construction, Chicago, IL.

[6] ASCE. 1971. *Plastic Design in Steel — A Guide and Commentary,* 2nd ed., American Society of Civil Engineers, New York.

[7] Astaneh-Asl, A. and Roberts, J. eds. 1996. *Seismic Design, Evaluation and Retrofit of Steel Bridges,* Proceedings of the Second U.S. Seminar, San Francisco, CA.

[8]   AWS. 1995. *Bridge Welding Code.* (ANSI/AASHTO/AWS D1.5-95), American Welding Society, Miami, FL.

[9]   Bazant , Z. P. and Cedolin, L. 1991. *Stability of Structures,* Oxford University Press, New York.

[10] Caltrans. 1993. *Design Criteria for I-880 Replacement,* California Department of Transportation, Sacramento, CA.

[11] Caltrans. 1994. *Design Criteria for SR-14/I-5 Replacement,* California Department of Transportation, Sacramento, CA.

[12] Caltrans. 1995. *Bridge Design Specifications,* California Department of Transportation, Sacramento, CA.

[13] Caltrans. 1995. *Bridge Memo to Designers (20-4),* California Department of Transportation, Sacramento, CA.

[14] Caltrans. 1996. *Guidelines for Generation of Response-Spectrum-Compatible Rock Motion Time History for Application to Caltrans Toll Bridge Seismic Retrofit Projects,* Caltrans Seismic Advisory Board, California Department of Transportation, Sacramento, CA.

[15] Caltrans. 1996. *Guidelines for Performing Site Response Analysis to Develop Seismic Ground Motions for Application to Caltrans Toll Bridge Seismic Retrofit Projects,* Caltrans Seismic Advisory Board, California Department of Transportation, Sacramento, CA.

[16] Caltrans. 1996. *Seismic Design Criteria for retrofit of the West Approach to the San Francisco-Oakland Bay Bridge,* Prepared by Keever, M., California Department of Transportation, Sacramento, CA.

[17] Caltrans. 1997. *San Francisco – Oakland Bay Bridge West Spans Seismic Retrofit Design Criteria,* Prepared by Reno, M. and Duan, L., edited by Duan, L., California Department of Transportation, Sacramento, CA.

[18] Caltrans. 1997. *Full Scale Isolation Bearing Testing Document* (Draft), Prepared by Mellon, D., California Department of Transportation, Sacramento, CA.

[19] Duan, L. and Chen, W.F. 1990. "A Yield Surface Equation for Doubly Symmetrical Section," *Structural Eng.,* 12(2), 114-119.
      CRC

[20] Duan, L. and Cooper, T. R. 1995. "Displacement Ductility Capacity of Reinforced Concrete Columns," *ACI Concrete Int.,* 17(11). 61-65.

[21] Duan, L. and Reno, M. 1995. "Section Properties of Latticed Members," *Research Report,* California Department of Transportation, Sacramento, CA.

[22] Duan, L., Reno, M., and Uang, C.M. 1997. "Buckling Model Interaction for Compression Built-up Members," *AISC Eng. J.* (in press).

[23] ENR. 1997. Seismic Superstar Billion Dollar California Freeway — Cover story: Rising form the Rubble, New Freeway Soars and Swirls Near Quake, Engineering News Records, Jan. 20, McGraw-Hill, New York.

[24] FHWA. 1987. *Seismic Design and Retrofit Manual for Highway Bridges,* Report No. FHWA-IP-87-6, Federal Highway Administration, Washington, D.C.

[25] FHWA. 1995. *Seismic Retrofitting Manual for Highway Bridges,* Publication No. FHWA-RD-94-052, Federal Highway Administration, Washington, D.C.

[26] Goyal, A. and Chopra, A.K. 1989. *Earthquake Analysis & Response of Intake-Outlet Towers,* EERC Report No. UCB/EERC-89/04, University of California, Berkeley, CA.

[27] Housner, G.W. 1990. *Competing Against Time,* Report to Governor George Deuknejian from The Governor's Broad of Inquiry on the 1989 Loma Prieta Earthquake, Sacramento, CA.

[28] Housner, G.W. 1994. *The Continuing Challenge — The Northridge Earthquake of January 17, 1994,* Report to Director, California Department of Transportation, Sacramento, CA.

[29] Institute of Industrial Science (IIS). 1995. *Incede Newsletter,* Special Issue, International Center for Disaster-Mitigation Engineering, Institute of Industrial Science, The University of Tokyo, Japan.

[30] Japan Ministry of Construction (JMC). 1994. *Manual of Menshin Design of Highway Bridges,* English Version: EERC, Report 94/10, University of California, Berkeley, CA.

[31] Kulak, G.L., Fisher, J.W., and Struik, J.H. 1987. *Guide to the Design Criteria for Bolted and Riveted Joints,* 2nd ed., John Wiley & Sons, New York.

[32] Liew, J.Y.R. 1992. "Advanced Analysis for Frame Design," Ph.D. Dissertation, Purdue University, West Lafayette, IN.

[33] McCormac, J. C. 1989. *Structural Steel Design, LRFD Method,* Harper & Row, New York.

[34] Park, R. and Paulay, T. 1975. Reinforced Concrete Structures, John Wiley & Sons, New York.

[35] Priestley, M.J.N., Seible, F., and Calvi, G.M. 1996. *Seismic Design and Retrofit of Bridges,* John Wiley & Sons, New York.

[36] Salmon, C. G. and Johnson, J. E. 1996. *Steel Structures: Design and Behavior, Emphasizing Load and Resistance Factor Design,* Fourth ed., HarperCollins College Publishers, New York.

[37] Uang, C. M. and Kleiser, M. 1997. "Cyclic Testing of Latticed Members for San Francisco-Oakland Bay Bridge," *Final Report,* Division of Structural Engineering, University of California at San Diego, La Jolla, CA.

[38] USS. 1968. *Steel Design Manual,* Brockenbrough, R.L. and Johnston, B.G., Eds., United States Steel Corporation, ADUSS 27-3400-02, Pittsburgh, PA.

[39] Xie, L. L. and Housner, G. W. 1996. *The Greater Tangshan Earthquake,* Vol. I and IV, California Institute of Technology, Pasadena, CA.

[40] Zelinski, R. 1994. *Seismic Design Momo Various Topics Preliminary Guidelines,* California Department of Transportation, Sacramento, CA.

Further Reading

[1] Chen, W.F. and Duan, L. 1998. *Handbook of Bridge Engineering,* (in press) CRC Press, Boca Raton, FL.

[2] Clough, R.W. and Penzien, J. 1993. *Dynamics of Structures,* 2nd ed., McGraw-Hill, New York.

[3] Fukumoto, Y. and Lee, G. C. 1992. *Stability and Ductility of Steel Structures under Cyclic Loading,* CRC Press, Boca Raton, FL.

[4] Gupta, A.K. 1992. *Response Spectrum Methods in Seismic Analysis and Design of Structures,* CRC Press, Boca Raton, FL.

Appendix A

16.A.1 Section Properties for Latticed Members

This section presents practical formulas proposed by Duan and Reno [21] for calculating section properties for latticed members.

Concept

It is generally assumed that section properties can be computed based on cross-sections of main components if the lacing bars and battens can assure integral action of the solid main components [33, 36]. To consider actual section integrity, reduction factors β_m for moment of inertia, and β_t for torsional constant are proposed depending on shear-flow transferring-capacity of lacing bars and connections. For clarity and simplicity, typical latticed members as shown in Figure 16.19 are discussed.

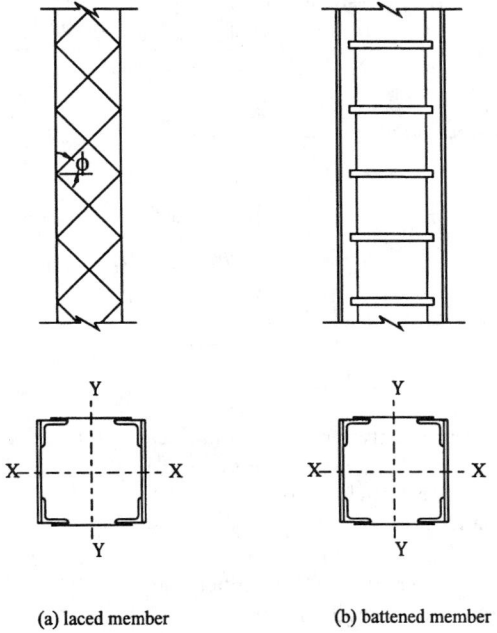

(a) laced member   (b) battened member

Figure 16.19   Typical latticed members.

Section Properties

1. Cross-sectional area — The contribution of lacing bars is assumed negligible. The cross-sectional area of latticed member is only based on main components.

$$A = \sum A_i \tag{16.26}$$

where A_i is cross-sectional area of individual component i.

2. Moment of inertia — For lacing bars or battens within web plane (bending about y-y axis in Figure 16.19)

$$I_{y-y} = \sum I_{(y-y)i} + \beta_m \sum A_i x_i^2 \tag{16.27}$$

where
I_{y-y} = moment of inertia of a section about y-y axis considering shear transferring capacity
I_i = moment of inertia of individual component i
x_i = distance between y-y axis and center of individual component i
β_m = reduction factor for moment of inertia and may be determined by the following formula:

For laced member (Figure 16.19a):

$$\beta_m = \dfrac{m \sin\phi \times \text{smaller of} \begin{cases} m_{\text{lacing}}\left(P_n^{\text{comp}} + P_n^{\text{ten}}\right) \\ m_{\text{lacing}} n_r A_r (0.6 F_u) \end{cases}}{F_{yf} A_f} \leq 1.0 \qquad (16.28a)$$

For battened member (Figure 16.19b):

$$\beta_m = \dfrac{m \times \text{smaller of} \begin{cases} m_{\text{batten}} A_b (0.6 F_{yw}) \\ m_{\text{batten}} (2 M_{p-\text{batten}}) / h \\ m_{\text{batten}} n_r A_r (0.6 F_u) \end{cases}}{F_{yf} A_f} \leq 1.0 \qquad (16.28b)$$

in which

ϕ	=	the angle between the diagonal lacing bar and the axis perpendicular to the member axis (see Figure 16.19)
A_b	=	cross-sectional area of batten plate
A_f	=	flange area
F_{yf}	=	yield strength of flange
F_{yw}	=	yield strength of web member (battens or lacing bars)
F_u	=	ultimate strength of rivets
m	=	number of panels between point of maximum moment to point of zero moment to either side (as an approximation, half of member length $L/2$ may be used)
m_{batten}	=	number of batten planes
m_{lacing}	=	number of lacing planes
n_r	=	number of rivets of connecting lacing bar and main component at one joint
h	=	depth of member in lacing plane
A_r	=	nominal area of rivet
$M_{p-\text{batten}}$	=	plastic moment of a batten plate about strong axis
P_n^{comp}	=	nominal compressive strength of lacing bar and can be determined by AISC-LRFD [4] column curve
P_n^{ten}	=	nominal tensile strength of lacing bar and can be determined by AISC-LRFD [4]

Since the section integrity mainly depends on the shear transference between various components, it is rational to introduce the β_m factor in Equation 16.27. As seen in Equations 16.28a and 16.28b, β_m is defined as the ratio of the shear capacity transferred by lacing bars/battens and connections to the shear-flow ($F_{yf} A_f$) required by the plastic bending moment of a fully integral section. For laced members, the shear transferring capacity is controlled by either lacing bars or connecting rivets, the smaller of the two values should be used in Equation 16.28a. For battened members, the shear transferring capacity is controlled by either pure shear strength of battens (0.6 $F_{yw} A_b$), or flexural strength of battens or connecting rivets, the smaller of the three values should be used in Equation 16.28b. It is important to point out that the limiting value unity for β_m implies a fully integral section when shear can be transferred fully by lacings and connections. For lacing bars or battens within flange plane (bending about x-x axis in Figure 16.19). The contribution of lacing bars is assumed negligible and only the main components are considered.

$$I_{x-x} = \sum I_{(x-x)i} + \sum A_i y_i^2 \qquad (16.29)$$

3. Elastic section modulus

$$S = \frac{I}{C} \qquad (16.30)$$

where
S = elastic section modulus of a section
C = distance from elastic neutral axis to extreme fiber

4. Plastic section modulus
   For lacing bars or battens within flange plane (bending about x-x axis in Figure 16.19)

$$Z_{x-x} = \sum y_i^* A_i^* \qquad (16.31)$$

For lacing bars or battens within web plane (bending about y-y axis in Figure 16.19)

$$Z_{y-y} = \beta_m \sum x_i^* A_i^* \qquad (16.32)$$

where
Z = plastic section modulus of a section about plastic neutral axis
x_i^* = distance between center of gravity of a section A_i^* and plastic neutral y-y axis
y_i^* = distance between center of gravity of a section A_i^* and plastic neutral x-x axis
A_i^* = cross-section area above or below plastic neutral axis

It should be pointed out that the plastic neutral axis is generally different from the elastic neutral axis. The plastic neutral axis is defined by equal plastic compression and tension forces for this section.

5. Torsional constant
   For a box-shaped section

$$J = \frac{4 (A_{\text{close}})^2}{\sum \frac{b_i}{t_i}} \qquad (16.33)$$

For an open thin-walled section

$$J = \sum \frac{b_i t_i^3}{3} \qquad (16.34)$$

where
A_{close} = area enclosed within mean dimension for a box
b_i = length of a particular segment of the section
t_i = average thickness of segment b_i
For determination of torsional constant of a latticed member, it is proposed that the lacing bars or batten plates be replaced by reduced equivalent thin-walled plates defined as:

$$A_{\text{equiv}} = \beta_t A_{\text{equiv}}^* \qquad (16.35)$$

For laced member (Figure 16.19a)

$$A^*_{\text{equiv}} = 3.12A_d \sin\phi \cos^2\phi \tag{16.36a}$$

For battened member (Figure 16.19b)

$$A^*_{\text{equiv}} = \frac{74.88}{\frac{2ah}{I_b} + \frac{a^2}{I_f}} \tag{16.36b}$$

$$t_{\text{equiv}} = \frac{A_{\text{equiv}}}{h} \tag{16.37}$$

where
a =   distance between two battens along member axis
A_{equiv} =   cross-section area of a thin-walled plate equivalent to lacing bars considering shear transferring capacity
A^*_{equiv} =   cross-section area of a thin-walled plate equivalent to lacing bars or battens assuming full section integrity
t_{equiv} =   thickness of equivalent thin-walled plate
A_d =   cross-sectional area of all diagonal lacings in one panel
I_b =   moment of inertia of a batten plate
I_f =   moment of inertia of a side solid flange about the weak axis
β_t =   reduction factor for torsion constant may be determined by the following formula:

For laced member (Figure 16.19a)

$$\beta_t = \frac{\cos\phi \times \text{smaller of} \begin{cases} P_n^{\text{comp}} + P_n^{\text{ten}} \\ n_r A_r (0.6F_u) \end{cases}}{0.6F_{yw}A^*_{\text{equiv}}} \leq 1.0 \tag{16.38a}$$

For battened member (Figure 16.19b)

$$\beta_t = \frac{\text{smaller of} \begin{cases} A_b(0.6F_{yw})h/a \\ 2M_{p-\text{batten}}/a \\ n_r A_r (0.6F_u)h/a \end{cases}}{0.6F_{yw}A^*_{\text{equiv}}} \leq 1.0 \tag{16.38b}$$

The torsional integrity is from lacings and battens. A reduction factor β_t, similar to that used for the moment of inertia, is introduced to consider section integrity when the lacing is weaker than the solid plate side of the section. β_t factor is defined as the ratio of the shear capacity transferred by lacing bars and connections to the shear-flow ($0.6F_{yw}A^*_{\text{equiv}}$) required by the equivalent thin-walled plate. It is seen that the limiting value of unity for β_t implies a fully integral section when shear in the equivalent thin-walled plate can be transferred fully by lacings and connections.

Based on the equal lateral stiffness principle, an equivalent thin-walled plate for a lacing plane, Equation 16.36a and 16.36b can be obtained by considering $E/G = 2.6$ for steel material and shape factor for shear $n = 1.2$ for a rectangular section.

6. Warping constant
   For a box-shaped section

$$C_w \approx 0 \tag{16.39}$$

For an I-shaped section

$$C_w = \frac{I_f h^2}{2} \tag{16.40}$$

where
I_f =   moment of inertia of one solid flange about the weak axis (perpendicular to the flange) of the cross-section
h =   distance between center of gravity of two flanges

16.A.2 Buckling Mode Interaction For Compression Built-up members

An important phenomenon of the behavior of these compressive **built-up members** is the interaction of buckling modes [9]; i.e., interaction between the individual (or local) component buckling (Figure 16.20a) and the global member buckling (Figure 16.20b). This section presents the practical approach proposed by Duan, Reno, and Uang [22] that may be used to determine the effects of interaction of buckling modes for capacity assessment of existing built-up members.

Buckling Mode Interaction Factor

To consider buckling mode interaction between the individual components and the global member, it is proposed that the usual effective length factor K of a built-up member be multiplied by a *buckling mode interaction factor*, γ_{LG}, that is,

$$K_{LG} = \gamma_{LG} K \tag{16.41}$$

where
K =   usual effective length factor of a built-up member
K_{LG} =   effective length factor considering buckling model interaction

Limiting Effective Slenderness Ratios

For practical design, the following two limiting effective slenderness ratios are suggested for consideration of buckling mode interaction:

$$K_a a / r_f = 1.1\sqrt{E/F} \tag{16.42}$$

$$K_a a / r_f = 0.75(KL/r) \tag{16.43}$$

where $(K_a a / r_f)$ is the largest effective slenderness ratio of individual components between connectors and (KL/r) is governing effective slenderness ratio of a built-up member.

The first limit Equation 16.42 is based on the argument that if an individual component is very short, the failure mode of the component would be material yielding (say 95% of the yield load), not member buckling. This implies that no interaction of buckling modes occurs when an individual component is very short.

The second limit Equation 16.43 is set forth by the current design specifications. Both the AISC-LRFD [4] and AASHTO-LRFD [1] imply that when the effective slenderness ratio $(K_a a / r_i)$ of each individual component between the connectors does not exceed 75% of the governing effective

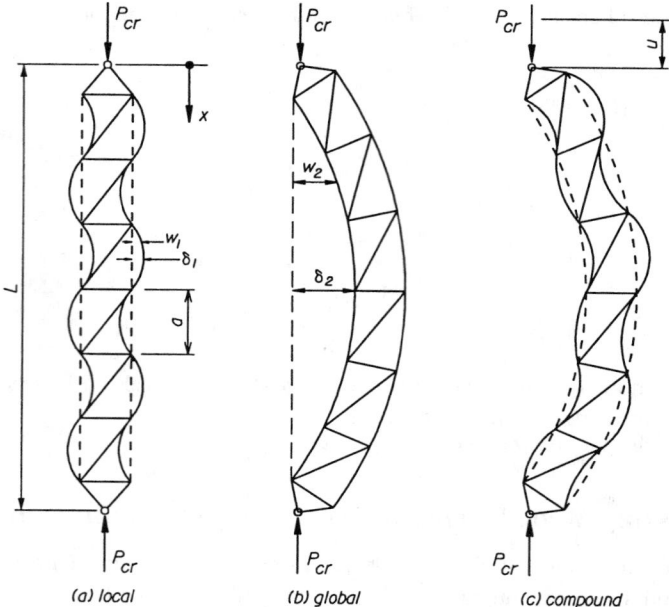

Figure 16.20 Buckling modes of built-up members.

slenderness ratio of the built-up member, no strength reduction due to interaction of buckling modes needs to be considered. The study reported by Duan, Reno, and Uang [22] has justified that the rule of $(K_a a/r_i) < 0.75(KL/r)$ is consistent with the theory.

Analytical Equation

The buckling mode interaction factor γ_{LG} is defined as:

$$\gamma_{LG} = \sqrt{\frac{P_G}{P_{cr}}} = \sqrt{\frac{\pi^2 EI}{(KL)^2 P_{cr}}} \tag{16.44}$$

where

P_{cr} = elastic buckling load considering buckling mode interaction
P_G = elastic buckling load without considering buckling mode interaction
L = unsupported member length
I = moment of inertia of a built-up member

γ_{LG} can be computed by the following equation:

$$\gamma_{LG} = \sqrt{\frac{1 + \alpha^2}{1 + \dfrac{\alpha^2}{1 + \dfrac{(\delta_o/a)^2 (K_a a/r_f)^2}{2\left[1 - \dfrac{(K_a a/r_f)^2}{(\gamma_{LG} KL/r)^2}\right]^3}}}} \tag{16.45}$$

where

(δ_o/a) = imperfection (out-of-straightness) parameter of individual component (see Figure 16.21)
α = separation as defined as:

$$\alpha = \frac{h}{2}\sqrt{\frac{A_f}{I_f}} = \frac{h}{2r_f} \qquad (16.46)$$

in which

I_f = moment inertia of one side individual components (see Figure 16.21)
A_f = cross-section area of one side individual components (see Figure 16.21)
h = depth of latticed member, distance between center of gravity of two flanges in lacing plane (see Figure 16.21)
r = radius of gyration of a built-in member
r_f = radius of gyration of individual component

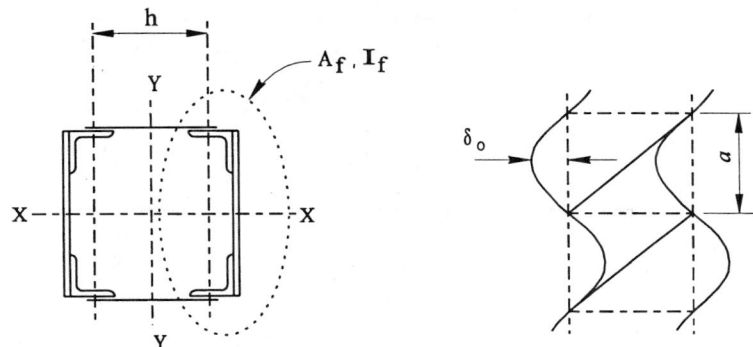

Figure 16.21 Typical cross-section and local components.

For widely separated built-up members with $\alpha \geq 2$, the buckling mode interaction factor γ_{LG} can be accurately estimated by the following equation on the conservative side:

$$\gamma_{LG} = \sqrt{1 + \frac{(\delta_o/a)^2(K_a a/r_j)^2}{2\left[1 - \frac{(K_a a/r_j)^2}{(\gamma_{LG}KL/r)^2}\right]^3}} \qquad (16.47)$$

Graphical Solution

Although γ_{LG} can be obtained by solving Equation 16.45, an iteration procedure must be used. For design purposes, solutions in chart forms are more desirable. Figures 16.22 to 16.24 provide engineers with alternative graphic solutions for widely separated built-up members with $\alpha \geq 2$. In these figures, the out-of-straightness ratios (δ_o/a) considered are 1/500, 1/1000, and 1/1500, and the effective slenderness ratios (KL/r) considered are 20, 40, 60, 100, and 140. In all these figures, the top line represents $KL/r = 20$ and the bottom line represents $KL/r = 140$.

16.A.3 Acceptable Force D/C Ratios and Limiting Values

Since it is uneconomical and impossible to design bridges to withstand seismic forces elastically, the non-linear inelastic responses of the bridges are expected. The performance-based criteria accepts certain seismic damage in some *other* components so that the *critical* components and the bridges will be kept essentially elastic and functional after the SEE and FEE. This section presents the concept of acceptable force D/C ratios, limiting member slenderness parameters and limiting width-thickness ratios, as well as expected ductility.

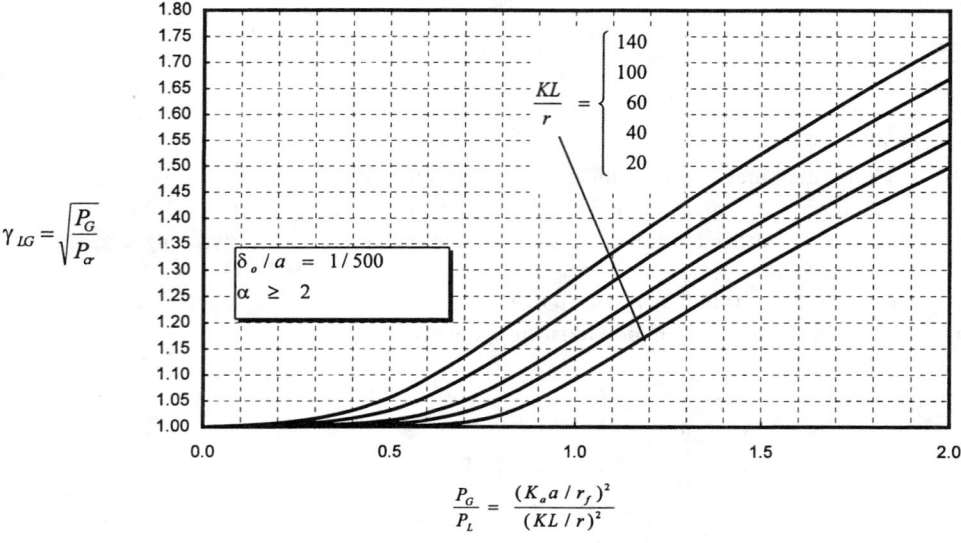

$$\gamma_{LG} = \sqrt{\frac{P_G}{P_\sigma}}$$

$$\frac{P_G}{P_L} = \frac{(K_a a / r_f)^2}{(KL / r)^2}$$

Figure 16.22   Buckling mode interaction factor γ_{LG} for $\delta_0/a = 1/500$.

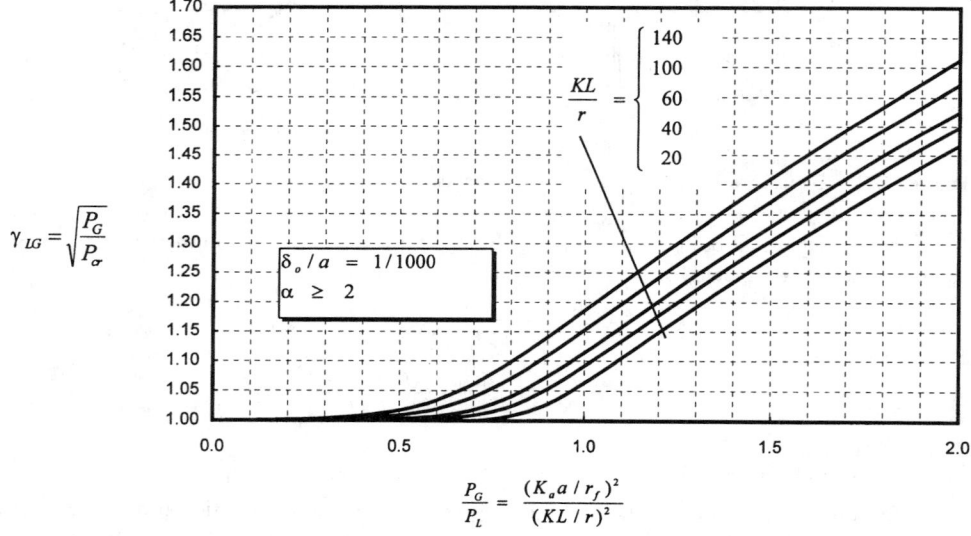

$$\gamma_{LG} = \sqrt{\frac{P_G}{P_\sigma}}$$

$$\frac{P_G}{P_L} = \frac{(K_a a / r_f)^2}{(KL / r)^2}$$

Figure 16.23   Buckling mode interaction factor γ_{LG} for $\delta_0/a = 1/1000$.

Definition of Force Demand/Capacity (D/C) Ratios

For members subjected to an individual load, force demand is defined as a factored individual force, such as factored moment, shear, axial force, etc. which shall be obtained by the nonlinear dynamic time history analysis – Level I specified in Section 16.5, and capacity is defined according to the provisions in Section 16.6.

For members subjected to combined loads, force D/C ratio is based on the force interaction. For example, for a member subjected to combined axial load and bending moment (Figure 16.25), force demand D is defined as the distance from the origin point O(0, 0) to the factored force point $d(P_u, M_u)$, and capacity C is defined as the distance from the origin point O(0, 0) to the point $c(P^*, M^*)$ on the specified interaction surface or curve (failure surface or curve).

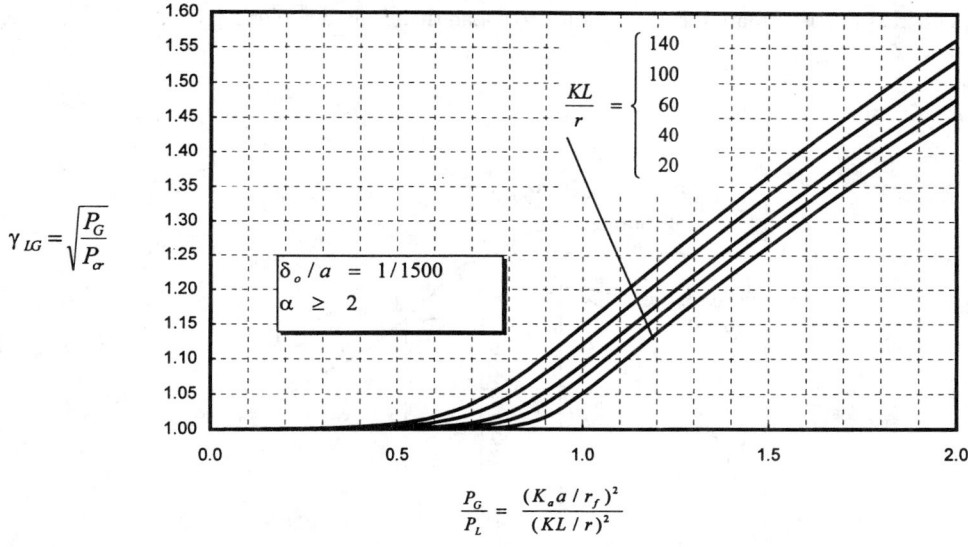

$$\gamma_{LG} = \sqrt{\frac{P_G}{P_\sigma}}$$

$$\frac{P_G}{P_L} = \frac{(K_a a / r_f)^2}{(KL / r)^2}$$

Figure 16.24 Buckling mode interaction factor γ_{LG} for $\delta_0/a = 1/1500$.

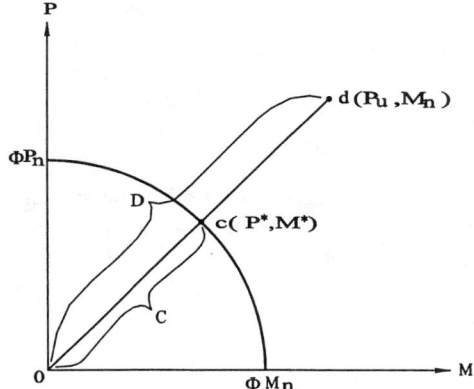

Figure 16.25 Definition of force D/C ratios for combined loadings.

Ductility and Load-Deformation Curves

"Ductility" is usually defined as a nondimensional factor, i.e., ratio of ultimate deformation to yield deformation [20, 34]. It is normally expressed by two forms:

1. curvature ductility ($\mu_\phi = \phi_u/\phi_y$)
2. displacement ductility ($\mu_\Delta = \Delta_u/\Delta_y$)

Representing section flexural behavior, *curvature ductility* is dependent on the section shape and material properties. It is based on the moment-curvature curve. Indicating structural system or member behavior, *displacement ductility* is related to both the structural configuration and its section behavior. It is based on the load-displacement curve.

A typical load-deformation curve including both the ascending and descending branches is shown in Figure 16.26. The **yield deformation** (Δ_y or ϕ_y) corresponds to a loading state beyond which the structure starts to respond inelastically. The **ultimate deformation** (Δ_u or ϕ_u) refers to the a loading state at which a structural system or a structural member can undergo without losing significant load-carrying capacity. It is proposed that the ultimate deformation (curvature or displacement) is defined as the deformation corresponding to a load that drops a maximum of 20% from the peak load.

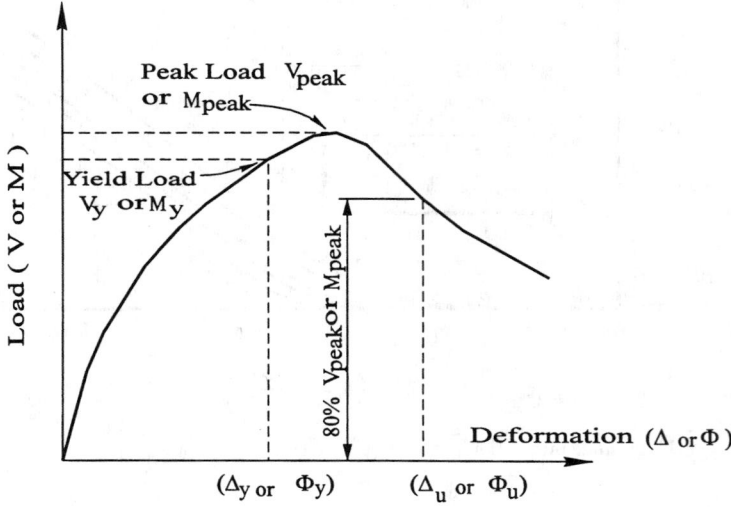

Figure 16.26   Load-deformation curves.

Force D/C Ratios and Ductility

The following discussion will give engineers a direct measure of seismic damage of structural components during an earthquake. Figure 16.27 shows a typical load-response curve for a single degree of freedom system. Displacement ductility is:

$$\mu_\Delta = \frac{\Delta_u}{\Delta_y} \qquad (16.48)$$

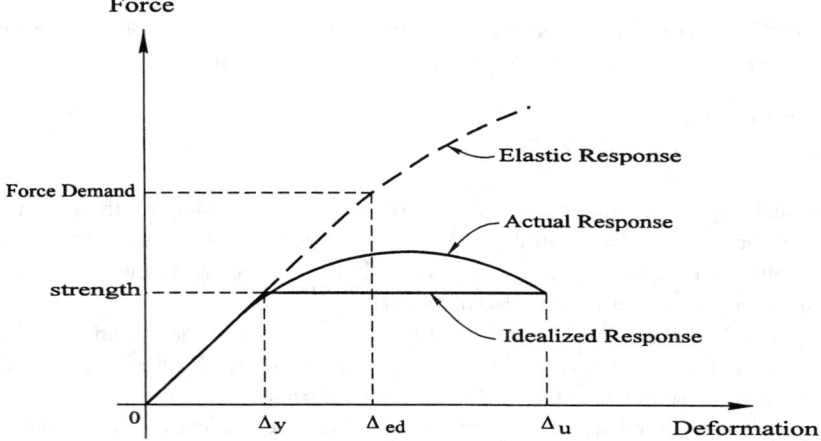

Figure 16.27   Response of a single degree of freedom of system.

A new term, *Damage Index,* is defined herein as the ratio of elastic displacement demand to ultimate displacement:

$$D_\Delta = \frac{\Delta_{ed}}{\Delta_u} \qquad (16.49)$$

When damage index $D_\Delta < 1/\mu_\Delta$ ($\Delta_{ed} < \Delta_y$), it implies that no damage occurs and the structure responds elastically; $1/\mu_\Delta < D_\Delta < 1.0$ indicates certain damages occur and the structure responds inelastically; $D_\Delta > 1.0$, however, means that a structural system collapses.

Based on the equal displacement principle, the following relationship is obtained as:

$$\frac{\text{Force Demand}}{\text{Force Capacity}} = \frac{\Delta_{ed}}{\Delta_y} = \mu_\Delta D_\Delta \qquad (16.50)$$

It is seen from Equation 16.50 that the force D/C ratio is related to both the structural characters in terms of ductility μ_Δ and the degree of damage in terms of damage index D_Δ. Table 16.4 shows detailed data for this relationship.

TABLE 16.4 Force D/C Ratio and Damage Index

Force D/C ratio	Damage index D_Δ	Expected system displacement ductility μ_Δ
1.0	No damage	No requirement
1.2	0.4	3.0
1.5	0.5	3.0
2.0	0.67	3.0
2.5	0.83	3.0

General Limiting Values

To ensure the important bridges have ductile load paths, general limiting slenderness parameters and width-thickness ratios are specified in Sections 16.7.3 and 16.7.4.

For steel members, λ_{cr} is the limiting member slenderness parameter for column elastic buckling and is taken as 1.5 from AISC-LRFD [4] and λ_{br} is the limiting member slenderness parameter for beam elastic torsional buckling and is calculated by AISC-LRFD [4]. For a *critical* member, a more strict requirement, 90% of those elastic buckling limits is proposed. Regardless of the force D/C ratios, all steel members must not exceed these limiting values. For existing steel members with D/C ratios less than one, this limit may be relaxed.

For concrete members, the general limiting parameter $KL/r = 60$ is proposed.

Acceptable Force D/C Ratios DC_{accept}

Acceptable force D/C ratios (DC_{accept}) depends on both the structural characteristics in terms of ductility μ_Δ and the degree of damage to the structure that can be accepted by practicing engineers in terms of damage index D_Δ.

To ensure a steel member has enough inelastic deformation capacity during an earthquake, it is necessary to limit both the member slenderness parameters and the section width-thickness ratios within the specified ranges so that the acceptable D/C ratios and the energy dissipation can be achieved.

Upper Bound Acceptable D/C Ratio DC_p

1. For *other* members, the large acceptable force D/C ratios ($DC_p = 2$ to 2.5) are proposed in Table 16.2. This implies that the damage index equals $0.67 \sim 0.83$ and more damage will occur at *other* members and large member ductility will be expected. To achieve this,

- the limiting width-thickness ratio was taken as $\lambda_{p-\text{Seismic}}$ from AISC-Seismic Provisions [5], which can provide flexural ductility 8 to 10.

- the limiting slenderness parameters were taken as λ_{bp} for flexural moment dominant members from AISC-LRFD [4], which can provide flexural ductility 8 to 10.

2. For *critical* members, small acceptable force D/C ratios ($DC_p = 1.2$ to 1.5) are proposed in Table 16.2, as the design purpose is to keep *critical* members essentially elastic and allow little damage (damage index equals to $0.4 \sim 0.5$). Thus, small member ductility is expected. To achieve this,

   - the limiting width-thickness ratio was taken as λ_p from AISC-LRFD [5], which can provide flexural ductility at least 4.

   - the limiting slenderness parameters were taken as λ_{bp} for flexural moment dominant members from AISC-LRFD [4], which can provide flexural ductility at least 4.

3. For axial load dominant members, the limiting slenderness parameters were taken as $\lambda_{cp} = 0.5$, corresponding to 90% of the axial yield load by the AISC-LRFD [4] column curve. This limit will provide the potential for axial load dominated members to develop expected inelastic deformation.

Lower Bound Acceptable D/C Ratio DC_r

The lower bound acceptable force D/C ratio $DC_{rc} = 1$ is proposed in Table 16.2. For $DC_{\text{accept}} = 1$, it is unnecessary to enforce more strict limiting values for members and sections. Therefore, the limiting slenderness parameters for elastic global buckling specified in Table 16.2 and the limiting width-thickness ratios specified in Table 16.3 for elastic local buckling are proposed.

Acceptable D/C Ratios Between Upper and Lower Bounds $DC_r < DC_{\text{accept}} < DC_p$

When acceptable force D/C ratios are between the upper and the lower bounds, $DC_r < DC_{\text{accept}} < DC_p$, a linear interpolation as shown in Figure 16.28 is proposed to determine the limiting slenderness parameters and width-thickness ratios. The following formulas can be used:

$$\lambda_{pr} = \begin{cases} \left[\lambda_p + (\lambda_r - \lambda_p)\left(\frac{DC_p - DC_{\text{accept}}}{DC_p - DC_r}\right)\right] & \text{for } \textit{critical} \text{ members} \\ \left[\lambda_{p-\text{Seismic}} + (\lambda_r - \lambda_{p-\text{Seismic}})\left(\frac{DC_p - DC_{\text{accept}}}{DC_p - DC_r}\right)\right] & \text{for } \textit{other} \text{ members} \end{cases} \quad (16.51)$$

For axial load dominant members ($P_u/P_n \geq M_u/M_n$)

$$\lambda_{cpr} = \begin{cases} \lambda_{cp} + (0.9\lambda_{cr} - \lambda_{cp})\left(\frac{DC_p - DC_{\text{accept}}}{DC_p - DC_r}\right) & \text{for } \textit{critical} \text{ members} \\ \lambda_{cp} + (\lambda_{cr} - \lambda_{cp})\left(\frac{DC_p - DC_{\text{accept}}}{DC_p - DC_r}\right) & \text{for } \textit{other} \text{ members} \end{cases} \quad (16.52)$$

For flexural moment dominant members ($M_u/M_n > P_u/P_n$)

$$\lambda_{bpr} = \begin{cases} \lambda_{bp} + (0.9\lambda_{br} - \lambda_{bp})\left(\frac{DC_p - DC_{\text{accept}}}{DC_p - DC_r}\right) & \text{for } \textit{critical} \text{ members} \\ \lambda_{bp} + (\lambda_{br} - \lambda_{bp})\left(\frac{DC_p - DC_{\text{accept}}}{DC_p - DC_r}\right) & \text{for } \textit{other} \text{ members} \end{cases} \quad (16.53)$$

where λ_r, λ_p, $\lambda_{p-\text{Seismic}}$ are limiting width-thickness ratios specified by Table 16.3.

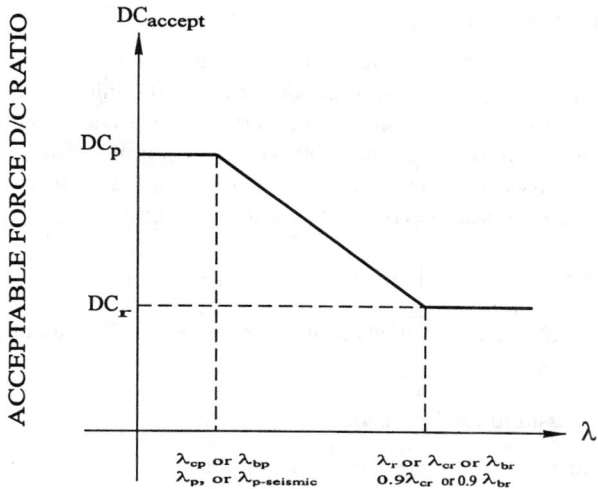

Figure 16.28   Acceptable D/C ratios and limiting slenderness parameters and width-thickness ratios.

Limiting Width-Thickness Ratios

The basic limiting width-thickness ratios λ_r, λ_p, $\lambda_{p-\text{Seismic}}$ specified in Table 16.3 are proposed for important bridges.

16.A.4 Inelastic Analysis Considerations

This section presents concepts and formulas of reduced material and section properties and yield surface for steel members for possible use in inelastic analysis.

Stiffness Reduction

Concepts of stiffness reduction — tangent modulus — have been used to calculate inelastic effective length factors by AISC-LRFD [4], and to account for both the effects of residual stresses and geometrical imperfection by Liew [32].

To consider inelasticity of a material, the tangent modulus of the material E_t may be used in analysis. For practical application, stiffness reduction factor (SRF) $= (E_t/E)$ can be taken as the ratio of the inelastic to elastic buckling load of the column:

$$SRF = \frac{E_t}{E} \approx \frac{(P_{cr})_{\text{inelastic}}}{(P_{cr})_{\text{elastic}}} \tag{16.54}$$

where $(P_{cr})_{\text{inelastic}}$ and $(P_{cr})_{\text{elastic}}$ can be calculated by AISC-LRFD [4] column equations:

$$(P_{cr})_{\text{inelastic}} = 0.658^{\lambda_c^2} A_s F_y \tag{16.55}$$

$$(P_{cr})_{\text{elastic}} = \left[\frac{0.877}{\lambda_c^2}\right] A_s F_y \tag{16.56}$$

in which A_s is the gross section area of the member and λ_c is the slenderness parameter. By utilizing the calculated axial compression load P, the tangent modulus E_t can be obtained as:

$$E_t = \begin{cases} E & \text{for } P/P_y \leq 0.39 \\ -3(P/P_y)\ln(P/P_y) & \text{for } P/P_y > 0.39 \end{cases} \tag{16.57}$$

Reduced Section Properties

In an initial structural analysis, the section properties based on a fully integral section of a latticed member may be used. If section forces obtained from this initial analysis are lower than the section strength controlled by the shear-flow transferring capacity, assumed fully integral section properties used in the analysis are rational. Otherwise, section properties considering a partially integral section, as discussed in Section 17.A.1, I_{equiv} and J_{equiv}, may be used in the further analysis. This concept is similar to "cracked section" analysis for reinforced concrete structures [13].

1. Moment of inertia — latticed members

    (a) For lacing bars or battens within web plane (bending about y-y axis in Figure 16.19)

The following assumptions are made:

- Moment-curvature curve (Figure 16.29) behaves bi-linearly until the section reaches its ultimate moment capacity.

- For moments less than $M_d = \beta_m M_u$, the moment at first stiffness degradation, the section can be considered fully integral.

- For moments larger than M_d and less than M_u, the ultimate moment capacity of the section, the section is considered as a partially integral one; that is, bending stiffness should be based on a reduced moment of inertia defined by Equation 16.27.

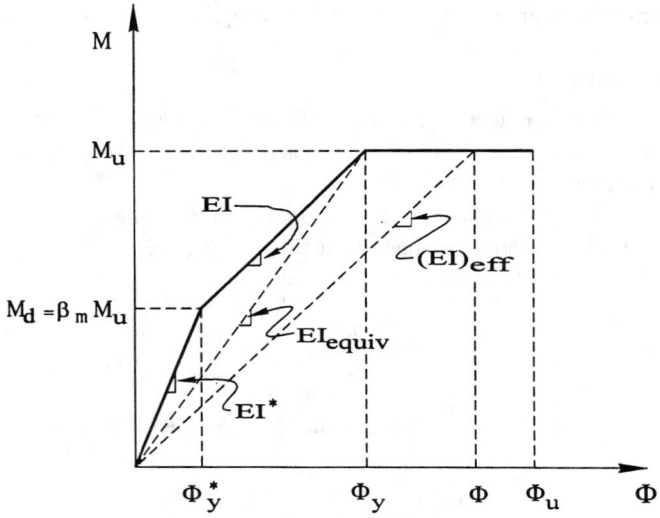

Figure 16.29   Idealized moment-curvature curve of a latticed member section.

An equivalent moment of inertia, I_{equiv}, based on the secant stiffness can be obtained as:

$$I_{equiv} = \frac{I^*_{y-y} I_{y-y}}{\beta_m I_{y-y} + (1 - \beta_m) I^*_{y-y}} \tag{16.58}$$

where

I_{y-y} = moment of inertia of a section about the y-y axis considering shear transfer-ring capacity

I^*_{y-y} = moment of inertia of a section about the y-y axis assuming full section integrity

β_m = reduction factor for the moment of inertia and may be determined by Equation 16.28

    (b) For lacing bars or battens within flange plane (bending about x-x axis in Figure 16.19)

Equation 16.29 is still valid for structural analysis.

2. Effective flexural stiffness — For steel members, when $\Phi_y < \Phi < \Phi_u$, the further reduced section property, effective flexural stiffness $(EI)_{\text{eff}}$ may be used in the analysis.

$$\frac{M_u}{\Phi_u} \leq (EI)_{\text{eff}} = \frac{M_u}{\Phi} \leq EI_{\text{equiv}} \tag{16.59}$$

3. Torsional constant — latticed members
Based on similar assumptions and principles used for moment of inertia, an equivalent torsional constant, J_{equiv}, is derived as follows:

$$J_{\text{equiv}} = \frac{J^* J}{\beta_t J + (1 - \beta_t) J^*} \tag{16.60}$$

where

J = torsional constant of a section considering shear transferring capacity (See Section 17.A.1)

J^* = torsional constant of a section assuming full section integrity

β_t = reduction factor for torsional constant may be determined by Equation 16.38

Yield Surface Equation for Doubly Symmetrical Sections

The yield or failure surface concept has been conveniently used in inelastic analysis to describe the full plastification of steel sections under action of axial force combined with biaxial bending. A four parameter yield surface equation for doubly symmetrical steel sections (I, thin-walled circular tube, thin-walled box, solid rectangular and circular sections), developed by Duan and Chen [19], is presented in this section for possible use in a nonlinear analysis.

The general shape of the yield surface for a doubly symmetrical steel section as shown in Figure 16.30 can be described approximately by the following general equation:

$$\left(\frac{M_x}{M_{pcx}}\right)^{\alpha_x} + \left(\frac{M_y}{M_{pcy}}\right)^{\alpha_y} = 1.0 \tag{16.61}$$

where M_{pcx} and M_{pcy} are the moment capacities about the respective axes, reduced for the presence of axial load; they can be obtained by the following formulas:

$$M_{pcx} = M_{px}\left[1 - \left(\frac{P}{P_y}\right)^{\beta_x}\right] \tag{16.62}$$

$$M_{pcy} = M_{py}\left[1 - \left(\frac{P}{P_y}\right)^{\beta_y}\right] \tag{16.63}$$

where

P = axial force

M_x = bending moment about the x-x principal axis

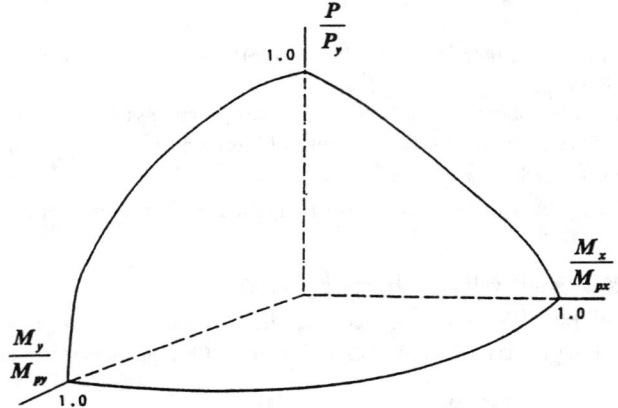

Figure 16.30   Typical yield surface for doubly symmetrical sections.

M_y =   bending moment about the y-y principal axis
M_{px} =   plastic moment about x-x principal axis
M_{py} =   plastic moment about y-y principal axis

The four parameters α_x, α_y, β_x, and β_y are dependent on sectional shapes and area distribution. It is seen that α_x and α_y represent biaxial loading behavior, while β_x and β_y describe uniaxial loading behavior. They are listed in Table 16.5:

TABLE 16.5   Parameters for Doubly Symmetrical Sections

Section types	α_x	α_y	β_x	β_y
Solid rectangular	$1.7 + 1.3\ (P/P_y)$	$1.7 + 1.3\ (P/P_y)$	2.0	2.0
Solid circular	2.0	2.0	2.1	2.1
I-shape	2.0	$1.2 + 2\ (P/P_y)$	1.3	$2 + 1.2\ (A_w/A_f)$
Thin-walled box	$1.7 + 1.5\ (P/P_y)$	$1.7 + 1.5\ (P/P_y)$	$2 - 0.5\ \overline{B} \geq 1.3$	$2 - 0.5\ \overline{B} \geq 1.3$
Thin-walled circular	2.0	2.0	1.75	1.75

Note: $\overline{B}$ is the ratio of width-to-depth of the box section with respect to the bending axis.

Equation 16.61 represents a smooth and convex surface in the three-dimensional stress-resultant space. It meets all special conditions and is easy to implement in a computer-based structural analysis.

17

Effective Length Factors of Compression Members

17.1 Introduction 17-1
17.2 Basic Concept 17-2
17.3 Isolated Columns 17-2
17.4 Framed Columns—Alignment Chart Method 17-3
   Alignment Chart Method • Requirements for Braced Frames •
   Simplified Equations to Alignment Charts
17.5 Modifications to Alignment Charts..................... 17-10
   Different Restraining Girder End Conditions • Different Restraining Column End Conditions • Column Restrained by Tapered Rectangular Girders • Unsymmetrical Frames • Effects of
   Axial Forces in Restraining Members in Braced Frames • Consideration of Partial Column Base Fixity • Inelastic K-factor
17.6 Framed Columns—Alternative Methods 17-25
   LeMessurier Method • Lui Method • Remarks
17.7 Unbraced Frames With Leaning Columns 17-29
   Rigid Columns • Leaning Columns • Remarks
17.8 Cross Bracing Systems 17-34
17.9 Latticed and Built-Up Members 17-35
   Laced Columns • Columns with Battens • Laced-Battened
   Columns • Columns with Perforated Cover Plates • Built-Up
   Members with Bolted and Welded Connectors
17.10 Tapered Columns 17-43
17.11 Crane Columns 17-43
17.12 Columns in Gable Frames............................. 17-48
17.13 Summary... 17-48
17.14 Defining Terms 17-49
References 17-50
Further Reading 17-53

Lian Duan
*Division of Structures,
California Department of
Transportation,
Sacramento, CA*

W.F. Chen
*School of Civil Engineering,
Purdue University,
West Lafayette, IN*

17.1 Introduction

The concept of the **effective length factors** of **columns** has been well established and widely used by practicing engineers and plays an important role in compression member design. The most structural design codes and specifications have provisions concerning the effective length factor. The aim of this chapter is to present a state-of-the-art engineering practice of the effective length factor for the design of columns in structures. In the first part of this chapter, the basic concept of the effective length factor is discussed. And then, the design implementation for isolated columns,

framed columns, crossing bracing systems, **latticed members, tapered columns, crane columns,** as well as columns in **gable frames** is presented. The determination of whether a frame is braced or unbraced is also addressed. Several detailed examples are given to illustrate the determination of effective length factors for different cases of engineering applications.

17.2 Basic Concept

Mathematically, the effective length factor or the *elastic K*-factor is defined as:

$$K = \sqrt{\frac{P_e}{P_{cr}}} = \sqrt{\frac{\pi^2 EI}{L^2 P_{cr}}} \tag{17.1}$$

where P_e is the Euler load, the elastic buckling load of a pin-ended column; P_{cr} is the elastic buckling load of an end-restrained framed column; E is the modulus of elasticity; I is the moment of inertia in the flexural buckling plane; and L is the unsupported length of column.

Physically, the K-factor is a factor that when multiplied by actual length of the end-restrained column (Figure 17.1a) gives the length of an equivalent pin-ended column (Figure 17.1b) whose buckling load is the same as that of the end-restrained column. It follows that effective length, KL,

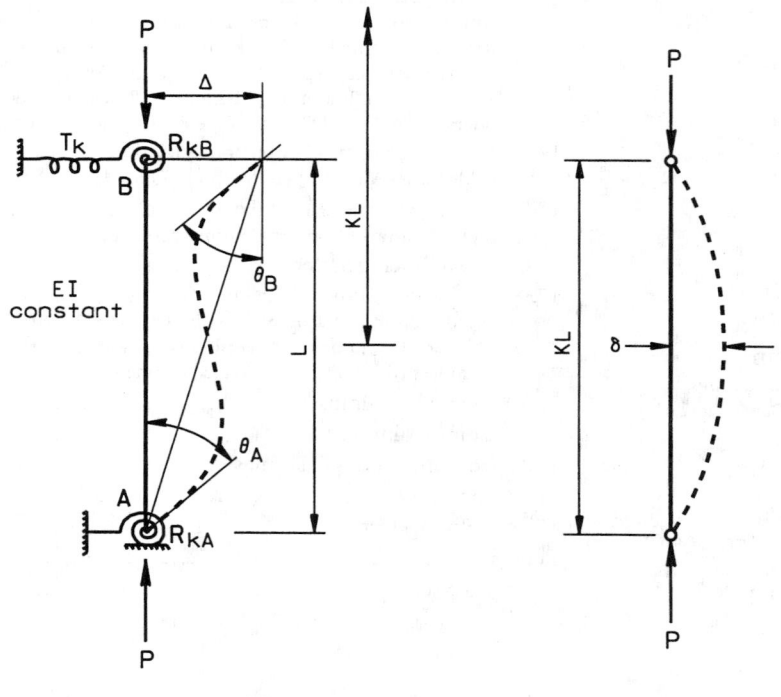

(a)   End-restrained Columns   (b)   Pin-ended Column

Figure 17.1   Isolated columns.

of an end-restrained column is the length between adjacent inflection points of its pure flexural buckling shape.

Specifications provide the resistance equations for pin-ended columns, while the resistance of framed columns can be estimated through the K-factor to the pin-ended columns strength equation.

Theoretical K-factor is determined from an elastic eigenvalue analysis of the entire structural system, while practical methods for the K-factor are based on an elastic eigenvalue analysis of selected subassemblages. The effective length concept is the only tool currently available for the design of compression members in engineering structures, and it is an essential part of analysis procedures.

17.3 Isolated Columns

From an eigenvalue analysis, the general K-factor equation of an end-restrained column as shown in Figure 17.1 is obtained as:

$$\det \begin{vmatrix} C + \frac{R_{kA}L}{EI} & S & -(C+S) \\ S & C + \frac{R_{kB}L}{EI} & -(C+S) \\ -(C+S) & -(C+S) & 2(C+S) - \left(\frac{\pi}{K}\right)^2 + \frac{T_kL^3}{EI} \end{vmatrix} = 0 \qquad (17.2)$$

where the stability functions C and S are defined as:

$$C = \frac{(\pi/K)\sin(\pi/K) - (\pi/K)^2 \cos(\pi/K)}{2 - 2\cos(\pi/K) - (\pi/K)\sin(\pi/K)} \qquad (17.3)$$

$$S = \frac{(\pi/K)^2 - (\pi/K)\sin(\pi/K)}{2 - 2\cos(\pi/K) - (\pi/K)\sin(\pi/K)} \qquad (17.4)$$

The largest value of K that satisfies Equation 17.2 gives the elastic buckling load of an end-restrained column.

Figure 17.2 [1, 3, 4] summarizes the theoretical K-factors for columns with some idealized end conditions. The recommended K-factors are also shown in Figure 17.2 for practical design applications. Since actual column conditions seldom comply fully with idealized conditions used in buckling analysis, the recommended K-factors are always equal to or greater than their theoretical counterparts.

17.4 Framed Columns—Alignment Chart Method

In theory, the **effective length factor** K for any column in a framed structure can be determined from a stability analysis of the entire structural analysis—eigenvalue analysis. Methods available for stability analysis include the slope-deflection method [17, 35, 71], three-moment equation method [13], and energy methods [42]. In practice, however, such analysis is not practical, and simple models are often used to determine the effective length factors for framed columns [38, 47, 55, 72]. One such practical procedure that provides an approximate value of the elastic K-factor is the **alignment chart** method [46]. This procedure has been adopted by the AISC [3, 4], ACI 318-95 [2], and AASHTO [1] specifications, among others. At present, most engineers use the **alignment chart** method in lieu of an actual stability analysis.

17.4.1 Alignment Chart Method

The structural models employed for determination of K-factor for framed columns in the alignment chart method are shown in Figure 17.3. The assumptions used in these models are [4, 17]:

1. All members have constant cross-section and behave elastically.
2. Axial forces in the girders are negligible.
3. All joints are rigid.
4. For **braced frames**, the rotations at the near and far ends of the girders are equal in magnitude and opposite in direction (i.e., girders are bent in single curvature).

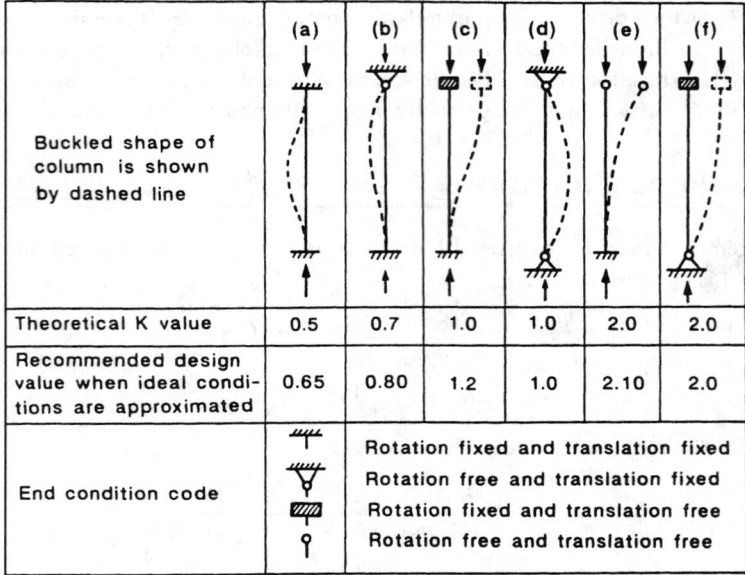

	(a)	(b)	(c)	(d)	(e)	(f)
Buckled shape of column is shown by dashed line						
Theoretical K value	0.5	0.7	1.0	1.0	2.0	2.0
Recommended design value when ideal conditions are approximated	0.65	0.80	1.2	1.0	2.10	2.0
End condition code		Rotation fixed and translation fixed				
		Rotation free and translation fixed				
		Rotation fixed and translation free				
		Rotation free and translation free				

Figure 17.2   Theoretical and recommended K-factors for isolated columns with idealized end conditions. (From American Institute of Steel Construction, *Load and Resistance Factor Design Specification for Structural Steel Buildings,* 2nd ed., Chicago, IL, 1993. With permission. Also from Johnston, B.G., Ed., *Structural Stability Research Council, Guide to Stability Design Criteria for Metal Structures,* 3rd ed., John Wiley & Sons, New York, 1976. With permission.)

5. For **unbraced frames**, the rotations at the near and far ends of the girders are equal in magnitude and direction (i.e., girders are bent in double curvature).
6. The stiffness parameters, $L\sqrt{P/EI}$, of all columns are equal.
7. All columns buckle simultaneously.

Using the slope-deflection equation method and stability functions, the effective length factor equations of framed columns are obtained as follows.

For columns in **braced frames:**

$$\frac{G_A G_B}{4}(\pi/K)^2 + \left(\frac{G_A + G_B}{2}\right)\left(1 - \frac{\pi/K}{\tan(\pi/K)}\right) + \frac{2\tan(\pi/2K)}{\pi/K} - 1 = 0 \qquad (17.5)$$

For columns in **unbraced frames:**

$$\frac{G_A G_B (\pi/K)^2 - 36^2}{6(G_A + G_B)} - \frac{\pi/K}{\tan(\pi/K)} = 0 \qquad (17.6)$$

where G_A and G_B are stiffness ratios of columns and girders at two end joints, A and B, of the column section being considered, respectively. They are defined by:

$$G_A = \frac{\sum_A (E_c I_c/L_c)}{\sum_A (E_g I_g/L_g)} \qquad (17.7)$$

$$G_B = \frac{\sum_B (E_c I_c/L_c)}{\sum_B (E_g I_g/L_g)} \qquad (17.8)$$

where $\sum$ indicates a summation of all members rigidly connected to the joint and lying in the plane in which buckling of column is being considered; subscripts c and g represent columns and girders, respectively.

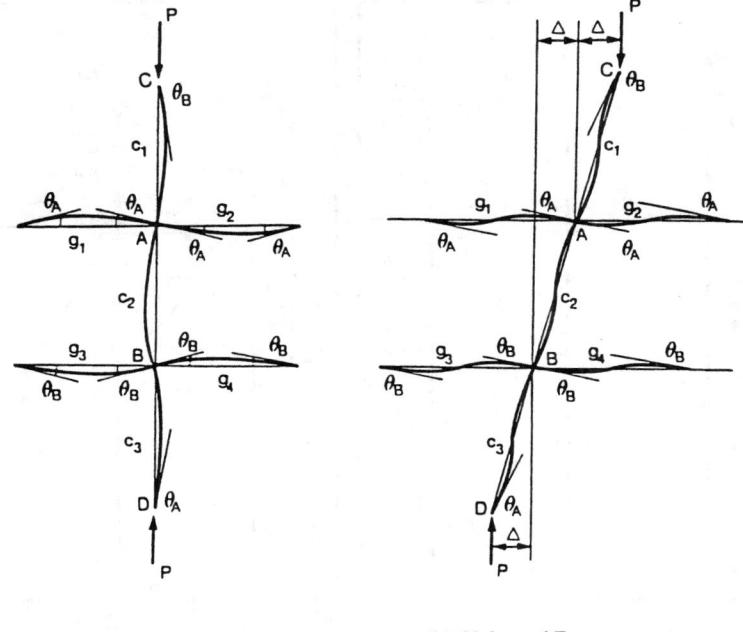

(a) Braced Frames   (b) Unbraced Frames

Figure 17.3   Subassemblage models for K-factors of framed columns.

Equations 17.5 and 17.6 can be expressed in the form of alignment charts, as shown in Figure 17.4. It is noted that for columns in braced frames, the range of K is $0.5 \leq K \leq 1.0$; for columns in unbraced frames, the range is $1.0 \leq K \leq \infty$. For column ends supported by but not rigidly connected to a footing or foundations, G is theoretically infinity, but, unless actually designed as a true friction free pin, may be taken as 10 for practical design. If the column end is rigidly attached to a properly designed footing, G may be taken as 1.0 [4].

EXAMPLE 17.1:

Given: A two-story steel frame is shown in Figure 17.5. Using the alignment chart, determine the K-factor for the elastic column DE. $E = 29,000$ ksi (200 GPa) and $Fy = 36$ ksi (248 MPa).

   Solution

1. For the given frame, section properties are

Members	Section	I_x in.^{4} (mm$^4 \times 10^8$)	L in. (mm)	I_x/L in.^{3} (mm^3)
AB and GH	W 10x22	118 (0.49)	180 (4,572)	0.656(10,750)
BC and HI	W10x22	118 (0.49)	144 (3,658)	0.819(13,412)
DE	W10x45	248 (1.03)	180 (4,572)	1.378(22,581)
EF	W10x45	248 (1.03)	144 (3,658)	1.722(28,219)
BE	W18x50	800 (3.33)	300 (7,620)	2.667(43,704)
EH	W18x86	1530 (6.37)	360 (9,144)	4.250(69,645)
CF	W16x40	518 (2.16)	300 (7,620)	1.727(28,300)
FI	W16x67	954 (3.97)	360 (9,144)	2.650(43,426)

2. Calculate G-factor for column DE:

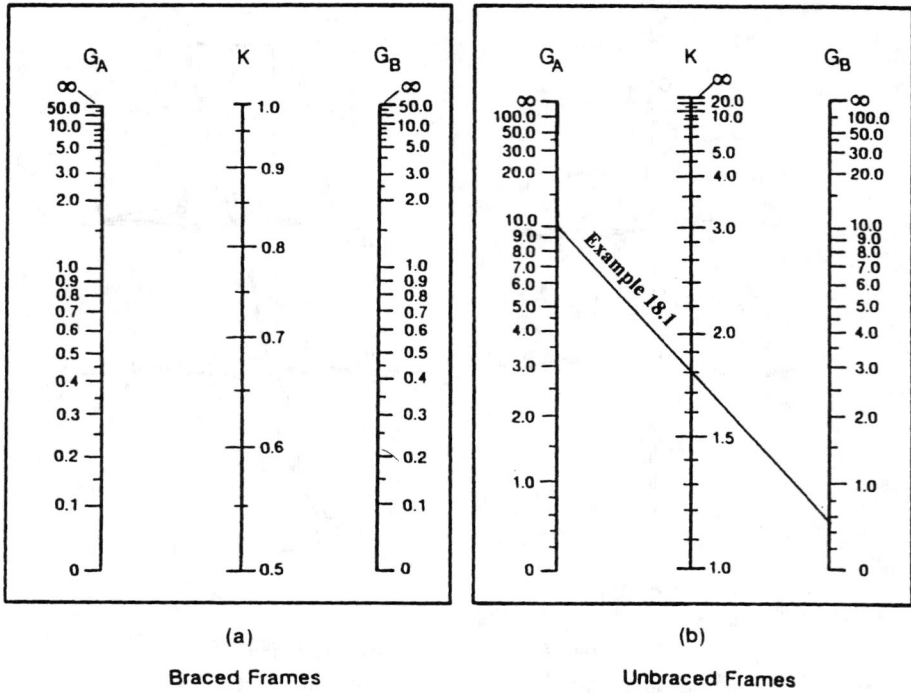

Figure 17.4 Alignment charts for effective length factors of framed columns. (From American Institute of Steel Construction, *Load and Resistance Factor Design Specification for Structural Steel Buildings,* 2nd ed., Chicago, IL, 1993. With permission. Also from Johnston, B.G., Ed., *Structural Stability Research Council, Guide to Stability Design Criteria for Metal Structures,* 3rd ed., John Wiley & Sons, New York, 1976. With permission.)

$$G_E = \frac{\sum_E (E_c I_c / L_c)}{\sum_E (E_g I_g / L_g)} = \frac{1.378 + 1.722}{2.667 + 4.250} = 0.448$$

$$G_D = 10 \quad \text{(AISC-LRFD, 1993)}$$

3. From the alignment chart in Figure 17.4b, $K = 1.8$ is obtained.

17.4.2 Requirements for Braced Frames

In stability design, one of the major decisions engineers have to make is the determination of whether a frame is braced or unbraced. The AISC-LRFD [4] states that a frame is braced when "lateral stability is provided by diagonal bracing, shear walls or equivalent means". However, there is no specific provision for the "amount of stiffness required to prevent sidesway buckling" in the AISC, AASHTO, and other specifications. In actual structures, a completely braced frame seldom exists. But in practice, some structures can be analyzed as braced frames as long as the lateral stiffness provided by the bracing system is large enough. The following brief discussion may provide engineers with the tools to make engineering decisions regarding the basic requirements for a braced frame.

1. **Lateral Stiffness Requirement**
   Galambos [34] presented a simple conservative procedure to evaluate minimum lateral stiffness provided by a bracing system so that the frame is considered braced.

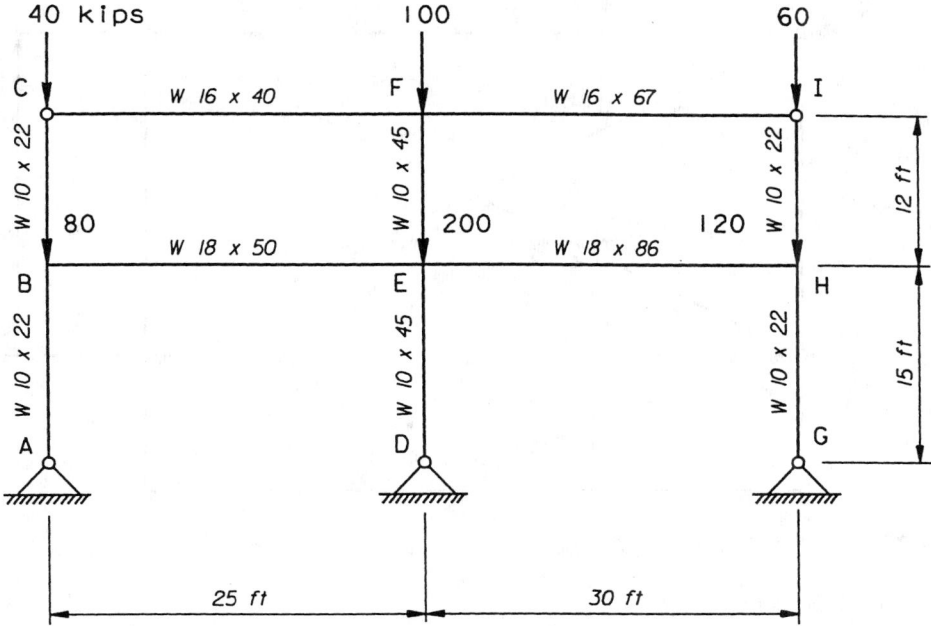

Figure 17.5 An unbraced two-story frame.

$$\text{Required lateral stiffness,}\quad T_k = \frac{\sum P_n}{L_c} \tag{17.9}$$

where $\sum$ represents the summation of all columns in one story, P_n is the nominal axial compression strength of a column using the effective length factor $K = 1$, and L_c is the unsupported length of a column.

2. **Bracing Size Requirement**

   Galambos [34] applied Equation 17.9 to a diagonal bracing (Figure 17.6) and obtained minimum requirements of diagonal bracing for a braced frame as

$$A_b = \frac{\left[1 + (L_b/L_c)^2\right]^{3/2} \sum P_n}{(L_b/L_c)^2\, E} \tag{17.10}$$

   where A_b is the cross-sectional area of diagonal bracing and L_b is the span length of the beam.

   A recent study by Aristizabal-Ochoa [8] indicates that the size of the diagonal bracing required for a totally braced frame is about 4.9 and 5.1% of the column cross-section for a "rigid frame" and "simple framing", respectively, and increases with the moment inertia of the column, the beam span, and the beam-to-column span ratio, L_b/L_c.

17.4.3   Simplified Equations to Alignment Charts

1. **ACI 318-95 Equations**

   The ACI Building Code [2] recommends the use of alignment charts as the primary design aid for estimating K-factors, following two sets of simplified K-factor equations as an alternative:

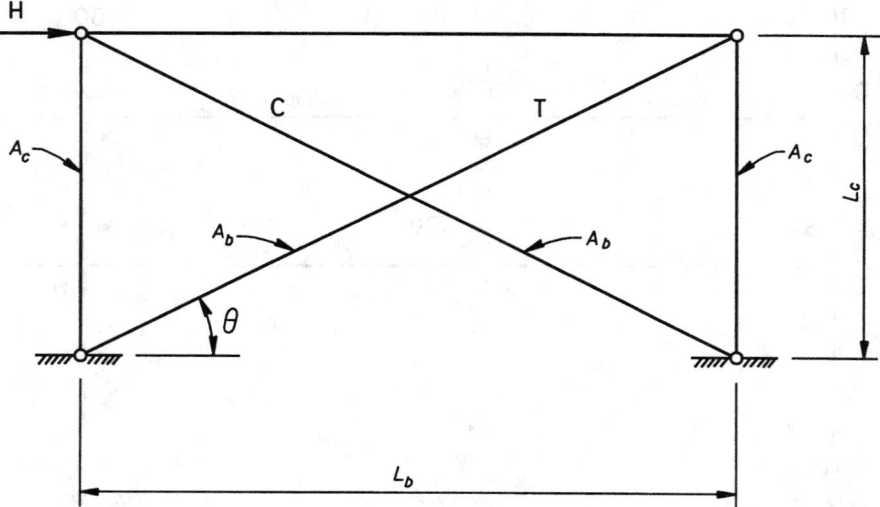

Figure 17.6   Diagonal cross bracing system.

For braced frames [19]:

$$K = 0.7 + 0.05 (G_A + G_B) \leq 1.0 \tag{17.11}$$
$$K = 0.85 + 0.05 G_{\min} \leq 1.0 \tag{17.12}$$

The smaller of the above two expressions provides an upper bound to the effective length factor for braced compression members.

For unbraced frames [32]:

For $G_m < 2$

$$K = \frac{20 - G_m}{20} \sqrt{1 + G_m} \tag{17.13}$$

For $G_m \geq 2$

$$K = 0.9 \sqrt{1 + G_m} \tag{17.14}$$

For columns hinged at one end

$$K = 2.0 + 0.3 G \tag{17.15}$$

where G_m is the average of G values at the two ends of the columns.

2. **Duan-King-Chen Equations**

A graphical alignment chart determination of the K-factor is easy to perform, while solving the chart Equations 17.5 and 17.6 always involves iteration. Although the ACI code provides simplified K-factor equations, generally, they may not lead to an economical design [40]. To achieve both accuracy and simplicity for design purposes, the following alternative K-factor equations were proposed by Duan, King and Chen [48].

For braced frames:

$$K = 1 - \frac{1}{5 + 9G_A} - \frac{1}{5 + 9G_B} - \frac{1}{10 + G_A G_B} \qquad (17.16)$$

For unbraced frames:

For $K < 2$

$$K = 4 - \frac{1}{1 + 0.2G_A} - \frac{1}{1 + 0.2G_B} - \frac{1}{1 + 0.01G_A G_B} \qquad (17.17)$$

For $K \geq 2$

$$K = \frac{2\pi a}{0.9 + \sqrt{0.81 + 4ab}} \qquad (17.18)$$

where

$$a = \frac{G_A G_B}{G_A + G_B} + 3 \qquad (17.19)$$

$$b = \frac{36}{G_A + G_B} + 6 \qquad (17.20)$$

3. **French Equations**
   For braced frames:

$$K = \frac{3G_A G_B + 1.4(G_A + G_B) + 0.64}{3G_A G_B + 2.0(G_A + G_B) + 1.28} \qquad (17.21)$$

For unbraced frames:

$$K = \sqrt{\frac{1.6G_A G_B + 4.0(G_A + G_B) + 7.5}{G_A + G_B + 7.5}} \qquad (17.22)$$

Equations 17.21 and 17.22 first appeared in the *French Design Rules for Steel Structure* [31] since 1966, and were later incorporated into the *European Recommendation for Steel Construction* [28]. They provide a good approximation to the alignment charts [26].

17.5 Modifications to Alignment Charts

In using the alignment charts in Figure 17.4 and Equations 17.5 and 17.6, engineers must always be aware of the assumptions used in the development of these charts. When actual structural conditions differ from these assumptions, unrealistic design may result [4, 43, 53]. The SSRC (Structural Stability Research Council) guide [43] provides methods that enable engineers to make simple modifications of the charts for some special conditions, such as unsymmetrical frames, column base conditions, girder far end conditions, and flexible conditions. A procedure that can be used to account for far ends of restraining columns being hinged or fixed was proposed by Duan and Chen [21, 22]. Consideration of effects of material inelasticity on the K-factor was developed originally by Yura [73] and expanded by Disque [20]. LeMessurier [52] presented an overview of unbraced frames with or without **leaning columns**. An approximate procedure is also suggested by AISC-LRFD [4]. Special attention should also be paid to calculation of the proper G values [10, 49] when partially restrained (PR) connections are used in frames. Several commonly used modifications are summarized in this section.

17.5.1   Different Restraining Girder End Conditions

When the end conditions of restraining girders are not rigidly jointed to columns, the girder stiffness (I_g/L_g) used in the calculation of G_A and G_B in Equations 17.7 and 17.8 should be multiplied by a modification factor, α_k, as:

$$G = \frac{\sum (E_c I_c/L_c)}{\sum \alpha_k \left(E_g I_g/L_g\right)} \tag{17.23}$$

where the modification factor, α_k, for braced frames developed by Duan and Lu [25] and for unbraced frames proposed by Kishi, Chen, and Goto [49] are given in Table 17.1 and 17.2. In these tables, R_{kN} and R_{kF} are elastic spring constants at the near and far ends of a restraining girder, respectively. R_{kN} and R_{kF} are the tangent stiffness of a semi-rigid connection at buckling.

TABLE 17.1   Modification Factor α_k for Braced Frames with Semi-Rigid Connections

End conditions of restraining girder		α_k
Near end	Far end	
Rigid	Rigid	1.0
Rigid	Hinged	1.5
Rigid	Semi-rigid	$\left(1 + \frac{6E_g I_g}{L_g R_{kF}}\right) / \left(1 + \frac{4E_g I_g}{L_g R_{kF}}\right)$
Rigid	Fixed	2.0
Semi-rigid	Rigid	$1/\left(1 + \frac{4E_g I_g}{L_g R_{kN}}\right)$
Semi-rigid	Hinged	$1.5/\left(1 + \frac{3E_g I_g}{L_g R_{kN}}\right)$
Semi-rigid	Semi-rigid	$\left(1 + \frac{6E_g I_g}{L_g R_{kF}}\right) / R^*$
Semi-rigid	Fixed	$2/\left(1 + \frac{4E_g I_g}{L_g R_{kN}}\right)$

Note: $R^* = \left(1 + \frac{4E_g I_g}{L_g R_{kN}}\right)\left(1 + \frac{4E_g I_g}{L_g R_{kF}}\right) - \left(\frac{E_g I_g}{L_g}\right)^2 \frac{4}{R_{kN} R_{kF}}$

EXAMPLE 17.2:

Given: A steel frame is shown in Figure 17.5. Using the alignment chart with the necessary modifications, determine the K-factor for elastic column EF. $E = 29{,}000$ ksi (200 GPa) and $F_y = 36$ ksi (248 MPa).

TABLE 17.2 Modification Factor, α_k, for Unbraced Frames with Semi-Rigid Connections

End conditions of restraining girder		α_k
Near end	Far end	
Rigid	Rigid	1
Rigid	Hinged	0.5
Rigid	Semi-rigid	$\left(1 + \frac{2E_g I_g}{L_g R_{kF}}\right) / \left(1 + \frac{4E_g I_g}{L_g R_{kF}}\right)$
Rigid	Fixed	2/3
Semi-rigid	Rigid	$1 / \left(1 + \frac{4E_g I_g}{L_g R_{kN}}\right)$
Semi-rigid	Hinged	$0.5 / \left(1 + \frac{3E_g I_g}{L_g R_{kN}}\right)$
Semi-rigid	Semi-rigid	$\left(1 + \frac{2E_g I_g}{L_g R_{kF}}\right) / R^*$
Semi-rigid	Fixed	$(2/3) / \left(1 + \frac{4E_g I_g}{L_g R_{kN}}\right)$

Note: $R^* = \left(1 + \frac{4E_g I_g}{L_g R_{kN}}\right)\left(1 + \frac{4E_g I_g}{L_g R_{kF}}\right) - \left(\frac{E_g I_g}{L_g}\right)^2 \frac{4}{R_{kN} R_{kF}}$

Solution

1. Calculate G-factor with modification for column EF.

   Since the far end of restraining girders are hinged, girder stiffness should be multiplied by 0.5 (see Table 17.2). Using section properties in Example 17.1, we obtain:

$$G_F = \frac{\sum (E_c I_c / L_c)}{\sum \alpha_k (E_g I_g / L_g)} = \frac{1.722}{0.5(1.727) + 0.5(2.650)} = 0.787$$

$$GE = 0.448$$

2. From the alignment chart in Figure 17.4b, $K = 1.22$ is obtained.

17.5.2 Different Restraining Column End Conditions

To consider different far end conditions of restraining columns, the general effective length factor equations for column $C2$ (Figure 17.3) were derived by Duan and Chen [21, 22, 23]. By assuming that the far ends of columns $C1$ and $C3$ are hinged and using the slope-deflection equation approach for the subassemblies shown in Figure 17.3, we obtain the following.

1. For a Braced Frame [21]:

$$C^2 - S^2 \left[G_{AC1} + G_{BC3} + G_{AC2} G_{BC2} + \frac{\frac{2G_{BC3}}{G_A} + \frac{2G_{AC1}}{G_B}}{C} - G_{AC1} G_{BC3}\left(\frac{S}{C}\right)^2 \right]$$

$$+ 2C\left(\frac{1}{G_A} + \frac{1}{G_B}\right) + \frac{4}{G_A G_B} = 0 \qquad (17.24)$$

where C and S are stability functions as defined by Equations 17.3 and 17.4; G_A and G_B are defined in Equations 17.7 and 17.8; G_{AC1}, G_{AC2}, G_{BC2}, and G_{BC3} are stiffness ratios of columns at A-th and B-th ends of the columns being considered, respectively. They are defined as:

$$G_{Ci} = \frac{E_{ci} I_{ci}/L_{ci}}{\sum (E_{ci} I_{ci}/L_{ci})} \tag{17.25}$$

where $\sum$ indicates a summation of all columns rigidly connected to the joint and lying in the plane in which buckling of the column is being considered.

Although Equation 17.24 was derived for the special case in which the far ends of both columns $C1$ and $C3$ are hinged, this equation is also applicable if adjustment to G_{Ci} is made as follows: (1) if the far end of column Ci ($C1$ or $C3$) is fixed, then take $G_{Ci} = 0$ (except for G_{C2}); (2) if the far end of the column Ci ($C1$ or $C3$) is rigidly connected, then take $G_{Ci} = 0$ and $G_{C2} = 1.0$. Therefore, Equation 17.24 can be specialized for the following conditions:

(a) If the far ends of both columns $C1$ and $C3$ are fixed, we have $G_{AC1} = G_{BC3} = 0$ and Equation 17.24 reduces to

$$C^2 - S^2 (G_{AC2}G_{BC2}) + 2C \left(\frac{1}{G_A} + \frac{1}{G_B}\right) + \frac{4}{G_A G_B} = 0 \tag{17.26}$$

(b) If the far end of column $C1$ is rigidly connected and the far end of column $C3$ is fixed, we have $G_{AC2} = 1.0$ and $G_{AC1} = G_{BC3} = 0$, and Equation 17.24 reduces to

$$C^2 - S^2 + G_{BC2} + 2C \left(\frac{1}{G_A} + \frac{1}{G_B}\right) + \frac{4}{G_A G_B} = 0 \tag{17.27}$$

(c) If the far end of column $C1$ is rigidly connected and the far end of column $C3$ is hinged, we have $G_{AC1} = 0$ and $G_{AC2} = 1.0$, and Equation 17.24 reduces to

$$C^2 - S^2 \left(G_{BC3} + G_{BC2} + \frac{2G_{BC3}}{G_A C}\right)$$
$$+ 2C \left(\frac{1}{G_A} + \frac{1}{G_B}\right) + \frac{4}{G_A G_B} = 0 \tag{17.28}$$

(d) If the far end of column $C1$ is hinged and the far end of column $C3$ is fixed, we have $G_{BC3} = 0$ and Equation 17.24 reduces to

$$C^2 - S^2 \left(G_{AC1} + G_{AC2}G_{BC2} + \frac{2G_{AC1}}{G_B C}\right)$$
$$+ 2C \left(\frac{1}{G_A} + \frac{1}{G_B}\right) + \frac{4}{G_A G_B} = 0 \tag{17.29}$$

(e) If the far ends of both columns $C1$ and $C3$ are rigidly connected (i.e., assumptions used in developing the alignment chart), we have $G_{C2} = 1.0$ and $G_{Ci} = 0$, and Equation 17.24 reduces to

$$C^2 - S^2 + 2C\left(\frac{1}{G_A} + \frac{1}{G_B}\right) + \frac{4}{G_A G_B} = 0 \tag{17.30}$$

which can be rewritten in the form of Equation 17.5.

2. For an Unbraced Frame [22, 23]:

$$\det \begin{vmatrix} a_{11} & a_{12} & a_{13} \\ a_{21} & a_{22} & a_{23} \\ a_{31} & a_{32} & a_{33} \end{vmatrix} = 0 \tag{17.31}$$

or

$$a_{11}a_{22}a_{33} + a_{21}a_{32}a_{13} + a_{31}a_{23}a_{12} - a_{31}a_{22}a_{13}$$
$$- a_{21}a_{12}a_{33} + a_{11}a_{23}a_{32} = 0 \tag{17.32}$$

where

$$a_{11} = C + \frac{6}{G_A} - G_{AC1}\frac{S^2}{C} \tag{17.33}$$

$$a_{22} = C + \frac{6}{G_B} - G_{BC3}\frac{S^2}{C} \tag{17.34}$$

$$a_{33} = -2\left[C + S - \frac{1}{2}\left(\frac{\pi}{K}\right)^2\right] \tag{17.35}$$

$$a_{12} = G_{AC2}S \tag{17.36}$$

$$a_{21} = G_{BC2}S \tag{17.37}$$

$$a_{31} = a_{32} = C + S \tag{17.38}$$

$$a_{13} = -(C + S) + G_{AC1}\left(S + \frac{S^2}{C}\right) \tag{17.39}$$

$$a_{23} = -(C + S) + G_{BC3}\left(S + \frac{S^2}{C}\right) \tag{17.40}$$

Although Equation 17.31 was derived for the special case in which the far ends of both columns $C1$ and $C3$ are hinged, it can be adjusted to account for the following cases: (1) if the far end of column Ci ($C1$ or $C3$) is fixed, then take $G_{Ci} = 0$ (except for G_{C2}); (2) if the far end of column Ci ($C1$ or $C3$) is rigidly connected, then take $G_{Ci} = 0$ and $G_{C2} = 1.0$. Therefore, Equation 17.31 can be used for the following conditions:

(a) If the far ends of both columns $C1$ and $C3$ are fixed, we take $G_{C1} = G_{C3} = 0$, and obtain from Equations 17.33, 17.34, 17.39, and 17.40,

$$a_{11} = C + \frac{6}{G_A} \tag{17.41}$$

$$a_{22} = C + \frac{6}{G_B} \tag{17.42}$$

$$a_{13} = a_{23} = -(C + S) \tag{17.43}$$

(b) If the far end of column $C1$ is rigidly connected and the far end of column $C3$ is fixed, we take $G_{AC2} = 1.0$ and $G_{AC1} = G_{BC3} = 0$, and obtain from Equations 17.33, 17.34, 17.36, 17.39, and 17.40,

$$a_{11} = C + \frac{6}{G_A} \tag{17.44}$$

$$a_{22} = C + \frac{6}{G_B} \tag{17.45}$$

$$a_{12} = S \tag{17.46}$$

$$a_{13} = a_{23} = -(C + S) \tag{17.47}$$

(c) If the far end of column $C1$ is rigidly connected and the far end of column $C3$ is hinged, we take $G_{AC1} = 0$ and $G_{AC2} = 1.0$, and obtain from Equations 17.33, 17.36, and 17.39,

$$a_{11} = C + \frac{6}{G_A} \tag{17.48}$$

$$a_{12} = S \tag{17.49}$$

$$a_{13} = -(C + S) \tag{17.50}$$

(d) If the far end of column $C1$ is hinged and the far end of column $C3$ is fixed, we have $G_{BC3} = 0.0$, and obtain from Equations 17.34 and 17.40,

$$a_{22} = C + \frac{6}{G_B} \tag{17.51}$$

$$a_{23} = -(C + S) \tag{17.52}$$

(e) If the far ends of both columns $C1$ and $C3$ are rigidly connected (i.e., assumptions used in developing the alignment chart, that is $\theta_C = \theta_B$ and $\theta_D = \theta_A$), we take $G_{C2} = 1.0$ and $G_{Ci} = 0$, and obtain from Equations 17.33 to 17.40,

$$a_{11} = C + \frac{6}{G_A} \tag{17.53}$$

$$a_{22} = C + \frac{6}{G_B} \tag{17.54}$$

$$a_{12} = a_{21} = S \tag{17.55}$$

$$a_{13} = a_{23} = -(C + S) \tag{17.56}$$

Equation 17.31 is reduced to the form of Equation 17.6.

The procedures to obtain the K-factor directly from the alignment charts without resorting to solve Equations 17.24 and 17.31 were also proposed by Duan and Chen [21, 22].

17.5.3   Column Restrained by Tapered Rectangular Girders

A modification factor, α_T, was developed by King et al. [48] for those framed columns restrained by tapered rectangular girders with different far end conditions. The following modified G-factor is introduced in connection with the use of alignment charts:

$$G = \frac{\sum (E_c I_c / L_c)}{\sum \alpha_T (E_g I_g / L_g)} \tag{17.57}$$

where I_g is the moment of inertia of the girder at near end. Both closed-form and approximate solutions for modification factor α_T were derived. It is found that the following two-parameter power function can describe the closed-form solutions very well:

$$\alpha_T = D (1 - r)^\beta \tag{17.58}$$

in which the parameter D is a constant depending on the far end conditions, and β is a function of the far end conditions and tapering factor, α and r, as defined in Figure 17.7.

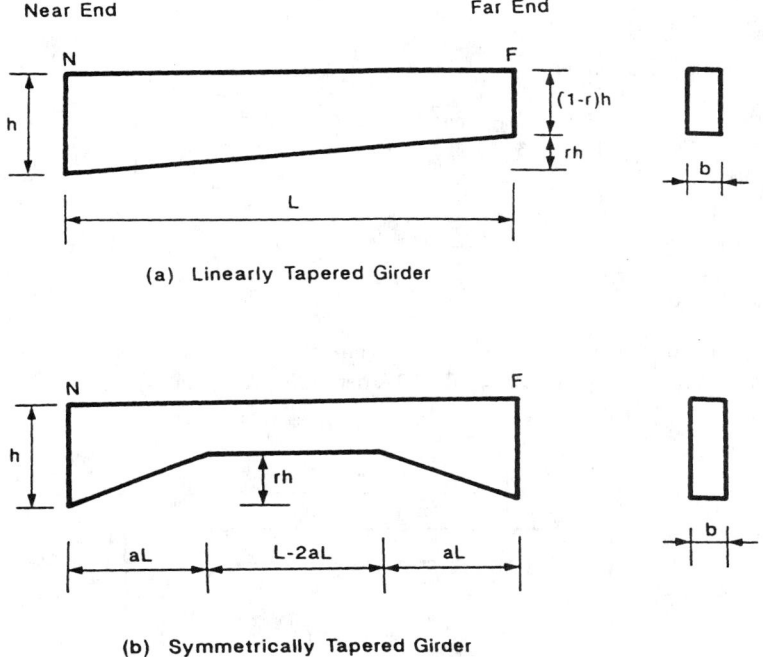

(a) **Linearly Tapered Girder**

(b) **Symmetrically Tapered Girder**

Figure 17.7   Tapered rectangular girders. (From King, W.S., Duan, L., et al., *Eng. Struct.*, 15(5), 369, 1993. With kind permission from Elsevier Science, Ltd, The Boulevard, Langford Lane, Kidlington OX5 IGB, UK.)

For a braced frame:

$$D = \begin{Bmatrix} 1.0 & \text{rigid far end} \\ 2.0 & \text{fixed far end} \\ 1.5 & \text{hinged far end} \end{Bmatrix} \tag{17.59}$$

For an unbraced frame:

$$D = \begin{Bmatrix} 1.0 & \text{rigid far end} \\ 2/3 & \text{fixed far end} \\ 0.5 & \text{hinged far end} \end{Bmatrix} \tag{17.60}$$

1. For a linearly tapered rectangular girder (Figure 17.7a)
   For a braced frame:

$$\beta = \begin{cases} 0.02 + 0.4r & \text{rigid far end} \\ 0.75 - 0.1r & \text{fixed far end} \\ 0.75 - 0.1r & \text{hinged far end} \end{cases} \tag{17.61}$$

For an unbraced frame:

$$\beta = \begin{cases} 0.95 & \text{rigid far end} \\ 0.70 & \text{fixed far end} \\ 0.70 & \text{hinged far end} \end{cases} \tag{17.62}$$

2. For a symmetrically tapered rectangular girder (Figure 17.7b)
   For a braced frame:

$$\beta = \begin{cases} 3 - 1.7a^2 - 2a & \text{rigid far end} \\ 3 + 2.5a^2 - 5.55a & \text{fixed far end} \\ 3 - a^2 - 2.7a & \text{hinged far end} \end{cases} \tag{17.63}$$

For an unbraced frame:

$$\beta = \begin{cases} 3 + 3.8a^2 - 6.5a & \text{rigid far end} \\ 3 + 2.3a^2 - 5.45a & \text{fixed far end} \\ 3 - 0.3a & \text{hinged far end} \end{cases} \tag{17.64}$$

EXAMPLE 17.3:

Given: A one-story frame with a symmetrically tapered rectangular girder is shown in Figure 17.8. Assuming $r = 0.5$, $a = 0.2$, and $I_g = 2I_c = 2I$, determine K-factor for column AB.

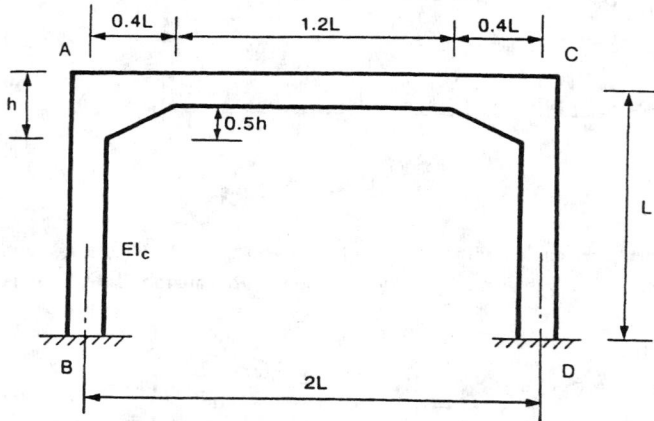

Figure 17.8 A simple frame with rectangular sections. (From King, W.S., Duan, L., et al., *Eng. Struct.*, 15(5), 369, 1993. With kind permission from Elsevier Science, Ltd, The Boulevard, Langford Lane, Kidlington OX5 IGB, UK.)

Solution

1. Using the alignment chart with modification

   For joint A, since the far end of the girder is rigid, use Equations 17.64 and 17.58,

$$\beta = 3 + 3.8\,(0.2)^2 - 6.5\,(0.2) = 1.852$$

$$\alpha_T = (1 - 0.5)^{1.852} = 0.277$$

$$G_A = \frac{\sum E_c I_c / L_c}{\sum \alpha_T E_g I_g / L_g} = \frac{EI/L}{0.277 E(2I)/2L} = 3.61$$

$$G_B = 1.0$$

$$(\text{AISC} - \text{LRFD } 1993)$$

   From the alignment chart in Figure 17.4b, $K = 1.59$ is obtained.

2. Using the alignment chart without modification

   A direct use of Equations 17.7 and 17.8 with an average section $(0.75\,h)$ results in:

$$I_g = 0.75^3 (2I) = 0.844I$$

$$G_A = \frac{EI/L}{0.844EI/2L} = 2.37 \quad G_B = 1.0$$

   From the alignment chart in Figure 17.4b $K = 1.49$, or $(1.49 - 1.59)/1.59 = -6\%$ in error on the less conservative side.

17.5.4 Unsymmetrical Frames

When the column sizes or column loads are not identical, adjustments to the alignment charts are necessary to obtain a correct K-factor. SSRC Guide [43] presents a set of curves as shown in Figure 17.9 for a modification factor, β, originally developed by Chu and Chow [18].

$$K_{\text{adjusted}} = \beta K_{\text{alignment chart}} \tag{17.65}$$

If the K-factor of the column under the load λP is desired, further modifications to K are necessary. Denoting K' as the effective length factor of the column with $I'_c = \alpha I_c$ subjected to the axial load, $P' = \lambda P$, as shown in Figure 17.9, then we have:

$$K' = K_{\text{adjusted}} \frac{L}{L'} \sqrt{\frac{\alpha}{\lambda}} \tag{17.66}$$

Equation 17.66 can be used to determine K-factors for columns in adjacent stories with different heights, L'.

17.5.5 Effects of Axial Forces in Restraining Members in Braced Frames

Bridge and Fraser [14] observed that K-factors of a column in a braced frame may be greater than unity due to "negative" restraining effects. Figure 17.10 shows the solutions obtained by considering both the "positive" and "negative" values of G-factors. The shaded portion of the graph corresponds to the alignment chart shown in Figure 17.4a when both G_A and G_B are positive.

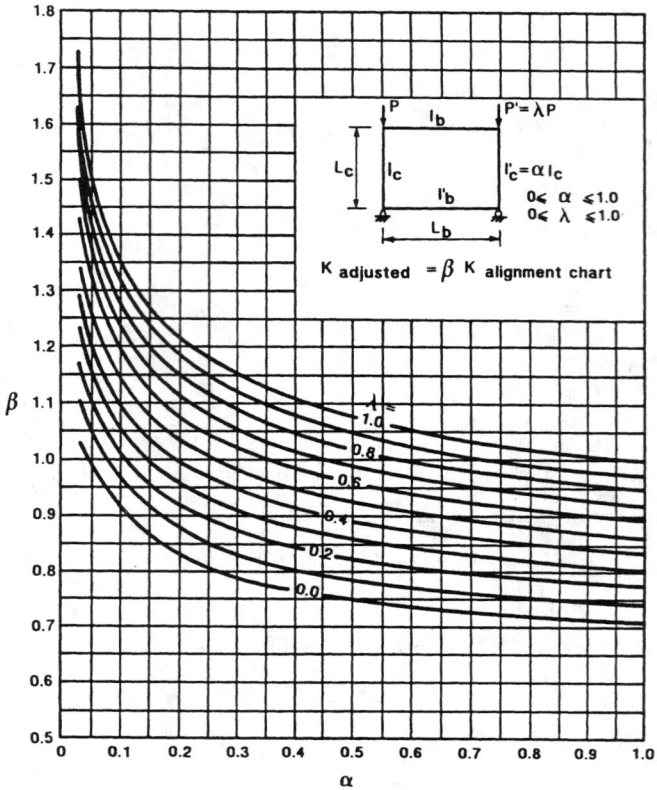

Figure 17.9   Chart for the modification factor β in an unsymmetrical frame. (from Johnston, B.G., Ed., *Structural Stability Research Council, Guide to Stability Design Criteria for Metal Structures*, 3rd ed., John Wiley & Sons, New York, 1976. With permission.)

To account for the effect of axial forces in the restraining members, Bridge and Fraser [14] proposed a more general expression for G-factor:

$$
\begin{aligned}
G &= \frac{(I/L)}{\sum_n (I/L)_n \gamma_n m_n} \\
  &= \frac{\text{stiffness of member } i \text{ under investigation}}{\text{stiffness of all rigidly connected members}}
\end{aligned}
\tag{17.67}
$$

where γ is a function of the stability functions S and C (Equations 17.3 and 17.4); m is a factor to account for the end conditions of the restraining member (see Figure 17.11); and subscript n represents the other members rigidly connected to member i. The summation in the denominator is for all members meeting at the joint.

Using Figures 17.10 and 17.11 and Equation 17.67, the effective length factor K_i for the i-th member can be determined by the following steps:

1. Sketch the buckled shape of the structure under consideration.
2. Assume a value of K_i for the member being investigated.
3. Calculate values of K_n for each of the other members that are rigidly connected to the i-th member using the equation

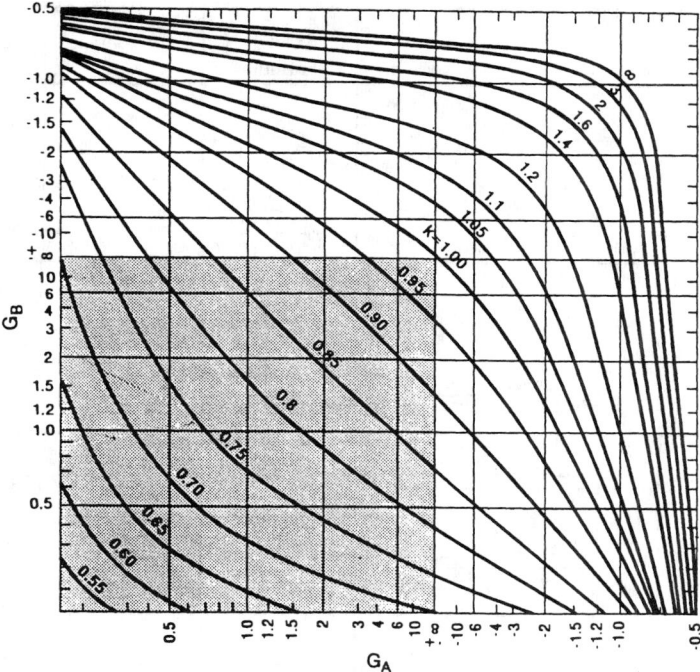

Figure 17.10   Effective length chart considering both positive and negative effects in braced frame. (From Bridge, R.Q. and Fraser, D.J., ASCE, *J. Struct. Eng.*, 113(6), 1341, 1987. With permission.)

$$K_n = K_i \frac{L_i}{L_n} \sqrt{\left(\frac{P_i}{P_n}\right)\left(\frac{I_n}{I_i}\right)} \tag{17.68}$$

4. Calculate γ and obtain m from Figure 17.11 for each member.
5. Calculate G_i for the i-th member using Equation 17.67.
6. Obtain K_i from Figure 17.10 and compare with the assumed K_i at Step 2.
7. Repeat the procedure by using the calculated K_i as the assumed K_i until K_i calculated at the end the cycle is approximately (say 10%) equal to the K_i at the beginning of the cycle.
8. Repeat steps 2 to 7 for other members of the frame.
9. The largest set of K values obtained is then used for design.

The above procedure has been illustrated [14] and verified [50] to provide a good elastic K-factor of columns in braced frames.

EXAMPLE 17.4:

Given: A braced column is shown in Figure 17.12. Consider axial force effects to determine K-factors for columns AB and BC.

    Solutions

1. Sketch the buckled shape as shown in Figure 17.12b

Axial Forces	Cases	Exact γ Formula	Approximate γ Formula	m
Tension	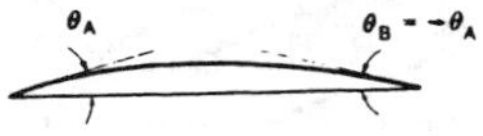 (diagram θ_A, $\theta_B=-\theta_A$)	$\dfrac{C-S}{2}$	$1 + \dfrac{1}{1.5\,K_n^2}$	1
	 (diagram θ_A, $M_B=0$)	$\dfrac{C-S^2/C}{3}$		1.5
	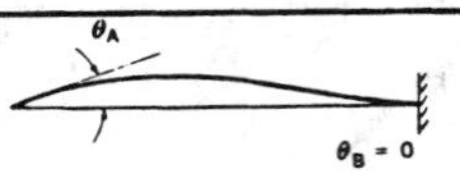 (diagram θ_A, $\theta_B=0$)	$\dfrac{C}{4}$	$1 + \dfrac{1}{4\,K_n^2}$	2
	 (diagram θ_A, $\theta_B=\theta_A$)	$\dfrac{C+S}{6}$		3
Compression	 (diagram θ_A, $\theta_B=-\theta_A$)	$\dfrac{C-S}{2}$	$\begin{cases} 1 - \dfrac{1}{K_n^2} & \text{for } K_n > 1.0 \\[2mm] 2 - \dfrac{2}{K_n^2} & \text{for } K_n \le 1.0 \end{cases}$	1
	 (diagram θ_A, $M_B=0$)	$\dfrac{C-S^2/C}{3}$		1.5
	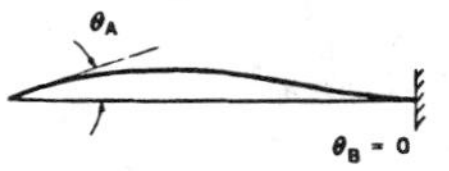 (diagram θ_A, $\theta_B=0$)	$\dfrac{C}{4}$	$\begin{cases} 1 - \dfrac{1}{2K_n^2} & \text{for } K_n > 0.7 \\[2mm] 2 - \dfrac{1}{K_n^2} & \text{for } K_n \le 0.7 \end{cases}$	2
	 (diagram θ_A, $\theta_B=\theta_A$)	$\dfrac{C+S}{6}$	$1 - \dfrac{1}{4\,K_n^2}$	3
Note	C and S are stability Equations (18.3.2) and (18.3.3)			

Figure 17.11   Values of γ and m to account for the effect of axial forces in the restraining members.

2. Assume $K_{AB} = 0.94$.

3. Calculate K_{BC} by Equation 17.68.

$$K_{BC} = K_{AB}\frac{L_{AB}}{L_{BC}} + \sqrt{\left(\frac{P_{AB}}{P_{BC}}\right)\left(\frac{I_{BC}}{I_{AB}}\right)} = 0.94\frac{L}{L}\sqrt{\frac{2PI}{P(1.2I)}}$$

$$= 1.22$$

4. Calculate γ and obtain m from Figure 17.11 for member BC.

Since $K_{BC} > 1.0$

$$\gamma_{BC} = 1 - \frac{1}{K_{BC}^2} = 1 - \frac{1}{1.22^2} = 0.33$$

Far end is pinned, $m_{BC} = 1.5$

5. Calculate G-factor for the member AB using Equation 17.67.

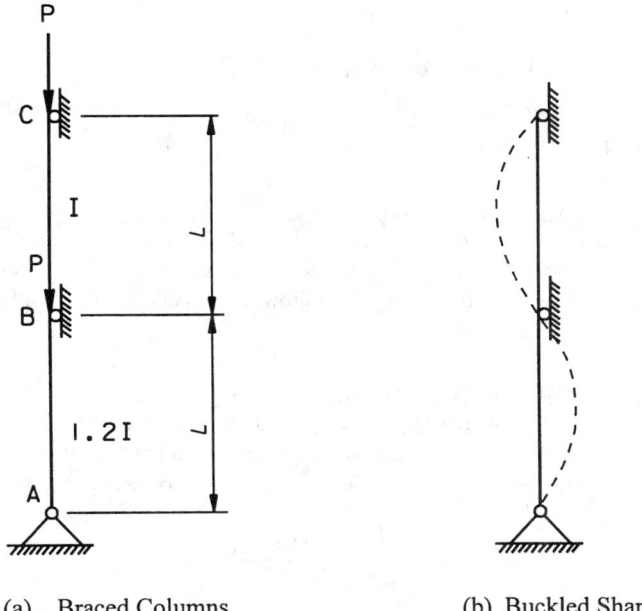

(a) Braced Columns   (b) Buckled Shape

Figure 17.12   Braced columns.

$$GB = \frac{(I/L)}{\sum_n (I/L)_n \gamma_n m_n} = \frac{(1.2I/L)}{(I/L)(0.33)(1.5)} = 2.42$$

$$G_A = \infty$$

6. From Figure 17.10, $K_{AB} = 0.93$.
    Comparing with the assumed $K_{AB} = 0.94$ O.K.
7. Repeat the above procedure for member BC.

> Assume $K_{BC} = 1.2$
>
> Calculate K_{AB} by Equation 17.68

$$K_{AB} = K_{BC} \frac{L_{BC}}{L_{AB}} \sqrt{\left(\frac{P_{BC}}{P_{AB}}\right)\left(\frac{I_{AB}}{I_{BC}}\right)} = 1.2 \frac{L}{L} \sqrt{\frac{P(1.2I)}{2PI}} = 0.93$$

> Calculate γ and obtain m from Figure 17.11 for member AB
>
> Since $K_{AB} < 1.0$

$$\gamma_{AB} = 2 - \frac{2}{K_{AB}^2} = 2 - \frac{2}{0.93^2} = -0.312$$

> Far end is pinned, $m_{AB} = 1.5$
>
> Calculate G-factor for the member BC using Equation 17.67.

$$GB = \frac{(I/L)}{\sum_n (I/L)_n \gamma_n m_n} = \frac{(I/L)}{(1.2I/L)(-0.312)(1.5)} = -1.78$$

$$G_A = \infty$$

Read Figure 17.10, $K_{BC} = 1.18$

Comparing with the assumed $K_{AB} = 1.20$ O.K.

8. It is seen that the largest set of K-factors is
$K_{AB} = 1.22$ and $K_{BC} = 0.93$

17.5.6   Consideration of Partial Column Base Fixity

In computing the effective length factor for monolithic connections, it is important to properly evaluate the degree of fixity in foundation. The following two approaches can be used to account for foundation fixity.

1. **Fictitious Restraining Beam Approach**
   Galambos [33] proposed that the effect of partial base fixity can be modelled as a fictitious beam. The approximate expression for the stiffness of the fictitious beam accounting for rotation of foundation in the soil has the form:

$$\frac{I_s}{L_B} = \frac{qBH^3}{72E_{\text{steel}}} \tag{17.69}$$

   where q is the modulus of subgrade reaction (varies from 50 to 400 lb/in.^{3}, 0.014 to 0.109 N/mm^3); B and H are the width and length (in bending plane) of the foundation; and E_{steel} is the modulus of elasticity of steel.

   Based on studies by Salmon, Schenker, and Johnston [65], the approximate expression for the stiffness of the fictitious beam accounting for the rotations between column ends and footing due to deformation of base plate, anchor bolts, and concrete can be written as:

$$\frac{I_s}{L_B} = \frac{bd^2}{72E_{\text{steel}}/E_{\text{concrete}}} \tag{17.70}$$

   where b and d are the width and length of the base plate, and subscripts *concrete* and *steel* represent concrete and steel, respectively. Galambos [33] suggested that the smaller of the stiffness calculated by Equations 17.69 and 17.70 be used in determining K-factors.

2. **AASHTO-LRFD Approach**
   The following values are suggested by AASHTO-LRFD [1]:

   $G = 1.5$ footing anchored on rock
   $G = 3.0$ footing not anchored on rock
   $G = 5.0$ footing on soil
   $G = 1.0$ footing on multiple rows of end bearing piles

17.5.7   Inelastic K-factor

The effect of material inelasticity and end restrain on the K-factors has been studied during the last two decades [12, 15, 20, 44, 45, 58, 64, 67, 68, 69, 73] The inelastic K-factor developed originally by Yura [73] and expanded by Disque [20] makes use of the alignment charts with simple modifications. To consider inelasticity of material, the G values as defined by Equations 17.7 and 17.8 are replaced by G^* [20] as follows:

$$G^* = SRF(G) = \frac{E_t}{E}G \tag{17.71}$$

in which E_t is the tangent modulus of the material. For practical application, stiffness reduction factor $(SRF) = (E_t/E)$ can be taken as the ratio of the inelastic to elastic buckling stress of the column

$$SRF = \frac{E_t}{E} \approx \frac{(F_{cr})_{\text{inelastic}}}{(F_{cr})_{\text{elastic}}} \approx \frac{P_u/A_g}{(F_{cr})_{\text{elastic}}} \tag{17.72}$$

where P_u is the factored axial load and A_g is the cross-sectional area of the member. $(F_{cr})_{\text{inelastic}}$ and $(F_{cr})_{\text{elastic}}$ can be calculated by AISC-LRFD [4] column equations:

$$(F_{cr})_{\text{inelastic}} = (0.658)^{\lambda_c^2} F_y \tag{17.73}$$

$$(F_{cr})_{\text{elastic}} = \left[\frac{0.877}{\lambda_c^2}\right] F_y \tag{17.74}$$

$$\lambda_c = \frac{KL}{r\pi}\sqrt{\frac{F_y}{E}} \tag{17.75}$$

in which K is the elastic effective length factor and r is the radius of gyration about the plane of buckling. Table 17.3 gives SRF values for different stress levels and slenderness parameters.

TABLE 17.3   Stiffness Reduction Factor (SRF) for G values

	$\left(\frac{KL}{r}\right)_{\text{elastic}}$			
	36 ksi	50 ksi		SRF
$\frac{P_u}{A_g F_y}$	(248 MPa)	(345 MPa)	λ_c	(Eq. 18.72)
1.00	0.0	0.0	0.155	0.000
0.95	31.2	26.5	0.350	0.133
0.90	44.7	38.0	0.502	0.258
0.85	55.6	47.1	0.623	0.376
0.80	65.1	55.2	0.730	0.486
0.75	73.9	62.7	0.829	0.588
0.70	82.3	69.8	0.923	0.680
0.65	90.5	76.8	1.015	0.763
0.60	98.5	83.6	1.105	0.835
0.55	106.6	90.4	1.195	0.896
0.50	114.7	97.4	1.287	0.944
0.45	123.2	104.5	1.381	0.979
0.40	131.9	111.9	1.480	0.998
0.39	133.7	113.5	1.500	1.000

EXAMPLE 17.5:

Given: A two-story steel frame is shown in Figure 17.5. Use the alignment chart to determine K-factor for inelastic column DE. $E = 29,000$ ksi (200 GPa) and $F_y = 36$ ksi (248 MPa).

Solution

1. Calculate the axial stress ratio:

$$\frac{P_u}{A_g F_y} = \frac{300}{13.3(36)} = 0.63$$

2. Obtain $SRF = 0.793$ from Table 17.3
3. Calculate modified G-factor.

$$G_E = 0.448 \text{ (Example 17.1)}$$
$$G_E^* = SRF(G_E) = 0.794(0.448) = 0.355$$
$$G_D = 10 \quad \text{(AISC-LRFD 1993)}$$

4. From the alignment chart in Figure 17.4b , we have

$$(K_{DE})_{\text{inelastic}} = 1.75$$

17.6 Framed Columns—Alternative Methods

17.6.1 LeMessurier Method

Considering that all columns in a story buckle simultaneously and strong columns will brace weak columns (Figure 17.13), a more accurate approach to calculate K-factors for columns in a sidesway frame was developed by LeMessurier [52]. The K_i value for the i-th column in a story can be obtained by the following expression:

$$K_i = \sqrt{\frac{\pi^2 E I_i}{L_i^2 P_i} \left(\frac{\sum P + \sum C_L P}{\sum PL} \right)} \tag{17.76}$$

where P_i is the axial compressive force for member i, subscript i represents the i-th column, and $\sum P$ is the sum of the axial force of all columns in a story.

$$P_L = \frac{\beta E I}{L^2} \tag{17.77}$$

$$\beta = \frac{6(G_A + G_B) + 36}{2(G_A + G_B) + G_A G_B + 3} \tag{17.78}$$

$$C_L = \left(\beta \frac{K_o^2}{\pi^2} - 1 \right) \tag{17.79}$$

in which K_o is the effective length factor obtained by the alignment chart for unbraced frames, and P_L is only for rigid columns which provide sidesway stiffness.

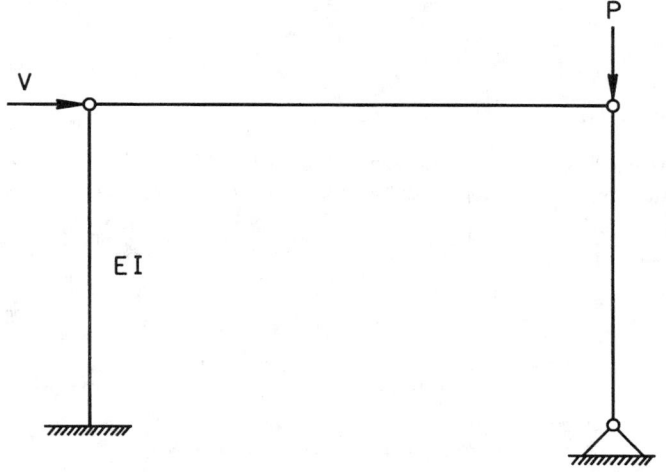

Figure 17.13   Subassemblage of the LeMessurier method.

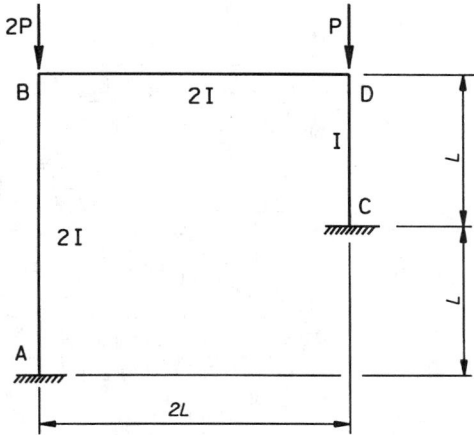

(a) Frame Dimensions and Loads

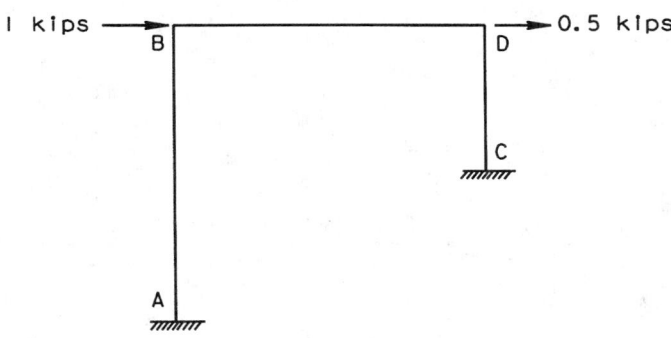

(b) Frame Subjected to Fictitious Lateral Loads

Figure 17.14   A frame with unequal columns.

EXAMPLE 17.6:

Given: A sway frame with columns of unequal height is shown in Figure 17.14a. Determine elastic K-factors for columns by using the LeMessurier method. Member properties are:

Member	A in.^{2}	(mm^2)	I in.^{4}	(mm$^4 \times 10^8$)	L in.	(mm)
AB	21.5	(13,871)	620	(2.58)	240	(6,096)
BD	21.5	(13,871)	620	(2.58)	240	(6,096)
CD	7.65	(4,935)	310	(1.29)	120	(3,048)

Solution

The detailed calculations are listed in Table 17.4.

Using Equation 17.76, we obtain:

$$K_{AB} = \sqrt{\frac{\pi^2 E I_{AB}}{L_{AB}^2 P_{AB}}\left(\frac{\sum P + \sum C_L P}{\sum P_L}\right)}$$

$$= \sqrt{\frac{\pi^2 E(620)}{(240)^2(2P)}\left(\frac{3P + 0.495P}{0.271E}\right)} = 0.83$$

$$K_{CD} = \sqrt{\frac{\pi^2 E I_{CD}}{L_{CD}^2 P_{CD}}\left(\frac{\sum P + \sum C_L P}{\sum P_L}\right)}$$

$$= \sqrt{\frac{\pi^2 E(310)}{(120)^2(P)}\left(\frac{3P + 0.495P}{0.271E}\right)} = 1.66$$

TABLE 17.4   Example 17.6—Detailed Calculation by the
LeMessurier Method

Members	AB	CD	Sum	Notes
I in.^{4}	620	310	—	
(mm$^4 \times 10^8$)	(2.58)	(1.29)	—	
L in.	240	120	—	
(mm)	(6,096)	(3,048)	—	
G_{top}	1.0	1.0	—	Eq. 18.7
G_{bottom}	0.0	0.0	—	Eq. 18.7
β	8.4	8.4	—	Eq. 18.78
K_{io}	1.17	1.17	—	Alignment Chart
C_L	0.165	0.165	—	Eq. 18.79
P_L	0.09E	0.181E	0.271E	Eq. 18.77
P	2P	P	3P	
$C_L P$	0.33P	0.165P	0.495P	

17.6.2 Lui Method

A simple and straightforward approach for determining the effective length factors for framed columns without the use of alignment charts and other charts was proposed by Lui [57]. The formulas take into account both the member instability and frame instability effects explicitly. The K-factor for the i-th column in a story was obtained in a simple form:

$$K_i = \sqrt{\left(\frac{\pi^2 E I_i}{P_i L_i^2}\right)\left[\left(\sum \frac{P}{L}\right)\left(\frac{1}{5\sum \eta} + \frac{\Delta_1}{\sum H}\right)\right]} \qquad (17.80)$$

where $\sum(P/L)$ represents the sum of the axial force-to-length ratio of all members in a story, $\sum H$ is the story lateral load producing Δ_1, Δ_1 is the first-order inter-story deflection, and η is the member stiffness index and can be calculated by

$$\eta = \frac{\left(3 + 4.8m + 4.2m^2\right)EI}{L^3} \qquad (17.81)$$

in which m is the ratio of the smaller to larger end moments of the member; it is taken as positive if the member bends in reverse curvature and negative for single curvature.

It is important to note that the term $\sum H$ used in Equation 17.80 is not the actual applied lateral load. Rather, it is a small disturbing or fictitious force (taken as a fraction of the story gravity loads) to be applied to each story of the frame. This fictitious force is applied in a direction such that the deformed configuration of the frame will resemble its buckled shape.

EXAMPLE 17.7:

Given: Determine K-factors by using the Lui method for the frame shown in Figure 17.14a.

$$E = 29,000 \text{ ksi } (200 \text{ GPa}).$$

Solution

Apply fictitious lateral forces at B and D (Figure 17.14b) and perform a first-order analysis. Detailed calculation is shown in Table 17.5.

Using Equation 17.80, we obtain:

$$K_{AB} = \sqrt{\left(\frac{\pi^2 E I_{AB}}{P_{AB} L_{AB}^2}\right)\left[\left(\sum \frac{P}{L}\right)\left(\frac{1}{5\sum \eta} + \frac{\Delta_1}{\sum H}\right)\right]}$$

$$= \sqrt{\left(\frac{\pi^2 (29,000)(620)}{(2P)(240)^2}\right)\left[\left(\frac{P}{60}\right)\left(\frac{1}{5(56.24)} + 0.019\right)\right]} = 0.76$$

$$K_{CD} = \sqrt{\left(\frac{\pi^2 E I_{CD}}{P_{CD} L_{CD}^2}\right)\left[\left(\sum \frac{P}{L}\right)\left(\frac{1}{5\sum \eta} + \frac{\Delta_1}{\sum H}\right)\right]}$$

$$= \sqrt{\left(\frac{\pi^2 (29,000)(310)}{(P)(120)^2}\right)\left[\left(\frac{P}{60}\right)\left(\frac{1}{5(56.24)} + 0.019\right)\right]} = 1.52$$

TABLE 17.5   Example 17.7—Detailed Calculation by the Lui Method

Members	AB	CD	Sum	Notes
I in.^{4}	620	310	—	
(mm$^4 \times 10^8$)	(2.58)	(1.29)		
L in.	240	120	—	
(mm)	(6096)	(3048)		
H kips	1.0	0.5	1.5	
(kN)	(4.448)	(2.224)	(6.672)	
Δ_1 in.	0.0286	0.0283	—	
(mm)	(0.7264)	(0.7188)		
$\Delta_1 / \sum H$ in./kips	—	—	0.019	Average
(mm/kN)			(0.108)	
M_{top} k-in.	−38.8	56.53	—	
(kN-m)	(−4.38)	(6.39)		
M_{bottom} k-in.	−46.2	81.18	—	
(kN-m)	(−5.22)	(9.17)		
m	0.84	0.69	—	
η kips/in.	13.00	43.24	56.24	Eq. 18.1
(kN/mm)	(2.28)	(7.57)	(9.85)	
P/L kips/in.	$P/120$	$P/120$	$P/60$	
(kN/mm)	$P/3048$	$P/3048$	$P/1524$	

17.6.3   Remarks

For comparison, Table 17.6 summarizes K-factors for the frame shown in Figure 17.14a obtained from the alignment chart, the LeMessurier and Lui methods, as well as an eigenvalue analysis. It is seen that errors in alignment chart results are rather significant in this case. Although K-factors predicted by Lui's and LeMessurier's formulas are identical in most cases, the simplicity and independence of any chart in the case of Lui's formulas make it more desirable for design office use [66].

TABLE 17.6   Comparison of K-Factors for the Frame in Figure 17.14a

Columns	Theoretical	Alignment chart	Lui Eq. 18.80	LeMessurier Eq. 18.76
AB	0.70	1.17	0.76	0.83
CD	1.40	1.17	1.52	1.67

17.7   Unbraced Frames With Leaning Columns

A column framed with simple connections is often called a **leaning column**. It has no lateral stiffness or sidesway resistance. A column framed with rigid moment-resisting connections is called a **rigid column**. It provides the lateral stiffness or sidesway resistance to the frame. When a frame system (Figure 17.15a) includes leaning columns, the effective length factors of rigid columns must be modified. Several approaches to account for the effect of "leaning columns" were reported in the literature [16, 52, 54, 73]. A detailed discussion about the leaning columns for practical applications was presented by Geschwindner [37].

17.7.1   Rigid Columns

1. **Yura Method**

   Yura [73] discussed frames with leaning columns and noted the behavior of stronger columns assisting weaker ones in resisting sidesway. He concluded that the alignment chart gives valid sidesway buckling solutions if the columns are in the elastic range and all columns in a story reach their individual buckling loads simultaneously. For columns that do not satisfy these two conditions, the alignment chart is generally overly conservative. Yura states that

   (a) The maximum load-carrying capacity of an individual column is limited to the load permitted on that column for braced case $K = 1.0$.

   (b) The total gravity loads that produce sidesway are distributed among the columns, which provides lateral stiffness in a story.

2. **Lim and McNamara Method**

   Based on the story buckling concept and using the stability functions, Lim and McNamara [54] presented the following formula to account for the leaning column effect.

$$K_n = K_o \sqrt{1 + \frac{\sum Q}{\sum P}\left(\frac{F_o}{F_n}\right)} \qquad (17.82)$$

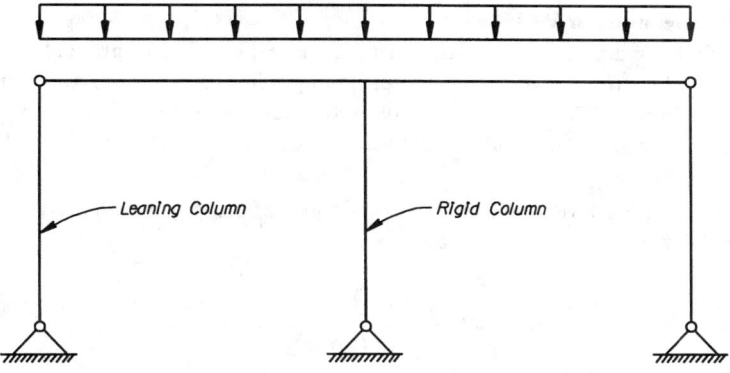

(a)   A Leaning-Column Frame

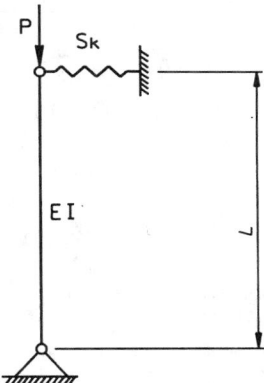

(b)   Model for a Leaning Column

Figure 17.15   A frame with leaning columns.

where K_n is the effective length factor accounting for the leaning columns; K_o is the effective length factor determined by the alignment chart (Figure 17.3b) not accounting for the leaning columns; $\sum P$ and $\sum Q$ are the loads on the restraining columns and on the leaning columns in a story, respectively; and F_o and F_n are the eigenvalue solutions for a frame without and with leaning columns, respectively. For normal column end conditions that fall somewhere between fixed and pinned, $F_o/F_n = 1$ provides a K-factor on the conservative side by less than 2% [37]. Using $F_o/F_n = 1$, Equation 17.82 becomes:

$$K_n = K_o\sqrt{1 + \frac{\sum Q}{\sum P}} \qquad (17.83)$$

Equation 17.83 gives the same K-factor as the modified Yura approach [37].

3. **LeMessurier and Lui Methods**

Equation 17.76 developed by LeMessurier [52] and Equation 17.80 proposed by Lui [57] can be used for frames with and without leaning columns. Since the K-factor expressions Equations 17.76 and 17.80 were derived for an entire story of the frame, they are applicable to frames with and without leaning columns.

4. **AISC-LRFD Method**

The current AISC-LRFD [4] commentary adopts the following modified effective length factor, K_i', for the i-th rigid column:

$$K_i' = \sqrt{\frac{\pi^2 E I_i}{L_i^2 P_{ui}} \left(\frac{\sum P_u}{\sum P_{e2}} \right)} \tag{17.84}$$

where $\sum P_{e2}$ is the Euler loads of all columns in a story providing lateral stiffness for the frame based on the effective length factor obtained from the alignment chart for an unbraced frame; P_{ui} is the required axial compressive strength for the i-th rigid column; and $\sum P_u$ is the required axial compressive strength of all columns in a story. When E and L^2 are constant for all columns in a story, AISC [4] suggested that:

$$K_i' = \sqrt{\frac{\sum P_u}{P_{ui}} \times \left(\frac{I_i}{\sum \frac{I_i}{K_{io}^2}} \right)} \tag{17.85}$$

except

$$K_i' \geq \sqrt{\frac{5}{8}} K_{io} \tag{17.86}$$

where K_{io} is the effective length factor of a rigid column based on the alignment chart for unbraced frames.

EXAMPLE 17.8:

Given: A frame with a leaning column is shown in Figure 17.16a [59]. Evaluate the K-factor for column AB using various methods. The bottom of column AB is assumed to be ideally pin-ended for comparison purposes. $E = 29,000$ ksi (200 GPa).

   Solution

1. **Alignment Chart Method**

$$
\begin{aligned}
G_A &= \infty \\
G_B &= \frac{\sum E_c I_c / L_c}{\sum \alpha_k E_g I_g / L_g} = \frac{EI/L}{0.5EI/L} = 2.0
\end{aligned}
$$

From Figure 17.3b, we have $K_{AB} = 2.6$.

2. **Lim and McNamara Method**

For this frame, $\sum P = \sum Q = P$ and $K_o = 2.6$. From Equation 17.83, we have

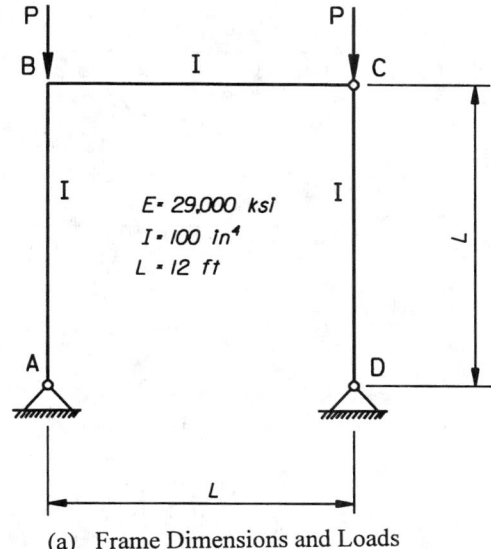

(a) Frame Dimensions and Loads

(b) Frame Subjected to Fictitious Lateral Loads

Figure 17.16 A leaning column frame.

$$K_{AB} = K_o\sqrt{1 + \frac{\sum Q}{\sum P}} = 2.6\sqrt{1+1} = 3.68$$

3. **LeMessurier Method**

For column AB, $G_A = \infty$ and $G_B = 2.0$; from the alignment chart, $K_o = 2.6$. According to Equations 17.76 to 17.79 we have,

$$\beta \,|_{G_A=\infty} = \frac{6(G_A + G_B) + 36}{2(G_A + G_B) + G_A G_B + 3}\,|_{G_A=\infty} = \frac{6}{2 + G_B} = \frac{6}{2 + 2} = 1.5$$

$$\sum P_L = (P_L)_{AB} = \frac{\beta EI}{L^2} = 1.5 \frac{EI}{L^2}$$

$$C_L = \left(\beta \frac{K_o^2}{\pi^2} - 1\right) = (1.5)\frac{2.6^2}{\pi^2} - 1 = 0.0274$$

$$K_{AB} = \sqrt{\frac{\pi^2 E I_{AB}}{L_{AB}^2 P_{AB}}\left(\frac{\sum P + \sum C_L P}{\sum P_L}\right)} = \sqrt{\frac{\pi^2 EI}{L^2 P}\left(\frac{2P + 0.0274P}{1.5EI/L^2}\right)}$$

$$= \sqrt{13.34} = 3.65$$

4. AISC-LRFD Method

Using Equation 17.85 for column AB:

$$K_{AB} = \sqrt{\frac{\sum P_u}{P_{AB}} \times \left(\frac{I_{AB}}{\sum \frac{I}{K_{io}^2}}\right)} = K_{io}\sqrt{2} = 3.68$$

5. Lui Method

(a) Apply a small lateral force, $H = 1$ kip, as shown in Figure 17.16b.

(b) Perform a first-order analysis and find $\Delta_1 = 0.687$ in. (17.45 mm).

(c) Calculate η factors from Equation 17.81.

Since column CD buckles in a single curvature, $m = -1$,

$$\eta_{CD} = \frac{(3 + 4.8m + 4.2m^2)EI}{L^3} = \frac{(3 - 4.8 + 4.2)EI}{L^3} = \frac{2.4EI}{L^3}$$

For column AB, $m = 0$,

$$\eta_{AB} = \frac{(3 + 4.8m + 4.2m^2)EI}{L^3} = \frac{3EI}{L^3}$$

$$\sum \eta = \frac{3EI}{L^3} + \frac{2.4EI}{L^3} = \frac{5.4(29,000)(100)}{(144)^3}$$

$$= 5.245 \text{ kips/in. } (0.918 \text{ kN/mm})$$

(d) Calculate the K-factor from Equation 17.80.

$$K_{AB} = \sqrt{\left(\frac{\pi^2 E I_{AB}}{P_{AB} L_{AB}^2}\right)\left[\left(\sum \frac{P}{L}\right)\left(\frac{1}{5\sum \eta} + \frac{\Delta_1}{\sum H}\right)\right]}$$

$$= \sqrt{\left(\frac{\pi^2(29,000)(100)}{P(144)^2}\right)\left[\left(\frac{2P}{144}\right)\left(\frac{1}{5(5.245)} + \frac{0.687}{1}\right)\right]} = 3.73$$

From an eigenvalue analysis, $K_{AB} = 3.69$ is obtained. It is seen that a direct use of the alignment chart leads to a significant error for this frame, and other approaches give good results. However, the LeMessurier approach requires the use of the alignment chart, and the Lui approach requires a first-order analysis subjected to a fictitious lateral loading.

17.7.2 Leaning Columns

Recognizing that a leaning column is being braced by rigid columns, Lui [57] proposed a model for the leaning column, as shown in Figure 17.15b. Rigid columns provide lateral stability to the whole structure and are represented by a translation spring with a spring stiffness, S_K. The K-factor for a leaning column can be obtained as:

$$K = \text{larger of} \begin{cases} 1 \\ \sqrt{\dfrac{\pi^2 EI}{S_K L^3}} \end{cases} \qquad (17.87)$$

For most commonly framed structures, the term $(\pi^2 EI / S_K L^3)$ normally does not exceed unity and so $K = 1$ often governs. AISC-LRFD [4] suggests that leaning columns with $K = 1$ may be used in unbraced frames provided that the lack of lateral stiffness from simple connections to the frame $(K = \infty)$ is included in the design of moment frame columns.

17.7.3 Remarks

Numerical studies by Geschwindner [37] found that the Yura approach gives overly conservative results for some conditions; Lim and McNamara's approach provides sufficiently accurate results for design, and the LeMessurier approach is the most accurate of the three. The Lim and McNamara approach could be appropriate for preliminary design while the LeMessurier and Lui approaches would be appropriate for final design.

17.8 Cross Bracing Systems

Diagonal bracing or X-bracing is commonly used in steel structures to resist horizontal loads. In the current practice, the design of this type of bracing system is based on the assumptions that the compression diagonal has negligible capacity and the tension diagonal resists the total load. The assumption that compression diagonal has a negligible capacity usually results in an overdesign [62, 63].

Picard and Beaulieu [62, 63] reported theoretical and experimental studies on double diagonal cross bracings (Figure 17.6) and found that

1. A general effective length factor equation (Figure 17.17) is given as

$$K = \sqrt{0.523 - \frac{0.428}{C/T}} \geq 0.50 \qquad (17.88)$$

   where C and T represent compression and tension forces obtained from an elastic analysis, respectively.

2. When the double diagonals are continuous and attached at their intersection point, the effective length of the compression diagonal is 0.5 times the diagonal length, i.e., $K = 0.5$, because the C/T ratio is usually smaller than 1.6.

EL-Tayem and Goel [27] reported a theoretical and experimental study about the X-bracing system made from single equal-leg angles. They concluded that:

1. Design of an X-bracing system should be based on an exclusive consideration of one half diagonal only.

2. For X-bracing systems made from single equal-leg angles, an effective length of 0.85 times the half diagonal length is reasonable, i.e., $K = 0.425$.

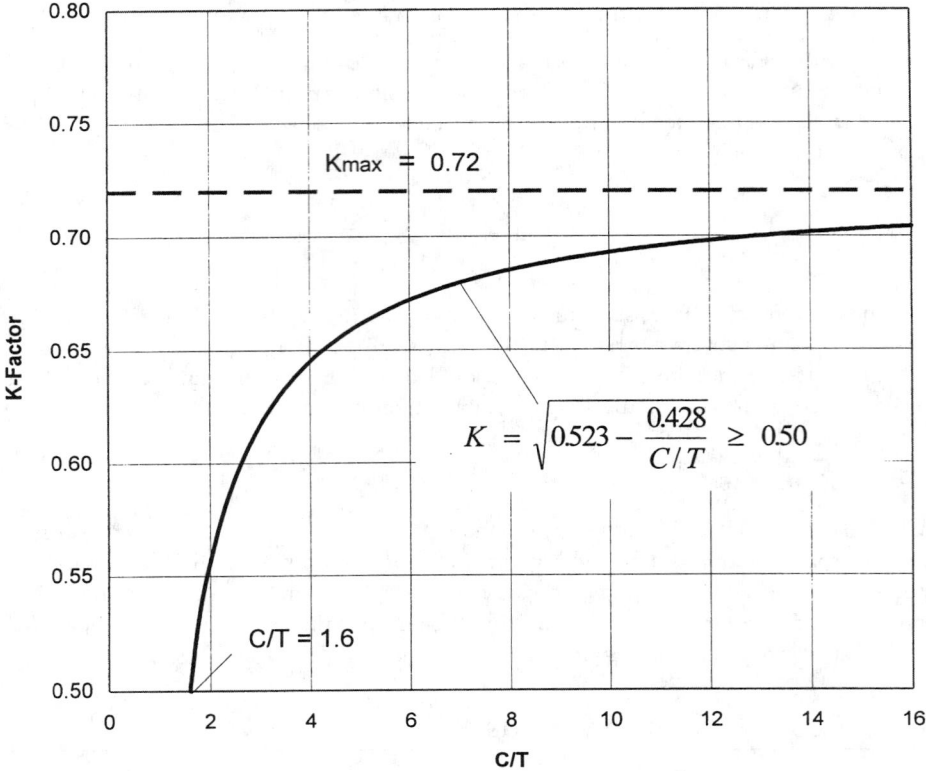

Figure 17.17 Effective length factor of compression diagonal. (From Picard, A. and Beaulieu, D., *AISC Eng. J.*, 24(3), 122, 1987. With permission.)

17.9 Latticed and Built-Up Members

The main difference of behavior between solid-webbed members, latticed members, and **built-up members** is the effect of shear deformation on their buckling strength. For **solid-webbed** members, shear deformation has a negligible effect on buckling strength. Whereas for latticed structural members using lacing bars and batten plates, shear deformation has a significant effect on buckling strength. It is a common practice that when a buckling model involves relative deformation produced by shear forces in the connectors, such as lacing bars and batten plates, between individual components, a modified effective length factor, K_m, is defined as follows:

$$K_m = \alpha_v K \qquad (17.89)$$

in which K is the usual effective length factor of a latticed member acting as a unit obtained from a structural analysis, and α_v is the shear factor to account for shear deformation on the buckling strength, or the modified effective slenderness ratio, $(KL/r)_m$ should be used in the determination of the compressive strength. Details of the development of the shear factor, α_v, can be found in textbooks by Bleich [13] and Timoshenko and Gere [70]. The following section briefly summarizes α_v formulas for various latticed members.

17.9.1 Laced Columns

For laced members as shown in Figure 17.18, by considering shear deformation due to the length-ening of diagonal lacing bars in each panel and assuming hinges at joints, the shear factor, α_v, has the form:

$$\alpha_v = \sqrt{1 + \frac{\pi^2 EI}{(KL^2)} \frac{1}{A_d E_d \sin \phi \cos^2 \phi}} \qquad (17.90)$$

where E_d is the modulus of elasticity of materials for the lacing bars, A_d is the cross-sectional area of all diagonals in one panel, and ϕ is the angle between the lacing diagonal and the axis that is perpendicular to the member axis.

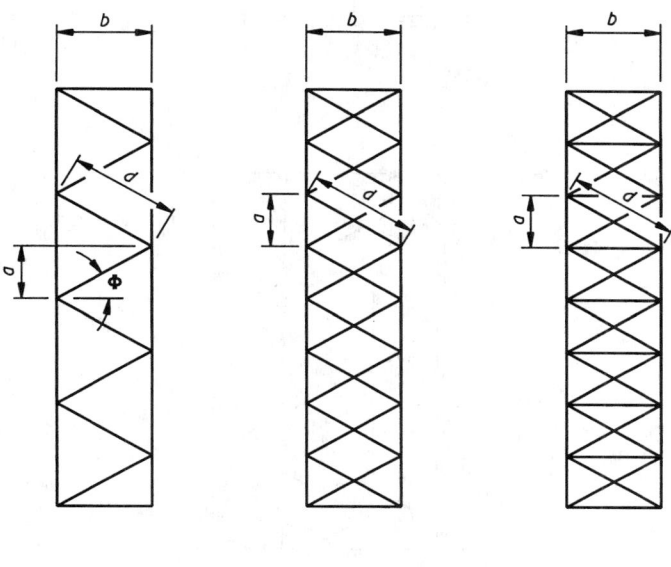

(a) Single Lacing (b) Double Lacing

Figure 17.18 Typical configurations of laced members.

If the length of the lacing bars is given (Figure 17.18), Equation 17.90 can be rewritten as:

$$\alpha_v = \sqrt{1 + \frac{\pi^2 EI}{(KL^2)} \frac{d^3}{A_d E_d ab^2}} \qquad (17.91)$$

where a, b, and d are the height of panel, depth of member, and length of diagonal, respectively.

The SSRC [36] suggested that a conservative estimate of the influence of 60 or 45° lacing, as generally specified in bridge design practice, can be made by modifying the overall effective length factor, K, by multiplying a factor, α_v, originally developed by Bleich [13] as follows:

For $\frac{KL}{r} > 40$,

$$\alpha_v = \sqrt{1 + 300/(KL/r)^2} \qquad (17.92)$$

For $\frac{KL}{r} \leq 40$,

$$\alpha_v = 1.1 \qquad (17.93)$$

EXAMPLE 17.9:

Given: A laced column with angles and cover plates is shown in Figure 17.19. $K_y = 1.25$, $L = 30$ ft (9144 mm). Determine the modified effective length factor, $(K_y)_m$, by considering the shear deformation effect.

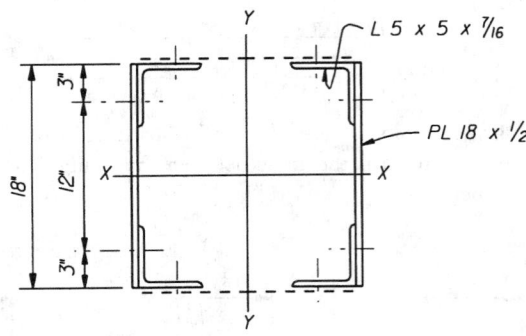

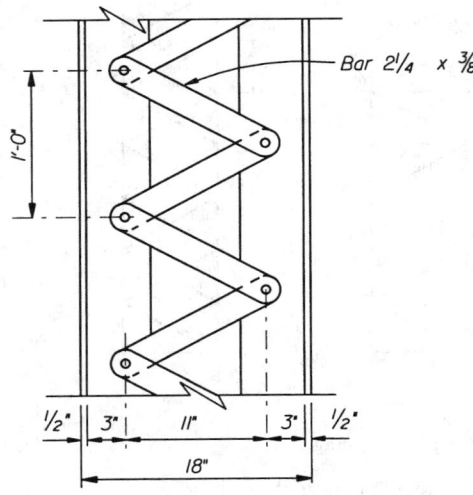

Figure 17.19   A laced column.

Section properties:

$$I_y = 2259 \text{ in.}^4 \ (9.4 \times 108 \text{ mm}^4)$$
$$E = E_d \qquad A_d = 1.69 \text{ in.}^2 \ (1090 \text{ mm}^2)$$
$$a = 6 \text{ in. (152 mm)} \qquad b = 11 \text{ in. (279 mm)}$$
$$d = 12.53 \text{ in. (318 mm)}$$

Solution

1. Calculate the shear factor, α_v, by Equation 17.91.

$$\alpha_v = \sqrt{1 + \frac{\pi^2 E I}{(KL)^2} \frac{d^3}{A_d E_d a b^2}}$$

$$= \sqrt{1 + \frac{\pi^2 E (2259)}{(1.25 \times 30 \times 12)^2} \frac{12.53^3}{1.69 E (6)(11)^2}} = 1.09$$

2. Calculate $(K_y)_m$ by Equation 17.89.

$$(K_y)_m = \alpha_v K_y = 1.09(1.25) = 1.36$$

17.9.2 Columns with Battens

The battened column has a greater shear flexibility than either the laced column or the column with perforated cover plates, hence the effect of shear distortion must be taken into account in calculating the effective length of a column [43]. For the battened members shown in Figure 17.20a, assuming that points of inflection in the battens are at the batten midpoints, and that points of inflection in the longitudinal element occur midway between the battens, the shear factor, α_v, is obtained as:

$$\alpha_v = \sqrt{1 + \frac{\pi^2 EI}{(KL)^2} \left(\frac{ab}{12E_b I_b} + \frac{a^2}{24EI_f} \right)} \tag{17.94}$$

where E_b is the modulus of elasticity of materials for the batten plates, I_b is the moment inertia of all the battens in one panel in the buckling plane, and I_f is the moment inertia of one side of the main components taken about the centroid axis of the flange in the buckling plane.

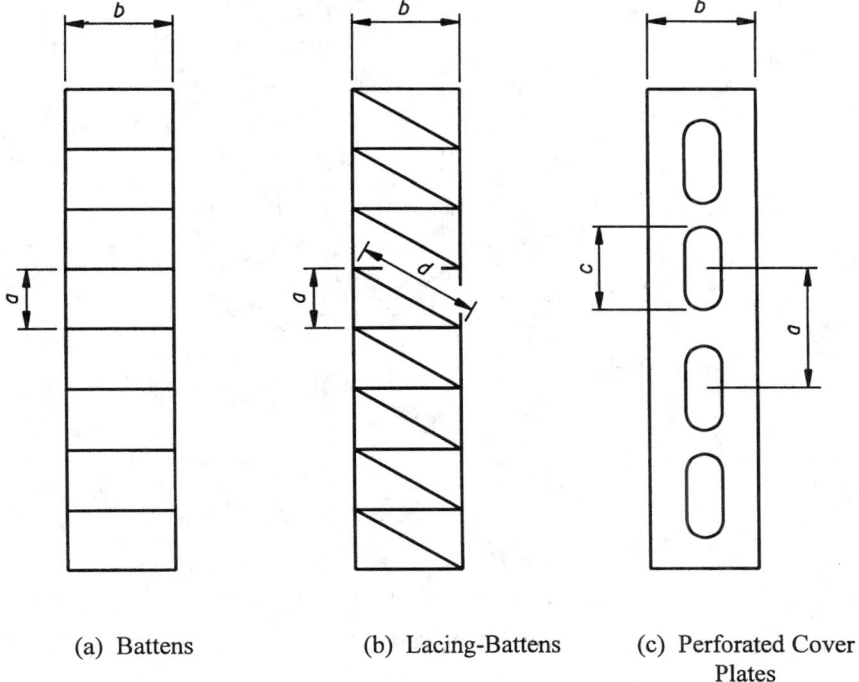

(a) Battens	(b) Lacing-Battens	(c) Perforated Cover Plates

Figure 17.20   Typical configurations of members with battens and with perforated cover plates.

EXAMPLE 17.10:

Given: A battened column is shown in Figure 17.21. $K_y = 0.8$, $L = 30$ ft (9144 mm). Determine the modified effective length factor, $(K_y)_m$, by considering the shear deformation effect.

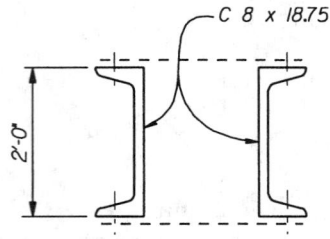

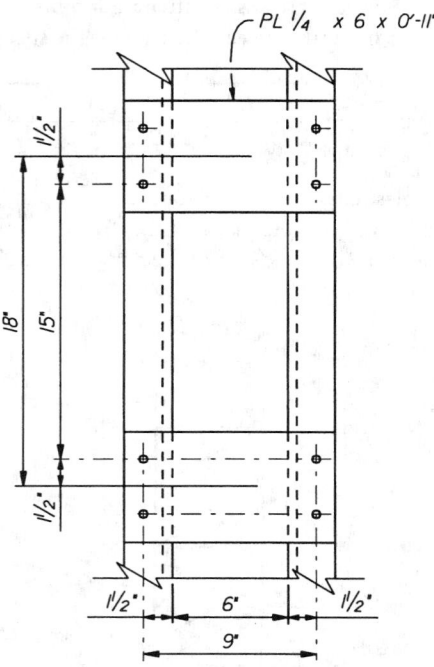

Figure 17.21   A battened column.

Section properties:

$$
\begin{aligned}
I_y &= 144 \text{ in.}^4 \ (6.0 \times 10^7 \text{ mm}^4) \\
E &= E_b \\
I_f &= 1.98 \text{ in.}^4 \ (8.24 \times 10^5 \text{ mm}^4) \\
a &= 15 \text{ in. } (381 \text{ mm}) \\
b &= 9 \text{ in. } (229 \text{ mm}) \\
I_b &= 9 \text{ in.}^4 \ (3.75 \times 10^6 \text{ mm}^4)
\end{aligned}
$$

Solution

1. Calculate the shear factor, α_v, by Equation 17.94.

$$\alpha_v = \sqrt{1 + \frac{\pi^2 EI}{(KL)^2}\left(\frac{ab}{12EI_b} + \frac{a^2}{24EI_f}\right)}$$

$$= \sqrt{1 + \frac{\pi^2 E(144)}{(0.8 \times 30 \times 12)^2}\left(\frac{15(9)}{12E(9)} + \frac{15^2}{24E(1.98)}\right)} = 1.05$$

2. Calculate $(K_y)_m$ by Equation 17.89.

$$\left(K_y\right)_m = \alpha_v K_y = 1.05(0.8) = 0.84$$

17.9.3 Laced-Battened Columns

For the laced-battened columns, as shown in Figure 17.20b, considering the shortening of the battens and the lengthening of the diagonal lacing bars in each panel, the shear factor, α_v, can be expressed as:

$$\alpha_v = \sqrt{1 + \frac{\pi^2 EI}{(KL)^2}\left(\frac{d^3}{A_d E_d ab^2} + \frac{b}{a A_b E_b}\right)} \tag{17.95}$$

where E_b is the modulus of elasticity of the materials for battens and A_b is the cross-sectional area of all battens in one panel.

17.9.4 Columns with Perforated Cover Plates

For members with perforated cover plates, shown in Figure 17.20c, considering the horizontal cross member as infinitely rigid, the shear factor, α_v, has the form:

$$\alpha_v = \sqrt{1 + \frac{\pi^2 EI}{(KL)^2}\left(\frac{9c^3}{64a E I_f}\right)} \tag{17.96}$$

where c is the length of a perforation.

It should be pointed out that the usual K-factor based on a solid member analysis is included in Equations 17.90 to 17.96. However, since the latticed members studied previously have pin-ended conditions, the K-factor of the member in the frame was not included in the second terms of the square root of the above equations in their original derivations [13, 70].

EXAMPLE 17.11:

Given: A column with perforated cover plates is shown in Figure 17.22. $K_y = 1.3$, $L = 25$ ft (7620 mm). Determine the modified effective length factor, $(K_y)_m$, by considering the shear deformation effect.

Section properties:

$$I_y = 2467 \text{ in.}^4 \ (1.03 \times 10^8 \text{ mm}^4)$$
$$I_f = 35.5 \text{ in.}^4 \ (1.48 \times 10^6 \text{ mm}^4)$$
$$a = 30 \text{ in. (762 mm)}$$
$$c = 14 \text{ in. (356 mm)}$$

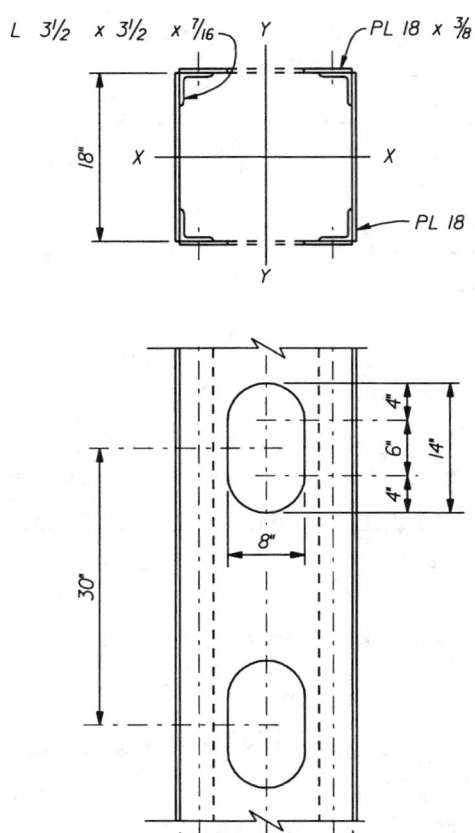

Figure 17.22   A column with perforate cover plates.

Solution

1. Calculate the shear factor, α_v, by Equation 17.96.

$$\alpha_v = \sqrt{1 + \frac{\pi^2 EI}{(KL)^2}\left(\frac{9c^3}{64a\,EI_f}\right)}$$

$$= \sqrt{1 + \frac{\pi^2 E(2467)}{(1.3 \times 25 \times 12)^2}\left(\frac{9(14)^3}{64(30)E(35.5)}\right)} = 1.03$$

2. Calculate $(K_y)_m$ by Equation 17.89.

$$\left(K_y\right)_m = \alpha_v K_y = 1.03(1.3) = 1.34$$

17.9.5   Built-Up Members with Bolted and Welded Connectors

AISC-LRFD [4] specifies that if the buckling of a built-up member produces shear forces in the connectors between individual component members, the usual slenderness ratio, KL/r, for compression members must be replaced by the modified slenderness ratio, $\left(\frac{KL}{r}\right)_m$, in determining the compressive strength.

1. For snug-tight bolted connectors:

$$\left(\frac{KL}{r}\right)_m = \sqrt{\left(\frac{KL}{r}\right)_o^2 + \left(\frac{a}{r_i}\right)^2}$$ (17.97)

2. For welded connectors and for fully tightened bolted connectors:

$$\left(\frac{KL}{r}\right)_m = \sqrt{\left(\frac{KL}{r}\right)_o^2 + 0.82\frac{\alpha^2}{(1+\alpha^2)}\left(\frac{a}{r_{ib}}\right)^2}$$ (17.98)

where $\left(\frac{KL}{r}\right)_o$ is the slenderness ratio of the built-up member acting as a unit, $\left(\frac{KL}{r}\right)_m$ is the modified slenderness ratio of the built-up member, $\frac{a}{r_i}$ is the largest slenderness ratio of the individual components, $\frac{a}{r_{ib}}$ is the slenderness ratio of the individual components relative to its centroidal axis parallel to the axis of buckling, a is the distance between connectors, r_i is the minimum radius of gyration of individual components, r_{ib} is the radius of gyration of individual components relative to its centroidal axis parallel to the member axis of buckling, α is the separation ratio $= h/2r_{ib}$, and h is the distance between centroids of individual components perpendicular to the member axis of buckling.

Equation 17.97 is the same as that used in the current Italian code as well as other European specifications, based on test results [74]. In the equation, the bending effect is considered in the first term in square root, and shear force effect is taken into account in the second term. Equation 17.98 was derived from elastic stability theory and verified by test data [9]. In both cases the end connectors must be welded or slip-critical bolted [9].

EXAMPLE 17.12:

Given: A built-up member with two back-to-back angles is shown in Figure 17.23. Determine the modified slenderness ratio, $(KL/r)_m$, in accordance with AISC-LRFD [4] and Equation 17.98.

$$
\begin{aligned}
r_{ib} &= 0.735 \text{ in. (19 mm)} \\
a &= 48 \text{ in. (1219 mm)} \\
h &= 1.603 \text{ in. (41 mm)} \\
(KL/r)_o &= 70
\end{aligned}
$$

Solution

1. Calculate the separation factor α.

$$\alpha = \frac{h}{2r_{ib}} = \frac{1.603}{2(0.735)} = 1.09$$

2. Calculate the modified slenderness ratio, $(KL/r)_m$, by Equation 17.98.

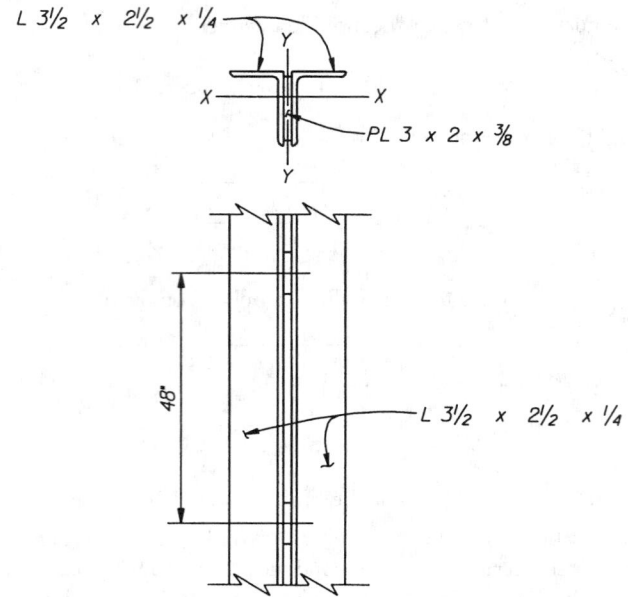

Figure 17.23   A built-up member with back-to-back angles.

$$\left(\frac{KL}{r}\right)_m = \sqrt{\left(\frac{KL}{r}\right)_o^2 + 0.82\frac{\alpha^2}{(1+\alpha^2)}\left(\frac{a}{r_{ib}}\right)^2}$$

$$= \sqrt{(70)^2 + 0.82\frac{1.09^2}{(1+1.09^2)}\left(\frac{48}{0.735}\right)^2}$$

$$= 82.5$$

17.10   Tapered Columns

The state-of-the-art design for tapered structural members was provided in the SSRC guide [36]. The charts shown in Figures 17.24 and 17.25 can be used to evaluate the effective length factors for tapered columns restrained by prismatic beams [36]. In these figures, I_T and I_B are the moment of inertia of the top and bottom beam, respectively; b and L are the length of beam and column, respectively; and γ is the tapering factor as defined by:

$$\gamma = \frac{d_1 - d_o}{d_o} \qquad\qquad (17.99)$$

where d_o and d_1 are the section depth of column at the smaller and larger end, respectively.

17.11   Crane Columns

The columns in mill buildings and warehouses are designed to support overhead crane loads. The cross-section of a crane column may be uniform or stepped (see Figure 17.26). Over the past two decades, a number of simplified procedures have been developed for evaluating the K-factors for crane columns [5, 6, 7, 11, 29, 30, 41, 51, 61]. Those procedures have limitations in terms of column geometry, loading, and boundary conditions. Most importantly, most of these studies ignored the

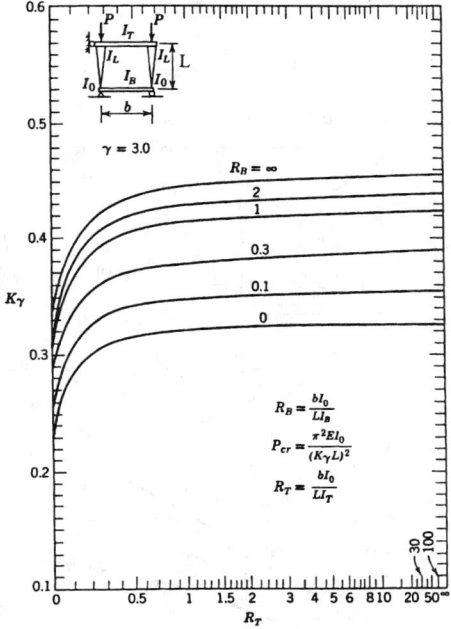

Figure 17.24 Effective length factor for tapered columns in braced frames. (From Galambos, T.V., Ed., *Structural Stability Research Council, Guide to Stability Design Criteria for Metal Structures*, 4th ed., John Wiley & Sons, New York, 1988. With permission.)

interaction effect between the left and right column of frames and were based on isolated member analyses [59]. Recently, a simple yet reasonably accurate procedure for calculating the K-factors for crane columns with any value of relative shaft length, moment of inertia, loading, and boundary conditions was developed by Lui and Sun [59]. Based on the story stiffness concept and accounting for both member and frame instability effects in the formulation, Lui and Sun [59] proposed the following procedure [see Figure 17.27]:

1. Apply the fictitious lateral loads, αP (α is an arbitrary factor; 0.001 may be used), in such a direction as to create a deflected geometry for the frame that closely approximates its actual buckled configuration.
2. Perform a first-order elastic analysis on the frame subjected to the fictitious lateral loads (Figure 17.27b). Calculate $\Delta_1 / \sum H$, where Δ_1 is the average lateral deflection at the intermediate load points (i.e., points B and F) of columns, and $\sum H$ is the sum of all fictitious lateral loads that act at and above the intermediate load points.
3. Calculate η using results obtained from a first-order elastic analysis for lower shafts (i.e., segments AB and FG), according to Equation 17.81.
4. Calculate the K-factor for the lower shafts using Equation 17.80.
5. Calculate the K-factor for upper shafts using the following formula:

$$K_U = K_L \left(\frac{L_L}{L_U}\right) \sqrt{\left(\frac{P_L + P_U}{P_U}\right)\left(\frac{I_U}{I_L}\right)} \tag{17.100}$$

where P is the applied load and subscripts U and L represent upper and lower shafts, respectively.

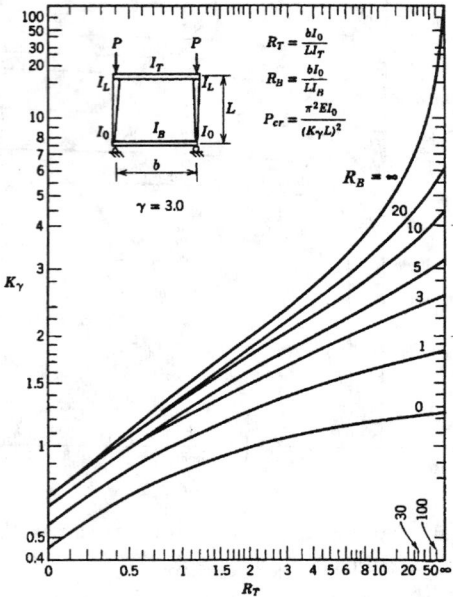

Figure 17.25   Effective length factor for tapered columns in unbraced frames. (From Galambos, T.V., Ed., *Structural Stability Research Council, Guide to Stability Design Criteria for Metal Structures,* 4th ed., John Wiley & Sons, New York, 1988. With permission.)

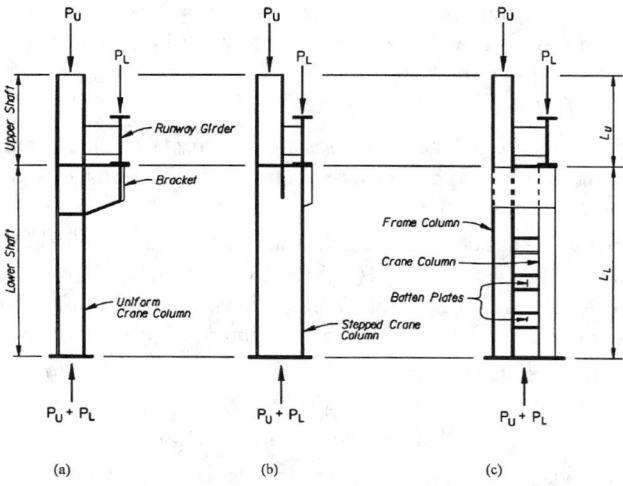

Figure 17.26   Typical crane columns. (From Lui, E.M. and Sun, M.Q., *AISC Eng. J.,* 32(2), 98, 1995. With permission.)

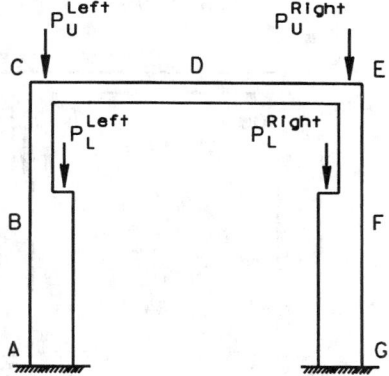

(a) Frame Subjected to Gravity Loads

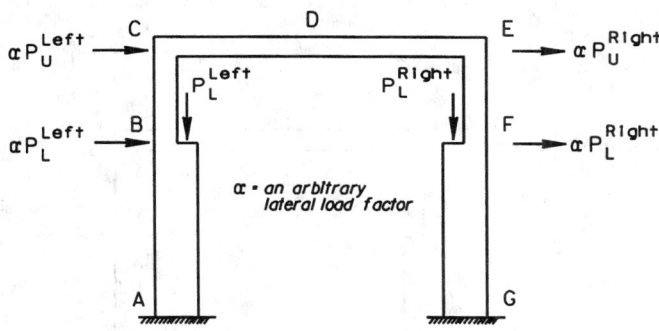

(b) Frame Subjected to Fictitious Lateral Loads

Figure 17.27   Crane column model for effective length factor computation. (From Lui, E.M. and Sun, M.Q., *AISC Eng. J.*, 32(2), 98, 1995. With permission.)

EXAMPLE 17.13:

Given: A stepped crane column is shown in Figure 17.28a. The example is the same frame as used by Fraser [30] and Lui and Sun [59]. Determine the effective length factors for all columns using the Lui approach. E = 29,000 ksi (200 GPa).

$$
\begin{aligned}
I_{AB} &= I_{FG} = I_L = 30,000 \text{ in.}^4 (1.25 \times 10^{10} \text{ mm}^4) \\
A_{AB} &= A_{FG} = A_L = 75 \text{ in.}^2 (48,387 \text{ mm}^2) \\
I_{BC} &= I_{EF} = I_{CE} = I_U = 5,420 \text{ in.}^4 (2.26 \times 10^9 \text{ mm}^4) \\
A_{BC} &= A_{EF} = A_{CE} = A_U = 34.14 \text{ in.}^2 (22,026 \text{ mm}^2)
\end{aligned}
$$

Solution

1. Apply a set of fictitious lateral forces with $\alpha = 0.001$ as shown in Figure 17.28b.

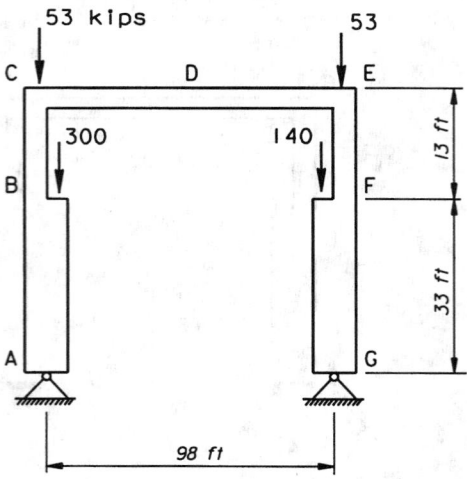

(a) Frame Subjected to Gravity Loads

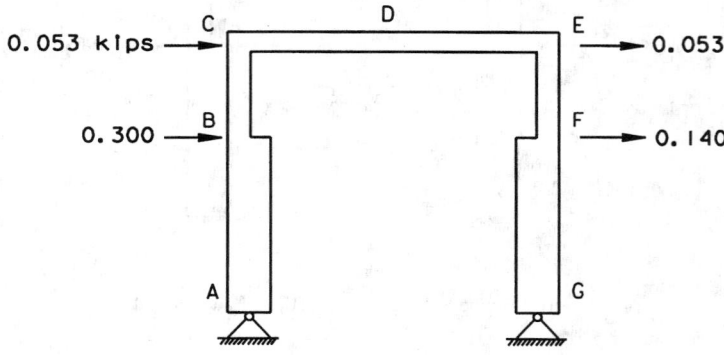

(b) Frame Subjected to Fictitious Lateral Loads

Figure 17.28   A pin-based stepped crane column. (From Lui, E.M. and Sun, M.Q., *AISC Eng. J.*, 32(2), 98, 1995. With permission.)

2. Perform a first-order analysis and find

$$(\Delta_1)_B = 0.1086 \text{ in.}(2.76 \text{ mm}) \quad \text{and} \quad (\Delta_1)_F = 0.1077 \text{ in.} \ (2.74 \text{ mm})$$

so,

$$\frac{\Delta_1}{\sum H} = \frac{(0.1086 + 0.1077)/2}{0.053 + 0.3 + 0.053 + 0.14} = 0.198 \text{ in./kips} \ (1.131 \text{ mm/kN})$$

3. Calculate η factors from Equation 17.81.
   Since the bottom of column AB and FG is pin-based, $m = 0$,

$$\eta_{AB} = \eta_{FG} = \frac{(3 + 4.8m + 4.2m^2)EI}{L^3} = \frac{3EI}{L^3}$$

$$= \frac{(3)(29,000)(30,000)}{(396)^3} = 42.03 \text{ kips/in. } (7.36 \text{ mm/kN})$$

$$\sum \eta = 42.03 + 42.03 = 84.06 \text{ kips/in. } (14.72 \text{ mm/kN})$$

4. Calculate the K-factors for columns AB and FG using Equation 17.80.

$$K_{AB} = \sqrt{\left(\frac{\pi^2(29,000)(30,000)}{(353)(396)^2}\right)\left[\left(\frac{353 + 193}{396}\right)\left(\frac{1}{5(84.06)} + 0.198\right)\right]}$$

$$= 6.55$$

$$K_{FG} = \sqrt{\left(\frac{\pi^2(29,000)(30,000)}{(193)(396)^2}\right)\left[\left(\frac{353 + 193}{396}\right)\left(\frac{1}{5(84.06)} + 0.198\right)\right]}$$

$$= 8.85$$

5. Calculate the K-factors for columns BC and EF using Equation 17.100.

$$K_{BC} = K_{AB}\left(\frac{L_{AB}}{L_{BC}}\right)\sqrt{\left(\frac{P_{AB} + P_{BC}}{P_{BC}}\right)\left(\frac{I_{BC}}{I_{AB}}\right)}$$

$$= 6.55\left(\frac{396}{156}\right)\sqrt{\left(\frac{353}{53}\right)\left(\frac{5420}{30,000}\right)} = 18.2$$

$$K_{EF} = K_{FG}\left(\frac{L_{FG}}{L_{EF}}\right)\sqrt{\left(\frac{P_{FG} + P_{EF}}{P_{EF}}\right)\left(\frac{I_{EF}}{I_{FG}}\right)}$$

$$= 8.85\left(\frac{396}{156}\right)\sqrt{\left(\frac{193}{53}\right)\left(\frac{5420}{30,000}\right)} = 18.2$$

The K-factors calculated above are in good agreement with the theoretical values reported by Lui and Sun [59].

17.12 Columns in Gable Frames

For a pin-based **gable frame** subjected to a uniformly distributed load on the rafter, as shown in Figure 17.29a, Lu [56] presented a graph (Figure 17.29b) to determine the effective length factors of columns. For frames having different member sizes for rafter and columns with (L/h) of (f/h) ratios not covered in Figure 17.29, an approximate method is available for determining K-factors of columns [39]. The method is to find an equivalent portal frame whose span length is equal to twice the rafter length, L_r (see Figure 17.29a). The K-factors can be determined form the alignment charts using $G_{top} = \frac{I_c/h}{I_r/2L_r}$ and corresponding G_{bottom}.

17.13 Summary

This chapter summarizes the state-of-the-art practice of the effective length factors for isolated columns, framed columns, diagonal bracing systems, latticed and built-up members, tapered

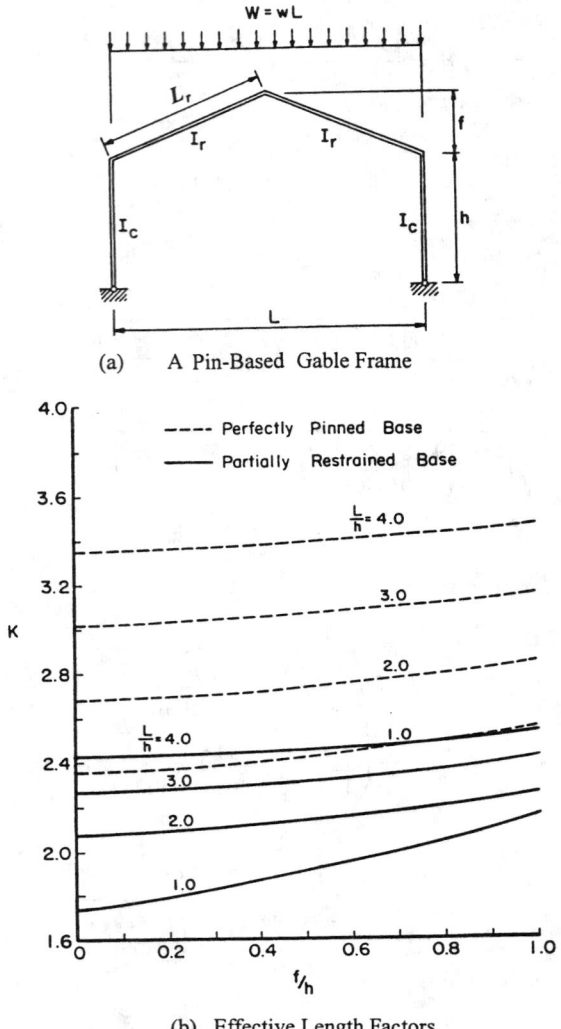

Figure 17.29 Effective length factors for columns in a pin-based gable frame. (From Lu, L.W., *AISC Eng. J.*, 2(2), 6, 1965. With permission.)

columns, crane columns, and columns in gable frames. Design implementation with formulas, charts, tables, various modification factors adopted in current codes and specifications, as well as those used in engineering practice are described. Several examples illustrate the steps of practical applications of various methods.

17.14 Defining Terms

Alignment chart: A monograph for determining the effective length factor K for some types of compression members.

Braced frame: A frame in which the resistance to lateral load or frame instability is primarily provided by diagonal bracing, shear walls, or equivalent means.

Build-up member: A member made of structural metal elements that are welded, bolted, and riveted together.

Column: A structural member whose primary function is to carry loads parallel to its longitudinal axis.

Crane column: A column that is designed to support overhead crane loads.

Effective length factor K**:** A factor that when multiplied by actual length of the end-restrained column gives the length of an equivalent pin-ended column whose elastic buckling load is the same as that of the end-restrained column.

Framed column: A column in a framed structure.

Gable frame: A frame with a gabled roof.

Latticed member: A member made of two or more rolled-shapes that are connected to one another by means of lacing bars, batten plates, or perforated plates.

Leaning column: A column that is connected to a frame with simple connections and does not provide lateral stiffness or sidesway resistance.

LRFD (Load and Resistance Factor Design): A method of proportioning structural components (members, connectors, connecting elements, and assemblages) such that no applicable limit state is exceeded when the structure is subjected to all appropriate load combinations.

Tapered column: A column that has a continuous reduction in section from top to bottom.

Unbraced frame: A frame in which the resistance to lateral loads is provided by the bending stiffness of frame members and their connections.

References

[1] American Association of State Highway and Transportation Officials. 1994. *LRFD Bridge Design Specifications,* 1st ed., AASHTO, Washington, D.C.

[2] American Concrete Institute. 1995. *Building Code Requirements for Structural Concrete (ACI 318-95) and Commentary (ACI 318R-95).* ACI, Farmington Hills, MI.

[3] American Institute of Steel Construction. 1989. *Allowable Stress Design Specification for Structural Steel Buildings,* 9th ed., AISC, Chicago, IL.

[4] American Institute of Steel Construction. 1993. *Load and Resistance Factor Design Specification for Structural Steel Buildings,* 2nd ed., AISC, Chicago, IL.

[5] Association of Iron and Steel Engineers. 1991. *Guide for the Design and Construction of Mill Buildings,* AISE, Technical Report, No. 13, Pittsburgh, PA.

[6] Anderson, J.P. and Woodward, J.H. 1972. Calculation of Effective Lengths and Effective Slenderness Ratios of Stepped Columns. *AISC Eng. J.,* 7(4):157-166.

[7] Agarwal, K.M. and Stafiej, A.P. 1980. Calculation of Effective Lengths of Stepped Columns. *AISC Eng. J.,* 15(4):96-105.

[8] Aristizabal-Ochoa, J.D. 1994. K-Factors for Columns in Any Type of Construction: Nonparadoxical Approach. *J. Struct. Eng.,* 120(4):1272-1290.

[9] Aslani, F. and Goel, S.C. 1991. An Analytical Criteria for Buckling Strength of Built-Up Compression Members. *AISC Eng. J.,* 28(4):159-168.

[10] Barakat, M. and Chen, W.F. 1991. Design Analysis of Semi-Rigid Frames: Evaluation and Implementation. *AISC Eng. J.,* 28(2):55-64.

[11] Bendapudi, K.V. 1994. Practical Approaches in Mill Building Columns Subjected to Heavy Crane Loads. *AISC Eng. J.,* 31(4):125-140.

[12] Bjorhovde, R. 1984. Effect of End Restraints on Column Strength—Practical Application, *AISC Eng. J.,* 21(1):1-13.

[13] Bleich, F. 1952. *Buckling Strength of Metal Structures.* McGraw-Hill, New York.

[14] Bridge, R.Q. and Fraser, D.J. 1987. Improved G-Factor Method for Evaluating Effective Length of Columns. *J. Struct. Eng.,* 113(6):1341-1356.

[15] Chapius, J. and Galambos, T.V. 1982. Restrained Crooked Aluminum Columns, *J. Struct. Div.,* 108(ST12):511-524.

[16] Cheong-Siat-Moy, F. 1986. *K*-Factor Paradox, *J. Struct. Eng.,* 112(8):1647-1760.

[17] Chen, W.F. and Lui, E.M. 1991. *Stability Design of Steel Frames,* CRC Press, Boca Raton, FL.

[18] Chu, K.H. and Chow, H.L. 1969. Effective Column Length in Unsymmetrical Frames, *Publ. Intl. Assoc. Bridge Struct. Eng.,* 29(1).

[19] Cranston, W.B. 1972. Analysis and Design of Reinforced Concrete Columns. *Research Report* No. 20, Paper 41.020, Cement and Concrete Association, London.

[20] Disque, R.O. 1973. Inelastic *K*-Factor in Design. *AISC Eng. J.,* 10(2):33-35.

[21] Duan, L. and Chen, W.F. 1988. Effective Length Factor for Columns in Braced Frames. *J. Struct. Eng.,* 114(10):2357-2370.

[22] Duan, L. and Chen, W.F. 1989. Effective Length Factor for Columns in Unbraced Frames. *J. Struct. Eng.,* 115(1):149-165.

[23] Duan, L. and Chen, W.F. 1996. Errata of Paper: Effective Length Factor for Columns in Unbraced Frames. *J. Struct. Eng.,* 122(1):224-225.

[24] Duan, L., King, W.S., and Chen, W.F. 1993. *K*-Factor Equation to Alignment Charts for Column Design. *ACI Struct. J.,* 90(3):242-248.

[25] Duan, L. and Lu, Z.G. 1996. A Modified *G*-Factor for Columns in Semi-Rigid Frames. *Research Report,* Division of Structures, California Department of Transportation, Sacramento, CA.

[26] Dumonteil, P. 1992. Simple Equations for Effective Length Factors. *AISC Eng. J.,* 29(3):111-115.

[27] El-Tayem, A.A. and Goel, S.C. 1986. Effective Length Factor for the Design of X-Bracing Systems. *AISC Eng. J.,* 23(4):41-45.

[28] European Convention for Constructional Steelwork. 1978. European Recommendations for Steel Construction, ECCS.

[29] Fraser, D.J. 1989. Uniform Pin-Based Crane Columns, Effective Length, *AISC Eng. J.,* 26(2):61-65.

[30] Fraser, D.J. 1990. The In-Plane Stability of a Frame Containing Pin-Based Stepped Column. *AISC Eng. J.,* 27(2):49-53.

[31] French. 1975. Regles de calcul des constructions en acier CM66, Eyrolles, Paris, France.

[32] Furlong, R.W. 1971. Column Slenderness and Charts for Design. *ACI Journal, Proceedings,* 68(1):9-18.

[33] Galambos, T.V. 1960. Influence of Partial Base Fixity on Frame Instability. *J. Struct. Div.,* 86(ST5):85-108.

[34] Galambos, T.V. 1964. Lateral Support for Tier Building Frames. *AISC Eng. J.,* 1(1):16-19.

[35] Galambos, T.V. 1968. *Structural Members and Frames.* Prentice-Hall International, London, U.K.

[36] Galambos, T.V., Ed. 1988. *Structural Stability Research Council, Guide to Stability Design Criteria for Metal Structures,* 4th ed., John Wiley & Sons, New York.

[37] Geschwindner, L.F. 1995. A Practical Approach to the "Leaning" Column, *AISC Eng. J.,* 32(2):63-72.

[38] Gurfinkel, G. and Robinson, A.R. 1965. Buckling of Elasticity Restrained Column. *J. Struct. Div.,* 91(ST6):159-183.

[39] Hansell, W.C. 1964. Single-Story Rigid Frames, in *Structural Steel Design,* Chapt. 20, Ronald Press, New York.

[40] Hu, X.Y., Zhou, R.G., King, W.S., Duan, L., and Chen, W.F. 1993. On Effective Length Factor of Framed Columns in ACI Code. *ACI Struct. J.*, 90(2):135-143.

[41] Huang, H.C. 1968. Determination of Slenderness Ratios for Design of Heavy Mill Building Stepped Columns. *Iron Steel Eng.*, 45(11):123.

[42] Johnson, D.E. 1960. Lateral Stability of Frames by Energy Method. *J. Eng. Mech.*, 95(4):23-41.

[43] Johnston, B.G., Ed. 1976. *Structural Stability Research Council, Guide to Stability Design Criteria for Metal Structures*, 3rd ed., John Wiley & Sons, New York.

[44] Jones, S.W., Kirby, P.A., and Nethercot, D.A. 1980. Effect of Semi-Rigid Connections on Steel Column Strength, *J. Constr. Steel Res.*, 1(1):38-46.

[45] Jones, S.W., Kirby, P.A., and Nethercot, D.A. 1982. Columns with Semi-Rigid Joints, *J. Struct. Div.*, 108(ST2):361-372

[46] Julian, O.G. and Lawrence, L.S. 1959. *Notes on J and L Nomograms for Determination of Effective Lengths*. Unpublished report.

[47] Kavanagh, T.C. 1962. Effective Length of Framed Column. *Trans. ASCE*, 127(II):81-101.

[48] King, W.S., Duan, L., Zhou, R.G., Hu, Y.X., and Chen, W.F. 1993. *K*-Factors of Framed Columns Restrained by Tapered Girders in US Codes. *Eng. Struct.*, 15(5):369-378.

[49] Kishi, N., Chen, W.F., and Goto, Y. 1995. Effective Length Factor of Columns in Semi-Rigid and Unbraced Frames. *Structural Engineering Report CE-STR-95-5*, School of Civil Engineering, Purdue University, West Lafayette, IN.

[50] Koo, B. 1988. Discussion of Paper "Improved *G*-Factor Method for Evaluating Effective Length of Columns" by Bridge and Fraser. *J. Struct. Eng.*, 114(12):2828-2830.

[51] Lay, M.G. 1973. Effective Length of Crane Columns, *Steel Const.*, 7(2):9-19.

[52] LeMessurier, W.J. 1977. A Practical Method of Second Order Analysis. Part 2—Rigid Frames. *AISC Eng. J.*, 14(2):49-67.

[53] Liew, J.Y.R., White, D.W., and Chen, W.F. 1991. Beam-Column Design in Steel Frameworks—Insight on Current Methods and Trends. *J. Const. Steel. Res.*, 18:269-308.

[54] Lim, L.C. and McNamara, R.J. 1972. Stability of Novel Building System, *Structural Design of Tall Steel Buildings, Vol. II-16, Proceedings*, ASCE-IABSE International Conference on the Planning and Design of Tall Buildings, Bethlehem, PA, 499-524.

[55] Lu, L.W. 1962. A Survey of Literature on the Stability of Frames. *Weld. Res. Conc. Bull.*, New York.

[56] Lu, L.W. 1965. Effective Length of Columns in Gable Frame, *AISC Eng. J.*, 2(2):6-7.

[57] Lui, E.M. 1992. A Novel Approach for *K*-Factor Determination. *AISC Eng. J.*, 29(4):150-159.

[58] Lui, E.M. and Chen, W.F. 1983. Strength of Columns with Small End Restraints, *J. Inst. Struct. Eng.*, 61B(1):17-26

[59] Lui, E.M. and Sun, M.Q. 1995. Effective Length of Uniform and Stepped Crane Columns, *AISC Eng. J.*, 32(2):98-106.

[60] Maquoi, R. and Jaspart, J.P. 1989. Contribution to the Design of Braced Framed with Semi-Rigid Connections. *Proc. 4th International Colloquium, Structural Stability Research Council*, 209-220. Lehigh University, Bethlehem, PA.

[61] Moore, W.E. II. 1986. A Programmable Solution for Stepped Crane Columns. *AISC Eng. J.*, 21(2):55-58.

[62] Picard, A. and Beaulieu, D. 1987. Design of Diagonal Cross Bracings. Part 1: Theoretical Study. *AISC Eng. J.*, 24(3):122-126.

[63] Picard, A. and Beaulieu, D. 1988. Design of Diagonal Cross Bracings. Part 2: Experimental Study. *AISC Eng. J.*, 25(4):156-160.

[64] Razzaq, Z. 1983. End Restraint Effect of Column Strength. *J. Struct. Div.*, 109(ST2):314-334.

[65] Salmon, C.G., Schenker, L., and Johnston, B.G. 1957. Moment-Rotation Characteristics of Column Anchorage. *Trans. ASCE,* 122:132-154.

[66] Shanmugam, N.E. and Chen, W.F. 1995. An Assessment of *K* Factor Formulas. *AISC Eng. J.,* 32(3):3-11.

[67] Sohal, I.S., Yong, Y.K., and Balagura, P.N. 1995. *K*-Factor in Plastic and SOIA for Design of Steel Frames, *Proceeding International Conference on Stability of Structures,* ICSS 95, June 7-9, Coimbatore, India, 411-421.

[68] Sugimoto, H. and Chen, W.F. 1982. Small End Restraint Effects on Strength of H-Columns, *J. Struct. Div.,* 108(ST3):661-681.

[69] Vinnakota, S. 1982. Planar Strength of Restrained Beam Columns. *J. Struct. Div.,* 108(ST11):2349-2516.

[70] Timoshenko, S.P. and Gere, J.M. 1961. *Theory of Elastic Stability,* 2nd ed., McGraw-Hill, New York.

[71] Winter, G. et al. 1948. Buckling of Trusses and Rigid Frames, Cornell Univ. Bull. No. 36, Engineering Experimental Station, Cornell University, Ithaca, NY.

[72] Wood, R.H. 1974. Effective Lengths of Columns in Multi-Storey Buildings. *Struct. Eng.,* 52(7,8,9):234-244, 295-302, 341-346.

[73] Yura, J.A. 1971. The Effective Length of Columns in Unbraced Frames. *AISC Eng. J.,* 8(2):37-42.

[74] Zandonini, R. 1985. Stability of Compact Built-Up Struts: Experimental Investigation and Numerical Simulation (in Italian). *Construzioni Metalliche,* No. 4.

Further Reading

[1] Chen, W.F. and Lui, E.M. 1987. *Structural Stability: Theory and Implementation,* Elsevier, New York.

[2] Chen, W.F., Goto, Y. and Liew, J.Y.R. 1996. *Stability Design of Semi-Rigid Frames,* John Wiley and Sons, New York.

[3] Chen, W.F. and Kim, S.E. 1997. *LRFD Steel Design Using Advanced Analysis,* CRC Press, Boca Raton, FL.

18

Stub Girder Floor Systems

18.1 Introduction 18-1
18.2 Description of the Stub Girder Floor System........... 18-2
18.3 Methods of Analysis and Modeling 18-6
      General Observations • Preliminary Design Procedure • Choice
      of Stub Girder Component Sizes • Modeling of the Stub Girder
18.4 Design Criteria For Stub Girders 18-13
      General Observations • Governing Sections of the Stub Girder
      • Design Checks for the Bottom Chord • Design Checks for the
      Concrete Slab • Design Checks for the Shear Transfer Regions
      • Design of Stubs for Shear and Axial Load • Design of Stud
      Shear Connectors • Design of Welds between Stub and Bottom
      Chord • Floor Beam Connections to Slab and Bottom Chord •
      Connection of Bottom Chord to Supports • Use of Stub Girder
      for Lateral Load System • Deflection Checks
18.5 Influence of Method of Construction 18-22
18.6 Defining Terms 18-23
References 18-23
Further Reading 18-24

Reidar Bjorhovde
Department of Civil and
Environmental Engineering,
University of Pittsburgh,
Pittsburgh, PA

18.1 Introduction

The stub girder system was developed in response to a need for new and innovative construction techniques that could be applied to certain parts of all multi-story steel-framed buildings. Originated in the early 1970s, the design concept aimed at providing construction economies through the integration of the electrical and mechanical service ducts into the part of the building volume that is occupied by the floor framing system [11, 12]. It was noted that the overall height of the floor system at times could be large, leading to significant increases in the overall height of the structure, and hence the steel tonnage for the project. At other times the height could be reduced, but only at the expense of having sizeable web penetrations for the ductwork to pass through. This solution was often accompanied by having to reinforce the web openings by stiffeners, increasing the construction cost even further.

The **composite** stub girder floor system subsequently was developed. Making extensive use of relatively simple shop fabrication techniques, basic elements with limited fabrication needs, simple connections between the main floor system elements and the structural columns, and composite action between the concrete floor slab and the steel load-carrying members, a floor system of significant strength, stiffness, and ductility was devised. This led to a reduction in the amount of structural steel that traditionally had been needed for the floor framing. When coupled with the use of continuous, composite transverse floor beams and the shorter erection time that was needed for the stub girder system, this yielded attractive cost savings.

0-8493-2674-5/97/$0.00+$.50

Since its introduction, the stub girder floor system has been used for a variety of steel-framed buildings in the U.S., Canada, and Mexico, ranging in height from 2 to 72 stories. Despite this relatively widespread usage, the analysis techniques and design criteria remain unknown to many designers. This chapter will offer examples of practical uses of the system, together with recommendations for suitable design and performance criteria.

18.2 Description of the Stub Girder Floor System

The main element of the system is a special girder, fabricated from standard hot-rolled wide-flange shapes, that serves as the primary framing element of the floor. Hot-rolled wide-flange shapes are also used as transverse floor beams, running in a direction perpendicular to the main girders. The girder and the beams are usually designed for composite action, although the system does not rely on having composite floor beams, and the latter are normally analyzed as continuous beams. As a result, the transverse floor beams normally use a smaller drop-in span within the positive moment region. This results in further economies for the floor beam design, since it takes advantage of continuous beam action.

Allowable stress design (ASD) or load and resistance factor design (LRFD) criteria are equally applicable for the design of stub girders, although LRFD is preferable, since it gives lower steel weights and simple connections. The costs that are associated with an LRFD-designed stub girder therefore tend to be lower.

Figure 18.1 shows the elevation of a typical stub girder. It is noted that the girder that is shown

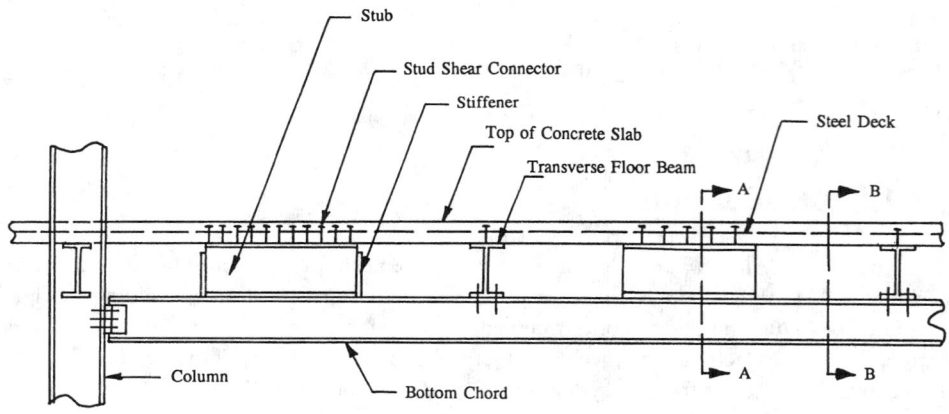

Figure 18.1 Elevation of a typical stub girder (one half of span is shown).

makes use of four stubs, oriented symmetrically with respect to the midspan of the member. The locations of the transverse floor beams are assumed to be the quarter points of the span, and the supports are simple. In practice many variations of this layout are used, to the extent that the girders may utilize any number of stubs. However, three to five stubs is the most common choice. The locations of the stubs may differ significantly from the symmetrical case, and the exterior (= end) stubs may have been placed at the very ends of the bottom chord. However, this is not difficult to address in the modeling of the girder, and the essential requirements are that the forces that develop as a result of the choice of girder geometry be accounted for in the design of the girder components and the adjacent structure. These actual forces are used in the design of the various elements, as distinguished from the simplified models that are currently used for many structural components.

The choices of elements, etc., are at the discretion of the design team, and depend on the service requirements of the building as seen from the architectural, structural, mechanical, and electrical viewpoints. Unique design considerations must be made by the structural engineer, for example, if it is decided to eliminate the exterior openings and connect the stubs to the columns in addition to the chord and the slab.

Figure 18.1 shows the main components of the stub girder, as follows:

1. Bottom chord
2. Exterior and interior stubs
3. Transverse floor beams
4. **Formed steel deck**
5. Concrete slab with longitudinal and transverse reinforcement
6. Stud shear connectors
7. Stub stiffeners
8. Beam-to-column connection

The bottom chord should preferably be a hot-rolled wide-flange shape of column-type proportions, most often in the W12 to W14 series of wide-flange shapes. Other chord cross-sections have been considered [19]; for example, T shapes and rectangular tubes have certain advantages as far as welded attachments and fire protection are concerned, respectively. However, these other shapes also have significant drawbacks. The rolled tube, for example, cannot accommodate the shear stresses that develop in certain regions of the bottom chord. Rather than using a T or a tube, therefore, a smaller W shape (in the W10 series, for example) is most likely the better choice under these conditions.

The steel grade for the bottom chord, in particular, is important, since several of the governing regions of the girder are located within this member, and tension is the primary stress resultant. It is therefore possible to take advantage of higher strength steels, and 50-ksi-yield stress steel has typically been the choice, although 65-ksi steel would be acceptable as well.

The floor beams and the stubs are mostly of the same size W shape, and are normally selected from the W16 and W18 series of shapes. This is directly influenced by the size(s) of the **HVAC** ducts that are to be used, and input from the mechanical engineer is essential at this stage. Although it is not strictly necessary that the floor beams and the stubs use identical shapes, it avoids a number of problems if such a choice is made. At the very least, these two components of the floor system should have the same height.

The concrete slab and the steel deck constitute the top chord of the stub girder. It is made either from **lightweight** or **normal weight** concrete, although if the former is available, even at a modest cost premium, it is preferred. The reason is the lower dead load of the floor, especially since the shores that will be used are strongly influenced by the concrete weight. Further, the shores must support several stories before they can be removed. In other words, the stub girders must be designed for shored construction, since the girder requires the slab to complete the system. In addition, the bending rigidity of the girder is substantial, and a major fraction is contributed by the bottom chord. The reduction in slab stiffness that is prompted by the lower value of the modulus of elasticity for the lightweight concrete is therefore not as important as it may be for other types of composite bending members.

Concrete strengths of 3000 to 4000 psi are most common, although the choice also depends on the limit state of the stud shear connectors. Apart from certain long-span girders, some local regions in the slab, and the desired mode of behavior of the slab-to-stub connection (which limits the maximum f_c' value that can be used), the strength of the stub girder is not controlled by the concrete. Consequently, there is little that can gained by using high-strength concrete.

The steel deck should be of the composite type, and a number of manufacturers produce suitable types. Normal deck heights are 2 and 3 in., but most floors are designed for the 3-in. deck. The deck ribs are run parallel to the longitudinal axis of the girder, since this gives better deck support on the transverse floor beams. It also increases the top chord area, which lends additional stiffness to a member that can span substantial distances. Finally, the parallel orientation provides a continuous rib trough directly above the girder centerline, improving the composite interaction of the slab and the girder.

Due to fire protection requirements, the thickness of the concrete cover over the top of the deck ribs is either 4-3/16 in. (normal weight concrete) or 3-1/4 in. (lightweight concrete). This eliminates the need for applying fire protective material to the underside of the steel deck.

Stud shear connectors are distributed uniformly along the length of the exterior and interior stubs, as well as on the floor beams. The number of connectors is determined on the basis of the computed shear forces that are developed between the slab and the stubs. This is in contrast to the current design practice for simple composite beams, which is based on the smaller of the ultimate axial load-carrying capacity of the slab and the steel beam [2, 3]. However, the simplified approach of current specifications is not applicable to members where the cross-section varies significantly along the length (nonprismatic beams). The computed shear force design approach also promotes connector economy, in the sense that a much smaller number of shear connectors is required in the interior shear transfer regions of the girder [5, 7, 21].

The stubs are welded to the top flange of the bottom chord with fillet welds. In the original uses of the system, the design called for all-around welds [11, 12]; subsequent studies demonstrated that the forces that are developed between the stubs and the bottom chord are concentrated toward the end of the stubs [5, 6, 21]. The welds should therefore be located in these regions.

The type and locations of the stub stiffeners that are indicated for the exterior stubs in Figure 18.1, as well as the lack of stiffeners for the interior stubs, represent one of the major improvements that were made to the original stub girder designs. Based on extensive research [5, 21], it was found that simple end-plate stiffeners were as efficient as the traditional fitted ones, and in many cases the stiffeners could be eliminated at no loss in strength and stiffness to the overall girder.

Figure 18.1 shows that a simple (shear) connection is used to attach the bottom chord of the stub girder to the adjacent structure (column, concrete building core, etc.). This is the most common solution, especially when a duct opening needs to be located at the exterior end of the girder. If the support is an exterior column, the slab will rest on an edge member; if it is an interior column, the slab will be continuous past the column and into the adjacent bay. This may or may not present problems in the form of slab cracking, depending on the reinforcement details that are used for the slab around the column.

The stub girder has sometimes been used as part of the lateral load-resisting system of steel-framed buildings [13, 17]. Although this has certain disadvantages insofar as column moments and the concrete slab reinforcement are concerned, the girder does provide significant lateral stiffness and ductility for the frame. As an example, the maintenance facility for Mexicana Airlines at the Mexico City International Airport, a structure utilizing stub girders in this fashion [17], survived the 1985 Mexico City earthquake with no structural damage.

Expanding on the details that are shown in Figure 18.1, Figure 18.2 illustrates the cross-section of a typical stub girder, and Figure 18.3 shows a complete girder assembly with lights, ducts, and suspended ceiling. Of particular note are the longitudinal reinforcing bars. They add flexural strength as well as ductility and stiffness to the girder, by helping the slab to extend its service range.

The longitudinal **rebars** are commonly placed in two layers, with the top one just below the heads of the stud shear connectors. The lower longitudinal rebars must be raised above the deck proper, using high chairs or other means. This assures that the bars are adequately confined.

The transverse rebars are important for adding shear strength to the slab, and they also help in the shear transfer from the connectors to the slab. The transverse bars also increase the overall ductility of the stub girder, and placing the bars in a herring bone pattern leads to a small improvement in the effective width of the slab.

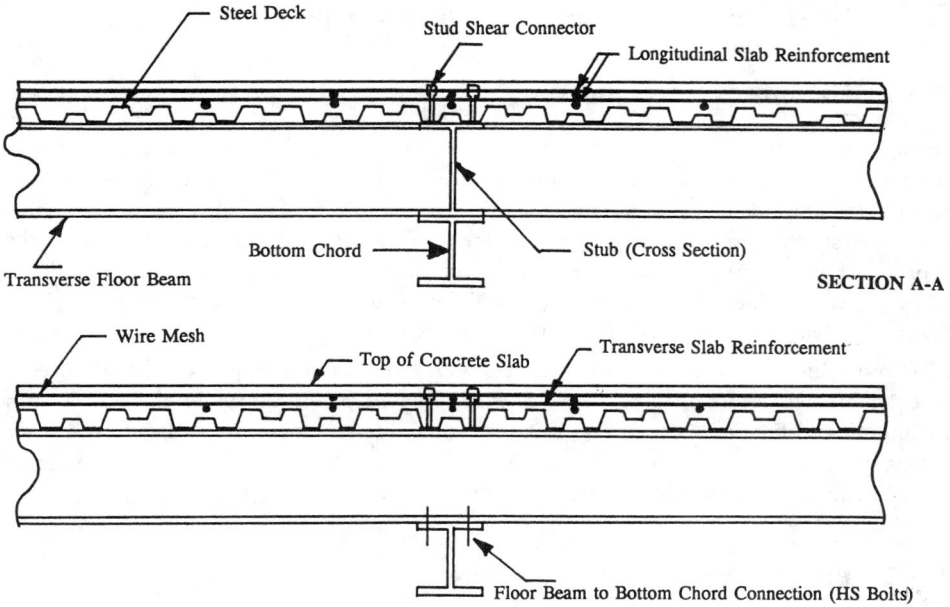

Figure 18.2 Cross-sections of a typical stub girder (refer to Figure 18.1 for section location).

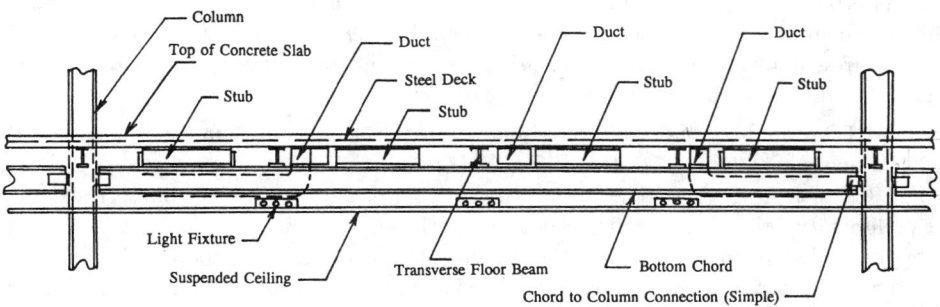

Figure 18.3 Elevation of a typical stub girder, complete with ductwork, lights, and suspended ceiling (duct sizes, etc., vary from system to system).

The common choices for stub girder floor systems have been 36- or 50-ksi-yield stress steel, with a preference for the latter, because of the smaller bottom chord size that can be used. Due to its function in the girder, there is no reason why steels such as ASTM A913 (65 ksi) cannot be used for the bottom chord. However, all detail materials (stiffeners, connection angles, etc.) are made from 36-ksi steel. Welding is usually done with 70-grade low hydrogen electrodes, using either the SMAW, FCAW, or GMAW process, and the stud shear connectors are welded in the normal fashion. All of the work is done in the fabricating shop, except for the shear connectors, which are applied in the field, where they are welded directly through the steel deck. The completed stub girders are then shipped to the construction site.

18.3   Methods of Analysis and Modeling

18.3.1   General Observations

In general, any number of methods of analysis may be used to determine the bending moments, shear forces, and axial forces throughout the components of the stub girder. However, it is essential to bear in mind that the modeling of the girder, or, in other words, how the actual girder is transformed into an idealized structural system, should reflect the relative stiffness of the elements. This means that it is important to establish realistic trial sizes of the components, through an appropriate preliminary design procedure. The subsequent modeling will then lead to stress resultants that are close to the magnitudes that can be expected in actual stub girders.

Based on this approach, the design that follows is likely to require relatively few changes, and those that are needed are often so small that they have no practical impact on the overall stiffness distribution and final member forces. The preliminary design procedure is therefore a very important step in the overall design. However, it will be shown that by using an LRFD approach, the process is simple, efficient, and accurate.

18.3.2   Preliminary Design Procedure

Using the LRFD approach for the preliminary design, it is not necessary to make any assumptions as regards the stress distribution over the depth of the girder, other than to adhere to the strength model that was developed for normal composite beams [3, 15]. The stress distribution will vary anyway along the span because of the openings.

The strength model of Hansell et al. [15] assumes that when the ultimate moment is reached, all or a portion of the slab is failing in compression, with a uniformly distributed stress of $0.85 f_c'$. The steel shape is simultaneously yielding in tension. Equilibrium is therefore maintained, and the internal stress resultants are determined using first principles. Tests have demonstrated excellent agreement with theoretical analyses that utilize this approach [5, 7, 15, 21].

The LRFD procedure uses load and resistance factors in accordance with the American Institute of Steel Construction (AISC) LRFD specification [3]. The applicable resistance factor is given by the AISC LRFD specification, Section D1, for the case of gross cross-section yielding. This is because the preliminary design is primarily needed to find the bottom chord size, and this component is primarily loaded in tension [5, 7, 10, 21]. The load factors of the LRFD specification are those of the American Society of Civil Engineers (ASCE) load standard [4], for the combination of dead plus live load.

The load computations follow the choice of the layout of the floor framing plan, whereby girder and floor beam spans are determined. This gives the tributary areas that are needed to calculate the dead and live loads. The load intensities are governed by local building code requirements or by the ASCE recommendations, in the absence of a local code.

Reduced live loads should be used wherever possible. This is especially advantageous for stub girder floor systems, since spans and tributary areas tend to be large. The ASCE load standard [4] makes use of a live load reduction factor, RF, that is significantly simpler to use, and also less conservative than that of earlier codes. The standard places some restrictions on the value of RF, to the effect that the reduced live load cannot be less than 50% of the nominal value for structural members that support only one floor. Similarly, it cannot be less than 40% of the nominal live load if two or more floors are involved.

Proceeding with the preliminary design, the stub girder and its floor beam locations determine the magnitudes of the concentrated loads that are to be applied at each of the latter locations. The following illustrative example demonstrates the steps of the solution.

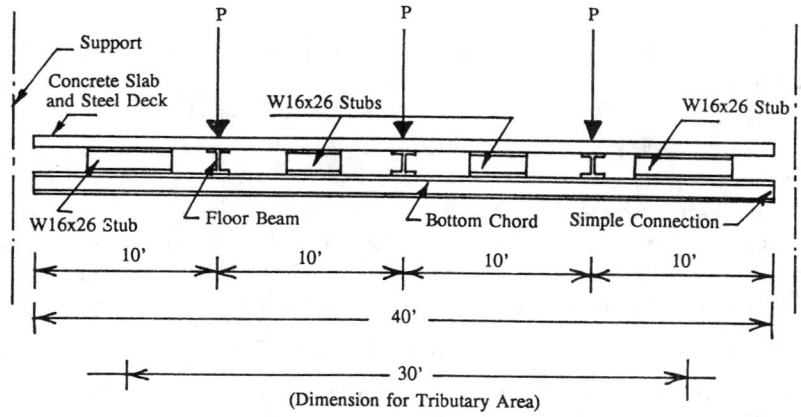

Figure 18.4 Stub girder layout used for preliminary design example.

EXAMPLE 18.1:

Given: Figure 18.4 shows the layout of the stub girder for which the preliminary sizes are needed. Other computations have already given the sizes of the floor beam, the slab, and the steel deck. The span of the girder is 40 ft, the distance between adjacent girders is 30 ft, and the floor beams are located at the quarter points. The steel grade remains to be chosen (36- and 50-ksi-yield stress steel are the most common); the concrete is lightweight, with $w_c = 120$ pcf and a compressive strength of $f'_c = 4000$ psi.

> ***Solution***

Loads:

> Estimated dead load $= 74$ psf
>
> Nominal live load $= 50$ psf

Live load reduction factor:

$$RF = 0.25 + 15/\sqrt{[2 \times (30 \times 30)]} = 0.60$$

Reduced live load:

$$RLL = 0.60 \times 50 = 30 \text{ psf}$$

Load factors (for D + L combination):

> For dead load: 1.2
>
> For live load: 1.6

Factored distributed loads:

> Dead Load, $DL = 74 \times 1.2 = 88.8$ psf
>
> Live Load, $LL = 30 \times 1.6 = 48.0$ psf
>
> Total $= 136.8$ psf

Concentrated factored load at each floor beam location:
Due to the locations of the floor beams and the spacing of the stub girders, the magnitude of each load, P, is:

$$P = 136.8 \times 30 \times 10 = 41.0 \text{ kips}$$

Maximum factored midspan moment:

The girder is symmetric about midspan, and the maximum moment therefore occurs at this location:

$$M_{max} = 1.5 \times P \times 20 - P \times 10 = 820 \text{ k-ft}$$

Estimated interior moment arm for full stub girder cross-section at midspan (refer to Figure 18.2 for typical details):

The interior moment arm (i.e., the distance between the compressive stress resultant in the concrete slab and the tensile stress resultant in the bottom chord) is set equal to the distance between the slab centroid and the bottom chord (wide-flange shape) centroid. This is simplified and conservative. In the example, the distance is estimated as

$$\text{Interior moment arm: } d = 27.5 \text{ in.}$$

This is based on having a 14 series W shape for the bottom chord, W16 floor beams and stubs, a 3-in.-high steel deck, and 3-1/4 in. of lightweight concrete over the top of the steel deck ribs (this allows the deck to be used without having sprayed-on fire protective material on the underside). These are common sizes of the components of a stub girder floor system.

In general, the interior moment arm varies between 24.5 and 29.5 in., depending on the heights of the bottom chord, floor beams/stubs, steel deck, and concrete slab.

Slab and bottom chord axial forces, F (these are the compressive and tensile stress resultants):

$$F = M_{max}/d = (820 \times 12)/27.5 = 357.9 \text{ kips}$$

Required cross-sectional area of bottom chord, A_s:

The required cross-sectional area of the bottom chord can now be found. Since the chord is loaded in tension, the ϕ value is 0.9.

It is also important to note that in the vierendeel analysis that is commonly used in the final evaluation of the stub girder, the member forces will be somewhat larger than those determined through the simplified preliminary procedure. It is therefore recommended that an allowance of some magnitude be given for the vierendeel action. This is done most easily by increasing the area, A_s, by a certain percentage. Based on experience [7, 10], an increase of one-third is suitable, and such has been done in the computations that follow.

On the basis of the data that have been developed, the required area of the bottom chord is:

$$A_s = \frac{(M_{max}/d)}{\phi \times F_y} \times \frac{4}{3} = \frac{F}{0.9 \times F_y} \times \frac{4}{3}$$

which gives A_s values for 36-ksi and 50-ksi steel of

$$A_s = \frac{357.9}{0.9 \times 36} \times \frac{4}{3} = 14.73 \text{ in.}^2 \quad (F_y = 36 \text{ ksi})$$

$$A_s = \frac{357.9}{0.9 \times 50} \times \frac{4}{3} = 10.60 \text{ in.}^2 \quad (F_y = 50 \text{ ksi})$$

Conclusions:

If 36-ksi steel is chosen for the bottom chord of the stub girder, the wide-flange shapes W12x50 and W14x53 will be suitable. If 50-ksi steel is the choice, the sections may be W12x40 or W14x38.

Obviously the final decision is up to the structural engineer. However, in view of the fact that the W12 series shapes will save approximately 2 in. in net floor system height, per story of the building, this would mean significant savings if the overall structure is 10 to 15 stories or more. The differences in stub girder strength and stiffness are not likely to play a role [7, 10, 14].

18.3.3   Choice of Stub Girder Component Sizes

Some examples have been given in the preceding for the choices of chord and floor beam sizes, deck height, and slab configuration. These were made primarily on the basis of acceptable geometries, deck size, and fire protection requirements, to mention some examples. However, construction economy is critical, and the following guidelines will assist the user. The data that are given are based on actual construction projects.

Economical span lengths for the stub girder range from 30 to 50 ft, although the preferable spans are 35 to 45 ft; 50-ft span girders are erectable, but these are close to the limit where the dead load becomes excessive, which has the effect of making the slab govern the overall design. This is usually not an economical solution. Spans shorter than 30 ft are known to have been used successfully; however, this depends on the load level and the type of structure, to mention the key considerations.

Depending on the type and configuration of steel deck that has been selected, the floor beam spacing should generally be maintained between 8 and 12 ft, although larger values have been used. The decisive factor is the ability of the deck to span the distance between the floor beams.

The performance of the stub girder is not particularly sensitive to the stub lengths that are used, as long as these are kept within reasonable limits. In this context it is important to observe that it is usually the exterior stub that controls the behavior of the stub girder. As a practical guideline, the exterior stubs are normally 5 to 7 ft long; the interior stubs are considerably shorter, normally around 3 ft, but components up to 5 ft long are known to have been used. When the stub lengths are chosen, it is necessary to bear in mind the actual purpose of the stubs and how they carry the loads on the stub girder. That is, the stubs are loaded primarily in shear, which explains why the interior stubs can be kept so much shorter than the exterior ones.

The shear connectors that are welded to the top flange of the stub, the stub web stiffeners, and the welds between the bottom flange of the stub and the top flange of the bottom chord are crucial to the function of the stub girder system. For example, the first application of stub girders utilized fitted stiffeners at the ends and sometimes at midlength of all of the stubs. Subsequent research demonstrated that the midlength stiffener did not perform any useful function, and that only the exterior stubs needed stiffeners in order to provide the requisite web stability and shear capacity [5, 21]. Regardless of the span of the girder, it was found that the interior stubs could be left unstiffened, even when they were made as short as 3 ft [7, 14].

Similar savings were realized for the welds and the shear connectors. In particular, in lieu of all-around fillet welds for the connection between the stub and the bottom chord, the studies showed that a significantly smaller amount of welding was needed, and often only in the vicinity of the stub ends. However, specific weld details must be based on appropriate analyses of the stub, considering overturning, weld capacity at the tension end of the stub, and adequate ability to transfer shear from the slab to the bottom chord.

18.3.4   Modeling of the Stub Girder

The original work of Colaco [11, 12] utilized a vierendeel modeling scheme for the stub girder to arrive at a set of stress resultants, which in turn were used to size the various components. Elastic finite element analyses were performed for some of the girders that had been tested, mostly to examine local stress distributions and the correlation between test and theory. However, the finite element solution is not a practical design tool.

Other studies have examined approaches such as non**prismatic beam** analysis [6, 21] and variations of the finite element method [16]. The nonprismatic beam solution is relatively simple to apply. On the other hand, it is not as accurate as the vierendeel approach, since it tends to overlook some important local effects and overstates the service load deflections [5, 21].

On the whole, therefore, the vierendeel modeling of the stub girder has been found to give the most accurate and consistent results, and the correlation with test results is good [5, 6, 11, 14, 21]. Finally, it offers the best physical similarity with actual girders; many designers have found this to be an important advantage.

There are no "simple" methods of analysis that can be used to find the bending moments, shear forces, and axial forces in **vierendeel girders.** Once the preliminary sizing has been accomplished, a computer solution is required for the girder. In general, all that is required for the vierendeel evaluation is a two-dimensional plane frame program for elastic structural analysis. This gives moments, shears, and axial forces, as well as deflections, joint rotations, and other displacement characteristics. The stress resultants are used to size the girder and its elements and connections; the displacements reflect the **serviceability** of the stub girder.

Once the stress resultants are known, the detailed design of the stub girder can proceed. A final run-through of the girder model should then be done, using the components that were chosen, to ascertain that the performance and strength are sufficient in all respects. Under normal circumstances no alterations are necessary at this stage.

As an illustration of the vierendeel modeling of a stub girder, the girder itself is shown in Figure 18.5a and the vierendeel model in Figure 18.5b. The girder is the same as the one used for the preliminary design example. It has four stubs and is symmetrical about midspan; therefore, only half is illustrated. The boundary conditions are shown in Figure 18.5b.

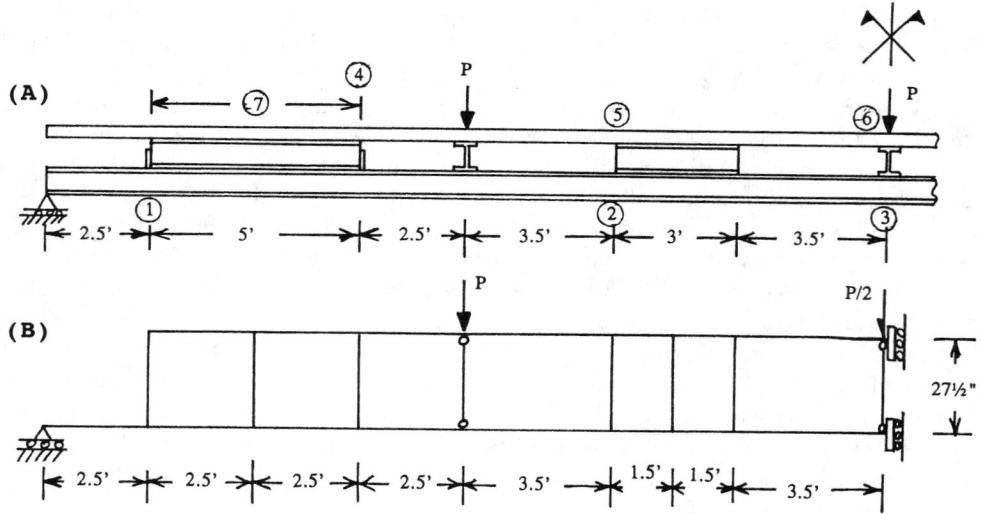

Figure 18.5   An actual stub girder and its vierendeel model (due to symmetry, only one half of the span is shown).

The bottom chord of the model is assigned a moment of inertia equal to the major axis I value, I_x, of the wide-flange shape that was chosen in the preliminary design. However, some analysts believe that since the stub is welded to the bottom chord, a portion of its flexural stiffness should be added to that of the moment of inertia of the wide-flange shape [5, 7, 14, 21] This approach is identical to treating the bottom chord W shape as if it has a cover plate on its top flange. The area of this cover plate is the same as the area of the bottom flange of the stub. This should be done only in the areas where the stubs are placed. In the regions of the interior and exterior stubs it is therefore realistic to increase the moment of inertia of the bottom chord by the parallel-axis value of $A_f \times d_f^2$, where A_f designates the area of the bottom flange of the stub and d_f is the distance between the centroids of the flange plate and the W shape. The contribution to the overall stub girder stiffness is generally small.

The bending stiffness of the top vierendeel chord equals that of the effective width portion of the slab. This should include the contributions of the steel deck as well as the reinforcing steel bars that are located within this width. In particular, the influence of the deck is important. The effective width is determined from the criteria in the AISC LRFD specification, Section I3.1 [3]. It is noted that these were originally developed on the basis of analyses and tests of prismatic composite beams. The approach has been found to give conservative results [5, 21], but should continue to be used until more accurate criteria are available.

In the computations for the slab, the cross-section is conveniently subdivided into simple geometrical shapes. The individual areas and moments of inertia are determined on the basis of the usual transformation from concrete to steel, using the modular ratio $n = E/E_c$, where E is the modulus of elasticity of the steel and E_c is that of concrete. The latter must reflect the density of the concrete that is used, and can be computed from [1]:

$$E_c = 33 \times w_c^{1.5} \times \sqrt{f_c'} \qquad (18.1)$$

The shear connectors used for the stub are required to develop 100% interaction, since the design is based on the computed shear forces, rather than the axial capacity of the steel beam or the concrete slab, as is used for prismatic beams in the AISC Specifications [2, 3]. However, it is neither common nor proper to add the moment of inertia contribution of the top flange of the stub to that of the slab, contrary to what is done for the bottom chord. The reason for this is that dissimilar materials are joined, and some local concrete cracking and/or crushing can be expected to take place around the shear connectors.

The discretization of the stubs into vertical vierendeel girder components is relatively straightforward. Considering the web of the stub and any stiffeners, if applicable (for exterior stubs, most commonly, since interior stubs usually can be left unstiffened), the moment of inertia about an axis that is perpendicular to the plane of the web is calculated. As an example, Figure 18.6 shows the

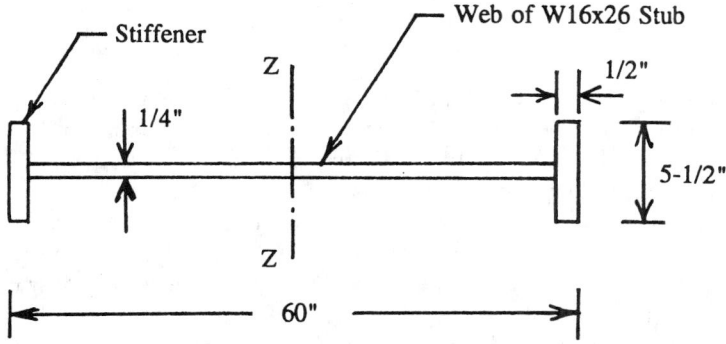

Figure 18.6 Horizontal cross-section of stub with stiffeners.

stub and stiffener configuration for a typical case. The stub is a 5-ft long W16x26 with 5-1/2x1/2-in. end-plate stiffeners. The computations give:

Moment of inertia about the $Z - Z$ axis:

$$I_{ZZ} = \left[0.25 \times (60)^3 \right]/12 + 2 \times 5.5 \times 0.5 \times (30)^2$$

$$= \quad 9450 \text{ in.}^4$$

Depending on the number of vierendeel truss members that will represent the stub in the model, the bending stiffness of each is taken as a fraction of the value of I_{ZZ}. For the girder shown in Figure 18.5, where the stub is discretized as three vertical members, the magnitude of I_{vert} is found as:

Moment of inertia of vertical member:

$$I_{vert} = I_{ZZ}/(\text{no. of verticals}) = 9450/3 = 3150 \text{ in.}^4$$

The cross-sectional area of the stub, including the stiffeners, is similarly divided between the verticals:

Area of vertical member:

$$\begin{aligned} A_{vert} &= \quad [A_{web} + 2 \times A_{st}]/(\text{no. of verticals}) \\ &= \quad [0.25 \times (60 - 2 \times 0.5) + 2 \times 5.5 \times 0.5]/3 \\ &= \quad 6.75 \text{ in.}^2 \end{aligned}$$

Several studies have aimed at finding the optimum number of vertical members to use for each stub. However, the strength and stiffness of the stub girder are only insignificantly affected by this choice, and a number between 3 and 7 is usually chosen. As a rule of thumb, it is advisable to have one vertical per foot length of stub, but this should serve only as a guideline.

The verticals are placed at uniform intervals along the length of the stub, usually with the outside members close to the stub ends. Figure 18.5 illustrates the approach. As for end conditions, these vertical members are assumed to be rigidly connected to the top and bottom chords of the vierendeel girder.

One vertical member is placed at each of the locations of the floor beams. This member is assumed to be pinned to the top and bottom chords, as shown in Figure 18.5, and its stiffness is conservatively set equal to the moment of inertia of a plate with a thickness equal to that of the web of the floor beam and a length equal to the beam depth. In the example, $t_w = 0.25$ in.; the beam depth is 15.69 in. This gives a moment of inertia of

$$\left(\left[15.69 \times 0.25^3 \right]/12 \right) = 0.02 \text{ in.}^4$$

and the cross-sectional area is

$$(15.69 \times 0.25) = 3.92 \text{ in.}^2$$

The vierendeel model shown in Figure 18.5b indicates that the portion of the slab that spans across the opening between the exterior end of the exterior stub and the support for the slab (a column, or a corbel of the core of the structural frame) has been neglected. This is a realistic simplification, considering the relatively low rigidity of the slab in negative bending.

Figure 18.5b also shows the support conditions that are used as input data for the computer analysis. In the example, the symmetrical layout of the girder and its loads make it necessary to analyze only one-half of the span. This cannot be done if there is any kind of asymmetry, and the entire girder must then be analyzed. For the girder that is shown, it is known that only vertical displacements can take place at midspan; horizontal displacements and end rotations are prevented at this location. At the far ends of the bottom chord only horizontal displacements are permitted, and end rotations are free to occur. The reactions that are found are used to size the support elements, including the bottom chord connections and the column.

The structural analysis results are shown in Figure 18.7, in terms of the overall bending moment, shear force, and axial force distributions of the vierendeel model given in Figure 18.5b. Figure 18.7d repeats the layout details of the stub girder, to help identify the locations of the key stress resultant magnitudes with the corresponding regions of the girder.

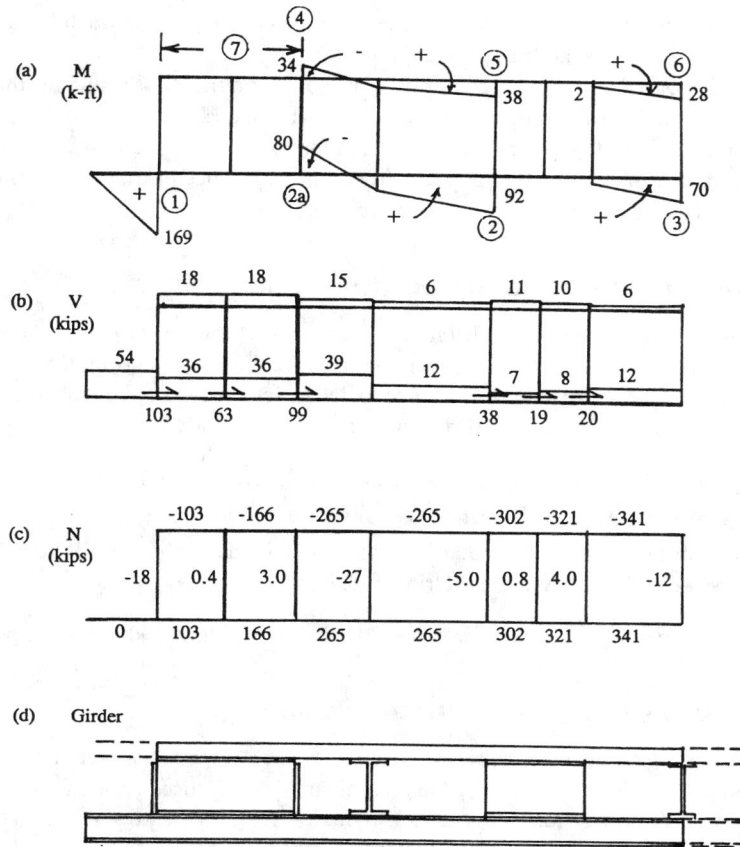

Figure 18.7 Distributions of bending moments, shear forces, and axial forces in a stub girder (see Figure 18.5) (dead load = 74 psf; nominal live load = 50 psf).

The design of the stub girder and its various components can now be done. This must also include deflection checks, even though research has demonstrated that the overall design will never be governed by deflection criteria [7, 14]. However, since the girder has to be built in the shored condition, the girder is often fabricated with a camber, approximately equal to the dead load deflection [7, 10].

18.4 Design Criteria For Stub Girders

18.4.1 General Observations

In general, the design of the stub girder and its components must consider overall member strength criteria as well as local checks. For most of these, the AISC Specifications [2, 3] give requirements that address the needs. Further, although LRFD and ASD are equally applicable in the design of the girder, it is recommended that LRFD be used exclusively. The more rational approach of this specification makes it the method of choice.

In several important areas there are no standardized rules that can be used in the design of the stub girder, and the designer must rely on rational engineering judgment to arrive at satisfactory solutions. This applies to the parts of the girder that have to be designed on the basis of computed forces, such as shear connectors, stiffeners, stub-to-chord welds, and slab reinforcement. The modeling and evaluation of the capacity of the central portion of the concrete slab are also subject to interpretation. However, the design recommendations that are given in the following are based on a wide variety of practical and successful applications.

It is again emphasized that the design throughout is based on the stress resultants that have been determined in the vierendeel or other analysis, rather than on idealized code criteria. However, the capacities of materials and fasteners, as well as the requirements for the stability and strength of tension and compression members, adhere strictly to the AISC Specifications. Any interpretations that have been made are always to the conservative side.

18.4.2 Governing Sections of the Stub Girder

Figures 18.5 and 18.7 show certain circled numbers at various locations throughout the span of the stub girder. These reflect the sections of the girder that are the most important, for one reason or another, and are the ones that must be examined to determine the required member size, etc. These are the governing sections of the stub girder and are itemized as follows:

1. Points 1, 2, and 3 indicate the critical sections for the bottom chord.
2. Points 4, 5, and 6 indicate the critical sections for the concrete slab.
3. Point 7, which is a region rather than a specific point, indicates the critical shear transfer region between the slab and the exterior stub.

The design checks that must be made for each of these areas are discussed in the following.

18.4.3 Design Checks for the Bottom Chord

The size of the bottom chord is almost always governed by the stress resultants at midspan, or point 3 in Figures 18.5 and 18.7. This is also why the preliminary design procedure focused almost entirely on determining the required chord cross-section at this location. As the stress resultant distributions in Figure 18.7 show, the bottom chord is subjected to combined positive bending moment and tensile force at point 3, and the design check must consider the beam-tension member behavior in this area. The design requirements are given in Section H1.1, Eqs. (H1-1a) and (H1-1b), of the AISC LRFD Specification [3].

Combined bending and tension must also be evaluated at point 2, the exterior end of the interior stub. The local bending moment in the chord is generally larger here than at midspan, but the axial force is smaller. Only a computation can confirm whether point 2 will govern in lieu of point 3. Further, although the location at the interior end of the exterior stub (point 2a) is rarely critical, the combination of negative moment and tensile force should be evaluated.

At point 1 of the bottom chord, which is located at the exterior end of the exterior stub, the axial force is equal to zero. At this location the bottom chord must therefore be checked for pure bending, as well as shear.

The preceding applies only to a girder with simple end supports. When it is part of the lateral load-resisting system, axial forces will exist in all parts of the chord. These must be resisted by the adjacent structural members.

18.4.4 Design Checks for the Concrete Slab

The top chord carries varying amounts of bending moment and axial force, as illustrated in Figure 18.7, but the most important areas are indicated as points 4 to 6. The axial forces are always compressive in the concrete slab; the bending moments are positive at points 5 and 6, but negative at point 4. As a result, this location is normally the one that governs the performance of the slab, not the least because the reinforcement in the positive moment region includes the substantial cross-sectional area of the steel deck.

The full effective width of the slab must be analyzed for combined bending and axial force at all of points 4 through 6. Either the composite beam-column criteria of the AISC LRFD specification [3] or the criteria of the reinforced concrete structures code of the American Concrete Institute (ACI) [1] may be used for this purpose.

18.4.5   Design Checks for the Shear Transfer Regions

Region 7 is the shear transfer region between the concrete slab and the exterior stub, and the combined shear and longitudinal compressive capacity of the slab in this area must be determined. The shear transfer region between the slab and the interior stub always has a smaller shear force.

Region 7 is critical, and several studies have shown that the slab in this area will fail in a combination of concrete crushing and shear [5, 6, 7, 21]. The shear failure zone usually extends from corner to corner of the steel deck, over the top of the shear connectors, as illustrated in Figure 18.8. This also

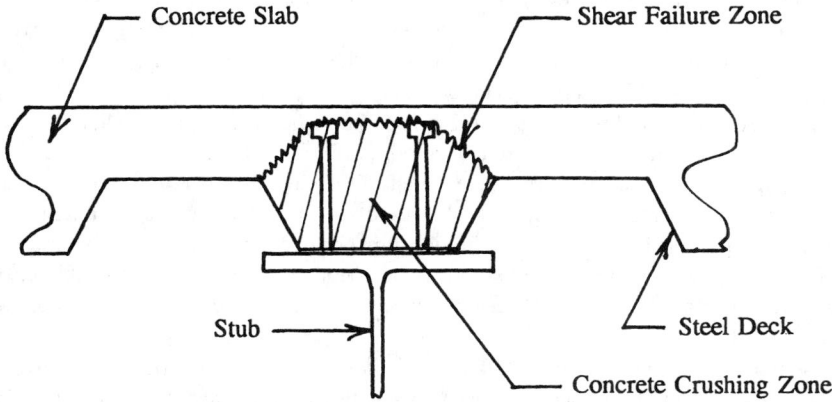

Figure 18.8   Shear and compression failure regions in the slab of the stub girder.

emphasizes why the placement of the longitudinal reinforcing steel bars in the central flute of the steel deck is important, as well as the location of the transverse bars: both groups should be placed just below the level of the top of the shear connectors (see Figure 18.2). The welded wire mesh reinforcement that is used as a matter of course, mostly to control shrinkage cracking in the slab, also assists in improving the strength and ductility of the slab in this region.

18.4.6   Design of Stubs for Shear and Axial Load

The shear and axial force distributions indicate the governing stress resultants for the stub members. It is important to note that since the vierendeel members are idealized from the real (i.e., continuous) stubs, bending is not a governing condition. Given the sizes and locations of the individual vertical members that make up the stubs, the design checks are easily made for axial load and shear. For example, referring to Figure 18.7, it is seen that the shear and axial forces in the exterior and interior stubs, and the axial forces in the verticals that represent the floor beams, are the following:

Exterior stub verticals:

$$
\begin{array}{llll}
\text{Shear forces:} & 103 \text{ kips} & 63 \text{ kips} & 99 \text{ kips} \\
\text{Axial forces:} & -18 \text{ kips} & 0.4 \text{ kips} & 3 \text{ kips}
\end{array}
$$

Interior stub verticals:

Shear forces:	38 kips	19 kips	20 kips
Axial forces:	−5 kips	0.8 kips	4 kips

Floor beam verticals:

Exterior:	Axial force = −39 kips
Interior:	Axial force = −12 kips

Shear forces are zero in these members.

The areas and moments of inertia of the verticals are known from the modeling of the stub girder. Figure 18.7 also shows the shear and axial forces in the bottom and top chords, but the design for these elements has been addressed earlier in this chapter.

The design checks that are made for the stub verticals will also indicate whether there is a need for stiffeners for the stubs, since the evaluations for axial load capacity should always first be made on the assumption that there are no stiffeners. However, experience has shown that the exterior stubs always must be stiffened; the interior stubs, on the other hand, will almost always be satisfactory without stiffeners, although exceptions can occur.

The axial forces that are shown for the stub verticals in the preceding are small, but typical, and it is clear that in all probability only the exterior end of the exterior stub really requires a stiffener. This was examined in one of the stub girder research studies, where it was found that a single stiffener would suffice, although the resulting lack of structural symmetry gave rise to a tensile failure in the unstiffened area of the stub [21]. Although this occurred at a very late stage in the test, the type of failure represents an undesirable mode of behavior, and the use of single stiffeners therefore was discarded. Further, by reason of ease of fabrication and erection, stiffeners should always be provided at both stub ends.

It is essential to bear in mind that if stiffeners are required, the purpose of such elements is to add to the area and moment of inertia of the web, to resist the axial load that is applied. There is no need to provide bearing stiffeners, since the load is not transmitted in this fashion. The most economical solution is to make use of end-plate stiffeners of the kind that is shown in Figure 18.1; extensive research evaluations showed that this was the most efficient and economical choice [5, 6, 21].

The vertical stub members are designed as columns, using the criteria of Section E1 of the AISC Specification [3]. For a conservative solution, an effective length factor of 1.0 may be used. However, it is more realistic to utilize a K value of 0.8 for the verticals of the stubs, recognizing the end restraint that is provided by the connections between the chords and the stubs. The K-factor for the floor beam verticals must be 1.0, due to the pinned ends that are assumed in the modeling of these components, as well as the flexibility of the floor beam itself in the direction of potential buckling of the vertical member.

18.4.7 Design of Stud Shear Connectors

The shear forces that must be transferred between the slab and the stubs are given by the vierendeel girder shear force diagram. These are the factored shear force values which are to be resisted by the connectors. The example shown in Figure 18.7 indicates the individual shear forces for the stub verticals, as listed in the preceding section. However, in the design of the overall shear connection, the total shear force that is to be transmitted to the stub is used, and the stud connectors are then distributed uniformly along the stub. The design strength of each connector is determined in accordance with Section I5.3 of the LRFD Specification [3], including any deck profile reduction factor (Section I3.5).

Analyzing the girder whose data are given in Figure 18.7, the following is known:

Exterior stub:

Total shear force = V_{es} = 103 + 63 + 99 = 265 kips

Interior stub:

Total shear force = V_{is} = 38 + 19 + 20 = 77 kips

The nominal strength, Q_n, of the stud shear connectors is given by Eq. (I5-1) in Section I5.3 of the LRFD Specification, thus:

$$Q_n = 0.5 \times A_{sc}\sqrt{f'_c \times E_c} \le A_{sc} \times F_u \tag{18.2}$$

where A_{sc} is the cross-sectional area of the stud shear connector, f'_c and E_c are the compressive strength and modulus of elasticity of the concrete, and F_u is the specified minimum tensile strength of the stud shear connector steel, or 60 ksi (ASTM A108).

In the equation for Q_n, the left-hand side reflects the ultimate limit state of shear yield failure of the connector; the right-hand side gives the ultimate limit state of tension fracture of the stud. Although shear almost always governs and is the desirable mode of behavior, a check has to be made to ensure that tension fracture will not take place. This as achieved by the appropriate value of E_c, setting $F_u = 60$ ksi, and solving for f'_c from Equation 18.2. The requirement that must be satisfied in order for the stud shear limit state to govern is given by Equation 18.3:

$$f'_c \le \frac{57,000}{w_c} \tag{18.3}$$

This gives the limiting values for concrete strength as related to the density; data are given in Table 18.1.

TABLE 18.1 Concrete Strength Limitations for Ductile Shear Connector Failure

Concrete density, w_c (pcf)	Maximum concrete strength, f'_c (psi)
145 (= NW)	4000
120	4800
110	5200
100	5700
90	6400

Note: NW = normal weight.

For concrete with $w_c = 120$ pcf and $f'_c = 4,000$ psi, as used in the design example, $E_c = 2,629,000$ psi. Using 3/4-in. diameter studs, the nominal shear capacity is:

$$Q_n = 0.5 \left[\pi (0.75)^2/4 \right] \sqrt{(4 \times 2,629)} \le \left[\pi (0.75)^2/4 \right] 60$$

which gives

$$Q_n = 22.7 \text{ kips} < 26.5 \text{ kips}$$

The LRFD Specification [3] does not give a resistance factor for shear connectors, on the premise that the ϕ value of 0.85 for the overall design of the composite member incorporates the stud strength variability. This is not satisfactory for composite members such as stub girders and composite trusses. However, a study was carried out to determine the resistance factors for the two ultimate limit states for stud shear connectors [20]. Briefly, on the basis of extensive analyses of test data from a variety of sources, and using the Q_n equation as the nominal strength expression, the values of the resistance factors that apply to the shear yield and tension fracture limit states, respectively, are:

Stud shear connector resistance factors:

Limit state of shear yielding: $\phi_{conn} = 0.90$

Limit state of tension fracture: $\phi_{conn} = 0.75$

The required number of shear connectors can now be found as follows, using the total stub shear forces, V_{es} and V_{is}, computed earlier in this section:

Exterior stub:

$$
\begin{aligned}
n_{es} &= V_{es}/(0.9 \times Q_n) = V_{es}/(\phi_{conn} Q_n) \\
&= 265/(0.9 \times 22.7) = 13.0
\end{aligned}
$$

i.e., use $n_{es} = 14$-3/4-in. diameter stud shear connectors, placed in pairs and distributed uniformly along the length of the top flange of each of the exterior stubs.

Interior stub:

$$
\begin{aligned}
n_{is} &= V_{is}/(0.9 \times Q_n) = V_{is}/(\phi_{conn} Q_n) \\
&= 77/(0.9 \times 22.7) = 3.8
\end{aligned}
$$

i.e., use $n_{is} = 4$-3/4-in. diameter stud shear connectors, placed singly and distributed uniformly along the length of the top flange of each of the interior stubs.

Considering the shear forces for the stub girder of Figures 18.5 and 18.7, the number of connectors for the exterior stub is approximately three times that for the interior one, as expected. Depending on span, loading, etc., there are instances when it will be difficult to fit the required number of studs on the exterior stub, since typical usage entails a double row, spaced as closely as permitted (four diameters in any direction [Section I5.6, AISC LRFD Specification [3]]). Several avenues may be followed to remedy such a problem; the easiest one is most likely to use a higher strength concrete, as long as the limit state requirements for Q_n and Table 18.1 are satisfied. This entails only minor reanalysis of the girder.

18.4.8   Design of Welds between Stub and Bottom Chord

The welds that are needed to fasten the stubs to the top flange of the bottom chord are primarily governed by the shear forces that are transferred between these components of the stub girder. The shear force distribution gives these stress resultants, which are equal to those that must be transferred between the slab and the stubs. Thus, the factored forces, V_{es} and V_{is}, that were developed in Section 18.4.7 are used to size the welds.

Axial loads also act between the stubs and the chord; these may be compressive or tensile. In Figure 18.7 it is seen that the only axial force of note occurs in the exterior vertical of the exterior stub (load = 18 kips); the other loads are very small compressive or tensile forces. Unless a significant tensile force is found in the analysis, it will be a safe simplification to ignore the presence of the axial forces insofar as the weld design is concerned.

The primary shear forces that have to be taken by the welds are developed in the outer regions of the stubs, although it is noted that in the case of Figure 18.5, the central vertical element in both stubs carries forces of some magnitude (63 and 19 kips, respectively). However, this distribution is a result of the modeling of the stubs; analyses of girders where many more verticals were used have confirmed that the major part of the shear is transferred at the ends [7, 10, 21]. The reason is that the stub is a full shear panel, where the internal moment is developed through stress resultants that act at points toward the ends, in a form of bending action. Tests have also verified this characteristic of the girder behavior [6, 21]. Finally, concentrating the welds at the stub ends will have significant economic impact [5, 7, 21].

In view of these observations, the most effective placement of the welds between the stubs and the bottom chord is to concentrate them across the ends of the stubs and along a short distance of both sides of the stub flanges. For ease of fabrication and structural symmetry, the same amount of welding should be placed at both ends, although the forces are always smaller at the interior ends of the stubs. Such U-shaped welds were used for a number of the full-size girders that were tested [5, 6, 21], with only highly localized yielding occurring in the welds. A typical detail is shown in Figure 18.9; this reflects what is recommended for use in practice.

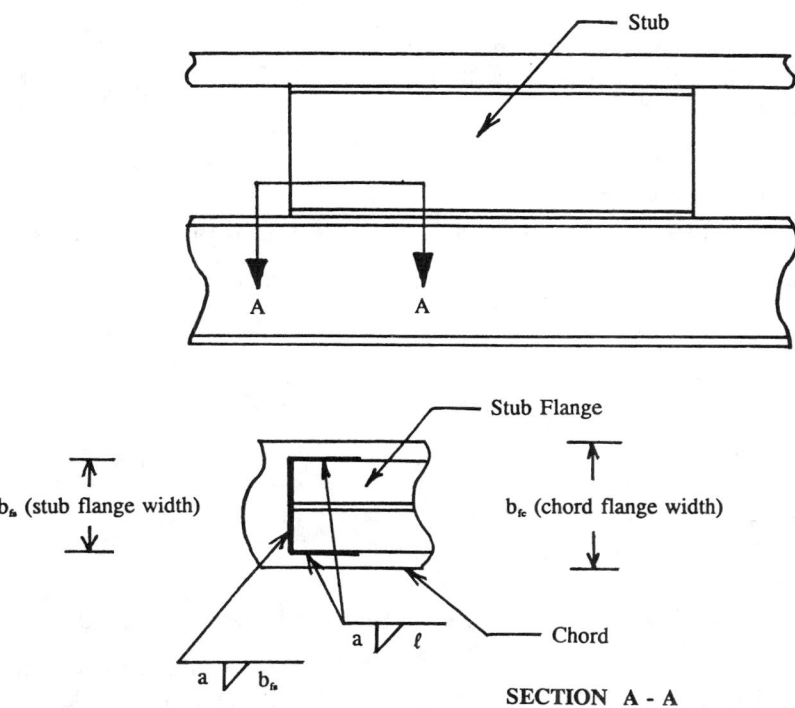

Figure 18.9   Placement of U-shaped fillet weld for attachment at each end of stub to bottom chord.

Prior to the research that led to the change of the welded joint design, the stubs were welded with all-around fillet welds for the exterior as well as the interior elements. The improved, U-shaped detail provided for weld metal savings of approximately 75% for interior stubs and around 50% for exterior stubs.

For the sample stub girder, W16x26 shapes are used for the stubs. The total forces to be taken by the welds are:

Exterior stub: $V_{es} = 265$ kips

Interior stub: $V_{is} = 77$ kips

Using E70XX electrodes and 5/16-in. fillet welds (the fillet weld size must be smaller than the thickness of the stub flange, which is 3/8 in. for the W16x26), the total weld length for each stub is L_w, given by (refer to Figure 18.9):

$$L_w = 2(b_{fs} + 2\ell)$$

since U-shaped welds of length $(b_{fs} + 2\ell)$ are placed at each stub end. The total weld lengths required for the stub girder in question are therefore:

Exterior stub:

$$(L_w)_{es} = V_{es}/(0.707a\phi_w F_w)$$
$$= 265/[0.707(5/16) \times 0.75(0.6 \times 70)] = 38.1 \text{ in.}$$

Interior stub:

$$(L_w)_{is} = V_{is}/(0.707a\phi_w F_w)$$
$$= 77/[0.707(5/16) \times 0.75(0.6 \times 70)] = 11.1 \text{ in.}$$

In the above expressions, $a = 5/16$ in. $=$ fillet weld size, $\phi_w = 0.75$, and $F_w = 0.6F_{EXX} = 0.6 \times 70 = 42$ ksi for E70XX electrodes (Table J2.3, AISC LRFD Specification [3]). The total U-weld lengths at each stub end are therefore:

Exterior stub: $L_{Ues} = 19.1$ in.
Interior stub: $L_{Uis} = 5.6$ in.

With a flange width for the W16x26 of 5.50 in., the above lengths can be simplified as:

$$L_{Ues} = 5.50 + 7.0 + 7.0$$

where ℓ_{es} is chosen as 7.0 in. For the interior stub:

$$L_{Uis} = 5.50 + 2.0 + 2.0$$

where ℓ_{is} is chosen as 2.0 in.

The details chosen are a matter of judgment. In the example, the interior stub for all practical purposes requires no weld other than the one across the flange, although at least a minimum weld return of 1/2 in. should be used.

18.4.9   Floor Beam Connections to Slab and Bottom Chord

In the vierendeel model, the floor beam is represented as a pinned-end compression member. It is designed using a K-factor of 1.0, and the floor beam web by itself is almost always sufficient to take the axial load. However, the floor beam must be checked for web crippling and web buckling under **shoring** conditions.

No shear is transferred from the beam to the slab or the bottom chord. In theory, therefore, any attachment device between the floor beam and the other components should not be needed. However, due to construction stability requirements, as well as the fact that the floor beam usually is designed for composite action normal to the girder, fasteners are needed. In practice, these are not actually designed; rather, one or two stud shear connectors are placed on the top flange of the beam, and two high-strength bolts attach the lower flange to the bottom chord.

18.4.10   Connection of Bottom Chord to Supports

In the traditional use of stub girders, the girder is supported as a simple beam, and the bottom chord end connections need to be able to transfer vertical reactions to the supports. The latter structural elements may be columns, or the girder may rest on corbels or other types of supports that are part of the concrete core of the building. For both of these cases the reactions that are to be carried to the adjacent structure are given by the analysis, and the response needs for the supports are clear.

Any shear-type beam connections may be used to connect the bottom chord to a column or a corbel or similar bracket. It is important to ascertain that the chord web shear capacity is sufficient, including block shear (Section J5 of the AISC LRFD Specification [3]).

Some designers prefer to use slotted holes for the connections, and to delay the final tightening of the bolts until after the shoring has been removed. This is done on the premise that the procedure will leave the slab essentially stress free from the construction loads, leading to less cracking in the slab during service. Other designers specify additional slab reinforcement to take care of any cracking problem. Experience has shown that both methods are suitable.

The slab may be supported on an edge beam or similar element at the exterior side of the floor system. There is no force transfer ability required of this support. In the interior of the building the slab will be continuously cast across other girders and around columns; this will almost always lead to some cracking, both in the vicinity of the columns as well as along beams and girders. With suitable placement of floor slab joints, this can be minimized, and appropriate transverse reinforcement for the slab will reduce, if not eliminate, the longitudinal cracks.

Data on the effects of various types of cracks in composite floor systems are scarce. Current opinion appears to be that the strength may not be influenced very much. In any case, the mechanics of the short- and long-term service response of composite beams is not well understood. Recent studies have developed models for the cracking mechanism and the crack propagation [18]; the correlation with a wide variety of laboratory tests is good. However, a comprehensive study of concrete cracking and its implications for structural service and strength needs to be undertaken.

18.4.11   Use of Stub Girder for Lateral Load System

The stub girder was originally conceived only as being part of the vertical load-carrying system of structural frames, and the use of simple connections, as discussed in Section 18.4.9, came from this development. However, because a deep, long-span member can be very effective as a part of the lateral load-resisting system for a structure, several attempts have been made to incorporate the stub girder into moment frames and similar systems. The projects of Colaco in Houston [13] and Martinez-Romero [17] in Mexico City were successful, although the designers noted that the cost premium could be substantial.

For the Colaco structure, his applications reduced drift, as expected, but gave much more complex beam-to-column connections and reinforcement details in the slab around the columns. Thus, the exterior stubs were moved to the far ends of the girders, and moment connections were designed for the full depth. For the Mexico City building, the added ductility was a prime factor in the survival of the structure during the 1985 earthquake.

The advantages of using the stub girders in moment frames are obvious. Some of the disadvantages have been outlined; in addition, it must be recognized that the lack of room for perimeter **HVAC** ducts may be undesirable. This can only be addressed by the mechanical engineering consultant. As a general rule, a designer who wishes to use stub girders as part of the lateral load-resisting system should examine all structural effects, but also incorporate nonstructural considerations such as are prompted by HVAC and electronic communication needs.

18.4.12   Deflection Checks

The service load deflections of the stub girder are needed for several purposes. First, the overall dead load deflection is used to assess the camber requirements. Due to the long spans of typical stub girders, as well as the flexibility of the framing members and the connections during construction, it is important to end up with a floor system that is as level as possible by the time the structure is ready to be occupied. Thus, the girders must be built in the shored condition, and the camber should be approximately equal to the dead load deflection.

Second, it is essential to bear in mind that each girder will be shored against a similar member at the level below the current construction floor. This member, in turn, is similarly shored, albeit against a girder whose stiffness is greater, due to the additional curing time of the concrete slab. This has a cumulative effect for the structure as a whole, and the dead load deflection computations must take this response into account.

In other words, the support for the shores is a flexible one, and deflections therefore will occur in the girder as a result of floor system movements of the structure at levels in addition to the one under consideration. Although this is not unique to the stub girder system, the span lengths and the interaction with the frame accentuate the influence on the girder design.

Depending on the structural system, it is also likely that the flexibility of the columns and the connections will add to the vertical displacements of the stub girders. The deflection calculations should incorporate these effects, preferably by utilizing realistic modified E_c values and determining displacements as they occur in the frame. Thus, the curing process for the concrete might be considered, since the strength development as a function of time is directly related to the value of E_c [1]. This is a subject that is open for study, although similar criteria have been incorporated in studies of the strength and behavior of composite frames [8, 9]. However, detailed evaluations of the influence of time-dependent stiffness still need to be made for a wide variety of floor systems and frames. The cumulative deflection effects can be significant for the construction of the building, and consequently also must enter into the contractor's planning. This subject is addressed briefly in Section 18.5.

Third, the live load deflections must be determined to assess the serviceability of the floor system under normal operating conditions. However, several studies have demonstrated that such displacements will be significantly smaller than the $L/360$ requirement that is normally associated with live load deflections [6, 7, 10, 14, 21]. It is therefore rarely possible to design a girder that meets the strength and the deflection criteria simultaneously [14]. In other words, strength governs the overall design.

Finally, although they rarely play a role in the overall response of the stub girder, the deflections and end rotations of the slab across the openings of the girder should also be checked. This is primarily done to assess the potential for local cracking, especially at the stub ends and at the floor beams. However, proper placement of the longitudinal girder reinforcement is usually sufficient to prevent problems of this kind, since the deformations tend to be small.

18.5 Influence of Method of Construction

A number of construction-related considerations have already been addressed in various sections of this chapter. The most important ones relate to the fact that the stub girders must be built in the shored condition. The placement and removal of the shores may have a significant impact on the performance of the member and the structure as a whole. In particular, too early shore removal may lead to excessive deflections in the girders at levels above the one where the shores were located. This is a direct result of the low stiffness of **"green"** concrete. It can also lead to "ponding" of the concrete slab, producing larger dead loads than accounted for in the original design. Finally, larger girder deflections can be translated into an "inward pulling" effect on the columns of the frame. However, this is clearly a function of the framing system.

On the other hand, the use of high early strength cement and similar products can reduce this effect significantly. Further, since the concrete usually is able to reach about 75% of the 28-day strength after 7 to 10 days, the problem is less severe than originally thought [5, 7, 10]. In any case, it is important for the structural engineer to interact with the general contractor, in order that the influence of the method of construction on the girders as well as the frame can be quantified, however simplistic the analysis procedure may be.

Due to the larger loads that can be expected for the shores, the latter must either be designed as structural members or at least be evaluated by the structural engineer. The size of the shores is also influenced by the number of floors that are to have these supports left in place. As a general rule, when stub girders are used for multi-story frames, the shores should be left in place for at least three floor levels. Some designers prefer a larger number; however, any choices of this kind should be based on computations for sizes and effects. Naturally, the more floors that are specified, the larger the shores will have to be.

18.6 Defining Terms

Composite: Steel and concrete acting in concert.

Formed steel deck: A thin sheet of steel shaped into peaks and valleys called corrugations.

Green concrete: concrete that has just been placed.

HVAC: Heating, ventilating, and air conditioning.

Lightweight: Refers to concrete with unit weights between 90 and 120 pcf.

Normal weight: Refers to concrete with unit weights of 145 lb per cubic foot (pcf).

Prismatic beam: A beam with a constant size cross-section over the full length.

Rebar: An abbreviated name for reinforcing steel bars.

Serviceability: The ability of a structure to function properly under normal operating conditions.

Shoring: Temporary support.

Vierendeel girder: A girder with top and bottom chords attached to each other through fully welded connections to vertical (generally) members.

References

[1] American Concrete Institute. 1995. *Building Code Requirements for Reinforced Concrete,* ACI Standard No. 318-95, ACI, Detroit, MI.

[2] American Institute of Steel Construction. 1989. *Specification for the Allowable Stress Design, Fabrication, and Erection of Structural Steel for Buildings,* 9th ed., AISC, Chicago, IL.

[3] American Institute of Steel Construction. 1993. *Specification for the Load and Resistance Factor Design, Fabrication, and Erection of Structural Steel for Buildings,* 2nd ed., AISC, Chicago, IL.

[4] American Society of Civil Engineers. 1995. *Minimum Design Loads for Buildings and Other Structures,* ASCE/ANSI Standard No. 7-95, ASCE, New York.

[5] Bjorhovde, R., and Zimmerman, T.J. 1980. Some Aspects of Stub Girder Design, *AISC Eng. J.,* 17(3), Third Quarter, September (pp. 54-69).

[6] Bjorhovde, R. 1981. *Full-Scale Test of a Stub Girder,* Report submitted to Dominion Bridge Company, Calgary, Alberta, Canada. Department of Civil Engineering, University of Alberta, Edmonton, Alberta, Canada, June.

[7] Bjorhovde, R. 1985. Behavior and Strength of Stub Girder Floor Systems, in *Composite and Mixed Construction,* ASCE Special Publication, ASCE, New York.

[8] Bjorhovde, R. 1987. *Design Considerations for Composite Frames,* Proceedings 2nd International and 5th Mexican National Symposium on Steel Structures, IMCA and SMIE, Morelia, Michoacan, Mexico, November 23-24.

[9] Bjorhovde, R. 1994. Concepts and Issues in Composite Frame Design, *Steel Structures,* Journal of the Singapore Society for Steel Structures, 5(1), December (pp. 3-14).

[10] Chien, E.Y.L. and Ritchie, J.K. 1984. *Design and Construction of Composite Floor Systems,* Canadian Institute of Steel Construction (CISC), Willowdale (Toronto), Ontario, Canada.

[11] Colaco, J.P. 1972. A Stub Girder System for High-Rise Buildings, *AISC Eng. J.*, 9(2), Second Quarter, July (pp. 89-95).

[12] Colaco, J. P. 1974. Partial Tube Concept for Mid-Rise Structures, *AISC Eng. J.*, 11(4), Fourth Quarter, December (pp. 81-85).

[13] Colaco, J.P. and Banavalkar, P.V. 1979. *Recent Uses of the Stub Girder System*, Proceedings 1979 National Engineering Conference, American Institute of Steel Construction, Chicago, IL, May.

[14] Griffis, T.C. 1983. *Stiffness Criteria for Stub Girder Floor Systems*, M.S. thesis, University of Arizona, Tucson, AZ.

[15] Hansell, W.C., Galambos, T.V., Ravindra, M.K., and Viest, I.M. 1978. Composite Beam Criteria in LRFD, *J. Structural Div.*, ASCE, 104(ST9), September (pp. 1409-1426).

[16] Hrabok, M.M. and Hosain, M.U. 1978. Analysis of Stub Girders Using Sub-Structuring, *Intl. J. Computers and Structures*, 8(5), 615-619.

[17] Martinez-Romero, E. 1983. *Continuous Stub Girder Structural System for Floor Decks*, Technical report, EMRSA, Mexico City, Mexico, February.

[18] Morcos, S.S. and Bjorhovde, R. 1995. Fracture Modeling of Concrete and Steel, *J. Structural Eng.*, ASCE, 121(7), 1125-1133.

[19] Wong, A.F. 1979. *Conventional and Unconventional Composite Floor Systems*, M.Eng. thesis, University of Alberta, Edmonton, Alberta, Canada.

[20] Zeitoun, L.A. 1984. *Development of Resistance Factors for Stud Shear Connectors*, M.S. thesis, University of Arizona, Tucson, AZ.

[21] Zimmerman, T.J. and Bjorhovde, R. 1981. *Analysis and Design of Stub Girders*, Structural Engineering Report No. 90, University of Alberta, Edmonton, Alberta, Canada, March.

Further Reading

The references that accompany this chapter are all-encompassing for the literature on stub girders. Primary references that should be studied in addition to this chapter are [5, 7, 10, 11], and [13].

19

Plate and Box Girders

19.1 Introduction 19-1
19.2 Stability of the Compression Flange 19-2
     Vertical Buckling • Lateral Buckling • Torsional Buckling • Compression Flange of a Box Girder
19.3 Web Buckling Due to In-Plane Bending 19-5
19.4 Nominal Moment Strength 19-6
19.5 Web Longitudinal Stiffeners for Bending Design 19-6
19.6 Ultimate Shear Capacity of the Web 19-7
19.7 Web Stiffeners for Shear Design 19-10
19.8 Flexure-Shear Interaction 19-11
19.9 Steel Plate Shear Walls 19-11
19.10 In-Plane Compressive Edge Loading 19-15
19.11 Eccentric Edge Loading 19-16
19.12 Load-Bearing Stiffeners 19-17
19.13 Web Openings 19-18
19.14 Girders with Corrugated Webs....................... 19-18
19.15 Defining Terms 19-21
References 19-21

Mohamed Elgaaly
Department of Civil &
Architectural Engineering,
Drexel University,
Philadelphia, PA

19.1 Introduction

Plate and box girders are used mostly in bridges and industrial buildings, where large loads and/or long spans are frequently encountered. The high torsional strength of box girders makes them ideal for girders curved in plan. Recently, thin steel plate **shear walls** have been effectively used in buildings. Such walls behave as vertical plate girders with the building columns as flanges and the floor beams as intermediate stiffeners. Although traditionally simply supported plate and box girders are built up to 150 ft span, several three-span continuous girder bridges have been built in the U.S. with center spans exceeding 400 ft.

In its simplest form a plate girder is made of two flange plates welded to a web plate to form an I section, and a box girder has two flanges and two webs for a single-cell box and more than two webs in multi-cell box girders (Figure 19.1). The designer has the freedom in proportioning the cross-section of the girder to achieve the most economical design and taking advantage of available high-strength steels. The larger dimensions of plate and box girders result in the use of slender webs and flanges, making buckling problems more relevant in design. Buckling of plates that are adequately supported along their boundaries is not synonymous with failure, and these plates exhibit post-buckling strength that can be several times their buckling strength, depending on the plate slenderness. Although plate buckling has not been the basis for design since the early 1960s, buckling strength is often required to calculate the post-buckling strength.

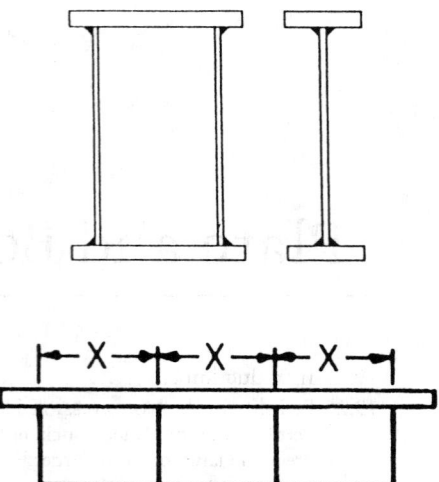

Figure 19.1   Plate and box girders.

The trend toward **limit state** format codes placed the emphasis on the development of new design approaches based on the ultimate strength of plate and box girders and their components. The post-buckling strength of plates subjected to shear is due to the diagonal tension field action. The post-buckling strength of plates subjected to uniaxial compression is due to the change in the **stress** distribution after buckling, higher near the supported edges. An **effective width** with a uniform stress, equal to the yield stress of the plate material, is used to calculate the post-buckling strength [40].

The flange in a box girder and the web in plate and box girders are often reinforced with stiffeners to allow for the use of thin plates. The designer has to find a combination of plate thickness and stiffener spacing that will optimize the weight and reduce the fabrication cost. The stiffeners in most cases are designed to divide the plate panel into subpanels, which are assumed to be supported along the stiffener lines. Recently, the use of **corrugated webs** resulted in employing thin webs without the need for stiffeners, thus reducing the fabrication cost and also improving the fatigue life of the girders.

The web of a girder and load-bearing diaphragms can be subjected to in-plane compressive patch loading. The ultimate capacity under this loading condition is controlled by **web crippling**, which can occur prior to or after local yielding. The presence of openings in plates subjected to in-plane loads is unavoidable in some cases, and the presence of openings affects the stability and ultimate strength of plates.

19.2   Stability of the Compression Flange

The compression flange of a plate girder subjected to bending usually fails in lateral buckling, local torsional buckling, or yielding; if the web is slender the compression flange can fail by vertical buckling into the web (Figure 19.2).

19.2.1   Vertical Buckling

The following limiting value for the **web slenderness ratio** to preclude this mode of failure [4] can be used,

$$h/t_w \leq \left[0.68E/\sqrt{F_{yf}\left(F_{yf} + F_r\right)}\right]\sqrt{A_w/A_f} \qquad (19.1)$$

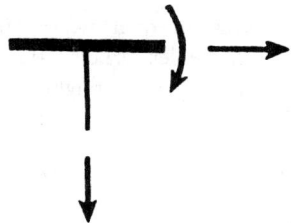

Figure 19.2 Compression flange modes of failure.

where h and t_w are the web height and thickness, respectively; A_w is the area of the web; A_f is the area of the flange; E is Young's modulus of elasticity; F_{yf} is the yield stress of the flange material; and F_r is the residual tension that must be overcome to achieve uniform yielding in compression. This limiting value may be too conservative since vertical buckling of the compression flange into the web occurs only after general yielding of the flange. This limiting value, however, can be helpful to avoid fatigue cracking under repeated loading due to out-of-plane flexing, and it also facilitates fabrication.

The American Institute of Steel Construction (**AISC**) specification [32] uses Equation 19.1 when the spacing between the vertical stiffeners, a, is more than 1.5 times the web depth, h ($a/h > 1.5$). In such a case the specification recommends that

$$h/t_w \leq 14,000/\sqrt{F_{yf}(F_{yf} + 16.5)} \qquad (19.2)$$

where a minimum value of $A_w/A_f = 0.5$ was assumed and the residual tension was taken to be 16.5 ksi. Furthermore, when a/h is less than or equal to 1.5, higher web slenderness is permitted, namely

$$h/t_w \leq 2000/\sqrt{F_{yf}} \qquad (19.3)$$

19.2.2 Lateral Buckling

When a flange is not adequately supported in the lateral direction, elastic lateral buckling can occur. The compression flange, together with an effective area of the web equal to $A_w/6$, can be treated as a column and the buckling stress can be calculated from the Euler equation [2]:

$$F_{cr} = \pi^2 E/(\lambda)^2 \qquad (19.4)$$

where λ is the slenderness ratio, which is equal to L_b/r_T; L_b is the length of the unbraced segment of the beam; and r_T is the radius of gyration of the compression flange plus one-third of the compression portion of the web.

The AISC specification adopted Equation 19.4, rounding $\pi^2 E$ to 286,000 and assuming that elastic buckling will occur when the slenderness ratio, λ, is greater than $\lambda_r (= 756/\sqrt{F_{yf}})$.

Furthermore, Equation 19.4 is based on uniform compression; in most cases the bending is not uniform within the length of the unbraced segment of the beam. To account for nonuniform bending, Equation 19.4 should be multiplied by a coefficient, C_b [25], where

$$C_b = 12.5M_{max}/(2.5M_{max} + 3M_A + 4M_B + 3M_C) \qquad (19.5)$$

where

M_{max}	=	absolute value of maximum moment in the unbraced beam segment
M_A	=	absolute value of moment at quarter point of the unbraced beam segment
M_B	=	absolute value of moment at centerline of the unbraced beam segment
M_C	=	absolute value of moment at three-quarter point of the unbraced beam segment

When the slenderness ratio, λ, is less than or equal to $\lambda_p (= 300/\sqrt{F_{yf}})$, the flange will yield before it buckles, and $F_{cr} = F_{yf}$. When the flange slenderness ratio, λ, is greater than λ_p and smaller than or equal to λ_r, inelastic buckling will occur and a straight line equation must be adopted between yielding ($\lambda \leq \lambda_p$) and elastic buckling ($\lambda > \lambda_r$) to calculate the inelastic buckling stress, namely

$$F_{cr} = C_b F_{yf} \left[1 - 0.5(\lambda - \lambda_p)/(\lambda_r - \lambda_p) \right] \leq F_{yf} \tag{19.6}$$

19.2.3 Torsional Buckling

If the outstanding width-to-thickness ratio of the flange is high, torsional buckling may occur. If one neglects any restraint provided by the web to the flange rotation, then the flange can be treated as a long plate, which is simply supported (hinged) at one edge and free at the other, subjected to uniaxial compression in the longitudinal direction. The elastic buckling stress under these conditions can be calculated from

$$F_{cr} = k_c \pi^2 E/12(1 - \mu^2)\lambda^2 \tag{19.7}$$

where k_c is a buckling coefficient equal to 0.425 for a long plate simply supported and free at its longitudinal edges; λ is equal to $b_f/2t_f$; b_f and t_f are the flange width and thickness, respectively; and E and μ are Young's modulus of elasticity and the Poisson ratio, respectively.

The AISC specification adopted Equation 19.7, rounding $\pi^2 E/12(1 - \mu^2)$ to 26,200 and assuming $k_c = 4\sqrt{h/t_w}$, where $0.35 \leq k_c \geq 0.763$. Furthermore, to allow for nonuniform bending, the buckling stress has to be multiplied by C_b, given by Equation 19.5. Elastic torsional buckling of the compression flange will occur if λ is greater than $\lambda_r (= 230/\sqrt{F_{yf}/k_c})$. When λ is less than or equal to $\lambda_p (= 65/\sqrt{F_{yf}})$, the flange will yield before it buckles, and $F_{cr} = F_{yf}$. When $\lambda_p < \lambda \leq \lambda_r$, inelastic buckling will occur and Equation 19.6 shall be used.

19.2.4 Compression Flange of a Box Girder

Lateral-torsional buckling does not govern the design of the compression flange in a box girder. Unstiffened flanges and flanges stiffened with longitudinal stiffeners can be treated as long plates supported along their longitudinal edges and subjected to uniaxial compression. In the **AASHTO** (American Association of State Highway and Transportation Officials) specification [1], the nominal flexural stress, F_n, for the compression flange is calculated as follows:
If $w/t \leq 0.57\sqrt{kE/F_{yf}}$, then the flange will yield before it buckles, and

$$F_n = F_{yf} \tag{19.8}$$

If $w/t > 1.23\sqrt{kE/F_{yf}}$, then the flange will elastically buckle, and

$$F_n = k\pi^2 E/12(1 - \mu^2)(w/t)^2$$

or

$$F_n = 26{,}200\, k(t/w)^2 \tag{19.9}$$

If $0.57\sqrt{kE/F_{yf}} < w/t \leq 1.23\sqrt{kE/F_{yf}}$, then the flange buckles inelastically, and

$$F_n = 0.592 F_{yf} [1 + 0.687 \sin(c\pi/2)] \tag{19.10}$$

In Equations 19.8 to 19.10,

w = the spacing between the longitudinal stiffeners, or the flange width for unstiffened flanges
c = $\left[1.23 - (w/t)\sqrt{F_{yf}/kE} \right]/0.66$
k = $\left(8I_s/wt^3 \right)^{1/3} \leq 4.0$, for $n = 1$
k = $\left(14.3 I_s/wt^3 n^4 \right)^{1/3} \leq 4.0$, for $n = 2, 3, 4,$ or 5

n =   number of equally spaced longitudinal stiffeners

I_s =   the moment of inertia of the longitudinal stiffener about an axis parallel to the flange and taken at the base of the stiffener

The nominal stress, F_n, shall be reduced for hybrid girders to account for the nonlinear variation of stresses caused by yielding of the lower strength steel in the web of a hybrid girder. Furthermore, another reduction is made for slender webs to account for the nonlinear variation of stresses caused by local bend buckling of the web. The reduction factors for hybrid girders and slender webs will be given in Section 19.3. The longitudinal stiffeners shall be equally spaced across the compression flange width and shall satisfy the following requirements [1].

The projecting width, b_s, of the stiffener shall satisfy:

$$b_s \leq 0.48t_s\sqrt{E/F_{yc}} \tag{19.11}$$

where

t_s =   thickness of the stiffener

F_{yc} =   specified minimum yield strength of the compression flange

The moment of inertia, I_s, of each stiffener about an axis parallel to the flange and taken at the base of the stiffener shall satisfy:

$$I_s \geq \Psi w t^3 \tag{19.12}$$

where

Ψ = $0.125k^3$ for $n = 1$

        = $0.07k^3n^4$ for $n = 2, 3, 4,$ or 5

n =   number of equally spaced longitudinal compression flange stiffeners

w =   larger of the width of compression flange between longitudinal stiffeners or the distance from a web to the nearest longitudinal stiffener

t =   compression flange thickness

k =   buckling coefficient as defined in connection with Equations 19.8 to 19.10

The presence of the in-plane compression in the flange magnifies the deflection and stresses in the flange from local bending due to traffic loading. The amplification factor, $1/(1 - \sigma_a/\sigma_{cr})$, can be used to increase the deflections and stresses due to local bending; where σ_a and σ_{cr} are the in-plane compressive and buckling stresses, respectively.

19.3   Web Buckling Due to In-Plane Bending

Buckling of the web due to in-plane bending does not exhaust its capacity; however, the distribution of the compressive bending stress changes in the post-buckling range and the web becomes less efficient. Only part of the compression portion of the web can be assumed effective after buckling. A reduction in the girder moment capacity to account for the web bend buckling can be used, and the following reduction factor [4] has been suggested:

$$R = 1 - 0.0005(A_w/A_f)(h/t - 5.7\sqrt{E/F_{yw}}) \tag{19.13}$$

It must be noted that when $h/t = 5.7\sqrt{E/F_{yw}}$, the web will yield before it buckles and there is no reduction in the moment capacity. This can be determined by equating the bend buckling stress to the web yield stress, i.e.,

$$k\pi^2 E/\left[12(1 - \mu^2)(h/t)^2\right] = F_{yw} \tag{19.14}$$

where k is the web bend buckling coefficient, which is equal to 23.9 if the flange simply supports the web and 39.6 if one assumes that the flange provides full fixity; the 5.7 in Equation 19.13 is based on a k value of 36.

The AISC specification replaces the reduction factor given in Equation 19.13 by

$$R_{PG} = 1 - [a_r/(1,200 + 300a_r)] \, (h/t - 970/\sqrt{F_{cr}}) \tag{19.15}$$

where a_r is equal to A_w/A_f and 970 is equal to $5.7\sqrt{29000}$; it must be noted that the yield stress in Equation 19.13 was replaced by the flange critical buckling stress, which can be equal to or less than the yield stress as discussed earlier. It must also be noted that in homogeneous girders the yield stresses of the web and flange materials are equal; in hybrid girders another reduction factor, R_e, [39] shall be used:

$$R_e = \left[12 + a_r(3m - m^3)\right]/(12 + 2a_r) \tag{19.16}$$

where a_r is equal to the ratio of the web area to the compression flange area (≤ 10) and m is the ratio of the web yield stress to the flange yield or buckling stress.

19.4 Nominal Moment Strength

The nominal moment strength can be calculated as follows.
  Based on tension flange yielding:

$$M_n = S_{xt} R_e F_{yt} \tag{19.17a}$$

or
  Based on compression flange buckling:

$$M_n = S_{xc} R_{PG} R_e F_{cr} \tag{19.17b}$$

where S_{xc} and S_{xt} are the section moduli referred to the compression and tension flanges, respectively; F_{yt} is the tension flange yield stress; F_{cr} is the compression flange buckling stress calculated according to Section 19.2; R_{PG} is the reduction factor calculated using Equation 19.15; and R_e is a reduction factor to be used in the case of hybrid girders and can be calculated using Equation 19.16.

19.5 Web Longitudinal Stiffeners for Bending Design

Longitudinal stiffeners can increase the bending strength of plate girders. This increase is due to the control of the web lateral deflection, which increases its flexural stress capacity. The presence of the stiffener also improves the bending resistance of the flange due to a greater web restraint. If one longitudinal stiffener is used, its optimum location is 0.20 times the web depth from the compression flange. In this case the web plate elastic bend buckling stress increases more than five times that without the stiffener. Tests [8] showed that an adequately proportioned longitudinal stiffener at 0.2h from the compression flange eliminates bend buckling in girders with web slenderness, h/t, as large as 450. Girders with larger slenderness will require two or more longitudinal stiffeners to eliminate web bend buckling. It must be noted that the increase in the bending strength of a longitudinally stiffened thin-web girder is usually small because the web contribution to the bending strength is small. However, longitudinal stiffeners can be important in a girder subjected to repeated loads because they reduce or eliminate the out-of-plane bending of the web, which increases resistance to fatigue cracking at the web-to-flange juncture and allows more slender webs to be used [42].
  The AISC specification does not address longitudinal stiffeners; on the other hand, the AASHTO specification states that longitudinal stiffeners should consist of either a plate welded longitudinally to one side of the web or a bolted angle, and shall be located at a distance of 0.4 D_c from the inner

surface of the compression flange, where D_c is the depth of the web in compression at the section with the maximum compressive flexural stress. Continuous longitudinal stiffeners placed on the opposite side of the web from the transverse intermediate stiffeners, as shown in Figure 19.3, are preferred. If longitudinal and transverse stiffeners must be placed on the same side of the web, it

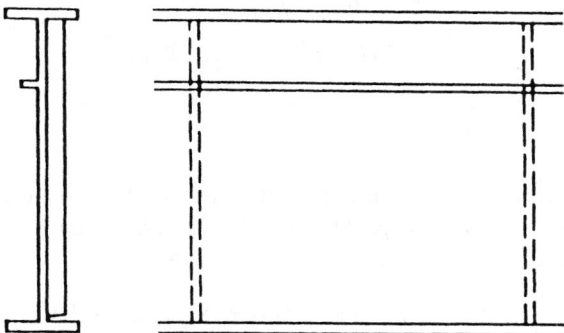

Figure 19.3   Longitudinal stiffener for flexure.

is preferable that the longitudinal stiffener not be interrupted for the transverse stiffener. Where the transverse stiffeners are interrupted, the interruptions must be carefully detailed with respect to fatigue.

To prevent local buckling, the projecting width, b_s of the stiffener shall satisfy the requirements of Equation 19.11. The section properties of the stiffener shall be based on an effective section consisting of the stiffener and a centrally located strip of the web not exceeding 18 times the web thickness. The moment of inertia of the longitudinal stiffener and the effective web strip about the edge in contact with the web, I_s, and the corresponding radius of gyration, r_s, shall satisfy the following requirements:

$$I_s \geq ht_w^3 \left[2.4(a/h)^2 - 0.13 \right] \tag{19.18}$$

and

$$r_s \geq 0.234a\sqrt{F_{yc}/E} \tag{19.19}$$

where

a =   spacing between transverse stiffeners

19.6   Ultimate Shear Capacity of the Web

As stated earlier, in most design codes buckling is not used as a basis for design. Minimum slenderness ratios, however, are specified to control out-of-plane deflection of the web. These ratios are derived to give a small factor of safety against buckling, which is conservative and in some cases extravagant.

Before the web reaches its theoretical **buckling load** the shear is taken by beam action and the shear stress can be resolved into diagonal tension and compression. After buckling, the diagonal compression ceases to increase and any additional loads will be carried by the diagonal tension. In very thin webs with stiff boundaries, the plate buckling load is very small and can be ignored and the shear is carried by a complete diagonal tension field action [41]. In welded plate and box girders the web is not very slender and the flanges are not very stiff; in such a case the shear is carried by beam action as well as incomplete tension field action.

Based on test results, the analytical model shown in Figure 19.4 can be used to calculate the ultimate shear capacity of the web of a welded plate girder [5]. The flanges are assumed to be too

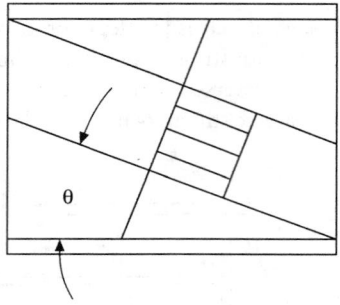

Figure 19.4   Tension field model by Basler.

flexible to support the vertical component from the tension field. The inclination and width of the tension field were defined by the angle Θ, which is chosen to maximize the shear strength. The ultimate shear capacity of the web, V_u, can be calculated from

$$V_u = \left[\tau_{cr} + 0.5\sigma_{yw}(1 - \tau_{cr}/\tau_{yw}) \sin \Theta_d\right] A_w \tag{19.20}$$

where
τ_{cr} = critical buckling stress in shear
τ_{yw} = yield stress in shear
σ_{yw} = web yield stress
Θ_d = angle of panel diagonal with flange
A_w = area of the web

   In Equation 19.20, if $\tau_{cr} \geq 0.8\tau_{yw}$, the buckling will be inelastic and

$$\tau_{cr} = \tau_{cri} = \sqrt{0.8\tau_{cr}\tau_{yw}} \tag{19.21}$$

   It was shown later [23] that Equation 19.20 gives the shear strength for a complete tension field instead of the limited band shown in Figure 19.4. The results obtained from the formula, however, were in good agreement with the test results, and the formula was adopted in the AISC specification.

   Many variations of this incomplete tension field model have been developed; are view can be found in the SSRC *Guide to Stability Design Criteria for Metal Structures* [22]. The model shown in Figure 19.5 [36, 38] gives better results and has been adopted in codes in Europe. In the model shown in Figure 19.5, near failure the tensile membrane stress, together with the buckling stress, causes yielding, and failure occurs when hinges form in the flanges to produce a combined mechanism that includes the yield zone ABCD. The vertical component of the tension field is added to the shear at buckling and combined with the frame action shear to calculate the ultimate shear strength. The ultimate shear strength is determined by adding the shear at buckling, the vertical component of the tension field, and the frame action shear, and is given by

$$V_u = \tau_{cr} A_w + \sigma_t A_w \left[(2c/h) + \cot \Theta - \cot \Theta_d\right] \sin^2 \Theta + 4M_p/c \tag{19.22}$$

where
σ_t = $-1.5\tau_{cr} \sin 2\Theta + \sqrt{\sigma_{yw}^2 + (2.25 \sin^2 2\Theta - 3)\tau_{cr}^2}$
c = $(2/\sin \Theta)\sqrt{Mp/(\sigma_t t_w)} \quad 0 \leq c \leq a$
M_p = plastic moment capacity of the flange with an effective depth of the web, b_e, given by
b_e = $30t_w[1 - 2(\tau_{cr}/\tau_{yw})]$

   where $(\tau_{cr}/\tau_{yw}) \leq 0.5$; reduction in M_p due to the effect of the flange axial compression shall be considered and when $\tau_{cr} > 0.8\tau_{yw}$,

$$\tau_{cr} = \tau_{cri} = \tau_{yw}[1 - 0.16(\tau_{yw}/\tau_{cr})]$$

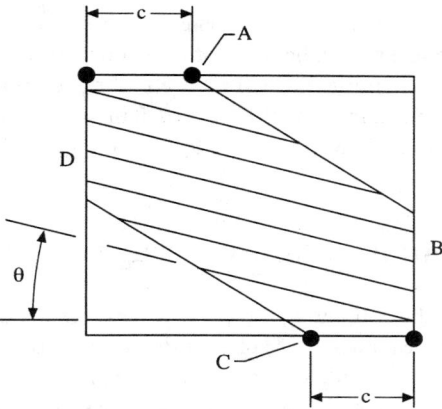

Figure 19.5 Tension field model by Rockey et al.

The maximum value of V_u must be found by trial; Θ is the only independent variable in Equation 19.22, and the optimum is not difficult to determine by trial since it is between $\Theta_d/2$ and 45 degrees, and V_u is not sensitive to small changes from the optimum Θ.

Recently [2, 33], it has been argued that the post-buckling strength arises not due to a diagonal tension field action, but by redistribution of shear stresses and local yielding in shear along the boundaries. A case in between is to model the web panel as a diagonal tension strip anchored by corner zones carrying shear stresses and act as gussets connecting the diagonal tension strip to the vertical stiffeners which are in compression [9]. On the basis of test results, it can be concluded that unstiffened webs possess a considerable reserve of post-buckling strength [16, 24]. The incomplete diagonal tension field approach, however, is only reasonably accurate up to a maximum aspect ratio (stiffeners spacing: web depth) equal to 6. Research is required to develop an appropriate method of predicting the post-buckling strength of unstiffened girders.

In the AISC specification, the shear capacity of a plate girder web can be calculated, using the model shown in Figure 19.4, as follows:

For $h/t_w \leq 187\sqrt{k_v/F_{yw}}$, the web yields before buckling, and

$$V_n = 0.6A_w F_{yw} \tag{19.23}$$

For $h/t_w > 187\sqrt{k_v/F_{yw}}$, the web will buckle and a tension field will develop, and

$$V_n = 0.6A_w F_{yw}\left[C_v + (1-C_v)/1.15\sqrt{1+(a/h)^2}\right] \tag{19.24}$$

where

k_v = buckling coefficient = $5 + 5/(a/h)^2$
    = 5, if $(a/h) > 3$ or $[260/(h/t_w)]^2$
C_v = ratio of the web buckling stress to the shear yield stress of the web
    = $187\sqrt{k_v/F_{yw}}/(h/t_w)$, for $187\sqrt{k_v/F_{yw}} \leq h/t_w \leq 234\sqrt{k_v/F_{yw}}$
    = $44{,}000\,k_v/[(h/t_w)^2 F_{yw}]$, for $h/t_w > 234\sqrt{k_v/F_{yw}}$

It must be noted that, in the above, the web buckling is elastic when $h/t_w > 234\sqrt{k_v/F_{yw}}$.

The AISC specification does not permit the consideration of tension field action in end panels, hybrid and web-tapered plate girders, and when a/h exceeds 3.0 or $[260/(h/tw)]^2$. This is contrary to the fact that a tension field can develop in all these cases; however, little or no research was conducted. Furthermore, tension field can be considered for end panels if the end stiffener is designed for this purpose. When neglecting the tension field action, the nominal shear capacity can be calculated from

$$V_n = 0.6A_w F_{yw} C_v \tag{19.25}$$

Care must be exercised in applying the tension field models developed primarily for welded plate girders to the webs of a box girder. The thin flange of a box girder can provide very little or no resistance against movements in the plane of the web. If the web of a box girder is transversely stiffened and if the model shown in Figure 19.4 is used, it may overpredict the web strength. Hence, it is advisable to use the model shown in Figure 19.5, assuming the plastic moment capacity of the flange to be negligible.

19.7 Web Stiffeners for Shear Design

Transverse stiffeners must be stiff enough to prevent out-of-plane displacement along the panel boundaries in computing shear buckling of plate girder webs. To provide the out-of-plane support an equation, developed for an infinitely long web with simply supported edges and equally spaced stiffeners, to calculate the required moment of inertia of the stiffeners, I_s, namely for $a \leq h$,

$$I_s = 2.5ht_w^3[(h/a) - 0.7(a/h)] \tag{19.26}$$

The AASHTO formula for load-factor design is

$$I_s = Jat_w^3 \tag{19.27}$$

where
$J \quad = \quad [2.5(h/a)^2 - 2] \geq 0.5$

Equation 19.27 is the same as Equation 19.26 except that the coefficient of (a/h) in the second term between brackets is 0.8 instead of 0.7. Equation 19.27 was adopted by the AISC specification as well. The moment of inertia of the transverse stiffener shall be taken about the edge in contact with the web for single-sided stiffeners and about the mid-thickness of the web for double-sided stiffeners.

To prevent local buckling of transverse stiffeners, the width, b_s, of each projecting stiffener element shall satisfy the requirements of Equation 19.11 using the yield stress of the stiffener material rather than that of the flange, as in Equation 19.11. Furthermore, b_s shall also satisfy the following requirements:

$$16.0t_s \geq b_s \leq 0.25b_f \tag{19.28}$$

where
$b_f \quad = \quad$ the full width of the flange

Transverse stiffeners shall consist of plates or angles welded or bolted to either one or both sides of the web. Stiffeners that are not used as connection plates shall be a tight fit at the compression flange, but need not be in bearing with the tension flange. The distance between the end of the web-to-stiffener weld and the near edge of the web-to-flange fillet weld shall not be less than $4t_w$ or more than $6t_w$. Stiffeners used as connecting plates for diaphragms or cross-frames shall be connected by welding or bolting to both flanges.

In girders with longitudinal stiffeners the transverse stiffener must also support the longitudinal stiffener as it forces a horizontal node in the bend buckling configuration of the web. In such a case it is recommended that the transverse stiffener section modulus, S_T, be equal to $S_L(h/a)$, where S_L is the section modulus of the longitudinal stiffener and h and a are the web depth and the spacing between the transverse stiffeners, respectively. In the AASHTO specification, the moment of inertia of transverse stiffeners used in conjunction with longitudinal stiffeners shall also satisfy

$$I_t \geq (b_t/b_l)(h/3a)I_l \tag{19.29}$$

where b_t and b_l = projecting width of transverse and longitudinal stiffeners, respectively, and I_t and I_l = moment of inertia of transverse and longitudinal stiffeners, respectively.

Transverse stiffeners in girders that rely on a tension field must also be designed for their role in the development of the diagonal tension. In this situation they are compression members, and so must be checked for local buckling. Furthermore, they must have cross-sectional area adequate for the axial force that develops. The axial force, F_s, can be calculated based on the analytical model [5] shown in Figure 19.4, and is given by

$$F_s = 0.5 F_{yw} a t_w \left[1 - \left(\tau_{cr}/\tau_{yw}\right)\right]\left(1 - \cos \Theta_d\right) \tag{19.30}$$

The AISC and AASHTO specifications assume that a width of the web equal to $18t_w$ acts with the stiffener and give the following formula for the cross-sectional area, A_s, of the stiffeners:

$$A_s \geq \left[0.15 B h t_w (1 - C_v) V_u / 0.9 V_n - 18 t_w^2\right] (F_{yw}/F_{ys}) \tag{19.31}$$

where the new notations are

$0.9 V_u$ = shear due to bf factored loads
B = 1.0 for double-sided stiffeners
= 1.8 for single-sided angle stiffeners
= 2.4 for single-sided plate stiffeners

If longitudinally stiffened girders are used, h in Equation 19.31 shall be taken as the depth of the web, since the tension field will occur between the flanges and the transverse stiffeners.

The optimum location of a longitudinal stiffener that is used to increase resistance to shear buckling is at the web mid-depth. In this case the two subpanels buckle simultaneously and the increase in the critical stress is substantial. To obtain the tension field shear resistance one can assume that only one tension field is developed between the flanges and transverse stiffeners even if longitudinal stiffeners are used.

19.8   Flexure-Shear Interaction

The shear capacity of a girder is independent of bending as long as the applied moment is less than the moment that can be taken by the flanges alone, $M_f = \sigma_{yf} A_f h$; any larger moment must be resisted in part by the web, which reduces the shear capacity of the girder. When the girder is subjected to pure bending with no shear, the maximum moment capacity is equal to the plastic moment capacity, M_p, due to yielding of the girder's entire cross-section. In view of the tension field action and based on test results [3], a simple conservative interaction equation is given by

$$(V/V_u)^2 + (M - M_f)/(M_p - M_f) = 1 \tag{19.32}$$

where M and V are the applied moment and shear, respectively.

In the AISC specification, plate girders with webs designed for tension field action and when the ultimate shear, V_u, is between 0.54 and $0.9V_n$, and the ultimate moment, M_u, is between 0.675 and $0.9M_n$, where V_n and M_n are the nominal shear and moment capacities in absence of one another; the following interaction equation must be satisfied,

$$M_u/0.9M_n + 0.625(V_u/0.9V_n) \leq 1.375 \tag{19.33}$$

19.9   Steel Plate Shear Walls

Although the post-buckling behavior of plates under monotonic loads has been under investigation for more than half a century, post-buckling strength of plates under cyclic loading has not been investigated until recently [7]. The results of this investigation indicate that plates can be subjected to few reversed cycles of loading in the post-buckling domain, without damage. In steel plate shear walls, the boundary members are stiff and the plate is relatively thin; in such cases a complete tension field can be developed. The plate can be modeled as a series of tension bars inclined at an angle, ϕ [27]. The angle of inclination, ϕ, is a function of the panel length and height, the plate thickness, the cross-sectional areas of the surrounding beams and columns, and the moment of inertia of the columns. It can be determined by applying the principle of least work and is given by

$$\tan^4 \alpha = \left[(2/t_w L) + (1/A_c) \right] / \left[(2/t_w L) + (2h/A_b L) + (h^4/180 I_c L^2) \right] \qquad (19.34)$$

where

α	=	angle of inclination of tension field with the vertical axis
L	=	panel length
h	=	panel height
t_w	=	wall thickness
A_b	=	cross-sectional area of beam
A_c	=	cross-sectional area of column
I_c	=	moment of inertia of column about axis perpendicular to the plane of the wall

Although this model can predict the ultimate capacity to a reasonable degree of accuracy it cannot depict the load-deflection characteristics to the same degree of accuracy. Based on test results and finite element analysis [17, 18], the stresses in the inclined tensile plate strips are not uniform but are higher near the supporting boundaries than the center of the plate, and yielding of these strips starts near their ends and propagates toward the midlength. The following method can be used to calculate the ultimate capacity and determine the load-deflection characteristics of a thin-steel-plate shear wall.

The plate in the shear wall is replaced by a series of truss elements in the diagonal tension direction, as shown in Figure 19.6. A minimum of four truss members shall be used to replace the plate panel

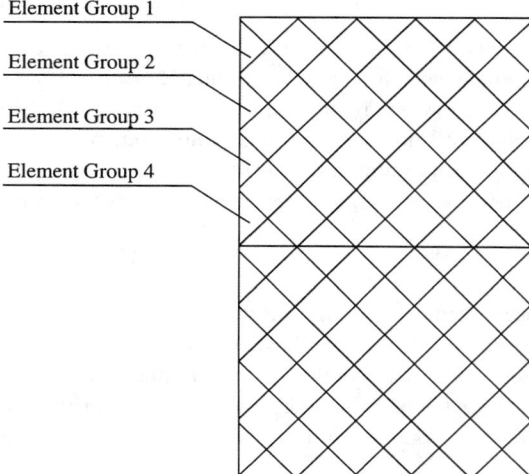

Figure 19.6   Steel plate shear wall analytical model.

in order to depict the panel behavior to a reasonable degree of accuracy; however, six members are recommended. The stress-strain relationship for the truss elements shall be assumed to be bilinearly elastic perfectly plastic, as shown in Figure 19.7, where E is Young's modulus of elasticity and σ_y is the tensile yield stress of the plate material. In Figure 19.7 the first slope represents the elastic

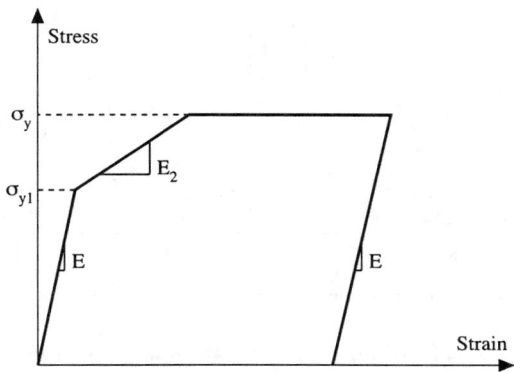

Figure 19.7   Stress-strain relationship for truss element.

response and the second represents the reduced stiffness caused by partial yielding; σ_{y1} and E_2 can be determined using a semi-empirical approach for welded as well as bolted shear walls, and can be calculated using the following equations:

$$\sigma_{y1} = (0.423 + 0.816 b_e/L)\sigma_y \tag{19.35}$$

where

$$b_e = \left[14.6\pi^2 E/12(1 - v^2)\tau_y\right]^{0.5} t \tag{19.36}$$

L = length of the strip
t = thickness of the plate
and

$$E_2 = \left[(\sigma_{y2} - \sigma_{y1})(3 - \sigma_{y1})/\sigma_{y2}(2 + \alpha)\right] E \tag{19.37}$$

where α is the ratio between the plastic strain, ε_p, and the strain at the initiation of yielding, ε_y. This ratio is in the range of 5 to 20, depending on the stiffness of the columns relative to the thickness of the plate; a value of 10 can be used in design. In the derivation of Equations 19.35 and 19.36 the inclination angle of the equivalent truss elements was assumed to be 45 degrees. The angle of inclination is usually in the range of 38 to 43 degrees and the effect of this assumption on the overall behavior of the wall is negligible.

In order to define the load displacement relationship of the truss elements in a bolted shear wall, the parameters P_{y1}, K_1, P_y, and K_2 shown in Figure 19.8 need to be calculated. For a bolted plate, the initial yielding can occur when the plate at the bolted connection starts slipping or when it locally yields near the boundaries as in the welded plate. The load due to slippage is controlled by the friction coefficient between the connected surfaces and the normal force applied by the bolts. In case the bolts were pre-tensioned to 70% of their ultimate tensile strength, slip will occur at a load equal to

$$P_{y1} = n(0.7\mu F_u^b A^b) \tag{19.38}$$

where n is the number of bolts at one end of the truss element; F_u^b and A^b are the ultimate tensile strength and cross-section area of the bolt, respectively; and μ is the friction coefficient between the connected surfaces. The load that causes initial yielding at the ends of the strip is the same as for the welded plate, and can be obtained from Equation 19.35 by multiplying the stress, σ_{y1}, by the strip cross-sectional area, A^p.

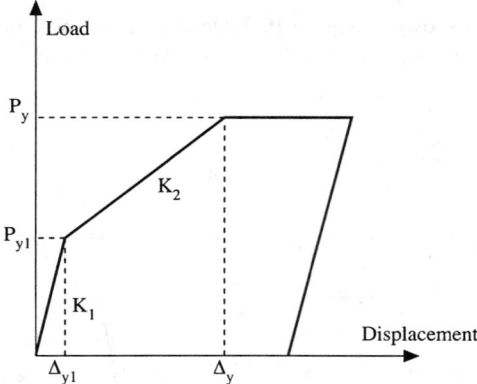

Figure 19.8   Load-displacement relationship for truss element.

The usable load is the smaller of the loads that causes slippage or initial yielding at the ends of the strip. The initial stiffness of the equivalent truss element can be calculated as follows:

$$K_1 = EA^P/L \tag{19.39}$$

The ultimate load is controlled by the total yielding of the plate strip, tearing at the bolt holes, or shearing of the bolts. The smallest failure load is the controlling ultimate capacity of the truss element. The ultimate load due to the strip yielding along its entire length,

$$P'_y = \sigma_y A^P \tag{19.40}$$

The bolted shear wall should be designed such that plate yielding controls.

Tearing at the bolt holes will occur when the edge distance is small or the bolt spacing is large, which should be avoided in design because it is a brittle failure. The tearing load, denoted as P''_y, can be calculated using the following formula:

$$P''_y = 1.4 n \sigma_u (L_e - D/2) t \tag{19.41}$$

where n is the number of bolts at the end of the truss element, σ_u is the ultimate tensile strength of the plate material, L_e is the distance from the edge of the plate to the centroid of the bolt hole, D is the diameter of the bolt, and t is the thickness of the plate. Note that in the above formula the ultimate strength of the plate material in shear was assumed to be 0.7, its ultimate strength in tension.

The shear failure of the bolts can occur if the shear strength of the bolts is small or the spacing between them is large. In such case the ultimate capacity of the truss element, which is denoted as P'''_y, can be calculated from one of the following formulas:

$$P'''_y = 0.45 n F^b_u A^b \tag{19.42a}$$
$$P'''_y = 0.60 n F^b_u A^b \tag{19.42b}$$

where n is the number of bolts and F^b_u and A^b are the ultimate strength and the gross cross-section area of the bolts, respectively. Equation 19.42a is used if the shear plane is within the threaded part of the bolt and Equation 19.42b is used if the shear plane is not within the threaded part of the bolt.

The ultimate load of the truss element is the smallest value of the plate total yielding capacity, the tearing capacity, and the bolt shearing capacity, i.e.,

$$P_y = \min(P'_y, P''_y, P'''_y) \tag{19.43}$$

As stated earlier, the plate yielding shall control and the designer must ensure that P_y' is the smallest of the three values.

As can be seen in Figure 19.8, in order to define the stiffness, K_2, the displacement of the truss element when the load reaches the ultimate capacity, P_y, needs to be determined. This ultimate displacement includes the stretching of the element as well as the slippage and the bearing deformation of the plate and the bolts at the connections. The plate strip represented by the truss element is stretched under load; the elongation includes both elastic and plastic deformations. As discussed earlier, due to the nonuniform strain distribution along the length of the strip, the plastic deformation will occur mostly near the ends. If one assumes that the strain distribution along the length of the strip is a second-degree parabola, and using a plastic deformation factor a, the elongation of the truss element due to the plate elastic and plastic deformations can be calculated using the following equation:

$$\Delta_{\text{def}} = (\sigma_y/3E)(2 + \alpha)L \qquad (19.44)$$

The elongation of the truss element due to slippage can be approximated by two times the hole clearance, taking into consideration the slippage at both ends of the element [26]. The local deformation at the bolt holes includes the effects of shearing, bending, and bearing deformations of the fastener as well as local deformation of the connected plates, and can be taken as 0.2 times the bolt diameter [21]. Having defined the ultimate elongation of the truss element, its reduced stiffness after slippage and/or initial yielding can be obtained using the following equation:

$$K_2 = (P_y - P_{y1})/\left[0.125 + 0.2D + \sigma_y(\alpha - 1)L/(3E)\right] \qquad (19.45)$$

19.10 In-Plane Compressive Edge Loading

Webs of plate and box girders and load-bearing diaphragms in box girders can be subjected to local in-plane compressive loads. Vertical (transverse) stiffeners can be provided at the location of the load to prevent web crippling; however, this is not always possible, such as in the case of a moving load, and it involves higher cost. Failure of the web under this loading is always due to crippling [10], as shown in Figure 19.9; in thin webs crippling occurs before yielding of the web and in stocky webs after yielding. The formula in the AISC specification predicts the crippling load, P_{cr}, to a reasonable

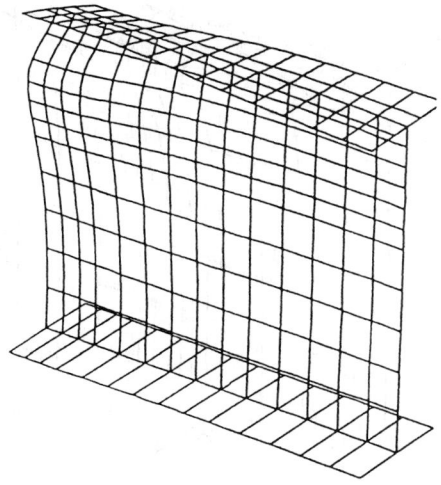

Figure 19.9   Deformed shape under in-plane edge loading (only half of the beam is shown).

degree of accuracy [37]; this formula is

$$P_{cr} = 135t_w^2 \left[1 + 3(N/d)\left(t_w/t_f\right)^{1.5}\right]\left(F_{yw}t_f/t_w\right)^{0.5} \tag{19.46}$$

The formula given by Equation 19.46 is applicable if the load is applied at a distance not less than half the member's depth from its end; if the load is at a distance less than half the member's depth, the following formulae shall be used [14]:

$$\text{For } N/d \leq 0.2, \ P_{cr} = 68t_w^2 \left[1 + 3(N/d)\left(t_w/t_f\right)^{1.5}\right]\left(F_{yw}t_f/t_w\right)^{0.5} \tag{19.47a}$$

$$\text{For } N/d > 0.2, \ P_{cr} = 68t_w^2 \left[1 + (4N/d - 0.2)(t_w/t_f)^{1.5}\right]\left(F_{yw}t_f/t_w\right)^{0.5} \tag{19.47b}$$

In addition to web crippling, the AISC specification requires a web yielding check; furthermore, when the relative lateral movement between the loaded compression flange and the tension flange is not restrained at the point of load application, sidesway buckling must be checked.

19.11 Eccentric Edge Loading

Eccentricities in loading with respect to the plane of the web are unavoidable, and it was found that there is a reduction in the web capacity due to the presence of an eccentricity [13, 14]; for example, in one case, an eccentricity of 0.5 in. reduced the web ultimate capacity to about half its capacity under in-plane load. Furthermore, it was found that the effect of the load eccentricity in reducing the ultimate capacity decreases as the ratio of the flange-to-web thickness increases. A deformed beam subjected to eccentric load near failure is shown in Figure 19.10. Web strength reduction factors for various eccentricities as a function of the flange width and for various flange-to-web thickness ratios are given in Figure 19.11.

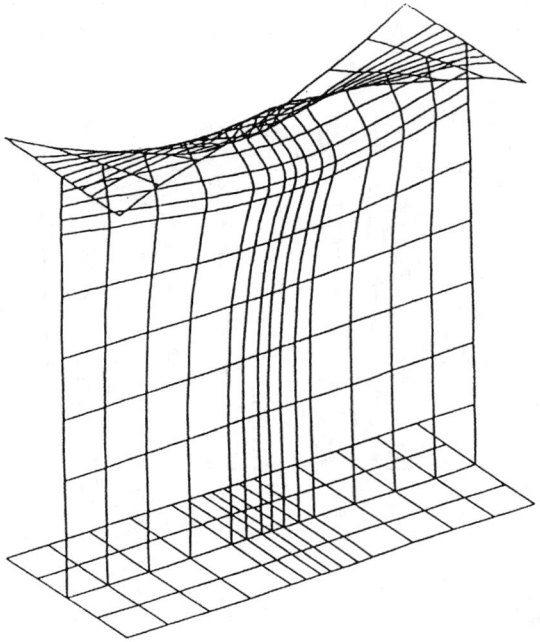

Figure 19.10   Deformed shape under eccentric edge loading.

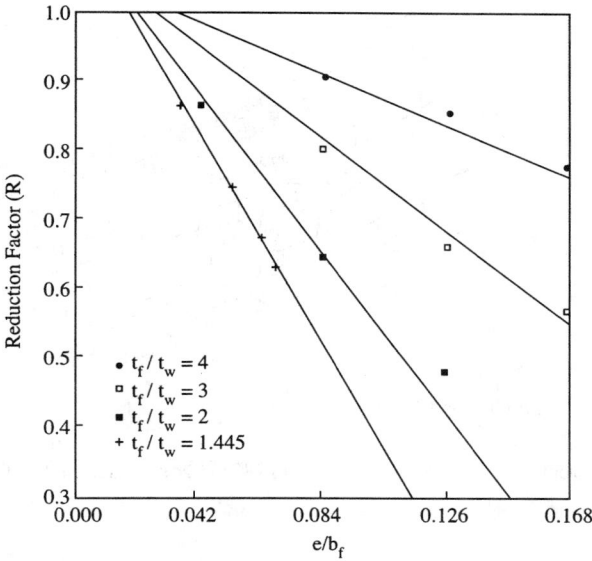

Figure 19.11 Eccentricity reduction factor.

The failure mechanism in the case of eccentric loading is different from that for in-plane loading. The flange twisting moment acting at the web-flange intersection can cause failure due to bending rather than crippling of the web, if the eccentricity is large enough. In most cases, however, the failure mode is due to a combination of web bending and crippling. Failure mechanisms were developed, and formulas to calculate the ultimate capacity of the web under eccentric edge loading were derived [15]. Currently, the AISC specification does not address the effect of the eccentricity on the reduction of the web crippling load. Eccentricities can also arise due to moments applied to the top flange in addition to vertical loads. An example would be a beam resting on the top flange of another beam and the two flanges are welded together. Rotation of the supported beam will impose a twisting moment in the flange of the supporting beam and bending of its web, which will reduce its crippling load.

19.12 Load-Bearing Stiffeners

Webs of girders are often strengthened with transverse stiffeners at points of concentrated loads and over intermediate and end supports. The AISC specification requires that these stiffeners be double sided, extend at least one-half the beam depth, and either bear on or be welded to the loaded flange. The specification, further, requires that they be designed as axially loaded members with an effective length equals to 0.75 times the web depth; and a strip of the web, with a width equal to 25 times its thickness for intermediate stiffeners and 12 times the thickness for end stiffeners, shall be considered in calculating the geometric properties of the stiffener.

The failure, in cases where the stiffener depth is less than 75% of the depth of the web, can be due to crippling of the web below the stiffener [14, 15], as shown in Figure 19.12. The failure, otherwise, is due to global buckling of the stiffener, provided that the thickness of the stiffener is adequate to prevent local buckling. The optimum depth of the stiffener is 0.75 times the web depth. The AISC specification does not account for factors such as the stiffener depth and load eccentricity.

In box girders intermediate diaphragms are provided to limit cross-sectional deformation and load-bearing diaphragms are used at the supports to transfer loads to the bridge bearings. Di-

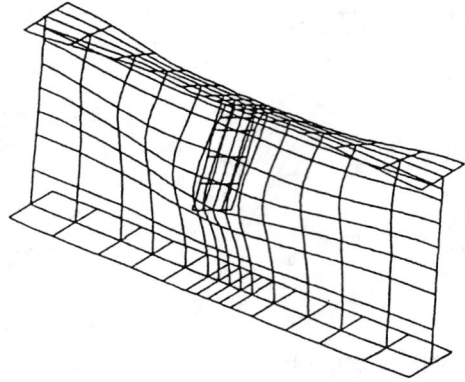

Figure 19.12   Web crippling below stiffener.

aphragm design is treated in the BS 5400: Part 3 (1983) [6] and discussed in Chapter 7 of the SSRC guide [22].

19.13   Web Openings

Openings are frequently encountered in the webs of plate and box girders. Research on the buckling and ultimate strength of plates with rectangular and circular openings subjected to in-plane loads has been performed by many investigators. The research has included reinforced and unreinforced openings. A theoretical method of predicting the ultimate capacity of slender webs containing circular and rectangular holes, and subjected to shear, has been developed [34, 35]. The solution is obtained by considering the equilibrium of two tension bands, one above and the other below the opening. These bands have been chosen to conform to the failure pattern observed in tested plate girders with holes. Experimental results showed that the method gives satisfactory and safe predictions. The calculated values were found to be between 5 and 30% below the test results.

Solutions for transversely stiffened webs subjected to shear and bending with centrally located holes are available [28, 30] and are applicable for webs with depth-to-thickness ratios of 120 to 360, panel aspect ratio between 0.7 and 1.5, hole depth greater than $1/10th$ of the web depth, and for circular, elongated circular, and rectangular holes.

19.14   Girders with Corrugated Webs

Corrugated webs can be used in an effort to decrease the weight of steel girders and reduce its fabrication cost. Studies have been conducted in Europe and Japan and girders with corrugated webs have been used in these countries [12]. The results of the studies indicate that the fatigue strength of girders with corrugated webs can be 50% higher compared to girders with flat stiffened webs. In addition to the improved fatigue life, the weight of girders with corrugated webs can be as much as 30 to 60% less than the weight of girders with flat webs and have the same capacity. Due to the weight savings, larger clear spans can be achieved. Beams and girders with corrugated webs are economical to use and can improve the aesthetics of the structure. Beams manufactured and used in Germany for buildings have a web thickness that varies between 2 and 5 mm, and the corresponding web height-to-thickness ratio is in the range of 150 to 260. The corrugated webs of two bridges built in France were 8 mm thick and the web height-to-thickness ratio was in the range of 220 to 375.

Failure in shear is usually due to buckling of the web and the failure in bending is due to yielding of the compression flange and its vertical buckling into the corrugated web, which buckles [19, 20]. The shear buckling mode is global for dense corrugation and local for course corrugation, as shown in Figure 19.13. The load-carrying capacity of the specimens drops after buckling, with some residual load-carrying capacity after failure. In the local buckling mode, the corrugated web acts as a series

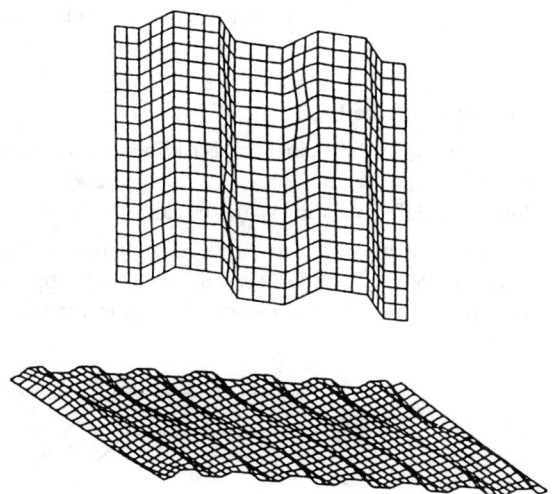

Figure 19.13   Local and global buckling.

of flat-plate subpanels that mutually support each other along their vertical (longer) edges and are supported by the flanges at their horizontal (shorter) edges. These flat-plate subpanels are subjected to shear, and the elastic buckling stress is given by

$$\tau_{cre} = k_s \left[\pi^2 E / 12(1 - \mu^2)(w/t)^2 \right] \qquad (19.48)$$

where
k_s = buckling coefficient, which is a function of the panel aspect ratio, h/w, and the boundary support conditions
h = the web depth
t = the web thickness
w = the flat-plate subpanel width — the horizontal or the inclined, whichever is bigger
E = Young's modulus of elasticity
μ = the Poisson ratio
   The buckling coefficient, ks, is given by
k_s = $5.34 + 2.31(w/h) - 3.44(w/h)^2 + 8.39(w/h)^3$, for the longer edges simply supported and the shorter edges clamped
k_s = $8.98 + 5.6(w/h)^2$, in the case where all edges are clamped
   An average local buckling stress, $\tau_{av} (= 0.5[\tau_{ssf} + \tau_{fx}])$, is recommended, and in the case of $\tau_{cre} \geq 0.8\tau_y$, inelastic buckling will occur and the inelastic buckling stress, τ_{cri}, can be calculated by $\tau_{cri} = (0.8 * \tau_{cre} * \tau_y)^{0.5}$, where $\tau_{cri} \leq \tau_y$.
   As stated earlier, the mode of failure is local and/or global buckling; when global buckling controls, the buckling stress can be calculated for the entire corrugated web panel, using orthotropic-plate buckling theory. The global elastic buckling stress, τ_{cre}, can be calculated from

$$\tau_{cre} = k_s \left[(D_x)^{0.25} (D_y)^{0.75} \right] / th^2 \qquad (19.49)$$

where

$$D_x = (q/s)Et^3/12$$

$$D_y = EI_y/q$$

$$I_y = 2bt(h_r/2)^2 + \{t(h_r)^3/6\sin\Theta\}$$

k_s = Buckling coefficient, equal to 31.6 for simply supported boundaries and 59.2 for clamped boundaries

t = corrugated plate thickness

b, h_r, q, s, and Θ are as shown in Figure 19.14.

In the aforementioned, when $\tau_{cre} \geq 0.8\tau_y$, inelastic buckling will occur and the inelastic buckling stress, τ_{cri}, can be calculated by $\tau_{cri} = (0.8 * \tau_{cre} * \tau_y)^{0.5}$, where $\tau cri \leq \tau_y$. For design, it is recommended that the local and global buckling values be calculated and the smaller value controls.

As stated earlier, the failure in bending is due to compression flange yielding and vertical buckling into the web, as shown in Figure 19.15. The failure is sudden, with no appreciable residual strength. The web offers negligible contribution to the moment carrying capacity of the beam, and for

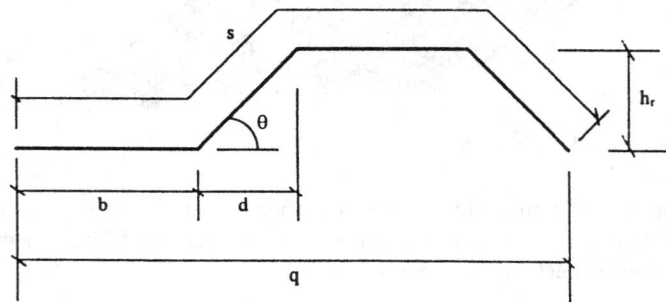

Figure 19.14   Dimensions of corrugation profile.

Figure 19.15   Bending failure of a beam with corrugate web.

design, the ultimate moment capacity can be calculated based on the flange yielding, ignoring any contribution from the web. The stresses in the web are equal to zero except near the flanges. This is because the corrugated web has no stiffness perpendicular to the direction of the corrugation, except for a very small distance which is adjacent to and restrained by the flanges, and the stresses are appreciable only within the horizontal folds of the corrugation.

It must be noted that the common practice is to fillet weld the web to the flanges from one side only; under static loading this welding detail was found to be adequate and there is no need to weld from both sides. Finally, the bracing requirements of the compression flange in beams and girders with corrugated webs are less severe compared to conventional beams and girders with flat webs. Lateral-torsional buckling of beams and girders with corrugated webs has been investigated [20].

19.15 Defining Terms

AASHTO: American Association of State Highway and Transportation Officials.

AISC: American Institute of Steel Construction.

Buckling load: The load at which a compressed element or member assumes a deflected position.

Effective width: Reduced flat width of a plate element due to buckling, the reduced width is termed the effective width.

Corrugated web: A web made of corrugated steel plates, where the corrugations are parallel to the depth of the girder.

Factored load: The nominal load multiplied by a load factor to account for unavoidable deviations of the actual load from the nominal load.

Limit state: A condition at which a structure or component becomes unsafe (strength limit state) or no longer useful for its intended function (serviceability limit state).

LRFD (Load and Resistance Factor Design): A method of proportioning structural components such that no applicable limit state is exceeded when the structure is subjected to all appropriate load combinations.

Shear wall: A wall in a building to carry lateral loads from wind and earthquakes.

Stress: Force per unit area.

Web crippling: Local buckling of the web plate under local loads.

Web slenderness ratio: The depth-to-thickness ratio of the web.

References

[1] American Association of State Highway and Transportation Officials. 1994. AASHTO LRFD Bridge Design Specifications, Washington, D.C.

[2] Ajam, W. and Marsh, C. 1991. Simple Model for Shear Capacity of Webs, *ASCE Struct. J.*, 117(2).

[3] Basler, K. 1961. Strength of Plate Girders Under Combined Bending and Shear, *ASCE J. Struct. Div.*, October, vol. 87.

[4] Basler K. and Thürlimann, B. 1963. Strength of Plate Girders in Bending, *Trans. ASCE*, 128.

[5] Basler, K. 1963. Strength of Plate Girders in Shear, *Trans. ASCE*, Vol. 128, Part II, 683.

[6] British Standards Institution. 1983. BS 5400: Part 3, Code of Practice for Design of Steel Bridges, BSI, London.

[7] Caccese, V., Elgaaly, M., and Chen, R. 1993. Experimental Study of Thin Steel-Plate Shear Walls Under Cyclic Load, *ASCE J. Struct. Eng.*, February.

[8] Cooper, P.B. 1967. Strength of Longitudinally Stiffened Plate Girders, *ASCE J. Struct. Div.,* 93(ST2), 419-452.

[9] Dubas, P. and Gehrin, E. 1986. Behavior and Design of Steel Plated Structures, ECCS Publ. No. 44, TWG 8.3, 110-112.

[10] Elgaaly, M. 1983. Web Design Under Compressive Edge Loads, *AISC Eng. J.,* Fourth Quarter.

[11] Elgaaly, M. and Nunan W. 1989. Behavior of Rolled Sections Webs Under Eccentric Edge Compressive Loads, *ASCE J. Struct. Eng.,* 115(7).

[12] Elgaaly, M. and Dagher, H. 1990. Beams and Girders with Corrugated Webs, Proceedings of the SSRC Annual Technical Session, St. Louis, MO.

[13] Elgaaly, M. and Salkar, R. 1990. Behavior of Webs Under Eccentric Compressive Edge Loads, Proceedings of IUTAM Symposium, Prague, Czechoslovakia.

[14] Elgaaly, M. and Salkar, R. 1991. Web Crippling Under Edge Loading, Proceedings of the AISC National Steel Construction Conference, Washington, D.C.

[15] Elgaaly, M., Salkar, R., and Eash, M. 1992. Unstiffened and Stiffened Webs Under Compressive Edge Loads, Proceedings of the SSRC Annual Technical Session, Pittsburgh, PA.

[16] Evans, H.R. and Mokhtari, A.R. 1992. Plate Girders with Unstiffened or Profiled Web Plates, *J. Singapore Struct. Steel Soc.,* 3(1), December.

[17] Elgaaly, M., Caccese, V., and Du, C. 1993. Postbuckling Behavior of Steel-Plate Shear Walls Under Cyclic Loads, *ASCE J. Struct. Eng.,* 119(2).

[18] Elgaaly, M., Liu, Y., Caccese, V., Du, C., Chen, R., and Martin, D. 1994. Non-Linear Behavior of Steel Plate Shear Walls, *Computational Structural Engineering for Practice,* edited by Papadrakakis and Topping, Civil-Comp Press, Edinburgh, UK.

[19] Elgaaly, M., Hamilton, R., and Seshadri, A. 1996. Shear Strength of Beams with Corrugated Webs, *J. Struct. Eng.,* 122(4).

[20] Elgaaly, M., Seshadri, A., and Hamilton, R. 1996. Bending Strength of Beams with Corrugated Webs, *J. Struct. Eng.,* 123(6).

[21] Fisher, J.W. 1965. Behavior of Fasteners and Plates with Holes, *J. Struct. Eng., ASCE,* 91(6).

[22] Galambos, T.V., Ed. 1988. *Guide to Stability Design Criteria for Metal Structures,* 4th ed., Wiley Interscience, New York.

[23] Gaylord, E.H. 1963. Discussion of K. Basler Strength of Plate Girders in Shear, *Trans. ASCE,* 128, Part II, 712.

[24] Hoglund, T. 1971. Behavior and Load Carrying Capacity of Thin Plate I-Girders, Division of Building Statics and Structural Engineering, Royal Institute of Technology, Bulletin No. 93, Stockholm.

[25] Kirby, P.A. and Nethercat, D.A. 1979. *Design for Structural Stability,* John Wiley & Sons, New York.

[26] Kulak, G.L., Fisher, J.W., and Struik, J.H.A. 1987. *Guide to Design Criteria for Bolted and Riveted Joints,* 2nd ed., Wiley Interscience, New York.

[27] Kulak, G.L. 1985. Behavior of Steel Plate Shear Walls, Proc. of the AISC Int. Eng. Symp. on Struct. Steel, American Institute of Steel Construction (AISC), Chicago, IL.

[28] Lee, M.M.K., Kamtekar, A.G., and Little, G.H. 1989. An Experimental Study of Perforated Steel Web Plates, *Struct. Engineer,* 67(2/24).

[29] Lee, M.M.K. 1990. Numerical Study of Plate Girder Webs with Holes, *Proc. Inst. Civ. Eng.,* Part 2.

[30] Lee, M.M.K. 1991. A Theoretical Model for Collapse of Plate Girders with Perforated Webs, *Struct. Eng.,* 68(4).

[31] Lindner, J. 1990. Lateral-Torsional Buckling of Beams with Trapezoidally Corrugated Webs, Proceedings of the 4th International Colloquium on Stability of Steel Structures, Budapest, Hungary.

[32] American Institute of Steel Construction. 1993. *Load and Resistance Factor Design Specification for Structural Steel Buildings,* AISC, Chicago.

[33] Marsh, C. 1985. Photoelastic Study of Postbuckled Shear Webs, *Canadian J. Civ. Eng.,* 12(2).

[34] Narayanan, R. and Der Avanessian, N.G.V. 1983. Strength of Webs Containing Cut-Outs, IABSE Proceedings P-64/83.

[35] Narayanan, R. and Der Avanessian, N.G.V. 1983. Equilibrium Solution for Predicting the Strength of Webs with Rectangular Holes, *Proc. ICE,* Part 2.

[36] Porter, D.M., Rockey, K.C., and Evans, H.R. 1975. The Collapse Behavior of Plate Girders Loaded in Shear, *Struct. Eng.,* 53(8), 313-325.

[37] Roberts, T.M. 1981. Slender Plate Girders Subjected to Edge Loading, *Proc. Inst. Civil Eng.,* Part 2, 71.

[38] Rockey, K.C. and Skaloud, M. 1972. The Ultimate Load Behavior of Plate Girders Loaded in Shear, *Struct. Eng.,* 50(1).

[39] Schilling, C.G. and Frost, R.W. 1964. Behavior of Hybrid Beams Subjected to Static Loads, *ASCE, J. Struct. Div.,* 90(ST3), 55-88.

[40] Von Karman, T., Sechler, E.F., and Donnell, L.H. 1932. The Strength of Thin Plates in Compression, *Trans. ASME,* 54(2).

[41] Wagner, H. 1931. Flat Sheet Metal Girder with Very Thin Metal Web, NACA Tech. Memo. Nos. 604, 605, 606.

[42] Yen, B.T. and Mueller, J.A. 1966. Fatigue Tests of Large-Size Welded Plate Girders, *Weld. Res. Counc. Bull.,* No. 118, November.

20

Steel Bridge Construction

20.1 Introduction 20-1
20.2 Construction Engineering in Relation to Design
      Engineering 20-2
20.3 Construction Engineering Can Be Critical 20-2
20.4 Premises and Objectives of Construction Engineering 20-3
20.5 Fabrication and Erection Information Shown on
      Design Plans 20-4
20.6 Erection Feasibility 20-4
20.7 Illustrations of Challenges in Construction
      Engineering 20-4
20.8 Obstacles to Effective Construction Engineering 20-5
20.9 Examples of Inadequate Construction Engineering
      Allowances and Effort 20-5
20.10 Considerations Governing Construction Engineering
      Practices... 20-6
20.11 Two General Approaches to Fabrication and
      Erection of Bridge Steelwork 20-7
20.12 Example of Arch Bridge Construction.................. 20-8
20.13 Which Construction Procedure Is To Be Preferred? ... 20-9
20.14 Example of Suspension Bridge Cable Construction ... 20-13
20.15 Example of Cable-Stayed Bridge Construction 20-17
20.16 Field Checking at Critical Erection Stages 20-19
20.17 Determination of Erection Strength Adequacy 20-20
20.18 Philosophy of the Erection Rating Factor 20-22
20.19 Minimum Erection Rating Factors 20-23
20.20 Deficiencies of Typical Construction Procedure
      Drawings and Instructions 20-24
20.21 Shop and Field Liaison by Construction Engineers 20-25
20.22 Construction Practices and Specifications—
      The Future 20-26
20.23 Concluding Comments 20-26
20.24 Further Illustrations 20-28
References... 20-60

Jackson Durkee
Consulting Structural Engineer,
Bethlehem, PA

20.1 Introduction

This chapter addresses some of the principles and practices applicable to the construction of medium- and long-span steel bridges — structures of such size and complexity that construction engineering becomes an important or even the governing factor in the successful fabrication and erection of the superstructure steelwork.

0-8493-2674-5/97/$0.00+$.50
© 1997 by CRC Press LLC

We begin with an explanation of the fundamental nature of construction engineering, then go on to explain some of the challenges and obstacles involved. Two general approaches to the fabrication and erection of bridge steelwork are described, with examples from experience with arch bridges, suspension bridges, and cable-stayed bridges.

The problem of erection-strength adequacy of trusswork under erection is considered, and a method of appraisal offered that is believed to be superior to the standard working-stress procedure.

Typical problems in respect to construction procedure drawings, specifications, and practices are reviewed, and methods for improvement suggested. Finally, we take a view ahead, to the future prospects for effective construction engineering in the U.S.

This chapter also contains a large number of illustrations showing a variety of erection methods for several types of steel bridges.

20.2   Construction Engineering in Relation to Design Engineering

With respect to bridge steelwork the differences between construction engineering and design engineering should be kept firmly in mind. Design engineering is of course a concept and process well known to structural engineers; it involves preparing a set of plans and specifications — known as the contract documents — that define the structure in its completed configuration, referred to as the geometric outline. Thus, the design drawings describe to the contractor the steel bridge superstructure that the owner wants to see in place when the project is completed. A considerable design engineering effort is required to prepare a good set of contract documents.

Construction engineering, however, is not so well known. It involves governing and guiding the fabrication and erection operations needed to produce the structural steel members to the proper cambered or "no-load" shape, and get them safely and efficiently "up in the air" in place in the structure, such that the completed structure under the deadload conditions and at normal temperature will meet the geometric and stress requirements stipulated on the design drawings.

Four key considerations may be noted: (1) design engineering is widely practiced and reasonably well understood, and is the subject of a steady stream of technical papers; (2) construction engineering is practiced on only a limited basis, is not as well understood, and is hardly ever discussed; (3) for medium- and long-span bridges, the construction engineering aspects are likely to be no less important than design engineering aspects; and (4) adequately staffed and experienced construction engineering offices are a rarity.

20.3   Construction Engineering Can Be Critical

The construction phase of the total life of a major steel bridge will probably be much more hazardous than the service-use phase. Experience shows that a large bridge is more likely to suffer failure during erection than after completion. Many decades ago, steel bridge design engineering had progressed to the stage where the chance of structural failure under service loadings became altogether remote. However, the erection phase for a large bridge is inherently less secure, primarily because of the prospect of inadequacies in construction engineering and its implementation at the job site. Indeed, the hazards associated with the erection of large steel bridges will be readily apparent from a review of the illustrations in this chapter.

For significant steel bridges the key to construction integrity lies in the proper planning and engineering of steelwork fabrication and erection. Conversely, failure to attend properly to construction engineering constitutes an invitation to disaster. In fact, this thesis is so compelling that whenever a steel bridge failure occurs during construction (see for example Figure 20.1), it is reasonable to assume that the construction engineering investigation was either inadequate, not properly implemented, or both.

Figure 20.1   Failure of a steel girder bridge during erection, 1995. Steel bridge failures such as this one invite suspicion that the construction engineering aspects were not properly attended to.

20.4   Premises and Objectives of Construction Engineering

Obviously, when the structure is in its completed configuration it is ready for the service loadings. However, during the erection sequences the various components of major steel bridges are subject to stresses that may be quite different from those provided for by the designer. For example, during construction there may be a derrick moving and working on the partially erected structure, and the structure may be cantilevered out some distance causing tension-designed members to be in compression and vice versa. Thus, the steelwork contractor needs to engineer the bridge members through their various construction loadings, and strengthen and stabilize them as may be necessary. Further, the contractor may need to provide temporary members to support and stabilize the structure as it passes through its successive erection configurations.

In addition to strength problems there are also geometric considerations. The steelwork contractor must engineer the construction sequences step by step to ensure that the structure will fit properly together as erection progresses, and that the final or closing members can be moved into position and connected. Finally, of course, the steelwork contractor must carry out the engineering studies needed to ensure that the geometry and stressing of the completed structure under normal temperature will be in accordance with the requirements of the design plans and specifications.

20.5   Fabrication and Erection Information Shown on Design Plans

Regrettably, the level of engineering effort required to accomplish safe and efficient fabrication and erection of steelwork superstructures is not widely understood or appreciated in bridge design offices, nor indeed by a good many steelwork contractors. It is only infrequently that we find a proper level of capability and effort in the engineering of construction.

The design drawings for an important bridge will sometimes display an erection scheme, even though most designers are not experienced in the practice of erection engineering and usually expend only a minimum or even superficial effort on erection studies. The scheme portrayed may not be practical, or may not be suitable in respect to the bidder or contractor's equipment and experience. Accordingly, the bidder or contractor may be making a serious mistake if he relies on an erection scheme portrayed on the design plans.

As an example of misplaced erection effort on the part of the designer, there have been cases where the design plans show cantilever erection by deck travelers, with the permanent members strengthened correspondingly to accommodate the erection loadings; but the successful bidder elected to use water-borne erection derricks with long booms, thereby obviating the necessity for most or all of the erection strengthening provided on the design plans. Further, even in those cases where the contractor would decide to erect by cantilevering as anticipated on the plans, there is hardly any way for the design engineer to know what will be the weight and dimensions of the contractor's erection travelers.

20.6   Erection Feasibility

Of course, the bridge designer does have a certain responsibility to his client and to the public in respect to the erection of the bridge steelwork. This responsibility includes (1) making certain, during the design stage, that there is a feasible and economical method to erect the steelwork; (2) setting forth in the contract documents any necessary erection guidelines and restrictions; and (3) reviewing the contractor's erection scheme, including any strengthening that may be needed, to verify its suitability. It may be noted that this latter review does not relieve the contractor from responsibility for the adequacy and safety of the field operations.

Bridge annals include a number of cases where the designing engineer failed to consider erection feasibility. In one notable instance the design plans showed the 1200-ft (366-m) main span for a long crossing over a wide river as an esthetically pleasing steel tied-arch. However, erection of such a span in the middle of the river was impractical; one bidder found that the tonnage of falsework required was about the same as the weight of the permanent steelwork. Following opening of the bids, the owner found the prices quoted to be well beyond the resources available, and the tied-arch main span was discarded in favor of a through-cantilever structure, for which erection falsework needs were minimal and practical.

It may be noted that designing engineers can stand clear of serious mistakes such as this one, by the simple expedient of conferring with prospective bidders during the preliminary design stage of a major bridge.

20.7   Illustrations of Challenges in Construction Engineering

Space does not permit comprehensive coverage of the numerous and difficult technical challenges that can confront the construction engineer in the course of the erection of various types of major steel bridges. However, some conception of the kinds of steelwork erection problems, the methods available to resolve them, and hazards involved can be conveyed by views of bridges in various stages of erection; refer to the illustrations in the text.

20.8   Obstacles to Effective Construction Engineering

There is an unfortunate tendency among designing engineers to view construction engineering as relatively unimportant. This view may be augmented by the fact that few designers have had any significant experience in the engineering of construction.

Further, managers in the construction industry must look critically at costs, and they can readily develop the attitude that their engineers are doing unnecessary theoretical studies and calculations, detached from the practical world. (And indeed, this may sometimes be the case.) Such management apprehension can constitute a serious obstacle to staff engineers who see the need to have enough money in the bridge tender to cover a proper construction engineering effort for the project. There is the tendency for steelwork construction company management to cut back the construction engineering allowance, partly because of this apprehension and partly because of the concern that other tenderers will not be allotting adequate money for construction engineering. This effort is often thought of by company management as "a necessary evil" at best — something they would prefer not to be bothered with or burdened with.

Accordingly, construction engineering tends to be a difficult area of endeavor. The way for staff engineers to gain the confidence of management is obvious — they need to conduct their investigations to a level of technical proficiency that will command management respect and support, and they must keep management informed as to what they are doing and why it is necessary. As for management's concern that other bridge tenderers will not be putting into their packages much money for construction engineering, this concern is no doubt usually justified, and it is difficult to see how responsible steelwork contractors can cope with this problem.

20.9   Examples of Inadequate Construction Engineering Allowances and Effort

Even with the best of intentions, the bidder's allocation of money to construction engineering can be inadequate. A case in point involved a very heavy, long-span cantilever truss bridge crossing a major river. The bridge superstructure carried a contract price of some $30 million, including an allowance of $150,000, or about one-half of 1%, for construction engineering of the permanent steelwork (i.e., not including such matters as design of erection equipment). As fabrication and erection progressed, many unanticipated technical problems came forward, including brittle-fracture aspects of certain grades of the high-strength structural steel, and aerodynamic instability of H-shaped vertical and diagonal truss members. In the end the contractor's construction engineering effort mounted to about $1.3 million, almost nine times the estimated cost.

Another significant example — this one in the domain of buildings — involved a design-and-construct project for airplane maintenance hangars at a prominent airport. There were two large and complicated buildings, each 100×150 m (328×492 ft) in plan and 37 m (121 ft) high with a 10-m (33-ft) deep space-frame roof. Each building contained about 2300 tons of structural steelwork. The design-and-construct steelwork contractor had submitted a bid of about $30 million, and included therein was the magnificent sum of $5000 for construction engineering, under the expectation that this work could be done on an incidental basis by the project engineer in his "spare time".

As the steelwork contract went forward it quickly became obvious that the construction engineering effort had been grossly underestimated. The contractor proceeded of staff-up appropriately and carried out in-depth studies, leading to a detailed erection procedure manual of some 270 pages showing such matters as erection equipment and its positioning and clearances; falsework requirements; lifting tackle and jacking facilities; stress, stability, and geometric studies for gravity and wind loads; step-by-step instructions for raising, entering, and connecting steelwork components; closing and swinging the roof structure and portal frame; and welding guidelines and procedures. This erection procedure manual turned out to be a key factor in the success of the fieldwork. The cost of this construction engineering effort amounted to ten times the estimate, but still came to a mere one-fifth of 1% of the total contract cost.

In yet another example a major steelwork general contractor was induced to sublet the erection of a long-span cantilever truss bridge to a reputable erection contractor, whose quoted price for the work was less than the general contractor's estimated cost. During the erection cycle the general contractor's engineers made some visits to the job site to observe progress, and were surprised and disconcerted to observe how little erection engineering and planning had been accomplished. For example, the erector had made no provision for installing jacks in the bottom-chord jacking points for closure of the main span; it was left up to the field forces to provide the jack bearing components inside the bottom-chord joints and to find the required jacks in the local market. When the job-built installations were tested it was discovered that they would not lift the cantilevered weight, and the job had to be shut down while the field engineer scouted around to find larger-capacity jacks. Further, certain compression members did not appear to be properly braced to carry the erection loadings; the erector had not engineered those members, but just assumed they were adequate. It became obvious that the erector had not appraised the bridge members for erection adequacy and had done little or no planning and engineering of the critical evolutions to be carried out in the field.

Many further examples of inadequate attention to construction engineering could be presented. Experience shows that the amounts of money and time allocated by steelwork contractors for the engineering of construction are frequently far less than desirable or necessary. Clearly, effort spent on construction engineering is worthwhile; it is obviously more efficient and cheaper, and certainly much safer, to plan and engineer steelwork construction in the office in advance of the work, rather than to leave these important matters for the field forces to work out. Just a few bad moves on site, with the corresponding waste of labor and equipment hours, will quickly use up sums of money much greater than those required for a proper construction engineering effort — not to mention the costs of any job accidents that might occur.

The obvious question is "Why is construction engineering not properly attended to?" Do not contractors learn, after a bad experience or two, that it is both necessary and cost effective to do a thorough job of planning and engineering the construction of important bridge projects? Experience and observation would seem to indicate that some steelwork contractors learn this lesson, while many do not. There is always pressure to reduce bid prices to the absolute minimum, and to add even a modest sum for construction engineering must inevitably reduce the chance of being the low bidder.

20.10   Considerations Governing Construction Engineering Practices

There are no textbooks or manuals that define how to accomplish a proper job of construction engineering. In bridge construction (and no doubt in building construction as well), the engineering of construction tends to be a matter of each firm's experience, expertise, policies and practices. Usually there is more than one way to build the structure, depending on the contractor's ingenuity and engineering skill, his risk appraisal and inclination to assume risk, the experience of his fabrication and erection work forces, his available equipment, and his personal preferences. Experience shows that each project is different; and although there will be similarities from one bridge of a given type to another, the construction engineering must be accomplished on an individual project basis. Many aspects of the project at hand will turn out to be different from those of previous similar jobs, and also there may be new engineering considerations and requirements for a given project that did not come forward on previous similar work.

During the estimating and bidding phase of the project the prudent, experienced bridge steelwork contractor will "start from scratch" and perform his own fabrication and erection studies, irrespective of any erection schemes and information that may be shown on the design plans. These studies can involve a considerable expenditure of both time and money, and thereby place that contractor at a disadvantage in respect to those bidders who are willing to rely on hasty, superficial studies, or — where the design engineer has shown an erection scheme — to simply assume that it has been engineered correctly and proceed to use it. The responsible contractor, on the other hand, will appraise the feasible construction methods and evaluate their costs and risks, and then make his selection.

After the contract has been executed the contractor will set forth how he intends to fabricate and erect, in detailed plans that could involve a large number of calculation sheets and drawings along with construction procedure documents. It is appropriate for the design engineer on behalf of his client to review the contractor's plans carefully, perform a check of construction considerations, and raise appropriate questions. Where the contractor does not agree with the designer's comments the two parties get together for review and discussion, and in the end they concur on essential factors such as fabrication and erection procedures and sequences, the weight and positioning of erection equipment, the design of falsework and other temporary components, erection stressing and strengthening of the permanent steelwork, erection stability and bracing of critical components, any erection check measurements that may be needed, and span closing and swinging operations.

The designing engineer's approval is needed for certain fabrication plans, such as the cambering of individual members; however, in most cases the designer should stand clear of actual *approval* of the contractor's construction plans since he is not in a position to accept construction responsibility, and too many things can happen during the field evolutions over which the designer has no control.

It should be emphasized that even though the designing engineer has usually had no significant experience in steelwork construction, the contractor should welcome his comments and evaluate them carefully and respectfully. In major bridge projects many matters can get out of control or can be improved upon, and the contractor should take advantage of every opportunity to improve his prospects and performance. The experienced contractor will make sure that he works constructively with the designing engineer, standing well clear of antagonistic or confrontational posturing.

20.11 Two General Approaches to Fabrication and Erection of Bridge Steelwork

As has been stated previously, the objective in steel bridge construction is to fabricate and erect the structure so that it will have the geometry and stressing designated on the design plans, under full dead load at normal temperature. This geometry is known as the geometric outline. In the case of steel bridges there have been, over the decades, two general procedures for achieving this objective:

1. *The "field adjustment" procedure* — Carry out a continuing program of field surveys and measurements, and perform certain adjustments of selected steelwork components in the field as erection progresses, in an attempt to discover fabrication and erection deficiencies and compensate for them.

2. *The "shop control" procedure* — Place total reliance on first-order surveying of span baselines and pier elevations, and on accurate steelwork fabrication and erection augmented by meticulous construction engineering; and proceed with erection without any field adjustments, on the basis that the resulting bridge deadload geometry and stressing will be as good as can possibly be achieved.

Bridge designers have a strong tendency to overestimate the capability of field forces to accomplish accurate measurements and effective adjustments of the partially erected structure, and at the same time they tend to underestimate the positive effects of precise steel bridgework fabrication and erection. As a result, we continue to find contract drawings for major steel bridges that call for field evolutions such as the following:

1. **Continuous trusses and girders** — At the designated stages, measure or "weigh" the reactions on each pier, compare them with calculated theoretical values, and add or remove bearing-shoe shims to bring measured values into agreement with calculated values.

2. **Arch bridges** — With the arch ribs erected to midspan and only the short, closing "crown sections" not yet in place, measure thrust and moment at the crown, compare them with calculated theoretical values, and then adjust the shape of the closing sections to correct for errors in span-length measurements and in bearing-surface angles at skewback supports, along with accumulated fabrication and erection errors.

3. **Suspension bridges** — Following erection of the first cable wire or strand across the spans from anchorage to anchorage, survey its sag in each span and adjust these sags to comport with calculated theoretical values.

4. **Arch bridges and suspension bridges** — Carry out a deck-profile survey along each side of the bridge under the steel-load-only condition, compare survey results with the theoretical profile, and shim the suspender sockets so as to render the bridge floorbeams level in the completed structure.

5. **Cable-stayed bridges** — At each deck-steelwork erection stage, adjust tensions in the newly erected cable stays so as to bring the surveyed deck profile and measured stay tensions into agreement with calculated theoretical data.

There are two prime obstacles to the success of "field adjustment" procedures of whatever type: (1) field determination of the actual geometric and stress conditions of the partially erected structure and its components will not necessarily be definitive, and (2) calculation of the corresponding "proper" or "target" theoretical geometric and stress conditions will most likely prove to be less than authoritative.

20.12 Example of Arch Bridge Construction

In the case of the arch bridge closing sections referred to heretofore, experience on the construction of two major fixed-arch bridges crossing the Niagara River gorge from the U.S. to Canada—the Rainbow and the Lewiston-Queenston arch bridges (see Figures 20.2 through 20.5)—has demonstrated the difficulty, and indeed the futility, of attempts to make field-measured geometric and stress conditions agree with calculated theoretical values. The broad intent for both structures was to make such adjustments in the shape of the arch-rib closing sections at the crown (which were nominally about 1 ft [0.3 m] long) as would bring the arch-rib actual crown moments and thrusts into agreement with the calculated theoretical values, thereby correcting for errors in span-length measurements, errors in bearing-surface angles at the skewback supports, and errors in fabrication and erection of the arch-rib sections.

Following extensive theoretical investigations and on-site measurements the steelwork contractor found, in the case of each Niagara arch bridge, that there were large percentage differences between the field-measured and the calculated theoretical values of arch-rib thrust, moment, and line-of-thrust position, and that the measurements could not be interpreted so as to indicate what corrections to the theoretical closing crown sections, if any, should be made. Accordingly, the contractor concluded that the best solution in each case was to abandon any attempts at correction and simply install the theoretical-shape closing crown sections. In each case, the contractor's recommendation was accepted by the designing engineer.

Points to be noted in respect to these field-closure evolutions for the two long-span arch bridges are that accurate jack-load closure measurements at the crown are difficult to obtain under field conditions; and calculation of corresponding theoretical crown thrusts and moments are likely to be questionable because of uncertainties in the dead loading, in the weights of erection equipment,

Figure 20.2　Erection of arch ribs, Rainbow Bridge, Niagara Falls, New York, 1941. Bridge span is 950 ft (290 m), with rise of 150 ft (46 m); box ribs are 3 × 12 ft (0.91 × 3.66 m). Tiebacks were attached starting at the end of the third tier and jumped forward as erection progressed (see Figure 20.3). Much permanent steelwork was used in tieback bents. Derricks on approaches load steelwork on material cars that travel up arch ribs. Travelers are shown erecting last full-length arch-rib sections, leaving only the short, closing crown sections to be erected. Canada is at right, the U.S. at left. (Courtesy of Bethlehem Steel Corporation.)

and in the steelwork temperature. Therefore, attempts to adjust the shape of the closing crown sections so as to bring the actual stress condition of the arch ribs closer to the theoretical condition are not likely to be either practical or successful.

It was concluded that for long, flexible arch ribs, the best construction philosophy and practice is (1) to achieve overall geometric control of the structure by performing all field survey work and steelwork fabrication and erection operations to a meticulous degree of accuracy, and then (2) to rely on that overall geometric control to produce a finished structure having the desired stressing and geometry. For the Rainbow arch bridge, these practical construction considerations were set forth definitively by the contractor in [2]. The contractor's experience for the Lewiston-Queenston arch bridge was similar to that on Rainbow, and was reported — although in considerably less detail — in [10].

20.13　Which Construction Procedure Is To Be Preferred?

The contractor's experience on the construction of the two long-span fixed-arch bridges is set forth at length since it illustrates a key construction theorem that is broadly applicable to the fabrication and erection of steel bridges of all types. This theorem holds that the contractor's best procedure for achieving, in the completed structure, the deadload geometry and stressing stipulated on the design plans, is generally as follows:

1. Determine deadload stress data for the structure, at its geometric outline and under normal temperature, based on accurately calculated weights for all components.

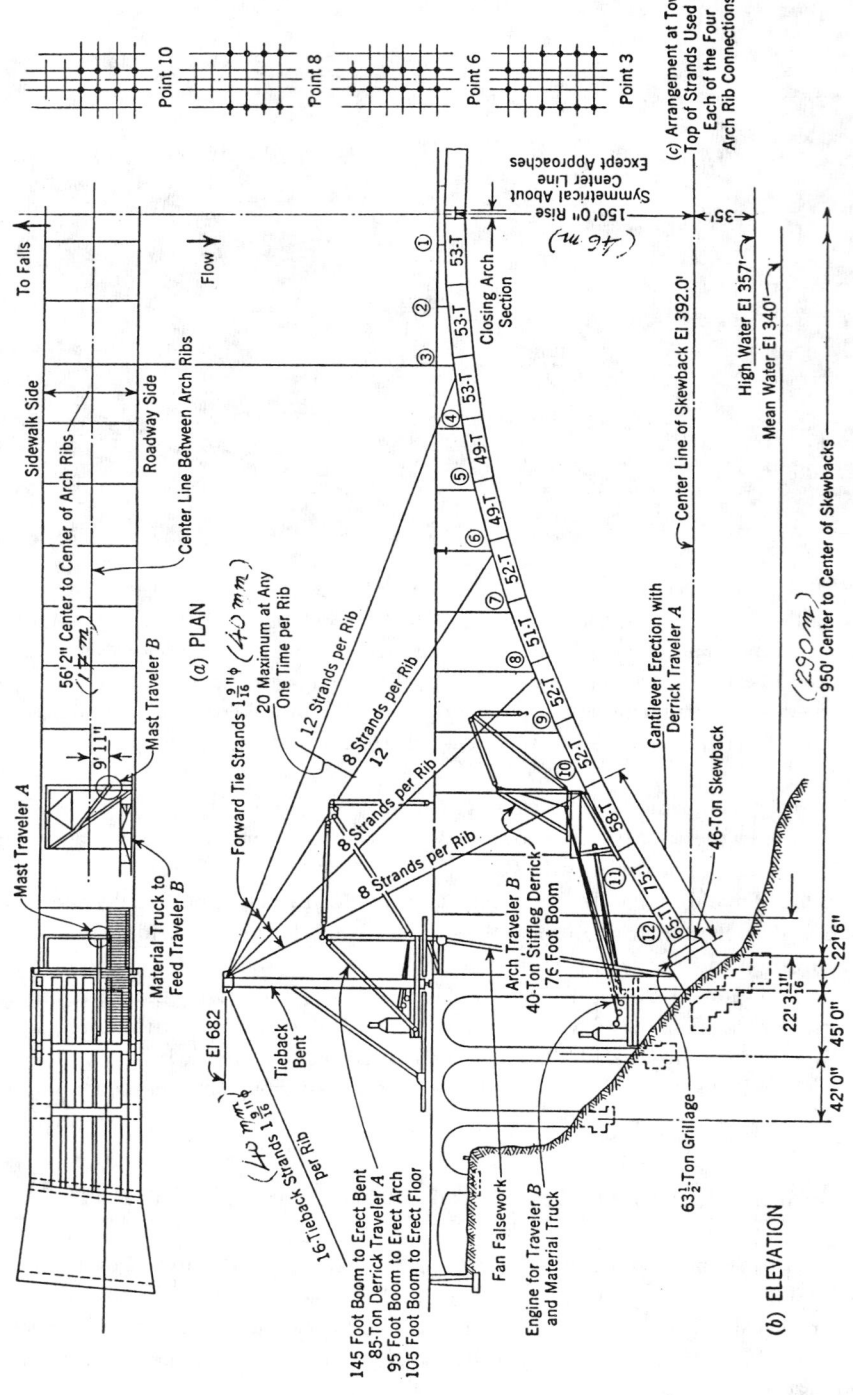

Figure 20.3 Rainbow Bridge, Niagara Falls, New York, showing successive arch tieback positions. Arch-rib erection geometry and stressing were controlled by means of measured tieback tensions in combination with surveyed arch-rib elevations.

Figure 20.4 Lewiston-Queenston arch bridge, near Niagara Falls, New York, 1962. The world's longest fixed-arch span, at 1000 ft (305 m); rise is 159 ft (48 m). Box arch-rib sections are typically about 3 × 13-1/2 ft (0.9 × 4.1 m) in cross-section and about 44-1/2 ft (13.6 m) long. Job was estimated using erection tiebacks (same as shown in Figure 20.3), but subsequent studies showed the long, sloping falsework bents to be more economical (even if less secure looking). Much permanent steelwork was used in the falsework bents. Derricks on approaches load steelwork onto material cars that travel up arch ribs. The 115-ton-capacity travelers are shown erecting the last full-length arch-rib sections, leaving only the short, closing crown sections to be erected. Canada is at left, the U.S. at right. (Courtesy of Bethlehem Steel Corporation.)

2. Determine the cambered (i.e., "no-load") dimensions of each component. This involves determining the change of shape of each component from the deadload geometry, as its deadload stressing is removed and its temperature is changed from normal to the "shop-tape" temperature.

3. Fabricate, with all due precision, each structural component to its proper no-load dimensions — except for certain flexible components such as wire rope and strand members, which may require special treatment.

4. Accomplish shop assembly of members and "reaming assembled" of holes in joints, as needed.

5. Carry out comprehensive engineering studies of the structure under erection at each key erection stage, determining corresponding stress and geometric data, and prepare a step-by-step erection procedure plan, incorporating any check measurements that may be necessary or desirable.

6. During the erection program, bring all members and joints to the designated alignment prior to bolting or welding.

7. Enter and connect the final or closing structural components, following the closing procedure plan, without attempting any field measurements thereof or adjustments thereto.

In summary, the key to construction success is to accomplish the field surveys of critical baselines and support elevations with all due precision, perform construction engineering studies comprehensively and shop fabrication accurately, and then carry the erection evolutions through in the field without any second guessing and ill-advised attempts at measurement and adjustment.

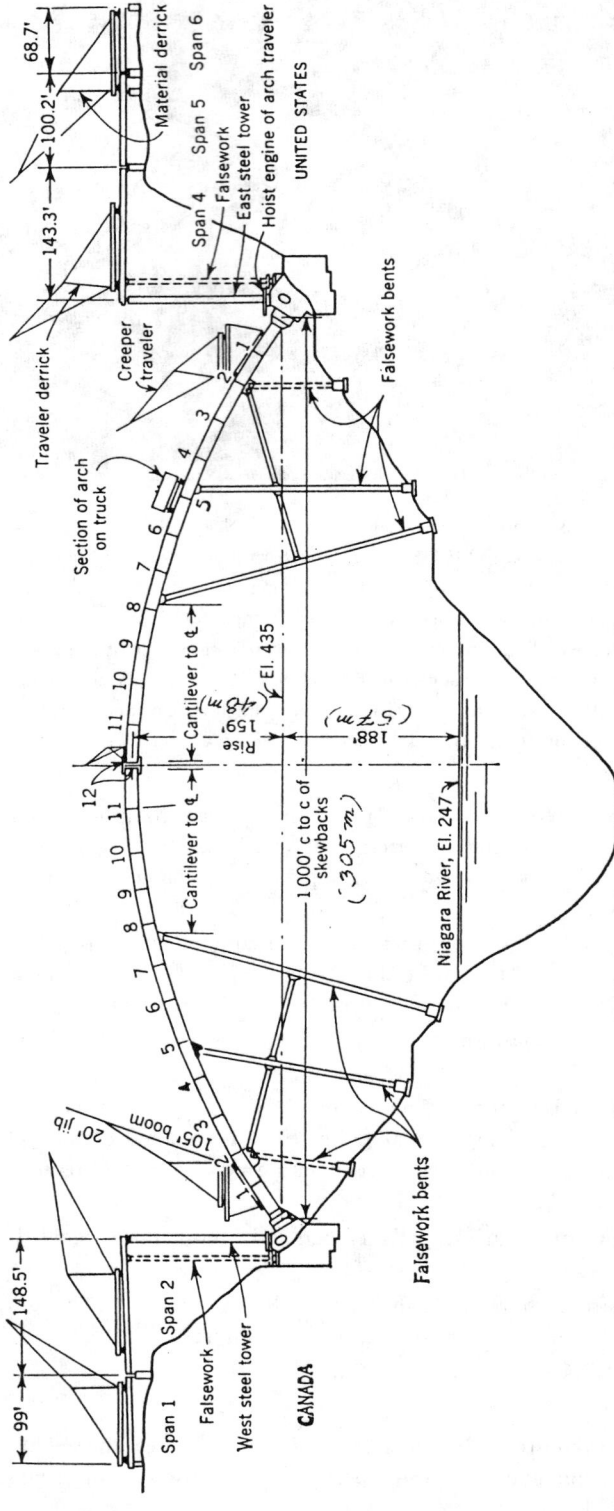

Figure 20.5 Lewiston-Queenston arch bridge, near Niagara Falls, New York. Crawler cranes erect steelwork for spans 1 and 6 and erect material derricks thereon. These derricks erect traveler derricks, which move forward and erect supporting falsework and spans 2, 5, and 4. Traveler derricks erect arch-rib sections 1 and 2 and supporting falsework at each skewback, then set up creeper derricks, which erect arches to midspan.

It may be noted that no special treatment is accorded to statically indeterminate members; they are fabricated and erected under the same governing considerations applicable to statically determinate members, as set forth above. It may be noted further that this general steel bridge construction philosophy does not rule out check measurements altogether, as erection goes forward; under certain special conditions, measurements of stressing and/or geometry at critical erection stages may be necessary or desirable in order to confirm structural integrity. However, before the erector calls for any such measurements he should make certain that they will prove to be practical and meaningful.

20.14 Example of Suspension Bridge Cable Construction

In order to illustrate the "shop control" construction philosophy further, its application to the main cables of the first Wm. Preston Lane, Jr., Memorial Bridge, crossing the Chesapeake Bay in Maryland, completed in 1952 (Figure 20.6), will be described. Suspension bridge cables constitute one of the most difficult bridge erection challenges. Up until "first Chesapeake" the cables of major suspension bridges had been adjusted to the correct position in each span by means of a sag survey of the first-erected cable wires or strands, using surveying instruments and target rods. However, on first Chesapeake, with its 1600-ft (488-m) main span, 661-ft (201-m) side spans, and 450-ft (l37-m) back spans, the steelwork contractor recommended abandoning the standard cable-sag survey and adopting the "setting-to-mark" procedure for positioning the guide strands — a significant new concept in suspension bridge cable construction.

Figure 20.6 Suspension spans of first Chesapeake Bay Bridge, Maryland, 1952. Deck steelwork is under erection and is about 50% complete. A typical four-panel through-truss deck section, weighing about 100 tons, is being picked in west side span, and also in east side span in distance. Main span is 1600 ft (488 m) and side spans are 661 ft (201 m); towers are 324 ft (99 m) high. Cables are 14 in. (356 mm) in diameter and are made up of 61 helical bridge strands each (see Figure 20.8).

The steelwork contractor's rationale for "setting to marks" was spelled out in a letter to the designing engineer (see Figure 20.7). (The complete letter is reproduced because it spells out significant construction philosophies.) This innovation was accepted by the designing engineer. It should be noted that the contractor's major argument was that setting to marks would lead to a more accurate cable placement than would the sag survey. The minor arguments, alluded to in the letter, were the resulting savings in preparatory office engineering work and in the field engineering effort, and most likely in construction time as well.

Each cable consisted of 61 standard helical-type bridge strands, as shown in Figure 20.8. To implement the setting-to-mark procedure each of three lower-layer "guide strands" of each cable (i.e., strands 1, 2, and 3) was accurately measured in the manufacturing shop under the simulated

July 6th, 1951
JJ:MM
C-1756

[To the designing engineer]

Gentlemen: Attention of Mr. _____
              Re: <u>Chesapeake Bay Bridge—Suspension Span Cables</u>

In our studies of the method of cable erection, we have arrived at the conclusion that setting of the guide strands to measured marks, instead of to surveyed sag, is a more satisfactory and more accurate method. Since such a procedure is not in accordance with the specifications, we wish to present for your consideration the reasoning which has led us to this conclusion, and to describe in outline form our proposed method of setting to marks.

On previous major suspension bridges, most of which have been built with parallel-wire instead of helical-strand cables, the though has evidently been that setting the guide wire or guide strand to a computed sag, varying with the temperature, would be the most accurate method. This is associated with the fact that guide wires were never measured and marked to length. These established methods were carried over when strand-type cables came into use. An added reason may have been the knowledge that a small error in length results in a relatively large error in sag; and on the present structure the length-error to sag-error rations are 1:2.4 and 1:1.5 for the main span and side spans, respectively.

However, the reading of the sag in the field is a very difficult operation because of the distances involved, the slopes of the side spans and backstays, the fact that even slight wind causes considerable motion to the guide strand, and for other practical reasons. We also believe that even though readings are made on cloudy days or at night, the actual temperature of all portions of the structure which will affect the sag cannot be accurately known. We are convinced setting the guide strands according to the length marks thereon, which are place under what amount to laboratory or ideal conditions at the manufacturing plant, will produce more accurate results than would field measurement of the sag.

To be specific, consider the case of field determination of sag in the main span, where it is necessary to establish accessible platforms, and an H.I. and a foresight somewhat below the desired sag elevation; and then to sight on the foresight and bring a target, hung from the guide strand, down to the line-of-sight. In the present case it is 1600 ft (488 m) to the foresight and 800 ft (244 m) to the target. Even if the line-of-sight were established just right, it would be only under perfect conditions of temperature and air—if indeed then—that such a survey would be precise. The difficulties are still greater in the side spans and back spans, where inclined lines-of-sight must be established by a series of offset measurements from distant bench marks. There is always the danger, particularly in the present location and at the time now scheduled, that days may be lost in waiting for the right conditions of weather to make an instrument survey feasible.

There is a second factor of doubt involved. The strand is measured under a known stress and at a known modulus, with "mechanical stretch" taken out. It is then reeled to a relatively small diameter and unreeled at the bridge site. Under its own weight, and until the full dead load has been applied, there is an indeterminable loss in mechanical set, or loss of modulus. A strand set to proper sag for the final modulus will accordingly be set too low, and the final cable will be below plan elevation. This possible error can only be on the side that is less desirable. Evidently, also, it could be on the order of 1 1/2 in. (40 mm) of sag increase for 1% of temporary reduction in modulus. If the strand were set to sag based on the assumed smaller modulus than will exist the fully loaded condition, we doubt whether this smaller modulus could be chosen closely enough to ensure that the final sag would be correct. We are assured, however, by our manufacturing plant, that even though the modulus under bare-cable weight may be subject to unknown variation, the modulus which existed at the pre-stressing bed under the measuring tension will be duplicated when this same tension is

Figure 20.7   Setting cable guide strands to marks.

reached under dead load. Therefore, if the guide strand is set to measured marks, the doubt as to modulus is eliminated.

A third source of error is temperature. In past practice the sag has been adjusted, by reference to a chart, in accordance with the existing temperature. Granted that the adjustment is made in the early morning (the fog having risen but the sun not), it is hard to conceive that the actual average temperature in 3955 ft (1205 m) of strand will be that recorded by any thermometer. The mainspan sag error is about 0.7 in. (18 mm) per deg C of temperature.

These conditions are all greatly improved at the strand pre-stressing bed. There seems to be no reason to doubt that the guide strands can be measured and marked to an insignificant degree of error, at a stipulated stress and under a well-soaked and determinable temperature. Any errors in sag level must result from something other than the measured length of the guide strand.

There is on indispensable condition, which however holds for either method of setting. That is, that the total distance from anchorage to anchorage, and the total calculated length of strand under its own-weight stress, must agree within the limits of shimming provided in the anchorages. Therefore, this distance in the field must be checked to close agreement. While the measured length of strand will be calculated with precision, it is interesting to note that in the measured length of strand will be calculated with precision, it is interesting to note that in this calculation, it is not essential that the modulus be known with exactness. The important factor is that the strand length under the final deadload stress will be calculated exactly; and since that length is measured under the corresponding average strand stress, knowledge of the modulus is not a consideration. If the modulus at deadload stress is not as assumed, the only effect will be a change of deflection under live load, and this is minor. We emphasize again that the strand length under dead load, and the length as measured in the prestressing bed, will be identical regardless of the modulus.

The calculation for the bare-cable position result in pulled-back positions for the tops of the towers and cable bents, in order to control the unbalanced forces tending to slip the strands in the saddles. These pullback distances may be slightly in error without the slipping forces overcoming friction and thereby becoming apparent. Such errors would affect the final sags of strands set to sag. However, they would have no effect on the final sags of strands set-to-mark at the saddles; these errors change the temporary strand sags only, and under final stress the sags and the shaft leans will be as called for by the design plans.

It sometimes has happened that a tower which at its base is square to the bridge axis, acquires a slight skew as it rises. The amount of this skew has never, so far as we know, been important. If it is disregarded and the guide strands are attached without any compensating change, then the final loading will, with virtual certainty, pull the tower square.

All sources of possible maladjustment have now been discussed except one—the errors in the several span lengths at the base of the towers and bents. The intention is to recognize and accept these, by performing the appropriate check measurements; and to correct for them by slipping the guide strands designated amounts through the saddles such that the center-of-saddle mark on the strand will be offset by that same amount from the centerline of the saddle.

If we have left unexplained herein any factor that seems to you to render our procedure questionable, we are anxious to know of it and discuss it with you in the near future; and we will be glad to come to your offices for this purpose. The detailed preparations for observing strand sags would require considerable time, and we are not now doing any work along those lines.

Yours very truly,

Chief Engineer

Figure 20.7 *(Continued)* Setting cable guide strands to marks.

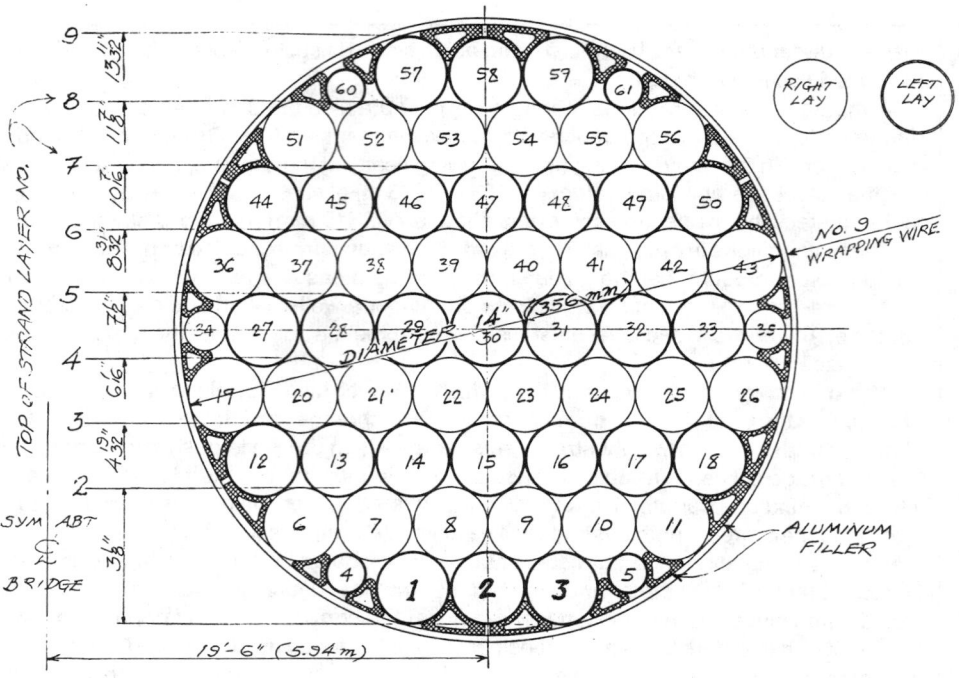

Figure 20.8 Main cable of first Chesapeake Bay suspension bridge, Maryland. Each cable consists of 61 helical-type bridge strands, 55 of 1-11/16 in. (43 mm) and 6 of 29/32 in. (23 mm) diameter. Strands 1, 2, and 3 were designated "guide strands" and were set to mark at each saddle and to normal shims at anchorages.

full-deadload tension, and circumferential marks were placed at the four center-of-saddle positions of each strand. Then, in the field, the guide strands (each about 3955 ft [1205 m] long) were erected and positioned according to the following procedure:

1. Place the three guide strands for each cable "on the mark" at each of the four saddles and set normal shims at each of the two anchorages.
2. Under conditions of uniform temperature and no wind, measure the sag differences among the three guide strands of each cable, at the center of each of the five spans.
3. Calculate the "center-of-gravity" position for each guide-strand group in each span.
4. Adjust the sag of each strand to bring it to the center-of-gravity position in each span. This position was considered to represent the correct theoretical guide-strand sag in each span.

The maximum "spread" from the highest to the lowest strand at the span center, prior to adjustment, was found to be 1-3/4 in. (44 mm) in the main span, 3-1/2 in. (89 mm) in the side spans, and 3-3/4 in. (95 mm) in the back spans. Further, the maximum change of perpendicular sag needed to bring the guide strands to the center-of-gravity position in each span was found to be 15/16 in. (24 mm) for the main span, 2-1/16 in. (52 mm) for the side spans, and 2-1/16 in. (52 mm) for the back spans. These small adjustments testify to the accuracy of strand fabrication and to the validity of the setting-to-mark strand adjustment procedure, which was declared to be a success by all parties concerned. It seems doubtful that such accuracy in cable positioning could have been achieved using the standard sag-survey procedure.

With the first-layer strands in proper position in each cable, the strands in the second and subsequent layers were positioned to hang correctly in relation to the first layer, as is customary and proper for suspension bridge cable construction.

This example provides good illustration that the construction engineering philosophy referred to as the shop-control procedure can be applied advantageously not only to typical rigid-type steel structures, such as continuous trusses and arches, but also to flexible-type structures, such as suspension bridges. There is, however, an important caveat: the steelwork contractor must be a firm of suitable caliber and experience.

20.15   Example of Cable-Stayed Bridge Construction

In the case cable-stayed bridges, the first of which were built in the 1950s, it appears that the governing construction engineering philosophy calls for field measurement and adjustment as the means for control of stay-cable and deck-structure geometry and stressing. For example, we have seen specifications calling for the completed bridge to meet the following geometric and stress requirements:

1. The deck elevation at midspan shall be within 12 in. (305 mm) of theoretical.
2. The deck profile at each cable attachment point shall be within 2 in. (50 mm) of a parabola passing through the actual (i.e., field-measured) midspan point.
3. Cable-stay tensions shall be within 5% of the "corrected theoretical" values.

Such specification requirements introduce a number of problems of interpretation, field measurement, calculation, and field correction procedure, such as the following:

1. Interpretation:
   - The specifications are silent with respect to transverse elevation differentials. Therefore, two deck-profile control parabolas are presumably needed, one for each side of the bridge.
2. Field measurement of actual deck profile:
   - The temperature will be neither constant nor uniform throughout the structure during the survey work.
   - The survey procedure itself will introduce some inherent error.
3. Field measurement of cable-stay tensions:
   - Hydraulic jacks, if used, are not likely to be accurate within 2%, perhaps even 5%; further, the exact point of "lift off" will be uncertain.
   - Other procedures for measuring cable tension, such as vibration or strain gaging, do not appear to define tensions within about 5%.
   - All cable tensions cannot be measured simultaneously; an extended period will be needed, during which conditions will vary and introduce additional errors.
4. Calculation of "actual" bridge profile and cable tensions:
   - Field-measured data must be transformed by calculation into "corrected actual" bridge profiles and cable tensions, at normal temperature and without erection loads.
   - Actual dead weights of structural components can differ by perhaps 2% from nominal weights, while temporary erection loads probably cannot be known within about 5%.

- The actual temperature of structural components will be uncertain and not uniform.
- The mathematical model itself will introduce additional error.

5. "Target condition" of bridge:

   - The "target condition" to be achieved by field adjustment will differ from the geometric condition, because of the absence of the deck wearing surface and other such components; it must therefore be calculated, introducing additional error.

6. Determining field corrections to be carried out by erector, to transform "corrected actual" bridge into "target condition" bridge:

   - The bridge structure is highly redundant, and changing any one cable tension will send geometric and cable-tension changes throughout the structure. Thus, an iterative correction procedure will be needed.

It seems likely that the total effect of all these practical factors could easily be sufficient to render ineffective the contractor's attempts to fine tune the geometry and stressing of the as-erected structure in order to bring it into agreement with the calculated bridge target condition. Further, there can be no assurance that the specifications requirements for the deck-profile geometry and cable-stay tensions are even compatible; it seems likely that *either* the deck geometry *or* the cable tensions may be achieved, but not *both*.

Specifications clauses of the type cited seem clearly to constitute unwarranted and unnecessary field-adjustment requirements. Such clauses are typically set forth by bridge designers who have great confidence in computer-generated calculations, but do not have a sufficient background in and understanding of the practical factors associated with steel bridge construction. Experience has shown that field procedures for major bridges developed unilaterally by design engineers should be reviewed carefully to determine whether they are practical and desirable and will in fact achieve the desired objectives.

In view of all these considerations, the question comes forward as to what design and construction principles should be followed to ensure that the deadload geometry and stressing of steel cable-stayed bridges will fall within acceptable limits. Consistent with the general construction-engineering procedures recommended for other types of bridges, we should abandon reliance on field measurements followed by adjustments of geometry and stressing, and instead place prime reliance on proper geometric control of bridge components during fabrication, followed by accurate erection evolutions as the work goes forward in the field.

Accordingly, the proper construction procedure for cable-stayed steel bridges can be summarized as follows:

1. Determine the actual bridge baseline lengths and pier-top elevations to a high degree of accuracy.
2. Fabricate the bridge towers, cables, and girders to a high degree of geometric precision.
3. Determine, in the fabricating shop, the final residual errors in critical fabricated dimensions, including cable-stay lengths after socketing, and positions of socket bearing surfaces or pinholes.
4. Determine "corrected theoretical" shims for each individual cable stay.
5. During erection, bring all tower and girder structural joints into shop-fabricated alignment, with fair holes, etc.
6. At the appropriate erection stages, install "corrected theoretical" shims for each cable stay.

7. With the structure in the all-steel-erected condition (or other appropriate designated condition), check it over carefully to determine whether any significant geometric or other discrepancies are in evidence. If there are none, declare conditions acceptable and continue with erection.

This construction engineering philosophy can be summarized by stating that if the steelwork fabrication and erection are properly engineered and carried out, the geometry and stressing of the completed structure will fall within acceptable limits; whereas, if the fabrication and erection are not properly done, corrective measurements and adjustments attempted in the field are not likely to improve the structure, or even to prove satisfactory. Accordingly, in constructing steel cable-stayed bridges we should place full reliance on accurate shop fabrication and on controlled field erection, just as is done on other types of steel bridges, rather than attempting to make measurements and adjustments in the field to compensate for inadequate fabrication and erection.

Figure 20.9 Cable-stayed orthotropic-steel-deck bridge over Mississippi River at Luling, La., 1982; view looking northeast. The main span is 1222 ft (372 m); the A-frame towers are 350 ft (107 m) high. A barge-mounted ringer derrick erected the main steelwork, using a 340-ft (104-m) boom with a 120-ft (37-m) jib to erect tower components weighing up to 183 tons, and using a shorter boom for deck components. Cable stays at the ends of projecting cross girders are permanent; others are temporary erection stays. Girder section 16-west of north portion of bridge, erected a few days previously, is projecting at left; companion girder section 16-east is on barge ready for erection (see Figure 20.10).

20.16   Field Checking at Critical Erection Stages   .

As has been stated previously, the best governing procedure for steel bridge construction is generally the shop control procedure, wherein full reliance is placed on accurate fabrication of the bridge

Figure 20.10   Luling Bridge deck steelwork erection, 1982; view looking northeast (refer to Figure 20.9) The twin box girders are 14 ft (4.3 m) deep; the deck plate is 7/16 in. (11 mm) thick. Girder section 16-east is being raised into position (lower right) and will be secured by large-pin hinge bars prior to fairing-up of joint holes and permanent bolting. Temporary erection stays are jumped forward as girder erection progresses.

components as the basis for the integrity of the completed structure. However, this philosophy does not rule out the desirability of certain checks in the field as erection goes forward, with the objective of providing assurance that the work is on target and no significant errors have been introduced.

It would be impossible to catalog those cases during steel bridge construction where a field check might be desirable; such cases will generally suggest themselves as the construction engineering studies progress. We will only comment that these field-check cases, and the procedures to be used, should be looked at carefully, and even skeptically, to make certain that the measurements will be both desirable and practical, producing meaningful information that can be used to augment job integrity.

20.17   Determination of Erection Strength Adequacy

Quite commonly, bridge member forces during the erection stages will be altogether different from those that will prevail in the completed structure. At each critical erection stage the bridge members must be reviewed for strength and stability, to ensure structural integrity as the work goes forward. Such a construction engineering review is typically the responsibility of the steelwork erector, who carries out thorough erection studies of the structure and calls for strengthening or stabilizing of members as needed. The erector submits the studies and recommendations to the designing engineer for review and comment, but normally the full responsibility for steelwork structural integrity during erection rests with the erector.

In the U.S., bridgework design specifications commonly require that stresses in steel structures under erection shall not exceed certain multiples of design allowable stresses. Although this type of erection stress limitation is probably safe for most steel structures under ordinary conditions, it is not necessarily adequate for the control of the erection stressing of large monumental-type bridges.

The key point to be understood here is that fundamentally, there is no logical fixed relationship between design allowable stresses, which are based upon somewhat uncertain long-term service loading requirements along with some degree of assumed structural deterioration, and stresses that are safe and economical during the bridge erection stages, where loads and their locations are normally well defined and the structural material is in new condition. Clearly, the basic premises of the two situations are significantly different, and "factored design stresses" must therefore be considered unreliable as a basis for evaluating erection safety.

There is yet a further problem with factored design stresses. Large truss-type bridges in various erection stages may undergo deflections and distortions that are substantial compared with those occurring under service conditions, thereby introducing apprehension regarding the effect of the secondary bending stresses that result from joint rigidity.

Recognizing these basic considerations, the engineering department of a major U.S. steelwork contractor went forward in the early 1970's to develop a logical philosophy for erection strength appraisal of large structural steel frameworks, with particular reference to long-span bridges, and implemented this philosophy with a stress analysis procedure. The effort was successful and the results were reported in a paper published by the American Society of Civil Engineers in 1977 [5]. This stress analysis procedure, designated the erection rating factor (ERF) procedure, is founded directly upon basic structural principles, rather than on bridge-member design specifications, which are essentially irrelevant to the problem of erection stressing.

It may be noted that a significant inducement toward development of the ERF procedure was the failure of the first Quebec cantilever bridge in 1907 (see Figures 20.11 and 20.12). It was quite obvious that evaluation of the structural safety of the Quebec bridge at advanced cantilever erection stages, such as that portrayed in Figure 20.11, by means of the factored design stress procedure, would inspire no confidence and would not be justifiable.

The erection rating factor (ERF) procedure can be summarized as follows:

1. Assume either (a) pin-ended members (no secondary bending), (b) plane-frame action (rigid truss joints, secondary bending in one plane), or (c) space-frame action (bracing-member joints also rigid, secondary bending in two planes), as engineering judgment dictates.

2. Determine, for each designated erection stage, the member primary forces (axial) and secondary forces (bending) attributable to gravity loads and wind loads.

3. Compute the member stresses induced by the combined erection axial forces and bending moments.

4. Compute the ERF for each member at three or five locations: at the middle of the member; at each joint, inside the gusset plates (usually at the first row of bolts); and, where upset member plates or gusset plates are used, at the stepped-down cross-section outside each joint.

5. Determine the minimum computed ERF for each member and compare it with the stipulated minimum value.

6. Where the computed minimum ERF equals or exceeds the stipulated minimum value, the member is considered satisfactory. Where it is less, the member may be inadequate; the critical part of it is reevaluated in greater detail and the ERF recalculated for further comparison with the stipulated minimum. (Initially calculated values can often be increased significantly.)

7. When the computed minimum ERF remains less than the stipulated minimum, the member must be strengthened as required.

Note that member forces attributable to wind are treated the same as those attributable to gravity loads. The old concept of "increased allowable stresses" for wind is not considered to be valid for erection conditions and is not used in the ERF procedure. Maximum acceptable ℓ/r and b/t

Figure 20.11   First Quebec railway cantilever bridge, 23 August 1907. Cantilever erection of south main span, 6 days before collapse. The tower traveler erected the anchor span (on falsework) and then the cantilever arm; then erected the top-chord traveler, which is shown erecting suspended span at end of cantilever arm. The main span of 1800 ft (549 m) was the world's longest of any type. The sidespan bottom chords second from pier (arrow) failed in compression because latticing connecting chord corner angles was deficient under secondary bending conditions.

values are included in the criteria. ERFs for members subjected to secondary bending moments are calculated using interaction equations.

20.18   Philosophy of the Erection Rating Factor

In order that the structural integrity or reliability of a steel framework can be maintained throughout the erection program, the minimum probable (or "minimum characteristic") strength value of each member must necessarily be no less than the maximum probable (or "maximum characteristic") force value, under the most adverse erection condition. In other words, the following relationship is required:

$$S - \Delta S \geq F + \Delta F \tag{20.1}$$

where

S =   computed or nominal strength value for the member
ΔS =   maximum probable member strength underrun from the computed or nominal value
F =   computed or nominal force value for the member
ΔF =   maximum probable member force overrun from the computed or nominal value

Equation 20.1 states that in the event the actual strength of the structural member is less than the nominal strength, S, by an amount ΔS, while at same time the actual force in the member is greater than the nominal force, F, by an amount ΔF, the member strength will still be no less than the member force, and so the member will not fail during erection. This equation provides a direct appraisal of erection realities, in contrast to the allowable-stress approach based on factored design stresses.

Figure 20.12 Wreckage of south anchor span of first Quebec railway cantilever bridge, 1907. View looking north from south shore a few days after collapse of 29 August 1907, the worst disaster in the history of bridge construction. About 20,000 tons of steelwork fell into the St. Lawrence River, and 75 workmen lost their lives.

Proceeding now to rearrange the terms in Equation 20.1, we find that

$$S\left(1 - \frac{\Delta S}{S}\right) \geq F\left(1 + \frac{\Delta F}{F}\right); \qquad \frac{S}{F} \geq \frac{1 + \frac{\Delta F}{F}}{1 - \frac{\Delta S}{S}} \tag{20.2}$$

The ERF is now defined as

$$\text{ERF} \equiv \frac{S}{F} \tag{20.3}$$

that is, the nominal strength value, S, of the member divided by its nominal force value, F. Thus, for erection structural integrity or reliability to be maintained, it is necessary that

$$\text{ERF} \geq \frac{1 + \frac{\Delta F}{F}}{1 - \frac{\Delta S}{S}} \tag{20.4}$$

20.19 Minimum Erection Rating Factors

In view of possible errors in (1) the assumed weight of permanent structural components, (2) the assumed weight and positioning of erection equipment, and (3) the mathematical models assumed for purposes of erection structural analysis, it is reasonable to assume that the actual member force for a given erection condition may exceed the computed force value by as much as 10%; that is, it is reasonable to take $\Delta F/F$ as equal to 0.10.

For tension members, uncertainties in (1) the area of the cross-section, (2) the strength of the material, and (3) the member workmanship, indicate that the actual member strength may be up to 15% less than the computed value; that is, $\Delta S/S$ can reasonably be taken as equal to 0.15. The additional uncertainties associated with compression member strength suggest that $\Delta S/S$ be taken as 0.25 for those members. Placing these values into Equation 20.4, we obtain the following minimum ERFs:

$$
\begin{aligned}
\text{Tension members:} \quad \text{ERF}_{t,\min} &= (1 + 0.10)/(1 - 0.15) \\
&= 1.294, \text{ say } 1.30 \\
\text{Compression members:} \quad \text{ERF}_{c,\min} &= (1 + 0.10)/(1 - 0.25) \\
&= 1.467, \text{ say } 1.45
\end{aligned}
$$

The proper interpretation of these expressions is that if, for a given tension (compression) member, the ERF is calculated as 1.30 (1.45) or more, the member can be declared safe for the particular erection condition. Note that higher, or lower, values of erection rating factors may be selected if conditions warrant.

The minimum ERFs determined as indicated are based on experience and judgment, guided by analysis and test results. They do not reflect any specific probabilities of failure and thus are not based on the concept of an acceptable risk of failure, which might be considered the key to a totally rational approach to structural safety. This possible shortcoming in the ERF procedure might be at least partially overcome by evaluating the parameters $\Delta F/F$ and $\Delta S/S$ on a statistical basis; however, this would involve a considerable effort, and it might not even produce significant results.

It is important to recognize that the ERF procedure for determining erection strength adequacy is based directly on fundamental strength and stability criteria, rather than being only indirectly related to such criteria through the medium of a design specification. Thus, the procedure gives uniform results for the erection rating of framed structural members irrespective of the specification that was used to design the members. Obviously, the end use of the completed structure is irrelevant to its strength adequacy during the erection configurations, and therefore the design specification should not be brought into the picture as the basis for erection appraisal.

Experience with application of the ERF procedure to long-span truss bridges has shown that it places the erection engineer in much better contact with the physical significance of the analysis than can be obtained by using the factored design stress procedure. Further, the ERF procedure takes account of secondary stresses, which have generally been neglected in erection stress analysis.

Although the ERF procedure was prepared for application to truss bridge members, the simple governing structural principle set forth by Equation 20.1 could readily be applied to bridge components of any type.

20.20   Deficiencies of Typical Construction Procedure Drawings and Instructions

At this stage of the review it is appropriate to bring forward a key problem in the realm of bridge construction engineering: the strong tendency for construction procedure drawings to be insufficiently clear, and for step-by-step instructions to be either lacking or less than definitive. As a result of these deficiencies it is not uncommon to find the contractor's shop and field evolutions to be going along under something less than suitable control.

Shop and field operations personnel who are in a position to speak frankly to construction engineers will sometimes let them know that procedure drawings and instructions often need to be clarified and upgraded. This is a pervasive problem, and it results from two prime causes: (1) the fabrication and erection engineers responsible for drawings and instructions do not have adequate on-the-job experience, and (2) they are not sufficiently skilled in the art of setting forth on the documents, clearly and concisely, exactly what is to be done by the operations forces—and, sometimes of equal importance, what *is not* to be done.

This matter of clear and concise construction procedure drawings and instructions may appear to be a pedestrian matter, but it is decidedly not. *It is a key issue of utmost importance to the success of steel bridge construction.*

20.21   Shop and Field Liaison by Construction Engineers

In addition to the need for well-prepared construction procedure drawings and instructions, it is essential for the staff engineers carrying out construction engineering to set up good working relations with the shop and field production forces, and to visit the work sites and establish effective communication with the personnel responsible for accomplishing what is shown on the documents (see Figure 20.13).

Figure 20.13   Visiting the work site. It is of first-order importance for bridge construction engineers to visit the site regularly and confer with the job superintendent and his foremen regarding practical considerations. Construction engineers have much to learn from the work forces in shop and field, and vice versa. (Courtesy of Bethlehem Steel Corporation.)

Construction engineers should review each projected operation in detail with the work forces, and upgrade the procedure drawings and instructions as necessary, as the work goes forward. Further, engineers should be present at the work sites during critical stages of fabrication and erection. As a component of these site visits, the engineers should organize special meetings of key production personnel to go over critical operations in detail—complete with slides and blackboard as needed—thereby providing the work forces with opportunities to ask questions and discuss procedures and potential problems, and providing engineers the opportunity to determine how well the work forces understand the operations to be carried out.

This matter of liaison between the office and the work sites—like the preceding issue of clear construction procedure documents—may appear to be somewhat prosaic; again, however, *it is a matter of paramount importance.* Failure to attend to these two key issues constitutes a serious problem in steel bridge construction, and opens the door to high costs and delays, and even to erection accidents.

20.22   Construction Practices and Specifications—The Future

The many existing differences of opinion and procedures in respect to proper governance of steel-work fabrication and erection for major steel bridges raises the question: How do proper bridge construction guidelines come into existence and find their way into practice and into bridge specifi-cations? Looking back over the period roughly from 1900 to 1975, we find that the major steelwork construction companies in the U.S. developed and maintained competent engineering departments that planned and engineered large bridges (and smaller ones as well) through the fabrication and erection processes with a high degree of proficiency. Traditionally, the steelwork contractor's engi-neers worked in cooperation with design-office engineers to develop the full range of bridgework technical factors, including construction procedures and practices.

However, times have changed during the last two decades; since 1970s major steel bridge con-tractors have all but disappeared in the U.S., and further, very few bridge design offices have on their staffs engineers experienced in fabrication and erection engineering. As a result, construction-engineering often receives less attention and effort than it needs and deserves, and this is not a good omen for the future of the design and construction of large bridges in the U.S.

Bridge construction engineering is not a subject that is or can be taught in the classroom; it must be learned on the job with major steelwork contractors. The best route for an aspiring young construction engineer is to spend significant amounts of time in the fabricating shop and at the job site, interspersed with time doing construction-engineering technical work in the office. It has been pointed out previously that although construction engineering and design engineering are related, they constitute different practices and require diverse backgrounds and experience. Design engineering can essentially be learned in the design office; construction engineering, however, cannot—it requires a background of experience at work sites. Such experience, it may be noted, is valuable also for design engineers; however, it is not as necessary for them as it is for construction engineers.

The training of future steelwork construction engineers in the U.S. will be handicapped by the demise of the "Big Two" steelwork contractors in the 1970s. Regrettably, it appears that surviving steelwork contractors in the U.S. generally do not have the resources for supporting strong engi-neering departments, and so there is some question as to where the next generation of steel bridge construction engineers in the U.S. will be coming from.

20.23   Concluding Comments

In closing this review of steel bridge construction it is appropriate to quote from the work of an illustrious British engineer, teacher, and author, the late Sir Alfred Pugsley [14]:

> A further crop of [bridge] accidents arose last century from overloading by traffic
> of various kinds, but as we have seen, engineers today concentrate much of their effort
> to ensure that a margin of strength is provided against this eventuality. But there is
> one type of collapse that occurs almost as frequently today as it has over the centuries:
> collapse at a late stage of erection.
>
>   The erection of a bridge has always presented its special perils and, in spite of
> ever-increasing care over the centuries, few great bridges have been built without loss
> of life. Quite apart from the vagaries of human error, with nearly all bridges there
> comes a critical time near completion when the success of the bridge hinges on some
> special operation. Among such are ... the fitting of a last section ... in a steel arch,

the insertion of the closing central [members] in a cantilever bridge, and the lifting of the roadway deck [structure] into position on a suspension bridge. And there have been major accidents in many such cases. It may be wondered why, if such critical circumstances are well known to arise, adequate care is not taken to prevent an accident. Special care is of course taken, but there are often reasons why there may still be "a slip betwixt cup and lip". Such operations commonly involve unusually close cooperation between constructors and designers, and between every grade of staff, from the laborers to the designers and directors concerned; and this may put a strain on the design skill, on detailed inspection, and on practical leadership that is enough to exhaust even a Brunel.

In such circumstances it does well to . . . recall [the] dictum . . . that "it is essential not to have faith in human nature. Such faith is a recent heresy and a very disastrous one." One must rely heavily on the lessons of past experience in the profession. Some of this experience is embodied in professional papers describing erection processes, often (and particularly to young engineers) superficially uninteresting. Some is crystallized in organizational habits, such as the appointment of resident engineers from both the contracting and [design] sides. And some in precautions I have myself endeavored to list . . . .

It is an easy matter to list such precautions and warnings, but quite another for the senior engineers responsible for the completion of a bridge to stand their ground in real life. This is an area of our subject that depends in a very real sense on the personal qualities of bridge engineers . . . . At bottom, the safety of our bridges depends heavily upon the integrity of our engineers, particularly the leading ones.

20.24   Further Illustrations of Bridges Under Construction, Showing Erection Methods

Figure 20.14   Royal Albert Bridge across River Tamar, Saltash, England, 1857. The two 455-ft (139-m) main spans, each weighing 1060 tons, were constructed on shore, floated out on pairs of barges, and hoisted about 100 ft (30 m) to their final position using hydraulic jacks. Pier masonry was built up after each 3-ft (1-m) lift.

Figure 20.15 Eads Bridge across the Mississippi River, St. Louis, Mo., 1873. The first important metal arch bridge in the U.S., supported by four planes of hingeless trussed arches having chrome-steel tubular chords. Spans are 502-520-502 ft (153-158-153 m). During erection, arch ribs were tied back by cables passing over temporary towers built on the piers. Arch ribs were packed in ice to effect closure.

GLASGOW STEEL BRIDGE,
CHICAGO AND ALTON RAILROAD.
April 8, 1879.

WM. SOOY SMITH,
Engineer.

Figure 20.16   Glasgow (Missouri) railway truss bridge, 1879.   Erection on full supporting falsework was commonplace in the 19th century.   The world's first all-steel bridge, with five 315-ft (96-m) through-truss simple spans, crossing the Missouri River.

Figure 20.17   Niagara River railway cantilever truss bridge, near Niagara Falls, New York 1883. Massive wood erection traveler constructed side span on falsework, then cantilevered half of main span to midspan. Erection of other half of bridge was similar. First modern-type cantilever bridge, with 470-ft (143-m) clear main span having a 120-ft (37-m) center suspended span.

The massive cantilevers of the Forth bridge, shown under erection, were conceived in the shadow of the Tay bridge disaster.

Figure 20.18   Construction of monumental Forth Bridge, Scotland, 1888. Numerous small movable booms were used, along with erection travelers for cantilevering the two 1710-ft (521-m) main spans. The main compression members are tubes 12 ft (3.65 m) in diameter; many other members are also tubular. Total steelwork weight is 51,000 tons. Records are not clear regarding such essentials as cambering and field fitting of individual members in this heavily redundant railway bridge. The Forth is arguably the world's greatest steel structure.

Figure 20.19   Pecos River railway viaduct, Texas, 1892. Erection by massive steam-powered wood traveler having many sets of falls and very long reach. Cantilever-truss main span has 185-ft (56-m) clear opening.

Figure 20.20 Raising of suspended span, Carquinez Strait Bridge, California, 1927. The 433-ft (132-m) suspended span, weighing 650 tons, was raised into position in 35 min., driven by four counterweight boxes having a total weight of 740 tons.

Figure 20.21 First Cooper River cantilever bridge, Charleston, S.C., 1929. Erection travelers constructed 450-ft (137-m) side spans on falsework, then went on to erect 1050-ft (320-m) main span (including 437.5-ft [133-m] suspended span) by cantilevering to midspan.

Figure 20.22   Erecting south tower of Golden Gate Bridge, San Francisco, 1935. A creeper traveler with two 90-ft (27-m) booms erects a tier of tower cells for each leg, then is jumped to the top of that tier and proceeds to erect the next tier. The tower legs are 90 ft (27 m) center-to-center and 690 ft (210 m) high. When the traveler completed the north tower (in background) it erected a Chicago boom on the west tower leg, which dismantled the creeper, erected tower-top bracing, and erected two small derricks (one shown) to service cable erection. Each tower contains 22,200 tons of steelwork.

Figure 20.23 Balanced-cantilever erection, Governor O.K. Allen railway/highway cantilever bridge, Baton Rouge, La., 1939. First use of long balanced-cantilever erection in the U.S. On each pier 650 ft (198 m) of steelwork, about 4000 tons, was balanced on the 40-ft (12-m) base formed by a sloping falsework bent. The compression load at the top of the falsework bent was measured at frequent intervals and adjusted by positioning a counterweight car running at bottom-chord level. The main spans are 848-650-848 ft (258-198-258 m); 650 ft span shown. (Courtesy of Bethlehem Steel Corporation.)

Figure 20.24   Tower erection, second Tacoma Narrows Bridge, Washington, 1949. This bridge replaced first Tacoma Narrows bridge, which blew down in a 40-mph (18-m/sec) wind in 1940. The tower legs are 60 ft (18 m) on centers and 462 ft (141 m) high. The creeper traveler is shown erecting the west tower, in background. On the east tower, the creeper erected a Chicago boom at the top of the south leg; this boom dismantled the creeper, then erected the tower-top bracing and a stiffleg derrick, which proceeded to dismantle the Chicago boom. The tower manhoist can be seen at the second-from-topmost landing platform. Riveting cages are approaching the top of the tower. Note tower-base erection kneebraces, required to ensure tower stability in free-standing condition (see Figure 20.27).

Figure 20.25 Aerial spinning of parallel-wire main cables, second Tacoma Narrows suspension bridge, Washington, 1949. Each main cable consists of 8702 parallel galvanized high-strength wires of 0.196-in. (4.98-mm) diameter, laid up as 19 strands of mostly 460 wires each. Following compaction the cable became a solid round mass of wires with a diameter of 20-1/4 in. (514 mm).

Figure 20.25a Tramway starts across from east anchorage carrying two wire loops. Three 460-wire strands have been spun, with two more under construction. Tramway spinning wheels pull wire loops across the three spans from east anchorage to west anchorage. Suspended footbridges provide access to cables. Spinning goes on 24 hours per day.

Figure 20.25b   Tramway arrives at west anchorage. Wire loops shown in Figure 20.25a are removed from spinning wheels and placed around strand shoes at west anchorage. This tramway then returns empty to east anchorage, while tramway for other "leg" of endless hauling rope brings two wire loops across for second strand that is under construction for this cable.

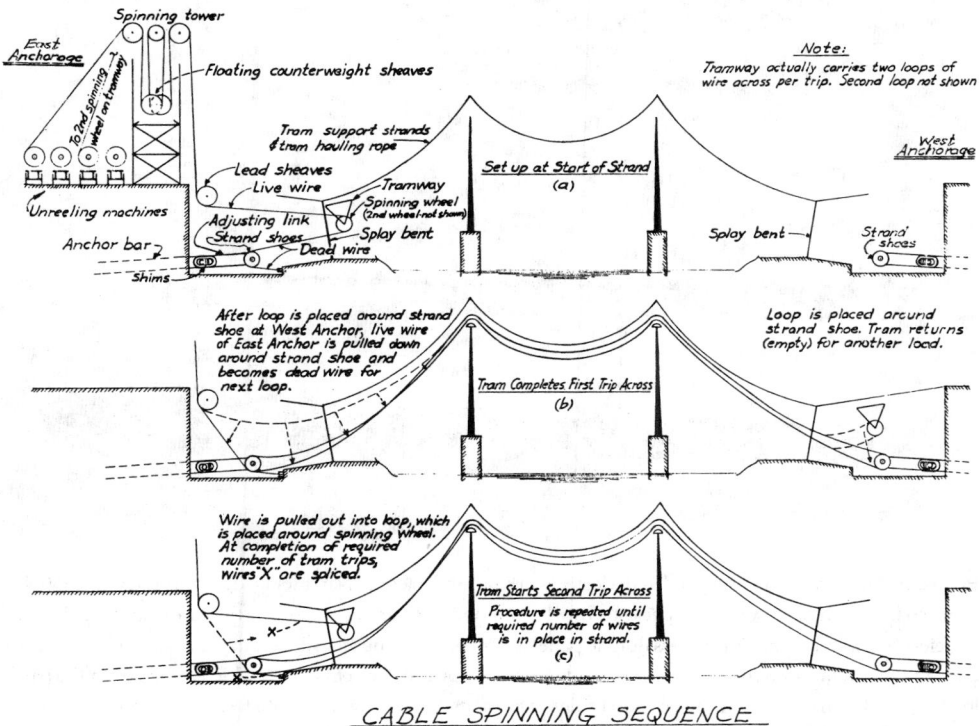

Figure 20.26a   Erection of individual wire loops.

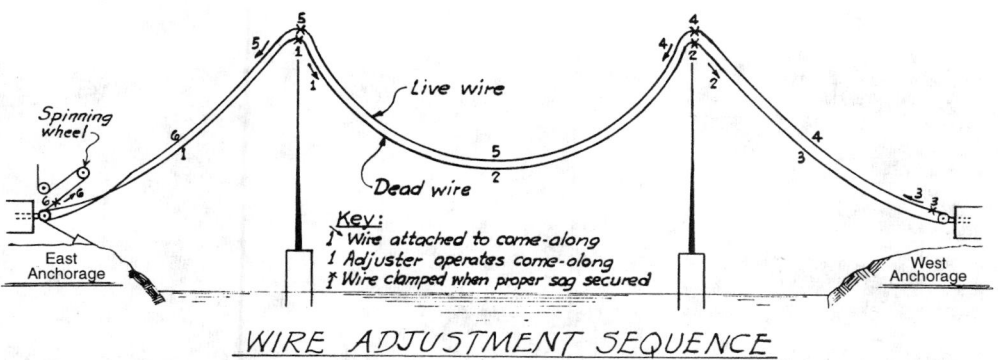

Figure 20.26b   Adjustment of individual wire loops.

Figure 20.26   Cable-spinning procedure for constructing suspension bridge parallel-wire main cables, showing details of aerial spinning method for forming individual 5-mm wires into strands containing 400 to 500 wires. Each wire loop is erected as shown in Figure 20.26a (refer to Figure 20.25), then adjusted to the correct sag as shown in Figure 20.26b. Each completed strand is banded with tape, then adjusted to the correct sag in each span. With all strands in place, they are compacted to form a solid round homogeneous mass of wires. The aerial spinning method was developed by John Roebling in the mid-19th century.

Figure 20.27   Erection of suspended deck steelwork, second Tacoma Narrows Bridge, Washington, 1950. The
Chicago boom on the tower raises deck steelwork components to deck level, where they are transported to deck
travelers by material cars. Each truss double panel is connected at top-chord level to previously erected trusses,
and left open at bottom-chord level to permit temporary upward deck curvature, which results from the partial
loading condition of the main suspension cables. The main span (at right) is 2800 ft (853 m), and side spans are
1100 ft (335 m). The stiffening trusses are 33 ft (10 m) deep and 60 ft (18 m) on centers. Tower-base kneebraces
(see Figure 20.24) show clearly here.

|(a)|(b)|(c)|

Figure 20.28   Moving deck traveler forward, second Tacoma Narrows Bridge, Washington, 1950. The traveler
pulling-falls leadline passes around the sheave beams at the forward end of the stringers, and is attached to the
front of the material car (at left). The material car is pulled back toward the tower, advancing the traveler two
panels to its new position at the end of the deck steelwork. Arrows show successive positions of material car.
(a) Traveler at star of move, (b) traveler advanced one panel, and (c) traveler at end of move.

Figure 20.29   Erecting closing girder sections of Passaic River Bridge, New Jersey Turnpike, 1951. Huge double-boom travelers, each weighing 270 tons, erect closing plate girders of the 375-ft (114-m) main span. The closing girders are 14 ft (4.3 m) deep and 115 ft (35 m) long and weigh 146 tons each. Sidewise entry was required (as shown) because of long projecting splice material. Longitudinal motion was provided at one pier, where girders were jacked to effect closure. Closing girders were laterally stable without floor steel fill-in, such that derrick falls could be released immediately. (Courtesy of Bethlehem Steel Corporation.)

(a)   (b)

Figure 20.30   Floating-in erection of a truss span, first Chesapeake Bay Bridge, Maryland, 1951. Erected 300-ft (91-m) deck-truss spans form erection dock, providing a work platform for two derrick travelers. A permanent deck-truss span serves as a falsework truss supported on barges and is shown carrying the 470-ft (143-m) anchor arm of the through-cantilever truss. This span is floated to its permanent position, then landed onto its piers by ballasting the barges. (a) Float leaves erection dock, and (b) float arrives at permanent position. (Courtesy of Bethlehem Steel Corporation.)

Figure 20.31   Floating-in erection of a truss span, first Chesapeake Bay Bridge, Maryland, 1952. A 480-ft (146-m) truss span, weighing 850 tons, supported on falsework consisting of a permanent deck-truss span along with temporary members, is being floated-in for landing onto its piers. Suspension bridge cables are under construction in background. (Courtesy of Bethlehem Steel Corporation.)

Figure 20.32   Erection of a truss span by hoisting, first Chesapeake Bay Bridge, Maryland, 1952. A 360-ft (110-m) truss span is floated into position on barges and picked clear using four sets of lifting falls. Suspension bridge deck is under construction at right. (Courtesy of Bethlehem Steel Corporation.)

Figure 20.33   Erection of suspension bridge deck structure, first Chesapeake Bay Bridge, Maryland, 1952. A typical four-panel through-truss deck section, weighing 99 tons, has been picked from the barge and is being raised into position using four sets of lifting falls attached to main suspension cables. The closing deck section is on the barge, ready to go up next. (Courtesy of Bethlehem Steel Corporation.)

Figure 20.34   Greater New Orleans cantilever bridge, Louisiana, 1957. Tall double-boom deck travelers started at ends of main bridge and erected anchor spans on falsework, then the 1575-ft (480-m) main span by cantilevering to midspan. (Courtesy of Bethlehem Steel Corporation.)

Figure 20.35 Tower erection, second Delaware Memorial Bridge, Wilmington, Del., 1966. The tower erection traveler has reached the topmost erecting position and swings into place the 23-ton closing top-strut section. The tower legs were jacked apart about 2 in. (50 mm) to provide entering clearance. The traveler jumping beams are in the topmost working position, above the cable saddles. The tower steelwork is about 418 ft (127 m) high. Cable anchorage pier is under construction at right. First Delaware Memorial Bridge (1951) is at left. The main span of both bridges is 2150 ft (655 m). (Courtesy of Bethlehem Steel Corporation.)

Figure 20.36   Erecting orthotropic-plate decking panel, Poplar Street Bridge, St. Louis, Mo., 1967. A five-span, 2165-ft (660-m) continuous box-girder bridge, main span 600 ft (183 m). Projecting box ribs are 5-1/2 × 17 ft (1.7 × 5.2 m) in cross-section, and decking section is 27 × 50 ft (8.2 × 15.2 m). Decking sections were field welded, while all other connections were field bolted. Box girders are cantilevered to falsework bents using overhead "positioning travelers" (triangular structure just visible above deck at left) for intermediate support. (Courtesy of Bethlehem Steel Corporation.)

Figure 20.37 Erection of parallel-wire-strand (PWS) cables, the Newport Bridge suspension spans, Narragansett Bay, R.I., 1968. Bridge engineering history was made at Newport with the development and application of shop-fabricated parallel-wire socketed strands for suspension bridge cables. Each Newport cable was formed of seventy-six 61-wire PWS, about 4512 ft (1375 m) long. Individual wires are 0.202 in. (5.13 mm) in diameter and are zinc coated. Parallel-wire cables can be constructed of PWS faster and at lower cost than by traditional air spinning of individual wires (see Figures 20.25 and 20.26). (Courtesy of Bethlehem Steel Corporation.)

Figure 20.37a Aerial tramway tows PWS from west anchorage up side span, then on across other spans to east anchorage. Strands are about 1-3/4 in. (44 mm) in diameter.

Figure 20.37b   Cable formers maintain strand alignment in cables prior to compaction. Each finished cable is about 15-1/4 in. (387 mm) in diameter. (Courtesy of Bethlehem Steel Corporation.)

Figure 20.38   Pipe-type anchorage for parallel-wire-strand (PWS) cables, the Newport Bridge suspension spans, Narragansett Bay, R.I., 1967. Pipe anchorages shown will be embedded in anchorage concrete. The socketed end of each PWS is pulled down its pipe from the upper end, then seated and shim-adjusted against the heavy bearing plate at the lower end. The pipe-type anchorage is much simpler and less costly than the standard anchor-bar type used with aerial-spun parallel-wire cables (see Figure 20.25b). (Courtesy of Bethlehem Steel Corporation.)

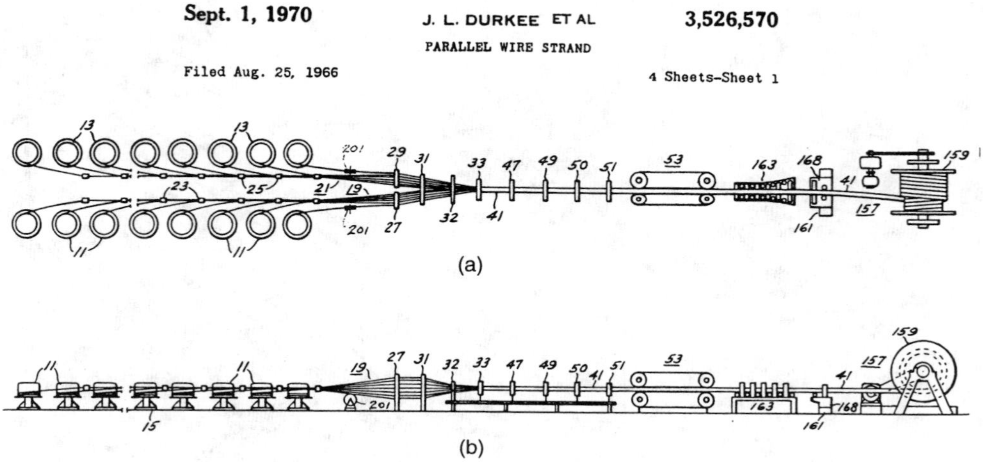

Sept. 1, 1970   J. L. DURKEE ET AL   3,526,570
                                 PARALLEL WIRE STRAND

Filed Aug. 25, 1966   4 Sheets—Sheet 1

(a)

(b)

Figure 20.39 Manufacturing facility for production of shop-fabricated parallel-wire strands (PWS). Prior to 1966, parallel-wire suspension bridge cables had to be constructed wire-by-wire in the field using the aerial spinning procedure developed by John Roebling in the mid-19th century (refer to Figures 20.25 and 20.26). In the early 1960s a major U.S. steelwork contractor originated and developed a procedure for manufacturing and reeling parallel-wire strands, as shown in these patent drawings. A PWS can contain up to 127 wires (see Figures 20.45 and 20.46). (a) Plan view of PWS facility. Turntables 11 contain "left-hand" coils of wire and turntables 13 contain "right-hand" coils, such that wire cast is balanced in the formed strand. Fairleads 23 and 25 guide the wires into half-layplates 27 and 29, followed by full layplates 31 and 32 whose guide holes delineate the hexagonal shape of final strand 41. (b) Elevation view of PWS facility. Hexagonal die 33 contains six spring-actuated rollers that form the wires into regular-hexagon shape; and similar roller dies 47, 49, 50, and 51 maintain the wires in this shape as PWS 41 is pulled along by hexagonal dynamic clamp 53. The PWS is bound manually with plastic tape at about 3-ft (1-m) intervals as it passes along between roller dies. The PWS passes across roller table 163, then across traverse carriage 168, which is operated by traverse mechanism 161 to direct the PWS properly onto reel 159. Finally, the reeled PWS is moved off-line for socketing. Note that wire measuring wheels (201) can be installed and used for control of strand length.

Figure 20.40 Suspended deck steelwork erection, the Newport Bridge suspension spans, Narragansett Bay, R.I., 1968. The closing mainspan deck section is being raised into position by two cable travelers, each made up of a pair of 36-in. (0.91-m) wide-flange rolled beams that ride the cables on wooden wheels. The closing section is 40-1/2 ft (12 m) long at top-chord level, 66 ft (20 m) wide and 16 ft (5 m) deep, and weighs about 140 tons. (Courtesy of Bethlehem Steel Corporation.)

Figure 20.41   Erection of Kansas City Southern Railway box-girder bridge, near Redland, Okla., by "launching", 1970. This nine-span continuous box-girder bridge is 2110 ft (643 m) long, with a main span of 330 ft (101 m). Box cross-section is 11 × 14.9 ft (3.35 × 4.54 m). The girders were launched in two "trains", one from the north end and one from the south end. A "launching nose" was used to carry the leading end of each girder train up onto the skidway supports as the train was pushed out onto successive piers. Closure was accomplished at center of main span. (Courtesy of Bethlehem Steel Corporation.)

Figure 20.41a   Leading end of north girder train moves across 250-ft (76-m) span 4, approaching pier 5. Main span is to right of pier 5.

Figure 20.41b   Launching nose rides up onto pier 5 skidway units, removing girder-train leading-end sag.

Figure 20.41c   Leading end of north girder train is now supported on pier 5.

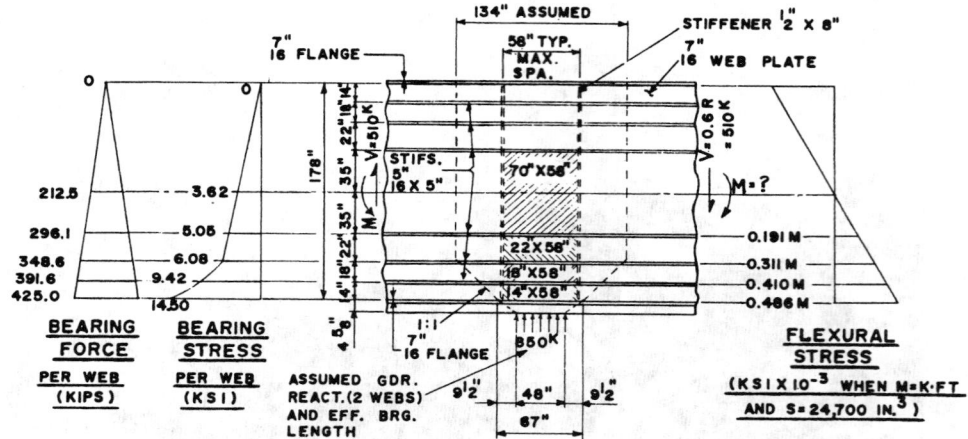

Figure 20.42a Typical assumed erection loading of box-girder web panels in combined moment, shear, and transverse compression.

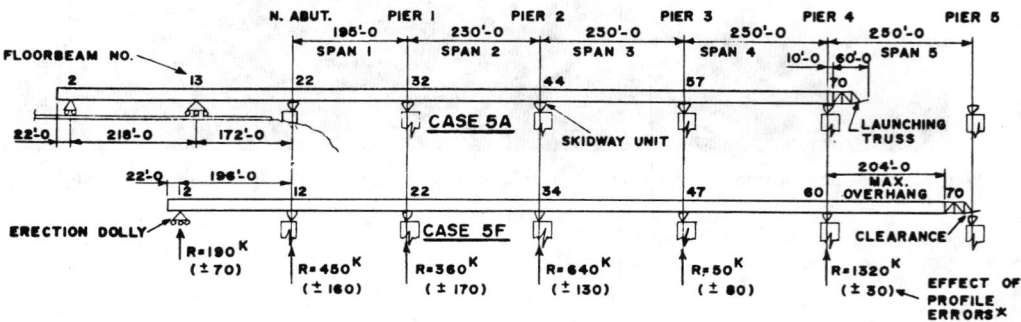

Figure 20.42b Launch of north girder train from pier 4 to pier 5.

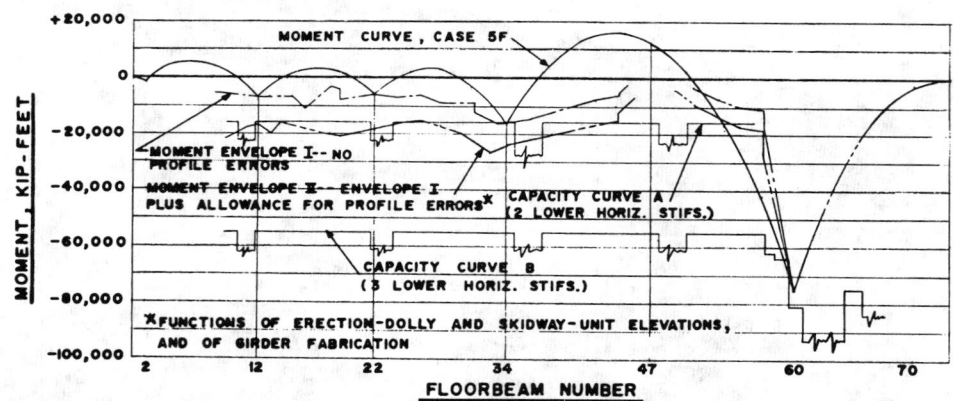

Figure 20.42c Negative-moment envelopes occurring simultaneously with reaction, for launch of north girder train to pier 5.

Figure 20.42 Erection strengthening to withstand launching, Kansas City Southern Railway box-girder bridge, near Redland, Okla. (see Figure 20.41).

Figure 20.43   Erection of west arch span of twin-arch Hernando de Soto Bridge, Memphis, Tenn., 1972. The two 900-ft (274-m) continuous-truss tied-arch spans were erected by a high-tower derrick boat incorporating a pair of barges. West-arch steelwork (shown) was cantilevered to midspan over two pile-supported falsework bents. Projecting east-arch steelwork (at right) was then cantilevered to midspan (without falsework) and closed with falsework-supported other half-arch. (Courtesy of Bethlehem Steel Corporation.)

Figure 20.44   Closure of east side span, Commodore John Barry cantilever truss bridge, Chester, Pa., 1973. A high-tower derrick boat (in background) started erection of trusses at both main piers, supported on falsework; then erected top-chord travelers for main and side spans. The sidespan traveler carried steelwork erection to closure, as shown, and the falsework bent was then removed. The mainspan traveler then cantilevered the steelwork (without falsework) to midspan, concurrently with erection by the west-half mainspan traveler, and the trusses were closed at midspan. Commodore Barry has a 1644-ft (501-m) main span, the longest cantilever span in the U.S., and 822-ft (251-m) side spans. (Courtesy of Bethlehem Steel Corporation.)

Figure 20.45   Reel of parallel-wire strand (PWS), Akashi Kaikyo suspension bridge, Kobe, Japan, 1994. Each socketed PWS is made up of 127 0.206-in. (5.23-mm) wires, is 13,360 ft (4073 m) long, and weighs 96 tons. Plastic-tape bindings secure the strand wires at 1-m intervals. Sockets can be seen on right side of reel. These PWS are the longest and heaviest ever manufactured. (Courtesy of Nippon Steel—Kobe Steel.)

Figure 20.46   Parallel-wire-strand main cable, Akashi Kaikyo suspension bridge, Kobe, Japan, 1994. The main span is 6529 ft (1990 m), by far the world's longest. The PWS at right is being towed across the spans, supported on rollers. The completed cable is made up of 290 PWS, making a total of 36,830 wires, and has a diameter of 44.2 in. (1122 mm) following compaction—the largest bridge cables built to date. Each 127-wire PWS is about 2-3/8 in. (60 mm) in diameter. (Courtesy of Nippon Steel—Kobe Steel.)

Figure 20.47 Artist's rendering of proposed Messina Strait suspension bridge, Italy. The Messina Strait crossing has been under discussion since about 1850, under investigation since about 1950, and under active design since about 1980. The enormous bridge shown would connect Sicily to mainland Italy with a single span of 10,827 ft (3300 m). Towers are 1250 ft (380 m) high. The bridge construction problems for such a span will be tremendously challenging. (Courtesy of Stretto di Messina, S.p.A.)

References

[1] Conditions of Contract and Forms of Tender, Agreement and Bond for Use in Connection with Works of Civil Engineering Construction, 6th ed. (commonly known as "ICE Conditions of Contract"), *Inst. Civil Engrs.* (U.K.), 1991.

[2] Copp, J.I., de Vries, K., Jameson, W.H., and Jones, J. 1945. Fabrication and Erection Controls, Rainbow Arch Bridge Over Niagara Gorge — A Symposium, *Transactions ASCE*, vol. 110.

[3] Durkee, E.L, 1945. Erection of Steel Superstructure, Rainbow Arch Bridge Over Niagara Gorge — A Symposium, *Transactions ASCE*, vol. 110.

[4] Durkee, J.L. 1972. Railway Box-Girder Bridge Erected by Launching, *J. Struct. Div., ASCE*, July.

[5] Durkee, J.L. and Thomaides, S.S. 1977. Erection Strength Adequacy of Long Truss Cantilevers, *J. Struct. Div., ASCE*, January.

[6] Durkee, J.L. 1977. Needed: U.S. Standard Conditions for Contracting, *J. Struct. Div., ASCE*, June.

[7] Durkee, J.L. 1982. Bridge Structural Innovation: A Firsthand Report, *J. Prof. Act., ASCE*, July.

[8] Durkee, J.L. 1966. Advancements in Suspension Bridge Cable Construction, Proceedings, International Symposium on Suspension Bridges, Laboratorio Nacional de Engenharia Civil, Lisbon.

[9] Enquiry into the Basis of Design and Methods of Erection of Steel Box Girder Bridges. Final Report of Committee, 4 vols. (commonly known as "The Merrison Report"), *HMSO* (London), 1973/4.

[10] Feidler, L.L., Jr. 1962. Erection of the Lewiston-Queenston Bridge, *Civil Engrg., ASCE*, November.

[11] Freudenthal, A.M., Ed. 1972. The Engineering Climatology of Structural Accidents, *Proceedings of the International Conference on Structural Safety and Reliability*, Pergamon Press, Elmsford, N.Y.

[12] Holgate, H., Kerry, J.G.G., and Galbraith, J. 1908. Royal Commission Quebec Bridge Inquiry Report, Sessional Paper No. 154, vols. I and II, S.E. Dawson, Ottawa, Canada.

[13] Petroski, H. 1993. Predicting Disaster, *American Scientist*, vol. 81, March.

[14] Pugsley, A. 1968. The Safety of Bridges, *The Structural Engineer*, U.K., July.

[15] Ratay, R.T., Ed. 1996. *Handbook of Temporary Structures in Construction*, 2nd ed., McGraw-Hill, New York.

[16] Schneider, C.C. 1905. The Evolution of the Practice of American Bridge Building, *Transactions ASCE*, vol.54.

[17] Sibly, P.G. and Walker, A.C. 1977. Structural Accidents and Their Causes, *Proc. Inst. Civil Engrs.* (U.K.), vol. 62(1), May.

[18] Smith, D.W. 1976. Bridge Failures, *Proc. Inst. Civil Engrs.* (U.K.), vol. 60(1), August.

21

Basic Principles of Shock Loading

21.1   Introduction 21-1
21.2   Requirements for Optimum Design 21-2
21.3   Absorbing Kinetic Energy 21-2
21.4   Material Properties for Optimum Design 21-7
21.5   Section Properties for Optimum Design 21-8
21.6   Detailing and Workmanship for Shock Loading 21-10
21.7   An Example of Shock Loading 21-11
21.8   Conclusions 21-12
21.9   Defining Terms 21-13
References 21-13
Further Reading 21-13

O.W. Blodgett and
D.K. Miller
*The Lincoln Electric Company,
Cleveland, OH*

21.1   Introduction

Shock loading presents an interesting set of problems to the design engineer. In the engineering community, design for **static loading** traditionally has been the most commonly used design procedure. Designing for shock loading, however, requires a change of thinking in several areas. The objectives of this discussion are to introduce the basic principles of shock loading and to consider their effect upon the integrity of structures.

To understand shock loading, one must first establish the various loading scheme definitions. There are four loading modes that are a function of the strain rate and the number of loading cycles experienced by the member. They are the following:

- static loading
- **fatigue loading**
- shock loading
- **shock/fatigue combination loading**

Static loading occurs when a force is slowly applied to a member. This is a slow or constant loading process that is equivalent to loading the member over a period of 1 min. Designing for static loading traditionally has been the approach used for a wide variety of components, such as buildings, water towers, dams, and smoke stacks.

Fatigue loading occurs when the member experiences alternating, repeated, or fluctuating stresses. Fatigue loading generally is classified by the number of loading cycles on the member and the stress level. Low cycle fatigue usually is associated with stress levels above the yield point, and fracture initiates in less than 1000 cycles. High cycle fatigue involves 10,000 cycles or more and overall stress levels that are below the yield point, although stress raisers from notch-like geometries and/or the residual stresses of welding result in localized plastic deformation. High cycle fatigue failures often occur with overall maximum stresses below the yield strength of the material.

In shock loading, an impact-type force is applied over a short instant of time. The yield and ultimate tensile strengths of a given member can be higher for loads with accelerated strain rates when compared to static loading (see Figure 21.1) [4]. The total elongation, however, remains constant at strain rates above approximately 10^{-4} in./in./s.

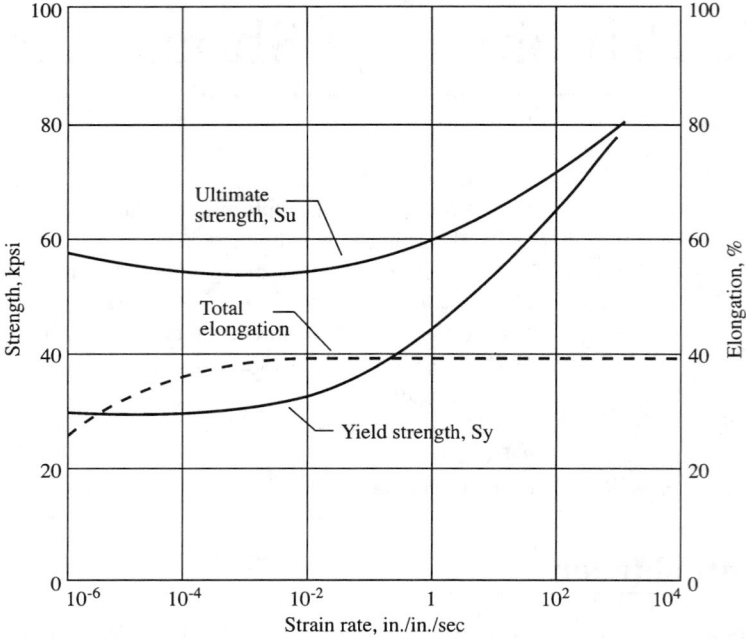

Figure 21.1   Influence of the strain rate on tensile properties. (From Manjoine, M.J., *J. Applied Mech.,* 66, A211, 1994. With permission.)

Shock/fatigue combination loading is equivalent to a shock load applied many times. It exhibits the problems of both fatigue loading and shock loading. Some examples of this loading mode are found in jack hammers, ore crushers, pile drivers, foundry "shake-outs", and landing gear on aircraft.

21.2   Requirements for Optimum Design

For an optimum design, shock loading requires that the entire volume of the member be stressed to the maximum (i.e., the yield point). This refers to both the entire length of the member and the entire cross-section of that member. In practice, however, most members do not have a geometry that allows for the member to be maximally loaded for its entire length, or stressed to the maximum across the entire cross-section.

21.3   Absorbing Kinetic Energy

In Figure 21.2, the member on the left acts as a cantilever beam. Under shock loading, the beam is deflected, which allows that force to be absorbed over a greater distance. The result is a decrease in the impact intensity applied to the beam. The member on the right, however, is hit straight on as a column, and is restricted to little deflection. Therefore, as the column absorbs the same amount of energy, the resulting force is extremely high.

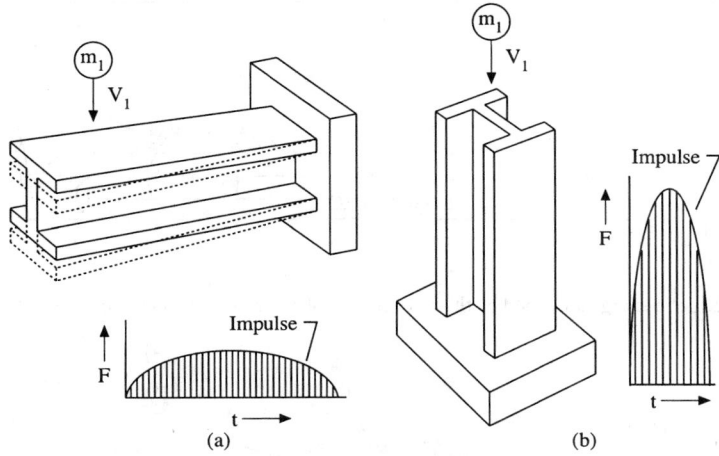

Figure 21.2 Absorbing kinetic energy. (From The Lincoln Electric Co. With permission.)

In shock loading, the energy of the applied force is ultimately absorbed, or transferred, to the structural component designed to resist the force. For example, a vehicle striking a column illustrates this transfer of energy, as shown in Figure 21.3. Prior to impact, the vehicle has a certain value of **kinetic energy** that is equal to $1/2mV^2$. The members that absorb the kinetic energy applied by the vehicle are illustrated in Figure 21.4, where the spring represents the absorbed **potential energy**, E_p, in "a". Figure 21.4b illustrates a vertical stop much like Figure 21.2a. The entire kinetic energy must be absorbed by the receiving member (less any insignificant conversion byproducts—heat, noise, etc.) and momentarily stored as elastic potential energy. This elastic potential energy is stored in the member due to its stressed or deflected condition.

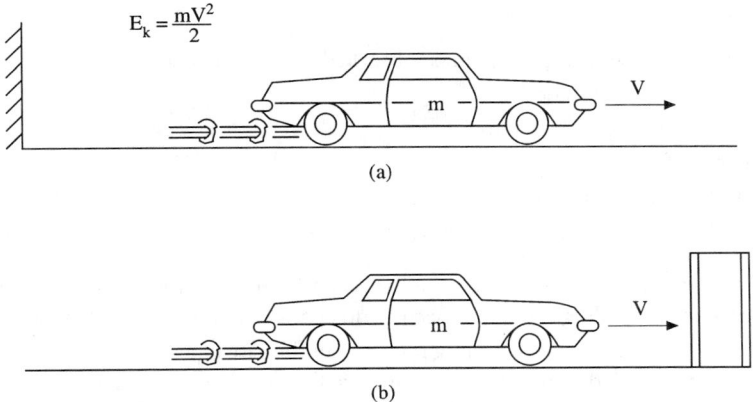

Figure 21.3 Kinetic energy prior to impact. (From The Lincoln Electric Co. With permission.)

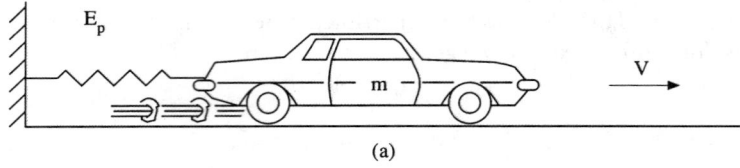

(a)

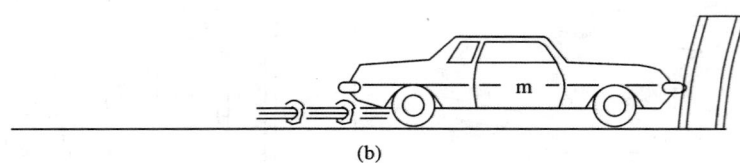

(b)

Figure 21.4   Kinetic energy absorbed by the recovery member. (From The Lincoln Electric Co. With permission.)

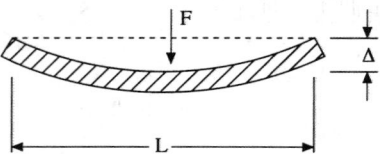

Figure 21.5   Maximum potential energy that can be absorbed by a simply supported beam. (From The Lincoln Electric Co. With permission.)

In Figure 21.5, the maximum potential energy (E_p) that can be elastically absorbed by a simply supported beam is derived. The maximum potential energy that may be stored in the loaded (deflected) beam when stressed to yield is

$$E_p = \frac{\sigma_Y^2 I L}{6 E C^2}$$
(21.1)

where

E_p = maximum **elastic capacity** (max. absorbed potential energy)
σ_Y = yield strength of material
I = moment of inertia of section
L = length of member
E = modulus of elasticity of material
C = distance to outer fiber from neutral axis of section

Figure 21.6 shows various equations that can be used to determine the amount of energy stored in a particular beam, depending upon the specific end conditions and point of load application.

Figure 21.7 addresses the shock loading potential in the case of a simply supported beam with a concentrated force applied in the center. To analyze this problem, two issues must be considered. First, the property of the material is σ_Y^2/E, and second, the property of the cross-section is I/C^2. The objective for optimum performance in shock loading is to ensure maximal energy absorption capability. Both contributing elements, material properties and section properties, will be examined later in this chapter.

Plotted in Figure 21.8 is the applied strain on the horizontal axis and the resisting stress on the vertical axis. If this member is strained up to the yield point stress, the area under that triangle

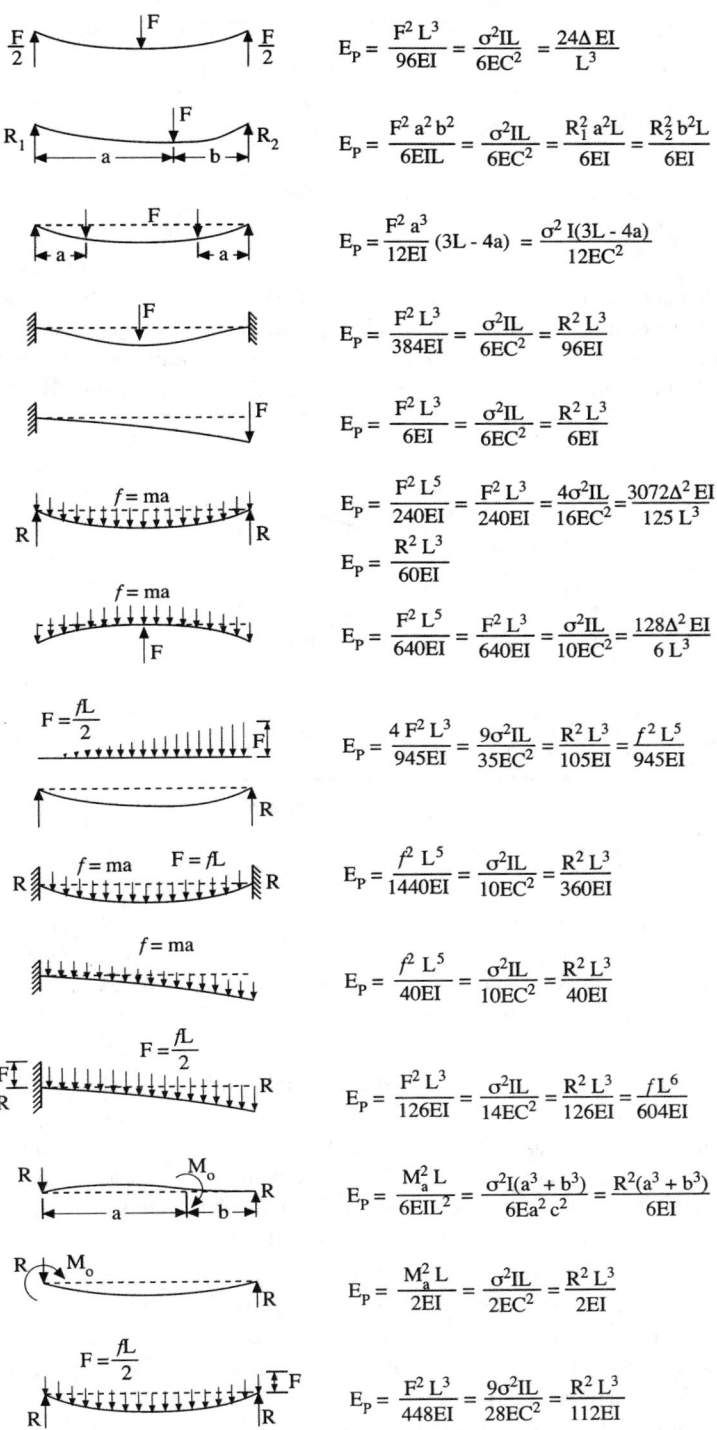

Figure 21.6 Energy storage capability for various end conditions and loading. (From The Lincoln Electric Co. With permission.)

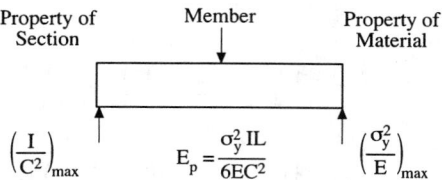

Figure 21.7   An example of shock loading for a simply supported beam. (From The Lincoln Electric Co. With permission.)

represents the maximum energy that can be absorbed elastically (E_p). When the load is removed, it will return to its original position. The remaining area, which is not cross-hatched, represents the plastic (or inelastic) energy if the member is strained. A tremendous amount of plastic energy can be absorbed, but of course, by that time the member is deformed (permanently set) and may be of no further use.

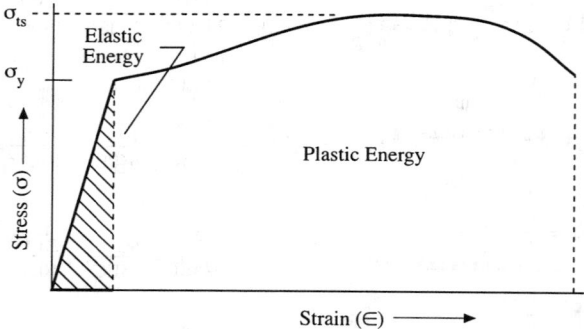

Figure 21.8   Stress-strain curve. (From The Lincoln Electric Co. With permission.)

The stress-strain diagram can help illustrate the total elastic potential capability of a material. The area of the triangular region is $\sigma_Y \varepsilon_Y / 2$. Since $E = \sigma_Y / \varepsilon_Y$ and $\sigma_Y = E \varepsilon_Y$, then this area is equal to $\sigma_Y^2 / 2E$. This gives the elastic energy potential per unit volume of material. As shown in Figure 21.7, the property of the material was σ_Y^2 / E (as defined by Equation 21.1), which is directly proportional to the triangular area in Figure 21.8.

The stress-strain curve for two members (one smooth and one notched) is illustrated in Figure 21.9. At the top of the illustration, the stress-strain curve is shown. Extending from the stress-strain curve is the length of the member. If the member is smooth, as on the lower left, it can be stressed up to the yield point, resulting in the highest strain that can be attained and the maximum elastic energy per unit volume of material. Applying this along the length of the member, the solid prism represents the total energy absorbed by that member.

On the right side of the illustration, note that the notch shown gives a stress raiser. If the stress raiser is assigned an arbitrary value of three-to-one, the area under the notch will display stresses three times greater than the average stress. Therefore, to keep the stress in the notched region below the yield stress, the applied stress must be reduced to 1/3 of the original amount. This will reduce the strain to 1/3, which corresponds to only 1/9 of the energy that will be absorbed. In other words, 1/9 of the elastic energy can be absorbed if the stress at the notch is kept to the yield stress. Notches and cracks in members subject to shock loading can become very dangerous, especially if the stress flow is normal to the notch. Therefore, structural members in shock loading require good workmanship and detailing.

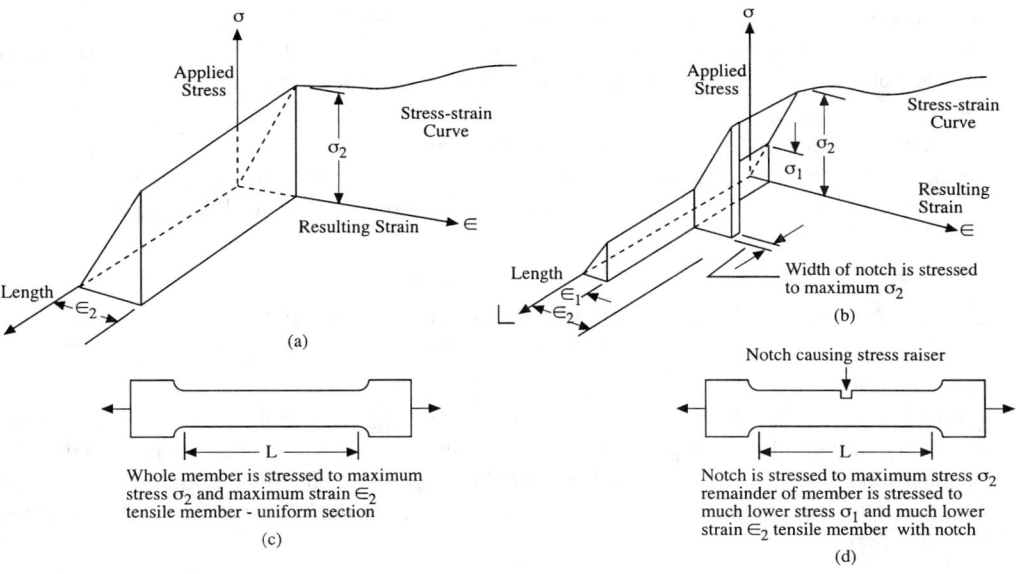

Figure 21.9 Smooth vs. notched. (From The Lincoln Electric Co. With permission.)

21.4 Material Properties for Optimum Design

As shown previously, in Figure 21.8, the elastic energy absorbed during a shock load is equal to the triangular-shaped area under the stress-strain curve up to the yield point. This energy is equal to $\sigma_Y^2/2E$, which is directly a function of the material. In order to maximize the amount of energy that can be absorbed (E_p), with regard to material properties, one must maximize this triangular-shaped area under the stress-strain curve.

In order to compare the energy absorption capability of various materials in shock loading, a list of materials and their corresponding mechanical properties is compiled in Table 21.1. For each

TABLE 21.1 Mechanical Properties of Common Design Materials

Material	Yield strength, σ_Y (ksi)	Ultimate tensile strength, σ_u (ksi)	Modulus of elasticity, E (psi)	Percent elongation (%)	Absorbed elastic energy, $\sigma_Y^2/2E$ (psi)
Mild steel	35	60	30×10^6	35	20.4
Medium carbon steel	45	85	30×10^6	25	33.7
High carbon steel	75	120	30×10^6	8	94.0
A514 Steel	100	115–135	30×10^6	18	166.7
Gray cast iron	6	20	15×10^6	5	1.2
Malleable cast iron	20	50	23×10^6	10	8.7
5056-H18 Aluminum alloy	59	63	10×10^6	10	174.1

material listed in column 1, typical yield and ultimate tensile strengths are tabulated in columns 2 and 3. The modulus of elasticity is recorded in column 4, and column 5 lists the typical percent elongation. Column 6 details the calculated value of $\sigma_Y^2/2E$ for each material.

Higher strength steel will provide a maximum value of $\sigma_Y^2/2E$, since the modulus of elasticity is approximately constant for all steels. However, if aluminum is an option, 5xxx series alloys will provide moderate-to-high strengths with high values of $\sigma_Y^2/2E$. In addition, the 5xxx series alloys usually have the highest welded strengths among aluminum alloys with good corrosion resistance [3].

Some materials, such as cast iron, should not be used for shock loading. Cast iron has a very low value of $\sigma_Y^2/2E$ and will absorb only a small amount of elastic energy. It is ironic that some engineers believe that cast iron is a better material for shock loading than steel or aluminum alloys; this is not the case.

When choosing a material for shock loading applications, the designer should also remember to consider other characteristics of the material such as machineability, weldability, corrosion resistance, etc. For example, very high strength steels are appropriate for shock loading applications; however, they are not very weldable.

21.5   Section Properties for Optimum Design

It was indicated earlier that most engineers are well versed in the principles of steady loading. To demonstrate the differences between steady and shock loading, an example is shown in Figure 21.10. On the left, a steady load is applied to a simply supported beam, and the maximum force that the member can carry for a given stress is $F = 4\sigma I/LC$. From this equation, it is important to realize that if the length of the beam (L) is doubled, then the beam would be able to carry only half of the load at the same stress.

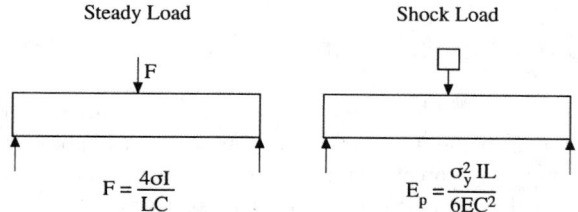

Figure 21.10   Steady load vs. shock load. (From The Lincoln Electric Co. With permission.)

With the shock loading equation, $E_p = \sigma_Y^2 I L/6EC^2$, notice that the length is in the numerator. Therefore, under shock loading, doubling the span of the beam will also double the amount of energy that it can absorb for the same stress level. *This is just the opposite of what happens under static loading.* Under static loading conditions with a fixed load, doubling the length of the beam also doubles the stress level, but with shock loading, doubling the length of the beam reduces the stress level to only 70% of what it was before.

In another example, consider the case of a variable-depth girder shown in the lower portion of Figure 21.11. On the top left there is a prismatic beam (a beam of the same cross-section throughout) under steady load. If, for instance, this is a crane beam in the shop, and there is concern about the weight of the beam, the beam depth can be reduced at the ends. This reduces the dead weight, but leaves the strength capacity of the beam unchanged under a static loading condition. The right-hand side of Figure 21.11 illustrates what happens under shock loading. The equations show that by reducing the beam depth, thus increasing the amount of bending stresses away from the center of the beam to a maximum, E_p is actually doubled.

In Table 21.2, the properties of two beams are analyzed. Beam A, a 12-in. deep, 65-lb beam, was first suggested. It has a **moment of inertia** of 533.4 in.[4]. The steady load section modulus will equal $S = I/C$, which results in a value of 88 in.[3]. Taking a more conventional (i.e., static loading) approach, beam B represents an attempt to create a better beam for shock loading. In this case, the beam size is increased to a depth of 24 in. and a weight of 76 lb. Its moment of inertia is now 2000 in.[4], or four times higher than that of beam A. However, in terms of shock loading alone, beam B has virtually no advantage over beam A because the value of I/C^2 has not changed.

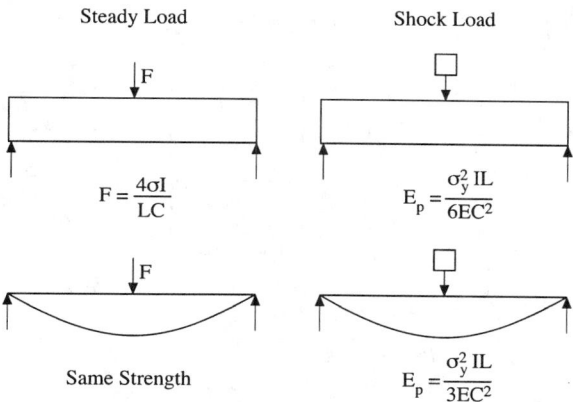

Figure 21.11 illustration with equations:

Steady Load

$$F = \frac{4\sigma I}{LC}$$

Same Strength

Shock Load

$$E_p = \frac{\sigma_y^2 \, IL}{6EC^2}$$

$$E_p = \frac{\sigma_y^2 \, IL}{3EC^2}$$

Figure 21.11   Steady load vs. shock load for a variable-depth beam. (From The Lincoln Electric Co. With permission.)

From Equation 21.1, I/C^2 is directly proportional to the amount of potential energy that can be absorbed, therefore, since I/C^2 has not changed, no additional energy can be absorbed by beam B as compared to beam A.

TABLE 21.2   Properties of Beam A and Beam B

Section property	Beam A 12 in. WF 65 lb	Beam B 24 in. WF 76 lb
Moment of inertia, I	533.4 in.^{4}	2096.4 in.^{4}
C	6.06 in.	11.96 in.
Steady load strength, $S = \frac{I}{C}$	$\frac{533.4}{6.06} = 88.2$ in.^{3}	$\frac{2096.4}{11.96} = 175$ in.^{3}
Shock load strength, $\frac{I}{C^2}$	$\frac{533.4}{(16.067)^2} = 14.5$ in.^{2}	$\frac{2096.4}{(11.96)^2} = 14.6$ in.^{2}

To determine the relative efficiency or economy of the configuration, it is possible to divide I/C^2 by the cross-sectional area (A). Since the material costs are directly related to the material weight (weight is equal to the cross-sectional area multiplied by the density and the length), I/C^2A is a good measure of the configuration's efficiency. Figure 21.12 shows that, for the same flange dimension and web thickness, changing the overall depth of section has little effect upon the shock load strength, I/C^2A. For this reason, when a beam is made deeper, it also becomes more rigid. When a shock load strikes it, the resultant force will be extremely high due to the small deflection. This example demonstrates the importance of discarding the preconceived ideas that come from experience with static loads, and embracing a new way of thinking that is appropriate for shock loading situations.

The effect of the beam web thickness is examined in Figure 21.13. For the same overall depth of section and flange size, decreasing the web thickness increases the shock load strength, I/C^2A. This example makes it clear that decreasing the web thickness actually increases the efficiency of shock load absorption for the same cross-sectional area.

Figure 21.14, on the other hand, demonstrates what happens when the flange width is considered a variable. For the same overall depth of section, increasing the flange width increases the shock load strength, I/C^2A. This puts the steel on the outer fibers, which is an improvement for shock loading. Two different section geometries are considered in Figure 21.15. The box-shaped section has a much higher moment of inertia than the H-shaped section; however, the distance to the outer

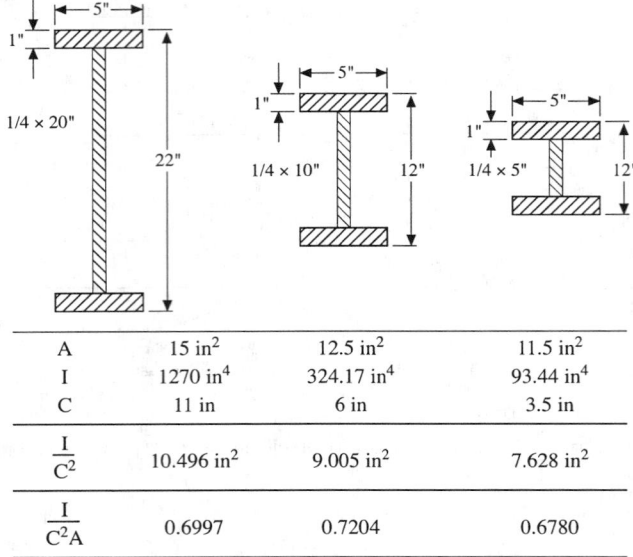

A	15 in²	12.5 in²	11.5 in²
I	1270 in⁴	324.17 in⁴	93.44 in⁴
C	11 in	6 in	3.5 in
$\dfrac{I}{C^2}$	10.496 in²	9.005 in²	7.628 in²
$\dfrac{I}{C^2 A}$	0.6997	0.7204	0.6780

Figure 21.12   Effect of depth on shock loading capacity. (From The Lincoln Electric Co. With permission.)

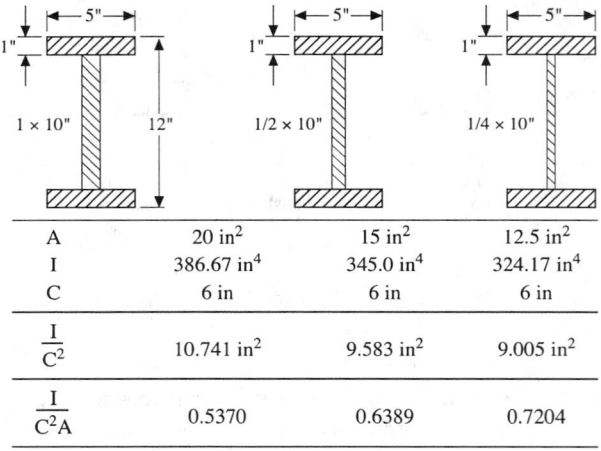

A	20 in²	15 in²	12.5 in²
I	386.67 in⁴	345.0 in⁴	324.17 in⁴
C	6 in	6 in	6 in
$\dfrac{I}{C^2}$	10.741 in²	9.583 in²	9.005 in²
$\dfrac{I}{C^2 A}$	0.5370	0.6389	0.7204

Figure 21.13   Effect of web thickness on shock loading capacity. (From The Lincoln Electric Co. With permission.)

fiber in both cases is equal. By implementing the box-shaped design, there is an increased capability to absorb a shock load over the H section. In the examples previously presented, it is important to realize that improvements were made by changing the section, not the material.

21.6   Detailing and Workmanship for Shock Loading

The fabrication of structures subjected to shock loading is generally no different from that required for slowly loaded structures, except that good workmanship is even more important. Wherever shock loading is a factor, the smallest fabrication problem can have negative, sometimes disastrous, consequences. For welded structures, postweld nondestructive inspection of these assemblies is recommended to verify the integrity of the welds.

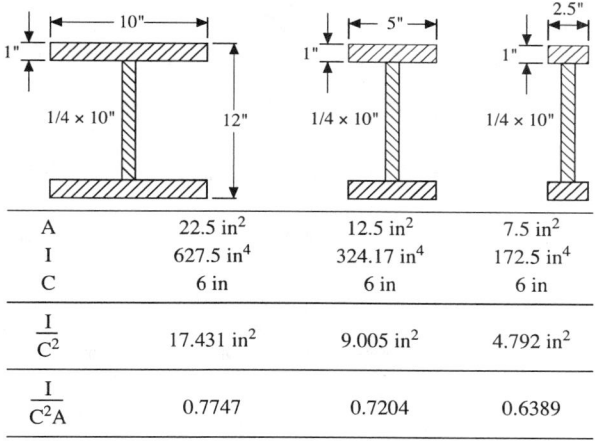

Figure 21.14 Effect of flange width on shock loading capacity. (From The Lincoln Electric Co. With permission.)

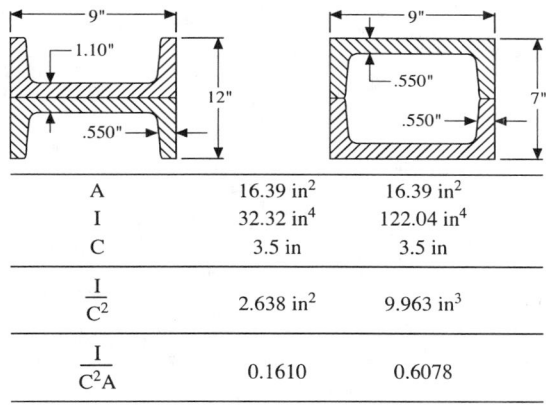

Figure 21.15 Effect of section geometry on shock loading capacity. (From The Lincoln Electric Co. With permission.)

21.7   An Example of Shock Loading

Figure 21.16 illustrates a real-world example of shock loading. The drawing shows ore falling into a primary gyratory crusher. When the ore comes down the chute, it strikes a bumper plate rather than falling with great impact directly into the crusher. This transforms some of the kinetic energy into the structure that supports the bumper. If the bumper plate is considered as a design problem, the falling ore has a normal velocity component to the surface of the bumper plate. Prior to impact, the normal velocity of the ore could equal 213 in./s. When the ore strikes the bumper plate, this normal velocity becomes zero.

If the force caused by the moving ore is found in terms of an incremental time (Δt), the "**impulse**" (force multiplied by incremental time) is equal to the change in **momentum** of the falling ore. After determining the force applied to the bumper plate, the plate size can be calculated. Equations 21.2 through 21.7 below outline the calculation procedure for this problem.

Mass of ore:

$$m = \frac{W}{g} = \frac{AV_1(\Delta t)\delta}{g} \tag{21.2}$$

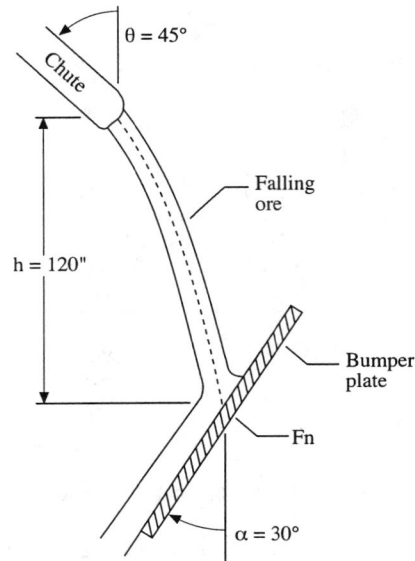

Figure 21.16   An example of shock loading. (From The Lincoln Electric Co. With permission.)

Impulse:

$$Imp = F(\Delta t) \tag{21.3}$$

Change in momentum:

$$\Delta M = mV_n - mV_0 \tag{21.4}$$

Impulse equals change in momentum:

$$Imp = \Delta M \tag{21.5}$$

$$F(\Delta t) = \Delta M = mV_n - mV_0 = \left[\frac{AV_1(\Delta t)\delta}{g}\right]V_n - mV_0 \tag{21.6}$$

$$F = \frac{AV_1\delta V_n}{g} \tag{21.7}$$

where
m = mass of ore
W = weight of ore
t = time ore falls
A = cross-sectional area of ore stream
δ = density of ore
g = acceleration of gravity
V_1 = ore velocity prior to impact
V_0 = ore velocity after impact (assume $V_0 = 0$)
V_n = normal velocity of ore at impact
F = force of impact

This force, F, can then be used to size the bumper plate and the supporting structure of the crusher.

21.8 Conclusions

In structural design, historically, the first tool an engineer utilizes is one applicable to static loading. However, in shock loading applications, new techniques are required to prevent failure and to utilize energy absorption. For example, optimum shock loading performance occurs when a member's complete volume is stressed to a maximum.

When sizing a structural member, remember to consider both the material properties and the section properties. The amount of absorbed kinetic energy is directly related to both of these properties. Detailing and workmanship are also crucial because flaws and cracks act as stress raisers that can reduce the member's ability to withstand shock loads.

21.9 Defining Terms

Elastic capacity: The maximum amount of potential energy that can be stored in a member.

Fatigue loading: Alternating, repeated, or fluctuating stresses that a member experiences over its operating life.

Impulse: The product of the force on a body and the time during which that force acts; equal to the change in momentum.

Kinetic energy: Energy associated with motion.

Moment of inertia: A measure of the resistance of a body to angular acceleration about a given axis that is equal to the sum of the products of each element of mass in the body and the square of the element's distance from the axis.

Momentum: A property of a moving body that the body has by virtue of its mass and motion and that is equal to the product of the body's mass and velocity.

Potential energy: Energy that a piece of matter has because of its position.

Shock/fatigue combination loading: The application of a shock load repeated many times.

Shock loading: The application of an extremely high force over a very short duration of time.

Static loading: Constant loading of a member.

References

[1] Blodgett, O.W. 1996. *Design of Welded Structures,* The James F. Lincoln Arc Welding Foundation.

[2] Blodgett, O.W. 1976. *Design of Weldments,* The James F. Lincoln Arc Welding Foundation.

[3] Kissell, J. Randolph. 1996. "Aluminum and Its Alloys", *Marks' Standard Handbook for Mechanical Engineers.* 10th ed., pp. 6.53–6.60, McGraw-Hill, New York.

[4] Manjoine, M.J. 1944. "Influence of Rate of Strain and Temperature on Yield Stresses of Mild Steel", *J. Appl. Mech.,* 66, A211–A218.

[5] Shigley, J.E. and Mischke, C.R. 1989. *Mechanical Engineering Design,* 5th ed., McGraw-Hill, New York.

[6] Stout, R.D. and Doty, W.D. 1971. *Weldability of Steels,* 2nd ed., Welding Research Council.

Further Reading

[1] *Fatigue Crack Propagation: A Symposium Presented at the Sixty-Ninth Annual Meeting.* 1967. ASTM Special Technical Publication No. 415.

[2] Harris, C.M. and Crede, C.E. 1976. *Shock and Vibration Handbook,* 2nd ed., McGraw-Hill, New York.

[3] Potter, J.M. and McHenry, H.I. 1990. *Fatigue and Fracture Testing of Weldments,* ASTM STP 1058.

[4] Sandor, B.I. 1972. *Fundamentals of Cyclic Stress and Strain,* The University of Wisconsin Press.

22

Welded Connections

22.1	Introduction	22-1
22.2	Joint and Weld Terminology	22-2
22.3	Determining Weld Size	22-6
22.4	Principles of Design	22-18
22.5	Welded Joint Details	22-22
22.6	Design Examples of Specific Components	22-27
22.7	Understanding Ductile Behavior	22-32
22.8	Special Considerations for Welded Structures	22-36
22.9	Materials	22-38
22.10	Connection Details	22-41
22.11	Achieving Ductile Behavior in Seismic Sections	22-45
22.12	Workmanship Requirements	22-52
22.13	Inspection	22-56
22.14	Post-Northridge Assessment	22-58
22.15	Defining Terms	22-60
	References	22-62
	Further Reading	22-63

O.W. Blodgett and
D. K. Miller
The Lincoln Electric Company,
Cleveland, OH

22.1 Introduction

Arc welding has become a popular, widely used method for making steel structures more economical. Although not a new process, welding is still often misunderstood. Perhaps some of the confusion results from the complexity of the technology. To effectively and economically design a building that is to be welded, the engineer should have a knowledge of metallurgy, fatigue, fracture control, weld design, welding processes, welding procedure variables, nondestructive testing, and welding economics. Fortunately, excellent references are readily available, and industry codes specify the minimum standards that are required to be met. Finally, the industry is relatively mature. Although new developments are made every year, the fundamentals of welding are well understood, and many experienced engineers may be consulted for assistance.

Welding is the only joining method that creates a truly one-piece member. All the components of a welded steel structure act in unison, efficiently and effectively transferring loads from one piece to another. Only a minimum amount of material is required when welding is used for joining. Alternative joining methods, such as bolting, are generally more expensive and require the use of lapped plates and angles, increasing the number of pieces required for construction. With welded construction, various materials with different tensile strengths may be mixed, and otherwise unattainable shapes can be achieved. Along with these advantages, however, comes one significant drawback: any problems experienced in one element of a member may be transferred to another.

0-8493-2674-5/97/$0.00+$.50

For example, a crack that exists in the flange of a beam may propagate through welds into a column flange. This means that, particularly in a dynamically loaded structure that is to be joined by welding, all details must be carefully controlled. Interrupted, non-continuous **backing** bars, tack welds, and even seemingly minor arc strikes have resulted in cracks propagating through primary members.

In order to best utilize the unique capabilities of welding, it is imperative to consider the entire design–fabrication–erection sequence. A properly designed welded connection not only transfers stresses safely, but also is economical to fabricate. Successful integration of design, welding processes, metallurgical considerations, inspection criteria, and in-service inspection depends upon mutual trust and free communication between the engineer and the fabricator.

22.2   Joint and Weld Terminology

A welded connection consists of two or more pieces of **base metal** joined by **weld metal**. Engineers determine joint type and generally specify weld type and the required throat dimension. Fabricators select the joint details to be used.

22.2.1   Joint Types

When pieces of steel are brought together to form a joint, they will assume one of the five configurations presented in Figure 22.1. Of the five, butt, tee, corner, and lap joints are common in construction. Coverplates on rolled beams, and angles to gusset plates would be examples of lap joints. Edge joints are more common for sheet metal applications. Joint types are merely descriptions of the relative positioning of the materials; the joint type does not imply a specific type of weld.

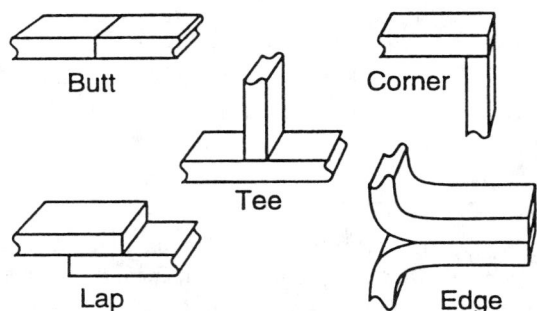

Figure 22.1   Joint types. (Courtesy of The Lincoln Electric Company. With permission.)

22.2.2   Weld Types

Welds may be placed into three major categories: groove welds, fillet welds, and plug or slot welds (see Figure 22.2). For groove welds, there are two subcategories: complete joint penetration (CJP) groove welds and partial joint penetration (PJP) groove welds (see Figure 22.3). Plug welds are commonly used to weld decking to structural supports. Groove and fillet welds are of prime interest for major structural connections.

In Figure 22.4, terminology associated with groove welds and fillet welds is illustrated. Of great interest to the designer is the dimension noted as the "throat." The throat is theoretically the weakest plane in the weld. This generally governs the strength of the welded connection.

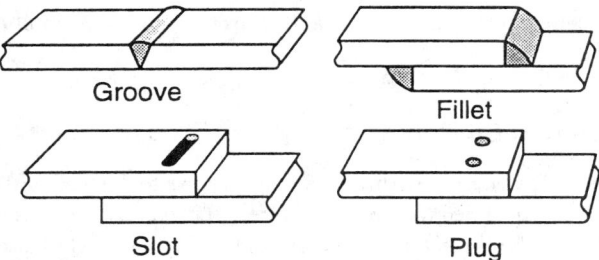

Figure 22.2 Major weld types. (Courtesy of The Lincoln Electric Company. With permission.)

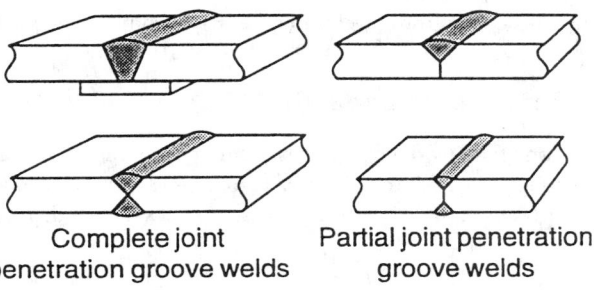

Figure 22.3 Types of groove welds. (Courtesy of The Lincoln Electric Company. With permission.)

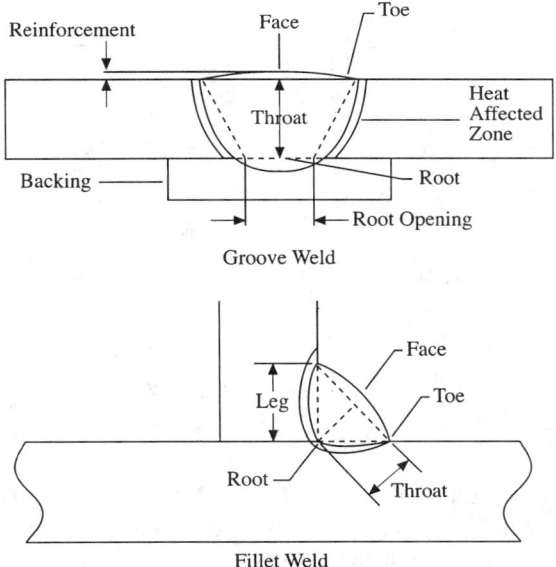

Figure 22.4 Weld terminology. (Courtesy of The Lincoln Electric Company. With permission.)

22.2.3 Fillet Welds

Fillet welds have a triangular cross-section and are applied to the surface of the materials they join. Fillet welds by themselves do not fully fuse the cross-sectional areas of parts they join, although it is still possible to develop full-strength connections with fillet welds.

The size of a fillet weld is usually determined by measuring the leg size, even though the weld is designed by determining the required throat size. For equal-legged, flat-faced fillet welds applied to plates that are oriented 90° apart, the throat dimension is found by multiplying the leg size by 0.707 (i.e., sine 45°).

22.2.4 Complete Joint Penetration (CJP) Groove Welds

By definition, CJP groove welds have a throat dimension equal to the thickness of the plate they join (see Figure 22.3). For prequalified welding procedure specifications, the American Welding Society (AWS) D1.1-96 [9] *Structural Welding Code* requires backing (see Weld Backing) if a CJP weld is made from one side, and **back gouging** if a CJP weld is made from both sides. This ensures complete fusion throughout the thickness of the material being joined. Otherwise, procedure qualification testing is required to prove that the full throat is developed. A special exception to this is applied to tubular connections whose CJP groove welds may be made from one side without backing.

22.2.5 Partial Joint Penetration (PJP) Groove Welds

A PJP groove weld is one that, by definition, has a throat dimension less than the thickness of the materials it joins (see Figure 22.3). An "**effective throat**" is associated with a PJP groove weld (see Figure 22.5). This term is used to delineate the difference between the depth of groove preparation

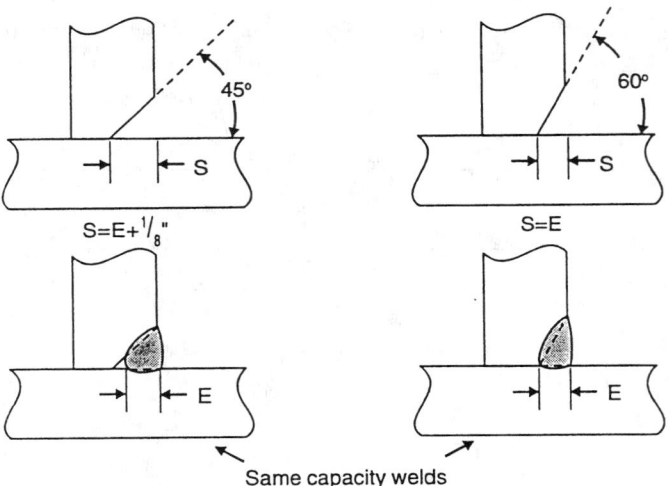

Figure 22.5 PJP groove welds: "E" vs. "S". (Courtesy of The Lincoln Electric Company. With permission.)

and the probable depth of fusion that will be achieved. When submerged arc welding (which has inherently deep penetration) is used, and the weld groove included angle is 60°, the D1.1-96 [9] code allows the designer to rely on the full depth of joint preparation to be used for delivering the required throat dimension. When other processes with less penetration are used, such as shielded metal arc welding, and when the groove angle is restricted to 45°, it is doubtful that fusion to the root of the joint will be obtained. Because of this, the D1.1-96 code assumes that 1/8 in. of the PJP joint may not be fused. Therefore, the effective throat is assumed to be 1/8 in. less than the depth of preparation. This means that for a given included angle, the depth of joint preparation must be increased to offset the loss of penetration.

The effective throat on a PJP groove weld is abbreviated utilizing a capital "E". The required depth of groove preparation is designated by a capital "S". Since the engineer does not normally know which welding process a fabricator will select, it is necessary for the engineer to specify only the dimension for E. The fabricator then selects the welding process, determines the position of welding, and thus specifies the appropriate S dimension, which will be shown on the shop drawings. In most cases, both the S and E dimensions will be contained on the welding symbols of shop drawings, the effective throat dimension showing up in parentheses.

22.2.6 Double-Sided Welds

Welds may be single or double. Double welds are made from both sides of the member (see Figure 22.6). Double-sided welds may require less weld metal to complete the joint. This, of course, has advantages with respect to cost and is of particular importance when joining thick members. However, double-sided joints necessitate access to both sides. If the double joint necessitates overhead welding, the economies of less weld metal may be lost because overhead welding bf deposition rates are inherently slower. For joints that can be repositioned, this is of little consequence. There are also distortion considerations, where the double-sided joints have some advantages in balancing weld shrinkage strains.

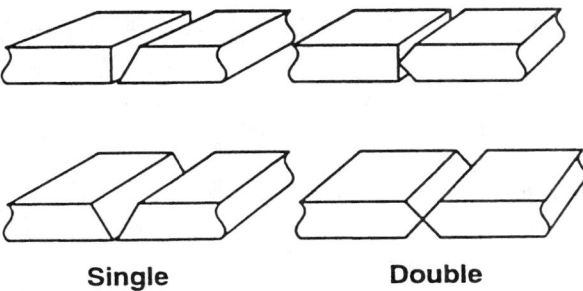

Single   **Double**

Figure 22.6   Single- vs. double-sided joints. (Courtesy of The Lincoln Electric Company. With permission.)

22.2.7 Groove Weld Preparations

Within the groove weld category, there are several types of preparations (see Figure 22.7). If the joint contains no preparation, it is known as a square groove. Except for thin sections, the square groove is rarely used. The bevel groove is characterized by one plate cut at a 90° angle and a second plate with a bevel cut. A vee groove is similar to a bevel, except both plates are bevel cut. A J-groove resembles a bevel, except the root has a radius, as opposed to a straight cut. A U-groove is similar to two J-grooves put together. For butt joints, vee and U-groove details are typically used when welding in the flat position since it is easier to achieve uniform fusion when welds are placed upon the inclined surfaces of these details versus the vertical edge of one side of the bevel or J-groove counterparts.

Properly made, any CJP groove preparation will yield a connection equal in strength to the connected material. The factors that separate the advantages of each type of preparation are largely fabrication related. Preparation costs of the various grooves differ. The flat surfaces of vee and bevel groove weld preparations are generally more economical to produce than the U and J counterparts, although less weld metal is usually required in the later examples. For a given plate thickness, the volume of weld metal required for the different types of grooves will vary, directly affecting fabrication costs. As the volume of weld metal cools, it generates **residual stresses** in the connection that have a direct effect on the extent of distortion and the probability of cracking or lamellar tearing. Reducing weld volume is generally advantageous in limiting these problems. The decision as to which

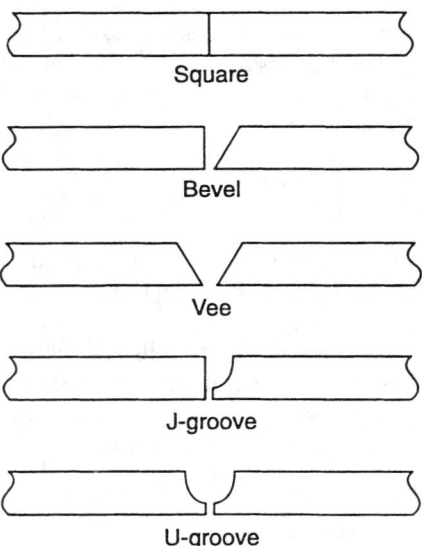

Figure 22.7   Groove weld preparation. (Courtesy of The Lincoln Electric Company. With permission.)

groove type will be used is usually left to the fabricator who, based on knowledge, experience, and available equipment, selects the type of groove that will generate the required quality at a reasonable cost. In fact, design engineers should not specify the type of groove detail to be used, but rather determine whether a weld should be a CJP or a PJP.

22.2.8   Interaction of Joint Type and Weld Type

Not every weld type can be applied to every type of joint. For example, butt joints can be joined only with groove welds. A fillet weld cannot be applied to a butt joint. Tee joints may be joined with fillet welds or groove welds. Similarly, corner joints may be joined with either groove welds or fillet welds. Lap joints would typically be joined with fillet welds or plug/slot welds. Table 22.1 illustrates possible combinations.

22.3   Determining Weld Size

22.3.1   Strength of Welded Connections

A welded connection can be designed and fabricated to have a strength that matches or exceeds that of the steel it joins. This is known as a full-strength connection and can be considered 100% efficient; that is, it has strength equivalent to that of the **base metal** it joins. Welded connections can be designed so that if loaded to destruction, failure would occur in the base material. Poor weld quality, however, may adversely affect weld strength.

A connection that duplicates the base metal capacity is not always necessary and when unwarranted, its specification unnecessarily increases fabrication costs. In the absence of design information, it is possible to specify welds that have strengths equivalent to the base metal capacity. Assuming the base metal thickness has been properly selected, a weld that duplicates the strength of the base metal will be adequate as well. This, however, is a very costly approach. Economical connections cannot be designed on this basis. Unfortunately, the overuse of the CJP detail and the requirement of "matching **filler metal**" (i.e., weld metal of a strength that is equal to that of the base metal) serves as evidence that this is often the case.

TABLE 22.1   Weld Type/Joint Type Interaction

	Fillet	Groove	Plug/Slot
Butt	N.A.		N.A.
Tee			N.A.
Corner			N.A.
Lap		N.A.	

Courtesy of Lincoln Electric Company. With permission.

22.3.2   Variables Affecting Welded Connection Strength

The strength of a welded connection is dependent on the weld metal strength and the area of weld that resists the load. Weld metal strength is a measure of the capacity of the deposited weld metal itself, measured in units such as ksi (kips per square inch). The connection strength reflects the combination of weld metal strength and cross-sectional area, and would be expressed as a unit of force, such as kips. If the product of area times the weld metal strength exceeds the loads applied, the weld should not fail in static service. For cyclic dynamic service, fatigue must be considered as well.

The area of weld metal that resists fracture is the product of the theoretical throat multiplied by the length. The **theoretical weld throat** is defined as the minimum distance from the root of the weld to its theoretical face. For a CJP groove weld, the theoretical throat is assumed to be equal to the thickness of the plate it joins. Theoretical throat dimensions of several types of welds are shown in Figure 22.8.

For fillet welds or partial joint penetration groove welds, using filler metal with strength levels equal to or less than the base metal, the theoretical failure plane is through the weld throat. When the same weld is made using filler metal with a strength level greater than that of the base metal, the failure plane may shift into the fusion boundary or **heat-affected zone**. Most designers will calculate the load capacity of the base metal, as well as the capacity of the weld throat. The fusion zone and its capacity is not generally checked, as this is unnecessary when matching or undermatching weld metal is used. When overmatching weld metal is specifically selected, and the required weld size is deliberately reduced to take advantage of the overmatched weld metal, the designer must check the capacity of the fusion zone (controlled by the base metal) to ensure adequate capacity in the connection.

Complete joint penetration groove welds that utilize weld metal with strength levels exactly equal to the base metal will theoretically fail in either the weld or the base metal. Even with matching weld

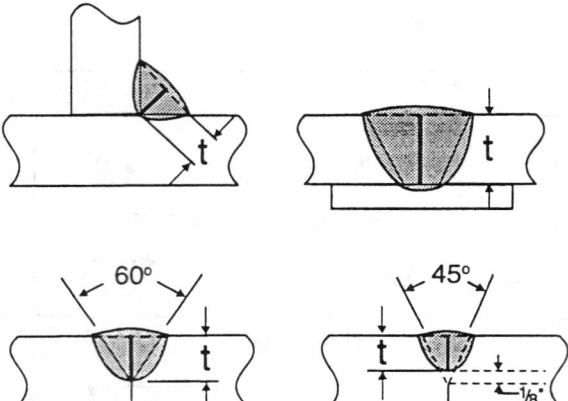

Figure 22.8   Theoretical throats. (Courtesy of The Lincoln Electric Company. With permission.)

metal, the weld metal is generally slightly higher in strength than the base metal, so the theoretical failure plane for transversely loaded connections is assumed to be in the base metal.

22.3.3   Determining Throat Size for Tension or Shear Loads

Connection strength is governed by three variables: weld metal strength, weld length, and weld throat. The weld length is often fixed, due to the geometry of the parts being joined, leaving one variable to be determined, namely, the throat dimension.

For tension or shear loads, the required capacity the weld must deliver is simply the force divided by the length of the weld. The result, in units of force per length (such as kips per inch) can be divided by the weld metal strength, in units of force per area (such as kips per square inch). The final result would be the required throat, in inches. Weld metal allowables that incorporate factors of safety can be used instead of the actual weld metal capacity. This directly generates the required throat size.

To determine the weld size, it is necessary to consider what type of weld is to be used. Assume the preceding calculation determined the need for a 1-in. throat size. If a single fillet weld is to be used, a throat of 1 in. would necessitate a leg size of 1.4 in., shown in Figure 22.9. For double-sided fillets, two 0.7-in. leg size fillets could be used. If a single PJP groove weld is used, the effective throat would have to be 1 in. The actual depth of preparation of the production joint would be 1 in. or greater, depending on the welding procedure and included angle used. A double PJP groove weld would require two effective throats of 0.5 in. each. A final option would be a combination of partial joint penetration groove welds and external fillet welds. As shown in Figure 22.9, a 60° included angle was utilized for the PJP groove weld and an unequal leg fillet weld was applied externally. This acts to shift the effective throat from the normal 45° angle location to a 30° throat.

If the plates being joined are 1 in. thick, a CJP groove weld is the only type of groove weld that will effectively transfer the stress, since the throat on a CJP weld is equal to the plate thickness. PJP groove welds would be incapable of developing adequate throat dimensions for this application, although the use of a combination PJP-fillet weld would be a possibility.

22.3.4   Determining Throat Size for Compressive Loads

When joints are subject only to compression, the unwelded portion of the joint may be milled-to-bear, reducing the required weld throat. Typical of these types of connections are column splices where PJP groove welds frequently are used for static structures.

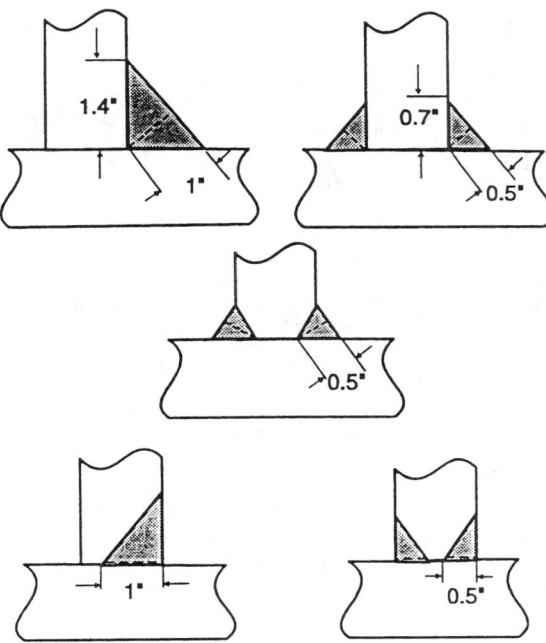

Figure 22.9   Weld combinations with equal throat dimensions. (Courtesy of The Lincoln Electric Company. With permission.)

22.3.5   Determining Throat Size for Bending or Torsional Loads

When a weld, or group of welds, is subject to bending or torsional loads, the weld(s) will not be uniformly loaded. In order to determine the stress on the weld(s), a weld size must be assumed and the resulting stress distribution calculated. An iterative approach may be used to optimize the weld size.

A simpler approach is to treat the weld as a line with no throat. Standard design formulas may be used to determine bending, vertical shear, torsion, etc. These formulas normally result in unit stresses. When applied to welds treated as a line, the formulas result in a force on the welds, measured in pounds per linear inch, from which the capacity of the weld metal, or applicable allowable values, may be used to determine the required throat size.

The following is a simple method used to determine the correct amount of welding required to provide adequate strength for either a bending or a torsional load. In this method, the weld is treated as a line, having no area but having a definite length and cross-section. This method offers the following advantages:

1. It is not necessary to consider throat areas.
2. Properties of the weld are easily found from a table without knowledge of weld leg size.
3. Forces are considered per unit length of weld, rather than converted to stresses. This facilitates dealing with combined-stress problems.
4. Actual values of welds are given as force per unit length of weld instead of unit stress on throat of weld.

Visualize the welded connection as a line (or lines), following the same outline as the connection but having no cross-sectional area. In Figure 22.10, the desired area of the welded connection, A_w, can be presented by just the length of the weld. The stress on the weld cannot be determined unless the weld size is assumed; but by following the proposed procedure, which treats the weld as a line,

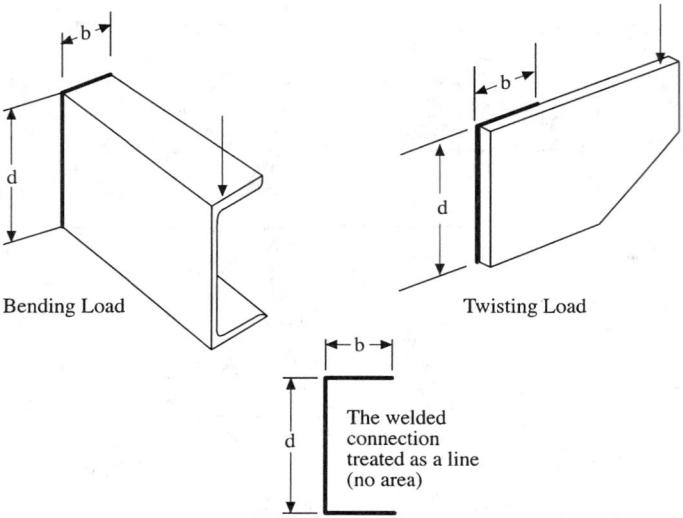

Figure 22.10   Treating the weld as a line for a twisting or bending load: A_w = length of weld (in.), Z_w = section modulus of weld (in.^{2}), J_w = polar moment of inertia of weld (in.^{3}). (Courtesy of The Lincoln Electric Company. With permission.)

the solution is more direct, is much simpler, and becomes basically one of determining the force on the weld(s).

22.3.6   Treating the Weld as a Line to Find Weld Size

By inserting this property of the welded connection into the standard design formula used for a particular type of load (Table 22.2), the unit force on the weld is found in terms of pounds per linear inch of weld.

Normally, use of these standard design formulas results in a unit stress, in pounds per square inch, but with the weld treated as a line, these formulas result in a unit force on the weld, in units of pounds per linear inch.

For problems involving bending or twisting loads, Table 22.3 is used. It contains the section modulus, S_w, and polar moment of inertia, J_w, of 13 typical welded connections with the weld treated as a line. For any given connection, two dimensions are needed: width, b, and depth, d. Section modulus, S_w, is used for welds subjected to bending; polar moment of inertia, J_w, for welds subjected to twisting. Section modulus, S_w, in Table 22.3 is shown for symmetric and asymmetric connections. For asymmetric connections, S_w values listed differentiate between top and bottom, and the forces derived therefrom are specific to location, depending on the value of S_w used.

When more than one load is applied to a welded connection, they are combined vectorially, but must occur at the same location on the welded joint.

22.3.7   Use Allowable Strength of Weld to Find Weld Size

Weld size is obtained by dividing the resulting unit force on the weld by the allowable strength of the particular type of weld used, obtained from Table 22.4 or 22.5. For a joint that has only a transverse load applied to the weld (either fillet or butt weld), the allowable transverse load may be used from the applicable table. If part of the load is applied parallel (even if there are transverse loads in addition), the allowable parallel load must be used.

TABLE 22.2 Standard Design Formulas Used for Determining Force on Weld

Type of Loading		Standard Design Formula Stress lbs/in^2	Treating the Weld as a Line Force lbs/in
		Primary Welds Transmit Entire Load at this Point	
	Tension or compression	$\sigma = \dfrac{P}{A}$	$f = \dfrac{P}{A_w}$
	Vertical Shear	$\tau = \dfrac{V}{A}$	$f = \dfrac{V}{A_w}$
	Bending	$\sigma = \dfrac{M}{Z}$	$f = \dfrac{M}{Z_w}$
	Twisting	$\tau = \dfrac{TC}{J}$	$f = \dfrac{TC}{J_w}$
		Secondary Welds Hold Section Together - Low Stress	
	Horizontal Shear	$\tau = \dfrac{VAy}{It}$	$f = \dfrac{VAy}{In}$
	Torsional Horizontal Shear	$\tau = \dfrac{TC}{J}$	$f = \dfrac{TCt}{J}$

Courtesy of The Lincoln Electric Company. With permission.

22.3.8 Applying the System to Any Welded Connection

1. Find the position on the welded connection where the combination of forces will be maximum. There may be more than one that must be considered.
2. Find the value of each of the forces on the welded connection at this point. Use Table 22.2 for the standard design formula to find the force on the weld. Use Table 22.3 to find the property of the weld treated as a line.
3. Combine (vectorially) all the forces on the weld at this point.
4. Determine the required weld size by dividing this value (step 3) by the allowable force in Table 22.4 or 22.5.

TABLE 22.3 Properties of Welded Connection; Treating Weld as a Line

Properties Of Weld Treated As A Line		
Ouline of Welded Joint b = width d=depth	Bending (about horizontal axis x - x)	Twisting
	$S_w = \dfrac{d^2}{6}$ In^2	$J_w = \dfrac{d^3}{12}$ In^3
	$S_w = \dfrac{d^2}{3}$	$J_w = \dfrac{d(3b^2 + d^2)}{6}$
	$S_w = bd$	$J_w = \dfrac{b^3 + 3bd^2}{6}$
$N_y = \dfrac{b^2}{2(b+d)}$ $N_x = \dfrac{d^3}{2(b+d)}$	$S_w = \dfrac{4bd + d^2}{6} = \dfrac{d^2(2b+d)}{3(b+d)}$ Top Bottom	$J_w = \dfrac{(b+d)^4 - 6b^2d^2}{12(b+d)}$
$N_y = \dfrac{b^2}{2(b+d)}$	$S_w = bd + \dfrac{d^2}{6}$	$J_w = \dfrac{(2b+d)^3}{12} - \dfrac{b^2(b+d)^2}{(2b+d)}$
$N_x = \dfrac{d^3}{2(b+d)}$	$S_w = \dfrac{2bd + d^2}{3} = \dfrac{d^2(2b+d)}{3(b+d)}$ Top Bottom	$J_w = \dfrac{(b+2d)^3}{12} - \dfrac{b^2(b+d)^2}{(b+2d)}$
	$S_w = bd + \dfrac{d^2}{3}$	$J_w = \dfrac{(b+d)^3}{6}$
$N_y = \dfrac{d^2}{b+2d}$	$S_w = \dfrac{2bd + d^2}{3} = \dfrac{d^2(2b+d)}{3(b+d)}$ Top Bottom	$J_w = \dfrac{(b+2d)^3}{12} - \dfrac{d^2(b+d)^2}{(b+2d)}$
$N_y = \dfrac{d^3}{2(b+d)}$	$S_w = \dfrac{4bd + d^2}{3} = \dfrac{4bd^2 + d^3}{6b+3d}$ Top Bottom	$J_w = \dfrac{d^3(4b+d)}{6(b+d)} + \dfrac{b^3}{6}$
	$S_w = bd + \dfrac{d^2}{3}$	$J_w = \dfrac{b^3 + 3bd^2 + d^3}{6}$
	$S_w = 2bd + \dfrac{d^2}{3}$	$J_w = \dfrac{2b^3 + 6bd^2 + d^3}{6}$
	$S_w = \dfrac{\pi d^2}{4}$	$J_w = \dfrac{\pi d^3}{4}$
	$S_w = \dfrac{\pi d^2}{2} + \pi D^2$	

Courtesy of The Lincoln Electric Company. With permission.

TABLE 22.4   Stress Allowables for Weld Metal

Type of weld	Stress in weld		Allowable connection stress	Required filler metal strength level
Complete joint penetration groove welds	Tension normal to the effective area		Same as base metal	Matching filler metal shall be used
	Compression normal to the effective area		Same as base metal	Filler metal with a strength level equal to or one classification (10 ksi [69 MPa]) less than matching filler metal may be used
	Tension or compression parallel to the axis of the weld		Same as base metal	
	Shear on the effective areas		0.30 × nominal tensile strength of filler metal, except shear stress on base metal shall not exceed 0.40 × yield strength of base metal	Filler metal with a strength level equal to or less than matching filler metal may be used
Partial joint penetration groove welds	Compression normal to effective area	Joint not designed to bear	0.50 × nominal tensile strength of filler metal, except stress on base metal shall not exceed 0.60 × yield strength of base metal	
		Joint designed to bear	Same as base metal	
	Tension or compression parallel to the axis of the weld[a]		Same as base metal	Filler metal with strength level equal to or less than matching filler metal may be used
	Shear parallel to axis of weld		0.30 × nominal tensile strength of filler metal, except shear stress on base metal shall not exceed 0.40 × yield strength of base metal	
	Tension normal to effective area		0.30 × nominal tensile strength of filler metal, except tensile stress on base metal shall not exceed 0.60 × yield strength of base metal	
Fillet weld	Shear on effective area		0.30 × nominal tensile strength of filler metal	
	Tension or compression parallel to axis of weld[a]		Same as base metal	Filler metal with a strength level equal to or less than matching filler metal may be used
Plug and slot welds	Shear parallel to faying surfaces (on effective area)		0.30 × nominal tensile strength of filler metal, except shear stress on base metal shall not exceed 0.40 × yield strength of base metal	Filler metal with a strength level equal to or less than matching filler metal may be used

[a] Fillet weld and partial joint penetration groove welds joining the component elements of built-up members, such as flange-to-web connections, may be designed without regard to the tensile or compressive stress in these elements parallel to the axis of the welds.

TABLE 22.5   AISC Fatigue Allowables

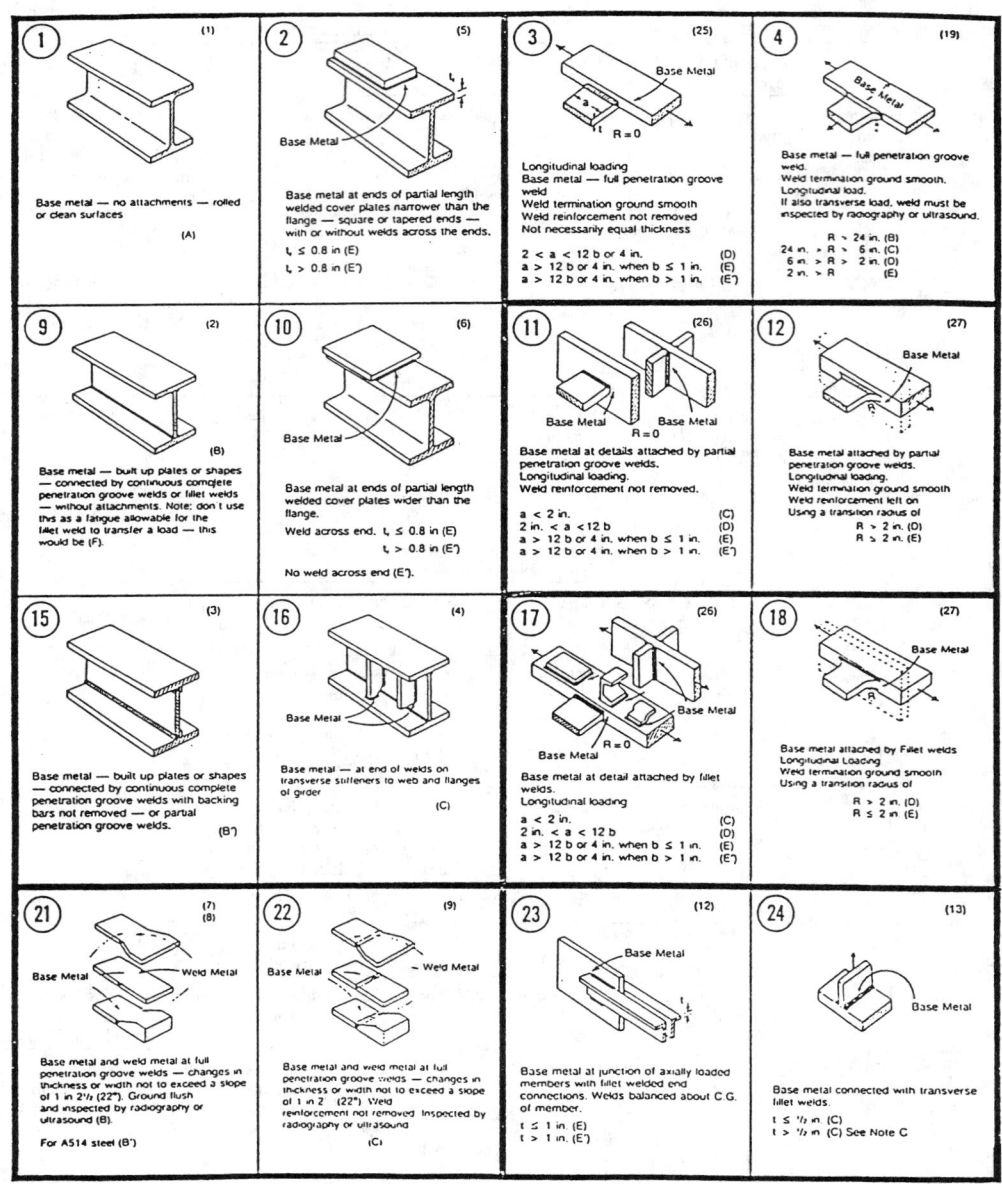

From American Institute of Steel Construction; American Welding Society. *Structural Welding Code: Steel: ANSI/AWS D1.1-96.* Miami, Florida, 1996. With permission.

TABLE 22.5 AISC Fatigue Allowables *(continued)*

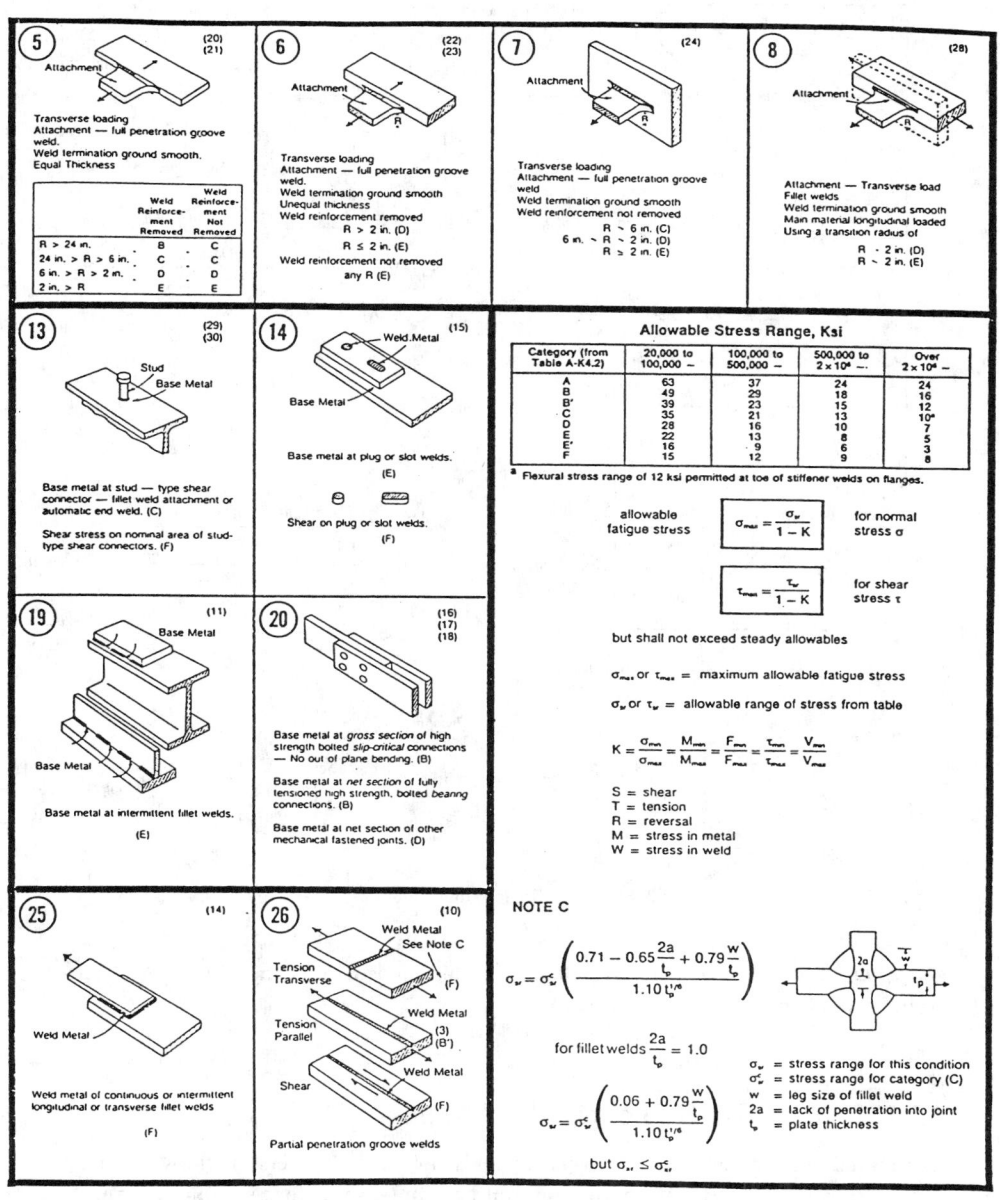

From American Institute of Steel Construction; American Welding Society. *Structural Welding Code: Steel: ANSI/AWS D1.1-96.* Miami, Florida, 1996. With permission.

22.3.9 Sample Calculations Using This System

The example in Figure 22.11 illustrates the application of this procedure.

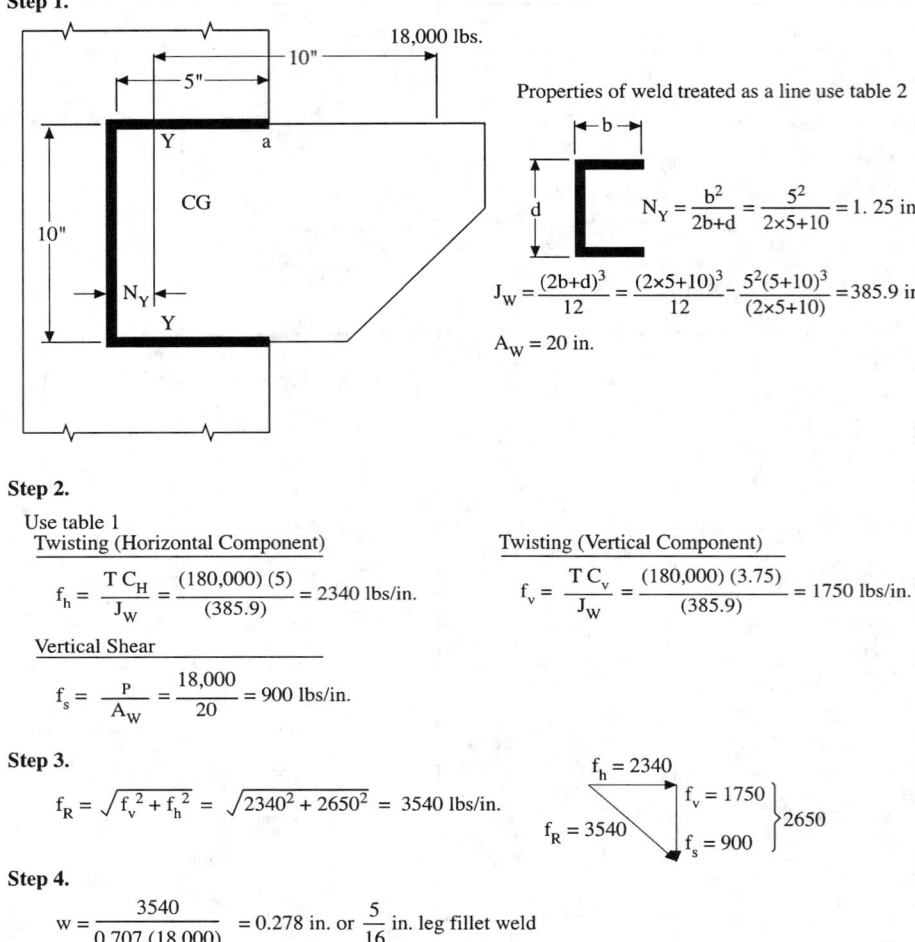

Step 1.

18,000 lbs.

Properties of weld treated as a line use table 2

$$N_Y = \frac{b^2}{2b+d} = \frac{5^2}{2\times5+10} = 1.25 \text{ in.}$$

$$J_W = \frac{(2b+d)^3}{12} = \frac{(2\times5+10)^3}{12} - \frac{5^2(5+10)^3}{(2\times5+10)} = 385.9 \text{ in.}^3$$

$$A_W = 20 \text{ in.}$$

Step 2.

Use table 1

Twisting (Horizontal Component)

$$f_h = \frac{T C_H}{J_W} = \frac{(180,000)(5)}{(385.9)} = 2340 \text{ lbs/in.}$$

Twisting (Vertical Component)

$$f_v = \frac{T C_v}{J_W} = \frac{(180,000)(3.75)}{(385.9)} = 1750 \text{ lbs/in.}$$

Vertical Shear

$$f_s = \frac{P}{A_W} = \frac{18,000}{20} = 900 \text{ lbs/in.}$$

Step 3.

$$f_R = \sqrt{f_v^2 + f_h^2} = \sqrt{2340^2 + 2650^2} = 3540 \text{ lbs/in.}$$

$f_h = 2340$

$f_v = 1750$

$f_R = 3540$

$f_s = 900$

$\left.\right\}2650$

Step 4.

$$w = \frac{3540}{0.707(18,000)} = 0.278 \text{ in. or } \frac{5}{16} \text{ in. leg fillet weld}$$

Figure 22.11 Sample problem using steps outlined in this approach to determine weld size. (Courtesy of The Lincoln Electric Company. With permission.)

22.3.10 Weld Size for Longitudinal Welds

Longitudinal welds include the web-to-flange welds on I-shaped girders and the welds on the corners of box girders. These welds primarily transmit horizontal shear forces resulting from the change in moment along the member. To determine the force between the members being joined, the following equation may be used:

$$f = \frac{Vay}{In}$$

where

f = force on weld per unit length

V = total shear on section at a given position along the beam
a = area of flange connected by the weld
y = distance from the neutral axis of the whole section to the center of gravity of the flange
I = moment of inertia of the whole section
n = number of welds joining the flange to webs per joint

The resulting force per unit length is then divided by the allowable stress in the weld metal and the weld throat is attained. This particular procedure is emphasized because the resultant value for the weld throat is nearly always less than the minimum allowable weld size. The minimum size then becomes the controlling factor.

22.3.11 Minimum Weld Size

Many codes specify minimum weld sizes that are a function of plate thickness. These are not design-related requirements, but rather reflect the inherent interaction of heat input and weld size.

22.3.12 Heat Input and Weld Size

Heat input and weld bead size (or cross-sectional area) are directly related. Heat input is typically calculated with the following equation:

$$H = \frac{60EI}{1000S}$$

where
H = heat input (kJ/in.)
E = arc volts
I = amperage
S = travel speed (in./min)

In order to create a larger weld in one pass, two approaches may be used: higher amperages (I) or slower travel speeds (S) must be employed. Notice that either procedure modification results in a higher heat input. Welding codes have specified minimum acceptable weld sizes with the primary purpose of dictating minimum heat input levels. For example, almost independent of the welding process used, a 1/4-in. fillet weld will require a heat input of approximately 20–30 kJ/in. By prescribing a minimum fillet weld size, these specifications have, in essence, specified a minimum heat input.

Understanding that the minimum fillet weld size is related to heat input, we must also note that there is an inherent interaction of preheat and heat input. The prescribed minimum fillet weld sizes assume the required preheats are also applied. If a situation arises where it is impossible to construct the minimum fillet weld size, it may be appropriate to increase the required preheat to compensate for the reduced energy of welding.

The minimum fillet weld size need never exceed the thickness of the thinner part. It is important to recognize the implications of this requirement. In some extreme circumstances, the connection might involve a very thin plate being joined to an extremely thick plate. The code requirements would dictate that the weld need not exceed the size of the thinner part. However, under these circumstances, additional preheat based upon the thicker material may be justified.

22.3.13 Required Weld vs. Minimum Weld Sizes

When welds are properly sized based upon the forces they are required to transfer, the appropriate weld size frequently is found to be surprisingly small. Even on bridge plate girders that may be 18 to 20 ft deep, with flange thicknesses exceeding 2 in., the required fillet weld size to transmit the horizontal shear forces may be in the range of a 3/32-in. continuous fillet. Intuition indicates that something would be wrong when trying to apply this small weld to join a flange that may be 2 in. thick to a web that is 3/4 in. thick. This is not to indicate a fault with the method used to determine weld size, but rather reveals the small shear forces involved. However, when attempts are made to fabricate this plate girder with these small weld sizes, extremely high travel speeds or very low currents would be required. This naturally would result in an extremely low heat input value. The cooling rates that would be experienced by the weld metal and the base material, specifically the heat-affected zone, would be exceedingly high. A brittle microstructure could be formed. To avoid this condition, the minimum weld size would dictate that a larger weld is required. This is frequently the case for longitudinal welds that resist shear. Any further increase in specified weld size is unnecessary and directly increases fabrication costs.

22.3.14 Single-Pass Minimum Sized Welds

Controlling the heat input by specifying the minimum fillet weld size necessitates that this minimum fillet weld be made in a single pass. If multiple passes are used to construct the minimum sized fillet weld, the intent of the requirement is circumvented. In the past, some recommendations included minimum fillet weld sizes of 3/8 in. and larger. A single-pass 3/8-in. fillet weld can be made only in the flat or vertical position. In the horizontal position, multiple passes are required, and the spirit of the requirement is invalidated. For this reason, the largest minimum fillet weld in Table 5.8 of the AWS D1.1-96 code [9] is 5/16-in. However, even this weld may necessitate multiple passes, depending on the particular welding process used. For example, a quality 5/16-in. fillet weld cannot be made in a single pass with the shielded metal arc welding process utilizing l/8-in.-diameter electrodes, except perhaps in the vertical plane.

It may not be possible to make the required minimum sized fillet weld in a single weld pass under all conditions. For example, it is impossible to make a 5/16-in. fillet weld in a single pass in the overhead position. Under these conditions, it is important to remember the principles that underlie the code requirements. For the preceding example, the overhead fillet weld would necessitate three weld passes. Each weld pass would be made with approximately one-third of the heat input normally associated with the 5/16-in. fillet weld. In order to ensure satisfactory results, it would be desirable to utilize additional preheat to offset the naturally resulting lower heat input that would result from each of these weld passes.

22.3.15 Minimum Sized Groove Welds

When CJP groove welds are made, there is no need to specify the minimum weld size, because the weld size will be the thickness of the base material being joined. This is not the case, however, for PJP, groove welds, so the various codes typically specify minimum PJP groove weld sizes as well. When making CJP groove welds, it is a good practice to make certain that the individual passes applied to the groove meet or exceed the minimum weld size for PJP groove welds.

22.4 Principles of Design

Many welding-related problems have at their root a violation of basic design principles. For dynamically loaded structures, attention to detail is particularly critical. This applies equally to high-cycle fatigue loading, short duration abrupt-impact loading, and seismic loading. The following constitutes a review of basic welding engineering principles that apply to all construction.

22.4.1 Transfer of Forces

Not all welds are evenly loaded. This applies to weld groups that are subject to bending as well as those subject to variable loads along their length. The situation is less obvious when steels of different geometries are joined by welding. A rule of thumb is to assume the transfer of force takes place from one member, through the weld, to the member that lies parallel to the force that is applied. Some examples are illustrated in Figure 22.12. For most simple static loading applications, redistribution of stress throughout the member accommodates the variable loading levels. For dynamically loaded members, however, this is an issue that must be carefully addressed in the design. The addition of stiffeners or continuity plates to column webs helps to unify the distribution of stress across the groove weld.

22.4.2 Minimize Weld Volumes

A good principle of welded design is to always use the smallest amount of weld metal possible for a given application. This not only has sound economic implications, but it reduces the level of residual stress in the connection due to the welding process. All heat-expanded metal will shrink as it cools, inducing residual stresses in the connection. These tendencies can be minimized by reducing the volume of weld metal. Details that will minimize weld volumes for groove welds generally involve minimum root openings, minimum included angles, and the use of double-sided joints.

22.4.3 Recognize Steel Properties

Steel is not a perfectly isotropic material. The best mechanical properties usually are obtained in the same orientation in which the steel was originally rolled, called the X axis. Perpendicular to the X axis is the width of the steel, or the Y axis. Through the thickness, or the Z axis, the steel will exhibit the least amount of ductility, lowest strength, and lowest toughness properties. It is always desirable, if possible, to allow the residual stresses of welding to elongate the steel in the X direction. Of particular concern are large welds placed on either side of the thickness of the steel where the weld shrinkage stress will act in the Z axis. This can result in lamellar tearing during fabrication, or under extreme loading conditions, can result in subsurface fracture.

22.4.4 Provide Ample Access for Welding

It is essential that the design provide adequate access for both welder and welding equipment, as well as good visibility for the welder. As a general rule, if the welder cannot see the joint, neither can the inspector; weld quality will naturally suffer. It is important that adequate access be provided for the proper placement of the welding electrode with respect to the joint. This is a function of the welding process. Gas-shielded processes, for example, must have ample access for insertion of the shielding gas nozzle into the weld joint. Overall access to the joint is a function of the configuration of the surrounding material. The prequalified groove weld details listed in AWS D1.1-96 [9] take these issues into consideration.

22.4.5 No Secondary Members in Welded Design

A fundamental premise of welding design is that there are no secondary members. Anything that is joined by welding can, and will, transfer stress between joined materials. For instance, segmented pieces of steel used for weld backing can result in a stress concentration at the interface of the backing. Attachments that are simply tack welded in place may become major load-carrying members, resulting in the initiation of fracture and propagation throughout the structure. These details must be considered in the design phase of every project, and also controlled during fabrication and erection.

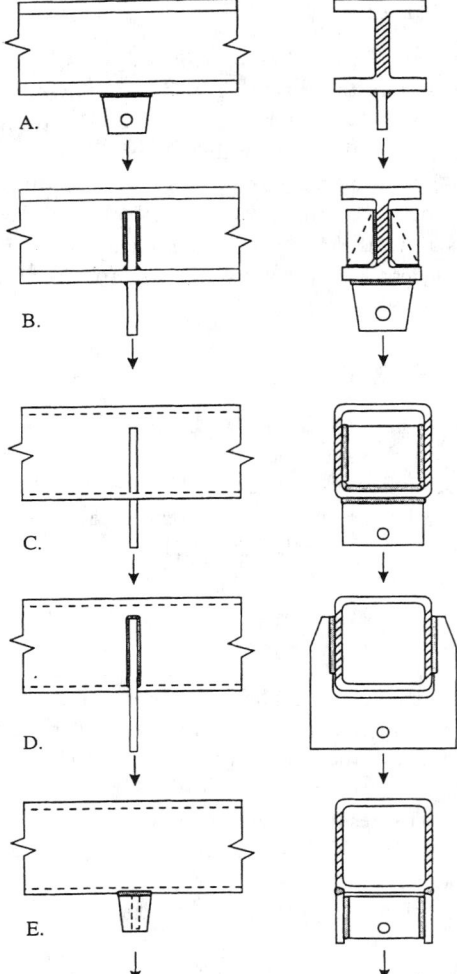

Figure 22.12   Examples of transfer of force. (a) The leg welded under the beam has direct force transfer when oriented parallel to, and directly under, the beam web. (b) The same leg rotated 90° will result in an uneven distribution of stress along the weld length, unless stiffeners are added. The stiffeners could be triangular in shape, since the purpose is to provide a path for force transfer into the weld. (c) For hollow box sections, a lug attached perpendicular to the beam's longitudinal axis results in an unevenly loaded weld until an internal diaphragm is added. (d) Wrapping the lug around the outside of the box section permits it to be directly welded to the section that is parallel to the load, i.e., the vertical sides. (e) Side plates are added to this lug in order to provide a path for force transfer to the vertical sides of the box section. (Courtesy of The Lincoln Electric Company. With permission.)

22.4.6   Residual Stresses in Welding

As heat-expanded weld metal and the surrounding base metal cool to room temperature, they shrink volumetrically. Under most conditions, this contraction is restrained or restricted by the surrounding material, which is relatively rigid and resists the shrinkage. This causes the weld to induce a residual stress pattern, where the weld metal is in residual tension and the surrounding base metal is in residual compression. The residual stress pattern is three dimensional since the metal shrinks volumetrically. The residual stress distribution becomes more complex when multiple-pass welding is performed. The final weld pass is always in residual tension, but subsequent passes will induce compression in previous weld beads that were formerly in tension.

For relatively flexible assemblages, these residual stresses induce distortion. As assemblages become more rigid, the same residual stresses can cause weld cracking, typically occurring shortly after fabrication. If distortion does not occur, or when cracking does not occur, the residual stresses do not relieve themselves, but are "locked in". Residual stresses are considered to be at the yield point of the material involved. Because any area that is subject to residual tensile stress is surrounded by a region of residual compressive stress, there is no loss in overall capacity of **as-welded** structures. However, this reduces the fatigue life for low-stress-range, high-cycle applications.

Small welded assemblies can be thermally stress relieved by heating the steel to 1150°F, holding it for a predetermined length of time (typically 1 h/in. of thickness), and allowing it to return to room temperature. Residual stresses can be reduced by this method, but they are never totally eliminated. This approach is not practical for large assemblies, and care must be exercised to ensure that the components being stress relieved have adequate support when at the elevated temperature, where the yield strength and the modulus of elasticity are greatly reduced, as opposed to room temperature properties. For most structural applications, residual stresses cause no particular problem to the performance of the system, and due to the complexity of stress relief activities, welded structures commonly are used in the as-welded condition.

When loads are applied to as-welded structures, there is some redistribution or gradual decrease in the residual stress patterns. Usually called "shake down", the thermal expansion and contraction experienced by a typical structure as it goes through a climatic season, as well as initial service loads applied to the building, result in a gradual reduction in the residual stresses from welding.

These residual stresses should be considered in any structural application. On a macro level, they will affect the erector's overall sequence of assembling of a building. On a micro level, they will dictate the most appropriate weld bead sequencing in a particular groove-welded joint. For welding applications involving repair, control of residual stresses is particularly important, since the degree of restraint associated with weld repair conditions is inevitably very high. Under these conditions, as well as applications involving heavy, highly restrained, very thick steel for new construction, the experience of a competent welding engineer can be helpful in avoiding the creation of unnecessarily high residual stresses.

22.4.7   Triaxial Stresses and Ductility

The commonly reported values for ductility of steel generally are obtained from uniaxial tensile coupons. The same degree of ductility cannot be achieved under biaxial or triaxial loading conditions. This is particularly significant since residual stresses are always present in any as-welded structure. A more detailed discussion on this subject is found in Section 22.7.

22.4.8   Flat Position Welding

Whenever possible, weld details should be oriented so that the welding can be performed in the flat position, taking advantage of gravity, which helps hold the molten weld metal in place. Flat position welds are made with a lower requirement for operator skill, and at the higher deposition rates that correspond to economical fabrication. This is not to say, however, that overhead welding should be avoided at all costs. An overhead weld may be advantageous if it allows for double-sided welding, with a corresponding reduction in the weld volume. High-quality welds can be made in the vertical plane, and, with the welding consumables available today, can be made at an economical rate.

22.5   Welded Joint Details

22.5.1   Selection of Fillet vs. PJP Groove Welds

For applications where either fillet welds or PJP groove welds are acceptable, the selection is usually based on cost. A variety of factors must be considered in order to determine the most economical weld type.

For welds with equal throat dimensions, the PJP configuration requires one-half the volume of weld metal required by the fillet weld. Alternatively, for equal weld metal volumes, the PJP option is approximately 40% stronger than the fillet weld. Additional factors must be considered, however.

For PJP welds, the bevel surface must be prepared prior to welding, increasing joint preparation cost. Typically achieved by flame cutting, this additional operation requires fuel gas, oxygen, and, most costly of all, labor.

In general, fillet welds are the easiest welds to produce. Access into the more narrow included angles of groove welds usually requires more careful control of welding parameters, commonly resulting in slower welding speeds. The root pass of a PJP groove weld, made into a joint with no root opening, necessitates sufficient included groove angles to avoid centerline cracking tendencies due to poor cross-sectional bead shape. Slag removal may be difficult in root passes as well. These problems do not exist in fillet welds when applied to 90° intersections of T joint members. Such issues can be of concern for skewed T joints, particularly when the acute angle side is less than 60°.

Typical shop practices have generated a general rule of thumb suggesting that fillet welds are the most cost-effective details for connections requiring throats of 1/2 in. or less, which equates to a leg size of 3/4 in. PJP groove welds are generally the best choice for throat sizes of 3/4 in. or greater. This would roughly equate to a 1-in. fillet weld. In general, fillet welds should not exceed 1-in., nor should PJP groove welds be specified for throat dimensions less than 1/2 in. Between these boundaries, specific shop practices will determine the most economical approach.

22.5.2   Weld Backing

When there is a gap between two members to be joined, it is difficult to bridge the space with weld metal. On the other hand, when two members are tightly abutted to each other, it is difficult to obtain complete fusion. To overcome these problems, weld backing is added behind the members to act as a support for the weld metal (Figure 22.13). Weld backing fits into one of two categories: fusible-permanent steel backing or removable backing.

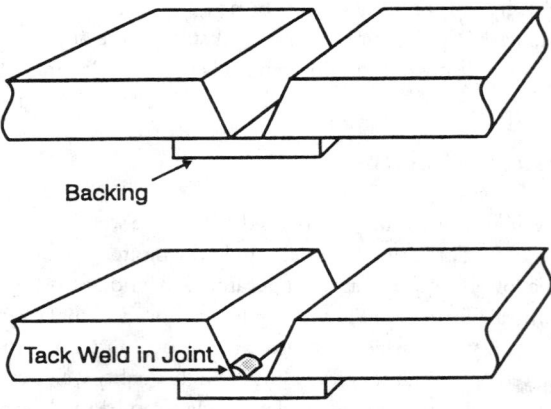

Figure 22.13   Weld backing. (Courtesy of The Lincoln Electric Company. With permission.)

22.5.3 Fusible Backing

Fusible steel backing, commonly known as backing bars, becomes part of the final structure when left in place, so steel that would meet quality requirements for primary members should be used for backing. In general, however, notch toughness properties are not specified for backing. The backing must be continuous for the length of the joint. If multiple pieces of steel backing are to be used in a single joint, they must be joined with CJP groove welds before being applied to the joint they are to back. Welds joining segments of backing bars should be inspected with radiography or ultrasonography to ensure soundness. Interrupted backing bars have been the source of fracture, as well as fatigue crack initiation, and are unacceptable.

For building construction, steel backing is frequently used to compensate for dimensional variations that inevitably occur under field conditions. To maintain plumb columns, there will be slight variations in the dimensions between the columns in a bay. Since the beams are cut to length before the exact dimensions are known, an oversized gap will often result between the beam and the column. Steel backing is inserted underneath this gap, and weld metal is used to bridge this space. It is important to remember, however, that the steel backing becomes part of the final structure if it is left in place.

22.5.4 Removable Backing

Removable backing includes fiberglass tapes, ceramic tiles, and fluxes attached to flexible tape. Removable backing generally is applied when the joint is to be welded with an open arc process such as flux core or shielded metal arc welding. Such backing is applied to the joint with some type of adhesive before the joint is welded. Upon completion of welding, the temporary backing is removed.

Removable backing may be less costly for the fabricator than using the alternatives of double-sided joints or fusible backing. A major obstacle in the use of many of these types of backing is the adhesive that holds the material in place. This is particularly a concern when preheat is required. In some situations, mechanical means have been used to assist in holding the backing in place. When supports are attached by tack welds, care should be exercised to ensure that appropriate techniques are employed.

22.5.5 Copper Backing

Another type of removable backing would be a copper chill bar placed under the joint. Because of the high thermal conductivity of copper, the large difference in melting points of copper versus steel, and physical and chemical differences between the metals, molten weld metal can be supported by copper and the two materials rarely fuse together. This makes copper an attractive material to use for weld backing.

However, this practice is discouraged or prohibited by many codes, because of the possibility of the arc impinging itself on the copper and drawing some of the melted copper into the weld metal. Copper promotes centerline cracking. This would, of course, be unacceptable. As a practical matter, fabricators avoid this practice simply because the copper backing is extremely expensive, and is rapidly ruined when the arc melts a portion of the copper. Copper backing can be used successfully under controlled conditions, which generally involve mechanized welding and joints that do not utilize root openings.

In some situations, the fabricator will mill a groove in a copper chill bar, and fill the groove with clean, dry submerged arc flux. The flux then acts as the backing, and ensures the arc does not melt any of the copper. This is an efficient method and does not have the same ramifications as welding directly against copper. To ensure tight fit of the copper to the back of the joint, pneumatic, mechanical, or hydraulic pressure may be applied to achieve close alignment. Any temporary welds made to attach the backing system to the structural member must employ appropriate welding techniques.

22.5.6 Weld Tabs

Weld tabs, commonly known as starting and run off tabs, are added to the ends of joints in order to facilitate quality welding for the full length of the joint. The start and finish ends of weld beads are known to be more defect prone than the continuous weld between these points. Under starting conditions, the **weld pool** must be established, adequate shielding developed, and thermal equilibrium established. At the termination of a weld, the crater experiences rapid cooling with the extinguishing arc. Shielding is reduced. Cracks and porosity are more likely to occur in craters than at other points of the weld. Starts and stops can be placed on these extension tabs and subsequently removed upon the completion of the weld (see Figure 22.14).

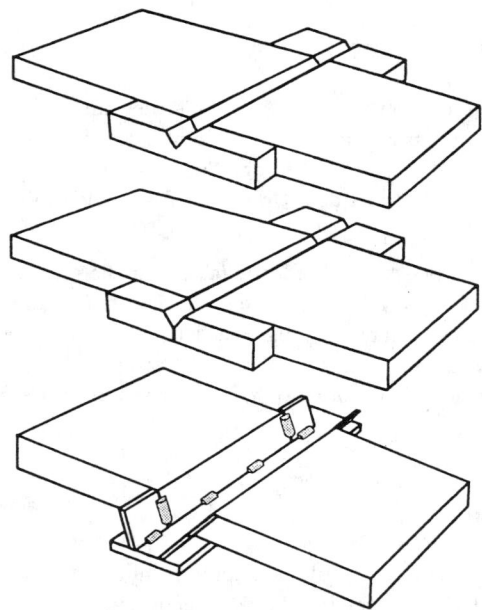

Figure 22.14   Examples of weld tabs. (Courtesy of The Lincoln Electric Company. With permission.)

It is preferable to attach the weld tabs by tack welding within the joint (in Figure 22.14, notice the tack welds in the third example). Preheat requirements must be met when attaching weld tabs, unless the production weld is made with the submerged arc welding process, which will remelt these zones. It is important for weld tabs to have the same geometry as the weld joint to ensure the full throat or plate thickness dimension is maintained at the ends of the weld joint.

When a weld tab containing weld metal of questionable quality is left in place, a fracture can initiate in these regions and propagate along the length of the weld. Weld tabs are removed for bridge fabrications, and since 1989, weld tab removal has been required by American Institute of Steel Construction (AISC) specifications when "jumbo" sections or heavy built-up sections are joined in tension applications by CJP groove welds.

22.5.7 Weld Access Holes

Weld access holes are provided in the web of beam sections to be joined to columns. The access hole in the upper flange connection permits the application of weld backing. The lower weld access hole permits access for the welder to make the bottom flange groove weld. AISC and AWS prescribe minimum weld access hole sizes for these connections ([9], para. 5.17, Figure 5.2). It must be emphasized that these minimum dimensions can be increased for specific requirements necessitated by the weld process, overall geometry, etc. However, the designer must be certain that the resultant section loss is acceptable.

In order to provide ample access for electrode placement, visibility of the joint, and effective cleaning of the weld bead, it is imperative to provide adequate access. In addition to offering access for welding operations, properly sized weld access holes provide an important secondary function: they prevent the interaction of the residual stress fields generated by the vertical weld associated with the web connection and the horizontal weld between the beam flange and column face. The weld access hole acts as a physical barrier to preclude the interaction of these residual stress fields, which can result in cracking. It is best for the weld access hole to terminate in an area of residual compressive stress [21]. More ductile behavior can be obtained under these conditions.

Weld access holes must be properly made. Nicks, gouges, and other geometric discontinuities can act as stress raisers, increasing local stress levels and acting as points of fracture initiation. AISC requires that weld access holes be ground to a bright finish on applications where tension splices are applied to heavy sections. Although not mandated by the codes, these requirements for tension members may be needed for successful fabrication of compression members when connection details typically associated with tension members are applied to compression members (e.g., CJP groove welds) [22].

22.5.8 Lamellar Tearing

Lamellar tearing is a welding-related type of cracking that occurs in the base metal. It is caused by the shrinkage strains of welding acting perpendicular to planes of weakness in the steel. These planes are the result of inclusions in the base metal that have been flattened into very thin plates that are roughly parallel to the surface of the steel. When stressed perpendicular to the direction of rolling, the metallurgical bonds across these plates can separate. Since the various plates are not on the same plane, a fracture may jump between the plates, resulting in a stair-stepped pattern of fractures, illustrated in Figure 22.15. This type of fracture generally occurs near the time of fabrication, and can be confused with underbead cracking.

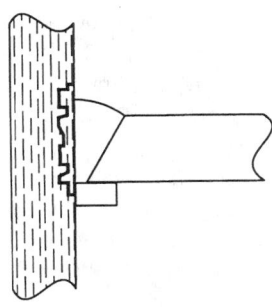

Figure 22.15   Lamellar tearing. (Courtesy of The Lincoln Electric Company. With permission.)

Several approaches can be taken to overcome lamellar tearing. The first variable is the steel itself. Lower levels of inclusions within the steel will help mitigate this tendency. This generally means lower sulfur levels, although the characteristics of the sulfide inclusion are also important. Manganese sulfide is relatively soft, and when the steel is rolled at hot working temperatures of 1600–2000°F, the sulfide inclusions flatten significantly. If steel is first treated to reduce the sulfur, and then calcium treated, for example, the resultant sulfide is harder than the surrounding steel, and during the rolling process, is more likely to remain spherical. This type of material will have much less of a tendency toward lamellar tearing.

Current developments in steel-making practice have helped to minimize lamellar tearing tendencies. With continuously cast steel, the degree of rolling after casting is diminished. The reduction in the amount of rolling has directly affected the degree to which these laminations are flattened, and has correspondingly reduced lamellar tearing tendencies.

The second variable involves the weld joint design. For a specific joint detail, it may be possible to alternate the weld joint to minimize lamellar tearing tendencies. For example, on corner joints it is preferred to bevel the member in which lamellar tearing would be expected, that is, the plate that will be strained in the through-thickness direction. This is illustrated in Figure 22.16.

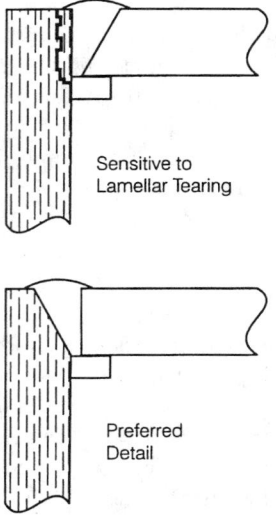

Sensitive to
Lamellar Tearing

Preferred
Detail

Figure 22.16   Lamellar tearing. (Courtesy of The Lincoln Electric Company. With permission.)

A reduction in the volume of weld metal used will help to reduce the stress that is imposed in the through-thickness direction. For example, a single bevel groove weld with a 3/8-in. root opening and 30° included angle will require approximately 22% less weld metal for a 1-1/2-in.-thick plate, compared to a 1/4-in. root opening and a 45° joint. The corresponding reduction in shrinkage stresses may be sufficient to eliminate lamellar tearing.

In extreme cases, it may be necessary to resort to special measures to minimize lamellar tearing, which may involve peening. This technique involves the mechanical deformation of the weld surface, which results in compressive residual stresses that minimize the magnitude of the residual tensile stresses that naturally occur after welding. In order for peening to be effective, it is generally performed when the weld metal is warm (above 300°F), and must cause plastic deformation of the weld surface. Peening is restricted from being applied to root passes (because the partially completed weld joint could easily crack), as well as final weld layers, because the peening can inhibit appropriate visual weld inspection and embrittle the weld metal, which will not be reheated ([9], para. 5.27).

Another specialized technique that can be used to overcome lamellar tearing tendencies is the "buttering layer" technique. With this approach, the surface of the steel where there might be a risk of lamellar tearing is milled to produce a slight cavity in which the butter layer can be applied. Individual weld beads are placed into this cavity. Since the weld beads are not constrained by being attached to a second surface, they solidify and cool, and thereby shrink, with a minimum level of applied stress to the material on which they are placed. After the butter layer is in place, it is possible to weld upon that surface with much less concern about lamellar tearing. This concept is illustrated in Figure 22.17.

Lamellar tearing tendencies are aggravated by the presence of hydrogen. When such tendencies are encountered, it is important to review the low hydrogen practice, examining the electrode selection, care of electrodes, application of preheat, and interpass temperature. Additional preheat can minimize lamellar tearing tendencies.

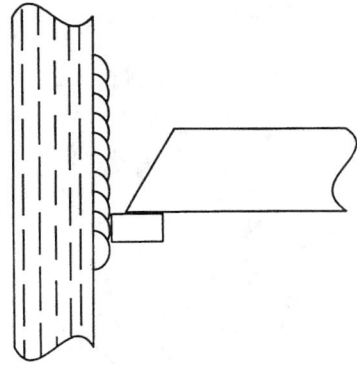

Figure 22.17 "Buttered" surface. (Courtesy of The Lincoln Electric Company. With permission.)

22.6 Design Examples of Specific Components

To demonstrate the design principles of welded connections, five examples are presented. The objective of each example is to determine either the weld leg size or the weld length. These are representative of several beam-to-column design concepts. For further details and examples consult [20].

22.6.1 Flexible Seat Angles

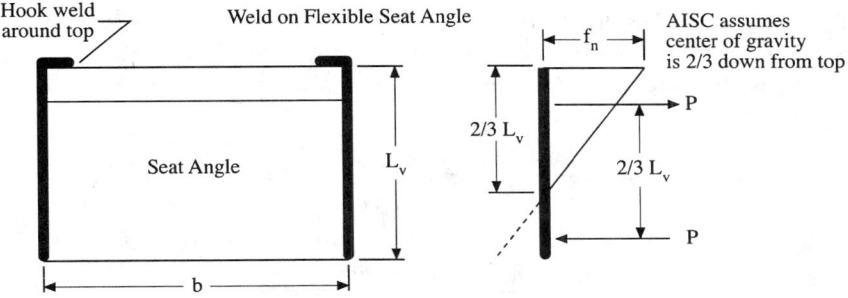

determine maximum unit *horizontal force on weld* (F_n)

$$M = \frac{R}{2}e_f = \frac{2}{3}L_v P \quad \text{also} \quad P = \frac{1}{2}F_n\frac{2}{3}L_v$$

from this

$$F_n = \frac{2.25Re_r}{L_v^2}$$

unit vertical force on weld

$$F_v = \frac{R}{2L_v}$$

resultant unit force on weld (at top)

$$f_r = \sqrt{f_n^2 + f_v^2} = \sqrt{\left(\frac{2.25Re_r}{L_v^2}\right)^2 + \left(\frac{R}{2L_v}\right)^2} = \frac{R}{2L_v^2}\sqrt{L_v^2 + 20.25e_f^2}$$

leg size of fillet weld

$$W = \frac{R\sqrt{L_v^2 + 20.25e_f^2}}{2L_v^2(.707)(.30EXX)}$$

22.6.2 Stiffened Seat Brackets

Beam is supported on a stiffened seat.

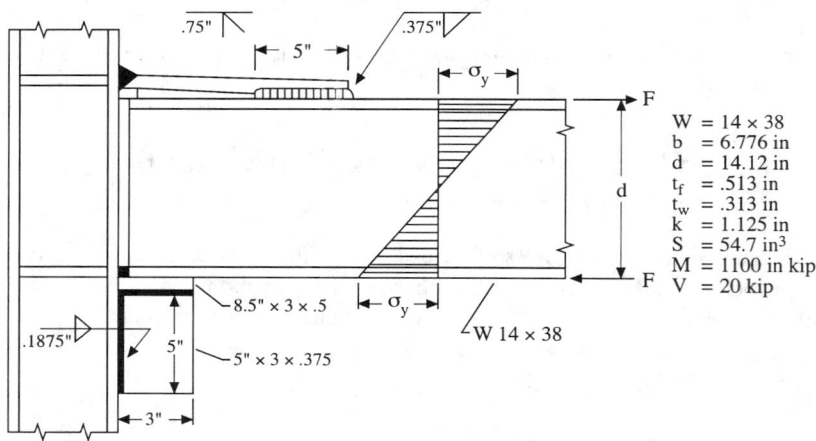

W	= 14 × 38
b	= 6.776 in
d	= 14.12 in
t_f	= .513 in
t_w	= .313 in
k	= 1.125 in
S	= 54.7 in³
M	= 1100 in kip
V	= 20 kip

In this particular connection, the shear reaction is taken as bearing through the lower flange of the beam. There is no welding directly on the web. For this reason it cannot be assumed that the web can be stressed up to its yield in bending throughout its full depth. Since full plastic moment cannot be assumed, the bending stress allowable is held to $\sigma = .60\sigma_y$, or 22 ksi. AISC Sect. 1.5.1.4.1.

Check the bending stress in the beam:

$$\sigma = \frac{M}{S} = \frac{1100 \text{ in.-kip}}{54.7 \text{ in.}} = 20.1 \text{ ksi} \quad < \quad .60\sigma_y \text{ or } 22 \text{ ksi} \quad \text{OK}$$

Bending force in the connection plate:

$$F = \frac{M}{d} = \frac{1100 \text{ in.-kip}}{14.12 \text{ in.}} = 78.0 \text{ kip}$$

Area of the top connection plate:

$$A_p = \frac{F}{\sigma} = \frac{78.0 \text{ kip}}{22 \text{ ksi}} = 3.54 \text{ in.}^2$$

or use a 5 × 3 in. plate which gives a value of $A_p = 3.75$ in.^{2} > 3.54 in.^{2} OK

If a 3/8 in. fillet weld is used to connect top plate to upper beam flange:

$$f_w = (.707)(3/8 \text{ in.})(21 \text{ ksi}) = 5.56 \text{ kips for linear inch of weld.}$$

Length of fillet weld:

$$L = \frac{F}{f_w} = \frac{78.0 \text{ kip}}{5.56 \text{ kip/in.}} = 14.1 \text{ in.}$$

or use 5 in. across the end of the plate, and 5 in. along each side, a total length of 15 in. > 14.1 in. OK

22.6.3 Web Framing Angles

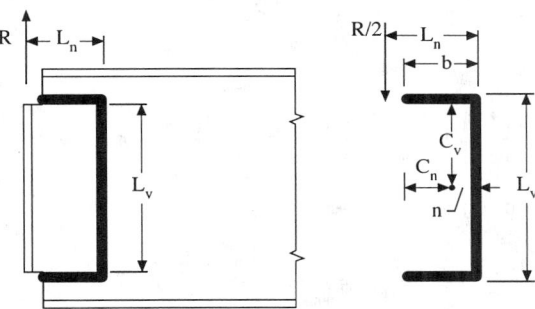

Web Framing Angles to Beam (usually shop weld)

Twisting

$$\text{(horizontal)} \quad F_n = \frac{J C_v}{J_w}$$

$$\text{(vertical)} \quad F_{v1} = \frac{J C_n}{J_w}$$

Shear

$$\text{(vertical)} \quad F_{v2} = \frac{R}{2(2b - L_v)}$$

Resultant force

$$F_r = \sqrt{F_n^2 - (F_{v1} + F_{v2})^2}$$

Leg size of fillet weld

$$W = \frac{F_r}{.707(.30 E X X)}$$

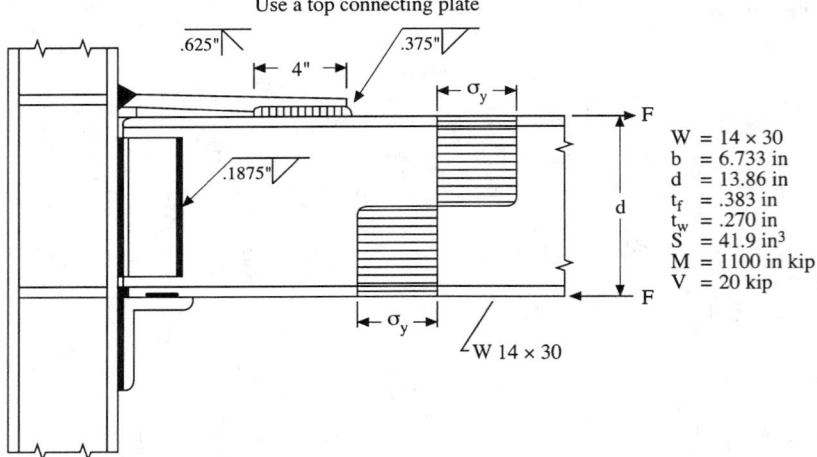

22.6.4   Top Plate Connections

The welding of the flanges and nearly full depth of the web would allow the beam to develop its full plastic moment. This will allow the "compact" beam to have a 10% higher bending allowable, or $\sigma = .66\sigma_y$. This also allows the end of the beam, and its welded connection, to be designed for 90% of end moment due to gravity loading. AISC Sect. 1.5.1.4.1.

Check the bending stress in the beam:

$$\sigma = \frac{.9M}{S} = \frac{.9(1100 \text{ in.-kip})}{41.9 \text{ in.}^3} = 23.6 \text{ ksi} < .66\sigma_y \text{ or } 24 \text{ ksi} \quad \text{OK}$$

Bending force in the top connecting plate:

$$F = \frac{.9M}{d} = \frac{.9(1100 \text{ in.-kip})}{13.86 \text{ in.}} = 71.5 \text{ kip}$$

Area of top connection plate:

$$A_p = \frac{F}{\sigma} = \frac{71.5 \text{ kip}}{24 \text{ ksi}} = 2.98 \text{ in.}^2$$

Or use a 5-1/2 in. by 5/8-in. plate, $A_p = 3.44 \text{ in.}^2 > 2.98 \text{ in.}^2$ OK

If a 3/8-in. fillet weld is used to connect the top plate to the upper beam flange:

$$f_w = (.707)(3/8 \text{ in.})(21 \text{ ksi}) = 5.56 \text{ kip per linear inch of weld}$$

Length of fillet weld

$$L = \frac{F}{f_w} = \frac{71.5 \text{ kip}}{5.56 \text{ kip/in.}} = 12.9 \text{ in.}$$

or use 5-1/2 in. across the end of the plate end 4 in. along each side, a total length of 13-1/2 in.

The lower flange of the beam is butt welded directly to the flange of the column. Since the web angle carries the shear reaction, no further work is required on this lower portion of the connection. The seat angle simply serves to provide temporary support for the beam during erection and a backing for the flange groove butt weld.

22.6.5   Directly Connected Beam-to-Column Connections

Design a fully welded beam-to-column connection for a W14x30 beam a W8x31 column to transfer an end moment of $M = 1000$ in.-kips, and a vertical shear of $V = 20$ kips. This example will be considered with several variations. Use A36 steel and E70 filler metal.

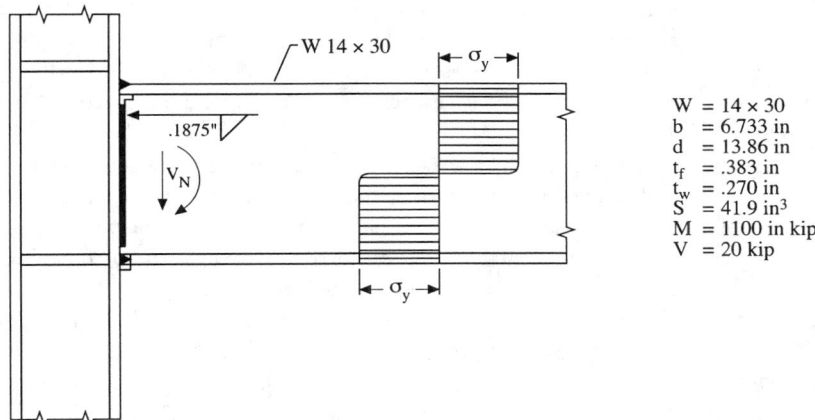

$$
\begin{aligned}
W &= 14 \times 30 \\
b &= 6.733 \text{ in} \\
d &= 13.86 \text{ in} \\
t_f &= .383 \text{ in} \\
t_w &= .270 \text{ in} \\
S &= 41.9 \text{ in}^3 \\
M &= 1100 \text{ in kip} \\
V &= 20 \text{ kip}
\end{aligned}
$$

The welding of the flanges and full depth of the web would allow the beam to develop its full plastic moment. This will allow the "compact" beam to have a 10% higher bending moment, or $\sigma = .66\sigma_y$. This also allows the end of the beam, and its welded connection, to be designed for 90% of the end moment due to gravity loading. AISC Sect. 1.5.1.4.1.

$$
\text{actual } \frac{b}{2t_f} = \frac{6.733}{2(.383)} = 8.79 \quad \text{AISC allowable } \frac{65}{\sqrt{\sigma_y}} = \frac{65}{\sqrt{36}} = 10.83 \quad \text{OK}
$$

$$
\text{actual } \frac{\sigma}{t} = \frac{13.86 - 2(.383)}{.290} = 48.5 \quad \text{AISC allowable } \frac{d}{t} = \frac{64D}{\sigma_y}\left(1 - 3.74\frac{\sigma_x}{\sigma_y}\right)
$$

$$
= 106.7 \quad \text{OK}
$$

hence this beam has a "compact" section.

$$
\sigma = \frac{.9M}{S} = \frac{.9(1100 \text{ in.-kip})}{41.9 \text{ in.}^3} = 23.63 \text{ ksi} < .66\sigma_y \text{ or } 24 \text{ ksi} \quad \text{OK}
$$

The weld on the web must be able to stress the web in bending to yield (σ_y) throughout its depth (see the bending stress distribution above).

Unit force this weld:

$$
f_w = \frac{V}{2L} = \frac{20 \text{ kip}}{2[13.86 - (2 \times .383)]} = .764 \text{ kip per linear inch}
$$

Leg size of fillet weld:

$$
w = \frac{.764 \text{ kip/in.}}{.707(21 \text{ ksi})} = .05 \text{ in.}
$$

However, this is welded to a .433-in.-thick flange of the column, so the minimum fillet weld size for this would be 3/16 in.

22.7   Understanding Ductile Behavior

"Ductility" can mean different things to different people. Materials such as cast iron are generally considered "brittle", while steel is called "ductile". A physical metallurgist may talk in terms of cleavage and ductile dimpling to define material behavior on a microscopic level. This is of little benefit to the structural engineer, who is more concerned with global deformation than microscopic behavior. Global deformation would include buckling, plastic hinge formation, stretching of members, and other inelastic behaviors that are visually observable. To achieve ductile behavior, the structural engineer, will select ductile materials for construction. To assume, however, that global deformation will occur simply because a ductile material has been selected can lead to unexpected brittle fracture of even ductile materials.

It is essential, therefore, that global behavior be separated from microscopic behavior. A material that fails by low-energy cleavage fracture cannot be made to function in a globally ductile manner, although it is possible for a structural element to fail with little or no deformation, and yet the fracture surface would exhibit the characteristics of ductile dimpling. Microscopic ductility and global ductility are separate issues, and the structural engineer must understand what conditions lead to global ductile behavior. This is particularly important where welding is applied, since welding introduces residual stresses and geometric influences that can affect the achievement of ductile behavior.

For global ductility to be possible, the following conditions must be achieved:

1. There must be a shear stress component (τ) that results from the applied load.
2. The shear stress must be of sufficient magnitude so as to exceed the critical shear stress of the material.
3. The shear stress must result in an inelastic shear strain that acts in a direction to relieve the particular stress that is applied.
4. There must be a sufficient length of unrestrained material to permit a reduction in the cross-sectional area (i.e., to allow for "necking" to occur).

These conditions are met in the specimens typically used to measure ductility of steels. As illustrated in Figure 22.18, the preceding four principles will be applied to the uniaxial tensile specimen. In Figure 22.18a, the specimen has been stretched to a point so that the resultant stresses, σ_1, are below the yield point, σ_y. The stress, σ_1, has caused a shear stress, τ_{1-2}, that acts on a 45° plane to the applied stress. Rather than focusing on σ_1 being less than σ_y, it is better to realize the resultant shear stress, τ_{1-2}, is less than the critical shear stress, τ_{cr}. A Mohr's Circle diagram assists in visualizing this behavior. Once τ_{cr} is exceeded, slippage along shear planes can occur, resulting in elongation. In Figure 22.18a, a shear stress has resulted from the applied stress (i.e., condition 1 from above was achieved), but the shear stress is not sufficient to exceed the critical value (i.e., condition 2 has not been achieved). All behavior under these conditions would be elastic, and although brittle fracture would not occur, neither would ductile behavior be achieved.

In Figure 22.18b, the load, F, has been increased so that the resultant shear, τ_{1-2}, exceeds τ_{cr}, resulting in slip on the plane oriented at 45°. This slip results in elongation, or stretching of the member. Global ductility is seen. This behavior occurs because condition 2 has been achieved. Figure 22.18c illustrates the continued application of force, resulting in slip occurring on multiple planes, eventually resulting in a reduction in cross-section, or necking. This is possible because all four conditions have been achieved.

A further increase in load, illustrated in Figure 22.18d, causes the critical tensile strength, σ_t, to be exceeded. The sample eventually breaks, and the final fracture surface exhibits little deformation. This occurs because, due to the localized deformations occurring in the necking region, the stresses in the other two principle directions are no longer zero. This is triaxial stress, and as illustrated in the

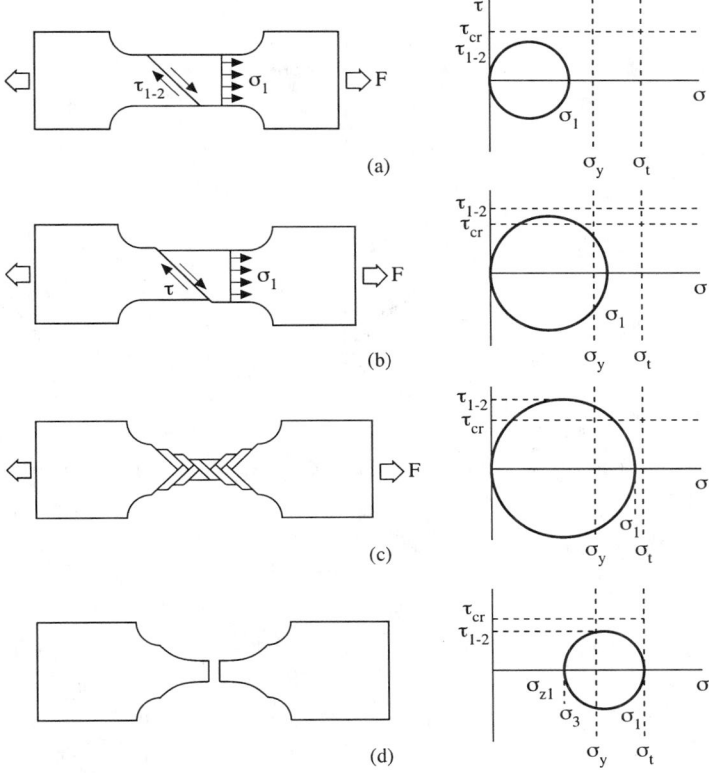

Figure 22.18 Shear stresses in simple, uniaxial tensile specimens. (Courtesy of The Lincoln Electric Company. With permission.)

Mohr's Circle, there is a resultant decrease in the shear stress. No longer is condition 1 maintained, and brittle fracture (in a global sense) occurs across the necked region.

These sample principles can be applied to various connection details. Consider, for example, the taper required for tension members that have thickness or width transitions. As seen in Figure 22.19, the sharp 90° transition results in a biaxial stress state near the transition. While ductile behavior could occur in the area where uniaxial stress exists, the second stress will reduce the shear stress, reducing its ductility capacity. The tapered transition allows for essentially uniaxial stresses to be maintained through the transition range, encouraging shear stresses capable of producing ductile behavior.

22.7.1 Two Residual Stresses Isolated

Figure 22.20 illustrates that two important residual stresses exist in the weld access hole's termination zone. This butt joint in the flange has a residual stress, σ_3, longitudinal to the length of the flange, as well as a stress transverse to the flange, σ_1. The longitudinal stress is tensile along the center line of the flange where the weld access hole terminates. It can be compared to tightening a steel cable lengthwise in the center in tension, with compression spread out on both sides. The transverse stress, σ_1, is positive (tensile) in the weld zone, as well as in an adjacent portion of the plate going through zero, and then compression. Beyond the adjacent plate, it becomes zero and then negative (compression). This transverse stress, σ_1, is also similar to tightening a steel cable.

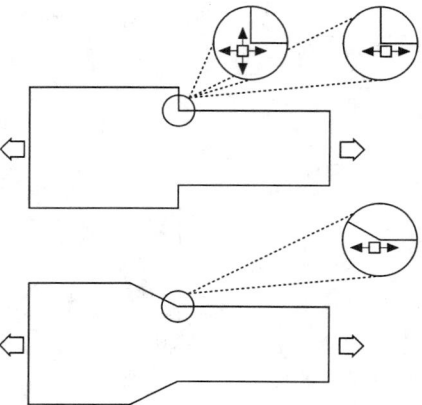

Figure 22.19   Stress state in transition connections. (Courtesy of The Lincoln Electric Company. With permission.)

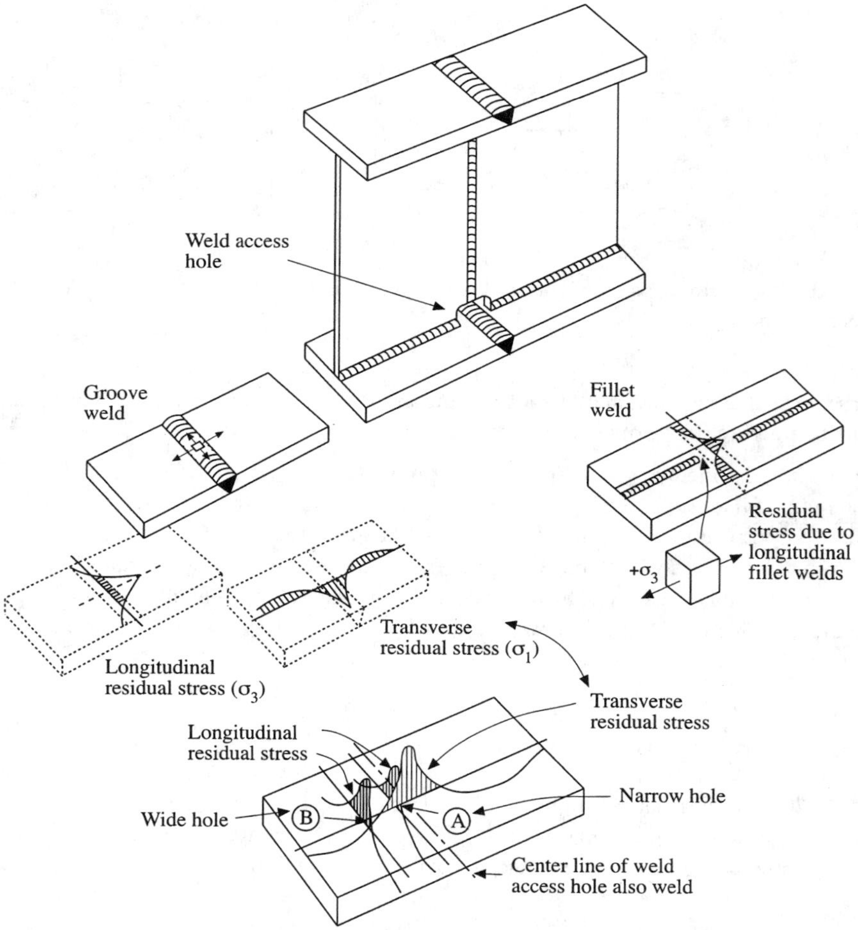

Figure 22.20   Resultant residual stress of welding. (Courtesy of The Lincoln Electric Company. With permission.)

22.7.2   Residual Stresses Applied

These residual stresses may be applied to a weld detail having a narrow weld access hole, as shown in Figure 22.21. This hole terminates at a point where σ_1 and σ_3 are in tension. Since the web at the edge of the weld access hole offers some restraint against movement in the through-thickness direction of the flange plate, stress in the σ_2 direction may have an appreciable value. All of the circles will be small. Neither τ_{2-3} nor τ_{1-3} will probably ever reach the critical shear stress value, and plastic strain or ductility will not occur, as the lower portion of Figure 22.21 illustrates.

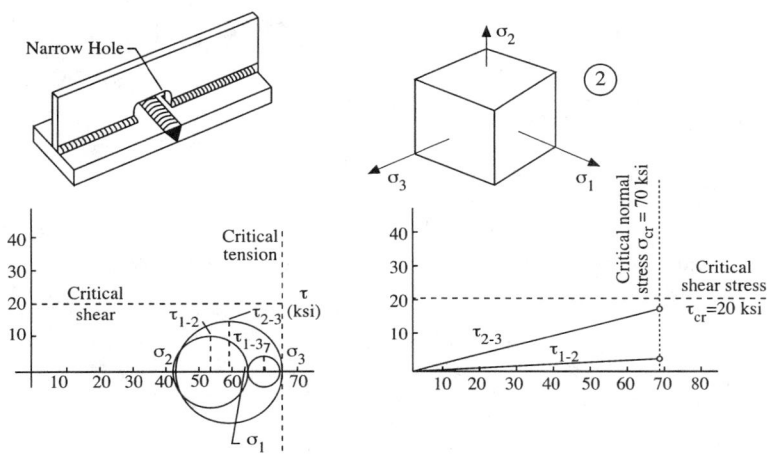

Figure 22.21   Mohr's circle of stress for element 2. (Courtesy of The Lincoln Electric Company. With permission.)

If the weld access hole can be cut with circular ends, sometimes called a pear-shaped opening, the stress, σ_2, in the through thickness of the flange plate will be greatly reduced, probably to zero in this critical section, as shown in Figure 22.22. This will produce a very large circle with σ_3 and the resulting shear stress, τ_{2-3}, will be very high — high enough to exceed the critical value well before σ_3 reaches its critical value for failure. This would result in a more ductile behavior.

If the weld access hole can be made wider, so that it terminates in a zone where the transverse residual stress, σ_1, is compressive (see Figure 22.23), then a more favorable stress condition will result in greater ductility in the σ_3 direction. In this case, shear stress, τ_{1-3}, will be high as shown on Mohr's Circle of stress, and the critical shear value will be reached at a much lower tensile stress or load value. This will produce more ductility in the σ_3 direction, greatly reducing the chance of a transverse crack in the flange at the termination of the weld access hole.

If a pear-shaped wide weld access hole is used, and the through-thickness stress, σ_2, becomes zero, it simply increases the shear stress τ_{2-3} and would seem to improve ductility (see Figure 22.24). However, looking at the resulting stress-strain curve of the flange plate at the termination of the weld access hole, it appears that rounding the ends of the wide access hole in this case does not appreciably increase the ductility. This is probably because the wide weld access hole already has excellent ductility.

Figure 22.25 shows stress-strain curves of the four different weld access hole details just discussed. The principles outlined herein can be applied to other details, evaluating the potential of biaxial or triaxial stresses and their effect on shear stress development. Consideration of these principles can assist in avoiding brittle fracture by encouraging ductile behavior.

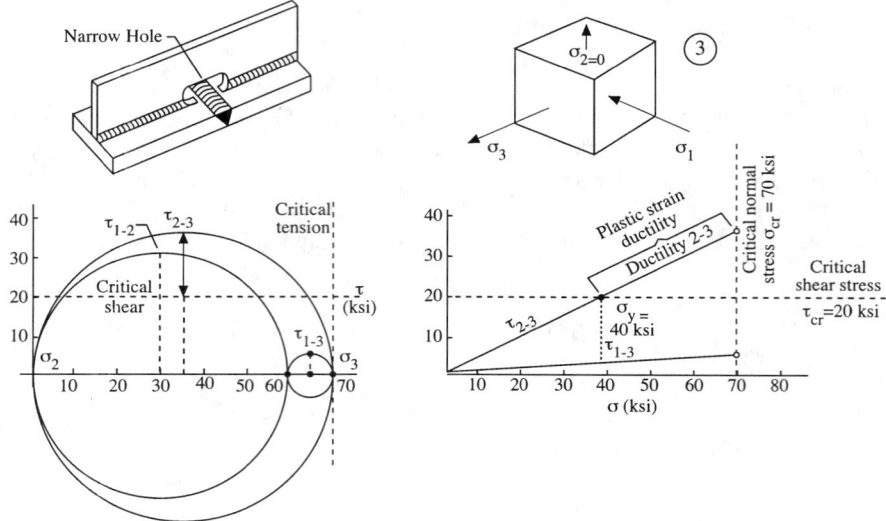

Figure 22.22   Mohr's circle of stress for element 3.   (Courtesy of The Lincoln Electric Company.   With permission.)

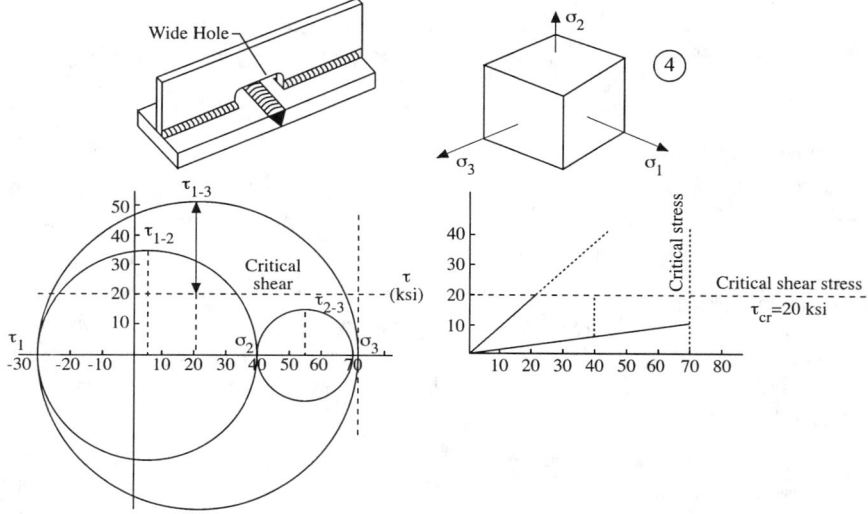

Figure 22.23   Mohr's circle of stress for element 4.   (Courtesy of The Lincoln Electric Company.   With permission.)

22.8   Special Considerations for Welded Structures Subject to Seismic Loading

22.8.1   Unique Aspects of Seismically Loaded Structures

Demands on Structural Systems

During an earthquake, even structures specifically designed for seismic resistance are subject to extreme demands. Any structure designed with a response modification factor, R_w, greater than unity will be loaded beyond the yield stress of the material. This is far more demanding than other anticipated types of loading. Due to the inherent ductility of steel, stress concentrations within a

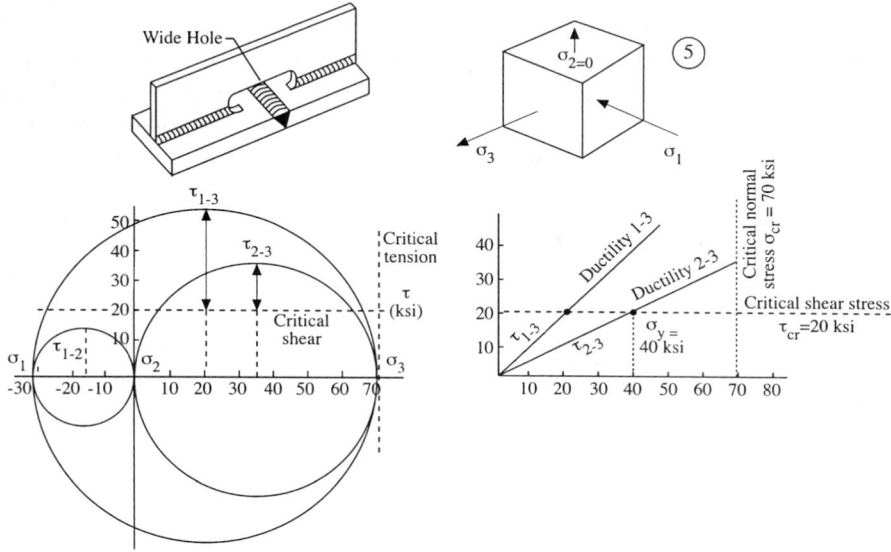

Figure 22.24   Mohr's circle of stress for element 5.   (Courtesy of The Lincoln Electric Company.   With permission.)

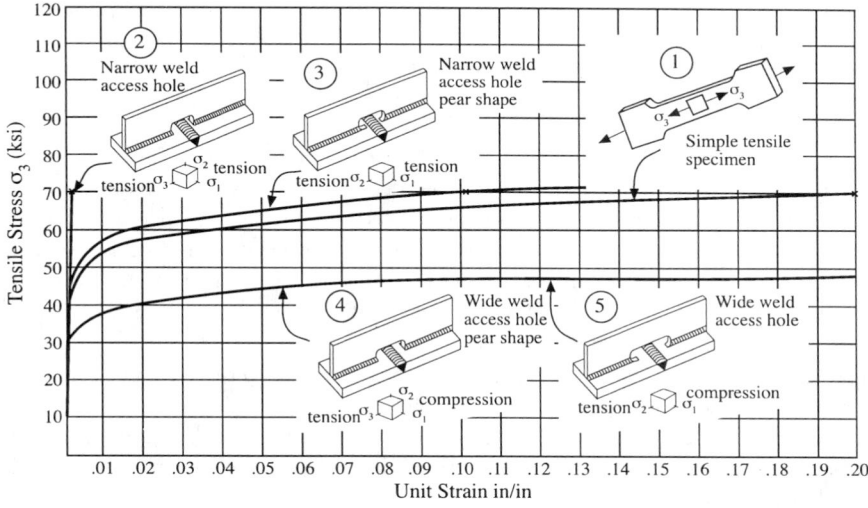

Figure 22.25   Stress-strain curves of four different weld access hole details. (Courtesy of The Lincoln Electric Company. With permission.)

steel structure are gradually distributed by plastic deformation. If the steel has a moderate degree of notch toughness, this redistribution eliminates localized areas of high stress, whether due to design, material, or fabrication irregularities. For statically loaded structures, the redistribution of stresses is relatively inconsequential. For cyclically loaded structures, repetition of this redistribution can lead to fatigue failure. In seismic loading, however, it is expected that portions of the structure will be loaded well beyond the elastic limit, resulting in plastic deformation. Localized areas of high stress will not simply be spread out over a larger region by plastic deformation. The resultant design, details, materials, fabrication, and erection must be carefully controlled in order to resist these extremely demanding loading conditions.

Demand for Ductility

Seismic designs have relied on ductility to protect structures during earthquakes. Unfortunately, much confusion exists regarding the measured property of ductility in steel, and ductility can be experienced in steel configured in various ways. It is essential that a fundamental understanding of ductility be achieved in order to ensure ductile behavior in the steel in general, and particularly in the welded connections.

Requirements for Efficient Welded Structures

Five elements are present in any efficient welded structure:

- Good overall design
- Good materials
- Good details
- Good workmanship
- Good inspection

Each element is important, and emphasis on one will not overcome deficiencies in others. Both the Northridge earthquake in 1994 and the Kobe earthquake in 1995 showed that deficiencies in one or more of the preceding areas may have contributed to the degradation in performance of Steel Moment-Resisting Frames (SMRFs).

22.9 Materials

22.9.1 Base Metal

Base metal properties are particularly important in structures subject to seismic loading. Unlike most static designs, seismically resistant structures depend on acceptable material behavior beyond the elastic limit. The basic premise of seismic design is to absorb seismic energies through yielding of the material. For static design, additional yield strength capacity in the steel may be desirable, but for applications where yielding is the desired method for achieving energy absorption, higher than expected yield strengths may have a dramatic negative effect. This is especially important as it relates to connections, both bolted and welded.

Figure 22.26 illustrates five material zones that occur near the groove weld in a beam-to-column connection. If it is assumed that the web is incapable of transferring any moment, it is essential that the plastic section modulus of the flanges (Z_f) times the tensile strength be greater than the entire plastic section property (Z) times the yield strength in the beam. All five material properties must be considered in order for the connection to behave satisfactorily. Note that this was the standard connection detail used for special moment-resisting frame (SMRF) systems prior to the Northridge earthquake.

Current American Society for Testing and Materials (ASTM) specifications do not place an upper limit on the yield strength for most structural steels, but specify only a minimum acceptable value. For instance, for ASTM A36 steel, the minimum acceptable yield strength is 36 ksi. This precludes a steel that has a yield strength of 35.5 ksi as being acceptable, but does nothing to prohibit the delivery of a 60-ksi steel. The tensile strength range is specified as 58–80 ksi. Although A36 is commonly specified for beams, columns are typically specified to be of ASTM A572 grade 50. With a 50-ksi minimum yield strength and a minimum tensile strength of 65 ksi, many designers were left with the false impression that the yield strength of the beam could naturally be less than that of the column. Due to the specification requirements, it is possible to produce steel that meets the requirements of both A36 and A572 grade 50. This material has been commercially promoted as "dual-certified" material. However, no matter what the material is called, it is critical for the connection illustrated

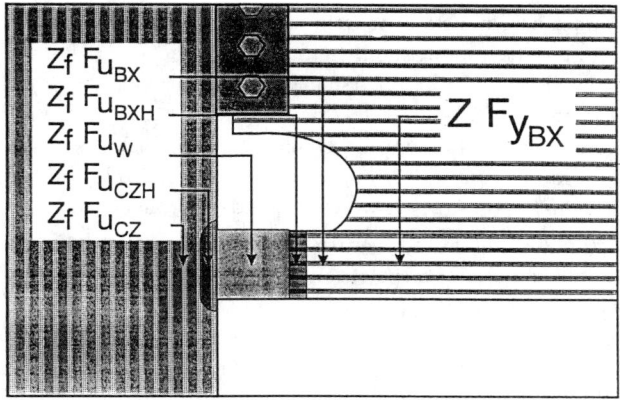

Figure 22.26   Five material zones that occur near the groove weld in a beam-to-column connection. (Courtesy of The Lincoln Electric Company. With permission.)

in Figure 22.26 to have controls on material properties that are more rigorous than the current ASTM standards impose.

Much of the focus of post-Northridge research has related to the beam yield-to-tensile ratio, commonly denoted as F_y/F_u . This is often compared to the ratio of Z_f/Z , with the desired relationship being

$$\frac{Z_f}{Z} > \frac{F_y}{F_u}$$

This suggests that not only is F_y (yield strength) important, but the ratio is important as well. For rolled W shapes, Z_f/Z ranges from 0.6 to 0.9. Based on ASTM *minimum* specified properties, F_y/F_u is as follows:

$$A36 \qquad 0.62$$
$$A572Gr50 \qquad 0.77$$

However, when actual properties of the steel are used, this ratio may increase. In the case of one building damaged in Northridge, mill test reports indicated the ratio to be 0.83.

ASTM steel specifications need further controls to limit the upper value of acceptable yield strengths for materials as well as the ratio of F_y/F_u . A new ASTM specification has been proposed to address these issues, although its approval will probably not be achieved before 1997.

In Figure 22.26, five zones have been identified in the area of the connection, with the sixth material property being located in the beam. Thus far, only two have been discussed: the beam yield strength and the beam ultimate strength. These are designated with the subscript X to indicate that these are the properties in the orientation of the longitudinal axis of the beam. When the beam is produced, the longitudinal direction is considered the "direction of rolling". In general, steel exhibits its best mechanical properties in this orientation. When this axis is designated as the X axis, the width of the beam would be known as the Y direction and the Z axis is through the thickness of the flanges. For beam properties, the X axis is the one of interest.

The properties of interest with respect to the column are oriented in the column Z axis, which will exhibit the least desirable mechanical properties. Current ASTM specifications do not require measurement of properties in this orientation. While there are ASTM standards for the measurement of through-thickness properties (ASTM 770), these are not normally applied for structural applications. It is this through-thickness strength, however, that is important to the performance of the connection.

Notch toughness is defined as the ability of a material to resist propagation of a preexisting crack-like flaw while under tensile stress. Pre-Northridge specifications did not include notch toughness requirements for either base materials or weld metals. When high loads are applied, and when notch-like details or imperfections exist, notch toughness is the material property that resists crack propagation from that discontinuity. Rolled shapes routinely produced today, specifically for lighter weight shapes in the group 1, 2, and 3 categories, generally are able to deliver a minimum notch toughness of 15 ft.-lb at 40°F. This is probably adequate toughness, although additional research should be performed in this area. For heavy columns made of group 4 and 5 shapes, this level of notch toughness may not be routinely achieved in standard production.

After Northridge, many engineers began to specify the supplemental requirements for notch toughness that are invoked by AISC specifications for welded tension splices in jumbo sections (group 4 and 5 rolled shapes). This requirement for 20 ft.-lb at 70°F is obtained from a Charpy V-notch specimen extracted from the web/flange interface, an area expected to have the lowest toughness in the cross-section of the shape. Since columns are not designed as tension members under most conditions, this requirement would not automatically be applied for column applications. However, as an interim specification, it seems reasonable to ensure minimum levels of notch toughness for heavy columns also.

22.9.2 Weld Metal Properties

Significant properties of weld metal are yield strength, tensile strength, toughness, and elongation. These properties usually may be obtained from data on the particular filler metal that will be employed to make the connection. The American Welding Society (AWS) filler metal classification system defines the minimum acceptable properties for the weld metal when deposited under very specific conditions. Most "70" series electrodes (e.g., E7018, E70T-1, E70T-6) have a minimum specified yield strength of 58 ksi and a minimum tensile strength of 70 ksi. As in the specifications for steel, there are no upper limits on the yield strength. However, in welded design, it is generally assumed that the weld metal properties will exceed those of the base metal, and any yielding that would occur in the connection should be concentrated in the base metal, not in the weld metal, since the base metal is assumed to be more homogeneous and more likely to be free of discontinuities than the weld. Most commercially available filler metals have a "70" classification, exceeding the minimum specified strength properties of the commonly used A36 and A572 grade 50.

These weld metal properties are obtained under very specific testing conditions prescribed by the AWS *A5 Filler Metal Specifications*. Weld metal properties are a function of many variables, including preheat and interpass temperatures, welding parameters, base metal composition, and joint design. Deviations in these conditions from those obtained for the test welds may result in differences in mechanical properties. Most of these changes will result in an increase in yield and tensile strength, along with a corresponding decrease in elongation and, in general, a decrease in toughness. When weld metal properties exceed those of the base metal, and when the connection is loaded into the inelastic range, plastic deformations would be expected to occur in the base metal, not in the weld metal itself. The increase in the strength of the weld metal compensates for the loss in ductility. The general trend to strength levels higher than those obtained under the testing conditions is of little consequence in actual fabrication.

There are conditions that may result in lower levels of strength, and the Northridge earthquake experience revealed that this may be more commonplace and more significant than originally thought. The interpass temperature is the temperature of the steel when the arc is initiated for subsequent welding. There are two aspects to the interpass temperature: the minimum level, which should always be the minimum preheat temperature, and the maximum level, beyond which welding should not be performed. Because of the relatively short length of beam-to-column flange welds, an operator may continue welding at a pace that will allow the temperature of the steel at the connection to increase to unacceptably high levels. After one or two weld passes, this temperature may approach the 1000°F range. In such a case, the strength of the weld deposit will be rapidly decreased.

Weld metal toughness is an area of particular interest in the post-Northridge specifications. Previous specifications did not include any requirement for minimum notch toughness levels in the weld deposits, allowing for the use of filler metals that have no minimum specified requirements. For connections that are subject to inelastic loading, it now appears that minimum levels of notch toughness must be specified. The actual limits on notch toughness have not been experimentally determined. With the AWS filler metal classifications in effect in 1996, electrodes are classified as either having no minimum specified notch toughness or having notch toughness values of 20 ft.-lb at a temperature of 0°F or lower. As an interim specification, 20 ft.-lb at 0°F or lower has been recommended. It should be noted that the more demanding notch toughness requirements impose several undesirable consequences upon fabrication, including increased cost of materials, lower deposition rates, less operator appeal, and greater difficulty in obtaining sound weld deposits. Therefore, ultra-conservative requirements imposed "just to be safe" may actually be unacceptable. Research will be conducted to determine precise toughness requirements. Until then, based upon practical issues of availability, 20 ft.-lb at 0°F is a reasonable specification.

22.9.3 Heat-Affected Zones

As illustrated in Figure 22.26, the base metal **heat-affected zones** (HAZs) represent material that may affect connection performance as well. The HAZ is that base metal that has been thermally changed due to the energy introduced into it by the welding process. In the small region immediately adjacent to the weld, the base metal has experienced a different thermal history than the rest of the base material. For most hot-rolled steels, the area of concern is a HAZ that is cooled too rapidly, resulting in a hardened HAZ. For quenched-and-tempered steels, the HAZ may be cooled too slowly, resulting in a softening of the area. In columns, the HAZ of interest is the Z direction area immediately adjacent to the groove weld. For the beam, these are oriented in the X direction.

Excessively high heat input can negatively affect HAZ properties by causing softening in these areas. Excessively low heat input can result in hardening of the HAZs. Weld metal properties may be negatively affected by extremely high heat input welding procedures, causing a decrease in both the yield strength and tensile strength, as well as the notch toughness of the weld deposit. Excessively low heat input may result in high-strength weld metal and also decrease the notch toughness of the weld deposit. Optimum mechanical properties are generally achieved if the heat input is maintained in the 30–80 kJ/in. range. Post-Northridge evaluation of fractured connections has revealed that excessively high heat input welding procedures were often used, confirmed by the presence of very large weld beads that sometimes exceeded the maximum limits prescribed by the D1.1-96 code. These may have had some corollary effects on weld metal and HAZ properties.

22.10 Connection Details

Since there are no secondary members in welded construction, any material connected by a weld participates in the structural system — positively or negatively. Unexpected load paths can be developed by the unintentional metallurgical path resulting from the one-component system created by welding. This phenomenon is particularly significant in detailing.

22.10.1 Weld Backing

Pre-Northridge specifications typically allowed steel backing to be left in place. Most of the fractures experienced in Northridge initiated immediately above the naturally occurring unfused region, between the backing and the column face. When this area experienced tensile loading due to lateral displacements, this region would result in a stress concentration and a notch that served as a crack initiation point (see Figure 22.27).

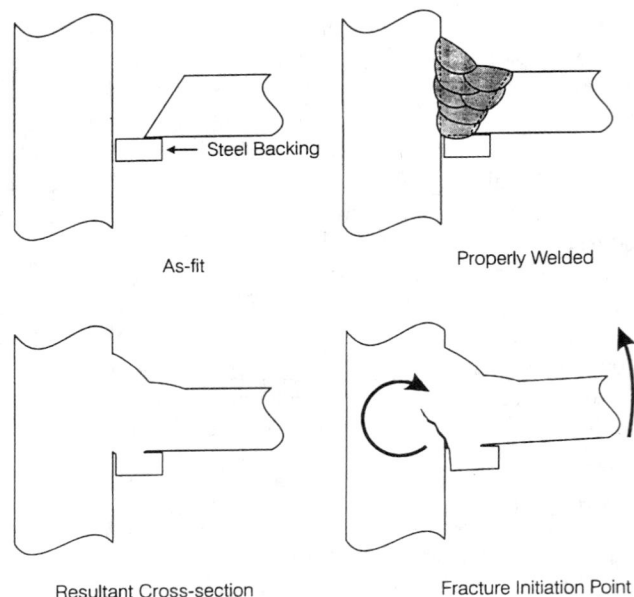

Figure 22.27 Weld backing and fracture initiation. (Courtesy of The Lincoln Electric Company. With permission.)

After Northridge, many specifications began to call for the removal of steel backing from the bottom beam-flange-to-column connection. This activity not only eliminates the notch-like condition, it permits gouging the weld root to sound metal, and allows for the depositing of a reinforcing fillet weld that provides a more gradual transition in the 90° interface between the beam and the column.

Not all backing is required to be removed. For welds subject to horizontal shear (such as corner joints in box columns), backing can be left in place. In butt joints, the degree of stress amplification that occurs due to backing left in place is much less severe than what occurs in T joints. Backing removal is expensive and, particularly when done in the overhead position, requires considerable welder skill. Some recommendations have not required removal of top beam-flange-to-column connection backing because the removal operation (gouge, clean, inspect, and reweld) must be performed through the weld access hole. This difficult operation may do more harm than good.

There is increased interest in ceramic backing. Welding procedures that employ ceramic backing must be qualified by test for work done in accordance with D1.1-96, para. 5.10. Welders must be trained in the proper use of these materials. Ceramic is nonconductive, requiring that the welder establish a "bridge" between the two steel members to be welded in order to maintain the electrical arc between the two members. While this can be accomplished fairly readily with small root opening dimensions (such as 1/4 in.), it becomes increasingly difficult with larger root openings (such as 1/2 in.). Wide, thin root passes on ceramic-backed joints may crack due to high shrinkage stresses imposed on small weld throats.

One benefit of the activity of fusible backing removal is that it permits the weld joint to be back gouged to sound material. The root of the weld joint is always the most problem-prone region. The center of the length of the bottom beam-flange-to-column weld is difficult to make, since the welder must work through the weld access hole. This is also one of the most difficult areas to inspect with confidence. In a typical beam-to-column connection, the bottom beam-flange-to-column weld must be interrupted midlength due to interference with the web. This area is particularly sensitive to workmanship problems, and is also a difficult region to inspect with ultrasonic testing. The back-gouging operations provide the opportunity for visual verification that sound weld metal has been obtained, particularly in the center of the joint length. This is similar to the D1.1-96 code requirement for back gouging of double-sided joints.

22.10.2 Weld Tabs

Weld tabs are auxiliary pieces of metal on which the welding arc may be started or stopped. For statically loaded structures, these are usually left in place. For seismic construction, weld tabs should be removed from critical connections that are subject to inelastic loading, because metal of questionable integrity may be deposited in the region of these weld tabs.

Weld tab removal is probably most important on beam-to-column connections where the column flange width is greater than the beam flange width. It is reasonable to expect that stress flow would take place through the left-in-place weld tab. However, for butt splices where the same width of material is joined, weld tabs extending beyond the width of the joint would not be expected to carry significant stress, making weld tab removal less critical. It is unlikely that tab removal from continuity plate welds would be justified.

For beam-to-column connections where columns are box shapes, the natural stress distribution causes the ends of the groove weld between the beam and column to be loaded to the greatest level, the same region as would contain the weld tab. Just the opposite condition exists when columns are composed of I-shaped members. The center of the weld is loaded most severely, causing the areas in which the weld tabs would be located to have the lowest stress level. For welds subject to high levels of stress, however, weld tabs should be removed.

22.10.3 Welds and Bolts Sharing Loads

Welding provides a continuous metallurgical path that relies upon the internal metallurgical structure of the fused metal to provide continuity and strength. Rivets and bolts rely on friction, shear of the fastening element, or bearing of the joint material to provide for transfer of loads between members. When mechanical fasteners such as bolts are combined with welds, caution must be exercised in assigning load-carrying capacity to each joining method.

Traditionally, it was thought that welds used in conjunction with bolts should be designed to carry the full load, assuming that the mechanical fasteners have no load-carrying capacity until the weld fails. The development of high-strength fasteners, however, created the assumption that loads can be shared equally between welds and fasteners. This has led to connection details that employ both joining systems. Specifically, the welded flange, bolted web detail used for many beam-to-column connections in SMRFs assumes that the bolted web is able to share loads equally with the welded flanges. Although most analyses suggest that vertical loads are transferred through the shear tab connection (bolted) and moments are transferred through the flanges (welded), the web does have some moment capacity. Depending on the particular rolled shape involved, the moment capacity of the web can be significant. Testing of specimens with the welded web detail, as compared to the bolted web detail, generally has yielded improved performance results. This has called into question the adequacy of the assumption of high-strength bolts sharing loads with welds when subject to inelastic loading. Post-Northridge research provides further evidence that the previously accepted assumptions may have been inadequate. Previous design rules regarding the capacity of bolted connections should be reexamined. This may necessitate additional fasteners, or larger sizes of shear tabs (both in thickness and in width). Stipulations regarding the addition of supplemental fillet welds on shear tabs, currently a function of the ratio of Z_f/Z, are probably also inadequate and will require revision.

Pending further research, the conservative approach is to utilize welded web details. This does not preclude the use of a bolted shear tab for erection purposes, but would rely on welds as a singular element connecting the web to the column.

22.10.4 Weld Access Holes

The performance of a connection during seismic loading can be limited by poorly made, or improperly sized, weld access holes. In the beam-to-column connection illustrated in Figure 22.28, a welded web connection has been assumed. As the flange groove weld shrinks volumetrically, a

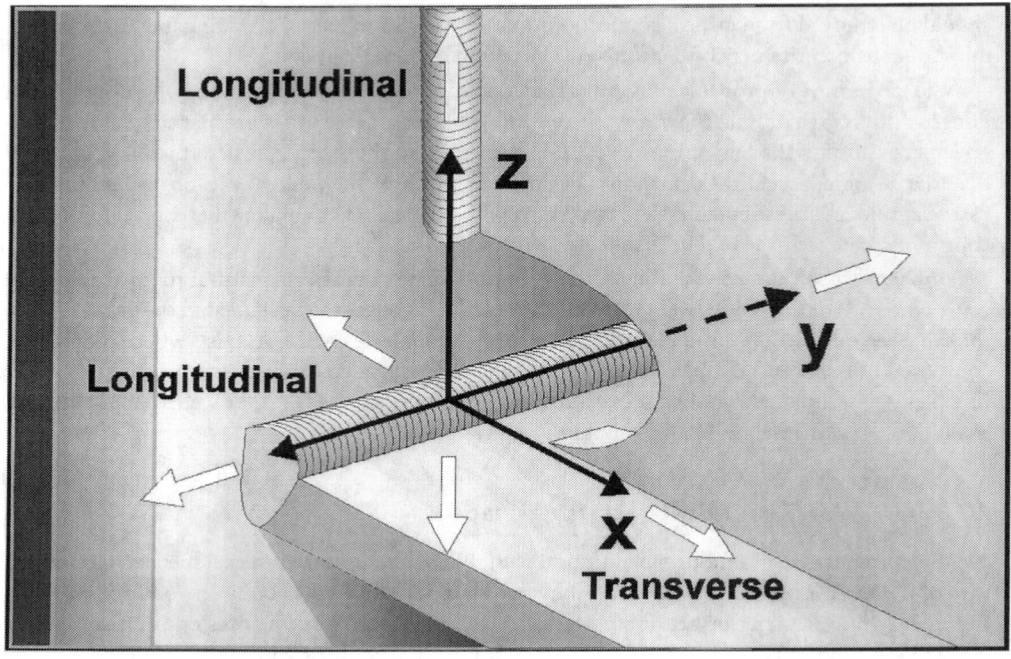

Figure 22.28   A generous weld access hole in this beam-to-column connection provides resistance to cracking. (Courtesy of The Lincoln Electric Company. With permission.)

residual stress field will develop perpendicular to the longitudinal axis of the weld, as illustrated in direction X in the figure. Concurrently, as the groove weld shrinks longitudinally, a residual stress pattern is established along the length of the weld, designated as direction Y. When the web weld is made, the longitudinal shrinkage of this weld results in a stress pattern in the Z direction. These three residual stress patterns meet at the intersection of the web and flange of the beam with the face of the column. When steel is loaded in all three orthogonal directions simultaneously, even the most ductile steel cannot exhibit ductility. At the intersection of these three welds, cracking tendencies would be significant. A generous weld access hole, however, will physically interrupt the interaction of the Z axis stress field and the biaxial (X and Y) stress field, thereby increasing the resistance to cracking during fabrication.

The quality of weld access holes may affect both resistance to fabrication-related cracking and resistance to cracking that may result from seismic events. Access holes usually are cut into the steel by a thermal cutting process, either oxy-fuel or plasma arc. Both processes rely on heating the steel to a high temperature and removing the heated material by pressurized gases. In the case of oxy-fuel cutting, oxidation of the steel is a key ingredient in this process. In either process, the steel on either side of the cut (called the "kerf") has been heated to an elevated temperature and rapidly cooled. In the case of oxy-fuel cutting, the surface may be enriched with carbon. For plasma cut surfaces, metallic compounds of oxygen and nitrogen may be present on this surface. The resultant surface may be hard and crack sensitive, depending on the combinations of the cutting procedure, base metal chemistry, and thickness of the materials involved. Under some conditions, the surface may contain small cracks, which can be the points of stress amplification that cause further cracking during fabrication or during seismic events.

Nicks or gouges may be introduced during the cutting process, particularly when the cutting torch is manually propelled during the formation of the access hole. These nicks may act as stress amplification points, increasing the possibility of cracking. To decrease the likelihood of notches and/or microcracks on thermally cut surfaces, AISC has specific provisions for making access holes in heavy group 4 and 5 rolled shapes. These provisions include the need for a preheat before cutting, requirements for grinding of these surfaces, and inspection of these surfaces with magnetic particle (MT) or dye penetrant (PT) inspection. Whether these provisions should be required for all connections that may be subject to seismic energies is unknown at this time. However, for connection details that impose high levels of stress on the connection, and specifically those that demand inelastic performance, it is apparent that every detail in the access hole region is a critical variable. In the Northridge earthquake, some cracking initiated from weld access holes.

22.11 Achieving Ductile Behavior in Seismic Sections

22.11.1 System Options

Several systems may be employed to achieve seismic resistance, including eccentrically braced frames (EBFs), concentrically braced frames (CBFs), SMRFs, and base isolation. Of the four mentioned, only base isolation is expected to reduce demand on the structure. The other three systems assume that at some point within the structure, plastic deformations will occur in members, thus absorbing seismic energy.

In a CBF, the brace member is expected to be subject to inelastic deformations. The welded connections at the termination of a brace are subject to significant tension or compression loads, although rotation demands in the connections are fairly low. Designing these connections requires the engineer to develop the capacity of the brace member in compression and tension. Recent experiences with CBF systems have reaffirmed the importance of the brace dimensions (b/t ratio), and the importance of good details in the connection itself. Problems appear to be associated with misplaced welds, undersized welds, missing welds, or welds of insufficient throat due to construction methods. In order to place the brace into the building frame, a gusset plate is usually welded into the corners of the frame. The brace is slit along its longitudinal axis and rotated into place. To maintain adequate dimensions for field assembly, the slot in the tube must be oversized, compared to the gusset, resulting in natural gaps between the tube and the gusset plate. When this dimension increases, as illustrated in Figure 22.29, it is important to consider the effect of the root opening on the strength of the fillet weld. For gaps exceeding 1/16 in., the D1.1-96 code requires that the weld leg size be increased by the amount of the gap, ensuring a constant actual throat dimension is maintained.

EBFs and SMRFs are significantly different structural systems, but some welding design principles apply equally to both systems. It is possible to design an EBF so that the "link" consists simply of a rolled steel member. In Figure 22.30, these examples are illustrated by the links designated as c1. In other EBF systems, however, the connection itself can be part of the link, as illustrated by c2. When this design method is used, the welded connections become critical since the expected loading on the connection is in the inelastic region. Much of the discussion under SMRF may be applied to these situations.

The SMRF system is commonly applied to low-rise structures. Advantages of this type of system include desirable architectural elements that leave the structure free of interrupting diagonal members. Extremely high demands for inelastic behavior in the connections are inherent to this system.

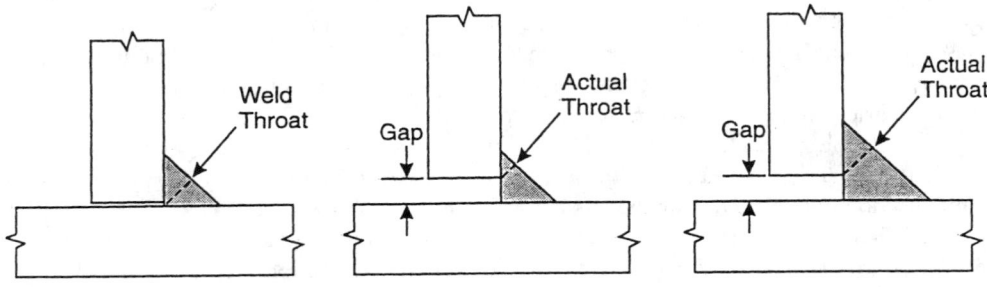

Figure 22.29   Effect of root openings (gaps) on fillet weld throat dimensions. (Courtesy of The Lincoln Electric Company. With permission.)

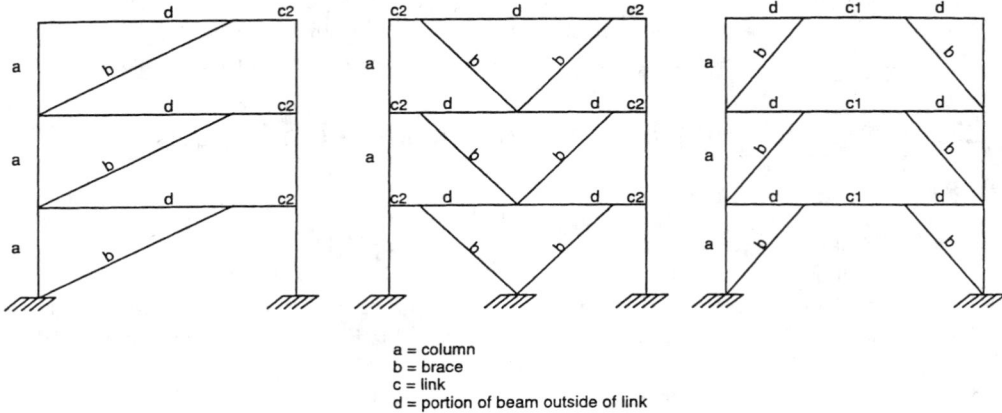

Figure 22.30   Examples of EBF systems. (From American Institute of Steel Construction. *Seismic Provisions for Steel Buildings.* 1992. With permission.)

When subject to lateral displacements, the structure assumes a shape as shown in Figure 22.31. Note that the highest moments shown in Figure 22.31 are applied at the connection. Figure 22.31 shows a plot of the section properties. Section properties are at their lowest value at the column face, because of the weld access holes that permit the deposition of the CJP beam-flange-to-column-flange welds. These section properties may be further reduced by deleting the beam web from the calculation of section properties. This is a reasonable assumption when the beam-web-to-column-shear tab is connected by the means of high-strength bolts. Greater capacity is achieved when the beam web is directly welded to the column flange with a CJP weld. The section properties at the end of the beam are least, precisely an area where the moment levels are the greatest, leading to the highest level of stresses. A plot of stress distribution is shown in Figure 22.31. The weld is therefore in the area of highest stress, making it critical to the performance of the connection. Details in either EBF or SMRF structures that place this type of demand on the weld require careful scrutiny.

22.11.2   Ductile Hinges in Connections

The SMRF concept is based on the premise that plastic hinges will form in the beams, absorbing seismically induced energies by inelastically stretching and deforming the steel. The connection is not expected to break. Following the Northridge earthquake, however, there was little or no evidence of hinge formation. Instead, the connections or portions of the connection experienced brittle fracture.

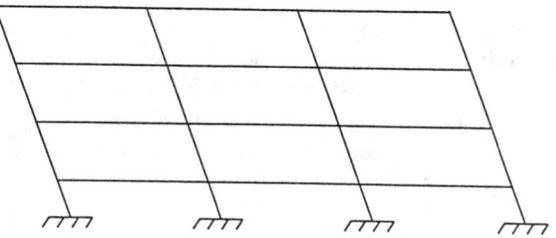

a.   SMRF Systems subject to lateral displacements.

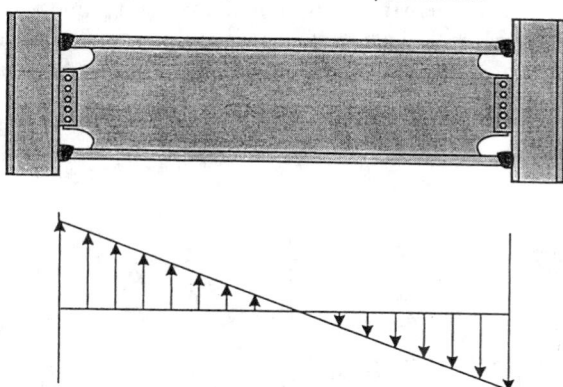

b.   Moment diagram of SMRF subject to lateral displacements.

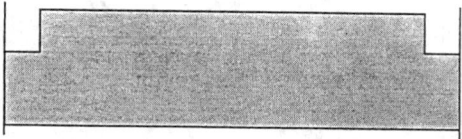

c.   Section propeties of SMRF subject to lateral displacements.

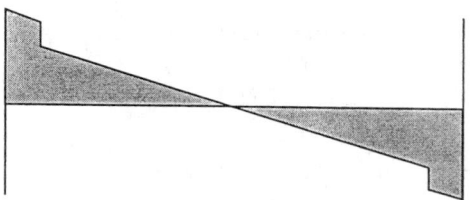

d . Stress distribution of SMRF subject to lateral displacements.

Figure 22.31   Analysis of SMRF behavior. (Courtesy of The Lincoln Electric Company. With permission.)

Most of the ductility data is obtained from smooth, slowly loaded, uniaxially loaded tensile specimens that are free to neck down. If a notch is placed in the specimen, perpendicular to the applied load, the specimen will be unable to exhibit its normal ductility, usually measured as elongation. The presence of notch-like conditions in the Northridge connections reduced the ductile behavior.

In 1994, initial research on SMRF connections attempted to eliminate the issues of notch-like conditions in the test specimens by removing weld backing and weld tabs and controlling weld soundness. Even with these changes, brittle fractures occurred when the standard details were tested. The testing program then evaluated several modified details with short cover plates, with better success. The beam-to-column connection will be examined with respect to the previously outlined conditions required for ductility (see Section 22.7).

Figure 22.32 shows two regions in question. Point A is at the weld joining the beam flange to the face of the column flange. Here there is restraint against strain (movement) across the width of the beam flange (ε_1) as well as through the thickness of the beam flange (ε_2). Point B is along the length of the beam flange away from the connecting weld. There is no restraint across the width of the flange or through its thickness.

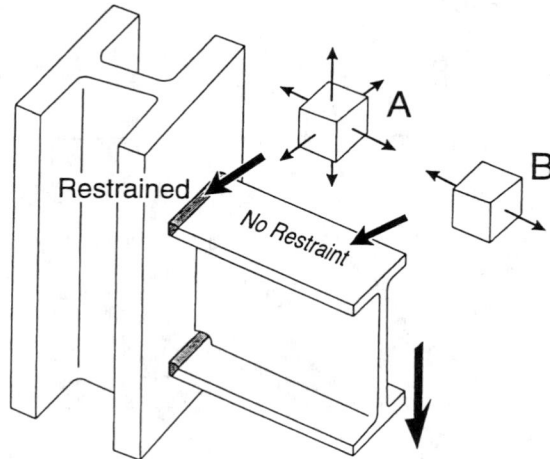

Figure 22.32   Regions to be analyzed relative to potential for ductile behavior. (Courtesy of The Lincoln Electric Company. With permission.)

The following equations can be found in most texts concerning strength of materials:

$$\varepsilon_3 = \frac{1}{E}(\sigma_3 - \mu\sigma_2 - \mu\sigma_1) \tag{22.1a}$$

$$\varepsilon_2 = \frac{1}{E}(-\mu\sigma_3 + \sigma_2 - \mu\sigma_1) \tag{22.1b}$$

$$\varepsilon_1 = \frac{1}{E}(-\mu\sigma_3 - \mu\sigma_2 + \sigma_1) \tag{22.1c}$$

It can be shown that:

$$\sigma_1 = \frac{E\left[\mu\varepsilon_3 + \mu\varepsilon_2 + (1-\mu)\varepsilon_1\right]}{(1+\mu)(1-2\mu)} \tag{22.2a}$$

$$\sigma_2 = \frac{E\left[\mu\varepsilon_3 + (1-\mu)\varepsilon_2 + \mu\varepsilon_1\right]}{(1+\mu)(1-2\mu)} \tag{22.2b}$$

$$\sigma_3 = \frac{E\left[(1-\mu)\varepsilon_3 + \mu\varepsilon_2 + \mu\varepsilon_1\right]}{(1+\mu)(1-2\mu)} \tag{22.2c}$$

The unit cube in Figure 22.33 is an element of the beam flange from point B in Figure 22.32. The applied force due to the moment is σ_3. Assuming strain in direction 3 to be + 0.001 in./in., and Poisson's ratio of $\mu = 0.3$ for steel, ε_2 and ε_3 can be found to be equal to -0.0003 in./in.

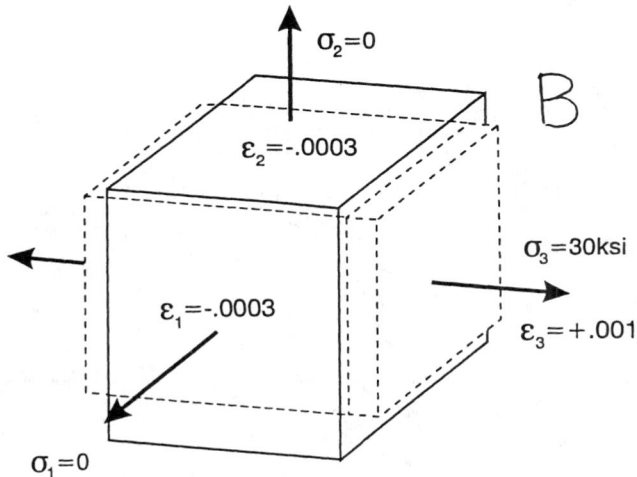

Figure 22.33 Unit cube showing applied stress from Figure 22.32. (Courtesy of The Lincoln Electric Company. With permission.)

Using these strains, from Equations 22.2a to 22.2c, it is found that

$$\sigma_1 = 0 \text{ ksi}$$
$$\sigma_2 = 0 \text{ ksi}$$
$$\sigma_3 = 30 \text{ ksi}$$

These stresses are plotted as a dotted circle on Figure 22.34. These values are then extrapolated to the point where fracture would occur, that is, where the net tensile strength is 70 ksi. The larger solid line circle is for a stress of 70 ksi or ultimate tensile stress. The resulting maximum shear stresses, τ_{1-3} and τ_{2-3}, are the radii of these two circles, or 35 ksi. The ratio of shear to tensile stress for steel is 0.5. Figure 22.35 plots this as line B. At a yield point of 55 ksi, the critical shear value is half of this, or 27.5 ksi. When this critical shear stress is reached, plastic straining takes place and ductile behavior will result up to the ultimate tensile strength, here 70 ksi. Figure 22.38 shows a predicated stress-strain curve indicating ample ductility.

Figure 22.36 shows an element from point A of Figure 22.32 at the junction of the beam and column flange. Whether weld metal or the material in the column or beam is considered makes

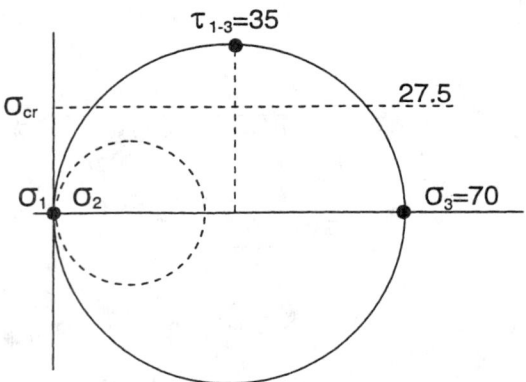

Figure 22.34   A plot of the tensile stress and shear stress from Figure 22.32. (Courtesy of The Lincoln Electric Company. With permission.)

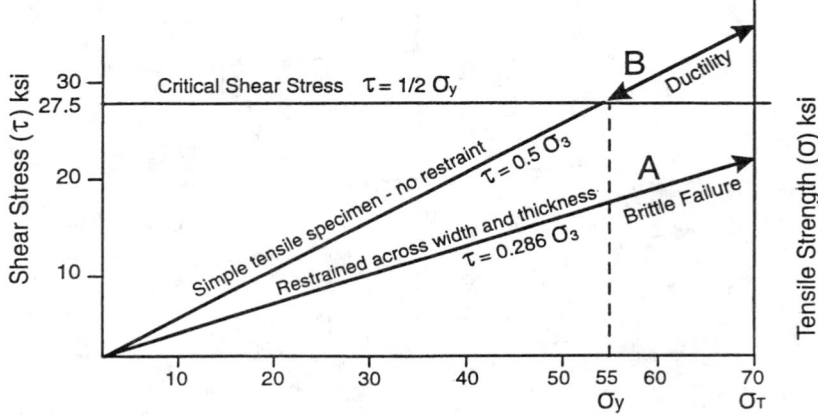

Figure 22.35   The ratio of shear to tensile stress. (Courtesy of The Lincoln Electric Company. With permission.)

little difference. This region is highly restrained. Suppose it is assumed:

$$\varepsilon_3 = +0.001 \text{ in./in. (as before)}$$
$$\varepsilon_2 = 0 \text{ (since it is highly restrained}$$
$$\varepsilon_1 = 0 \text{ with little strain possible)}$$

Then, from the given equations, the following stresses are found:

$$\sigma_1 = 17.31 \text{ ksi}$$
$$\sigma_2 = 17.31 \text{ ksi}$$
$$\sigma_3 = 40.38 \text{ ksi}$$

The stresses are plotted as a dotted circle in Figure 22.37.
If these stresses are increased to the ultimate tensile strength, it is found that

$$\sigma_1 = 30.0 \text{ ksi}$$
$$\sigma_2 = 30.0 \text{ ksi}$$
$$\sigma_3 = 70.0 \text{ ksi}$$

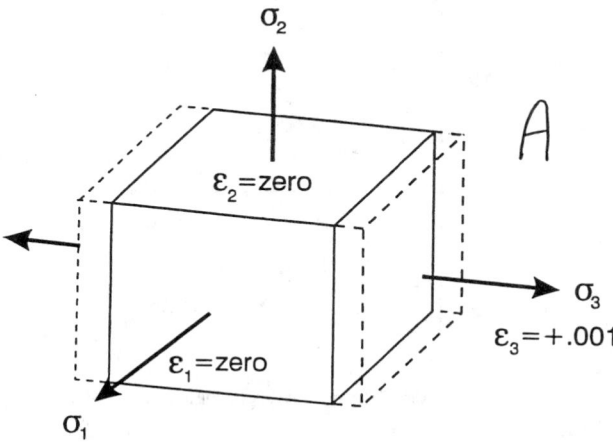

Figure 22.36 The highly restrained region at the junction of the beam and column flange shown in Figure 22.32. (Courtesy of The Lincoln Electric Company. With permission.)

The solid line circle in Figure 22.37 is a plot of stresses for this condition. The maximum shear stresses are $\tau_{1-3} = \tau_{2-3} = 20$ ksi. Since these are less than the critical shear stress (27.5 ksi), no plastic movement, or ductility, would be expected.

In this case, the ratio of shear to tensile stress is 0.286. In Figure 22.35, this condition is plotted as line A. It never exceeds the value of the critical shear stress (27.5 ksi); therefore, there will be no plastic strain or movement, and it will behave as a brittle material. Figure 22.38 shows a predicated stress-strain curve going upward as a straight line (A) (elastic) until the ultimate tensile stress is reached in a brittle manner. It would therefore be expected that, at the column face or in the weld where high restraint exists, little ductility would result. This is where brittle fractures have occurred, both in the laboratory and in actual Northridge structures.

In the SMRF system, the greatest moment (due to lateral forces) will occur at the column face. This moment must be resisted by the beam's section properties, which because of weld access holes are lowest at the column face. Thus, the highest stresses occur at this point, the point where analysis shows ductility to be impossible.

In Figure 22.32, material at point B was expected to behave as shown in Figure 22.34a, and as line B in Figure 22.35, and curve B in Figure 22.38; that is, with ample ductility. Plastic hinges must be forced to occur in this region.

Several post-Northridge designs have employed details that encourage use of this potential ductility. The coverplated design illustrated in Figure 22.39 accomplishes two important things: first, the stress level at point A is reduced as a result of the increased cross-section at the weld. This region, incapable of ductility, must be kept below the critical tensile stress and the increase in area accomplishes this goal. Second, and most significant, the most highly stressed region is now at point B, the region of the beam that is capable of exhibiting ductility.

The real success of this connection will depend upon getting the adjacent beam to plastically bend before this critical section cracks. The way in which a designer selects structural details under particular load conditions greatly influences whether the condition provides enough shear stress component so that the critical shear value may be exceeded first, producing sufficient plastic movement before the critical normal stress value is exceeded. This will result in a ductile detail and minimize the chances of cracking.

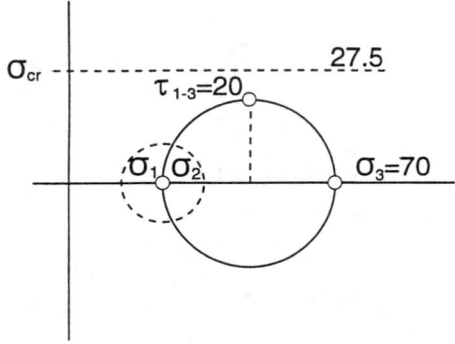

Figure 22.37   A plot of the tensile stress and shear stress from Figure 22.35. (Courtesy of The Lincoln Electric Company. With permission.)

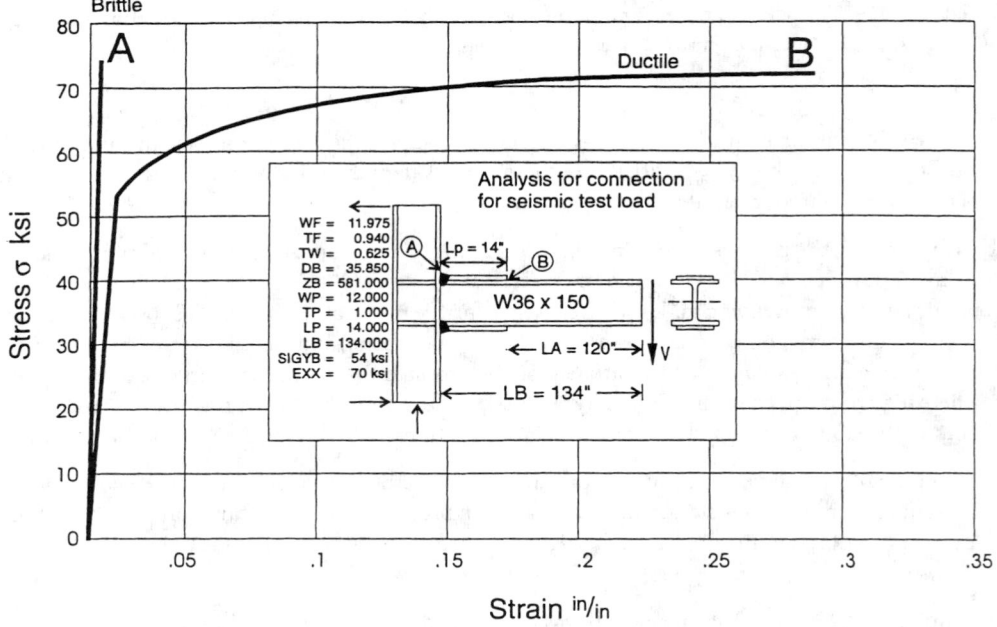

Figure 22.38   Stress-strain curve. (Courtesy of The Lincoln Electric Company. With permission.)

22.12   Workmanship Requirements

In welded construction, the performance of the structural system often depends on the ability of skilled welders to deposit sound weld metal. As the level of loading increases, dependence on high-quality fabrication increases. For severely loaded connections, good workmanship is a key contributor to acceptable performance.

Design and fabrication specifications such as the AISC *Manual of Steel Construction* and the AWS D1.1 *Structural Welding Code: Steel* [9] communicate minimum acceptable practices. It is impossible for any code to cover every situation that will ever be contemplated. It is the responsibility of the engineer to specify any additional requirements that supersede minimum acceptable standards.

The D1.1-96 code does not specifically address seismic issues, but does establish a minimum level of quality that must be achieved in seismic applications. Additional requirements are probably

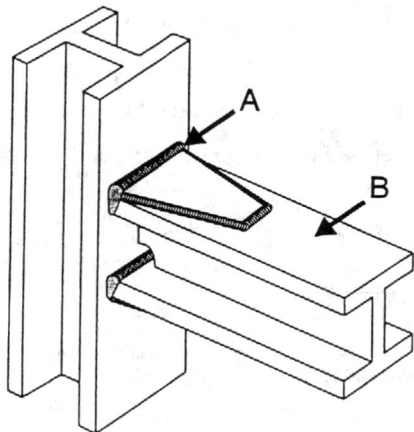

Figure 22.39 Coverplate detail takes advantage of the region where ductility is possible. (Courtesy of The Lincoln Electric Company. With permission.)

warranted. These would include requirements for nondestructive testing, notch tough weld deposits, and additional requirements for in-process verification inspection.

22.12.1 Purpose of the Welding Procedure Specification

The welding procedure specification (WPS) is somewhat analogous to a cook's recipe. It outlines the steps required to make a good-quality weld under specific conditions. It is the primary method to ensure the use of welding variables essential to weld quality. In addition, it permits inspectors and supervisors to verify that the actual welding is performed in conformance with the constraints of the WPS. Examples of WPSs are shown in Figures 22.40 and 22.41.

WPSs typically are submitted to the inspector for review prior to the start of welding. For critical projects, the services of welding engineers may be needed. WPSs are intended to be communication tools for maintenance of weld quality. All parties involved with the fabrication sequence must have access to these documents to ensure conformance to their requirements.

22.12.2 Effect of Welding Variables

Specific welding variables that determine the quality of the deposited weld metal are a function of the particular welding process being used, but the general trends outlined below are applicable to all welding processes.

Amperage is a measure of the amount of current flowing through the electrode and the work. An increase in amperage generally means higher deposition rates, deeper penetration, and more melting of base metal. The role of amperage is best understood in the context of heat input and current density, which are described below.

Arc voltage is directly related to arc length. As the voltage increases, the arc length increases. Excessively high voltages may result in weld metal porosity, while extremely low voltages will produce poor weld bead shapes. In an electrical circuit, the voltage is not constant, but is composed of a series of voltage drops. Therefore, it is important to monitor voltage near the arc.

Travel speed is the rate at which the electrode is moved relative to the joint. Travel speed, which has an inverse effect on the size of weld beads, is a key variable used in determining heat input.

Polarity is a definition of the direction of current flow. Positive (or reverse) polarity is achieved when the electrode lead is connected to the positive terminal of the direct current power supply. The work lead would be connected to the negative terminal. Negative (or straight) polarity occurs

WELDING PROCEDURE SPECIFICATION (WPS) Yes ☒
PREQUALIFIED __X__ QUALIFIED BY TESTING _____
or PROCEDURE QUALIFICATION RECORDS (PQR) Yes ☐

Identification # __96-008__
Revision __1__ Date __4-96__ By __Y. LYE__
Authorized by __O. KAY__ Date _____
Type—Manual ☐ Semi-Automatic ☒
Machine ☐ Automatic ☐

Company Name __ABC Fabricating__
Welding Process(es) __FCAW__
Supporting PQR No.(s) __N.A.__

JOINT DESIGN USED
Type: __T-joint, Fillet Weld 5/16"__
Single ☐ Double Weld ☒
Backing: Yes ☐ No ☒
Backing Material:
Root Opening __—__ Root Face Dimension __—__
Groove Angle: __—__ Radius (J–U) __—__
Back Gouging: Yes ☐ No ☒ Method __—__

POSITION
Position of Groove: __—__ Fillet: __HORIZ. (2F)__
Vertical Progression: Up ☐ Down ☐ __N.A.__

ELECTRICAL CHARACTERISTICS

Transfer Mode (GMAW) Short-Circuiting ☐
Globular ☐ Spray ☐
Current: AC ☐ DCEP ☒ DCEN ☐ Pulsed ☐
Other __—__

BASE METALS
Material Spec. __ASTM A572 / A36__
Type or Grade __Gr 50 / —__
Thickness: Groove __—__ Fillet __1/2" → 2 1/4"__
Diameter (Pipe) __—__

Tungsten Electrode (GTAW)
Size: __—__
Type: _____

FILLER METALS
AWS Specification __A 5.20__
AWS Classification __E 70T-1__

TECHNIQUE
(Stringer) or Weave Bead: _____
Multi-pass or (Single Pass) (per side) _____
Number of Electrodes __1__
Electrode Spacing Longitudinal __—__
Lateral __—__
Angle __—__

SHIELDING
Flux __—__ Gas __X__
Composition __100% CO₂__
Electrode-Flux (Class) _____ Flow Rate __45 CFH__
Gas Cup Size __#4__

Contact Tube to Work Distance __1 1/8" ± 1/4"__
Peening __—__
Interpass Cleaning: __N.A.__

PREHEAT
Preheat Temp., Min __150°F__
Interpass Temp., Min __150°F__ Max __—__

POSTWELD HEAT TREATMENT
Temp. __—__
Time __—__

WELDING PROCEDURE

Pass or Weld Layer(s)	Process	Filler Metals		Current		Volts	Travel Speed	Joint Details
		Class	Diam.	Type & Polarity	Amps or Wire Feed Speed			
1	FCAW	E70T-1	3/32"	DC+	460A	31 V	15-17 ipm	

Form E-1 (Front)

Figure 22.40 Example of Welding Procedure Specification (WPS). (From American Welding Society. *Structural Welding Code: Steel: ANSI/AWS D1.1-96.* Miami, Florida, 1996. With permission.)

WELDING PROCEDURE SPECIFICATION (WPS) Yes ☒
PREQUALIFIED ___X___ **QUALIFIED BY TESTING** _____
or PROCEDURE QUALIFICATION RECORDS (PQR) Yes ☐

Identification # __12817__
Revision __2__ Date _3-89_ By _I. See_
Authorized by _J. Walk_ Date _4-89_
Company Name _XYZ Erectors_
Welding Process(es) _FCAW_
Supporting PQR No.(s) _N.A._
Type—Manual ☐ Semi-Automatic ☒
Machine ☐ Automatic ☐

JOINT DESIGN USED
Type: Lap Joint - Fillet Weld 5/16"
Single ☒ Double Weld ☐
Backing: Yes ☐ No ☒
Backing Material:
Root Opening _--_ Root Face Dimension _--_
Groove Angle: _--_ Radius (J–U) _--_
Back Gouging: Yes ☐ No ☒ Method _--_

BASE METALS
Material Spec. _ASTM A36/A36_
Type or Grade _---_
Thickness: Groove _---_ Fillet 1/2" - 5/8"
Diameter (Pipe) _---_

FILLER METALS
AWS Specification _S.20_
AWS Classification _E71T-8_

SHIELDING
Flux _---_ Gas _None_
Composition _--_
Electrode-Flux (Class) _--_ Flow Rate _--_
Gas Cup Size _--_

PREHEAT
Preheat Temp., Min _None (70°F min.)_
Interpass Temp., Min _None_ Max_____

POSITION
Position of Groove:_____ Fillet: Vertical (3F)
Vertical Progression: Up ☒ Down ☐

ELECTRICAL CHARACTERISTICS
Transfer Mode (GMAW) Short-Circuiting ☐
Globular ☐ Spray ☐
Current: AC ☐ DCEP ☐ DCEN ☒ Pulsed ☐
Other _----_
Tungsten Electrode (GTAW)
Size: _---_
Type: _---_

TECHNIQUE
Stringer or Weave Bead: _Weave_
Multi-pass or Single Pass (per side) _single_
Number of Electrodes _1_
Electrode Spacing Longitudinal _---_
Lateral _---_
Angle _---_
Contact Tube to Work Distance 1/2" - 3/4"
Peening _---_
Interpass Cleaning: _N.A._

POSTWELD HEAT TREATMENT
Temp. _---_
Time _---_

WELDING PROCEDURE

| Pass or Weld Layer(s) | Process | Filler Metals | | Current | | Volts | Travel Speed | Joint Details |
		Class	Diam.	Type & Polarity	Amps or Wire Feed Speed			
1	FCAW	E71T-8	0.068"	DC-	250A (150 ipm)	19-21	5.5 - 6.5ipm	 3F

Form E-1 (Front)

Figure 22.41 Example of Welding Procedure Specification (WPS). (From American Welding Society. *Structural Welding Code: Steel: ANSI/AWS D1.1-96.* Miami, Florida, 1996. With permission.)

when the electrode is connected to the negative terminal. For most welding processes, the required electrode polarity is a function of the design of the electrode. For submerged arc welding, either polarity could be utilized.

Current density is determined by dividing the welding amperage by the cross-sectional area of the electrode. The current density is therefore proportional to I/d^2. As the current density increases, both deposition rates and penetration increase.

Preheat and interpass temperatures are used to control cracking tendencies, typically in the base material. Excessively high preheat and interpass temperatures will reduce the yield and tensile strength of the weld metal as well as the toughness. When base metals receive little or no preheat, the resultant rapid cooling can promote cracking as well as excessively high yield and tensile properties in the weld metal, and a corresponding reduction in toughness and elongation.

The WPS defines and controls all of the preceding variables. Conformance to the WPS is particularly important in the case of seismically loaded structures, because of the high demand placed on welded connections under these situations.

22.12.3   Fit-Up

The orientation of the various pieces prior to welding is known as "fit-up". The AWS D1.1-96 code [9] contains specific tolerances that are applied to the as-fit dimensions of a joint prior to welding. There must be ample access to the root of the joint to ensure good, uniform fusion between the members being joined. Excessively small root openings or included angles in groove welds do not permit uniform fusion. Excessively large root openings or included angles result in the need for greater volumes of weld metal, with corresponding increases in shrinkage stresses, which in turn increases distortion and cracking tendencies. The D1.1-96 tolerances for fit-up are generally measured in 1/16-in. increments.

22.12.4   Field vs. Shop Welding

Many believe that the highest quality welding is obtained under shop welding conditions. The greatest differences between field and shop welding are related to control. For shop fabrication, the work force is generally more stable. Supervision practices are well understood and communication is generally more efficient. Under field welding conditions, control of a project seems to be more difficult. While there are environmental challenges to field conditions, including temperature, wind, and moisture, these seem to pose fewer problems than do the management issues.

For field welding, the gasless welding processes such as self-shielded flux cored welding and shielded metal arc welding usually are preferred. Gas metal arc, gas tungsten arc, and gas-shielded flux cored arc welding are all limited due to their sensitivity to wind-related gas disturbances. It is imperative that field welding conditions receive an appropriate increase in the monitoring and control area to ensure consistent quality. D1.1-96 imposes the same requirements on field welding as on shop welding. This includes qualification of welders, the use of welding procedures, and the resultant quality requirements.

22.13   Inspection

The AWS D1.1-96 code requires that all welds be inspected, specifically by means of visual inspection. In addition, at the engineer's discretion and as identified in contract documents, nondestructive testing may be required for finished **weldments**. This enables the engineer with a knowledge of the complexity of the project to specify additional inspection methodologies commensurate with the degree of confidence required for a particular project. In the case of seismically loaded structures, and connections subject to high stress levels, the need for inspection increases.

22.13.1 In-Process Visual Inspection

D1.1-96 mandates the use of in-process visual inspection. Before welding, the inspector reviews welder qualification records, welding procedure specifications, and the contract documents to confirm that applicable requirements are met. Before welding is performed, the inspector verifies fit-up and joint cleanliness, examines the welding equipment to ensure it is in proper working order, verifies that the materials involved meet the various requirements, and confirms that the required levels of preheat have been properly applied. During welding, the inspector confirms that the WPS is being carried out and that the intermediate weld passes meet the various requirements. After welding is finished, final bead shapes and welding integrity can be visually confirmed. Effective visual inspection is a critical component for ensuring consistent weld quality.

22.13.2 Nondestructive Testing

Four major nondestructive testing methods may be used to verify weld integrity after welding operations are completed. Each should be used in conjunction with effective visual inspection. No process is 100% capable of detecting all discontinuities in a weld.

Dye penetrant (PT) inspection involves the application of a liquid that is drawn into a surface-breaking discontinuity, such as a crack or porosity, by capillary action. When the excess residual dye is removed from the surface, a developer is applied that will absorb the penetrant contained within the discontinuity. The result is a stain in the developer that shows that a discontinuity is present. PT testing is limited to surface-breaking discontinuities. It cannot read subsurface discontinuities, but it is highly effective in accenting very small discontinuities.

Magnetic particle (MT) inspection utilizes the change in magnetic flux that occurs when a magnetic field is present in the vicinity of a discontinuity. The change will show up as a different pattern when magnetic dust-like particles are applied to the surface of the part. The process is highly effective in locating discontinuities that are on the surface or slightly subsurface. The magnetic field can be created in the material in one of two ways: the current is directly passed through the material or the magnetic field is induced through a coil on a yoke. Since the process is most sensitive to discontinuities that lie perpendicular to the magnetic flux path, it is necessary to energize the part in two directions in order to fully inspect the component.

Radiographic (RT) inspection uses X-rays or gamma rays that are passed through the weld to expose a photographic film on the opposite side of the joint. High-voltage generators produce X-rays, while gamma rays are created by atomic disintegration of radioisotopes. Whenever radiographic inspection is employed, workers must be protected from exposure to excessive radiation. RT relies on the ability of the material to pass some of the radiation through, while absorbing part of this energy within the material. Different materials have different absorption rates. As the different levels of radiation are passed through the material, portions of the film are exposed to a greater or lesser degree. When this film is developed, the resulting radiograph will bear the image of the cross-section of the part. The radiograph is actually a negative. The darkest regions are those that were most exposed when the material being inspected absorbed the least amount of radiation. Porosity will show up as small dark round circles. Slag is generally dark and will look similar to porosity, but will have irregular shapes. Cracks appear as dark lines. Excessive reinforcement will result in a light region.

A radiographic test is most effective for detecting volumetric discontinuities such as slag or porosity. When cracks are oriented perpendicular to the direction of a radiographic source, they may be missed with the RT method. Therefore, RT inspection is most appropriate for butt joints and is generally not appropriate for inspection of corner or T joints. Radiographic testing has the advantage of generating a permanent record for future reference.

In ultrasonic (UT) inspection, solid discontinuity-free materials will transmit high-frequency sound waves throughout the part in an uninterrupted manner. A receiver "hears" the sound reflected off of the back surface of the part being inspected. If there is a discontinuity between the transmitter and the back of the part, an intermediate signal will be sent to the receiver indicating its presence. The pulses are read on a CRT screen. The magnitude of the signal received from the discontinuity indicates its size. UT is most sensitive to planar discontinuities, i.e., cracks. UT effectiveness is dependent on the operator's skill, so UT technician training and certification is essential. With currently available technology, UT is capable of reading a variety of discontinuities that would be acceptable for many applications. Acceptance criteria must be clearly communicated to the inspection technicians so unnecessary repairs are avoided.

22.13.3 Applications for Nondestructive Testing Methods

Visual inspection is the most comprehensive method available to verify conformance with the wide variety of issues that can affect weld quality and should be thoroughly applied on every welding project. To augment visual inspection, nondestructive testing can be specified to verify the integrity of the deposited weld metal. The selection of the inspection method should reflect the probable discontinuities that would be encountered, and the consequences of undetected discontinuities. Consideration must be given to the conditions under which the inspection would be performed, such as field vs. shop conditions. The nature of the joint detail (butt, T, corner, etc.) and the weld type (CJP, PJP, fillet weld) will determine the choice of the inspection process in many situations. MT inspection is usually preferred over PT inspection because of its relative simplicity. Cleanup is easy, and the process is sensitive. PT is normally reserved for applications where the material is nonmagnetic, and MT would not be applicable. While MT is suitable for detection of surface or slightly subsurface discontinuities only, it is in these areas that many welding defects are located. It is very effective in crack detection, and can be utilized to ensure complete crack removal before subsequent welding is performed on damaged structures.

UT inspection has become the primary nondestructive testing method used for most building applications. It can be utilized to inspect butt, T, and corner joints, is relatively portable, and is free from the radiation concerns associated with RT inspection. UT is especially sensitive to the identification of cracks, the most significant defect in a structural system. Although it may not detect spherical or cylindrical voids such as porosity, nondetection of these types of discontinuities has fewer consequences.

22.14   Post-Northridge Assessment

Prior to the Northridge earthquake, the SMRF with the "pre-Northridge" beam-to-column detail was unchallenged regarding its ability to perform as expected. This confidence existed in spite of a fairly significant failure rate when these connections had been tested in previous research. The pre-Northridge detail consisted of the following:

- CJP groove welds of the beam flanges to the column face, with weld backing and weld tabs left in place.
- No specific requirement for minimum notch toughness properties in the weld deposit.
- A bolted web connection with or without supplemental fillet welds of the shear tab to the beam web.
- Standard ASTM A36 steel for the beam and ASTM 572 grade 50 for the column (i.e., no specific limits on yield strength or the F_y/F_u ratio).

As a result of the Northridge earthquake, and research performed immediately afterward, confidence in this detail has been severely shaken. Whether any variation of this detail will be suitable for use in the future is currently unknown. More research must be done, but one can speculate that, with the possible exception of small-sized members, some modification of this detail will be required.

Although testing of this configuration had a fairly high failure rate in pre-Northridge tests, many successful results were obtained. Further research will determine which variables are the most significant in predicting performance success. Some changes in materials and design practice also should be considered. In recent years, recycling of steel has become a more predominant method of manufacture. This is not only environmentally responsible, it is economical. However, in recycling, residual alloys can accumulate in the scrap charge, inadvertently increasing steel strength levels. In the past 20 years, the average yield strength of ASTM A36 steel has increased approximately 15%. Testing done with lower yield strength steel would be expected to exhibit different behavior than test specimens made of today's higher strength steels (in spite of the same ASTM designation). For practical reasons, laboratory specimens tend to be small in size. Success in small-sized specimens was extrapolated to apply to very large connection assemblies in actual structures. The design philosophy that led to fewer SMRFs throughout a structure required that each of the remaining frames be larger in size. This corresponded to heavier and deeper beams, and much heavier columns, with an increase in the size of the weld between the two rolled members. The effect of size on restraint and triaxial stresses was not researched, resulting in some new discoveries about the behavior of the large-sized assemblages during the Northridge earthquake.

The engineering community generally agrees that the pre-Northridge connection (as defined above) is no longer adequate and some modification will be required. Any deviation from the definition above will constitute a modification for the purposes of this discussion.

22.14.1 Minor Modifications to the SMRF Connection

With the benefit of hindsight, several aspects of the pre-Northridge connection detail appear to be obviously deficient. Weld backing left in place in a connection subject to both positive and negative moments where the root of the flange weld can be put into tension creates high-stress concentrations that may result in cracking. Failure to specify minimum toughness levels for weld metal for heavily loaded connections is another deficiency. The superior performance of the welded web vs. the bolted web in past testing draws into question the assumption of load sharing between welds and bolts. Now it seems that tighter control of the strength properties of the beam steel and the relationship to the column is essential.

Some preliminary tests suggest that tightly controlling all of these variables may result in acceptable performance. However, the authors know of no test of unmodified beam-to-column connections where the connection zone has remained crack free when acceptable rotation limits were achieved. For smaller sized members, this approach may be technically possible, although the degree of control necessary on both the material properties and the welding operations may make it impractical.

22.14.2 Coverplated Designs

This concept uses short coverplates that are added to the top and bottom flanges of the beam. Fillet welds transfer the coverplate forces to the beam flanges. The bottom flange coverplate is shop welded to the column flange, and the bottom beam flange is field welded to the column flange and to the coverplate. Both the top flange and the top flange coverplate are field welded to the column flange with a common weld. The web connection may be welded or high-strength bolted. These connections have been tested to a limited extent, with generally favorable results.

Following Northridge, the coverplate approach received significant attention because it offered early promise of a viable solution. Other methods may prove to be superior as time passes. While the coverplate solution treats the beam in the same way as other approaches (i.e., it moves the plastic hinge into a region where ductility can be demonstrated), it concentrates all the loading to the column into a relatively short distance. Other alternatives may treat the column in a more gentle manner.

22.14.3   Flange Rib Connections

This concept utilizes one or two vertical ribs attached between the beam flanges and column face. The intent of the rib plates is to reduce the demand on the weld at the column flange and to shift the plastic hinge from the column face. In limited testing, flange rib connections have demonstrated acceptable levels of plastic rotation provided that the girder flange welding is done correctly.

Vertical ribs appear to function very similarly to the coverplated designs, but offer the additional advantage of spreading the load over a greater portion of the column. The single-rib designs appear to be better than the twin-rib approaches because the stiffening device is in alignment with the column web (for I-shaped columns) and facilitates easy access to either side of the device for welding. It is doubtful that the rib design would be appropriate for box column applications.

22.14.4   Top and Bottom Haunch Connections

In this configuration, haunches are placed on both the top and bottom flanges. In two tests of the top and bottom haunch connection, it has exhibited extremely ductile behavior, achieving plastic rotations as great as 0.07 rad. Tests of single, haunched beam–column connections have not been as conclusive; further tests are planned.

Although they are costly, haunches appear to be the most straightforward approach to obtaining the desired behavior out of the connection. The treatment to the column is particularly desirable, greatly increasing the portion of the column participating in the transfer of moment.

22.14.5   Reduced Beam Section Connections

In this configuration, the cross-section of the beam is deliberately reduced within a segment to produce a plastic hinge within the beam span, away from the column face. A variant of this approach produces the so-called "dog bone" profile.

Reduced section details offer the prospect of a low-cost connection and increased performance out of detailing that is very similar to the pre-Northridge connection. Control of material properties of the beam will still be a major variable if this detail is used. Lateral bracing will probably be required in the area of the reduced section to prevent buckling, particularly at the bottom flange when loaded in compression.

22.14.6   Partially Restrained Connections

Some have suggested that partially restrained (PR) connection details will offer a performance advantage over the SMRF. The relative merits of a PR frame vs. a rigid frame are beyond the scope of this work. However, many engineers immediately think of bolted PR connections when it is possible to utilize welded connections for PR performance as well.

Illustrated in Figure 22.42 is a detail that can be employed utilizing the PR concept. Detailing rules must be developed, and tests done, before these details are employed. They are supplied to offer welded alternatives to bolted PR connections.

22.15   Defining Terms

As-welded: The condition of weld metal, weld joints, and weldments after welding, but prior to any subsequent thermal, mechanical, or chemical treatments.

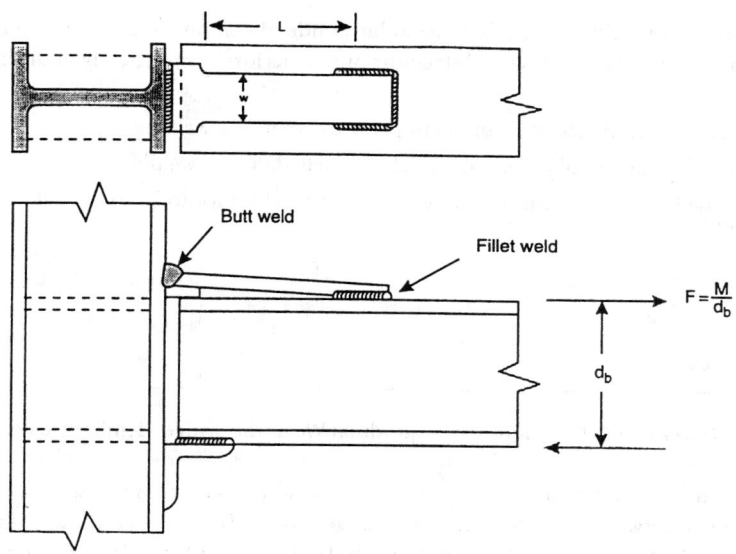

Figure 22.42 A partially restrained welded connection detail. (Courtesy of The Lincoln Electric Company. With permission.)

Autogenous weld: A fusion weld made without the addition of filler metal.

Back gouging: The removal of weld metal and base metal from the other side of a partially welded joint to facilitate complete fusion and complete joint penetration upon subsequent welding from that side.

Backing: A material or device placed against the back side of the joint, or at both sides of a weld in electroslag and electrogas welding, to support and retain molten weld metal. The material may be partially fused or remain unfused during welding and may be either metal or nonmetal.

Base metal: The material to be welded, brazed, soldered, or cut.

Deposition rate: The weight of material deposited in a unit of time.

Effective throat: The minimum distance minus any convexity between the weld root and the face of a fillet weld.

Filler metal: The metal to be added in making a welded, brazed, or soldered joint.

Heat affected zone (HAZ): That portion of the base metal that has not been melted, but whose mechanical properties or microstructure have been altered by the heat of welding, brazing, soldering, or cutting.

Nugget: The weld metal joining the workpieces in spot, roll spot, seam, or projection welds.

Postheating: The application of heat to an assembly after welding, brazing, soldering, thermal spraying, or thermal cutting.

Preheating: The application of heat to the base metal immediately before welding, brazing, soldering, thermal spraying, or cutting.

Residual stress: Stress present in a member that is free of external forces or thermal gradients.

Theoretical weld throat: The distance from the beginning of the joint root perpendicular to the hypotenuse of the largest right triangle that can be inscribed within the cross-section of a fillet weld. This dimension is based on the assumption that the root opening is equal to zero.

Weldability: The capacity of material to be welded under the imposed fabrication conditions into a specific, suitable designed structure and to perform satisfactorily in the intended service.

Weldment: An assembly whose component parts are joined by welding.

Weld metal: That portion of a weld that has been melted during welding.

Weld pool: The localized volume of molten metal in a weld prior to its solidification as weld metal.

References

[1] Alexander, W.G. 1991. Designing Longitudinal Welds for Bridge Members, *Eng. J.*, 28(1), 29-36.

[2] American Association of State Highway and Transportation Officials and American Welding Society. 1995. *Bridge Welding Code: ANSI/AASHTO/AWS D1.5-95.* Miami, FL.

[3] American Institute of Steel Construction. 1994. *Manual of Steel Construction: Load & Resistance Factor Design*, 2nd ed., Chicago, IL.

[4] American Society for Metals. 1983. *Metals Handbook, Ninth Edition, Volume 6: Welding, Brazing, and Soldering*, Metals Park, OH.

[5] American Welding Society. 1988. *Guide for the Visual Inspection of Welds: ANSI/AWS B1.11-88*, Miami, FL.

[6] American Welding Society. 1991. *Standard Symbols for Welding, Brazing and Nondestructive Examination: ANSI/AWS A2.4-93*, Miami, FL.

[7] American Welding Society. 1989. *Standard Welding Terms and Definitions: ANSI/AWS A3.0-89*, Miami, FL.

[8] American Welding Society. 1989. *Structural Welding Code: Sheet Steel: ANSI/AWS D1.3-89*, Miami, FL.

[9] American Welding Society. 1996. *Structural Welding Code: Steel: ANSI/AWS D1.1-96*, Miami, FL.

[10] American Welding Society. 1995. Structural Welding Committee Position Statement on Northridge Earthquake Welding Issues, Miami, FL.

[11] American Welding Society. 1987. *Welding Handbook, Eighth Edition, Volume 1: Welding Technology*, L.P. Conner, Ed., Miami, FL.

[12] American Welding Society. 1991. *Welding Handbook, Eighth Edition, Volume 2: Welding Processes*, R.L. O'Brien, Ed., Miami, FL.

[13] American Welding Society. 1996. *Welding Handbook, Eighth Edition, Volume 3: Materials and Applications, Part 1*, W.R. Oates, Ed., Miami, FL.

[14] American Welding Society. 1976. *Welding Handbook: Volume One, Seventh Edition: Fundamentals of Welding*, C. Weisman, Ed., Miami, FL.

[15] American Welding Society. 1978. *Welding Handbook: Volume Two, Seventh Edition: Welding Processes—Arc and Gas Welding and Cutting, Brazing, and Soldering*, W.H. Kearns, Ed., Miami, FL.

[16] American Welding Society. 1980. *Welding Handbook: Volume Three, Seventh Edition: Welding Processes—Resistance and Solid-State Welding and Other Joining Processes*, W.H. Kearns, Ed., Miami, FL.

[17] American Welding Society. 1982. *Welding Handbook: Volume Four, Seventh Edition: Metals and Their Weldability*, W.H. Kearns, Ed., Miami, FL.

[18] American Welding Society. 1984. *Welding Handbook: Volume Five, Seventh Edition: Engineering, Costs, Quality and Safety*, W.H. Kearns, Ed., Miami, FL.

[19] Barsom, J.M. and Rolfe, S.T. 1987. *Fracture and Fatigue Control in Structures: Applications of Fracture Mechanics,* 2nd ed., Prentice-Hall, Englewood Cliffs, NJ.

[20] Blodgett, O.W. 1966. *Design of Welded Structures,* The James F. Lincoln Arc Welding Foundation, Cleveland, OH.

[21] Blodgett, O.W. 1995. Details to Increase Ductility in SMRF Connections, *The Welding Innovation Quarterly,* XII(2).

[22] Blodgett, O.W. 1993. The Challenge of Welding Jumbo Shapes, Part II: Increasing Ductility of Connections, *The Welding Innovation Quarterly,* X(1).

[23] Lincoln Electric Company. 1995. *The Procedure Handbook of Arc Welding,* 13th ed., Cleveland, OH.

[24] Miller, D.K. 1996. Ensuring Weld Quality in Structural Applications, Part I: The Roles of Engineers, Fabricators & Inspectors, *The Welding Innovation Quarterly,* XIII(2).

[25] Miller, D.K. 1996. Ensuring Weld Quality in Structural Applications, Part II: Effective Visual Inspection, *The Welding Innovation Quarterly,* XIII(3).

[26] Miller, D.K. 1994. Northridge: The Role of Welding Clarified, *The Welding Innovation Quarterly,* XI(2).

[27] Miller, D.K. 1996. Northridge: An Update, *The Welding Innovation Quarterly,* XIII(1).

[28] Miller, D.K. 1994. *Welding of Steel Bridges,* The James F. Lincoln Arc Welding Foundation, Cleveland, OH.

[29] Miller, D.K. 1988. What Structural Engineers Need to Know About Weld Metal, *1988 National Steel Construction Conference Proceedings:* 35.1-35.15. American Institute of Steel Construction, Chicago, IL.

Further Reading

[1] American Association of State Highway and Transportation Officials. 1978. *Guide Specifications for Fracture Critical Non-Redundant Steel Bridge Members, 1978.* (As revised by Interim Specifications for Bridges, 1981, 1983, 1984, 1985, 1986, and 1991.) Washington, D.C.

[2] American Institute of Steel Construction. 1994. *Northridge Steel Update I,* Chicago, IL.

[3] American Society for Metals. 1994. *Hydrogen Embrittlement and Stress Corrosion Cracking,* R. Gibala and R.F. Hehemann, Eds., Metals Park, OH.

[4] American Welding Society. various dates. *A5 Filler Metal Specifications,* Miami, FL.

[5] American Welding Society. 1993. *Standard Methods for Determination of the Diffusible Hydrogen Content of Martensitic, Bainitic, and Ferritic Steel Weld Metal Produced by Arc Welding: ANSI/AWS A4.3-93,* Miami, FL.

[6] Bailey, N., Coe, F.R., Gooch, T.G., Hart, P.H.M., Jenkins, N., and Pargeter, R.J. 1973. *Welding Steels Without Hydrogen Cracking,* 2nd ed., Abington Publishing, Cambridge, England.

[7] Boniszewski, T. 1992. *Self-Shielded Arc Welding.* Abington Press, Cambridge, England.

[8] Masubuchi, K. 1980. *Analysis of Welded Structures,* 1st ed. Pergamon Press, Oxford, England.

[9] Roeder, C.W. 1985. *Use of Thermal Stresses for Repair of Seismic Damage to Steel Structures.* University of Washington, Seattle, WA.

[10] SAC Joint Venture. 1995. *Interim Guidelines: Evaluation, Repair, Modification and Design of Welded Steel Moment Frame Structures,* Report No. SAC-95-02 (FEMA 267), Sacramento, CA.

[11] Shanafelt, G.O. and Horn, W.B. 1984. *Guidelines for Evaluation and Repair of Damaged Steel Bridge Members,* National Cooperative Highway Research Program Report 271, Transportation Research Board, Washington, D.C.

[12] Stout, R.D. 1987. *Weldability of Steels,* 4th ed., Welding Research Council, New York.

[13] Wilson, A.D. 1990. Hardness Testing of Thermal Cut Edges of Steel, *Eng. J.,* 27(3), 98-105.

23

Composite Connections

23.1 Introduction 23-1
23.2 Connection Behavior Classification 23-2
23.3 PR Composite Connections 23-5
23.4 Moment-Rotation (M-θ) Curves 23-9
23.5 Design of Composite Connections in Braced Frames.. 23-12
23.6 Design for Unbraced Frames 23-20
References 23-27

Roberto Leon
School of Civil and Environmental Engineering, Georgia Institute of Technology, Atlanta, GA

23.1 Introduction

The vast majority of steel buildings built today incorporate a floor system consisting of composite beams, composite joists or trusses, stub girders, or some combination thereof [29]. Traditionally the strength and stiffness of the floor slabs have only been used for the design of simply-supported flexural members under gravity loads, i.e., for members bent in single curvature about the strong axis of the section. In this case the members are assumed to be pin-ended, the cross-section is assumed to be prismatic, and the effective width of the slab is approximated by simple rules. These assumptions allow for a member-by-member design procedure and considerably simplify the checks needed for strength and serviceability limit states. Although most structural engineers recognize that there is some degree of continuity in the floor system because of the presence of reinforcement to control crack widths over column lines, this effect is considered difficult to quantify and thus ignored in design.

The effect of the floor slabs has also been neglected when assessing the strength and stiffness of frames subjected to lateral loads for four principal reasons. First, it has been assumed that neglecting the additional strength and stiffness provided by the floor slabs always results in a conservative design. Second, a sound methodology for determining the M-θ curves for these connections is a prerequisite if their effect is going to be incorporated into the analysis. However, there is scant data available in order to formulate reliable moment-rotation (M-θ) curves for composite connections, which fall typically into the partially restrained (PR) and partial strength (PS) category. Third, it is difficult to incorporate into the analysis the non-prismatic composite cross-section that results when the member is subjected to double curvature as would occur under lateral loads. Finally, the degree of composite interaction in floor members that are part of lateral-load resisting systems in seismic areas is low, with most having only enough shear transfer capacity to satisfy diaphragm action.

Research during the past 10 years [25] and damage to steel frames during recent earthquakes [22] have pointed out, however, that there is a need to reevaluate the effect of composite action in modern frames. The latter are characterized by the use of few bents to resist lateral loads, with the ratio of number of gravity to moment-resisting columns often as high as 6 or more. In these cases the aggregate effect of many PR/PS connections can often add up to a significant portion of the lateral resistance of a frame. For example, many connections that were considered as pins in the analysis (i.e., connections to columns in the gravity load system) provided considerable lateral strength and stiffness to steel moment-resisting frames (MRFs) damaged during the Northridge earthquake. In these cases many of the fully restrained (FR) welded connections failed early in the load history, but the frames generally performed well. It has been speculated that the reason for the satisfactory performance was that the numerous PR/PS connections in the gravity load system were able to provide the required resistance since the input base shear decreased as the structure softened. In these PR/PS connections, much of the additional capacity arises from the presence of the floor slab which provides a moment transfer mechanism not accounted for in design.

In this chapter general design considerations for a particular type of composite PR/PS connection will be given and illustrated with examples for connections in braced and unbraced frames. Information on design of other types of bolted and composite PR connections is given elsewhere [22], (Chapter 6 of [29]). The chapter begins with discussions of both the development of M-θ curves and the effect of PR connections on frame analysis and design. A clear understanding of these two topics is essential to the implementation of the design provisions that have been proposed for this type of construction [26] and which will be illustrated herein.

23.2   Connection Behavior Classification

The first step in the design of a building frame, after the general topology, the external loads, the materials, and preliminary sizes have been selected, is to carry out an analysis to determine member forces and displacements. The results of this analysis depend strongly on the assumptions made in constructing the structural model. Until recently most computer programs available to practicing engineers provided only two choices (rigid or pinned) for defining the connections stiffness. In reality connections are very complex structural elements and their behavior is best characterized by M-θ curves such as those given in Figure 23.1 for typical steel connections to an A36 W24x55 beam ($M_{p,\text{beam}} = 4824$ kip-in.). In Figure 23.1, M_{conn} corresponds to the moment at the column face, while θ_{conn} corresponds to the total rotation of the connection and a portion of the beam generally taken as equal to the beam depth. These curves are shown for illustrative purposes only, so that the different connection types can be contrasted. For each of the connection types shown, the curves can be shifted through a wide range by changing the connection details, i.e., the thickness of the angles in the top and seat angle case.

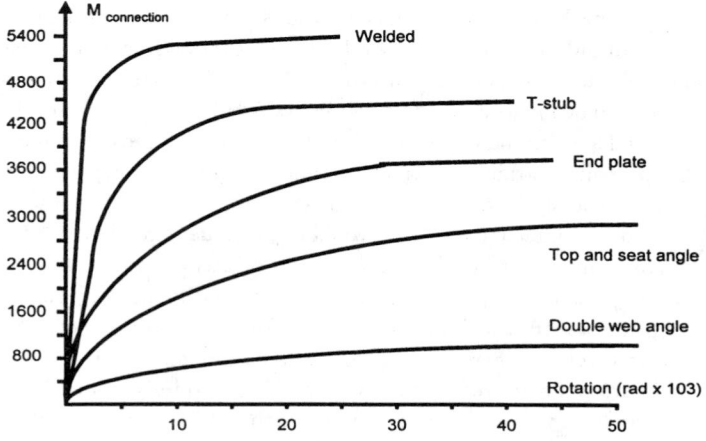

Figure 23.1   Typical moment-rotation curves for steel connections.

While the M-θ curves are highly non-linear, at least three key properties for design can be obtained from such data. Figure 23.2 illustrates the following properties, as well as other relevant connection characteristics, for a composite connection:

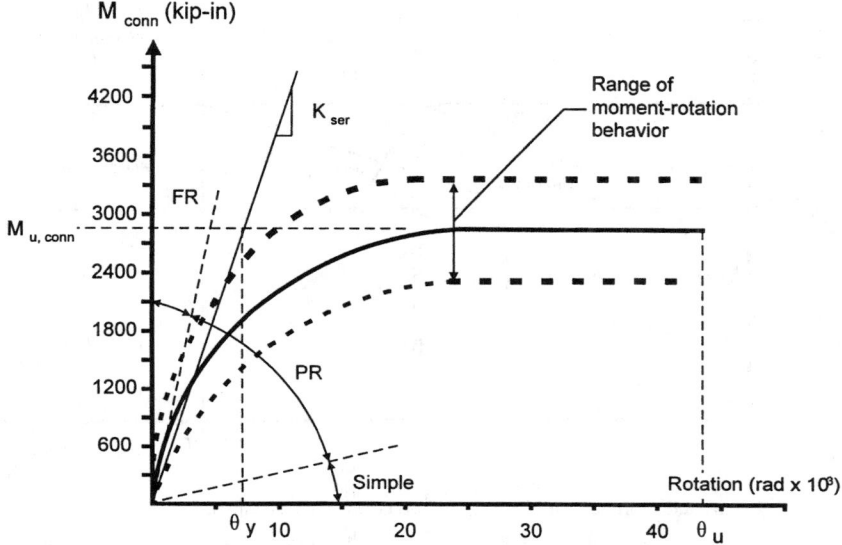

Figure 23.2   Definition of connection properties for PR connections.

1. Initial stiffness (k_{ser}), which will be used in calculating deflection and vibration performance under service loads. In these analysis the connection will be represented by a linear rotational spring. Since the curves are non-linear from the beginning, and k_{ser} will be assumed constant, the latter needs to be defined as the secant stiffness to some predetermined rotation.

2. Ultimate strength ($M_{u,conn}$), which will be used in assessing the ultimate strength of the frame. The strength is controlled either by the strength of the connection itself or that of the framing beam. In the former case the connection is defined as partial strength (PS) and in the latter as full strength (FS).

3. Maximum available rotation (θ_u), which will be used in checking both the redistribution capacity under factored gravity loads and the drift under earthquake loads. The required rotational capacity depends on the design assumptions and the redundancy of the structure.

It is often useful also to define a fourth quantity, the ductility (μ) of the connection. This is defined as the ratio of the ultimate rotation capacity (θ_u) to some nominal "yield" rotation (θ_y). It should be understood that the definition of θ_y is subjective and needs to account for the shape of the curve (i.e., how sharp is the transition from the service to the yield level — the sharper the transition the more valid the definition shown in Figure 23.2). In the design procedure to be discussed in this chapter, the initial stiffness, ultimate strength, maximum rotation, and ductility are properties that will need to be check by the structural engineer.

Figure 23.2 schematically shows that there can be a considerable range of strength and stiffness for these connections. The range depends on the specific details of the connection, as well as the normal variability expected in materials and construction practices. Figure 23.2 also shows that certain ranges of initial stiffness can be used to categorize the initial connection stiffness as either fully restrained (FR), partially restrained (PR), or simple. Because the connection behavior is strongly influenced by the strength and stiffness of the framing members, it is best to non-dimensionalize M-θ curves as shown in Figure 23.3.

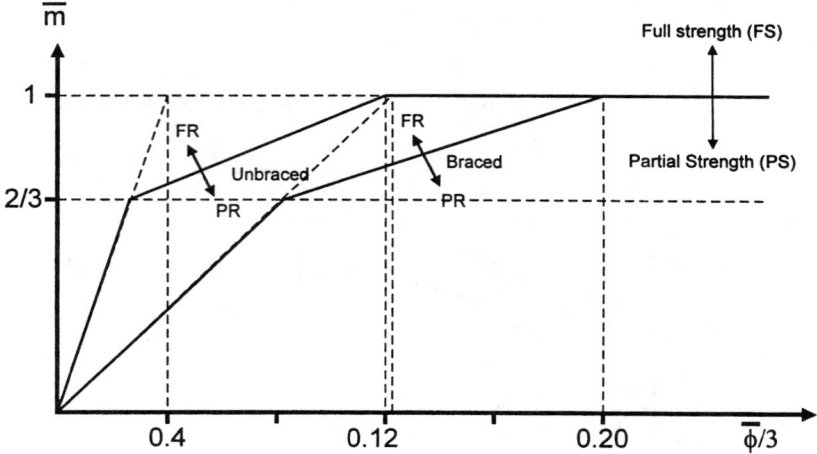

Figure 23.3   Normalized moment-rotation curves and connection classification. (After Eurocode 3, Design of Steel Structures, Part 1: General Rules and Rules for Buildings, ENV 1993-1-1: 1992, Comite Europeen de Normalisation (CEN), Brussels, 1992.)

In Figure 23.3, the vertical axis represents the ratio ($\overline{m}$) of the moment capacity of the connection ($M_{u,\text{conn}}$) to the nominal plastic moment capacity ($M_{p,\text{beam}} = Z_x F_y$) of the steel beam framing into it. As noted above, if this ratio is less than one then the connection is considered partial strength (PS); if it is equal or greater than one, then it is classified as a full strength (FS) connection. The horizontal axis is normalized to the end rotation of the framing beam assuming simple supports at the beam ends (θ_{ss}). This rotation depends, of course, on the loading configuration and the level of loading. Generally a factored distributed gravity load (w_u) and linear elastic behavior up to the full plastic capacity are assumed ($\theta_{ss} = w_u L_{\text{beam}}^3 / 24 E I_{\text{beam}}$). The resulting reference rotation ($\overline{\phi} = M_p L / E I$), based on a M_p of $w_u L^2 / 8$, is $M_p L / (3 E I) = \overline{\phi}/3$. It should be noted that the connection rotation is normalized with respect to the properties of the beam and not the column and that this normalization is meaningful only in the context of gravity loads. The column is assumed to be continuous and part of a strong column–weak beam system. For gravity loads its stiffness and strength are considered to contribute little to the connection behavior. This assumption, of course, does not account for panel zone flexibility which is important in many types of FS connections.

The non-dimensional format of Figure 23.3 is important because the terms partially restrained (PR) and full restraint (FR) can only be defined with respect to the stiffness of the framing members. Thus, a FR connection is defined as one in which the ratio (α) of the connection stiffness (k_{ser}) to the stiffness of the framing beam ($E I_{\text{beam}} / L_{\text{beam}}$) is greater than some value. For unbraced frames the recommended value ranges from 18 to 25, while for braced frames they range from 8 to 12. Figure 23.3 shows the limits chosen by the Eurocode, which are 25 for the unbraced case and 8 for the braced case [15]. These ranges have been selected based on stability studies that indicate that the global buckling load of a frame with PR connections with stiffnesses above these limits is decreased by less than 5% over the case of a similar frame with rigid connections. The large difference between the braced and unbraced values stems from the P-Δ and P-δ effects on the latter. PR connections are defined as those having α ranging from about 2 up to the FR limit. Connections with α less than 2 are regarded as pinned.

23.3   PR Composite Connections

Conventional steel design in the U.S. separates the design of the gravity and lateral load resisting systems. For gravity loads the floor beams are assumed to be simply supported and their section properties are based on assumed effective widths for the slab (AISC Specification I3.1 [2]) and a simplified definition of the degree of interaction (Lower Bound Moment of Inertia, Part 5 [3]). The simple supports generally represent double angle connections or single plate shear connections to the column flange. For typical floor beam sizes, these connections, tested without slabs, have shown low initial stiffness ($\alpha < 4$) and moment capacity ($M_{u,\text{conn}} < 0.1M_{p,\text{beam}}$) such that their effect on frame strength and stiffness can be characterized as negligible. In reality when live loads are applied, the floor slab will contribute to the force transfer at the connection if any slab reinforcement is present around the column. This reinforcement is often specified to control crack widths over the floor girders and column lines and to provide structural integrity. This results in a weak composite connection as shown in Figure 23.4. The effect of a weak PR composite connection on the behavior under gravity loads is shown in Example 23.1.

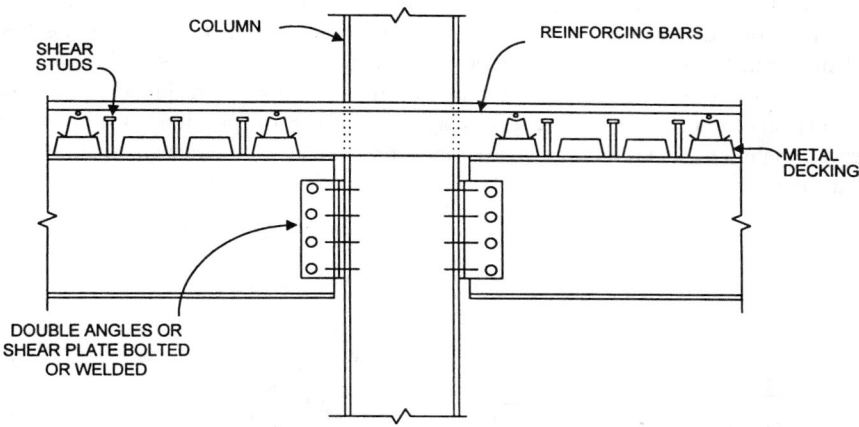

Figure 23.4   Weak PR composite connection.

EXAMPLE 23.1:   Effect of a Weak Composite Connection

Consider the design of a simply-supported composite beam for a DL $= 100$ psf and a LL $= 80$ psf. The span is 30 ft and the tributary width is 10 ft. For this case the factored design moment (M_u) is 3348 kip-in. and the required nominal moment (M_n) is 3720 kip-in. From the AISC LRFD Manual [3] one can select an A36 W18x35 composite beam with 92% interaction (PNA $= 3$, $\phi M_p = 3720$ kip-in., and $I_{LB} = 1240$ in.^{4}). The W18x35 was selected based on optimizing the section for the construction loads, including a construction LL allowance of 20 psf. The deflection under the full live load for this beam is 0.4 in., well below the 1 in. allowed by the L/360 criterion. Thus, this section looks fine until one starts to check stresses. If we assume that all the dead load stresses from 1.2DL, which are likely to be present after the construction period, are carried by the steel beam alone, then:

$$\sigma_{DL,\text{steel alone}} = M_{DL}/S_x = 1620 \text{ kip-in. }/57.6 \text{ in.}^3 = 28.1 \text{ ksi}$$

The stresses from live loads are then superimposed, but on the composite section. For this section $S_{eff} = 91.9$ in.^{3}, so the additional stress due to the arbitrary point-in-time (APT) live load (0.5LL) is:

$$\sigma_{LL(APT)} = M_{LL(APT)}/S_{eff} = 540 \text{ kip-in.}/91.9 \text{ in.}^3 = 5.9 \text{ ksi}$$

Thus, the total stress (σ_{APTl}) under the APT live load is:

$$\sigma APT = \sigma_{DL,\text{steel alone}} + \sigma_{LL(APT)} = 28.1 + 5.9 = 34.0 \text{ ksi}$$

Under the full live load ($1.0LL$), the stresses are:

$$\sigma_{APT} = \sigma_{DL,\text{steel alone}} + 2\sigma_{LL(APT)} = 28.1 + 11.8 = 39.9 \text{ ksi} > F_y = 36 \text{ ksi}$$

Thus, the beam has yielded under the full live loads even though the deflection check seemed to imply that there were no problems at this level. The current LRFD provisions do not include this check, which can govern often if the steel section is optimized for the construction loads.

Let us investigate next what the effect of a weak PR connection, similar to that shown in Figure 23.3, will be on the service performance of this beam. Assume that the beam frames into a column with double web angles connection and that four #3 Grade 60 bars have been specified on the slab to control cracking. These bars are located close enough to the column so that they can be considered part of the section under negative moment. The connection will be studied using the very simple model shown in Figure 23.5. In this model all deformations are assumed to be concentrated in an area very close to the connection, with the beam and column behaving as rigid bodies. The reinforcing bars are treated as a single spring (K_{bars}) while the contribution to the bending stiffness of the web angles (K_{shear}) is ignored. The connection is assumed to rotate about a point about 2/3 of the depth of the beam.

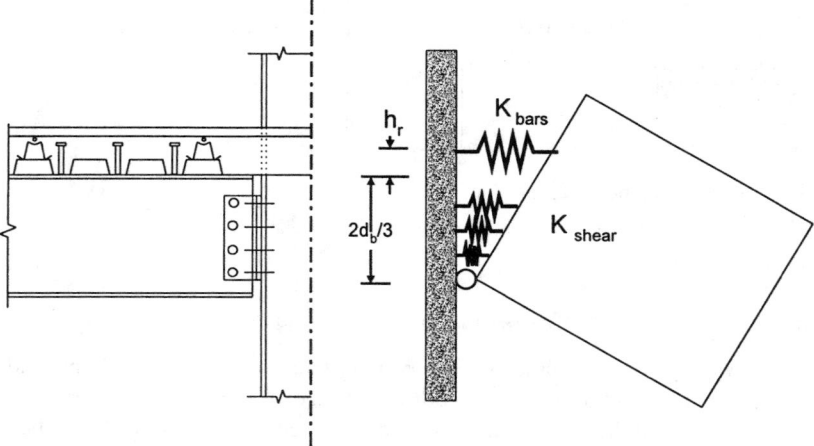

Figure 23.5 Simple mechanistic connection model.

Assuming that the angles and bolts can carry a combination of compression and shear forces without failing, at ultimate the yielding of the slab reinforcement will provide a tensile force (T) equal to:

$$T = \left(4 \text{ bars} * 0.11 \text{ in.}^2/ \text{ bar} * 60 \text{ ksi}\right) = 26.4 \text{ kips}$$

This force acts an eccentricity (e) of at least:

$$e = \text{two-thirds of the beam depth} + \text{deck rib height} = 12\text{in.} + 3 \text{ in.} = 15 \text{ in.}$$

This results in a moment capacity for the connection ($M_{u,\text{conn}}$) equal to:

$$M_{u,\text{conn}} = T * e = 26.4 * 15 = 396 \text{ kip-in.}$$

The capacity of the beam ($M_{p,\text{beam}}$) is:

$$M_{p,\text{beam}} = Z_x * F_y = 66.5 \text{ in.}^3 * 36 \text{ ksi} = 2394 \text{ kip-in.}$$

Thus, the ratio ($\overline{m}$) of the connection capacity to the steel beam capacity is:

$$\overline{m} = 396/2394 * 100 \approx 17\%$$

If we assume that (1) the bars yield and transfer most of their force over a development length of 24 bar diameters from the point of inflection, (2) the strain varies linearly, and (3) the connection region extends for a length equal to the beam depth (18 in.), then the slab reinforcement can be modeled by a spring (K_{bars}) equal to:

$$K_{\text{bars}} = EA/L = \left(30{,}000 \text{ ksi} * 0.44 \text{ in.}^2 \right) / (18 \text{ in.}) = 733.3 \text{ kips/in.}$$

Yield will be achieved at a rotation (θ_y) equal to:

$$
\begin{aligned}
\theta_y &= (T/(K_{\text{bars}} * e)) = \left(26.4 \text{ kips}/ \left(733 \text{ kips/in.} \times 15 \text{ in.} \right) \right) \\
&= 0.0024 \text{ radians or 2.4 milliradians}
\end{aligned}
$$

The connection stiffness (K_{ser}) can be approximated as:

$$K_{\text{ser}} = M_{u,\text{conn}}/\theta_y = 396 \text{ kip-in.} /0.0024 \text{ radians} = 165{,}000 \text{ kip-in./radian}$$

Assuming that the beam spans 30 ft, the beam stiffness is:

$$K_{\text{beam}} = EI_{\text{beam}} /L_{\text{beam}} = \left(30{,}000 \text{ ksi} * 510 \text{ in.}^4 /360 \text{ in.} \right) = 42{,}500 \text{ kip-in./radian}$$

Thus, the ratio of connection to beam stiffness (α) is:

$$\alpha = K_{\text{ser}}/K_{\text{beam}} = 165{,}000/42{,}500 = 3.9$$

The relatively low values of α and $\overline{m}$ obtained for this connection, even assuming the non-composite properties in order to maximize α and $\overline{m}$, would seem to indicate that this connection will have little effect on the behavior of the floor system. This is incorrect for two reasons. First, the rotations (0.0024 radian) at which the connection strength is achieved are within the service range, and thus much of the connection strength is activated earlier than for a steel connection. Second, the composite connections only work for live loads and thus provide substantial reserve capacity to the system. The moments at the supports ($M_{PR \text{ conn}}$) due to the presence of these weak connections for the case of a uniformly distributed load (w) are:

$$M_{PR \text{ conn}} = wL^2/12 * 1/(1+2/\alpha) = wL^2/18.2$$

For the case of w being the APT live load, the moment is 238 kip-in., while for the case of the full live load it is 476 kip-in. This reduces the moments at the centerline from 540 kip-in. to 302 kip-in. for the APT live load and from 1080 kip-in. to 604 kip-in. for the full live load. The maximum additional stress is 6.6 ksi under full LL loads, so no yielding will occur. Thus, if a significant portion of the beam's capacity has been used up by the dead loads, a weak composite connection can prevent excessive deflections at the service level.

The connection illustrated in Figure 23.4 is one of the weakest variations possible when activating composite action. Figures 23.6 through 23.8 show three other variations, one with a seat angle, one with an end plate (partial or full), and one with a welded plate as the bottom connection. As compared with the simple connection in Figure 23.4, both the moment capacity and the initial stiffness of these latter connections can be increased by more slab steel, thicker web angles or end plates, and friction bolts in the seat and web connections. The selection of a bolted seat angle, end plate, or welded plate will depend on the amount of force that the designer wants to transfer at the connection and on local construction practices.

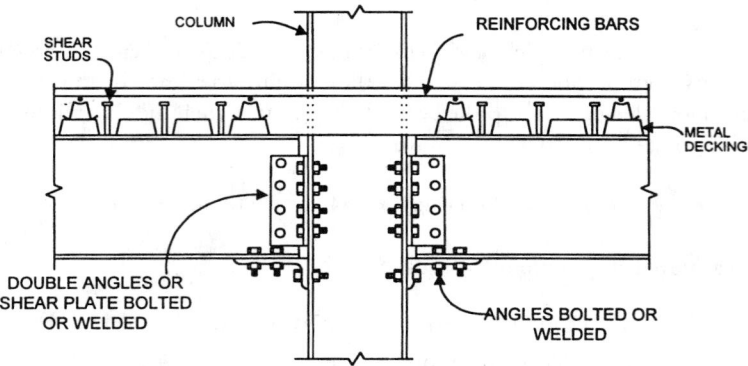

Figure 23.6   Seat angle composite connection.

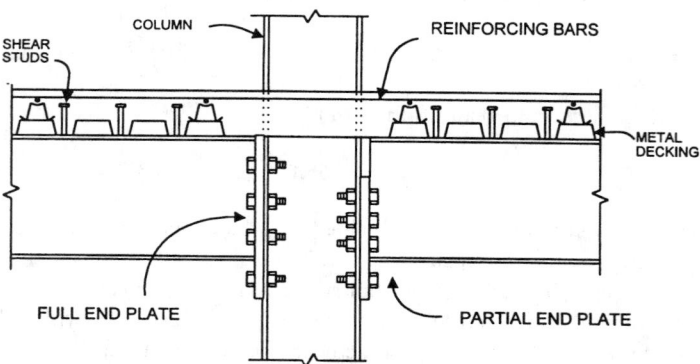

Figure 23.7   End plate composite connection.

The behavior of these connections under gravity loads (negative moments) should be governed by gradual yielding of the reinforcing bars, and not by some brittle or semi-ductile failure mode. Examples of these latter modes are shear of the bolts and local buckling of the bottom beam flange. Both modes of failure are difficult to eliminate at large deformations due to the strength increases resulting from strain hardening of the connecting elements. The design procedures to be proposed here for composite PR connections intend to insure very ductile behavior of the connection to allow redistribution of forces and deformations consistent with a plastic design approach. Therefore, the intent in design will be to delay but not eliminate all brittle and semi-brittle modes of failure through a capacity design philosophy [22].

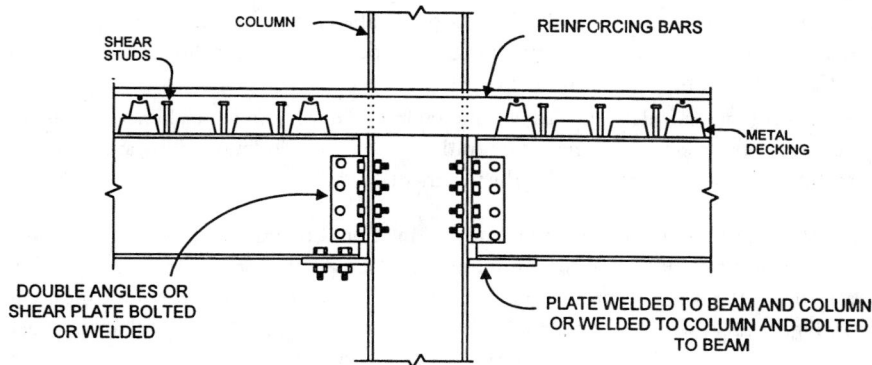

Figure 23.8   Welded bottom plate composite connection.

For the connections shown in Figures 23.6 through 23.8, if the force in the slab steel at yielding is moderate, it is likely that the bolts in a seat angle or a partial end plate will be able to handle the shear transfer between the column and the beam flanges. If the forces are high, an oversized plate with fillet welds can be used to transfer these forces. The connections in Figures 23.6 and 23.7 will probably be true PR/PS connections, while that in Figure 23.8 will likely be a PR/FS connection. In the latter case it is easy to see that considerable strength and stiffness can be obtained, but there are potential problems. These include the possibility of activating other less desirable failure mechanisms such as web crippling of the column panel zone or weld fracture.

The behavior of these connections under lateral loads that induce moment reversals (positive moments) at the connections should be governed by gradual yielding of the bottom connection element (angle, partial end plate, or welded plate). Under these conditions the slab can transfer very large forces to the column by bearing if the slab contains reinforcement around the column in the two principal directions. In this case, brittle failure modes to avoid include crushing of the concrete and buckling of the slab reinforcement.

The composite connections discussed here provide substantial strength reserve capacity, reliable force redistribution mechanisms (i.e., structural integrity), and ductility to frames. In addition, they provide benefits at the service load level by reducing deflection and vibration problems. Issues related to serviceability of structure with PR frames will be treated in the section on design of composite connections in braced frames.

23.4   Moment-Rotation (M-θ) Curves

As noted earlier, a prerequisite for design of frames incorporating PR connections is a reliable knowledge of the M-θ curves for the connections being used. There are at least four ways of obtaining them:

1. From experiments on full-scale specimens that represent reasonably well the connection configuration in the real structure [21]. This is expensive, time-consuming, and not practical for everyday design unless the connections are going to be reused in many projects.

2. From catalogs of M-θ curves that are available in the open literature [6, 16, 20, 27]. As discussed elsewhere [7, 22], extreme care should be used in extrapolating from the equations in these databases since they are based mostly on tests on small specimens that do not properly model the boundary conditions.

3. From advanced analysis, based primarily on detailed finite element models of the connection, that incorporate all pertinent failure modes and the non-linear material properties of the connection components.

4. From simplified models, such as that shown in Figure 23.5, in which behavioral aspects are lumped into simple spring configurations and other modes of failure are eliminated by establishing proper ranges for the pertinent variables.

Ideally M-θ curves for a new type of connection should be obtained by a combination of experimentation and advanced analysis. Simplified models can then be constructed and calibrated to other tests for similar types of connections available in the literature. For the composite connections shown in Figure 23.6, which will be labeled PR-CC, Leon et al. [23] followed that approach. They developed the following M-θ equation for these connections under negative moment for rotations less than 20 milliradians:

$$M^- = C1 * \left(1 - e^{(-C2*\theta)}\right) + C3 * \theta \tag{23.1}$$

where

$$
\begin{aligned}
C1 &= 0.1800 * \left[\left(4 * A_{rb} * F_{yrb}\right) + \left(0.857 * A_{sL} * F_{yL}\right)\right] * (d + Y3) \\
C2 &= 0.7750 \\
C3 &= 0.0070 * (A_{sL} + A_{wL}) * (d + Y3) * F_{yL} \\
\theta &= \text{relative rotation (milliradians)} \\
A_{wL} &= \text{area of web angles resisting shear (in.}^2) \\
A_{sL} &= \text{area of seat angle leg (in.}^2) \\
A_{rb} &= \text{effective area of slab reinforcement (in.}^2) \\
d &= \text{depth of steel beam (in.)} \\
Y3 &= \text{distance from top of steel shape to center of slab force (in.)} \\
F_{yL} &= \text{yield stress of seat and web angles (ksi)} \\
F_{yrb} &= \text{yield stress of slab reinforcement (ksi)}
\end{aligned}
$$

Since these connections will have unsymmetric M-θ characteristics due to presence of the concrete slab, the following equation was developed for these connections under positive moments for rotations less than 10 milliradians:

$$M^+ = C1 * \left(1 - e^{(-C2*\theta)}\right) + (C3 + C4) * \theta \tag{23.2}$$

where

$$
\begin{aligned}
C1 &= 0.2400 = * \left[(0.48 * A_{Wl}) + A_{Sl}\right] * (d + Y3) * F_{yl} \\
C2 &= 0.0210 * (d + Y3/2) \\
C3 &= 0.0100 * (A_{wL} + A_{sL}) * (d + Y3) * F_{yL} \\
C4 &= 0.0065 * A_{wL} * (d + Y3) * F_{yL}
\end{aligned}
$$

For preliminary design it may be necessary to model the connections as bi-linear springs only, characterized by a service stiffness (k_{conn}), an ultimate strength ($M_{u,conn}$), and hardening stiffness (k_{ult}). Simplified expressions for these are as follows:

$$K_{conn} = 85 \left[\left(4A_{rb}F_{yr}\right) + \left(A_{wL}F_{yL}\right)\right] (d + Y3) \tag{23.3}$$

$$M_{u,conn} = 0.245 \left[\left(4A_{rb}F_{yr}\right) + \left(A_{wL}F_{yL}\right)\right] (d + Y3) \tag{23.4}$$

$$K_{ult} = 12.2 \left[\left(4A_{rb}F_{yr}\right) + \left(A_{wL}F_{yL}\right)\right] (d + Y3) \tag{23.5}$$

For a final check, it is desirable to model the entire response using Equations 23.1 and 23.2 or some piecewise linear version of them. The author has proposed a tri-linear version for Equation 23.1 for which the three breakpoints are defined as [5]:

$\theta 1$ = the rotation at which the tangent stiffness reaches 80% of its original value

$M1$ = moment corresponding to $\theta 1$

$\theta 2$ = the rotation at which the exponential term of the connection equations $\left(e^{-C2*q}\right)$ is equal to 0.10

$M2$ = moment corresponding to $\theta 2$

$\theta 3$ = equal to 0.020 radians, close to the maximum rotation required for this type of connection

$M3$ = moment corresponding to $\theta 3$

It is necessary in this case to differentiate Equation 23.1 and set θ equal to zero to find an initial stiffness, and then backsolve for the rotation corresponding to 80% of that initial stiffness. All the examples in this chapter are worked out in English units because metric versions of Equations 23.1 through 23.5 have not yet been properly tested.

EXAMPLE 23.2: Moment-Rotation Curves

Figure 23.9b shows the complete M-θ curve for the composite PR connection shown in Figure 23.9a. The values shown in Figure 23.9b were taken directly from substituting into Equations 23.1 through 23.4. The shaded squares show the breakpoints for the trilinear curves described in the previous section. The trilinear curve for positive moment was derived by using the same definitions as for negative moments but limiting the rotations to 10 milliradians, the limit of applicability of Equation 23.2. Tables for the preliminary and final design of this type of connection are given in a recently issued design guide [26].

The M-θ curves shown in Figure 23.9b are predicated on a certain level of detailing and some assumptions regarding Equations 23.1 through 23.5, including the following:

1. In Equations 23.1 and 23.2, the area of the seat angles (A_{sL}) shall not be taken as more than 1.5 times that of the reinforcing bars (A_{rb}).

2. In Equations 23.1 and 23.2, the area of the web angles (A_{wL}) resisting shear shall not be taken as more than 1.5 times that of one leg of the seat angle (A_{sL}) for A572 Grade 50 steel and 2.0 for Grade A36.

3. The studs shall be designed for full interaction and all provisions of Chapter I of the LRFD Specification [2] shall be met.

4. All bolts, including those to the beam web, shall be slip-critical and only standard and short-slotted holes are permitted.

5. Maximum nominal steel yield strength shall be taken as 50 ksi for the beam and 60 ksi for the reinforcing bars. Maximum concrete strength shall be taken as 5 ksi.

6. The slab reinforcement should consist of at least six longitudinal bars placed symmetrically within a total effective width of seven column flange widths. For edge beams the steel should be distributed as symmetrically as possible, with at least 1/3 of the total on the edge side.

7. Transverse reinforcement, consistent with a strut-and-tie model, shall be provided. In the limit the amount of transverse reinforcement will be equal to that of the longitudinal reinforcement.

8. The maximum bar size allowed is #6 and the transverse reinforcement should be placed below the top of the studs whenever possible.

9. The slab steel should extend for a distance given by the longest of $L_b/4$ or 24 bar diameters past the assumed inflection point. At least two bars should be carried continuously across the span.

10. All splices and reinforcement details shall be designed in accordance with ACI 318-95 [1].

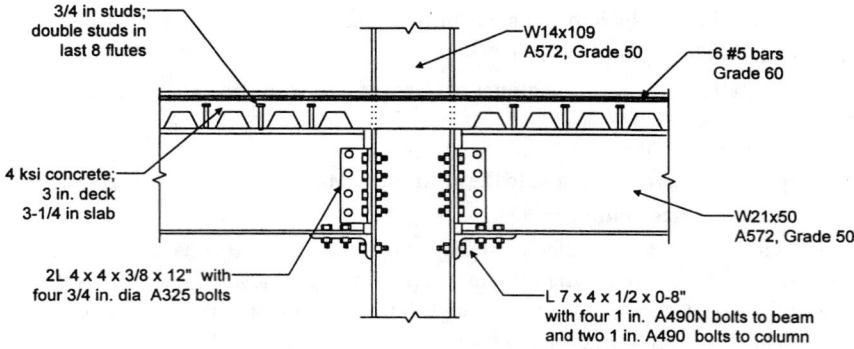

(a) Details of Type I PR-CC

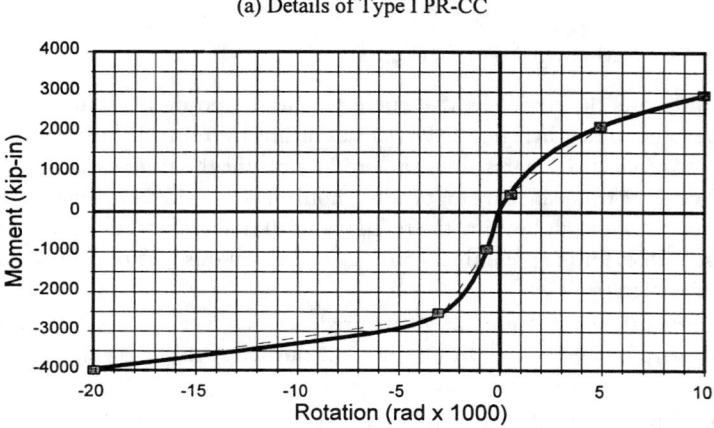

(b) Moment-rotation curve

Figure 23.9   Typical PR-CC connection and its moment-rotation curves.

11. Whenever possible the space between the column flanges shall be filled with concrete. This aids in transferring the forces and reduces stability problems in the column flanges and web.

These detailing requirements must be met because the analytical studies used to derive Equations 23.1 and 23.2 assumed this level of detailing and material performance. Only Item 11 is optional but strongly encouraged for unbraced applications Compliance with these requirements means that extensive checks for the ultimate rotation capacity will not be needed.

23.5   Design of Composite Connections in Braced Frames

The design of PR-CCs requires that the designer carefully understand the interaction between the detailing of the connection and the design forces. Figure 23.10 shows the moments at the end and centerline, as well as the centerline deflection, for the case of a prismatic beam under a distributed load with two equal PR connections at its ends. The graph shows three distinct, almost linear zones for each line; two horizontal zones at either end and a steep transition zone between α of 0.2 and 20. Note that the horizontal axis, which represents the ratio of the connection to the beam stiffness, is logarithmic. This means that relatively large changes in the stiffness of the connection have a relatively minor effect. For example, consider the case of a beam with PR-CCs with a nominal α of 10. This gives moments of $wL^2/13.2$ at the end and $wL^2/20.3$ at centerline, with a corresponding deflection of 1.67 $wL^4/384EI$. If the service stiffness (k_{ser}) for this connection is underestimated by 25% ($\alpha = 7.5$) these values change to $wL^2/13.6$, $wL^2/19.4$, and $1.84wL^4/384EI$. These represent changes of 3.0%, 4.4%, and 10%, respectively, and will not affect the service or ultimate performance of the system significantly. This is why the relatively large range of moment-rotation behavior, typical of PR connections and shown schematically in Figure 23.2, does not pose an insurmountable problem from the design standpoint.

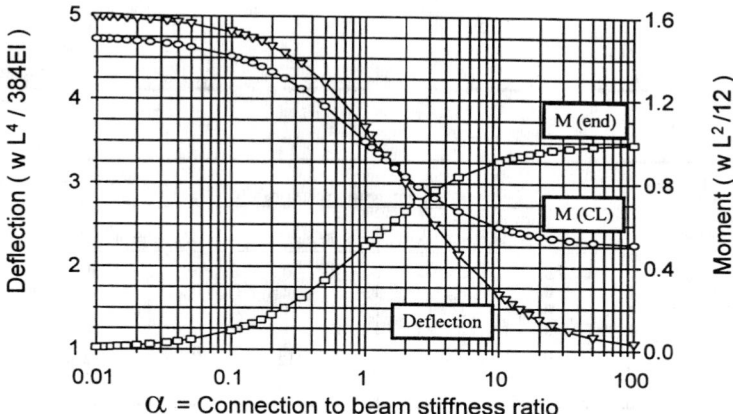

Figure 23.10   Moments and deflections for a prismatic beam with PR connections under a distributed load.

For continuous composite floors in braced frames, where the floor system does not participate in resisting lateral loads, the design for ultimate strength can be based on elastic analysis such as that shown in Figure 23.10 or on plastic collapse mechanisms. If elastic analysis is used, it is important to recognize that both the bending resistance and the moments of inertia change from regions of negative to positive moments. The latter effect, which would be important in elastic analysis, is not considered in the calculations for Figure 23.10. In the case of the fixed ended beam with full strength connections (FR/FS), elastic analysis ($\alpha = \infty$ in Figure 23.10) results in the maximum force corresponding to the area of lesser resistance. This is why it would be inefficient to design continuous composite beams with FR connections from the strength standpoint. As the connection stiffness is reduced, the ratio of the moment at the end to the centerline begins to decrease. From Figure 23.10, for a prismatic beam, the optimum connection stiffness is found to be around $\alpha = 3$, where the moments at the ends and middle are equal ($wL^2/16$). This indicates that it takes relatively little restraint to get a favorable distribution of the loads. If the effect of the changing moments of inertia is included, as it should for the case of composite beams, the sloping portions of the moment curves in Figure 23.10 will move to the right. For this case, the optimum solution will not be at the intersection of the M(end) and M (CL) lines but at the location where the ratio of $M_{p,ci}/M_{p,b}$ equals M(end)/ M (CL). Preliminary studies indicate that the optimum connection stiffness for composite beams is generally found to be still around α of 3 to 6. This indicates that it takes relatively little restraint to get a favorable distribution of the loads. This type of simple elastic analysis, however, cannot account for the fact that the connection M-θ curves are non-linear and thus will not be useful in the analysis of PR/PS connections such as PR-CCs.

Design of continuous beams with PR connections can be carried out efficiently by using plastic analysis. The collapse load factor for a beam (λ_b) with a plastic moment capacity $M_{p,b}$ at the center, and connection capacities $M_{p,c1}$, and $M_{p,c2}(M_{p,c1} > M_{p,c2})$ at its ends, can be written as:

$$\lambda_b = \frac{d}{(PL \text{ or } wL^2)} \left(aM_{p,c1} + bM_{p,c2} + cM_{p,b} \right) \qquad (23.6)$$

where the coefficients a, b, c, and d are given in Table 23.1, P and w are the point and distributed loads, and L is the beam length, respectively. For Load Cases 1 through 4, the spacing between the loads is assumed equal.

TABLE 23.1   Values of Constants in Equation 23.7 for Different Loading Configurations

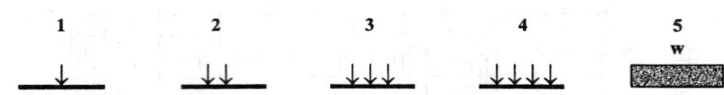

	Connection relationship										
	$M_{p,c1} = M_{p,c2}$				$M_{p,c1} > M_{p,c2}$				$M_{p,c2} = 0$		
	a	b	c	d	a	b	c	d	a	b	c	d
1	1	0	1	4	1	1	2	2	1	0	2	2
2	1	0	1	3	1	2	3	1	1	0	3	1
3	1	0	1	2	1	1	2	1	1	0	2	1
4	1	0	1	$\frac{5}{3}$	2	3	5	$\frac{5}{12}$	2	0	5	$\frac{5}{12}$
5	1	0	1	8					1	0	$\frac{L}{x}$	$\frac{2L}{L-x}$

For the case of a distributed load (Load Case 5) with unequal end connections ($M_{p,c1} > M_{p,c2}$), it is not possible to write a simple expression in the form of Equation 23.6 because the solution requires locating the position of the center hinge. For the case of $M_{p,c2} = 0$, the position can be calculated by:

$$x = \frac{M_{p,b}}{M_{p,c1}} L \left\{ \sqrt{1 + \frac{M_{p,c1}}{M_{p,b}}} - 1 \right\}$$

(23.7)

If plastic analysis is used, it is important to recognize that the flexural strength changes from the area of negative ($M_{p,c1}$ and $M_{p,c2}$) to positive moment ($M_{p,b}$), and that the ratio of $M_{p,ci}/M_{p,b}$ will often be 0.6 or less.

For the service limit state, it is important again to recognize that the results shown in Figure 23.10 are valid only for a prismatic beam. In reality a continuous composite beam will be non-prismatic, with the positive moment of inertia of the cross-section (I_{pos}) often being 1.5 to 2.0 times greater than the negative one (I_{neg}). It has been suggested that an equivalent inertia (I_{eq}), representing a weighted average, should be used [5]:

$$I_{eq} = 0.4 I_{neg} + 0.6 I_{pos}$$

(23.8)

The effect of accounting for the non-prismatic characteristics of the beam is far more important in calculating deflections than in calculating the required flexural resistance. For calculating deflections of beams with equal PR connections at both ends, the following expression has been proposed [5]:

$$\delta_{PR} = \delta_{FR} + \frac{C_\theta \theta_{sym} L}{4}$$

(23.9)

where

δ_{PR} = the deflection of the beam with partially restrained connections
δ_{FR} = the deflection of the beam with fixed-fixed connections
C_θ = a deflection coefficient

θ_{sym} = the service load rotation corresponding to a beam with both connections equal to the stiffest connection present

When the beam has equal connection stiffnesses, C_θ equals one. Values for the constant C_θ in Equation 23.9 are given in Table 23.2 for some common loading cases.

TABLE 23.2 Constants for Deflection Calculations by Equation 23.9

K_b/K_a	$1/(1+\alpha)$								
	0.9	0.8	0.7	0.6	0.5	0.4	0.3	0.2	0.1
1	1	1	1	1	1	1	1	1	1
0.9	1.05	1.04	1.04	1.03	1.03	1.02	1.02	1.01	1.01
0.8	1.11	1.09	1.08	1.07	1.05	1.04	1.03	1.02	1.01
0.7	1.18	1.15	1.13	1.11	1.09	1.07	1.05	1.03	1.02
0.6	1.27	1.22	1.18	1.15	1.12	1.09	1.07	1.04	1.02
0.5	1.39	1.31	1.25	1.20	1.16	1.12	1.08	1.05	1.03
0.4	1.54	1.41	1.32	1.25	1.20	1.15	1.10	1.07	1.03
0.3	1.76	1.55	1.41	1.32	1.24	1.18	1.12	1.08	1.04
0.2	2.09	1.72	1.52	1.39	1.29	1.21	1.15	1.09	1.04
0.1	2.63	1.97	1.66	1.47	1.34	1.25	1.17	1.10	1.05
0	3.70	2.32	1.83	1.57	1.40	1.28	1.19	1.12	1.05

Note: K_b = stiffness of the less stiff connection; K_a = stiffness of the stiffer connection; $1/(1+\alpha/2) = M_{conn,PR}/M_{conn,fixed}$ and; $\alpha = EI/(K_aL)$.

The value of θ_{symm} is given by:

$$\theta_{symm} = \frac{M_{FEM}}{K_{ser} + \left(1 + \frac{2}{\alpha}\right)}$$

where

M_{FEM} = the fixed end moment

K_{conn} = the stiffness of the connection

α = the ratio of the connection to the beam stiffness

The effect of partially restrained connections on floor vibrations is an area that has received comparatively little attention. Figure 23.11 shows the changes in natural frequency for a prismatic beam with a distributed load as the stiffness of the end connections change. The connections at both ends are assumed equal and the connection stiffness is assumed to be linear. The natural frequency (f_n, Hz) is given by:

$$f_n = \frac{K_n^2}{2\pi}\sqrt{\frac{EI}{mL^4}} \qquad (23.10)$$

where m is the mass per unit length, L is the length, and EI is the stiffness of the beam.

Generally m is taken as the distributed load (w) given by the dead plus 25% of the live loads and divided by the acceleration of gravity ($g = 386$ in./s^2). Limit values of K_n range from π^2 for the simply supported case to $(1.5\pi)^2$ for the fixed case.

EXAMPLE 23.3:

Design a continuous floor system in a braced frame. The system will consist of a three-span girder with a total length of 96 ft, and will be designed for dead loads of 80 psf and live loads of 100 psf. The reduced live loads will be taken as 60 psf. This girder supports floor beams spanning 28 ft in the perpendicular direction every 8 ft, for a total of three point loads per span. In addition to the distributed loads described above, the interior span will support equipment weighing 15 kips, to be installed before the slab is cast (Figure 23.12). Cambering will be provided to offset all dead loads, including the equipment. The connections to the exterior columns will be assumed as pinned since

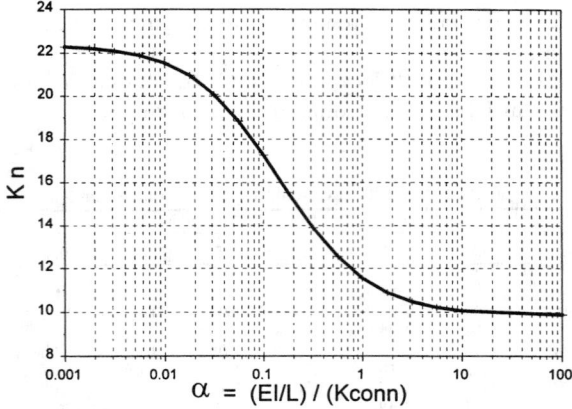

Figure 23.11   First natural frequency of vibration for a beam with PR connections.

an overhang would be required to anchor the slab reinforcement. The steel will be A572 Grade 50 and a 3-1/4 in. lightweight concrete slab ($f'_c = 4$ ksi) on 3 in. metal deck (Y2 = 4.5 in.) will be assumed.

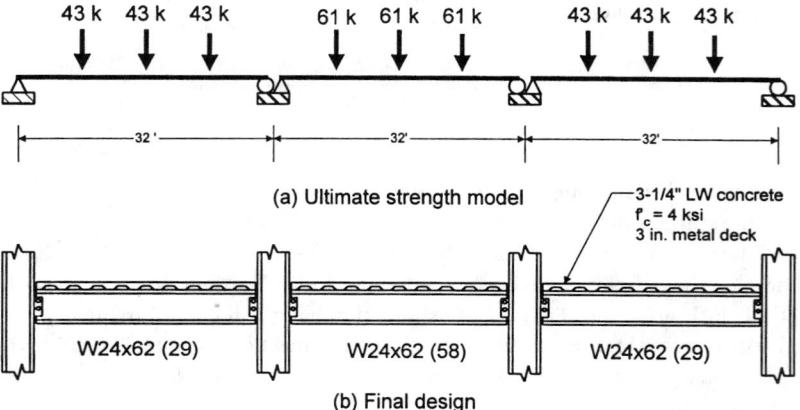

Figure 23.12   Design of composite floor system as simply supported beams (numbers in parenthesis are the number of shear studs).

The construction dead loads are assumed as 60 psf and the construction live loads are taken as 15 psf. The design construction load, assuming distributed loads, is:

$$w_{u,const} = [1.2(0.06) + 1.6(0.015)] (28 \text{ ft.}) = 2.69 \text{ k/ft}$$
$$M_{u,const} = wL^2/8 = (2.69)(32)^2/8 = 344 \text{ k-ft} = 4129 \text{ kip-in.}$$
$$Z_x = 4129/(0.9 \times 50) = 91.8 \text{ in.}^3$$

Assuming that the beam will be supported laterally during the construction phase, the most economical steel section would be a W21x44 ($Z_x = 95.4$ in.^{3}). For the ultimate strength limit state, assuming three point loads at the location of the floor beams, for the interior span:

$$P_{u,} = [1.2(0.08) + 1.6(0.06)] (28 \text{ ft.})(8 \text{ ft.}) + 1.2(15 \text{ kips}) = 61.0 \text{ kips}$$
$$\phi M_{u,} = 15P_uL/32 = 15(61.0)(32)/32 = 915 \text{ k-ft} = 10,980 \text{ kip-in.}$$

For the ultimate strength limit state in the exterior spans:

$$P_{u,} = [1.2(0.08) + 1.6(0.06)] \, (28 \text{ ft.})(8 \text{ ft.}) = 43.0 \text{ kips}$$
$$\phi M_{u,} = 15 P_u L/32 = 15(61.0)(32)/32 = 645 \text{ k-ft} = 7{,}742 \text{ kip-in.}$$

If we assume typical current construction practice and design these girders as simply supported composite beams, for the ultimate load condition the section required will be a fully composite W24x55 (Y2 = 4.5 in. and $\sum Q_n = 810$ kips). Assuming $f'_c = 4$ ksi and 3/4 in. headed studs, 38 shear studs per half-span, or more than two studs per flute, will be needed. This is not a very efficient design, and thus a partially composite W24x62 will be a better choice ($\phi M_p = 930$ kip-ft with Y2 = 4.0 in. and $\sum Q_n = 598$ kips). This results in 29 studs per half-span or roughly two studs per flute.

The service load deflection in this case would be:

$$\delta = 19 P L^3/384 E I = \left[19 \times (0.06 \times 28 \times 8) \times (32 \times 12)^3 \right] / $$
$$[384 \times 29000 \times 2180] = 0.595 \text{ in.} \approx L/640$$

For the exterior spans, a W24x62 with the minimum amount of interaction (25%, or $M_p = 755$ kip-ft with Y2 = 4.0 in. and $\sum Q_n = 228$ kips) and 21 studs total will suffice.

If we were to provide a PR-CC such as the one shown in Figure 23.9, one could calculate its ultimate strength ($M_{u,\text{conn}}$), from Equation 23.4 as:

$$M_{u,\text{conn}} = 0.245 \left[\left(4 \times \left(6 \times 0.31 \text{ in.}^2 \right) \times 60 \text{ ksi} \right) + (4.00 \times 50) \right] \times (21 + 4) = 3{,}959 \text{ kip-in.}$$

Note that the nominal capacity of the connection ($M_{u,\text{conn}} = 3{,}959$ kip-in.) has to be less than or equal to that of the steel beam $\phi(M_{p,b} = 4{,}293$ kip-in.) in order to insure that the hinging will not occur in the beam. The author has suggested [22] that a good starting point for the strength design is to assume that the connection will carry about 70 to 80% of $M_{p,b}$. For our case the ratio is 3,959/6,888 = 0.58 which is somewhat lower but reasonable because of the heavy dead loads.

For the interior span, from Equation 23.6 and assuming that $M_{p,c1} = M_{p,c2} = \phi M_{u,\text{conn}} = (0.9 \times 3959) = 3563$ kip-in., for a collapse load factor (λ_p) of 1.00:

$$1.00 = \frac{(2)}{(61 \times 32 \times 12)} \left(3563 + \phi M_{p,b} \right)$$
$$\phi M_{p,b} = 8149 \text{ kip-in.} = 679 \text{ kip-ft}$$

For the exterior span, from Equation 23.6 and assuming that $M_{p,c1} = 0$ and $M_{p,c2} = \phi M_{u,\text{conn}} = (0.9 \times 3959) = 3563$ kip-in., for a collapse load factor (λ_p) of 1.00:

$$1.00 = \frac{(1)}{(43 \times 32 \times 12)} \left(3563 + 2\phi M_{p,b} \right)$$
$$\phi M_{p,b} = 6474 \text{ kip-in.} = 540 \text{ kip-ft}$$

The required strength can now be provided by a fully composite W21x44 ($\phi M_n = 683$ kip-in., and $\sum Q_n = 650$ kips or two studs per flute) and by a partially composite W21x44 ($\phi M_n = 564$ kip-in., and $\sum Q_n = 260$ kips or one studs per flute). Figure 23.13 shows the analysis model and the final design for this case, as well as the moment diagram for the case of DL + LL.

Figure 23.13c shows that the dead load moments are calculated on the simply supported structure (SS), while the live load ones are calculated on the continuous structure (PR). For calculation purposes, the moments of inertia were taken as 1699 in.[4] for the interior span and 1399 in.[4] for the exterior span, as per Equation 23.8. Figure 23.13c indicates that the maximum moment in the interior span at full service load is 647 kip-ft. This is close to the factored capacity of the section

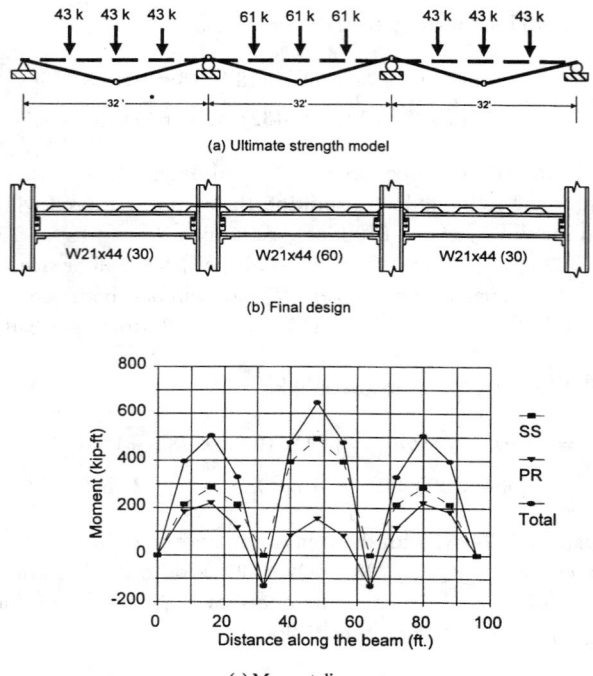

Figure 23.13   Continuous beam design with PR connections.

($\phi M_n = 683$ kip-ft). Thus, careful attention should be paid to the stresses and deflections at service loads when using a plastic design approach since the latter does not consider construction sequence or the onset of yielding. In this case perhaps a W21x50 section, with the same number of studs, would be a more prudent design.

In computing the forces for the case of the PR system, the connection stiffness was calculated directly as a secant stiffness at 0.002 radian from Equation 23.1. The stiffness was 1.135×10^6 kip-in./rad, which is slightly lower than the 1.398×10^6 kip-in./rad given by Equation 23.3. The α for this connection is:

$$\alpha = \frac{K_{\text{ser}} L}{EI} = \frac{(1.135 \times 10^6)(32 \times 12)}{(29000)(1699)} = 8.84$$

This puts this PR connection near the middle of the PR range for unbraced frames and near the rigid case for the case of braced frames.

The deflection of the center span under the full live load is, from Equation 23.9:

$$\delta_{PR} = \frac{PL^3}{96EI} + \frac{C_\theta M_{FF} L}{4 K_{\text{conn}} \left(1 + \frac{2}{\alpha}\right)} = 0.161 + 0.109 = 0.270 \text{ in.} \approx L/1500$$

This deflection is considerably less than that computed for the simply supported case even when a much larger section (W24x62) was used in the latter case. An idea of the effect of this PR connection can be gleaned from inspecting Figure 23.10. Although Figure 23.10 corresponds to a different case, the moment diagrams are not substantially different and thus a meaningful comparison can be made for the elastic case. From Figure 23.10, the difference in deflection between a simple support and a PR connection with $\alpha = 8.84$ is roughly a factor of 2.8 (5/1.8), while the difference in moment of inertia is only 1.28 (2180 in.^{4} / 1699 in.^{4}). In this example, the design was governed by strength and not deflections. However, this example clearly shows the impact of a PR connection in reducing floor deflections.

In addition to the strength calculation above, the design procedure requires that the following limit states and design criteria be satisfied (refer to Figure 23.9a for details):

1. Shear strength of the bolts attaching the seat angle to the beam (ϕV_{bolts}): The bolts have to be designed to transfer, through shear, a compressive force corresponding to 1.25 of the force (T_{slab}) in the slab reinforcement. The 1.25 factor accounts for the typical overstrength of the reinforcement, and intends to insure that the bolts will be able to carry a force consistent with first yielding of the slab steel. Assuming 1 in. diameter A490N bolts:

$$(\phi V_{bolts}) = 1.25T_{slab} = 1.25F_y A_{bars} = 139.5 \text{ kips}$$
$$N_{bolts} = (\phi V_{bolts})/35.3 = 3.95 \cong 4 \text{ bolts (O.K.)}$$

2. Bearing strength at the bolt holes (ϕR_n): The thickness of the angle will be governed by the required flexural resistance of the angle leg connecting to the beam flange in the case of a connection in an unbraced frame, where tensile forces at the bottom of the connection are possible. It will be governed by either bearing of the bolts or compressive yielding of the angle leg in the case of a connection in a braced frame. In this case:

$$\phi R_n = \phi(2.4dt F_u) = 0.75(2.4 \times 0.875 \times 0.5 \times 65)$$
$$= 51.2 \text{ kips/bolt} > \phi V_{bolts} , \text{ O.K.}$$

3. Tension yield and rupture of the seat angle: This limit state is strictly applicable to the case of unbraced frames where pull-out of the angle under positive moments is possible. For the case of a connection in a braced frame, it is prudent to check the angle for yielding under compressive forces (ϕC_n) and possible buckling. The latter is never a problem given the short gage lengths, while the former is:

$$\phi C_n = \phi \left(A_g F_y \right) = 0.9 \times (8 \times 0.5) \times 50 = 180 \text{ kips} > \phi V_{bolts} , \text{ O.K.}$$

4. Number and distribution of slab bars, including transverse reinforcement, to insure a proper strut-and-tie action at ultimate (see section following Example 23.2 for details).
5. Number and distribution of shear studs to provide adequate composite action (checked above as part of the flexural design).
6. Tension strength, including prying action, for the bolts connecting the beam to the column.
7. Shear capacity of the web angles.
8. Block shear capacity of the web angles.
9. Check for the need for column stiffeners

Limit states (6) through (9) can be checked following the current LRFD provisions, and the details will not be provided here. However, it should be clear from the few calculations shown above that the shear capacity of the bolts is the primary mechanism limiting the forces in the connection.

The structural benefits of using a PR-CC connection are clear from the results of this example. From the economic standpoint, for a PR-CC to be beneficial, the cost of the additional reinforcing bars and seat angle bars has to be offset by that of the additional studs and larger sections required for the simply supported case. In some instances the benefits may not be there from the economic standpoint, but the designer may choose to use PR-CCs anyway because of their additional redundancy and toughness.

In Example 23.3, the design was controlled by strength and thus it was relatively simple to calculate forces based on plastic analysis and proportion the connection based on a simplified model similar to that shown in Figure 23.5. Since deflections did not control the design, the connection stiffness did not play an appreciable role in the preliminary design. If serviceability criteria control the design, then the proportioning of the connection can start from Equation 23.3. In this case the analysis has to be iterative, since the value of the connection stiffness will affect the moment diagram and the deflection. For applications in braced frames, however, experience indicates that it is strength and not stiffness that governs the design. This is because the steel beam size is controlled by the construction loads if the typical unshored construction process is used. In general, the steel beam selected is capable of providing the required stiffness even if it is the minimum amount of interaction (25% is recommended by AISC and 50% by this author).

23.6   Design for Unbraced Frames

As noted earlier, the design of frames with PR connections requires that the effects of the non-linear stiffness and partial strength characteristics of the connections be incorporated into the analysis. From the practical standpoint, the main difference between the design of unbraced FR and PR frames is the contribution of the connections to the lateral drift. The designer thus needs to balance not just the stiffness of the columns and beams to satisfy drift requirements, but account for the additional contribution of the concentrated rotations at the connections. There are no established practical rules on the best distribution of resistance to drift between columns, beams, and connections for PR-CCs. Trial designs indicate that distributing them about equally is reasonable (i.e., 33% to the beams, columns, and connections, respectively), and that it may be advantageous in low-rise frames to count on the columns to carry the majority of the resistance to drift (say 40 to 45% to columns, and the rest divided about equally between the beams and connections). The use of fixed column bases is imperative in the design of PR-CC frames, just as it is in the design of almost all unbraced FR frames, in order to limit drifts. Thus, designers should pay careful attention to the detailing of the foundations and the column bases.

The required level of analysis for the design of unbraced frames with PR connections is currently not covered in any detail by design codes. The AISC LRFD specification [2] allows for the use of such connections by requiring that the designer provide a reliable amount of end restraint for the connections by means of tests, advanced analysis, or documented satisfactory performance. The AISC LRFD specification, however, does not provide any guidance on the analysis requirements except to note that the influence of PR connections on stability and P-Δ effects need to be incorporated into the design. The new NEHRP provisions and AISC seismic provisions [4, 28] will contain generic design requirements for frames with PR connections for use in intermediate and ordinary moment frames (IMF and OMF). In addition, it will contain some specific requirements for some specific types of connections, such as the PR-CCs described in this chapter. It is unlikely that there will be an attempt in the near future to codify the analysis and design of PR frames since it would be difficult to develop guidelines to cover the vast array of connection types available (Figure 23.1). Thus, design of PR frames will remain essentially the responsibility of the structural engineer with guidance, for particular types of connections, from design guidelines [8, 26], books [12, 13, 14], and other technical publications. The proposed procedures to be described next remain, therefore, only a suggestion for proportioning the entire system. Only the detailing of the connections, including checking all pertinent failure modes, should be regarded as a requirement.

The design procedure to be discussed is divided into two distinct parts. For the service limit states (deflections, drift, and vibrations) the design will use a linear elastic model with elastic rotational springs at the beam ends to simulate the influence of the PR connections. For the ultimate limit states (strength and stability), a modified, second-order plastic analysis approach will be used. In this case the connections will be modeled as elastic-perfectly plastic hinges and the stability effects will be modeled through a simplified second-order approach [26]. Because the latter was calibrated to a population of regular frames with PR-CCs, the approach is only usable for PR-CC frames. The design process will be illustrated with calculations for the frame shown in Figure 23.14. For a complete design example, including all intermediate steps and design aids, the reader is referred to [26].

EXAMPLE 23.4:

Conduct the preliminary design for the frame shown in Figure 23.14. The frame is a typical interior frame, has a tributary width of 30 ft, and will be designed for an 80 mph design wind and for forces consistent with UBC 1994 seismic zone 2A. The dead loads are 55 psf for the slab and framing and 30 psf for partitions, mechanical, and miscellaneous. The weight of the facade is estimated as 700 plf. The live loads are 50 psf and 125 psf in the exterior and interior bays, respectively, and will be reduced as per ASCE 7-95. The roof dead and live loads are 30 psf and 20 psf, respectively. The floor slab will consist of a 3-1/4 in. lightweight slab on a 3 in. metal deck, resulting is a typical Y2 for the slab of 4.5 in. The design of the entire frame is beyond the scope of this chapter, so calculations for only a few key steps will be given.

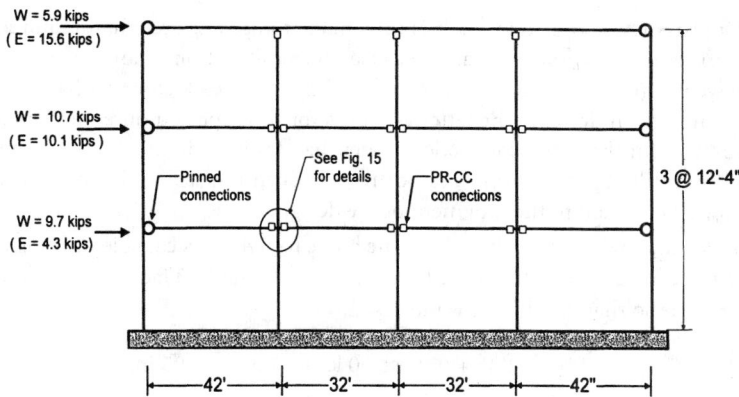

Figure 23.14 Frame for Example 23.4.

Part 1: Select beams and determine desired moments at the connections:

Step 1: Select the beam sizes based on the factored construction loads, as illustrated in Example 23.3. For this case the exterior bays require a W21x50, while the interior bays require a W21x44.

Step 2: Select moment capacity desired at the supports (M_{us}) based on the live loads. A good starting point is 75% of the M_p of the steel beam selected in Step 1, but the choice is left to the designer. Once M_{us} has been chosen, the factored moment at the center of the span (M_{uc}) can be computed as the difference between the ultimate simply supported factored moment (M_u, static moment) and M_{us}. For the interior span, $w_u = 6.66$ kip/ft and $M_u = 1020$ kip-ft of which roughly 55% corresponds to the dead loads and 45% to the live loads. Thus, select a connection capable of carrying:

$$M_{us} = 1020 \times .45 \times 0.75 = 330 \text{ kip-ft} = 3965 \text{ kip-in.}$$
$$M_{uc} = 1020 - 330 = 690 \text{ kip-ft} = 8,280 \text{ kip-in.}$$

Step 3: Select a composite beam to carry M_{uc} and check that it can carry the unfactored service loads without yielding of the slab reinforcement. Assume that full composite action will be required to limit vertical deflections and lateral drifts. Using the steel beams from Step 1 and following the procedure from Example 23.3, the exterior bays require a W21x50 with 58 3/4 in. diameter studs, while the interior bays require a W21x44 with 52 studs. The design procedure for lateral loads was derived assuming that the beams were fully composite. In this case that means increasing the number of studs to 66 and 58, respectively, which is a very small increase. The moments of inertia computed from Equation 23.4, and including the contribution of the reinforcement are 1843 in.^{4} for the W21x44 and 1899 in.^{4} for the W21x50.

Part 2: Preliminary connection design:

Step 4: Compute the amount of slab reinforcement (A_{rb}) required to carry M_{us}. Assume that the moment arm is equal to the beam depth plus the deck rib height plus 0.5 in. The nominal required moment capacity is:

$$M_n = M_{us}/\phi = 3950/0.9 = 4388 \text{ kip-in.}$$
$$A_{rb} = 4388/(60 \text{ ksi} \times (21 + 3 + 0.5)) = 2.98 \text{ in.}^2$$

Try 8 #5 bars ($A_{rb} = 2.48 \text{ in.}^2$). It is reasonable to use less area than required by the equations above ($A_{rb} = 2.98 \text{ in.}^2$) because those calculation ignore the contribution of the web angles to the ultimate capacity and the $\phi = 0.9$ factor that has been added to the connection design. The latter accounts for the expected differences in stiffness and strength for the entire connection rather than for its individual components. Currently the LRFD Specification does not require such a factor and thus its use, while recommended, is left to the judgment of the designer.

Step 5: Choose a seat angle so that the area of the angle leg (A_{sL}) is capable of transmitting a tensile force equal to 1.33 times the force in the slab. The 1.33 factor is used to obtain a thicker angle so that its stiffness is increased.

$$A_{sL} = 2.48 * (60 \text{ ksi}/50 \text{ ksi}) * 1.33 = 3.95 \text{ in.}^2$$

Try a L7x4x1/2x8" ($A_{sL} = 4.00 \text{ in.}^2$).

Step 6: For transferring the shear force consistent with the rebar reaching $1.25 F_y$, the bolt shear capacity required is:

$$V_{bolt} = 2.48 \text{ in.}^2 \times 60 \text{ ksi} \times 1.25 = 186 \text{ kips}$$

This requires four 1-in. A490X bolts. Note that if the number of bolts is taken greater than 4, they would be difficult to fit into the commonly available angle shapes. In general the number and size of bolts required to carry the shear at the bottom of the connection is the governing parameter in design. Thus, another possible way of selecting the amount of moment desired at the connection (see Step 2) is to select the size and number of bolts and determine M_{us} as:

$$M_{us} = V_{bolt} \left(\text{beam depth } + \text{ deck height } + 0.5 \text{ in.} \right)$$

Step 7: Determine the number and size of bolts required for the connection to the column flange. From typical tension capacity calculations, including prying action, two 1-in. A490X bolts are required for the connection to the column. In general, and for ease of construction, these bolts should be the same size as those determined from Step 6.

Step 8: Select web angles (A_{wL}) assuming a bearing connection. Check bearing and block shear capacity. The factored shear (V_u) is:

$$V_u = 6.6(35/2) = 115.5 \text{ kips}$$

The factored shear from lateral loads is based on assuming the formation of a sidesway mechanism in which one end of the beam reaches its positive moment and the other its negative moment capacity. Since the connection has not been completely designed, assume that the negative and positive moment capacities are the same. This is conservative since the positive capacity will generally be smaller than the negative one.

$$V_u = 2M_{n,\text{conn}}/L = (4{,}388 \text{ kip-in.} + 4{,}388 \text{ kip-in.})/(35 \times 12) = 18.7 \text{ kips}$$

From Tables 9-2 in the LRFD Manual, four 3/4 in.-diameter A325N bolts, with a pair of L4x4x1/4x12" can carry 117 kips. Note that for calculation purposes, the area of the web angles (A_{wl}) in Equations 23.1 and 23.2 is limited to the smallest of the gross shear area of the angles ($2 \times 12 \times 1/4 = 6.00$ in.^{2}) or 1.5 times the area of the seat angle ($1.5 A_{Sl} = 1.5 \times 8 \times 1/2 = 6.00$ in.^{2}). This is required because Equations 23.1 and 23.2 were derived with this limit as an assumption.

Step 9: Determine connection strengths and stiffness for preliminary lateral load design. From Equations 23.3 through 23.5:

$$
\begin{aligned}
k_{\text{conn}} &= 85\left[(4 * 2.48 * 60) + (6 * 50)\right](21 + 3.5) = 1.864 \times 10^6 \text{ kip-in./rad} \\
M_{u,\text{conn}} &= 0.245\left[(4 * 2.48 * 60) + (6 * 50)\right](21 + 3.5) = 5373 \text{ kip-in.} \\
k_{\text{ult}} &= 12\left[(4 * 2.48 * 60) + (6 * 50)\right](21 + 3.5) = 263.2 \times 10^3 \text{ kip-in./rad}
\end{aligned}
$$

From the more complex Equation 23.1, the ultimate moment at 0.02 radians is 5040 kip-in., the secant stiffness to 0.002 radians is 1.403×10^6 kip-in./rad, and the ultimate secant stiffness is 252×10^3 kip-in./rad. Thus, the approximate formulas seem to provide a good preliminary estimate. Whenever possible, the use of Equations 23.1 and 23.2 is recommended.

The stiffness ratio for this connection is:

$$\alpha = 1.403 \times 10^6 \times (35 \times 12)/(29000 * 1699) = 11.95$$

Step 10: Check deflections under live load based on the service stiffness computed in Step 9. As for Example 23.3, the centerline deflection under full live loads is small since the α is large.

The connection designed in Steps 4 through 9 is shown in Figure 23.15. In the next steps, the adequacy of the connections, designed for gravity loads, to handle the design lateral loads will be checked.

Part 3: Preliminary lateral load design:

Step 11: Determine column sizes based on drift requirements and/or gravity load requirements. From the gravity loads, and making a 10% allowance for second order effects, a W14x74 was selected for the exterior leaner columns and a W14x132 for the interior columns. The selection of the interior columns was checked by satisfying the interaction equations (Equations H1-1a,b in the LRFD Specification) assuming that (a) the required moment capacity will be given by the summation of the moment capacities on either side of the connection ($M_{b,\text{conn}}^- = 4193$ kip-in. and $M_{p,\text{conn}}^+ = 3655$ kip-in. from Equations 23.1 and 23.2); (b) the axial load is given by 1.2DL + 0.5LL ($P_u = 365$ kips, including live load reductions); and (c) B1 = 1.0 and B2 = 1.1.

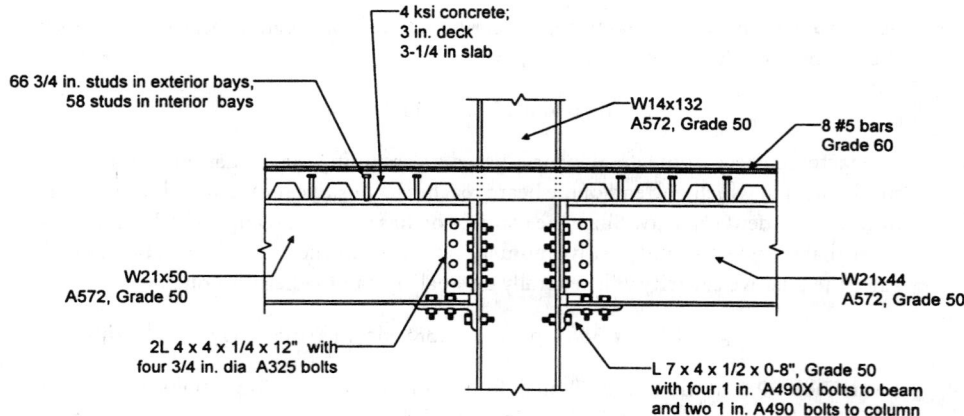

Figure 23.15   Connection details for Example 23.4.

The total story drift (Δ) can be calculated, for preliminary design purposes as:

$$\Delta = VH^2 \left[\frac{1}{\sum K_c} + \frac{1}{\sum K_g} + \frac{1}{\sum K_{\text{conn}}} \right] \tag{23.11}$$

$$\sum K_b = \sum \left(\frac{12EI_{\text{eq}}}{L_b} \right) = \frac{(12)(29000)(2 * 1843 + 1898)}{(420)} = 4.635 \times 10^6$$

$$\sum K_{\text{conn}} = 4(1.403 \times 10^6) = 5.612 \times 10^6$$

$$\sum K_c = \sum \left(\frac{12EI_c}{H} \right) = \frac{(2)(12)(29000)(1530)}{(148)} = 7.195 \times 10^6$$

where

I_c and I_{eq}	=	the moments of inertia of the columns and beams
L_b	=	the girder length
H	=	the story height
V	=	the story shear
k_{conn}	=	the connection stiffness

For the girders the effective stiffness of the exterior girders, which are pin-connected at one end and have a PR connection at the other, was taken as equal to that of one girder. For the columns, only the two interior ones are used since the exterior ones are leaners. The summations are over all beams, columns, and connections participating in the lateral load resisting system. Note that for unbraced frames subjected to lateral loads one connection at each column line will be loading and one will be unloading. Thus, their k_{conn} should be different on either side of a column; however, for preliminary design it is sufficient to assume the k_{conn} for negative moments. For the critical first story, the shear (H) due to the wind loads is 26.3 kips. For drift design, this value will be checked against an allowable drift of 0.25%. From Equation 23.11, the interstory drift is 0.31 in. or $H/482$ which is well within the $H/400$ normally allowed. Note from the calculations of stiffness in Equation 23.11 that the connections actually provide about 32% of the lateral resistance. The beams provide about 27% and the columns provide the remaining 41%. This represents a well-balanced distribution of stiffness. Note again that the contribution of the columns, which is intimately tied to the assumption of base fixity, is the key to limiting drifts. The drift under seismic loads ($H = 34$ kips) can be checked roughly by calculating the elastic drift and multiplying it by an amplification factor (C_d). The new ASIC Seismic and NEHRP provisions give $C_d = 5.5$ for PR-CCs, and allow a maximum of 1.5% drift for this type of structure. Thus:

$$\Delta = (0.31 \text{ in.} \times (34/26.3) \times 5.5) = 2.20 \text{ in.} \rightarrow (2.20/148) \times 100 = 1.5\%, \text{ O.K.}$$

Although this frame barely meets the displacement criteria, a more refined non-linear analysis should be carried out to determine the actual drift.

Step 12: The strength of a frame can be calculated based on a sidesway, plastic collapse mechanism (Figure 23.16). The first-order, rigid-plastic collapse load factor (λ_p) for this type of structure is given by [26]:

$$\lambda_p = \frac{(N+1)M_{p,\text{col}} + ((N-1)*S)\left(M^+_{p,\text{conn}} + M^-_{p,\text{conn}}\right)_{\text{int}} + (S)\left(M^+_{p,\text{conn}} + M^-_{p,\text{conn}}\right)_{\text{ext}}}{\sum(V_i * H_i)} \tag{23.12}$$

where

N	=	the number of bays
S	=	the number of stories
V_i and H_i	=	the loads and heights at each story
$M_{p,\text{col}}$	=	the column plastic capacity at the base
$M_{p,\text{conn}}$	=	the connection capacity at 10 milliradians

ext and int refer to the exterior and interior connections
$+$ and $-$ refer to the positive and negative moment capacities.

For our case:

$$\lambda_p = \frac{(5)(10,530) + ((3)*2)(4193 + 3655)}{\sum(15.6 * 444 + 10.1 * 296 + 4.3 * 148)} = 9.45$$

This is apparently a very large collapse load factor, but there is a substantial reduction in that capacity due to second-order effects (Figure 23.17). Consideration of the P-Δ effects results in a second-order collapse load factor λ_k which is a function of the rigid plastic collapse load factor (λ_p) and the ratio (S_p) of the lateral displacement at collapse (Δ_k) to the displacement at a load factor of one (Δ_w). For proportional loading this leads to a second-order collapse load factor (λ_k) equal to [18]:

$$\lambda_k = \frac{\lambda_p}{1 + S_p\lambda_p^2\left(\dfrac{\sum P\theta\delta}{\sum M_p\phi}\right)} \tag{23.13}$$

where

P	=	the axial loads
θ	=	the story rotation
δ	=	the elastic interstory drift
ϕ	=	the rotation of the plastic hinges
M_p	=	the moments at the hinges (Figure 23.16)

Values of S_p for various numbers of stories and story heights are given in Table 23.3. The ratio of θ/ϕ in the denominator of Equation 23.13 is equal to 1.0 when the member rotations (θ) are equal to the plastic hinge rotations (ϕ) as would be the case in the weak beam–strong column sway mechanism envisioned here. This results in $\lambda_k = 2.99$, which is a reasonable collapse load factor [17].

Step 13: Check strong column–weak beam behavior by requiring that:

$$1.25\left[(M^-)_{p,\text{conn}} + (M^+)_{p,\text{conn}}\right] \le \sum\left(M_{p,\text{col}}\left[1 - \frac{P}{P_{\max}}\right]\right) \tag{23.14}$$

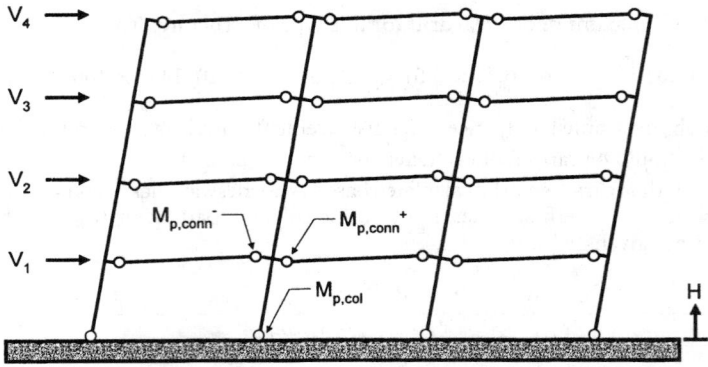

Figure 23.16   Plastic collapse mechanism and second order effects.

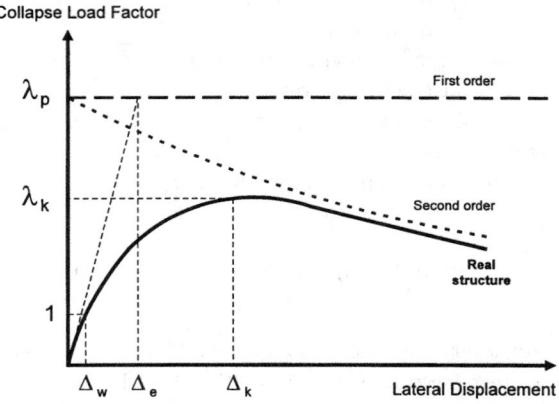

Figure 23.17   Simplified computation of plastic second-order effects. (After Horne, M.R. and Morris, L.J. *Plastic Design of Low-Rise Frames,* The MIT Press, Cambridge, MA, 1982.)

TABLE 23.3   Values of S_p

Number of stories	Story height (ft)	Proportional loading
	12	8
4	14	7.2
	16	5.4
	12	5.9
6	14	4.5
	16	3.9
	12	3.6
8	14	2.8
	16	2

From Hoffmann, J.J., Design Procedures and Analysis Tools for Semi-Rigid Composite Members and Frames, M.S. Thesis, University of Minnesota, Minneapolis. With permission.

Step 14: Check stability of the columns by using AISC LRFD [2] Equation H1-1(a) and (b). Assume that lateral loads will control and that the maximum moments are equal to $M^+_{p,\text{conn}}$ plus $M^+_{p,\text{conn}}$. These $M_{p,\text{conn}}$ shall be based on a 1.25 overstrength factor for both rebar and steel angle, and a $\phi = 1.00$. For calculating G factors, assume that the effective moment of inertia for the beams is:

$$I_{\text{eff}} = I_{\text{eq}} \left(\frac{1}{1 + \frac{6}{\alpha}} \right) \tag{23.15}$$

Procedures for determining the stability of PR frames are still under development (see Chapter 3 of ASCE 1997 for more details).

Once a preliminary design has been completed, the final checks need to be made with advanced analysis tools unless the frame is very regular and seismic forces are not a concern. The level of modeling required is left to the discretion of the designer, but should, at a minimum, include tri-linear springs to model the connections and include second-order effects.

The design procedures illustrated here are limited to only one type of connection. To the author's knowledge only two other types of composite connections have received a similar level of development: those between steel beams and concrete columns and those for composite end plates. An extensive treatment of the general topic of connection design is given in Chapter 6 of Viest et al. [29]. The latest information on design of composite and PR connections can also be found in the proceedings of several international conferences [9, 10, 11, 19].

References

[1] ACI 318-95. 1995. Building Code Requirements for Reinforced Concrete (ACI 318-95), American Concrete Institute, Detroit, MI.

[2] AISC. 1993. *Load and Resistance Factor Design Specification for Structural Steel Buildings,* 2nd. ed., American Institute of Steel Construction, Chicago, IL.

[3] AISC. 1994. *Manual of Steel Construction—Load and Resistance Factor Design,* 2nd. ed., American Institute of Steel Construction, Chicago. IL.

[4] AISC. 1997. *Seismic Provisions for Structural Steel Building, Load and Resistance Factor Design,* American Institute of Steel Construction, Chicago, IL.

[5] Ammerman, D.J. and Leon, R.T. 1990. Unbraced Frames with Semi-Rigid Connections, *AISC Eng. J.,* 27(1), 12-21.

[6] Ang, K. M. and Morris, G. A. 1984. Analysis of Three-Dimensional Frames with Flexible Beam-Column Connections, *Can. J. Civ. Eng.,* 11, 245-254.

[7] ASCE. 1997. *Effective Length and Notional Load Approaches for assessing Frame Stability: Implications for American Steel Design,* 1st ed., ASCE, New York.

[8] ASCE Task Committee on Design Criteria for Composite Structures in Steel and Concrete. 1998, Design Guide for Partially-Restrained Composite Connections (PR-CC), *ASCE J. Struc. Eng.,* to appear in 1998.

[9] Bjorhovde, R., Colson, A., and Zandonini, R., Eds. 1996. *Connections in Steel Structures III: Behaviour, Strength and Design,* Proceedings of the Third International Workshop on Connections held at Trento, Italy, May, 1995, Pergamon Press, London.

[10] Bjorhovde, R., Colson, A., Haaijer, G., and Stark, J.W.B., Eds. 1992. *Connections in Steel Structures II: Behaviour, Strength and Design,* Proceedings of the Second Workshop on Connections held at Pittsburgh, April 1991, AISC, Chicago, IL.

[11] Bjorhovde, R., Brozzetti, J., and Colson, A., Eds. 1988. *Connections in Steel Structures: Behaviour, Strength and Design,* Proceedings of the Workshop on Connections held at the Ecole Normale Superiere, Cachan, France, May 1987, Elsevier Applied Science, London.

[12] Chen, W.F. and Toma, S. 1994. *Advanced Analysis of Steel Frames,* CRC Press, Boca Raton, FL.

[13] Chen, W.F. and Lui, E. 1991. *Stability Design of Steel Frames,* CRC Press, Boca Raton, FL.

[14] Chen, W.F., Goto, Y., and Liew, J. Y. R. 1995. *Stability Design of Semi-Rigid Frames,* John Wiley & Sons, New York.

[15] Eurocode 3. 1992. Design of Steel Structures, Part 1: General Rules and Rules for Buildings, ENV 1993-1-1:1992, Comite Europeen de Normalisation (CEN), Brussels.

[16] Goverdhan, A.V. 1984. A Collection of Experimental Moment Rotation Curves Evaluation of Predicting Equations for Semi-Rigid Connections, M.Sc. Thesis, Vanderbilt University, Nashville, TN.

[17] Hoffman, J. J. 1994. Design Procedures and Analysis Tools for Semi-Rigid Composite Members and Frames, M.S. Thesis, University of Minnesota, Minneapolis.

[18] Horne, M.R. and Morris, L.J. 1982. *Plastic Design of Low-Rise Frames,* The MIT Press, Cambridge, MA.

[19] IABSE. 1989. Bolted and Special Connections, Proceedings of the International Colloquium held in Moscow, USSR, May 1989, VNIPIP, 4 Vols., IABSE, Zurich.

[20] Kishi, N. and Chen, W. F. 1986. Data Base of Steel Beam-to-Column Connections, Vol. 1 & 2, Structural Engineering Report No. CE-STR-86-26, School of Civil Engineering, Purdue University, West Lafayette, IN.

[21] Leon, R.T. and Deierlein, G.G. 1996. Considerations for the Use of Quasi-Static Testing, *Earthquake Spectra,* 12(1), 87-110.

[22] Leon, R.T. 1996. Seismic Performance of Bolted and Riveted Connections, in *Background Reports on Metallurgy, Fracture Mechanics, Welding, Moment Connections and Frame System Behavior,* SAC Report 95-09, SAC Joint Venture, Sacramento, CA.

[23] Leon, R.T., Ammerman, D., Lin, J., and McCauley, R. 1987. Semi-Rigid Composite Steel Frames, *AISC Eng. J.,* 24(4), 147-156.

[24] Leon, R.T. and Hajjar, J.F. 1996. Effect of Floor Slabs on the Performance of Steel Moment Connections, Proceedings of the 11WCEE, Elsevier, London.

[25] Leon, R.T. and Zandonini, R. 1992. Composite Connections, in *Steel Design: An International Guide,* R. Bjorhovde and P. Dowling, Eds., Elsevier Publishers, London.

[26] Leon, R.T., Hoffman, J., and Staeger, T. 1996. Design of Partially-Restrained Composite Connections, AISC Design Guide 9, American Institute of Steel Construction, Chicago, IL.

[27] Nethercot, D.A. 1985. Steel Beam to Column Connections — A Review of Test Data and Their Applicability to the Evaluation of the Joint Behaviour of the Performance of Steel Frames, CIRIA, London.

[28] HEHRP. 1997. NEHRP Recommended Provisions for Seismic Regulations for New Buildings, BSSC, Washington, D.C.

[29] Viest, I.M., Colaco, J.P., Furlong, R.W., Griffis, L.G., Leon, R.T., and Wyllie, L.A. 1996. *Composite Construction Design for Buildings,* McGraw-Hill, New York.

24

Fatigue and Fracture

24.1 Introduction 24-1
24.2 Design and Evaluation of Structures for Fatigue 24-7
      Classification of Structural Details for Fatigue • Scale Effects in
      Fatigue • Distortion and Multiaxial Loading Effects in Fatigue
      • The Effective Stress Range for Variable-Amplitude Loading •
      Low-Cycle Fatigue Due to Seismic Loading
24.3 Evaluation of Structural Details for Fracture 24-18
      Specification of Steel and Filler Metal • Fracture Mechanics
      Analysis
24.4 Summary 24-26
24.5 Defining Terms 24-26
References 24-27
Further Reading 24-30

Robert J. Dexter and
John W. Fisher
Department of Civil Engineering,
Lehigh University,
Bethlehem, PA

24.1 Introduction

This chapter provides an overview of aspects of fatigue and fracture that are relevant to design or assessment of structural components made of concrete, steel, and aluminum. This chapter is intended for practicing civil and structural engineers engaged in regulation, design, inspection, repair, and retrofit of a variety of structures, including buildings; bridges; sign, signal, and luminaire support structures; chimneys; transmission towers; etc. Established procedures are explained for design and in-service assessment to ensure that structures are resistant to fatigue and fracture. This chapter is not intended as a comprehensive review of the latest research results in the subject area; therefore, many interesting aspects of fatigue and fracture are not discussed.

The design and assessment procedures outlined in this chapter may be applied to other similar structures, even outside the traditional domain of civil engineers, including offshore structures, cranes, heavy vehicle frames, and ships. The mechanical engineering approach, which works well for smooth machine parts, gives an overly optimistic assessment of the fatigue strength of structural details. There are many cases of failures of these types of structures, such as the crane in Figure 24.1 or the vehicle frame in Figure 24.2, which would have been predicted had the structural engineering approach been applied.

The possibility of fatigue must be checked for any structural member that is subjected to cyclic loading. Among the few cases where cracking has occurred in structures, the cracks are usually only a nuisance and may even go unnoticed. Only in certain truly non-redundant structural systems can cracking lead to structural collapse. The loading for most structures is essentially under fixed-load or load-control boundary conditions. On a local scale, however, most individual members and connections in redundant structures are essentially under displacement-control boundary conditions. In other words, because of the stiffness of the surrounding structure, the ends of the member

Figure 24.1   Fatigue cracking at welded detail in crane boom.

have to deform in a way that is compatible with nearby members. Under displacement control, a member can continue to provide integrity (e.g., transfer shear) after it has reached ultimate strength and is in the descending branch of the load-displacement curve. This behavior under displacement control is referred to as load shedding. In order for load shedding to be fully effective, individual critical members in tension must elongate to several times the yield strain locally without completely fracturing.

Good short-term performance should not lead to complacency, because fatigue and stress-corrosion cracking may take decades to manifest. Corrosion and other structural damage can precipitate and accelerate fatigue and fracture. Also, fabrication cracks may be built into a structure and never discovered. These dormant cracks can fracture if the structure is ever loaded into the inelastic range, such as in an earthquake.

Fatigue cracking in steel bridges in the U.S. has become a more frequent occurrence since the 1970s. Figure 24.3 shows a large crack that was discovered in 1970 at the end of a coverplate in one of the Yellow Mill Pond multibeam structures located at Bridgeport, Connecticut. Between 1970 and 1981, numerous fatigue cracks were discovered at the ends of coverplates in this bridge [19].

Fatigue cracking in bridges, such as shown in Figure 24.3, resulted from an inadequate experimental base and overly optimistic specification provisions developed from the experimental data in

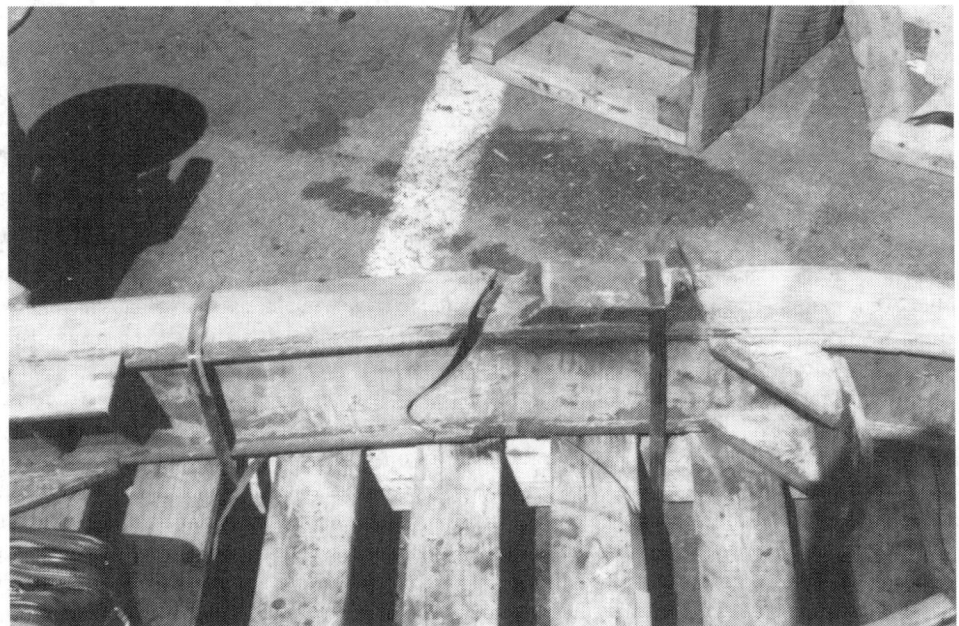

Figure 24.2 Fatigue cracking at welded detail in vehicle frame.

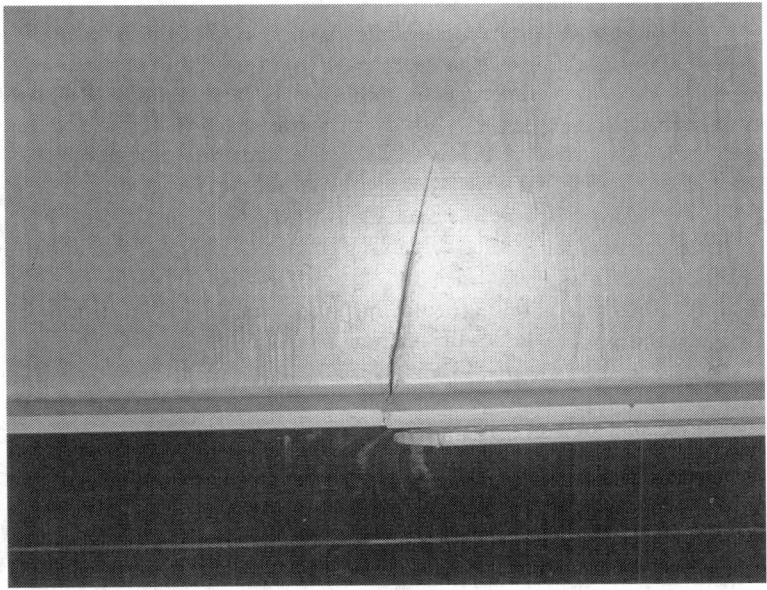

Figure 24.3 Fatigue crack originating from the weld toe of a coverplate end detail in one of the Yellow Mill Pond structures.

the 1960s. The assumption of a fatigue limit at two million cycles proved to be incorrect. As a result of extensive large-scale fatigue testing, it is now possible to clearly identify and avoid details that are expected to have low fatigue strength. The fatigue problems with the older bridges can be avoided in new construction. Fortunately, it is also possible to retrofit or upgrade the fatigue strength of existing bridges with poor details.

Low-cycle fatigue is a possible failure mode for structural members or connections that are cycled into the inelastic region for a small number of cycles. For example, bracing members in a braced frame or beam-to-column connections in a welded special-moment frame (WSMF) may be subjected to low-cycle fatigue in an earthquake. In sections that are cyclically buckling, the low-cycle fatigue is linked to the buckling behavior. This emerging area of research is briefly discussed in Section 24.2.5.

The primary emphasis in this chapter is on high-cycle fatigue. Truck traffic causes high-cycle fatigue of bridges. Fatigue cracking may occur in industrial buildings subjected to loads from cranes or other equipment or machinery. Although it has not been a problem in the past, fatigue cracking could occur in high-rise buildings frequently subjected to large wind loads. Wind loads have caused numerous fatigue problems in sign, signal, and luminaire support structures [32], transmission towers, and chimneys.

Although cracks can form in structures cycled in compression, they arrest and are not structurally significant. Therefore, only members or connections for which the stress cycle is at least partially in tension need to be assessed. If a fatigue crack forms in one element of a bolted or riveted built-up structural member, the crack cannot propagate directly into neighboring elements. Usually, a riveted member will not fail until a second crack forms in another element. Therefore, riveted built-up structural members are inherently redundant. Once a fatigue crack forms, it can propagate directly into all elements of a continuous welded member and cause failure at service loads. The lack of inherent redundancy in welded members is one reason that fatigue and fracture changed from a nuisance to a significant structural integrity problem as welding became widespread in the 1940s. Welded structures are not inferior to bolted or riveted structures; they just require more attention to design, detailing, and quality.

In structures such as bridges and ships, the ratio of the fatigue-design load to the strength-design loads is large enough that fatigue may control the design of much of the structure. In long-span bridges, the load on much of the superstructure is dominated by the dead load, with the fluctuating live load relatively small. These members will not be sensitive to fatigue. However, the deck, stringers, and floorbeams of bridges are subjected to primarily live load and therefore may be controlled by fatigue.

In structures controlled by fatigue, fracture is almost always preceded by fatigue cracking; therefore, the primary emphasis should be on preventing fatigue. Usually, the steel and filler metal have minimum specified toughness values (such as a Charpy V-Notch [CVN] test requirement). In this case, the cracks can grow to be quite long before fracture occurs. Fatigue cracks grow at an exponentially increasing rate; therefore, most of the life transpires while the crack is very small. Additional fracture toughness, greater than the minimum specified values, will allow the crack to grow to a larger size before sudden fracture occurs. However, the crack is growing so rapidly at the end of life that the additional toughness may increase the life only insignificantly.

However, fracture is possible for buildings that are not subjected to cyclic loading. Several large tension chords of long-span trusses fractured while under construction in the 1980s. The tension chords consisted of welded jumbo shapes, i.e., shapes in groups 4 and 5, as shown in Figure 24.4 [22]. These jumbo shapes are normally used for columns, where they are not subjected to tensile stress. These sections often have low fracture toughness, particularly in the core region of the web and flange junction. The low toughness has been attributed to the relatively low rolling deformation and slow cooling in these thick shapes. The low toughness is of little consequence if the section is used as a column and remains in compression. The fractures of jumbo tension chords occurred at welded splices at groove welds or at flame-cut edges of cope holes, as shown in Figure 24.4. In both cases the cracks formed at cope holes in the hard layer formed from thermal cutting. These cracks propagated in the core region of these jumbo sections, which has very low toughness. As a consequence of these brittle fractures, AISC (American Institute of Steel Construction) specifications now have a supplemental CVN notch toughness requirement for shapes in groups 4 and 5 and (for the same reasons) plates greater than 51 mm thick, when these are welded and subject to primary

Figure 24.4 (a) View of jumbo section used as tension chord in a roof truss and (b) closeup view of fracture in web originating from weld access holes at welded splice.

tensile stress from axial load or bending. Poorly prepared cope holes have resulted in cracks and fractures in lighter shapes as well.

The detailing rules that are used to prevent fatigue are intended to avoid notches and other stress concentrations. These detailing rules are useful for the avoidance of brittle fracture as well as fatigue. For example, the detailing rules in AASHTO (American Association of State Highway Transportation Officials) bridge design specifications would not permit a backing bar to be left in place because of the unfused notch perpendicular to the tensile stress in the flange. Along with low-toughness weld metal, this type of backing bar notch was a significant factor in the brittle fracture of WSMF connections in the Northridge earthquake [33, 53, 55]. Figure 24.5 shows a cross-section of a beam-flange-to-column weld from a building that experienced such a fracture. It is clear that the crack emanated from the notch created by the backing bar.

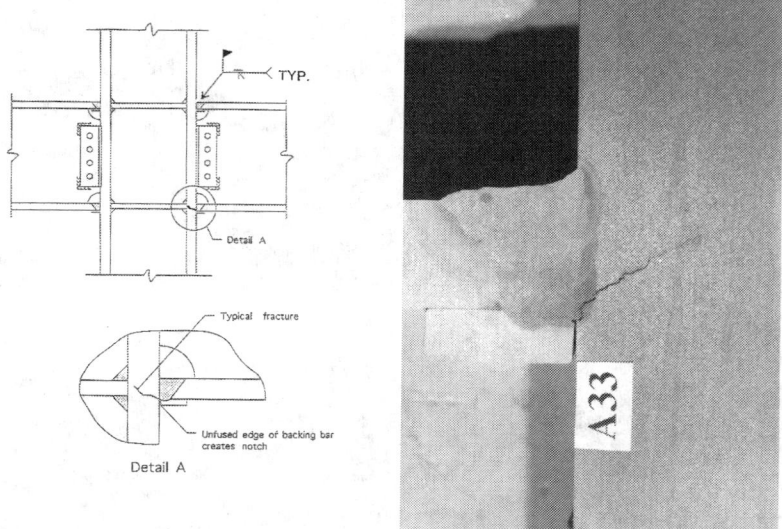

Figure 24.5 Welded steel moment frame (WSMF) connection showing (a) location of typical fractures and (b) typical crack, which originated at the backing bar notch and propagated into the column flange.

Detailing rules similar to the AASHTO detailing rules are included in American Welding Society (AWS) D1.1 *Structural Welding Code—Steel* for dynamically loaded structures. Dynamically loaded has been interpreted to mean fatigue loaded. Unfortunately, most seismically loaded building frames have not been required to be detailed in accordance with these rules. Even though it is not required, it might be prudent in seismic design to follow the AWS D1.1 detailing rules for all dynamically loaded structures.

Design for fracture resistance in the event of an extreme load is more qualitative than fatigue design, and usually does not involve specific loads. Details are selected to maximize the strength and ductility without increasing the basic section sizes required to satisfy strength requirements. The objective is to get the yielding to spread across the cross-section and develop the reserve capacity of the structural system without allowing premature failure of an individual component to precipitate total failure of the structure. The process of design for fracture resistance involves (1) predicting conceivable failure modes due to extreme loading, then (2) correctly selecting materials for and detailing the "critical" members and connections involved in each failure mode to achieve maximum ductility. Critical members and connections are those that are required to yield, elongate, or form a plastic hinge before the ultimate strength can be achieved for these conceived failure modes. Usually, the cost to upgrade a design meeting strength criteria to also be resistant to fatigue and fracture is very reasonable. The cost may increase due to (1) details that are more expensive to fabricate, (2) more expensive welding procedures, and (3) more expensive materials.

Quantitative means for assessing fracture are presented. Because of several factors, there is at best only about ± 30% accuracy in these fracture predictions, however. These factors include (1) variability of material properties; (2) changes in apparent toughness values with changes in test specimen size and geometry; (3) differences in toughness and strength of the weld zone; (4) complex residual stresses; (5) high gradients of stress in the vicinity of the crack due to stress concentrations; and (6) the behavior of cracks in complex structures of welded intersecting plates.

24.2   Design and Evaluation of Structures for Fatigue

Testing on full-scale welded members has indicated that the primary effect of constant amplitude loading can be accounted for in the live-load stress range [15, 20, 21, 34]; that is, the mean stress is not significant. The reason that the dead load has little effect on the lower bound of the results is that, locally, there are very high residual stresses. In details that are not welded, such as anchor bolts, there is a strong mean stress effect [54]. A worst-case conservative assumption (i.e., a high-tensile mean stress) is made in the testing and design of these nonwelded details.

The strength and type of steel have only a negligible effect on the fatigue resistance expected for a particular detail. The welding process also does not typically have an effect on the fatigue resistance. The independence of the fatigue resistance from the type of steel greatly simplifies the development of design rules for fatigue since it eliminates the need to generate data for every type of steel.

The established approach for fatigue design and assessment of metal structures is based on the S-N curve. Typically, small-scale specimen tests will result in longer apparent fatigue lives. Therefore, the S-N curve must be based on tests of full-size structural components such as girders. The reasons for these scale effects are discussed in Section 24.2.2. When information about a specific crack is available, a fracture mechanics crack growth rate analysis should be used to calculate remaining life [9, 10]. However, in the design stage, without specific initial crack size data, the fracture mechanics approach is not any more accurate than the S-N curve approach [35]. Therefore, the fracture mechanics crack growth analysis will not be discussed further.

Welded and bolted details for bridges and buildings are designed based on the nominal stress range rather than the local "concentrated" stress at the weld detail. The nominal stress is usually obtained from standard design equations for bending and axial stress and does not include the effect of stress concentrations of welds and attachments. Usually, the nominal stress in the members can be easily calculated without excessive error. However, the proper definition of the nominal stresses may become a problem in regions of high stress gradients.

The lower-bound S-N curves for steel in the AASHTO, AISC, AWS, and the American Railway Engineers Association (AREA) provisions are shown in Figure 24.6. These S-N curves are based on a lower bound with a 97.5% survival limit. S-N curves are presented for seven categories (A through E′) of weld details. The effect of the welds and other stress concentrations is reflected in the ordinate of the S-N curves for the various detail categories. The slope of the regression line fit to the test data for welded details is typically in the range 2.9 to 3.1 [34]. Therefore, in the AISC and AASHTO codes as well as in Eurocode 3 [18], the slopes have been standardized at 3.0.

Figure 24.6 shows the constant-amplitude fatigue limits (CAFLs) for each category as horizontal dashed lines. The CAFLs in Figure 24.6 were determined from the full-scale test data. When constant-amplitude tests are performed at stress ranges below the CAFL, noticeable cracking does not occur. Note that for all but category A, the fatigue limits occur at numbers of cycles much greater than two million, and therefore the CAFL should not be confused with the fatigue strength. Fatigue strength is a term representing the nominal stress range corresponding to the lower-bound S-N curve at a particular number of cycles, usually two million cycles. Most structures experience what

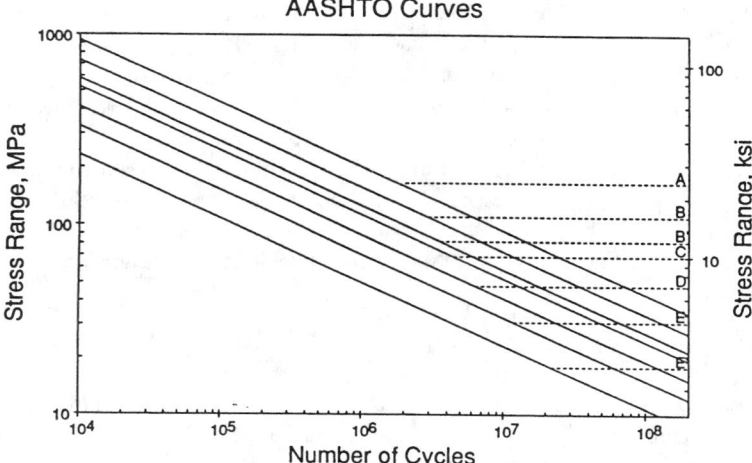

Figure 24.6   The AASHTO/AISC S-N curves. Dashed lines are the constant-amplitude fatigue limits and indicate the detail category.

is known as long-life variable-amplitude loading, i.e., very large numbers of random-amplitude cycles greater than the number of cycles associated with the CAFL. For example, a structure loaded continuously at an average rate of three times per minute (0.05 Hz) would accumulate 10 million cycles in only 6 years. The CAFL is the only important property of the S-N curve for long-life variable-amplitude loading, as discussed further in Section 24.2.4.

Similar S-N curves have been proposed by the Aluminum Association for welded aluminum structures. Table 24.1 summarizes the CAFLs for steel and aluminum for categories A through E'. The design procedures are based on associating weld details with specific categories. For both steel and aluminum, the separation of details into categories is approximately the same. Since fatigue is typically only a serviceability problem, fatigue design is carried out using service loads.

TABLE 24.1   Constant-Amplitude Fatigue Limits for AASHTO and Aluminum Association S-N Curves

Detail category	CAFL for steel (MPa)	CAFL for aluminum (MPa)
A	165	70
B	110	41
B′	83	32
C	69	28
D	48	17
E	31	13
E′	18	7

The nominal stress approach is simple and sufficiently accurate, and therefore is preferred when applicable. However, for details not covered by the standard categories, or for details in the presence of secondary stresses or high-stress gradients, the "hot-spot" stress range approach may be the only alternative. The hot-spot stress range is the stress range in a plate normal to the weld axis at some small distance from the weld toe. The hot-spot stress may be determined by strain gage measurement, finite element analysis, or empirical formulas. Unfortunately, methods and locations for measuring or calculating hot-spot stress as well as the associated S-N curve vary depending on which code or recommendation is followed [56].

In U.S. practice (i.e., the hot-spot method that has been used with the American Petroleum Institute's API RP-2A and AWS D1.1) the hot-spot stress is determined with a strain gage located nominally 5mm from the weld toe [16]. Actually, the 5-mm distance was not specifically selected. Rather, this distance was just the closest to the weld toe that a 3-mm strain gage could be placed. This definition of hot-spot stress originated from early experimental work on pressure vessels and tubular joints and has been the working definition of hot-spot stress in the U.S. offshore industry [40]. This approach is also used for other welded tubular joints and for details in ships and other marine structures.

The S-N curve used with the hot-spot stress approach is essentially the same as the nominal stress S-N curve (category C) for a transverse butt or fillet weld in a nominal membrane stress field (i.e., a stress field without any global stress concentration). The geometrical stress concentration and discontinuities associated with the local weld toe geometry are built into the S-N curve, while the global stress concentration is included in the hot-spot stress range.

24.2.1 Classification of Structural Details for Fatigue

It is standard practice in fatigue design of welded structures to separate the weld details into categories having similar fatigue resistance in terms of the nominal stress. Most common details can be idealized as analogous to one of the drawings in the specifications. The categories in Figure 24.6 range from A to E′ in order of decreasing fatigue strength. There is an eighth category, F, in the specifications, which applies to fillet welds loaded in shear. However, there have been very few if any failures related to shear, and the stress ranges are typically very low such that fatigue rarely would control the design. Therefore, the shear stress category F will not be discussed further.

In fact there have been very few if any failures attributed to details that have a fatigue strength greater than category C. Most structures have many more severe details, and these will generally govern the fatigue design. Therefore, unless all connections in highly stressed elements of the structure are high-strength bolted connections rather than welded, it is usually a waste of time to check category C and better details. Therefore, only category C and more severe details will be discussed in this section.

Severely corroded members should be evaluated to determine the stress range with respect to the reduced thickness and loss of section. Corrosion notches and pits may lead to fatigue cracks and should be specially evaluated. Otherwise severely corroded members may be treated as category E [44].

In addition to being used by AISC and AASHTO specifications, the S-N curves in Figure 24.6 and detail categories are essentially the same as those adopted by the AREA and AWS *Structural Welding Code* D1.1. The AASHTO/AISC S-N curves are also the same as 7 of the 11 S-N curves in the Eurocode 3. The British Standard (BS) 7608 has slightly different S-N curves, but these can be correlated to the nearest AISC S-N curve for comparison.

The following is a brief simplified overview of the categorization of fatigue details. In all cases, the applicable specifications should also be checked. Several reports have been published that show a large number of illustrations of details and their categories in addition to those in AISC and AASHTO specifications [14, 57]. Also, the Eurocode 3 and the BS 7608 have more detailed illustrations for their categorization than does the AISC or AASHTO specifications. Maddox [38] discusses categorization of many details in accordance with BS 7608, from which roughly equivalent AISC categories can be inferred.

In most cases, the fatigue strength recommended in these European standards is similar to the fatigue strength in the AISC and AASHTO specifications. However, there are several cases where the fatigue strength is significantly different; usually the European specifications are more conservative. Some of these cases are discussed in the following, as well as the fatigue strength for details that are not found in the specifications.

Mechanically Fastened Joints

Small holes are considered category D details. Therefore, rivetted and mechanically fastened joints (other than high-strength bolted joints) loaded in shear are evaluated as category D in terms of the net-section nominal stress. Pin plates and eyebars are designed as category E details in terms of the stress on the net section. In the AISC specifications, bolted joints loaded in direct tension are evaluated in terms of the maximum unfactored tensile load, including any prying load. Typically, these provisions are applied to hanger-type or bolted flange connections where the bolts are tensioned against the plies. If the number of cycles exceeds 20,000, the allowable load is reduced relative to the allowable load for static loading. Prying is very detrimental to fatigue, so if the number of cycles exceeds 20,000, it is advisable to minimize prying forces.

When bolts are tensioned against the plies, the total fluctuating load is resisted by the whole area of the precompressed plies, so that the bolts are subjected to only a fraction of the total load [37]. The analysis to determine this fraction is difficult, and this is one reason that the bolts are designed in terms of the maximum load rather than a stress range in the AISC specifications. In BS 7608, a slightly different approach is used for bolts in tension that achieves approximately the same result as the AISC specification for high-strength bolts. The stress range, on the tensile stress area of the bolt, is taken as 20% of the total applied load, regardless of the fluctuating part of the total load. The S-N curve for bolts is proportional to F_u, so that for high-strength bolts the result is an S-N curve between category E and E' for cycles less than two million. The tensile stress area, A_t, is given by

$$A_t = \frac{\pi}{4} \left(d_b - \frac{0.9743}{n} \right)^2 \tag{24.1}$$

where
d_b = the nominal diameter (the body or shank diameter)
n = threads per inch
(Note that the constant would be different if SI units were used.)

In the Eurocode 3, the fatigue strength of bolts is given in terms of the actual stress range in the bolts, although it is not clear how to calculate this for pretensioned connections. The recommended fatigue strength is given in terms of the tensile stress area of the bolt and does not depend on tensile strength. The design S-N curve from Eurocode 3 is about the same as category E', which is consistent with BS 7608 for high-strength bolts.

Anchor bolts in concrete cannot be adequately pretensioned and therefore do not behave like hanger-type or bolted flange connections. At best they are pretensioned between nuts on either side of the column base plate and the part below the bottom nut is still exposed to the full load range. Some additional test data was recently generated at the ATLSS Center at Lehigh University [54] for grade 55 and grade 105 anchor bolts. When combined with the existing data [27], the data show that the fatigue strength for anchor bolts is slightly greater than category E' in terms of the stress range on the tensile stress area of the bolt. Some of the bolts were tested with an intentional misalignment of 1:40, and these had only slightly lower fatigue strength than the aligned bolts, bringing the lower bound of the data closer to the category E' S-N curve.

The ATLSS data show the CAFL for all anchor bolts is slightly greater than the category D CAFL (48 MPa). The ATLSS data and Karl Frank's data show that proper tightening between the double nuts had a slight beneficial effect on the CAFL, but not enough to increase it by one category. In summary, for all types of bolts, if the actual stress range on the tensile stress area can be determined, it is recommended that (1) for finite life, the category E' S-N curve be used, and (2) for infinite life, the CAFL equivalent to that for category D be used (48 MPa).

Welded Joints

Welded joints are considered longitudinal if the axis of the weld is parallel to the primary stress range. Continuous longitudinal welds are category B or B' details. However, the termination of longitudinal fillet welds is more severe (category E). (The termination of full-penetration groove longitudinal welds requires a ground transition radius but gives a greater fatigue strength, depending on the radius.) If longitudinal welds must be terminated, it is better to terminate at a location where the stress ranges are less severe.

Category C includes transverse full-penetration groove welds (butt joints) subjected to nondestructive evaluation (NDE). Experiments conducted at Lehigh University showed that groove welds containing large internal discontinuities that were not screened out by NDE had a fatigue strength comparable to category E. The BS 7608 and the British Standards Institute published document PD 6493 [12] have reduced fatigue strength curves for groove welds with defects that are generally in agreement with these experimental data. Transverse groove welds with a permanent backing bar are reduced to category D [38]. One-sided welds with melt through (without backing bars) are also classified as category D.

Cope holes for weld access and to avoid intersecting welds, with edges conforming to the ANSI (American National Standards Institute) smoothness of 1000, may be considered a category D detail. Poorly executed cope holes must be treated as a category E detail. In some cases small cracks have occurred from the thermal-cut edges if martensite is developed. In those cases, crack extension will occur at lower stress ranges. Testing performed at ATLSS as well as at TNO in the Netherlands [17] has shown that the cope hole has lower fatigue strength than overlapping welds, which are less expensive but have traditionally been avoided because of the discontinuity at the overlap.

There have been many fatigue-cracking problems in structures at miscellaneous and seemingly unimportant attachments to the structure for such things as racks and hand rails. Attachments are a "hard spot" on the strength member that create a stress concentration at the weld. Often, it is not realized that such secondary members become part of the girder, i.e., that these secondary members stretch with the girder and therefore are subject to large stress ranges. Consequently, problems have occurred with fatigue of such secondary members.

Attachments normal to flanges or plates that do not carry significant load are rated category C if less than 51 mm long in the direction of the primary stress range, D if between 51 and 101 mm long, and E if greater than 101 mm long. (The 101-mm limit may be smaller for plates thinner than 9 mm.) If there is not at least 10 mm edge distance, then category E applies for an attachment of any length. The category E$'$, slightly worse than category E, applies if the attachment plates or the flanges exceed 25 mm in thickness. Transverse stiffeners are treated as short attachments (category C). Note that the attachment to the round tube in the crane boom in Figure 24.1, the transverse attachment in the vehicle frame in Figure 24.2, and the coverplate end detail in Figure 24.3 are all category E and E$'$ attachments.

The cruciform joint where the load-carrying member is discontinuous is considered a category C detail because it is assumed that the plate transverse to the load-carrying member does not have any stress range. A special reduction factor for the fatigue strength is provided when the load-carrying plate exceeds 13 mm in thickness. This factor accounts for the possible crack initiation from the unfused area at the root of the fillet welds (as opposed to the typical crack initiation at the weld toe for thinner plates) [26]. An example of cracking through the fillet weld throat of an attachment plate is shown in Figure 24.7.

Transverse stiffeners that are used for cross-bracing or diaphragms are also treated as category C details with respect to the stress in the main member. In most cases, the stress range in the stiffener from the diaphragm loads is not considered because these loads are typically unpredictable. In any case, the stiffener must be attached to the flanges, so even if the transverse loads were significant, most of the load would be transferred in shear to the flanges. (The web has very little out-of-plane stiffness.) In theory, the shear stress range in the fillet welds to the flanges should be checked, but shear stress ranges rarely govern design.

In most other types of load-carrying attachments, there is interaction between the stress range in the transverse load-carrying attachment and the stress range in the main member. In practice, each of these stress ranges is checked separately. The attachment is evaluated with respect to the stress

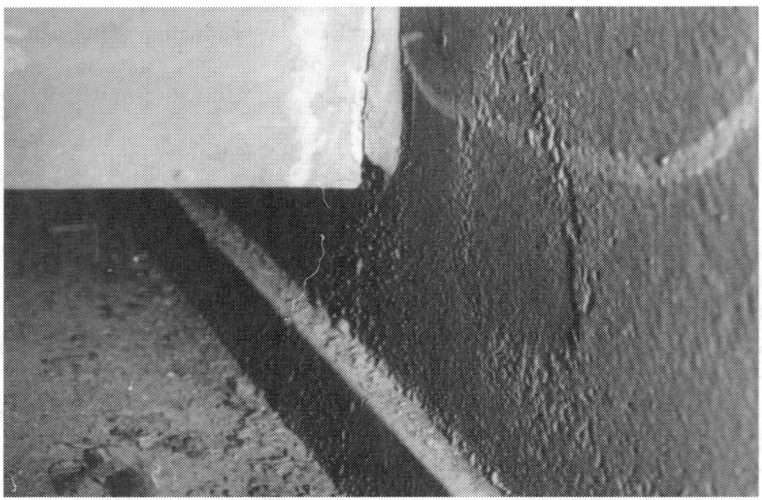

Figure 24.7   Cracking through the throat of fillet welds on an attachment plate.

range in the main member and then it is separately evaluated with respect to the transverse stress range. The combined multiaxial effect of the two stress ranges is taken into account by a decrease in the fatigue strength; that is, most load-carrying attachments are considered category E details. Multiaxial effects are discussed in greater detail in Section 24.2.3.

If the fillet or groove weld ends of a longitudinal attachment (load bearing or not) are ground smooth to a transition radius greater than 50 mm, the attachment can be considered category D (load bearing or not). If the transition radius of a groove-welded longitudinal attachment is increased to greater than 152 mm (with the groove-weld ends ground smooth), the detail (load bearing or not) can be considered category C.

Misalignment is a primary factor in susceptibility to cracking. The misalignment causes eccentric loading, local bending, and stress concentration. If the ends of a member with a misaligned connection are essentially fixed, the stress concentration factor (SCF) associated with misalignment is

$$\text{SCF} = 1.0 + 3e/t \tag{24.2}$$

where e is the eccentricity and t is the smaller of the thicknesses of two opposing loaded members. The nominal stress times the SCF should then be compared to the appropriate category. Generally, such misalignment should be avoided at fatigue critical locations. Equation 24.2 can also be used where e is the distance that the weld is displaced out of plane due to angular distortion. In either case, if the ends are pinned, the SCF is twice as large. A thorough guide to the SCF for various types of misalignment and distortion, including plates of unequal thickness, can be found in the British Standards Institute published document PD 6493 [12].

Reinforced and Prestressed Concrete and Bridge Stay Cables

Concrete structures are typically less sensitive to fatigue than welded steel and aluminum structures. However, fatigue may govern the design when impact loading is involved, such as for pavement, bridge decks, and rail ties. Also, as the age of concrete girders in service increases, and as the applied stress ranges increase with increasing strength of concrete, the concern for fatigue in concrete structural members has also increased.

According to ACI (American Concrete Institute) Committee Report 215R-74 in the *Manual of Standard Practice* [2], the fatigue strength of plain concrete at 10 million cycles is approximately 55% of the ultimate strength. However, even if failure does not occur, repeated loading may contribute to premature cracking of the concrete, such as inclined cracking in prestressed beams. This cracking can then lead to localized corrosion and fatigue of the reinforcement [30].

The fatigue strength of straight, unwelded reinforcing bars and prestressing strand can be described (in terms of the categories for steel details) with the category B S-N curve. The lowest stress range that has been known to cause a fatigue crack in a straight reinforcing bar is 145 MPa, which occurred after more than a million cycles. As expected, based on the results for steel details, minimum stress and yield strength had minimal effect on the fatigue strength of reinforcing bars. Bar size, geometry, and deformations also had minimal effect. ACI Committee 215 [2] suggested that members be designed to limit the stress range in the reinforcing bar to 138 MPa for high levels of minimum stress (possibly increasing to 161 MPa for less minimum stress). Fatigue tests show that previously bent bars had only about half the fatigue strength of straight bars, and failures have occurred down to 113 MPa [47]. Committee 215 recommends that half of the stress range for straight bars be used (i.e., 69 MPa) for the worst-case minimum stress. Equating this recommendation to the S-N curves for steel details, bent reinforcement may be treated as a category D detail.

Provided the quality is good, butt welds in straight reinforcing bars do not significantly lower the fatigue strength. However, tack welds reduce the fatigue strength of straight bars about 33%, with failures occurring as low as 138 MPa. Fatigue failures have been reported in welded wire fabric and bar mats [51].

If prestressed members are designed with sufficient precompression that the section remains uncracked, there is not likely to be any problem with fatigue. This is because the entire section is resisting the load ranges and the stress range in the prestessing strand is minimal. Similarly, for unbonded prestressed members, the stress ranges will be very small. Although the fatigue strength of prestressing strand in air is about equal to category B, when the anchorages are tested as well, the fatigue strength of the system is as low as half the fatigue strength of the wire alone (i.e., about category E). However, there is reason to be concerned for bonded prestressing at cracked sections because the stress range increases locally. The concern for cracked sections is even greater if corrosion is involved. The pitting from corrosive attack can dramatically lower the fatigue strength of reinforcement [30].

The above data were generated in tests of the prestressing systems in air. When actual beams are tested, the situation is very complex, but it is clear that much lower fatigue strength can be obtained [45, 48]. Committee 215 has recommended the following for prestressed beams:

1. The stress range in prestressed reinforcement, determined from an analysis considering the section to be cracked, shall not exceed 6% of the tensile strength of the reinforcement. (*Note*: This is approximately equivalent to category C.)
2. Without specific experimental data, the fatigue strength of unbonded reinforcement and their anchorages shall be taken as half of the fatigue strength of the prestressing steel. (*Note*: This is approximately equivalent to Category E.) Lesser values shall be used at anchorages with multiple elements.

The Post-Tensioning Institute (PTI) has issued "Recommendations for Stay Cable Design and Testing". The PTI recommends that uncoupled bar stay cables are category B details, while coupled (glued) bar stay cables are category D. The fatigue strengths of stay cables are verified through fatigue testing. Two types of tests are performed: (1) fatigue testing of the strand and (2) testing of relatively short lengths of the assembled cable with anchorages. The recommended test of the system is two million cycles at a stress range (158 MPa) that is 35 MPa greater than the fatigue allowable for category B at two million cycles. This test should pass with less than 2% wire breaks. A subsequent proof test must achieve 95% of the actual ultimate tensile strength of the tendons.

24.2.2   Scale Effects in Fatigue

As previously mentioned, fatigue tests on small-scale specimens will give higher apparent fatigue strength and are therefore unconservative [39, 41, 46]. There are several possible reasons for the observed scale effects. First, there is a well-known thickness effect in fatigue. This thickness effect is reflected in many places in AASHTO and AISC specifications where the fatigue strength is reduced for details with plate thickness greater than 20 or 25 mm in certain cases. For example, when coverplates exceed 25 mm in thickness or are wider than the flange, category E′ applies rather than category E. However, there may be cases where the coverplate is both wider than the flange and thicker than 25 mm. The fatigue strength in this case may be even less than category E′. One such case is shown in Figure 24.8, which is a wind-bracing gusset attached to the bottom of a floorbeam flange. The fatigue crack began at the termination of the fillet weld (along the top weld toe) where the plates overlap gusset laps.

Figure 24.8   Fatigue crack originating from the upper weld toe of a fillet weld where the fillet weld terminates near the overlap of thick plates.

In BS 7608, the fatigue strength of many details are keyed to plates with thickness 16 mm and less. For plates exceeding 16 mm, an equation is given that reduces the fatigue strength for thicker plates. A similar equation is used in Eurocode 3 for plates greater than 25 mm thick. These equations produce reductions in fatigue strength proportional to the 1/4 power of the ratio of the thickness to the base thickness (i.e., 16 or 25 mm).

Another effect is that the applied stress range may be different in small-scale specimens. For example, the stress concentration associated with welded attachments varies with the length of the attachment in the direction of the stresses. Also, in large-scale specimens, even though the nominal stress state is uniaxial or bending, unique local multiaxial stress states may develop naturally in complex details from random stress concentrations (e.g., poor workmanship and weld shape) and eccentricities (e.g., asymmetry of the design, tolerances, misalignment, distortion from welding). These complex natural stress states may be difficult to simulate in small-scale specimens and are difficult if not impossible to simulate analytically.

The state of residual stress from welding may be significantly different for small specimens due to the lack of constraint. Even if the specimens are cut from large-scale members, the residual stress will be altered. Finally, the volume of weld metal in full-scale members is sufficient to contain a structurally relevant representative sample of discontinuities (e.g., microcracks, pores, slag inclusions, hydrogen cracks, tack welds, and other notches).

24.2.3 Distortion and Multiaxial Loading Effects in Fatigue

In the AASHTO/AISC fatigue design provisions, the loading is assumed to be simple uniaxial loading. However, the loading may often be more complex than is commonly assumed in design. For example, fatigue design is based on the primary tension and bending stress ranges. Torsion, racking, transverse bending, and membrane action in plating are considered secondary loads and are typically not considered in fatigue analysis.

However, it is clear from the type of cracks that occur in bridges that a significant proportion of the cracking is due to distortion resulting from such secondary loading [24]. The solution to the problem of fatigue cracking due to secondary loading usually relies on the qualitative art of good detailing. Often, the best solution to distortion cracking problems may be to stiffen the structure. Typically, the better connections are more rigid. For example, transverse bracing and floorbeam attachment plates on welded girders should be welded directly to both flanges as well as the web. Numerous fatigue cracks have occurred due to distortion in the "web gap", i.e., the narrow gap between the stiffener and the flange. Figure 24.9 shows an example of a longitudinal crack formed along the longitudinal fillet weld that originated in the web gap (between the top of the stiffener and the flange) when such attachment plates are not welded to the flange. There has been a tendency to avoid welding to the tension flange due to an unfounded concern about brittle fracture.

In some cases a better solution is to allow the distortion to take place over a greater area so that lower stresses are created; that is, the detail should be made more flexible. For example, if a transverse stiffener is not welded to a flange, it is important to ensure that the gap between the flange and the end of the stiffener is sufficiently large, between four and six times the thickness of the web [23]. Another example where the best details are more flexible is connection angles for simply supported beams. Despite our assumptions, such simple connections transmit up to 40% of the theoretical fixed-end moment, even though they are designed to transmit only shear forces. For a given load, the moment in the connection decreases significantly as the rotational stiffness of the connection decreases. The increased flexibility of connection angles allows the limited amount of end rotation to take place with reduced bending stresses. A criterion has been developed for the design of these angles to provide sufficient flexibility [58]. The criterion states that the angle thickness (t) must be

$$t < 12 \left(g^2 / L \right) \tag{24.3}$$

Figure 24.9   Typical distortion-induced cracking in the web gap of an attachment plate.

where g is the gage in inches and L is the span length in inches. For example, for connection angles with a gage of 76 mm and a beam span of 7 m, the angle thickness should be just less than 10 mm. To solve a connection-angle cracking problem in service, the topmost rivet or bolt may be removed and replaced with a loose bolt to ensure the shear capacity. For loose bolts, steps are required to ensure that the nuts do not back off.

Significant stresses from secondary loading are often in a different direction than the primary stresses. Fortunately, experience with multiaxial loading experiments on large-scale welded structural details indicates the loading perpendicular to the local notch or the weld toe dominates the fatigue life. The cyclic stress in the other direction has no effect if the stress range is below 83 MPa and only a small influence above 83 MPa [16, 24].

The recommended approach for multiaxial loads is

1. Decide which loading (primary or secondary) dominates the fatigue cracking problem (typically the loading perpendicular to the weld axis or perpendicular to where cracks have previously occurred in similar details).
2. Perform the fatigue analysis using the stress range in this direction (i.e., ignore the stresses in the orthogonal directions).

24.2.4   The Effective Stress Range for Variable-Amplitude Loading

An actual service load history is likely to consist of cycles with a variety of different load ranges, i.e., variable-amplitude loading [25]. However, the fatigue design provisions are based on constant-amplitude loading and do not give any guidance for variable-amplitude loading. A procedure is shown below to convert variable stress ranges to an equivalent constant-amplitude stress range with the same number of cycles. This procedure is based on the damage summation rule jointly credited to Palmgren and Miner (referred to as Miner's rule) [42]. If the slope of the S-N curve is equal to 3, then the relative damage of stress ranges is proportional to the cube of the stress range. Therefore, the effective stress range is equal to the cube root of the mean cube of the stress ranges, i.e.,

$$S_{\text{effective}} = \left[(n_i / N_{\text{total}}) \, S_i^3 \right]^{1/3} \qquad (24.4)$$

The LRFD (load and resistance factor design) version of the AASHTO specification implies such an effective stress range using the straight line extension of the constant-amplitude curve. This is essentially the approach for variable-amplitude loading in BS 7608. Eurocode 3 also uses the effective stress range concept.

Research on such high-cycle variable-amplitude fatigue has shown that if all but 0.01% of the stress ranges are below the CAFLs, fatigue cracking does not occur [25]. The simplified fatigue-design procedure in the *AASHTO LRFD Bridge Design Specifications* [1] for structures with very large numbers of cycles is based on this observation. The objective of the AASHTO fatigue-design procedure is to ensure that the stress ranges at critical details due to a fatigue limit state load range are less than the CAFL for the particular details. The fatigue limit state load range is defined as having a probability of exceedence over the lifetime of the structure of 0.01%. A structure with millions of cycles is likely to see load ranges with this magnitude or greater hundreds of times; therefore, the fatigue limit state load range is not as large as the extreme loads used to check ultimate strength.

24.2.5 Low-Cycle Fatigue Due to Seismic Loading

Steel-braced frames and moment-resisting frames are expected to withstand cyclic plastic deformation without cracking in a large earthquake. If brittle fracture of these moment frame connections is suppressed, the connections can be cyclically deformed into the plastic range and will eventually fail by tearing at a location of strain concentration. This failure mode can be characterized as low-cycle fatigue. Low-cycle fatigue has been studied for pressure vessels and some other types of mechanical engineering structures. Since low-cycle fatigue is an inelastic phenomenon, the strain range is the key parameter rather than the stress range. However, at this time very little is understood about low-cycle fatigue in structures. For example, it is a very difficult task just to predict accurately the local strain range at a location of cyclic buckling.

Research performed to date indicates the feasibility of predicting curves for low-cycle fatigue from strain range vs. number of cycles in a manner analogous to high-cycle fatigue design using stress-range-based S-N curves. For example, low-cycle fatigue experiments were performed on specimens that would buckle as well as compact specimens that would not buckle but rather would fail from cracking at the welds [36]. These tests showed that the number of cycles to failure by low-cycle fatigue of welded connections could be predicted by the local strain range in a power law that is analogous to the power law (with stress range) represented by an S-N curve. They also showed that Miner's rule could be used to predict the number of variable-amplitude cycles to failure based on constant-amplitude test data.

More recently, Castiglioni [8, 13] has conducted similar experiments and plotted the results in terms of a fictitious elastic stress range that is equal to the strain range times the modulus of elasticity. In this manner he has shown that the low-cycle fatigue data plot along the same S-N curves from the Eurocode (similar to the AASHTO S-N curves) that are normally used for high-cycle fatigue. Castiglioni has equated the slenderness of the flanges with different fatigue categories, in effect treating the propensity for buckling like a "notch".

It can be hypothesized from these preliminary data that the same model used for high-cycle fatigue design, i.e., the S-N curves (converted to strain), can be used to predict fatigue behavior in the very-low-cycle regime characteristic of earthquake loading. Such a model could be very useful in seismic design of welded and bolted steel connections. Just by inspection, alternative details for a connection can be ranked in accord with their expected fatigue strength, i.e., the expected strain range that would cause cracking after a certain minimum number of cycles. After some limited verification through very-low-cycle inelastic experiments, these comparisons could rely on the existing knowledge base for the relative fatigue strength of various details in high-cycle fatigue.

The detailing rules that are used to prevent high-cycle fatigue are intended to avoid notches and other stress concentrations. These detailing rules could also be useful for preventing brittle fracture and premature low-cycle fatigue cracking. The relative fatigue strength is given by the detail category and the corresponding S-N curve.

24.3   Evaluation of Structural Details for Fracture

Unlike fatigue, fracture behavior depends strongly on the type and strength level of the steel or filler metal. In general, fracture toughness has been found to decrease with increasing yield strength of a material, suggesting an inverse relationship between the two properties. In practice, however, fracture toughness is more complex than implied by this simple relationship since steels with similar strength levels can have widely varying levels of fracture toughness.

Steel exhibits a transition from brittle to ductile fracture behavior as the temperature increases. For example, Figure 24.10 shows a plot of the energy required to fracture CVN impact test specimens of A588 structural steel at various temperatures. These results are typical for ordinary hot-rolled

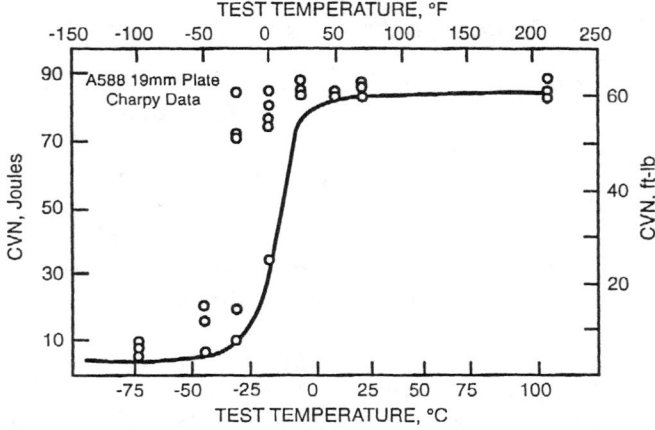

Figure 24.10   Charpy energy transition curve for A588 grade 50 (350-MPa yield strength) structural steel.

structural steel. The transition phenomena shown in Figure 24.10 is a result of changes in the underlying microstructural fracture mode. There are really at least three distinct types of fracture with distinctly different behavior.

1. *Brittle fracture* is associated with cleavage, which is transgranular fracture on select crystallographic planes on a microscopic scale. This type of fracture occurs at the lower end of the temperature range, although the brittle behavior can persist up to the boiling point of water in some low-toughness materials. This part of the temperature range is called the lower shelf because the minimum toughness is fairly constant up to the transition temperature. Brittle fracture is sometimes called elastic fracture because the plasticity that occurs is negligible and consequently the energy absorbed in the fracture process is also negligible.

2. *Transition-range fracture* occurs at temperatures between the lower shelf and the upper shelf and is associated with a mixture of cleavage and fibrous fracture on a microstructural scale. Because of the mixture of micromechanisms, transition-range fracture is characterized by extremely large variability. Fracture in the transition region is sometimes referred to as elastic-plastic fracture because the plasticity is limited in extent but has a significant impact on the toughness.

3. *Ductile fracture* is associated with a process of void initiation, growth, and coalescence on a microstructural scale, a process requiring substantial energy, and occurs at the higher end of the temperature range. This part of the temperature range is referred to as the upper shelf because the toughness levels off and is essentially constant for

higher temperatures. Ductile fracture is sometimes called fully plastic fracture because there is substantial plasticity across most of the remaining cross-section ahead of a crack. Ductile fracture is also called fibrous fracture due to the fibrous appearance of the fracture surface, or shear fracture due to the usually large slanted shear lips on the fracture surface.

Ordinary structural steel such as A36 or A572 is typically only hot rolled. To achieve very high-toughness, steels must be controlled rolled, i.e., rolled at lower temperatures, or must receive some auxiliary heat treatment such as normalization. In contrast to the weld metal, the cost of the steel is a major part of total costs. The expense of the high-toughness steels has not been found to be warranted for most building and bridges, whereas the cost of high-toughness filler metal is easily justifiable. Hot-rolled steels, which fracture in the transition region at the lowest service temperatures, have sufficient toughness for the required performance of most welded buildings and bridges.

24.3.1 Specification of Steel and Filler Metal

ASTM (American Society for Testing and Materials) specifications for bridge steel (A709) and ship steel (A131) provide for minimum CVN impact test energy levels. Structural steel specified by A36, A572, or A588, without supplemental specifications, does not require the Charpy test to be performed. If there is concern about brittle fracture and either (1) high ductility demand, (2) concern with low-temperature exposed structures, or (3) dynamic loading, then the CVN impact test should be specified by the purchaser of steel as a supplemental requirement. The results of the CVN test, impact energies, are often referred to as "notch-toughness" values.

Because the Charpy test is relatively easy to perform, it will likely continue to be the measure of toughness used in steel specifications. In the range of temperatures called the brittle region, the Charpy notch toughness is approximately correlated with K_c from fracture mechanics tests. The CVN specification works by ensuring that the transition from brittle to ductile fracture occurs at some temperature less than service temperature. The notch toughness requirement ensures that brittle fracture will not occur as long as large cracks do not develop. Often 34 J (25 ft-lb), 27 J (20 ft-lb), or 20 J (15 ft-lb) are specified at a particular temperature. The intent of specifying any of these numbers is the same, i.e., to make sure that the transition starts below this temperature.

Charpy toughness requirements for steel and weld metal for bridges and buildings are compared in Table 24.2. This table is simplified and does not include all the requirements.

TABLE 24.2 Minimum Charpy Impact Test Requirements for Bridges and Buildings

| | Minimum service temperature | | |
| | −18°C | −34°C | −51°C |
Material	Joules@°C	Joules@°C	Joules@°C
Steel: nonfracture critical members[a,b]	20@21	20@4	20@−12
Steel: fracture critical members[a,b]	34@21	34@4	34@−12
Weld metal for nonfracture critical[a]	27@−18	27@−18	27@−29
Weld metal for fracture critical[a,b]	34@−29°C for all service temperatures		
AISC: Jumbo sections and plates thicker than 50 mm[b]	27@21°C for all service temperature		

[a] These requirements are for welded steel with minimum specified yield strength up to 350 MPa up to 38 mm thick. Fracture critical members are defined as those that if fractured would result in collapse of the bridge.

[b] The requirements pertain only to members subjected to tension or tension due to bending.

Note that the bridge steel specifications require a CVN at a temperature 38°C *greater* than the minimum service temperature. This temperature shift accounts for the effect of strain rates, which are lower in the service loading of bridges (on the order of 10^{-3}) than in the Charpy test (greater than 10^1). It is possible to measure the toughness using a Charpy specimen loaded at a strain rate characteristic of bridges, called an intermediate strain rate, although the test is more difficult and the results are more variable. When the CVN energies from an intermediate strain rate are plotted as a function of temperature, the transition occurs at a temperature about 38°C lower for materials with yield strength up to 450 MPa.

As shown in Table 24.2, the AWS D1.5 *Bridge Welding Code* specifications for weld metal toughness are more demanding than the specifications for base metal. This is reasonable because the weld metal is always the location of discontinuities and high tensile residual stresses. However, there are no requirements for weld metal toughness in AWS D1.1. This lack of requirements was rationalized because typically the weld deposits are of higher toughness than the base metal. However, this is not always the case, e.g., the self-shielded flux-cored arc welds (FCAW-S) used in many of the WSMFs that fractured in the Northridge earthquake were reported to have very low toughness. The commentary in the AISC manual does warn that for "dynamic loading, the engineer may require the filler metals used to deliver notch-tough weld deposits".

ASTM A673 has specifications for the frequency of Charpy testing. The *H* frequency requires a set of three CVN specimens to be tested from one location for each heat or about 50 tons. These tests can be taken from a plate with thickness up to 9 mm different from the product thickness if it is rolled from the same heat. The P frequency requires a set of three specimens to be tested from one end of every plate, or from one shape in every 15 tons of that shape. For bridge steel, the AASHTO code requires CVN tests at the H frequency as a minimum. For fracture critical members, the guide specifications require CVN testing at the P frequency. In the AISC code, CVN tests are required at the P frequency for thick plates and jumbo sections. A special test location in the core of the jumbo section is specified, as well as a requirement that the section tested be produced from the top of the ingot.

Even the P testing frequency may be insufficient for as-rolled structural steel. In a recent report for the National Cooperative Highway Research Program (NCHRP) [28], CVN data were obtained from various locations on bridge steel plates. The data showed that because of extreme variability in CVN across as-rolled plates, it would be possible to miss potentially brittle areas of plates if only one location per plate is sampled. For plates that were given a normalizing heat treatment, the excessive variability was eliminated.

24.3.2 Fracture Mechanics Analysis

Fracture mechanics is based on the mathematical analysis of solids with notches or cracks. Relationships between the material toughness, the crack size, and the stress or displacement will be derived below using fracture mechanics. The objective of a fracture mechanics analysis (as outlined herein) is to ensure that brittle fracture does not occur. Even if ductile fracture occurs before local buckling or another failure mode, ductile fracture is considered to give acceptable ductility. Brittle fracture occurs with nominal net-section stresses below or just slightly above the yield point. Therefore, the relatively simple principles of linear-elastic fracture mechanics (LEFM) can be used to conservatively assess whether a welded joint is likely to fail by brittle fracture rather than in a ductile manner.

It significantly simplifies the presentation and practical use of fracture mechanics if the discussion is confined to brittle fracture only. Worst-case assumptions are made regarding numerous factors that can enhance fracture toughness, e.g., temperature, strain rate, constraint, and notch acuity or sharpness. These assumptions eliminate the need for extensive discussion of these effects.

If necessary, these effects can be considered and more advanced principles of fracture mechanics can be used to estimate the maximum monotonic or cyclic rotation before ductile tearing failure. Fracture mechanics can also be used to predict the subcritical propagation of cracks due to fatigue and/or stress-corrosion cracking that may precede fracture. In order to present a thorough discussion of the brittle fracture problem here, we cannot provide a detailed discussion of many other interesting fracture mechanics topics. There are several excellent books on fracture mechanics that cover these topics in detail [6, 9, 10].

Although cracks can be loaded by shear, experience shows that only the tensile stress normal to the crack is important in causing fatigue or fracture in steel structures. This tensile loading is referred to as Mode I. When the plane of the crack is not normal to the maximum principal stress, a crack that propagates subcritically or in a stable manner will generally turn as it extends, such that it becomes normal to the principal tensile stress. Therefore, it is typically recommended that a welding defect or crack-like notch that is not oriented normal to the primary stresses can be idealized as an equivalent crack with a size equal to the projection of the actual crack area on a plane that is normal to the primary stresses (see [12], for example).

Brittle fracture occurs with nominal net-section stresses below or just slightly above the yield point. Therefore, the relatively simple principles of LEFM can be used to conservatively assess whether a welded joint is likely to fail by brittle fracture rather than in a ductile manner. LEFM gives a relatively straightforward method for predicting fracture, based on a parameter called the stress-intensity factor (K), which characterizes the stresses at notches or cracks [31]. The applied K is determined by the size of the crack (or crack-like notch) and the nominal cross-section stress remote from the crack. Crack-like notches and weld defects are idealized as cracks, which include crack-like notches and weld defects as well. In the case of linear elasticity, the stress-intensity factor can be considered a measure of the magnitude of the crack tip stress and strain fields. Solutions for the applied stress-intensity factor, K, for a variety of geometries can be found in handbooks [43, 49, 50, 52]. Most of the solutions are variations on standard test specimens that have been studied extensively. The following discussion presents a few useful solutions and examples of their application to welded joints.

In general, the applied stress-intensity factor is given as

$$K = F_c * F_s * F_w * F_g * \sigma \sqrt{\pi a} \qquad (24.5)$$

where the F terms are modifiers on the order of 1.0, specifically:

F_c = the factor for the effect of crack shape

F_s = the factor, equal to 1.12, that is used if a crack originates at a free surface

F_w = a correction for finite-width, which is necessary because the basic solutions were generally derived for infinite or semi-infinite bodies

F_g = a factor for the effect of nonuniform stresses, such as bending stress gradient

An SCF is defined as the ratio of the peak stress near the stress raiser to the nominal cross-section stress remote from the stress raiser. SCFs are often used in fracture assessments when the crack is located near a stress raiser. For example, a crack may be located at a plate edge that is badly corroded. Any SCF would also be included in F_g.

The stress-intensity factor has the unusual units of MPa-m$^{1/2}$ or ksi-in.$^{1/2}$. The material fracture toughness is characterized in terms of the applied K at the onset of fracture in simplified small test specimens, called K "critical" or K_c. The fracture toughness (K_c) is considered a transferable material property; i.e., fracture of structural details is predicted if the value of the applied K in the detail exceeds K_c. Equation 24.5 relates the important factors that influence fracture: K_c represents the material, σ represents the design, and a represents the fabrication and inspection.

In this section, K_c is used as any type of critical K associated with a quasi-static strain rate, derived from any one of a variety of test methods. One measure of K_c is the plane-strain fracture toughness, which is given the special subscript "I" for plane strain, K_{Ic}. K_{Ic} must be measured in specimens that are very thick and approximate plane strain. If the fracture toughness is measured in an impact test, the special designation K_d is used, where the subscript "d" is for dynamic. In practice, K_c is often estimated from correlations with the result from a CVN test because the CVN is much cheaper to perform and requires less material than a fracture mechanics test, and all test laboratories are equipped for the CVN test. A widely accepted correlation for the lower shelf and lower transition region between K_d and CVN [9]:

$$K_d = 11.5 * \sqrt{CVN} \tag{24.6}$$

where CVN is given in joules and K_d is given in MPa-m$^{1/2}$. A different constant is used for English units. This correlation is used to construct the lower part of the curve for dynamic fracture toughness (K_d) as a function of temperature directly from the curve of CVN vs. temperature. There is a temperature shift between the intermediate load rate values of K_c and the impact load rate values of K_d that is approximately equal to the temperature shift that occurs for CVN data. Therefore, K_c values for structural steel are obtained by shifting the K_d curve to a temperature that is 38°C lower. However, for brittle materials there is essentially no temperature shift and therefore K_c is approximately equal to K_d.

Prior to the 1994 Northridge earthquake, the welds in the WSMF connections were commonly made with the FCAW-S process using an E7XT-4 weld wire. For the connections fractured in the Northridge earthquake that have been investigated so far, the weld metal CVN is plotted in Figure 24.11. The lower-bound impact energy is between 4 and 14 J for temperatures up to 50°C. If recommended weld procedures are followed, the fracture toughness increases slightly but remains inadequate.

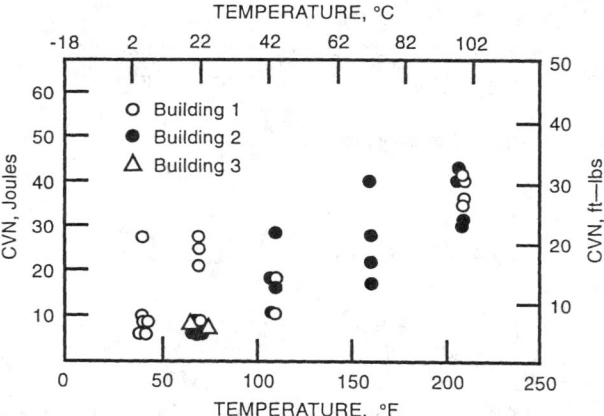

Figure 24.11 Typical Charpy impact energy from E7XT-4 self-shielded flux-cored arc welding weld metal from Northridge WSMF connections.

The lower bounds of the CVN and K_c data for the E7XT-4 weld metal is similar to the lower bounds from other brittle materials, such as the core region of the jumbo sections shown in Figure 24.4. This similarity in the data suggests that a there may be a lower-bound value of the fracture toughness that can be assumed for brittle ferritic weld metal, structural steel, and the heat-affected zone (HAZ). The lower-bound fracture toughness reflects the worst effects of temperature and strain rate. For these materials, the lower-bound fracture toughness was between 45 and 50 MPa-m$^{1/2}$. This concept of a lower-bound fracture toughness is very useful for fracture assessment.

As a consequence of the brittle fractures in jumbo sections, AISC specifications now have a supplemental Charpy requirement for shapes in groups 4 and 5 and (for the same reasons) plates greater than 51 mm thick, when these are welded and subject to primary tensile stress from axial load or bending. These jumbo shapes and thick plates must exhibit 27 J at 21°C. Using the correlation of Equation 24.6, 27 J will give a K_c of 60 MPa-m$^{1/2}$.

There are size and constraint effects and other complications that make the LEFM fracture toughness, K_c, less than perfect as a material property. This is especially true when K_c is estimated based only on a correlation to CVN. Nevertheless, as illustrated in the following, the conservative lower-bound value of K_c can be used by structural engineers to avoid brittle fracture.

Center Crack

The K solution for an infinitely wide plate with a through crack subject to uniform tensile membrane stress is

$$K = \sigma\sqrt{\pi a} \tag{24.7}$$

where

σ = nominal cross-section stress remote from the crack
$2a$ = the total overall crack length

If the total width of the panel is given as 2W, F_w for this crack geometry can be approximated by the Fedderson or secant formula:

$$F_w = \sqrt{\sec\frac{\pi a}{2W}} \tag{24.8}$$

This formula gives a value that is close to 1.0 and can be ignored for a/W less than a third. For a/W of about 0.5, the secant formula gives F_w of about 1.2. However, the values from the secant equation go to infinity as a approaches W. The secant formula is reasonably accurate for a/W up to 0.85. The F_w may be used for other crack geometries as well.

Many common buried defects and notches in welded joints can be idealized as a center crack in tension. For example, Figure 24.12 shows a backing bar with a fillet that is idealized as a center crack. The unfused area of the backing bar creates a crack-like notch with one tip in the root of

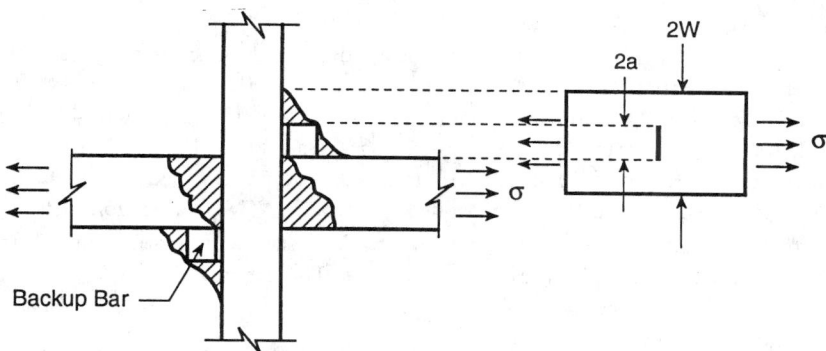

Figure 24.12 Cross-section of a one-sided groove-welded cruciform-type connection with loaded plate discontinuous and idealization of the notches from a backing bar with a fillet weld as a center-cracked tension panel.

the fillet weld and one tip at the root of the groove weld. The crack is asymmetrical but since the connection is subjected to uniform tension, the crack can be analyzed as if it were in a symmetric center-cracked panel. Of course, the applied K is higher on the crack tip that is in the root of the fillet weld because there is a high F_w for this side. Assuming the weld metal of the groove weld and the fillet weld have comparable toughness, the fillet weld side of the backing bar will govern the fracture limit state. Therefore, the panel is idealized as being symmetric with respect to the center of the backing bar.

Assuming negligible weld root penetration, the crack size ($2a$) is taken as being equal to the backing bar thickness, say 13 mm. For a/W of 0.5, Equation 24.8 gives F_w equal to 1.2. Although this idealization seems like a gross approximation at best, the validity of the K solution for this particular weld joint was verified based on observed fatigue crack propagation rates.

If this is a grade 50 steel, the yield strength could be up to 450 MPa. The notch tip could be subjected to full tensile residual stress. Therefore, Equation 24.5 is solved with the cross-section stress equal to 450 MPa, with the F_w factor of 1.2, and a of 6 mm, giving 74 MPa-m$^{1/2}$. It can be seen that this configuration could cause a brittle fracture for very brittle materials. However, weld metal and base steel with modest toughness could easily withstand this defect.

Edge Crack

The stress-intensity factor for an edge crack in an infinitely wide plate is

$$K = 1.12\sigma\sqrt{\pi a} \qquad (24.9)$$

where σ is the remote cross-section nominal stress and a is the depth of the edge crack or crack-like notch. It can be seen that the edge crack equation is treated like half of a center crack, i.e., Equation 24.7, where a and W are the total length and width, respectively, for the edge crack. The F_s of 1.12 is applied to account for the free edge, which is not restrained as it is in the center-cracked geometry. Equation 24.9 is also modified with the F_w as in Equation 24.8.

Figure 24.5 showed a cross-section near the crack origin for a WSMF beam-flange-to-column weld that fractured in the Northridge earthquake. The fracture surfaces, such as shown in Figure 24.13, indicate that the fractures originate in the root of the weld, typically at a lack-of-fusion defect. This lack-of-fusion defect is difficult to avoid when the weld must be stopped on one side of the web and started on the other side. The weld fracture surface in Figure 24.13 shows the crack-like notch formed by the combination of a lack-of-fusion defect and the unfused edge of the backing bar. On a cross-section at the deepest point of the lack-of-fusion defect, the total depth of the notch, including the unfused edge of the backing bar, is from 13 to 19 mm.

The value of 45 MPa-m$^{1/2}$ can be used as a lower bound to the fracture toughness of structural steel or weld metal. Equation 24.9 may be used to predict brittle fracture for the WSMF connection welds when K exceeds 45 MPa-m$^{1/2}$. For a notch depth of 13 to 19 mm, Equation 24.9 would predict that brittle fracture is likely to occur for cross-section stress between 160 and 200 MPa, well below the yield point. These types of LEFM calculations, had they been performed prior to the earthquake, would have predicted that brittle fracture would occur in the WSMF connections before yielding.

The propagation path of the unstable dynamic crack is seemingly chaotic, as it is influenced by dynamic stress waves and complex residual stress fields. The critical event was the initiation of the unstable crack in the brittle weld. There is little significance to whether the crack propagated in the weld or turned and entered the column.

The solution to the WSMF cracking problem is elimination of the possibility of such large defects (i.e., $\approx$ 19 mm) that result from the backing bar and any lack-of-fusion defect. The backing bar removal and back gouging to minimize the lack of fusion will eliminate most large crack-like conditions. However, this is not enough to eliminate the problem of brittle fracture. Appropriate weld notch toughness requirements must be specified to avoid the use of low-toughness weld metal.

There are certainly many other equally important design issues that influenced these fractures. The overall lack of redundancy, i.e., the reliance on only one or two massive WSMFs to resist lateral load in each direction, contributes to large forces and increases the thickness of the members and the high constraint of the connections. Even if brittle fracture is avoided, welds will typically fail at a lower level of plastic strain than base metal. Therefore, it can also be argued that it is imprudent to rely on welds for extensive plastic deformation. Several improved WSMF connections have been proposed, most of which are designed such that the plastic hinge develops in the span away from the connection. Nevertheless, in the event of unexpected loading, it is still desirable that these weld joints have a ductile failure mode. Therefore, while these improved connection designs may be

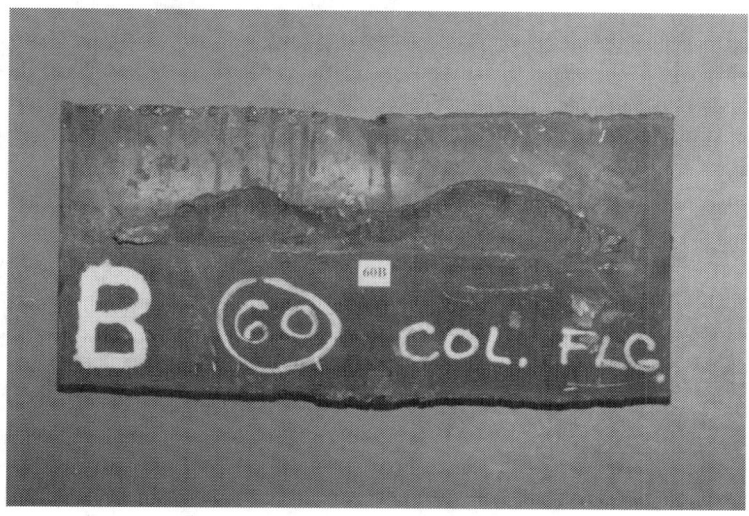

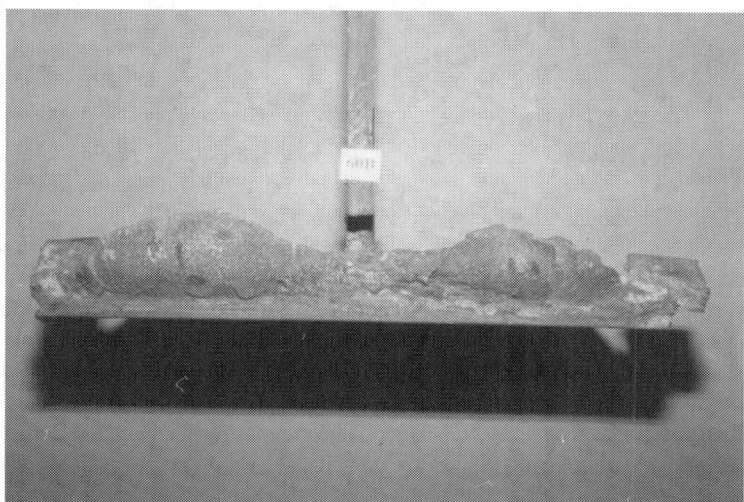

Figure 24.13 Fracture surfaces of a typical beam-flange-to-column joint of a WSMF that fractured in the Northridge earthquake, showing the lack-of-fusion defect in the center of the weld.

worthwhile, the low-toughness weld metal and joint design with a built-in notch should still not be used under most circumstances.

Buried Penny-Shaped Crack

Many internal weld defects are idealized as an ellipse or a circle that is circumscribed around the projection of the weld defect on a plane perpendicular to the stresses. Often, the increased accuracy accrued by using the relatively complex elliptical formula is not worth the effort, and the circumscribed penny-shaped or circular crack is always conservative. The stress-intensity factor for the penny-shaped crack in an infinite body is given as

$$K = \frac{2}{\pi}\sigma\sqrt{\pi a} \qquad (24.10)$$

where a is the radius of the circular crack. As for the other types of cracks, the F_w can be calculated using Equation 24.10. In terms of Equation 24.8, the crack shape factor F_c in this case is $2/\pi$ or 0.64. Using (1) the lower-bound fracture toughness of 45 MPa-m$^{1/2}$ and (2) an upper-bound residual stress plus applied stress equal to the upper-bound yield strength for grade 50 steel (450 MPa), Equation 24.10 shows that a penny-shaped crack would have to have a radius exceeding 8 mm to be critical, i.e., the diameter of the allowable welding defect would be 15 mm (providing that fatigue is not a potential problem).

The crack shape factor F_c is more favorable (0.64) for buried cracks as opposed to F_s of 1.12 for edge cracks, and the defect size is equal to $2a$ for the buried crack and only a for the edge crack. These factors explain why edge cracks of a given size are much more dangerous than buried cracks of the same size.

24.4 Summary

Structural elements for which the live load is a large percentage of the total load are potentially susceptible to fatigue. Many factors in fabrication can increase the potential for fatigue including notches, misalignment, and other geometrical discontinuities, thermal cutting, weld joint design (particularly backing bars), residual stress, nondestructive evaluation and weld defects, intersecting welds, and inadequate weld access holes. The fatigue design procedures in the AASHTO and AISC specifications are based on control of the stress range and knowledge of the fatigue strength of the various details. Using these specifications, it is possible to identify and avoid details that are expected to have low fatigue strength.

Welded connections and thermal-cut hole copes, blocks, or cuts are potentially susceptible to brittle fracture. Many interrelated design variables can increase the potential for brittle fracture including lack of redundancy, large forces and moments with dynamic loading rates, thick members, geometrical discontinuities, and high constraint of the connections. Low temperature can be a factor for exposed structures. The factors mentioned above that influence the potential for fatigue have a similar effect on the potential for fracture. In addition, cold work, flame straightening, weld heat input, and weld sequence can also affect the potential for fracture. The AASHTO specifications [1] require a minimum CVN notch toughness at a specified temperature for the base metal and the weld metal of members loaded in tension or tension due to bending. Almost two decades of experience with these bridge specifications has proven that they are successful in significantly reducing the number of brittle fractures. Simple, LEFM concepts can be used to predict the potential for brittle fracture in buildings.

24.5 Defining Terms

a: Crack length.

A_t: The tensile stress area of a bolt.

AASHTO: American Association of State Highway Transportation Officials.

ACI: American Concrete Institute.

AISC: American Institute of Steel Construction.

ASTM: American Society for Testing and Materials.

AWS: American Welding Society.

BSI: British Standards Institute.

CAFL: Constant amplitude fatigue limit, a level of stress range below which noticeable crack-ing does not occur in constant amplitude fatigue tests.

CVN: Charpy V-Notch impact test energy.

d_b: The nominal diameter of a bolt (the body or shank diameter).

F_c: Factor for the effect of crack shape.

F_s: Factor, equal to 1.12, that is used if a crack originates at a free surface.

F_w: Correction for finite-width.

F_g: Factor for the effect of non-uniform stresses, such as bending stress gradient.

F_u: Ultimate tensile strength.

K: Stress-intensity factor.

K_c: Fracture toughness.

K_d: Dynamic fracture toughness.

L: Span length.

LFRD: Load and resistance factor design.

n: Threads per inch in bolts.

n_i: Number of cycles in an interval of stress range i.

N_{total}: Total number of stress ranges.

PTI: Post-Tensioning Institute.

$S_{\text{effective}}$: The effective stress range which is equal to the cube root of the mean cube of the all stress ranges.

S_i: Stress range for interval i.

SCF: Stress concentration factor.

t: Plate thickness.

g: Gage.

References

[1] American Association of State Highway Transportation Officials. 1994. *AASHTO LRFD Bridge Design Specifications,* First edition, Washington, D.C.

[2] American Concrete Institute Committee 215. 1996. Considerations for Design of Concrete Structures Subjected to Fatigue Loading, ACI 215R-74 (Revised 1992), *ACI Manual of Standard Practice,* Vol. 1.

[3] Ad-hoc Committee on Cable-Stayed Bridges. 1986. *Recommendations for Stay Cable Design and Testing,* Post-Tensioning Institute, Phoenix, AZ.

[4] American Institute of Steel Construction. 1994. *Load and Resistance Factor Design Specification for Structural Steel Buildings,* Second edition, Chicago, IL.

[5] American Petroleum Institute. 1989. *Recommended Practice for Planning, Designing, and Constructing Fixed Offshore Platforms,* API RP2A, 18th edition.

[6] Anderson, T.L. 1995. *Fracture Mechanics—Fundamentals and Applications,* Second edition, CRC Press, Boca Raton, FL.

[7] American Welding Society. 1996. *Structural Welding Code—Steel,* ANSI/AWS D1.1, Miami, FL.

[8] Ballio, G. and Castiglioni, C.A. 1994. A Unified Approach for the Design of Steel Structures Under Low and/or High Cycle Fatigue, *J. Constructional Steel Res.*

[9] Barsom, J.M. and Rolfe, S.T. 1987. *Fracture and Fatigue Control in Structures,* Second edition, Prentice-Hall, Englewood Cliffs, NJ.

[10] Broek, D. 1987. *Elementary Fracture Mechanics,* Fourth edition, Martinus Nijhoff Publishers, Dordrecht, the Netherlands.

[11] British Standards Institute. 1994. *Code of Practice for Fatigue Design and Assessment of Steel Structures,* BS 7608, London.

[12] British Standards Institute. 1991. *Guidance on Some Methods for the Derivation of Acceptance Levels for Defects in Fusion Welded Joints,* BSI PD 6493, London.

[13] Castiglioni, C.A. 1995. Cumulative Damage Assessment in Structural Steel Details, *Extending the Lifespan of Structures,* IABSE Symposium, San Francisco, CA, pp. 1061-1066.

[14] Demers, C. and Fisher, J.W. 1990. *Fatigue Cracking of Steel Bridge Structures, Volume I: A Survey of Localized Cracking in Steel Bridges—1981 to 1988,* Report No. FHWA-RD-89-166; *Volume II, A Commentary and Guide for Design, Evaluation, and Investigating Cracking,* Report No. FHWA-RD-89-167, Federal Highway Administration, McLean, VA.

[15] Dexter, R.J., Fisher, J.W., and Beach, J.E. 1993. Fatigue Behavior of Welded HSLA-80 Members, *Proceedings, 12th International Conference on Offshore Mechanics and Arctic Engineering,* Vol. III, Part A, Materials Engineering, American Society of Mechanical Engineers, New York, pp. 493-502.

[16] Dexter, R.J., Tarquinio, J.E., and Fisher, J.W. 1994. An Application of Hot-Spot Stress Fatigue Analysis to Attachments on Flexible Plate, *Proceedings of the 13th International Conference on Offshore Mechanics and Arctic Engineering,* American Society of Mechanical Engineers, New York.

[17] Djikstra, O.D., Wardenier, J., and Hertogs, A.A. 1988. The Fatigue Behavior of Welded Splices With and Without Mouse Holes in IPE 400 and HEM 320 Beams, *Proceedings of the International Conference on Weld Failures—Weldtech '88,* The Welding Institute, Abington Hall, Cambridge, U.K.

[18] European Committee for Standardization. 1992. *Eurocode 3: Design of Steel Structures—Part 1.1: General Rules and Rules for Buildings,* ENV 1993-1-1, Brussels.

[19] Fisher, J.W. 1984. *Fatigue and Fracture in Steel Bridges,* ISBNO-471-80469-X, John Wiley & Sons, New York.

[20] Fisher, J.W., Frank, K.H., Hirt, M.A., and McNamee, B.M. 1970. *Effect of Weldments on the Fatigue Strength of Steel Beams,* National Cooperative Highway Research Program Report 102, Highway Research Board, Washington, D.C.

[21] Fisher, J.W., Albrecht, P.A., Yen, B.T., Klingerman, D.J., and McNamee, B.M. 1974. *Fatigue Strength of Steel Beams with Welded Stiffeners and Attachments,* National Cooperative Highway Research Program Report 147, Transportation Research Board, Washington, D.C.

[22] Fisher, J.W. and Pense, A.W. 1987. Experience with Use of Heavy W Shapes in Tension, *Eng. J.,* American Institute of Steel Construction, 24(2).

[23] Fisher, J.W. and Keating, P.B. 1989. Distortion-Induced Fatigue Cracking of Bridge Details with Web Gaps, *J. Constructional Steel Res.,* (12), 215-228.

[24] Fisher, J.W., Jian, J., Wagner, D.C., and Yen, B.T. 1990. *Distortion-Induced Fatigue Cracking in Steel Bridges,* National Cooperative Highway Research Program Report 336, Transportation Research Board, Washington, D.C.

[25] Fisher, J.W., et al. 1993. *Resistance of Welded Details Under Variable Amplitude Long-Life Fatigue Loading,* National Cooperative Highway Research Program Report 354, Transportation Research Board, Washington, D.C.

[26] Frank, K.H. and Fisher, J.W. 1979. Fatigue Strength of Fillet Welded Cruciform Joints, *J. Structural Div.,* ASCE, 105(ST9), 1727-1740.

[27] Frank, K.H. 1980. Fatigue Strength of Anchor Bolts, *J. Structural Div.,* ASCE, 106 (ST).

[28] Frank, K.H., et al. 1993. *Notch Toughness Variability in Bridge Steel Plates,* National Cooperative Highway Research Program Report 355, Washington, D.C.

[29] American Association of State Highway and Transportation Officials. 1978. *Guide Specifications for Fracture Critical Non-Redundant Steel Bridge Members,* Washington, D.C.

[30] Hahin, C. 1994. *Effects of Corrosion and Fatigue on the Load-Carrying Capacity of Structural Steel and Reinforcing Steel,* Illinois Physical Research Report No. 108, Illinois Department of Transportation, Springfield, IL.

[31] Irwin, G.R. 1957. Analysis of Stresses and Strains Near the End of Crack Traversing a Plate, *Trans., ASME, J. Appl. Mech.,* Vol. 24. Also reprinted in ASTM volume on classic papers.

[32] Kaczinski, M.R., Dexter, R.J., and Van Dien, J.P. 1996. *Fatigue-Resistant Design of Cantilevered Signal, Sign, and Light Supports,* Final Report for NCHRP Project 10-38, ATLSS Engineering Research Center, Lehigh University.

[33] Kaufmann, E.J., Xue, M., Lu, L.-W., and Fisher, J.W. 1996. Achieving Ductile Behavior of Moment Connections, *Modern Steel Construction,* 36(1), 30-39.

[34] Keating, P.B. and Fisher, J.W. 1986. *Evaluation of Fatigue Tests and Design Criteria on Welded Details,* National Cooperative Highway Research Program, Report 286, Washington, D.C.

[35] Kober, Dexter, R.J., Kaufmann, E.J., Yen, B.T., and Fisher, J.W. 1994. The Effect of Welding Discontinuities on the Variability of Fatigue Life, *Fracture Mech.,* Twenty-Fifth Volume, ASTM STP 1220, F. Erdogan and Ronald J. Hartranft, Eds., American Society for Testing and Materials, Philadelphia, PA.

[36] Krawinkler, H. and Zohrei, M. 1983. Cumulative Damage in Steel Structures Subjected to Earthquake Ground Motion, *Computers and Structures,* 16(1-4), 531-541.

[37] Kulak, G.L., Fisher, J.W., and Struick, J.H. 1987. *Guide to Design Criteria for Bolted and Riveted Joints,* Second Edition, Prentice-Hall, Englewood Cliffs, NJ.

[38] Maddox, S.J. 1991. *Fatigue Strength of Welded Structures,* Second Edition, Abington Publishing, Cambridge, UK.

[39] Marsh, K.J., Ed. 1988. *Full-Scale Fatigue Testing of Components and Structures,* Butterworths, London.

[40] Marshall, P.W. 1992. *Design of Welded Tubular Connections,* Elsevier, New York.

[41] Miki, C., Nishimura, T., Tajima, J., and Okukawa, A. 1980. Fatigue Strength of Steel Members Having Longitudinal Single-Bevel Groove Welds, *Trans. Japan Weld. Soc.,* 11(1), 43-56.

[42] Miner, M.A. 1945. Cumulative Damage in Fatigue, *J. Appl. Mech.,* 12, A-159.

[43] Murakami, Y., et al., Eds. 1987. *Stress Intensity Factors Handbook,* Vols. 1 and 2, Pergamon Press, Oxford, U.K.

[44] Outt, J.M.M., Fisher, J.W., and Yen, B.T. 1984. *Fatigue Strength of Weathered and Deteriorated Riveted Members,* Report DOT/OST/P-34/85/016, Department of Transportation, Federal Highway Administration, Washington, D.C.

[45] Overnman, T.R., Breen, J.E., and Frank, K.H. 1984. *Fatigue Behavior of Pretensioned Concrete Girders,* Research Report 300-2F, Center for Transportation Research, The University of Texas at Austin.

[46] Petershagen, H. 1986. Fatigue Problems in Ship Structures, *Advances in Marine Structures,* Elsevier Applied Science, London, pp. 281-304.

[47] Pfister, J.F. and Hognestad, E. 1964. High Strength Bars as Concrete Reinforcement, Part 6, Fatigue Tests, *J. PCA Res. Dev. Lab.,* 6(1), 65-84.

[48] Rabbat, B.G., et al. 1979. Fatigue Tests of Pretensioned Girders with Blanketed and Draped Strands, *J. Prestressed Concrete Inst.,* 24(4), 88-115.

[49] Rooke, D.P. and Cartwright, D.J. 1974. *Compendium of Stress Intensity Factors,* Her Majesty's Stationery Office, London.

[50] Rooke, D.P. 1986. *Compounding Stress Intensity Factors,* Research Reports in Materials Science (Series One), The Parthenon Press, Lancashire, U.K.

[51] Sternberg, F. 1969. *Performance of Continuously Reinforced Concrete Pavement, I-84 Southington,* Connecticut State Highway Department.

[52] Tada, H. 1985. *The Stress Analysis of Cracks Handbook,* Paris Productions, Inc., St. Louis, MO.

[53] Tide, R.H.R., Fisher, J.W., and Kaufmann, E.J. 1996. Substandard Welding Quality Exposed: Northridge, California Earthquake, January 17, 1994, IIW Asian Pacific Welding Congress, Auckland, New Zealand, February 4-9.

[54] VanDien, J.P., Kaczinski, M.R., and Dexter, R.J. 1996. Fatigue Testing of Anchor Bolts, *Building an International Community of Structural Engineers,* Vol. 1, Proceedings of Structures Congress XIV, Chicago, pp. 337-344.

[55] Xue, M., Kaufmann, E.J., Lu, L.W., and Fisher, J.W. 1996. Achieving Ductile Behavior of Moment Connections—Part II, *Modern Steel Construction,* 36(6), 38-42.

[56] Yagi, J., Machida, S., Tomita, Y., Matoba, M., and Kawasaki, T. 1991. *Definition of Hot-Spot Stress in Welded Plate Type Structure for Fatigue Assessment,* International Institute of Welding, IIW-XIII-1414-91.

[57] Yen, B.T., Huang, T., Lai, L.Y., and Fisher, J.W. 1990. *Manual for Inspecting Bridges for Fatigue Damage Conditions,* Report No. FHWA-PA-89-022 + 85-02, Fritz Engineering Laboratory Report No. 511.1, Pennsylvania Department of Transportation, Harrisburg, PA.

[58] Yen, B.T., et al. 1991. Fatigue Behavior of Stringer-Floorbeam Connections, *Proceedings of the Eighth International Bridge Conference,* IBC-91-19, Engineers' Society of Western Pennsylvania, pp. 149-155.

Further Reading

[1] Anderson, T.L. 1995. *Fracture Mechanics — Fundamentals and Applications,* 2nd ed., CRC Press, Boca Raton, FL.

[2] Barsom, J.M. and Rolfe, S.T. 1987. *Fracture and Fatigue Control in Structures,* 2nd ed., Prentice-Hall, Englewood Cliffs, NJ.

[3] Broek, D. 1987. *Elementary Fracture Mechanics,* 4th ed., Martinus Nijhoff Publishers, Dordrecht, Netherlands.

[4] Fisher, J.W. 1984. *Fatigue and Fracture in Steel Bridges,* John Wiley & Sons, New York.

[5] Maddox, S.J. 1991. *Fatigue Strength of Welded Structures,* 2nd ed., Abington Publishing, Cambridge, U.K.

25

Underground Pipe

25.1   Introduction 25-1
25.2   External Loads 25-2
      Overburden • Surcharge at Grade • Live Loads • Seismic Loads
25.3   Internal Loads 25-5
      Internal Pressure and Vacuum • Pipe and Contents
25.4   Design Methods 25-5
      General • Flexible Design • Rigid Design
25.5   Joints 25-16
      General • Joint Types • Hydrostatic Testing
25.6   Corrosion Protection 25-18
      Coatings • Cathodic Protection
References 25-20

J. M. Doyle and
S.J. Fang
Sargent & Lundy, LLC
Chicago, IL

25.1   Introduction

Throughout recorded history, works have been constructed for conveying water from one place to another. The Roman aqueducts are often mentioned as examples of great technical achievement; indeed, some of these early structures are still in use today. Although most of the early water carrying structures were open channels, conduits and pipes of various materials were also used in Roman times. It appears, though, that the effectiveness of the early pipes was limited because their materials were weak in tensile capacity. Therefore, the pipes could not carry fluid under any appreciable pressure. Beginning in the 17th century, wood and cast iron were used in water pipe applications in order to carry water under pressure from pumping, which was introduced about the same time. Since then, many materials have evolved for use in pipes. As a general rule, the goals for new pipe material development has been increased tensile strength, reduced weight, and, of course, reduced cost.

    Pipe that is buried underground must sustain other loads besides the internal fluid pressure. That is, it must support the soil overburden, groundwater, loads applied at the ground surface, such as vehicular traffic, and forces induced by seismic motion. Buried pipe is, therefore, a structure as well as a conduit for conveying fluid. That being the case, special design procedures are required to insure that both functions are fulfilled. It is the purpose of this chapter to present techniques that are currently in use for the design of underground pipelines. Such lines are used for public water systems, sewers, drainage facilities, and many industrial processes. Pipe materials to be considered include steel, concrete, and fiberglass reinforced plastic. This selection provides examples of both flexible and rigid behavior. The methodologies presented here can be applied to other materials as well. Design procedures given are, for the most part, based on material contained in U.S. national standards or recommended practices developed by industry organizations. It is our intention to provide an exposition of the essential elements of the various design procedures. No claim is made to total inclusiveness for the methodologies discussed. Readers interested in the full range of refinements and subtleties of any of the approaches are encouraged to consult the cited works. For convenience when comparing references, the notations used in work by others will be maintained here. Attention is focused on large-diameter lines, generally greater than 24 in. Worked sample problems are included to illustrate the material presented.

25.2   External Loads

25.2.1   Overburden

The vertical load that the pipe supports consists of a block of soil extending from the ground surface to the top of the pipe plus (or minus) shear forces along the edges of the block. The shear forces are developed when the soil prism above the pipe or the soil surrounding the prism settle relative to each other. For example, the soil prism above the pipe in an excavated trench would tend to settle relative to the surrounding soil. The shear forces between the backfill and the undisturbed soil would resist the settlement, thus reducing the prism load to be carried by the pipe. For a pipe placed on the ground and covered by a new fill, the effect may be the same or opposite, in which case the load to be supported by the pipe would be greater than the soil prism. The difference in behavior depends on the difference in settlement between the pipe itself and the fill material. Sketches of typical methods of buried pipe installation are shown in Figure 25.1.

Methods developed by Marston and Spangler, and their co-workers, at Iowa State University [28, 29, 34, 35, 36, 39] over a period of about 50 years, are the accepted tools for evaluating overburden loads on buried conduits and are widely used in design practice. The general form of the expression, developed by this group, used to calculate the overburden load carried by the pipe is given as

$$W_c = CwB^2 \tag{25.1}$$

where:

W_c = total load on pipe, per unit of length
C = load coefficient, dependent on type of installation, trench or fill, on the soil type, and on relative rates of settlement of the pipe and surrounding soil
w = unit weight of soil supported by pipe
B = width of trench of outer diameter of pipe

Values for the load coefficient, C, for varying conditions of installation, are given in several standard references (see, e.g., [20]).

The American Water Works Association (AWWA) [21], in its design manual for steel pipe, recommends that the total overburden load on buried steel pipes be assumed equal to a soil prism with width equal to the outer diameter of the pipe and height equal to the cover depth. That is,

$$W_c = wB_ch \tag{25.2}$$

where

B_c = external pipe diameter
h = depth from ground surface to top of pipe

25.2.2   Surcharge at Grade

Besides the direct loads imposed by the soil overburden, underground pipes must also sustain loads applied on the ground surface. Typically, such loads occur as a result of vehicular traffic passing over the route of the pipe. However, they can be caused by static objects placed directly, or nearly so, above the pipe as well.

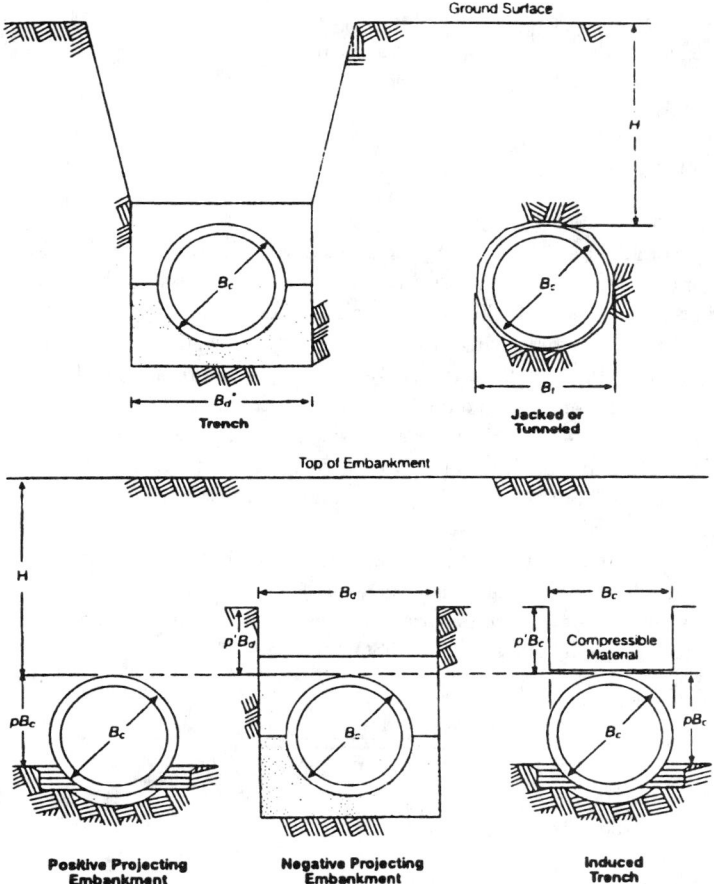

Figure 25.1 Typical underground pipe installations. (Reprinted from *Concrete Pressure Pipe*, M9, by permission. Copyright©1995, American Water Works Association.)

Experimental results, by the Iowa State University researchers and others [33, 37], have confirmed that the load intensity at the pipe depth, due to surface loads, can be predicted on the basis of the theory of elasticity. The effects of an arbitrary spatial distribution of surface load can be obtained by utilizing the well-known Boussinesq solution [41], for a point load on an elastic half space, as an influence function.

Since the Boussinesq solution provides a stress distribution for which magnitudes decay with distance from the load, it follows that the intensity of surface loads decreases with increased depth. Therefore, the consequence of traffic, or other surface loads, on deeply buried pipes is relatively minor. Conversely, surface loads applied over pipes with shallow cover can be quite serious. For this reason, a minimum cover is usually required in any place where vehicular traffic will operate over underground conduits.

Prior to development of present day computational tools, the evaluation of the Boussinesq equations to determine the total load on a buried pipe due to an arbitrary surface load was beyond the capability of most practitioners. For that reason, tables were developed, based on simple surface load distributions, and have been included in most design literature for buried pipe for many years. See, for example, the tables of values in the AWWA Manual M11 [21]. Loading configurations not covered by the previously developed tables can be investigated using available software programs.

Mathcad [30], for example, can be utilized to carry out the analysis necessary to evaluate the effect of arbitrary surface loads on buried structures, including pipes.

25.2.3 Live Loads

The main source of design live loads on buried pipes is wheeled traffic from highway trucks, railroad locomotives, and aircraft. Loads transmitted to buried structures by the standard HS-20 truck loading [1] and the Cooper E-80 railroad loading have been evaluated using the Boussinesq solution and engineering judgment, for varying depths of cover, and are available, in different forms, in several publications (see, e.g., [6, 20]). Due to the wide variation in aircraft wheel loadings, it is usually necessary to evaluate each case separately. FAA Advisory Circular 150/5320-5B provides information on aircraft wheel loads. The load intensity at the depth of the pipe has been reported in numerous references. Simple load intensities for the HS-20 truck loads and for the Cooper E-80 locomotive loads, at varying depths, are given in Tables 25.1 and 25.2, respectively [6]. More comprehensive tables for truck and railroad loads have been published [20, 27]. In general, the intensities given in Tables 25.1 and 25.2 are close to the intensities given in the other tables, though some differences do exist. For examples in this chapter, live loads will be based on the intensities given in Tables 25.1 and 25.2. In case of doubt as to appropriate live load values to use in design of buried pipe, the advice of a geotechnical engineer should be obtained.

TABLE 25.1   HS-20 Live Load

Height of cover, ft	Live load, lb/ft^2
1	1800
2	800
3	600
4	400
5	250
6	200
7	175
8	100
Over 8	Neglect

From American Society for Testing and Materials. 1994. A796. *Standard Practice for Structural Design of Corrugated Steel Pipe, Pipe-Arches and Arches for Storm and Sanitary Sewers and Other Buried Applications.* With permission.

TABLE 25.2   Cooper E-80 Live Load

Height of cover, ft	Live load, lb/ft^2
2	3800
5	2400
8	1600
10	1100
12	800
15	600
20	300
30	100
Over 30	Neglect

From American Society for Testing and Materials. 1994. A796. *Standard Practice for Structural Design of Corrugated Steel Pipe, Pipe-Arches and Arches for Storm and Sanitary Sewers and Other Buried Applications.* With permission.

25.2.4 Seismic Loads

In zones of high seismicity, buried conduits must be designed for the stresses imposed by earthquake ground motions. The American Society of Civil Engineers (ASCE) has developed procedures for evaluation of the magnitude of axial and flexural strains induced in underground lines by seismic motions [24]. The document reflects the research efforts of many of the leading seismic engineers in the country and the methodology is widely used for design of underground conduits of all kinds.

As a general rule, the stresses in pipe walls due to seismic motion–induced strains are quite small and do not adversely affect the design. Since most design codes allow for an increase in allowable stress, or a decrease in load factors, when seismic loads are included in a load combination, buried pipes that are sized to sustain other design loads usually have sufficient strength to resist seismic-imposed stresses.

Consequently, the major consideration to be addressed in design of underground pipe is not strength but excessive relative movement. Unrestrained slip joints in buried pipe may be subject to relative movement, between the two segments meeting at a joint, that exceeds the limit of the joint's capacity to function. For that reason, slip joint pipe must be investigated for maximum relative movement when subject to seismic motion. Types of pipe commonly utilizing slip joints include ductile iron, reinforced and prestressed concrete, and fiberglass reinforced plastic.

25.3 Internal Loads

25.3.1 Internal Pressure and Vacuum

Underground pipe systems operate under varying levels of internal pressure. Gravity sewer lines normally operate under fairly low internal pressure whereas water supply mains and industrial process pipes may be subject to internal pressures of several hundred pounds per square inch. High-pressure pipelines are often designed for a continuous operating pressure and for a short-term transient pressure.

Certain operational events may cause a temporary vacuum in buried conduits. In most cases the duration of application of vacuum loading is extremely short and its effects can usually be examined separately from other live loads. For design, a hydraulic analysis of the system may be used to predict the magnitude and time variation of transients in both the positive and negative internal pressure.

25.3.2 Pipe and Contents

The effects of dead weight of the pipe wall and the fluid carried must be resisted by the structural capacity of the pipe. Neither of these loads contribute significantly to the overall stress state in most circumstances. In practice, loads from these two sources are often neglected in design of steel or plastic pipe, but they are usually included in design of prestressed and reinforced concrete pressure pipe and can be included in design of concrete nonpressure pipe as well. Formulas for determination of pipe wall bending moments and thrust forces, due to self-weight and fluid loads, are available in standard stress analysis references [43]. Since these loads are usually small compared to the overburden, they can be added to the vertical soil loads for simplicity and with conservatism.

25.4 Design Methods

25.4.1 General

The principal structural consideration in design of buried pipe is the ability to support all imposed loads. Other important items include the type of joints to be used and protection against environmental exposure. There are two fundamental approaches to design of buried pipe, based on the pipe's behavior under load [32, 40]. Pipe that undergoes relatively large deformations under its gravity loads, and obtains a large part of its supporting capacity from the passive pressure of the surrounding soil, is referred to as "flexible". As will be observed, the evaluation of the contribution of the soil to pipe strength is difficult due to varying conditions of pie installation. For that reason, prudence in design must be followed. However, as with most design problems, the engineer must, ultimately, balance conservatism with economic considerations.

Pipes with stiffer walls that resist most of the imposed load without much benefit of engagement of passive soil pressure, because deformation under load is restricted, are called "rigid". Steel, both corrugated and plain plate, ductile iron, and fiberglass reinforced plastic pipes are considered flexible; concrete pipe is considered rigid. Different methodologies are employed in assessing the strength of each type.

25.4.2   Flexible Design

Plain Steel

The structural capacity of flexible pipes is evaluated on the basis of resistance to buckling (compressive yield) and vertical diametrical deflection under load. Additionally, for flexible pipes, a nonstructural requirement in the form of a minimum stiffness to ensure that the pipe is not damaged during shipping and handling is normally imposed. In the case of steel pipes designed according to the recommendations of AWWA Manual M11 [21], the following two equations are used to choose pipe wall thickness sufficient to satisfy the handling requirement:

$$t \quad \geq \frac{D}{288} \qquad \text{for diameter up to 54 in.}$$

$$t \quad \geq \frac{D+20}{400} \quad \text{for diameter greater than 54 in.} \tag{25.3}$$

It is of interest to note that for many years, a minimum thickness of $D/200$ was used by pipe designers. In our experience, wall thicknesses meeting this ratio will usually result in designs that also satisfy the strength and deflection criteria discussed below. Tensile stresses due to internal pressure must be limited to a fraction of the tensile yield of the material. AWWA recommends limiting the tensile stress to 50% of yield.

Collapse, or buckling, of flexible pipes is difficult to predict theoretically because of the indeterminate nature of the load pattern. AWWA has published an expression for the determination of capacity of a given pipe to support imposed loads. The equation, given as Equation 6-7 in AWWA Manual M11 [21], incorporates the effects of the passive soil resistance, the buoyant effect of groundwater, and the stiffness of the pipe itself. Allowable buckling pressure is given by:

$$q_a = \left(\frac{1}{FS}\right)\left(32 R_w B' E' \frac{EI}{D^3}\right)^{1/2} \tag{25.4}$$

where

q_a	=	allowable buckling pressure (psi)
FS	=	factor of safety
	=	2.5 for $(12h/D) \geq 2$ and
	=	3.0 for $(12h/D) < 2$
R_w	=	water buoyancy factor
	=	1-0.33 (h_w/h)
h	=	height of ground surface above top of pipe (ft)
h_w	=	height of groundwater surface above top of pipe (ft)
D	=	diameter of pipe (in.)
B'	=	coefficient of elastic support
	=	$\frac{1}{1+4e^{-0.065h}}$

E' = modulus of soil reaction (psi)
E = modulus of elasticity of pipe wall (psi)
I = moment of inertia per inch length of pipe wall (in.^{3})

In case vacuum load and surface live load are both included in the design conditions, AWWA recommends that separate load combinations be considered for each. That is because vacuum loads usually occur only for a short time and the probability of vacuum and maximum surface load occurring simultaneously is very small. In particular the following two load cases should be considered. For traffic live load:

$$q_a \geq \gamma_w h_w + R_w \frac{W_c}{D} + \frac{W_L}{D} \tag{25.5}$$

where

γ_w = specific weight of water (0.0361 lb/in.^{3})
W_L = live load on pipe (lb/in. length of pipe)
W_c = vertical soil load on pipe (lb/in. length of pipe)

For vacuum load:

$$q_a \geq \gamma_w h_w + R_w \frac{W_c}{D} + P_v \tag{25.6}$$

where

P_v = internal vacuum pressure (psi)

Deflection is determined by the Spangler formula:

$$\Delta y = D_l \left(\frac{K W_c r^3}{EI + 0.061 E' r^3} \right) \tag{25.7}$$

where

Δy = deflection of pipe (in.)
D_l = deflection lag factor (1.0 to 1.5)
K = bedding constant (0.1)
r = pipe radius (in.)

This form of the deflection equation was obtained by ordinary bending theory of a ring subject to an assumed pattern of applied vertical load, width of vertical reaction, and distribution of horizontal passive pressure [38, 42] and has been used in pipe design for over 50 years. According to the formula, deflection is limited by the stiffness of the pipe wall itself and by the effect of the passive pressure. It is significant to note that in the sizes of steel pipes often encountered, the ratio of the two components of resistance is on the order of 1:20, with the pipe wall stiffness being the smaller. Therefore, it is obvious that the passive resistance, which is closely related to the type of backfill and its degree of compaction, is the dominant influence on the vertical deflection of flexible pipes. That being the case, it becomes apparent that increasing the strength of a flexible pipe will probably be an inefficient way to properly limit deflection of underground pipe in most cases. The pipe installation must be completed as specified in order for this to be achieved.

Efforts to quantify the modulus of soil reaction, E', have continued since the initial development of the deflection equation. Suggested values are published in numerous references, including AWWA Manual M11 [21]. Values given there range from 200 to 3000 psi. The values depend on the type and level of compaction of the surrounding soil. Since pipe designers often have little control over the installation of pipe, historically, a value of E' in the range of 700 to 1000 psi has been assumed representative of average installations for estimating deflection at time of design.

In a recent work, engineers at the U.S. Bureau of Reclamation addressed the question of deflection of flexible pipe [27]. Their work, which is based on the wide experience of the Bureau of Reclamation in construction of all kinds of underground pipes, discusses appropriate values of E' based on not only the backfill and compaction used, but also the native soil. In addition to the soil modulus values, the authors also give a modified form of the deflection equation that includes a factor to account for long-term deflection, T_f (which replaces the factor D_l in Equation 25.7), and an additional multiplier on the soil modulus, called a design factor, F_d, with values ranging from 0.3 to 1.0. The combined effect of these two changes is, generally, to predict larger deflections than with the original Spangler equation. The revised equation becomes:

$$\Delta y = T_f \left(\frac{K W_c r^3}{EI + 0.061 F_d E' r^3} \right) \tag{25.8}$$

where
T_f =   time lag factor
F_d =   design factor

Values for Spangler's deflection lag factor, D_l, of 1.0 to 1.5 are recommended; designers usually use the 1.5 value for conservatism. Since the minimum recommended value for T_f is 1.5, the deflections by the modified equation will be higher. Values of the design factor, F_d, are presented for three cases, A, B, and C. The value for case A is 1.0; case B values, which are recommended for design, vary from 0.5 to 1.0; and case C values, which are recommended for designs in which deflection is critical, range from 0.3 to 0.75. In all cases the values of F_d increase with quality and level of compaction of the backfill.

It follows that control of bedding and backfill of flexible pipes during construction is critical to their performance. The required passive pressure can be developed only in high-quality fill material, compacted to the proper density. The material surrounding the pipe and extending above the pipe for at least 12 in. should be a well-graded granular stone. Coarse-grained material provides much higher passive resistance and, therefore, limits pipe deflection, in flexible pipe systems, more than fine-grained soil types. Compaction in the lower levels of the pipe is critical. Hand tampers or similar equipment are necessary to ensure that adequate density is obtained in the region below the lower haunches of the pipe. Historically, failure to achieve the proper level of compaction in this area of difficult accessibility has been identified as a major contributing cause to excessive deformations in flexible pipe construction.

It is common practice to limit the final vertical deflection of unlined pipes to less than 5% of the diameter. Deflection of pipes with cement mortar coatings should be limited to 2% of the diameter. Field observations of steel pipes in service indicate that once the deflection reaches 20% of the diameter, collapse is imminent.

EXAMPLE 25.1:

A 96-in.-diameter steel pipe with a 1/2-in. wall is installed with its top 15 ft below the ground surface. The local water table is located 7 ft below the surface. Assume that the soil has a modulus of reaction, E', of 1000 psi, and that it has a unit weight of 120 pcf.

1. Verify that the pipe will satisfy the buckling and deflection criteria given in AWWA Manual M11 [21].
2. Determine the amount of vacuum load that can be supported by the pipe.

   Solution

1. The weight of soil bearing on the pipe is calculated from the prism of soil from the top of the pipe to the ground surface:

$$\begin{aligned} W_c &= \gamma_s h D \\ &= 120 \times 15 \times (96/12) = 14,400 \text{ lb/ft} \end{aligned}$$

Determine the h/D ratio to obtain the appropriate factor of safety:

$$\frac{h}{D} = \frac{15}{8} = 1.875 < 2; \qquad \text{therefore} \ \ FS = 3$$

The groundwater surface is $15 - 7 = 8$ ft above the top of the pie. The water buoyancy factor (R_w) and the coefficient of elastic support (E') are calculated on the basis of the depth of cover and groundwater:

$$
\begin{aligned}
h_w &= 8 \, ft \\
R_w &= 1 - 0.33 \frac{h_w}{h} = 1 - 0.33 \times \frac{8}{15} = 0.824 \\
B' &= \frac{1}{1 + 4e^{-0.065 \times 15}} = 0.399
\end{aligned}
$$

The modulus of elasticity for steel is 29×10^6 psi; the moment of inertia per inch length of pipe is

$$I = \frac{t^3}{12} = \frac{0.5^3}{12} = 0.0104 \ \text{in.}^3; \qquad \text{hence the product} \ EI = 302{,}083 \ \text{in.-lb}$$

Therefore, by Equation 25.4, the allowable buckling pressure is

$$q_a = \left(\frac{1}{3}\right)\left(32 \times 0.824 \times 0.399 \times 1{,}000 \times \frac{302{,}083}{96^3}\right)^{1/2} = 19.968 \ \text{psi}$$

The total applied load intensity, Q, is given by

$$Q = \gamma_w h_w + R_w \frac{W_c}{D} + \frac{W_L}{D} = 0.0361 \times 96 + 0.824 \times \frac{14{,}400}{12 \times 96} + 0 = 13.766 \ \text{psi}$$

Since $Q < q_a$, the pipe is safe against buckling. Check deflection:

$$\Delta y = 1.5 \left(\frac{0.1 \times \frac{14{,}400}{12} \times 48^3}{302{,}083 + 0.061 \times 1{,}000 \times 48^3} \right) = 2.824 \ \text{in.}$$

The calculated deflection is approximately 3% of the diameter, less than the 5% usually specified as the limit for unlined pipe.

2. The vacuum pressure that can be supported within the buckling capacity of the pipe is the difference between the calculated critical buckling capacity, q_a, and the applied load intensity, Q:

$$P_v = q_a - Q = 19.968 - 13.766 = 6.202 \ \text{psi}$$

Corrugated Steel

Corrugated steel has the advantage of greater flexural strength per unit weight of material than plain steel, and has been widely used in surface drainage systems and to a lesser extent in process water systems. In this form, the pipe is assembled from corrugated sheets, rolled to radius and bolted or riveted together.

Corrugated steel pipes can be designed according to ASTM standard practice A796 (American Society for Testing and Materials). The practice covers both curve and tangent ("sinusoidal") walls and smooth walls with helical ribs of rectangular section at regular intervals for increased strength.

As with plain steel pipe, this design procedure requires a minimum stiffness in the pipe wall for shipping and handling. To make a quantitative evaluation of the degree of stiffness, a flexibility factor, defined for all wall configurations as

$$FF = \frac{D^2}{EI} \tag{25.9}$$

where

FF = flexibility factor (in.-lb^{-1})
D = pipe diameter (in.)
E = modulus of elasticity (psi)
I = moment of inertia of wall cross-section per inch (in.^{3})

is subject to limits depending on the corrugation configuration and the type of installation. For example, in configurations of sinusoidal corrugations, specified in ASTM A760 and A761 [4, 5], values of the flexibility factor are restricted to 0.020 to 0.060.

The phenomenon of buckling of buried corrugated pipes has been investigated, through prototype testing, by Watkins [3]. Design curves utilizing the results of that research were originally published in an American Iron and Steel Institute (AISI) design manual [3] and have been continued into the current edition of the book. The curves provide buckling loads for corrugated steel–walled pipes as a function of diameter-to-radius of gyration ratio. The design equations given in ASTM A796 [6] are of the same general form as the design curves developed by AISI. That is, there are three ranges of behavior—elastic buckling, inelastic buckling, and yield—and the dependence of the expressions on the independent variable, D/r, is the same in the two regimes of the formulas in both documents. The principal difference between the two approaches is the inclusion of an explicit dependence on soil stiffness in the ASTM A796 equations. The AISI formulas, on the other hand, account for soil stiffness by reduction in applied load for well-compacted backfills.

The applicable formulas for critical buckling stress, as given in ASTM A796 [6], and their applicable ranges of diameter-to-radius of gyration ratio are given below:

$$f_c = f_u - \frac{f_u^2}{48E}\left(\frac{kD}{r}\right)^2 \quad \text{for} \quad \frac{kD}{r} \le \sqrt{\frac{24E}{f_u}}$$

$$f_c = \frac{12E}{\left(\frac{kD}{r}\right)^2} \quad \text{for} \quad \frac{kD}{r} > \sqrt{\frac{24E}{f_u}} \tag{25.10}$$

subject to the provision that:

$$f_c \le f_y \tag{25.11}$$

where

f_c = critical buckling stress (psi)
f_y = specified minimum yield stress (psi)
f_u = specified minimum ultimate stress (psi)
k = soil stiffness factor = 0.22 for material at 90% density
E = modulus of elasticity (psi)
D = pipe diameter (in.)
r = radius of gyration of pipe wall (in.)

The buckling formulas in ASTM A796 [6] also are given in the AASHTO (American Association of State Highway and Transportation Officials) specification for design of highway bridges [1].

The corresponding equations provided in the AISI handbook [3] are

$$f_c = 40,000 - 0.081 \left(\frac{D}{r} \right)^2 \qquad \text{for } 294 < \frac{D}{r} < 500$$

$$f_c = \frac{4.93 \times 10^9}{\left(\frac{D}{r} \right)^2} \qquad \text{for } \frac{D}{r} > 500 \qquad (25.12)$$

with, again, the provision that the critical buckling stress cannot exceed the yield stress, f_y.

A comparison of the two sets of formulas can be made to determine the maximum variation. The soil parameter, k, obviously affects the results. In ASTM A796 [6], a value of $k = 0.22$ is recommended for good fill material compacted to 90% of standard density; no suggestions are provided for other backfill conditions. The AISI expressions for critical buckling stress (Equation 25.12) do not contain any dependence on the degree of compaction of the surrounding soil. However, the handbook does recommend load factors, which multiply the applied loads, that are related to the degree of compaction. For example, the recommended load factor for 90% compaction is 0.75. Use of a load factor of 0.75 has the same effect as increasing the allowable stress by 1.33. If the results from the ASTM and AISI equations are compared on that basis, the values are within 10%. On the other hand, if a load factor of 1.0, which corresponds to a density of only 80% standard, is used, the soil stiffness, k, must be increased to 0.26 for the two sets of formulas to give approximately the same results. Clearly, use of the higher value for k results in a slightly more conservative design, and that may be desirable, in view of the normally unknown character of the actual installed backfill.

In either case, the appropriate wall cross-section must be selected to satisfy

$$\frac{W_c}{2A} \le \frac{f_c}{(SF)} \qquad (25.13)$$

where

A = cross-section area of wall per unit length

SF = safety factor = 2

W_c = vertical load per unit length of pipe (*Note:* W_c must be multiplied by the appropriate load factor if the AISI equations are used)

It is of interest to note the D/r ratios that form the transition between elastic and inelastic and between buckling and yield behavior in the ASTM A796 equations. For pipe meeting ASTM A760, maximum thickness of 0.168 in., the specified minimum yield and ultimate stress are 33 ksi and 45 ksi, respectively. For those values, elastic buckling controls design for D/r ratios greater than 478 and yield controls for D/r ratios less than 350, for k equal to 0.26. These values correspond fairly closely with the AISI limits of 500 and 294, respectively.

EXAMPLE 25.2:

Consider the pipe in Example 25.1. Determine the minimum wall thickness of 3 in. x 1 in. corrugated pipe that will satisfy the buckling expressions of Equation 25.10 and the handling requirement of Equation 25.9 with FF limited to a maximum value of 0.033. Assume a value of k of 0.26. Also, the minimum specified yield (f_y) and ultimate (f_u) stresses for the material are 33,000 and 45,000 psi, respectively.

Solution   The radius of gyration for all thicknesses in the 3 × 1 series range from 0.3410 to 0.3499 in. (see Table 25.3).

Therefore, in the calculations use

$$r = 0.34 \text{ in.}$$

It follows that

$$\frac{kD}{r} = \frac{0.26 \times 96}{0.34} = 52.8 \qquad \sqrt{\frac{24E}{f_u}} = \sqrt{\frac{24 \times 29 \times 10^6}{45,000}} = 124.4$$

Since $\frac{kD}{r} < \sqrt{\frac{24E}{f_u}}$, the first of Equation 25.10 must be used:

$$
\begin{aligned}
f_c &= f_u - \frac{f_u^2}{48E}\left(\frac{kD}{r}\right)^2 = 45,000 - \frac{45,000^2}{48 \times 29 \times 10^6}\left(\frac{0.26 \times 96}{0.34}\right)^2 \\
&= 37,160 \text{ psi} > f_y
\end{aligned}
$$

Since the calculated buckling stress exceeds the yield, the critical stress is the yield stress, 33,000 psi. The required wall area per foot of length can be determined by rearrangement of Equation 25.13:

$$
A = \frac{W_c}{2\frac{f_c}{(SF)}} = \frac{14,400}{2 \times \frac{33,000}{2}} = 0.437 \text{ in.}^2
$$

The lightest section, thickness of 0.052 in., will satisfy the stress requirement. Check the handling requirement. The moment of inertia per inch of wall is 6.892×10^{-3} in.^{3}:

$$
FF = \frac{D^2}{EI} = \frac{96^2}{(29 \times 10^6) \times (6.892 \times 10^{-3})} = 0.046 > 0.033
$$

Since the flexibility factor is too large, try the next section in the series, the 0.064-in. thickness, and $I = 8.658 \times 10^{-3}$ in.^{3}:

$$
FF = \frac{D^2}{EI} = \frac{96^2}{(29 \times 10^6) \times (8.658 \times 10^{-3})} = 0.037 > 0.033
$$

Since the flexibility is still too great, the next section in the series, with a thickness of 0.079 in. and flexibility factor of 0.029 is chosen. This example demonstrates a condition that occurs quite often in the design of flexible pipes; if the handling and installation minimum stiffness requirements are met, the strength requirements are automatically taken care of.

TABLE 25.3   Sectional Properties for 3 × 1 in. Corrugated Sheets

Thickness in.	Area in.^{2}/ft	Moment of inertia in.$^3 \times 10^{-3}$	Radius of gyration in.
0.052	0.711	6.892	0.3410
0.064	0.890	8.658	0.3417
0.079	1.113	10.883	0.3427
0.109	1.560	15.458	0.3448
0.138	2.008	20.175	0.3472
0.168	2.458	25.083	0.3499

From American Society for Testing and Materials. 1994. A796. *Standard Practice for Structural Design of Corrugated Steel Pipe, Pipe-Arches and Arches for Storm and Sanitary Sewers and Other Buried Applications.* With permission.

Fiberglass Reinforced Plastic

Fiberglass reinforced plastic (FRP) pipe is fabricated by winding glass strands into a matrix of organic resin on a mandrel of the desired diameter. A variation on the fiberglass-resin matrix utilizes cement of polymer mortar incorporated into the structure to add stiffness and reduce cost of materials. ASTM standards D3262 [9], D3517 [10], and D3754 [11], and AWWA standard C950 [19] provide requirements for manufacture of both the fiberglass-resin and the mortar pipe in diameters up to 144 in. Structural strength and rigidity against external loads for this type of pipe are established by load tests performed as specified by ASTM D2412 [8]. In the load test, equal and opposite concentrated loads are applied on opposite ends of a diameter. Load deflection data are obtained from which stiffness and related buckling strength of the pipe can be determined.

Each of the mentioned pipe specifications provides for levels of pipe stiffness (PS) of 9, 18, 36, and 72 psi. These values represent applied force per unit length of pipe divided by deflection. Use of the pipe stiffness and the formula for deflection of a point-loaded circular ring allows determination of the product, EI, of the composite pipe wall. In FRP construction, the modulus of elasticity (E) depends on several variables: the moduli of the resin and the glass reinforcement, the relative amounts of glass 20 and resin, and the angle of the filament winding. For that reason, it is convenient to utilize the experimentally determined overall pipe stiffness in design rather than to base calculations on the composite modulus of elasticity of the material.

In particular, the buckling formula (Equation 25.4) and Spangler's equation for deflection (Equation 25.7) can be recast in terms of the pipe stiffness, as shown in the following steps. The formula for deflection of a concentrated loaded pipe can be rearranged to provide an expression for EI:

$$EI = 0.149\frac{F}{\Delta}r^3 \tag{25.14}$$

where
F = concentrated load per inch of pipe (lb/in.)
D = pipe deflection (in.)
r = pipe radius (in.)

Since the pipe stiffness is defined as

$$PS = \frac{F}{\Delta} \tag{25.15}$$

the relationship between EI and PS becomes

$$EI = 0.149(PS)r^3 \tag{25.16}$$

The allowable buckling stress expression, therefore, can be rewritten:

$$q_a = \left(\frac{1}{FS}\right)\left(0.596R_w B' E' PS\right)^{1/2} \tag{25.17}$$

and the deflection formula is

$$\Delta y = D_l \left(\frac{KW_c}{0.149PS + 0.061E'}\right) \tag{25.18}$$

For underground installations, many fiberglass pipe manufacturers recommend a minimum pipe stiffness of 36 psi in order to ensure sufficient stiffness to perform backfilling properly. Deflections are normally restricted to 5% of the diameter. In contrast to design of steel pipe, it is normal practice to consider the bending stresses induced in the wall by deflection of the pipe. Methods for evaluating these stresses in combination with the stresses due to internal pressure have been developed by a committee of AWWA and will be included in one chapter of a manual on design of fiberglass reinforced water pipe, scheduled for publication in the near future [22].

25.4.3 Rigid Design

Concrete Pressure Pipe

Concrete pipe can be used for both pressure and nonpressure applications. It offers the advantage of being corrosion resistant in conditions where steel might be attacked, and in some instances it may be a more cost-effective solution than steel or plastic.

When concrete pipe is used in high-pressure systems, prestressed concrete pipe is the type most often selected. The pipe is manufactured in diameters from 24 to 144 in. and is fabricated with walls from 4 to 12 in. thick. A steel cylinder, usually of 16-gal thickness, is embedded in the wall for leak protection. The outer surface of the wall is wrapped with high-strength wire under tensile stresses in the 175–200 ksi range. The prestressing places the concrete wall into compression of sufficient magnitude so that it will not be fully relieved under design internal pressure loadings. Finally, a coat of sand-cement mortar is applied over the prestressing wires to provide corrosion protection. A comprehensive design procedure for this type of pipe is contained in AWWA C304 [17]. Prestressed pipe is normally designed only by pipe manufacturers. The design provisions meet AWWA C304 and the actual pipe is fabricated according to AWWA C301 [14]. Pipe purchasers must indicate the design pressures, including transients, installation conditions, and surface loads.

Reinforced concrete pipe can be designed to sustain internal pressure loads, but the maximum pressures that can be carried are significantly less than with prestressed pipe and its use in such applications is limited. AWWA C300, C302, and C303 [13, 15, 16] are all specifications covering the design and fabrication of reinforced concrete pressure pipe in differing configurations of reinforcing and with or without the steel cylinder pressure boundary. Design procedures for all three specifications are presented in AWWA Manual M9 [20] and the interested reader is encouraged to review that manual for details. As with prestressed pipe, the pipe specifier usually supplies only the performance attributes and the pipe fabricator performs the design to meet the appropriate specification.

Concrete Nonpressure Pipe

ASTM C76 [7] contains specification requirements for reinforced concrete pipe not intended for pressure applications. Five classes of pipe, classes I–V, respectively, representing five levels of structural strength, are specified. The strength is characterized by the concentrated load required to cause a crack of 0.01 in. width and the ultimate concentrated load. Load values are determined experimentally by the three-edge-bearing test. The test simulates concentrated loads applied at opposite ends of a pipe diameter. These loads are referred to as D-loads ($D_{0.01}$ and D_{ult}): the concentrated force per unit length of pipe per unit length of diameter necessary to cause either the 10-mil-width crack or ultimate failure of the pipe. D-load values for the five pipe classes included in ASTM C76 are shown in Table 25.4.

In determination of the strength required to resist external loads, the total pipe load is estimated by standard methods. Bedding factors, based on the type of installation, the soil type, and its level of compaction, have been developed by the American Concrete Pipe Association [2]. These factors represent the ratio of the maximum bending moment due to a concentrated load to the moment caused by the actual live and dead load of the same magnitude as the concentrated load.

TABLE 25.4 D-Loads for ASTM C76 Concrete Pipe

Pipe class	$D_{0.01}$ load	D_{ult} load
I	800	1200
II	1000	1500
III	1350	2000
IV	2000	3000
V	3000	3750

From American Society for Testing and Materials. 1994. C76. *Standard Specification for Reinforced Concrete Culvert, Storm Drain, and Sewer Pipe.* With permission.

ACPA has defined four standard installation types, for which relevant information is shown in Tables 25.5 and 25.6. Bedding factors for embankment installations are given in Table 25.7. Other bedding factors, for trench installations and for live load effects, have also been obtained by ACPA but are not reproduced here. It is noted that ACPA recommends using the dead load factor for live load contributions as well, if the tabulated live load factor is larger than the dead load factor. The calculation methodology used to obtain the various factors is described in the ACPA design data [2].

TABLE 25.5 Equivalent USCS and AASHTO Soil Classifications for SIDD Soil Designations (ACPA)

	Representative Soil Types	
SIDD[a] soil	USCS	AASHTO
Gravelly sand (category I)	SW, SP, GW, GP	A1, A3
Sandy silt (category II)	GM, SM, ML Also GC, SC with less then 20% passing #200 sieve	A2, A4
Silty clay (category III)	CL, MH, GC, SC CH	A5, A6 A7

[a] Standard Installations Direct Design, ACPA
From American Concrete Pipe Association. 1995. *Design Data 40. Standard Installations and Bedding Factors for the Indirect Design Method.* With permission.

TABLE 25.6 Standard Embankment Installation Soils and Minimum Compaction Requirements (ACPA)

Installation type	Haunch and outer bedding	Lower side
Type 1	95% Category I	90% Category I, 95% category II, or 100% category III
Type 2	90% Category I or 95% category II	85% Category I, 90% category II, or 95% category III
Type 3	85% Category I, 90% category II, or 95% category III	85% Category I, 90% category II, or 95% category III
Type 4	No compaction required, except if category III, use 85% category III	No compaction required except if category III, use 85% category III

Note: Bedding thickness for all types: $D_o/24$ minimum, not less than 3 in. If rock foundation, use $D_o/12$ minimum, not less than 6 in. Compaction is standard Proctor.
From American Concrete Pipe Association. 1995. *Design Data 40. Standard Installations and Bedding Factors for the Indirect Design Method.* With permission.

TABLE 25.7 Bedding Factors, Embankment Condition, B_{fe} (ACPA)

Pipe diameter, (in.)	Standard installation			
	Type 2	Type 1	Type 3	Type 4
12	3.2	4.4	2.5	1.7
24	3.0	4.2	2.4	1.7
36	2.9	4.0	2.3	1.7
72	2.8	3.8	2.2	1.7
144	2.8	3.6	2.2	1.7

From American Concrete Pipe Association. 1995. *Design Data 40. Standard Installations and Bedding Factors for the Indirect Design Method.* With permission.

Use of the appropriate bedding factor allows the conversion of the actual load to an equivalent point load. Comparison of that equivalent load with standard D-loads is used to establish the appropriate class of pipe with sufficient capacity to support the design loads. Normal procedure is to utilize a design factor of safety of 1.0 against the D-load required to cause a 0.01-in. crack. Details of the procedure are illustrated in the following example problem for embankment installation. The design procedure is similar for trench installation.

EXAMPLE 25.3:

Consider a 48-in.-diameter reinforced concrete pipe to be installed beneath a railroad for surface drainage. The pipe is to be installed in an embankment with a depth of cover of 5 ft. For the purpose of this example, assume that the overburden load is equal to the prism of soil above the pipe. Soil unit weight is 120 pcf and the backfill conditions are such that a standard installation type 3 exists.

Solution   For a 48-in. pipe, the wall thickness of a pipe meeting ASTM C76 is 5 in. The soil dead weight is given by

$$W_E = whB_c = 120 \times 5 \times \frac{48 + 10}{12} = 2900 \text{ lb/ft}$$

The live load intensity is obtained from Table 25.2:

$$W_L = w_{LL}B_c = 2400 \times \left(\frac{48 + 10}{12}\right) = 11{,}600 \text{ lb/ft}$$

The total overburden plus live load is

$$W_E + W_L = 2{,}900 + 11{,}600 = 14{,}500 \text{ lb/ft}$$

From Table 25.7, the bedding factor, B_{fe}, is found to be 2.2; the live load bedding factor (not tabulated here) is also 2.2 for this installation. Therefore, use a bedding factor of 2.2 for the total load:

$$\frac{\text{Total load}}{\text{Bedding factor}} = \frac{14{,}500}{2.2} = 6{,}591$$

To obtain required D-load, divide this result by the pipe diameter:

$$D_{0.01} \text{ required} = \frac{6591}{4} = 1648$$

Using class IV pipe, $D_{0.01} = 2000$, $D_{\text{ult}} = 3000$.

25.5   Joints

25.5.1   General

In order to form a continuous conduit from the individual pipe sections, it is necessary to connect the sections together in such a way that the pressure-containing and load-resisting capability is preserved in the completed assembly. Each type of pipe discussed previously utilizes special types of joints as explained in the following.

25.5.2   Joint Types

Plain Steel

In plain steel plate pipe, the individual pipe sections are fabricated from plate, rolled to the proper radius, and welded together. Joints, in fabricated sections, are either continuous helical or longitudinal. When installed, the sections are welded together using either bell-and-spigot or butt joints.

In plain steel pipe, full penetration butt welds are used extensively for field joints. In water works construction, welding of pipelines is covered by AWWA C206 [12]. That standard requires that welding procedures and welding operators be prequalified before use on a job. In addition, tolerances on fit-up are specified and inspection requirements are set out. Strict adherence to project specifications is necessary to guarantee that the desired continuity is obtained at the junction.

Although connecting pipe segments by use of full-penetration butt welds has enjoyed wide acceptance in pipeline construction, bell-and-spigot joints, fillet welded, may also be used. These joints require only fillet welds and are generally considered to be less expensive to install than the full-penetration butt weld. However, due to the inherent eccentricity in such joints, a potential for failure exists under certain temperature conditions when longitudinal tensile stresses are developed. Some failures of welded bell-and-spigot joints were reported in the technical literature a few years ago [26, 31]. Since that time, requirements for welding of bell-and-spigot joints in steel pipe in AWWA C206 [12] have been revised to minimize the potential for failure in this type of joint.

Corrugated Steel

The field joints used in corrugated steel are usually made by bolting, either in lap joints or with coupling bands that fit over two adjacent sections. In most cases, gaskets should be used at joints to provide leak tightness.

Fiberglass Reinforced Plastic

Several types of joints are used in fiberglass pipe. Coupling or bell-and-spigot joints with O-ring gaskets (see Figure 25.2) and mechanical couplings, for unrestrained joints, are specified by the ASTM fiberglass pipe specifications mentioned previously. These joints can be used with restraining devices, such as tie rods, if necessary. In addition, continuous hand lay-up joints consisting of alternating layers of glass fabric and resin or adhesive-bonded bell-and-spigot joints are used for joints that must resist longitudinal force as well as contain the pressure exerted by the fluid carried.

Figure 25.2   Bell-and-spigot and coupling joints for fiberglass pipe. (From American Society for Testing and Materials. 1991. D3517. *Standard Specification for "Fiberglass" Glass-Fiber-Reinforced Thermosetting-Resin Pressure Pipe.* With permission.)

Prestressed Concrete

In straight runs of prestressed concrete pressure pipe, the most common joint type is the bell-and-spigot slip-on joint with a rubber O-ring gasket (see Figure 25.3). When making the joint, care should be used to ensure that the gasket is in its proper place and that the mating ends are properly located with respect to each other. The exterior of the joint should be filled with flowable sand-cement grout, contained by a suitable appliance. Grouting of the inside joint gap may be required, depending on water chemistry. When it is, the grouting should be completed after the backfill is compacted. The interior surface of each joint must be smoothed to allow unrestricted flow. When axial tension forces must be transmitted across a joint, locking variations of the basic slip-on joint are available.

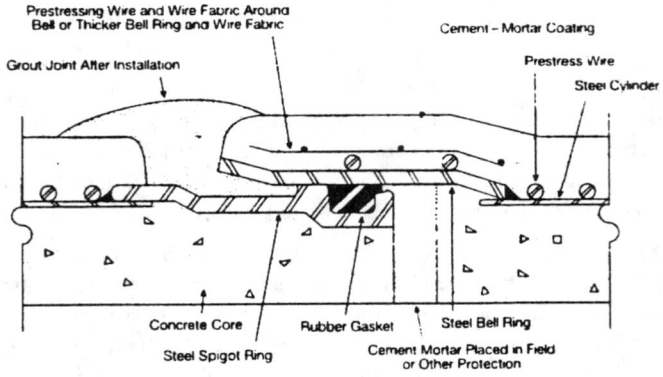

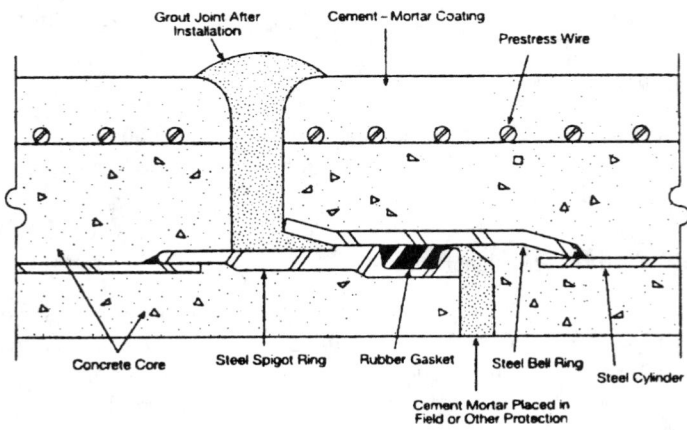

Figure 25.3   Prestressed concrete joint details. (Reprinted from *Concrete Pressure Pipe*, M9, by permission. Copyright©1995, American Water Works Association.)

Reinforced Concrete

Typical joints for reinforced concrete pressure pipe are shown in Figure 25.4. Joints for concrete nonpressure pipe are similar to the concrete-only joints in Figure 25.4. In some cases, gaskets are not used in nonpressure pipe.

25.5.3 Hydrostatic Testing

A field hydrostatic test is usually performed to verify that all joints are watertight. Test pressure should exceed the maximum design pressure, including transients, by at least 25%. Leakage through welded joints should be virtually nonexistent. It is common to allow a slight leakage rate for O-ring gasketed joints (AWWA, C600).

25.6   Corrosion Protection

There are three environmental agents that exert strong influence on corrosion of the pipe wall material in buried installations. These are the water, or other fluid, carried, the soil in contact with the pipe, and the groundwater. In the case of certain process water systems, such as power

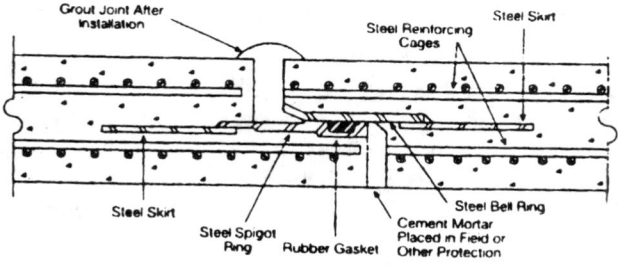

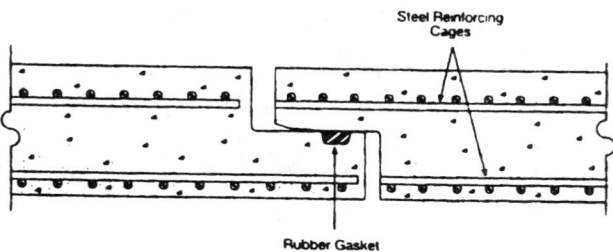

Figure 25.4   Joint details for reinforced concrete pipe. (Reprinted from *Concrete Pressure Pipe*, M9, by permission. Copyright©1995, American Water Works Association.)

plant condenser cooling systems, the water may be circulated continuously within a closed loop using cooling towers, lakes, or other means of exhausting heat. When closed systems are used, even in fresh water environments, the concentrations of certain compounds in the water may increase and cause elevated corrosion rates in steel pipes. Chlorides are generally believed to be the most aggressive compounds, normally found in water sources, in regard to corrosion of steel. Chlorides can also be harmful to concrete pipes, posing threats to the concrete itself and to steel reinforcing. Sulfates are not usually associated with steel corrosion, but they can be detrimental to concrete.

Once-through systems, on the other hand, are usually less corrosive for steel pipes and less harmful to concrete than the closed-cycle systems. When brackish water is used for cooling, positive steps must be taken to ensure that corrosion is controlled.

While the process water carried may promote corrosion or other damage on the inside of the pipe, the outside surface may be attacked by the surrounding soil, the groundwater, or both. Soils with low electrical resistivity may help advance corrosion in steel. Soils that have sulfate compounds above certain critical levels can cause damage to concrete pipe. The groundwater can have the same effects on the exterior of the pipe as the process fluid has on the inside. Specifically, groundwater with high chloride or sulfate contents may be harmful to the pipe material.

Because of the wide range of possibilities for the existence of detrimental chemical action, it is essential that the nature of the external and internal environments of the pipe be evaluated in the design process. Chemical analyses of the process water and the groundwater are essential. Also, chemical analysis of the soil and resistivity survey results must be available in order to make the best choice of pipe system to withstand the exposure throughout the design life of the facility.

25.6.1   Coatings

Coatings can be used to inhibit corrosion or other forms of deterioration in both concrete and steel pipes. Type and extent of coatings depends on the service environment. Steel pipes are almost always coated externally. Coal tar enamel and wrapping has been used successfully in the U.S. for decades. Epoxies and urethanes, among others, have become popular in more recent times.

Cement mortar coatings may be applied to both the interior and exterior of steel pipes. This type of coating offers several advantages and has an extensive record of satisfactory service. When exterior coating of prestressed concrete pipes is desired, certain epoxies are acceptable. Because prevention of corrosion in the prestressing wires is so important, pipe designers sometimes specify an additional coating to supplement the protection furnished by the cement mortar coating.

25.6.2 Cathodic Protection

Protection against corrosion may, in certain circumstances, require a cathodic protection system. For example, cathodic protection has proven to be very successful in providing leak-free high-pressure oil and natural gas pipelines throughout the U.S. Power plant sites have widely dispersed grounding systems, which can cause unpredictable stray currents that may promote steel corrosion. Cathodic protection must be designed by competent engineers based on information regarding the extent of buried facilities, the soil resistivity measurements, and the plant grounding system. Electrical continuity should be provided on prestressed concrete pipe if present or future installation of cathodic protection is a possibility.

References

[1] American Association of State Highway and Transportation Officials (AASHTO). 1992. *Standard Specifications for Highway Bridges.* 15th ed.

[2] American Concrete Pipe Association (ACPA). 1995. Design Data 40. *Standard Installations and Bedding Factors for the Indirect Design Method.*

[3] American Iron and Steel Institute (AISI). 1977. *Handbook of Steel Drainage & Highway Construction Products.*

[4] American Society for Testing and Materials (ASTM). 1993. A760. *Standard Specification for Corrugated Steel Pipe, Metallic Coated for Sewers and Drains.*

[5] American Society for Testing and Materials (ASTM). 1990. A761. *Standard Specification for Corrugated Steel Structural Plate, Zinc-Coated, for Field-Bolted Pipe, Pipe-Arches, and Arches.*

[6] American Society for Testing and Materials (ASTM). 1994. A796. *Standard Practice for Structural Design of Corrugated Steel Pipe, Pipe-Arches and Arches for Storm and Sanitary Sewers and Other Buried Applications.*

[7] American Society for Testing and Materials (ASTM). 1994. C76. *Standard Specification for Reinforced Concrete Culvert, Storm Drain, and Sewer Pipe.*

[8] American Society for Testing and Materials (ASTM). 1993. C2412. *Standard Test Method for Determination of External Loading Characteristics of Plastic Pipe by Parallel-Plate Loading.*

[9] American Society for Testing and Materials (ASTM). 1993. D3262. *Standard Specification for "Fiberglass" (Glass-Fiber-Reinforced Thermosetting-Resin Sewer Pipe).*

[10] American Society for Testing and Materials (ASTM). 1991. D3517. *Standard Specification for "Fiberglass" (Glass-Fiber-Reinforced Thermosetting-Resin Pressure Pipe).*

[11] American Society for Testing and Materials (ASTM). 1991. D3754. *Standard Specification for "Fiberglass" (Glass-Fiber-Reinforced Thermosetting-Resin Sewer and Industrial Pressure Pipe).*

[12] American Water Works Association (AWWA). 1991. C206. *Standard for Field Welding of Steel Water Pipe.*

[13] American Water Works Association (AWWA). 1989. C300. *Standard for Reinforced Concrete Pressure Pipe, Steel-Cylinder Type, for Water and Other Liquids.*

[14] American Water Works Association (AWWA). 1992. C301. *Standard for Prestressed Concrete Pressure Pipe, Steel-Cylinder Type, for Water and Other Liquids.*

[15] American Water Works Association (AWWA). 1987. C302. *Standard for Reinforced Concrete Pressure Pipe, Noncylinder Type, for Water and Other Liquids.*

[16] American Water Works Association (AWWA). 1987. C303. *Standard for Reinforced Concrete Pressure Pipe, Steel Cylinder Type, Pretensioned, for Water and Other Liquids.*

[17] American Water Works Association (AWWA). 1992. C304. *Standard for Design of Prestressed Concrete Cylinder Pipe.*

[18] American Water Works Association (AWWA). 1987. C303. *Standard for Installation of Ductile-Iron Water Mains and Their Appurtenances.*

[19] American Water Works Association (AWWA). 1988. C950. *Standard for Fiberglass Pressure Pipe.*

[20] American Water Works Association (AWWA). 1995. Manual M9. *Concrete Pressure Pipe.*

[21] American Water Works Association (AWWA). 1989. Manual M11. *Steel Pipe-A Guide for Design and Installation.*

[22] American Water Works Association (AWWA). 1997 (projected). Manual M45. *Fiberglass Pressure Pipe.*

[23] Anton, W.F., J.E. Herold, R.T. Dailey, and W.J. Cichanski. 1990. *Investigation & Rehabilitation of Seattle's Tolt Pipeline.* Proceedings of the International Conference on Pipeline Design and Installation. ASCE. Las Vegas. pp. 213–229.

[24] Committee on Seismic Analysis of the ASCE Structural Division Committee on Nuclear Structures and Materials. 1983. *Seismic Response of Buried Pipes and Structural Components.* American Society of Civil Engineers (ASCE).

[25] Doyle, J.M. and S.L. Chu. 1968. Plastic Design of Flexible Conduits. *J. of Structural Division (ASCE),* 94, 1935–1944.

[26] Eberhardt, A. 1990. 108-in. Diameter Steel Water Conduit Failure and Assessment of AWWA Practice. *J. of Performance of Constructed Facilities (ASCE),* 4, 30–50.

[27] Howard, A.K., L.A. Kinney, and R.P. Fuerst. 1995. *Method for Prediction of Flexible Pipe Deflection.* Report M-25 (M0250000.995). U.S. Bureau of Reclamation. Denver, CO.

[28] Marston, A. and A.O. Anderson. 1913. The Theory of Loads on Pipes in Ditches and Tests of Cement and Clay Drain Tile and Sewer Pipe. Bull. 31. Iowa Engineering Experiment Station. Ames, IA.

[29] Marston, A. 1930. The Theory of External Loads on Closed Conduits in the Light of the Latest Experiments. Bull. 96. Iowa Engineering Experiment Station. Ames, IA.

[30] MathSoft Inc. 1994. *Mathcad PLUS 5.0.* MathSoft Inc. Cambridge, MA.

[31] Moncarz, P.D., J.C. Shyne, and G.K. Derbalian. 1987. Failures of 108-inch Steel Pipe Water Main. *J. of Performance of Constructed Facilities (ASCE),* 1, 168–187.

[32] Netzel, R.J. and J.M. Doyle. 1983. Design of Underground Pipelines. *Proc. American Power Conference,* 45, 792.

[33] Newmark, N.M. 1935. Simplified Computation of Vertical Pressures in Elastic Foundations. Circular No. 24. Engineering Experiment Station. University of Illinois. Urbana, IL.

[34] Schlick, W.J. 1932. Loads on Pipe in Wide Ditches. Bull. 108. Iowa Engineering Station. Ames, IA.

[35] Schlick, W.J. 1952. Loads on Negative Projecting Conduits. *Proc. Highway Research Board,* 31, 308.

[36] Spangler, M.G., R. Winfrey, and C. Mason. 1926. Experimental Determination of Static and Impact Loads Transmitted to Culverts. Bull. 76. Iowa Engineering Experiment Station. Ames, IA.

[37] Spangler, M.G. and R.L. Hennessy. 1946. A Method of Computing Live Loads Transmitted to Underground Conduits. *Proc. Highway Research Board,* 26, 83.

[38] Spangler, M.G. 1941. The Structural Design of Flexible Pipe Culverts. Bull. 153. Iowa Engineering Experiment Station. Ames, IA.

[39] Spangler, M.G. 1950. A Theory of Loads on Negative Projecting Conduits. *Proc. Highway Research Board,* 29, 153.

[40] Spangler, M.G. and R.L. Handy. 1982. *Soil Engineering.* 4th ed. Harper & Row.

[41] Timoshenko, S. and J.N. Goodier. 1951. *Theory of Elasticity.* McGraw-Hill, New York.

[42] Watkins, R.K. and M.G. Spangler. 1958. Some Characteristics of the Modulus of Passive Resistance of Soil: A Study in Similitude. *Proc. Highway Research Board,* 37, 576.

[43] Young, W.C. 1989. *Roark's Formulas for Stress & Strain.* 6th ed. McGraw-Hill, New York.

26

Structural Reliability[1]

26.1 Introduction 26-1
Definition of Reliability • Introduction to Reliability-Based
Design Concepts

26.2 Basic Probability Concepts 26-3
Random Variables and Distributions • Moments • Concept of
Independence • Examples • Approximate Analysis of Moments
• Statistical Estimation and Distribution Fitting

26.3 Basic Reliability Problem 26-15
Basic $R - S$ Problem • More Complicated Limit State Functions
Reducible to $R - S$ Form • Examples

26.4 Generalized Reliability Problem 26-21
Introduction • FORM/SORM Techniques • Monte Carlo Simulation

26.5 System Reliability 26-27
Introduction • Basic Systems • Introduction to Classical System
Reliability Theory • Redundant Systems • Examples

26.6 Reliability-Based Design (Codes) 26-31
Introduction • Calibration and Selection of Target Reliabilities •
Material Properties and Design Values • Design Loads and Load
Combinations • Evaluation of Load and Resistance Factors

26.7 Defining Terms 26-35

Acknowledgments 26-36

References 26-36

Further Reading 26-38

Appendix 26-38

D. V. Rosowsky
Department of Civil Engineering,
Clemson University,
Clemson, SC

26.1 Introduction

26.1.1 Definition of Reliability

Reliability and **reliability-based design** (RBD) are terms that are being associated increasingly with the design of civil engineering structures. While the subject of reliability may not be treated explicitly in the civil engineering curriculum, either at the graduate or undergraduate levels, some basic knowledge of the concepts of structural reliability can be useful in understanding the development and bases for many modern design codes (including those of the American Institute of Steel

---

[1]Parts of this chapter were previously published by CRC Press in *The Civil Engineering Handbook*, W.F. Chen, Ed., 1995.

0-8493-2674-5/97/$0.00+$.50
© 1997 by CRC Press LLC

Construction [AISC], the American Concrete Institute [ACI], the American Association of State Highway Transportation Officials [AASHTO], and others).

Reliability simply refers to some probabilistic measure of satisfactory (or safe) performance, and as such, may be viewed as a complementary function of the probability of failure.

$$\text{Reliability} = fcn\,(1 - P_{\text{failure}}) \tag{26.1}$$

When we talk about the reliability of a structure (or member or system), we are referring to the probability of safe performance for a particular **limit state**. A limit state can refer to ultimate failure (such as collapse) or a condition of unserviceability (such as excessive vibration, deflection, or cracking). The treatment of structural loads and resistances using probability (or reliability) theory, and of course the theories of structural analysis and mechanics, has led to the development of the latest generation of probability-based, reliability-based, or **limit states design** codes.

If the subject of structural reliability is generally not treated in the undergraduate civil engineering curriculum, and only a relatively small number of universities offer graduate courses in structural reliability, why include a basic (introductory) treatment in this handbook? Besides providing some insight into the bases for modern codes, it is likely that future generations of structural codes and specifications will rely more and more on probabilistic methods and reliability analyses. The treatment of (1) structural analysis, (2) structural design, and (3) probability and statistics in most civil engineering curricula permits this introduction to structural reliability without the need for more advanced study. This section by no means contains a complete treatment of the subject, nor does it contain a complete review of probability theory. At this point in time, structural reliability is usually only treated at the graduate level. However, it is likely that as RBD becomes more accepted and more prevalent, additional material will appear in both the graduate and undergraduate curricula.

26.1.2 Introduction to Reliability-Based Design Concepts

The concept of RBD is most easily illustrated in Figure 26.1. As shown in that figure, we consider the

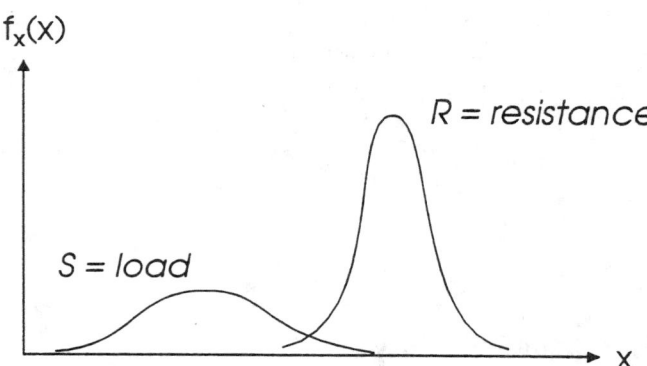

Figure 26.1   Basic concept of structural reliability.

acting load and the structural resistance to be random variables. Also as the figure illustrates, there is the possibility of a resistance (or strength) that is inadequate for the acting load (or conversely, that the load exceeds the available strength). This possibility is indicated by the region of overlap on Figure 26.1 in which realizations of the load and resistance variables lead to **failure**. The objective of RBD is to ensure the probability of this condition is acceptably small. Of course, the load can refer to any appropriate structural, service, or environmental loading (actually, its effect), and the resistance can refer to any limit state capacity (i.e., flexural strength, bending stiffness, maximum tolerable deflection, etc.). If we formulate the simplest expression for the probability of failure (P_f) as

$$P_f = P\left[(R - S) < 0\right] \tag{26.2}$$

we need only ensure that the units of the resistance (R) and the load (S) are consistent. We can then use probability theory to estimate these limit state probabilities.

Since RBD is intended to provide (or ensure) uniform and acceptably small failure probabilities for similar designs (limit states, materials, occupancy, etc.), these acceptable levels must be predetermined. This is the responsibility of code development groups and is based largely on previous experience (i.e., calibration to previous design philosophies such as **allowable stress design** [ASD] for steel) and engineering judgment. Finally, with information describing the statistical variability of the loads and resistances, and the target probability of failure (or target reliability) established, factors for codified design can be evaluated for the relevant load and resistance quantities (again, for the particular limit state being considered). This results, for instance, in the familiar form of design checking equations:

$$\phi R_n \geq \sum_i \gamma_i Q_{n,i} \tag{26.3}$$

referred to as **load and resistance factor design** (LRFD) in the U.S., and in which R_n is the nominal (or design) resistance and Q_n are the **nominal load effects**. The factors γ_i and ϕ in Equation 26.3 are the load and resistance factors, respectively. This will be described in more detail in later sections. Additional information on this subject may be found in a number of available texts [3, 21].

26.2 Basic Probability Concepts

This section presents an introduction to basic probability and statistics concepts. Only a sufficient presentation of topics to permit the discussion of reliability theory and applications that follows is included herein. For additional information and a more detailed presentation, the reader is referred to a number of widely used textbooks (i.e., [2, 5]).

26.2.1 Random Variables and Distributions

Random variables can be classified as being either discrete or continuous. Discrete random variables can assume only discrete values, whereas continuous random variables can assume any value within a range (which may or may not be bounded from above or below). In general, the random variables considered in structural reliability analyses are continuous, though some important cases exist where one or more variables are discrete (i.e., the number of earthquakes in a region). A brief discussion of both discrete and continuous random variables is presented here; however, the reliability analysis (theory and applications) sections that follow will focus mainly on continuous random variables.

The relative frequency of a variable is described by its probability mass function (PMF), denoted $p_X(x)$, if it is discrete, or its probability density function (PDF), denoted $f_X(x)$, if it is continuous. (A histogram is an example of a PMF, whereas its continuous analog, a smooth function, would represent a PDF.) The cumulative frequency (for either a discrete or continuous random variable) is described by its cumulative distribution function (CDF), denoted $F_X(x)$. (See Figure 26.2.)

There are three basic axioms of probability that serve to define valid probability assignments and provide the basis for probability theory.

1. The probability of an event is bounded by zero and one (corresponding to the cases of zero probability and certainty, respectively).
2. The sum of all possible outcomes in a sample space must equal one (a statement of collectively exhaustive events).

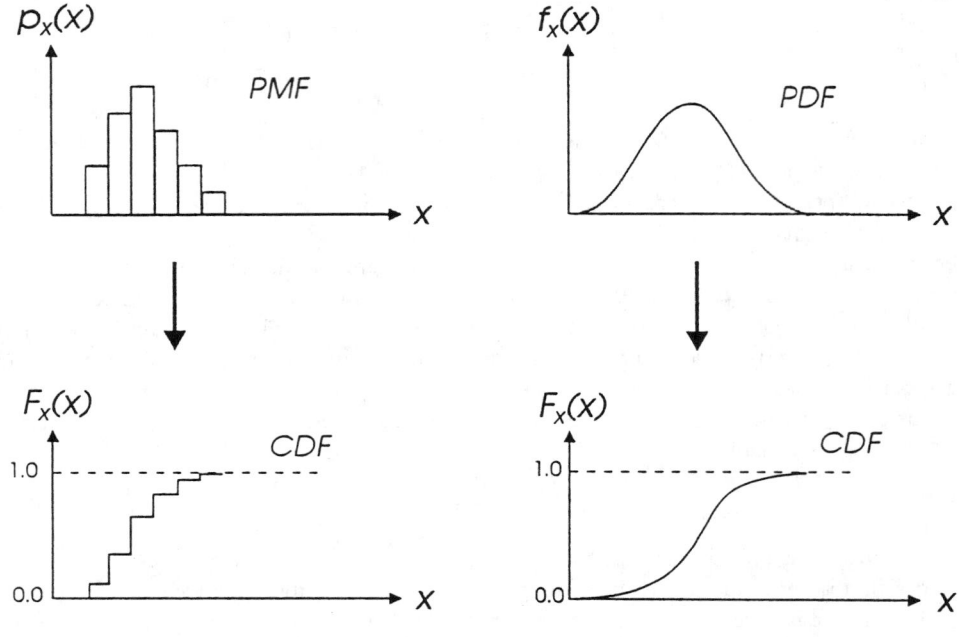

(a) Discrete random variable   (b) Continuous random variable

Figure 26.2   Sample probability functions.

3. The probability of the union of two mutually exclusive events is the sum of the two individual event probabilities, $P[A \cup B] = P[A] + P[B]$.

The PMF or PDF, describing the relative frequency of the random variable, can be used to evaluate the probability that a variable takes on a value within some range.

$$P[a < X_{\text{discr}} \le b] = \sum_{a}^{b} p_X(x) \tag{26.4}$$

$$P[a < X_{\text{cts}} \le b] = \int_{a}^{b} f_X(x)dx \tag{26.5}$$

The CDF is used to describe the probability that a random variable is less than or equal to some value. Thus, there exists a simple integral relationship between the PDF and the CDF. For example, for a continuous random variable,

$$F_X(a) = P[X \le a] = \int_{-\infty}^{a} f_X(x)dx \tag{26.6}$$

There are a number of common distribution forms. The probability functions for these distribution forms are given in Table 26.1.

An important class of distributions for reliability analysis is based on the statistical theory of extreme values. Extreme value distributions are used to describe the distribution of the largest or smallest of a set of independent and identically distributed random variables. This has obvious implications for reliability problems in which we may be concerned with the largest of a set of 50 annual-extreme snow loads or the smallest (lowest) concrete strength from a set of 100 cylinder tests, for example. There are three important extreme value distributions (referred to as Type I, II, and III,

TABLE 26.1 Common Distribution Forms and Their Parameters

Distribution	PMF or PDF	Parameters	Mean and variance
Binomial	$p_X(x) = \binom{n}{x} p^x (1-p)^{n-x}$ $x = 0, 1, 2, \ldots, n$	p	$E[X] = np$ $\text{Var}[X] = np(1-p)$
Geometric	$p_X(x) = p(1-p)^{x-1}$ $x = 0, 1, 2, \ldots$	p	$E[X] = 1/p$ $\text{Var}[X] = (1-p)/p^2$
Poisson	$p_X(x) = \frac{(\upsilon t)^x}{x!} e^{-\upsilon t}$ $x = 0, 1, 2, \ldots$	υ	$E[X] = \upsilon t$ $\text{Var}[X] = \upsilon t$
Exponential	$f_X(x) = \lambda e^{-\lambda x}$ $x \geq 0$	λ	$E[X] = 1/\lambda$ $\text{Var}[X] = 1/\lambda^2$
Gamma	$f_X(x) = \frac{\upsilon(\upsilon x)^{k-1} e^{-\upsilon x}}{\Gamma(k)}$ $x \geq 0$	υ, k	$E[X] = k/\upsilon$ $\text{Var}[X] = k/\upsilon^2$
Normal	$f_X(x) = \frac{1}{\sqrt{2\pi}\sigma} \exp\left[-\frac{1}{2}\left(\frac{x-\mu}{\sigma}\right)^2\right]$ $-\infty < x < \infty$	μ, σ	$E[X] = \mu$ $\text{Var}[X] = \sigma^2$
Lognormal	$f_X(x) = \frac{1}{\sqrt{2\pi}\zeta x} \exp\left[-\frac{1}{2}\left(\frac{\ln x - \lambda}{\zeta}\right)^2\right]$ $x \geq 0$	λ, ζ	$E[X] = \exp\left(\lambda + \frac{1}{2}\zeta^2\right)$ $\text{Var}[X] = E^2[X]\left(\exp\left(\zeta^2\right) - 1\right)$
Uniform	$f_X(x) = \frac{1}{b-a}$ $a < x < b$	a, b	$E[X] = \frac{(a+b)}{2}$ $\text{Var}[X] = \frac{1}{12}(b-a)^2$
Extreme Type I (largest)	$f_X(x) = \alpha \exp\left[-\alpha(x-u) - e^{-\alpha(x-u)}\right]$ $-\infty < x < \infty$	α, u	$E[X] = u + \frac{\gamma}{\alpha}$ $(\gamma \cong 0.5772)$ $\text{Var}[X] = \frac{\pi^2}{6\alpha^2}$
Extreme Type II (largest)	$f_X(x) = \frac{k}{x}\left(\frac{u}{x}\right)^k e^{-\left(\frac{u}{x}\right)^k}$ $x \geq 0$	k, u	$E[X] = u\Gamma\left(1 - \frac{1}{k}\right)$ $(k > 1)$ $\text{Var}[X] = u^2\left[\Gamma\left(1-\frac{2}{k}\right) - \Gamma^2\left(1-\frac{1}{k}\right)\right]$ $(k > 2)$
Extreme Type III (smallest)	$f_X(x) = \frac{k}{w-\varepsilon}\left(\frac{x-\varepsilon}{w-\varepsilon}\right)^{k-1} \exp\left[-\left(\frac{x-\varepsilon}{w-\varepsilon}\right)^k\right]$ $x \geq \varepsilon$	k, w, ε	$E[X] = \varepsilon + (u-\varepsilon)\Gamma\left(1+\frac{1}{k}\right)$ $\text{Var}[X] = (u-\varepsilon)^2\left[\Gamma\left(1+\frac{2}{k}\right) - \Gamma^2\left(1+\frac{1}{k}\right)\right]$

respectively), which are also included in Table 26.1. Additional information on the derivation and application of extreme value distributions may be found in various texts (e.g., [3, 21]).

In most cases, the solution to the integral of the probability function (see Equations 26.5 and 26.6) is available in closed form. The exceptions are two of the more common distributions, the normal and lognormal distributions. For these cases, tables are available (i.e., [2, 5, 21]) to evaluate the integrals. To simplify the matter, and eliminate the need for multiple tables, the standard normal distribution is most often tabulated. In the case of the normal distribution, the probability is evaluated:

$$P\left[a < X \leq b\right] = F_X(b) - F_X(a) = \Phi\left(\frac{b - \mu_x}{\sigma_x}\right) - \Phi\left(\frac{a - \mu_x}{\sigma_x}\right) \tag{26.7}$$

where $F_X(\cdot)$ = the particular normal distribution, $\Phi(\cdot)$ = the standard normal CDF, μ_x = mean of random variable X, and σ_x = standard deviation of random variable X. Since the standard normal variate is therefore the variate minus its mean, divided by its standard deviation, it too is a normal random variable with mean equal to zero and standard deviation equal to one. Table 26.2 presents the standard normal CDF in tabulated form.

In the case of the lognormal distribution, the probability is evaluated (also using the standard normal probability tables):

$$P[a < Y \leq b] = F_y(b) - F_Y(a) = \Phi\left(\frac{\ln b - \lambda_y}{\xi_y}\right) - \Phi\left(\frac{\ln a - \lambda_y}{\xi_y}\right) \tag{26.8}$$

where $F_Y(\cdot)$ = the particular lognormal distribution, $\Phi(\cdot)$ = the standard normal CDF, and λ_y and ξ_y are the lognormal distribution parameters related to μ_y = mean of random variable Y and V_y = coefficient of variation (COV) of random variable Y, by the following:

$$\lambda_y = \ln \mu_y - \frac{1}{2}\xi_y^2 \tag{26.9}$$

$$\xi_y^2 = \ln\left(V_y^2 + 1\right) \tag{26.10}$$

Note that for relatively low coefficients of variation ($V_y \approx 0.3$ or less), Equation 26.10 suggests the approximation, $\xi \approx V_y$.

26.2.2 Moments

Random variables are characterized by their distribution form (i.e., probability function) and their moments. These values may be thought of as shifts and scales for the distribution and serve to uniquely define the probability function. In the case of the familiar normal distribution, there are two moments: the mean and the standard deviation. The mean describes the central tendency of the distribution (the normal distribution is a symmetric distribution), while the standard deviation is a measure of the dispersion about the mean value. Given a set of n data points, the sample mean and the sample variance (which is the square of the sample standard deviation) are computed as

$$m_x = \frac{1}{n}\sum_i X_i \tag{26.11}$$

$$\hat{\sigma}_x^2 = \frac{1}{n-1}\sum_i (X_i - m_x)^2 \tag{26.12}$$

Many common distributions are two-parameter distributions and, while not necessarily symmetric, are completely characterized by their first two moments (see Table 26.1). The population mean, or first moment of a continuous random variable, is computed as

$$\mu_x = E[X] = \int_{-\infty}^{+\infty} x f_X(x)dx \tag{26.13}$$

where $E[X]$ is referred to as the expected value of X. The population variance (the square of the population standard deviation) of a continuous random variable is computed as

$$\sigma_x^2 = \text{Var}[X] = E\left[(X - \mu_x)^2\right] = \int_{-\infty}^{+\infty} (x - \mu_x)^2 f_X(x)dx \tag{26.14}$$

The population variance can also be expressed in terms of expectations as

$$\sigma_x^2 = E[X^2] - E^2[X] = \int_{-\infty}^{+\infty} x^2 f_X(x)dx - \left(\int_{-\infty}^{+\infty} x f_X(x)dx\right)^2 \tag{26.15}$$

TABLE 26.2 Complementary Standard Normal Table, $\Phi(-\beta) = 1 - \Phi(\beta)$

β	$\Phi(-\beta)$	β	$\Phi(-\beta)$	β	$\Phi(-\beta)$
.00	.50000 + 00	.47	.3192E + 00	.94	.1736E + 00
.01	.4960E + 00	.48	.3156E + 00	.95	.1711E + 00
.02	.4920E + 00	.49	.3121E + 00	.96	.1685E + 00
.03	.4880E + 00	.50	.3085E + 00	.97	.1660E + 00
.04	.4840E + 00	.51	.3050E + 00	.98	.1635E + 00
.05	.4801E + 00	.52	.3015E + 00	.99	.1611E + 00
.06	.4761E + 00	.53	.2981E + 00	1.00	.1587E + 00
.07	.4721E + 00	.54	.2946E + 00	1.01	.1562E + 00
.08	.4681E + 00	.55	.2912E + 00	1.02	.1539E + 00
.09	.4641E + 00	.56	.2877E + 00	1.03	.1515E + 00
.10	.4602E + 00	.57	.2843E + 00	1.04	.1492E + 00
.11	.4562E + 00	.58	.2810E + 00	1.05	.1469E + 00
.12	.4522E + 00	.59	.2776E + 00	1.06	.1446E + 00
.13	.4483E + 00	.60	.2743E + 00	1.07	.1423E + 00
.14	.4443E + 00	.61	.2709E + 00	1.08	.1401E + 00
.15	.4404E + 00	.62	.2676E + 00	1.09	.1379E + 00
.16	.4364E + 00	.63	.2643E + 00	1.10	.1357E + 00
.17	.4325E + 00	.64	.2611E + 00	1.11	.1335E + 00
.18	.4286E + 00	.65	.2578E + 00	1.12	.1314E + 00
.19	.4247E + 00	.66	.2546E + 00	1.13	.1292E + 00
.20	.4207E + 00	.67	.2514E + 00	1.14	.1271E + 00
.21	.4168E + 00	.68	.2483E + 00	1.15	.1251E + 00
.22	.4129E + 00	.69	.2451E + 00	1.16	.1230E + 00
.23	.4090E + 00	.70	.2420E + 00	1.17	.1210E + 00
.24	.4052E + 00	.71	.2389E + 00	1.18	.1190E + 00
.25	.4013E + 00	.72	.2358E + 00	1.19	.1170E + 00
.26	.3974E + 00	.73	.2327E + 00	1.20	.1151E + 00
.27	.3936E + 00	.74	.2297E + 00	1.21	.1131E + 00
.28	.3897E + 00	.75	.2266E + 00	1.22	.1112E + 00
.29	.3859E + 00	.76	.2236E + 00	1.23	.1093E + 00
.30	.3821E + 00	.77	.2207E + 00	1.24	.1075E + 00
.31	.3783E + 00	.78	.2177E + 00	1.25	.1056E + 00
.32	.3745E + 00	.79	.2148E + 00	1.26	.1038E + 00
.33	.3707E + 00	.80	.2119E + 00	1.27	.1020E + 00
.34	.3669E + 00	.81	.2090E + 00	1.28	.1003E + 00
.35	.3632E + 00	.82	.2061E + 00	1.29	.9853E − 01
.36	.3594E + 00	.83	.2033E + 00	1.30	.9680E − 01
.37	.3557E + 00	.84	.2005E + 00	1.31	.9510E − 01
.38	.3520E + 00	.85	.1977E + 00	1.32	.9342E − 01
.39	.3483E + 00	.86	.1949E + 00	1.33	.9176E − 01
.40	.3446E + 00	.87	.1922E + 00	1.34	.9012E − 01
.41	.3409E + 00	.88	.1894E + 00	1.35	.8851E − 01
.42	.3372E + 00	.89	.1867E + 00	1.36	.8691E − 01
.43	.3336E + 00	.90	.1841E + 00	1.37	.8534E − 01
.44	.3300E + 00	.91	.1814E + 00	1.38	.8379E − 01
.45	.3264E + 00	.92	.1788E + 00	1.39	.8226E − 01
.46	.3228E + 00	.93	.1762E + 00	1.40	.8076E − 01
1.41	.7927E − 01	1.88	.3005E − 01	2.35	.9387E − 02
1.42	.7780E − 01	1.89	.2938E − 01	2.36	.9138E − 02
1.43	.7636E − 01	1.90	.2872E − 01	2.37	.8894E − 02
1.44	.7493E − 01	1.91	.2807E − 01	2.38	.8656E − 02
1.45	.7353E − 01	1.92	.2743E − 01	2.39	.8424E − 02
1.46	.7215E − 01	1.93	.2680E − 01	2.40	.8198E − 02
1.47	.7078E − 01	1.94	.2619E − 01	2.41	.7976E − 02
1.48	.6944E − 01	1.95	.2559E − 01	2.42	.7760E − 02
1.49	.6811E − 01	1.96	.2500E − 01	2.43	.7549E − 02
1.50	.6681E − 01	1.97	.2442E − 01	2.44	.7344E − 02
1.51	.6552E − 01	1.98	.2385E − 01	2.45	.7143E − 02
1.52	.6426E − 01	1.99	.2330E − 01	2.46	.6947E − 02
1.53	.6301E − 01	2.00	.2275E − 01	2.47	.6756E − 02
1.54	.6178E − 01	2.01	.2222E − 01	2.48	.6569E − 02
1.55	.6057E − 01	2.02	.2169E − 01	2.49	.6387E − 02
1.56	.5938E − 01	2.03	.2118E − 01	2.50	.6210E − 02
1.57	.5821E − 01	2.04	.2068E − 01	2.51	.6037E − 02
1.58	.5705E − 01	2.05	.2018E − 01	2.52	.5868E − 02
1.59	.5592E − 01	2.06	.1970E − 01	2.53	.5703E − 02
1.60	.5480E − 01	2.07	.1923E − 01	2.54	.5543E − 02
1.61	.5370E − 01	2.08	.1876E − 01	2.55	.5386E − 02
1.62	.5262E − 01	2.09	.1831E − 01	2.56	.5234E − 02
1.63	.5155E − 01	2.10	.1786E − 01	2.57	.5085E − 02

TABLE 26.2   Complementary Standard Normal Table,
$\Phi(-\beta) = 1 - \Phi(\beta)$ *(continued)*

β	$\Phi(-\beta)$	β	$\Phi(-\beta)$	β	$\Phi(-\beta)$
1.64	$.5050E - 01$	2.11	$.1743E - 01$	2.58	$.4940E - 02$
1.65	$.4947E - 01$	2.12	$.1700E - 01$	2.59	$.4799E - 02$
1.66	$.4846E - 01$	2.13	$.1659E - 01$	2.60	$.4661E - 02$
1.67	$.4746E - 01$	2.14	$.1618E - 01$	2.61	$.4527E - 02$
1.68	$.4648E - 01$	2.15	$.1578E - 01$	2.62	$.4396E - 02$
1.69	$.4551E - 01$	2.16	$.1539E - 01$	2.63	$.4269E - 02$
1.70	$.4457E - 01$	2.17	$.1500E - 01$	2.64	$.4145E - 02$
1.71	$.4363E - 01$	2.18	$.1463E - 01$	2.65	$.4024E - 02$
1.72	$.4272E - 01$	2.19	$.1426E - 01$	2.66	$.3907E - 02$
1.73	$.4182E - 01$	2.20	$.1390E - 01$	2.67	$.3792E - 02$
1.74	$.4093E - 01$	2.21	$.1355E - 01$	2.68	$.3681E - 02$
1.75	$.4006E - 01$	2.22	$.1321E - 01$	2.69	$.3572E - 02$
1.76	$.3920E - 01$	2.23	$.1287E - 01$	2.70	$.3467E - 02$
1.77	$.3836E - 01$	2.24	$.1255E - 01$	2.71	$.3364E - 02$
1.78	$.3754E - 01$	2.25	$.1222E - 01$	2.72	$.3264E - 02$
1.79	$.3673E - 01$	2.26	$.1191E - 01$	2.73	$.3167E - 02$
1.80	$.3593E - 01$	2.27	$.1160E - 01$	2.74	$.3072E - 02$
1.81	$.3515E - 01$	2.28	$.1130E - 01$	2.75	$.2980E - 02$
1.82	$.3438E - 01$	2.29	$.1101E - 01$	2.76	$.2890E - 02$
1.83	$.3363E - 01$	2.30	$.1072E - 01$	2.77	$.2803E - 02$
1.84	$.3288E - 01$	2.31	$.1044E - 01$	2.78	$.2718E - 02$
1.85	$.3216E - 01$	2.32	$.1017E - 01$	2.79	$.2635E - 02$
1.86	$.3144E - 01$	2.33	$.9903E - 02$	2.80	$.2555E - 02$
1.87	$.3074E - 01$	2.34	$.9642E - 02$	2.81	$.2477E - 02$
2.82	$.2401E - 02$	3.29	$.5009E - 03$	3.76	$.8491E - 04$
2.83	$.2327E - 02$	3.30	$.4834E - 03$	3.77	$.8157E - 04$
2.84	$.2256E - 02$	3.31	$.4664E - 03$	3.78	$.7836E - 04$
2.85	$.2186E - 02$	3.32	$.4500E - 03$	3.79	$.7527E - 04$
2.86	$.2118E - 02$	3.33	$.4342E - 03$	3.80	$.7230E - 04$
2.87	$.2052E - 02$	3.34	$.4189E - 03$	3.81	$.6943E - 04$
2.88	$.1988E - 02$	3.35	$.4040E - 03$	3.82	$.6667E - 04$
2.89	$.1926E - 02$	3.36	$.3897E - 03$	3.83	$.6402E - 04$
2.90	$.1866E - 02$	3.37	$.3758E - 03$	3.84	$.6147E - 04$
2.91	$.1807E - 02$	3.38	$.3624E - 03$	3.85	$.5901E - 04$
2.92	$.1750E - 02$	3.39	$.3494E - 03$	3.86	$.5664E - 04$
2.93	$.1695E - 02$	3.40	$.3369E - 03$	3.87	$.5437E - 04$
2.94	$.1641E - 02$	3.41	$.3248E - 03$	3.88	$.5218E - 04$
2.95	$.1589E - 02$	3.42	$.3131E - 03$	3.89	$.5007E - 04$
2.96	$.1538E - 02$	3.43	$.3017E - 03$	3.90	$.4804E - 04$
2.97	$.1489E - 02$	3.44	$.2908E - 03$	3.91	$.4610E - 04$
2.98	$.1441E - 02$	3.45	$.2802E - 03$	3.92	$.4422E - 04$
2.99	$.1395E - 02$	3.46	$.2700E - 03$	3.93	$.4242E - 04$
3.00	$.1350E - 02$	3.47	$.2602E - 03$	3.94	$.4069E - 04$
3.01	$.1306E - 02$	3.48	$.2507E - 03$	3.95	$.3902E - 04$
3.02	$.1264E - 02$	3.49	$.2415E - 03$	3.96	$.3742E - 04$
3.03	$.1223E - 02$	3.50	$.2326E - 03$	3.97	$.3588E - 04$
3.04	$.1183E - 02$	3.51	$.2240E - 03$	3.98	$.3441E - 04$
3.05	$.1144E - 02$	3.52	$.2157E - 03$	3.99	$.3298E - 04$
3.06	$.1107E - 02$	3.53	$.2077E - 03$	4.00	$.3162E - 04$
3.07	$.1070E - 02$	3.54	$.2000E - 03$	4.10	$.2062E - 04$
3.08	$.1035E - 02$	3.55	$.1926E - 03$	4.20	$.1332E - 04$
3.09	$.1001E - 02$	3.56	$.1854E - 03$	4.30	$.8524E - 05$
3.10	$.9676E - 03$	3.57	$.1784E - 03$	4.40	$.5402E - 05$
3.11	$.9354E - 03$	3.58	$.1717E - 03$	4.50	$.3391E - 05$
3.12	$.9042E - 03$	3.59	$.1653E - 03$	4.60	$.2108E - 05$
3.13	$.8740E - 03$	3.60	$.1591E - 03$	4.70	$.1298E - 05$
3.14	$.8447E - 03$	3.61	$.1531E - 03$	4.80	$.7914E - 06$
3.15	$.8163E - 03$	3.62	$.1473E - 03$	4.90	$.4780E - 06$
3.16	$.7888E - 03$	3.63	$.1417E - 03$	5.00	$.2859E - 06$
3.17	$.7622E - 03$	3.64	$.1363E - 03$	5.10	$.1694E - 06$
3.18	$.7363E - 03$	3.65	$.1311E - 03$	5.20	$.9935E - 07$
3.19	$.7113E - 03$	3.66	$.1261E - 03$	5.30	$.5772E - 07$
3.20	$.6871E - 03$	3.67	$.1212E - 03$	5.40	$.3321E - 07$
3.21	$.6636E - 03$	3.68	$.1166E - 03$	5.50	$.1892E - 07$
3.22	$.6409E - 03$	3.69	$.1121E - 03$	6.00	$.9716E - 09$
3.23	$.6189E - 03$	3.70	$.1077E - 03$	6.50	$.3945E - 10$
3.24	$.5976E - 03$	3.71	$.1036E - 03$	7.00	$.1254E - 11$
3.25	$.5770E - 03$	3.72	$.9956E - 04$	7.50	$.3116E - 13$
3.26	$.5570E - 03$	3.73	$.9569E - 04$	8.00	$.6056E - 15$
3.27	$.5377E - 03$	3.74	$.9196E - 04$	8.50	$.9197E - 17$
3.28	$.5190E - 03$	3.75	$.8837E - 04$	9.00	$.1091E - 18$

The COV is defined as the ratio of the standard deviation to the mean, and therefore serves as a nondimensional measure of variability.

$$\text{COV} = V_X = \frac{\sigma_x}{\mu_x} \qquad (26.16)$$

In some cases, higher order (> 2) moments exist, and these may be computed similarly as

$$\mu_x^{(n)} = E\left[(X - \mu_x)^n\right] = \int_{-\infty}^{+\infty} (x - \mu_x)^n \, f_X(x)dx \qquad (26.17)$$

where $\mu_x^{(n)}$ = the nth central moment of random variable X. Often, it is more convenient to define the probability distribution in terms of its parameters. These parameters can be expressed as functions of the moments (see Table 26.1).

26.2.3   Concept of Independence

The concept of statistical independence is very important in structural reliability as it often permits great simplification of the problem. While not all random quantities in a reliability analysis may be assumed independent, it is certainly reasonable to assume (in most cases) that loads and resistances are statistically independent. Often, the assumption of independent loads (actions) can be made as well.

Two events, A and B, are statistically independent if the outcome of one in no way affects the outcome of the other. Therefore, two random variables, X and Y, are statistically independent if information on one variable's probability of taking on some value in no way affects the probability of the other random variable taking on some value. One of the most significant consequences of this statement of independence is that the joint probability of occurrence of two (or more) random variables can be written as the product of the individual marginal probabilities. Therefore, if we consider two events (A = probability that an earthquake occurs and B = probability that a hurricane occurs), and we assume these occurrences are statistically independent in a particular region, the probability of both an earthquake and a hurricane occurring is simply the product of the two probabilities:

$$P\left[A \text{ ``and'' } B\right] = P\left[A \cap B\right] = P[A]P[B] \qquad (26.18)$$

Similarly, if we consider resistance (R) and load (S) to be continuous random variables, and assume independence, we can write the probability of R being less than or equal to some value r and the probability that S exceeds some value s (i.e., failure) as

$$\begin{aligned} P[R \le r \cap S > s] &= P[R \le r]P[S > s] \\ &= P[R \le r](1 - P[S \le s]) = F_R(r)(1 - F_S(s)) \qquad (26.19) \end{aligned}$$

Additional implications of statistical independence will be discussed in later sections. The treatments of dependent random variables, including issues of correlation, joint probability, and conditional probability are beyond the scope of this introduction, but may be found in any elementary text (e.g., [2, 5]).

26.2.4   Examples

Three relatively simple examples are presented here. These examples serve to illustrate some important elements of probability theory and introduce the reader to some basic reliability concepts in structural engineering and design.

EXAMPLE 26.1:

The Richter magnitude of an earthquake, given that it has occurred, is assumed to be exponentially distributed. For a particular region in Southern California, the exponential distribution parameter (λ) has been estimated to be 2.23. What is the probability that a given earthquake will have a magnitude greater than 5.5?

$$
\begin{aligned}
P[M > 5.5] &= 1 - P[M \le 5.5] = 1 - F_X(5.5) \\
&= 1 - \left[1 - e^{-5.5\lambda}\right] \\
&= e^{-2.23 \times 5.5} = e^{-12.265} \\
&\approx 4.71 \times 10^{-6}
\end{aligned}
$$

Given that two earthquakes have occurred in this region, what is the probability that both of their magnitudes were greater than 5.5?

$$
\begin{aligned}
P[M_1 > 5.5 \cap M_2 > 5.5] &= P[M_1 > 5.5]P[M_2 > 5.5] \ \text{(assumed independence)} \\
&= (P[M > 5.5])^2 \ \text{(identically distributed)} \\
&= \left(4.71 \times 10^{-6}\right)^2 \\
&\approx 2.22 \times 10^{-11} \ \text{(very small!)}
\end{aligned}
$$

EXAMPLE 26.2:

Consider the cross-section of a reinforced concrete column with 12 reinforcing bars. Assume the load-carrying capacity of each of the 12 reinforcing bars (R_i) is normally distributed with mean of 100 kN and standard deviation of 20 kN. Further assume that the load-carrying capacity of the concrete itself is $r_c = 500$ kN (deterministic) and that the column is subjected to a known load of 1500 kN. What is the probability that this column will fail?

First, we can compute the mean and standard deviation of the column's total load-carrying capacity.

$$
E[R] = m_R = r_c + \sum_{i=1}^{12} E[R_i] = 500 + 12(100) = 1700 \ \text{kN}
$$

$$
\text{Var}[R] = \sigma_R^2 = \sum_{i=1}^{12} \sigma_{R_i}^2 = 12\,(20)^2 = 4800 \ \text{kN}^2 \quad \therefore \sigma_R = 69.28 \ \text{kN}
$$

Since the total capacity is the sum of a number of normal variables, it too is a normal variable (central limit theorem). Therefore, we can compute the probability of failure as the probability that the load-carrying capacity, R, is less than the load of 1500 kN.

$$
P[R < 1500] = F_R(1500) = \Phi\left(\frac{1500 - 1700}{69.28}\right) = \Phi(-2.89) \approx 0.00193
$$

EXAMPLE 26.3:

The moment capacity (M) of the simply supported beam ($l = 10$ ft) shown in Figure 26.3 is assumed to be normally distributed with mean of 25 ft-kips and COV of 0.20. Failure occurs if the maximum moment exceeds the moment capacity. If only a concentrated load $P = 4$ kips is applied

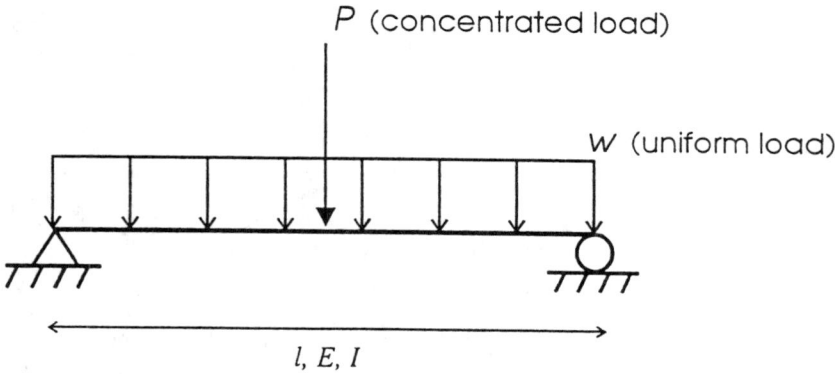

Figure 26.3 Simply supported beam (for Example 26.3).

at midspan, what is the failure probability?

$$M_{max} = \frac{Pl}{4} = \frac{4\left(10'\right)}{4} = 10 \text{ ft-kips}$$

$$P_f = P\left[M < M_{max}\right] = F_M(10) = \Phi\left(\frac{10 - 25}{5}\right) = \Phi(-3.0) \approx 0.00135$$

If only a uniform load $w = 1$ kip/ft is applied along the entire length of the beam, what is the failure probability?

$$M_{max} = \frac{wl^2}{8} = \frac{1\left(10'\right)^2}{8} = 12.5 \text{ ft-kips}$$

$$P_f = P[M < M_{max}] = F_M(12.5) = \Phi\left(\frac{12.5 - 25}{5}\right) = \Phi(-2.5) \approx 0.00621$$

If the beam is subjected to both P and w simultaneously, what is the probability the beam performs safely?

$$M_{max} = \frac{Pl}{4} + \frac{wl^2}{8} = 10 + 12.5 = 22.5 \text{ ft-kips}$$

$$P_f = P[M < M_{max}] = F_M(22.5) = \Phi\left(\frac{22.5 - 25}{5}\right) = \Phi(-0.5) \approx 0.3085$$

$$\therefore \quad P\left(\text{“safety”}\right) = P_S = (1 - P_f) = 0.692$$

Note that this failure probability is not simply the sum of the two individual failure probabilities computed previously. Finally, for design purposes, suppose we want a probability of safe performance $P_s = 99.9\%$, for the case of the beam subjected to the uniform load (w) only. What value of w_{max} (i.e., maximum allowable uniform load for design) should we specify?

$$M_{allow.} = \frac{w_{max}\left(l^2\right)}{8} = w_{max}\left(\frac{10^2}{8}\right) = 12.5\,(w_{max})$$

$$goal : P\left[M > 12.5 w_{max}\right] = 0.999$$

$$1 - F_M\left(12.5 w_{max}\right) = 0.999$$

$$1 - \Phi\left(\frac{12.5 w_{max} - 25}{5}\right) = 0.999$$

$$\therefore \Phi^{-1}(1.0 - 0999) = \frac{12.5 w_{max} - 25}{5}$$

$$\therefore w_{max} = \frac{(-3.09)(5) + 25}{12.5} \approx 0.76 \text{ kips/ft}$$

EXAMPLE 26.4:

The total annual snowfall for a particular location is modeled as a normal random variable with mean of 60 in. and standard deviation of 15 in. What is the probability that in any given year the total snowfall in that location is between 45 and 65 in.?

$$
\begin{aligned}
P[45 < S \le 65] \quad &= \quad F_S(65) - F_S(45) = \Phi\left(\frac{65-60}{15}\right) - \Phi\left(\frac{45-60}{15}\right) \\
&= \quad \Phi(0.33) - \Phi(-1.00) = \Phi(0.33) - [1 - \Phi(1.00)] \\
&= \quad 0.629 - (1 - 0.841) \approx 0.47 \ \ \text{(about 47\%)}
\end{aligned}
$$

What is the probability the total annual snowfall is at least 30 in. in this location?

$$
\begin{aligned}
1 - F_S(30) \quad &= \quad 1 - \Phi\left(\frac{30-60}{15}\right) \\
&= \quad 1 - \Phi(-2.0) = 1 - [1 - \Phi(2.0)] \\
&= \quad \Phi(2.0) \approx 0.977 \ \ \text{(about 98\%)}
\end{aligned}
$$

Suppose for design we want to specify the 95th percentile snowfall value (i.e., a value that has a 5% exceedence probability). Estimate the value of $S_{.95}$.

$$
P[S > S_{.95}] \equiv 0.05 \qquad P[S < S_{.95}] = .95
$$

$$
\Phi\left(\frac{S_{.95} - 60}{15}\right) = 0.95
$$

$$
\therefore S_{.95} = \left[15 \times \Phi^{-1}(.95)\right] + 60
$$

$$
= (15)(1.64) + 60 = 84.6 \ \text{in.}
$$

$$
\text{(so, specify 85 in.)}
$$

Now, assume the total annual snowfall is lognormally distributed (rather than normally) with the same mean and standard deviation as before. Recompute $P[45 \ \text{in.} \le S \le 65 \ \text{in.}]$. First, we obtain the lognormal distribution parameters:

$$
\xi^2 \quad = \quad \ln(V_S^2 + 1) = \ln\left(\left(\frac{15}{60}\right)^2 + 1\right) = 0.061
$$

$$
\xi \quad = \quad 0.246 \ (\approx 0.25 = V_S; \qquad \text{o.k. for } V \approx 0.3 \ \text{or less})
$$

$$
\lambda \quad = \quad \ln(m_S) - 0.5\xi^2 = \ln(60) - 0.5(0.61) = 4.064
$$

Now, using these parameters, recompute the probability:

$$
\begin{aligned}
P[45 < S_{LN} \le 65] \quad &= \quad F_S(65) - F_S(45) = \Phi\left(\frac{\ln(65) - 4.06}{0.25}\right) - \Phi\left(\frac{\ln(45) - 4.06}{0.25}\right) \\
&= \quad \Phi(0.46) - \Phi(-1.01) = \Phi(0.46) - [1 - \Phi(1.01)] \\
&= \quad 0.677 - (1 - 0.844) \approx 0.52 \ \ \text{(about 52\%)}
\end{aligned}
$$

Note that this is slightly higher than the value obtained assuming the snowfall was normally distributed (47%). Finally, again assuming the total annual snowfall to be lognormally distributed, recompute the 5% exceedence limit (i.e., the 95th percentile value):

$$
P[S < S_{.95}] = .95
$$

$$
\Phi\left(\frac{\ln(S_{.95}) - 4.06}{0.25}\right) = 0.95
$$

$$\therefore \ln(S_{.95}) = \left[.25 \times \Phi^{-1}(.95)\right] + 4.06$$
$$= (.25)(1.64) + 4.06 = 4.47 \therefore S_{.95} = \exp(4.47) \approx 87.4 \text{ in.}$$
$$\text{(specify 88 in.)}$$

Again, this value is slightly higher than the value obtained assuming the total snowfall was normally distributed (about 85 in.).

26.2.5 Approximate Analysis of Moments

In some cases, it may be desired to estimate approximately the statistical moments of a function of random variables. For a function given by

$$Y = g(X_1, X_2, \ldots, X_n) \tag{26.20}$$

approximate estimates for the moments can be obtained using a first-order Taylor series expansion of the function about the vector of mean values. Keeping only the 0th- and 1st-order terms results in an approximate mean

$$E[Y] \approx g(\mu_1, \mu_2, \ldots, \mu_n) \tag{26.21}$$

in which μ_i = mean of random variable X_i, and an approximate variance

$$\text{Var}[Y] \approx \sum_{i=1}^{n} c_i^2 \text{Var}[X_i] + \sum_{i \neq j}^{n} \sum^{n} c_i c_j \text{Cov}[X_i, X_j] \tag{26.22}$$

in which c_i and c_j are the values of the partial derivatives $\partial g / \partial X_i$ and $\partial g / \partial X_j$, respectively, evaluated at the vector of mean values $(\mu_1, \mu_2, \ldots, \mu_n)$, and $\text{Cov}[X_i, X_j]$ = covariance function of X_i and X_j. If all random variables X_i and X_j are mutually uncorrelated (statistically independent), the approximate variance reduces to

$$\text{Var}[Y] \approx \sum_{i=1}^{n} c_i^2 \text{Var}[X_i] \tag{26.23}$$

These approximations can be shown to be valid for reasonably linear functions $g(X)$. For non-linear functions, the approximations are still reasonable if the variances of the individual random variables, X_i, are relatively small.

The estimates of the moments can be improved if the second-order terms from the Taylor series expansions are included in the approximation. The resulting second-order approximation for the mean assuming all X_i, X_j uncorrelated is

$$E[Y] \approx g(\mu_1, \mu_2, \ldots, \mu_n) + \frac{1}{2} \sum_{i=1}^{n} \left(\frac{\partial^2 g}{\partial X_i^2} \right) \text{Var}[X_i] \tag{26.24}$$

For uncorrelated X_i, X_j, however, there is no improvement over Equation 26.23 for the approximate variance. Therefore, while the second-order analysis provides additional information for estimating the mean, the variance estimate may still be inadequate for nonlinear functions.

26.2.6 Statistical Estimation and Distribution Fitting

There are two general classes of techniques for estimating statistical moments: point-estimate methods and interval-estimate methods. The method of moments is an example of a point-estimate method, while confidence intervals and hypothesis testing are examples of interval-estimate techniques. These topics are treated generally in an introductory statistics course and therefore are not covered in this chapter. However, the topics are treated in detail in Ang and Tang [2] and Benjamin and Cornell [5], as well as many other texts.

The most commonly used tests for goodness-of-fit of distributions are the Chi-Squared (χ^2) test and the Kolmogorov-Smirnov (K-S) test. Again, while not presented in detail herein, these tests are described in most introductory statistics texts. The χ^2 test compares the observed relative frequency histogram with an assumed, or theoretical, PDF. The K-S test compares the observed cumulative frequency plot with the assumed, or theoretical, CDF. While these tests are widely used, they are both limited by (1) often having only limited data in the tail regions of the distribution (the region most often of interest in reliability analyses), and (2) not allowing evaluation of goodness-of-fit in specific regions of the distribution. These methods do provide established and effective (as well as statistically robust) means of evaluating the relative goodness-of-fit of various distributions over the entire range of values. However, when it becomes necessary to assure a fit in a particular region of the distribution of values, such as an upper or lower tail, other methods must be employed. One such method, sometimes called the inverse CDF method, is described here. The inverse CDF method is a simple, graphical technique similar to that of using probability paper to evaluate goodness-of-fit.

It can be shown using the theory of order statistics [5] that

$$E\left[F_X(y_i)\right] = \frac{i}{n+1} \tag{26.25}$$

where $F_X(\cdot)$ = cumulative distribution function, y_i = mean of the ith order statistic, and n = number of independent samples. Hence, the term $i/(n+1)$ is referred to as the ith rank mean plotting position. This well-known plotting position has the properties of being nonparametric (i.e., distribution independent), unbiased, and easy to compute. With a sufficiently large number of observations, n, a cumulative frequency plot is obtained by plotting the rank-ordered observation x_i versus the quantity $i/(n+1)$. As n becomes large, this observed cumulative frequency plot approaches the true CDF of the underlying phenomenon. Therefore, the plotting position is taken to approximate the CDF evaluated at x_i:

$$F_X(x_i) \approx \left(\frac{i}{n+1}\right) \qquad i = 1, \ldots, n \tag{26.26}$$

Simply examining the resulting estimate for the CDF is limited as discussed previously. That is, assessing goodness-of-fit in the tail regions can be difficult. Furthermore, relative goodness-of-fit over all regions of the CDF is essentially impossible. To address this shortcoming, the inverse CDF is considered. For example, taking the inverse CDF of both sides of Equation (26.26) yields

$$F_X^{-1}\left[F_X(x_i)\right] \approx F_X^{-1}\left[\left(\frac{i}{n+1}\right)\right] \tag{26.27}$$

where the left-hand side simply reduces to x_i. Therefore, an estimate for the ith observation can be obtained provided the inverse of the assumed underlying CDF exists (see Table 26.5). Finally, if the ith (rank-ordered) observation is plotted against the inverse CDF of the rank mean plotting position, which serves as an estimate of the ith observation, the relative goodness-of-fit can be evaluated over the entire range of observations. Essentially, therefore, one is seeking a close fit to the 1:1 line. The better this fit, the better the assumed underlying distribution $F_X(\cdot)$. Figure 26.4 presents an example of a relatively good fit of an Extreme Type I largest (Gumbel) distribution to annual maximum wind speed data from Boston, Massachusetts.

Caution must be exercised in interpreting goodness-of-fit using this method. Clearly, a perfect fit will not be possible, unless the phenomenon itself corresponds directly to a single underlying distribution. Furthermore, care must be taken in evaluating goodness-of-fit in the tail regions, as often limited data exists in these regions. A poor fit in the upper tail, for instance, may not necessarily mean that the distribution should be rejected. This method does have the advantage, however, of

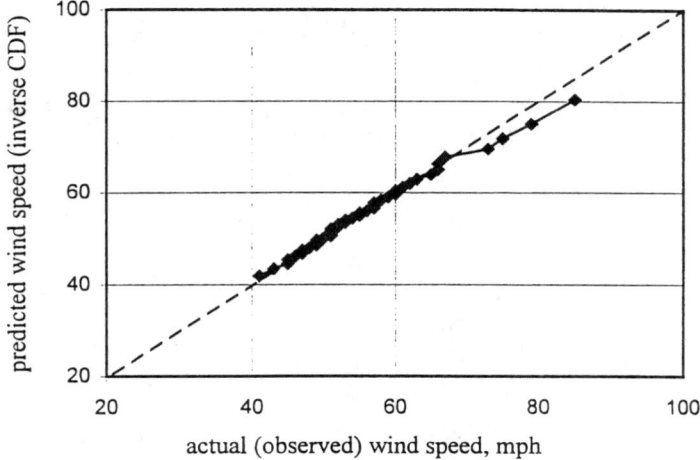

Figure 26.4   Inverse CDF (Extreme Type I largest) of annual maximum wind speeds, Boston, MA (1936–1977).

permitting an evaluation over specific ranges of values corresponding to specific regions of the distribution. While this evaluation is essentially qualitative, as described herein, it is a relatively simple extension to quantify the relative goodness-of-fit using some measure of correlation, for example.

Finally, the inverse CDF method has advantages over the use of probability paper in that (1) the method can be generalized for any distribution form without the need for specific types of plotting paper, and (2) the method can be easily programmed.

26.3   Basic Reliability Problem

A complete treatment of structural reliability theory is not included in this section. However, a number of texts are available (in varying degrees of difficulty) on this subject [3, 10, 21, 23]. For the purpose of an introduction, an elementary treatment of the basic (two-variable) reliability problem is provided in the following sections.

26.3.1   Basic $R - S$ Problem

As described previously, the simplest formulation of the failure probability problem may be written:

$$P_f = P[R < S] = P[R - S < 0] \tag{26.28}$$

in which R = resistance and S = load. The simple function, $g(X) = R - S$, where X = vector of basic random variables, is termed the limit state function. It is customary to formulate this limit state function such that the condition $g(X) < 0$ corresponds to failure, while $g(X) > 0$ corresponds to a condition of safety. The limit state surface corresponds to points where $g(X) = 0$ (where the term "surface" implies it is possible to have problems involving more than two random variables). For the simple two-variable case, if the assumption can be made that the load and resistance quantities are statistically independent, and that the population statistics can be estimated by the sample statistics, the failure probabilities for the cases of normal or lognormal variates (R, S) are given by

$$P_{f(N)} \quad = \quad \Phi\left(\frac{0 - m_M}{\hat{\sigma}_M}\right) = \Phi\left(\frac{m_S - m_R}{\sqrt{\hat{\sigma}_S^2 + \hat{\sigma}_R^2}}\right) \tag{26.29}$$

$$P_{f(LN)} = \Phi\left(\frac{0 - m_M}{\hat{\sigma}_M}\right) = \Phi\left(\frac{\lambda_S - \lambda_R}{\sqrt{\xi_S^2 + \xi_R^2}}\right) \qquad (26.30)$$

where $M = R - S$ is the safety margin (or limit state function). The concept of a safety margin and the reliability index, β, is illustrated in Figure 26.5. Here, it can be seen that the reliability index,

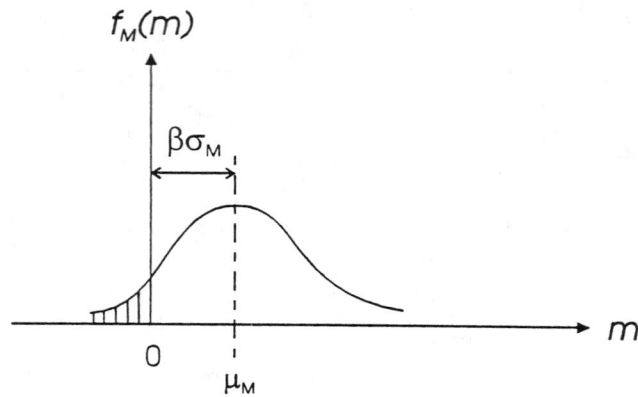

Figure 26.5   Safety margin concept, $M = R - S$.

β, corresponds to the distance (specifically, the number of standard deviations) the mean of the safety margin is away from the origin (recall, $M = 0$ corresponds to failure). The most common, generalized definition of reliability is the second-moment reliability index, β, which derives from this simple two-dimensional case, and is related (approximately) to the failure probability by

$$\beta \approx \Phi^{-1}(1 - P_f) \qquad (26.31)$$

where $\Phi^{-1}(\cdot) =$ inverse standard normal CDF. Table 26.2 can also be used to evaluate this function. (In the case of normal variates, Equation 26.31 is exact. Additional discussion of the reliability index, β, may be found in any of the texts cited previously.) To gain a feel for relative values of the reliability index, β, the corresponding failure probabilities are shown in Table 26.3. Based on the above discussion (Equations 26.29 through 26.31), for the case of R and S both distributed normal or lognormal, expressions for the reliability index are given by

$$\beta_{(N)} = \frac{m_M}{\hat{\sigma}_M} = \frac{m_R - m_S}{\sqrt{\hat{\sigma}_R^2 + \hat{\sigma}_S^2}} \qquad (26.32)$$

$$\beta_{(LN)} = \frac{m_M}{\hat{\sigma}_M} = \frac{\lambda_R - \lambda_S}{\sqrt{\xi_R^2 + \xi_S^2}} \qquad (26.33)$$

For the less generalized case where R and S are not necessarily both distributed normal or lognormal (but are still independent), the failure probability may be evaluated by solving the convolution integral shown in Equation 26.34a or 26.34b either numerically or by simulation:

$$P_f = P[R < S] = \int_{-\infty}^{+\infty} F_R(x)f_S(x)dx \qquad (26.34a)$$

$$P_f = P[R < S] = \int_{-\infty}^{+\infty} [1 - F_S(x)]\,f_R(x)dx \qquad (26.34b)$$

TABLE 26.3   Failure Probabilities and
Corresponding Reliability Values

Probability of failure, P_f	Reliability index, β
.5	0.00
.1	1.28
.01	2.32
.001	3.09
10^{-4}	3.71
10^{-5}	4.75
10^{-6}	5.60

Again, the second-moment reliability is approximated as $\beta = \Phi^{-1}(1 - P_f)$. Additional methods for evaluating β (for the case of multiple random variables and more complicated limit state functions) are presented in subsequent sections.

26.3.2   More Complicated Limit State Functions Reducible to $R - S$ Form

It may be possible that what appears to be a more complicated limit state function (i.e., more than two random variables) can be reduced, or simplified, to the basic $R - S$ form. Three points may be useful in this regard:

1. If the COV of one random variable is very small relative to the other random variables, it may be able to be treated as a it deterministic quantity.

2. If multiple, statistically independent random variables (X_i) are taken in a summation function ($Z = aX_1 + bX_2 + \ldots$), and the random variables are assumed to be normal, the summation can be replaced with a single normal random variable (Z) with moments:

$$E[Z] = aE[X_1] + bE[X_2] + \ldots \tag{26.35}$$

$$\mathrm{Var}[Z] = \sigma_Z^2 = a^2\sigma_{x_1}^2 + b^2\sigma_{x_2}^2 + \ldots \tag{26.36}$$

3. If multiple, statistically independent random variables (Y_i) are taken in a product function ($Z' = Y_1 Y_2 \ldots$), and the random variables are assumed to be lognormal, the product can be replaced with a single lognormal random variable (Z') with moments (shown here for the case of the product of two variables):

$$E[Z'] = E[Y_1]E[Y_2] \tag{26.37}$$

$$\mathrm{Var}[Z'] = \mu_{Y_1}^2\sigma_{Y_2}^2 + \mu_{Y_2}^2\sigma_{Y_1}^2 + \sigma_{Y_1}^2\sigma_{Y_2}^2 \tag{26.38}$$

Note that the last term in Equation 26.38 is very small if the coefficients of variation are small. In this case, and more generally, for the product of n random variables, the COV of the product may be expressed:

$$V_Z \approx \sqrt{V_{Y_1}^2 + V_{Y_2}^2 + \ldots + V_{Y_n}^2} \tag{26.39}$$

When it is not possible to reduce the limit state function to the simple $R - S$ form, and/or when the random variables are not both normal or lognormal, more advanced methods for the evaluation of the failure probability (and hence the reliability) must be employed. Some of these methods will be described in the next section after some illustrative examples.

26.3.3 Examples

The following examples all contain limit state functions that are in, or can be reduced to, the form of the basic $R - S$ problem. Note that in all cases the random variables are all either normal or lognormal. Additional information suggesting when such distribution assumptions may be reasonable (or acceptable) is also provided in these examples.

EXAMPLE 26.5:

Consider the statically indeterminate beam shown in Figure 26.6, subjected to a concentrated load, P. The moment capacity, M_{cap}, is a random variable with mean of 20 ft-kips and standard deviation

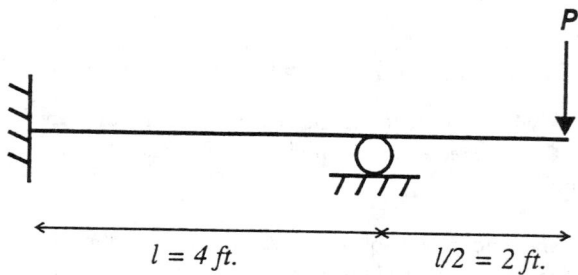

Figure 26.6   Cantilever beam subject to point load (Example 26.5).

of 4 ft-kips. The load, P, is a random variable with mean of 4 kips and standard deviation of 1 kip. Compute the second-moment reliability index assuming P and M_{cap} are normally distributed and statistically independent.

$$M_{max} = \frac{Pl}{2}$$

$$P_f = P\left[M_{cap} < \frac{Pl}{2}\right] = P\left[M_{cap} - \frac{Pl}{2} < 0\right] = P\left[M_{cap} - 2P < 0\right]$$

Here, the failure probability is expressed in terms of $R - S$, where $R = M_{cap}$ and $S = 2P$. Now, we compute the moments of the safety margin given by $M = R - S$:

$$m_M = E[M] = E[R - S] = E[R] - E[S] = m_{M_{cap}} - 2m_P = 20 - 2(4) = 12 \text{ ft-kips}$$

$$\hat{\sigma}_M^2 = \text{Var}[M] = \text{Var}[R] + \text{Var}[S] = \hat{\sigma}_{M_{cap}}^2 + (2)^2\hat{\sigma}_P^2 = (4)^2 + 4(1)^2 = 20 \text{ (ft-kips)}^2$$

Finally, we can compute the second-moment reliability index, β, as

$$\beta = \frac{m_M}{\hat{\sigma}_M} = \frac{m_R - m_S}{\sqrt{\hat{\sigma}_R^2 + \hat{\sigma}_S^2}} = \frac{12}{\sqrt{20}} \approx 2.68$$

(The corresponding failure probability is therefore $P_f \approx \Phi(-\beta) = \Phi(-2.68) \approx 0.00368$.)

EXAMPLE 26.6:

When designing a building, the total force acting on the columns must be considered. For a particular design situation, the total column force may consist of components of dead load (self-weight), live load (occupancy), and wind load, denoted D, L, and W, respectively. It is reasonable to assume these variables are statistically independent, and here we will further assume them to be normally distributed with the following moments:

Variable	Mean(m)	SD(σ)
D	4.0 kips	0.4 kips
L	8.0 kips	2.0 kips
W	3.4 kips	0.7 kips

If the column has a strength that is assumed to be deterministic, $R = 20$ kips, what is the probability of failure and the corresponding second-moment reliability index, β?

First, we compute the moments of the combined load, $S = D + L + W$:

$$
\begin{aligned}
m_S &= m_D + m_L + m_W = 4.0 + 8.0 + 3.4 = 15.4 \text{ kips} \\
\hat{\sigma}_S &= \sqrt{\hat{\sigma}_D^2 + \hat{\sigma}_L^2 + \hat{\sigma}_W^2} = \sqrt{(0.4)^2 + (2.0)^2 + (0.7)^2} = 2.16 \text{ kips}
\end{aligned}
$$

Since S is the sum of a number of normal random variables, it is itself a normal variable. Now, since the resistance is assumed to be deterministic, we can simply compute the failure probability directly in terms of the standard normal CDF (rather than formulating the limit state function).

$$
\begin{aligned}
P_f &= P[S > R] = 1 - P[S < R] = 1 - F_S(20) \\
&= 1 - \Phi\left(\frac{20 - 15.4}{2.16}\right) = 1 - \Phi(2.13) \approx 1 - (.9834) = .0166 \\
&(\therefore \beta = 2.13)
\end{aligned}
$$

If we were to formulate this in terms of a limit state function (of course, the same result would be obtained), we would have $g(X) = R - S$, where the moments of S are given above and the moments of R would be $m_R = 20$ kips and $\sigma_R = 0$. Now, if we assume the resistance, R, is a random variable (rather than being deterministic), with mean and standard deviation given by $m_R = 20$ kips and $\sigma_R = 2$ kips (i.e., COV = 0.10), how would this additional uncertainty affect the probability of failure (and the reliability)? To answer this, we analyze this as a basic $R - S$ problem, assuming normal variables, and making the reasonable assumption that the loads and resistance are independent quantities. Therefore, from Equation 26.29:

$$
\begin{aligned}
P_f &= P[R - S < 0] = \Phi\left(\frac{m_S - m_R}{\sqrt{\hat{\sigma}_S^2 + \hat{\sigma}_R^2}}\right) = \Phi\left(\frac{15.4 - 20}{\sqrt{(2.16)^2 + (2)^2}}\right) \\
&= \Phi\left(\frac{-4.6}{\sqrt{8.67}}\right) = \Phi(-1.56) \approx 0.0594 \\
&(\therefore \beta = 1.56)
\end{aligned}
$$

As one would expect, the uncertainty in the resistance serves to increase the failure probability (in this case, fairly significantly), thereby decreasing the reliability.

EXAMPLE 26.7:

The fully plastic flexural capacity of a steel beam section is given by the product YZ, where $Y =$ steel yield strength and $Z =$ section modulus. Therefore, for an applied moment, M, we can express the limit state function as $g(X) = YZ - M$, where failure corresponds to the condition $g(X) < 0$. Given the statistics shown below and assuming all random variables are lognormally distributed (this ensures non-negativity of the load and resistance variables), reduce this to the simple $R - S$ form and estimate the second-moment reliability index.

Variable	Distribution	Mean	COV
Y	Lognormal	40 ksi	0.10
Z	Lognormal	50 in.^{3}	0.05
M	Lognormal	1000 in.-kip	0.20

First, we obtain the moments of R and S as follows:
"R" $= YZ$:

$$E[R] = m_R = m_Y m_Z = (40)(50) = 2000 \text{ in.-kips}$$

$$V_R = \text{COV} \approx \sqrt{V_Y^2 + V_Z^2} = 0.112 \text{ (since COVs are "small")}$$

"S" $= M$:

$$E[S] = m_M = 1000 \text{ in.-kips}$$

$$V_S = \text{COV} = V_M = 0.20$$

Now, we can compute the lognormal parameters (λ and ξ) for R and S:

$$\xi_R \approx V_R = 0.112 \text{ (since small COV)}$$

$$\lambda_R = \ln m_R - \frac{1}{2}\xi_R^2 = \ln(2000) - \frac{1}{2}(.112)^2 = 7.595$$

$$\xi_S \approx V_S = 0.20 \text{ (since small COV)}$$

$$\lambda_S = \ln m_S - \frac{1}{2}\xi_S^2 = \ln(1000) - \frac{1}{2}(.2)^2 = 6.888$$

Finally, the second-moment reliability index, β, is computed:

$$\beta_{LN} = \frac{\lambda_R - \lambda_S}{\sqrt{\xi_R^2 + \xi_S^2}} = \frac{7.595 - 6.888}{\sqrt{(.112)^2 + (.2)^2}} \approx 3.08$$

Since the variability in the section modulus, Z, is very small ($V_Z = 0.05$), we could choose to neglect it in the reliability analysis (i.e., assume Z deterministic). Still assuming variables Y and M to be lognormally distributed, and using Equation 26.33 to evaluate the reliability index, we obtain $\beta = 3.17$. If we further assumed Y and M to be normal (instead of lognormal) random variables, the reliability index computed using Equation 26.32 would be $\beta = 3.54$. This illustrates the relative error one might expect from (a) assuming certain variables with low COVs to be essentially deterministic (i.e., 3.17 vs. 3.08), and (b) assuming the incorrect distributions, or simply using the normal distribution when more statistical information is available suggesting another distribution form (i.e., 3.54 vs. 3.08).

EXAMPLE 26.8:

Consider again the simply supported beam shown in Figure 26.3, subjected to a uniform load, w (only), along its entire length. Assume that, in addition to w being a random variable, the member properties E and I are also random variables. (The length, however, may be assumed to be deterministic.) Formulate the limit state function for excessive deflection (assume a maximum allowable deflection of $l/360$, where l = length of the beam) and then reduce it to the simple $R - S$ form. (Set-up only.)

$$\delta_{\max} = \frac{5wl^4}{384EI}$$

$$P_f = P[\delta_{\text{allow.}} - \delta_{\max} < 0]$$

The failure probability is in the $R - S$ form ($R = \delta_{\text{allow.}}$ and $S = \delta_{\max}$); however, we still must express the limit state function in terms of the basic variables.

$$
\begin{aligned}
g(X) &= \frac{l}{360} - \frac{5wl^4}{384EI} < 0 \ \text{(for failure)} \\
&= \frac{EI}{360} - \frac{5wl^3}{384} < 0 \\
&= \frac{384}{360}(EI) - 5wl^3 < 0 \\
&= 1.067(EI) - 5l^3(w) < 0
\end{aligned}
$$

Note that the limit state function is now expressed in the simple $R - S$ form, with $R = EI$ and $S = w$. If E and I are assumed to be lognormally distributed, their product, EI, is also a lognormal random variable, and the moments can be computed as was done in the previous example. Finally, if the uniform load, w, can be assumed lognormal as well, the second-moment reliability index could be computed (also as done in the previous example).

26.4   Generalized Reliability Problem

26.4.1   Introduction

As discussed previously for the simple two-variable $(R - S)$ case in which R and S are assumed to be independent, identically distributed (normal or lognormal) random variables permits a closed-form solution to the failure probability. However, such a two-variable simplification of the limit state is often not possible for many structural reliability problems. Furthermore, the joint probability function for the random variables in the limit state equation is seldom known precisely, due to limited data. Even if the basic variables are mutually statistically independent and all marginal density functions are known, it is often impractical (or impossible) to perform the numerical integration of the multidimensional convolution integral over the failure domain. In this section, a number of widely used techniques for evaluating structural reliability under general conditions are presented.

26.4.2   FORM/SORM Techniques

First-order second-moment (FOSM) methods were the first techniques used to evaluate structural reliability. The name refers to the way in which the limit state is linearized (first-order) and the way in which the method characterizes the basic variables (second moment). Later, more advanced methods were developed to include information about the complete distributions characterizing the random variables. These advanced FOSM techniques became known as **first-order reliability methods (FORM)**. Finally, among the most recent developments has been the refined curve-fitting of the limit state surface in the analysis, giving rise to the so-called **second-order reliability methods (SORM)**. Details of these reliability analysis techniques may be found in the literature [3, 21, 23].

When the simple limit state (safety margin) is defined by $M = R - S$, we have already seen that the reliability index, β, can be expressed (see Figure 26.5):

$$\beta = \frac{\mu_M}{\sigma_M} = \frac{E[g(X)]}{SD[g(X)]} \tag{26.40}$$

where $E[g(X)]$ and $SD[g(X)]$ are the mean and standard deviation of the limit state function, respectively. Therefore, for the simple $R - S$ case, β is the distance from the mean of the safety margin ($\mu_M = \mu_R - \mu_S$) to the origin in units of standard deviations of M. This is illustrated in Figure 26.1. In this simple second-moment formulation, no mention is made of the underlying probability distributions. The reliability index, β, depends only on measures of central tendency and dispersion of the margin of safety, M, for the limit state function.

For the more general case where the number of random variables may be greater than two, the limit state surface may be nonlinear, and the random variables may not be normal, a number of iterative solution techniques have been developed. These techniques are all very similar, and differ mainly in the approach taken to a minimization problem. One general procedure is presented at the end of this section. Other approaches may be found in the literature [3, 12, 21, 23]. What follows is a summary of the mathematics behind the formulation of FORM techniques. It is not necessary to fully understand the development of these methods, and those wishing to skip this material can go directly to the algorithm provided later in this section.

To simplify the presentation herein, the basic variables, X_i, are assumed to be statistically independent and therefore uncorrelated. This assumption, as discussed earlier, is often reasonable for many structural reliability problems. Further, it can be shown that weak correlation (i.e., $\rho < 0.2$, where ρ is the correlation coefficient) can generally be neglected and that strong correlation (i.e., $\rho > 0.8$) can be considered to imply fully dependent variables. Additional discussion of correlated variables in FORM/SORM may be found in the references [21, 23]. The limit state function, expressed in terms of the basic variables, X_i, is first transformed to reduced variables, u_i, having zero mean and unit standard deviation:

$$u_i = \frac{X_i - \mu_{X_i}}{\sigma_{X_i}} \tag{26.41}$$

A transformed limit state function can then be expressed in terms of the reduced variables:

$$g_1(u_1, \ldots, u_n) = 0 \tag{26.42}$$

with failure now being defined as $g_1(u) < 0$. The space corresponding to the reduced variables can be shown to have rotational symmetry, as indicated by the concentric circles of equiprobability shown on Figure 26.7. The reliability index, β, is now defined as the shortest distance between the limit state surface, $g_1(u) = 0$, and the origin in reduced variable space (see Figure 26.7). The point $(u_1^*, \ldots, u_n^*)$ on the limit state surface that corresponds to this minimum distance is referred to as the checking (or design) point and can be determined by simultaneously solving the set of equations:

$$\alpha_i = \frac{\frac{\partial g_1}{\partial u_i}}{\sqrt{\sum_i \left(\frac{\partial g_1}{\partial u_i} \right)^2}} \tag{26.43}$$

$$u_i^* = -\alpha_i \beta \tag{26.44}$$

$$g_1 \left(u_1^*, \ldots, u_n^* \right) = 0 \tag{26.45}$$

and searching for the direction cosines, α_i, that minimize β. The partial derivatives in Equation 26.43 are evaluated at the reduced space design point $(u_1^*, \ldots, u_n^*)$. This procedure, and Equations 26.43 through 26.45, result from linearizing the limit state surface (in reduced space) and computing the

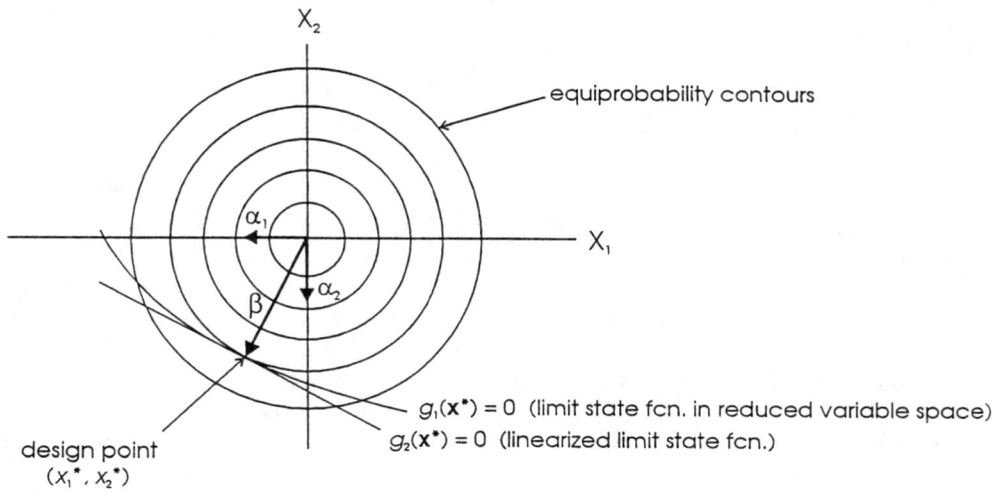

Figure 26.7 Formulation of reliability analysis in reduced variable space. (Adapted from Ellingwood, B., Galambos, T.V., MacGregor, J. G. and Cornell, C. A. 1980. Development of a Probability Based Load Criterion for American National Standard A58, NBS Special Publication SP577, National Bureau of Standards, Washington, D.C.)

reliability as the shortest distance from the origin in reduced space to the limit state hyperplane. It may be useful at this point to compare Figures 26.5 and 26.7 to gain some additional insight into this technique.

Once the convergent solution is obtained, it can be shown that the checking point in the original random variable space corresponds to the points:

$$X_i^* = \mu_{X_i} \left(1 - \alpha_i \beta V_{X_i} \right) \tag{26.46}$$

such that $g(X_1^*, \ldots, X_n^*) = 0$. These variables will correspond to values in the upper tails of the probability distributions for load variables and the lower tails for resistance (or geometric) variables.

The formulation described above provides an exact estimate of the reliability index, β, for cases in which the basic variables are normal and in which the limit state function is linear. In other cases, the results are only approximate. As many structural load and resistance quantities are known to be non-normal, it seems reasonable that information on distribution type be incorporated into the reliability analysis. This is especially true since the limit state probabilities can be affected significantly by different distributions' tail behaviors. Methods that include distribution information are known as full-distribution methods or advanced FOSM methods. One commonly used technique is described below.

Because of the ease of working with normal variables, the objective here is to transform the non-normal random variables into equivalent normal variables, and then to perform the analysis for a solution of the reliability index, as described previously. This transformation is accomplished by approximating the true distribution by a normal distribution at the value corresponding to the design point on the failure surface. By fitting an equivalent normal distribution at this point, we are forcing the best approximation to be in the tail of interest of the particular random variable. The fitting is accomplished by determining the mean and standard deviation of the equivalent normal variable such that, at the value corresponding to the design point, the cumulative probability and the probability density of the actual (non-normal) and the equivalent normal variable are equal. (This is the basis for the so-called Rackwitz-Fiessler algorithm.) These moments of the equivalent normal variable are given by

$$\sigma_i^N = \frac{\phi\left(\Phi^{-1}\left(F_i\left(X_i^*\right)\right)\right)}{f_i\left(X_i^*\right)} \qquad (26.47)$$

$$\mu_i^N = X_i^* - \Phi^{-1}\left(F_i\left(X_i^*\right)\right)\sigma_i^N \qquad (26.48)$$

in which $F_i(\cdot)$ and $f_i(\cdot)$ are the non-normal CDF and PDF, respectively, $\phi(\cdot)$ = standard normal PDF, and $\Phi^{-1}(\cdot)$ = inverse standard normal CDF. Once the equivalent normal mean and standard deviation given by Equations 26.47 and 26.48 are determined, the solution proceeds exactly as described previously. Since the checking point, X_i^*, is updated at each iteration, the equivalent normal mean and standard deviation must be updated at each iteration cycle as well. While this can be rather laborious by hand, the computer handles this quite efficiently. Only in the case of highly nonlinear limit state functions does this procedure yield results that may be in error.

One possible procedure for computing the reliability index, β, for a limit state with non-normal basic variables is shown below:

1. Define the appropriate limit state function.
2. Make an initial guess at the reliability index, β.
3. Set the initial checking point values, $X_i^* = \mu_i$, for all i variables.
4. Compute the equivalent normal mean and standard deviation for non-normal variables.
5. Compute the partial derivatives $(\partial g/\partial X_i)$ evaluated at the design point X_i^*.
6. Compute the direction cosines, α_i, as

$$\alpha_i = \frac{\frac{\partial g}{\partial X_i}\sigma_i^N}{\sqrt{\sum_i \left(\frac{\partial g}{\partial X_i}\sigma_i^N\right)^2}} \qquad (26.49)$$

7. Compute the new values of design point X_i^* as

$$X_i^* = \mu_i^N - \alpha_i\beta\sigma_i^N \qquad (26.50)$$

8. Repeat steps 4 through 7 until estimates of α_i stabilize (usually fast).
9. Compute the value of β such that $g(X_1^*, \ldots, X_n^*) = 0$.
10. Repeat steps 4 through 9 until the value for β converges. (This normally occurs within five cycles or less, depending on the nonlinearity of the limit state function.)

As with the previous procedure, this method is easily programmed on the computer. Many spreadsheet programs and other numerical analysis software packages also have considerable statistical capabilities, and therefore can be used to perform these types of analyses. This procedure can also be modified to estimate a design parameter (i.e., a section modulus) such that a specific target reliability is achieved. Other procedures are presented elsewhere in the literature [3, 12, 21, 23] including a somewhat different technique in which the equivalent normal mean and standard deviation are used directly in the reduction of the variables to standard normal form (i.e., u_i space). Additional information on SORM techniques may be found in the literature [8, 9].

26.4.3 Monte Carlo Simulation

An alternative to integration of the relevant joint probability equation over the domain of random variables corresponding to failure is to use Monte Carlo simulation (MCS). While FORM/SORM techniques are approximate in the case of nonlinear limit state functions, or with non-normal random variables (even when advanced FORM/SORM techniques are used), MCS offers the advantage of providing an exact solution to the failure probability. The potential disadvantage of MCS is the amount of computing time needed, especially when very small probabilities of failure are being estimated. Still, as computing power continues to increase and with the development and refinement of variance reduction techniques (VRTs) MCS is becoming more accepted and more utilized, especially for the analysis of increasingly complicated structural systems. VRTs such as importance sampling, stratified sampling, and Latin hypercube sampling can often be used to significantly reduce the number of simulations required to obtain reliable estimates of the failure probability.

A brief description of MCS is presented here. Additional information may be found elsewhere [21, 22]. The concept behind MCS is to generate sets of realizations of the random variables in the limit state function (with the assumed known probability distributions) and to record the number of times the resulting limit state function is less than zero (i.e., failure). The estimate of the probability of failure (P_f) then is simply the number of failures divided by the total number of simulations (N). Clearly, the accuracy of this estimate increases as N increases, and a larger number of simulations are required to reliably estimate smaller failure probabilities. Table 26.4 presents the number of simulations required to obtain three different confidence intervals on the estimate of P_f for some typical values in structural reliability analyses.

TABLE 26.4   Approximate Number of Simulations Required for Given Confidence Intervals ($\alpha \times 100\%$) on Reliability Index

$\beta \pm \varepsilon$	$\alpha = 0.90$ ($k = 1.64$)	$\alpha = 0.95$ ($k = 1.96$)	$\alpha = 0.99$ ($k = 2.58$)
$1.5 \pm .10$	1,000	1,400	2,500
$1.5 \pm .05$	4,000	5,700	9,800
$1.5 \pm .01$	100,000	142,000	246,000
$2.0 \pm .10$	2,000	3,000	5,100
$2.0 \pm .05$	8,200	12,000	20,500
$2.0 \pm .01$	240,000	342,000	592,000
$3.0 \pm .10$	18,000	25,600	44,300
$3.0 \pm .05$	75,000	107,000	186,000
$3.0 \pm .01$	2,270,000	3,240,000	5,610,000

The generation of random variates is a relatively simple task (provided the random variables may be assumed independent) and requires only (1) that the relevant CDF is invertable (or in the case of normal and lognormal variates, numerical approximations exist for the inverse CDF), and (2) that a uniform random number generator is available. (See the Appendix for two examples of uniform random number generators. Random number generators for other distributions may be available to you, and would further simplify the simulation analysis.) The generation of correlated variates is not described here, but information may be found in the literature [9, 21, 23] As shown in Figure 26.8, the value of the CDF for random variable X is (by definition) uniformly distributed on {0, 1}. Therefore, if we generate a uniform {0, 1} deviate and substitute this into the inverse of the CDF of interest (with the relevant parameters or moments), we obtain a realization of a variate with this CDF. For example, consider the generation of an exponential variate with parameter λ. The CDF is expressed:

$$F_X(x) = 1 - \exp(-\lambda x) \tag{26.51}$$

If we substitute u_i (a uniform {0, 1} deviate; see the Appendix) for $F_X(x)$ and invert the CDF to solve for x_i, we obtain

$$x_i = -\frac{1}{\lambda} \ln(1 - u_i) \tag{26.52}$$

Here, x_i is an exponential variate with parameter λ. As another example, consider the normal

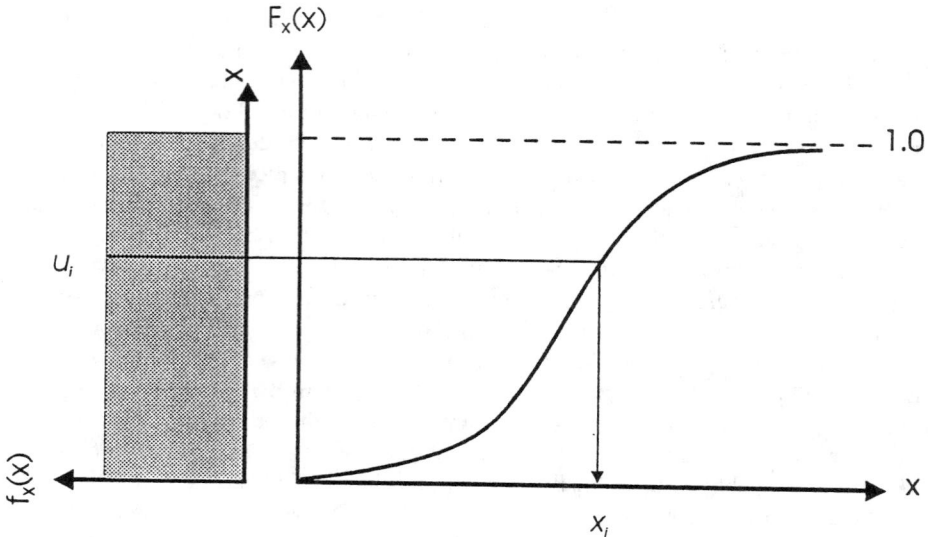

Figure 26.8   Random variable simulation.

distribution, for which no closed-form expression exists for the CDF or its inverse. The generalized normal CDF can be written as a function of the standard normal CDF as

$$F_X(x) = \Phi\left(\frac{x - \mu_x}{\sigma_x}\right) \tag{26.53}$$

Therefore, an expression for a generalized normal variate would be:

$$x_i = \mu_x + \sigma_x \Phi^{-1}(u_i) \tag{26.54}$$

where μ_x and σ_x are the mean and standard deviation, respectively, $u_i =$ uniform {0, 1} deviate, and $\Phi^{-1}(\cdot) =$ inverse standard normal CDF. While not available in closed form, numerical approximations for $\Phi^{-1}(\cdot)$ (i.e., in the form of algorithms or subroutines) are available (e.g., [12]). The Appendix presents approximate functions for both $\Phi(\cdot)$ and $\Phi^{-1}(\cdot)$. Table 26.5 presents the inverse CDFs for a number of common distribution types.

TABLE 26.5   Common Distributions, CDFs, and Inverse CDFs

Distribution	CDF ($= u_i$)	Inverse CDF
Normal	$F_X(x) = \Phi\left(\frac{x-\mu}{\sigma}\right)$	$x_i = \left(\Phi^{-1}(u_i) \times \sigma\right) + \mu$
Lognormal	$F_X(x) = \Phi\left(\frac{\ln x - \lambda}{\xi}\right)$	$x_i = \exp\left[\left(\Phi^{-1}(u_i) \times \xi\right) + \lambda\right]$
Uniform	$F_X(x) = \frac{x-a}{b-a}$	$x_i = a + (b-a)u_i$
Exponential	$F_X(x) = 1 - \exp(-\lambda x)$	$x_i = -\frac{1}{\lambda}\ln(1 - u_i)$
Extreme Type I (largest), "Gumbel"	$F_X(x) = \exp\left(-\exp\left(-\alpha(x - u)\right)\right)$	$x_i = -\frac{1}{\alpha}\ln(-\ln u_i) + u$
Extreme Type II (largest)	$F_X(x) = \exp\left(-(u/x)^k\right)$	$x_i = u\left(-\ln u_i\right)^{-1/k}$
Extreme Type III (smallest), "Weibull"	$F_X(x) = 1 - \exp\left[-\left(\frac{x-\varepsilon}{w-\varepsilon}\right)^k\right]$	$x_i = (-\ln(1 - u_i))^{1/k}(w - \varepsilon) + \varepsilon$

MCS can provide a very powerful tool for the solution of a wide variety of problems. Improvements in efficiency over crude or direct MCS can be realized by improved algorithmic design (programming) and by the utilization of VRTs. Monte Carlo techniques can also be used for the simulation of discrete and continuous random processes.

26.5 System Reliability

26.5.1 Introduction

While most structural codes in the U.S. treat design on a member-by-member basis, most elements within a structure are actually performing as part of an often complicated structural system. Interest in characterizing the performance and safety of structural systems has led to an increased interest in the area of system reliability. The classical theories of series and parallel system reliability are well developed and have been applied to the analysis of such complicated structural systems as nuclear power plants and offshore structures. In the following sections, a brief introduction to system reliability is presented along with some examples. This subject within the broad field of structural reliability is relatively new, and advances both in the theory and application of system reliability concepts to civil engineering design can be expected in the coming years.

26.5.2 Basic Systems

The two types of systems in classical theory are the series (or weakest link) system and the parallel system. The literature is replete with formulations for the reliability of these systems, including the possibility of correlated element strengths (e.g., [23, 24]). The relevant limit state is defined by the system type. For a series system, the system limit state is taken by definition to correspond to the first member failure, hence the name "weakest link." In the case of the strictly parallel system, the system limit state is taken by definition to correspond to the failure of all members. Formulations for the system reliability of a parallel system in which the load-deformation behavior of the members is assumed to be ductile or brittle are both well developed and presented in the literature (see [24], for example). In all cases, the system reliabilities are expressed in terms of the component (or member) reliabilities.

Classical system reliability theory has been able to be extended somewhat to model more complicated systems using combinations of series and parallel systems. These formulations, however, are still subject to limitations with regard to possible load sharing (distribution of load among components of the system) and time-dependent effects, such as degrading member resistances.

26.5.3 Introduction to Classical System Reliability Theory

For a system limit state defined by $g(x_1, \ldots, x_m) = 0$, where x_i are the basic variables, the failure probability is computed as the integral over the failure domain $(g(X) < 0)$ of the joint probability density function of X. In general, the failure of any system can be expressed as a union and/or intersection of events. For example, the failure of an ideal series (or weakest link) system may be expressed,

$$F_{\text{sys}} = F_1 \cup F_2 \cup \ldots \cup F_m \qquad (26.55)$$

in which $\cup$ denotes the Boolean OR operator and $F_i = i$th component (element) failure event. A statically determinate truss is modeled as a series system since the failure of the truss corresponds to the failure of any single member. Both first-order and second-order (which includes information on the joint probability behavior) bounds have been developed to express the system failure probability as a function of the individual element failure probabilities. These formulations are well developed and presented in the literature [3, 10, 21, 24].

The failure of a strictly parallel system may be expressed,

$$F_{\text{sys}} = F_1 \cap F_2 \cap \ldots \cap F_m \tag{26.56}$$

in which $\cap$ denotes the Boolean AND operator. Such is the case for the classical "Daniels" system of parallel, ductile rods or cables subject to equal deformation. In this case, system failure corresponds to the failure of all members or elements. First- and second-order bounds are also available for this system idealization (e.g., [17]). Furthermore, bounds that account for possible dependence of failure modes (modal correlation) have been developed [3]. If the parallel system is composed of brittle elements, the analysis may be further complicated by having to account for load redistribution following member failure. This total failure may therefore be the result of progressive element failures.

Returning again to the two fundamental system types, series and parallel, we can examine the probability distributions for the strength of these systems as functions of the distributions of the strengths of the individual members (elements). In the simple structural idealization of a series system of n elements (for which the characterization of the member failures as brittle or ductile is irrelevant since system failure corresponds to first-member failure), the distribution function for the system strength, R_{sys}, can be expressed:

$$F_{R_{\text{sys}}}(r) = 1 - \prod_{i=1}^{n} \left(1 - F_{R_i}(r)\right) \tag{26.57}$$

where the individual member strengths are assumed independent. In Equation 26.57, $F_{R_i}(r) =$ distribution function (CDF) for the individual member resistance. If the n individual member strengths are also identically distributed (i.e., have the same parent distribution, $F_R(r)$, with the same moments), Equation 26.57 can be simplified to

$$F_{R_{\text{sys}}}(r) = 1 - (1 - F_R(r))^n \tag{26.58}$$

In the case of the idealized parallel system of n elements, the system failure is dependent on whether the member behavior is perfectly brittle or perfectly ductile. In the simple case of the parallel system with n perfectly ductile elements, the system strength is given by

$$R_{\text{sys}} = \sum_{i=1}^{n} R_i \tag{26.59}$$

where $R_i =$ strength of element i. The central limit theorem (see [2, 5]) suggests that as the number of members in this system gets large, the system strength approaches a normal random variable, regardless of the distributions of the individual member strengths. When the member behavior is perfectly brittle, the system behavior is dependent on the degree of indeterminacy (redundancy) of the system and the ability of the system to redistribute loads to other members. For some applications, it may be reasonable to model structures idealized as parallel systems with brittle members as series systems, if the brittle failure of one member is likely to overload the remaining members. The issue of correlated member strengths (and correlated failure modes) is beyond the scope of this introduction, but information may be found in [3, 23, 24].

It is appropriate at this point to present the simple first-order bounds for the two fundamental systems. Additional information on the development and application of these as well as the second-order bounds may be found in the literature cited previously. The first-order bounds for a series system are given by

$$\max_{i=1}^{n} \left\{ P_{f_i} \right\} \leq P_{f_{\text{sys}}} \leq 1 - \left(\prod_{i=1}^{n} (1 - P_{f_i}) \right) \tag{26.60}$$

where P_{f_i} = failure probability for member (element) i. The first-order bounds for a parallel system are given by

$$\prod_{i=1}^{n} P_{f_i} \leq P_{f_{sys}} \leq \min_{i=1}^{n} \left\{ P_{f_i} \right\} \qquad (26.61)$$

Improved (second-order) bounds (the first-order bounds are often too broad to be of practical use) that include information on the joint probability behavior (i.e., member or modal correlation) have been developed and are described in the literature (e.g., [3, 10, 24]).

Classical system reliability theory, as briefly introduced above, is limited in that it cannot account for more complicated load-deformation behavior and the time dependencies associated with load redistribution following (brittle) member failure. Generalized formulations for the reliability of systems that are neither strictly series nor strictly parallel type systems are not available. Analyses of these systems are often based on combined series and parallel system models in which the complete system is modeled as some arrangement of these classical subsystems. These solutions tend to be problem specific and still do not address any possible time-dependent or load-sharing issues.

26.5.4 Redundant Systems

A redundant (indeterminate) system may be defined as having some overload capacity following the failure of an element. The level of redundancy (or degree of indeterminacy) refers to the number of element failures that can be tolerated without the system failing. The reliability of such a structure is dependent on the nature (type) of redundancy. The level of redundancy dictates how many members can fail prior to collapse, and therefore answers the question, "Would the failure of member j lead to impending collapse?" Furthermore, load-deformation behavior of the individual members specifies whether or not the limit states are load-path dependent. For ductile element behavior (i.e., the Daniels system), the limit state is effectively load-path independent, implying the order of member failures is not significant. For a system of brittle elements, however, the limit state may be load-path dependent. In this case, the performance of the system is related to the load redistribution behavior following member failure, and hence the order (or relative position) of member failures becomes important. The parallel-member system model with brittle elements (i.e., perfectly elastic load-deformation behavior) is appropriate for (and has been used to model) a wide range of redundant structural systems including floors, roofs, and wall systems.

26.5.5 Examples

Three examples are described in this section. The first example considers a series system in which the elements are considered to represent different modes of failure. Modal failure analysis is often treated using the concepts of system reliability (i.e [3]). Here, the structure being considered (actually, the simply supported beam element, i.e., Figure 26.3) may fail in any one of three different modes: flexure, shear, and excessive deflection. (The last mode corresponds to a serviceability-type limit state rather than an ultimate strength type.) The "failure" of the structural element is assumed to occur when any of these limit states is violated. For simplicity, the modal failure probabilities are assumed to be uncorrelated. (For information on handling correlated failure modes, see [3]). In other words, the element (system) fails when it fails in flexure, or it fails in shear, or it experiences excessive deflection:

$$F_{sys} = F_M \cup F_V \cup F_\delta \qquad (26.62)$$

If, for example, the probabilities of moment, shear, and deflection failure, respectively, are given by $F_M = 0.0015$, $F_V = 0.002$, and $F_\delta = 0.005$, the first-order bounds shown in Equation 26.60 result in

$$0.005 \leq P_{f_{sys}} \leq 1 - (1 - 0.0015)(1 - 0.002)(1 - 0.005)$$

$$0.005 \leq P_{f_{sys}} \leq 0.0085 \tag{26.63}$$

This corresponds to a range for β of $2.39 \leq \beta_{sys} \leq 2.58$.

The second example considers a strictly parallel system of five cables supporting a load (see Figure 26.9). In this case, the system failure corresponds to the condition where the cable system

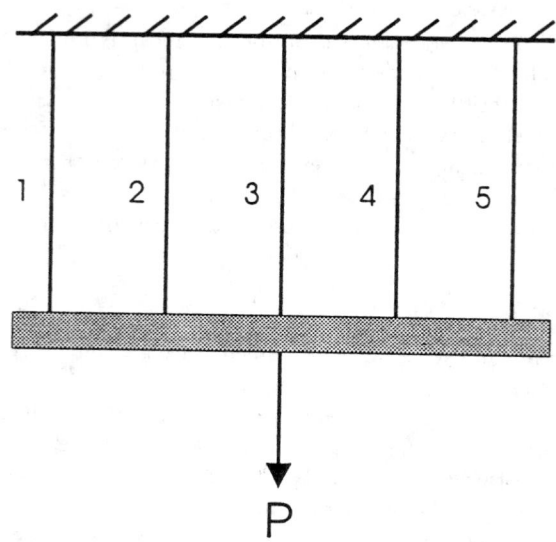

Figure 26.9   Five-element parallel system.

can no longer carry any load. Therefore, all of the cables must have failed for the system to have failed. In this simple example, the issue of load redistribution following the failure of one of the cables is not addressed; however, this problem has been studied extensively (e.g., [19]). Here, the five cable strengths are assumed to be statistically independent, and the system failure probability is the probability that P is large enough to fail all of the cables simultaneously:

$$F_{sys} = F_1 \cap F_2 \cap \ldots \cap F_5 \tag{26.64}$$

If, for example, the probability of failure of an individual cable is 0.001, and the cable strengths are assumed to be independent, identically distributed random variables, the first-order bounds on the system failure probability given by Equation 26.61 become

$$(0.001)^5 \leq P_{f_{sys}} \leq 0.001 \tag{26.65}$$

Here, the lower bound corresponds to the case of perfectly uncorrelated member strengths (i.e., independent cable failures), while the upper bound corresponds to the case of perfect correlation. These first-order bounds, as indicated by Equation 26.65, become very wide with increasing n. Here, information on correlation can be important in computing narrower and more useful bounds.

Finally, as a third example, a combined (series and parallel) system is considered. In this case, the event probabilities correspond to the failure probabilities of different components required for a safe shutdown of a nuclear power plant. While these events are assumed to be independent, their arrangement describing safe system performance (see Figure 26.10) forms a combined series-parallel system. In this case, the three subsystems are arranged in series: subsystem A is a series system and

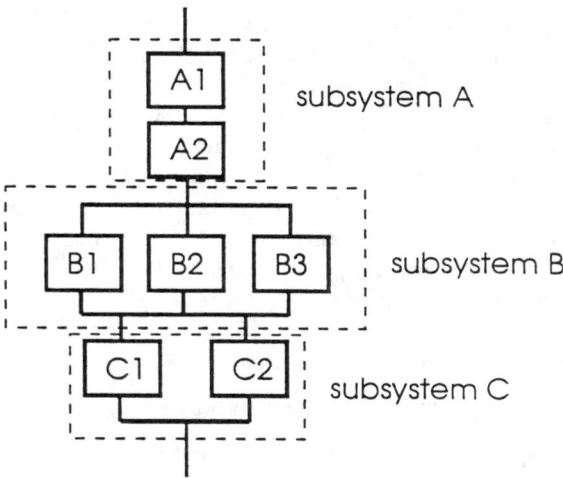

Figure 26.10 Safe shutdown of a nuclear power plant.

subsystems B and C are parallel systems. In this case, the system failure probability is given by

$$F_{sys} = F_A \cup F_B \cup F_C \qquad (26.66)$$

or, expressed in terms of the individual component failure probabilities:

$$F_{sys} = \left[F_{A_1} \cup F_{A_2} \right] \cup \left[F_{B_1} \cap F_{B_2} \cap F_{B_3} \right] \cup \left[F_{C_1} \cap F_{C_2} \right] \qquad (26.67)$$

26.6 Reliability-Based Design (Codes)

26.6.1 Introduction

This section will provide a brief introduction to reliability-based design concepts in civil engineering, with specific emphasis on structural engineering design. Since the 1970s, the theories of probability and statistics and reliability have provided the bases for modern structural design codes and specifications. Thus, probabilistic codes have been replacing previous deterministic-format codes in recent years. RBD procedures are intended to provide more predictable levels of safety and more risk-consistent (i.e., design-to-design) structures, while utilizing the most up-to-date statistical information on material strengths, as well as structural and environmental loads. An excellent discussion of RBD in the U.S. as well as other countries is presented in [21]. Other references are also available that deal specifically with probabilistic code development in the U.S. [12, 13, 15]. The following sections provide some basic information on the application of reliability theory to aspects of RBD.

26.6.2 Calibration and Selection of Target Reliabilities

Calibration refers to the linking of new design procedures to previous existing design philosophies. Much of the need for calibration arises from making any new code changes acceptable to the engineering and design communities. For purely practical reasons, it is undesirable to make drastic changes in the procedures for estimating design values, for example, or in the overall formats of design checking equations. If such changes are to be made, it is impractical and uneconomical to make them often. Hence, code development is an often slow process, involving many years and many revisions. The other justification for code calibration has been the notion that previous design philosophies (i.e., ASD) have resulted in safe designs (or designs with acceptable levels of performance), and that therefore these previous levels of safety should serve as benchmarks in the development of new specifications or procedures (i.e., LRFD).

The actual process of calibration is relatively simple. For a given design procedure (i.e., ASD for steel beams in flexure), estimate the reliability based on the available statistical information on the loads and resistances and the governing checking equation. This becomes the target reliability and is used to develop the appropriate load and resistance factors, for example, for the new procedure (i.e., LRFD). In the development of LRFD for both steel and wood, for example, the calibration process revealed an inconsistency in the reliability levels for different load combinations. As this was undesirable, a single target reliability was selected and the new LRFD procedures were able to correct this problem. For more information on code calibration, the reader is referred to the literature [12, 15, 21].

26.6.3 Material Properties and Design Values

The basis for many design values encountered in structural engineering design is now probabilistic. Earlier design values were often based on mean values of member strength, for example, with the factor of safety intended to account for all forms of uncertainty, including material property variability. Later, as more statistical information became available, as people became more aware of the concept of relative uncertainty, and with the use of probabilistic methods in code development, characteristic values were selected for use in design. The characteristic values (referred to as nominal or design values in most specifications) are generally selected from the lower tail of the distribution describing the material property (see Figure 26.11). Typically, the 5th percentile value (that value below which 5% of the probability density lies) is selected as the **nominal resistance** (i.e., nominal strength), though in some cases, a different percentile value may be selected. While this value may serve as the starting point for establishing the design value, modifications are often needed to account for such things as size effects, system effects, or (in the case of wood) moisture content effects, etc. The bases for the design resistance values for specifications in the U.S. are described in the literature (e.g., [12, 16, 20]). An excellent review of resistance modeling and a summary of statistical properties for structural elements is presented in [21]. Table 26.6 presents some typical resistance statistics for concrete and steel members. Additional statistics are available, along with statistics for masonry, aluminum, and wood members in [12] as well. The mean values are presented in ratio to their nominal (or design) values, m_R/R_n. In addition, the coefficient of variation, V_R, and the PDF are listed in Table 26.6.

26.6.4 Design Loads and Load Combinations

The selection of design load values, such as those found in the ASCE 7-95 standard [4] (formerly the ANSI A58.1 standard), *Minimum Design Loads for Buildings and Other Structures*, is also largely probability based. Though somewhat more complicated than the selection of design resistance values as described above, the concept is quite similar. Of course, greater complexity is introduced since we may be concerned with both spatial and temporal variations in the load effects. In addition, because of the difficulties in conducting load surveys, and the large amount of variability associated with naturally occurring phenomena giving rise to many structural and environmental loadings, there is a high degree of uncertainty associated with these quantities. A number of load surveys have been conducted, and the valuable data collected have formed the basis for many of our design values (e.g., [7, 11, 14, 18]). When needed, such as in the case where data simply are not available or able to be collected with any reasonable amount of effort, this information is supplemented by engineering judgment and expert opinion. Therefore, design load values are based on (1) statistical information, such as load survey data, and (2) engineering judgment, including past experience, and scenario analysis. As shown in Figure 26.11, the design load value can be visualized as some characteristic value in the upper tail of the distribution describing the load. For example, the 95th

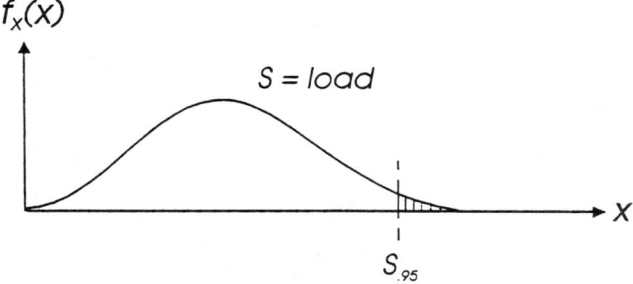

(a) design load (e.g., 5% exceedence probability)

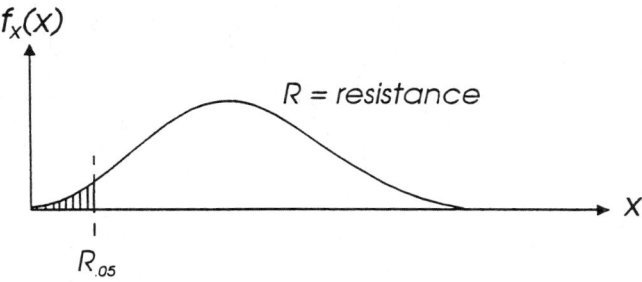

(b) nominal resistance (e.g., 5th-percentile)

Figure 26.11 Typical specification of design (nominal) load and resistance values.

percentile wind speed is that value of wind speed that has a 5% $(1 - .95)$ exceedence probability. Probabilistic load modeling represents an extensive area of research, and a significant amount of work is reported on in the literature [21, 28]. A summary of load statistics is presented in Table 26.7.

In most codes, a number of different load combinations are suggested for use in the appropriate checking equation format. For example, the ASCE 7-95 standard recommends the following load combinations [4]:

$$
\begin{aligned}
U &= 1.4D_n \\
U &= 1.2D_n + 1.6L_n \\
U &= 1.2D_n + 1.6S_n + (0.5L_n \text{ or } 0.8W_n) \\
U &= 1.2D_n + 1.3W_n + 0.5L_n \\
U &= 1.2D_n + 1.0E_n + 0.5L_n + 0.2S_n \\
U &= 0.9D_n + (-1.3W_n \text{ or } 1.0E_n)
\end{aligned}
\tag{26.68}
$$

where D_n, L_n, S_n, W_n, and E_n are the nominal (design) values for dead load, live load, snow load, wind load, and earthquake load, respectively. A similar set of load combinations may be found in both the ACI and AISC specifications, though in the case of the ACI code the load factors (developed earlier) are slightly different. These load combinations were developed in order to ensure essentially equal exceedence probabilities for all combinations, U. A discussion of the bases for these load

TABLE 26.6   Typical Resistance Statistics for Concrete and Steel Members

	Type of member	m_R/R_n	V_R
Concrete elements			
Flexure, reinforced concrete	Continuous one-way slabs	1.22	0.16
	Two-way slabs	1.12-1.16	0.15
	One-way pan joists	1.13	0.14
	Beams, grade 40, $f'_c = 5$ ksi	1.14-118	0.14
	Beams, grade 60, $f'_c = 5$ ksi	1.01-1.09	0.08-0.12
	Overall values	1.05	0.11
Flexure, prestressed concrete	Plant precast pretensioned	1.06	0.08
	Cast-in-place post-tensioned	1.04	0.10
Axial load and flexure	Short columns, compression	0.95-1.05	0.14-0.16
	Short columns, tension	1.05	0.12
	Slender columns, compression	1.10	0.17
	Slender columns, tension	0.95	0.12
Shear	Beams with $a/d < 2.5$, $\rho_w = 0.008$:		
	No stirrups	0.93	0.21
	Minimum stirrups	1.00	0.19
	Moderate stirrups	1.09	0.17
Hot-rolled steel elements			
	Tension member, yield	1.05	0.11
	Tension member, ultimate	1.10	0.11
	Compact beam, uniform moment	1.07	0.13
	Compact beam, continuous	1.11	0.13
	Elastic beam, LTB	1.03	0.12
	Inelastic beam, LTB	1.11	0.14
	Beam columns	1.07	0.15
	Plate-girders, flexure	1.08	0.12
	Plate girders, shear	1.14	0.16
	Compact composite beams	1.04	0.14
	Fillet welds	0.88	0.18
	ASS bolts in tension, A325	1.20	0.09
	ASS bolts in tension, A490	1.07	0.05
	HSS bolts in shear, A325	0.60	0.10
	HSS bolts in shear, A490	0.52	0.07

Adapted from Ellingwood, B., Galambos, T.V., MacGregor, J.G., and Cornell, C.A. 1980. "Development of a Probability Based Load Criterion for American National Standard A58," NBS Special Publication SP577, National Bureau of Standards, Washington, D.C.

TABLE 26.7   Typical Load Statistics

Load type	Mean-to-nominal	COV	Distribution
Dead load	1.05	0.10	Normal
Live load			
Sustained component	0.30	0.60	Gamma
Extraordinary component	0.50	0.87	Gamma
Total (max., 50 years)	1.00	0.25	Type I
Snow load (annual max.)			
General site (northeast U.S.)	0.20	0.87	Lognormal
Wind load			
50-year maximum	0.78	0.37	Type I
Annual maximum	0.33	0.59	Type I
Earthquake load	0.5-1.0	0.5-1.4	Type II

combinations may be found in [12]. A comparison of LRFD with other countries' codes may be found in [21].

One important tool used in the development of the load combinations is known as Turkstra's Rule [25, 26], developed as an alternative to more complicated load combination analysis. This rule states that, in effect, the maximum of a combination of two or more load effects will occur when one of the loads is at its maximum value while the other loads take on their instantaneous or arbitrary point-in-time values. Therefore, if n time-varying loads are being considered, there are at least n corresponding load combinations that would need to be considered. This rule may be written generally as

$$\max \{Z\} = \max_i \left[\max_T X_i(t) + \sum_{\substack{j=1 \\ j \neq i}}^{n} X_j(t) \right] \tag{26.69}$$

where $\max \{Z\}$ = maximum combined load, $X_i(t)$, $i = 1, \ldots, n$ are the time-varying loads being considered in combination, and t = time. In the equation above, the first term in the brackets represents the maximum in the lifetime (T) of load X_i, while the second term is the sum of all other loads at their point-in-time values. This approximation may be unconservative in some cases where the maximum load effect occurs as a result of the combination of multiple loads at near maximum values. However, in most cases, the probability of this occurring is small, and thus Turkstra's Rule has been shown to be a good approximation for most structural load combinations [27].

26.6.5 Evaluation of Load and Resistance Factors

Recall that for the generalized case of non-normal random variables, the following expression was developed (see Equation 26.50):

$$X_i^* = \mu_i^N - \alpha_i \beta \sigma_i^N \tag{26.70}$$

If we further define the design point value X_i^* in terms of a nominal (design) value X_n:

$$X_i^* = \gamma_i X_n \tag{26.71}$$

where γ_i = partial factor on load X_i (or the inverse of the resistance factor). Therefore, for the popular LRFD format in the U.S. in which the design equation has the form

$$\phi R_n \geq \sum_i \gamma_i X_{n,i} \tag{26.72}$$

the **load factors** may be computed as

$$\gamma_i = \frac{\mu_i^N - \alpha_i \hat{\beta} \sigma_i^N}{X_{n,i}} \tag{26.73}$$

and the **resistance factor** is given by

$$\phi = \frac{R_n}{\mu_i^N - \alpha_i \hat{\beta} \sigma_i^N} \tag{26.74}$$

In Equations 26.73 and 26.74, α_i = direction cosine from the convergent iterative solution for random variable i, β =convergent reliability index (i.e., the target reliability), and $X_{n,i}$ and R_n are the nominal load and resistance values, respectively.

Additional information on the evaluation of load and resistance factors based on FORM/SORM techniques, as well as comparisons between different code formats, may be found in the literature [3, 12, 21].

26.7 Defining Terms[2]

Allowable stress design (or working stress design): A method of proportioning structures such that the computed elastic stress does not exceed a specified limiting stress.

---

[2]Selected terms taken from [12].

Calibration: A process of adjusting the parameters in a new standard to achieve approximately the same reliability as exists in a current standard or specification.

Factor of safety: A factor by which a designated limit state force or stress is divided to obtain a specified limiting value.

Failure: A condition where a limit state is reached.

FORM/SORM (FOSM): First- and Second-order reliability methods (first-order second-moment reliability methods). Methods that involve (1) a first- or second-order Taylor series expansion of the limit state surface, and (2) computing a notional reliability measure that is a function only of the means and variances (first two moments) of the random variables. (Advanced FOSM includes full distribution information as well as any possible correlations of random variables.)

Limit state: A criterion beyond which a structure or structural element is judged to be no longer useful for its intended function (serviceability limit state) or beyond which it is judged to be unsafe (ultimate limit state).

Limit states design: A design method that aims at providing safety against a structure or structural element being rendered unfit for use.

Load factor: A factor by which a nominal load effect is multiplied to account for the uncertainties inherent in the determination of the load effect.

LRFD: Load and resistance factor design. A design method that uses load factors and resistance factors in the design format.

Nominal load effect: Calculated using a nominal load; the nominal load frequently is determined with reference to a probability level; e.g., 50-year mean recurrence interval wind speed used in calculating the wind load for design.

Nominal resistance: Calculated using nominal material and cross-sectional properties and a rationally developed formula based on an analytical and/or experimental model of limit state behavior.

Reliability: A measure of relative safety of a structure or structural element.

Reliability-based design (RBD): A design method that uses reliability (probability) theory in the safety checking process.

Resistance factor: A factor by which the nominal resistance is multiplied to account for the uncertainties inherent in its determination.

Acknowledgments

The author is grateful for the comments and suggestions provided by Professor James T. P. Yao at Texas A&M University and Professor Theodore V. Galambos at the University of Minnesota. In addition, discussions with Professor Bruce Ellingwood at Johns Hopkins University were very helpful in preparing this chapter.

References

[1] Abramowitz, M. and Stegun, I.A., Eds. 1966. *Handbook of Mathematical Functions*, Applied Mathematics Series No. 55, National Bureau of Standards, Washington, D.C.
[2] Ang, A.H.-S. and Tang, W.H. 1975. *Probability Concepts in Engineering Planning and Design, Volume I: Basic Principles*, John Wiley & Sons, New York.
[3] Ang, A.H.-S. and Tang, W.H. 1975. *Probability Concepts in Engineering Planning and Design, Volume II: Decision, Risk, and Reliability*, John Wiley & Sons, New York.

[4] American Society of Civil Engineers. 1996. *Minimum Design Loads for Buildings and Other Structures*, ASCE 7-95, New York.

[5] Benjamin, J.R. and Cornell, C.A. 1970. *Probability, Statistics, and Decision for Civil Engineers*, McGraw-Hill, New York.

[6] Bratley, P., Fox, B.L., and Schrage, L.E. 1987. *A Guide to Simulation*, Second Edition, Springer-Verlag, New York.

[7] Chalk, P. and Corotis, R.B. 1980. A Probability Model for Design Live Loads, *J. Struct. Div.*, ASCE, 106(10):2017-2033.

[8] Chen, X. and Lind, N.C. 1983. Fast Probability Integration by Three-Parameter Normal Tail Approximation, *Structural Safety*, 1(4):269-276.

[9] Der Kiureghian, A. and Liu, P.L. 1986. Structural Reliability Under Incomplete Probability Information, *J. Eng. Mech.*, ASCE, 112(1):85-104.

[10] Ditlevsen, O. 1981. *Uncertainty Modelling*, McGraw-Hill, New York.

[11] Ellingwood, B. and Culver, C.G. 1977. Analysis of Live Loads in Office Buildings, *J. Struct. Div.*, ASCE, 103(8):1551-1560.

[12] Ellingwood, B., Galambos, T.V., MacGregor, J.G., and Cornell, C.A. 1980. Development of a Probability Based Load Criterion for American National Standard A58, NBS Special Publication *SP577*, National Bureau of Standards, Washington, D.C.

[13] Ellingwood, B., MacGregor, J.G., Galambos, T.V. and Cornell, C.A. 1982. Probability Based Load Criteria: Load Factors and Load Combinations, *J. Struct. Div.*, ASCE, 108(5):978-997.

[14] Ellingwood, B. and Redfield, R. 1982. Ground Snow Loads for Structural Design, *J. Struct. Eng.*, ASCE, 109(4):950-964.

[15] Galambos, T.V., Ellingwood, B., MacGregor, J.G., and Cornell, C.A. 1982. Probability Based Load Criteria: Assessment of Current Design Practice, *J. Struct. Div.*, ASCE, 108(5):959-977.

[16] Galambos, T.V. and Ravindra, M.K. 1978. Properties of Steel for Use in LRFD, *J. Struct. Div.*, ASCE, 104(9):1459-1468.

[17] Grigoriu, M. 1989. Reliability of Daniels Systems Subject to Gaussian Load Processes, *Structural Safety*, 6(2-4):303-309.

[18] Harris, M.E., Corotis, R.B., and Bova, C.J. 1981. Area-Dependent Processes for Structural Live Loads, *J. Struct. Div.*, ASCE, 107(5):857-872.

[19] Hohenbichler, M. and Rackwitz, R. 1983. Reliability of Parallel Systems Under Imposed Uniform Strain, *J. Eng. Mech. Div.*, ASCE, 109(3):896-907.

[20] MacGregor, J.G., Mirza, S.A., and Ellingwood, B. 1983. Statistical Analysis of Resistance of Reinforced and Prestressed Concrete Members, *ACI J.*, 80(3):167-176.

[21] Melchers, R.E. 1987. *Structural Reliability: Analysis and Prediction*, Ellis Horwood Limited, distributed by John Wiley & Sons, New York.

[22] Rubinstein, R.Y. 1981. *Simulation and the Monte Carlo Method*, John Wiley & Sons, New York.

[23] Thoft-Christensen, P. and Baker, M.J. 1982. *Structural Reliability Theory and Its Applications*, Springer-Verlag, Berlin.

[24] Thoft-Christensen, P. and Murotsu, Y. 1986. *Application of Structural Systems Reliability Theory*, Springer-Verlag, Berlin.

[25] Turkstra, C.J. 1972. Theory of Structural Design Decisions, *Solid Mech. Study No. 2*, University of Waterloo, Ontario, Canada.

[26] Turkstra, C.J. and Madsen, H.O. 1980. Load Combinations in Codified Structural Design, *J. Struct. Div.*, ASCE, 106(12):2527-2543.

[27] Wen, Y.-K. 1977. Statistical Combinations of Extreme Loads, *J. Struct. Div.*, ASCE, 103(6):1079-1095.

[28] Wen, Y.-K. 1990. *Structural Load Modeling and Combination for Performance and Safety Evaluation*, Elsevier, Amsterdam.

Further Reading

Melchers [21] provides one of the best overall presentations of structural reliability, both its theory and applications. Ang and Tang [3] also provides a good summary. For a more advanced treatment, refer to Ditlevsen [10], Thoft-Christensen and Baker [23], or Thoft-Christensen and Murotsu [24].

The International Conference on Structural Safety and Reliability (ICOSSAR) and the International Conference on the Application of Statistics and Probability in Civil Engineering (ICASP) are each held every 4 years. The proceedings from these conferences include short papers on a variety of state-of-the-art topics in structural reliability. The conference proceedings may be found in the engineering libraries at most universities. A number of other conferences, including periodic specialty conferences cosponsored by ASCE, also include sessions pertaining to reliability.

Appendix

Some Useful Functions for Simulation

1. $\Phi(\cdot) = $ standard normal cumulative distribution function
   Approximate algorithm [1]:

$$\Phi(x) = 1 - \tfrac{1}{2}\left(1 + c_1 x + c_2 x^2 + c_3 x^3 + c_4 x^4\right)^{-4} + \varepsilon(x) \quad ; \quad x \geq 0$$

$$|\varepsilon(x)| < 2.5 \times 10^{-4}$$

$$c_1 = 0.196854 \quad c_2 = 0.115194 \quad c_3 = 0.000344 \quad c_4 = 0.019527$$

Note: $\Phi(-x) = 1 - \Phi(x)$

2. $\Phi^{-1}(\cdot) = $ inverse standard normal cumulative distribution function.
   Simple approximate routine for inverse standard normal CDF:

   Input: CDF

   Output: Z (also: CDF is returned unchanged)

   Format: $Z = \Phi^{-1}(\text{CDF})$

   Usage: `call invcdf(arg1, arg2)`

```
subroutine invcdf(cdf,z)
data a0,a1,a2 / 2.515517, 0.802853, 0.010328/
data b1,b2,b3 / 1.432788, 0.189269, 0.001308/
y = cdf
if(cdf .gt. 0.5) y = 1.0 - cdf
if(y .le. 0.0) y = 3.0E-39     (note: machine-dependent)
v = sqrt(-2.0*alog(y))
z = v-(a0+v*(a1+v*a2))/(1.0+v*(b1+v*(b2+v*b3)))
if(cdf .lt. 0.5) z = -z
return
end
```

3. RNUM = uniform {0, 1} random number generator.

   Simple routine to generate an array "`arrcdf`" of "`inum`" uniform {0, 1} random deviates. (The parameter "`dseed`" is just a starting random seed.)

```
subroutine RNUM(dseed,inum,arrcdf)
dimension arrcdf(50000)
double precision dseed
k = 5701
j = 3612
m = 566927
rm = 566927.0
do 100 i = 1,inum
   ix = int(dseed* rm)
   irand = mod (j* ix + k, m)
   arrcdf(i) = (real(irand) + 0.5)/rm
   dseed = arrcdf(i)
100   continue
return
end
```

4. UNIF = uniform random deviate between 0 and 1.

   Recursive uniform {0, 1} random number generator [6]:

   INPUT: I_X = any integer between 0 and 2147483647

   OUTPUT: new pseudorandom value, and uniform variate between 0 and 1

$$K_1 = \frac{I_X}{127773}$$

$$I_X = 16807\left(I_X - \frac{K_1}{127773}\right) - 2836(K_1)$$

   if $(I_X < 0)$ then $(I_X = I_X + 2147483647)$

   $\text{UNIF} = I_X \times 4.656612875E^{-10}$

27

Passive Energy Dissipation and Active Control

27.1 Introduction 27-1
27.2 Basic Principles and Methods of Analysis 27-3
Single-Degree-of-Freedom Structural Systems • Multi-Degree-of-Freedom Structural Systems • Energy Formulations • Energy-Based Design
27.3 Recent Development and Applications 27-12
Passive Energy Dissipation • Active Control
27.4 Code Development 27-22
27.5 Concluding Remarks 27-23
References 27-23

T.T. Soong and
G.F. Dargush
Department of Civil Engineering,
State University of New York
at Buffalo, Buffalo, NY

27.1 Introduction

In recent years, innovative means of enhancing structural functionality and safety against natural and man-made hazards have been in various stages of research and development. By and large, they can be grouped into three broad areas, as shown in Table 27.1: (1) base isolation, (2) passive energy dissipation, and (3) active control. Of the three, base isolation can now be considered a more mature technology with wider applications as compared with the other two [2].

TABLE 27.1 Structural Protective Systems

Seismic isolation	Passive energy dissipation	Active control
Elastomeric bearings	Metallic dampers Friction dampers	Active bracing systems Active mass dampers
Lead rubber bearings	Viscoelastic dampers	Variable stiffness or damping systems
	Viscous fluid dampers	
Sliding friction pendulum	Tuned mass dampers Tuned liquid dampers	

Passive energy dissipation systems encompass a range of materials and devices for enhancing damping, stiffness, and strength, and can be used both for natural hazard mitigation and for rehabilitation of aging or deficient structures [46]. In recent years, serious efforts have been undertaken to develop the concept of energy dissipation, or supplemental damping, into a workable technology, and a number of these devices have been installed in structures throughout the world. In general,

0-8493-2674-5/97/$0.00+$.50

such systems are characterized by a capability to enhance energy dissipation in the structural systems in which they are installed. This effect may be achieved either by conversion of kinetic energy to heat or by transferring of energy among vibrating modes. The first method includes devices that operate on principles such as frictional sliding, yielding of metals, phase transformation in metals, deformation of viscoelastic solids or fluids, and fluid orificing. The latter method includes supplemental oscillators that act as dynamic absorbers. A list of such devices that have found applications is given in Table 27.1.

Among the current passive energy dissipation systems, those based on deformation of viscoelastic polymers and on fluid orificing represent technologies in which the U.S. industry has a worldwide lead. Originally developed for industrial and military applications, these technologies have found recent applications in natural hazard mitigation in the form of either energy dissipation or elements of seismic isolation systems.

The possible use of active control systems and some combinations of passive and active systems, so-called hybrid systems, as a means of structural protection against wind and seismic loads has also received considerable attention in recent years. Active/hybrid control systems are force delivery devices integrated with real-time processing evaluators/controllers and sensors within the structure. They must react simultaneously with the hazardous excitation to provide enhanced structural behavior for improved service and safety. Figure 27.1 is a block diagram of the active structural control problem. The basic task is to find a control strategy that uses the measured structural responses

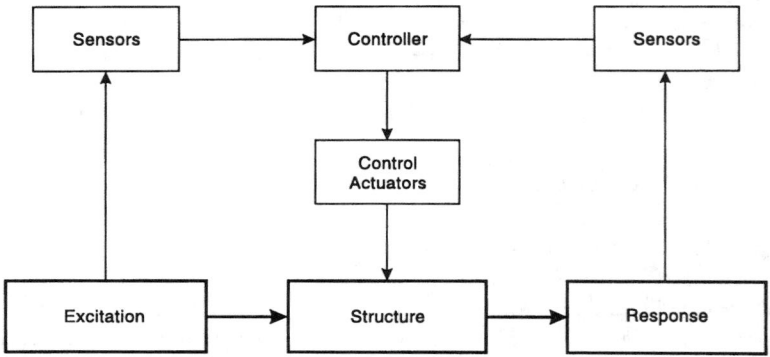

Figure 27.1   Block diagram of active structural control.

to calculate the control signal that is appropriate to send to the actuator. Structural control for civil engineering applications has a number of distinctive features, largely due to implementability issues, that set it apart from the general field of feedback control. First of all, when addressing civil structures, there is considerable uncertainty, including nonlinearity, associated with both physical properties and disturbances such as earthquakes and wind. Additionally, the scale of the forces involved is quite large, there are a limited number of sensors and actuators, the dynamics of the actuators can be quite complex, and the systems must be fail safe [10, 11, 23, 24, 27, 44].

Nonetheless, remarkable progress has been made over the last 20 years in research on using active and hybrid systems as a means of structural protection against wind, earthquakes, and other hazards [45, 47]. Research to date has reached the stage where active systems such as those listed in Table 27.1 have been installed in full-scale structures. Some active systems are also used temporarily in construction of bridges or large span structures (e.g., lifelines, roofs) where no other means can provide adequate protection. Additionally, most of the full-scale systems have been subjected to actual wind forces and ground motions, and their observed performances provide invaluable information in terms of (1) validating analytical and simulation procedures used to predict system performance, (2) verifying complex electronic-digital-servohydraulic systems under actual loading conditions, and (3) verifying the capability of these systems to operate or shutdown under prescribed conditions.

The focus of this chapter is on passive energy dissipation and active control systems. Their basic operating principles and methods of analysis are given in Section 27.2, followed by a review in Section 27.3 of recent development and applications. Code development is summarized in Section 27.4, and some comments on possible future directions in this emerging technological area are advanced in Section 27.5. In the following subsections, we shall use the term *structural protective systems* to represent either passive energy dissipation systems or active control systems.

27.2 Basic Principles and Methods of Analysis

With recent development and implementation of modern structural protective systems, the entire structural engineering discipline is now undergoing a major change. The traditional idealization of a building or bridge as a static entity is no longer adequate. Instead, structures must be analyzed and designed by considering their dynamic behavior. It is with this in mind that we present some basic concepts related to topics that are of primary importance in understanding, analyzing, and designing structures that incorporate structural protective systems.

In what follows, a simple single-degree-of-freedom (SDOF) structural model is discussed. This represents the prototype for dynamic behavior. Particular emphasis is given to the effect of damping. As we shall see, increased damping can significantly reduce system response to time-varying disturbances. While this model is useful for developing an understanding of dynamic behavior, it is not sufficient for representing real structures. We must include more detail. Consequently, a multi-degree-of-freedom (MDOF) model is then introduced, and several numerical procedures are outlined for general dynamic analysis. A discussion comparing typical damping characteristics in traditional and control-augmented structures is also included. Finally, a treatment of energy formulations is provided. Essentially one can envision an environmental disturbance as an injection of energy into a structure. Design then focuses on the management of that energy. As we shall see, these energy concepts are particularly relevant in the discussion of passively or actively damped structures.

27.2.1 Single-Degree-of-Freedom Structural Systems

Consider the lateral motion of the basic SDOF model, shown in Figure 27.2, consisting of a mass, m, supported by springs with total linear elastic stiffness, k, and a damper with linear viscosity, c. This SDOF system is then subjected to an external disturbance, characterized by $f(t)$. The excited model responds with a lateral displacement, $x(t)$, relative to the ground, which satisfies the equation of motion:

$$m\ddot{x} + c\dot{x} + kx = f(t) \tag{27.1}$$

in which a superposed dot represents differentiation with respect to time. For a specified input, $f(t)$, and with known structural parameters, the solution of this equation can be readily obtained.

In the above, $f(t)$ represents an arbitrary environmental disturbance such as wind or an earthquake. In the case of an earthquake load,

$$f(t) = -m\ddot{x}_g(t) \tag{27.2}$$

where $\ddot{x}_g(t)$ is ground acceleration.

Consider now the addition of a generic passive or active control element into the SDOF model, as indicated in Figure 27.3. The response of the system is now influenced by this additional element. The symbol Γ in Figure 27.3 represents a generic integrodifferential operator, such that the force

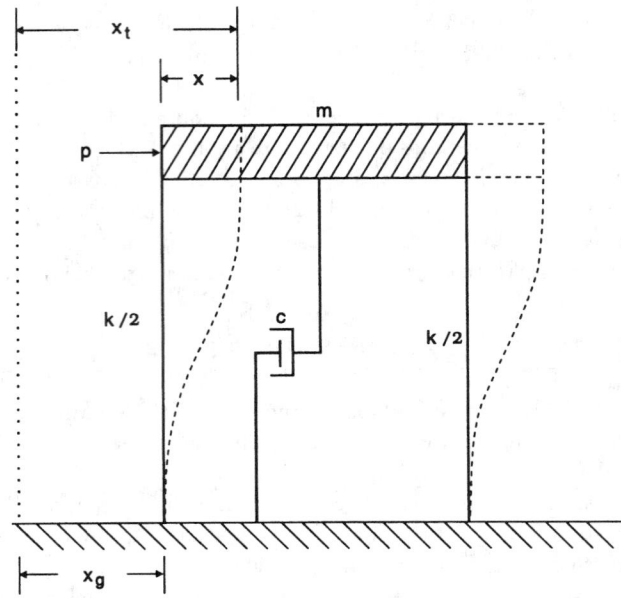

Figure 27.2   SDOF model.

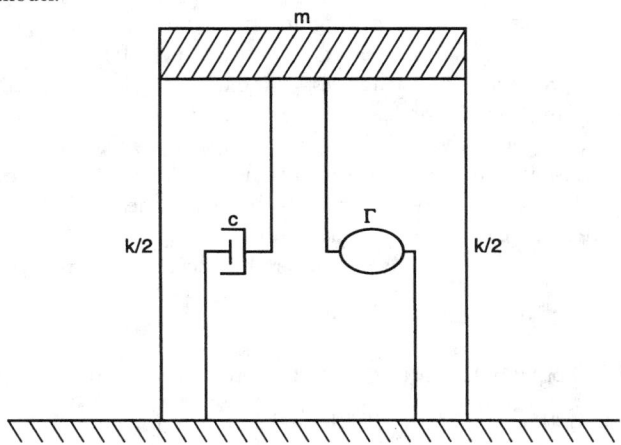

Figure 27.3   SDOF model with passive or active control element.

corresponding to the control device is written simply as Γx. This permits quite general response characteristics, including displacement, velocity, or acceleration-dependent contributions, as well as hereditary effects. The equation of motion for the extended SDOF model then becomes, in the case of an earthquake load,

$$m\ddot{x} + c\dot{x} + kx + \Gamma x = -(m + \overline{m})\ddot{x}_g \qquad (27.3)$$

with $\overline{m}$ representing the mass of the control element.

The specific form of Γx needs to be specified before Equation 27.3 can be analyzed, which is necessarily highly dependent on the device type. For passive energy dissipation systems, it can be represented by a force-displacement relationship such as the one shown in Figure 27.4, representing a rate-independent elastic-perfectly plastic element. For an active control system, the form of Γx is governed by the control law chosen for a given application. Let us first note that, denoting the control force applied to the structure in Figure 27.1 by $u(t)$, the resulting dynamical behavior of the

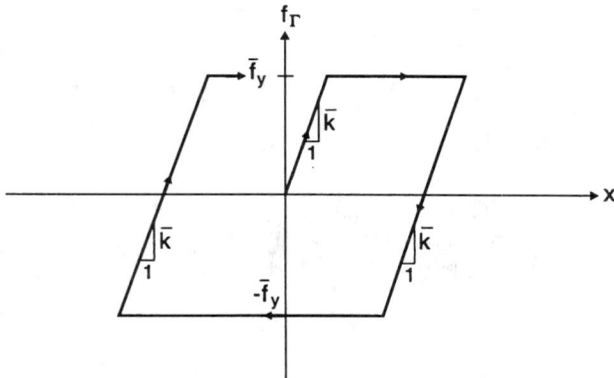

Figure 27.4   Force-displacement model for elastic-perfectly plastic passive element.

structure is governed by Equation 27.3 with

$$\Gamma x = -u(t) \tag{27.4}$$

Suppose that a feedback configuration is used in which the control force, $u(t)$, is designed to be a linear function of measured displacement, $x(t)$, and measured velocity, $\dot{x}(t)$. The control force, $u(t)$, takes the form

$$u(t) = g_1 x(t) + g_2 \dot{x}(t) \tag{27.5}$$

In view of Equation 27.4, we have

$$\Gamma x = -[g_1 + g_2 d/dt] x \tag{27.6}$$

The control law is, of course, not necessarily linear in $x(t)$ and $\dot{x}(t)$ as given by Equation 27.5. In fact, nonlinear control laws may be more desirable for civil engineering applications [61]. Thus, for both passive and active control cases, the resulting Equation 27.3 can be highly nonlinear.

Assume for illustrative purposes that the base structure has a viscous damping ratio $\zeta = 0.05$ and that a simple massless yielding device is added to serve as a passive element. The force-displacement relationship for this element, depicted in Figure 27.4, is defined in terms of an initial stiffness, $\bar{k}$, and a yield force, $\bar{f}_y$. Consider the case where the passively damped SDOF model is subjected to the 1940 El Centro S00E ground motion as shown in Figure 27.5. The initial stiffness of the elastoplastic passive device is specified as $\bar{k} = k$, while the yield force, $\overline{f}_y$, is equal to 20% of the maximum applied ground force. That is,

$$\overline{f}_y = 0.20 \, \text{Max} \left\{ m \left| \ddot{x}_g \right| \right\} \tag{27.7}$$

The resulting relative displacement and total acceleration time histories are presented in Figure 27.6. There is significant reduction in response compared to that of the base structure without the control element, as shown in Figure 27.7. Force-displacement loops for the viscous and passive elements are displayed in Figure 27.8. In this case, the size of these loops indicates that a significant portion of the energy is dissipated in the control device. This tends to reduce the forces and displacements in the primary structural elements, which of course is the purpose of adding the control device.

27.2.2   Multi-Degree-of-Freedom Structural Systems

In light of the preceding arguments, it becomes imperative to accurately characterize the behavior of any control device by constructing a suitable model under time-dependent loading. Multiaxial representations may be required. Once that model is established for a device, it must be properly

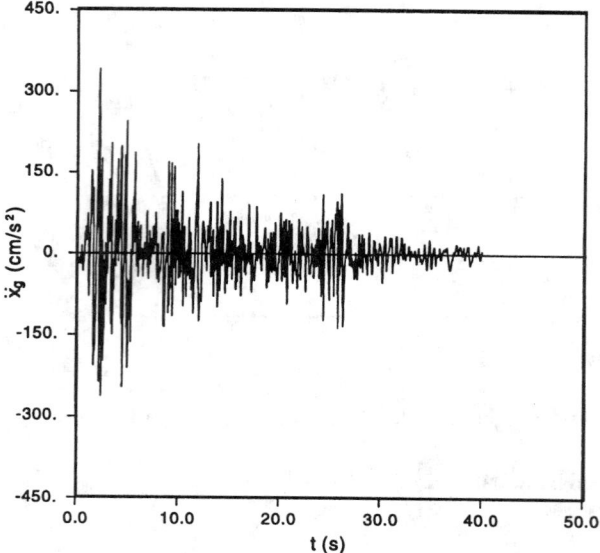

Figure 27.5   1940 El Centro S00E accelerogram.

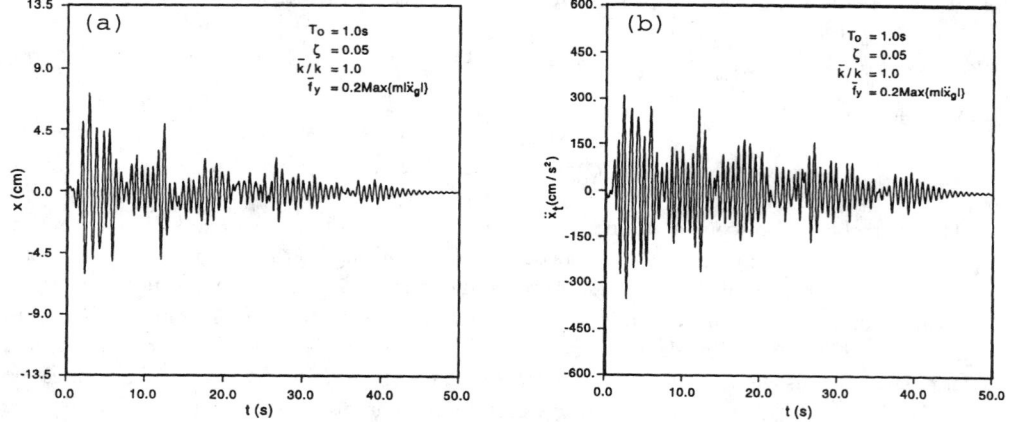

Figure 27.6   1940 El Centro time history response for SDOF with passive element: (a) displacement, (b) acceleration.

incorporated into a mathematical idealization of the overall structure. Seldom is it sufficient to employ an SDOF idealization for an actual structure. Thus, in the present subsection, the formulation for dynamic analysis is extended to an MDOF representation.

The finite element method (FEM) (e.g., [63]) currently provides the most suitable basis for this formulation. From a purely physical viewpoint, each individual structural member is represented mathematically by one or more finite elements having the same mass, stiffness, and damping characteristics as the original member. Beams and columns are represented by one-dimensional elements, while shear walls and floor slabs are idealized by employing two-dimensional finite elements. For more complicated or critical structural components, complete three-dimensional models can be developed and incorporated into the overall structural model in a straightforward manner via substructuring techniques.

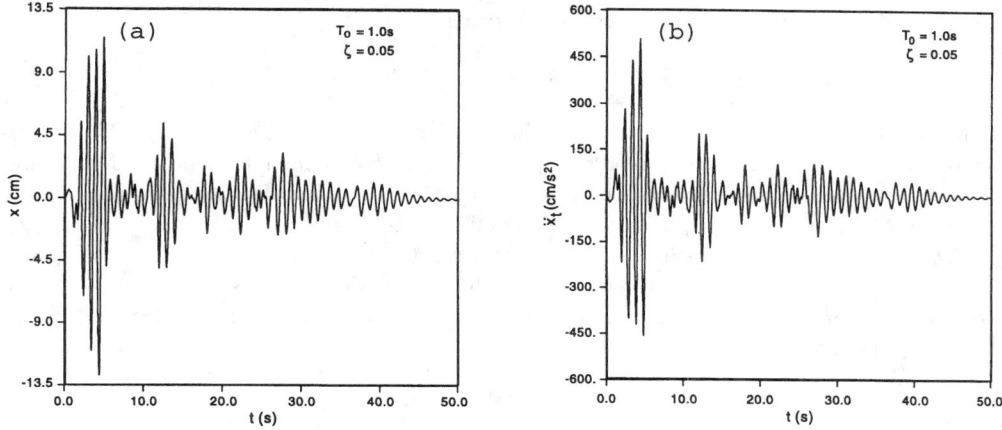

Figure 27.7 1940 El Centro SDOF time history response: (a) displacement, (b) acceleration.

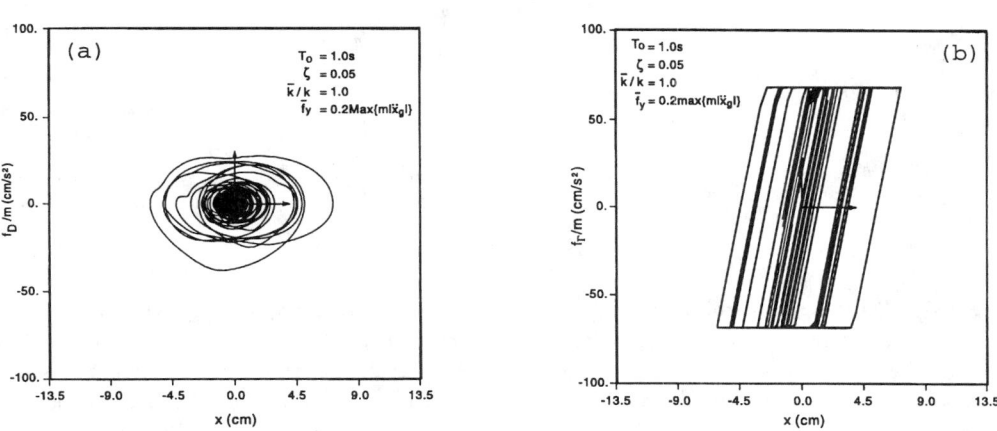

Figure 27.8 1940 El Centro SDOF force-displacement response for SDOF with passive element: (a) viscous element, (b) passive element.

The FEM actually was developed largely by civil engineers in the 1960s from this physical perspective. However, during the ensuing decades the method has also been given a rigorous mathematical foundation, thus permitting the calculation of error estimates and the utilization of adaptive solution strategies (e.g., [49]). Additionally, FEM formulations can now be derived from variational principles or Galerkin weighted residual procedures. Details of these formulations are beyond our scope. However, it should be noted that numerous general-purpose finite element software packages currently exist to solve the structural dynamics problem, including ABAQUS, ADINA, ANSYS, and MSC/NASTRAN. While none of these programs specifically addresses the special formulations needed to characterize structural protective systems, most permit generic user-defined elements. Alternatively, one can utilize packages geared exclusively toward civil engineering structures, such as ETABS, DRAIN, and IDARC, which in some cases can already accommodate typical passive elements.

Via any of the above-mentioned methods and programs, the displacement response of the structure is ultimately represented by a discrete set of variables, which can be considered the components of a generalized relative displacement vector, $x(t)$, of dimension N. Then, in analogy with Equation 27.3, the N equations of motion for the discretized structural system, subjected to uniform base excitation and time varying forces, can be written:

$$M\ddot{x} + C\dot{x} + Kx + \Gamma x = -(M + \overline{M})\ddot{x}_g \tag{27.8}$$

where M, C, and K represent the mass, damping, and stiffness matrices, respectively, while Γ symbolizes a matrix of operators that model the protective system present in the structure. Meanwhile, the vector $\ddot{x}_g$ contains the rigid body contribution of the seismic ground displacement to each degree of freedom. The matrix $\overline{M}$ represents the mass of the protective system.

There are several approaches that can be taken to solve Equation 27.8. The preferred approach, in terms of accuracy and efficiency, depends upon the form of the various terms in that equation. Let us first suppose that the protective device can be modeled as direct linear functions of the acceleration, velocity, and displacement vectors. That is,

$$\Gamma x = \overline{M}\ddot{x} + \overline{C}\dot{x} + \overline{K}x \tag{27.9}$$

Then, Equation 27.8 can be rewritten as

$$\hat{M}\ddot{x} + \hat{C}\dot{x} + \hat{K}x = -\hat{M}\ddot{x}_g \tag{27.10}$$

in which

$$\hat{M} = M + \overline{M} \tag{27.11a}$$
$$\hat{C} = C + \overline{C} \tag{27.11b}$$
$$\hat{K} = K + \overline{K} \tag{27.11c}$$

Equation 27.10 is now in the form of the classical matrix structural dynamic analysis problem. In the simplest case, which we will now assume, all of the matrix coefficients associated with the primary structure and the passive elements are constant. As a result, Equation 27.10 represents a set of N linear second-order ordinary differential equations with constant coefficients. These equations are, in general, coupled. Thus, depending upon N, the solution of Equation 27.10 throughout the time range of interest could become computationally demanding. This required effort can be reduced considerably if the equation can be uncoupled via a transformation; that is, if $\hat{M}$, $\hat{C}$, and $\hat{K}$ can be diagonalized. Unfortunately, this is not possible for arbitrary matrices $\hat{M}$, $\hat{C}$, and $\hat{K}$. However, with certain restrictions on the damping matrix, $\hat{C}$, the transformation to modal coordinates accomplishes the objective via the modal superposition method (see, e.g., [7]).

As mentioned earlier, it is more common having Γx in Equation 27.9 nonlinear in x for a variety of passive and active control elements. Consequently, it is important to develop alternative numerical approaches and design methodologies applicable to more generic passively or actively damped structural systems governed by Equation 27.8. Direct time-domain numerical integration algorithms are most useful in that regard. The Newmark beta algorithm, for example, is one of these algorithms and is used extensively in structural dynamics.

27.2.3 Energy Formulations

In the previous two subsections, we have considered SDOF and MDOF structural systems. The primary thrust of our analysis procedures has been the determination of displacements, velocities, accelerations, and forces. These are the quantities that, historically, have been of most interest. However, with the advent of innovative concepts for structural design, including structural protective systems, it is important to rethink current analysis and design methodologies. In particular, a focus on energy as a design criterion is conceptually very appealing. With this approach, the engineer is concerned, not so much with the resistance to lateral loads but rather, with the need to dissipate the energy input into the structure from environmental disturbances. Actually, this energy concept is not new. Housner [21] suggested an energy-based design approach even for more traditional structures several decades ago. The resulting formulation is quite appropriate for a general discussion of energy dissipation in structures equipped with structural protective systems.

In what follows, an energy formulation is developed for an idealized structural system, which may include one or more control devices. The energy concept is ideally suited for application to non-traditional structures employing control elements, since for these systems proper energy management is a key to successful design. To conserve space, only SDOF structural systems are considered, which can be easily generalized to MDOF systems.

Consider once again the SDOF oscillator shown in Figure 27.2 and governed by the equation of motion defined in Equation 27.1. An energy representation can be formed by integrating the individual force terms in Equation 27.1 over the entire relative displacement history. The result becomes

$$E_K + E_D + E_S = E_I \qquad (27.12)$$

where

$$E_K = \int m\ddot{x}dx = \frac{m\dot{x}^2}{2} \qquad (27.13a)$$

$$E_D = \int c\dot{x}dx = \int c\dot{x}^2dt \qquad (27.13b)$$

$$E_S = \int kxdx = \frac{kx^2}{2} \qquad (27.13c)$$

$$E_I = \int fdx \qquad (27.13d)$$

The individual contributions included on the left-hand side of Equation 27.12 represent the relative kinetic energy of the mass (E_K), the dissipative energy caused by inherent damping within the structure (E_D), and the elastic strain energy (E_S). The summation of these energies must balance the input energy (E_I) imposed on the structure by the external disturbance. Note that each of the energy terms is actually a function of time, and that the energy balance is required at each instant throughout the duration of the loading.

Consider aseismic design as a more representative case. It is unrealistic to expect that a traditionally designed structure will remain entirely elastic during a major seismic disturbance. Instead, inherent ductility of structures is relied upon to prevent catastrophic failure, while accepting the fact that some damage may occur. In such a case, the energy input (E_I) from the earthquake simply exceeds the capacity of the structure to store and dissipate energy by the mechanisms specified in Equations 27.13a–c. Once this capacity is surpassed, portions of the structure typically yield or crack. The stiffness is then no longer a constant, and the spring force in Equation 27.1 must be replaced by a more general functional relation, $g_S(x)$, which will commonly incorporate hysteretic effects. In general, Equation 27.13c is redefined as follows for inelastic response:

$$E_S = \int g_S(x)dx = E_{S_e} + E_{S_p} \qquad (27.14)$$

in which E_S is assumed separable into additive contributions E_{S_e} and E_{S_p}, representing the fully recoverable elastic strain energy and the dissipative plastic strain energy, respectively.

Figure 27.9a provides the energy response of a 0.3-scale, six-story concentrically braced steel structure as measured by Uang and Bertero [54]. The seismic input consisted of the 1978 Miyagi-Ken-Oki Earthquake signal scaled to produce a peak shaking table acceleration of 0.33 g, which was deemed to represent the damageability limit state of the model. At this level of loading, a significant

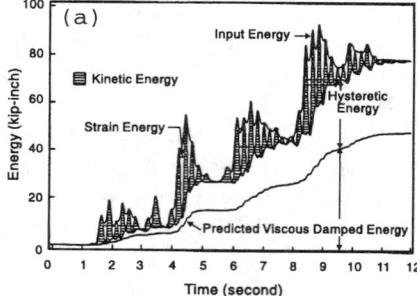 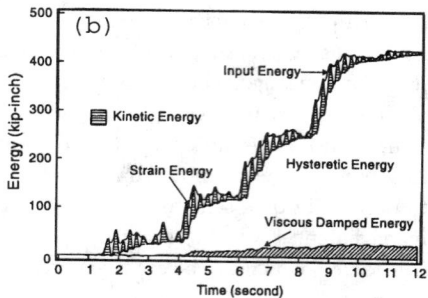

Figure 27.9   Energy response of a traditional structure: (a) damageability limit state, (b) collapse limit state. (From Uang, C.M. and Bertero, V.V. 1986. *Earthquake Simulation Tests and Associated Studies of a 0.3 Scale Model of a Six-Story Concentrically Braced Steel Structure.* Report No. UCB/ EERC - 86/10, Earthquake Engineering Research Center, Berkeley, CA. With permission.)

portion of the energy input to the structure is dissipated, with both viscous damping and inelastic hysteretic mechanisms having substantial contributions. If the intensity of the signal is elevated, an even greater share of the energy is dissipated via inelastic deformation. Finally, for the collapse limit state of this model structure at 0.65 g peak table acceleration, approximately 90% of the energy is consumed by hysteretic phenomena, as shown in Figure 27.9b. Evidently, the consumption of this quantity of energy has destroyed the structure.

From an energy perspective, then, for proper aseismic design, one must attempt to minimize the amount of hysteretic energy dissipated by the structure. There are basically two viable approaches available. The first involves designs that result in a reduction in the amount of energy input to the structure. Base isolation systems and some active control systems, for example, fall into that category. The second approach, as in the passive and active control system cases, focuses on the introduction of additional energy-dissipating mechanisms into the structure. These devices are designed to consume a portion of the input energy, thereby reducing damage to the main structure caused by hysteretic dissipation. Naturally, for a large earthquake, the devices must dissipate enormous amounts of energy.

The SDOF system with a control element is displayed in Figure 27.3, while the governing integrodifferential equation is provided in Equation 27.3. After integrating with respect to x, an energy balance equation can be written:

$$E_K + E_D + E_{S_e} + E_{S_p} + E_C = E_I \tag{27.15}$$

where the energy associated with the control element is

$$E_C = \int \Gamma x \, dx \tag{27.16}$$

and the other terms are as previously defined.

As an example of the effects of control devices on the energy response of a structure, consider the tests, of a one-third scale three-story lightly reinforced concrete framed building, conducted by Lobo et al. [30]. Figure 27.10a displays the measured response of the structure due to the scaled 1952 Taft N21E earthquake signal normalized for peak ground accelerations of 0.20 g. A considerable portion of the input energy is dissipated via hysteretic mechanisms, which tend to damage the primary structure through cracking and the formation of plastic hinges. On the other hand, damage is minimal with the addition of a set of viscoelastic braced dampers. The energy response of the braced structure, due to the same seismic signal, is shown in Figure 27.10b. Notice that although the input energy has increased slightly, the dampers consume a significant portion of the total, thus protecting the primary structure.

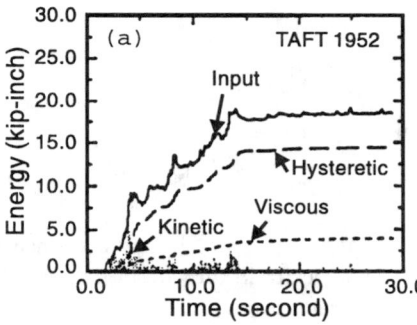
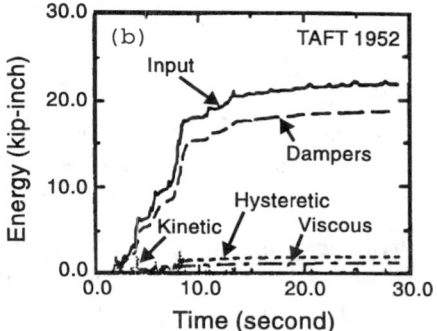

Figure 27.10   Energy response of test structure: (a) without passive devices, (b) with passive devices. (From Lobo, R.F., Bracci, J.M., Shen, K.L., Reinhorn, A.M., and Soong., T.T. 1993. *Earthquake Spectra*, 9(3), 419–446. With permission.)

27.2.4   Energy-Based Design

While the energy concept, as outlined briefly above, does not currently provide the basis for aseismic design codes, there is a considerable body of knowledge that has been developed from its application to traditional structures. Housner [21, 22] was the first to propose an energy-based philosophy for earthquake-resistant design. In particular, he was concerned with limit design methods aimed toward preventing collapse of structures in seismically active regions. Housner assumed that the energy input calculated for an undamped, elastic idealization of a structure provided a reasonable upper bound to that for the actual inelastic structure.

Berg and Thomaides [4] examined the energy consumption in SDOF elastoplastic structures via numerical computation and developed energy input spectra for several strong-motion earthquakes. These spectra indicate that the amount of energy, E_I, imparted to a structure from a given seismic event is quite dependent upon the structure itself. The mass, the natural period of vibration, the critical damping ratio, and yield force level were all found to be important characteristics.

On the other hand, their results did suggest that the establishment of upper bounds for E_I might be possible, and thus provided support for the approach introduced by Housner. However, the energy approach was largely ignored for a number of years. Instead, limit state design methodologies were developed which utilized the concept of displacement ductility to construct inelastic response spectra as proposed initially by Veletsos and Newmark [56].

More recently, there has been a resurgence of interest in energy-based concepts. For example, Zahrah and Hall [62] developed an MDOF energy formulation and conducted an extensive parametric study of energy absorption in simple structural frames. Their numerical work included a comparison between energy-based and displacement ductility-based assessments of damage, but the authors stopped short of issuing a general recommendation.

A critical assessment of the energy concept as a basis for design was provided by Uang and Bertero [55]. The authors initially contrast two alternative definitions of the seismic input energy. The quantity specified in Equation 27.13d is labeled the relative input energy, while the absolute input energy (E_{I_a}) is defined by

$$E_{I_a} = \int m\ddot{x}_t dx_g \qquad (27.17a)$$

In conjunction with this latter quantity, an absolute kinetic energy (E_{K_a}) is also required, where

$$E_{K_a} = \frac{m\dot{x}_t^2}{2} \tag{27.17b}$$

The absolute energy equation corresponding to Equation 27.15 then becomes

$$E_{K_a} + E_D + E_{S_e} + E_{S_p} + E_C = E_{I_a} \tag{27.18}$$

Based upon the development of input energy spectra for an SDOF system, the authors conclude that, while both measures produce approximately equivalent spectra in the intermediate period range, E_{I_a} should be used as a damage index for short period structures and E_I is more suitable for long period structures. Furthermore, an investigation revealed that the assumption of Housner to employ the idealized elastic strain energy, as an estimate of the actual input energy, is not necessarily conservative. Uang and Bertero [55] also studied an MDOF structure, and concluded that the input energy spectra for an SDOF can be used to predict the input energy demand for that type of building. In a second portion of the report, an investigation was conducted on the validity of the assumption that energy dissipation capacity can be used as a measure of damage. In testing cantilever steel beams, reinforced concrete shear walls, and composite beams the authors found that damage depends upon the load path.

The last observation should come as no surprise to anyone familiar with classical failure criteria. However, it does highlight a serious shortcoming for the use of the energy concept for limit design of traditional structures. As was noted above, in these structures, a major portion of the input energy must be dissipated via inelastic deformation, but damage to the structure is not determined simply by the magnitude of the dissipated energy. On the other hand, in non-traditional structures incorporating passive damping mechanisms, the energy concept is much more appropriate. The emphasis in design is directly on energy dissipation. Furthermore, since an attempt is made to minimize the damage to the primary structure, the selection of a proper failure criterion is less important.

27.3   Recent Development and Applications

As a result of serious efforts that have been undertaken in recent years to develop and implement the concept of passive energy dissipation and active control, a number of these devices have been installed in structures throughout the world, including Japan, New Zealand, Italy, Mexico, Canada, and the U.S. In what follows, advances in terms of their development and applications are summarized.

27.3.1   Passive Energy Dissipation

As alluded to in Section 27.1 and Table 27.1, a number of passive energy dissipation devices have been developed and installed in structures for performance enhancement under wind or earthquake loads. Discussions presented below are centered around some of the more common devices that have found applications in these areas.

Metallic Yield Dampers

One of the effective mechanisms available for the dissipation of energy input to a structure from an earthquake is through inelastic deformation of metals. The idea of utilizing added metallic energy dissipators within a structure to absorb a large portion of the seismic energy began with the conceptual and experimental work of Kelly et al. [26] and Skinner et al. [42]. Several of the devices considered included torsional beams, flexural beams, and U-strip energy dissipators. During the ensuing years, a wide variety of such devices have been studied or tested [5, 52, 53, 59]. Many of these devices use mild steel plates with triangular or X shapes so that yielding is spread almost uniformly throughout the material. A typical *X*-shaped plate damper or ADAS (added damping and stiffness) device is shown in Figure 27.11. Other materials, such as lead and shape-memory alloys, have also been evaluated [1]. Some particularly desirable features of these devices are their stable hysteretic behavior, low-cycle fatigue property, long-term reliability, and relative insensitivity to environmental temperature. Hence, numerous analytical and experimental investigations have been conducted to determine these characteristics of individual devices.

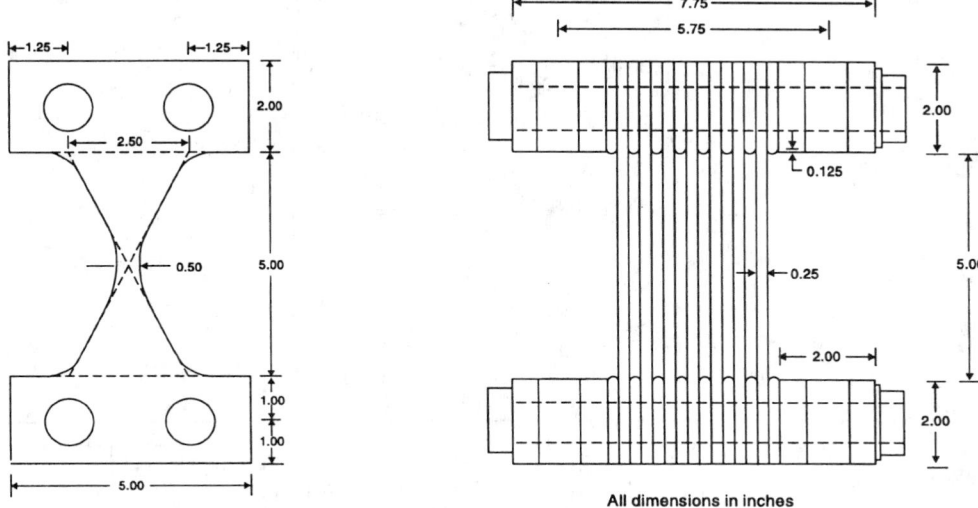

Figure 27.11 Added damping and stiffness (ADAS) device. (From Whittaker, A.S., Bertero, V.V., Thompson, C.L., and Alonso, L.J. 1991. *Earthquake Spectra*, 7(4), 563-604. With permission.)

After gaining confidence in their performance based primarily on experimental evidence, implementation of metallic devices in full-scale structures has taken place. The earliest implementations of metallic dampers in structural systems occurred in New Zealand and Japan. A number of these interesting applications are reported in Skinner et al. [43] and Fujita [17]. More recent applications include the use of ADAS dampers in seismic upgrade of existing buildings in Mexico [31] and in the U.S. [36]. The seismic upgrade project discussed in Perry et al. [36] involves the retrofit of the Wells Fargo Bank building in San Francisco, California. The building is a two-story nonductile concrete frame structure originally constructed in 1967 and subsequently damaged in the 1989 Loma Prieta earthquake. The voluntary upgrade by Wells Fargo utilized chevron braces and ADAS damping elements. More conventional retrofit schemes were rejected due to an inability to meet the performance objectives while avoiding foundation work. A plan view of the second floor including upgrade details is provided in Figure 27.12. A total of seven ADAS devices were employed, each with a yield force of 150 kps. Both linear and nonlinear analyses were used in the retrofit design process. Further three-dimensional response spectrum analyses, using an approximate equivalent linear representation for the ADAS elements, furnished a basis for the redesign effort. The final design was verified with DRAIN-2D nonlinear time history analyses. A comparison of computed response before and after the upgrade is contained in Figure 27.13. The numerical results indicated that the revised design was stable and that all criteria were met. In addition to the introduction of the bracing and ADAS dampers, several interior columns and a shear wall were strengthened.

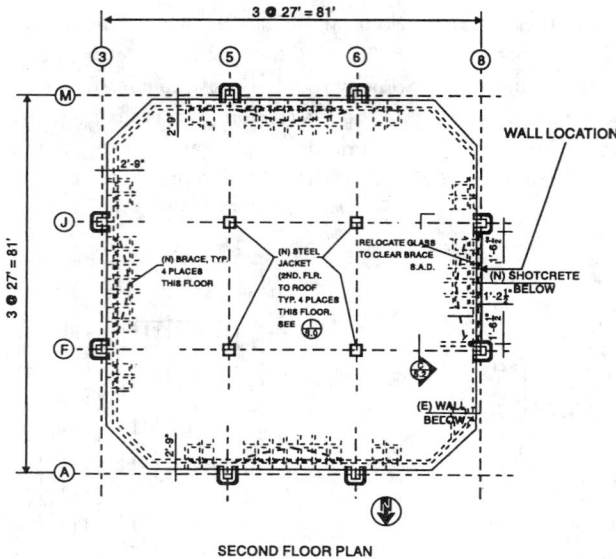

SECOND FLOOR PLAN

Figure 27.12 Wells Fargo Bank building retrofit details. (From Perry, C.L., Fierro, E.A., Sedarat, H., and Scholl, R.E. 1993. *Earthquake Spectra*, 9(3), 559-579. With permission.)

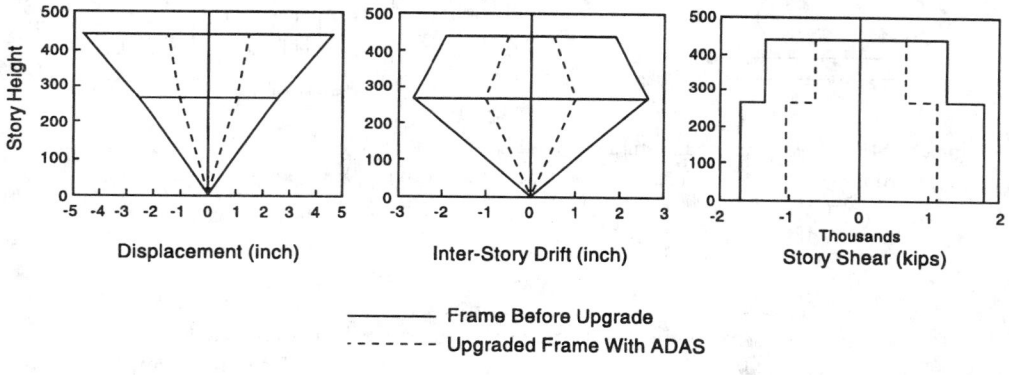

——— Frame Before Upgrade

- - - - - Upgraded Frame With ADAS

Figure 27.13 Comparison of computed results for Wells Fargo Bank building—envelope of response values in x direction. (From Perry, C.L., Fierro, E.A., Sedarat, H., and Scholl, R.E. 1993. *Earthquake Spectra*, 9(3), 559-579. With permission.)

Friction Dampers

Friction dampers utilize the mechanism of solid friction that develops between two solid bodies sliding relative to one another to provide the desired energy dissipation. Several types of friction dampers have been developed for the purpose of improving seismic response of structures. A simple brake lining frictional system was studied by Pall et al. [34]; however, a special damper mechanism, devised by Pall and Marsh [33], and depicted in Figure 27.14, permits much more effective operation. During cyclic loading, the mechanism tends to straighten buckled braces and also enforces slippage in both tensile and compressive directions.

Several alternative friction damper designs have also been proposed in the recent literature. For example, Roik et al. [39] discuss the use of three-stage friction-grip elements. A simple conceptual design, the slotted bolted connection (SBC), was investigated by FitzGerald et al. [15] and Grigorian et al. [18]. Another design of a friction damper is the energy dissipating restraint (EDR) manu-

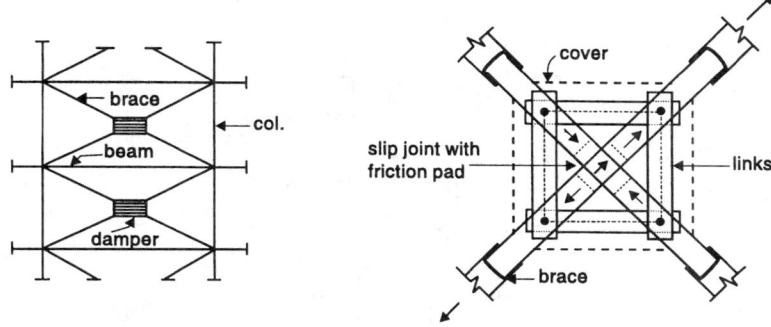

Figure 27.14   *X*-braced friction damper. (From Pall, A.S. and Marsh, C. 1982. *J. Struct. Div.*, ASCE, 1208(ST6), 1313-1323. With permission.)

factured by Fluor Daniel, Inc. There are several novel aspects of the EDR that combine to produce very different response characteristics. A detailed presentation of the design and its performance is provided in Nims et al. [32].

In recent years, there have been several commercial applications of friction dampers aimed at providing enhanced seismic protection of new and retrofitted structures. This activity in North America is primarily associated with the use of Pall friction devices in Canada. For example, the applications of friction dampers to the McConnel Library of the Concordia University in Montreal, Canada, is discussed in Pall and Pall [35]. A total of 143 dampers were employed in this case. Interestingly, the architects chose to expose 60 of the dampers to view due to their aesthetic appeal. A series of nonlinear DRAIN-TABS [19] analyses were utilized to establish the optimum slip load for the devices, which ranges from 600 to 700 kN, depending upon the location within the structure. For the three-dimensional time-history analyses, artificial aseismic signals were generated with a wide range of frequency contents and a peak ground acceleration scaled to 0.18 g to represent expected ground motion in Montreal. Under this level of excitation, an estimate of the equivalent damping ratio for the structure with frictional devices is approximately 50%. In addition, for this library complex, the use of the friction dampers resulted in a net savings of 1.5% of the total building cost. The authors noted that higher savings would be expected in more seismically vulnerable regions.

Viscoelastic Dampers

Viscoelastic materials used in structural applications are usually copolymers or glassy substances that dissipate energy through shear deformation. A typical viscoelastic (VE) damper, which consists of viscoelastic layers bonded with steel plates, is shown in Figure 27.15. When mounted in a structure, shear deformation and hence energy dissipation takes place when structural vibration induces relative motion between the outer steel flanges and the center plate. Significant advances in research and development of VE dampers, particularly for seismic applications, have been made in recent years through analyses and experimental tests (e.g., [6, 29, 41]).

The first applications of VE dampers to structures were for reducing acceleration levels, or increasing human comfort, due to wind. In 1969, VE dampers were installed in the twin towers of the World Trade Center in New York as an integral part of the structural system. There are about 10,000 VE dampers in each tower, evenly distributed throughout the structure from the 10th to the 110th floor. The towers have experienced a number of moderate to severe wind storms over the last 25 years. The observed performance of the VE dampers has been found to agree well with theoretical values. In 1982, VE dampers were incorporated into the 76-story Columbia SeaFirst Building in Seattle, Washington, to protect against wind-induced vibrations [25]. To reduce the wind-induced vibration, the design called for 260 dampers to be located alongside the main diagonal members

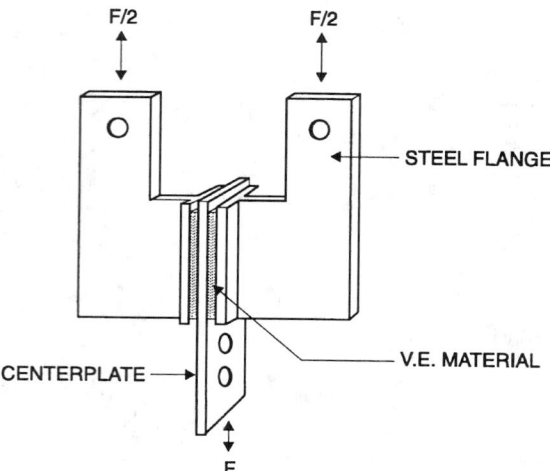

Figure 27.15   Typical viscoelastic damper configuration.

in the building core. The addition of VE dampers to this building was calculated to increase its damping ratio in the fundamental mode from 0.8 to 6.4% for frequent storms and to 3.2% at design wind. Similar applications of VE dampers were made to the Two Union Square Building in Seattle in 1988. In this case, 16 large VE dampers were installed parallel to four columns in one floor.

Seismic applications of VE dampers to structures began only recently. A seismic retrofit project using VE dampers began in 1993 for the 13-story Santa Clara County building in San Jose, California [8]. Situated in a high seismic risk region, the building was built in 1976. It is approximately 64 m in height and nearly square in plan, with 51 × 51 m on typical upper floors. The exterior cladding consists of full-height glazing on two sides and metal siding on the other two sides. The exterior cladding, however, provides little resistance to structural drift. The equivalent viscous damping in the fundamental mode is less than 1% of critical.

The building has been extensively instrumented, providing invaluable response data obtained during a number of past earthquakes. A plan for seismic upgrade of the building was developed, in part, when the response data indicated large and long-duration response, including torsional coupling, to even moderate earthquakes. The final design called for installation of two dampers per building face per floor level, as shown in Figure 27.16, which would increase the equivalent damping in the fundamental mode of the building to about 17% of critical, providing substantial reductions to building response under all levels of ground shaking. A typical damper configuration is shown in Figure 27.17.

Viscous Fluid Dampers

Damping devices based on the operating principle of high-velocity fluid flow through orifices have found numerous applications in shock and vibration isolation of aerospace and defense systems. In recent years, research and development of viscous fluid (VF) dampers for seismic applications to civil engineering structures have been performed to accomplish three major objectives. The first was to demonstrate by analysis and experiment that viscous fluid dampers can improve seismic capacity of a structure by reducing damage and displacements and without increasing stresses. The second was to develop mathematical models for these devices and demonstrate how these models can be incorporated into existing structural engineering software codes. Finally, the third was to evaluate reliability and environmental stability of the dampers for structural engineering applications.

As a result, VF dampers have in recent years been incorporated into civil engineering structures. In several applications, they were used in combination with seismic isolation systems. For example, VF dampers were incorporated into base isolation systems for five buildings of the new San Bernardino

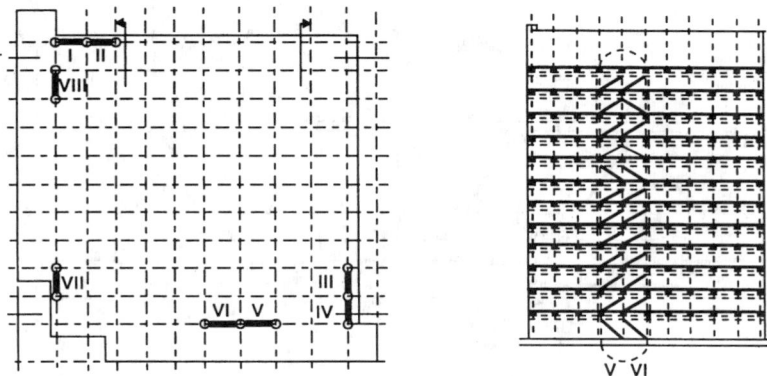

Figure 27.16   Location of viscoelastic dampers in Santa Clara County building.

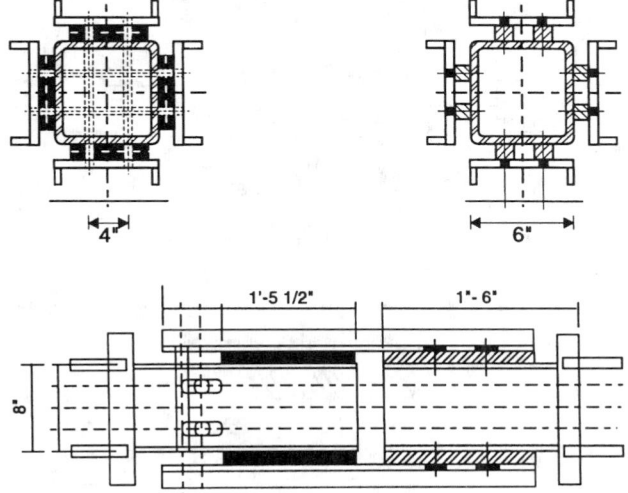

Figure 27.17   Santa Clara County building viscoelastic damper configuration.

County Medical Center, located close to two major fault lines, in 1995. The five buildings required a total of 233 dampers, each having an output force of 320,000 lb and generating an energy dissipation level of 3,000 hp at a speed of 60 in./s. A layout of the damper-isolation system assembly is shown in Figures 27.18 and 27.19 gives the dimensions of the viscous dampers employed.

Tuned Mass Dampers

The modern concept of tuned mass dampers (TMDs) for structural applications has its roots in dynamic vibration absorbers, studied as early as 1909 by Frahm [9]. A schematic representation of Frahm's absorber is shown in Figure 27.20, which consists of a small mass, m, and a spring with spring stiffness k attached to the main mass, M, with spring stiffness K. Under a simple harmonic load, one can show that the main mass, M, can be kept completely stationary when the natural frequency, $(\sqrt{k/m})$, of the attached absorber is chosen to be (or tuned to) the excitation frequency.

As in the case of VE dampers, early applications of TMDs have been directed toward mitigation of wind-induced excitations. It appears that the first structure in which a TMD was installed is the Centerpoint Tower in Sydney, Australia [12, 28]. One of only two buildings in the U.S. equipped with a TMD is the 960-ft Citicorp Center in New York, in which the TMD is situated on the 63rd floor. At this elevation, the building can be represented by a simple modal mass of approximately

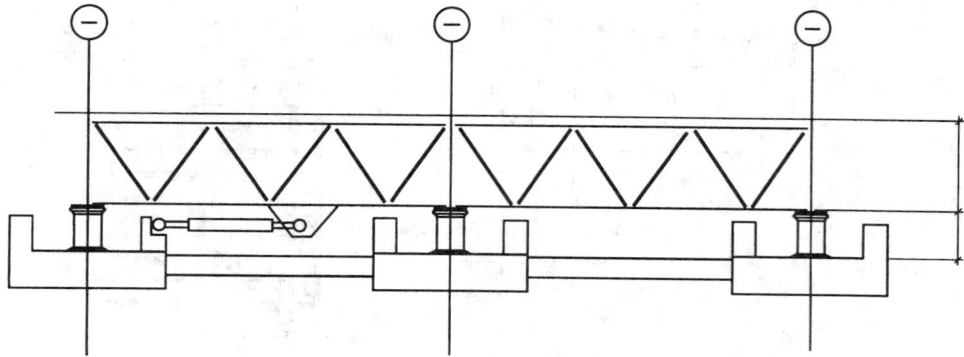

Figure 27.18 San Bernardino County Medical Center damper-base isolation system assembly.

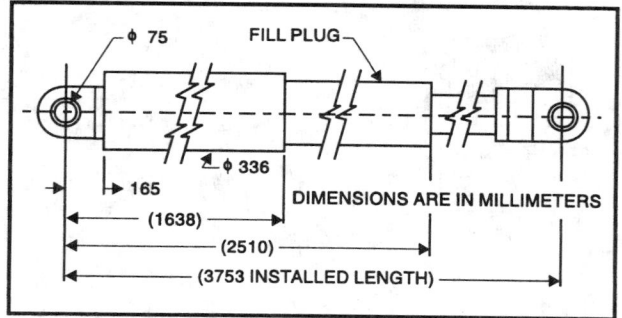

Figure 27.19 Dimensions of viscous fluid damper for San Bernardino County Medical Center.

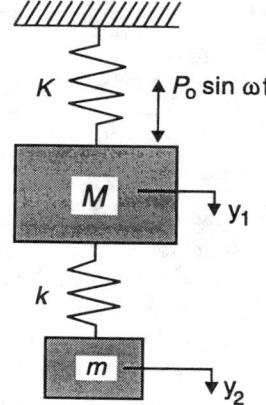

Figure 27.20 Undamped absorber and main mass subject to harmonic excitation (Frahm's absorber).

20,000 tons, to which the TMD is attached to form a two-DOF system. Tests and actual observations have shown that the TMD produces an approximate effective damping of 4% as compared to the 1% original structural damping, which can reduce the building acceleration level by about 50% [37, 38]. The same design principles were followed in the development of the TMD for installation in the John Hancock Tower, Boston, Massachusetts [13]. In this case, however, the TMD consists of two 300-ton mass blocks. They move in phase to provide lateral response control and out of phase for torsional control.

Recently, numerical and experimental studies have been carried out to examine the effectiveness of TMDs in reducing seismic response of structures. It is noted that a passive TMD can only be tuned to a single structural frequency. While the first-mode response of an MDOF structure with TMD can be substantially reduced, the higher mode response may in fact increase as the number of stories increases. For earthquake-type excitations, it has been demonstrated that, for shear structures up to 12 floors, the first mode response contributes more than 80% to the total motion [60]. However, for a taller building on a firm ground, higher modal response may be a problem that needs further study. Villaverde [57] studied three structures—a two-dimensional ten-story shear building, three-dimensional one-story frame building, and a three-dimensional cable-stayed bridge—using nine kinds of real earthquake records. Numerical and experimental results show that the effectiveness of TMDs on reducing the response of the same structure under different earthquakes or of different structures under the same earthquake is significantly different; some cases give good performance and some have little or even no effect. This implies that there is a dependency of the attained reduction in response on the characteristics of the ground motion that excites the structure. This response reduction is large for resonant ground motions and diminishes as the dominant frequency of the ground motion gets further away from the structure's natural frequency to which the TMD is tuned.

It is also noted that the interest in using TMDs for vibration control of structures under earthquake loads has resulted in some innovative developments. An interesting approach is the use of a TMD with active capability, so-called active mass damper (AMD) or active tuned mass damper (ATMD). Systems of this type have been implemented in a number of tall buildings in recent years in Japan [48]. Some examples of such systems will be discussed in Section 27.3.2.

Tuned Liquid Dampers

The basic principles involved in applying a tuned liquid damper (TLD) to reduce the dynamic response of structures are quite similar to those discussed above for the TMD. In effect, a secondary mass in the form of a body of liquid is introduced into the structural system and tuned to act as a dynamic vibration absorber. However, in the case of TLDs, the response of the secondary system is highly nonlinear due either to liquid sloshing or the presence of orifices. TLDs have also been used for suppressing wind-induced vibrations of tall structures. In comparison with TMDs, the advantages associated with TLDs include low initial cost, virtually free maintenance, and ease of frequency tuning.

It appears that TLD applications have taken place primarily in Japan. Examples of TLD-controlled structures include the Nagasaki Airport Tower, installed in 1987, the Yokohama Marine Tower, also installed in 1987, the Shin-Yokohama Prince Hotel, installed in 1992, and the Tokyo International Airport Tower, installed in 1993 [50, 51]. The TLD installed in the 77.6-m Tokyo Airport Tower, for example, consists of about 1400 vessels containing water, floating particles, and a small amount of preservatives. The vessels, shallow circular cylinders 0.6 m in diameter and 0.125 m in height, are stacked in six layers on steel-framed shelves. The total mass of the TLD is approximately 3.5% of the first-mode generalized mass of the tower and its sloshing frequency is optimized at 0.743 Hz. Floating hollow cylindrical polyethylene particles were added in order to optimize energy dissipation through an increase in surface area together with collisions between particles.

The performance of the TLD has been observed during several storm episodes. In one such episode, with a maximum instantaneous wind speed of 25 m/s, the observed results show that the TLD reduced the acceleration response in the cross-wind direction to about 60% of its value without the TLD.

27.3.2 Active Control

As mentioned in Section 27.1, the development of active or hybrid control systems has reached the stage of full-scale applications to actual structures. Since 1989, more than 20 active or hybrid systems have been installed in building structures in Japan, the only country in which these applications have taken place. In addition, 14 bridge towers have employed active systems during erection [16]. Described briefly below are two of these systems and their observed performance. The performance of these systems under recent wind and earthquake episodes is summarized in this section. More details of these applications can be found in [48].

Sendagaya INTES Building

An AMD system was installed in the Sendagaya INTES building in Tokyo in 1991. As shown in Figure 27.21, the AMD was installed atop the 11th floor and consists of two masses to control transverse and torsional motions of the structure while hydraulic actuators provide the active control capabilities. The top view of the control system is shown in Figure 27.22, where ice thermal storage tanks are used as mass blocks so that no extra mass is introduced. The masses are supported by multistage rubber bearings intended to reduce the control energy consumed in the AMD and for ensuring smooth mass movements [20].

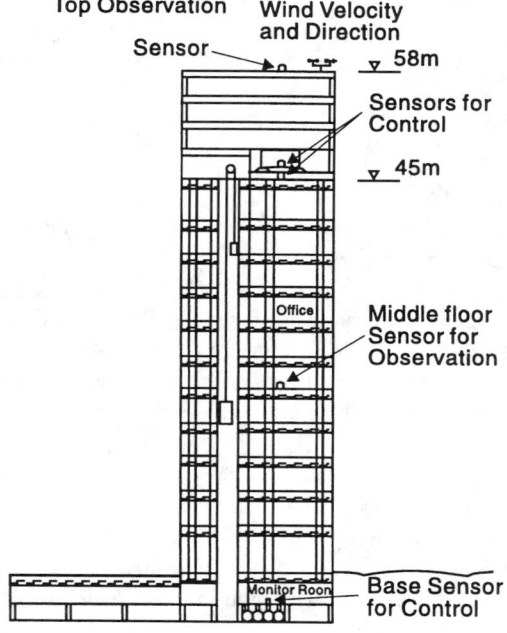

Figure 27.21   Sendagaya INTES building.

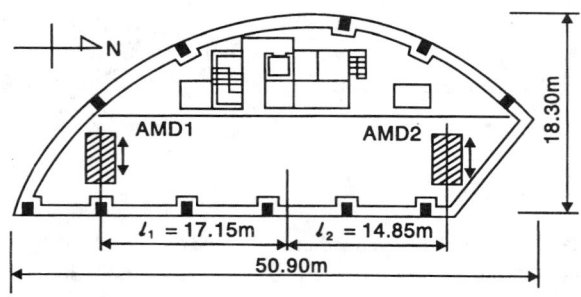

Figure 27.22   Top view of the active mass damper (AMD) in Sendagaya building.

Sufficient data were obtained for evaluation of the AMD performance when the building was subjected to strong wind on March 29, 1993, with peak instantaneous wind speed of 30.6 m/s. An example of the response Fourier spectra using samples of 30-s duration is shown in Figure 27.23, showing good performance in the low frequency range. The response of the fundamental mode was reduced by 18 and 28% for translation and torsion, respectively. Similar performance characteristics were observed during a series of earthquakes recorded between May 1992 and February 1993.

Hankyu Chayamachi Building

In 1992, an AMD system was installed in the 160-m, 34-story Hankyu Chayamachi building (shown in Figure 27.24), located in Osaka, Japan, for the primary purpose of occupant comfort control. In this case, the heliport at the roof top is utilized as the moving mass of the AMD, which

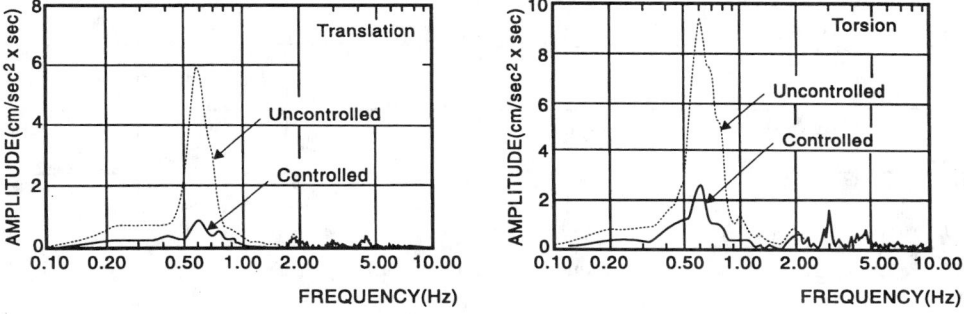

Figure 27.23 Sendgaya building—response Fourier spectra (March 29, 1993).

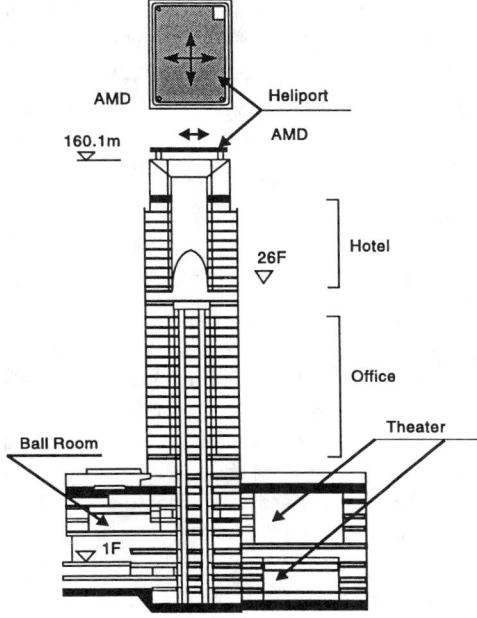

Figure 27.24 Hankyu Chayamachi building.

weighs 480 tons and is about 3.5% of the weight of the tower portion. The heliport is supported
by six multi-stage rubber bearings. The natural period of rubber and heliport system was set to
3.6 s, slightly lower than that of the building (3.8 s). The AMD mechanism used here has the same
architecture as that of Sendagaya INTES, namely, scheme of the digital controller, servomechanism,
and the hydraulic design, except that two actuators of 5-ton thrusts are attached in horizontal
orthogonal directions. Torsional control is not considered here.

Acceleration Fourier spectra during a recent typhoon are shown in Figure 27.25. Since the building
in this case oscillated primarily in its fundamental mode, significant reductions in acceleration levels
were observed.

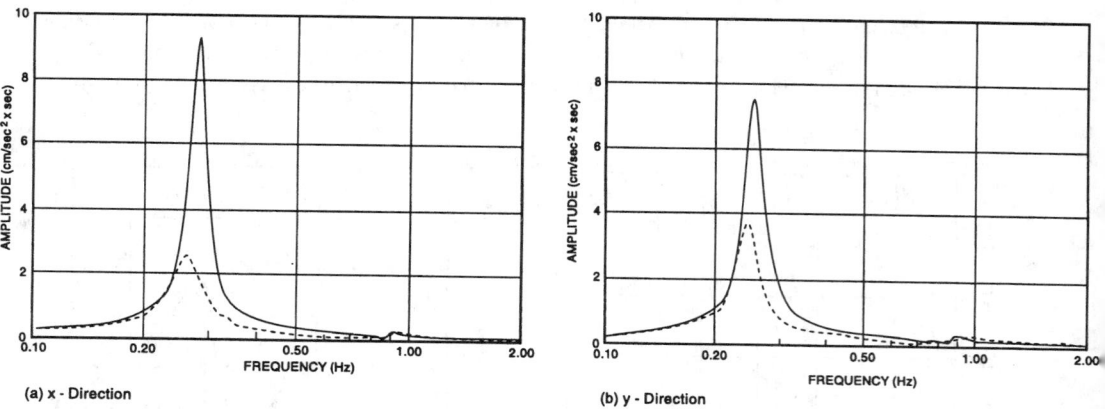

Figure 27.25   Hankyu Chayamachi building—acceleration Fourier spectra.

An observation to be made in the performance of control systems such as those described above
is that efficient active control systems can be implemented with existing technology under practical
constraints such as power requirements and stringent demand of reliability. Thus, significant strides
have been made considering that serious implementational efforts began less than ten years ago. On
the other hand, the active dampers developed for the Sendagaya INTES and Hankyu Chayamachi
buildings were designed primarily for response control due to wind and moderate earthquakes. In
order to reach the next level in active/hybrid control technology, an outstanding issue that needs to
be addressed is whether such systems, with limited control resources and practical constraints such
as mass excursions, can be made effective under strong earthquakes.

27.4   Code Development

At present, design of active control systems for structural applications is not addressed in any
model code in the U.S. However, extensive efforts in the field of passive energy dissipation and
the increased interest of the engineering profession in this area has resulted in the development of
tentative requirements for the design and implementation of passive energy dissipation devices. The
Energy Dissipation Working Group of the Base Isolation Subcommittee of the Structural Engineers
Association of Northern California (SEAONC) has developed a document addressing these tentative
requirements that provides design guidelines applicable to a wide range of system hardware [58].
The scope includes metallic, friction, viscoelastic, and viscous devices. On the other hand, TMDs
and TLDs are not addressed.

The general philosophy of that document is to confine inelastic deformation primarily to the
energy dissipators, while the main structure remains elastic for the design basis earthquake. Fur-

thermore, since passive energy dissipation technology is still relatively new, a conservative approach is taken on many issues. For example, an experienced independent engineering review panel must be formed to conduct a review of the energy dissipation system and testing programs.

According to the April 1993 version of the tentative requirements, static lateral force analysis cannot be used for design of structures incorporating energy dissipation devices. Dynamic analysis is mandatory. For rate-dependent devices (i.e., VE and VF), response spectrum analysis may be used provided that the remainder of the structure operates in the elastic range during the design basis earthquake. For all rate-independent devices (e.g., metallic and friction dampers) and for any case involving an inelastic structure, the document requires the use of nonlinear time-history analysis.

Prototype testing of energy dissipating devices is also specified. The program included 200 cycles at the design wind force, 50 cycles at one-half the device design displacement, 50 cycles at the device design displacement, and 10 cycles at the maximum device displacement. This program must be repeated at various frequencies for rate-dependent dampers. In addition, general statements are included to indicate that consideration should be given during design to other environmental factors, such as operating temperature, moisture, and creep.

The 1994 edition of the *National Earthquake Hazard Reduction Program Recommended Provisions for Seismic Regulations for New Buildings* [14] contains an appendix on passive energy dissipation systems, which is similar to the SEAONC document in both scope and philosophy. Also under development is a document on "Guidelines and Commentary for Seismic Rehabilitation of Buildings" by the Applied Technology Council for the Building Seismic Safety Council [3]. It is expected to include a section on guidelines and commentary for energy dissipation systems when completed in 1997.

27.5   Concluding Remarks

We attempted to introduce the basic concepts of passive energy dissipation and active control, and to present up-to-date current development, structural applications, and code-related activities in this exciting and fast expanding field. While significant strides have been made in terms of implementation of these concepts to structural design and retrofit, it should be emphasized that this entire technology is still evolving. Significant improvements in both hardware and design procedures will certainly continue for a number of years to come.

The acceptance of innovative systems in structural engineering is based on a combination of performance enhancement versus construction costs and long-term effects. Continuing efforts are needed in order to facilitate wider and speedier implementation. These include effective system integration and further development of analytical and experimental techniques by which performances of these systems can be realistically assessed. Structural systems are complex combinations of individual structural components. New innovative devices need to be integrated into these complex systems, with realistic evaluation of their performance and impact on the structural system, as well as verification of their ability for long-term operation. Additionally, innovative ideas of devices require exploration through experimentation and adequate basic modeling. A series of standardized benchmark structural models representing large buildings, bridges, towers, lifelines, etc., with standardized realistically scaled-down excitations representing natural hazards, will be of significant value in helping to provide an experimental and analytical testbed for proof-of-concept of existing and new devices.

References

[1] Aiken, I.D. and Kelly, J.M. 1992. Comparative Study of Four Passive Energy Dissipation Systems. *Bull. N.Z. Nat. Soc. Earthquake Eng.*, 25(3), 175-192.

[2]   Applied Technology Council. 1993. *Proceedings on Seismic Isolation, Passive Energy Dis-sipation, and Active Control*, ATC 17-1, Redwood City, CA.

[3]   Applied Technology Council. 1995. *Guidelines and Commentary for Seismic Rehabilitation of Buildings*, ATC-33, 75% Complete Draft, Redwood City, CA.

[4]   Berg, G.V. and Thomaides, S.S. 1960. Energy Consumption by Structures in Strong Motion Earthquakes, *Proc. 2nd World Conf. Earthquake Eng.*, II, 681-697, Tokyo.

[5]   Bergman, D.M. and Goel, S.C. 1987. *Evaluation of Cyclic Testing of Steel-Plate Devices for Added Damping and Stiffness*, Report No. UMCE 87-10, University of Michigan, Ann Arbor, MI.

[6]   Chang, K.C., Shen, K.L., Soong, T.T. and Lai, M.L. 1994. Seismic Retrofit of a Concrete Frame with Added Viscoelastic Dampers, *5th Nat. Conf. Earthquake Eng.*, Chicago, IL.

[7]   Clough, R.W. and Penzien, J. 1975. *Dynamics of Structures*, McGraw-Hill, NY.

[8]   Crosby, P., Kelly, J.M. and Singh, J. 1994. Utilizing Viscoelastic Dampers in the Seismic Retrofit of a Thirteen Story Steel Frame Building, *Struct. Congress XII*, Atlanta, GA, pp. 1286-1291.

[9]   Den Hartog, J.P. 1956. *Mechanical Vibrations*, 4th ed., McGraw-Hill, New York.

[10] Dyke, S.J., Spencer Jr., B.F., Quast P., Sain, M.K., Kaspari Jr., D.C. and Soong, T.T. 1994. *Experimental Verification of Acceleration Feedback Control Strategies for an Active Tendon System*, NCEER-94-0024, Buffalo, NY.

[11] Dyke, S.J., Spencer Jr., B.F., Quast, P. and Sain, M.K. 1995. The Role of Control-Structure Interaction in Protective System Design, *J. Eng. Mech., ASCE*, 121(2), 322-338.

[12] ENR. 1971. Tower Cables Handle Wind, Water Tank Dampens It, *Eng. News-Record*, p.23, Dec. 9.

[13] ENR. 1975. Hancock Tower Now to Get Dampers, *Eng. News-Record*, p.11, October 30.

[14] Federal Emergency Management Association. 1995. *1994 NEHRP Recommended Provisions for Seismic Regulations for New Buildings*, Report FEMA 222A, Washington, D.C.

[15] FitzGerald, T.F., Anagnos, T., Goodson, M. and Zsutty, T. 1989. Slotted Bolted Connections in Aseismic Design for Concentrically Braced Connections, *Earthquake Spectra*, 5(2), 383-391.

[16] Fujino, Y. 1994. Recent Research and Developments on Control of Bridges under Wind and Traffic Excitations in Japan, *Proc. Int. Workshop on Struct. Control*, pp. 144-150.

[17] Fujita, T. (Ed.). 1991. Seismic Isolation and Response Control for Nuclear and Non-Nuclear Structures, Special Issue for the Exhibition of *11th Int. Conf. on SMiRT*, Tokyo, Japan.

[18] Grigorian, C.E., Yang, T.S. and Popov, E.P. 1993. Slotted Bolted Connection Energy Dissipators, *Earthquake Spectra*, 9(3), 491-504.

[19] Guendeman-Israel, R. and Powell, G.H. 1977. *DRAIN-TABS—A Computerized Program for Inelastic Earthquake Response of Three Dimensional Buildings*, Report No. UCB/EERC 77-08, University of California, Berkeley, CA.

[20] Higashino, M. and Aizawa, S. 1993. Application of Active Mass Damper System in Actual Buildings, *Proc. Int. Workshop on Structural Control*, G.W. Housner and S.F. Masri (Eds.), Los Angeles, CA, pp. 194-205.

[21] Housner, G.W. 1956. Limit Design of Structures to Resist Earthquakes, *Proc. 1st World Conf. on Earthquake Eng.*, pp. 5-1 – 5-13, Earthquake Eng. Research Center, Berkeley, CA.

[22] Housner, G.W. 1959. Behavior of Structures During Earthquakes, *J. Eng. Mech. Div., ASCE*, 85(EM4), 109-129.

[23] Housner, G.W, Soong, T.T. and Masri, S.F. 1994. Second Generation of Active Structural Control in Civil Engineering, *Proc. 1st World Conf. on Struct. Control*, Los Angeles, CA, FA2, 3-18.

[24] Housner, G.W., Masri, S.F. and Chassiakos, A.G. (Eds.). 1994. *Proc. 1st World Conf. on Struct. Control*, Los Angeles, CA.

[25] Keel, C.J. and Mahmoodi, P. 1986. Designing of Viscoelastic Dampers for Columbia Center Building, *Building Motion in Wind*, N. Isyumov and T. Tschanz (Eds.), ASCE, New York, pp. 66-82.

[26] Kelly, J.M., Skinner, R.I. and Heine, A.J. 1972. Mechanisms of Energy Absorption in Special Devices for Use in Earthquake Resistant Structures, *Bull. N.Z. Nat. Soc. Earthquake Eng.*, 5, 63-88.

[27] Kobori, T. 1994. Future Direction on Research and Development of Seismic-Response-Controlled Structure, *Proc. 1st World Conf. on Struct. Control*, Los Angeles, CA, Panel, 19-31.

[28] Kwok, K.C.S. and MacDonald, P.A. 1987. Wind-Induced Response of Sydney Tower, *Proc. 1st Nat. Struct. Eng. Conf.*, pp. 19-24.

[29] Lai, M.L., Chang, K.C., Soong, T.T., Hao, D.S. and Yeh, Y.C. 1995. Full-scale Viscoelastically Damped Steel Frame, *ASCE J. Struct. Eng.*, 121(10), 1443-1447.

[30] Lobo, R.F., Bracci, J.M., Shen, K.L., Reinhorn, A.M. and Soong., T.T. 1993. Inelastic Response of R/C Structures with Viscoelastic Braces, *Earthquake Spectra*, 9(3), 419-446.

[31] Martinez-Romero, E. 1993. Experiences on the Use of Supplemental Energy Dissipators on Building Structures, *Earthquake Spectra*, 9(3), 581-624.

[32] Nims, D.K., Richter, P.J. and Bachman, R.E. 1993. The Use of the Energy Dissipating Restraint for Seismic Hazard Mitigation, *Earthquake Spectra*, 9(3), 467-489.

[33] Pall, A.S. and Marsh, C. 1982. Response of Friction Damped Braced Frames, *J. Struct. Div.*, ASCE, 1208(ST6), 1313-1323.

[34] Pall, A.S., Marsh, C. and Fazio, P. 1980. Friction Joints for Seismic Control of Large Panel Structures, *J. Prestressed Concrete Inst.*, 25(6), 38-61.

[35] Pall, A.S. and Pall, R. 1993. Friction-Dampers Used for Seismic Control of New and Existing Building in Canada, *Proc. ATC 17-1 Seminar on Isolation, Energy Dissipation and Active Control*, San Francisco, CA, 2, 675-686.

[36] Perry, C.L., Fierro, E.A., Sedarat, H. and Scholl, R.E. 1993. Seismic Upgrade in San Francisco Using Energy Dissipation Devices, *Earthquake Spectra*, 9(3), 559-579.

[37] Petersen, N.R. 1980. Design of Large Scale TMD, *Struct. Control*, North Holland, Amsterdam, pp. 581-596.

[38] Petersen, N.R. 1981. Using Servohydraulics to Control High-Rise Building Motion, *Proc. Nat. Convention Fluid Power*, Chicago, pp. 209-213.

[39] Roik, K., Dorka, U. and Dechent, P. 1988. Vibration Control of Structures under Earthquake Loading by Three-Stage Friction-Grip Elements, *Earthquake Eng. Struct. Dyn.*, 16, 501-521.

[40] Sakurai, T., Shibata, K., Watanabe, S., Endoh, A., Yamada, K., Tanaka, N. and Kobayashi, H. 1992. Application of Joint Damper to Thermal Power Plant Buildings, *Proc. 10th World Conf. Earthquake Eng.*, Madrid, Spain, 7, 4149-4154.

[41] Shen, K.L., Soong, T.T., Chang, K.C. and Lai, M.L. 1995. Seismic Behavior of Reinforced Concrete Frame with Added Viscoelastic Dampers, *Eng. Structures*, 17(5), 372-380.

[42] Skinner, R.I., Kelly, J.M. and Heine, A.J. 1975. Hysteresis Dampers for Earthquake-Resistant Structures, *Earthquake Eng. Struct. Dyn.*, 3, 287-296.

[43] Skinner, R.I., Tyler, R.G., Heine, A.J. and Robinson, W.H. 1980. Hysteretic Dampers for the Protection of Structures from Earthquakes, *Bull. N.Z. Nat. Soc. Earthquake Eng.*, 13(1), 22-36.

[44] Soong, T.T. 1990. *Active Structural Control: Theory and Practice*, Longman Scientific and Technical, Essex, England, and Wiley, New York.

[45] Soong, T.T. and Constantinou, M.C. (Eds.). 1994. *Passive and Active Structural Vibration Control in Civil Engineering*, Springer-Verlag, Wien and New York.

[46] Soong, T.T. and Dargush, G.F. 1997. *Passive Energy Dissipation Systems in Structural Engineering*, Wiley, London.

[47] Soong, T.T. and Reinhorn, A.M. 1993. An Overview of Active and Hybrid Structural Control Research in the U.S., *J. Struct. Design Tall Bldg.*, 2, 193-209.

[48] Soong, T.T., Reinhorn, A.M., Aizawa, S. and Higashino, M. 1994. Recent Structural Applications of Active Control Technology, *J. Struct. Control*, 1(2), 5-21.

[49] Szabó, B. and Babuška, I. 1991. *Finite Element Analysis*, John Wiley & Sons, New York.

[50] Tamura, Y., Shimada, K., Sasaki, A., Kohsaka, R. and Fuji, K. 1994. Variation of Structural Damping Ratios and Natural Frequencies of Tall Buildings During Strong Winds, *Proc. 9th Int. Conf. Wind Eng.*, New Delhi, India, 3, 1396-1407.

[51] Tamura, Y., Fujii, K., Ohtsuki, T., Wakahara, T. and Kohsaka, R. 1995. Effectiveness of Tuned Liquid Dampers and Wind Excitations, *Eng. Struct.*, 17(9), 609-621.

[52] Tsai, K.C., Chen, H.W., Hong, C.P. and Su, Y.F. 1993. Design of Steel Triangular Plate Energy Absorbers for Seismic-Resistant Construction, *Earthquake Spectra*, 9(3), 505-528.

[53] Tyler, R.G. 1985. Test on a Brake Lining Damper for Structures, *Bull. N.Z. Nat. Soc. Earthquake Eng.*, 18(3), 280-284.

[54] Uang, C.M. and Bertero, V.V. 1986. *Earthquake Simulation Tests and Associated Studies of a 0.3 Scale Model of a Six-Story Concentrically Braced Steel Structure*, Report No. UCB/EERC-86/10, Earthquake Engineering Research Center, Berkeley, CA.

[55] Uang, C.M. and Bertero, V.V. 1988. *Use of Energy as a Design Criterion in Earthquake Resistant Design*, Report No. UCB/EERC-88/18, Earthquake Engineering Research Center, Berkeley, CA.

[56] Veletsos, A.S. and Newmark, N.M. 1960. Effect of Inelastic Behavior on the Response of Simple Systems to Earthquake Motions, *Proc. 2nd World Conf. Earthquake Eng.*, II, 895-912, Tokyo.

[57] Villaverde, R. 1994. Seismic-Control of Structures with Damped Resonant Appendages, *Proc. 1st World Conf. Struct. Control*, Los Angeles, CA, 1, WP4-113 - WP4-122.

[58] Whittaker, A.S., Aiken, I., Bergman, P., Clark, J., Cohen, J., Kelly, J.M. and Scholl, R. 1993. Code Requirements for the Design and Implementation of Passive Energy Dissipation Systems, *Proc. ATC 17-1 Seminar Seismic Isolation, Passive Energy Dissipation and Active Control*, San Francisco, 2, 497-508.

[59] Whittaker, A.S., Bertero, V.V., Thompson, C.L. and Alonso, L.J. 1991. Seismic Testing of Steel Plate Energy Dissipation Devices, *Earthquake Spectra*, 7(4), 563-604.

[60] Wirsching, P.H. and Yao, J.T.P. 1970. Modal Response of Structures, *J. Struct. Div.*, ASCE, 96(4), 879-883.

[61] Wu, Z. and Soong, T.T. 1995. Nonlinear Feedback Control for Improved Peak Response Reduction, *J. Smart Mater. Struct.*, 4, A140-A147.

[62] Zahrah, T.F. and Hall, W.J. 1982. *Seismic Energy Absorption in Simple Structures*, Structural Research Series No. 501, University of Illinois, Urbana, IL.

[63] Zienkiewicz, O.C. and Taylor, R.L. 1989. *The Finite Element Method*, Vols. 1 and 2, 4 ed., McGraw-Hill, London.

28

An Innnovative Design For Steel Frame Using Advanced Analysis[1]

28.1 Introduction 28-1
28.2 Practical Advanced Analysis 28-5
   Second-Order Refined Plastic Hinge Analysis • Analysis of Semi-Rigid Frames • Geometric Imperfection Methods • Numerical Implementation
28.3 Verifications 28-18
   Axially Loaded Columns • Portal Frame • Six-Story Frame • Semi-Rigid Frame
28.4 Analysis and Design Principles 28-28
   Design Format • Loads • Load Combinations • Resistance Factors • Section Application • Modeling of Structural Members • Modeling of Geometric Imperfection • Load Application • Analysis • Load-Carrying Capacity • Serviceability Limits • Ductility Requirements • Adjustment of Member Sizes
28.5 Computer Program 28-34
   Program Overview • Hardware Requirements • Execution of Program • Users' Manual
28.6 Design Examples 28-41
   Roof Truss • Unbraced Eight-Story Frame • Two-Story Four-Bay Semi-Rigid Frame
28.7 Defining Terms 28-53
References 28-54
Further Reading 28-55

Seung-Eock Kim
Department of Civil Engineering, Sejong University, Seoul, South Korea

W. F. Chen
School of Civil Engineering, Purdue University, West Lafayette, IN

28.1 Introduction

The steel design methods used in the U.S. are allowable stress design (**ASD**), plastic design (**PD**), and load and **resistance factor** design (**LRFD**). In ASD, the stress computation is based on a first-order elastic analysis, and the geometric nonlinear effects are implicitly accounted for in the member design equations. In PD, a first-order plastic-hinge analysis is used in the structural analysis. PD allows inelastic force redistribution throughout the structural system. Since geometric nonlinearity and gradual yielding effects are not accounted for in the analysis of plastic design, they are approximated in member design equations. In LRFD, a first-order elastic analysis with amplification factors or a

---

[1]The material in this chapter was previously published by CRC Press in *LRFD Steel Design Using Advanced Analysis*, W. F. Chen and Seung-Eock Kim, 1997.

direct second-order elastic analysis is used to account for geometric nonlinearity, and the ultimate strength of **beam-column** members is implicitly reflected in the design interaction equations. All three design methods require separate member capacity checks including the calculation of the K factor. In the following, the characteristics of the LRFD method are briefly described.

The strength and stability of a structural system and its members are related, but the interaction is treated separately in the current American Institute of Steel Construction (AISC)-LRFD specification [2]. In current practice, the interaction between the structural system and its members is represented by the effective length factor. This aspect is described in the following excerpt from SSRC Technical Memorandum No. 5 [28]:

> Although the maximum strength of frames and the maximum strength of component members are interdependent (but not necessarily coexistent), it is recognized that in many structures it is not practical to take this interdependence into account rigorously. At the same time, it is known that difficulties are encountered in complex frameworks when attempting to compensate automatically in **column** design for the instability of the entire frame (for example, by adjustment of column effective length). Therefore, SSRC recommends that, in design practice, the two aspects, stability of separate members and elements of the structure and stability of the structure as a whole, be considered separately.

This design approach is marked in Figure 28.1 as the indirect analysis and design method.

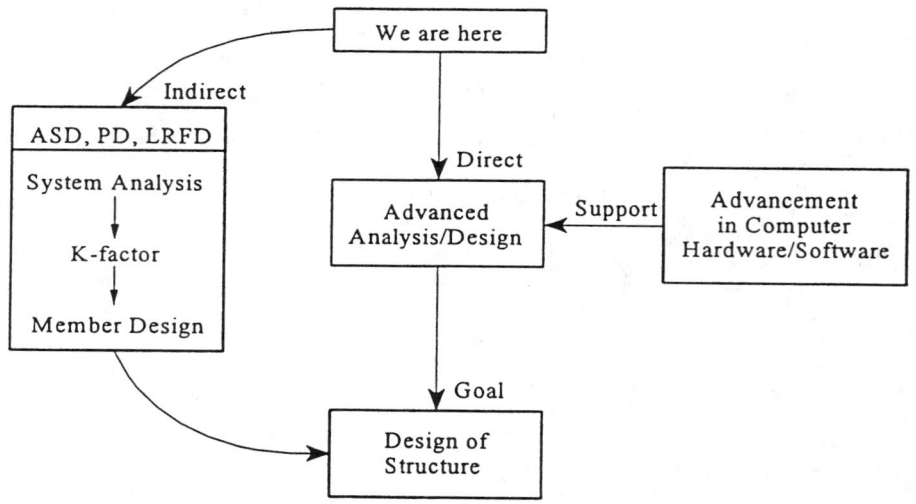

Figure 28.1   Analysis and design methods.

In the current AISC-LRFD specification [2], first-order elastic analysis or second-order elastic analysis is used to analyze a structural system. In using first-order elastic analysis, the first-order moment is amplified by B_1 and B_2 factors to account for second-order effects. In the specification, the members are isolated from a structural system, and they are then designed by the member strength curves and interaction equations as given by the specifications, which implicitly account for second-order effects, inelasticity, residual stresses, and **geometric imperfections** [8]. The column curve and beam curve were developed by a curve-fit to both theoretical solutions and experimental data, while the beam-column interaction equations were determined by a curve-fit to the so-called "exact" **plastic-zone** solutions generated by Kanchanalai [14].

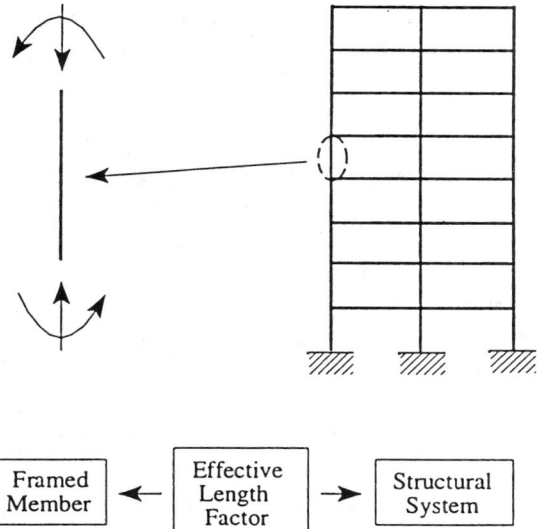

| Framed Member | ← | Effective Length Factor | → | Structural System |

Figure 28.2 Interaction between a structural system and its component members.

In order to account for the influence of a structural system on the strength of individual members, the effective length factor is used, as illustrated in Figure 28.2. The effective length method generally provides a good design of framed structures. However, several difficulties are associated with the use of the effective length method, as follows:

1. The effective length approach cannot accurately account for the interaction between the structural system and its members. This is because the interaction in a large structural system is too complex to be represented by the simple effective length factor K. As a result, this method cannot accurately predict the actual required strengths of its framed members.

2. The effective length method cannot capture the inelastic redistributions of internal forces in a structural system, since the first-order elastic analysis with B_1 and B_2 factors accounts only for second-order effects but not the inelastic redistribution of internal forces. The effective length method provides a conservative estimation of the ultimate load-carrying capacity of a large structural system.

3. The effective length method cannot predict the failure modes of a structural system subject to a given load. This is because the LRFD interaction equation does not provide any information about failure modes of a structural system at the **factored loads**.

4. The effective length method is not user friendly for a computer-based design.

5. The effective length method requires a time-consuming process of separate member capacity checks involving the calculation of K factors.

With the development of computer technology, two aspects, the stability of separate members and the stability of the structure as a whole, can be treated rigorously for the determination of the maximum strength of the structures. This design approach is marked in Figure 28.1 as the direct analysis and design method. The development of the direct approach to design is called **advanced analysis**, or more specifically, second-order inelastic analysis for frame design. In this direct approach, there is no need to compute the effective length factor, since separate member capacity checks encompassed by the specification equations are not required. With the current available computing technology, it is feasible to employ advanced analysis techniques for direct frame design. This method has been considered impractical for design office use in the past. The purpose of this chapter is to present a practical, direct method of steel frame design, using advanced analysis, that will produce almost identical member sizes as those of the LRFD method.

The advantages of advanced analysis in design use are outlined as follows:

1. Advanced analysis is another tool for structural engineers to use in steel design, and its adoption is not mandatory but will provide a flexibility of options to the designer.
2. Advanced analysis captures the **limit state** strength and stability of a structural system and its individual members directly, so separate member capacity checks encompassed by the specification equations are not required.
3. Compared to the LRFD and ASD, advanced analysis provides more information of structural behavior by direct inelastic **second-order analysis**.
4. Advanced analysis overcomes the difficulties due to incompatibility between the elastic global analysis and the limit state member design in the conventional LRFD method.
5. Advanced analysis is user friendly for a computer-based design, but the LRFD and ASD are not, since they require the calculation of K factor on the way from their analysis to separate member capacity checks.
6. Advanced analysis captures the inelastic redistribution of internal forces throughout a structural system, and allows an economic use of material for highly indeterminate steel frames.
7. It is now feasible to employ advanced analysis techniques that have been considered impractical for design office use in the past, since the power of personal computers and engineering workstations is rapidly increasing.
8. Member sizes determined by advanced analysis are close to those determined by the LRFD method, since the advanced analysis method is calibrated against the LRFD column curve and beam-column interaction equations. As a result, advanced analysis provides an alternative to the LRFD.
9. Advanced analysis is time effective since it completely eliminates tedious and often confused member capacity checks, including the calculation of K factors in the LRFD and ASD.

Among various advanced analyses, including plastic-zone, quasi-plastic hinge, elastic-plastic hinge, notional-load plastic-hinge, and refined **plastic hinge** methods, the refined plastic hinge method is recommended, since it retains the efficiency and simplicity of computation and accuracy for practical use. The method is developed by imposing simple modifications on the conventional elastic-plastic hinge method. These include a simple modification to account for the gradual sectional **stiffness** degradation at the plastic hinge locations and to include the gradual member stiffness degradation between two plastic hinges.

The key considerations of the conventional LRFD method and the practical advanced analysis method are compared in Table 28.1. While the LRFD method does account for key behavioral effects implicitly in its column strength and beam-column interaction equations, the advanced analysis method accounts for these effects explicitly through **stability functions**, stiffness degradation functions, and geometric imperfections, to be discussed in detail in Section 28.2.

Advanced analysis holds many answers to real behavior of steel structures and, as such, we recommend the proposed design method to engineers seeking to perform frame design in efficiency and rationality, yet consistent with the present LRFD specification. In the following sections, we will present a practical advanced analysis method for the design of steel frame structures with LRFD. The validity of the approach will be demonstrated by comparing case studies of actual members and

TABLE 28.1   Key Considerations of Load and Resistance Factor Design (LRFD) and Proposed Methods

Key consideration	LRFD	Proposed method
Second-order effects	Column curve B_1, B_2 factor	Stability function
Geometric imperfection	Column curve	Explicit imperfection modeling method $\psi = 1/500$ for unbraced frame $\delta_c = L_c/1000$ for braced frame Equivalent notional load method $\alpha = 0.002$ for unbraced frame $\alpha = 0.004$ for braced frame Further reduced tangent modulus method $E_t' = 0.85E_t$
Stiffness degradation associated with residual stresses	Column curve	CRC tangent modulus
Stiffness degradation associated with flexure	Column curve Interaction equations	Parabolic degradation function
Connection nonlinearity	No procedure	Power model/rotational spring

frames with the results of analysis/design based on exact plastic-zone solutions and LRFD designs. The wide range of case studies and comparisons should confirm the validity of this advanced method.

28.2   Practical Advanced Analysis

This section presents a practical advanced analysis method for the direct design of steel frames by eliminating separate member capacity checks by the specification. The refined plastic hinge method was developed and refined by simply modifying the conventional elastic-plastic hinge method to achieve both simplicity and a realistic representation of actual behavior [15, 25]. Verification of the method will be given in the next section to provide final confirmation of the validity of the method.

Connection flexibility can be accounted for in advanced analysis. Conventional analysis and design of steel structures are usually carried out under the assumption that beam-to-column connections are either fully rigid or ideally pinned. However, most connections in practice are semi-rigid and their behavior lies between these two extreme cases. In the AISC-LRFD specification [2], two types of construction are designated: Type FR (fully restrained) construction and Type PR (partially restrained) construction. The LRFD specification permits the evaluation of the flexibility of connections by "rational means".

Connection behavior is represented by its moment-rotation relationship. Extensive experimental work on connections has been performed, and a large body of moment-rotation data collected. With this data base, researchers have developed several connection models, including linear, polynomial, B-spline, power, and exponential. Herein, the three-parameter power model proposed by Kishi and Chen [21] is adopted.

Geometric imperfections should be modeled in frame members when using advanced analysis. Geometric imperfections result from unavoidable error during fabrication or erection. For structural members in building frames, the types of geometric imperfections are out-of-straightness and out-of-plumbness. Explicit modeling and equivalent **notional loads** have been used to account for geometric imperfections by previous researchers. In this section, a new method based on further reduction of the tangent stiffness of members is developed [15, 16]. This method provides a simple means to account for the effect of imperfection without inputting notional loads or explicit geometric imperfections.

The practical advanced analysis method described in this section is limited to two-dimensional braced, unbraced, and semi-rigid frames subject to static loads. The spatial behavior of frames is not considered, and lateral torsional buckling is assumed to be prevented by adequate lateral bracing. A compact W section is assumed so sections can develop full plastic moment capacity without local buckling. Both strong- and weak-axis bending of wide flange sections have been studied using the practical advanced analysis method [15]. The method may be considered an interim analysis/design procedure between the conventional LRFD method widely used now and a more rigorous advanced analysis/design method such as the plastic-zone method to be developed in the future for practical use.

28.2.1   Second-Order Refined Plastic Hinge Analysis

In this section, a method called the refined plastic hinge approach is presented. This method is comparable to the elastic-plastic hinge analysis in efficiency and simplicity, but without its limitations. In this analysis, stability functions are used to predict second-order effects. The benefit of stability functions is that they make the analysis method practical by using only one element per beam-column. The **refined plastic hinge analysis** uses a two-surface yield model and an effective tangent modulus to account for stiffness degradation due to distributed plasticity in framed members. The member stiffness is assumed to degrade gradually as the second-order forces at critical locations approach the cross-section plastic strength. Column tangent modulus is used to represent the effective stiffness of the member when it is loaded with a high axial load. Thus, the refined plastic hinge model approximates the effect of distributed plasticity along the element length caused by initial imperfections and large bending and axial force actions. In fact, research by Liew et al. [25, 26], Kim and Chen [16], and Kim [15] has shown that refined plastic hinge analysis captures the interaction of strength and stability of structural systems and that of their component elements. This type of analysis method may, therefore, be classified as an advanced analysis and separate specification member capacity checks are not required.

Stability Function

To capture second-order effects, stability functions are recommended since they lead to large savings in modeling and solution efforts by using one or two elements per member. The simplified stability functions reported by Chen and Lui [7] or an alternative may be used. Considering the prismatic beam-column element, the incremental force-displacement relationship of this element may be written as

$$\begin{bmatrix} \dot{M}_A \\ \dot{M}_B \\ \dot{P} \end{bmatrix} = \frac{EI}{L} \begin{bmatrix} S_1 & S_2 & 0 \\ S_2 & S_1 & 0 \\ 0 & 0 & A/I \end{bmatrix} \begin{bmatrix} \dot{\theta}_A \\ \dot{\theta}_B \\ \dot{e} \end{bmatrix} \qquad (28.1)$$

where

S_1, S_2	=	stability functions
$\dot{M}_A, \dot{M}_B$	=	incremental end moment
$\dot{P}$	=	incremental axial force
$\dot{\theta}_A, \dot{\theta}_B$	=	incremental joint rotation
$\dot{e}$	=	incremental axial displacement
A, I, L	=	area, moment of inertia, and length of beam-column element
E	=	modulus of elasticity.

In this formulation, all members are assumed to be adequately braced to prevent out-of-plane buckling, and their cross-sections are compact to avoid local buckling.

Cross-Section Plastic Strength

Based on the AISC-LRFD bilinear interaction equations [2], the cross-section plastic strength may be expressed as Equation 28.2. These AISC-LRFD cross-section plastic strength curves may be adopted for both strong- and weak-axis bending (Figure 28.3).

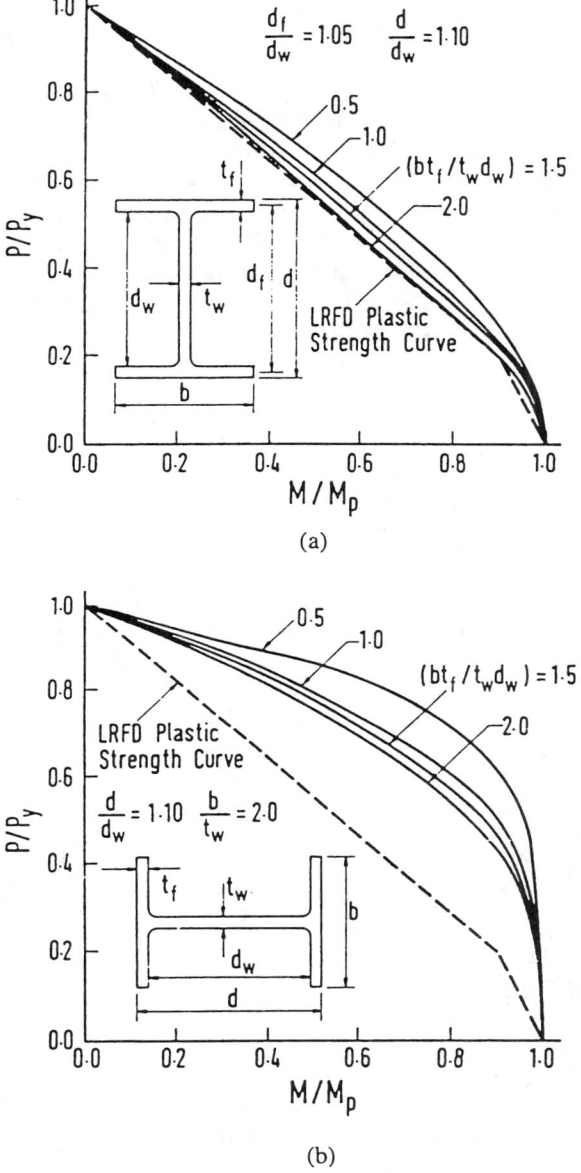

Figure 28.3 Strength interaction curves for wide-flange sections.

$$\frac{P}{P_y} + \frac{8}{9}\frac{M}{M_p} = 1.0 \quad \text{for} \quad \frac{P}{P_y} \geq 0.2 \qquad (28.2a)$$

$$\frac{P}{P_y} + \frac{M}{M_p} = 1.0 \quad \text{for} \quad \frac{P}{P_y} < 0.2 \qquad (28.2b)$$

where
P, M = second-order axial force and bending moment
P_y = squash load
M_p = plastic moment capacity

CRC Tangent Modulus

The **CRC** tangent modulus concept is employed to account for the gradual yielding effect due to residual stresses along the length of members under axial loads between two plastic hinges. In this concept, the elastic modulus, E, instead of moment of inertia, I, is reduced to account for the reduction of the elastic portion of the cross-section since the reduction of elastic modulus is easier to implement than that of moment of inertia for different sections. The reduction rate in stiffness between the weak and strong axis is different, but this is not considered here. This is because rapid degradation in stiffness in the weak-axis strength is compensated well by the stronger weak-axis plastic strength. As a result, this simplicity will make the present methods practical. From Chen and Lui [7], the CRC E_t is written as (Figure 28.4):

$$E_t \quad = \quad 1.0E \ \text{ for } \ P \le 0.5P_y \tag{28.3a}$$

$$E_t \quad = \quad 4\frac{P}{P_y}E\left(1 - \frac{P}{P_y}\right) \ \text{ for } \ P > 0.5P_y \tag{28.3b}$$

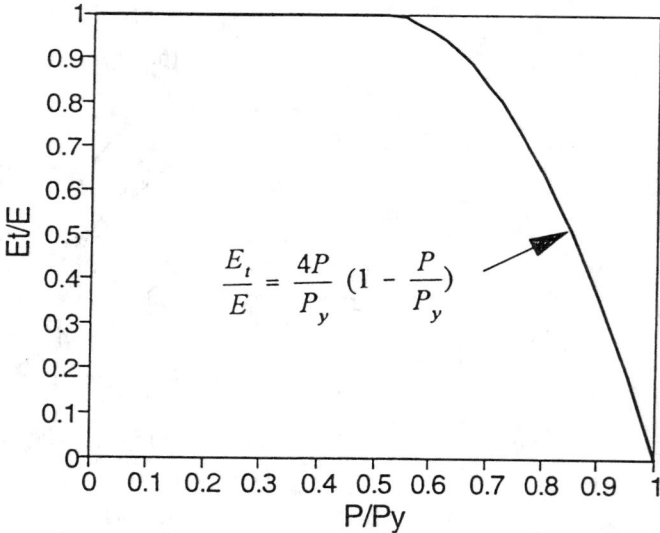

Figure 28.4 Member tangent stiffness degradation derived from the CRC column curve.

Parabolic Function

The tangent modulus model in Equation 28.3 is suitable for $P/P_y > 0.5$, but it is not sufficient to represent the stiffness degradation for cases with small axial forces and large bending moments. A gradual stiffness degradation of plastic hinge is required to represent the distributed plasticity effects associated with bending actions. We shall introduce the hardening plastic hinge model to represent the gradual transition from elastic stiffness to zero stiffness associated with a fully developed plastic hinge. When the hardening plastic hinges are present at both ends of an element, the incremental force-displacement relationship may be expressed as [24]:

$$
\begin{bmatrix} \dot{M}_A \\ \dot{M}_B \\ \dot{P} \end{bmatrix} = \frac{E_t I}{L} \begin{bmatrix} \eta_A \left[S_1 - \frac{S_2^2}{S_1}(1 - \eta_B) \right] & \eta_A \eta_B S_2 & 0 \\ \eta_A \eta_B S_2 & \eta_B \left[S_1 - \frac{S_2^2}{S_1}(1 - \eta_A) \right] & 0 \\ 0 & 0 & A/I \end{bmatrix} \begin{bmatrix} \dot{\theta}_A \\ \dot{\theta}_B \\ \dot{e} \end{bmatrix}
$$

$$(28.4)$$

where

$\dot{M}_A, \dot{M}_B, \dot{P}$ = incremental end moments and axial force, respectively
S_1, S_2 = stability functions
E_t = tangent modulus
η_A, η_B = element stiffness parameters

The parameter η represents a gradual stiffness reduction associated with flexure at sections. The partial plastification at cross-sections in the end of elements is denoted by $0 < \eta < 1$. The η may be assumed to vary according to the parabolic expression (Figure 28.5):

$$\eta = 4\alpha(1 - \alpha) \text{ for } \alpha > 0.5 \qquad (28.5)$$

where α is the force state parameter obtained from the limit state surface corresponding to the element end (Figure 28.6):

$$
\alpha = \frac{P}{P_y} + \frac{8}{9}\frac{M}{M_p} \text{ for } \frac{P}{P_y} \geq \frac{2}{9}\frac{M}{M_p} \qquad (28.6a)
$$

$$
\alpha = \frac{P}{2P_y} + \frac{M}{M_p} \text{ for } \frac{P}{P_y} < \frac{2}{9}\frac{M}{M_p} \qquad (28.6b)
$$

where

P, M = second-order axial force and bending moment at the cross-section
M_p = plastic moment capacity

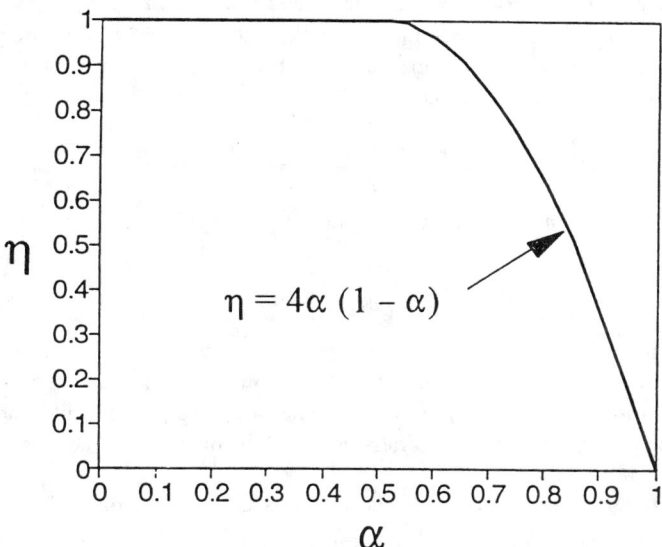

Figure 28.5 Parabolic plastic hinge stiffness degradation function with $\alpha_0 = 0.5$ based on the load and resistance factor design sectional strength equation.

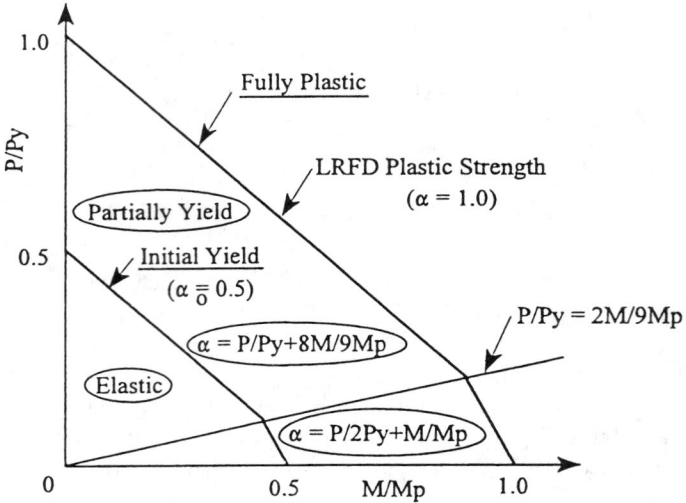

Figure 28.6   Smooth stiffness degradation for a work-hardening plastic hinge based on the load and resistance factor design sectional strength curve.

28.2.2   Analysis of Semi-Rigid Frames

Practical Connection Modeling

The three-parameter power model contains three parameters: initial connection stiffness, R_{ki}, ultimate connection moment capacity, M_u, and shape parameter, n. The power model may be written as (Figure 28.7):

$$m = \frac{\theta}{(1+\theta^n)^{1/n}} \quad \text{for} \quad \theta > 0, \quad m > 0 \tag{28.7}$$

where $m = M/M_u$, $\theta = \theta_r/\theta_o$, θ_o = reference plastic rotation, M_u/R_{ki}, M_u = ultimate moment capacity of the connection, R_{ki} = initial connection stiffness, and n = shape parameter. When the connection is loaded, the connection tangent stiffness, R_{kt}, at an arbitrary rotation, θ_r, can be derived by simply differentiating Equation 28.7 as:

$$R_{kt} = \frac{dM}{d\,|\theta_r|} = \frac{M_u}{\theta_o\,(1+\theta^n)^{1+1/n}} \tag{28.8}$$

When the connection is unloaded, the tangent stiffness is equal to the initial stiffness as:

$$R_{kt} = \frac{dM}{d\,|\theta_r|} = \frac{M_u}{\theta_o} = R_{ki} \tag{28.9}$$

It is observed that a small value of the power index, n, makes a smooth transition curve from the initial stiffness, R_{kt}, to the ultimate moment, M_u. On the contrary, a large value of the index, n, makes the transition more abruptly. In the extreme case, when n is infinity, the curve becomes a bilinear line consisting of the initial stiffness, R_{ki}, and the ultimate moment capacity, M_u.

Practical Estimation of Three Parameters Using Computer Program

An important task for practical use of the power model is to determine the three parameters for a given connection configuration. One difficulty in determining the three parameters is the need for numerical iteration, especially to estimate the ultimate moment, M_u. A set of nomographs

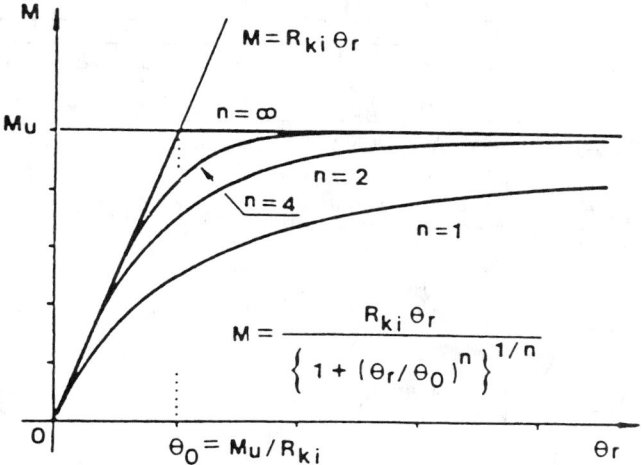

Figure 28.7 Moment-rotation behavior of the three-parameter model.

was proposed by Kishi et al. [22] to overcome the difficulty. Even though the purpose of these nomographs is to allow the engineer to rapidly determine the three parameters for a given connection configuration, the nomographs require other efforts for engineers to know how to use them, and the values of the nomographs are approximate.

Herein, one simple way to avoid the difficulties described above is presented. A direct and easy estimation of the three parameters may be achieved by use of a simple computer program 3PARA.f. The operating procedure of the program is shown in Figure 28.8. The input data, CONN.DAT, may be easily generated corresponding to the input format listed in Table 28.2.

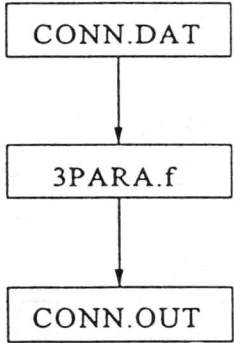

Figure 28.8 Operating procedure of computer program estimating the three parameters.

As for the shape parameter, n, the equations developed by Kishi et al. [22] are implemented here. Using a statistical technique for n values, empirical equations of n are determined as a linear function of $\log_{10} \theta_0$, shown in Table 28.3. This n value may be calculated using 3PARA.f.

Load-Displacement Relationship Accounting for Semi-Rigid Connection

The connection may be modeled as a rotational spring in the moment-rotation relationship represented by Equation 28.10. Figure 28.9 shows a beam-column element with **semi-rigid connections** at both ends. If the effect of connection flexibility is incorporated into the member stiffness,

TABLE 28.2 Input Format

Line	Input data						Remark
1	ITYPE	F_y	E				Connection type and material properties
2	l_t	t_t	k_t	g_t	W	d	Top/ seat-angle data
3	l_a	t_a	k_a	g_a			Web-angle data

ITYPE	=	Connection type (1 = top and seat-angle connection, 2 = with web-angle connection)
F_y	=	yield strength of angle
E	=	Young's modulus (= 29, 000 ksi)
l_t	=	length of top angle
t_t	=	thickness of top angle
k_t	=	k value of top angle
g_t	=	gauge of top angle(= 2.5 in., typical)
W	=	width of nut (W = 1.25 in. for 3/4D bolt, W = 1.4375 in. for 7/8D bolt)
d	=	depth of beam
l_a	=	length of web angle
t_a	=	thickness of web angle
k_a	=	k value of web angle
g_a	=	gauge of web angle

Note:
(1) Top- and seat-angle connections need lines 1 and 2 for input data, and top and seat angle with web-angle connections need lines 1, 2, and 3.
(2) All input data are in free format.
(3) Top- and seat-angle sizes are assumed to be the same.
(4) Bolt sizes of top angle, seat angle, and web angle are assumed to be the same.

TABLE 28.3 Empirical Equations for Shape Parameter, n

Connection type	n	
Single web-angle connection	$0.520 \log_{10} \theta_o + 2.291$	for $\log_{10} \theta_o > -3.073$
	0.695	for $\log_{10} \theta_o < -3.073$
Double web-angle connection	$1.322 \log_{10} \theta_o + 3.952$	for $\log_{10} \theta_o > -2.582$
	0.573	for $log_{10}\theta_o < -2.582$
Top- and seat-angle connection	$2.003 \log_{10} \theta_o + 6.070$	for $\log_{10} \theta_o > -2.880$
	0.302	for $\log_{10} \theta_o < -2.880$
Top- and seat-angle connection with double web angle	$1.398 \log_{10} \theta_o + 4.631$	for $\log_{10} \theta_o > -2.721$
	0.827	for $\log_{10} \theta_o < -2.721$

From Kishi, N., Goto, Y., Chen, W. F., and Matsuoka, K. G. 1993. *Eng. J.*, AISC, pp. 90-107. With permission.

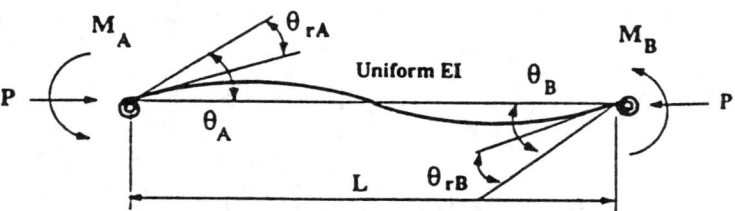

Figure 28.9 Beam-column element with semi-rigid connections.

the incremental element force-displacement relationship of Equation 28.1 is modified as [24]:

$$\begin{bmatrix} \dot{M}_A \\ \dot{M}_B \\ \dot{P} \end{bmatrix} = \frac{E_t I}{L} \begin{bmatrix} S_{ii}^* & S_{ij}^* & 0 \\ S_{ij}^* & S_{jj}^* & 0 \\ 0 & 0 & A/I \end{bmatrix} \begin{bmatrix} \dot{\theta}_A \\ \dot{\theta}_B \\ \dot{e} \end{bmatrix} \qquad (28.10)$$

where

$$S_{ii}^* = \left(S_{ij} + \frac{E_t I S_{ii} S_{jj}}{L R_{kt B}} - \frac{E_t I S_{ij}^2}{L R_{kt B}} \right) / R^* \qquad (28.11a)$$

$$S_{jj}^* = \left(S_{jj} + \frac{E_t I S_{ii} S_{jj}}{L R_{kt A}} - \frac{E_t I S_{ij}^2}{L R_{kt A}} \right) / R^* \qquad (28.11b)$$

$$S_{ij}^* = S_{ij} / R^* \qquad (28.11c)$$

$$R^* = \left(1 + \frac{E_t I S_{ii}}{L R_{kt A}} \right) \left(1 + \frac{E_t I S_{jj}}{L R_{kt B}} \right) - \left(\frac{E_t I}{L} \right)^2 \frac{S_{ij}^2}{R_{kt A} R_{kt B}} \qquad (28.11d)$$

where $R_{kt A}$, $R_{kt B}$ = tangent stiffness of connections A and B, respectively; S_{ii} S_{ij} = generalized stability functions; and S_{ii}^*, S_{jj}^* = modified stability functions that account for the presence of end connections. The tangent stiffness ($R_{kt A}$, $R_{kt B}$) accounts for the different types of semi-rigid connections (see Equation 28.8).

28.2.3 Geometric Imperfection Methods

Geometric imperfection modeling combined with the CRC tangent modulus model is discussed in what follows. There are three: the explicit imperfection modeling method, the equivalent notional load method, and the further reduced tangent modulus method.

Explicit Imperfection Modeling Method

Braced Frame

The refined plastic hinge analysis implicitly accounts for the effects of both residual stresses and spread of yielded zones. To this end, refined plastic hinge analysis may be regarded as equivalent to the plastic-zone analysis. As a result, geometric imperfections are necessary only to consider fabrication error. For braced frames, member out-of-straightness, rather than frame out-of-plumbness, needs to be used for geometric imperfections. This is because the $P - \Delta$ effect due to the frame out-of-plumbness is diminished by braces. The ECCS [10, 11], AS [30], and Canadian Standard Association (CSA) [4, 5] specifications recommend an initial crookedness of column equal to 1/1000 times the column length. The AISC code recommends the same maximum fabrication tolerance of $L_c/1000$ for member out-of-straightness. In this study, a geometric imperfection of $L_c/1000$ is adopted.

The ECCS [10, 11], AS [30], and CSA [4, 5] specifications recommend the out-of-straightness varying parabolically with a maximum in-plane deflection at the midheight. They do not, however, describe how the parabolic imperfection should be modeled in analysis. Ideally, many elements are needed to model the parabolic out-of-straightness of a beam-column member, but it is not practical. In this study, two elements with a maximum initial deflection at the midheight of a member are found adequate for capturing the imperfection. Figure 28.10 shows the out-of-straightness modeling for a braced beam-column member. It may be observed that the out-of-plumbness is equal to 1/500 when the half segment of the member is considered. This value is identical to that of sway frames as discussed in recent papers by Kim and Chen [16, 17, 18]. Thus, it may be stated that the imperfection values are essentially identical for both sway and braced frames. It is noted that this explicit modeling method in braced frames requires the inconvenient imperfection modeling at the center of columns although the inconvenience is much lighter than that of the conventional LRFD method for frame design.

Unbraced Frame

The CSA [4, 5] and the AISC codes of standard practice [2] set the limit of erection out-of-plumbness at $L_c/500$. The maximum erection tolerances in the AISC are limited to 1 in. toward

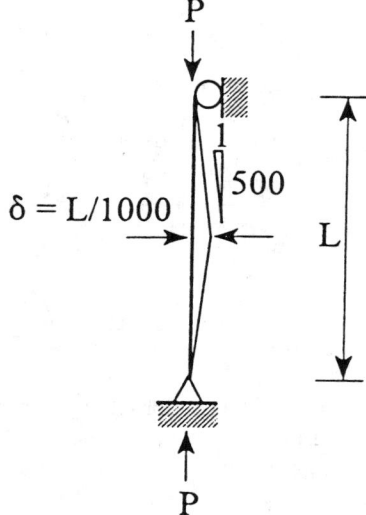

Figure 28.10   Explicit imperfection modeling of a braced member.

the exterior of buildings and 2 in. toward the interior of buildings less than 20 stories. Considering the maximum permitted average lean of 1.5 in. in the same direction of a story, the geometric imperfection of $L_c/500$ can be used for buildings up to six stories with each story approximately 10 ft high. For taller buildings, this imperfection value of $L_c/500$ is conservative since the accumulated geometric imperfection calculated by 1/500 times building height is greater than the maximum permitted erection tolerance.

In this study, we shall use $L_c/500$ for the out-of-plumbness without any modification because the system strength is often governed by a weak story that has an out-of-plumbness equal to $L_c/500$ [27] and a constant imperfection has the benefit of simplicity in practical design. The explicit geometric imperfection modeling for an unbraced frame is illustrated in Figure 28.11.

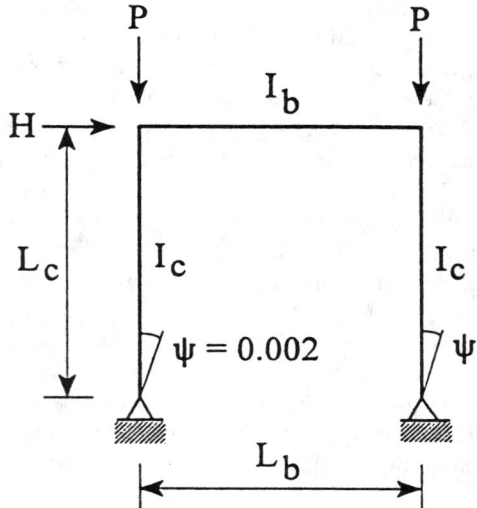

Figure 28.11   Explicit imperfection modeling of an unbraced frame.

Equivalent Notional Load Method

Braced Frame

The ECCS [10, 11] and the CSA [4, 5] introduced the equivalent load concept, which accounted for the geometric imperfections in unbraced frames, but not in braced frames. The notional load approach for braced frames is also necessary to use the proposed methods for braced frames.

For braced frames, an equivalent notional load may be applied at midheight of a column since the ends of the column are braced. An equivalent notional **load factor** equal to 0.004 is proposed here, and it is equivalent to the out-of-straightness of $L_c/1000$. When the free body of the column shown in Figure 28.12 is considered, the notional load factor, α, results in 0.002 with respect to one-half of the member length. Here, as in explicit imperfection modeling, the equivalent notional load factor is the same in concept for both sway and braced frames.

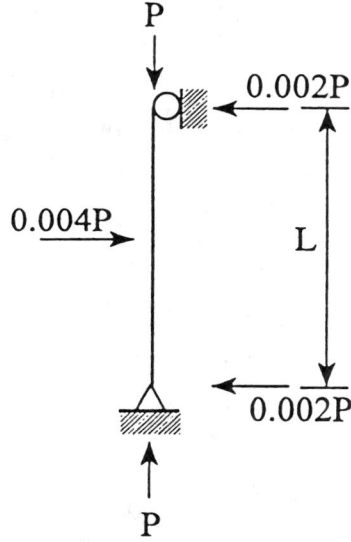

Figure 28.12   Equivalent notional load modeling for geometric imperfection of a braced member.

One drawback of this method for braced frames is that it requires tedious input of notional loads at the center of each column. Another is the axial force in the columns must be known in advance to determine the notional loads before analysis, but these are often difficult to calculate for large structures subject to lateral wind loads. To avoid this difficulty, it is recommended that either the explicit imperfection modeling method or the further reduced tangent modulus method be used.

Unbraced Frame

The geometric imperfections of a frame may be replaced by the equivalent notional lateral loads expressed as a fraction of the gravity loads acting on the story. Herein, the equivalent notional load factor of 0.002 is used. The notional load should be applied laterally at the top of each story. For sway frames subject to combined gravity and lateral loads, the notional loads should be added to the lateral loads. Figure 28.13 shows an illustration of the equivalent notional load for a portal frame.

Further Reduced Tangent Modulus Method

Braced Frame

The idea of using the reduced tangent modulus concept is to further reduce the tangent modulus, E_t, to account for further stiffness degradation due to geometrical imperfections. The degradation of member stiffness due to geometric imperfections may be simulated by an equivalent reduction

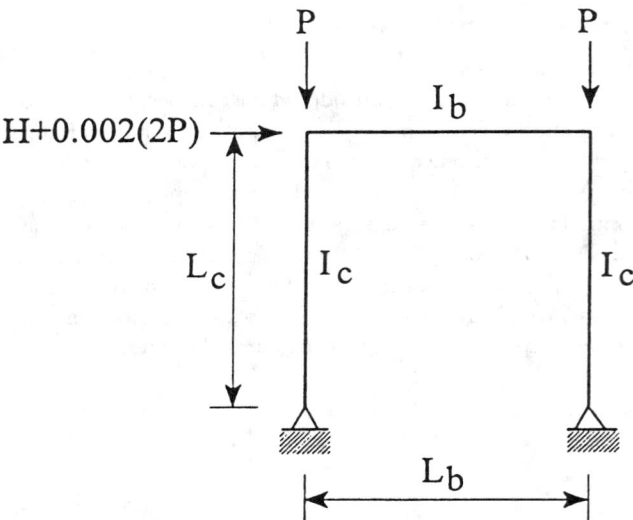

Figure 28.13   Equivalent notional load modeling for geometric imperfection of an unbraced frame.

of member stiffness. This may be achieved by a further reduction of tangent modulus as [15, 16]:

$$E'_t = 4\frac{P}{P_y}\left(1 - \frac{P}{P_y}\right)E\xi_i \text{ for } P > 0.5P_y \tag{28.12a}$$

$$E'_t = E\xi_i \text{ for } P \leq 0.5P_y \tag{28.12b}$$

where

E'_t = reduced E_t

ξ_i = reduction factor for geometric imperfection

   Herein, the reduction factor of 0.85 is used, and the further reduced tangent modulus curves for the CRC E_t with geometric imperfections are shown in Figure 28.14. The further reduced tangent modulus concept satisfies one of the requirements for advanced analysis recommended by the SSRC task force report [29], that is: "The geometric imperfections should be accommodated implicitly within the element model. This would parallel the philosophy behind the development of most modern column strength expressions. That is, the column strength expressions in specifications such as the AISC-LRFD implicitly include the effects of residual stresses and out-of-straightness."

   The advantage of this method over the other two methods is its convenience for design use, because it eliminates the inconvenience of explicit imperfection modeling or equivalent notional loads. Another benefit of this method is that it does not require the determination of the direction of geometric imperfections, often difficult to determine in a large system. On the other hand, in other two methods, the direction of geometric imperfections must be taken correctly in coincidence with the deflection direction caused by bending moments, otherwise the wrong direction of geometric imperfection in braced frames may help the bending stiffness of columns rather than reduce it.

Unbraced Frame

   The idea of the further reduced tangent modulus concept may also be used in the analysis of unbraced frames. Herein, as in the braced frame case, an appropriate reduction factor of 0.85 to E_t can be used [18, 19, 20]. The advantage of this approach over the other two methods is its convenience and simplicity because it completely eliminates the inconvenience of explicit imperfection modeling or the notional load input.

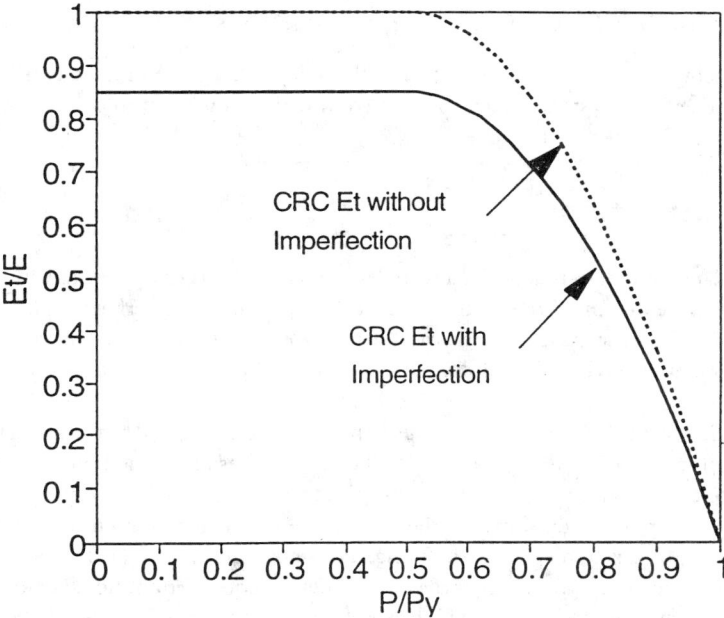

Figure 28.14   Further reduced CRC tangent modulus for members with geometric imperfections.

28.2.4   Numerical Implementation

The nonlinear global solution methods may be divided into two subgroups: (1) iterative methods and (2) simple incremental method. Iterative methods such as Newton-Raphson, modified Newton-Raphson, and quasi-Newton satisfy equilibrium equations at specific external loads. In these methods, the equilibrium out-of-balance present following the linear load step is eliminated (within tolerance) by taking corrective steps. The iterative methods possess the advantage of providing the exact load-displacement frame; however, they are inefficient, especially for practical purposes, in the trace of the hinge-by-hinge formation due to the requirement of the numerical iteration process.

The simple incremental method is a direct nonlinear solution technique. This numerical procedure is straightforward in concept and implementation. The advantage of this method is its computational efficiency. This is especially true when the structure is loaded into the inelastic region since tracing the hinge-by-hinge formation is required in the element stiffness formulation. For a finite increment size, this approach approximates only the nonlinear structural response, and equilibrium between the external applied loads and the internal element forces is not satisfied. To avoid this, an improved incremental method is used in this program. The applied load increment is automatically reduced to minimize the error when the change in the element stiffness parameter ($\Delta\eta$) exceeds a defined tolerance. To prevent plastic hinges from forming within a constant-stiffness load increment, load step sizes less than or equal to the specified increment magnitude are internally computed so plastic hinges form only after the load increment. Subsequent element stiffness formations account for the stiffness reduction due to the presence of the plastic hinges. For elements partially yielded at their ends, a limit is placed on the magnitude of the increment in the element end forces.

The applied load increment in the above solution procedure may be reduced for any of the following reasons:

1. Formation of new plastic hinge(s) prior to the full application of incremental loads.
2. The increment in the element nodal forces at plastic hinges is excessive.

3. Nonpositive definiteness of the structural stiffness matrix.

As the stability limit point is approached in the analysis, large step increments may overstep a limit point. Therefore, a smaller step size is used near the limit point to obtain accurate collapse displacements and second-order forces.

28.3 Verifications

In the previous section, a practical advanced analysis method was presented for a direct two-dimensional frame design. The practical approach of geometric imperfections and of semi-rigid connections was also discussed together with the advanced analysis method. The practical advanced analysis method was developed using simple modifications to the conventional elastic-plastic hinge analysis.

In this section, the practical advanced analysis method will be verified by the use of several benchmark problems available in the literature. Verification studies are carried out by comparing with the plastic-zone solutions as well as the conventional LRFD solutions. The strength predictions and the load-displacement relationships are checked for a wide range of steel frames including axially loaded columns, portal frame, six-story frame, and semi-rigid frames [15]. The three imperfection modelings, including explicit imperfection modeling, equivalent notional load modeling, and further reduced tangent modulus modeling, are also verified for a wide range of steel frames [15]).

28.3.1 Axially Loaded Columns

The AISC-LRFD column strength curve is used for the calibration since it properly accounts for second-order effects, residual stresses, and geometric imperfections in a practical manner. In this study, the column strength of proposed methods is evaluated for columns with slenderness parameters, $\left[\lambda_c = \frac{KL}{r}\sqrt{F_y/(\pi^2 E)}\right]$, varying from 0 to 2, which is equivalent to slenderness ratios (L/r) from 0 to 180 when the yield stress is equal to 36 ksi.

In explicit imperfection modeling, the two-element column is assumed to have an initial geometric imperfection equal to $L_c/1000$ at column midheight. The predicted column strengths are compared with the LRFD curve in Figure 28.15. The errors are found to be less than 5% for slenderness ratios up to 140 (or λ_c up to 1.57). This range includes most columns used in engineering practice.

In the equivalent notional load method, notional loads equal to 0.004 times the gravity loads are applied midheight to the column. The strength predictions are the same as those of the explicit imperfection model (Figure 28.16).

In the further reduced tangent modulus method, the reduced tangent modulus factor equal to 0.85 results in an excellent fit to the LRFD column strengths. The errors are less than 5% for columns of all slenderness ratios. These comparisons are shown in Figure 28.17.

28.3.2 Portal Frame

Kanchanalai [14] performed extensive analyses of portal and leaning column frames, and developed exact interaction curves based on plastic-zone analyses of simple sway frames. Note that the simple frames are more sensitive in their behavior than the highly redundant frames. His studies formed the basis of the interaction equations in the AISC-LRFD design specifications [2, 3]. In his studies, the stress-strain relationship was assumed elastic-perfectly plastic with a 36-ksi yield stress and a 29,000-ksi elastic modulus. The members were assumed to have a maximum compressive residual stress of $0.3F_y$. Initial geometric imperfections were not considered, and thus an adjustment of his interaction curves is made to account for this. Kanchanalai further performed experimental work to verify his analyses, which covered a wide range of portal and leaning column frames with

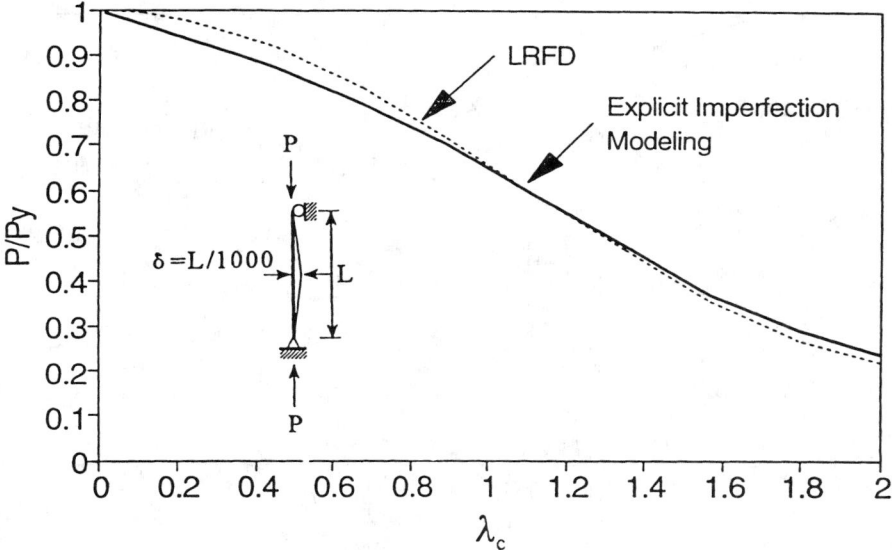

Figure 28.15 Comparison of strength curves for an axially loaded pin-ended column (explicit imperfection modeling method).

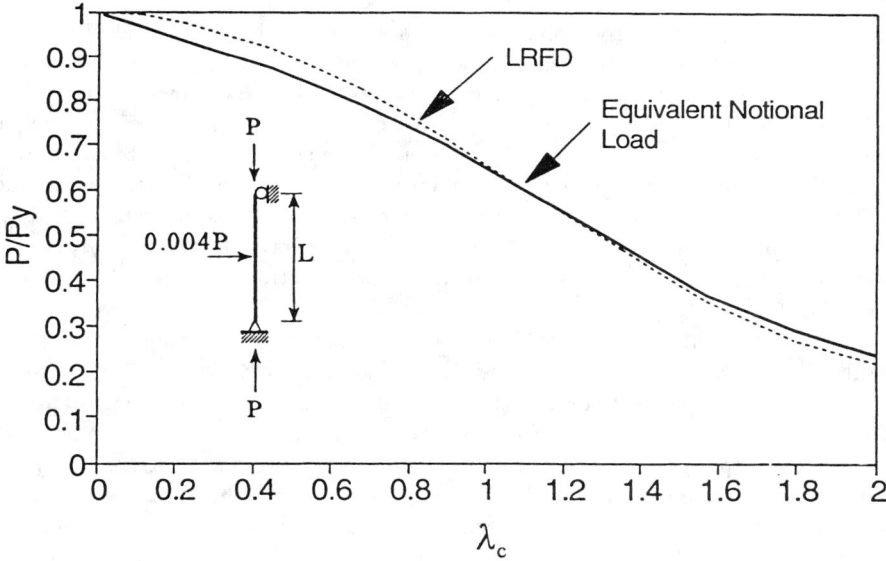

Figure 28.16 Comparison of strength curves for an axially loaded pin-ended column (equivalent notional load method).

slenderness ratios of 20, 30, 40, 50, 60, 70, and 80 and relative stiffness ratios (G) of 0, 3, and 4. The ultimate strength of each frame was presented in the form of interaction curves consisting of the nondimensional first-order moment ($HL_c/2M_p$ in portal frames or HL_c/M_p in leaning column frames in the x axis) and the nondimensional axial load (P/P_y in the y axis).

In this study, the AISC-LRFD interaction curves are used for strength comparisons. The strength calculations are based on the LeMessurier K factor method [23] since it accounts for story buckling and results in more accurate predictions. The inelastic stiffness reduction factor, τ [2], is used to

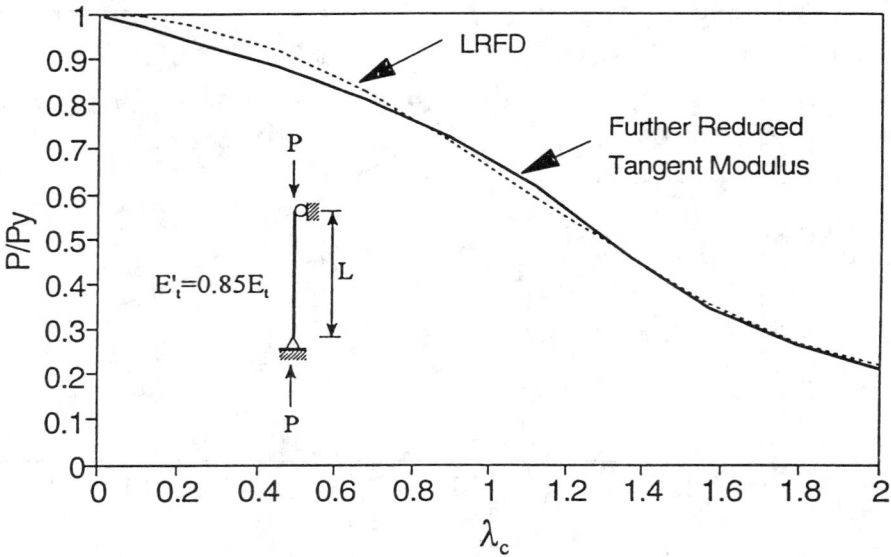

Figure 28.17 Comparison of strength curves for an axially loaded pin-ended column (further reduced tangent modulus method).

calculate K in LeMessurier's procedure. The resistance factors ϕ_b and ϕ_c in the LRFD equations are taken as 1.0 to obtain the nominal strength. The interaction curves are obtained by the accumulation of a set of moments and axial forces which result in unity on the value of the interaction equation.

When a geometric imperfection of $L_c/500$ is used for unbraced frames, including leaning column frames, most of the strength curves fall within an area bounded by the plastic-zone curves and the LRFD curves. In portal frames, the conservative errors are less than 5%, an improvement on the LRFD error of 11%, and the maximum unconservative error is not more than 1%, shown in Figure 28.18. In leaning column frames, the conservative errors are less than 12%, as opposed to the 17% error of the LRFD, and the maximum unconservative error is not more than 5%, as shown in Figure 28.19.

When a notional load factor of 0.002 is used, the strengths predicted by this method are close to those given by the explicit imperfection modeling method (Figures 28.20 and 28.21).

When the reduced tangent modulus factor of 0.85 is used for portal and leaning column frames, the interaction curves generally fall between the plastic-zone and LRFD curves. In portal frames, the conservative error is less than 8% (better than the 11% error of the LRFD) and the maximum unconservative error is not more than 5% (Figure 28.22). In leaning column frames, the conservative error is less than 7% (better than the 17% error of the LRFD) and the maximum unconservative error is not more than 5% (Figure 28.23).

28.3.3 Six-Story Frame

Vogel [32] presented the load-displacement relationships of a six-story frame using plastic-zone analysis. The frame is shown in Figure 28.24. Based on ECCS recommendations, the maximum compressive residual stress is $0.3F_y$ when the ratio of depth to width (d/b) is greater than 1.2, and is $0.5F_y$ when the d/b ratio is less than 1.2 (Figure 28.25). The stress-strain relationship is elastic-plastic with strain hardening as shown in Figure 28.26. The geometric imperfections are $L_c/450$.

For comparison, the out-of-plumbness of $L_c/450$ is used in the explicit modeling method. The notional load factor of 1/450 and the reduced tangent modulus factor of 0.85 are used. The further

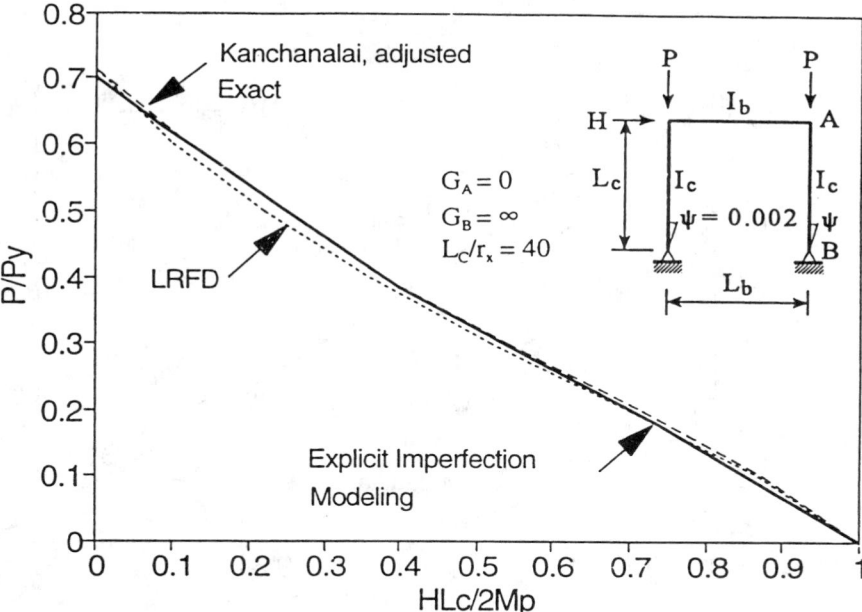

Figure 28.18 Comparison of strength curves for a portal frame subject to strong-axis bending with $L_c/r_x = 40$, $G_A = 0$ (explicit imperfection modeling method).

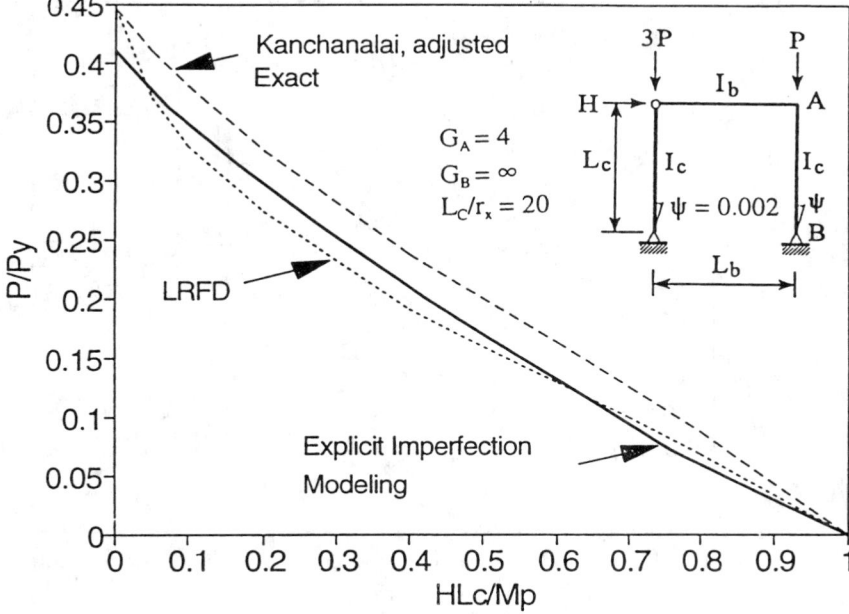

Figure 28.19 Comparison of strength curves for a leaning column frame subject to strong-axis bending with $L_c/r_x = 20$, $G_A = 4$ (explicit imperfection modeling method).

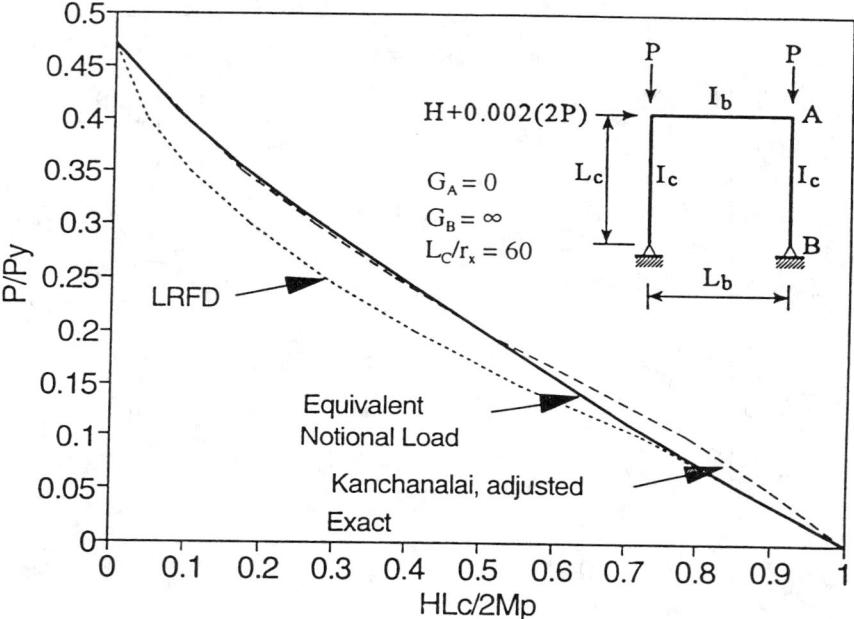

Figure 28.20   Comparison of strength curves for a portal frame subject to strong-axis bending with $L_c/r_x = 60$, $G_A = 0$ (equivalent notional load method).

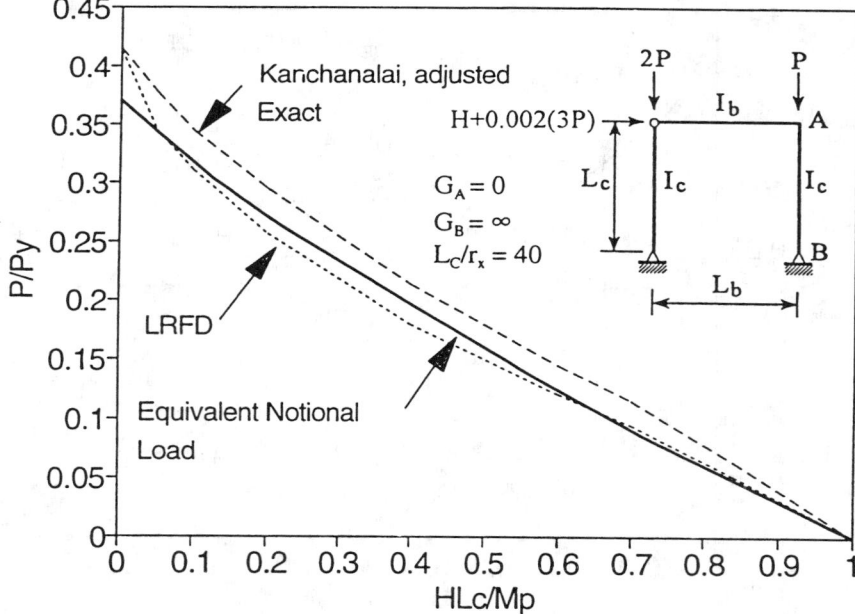

Figure 28.21   Comparison of strength curves for a leaning column frame subject to strong-axis bending with $L_c/r_x = 40$, $G_A = 0$ (equivalent notional load method).

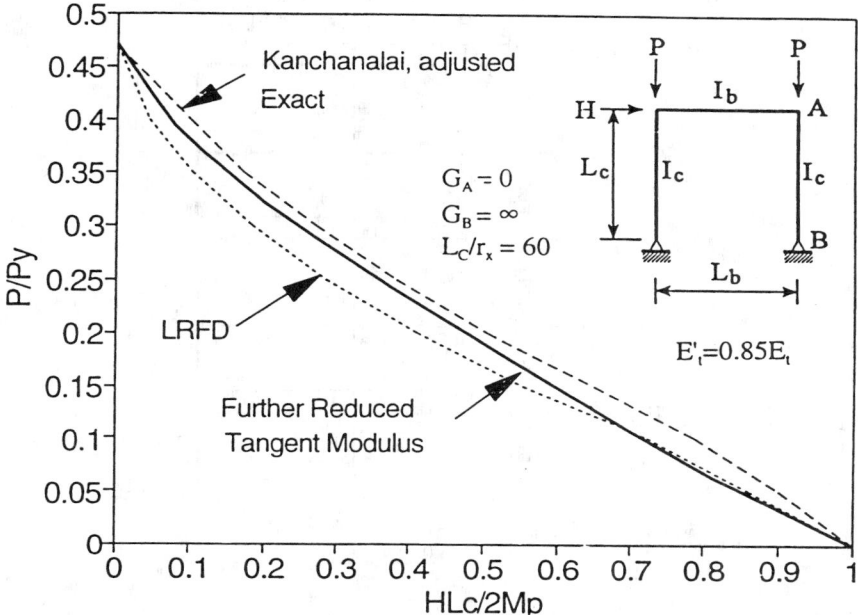

Figure 28.22 Comparison of strength curves for a portal frame subject to strong-axis bending with $L_c/r_x = 60$, $G_A = 0$ (further reduced tangent modulus method).

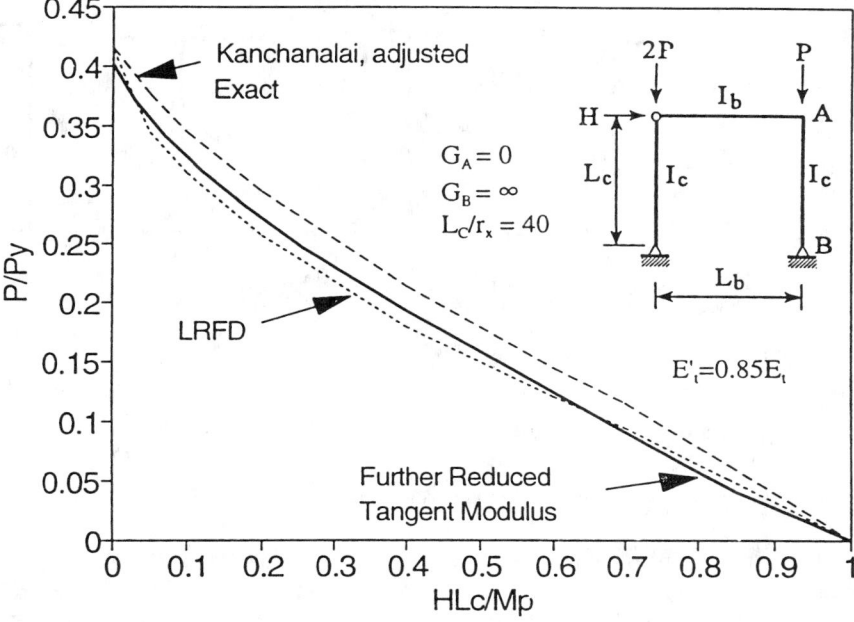

Figure 28.23 Comparison of strength curves for a leaning column frame subject to strong-axis bending with $L_c/r_x = 40$, $G_A = 0$ (further reduced tangent modulus method).

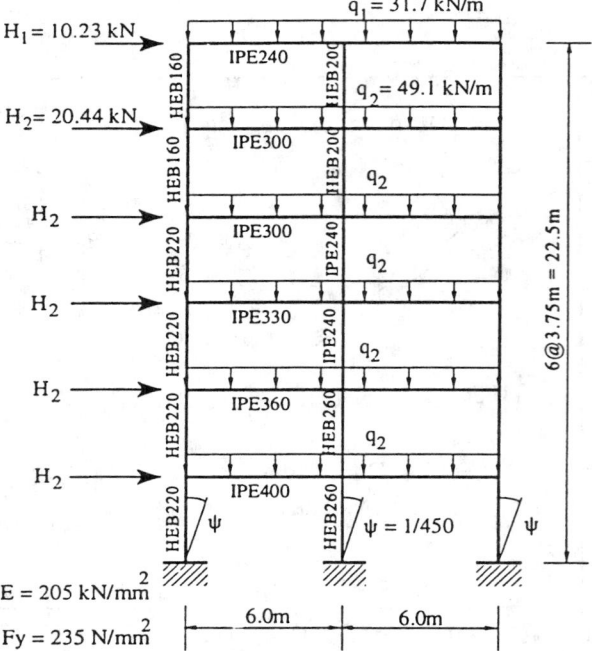

Figure 28.24   Configuration and load condition of Vogel's six-story frame for verification study.

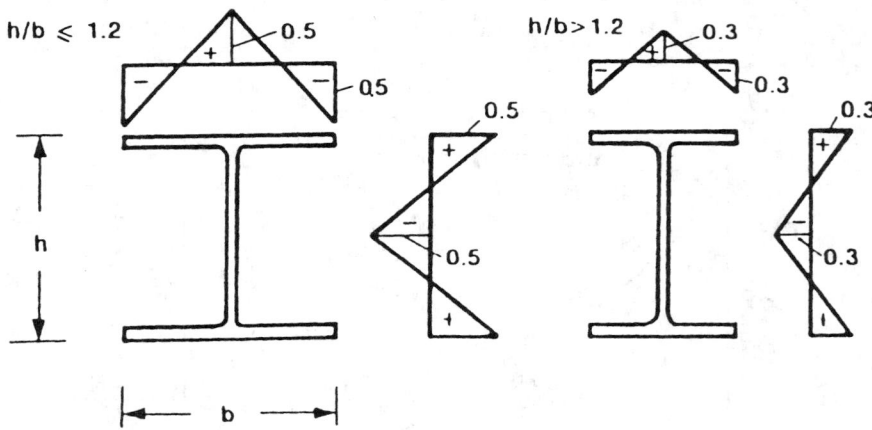

$$\text{Residual stress distributions} \quad \left(\overline{\sigma}_{res} = \frac{\sigma_{res}}{235 \text{N/mm}^2} \right)$$

Figure 28.25   Residual stresses of cross-section for Vogel's frame.

reduced tangent modulus is equivalent to the geometric imperfection of $L_c/500$. Thus, the geometric imperfection of $L_c/4500$ is additionally modeled in the further reduced tangent modulus method, where $L_c/4500$ is the difference between the Vogel's geometric imperfection of $L_c/450$ and the proposed geometric imperfection of $L_c/500$.

The load-displacement curves for the proposed methods together with the Vogel's plastic-zone analysis are compared in Figure 28.27. The errors in strength prediction by the proposed methods are less than 1%. Explicit imperfection modeling and the equivalent notional load method un-

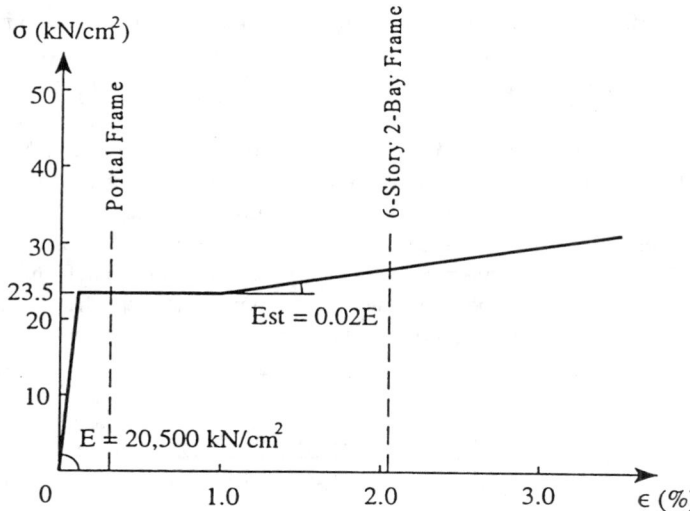

Figure 28.26 Stress-strain relationships for Vogel's frame.

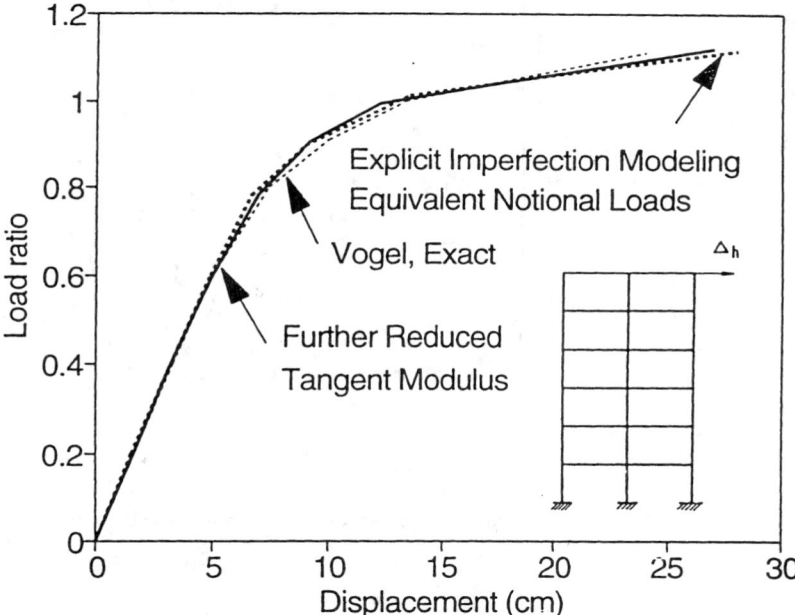

Figure 28.27 Comparison of displacements for Vogel's six-story frame.

derpredict lateral displacements by 3%, and the further reduced tangent modulus method shows a good agreement in displacement with Vogel's exact solution. Vogel's frame is a good example of how the reduced tangent modulus method predicts lateral displacement well under reasonable load combinations.

28.3.4 Semi-Rigid Frame

In the open literature, no benchmark problems solving semi-rigid frames with geometric imperfections are available for a verification study. An alternative is to separate the effects of semi-rigid connections and geometric imperfections. In previous sections, the geometric imperfections were studied and comparisons between proposed methods, plastic-zone analyses, and conventional LRFD methods were made. Herein, the effect of semi-rigid connections will be verified by comparing analytical and experimental results.

Stelmack [31] studied the experimental response of two flexibly connected steel frames. A two-story, one-bay frame in his study is selected as a benchmark for the present study. The frame was fabricated from A36 W5x16 sections, with pinned base supports (Figure 28.28). The connections were bolted top and seat angles (L4x4x1/2) made of A36 steel and A325 3/4-in.-diameter bolts (Figure 28.29). The experimental moment-rotation relationship is shown in Figure 28.30. A gravity load of 2.4 kips was applied at third points along the beam at the first level, followed by a lateral load application. The lateral load-displacement relationship was provided by Stelmack.

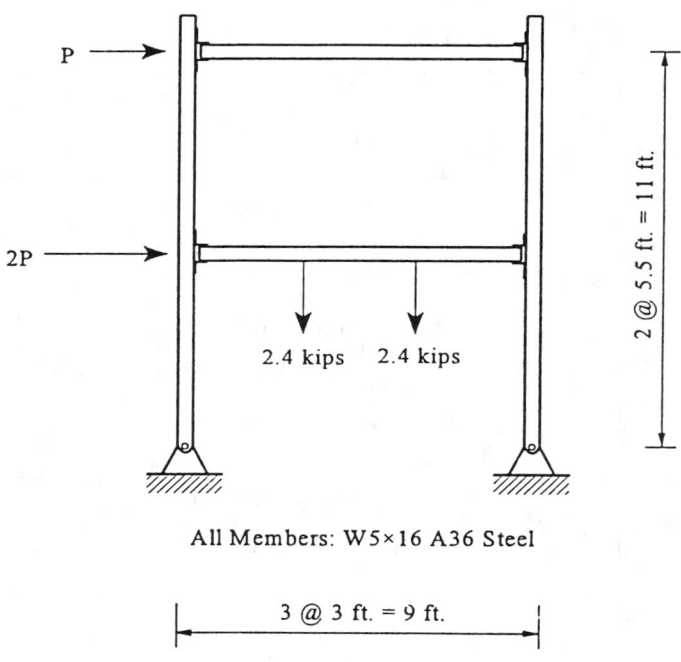

Figure 28.28   Configuration and load condition of Stelmack's two-story semi-rigid frame.

Herein, the three parameters of the power model are determined by curve-fitting and the program 3PARA.f is presented in Section 28.2.2. The three parameters obtained by the curve-fit are $R_{ki} = 40,000$ k-in./rad, $M_u = 220$ k-in., and $n = 0.91$. We obtain three parameters of $R_{ki} = 29,855$ kips/rad $M_u = 185$ k-in. and $n = 1.646$ with 3PARA.f.

The moment-rotation curves given by experiment and curve-fitting show good agreement (Figure 28.30). The parameters given by the Kishi-Chen equations and by experiment show some deviation (Figure 28.30). In spite of this difference, the Kishi-Chen equations, using the computer program (3PARA.f), are a more practical alternative in design since experimental moment-rotation curves are not usually available [19]. In the analysis, the gravity load is first applied, then the lateral load. The lateral displacements given by the proposed methods and by the experimental method compare well (Figure 28.31). The proposed method adequately predicts the behavior and strength of semi-rigid connections.

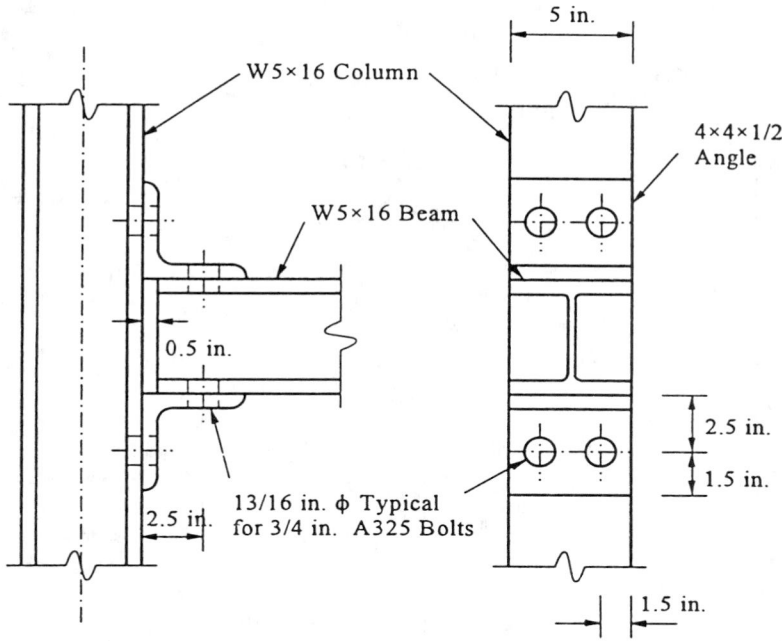

Figure 28.29   Top and seat angle connection details.

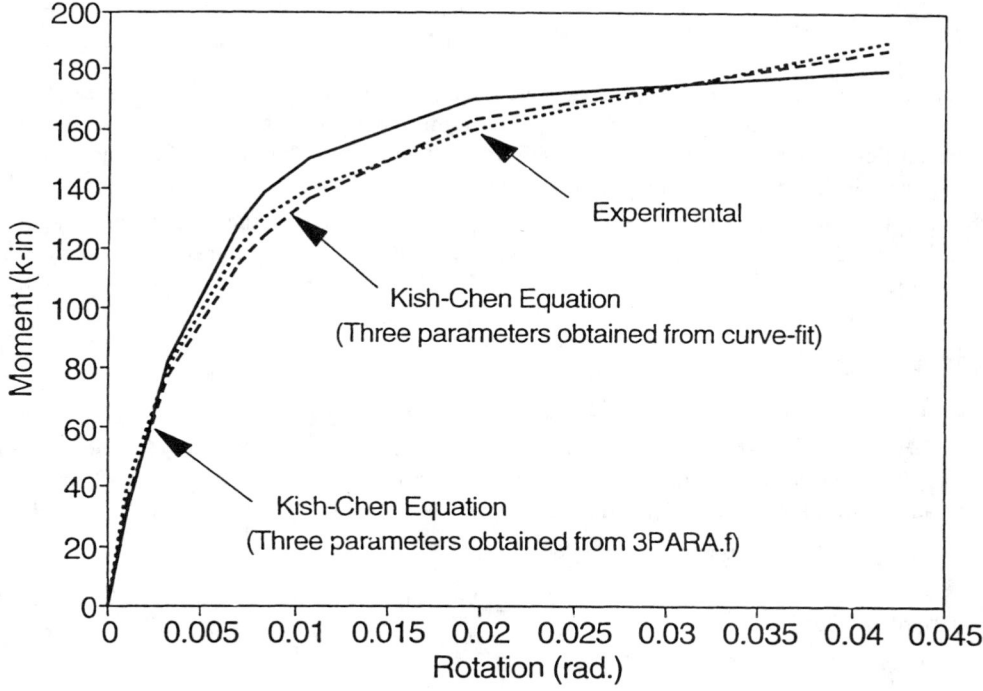

Figure 28.30   Comparison of moment-rotation relationships of semi-rigid connection by experiment and the Kishi-Chen equation.

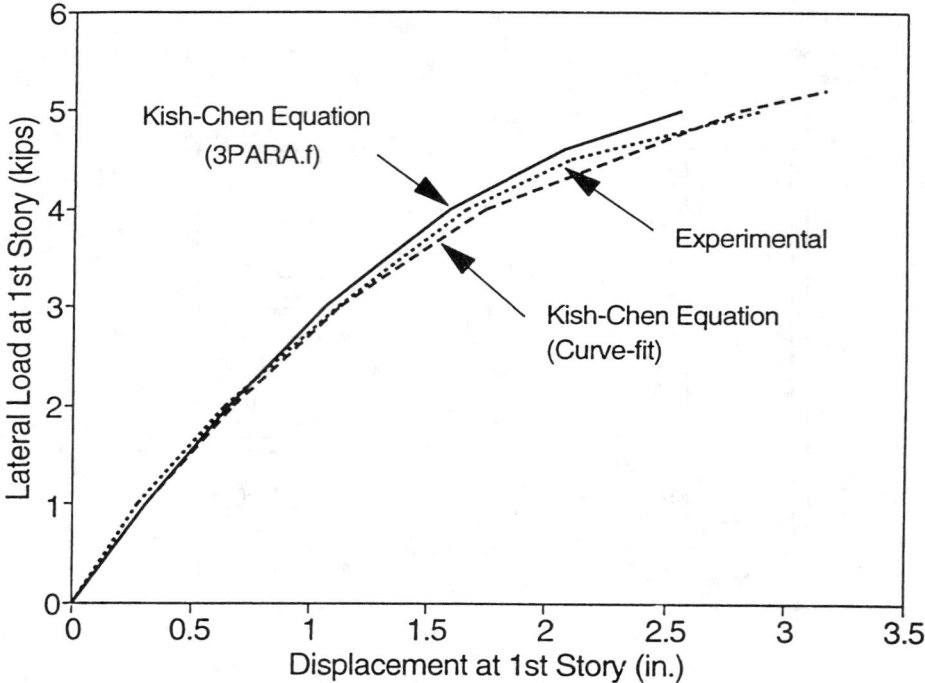

Figure 28.31   Comparison of displacements of Stelmack's two-story semi-rigid frame.

28.4   Analysis and Design Principles

In the preceding section, the proposed advanced analysis method was verified using several bench-mark problems available in the literature. Verification studies were carried out by comparing it to the plastic-zone and conventional LRFD solutions. It was shown that practical advanced analysis predicted the behavior and failure mode of a structural system with reliable accuracy.

In this section, analysis and design principles are summarized for the practical application of the advanced analysis method. Step-by-step analysis and design procedures for the method are presented.

28.4.1   Design Format

Advanced analysis follows the format of LRFD. In LRFD, the factored load effect does not exceed the factored nominal resistance of the structure. Two safety factors are used: one is applied to loads, the other to resistances. This approach is an improvement on other models (e.g., ASD and PD) because both the loads and the resistances have unique factors for unique uncertainties. LRFD has the format

$$\phi R_n \geq \sum_{i=1}^{m} \gamma_i Q_{ni} \qquad (28.13)$$

where

R_n =   nominal resistance of the structural member
Q_n =   nominal load effect (e.g., axial force, shear force, bending moment)
ϕ =   resistance factor (≤ 1.0) (e.g., 0.9 for beams, 0.85 for columns)

γ_i = load factor (usually > 1.0) corresponding to Q_{ni} (e.g., $1.4D$ and $1.2D + 1.6L + 0.5S$)

i = type of load (e.g., D = dead load, L = live load, S = snow load)

m = number of load type

Note that the LRFD [2] uses separate factors for each load and therefore reflects the uncertainty of different loads and combinations of loads. As a result, a relatively uniform reliability is achieved.

The main difference between conventional LRFD methods and advanced analysis methods is that the left side of Equation 28.13 (ϕR_n) in the LRFD method is the resistance or strength of the component of a structural system, but in the advanced analysis method, it represents the resistance or the load-carrying capacity of the whole structural system.

28.4.2   Loads

Structures are subjected to various loads, including dead, live, impact, snow, rain, wind, and earthquake loads. Structures must be designed to prevent failure and limit excessive deformation; thus, an engineer must anticipate the loads a structure may experience over its service life with reliability.

Loads may be classified as static or dynamic. Dead loads are typical of static loads, and wind or earthquake loads are dynamic. Dynamic loads are usually converted to equivalent static loads in conventional design procedures, and it may be adopted in advanced analysis as well.

28.4.3   Load Combinations

The load combinations in advanced analysis methods are based on the LRFD combinations [2]. Six factored combinations are provided by the LRFD specification. The one must be used to determine member sizes. Probability methods were used to determine the load combinations listed in the LRFD specification (LRFD-A4). Each factored load combination is based on the load corresponding to the 50-year recurrence, as follows:

$$(1)\ 1.4D \tag{28.14a}$$

$$(2)\ 1.2D + 1.6L + 0.5(L_r \text{ or } S \text{ or } R) \tag{28.14b}$$

$$(3)\ 1.2D + 1.6(L_r \text{ or } S \text{ or } R) + (0.5L \text{ or } 0.8W) \tag{28.14c}$$

$$(4)\ 1.2D + 1.3W + 0.5L + 0.5(L_r \text{ or } S \text{ or } R) \tag{28.14d}$$

$$(5)\ 1.2D \pm 1.0E + 0.5L + 0.2S \tag{28.14e}$$

$$(6)\ 0.9D \pm (1.3W \text{ or } 1.0E) \tag{28.14f}$$

where

D = dead load (the weight of the structural elements and the permanent features on the structure)

L = live load (occupancy and movable equipment)

L_r = roof live load

W = wind load

S = snow load

E = earthquake load

R = rainwater or ice load.

The LRFD specification specifies an exception that the load factor on live load, L, in combinations (3)–(5) must be 1.0 for garages, areas designated for public assembly, and all areas where the live load is greater than 100 psf.

28.4.4   Resistance Factors

The AISC-LRFD cross-section strength equations may be written as

$$\frac{P}{\phi_c P_y} + \frac{8}{9}\frac{M}{\phi_b M_p} = 1.0 \quad \text{for} \quad \frac{P}{\phi_c P_y} \geq 0.2 \qquad (28.15\text{a})$$

$$\frac{P}{2\phi_c P_y} + \frac{M}{\phi_b M_p} = 1.0 \quad \text{for} \quad \frac{P}{\phi_c P_y} < 0.2 \qquad (28.15\text{b})$$

where

P, M = second-order axial force and bending moment, respectively
P_y = squash load
M_p = plastic moment capacity
ϕ_c, ϕ_b = resistance factors for axial strength and flexural strength, respectively

Figure 28.32 shows the cross-section strength including the resistance factors, ϕ_c and ϕ_b. The

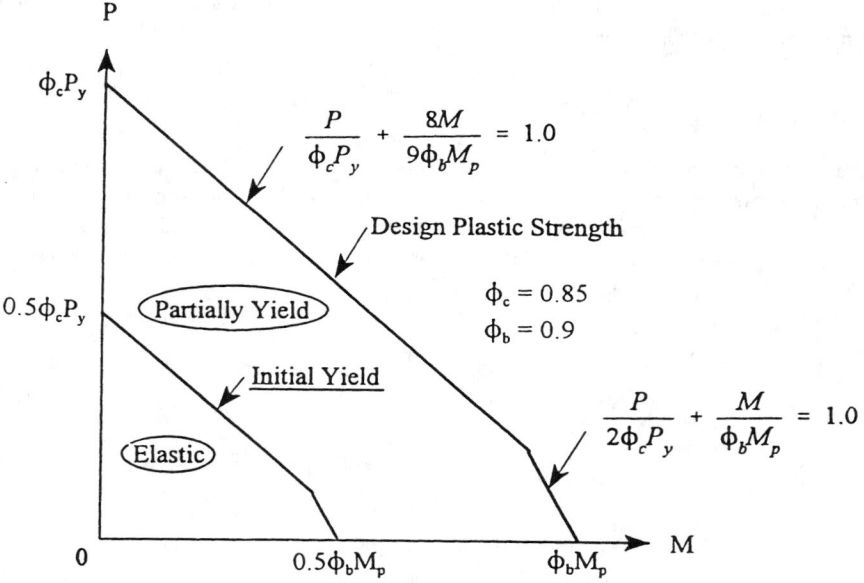

Figure 28.32   Stiffness degradation model including reduction factors.

reduction factors, ϕ_c and ϕ_b, are built into the analysis program and are thus automatically included in the calculation of the load-carrying capacity. The reduction factors are 0.85 for axial strength and 0.9 for flexural strength, corresponding to AISC-LRFD specification [2]. For connections, the ultimate moment, M_u, is reduced by the reduction factor 0.9.

28.4.5   Section Application

The AISC-LRFD specification uses only one column curve for rolled and welded sections of W, WT, and HP shapes, pipe, and structural tubing. The specification also uses some interaction equations for doubly and singly symmetric members, including W, WT, and HP shapes, pipe, and structural tubing, even though the interaction equations were developed on the basis of W shapes by Kanchanalai [14].

The present advanced analysis method was developed by calibration with the LRFD column curve and interaction equations described in Section 28.3. To this end, it is concluded that the proposed methods can be used for various rolled and welded sections, including W, WT, and HP shapes, pipe, and structural tubing without further modifications.

28.4.6 Modeling of Structural Members

Different types of advanced analysis are (1) plastic-zone method, (2) quasi-plastic hinge method, (3) elastic-plastic hinge method, and (4) refined plastic hinge method. An important consideration in making these advanced analyses practical is the required number of elements for a member in order to predict realistically the behavior of frames.

A sensitivity study of advanced analysis is performed on the required number of elements for a beam member subject to distributed transverse loads. A two-element model adequately predicts the strength of a member. To model parabolic out-of-straightness in a beam-column, a two-element model with a maximum initial deflection at the midheight of a member adequately captures imperfection effects. The required number of elements in modeling each member to provide accurate predictions of the strengths is summarized in Table 28.4. It is concluded that practical advanced analysis is computationally efficient.

TABLE 28.4 Necessary Number of Elements

Member	Number of elements
Beam member subject to uniform loads	2
Column member of braced frame	2
Column member of unbraced frame	1

28.4.7 Modeling of Geometric Imperfection

Geometric imperfection modeling is required to account for fabrication and erection tolerances. The imperfection modeling methods used here are the explicit imperfection, the equivalent notional load, and the further reduced tangent modulus models. Users may choose one of these three models in an advanced analysis. The magnitude of geometric imperfections is listed in Table 28.5.

TABLE 28.5 Magnitude of Geometric Imperfection

Geometric imperfection method	Magnitude
Explicit imperfection modeling method	$\psi = 2/1000$ for unbraced frames $\psi = 1/1000$ for braced frames
Equivalent notional load method	$\alpha = 2/1000$ for unbraced frames $\alpha = 4/1000$ for braced frames
Further reduced tangent modulus method	$E_t' = 0.85E_t$

Geometric imperfection modeling is required for a frame but not a truss element, since the program computes the axial strength of a truss member using the LRFD column strength equations, which account for geometric imperfections.

28.4.8 Load Application

It is necessary, in an advanced analysis, to input proportional increment load (not the total loads) to trace nonlinear load-displacement behavior. The incremental loading process can be achieved by scaling down the combined factored loads by a number between 10 and 50. For a highly redundant structure (such as one greater than six stories), dividing by about 10 is recommended, and for a nearly statically determinate structure (such as a portal frame), the incremental load may be factored down by 50. One may choose a number between 10 and 50 to reflect the redundancy of a particular structure. Since a highly redundant structure has the potential to form many plastic hinges and the applied load increment is automatically reduced as new plastic hinges form, the larger incremental load (i.e., the smaller scaling number) may be used.

28.4.9 Analysis

Analysis is important in the proposed design procedures, since the advanced analysis method captures key behaviors including second-order and inelasticity in its analysis program. Advanced analysis does not require separate member capacity checks by the specification equations. On the other hand, the conventional LRFD method accounts for inelastic second-order effects in its design equations (not in analysis). The LRFD method requires tedious separate member capacity checks. Input data used for advanced analysis is easily accessible to users, and the input format is similar to the conventional linear elastic analysis. The format will be described in detail in Section 28.5. Analyses can be simply carried out by executing the program described in Section 28.5. This program continues to analyze with increased loads and stops when a structural system reaches its ultimate state.

28.4.10 Load-Carrying Capacity

Because consideration at moment redistribution may not always be desirable, the two approaches (including and excluding inelastic moment redistribution) are presented. First, the load-carrying capacity, including the effect of inelastic moment redistribution, is obtained from the final loading step (limit state) given by the computer program. Second, the load-carrying capacity without the inelastic moment redistribution is obtained by extracting that force sustained when the first plastic hinge formed. Generally, advanced analysis predicts the same member size as the LRFD method when moment redistribution is not considered. Further illustrations on these two choices will be presented in Section 28.6.

28.4.11 Serviceability Limits

The serviceability conditions specified by the LRFD consist of five limit states: (1) deflection, vibration, and **drift**; (2) thermal expansion and contraction; (3) connection slip; (4) camber; and (5) corrosion. The most common parameter affecting the design serviceability of steel frames is deflection.

Based on the studies by the Ad Hoc Committee [1] and by Ellingwood [12], the deflection limits recommended (Table 28.6) were proposed for general use. At **service load** levels, no plastic hinges are permitted anywhere in the structure to avoid permanent deformation under service loads.

TABLE 28.6   Deflection Limitations of Frame

Item	Deflection ratio
Floor girder deflection for service live load	$L/360$
Roof girder deflection	$L/240$
Lateral drift for service wind load	$H/400$
Interstory drift for service wind load	$H/300$

28.4.12 Ductility Requirements

Adequate inelastic rotation capacity is required for members in order to develop their full plastic moment capacity. The required rotation capacity may be achieved when members are adequately braced and their cross-sections are compact. The limitations of compact sections and lateral unbraced length in what follows leads to an inelastic rotation capacity of at least three and seven times the elastic rotation corresponding to the onset of the plastic moment for non-seismic and seismic regions, respectively.

Compact sections are capable of developing the full plastic moment capacity, M_p, and sustaining large hinge rotation before the onset of local buckling. The compact section in the LRFD specification is defined as:

1. Flange

   - For non-seismic region

$$\frac{b_f}{2t_f} \leq \frac{65}{\sqrt{F_y}} \tag{28.16}$$

   - For seismic region

$$\frac{b_f}{2t_f} \leq \frac{52}{\sqrt{F_y}} \tag{28.17}$$

   where
   b_f = width of flange
   t_f = thickness of flange
   F_y = yield stress in ksi

2. Web

   - For non-seismic region

$$\frac{h}{t_w} \leq \frac{640}{\sqrt{F_y}} \left(1 - \frac{2.75 P_u}{\phi_b P_y}\right) \quad \text{for} \quad \frac{P}{\phi_b P_y} \leq 0.125 \tag{28.18a}$$

$$\frac{h}{t_w} \leq \frac{191}{\sqrt{F_y}} \left(2.33 - \frac{P_u}{\phi_b P_y}\right) \geq \frac{253}{\sqrt{F_y}} \quad \text{for} \quad \frac{P}{\phi_b P_y} > 0.125 \tag{28.18b}$$

   - For seismic region

$$\frac{h}{t_w} \leq \frac{520}{\sqrt{F_y}} \left(1 - \frac{1.54 P_u}{\phi_b P_y}\right) \quad \text{for} \quad \frac{P}{\phi_b P_y} \leq 0.125 \tag{28.19a}$$

$$\frac{h}{t_w} \leq \frac{191}{\sqrt{F_y}} \left(2.33 - \frac{P_u}{\phi_b P_y}\right) \geq \frac{253}{\sqrt{F_y}} \quad \text{for} \quad \frac{P}{\phi_b P_y} > 0.125 \tag{28.19b}$$

   where
   h = clear distance between flanges
   t_w = thickness of web
   F_y = yield strength in ksi

In addition to the compactness of section, the lateral unbraced length of beam members is also a limiting factor for the development of the full plastic moment capacity of members. The LRFD provisions provide the limit on spacing of braces for beam as:

- For non-seismic region

$$L_{pd} \leq \frac{[3,600 + 2,200(M_1/M_2)] \, r_y}{F_y} \qquad (28.20a)$$

- For seismic region

$$L_{pd} \leq \frac{2,500r_y}{F_y} \qquad (28.20b)$$

where

L_{pd} = unbraced length
r_y = radius of gyration about y axis
F_y = yield strength in ksi
M_1, M_2 = smaller and larger end moment, respectively
M_1/M_2 = positive in double curvature bending

   The AISC-LRFD specification explicitly specifies the limitations for beam members as described above, but not for beam-column members. More studies are necessary to determine the reasonable limits leading to adequate rotation capacity of beam-column members. Based on White's study [33], the limitations for beam members seem to be used for beam-column members until the specification provides the specific values for beam-column members.

28.4.13   Adjustment of Member Sizes

If one of three conditions — strength, serviceability, or **ductility** — is not satisfied, appropriate adjustments of the member sizes should be made. This can be done by referring to the sequence of plastic hinge formation shown in the P.OUT. For example, if the load-carrying capacity of a structural system is less than the factored load effect, the member with the first plastic hinge should be replaced with a stronger member. On the other hand, if the load-carrying capacity exceeds the factored load effect significantly, members without plastic hinges may be replaced with lighter members. If lateral drift exceeds drift requirements, columns or beams should be sized up, or a braced structural system should be considered instead to meet this serviceability limit.

   In semi-rigid frames, behavior is influenced by the combined effects of members and connections. As an illustration, if an excessive lateral drift occurs in a structural system, the drift may be reduced by increasing member sizes or using more rigid connections. If the strength of a beam exceeds the required strength, it may be adjusted by reducing the beam size or using more flexible connections. Once the member and connection sizes are adjusted, the iteration leads to an optimum design. Figure 28.33 shows a flow chart of analysis and design procedure in the use of advanced analysis.

28.5   Computer Program

This section describes the Practical Advanced Analysis Program (PAAP) for two-dimensional steel frame design [15, 24]. The program integrates the methods and techniques developed in Sections 28.2 and 28.3. The names of variables and arrays correspond as closely as possible to those used in theoretical derivations. The main objective of this section is to present an educational version of software to enable engineers and graduate students to perform planar frame analysis for a more realistic prediction of the strength and behavior of structural systems.

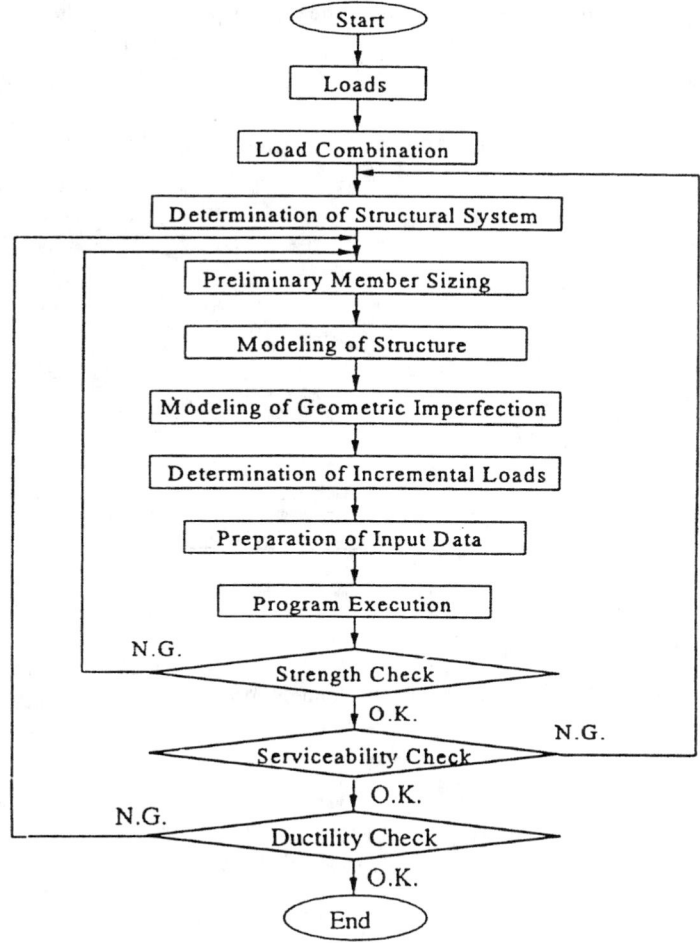

Figure 28.33   Analysis and design procedure.

The instructions necessary for user input into PAAP are presented in Section 28.5.4. Except for the requirement to input geometric imperfections and incremental loads, the input data format of the program is basically the same as that of the usual linear elastic analysis program. The user is advised to read all the instructions, paying particular attention to the appended notes, to achieve an overall view of the data required for a specific PAAP analysis. The reader should recognize that no system of units is assumed in the program, and take the responsibility to make all units consistent. Mistaken unit conversion and input are a common source of erroneous results.

28.5.1   Program Overview

This FORTRAN program is divided in three: DATAGEN, INPUT, and PAAP. The first program, DATAGEN, reads an input data file, P.DAT, and generates a modified data file, INFILE. The second program, INPUT, rearranges INFILE into three working data files: DATA0, DATA1, and DATA2. The third program, PAAP, reads the working data files and provides two output files named P.OUT1 and P.OUT2. P.OUT1 contains an echo of the information from the input data file, P.DAT. This file may be used to check for numerical and incompatibility errors in input data. P.OUT2 contains the load and displacement information for various joints in the structure as well as the element joint forces for all types of elements at every load step. The load-displacement results are presented at the end of every load increment. The sign conventions for loads and displacements should follow the frame degrees of freedom, as shown in Figures 28.34 and 28.35.

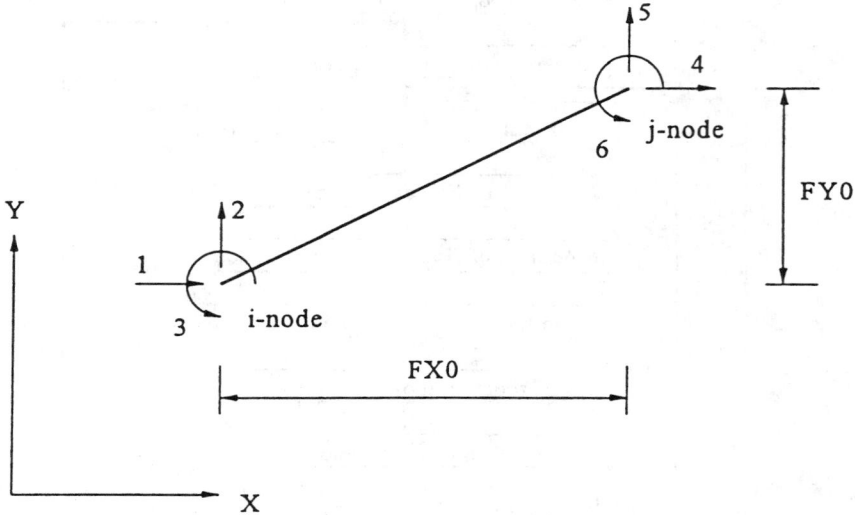

Figure 28.34   Degrees of freedom numbering for the frame element.

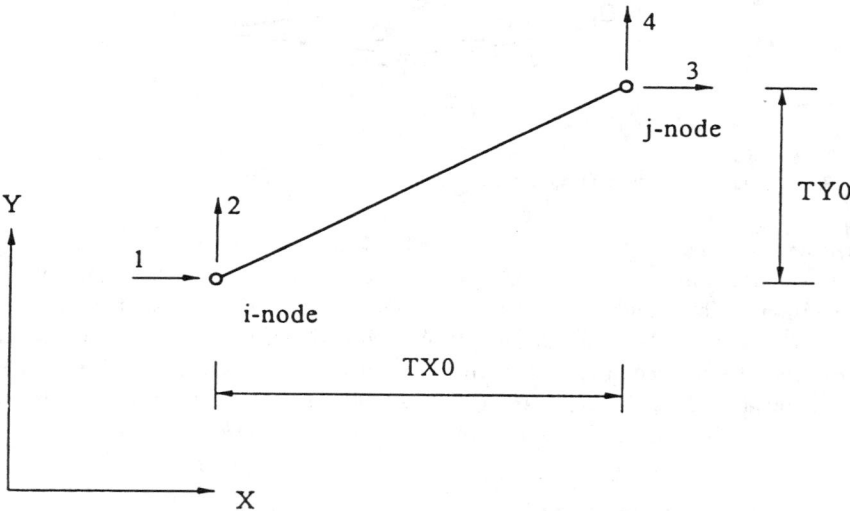

Figure 28.35   Degrees of freedom numbering for the truss element.

The element joint forces are obtained by summing the product of the element incremental displacements at every load step. The element joint forces act in the global coordinate system and must be in equilibrium with applied forces. After the output files are generated, the user can view these files on the screen or print them with the MS-DOS PRINT command. The schematic diagram in Figure 28.36 sets out the operation procedure used by PAAP and its supporting programs [15].

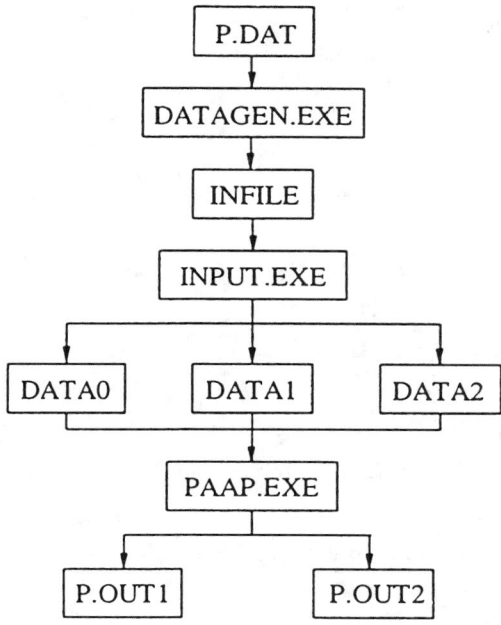

Figure 28.36 Operating procedures of the proposed program.

28.5.2 Hardware Requirements

This program has been tested in two computer processors. First it was tested on an IBM 486 or equivalent personal computer system using Microsoft's FORTRAN 77 compiler v1.00 and Lahey's FORTRAN 77 compiler v5.01. Second, its performance in the workstation environment was tested on a Sun 5 using a Sun FORTRAN 77 compiler. The program sizes of DATAGEN, INPUT, and PAAP are 8 kB, 9 kB, and 94 kB, respectively. The total size of the three programs is small, 111 kB (= 0.111 MB), and so a 3.5-in. high-density diskette (1.44 MB) can accommodate the three programs and several example problems.

The memory required to run the program depends on the size of the problem. A computer with a minimum 640 K of memory and a 30-MB hard disk is generally required. For the PC applications, the array sizes are restricted as follows:

1. Maximum total degrees of freedom, MAXDOF = 300
2. Maximum translational degrees of freedom, MAXTOF = 300
3. Maximum rotational degrees of freedom, MAXROF = 100
4. Maximum number of truss elements, MAXTRS = 150
5. Maximum number of connections, MAXCNT = 150

It is possible to run bigger jobs in UNIX workstations by modifying the above values in the PARAMETER and COMMON statements in the source code.

28.5.3 Execution of Program

A computer diskette is provided in *LRFD Steel Design Using Advanced Analysis*, by W. F. Chen and S. E. Kim, [6], containing four directories with the following files, respectively.

1. Directory PSOURCE
   - DATAGEN.FOR

- `INPUT.FOR`
- `PAAP.FOR`

2. Directory `PTEST`
   - `DATAGEN.EXE`
   - `INPUT.EXE`
   - `PAAP.EXE`
   - `RUN.BAT` (batch file)
   - `P.DAT` (input data for a test run)
   - `P.OUT1` (output for a test run)
   - `P.OUT2` (output for a test run)

3. Directory `PEXAMPLE`
   - All input data for the example problems presented in Section 28.6

4. Directory `CONNECT`
   - `3PARA.FOR` (program for semi-rigid connection parameters)
   - `3PARA.EXE`
   - `CONN.DAT` (input data)
   - `CONN.OUT` (output for three parameters)

To execute the programs, one must first copy them onto the hard disk (i.e., copy `DATAGEN.EXE`, `INPUT.EXE`, `PAAP.EXE`, `RUN.BAT`, and `P.DAT` from the directory `PTEST` on the diskette to the hard disk). Before launching the program, the user should test the system by running the sample example provided in the directory. The programs are executed by issuing the command `RUN`. The batch file `RUN.BAT` executes `DATAGEN`, `INPUT`, and `PAAP` in sequence. The output files produced are `P.OUT1` and `P.OUT2`. When the compilers are different between the authors' and the user's, the program (`PAAP.EXE`) may not be executed. This problem may be easily solved by recompiling the source programs in the directory `PSOURCE`.

The input data for all the problems in Section 28.6 are provided in the directory `PEXAMPLE`. The user may use the input data for his or her reference and confirmation of the results presented in Section 28.6. It should be noted that `RUN` is a batch command to facilitate the execution of `PAAP`. Entering the command `RUN` will write the new files, including `DATA0`, `DATA1`, `DATA2`, `P.OUT1`, and `P.OUT2`, over the old ones. Therefore, output files should be renamed before running a new problem.

The program can generate the output files `P.OUT1` and `P.OUT2` in a reasonable time period, described in the following. The run time on an IBM 486 PC with memory of 640 K to get the output files for the eight-story frame shown in Figure 28.37 is taken as 4 min 10 s and 2 min 30 s in real time rather than CPU time by using Microsoft FORTRAN and Lahey FORTRAN, respectively. In the Sun 5, the run time varies approximately 2–3 min depending on the degree of occupancy by users.

The directory `CONNECT` contains the program that computes the three parameters needed for semi-rigid connections. The operation procedure of the program, the input data format, and two examples were presented in Section 28.2.2.

28.5.4 Users' Manual

Analysis Options

PAAP was developed on the basis of the theory presented in Section 28.2. While the purpose of the program is basically for advanced analysis using the second-order inelastic concept, the program

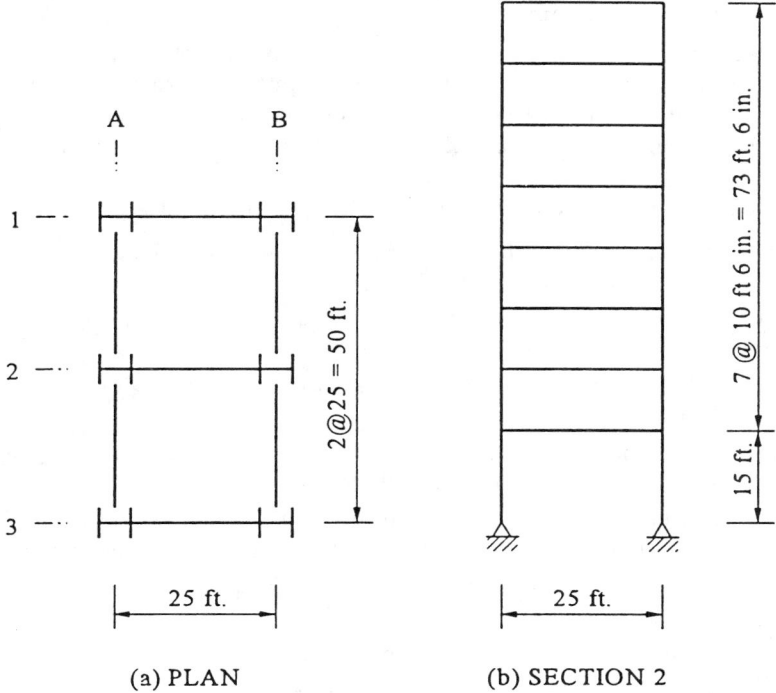

Figure 28.37   Configuration of the unbraced eight-story frame.

can be used also for first- and second-order elastic analyses. For a first-order elastic analysis, the total factored load should be applied in one load increment to suppress numerical iteration in the nonlinear analysis algorithm. For a second-order elastic analysis, a yield strength of an arbitrarily large value should be assumed for all members to prevent yielding.

Coordinate System

A two-dimensional (x, y) global coordinate system is used for the generation of all the input and output data associated with the joints. The following input and output data are prepared with respect to the global coordinate system.

1. Input data

    - joint coordinates
    - joint restraints
    - joint load

2. Output data

    - joint displacement
    - member forces

Type of Elements

The analysis library consists of three elements: a plane frame, a plane truss, and a connection. The connection is represented by a zero-length rotational spring element with a user-specified nonlinear moment-rotation curve. Loading is allowed only at nodal points. Geometric and material nonlinearities can be accounted for by using an iterative load-increment scheme. Zero-length plastic hinges are lumped at the element ends.

Locations of Nodal Points

The geometric dimensions of the structures are established by placing joints (or nodal points) on the structures. Each joint is given an identification number and is located in a plane associated with a global two-dimensional coordinate system. The structural geometry is completed by connecting the predefined joints with structural elements, which may be a frame, a truss, or a connection. Each element also has an identification number.

The following are some of the factors that need to be considered in placing joints in a structure:

1. The number of joints should be sufficient to describe the initial geometry and the response behavior of the structures.

2. Joints need to be located at points and lines of discontinuity (e.g., at changes in material properties or section properties).

3. Joints should be located at points on the structure where forces and displacements need to be evaluated.

4. Joints should be located at points where concentrated loads will be applied. The applied loads should be concentrated and act on the joints.

5. Joints should be located at all support points. Support conditions are represented in the structural model by restricting the movement of the specific joints in specific directions.

6. Second-order inelastic behavior can be captured by the use of one or two elements per member, corresponding to the following guidelines:

   - Beam member subjected to uniform loads: two elements
   - Column member of braced frame: two elements
   - Column member of unbraced frame: one element

Degrees of Freedom

A two-joint frame element has six displacement components, as shown in Figure 28.34. Each joint can translate in the global x and y directions, and rotate about the global z axis. The directions associated with these displacement components are known as degrees of freedom of the joint. A two-joint truss element has four degrees of freedom, as shown in Figure 28.35. Each join has two translational degrees of freedom and no rotational component.

If the displacement of a joint corresponding to any one of its degrees of freedom is known to be zero (such as at a support), then it is labeled an inactive degree of freedom. Degrees of freedom where the displacements are not known are termed active degree of freedoms. In general, the displacement of an inactive degree of freedom is usually known, and the purpose of the analysis is to find the reaction in that direction. For an active degree of freedom, the applied load is known (it could be zero), and the purpose of the analysis is to find the corresponding displacement.

Units

There are no built-in units in PAAP. The user must prepare the input in a consistent set of units. The output produced by the program will conform to the same set of units. Therefore, if the user chooses to use kips and inches as the input units, all the dimensions of the structure must be entered in inches and all the loads in kips. The material properties should also conform to these units. The output units will then be in kips and inches, so that the frame member axial force will be in kips, bending moments will be in kip-inches, and displacements will be in inches. Joint rotations, however, are in radians, irrespective of units.

Input Instructions

Described here is the input sequence and data structure used to create an input file called P . DAT. The analysis program, PAAP, can analyze any structure with up to 300 degrees of freedom, but it is possible to recompile the source code to accommodate more degrees of freedom by changing the size of the arrays in the `PARAMETER` and `COMMON` statements. The limitation of degrees of freedom can be solved by using dynamic storage allocation. This procedure is common in finite element programs [13, 9], and it will be used in the next release of the program.

The input data file is prepared in a specific format. The input data consists of 13 data sets, including five control data, three section property data, three element data, one boundary condition, and one load data set as follows:

1. Title
2. Analysis and design control
3. Job control
4. Total number of element types
5. Total number of elements
6. Connection properties
7. Frame element properties
8. Truss element properties
9. Connection element data
10. Frame element data
11. Truss element data
12. Boundary conditions
13. Incremental loads

Input of all data sets is mandatory, but some of the data associated with elements (data sets 6–11) may be skipped, depending on the use of the element. The order of data sets in the input file must be strictly maintained. Instructions for inputting data are summarized in Table 28.7

28.6 Design Examples

In previous sections, the concept, verifications, and computer program of the practical advanced analysis method for steel frame design have been presented. The present advanced analysis method has been developed and refined to achieve both simplicity in use and, as far as possible, a realistic representation of behavior and strength. The advanced analysis method captures the limit state strength and stability of a structural system and its individual members. As a result, the method can be used for practical frame design without the tedious separate member capacity checks, including the calculation of K factor.

The aim of this section is to provide further confirmation of the validity of the LRFD-based advanced analysis methods for practical frame design. The comparative design examples in this section show the detailed design procedure for advanced and LRFD design procedures [15]. The design procedures conform to those described in Section 28.4 and may be grouped into four basic steps: (1) load condition, (2) structural modeling, (3) analysis, and (4) limit state check. The design examples cover simple structures, truss structures, braced frames, unbraced frames, and semi-rigid frames. The three practical models — explicit imperfection, equivalent notional load, and further reduced tangent modulus — are used for the design examples. Member sizes determined by advanced procedures are compared with those determined by the LRFD, and good agreement is generally observed.

The design examples are limited to two-dimensional steel frames, so that the spatial behavior is not considered. Lateral torsional buckling is assumed to be prevented by adequate lateral braces. Compact W sections are assumed so that sections can develop their full plastic moment capacity without buckling locally. All loads are statically applied.

TABLE 28.7   Input Data Format for the Program PAAP

Data set	Column	Variable	Description
Title	A70	—	Job title and general comments
Analysis and design control	1–5	IGEOIM	Geometric imperfection method 0: No geometric imperfection (default) 1: Explicit imperfection modeling 2: Equivalent notional load 3: Further reduced tangent modulus
	6–10	ILRFD	Strength reduction factor, $\phi_c = 0.85$, $\phi_b = 0.9$ 0: No reduction factors considered (default) 1: Reduction factors considered
Job control	1–5	NNODE	Total number of nodal points of the structure
	6–10	NBOUND	Total number of supports
	11–15	NINCRE	Allowable number of load increments (default = 100); at least two or three times larger than the scaling number
Total number of element types	1–5	NCTYPE	Number of connection types (1–30)
	6–10	NFTYPE	Number of frame types (1–30)
	11–15	NTTYPE	Number of truss types (1–30)
Total number of elements	1–5	NUMCNT	Number of connection elements (1–150)
	6–10	NUMFRM	Number of frame elements (1–100)
	11–15	NUMTRS	Number of truss elements (1–150)
Connection properties	1-5	ICTYPE	Connection type number
	6–15*	M_u	Ultimate moment capacity of connection
	16–25*	R_{ki}	Initial stiffness of connection
	26–35*	N	Shape parameter of connection
Frame element properties	1-5	IFTYPE	Frame type number
	6-15*	A	Cross-section area
	6-25*	I	Moment of inertia
	26-35*	Z	Plastic section modulus
	36-45*	E	Modulus of elasticity
	46-55*	FY	Yield stress
	55-60	IFCOL	Identification of column member, IFCOL = 1 for column (default = 0)
Truss element properties	1-5	ITYPE	Truss type number
	6-15*	A	Cross-section area
	16-25*	I	Moment of inertia
	26-35*	E	Modulus of elasticity
	36-45*	FY	Yield stress
	46-50	ITCOL	Identification of column member, ITCOL = 1 for column (default = 0)
Connection element data	1-5	LCNT	Connection element number
	6-10	IFMCNT	Frame element number containing the connection
	11-15	IEND	Identification of element ends containing the connection 1: Connection attached at element end i 2: Connection attached at element end j
	16-20	JDCNT	Connection type number
	21-25	NOSMCN	Number of same elements for automatic generation (default = 1)
	26-30	NELINC	Element number (IFMCNT) increment of automatically generated elements (default = 1)
Frame element data	1-5	LFRM	Frame element number
	6-15*	FXO	Horizontal projected length; positive for i-j direction in global x direction
	16-25*	FYO	Vertical projected length; positive for i-j direction in global y direction
	26-30	JDFRM	Frame type number
	31-35	IFNODE	Number of node i
	36-40	JFNODE	Number of node j
	41-45	NOSMFE	Number of same elements for automatic generation (default = 1)
	46-50	NODINC	Node number increment of automatically generated elements (default = 1)

TABLE 28.7 Input Data Format for the Program PAAP *(continued)*

Data set	Column	Variable	Description
Truss	1-5	LTRS	Truss element number
element	6-15*	TXO	Horizontal projected length; positive for i-j direction in
data			global x direction
	16-25*	TYO	Vertical projected length; positive for i-j direction in
			global y direction
	26-30	JDTRS	Truss type number
	31-35	ITNODE	Number of node i
	36-40	JTNODE	Number of node j
	51-55	NOSMTE	Number of same elements for automatic generation
			(default = 1)
	56-60	NODINC	Node number increment of automatically generated
			elements (default = 1)
Boundary	1-5	NODE	Node number of support
conditions	6-10	XFIX	XFIX = 1 for restrained in global x direction (default = 0)
	11-15	YFIX	YFIX = 1 for restrained in global y direction (default = 0)
	16-20	RFIX	RFIX = 1 for restrained in rotation (default = 0)
	21-25	NOSMBD	Number of same boundary condition for automatic
			generation (default = 1)
	26-30	NODINC	Node number increment of automatically generated
			supports (default = 1)
	1-5	NODE	Node number where a load applied
Incremental	6-15*	XLOAD	Incremental load in global x direction (default = 0)
loads	16-25*	YLOAD	Incremental load in global y direction (default = 0)
	26-35*	RLOAD	Incremental moment in global θ direction (default = 0)
	36-40	NOSMLD	Number of same loads for automatic generation
			(default = 1)
	41-45	NODINC	Node number increment of automatically generated loads
			(default = 1)

* indicates that the real value (F or E format) should be entered; otherwise input the integer value (I format).

28.6.1 Roof Truss

Figure 28.38 shows a hinged-jointed roof truss subject to gravity loads of 20 kips at the joints. A36 steel pipe is used. All member sizes are assumed identical.

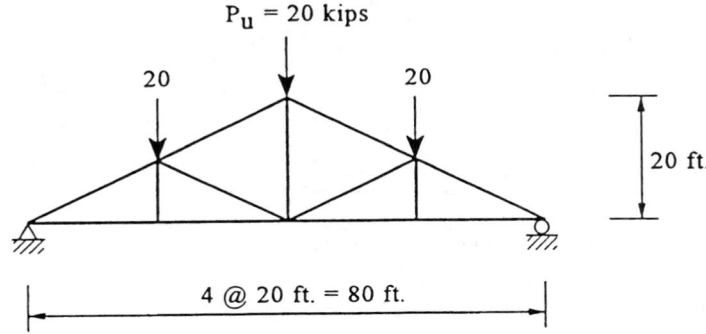

Figure 28.38 Configuration and load condition of the hinged-jointed roof truss.

Design by Advanced Analysis

Step 1: Load Condition and Preliminary Member Sizing

The critical factored load condition is shown in Figure 28.38. The member forces of the truss may be obtained (Figure 28.39) using equilibrium conditions. The maximum compressive force is 67.1 kips. The effective length is the same as the actual length (22.4 ft) since K is 1.0. The preliminary member size of steel pipe is 6 in. in diameter with 0.28 in. thickness ($\phi P_n = 81$ kips), obtained using the column design table in the LRFD specification.

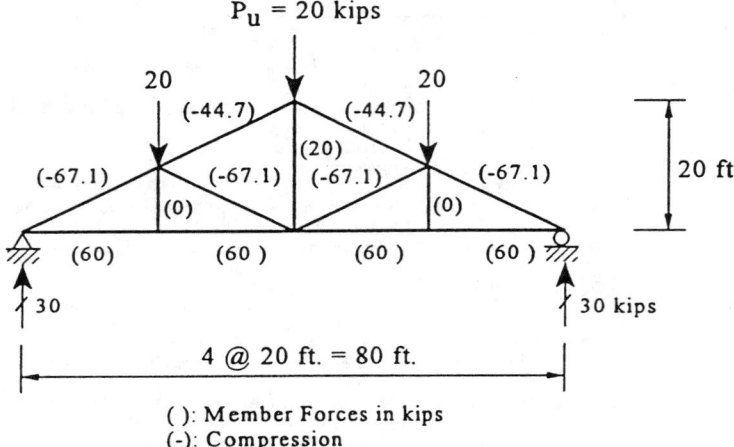

Figure 28.39   Member forces of the hinged-jointed roof truss.

Step 2: Structural Modeling

Each member is modeled with one truss element without geometric imperfection since the program computes the axial strength of the truss member with the LRFD column strength equations, which indirectly account for geometric imperfections. An incremental load of 0.5 kips is determined by dividing the factored load of 20 kips by a scaling factor of 40, as shown in Figure 28.40.

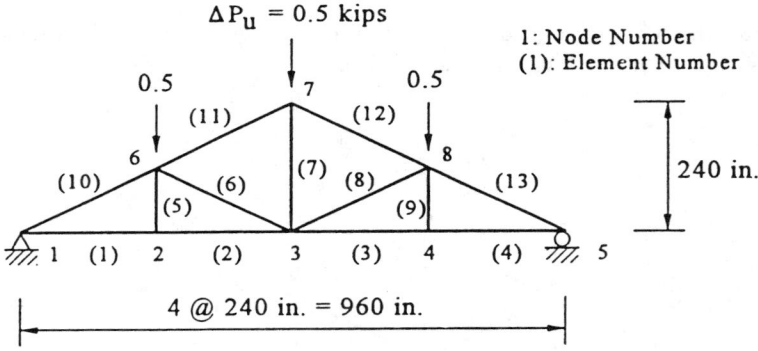

Figure 28.40   Modeling of the hinged-jointed roof truss.

Step 3: Analysis

Referring to the input instructions described in Section 28.5.4, the input data may be easily generated, as listed in Table 28.8. Note that the total number of supports (NBOUND) in the hinged-jointed truss must be equal to the total number of nodal points, since the nodes of a truss element are restrained against rotation. Programs DATAGEN, INPUT, and PAAP are executed in sequence by entering the batch file command RUN on the screen.

Step 4: Check of Load-Carrying Capacity

Truss elements 10 and 13 fail at load step 48, with loads at nodes 6, 7, and 8 being 24 kips. Since this truss is statically determinant, failure of one member leads to failure of the whole system. Load step 49 shows a sharp increase in displacement and indicates a system failure. The member force of element 10 is 80.4 kips ($F_x = 72.0$ kips, $F_y = 35.7$ kips). Since the load-carrying capacity of 24 kips at nodes 6, 7, and 8 is greater than the applied load of 20 kips, the member size is adequate.

TABLE 28.8 Input Data, P . DAT, of the Hinged-Jointed Roof Truss

```
Roof truss
   0      1
   8      8    100
   0      0     1
   0      0    13
   1         5.58         28.1     29000.0          36.0
   1       240.0           0.0     1     1     2     4
   5         0.0         120.0     1     2     6
   6      -240.0         120.0     2     3     6
   7         0.0         240.0     1     3     7
   8       240.0         120.0     2     3     8
   9         0.0         120.0     1     4     8
  10       240.0         120.0     2     1     6
  11       240.0         120.0     2     6     7
  12       240.0        -120.0     2     7     8
  13       240.0        -120.0     2     8     5
   1      1      1      1
   2                    1     3
   5             1      1
   6                    1     3
   6                   -0.5                        3
```

Step 5: Check of Serviceability

Referring to P.OUT2, the deflection at node 3 corresponding to load step 1 is equal to 0.02 in. This deflection may be considered elastic since the behavior of the beam is linear and elastic under small loads. The total deflection of 0.8 in. is obtained by multiplying the deflection of 0.02 in. by the scaling factor of 40. The deflection ratio over the span length is 1/1200, which meets the limitation 1/360. The deflection at the service load will be smaller than that above since the factored load is used for the calculation of deflection above.

Comparison of Results

The advanced analysis and LRFD methods predict the same member size of steel pipe with 6 in. diameter and 0.28 in. thickness. The load-carrying capacities of element 10 predicted by these two methods are the same, 80.5 kips. This is because the truss system is statically determinant, rendering inelastic moment redistribution of little or no benefit.

28.6.2 Unbraced Eight-Story Frame

Figure 28.37 shows an unbraced eight-story one-bay frame with hinged supports. All beams are rigidly connected to the columns. The column and beam sizes are the same. All beams are continuously braced about their weak axis. Bending is primarily about the strong axis at the column. A36 steel is used for all members.

Design by Advanced Analysis

Step 1: Load Condition and Preliminary Member Sizing

The uniform gravity loads are converted to equivalent concentrated loads in Figure 28.41. The preliminary column and beam sizes are selected as W33x130 and W21x50.

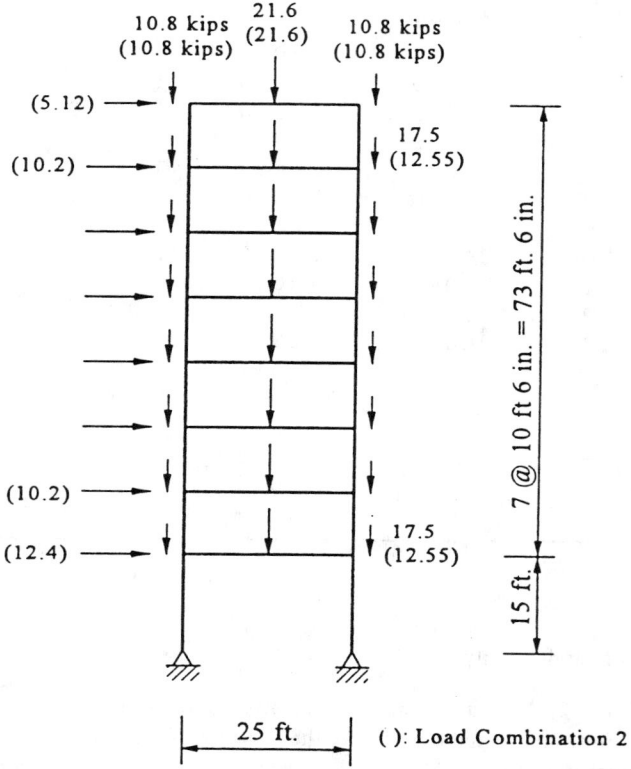

Figure 28.41   Concentrated load condition converted from the distributed load for the two-story frame.

Step 2: Structural Modeling

Each column is modeled with one element since the frame is unbraced and the maximum moment in the member occurs at the ends. Each beam is modeled with two elements.

The explicit imperfection and the further reduced tangent modulus models are used in this example, since they are easier in preparing the input data than equivalent notional load models. Figure 28.42 shows the model for the eight-story frame. The explicit imperfection model uses an out-of-plumbness of 0.2%, and in the further reduced tangent modulus model, $0.85\,E_t$ is used.

Herein, a scaling factor of 10 is used due to the high indeterminacy. The load increment is automatically reduced if the element stiffness parameter, η, exceeds the predefined value 0.1. The 54 load steps required to converge on the solution are given in P.OUT2.

Step 3: Analysis

The input data may be easily generated, as listed in Table 28.9a and 28.9b. Programs are executed in sequence by typing the batch file command RUN.

Step 4: Check of Load-Carrying Capacity

From the output file, P.OUT, the ultimate load-carrying capacity of the structure is obtained as 5.24 and 5.18 kips with respect to the lateral load at roof in load combination 2 by the imperfection

Table 28.9a Input Data, P.DAT, of the Explicit Imperfection Modeling for the Unbraced Eight-Story Frame

```
Unbraced eight-story frame, explicit imperfection modeling
    1     1
   26     2
    0     2     0
    0    32     0
    1        38.30      6710.00       467.00    29000.00        36.00
    2        14.70       984.00       110.00    29000.00        36.00
    1         0.252       126.00     1     2     1     7
    8         0.360       180.00     1     9     8
    9         0.252       126.00     1    19    18     7
   16         0.360       180.00     1    26    25
   17       150.00          0.00     2     1    10     8
   25       150.00          0.00     2    10    18     8
    9     1     1
   26     1     1
    1         0.512        -1.080
    2         1.020        -1.255                              6
    8         1.240        -1.255
   10                      -2.160
   11                      -2.510                              7
   18                      -1.080
   19                      -1.255                              7
```

Table 28.9b Input Data, P.DAT, of the Explicit Imperfection Modeling for the Unbraced Eight-Story Frame

```
Unbraced eight-story frame, further reduced tangent modulus
    3     1
   26     2
    0     2     0
    0    32     0
    1        38.30      6710.00       467.00    29000.00        36.00     1
    2        14.70       984.00       110.00    29000.00        36.00
    1         0.         126.00     1     2     1     7
    8         0.         180.00     1     9     8
    9         0.         126.00     1    19    18     7
   16         0.         180.00     1    26    25
   17       150.00          0.00     2     1    10     8
   25       150.00          0.00     2    10    18     8
    9     1     1
   26     1     1
    1         0.512        -1.080
    2         1.020        -1.255                              6
    8         1.240        -1.255
   10                      -2.160
   11                      -2.510                              7
   18                      -1.080
   19                      -1.255                              7
```

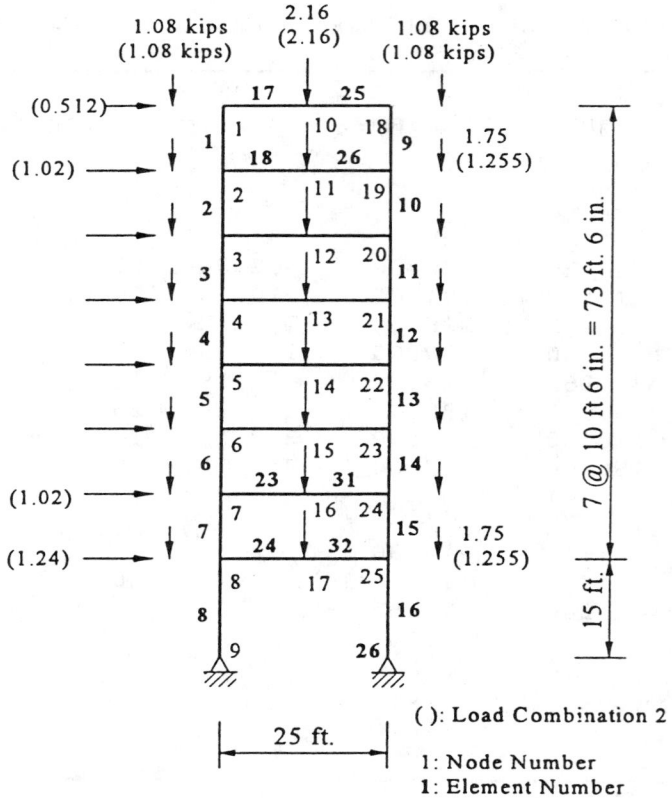

Figure 28.42   Structural modeling of the eight-story frame.

method and the reduced tangent modulus method, respectively. This load-carrying capacity is 3 and 2% greater, respectively, than the applied factored load of 5.12 kips. As a result, the preliminary member sizes are satisfactory.

Step 5: Check of Serviceability

The lateral drift at the roof level by the wind load (1.0W) is 5.37 in. and the drift ratio is 1/198, which does not satisfy the drift limit of 1/400. When W40x174 and W24x76 are used for column and beam members, the lateral drift is reduced to 2.64 in. and the drift ratio is 1/402, which satisfies the limit 1/400. The design of this frame is thus governed by serviceability rather than strength.

Comparison of Results

The sizes predicted by the proposed methods are W33x130 columns and W21x50 beams. They do not, however, meet serviceability conditions and must therefore be increased to W40x174 and W24x76 members. The LRFD method results in the same (W40x174) column but a larger (W27x84) beam (Figure 28.43).

28.6.3 Two-Story Four-Bay Semi-Rigid Frame

Figure 28.44 shows a two-story four-bay semi-rigid frame. The height of each story is 12 ft and it is 25 ft wide. The spacing of the frames is 25 ft. The frame is subjected to a distributed gravity and concentrated lateral loads. The roof beam connections are the top and seat angles of L6x4.0x3/8x7 with double web angles of L4x3.5x1/4x5.5 made of A36 steel. The floor beam connections are the

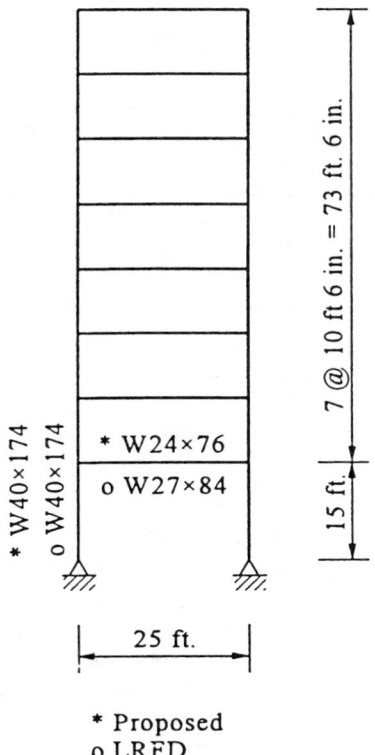

Figure 28.43   Comparison of member sizes of the eight-story frame.

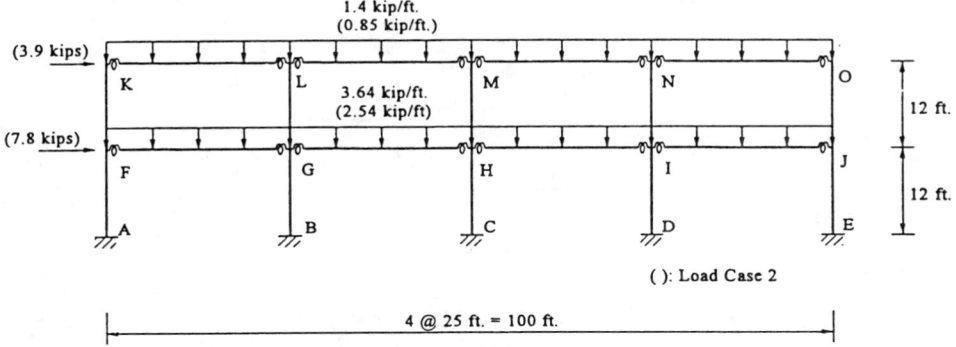

Figure 28.44   Configuration and load condition of two-story four-bay semi-rigid frame.

top and seat angles of L6x4x9/16x7 with double web angles of L4x3.5x5/16x8.5. All fasteners are A325 3/4-in.-diameter bolts. All members are assumed to be continuously braced laterally.

Design by Advanced Analysis

Step 1: Load Condition and Preliminary Member Size

The load conditions are shown in Figure 28.44. The initial member sizes are selected as W8x21, W12x22, and W16x40 for the columns, the roof beams, and the floor beams, respectively.

Step 2: Structural Modeling

Each column is modeled with one element and beam with two elements. The distributed gravity loads are converted to equivalent concentrated loads on the beam, shown in Figure 28.45. In explicit

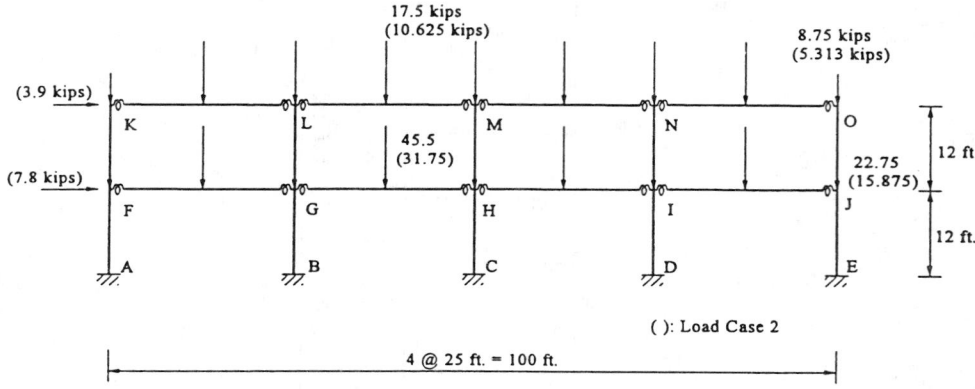

Figure 28.45 Concentrated load condition converted from the distributed load for the two-story one-bay semi-rigid frame.

imperfection modeling, the geometric imperfection is obtained by multiplying the column height by 0.002. In the equivalent notional load method, the notional load is 0.002 times the total gravity load plus the lateral load. In the further reduced tangent modulus method, the program automatically accounts for geometric imperfection effects. Although users can choose any of these three models, the further reduced tangent modulus model is the only one presented herein. The incremental loads are computed by dividing the concentrated load by the scaling factor of 20.

Step 3: Analysis

The three parameters of the connections can be computed by use of the computer program 3PARA. Corresponding to the input format in Table 28.2, the input data, CONN.DAT, may be generated, as shown in Tables 28.10a and 28.10b.

Table 28.10a Input Data, CONN.DAT, of Connection for Roof Beam

2	36.0	29000			
7	0.375	0.875	2.5	1.25	12
5.5	0.25	0.6875	2.5		

Table 28.10b Input Data, CONN.DAT, of Connection for Floor Beam

2	36.0	29000			
7	0.5625	1.0625	2.5	1.25	16
8.5	0.3125	0.75	2.5		

Referring to the input instructions (Section 28.5.4), the input data is written in the form shown in Tables 28.11a and 28.11b. Programs DATAGEN, INPUT, and PAAP are executed sequentially by typing "RUN." The program will continue to analyze with increasing load steps up to the ultimate state.

Step 4: Check of Load-Carrying Capacity

As shown in output file, P.OUT2, the ultimate load-carrying capacities of the load combinations 1 and 2 are 46.2 and 42.9 kips, respectively, at nodes 7–13 (Figure 28.45). Compared to the applied loads, 45.5 and 31.75 kips, the initial member sizes are adequate.

Table 28.11a   Input Data, P.DAT, of the Four-Bay Two-Story Semi-Rigid Frame

```
Four-bay two-story semi-rigid frame (Load case 1)
      3    1
     23    5  100
      2    3    0
     16   26    0
      1    1361.0    607384.0      0.927
      2     446.0     90887.0      1.403
      1      6.16        75.3      20.4    29000.0      36.0    1
      2      5.48       156.0      29.3    29000.0      36.0
      3     11.8        518.0      72.9    29000.0      36.0
      1   11    1    1
      2   12    2    1
      3   13    1    1
      4   14    2    1
      5   15    1    1
      6   16    2    1
      7   17    1    1
      8   18    2    1
      9   19    1    2
     10   20    2    2
     11   21    1    2
     12   22    2    2
     13   23    1    2
     14   24    2    2
     15   25    1    2
     16   26    2    2
      1      0.0       144.0    1    1     6
      2      0.0       144.0    1    2     8
      3      0.0       144.0    1    3    10
      4      0.0       144.0    1    4    12
      5      0.0       144.0    1    5    14
      6      0.0       144.0    1    6    15
      7      0.0       144.0    1    8    17    3    2
     10      0.0       144.0    1   14    23
     1J    150.0         0.0    3    6     7    8
     19    150.0         0.0    2   15    16    8
      1    1    1    1    5
      6            -1.1375
      7            -2.2750                  7
     14            -1.1375
     15            -0.4375
     16            -0.8750                  7
     23            -0.4375
```

Table 28.11b   Input Data, P . DAT, of the Four-Bay Two-Story Semi-Rigid Frame

```
Four-bay two-story semi-rigid frame (Load case 2)
    3     1
   23     5   100
    2     3     0
   16    26     0
    1    1361.0    607384.0      0.927
    2     446.0     90887.0      1.403
    1       6.16        75.3       20.4     29000.0        36.0       1
    2       6.48       156.0       29.3     29000.0        36.0
    3      11.8        518.0       72.9     29000.0        36.0
    1    11    1    1
    2    12    2    1
    3    13    1    1
    4    14    2    1
    5    15    1    1
    6    16    2    1
    7    15    1    1
    8    16    2    1
    9    19    1    2
   10    20    2    2
   11    21    1    2
   12    22    2    2
   13    23    1    2
   14    24    2    2
   15    25    1    2
   16    26    2    2
    1       0.0       144.0    1    1     6
    2       0.0       144.0    1    2     8
    3       0.0       144.0    1    3    10
    4       0.0       144.0    1    4    12
    5       0.0       144.0    1    5    14
    6       0.0       144.0    1    6    15
    7       0.0       144.0    1    8    17    3    2
   10       0.0       144.0    1   14    23
   11     150.0         0.0    3    6     7    8
   19     150.0         0.0    2   15    16    8
    1    1    1    1     5
    6       0.39    -0.7938
    7               -1.5880                 7
   14               -0.7938
   15       0.195   -0.2657
   16               -0.5313                 7
   23               -0.2657
```

Step 5: Check of Serviceability

The lateral displacement at roof level corresponding to 1.0W is computed as 0.51 in. from the computer output, P.OUT2. The drift ratio is 1/565, which satisfies the limitation 1/400. The preliminary member sizes are satisfactory.

Comparison of Results

The member sizes by the advanced analysis and LRFD methods are compared in Figure 28.46. The beam members are one size larger in the advanced analysis method, and the interior columns are one size smaller.

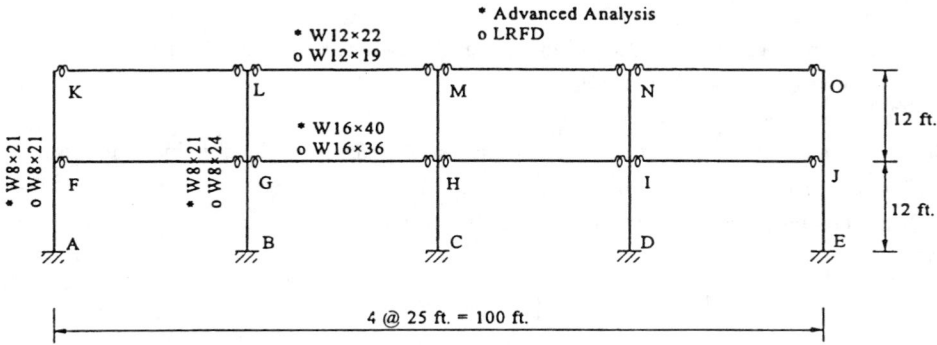

Figure 28.46   Comparison of member sizes of the two-story four-bay semi-rigid frame.

28.7   Defining Terms

Advanced analysis: Analysis predicting directly the stability of a structural system and its component members and not needing separate member capacity checks.

ASD: Acronym for Allowable Stress Design.

Beam-columns: Structural members whose primary function is to carry axial force and bending moment.

Braced frame: Frame in which lateral deflection is prevented by braces or a shear walls.

CRC: Acronym for Column Research Council.

Column: Structural member whose primary function is to carry axial force.

Drift: Lateral deflection of a building.

Ductility: Ability of a material to undergo a large deformation without a significant loss in strength.

Factored load: The product of the nominal load and a load factor.

Flexural member: Structural member whose primary function is to carry bending moment.

Geometric imperfection: Unavoidable geometric error during fabrication and erection.

Limit state: A condition in which a structural or structural component becomes unsafe (strength limit state) or unfit for its intended function (serviceability limit state).

Load factor: A factor to account for the unavoidable deviations of the actual load from its nominal value and uncertainties in structural analysis.

LRFD: Acronym for Load Resistance Factor Design.

Notional load: Load equivalent to geometric imperfection.

PD: Acronym for Plastic Design.

Plastic hinge: A yield section of a structural member in which the internal moment is equal to the plastic moment of the cross-section.

Plastic zone: A yield zone of a structural member in which the stress of a fiber is equal to the yield stress.

Refined plastic hinge analysis: Modified plastic hinge analysis accounting for gradual yielding of a structural member.

Resistance factors: A factor to account for the unavoidable deviations of the actual resistance of a member or a structural system from its nominal value.

Second-order analysis: Analysis to use equilibrium equations based on the deformed geometry of a structure under load.

Semi-rigid connection: Beam-to-column connection whose behavior lies between fully rigid and ideally pinned connection.

Service load: Nominal load under normal usage.

Stability function: Function to account for the bending stiffness reduction due to axial force.

Stiffness: Force required to produce unit displacement.

Unbraced frame: Frame in which lateral deflections are not prevented by braces or shear walls.

References

[1] Ad Hoc Committee on Serviceability. 1986. Structural Serviceability: a critical appraisal and research needs, *J. Struct. Eng.*, ASCE, 112(12), 2646-2664.

[2] American Institute of Steel Construction. 1994. *Load and Resistance Factor Design Specification,* 2nd ed., Chicago.

[3] American Institute of Steel Construction. 1986. *Load and Resistance Factor Design Specification for Structural Steel Buildings,* Chicago.

[4] Canadian Standard Association. 1994. *Limit States Design of Steel Structures,* CAN/CSA-S16.1-M94.

[5] Canadian Standard Association. 1989. *Limit States Design of Steel Structures,* CAN/CSA-S16.1-M89.

[6] Chen, W. F. and Kim, S. E. 1997. *LRFD Steel Design Using Advanced Analysis,* CRC Press, Boca Raton, FL.

[7] Chen, W. F. and Lui, E. M. 1992. *Stability Design of Steel Frames,* CRC Press, Boca Raton, FL.

[8] Chen, W.F. and Lui, E.M. 1986. *Structural Stability—Theory and Implementation,* Elsevier, New York.

[9] Cook, R.D., Malkus, D.S., and Plesha, M.E. 1989. *Concepts and Applications of Finite Element Analysis,* 3rd ed., John Wiley & Sons, New York.

[10] ECCS. 1991. *Essentials of Eurocode 3 Design Manual for Steel Structures in Building,* ECCS-Advisory Committee 5, No. 65.

[11] ECCS. 1984. *Ultimate Limit State Calculation of Sway Frames with Rigid Joints,* Technical Committee 8—Structural Stability Technical Working Group 8.2— System, Publication No. 33.

[12] Ellingwood. 1989. Serviceability Guidelines for Steel Structures, *Eng. J.*, AISC, 26, 1st Quarter, pp. 1-8.

[13] Hughes, T. J. R. 1987. *The Finite Element Method: Linear Static and Dynamic Finite Element Analysis,* Prentice-Hall, Englewood Cliffs, NJ.

[14] Kanchanalai, T. 1977. The Design and Behavior of Beam-Columns in Unbraced Steel Frames, AISI Project No. 189, Report No. 2, Civil Engineering/Structures Research Lab., University of Texas at Austin.

[15] Kim, S.E. 1996. Practical Advanced Analysis for Steel Frame Design, Ph.D. thesis, School of Civil Engineering, Purdue University, West Lafayette, IN.

[16] Kim, S. E. and Chen, W. F. 1996. Practical advanced analysis for steel frame design, *ASCE Structural Congress XIV*, Chicago, Special Proceeding Volume on Analysis and Computation, April, pp. 19-30.

[17] Kim, S. E. and Chen, W. F. 1996. Practical advanced analysis for braced steel frame design, *J. Struct. Eng.*, ASCE, 122(11), 1266-1274.

[18] Kim, S. E. and Chen, W. F. 1996. Practical advanced analysis for unbraced steel frame design, *J. Struct. Eng.*, ASCE, 122(11), 1259-1265.

[19] Kim, S. E. and Chen, W. F. 1996. Practical advanced analysis for semi-rigid frame design, *AISC Eng. J.*, 33(4), 129-141.

[20] Kim, S. E. and Chen, W. F. 1996. Practical advanced analysis for frame design—Case study, *SSSS J.*, 6(1), 61-73.

[21] Kish, N. and Chen, W. F. 1990. Moment-rotation relations of semi-rigid connections with angles, *J. Struct. Eng.*, ASCE, 116(7), 1813-1834.

[22] Kishi, N., Goto, Y., Chen, W. F., and Matsuoka, K. G. 1993. Design aid of semi-rigid connections for frame analysis, *Eng. J.*, AISC, 4th quarter, pp. 90-107.

[23] LeMessurier, W. J. 1977. A practical method of second order analysis, Part 2—Rigid frames. *AISC Eng. J.*, 14(2), 49-67.

[24] Liew, J. Y. R. 1992. Advanced Analysis for Frame Design, Ph.D. thesis, School of Civil Engineering, Purdue University, West Lafayette, IN.

[25] Liew, J. Y. R., White, D. W., and Chen, W. F. 1993. Second-order refined plastic hinge analysis of frame design: Part I, *J. Struct. Eng.*, ASCE, 119(11), 3196-3216.

[26] Liew, J. Y. R., White, D. W., and Chen, W. F. 1993. Second-order refined plastic-hinge analysis for frame design: Part 2, *J. Struct. Eng.*, ASCE, 119(11), 3217-3237.

[27] Maleck, A. E., White, D. W., and Chen, W. F. 1995. Practical application of advanced analysis in steel design, *Proc. 4th Pacific Structural Steel Conf.*, Vol. 1, Steel Structures, pp. 119-126.

[28] SSRC 1981. General principles for the stability design of metal structures, Technical Memorandum No. 5, Civil Engineering, ASCE, February, 53-54.

[29] White, D. W. and Chen, W. F., Eds., 1993. *Plastic Hinge Based Methods for Advanced Analysis and Design of Steel Frames: An Assessment of the State-of-the-art*, SSRC, Lehigh University, Bethlehem, PA.

[30] Standards Australia. 1990. *AS4100-1990, Steel Structures*, Sydney, Australia.

[31] Stelmack T. W. 1982. Analytical and Experimental Response of Flexibly-Connected Steel Frames, M.S. dissertation, Department of Civil, Environmental, and Architectural Engineering, University of Colorado.

[32] Vogel, U. 1985. Calibrating frames, *Stahlbau*, 10, 1-7.

[33] White, D. W. 1993. Plastic hinge methods for advanced analysis of steel frames, *J. Constr. Steel Res.*, 24(2), 121-152.

Further Reading

[1] Chen, W. F. and Kim, S. E. 1997. *LRFD Steel Design Using Advanced Analysis*, CRC Press, Boca Raton, FL.

[2] Chen, W. F. and Lui, E. M. 1992. *Stability Design of Steel Frames*, CRC Press, Boca Raton, FL.

[3] Chen, W. F. and Toma, S. 1994. *Advanced Analysis of Steel Frames*, CRC Press, Boca Raton, FL.

[4] Chen, W. F. and Sohal, I. 1995. *Plastic Design and Second-Order Analysis of Steel Frames*, Springer-Verlag, New York.

[5] Chen, W. F., Goto, Y., and Liew, J. Y. R. 1996. *Stability Design of Semi-Rigid Frames*, John Wiley & Sons, New York.

29

Welded Tubular Connections—CHS Trusses

29.1   Introduction 29-1
29.2   Architecture... 29-1
29.3   Characteristics of Tubular Connections 29-2
29.4   Nomenclature... 29-2
29.5   Failure Modes.. 29-2
        Local Failure • General Collapse • Unzipping or Progressive
        Failure • Materials Problems • Fatigue
29.6   Reserve Strength 29-7
29.7   Empirical Formulations 29-8
29.8   Design Charts... 29-9
        Joint Efficiency • Derating Factor
29.9   Application 29-11
29.10 Summary and Conclusions 29-14
References 29-14

Peter W. Marshall
MHP Systems Engineering,
Houston, Texas

29.1   Introduction

Truss connections in circular hollow sections (CHS) present unique design challenges. This chapter discusses the following elements of the subject: Architecture, Characteristics of Tubular Connections, Nomenclature, Failure Modes, Reserve Strength, Empirical Formulations, Design Charts, Application, and Summary and Conclusions.

29.2   Architecture

"Architecture" is defined as the art and science of designing and successfully executing structures in accordance with aesthetic considerations and the laws of physics, as well as practical and material considerations. Where tubular structures are exposed for dramatic effect, it is often disappointing to see grand concepts fail in execution due to problems in the structural connections of tubes. Such "failures" range from awkward ugly detailing, to learning curve problems during fabrication, to excessive deflections or even collapse. Such failures are unnecessary, as the art and science of welded tubular connections has been codified in the AWS Structural Welding Code [1].

A well-engineered structure requires that a number of factors be in reasonable balance. Factors to be considered in relation to economics and risk in the design of welded tubular structures and their connections include: (1) static strength, (2) fatigue resistance, (3) fracture control, and (4) weldability. Static strength considerations are so important that they often dictate the very architecture and layout of the structure; certainly they dominate the design process and are the focus of this chapter. Many of the other factors also require early attention in design, and arise again in setting up QC/QA programs during construction; these are discussed further in sections of the Code dealing with materials, welding technique, qualification, and inspection.

29.3 Characteristics of Tubular Connections

Tubular members benefit from an efficient distribution of their material, particularly in regard to beam bending or column buckling about multiple axes. However, their resistance to concentrated radial loads are more problematic. For architecturally exposed applications, the clean lines of a closed section are esthetically pleasing and they minimize the amount of surface area for dirt, corrosion, or other fouling. Simple welded tubular joints can extend these clean lines to include the structural connections.

Although many different schemes for stiffening tubular connections have been devised [3], the most practical connection is made by simply welding the branch member to the outside surface of the main member (or chord). Where the main member is relatively compact (D/T less than 15 or 20), the branch member thickness is limited to 50 or 60% of the main member thickness, and a prequalified weld detail is used, the connection can develop the full static capacity of the members joined. Where the foregoing conditions are not met, e.g., with large diameter tubes, a short length of heavier material (or joint can) is inserted into the chord to locally reinforce the connection area. Here, the design problem reduces to one of selecting the right combination of thickness, yield strength, and notch toughness for the chord or joint can. The detailed considerations involved in this design process are the subject of this chapter.

29.4 Nomenclature

Non-dimensional parameters for describing the geometry of a tubular connection are given in the following list. Beta, eta, theta, and zeta describe the surface topology. Gamma and tau are two very important thickness parameters. Alpha (not shown) is an ovalizing parameter, depending on load pattern (it was formerly used for span length in beams loaded via tee connections).

β (beta)	d/D, branch diameter/main diameter
η (eta)	branch footprint length/main diameter
θ (theta)	angle between branch and main member axes
ζ (zeta)	g/D, gap/diameter (between balancing branches of a K-connection)
γ (gamma)	R/T, main member radius/thickness ratio
τ (tau)	t/T, branch thickness/main thickness

In AWS D1.1 [1], the term "T-, Y-, and K-connection" is used generically to describe simple structural connections or nodes, as opposed to co-axial butt and lap joints. A letter of the alphabet (T, Y, K, X) is used to evoke a picture of what the node subassemblage looks like.

29.5 Failure Modes

A number of unique failure modes are possible in tubular connections. In addition to the usual checks on weld stress, provided for in most design codes, the designer must check for the following failure modes, listed together with the relevant AWS D1.1-96 [1] code sections:

Local failure (punching shear)	2.40.1.1
General collapse	2.40.1.2
Unzipping (progressive weld failure)	2.40.1.3
Materials problems	
(fracture and delamination)	2.42, C4.12.4.4, and 2.1.3
Fatigue	2.36.6

29.5.1 Local Failure

AWS design criteria for this failure mode have traditionally been formulated in terms of punching shear. The main member acts as a cylindrical shell in resisting the concentrated radial line loads (kips/in.) delivered to it at the branch member footprint. Although the resulting localized shell stresses in the main member are quite complex, a simplified but still quite useful representation can be given in terms of punching shear stress, v_p:

$$\text{acting } v_p = f_n \tau \sin \theta \tag{29.1}$$

where f_n is the nominal stress at the end of the branch member, either axial or bending, which are treated separately. Punching shear is the notional stress on the potential failure surface, as illustrated in Figure 29.1. The overriding importance of chord thickness is reflected in tau, while $\sin \theta$ indicates that it is the radial component of load that causes all the mischief.

The allowable punching shear stress is given in the Code as:

$$\text{allowable } v_p = \frac{F_{yo}}{0.6\gamma} \cdot Q_q \cdot Q_f \tag{29.2}$$

We see that the allowable punching shear stress is primarily a function of main member yield strength (F_{yo}) and gamma ratio (main member radius/thickness), with some trailing terms that tend towards unity. The term Q_q reflects the considerable influence of connection type, geometry, and load pattern, while interactions between branch and chord loads are covered by the reduction factor Q_f. Interactions between brace axial load and bending moments are treated analogous to those for a fully plastic section.

Since 1992, the AWS code also includes tubular connection design criteria in total load ultimate strength format, compatible with an LRFD design code formulation. This was derived from, and intended to be comparable to, the original punching shear criteria.

29.5.2 General Collapse

In addition to local failure of the main member in the vicinity of the branch member, a more widespread mode of collapse may occur, e.g., general ovalizing plastic failure in the cylindrical shell of the main member. To a large extent, this is now covered by strength criteria that are specialized by connection type and load pattern, as reflected in the Q_q factor.

For balanced K-connections, the inward radial loads from one branch member is compensated by outward loads on the other, ovalizing is minimized, and capacity approaches the local punching shear limit. For T and Y connections, the radial load from the single branch member is reacted by beam shear in the main member or chord, and the resulting ovalizing leads to lower capacity. For cross or X connections, the load from one branch is reacted by the opposite branch, and the resulting double dose of ovalizing in the main member leads to still further reductions in capacity. The Q_q term also reflects reduced ovalizing and increased capacity, as the branch member diameter approaches that of the main member.

Thus, for design purposes, tubular connections are classified according to their configuration (T, Y, K, X, etc.). For these "alphabet" connections, different design strength formulae are often applied

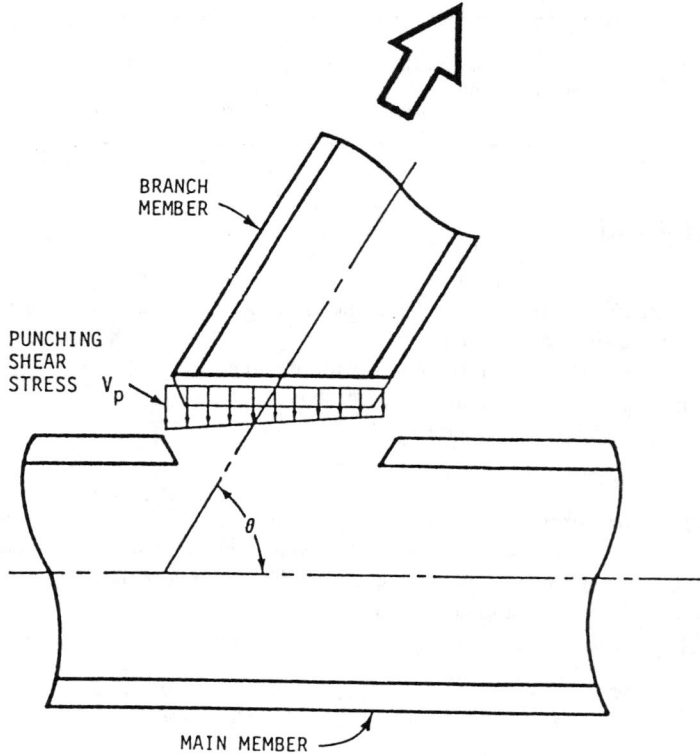

Figure 29.1   Local failure mode and punching shear V_p.

to each different type. Until recently, the research, testing, and analysis leading to these criteria dealt only with connections having their members in a single plane, as in a roof truss or girder.

Many tubular space frames have bracing in multiple planes. For some loading conditions, these different planes interact. When they do, criteria for the "alphabet" joints are no longer satisfactory. In AWS, an "ovalizing parameter" (alpha, Appendix L) may be used to estimate the beneficial or deleterious effect of various branch member loading combinations on main member ovalizing. This reproduces the trend of increasingly severe ovalizing in going from K to T/Y to X-connections, and has been shown to provide useful guidance in a number of more adverse planar (e.g., all-tension double-K [9]) and multi-planar (e.g., hub [11]) situations. However, for similarly loaded members in adjacent planes, e.g., paired KK connections in delta trusses, Japanese data indicate that no increase in capacity over the corresponding uniplanar connections should be taken [2].

The effect of a short joint can (less than 2.5 diameters) in reducing the ovalizing or crushing capacity of cross connections is addressed in AWS section 2.40.1.2(2) [1]. Since ovalizing is less severe in K-connections, the rule of thumb is that the joint can need only extend 0.25 to 0.4 diameters beyond the branch member footprints to avoid a short-can penalty. Intermediate behavior would apply to T/Y connections.

A more exhaustive discussion would also consider the following modes of general collapse in addition to ovalizing: beam bending of the chord (in T-connection tests), beam shear (in the gap of K-connections), transverse crippling of the main member sidewall, and local buckling due to uneven load transfer (either brace or chord). These are illustrated in Figure 29.2.

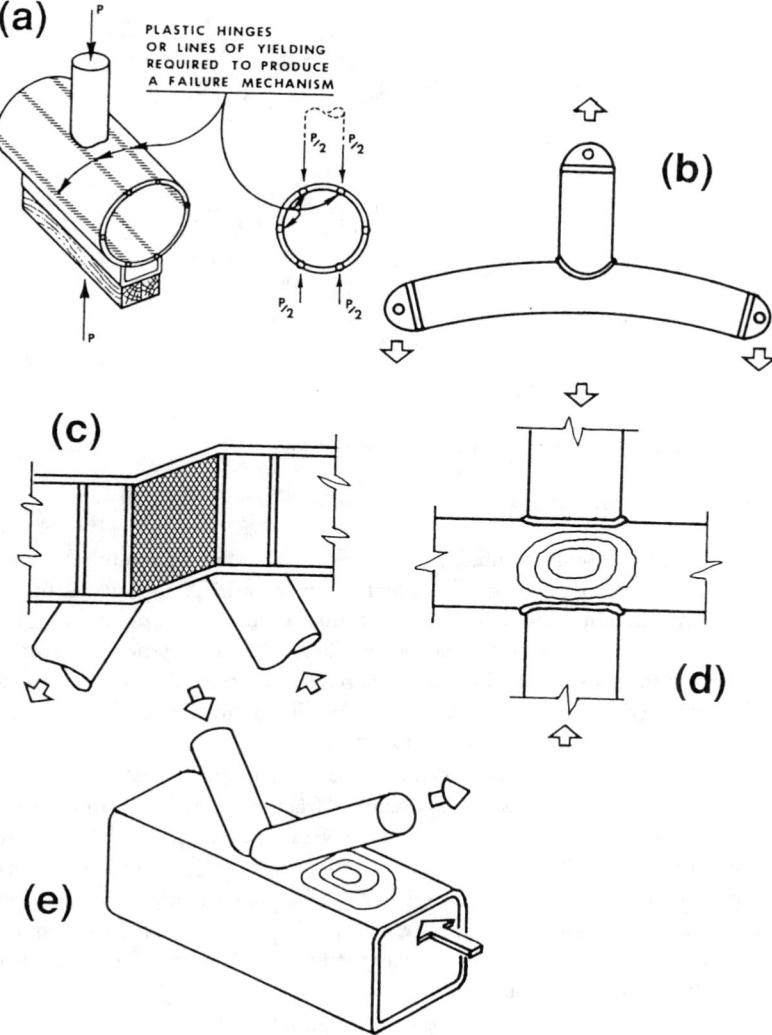

Figure 29.2 Failure modes — general colapse. (a) Ovalizing, (b) beam bending, (c) beam shear in the gap, (d) sidewall (web) crippling, and (e) local buckling due to uneven distribution of axial load.

29.5.3 Unzipping or Progressive Failure

The initial elastic distribution of load transfer across the weld in a tubular connection is highly non-uniform, as illustrated in Figure 29.3, with the peak line load often being a factor of two higher than that indicated on the basis of nominal sections, geometry, and statics. Some local yielding is required for tubular connections to redistribute this and reach their design capacity. If the weld is a weak link in the system, it may "unzip" before this redistribution can happen. Criteria given in the AWS code are intended to prevent this unzipping, taking advantage of the higher reserve strength in weld allowable stresses than is the norm elsewhere. For mild steel tubes and overmatched E70 weld metal, weld effective throats as small as 70% of the branch member thickness are permitted.

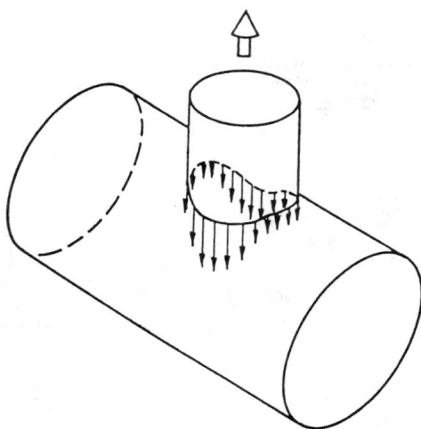

Figure 29.3   Uneven distribution of load across the weld.

29.5.4   Materials Problems

Most fracture control problems in tubular structures occur in the welded tubular connections, or nodes. These require plastic deformation in order to reach their design capacity. Fatigue and fracture problems for many different node geometries are brought into a common focus by use of the "hot spot" stress, as would be measured by a strain gauge, adjacent to and perpendicular to the toe of the weld joining branch to main member, in the worst region of localized plastic deformation (usually in the chord). Hot spot stress has the advantage of placing many different connection geometries on a common basis with regard to fatigue and fracture.

Charpy impact testing is a method for qualitative assessment of material toughness. The method has been, and continues to be, a reasonable measure of fracture safety when employed with a definitive program of nondestructive testing to eliminate weld area flaws. The AWS recommendations for material selection (C2.42.2.2) and weld metal impact testing (C4.12.4.4) are based on practices that have provided satisfactory fracture experience in offshore structures located in moderate temperature environments, i.e., 40°F (+5°C) water and 14°F (−10°C) air exposure. For environments that are either more or less hostile, impact testing temperatures should be reconsidered based on LAST (lowest anticipated service temperature).

In addition to weld metal toughness, consideration should be given to controlling the properties of the heat affected zone (HAZ). Although the heat cycle of welding sometimes improves hot rolled base metals of low toughness, this region will more often have degraded toughness properties. A number of early failures in welded tubular connections involved fractures that either initiated in or propagated through the HAZ, often obscuring the identification of other design deficiencies, e.g., inadequate static strength.

A more rigorous approach to fatigue and fracture problems in welded tubular connections has been taken by using fracture mechanics [5]. The CTOD (crack tip opening displacement) test is used to characterize materials that are tough enough to undergo some plasticity before fracture.

Underneath the branch member footprint, the main member is subjected to stresses in the thru-thickness or short transverse direction. Where these stresses are tensile, due to weld shrinkage or applied loading, delamination may occur — either by opening up pre-existing laminations or by laminar tearing in which microscopic inclusions link up to give a fracture having a woody appearance, usually in or near the HAZ. These problems are addressed in API joint can steel specifications 2H, 2W, and 2Y. Pre-existing laminations are detected with ultrasonic testing. Microscopic inclusions are prevented by restricting sulfur to very low levels (< 60 ppm) and by inclusion shape control metallurgy in the steel-making ladle. As a practical matter, weldments that survive the weld shrinkage phase usually perform satisfactorily in ordinary service.

Joint can steel specifications also seek to enhance weldability with limitations on carbon and other alloying elements, as expressed by carbon equivalent or P_{cm} formulae. Such controls are increasingly important as residual elements accumulate in steel made from scrap. In AWS Appendix XI [1], the preheat required to avoid HAZ cracking is related to carbon equivalent, base metal thickness, hydrogen level (from welding consumables), and degree of restraint.

29.5.5 Fatigue

This failure mode has been observed in tubular joints in offshore platforms, dragline booms, drilling derricks, radio masts, crane runways, and bridges. The nominal stress, or detail classification approach, used for non-tubular structures fails to recognize the wide range of connection efficiencies and stress concentration factors that can occur in tubular structures. Thus, fatigue design criteria based on either punching shear or hot spot stress appear in the AWS Code. The subject is also summarized in recent papers on tubular offshore structures [7, 8].

29.6 Reserve Strength

While the elastic behavior of tubular joints is well predicted by shell theory and finite element analysis, there is considerable reserve strength beyond theoretical yielding due to triaxiality, plasticity, large deflection effects, and load redistribution. Practical design criteria make use of this reserve strength, placing considerable demands on the notch toughness of joint-can materials. Through joint classification (API) or an ovalizing parameter (AWS), they incorporate elements of general collapse as well as local failure. The resulting criteria may be compared against the supporting data base of test results to ferret out bias and uncertainty as measures of structural reliability. Data for K, T/Y, and X joints in compression show a bias on the safe side of 1.35, beyond the nominal safety factor of 1.8, as shown in Figure 29.4. Tension joints appear to show a larger bias of 2.85; however, this reduces to 2.05 for joints over 0.12 in. thick, and 1.22 over 0.5 in., suggesting a thickness effect for tests that end in fracture.

For overload analysis of tubular structures (e.g., earthquake), we need not only ultimate strength, but also the load-deflection behavior. Early tests showed ultimate deflections of 0.03 to 0.07 chord diameters, giving a typical ductility of 0.10 diameters for a brace with weak joints at both ends. As more different types of joints were tested, a wider variety of load-deflection behaviors emerged, making such generalizations tenuous.

Cyclic overload raises additional considerations. One issue is whether the joint will experience a ratcheting or progressive collapse failure, or will achieve stable behavior with plasticity contained at local hotshots, a process called "shakedown" (as in shakedown cruise). While tubular connections have withstood 60 to several hundred repetitions of load in excess of their nominal capacity, a conservative analytical treatment is to consider that the cumulative plastic deformation or energy absorption to failure remains constant.

When tubular joints and members are incorporated into a space frame, the question arises as to whether computed bending moments are primary (i.e., necessary for structural stability, as in a sidesway portal situation, and must be designed for) or secondary (i.e., an unwanted side effect of deflection which may be safely ignored or reduced). When proportional loading is imposed, with both axial load and bending moment being maintained regardless of deflection, the joint simply fails when it reaches its failure envelope. However, when moments are due to imposed lateral deflection, and then axial load is imposed, the load path skirts along the failure envelope, shedding the moment and sustaining further increases in axial load.

Another area of interaction between joint behavior and frame action is the influence of brace bending/rotation on the strength of gap K-connections. If rotation is prevented, bending moments develop which permit the gap region to transfer additional load. If the loads remain strictly axial,

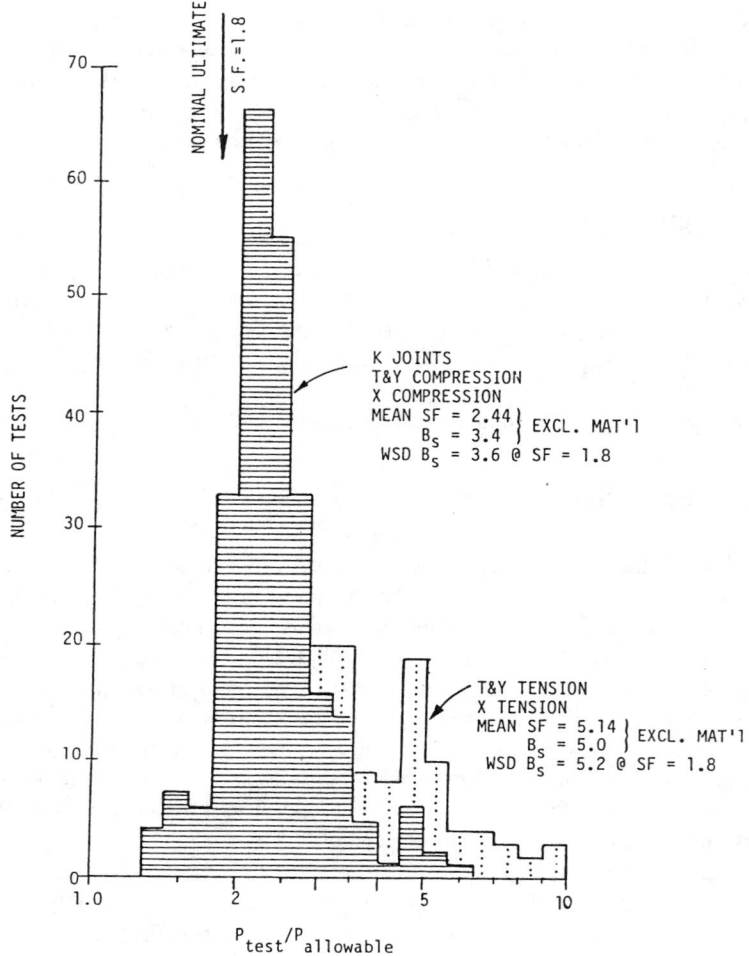

Figure 29.4   Comparison of AWS design criteria with the WRC database.

brace end rotation occurs in the absence of restraining moments, and a lower joint capacity is found. These problems arise for circular tubes as well as box connections, and a recent trend has been to conduct joint-in-frame tests to achieve a realistic balance between the two limiting conditions. Loads that maintain their original direction (as in an inelastic finite element analysis) or, worse yet, follow the deflection (as in testing arrangements with a two-hinge jack), result in a plastic instability of the compression brace stub which grossly understates the actual joint strength. Existing data bases may need to be screened for this problem.

29.7   Empirical Formulations

Because of the foregoing reserve strength issues, AWS design criteria have been derived from a database of ultimate strength tubular joint tests. Comparison with the database (Figure 29.4) indicates a safety index of 3.6 against known static loads for the AWS punching shear criteria. Safety index is the safety margin, including hidden bias, expressed in standard deviations of total uncertainty. Since these criteria are used to select the main member chord or joint can, the choice of safety index is similar to that used for sizing other structural members, rather than the higher safety margins used for workmanship-sensitive connection items such as welds or bolts.

When the ultimate axial load is used in the context of AISC-LRFD, with a resistance factor of 0.8, AWS ultimate strength is nominally equivalent to punching shear allowable stress design (ASD) for structures having 40% dead load and 60% live load. LRFD falls on the safe side of ASD for structures having a lower proportion of dead load. AISC criteria for tension and compression members appear to have made the equivalency trade-off at 25% dead load; thus, the LRFD criteria given by AWS would appear to be conservative for a larger part of the population of structures.

In Canada, using these resistance factors with slightly different load factors, a 4.2% difference in overall safety factor results — within calibration accuracy [10].

29.8   Design Charts

Research, testing, and applications have progressed to the point where tubular connections are about as reliable as the other structural elements with which designers deal. One of the principal barriers to more widespread use seems to be unfamiliarity. To alleviate this problem, design charts have been presented in "Designing Tubular Connections with AWS D1.1", by P. W. Marshall [4].

The capacity of simple, direct, welded, tubular connections is given in terms of punching shear efficiency, E_v, where

$$E_v = \frac{\text{allowable punching shear stress}}{\text{main member allowable tension stress}} \tag{29.3}$$

Charts for punching shear efficiency for axial load, in-plane bending, and out-of-plane bending appear as Figures 29.5 through 29.9. Note that for axial load, separate charts are given for K-connections, T/Y connections, and X connections, reflecting their different load patterns and different values of the ovalizing parameter (alpha). Within each connection or load type, punching shear efficiency is a function of the geometry parameters, diameter ratio (beta) and chord radius/thickness (gamma), as defined earlier. For K-connections, the gap, g, between braces (of diameter d) is also significant, with the behavior reverting to that of T/Y connections for very large gap. Punching shear efficiency cannot exceed a value of 0.67, the material limit for shear.

29.8.1   Joint Efficiency

The importance of branch/chord thickness ratio tau (t/T) and of angle $(\sin \theta)$ becomes apparent in the expression for joint efficiency, E_j, given by:

$$E_j = \frac{E_v \quad \cdot \quad Q_f}{(t/T)\sin \theta} \quad \cdot \quad \frac{F_{yo}(\text{chord})}{F_y(\text{branch})} \tag{29.4}$$

where Q_f is the derating factor to account for chord utilization (described in the next section), and the ratio of specified minimum yield strengths F_{yo}/F_y drops out if chord and branch are of the same material. In LRFD, joint efficiency is the characteristic ultimate capacity of the tubular connection, as a fraction of the branch member yield capacity. In ASD, joint efficiency is the branch member nominal stress (as a fraction of tension allowable) at which the tubular connection reaches its allowable punching shear. Connections with 100% joint efficiency develop the full yield capacity of the attached branch member, in either design format.

29.8.2   Derating Factor

In most structures, the main member (chord) at tubular connections must do double duty, carrying loads of its own (axial stress f_a and bending f_b) in addition to the localized loadings (punching

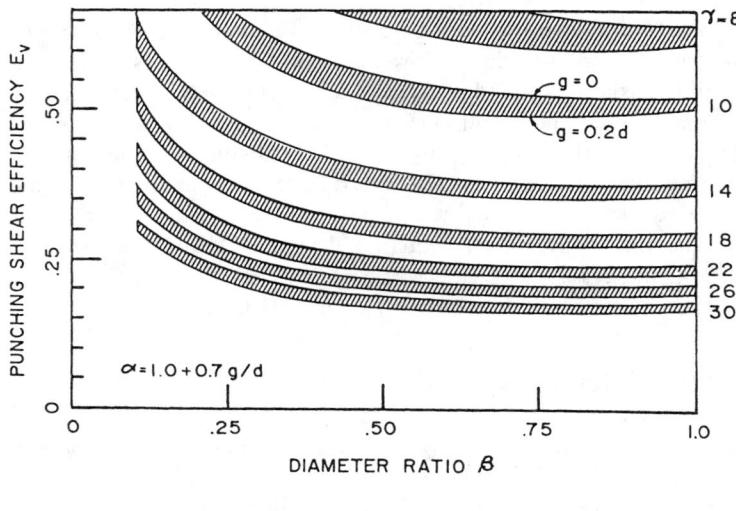

CIRCULAR K-JOINTS, SMALL GAP

$(g \leq 0.2d)$

Figure 29.5   Values of Q_q for axial load in K-connections.

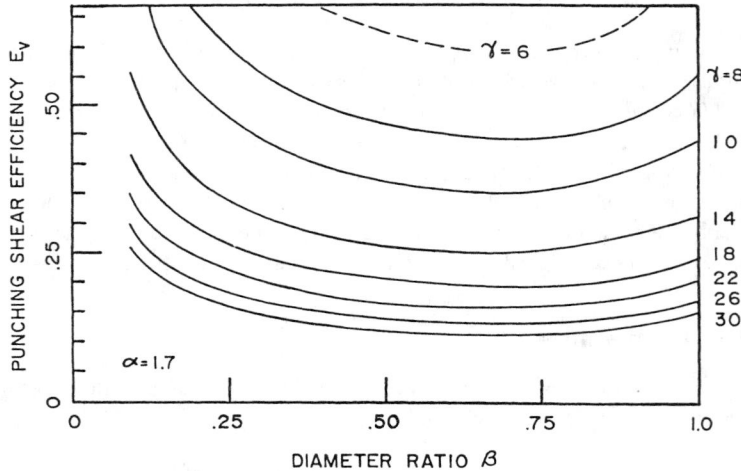

Figure 29.6   Values of Q_q for axial load in T- and Y-connections.

shear) imposed by the branch members. Interaction between these two causes a reduction in the punching shear capacity, as reflected in the Q_f derating factor, shown in Figure 29.10.

In-plane bending experiences the most severe interaction, as localized shell bending stresses at the tubular intersection are in the same direction and directly additive to the chord's own nominal stresses over a large part of the cross-section. For chords with very high R/T (gamma) and high nominal compressive stresses, buckling tendencies further reduce the capacity for localized shell stresses. Out-of-plane bending is less vulnerable to both these sources of interaction, as high shell stresses only occupy a localized part of the cross-section, and are transverse to P-Δ effects. Axially loaded connections of the types tested thus far exhibit intermediate behavior (although the gap region in K-connections might be expected to behave more like in-plane bending).

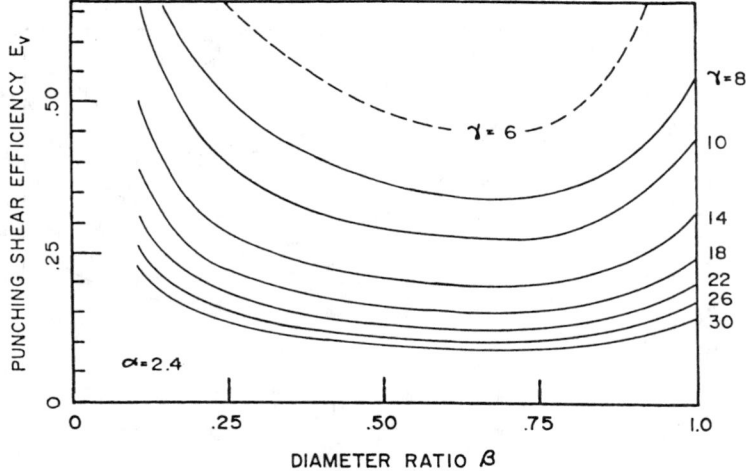

Figure 29.7 Values of Q_q for axial load in X-connections and other configurations subject to crushing.

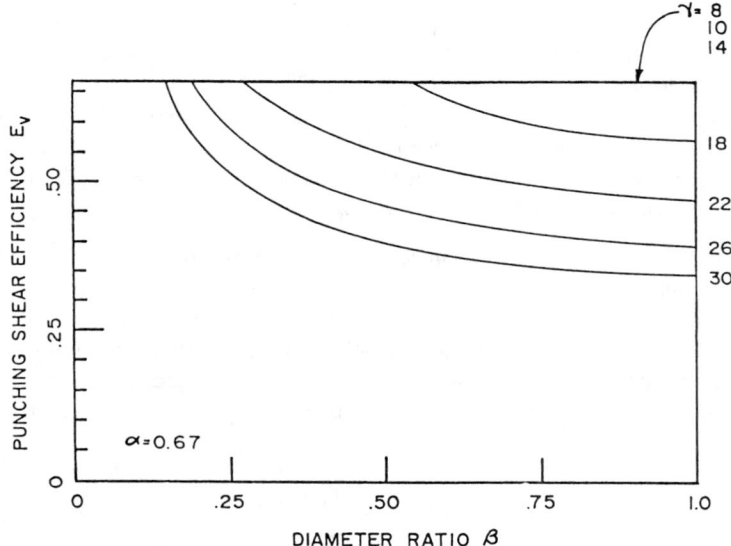

Figure 29.8 Values of Q_q for in-plane bending.

29.9 Application

What follows is a step-by-step design procedure for simple tubular trusses, applying the charts presented in the foregoing section.

Step 1. Lay out the truss and calculate member forces using statically determinate pin-end assumptions. Flexibility of the connections results in secondary bending moments being lower than given by typical rigid-joint computer frame analyses.

Step 2. Select members to carry these axial loads, using the appropriate governing design code, e.g., AISC. While doing this, consider the architecture of the connections along the following guidelines:

1. Keep compact members, especially low D/T for the main member (chord).
2. Keep branch/main thickness ratio (tau) less than unity, preferably about 0.5.

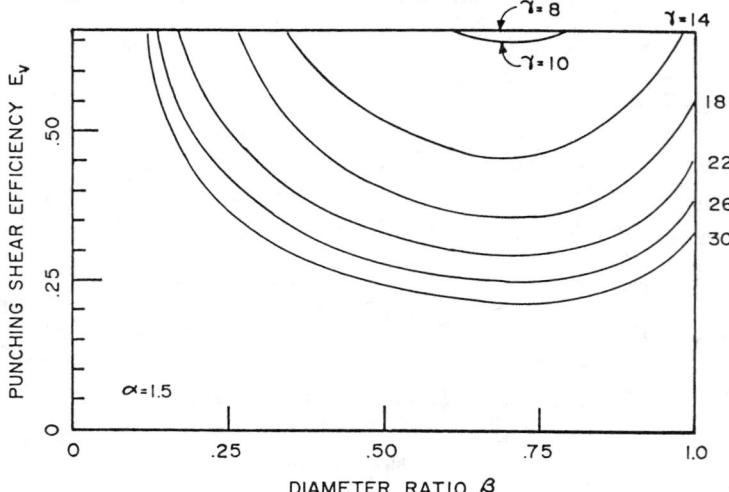

Figure 29.9 Values of Q_q for out-of-plane bending.

3. Select branch members to aim for large beta (branch/main diameter ratio), subject to avoidance of large eccentricity moments.

4. In K-connections, use a minimum gap of 2 in. between the braces for welding access. For small tubes, this may be reduced to 20% of the branch member diameter. Connection eccentricities up to 25% of the chord diameter may be used to accomplish this. Reconsider truss layout if this gets awkward.

Step 3. Calculate and distribute eccentricity moments and moments due to loads applied in-between panel points. These are not secondary moments, and must be provided for. They may be allocated entirely to the chord, for connection eccentricities less than 25% of the chord diameter, but should be distributed to both chord and braces for larger eccentricities, portal frames, or Vierendeel type trusses. Recheck members for these moments and resize as necessary.

Step 4. For each branch member, calculate A_y, utilization against member-end yield at the joint. For allowable stress design,

$$A_y = \frac{|f_a|}{0.6F_y} \text{ or } \frac{|f_b|}{0.6F_y} \tag{29.5}$$

where

f_a = nominal axial stress
f_b = bending in the branch

Where used, the 1/3 increase is applicable to the denominator.

Step 5. Also calculate chord utilization, using the formula in Figure 29.10 with chord nominal stresses and specified minimum yield strength. Use the appropriate chart in the figure to determine the derating factor Q_f. At heavily sheared gap K-connections and at eccentric bearing shoes, it may (rarely) also be necessary to check beam shear in the main member, and its interaction with other chord stresses, e.g., using AISC criteria. For circular sections, the effective area for beam shear is half the gross area.

Step 6. For each end of each branch member, calculate the joint efficiency E_j using Equation 29.4 and the appropriate charts for punching shear efficiency E_v. Joint efficiencies less than 0.5 are sometimes considered poor practice, rendering the structure vulnerable to incidental loads which the members could resist, but not the weaker joints.

Step 7. For axial loading alone, or bending alone, the connection is satisfactory if member-end utilization is less than joint efficiency, i.e., $A_y/E_j \leq 1.0$. For the general case, with combinations

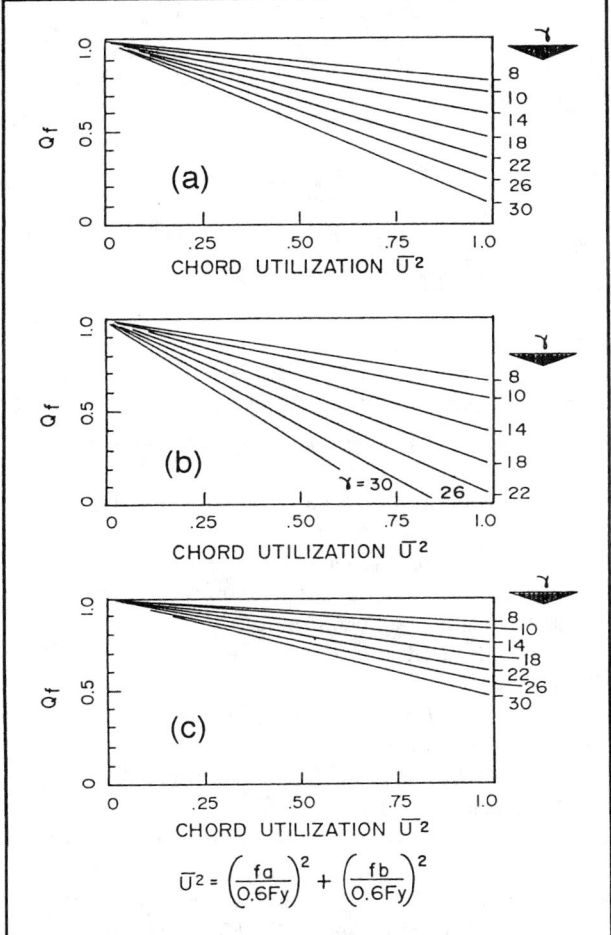

Figure 29.10 Derating factor Q_f for (a) axial loads in branch, (b) in-plane bending, and (c) out-of-plane bending.

of axial load and bending, the connection must satisfy the following interaction formula:

$$(A_y/E_j)^{1.75}_{\text{axial}} + (A_y/E_j)_{\text{bending}} \le 1.0 \tag{29.6}$$

Step 8. To redesign unsatisfactory connections, go back to Step 2 and

1. increase the chord thickness, or
2. increase the branch diameter, or
3. both of the above.

Consider overlapped connections (AWS section 2.40.1.6) or stiffened connections only as a last resort. Overlapped connections increase the complexity of fabrication, but can result in substantial reductions in the required chord wall thickness.

Step 9. When the designer thinks he is done, he should talk to potential fabricators and erectors. Their feedback could be valuable for avoiding unnecessary, difficult, and expensive construction headaches. Also make sure they are familiar with, and prepared to follow, AWS Code requirements for special welder qualifications, and that they are capable of coping the brace ends with sufficient accuracy to apply AWS prequalified procedures. Considerable savings can be realized by specifying partial joint penetration welds for tubular T-, Y-, and K-connections with no root access, where these are appropriate to service requirements. Fabrication and inspection practices for welded tubular connections have been addressed by Post [12].

29.10   Summary and Conclusions

This chapter has served as a brief introduction to the subject of designing welded tubular connections for circular hollow sections. More detail on the background and use of AWS D1.1 in this area can be found in [6].

References

[1] AWS D1.1-96. 1996. *Structural Welding Code — Steel,* American Welding Society, Miami, FL.

[2] Kurobane, Y. 1995. Comparison of AWS vs. International Criteria, ASCE Structures Congress, Atlanta, GA.

[3] Marshall, P.W. 1986. Design of Internally Stiffened Tubular Joints, *Proc. IIW/AIJ Intl. Conf. on Safety Criteria in the Design of Tubular Structures,* Tokyo.

[4] Marshall, P.W. 1989. Designing Tubular Connections, *Welding J.*

[5] Marshall, P.W. 1990. Advanced Fracture Control Procedures for Deepwater Offshore Towers, *Welding J.*

[6] Marshall, P.W. 1992. *Design of Welded Tubular Connections: Basis and Use of AWS D.1.1,* Elsevier Science Publishers, Amsterdam.

[7] Marshall, P.W. 1993. API Provisions for SCF, S-N and Size-Profile Effects, *Proc Offshore Tech. Conf.,* OTC 7155, Houston, TX.

[8] Marshall, P.W. 1996. Offshore Tubular Structures, *Proc. AWS Intl. Conf. on Tubular Structures,* Vancouver.

[9] Marshall, P.W. and Luyties, W.H. 1982. Allowable Stresses for Fatigue Design, *Proc. Intl. Conf. on Behavior of Off-Shore Structures,* BOSS-82 at MIT, McGraw-Hill, New York.

[10] Packer, J.A. et al. 1984. Canadian Implementation of CIDECT Monograph 6, IIW Doc. XV-E-84-072.

[11] Paul, J.C. 1992. The Ultimate Behavior of Multiplanar TT- and KK-Joints Made of Circular Hollow Sections, Ph.D. thesis, Kumamoto University, Japan.

[12] Post, J.W. 1996. Fabrication and Inspection Practices for Welded Tubular Connections, *Proc. AWS Intl. Conf. on Tubular Structures,* Vancouver.

[13] Wardenier, J. 1987. Design and Calculation of Predominantly Statically Loaded Joints Between Round (Circular) Hollow Sections, van Leeuwen Technical Information No. 7.

Index

3PARA.f computer program, **28**-11

AASHTO-LRFD approach, **17**-22
absorbers, dynamic vibration, **27**-17
abutments, concrete substructure, **10**-26
acceleration, earthquake, **5**-18
acceptable D/C ratios, **16**-50
acceptable force D/C ratios and limiting values, **16**-45
acceptable force D/C ratios DC_{accept}, **16**-49
acceptable force D/C ratios and limiting values, **16**-29
acceptance criteria, **16**-34
access holes, weld, **22**-25, **22**-44
ACI **318**-95 equations, **17**-7
ACI moment coefficients, **4**-38
active control, **27**-1, **27**-12
active mass damper system, **27**-20
ADAS damping elements, **27**-13
adjustment of
 design values, **9**-6
 member sizes, **28**-34
admixture, **4**-7
advanced analysis design format, **28**-28
advanced analysis load, **28**-29
advanced analysis method, examples, **28**-41
advantages of advanced analysis, **28**-4
aeroelastic model tests, **12**-69
aircraft wheel load, **25**-4
AISC-LRFD method, **17**-30
alignment chart method, **17**-3
allowable bending stress, **3**-50
allowable compressive stress for cones, **11**-14
allowable in-plane shear stress, **11**-15
allowable shear stress, **3**-51
allowable strength of weld to find weld size, **22**-10
allowable stress design (ASD), **3**-8, **7**-4, **18**-2, **28**-1
alloy characteristics, **8**-2
alloy selection, aluminum, **8**-20
aluminum
 adhesive bonded connections, **8**-19
 alloy selection, **8**-20
 buckling curves for alloys, **8**-4
 codes and specifications, **8**-3
 compared to steel, **8**-4
 definition of terms, **8**-30
 design adhesive joint problems, **8**-19
 design considerations, **8**-20
 design general considerations, **8**-20
 design studies, **8**-21
 determining components strength, **8**-6
 economics of design, **8**-29
 effects of joining, **8**-9
 effects of strain rate, **8**-6
 effects of temperature, **8**-6
 effects of welding, **8**-5
 equations, **8**-4
 fatigue behavior, **8**-19

 lateral buckling, **8**-10
 material background, **8**-1
 mechanical connections, **8**-17
 members under combined bending and axial
  loads, **8**-10
 metal working, **8**-21
 nominal shape factors, **8**-9
 physical properties, **8**-2
 product selection, **8**-20
 safety and resistance factors, **8**-4
 stiffeners for flat plates, **8**-14
 structural behavior, **8**-4
 structures design, **8**-4
 structures finishing, **8**-21
 thin elements of columns and beams, **8**-12
 thin roofing and siding products, **8**-18
 web crushing, **8**-13
 welded connections, **8**-18
 welded structures, **24**-8
amplitude loading, variable, **24**-16
analysis, **28**-32
 and design, frame, **23**-2
 and design of frames, **12**-25
 and design of unbraced frames, **12**-36
 and design procedures, **28**-28
 buckling discrete, **13**-38
 charts for gable frames, **2**-136
 ductility, inelastic, **2**-119
 dynamic response spectrum, **16**-14
 dynamic time history, **16**-14
 elastic, **6**-12, **6**-28, **23**-13
 elastic buckling, **2**-143
 elastic flexural, **4**-12
 first-order elastic, **12**-37
 first-order hinge-by-hinge, **2**-142
 Fourier, **2**-170
 fracture mechanics, **24**-20
 input data for the computer, **18**-12
 linear, **6**-38
 linear buckling, **1**-12
 load, advanced, **28**-29
 load application in advanced, **28**-31
 load combination in advanced, **28**-29
 method, **16**-14
 method of, **27**-3
 modal failure, **26**-29
 non-linear, **6**-37
 non-prismatic beam, **18**-10
 numerical, **2**-175
 of an isolated sub-frame, **12**-37
 of beam with tension compression reinforcement, **4**-11
 of frames, **4**-37
 of moments, approximate, **26**-13
 of prismatic composite beams, **18**-11
 of rectangular beams with tension reinforcement only, **4**-9
 of semi-rigid frames, **28**-10
 plastic, **6**-15, **6**-36, **23**-14

plastic hinge, 2-162
reliability, 26-9
resistance factors in advanced, 28-30
second-order elastic, 2-143, 2-159
second-order inelastic, 2-144
section application in advanced, 28-30
static linear, 16-14
structural, 2-1, 28-1
vierendeel, 18-8
analytical equation, 16-44
analytical model, 2-1
anchor bolts, 3-90
anchor bolts (rods) and anchorage assemblies, 16-23
anchors and hoops in solid slabs, 6-54
anchors, transmission structure, 15-25
angle connectors, 6-53
angle, rupture of the seat, 23-19
angles, web, 23-19
application of friction dampers, 27-15
application of slope deflection method to frames, 2-28
approximate analysis, 4-38
approximate analysis of moments, 26-13
arc, weld, 7-43
arch bridge, 10-47
    Langer bridge, 10-48
    Lohse bridge, 10-48
    structural analysis, 10-47
    structural features, 10-47
    trussed and Nielsen, 10-49
    type of arch, 10-47
arches, wood, 9-35
architecture, 29-1
area, criterion, 3-51
area method, moment, 2-15
area, stiffener, 3-55
ASD, 3-8
    method, 7-38
aseismic design, 27-9
AWS code, 29-3, 29-13
axial compression,
    uniform, 11-14
    wood, 9-19
axial force, 2-5
axially loaded
    column, 28-18
    connection, 29-10
    member, 2-86
axisymmetric circular plate, 1-10

background, aluminum material, 8-1
background of criteria development, 16-10
backing,
    weld, 22-22, 22-41
    weld ceramic, 22-42
    weld copper, 22-23
    weld fusible, 22-23
    weld removable, 22-23
bars, slab, 23-19
base isolation, 27-1
base metal properties, 22-38

base plate
    axially loaded, 3-85
    column, 3-84
    with moments, 3-87
    with shear, 3-89
basement walls, 4-62
basic axioms of probability, 26-3
basic concepts of stability, 1-11
basic modeling, 27-23
basic probability concepts, 26-3
basic $R - S$ problem, 26-18
basic reliability concepts, 26-9
basic reliability problem, 26-15
basic systems, 26-27
batens, column with, 17-37
beam, 2-1
    and column splices, 3-82
    bearing plates, 3-91
    bridge, strutted, 10-46
    column stability, 2-149
    column subjected to end moments, 2-153
    column subjected to transverse loading, 2-152
    columns, plastic design of, 3-99
    composite, 3-94
    connection, stiffness, 23-15
    continuous, 2-11, 2-74, 6-25
    curved, 9-34
    deflection, 2-14
    deflections of, 23-14
    design of wood, 9-8
    element, finite element method, 2-111
    fix-ended, 2-7
    flexural, 27-12
    prestressed, 24-13
    response, 6-9
    simply supported, 6-9
    torsional, 27-12
    webs, shear strength, 7-32
    yield-to-tensile ratio, 22-39
beam-column (bent cap) joints, 16-32
bearing, bridge, 10-31
bearing capacity, wood, 9-13
bearing plates, beam, 3-91
bearing stiffeners, 3-54
bearing stiffeners, load, 19-17
behavior classification, connection, 23-2
behavior, fracture, 24-18
behavior of structural steel, 3-2
behavior of the members, 6-75
behavior, steel structures, 28-4
bending
    and axial tension, 9-19
    axial compression due to, 11-14
    biaxial, 3-47
    capacity of column at base, 4-51
    load flange, 3-30
    local flange, 3-37
    moment, 2-5, 2-71
    moment, axial compression due to, 11-11
    moment diagram, 2-6
    of circular plates, 2-50
    of simply supported rectangular plates, 2-43
    of thin plates, 2-39
    round tubes and curved panels, 8-16

strain energy in, 2-76
stress, allowable, 3-50
wood biaxial, 9-19
wood combined, 9-19
between nodes, loading, 2-95
biaxial bending, 3-47
    allowable stress design, 3-47
    load and resistance factor design, 3-48
    wood, 9-19
block connectors, 6-54
block node (type BK), 13-46
body waves, 5-3
bolt, 3-6, 23-19
    anchor, 3-90
    and holes, 16-24
    holes, 3-59, 23-19
    loaded in combined tension and shear, 3-61
    loaded in shear, 3-60
    loaded in tension, 3-60
    stress range, 24-10
    wood, 9-26
bolted
    bracket type connections, 3-67
    connection, 3-59, 7-43, 15-14
    connectors, 17-40
    hanger type connections, 3-65
    moment resistant connections, 3-69
    shear connections, 3-69
bolting, steel bridge, 10-11
boundary conditions, 2-2, 16-15
boundary conditions, plate, 2-42
bounds, simple first-order, 26-28
Boussinesq solution, 25-3
bowl node (type NK), 13-44
braced dampers, viscoelastic, 27-10
braced frame, 3-49, 17-6, 23-12, 23-15
braced, frame, 28-13, 28-15
braces, chevron, 27-13
bracing size requirement for braced frame, 17-7
bracing system, crossing, 17-33
brackets and corbels, 4-43
Bresler load countour method, 4-18
Bresler reciprocal method, 4-18
bridge
    arch, 10-47, 20-2, 20-8
    bearing, 10-31
    cable-stayed, 10-49, 20-2, 20-8, 20-17
    cambering of individual members, 20-7
    classification
        by materials, 10-3
        by objectives, 10-3
        by structural system, 10-3
        by support condition, 10-6
    components, 5-70
    concrete, 10-13
    concrete substructure, 10-25
    construction, 2-26
        example, 20-13, 20-17
        illustrations, 20-28
        philosophy, 20-9
        procedure, 20-9
    conventional, 5-70
    critical erection stages, 20-19
    deck, orthotropic, 8-26

design, 10-7
design analysis, 10-47
design, loads in, 10-7
erection methods, 20-28
erection rating factor, 20-21
erection strength adequacy, 20-20
erection studies, 20-6
example of arch bridge construction, 20-8
expansion joints, 10-33
fabrication and erection, 20-7
fabrication studies, 20-6
failure, 20-2
falsework, 20-4
fatigue cracking in, 24-2
field adjustment procedure, 20-7
field checking, 20-19
flexible, 10-58
flexible arch ribs, 20-9
floor system, 10-28
future of construction practices, 20-26
general approaches for major, 20-7
geometric outline, 20-7
girder, 10-34
girder bridge design example, 10-61, 10-77
guard railings, 10-34
influence lines, 10-8
Niagara Falls, 20-8
permanent steelwork, 20-4
plan factors, 10-6
proper construction procedure, 20-18
Rahmen bridges, 10-45
Rainbow arch, 20-9
rigid frame bridges, 10-45
shop and field liaison, 20-25
shop control procedure, 20-7, 20-13
specification requirements problems, 20-17
stay cables, 24-12
steel, 10-8
steel, construction, 20-1
structure, 10-1
strutted beam, 10-46
suspension, 10-53, 20-2, 20-8
truss, 10-42
types of bearings, 10-31
under construction, 20-28
brittle fracture, 24-18
brittle fractures in jumbo sections, 24-22
brittle materials, 24-22
buckling
    aluminum lateral, 8-10
    analysis,
        elastic, 2-143
        linear, 1-12
        behavior methods of, 13-38
    column, 11-11
    continuum analogy analysis, 13-38
    discrete analysis, 13-38
    flexural, 3-17, 7-36
    lateral, 1-23
    lateral torsional, 3-27, 6-38
    local, 3-55, 6-27, 11-10
    local stiffener, 11-13
    of buried corrugated pipes, 25-10
    of plates, 19-1

of shell, 2-70
of thin plate, 2-58
overall column, 7-36
sidesway web, 3-31, 3-38
torsional, 3-17, 7-36
torsional-flexural, 1-21
buckling mode interaction factor, 16-43
buckling mode interaction for compression built-
    up-members, 16-43
buildings,
    earthquake concrete, 5-66
    earthquake masonry, 5-68
    metal, 7-47
built-up compression members, 3-22
bundled bars, 4-26, 4-27
buoyancy and hydrodynamic mass, 16-12
buried penny-shape crack, 24-25

cable stayed bridge
    structural analysis, 10-50
    structural features, 10-49
    tension in cable, 10-52
    type of, 10-50
cable tension general principles, 10-52
cables, bridge stay, 24-12
cables, types of, 10-54
CADPRO program, 15-29
calibration and selection of target reliabilities, 26-31
calibration,
    code, 26-31
    need for, 26-31
    process of, 26-32
cantilever column, 1-15
capacity design philosophy, 23-8
capacity,
    inelastic rotation, 28-33
    wood bearing, 9-13
    wood moment, 9-9
    wood shear, 9-11
carbon steel, 3-3
carry-over factor, 2-32
Castigliano's theorem, 2-79
categorization of analysis method, 2-142
causes of earthquakes and faulting, 5-3
center crack, 24-23
central difference method, 2-178
ceramic backing, weld, 22-42
challenges in construction engineering, 20-4
channel connectors, 6-53
characteristics of tubular connections, 29-2
Charpy impact testing method, 29-6
chevron braces, 27-13
choice of displacement function, 2-101
circular hollow section (CHS), 29-1
circular plate, 1-7, 2-60
circumferential, stiffener, 14-27
circumferential stresses, 2-18
civil engineering structure, 27-16
CJP, welded connections, 22-4
clamped edge, plate, 2-42
Clapeyron's theorem, 2-11
classical system reliability theory, 26-27, 26-29

classification,
    connection behavior, 23-2
    multistory frames, 12-1
    cross-sections, 6-27
    of structural details for fatigue, 24-9
    of tall building frames, 12-6
coatings, underground pipe, 25-19
code calibration, 26-31
code development of active control systems, 27-22
code, seismic design, 5-57
code, selected seismic, 5-57
coefficient, ELF seismic, 5-57
coefficient of variation (COV), 26-17
cold-formed
    steel design, 7-4
    steel member, 7-1
    steel members, effective width, 7-12
    steel members, types of, 7-2
    steel sections, 7-2
    steel sections, properties of, 7-9
    steel tension members, 7-20
collapse load factor, 23-25
collapse mechanism, 2-130
collapse mechanism, plastic, 2-118, 23-13
collapse of flexible pipes, 25-6
column, 4-40
    and beams, plastic design of, 3-98
    axially loaded, 28-18
    base fixity, partial, 17-22
    base plate, 3-84
    buckling, 11-11
    buckling, overall, 7-36
    cantilever, 1-15
    composite, 3-93, 6-69
    crane, 17-42
    design, wood, 9-15
    effective length factor, 2-149
    end conditions, different restraining, 17-11
    framed, 17-3
    imperfect, 1-20
    in gable frames, 17-47
    isolated, 17-3
    laced-battened, 17-39
    leaning, 17-33
    member laced, 17-34
    restrained by tapered rectangular girders,
        17-15
    rigid, 17-28
    simply supported, 1-13
    stability, 2-145
    stiffeners, 23-19
    tapered, 17-42
    unbraced frames with leaning, 17-28
    under axial load and biaxial bending, 4-17
    under bending and axial load, 4-14
    under flexural-torsional buckling, 8-8
    with batens, 17-37
    with one end fixed and one end free, 2-148
    with perforated cover plates, 17-39
    with pinned ends, 2-146
    wood solid, 9-16
    wood spaced, 9-18
combination, structural load, 26-35

combined
    axial load and bending, 7-40
    bending and shear, 7-33
    bending and web crippling, 7-34
    bending, torsion and axial force, 3-48
    combined bending, wood, 9-19
    compressive axial load and bending, 7-40
    footings, 4-54
    shear and flexure, 2-172
    shear, moment, and axial load, 4-20
    shear, moment, and torsion, 4-20
    tensile axial load and bending, 7-40
    torsion and axial load, 4-21
combined flexure and axial force, 3-44
    allowable stress design, 3-44
    load and resistance factor design, 3-45
compact section, 3-27, **28**-33
compact section member, 3-30
compaction piles, 5-53
comparison of results, **28**-45, **28**-48, **28**-53
complete element, 2-101
complicated limit state function, **26**-17
components, bridge, 5-70
composite
    beam-columns, 3-95
    beams, 3-94, 6-8, 6-59
    column, 3-93,6-69
    connection, **23**-1, **23**-9
        design, **23**-12, **23**-20
        PR, **23**-5
        weak, **23**-5
    construction, 6-2, 7-52
    decking, 6-8
    floor slabs, 3-95
    floor systems, **12**-8
    members, 3-92
    slab, 6-91, 6-93
composite systems
    braced composite frames, **12**-44
    steel-concrete, **12**-43
    unbraced composite frames, **12**-45
compound trusses, 2-26
compression buckling of the web, 3-31, 3-38
compression flange of a box girder, **19**-4
compression lap splices, 4-26
compression members, 3-17
    allowable stress design, 3-19
    built-up, 3-22
    effective length of, **16**-16
    load and resistance factor design, 3-19
compression, uniform axial, **11**-14
compression, wood axial, 9-19
computer program
    3PARA.f, **28**-11
    PAAP, **28**-34
    SAP, **13**-31
    SAP-90, **15**-23
    STRUDL, **13**-31
    using, **28**-10
computer solution for the girder, **18**-10
computing the reliability index, **26**-24
concentrated load criteria, 3-37
concentrically loaded compression members, 7-34
concept of a safety margin, **26**-16

concept of equivalent lateral force (ELF), 5-56
concept of statistical independence, **26**-9
concepts,
    basic probability, **26**-3
    basic reliability, **26**-9
    statistics, **26**-3
concrete, 6-6
concrete bridge, **10**-13
    box girder, **10**-17
    cast-in-place box girder, **10**-14
    deck girder, **10**-14
    design consideration, **10**-15, **10**-18
    precast I girder, **10**-16
    prestressed, **10**-16
    reinforced, **10**-14
    segmental bridge, **10**-17
    slab, **10**-14, **10**-16
concrete buildings, earthquake, 5-66
concrete pipe, prestressed, **25**-14
concrete pipe, reinforced, **25**-16
concrete properties for cooling towers, **14**-31
concrete, reinforced, **14**-19
concrete structures, **16**-32
concrete substructure
    abutments, **10**-26
    bents and piers, **10**-25
    bridge, **10**-25
    design consideration, **10**-27
conditions,
    boundary, 2-2
    moving load, 2-71
    plate boundary, 2-42
conditions for global ductility, **22**-32
cone-cylinder junction rings, **11**-14
cone, failure, 3-91
conical shell, 2-64
connection, 3-58
    aluminum mechanical, 8-17
    aluminum welded, 8-18
    and floor vibrations, **23**-15
    and joints, 7-42
    axially loaded, **29**-10
    beam and column splices, 3-82
    behavior classification, **23**-2
    bolted, 3-59, 7-43, **15**-14
    bolted bracket type, 3-67
    bolted hanger type, 3-65
    bolted moment resistant, 3-69
    bolted shear, 3-69
    characteristics of tubular, **29**-2
    composite, **23**-1, **23**-9
    design, composite, **23**-12, **23**-20
    design of moment resistant, 3-69
    flange rib, **22**-60
    for tension members, welded, 3-77
    full strength, **23**-4
    lateral resistance, **23**-24
    minor modifications to the SMRF, **22**-59
    modeling, practical, **28**-10
    nomenclature of tubular, **29**-2
    of bottom chord to supports, **18**-20
    partial strength, **23**-4
    partially restrained, **22**-60
    PR composite, **23**-5

PR/FS, **23**-9
reduced beam section, **22**-60
rigid, **3**-69
screw, **7**-43
semi-rigid end, **2**-96
shear, **6**-44
shop welded-field bolted, **3**-82
simple shear, **12**-27
splices, **16**-22
stiffness, **23**-4, **23**-18
stiffness, beam, **23**-15
to slab and bottom chord, **18**-20
top and bottom haunch, **22**-60
transmission structure, **15**-14
type of, **3**-58
weak composite, **23**-5
weak PR, **23**-6
welded, **3**-73, **3**-78, **7**-42, **22**-1, **24**-26
welded bracket type, **3**-78
welded moment resistant, **3**-80
welded shear, **3**-80
welded tubular, **29**-1
connectors, **6**-53
connectors,
bolted, **17**-40
stud shear, **18**-4
welded, **17**-40
consistent deformations, **2**-36
constant amplitude fatigue limits (CAFL), **24**-7,
24-17
construction,
bridge, **2**-26
composite, **7**-52
construction engineering
and design engineering, **20**-2
can be critical, **20**-2
challenges, **20**-4
effort, **20**-5
experience, **20**-26
objectives, **20**-3
practices, **20**-6
cost of, **20**-6
inadequate, **20**-5
obstacles to effective, **20**-5
construction loads on cooling towers, **14**-12
construction of response spectra, **5**-40
construction, residential, **7**-52
construction,
steel bridge, **20**-1
weld seismic, **22**-43
continuous beams, **2**-11, **6**-25
continuous composite slabs, **6**-101
continuous floor system, **23**-15
continuous structure, **23**-17
contract documents, **20**-2
control, active, **27**-1
control of bedding and backfill, **25**-8
control of cracking, **6**-35
convention, sign, **2**-17, **2**-31
conventional bridges, **5**-70
cooling tower
circumferential stiffeners, **14**-27
components of a, **14**-2
concrete strength, **14**-30

construction loads on, **14**-12
cracking load, **14**-21
damage and failures of, **14**-4
dead load, **14**-8
design and detailing of components, **14**-24
during construction, **14**-29
earthquake loading on, **14**-11
finite element method, **14**-13
foundations, **14**-27
geometry of, **14**-6
hyperbolic, **14**-18
internal conditions, **14**-10
load conditions of, **14**-7
method of analysis, **14**-12
modeling, **14**-13
seismic analysis, **14**-23
settlement, **14**-12
shell wall thickness, **14**-24
spacing of, **14**-11
structure, **14**-1
supporting columns, **14**-26
temperature variations on, **14**-11
wall thickness, **14**-29
wind load, **14**-8
cope holes, **24**-11
copper backing, weld, **22**-23
correct collapse mechanism, **2**-124
correct limit load, **2**-124
corroded member, severely, **24**-9
corrosion cracking, stress, **24**-2
corrosion protection of steel, **3**-5
corrosion resistant high strength low alloy steel, **3**-3
corrugated web, **8**-15
cost of construction engineering, **20**-6
coverplated, design, **22**-59
crack, center, **24**-23
crack, edge, **24**-24
cracking of concrete, **6**-103
cracking, stress corrosion, **24**-2
crane column, **17**-42
CRC tangent modulus, **28**-8
crippling, web, **3**-31, **3**-37, **7**-33
criteria
acceptance, **16**-34
concentrated load, **3**-37
development, background of, **16**-10
for concentrated load, **3**-30
performance acceptance, **16**-26
structural design, **3**-7
transverse stiffeners, **3**-51
criterion
area, **3**-51
deflection, **3**-31, **3**-39
flexural strength, **3**-33, **3**-52
moment of inertial, **3**-51
shear strength, **3**-30, **3**-36, **3**-53
shear transfer, **3**-52
critical damping, **5**-2
critical damping ratio, **2**-167
critical erection stages, bridge, **20**-19
cross-section plastic strength, **28**-6
crossing bracing system, **17**-33
cruciform joint, **24**-11
crushing, aluminum web, **8**-13

crushing, web, **8**-13
CTOD test, **29**-6
curve, trilinear, **23**-11
curved beams and arches, **9**-34
curved flexural members, **2**-17
curves, moment rotation, **23**-1, **23**-9, **23**-11
cylinder node (type ZK), **13**-46
cylinders and cones under combined loads, **11**-16
cylindrical, shell structure, **11**-10
cylindrical tubular member, 7-42

damageability limit state, **27**-9
damped free vibration, **2**-167
dampers,
    friction, **27**-14
    metallic yield, **27**-12
    tuned liquid, **27**-19
    tuned mass, **27**-17
    VE, **27**-15
    viscoelastic, **27**-15
    viscoelastic braced, **27**-10
    viscous fluid, **27**-16
damping, **2**-175
    critical, **5**-2
    modal, **2**-175
dead load analysis method, **14**-15
dead load moments, **23**-17
deck ribs oriented parallel to steel beams, **6**-51
deck ribs oriented perpendicular to steel beam,
    6-51
deficiencies of construction procedures, **20**-24
deflection, **6**-34, **6**-63, **6**-103, **23**-20
deflection, beam, **2**-14
deflection checks of the stub girder, **18**-21
deflection criterion, **3**-31, **3**-39
deflection, design estimating, **25**-7
deflection equation, **25**-7
deflection equations, slope, **2**-154
deflection method, slope, **2**-28
deflections of beams, **23**-14
deflections, reducing floor, **23**-18
deflections, wood, **9**-37
derating factor, **29**-9
description of Monte Carlo simulation, **26**-25
design
    advantage of advanced analysis, **28**-4
    and impact loading, **24**-12
    aseismic, **27**-9
    based on the S-N curve, **24**-7
    bases, **7**-4
    basis earthquake, **27**-23
    braced or unbraced frame, **17**-6
    bridge, **10**-7
    built-up highway girder, **8**-23
    by advanced analysis, **28**-43, **28**-46, **28**-49
    charts, **29**-9
    checks for shear transfer regions, **18**-15
    checks for the bottom chord, **18**-14
    checks for the concrete slab, **18**-14
    code, seismic, **5**-57
    codes, **6**-4
    composite connection, **23**-12, **23**-20
    considerations, aluminum, **8**-20

considerations for footings, **4**-46
considerations for wood, **9**-5
considerations of bracing frames, **12**-31
coverplated, **22**-59
criteria for bridges, seismic, **16**-5
criteria for stub girders, **18**-13
criteria, no-collapse-based, **16**-6
criteria, performance-based, **16**-8
criteria, structural, **3**-7
earthquake resistant, **27**-11
energy-based, **27**-11
engineering, **20**-2
equivalent static method, **5**-56
estimating deflection, **25**-7
examples, **28**-41
examples of welded connections, **22**-27
fastener and connection, **9**-22
flexural members, **3**-27
for fracture resistance, **24**-6
for seismic loading, **4**-40
for unbraced frames, **23**-20
formats, steel, **3**-8
frame analysis and, **23**-2
latticed tower or space frame, **8**-28
lighting standard, **8**-21
loads and load combinations, **26**-32
of beams, **4**-20, **4**-21
of cold-formed steel structure, **7**-4
of colded-formed structural members, **7**-19
of column web stiffeners, **3**-72
of corrugated steel pipes, **25**-9
of deep beams, **4**-22
of flexural members, **7**-20
of guys, **15**-21
of stirrup reinforcement, **4**-21
of structures, earthquake effects and, **5**-61
of stubs for shear and axial load, **18**-15
of stud shear connectors, **18**-16
of the stub girder, **18**-13
of underground pipe, **25**-5
of welds between stub and bottom chord,
    18-18
orthotropic bridge deck, **8**-26
overhead sign truss, **8**-22
(PD), plastic, **3**-8, **28**-1
philosophy, **3**-7
philosophy, capacity, **23**-8
plastic, **3**-98
procedure for concrete pipe, **25**-14
procedures, **4**-29
procedures for plate girders, **10**-36
recommendations for stay cable, **24**-13
reliability-based, **26**-1
reliability-based (codes), **26**-31
resources for arche, **9**-37
resources for curved beam, **9**-37
roofing or siding for a building, **8**-25
seismic, **5**-58, **24**-6
ship hull, **8**-27
stresses, **20**-21
structural, **2**-1
structural engineering, **26**-31
studies, aluminum, **8**-21
tension member, **9**-14

tubular trusses, **29**-11
values, wood, **9**-6
web stiffeners for shear, **19**-10
wood beam, **9**-8
wood column, **9**-15
design earthquake
    code approach, **5**-35
    probabilistic seismic hazard analysis approach,
        5-36
    upper-bound approach, **5**-36
design load
    construction loads, **13**-31
    dead load, **13**-25
    live load, snow or rain, **13**-26
    temperature effect, **13**-29
    wind load, **13**-27
details, welded joint, **22**-22
details, welded web, **22**-43
determinacy, stability and, **2**-27
determinate externally, statically, **2**-4
determination of capacities, **16**-16
determination of demands, **16**-14
determination of SRMDs properties, **16**-26
determining throat size for bending or torsional
    loads, **22**-9
determining throat size for compressive loads, **22**-8
determining throat size for tension or shear loads,
    22-8
development of bars in compression, **4**-25
development length of reinforcing bars, **4**-52
development of bars in tension, **4**-25
development of hooks in tension, **4**-26
development of reinforcement, **4**-25
diagram, bending moment, **2**-6
diagram, moment, **23**-20
diagram, shear, **2**-6
diaphragms, shear, **7**-48
difference method, central, **2**-178
different restraining column end conditions, **17**-11
different restraining girder end conditions, **17**-10
differential settlement, **4**-52
direct design method, **4**-32
direct design of steel frames, **28**-5
disc node (type TK), **13**-43
discrete analysis, buckling, **13**-38
dissipation, passive energy, **27**-1
distance, minimum edge, **3**-64
distortion and multiaxial loading effects in fatigue,
    24-15
distributed mass systems, **2**-171
distribution,
    lognormal, **26**-6
    moment, **2**-32
    normal, **26**-6
distribution method, moment, **2**-31
distribution of seismicity, **5**-5
dome,
    spherical, **2**-63
    geodesic, **13**-21
    Kiewitt, **13**-21
    lamella, **13**-21
    Ribbed, **13**-21
    Schwedler, **13**-21
    three-way grid, **13**-21

double layer grids, **13**-33
    basic elements, **13**-6
    cambering and slope, **13**-15
    composed of latticed trusses, **13**-7
    composed of square pyramids, **13**-7
    composed of triangular pyramids, **13**-8
    depth and module, **13**-13
    design parameters, **13**-13
    free in one side, **13**-12
    method of support, **13**-12
    multi-column supports, **13**-12
    optimum design, **13**-14
    support along perimeters, **13**-12
    type choosing, **13**-8
    types and geometry, **13**-6
    types of, **13**-7
    types of cambering, **13**-15
doubly curved thin shells, **14**-12
dowels on footings, **4**-51
dowels, wood, **9**-26
drift assessment of a frame, **12**-30
Duan-King-Chen equations, **17**-8
ductibility, steel, **7**-10
ductile fracture, **24**-18
ductile tearing failure, **24**-20
ductility and load-deformation curves, **16**-47
ductility, inelastic analysis, **2**-119
ductility requirements, **28**-33
dynamic behavior, prototype for, **27**-3
dynamic response spectrum analysis, **16**-14
dynamic, structural, **2**-165
dynamic time history analysis, **16**-14
dynamic vibration absorbers, **27**-17

earthquake, **5**-1
    acceleration, **5**-18
    acceleration and duration, **5**-47
    and cyclic plastic deformation, **24**-17
    and faulting, causes of, **5**-3
    body wave magnitudes, **5**-12
    bridge behavior during an, **5**-70
    closest horizontal distance, **5**-25
    code history and development, **5**-55
    concrete buildings, **5**-66
    concrete buildings typical problems, **5**-67
    damages, **16**-5
    damages in substructures, **10**-27
    demand for ductility, **22**-38
    demands of structural systems, **22**-36
    design basis, **27**-23
    effects and design of structures, **5**-61
    elastic response spectra, **5**-19
    engineering intensity scale (EIS), **5**-24
    epicentral distance, **5**-25
    estimation of forces, **5**-63
    fault types, **5**-3
    faults, **5**-1, **5**-3
    FRISK code, **5**-38
    functionality evaluation, **16**-10
    Great Kanto, **5**-9
    ground motion, **5**-2, **5**-24
    ground motion and damage, **5**-42
    house over garage, **5**-65

hypocentral distance, **5**-25
inelastic response spectra, **5**-21
intensity, **5**-24
large, **5**-32
largest in North America, **5**-3
lateral loading, **5**-63
level of damage, fragility, **5**-2
load, **2**-2, **16**-12
loading, **24**-17
loading on cooling towers, **14**-11
magnitude, **5**-11, **5**-37
magnitude on the basis geological evidence,
        5-38
masonry buildings, **5**-68
measurement of, **5**-11
modified mercalli intensity (MMI), **5**-13
most hazardous structure, **5**-68
non building structures, **5**-70
Northridge, **22**-58
Pacific plate, **5**-3
peak ground acceleration (PGA), **5**-18
peak ground displacement (PGD), **5**-18
peak ground velocity (PGV), **5**-18
performance of buildings, **5**-61
post-Northridge assessment, **22**-58
pounding, **5**-63
PSHA, **5**-36
resistance of forces, **5**-61
resistant design, **27**-11
resistiveness, **5**-62
response spectra, **5**-21
response spectrum intensity, **5**-23
safety evaluation, **16**-10
seismic design codes, **5**-54
seismic hazard and design, **5**-35
seismic moment, **5**-12
seismicity on the basis geological evidence,
        5-38
sophisticated models of, occurrence, **5**-39
space frame resistance, **13**-33
special moment resisting frame, **22**-58
steel frame buildings, **5**-65
strong motion duration, **5**-34
tectonic related, **5**-1
types of building and performance, **5**-64
wood frame, **5**-64
eccentric brace frame (EBF), **5**-65
eccentric edge loading, **19**-16
edge crack, **24**-24
edge distance, minimum, **3**-64
edge loading, eccentric, **19**-16
edge, plate clamped, **2**-42
edge, plate free, **2**-43
effect of soils on ground motion, **5**-42
effective area of welds, **3**-73
effective length, **1**-18
effective length factor, **18**-16
effective length factors of columns, **17**-1
effective length method difficulties, **28**-3
effective length of compression members, **16**-16
effective stress range, **24**-16
effective width, **6**-26
effective width of concrete flange, **6**-11

effective width of uniformly compressed elements
        with stiffeners, **7**-18
effects of axial forces, **17**-17
effects of welding, **8**-5
effects, plastic hinge, **2**-160
eigenvectors, **2**-171
elastic analysis, **6**-12, **23**-13
elastic analysis of the cross-section, **6**-28
elastic analysis, second-order, **2**-143, **2**-159
elastic behavior of the section, **6**-71
elastic buckling analysis, **2**-143
elastic flexural analysis, **4**-12
elastic K-factor, **17**-2
elasticity, modulus of, **7**-9
elasticity, theory of, **25**-3
element
        complete, **2**-101
        isoparametric, **2**-103
        section, slender, **3**-27
        shape function, **2**-104
        shapes and discretization, **2**-100
        stiffness matrix, **2**-86
        strength, steel, **7**-11
        three-dimensional frame, **2**-163
ELF seismic coefficient, **5**-57
empirical formulations, **29**-8
end-bearing splices, **4**-26
end-restrained column, **17**-3
energy-based design, **27**-11
energy dissipation, passive, **27**-1
energy formulations, **27**-8
energy, kinetic, **27**-2
energy methods in structural analysis, **2**-75
energy relations in structural analysis, **2**-77
energy, seismic input, **27**-11
engineering, design, **20**-2
engineering design, structural, **26**-31
epicentral distance, earthquake, **5**-25
equation, deflection, **25**-7
equation of motion, **2**-165
equations, slope deflection, **2**-154
equilateral triangular plate, **2**-55
equilibrium, **2**-122
equilibrium, state of, **1**-11, **2**-2
equivalent frame method, **4**-32
erection feasibility, **20**-4
erection of steelwork superstructures, **20**-4
erection of the bridge steelwork, **20**-4
erection rating factor, **20**-21
erection rating factors advantages, **20**-24
erection strength adequacy, bridge, **20**-20
estimating deflection, design, **25**-7
Euler load, **2**-59
evaluation of load and resistance factors, **26**-35
evaluation of structural details for fracture, **24**-18
exact estimate of the reliability index, **26**-23
exact plastic analysis solution, **2**-122
example of combined system, **26**-30
examples of advanced analysis method, **28**-41
examples of inadequate construction engineering,
        20-5
examples of limit state functions, **26**-18
examples of probability theory, **26**-9
existing materials, **16**-13

expansion joints, bridge, **10**-33
explicit imperfection modeling method, **28**-13
externally, statically determinate, **2**-4
externally, statically indeterminate, **2**-4
externally, unstable, **2**-4

fabrication of steelwork superstructures, **20**-4
factor, collapse load, **23**-25
factor, derating, **29**-9
factored design stresses, **20**-21
factors, bridge plan, **10**-6
failure analysis, modal, **26**-29
failure cone, **3**-91
failure, ductile tearing, **24**-20
failure mechanisms, **6**-70
failure modes, **6**-9, **29**-2
failure of latticed structures, **13**-35
failure, probability of, **26**-30
failure probability problem, **26**-15
Fast Fourier Transform, **5**-20
fastener spacing, maximum, **3**-63
fastners, structural, **3**-6
fatigue and fracture, **24**-1
    design evaluation of structures, **24**-7
    evaluation of structural details, **24**-18
    high-cycle, **24**-4
    lowest stress range, **24**-13
    of secondary members, **24**-11
    tension chords, **24**-4
fatigue behavior, aluminum, **8**-19
fatigue of tubular connections, **29**-7
fatigue resistance, **29**-1
fatigue, strength, **24**-9
fault, earthquake, **5**-1, **5**-3
fault types, earthquake, **5**-3
fictitious restraining beam approach, **17**-22
field checking, bridge, **20**-19
fillet limitations, weld, **3**-76
finishing, aluminum structures, **8**-21
finite element method, **14**-13
    basic concept, **2**-98
    beam element, **2**-111
    choice of displacement function, **2**-101
    element shape function, **2**-104
    element shapes and discretization, **2**-100
    formulation of stiffness matrix, **2**-104
    isoparametric elements, **2**-103
    isoparametric families of elements, **2**-103
    nodal degrees of freedom, **2**-103
    plane strain, **2**-99
    plane stress, **2**-99
    plates in bendings, **2**-113
    plates subjected to in-plane forces, **2**-109
finite element method (FEM), **27**-7
finite element software packages, **27**-7
fire protection requirements, **18**-4
fireproofing of steel, **3**-4
first-order hinge-by-hinge analysis, **2**-142
fix-ended beams, **2**-7
fixed-based gable frame, **2**-138
fixed-end beams, **2**-31
fixed support, **2**-2
flange bending, load, **3**-30

flange bending, local, **3**-37
flange rib connection, **22**-60
flexibility method, **2**-84
flexibility of connection angles, **24**-15
flexible, bridge, **10**-58
flexural beam, **27**-12
flexural buckling, **3**-17, **7**-36
flexural capacity, **6**-96
flexural design of beams and one-way slabs, **4**-8
flexural members, **2**-5, **2**-87, **3**-26, **3**-50, **4**-40
    allowable stress design, **3**-27
    bending moment, **7**-15
    bending strengths, **7**-21
    continuous beams, **3**-39
    curved, **2**-17
    design, **3**-27
    design of, **7**-20
    lateral bracing of beams, **3**-42
    load and resistance factor design, **3**-33
flexural reinforcement, **4**-49
flexural strength, **4**-13, **23**-14
flexural strength criterion, **3**-33, **3**-52
flexural-torsional buckling, **3**-17
flexure-shear interaction, **3**-54, **19**-11
floor beam connections to slab and bottom chord,
    18-20
floor deflections, reducing, **23**-18
floor slabs, composite, **3**-95
floor system
    bridge, **10**-28
    composite, **12**-8
    concrete deck, **10**-29
    continuous, **23**-15
    floor beams, **10**-30
    pavement, **10**-29
    sizes of the components of a stub girder, **18**-8
    steel deck, **10**-29
    stringers, **10**-29
floor vibrations, connections and, **23**-15
floor vibrations, wood, **9**-38
fluid dampers, viscous, **27**-16
footing reinforcement, **4**-49
footings, **4**-45
footings on piles, **4**-59
force, axial, **2**-5
force D/C ratios and ductility, **16**-48
force, shear, **2**-5
forced vibration, **2**-168
forces, seismic, **23**-27
FORM/SORM techniques, **26**-21, **26**-25, **26**-35
formats, steel design, **3**-8
formulation of stiffness matrix, **2**-104
formulations, energy, **27**-8
formwork and scaffolding, **14**-29
foundations, cooling tower, **14**-27
foundations, transmission structure, **15**-23
Fourier analysis, **2**-170
fracture behavior, **24**-18
fracture, brittle, **24**-18
fracture control, **29**-1
fracture, ductile, **24**-18
fracture, fatigue and, **24**-1
fracture mechanics analysis, **24**-20
fracture predictions, **24**-7

fracture, transition range, **24**-18
fracture, types of, **24**-18
frame, **2**-1, **2**-27, **3**-49, **4**-37
    analysis and design, **23**-2
    and lateral loads, **23**-1
    application of slope deflection method, **2**-28
    braced, **3**-49, **17**-6, **23**-12, **23**-15, **28**-13,
        28-15
    bracing size requirement for braced, **17**-7
    classification of tall building, **12**-6
    column in gable, **17**-47
    design for unbraced, **23**-20
    earthquake wood, **5**-64
    effects of axial forces, **17**-17
    element, three-dimensional, **2**-163
    fixed-based gable, **2**-138
    gable, **2**-133
    interior, **23**-21
    intermediate moment, **23**-20
    lateral stiffness requirement for braced, **17**-6
    moment resisting, **23**-2
    ordinary moment, **23**-20
    pinned-base gable, **2**-136
    portal, **2**-127, **2**-175, **28**-18
    rectangular, **2**-131
    semi-rigid, **28**-25
    six-story, **28**-20
    stability, **2**-142
    stability of PR, **23**-27
    unbraced, **3**-49, **28**-13, **28**-15
    unbraced eight-story, **28**-45
    unsymmetrical, **17**-17
    with leaning columns, unbraced, **17**-28
framed column, **17**-3
framed columns alternative methods, **17**-24
free edge, plate, **2**-43
free vibration, **2**-166, **2**-167
french equations, **17**-9
friction dampers, **27**-14
friction grip bolts, **6**-54
FRISK code, earthquake, **5**-38
full-scale systems, **27**-2
full strength connection, **23**-4
fully and partially composite beams, **6**-59
fully composite beams, **6**-59
function, element shape, **2**-104
function, stability, **2**-155, **28**-6
functionality evaluation earthquake, **16**-10
functionality, structural, **27**-1
functions, probability, **26**-4
fusible backing, weld, **22**-23

gable frame, **2**-133
    analysis charts for, **2**-136
    fixed-based, **2**-138
    pinned-base, **2**-136
Gaussian curvature, **2**-54
general collapse, **29**-3
general design procedure, **16**-26
general limiting slenderness parameters and width-
    thickness ratios, **16**-27
general limiting values, **16**-49
generalized displacement amplitudes, **2**-101

generalized reliability problem, **26**-21
geodesic domes, **13**-21
geometric imperfection method, **28**-13
geometric instability, **2**-4
geometries, shell structure, **11**-5
girder bridge, **10**-34
    box girder, **10**-41
    composite girder, **10**-39
    grillage girder, **10**-41
    important factors, **10**-38
    plate girder, **10**-35
    structural features, **10**-34
girder,
    plate, **3**-50
    Vierendeel, **2**-140
    welded plate, **19**-7
girders with corrugated webs, **19**-18
goodness-of-fit, **26**-14
governing equation, **1**-24
governing sections of the stub girder, **18**-14
graphical solution, **16**-45
grids, double layer, **13**-33
grillages, **2**-90, **2**-139
ground motion, earthquake, **5**-2, **5**-24
ground motion, effect of soils on, **5**-42
guard railings, bridge, **10**-34
guide strands, **20**-16
gusset plate connections, **16**-18
guys, design of, **15**-21

Hankyu Chayamachi building, **27**-21
hardware requirements, **28**-37
hazards, seismic, **5**-1
(HAZ), **29**-6
heat input and weld size, **22**-17
heavyweight concrete, **4**-4
high-cycle fatigue, **24**-4
high early strength cement, **18**-22
high strength concrete, **4**-4
high strength low alloy steel, **3**-3
high-order differential equation, **1**-17
hinge, **2**-2
hinge analysis, plastic, **2**-162
hinge-by-hinge analysis, first-order, **2**-142
hinge effects, plastic, **2**-160
historic seismicity record, **5**-39
holes, bolt, **3**-59, **23**-19
holes, cope, **24**-11
holes, weld access, **22**-25, **22**-44
hoop tension, **11**-14
hull, ship, **8**-27
hydrostatic testing for joints, **25**-18
hyperbolic cooling towers, **14**-18
hypocentral distance, earthquake, **5**-25

idealized models, **2**-4
impact loads, **2**-2
imperfect columns, **1**-20
imperfection method, geometric, **28**-13
implementation, numerical, **28**-17
improved second-order bounds, **26**-29
in-plane compressive edge loading, **19**-15

inadequate construction engineering, **20**-5
indeterminate externally, statically, **2**-4
index, reliability, **26**-16
inelastic analysis
    considerations, **16**-51
    ductility, **2**-119
    equilibrium method, **2**-125
    gable frames, **2**-133
    mechanism method, **2**-129
    overall view, **2**-118
    plastic hinge, **2**-120
    plastic moment, **2**-121
    redistribution of forces, **2**-119
    second-order, **2**-144
    theory of plastic analysis, **2**-122
inelastic K-factor, **17**-22
inelastic rotation capacity, **28**-33
influence lines, **2**-70
    bending moment, **2**-74
    bridge, **10**-8
    for bending moment in simple beams, **2**-71
    for continuous beams, **2**-74
    for shear in simple beams, **2**-71
    for trusses, **2**-72
    qualitative, **2**-73
    shear, **2**-74
    support reaction, **2**-73
influence of local buckling, **6**-80
influence of method of construction, **18**-22
input energy, seismic, **27**-11
inspection of welding, **10**-10
inspection,
    weld radiographic, **22**-57
    weld ultrasonic, **22**-57
    welded connections, **22**-56
    weld, **3**-6
instability, geometric, **2**-4
instability, structural, **1**-12
intensity, earthquake, **5**-24
intensity, seismic, **5**-13
interior frame, **23**-21
intermediate moment frame, **23**-20
intermediate stiffeners, **3**-55
inverse CDF method, **26**-14
isolated column, **17**-3
isolation, base, **27**-1
isoparametric elements, **2**-103
isoparametric families of elements, **2**-103

joint and weld terminology, **22**-2
joint, cruciform, **24**-11
joint details, welded, **22**-22
joint efficiency, **29**-9
jointing system, **13**-39
    design requirements, **13**-40
    main groups of, **13**-40
    mero connector, **13**-40
    nodus system, **13**-50
    NS space truss system, **13**-52
    oktaplatte system, **13**-48
    space deck, **13**-46
    triodetic system, **13**-47
    types of, **13**-40

unibat system, **13**-50
unistrut system, **13**-47
joints, **2**-1
    bridge expansion, **10**-33
    connection and, **7**-42
    for fiberglass reinforced plastic pipe, **25**-17
    for reinforced concrete pipe, **25**-18
    method of, **2**-24
    of frames, **4**-42
    underground pipes, **25**-16
    unrestrained slip, **25**-5
    welded, **24**-10, **25**-18
jumbo shapes, welded, **24**-4

K-factor, **18**-16
Kiewitt domes, **13**-21
kinetic energy, **27**-2

laced-battened column, **17**-39
laced column, member, **17**-34
lamella domes, **13**-21
lamellar tearing, weld, **22**-25
Langer bridge, **10**-48
lateral buckling, **1**-23
lateral buckling, aluminum, **8**-10
lateral buckling of the flange, **19**-3
lateral force resisting systems (LFRS), **5**-61
lateral loading, earthquake, **5**-63
lateral loads, frames and, **23**-1
lateral PGD, **5**-51
lateral resistance, connection, **23**-24
lateral stiffness requirement for braced frame, **17**-6
lateral torsional buckling, **3**-27, **6**-38
latticed and built-up members, **17**-34
latticed shell
    braced barrel vaults, **13**-18
    braced domes, **13**-20
    buckling analysis of, **13**-35
    difference from double layer grids, **13**-17
    form and layer, **13**-17
    hyperbolic paraboloid shell, **13**-23
    intersection and combination, **13**-24
    types of buckling, **13**-35
law of conservation of energy, **2**-78
layout, structural, **12**-25
leaning column, **17**-33
LeMessurier and Lui method, **17**-30
LeMessurier method, **17**-24
length, effective, **1**-18
length, stiffener, **3**-55
level of redundancy, **26**-29
lightweight concrete, **4**-4
Lim and McNamara method, **17**-28
limit analysis, **4**-39
limit state, **6**-36
limit state, damageability, **27**-9
limit state of serviceability, **7**-5
limit state of strength, **7**-5
limit state, service, **23**-14, **23**-20
limit states, **16**-16
limit states and resistance factors, **16**-16
limit states, serviceability, **28**-32

limit states, ultimate, **23**-20
limitations, shell structure, **11**-2
limitations, weld fillet, **3**-76
limiting effective slenderness ratios, **16**-43
limiting width-thickness ratios, **16**-51
linear analysis with redistribution, **6**-38
linear buckling analysis, **1**-12
linear elastic fracture mechanics (LEFM), **24**-21
lines,
    bridge influence, **10**-8
    influence, **2**-70
    qualitative influence, **2**-73
liquefaction, mitigation of, **5**-53
liquefaction, soil's capacity against, **5**-47
liquid dampers, tuned, **27**-19
load
    advanced analysis, **28**-29
    aircraft wheel, **25**-4
    and reactions, **2**-2
    and resistance factor design (LRFD), **3**-9, **7**-4,
        18-2, **26**-3, **28**-1
    application in advanced analysis, **28**-31
    bearing stiffeners, **19**-17
    load-carrying capacity, **28**-32
    combination in advanced analysis, **28**-29
    combination, structural, **26**-35
    conditions, moving, **2**-71
    criteria, concentrated, **3**-30, **3**-37
    design of stubs for shear and axial, **18**-15
    earthquake, **2**-2, **16**-12
    eccentric edge, **19**-16
    Euler, **2**-59
    factor, collapse, **23**-25
    factors and combinations, **16**-12
    flange bending, **3**-30
    frames and lateral, **23**-1
    impact, **2**-2
    introduction region, **6**-81
    live, **24**-26
    member with transverse, **2**-157
    method, unit, **2**-82
    moments, dead, **23**-17
    multiaxial, **24**-16
    permanent, **2**-2
    ponding, **2**-2
    response to suddenly applied, **2**-169
    shedding, **24**-2
    system, stub girder for lateral, **18**-21
    total overburden, **25**-2
    underground pipe, **25**-1
    variable, **2**-2
    wind, **2**-2, **16**-12
loaded column, axially, **28**-18
loaded member, axially, **2**-86
loading between nodes, **2**-95
loading, earthquake, **24**-17
loading, earthquake lateral, **5**-63
loading, variable amplitude, **24**-16
loads and load combinations, **16**-12
local buckling, **3**-55, **6**-27, **6**-80, **11**-10
local buckling of tubes and curved panels, **8**-16
local flange bending, **3**-37
local stiffener, buckling, **11**-13
local web yielding, **3**-31, **3**-37

lock-in effects, **12**-63
lognormal distribution, **26**-6
Lohse bridge, **10**-48
longitudinal shear capacity, **6**-94
low cycle fatigue due to seismic loading, **24**-17
low cycle fatigue in structures, **24**-17
lower bound acceptable D/C ratio DC_r, **16**-50
lower bound theorem, **2**-124
LRFD, **3**-9, **29**-9
LRFD, method, **7**-40
Lui method, **17**-26

magnitude, earthquake, **5**-11, **5**-37
main components of the stub girder, **18**-3
masonry buildings, earthquake, **5**-68
mass dampers, tuned, **27**-17
mass systems, distributed, **2**-171
material background, aluminum, **8**-1
material problems, **29**-6
material properties and design values, **26**-32
materials, **3**-2, **6**-5
    and mechanical properties, **7**-7
    brittle, **24**-22
    shell structure, **11**-3
    underground pipe, **25**-1
    welded connections, **22**-38
mathcad program, **25**-4
matrix analysis, **4**-38
matrix, element stiffness, **2**-86
matrix method, **2**-83
matrix, structure stiffness, **2**-91
maximum fastener spacing, **3**-63
maximum moment, **23**-17
mean and standard deviation, **26**-6
measurement of earthquake, **5**-11
mechanical connections, **4**-26
mechanical connections, aluminum, **8**-17
mechanically fastened joints, **24**-10
mechanics analysis, fracture, **24**-20
mechanism, **2**-122
mechanism, collapse, **2**-130
mechanism method, **2**-129
mechanism, plastic collapse, **2**-118, **23**-13
mechanism, types of, **2**-130
meizoseismal strong ground motion, **5**-30
member
    axially loaded, **2**-86
    bent in double curvature, **2**-159
    bent in single curvature, **2**-159
    compact section, **3**-30
    composite, **3**-92
    compression, **3**-17
    concentrically loaded compression, **7**-34
    curved flexural, **2**-17
    cylindrical tubular, **7**-42
    design tension, **9**-14
    flexural, **2**-5, **2**-87, **3**-26, **3**-50
    laced column, **17**-34
    latticed and built-up, **17**-34
    section properties for latticed, **16**-38
    sectional properties, **7**-19
    severely corroded, **24**-9
    size adjustment, **28**-34

subjected to side sway, 2-156
tension, 3-9
thin-walled, 1-21
with a hinge at one end, 2-156
with bolted and welded connectors, 17-40
with end restraints, 2-157
with tensile axial force, 2-159
with transverse loading, 2-157
membrane theory of cylindrical shells, 2-66
membrane theory of shells of revolution, 2-63
metal buildings, 7-47
metal properties, base, 22-38
metal properties, weld, 22-40
metal toughness, weld, 22-41
metal working, aluminum, 8-21
metallic yield dampers, 27-12
method
     alignment chart, 17-3
     analysis, 16-14
     ASD, 7-38
     bridge, erection, 20-28
     central difference, 2-178
     Charpy impact method, 29-6
     equilibrium, 2-125
     equivalent notional load, 28-15
     explicit imperfection modeling, 28-13
     finite element, 14-13
     finite element (FEM), 27-7
     flexibility, 2-84
     for computing sectional properties, 7-19
     for determination of K-factor, 17-3
     for evaluating overburden loads, 25-2
     for ground motion modeling, 5-30
     for quantification of PGD, 5-51
     further reduced tangent modulus, 28-15
     geometric imperfection, 28-13
     inverse CDF, 26-14
     LeMessurier, 17-24
     LeMessurier and Lui, 17-30
     Lim and McNamara, 17-28
     linear static analysis, 14-15
     LRFD, 7-40
     Lui, 17-26
     matrix, 2-83
     mechanism, 2-129
     moment area, 2-15
     moment distribution, 2-31
     of analysis, 27-3
     of analysis and modeling stub girders, 18-6
     of analysis for cooling towers, 14-12
     of consistent deformations, 2-36
     of construction and its influence, 18-22
     of joints, 2-24
     of moments, 26-13
     of sections, 2-26
     practical advanced analysis, 28-4, 28-18
     refined plastic hinge, 28-4
     response spectrum, 14-22
     slope deflection, 2-28
     slope-deflection equation, 17-4
     stiffness, 2-86
     to determine the correct amount of welding,
        22-9
     unit load, 2-82
     Yura, 17-28
minimum edge distance, 3-64
minimum erection rating factors, 20-23
minimum pipe stiffness, 25-13
minimum reinforcement, 4-21, 4-24, 6-35
minimum sized groove welds, 22-18
minimum slab thickness and reinforcement, 4-31
minimum weld size, 22-17
misplaced erection effort of steelwork, 20-4
mitigation of liquefaction, 5-53
mixing concrete, 4-8
modal damping, 2-175
modal failure analysis, 26-29
model,
     analytical, 2-1
     idealized, 2-4
     SDOF, 27-3
     structural, 2-1
model tests, aeroelastic, 12-69
modeling, basic, 27-23
modeling considerations, 16-15
modeling cooling tower, 14-13
modeling of geometric imperfection, 28-31
modeling of structural members, 28-31
modeling of the stub girder, 18-9
modeling, practical connection, 28-10
modification for end connections, 2-161
modifications to account for plastic hinge effects,
        2-160
modifications to alignment charts, 17-9
modulus of elasticity, 7-9
modulus of soil reaction, 25-7
modulus, shear, 7-9
modulus, tangent, 7-9
moment
     area method, 2-15
     bending, 2-5, 2-71
     capacity, 9-9
     carry-over, 2-31
     connections welded or bolted, 12-35
     dead load, 23-17
     diagram, 23-20
     diagram, bending, 2-6
     distributed, 2-31
     distribution, 2-32, 4-38
     distribution for frames, 2-33
     distribution method, 2-31
     earthquake seismic, 5-12
     fixed-end, 2-31
     frame, intermediate, 23-20
     frame, ordinary, 23-20
     maximum, 23-17
     method of, 26-13
     resistance, 6-61
     resisting frames, 23-2
     resisting frames (MRF), 5-65
     rotation curves, 23-1, 23-9, 23-11
     strength, nominal, 19-6
     unbalanced, 2-31
Monte Carlo simulation, 26-24
Monte-Carlo simulation (MCS), 26-27
motion, earthquake ground, 5-24
motion, equation of, 2-165

movements, tower foundation, **15**-27
moving load conditions, **2**-71
multi-degree-of-freedom structural systems, **27**-5
multiaxial load, **24**-16
multiple degree systems, **2**-170
multistory frames
    beams with web openings, **12**-12
    braced versus unbraced, **12**-3
    bracing systems, **12**-3, **12**-28
    classification, **12**-1
    comparison of floor spanning systems, **12**-19
    composite beams and girders, **12**-10
    composite trusses, **12**-15
    design concepts, **12**-23
    designing gravity frames, **12**-24
    drift assessment, **12**-34
    fabricated tapered beams, **12**-13
    floor diaphragms, **12**-19
    floor structures, **12**-8
    haunched beams, **12**-13
    long-span flooring systems, **12**-12
    moment resistant frames, **12**-33
    parallel beam system, **12**-14
    prestressed composite beams, **12**-17
    rigid frames, **12**-2
    simple frames, **12**-2
    structural schemes, **12**-23
    stud girder system, **12**-17
    sway versus non-sway, **12**-5

nails, wood, **9**-23
Navier solution, **2**-45
NDS® provisions, **9**-18, **9**-21, **9**-30, **9**-38
need for calibration, **26**-31
new materials, **16**-13
Newmark beta algorithm, **27**-8
no-collapse based design criteria, **16**-6
nodal degrees of freedom, **2**-103
nodes, loading between, **2**-95
nomenclature of tubular connections, **29**-2
nomenclature, shell structure, **11**-8
nominal moment strength, **16**-24, **19**-6
nominal shear strength, **16**-24
nominal strength of concrete structures, **16**-24
nominal strength of steel structures, **16**-16
non-linear analysis, **6**-37
non-normal random variables, **26**-23
noncompact section, **3**-27
nondestructive testing, weld, **22**-57, **22**-58
normal distribution, **26**-6
normal variables, **26**-23
Northridge, earthquake, **22**-58
number of independent mechanisms, **2**-129
numerical analysis, **2**-175
numerical implementation, **28**-17

objectives of construction engineering, **20**-3
objectives of seismic design, **16**-10
obstacles to effective construction engineering, **20**-5
one-way shear, **4**-49
one-way slabs strength design, **4**-19
ordinary moment frame, **23**-20

orthotropic plate, **2**-56
oscillator, SDOF, **27**-9
ovalizing parameter, **29**-4
overall column buckling, **7**-36
overburden load, total, **25**-2
overburden, underground pipe, **25**-2
overlapping weld, **24**-11
overview, shell structure, **11**-1

PAAP
    analysis option, **28**-38
    coordinate system, **28**-39
    degrees of freedom, **28**-40
    hardware requirements, **28**-37
    locations of nodal points, **28**-40
    program execution, **28**-37
    type of elements, **28**-39
    units, **28**-40
    users' manual, **28**-38
Pacific plate, earthquake, **5**-3
packages, software, **27**-7
painting, steel bridge, **10**-13
panel, curtain, and bearing walls, **4**-60
panels, wood structural, **9**-30
partial strength connection, **23**-4
partially composite beams, **6**-60
partially restrained connection, **22**-60
partition walls, **4**-62
passive energy dissipation, **27**-1
passive energy dissipation recent development, **27**-12
pavement, floor system, **10**-29
PCA load contour method, **4**-18
PD, **3**-8, **28**-1
performance acceptance criteria, **16**-26
performance-based design criteria, **16**-8
performance of control systems, **27**-22
performance requirements, **16**-10
permanent loads, **2**-2
PGD, lateral, **5**-51
philosophy, capacity design, **23**-8
philosophy, design, **3**-7
philosophy of the erection rating factor, **20**-22
physical properties, aluminum, **8**-2
piles, compaction, **5**-53
piles, tower foundation, **15**-25
pinned-base gable frame, **2**-136
pipe, prestressed concrete, **25**-14
pipe, reinforced concrete, **25**-16
pipe stiffness, minimum, **25**-13
pipe, underground, **25**-1
pipes, steel, **25**-6
PJP, welded connections, **22**-4
plan factors, bridge, **10**-6
plane strain, finite element method, **2**-99
plane stress, finite element method, **2**-99
plane structures, **2**-4
plastic analysis, **6**-15, **6**-36, **23**-14
plastic behavior of the section, **6**-72
plastic collapse mechanism, **2**-118, **23**-13
plastic design (PD), **3**-8, **3**-98, **28**-1
plastic design of beam-columns, **3**-99
plastic design of columns and beams, **3**-98
plastic hinge analysis, **2**-162

plastic hinge effects, 2-160
plastic hinge length, 2-121
plastic moment, 2-122
plastic resistance of the cross-section, 6-29
plate, 1-24, 2-39
    and box girders, 19-1
    axisymmetric circular, 1-10
    basic assumptions, 1-1
    beam bearing, 3-91
    bending of circular, 2-50
    bending of simply supported rectangular,
        2-43
    bending of thin, 2-39
    boundary conditions, 1-6, 2-42
    buckling of, 19-1
    buckling of thin, 2-58
    circular, 1-7, 2-60
    clamped edge, 2-42
    column base, 3-84
    columns with perforated cover, 17-39
    constitutive equations, 1-4
    equilateral triangular, 2-55
    equilibrium equations, 1-2
    examples of bending problems, 1-9
    free edge, 2-43
    governing equation, 1-24
    governing equations, 1-2
    in bendings rectangular element, 2-113
    in-plane and out-of-plane problems, 1-4
    of various shapes and boundary conditions,
        2-54
    orthotropic, 2-56
    rectangular, 2-58
    simply supported, 1-24
    simply supported edge, 2-43
    simply supported rectangular, 1-9
    strain-displacement relationships, 1-2
    strain energy of simple, 2-51
    subjected to in-plane forces, 2-109
    supported at corners, rectangular, 2-56
    theory, 2-54
plate girders, 3-50
    allowable stress design, 3-50
    bending strength, 19-6
    design procedures, 10-36
    load and resistance factor design, 3-52
    welded, 19-7
plated structures, 2-4
poles, transmission, 15-17
ponding load, 2-2
portal frame, 2-127, 2-175
portal, frame, 28-18
pounding, earthquake, 5-63
PR composite connections, 23-5
PR/FS connection, 23-9
practical advanced analysis method, 28-4
practical advanced analysis program (PAAP), 28-34
practical connection modeling, 28-10
practical estimation of three-parameters, 28-10
predictions, fracture, 24-7
preliminary design procedure of girders, 18-6
pressure, velocity, 14-10
prestressed, beam, 24-13

prestressed concrete beams and one-way slab strength
        design, 4-24
prestressed, concrete bridge, 10-16
prestressed concrete pipe, 25-14
prestressed concrete strength design, 4-12
principle of superposition, 2-4
principle of virtual work, 2-124
probability, 26-3
probability concepts, basic, 26-3
probability functions, 26-4
probability of failure, 26-30
probability problem, failure, 26-15
problem, basic reliability, 26-15
problem, failure probability, 26-15
problem, generalized reliability, 26-21
process of calibration, 26-32
product selection, aluminum, 8-20
program,
    CADPRO, 15-29
    overview, 28-35
    practical advanced analysis, 28-34
    SHAKE, 5-48
    using computer, 28-10
properties,
    base metal, 22-38
    member sectional, 7-19
    of concrete, 4-2
    of concrete and reinforcing steel, 4-2
    of wood, 9-3
    weld metal, 22-40
proportioning and mixing concrete, 4-6
proportioning concrete mix, 4-6
protection requirements, fire, 18-4
protection, structural, 27-2
protective systems, structural, 27-3, 27-8
prototype for dynamic behavior, 27-3
prying action, 16-24
PSHA, earthquake, 5-36
punching shear, 4-48
punching shear and two-way action, 6-102

qualitative influence lines, 2-73
quenched and tempered alloy steel, 3-3

radiographic inspection, weld, 22-57
Rahmen bridges, 10-45
railings, bridge guard, 10-34
random variables and distributions, 26-3
range, bolt stress, 24-10
range, effective stress, 24-16
range fracture, transition, 24-18
ratio of the fatigue-design to the strength-design
        loads, 24-4
reactions, loads and, 2-2
recommendations for stay cable design, 24-13
rectangular frame, 2-131
rectangular plate, 2-58
reduced section properties, 16-52
reducing floor deflections, 23-18
redundancy, level of, 26-29
redundant systems, 26-29
redundants, 2-36

refined plastic hinge approach, **2**-162
regions, shear transfer, **18**-15
reinforced and prestressed concrete, **24**-12
reinforced concrete, **14**-19
reinforced concrete beam, **4**-19
reinforced, concrete bridge, **10**-14
reinforced concrete pipe, **25**-16
reinforced concrete strength design, **4**-8
reinforced concrete structures, **16**-25
reinforcement ratios, **4**-10, **4**-12, **4**-14
reinforcing steel, **4**-5, **6**-7
reliability analysis, **26**-9
reliability-based design, **26**-1, **26**-31
reliability-based design concepts, **26**-2
reliability concepts, basic, **26**-9
reliability, definition of, **26**-1
reliability index, **26**-16, **26**-23
reliability of the dampers, **27**-16
reliability problem,
        basic, **26**-15
        generalized, **26**-21
reliability, structural, **26**-9
reliability, system, **26**-27
removable backing, weld, **22**-23
required weld vs. minimum weld sizes, **22**-17
requirements, ductility, **28**-33
requirements, fire protection, **18**-4
requirements for braced frames, **17**-6
requirements, hardware, **28**-37
reserve strength, **29**-7
residential construction, **7**-52
resistance, connection lateral, **23**-24
resistance factors, **16**-16
resistance factors in advanced analysis, **28**-30
resistance of a section to combined compression
        and bending, **6**-74
resistance of members to combined compression
        and bending, **6**-78
resistance of members to compression, **6**-75
resistance of the section under compression, **6**-72
resistance, weld, **7**-43
resistant design, earthquake, **27**-11
resisting frames, moment, **23**-2
resistiveness, earthquake, **5**-62
resonance, **2**-169
response spectra, earthquake, **5**-21
response spectrum method, **14**-22
response to suddenly applied load, **2**-169
response to time varying-load, **2**-170
restrained connection, partially, **22**-60
restraining members in braced frames, **17**-17
restrictions for application of design methods, **6**-81
results, comparison of, **28**-45, **28**-48, **28**-53
Ribbed domes, **13**-21
rigid column, **17**-28
rigid connection, **3**-69
rigid frame bridge
        p-Rahmen, **10**-46
        portal frame, **10**-45
        structural features, **10**-45
        Vierendeel bridge, **10**-46
ring stiffeners, **11**-12
rings, stiffening, **11**-14
risk, seismic, **5**-1

rivets and holes, **16**-23
roller, **2**-2
roof structure, shell, **7**-49
roof truss, **28**-43
rotation capacity, inelastic, **28**-33
rotation curves, moment, **23**-1, **23**-9, **23**-11
rotational spring, **2**-2
rule, Turkstra's, **26**-34
rules to prevent fatigue, **24**-6
rupture of the seat angle, **23**-19

safety evaluation earthquake, **16**-10
scaffolding, formwork and, **14**-29
scale effects in fatigue, **24**-14
scheme drawings for multistory building design,
        12-24
Schwedler domes, **13**-21
scope, **6**-4
scope, shell structure, **11**-2
screw, connection, **7**-43
screws, wood, **9**-23
screws, wood lag, **9**-26
SDOF
        model, **27**-3
        oscillator, **27**-9
        structural systems, **27**-9
        system, **2**-166, **27**-12
seat angle, rupture of the, **23**-19
second-order elastic analysis, **2**-143, **2**-159
second-order inelastic analysis, **2**-144
second-order refined plastic hinge analysis, **2**-162,
        28-6
section
        application in advanced analysis, **28**-30
        circular hollow, **29**-1
        compact, **3**-27, **28**-33
        member, compact, **3**-30
        method of, **2**-26
        noncompact, **3**-27
        properties, **16**-39
        properties for latticed members, **16**-38
        slender element, **3**-27
section properties of latticed members, **16**-15
sectional properties, member, **7**-19
seismic
        applications of VE dampers, **27**-16
        code, selected, **5**-57
        coefficient, ELF, **5**-57
        construction, weld, **22**-43
        design, **5**-58, **24**-6
        design code, **5**-57
        design criteria for bridges, **16**-5
        forces, **23**-27
        hazard analysis, full probabilistic, **5**-38
        hazard and design earthquake, **5**-35
        hazards, **5**-1
        input energy, **27**-11
        intensity, **5**-13
        liquefaction of soils, **5**-44
        seismic moment, earthquake, **5**-12
        response modification devices, **16**-15, **16**-25,
                16-34
        risk, **5**-1

seismicity, distribution of, **5**-5
seismicity record, historic, **5**-39
seismological model-based relations, **5**-26
selected seismic code, **5**-57
selection, aluminum alloy, **8**-20
selection, aluminum product, **8**-20
semi-rigid connection, **28**-11
semi-rigid end connection, **2**-96
semi-rigid frame, **28**-25
Sendagaya INTES building, **27**-20
service limit state, **23**-14, **23**-20
serviceability considerations, wood, **9**-37
serviceability limit states, **6**-18, **6**-31, **6**-103, **28**-32
setting-to-mark-strand-adjustment procedure, **20**-16
settlement, cooling tower, **14**-12
severely corroded member, **24**-9
SHAKE program, **5**-48
shakedown process, **29**-7
shape function, element, **2**-104
shapes, structural steel, **3**-5
shear
        and flexure, combined, **2**-172
        and torsion, **4**-19
        and torsion in curved panels, **8**-17
        and torsion in round tubes, **8**-17
        capacity of the web, **19**-7
        capacity, wood, **9**-11
        connection, **6**-44, **6**-59
        connection detailing, **6**-57
        connection simple, **12**-27
        connection, welded, **3**-80
        connections, bolted, **3**-69
        connector, **6**-8, **6**-48
        connectors for the stub, **18**-11
        connectors spacing, **6**-56
        connectors, stud, **18**-4
        design, web stiffeners for, **19**-10
        diagram, **2**-6
        diaphragms, 7-48
        effects, **6**-81
        flexibility greater, **17**-37
        force, **2**-5
        in simple beams, **2**-71
        influence lines, **2**-74
        modulus, 7-9
        reinforcement, **4**-21, **4**-25
        strength, beam webs, 7-32
        strength criterion, **3**-30, **3**-36, **3**-53
        strength of mechanical shear connectors, **6**-48
        strength provided by concrete, **4**-20, **4**-24
        strength provided by the shear reinforcement, **4**-24
        stress, allowable, **3**-51
        stress for a cylindrical shell, **11**-12
        studs, **23**-19
        transfer, criterion, **3**-52
        transfer mechanisms, **6**-45
        transfer regions, **18**-15
        vertical, **6**-17, **6**-102
        vibration of beams, **2**-172
        wall and diaphragm resistance, **9**-33
        wall design resources, **9**-33
        wall required resistance, **9**-33
        walls, **4**-62
        walls and diaphragms, **9**-32
shedding, load, **24**-2
shell
        buckling of, **2**-70
        conical, **2**-64
        doubly curved thin, **14**-12
        element stress resultants in, **2**-61
        membrane theory of cylindrical, **2**-66
        of revolution, membrane theory, **2**-63
        of revolution subjected to unsymmetrical
                loading, **2**-66
        roof structure, 7-49
        spherical with equal biaxial stresses, **11**-20
        spherical with unequal biaxial stresses, **11**-20
        structural, **2**-61
        subjected to external pressure, **11**-19
        subjected to shear, **11**-20
        symmetrically loaded circular cylindrical, **2**-67
        thin, **2**-61
        tolerances for cylindrical and conical, **11**-18
        tolerances for formed heads, **11**-21
        toroidal and ellipsoidal heads, **11**-21
        under uniform axial compression, **11**-10
        wall thickness on cooling towers, **14**-24
shell structure
        buckling design method, **11**-7
        design rules, **11**-2
        failure modes, **11**-6
        geometries, **11**-5
        limitations, **11**-2
        loads and load combination, **11**-7
        materials, **11**-3
        maximum allowable stress for cylindrical,
                11-10
        nomenclature, **11**-8
        other materials, **11**-3
        overview, **11**-1
        production practice, **11**-2
        scope, **11**-2
        stability analysis and design, **11**-3
        steel, **11**-3
        stress components, **11**-3
        stress factor, **11**-7
shock and vibration isolation, **27**-16
shock loading
        absorbing kinetic energy, **21**-2
        an example of, **21**-11
        basic principles, **21**-1
        detailing and workmanship, **21**-10
        loading scheme definitions, **21**-1
        material properties for optimum design, **21**-7
        requirements for optimum design, **21**-2
        section properties for optimum design, **21**-8
shop control procedure, **20**-17
shop welded-field bolted connections, **3**-82
short columns under axial load and bending, **4**-15
short columns under minimum eccentricity, **4**-14
side sway, member subjected to, **2**-156
sidesway web buckling, **3**-31, **3**-38
sign convention, **2**-17, **2**-31
simple beam
        influence lines for bending moment in, **2**-71
        influence lines for shear in, **2**-71

simple first-order bounds, **26**-28
simple shear, connection, **12**-27
simplified equations to alignment charts, **17**-7
simply supported column, **1**-13
simply supported composite beams, **6**-8
simply supported composite slabs, **6**-93
simply supported plate, **1**-24
simply supported rectangular plate subjected to uniform load, **1**-9
simply supported structure, **23**-17
single-column spread footings, **4**-53
single degree of freedom (SDOF), **5**-2
single-degree-of-freedom structural systems, **27**-3
single pass minimum sized welds, **22**-18
six-story frame, **28**-20
size adjustment, member, **28**-34
size and length limitations of welds, **3**-76
size, minimum weld, **22**-17
size of footings, **4**-47
skeletal structure, **2**-4
slab bars, **23**-19
slab, concrete bridge, **10**-14, **10**-16
slab design based on partial interaction theory, **6**-99
slabs, composite floor, **3**-95
slender element section, **3**-27
slenderness effects, **4**-16
slip joints, unrestrained, **25**-5
slope deflection, **4**-37
    equation method, **17**-4
    equations, **2**-154
    method, **2**-28
software packages, **27**-7
soil capacity against liquefaction, **5**-47
solid columns, wood, **9**-16
solid structure, three-dimensional, **2**-4
solution, Boussinesq, **25**-3
solution, Navier, **2**-45
space frame
    advantages of, **13**-4
    assembly by sliding element in the air, **13**-17
    assembly of elements in the air, **13**-17
    basic concepts, **13**-3
    bearing joints, **13**-53
    computers and design of, **13**-31
    defining terms, **13**-54
    definition of, **13**-2
    double-layer grids, **13**-6
    earthquake resistance, **13**-33
    erection by strips or blocks, **13**-17
    hoisting by Derrick masts or cranes, **13**-17
    introduction to, **13**-1
    jacking up, **13**-17
    jointing systems, **13**-39
    lifting up, **13**-17
    methods of erection, **13**-16
    preliminary planning guidelines, **13**-5
    stability, **13**-35
    static analysis, **13**-31, **13**-32
    stiffness and flexibility method, **13**-31
    structural analysis, **13**-25
spaced columns, wood, **9**-18
spacing limitation for shear reinforcement, **4**-25
spacing, maximum fastener, **3**-63
spacing of shear connectors, **3**-97

spacing, stiffener, **3**-55
spatial structures, **2**-4
specification of steel and filler metal, **24**-19
spectra, earthquake response, **5**-21
spectrum method, response, **14**-22
spherical dome, **2**-63
spikes, wood, **9**-23
splices, **4**-26
spring, rotational, **2**-2
spring, translational, **2**-2
stability and determinacy, **2**-27
stability
    basic concepts of, **1**-11
    beam-column, **2**-149
    column, **2**-145
    frame, **2**-142
    stability equation of a column, **2**-145
    space frame, **13**-35
stability function, **2**-155, **28**-6
stability of guyed poles, **15**-21
stability of PR frames, **23**-27
stability of the compression flange, **19**-2
state, damageability limit, **27**-9
state function, complicated limit, **26**-17
state of equilibrium, **2**-2
state, service limit, **23**-14, **23**-20
states of equilibrium, **1**-11
states, serviceability limit, **28**-32
states, ultimate limit, **23**-20
static linear analysis, **16**-14
static strength, **29**-1
statically determinate externally, **2**-4
statically indeterminate externally, **2**-4
statistical estimation and distribution fitting, **26**-13
statistical theory of extreme values, **26**-4
statistics concepts, **26**-3
stay cables, bridge, **24**-12
steel
    bridge construction, **20**-1
    carbon, **3**-3
    concrete interface separation, **6**-55
    corrosion protection of, **3**-5
    corrosion resistant high strength low alloy, **3**-3
    deck, **6**-91
    decking, **6**-8
    design, bracing requirements, **7**-34
    design formats, **3**-8
    design, maximum flat-width-to-thickness ratios, **7**-11
    dual certified material, **22**-38
    ductibility, **7**-10
    element strength, **7**-11
    fireproofing of, **3**-4
    frame design, **28**-34
    framed buildings, multi-story, **18**-1
    high strength low alloy, **3**-3
    pipes, **25**-6
    plate shear walls, **19**-1, **19**-11
    quenched and tempered alloy, **3**-3
    reinforcing, **6**-7
    shapes, structural, **3**-5
    shell structure , **11**-3
    structural, **6**-8, **16**-13

structure, 16-25, 16-26
structure to resist horizontal loads, 17-33
structures behavior, 28-4
TMCP, 3-4
types of, 3-3
weldability of, 3-7
work contractor, 20-3
steel bridge, 10-8
    bolting, 10-11
    construction on site, 10-12
    fabrication in shop, 10-12
    painting, 10-13
    welding, 10-8
steelwork fabrication and erection, 20-2
steelwork, superstructure, 20-1
stiffened elements under uniform compression,
        7-12
stiffened elements with stress gradient, 7-15
stiffener
    area, 3-55
    bearing, 3-54
    buckling local, 11-13
    circumferential, 14-27
    criteria transverse, 3-51
    intermediate, 3-55
    length, 3-55
    load bearing, 19-17
    moment of inertia, 3-55
    ring, 11-12
    spacing, 3-55
    transverse, 24-11
stiffeners, column, 23-19
stiffening rings, 11-14
stiffness
    beam connection, 23-15
    connection, 23-4, 23-18
    matrix, element, 2-86
    matrix, structure, 2-91
    method, 2-86
    minimum pipe, 25-13
    reduction, 16-51
strain energy, 2-51
    due to uniaxial stress, 2-75
    in bending, 2-76
    in shear, 2-77
strength
    and deflection criteria, 25-6
    connection, full, 23-4
    connection, partial, 23-4
    criterion, flexural, 3-33, 3-52
    criterion, shear, 3-30, 3-36, 3-53
    fatigue, 24-9
    flexural, 23-14
    increase from cold work of forming, 7-9
    nominal moment, 19-6
    of welded connections, 22-6
    of welds, 3-73
    reserve, 29-7
    static, 29-1
    tensile, 7-9
    tension, 23-19
stress
    allowable bending, 3-50
    allowable shear, 3-51

circumferential, 2-18
concentration factor (SCF), 24-21
corrosion cracking, 24-2
for a cylinder under external pressure, 11-11
gradient, stiffened elements with, 7-15
limitation, 6-33
range, bolt, 24-10
range effect in fatigue, 24-15
range, effective, 24-16
resultants in shell element, 2-61
strain, 7-9
strain behavior of structural steel, 3-2
strain energy due to uniaxial, 2-75
stringers, floor system, 10-29
strip, grid and mat foundations, 4-57
structural
    analysis, 2-1, 10-47, 28-1
    analysis energy methods in, 2-75
    analysis, energy relations in, 2-77
    behavior, aluminum, 8-4
    component classification, 16-26
    concrete, 16-13
    concrete design, 4-2
    design, 2-1
    design criteria, 3-7
    design history, 6-1
    dynamic, 2-165
    engineering, 27-23
    engineering design, 26-31
    fasteners, 3-6
    form of vertical truss, 12-29
    functionality, 27-1
    instability, 1-12
    layout, 12-25
    load combination, 26-35
    material, 16-13
    model, 2-1
    panels, wood, 9-30
    protection, 27-2
    protective systems, 27-3, 27-8
    reliability, 26-9
    scheme for multistory frames, 12-24
    shell, 2-61
    steel, 6-8, 16-13
    steel, behavior of, 3-2
    steel shapes, 3-5
    systems and assemblies, 7-46
structural deformation capacity, 16-25
structure
    aluminum welded, 24-8
    behavior, steel, 28-4
    bridge, 10-1
    civil engineering, 27-16
    continuous, 23-17
    cooling tower, 14-1
    design, aluminum, 8-4
    finishing, aluminum, 8-21
    low cycle fatigue in, 24-17
    plane, 2-4
    plated, 2-4
    shell roof, 7-49
    simply supported, 23-17
    skeletal, 2-4
    spatial, 2-4

steel, **16**-25, **16**-26
stiffness matrix, **2**-91
three-dimensional solid, **2**-4
two dimensional, **2**-1
under earthquake load, **27**-19
with several redundants, **2**-37
strutted beam bridge, **10**-46
stub girder component sizes, **18**-9
stub girder floor system description, **18**-2
stub girder floor systems, **18**-1, **18**-5
stub girder for lateral load system, **18**-21
stub girders built in the shore condition, **18**-22
stud assemblies, wall, **7**-50
stud connectors used in solid concrete slabs, **6**-49
stud connectors used with profiled steel decking, **6**-50
stud shear connectors, **18**-4
studies, aluminum design, **8**-21
studs, shear, **23**-19
substructure, bridge concrete, **10**-25
superposition, principle of, **2**-4
superstructure steelwork, **20**-1
support, fixed, **2**-2
supported structure, simply, **23**-17
surface waves, **5**-3
suspension bridge, **10**-53
    cable design, **10**-54
    stability for wind, **10**-58
    stiffening girder, **10**-57
    structural analysis, **10**-54
    structural features of, **10**-53
    tower, **10**-57
    type of, **10**-54
symbols, welding, **3**-73
symmetrically loaded circular cylindrical shells, **2**-67
system
    basic, **26**-27
    bridge floor, **10**-28
    composite floor, **12**-8
    continuous floor, **23**-15
    crossing bracing, **17**-33
    distributed mass, **2**-171
    jointing, **13**-39
    multiple degree, **2**-170
    redundant, **26**-29
    reliability, **26**-27
    SDOF, **27**-12
    structural protective, **27**-3, **27**-8
systems to achieve seismic resistance, **22**-45

tabs, weld, **22**-24, **22**-43
tall building
    across wind response, **12**-63
    core braced systems, **12**-39
    frame tube systems, **12**-42
    framing systems, **12**-38
    moment-truss systems, **12**-40
    outrigger and belt truss systems, **12**-41
    wind effect, **12**-57
tangent modulus, **7**-9
tapered column, **17**-42
tearing failure, ductile, **24**-20

tearing, weld lamellar, **22**-25
techniques to limit serviceability problems, **6**-35
tectonic related, earthquake, **5**-1
temperature variations on cooling towers, **14**-11
tensile strength, **7**-9
tension, hoop, **11**-14
tension lap splices, **4**-26
tension members, **3**-9
    allowable stress design, **3**-10
    design, **9**-14
    load and resistance factor design, **3**-12
    pin-connected members, **3**-15
    threaded rods, **3**-17
tension strength, **23**-19
tension yield, **23**-19
terms, defining, **3**-99, **7**-53
testing, tower foundation, **15**-27
testing, weld nondestructive, **22**-57, **22**-58
tests, aeroelastic model, **12**-69
theorem
    Castigliano's, **2**-79
    Clapeyron's, **2**-11
    key construction, **20**-9
    lower bound, **2**-124
    of minimum potential energy, **2**-79
    of virtual work, **2**-78
    uniqueness, **2**-124
    upper bound, **2**-124
theory of elasticity, **25**-3
theory of order statistics, **26**-14
theory, plate, **2**-54
theory, system reliability, **26**-27
thickness effect in fatigue, **24**-14
thin shell, **2**-61
thin-walled members, **1**-21
three-dimensional frame element, **2**-163
three-dimensional solid structure, **2**-4
three-way grid domes, **13**-21
ties, **2**-1
TMCP steel, **3**-4
torsion reinforcement, **4**-22
torsional beam, **27**-12
torsional buckling, **3**-17, **7**-36
torsional buckling, lateral, **3**-27
torsional buckling of the flange, **19**-4
torsional-flexural buckling, **1**-21, **7**-37
torsional strength provided by concrete, **4**-21
total overburden load, **25**-2
toughness, weld metal, **22**-41
tower foundation, **15**-23
    cast in place concrete, **15**-23
    design examples, **15**-27
    direct embedment, **15**-25
    drilled concrete shafts, **15**-24
    example of drilled shaft, **15**-29
    example of spread footing, **15**-27
    geotechnical parameters, **15**-23
    movements, **15**-27
    piles, **15**-25
    rock foundation, **15**-25
    steel grillages, **15**-23
    testing, **15**-27
    types and selection design, **15**-23
    vibratory shells, **15**-25

tower, modeling cooling, **14**-13
tower, suspension bridge, **10**-57
transfer, criterion shear, **3**-52
transfer regions, shear, **18**-15
transition range fracture, **24**-18
translational spring, **2**-2
transmission poles, **15**-17
    concrete poles, **15**-20
    guyed poles, **15**-20
    stress analysis, **15**-18
    tubular steel poles, **15**-18
    wood poles, **15**-19
transmission structure
    analysis and design methodology, **15**-12
    anchor structures, **15**-6
    anchors, **15**-25
    angle towers, **15**-4
    application, **15**-1
    bolted connections, **15**-14
    calculation of loads using ASCE, **15**-6
    calculation of loads using NESC, **15**-5
    configuration and material, **15**-3
    connection, **15**-14
    constructibility, **15**-3
    construction and maintenance loads, **15**-7
    construction and other considerations, **15**-26
    deadend structures, **15**-4, **15**-8
    design for broken conductors, **15**-6
    design of steel lattice tower, **15**-10
    foundations, **15**-23
    loads on, **15**-5
    loads on structure, **15**-7
    longitudinal loading, **15**-8
    longitudinal unbalanced load, **15**-9
    maintenance considerations, **15**-4
    sectionalizing a long line, **15**-4
    security loads, **15**-6
    special loads, **15**-6
    state of the art review, **15**-4
    structure families, **15**-4
    tower geometry, **15**-10
    tower members, **15**-12
    tower testing, **15**-15
    transmission pole, **15**-17
    transverse load due to line angle, **15**-8
    transverse loads, **15**-8
    vertical loads, **15**-7
    wind load on structures, **15**-8
    wind load on wires, **15**-8
transverse reinforcement, **6**-57
transverse, stiffener, **24**-11
transverse stiffeners, criteria, **3**-51
treating the weld as a line to find weld size, **22**-10
treatment of energy formulations, **27**-3
triangular plate, equilateral, **2**-55
triaxial stresses and ductility, **22**-21
trilinear curve, **23**-11
truss bridge, **10**-42
    gerber truss bridge, **10**-45
    structural analysis and secondary stress, **10**-43
    structural features, **10**-42
    type of truss, **10**-43
truss connection, **29**-1
truss, roof, **28**-43

trusses, **2**-1, **2**-23
    compound, **2**-26
    wood, **9**-33
tubular member, cylindrical, **7**-42
tuned liquid dampers, **27**-19
tuned mass dampers, **27**-17
turbulence, wind effects, **12**-51
Turkstra's rule, **26**-34
two-column footings, **4**-55
two dimensional structure, **2**-1
two-story four-bay semi-rigid frame, **28**-48
two-way systems, **4**-28
type of connections, **3**-58
types of braced domes, **13**-21
types of cables, **10**-54
types of footing, **4**-46
types of fracture, **24**-18
types of mechanism, **2**-130
types of sections and advantages, **6**-69
types of steel, **3**-3
types of welded joints, **10**-10
types of wood products, **9**-2

U-strip energy dissipators, **27**-12
ultimate limit state, **6**-36
ultimate shear capacity of the web, **19**-7
ultrasonic inspection, weld, **22**-57
unbraced eight-story frame, **28**-45
unbraced, frame, **3**-49, **28**-13, **28**-15
unbraced frames, design for, **23**-20
unbraced frames with leaning columns, **17**-28
undamped free vibration, **2**-166
underground pipe, **25**-1
    cathodic protection, **25**-20
    coatings, **25**-19
    concrete non-pressure, **25**-14
    concrete pressure, **25**-13
    corrugated steel, **25**-9
    dead weight of the, **25**-5
    design methods, **25**-5
    example problem, **25**-8, **25**-11, **25**-16
    external loads, **25**-2
    fiberglass reinforced plastic (FRP), **25**-12
    flexible design, **25**-6
    internal load, **25**-5
    internal pressure and vacuum, **25**-5
    joints, **25**-16
    live loads, **25**-4
    load, **25**-1
    materials, **25**-1
    overburden, **25**-2
    seismic load, **25**-5
    surcharge at grade, **25**-2
uniaxial stress, **2**-75
uniform axial compression, **11**-14
uniformly compressed elements with an edge stiffener, **7**-16
uniqueness theorem, **2**-124
unit load method, **2**-82
unrestrained slip joints, **25**-5
unstable externally, **2**-4
unstiffened cone-cylinder junctions, **11**-14

unstiffened elements under uniform compression, 7-16
unsymmetrical frame, **17**-17
unsymmetrical loading, shells of revolution subjected to, **2**-66
upper bound acceptable D/C ratio DC_p, **16**-49
upper bound theorem, **2**-124
using computer program, **28**-10

values, wood design, **9**-6
variable amplitude loading, **24**-16
variable loads, **2**-2
variables affecting welded connection strength, **22**-7
variables, normal, **26**-23
variance reduction techniques (VRTs), **26**-25
VE dampers, **27**-15
velocity pressure, **14**-10
vertical buckling of the compression flange, **19**-2
vertical shear, **6**-17, **6**-102
vertical vierendeel girder components, **18**-11
vibration
    absorbers, dynamic, **27**-17
    connections and floor, **23**-15
    damped free, **2**-167
    forced, **2**-168
    free, **2**-166
    undamped free, **2**-166
    wood floor, **9**-38
vibro-replacement, **5**-53
vibrostabilization, **5**-53
vierendeel, analysis, **18**-8
Vierendeel girders, **2**-140
vierendeel modeling scheme for the stub girder, **18**-9
viscoelastic braced dampers, **27**-10
viscoelastic dampers, **27**-15
viscous fluid dampers, **27**-16
viscous fluid dampers for seismic applications, **27**-16

wall stud assemblies, **7**-50
walls, **4**-60
waves, body, **5**-3
waves, surface, **5**-3
weak composite connection, **23**-5
weak PR connection, **23**-6
web
    angles, **23**-19
    buckling due to in-plane bending, **19**-5
    corrugated, **8**-15
    crippling, **3**-31, **3**-37, **7**-33
    crushing, **8**-13
    longitudinal stiffeners for bending design, **19**-6
    openings in box girders, **19**-18
    openings in plate girders, **19**-18
    reinforcement, **4**-26, **4**-27
    stiffeners for shear design, **19**-10
weld, **3**-6
    access holes, **22**-25, **22**-44
    and bolts sharing loads, **22**-43
    arc, **7**-43

backing, **22**-22, **22**-41
buttering layer technique, **22**-26
ceramic backing, **22**-42
copper backing, **22**-23
dye penetrant inspection, **22**-57
effect of welding variables, **22**-53
effective area of, **3**-73
field vs. shop, **22**-56
fillet limitations, **3**-76
fit-up, **22**-56
fusible backing, **22**-23
fusible backing removal, **22**-42
heat affected zones, **22**-41, **29**-6
in-process visual inspection, **22**-57
inspections, **3**-6
lamellar tearing, **22**-25
magnetic particle inspection, **22**-57
maximum effective length of, **3**-77
maximum leg size, **3**-77
metal properties, **22**-40
metal toughness, **22**-41
minimize weld volumes, **22**-19
minimum effective length of, **3**-77
minimum leg size of fillet, **3**-76
nondestructive testing, **22**-57, **22**-58
overlapping, **24**-11
provide ample access for welding, **22**-19
purpose of the WPS, **22**-53
radiographic inspection, **22**-57
recognize steel properties, **22**-19
related type of cracking, **22**-25
removable backing, **22**-23
requirements for efficient welded structures, **22**-38
residual stresses applied, **22**-35
residual stresses in welding, **22**-20
resistance, **7**-43
seismic construction, **22**-43
selection of fillets vs. PJP groove, **22**-22
size and length limitations of, **3**-76
size, minimum, **22**-17
strength of, **3**-73
subjected to combined shear and flexure, **3**-78
tabs, **22**-24, **22**-43
temperature of the steel, **22**-40
transfer of forces, **22**-19
two residual stresses isolated, **22**-33
ultrasonic inspection, **22**-57
weldability, **29**-1
weldability of steel, **3**-7
welded
    bracket type connections, **3**-78
    connection, **3**-73, **3**-78, **24**-26
    connections, aluminum, **8**-18
    connections for tension members, **3**-77
    connectors, **17**-40
    joint details, **22**-22
    joints, **24**-10, **25**-18
    jumbo shapes, **24**-4
    moment resistant connection, **3**-80
    plate girder, **19**-7
    shear connection, **3**-80
    splices, **4**-26

structures, aluminum, **24**-8
tension splices specifications, **22**-40
tubular connection, **29**-1
tubular connection problems, **29**-6
web details, **22**-43
welded connections, **7**-42, **22**-1
  CJP, **22**-4
  connection details, **22**-41
  design examples, **22**-27
  determining weld size, **22**-6
  double-side welds, **22**-5
  ductile behavior in seismic sections, **22**-45
  ductile hinges in connections, **22**-46
  fillet welds, **22**-3
  flat position welding, **22**-21
  groove weld preparations, **22**-5
  inspection, **22**-56
  interaction of joint and weld, **22**-6
  joint and weld terminology, **22**-2
  joint types, **22**-2
  materials, **22**-38
  no secondary members in design, **22**-19
  PJP, **22**-4
  post-Northridge assessment, **22**-58
  principles of design, **22**-18
  subject to seismic loading, **22**-36
  triaxial stresses and ductility, **22**-21
  understanding ductile behavior, **22**-32
  weld types, **22**-2
  welded joint details, **22**-22
  workmanship requirements, **22**-52
welding,
  effects of, **8**-5
  inspection of, **10**-10
  steel bridge, **10**-8
  symbols, **3**-73
wheel load, aircraft, **25**-4
wind effects
  characteristics of wind, **12**-50
  cross-spectrum of turbulence, **12**-53
  forces due to turbulent flow, **12**-56
  forces due to uniform flow, **12**-54
  induced dynamic forces, **12**-54
  integral scales of turbulence, **12**-53
  mean wind speed, **12**-50
  on buildings, **12**-48, **12**-69
  response by wind tunnel tests, **12**-69
  response due to across wind, **12**-63
  response due to along wind, **12**-57
  spectrum of turbulence, **12**-53
  torsional response, **12**-67
  turbulence, **12**-51
wind-induced vibrations, **27**-15
wind load, **2**-2, **16**-12
  dynamics, **16**-12
  factor estimation, **14**-19

  on live load, **16**-12
  on structures, **16**-12
wind, stability for, **10**-58
wood
  arches, **9**-35
  axial compression, **9**-19
  beam design, **9**-8
  bearing capacity, **9**-13
  biaxial bending, **9**-19
  bolts, **9**-26
  built-up column, **9**-18
  column design, **9**-15
  combined bending, **9**-19
  combined bending and axial tension, **9**-19
  combined load design, **9**-19
  curved beams and arches, **9**-34
  deflections, **9**-37
  design values, **9**-6
  dowels, **9**-26
  fastener and connection design, **9**-22
  floor vibrations, **9**-38
  frame, earthquake, **5**-64
  lag screws, **9**-26
  loads and load combinations, **9**-5
  moment capacity, **9**-9
  nails, **9**-23
  NDS®, **9**-13
  non-structural performance, **9**-38
  other types of connections, **9**-30
  panel design resources, **9**-32
  panel design values, **9**-31
  panel section properties, **9**-31
  preliminary design considerations, **9**-5
  properties of, **9**-3
  screws, **9**-23
  serviceability considerations, **9**-37
  shear capacity, **9**-11
  shear walls and diaphragms, **9**-32
  solid columns, **9**-16
  spaced columns, **9**-18
  spikes, **9**-23
  structural panels, **9**-30
  structures, **9**-1
  tension member design, **9**-14
  trusses, **9**-33
worked examples, **6**-20

yield dampers, metallic, **27**-12
yield point, **7**-9
yield surface equation for doubly symmetrical sections, **16**-53
yield, tension, **23**-19
yielding, **7**-36
yielding, local web, **3**-31, **3**-37
Yura method, **17**-28